CALCULUS

CALCULUS

James F. Hurley
University of Connecticut

Wadsworth Publishing Company
Belmont, California
A Division of Wadsworth, Inc.

Mathematics Editor: Kevin Howat
Production Editor: Harold Humphrey
Cover and Text Designer: Lisa S. Mirski
Print Buyer: Karen Hunt
Art Editor: Marta Kongsle
Assistant Editor: Anne Scanlan-Rohrer
Editorial Assistants: Catherine Read, Ruth Singer
Copy Editor: Yvonne Howell
Technical Illustrators: Scientific Illustrators
Signing Representatives: Rich Giggey, Ira Zukerman
Cover: Computer graphic created by Tim Alt and Ken Weiss of Digital Art, Los Angeles, based on a design by Lisa S. Mirski. Two images were combined, using a posterized block pixel process.

Printed in the United States of America

3 4 5 6 7 8 9 10——91 90 89 88

Library of Congress Cataloging-in-Publication Data

Hurley, James Frederick, 1941–
Calculus.

Includes indexes.
1. Calculus. I. Title.
QA303.H859 1987 515 86-10987
ISBN 0-534-05592-3

Students:

A *Study Guide* and a *Student Solutions Manual* have been specially created to help you master the concepts presented in this textbook. Information about these can be obtained from your bookstore. Also, a *computer software supplement* specially designed to enhance your understanding of selected topics in this text is available for purchase through your bookstore or through the special order form in the back of this book.

To Cecile, JoJo, and Gigi

Contents

* Optional section

* Optional section

* Optional section

Preface

The image of Stonehenge on the cover represents many of the themes that motivated the writing of this book. Seen from this angle, the arrangement of stones suggests a gateway, which calculus surely is—not only to modern mathematics, but also to science, engineering, and the quantitative sophistication a liberal education seeks to impart. A major goal of the text is to convey to student readers the power of calculus as a tool for understanding the world around them. It is hoped that this will motivate them not merely to pass through this gateway, but to work their way through it actively. They need to emerge from calculus with solid mastery both of its computational procedures and its concepts, as well as confidence in using them to analyze and solve problems in other fields.

To foster those goals, there is a significant effort in the text to relate calculus to topics studied in other lower-division college courses—physics, chemistry, biology, economics, engineering, etc.—as well as to familiar occurrences in everyday life. Such topics are used in some cases to motivate new mathematical ideas and in others to illustrate extra-mathematical applications of calculus. Sometimes, they are used to do both. For instance, limits at infinity are motivated by the question of why new editions of textbooks appear periodically, an occurrence that often prevents students from reselling their texts. One-sided limits are related to change-of-phase in chemistry. And improper integrals are motivated by, and used to compute, the work done in sending a deep space probe out of the solar system. This kind of motivation and application is thoroughly integrated into the text, rather than being confined to isolated sections or subsections.

The current generation of students, whose background is so rich in pictorial experience, seems to learn best when given ample visual aids. The power of Stonehenge to evoke mental images in its viewers suggests the strong visual emphasis throughout the text. In addition to approximately 800 figures in the text itself, tables are frequently used to summarize the results of repeated calculations. And the answer section includes graphs whenever they are asked for in the exercises. Nearly all the art for the text was computer-generated at least twice to ensure accuracy and the most illuminating images possible.

The erection of Stonehenge was a triumph both of insight (in designing it) and persistent hard work (in completing it). Calculus also requires insight to understand its ideas and persistence to really master it, and the text makes no attempt to mislead its readers about that. To be sure, the primary thrust is toward helping the student develop the conceptual understanding and insight that must be the foundation of true mastery of calculus. But there is also emphasis on building the solid computational skills and the confidence in using them that are vital to successful advanced study and work in mathematics, science, and engineering. Many examples show the computational persistence needed to complete some of the more demanding exercises and to fulfill the expectations of later courses. There are discussions of the significance of central concepts, including the necessity as well as the impact of the hypotheses of some important theorems. There are counterexamples as well as examples, and the last few exercises in most sets challenge the students to think in some depth about the material of the section.

Because the most efficient computational implementation of mathematical ideas is usually algorithmic, the text provides suggested algorithms for solving problems in such areas as optimization, related rates, change of variable in integration, integration by parts, and convergence tests for infinite series. It is hoped that working with these can help the students develop the habit of thinking algorithmically and looking at problems as opportunities to apply general methods.

The book provides a wide range of mathematical proofs—from those essential to understanding the concepts and performing the calculations to more subtle ones that yield deeper theoretical insights. These proofs can be omitted without breaking the continuity of the text; instructors who wish to treat the theory more thoroughly, however, may find some of the more conceptual discussions of special interest. The section on differentials, for example, develops the connection between differentiation and linear approximation in some detail, which the instructor can assign according to students' needs and abilities.

There is enough material for a three-semester or four- or five-quarter course meeting four or five hours per week. (See below for information about the *Instructor's Manual,* which contains specific suggestions about outlines and pacing.) The text is divided into chapters that correspond to natural learning units, so that testing may be keyed to the chapter organization. To help students digest and organize the material of each chapter, there is a review section called *Looking Back* for self study. It summarizes the most important material in the chapter, presents a *Chapter Checklist* of key terms from each section, and contains a set of review exercises.

The trigonometric functions appear early, in Chapter 2, and so are available during the first term for applications and illustrations of the full chain rule (rather than just the general power rule). Besides the traditional logarithmic and exponential topics, Chapter 6 includes an optional section on log-log and semi-log graphing, techniques widely used in many fields but whose mathematical foundation most students never see. This material provides an easy but significant application of logarithmic functions and their properties, one that complements the more substantial applications presented earlier in the chapter. In keeping with the theme of the usefulness of calculus, elementary differential equations are also introduced early (Section 3.10) and are used in Chapter 6 in connection with some applications of exponential, logarithmic, and other transcendental functions. Chapter 15, which continues the study of differential equations through second-order linear equations and their applications, can be covered at any point after the completion of Chapter 6. The discussion of infinite series in Chapter 9 emphasizes the idea of finite approximation of irrational quantities. The latter theme, first discussed in the opening section, culminates in the development of Taylor polynomials and Taylor series.

The material on three-dimensional analytic geometry and calculus of several variables is approached from the vector point of view. This parallels the notation students see in concurrent science and engineering courses and also puts into sharp focus the analogies between single-variable and multivariable calculus. For example, the gradient is introduced as the multidimensional version of the ordinary derivative of a one-variable function. By looking for such similarities to single-variable calculus, the student is constantly reminded of the latter and thereby helped to deepen his or her mastery of it. The primary emphasis in Chapters 10 through 14 is on computation rather than theory, however. Sophisticated concepts like vector spaces, linear functions and transformations, and Hessian matrices are not introduced.

The computer-generated sky on the cover serves as a backdrop to highlight Stonehenge. It suggests the illuminating potential of computers for bringing many of the key ideas of calculus into sharper focus. The text tries to realize this potential by using computer output to illustrate topics like limits, differentiation, root finding, optimization, integration, Euler's method, calculation of values of the natural logarithm function, Taylor polynomials, and others. Throughout the book, there are optional exercises marked PC. They are designed for programmable calculator or personal computer solution, but are also suitable for a mainframe system or, in many cases, for a nonprogrammable calculator. In addition, there are a number of optional subsections that discuss numerically-oriented topics, and there are comprehensive discussions of numerical integration and polynomial approximation of transcendental functions. These features have been included because of the greater prominence numerical methods have assumed with the spread of affordable programmable calculators and microcomputers. But they are optional enhancements which can be covered in varying degrees of detail or omitted altogether without affecting later work. Indeed, the text has been successfully class-tested in a traditional course that placed little emphasis on numerical matters, in another that integrated programmable calculators, and in one given in conjunction with a computer laboratory section. (Ancillaries designed for use with microcomputers are described below.)

Stonehenge is one of the oldest surviving monuments to humanity's quest to understand our world. The persons who designed and erected its monoliths, and the mathematics they used, are unknown to us. But not so the great mathematicians who have built calculus, and their mathematical ideas. The book contains numerous historical notes to acknowledge their contributions, so that the reader can perceive the continuing evolution of mathematics as an area of human knowledge.

Ancillaries

There are several supplements available to meet special needs and widen the possible uses of the text.

- To assist students in fully mastering the text, a ***Study Guide*** has been prepared by George F. Feissner of the State University of New York College at Cortland. It outlines the content of each section in the form of statements to be filled in by students, to help them organize the material. It also provides additional worked-out examples that isolate the individual skills needed to successfully work the problems in the text. Those worked-out problems are annotated with suggestions on how to analyze, set up, and solve calculus problems. In addition, there are practice problems for the students to test their understanding as they study, and each chapter concludes with a *Self Test* designed to help them gauge their readiness for a chapter test.

- The ***Instructor's Manual*** contains sample outlines and examinations and offers section-by-section suggestions on efficient use of the text. Also included are sample BASIC programs for the numerical methods mentioned in the text. The guidance it provides about choosing material for class discussion is intended especially to assist teaching assistants, part-time instructors, visiting professors, and others new to the calculus classroom.

- A microcomputer diskette that is entitled ***Calculus (Kemeny/Hurley)*** has been specially prepared to accompany the text by John G. Kemeny of Dartmouth College and is available from True BASIC, Inc., 39 South Main Street, Hanover, New Hampshire 03755 (telephone: (800)TR-BASIC or, in New Hampshire, (603) 643-3882). It features several powerful general programs for calculus that provide striking illustrations of such things as tabulating and graphing functions, computing their limits, finding formulas for and plotting their tangents, finding extreme values over intervals, integrating them, using L'Hôpital's rule to evaluate indeterminate limits, and finding formulas for Taylor polynomials and the solution of second-order linear homogeneous differential equations. It also contains custom-designed programs for use in solving the text's PC problems.

- An ***Instructor's Solution Manual*** has been prepared by John T. Hardy, Jr., of the University of Houston and Jeffrey J. Morgan of Texas A & M University in collaboration with Paul R. Fallone, Jr., of the University of Connecticut at Hartford. It contains worked-out solutions of all the exercises and is available on a complimentary basis to those who adopt the text. The ***Student's Solution Manual,*** a shortened version consisting of the solutions to every fourth exercise, is available for sale to students if authorized by a professor who adopts the text.

- The forthcoming paperback ***Computer Laboratory Manual for Calculus*** discusses construction of structured computer programs for calculus and contains some sample BASIC programs for the numerical methods mentioned in the text and exercises. It is organized into chapters corresponding to material that might be presented throughout the year in a computer laboratory section of an introductory calculus course.

In closing, it is a happy duty to express my appreciation to my editors, Heather Bennett and Kevin Howat, and their colleagues, Richard Giggey, Richard Greenberg, Hal Humphrey, Marta Kongsle, Lisa Mirski, Catherine Read, Stephen Rutter, Anne Scanlan-Rohrer, Ruth Singer, and Ira Zukerman of Wadsworth Publishing Company for the support, encouragement, and hard work they have contributed to the project that culminated in publication of this text. The uniformly high integrity, professional skill, and personal dedication of the people at Wadsworth—who truly live their professed desire to be partners with their authors in enhancing higher education—have benefitted both me and this book in no small degree.

James F. Hurley

Storrs, Connecticut
July, 1986

Acknowledgements

It is a pleasure to acknowledge the help and support of the many individuals who have contributed so much to the publication of this text. Cecile N. Hurley supplied and assisted with the writing of numerous applications, and also patiently and faithfully typed the entire manuscript and its revisions, often from copy whose legibility was at best marginal. Joseph Hurley compiled and typed all the answers, and made computer plots of many of the graphs in the answer sections. Regina Hurley provided much clerical and collating assistance. My entire family cheerfully bore the many burdens that grew out of the time I so often devoted to the text instead of to them.

Since 1962, I have taught calculus from texts authored by many mathematicians, and have consulted numerous others while preparing lectures. All of these have helped me shape my view of the material presented here, and I am indebted to them all.

The students at the University of Connecticut and the University of Michigan at Flint who class-tested the text in manuscript form deserve thanks for the valuable feedback they provided to improve the book. The encouragement of my students over the years has played a major role in my decision to start, and persevere with, the writing of my texts.

Charles Paskewitz developed a customized three-dimensional surface plotting program for the IBM Personal Computer and Hewlett-Packard HP-7475A plotter. That was used to generate most of the preliminary illustrations in Chapters 11 and 12 and the corresponding part of the answer section. I am very grateful for his dedicated work, whose high caliber materially improved the visual quality of the manuscript. The final art for the text was produced from a Hewlett-Packard HP-86 computer driving a Hewlett-Packard HP-7550 plotter. The software used was developed by George E. Morris of Scientific Illustrators of Champaign, Illinois. Accuracy of the plots is to within a tolerance of approximately 0.001 in. As will be apparent upon examining the art, Mr. Morris's skills have significantly enhanced the appearance and quality of the text.

The accuracy reviewers—Paul R. Fallone, Jr., of the University of Connecticut, Hartford; Stuart Goldenberg of the California State Polytechnic University,

San Luis Obispo; John T. Hardy, Jr., and Jeffrey J. Morgan of the University of Houston; Eleanor Killam of the University of Massachusetts; Thomas J. Kyrouz of Salem State College; Robert McFadden of Northern Illinois University; Jerry M. Metzger of the University of North Dakota; Karl Seydel of Stanford University; Ken Seydel of Skyline College; Donald R. Snow of Brigham Young University; Michael R. Stein of Northwestern University; and Joseph Hurley—are responsible for the elimination of many errors. Jay Freedman of Saunders College Publishing also contributed a number of improvements. I am grateful to Joseph Hurley, Maripaz Nespral, Miriam Nespral, and Elliott Derman for their assistance in compiling and editing the index. All the reviewers and class-testers listed below deserve special thanks for the many suggestions and corrections they contributed. Remaining shortcomings are, of course, solely my responsibility.

Robert A. Bix
University of Michigan, Flint

Daniel J. Britten
University of Windsor

Sam Councilman
California State University, Long Beach

John P. D'Angelo
University of Illinois, Urbana-Champaign

George F. Feissner
State University of New York, Cortland

Francis G. Florey
University of Wisconsin, Superior

Norman J. Frisch
University of Wisconsin, Oshkosh

Frederick S. Gass
Miami University

Colin C. Graham
Northwestern University

Charles W. Groetsch
University of Cincinnati

Nathaniel Grossman
University of California, Los Angeles

Shelby K. Hildebrand
Texas Tech University

Daniel S. Kahn
Northwestern University

Hudson V. Kronk
State University of New York, Binghamton

Eleanor Killam
University of Massachusetts

Lawrence D. Kugler
University of Michigan, Flint

H. E. Lacey
Texas A & M University

Robert H. Lohman
Kent State University

Robert McFadden
Northern Illinois University

Renate McLaughlin
University of Michigan, Flint

Weston I. Nathanson
California State University, Northridge

Bruce P. Palka
University of Texas, Austin

Anthony L. Peressini
University of Illinois, Urbana-Champaign

Richard C. Randell
University of Iowa

Thomas W. Rishel
Cornell University

John T. Scheick
Ohio State University

Keith W. Schrader
University of Missouri, Columbia

Mark Smiley
Auburn University, Montgomery

Harvey A. Smith
Arizona State University

B. David Stacy
Oregon State University

Michael R. Stein
Northwestern University

Harvey B. Keynes
University of Minnesota

Richard B. Thompson
University of Arizona

Antoinette Trembinska
Northwestern University

Clifford E. Weil
Michigan State University

Harold N. Ward
University of Virginia

Beverly H. West
Cornell University

I am also grateful to the following colleagues at the University of Connecticut, who contributed many helpful suggestions for improving the presentation of the calculus of several variables, based on their teaching experience from my text *Intermediate Calculus.*

Kinetsu Abe
Richard K. Beatson
Kathy M. Bradunas
M. Esperanza S. Cayco
Gary Cornell
Louis J. DeLuca
Thomas J. Dunion, Jr.
Paul R. Fallone, Jr.
Andrew H. Haas
James S. Hefferon
John Kaliongis
Soon-kyu Kim
Gerald M. Leibowitz
Manuel Lerman
Oscar I. Litoff
Andrew Miller
Jerome H. Neuwirth
James H. Schmerl
Stuart J. Sidney
Jeffrey L. Tollefson
Edmond C. Tomastik
Charles I. Vinsonhaler
Murray Wachman
William J. Wickless
Elliot S. Wolk

Finally, it is a pleasure to thank the following users of *Intermediate Calculus* and *Multivariable Calculus* who kindly sent me suggestions for improving certain exposition.

Betty Kennedy
University of Victoria

James Riddell
University of Victoria

David J. Leeming
University of Victoria

Yong-tae Shin
Keimyung University

Francisco Parisi-Presicce
University of Southern California

James D. Stewart
McMaster University

Foreword to the Student

This book is written for you to read and learn from. This *Foreword* offers some suggestions on how to get the most out of the features that have been included to help you as you learn calculus. When you have completed your calculus study, you may find the text useful to keep for reference. Many students find it easiest to review a topic directly from the source where they first learned it.

Examples and Exercises

The *essential* ingredient in mastering calculus is working as many of the exercises as possible. Don't kid yourself. Anybody who has passed a precalculus course can *understand* the concepts of calculus and follow the solutions of the examples in the text. In fact, a careful effort has been made to avoid omitting any computational steps in those solutions. But that is ***not*** the same as learning to do calculus yourself, which can come only through *extensive practice with exercises.* Such practice is needed not only to learn the mechanical processes, but also to become proficient enough in using them so that you can finish examinations in the allotted time.

The examples worked out in the text illustrate how the mathematics presented in each section is applied to solve concrete problems, in calculus and in other fields of study. The examples are designed to prepare you to work the exercises, because most of those resemble the examples or are based on ideas discussed there. To build the skill and confidence needed for quizzes and tests, try to do as many of the exercises as you can without referring back to the worked examples.

Some exercises, especially those toward the end of each list, are more challenging. They are designed to let you dig beneath the surface: to think about the material more deeply and thereby to improve your mastery of it. Often hints are given for those problems. Whether or not they are formally assigned, you should try some of them in order to achieve the best proficiency you can in calculus.

Organization

The text is organized into chapters and sections. The first section of each chapter gives a general introduction to the upcoming topics. Each later section typically begins with one or two paragraphs that set the scene for what follows. They may pose a problem whose solution requires new methods or they may focus on a previous topic that will be explored further in the section.

Next come any new definitions that may be necessary to precisely state the results or procedures to be presented. Those statements specify exactly when the techniques under development can be used. Pay careful attention to them, to avoid the frustration many students experience when they try to apply a technique to a case where it is inapplicable!

Optional material is often separated from the main part of the text in subsections. Your instructor will indicate which of those subsections should be studied in detail.

Proofs Mathematics is a deductive science, whose results are provable from basic axioms and earlier theorems by using the rules of logical inference. Throughout the text, proofs of theorems are interspersed with the definitions, computational rules, and examples. Their primary purpose is to indicate how key results are derivable from previously established or assumed facts. They are written in enough detail for you to follow them step-by-step should your instructor advise you to study them carefully. Mastery of the proofs, however, is not a prerequisite to doing most of the exercises or to learning the subsequent computational methods. While reading them and understanding them can help you build a more solid foundation on which to learn computational calculus, many first courses do not put major emphasis on them.

Answers Answers to all odd-numbered computational problems appear at the end of the text, in Appendix IV. If one of your answers differs from that in the answer section, don't simply assume that you are wrong and need help to proceed. Instead, carefully check your work as you would if there were a discrepancy in answers with a fellow student. Such checking is one of the most effective ways of thoroughly learning calculus.

Review Sections Each chapter ends with a review section called *Looking Back*, a short summary of the chapter that puts its contents into perspective. Included are a *Chapter Checklist*, which gives the principal terms from the chapter, and a set of *Review Exercises* to help you judge how well you have learned the major problem-solving techniques of the chapter.

Inside Covers The inside front cover contains lists of key formulas from algebra, geometry, and trigonometry that are needed in calculus. Refer to it whenever an example uses an algebraic technique or geometric or trigonometric formula that isn't familiar. There is also a list of common algebraic errors, titled *Algebraic Pitfalls*. In checking your work for possible mistakes, it may be helpful to refer to this list to make sure that you have not fallen victim to one of those pitfalls, which are often very difficult for students to spot in their work. The inside back cover contains an index of symbols and notation used throughout the book, arranged by first page of occurrence of the symbols.

Format Definitions, lemmas, theorems, and corollaries are numbered sequentially within each section, with a double number that gives the section and item number of each numbered statement. For example, Definition 4.3 is the third numbered item in Section 4. The most important statements and formulas are boxed or printed in color. When new terms are introduced, they are printed in ***bold print.*** Examples are numbered separately, starting from Example 1 in each section. When two or more steps are combined on a single line of an example's solution, the steps are linked by a colored arrow →. A colored box ■ marks the end of the solution to each example. Proofs end with the symbol QED. That is the abbreviation of the Latin phrase *quod erat demonstrandum*, meaning *which was to be proved.* Until the 1800s, most mathematics texts were written in Latin, and that phrase is the Latin translation of the Greek phrase ὅπερ ἔδει δεῖξαι, with which Euclid ended his proofs more than 2000 years ago.

Mathematics is a very old subject rich in impressive discoveries, brilliant breakthroughs, and illustrious people. The cover of the text shows a digitized image of Stonehenge, where the ancient monoliths arranged with precise attention to key cyclical events have long intrigued those who have gazed upon this scene. The design of that arrangement constitutes an ancient application of mathematics to better understand the world around us. The digitized image suggests the contemporary interaction between calculus and the modern world that you will learn of in this text as you read and study the applications of calculus and the algorithms for performing some of its calculations on electronic devices. The *Historical Notes* you will find throughout the text present information about some of the men and women who contributed to the development of calculus and related areas, a process that, as we suggested, continues to this day.

Study Aids

To provide additional help in learning from the text, supplementary books and software are available.

- A ***Study Guide***, prepared by Professor George F. Feissner of the State University of New York College at Cortland, spotlights each section's major concepts and also provides strategic analysis of the steps needed to work the exercises for that section. Its suggestions about the exercises are followed by practice problems that focus on the specific skills needed to become proficient at solving calculus problems. Each chapter ends with a *Self Test*, a multiple-choice examination to help you gauge your readiness to take an examination on the material of that chapter.

- A ***Student Solutions Manual*** has been prepared by Professors John T. Hardy, Jr., and Jeffrey J. Morgan of the University of Houston. It contains solutions to every fourth problem and can be stocked by your college bookstore on your instructor's request.

- A disk entitled ***Calculus (Kemeny/Hurley)*** has been prepared by Professor John Kemeny of Dartmouth College, one of the inventors of the BASIC programming language. It contains a number of powerful general programs that illustrate major concepts of calculus quickly and clearly on a microcomputer screen, as well as specially designed programs for use in working the PC problems in the text. Those are exercises intended for solution on personal computers or programmable calculators, which extend the scope of calculus beyond hand calculation. Many problems in science and industry that are solved using calculus are too complex for hand solution, and the PC problems give you a chance to use modern electronic devices in the way they are actually used to apply calculus in contemporary work. Versions of this software are available for the IBM PC and computers that are compatible with it (most of those that use the MS-DOS operating system), the Apple Macintosh, and the Commodore Amiga.

Finally, a word about how to read a mathematics text: *actively!* Read with a pencil in your hand and paper at your side. Whenever the text does a calculation, work through it on your own, checking that you understand and can reproduce the steps shown in the book. Don't expect to see everything immediately without any thought. If calculus were *that* easy, then calculus courses would be short and calculus books thin. However, this book is written so that by performing routine algebraic calculations you can follow derivations presented or solutions given in the text. Such active reading is the best way to understand calculus and build confidence in using it. It also sharpens your algebraic skills, which are indispensable in mastering this subject.

Many students find it helpful to read through each section before it is discussed in lecture. That helps them get an idea of what the lecture will cover and often reveals points they may want to ask questions about during class as their instructor discusses the material. Other students prefer to let their instructor introduce them to the material, and then study the text and their notes after class. They find that combination often helps them understand points they may not have fully grasped the first time they saw them. Both approaches involve at least one careful reading of the text. Try each one to find which works better for you.

Do you have a willing mind, a sharp pencil, and a blank pad of paper ready? Then let's get started!

CALCULUS

PROLOGUE What Is Calculus?

Calculus is a part of mathematics that was developed to analyze change, the fundamental fact of life. Precalculus mathematics studies fairly static relationships. For instance, only elementary arithmetic is needed to compute the distance traveled by a car in 3 hours if its speed is constantly 50 miles per hour. Elementary algebra is enough to find how long it will take that car to overtake a second one traveling at a constant speed of 30 miles per hour, if the second car started its trip one hour earlier from the same point as the first. Similarly, it is easy to find the average speed of a car that travels 150 miles in 3 hours, but how is that average related to the car's speedometer reading at any particular time during the trip? Calculus provides a simple answer. Its very origin lies in the study of motion, whose description involves one of the principal concepts of calculus, the *derivative*. The derivative will enable us to solve all of the following problems.

1. An apartment building has 250 units. At a monthly rent of \$200.00 per unit, it has no vacancies. Management estimates that, on average, for each \$5.00 increase in rent, one unit becomes vacant. What rent should be charged to maximize monthly revenue?
2. You are challenged to blow up a balloon with enough force to increase its radius at the rate of 1/2 inch per second. How much air per second must your lungs pump into the balloon to maintain that rate of radial increase when the radius is 4 inches?
3. Two quantities x and y are related by the equation

   $$y^7 - 11x^5y^6 - x^{15} + 11 = 0.$$

   Noting that the point $(x, y) = (1, 1)$ satisfies this equation, how can you estimate the value of y when $x = 1.002$?
4. The function $f(x) = x^5 - 2x - 1$ has value 0 at some point a between $x = 1$ and $x = 2$. Find an approximation to a that is correct to five decimal places.

FIGURE 1.1

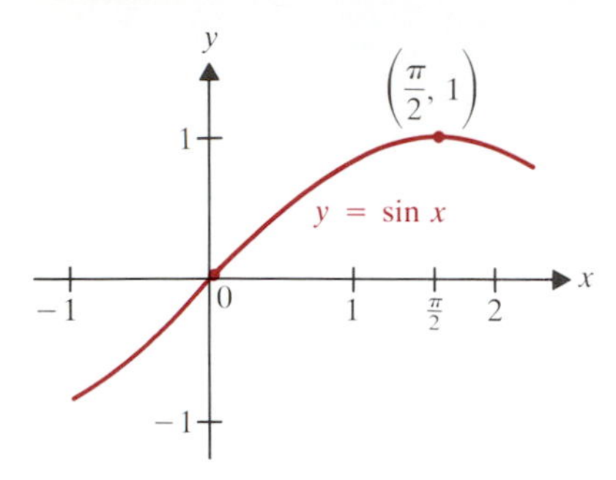

FIGURE 1.2

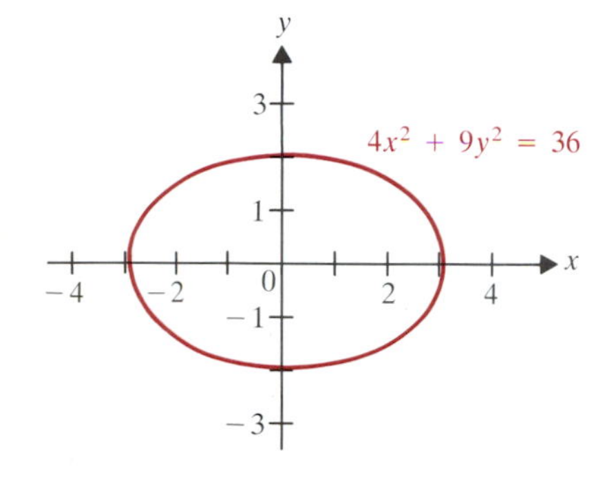

FIGURE 1.3

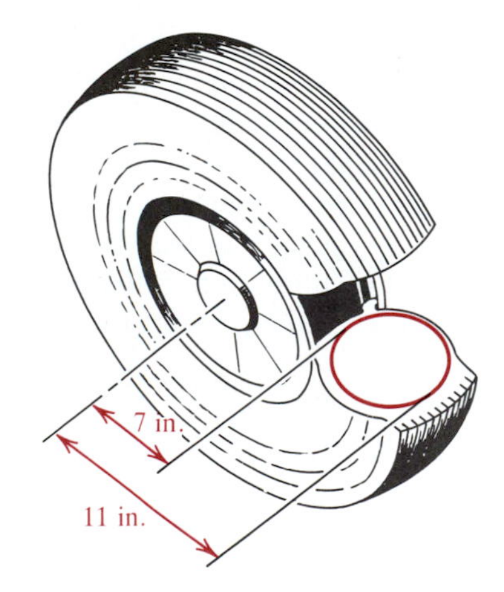

5. A driver in a car traveling 50 miles per hour applies the brakes, producing a steady deceleration of 20 feet per second each second. How long will it take the car to stop, and how far will it travel before coming to a stop?

Before getting to the derivative, we lay some groundwork in Chapter 1, which reviews background material from precalculus and introduces the fundamental concept of *limit*. (The derivative is a particular kind of limit.)

After derivatives have been explored, we turn in Chapters 4, 5, and 7 to another major topic of calculus, the *integral*, which is another kind of limit. We will use the integral to solve the following problems.

6. The rate of population increase in the United States is slackening. How can census data be used to estimate the long-term trend for the population?
7. A particle moves along the graph of $y = \sin x$ from $x = 0$ to $x = \pi/2$. What is the length of the particle's path? (**Figure 1.1.**)
8. An office is maintained at a constant temperature of 21°C. At 2:00 PM, the body of a murder victim is found in the office. The coroner arrives at 3:00 PM and measures the body's temperature as 31°C. An hour later, its temperature has dropped to 29°C. If the body temperature was 37°C at the time of death, then determine when the murder took place.
9. Find the area enclosed by the ellipse $4x^2 + 9y^2 = 36$. (**Figure 1.2.**)
10. A tire on a subcompact car has inner radius 7 inches and outer radius 11 inches. Assuming that the cross sections of the tire are circular, estimate the volume enclosed by the tire. (**Figure 1.3.**)

In Chapter 6, the previously developed calculus is extended to wider classes of functions: logarithmic, exponential, and inverse trigonometric functions.

Chapter 8 considers some analytic geometry in the plane, including an important alternative to the rectangular Cartesian coordinate system. In Chapter 9, the problem of adding infinitely many real numbers is explored. This surprising concept turns out to be closely related to an important procedure for approximating complicated functions by those with the simplest calculus—polynomial functions.

The rest of calculus considers functions of more than one variable, and their differentiation and integration. The geometric backdrop of the plane, which is used in the calculus of functions of one variable, is replaced by the three-dimensional coordinate space of ordered triples of real numbers. The relationship between derivatives of functions and the functions themselves, already introduced in elementary calculus via integration, is studied more thoroughly in differential equations.

Calculus has a long and rich history. As we proceed, we will include a number of notes about its development. A more detailed discussion is given in *The Historical Development of the Calculus* by C. H. Edwards, Jr. (Springer–Verlag, New York, 1979).

CHAPTER 1 Functions and Limits

1.0 Introduction

This chapter begins with a review of absolute value and inequalities, the Cartesian coordinate system in the plane, and the various forms of the equation of a straight line. The central concept in studying lines is *slope*, which is intimately related to differential calculus. Lines are often used to approximate more complicated curves, and the derivative calculates the slope of an approximating line to a curve. We complete our review of precalculus in Section 3 by considering functions and their graphs.

The study of calculus begins in Section 4, where the idea of limit is introduced. It describes the behavior of the values $f(x)$ of a function f as x varies and so is well suited to discussing the phenomenon of change. Section 5 presents several useful properties of limits that help simplify the calculation of limits of elementary functions. Section 6 broadens the meaning of limit to make it apply to a wider class of functions. The final section introduces continuous functions, which comprise one of the principal classes of functions encountered in calculus and its applications. Continuity is defined and studied using the facts about limits presented in the preceding three sections.

Section 8 is the review section of the chapter. Its chapter summary and list of major topics provide an overview of the chapter. The review exercises are study aids to help you measure your mastery of the chapter's topics.

Besides the material reviewed in the first three sections, successful study of calculus requires manipulative skills from precalculus such as factoring, solving equations using the quadratic formula, and performing algebraic operations with polynomials, radicals, and exponents. If you are rusty in any of those areas, then you should have a precalculus text available for reference and review as needed.

1.1 Absolute Value and Inequalities

Precalculus requires the set **R** of real numbers—for instance, to solve a quadratic equation as simple as

$$x^2 - 2 = 0. \tag{1}$$

Its roots are the two irrational numbers $\sqrt{2}$ and $-\sqrt{2}$. (As you probably recall, they cannot be expressed as the quotient m/n of two integers and so are not *rational* numbers.)

When you evaluate $\sqrt{2}$ on your calculator, however, you see something like 1.4142136, which *is* a rational number. Indeed,

$$1.4142136 = \frac{14142136}{10000000}.$$

Recall that the decimal expansion of any rational number either is *finite* or is an infinite *repeating* decimal. For instance,

$$\frac{1}{2} = 0.5 = 0.5000000\ldots, \qquad \frac{3}{11} = 0.272727\ldots, \qquad \frac{11}{24} = 0.4583333\ldots.$$

If you evaluate 3/11 on your calculator, you will see something like 0.2727273 or 0.27272727, which are 7- or 8-place *approximations* to 3/11. By contrast, irrational numbers like $\sqrt{2}$, $\sqrt{3}$, $\sqrt{5}$, and π have infinite *nonrepeating* decimal expressions. For instance,

$$\sqrt{2} = 1.41421356\ldots,$$

where no sequence of repeating digits occurs. The calculator value 1.4142136 is simply a 7-place approximation to $\sqrt{2}$: a *rational approximation* to an *irrational* number. Such approximations enable the calculator (or a person) to do decimal calculations involving irrational numbers. Keep in mind that many answers obtained by hand or by calculator in decimal form are not exact, but rather are approximate and subject to error in their last few decimal places.

FIGURE 1.1

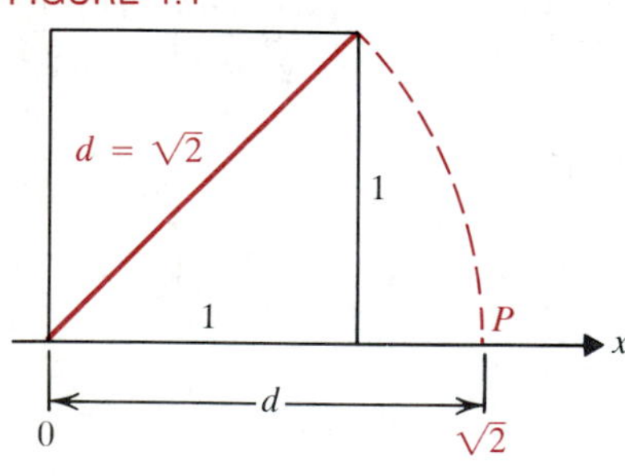

For our purposes, a matter of greater importance is that the rational numbers cannot be used as the coordinates of all the points on a line. If we choose an origin 0 and a unit distance, then there are points, like P in **Figure 1.1,** on the line *without* rational coordinates. But the set **R** of real numbers, which includes the irrationals like $\sqrt{2}$ as well as the rationals, is in *one-to-one correspondence with the points on a line.* This means that every real number x corresponds to one and only one point on the line, and every point on the line corresponds to exactly one real number x. For this reason, we label points on the line by the real numbers corresponding to them, as in **Figure 1.2.**

FIGURE 1.2

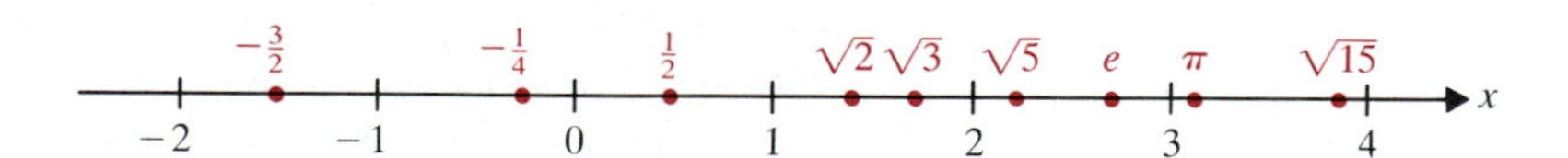

In calculus, we sometimes have to analyze inequalities. To review them, we begin by recalling that the inequality

$$a < b \qquad (\text{or } b > a)$$

between the real numbers a and b means that $b - a$ is a positive real number. We thus write $c > 0$ to express the fact that a certain number c is positive.

FIGURE 1.3

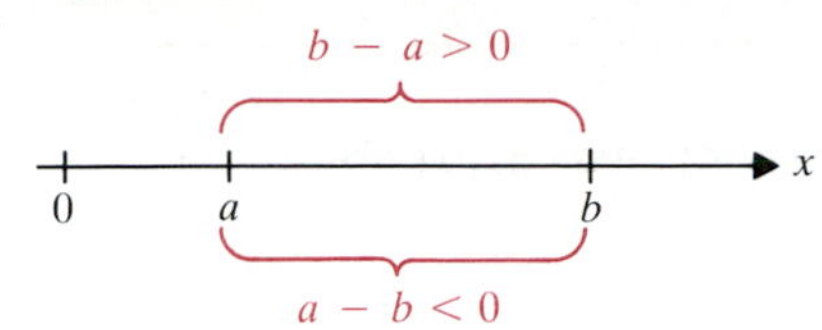

Remember that

$$a \leq b$$

means that either $a < b$ or $a = b$. (This is also written $b \geq a$.) Recall also that $a < b$ simply means geometrically that a is to the *left* of b on the number line, as illustrated in **Figure 1.3.** The basic properties of the relation *less than* are given by the following familiar rules.

> RULES OF INEQUALITY
>
> For all real numbers a, b, c, and d:
>
> ***Transitive Rule.*** If $a < b$ and $b < c$, then $a < c$.
>
> ***Additive Rules.***
>
> 1. If $a < b$, then $a + c < b + c$.
> 2. If $a < b$ and $c < d$, then $a + c < b + d$.
>
> ***Multiplicative Rules.***
>
> 1. If $a < b$ and $c > 0$, then $ac < bc$.
> 2. If $a < b$ and $c < 0$, then $ac > bc$.
> 3. If $a < b$ and $c < d$, and $a, b, c, d > 0$, then $ac < bd$.

Warning. Among these rules, the second multiplicative rule is often forgotten, especially when the multiplying factor may be either positive or negative. *Beware!* Multiplication by a negative number *reverses* the direction of an inequality. For instance,

$$2 < 3 \qquad \text{but} \qquad (-1)2 > (-1)3$$

since $-2 > -3$.

Our first example illustrates the use of the basic rules to solve linear inequalities.

Example 1

Solve the inequality $3 - 3x < 5 - x$.

FIGURE 1.4

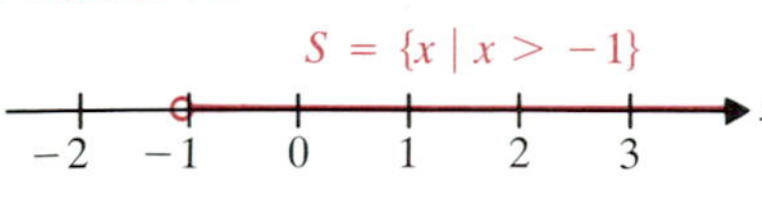

Solution. We add -3 and x to each side, and by the second additive rule get

$$-2x < 2.$$

Then we multiply each side by $-1/2$ and, remembering to reverse the inequality sign, obtain

$$x > -1.$$

Consequently, the solution set S consists of all real numbers to the right of -1 on the number line, that is, $S = \{x \mid x > -1\}$. (See **Figure 1.4.**) ■

The solution set S in Example 1 is more commonly written as the open interval $(-1, +\infty)$. The following definition and **Figure 1.5** review the various kinds of intervals on the real line.

1.1 DEFINITION

(a) The ***closed interval*** $[a, b] = \{x | a \leq x \leq b\}$.

(b) The ***open interval*** $(a, b) = \{x | a < x < b\}$.

(c) The ***half-open*** (or ***half-closed***) intervals $[a, b)$ and $(a, b]$ are

$$[a, b) = \{x | a \leq x < b\} \quad \text{and} \quad (a, b] = \{x | a < x \leq b\}.$$

(d) The ***open half-lines*** $(-\infty, b)$ and $(a, +\infty)$ are

$$(-\infty, b) = \{x | x < b\} \quad \text{and} \quad (a, +\infty) = \{x | x > a\}.$$

(e) The ***closed half-lines*** $(-\infty, b]$ and $[a, +\infty)$ are

$$(-\infty, b] = \{x | x \leq b\} \quad \text{and} \quad [a, +\infty) = \{x | x \geq a\}.$$

(f) $(-\infty, +\infty) = \mathbf{R}$, the entire ***real line.***

FIGURE 1.5

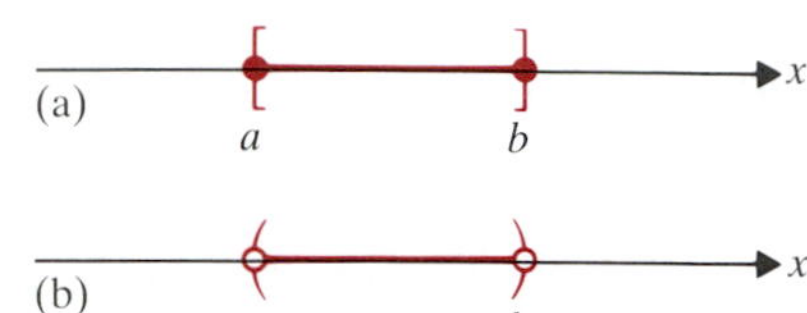

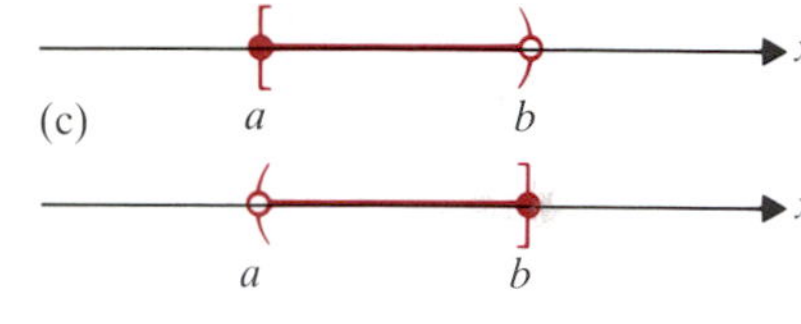

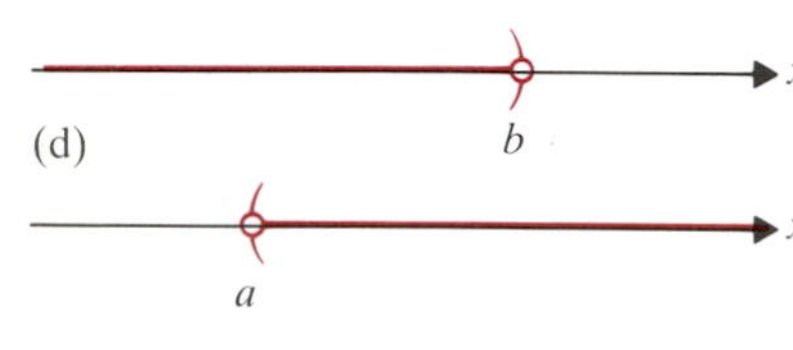

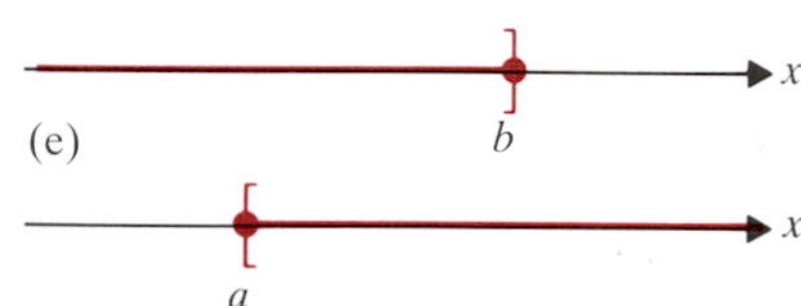

The points a and b are called ***endpoints*** of the intervals in Definition 1.1. Open intervals do not contain their endpoints, but closed intervals do. Take particular note that **the symbols $+\infty$** (read "plus infinity") **and $-\infty$** (read "minus infinity") **do not stand for real numbers.** Their use merely signifies that the corresponding interval is *unbounded*, that is, either a half line or the entire real line **R**. Note that -1 is not in the unbounded open interval $(-1, +\infty)$, the solution set S of Example 1.

In Definition 1.1 we have used double inequalities to describe the two simultaneous conditions on x in parts (a), (b), and (c). For instance,

$$a \leq x \leq b$$

means that x simultaneously satisfies $a \leq x$ and $x \leq b$. This is a symbolic way of saying that x lies between a and b, inclusive.

The notion of ***absolute value*** is very useful in working with inequalities. Geometrically, $|x|$ is simply the distance from the point x on the number line to the origin 0 and so is always a nonnegative number. This condition is expressed algebraically as

$$|x| = \begin{cases} x & \text{if } x \text{ is nonnegative} \\ -x & \text{if } x \text{ is negative.} \end{cases}$$

FIGURE 1.6

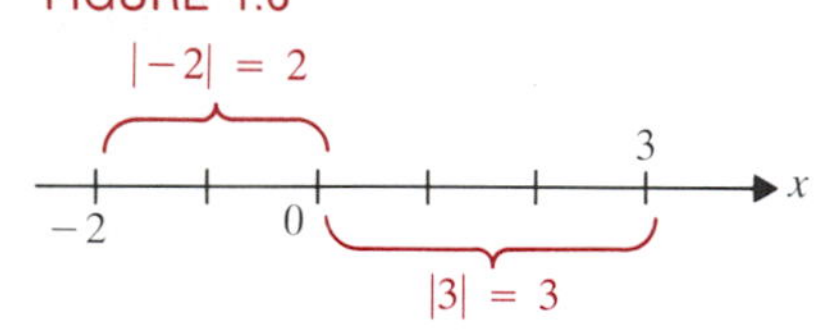

There is also a single defining equation for $|x|$. This involves $\sqrt{a}$ (read "radical a"), the *principal* (or *nonnegative*) *square root* of a. The facts that $\sqrt{3^2} = 3$ and $\sqrt{(-2)^2} = 2$ (see **Figure 1.6**) are instances of the important general formula

$$\sqrt{x^2} = |x| \tag{2}$$

Warning. Keep (2) firmly in mind. Calculus students sometimes mistakenly assume $\sqrt{x^2}$ is x in situations where x could be negative. But if $x = -2$, then

$$\sqrt{x^2} = \sqrt{4} = 2 = -x,$$

not x!

FIGURE 1.7

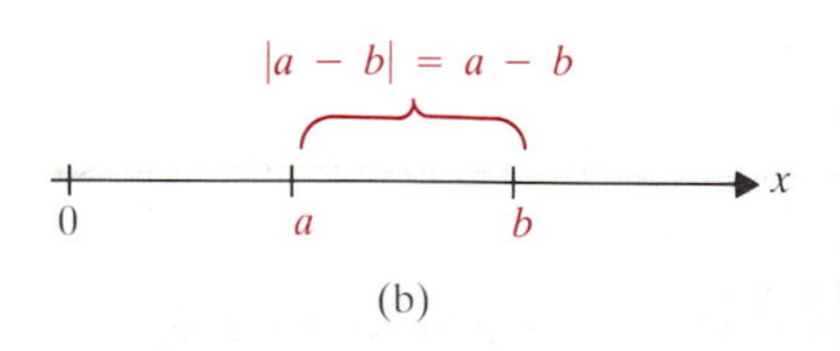

As **Figure 1.7** illustrates, $|a - b|$ is the distance between the points a and b on the real line. The most important algebraic properties of absolute value are given in the following theorem.

1.2 THEOREM

Let a and b be real numbers, and let $d > 0$. Then

(a) $|x| < d$ is equivalent to $-d < x < d$.

(b) $|x| > d$ is equivalent to $x > d$ or $x < -d$.

(c) $|x - a| < d$ is equivalent to $a - d < x < a + d$.

(d) ***Triangle Inequality.*** $|a + b| \le |a| + |b|$.
(e) $|ab| = |a|\,|b|$.
(f) $\left|\dfrac{a}{b}\right| = \dfrac{|a|}{|b|}$, if $b \ne 0$.
(g) $|a| = 0$ if and only if $a = 0$; $|a| > 0$ otherwise.
(h) $|a - b| = |b - a|$.

Partial Proof. (a) Since $|x|$ is the distance from the point x on the number line to the origin 0, $|x| < d$ is equivalent to x lying less than d units from the origin. Thus, $|x| < d$ is equivalent to x lying in the interval $(-d, d)$, that is, to $-d < x < d$.

(d) Since $|x|$ is always nonnegative, $x \le |x|$ and $x \ge -|x|$ for any real number x. Thus,

$$-|a| \le a \le |a| \qquad \text{and} \qquad -|b| \le b \le |b|.$$

Adding these inequalities term-by-term and using the second additive rule given before Example 1, we get

$$-(|a| + |b|) \le a + b \le |a| + |b|.$$

Then by (a),

$$|a + b| \le |a| + |b|.$$

(f) $$\left|\frac{a}{b}\right| = \sqrt{\frac{a^2}{b^2}} = \frac{\sqrt{a^2}}{\sqrt{b^2}} = \frac{|a|}{|b|}.$$

We leave the remaining parts of the proof as exercises (see Exercises 25–29). QED

The next example illustrates the usefulness of Theorem 1.2 in solving inequalities.

Example 2

Find the solution set of the given inequality.
(a) $|x^2 - 3| < 6$;   (b) $|3x + 1| \ge 10$.

Solution. (a) First, Theorem 1.2(a) says that $|x^2 - 3| < 6$ is equivalent to $-6 < x^2 - 3 < 6$. By the second additive rule, this in turn is equivalent to $-3 < x^2 < 9$. Since x^2 is always nonnegative, the condition $-3 < x^2$ is redundant. Dropping it and extracting principal square roots in the remaining half ($x^2 < 9$) of the inequality, we get

$$|x| < 3 \rightarrow -3 < x < 3,$$

by Exercise 27 and Theorem 1.2(a). The solution set is therefore $(-3, 3)$.

(b) From Theorem 1.2(b) and the additive and multiplicative rules, we have

$$\begin{aligned} 3x + 1 &\ge 10 \quad &\text{or}\quad 3x + 1 &\le -10,\\ 3x &\ge 9 \quad &\text{or}\quad 3x &\le -11,\\ x &\ge 3 \quad &\text{or}\quad x &\le -\frac{11}{3}. \end{aligned}$$

Thus the solution set consists of two half-lines that have no points in common, $[3, +\infty) \cup (-\infty, -11/3]$* and is shown in **Figure 1.8.** ■

FIGURE 1.8

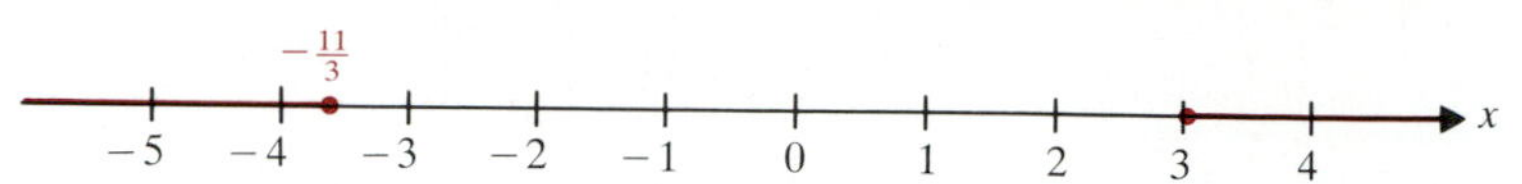

The most efficient way to solve nonlinear inequalities is to use factoring to reduce to several simultaneous linear inequalities. The strategy is to first use the rules to reduce the given inequality to one of the forms

$$r(x) \geq 0 \qquad \text{or} \qquad r(x) \leq 0,$$

where $r(x)$ is a rational expression, that is, a polynomial or quotient of polynomials. Then we factor $r(x)$ fully and *study the signs of the individual factors, which completely determine the sign of* $r(x)$. Our experience with linear inequalities of the form

(3) $$ax + b > 0,$$

shows that the sign of $ax + b$ is determined by whether x is to the left or to the right of $-b/a$ on the number line. This is illustrated in **Figure 1.9.** To obtain Figure 1.9(a), for which $a > 0$, note first that by the first additive rule, (3) is equivalent to

(4) $$ax > -b.$$

FIGURE 1.9

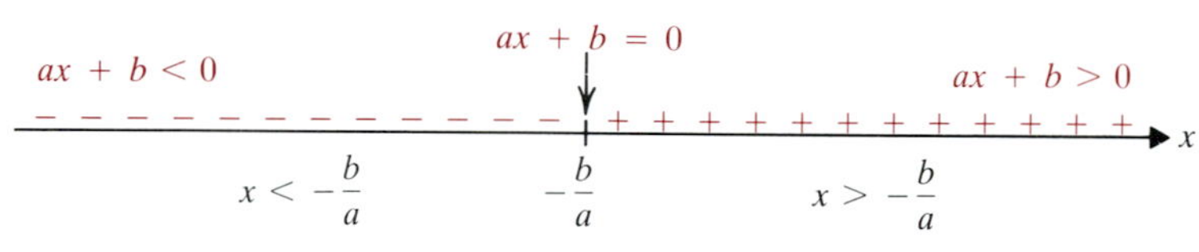

(a) The case $a > 0$

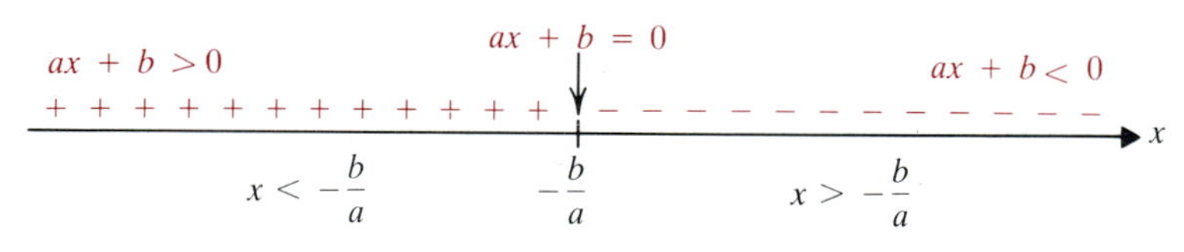

(b) The case $a < 0$

Then, because $a > 0$, the first multiplicative rule says that (4) is equivalent to

$$x > -\frac{b}{a}.$$

Hence, $x > -b/a$ is equivalent to $ax + b > 0$. Similar reasoning produces Figure 1.9(b). This discussion establishes the following fundamental property, which we use to solve inequalities.

> A linear factor $ax + b$ changes sign only at $x = -\dfrac{b}{a}$, where it takes on the value 0.

* The *union* $A \cup B$ of two sets A and B of real numbers consists of all real numbers x that belong to either set: $A \cup B = \{x | x \in A \text{ or } x \in B\}$.

The next two examples show how to exploit this fact to solve nonlinear inequalities.

Example 3

Solve for x:

$$3x^2 - 3x \geq 6 - 10x. \tag{5}$$

Solution. Adding $-6 + 10x$ to each side of (5), we get

$$3x^2 + 7x - 6 \geq 0.$$

This factors to give

$$(3x - 2)(x + 3) \geq 0. \tag{6}$$

Hence, (5) is equivalent to (6). We solve (6) by analyzing the signs of the linear factors $3x - 2$ and $x + 3$ over the complete real line. We split the real line at the key numbers $x = 2/3$ and $x = -3$, which are the only places where $(3x - 2)(x + 3)$ can change sign. The analysis can be conveniently carried out using the format of Table 1.

TABLE 1

x:	$(-\infty, -3)$	-3	$(-3, 2/3)$	$2/3$	$(2/3, +\infty)$ $\to$ **R**
$3x - 2$:	$-$		$-$		$+$
$x + 3$:	$-$		$+$		$+$
$(3x - 2)(x + 3)$:	$+$	0	$-$	0	$+$

Each row of the table represents the real line, split into the intervals $(-\infty, -3)$, $(-3, 2/3)$, and $(2/3, +\infty)$. In each of these intervals $(3x - 2)(x + 3)$ has constant sign, since change-of-sign occurs *only* at $x = 2/3$ and $x = -3$. To determine the sign of the factors on each interval, simply substitute a convenient number in the interval for x, such as -4 from $(-\infty, -3)$, 0 from $(-3, 2/3)$, and 1 from $(2/3, +\infty)$. In this way, we fill in the signs of the factors in Table 1 and then determine the sign of $(3x - 2)(x + 3)$ in the bottom row. The solution set for (6) consists of all points on the real line for which $(3x - 2)(x + 3)$ has either a + sign or is 0. We see from the table that the intervals corresponding to + are $(-\infty, -3)$ and $(2/3, +\infty)$. The solution set also includes the points $x = -3$ and $x = 2/3$, where $(3x - 2)(x + 3) = 0$. Thus the solution set is $(-\infty, -3] \cup [2/3, +\infty)$. ■

Table 1 not only gives the solution set for (6) but also gives a full report of the sign of $(3x - 2)(x + 3)$ over the entire real line. From it, for example, we also see that the solution set for $3x^2 - 3x < 6 - 10x$ is $(-3, 2/3)$. We will see in Chapter 3 that such tables can be very useful in calculus.

We next consider an inequality that involves quotients of polynomials.

Example 4

Solve for x:

$$\frac{x}{x + 2} > \frac{3}{x - 2}. \tag{7}$$

Solution. The first step is to reduce (7) to an inequality that involves just one rational expression. This is accomplished by the string of equivalent inequalities

$$\frac{x}{x+2} - \frac{3}{x-2} > 0, \qquad \frac{x(x-2)}{(x+2)(x-2)} - \frac{3(x+2)}{(x+2)(x-2)} > 0,$$

$$\frac{x^2 - 2x - 3x - 6}{(x+2)(x-2)} > 0, \qquad \frac{x^2 - 5x - 6}{(x+2)(x-2)} > 0,$$

$$\frac{(x-6)(x+1)}{(x+2)(x-2)} > 0. \tag{8}$$

To analyze (8) we make Table 2, which gives the sign of each linear factor over the entire real line. Here the key numbers are again the places where a linear factor changes sign: $x = 6$, $x = -1$, $x = -2$, and $x = 2$.

TABLE 2

x:	$(-\infty, -2)$	-2	$(-2, -1)$	-1	$(-1, 2)$	2	$(2, 6)$	6	$(6, +\infty)$ R
$x - 6$:	$-$		$-$		$-$		$-$		$+$
$x + 1$:	$-$		$-$		$+$		$+$		$+$
$x + 2$:	$-$		$+$		$+$		$+$		$+$
$x - 2$:	$-$		$-$		$-$		$+$		$+$
$\frac{(x-6)(x+1)}{(x+2)(x-2)}$:	$+$	undefined	$-$	0	$+$	undefined	$-$	0	$+$

The solution set of (8) is read from the bottom row of the table:

$$(-\infty, -2) \cup (-1, 2) \cup (6, +\infty). \quad ■$$

Take note in Example 4 that the points where sign changes occur are not solution points, because (7) is a strict inequality. At such points the rational expression either is 0 ($x = 6$, $x = -1$) or is not defined ($x = 2$, $x = -2$).

Exercises 1.1

In Exercises 1–22, solve the given inequality, expressing your answer in interval notation.

1. $2x - 3 < 5x - 6$
2. $4x + 7 > 2 + x$
3. $2 - x \le x + 5$
4. $2 - 3x \le 2x - 8$
5. $3 - 2x \le 8 - 3x$
6. $5 - 2x \ge x - 4$
7. $7 - 3x > x - 1$
8. $x + 3 < 3 - 2x$
9. $|3x - 1| < 5$
10. $|1 - 4x| \le 7$
11. $|2x + 11| \ge 9$
12. $|3x - 2| > 7$
13. $x^2 + 3x > 2x + 2$
14. $x^2 + 3x > 7x + 12$
15. $x^3 - 4x \ge 4 - x^2$
16. $9x - x^3 \le 45 - 5x^2$
17. $\dfrac{1}{x} > \dfrac{x+1}{x+2}$
18. $\dfrac{2}{x-1} < \dfrac{x}{x+2}$
19. $\dfrac{4x-3}{x+3} \ge \dfrac{2x-5}{x+1}$
20. $\dfrac{2x-3}{x+4} \le \dfrac{2x-1}{x-3}$
21. $|x^2 - 7| < 2$
22. $|x^2 - 4| \le 5$
23. For what values of x is $\sqrt{3x^2 + x - 2}$ a real number?
24. Repeat Exercise 23 for $\sqrt{\dfrac{2x-1}{x+5}}$.
25. **(a)** Prove Theorem 1.2(b). **(b)** Prove Theorem 1.2(c).
26. Show that if $0 < a < b$, then $a^2 < b^2$.
27. Show that if $a^2 < b^2$, then $|a| < |b|$.
28. Prove Theorem 1.2(e).
29. Prove Theorem 1.2(g).
30. Show that $\sqrt{2}$ is not rational. (*Hint:* Assume that $\sqrt{2} = m/n$ is in lowest terms. Obtain a contradiction by showing that 2 must divide both m and n.)
31. For all real numbers a and b, show that $a^2 + b^2 \ge 2ab$.
32. For all real numbers a and b, show that $a^2 + b^2 \ge -2ab$.
33. What can be concluded from Exercises 31 and 32 about $|ab|$?

Exercises 34–38 deal with the *Gibbs–Helmholtz equation*, an important tool in studying chemical reactions. It gives the free-energy change G' (the maximum amount of work that can be accomplished by a reaction at constant temperature and pressure). This depends on H', the change in heat content, on S', the change in entropy (randomness or disorder, such as that produced by the motion of molecules when changing from solid to liquid state), and on the Kelvin temperature T. The equation is

$$G' = H' - TS'. \tag{9}$$

34. In all *spontaneous reactions*, such as the rusting of iron, $G' < 0$. If $G' > 0$, energy must be added for the reaction to take place. What signs should H' and S' have so that a reaction is spontaneous at *any* temperature T? (Temperature measured on the Kelvin scale is *always* positive.)

35. Refer to Exercise 34. What signs should H' and S' have so that a reaction is nonspontaneous at *any* temperature T?

36. Refer to Exercise 34. What signs must H' and S' have for spontaneity to depend on the value of T?

37. Refer to Exercise 34. Suppose that $H' > 0$ and $S' > 0$. Give a condition on T that will make the reaction spontaneous.

38. Refer to Exercise 34. Suppose that $H' < 0$ and $S' < 0$, a situation that occurs rarely. Give a condition on T that will make the reaction spontaneous.

HISTORICAL NOTE

J. Willard Gibbs (1839–1903) was one of the first Americans to achieve prominence in the international scientific community. He was an important contributor to the theoretical development of thermodynamics, as well as chemistry, astronomy, and mathematical physics. The Gibbs–Helmholtz equation was contained in a long paper published during 1876–1878 that laid the foundations of chemical thermodynamics. He also contributed to the theory of infinite series and to the spread of vectors as important tools in mathematics and physics.

Hermann L. F. von Helmholtz (1821–1894) was a German physicist and physiologist who made important contributions to mechanics, electricity and nerve action, and hearing. He is credited with having formulated the principle of energy conservation in 1847. Unaware of Gibbs' slightly earlier work, he formulated the Gibbs–Helmholtz equation in 1882.

1.2 Coordinates, Lines, and Circles

The ordinary distance between two points P and Q on the real line with respective coordinates x_1 and x_2 is

$$d(P, Q) = |x_1 - x_2| = |x_2 - x_1|. \tag{1}$$

The ***directed distance*** from x_1 to x_2 is defined to be $x_2 - x_1$. This is also called ***increment*** (or *incremental change*) Δx in the variable x with initial value x_1 and subsequent value x_2:

$$\Delta x = x_2 - x_1 = \text{ending value of } x \text{ minus beginning value of } x.$$

Note: Δx is read "delta x" and is a *single symbol*. Do **not** confuse it with a product: the symbol Δ *never* stands for a mathematical quantity in this text.

As an illustration, figures compiled by the United States Department of Commerce show that during the first half of 1980, the annual rate R of retail sales in the United States (in billions of dollars) declined from $R_1 = 46$ to $R_2 = 42$ (approximately). The incremental change ΔR for that half-year was therefore

$$\Delta R = R_2 - R_1 = 42 - 46 = -4.$$

During the first quarter of 1981, R changed from approximately $R_3 = 43$ to approximately $R_4 = 45$. Thus the corresponding increment in R over that period was

$$\Delta R = R_4 - R_3 = 45 - 43 = 2.$$

As these illustrations show, Δx is positive precisely when x_2 is greater than x_1 (i.e., when x increases from its initial value). When x_2 is less than x_1 (i.e., when x decreases from its initial value), Δx is negative. When x_2 coincides with x_1 (i.e., when x does not change from its initial value), Δx is 0.

Just as real numbers are used to assign coordinates to points on a line, so every point P in the plane has one and only one pair of coordinates, and each pair of coordinates corresponds to exactly one point in the plane. Thus there *is a one-to-one correspondence between the set of all points P in the plane and the set of all ordered pairs* (x, y) *of real numbers.* This scheme for identifying points by their coordinates is called the ***Cartesian*** (or *rectangular*) ***coordinate system*** for the plane. We often refer to the coordinate plane as $\mathbf{R}^2$, the set of ordered pairs of real numbers. (See **Figure 2.1.**) Recall that the ***graph*** of an equation involving the variables x and y is the set of all points $P(x, y)$ whose coordinates satisfy the given equation. **Figure 2.2** shows the graph of the simple coordinate equations $x = c$ and $y = d$: lines parallel to the coordinate axes.

FIGURE 2.1

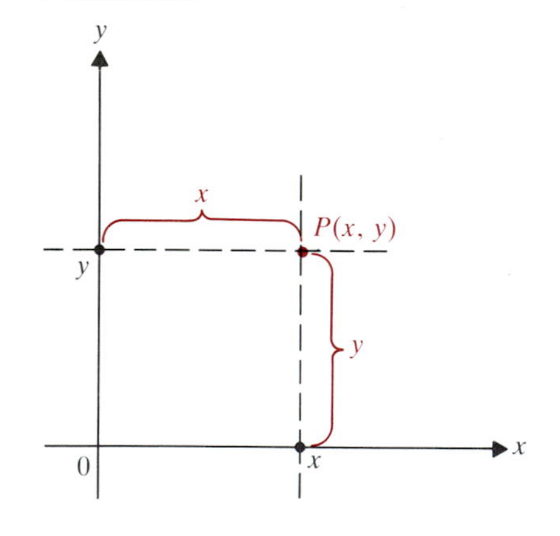

FIGURE 2.2

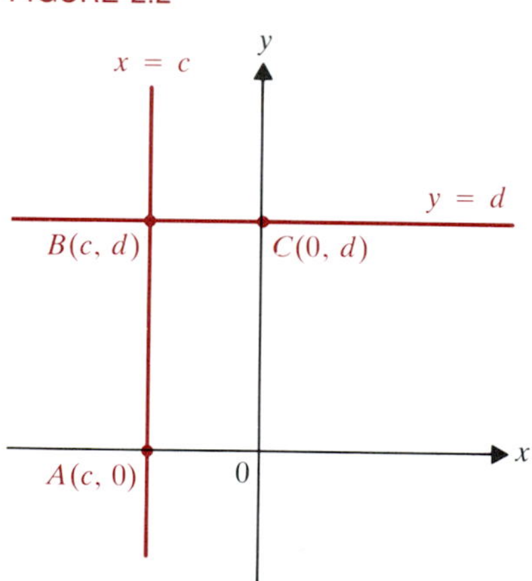

FIGURE 2.3

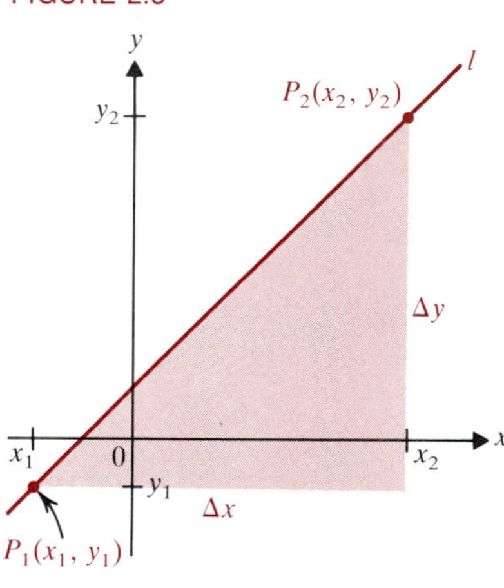

The most important concept associated with nonvertical lines is their ***slope.*** (See **Figure 2.3.**) If $P_1(x_1, y_1)$ and $P_2(x_2, y_2)$ are any two (distinct) points on a nonvertical line l, then recall that the *slope* of l is

$$m_l = \frac{\Delta y}{\Delta x} = \frac{y_2 - y_1}{x_2 - x_1}. \tag{2}$$

FIGURE 2.4

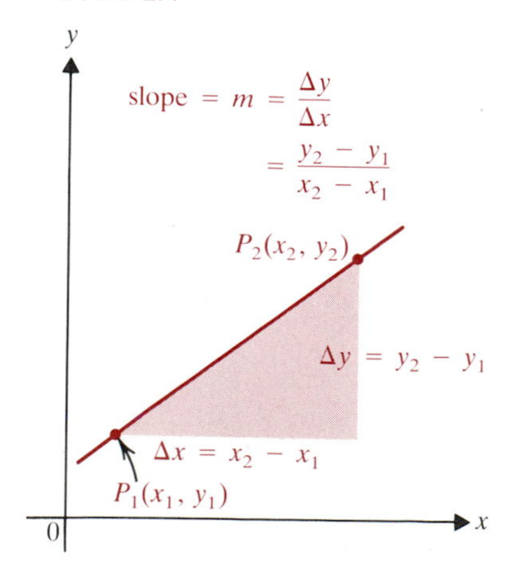

Example 1

Find the slope of the line through the points $(5, -2)$ and $(-1, 1)$.

Solution. $m = \dfrac{\Delta y}{\Delta x} = \dfrac{1 - (-2)}{-1 - 5} = \dfrac{3}{-6} = -\dfrac{1}{2}.$ ■

Equation (2) says that the slope of a line segment P_1P_2 is the ratio of the change in y to the change in x from P_1 to P_2. This is called the *average rate of change of y per unit change in x* over the interval $[x_1, x_2]$. (Indeed, *rate* is simply the anglicized version of the Latin *ratio.*) See **Figure 2.4.** Such an interpretation is frequently important in science. For instance, Celsius and Fahrenheit temperatures are related by a *linear equation*, that is, one whose graph is a straight line. As Celsius temperature changes from 0° to 100°, the Fahrenheit readings change from 32° to 212°.

Example 2

Find the average rate of change of Fahrenheit temperature F with respect to Celsius C by finding the slope of the corresponding line segment.

FIGURE 2.5

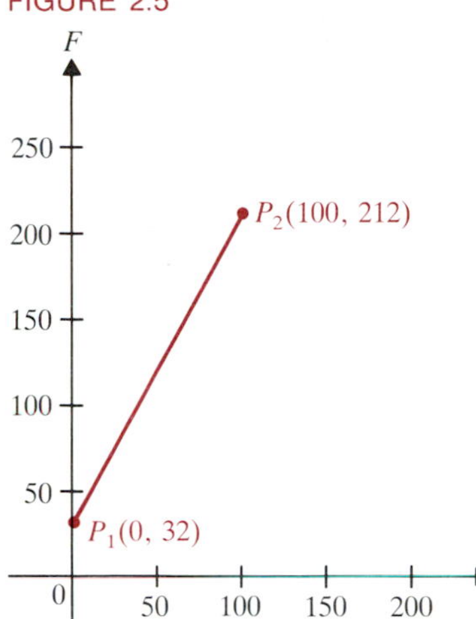

Solution. The line segment referred to is the one in **Figure 2.5** that joins $P_1 = (C_1, F_1) = (0, 32)$ to $P_2 = (C_2, F_2) = (100, 212)$. Its slope is

$$m = \frac{F_2 - F_1}{C_2 - C_1} = \frac{212 - 32}{100 - 0} = \frac{180}{100} = \frac{9}{5}.$$

Hence, the average rate of change of F relative to C over the interval from $C = 0$ to $C = 100$ is 9/5. This means that for an increase of one degree on the Celsius scale, there is an increase of slightly under 2°F. ■

From this, it is easy (Exercise 9) to derive the formula for converting from one temperature to another. Many students, in fact, find that formula easier to derive than to remember. Recall that if a point $P_0(x_0, y_0)$ lies on a line of slope m, then the line has the ***point-slope equation***

$$y - y_0 = m(x - x_0). \tag{3}$$

Example 3

Find the equation of the line in Example 1.

Solution. We have $m = -1/2$, and for P_0 we can use either $(5, -2)$ or $(-1, 1)$. The first choice gives

$$y - (-2) = -\frac{1}{2}(x - 5) \rightarrow y + 2 = -\frac{1}{2}x + \frac{5}{2},$$

$$2y + 4 = -x + 5 \rightarrow x + 2y = 1.$$

The second choice gives the same equation:

$$y - 1 = -\frac{1}{2}(x + 1) \rightarrow 2y - 2 = -x - 1 \rightarrow x + 2y = 1. \quad ■$$

To recognize that the two different choices of P_0 led to the same equation in Example 3, we put each equation into the ***general form*** of the equation of a line. That is an equation of the form

$$Ax + By = C. \tag{4}$$

It is easy to derive another useful form of the equation of a straight line from (3). Solving for y, we get

$$y = mx - mx_0 + y_0 = mx + (y_0 - mx_0).$$

FIGURE 2.6

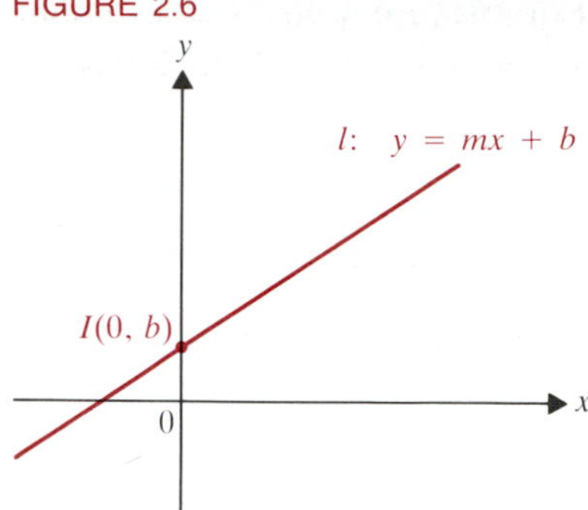

Letting $y_0 - mx_0 = b$ (a constant because y_0, m, and x_0 are constants) we get the ***slope-intercept form*** of the equation of l:

$$y = mx + b. \tag{5}$$

When $x = 0$, we see from (5) that $y = b$. Hence, $I(0, b)$ is the point where the line l crosses the y-axis. See **Figure 2.6.** The number b is called the ***y-intercept*** of l.

For instance, in Example 3 we have

$$x + 2y = 1 \rightarrow 2y = -x + 1 \rightarrow y = -\frac{1}{2}x + \frac{1}{2}.$$

Hence, the line has slope $-1/2$ and y-intercept $b = 1/2$.

Equation (5) is often useful in analyzing laboratory data.

FIGURE 2.7

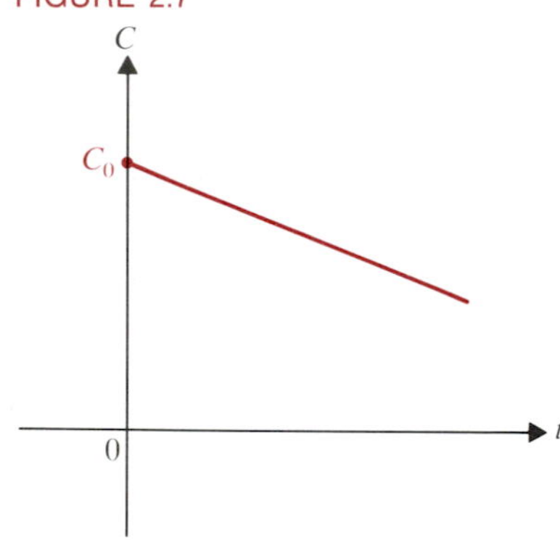

Example 4

Figure 2.7 shows a plot of the concentration C of a substance A involved in a chemical reaction over time. The slope is measured experimentally to be $-1/k$, where k is a positive constant associated with the reaction. What is the equation of the line?

Solution. By (5), the equation of the line in Figure 2.7 is

$$C = mt + b,$$

where b is C_0, the initial concentration of A. Since $m = -1/k$, we find

$$C = -\frac{1}{k}t + C_0 = C_0 - \frac{t}{k}. \blacksquare$$

Sometimes, as in the next example, it is possible to use the defining equation (2) of slope directly to obtain information about a line.

Example 5

A particle starts moving along the line in Example 1 at the point $P_1(3, -1)$. If it moves 2 units to the right to P_2, then what is its new position?

Solution. In moving from P_1 to P_2 on the line l, $\Delta x = 2$. Recalling from Example 1 that $m = -1/2$, we have from (2),

$$\Delta y = -\frac{1}{2}\Delta x = -\frac{1}{2} \cdot 2 = -1.$$

FIGURE 2.8

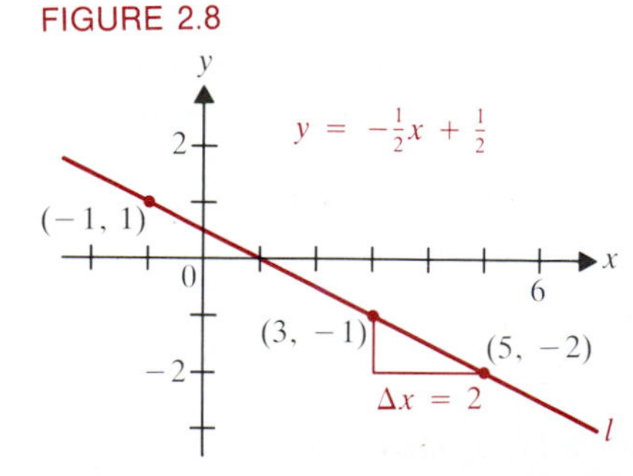

Hence (see Figure 2.8),

$$P_2 = (x_2, y_2) = (x_1 + \Delta x, y_1 + \Delta y) = (3 + 2, -1 - 1) = (5, -2). \blacksquare$$

Obviously, any two vertical lines $x = c$ and $x = d$ are parallel. The next result says that nonvertical lines are parallel precisely when they rise (or fall) at the same rate. (For the proof, see Exercises 60 and 61.)

2.1 THEOREM Two distinct nonvertical lines l_1 and l_2 with respective equations $y = m_1x + b_1$ and $y = m_2x + b_2$ are parallel if and only if $m_1 = m_2$ (that is, their slopes are the same).

Example 6

Find an equation for the line through $(5, -1)$ that is parallel to the line l with equation $x - 2y = 4$.

FIGURE 2.9

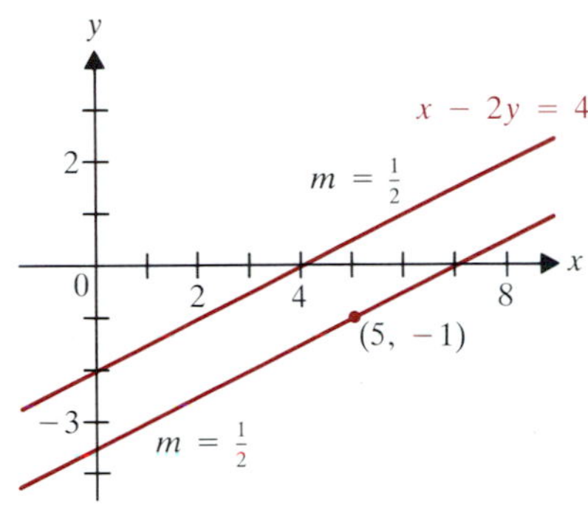

Solution. First, we find the slope of l. We have

$$-2y = -x + 4 \rightarrow y = \frac{1}{2}x - 2.$$

Thus $m = 1/2$. See **Figure 2.9.** This must be the slope of the line whose equation we are to find. From the point-slope formula, we get

$$y + 1 = \frac{1}{2}(x - 5) \rightarrow 2y + 2 = x - 5 \rightarrow x - 2y = 7. \quad ■$$

The next theorem gives a simple criterion for perpendicularity of nonvertical lines. (Note that any vertical line $x = c$ is perpendicular to any horizontal line $y = d$.) We use the notation $l_1 \perp l_2$ to stand for the statement that a line l_1 is perpendicular to a line l_2.

2.2 THEOREM

Two nonvertical lines l_1 and l_2 are perpendicular if and only if $m_1m_2 = -1$, that is, if and only if $m_1 = -1/m_2$.

Proof. Note that $l_1 \perp l_2$ if and only if $\bar{l}_1 \perp \bar{l}_2$, where $\bar{l}_1$ is the line through the origin parallel to l_1 and $\bar{l}_2$ is the line through the origin parallel to l_2. See **Figure 2.10.** By Theorem 2.1, $\bar{l}_1$ has slope m_1 and $\bar{l}_2$ has slope m_2. The y-intercept of both $\bar{l}_1$ and $\bar{l}_2$ is 0, because each line goes through (0, 0). Consider the points

FIGURE 2.10

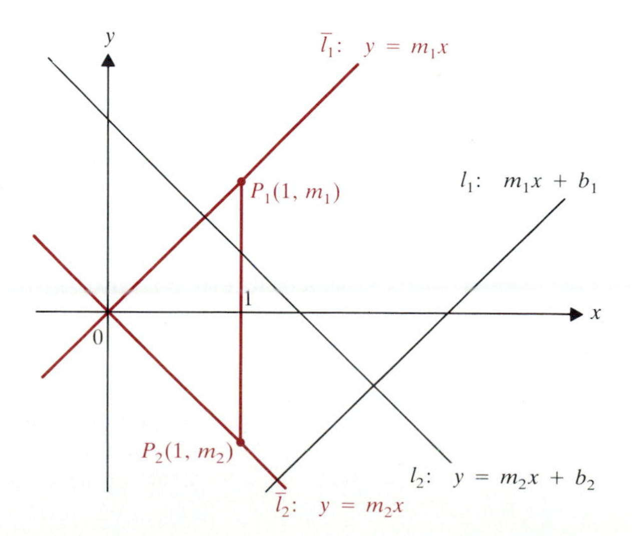

P_1 on $\bar{l}_1$ and P_2 on $\bar{l}_2$, where $x = 1$. The y-coordinate of P_1 is $y = m_1 1 + 0 = m_1$, and the y-coordinate of P_2 is $y = m_2 1 + 0 = m_2$. Now $\bar{l}_1 \perp \bar{l}_2$ holds if and only if angle $P_2 0 P_1$ is a right angle. By the Pythagorean theorem, this in turn is equivalent to the successive equations:

$$d(0, P_1)^2 + d(0, P_2)^2 = d(P_1, P_2)^2,$$
$$(1 + m_1{}^2) + (1 + m_2{}^2) = (1 - 1)^2 + (m_2 - m_1)^2,$$
$$2 + m_1{}^2 + m_2{}^2 = m_2{}^2 - 2m_1m_2 + m_1{}^2,$$
$$2 = -2m_1m_2,$$
$$m_1m_2 = -1. \quad \boxed{\text{QED}}$$

Example 7

Find an equation of the line l_2 through $(5, -1)$ that is perpendicular to the line l_1 with equation $x - 2y = 4$.

Solution. From our work in Example 6, $m_1 = 1/2$. By Theorem 2.2 then, the line we are seeking has slope m_2 satisfying

$$m_1m_2 = -1 \rightarrow m_2 = -\frac{1}{m_1} = -\frac{1}{1/2} = -2.$$

From the point-slope form, the equation of l_2 is therefore

$$y + 1 = -2(x - 5) \rightarrow y + 1 = -2x + 10 \rightarrow 2x + y = 9. \quad ■$$

FIGURE 2.11

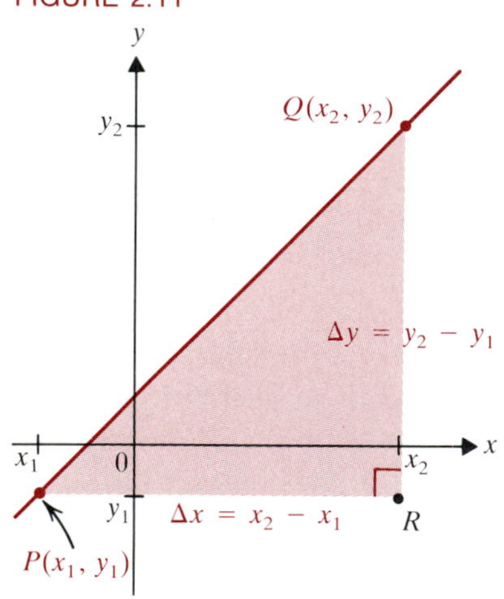

The interpretation of the slope of the line segment joining $P(x_1, y_1)$ to $Q(x_2, y_2)$ as the vertical rise $\Delta y = y_2 - y_1$ divided by the horizontal run $\Delta x = x_2 - x_1$ provides an easy derivation of the ***distance formula,***

$$(6) \qquad d(P, Q) = \sqrt{(x_2 - x_1)^2 + (y_2 - y_1)^2} = \sqrt{(x_1 - x_2)^2 + (y_1 - y_2)^2}.$$

From **Figure 2.11,** we see in fact that

$$d(P, Q) = \sqrt{(\Delta x)^2 + (\Delta y)^2} = \sqrt{(x_2 - x_1)^2 + (y_2 - y_1)^2},$$

since triangle PRQ is a right triangle.

The equation of the circle of radius r centered at $C(h, k)$ is easy to derive by using (6). See **Figure 2.12.** For a point $P(x, y)$ is on the circle if and only if

$$d(C, P) = \sqrt{(x - h)^2 + (y - k)^2} = r,$$

which in turn is equivalent to

$$(7) \qquad (x - h)^2 + (y - k)^2 = r^2.$$

FIGURE 2.12

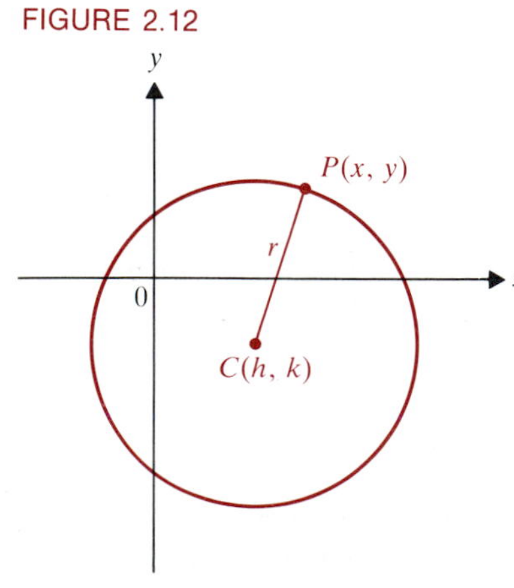

If the circle is centered at the origin, then $h = k = 0$; so (7) becomes

$$x^2 + y^2 = r^2.$$

Example 8

Find the equation of the circle centered at $(2, -3)$ with radius 4.

Solution. Here $h = 2$ and $k = -3$. Thus $x - h = x - 2$ and $y - k = y - (-3) = y + 3$. So (7) gives

$$(x - 2)^2 + (y + 3)^2 = 16. \quad \blacksquare$$

It turns out that the graph of *every* second-degree equation of the form

$$Ax^2 + By^2 + Dx + Ey + F = 0,$$

where $A = B \neq 0$, is either a circle, a single point, or the empty set. To find out which possibility holds in a given example, we *complete the square* in x and y. Recall that this procedure involves adding to each side of the equation the square of half the coefficient of x and the square of half the coefficient of y.

Example 9

Determine the graphs of the equations

(a) $2x^2 + 2y^2 - 4x + 12y - 7 = 0$; (b) $2x^2 + 2y^2 - 4x + 12y + 22 = 0$;
(c) $2x^2 + 2y^2 - 4x + 12y + 20 = 0$.

Solution. (a) First, divide the equation through by the common coefficient 2 of x^2 and y^2. That gives

$$x^2 + y^2 - 2x + 6y - \frac{7}{2} = 0.$$

Next, collect together the terms involving x and y, and then complete the square in each variable:

$$(x^2 - 2x \qquad) + (y^2 + 6y \qquad) = \frac{7}{2}, \tag{8}$$

$$(x^2 - 2x + 1) + (y^2 + 6y + 9) = \frac{7}{2} + 1 + 9, \tag{9}$$

$$(x - 1)^2 + (y + 3)^2 = \frac{27}{2}.$$

This equation is of the form (7) with $h = 1$, $k = -3$. So the equation is that of a circle with center $(1, -3)$ and radius $\sqrt{27/2}$.

(b) If we proceed exactly as in (a), we obtain the left side of (8) but with -11 on the right side instead of $7/2$. Thus, in place of (9) we get

$$(x^2 - 2x + 1) + (y^2 + 6y + 9) = -11 + 1 + 9,$$
$$(x - 1)^2 + (y + 3)^2 = -1.$$

This has the empty set as its graph, since the left-hand side is a sum of squares, which must be nonnegative.

(c) This time, in place of (8) we obtain an equation with -10 on the right side, and in place of (9) we have

$$(x^2 - 2x + 1) + (y^2 + 6y + 9) = -10 + 1 + 9,$$
$$(x - 1)^2 + (y + 3)^2 = 0.$$

This is a complicated equation for the single point $(1, -3)$, viewed as a circle of radius 0 centered at $(1, -3)$. ■

It is sometimes useful to be able to find the midpoint $M(c, d)$ of the line segment joining two given points $P_1(x_1, y_1)$ and $P_2(x_2, y_2)$. See **Figure 2.13,** where we have drawn the vertical lines $x = x_1$, $x = x_2$, and $x = c$, and the horizontal lines $y = y_1$ and $y = y_2$. Using the figure, one can derive the following familiar formulas for the coordinates of M (see Exercise 39):

$$c = \frac{x_1 + x_2}{2}, \qquad d = \frac{y_1 + y_2}{2}, \tag{10}$$

the arithmetic means of the coordinates of P_1 and P_2.

FIGURE 2.13

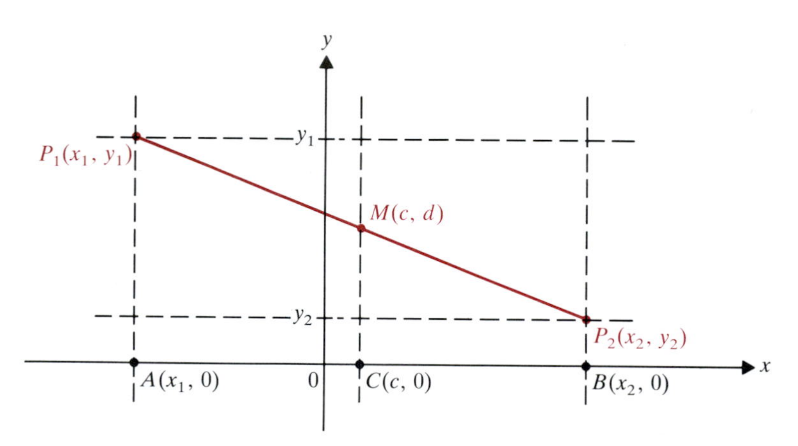

Example 10

If $A(6, 3)$ and $B(-2, -5)$ are the endpoints of a diameter of a circle, then find its equation.

Solution. The center of the circle is the midpoint $C(h, k)$ of AB. So, by (10), we have

$$h = \frac{6 - 2}{2} = 2, \qquad k = \frac{3 - 5}{2} = -1.$$

The radius r is half the diameter. Thus

$$r = \frac{1}{2}d(A, B) = \frac{1}{2}\sqrt{(6 + 2)^2 + (3 + 5)^2} = \frac{1}{2}\sqrt{64 + 64} = \frac{1}{2}\sqrt{64}\sqrt{2} = 4\sqrt{2}.$$

Therefore, the equation of the circle is

$$(x - 2)^2 + (y + 1)^2 = 32.$$

It is shown in **Figure 2.14.** ■

FIGURE 2.14

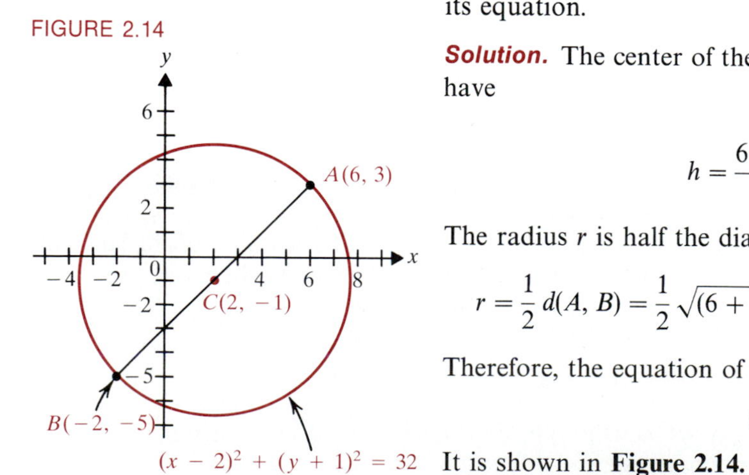

HISTORICAL NOTE

In 1637, René Descartes (1596–1650) published the first article that used algebraic equations to describe curves. It was an appendix to his noted philosophical treatise. *Discourse on the Method,* from which his famous dictum "*Cogito ergo sum*—I think, therefore I am" comes. This was the philosophical basis for his belief in his own existence. At about the same time, the French mathematician Pierre de Fermat (1601–1665) considered similar geometric questions in private correspondence; however, that work was not published until 1679. These two men are generally regarded as the founders of analytic geometry.

It is interesting that neither Descartes nor Fermat used the now-standard coordinate system, which is named for Descartes. That was first introduced a full century later by the Swiss mathematician Gabriel Cramer (1704–1752), who is more famous for his rule involving the solution of linear equations by determinants. But if history had exerted more influence on terminology than did fame, we might today be discussing "Cramerian coordinates" for the plane.

Exercises 1.2

In Exercises 1 and 2, find the slope of the line segment joining the given points.

1. **(a)** $(-1, 2), (2, -1)$
(b) $(-1, 3), (2, 7)$

2. **(a)** $(2, -3), (-3, 2)$
(b) $(5, -1), (7, 2)$

In Exercises 3–8, find the general form equation of the line satisfying the given conditions.

3. **(a)** Through the points in Exercise 1(a).
(b) Through the points in Exercise 1(b).

4. **(a)** Through $(-5, 2)$; slope 6.
(b) Through $(2, -3)$; slope $1/2$.

5. **(a)** Through $(-1, -3)$; y-intercept 3.
(b) Through $(2, -2)$; y-intercept 5.

6. **(a)** x-intercept 3, y-intercept 6.
(b) x-intercept -2, y-intercept 1.

7. **(a)** Through $(2, -1)$ parallel to $2x - 3y = 8$.
(b) Through $(-1, 3)$ parallel to $5x - y = 2$.

8. **(a)** Through $(2, -1)$ perpendicular to $2x - 3y = 8$.
(b) Through $(-1, 3)$ perpendicular to $5x - y = 2$.

9. Given that Fahrenheit and Celsius temperatures are related by a linear equation in F and C, derive a formula for F in terms of C. (See Example 2.)

10. The freezing point of mercury is $-39°$C, and the melting point of cadmium is 321°C. Suppose you devise a new temperature scale with 0°N the freezing point of mercury and 100°N the melting point of cadmium. Derive a formula to convert from New to Celsius temperature. (See Example 2 and Exercise 9.)

In Exercises 11 and 12, find the distance between the given pair of points and the midpoint of the line segment joining them.

11. **(a)** $(5, 1), (1, 4)$
(b) $(-2, -3), (3, 9)$

12. **(a)** $(3, -1), (-2, 3)$
(b) $(-1, 4), (4, -8)$

13. What is the equation of the x-axis?

14. What is the equation of the y-axis?

In Exercises 15–18, find the equation of the circle satisfying the given conditions.

15. **(a)** Center $(-1, 3)$; radius 2.
(b) Center $(2, -1)$; radius 3.

16. **(a)** Center $(4, 1)$; contains the point $(1, -3)$.
(b) Center $(3, -3)$; contains the point $(-2, 9)$.

17. **(a)** Center $(5, -2)$; tangent to the x-axis.
(b) Center $(-3, 2)$; tangent to the y-axis.

18. **(a)** Endpoints of diameter $(2, -3)$ and $(-4, 5)$.
(b) Endpoints of diameter $(-3, 1)$ and $(2, 4)$.

In Exercises 19–26, determine the nature of the graph of the given equation.

19. $x^2 + y^2 - 8x + 2y - 8 = 0$

20. $x^2 + y^2 - 4x - 6y - 3 = 0$

21. $3x^2 + 3y^2 - 6x - 18y + 27 = 0$

22. $5x^2 + 5y^2 + 20x - 10y + 120 = 0$

23. $x^2 + y^2 + 6x - 8y + 25 = 0$

24. $x^2 + y^2 - 10x + 6y + 34 = 0$

25. $x^2 + y^2 + 2x - 4y + 11 = 0$

26. $x^2 + y^2 + 4x - 2y + 6 = 0$

27. The velocity v of a falling body is related to the time t elapsed since it began to fall by a linear equation in v and t. If every second of fall, v increases by 9.8 m/sec, find a formula for v in terms of t and the initial velocity v_0.

28. The speed s of a car that is braking at a constant rate is related to the time t elapsed since braking began by a first-degree equation in s and t. If every second, s decreases by 15 ft/sec, find a formula for s in terms of t if $s = 44$ ft/sec when the brakes were applied. How long will it take the car to stop?

29. In business, total costs T of producing x units of a product are usually expressed as the sum of fixed costs F (overhead) and variable costs V related to manufacturing costs for x units of the product.

(a) If each unit costs \$3.00 to manufacture and fixed costs are \$1000 per day, give an equation for T on a day for which x units are produced.

(b) Under the assumptions in part (a), how much does it cost to produce 1000 units of the product?

(c) Interpret the \$3.00 unit cost in terms of the graph of T.

30. Suppose your home has a 1000-gallon oil tank and in the winter months your consumption of oil averages 5 gallons per day. Write an equation for the amount A of oil left in the tank n days after it is filled at the beginning of winter.

31. Suppose demand D for a certain agricultural product is 0 at a price of \$1.00 per unit and increases by 1000 units for every penny the price drops. Suppose market supply S is 0 at a price of 50¢ per unit and increases by 4000 units for every penny the price rises.

(a) Write equations for S and D in terms of the price p.

(b) The *law of supply and demand* in economics predicts that the price will adjust to that which just balances supply and demand. Use this law to predict the free market price of the product.

(c) How many units will be bought and sold at the price level found in (b)?

32. One of the recognized forms of depreciation for income tax purposes is *straight-line* depreciation. Suppose a business buys a one million dollar computer that has a useful life of ten years. At the end of ten years it can dispose of the computer for \$100,000.

(a) Find the equation relating the time t since purchase (in years) and the value V of the computer, if they are related by a linear equation.

(b) Find how much the value of the computer declines each year. (This is the *depreciation*.)

33. Measurements of heat loss through an insulated wall suggest that the temperature drop through the insulation is described by a linear equation. Suppose on the inside of a 9 cm thick batt of fiberglass insulation the temperature is 20°C, and on the outside of the batt the temperature is -10°C.

(a) Plot temperature T vs. distance D from the inside edge of the batt for the two measurements given.

(b) Write the equation of the line through these two points.

(c) Use your equation in (b) to compute the temperature inside the insulation at a point 3 cm from its inside edge.

34. Show that the lines $Ax + By = C$ and $Bx - Ay = D$ are perpendicular if A and B are nonzero.

35. In the Gibbs–Helmholtz equation (see Exercise 34, Section 1.1), G', H', and S' really represent the respective incremental changes in the free energy G, the heat content H, and the entropy S. Rewrite equation (9) of the last section in Δ-notation for increments. (Your result is the usual form of the Gibbs–Helmholtz equation.)

36. If a particle starts at $P_1(-3, 2)$ and moves with increments $\Delta x = -1$ and $\Delta y = -3$, then what are the coordinates of P_2?

37. A particle starts at $P_1(x_1, y_1)$ and moves to $P_2(-1, 1)$. If $\Delta x = -2$ and $\Delta y = 3$, find x_1 and y_1.

38. A particle moves from $P_1(3, 5)$ in such a way that $\Delta y = 2\,\Delta x$. If $y_2 = 1$, find x_2.

39. Prove the midpoint formula (10).

40. To find the intersection point of two nonvertical lines, we can first put the equations in slope-intercept form,

$$y_1 = m_1 x + b_1 \qquad \text{and} \qquad y_2 = m_2 x + b_2,$$

and we can then equate the expressions for y_1 and y_2 to find the x-coordinate of the point of intersection. Show that this gives

$$x_0 = \frac{b_2 - b_1}{m_1 - m_2} \qquad \text{if } m_1 \neq m_2.$$

41. Use the approach of Exercise 40 to find the point of intersection of $-x + 3y = 6$ and $x - y = 3$.

42. Use the approach of Exercise 40 to find the point of intersection of $x - 2y = 5$ and $2x + y = 3$.

43. If a line has x-intercept a and y-intercept b, then show that

$$\frac{x}{a} + \frac{y}{b} = 1$$

is an equation of the line. This is called the *intercept form* of the equation of the line.

44. Write the intercept form of the equation of the line through $(5, 0)$ and $(0, -2)$.

45. Write the intercept form of the equation of the line through $(0, 2)$ and $(-1, 0)$.

46. Are the points $A(3, 1)$, $B(-1, 5)$, and $C(2, 0)$ the vertices of a right triangle? If so, find its area.

47. Show that the points $A(1, 5)$, $B(-1, 3)$ and $C(4, 2)$ are the vertices of a right triangle.

48. Are the points $A(1, 3)$, $B(-2, 7)$, and $C(5, 6)$ the vertices of a right triangle? If so, find its area.

49. Find the equation of the set of points $P(x, y)$ that are equidistant from $A(1, 3)$ and $B(-1, 1)$.

50. Find the equation of the set of points $P(x, y)$ that are equidistant from $A(-1, 2)$ and $B(2, -1)$.

51. Find k so that $kx + 2y = 5$ is
(a) parallel to, **(b)** perpendicular to

$$-3x + y = 3.$$

52. Find k so that $3x - ky = 2$ is
(a) parallel to, **(b)** perpendicular to

$$x + 2y = 5.$$

53. Three consecutive vertices of a parallelogram are (2, 5), (4, 2), and (6, 6). Find the fourth vertex.

54. Show that (1, 5), (−1, 3), (4, 2), and (2, 0) are the vertices of a rectangle.

55. Draw a rectangle. Then introduce coordinate axes so that the x-axis is along one side and the y-axis is along an adjacent side. Suppose that one of those sides has length a and the other has length b.
(a) Find the midpoint of each side of the rectangle.
(b) Show that the quadrilateral obtained by joining midpoints of adjacent sides of the rectangle has all four sides of equal length.

56. Use the approach of Exercise 55 to show that the diagonals of any rectangle are of equal length.

57. Let $P(x_0, y_0)$ be a point not on the line l: $ax + by = c$, where a and b are nonzero.
(a) Find the equation of the line l' through P that is perpendicular to l.
(b) Find the point $P_1(x_1, y_1)$ where l' intersects l.
(c) Use the result of (b) to show that

$$d(P, P_0) = \frac{|ax_0 + by_0 - c|}{\sqrt{a^2 + b^2}}$$

is the distance from P_0 to the line l.

58. Use the formula in Exercise 57(c) to find the distance from the point $(-3, 2)$ to the line $-2x + y = 3$.

59. Repeat Exercise 58 for the point (5, 1) and the line $3x - 4y = 6$.

The following two exercises prove Theorem 2.1.

60. Suppose that (x_0, y_0) is a point of intersection of the lines $y = m_1x + b_1$ and $y = m_1x + b_2$. Then show that $b_1 = b_2$, so that the lines are identical. Conclude that *distinct* lines $y = m_1x + b_1$ and $y = m_1x + b_2$ must be parallel.

61. Suppose that the lines l_1: $y = m_1x + b_1$ and l_2: $y = m_2x + b_2$ are parallel. Use Exercise 40 to show that if $m_1 \neq m_2$, then the lines intersect. Hence, conclude that m_1 must equal m_2.

1.3 Functions and Graphs

The rest of this chapter, and much of the rest of the book, will deal with functions. Why are functions so central? As we mentioned in the Prologue, the fundamental nature of life itself involves change. The mathematician's way of dealing with change is to study it through functions, which describe how one quantity, say x, determines another, the value of a function f at x. We begin our review of functions by recalling the definition.

3.1 DEFINITION

A ***real-valued function f of a real variable x*** is a rule that associates to each number x in some set D of real numbers exactly one corresponding real number $y = f(x)$.

The number $y = f(x)$ (read "f of x") is called the ***value of f at x*** or the *image of x under f*. The set D is called the ***domain*** of f. For example, the rule $f(x) = 3x^2 + 1$ defines a function f whose domain is the entire set **R** of real numbers. However, the equation

$$x^2 + y^2 = 1$$

does *not* define y as a function of x. The problem is that for each real number x in the interval $D = [-1, 1]$, this rule defines *two* real numbers y, namely,

$$y_1 = \sqrt{1 - x^2} \qquad \text{and} \qquad y_2 = -\sqrt{1 - x^2}.$$

Definition 3.1 requires that for each x in D, *exactly one* real number y be determined by the rule.

Unless it is explicitly given, D is taken to be the set of all real numbers x for which the rule for calculating $f(x)$ is meaningful. For instance,

$$f(x) = \frac{1}{x}$$

defines a function that associates to each real number (except 0) its reciprocal. Since division by 0 is meaningless, the domain D of f is

$$D = (-\infty, 0) \cup (0, +\infty),$$

the entire real line except for 0. Your calculator may have a key to compute $1/x$ from an entry x. If you enter 0 and then press this key, you will probably be met by a blinking display of a very large number, or perhaps by the word "ERROR" in the display. Calculators are very efficient function evaluators but balk at trying to evaluate a function at points x outside the domain. Our first example illustrates how the techniques of Section 1 can be used to determine the domain of a given function.

Example 1

Find the domain of the function g if $g(x) = \sqrt{x^2 + 6x + 5}$.

Solution. Factoring the expression under the radical gives $g(x) = \sqrt{(x + 5)(x + 1)}$, which is defined for all numbers x such that $(x + 5)(x + 1) \geq 0$. From the following table, we see that $D = (-\infty, -5] \cup [-1, +\infty)$. ■

x:	$(-\infty, -5)$	-5	$(-5, -1)$	-1	$(-1, +\infty)$
$x + 5$:	$-$		$+$		$+$
$x + 1$:	$-$		$-$		$+$
$(x + 5)(x + 1)$:	$+$	0	$-$	0	$+$

Even though your calculator doesn't have a key for computing $g(x)$ directly, you can still think of its formula as specifying a sequence of calculator keys to press in order to compute $g(x)$. (If your calculator is programmable, you can actually program it to evaluate $g(x)$ for any x you enter from the set D.)

For each point a in the domain D of a given function f, the value $f(a)$ of f corresponding to a is the number that results from substituting a for x in the formula for f. In the function g in Example 1, we have

$$g(-8) = \sqrt{(-8)^2 + 6(-8) + 5} = \sqrt{64 - 48 + 5} = \sqrt{21} = g(2),$$

$$g(a) = \sqrt{a^2 + 6a + 5},$$

$$g(2a) = \sqrt{(2a)^2 + 6(2a) + 5} = \sqrt{4a^2 + 12a + 5},$$

$$g(a + b) = \sqrt{(a + b)^2 + 6(a + b) + 5} = \sqrt{a^2 + 2ab + b^2 + 6a + 6b + 5},$$

and so forth.

FIGURE 3.1

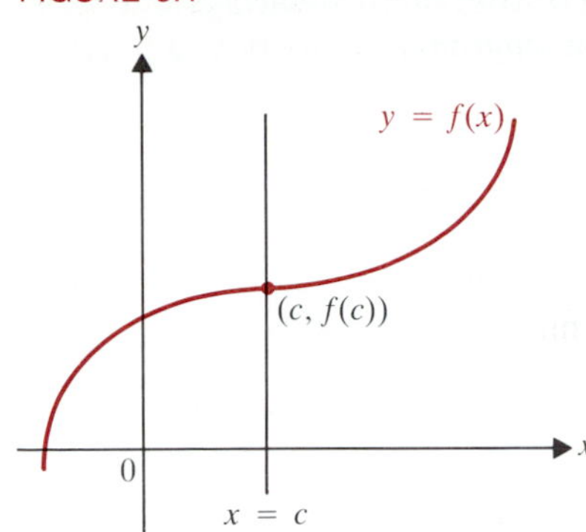

Occasionally, we will need to consider the set of all images $f(x)$ of elements x in D under the rule f. This set is called the ***range*** R of f,

$$R = \{y \mid y = f(x) \text{ for some } x \text{ in } D\}.$$

For example, if $f(x) = x^2$, then the domain D of f is the entire real line $\mathbf{R}$, because the square of any real number exists. Since $f(x) = x^2$ is nonnegative for any real number x, $R \subseteq [0, +\infty)$, the set of all nonnegative real numbers. It is not hard to see that R actually *equals* $[0, +\infty)$. For if $y \in [0, +\infty)$, then y has principal square root $x = \sqrt{y}$, and $y = (\sqrt{y})^2 = f(x)$ belongs to R.

Recall that the ***graph*** of a function f is the set of all ordered pairs (x, y) in the real plane such that $y = f(x)$. The condition in Definition 3.1, that to each x in the domain D of f there should correspond *one and only one* value $y = f(x)$, has a simple graphical interpretation. Namely, for each x in D, there is exactly one ordered pair $(x, f(x))$ on the graph. That is, *any vertical line $x = c$, where $c \in D$, must intersect the graph in the single point $(c, f(c))$*. See **Figure 3.1.** As a consequence, it follows immediately that the circle $x^2 + y^2 = 1$ in **Figure 3.2** is *not* the graph of a function, since, for instance, the y-axis $(x = 0)$ intersects the graph in the two points $(0, 1)$ and $(0, -1)$.

From the last section, the graph of a *linear function* f given by

$$f(x) = mx + b$$

for constants m and b is a straight line of slope m and y-intercept b. See **Figure 3.3.** The line rises as x increases in case $m > 0$, and falls as x increases if $m < 0$.

FIGURE 3.2

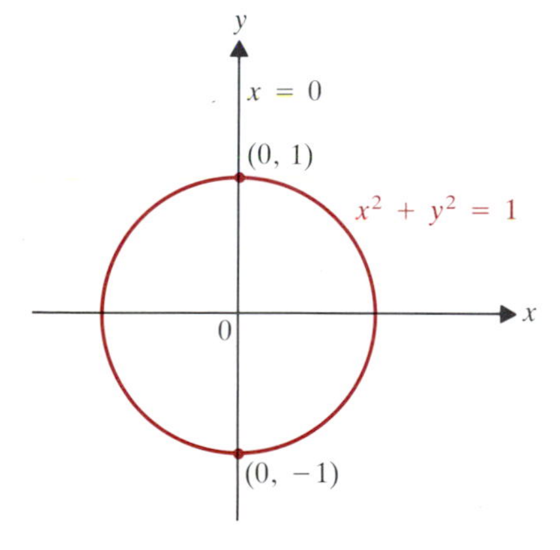

FIGURE 3.3

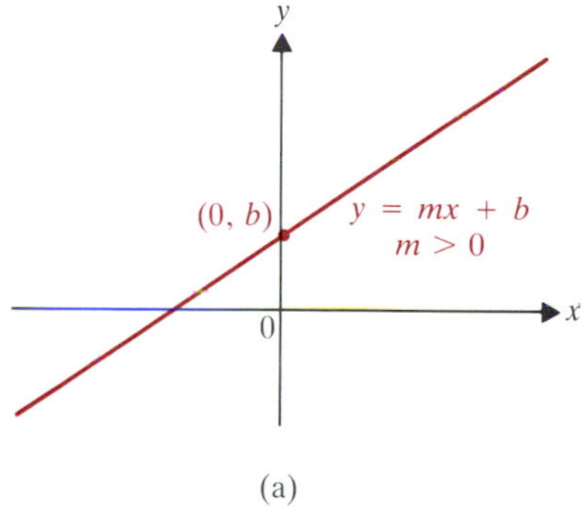

(a)

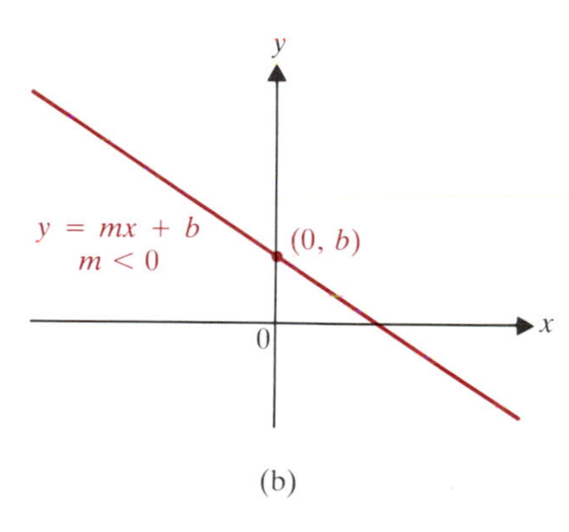

(b)

Linear functions with $b = 0$ are used to model proportionality relationships. If $y = mx$, then y is said to be ***proportional*** to x, or to *vary directly* as x. The slope m is called the *constant of proportionality*. Proportionality is a common relationship. For instance, in growing children, hormone dosage level D is usually proportional to the body surface area S, which physicians approximate using a formula based on the height and weight of the patient. The following example is based on an actual medical case.

Example 2

A cortisone-deficient child, aged 9 years, with body surface area S of approximately $1.00\ \text{m}^2$, was given a daily cortisone dose of 1.25 milligram (mg). At age

FIGURE 3.4

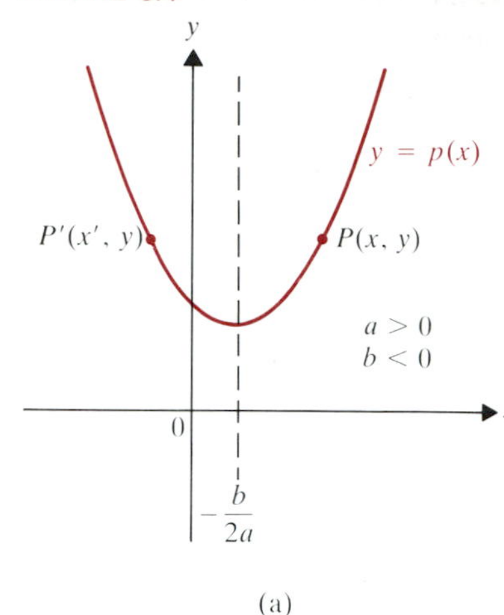

(a)

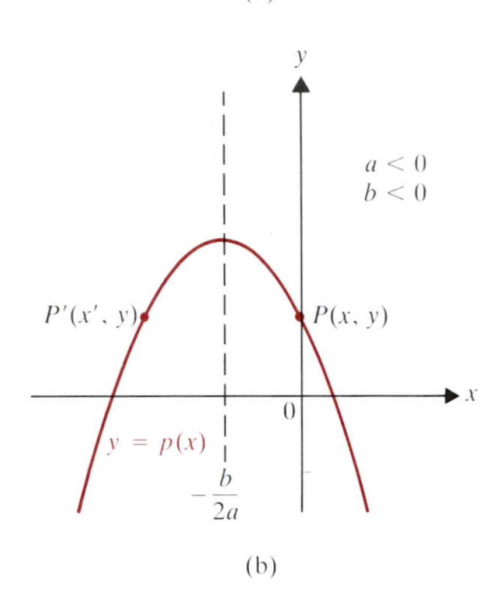

(b)

16, the child had grown so that $S \approx 1.80\ \mathrm{m}^2$. To what level should the cortisone dose have been raised? (The symbol $\approx$ means *approximately equals.*)

Solution. The dosage D is given by $D = mS$. Since $D = 1.25$ when $S = 1.00$, we have $1.25 = m$. When $S = 1.80$, then

$$d = mS = (1.25)(1.80) = 2.25 \text{ mg.} \quad ■$$

After linear functions, the next simplest functions are ***quadratic functions.***

$$q(x) = ax^2 + bx + c, \qquad a \neq 0. \tag{1}$$

The graphs of quadratic functions are always ***parabolas,*** which open *upward* if $a > 0$, *downward* if $a < 0$. The vertex of a parabola is found by completing the square on the right side of (1).

$$\begin{aligned} y &= a\left(x^2 + \frac{b}{a}x\right) + c = a\left(x^2 + \frac{b}{a}x + \frac{b^2}{4a^2}\right) + c - a\left(\frac{b^2}{4a^2}\right) \\ &= a\left(x + \frac{b}{2a}\right)^2 + c - \frac{b^2}{4a} = a\left(x + \frac{b}{2a}\right)^2 + \frac{4ac - b^2}{4a}. \end{aligned}$$

The ***vertex*** of the parabola is at the point $(x, q(x))$, where $x = -b/2a$. The vertex is then the lowest (if $a > 0$) or the highest (if $a < 0$) point on the graph. See **Figure 3.4.** The line $x = -b/2a$ is the ***axis*** of the parabola. The graph is ***symmetric*** about this axis, meaning that every point $P(x, y)$ on the parabola has a mirror-image point $P'(x', y)$ on the parabola, where P and P' are at the same distance from the axis. In graphing (1), it is convenient to start with $x = -b/2a$ and then add and subtract the same quantities to this value to get symmetric points on the graph.

Example 3

Graph the function $y = -2x^2 + 2x - 3$.

FIGURE 3.5

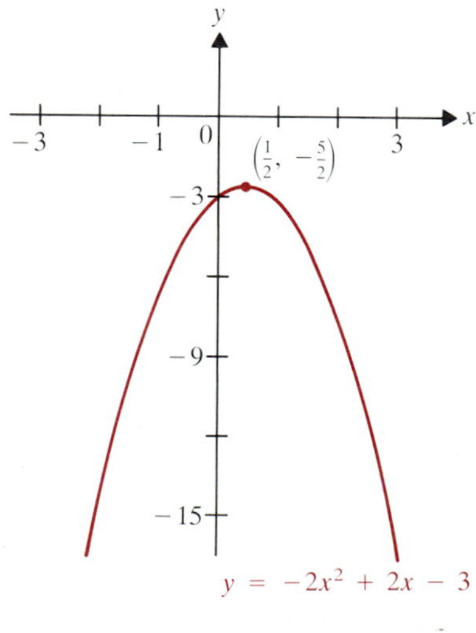

Solution. First, we complete the square in x:

$$\begin{aligned} y &= -2(x^2 - x) - 3 = -2\left(x^2 - x + \frac{1}{4}\right) - 3 + 2 \cdot \frac{1}{4} \\ &= -2\left(x - \frac{1}{2}\right)^2 - \frac{5}{2}. \end{aligned}$$

The factor $(x - \frac{1}{2})^2$ is smallest when $x = \frac{1}{2}$. Thus the vertex of the graph is at the point $(\frac{1}{2}, -\frac{5}{2})$. So the graph is symmetric with respect to the axis $x = \frac{1}{2}$. The graph opens downward since $a < 0$. It is easier to evaluate the function at integers, so in our table we use integers for all points except the vertex. The graph is shown in **Figure 3.5.**

x	$\frac{1}{2}$	0	1	−1	2	−2	3
y	$-\frac{5}{2}$	−3	−3	−7	−7	−15	−15

■

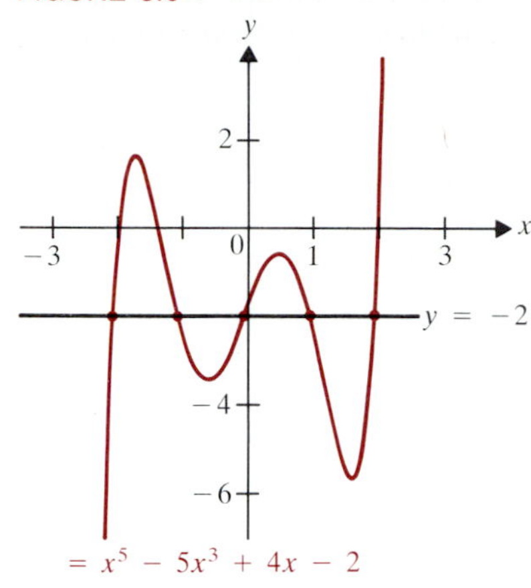

FIGURE 3.6

We will develop techniques to graph ***polynomial functions*** of degree 3 and higher in Chapter 3. Polynomial functions of degree n have the form

$$p(x) = a_n x^n + a_{n-1}x^{n-1} + \ldots + a_2 x^2 + a_1 x + a_0,$$

where the coefficients $a_0, a_1, a_2, \ldots, a_{n-1}, a_n$ are constants, with $a_n \neq 0$. To see that some care must be exercised in graphing higher-degree polynomials, consider the function

$$y = x^5 - 5x^3 + 4x - 2.$$

The following table of values is correct, but if the corresponding points were simply joined by the line $y = -2$, the resulting graph would be drastically different from the actual graph shown in **Figure 3.6**!

x	-2	-1	0	1	2
y	-2	-2	-2	-2	-2

Two important types of functions are *even* functions and *odd* functions. For both, x is in the domain D if and only if $-x$ is in the domain. Such a function is ***even*** if

$$f(x) = f(-x)$$

for every x in D. It is ***odd*** if

$$f(x) = -f(-x)$$

for every x in D. Any even power function $f(x) = x^{2k}$, for k an integer, is an even function, because

$$f(-x) = (-x)^{2k} = [(-x)^2]^k = [x^2]^k = x^{2k} = f(x).$$

The graph of such a function is symmetric in the y-axis. Any odd power function $f(x) = x^{2k-1}$, for k an integer, is an odd function, because

$$f(-x) = (-x)^{2k-1} = (-1)^{2k-1}x^{2k-1} = -x^{2k-1} = -f(x).$$

Operations on Functions

There are several ways to build new functions from given functions. The simplest operations correspond to basic arithmetic operations. The ***sum, difference,*** and ***product*** of two functions f and g are defined by the following rules on the domain $D_f \cap D_g$.*

$$(f + g)(x) = f(x) + g(x),$$
$$(f - g)(x) = f(x) - g(x),$$
$$(f \cdot g)(x) = f(x) \cdot g(x).$$

The ***quotient function*** is defined by

$$(f/g)(x) = \frac{f(x)}{g(x)}$$

* $D_f \cap D_g$ is the *intersection* of the domains D_f and D_g of f and g, respectively: $D_f \cap D_g = \{x \mid x \in D_f \text{ and } x \in D_g\}$, the set of points common to both domains.

on the domain $D_{f/g} = \{x \in D_f \cap D_g | g(x) \neq 0\}$. These functions are thus defined at a real number x *only* if both $f(x)$ and $g(x)$ are defined. In addition, $(f/g)(x)$ is defined only in case $g(x) \neq 0$. As an illustration, if

$$f(x) = \sqrt{x^2 - 1} \qquad \text{and} \qquad g(x) = \frac{1}{x + 2},$$

then

$$(f \cdot g)(x) = \frac{\sqrt{x^2 - 1}}{x + 2}$$

has domain $D = D_f \cap D_g$, where $D_f = \{x | x^2 - 1 \geq 0\} = (-\infty, -1] \cup [1, +\infty)$ and $D_g = (-\infty, -2) \cup (-2, +\infty)$. The intersection of D_f with D_g is obtained by deleting -2 from D_f, so $D_f \cap D_g$ is $(-\infty, -2) \cup (-2, -1] \cup [1, +\infty)$.

FIGURE 3.7

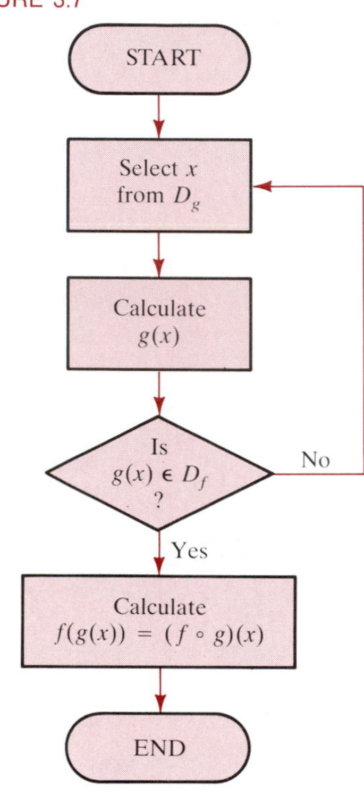

One of the most useful ways to combine two functions is by forming their *composition*. In mathematics itself, but perhaps even more so in applied fields, it is common to analyze a complicated functional relationship in stages. A case in point is the idea of a food chain in life science. Consider a large lake in which predator fish feed on smaller fish. The population N of the predator fish at any time t is a function of the size of its food source:

$$N = f(S), \tag{2}$$

where S is the population of the small fish. The population of prey will change in response to the size of its food source, say, algae. So we can consider it to be a function of the volume A of algae in the lake,

$$S = g(A), \tag{3}$$

for some function g. Equations (2) and (3) focus on the relationships between the *adjacent* members of the food chain. From these we see that the population N of predator fish is ultimately a function of the volume A of algae in the lake. We formalize this in the following definition.

3.2 DEFINITION

The ***composite function*** $f \circ g$ of two functions is defined by

$$(f \circ g)(x) = f(g(x)) \tag{4}$$

on the domain

$$D_{f \circ g} = \{x \in D_g | g(x) \in D_f\}. \tag{5}$$

We read "$f \circ g$" as "f composed with g" or "f following g."

In the discussion of the food chain, the population N of large fish is the composite of the two functions f and g:

$$N = f(S) = f(g(A)) = (f \circ g)(A).$$

The process of composition is illustrated schematically in **Figure 3.7.** The restriction (5) on the domain of $f \circ g$ is made to ensure that (4) can be applied. $D_{f \circ g}$ is entirely included in D_g, because the first step in evaluating $f(g(x))$ is calculation of $g(x)$. To compute f at $g(x)$, $g(x)$ must be in D_f.

Example 4

Suppose $f(x) = \sqrt{x^2 - 9}$ and $g(x) = 2x - 3$. Find $f \circ g$.

Solution. We first work out the formula for the composite function, and then we find $D_{f \circ g}$. Using (4), we have

$$(f \circ g)(x) = f(g(x)) = \sqrt{(g(x))^2 - 9} = \sqrt{(2x - 3)^2 - 9} = \sqrt{4x^2 - 12x + 9 - 9}$$
$$= \sqrt{4x^2 - 12x} = 2\sqrt{x^2 - 3x}.$$

Here, $D_g = \mathbf{R}$, and $D_f = \{x | x^2 - 9 \geq 0\} = (-\infty, -3] \cup [3, +\infty)$. Hence, $g(x) \in D_f$ is equivalent to $2x - 3 \in (-\infty, -3] \cup [3, +\infty)$. This in turn is equivalent to the successive conditions

$$2x - 3 \leq -3 \quad \text{or} \quad 2x - 3 \geq 3,$$
$$2x \leq 0 \quad \text{or} \quad 2x \geq 6,$$
$$x \leq 0 \quad \text{or} \quad x \geq 3.$$

Thus from (5), $D_{f \circ g} = (-\infty, 0] \cup [3, +\infty)$. ■

Exercises 1.3

In Exercises 1–10, find the domain of the given function.

1. $f(x) = \dfrac{1}{x + 2}$

2. $h(x) = \sqrt{x - 3}$

3. $s(x) = \dfrac{1}{\sqrt{2 - x}}$

4. $u(x) = \sqrt{\dfrac{x - 1}{x + 3}}$

5. $w(x) = \dfrac{\sqrt{x - 1}}{\sqrt{x + 3}}$

6. $q(x) = \dfrac{\sqrt{x - 3}}{x^2 - 3x - 18}$

7. $r(x) = \dfrac{\sqrt{x - 3}}{x^2 - 10x + 24}$

8. $k(x) = \sqrt{x^2 - 5x + 6}$

9. $m(x) = \sqrt{1 - x^2}$

10. $p(x) = \dfrac{3 - x}{9 - x^2}$

In Exercises 11–24, draw the graph of the given function.

11. $f(x) = 1/x$

12. $f(x) = |x|$

13. $f(x) = \sqrt{x}$

14. $f(x) = \sqrt{4 - x^2}$

15. $f(x) = \dfrac{1}{\sqrt{x - 3}}$

16. $f(x) = \begin{cases} 2 - x^2 & \text{for } x < 0 \\ 2 + x & \text{for } x \geq 0 \end{cases}$

17. $g(x) = \begin{cases} 2x + 1 & \text{for } x \leq 0 \\ 2x^2 + 1 & \text{for } x > 0 \end{cases}$

18. $f(x) = \begin{cases} 3x + 1 & \text{for } x < 1 \\ 2 - x^2 & \text{for } x \geq 1 \end{cases}$

19. $h(x) = \begin{cases} 2x + 5 & \text{for } x < -1 \\ x^2 + 1 & \text{for } x \geq -1 \end{cases}$

20. $f(x) = x^2 + 4x - 1$

21. $f(x) = 2x^2 + 6x - 2$

22. $f(x) = 2 - 3x - x^2$

23. $f(x) = -2x^2 + 4x + 5$

24. $c(x) = -x^3$

25. If

$$f(x) = \frac{2x - 1}{x^2 + x - 1},$$

find **(a)** $f(2)$, **(b)** $f(-2)$, **(c)** $f(2a)$, **(d)** $f(1/a)$, **(e)** $f(a + h)$, **(f)** $f(\sqrt{a})$, **(g)** $f(a^2)$.

26. Repeat Exercise 25 for $f(x) = \sqrt{x^2 + x + 1}$.

27. If $f(x) = x^2 - 3x + 2$, find **(a)** $f(a)$, **(b)** $f(a + h)$, **(c)** $\dfrac{f(a + h) - f(a)}{h}$.

28. Does the given equation define y as a function of x? Explain.

(a) $x + y^2 = 3$ **(b)** $x^2 + 2y^2 = 8$

In Exercises 29–34, find (a) $f + g$, (b) $f \cdot g$, (c) $f \div g$, (d) $g \div f$, (e) $f \circ g$, and (f) $g \circ f$ for the given functions. Specify the domain.

29. $f(x) = x^2 - 1$, $g(x) = 2x - 1$

30. $f(x) = \dfrac{1}{\sqrt{x^2 - 4}}$, $g(x) = x - 1$

31. $f(x) = \sqrt{x^2 - 4}$, $g(x) = \dfrac{1}{x + 1}$

32. $f(x) = \sqrt{x}$, $g(x) = 1 - x^2$

33. $f(x) = \dfrac{1}{x - 2}$, $g(x) = \dfrac{x + 1}{x + 3}$

34. $f(x) = x^3 + 1$, $g(x) = \sqrt[3]{x - 1}$

In Exercises 35–38, find the domain of the given function.

35. $f(x) = \sqrt{x^2 - 5x + 6} + \dfrac{1}{x - 2}$

36. $f(x) = \dfrac{1}{x - 5} + \sqrt{x - 4}$

37. $m(x) = \dfrac{3}{|x-2|} + \dfrac{1}{x^2}$

38. $n(x) = \sqrt{5x-3} + \dfrac{1}{3x-5}$

In Exercises 39–42, write the function f as the composite of two functions g and h.

39. $f(x) = \sqrt{x^2 - 5x + 7}$

40. $f(x) = \dfrac{1}{3x^2 + 5x - 1}$

41. $f(x) = \left(\dfrac{1}{2x-3}\right)^3$

42. $f(x) = (3x^2 - 4x + 1)^{3/2}$

43. Charles's law in physics and chemistry says that at constant pressure, the volume (in liters) of a fixed mass of gas is proportional to the Kelvin temperature (K). At 200 K, the volume of a sample of hydrogen is 15 liters. What will be the volume of the sample at 300 K if the pressure remains constant?

44. When a muscle exercises, it generates heat, which the body dissipates mostly through the skin (through perspiration). Heat lost through the skin, ΔH_l, is proportional to body surface area S. Heat produced during exercise ΔH_p is proportional to body mass M. Since the body must maintain an equilibrium between heat produced and heat lost, show that an individual's body surface area S must be proportional to his or her mass M. (For example, as an individual gains weight, fat deposits stretch the skin to increase its surface area; people are never shaped like boxes or balls.)

45. Dosage levels of some drugs are prescribed to be proportional to body mass. Use Exercise 44 to show that this is *equivalent* to prescribing dosage levels to be proportional to body surface area, as in Example 2 for cortisone.

46. A chemical reaction in which one molecule of a substance S reacts with one molecule of a substance T to produce one molecule of a substance U is written symbolically $S + T \to U$. The ***law of mass action*** states that the rate r of formation of U is proportional to the concentrations of S and T (in moles per liter) at any point of the reaction. Suppose the initial concentrations of S and T are S_0 and T_0. Here, one mole of U is produced through the reaction of one mole each of S and T.

(a) Express the concentrations of S and T after x moles of U have been produced.
(b) Write an equation for r as a function of x. What sort of function is this?
(c) Find the value of x for which your function in (b) is minimized.
(d) Is your answer in (c) meaningful in the context of the reaction?

47. A patient in a hospital receives an intravenous glucose solution from a cylindrical bottle of radius 6 cm and height 25 cm. The bottle has a control, adjusted so that the fluid level drops $\frac{1}{4}$ cm per minute. Recalling that the volume of a right circular cylinder of radius r and height h is $\pi r^2 h$, give a formula for

(a) the amount S of solution in cubic centimeters (cm^3) that has entered the patient's vein when the fluid height is h cm;
(b) the height of the fluid (in cm) t minutes after the bottle is hooked up to the patient (assume the bottle is full initially);
(c) the amount S as a function of t.
(d) How long does it take for all the fluid to enter the patient's vein?

48. Fluid motion in a vessel with circular cross section is governed by a law formulated in 1843 by the French physiologist Jean L. M. Poiseuille (1799–1869). That law states that $v = C(R^2 - r^2)$, where v is the velocity of the flow at distance r from the center of a vessel of radius R, and C is a constant that is determined by the viscosity of the fluid, the fluid pressure, and the length of the vessel. See **Figure 3.8.**

(a) When r is close to R, what can you say about v?
(b) For what value of r is v largest?
(c) How do (a) and (b) account for the possibility of fat globules being deposited on arterial walls (arteriosclerosis)?

FIGURE 3.8

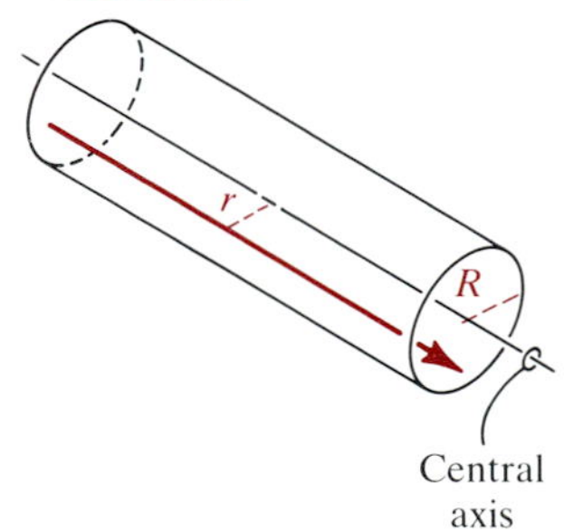

49. Classify the following functions as even, odd, or neither.

(a) $f(x) = a$ (constant function)
(b) $f(x) = x^2 + 1$
(c) $f(x) = x^3 - x$
(d) $f(x) = 3x^4 - 2x^2 + 1$
(e) $f(x) = |x|$
(f) $f(x) = (x-3)^5$

50. If f and g are even, then show that

(a) $f + g$ is even.
(b) $f \cdot g$ is even.
(c) $f \circ g$ is even.

51. If f and g are odd, then show that

(a) $f + g$ is odd.
(b) $f \cdot g$ is even.

52. If $f(x) = 1/x$, for $x \neq 0$, then identify the function $f \circ f$.

HISTORICAL NOTE

Charles's law in Exercise 43 was formulated in 1787 by Jacques A. C. Charles (1746–1823), a French physicist and pioneer balloonist. He made the first ascent in a hydrogen-filled balloon in 1783, ten days after the first ascent by a hot-air balloon had occurred. His contributions to ballooning rather than his gas law, or his calculus text published in 1784, are credited with having helped him survive the French revolution. In fact, his gas law was never publicized by him and seems to have first become widely known through its rediscovery in 1802 by Joseph L. Gay–Lussac (1778–1850), a French chemist and another pioneer balloonist. (In chemistry, Charles's law is often called the law of Charles and Gay–Lussac.) Gay–Lussac set an altitude record of 23,000 ft in a hydrogen balloon in 1804, a record that endured for half a century. Both men's interest in gases seems to have stemmed from ballooning.

1.4 Limits: A Measure of Change

In the Prologue, we said that calculus provides mathematical tools for studying change. As a start, we introduced incremental change in Section 1.2. A familiar example involving increments is the calculation of the average speed of an automobile over some time period. Let the distance (in miles or kilometers) traveled after t hours be

$$y = f(t).$$

Then the car travels distance

$$\Delta y = f(t_0 + \Delta t) - f(t_0) \tag{1}$$

during a time period of length Δt beginning at $t = t_0$. Its *average velocity* during this period is obtained by dividing Δy by Δt. From (1), the formula for average velocity is therefore

$$\frac{\Delta y}{\Delta t} = \frac{f(t_0 + \Delta t) - f(t_0)}{\Delta t}. \tag{2}$$

This quantity is of considerable interest, but its usefulness is somewhat limited. If, for example, a police radar unit records the instantaneous speed of your car as 80 in a zone with a speed limit of 50, then it will do you little good to make the computation in (2), even if your average velocity during the last hour was only 40!

This brings us back to a question raised in the Prologue. Namely, what is the relationship between the average velocity computed from (2) over the time interval Δt and the *instantaneous* velocity v_0 at time t_0? To answer that, consider $\Delta y/\Delta t$ for smaller and smaller elapsed times Δt. It seems reasonable that for very small time increments Δt, the average velocity $\Delta y/\Delta t$ over the interval $(t_0, t_0 + \Delta t)$ should be close to the instantaneous velocity v_0 at t_0. Furthermore, as Δt gets smaller and smaller, the average velocity $\Delta y/\Delta t$ should be closer and closer to the speedometer reading at $t = t_0$. This is the approach used in physics to define the instantaneous velocity at $t = t_0$ as the *limit* of the average velocity

$\Delta y/\Delta t$ as Δt approaches 0. To illustrate this idea, we consider a simpler, but no less exciting, type of motion.

Example 1

A sky diver experiences free fall before opening his parachute. The number y of feet fallen after t seconds is closely approximated by the function

$$y = 16t^2 = f(t).$$

Find the sky diver's velocity 4 seconds after beginning his fall.

Solution. We first use (2) to investigate the average velocity. We have

$$\frac{\Delta y}{\Delta t} = \frac{f(4 + \Delta t) - f(4)}{\Delta t} = \frac{16(4 + \Delta t)^2 - (16)(4)^2}{\Delta t},$$

$$= 16 \cdot \frac{\cancel{16} + 8\,\Delta t + (\Delta t)^2 - \cancel{16}}{\Delta t} = 16 \cdot \frac{\cancel{\Delta t}(8 + \Delta t)}{\cancel{\Delta t}},$$

$$\frac{\Delta y}{\Delta t} = 16(8 + \Delta t) = 128 + 16\,\Delta t, \tag{3}$$

for small (but nonzero) values of Δt. From (3), it seems clear that as Δt becomes smaller and smaller, the average velocity gets closer and closer to 128. We say then that the *limit* of the average velocity, as Δt approaches 0, is 128. By definition then, the instantaneous velocity at $t_0 = 4$ is 128. ■

FIGURE 4.1

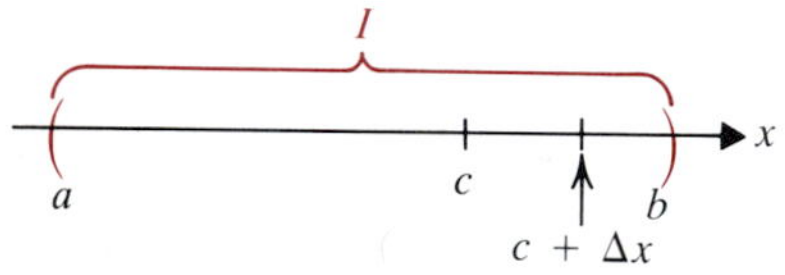

The above ideas are not confined to motion, but rather occur widely in mathematics and its applications. In general, suppose a function f is defined on an open interval I containing a point c as in **Figure 4.1.** Suppose x moves from $x = c$ to $x = c + \Delta x$, another value in I. Then we call

$$\Delta y = f(c + \Delta x) - f(c)$$

the corresponding incremental change in y, as illustrated in **Figure 4.2.** The ***average rate of change of y = f(x) with respect to x*** over the interval with endpoints c and $c + \Delta x$ is

$$\frac{\Delta y}{\Delta x} = \frac{f(c + \Delta x) - f(c)}{\Delta x}. \tag{2'}$$

FIGURE 4.2

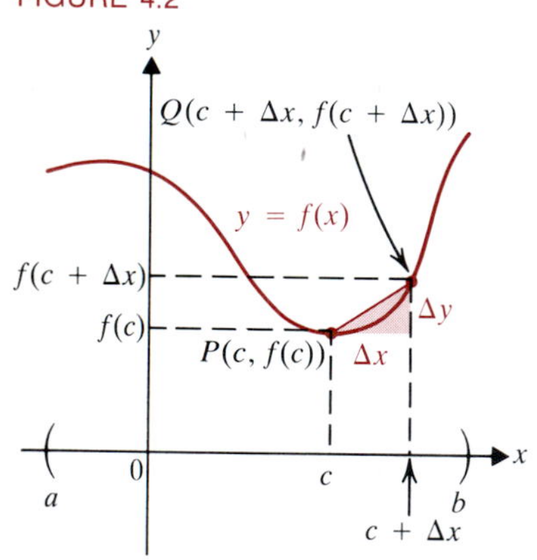

Notice that this is just the slope of the *secant PQ*, the segment in Figure 4.2 joining the points $P(c, f(c))$ and $Q(c + \Delta x, f(c) + \Delta y)$ on the graph of $y = f(x)$. As in Example 1, the instantaneous rate of change of y with respect to x is defined to be the limit of the average rate of change, as Δx approaches 0.

Example 2

Investigate the instantaneous rate of change of y with respect to x at $x = 1$ if $y = \sqrt{x}$.

Solution. If x changes from 1 to $1 + \Delta x$, then the corresponding change in y is

$$\Delta y = \sqrt{1 + \Delta x} - \sqrt{1} = \sqrt{x} - 1.$$

Thus values of x near 1 produce small values of both Δx and Δy, and

$$(4) \qquad \frac{\Delta y}{\Delta x} = \frac{\sqrt{1 + \Delta x} - 1}{\Delta x} = \frac{\sqrt{x} - 1}{x - 1}.$$

We can investigate this as Δx gets closer and closer to 0 by computing

$$g(x) = \frac{\sqrt{x} - 1}{x - 1}$$

as x gets closer and closer to 1. A calculator was used to compile Table 1.

TABLE 1

x	0.9	0.99	0.999	0.9995	0.9999	1.0001	1.001	1.01	1.1
$g(x)$	0.513167	0.501256	0.500126	0.500064	0.50002	0.49998	0.499875	0.4987562	0.4880885

From this it *seems* that $\Delta y/\Delta x$ has limit 0.5 as x approaches 1, that is, as Δx approaches 0. But this is not clear from (4), as the limit 128 was from (3) in Example 1. The problem is that as x gets close to 1, the numerator and denominator of (4) both approach 0, so the trend for their quotient is not evident. To get around this, we use the familiar algebraic formula $a^2 - b^2 = (a - b)(a + b)$ to rationalize the numerator, with $\sqrt{x}$ playing the role of a and 1 playing the role of b. If we multiply the numerator and denominator of (4) by $a + b = \sqrt{x} + 1$, then we get

$$(5) \qquad \frac{\Delta y}{\Delta x} = \frac{\sqrt{x} - 1}{x - 1} \cdot \frac{\sqrt{x} + 1}{\sqrt{x} + 1} = \frac{x - 1}{(x - 1)(\sqrt{x} + 1)}.$$

Now, as x *approaches* 1, it is *unequal* to 1. This permits the nonzero factor $x - 1$ to be cancelled from the numerator and denominator of (5), which gives

$$(6) \qquad \frac{\Delta y}{\Delta x} = \frac{\cancel{x - 1}}{(\cancel{x - 1})(\sqrt{x} + 1)} = \frac{1}{\sqrt{x} + 1}.$$

As x approaches 1, it seems evident that $\sqrt{x} \to 1$ also. Thus from (6), it now seems clear that as x approaches 1, $\Delta y/\Delta x$ approaches $0.5 = 1/(1 + 1)$. ■

Examples 1 and 2 seem to suggest that it is always possible to work out limits by doing some algebraic manipulation. But consider the limit of

$$(7) \qquad f(x) = \frac{\sin x}{x}$$

as x approaches 0. A calculator in radian mode produced Table 2.

TABLE 2

x	0.1, −0.1	0.01, −0.01	0.001, −0.001	0.0001, −0.0001
$f(x)$	0.9983342	0.9999833	0.9999998	0.999999996

Table 2 makes it appear that $f(x)$ has limit 1 as x approaches 0, but no amount of algebraic cleverness will make that obvious from the formula for $f(x)$. Indeed, a considerable part of Section 2.4 is devoted to showing that this limit really *is* 1.

The function g defined by

$$g(x) = \frac{\sqrt{x} - 1}{x - 1}$$

in Example 2 has a limit as x approaches 1, even though g is not defined at $x = 1$. The same is true of $f(x)$ in (7) for $x = 0$, but algebraic manipulation is ineffective in calculating its limit. As this suggests, we cannot define the concept of limit in terms of algebra. Indeed, the formulation of the idea of limit was a major advance in mathematics. (See the historical note at the end of this section.) The basic idea is that f has limit L as x approaches c if the *values of $f(x)$ become closer and closer to L as x gets closer and closer to c*, as illustrated by Tables 1 and 2 above. **Figure 4.3** shows this graphically for a function f not defined at $x = c$ but having limit L as x approaches c. As x moves along the x-axis toward c, $f(x)$ gets closer and closer to the value L.

FIGURE 4.3

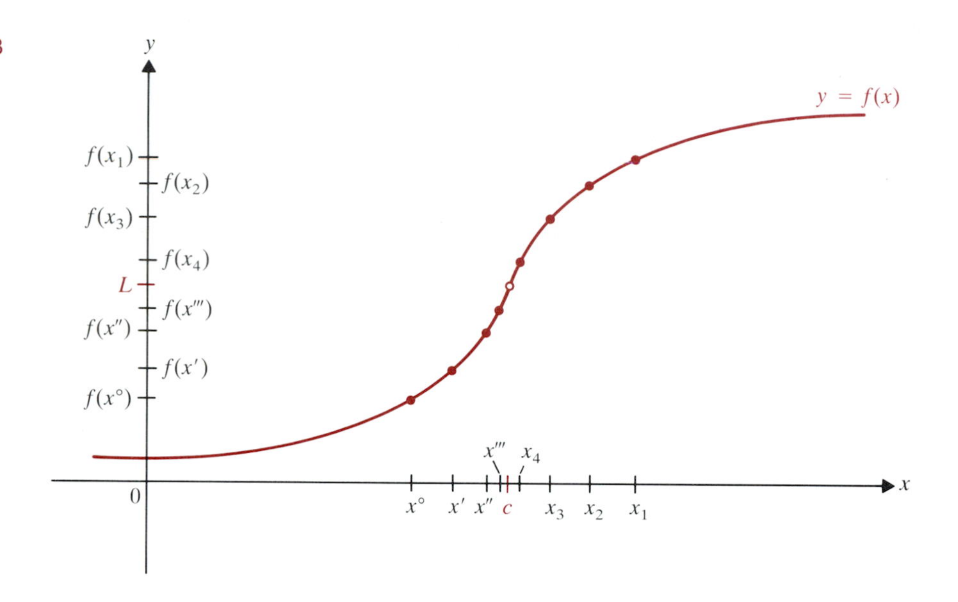

To express this more precisely, we use the fact that *two real numbers are close if and only if the distance between them on the real line is small.* That in turn is equivalent to the absolute value of their difference being small. This is brought out by **Figure 4.4,** which plots $|x - c|$ vs. x for the points along the x-axis in Figure 4.3. We can thus express the idea of limit as follows, where, to simplify notation, we introduce the symbol $\to$ to stand for *approaches.*

FIGURE 4.4

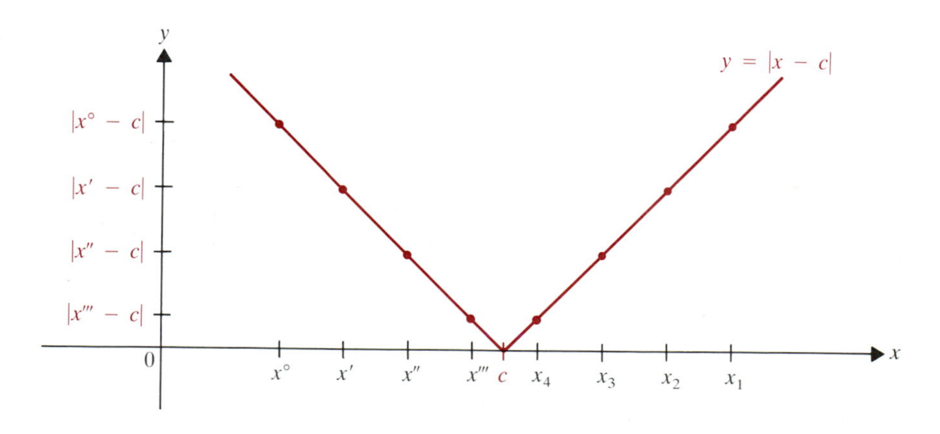

> The statement *the limit of* $f(x)$ *as* $x \to c$ *is* L means that $|f(x) - L|$ becomes and stays arbitrarily small (that is, as small as we like) when $|x - c|$ becomes sufficiently small.

The essential idea here is *change*: $|f(x) - L|$ changes in response to $|x - c|$ changing. We are not so much interested in what happens to f *at* the point $x = c$ as we are in the behavior of $f(x)$ as x *approaches* c through values close but unequal to c. Referring to Table 2 (where $L = 1$ and $c = 0$), for instance, it seems that *given any positive tolerance*, such as 0.01, 0.001, and 0.0001, *we can make* $|f(x) - L|$ *smaller than that tolerance by making* $|x - c|$ *sufficiently close to 0.* For example, the second and later columns of Table 2 suggest that we can make

$$\left|\frac{\sin x}{x} - 1\right| < 0.0001$$

by taking $|x - 0| < 0.01$. This is the crux of the formal definition of limit.

4.1 DEFINITION

Let f be defined at all points on an open interval I containing c, except possibly at c itself. Then the statement ***the limit of*** $\boldsymbol{f(x)}$ ***as*** $\boldsymbol{x}$ ***approaches*** $\boldsymbol{c}$ ***is*** $\boldsymbol{L}$, written

$$\lim_{x \to c} f(x) = L, \tag{8}$$

means that for any positive number ε, there is some positive number δ such that

$$|f(x) - L| < \varepsilon \qquad \text{whenever } 0 < |x - c| < \delta. \tag{9}$$

This is the fundamental definition of calculus and so deserves further comment. First, to emphasize that the limit concept is concerned with how the values of f *change* as $x \to c$, *we don't even require that f be defined at c.* Thus in (9), the condition $0 < |x - c|$ is present because we may not be able to evaluate $f(c)$, so we can't require $|f(c) - L| < \varepsilon$. Furthermore, even if $f(c)$ *is* defined, we really are interested in the behavior of f as x *approaches* c. In Example 2, we were able to compute

$$\lim_{x \to 1} \frac{\sqrt{x} - 1}{x - 1}$$

even though the function g defined by $(\sqrt{x} - 1)/(x - 1)$ *isn't* defined at $x = 1$.

FIGURE 4.5

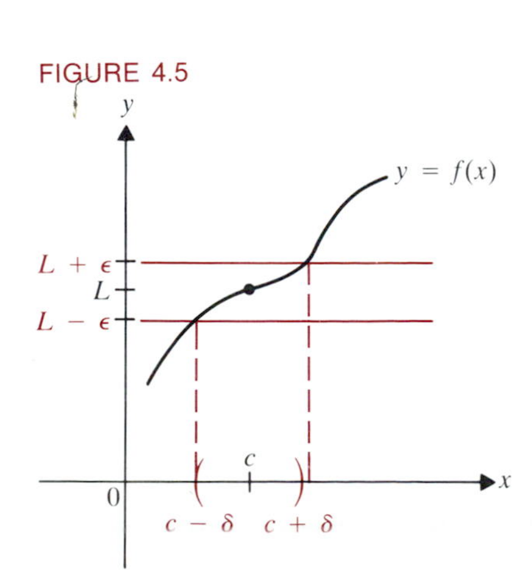

The Greek letters ε (epsilon) and δ (delta) are used in Definition 4.1 to convey the idea that we can make $|f(x) - L|$ *arbitrarily* small (less than *any* prescribed tolerance $\varepsilon > 0$) by making $|x - c|$ *sufficiently* small (less than some $\delta > 0$).

Geometrically, (9) says that the graph of $y = f(x)$ will lie within a strip between $y = L + \varepsilon$ and $y = L - \varepsilon$ if $x \neq c$ is between $c - \delta$ and $c + \delta$ (**Figure 4.5**). For, by Theorem 1.2(c), we can rewrite (9) in the form

$$L - \varepsilon < f(x) < L + \varepsilon \qquad \text{whenever } c - \delta < x < c + \delta,\ x \neq c$$

Given a tolerance ε and a reasonably simple function f, it is usually not hard to find a suitable δ in Definition 4.1. The next example illustrates this for $\lim_{x \to 2} x^2$, which pretty clearly must be 4.

Example 3

If $\varepsilon = 0.0001$, then find δ so that (9) is satisfied for $f(x) = x^2$, $c = 2$, and $L = 4$.

Solution. We are to find δ so that

$$|x^2 - 4| < 0.0001 \qquad \text{whenever } 0 < |x - 2| < \delta.$$

By Theorem 1.2(a), the first condition is equivalent to $-0.0001 < x^2 - 4 < 0.0001$, that is, to

$$(10) \qquad 3.9999 < x^2 < 4.0001.$$

Taking positive square roots, we find that (10) will hold if

$$(11) \qquad 1.999976 < x < 2.000024.$$

On subtracting 2 in (11), we obtain

$$-0.000024 < x - 2 < 0.000024,$$

which is equivalent to $|x - 2| < 0.000024$. We can thus use 0.000024 for δ. (Naturally, any *smaller* positive number, such as 0.00001, would also do for δ.) ■

It is not always as simple to find $\lim_{x \to c} f(x)$ as it was in Example 3. The next theorem is very helpful in computing limits of functions when substitution of $x = c$ into the formula for f fails to give useful information. It is based on the crucial fact mentioned above that $\lim_{x \to c} f(x)$ depends on the behavior of the function f *near* the limit point $x = c$.

4.2 THEOREM

Suppose that $f(x) = g(x)$ for all x in some interval I containing c, *except possibly at c itself*. If $\lim_{x \to c} g(x) = L$, then $\lim_{x \to c} f(x) = L$ also.

Proof. Let $\varepsilon > 0$ be given. Since $g(x)$ has limit L at c, there is some $\delta > 0$ such that

$$(12) \qquad |g(x) - L| < \varepsilon \qquad \text{whenever } 0 < |x - c| < \delta.$$

But for $0 < |x - c| < \delta$, we are told that $f(x) = g(x)$. Thus (12) says

$$|f(x) - L| < \varepsilon \qquad \text{whenever } 0 < |x - c| < \delta.$$

Hence, $\lim_{x \to c} f(x) = L$. QED

The next example suggests the usefulness of Theorem 4.2.

Example 4

If $f(x) = \dfrac{x^2 - x - 6}{x + 2}$, then find $\lim_{x \to -2} f(x)$.

Solution. As x gets close to -2, both $x^2 - x - 6 = (x - 3)(x + 2)$ and $x + 2$ become closer and closer to 0. So it is not clear what happens to their quotient $f(x)$. But since

$$f(x) = \frac{(x - 3)\cancel{(x + 2)}}{\cancel{x + 2}} = x - 3 = g(x),$$

FIGURE 4.6

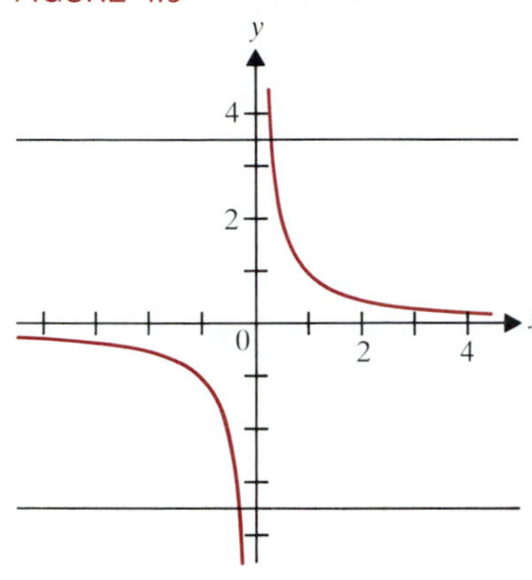

for all x near but unequal to -2, Theorem 4.2 says that

$$\lim_{x \to -2} f(x) = \lim_{x \to -2} g(x).$$

Now, as $x \to -2$, it is evident that $g(x) = x - 3$ approaches -5. Thus $\lim_{x \to -2} f(x) = -5$. ■

Not all functions have a limit at every point c. Consider for instance $f(x) = 1/x$ at $c = 0$. This function is not defined at $x = 0$, but that alone does not conflict with Definition 4.1. However, a glance at the graph in **Figure 4.6** shows the problem: f behaves far too wildly near $c = 0$ to have a limit. Indeed, for x positive and close to 0, $1/x$ is an enormous positive number, while for x negative and close to 0, $1/x$ is an enormous negative number. Thus the graph is not confined to *any* strip between $y = L + \varepsilon$ and $y = L - \varepsilon$ for x near 0: Figure 4.6 is clearly incompatible with the geometric criterion for a limit. (Compare Figures 4.5 and 4.6.)

Proving that $\lim_{x \to c} f(x) = L$*

It is possible to prove rigorously that limits of many elementary functions exist and have the expected values. To do so, we must show that condition (9) of Definition 4.1 holds. That involves finding, for any given $\varepsilon > 0$, a corresponding sufficiently small positive number δ such that

(9) $$|f(x) - L| < \varepsilon \qquad \text{whenever } 0 < |x - c| < \delta.$$

Since for smaller values of ε we usually need to use smaller values of δ, we can think of $\delta = \delta(\varepsilon)$ as depending on epsilon. This procedure can become rather involved, so we discuss it only for a simple case.

Example 5

Show that $\lim_{x \to 2} (5x - 3) = 7$.

Solution. Let $\varepsilon > 0$ be any positive real number. We want to find $\delta > 0$ so that (9) holds. Here, $f(x) = 5x - 3$, $c = 2$, and $L = 7$. Thus

(13) $$|f(x) - L| = |5x - 3 - 7| = |5x - 10| = 5|x - 2|.$$

We can then be sure that

$$|f(x) - L| < \varepsilon$$

if we have

(14) $$5|x - 2| < \varepsilon,$$

that is, if

(15) $$|x - 2| < \frac{\varepsilon}{5}.$$

Accordingly, we can let $\delta = \varepsilon/5$ (or $\varepsilon/10$ or any *smaller* positive number). Equations (13), (14), and (15) say that $|f(x) - L| < \varepsilon$ whenever $0 < |x - 2| < \delta$. ■

* Optional

HISTORICAL NOTE

Though the details of their theories were very different, the cofounders of calculus, Isaac Newton (1642–1727) of England and Gottfried W. Leibniz (1646–1716) of Germany, both tried to define limits in a purely algebraic way. They did so by introducing *infinitesimals,* quantities that were "infinitely small"—smaller than any positive real number but still greater than zero. Whatever infinitesimals were, then, they were not real numbers.

This approach laid calculus open to savage criticism, such as that leveled by Bishop George Berkeley (1685–1753). Enraged by criticism from mathematicians of allegedly imprecise *religious* thinking, this Anglican prelate authored a scathing essay, *The Analyst*, in 1734, in which he derisively referred to infinitesimals as the "ghosts of departed quantities." Over the next century, several mathematicians tried unsuccessfully to formulate a better theory of limits, but it was not until the German mathematician Karl Weierstrass (1815–1897) that the theory of limits was put on a logically sound foundation. Weierstrass managed the seemingly impossible: giving a definition of limit (essentially Definition 4.1) that avoided all reference to change, motion, geometry, or infinitesimals. His definition involves only real numbers, absolute value, and inequalities—all treated in precalculus!

One of the great achievements of modern mathematical logic occurred in 1960 when Abraham Robinson (1918–1974), then of the Hebrew University in Jerusalem and later of the University of California, Los Angeles, and Yale University, showed that there *is* a mathematically rigorous way to develop calculus using infinitesimals. To do so, one has to work with a larger system than the real line **R**, namely, the *hyperreal* number system ***R**. This set contains infinitesimals, that is, elements i satisfying $0 < i < p$ for every positive real number p. One can then define $f(x)$ to be *infinitely close* to L in case $L - f(x)$ is an infinitesimal, and this permits the definition of $\lim_{x\to c} f(x) = L$ to be that $f(x)$ is infinitely close to L whenever x is infinitely close to c! For an elementary account for this nonstandard approach to calculus, see *Elementary Calculus* by H. Jerome Keisler (Prindle, Weber & Schmidt, Boston, 1986).

Exercises 1.4

In Exercises 1–18, determine the value of the limit.

1. $\lim_{x\to 3} (x^3 - 5x^2 + x + 1)$

2. $\lim_{x\to -2} (4x^2 + 7x - 2)$

3. $\lim_{x\to 3} \dfrac{\sqrt[3]{x^2 - 5x - 2}}{2x - 4}$

4. $\lim_{x\to -2} \dfrac{\sqrt{x^3 - 8x + 1}}{5x + 1}$

5. $\lim_{x\to 0} |x^2 - 3x - 2|$

6. $\lim_{x\to 2} \dfrac{|2x - 4|}{3x - 4}$

7. $\lim_{x\to 3} \dfrac{1}{3x^2 - 2x - 8}$

8. $\lim_{x\to -3} \sqrt{5 - 2x - x^2}$

9. $\lim_{x\to 2} \dfrac{\dfrac{1}{x+1} - \dfrac{1}{3}}{x - 2}$

10. $\lim_{x\to 4} \dfrac{\dfrac{1}{\sqrt{x}} - \dfrac{1}{2}}{x - 4}$

11. $\lim_{x\to 3} \dfrac{x^2 - 9}{x - 3}$

12. $\lim_{x\to -1} \dfrac{x^4 - 1}{x + 1}$

13. $\lim_{x\to 16} \dfrac{\sqrt{x} - 4}{x - 16}$

14. $\lim_{x\to 8} \dfrac{\sqrt[3]{x} - 2}{x - 8}$

15. $\lim_{x\to -3} \dfrac{x^2 - x - 12}{x + 3}$

16. $\lim_{x\to -2} \dfrac{x^2 - 4x - 12}{x + 2}$

17. $\lim_{x\to -1} \dfrac{x^3 + 1}{x + 1}$

18. $\lim_{x\to -2} \dfrac{x^5 + 32}{x + 2}$

In Exercises 19–24, find a value δ corresponding to the given ε so that (9) of Definition 4.1 is satisfied.

19. $\lim_{x\to 2} (-3x + 1) = -5$, $\varepsilon = 0.01$

20. $\lim_{x\to -1} (2x + 3) = 1$, $\varepsilon = 0.01$

21. $\lim_{x\to -1} x^2 = 1$, $\varepsilon = 0.001$

22. $\lim_{x\to 2} (1 - x^2) = -3$, $\varepsilon = 0.001$

23. $\lim_{x\to 7} \sqrt{x-3} = 2$, $\varepsilon = 0.001$

24. $\lim_{x\to 1} \dfrac{1}{x+3} = \dfrac{1}{4}$, $\varepsilon = 0.001$

In Exercises 25–32, use Definition 4.1 to show the limits have the values given.

25. $\lim_{x\to 3} (3x-5) = 4$

26. $\lim_{x\to -2} (2x+1) = -3$

27. $\lim_{x\to -3} (-x+1) = 4$

28. $\lim_{x\to 5} (2x-3) = 7$

29. $\lim_{x\to 0} x^2 = 0$

30. $\lim_{x\to 0} x^3 = 0$

31. $\lim_{x\to 4} \sqrt{x} = 2$ (*Hint:* See Example 2.)

32. $\lim_{x\to 25} \sqrt{x-16} = 3$

In Exercises 33–36, explain why the limit fails to exist.

33. $\lim_{x\to 2} \dfrac{1}{x-2}$

34. $\lim_{x\to -1} \dfrac{1}{x+1}$

35. $\lim_{x\to 0} \dfrac{|x|}{x}$

36. $\lim_{x\to 1} \dfrac{x-1}{|x-1|}$

37. The ***Dirichlet function*** is named for the German mathematician P. G. Lejeune Dirichlet (1805–1859) who made many important contributions to mathematics and mathematical physics. In 1829, he defined a function g by the formula

$$g(x) = \begin{cases} 1 & \text{if } x \text{ is rational} \\ 0 & \text{if } x \text{ is irrational} \end{cases}$$

Does $\lim_{x\to c} g(x)$ exist for *any* c? Explain.

38. Let $f(x) = [x]$ be the greatest integer n such that $n \le x$. For which numbers c does $\lim_{x\to c} f(x)$ exist? Explain.

39. Use Definition 4.1 to show that if $\lim_{x\to c} f(x)$ exists, then it is unique. Do this by showing that if L_1 and L_2 are two limits for $f(x)$ as $x \to c$, then $L_1 = L_2$. (*Hint:* Assume that $L_1 \ne L_2$ and let $\varepsilon = \frac{1}{2}|L_1 - L_2|$. Then derive a contradiction.)

40. Definition 4.1 gives the condition for existence of $\lim_{x\to c} f(x)$. Negate that condition to obtain a condition for $\lim_{x\to c} f(x)$ to fail to exist.

1.5 Properties of Limits

This section contains several results that can be used to evaluate many limits without recourse to Definition 4.1. Our emphasis is on using, rather than deriving, these basic properties. Thus most of the proofs are deferred to Appendix II. Throughout, we use the symbol c for any limit point on the real line. The first result deals with some simple functions and the simplest operations on functions. A partial proof can be found at the end of the section.

5.1 THEOREM

(a) If $f(x) = a$ for all x, then $\lim_{x\to c} f(x) = a$. That is, *the limit of a constant function at any point is its constant value.*

(b) If $f(x) = x$ for all x, then $\lim_{x\to c} f(x) = c$.

(c) Let a be any fixed real number. If $\lim_{x\to c} f(x) = L$, then $\lim_{x\to c} af(x) = aL$. That is, *the limit of a constant times a function is the constant times the limit of the function, if that limit exists.*

(d) If $\lim_{x\to c} f(x) = L$ and $\lim_{x\to c} g(x) = M$, then

$$\lim_{x\to c} (f+g)(x) = L + M \qquad \text{and} \qquad \lim_{x\to c} (f-g)(x) = L - M.$$

That is, *the limit of the sum (or difference) of two functions is the sum (or difference) of the limits,* if those limits exist.

Theorem 5.1 makes it easy to evaluate the limit of any linear function.

Example 1

Find (a) $\lim_{x \to 2} (3x - 5)$ (b) $\lim_{x \to c} (mx + b)$.

Solution. (a) $\lim_{x \to 2} (3x - 5) = \lim_{x \to 2} 3x - \lim_{x \to 2} 5$ *by Theorem 5.1(d)*

$= 3 \lim_{x \to 2} x - 5$ *by Theorem 5.1 (c, a)*

$= 3 \cdot 2 - 5$ *by Theorem 5.1(b)*

$= 6 - 5 = 1.$

(b) Using exactly the same reasoning as in (a), we have

$$\lim_{x \to c} (mx + b) = \lim_{x \to c} mx + \lim_{x \to c} b$$

$$= m \lim_{x \to c} x + b = mc + b. \blacksquare$$

By repeatedly applying it or using mathematical induction (Appendix III), we can extend Theorem 5.1(d) to cover the limit of the sum of *any* number of functions. (The proof is asked for in Exercise 30.)

If $\lim_{x \to c} f_i(x) = L_i$ for $i = 1, 2, \ldots, n$, then

$$(1) \qquad \lim_{x \to c} [f_1(x) + f_2(x) + \ldots + f_n(x)] = L_1 + L_2 + \ldots + L_n.$$

The next result concerns the limits of two particularly important functions and is helpful in deriving more general rules for computing limits. The proof is given in Appendix II.

5.2 THEOREM

(a) $\lim_{x \to c} x^2 = c^2$ (b) $\lim_{x \to c} \frac{1}{x} = \frac{1}{c}$ if $c \neq 0$.

Example 2

Calculate $\lim_{x \to 3} \left(\frac{6}{x} - 3x^2 + 2x\right)$.

Solution.

$$\lim_{x \to 3}\left(\frac{6}{x} - 3x^2 + 2x\right) = \lim_{x \to 3}\left(6 \cdot \frac{1}{x} - 3x^2 + 2x\right)$$

$= 6 \lim_{x \to 3} \frac{1}{x} - 3 \lim_{x \to 3} x^2 + 2 \lim_{x \to 3} x$ *by Theorem 5.1(c) and (1)*

$= 6 \cdot \frac{1}{3} - 3 \cdot 9 + 2 \cdot 3$ *by Theorems 5.2 and 5.1(b)*

$= -19. \blacksquare$

In Theorem 5.2, both limits can be evaluated by *substituting* c for x in the formula for the function. That is,

$$\lim_{x \to c} f(x) = f(c) \tag{2}$$

in case $f(x) = x^2$ or $f(x) = 1/x$. In Section 7, we will see that many important functions satisfy (2). The next result gives a powerful means of evaluating limits of composite functions $h \circ g$ (see Definition 3.2) if h satisfies (2). Its proof is also found in Appendix II.

5.3 THEOREM

Suppose that $\lim_{x \to c} g(x) = M$ and $\lim_{t \to M} h(t) = h(M)$. Then

$$\lim_{x \to c} h(g(x)) = h(M) = h\left(\lim_{x \to c} g(x)\right).$$

The next example illustrates the use of Theorem 5.3.

Example 3

Find $\lim_{x \to 1} (2x - 3)^2$.

Solution. We can use Theorem 5.3 with $h(t) = t^2$ and $g(x) = 2x - 3$. Example 1(b) gives

$$\lim_{x \to 1} g(x) = \lim_{x \to 1} (2x - 3) = 2 \cdot 1 - 3 = -1 = M.$$

Hence,

$$\lim_{x \to 1} (2x - 3)^2 = h(M) = h(-1) = (-1)^2 = 1. \blacksquare$$

We claimed that Theorem 5.3 is a powerful result. To illustrate its power, we next show how it can be used to establish analogues of Theorem 5.1(d) for multiplication and division without recourse to Definition 4.1.

5.4 THEOREM

Suppose that $\lim_{x \to c} f(x) = L$ and $\lim_{x \to c} g(x) = M$. Then

(a) $\lim_{x \to c} (f \cdot g)(x) = LM$, (b) $\lim_{x \to c} \dfrac{1}{g(x)} = \dfrac{1}{M}$ if $M \neq 0$,

(c) $\lim_{x \to c} (f/g)(x) = \dfrac{L}{M}$ if $M \neq 0$.

Partial Proof. We prove (a) and leave the other parts for Exercise 31.

(a) First, consider the case $g = f$. We need to show that $\lim_{x \to c} g(x)^2 = M^2$, where $g(x)^2 = g(x) \cdot g(x)$. We apply Theorem 5.3 to $h(t) = t^2$. Observe that $g(x)^2 = h(g(x))$. By Theorem 5.2(a), $\lim_{t \to M} h(t) = M^2 = h(M)$. By the hypothesis of this theorem, $\lim_{x \to c} g(x) = M$. So Theorem 5.3 gives

$$\lim_{x \to c} g(x)^2 = \lim_{x \to c} h(g(x)) = h\left(\lim_{x \to c} g(x)\right) = h(M) = M^2. \tag{3}$$

For the general case, we write $(f \cdot g)(x) = f(x)g(x)$ in the form

(4) $$f(x)g(x) = \tfrac{1}{4}[(f(x) + g(x))^2 - (f(x) - g(x))^2].$$

[You should verify (4) by expanding and simplifying the right-hand side.] Now we can use (3) on the squared functions. We know from Theorem 5.2(d) that $\lim_{x \to c}[f(x) + g(x)] = L + M$ and $\lim_{x \to c}[f(x) - g(x)] = L - M$. Therefore, by (3), we have

$$\lim_{x \to c}[f(x) + g(x)]^2 = (L + M)^2 = L^2 + 2LM + M^2,$$

$$\lim_{x \to c}[f(x) - g(x)]^2 = (L - M)^2 = L^2 - 2LM + M^2.$$

Then (4) and Theorem 5.1(c) with $a = \tfrac{1}{4}$ give

$$\lim_{x \to c} f(x) = \frac{1}{4}[(L^2 + 2LM + M^2) - (L^2 - 2LM + M^2)] = LM. \quad \boxed{\text{QED}}$$

The next example illustrates how Theorem 5.4 can help work out limits.

Example 4

Find $\displaystyle\lim_{x \to 3} \frac{x^2 - 3x + 1}{3x^2 + 2}$.

Solution.

$$\lim_{x \to c} \frac{x^2 - 3x + 1}{3x^2 + 2} = \frac{\lim_{x \to 3}(x^2 - 3x + 1)}{\lim_{x \to 3}(3x^2 + 2)} \qquad \textit{by Theorem 5.4(c)}$$

$$= \frac{\lim_{x \to 3} x^2 - \lim_{x \to 3} 3x + \lim_{x \to 3} 1}{3\lim_{x \to 3} x^2 + \lim_{x \to 3} 2} \qquad \textit{by Theorem 5.1}$$

$$= \frac{3^2 - 3 \cdot 3 + 1}{3 \cdot 3^2 + 2} \qquad \textit{by Theorems 5.2(a) and 5.1}$$

$$= \frac{1}{29}. \quad \blacksquare$$

With practice, you will soon be able to shorten the work shown in Example 4. In fact, we now have enough tools to quickly evaluate the limits of many common functions. First, we can extend Theorem 5.4(a) as we did Theorem 5.1(d). (See Exercise 32.)

> If $\lim_{x \to c} f_i(x) = L_i$ for $i = 1, 2, \ldots, n$, then
>
> (5) $$\lim_{x \to c}[f_1(x) \cdot f_2(x) \ldots f_n(x)] = L_1 \cdot L_2 \ldots L_n.$$

A special case of this, for the functions $f_1(x) = f_2(x) = x$ and $n = 2$, is Theorem 5.2(a). We now see that for any positive integral power n,

(6) $$\lim_{x \to c} x^n = c^n$$

Moreover, it follows at once from (6) and (1) that for any polynomial function,

$$p(x) = a_0 + a_1x + a_2x^2 + \dots + a_nx^n,$$

we have $\lim_{x \to c} p(x) = p(c)$. Consequently, these functions also satisfy condition (2) following Example 2. This fact and Theorem 5.4(c) say that (2) also holds for *rational functions*—quotients of polynomial functions:

If $f(x) = \dfrac{p(x)}{q(x)}$, where $p(x)$ and $q(x)$ are polynomial functions, then

(7) $$\lim_{x \to c} f(x) = f(c) \text{ provided that } q(c) \neq 0.$$

Thus we can evaluate many limits (like the one in Example 4) by simple substitution.

Example 5

Find $\displaystyle\lim_{x \to 2} \frac{x^3 - 3x^2 + 4x + 1}{x^5 - 2x^2 - 5x - 1}$.

Solution. Here $q(c) = q(2) = 2^5 - 2(2)^2 - 5(2) - 1 = 13 \neq 0$. So by (7), the given function has limit $p(2)/q(2)$, where $p(2) = (2)^3 - 3(2)^2 + 4(2) + 1 = 5$. Hence, the limit is $5/13$. ■

The next result further extends the class of functions whose limits we can evaluate without using Definition 4.1. The proof is given in Appendix II.

5.5 THEOREM

Let c be any real number if n is odd, a positive real number if n is even. Then

$$\lim_{x \to c} \sqrt[n]{x} = \sqrt[n]{c}$$

Example 6

Find $\displaystyle\lim_{x \to 8} \frac{x^{2/3} - 5x^{1/3}}{(2x)^{1/2} + 1}$.

Solution. We use Theorems 5.4(c), 5.1, and 5.5 to get

$$\lim_{x \to 8} \frac{x^{2/3} - 5x^{1/3}}{(2x)^{1/2} + 1} = \frac{\lim\limits_{x \to 8} (x^{2/3} - 5x^{1/3})}{\lim\limits_{x \to 8} (2^{1/2}x^{1/2} + 1)}$$

$$= \frac{\lim\limits_{x \to 8} (x^{1/3} \cdot x^{1/3}) - 5 \lim\limits_{x \to 8} x^{1/3}}{2^{1/2} \lim\limits_{x \to 8} x^{1/2} + \lim\limits_{x \to 8} 1}$$

$$= \frac{\lim\limits_{x \to 8} x^{1/3} \cdot \lim\limits_{x \to 8} x^{1/3} - 5(8^{1/3})}{(2^{1/2})(8^{1/2}) + 1}$$

$$= \frac{(2)(2) - 5(2)}{4 + 1} = -\frac{6}{5}. \quad ■$$

From Theorems 5.5 and 5.3 we can obtain the following stronger version of Theorem 5.5.

5.6 COROLLARY

Suppose that $\lim_{x\to c} g(x) = M$, where $M > 0$ if n is even. Then

$$\lim_{x\to c} \sqrt[n]{g(x)} = \sqrt[n]{M} = \sqrt[n]{\lim_{x\to c} g(x)}$$

Proof. We apply Theorem 5.3 with $h(x) = \sqrt[n]{x} = x^{1/n}$. By Theorem 5.5, we have $\lim_{x\to c} h(x) = \sqrt[n]{c} = h(c)$. Then Theorem 5.3 says that

$$\lim_{x\to c} h(g(x)) = h\left(\lim_{x\to c} g(x)\right) = h(M) = M^{1/n}. \quad \boxed{\text{QED}}$$

The requirement that $M > 0$ in case n is even ensures that the nth-root function is defined on an open interval about c, so that Definition 4.1 can apply to $\sqrt[n]{g(x)}$ at $x = c$. Corollary 5.6 is even easier to use than it was to derive, as the next example illustrates.

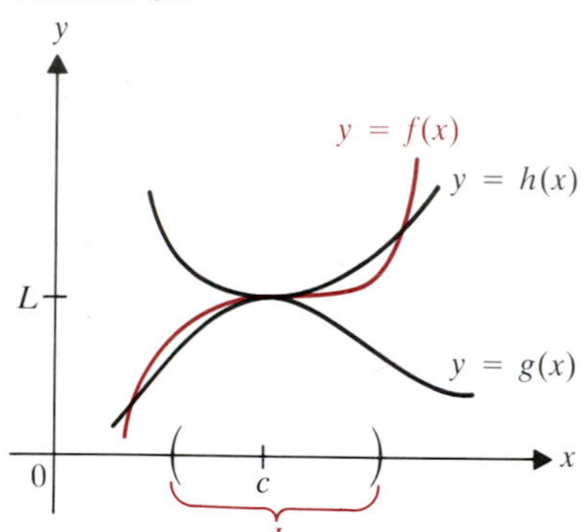

FIGURE 5.1

Example 7

Find $\lim_{x\to 3} \sqrt[3]{x^3 - 5x - 4}$.

Solution. $\lim_{x\to 3} \sqrt[3]{x^3 - 5x - 4} = \sqrt[3]{27 - 15 - 4} = \sqrt[3]{8} = 2.$ ■

The next result can often be used to find limits that are awkward to evaluate directly. If we can sandwich $f(x)$ between the values of two functions g and h that have common limit L as $x \to c$, then f also has limit L as $x \to c$. This is illustrated in **Figure 5.1.** The proof is left for Appendix II.

5.7 THEOREM

> ***Sandwich Theorem.*** Suppose that for all $x \neq c$ in an open interval I containing c,
>
> $$g(x) \le f(x) \le h(x).$$
>
> If $\lim_{x\to c} g(x) = \lim_{x\to c} h(x) = L$, then $\lim_{x\to c} f(x) = L$ also.

Notice that $g(x) \le f(x) \le h(x)$ is required only in some interval near c, not for *all* x. (Look at Figure 5.1 again.)

Example 8

Find $\lim_{x\to 0} x\sin(1/x)$.

Solution. Recall from trigonometry that all the values of the sine function lie between -1 and 1. Hence,

$$\left|\sin\frac{1}{x}\right| \le 1$$

for all x in the domain of $\sin(1/x)$. Multiplying by $|x| > 0$, we obtain

$$\left|x\sin\frac{1}{x}\right| \le |x|.$$

Then Theorem 1.2(a) gives

$$-|x| \leq x \sin \frac{1}{x} \leq |x|.$$

We can therefore apply Theorem 5.7 to $f(x) = x \sin(1/x)$ by letting $g(x) = -|x|$ and $h(x) = |x|$. It is not hard to see (Exercise 24) that $\lim_{x \to 0} |x| = 0$, so $\lim_{x \to 0} h(x) = 0$. Then by Theorem 5.1(c) with $a = -1$, we also have

$$\lim_{x \to 0} g(x) = -1 \cdot 0 = 0.$$

By the sandwich theorem, therefore, $\lim_{x \to 0} x \sin(1/x) = 0$. ■

Our final result, whose proof is also found in Appendix II, states that the limit of a nonnegative (or nonpositive) function cannot be a number lying on the other side of 0.

5.8 THEOREM

Let I be an open interval containing c. Suppose that $\lim_{x \to c} f(x) = L$.

(a) If $f(x) \geq 0$ for all $x \neq c$ in I, then $L \geq 0$.

(b) If $f(x) \leq 0$ for all $x \neq c$ in I, then $L \leq 0$.

For example, from the fact that $\cos x$ is positive for all $x \neq 0$ lying near 0, we can conclude from Theorem 5.8(a) that $\lim_{x \to 0} \cos x \geq 0$. (In Section 2.3, we will show that this limit is in fact $\cos 0 = 1$.)

Partial Proof of Theorem 5.1*

We leave the proofs of parts (a) and (b) for Exercises 27 and 28.

(c) If $\lim_{x \to c} f(x) = L$, then for any real number a, $\lim_{x \to c} af(x) = aL$.

Proof. If $a = 0$, this follows from (a) since $af(x) = 0$ for all x. Suppose then that $a \neq 0$. Let $\varepsilon > 0$ be given. We want to make

$$|af(x) - aL| = |a|\,|f(x) - L| < \varepsilon$$

by making $|x - c|$ sufficiently small. Since $\lim_{x \to c} f(x) = L$ and $\varepsilon/|a| > 0$, Definition 4.1 says that there is a $\delta > 0$ such that

$$|f(x) - L| < \frac{\varepsilon}{|a|} \qquad \text{whenever } 0 < |x - c| < \delta.$$

Then, whenever $0 < |x - c| < \delta$, we have

$$|af(x) - aL| = |a|\,|f(x) - L| < |a| \frac{\varepsilon}{|a|} = \varepsilon,$$

as required.

(d) If $\lim_{x \to c} f(x) = L$ and $\lim_{x \to c} g(x) = M$, then $\lim_{x \to c} (f + g)(x) = L + M$ and $\lim_{x \to c} (f - g)(x) = L - M$.

* Optional

Proof. We prove the first statement and leave the second for Exercise 29. Let $\varepsilon > 0$ be given. Then $\frac{1}{2}\varepsilon > 0$, so there are $\delta_1 > 0$ and $\delta_2 > 0$ such that

$$|f(x) - L| < \frac{1}{2}\varepsilon \qquad \text{if } 0 < |x - c| < \delta_1,$$

and

$$|g(x) - M| < \frac{1}{2}\varepsilon \qquad \text{if } 0 < |x - c| < \delta_2.$$

Let δ be the smaller of δ_1 and δ_2. Then if $0 < |x - c| < \delta$, the triangle inequality [Theorem 1.2(d)] says that

$$|(f + g)(x) - (L + M)| = |f(x) + g(x) - L - M| = |f(x) - L + g(x) - M|$$
$$\le |f(x) - L| + |g(x) - M| < \frac{1}{2}\varepsilon + \frac{1}{2}\varepsilon = \varepsilon. \quad \text{QED}$$

Exercises 1.5

In Exercises 1–23, evaluate the given limit using theorems of this section.

1. $\lim_{x\to 2} (3x - 1)$
2. $\lim_{x\to -1} (5 - 3x)$
3. $\lim_{x\to 1} (2x^2 - 7x + 4)$
4. $\lim_{x\to -1} (x^2 + 4x - 2)$
5. $\lim_{x\to 2} \dfrac{1}{8x - 3}$
6. $\lim_{x\to -3} \dfrac{1}{2 - 2x}$
7. $\lim_{x\to 3} \left(\dfrac{3}{x^2} - \dfrac{2}{x} + x^2 - 6\right)$
8. $\lim_{x\to -1} \left(\dfrac{5}{x^3} - \dfrac{3}{x} + 2x^2 - 4x + 1\right)$
9. $\lim_{x\to 2} \dfrac{3x^2 - 5x + 1}{x^3 - 3x + 1}$
10. $\lim_{x\to -2} \dfrac{x^2 - 2x + 3}{x^3 - x + 1}$
11. $\lim_{x\to 4} \dfrac{3\sqrt{x} - x^{3/2}}{x^{5/2} - 3x^{3/2}}$
12. $\lim_{x\to -8} \dfrac{2x^{1/3} + \sqrt{2}x^{2/3}}{x^{4/3} - x^{1/3}}$
13. $\lim_{x\to 3} \sqrt[3]{\dfrac{5x^3 - 2x^2 - 4x - 9}{x^2 + 2x - 3}}$
14. $\lim_{x\to -1} \sqrt{\dfrac{5x^{1/3} - 6x}{2x^2 + 3x + 2}}$
15. $\lim_{x\to 4} \dfrac{x^2 - 16}{x - 4}$
16. $\lim_{x\to -1} \dfrac{x^2 - 1}{x + 1}$
17. $\lim_{x\to 2} \dfrac{1/x - 1/2}{x - 2}$
18. $\lim_{x\to 3} \dfrac{1/(x + 1) - 1/4}{x - 3}$
19. $\lim_{x\to 0} \dfrac{3x^{8/3} - 27x^8 + 6}{4x^{5/3} - 17x^{17} - 2}$
20. $\lim_{x\to 1} \dfrac{15x^{17/2} - 8x^{11/9} - 18x + 2}{7x^{15} - 3x^{5/2} - 6x^5 + 8}$
21. $\lim_{x\to 0} x^2 \sin(1/x)$
22. $\lim_{x\to 0} x \cos(1/x)$
23. $\lim_{x\to 0} \left(x \sin \dfrac{1}{x^2} + x^2 \cos \dfrac{1}{x}\right)$.
24. Show that $\lim_{x\to 0} |x| = 0$.
25. If $0 \le f(x) \le k$ (a fixed constant) for all $x \ne 0$ in some interval I containing 0, then show that $\lim_{x\to 0} xf(x) = 0$.
26. Under the assumptions of Exercise 25, show $\lim_{x\to 0} x^n f(x) = 0$ for every positive integer n. What happens if n is a *negative* integer?
27. Prove Theorem 5.1(a).
28. Prove Theorem 5.1(b).
29. If $\lim_{x\to c} f(x) = L$ and $\lim_{x\to c} g(x) = M$, then prove that $\lim_{x\to c} (f - g)(x) = L - M$.
30. Prove the extended version (1) of Theorem 5.1(d).
31. (a) Prove Theorem 5.4(b). (b) Prove Theorem 5.4(c).
32. Prove the extended version (5) of Theorem 5.4(a).
33. **(a)** Show that if $\lim_{x\to c} f(x) = L$, then $\lim_{x\to c} [f(x)]^n = L^n$, for any positive integer n. [*Hint:* Use (5).]
    (b) Use Theorem 5.4(b) and (5) to show that the result in (a) also holds for negative integers n, provided that $L \ne 0$.

The next result is used in Section 6.3.

34. Suppose that $r = m/n$ is a rational number, where m and n are integers, and $n > 0$. If $\lim_{x\to c} f(x) = L$, then use Theorem 5.3, Corollary 5.6, and Exercise 33 to show that $\lim_{x\to c} [f(x)]^{m/n} = L^{m/n}$. (Assume that $L > 0$ if n is even and that $L \ne 0$ if $m < 0$.)
35. Suppose that for some fixed constant k, the function f satisfies $|f(x)| \le k$ for all $x \ne c$ in some interval I containing c. Show that if $\lim_{x\to c} g(x) = 0$, then $\lim_{x\to c} f(x) \cdot g(x) = 0$.

36. Show that $\lim_{x\to c} f(x) = L$ if and only if $\lim_{x\to c} (f(x) - L) = 0$.

37. Suppose that for some fixed constant l, the function f satisfies $|f(x) - f(c)| \leq l|x - c|$ for all $x \neq c$ in some interval I containing c. Then what is $\lim_{x\to c} f(x)$? Prove that your answer is correct.

38. If in Exercise 37 we know that $|f(x) - f(c)| \leq l|x - c|^n$ for some positive integer n, then what is $\lim_{x\to c} f(x)$?

39. Use the triangle inequality to show that
(a) $|a| - |b| \leq |a - b|$ (*Hint*: Use the fact that $a = a - b + b$);
(b) $|b| - |a| \leq |b - a|$; **(c)** $||a| - |b|| \leq |a - b|$.

40. Use Exercise 39 to show that if $\lim_{x\to c} f(x) = L$, then $\lim_{x\to c} |f(x)| = |L|$. (This generalizes Exercise 24.)

41. Suppose that $\lim_{x\to c} |f(x)| = |L|$. Then must $\lim_{x\to c} f(x) = L$? If so, then prove it. If not, then give an example in which $\lim_{x\to c} f(x) \neq L$ even though $\lim_{x\to c} |f(x)| = |L|$.

42. Suppose that $f(x) > 0$ for all $x \neq c$ in an open interval I containing c. If $\lim_{x\to c} f(x) = L$, then must L be positive? If so, then prove it. If not, then give an example in which L is not positive even though $f(x)$ is positive for all $x \neq c$ in I.

1.6 One-Sided Limits

In this section, we extend the concept of limit to a wider class of functions than those covered by Definition 4.1. Consider, for instance, the function f defined by

$$f(x) = \sqrt{x - 4}.$$

FIGURE 6.1

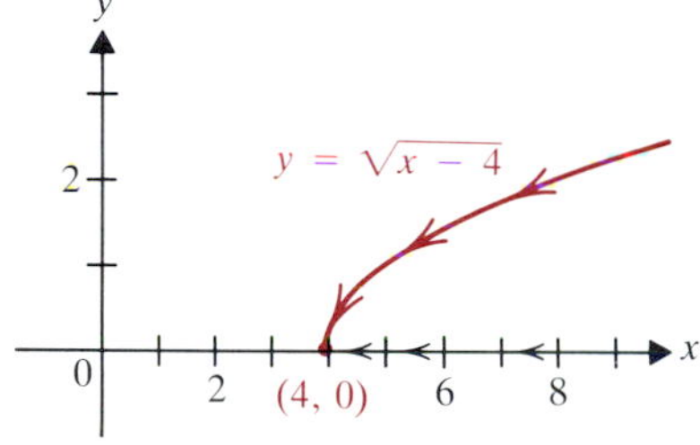

Its graph is shown in **Figure 6.1.** For every point $c > 4$, we have

$$\lim_{x\to c} f(x) = f(c) = \sqrt{c - 4},$$

by Corollary 5.6. There is no limit at $c = 4$, however. For $\lim_{x\to c} f(x)$ to exist, Definition 4.1 requires the function f to be defined throughout an entire open interval $(c - \delta, c + \delta)$ about c (except possibly at c itself). Here $f(x) = \sqrt{x - 4}$ is defined *only* for $x \geq 4$, so it is defined at no points $x < c$. It is natural, though, to say that $f(x)$ has *one-sided limit* 0 *as x approaches* 4 *from the right* (notation: $x \to 4^+$), or *from values greater than* 4. The following formal definition expresses this idea, and the corresponding notion of left-hand limit, in the language of Definition 4.1.

6.1 DEFINITION

Suppose that f is defined on an open interval (a, b). The ***one-sided limits*** of f are defined by

(a) $\lim_{x\to b^-} f(x) = L$ means for that any $\varepsilon > 0$, there is a $\delta > 0$ such that

$$|f(x) - L| < \varepsilon \qquad \text{whenever } 0 < |x - b| < \delta \text{ and } x < b,$$

that is, $x \in (b - \delta, b)$: $b - \delta < x < b$.

(b) $\lim_{x\to a^+} f(x) = R$ means that for any $\varepsilon > 0$, there is a $\delta > 0$ such that

$$|f(x) - R| < \varepsilon \qquad \text{whenever } 0 < |x - a| < \delta \text{ and } x > a,$$

that is, $x \in (a, a + \delta)$: $a < x < a + \delta$.

This is illustrated in **Figure 6.2.** The statement *the left-hand limit of* $f(x)$ *as* $x \to b^-$ *is L* means that for any $\varepsilon > 0$, there is a $\delta > 0$ such that the graph of $y = f(x)$ lies entirely within a strip between $y = L + \varepsilon$ and $y = L - \varepsilon$ when $x \in (b - \delta, b)$, as in Figure 6.2(a). There is a similar interpretation for the right-hand limit of $f(x)$ as $x \to a^+$, as shown in Figure 6.2(b).

As the next example illustrates, Definition 6.1 allows us to discuss limits of some simple functions that don't have limits according to Definition 4.1.

FIGURE 6.2

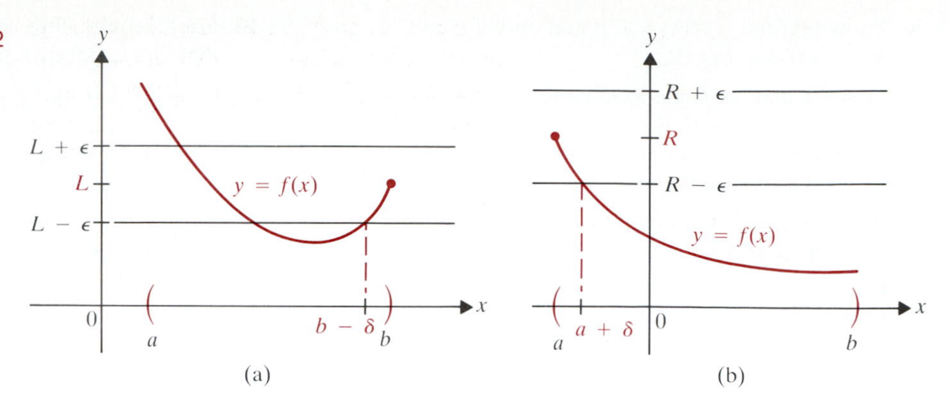

Example 1

Find the following one-sided limits.

(a) $\lim_{x \to 1^-} (3x - \sqrt{1 - x^2})$ (b) $\lim_{x \to 4^+} (\sqrt[4]{x^2 - 16} - x + 1)$

Solution. (a) Note that $1 - x^2 > 0$ on $(-1, 1)$. Thus the function $3x - \sqrt{1 - x^2}$ is defined as x approaches 1 from the left. As $x \to 1^-$, $3x \to 3$ and $\sqrt{1 - x^2} \to 0$. The analogues for one-sided limits of Corollary 5.6 and Theorem 5.1(d) give

$$\lim_{x \to 1^-} (3x - \sqrt{1 - x^2}) = 3 - 0 = 3.$$

(b) As $x \to 4^+$, $x^2 - 16 \to 0^+$ and $-x + 1 \to -3$, so we have

$$\lim_{x \to 4^+} (\sqrt[4]{x^2 - 16} - x + 1) = -3,$$

again using the analogues for one-sided limits of Corollary 5.6 and Theorem 5.1(d). ■

As in Example 1, we will often use the analogues of the theorems of the last section, which also hold for one-sided limits. For instance, if $\lim_{x \to c^+} f(x) = L$ and $\lim_{x \to c^+} g(x) = M$, then

$$\lim_{x \to c^+} [f(x) + g(x)] = L + M.$$

The proofs of the theorems of Section 5 carry over nearly verbatim to one-sided limits. (See Exercises 27–29.)

One-sided limits are helpful in dealing with functions at endpoints of their domains, as Example 1 illustrates. They are also useful in studying *schizophrenic functions*. We give this name to *functions that have multiple formulas:* one for x less than some number a, and one or more different ones for $x > a$. For instance, consider the function f given by

$$f(x) = \begin{cases} 110 + \frac{1}{2}x & \text{for } x \leq 0 \\ 190 + \frac{11}{10}x & \text{for } x \in (0, 100] \\ 844 + \frac{2}{5}(x - 110) & \text{for } x > 100 \end{cases}$$

Its graph is shown in **Figure 6.3.** Although it might look contrived, it actually models the *enthalpy H* (which may be thought of as the heat content) of water at various temperatures x measured in degrees Celsius. Enthalpy is an important concept in thermodynamics. The function f is of interest to chemists mainly at $x = 0$ and $x = 100$, where water undergoes a *change of phase*. The interest lies in the value of $\Delta y = \Delta H$ at these points. (An important application of ΔH is in determining the spontaneity of a chemical reaction. See Exercises 34-38 of Section 1 on the Gibbs-Helmoltz equation.) One-sided limits can be used to compute ΔH from the formulas for f.

FIGURE 6.3

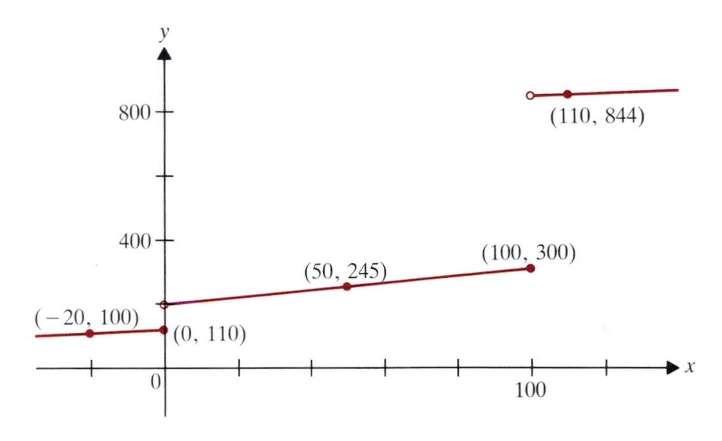

Example 2

Find the enthalpy change Δy at $x = 0$ and $x = 100$, the freezing and boiling points of water. (An appropriate unit for Δy here is calories per gram.)

Solution. Note that Δy is the size of the jump at $x = 0$ and $x = 100$. Thus at $x = 0$,

$$\begin{aligned}\Delta y &= \lim_{x\to 0^+} f(x) - \lim_{x\to 0^-} f(x)\\ &= \lim_{x\to 0^+}\left(190 + \frac{11}{10}x\right) - \lim_{x\to 0^-}\left(110 + \frac{1}{2}x\right)\\ &= 190 - 110 = 80 \text{ cal/g.}\end{aligned}$$

(In thermodynamics, this Δy is called ΔH_{fusion} for water.) At $x = 100$, we have

$$\begin{aligned}\Delta y &= \lim_{x\to 100^+} f(x) - \lim_{x\to 100^-} f(x)\\ &= \lim_{x\to 100^+}\left[844 + \frac{2}{5}(x - 110)\right] - \lim_{x\to 100^-}\left(190 + \frac{11}{10}x\right)\\ &= 844 - 4 - 300 = 540 \text{ cal/g.}\end{aligned}$$

(This is called $\Delta H_{\text{vaporization}}$ for water.) ■

We began this section by introducing one-sided limits as a device to discuss limiting behavior of functions that don't have limits in the sense of Definition 4.1. One-sided limits are also helpful in determining whether schizophrenic functions have limits at the points where they change formulas. The next result makes that possible.

6.2 THEOREM

Let f be defined on an open interval I containing c, except possibly at c. Then $\lim_{x \to c} f(x)$ exists if and only if

$$\lim_{x \to c^+} f(x) = \lim_{x \to c^-} f(x) = L,$$

in which case $\lim_{x \to c} f(x) = L$.

Proof. (a) Suppose first that $\lim_{x \to c} f(x)$ exists and is L. Then for any $\varepsilon > 0$, there is a $\delta > 0$ such that $|f(x) - L| < \varepsilon$ whenever $0 < |x - c| < \delta$. In particular, for $c - \delta < x < c$, we have $|f(x) - L| < \varepsilon$. So $\lim_{x \to c^-} f(x) = L$. Similarly, if $c < x < c + \delta$, then $|f(x) - L| < \varepsilon$, that is, $\lim_{x \to c^+} f(x) = L$.

(b) Conversely, suppose that $\lim_{x \to c^+} f(x) = L = \lim_{x \to c^-} f(x)$. Let $\varepsilon > 0$ be given. Then by Definition 6.1 there exist $\delta_1 > 0$ and $\delta_2 > 0$ such that

$$|f(x) - L| < \varepsilon \qquad \text{if } c - \delta_1 < x < c$$

and

$$|f(x) - L| < \varepsilon \qquad \text{if } c < x < c + \delta_2.$$

Let δ be the smaller of δ_1 and δ_2. Then, whenever $x \neq c$ is in $(c - \delta, c + \delta)$, we have $|f(x) - L| < \varepsilon$. That is, $|f(x) - L| < \varepsilon$ whenever $0 < |x - c| < \delta$. So by Definition 4.1, $\lim_{x \to c} f(x) = L$. QED

The next example illustrates the use of Theorem 6.2.

Example 3

Determine whether f and g have limits at c.

(a) $f(x) = \begin{cases} x^2 - x + 1 & \text{for } x < 1 \\ 2x - 1 & \text{for } x \geq 1 \end{cases}, \quad c = 1$

(b) $g(x) = \begin{cases} \dfrac{1}{1 + x} & \text{for } x < 0 \\ \dfrac{3}{x - 3} & \text{for } x \geq 0 \end{cases}, \quad c = 0.$

Solution. (a) To apply Theorem 6.2, we have to use the first formula for f as $x \to 1^-$ (since $x < 1$ then) and the second formula for f as $x \to 1^+$ (since $x > 1$ then). We find

$$\lim_{x \to 1^-} f(x) = \lim_{x \to 1^-} (x^2 - x + 1) = 1^2 - 1 + 1 = 1,$$

$$\lim_{x \to 1^+} f(x) = \lim_{x \to 1^+} (2x - 1) = 2(1) - 1 = 1.$$

Therefore $\lim_{x \to 1} f(x) = 1$ by Theorem 6.2.

(b) Proceeding as in (a), we have

$$\lim_{x\to 0^-} g(x) = \lim_{x\to 0^-} \frac{1}{1+x} = \frac{1}{1} = 1,$$

$$\lim_{x\to 0^+} g(x) = \lim_{x\to 0^+} \frac{3}{x-3} = \frac{3}{-3} = -1.$$

Since $\lim_{x\to 0^-} g(x) \neq \lim_{x\to 0^+} g(x)$, the function g does not have a limit at c. ■

The graphs of the functions f and g shown in **Figure 6.4** reflect the fact that f has a limit at $x = 1$, but g does not have a limit at $x = 0$.

FIGURE 6.4

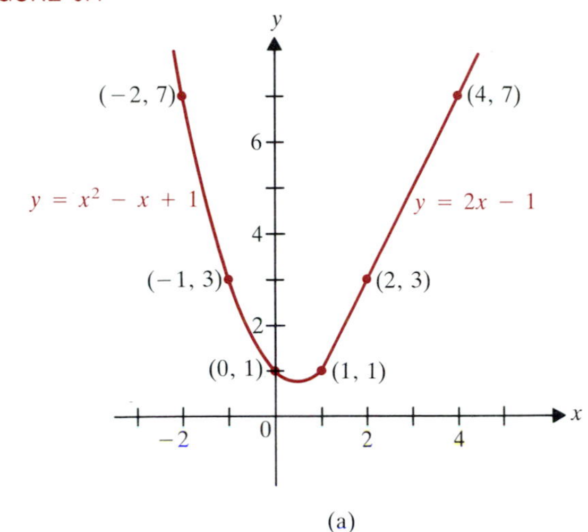

(a)

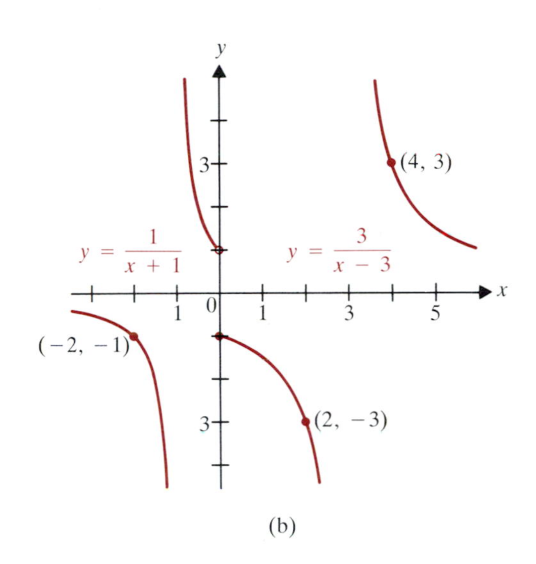

(b)

FIGURE 6.5

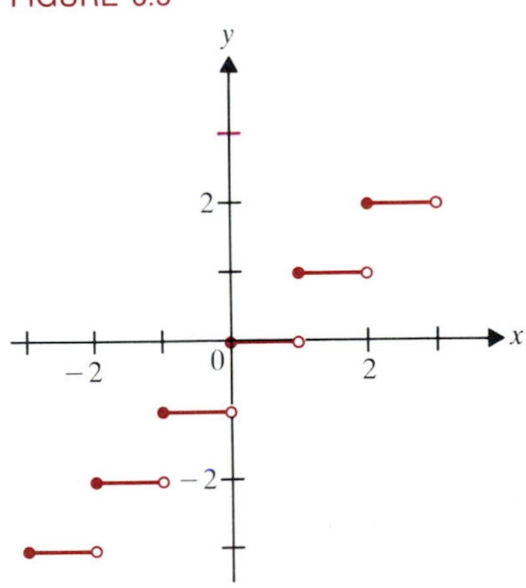

Our final example considers the *greatest integer function* f defined by

$$f(x) = [x] = \text{the greatest integer } n \leq x.$$

At every number c that is not an integer, f has a limit because

$$\lim_{x\to c} [x] = [c].$$

The graph of $y = [x]$ in **Figure 6.5** indicates that the function does not have a limit at any integer c. Theorem 6.2 makes that clear.

Example 4

Show that the greatest integer function has no limit at any integer n.

Solution. As $x \to n^+$, $x > n$, so $[x] = n$ for x close to but greater than n. Thus

$$\lim_{x\to n^+} [x] = [n] = n.$$

However, as $x \to n^-$, $x < n$, so $[x] = n - 1$ for x close to but less than n. Therefore

$$\lim_{x\to n^-} [x] = n - 1.$$

Since $\lim_{x\to n^+} [x] \neq \lim_{x\to n^-} [x]$, Theorem 6.2 says that $\lim_{x\to n} [x]$ does not exist. ■

Exercises 1.6

In Exercises 1–10, find the limits, if they exist, at the given points.

1. $\lim_{x\to 0^+} \sqrt{4-4x}$
2. $\lim_{x\to 3^-} \sqrt{3x-x^2}$
3. $\lim_{x\to 1^-} \sqrt{4-4x}$
4. $\lim_{x\to 0^+} \sqrt{3x-x^2}$
5. $\lim_{x\to 4^+} \sqrt{4-x}$
6. $\lim_{x\to 3^+} \sqrt{3x-x^2}$
7. $\lim_{x\to 1^+} \sqrt{4-4x}$
8. $\lim_{x\to 0^-} \sqrt{3x-x^2}$
9. (a) $\lim_{x\to 0^+} \frac{|x|}{x}$, (b) $\lim_{x\to 0^-} \frac{|x|}{x}$, (c) $\lim_{x\to 0} \frac{|x|}{x}$
10. (a) $\lim_{x\to 1^+} \frac{|x-1|}{x-1}$, (b) $\lim_{x\to 1^-} \frac{|x-1|}{x-1}$, (c) $\lim_{x\to 1} \frac{|x-1|}{x-1}$

In Exercises 11–20, find (a) $\lim_{x\to c^-} f(x)$, (b) $\lim_{x\to c^+} f(x)$ and (c) $\lim_{x\to c} f(x)$ (if they exist).

11. $c = 0,\ f(x) = \begin{cases} x^2 - 3x + 1 & \text{for } x \le 0 \\ 1 + 5x - x^2 & \text{for } x > 0 \end{cases}$
12. $c = 1,\ f(x) = \begin{cases} x^3 - x - 2 & \text{for } x \le 1 \\ 2 - 3x - x^2 & \text{for } x > 1 \end{cases}$
13. $c = -2,\ f(x) = \begin{cases} \dfrac{-1}{4 - x + x^2} & \text{for } x \le -2 \\ \dfrac{2x-3}{8 - 3x + 5x^2 + 2x^3} & \text{for } x > -2 \end{cases}$
14. $c = -1,\ f(x) = \begin{cases} \sqrt{7 + 2x^2}, & x \le -1 \\ \dfrac{x^2 - x + 1}{4x + 5}, & x > -1 \end{cases}$
15. $c = -1,\ f(x) = \begin{cases} \dfrac{x^2 - 3x - 2}{x + 1}, & x < -1 \\ \dfrac{5}{x + 2}, & x \ge -1 \end{cases}$
16. $c = 1,\ f(x) = \begin{cases} \sqrt{\dfrac{x-1}{x+1}}, & x < -1, x > 1 \\ \dfrac{2x-3}{x-1}, & -1 < x < 1 \end{cases}$
17. $c = 2,\ f(x) = \begin{cases} -x^2 + 3x + 5, & x \le 2 \\ \dfrac{2}{3 + x}, & x > 2 \end{cases}$
18. $c = -3,\ f(x) = \begin{cases} \dfrac{6}{2 - x}, & x < -3 \\ \dfrac{2x - 3}{5x^2 + 8x - 11}, & x \ge -3 \end{cases}$
19. $c = 1,\ f(x) = \begin{cases} \dfrac{x^3 - 1}{x - 1}, & x < 1 \\ \dfrac{x^2 - 1}{x - 1}, & x > 1 \end{cases}$
20. $c = -2,\ f(x) = \begin{cases} \dfrac{x^3 + 8}{x + 2}, & x < -2 \\ \dfrac{x^2 - 4}{x + 2}, & x > -2 \end{cases}$
21. If $g(x) = x - [x]$, then find $\lim_{x\to n^-} g(x)$ and $\lim_{x\to n^+} g(x)$ for any integer n.
22. Repeat Exercise 21 for $g(x) = [x]/x$, if $n \neq 0$.
23. *Entropy*, which may be thought of as a measurement of disorder or randomness in a physical system, is measured in calories per Kelvin. (To convert from degrees Celsius to Kelvin, add 273.) An approximate modeling function for the entropy of one mole of water under one atmosphere pressure is given by
$$S = \begin{cases} 10 + 0.04\,T & \text{for } T \le 0 \\ 15 + 0.05\,T & \text{for } 0 < T \le 100 \\ 44 + 0.03\,T & \text{for } T > 100 \end{cases}$$
where T is the temperature in degrees *Celsius*, and S is in calories per Kelvin. Compute ΔS at $T = 0$ and $T = 100$.
24. When water is in the liquid state, its volume is essentially constant as the pressure varies. But if the pressure is reduced to a sufficiently small level, the water vaporizes. Consider a closed cylinder fitted with a piston, that initially contains a 10-ml sample of water. As the piston is moved to increase the volume in the cylinder, the pressure falls to 24 mm of mercury. At that point, the water vaporizes and fills the volume, which is 1000 ml (1 liter). As the piston continues to move, the volume increases and the pressure decreases.
    - **(a)** Write a formula for the volume V of the water as a function of P, for $P > 24$.
    - **(b)** Use *Boyle's law*, which states that at constant temperature the product of the pressure and the volume of a given amount of gas is constant ($PV = C$), to give a formula for the volume V as a function of P for $P < 24$.
    - **(c)** Does $\lim_{P\to 0^+} V$ exist? If so, what is it?
    - **(d)** Find $\lim_{P\to 24^+} V$ and $\lim_{P\to 24^-} V$. Does $\lim_{P\to 24} V$ exist?

25. For the Dirichlet function g of Exercise 37, Section 1.4, does $\lim_{x\to c^+} g(x)$ exist for any c? How about $\lim_{x\to c^-} g(x)$?

26. Show that if $f(x)$ has a left-hand limit as $x \to c$, then that limit is unique (cf. Exercise 39 of Section 1.4).

27. Prove the analogue of Theorem 5.1(c) for left-hand limits.

28. Prove the analogue of Theorem 5.1(d) for right-hand limits.

29. Prove the analogue of Theorem 5.1(a, b) for left-hand limits.

30. In Example 3, state exactly which analogues of limit theorems from the last section were used, and where they were used in the solution.

HISTORICAL NOTE

Boyle's Law in Exercise 24 was formulated about 1660 by Robert Boyle (1627–1691), the Englishman who is often called the father of chemistry. He was held in the same high esteem in the British scientific community as Newton. Devoted to religion as well as to science, he provided in his will for an annual series of lectures in defense of Christianity. One of his principal themes was that science and religion were integrally and harmoniously related.

1.7 Continuous Functions

FIGURE 7.1

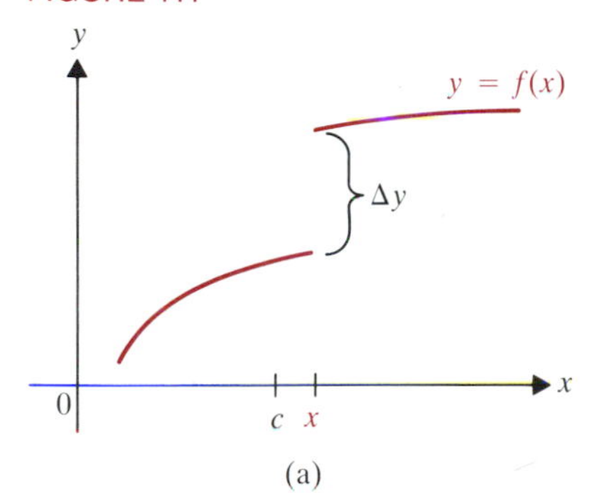

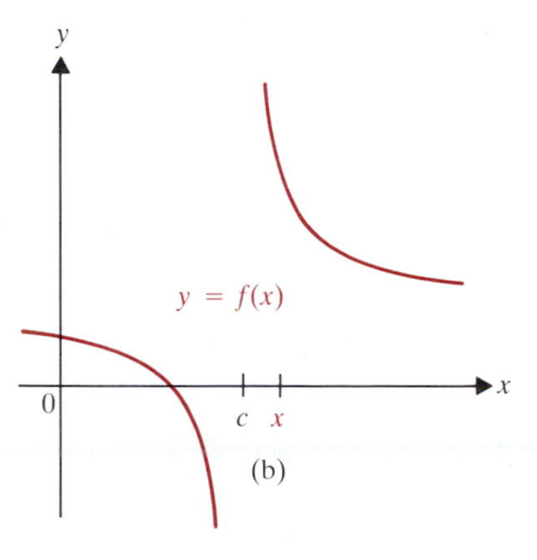

One of the main classes of functions that occur in calculus and its applications is made up of the *continuous* functions. For example, Einstein's equation

$$m = E/c^2$$

says that the mass m of a body is proportional to its energy E. An important feature of this relationship is that a small change in E produces only a relatively small change in m. Often in science, when a quantity y is a function of another quantity x,

$$y = f(x),$$

it is similarly true that small changes in x produce correspondingly small changes in y. In such circumstances, scientists say that y is a *continuous function* of x. Calculus makes it possible to formulate this idea in a precise mathematical way.

Let c be any point in the domain of f. Since y cannot change abruptly near $x = c$,

$$\Delta y = f(x) - f(c)$$

must be arbitrarily small when

$$\Delta x = x - c$$

is sufficiently small. For otherwise, perhaps under a microscope, the graph of $y = f(x)$ near $x = c$ would appear like that in **Figure 7.1.** That graph shows an abrupt change in y corresponding to a very small change in x near $x = c$. The above condition involving Δy and Δx means that $f(x)$ is arbitrarily close to $f(c)$ when x is close enough to c. That is, $\lim_{x\to c} f(x) = f(c)$. We thus define continuity of f at c as follows.

7.1 DEFINITION

The function f is ***continuous at*** $\boldsymbol{x = c}$ if

(1) $$\lim_{x\to c} f(x) = f(c).$$

We say that f is ***continuous on an open interval I*** in case f is continuous at every point c in I. If f is not continuous at $x = c$, it is said to be ***discontinuous*** there.

FIGURE 7.2

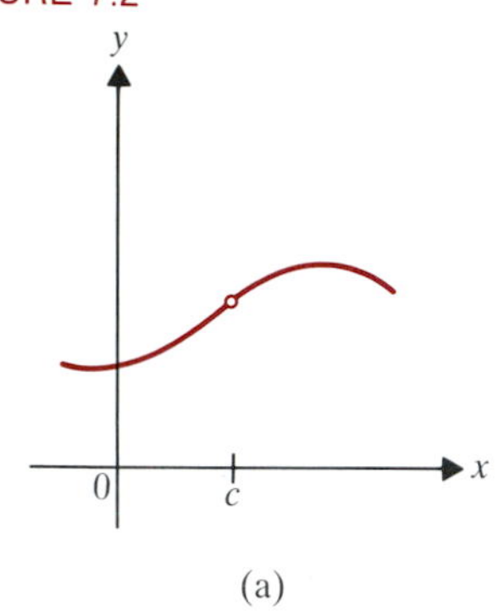

(a)

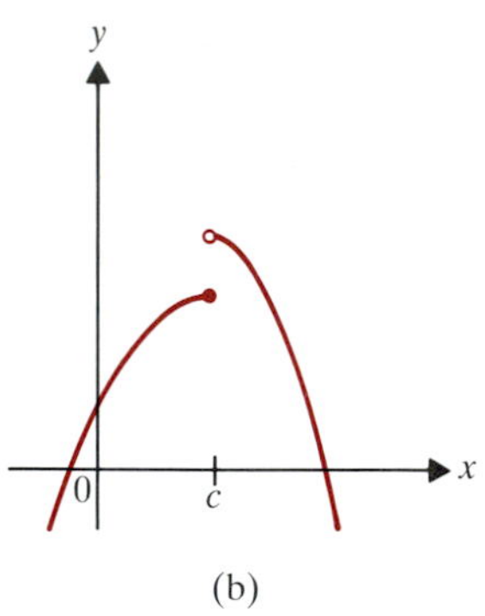

(b)

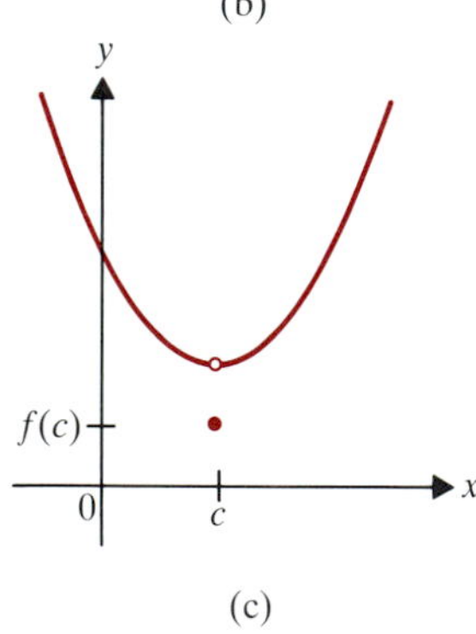

(c)

Example 1

Show that f is continuous at $x = 1$ if $f(x) = \dfrac{x^2 - x + 3}{x^2 + 1}$.

Solution. Using basic limit theorems from Section 5, we find

$$\lim_{x\to 1} f(x) = \frac{\lim_{x\to 1}(x^2 - x + 3)}{\lim_{x\to 1}(x^2 + 1)} = \frac{1^2 - 1 + 3}{1^2 + 1} = \frac{3}{2}$$
$$= f(1).$$

So according to Definition 7.1, f is continuous at $x = 1$. ■

In the language of Definition 4.1, (1) says that for any $\varepsilon > 0$, there is a $\delta > 0$ such that

$$|f(x) - f(c)| < \varepsilon \qquad \text{whenever } |x - c| < \delta.$$

Here we do not need the additional restriction $0 < |x - c|$ present in the definition of limit, since we are now interested in the behavior of f *at* as well as *near* $x = c$. Indeed, (1) requires f to be defined at c and to have limit $f(c)$ as $x \to c$. Continuous functions thus provide a link between algebra and calculus: for them, $\lim_{x\to c} f(x)$ can be correctly calculated by finding $f(c)$. Most people find substitution of c for x the natural way to evaluate $\lim_{x\to c} f(x)$, so continuous functions are inherently appealing and natural.

It is worth emphasizing that **Definition 7.1 requires *three conditions* to be met** in (1). These are

- c is in the domain of f, that is, $f(c)$ is defined;
- $\lim_{x\to c} f(x)$ exists, that is, f has a limit as $x \to c$;
- and the two quantities above are equal, that is, $\lim_{x\to c} f(x) = f(c)$.

Thus a function f fails to be continuous at $x = c$ if *any one* of the conditions

- f is undefined at $x = c$,
- $\lim_{x\to c} f(x)$ fails to exist,
- $\lim_{x\to c} f(x)$ exists but is not equal to $f(c)$

is true. Graphically, the presence of any of these three problems means that there is a *break* in the graph of the function f at $x = c$. **Figure 7.2**(a) shows the graph of a function that is undefined at $x = c$: the open circle indicates that there is a point missing. Figure 7.2(b) shows the graph of a function that has no limit at $x = c$. Figure 7.2(c) shows the graph of a function whose limit at $x = c$ is not equal to $f(c)$. These graphs suggest the origin of the name *continuous:* a function f is continuous on an open interval I if its graph has no breaks over I. Thus the graph of a continuous function over an open interval can be drawn in one continuous stroke. For example, the function $f(x) = |x|$ is continuous on $I = (-\infty, +\infty) = \mathbf{R}$: its graph has no breaks. See **Figure 7.3.**

FIGURE 7.3

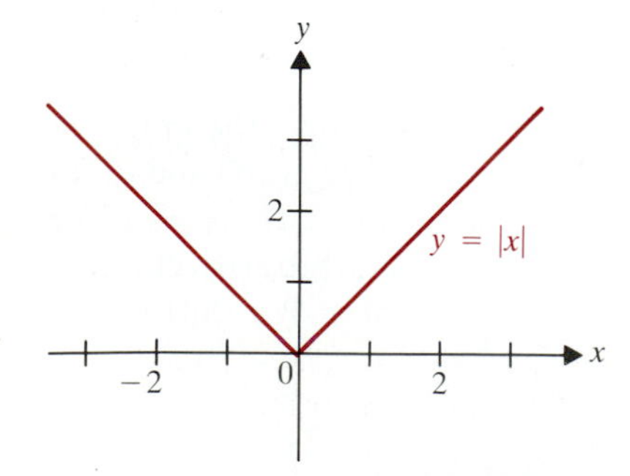

FIGURE 7.4

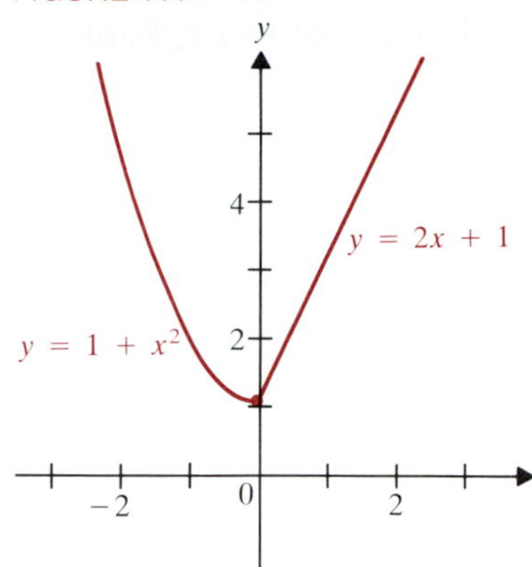

Example 2

Determine whether the following functions are continuous.

(a) $f(x) = \begin{cases} 1 + x^2 & \text{for } x \le 0 \\ 2x + 1 & \text{for } x > 0 \end{cases}$ at the point $c = 0$.

(b) $g(x) = 6x^2 - 8x + 5$ on the interval $(-\infty, +\infty)$.

Solution. (a) The function f is defined at 0: $f(0) = 1 + 0^2 = 1$. Also,

$$\lim_{x \to 0^-} f(x) = \lim_{x \to 0^-} (1 + x^2) = 1,$$

and

$$\lim_{x \to 0^+} f(x) = \lim_{x \to 0^+} (2x + 1) = 1.$$

Hence, by Theorem 6.2, $\lim_{x \to 0} f(x) = 1 = f(0)$. Thus f is continuous at c. See **Figure 7.4,** which shows the unbroken graph of f.

(b) For any real number c, we have from Equation (7) of Section 5

$$\lim_{x \to c} g(x) = 6c^2 - 8c - 5 = g(c).$$

Therefore g is continuous at c. Consequently, g is continuous on all of $\mathbf{R} = (-\infty, +\infty)$. ■

FIGURE 7.5

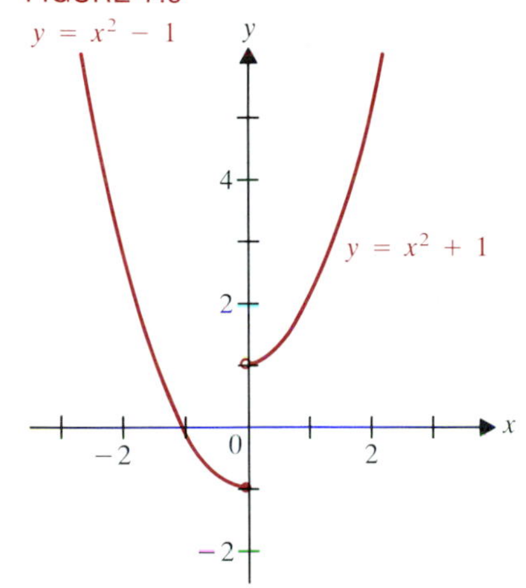

Example 3

Determine whether the following functions are continuous.

(a) $h(x) = \begin{cases} x^2 - 1 & \text{for } x \le 0 \\ x^2 + 1 & \text{for } x > 0 \end{cases}$ at $c = 0$.

(b) $k(x) = \dfrac{x^2 - 1}{x - 1}$ at $c = 1$.

Solution. (a) As in Example 2(a), this function is defined at $c = 0$. But this time we find that

$$\lim_{x \to 0^-} h(x) = \lim_{x \to 0^-} (x^2 - 1) = 0^2 - 1 = -1,$$

but

$$\lim_{x \to 0^+} h(x) = \lim_{x \to 0^+} (x^2 + 1) = 0^2 + 1 = 1.$$

Hence, by Theorem 6.2, h does not have a limit as $x \to 0$. Therefore, h is not continuous at $x = 0$. See **Figure 7.5,** which shows a break in the graph of h at $x = 0$.

(b) This function is not continuous at $c = 1$, because it is not defined there. (Division by 0 would occur at $c = 1$.) Note that for all $x \neq 1$, however,

$$k(x) = x + 1,$$

so $\lim_{x \to 1} k(x) = 2$. This is a function that has a limit at $x = c$, but is not continuous there. See **Figure 7.6.** Thus, being continuous at a point $x = c$ is a *stronger* requirement than having a limit there. ■

FIGURE 7.6

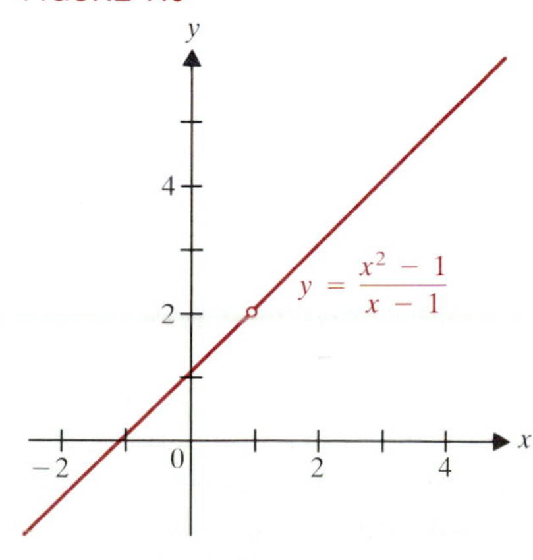

The real problem in Example 3(b) seems to lie with the defining formula for $k(x)$ rather than with the function itself. Indeed, the function g defined by

$$g(x) = \begin{cases} \dfrac{x^2 - 1}{x - 1} & \text{if } x \neq 1 \\ 2 & \text{if } x = 1 \end{cases}$$

is continuous at $c = 1$. This is so because, first of all, g is now defined at $x = 1$. Also, the reasoning in Example 3 shows that

$$\lim_{x \to 1} g(x) = 2.$$

Therefore g has a limit at $x = 1$ that coincides with $g(1)$. Consequently, g is continuous at $c = 1$. The only difference between g and k is that we have *enlarged the domain* by defining g at $c = 1$. We have done so, moreover, by defining the value of g at $c = 1$ to be $\lim_{x \to 1} g(x)$. That fills in the break in the graph in Figure 7.6.

In general, we say that a function f has a ***removable discontinuity*** at $x = c$ if c is not in the domain of f but $\lim_{x \to c} f(x) = L$ exists. In such a case, if we enlarge the domain by defining $f(c) = L$, then the reasoning used on $g(x)$ above shows that f is continuous at c.

By contrast, things are beyond repair when $\lim_{x \to c} f(x)$ fails to exist. In that case, f is said to have an ***essential discontinuity*** at $x = c$. For instance, the function h in Example 3(a) has an essential discontinuity at $c = 0$.

Most functions encountered in elementary calculus are continuous. For example, the remarks in Section 5 following Equation (6) (p. 41) say that any polynomial function p is continuous at every real number c. The next result extends this further. The first part says that a rational function is continuous at every point c except where its denominator is 0. The second part says that the nth-root function is continuous virtually everywhere that it is defined.

7.2 THEOREM

(a) Every rational function

$$f(x) = \frac{p(x)}{q(x)}$$

is continuous at every point c in its domain D.

(b) The function $f(x) = \sqrt[n]{x}$ is continuous at every real number c if n is odd, and at every nonnegative number c if n is even.

Proof. (a) From Equation (7) of Section 5, $\lim_{x \to c} f(x) = f(x)$, because $c \in D$ means that $q(c) \neq 0$. Thus f is continuous at c.

(b) From Theorem 5.5, $\lim_{x \to c} f(x) = \sqrt[n]{c} = f(c)$. QED

Similarly, the following facts are consequences of the theorems of Section 5.

7.3 THEOREM

Suppose that f and g are continuous at c. Then so are (a) $f + g$, (b) $f - g$, (c) $f \cdot g$, (d) f/g if $g(c) \neq 0$, (e) $1/g$ if $g(c) \neq 0$, and (f) $\sqrt[n]{g}$ for any positive integer n (where, in case n is even, we assume that $g(c) > 0$).

The preceding two theorems say that most simple functions of algebra are continuous. In particular, all polynomial functions, all rational functions, all square roots, cube roots, and so on are continuous at almost all points of their domains. What about a point like $c = 0$ for the function $f(x) = \sqrt{x}$? According to Definition 7.1, this function isn't continuous, since f has only a *one-sided* limit at $c = 0$, from the right. It seems inappropriate to call this function discontinuous at $c = 0$, though, because the graph is unbroken (**Figure 7.7**). It simply *ends* at $c = 0$. We thus widen the meaning of continuity as follows.

7.4 DEFINITION

The function f is ***continuous on a closed interval*** $[a, b]$ if f is continuous at every point of the open interval (a, b) in the sense of Definition 7.1 and also

$$\lim_{x \to a^+} f(x) = f(a) \qquad \text{and} \qquad \lim_{x \to b^-} f(x) = f(b).$$

Similarly, f is continuous on the closed half-line $[c, +\infty)$ or $(-\infty, c]$ if it is continuous on the corresponding open half-line $(c, +\infty)$ or $(-\infty, c)$ according to Definition 7.1 and $\lim_{x \to c^+} f(x) = f(c)$ or $\lim_{x \to c^-} f(x) = f(c)$.

Thus $f(x) = \sqrt{x}$ is continuous on $[0, +\infty)$. The following basic theorem on the continuity of composite functions follows easily from Theorem 5.3.

7.5 THEOREM

Suppose that g is continuous at x_0 and h is continuous at $t_0 = g(x_0)$. Then the composite function $h \circ g$ is continuous at x_0.

FIGURE 7.7

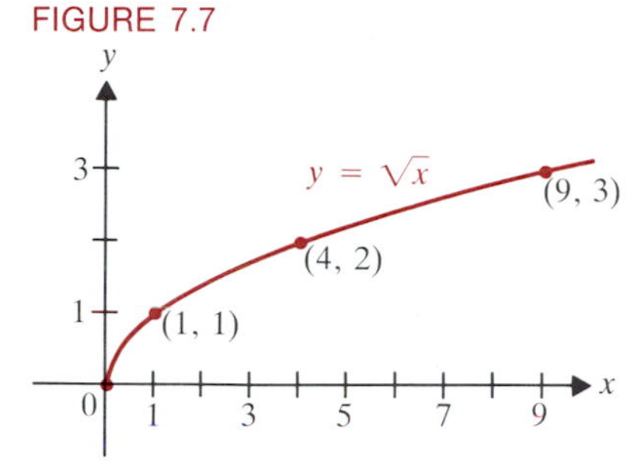

Proof. If g is continuous at x_0, then we have $\lim_{x \to x_0} g(x) = g(x_0) = M$ in Theorem 5.3. If h is continuous at $t_0 = g(x_0)$, then we also have $\lim_{t \to t_0} h(t) = h(t_0)$. Therefore

$$\lim_{x \to x_0} (h \circ g)(x) = h(M) = h(g(x_0)) = (h \circ g)(x_0).$$

Hence, $h \circ g$ is continuous at x_0. QED

This result guarantees that if we build up more complicated functions from simple continuous functions by composition, then the resulting functions will also be continuous. For instance, since $f(x) = \sqrt[n]{x}$ is continuous by Theorem 7.2(b), and $g(x) = (x^2 + 1)/(x^2 + 4)$ is continuous by Theorem 7.2(a), it follows that

$$(f \circ g)(x) = \sqrt{\frac{x^2 + 1}{x^2 + 4}} = f(g(x))$$

is continuous at all numbers in its domain, which happens to be all of **R** since both $x^2 + 1$ and $x^2 + 4$ are always positive. The next example shows how we can analyze the continuity of a function on **R**.

Example 4

Determine for which x the function $h(x) = \dfrac{\sqrt{x^2 - 1}}{x + 2}$ is continuous.

Solution. On p. 26 we saw that h has domain $(-\infty, -2) \cup (-2, -1] \cup [1, +\infty)$. The function is discontinuous at $x = -2$, since its denominator is 0 there. This is an essential discontinuity, because as $x \to -2^+$, for instance, $h(x)$ grows larger and larger without bound since we are dividing a number near $\sqrt{3}$ by a tiny positive number. Thus h has no limit as $x \to 2$. Since

$$\lim_{x \to -1^-} h(x) = 0 = h(-1) \qquad \text{and} \qquad \lim_{x \to 1^+} h(x) = 0 = h(1),$$

h is continuous on $(-2, -1]$ and $[1, +\infty)$ by Definition 7.4. Hence, h is continuous on its entire domain. ■

We close this section with two fundamental theorems about continuous functions, whose proofs are part of advanced calculus. The first seems obvious geometrically, since a continuous function has no breaks in its graph. It says that $f(x)$ cannot jump over any value y_1 lying between $f(a)$ and $f(b)$. Proving this directly from Definitions 7.1 and 7.4, without using pictures like **Figure 7.8,** however, involves considerable intricacy.

FIGURE 7.8

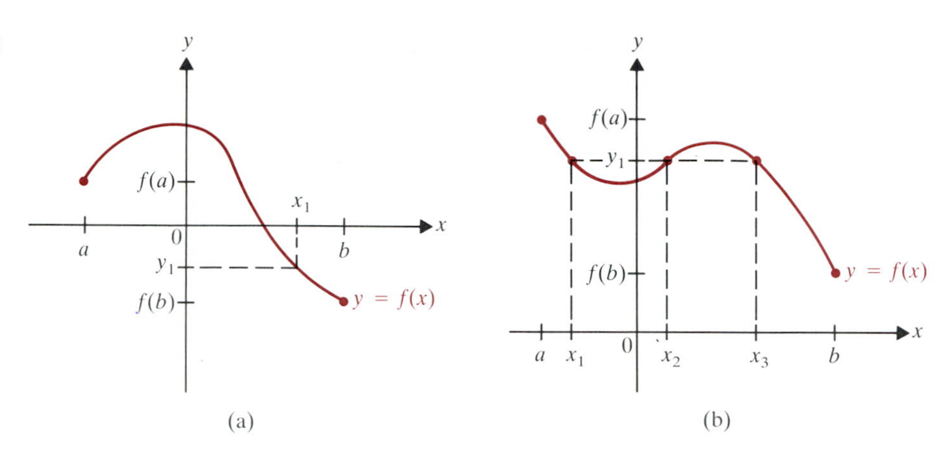

7.6 THEOREM

> ***Intermediate Value Theorem.*** If f is continuous on $[a, b]$ and $f(a) \neq f(b)$, then for any number y_1 between $f(a)$ and $f(b)$, we have
>
> $$f(x_1) = y_1$$
>
> for at least one number x_1 between a and b. That is, f takes on every intermediate value between $f(a)$ and $f(b)$.

FIGURE 7.9

As an illustration, consider $y = f(x) = x^2$ on the interval $[0, 2]$. The graph is shown in **Figure 7.9.** Note that $y_1 = 1$ lies between $f(0) = 0$ and $f(2) = 4$. For $y_1 = 1$, we have $x^2 = 1$ when $x_1 = 1$.

Theorem 7.6 can be used to locate intervals containing a root of an equation $f(x) = 0$, if f is continuous. If $[a, b]$ is an interval on which $f(a)$ and $f(b)$ have *opposite* signs, then Theorem 7.6 guarantees that there must be some number x_1 between a and b such that $f(x_1) = 0$.

FIGURE 7.10

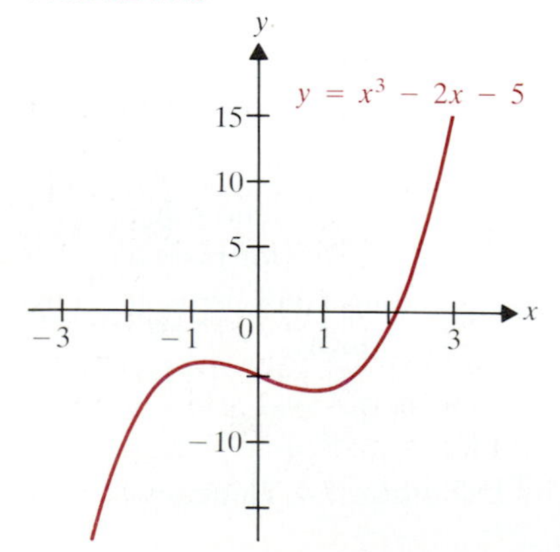

Example 5

Find an interval with integer endpoints that contains a root of the equation $x^3 - 2x - 5 = 0$.

Solution. The following table of values shows that the function $f(x) = x^3 - 2x - 5$ has opposite signs at $x = 2$ and $x = 3$.

x	-5	-4	-3	-2	-1	0	1	2	3	4	5
$f(x)$	-120	-61	-26	-9	-4	-5	-6	-1	16	51	111

Therefore there must be a root somewhere between $x = 2$ and $x = 3$. See **Figure 7.10.** ■

We can zero in on the root in Example 5 as follows. First, $f(2.5) = 5.625 > 0$; thus f changes sign on $[2, 2.5]$. So there must be a root in that interval. Similarly, $f(2.25) = 1.890625 > 0$, so there is a root in the interval $[2, 2.25]$. Continuing, we can evaluate f at the midpoint of each successive interval in which it changes sign. We find

$$f(2.125) = 0.345703125 > 0, \qquad \text{so there is a root in } [2, 2.125],$$

$$f(2.0625) = -0.351318359 < 0, \qquad \text{so there is a root in } [2.0625, 2.125],$$

and so on. Continuing in this way, we eventually come to the approximation $c = 2.09455148$ for the root. (As you can check if your calculator has enough digits, $f(c) \approx -1.8 \times 10^{-8}$, so this is *only* an approximation to the root.)

The method described above for approximating a zero of a continuous function f is called the ***bisection method.*** (See Exercises 35–39.) It is well suited to implementation on a computer or programmable calculator.

Theorem 7.6 is a basic theorem for continuous functions. It fails to hold more generally, because discontinuous functions sometimes *do* jump over values. Consider, for instance, the function h in Example 3(a),

$$h(x) = \begin{cases} x^2 - 1 & \text{for } x \leq 0 \\ x^2 + 1 & \text{for } x > 0 \end{cases}$$

on the interval $[-1/2, 1/2]$. We have $h(a) = h(-1/2) = 1/4 - 1 = -3/4$ and $h(b) = h(1/2) = 1/4 + 1 = 5/4$. But for the value $y_1 = 0$ between $h(a)$ and $h(b)$, there is *no* x_1 in the interval $[-1/2, 1/2]$ such that $h(x_1) = 0$, as a glance at Figure 7.5 confirms.

Our final result about continuous functions provides the theoretical underpinning for much of Chapter 3.

7.7 THEOREM

Extreme Value Theorem. If f is continuous on a closed interval $[a, b]$, then f has a maximum value M and a minimum value m on $[a, b]$. That is, there exist x_1 and x_2 in $[a, b]$ such that

$$f(x_1) = m \leq f(x) \qquad \text{for all } x \in [a, b],$$

and

$$f(x_2) = M \geq f(x) \qquad \text{for all } x \in [a, b].$$

Referring again to $y = f(x) = x^2$ on $[0, 2]$, we see from Figure 7.9 that the maximum value on $[0, 2]$ is 4 at $x_2 = 2$. The minimum value is 0 at $x_1 = 0$. *Some* discontinuous functions also have a maximum and a minimum on a closed interval. As Figure 7.5 shows, the function h in Example 3(a) has a maximum on $[-1/2, 1/2]$, namely, $h(1/2) = 5/4$, and has minimum value $-1 = h(0)$ on $[-1/2, 1/2]$. On the other hand, $y = 1/x$ certainly has no maximum or minimum on $[-1, 1]$, since as $x \to 0$, the function takes on arbitrarily large positive and negative values.

On an *open* interval, neither continuous nor discontinuous functions need have extreme values. On the open interval $(-2, 2)$, $f(x) = x^2$ does have a minimum, $0 = f(0)$. But there is no maximum since the points -2 and 2 are not in $(-2, 2)$. [For every number $2 - \delta$ close to but less than 2, $2 - \frac{1}{2}\delta$ is closer to 2 and f has a *larger* value at $2 - \frac{1}{2}\delta$ than at $2 - \delta$. So f has no maximum on $(-2, 2)$.]

HISTORICAL NOTE

The term *continuous function* goes back at least to 1747, when the French mathematician Jean d'Alembert (1717–1783) used it for functions defined by a single formula. The idea that small changes in x should produce small changes in $f(x)$ may have originated with Louis Arbogast (1759–1803) in 1791, who also identified the intermediate-value property of Theorem 7.6 as an equivalent notion. Condition (1) was used as the working definition of continuity by the Czech priest Bernhard Bolzano (1781–1848), who used it to prove Theorem 7.6 and many other fundamental properties of continuous functions. Bolzano's work was not published, and it appears that the famous French mathematician Augustin Louis Cauchy (1789–1857) independently arrived at (1) as the proper definition of continuity, and used it to give the proof of Theorem 7.6 found in most advanced calculus texts. The formulation we have given of Definition 7.1 is due to Weierstrass, about 1860.

Exercises 1.7

In Exercises 1–10, determine whether the function defined in the given exercise from Exercises 1.6 is continuous at the point c given in that exercise.

1. Exercise 11 **2.** Exercise 12

3. Exercise 13 **4.** Exercise 14

5. Exercise 15 **6.** Exercise 16

7. Exercise 17 **8.** Exercise 18

9. Exercise 19 **10.** Exercise 20

In Exercises 11–20 classify the given point as one of continuity, removable discontinuity, or essential discontinuity for the given function. In the case of a removable discontinuity, how should $f(c)$ be defined to make f continuous at c?

11. $f(x) = \sqrt{9 - 6x + x^2}$, $c = 4$

12. $f(x) = \dfrac{x^2 - 3x + 2}{x^2 - 6x + 5}$, $c = 1$

13. $f(x) = \dfrac{x^2 - 3x - 4}{x^2 + 2x + 1}$, $c = 1$

14. $f(x) = \dfrac{x^2 - 8x + 12}{x^2 - 4x + 4}$, $c = 2$

15. $f(x) = \sqrt{\dfrac{x^2 - 1}{x^2 + 2x + 1}}$, $c = -1$

16. $f(x) = \sqrt{\dfrac{x^2 - 1}{x^2 - 2x + 1}}$, $c = 1$

17. $f(x) = \begin{cases} x^2 - 2x + 1, & x \le 2 \\ 5 - 4x + x^2, & x > 2 \end{cases}$, $c = 2$

18. $f(x) = \begin{cases} \dfrac{x^2 - 4}{x - 2}, & x < 2 \\ -x^2 + 4x - 2, & x \ge 2 \end{cases}$, $c = 2$

19. $f(x) = \begin{cases} \dfrac{\sqrt{x} - 2}{x - 4}, & x < 4 \\ \dfrac{1}{8}x^2 - \dfrac{1}{4}x - \dfrac{3}{4}, & x > 4 \end{cases}$, $c = 4$

20. $f(x) = \begin{cases} \dfrac{\sqrt{x} - 3}{x - 9}, & x < 9 \\ \dfrac{1}{6}x^2 - \dfrac{3}{2}x + \dfrac{1}{3}, & x > 9 \end{cases}$, $c = 9$

In Exercises 21–30, use the definitions and theorems of this section to determine where the given functions are continuous.

21. $g(x) = \dfrac{1}{x - 2} + \sqrt{x - 1}$ **22.** $k(x) = \dfrac{3x - 1}{3x + 1}$

23. $p(x) = \dfrac{5x^2 - 3x + 1}{1 - x^2}$ **24.** $q(x) = \dfrac{-x^2 - 2x + 7}{5 - x^2}$

25. $r(x) = \sqrt[4]{x^4 - 16}$ **26.** $s(x) = \sqrt{x^2 - 8x + 12}$

27. $t(x) = \dfrac{2x - 8}{x^3 - x^2 - 4x + 4}$ **28.** $u(x) = \dfrac{6x - 1}{x^3 - 4x^2 - x + 4}$

29. $v(x) = \dfrac{\sqrt{x^2 - 9}}{x^3 - x^2 - 9x + 9}$ **30.** $w(x) = \dfrac{\sqrt{x^2 - 1}}{x^3 - 3x^2 - x + 3}$

In Exercises 31–34, show that the given equation has a solution between the two given points.

31. $x^3 - x - 1 = 0$, between $a = 1$ and $b = 2$

32. $x^3 + x - 1 = 0$, between $a = 0$ and $b = 1$

33. $x^4 - 4x^2 + x + 1 = 0$, between $a = 0$ and $b = 1$

34. $x^4 - x^3 - 2x - 3 = 0$, between $a = 1$ and $b = 2$

35. With reference to the bisection method described after Example 5, suppose that f is continuous and $f(a) \cdot f(b) < 0$.

(a) Explain why after n steps the method finds an interval of length $(b - a)/2^n$ that contains a root x^* of $f(x) = 0$.

(b) An approximation r to a quantity x^* is said to be ***accurate to k decimal places*** if $|x^* - r| < \frac{1}{2} \cdot 10^{-k}$. Explain why after eight iterations of the bisection method on an interval $[n, n + 1]$ when n is an integer, you have an approximation r of the root x^* that is accurate to at least two decimal places.

In Exercises 36–39, use Exercise 35(b) and a computer or programmable calculator program to find a 4-decimal-place approximation to the root x^* between a and b of the given exercise. What do you obtain when r is substituted into the given equation?

PC **36.** Exercise 32

PC **37.** Exercise 31

PC **38.** Exercise 34

PC **39.** Exercise 33

40. Prove that f is continuous at c if and only if $\lim_{h \to 0} f(c + h) = f(c)$.

41. (a) Use the fact that $f(x) = x^2$ is continuous on $\mathbf{R}$ to show that every nonnegative real number y has a real square root. (*Hint:* If n is an integer, then for some n, $n^2 \geq y$.)

(b) Show as in part (a) that every nonnegative real number has a real nth root if n is even.

42. Use the fact that $f(x) = x^n$ is continuous on $\mathbf{R}$ to show that every real number has a real nth root if n is odd.

43. For which real numbers x is the Dirichlet function (Exercise 37, Section 4) continuous?

44. For which real numbers x is the greatest integer function $f(x) = [x]$ continuous?

45. If f is continuous at c, but g is discontinuous at c, what can you say about the continuity of $f + g$ at c?

46. If f and g are discontinuous at c, then can $f + g$ be continuous at c?

47. If f is continuous at c, but g is discontinuous at c, then $f \cdot g$ may be continuous or discontinuous at c. Give examples to illustrate this.

48. If k is a constant such that $|f(x) - f(c)| \leq k|x - c|$ for all x in some interval I containing c, then prove that f is continuous at c.

49. Use Exercise 48 and Exercise 40 of Section 1.5 to prove that $f(x) = |x|$ is continuous for all real numbers x.

50. If f is continuous on $[a, b]$ and on $(b, c]$, then it is not necessarily true that f is continuous on $[a, c]$, the union of $[a, b]$ and $(b, c]$. Give an example to illustrate this.

1.8 Looking Back

This first chapter is made up of two distinct parts. In the first three sections we reviewed prerequisite material from precalculus: absolute value, inequalities, graphing, linear and more general functions, and combining functions in various ways. In the ensuing four sections we introduced and studied the fundamental operation of calculus: taking limits. The main use of Definition 4.1 was in deriving rules—like those in Section 5—that allow us to calculate limits mechanically. In reviewing this chapter, a good first goal is sharpening your skills in computing both two-sided and one-sided limits.

In Section 7, continuity was introduced. A working knowledge of what it means and how to test functions for continuity will be assumed in later chapters. One-sided limits are integrally involved with analyzing the continuity of functions that have more than one formula.

CHAPTER CHECKLIST

Section 1.1: rational and irrational numbers; greater than, less than; transitive, additive, multiplicative rules for inequalities; closed, open, half-open, half-closed intervals, half-lines; radical of a number; absolute value and its properties; triangle inequality; solving inequalities by factoring and tables.

Section 1.2: increments; Cartesian coordinates; graph of an equation; slope, average rate of change; point-slope equation, general form, slope-intercept equation of a line; criteria for parallel and perpendicular lines; distance formula; equation of a circle; completing the square.

Section 1.3: definitions of function, domain, graph; linear function, proportionality; quadratic functions, parabolas; polynomial functions; even and odd functions; composition of functions.

Section 1.4: average and instantaneous rates of change; limits; tolerances; $\varepsilon - \delta$ formulation and proof of limits.

Section 1.5: limit theorems for constant functions, constant multiples of functions, sum, difference, composition, product, reciprocal, and quotient of functions; limits of polynomials, rational functions, and nth roots of functions; sandwich theorem.

Section 1.6: left-hand limit; right-hand limit; schizophrenic functions; relation of one-sided limits to ordinary limits.

Section 1.7: continuity at a point and on an interval; threefold requirement for continuity at a point; removable discontinuity, essential discontinuity; continuity of rational functions and composite functions; intermediate value theorem; bisection method.

REVIEW EXERCISES 1.8

In Exercises 1–4, solve the given inequality.

1. $|2x - 3| < 5$

2. $x^2 - 4x < x - 2$

3. $\dfrac{1}{x+1} > \dfrac{1}{x}$

4. $\dfrac{3}{x-2} < \dfrac{x}{x+2}$

In Exercises 5–8, find the equation of the specified line.

5. Through (2, 5) and (−1, 4).

6. Through (−2, 1) and (−2, 3).

7. Through (2, 3) parallel to $x - 3y = 6$.

8. Through (2, 3) perpendicular to $x - 3y = 6$.

9. It costs a book publisher $50,000 to develop a manuscript for publication. Plant costs for printing of up to 7000 copies are $5.00 for each copy.
(a) Give the equation for the total cost of printing x copies, where $x \leq 7000$.
(b) Suppose from market research that the publisher expects to sell 7000 copies in the first year of publication. What is the minimum price that the company can charge bookstores if it is to recover its publication costs the first year?

10. Under proper conditions, the gas nitrous oxide (N_2O) decomposes into nitrogen (N_2) gas and oxygen (O_2) gas, in what is called a *zero order chemical reaction*, one in which the rate k of the reaction is constant. This reaction rate is the rate of change of the concentration C of N_2O per minute. (See **Figure 8.1** for a typical graph of C vs. t for zero order reactions.) Denote the initial concentration of N_2O by C_0. Write an equation for C in terms of t.

FIGURE 8.1

11. Find the equation of the circle centered at (1, −3) that passes through the point (4, 2).

12. If (1, 3) and (5, −1) are the endpoints of a diameter of a circle, then give its equation.

13. What is the graph of $2x^2 + 2y^2 - 8x + 6y + 10 = 0$?

14. What happens in Exercise 13 if the constant 10 is replaced by **(a)** 25/2 **(b)** 13?

In Exercises 15–18, find the domain of the given function.

15. $q(x) = \dfrac{\sqrt{x} - 3}{x - 9}$

16. $r(x) = \dfrac{1}{\sqrt{4x - 1}} + \sqrt{1 - x^2}$

17. $s(x) = \dfrac{1}{x + 5} + \sqrt{x^2 + 8x + 12}$

18. $t(x) = \dfrac{1}{x \cos x}$

In Exercises 19 and 20, graph the given functions.

19. $h(x) = 1 - 2x - x^2$

20. $k(x) = 2x^2 + 6x - 3$

21. Let $f(x) = [x]$ be the greatest integer $n \leq x$.
(a) Is the function f even, odd, or neither?
(b) Is $g(x) = x - [x]$ even, odd, or neither?

22. If $f(x) = \sqrt{x + 5}$ and $g(x) = x^2 + 1$, then find $f + g$, $f \cdot g$, f/g, $f \circ g$, and $g \circ f$.

In Exercises 23–32, evaluate the given limits if they exist.

23. $\displaystyle\lim_{t \to 5} \frac{t^2 - 4t - 5}{t^2 - 3t - 10}$

24. $\displaystyle\lim_{t \to -3} \frac{t^2 + 2t - 3}{t^2 + 5t + 6}$

25. $\displaystyle\lim_{x \to 2} \frac{|x - 2|}{x - 2}$

26. $\displaystyle\lim_{x \to 1} \frac{\sqrt{x^2 - 2x + 1}}{x - 1}$

27. $\lim_{x\to 2} \sqrt{\dfrac{5x^2 - 3x + 1}{x^3 - 2x + 1}}$

28. $\lim_{x\to -1} \sqrt{\dfrac{17x^2 + 7x - 2}{3x^2 + 3x - 1}}$

29. $\lim_{x\to 0} x \cos \dfrac{1}{x^2}$

30. $\lim_{x\to 1} (x - 1)^2 \sin \dfrac{1}{x - 1}$

31. $\lim_{x\to 2^+} \sqrt{x^2 - 4x + 4}$

32. $\lim_{x\to 3^-} \sqrt{x^2 - 2x - 3}$

In Exercises 33–40, determine whether the given function is continuous at the given point. If not, is the discontinuity removable?

33. $f(x) = \begin{cases} 1 - 3x, & x \le 0 \\ 2x^2 - 5x + 1, & x > 0 \end{cases}, \quad x = 0$

34. $g(x) = \begin{cases} x^2 - 1, & x \le 1 \\ x^2 - 3x + 2, & x > 1 \end{cases}, \quad x = 1$

35. $h(x) = \dfrac{x^2 - 5x + 4}{x^2 + x - 2}, \quad x = 1$

36. $k(x) = \dfrac{x^2 - x - 6}{x^2 - 2x - 8}, \quad x = -2$

37. $m(x) = \dfrac{x^2 - x - 6}{x^2 + 3x + 2}, \quad x = -1$

38. $n(x) = \dfrac{x^2 - 4x + 3}{x^2 - 2x - 3}, \quad x = -3$

39. $f(x) = \begin{cases} 2 - 3x + x^2, & x \le 1 \\ \dfrac{1}{2 - 3x + 2x^2}, & x > 1 \end{cases}, \quad x = 1$

40. $g(x) = \begin{cases} \dfrac{1}{x - 3}, & x < 3 \\ x - 3, & x \ge 3 \end{cases}, \quad x = 3$

In Exercises 41–44, determine where the given functions are continuous.

41. $f(x) = \begin{cases} 3 - 2x + x^2, & x \le 1 \\ 5x - 3, & 1 < x < 3 \\ x^2 - 4x + 1, & x \ge 3 \end{cases}$

42. $g(x) = \begin{cases} 2 - 3x, & x \le 2 \\ x^2 + x - 5, & 2 < x < 3 \\ 4x - 5, & x \ge 3 \end{cases}$

43. $h(x) = \dfrac{3}{x - 2} + \dfrac{1}{\sqrt{x - 1}}$

44. $j(x) = \dfrac{\sqrt{x^2 - 1}}{x^2 + 4x + 3}$

45. Show that $f(x) = \dfrac{1}{x^2 + 1}$ is continuous on $\mathbf{R}$.

46. If $f(a) > 0$ and f is continuous at $x = a$, then show that $f(x) > 0$ for all x in some interval $(a - \delta, a + \delta)$.

47. Show that f is continuous at c if and only if $\lim_{h\to 0} f(c - h) = f(c)$.

48. Prove from the definition that $\lim_{x\to -2} (-3x + 5) = 11$.

49. Prove that for any two positive real numbers a and b, their geometric mean $\sqrt{ab}$ is less than or equal to their arithmetic mean $\frac{1}{2}(a + b)$. When does equality hold?

50. Show that $x^4 - 3x^3 + 4x^2 - 5x + 1 = 0$ has a root between $x = 0$ and $x = 1$.

51. Does the function f in Exercise 41 satisfy the intermediate-value theorem on the interval $[2, 4]$?

52. If $f(x) = 0$ when x is an integer, and otherwise $f(x) = 1$, then for which x is f continuous?

53. Let $f(x)$ be the number of cents postage required for a first class letter of weight x ounces. At which points c is f discontinuous? What type of discontinuity does f have at each such point?

54. A buffalo herd in a game preserve has population P, which is a function of the amount of grass cover available for the herd. Grass yield Y is estimated in pounds per square foot by weighing the harvest from a 20-square-foot test plot. The number of grasshoppers affects this yield. An approximation for Y is given by the function

$$g(N) = 1 - \frac{1}{20} N\left(\frac{1}{20} N - 1\right),$$

where N is the number of grasshoppers counted on the test plot on July 1. An approximation for the population of the buffalo herd on October 1 is

$$P = f(Y) = 1000\sqrt{Y^2 + 3Y + 1}.$$

Find a formula for P as a function of N.

CHAPTER 2 Derivatives

2.0 Introduction

Our study of limits began with a discussion of the change in the value of $y = f(x)$ in response to change in x near a point c. Now that we have laid the foundation of limits and their properties, we return to the special type of limit used to define and calculate instantaneous rates of change. It is called the *derivative* of y (or f) with respect to x.

In the first section of this chapter, we see that the instantaneous rate of change of a function f at a point c is closely related to the tangent line to the graph of f at the point $(c, f(c))$. The next section derives formulas for calculating derivatives of many elementary functions. In Section 3, we pause to review the trigonometric functions, and then consider their derivatives in the next section.

Section 5 exploits the connection between the derivative of f and the tangent line to its graph to approximate values of f near $x = c$. This leads to a useful device called the *differential*.

In Section 6, we present the most important rule for computing derivatives, the *chain rule*, which is used to find the derivative of composite functions. Its importance is brought out in the next two sections, which apply the chain rule to find derivatives of inverse functions and functions defined by equations that relate x and y.

The final two sections discuss topics that will be useful later. Section 9 considers the derivative of the derivative function, which is called the *second derivative*. This is the main tool in Section 3.4 and also underlies much of Chapter 9. The last section presents the mean value theorem for derivatives, a key result that will be used in Section 3.3 (and later) to develop tools for studying the behavior of differentiable functions.

2.1 Instantaneous Rates of Change and Tangent Lines

FIGURE 1.1

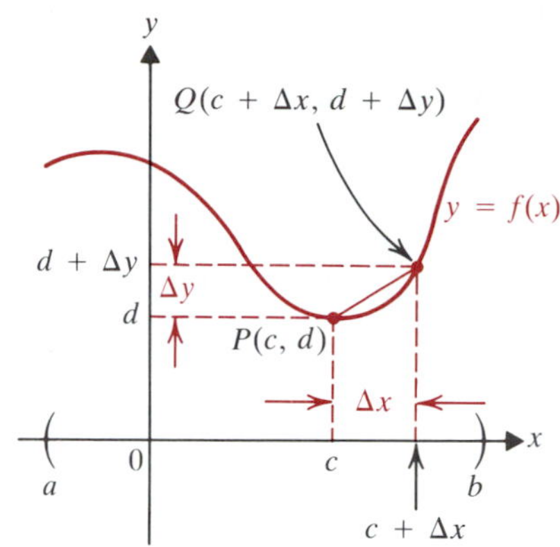

We began our discussion of limits in Section 1.4 by analyzing the rate of change of a quantity y per unit change in a variable x, where $y = f(x)$. Recall that the *average rate of change* of y with respect to x over the interval with endpoints c and $c + \Delta x$ is

$$\frac{\Delta y}{\Delta x} = \frac{f(c + \Delta x) - f(c)}{\Delta x}. \tag{1}$$

This is the slope of the *secant* joining the points $P(c, d)$ and $Q(c + \Delta x, d + \Delta y)$ of the graph of f, where $d = f(c)$ and $d + \Delta y = f(c + \Delta x)$. Refer to **Figure 1.1.**

As $\Delta x \to 0$, the average rate of change $\Delta y/\Delta x$ in (1) ought to approximate more and more closely the *instantaneous rate of change* of y with respect to x at $x = c$. This corresponds to approximating one's instantaneous speed, as shown on a car's speedometer, say, by computing the average speed over smaller and smaller time intervals of length $\Delta t \to 0$. (See Examples 1 and 2 of Section 1.4.) The following definition sums up these ideas.

1.1 DEFINITION

If $y = f(x)$, then the ***derivative of y with respect to x at x = c*** is

$$f'(c) = \left.\frac{dy}{dx}\right|_c = \lim_{\Delta x \to 0} \frac{\Delta y}{\Delta x} = \lim_{\Delta x \to 0} \frac{f(c + \Delta x) - f(c)}{\Delta x}, \tag{2}$$

if this limit exists. When the limit does exist, it is called the ***instantaneous rate of change of y with respect to x at x = c.***

The symbol

$$\left.\frac{dy}{dx}\right|_c$$

is read "$dy\,dx$ at c". **It is a single symbol, not a quotient.** It is intended to remind you that

$$\left.\frac{dy}{dx}\right|_c$$

is the limit of the difference quotient $\Delta y/\Delta x$ as $x \to 0$.

Example 1

If $f(x) = x^2$, then find (a) $f'(-1)$, (b) $f'(2)$.

Solution. (a) Using (2) with $c = -1$, we find

$$f'(-1) = \lim_{\Delta x \to 0} \frac{(-1 + \Delta x)^2 - (-1)^2}{\Delta x}$$

$$= \lim_{\Delta x \to 0} \frac{1 - 2\Delta x + (\Delta x)^2 - 1}{\Delta x} = \lim_{\Delta x \to 0} \frac{\Delta x(-2 + \Delta x)}{\Delta x}$$

$$= \lim_{\Delta x \to 0} (-2 + \Delta x) = -2.$$

(b) With $c = 2$, (2) gives

$$f'(2) = \lim_{\Delta x \to 0} \frac{(2 + \Delta x)^2 - 2^2}{\Delta x} = \lim_{\Delta x \to 0} \frac{4 + 4\Delta x + (\Delta x)^2 - 4}{\Delta x}$$

$$= \lim_{\Delta x \to 0} \frac{\Delta x(4 + \Delta x)}{\Delta x} = \lim_{\Delta x \to 0} (4 + \Delta x) = 4. \blacksquare$$

For most functions—including f in Example 1—as the point c varies, so does the value of $f'(c)$. We thus obtain the ***derivative function*** f' whose domain is the set of all points c for which the limit in (2) exists. The function f is said to be ***differentiable*** at $x = c$ when that limit does exist, and at such a point the value of the derivative function f' is simply the value $f'(c)$ of the limit in (2). The process of calculating $f'(c)$ is called *differentiation* of f at $x = c$. We say that we *differentiate* f at c when we compute $f'(c)$.

Example 2

If $f(x) = x^2$, then find the formula for $f'(x)$.

Solution. We use (2) with c any real number. That gives

$$f'(c) = \lim_{\Delta x \to 0} \frac{(c + \Delta x)^2 - c^2}{\Delta x} = \lim_{\Delta x \to 0} \frac{c^2 + 2c\,\Delta x + (\Delta x)^2 - c^2}{\Delta x}$$

$$= \lim_{\Delta x \to 0} \frac{\Delta x(2c + \Delta x)}{\Delta x} = \lim_{\Delta x \to 0} (2c + \Delta x) = 2c.$$

Thus the derivative function f' has the formula $f'(x) = 2x$. $\blacksquare$

Besides $f'(x)$ and dy/dx, other notations for the derivative function include $D_x f(x)$, $Df(x)$, and df/dx. We will sometimes use these, especially $D_x f(x)$. In place of $f'(x)$, we will frequently write dy/dx.

Just as there is more than one notation for the left side of (2), so there is more than one notation used in writing the limit. The most common variant of the notation in (2) is obtained by letting $x = c + \Delta x$. Then $\Delta x = x - c$, and so $\Delta x \to 0$ is equivalent to $x - c \to 0$, that is, to $x \to c$. This gives the following alternative version of (2).

$$f'(c) = \lim_{x \to c} \frac{f(x) - f(c)}{x - c}. \tag{3}$$

This form of the definition is sometimes more convenient to use than (2), as we will see later. Because h is simpler to write than Δx, mathematicians tend to prefer it to Δx in (2). The next example illustrates that.

Example 3

A sky diver in free fall drops y feet during the first x seconds, where $y = 16x^2$ (approximately). What is her speed of fall when $x = c$?

FIGURE 1.2

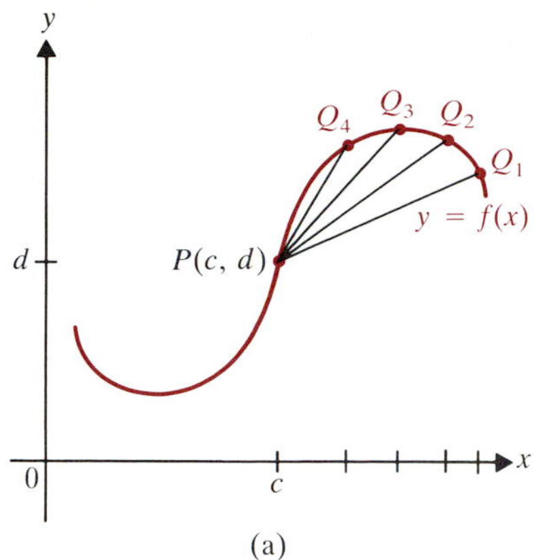

(a)

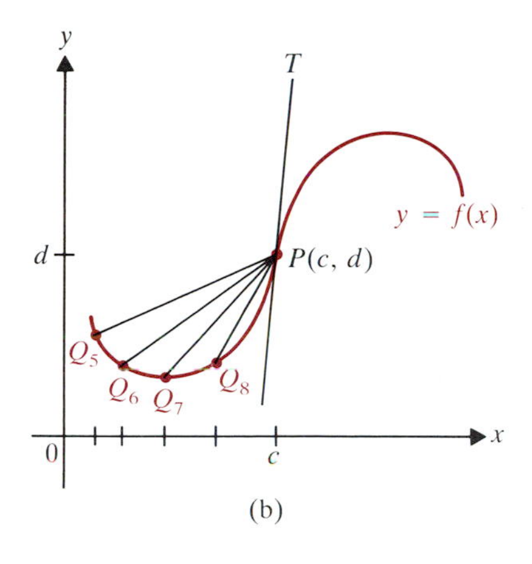

(b)

Solution. We are asked to compute dy/dx where $y = f(x) = 16x^2$ and $x = c$. We use (2), with h in place of Δx, getting

$$\left.\frac{dy}{dx}\right|_c = \lim_{h\to 0}\frac{f(c+h)-f(c)}{h} = \lim_{h\to 0}\frac{16(c+h)^2-16c^2}{h}$$

$$= \lim_{h\to 0} 16\,\frac{\cancel{c^2}+2ch+h^2-\cancel{c^2}}{h}$$

$$= 16\lim_{h\to 0}\frac{2ch+h^2}{h} = 16\lim_{h\to 0}\frac{\cancel{h}(2c+h)}{\cancel{h}}$$

$$= 16\lim_{h\to 0}(2c+h) = 16\cdot 2c = 32c. \quad \blacksquare$$

When $c = 4$, we get the result of Example 1 of Section 1.4: $dy/dx = 128$ when $x = 4$. *The advantage of the foregoing solution is that from it we can compute dy/dx at any point c.* For instance, $dy/dx = 32$ when $c = 1$, $dy/dx = 48$ when $c = 1.5$, and so forth.

Since the average rate of change $\Delta y/\Delta x$ is the slope of the line through the points $P(c, d)$ and $Q(c + \Delta x, d + \Delta y)$ on the graph of $y = f(x)$, we may ask what graphical interpretation there is for the instantaneous rate of change dy/dx. **Figure 1.2** suggests what happens as $\Delta x \to 0^+$ and as $\Delta x \to 0^-$. In Figure 1.2(a), as $\Delta x \to 0^+$, the point $Q(c + \Delta x, d + \Delta y)$ moves along the graph of $y = f(x)$ toward the point $P(c, d)$. The secant PQ changes in a corresponding way. For a small value of Δx, like Δx_4 in Figure 1.2(a), the secant seems to be almost *tangent* to the curve at P. In Figure 1.2(b), as $\Delta x \to 0$, the secant PQ seems to *more and more closely* approximate what looks like the tangent line to the curve at $P(c, d)$. This is the basis for the definition of the tangent line in calculus.

The last paragraph seems to suggest defining the tangent line as the limit of the secant lines as $\Delta x \to 0$. We only know how to define limits of *functions*, however. So our approach is to define the *slope m* of the tangent line at $P(c, d)$. Once m is known, we can write down the equation of the tangent line: $y - d = m(x - c)$. Figure 1.2 suggests that the slope of the tangent line should be the limit of the slope

$$\frac{\Delta y}{\Delta x} = \frac{f(c+\Delta x)-f(c)}{\Delta x}$$

of the secants as $\Delta x \to 0$. By Definition 1.1, if this difference quotient has a limit, then that limit is

$$f'(c) = \left.\frac{dy}{dx}\right|_c.$$

We thus give the following definition.

1.2 DEFINITION

The ***slope of the tangent line*** to the graph of $y = f(x)$ at $P(c, f(c))$ is the derivative

(4) $$f'(c) = \lim_{h\to 0}\frac{f(c+h)-f(c)}{h},$$

if f is differentiable at $x = c$. [If c is an endpoint of the domain of f, then the appropriate one-sided limit is used in (4).]

Example 4

Find the slope of the tangent line to the graph of $y = \sqrt{x}$ at (a) the point $(c, \sqrt{c})$, where $c > 0$; (b) the point $(0, 0)$.

Solution. (a) The slope of the tangent line is

$$\lim_{h \to 0} \frac{f(c+h) - f(c)}{h} = \lim_{h \to 0} \frac{\sqrt{c+h} - \sqrt{c}}{h}.$$

As h approaches 0, both the numerator and denominator of the difference quotient approach 0. To evaluate the limit, we use the same algebraic trick as in Example 2 of Section 1.4. We multiply numerator and denominator by $\sqrt{c+h} + \sqrt{c}$. This rationalizing factor permits cancellation of h as a common factor of the numerator and denominator:

$$\begin{aligned} \frac{\sqrt{c+h} - \sqrt{c}}{h} &= \frac{\sqrt{c+h} - \sqrt{c}}{h} \cdot \frac{\sqrt{c+h} + \sqrt{c}}{\sqrt{c+h} + \sqrt{c}} \\ &= \frac{(\cancel{c} + h) - \cancel{c}}{h(\sqrt{c+h} + \sqrt{c})} = \frac{\cancel{h}}{\cancel{h}(\sqrt{c+h} + \sqrt{c})} \\ &= \frac{1}{\sqrt{c+h} + \sqrt{c}}. \end{aligned}$$

Hence, the slope of the tangent is

$$\lim_{h \to 0} \frac{1}{\sqrt{c+h} + \sqrt{c}} = \frac{1}{\sqrt{c} + \sqrt{c}},$$

$$\left.\frac{dy}{dx}\right|_c = \frac{1}{2\sqrt{c}}. \tag{5}$$

(b) All of the calculations made before we took limits in (a) remain valid if $c = 0$. They show that if $c = 0$, then the slope of the tangent line would be

$$\lim_{h \to 0^+} \frac{1}{\sqrt{0+h} - \sqrt{0}} = \lim_{h \to 0^+} \frac{1}{\sqrt{h}}.$$

Thus there is no slope of the tangent line, since dy/dx *fails to exist* when $c = 0$. ■

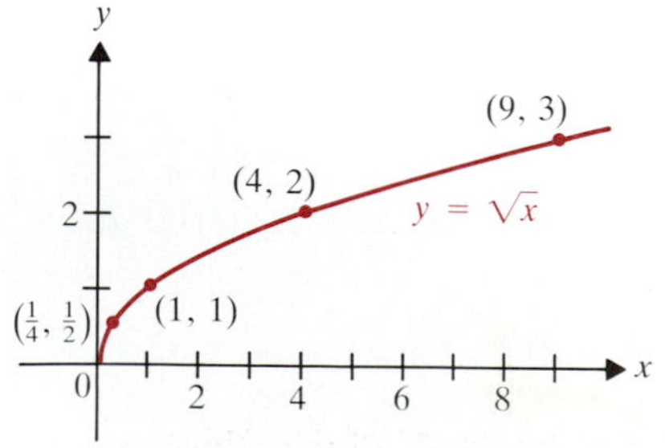

FIGURE 1.3

Figure 1.3 suggests what happened in Example 2(b). The graph of $y = \sqrt{x}$ appears to be tangent to the *vertical line* $x = 0$ (the y-axis) at $(0, 0)$. Since a vertical line does not have a slope, the fact that dy/dx fails to be the slope of the tangent line is perfectly appropriate.

Working out the limit in (4) at a general point c saved some work in Example 2(b). The solution of the next example further illustrates the advantage of such a general limit computation.

Example 5

Find the equation of the tangent line to the graph of $y = x^2 + 2x - 3$ at (a) $c = 2$, (b) $c = -2$.

Solution. (a) Using Definition 1.1, we have at any point $x = c$

$$\begin{aligned}\frac{dy}{dx} &= \lim_{h\to 0}\frac{f(c+h)-f(c)}{h}\\ &= \lim_{h\to 0}\frac{(c+h)^2+2(c+h)-3-(c^2+2c-3)}{h}\\ &= \lim_{h\to 0}\frac{c^2+2ch+h^2+2c+2h-3-c^2-2c+3}{h}\\ &= \lim_{h\to 0}\frac{2ch+h^2+2h}{h} = \lim_{h\to 0}\frac{h(2c+h+2)}{h}\\ &= \lim_{h\to 0}(2c+h+2) = 2c+2.\end{aligned}$$

When $c = 2$, we have $dy/dx = 2\cdot 2 + 2 = 6$. Note that $f(2) = 2^2 + 2\cdot 2 - 3 = 5$. Hence, the tangent line at $x = 2$ is

$$y - 5 = 6(x-2) \rightarrow y - 5 = 6x - 12 \rightarrow y = 6x - 7.$$

(b) Here $dy/dx = 2c + 2 = -4 + 2 = -2$. Also, $f(-2) = (-2)^2 + 2(-2) - 3 = -3$. So the tangent line is

$$y + 3 = -2(x+2) \rightarrow y + 3 = -2x - 4 \rightarrow y = -2x - 7.$$

We show the tangent lines and the graph in **Figure 1.4,** which seems to fit our idea about tangency at the points $P_1(2, 5)$ and $P_2(-2, -3)$ quite well. ■

FIGURE 1.4

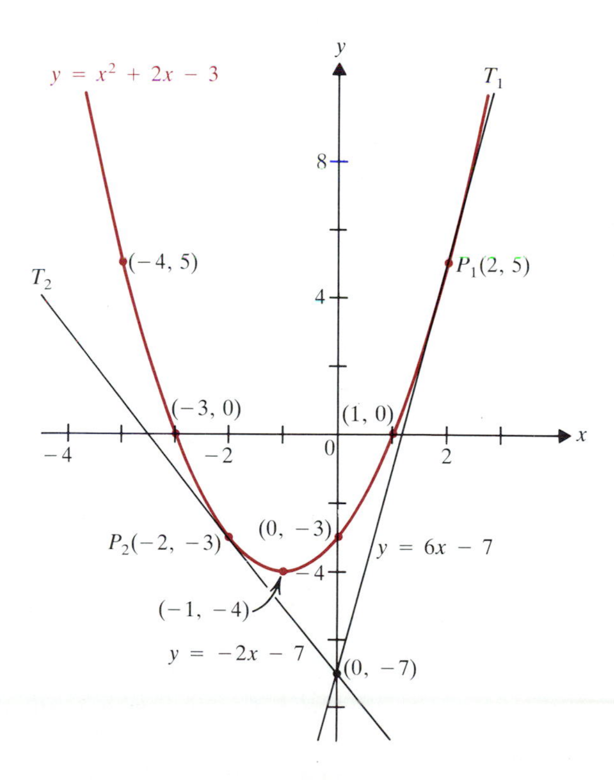

Recall that being continuous at a point $x = c$ is a more demanding condition than having a limit at c. Having a derivative is a still more demanding condition, as the next theorem indicates.

1.3 THEOREM

> If f is differentiable at c, then f is continuous at c.

Proof. Suppose that f is differentiable at $x = c$. Then f is defined at $x = c$, since the difference quotient in (2) involves $f(c)$. Moreover, using (3) and limit theorems, we have

$$\begin{aligned}\lim_{x \to c} [f(x) - f(c)] &= \lim_{x \to c} \left[\frac{f(x) - f(c)}{x - c} \cdot (x - c)\right] \\ &= \lim_{x \to c} \frac{f(x) - f(c)}{x - c} \cdot \lim_{x \to c} (x - c) \\ &= f'(c) \cdot 0 = 0.\end{aligned}$$

Hence, $\lim_{x \to c} f(x) = f(c)$. QED

Theorem 1.3 says that every differentiable function must be continuous. But continuous functions need not always be differentiable, as our concluding example shows.

Example 6

Show that the absolute value function f, given by

$$f(x) = |x|,$$

is continuous but not differentiable at $x = 0$.

Solution. First, f is continuous at 0 since

$$f(x) = \begin{cases} -x & \text{for } x < 0 \\ x & \text{for } x > 0. \end{cases}$$

Hence, $\lim_{x \to 0^+} f(x) = \lim_{x \to 0^+} x = 0$ and $\lim_{x \to 0^-} f(x) = \lim_{x \to 0^-} (-x) = 0$, so that

$$\lim_{x \to 0} f(x) = 0 = f(0).$$

But we have

$$\frac{f(0 + h) - f(0)}{h} = \frac{|0 + h| - |0|}{h} = \frac{|h|}{h}.$$

Therefore

$$\lim_{h \to 0^+} \frac{f(0 + h) - f(0)}{h} = \lim_{h \to 0^+} \frac{h}{h} = 1,$$

and

$$\lim_{h \to 0^-} \frac{f(0 + h) - f(0)}{h} = \lim_{h \to 0^-} \frac{-h}{h} = -1.$$

Hence, by Theorem 6.2 of the last chapter,

$$\lim_{h\to 0} \frac{f(c+h)-f(c)}{h}$$

fails to exist when $c = 0$. Thus f is not differentiable at $x = 0$. ■

In order of increasing strength, the hierarchy of properties of a function is

having a limit, being continuous, being differentiable

at a point. Most functions we consider will be differentiable, but some, like the absolute value function, may fail to be differentiable at certain points of their domains even though they may be continuous there.

HISTORICAL NOTE

The notation dy/dx is due to Gottfried W. Leibniz (1646–1716), the great German mathematician, scientist, and philosopher who developed calculus independently of but at about the same time as Newton. Leibniz was particularly adept at developing useful notation, some of which we will see later. He seems to have introduced, or popularized, many now-standard notational devices, including the equal sign, a raised dot for multiplication, and numerical superscripts to indicate powers. (Descartes originated the latter notation in 1637.)

In 1665, Newton gave the essence of Definition 1.2 for tangent lines. The precise limit definition (1) was given by the French mathematician Augustin L. Cauchy (1789–1857), who in 1821 published a calculus text containing many of the derivations and proofs still given in modern books. The notation $f'(x)$ for the derivative function was introduced by the French mathematician Joseph L. Lagrange (1736–1813) in a textbook published in 1797. Lagrange's text was the first to attempt a logically sound development of calculus free of infinitesimals. It fell short of modern standards of rigor, however. He based his development on power series, which we will discuss only in Chapter 9, after developing most of single-variable calculus.

Exercises 2.1

In Exercises 1–8, use Definition 1.2 to find the equation of the tangent line to the graph of the given function at the given points. Draw a figure showing the curve and the tangent lines.

1. $y = x^2 + 1$ at **(a)** $c = 1$, **(b)** $c = -2$.
2. $y = 1 - 2x^2$ at **(a)** $c = -1$, **(b)** $c = 2$.
3. $y = 1/x$ at **(a)** $c = 1$, **(b)** $c = -2$.
4. $y = x^2 - x - 1$ at **(a)** $c = 1$, **(b)** $c = -2$.
5. $y = 4 - x^2$ at **(a)** $c = 1$, **(b)** $c = -2$.
6. $y = \sqrt{x+2}$ at **(a)** $c = 2$, **(b)** $c = -1$.
7. $y = \sqrt{x+2}$ at $c = -2$.
8. $y = \sqrt{1-x^2}$ at **(a)** $c = 1$, **(b)** $c = 0$.

In Exercises 9–14, use Definition 1.1 to find the formula for $D_x f(x)$. Also give the domain of f'.

9. $f(x) = -11$
10. $f(x) = 5x - 2$
11. $f(x) = x^3 + 1$
12. $f(x) = \dfrac{1}{x^2+1}$
13. $f(x) = 1/x^2$
14. $f(x) = \sqrt{3x-2}$

In Exercises 15–20, use Definition 1.1 to find the velocity at $t = t_0$ of a body whose position at time t is s.

15. $s = 980t^2$ at **(a)** $t_0 = 1$, **(b)** $t_0 = 2$.
16. $s = 9.8t^2$ at **(a)** $t_0 = 0$, **(b)** $t_0 = 3$.

17. $s = t^2 - 4t + 6$ at **(a)** $t_0 = 1$, **(b)** $t_0 = 2$, **(c)** $t_0 = 3$.
(d) If positive velocity indicates upward motion, and negative velocity indicates downward motion, then analyze the direction of motion.

18. $s = 3t^2 - 6t + 2$ at **(a)** $t_0 = 0$, **(b)** $t_0 = 1$, **(c)** $t_0 = 2$. **(d)** Repeat Exercise 17(d).

19. $s = 5 + 4t - t^2$ at **(a)** $t_0 = 1$, **(b)** $t_0 = 2$, **(c)** $t_0 = 3$. **(d)** Repeat Exercise 17(d).

20. $s = -5 + 6t - t^2$ at **(a)** $t_0 = 2$, **(b)** $t_0 = 3$, **(c)** $t_0 = 4$. **(d)** Repeat Exercise 17(d).

The *normal line* to the graph of $y = f(x)$ at $P(c, f(c))$ is defined to be the line through P perpendicular to the tangent line. In Exercises 21–24, find the normal line to the graph of the function in the given exercise at the points specified in that exercise.

21. Exercise 1

22. Exercise 2

23. Exercise 3

24. Exercise 7

25. Show that the tangent lines to the graph of $y = x^2$ at $x_0 = 2$ and $x_1 = -2$ intersect on the y-axis.

26. Show that the tangents to $y = x^2$ at the two points (c, c^2) and $(-c, c^2)$ always intersect on the y-axis.

27. Show that the tangents to $y = x^2$ at the points (c, c^2) and (d, d^2) intersect at the point $(\frac{1}{2}(c + d), cd)$.

28. Let T be the tangent line to the graph of $y = x^2$ at any point $P(c, c^2)$. Show that T intersects the x-axis at $(\frac{1}{2}c, 0)$.

In Exercises 29–36, use Theorem 1.3 to show that f is *not* differentiable at the given c, by showing it is not continuous there.

29. $f(x) = 1/x, \quad c = 0$

30. $f(x) = \dfrac{1}{x - 2}, \quad c = 2$

31. $f(x) = \begin{cases} 1 - x^2, & x \le 0 \\ x^2, & x > 0 \end{cases}, \quad c = 0$

32. $f(x) = \begin{cases} \dfrac{1}{x + 2}, & x \le -3 \\ 1 + x, & x > -3 \end{cases}, \quad c = -3$

33. $f(x) = [x], \quad c = 3$

34. $f(x) = 1 - [x], \quad c = 2$

35. $f(x) = d(x)$, the Dirichlet function of Exercise 37, Section 1.4, at any point c.

36. $f(x) = 1/\sqrt{x}, \quad c = 0.$

37. Find the domain of f' if $f(x) = |x - 3|$.

38. Find the domain of f' if $f(x) = |5 - 3x|$.

39. If f is an even function that is differentiable on all of **R**, then show that f' is an odd function.

40. If f is an odd function differentiable on all of **R**, then show that f' is an even function.

PC **41.** Use a calculator or computer to estimate $f'(c)$ for $c = 0$, $\pi/6$, $\pi/4$, $\pi/3$, and $\pi/2$ if $f(x) = \sin x$. Do this by computing

$$\frac{f(c + h) - f(c)}{h}$$

for **(a)** $h = 0.1$ **(b)** $h = 0.01$ **(c)** $h = 0.001$ **(d)** $h = 0.0001$. On the basis of your results, what do you expect $f'(c)$ to be?

PC **42.** Repeat Exercise 41 for $f(x) = \cos x$ and $c = 0, \pi/6, \pi/4, \pi/3$, and $\pi/2$.

2.2 Basic Rules of Differentiation

So far, we have used Definition 1.1 to calculate $f'(c)$. Since that is a bit of a chore, in this section we derive *formulas* for $f'(x)$ that will apply directly to most elementary functions. The first rule is that *the derivative of a constant function is zero.* (For example, if a body's position $y(t)$ is constant, then its speed dy/dt is always zero.) To see this, suppose that $f(x) = c$ for all x. Then

$$\lim_{h \to 0} \frac{f(x + h) - f(x)}{h} = \lim_{h \to 0} \frac{\not{c} - \not{c}}{h} = \lim_{h \to 0} 0 = 0.$$

We state this as the following rule, which says that a horizontal line has slope zero at every point.

> If $f(x) = b$ for every real number x, then f is differentiable on all of **R** and
>
> (1) $$f'(x) = 0. \qquad \text{That is, } D_x(b) = 0.$$

The next simplest type of function f is linear, $f(x) = mx + b$. Since $f'(x)$ is the slope of the tangent line to the graph of f, we might expect $f'(x)$ to be m. Sure enough, we have

$$\lim_{h\to 0}\frac{f(x+h)-f(x)}{h} = \lim_{h\to 0}\frac{m(x+h)+b-(mx+b)}{h}$$

$$= \lim_{h\to 0}\frac{mx+mh+b-mx-b}{h}$$

$$= \lim_{h\to 0}\frac{mh}{h} = m.$$

This establishes the next rule.

> If $f(x) = mx + b$, then for every real number x,
>
> (2) $f'(x) = m$. That is, $D_x(mx + b) = m$.

We next turn to powers of x. First, consider $f(x) = x^2$. We have

$$D_x(x^2) = \lim_{h\to 0}\frac{(x+h)^2 - x^2}{h} = \lim_{h\to 0}\frac{x^2 + 2xh + h^2 - x^2}{h}$$

$$= \lim_{h\to 0}\frac{h(2x+h)}{h} = \lim_{h\to 0}(2x+h) = 2x.$$

Thus

(3) $$D_x(x^2) = 2x.$$

We continue by investigating the function $g(x) = x^3$. This time, the difference quotient will involve the cube of $x + h$, so we use the formula

$$(x+h)^3 = x^3 + 3x^2h + 3xh^2 + h^3$$

to obtain

$$g'(x) = \lim_{h\to 0}\frac{(x+h)^3 - x^3}{h}$$

$$= \lim_{h\to 0}\frac{x^3 + 3x^2h + 3xh^2 + h^3 - x^3}{h}$$

$$= \lim_{h\to 0}\frac{h(3x^2 + 3xh + h^2)}{h} = \lim_{h\to 0}(3x^2 + 3xh + h^2)$$

$$= 3x^2.$$

This result, together with (3), suggests the following general formula.

2.1 THEOREM

> ***Power Rule.*** If n is any positive integer and $f(x) = x^n$, then
>
> (4) $f'(x) = nx^{n-1}$. That is, $D_x(x^n) = nx^{n-1}$.

Proof. We remind you of another formula from elementary algebra, namely, the factor formula for the difference of two nth powers,

(5) $$a^n - b^n = (a-b)(a^{n-1} + a^{n-2}b + a^{n-3}b^2 + \ldots + ab^{n-2} + b^{n-1}).$$

(This is easy to verify by multiplying out the right side.) We will use (5) with $a = x + h$ and $b = x$. Notice that there are n terms in the second parentheses on the right side of (5), namely, a^{n-1} and $n - 1$ terms containing $b, b^2, \ldots, b^{n-1}$. So we have

$$f'(x) = \lim_{h \to 0} \frac{(x+h)^n - x^n}{h}$$

$$= \lim_{h \to 0} \frac{(\not{x} + h - \not{x})[(x+h)^{n-1} + (x+h)^{n-2} \cdot x + \ldots + (x+h)x^{n-2} + x^{n-1}]}{h}$$

$$= \lim_{h \to 0} \frac{\not{h}[(x+h)^{n-1} + (x+h)^{n-2} \cdot x + \ldots + (x+h)x^{n-2} + x^{n-1}]}{\not{h}}$$

$$= \lim_{h \to 0} [(x+h)^{n-1} + (x+h)^{n-2} \cdot x + \ldots + (x+h)x^{n-2} + x^{n-1}]$$

$$= x^{n-1} + x^{n-2} \cdot x + \ldots + x \cdot x^{n-2} + x^{n-1}$$

$$= \underbrace{x^{n-1} + x^{n-1} + \ldots + x^{n-1} + x^{n-1}}_{n \text{ terms}} = nx^{n-1}. \quad \text{QED}$$

Example 1

Find $f'(x)$ if
(a) $f(x) = x$ (b) $f(x) = x^4$.

Solution. (a) From Rule (2) with $m = 1$ and $b = 0$, we have $f'(x) = 1$.
(b) From Theorem 2.1 with $n = 4$, we get $f'(x) = 4x^3$. ■

To differentiate polynomial functions, we will apply Theorem 2.1 to constant multiples and sums of powers of x. So we need to know how to differentiate a constant multiple of any function f. The definition of $D_x[cf(x)]$ and Theorem 5.1(c) of Chapter 1 give

$$D_x[cf(x)] = \lim_{h \to 0} \frac{cf(x+h) - cf(x)}{h}$$

$$= \lim_{h \to 0} c \cdot \frac{f(x+h) - f(x)}{h}$$

$$= c \cdot \lim_{h \to 0} \frac{f(x+h) - f(x)}{h} = cf'(x).$$

This establishes the first part of the next result.

2.2 THEOREM

If f and g are differentiable at the real number x and c is any constant, then

(a) cf is differentiable at x and

$$D_x(cf(x)) = cf'(x); \tag{6}$$

(b) $f + g$ and $f - g$ are differentiable at x and

$$D_x[(f+g)(x)] = f'(x) + g'(x), \tag{7}$$

$$D_x[(f-g)(x)] = f'(x) - g'(x). \tag{8}$$

Proof. The discussion above establishes (6). We derive (7) and leave (8) for you as Exercise 29. We have

$$
\begin{aligned}
D_x[(f+g)(x)] &= \lim_{h\to 0}\frac{(f+g)(x+h)-(f+g)(x)}{h} \\
&= \lim_{h\to 0}\frac{f(x+h)+g(x+h)-f(x)-g(x)}{h} \\
&= \lim_{h\to 0}\left(\frac{f(x+h)-f(x)}{h}+\frac{g(x+h)-g(x)}{h}\right) \\
&= \lim_{h\to 0}\frac{f(x+h)-f(x)}{h}+\lim_{h\to 0}\frac{g(x+h)-g(x)}{h} \\
&= f'(x)+g'(x). \quad \boxed{\text{QED}}
\end{aligned}
$$

By repeated application of (7) and (8), we can extend them to sums or differences of any finite number of functions, that is,

$$
(9) \qquad D_x[f_1(x) \pm f_2(x) \pm \ldots \pm f_n(x)] = f'_1(x) \pm f'_2(x) \pm \ldots \pm f'_n(x),
$$

where for each $j = 1, 2, \ldots, n$, $f'_j(x)$ on the right has the same sign prefixed before it as $f_j(x)$ has on the left. As suggested after Example 1, we can use (9) and (6) to differentiate any polynomial function.

Example 2

Differentiate

(a) $f(x) = 5x^3 - 7x^2 - 2x + 3$

(b) $p(x) = a_n x^n + a_{n-1}x^{n-1} + \ldots + a_2x^2 + a_1x + a_0$.

Solution. (a)

$$
\begin{aligned}
f'(x) &= \frac{d}{dx}(5x^3) - \frac{d}{dx}(7x^2) - \frac{d}{dx}(2x) + \frac{d}{dx}(3) && \textit{by (7)} \\
&= 5\frac{d}{dx}(x^3) - 7\frac{d}{dx}(x^2) - 2\frac{d}{dx}(x) + 0 && \textit{by (6)} \\
&= 5(3x^2) - 7(2x) - 2 && \textit{by Theorem 2.1} \\
&= 15x^2 - 14x - 2.
\end{aligned}
$$

(b) $p'(x) = na_nx^{n-1} + (n-1)a_{n-1}x^{n-2} + \ldots + 2a_2x + a_1$. ■

Qualitatively, Theorem 2.2 says that the operation of taking derivatives, which we denote by d/dx, preserves sums, differences, and constant multiples in the sense that

- the derivative of a sum is the sum of the derivatives:

$$\frac{d}{dx}[f(x) + g(x)] = f'(x) + g'(x);$$

- the derivative of a difference is the difference of the derivatives:

$$\frac{d}{dx}[f(x) - g(x)] = f'(x) - g'(x);$$

- the derivative of a constant times a function is the constant times the derivative of the function:

$$\frac{d}{dx}[cf(x)] = cf'(x).$$

In view of the fact that the limit of a product of two functions is the product of the limits, it is natural to guess that the derivative of a product will be the product of the derivatives. This reasonable expectation is, however, ***false.*** For instance, consider the functions $f(x) = x^2$ and $g(x) = x^3$. We know from (4) that $f'(x) = 2x$ and $g'(x) = 3x^2$, so that

$$f'(x) \cdot g'(x) = 2x \cdot 3x^2 = 6x^3.$$

But $f(x) \cdot g(x) = x^2 \cdot x^3 = x^5$, so that we can also use (4) to find

$$\frac{d}{dx}[f(x) \cdot g(x)] = \frac{d}{dx}(x^5) = 5x^4,$$

which is *not* the product $f'(x) \cdot g'(x)$ of the derivatives. Over a period of more than a week in 1677, Leibniz rejected this and some other guesses before discovering the correct formula for the derivative of a product. Instances of the following rule had also been used by Newton as much as a decade earlier.

2.3 THEOREM

Product Rule. If f and g are differentiable at c, then so is $f \cdot g$ and

(10) $$(f \cdot g)'(c) = f(c)g'(c) + f'(c)g(c).$$

Proof. To transform the difference quotient

$$\frac{f(c+h)g(c+h) - f(c)g(c)}{h}$$

into a form whose limit we can evaluate, we subtract and add the term $f(c)g(c+h)$ in the numerator. This allows us to break the difference quotient into two terms, whose limits we can recognize. That is how we get

$$\begin{aligned}
(f \cdot g)'(c) &= \lim_{h\to 0} \frac{(f \cdot g)(c+h) - (f \cdot g)(c)}{h} \\
&= \lim_{h\to 0} \frac{f(c+h)g(c+h) - f(c)g(c)}{h} \\
&= \lim_{h\to 0} \frac{f(c+h)g(c+h) - f(c)g(c+h) + f(c)g(c+h) - f(c)g(c)}{h} \\
&= \lim_{h\to 0} \left(\frac{f(c+h) - f(c)}{h} \cdot g(c+h) + f(c) \cdot \frac{g(c+h) - g(c)}{h} \right) \\
&= \lim_{h\to 0} \left(\frac{f(c+h) - f(c)}{h} g(c+h) \right) + \lim_{h\to 0} \left(f(c) \cdot \frac{g(c+h) - g(c)}{h} \right) \\
&\qquad \text{by Theorem 5.1(d) of Chapter 1} \\
&= \lim_{h\to 0} \frac{f(c+h) - f(c)}{h} \lim_{h\to 0} g(c+h) + f(c) \lim_{h\to 0} \frac{g(c+h) - g(c)}{h} \\
&\qquad \text{by Theorems 5.4 and 5.1(c) of Chapter 1} \\
&= f'(c)g(c) + f(c)g'(c) \qquad \text{by Theorem 1.3} \\
&= f(c)g'(c) + f'(c)g(c). \quad \boxed{\text{QED}}
\end{aligned}$$

In words, (10) says that

> *the derivative of a product is the first function times the derivative of the second plus the derivative of the first function times the second:*
>
> $$\frac{d}{dx}[f(x) \cdot g(x)] = f(x)g'(x) + f'(x)g(x).$$

As an illustration of the product rule, we have

$$D_x[(x^3 - 3x^2 + x - 2)(x^4 - 5)]$$
$$= (x^3 - 3x^2 + x - 2) \cdot 4x^3 + (3x^2 - 6x + 1)(x^4 - 5).$$

If one of the functions f or g is a constant function, then the product rule reduces to (6) in Theorem 2.2. For instance, if $g(x) = c$, we have by the product rule

$$D_x[cf(x)] = cf'(x) + c'f(x)$$
$$= cf'(x) + 0 \cdot f(x) = cf'(x). \qquad \textit{by (1)}$$

The product rule can be used on products involving more than two factors, by repeated application. For example, we have by (10)

$$D_x[f(x) \cdot g(x) \cdot h(x)] = f(x)g(x) \cdot h'(x) + [f(x)g(x)]' \cdot h(x)$$
$$= f(x)g(x)h'(x) + [f(x)g'(x) + f'(x)g(x)] \cdot h(x)$$
$$= f(x)g(x)h'(x) + f(x)g'(x)h(x) + f'(x)g(x)h(x).$$

In general, the derivative of a product of n functions is the sum of n products formed by differentiating one factor and multiplying it by the product of the other factors.

Example 3

The *ideal gas law* relates pressure P (in atmospheres), volume V (in liters), and the Kelvin temperature T of a gas according to $PV = nRT$, where R is the gas constant 0.0821 liter·atm/K·mol, and n is the number of moles of the gas. Suppose at a certain instant that $P = 10.0$ atm and is increasing at a rate of 8.00×10^{-2} atm/min and that $V = 15.0$ liters and is decreasing at the rate of 0.100 liters/min. What is the rate of change of T per minute if $n = 20.0$ mol?

Solution. Let t stand for time, measured in minutes. We are given $P_0 = 10.0$, $V_0 = 15.0$, $dP/dt = 0.080$, and $dV/dt = -0.100$ at the time in question. Here, P, V, and T all change over time and so are functions of t. From

$$T = \frac{1}{nR} PV$$

we get

$$\frac{dT}{dt} = \frac{1}{nR}\frac{d}{dt}(PV) \qquad \textit{by Theorem 2.2(a)}$$

$$= \frac{1}{nR}\left(P\frac{dV}{dt} + \frac{dP}{dt} \cdot V\right) \qquad \textit{by the product rule (10)}$$

$$= \frac{1}{nR}[10.0(-0.100) + (0.080)(15.0)] = \frac{0.200}{nR}.$$

Since $R = 8.21 \times 10^{-2}$ and $n = 20.0$, we have

$$\frac{dT}{dt} = \frac{0.200}{(20.0)(8.21 \times 10^{-2})} \approx 0.122.$$

Thus T is rising at about 0.12 K/min. ■

We turn now to the question of finding the derivative of a quotient f/g of two functions f and g, each of which is differentiable. Note that

$$(f/g)(x) = f(x) \cdot \frac{1}{g(x)},$$

so that we could apply the product rule to differentiate f/g if we knew how to differentiate the function

$$\frac{1}{g(x)}.$$

FIGURE 2.1

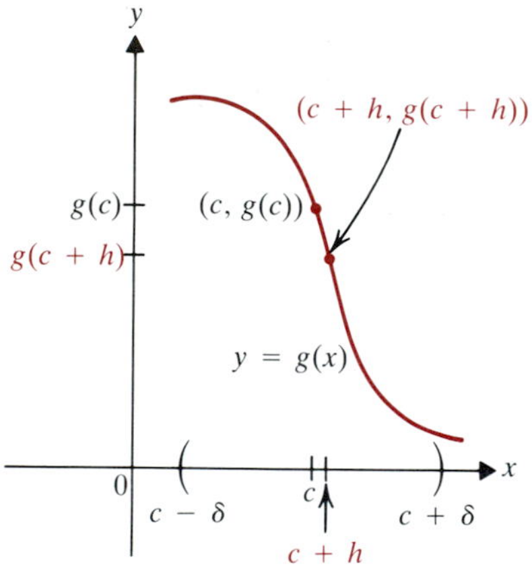

So we first derive a formula for $(1/g)'(x)$. Suppose that g is differentiable at c and that $g(c) \neq 0$. Then by Theorem 1.3, g is continuous at $x = c$. So, as **Figure 2.1** illustrates, for small values of h (less than some number $\delta > 0$), $g(c + h) \neq 0$. This is so because g cannot jump from its nonzero value at c to the value zero at $c + h$ if h is sufficiently small. Then we have

$$\left(\frac{1}{g}\right)'(c) = \lim_{h \to 0} \frac{\dfrac{1}{g(c+h)} - \dfrac{1}{g(c)}}{h} = \lim_{h \to 0} \frac{\dfrac{g(c) - g(c+h)}{g(c)g(c+h)}}{h}$$

$$= \lim_{h \to 0} \frac{g(c) - g(c+h)}{h} \cdot \frac{1}{g(c)g(c+h)}$$

$$= -\lim_{h \to 0} \frac{g(c+h) - g(c)}{h} \cdot \lim_{h \to 0} \frac{1}{g(c)g(c+h)}$$

$$= -g'(c) \cdot \frac{1}{g(c)^2} \qquad \textit{by Theorem 1.3}$$

(11)
$$\left(\frac{1}{g}\right)'(c) = -\frac{g'(c)}{g(c)^2}$$

In words, (11) says that

> *the derivative of the reciprocal of a function g is the negative of the derivative of the function divided by the square of the function:*
>
> $$D_x\left(\frac{1}{g(x)}\right) = -\frac{g'(x)}{g(x)^2}.$$

[Notice that we write $[g(x)]^2$ as $g(x)^2$.]

We can now derive the formula for the derivative of a quotient of two differentiable functions.

2.4
THEOREM

Quotient Rule. Suppose that f and g are differentiable at $x = c$ and that $g(c) \neq 0$. Then f/g is differentiable at $x = c$, and

$$(f/g)'(c) = \frac{g(c)f'(c) - f(c)g'(c)}{g(c)^2}. \tag{12}$$

Proof. Since $(f/g)(x) = f(x) \cdot \dfrac{1}{g(x)}$, we use the product rule and (11) to get

$$\begin{aligned}
(f/g)'(c) &= \left(f \cdot \frac{1}{g}\right)'(c) \\
&= f(c) \cdot \left(\frac{1}{g}\right)'(c) + f'(c) \cdot \left(\frac{1}{g}\right)(c) \\
&= f(c) \cdot \frac{(-g'(c))}{g(c)^2} + f'(c) \cdot \frac{1}{g(c)} && \textit{by (11)} \\
&= \frac{-f(c)g'(c) + f'(c)g(c)}{g(c)^2} \\
&= \frac{g(c)f'(c) - f(c)g'(c)}{g(c)^2}. \quad \boxed{\text{QED}}
\end{aligned}$$

The verbal description of (12) is:

the derivative of a quotient f/g is the denominator times the derivative of the numerator minus the numerator times the derivative of the denominator, all divided by the square of the denominator:

$$D_x\left(\frac{f(x)}{g(x)}\right) = \frac{g(x)f'(x) - f(x)g'(x)}{g(x)^2}.$$

With the aid of this rule, it is easy to find the derivative of any rational function.

Example 4

Find $f'(x)$ if $f(x) = \dfrac{x^3 - x + 2}{x^2 + 1}$.

Solution. Theorem 2.4 gives

$$\begin{aligned}
f'(x) &= \frac{(x^2 + 1) \cdot (3x^2 - 1) - (x^3 - x + 2) \cdot (2x)}{(x^2 + 1)^2} \\
&= \frac{3x^4 + 2x^2 - 1 - 2x^4 + 2x^2 - 4x}{(x^2 + 1)^2} \\
&= \frac{x^4 + 4x^2 - 4x - 1}{(x^2 + 1)^2}. \quad \blacksquare
\end{aligned}$$

We used (11) to derive the quotient rule. We can also use it to extend the power rule (Theorem 2.1) to negative integers. Recall that if n is a positive integer, then the power rule says that

$$D_x(x^n) = nx^{n-1}. \tag{4}$$

Suppose that n is a negative integer. Then $n = -k$ for some positive integer k, and the function $p(x) = x^n$ has formula

$$p(x) = x^{-k} = \frac{1}{x^k}.$$

Then (11) says that

$$p'(x) = -\frac{kx^{k-1}}{(x^k)^2} = -\frac{kx^{k-1}}{x^{2k}} = -\frac{k}{x^{2k-(k-1)}}$$
$$= -\frac{k}{x^{k+1}} = -kx^{-(k+1)},$$

that is,

$$\frac{d}{dx}(x^n) = nx^{n-1}. \tag{13}$$

Equation (13) is simply the assertion that the power rule (4) also holds if n is a negative integer. Henceforth then, feel free to use the power rule to differentiate *any* integer power of x.

Example 5

If $f(x) = \dfrac{1}{8x^3}$, then find $f'(x)$.

Solution. Since

$$f(x) = \frac{1}{8}x^{-3},$$

we get immediately from (13) and Theorem 2.2(a)

$$f'(x) = \frac{1}{8}\cdot(-3x^{-3-1}) = -\frac{3}{8}x^{-4} = -\frac{3}{8x^4}. \blacksquare$$

We close the section by giving a far-reaching extension of the power rule. If in Theorem 2.3 we let $g = f$, then (10) gives

$$D_x[f(x)^2] = f(x)f'(x) + f'(x)f(x) = 2f(x)f'(x). \tag{14}$$

In the same way, using (14) and (10) on $f(x)^3 = f(x)^2 \cdot f(x)$, we get

$$D_x(f(x)^3) = f(x)D_x(f(x)^2) + f'(x)f(x)^2$$
$$= f(x)\cdot 2f(x)f'(x) + f'(x)f(x)^2$$
$$= 3f(x)^2f'(x).$$

In this way, it can be shown (Exercise 31) that

$$D_x[f(x)^n] = nf(x)^{n-1}\cdot f'(x) \tag{15}$$

for any positive integer n. The reasoning used in deriving (13) will also show that (15) holds for negative integers as well.

2.5
THEOREM

General Power Rule. If n is any integer, then

(15) $$D_x[f(x)^n] = nf(x)^{n-1} \cdot f'(x).$$

Example 6

Find $D_x[(3x^3 - 2x^2 + 8)^5]$.

Solution. From (15), with $f(x) = 3x^3 - 2x^2 + 8$ and $n = 5$, we obtain

$$D_x[(3x^3 - 2x^2 + 8)^5] = 5(3x^3 - 2x^2 + 8)^4 \cdot (9x^2 - 4x). \quad ■$$

Example 7

If $f(x) = \dfrac{1}{(x^3 + 2x - 3)^2}$, then find $f'(x)$.

Solution. Here $f(x) = (x^3 + 2x - 3)^{-2}$. So (15) gives

$$f'(x) = -2(x^3 + 2x - 3)^{-3} \cdot D_x(x^3 + 2x - 3)$$

$$= -\frac{2}{(x^3 + 2x - 3)^3} \cdot (3x^2 + 2) = -\frac{6x^2 + 4}{(x^3 + 2x - 3)^3}. \quad ■$$

Exercises 2.2

In Exercises 1–28, use theorems of this section to differentiate the given function.

1. $f(x) = 2x - 8$
2. $g(x) = ax^4 + bx^3 + cx^2 + dx + e$
3. $f(x) = 5x^2 - 8x + 2$
4. $h(x) = \dfrac{a}{x^3} + \dfrac{b}{x^2} + \dfrac{c}{x} + d$
5. $f(x) = 7x^5 - 3x^4 + 7x^2 - 8$
6. $k(x) = 11x^6 - 2x^4 - 3x^3 + 3x - 5$
7. $f(x) = 4x^2 - 3x - 2 + \dfrac{1}{x} - \dfrac{3}{x^2}$
8. $l(x) = 5x^3 - 2x^2 - \dfrac{5}{x^2} + \dfrac{3}{x^3}$
9. $f(x) = \dfrac{1}{5x^2 - 8x + 11}$
10. $m(x) = \dfrac{1}{7x^3 + 5x^2 + 6}$
11. $f(x) = (3x^9 - 15x^7 + 11x^2 - 5x - 2)(7x^3 - 3x^2 + x + 1)$
12. $h(x) = (2x^5 - 14x^4 + 8x^3 - 2x^2 + 5) \cdot (x^{11} - 8x^5 + 4x^3 - 3x^2 + 7)$
13. $f(x) = (3x^2 - 2x + 1)^8$
14. $p(x) = (7x^3 - 3x^2 + 4)^5$
15. $f(x) = \dfrac{1}{(2x^2 - 8)^6}$
16. $q(x) = \dfrac{1}{(x^2 + 3x - 2)^7}$
17. $f(x) = \dfrac{x^2 + 1}{x^2 + 5}$
18. $s(x) = \dfrac{2x^2 + 1}{3x^4 + 2}$
19. $f(x) = \dfrac{3 + 2x}{3 - 2x}$
20. $t(x) = \dfrac{2x - 1}{2x + 1}$
21. $f(x) = (2x - 3)^5 - \dfrac{1}{(5x^2 + 1)^7}$
22. $u(x) = (x^2 + 1)^7 - \dfrac{1}{(x^4 + 1)^5}$
23. $f(x) = \left(\dfrac{x^2 + 1}{x^2 - 1}\right)^2$
24. $v(x) = \left(\dfrac{2 - x^3}{2 + x^3}\right)^2$
25. $f(x) = \dfrac{ax + b}{cx^2 + dx + e}$
26. $w(x) = \dfrac{x(x + 1)(x + 2)}{x + 3}$
27. $f(x) = [(x^2 + 1)^2 + 5]^2$
28. $y(x) = \dfrac{(1 + x^2)^3}{(1 + x^3)^2}$

29. Derive (8).

30. Derive (15) for the case $n = 4$.

31. Use mathematical induction to derive (15) for any positive integer n.

32. Derive (15) for any negative integer n.

33. In thermodynamics, the *enthalpy* H of a system (see Example 2 of Section 1.6) is defined to be $E + PV$, where E is the internal energy, P is pressure, and V is volume. Suppose that a reaction takes place over time, at constant pressure P. Regarding E and V as functions of time, give a formula for dH/dt, the rate of change of enthalpy, in terms of dE/dt and dV/dt. (This is one form of the *first law of thermodynamics*.)

34. While the ideal gas law is fundamental in studying the physical behavior of gases, all real gases deviate somewhat from it as the pressure and temperature vary. One refinement of the ideal gas law is the law of *van der Waals*, which applies to one mole of gas. It states that

$$P = \frac{RT}{V - b} - \frac{a}{V^2},$$

where a and b are constants that depend on the particular gas being studied, and P, V, R, and T are as in the ideal gas law. Find dP/dV in this law, assuming constant temperature.

35. In Example 3, suppose that $P = 8.0$ atm and $dP/dt = 5.0 \times 10^{-2}$ atm/min when $V = 12$ liters and $dV/dt = -0.05$ liters/min. Find dT/dt if $n = 15$ mol.

36. A sample of air is kept at 25°C while the pressure is raised. At a certain time, its volume V is 0.1 liter, and its pressure P is 1 atm and is rising at the rate of 0.1 atm/min. Use Boyle's law (Exercise 24, Section 1.6) to find dV/dt at this time.

37. The gravitational force between two bodies of mass m_1 and m_2 whose centers are at a distance r apart is given by *Newton's law of gravitation* as

$$F = G \cdot \frac{m_1 m_2}{r^2}$$

for G a constant, approximately equal to 6.674×10^{-17} $\text{N}\cdot\text{km}^2/\text{kg}^2$. ($N$ stands for *newton*, a metric unit of force.) The earth has mass $m_1 \approx 5.993 \times 10^{24}$ kg and radius about 6.400×10^3 km. Consider a meteor of negligible radius and mass $m_2 = 10^5$ kg. If the meteor is $R = 200$ km above the earth's surface and is moving toward the earth at 1 km/s, find dF/dt. [*Hint:* Use (15).]

38. As the meteor in Exercise 37 falls, it partially burns up. Suppose that at $R = 100$ km its mass is down to 10^4 kg, but its speed is up to 5 km/s. How does F compare to its value at $R = 200$ km? What is dF/dt at $R = 100$ km? (Treat m_2 as the constant 10^4 near $R = 100$.)

39. An extensively used relation in biochemistry is the *Michaelis–Menten* equation, which describes the rate of reaction involving an enzyme. The equation is

$$v = \frac{S \cdot V_{\text{max}}}{K_m + S},$$

where K_m is a constant, S is the concentration of the substance on which the enzyme acts, v is the rate of the reaction, and V_{max} is the *limiting* (or *maximal*) rate of the reaction.
(a) Find dv/dS; **(b)** use (1) to find dv/dt if $dS/dt = 2$.

40. Poiseuille's law (see Exercise 48, Section 1.3) states that the velocity v in cm/min of blood flow r units from the center of an artery of radius R is

$$v = c(R^2 - r^2)$$

where c is a constant.
(a) Find dv/dR if r is fixed. **(b)** Find dv/dr if R is fixed.
(c) If $R = 0.050$ cm, $c = 1$, and $dR/dt = 0.001$ cm/min, then find dv/dt at a fixed value of r. Use (15).

41. Let x be the number of workers employed in a plant and $p(x)$ the total value of the production of the plant. The quantity

$$\frac{p(x)}{x}$$

is called the *average productivity* of the work force at the plant. Find a formula for

$$\frac{d}{dx}\left(\frac{p(x)}{x}\right).$$

Show that this is a positive number if and only if $p'(x)$ is greater than the average productivity.

42. In Exercise 41, explain why it is advantageous to the company to hire more workers when

$$\frac{d}{dx}\left(\frac{p(x)}{x}\right) > 0.$$

What is likely to happen if

$$\frac{d}{dx}\left(\frac{p(x)}{x}\right)$$

is negative?

43. Show that if f is differentiable at c and $f \cdot g$ is differentiable at c, then g is differentiable at c if $f(c) \neq 0$.

44. Show that $f(x)^2$ may be differentiable at $x = c$ even if f is not differentiable at c. Why is your example not in conflict with Exercise 43?

45. If $f(x) = (x - 1)(x - 2)(x - 3)$, then show for $x \neq 1, 2,$ or 3 that

$$\frac{f'(x)}{f(x)} = \frac{1}{x - 1} + \frac{1}{x - 2} + \frac{1}{x - 3}.$$

46. Does Theorem 2.3 apply to $f(x) = x \cdot |x|$ at $x = 0$? What is $f'(0)$?

HISTORICAL NOTE

The Michaelis–Menten equation in Exercise 39 was developed in 1913 by the German chemist Leonor Michaelis (1875–1949) and his research assistant, Maud L. Menten. After World War I, Michaelis emigrated to the United States. One of his discoveries—that the protein keratin found in human hair is soluble in thioglycolic acid—made possible the development of home permanents for hair.

2.3 Trigonometric Functions and Limits*

FIGURE 3.1

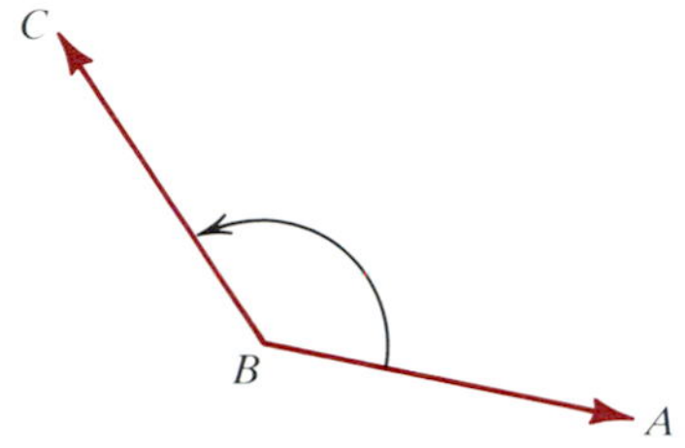

Trigonometric functions occur widely in applications, so we need to derive formulas for differentiating them. That is done in the next section. Here we remind you of the basic properties of these functions and work out some important trigonometric limits.

We begin with a review of radian measure for angles, which is better suited to calculus than is degree measure. **Figure 3.1** shows an angle ABC formed by two rays. This angle is congruent to the central angle DOE of the unit circle $x^2 + y^2 = 1$ shown in **Figure 3.2.** Angle DOE has its *initial side* OD along the positive x-axis, and is said to be in *standard position.* Side OE is the *terminal side* of the angle. The *radian measure* of angle DOE is defined as follows.

3.1 DEFINITION

The ***radian measure*** of an angle is s radians if the angle cuts off an arc of length s on the unit circle $x^2 + y^2 = 1$. Angles measured *counterclockwise* from the positive x-axis have *positive* radian measure. Those measured *clockwise* from the positive x-axis have *negative* radian measure.

FIGURE 3.2

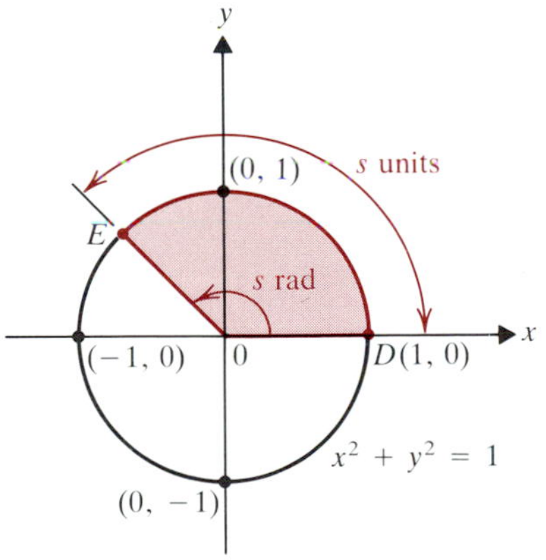

Since the circumference of the entire unit circle is 2π, we see that

$$360° = 2\pi \text{ rad},$$

where *rad* is the abbreviation of *radian.* We thus have the basic conversion formulas

(1) $$1° = \frac{\pi}{180} \text{ rad} \quad \text{and} \quad 1 \text{ rad} = \frac{180°}{\pi}.$$

Thus one radian is equivalent to approximately 57.3°. Most calculators are in degree mode when first turned on. To avoid error, develop the habit of switching to radian mode as soon as you turn on your calculator. From now on, we will follow the standard practice of *omitting* the unit *rad* when discussing angles. In calculus, *all* angles are measured in radians unless degrees are specifically mentioned.

The two basic trigonometric functions are *sine* and *cosine.*

3.2 DEFINITION

Let t be a real number and AOP an angle in standard position of radian measure t as in **Figure 3.3.** Then the ***sine function*** is defined by

(2) $$\sin t = y,$$

* The first part of this section, up to "Some Basic Limits," is a brief review of trigonometry, which can be skipped or gone over quickly.

FIGURE 3.3

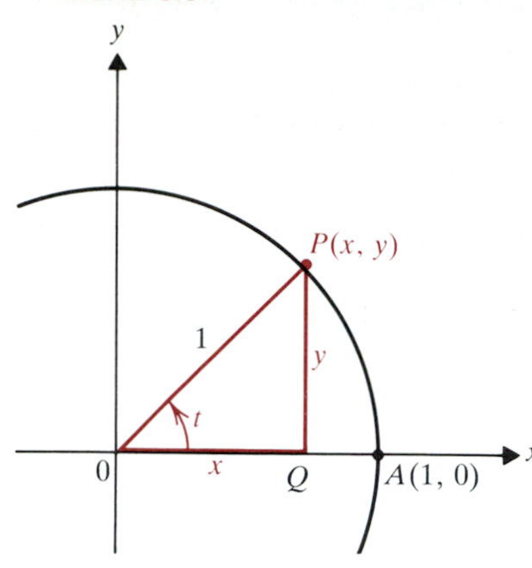

and the ***cosine function*** is defined by

$$\cos t = x. \tag{3}$$

The domain of both the sine and cosine functions is the entire real line, since for any real number t there is an angle with radian measure t. If, as in Figure 3.3, angle AOP is acute (i.e., its radian measure is less than $\pi/2$), then in the triangle QOP

$$\sin t = \frac{y}{1} = \frac{\text{length of side opposite } t}{\text{length of hypotenuse}}, \tag{4}$$

$$\cos t = \frac{x}{1} = \frac{\text{length of side adjacent to } t}{\text{length of hypotenuse}}. \tag{5}$$

Since the point $P(x, y)$ is on the unit circle, $x^2 + y^2 = 1$, equations (2) and (3) give the fundamental identity

$$\cos^2 t + \sin^2 t = 1. \tag{6}$$

Table 1 gives the sine and cosine of many commonly encountered angles between 0° and 180°. This table is easily constructed using (4) and (5) and the 45°–45°–90° and 30°–60°–90° right triangles shown in **Figure 3.4.** Be sure you can produce the entries from that sort of sketch. These are values you should *not* have to get from a calculator.

TABLE 1

t *(degrees)*	t *(radians)*	$\sin t$	$\cos t$
0	0	0	1
30	$\pi/6$	$1/2$	$\sqrt{3}/2$
45	$\pi/4$	$1/\sqrt{2}$	$1/\sqrt{2}$
60	$\pi/3$	$\sqrt{3}/2$	$1/2$
90	$\pi/2$	1	0
120	$2\pi/3$	$\sqrt{3}/2$	$-1/2$
135	$3\pi/4$	$1/\sqrt{2}$	$-1/\sqrt{2}$
150	$5\pi/6$	$1/2$	$-\sqrt{3}/2$
180	π	0	-1

FIGURE 3.4

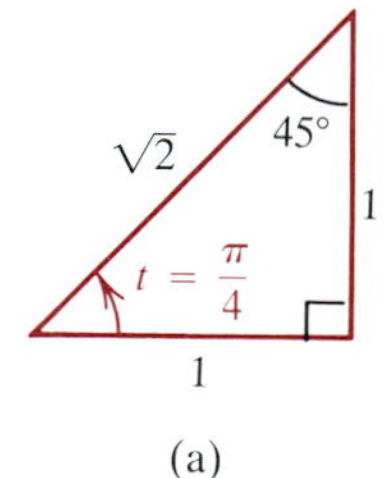

(a)

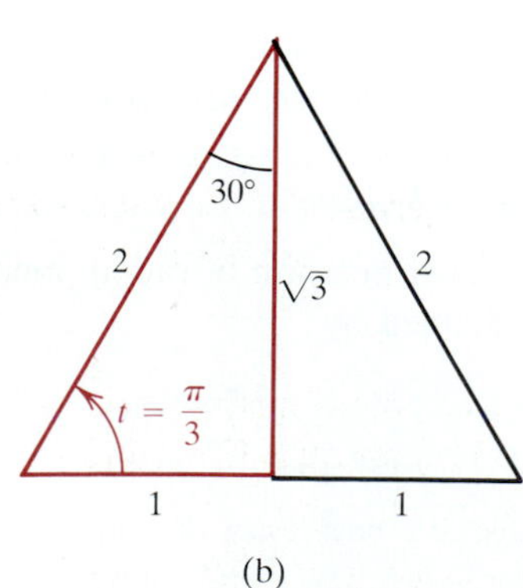

(b)

Sine is an odd function and cosine is an even function, that is,

$$\sin(-t) = -\sin t, \tag{7}$$

$$\cos(-t) = \cos t \tag{8}$$

for every real number t.

Since an angle of radian measure 2π corresponds to a complete revolution around the unit circle, angles of radian measure t and $t + 2\pi$ in standard position have the same terminal sides. Hence, from Definition 3.2,

$$\cos(t + 2\pi) = \cos t, \tag{9}$$

and

$$\sin(t + 2\pi) = \sin t. \tag{10}$$

We call this behavior *periodic*.

3.3 DEFINITION

A nonconstant function f is ***periodic*** if there is some number p such that

$$f(x + p) = f(x) \tag{11}$$

for every real number x. If there is a least such positive number p, then it is called the ***period*** of f.

From (9) and (10), cosine and sine are periodic functions, and it is not hard to see that they both have period 2π.

The four remaining trigonometric functions are defined as quotients and reciprocals of the sine and cosine functions. For each of them, we must exclude from the domain the points where the denominator is zero. (Recall that $\cos x = 0$ for odd multiples $(2n + 1)\pi/2$ of $\pi/2$, and $\sin x = 0$ for any integer multiple $n\pi$ of π, $n = 0, \pm 1, \pm 2, \pm 3, \ldots$.)

3.4 DEFINITION

The ***tangent function*** is defined by

$$\tan x = \frac{\sin x}{\cos x}, \qquad \text{for } x \neq (2n + 1)\frac{\pi}{2}.$$

The ***cotangent function*** is the reciprocal of the tangent function,

$$\cot x = \frac{1}{\tan x} = \frac{\cos x}{\sin x}, \qquad \text{for } x \neq n\pi.$$

The ***secant function*** is the reciprocal of the cosine function,

$$\sec x = \frac{1}{\cos x}, \qquad \text{for } x \neq (2n + 1)\frac{\pi}{2}.$$

The ***cosecant function*** is the reciprocal of the sine function,

$$\csc x = \frac{1}{\sin x}, \qquad \text{for } x \neq n\pi.$$

Example 1

Write the values of the six trigonometric functions at $x = \pi/6$.

Solution. From Figure 3.4(b) we obtain the entries $\sin \pi/6 = 1/2$ and $\cos \pi/6 = \sqrt{3}/2$ found in Table 1. Then directly from Definition 3.4 we find

$$\tan \pi/6 = \frac{1/2}{\sqrt{3}/2} = \frac{1}{\sqrt{3}} = \frac{\sqrt{3}}{3}, \qquad \cot \pi/6 = \frac{1}{\tan \pi/6} = \sqrt{3},$$

$$\sec \pi/6 = \frac{1}{\cos \pi/6} = \frac{2}{\sqrt{3}} = \frac{2\sqrt{3}}{3}, \qquad \csc \pi/6 = \frac{1}{\sin \pi/6} = 2. \quad \blacksquare$$

We next remind you of the fundamental formula for the cosine of the sum of two angles. (The details of its derivation are not germane to calculus, so we do not reprove the formula here.) If t and u are real numbers, then

$$\cos(t + u) = \cos t \cos u - \sin t \sin u. \tag{12}$$

By making appropriate substitutions in (12), we can obtain further formulas of this type. In particular, if we replace u by $-u$ in (12) and use (7) and (8), then we get

$$\cos(t - u) = \cos t \cos(-u) - \sin t \sin(-u) = \cos t \cos u + \sin t \sin u. \tag{13}$$

From (12) and (13) and the facts that $\cos\pi = -1$ and $\sin\pi = 0$, we have

(14) $$\cos(t+\pi) = -\cos t \quad \text{and} \quad \cos(\pi - t) = -\cos t.$$

Finally, (12) with $t = \frac{1}{2}\pi$ and $u = -t$ gives

(15) $$\cos(\tfrac{1}{2}\pi - t) = \cos\tfrac{1}{2}\pi\cos(-t) + \sin\tfrac{1}{2}\pi\sin t$$
$$= 0 + 1\cdot\sin t = \sin t,$$

and (15) with $u = \frac{1}{2}\pi - t$, that is, $t = \frac{1}{2}\pi - u$, gives

(16) $$\cos u = \sin(\tfrac{1}{2}\pi - u).$$

Formulas (15) and (16) account for the name *cosine.* The angles t and $\frac{1}{2}\pi - t$ are *complementary*, that is, their sum is a right angle. Since the sine of each of these angles is the cosine of its complement, sine and cosine are called *complementary functions*, or *cofunctions* for short.

We next derive the formulas for the sine function corresponding to Formulas (12)–(16). We have from (15), (12), and (16)

$$\sin(t+u) = \cos[\tfrac{1}{2}\pi - (t+u)] = \cos[(\tfrac{1}{2}\pi - t) - u]$$
$$= \cos(\tfrac{1}{2}\pi - t)\cos u + \sin(\tfrac{1}{2}\pi - t)\sin u$$
$$= \sin t\cos u + \cos t\sin u.$$

Thus

(17) $$\sin(t+u) = \sin t\cos u + \cos t\sin u.$$

If we replace u by $-u$ in (17), and use (7) and (8), then we get

(18) $$\sin(t-u) = \sin t\cos u - \cos t\sin u.$$

Reasoning like that used in deriving (14) then gives

(19) $$\sin(\pi - t) = \sin t \quad \text{and} \quad \sin(\pi + t) = -\sin t.$$

The *double-angle formulas* now follow immediately.

3.5 COROLLARY

(a) $\sin 2t = 2\sin t\cos t$

(b) $\cos 2t = \cos^2 t - \sin^2 t = 2\cos^2 t - 1 = 1 - 2\sin^2 t.$

Proof. (a) Let $u = t$ in (17). Then we have

$$\sin 2t = \sin t\cos t + \cos t\sin t = 2\sin t\cos t.$$

(b) Let $u = t$ in (12). We then obtain

$$\cos 2t = \cos^2 t - \sin^2 t.$$

Then (4) gives

$$\cos 2t = \cos^2 t - (1 - \cos^2 t) = 2\cos^2 t - 1,$$

and also

$$\cos 2t = (1 - \sin^2 t) - \sin^2 t = 1 - \sin^2 t.$$ QED

Corollary 3.5 leads to two identities that are very useful in the calculus of sine and cosine. These are sometimes called *half-angle* formulas.

3.6
COROLLARY

(a) $\cos^2 t = \dfrac{1 + \cos 2t}{2}$;

(b) $\sin^2 t = \dfrac{1 - \cos 2t}{2}$.

Proof. We simply solve for $\cos^2 t$ and $\sin^2 t$ in the second and third formulas of Corollary 3.5(b). We have

$$\cos 2t = 2\cos^2 t - 1 \to 1 + \cos 2t = 2\cos^2 t \to \cos^2 t = \frac{1 + \cos 2t}{2},$$

and similarly, $\sin^2 t = (1 - \cos 2t)/2$. QED

All six trigonometric functions are periodic. We saw above that sine and cosine have period 2π. So then do their reciprocals, cosecant and secant. However, tangent and cotangent have period π.

Example 2

Verify that the tangent function is periodic by showing that (11) holds for $p = \pi$.

Solution. Formulas (14) and (19) give

$$\tan(x + \pi) = \frac{\sin(x + \pi)}{\cos(x + \pi)} = \frac{-\sin x}{-\cos x} = \tan x. \quad \blacksquare$$

We can divide the basic identity

$$\cos^2 t + \sin^2 t = 1 \tag{6}$$

through by $\cos^2 t$ and $\sin^2 t$ to obtain basic identities involving $\tan t$ and $\cot t$. If we divide (6) through by $\cos^2 t$, we get

$$1 + \tan^2 t = \sec^2 t, \tag{20}$$

provided that t is not an odd multiple of $\pi/2$ (when $\cos^2 t = 0$). Division of (6) by $\sin^2 t$ produces

$$\cot^2 t + 1 = \csc^2 t, \tag{21}$$

which is valid for all t except integer multiples of π (where $\sin^2 t = 0$).

We close this review of the trigonometric functions by giving the graphs of the sine and cosine functions. (The graphs of the others are best drawn using methods of the next chapter.) Using (9), (10), (14), and (19), we can extend Table 1 to give the values of $\sin x$ and $\cos x$ for any integral multiple x of $\pi/6$ or $\pi/4$. Alternatively, we can use a calculator *in radian mode* to find values of $\sin x$ and $\cos x$ for x in $[0, 2\pi]$ and then use (7) and (8). In any case, we get the graphs shown in **Figures 3.5** and **3.6.**

Some Basic Trigonometric Limits

To differentiate the trigonometric functions, we need to know some basic facts about limits and continuity for the sine and cosine functions. To establish their continuity, we need the following preliminary result, whose proof is given at

FIGURE 3.5

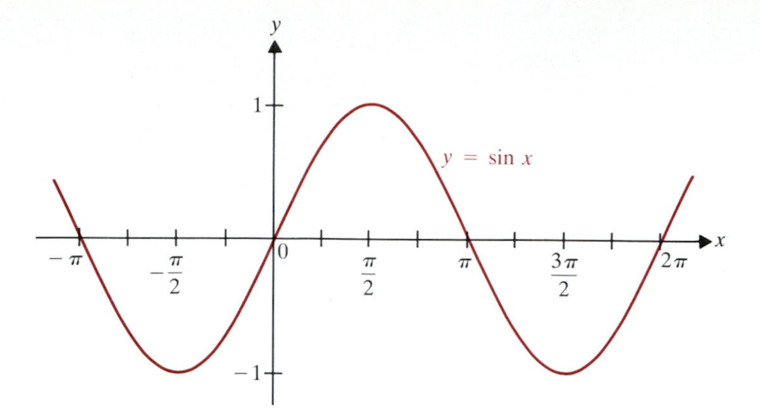

FIGURE 3.6

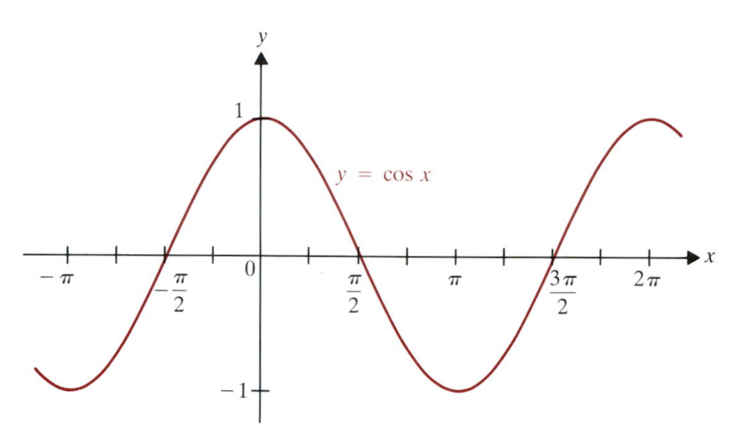

the end of this section. (A *lemma* is a result whose primary importance lies in deriving a more interesting fact, in this case, Theorem 3.8.)

3.7 LEMMA The functions $\sin x$ and $\cos x$ satisfy

$$|\sin t - \sin t_0| \leq |t - t_0|, \tag{22}$$

$$|\cos t - \cos t_0| \leq |t - t_0|. \tag{23}$$

It follows easily from the lemma that sine and cosine are continuous on the entire real line.

3.8 THEOREM The sine and cosine functions are continuous on $(-\infty, +\infty)$.

Proof. We use the sandwich theorem (Theorem 5.7 of Chapter 1). For $f(x) = \sin x$ or $f(x) = \cos x$, Lemma 3.7 says that

$$|f(x) - f(x_0)| \leq |x - x_0|.$$

By Theorem 1.2(c) of Chapter 1, this is equivalent to

$$-|x - x_0| \leq f(x) - f(x_0) \leq |x - x_0|.$$

Thus, by the sandwich theorem,

$$0 = \lim_{x \to x_0} (-|x - x_0|) \leq \lim_{x \to x_0} [f(x) - f(x_0)] \leq \lim_{x \to x_0} |x - x_0| = 0,$$

that is, $0 = \lim_{x \to x_0} [f(x) - f(x_0)]$. Hence, $\lim_{x \to x_0} f(x) = f(x_0)$. QED

This fact is not surprising, since the graphs of $\sin x$ and $\cos x$ in Figures 3.5 and 3.6 have no breaks. But we will now use it to complete the discussion of

$\lim_{x\to 0}(\sin x/x)$, which we first considered after Example 2 in Section 1.4. Table 2 on p. 31 made it seem that the limit is 1, but as yet we haven't shown that this is correct. To do that, we also need the following result from Euclidean geometry about the area of a circular sector.

3.9 LEMMA The area of a sector of a circle with radius r and central angle θ (measured in radians) is

$$A = \frac{1}{2}r^2\theta. \tag{24}$$

FIGURE 3.7

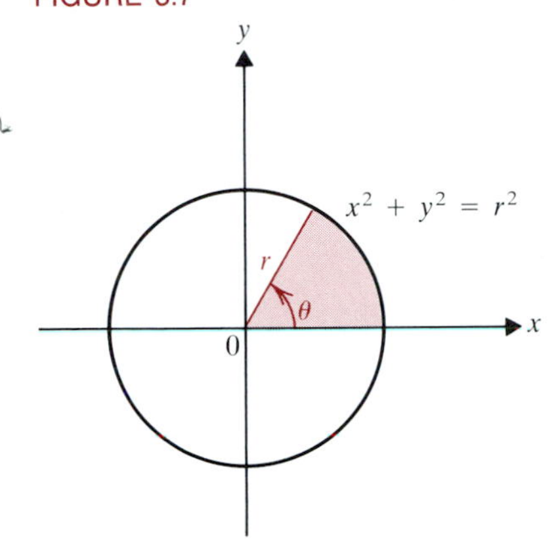

Proof. As θ increases from 0 to 2π, A increases from 0 up to πr^2. For any $\theta < 2\pi$, the piece of pie in **Figure 3.7** will cover the fraction $\theta/(2\pi)$ of the entire circular area. Hence,

$$A = \frac{\theta}{2\pi}\cdot \pi r^2 = \frac{1}{2}r^2\theta. \quad \boxed{\text{QED}}$$

Returning now to $\lim_{t\to 0}(\sin t/t)$, we can see trigonometric evidence in **Figure 3.8** that the limit should be 1. Notice that $\sin t$ is the y-coordinate of the point P, that is, $\sin t = d(P, Q)$. Also, t is the length PR of the arc PR on the unit circle, by the definition of radian measure. As $t \to 0$, we expect the ratio of $d(P, Q)$ to PR to approach 1, because the unit circle has a vertical tangent at the point $R(1, 0)$. We now proceed to prove that this limit is indeed 1.

3.10 THEOREM

$$\lim_{t\to 0}\frac{\sin t}{t} = 1.$$

FIGURE 3.8

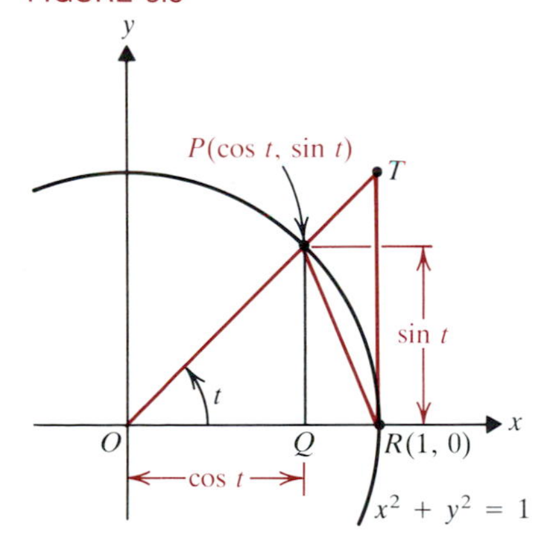

Proof. First, suppose $t \in (0, \pi/2)$. In Figure 3.8, the circular sector ORP lies between two triangles ORP and ORT, where RT is vertical. Thus the area of sector ORP lies between the areas of the two triangles—which are half the base times the height. Lemma 3.9 then gives

$$\frac{1}{2}d(O, R)\cdot d(P, Q) < \frac{1}{2}\cdot 1^2 \cdot t < \frac{1}{2}d(O, R)\cdot d(R, T),$$

$$\frac{1}{2}\cdot 1\cdot \sin t < \frac{1}{2}t < \frac{1}{2}\cdot 1 \cdot d(R, T), \tag{25}$$

$$\sin t < t < d(R, T).$$

Also,

$$d(R, T) = \frac{d(R, T)}{1} = \frac{d(R, T)}{d(O, R)} = \frac{\sin t}{\cos t},$$

since triangles OQP and ORT are similar. Substituting this into (25), we get

$$\sin t < t < \frac{\sin t}{\cos t}. \tag{26}$$

If we divide (26) through by $\sin t$, which is positive since $t \in (0, \pi/2)$, then we obtain

$$1 < \frac{t}{\sin t} < \frac{1}{\cos t}. \tag{27}$$

Now suppose $t \in (-\frac{1}{2}\pi, 0)$. Then $-t \in (0, \frac{1}{2}\pi)$, so (27) gives

$$1 < \frac{-t}{\sin(-t)} < \frac{1}{\cos(-t)}. \tag{28}$$

Since $\sin(-t) = -\sin t$ and $\cos(-t) = \cos t$, (28) simplifies to

$$1 < \frac{t}{\sin t} < \frac{1}{\cos t}.$$

Thus (27) holds for all $t \in (-\frac{1}{2}\pi, 0) \cup (0, \frac{1}{2}\pi)$. We want to apply the sandwich theorem (Theorem 5.7 of Chapter 1) to (27) and the interval $I = (-\frac{1}{2}\pi, \frac{1}{2}\pi)$. By Theorem 3.8, $\cos t$ is continuous at $t = 0$. So Theorem 5.4(b) of Chapter 1 says that as $t \to 0$, we have $1/\cos t \to 1/\cos 0 = 1$. Thus (27) shows that $t/\sin t$ is sandwiched between two functions that both have limit 1 as $t \to 0$. Therefore, according to the sandwich theorem, we have

$$\lim_{t\to 0} \frac{t}{\sin t} = 1. \tag{29}$$

Finally, since

$$\frac{\sin t}{t} = \frac{1}{\dfrac{t}{\sin t}},$$

Theorem 5.4(b) of Chapter 1 gives

$$\lim_{t\to 0} \frac{\sin t}{t} = \frac{1}{\lim\limits_{t\to 0} \dfrac{t}{\sin t}} = \frac{1}{1} = 1. \quad \boxed{\text{QED}}$$

Theorem 3.10 can be used on many functions that resemble $\sin x/x$.

Example 3

Find $\lim\limits_{x\to 0} \dfrac{3x}{\sin 4x}$.

Solution. We need to put $3x/\sin 4x$ into the form $\sin t/t$. For that, write

$$\frac{3x}{\sin 4x} = \frac{3}{4} \cdot \frac{4x}{\sin 4x},$$

and let $t = 4x$. Then $x \to 0$ is equivalent to $t \to 0$. Hence, by (29),

$$\lim_{x\to 0} \frac{3x}{\sin 4x} = \frac{3}{4} \lim_{x\to 0} \frac{4x}{\sin 4x} = \frac{3}{4} \lim_{t\to 0} \frac{t}{\sin t} = \frac{3}{4}. \quad \blacksquare$$

The cosine function has a limit corresponding to the one in Theorem 3.10. Referring back to Figure 3.8, notice that

$$\frac{1 - \cos t}{t} = \frac{1 - d(O, Q)}{PR} = \frac{d(Q, R)}{PR}.$$

As $t \to 0$, we would expect that

$$\frac{1 - \cos t}{t}$$

also approaches 0, since there is a vertical tangent at (1, 0). This indeed is what happens.

3.11 THEOREM

$$\lim_{t \to 0} \frac{1 - \cos t}{t} = 0.$$

Proof. We use the identity $\sin^2 t + \cos^2 t = 1$ in the form $1 - \cos^2 t = \sin^2 t$. Note that

$$\frac{1 - \cos t}{t} = \frac{1 - \cos t}{t} \cdot \frac{1 + \cos t}{1 + \cos t} = \frac{1 - \cos^2 t}{t(1 + \cos t)}$$

$$= \frac{\sin^2 t}{t(1 + \cos t)} = \frac{\sin t}{t} \cdot \frac{\sin t}{1 + \cos t}.$$

Therefore,

$$\lim_{t \to 0} \frac{1 - \cos t}{t} = \lim_{t \to 0} \frac{\sin t}{t} \cdot \lim_{t \to 0} \frac{\sin t}{1 + \cos t}$$

$$= 1 \cdot \frac{0}{1 + 1} = 0. \quad \text{QED}$$

This result also is useful in working out limits of functions that involve sine and cosine.

Example 4

Find $\lim_{x \to 0} \dfrac{\cos^2 x - 2\cos x + 1}{x^2}$.

Solution. Note $\cos^2 x - 2\cos x + 1 = (\cos x - 1)^2 = (1 - \cos x)^2$. So we have

$$\lim_{x \to 0} \frac{\cos^2 x - 2\cos x + 1}{x^2} = \lim_{x \to 0} \frac{1 - \cos x}{x} \cdot \lim_{x \to 0} \frac{1 - \cos x}{x}$$

$$= 0 \cdot 0 = 0. \quad \blacksquare$$

Proof of Lemma 3.7*

3.7 LEMMA

The functions $\sin x$ and $\cos x$ satisfy

(22) $$|\sin t - \sin t_0| \leq |t - t_0|,$$

and

(23) $$|\cos t - \cos t_0| \leq |t - t_0|$$

for any t, t_0 in any interval I.

Proof. Refer to **Figure 3.9**, where we show the points $P(\cos t_0, \sin t_0)$ and $Q(\cos t, \sin t)$ for a first-quadrant angle t_0 and a second-quadrant angle t. A straight line is the shortest distance between two points, so chord PQ is no longer than arc $\widehat{PQ}$. Since we are using radian measure, and P and Q are on the unit

* Optional

FIGURE 3.10

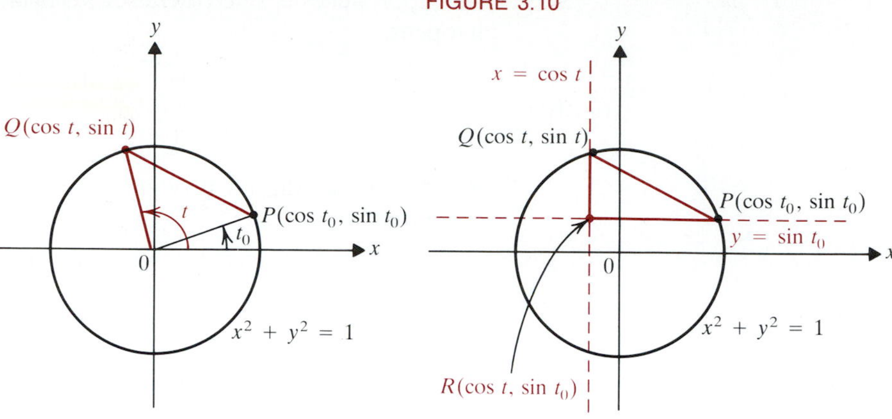

circle, arc $\widehat{PQ}$ has length $|t - t_0|$. Therefore,

(30) $$d(P, Q) \le |t - t_0|.$$

In **Figure 3.10** we draw the vertical line $x = \cos t$ through Q and the horizontal line $y = \sin t_0$ through P, and label their point of intersection R, so that the coordinates of R are $(\cos t, \sin t_0)$. Triangle PRQ is a right triangle, so the length of its hypotenuse is greater than the length of the other sides. That is,

(31) $$d(R, Q) = |\sin t - \sin t_0| < d(P, Q)$$

and

(32) $$d(R, P) = |\cos t - \cos t_0| < d(P, Q).$$

Combining (31) and (32) with (30), we obtain (22) and (23). QED

Exercises 2.3

In Exercises 1 and 2, express the given degree measure in radian measure.

1. **(a)** 135° **(b)** 45° **(c)** 225° **(d)** 300° **(e)** 315°
2. **(a)** 150° **(b)** 120° **(c)** 240° **(d)** 330° **(e)** 210°
3. Sine and cosine are both positive in the first quadrant, that is, for angles with radian measure $t \in (0, \frac{1}{2}\pi)$. What are their signs in the other three quadrants?

FIGURE 3.11

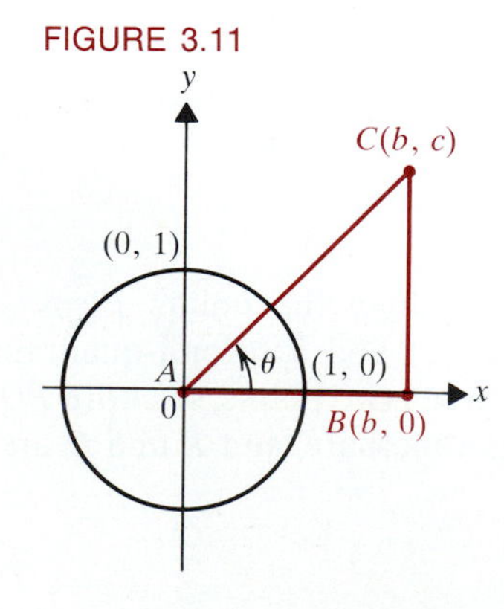

4. In triangle ABC in **Figure 3.11,** one says that

$$\sin\theta = \frac{d(B, C)}{d(A, C)} \quad \text{and} \quad \cos\theta = \frac{d(A, B)}{d(A, C)}.$$

By using similar triangles, show that this agrees with Definition 3.2.

In Exercises 5–8, use the right triangles in Figure 3.4, Equations (5)–(8) and (13)–(19), and Corollary 3.5 to find the given function value.

5. **(a)** $\sin\frac{7\pi}{6}$ **(b)** $\sin\frac{5\pi}{4}$ **(c)** $\sin\frac{3\pi}{2}$
   (d) $\sin\frac{11\pi}{6}$ **(e)** $\sin\frac{23\pi}{6}$ **(f)** $\sin\left(-\frac{5\pi}{6}\right)$
   (g) $\sin\left(-\frac{5\pi}{4}\right)$
6. **(a)** $\cos\frac{4\pi}{3}$ **(b)** $\cos\frac{3\pi}{2}$ **(c)** $\cos\frac{5\pi}{3}$
   (d) $\cos\frac{7\pi}{4}$ **(e)** $\cos\frac{17\pi}{6}$ **(f)** $\cos\left(-\frac{3\pi}{4}\right)$

(g) $\cos\left(-\frac{4\pi}{3}\right)$

7. (a) $\tan\frac{\pi}{3}$ (b) $\sec\frac{2\pi}{3}$ (c) $\cot\frac{3\pi}{4}$

(d) $\csc\frac{5\pi}{6}$ (e) $\tan\frac{11\pi}{6}$ (f) $\cot\frac{4\pi}{3}$

(g) $\sec\frac{5\pi}{3}$

8. (a) $\tan\frac{\pi}{4}$ (b) $\sec\frac{3\pi}{4}$ (c) $\cot\frac{2\pi}{3}$

(d) $\csc\frac{7\pi}{6}$ (e) $\tan\frac{4\pi}{3}$ (f) $\cot\frac{5\pi}{4}$

(g) $\sec\frac{11\pi}{6}$

9. Use Corollary 3.6 to find

(a) $\sin\frac{\pi}{12}$ (b) $\cos\frac{5\pi}{12}$ (c) $\sin\frac{5\pi}{12}$ (d) $\cos\frac{\pi}{12}$

10. Find

(a) $\sin\frac{7\pi}{12}$ (b) $\cos\frac{11\pi}{12}$ (c) $\sin\frac{7\pi}{24}$ (d) $\cos\frac{11\pi}{24}$

11. Show that

(a) $\sin(2\pi - t) = -\sin t$ (b) $\cos(2\pi - t) = \cos t$
(c) $\sin(t + \pi) = -\sin t$ (d) $\cos(t + \pi) = -\cos t$
(e) $\sin(t + \frac{1}{2}\pi) = \cos t$ (f) $\cos(t + \frac{1}{2}\pi) = -\sin t$

12. (a) Show that if $t \in [0, 2\pi]$, then $\sin\frac{t}{2} = \sqrt{\frac{1-\cos t}{2}}$.

What if $t \in [2\pi, 4\pi]$?

(b) Show that if $t \in [0, \pi]$, then $\cos\frac{t}{2} = \sqrt{\frac{1+\cos t}{2}}$.

What if $t \in [\pi, 2\pi]$?

In Exercises 13–14, find all the values $t \in [0, 2\pi]$ that satisfy the given equation.

13. (a) $2\sin^3 t - \sin t = 0$ (b) $\sin 2t = \cos t$
(c) $\sin 2t - 2\sin t - \cos t + 1 = 0$

14. (a) $2\cos^2 t - \cos t - 3 = 0$ (b) $\sin 2t = \sin t$
(c) $\cos 2t - 3\cos t = 0$

15. Show that $\sin 3t = 3\sin t - 4\sin^3 t$.

16. Show that $\cos 3t = 4\cos^3 t - 3\cos t$.

17. Is

$$f(x) = \frac{\sin x}{x}$$

continuous at $x = 0$? If not, what sort of discontinuity is there at $x = 0$?

18. Repeat Exercise 17 for $f(x) = \frac{1 - \cos x}{x}$.

In Exercises 19–36, evaluate the given limits.

19. $\lim_{x\to 0}\frac{\sin 2x}{x}$

20. $\lim_{x\to 0}\frac{x}{1-\cos x}$

21. $\lim_{x\to 0}\frac{4x}{\sin 3x}$

22. $\lim_{x\to 0}\frac{x^2}{1-\cos^3 x}$

23. $\lim_{x\to 0}\frac{\sin 2x}{\sin 3x}$

24. $\lim_{x\to 0}\frac{\sin^2 x + \sin x}{x}$

25. $\lim_{x\to 0}\frac{1-\cos x}{\sin x}$

26. $\lim_{x\to 0}\frac{1-\cos^4 x}{x^2}$

27. $\lim_{x\to \pi}\frac{\sin x}{\pi - x}$

28. $\lim_{x\to 1/2\pi}\frac{\sin x - 1}{x - \frac{1}{2}\pi}$

29. $\lim_{x\to 0}\frac{1-\cos 2x}{3x}$

30. $\lim_{x\to 0}\frac{1-\cos\sqrt{x}}{x}$

31. $\lim_{x\to 0^+}\frac{\sin x}{x^2}$

32. $\lim_{x\to 0}\frac{x}{x+\sin x}$

33. $\lim_{x\to 0}\frac{\sin^4 x}{x^2}$

34. $\lim_{x\to 0}\frac{x}{x\cos x + \sin x}$

35. $\lim_{x\to 0}\frac{\sin^4 2x}{3x^4}$

36. $\lim_{x\to 0}\frac{x\cos x - \sin x}{x}$

37. Set your calculator in degree mode and try to estimate

$$\lim_{\theta\to 0^\circ}\frac{\sin\theta^\circ}{\theta^\circ}.$$

What do you get? Using the fact that $\theta^\circ = \pi\theta/180$ rad, find

$$\lim_{\theta\to 0^\circ}\frac{\sin\theta^\circ}{\theta^\circ}.$$

38. In degree mode, try to estimate

$$\lim_{\theta\to 0^\circ}\frac{1-\cos\theta^\circ}{\theta^\circ}.$$

How does it compare to the result of Theorem 3.11? Using the fact that $\theta^\circ = \pi\theta/180$ rad, find

$$\lim_{\theta\to 0^\circ}\frac{1-\cos\theta^\circ}{\theta^\circ}.$$

In Exercises 39–42, find the limits.

39. (a) $\lim_{x\to 0}\frac{\sin ax}{bx}$ for any real numbers a and b, $b \neq 0$.

(b) $\lim_{x\to 0}\frac{\sin(\sin x)}{\sin x}$.

40. (a) $\lim_{x\to 0}\frac{\sin ax}{\sin bx}$ for any real numbers a and b, $b \neq 0$.

(b) $\lim_{x\to 0}\frac{\sin(\sin x)}{x}$

41. Find $\lim_{x\to 0}\frac{1-\cos x\sin^2 x - \cos^2 x}{x^3}$

42. Find $\lim_{x \to 0} \dfrac{\cos^3 x - 2\cos^2 x - \cos x + 1}{x^3}$

43. Graph the function $y = \sin 2x$. What is the period? What is the period of $y = \sin kx$ if k is *any* positive integer?

44. Graph the function $y = \cos \frac{1}{2}x$. What is the period? What is the period of $y = \cos \dfrac{x}{k}$ if k is *any* positive integer?

45. **(a)** Find a sine function with period 12.
(b) Find a cosine function with period p where p is a positive integer.

46. The *amplitude* of a sine or cosine function is defined to be half the difference between the maximum and minimum values of the function.
(a) What is the amplitude of $y = \sin x$ and $y = \cos x$?
(b) What is the amplitude of $y = 5 + 2\sin x$?
(c) What is the amplitude of $y = -3 + 2\cos x$?

47. Periodic phenomena are often modeled by a sine or cosine function. Consider the price of tomatoes over a one-year period in a northeastern city. The price varies from a high of \$1.40 per pound on March 1 to a low of \$0.60 per pound on September 1. Let t be the number of months elapsed since March 1. Consider the modeling function

$$P = 100 + 40\cos\frac{\pi}{6}t.$$

(a) What are the values of P at $t = 0$ and $t = 6$?
(b) What are the values of P at $t = 3$ and $t = 9$?
(c) Do the results of (a) and (b) suggest that P is a reasonable model for the price of tomatoes in cents?

48. Refer to Exercise 47. Suppose the price of lettuce in the same northeastern city varies from a high of \$1.00 a head on January 1 to a low of \$0.40 a head on July 1. Construct a modeling function for the price of lettuce over a 12-month period.

49. The *basal metabolism rate* of an animal is defined to be the number of kilocalories per hour (kcal/h) of heat given off by the animal's body. For most animals, this rate follows a *diurnal cycle* of 24 h, generally rising at night during periods of lower temperature. Suppose the highest rate for a human is about 0.6 kcal/h at 6:00 AM and the lowest rate is about 0.1 kcal/h at 6:00 PM. Use a cosine function to model this rate as a function of the number of hours since 6:00 AM. (See Exercises 45 and 47.)

50. The human heart beats an average of about 72 times per minute. As it contracts, it squeezes blood into the arterial system, causing blood pressure to rise (systolic phase). Afterwards, it relaxes to allow the heart to fill with blood (diastolic phase). The normal systolic blood pressure is about 120 mm of mercury. The normal diastolic blood pressure is about 80 mm of mercury. These are the extremes that are recorded when a person's blood pressure is measured.
(a) If we want to model blood pressure by a sine function of time t measured in minutes, then what should its period be?
(b) Construct a modeling function if we measure time from the midpoint between systolic and diastolic phases.

51. The height of a tall tree, such as a giant redwood of the northwestern United States, is calculated by using trigonometry. An observer 150 ft from the base of such a tree measures the angle of elevation to the top as 60°. How tall is the tree?

52. A navigator on a ship's bridge is 10 m above sea level. He measures the angle of elevation to the top of a mountain to be 30°. If his map shows the mountain peak to be 1000 m above sea level, how far is the ship from the mountain?

53. Derive the identity

(a) $\tan(x + y) = \dfrac{\tan x + \tan y}{1 - \tan x \tan y}$

(b) $\tan(x - y) = \dfrac{\tan x - \tan y}{1 + \tan x \tan y}$

54. **(a)** Show that the secant function is an even function.
(b) Show that the tangent, cosecant, and cotangent functions are odd functions.

HISTORICAL NOTE

The great eighteenth-century mathematician Leonhard Euler (pronounced *oiler*, 1707–1783) in 1755 formulated a definition of function that has over the years evolved into the modern definition (Definition 3.1, Chapter 1). Even earlier, in 1748, Euler, then a professor at the Royal Academy in Berlin, had given the definition of the two basic trigonometric functions, sine and cosine. So insightful was his approach to these functions that it has persisted, virtually unchanged, to the present day. Indeed, most of what we present in this section goes back to Euler, in some cases implicitly but in many cases explicitly part of his book *Introduction to the Analysis of Infinite Processes*, which is regarded as the first calculus text to embody modern ideas.

2.4 Differentiation of Trigonometric Functions

In this section, we use results of the last section to obtain formulas for the derivatives of the sine and cosine functions. We then use the reciprocal rule and the quotient rule of Section 2 to differentiate the other trigonometric functions.

If you did Exercise 41 of Section 1, then you saw that the difference quotient for the sine function,

$$\frac{\Delta y}{\Delta x} = \frac{\sin(c + h) - \sin c}{h},$$

seems to approach $\cos c$ as $h \to 0$ when $c = 0, \pi/6, \pi/4, \pi/3$, and $\pi/2$. For instance, when $c = \pi/3$, the values of $\Delta y/\Delta x$ shown in Table 1 appear to approach $\cos \pi/3 = 1/2$ as $\Delta x(=h) \to 0$. The following theorem states that, in fact, $D_x \sin x = \cos x$ for *all* x.

TABLE 1

Δx	$\Delta y/\Delta x$
0.1	0.4559019
0.01	0.4956610
0.001	0.4995668
0.0001	0.4999565

4.1 THEOREM For every real number x,

(1) $$\frac{d}{dx}(\sin x) = \cos x.$$

Proof. For any x, we have from Formula (17) of the last section

$$\frac{d}{dx}(\sin x) = \lim_{h \to 0} \frac{\sin(x + h) - \sin x}{h}$$

$$= \lim_{h \to 0} \frac{\sin x \cos h + \cos x \sin h - \sin x}{h}$$

$$= \lim_{h \to 0} \frac{\sin x \cos h - \sin x + \cos x \sin h}{h}$$

$$= \lim_{h \to 0} \frac{(\sin x)(\cos h - 1) + \cos x \sin h}{h}$$

$$= \lim_{h \to 0} \left((-\sin x) \cdot \frac{1 - \cos h}{h} + \cos x \cdot \frac{\sin h}{h} \right)$$

$$= \lim_{h \to 0} (-\sin x) \cdot \frac{1 - \cos h}{h} + \lim_{h \to 0} \cos x \cdot \frac{\sin h}{h}$$ *by Theorem 5.1(d) of Chapter 1*

$$= (-\sin x) \cdot \lim_{h \to 0} \frac{1 - \cos h}{h} + (\cos x) \cdot \lim_{h \to 0} \frac{\sin h}{h}$$ *by Theorem 5.1(c) of Chapter 1 ($\sin x$ is constant as $h \to 0$)*

$$= (-\sin x) \cdot 0 + (\cos x) \cdot 1$$ *by Theorems 3.10 and 3.11*

$$= \cos x.$$ QED

Example 1

Find $D_x\left(\dfrac{\sin x}{x^2 + x + 1}\right)$.

Solution. Using (1) and the quotient rule (Theorem 2.4), we get

$$D_x\left(\frac{\sin x}{x^2+x+1}\right) = \frac{(x^2+x+1)D_x(\sin x) - (\sin x)D_x(x^2+x+1)}{(x^2+x+1)^2}$$
$$= \frac{(x^2+x+1)\cdot\cos x - (\sin x)(2x+1)}{(x^2+x+1)^2}$$
$$= \frac{x^2\cos x + x(\cos x - 2\sin x) + \cos x - \sin x}{(x^2+x+1)^2}. \blacksquare$$

We next derive the formula for $D_x(\cos x)$.

4.2 THEOREM For every real number x,

$$\frac{d}{dx}(\cos x) = -\sin x. \tag{2}$$

Proof. For any x, we have from Formula (12) of the last section

$$\frac{d}{dx}(\cos x) = \lim_{h\to 0}\frac{\cos(x+h)-\cos x}{h}$$
$$= \lim_{h\to 0}\frac{\cos x\cos h - \sin x\sin h - \cos x}{h}$$
$$= \lim_{h\to 0}\frac{(\cos x)(\cos h - 1) - \sin x\sin h}{h}$$
$$= \lim_{h\to 0}(-\cos x)\cdot\frac{1-\cos h}{h} - \lim_{h\to 0}\sin x\cdot\frac{\sin h}{h}$$
$$= (-\cos x)\cdot 0 - (\sin x)\cdot 1 = -\sin x. \quad \textit{by Theorems 3.10 and 3.11}$$

QED

You should commit formulas (1) and (2) to memory, taking care to remember that the minus sign appears in, and only in, (2).

Example 2

Find $D_x\left(\dfrac{\sin x}{1+\cos x}\right)$.

Solution. From (1), (2), and the quotient rule, we have

$$D_x\left(\frac{\sin x}{1+\cos x}\right) = \frac{(1+\cos x)D_x(\sin x) - (\sin x)D_x(1+\cos x)}{(1+\cos x)^2}$$
$$= \frac{(1+\cos x)(\cos x) - (\sin x)(-\sin x)}{(1+\cos x)^2}$$
$$= \frac{\cos x + \cos^2 x + \sin^2 x}{(1+\cos x)^2}$$
$$= \frac{\cos x + 1}{(1+\cos x)^2} = \frac{1}{1+\cos x}. \blacksquare$$

Using the general power rule (Theorem 2.5), we can extend (1) and (2) to any positive or negative integer n:

(3) $$\frac{d}{dx}[\sin^n x] = n\sin^{n-1} x\cos x,$$

(4) $$\frac{d}{dx}[\cos^n x] = -n\cos^{n-1} x\sin x.$$

Example 3

If $f(x) = x^2\sin^2 x + \cos^2 x\sin^3 x$, then find $f'(x)$.

Solution. We use (3), (4), and the product rule (Theorem 2.3) to get

$$\begin{aligned} f'(x) &= 2x\sin^2 x + x^2\cdot 2\sin x\cdot\cos x + 2\cos x(-\sin x)\cdot\sin^3 x \\ &\quad + (\cos^2 x)\cdot(3\sin^2 x\cos x) \\ &= 2x\sin^2 x + 2x^2\sin x\cos x - 2\sin^4 x\cos x + 3\sin^2 x\cos^3 x. \end{aligned}$$ ■

As promised, Theorems 4.1 and 4.2 and the basic rules of differentiation from Section 2 are all we need to differentiate the remaining trigonometric functions. For example, Theorem 4.2 and the reciprocal rule or the general power rule from Section 2 give

$$\begin{aligned} D_x\sec x &= \frac{d}{dx}[(\cos x)^{-1}] = -(\cos x)^{-2}\cdot\frac{d}{dx}(\cos x) \\ &= -\frac{1}{\cos^2 x}(-\sin x) = \frac{1}{\cos x}\cdot\frac{\sin x}{\cos x}. \end{aligned}$$

Hence,

(5) $$D_x\sec x = \sec x\tan x.$$

Similarly, the quotient rule gives

$$\begin{aligned} D_x\tan x &= D_x\frac{\sin x}{\cos x} = \frac{\cos x\cdot\cos x - (\sin x)\cdot(-\sin x)}{\cos^2 x} \\ &= \frac{\cos^2 x + \sin^2 x}{\cos^2 x} = \frac{1}{\cos^2 x}. \end{aligned}$$

We therefore have

(6) $$D_x\tan x = \sec^2 x.$$

Example 4

Calculate (a) $D_x(\tan^2 x\sec x)$ and (b) $D_x\cot x$.

Solution. (a) Using the product and general power rules in conjunction with (5) and (6), we get

$$\begin{aligned} D_x(\tan^2 x\sec x) &= D_x(\tan^2 x)\cdot\sec x + \tan^2 x\cdot D_x(\sec x) \\ &= 2(\tan x)(\sec^2 x)\cdot\sec x + \tan^2 x\cdot\sec x\tan x \\ &= 2\tan x\sec^3 x + \sec x\tan^3 x. \end{aligned}$$

(b) Since $\cot x = (\tan x)^{-1}$, we can use the general power rule and (6) to obtain

$$D_x(\cot x) = D_x[(\tan x)^{-1}] = -(\tan x)^{-2} \cdot D_x(\tan x)$$
$$= -\frac{1}{\tan^2 x} \cdot \sec^2 x = -\frac{\cos^2 x}{\sin^2 x} \cdot \frac{1}{\cos^2 x}$$
$$= -\frac{1}{\sin^2 x} = -\csc^2 x. \quad \blacksquare$$

The result of Example 4(b) provides another differentiation rule:

(7) $$D_x(\cot x) = -\csc^2 x.$$

The last trigonometric function is cosecant, the reciprocal of the sine function. Again, we can use the general power rule to get

$$D_x(\csc x) = D_x[(\sin x)^{-1}] = -(\sin x)^{-2} \cdot D_x(\sin x)$$
$$= -\frac{1}{\sin^2 x} \cdot \cos x = -\frac{1}{\sin x} \cdot \frac{\cos x}{\sin x}.$$

Hence,

(8) $$D_x(\csc x) = -\csc x \cot x.$$

To keep the signs straight in the differentiation formulas for the trigonometric functions, observe from (2), (7), and (8) that the derivatives of functions whose names start with *co* (*co*sine, *co*tangent, and *co*secant) all have *negative* signs. The other derivative formulas do not involve minus signs. It suffices to memorize (1), (2), (5), and (6), and to notice that (7) and (8) follow mechanically from (6) and (5), respectively. To get (8) from (5), for example, merely insert the prefix *co*- throughout and add a minus sign. The same procedure also produces (7) from (6).

Example 5

Calculate $\frac{d}{dx}(\sec^2 x + \csc^2 x)$.

Solution. Using the sum and general power rules and Formulas (5) and (8), we get

$$\frac{d}{dx}(\sec^2 x + \csc^2 x) = \frac{d}{dx}(\sec^2 x) + \frac{d}{dx}(\csc^2 x)$$
$$= 2 \sec x \cdot \sec x \tan x + 2 \csc x(-\csc x \cot x)$$
$$= 2 \sec^2 x \tan x - 2 \csc^2 x \cot x. \quad \blacksquare$$

FIGURE 4.1

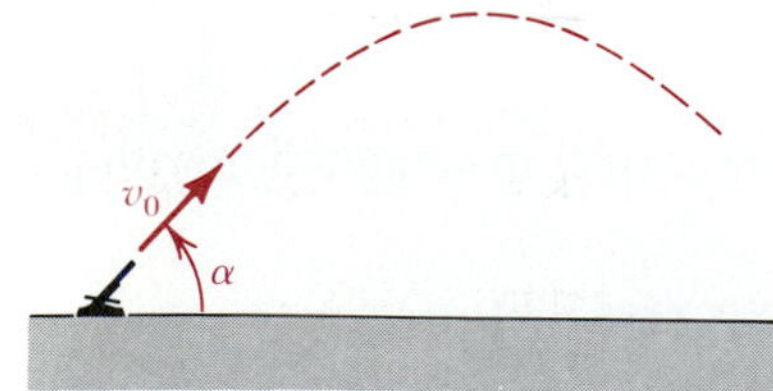

Our concluding example suggests the role of differentiation of trigonometric functions in applications to physics and engineering.

Example 6

When a projectile is launched with angle of elevation α and initial speed v_0 (see **Figure 4.1**), its horizontal displacement at time t is

(9) $$x = v_0 t \cos \alpha.$$

Find a formula for $dx/d\alpha$ and interpret. Show that $dx/d\alpha < 0$ for all $0 < \alpha < \pi/2$. Interpret this fact physically also.

Solution. From (9) and Theorem 4.2, we have

$$\frac{dx}{d\alpha} = -v_0 t \sin \alpha. \tag{10}$$

This is the rate of change in x per unit change in α and represents the rate at which x increases or decreases as the angle α is increased. Since $\sin \alpha > 0$ for $0 < \alpha < \pi/2$, (10) says that x *decreases* as α is increased. ■

Exercise 2.4

In Exercises 1–20, find $f'(x)$ for the given function f.

1. $f(x) = 5 \sin x - 2 \cos x$
2. $f(x) = 7 \sec x - 2 \cot x$
3. $f(x) = 3x^2 \sin x + 5x \cos x - \sin x \cos x$
4. $f(x) = x^3 \cot x - x^2 \sin x + 3 \sec x \tan x$
5. $f(x) = \tan^3 x + \sec^3 x$
6. $f(x) = (1 + \sec^2 x)^3$
7. $f(x) = (1 + \cos^2 x)^2(1 + \sin^2 x)^3$
8. $f(x) = (x^2 + x - 1)^2(\sec^2 x + \csc^2 x)$
9. $f(x) = (1 + \cos^2 x)(\sin^3 x + \sin^3 x \cos^2 x)$
10. $f(x) = (1 + \sec^2 x)^3(\sin^2 x \cos^2 x + x \cot x)$
11. $f(x) = \dfrac{\cos^2 x}{1 + \sin x}$
12. $f(x) = \dfrac{1}{\sin x + \cos x}$
13. $f(x) = \dfrac{\sin x + 1}{\sin x - 1}$
14. $f(x) = \dfrac{1 + \tan x}{1 - \tan x}$
15. $f(x) = \dfrac{\sin x}{x}$
16. $f(x) = \dfrac{\cos x}{x^2 + 1}$
17. $f(x) = \dfrac{\sin^3 x + \cos^3 x}{\sin x + 2 \cos x}$
18. $f(x) = \dfrac{\sin^2 x \cos x + \cos^2 x \sin x}{\sin x - \cos x}$
19. $f(x) = \sin 2x$ (*Hint:* See Corollary 3.5.)
20. $f(x) = \cos 2x$ (*Hint:* See Corollary 3.5.)
21. Use the results of Exercises 19 and 20 and formula (17) of the last section to find $D_x(\sin 3x)$.
22. Use the results of Exercises 19 and 20 and formula (12) of the last section to find $D_x(\cos 3x)$.
23. **(a)** Find the equation of the tangent line to $y = \cos x$ at $x = \pi/3$.
    (b) Find the equation of the tangent line to $y = \tan x$ at $x = 5\pi/6$.
24. In Example 6, the vertical displacement of the projectile at time t is

$$y = -\frac{1}{2} g t^2 + (v_0 \sin \alpha) t, \tag{11}$$

where g is the acceleration due to gravity.

(a) Calculate dy/dx. Interpret.
(b) What is the sign of dy/dx for $0 < \alpha < \pi/2$? Interpret.

25. In Example 6 and Exercise 24, the angle α is usually fixed so as to determine the horizontal range. Assuming that α is fixed and $0 < \alpha < \pi/2$, calculate dx/dt and dy/dt and interpret. For what values of t is the projectile rising? Falling?
26. In Example 6, the point of impact of the projectile (the range of the gun) is given by

$$x_f = \frac{v_0^2 \sin \alpha \cos \alpha}{2g},$$

where g is the acceleration due to gravity.
(a) Find $dx_f/d\alpha$. What does it mean when this is positive? Negative?
(b) For what values of α is $dx_f/d\alpha$ positive? Negative?
(c) For which values of α is $dx_f/d\alpha$ zero? Interpret the significance of such α.

27. A block of mass m is placed on an inclined plane of angle θ. (See **Figure 4.2.**) The tangential and normal forces on the block are given by

$$F_t = mg \sin \theta \qquad \text{and} \qquad F_n = mg \cos \theta$$

(a) Find $dF_t/d\theta$ and $dF_n/d\theta$.
(b) Assuming $\theta \in (0, \frac{1}{2}\pi)$, discuss the change in F_t and F_n as θ increases.

FIGURE 4.2

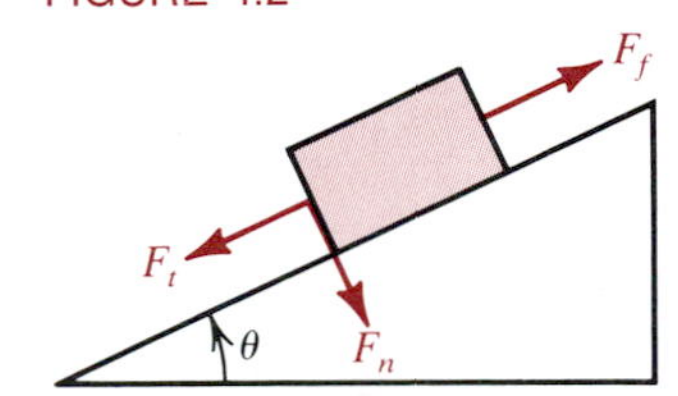

28. In Exercise 27, if the coefficient of friction is μ, the frictional force F_f is $F_f = \mu F_n$. (The block will slide down the slope if $F_t > F_f$.)
(a) Express the resultant tangential force F on the block due to F_t and F_f. (Take the positive direction downward.)
(b) Find $dF/d\theta$.

(c) Use your result in (b) to show that F increases as θ is increased in the interval $(0, \frac{1}{2}\pi)$.

29. If you drag an object of mass m along a horizontal sidewalk by pulling on it along an angle θ inclined to the sidewalk, as in **Figure 4.3**, then the force F needed to move the object is

$$F = \frac{mg\mu}{\mu \sin\theta + \cos\theta},$$

where μ is the coefficient of friction. Find $dF/d\theta$. For which θ is $dF/d\theta$ positive? Negative?

FIGURE 4.3

30. In Exercise 29,

(a) show that $F = mg\mu$ when $\theta = 0$, and $F = mg$ when $\theta = \frac{1}{2}\pi$;

(b) find $dF/d\theta$ for $\theta = 0$ and $\theta = \frac{1}{2}\pi$; show $dF/d\theta < 0$ for $\tan\theta < \mu$, and $dF/d\theta > 0$ for $\tan\theta > \mu$;

(c) from your results in (b), show that F decreases from $\theta = 0$, reaching its low point when $\tan\theta = \mu$.

2.5 The Tangent Approximation and Differentials

Definition 1.2 says that the slope of the tangent line to the graph of the differentiable function f at the point $P(c, f(c))$ is $f'(c)$. In Example 5 of Section 1, we found the tangent lines to the graph of $y = x^2 + 2x - 3$ at $c = 2$ and $c = -2$. Figure 1.4—reproduced here as **Figure 5.1**—shows that the tangent lines T_1 and T_2 almost coincide with the graph of $y = f(x)$ near the points of tangency. Near $c = 2$ for instance, Table 1 shows values of $y = f(x)$ and $y = T(x) = 6x - 7$ for values of x near $c = 2$. As the table shows, $T(x)$ closely approximates $f(x)$ near $x = c$.

FIGURE 5.1

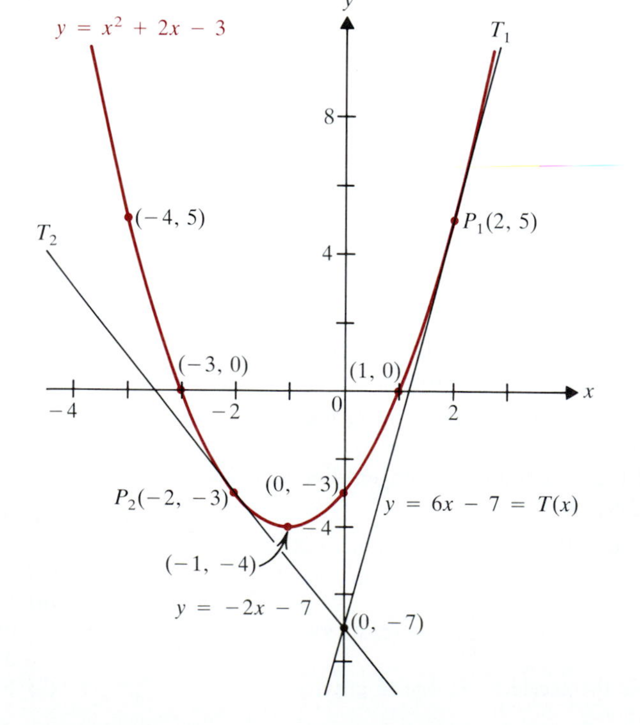

TABLE 1

x	$f(x)$	$T(x)$
1.9	4.41	4.4
1.99	4.9401	4.94
1.999	4.994001	4.994
1.9999	4.99940001	4.9994
2	5	5
2.0001	5.00060001	5.0006
2.001	5.006001	5.006
2.01	5.0601	5.06
2.1	5.61	5.6

More generally, near any point $P(c, f(c))$ on the graph of an arbitrary differentiable function f the value of $f(x)$ is close to the value of the ***tangent approximation:***

(1) $$f(x) \approx T(x) = f(c) + f'(c)(x - c).$$

The expression (1) is the approximation of f by the tangent-line function T obtained from the equation

$$y - f(c) = f'(c)(x - c)$$

of the tangent line to the graph of $y = f(x)$ at $P(c, f(c))$. For instance, in Example 5 of Section 1, $f(x) = x^2 + 2x - 3$ and $c = 2$. Thus

$$f'(x) = 2x + 2 = 2 \cdot 2 + 2 = 6$$

at $x = c$. Since $f(2) = 2^2 + 2 \cdot 2 - 3 = 5$, the equation of the tangent line to the graph of f at $x = 2$ is

$$y - 5 = 6(x - 2) \rightarrow y = 6x - 12 + 5 = 6x - 7 = T(x).$$

Table 1 illustrates the idea behind (1): for x near c, the value of $y = f(x)$ on the curve does not differ substantially from the value $y = T(x) = f(c) + f'(c)(x - c)$ on the tangent line. See **Figure 5.2.**

FIGURE 5.2

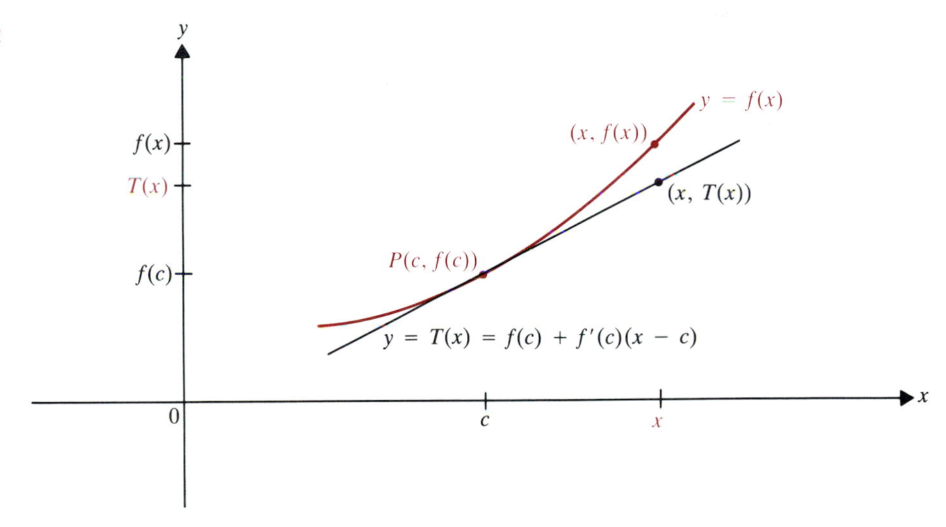

An attractive conceptual feature of (1) is that it provides a way to approximate (perhaps complicated) differentiable functions f by simple *linear functions* T. The following example illustrates this.

Example 1

Use the tangent approximation to estimate $\sqrt{9.006}$.

Solution. Observe that 9.006 is very near $c = 9$, where we know the exact value of the square-root function. Let $f(x) = \sqrt{x}$. Then formula (5) of Example 4 in Section 1 gives

$$\frac{dy}{dx} = \frac{1}{2\sqrt{c}} = \frac{1}{2\sqrt{9}} = \frac{1}{6}.$$

We want to approximate $f(x)$ for $x = 9.006$. Here then $x - c = 0.006$, and

$$f(x) \approx T(x) = f(c) + f'(c)(x - c) = 3 + \frac{1}{6}(9.006 - 9)$$
$$= 3 + 0.001 = 3.001. \blacksquare$$

To see how good the approximation just obtained is, enter 9.006 on your calculator and press the square root key. You should see 3.000999833 or a number very close to it. This rounds off to 3.001 to three decimal places, the number of places in 9.006.

An important application of the tangent approximation is to estimate increments Δy of a function $y = f(x)$ near a point where f is differentiable. As **Figure 5.3** illustrates,

$$\Delta y = f(x) - f(c) \approx f'(c)\,\Delta x. \tag{2}$$

This is just (1) rewritten with $\Delta x = x - c$. The right side of (2) is a constant multiple of Δx, so is a linear function of Δx. The linear relationship in (2) embodies the essence of Leibniz's calculus of differentials, and leads to his useful differential notation.

FIGURE 5.3

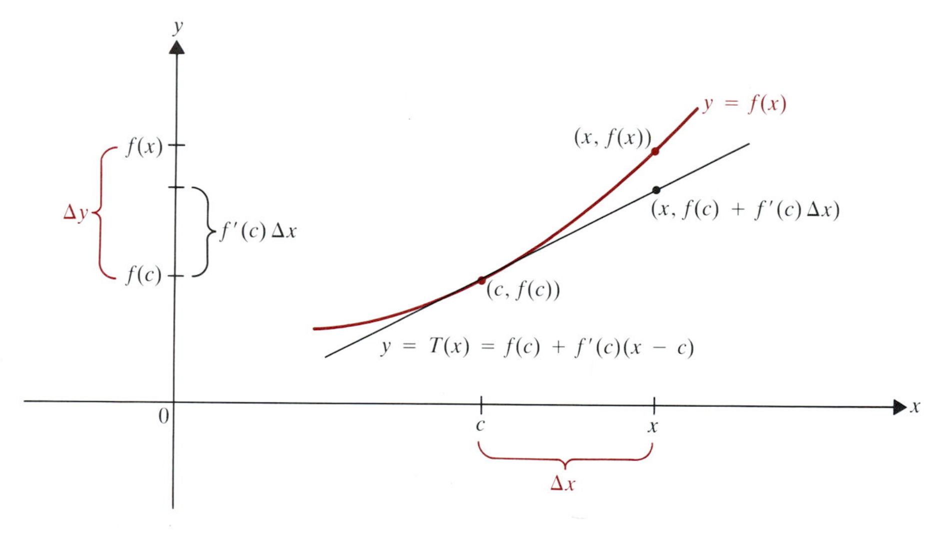

5.1 DEFINITION Suppose that f is differentiable at $x = c$ and is defined on an open interval I containing c. For $x \in I$, the ***differential of x at c*** is

$$dx = \Delta x = x - c. \tag{3}$$

The ***differential of y*** (or f) ***at c*** is

$$dy = \left.\frac{dy}{dx}\right|_c dx = f'(c)\,dx = f'(c) \cdot (x - c). \tag{4}$$

Note that in (3) and (4) *the quantity dy is a linear function of the quantity dx*. Together, (2) and (4) say that **the increment Δy is approximated by the differential dy.**

Many apparently exact calculations are really only approximate. For instance, although use of the y^x key on a calculator will usually give the answer 8 for 2^3, subtraction of 8 from this number may not produce a 0 in the calculator display! In fact, many popular calculators give the small positive number 0.0000000164 when this experiment is tried. The reason is that the y^x key only approximates y^x. In view of this, it is helpful to be able to estimate accuracy of computed answers. The next example indicates how the differential and tangent approximation can be used to do so.

Example 2

FIGURE 5.4

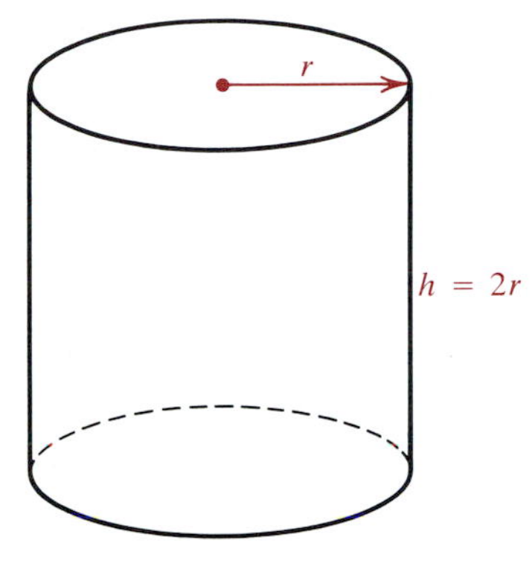

A certain cell is cylindrical, with radius equal to half its height, as in **Figure 5.4.** Suppose the radius is measured by an electron microscope to be 1.00×10^2 Å (Å is the *angstrom*, 10^{-10} m), with a possible error of 1%. Estimate the error in the computed volume of the cell.

Solution. The 1% error means that the true radius lies somewhere between 100 − (0.01)(100) and 100 + (0.01)(100) angstroms, that is, between 99 Å and 101 Å. Recall that the formula for the volume of a right circular cylinder of radius r and height h is

$$V = \pi r^2 h.$$

Since $r = \frac{1}{2}h$ for this cell, we have $h = 2r$. Thus

$$V = \pi r^2 \cdot 2r = 2\pi r^3 = f(r).$$

The calculated volume V_0 when $r_0 = 100$ Å $= 10^2$ Å is therefore

$$V_0 = 2\pi \cdot (10^2)^3 \text{ Å}^3 = 2\pi \cdot 10^6 \text{ Å}^3. \tag{5}$$

The maximum error in the volume occurs when $dr = \Delta r = \pm 1$ Å, the maximum possible error in the radius. From (2) and $f'(r) = (2\pi)(3r^2) = 6\pi r^2$, we get

$$\Delta V \approx dV \approx 6\pi {r_0}^2\, dr \approx \pm 6\pi \cdot (10^2)^2 \cdot 1 \text{ Å}^3,$$

$$\Delta V \approx \pm 6\pi \cdot 10^4 \text{ Å}^3. \tag{6}$$

The fact that the angstrom is such a small unit makes (6) difficult to interpret. But we can estimate the *relative error* by dividing ΔV by the computed volume V_0 from (5). That gives

$$\text{relative error} \approx \pm\frac{6\pi \cdot 10^4}{2\pi \cdot 10^6} \approx \pm 0.03.$$

We therefore expect an approximate relative error of up to 3% in the computed volume, *triple* the relative error in measuring r. ■

Example 2 illustrates how errors in measurement are often *multiplied* by computations with measured data. In approximating (5), since r is measured only to three significant figures, a scientist would express the volume to three significant figures as 6.28×10^6 Å^3. The maximum error of 3% would be less than 0.19×10^6 Å^3, meaning that the true volume would lie somewhere between 6.09×10^6 Å^3 and 6.47×10^6 Å^3. Thus the computed answer might not be accurate even to two significant figures! This helps explain the wording used in science in rules governing calculations with significant figures: An answer contains *no more* significant figures than the least number of significant figures of any factor. It may contain *less* than that many.

The differential provides a link between the derivative $f'(c)$ as the rate of change $y = f(x)$ with respect to x at $x = c$ and the derivative as the slope of the tangent line to $y = f(x)$ at $x = c$. This arises from the fact that the equation of the tangent line at $x = c$ is

$$y - f(c) = f'(c)(x - c) = f'(c)\,dx = dy. \tag{7}$$

In (4), if we divide by dx we get

$$\left.\frac{dy}{dx}\right|_c = f'(c), \tag{8}$$

the rate of change of y with respect to x at $x = c$. Equations (7) and (8) express *both* the equation of the tangent line and the rate of change of y with respect to x in terms of differentials. An added bonus from (8) is that it gives an interpretation of the Leibniz notation dy/dx for the derivative as an *actual*, not just *symbolic*, quotient. We can think of dy/dx as the quotient of differentials: dy by dx. See **Figure 5.5,** which shows dy and dx at $x = c$. Because of this, the part of calculus that deals with differentiation is called *differential calculus.*

FIGURE 5.5

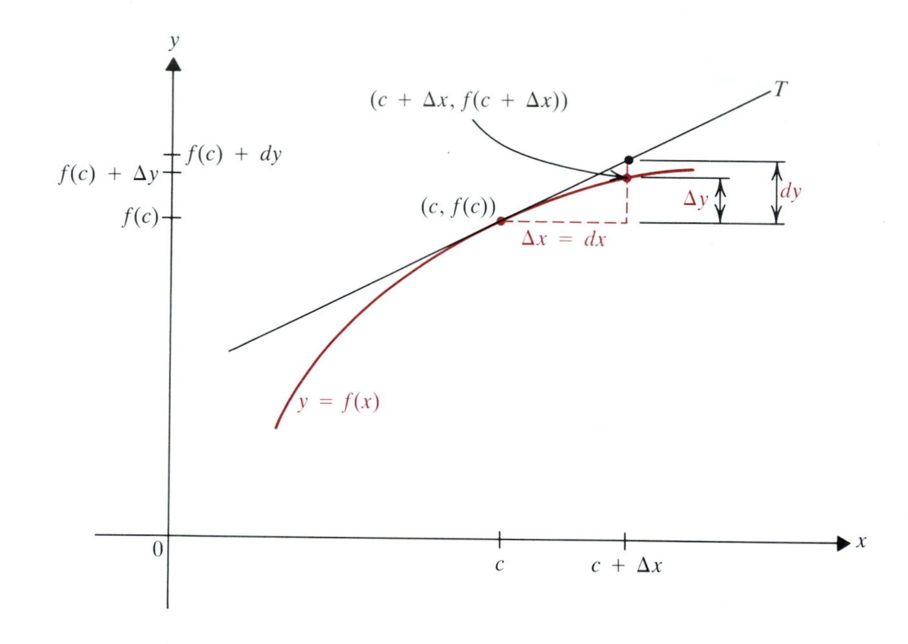

Differentiability and the Tangent Approximation*

We have seen that the tangent approximation $T(x)$ to $f(x)$ appears to be quite good (cf. Table 1) when x is close to c. In the rest of this section, we analyze this more carefully. We begin by introducing a function $e(x)$ to measure the *error* made in the approximation (1) of $f(x)$ by $T(x)$:

$$e(x) = f(x) - T(x) = f(x) - f(c) - f'(c)(x - c). \tag{9}$$

Since f is differentiable at c, it is continuous there by Theorem 1.3. Hence,

$$\begin{aligned}\lim_{x\to c} e(x) &= \lim_{x\to c} [f(x) - f(c)] - \lim_{x\to c} [f'(c)(x - c)]\\ &= [f(c) - f(c)] - f'(c)\cdot 0 = 0.\end{aligned}$$

* Optional

Still more is true. The error $e(x)$ in (9) actually has a *smaller order of magnitude* than $x - c$. By this we mean that $e(x)$ approaches 0 as $x \to c$ so much *faster* than x approaches c that $e(x)/(x - c)$ even approaches 0.

5.2 THEOREM

If f is differentiable at $x = c$ and $e(x)$ is defined by (9), then

$$\lim_{x \to c} \frac{e(x)}{x - c} = 0. \tag{10}$$

Proof. We have from (9)

$$\begin{aligned}\lim_{x \to c} \frac{e(x)}{x - c} &= \lim_{x \to c} \frac{f(x) - f(c) - f'(c)(x - c)}{x - c} \\ &= \lim_{x \to c} \frac{f(x) - f(c)}{x - c} - \lim_{x \to c} \frac{f'(c)(x - c)}{x - c} \\ &= f'(c) - f'(c) = 0. \quad \boxed{\text{QED}}\end{aligned}$$

by Theorem 5.1(d) of Chapter 1

Example 3

Verify (10) for $f(x) = 1/x$ and $c = 2$.

Solution. Since $f(x) = x^{-1}$, the power rule says that $f'(x) = -x^{-2}$. Hence, $f'(2) = -1/4$. We also have $f(c) = f(2) = 1/2$. Therefore

$$T(x) = f(c) + f'(c)(x - c) = \frac{1}{2} - \frac{1}{4}(x - 2).$$

Then from (9),

$$\begin{aligned}e(x) &= \frac{1}{x} - \frac{1}{2} - \left(-\frac{1}{4}\right)(x - 2) = \frac{1}{x} - \frac{1}{2} + \frac{1}{4}x - \frac{1}{2} \\ &= \frac{1}{x} + \frac{1}{4}x - 1 = \frac{4 + x^2 - 4x}{4x} = \frac{(x - 2)^2}{4x}.\end{aligned}$$

Thus

$$\lim_{x \to 2} \frac{e(x)}{x - 2} = \lim_{x \to 2} \frac{x - 2}{4x} = 0.$$

Hence, (10) is satisfied for $c = 2$. ■

Theorem 5.2 guarantees that for x close to c, the tangent approximation $T(x)$ is very close to $f(x)$. But its importance goes beyond that. It provides a means of *characterizing* the differentiability of f at $x = c$ without direct reference to the difference quotient. This characterization is the basis for our proof in the next section of the most important differentiation rule: the chain rule. The characterization essentially says that f differentiable at $x = c$ is equivalent to $f(x)$ being approximable near $x = c$ by a linear function

$$y = f(c) + m(x - c) = L(x)$$

so closely that the error $e(x) = f(x) - L(x)$ has a smaller order of magnitude than $x - c$. The precise statement takes the following form.

5.3 THEOREM

Suppose that f is defined on an open interval I containing c. Then f is differentiable at $x = c$ if and only if there is some real number m such that the function

$$e(x) = f(x) - f(c) - m \cdot (x - c) \tag{11}$$

satisfies

$$\lim_{x \to c} \frac{e(x)}{x - c} = 0. \tag{10}$$

Moreover, $m = f'(c)$.

Proof. Theorem 5.2 showed the "if part": If f is differentiable at $x = c$, then (10) does hold if we take for m in (11) the number $f'(c)$. We thus have to prove the "only-if part": If there is a real number m such that $e(x)$ in (11) satisfies (10), then f is differentiable at $x = c$ and the number m is actually $f'(c)$. We do this by first noting from (11) that

$$f(x) - f(c) = m(x - c) + e(x).$$

Therefore,

$$\begin{aligned} \lim_{x \to c} \frac{f(x) - f(c)}{x - c} &= \lim_{x \to c} \frac{m(x - c) + e(x)}{x - c} \\ &= \lim_{x \to c} \frac{m \cdot (x - c)}{x - c} + \lim_{x \to c} \frac{e(x)}{x - c} \\ &= m + 0 = m. \qquad \textit{by (10)} \end{aligned}$$

Equation (3) of Section 1 then tells us that f is differentiable at $x = c$ and $f'(c) = m$. QED

We emphasize that Theorem 5.3 is of conceptual and theoretical importance, but it is not very useful computationally, since it requires calculation of the number m $(=f'(c))$. Our use of it will be almost exclusively to derive other results that *will* have computational significance.

Exercises 2.5

In Exercises 1–4, (a) find the equation of the tangent line to $y = f(x)$ at $x = c$, (b) use the tangent approximation to estimate $f(x)$, and (c) compare your estimate in (b) to the value obtained from a calculator.

1. $y = \dfrac{1}{x^2 + 1}$, $c = 1$, $x = 1.002$

2. $y = \dfrac{1 - x}{x^2 + 1}$, $c = 0$, $x = 0.001$

3. $y = \cos x$, $c = \pi/2$, $x = 91°$

4. $y = \tan x$, $c = \pi/4$, $x = 44°$

In Exercises 5–12, find Δy exactly for the given $y = f(x)$, c, and $\Delta x = x - c$. Compare with dy.

5. $f(x) = x^2 + 3x - 1$, $c = 1$, $\Delta x = 0.1$

6. $f(x) = 1 - 2x - 3x^2$, $c = -1$, $\Delta x = 0.1$

7. $f(x) = x^3 - 2x^2 + 3$, $c = 2$, $\Delta x = -0.01$

8. $f(x) = x^3 + 3x^2 - 1$, $c = -2$, $\Delta x = -0.01$

9. $f(x) = 1/x^2$, $c = 2$, $x = 2.01$

10. $f(x) = \dfrac{3}{2x^2 - 1}$, $c = 1$, $x = 1.01$

11. $f(x) = \dfrac{x^2 + x + 1}{x^2 - x + 1}$, $c = 1$, $x = 0.99$

12. $f(x) = \dfrac{3 - x^2}{3 + x^2}$, $c = 1$, $x = 0.99$

In Exercises 13–18, use differentials to estimate the quantities.

13. The volume of a spherical shell of inner radius 10 and thickness 0.3. [Recall that the volume enclosed by sphere of radius a is $\frac{4}{3}\pi a^3$.]

14. The volume of a cylindrical shell of inner height and inner radius $r = 3$ and thickness 0.1.

15. The volume of paint needed to paint the walls of a cubical room of size 8 by 8 by 8 ft if the paint is to be applied in two coats each 1/128 in. thick.

16. The volume of paint needed to paint the outer curved surface of a cylindrical water tank of height 10 ft and inner radius 5 ft if two coats are applied, each 1/16 in. thick.

17. $\sin 31°$

18. $\cot 119°$

In Exercises 19–26, use differentials to estimate the error and relative error in the computed quantity attributable to the error in the measured quantity.

19. The volume of a spherical ball of radius $r = 10$ cm with error of ± 0.1 cm. (See Exercise 13.)

20. The volume of a cube of edge $x = 5$ cm with error of ± 0.1 cm.

21. The area of a square floor of side 10 ft if the side length is measured to within 1/4 in.

22. The area of a circular hole for a stained glass window of radius $r = 2$ ft if the hole's radius is measured accurately to the nearest 1/8 in.

23. The resistance $R = 1/d^2$ of a length of wire of diameter $d = 0.1$ in. measured to within a tolerance of $\pm 1\%$.

24. The current $I = 115/R$ flowing in an electrical circuit of resistance R ohms if R is measured to within a tolerance of $\pm 5\%$.

25. The force

$$F = \frac{1}{4\pi\varepsilon_0} \cdot \frac{1}{r^2}$$

of repulsion between two oppositely charged particles carrying unit charges at distance r cm if r is measured to within $\pm 1\%$. (Here treat ε_0, the *permittivity of a vacuum*, as an exact constant.)

26. The gravitational force $F = G/R^2$ between two 1-kg masses at distance R measured to within $\pm 2\%$. (Here treat the gravitational constant G as exact.)

In Exercises 27–36, verify (10) in Theorem 5.2 for the given function f at the given point c. (First find $f'(c)$.)

27. $f(x) = 2x + 3$, $c = 1$

28. $f(x) = -3x + 1$, $c = 2$

29. $f(x) = x^2$, $c = 1$

30. $f(x) = x^2 - 3$, $c = 1$

31. $f(x) = \sqrt{x}$, $c = 4$

32. $f(x) = \sqrt{x + 3}$, $c = 6$

33. $f(x) = \dfrac{1}{x + 1}$, $c = 1$

34. $f(x) = \dfrac{1}{x - 1}$, $c = 0$

35. $f(x) = x^3$, $c = 1$

36. $f(x) = x^3 - x + 1$, $c = 0$

In Exercises 37 and 38, use theorems about derivatives to prove the formula for differentials. Here $u = f(x)$ and $v = g(x)$, where f and g are differentiable functions.

37. **(a)** $dc = 0$ if c is any constant.
(b) $d(u + v) = du + dv$ **(c)** $d(u - v) = du - dv$
(d) $d(cu) = c\,du$ for any constant c.

38. **(a)** $d(u \cdot v) = u \cdot dv + v \cdot du$
(b) $d\left(\dfrac{1}{v}\right) = -\dfrac{dv}{v^2}$
(c) $d\left(\dfrac{u}{v}\right) = \dfrac{v\,du - u\,dv}{v^2}$
(d) $d(u^n) = nu^{n-1}\,du$ for n an integer.

39. For Boyle's law $PV = C$, show that $dP/P + dV/V = 0$.

40. For the ideal gas law $PV = nRT$ (n and R constants), use differentials to show that

$$P\frac{dV}{dT} + \frac{dP}{dT}V$$

is constant.

41. For systems under constant pressure P, the *first law of thermodynamics in differential form* is

$$dQ = dE + P\,dV,$$

where dQ is thought of as the heat supplied to the system, dE is the change in internal energy, and dV is the change in volume. The *heat capacity* C_p is defined as dQ/dT. Use differentials to find a formula for C_p.

42. In Exercise 41, if the volume of the system is kept constant, show that the heat capacity is exactly equal to the rate of change of internal energy.

43. If there is some number B such that $|e(x)| \le B(x - c)^2$ for all x in some open interval I containing c, then show that f is differentiable at c.

44. Repeat Exercise 43 if $|e(x)| \le B|x - c|^{3/2}$.

2.6 The Chain Rule

As on p. 26, consider a lake that supports a population S of small fish. If these fish feed on algae, then S is a function of the volume A of algae in the lake, say $S = g(A)$. In turn, since the amount of algae varies over time, $A = h(t)$ for some function h. It would seem that from the rate of growth (or decline) dA/dt of the volume of algae, and knowledge of g, we should be able to predict the rate of change dS/dt of the small fish population at any time t_0.

This expectation can indeed be realized. The tangent approximation for A near t_0 gives

$$(1) \qquad \Delta A = A - A_0 = h(t) - h(t_0) \approx h'(t_0) \cdot (t - t_0) = h'(t_0)\,\Delta t,$$

if we assume that h is a differentiable function of t with $A_0 = h(t_0)$. We similarly can approximate ΔS for A near A_0:

$$(2) \qquad \Delta S = S - g(A_0) = g(A) - g(A_0) \approx g'(A_0) \cdot (A - A_0) = g'(A_0)\,\Delta A,$$

if g is a differentiable function of A. Putting (1) and (2) together, we obtain

$$\Delta S \approx g'(A_0) \cdot \Delta A \approx g'(h(t_0)) \cdot h'(t_0)\,\Delta t.$$

Thus

$$(3) \qquad \frac{\Delta S}{\Delta t} \approx g'(h(t_0)) \cdot h'(t_0).$$

Equation (3) suggests that

$$(4) \qquad \left.\frac{dS}{dt}\right|_{t_0} = g'(h(t_0)) \cdot h'(t_0) = \left.\frac{dS}{dA}\right|_{A_0} \cdot \left.\frac{dA}{dt}\right|_{t_0}.$$

This is correct and is an instance of the single most important rule of differential calculus: the *chain rule*. It carries that name because it is a formula for differentiating a chain of composed functions, such as $S = g(A) = g(h(t)) = (g \circ h)(t)$ above. **Thorough mastery of the rule and its use is essential for success in learning calculus.** We proceed to give the formal statement of the chain rule and several examples of how it is used. The proof, which consists of formalizing the ideas leading to (4), is found at the end of the section.

6.1 THEOREM

Chain Rule. If g is differentiable at x_0 and f is differentiable at $u_0 = g(x_0)$, then $f \circ g$ is differentiable at x_0 and

$$(5) \qquad (f \circ g)'(x_0) = f'(g(x_0)) \cdot g'(x_0).$$

If $u = g(x)$ and $y = f(u) = (f \circ g)(x)$, then in Leibniz notation

$$(6) \qquad \left.\frac{dy}{dx}\right|_{x_0} = \left.\frac{dy}{du}\right|_{u_0} \cdot \left.\frac{du}{dx}\right|_{x_0}.$$

It is worth noting that the general power rule (Theorem 2.5) is a special case of the chain rule. If we apply (5) to the function $f(u) = u^n$, then we get

$$\begin{aligned} D_x[(f \circ g)(x_0)] &= D_x[g(x)^n]|_{x=x_0} = f'(g(x_0)) \cdot g'(x_0) \\ &= ng(x_0)^{n-1} \cdot g'(x_0), \end{aligned}$$

since $df/du = nu^{n-1}$. This is precisely what the general power rule asserts at any point x_0 where $g(x)$ is differentiable.

Example 1

If $y = \sin^5 x$, then find dy/dx at $x = \pi/6$.

Solution. Here $y = (\sin x)^5$, so (5) or the general power rule gives

$$\left.\frac{dy}{dx}\right|_{\pi/6} = 5(\sin x)^4 \cdot \cos x\Big|_{\pi/6} = 5 \cdot \left(\frac{1}{2}\right)^4 \cdot \frac{\sqrt{3}}{2} = \frac{5\sqrt{3}}{32}. \quad ■$$

The formulation (6) makes the chain rule easy to remember, because we can think of the du terms as *cancelling* on the right. This helps account for the enduring popularity of Leibniz's notation dy/dx for the derivative. To get a general formula for the derivative function $(f \circ g)'(x)$, we simply replace x_0 and u_0 in (5) and (6) by x and u. That produces the following ***general chain rule.***

> If f and g are differentiable functions, then
>
> (7) $$(f \circ g)'(x) = f'(g(x)) \cdot g'(x),$$
>
> that is,
>
> (8) $$\frac{dy}{dx} = \frac{dy}{du}\frac{du}{dx}$$
>
> if $u = g(x)$ and $y = f(u)$.

In applying the chain rule, it may help at the start to follow the procedure suggested by (6) and (8): introduce an intermediate variable u. This helps to identify the functions f and g and to avoid a problem that often plagues beginners. Many students forget to multiply $f'(g(x))$ by $g'(x)$.

Example 2

Differentiate $h(x) = \sqrt{5x^2 + 3x - 1}$.

Solution. We can let $u = 5x^2 + 3x - 1$ and $y = \sqrt{u}$. Recall from Example 4 of Section 1 that

$$\frac{dy}{du} = \frac{1}{2\sqrt{u}}.$$

Then (8) gives

$$h'(x) = \frac{dy}{dx} = \frac{dy}{du}\frac{du}{dx} = \frac{1}{2\sqrt{u}} \cdot (10x + 3) = \frac{10x + 3}{2\sqrt{5x^2 + 3x - 1}}. \quad ■$$

Note that the intermediate variable u does not appear in the final formula for $h'(x)$, since it did not appear in the formula for $h(x)$. We introduced it *solely* to avoid the sometimes tempting mistake of calculating $D_x\sqrt{5x^2 + 3x - 1}$ to be $1/(2\sqrt{5x^2 + 3x - 1})$. The next example, while very simple, underscores the fact that $D_x(\sin kx)$ is *not* simply $\cos kx$, if k is a constant.

Example 3

Differentiate $s(x) = \sin 3x$.

cos 3x(3)

Solution. Exercise 21 of Section 4 outlined one way to find $s'(x)$, but the chain rule provides a much easier solution. We let $u = 3x$ and $y = \sin u$ and apply (8). Since $du/dx = 3$, we get

$$\frac{dy}{dx} = \frac{dy}{du}\frac{du}{dx} = (\cos u) \cdot 3 = 3\cos 3x. \blacksquare$$

The intermediate variable u should be considered a *temporary* crutch, to be discarded as soon as you develop facility with the chain rule. Notice that while we *evaluate* composite functions $f(g(x))$ "from the inside out," we *differentiate* them "from the outside in." In Example 2, for instance, we first differentiated the square-root function and then *multiplied by the derivative of the next* (inside) *function*, $5x^2 + 3x - 1$. This is the mental process the substitution scheme is intended to make habitual. In Example 3, we differentiate $\sin 3x$ by first differentiating the sine function, getting $\cos 3x$, and then multiplying by the derivative, 3, of the inside function $3x$. As you work exercises, check each time, whether you introduce an intermediate variable or not, that you have followed this process.

Example 4

Find $\dfrac{dy}{dx}$ if $y = \dfrac{\sqrt{x^2 + 4x + 3}}{x^2 + 5}$.

Solution. Here y is a quotient, and so we use the quotient rule. In differentiating the numerator, we use the chain rule to get

$$\frac{dy}{dx} = \frac{(x^2 + 5) \cdot \dfrac{2x + 4}{2\sqrt{x^2 + 4x + 3}} - \sqrt{x^2 + 4x + 3} \cdot 2x}{(x^2 + 5)^2}$$

$$= \frac{(x^2 + 5)(x + 2) - 2x(x^2 + 4x + 3)}{(x^2 + 5)^2\sqrt{x^2 + 4x + 3}} \qquad \text{multiplying numerator and denominator by } \sqrt{x^2 + 4x + 3}$$

$$= \frac{x^3 + 2x^2 + 5x + 10 - 2x^3 - 8x^2 - 6x}{(x^2 + 5)^2\sqrt{x^2 + 4x + 3}}$$

$$= \frac{-x^3 - 6x^2 - x + 10}{(x^2 + 5)^2\sqrt{x^2 + 4x + 3}}. \blacksquare$$

The chain rule in the form (5) or (6) is useful in obtaining tangent approximations for complicated functions.

Example 5

Find an approximation for $h(1.98)$ if $h(x) = \sqrt{5x^2 + 3x - 1}$.

Solution. We could use the formula for $h(x)$ in Example 2, but it is usually easier in this type of problem to use (6) directly. Recall the substitutions:

$$y = \sqrt{u} \qquad \text{and} \qquad u = 5x^2 + 3x - 1.$$

For $x_0 = 2$, we have $u_0 = 25$, and so $y_0 = \sqrt{25} = 5$. Since

$$\frac{dy}{du} = \frac{1}{2\sqrt{u}} \quad \text{and} \quad \frac{du}{dx} = 10x + 3,$$

we obtain

$$\left.\frac{dy}{du}\right|_{u_0} = \frac{1}{2\sqrt{25}} = \frac{1}{10} \quad \text{and} \quad \left.\frac{du}{dx}\right|_{x_0} = 23.$$

So by (6),

$$\left.\frac{dy}{dx}\right|_{x_0} = \frac{1}{10} \cdot 23 = 2.3.$$

Therefore, for x near $x_0 = 2$,

$$h(x) \approx h(x_0) + h'(x_0)(x - x_0) = y_0 + \left.\frac{dy}{dx}\right|_{x_0} \cdot (x - x_0).$$

Thus

$$h(1.98) \approx 5 + 2.3(1.98 - 2) \approx 4.954.$$

By comparison, a calculator gives $h(1.98) \approx 4.953988$. ■

The chain rule extends to composite functions involving three, four, or any finite number of functions. As an illustration, suppose we want to differentiate the composite $f \circ g \circ h$ of three functions:

$$y = (f \circ g \circ h)(t). \tag{9}$$

Letting $x = h(t)$ and $u = g(x)$, we can rewrite (9) in the form

$$\begin{aligned} y &= f(g(h(t))) = f(g(x)), \qquad \text{where } x = h(t), \\ &= f(u), \qquad \text{where } u = g(x) \text{ and } x = h(t). \end{aligned}$$

Then we can apply (6) twice to get

$$\frac{dy}{dt} = \frac{dy}{du} \cdot \frac{du}{dt},$$

$$\frac{dy}{dt} = \frac{dy}{du} \cdot \frac{du}{dx} \cdot \frac{dx}{dt}. \tag{10}$$

Example 6

Find dy/dx if $y = \sin^3 x^5$.

Solution. Here $y = (f \circ g \circ h)(x)$, where $h(x) = x^5$, g is the sine function, and f is the function that cubes its variable:

$$y = v^3, \qquad \text{where } v = \sin u \text{ and } u = x^5.$$

Thus, using the reasoning that gave (10), we obtain

$$\begin{aligned} \frac{dy}{dx} &= 3v^2 \frac{dv}{dx} = (3\sin^2 x^5) \cdot \frac{d}{dx}(\sin x^5) \\ &= 3\sin^2 x^5 \cdot \cos x^5 \cdot \frac{d}{dx}(x^5) = 15x^4 \sin^2 x^5 \cos x^5. \end{aligned}$$ ■

To illustrate how (10) is applied to other fields, we return to the food chain described at the beginning of the section. Recall that there is a lake with a population S of small fish that feed on algae. If A is the volume of algae in the lake at time t, then

$$S = g(A), \qquad \text{where } A = h(t).$$

Suppose now that there is a population N of large fish that depend upon the small fish as a food source.

Example 7

Suppose that N is modeled by the function

$$N = f(S) = \frac{1}{4} S^2 + 100,$$

where S is the population of small fish in hundreds. Suppose that in turn S is given by

$$S = g(A) = 25 - \frac{1}{4}(A - 10)^2,$$

where A is the volume of algae in the lake in cubic meters. If $A = 8$ and is increasing at the rate of $\frac{1}{2}m^3$ per year, then find the rate of change of N.

Solution. We are given $dA/dt = 1/2$ and asked to find dN/dt when $A = 8$. From (10), we have

$$\frac{dN}{dt} = \frac{dN}{dS}\frac{dS}{dt} = \frac{dN}{dS}\frac{dS}{dA}\frac{dA}{dt} = \frac{1}{2}\frac{dN}{dS}\frac{dS}{dA}. \tag{11}$$

Since $S = 25 - \frac{1}{4}(A - 10)^2$, we have for $A = 8$

$$\frac{dS}{dA} = -\frac{1}{4} \cdot 2(A - 10)^1 \cdot 1 = -\frac{1}{2}(8 - 10) = 1,$$

and

$$S = 25 - \frac{1}{4}(8 - 10)^2 = 25 - \frac{1}{4} \cdot 4 = 24.$$

The formula $N = S^2/4 + 100$ gives $dN/dS = S/2 = 24/2 = 12$ when $A = 8$. Consequently, when $A = 8$ we obtain from (11)

$$\frac{dN}{dt} = \frac{1}{2}\frac{dN}{dS}\frac{dS}{dA} = \frac{1}{2} \cdot 12 \cdot 1 = 6.$$

Hence, we would expect an increase of 6 large fish per year in response to the algae growth. (Exercises 37–42 explore this model further.) ■

Proof of the Chain Rule (Theorem 6.1)*

As indicated in deriving (4), our proof of the chain rule is based on the tangent approximation. Suppose that g is differentiable at $x = x_0$ and that f is dif-

* Optional

ferentiable at $u_0 = g(x_0)$. Then Theorem 5.2 says that

$$e(u) = f(u) - f(u_0) - f'(u_0) \cdot (u - u_0) \tag{12}$$

satisfies

$$\lim_{u \to u_0} \frac{e(u)}{u - u_0} = 0. \tag{13}$$

Note that $e(u_0) = 0$. We also have from (12)

$$f(u) - f(u_0) = f'(u_0) \cdot (u - u_0) + e(u),$$

$$f(u) - f(u_0) = f'(u_0) \cdot (u - u_0) + E(u) \cdot (u - u_0). \tag{14}$$

Here we have introduced the function $E(u)$, which is defined by

$$E(u) = \begin{cases} \dfrac{e(u)}{u - u_0}, & \text{for } u \neq u_0, \\ 0, & \text{when } u = u_0, \end{cases}$$

so that $e(u) = E(u) \cdot (u - u_0)$ holds for all u. (Both are 0 at u_0.) Notice further that, by (13), $E(u)$ approaches 0 as $u \to u_0$. If we let $u = g(x)$, then (14) becomes

$$f(g(x)) - f(g(x_0)) = f'(g(x_0)) \cdot (g(x) - g(x_0)) + E(u) \cdot (g(x) - g(x_0)).$$

Division by $x - x_0$ then gives

$$\frac{f(g(x)) - f(g(x_0))}{x - x_0} = f'(g(x_0)) \cdot \frac{g(x) - g(x_0)}{x - x_0} + E(u) \cdot \frac{g(x) - g(x_0)}{x - x_0}. \tag{15}$$

As $x \to x_0$, $u = g(x) \to u_0 = g(x_0)$, since the differentiable function g is continuous at x_0 by Theorem 1.3. Letting $x \to x_0$ in (15), we have from (3) of Section 2.1 and the fact that $E(u) \to 0$ as $u \to u_0$

$$\begin{aligned} (f \circ g)'(x_0) &= f'(g(x_0)) \cdot g'(x_0) + 0 \cdot g'(x_0) \\ &= f'(g(x_0)) \cdot g'(x_0). \quad \boxed{\text{QED}} \end{aligned}$$

Exercises 2.6

In Exercises 1–26, differentiate the given functions.

1. $f(x) = (x^2 - 4x + 7)^3$

2. $g(x) = (6 - 8x^2)^5$

3. $h(x) = (5x^3 - 2x^2 + 3x - 2)^{-2}$

4. $m(x) = (x^2 - 1)^2(5x^3 - x + 1)^3$

5. $f(x) = (x^2 - 5x + 3)^{-2}(x^3 + x + 1)^3$

6. $g(x) = \left(\dfrac{x-1}{x+1}\right)^3 + 2\left(\dfrac{x-1}{x+1}\right)^2 + 5\sqrt{\dfrac{x-1}{x+1}}$

7. $p(x) = \dfrac{(x^2 + 1)^3}{(x^2 + x - 1)^2}$

8. $r(x) = \dfrac{5x - 2}{\sqrt{x^2 + 1}}$

9. $t(x) = \dfrac{\sqrt{x^2 - x + 2}}{x^2 + 3}$

10. $v(x) = \sin x^2$

11. $f(x) = 5(\sin 3x)(\cos 2x)$

12. $h(x) = \sqrt{\sin^2 x + \cos^3 x}$

13. $t(x) = \tan(3x^2 + 1)$

14. $c(x) = \cot\sqrt{1 - x^2}$

15. $g(t) = \dfrac{\sin 2t}{\cos 2t - \sin 2t}$

16. $k(t) = \sin^2(1 - t^2)$

17. $m(u) = \sin(\cos 2u)$

18. $p(u) = \cos(u^2 + \sin u)$

19. $q(x) = \sin\sqrt{1 + x^2}$

20. $r(x) = \cos^3(1/x^2)$

21. $s(x) = \sqrt{\sec(3x - 2)}$

22. $t(x) = \sqrt{\tan(5x^2 + 1/x)}$

23. $m(x) = \sin^2 5x^2$

24. $p(x) = \left(\dfrac{6x + 1}{x^2 + 3x + 4}\right)^2$

25. $r(x) = \sqrt{x^2 + \sqrt{1 + 2x^2}}$

26. $s(x) = \sqrt{(5x^2 + 2)^3 + (3x^3 + 1)^4}$

27. Find dy/dx if $y = f(x^4)$ and $f'(x) = 1/(2x^2 + 5x)$.

28. Find dy/dx if $y = f(\sqrt{x})$ and $f'(x) = \cos x$.

29. Find $\left.\dfrac{dy}{dx}\right|_{x_0}$ if $y = f(x)^4$, $f(x_0) = 2$, and $f'(x_0) = -1$.

30. Find $\left.\dfrac{dy}{dx}\right|_{x_0}$ if $y = \sqrt{f(x)}$, $f(x_0) = 4$, and $f'(x_0) = \dfrac{1}{2}$.

In Exercises 31–36, use the tangent approximation to estimate $f(x)$ for the given x. Compare with a calculator value.

31. $f(x) = \sqrt{x^2 + x + 3}$ at $x = 2.03$.

32. $f(x) = \sqrt{x^2 + 3x + 5}$ at $x = 0.97$.

33. $f(x) = 1/\sqrt{2x^2 + 1}$ at $x = 2.003$.

34. $f(x) = 1/\sqrt{x^2 - x + 4}$ at $x = 0.997$.

35. $f(x) = \cos\left(\frac{\pi}{6}x^2\right)$ at $x = 1.06$.

36. $f(x) = \sin\left(\frac{\pi}{3}x^2\right)$ at $x = 0.97$.

Exercises 37–42 refer to Example 7. Suppose the algae volume continues growing at $1/2\ m^3$ per year.

37. What are S, N, and dN/dt one year after the time t_0 of the example?

38. What are S, N, and dN/dt two years after the time t_0 of the example?

39. What are S, N, and dN/dt four years after t_0?

40. What are S, N, and dN/dt six years after t_0?

41. What algae level seems to support the largest population of small fish? Is this the algae level that also corresponds to the largest population of large fish?

42. What does this model predict about the long-term effects of indefinite algae growth at the rate of $1/2\ m^3$ per year?

43. In Exercise 47 of Section 3, we modeled the price (in cents per pound) of tomatoes t months after March 1 by

$$p = 100 + 40\cos\frac{\pi}{6}t, \qquad t \in [0, 12]$$

in a certain city in the northeastern United States. Find dp/dt. When is the price rising? When does the price fall?

44. In Exercise 48 of Section 3, we modeled the price of lettuce t months after January 1 by

$$p = 70 + 30\cos\frac{\pi}{6}t, \qquad t \in [0, 12].$$

Find dp/dt. When is the price rising? Falling?

45. A certain parasite is used to control a pest population. Suppose that N parasites per square mile result in a pest population P of $P = f(N) = 160{,}000 - N^2$ per square mile. Suppose further that the parasites can live only between temperatures of 50°F and 90°F with N modeled by the function $N = (T - 50)(90 - T)$, where T is the Fahrenheit temperature. Find whether the pest population is increasing or decreasing at **(a)** $T = 60$°F, **(b)** $T = 70$°F, and **(c)** $T = 80$°F.

46. In Exercise 54 of Section 1.8, find dP/dY and dY/dN to find dP/dN when $N = 30$.

47. The ***wind-chill index*** is defined by

$$W_C = (10\sqrt{s} + 10.45 - s)(33 - T),$$

where s is the wind speed in m/s (meters per second) and T is the Celsius temperature. (The normal human skin temperature is 33°C.) The *wind-chill corrected temperature* is the temperature T_C that for $s = 2.2$ m/s (5 miles/h) yields the *same* value for W_C as do T and s.

(a) Determine W_C if $T = 0$°C and $s = 16$ m/s (about 35 miles/h). What is T_C?

(b) Give the general formula for T_C in terms of s and T.

(c) Find dT_C/ds if T is fixed. What is the value at $T = 0$, $s = 16$?

48. Refer to Exercise 47.

(a) Find dT_C/dT if s is fixed. What is the value at $T = 0$, $s = 16$?

(b) Interpret your answers in part (a) and Exercise 47(c) in terms of a 1 m/s increase in wind speed or a 1°C drop in air temperature when $T = 0$, $s = 16$.

49. Use $\cos(\frac{1}{2}\pi - x) = \sin x$ and $\sin(\frac{1}{2}\pi - x) = \cos x$ to show that $D_x \cos x = -\sin x$ follows from Theorems 4.1 and 6.1.

50. **(a)** Show that the general power rule (Theorem 2.5) is a special case of the chain rule.

(b) Use Theorem 6.1 to rework Exercise 39 of Section 1.

(c) Use Theorem 6.1 to rework Exercise 40 of Section 1.

51. Some texts offer the following simple proof of (6) in Theorem 6.1, first given by Cauchy in 1821.

$$\left.\frac{dy}{dx}\right|_{x_0} = \lim_{\Delta x \to 0}\frac{\Delta y}{\Delta x} = \lim_{\Delta x \to 0}\left(\frac{\Delta y}{\Delta u}\cdot\frac{\Delta u}{\Delta x}\right) = \left.\frac{dy}{du}\right|_{u_0}\cdot\left.\frac{du}{dx}\right|_{x_0}.$$

(a) For this proof to be valid, what must be true of Δu when Δx is small but unequal to 0?

(b) If $g(x) = c$ for any constant c, show that the condition in (a) *fails* but the chain rule *still* is true for $f \circ g$ if f is differentiable at $g(x_0)$. Hence, show that Cauchy's proof is not valid in general.

52. **(a)** Use form (3) of the definition of derivative (p. 64) and the reasoning of Example 8 of Section 1.5 to show that

$$h(x) = x^2\sin^2\frac{1}{x}$$

is differentiable at $x = 0$.

(b) Conclude from (a) and the chain rule that

$$k(x) = x^4\sin^4\frac{1}{x}$$

is differentiable at $x = 0$. What is $k'(0)$?

(c) Let

$$y = u^2 = f(u) \qquad \text{and} \qquad u = x^2\sin^2\frac{1}{x} = g(x).$$

Show for $x_0 = 0$ that the condition in Exercise 51(a) fails but that the chain rule still is true for $f \circ g$ at $x = 0$. Hence, show that Cauchy's proof is not valid even when constant functions are excluded from consideration.

2.7 Differentiation of Inverse Functions and Roots

FIGURE 7.1

The chain rule lets us differentiate a much wider class of functions than we could formerly. In Example 2 of the last section, for instance, we were able to use the chain rule and the formula for the derivative of the square-root function to differentiate $h(x) = \sqrt{5x^2 + 3x - 1}$. However, there are still many root functions that we cannot differentiate, such as $k(x) = \sqrt[3]{5x^2 + 3x - 1}$. The obstacle is that we can differentiate $f(u) = \sqrt[n]{u}$ only for $n = 2$ at this point. This section will show how to differentiate these functions for any positive integer n. We approach the problem from the point of view of *inverse functions.*

As the term itself is intended to suggest, the inverse of a function f is a function g whose rule *undoes* the effect of applying the rule for f. This means that for any x in the domain of f, if we apply g to $f(x)$, then we get back to x again. That is, $g(f(x)) = x$. In **Figure 7.1,** this is accomplished graphically by drawing a horizontal line segment out to the curve, and then drawing a vertical line segment to the x-axis. Algebraically, getting back to x from a given value of y is accomplished by solving the equation $y = f(x)$ for $x = g(y)$ as a function of y.

Consider, for example, the squaring function $f(x) = x^2$ with domain restricted to the set $[0, +\infty)$ of nonnegative real numbers. To undo the action of f, we solve the equation $y = x^2$ for $x \in [0, +\infty)$:

(1) $$x = \sqrt{y} = g(y).$$

This says that the inverse of the squaring function f defined on $[0, +\infty)$ is the principal square-root function g on $[0, +\infty)$. Since $\sqrt{x^2} = x$ for any nonnegative real number x, we have

(2) $$(g \circ f)(x) = x \qquad \text{for all } x \text{ in the domain of } f.$$

Likewise, $(\sqrt{y})^2 = y$ for any nonnegative real number y, so

(3) $$(f \circ g)(y) = y \qquad \text{for all } y \text{ in the range of } f,$$

since the range of f is also the set of nonnegative real numbers.

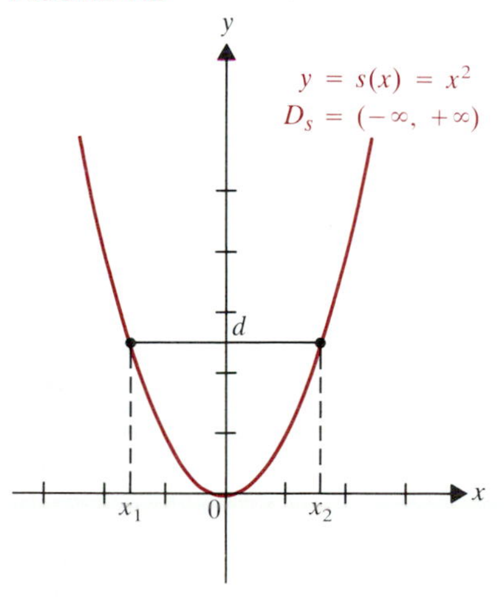

FIGURE 7.2

We have no difficulty solving $y = x^2$ for x as a function of y to get (1) if we restrict the domain of the squaring function to the set of nonnegative real numbers. As **Figure 7.2** shows, however, if we consider the squaring function $s(x) = x^2$ with domain the entire real line, we can no longer solve $y = x^2$ to get an inverse function $x = t(y)$. The problem is that for each positive real number y, there are *two* real numbers x such that $y = x^2$: the positive and negative square roots of y. Hence, the equation $y = x^2$ does not define x as a function of y. [Remember that to each number y in the domain of the function t, there must correspond one *and only one* real number $x = t(y)$.] This brings up the question: Under what circumstances does a function f have an inverse function? That is, when can we solve $y = f(x)$ for x as a function of y? Geometrically, as Figures 7.1 and 7.2 illustrate, *for every d in the domain of f, the horizontal line $y = d$ must meet the graph exactly once.* That is, every y in the range must be the image of one and only one x in the domain. Such functions have a special name.

7.1 DEFINITION

A function f is ***one-to-one*** if for every y in its range, $y = f(x)$ for exactly one x in the domain. That is, for all x_1 and x_2 in the domain f,

(4) $$\text{if } x_1 \neq x_2, \text{ then } f(x_1) \neq f(x_2).$$

Another common way of expressing the one-to-one condition (4) is

$$f(x_1) = f(x_2) \text{ only when } x_1 = x_2, \tag{5}$$

for any x_1 and x_2 in the domain of f. The next example shows how (4) and (5) can be tested in practice.

Example 1

Decide whether the given function f is one-to-one. If it is, then solve the equation $y = f(x)$ for x as a function of y.

(a) $f(x) = x^2 + 3x - 1$ (b) $f(x) = \dfrac{x+2}{3x-1}$.

FIGURE 7.3

$y = x^2 + 3x - 1$

$= (x + \frac{3}{2})^2 - \frac{13}{4}$

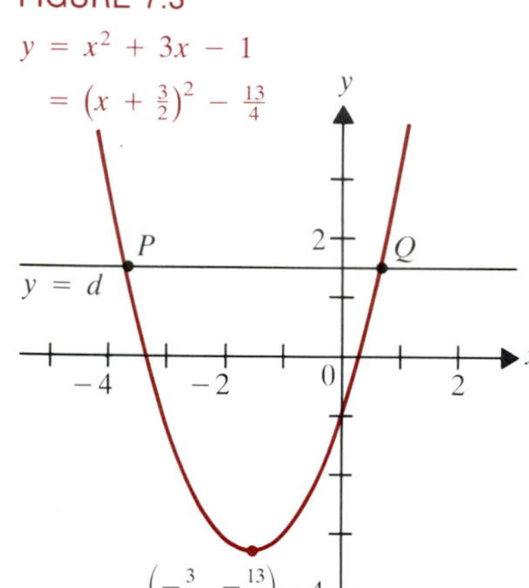

Solution. (a) In **Figure 7.3,** we show the graph of

$$f(x) = \left(x + \frac{3}{2}\right)^2 - \frac{13}{4}.$$

The figure shows that the horizontal-line test mentioned before Definition 7.1 fails for this function. For any $d > -13/4$, the line $y = d$ hits the graph in *two* points, labeled P and Q in the figure. Thus f is not a one-to-one function: $f(1) = f(-4) = 3$ shows that (4) fails for f.

(b) In this case, we use (5) to investigate the function without bothering to draw its graph. Suppose that $f(x_1) = f(x_2)$. To see whether this can happen only when $x_1 = x_2$, we proceed as follows.

$$\frac{x_1 + 2}{3x_1 - 1} = \frac{x_2 + 2}{3x_2 - 1} \to (x_1 + 2)(3x_2 - 1) = (x_2 + 2)(3x_1 - 1),$$

$$\cancel{3x_1x_2} + 6x_2 - x_1 - \cancel{2} = \cancel{3x_1x_2} + 6x_1 - x_2 - \cancel{2},$$

$$7x_2 = 7x_1 \to x_1 = x_2.$$

Hence, this f is one-to-one. To solve

$$y = \frac{x+2}{3x-1}$$

for x as a function of y, we first multiply the equation through by $3x - 1$. That leads to

$$3xy - y = x + 2 \to 3xy - x = y + 2 \to x(3y - 1) = y + 2,$$

$$x = \frac{y+2}{3y-1} = g(y),$$

if $y \neq 1/3$. ■

We can now give the formal definition of the inverse of a one-to-one function f.

7.2 DEFINITION

Let f be a one-to-one function with domain D and range R. The ***inverse function*** of f is the function f^{-1} with domain R and range D such that for all x in D

$$y = f(x) \text{ holds if and only if } x = f^{-1}(y). \tag{6}$$

Definition 7.2 introduces the standard notation f^{-1} for the inverse function, which we have been denoting by g up to now. To avoid confusion, we *never* use $f^{-1}(y)$ to denote the value of the reciprocal of the function f at a point y. $\left(\text{Instead, for the reciprocal we use the notation } \frac{1}{f(y)}.\right)$ We find f^{-1} as in Example 1(b):

To find f^{-1} for a given function f, try to solve the equation

$$y = f(x), \tag{7}$$

for x as a function of y. If that can be done to produce a formula

$$x = g(y), \tag{8}$$

then f^{-1} is the function g. For instance, in Example 1(b),

$$f^{-1}(y) = \frac{y+2}{3y-1}. \tag{9}$$

It is traditional to write x as the independent variable of any function, and y as the dependent variable. Thus we would normally write the formula for f^{-1} in Example 1(b) in the form

$$f^{-1}(x) = \frac{x+2}{3x-1},$$

rather than as in (9). In other words, once we solve $y = f(x)$ for x to get $x = f^{-1}(y)$, we then *interchange* x and y and write

$$y = f^{-1}(x).$$

Graphically, an interchange of x and y is equivalent to reflecting the xy-plane in the line $y = x$, since that reflection exchanges the x-axis with the y-axis (see **Figure 7.4**). Accordingly, we have the following basic fact about the graphs of $y = f(x)$ and $y = f^{-1}(x)$.

FIGURE 7.4

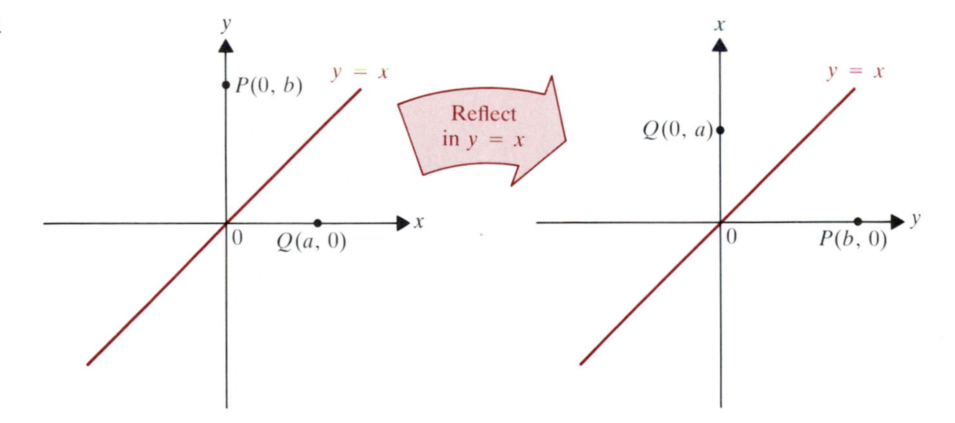

7.3 THEOREM

> The graph of $y = f^{-1}(x)$ is the reflection of the graph of $y = f(x)$ across the line $y = x$.

This means that the point (u, v) lies on the graph of $y = f(x)$ if and only if the point (v, u) lies on the graph of $y = f^{-1}(x)$. This property is illustrated in **Figure 7.5,** which shows the graphs of $y = x^2$ and $y = \sqrt{x}$.

FIGURE 7.5

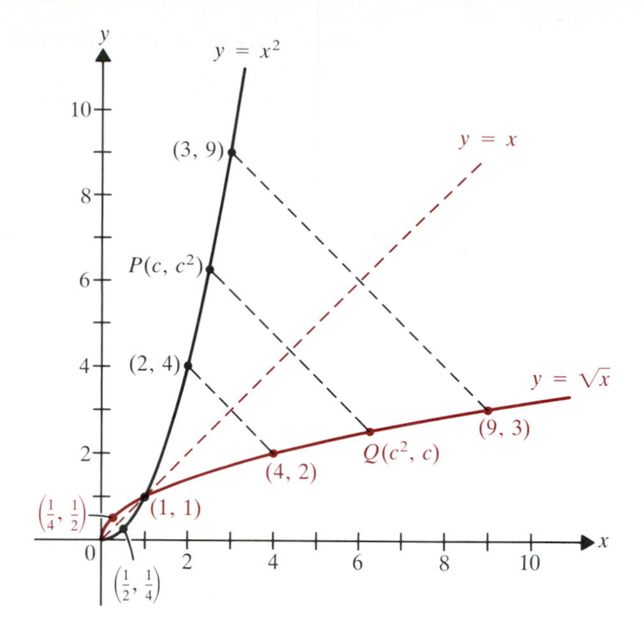

FIGURE 7.6

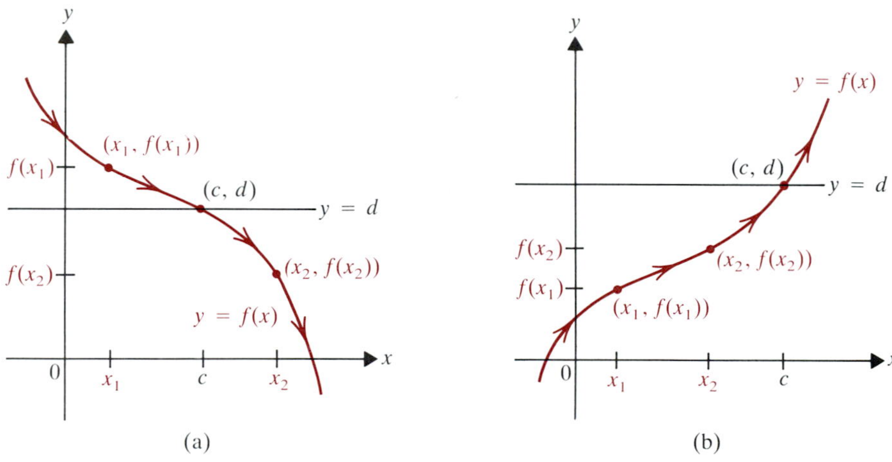

Figure 7.6 shows the graphs of two general types of functions that always have inverses, namely, *increasing* functions and *decreasing* functions. These are defined as follows.

7.4 DEFINITION A function f is called ***increasing*** (respectively, ***decreasing***) on a subset S of its domain if

$$\text{whenever } x_1 < x_2 \text{ in } S, \text{ then } f(x_1) < f(x_2) \qquad (\text{respectively, } f(x_1) > f(x_2)).$$

A function f is ***monotonic*** on S if it is either increasing on S or decreasing on S.

Definition 7.4 conforms to normal usage. When we speak of a declining unemployment rate, we expect to see a graph like that in Figure 7.6(a). If we hear that the stock market is rising, then we expect the graph of its index to be like the one in Figure 7.6(b). If f is an increasing function and we superimpose an arrowhead pointing toward the right on its graph, then the arrowhead will also point upward, as in Figure 7.6(b). For a decreasing function f, a similarly drawn arrowhead will point downward, as in Figure 7.6(a).

Note that if f is monotonic on its domain, then f is automatically one-to-one, since if $x_1 \neq x_2$, then either $x_1 < x_2$ or $x_1 > x_2$. Hence, $f(x_1) < f(x_2)$ or

$f(x_1) > f(x_2)$, so $f(x_1) \neq f(x_2)$. Thus, every function f that is monotonic on its entire domain has an inverse function.

Notice in Figure 7.5 that $f(x) = x^2$ is a continuous increasing function on its domain $[0, +\infty)$, and so is its inverse $g(x) = \sqrt{x}$. A result of advanced calculus says that *the inverse of a continuous monotonic function f is always continuous and monotonic in the same sense* (that is, increasing if f is increasing and decreasing if f is decreasing). Although we do not attempt to prove this, we will use it as needed.

Consider again the function $y = f(x) = x^2$ for $x \geq 0$ and its inverse function, $x = f^{-1}(y) = \sqrt{y}$. Note that for any x_0, $f'(x_0) = 2x_0$. Also, for $y_0 = f(x_0) = x_0{}^2$, we have

$$(D_y f^{-1})(y_0) = \frac{1}{2\sqrt{y_0}} = \frac{1}{2\sqrt{x_0{}^2}} = \frac{1}{2x_0},$$

since $x_0 \geq 0$. Thus

$$(f^{-1})'(y_0) = \frac{1}{f'(x_0)}. \tag{10}$$

The next result is that (10) is always true for a differentiable monotonic function f when $f'(x_0) \neq 0$.

7.5 THEOREM

Inverse Function Differentiation Rule. Let f be monotonic and differentiable on an interval I containing x_0, where x_0 is not an endpoint of I. If $f'(x_0) \neq 0$, then f^{-1} is differentiable at $y_0 = f(x_0)$ and

$$(f^{-1})'(y_0) = \frac{1}{f'(x_0)}. \tag{11}$$

Letting $y = f(x)$, we have in Leibniz notation

$$\left.\frac{dx}{dy}\right|_{y_0} = \frac{1}{\left.\dfrac{dy}{dx}\right|_{x_0}}. \tag{12}$$

Proof. Since f is differentiable on I, it is continuous there. In particular, f is continuous at x_0. Therefore, as $\Delta x \to 0$, so does $\Delta y = f(x_0 + \Delta x) - f(x_0)$. The above-quoted advanced calculus theorem tells us that f^{-1} is continuous at $y_0 = f(x_0)$. Therefore, as Δy approaches 0, so does $\Delta x = f^{-1}(y_0 + \Delta y) - f^{-1}(y_0)$. This shows that $\Delta x \to 0$ is equivalent to $\Delta y \to 0$. Note also that since f is monotonic, it is one-to-one. Thus $f(x_0 + \Delta x) \neq f(x_0)$ if $\Delta x \neq 0$. That is, $\Delta y = f(x_0 + \Delta x) - f(x_0) \neq 0$ if $\Delta x \neq 0$. We therefore have

$$(f^{-1})'(y_0) = \lim_{\Delta y \to 0} \frac{f^{-1}(y_0 + \Delta y) - f^{-1}(y_0)}{\Delta y} = \lim_{\Delta y \to 0} \frac{\Delta x}{\Delta y}$$

$$= \frac{1}{\displaystyle\lim_{\Delta y \to 0} \frac{\Delta y}{\Delta x}} \qquad \textit{by Theorem 5.4(b) of Chapter 1}$$

$$= \frac{1}{\displaystyle\lim_{\Delta x \to 0} \frac{\Delta y}{\Delta x}} = \frac{1}{f'(x_0)}. \quad \text{QED}$$

A real advantage of Theorem 7.5 is that if we just want the derivative of the inverse f^{-1} of a function f at a point y_0, then we don't need to solve $y = f(x)$ for x and differentiate the resulting formula. In fact, we may be unable to solve for x as a function of y and yet still be able to find $\left.\frac{dx}{dy}\right|_{y_0}$! The next example illustrates this.

Example 2

Find $\left.\frac{dx}{dy}\right|_{y_0=-9}$ if $y = x^5 + 3x^3 - 2x^2 + x - 2 = f(x)$.

Solution. We need only the *single* value x_0 such that $f(x_0) = -9$. A table of values for f can help locate such an x_0. From the table

x	-3	-2	-1	0	1	2	3
y	-347	-68	-9	-2	1	48	307

it follows that $x_0 = -1$. We have

$$\left.\frac{dy}{dx}\right|_{-1} = \left.(5x^4 + 9x^2 - 4x + 1)\right|_{x=-1} = 19.$$

Hence,

$$\left.\frac{dx}{dy}\right|_{y_0=-9} = \frac{1}{19}. \qquad \text{by (12)} \blacksquare$$

As promised at the beginning of the section, we can use Theorem 7.5 to extend the power rule for differentiation (Theorem 2.1),

$$D_x(x^n) = nx^{n-1},$$

to any rational number. In Section 1, we showed that it holds for any positive or negative integer n. Using Theorem 2.1, we can extend it to any nth-root function $f(x) = x^{1/n} = \sqrt[n]{x}$. First, notice that this is the inverse of the nth-power function $y = g(x) = x^n$, which is an increasing function on $(0, +\infty)$. If n is odd, then as **Figure 7.7** illustrates, the function g is also increasing on $(-\infty, 0)$. If n

FIGURE 7.7

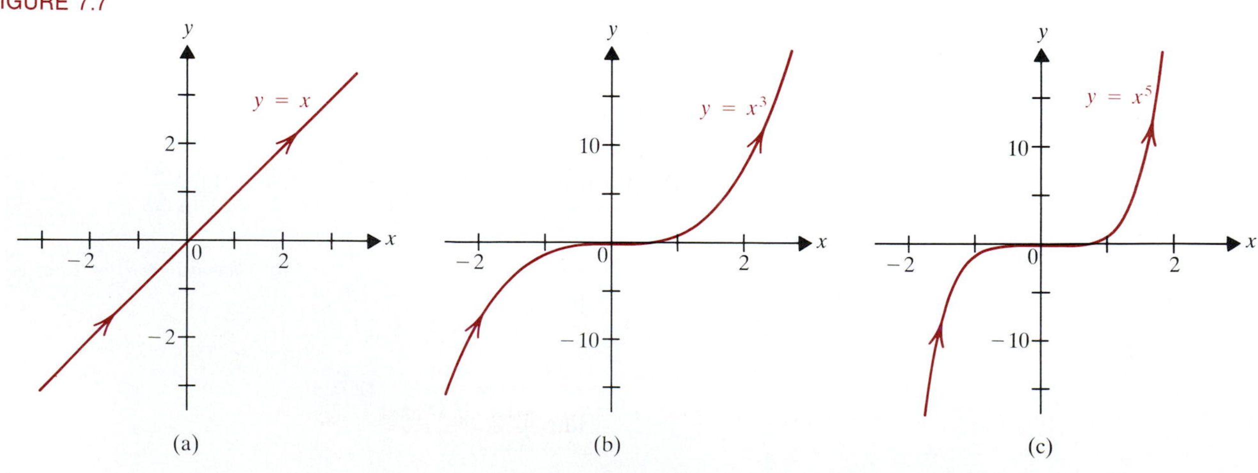

FIGURE 7.8

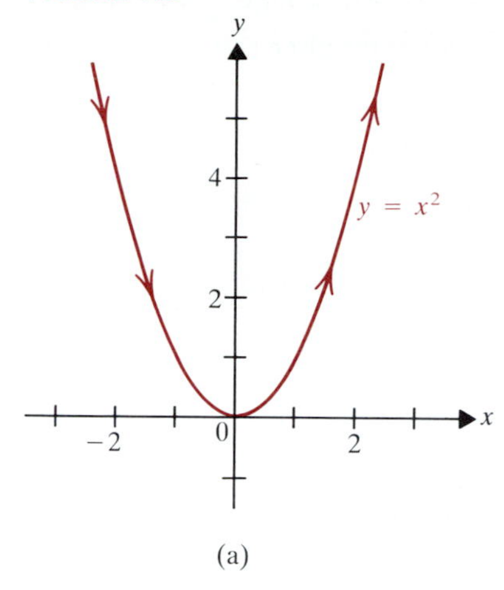

(a)

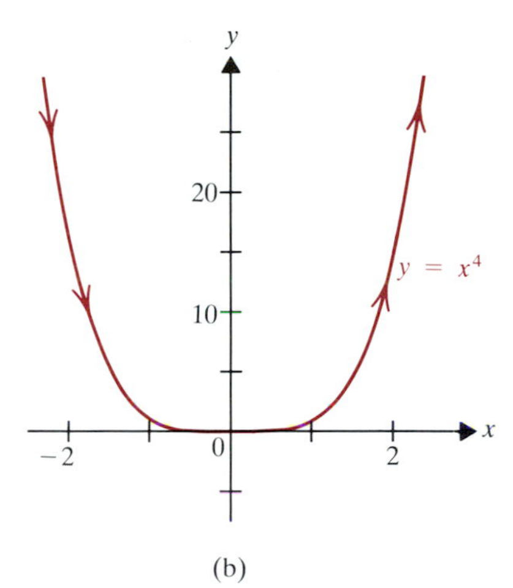

(b)

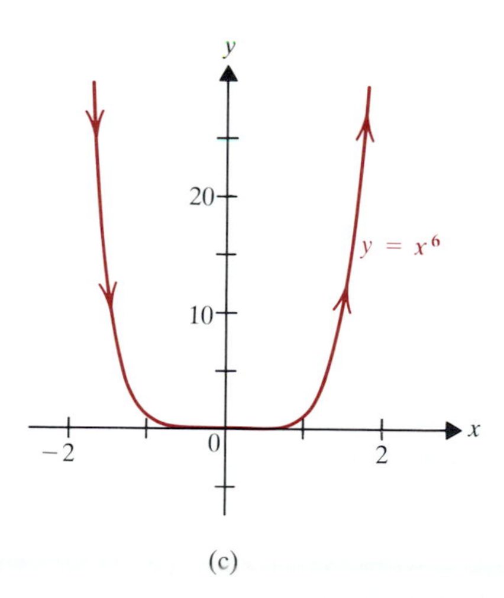

(c)

is even, then as **Figure 7.8** illustrates, g is decreasing on $(-\infty, 0)$. It follows then that g is monotonic on any interval I that does not include 0. On any such I, the power rule for integers applied to $x = g(y) = y^n$ gives

$$\frac{dx}{dy} = ny^{n-1}.$$

Hence, by (12) of Theorem 7.5 (read from right to left):

$$\frac{dy}{dx} = \frac{1}{\dfrac{dx}{dy}} = \frac{1}{ny^{n-1}}.$$

Making the substitution $y = x^{1/n}$, we obtain

$$\text{(13)} \qquad \frac{dy}{dx} = \frac{1}{n \cdot (x^{1/n})^{n-1}} = \frac{1}{n} \cdot \frac{1}{x^{1-1/n}} = \frac{1}{n} x^{1/n-1} \qquad \text{on } I.$$

Equation (13) says that the power rule holds for the function $f(x) = x^{1/n}$. We can extend the power rule to any rational power of x, $y = x^{m/n}$, easily now. For we can write

$$y = (x^{1/n})^m,$$

and apply the chain rule, the power rule for integers, and (13) to get

$$\frac{dy}{dx} = m(x^{1/n})^{m-1} \cdot \frac{d}{dx}(x^{1/n}) = mx^{m/n-1/n} \cdot \frac{1}{n} x^{1/n-1}$$

$$= \frac{m}{n} x^{m/n-1}.$$

We thus have extended the power rule to rational powers:

$$\text{(14)} \qquad D_x(x^{m/n}) = \frac{m}{n} x^{m/n-1}, \qquad \text{for any integers } m \text{ and } n,\ n \neq 0.$$

Strictly speaking, our derivation of (14) applies only to intervals I not containing 0, but in fact (14) is true for all x for which both $x^{m/n}$ and $x^{m/n-1}$ are defined, and we will freely use it without bothering to specify an interval I.

We can use (14) to differentiate the function k mentioned at the start of this section.

Example 3

(a) Find $k'(x)$ if $k(x) = \sqrt[3]{5x^2 + 3x - 1}$.

(b) Find $m'(x)$ if $m(x) = (7x^3 + 4x^2 - 3)^{3/2}$.

Solution. (a) We use the chain rule and (14), after writing

$$k(x) = (5x^2 + 3x - 1)^{1/3}.$$

We have

$$k'(x) = \frac{1}{3}(5x^2 + 3x - 1)^{-2/3}(10x + 3) = \frac{10x + 3}{3(5x^2 + 3x - 1)^{2/3}}$$

(b) $m'(x) = \dfrac{3}{2}(7x^3 + 4x^2 - 3)^{1/2}(21x^2 + 8x)$. ■

Exercises 2.7

In Exercises 1–10, determine whether the given function f is one-to-one. If so, find formulas for f^{-1} and $(f^{-1})'$ and give the domain of f^{-1}.

1. $f(x) = x^3 - 1$
2. $f(x) = 3x + 7$
3. $f(x) = 1/x$
4. $f(x) = \sqrt{4 - x^2}$, $x \in [0, 2]$
5. $f(x) = x^2 + 3x - 4$, $x \in [-4, 1]$
6. $f(x) = x^2 - 4x - 5$, $x \in [-1, 5]$
7. $f(x) = \dfrac{x-1}{x+1}$
8. $f(x) = \dfrac{2x+1}{x+3}$
9. $f(x) = \sqrt{1 - x^2}$, $x \in [-1, 1]$
10. $f(x) = 1 - x^2$, $x \in [-1, 1]$.

In Exercises 11–14, restrict the domain of f so that f has an inverse and find a formula for f^{-1}. Also give the domain of f^{-1} and a formula for $(f^{-1})'$.

11. $f(x) = x^2 + 1$
12. $f(x) = x^2 - 7$
13. $f(x) = \sqrt{16 - x^2}$
14. $f(x) = \sqrt{x^2 - 4}$

In Exercises 15–24, find $f'(x)$.

15. $f(x) = (x^2 + 3x - 1)^{3/2}$
16. $f(x) = (1 - 2x - x^2)^{4/3}$
17. $f(x) = (5x^3 - 3x + 7)^{2/3}$
18. $f(x) = (3 - 2x - x^3)^{2/5}$
19. $f(x) = \sqrt[3]{\dfrac{x-1}{x+1}}$
20. $f(x) = (\sin^2 x + \tan x)^{3/2}$
21. $f(x) = (\cos^3 2x - x^3)^{2/3}$
22. $f(x) = [1 + \sin(6x^2 + 1)]^{3/4}$
23. $f(x) = (5x^2 + \sqrt[3]{x^2 + 1})^{3/2}$
24. $f(x) = (x^3 - (2x^2 + 1)^{3/2})^{4/3}$

In Exercises 25–34, find dx/dy at the given point.

25. $y = x^3 - 3x^2 + 2x + 1$ at $(1, 1)$
26. $y = x^5 + 3x^3 - 2x^2 + x - 2$ at $(1, 1)$
27. $y = \sin x$ at $x = \pi/4$. How do you know there is an inverse function for $x \in [0, \pi/2]$?
28. $y = \cos x$ at $x = \pi/6$. How do you know there is an inverse function for $x \in [0, \pi/2]$?
29. $y = x^3 - 1$ at $y_0 = 7$.
30. $y = \sqrt{x^2 - 7}$, $x \in [\sqrt{7}, +\infty)$ at $y_0 = 3$.
31. $y = x^2 - 5x + 5$, $x \in [3, +\infty)$, $y_0 = -1$.
32. $y = 1 + x - x^2$, $x \in [2, +\infty)$, $y_0 = -1$.
33. $y = x^3 - 2x + 1$, $y_0 = -3$.
34. $y = 1 + 3x + x^3$, $y_0 = -3$.

35. The function $i(x) = x$ is called the *identity function* for composition. Show that $f \circ i = i \circ f = f$ for every function f.
36. Show that if f is an invertible function with domain D and range R, then $f^{-1} \circ f = i$ with restricted domain D and $f \circ f^{-1} = i$ with restricted domain R.
37. Use the result of Exercise 36 and the chain rule to show that if f^{-1} is defined at $y_0 = f(x_0)$, if $f'(x_0) \neq 0$, and if $(f^{-1})'(y_0)$ exists, then it must be given by (11).
38. Use the result of Exercise 36 and the chain rule to show that if $f'(x_0) = 0$ and f^{-1} exists, then f^{-1} is *not* differentiable at $y_0 = f(x_0)$. Give an example of such a function f.

2.8 Implicit Differentiation

So far we have dealt with functions explicitly given by a formula $y = f(x)$ for x in some set D. Sometimes two variables are related by a *functional equation* $F(x, y) = 0$. Often such an equation determines one or more differentiable functions $y = f(x)$. Occasionally, it may not determine any.

Example 1

Do the following equations determine y as one or more differentiable functions of x?

(a) $x^2 + y^2 + 1 = 0$ (b) $x^2 + y^3 + 1 = 0$ (c) $x^2 + y^2 - 1 = 0$

Solution. (a) For any real numbers x and y, $x^2 + y^2 \geq 0$. Thus $x^2 + y^2 + 1$ *cannot* be 0. Hence, this equation, which has the empty graph, cannot determine *any* functional relationships.

FIGURE 8.1

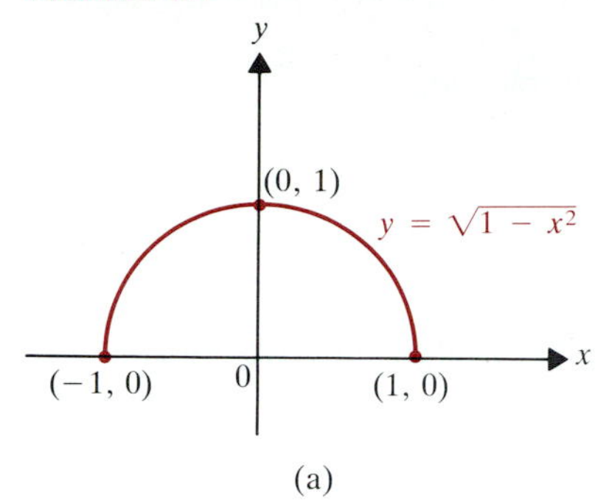

(a)

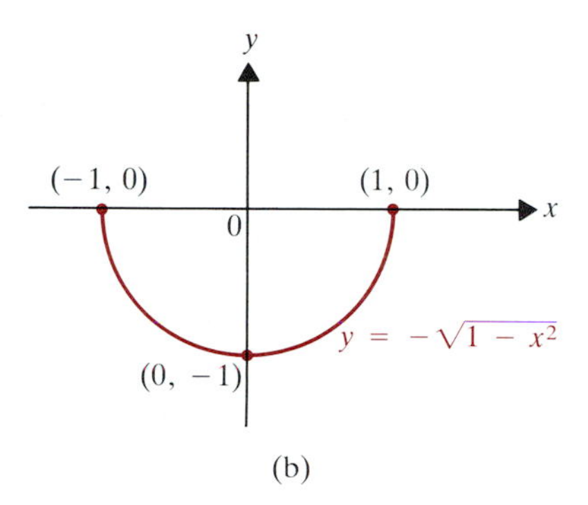

(b)

(b) Here, $y^3 = -1 - x^2$. Since every real number has exactly one real cube root, this defines y as a function of x:

$$y = (-1 - x^2)^{1/3} = -(1 + x^2)^{1/3} = f(x).$$

The domain D of f is the entire set $\mathbf{R}$ of real numbers, and f is differentiable on all of $\mathbf{R}$.

(c) In this case, $y^2 = 1 - x^2$. We cannot solve that equation uniquely for y as a single function of x, because there are two possibilities for any $x \in [-1, 1]$. Namely, $y = \sqrt{1 - x^2}$ or $y = -\sqrt{1 - x^2}$. However, each of these possibilities *does* define a function with domain $[-1, 1]$. Each of those functions is differentiable on $(-1, 1)$. The respective graphs are the upper half and the lower half of the circle $x^2 + y^2 = 1$, as shown in **Figure 8.1.** ■

In general, given a functional equation $F(x, y) = 0$, it may not be a simple matter to decide whether it does define y as one or more differentiable functions of x. That must be left for later courses. What we do in this section is show how to find a formula for dy/dx that holds if $F(x, y) = 0$ *does* define y as at least one differentiable function of x. If that functional equation happens to define y as more than one differentiable function of x, then our technique will produce a *single* formula for dy/dx that is true for *each* such function. How can that be? To see, we return to Example 1(c). We can differentiate each function there by the chain rule for any $x \in (-1, 1)$. If we do so, we get

$$y = (1 - x^2)^{1/2}, \qquad\qquad y = -(1 - x^2)^{1/2},$$

$$\frac{dy}{dx} = \frac{1}{2}(1 - x^2)^{-1/2}(-2x), \qquad \frac{dy}{dx} = -\frac{1}{2}(1 - x^2)^{-1/2}(-2x),$$

$$= -\frac{x}{(1 - x^2)^{1/2}}, \qquad\qquad = -\frac{x}{-(1 - x^2)^{1/2}},$$

$$= -\frac{x}{y}, \qquad\qquad = -\frac{x}{y}.$$

Thus for each of the two possible functions

$$y = y_1(x) = \sqrt{1 - x^2} \qquad \text{and} \qquad y = y_2(x) = -\sqrt{1 - x^2}$$

defined by the functional equation

$$x^2 + y^2 - 1 = 0,$$

it turns out that the derivative is given by the formula

(1) $$\frac{dy}{dx} = -\frac{x}{y}.$$

There is an easier way—called the ***method of implicit differentiation***—to find Formula (1). *If* $F(x, y) = 0$ *does define* $y = f(x)$ *as a differentiable function of* x, then we have

(2) $$F(x, f(x)) = 0.$$

We can regard the left side of (2) as a (perhaps complicated) function of x. Application of the differential operator D_x to each side of (1) gives

(3) $$D_x[F(x, f(x))] = D_x(0) = 0.$$

Since the left side of (1) is a function of x, we can use the chain rule to compute its derivative with respect to x, introducing some factors $f'(x)$ as we proceed.

We then get from (3) an equation involving x, $y = f(x)$, and $f'(x)$, which we can solve for $f'(x)$.

To illustrate the method, we return to Example 1(c).

Example 2

Find dy/dx if $x^2 + y^2 - 1 = 0$.

Solution. If this equation defines y as one or more differentiable functions of x, say $y = f(x)$, then the equation is

$$x^2 + f(x)^2 - 1 = 0.$$

We apply D_x to each side and use the chain rule on the function $y^2 = [f(x)]^2$ to get $D_x[f(x)^2] = 2f(x)^1 \cdot f'(x)$. So we obtain

$$D_x[x^2 + f(x)^2 - 1] = D_x(0) \rightarrow 2x + 2f(x)f'(x) - 0 = 0,$$

$$2f(x)f'(x) = -2x \rightarrow f'(x) = -\frac{x}{f(x)} = -\frac{x}{y},$$

if $f(x) \neq 0$. ■

This is *exactly* what we got before by first solving $x^2 + y^2 - 1 = 0$ for y as a function of x. Notice that we *must* exclude the places where $y = 0$, that is, where $x = \pm 1$. From Figure 8.1, those are places where the graphs of $y = y_1(x)$ and $y = y_2(x)$ have vertical tangents, so that y is not a differentiable function of x at such points.

If we apply the same sequence of steps to the equation in Example 1(a), then we come up with the same formula for dy/dx even though y is *not* a function of x! We emphasize that the method of implicit differentiation produces a formula for dy/dx that is correct **if** $F(x, y) = 0$ defines y as one or more differentiable functions of x. Should $F(x, y) = 0$ fail to define y as a differentiable function of x, then the formula for dy/dx is meaningless.

We next consider a more complicated functional equation that the method of implicit differentiation still handles quite well.

Example 3

If $y^7 - 11x^5y^6 - x^{15} + 11 = 0$ defines y as a differentiable function of x, then find dy/dx.

Solution. We proceed as in Example 2, regarding $y = f(x)$. Then

$$\frac{d}{dx}(y^7 - 11x^5y^6 - x^{15} + 11) = \frac{d}{dx}(0),$$

$$7y^6\frac{dy}{dx} - 55x^4y^6 - 11x^5\left(6y^5\frac{dy}{dx}\right) - 15x^{14} + 0 = 0, \qquad \textit{using the product rule and the chain rule}$$

$$\frac{dy}{dx} \cdot (7y^6 - 66x^5y^5) = 55x^4y^6 + 15x^{14},$$

$$\frac{dy}{dx} = \frac{55x^4y^6 + 15x^{14}}{7y^6 - 66x^5y^5}, \tag{4}$$

if $7y^6 - 66x^5y^5 \neq 0$. ■

The real value of the technique of implicit differentiation is that it may provide the *only* feasible way for us to estimate the value of y in a complicated functional equation $F(x, y) = 0$ near a known point (c, d) that satisfies the equation. In Example 3, for instance, it is not feasible to solve the equation explicitly for y as a function of x. (Indeed, there is no general formula, like the quadratic formula, that allows us to solve all polynomial equations $P(x, y) = 0$ of degree 5 or higher for y as a function of x.) But we *can* use the tangent approximation to estimate y for x near a point $x = c$ where we know

$$y = d \qquad \text{and} \qquad \left.\frac{dy}{dx}\right|_c .$$

This provides a solution to Problem 3 of the Prologue.

Example 4

Estimate the value of y when $x = 1.002$ if $y^7 - 11x^5y^6 - x^{15} + 11 = 0$.

Solution. We observe that (1, 1) satisfies the given equation. From (4) in Example 3, when $x = 1$ and $y = 1$, we have

$$\frac{dy}{dx} = \frac{55 + 15}{7 - 66} = -\frac{70}{59} \approx -1.1864.$$

Since $y = 1$ when $x = 1$, we have $f(1) = 1$. So the tangent approximation with $x = 1.002$ and $c = 1$ gives

$$\begin{aligned} f(x) &\approx f(c) + f'(c)(x - c) = 1 + (-1.1864)(0.002) \\ &\approx 1 - 0.0024 = 0.9976. \end{aligned} \tag{5}$$

Hence, when $x = 1.002$, $y \approx 0.998$. ■

If we need to continue approximating y as x varies from 1.002 to 1.004, 1.006, 1.008, 1.01, . . . , we can do so by using approximately obtained values for x and y in (4) with the tangent approximation (5). The arithmetic is well suited to a calculator (especially a programmable one) or a computer but is rather tedious for hand calculation. It provides a way of generating an *approximate* graph of a complicated equation like that in Example 4.

Example 5

In Example 4, estimate the value of y at $x = 1.002$, 1.004, 1.006, 1.008, and 1.01.

Solution. We use (4) and a calculator to compute dy/dx from (3) with $c = 1.002$, $d = 0.998$. We get $dy/dx \approx -1.1883$, and so

$$f(1.004) \approx 0.998 + (-1.1883)(0.002) \approx 0.996.$$

Using $c = 1.004$ and $d = 0.996$, we get $dy/dx \approx -1.1903$, and

$$f(1.006) \approx 0.996 + (-1.1903)(0.002) \approx 0.994.$$

Then $c = 1.006$ and $d = 0.994$ give $dy/dx \approx -1.1926$, and

$$f(1.008) \approx 0.994 + (-1.1926)(0.002) \approx 0.992.$$

Finally, $c = 1.008$ and $d = 0.992$ give $dy/dx \approx -1.1951$, and $f(1.01) \approx 0.990$. ■

Note: Since the input data for x and y was expressed only to three decimal places, we rounded all our answers to three decimal places. Table 1, which summarizes the results of Example 5, was generated by a computer.

TABLE 1

x	y	dy/dx	New y from (4)
1.	1.	−1.18644	.997627
1.002	.997627	−1.18835	.99525
1.004	.99525	−1.19052	.992869
1.006	.99287	−1.19296	.990483
1.008	.990483	−1.19568	.988092
1.01	.988092	−1.19869	.985695

Exercises 2.8

In Exercises 1–20, find a formula for dy/dx that holds in case the given equation defines y as a differentiable function of x.

1. $x^2 - y^2 = 9$

2. $4x^2 + 9y^2 = 36$

3. $x^2y + y^2x = 1$

4. $x^3y + y^3x = 3$

5. $x^3 + 2x^2y + 3xy^2 - y^3 = 8$

6. $x^4 - 5x^2y^2 + y^4 = 11$

7. $\sqrt{x} + \sqrt{y} = 5$

8. $\sqrt[3]{x} - \sqrt[3]{y} = 3$

9. $3x^2y^2 = y^3 - x^3 + 1$

10. $5x^2y^3 = xy^4 - x^2y + 2$

11. $\dfrac{2}{x} - \dfrac{3}{y} = xy + 1$

12. $\dfrac{5}{\sqrt{x}} - \dfrac{2}{\sqrt{y}} = x^2y^2 - 1$

13. $(xy^2 - x^2y)^2 = x^3 - y^3$

14. $(x^3y + xy^3)^2 = x^2y - xy$

15. $\sqrt{xy + 1} + \sqrt[3]{xy - 1} = xy - 1$

16. $\sqrt[3]{x^2y + x} = \sqrt{xy + 3}$

17. $y \sin^3 x + x \cos^2 y = 3$

18. $x^2 \tan^2 x - y^2 \sec^2 y = 1$

19. $\cos^2(x^3y) - \sin^3(xy^2) = x^2y^2$

20. $\sin\sqrt{xy} - \cos^2\sqrt{xy} = 8xy^2$

In Exercises 21–30, use the tangent approximation to approximate y for the given value of x.

21. $x^2y^2 - 3x - y^3 = 19$ at $x = 2.984$.
[Note that (3, 2) satisfies the functional equation.]

22. $x^2y^3 - x^2 + y^2x = 11$ at $x = 0.992$.
[Note that (1, 2) satisfies the equation.]

23. The equation in Example 4 at 1.0059.

24. $y^6 + 3x^2y^4 + 3x^4y^2 + x^6 - 8 = 0$ at $x = 0.97$.
[Note that (1, 1) satisfies the equation.]

25. $xy + x^3 + y^3 = 13$ at $x = -2.005$.
[Note that (−2, 3) satisfies the equation.]

26. $x^2y^2 + x^2y + x^3y = 3$ at $x = 0.98$.
[Note that (1, 1) satisfies the equation.]

27. $x^2 + 2xy + 3y^2 = 6$ at $x = 1.04$.
[Note that (1, 1) satisfies the equation.]

28. $x^2 - xy + 3y^2 = 15$ at $x = -1.013$.
[Note that (−1, 2) satisfies the equation.]

29. $5x + 4x^3 - y^4 - 3y + 1 = 0$ at $x = 0.971$.
[Note that (1, −2) satisfies the equation.]

30. $x^3 - 3xy^2 + y^3 = 1$ at $x = 1.97$.
[Note that (2, −1) satisfies the equation.]

In Exercises 31–34, use the approach of Example 5 to estimate the value of y at the given values of x.

31. $x = 2.968, 2.952, 2.936, 2.92$ in Exercise 21.

32. $x = 0.984, 0.976, 0.968, 0.96$ in Exercise 22.

33. $x = -2.010, -2.015, -2.020, -2.025$ in Exercise 25.

34. $x = 1.08, 1.12, 1.16, 1.20$ in Exercise 27.

35. The *hypocycloid of four cusps* is the graph of the equation $x^{2/3} + y^{2/3} = a^{2/3}$. Show that this curve is tangent to the x-axis at $x = a$ and $x = -a$.

36. The graph of $(x^2 + y^2)^2 = 4(x^2 - y^2)$ is called a *lemniscate*. Find a formula for the slope of the tangent to this curve at (c, d).

37. Find a formula for the tangent line to the ellipse

$$\frac{x^2}{a^2} + \frac{y^2}{b^2} = 1$$

at (c, d).

38. Find a formula for the tangent line to the hyperbola

$$\frac{x^2}{a^2} - \frac{y^2}{b^2} = 1$$

at (c, d).

39. Show that the circles $(x - 2)^2 + y^2 = 2$ and $x^2 + (y - 2)^2 = 2$ are tangent at the point (1, 1).

40. Show that the circles $(x - 4)^2 + y^2 = 8$ and $x^2 + (y - 4)^2 = 8$ are tangent at the point (2, 2).

41. Show that the circle $x^2 + y^2 = 2$ and the hyperbola $xy = 1$ are tangent at the points where they intersect.

42. Show that the parabolas $2x + y^2 = 1$ and $y^2 - 2x = 1$ intersect at right angles.

43. For which points on the ellipse $4x^2 + y^2 = 1$ does the tangent line have slope 1?

44. For which points on the ellipse $4x^2 + y^2 = 1$ does the tangent line pass through the point (3, 0)?

2.9 Repeated Differentiation

The derivative function f' is obtained by differentiating a function f. Its domain consists of all points c where f is differentiable. Since f' is itself a function, we can differentiate it as well.

9.1 DEFINITION The ***second derivative*** f'' of a function f is defined by $f''(x) = (f')'(x)$ at every point x where f' is differentiable. That is, for a real number c in the domain of f',

$$f''(c) = \lim_{h \to 0} \frac{f'(c+h) - f'(c)}{h}, \tag{1}$$

if this limit exists.

For example if $f(x) = \sin x$, then $f'(x) = \cos x$. Thus $f''(x) = D_x(\cos x) = -\sin x$.

There is a Leibniz notation for the second derivative function f'', namely,

$$\frac{d^2y}{dx^2}, \tag{2}$$

which is read "d second $y\,dx$ (squared)" or "d two $y\,dx$ two". The odd placement of the superscripts comes from the symbolic calculation

$$\frac{d^2y}{dx^2} = \frac{d}{dx}\left(\frac{dy}{dx}\right) = \frac{d(dy)}{dx\,dx} = \frac{d^2y}{dx^2}.$$

This time, however, we do **not** give any independent meaning to d^2y or dx^2: Regard (2) as a single, indivisible symbol. Other notations for $f''(x)$ include

$$D_x{}^2(f(x)), \qquad D^2f(x), \qquad \text{and} \qquad \frac{d^2}{dx^2}(f(x)).$$

We will usually use either $f''(x)$ or d^2y/dx^2.

No new theory or formulas are needed to calculate second derivatives.

Example 1

If $f(x) = x^2 \sin x$, then find $f''(x)$.

Solution. Since $f'(x) = 2x \sin x + x^2 \cos x$, we have

$$\begin{aligned} f''(x) &= (2 \sin x + 2x \cos x) + (2x \cos x - x^2 \sin x) \\ &= 2 \sin x + 4x \cos x - x^2 \sin x. \ \blacksquare \end{aligned}$$

We can find the second derivative of an implicitly defined function by differentiating the expression for dy/dx that we get from the method of implicit differentiation. The next example illustrates how the result can often be simplified.

Example 2

If $x^2 + y^2 - 1 = 0$ defines y as a differentiable function of x, then find d^2y/dx^2.

Solution. In Example 2 of the last section, we found $dy/dx = -x/y$. Differentiating this, we obtain

$$\frac{d^2y}{dx^2} = \frac{d}{dx}\left(-\frac{x}{y}\right) = -\frac{y \cdot 1 - x \cdot \dfrac{dy}{dx}}{y^2} = -\frac{y - x\left(-\dfrac{x}{y}\right)}{y^2}$$

$$= -\frac{y^2 + x^2}{y^3} = -\frac{1}{y^3}.$$

(The last step results from the defining equation, which says $x^2 + y^2 = 1$.) ■

There is an application of the second derivative that goes back to Newton and is still of considerable importance in physics. Consider a body whose position y at time t is given by $y = f(t)$. Recall that we defined instantaneous velocity at $t = t_0$ to be

$$v = \left.\frac{dy}{dt}\right|_{t_0} = f'(t_0). \tag{3}$$

In physics, the *acceleration* of a moving body at time $t = t_0$ is defined to be the instantaneous rate of change of velocity with respect to time.

9.2 DEFINITION If the position of a moving body at time t is given by $y = f(t)$, then the ***acceleration*** on the body at $t = t_0$ is

$$a(t_0) = \left.\frac{dv}{dt}\right|_{t_0} = \left.\frac{d^2y}{dt^2}\right|_{t_0} = f''(t_0). \tag{4}$$

The acceleration is the rate of change of dv/dt, and so measures the rate at which the velocity is increasing or decreasing. It thus shows how fast the moving body is speeding up or slowing down.

Newton's *second law of motion* describes the change in motion produced by a force F acting on a body of mass m. If the motion is along a straight line, then the law takes the form

$$F = ma = m\frac{d^2y}{dt^2}. \tag{5}$$

Example 3

A body falling freely from rest falls a distance $y = \frac{1}{2}gt^2$ in t seconds, where $g \approx 9.80 \text{ m/s}^2 \approx 32.2 \text{ ft/s}^2$ is the acceleration due to gravity. What is the force acting on the body?

Solution. The velocity of fall after t seconds is $v = y' = \frac{1}{2}g \cdot 2t = gt$. Thus

$$a = y'' = v' = g.$$

The acceleration of a freely falling body is just the acceleration due to gravity. Thus by (5) the force acting on the body is

$$F = ma = mg,$$

which is just its *weight*. (The term *free fall* comes from the fact that the body is free of all forces except its own weight.) ■

Velocity and acceleration are also used to analyze more complex straight-line motion, as the next example illustrates.

Example 4

Discuss the motion of a body moving along the x-axis whose directed distance from the origin at time t is

$$x = t^3 - 2t^2 + t - 3. \tag{6}$$

Solution. We have

$$v = \frac{dx}{dt} = 3t^2 - 4t + 1, \tag{7}$$

and

$$a = \frac{d^2x}{dt^2} = 6t - 4. \tag{8}$$

Table 1 shows the position of the body as computed from (6), starting from $t = 0$. It shows that the body starts moving toward the origin, reverses itself briefly by returning to -3, and then moves steadily to the right.

TABLE 1

t	0	1/2	1	2	3	4	5	10
x	-3	-2.875	-3	-1	9	33	77	807

To gain more insight into the motion, we use (7):

$$v = 3t^2 - 4t + 1 = (3t - 1)(t - 1).$$

Positive values of v indicate motion to the right; negative values mean motion to the left. Table 2 refines the data from Table 1 by specifying the direction of motion at any time. It also shows exactly when the direction of motion changes: at $t = 1/3$ and $t = 1$. **Figure 9.1** is a schematic representation of the information

TABLE 2

t:	$(-\infty, 1/3)$	1/3	$(1/3, 1)$	1	$(1, +\infty)$ $\rightarrow$ **R**
$3t - 1$:	$-$		$+$		$+$
$t - 1$:	$-$		$-$		$+$
v:	$+$	0	$-$	0	$+$
Conclusion:	Motion toward right	Motion stops	Motion toward left	Motion stops	Motion toward right

FIGURE 9.1

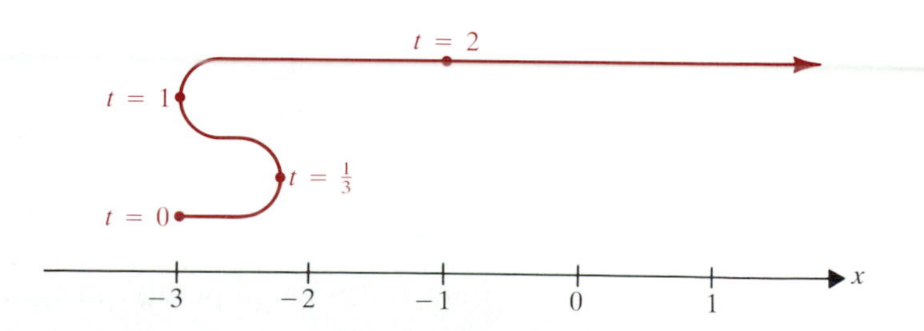

in Tables 1 and 2. It shows the direction of motion and the location of the body at various times.

Finally, (8) says that the acceleration is positive when $6t - 4 > 0$, that is, for $t > 2/3$, and is negative when $t < 2/3$. Thus the velocity of the body is decreasing up to $t = 2/3$. (The velocity becomes 0 at $t = 1/3$ and then becomes more negative, indicating motion toward the left.) After $t = 2/3$, the velocity begins to increase, reaching 0 at $t = 1$ (from Table 2) and then becoming more and more positive. ■

Just as we defined the second derivative f'' to be the derivative of f', so we can define third derivatives, fourth derivatives, and so on. In Chapter 9, we will see how these are used to obtain polynomial approximations to differentiable functions. For now, we define them just as in Definition 9.1.

9.3 DEFINITION

If $f''(x)$ exists on an interval I containing c, then the ***third derivative of f at c*** is

$$f'''(c) = f^{(3)}(c) = \left.\frac{d^3y}{dx^3}\right|_c = \lim_{h\to 0}\frac{f''(c+h) - f''(c)}{h}, \tag{9}$$

if this limit exists. We write $f'''(x)$ for the third derivative function. If the nth derivative $f^{(n)}(x)$ exists on an interval I containing c, for some integer $n \geq 3$, then the ***(n + 1)-st derivative of f at c*** is

$$f^{(n+1)}(c) = \left.\frac{d^{n+1}y}{dx^{n+1}}\right|_c = \lim_{h\to 0}\frac{f^{(n)}(c+h) - f^{(n)}(c)}{h}, \tag{10}$$

provided that this limit exists.

The prime notation of (9) quickly becomes unwieldy. An alternative is given in (10): superscripted arabic numerals in parentheses. There is no hard and fast rule for when the prime notation is dropped. You may see $f''''(c)$, for instance, but almost certainly will see $f^{(5)}(c)$ instead of $f'''''(c)$.

For polynomial functions, things wind down to 0 by the time you differentiate more times than the degree of the polynomial.

Example 5

Find $f^{(n)}(x)$ for all positive integers n if $f(x) = x^3 - 2x^2 + x - 3$.

Solution. We have

$$f'(x) = 3x^2 - 4x + 1, f''(x) = 6x - 4$$
$$f'''(x) = 6, f^{(4)}(x) = 0.$$

Thus for all $n \geq 4$, $f^{(n)}(x) = 0$. ■

By contrast, transcendental functions can be differentiated arbitrarily often without obtaining the identically zero function.

Example 6

Find $f^{(n)}(x)$ for all positive integers n if $f(x) = \cos x$.

Solution. Here $f'(x) = -\sin x$, so $f''(x) = -\cos x$, $f'''(x) = \sin x$, $f^{(4)}(x) = \cos x$. Since $f^{(4)}(x) = f(x) = \cos x$, no matter how many times we differentiate f, we get one of the four functions f, f', f'', or f''', which repeat in cycles. We therefore have

$$f(x) = f^{(4)}(x) = f^{(8)}(x) = f^{(12)}(x) = \cdots = \cos x,$$
$$f'(x) = f^{(5)}(x) = f^{(9)}(x) = f^{(13)}(x) = \cdots = -\sin x,$$
$$f''(x) = f^{(6)}(x) = f^{(10)}(x) = f^{(14)}(x) = \cdots = -\cos x,$$
$$f'''(x) = f^{(7)}(x) = f^{(11)}(x) = f^{(15)}(x) = \cdots = \sin x. \quad ■$$

HISTORICAL NOTE

Leibniz actually did define second differentials but was not able to do so in a way that proved satisfactory. He said d^2y was the *differential* of dy. That is, $d(dy) = d(f'(x)\,dx)$. While he was not able to explain why this should yield $f''(x)\,dx^2$, we can do so as follows. First, define d^2x to be 0. Writing $(dx)^2$ as dx^2, we then have

$$\begin{aligned} d(f'(x)\cdot dx) &= f'(x)\,d(dx) + d(f'(x))\,dx && \textit{by Exercise 38(a) of Section 5} \\ &= f'(x)\,d^2x + (f''(x)\,dx)\,dx = 0 + f''(x)(dx)^2. \end{aligned}$$

That is,

$$d^2y = f''(x)\,dx^2.$$

Hence, dividing by dx^2, we can get $f''(x)$ to be an *actual quotient* d^2y/dx^2. The real reason for not giving d^2y and dx^2 independent meanings in (2) is that, unlike dy and dx, which are used in discussing incremental changes, d^2y and dx^2 do not possess useful interpretations.

Exercises 2.9

In Exercises 1–14, find the second derivative of the given function.

1. $f(x) = \frac{1}{2}x^4 - \frac{3}{2}x^3 + 5x^2 - 4x + \pi$

2. $f(x) = ax^3 + bx^2 + cx + d$

3. $f(x) = \tan x$

4. $g(x) = \sec x$

5. $f(x) = (x^2 - 5x + 1)^3$

6. $g(x) = (2x^2 + 3x - 4)^3$

7. $h(x) = x^{3/2} - \dfrac{2}{\sqrt{x}}$

8. $k(x) = \sqrt{x} = \dfrac{1}{\sqrt{x}}$

9. $m(x) = \sqrt{x^2 - x + 2}$

10. $n(x) = (x^2 + 3x - 2)^{2/3}$

11. $p(x) = x^2 \sin 3x$

12. $q(x) = x^2 \cos 2x$

13. $r(x) = \sin^3(x^2 + 1)$

14. $s(x) = \cos^3(x^2 - 1)$

In Exercises 15–22, find d^2y/dx^2 if the given equation defines y as a differentiable function of x.

15. $x^2 - y^2 = 4$

16. $y^2 - x^2 = 9$

17. $x^3 + y^3 = 8$

18. $x^3 - y^3 = 27$

19. $x^2 + xy + y^2 = 1$

20. $x^2 - 2xy - y^2 = 4$

21. $x^{2/3} + y^{2/3} = 1$

22. $x^{2/3} + y^{2/3} = 4$

In Exercises 23–32, discuss the motion of a body along the x-axis if its directed distance x from the origin is given by the function f.

23. $f(t) = t^2 - 8t + 5$

24. $f(t) = 4t - t^2 - 1$

25. $f(t) = t^3 - 3t^2 + 2$

26. $f(t) = 3t^2 - t^3 + 3$

27. $f(t) = \frac{2}{3}t^3 - \frac{7}{2}t^2 + 3t + 4$

28. $f(t) = t^3 - 12t^2 + 36t - 10$

29. $f(t) = t^4 - 8t^2 + 5$

30. $f(t) = t^4 - 6t^2 + 5$

31. $f(t) = \sin t$

32. $f(t) = \cos 2t$

In Exercises 33–38, find all derivatives of the given function.

33. $f(x) = 5x^3 - 8x^2 + x + 3$

34. $g(x) = 8x^4 - 5x^2 + 6$

35. $h(x) = \sin x$

36. $k(x) = \cos 2x$

37. $m(x) = \dfrac{1}{x}$

38. $n(x) = \sqrt{x}$

39. If $f(x) = g(x) \cdot h(x)$, then give a formula for $f''(x)$.

40. In Exercise 39, give a formula for $f'''(x)$.

41. In Exercise 39, give a formula for $f^{(4)}(x)$.

42. From your results in Exercises 39–41, construct a formula for $f^{(n)}(x)$ in Exercise 39, in terms of the derivatives of g and h. (The result is known as *Leibniz's rule.* It can be proved by mathematical induction in the same way that the binomial theorem of algebra is proved.)

43. If $f(x) = (g \circ h)(x)$, then use the chain rule to give a formula for $f''(x)$.

44. If $f(x) = x^n$ for any positive integer n, then show that $f^{(n)}(x) = n! = n(n-1)(n-2) \ldots 3 \cdot 2 \cdot 1$.

45. If f is an even function such that f'' exists, show f'' is also an even function. (See Exercise 39 of Section 1.)

46. If f is an odd function such that f'' exists, then show that f'' is also an odd function. (See Exercise 40 of Section 1.)

47. *Simple harmonic motion* is motion for which the acceleration satisfies

$$a = \frac{d^2x}{dt^2} = -\omega^2 x, \tag{11}$$

where ω is a constant (called the *frequency*) and x is the displacement (distance of the moving body from its initial position). Notice that the force experienced is in the opposite direction from the displacement and is a linear function of the displacement. Show that any function of the form $x = A\cos\omega t + B\sin\omega t$ where A and B are constants satisfies (11). Here, ω, ϕ_0, and ϕ_1 are constants.

48. In Exercise 47, show that any function of the form **(a)** $x = a\cos(\omega t - \phi_0)$ or **(b)** $x = a\sin(\omega t - \phi_1)$ satisfies (11).

49. Show that $f(x) = x^{3/2}$ is differentiable at $c = 0$ but has no second derivative there.

50. Repeat Exercise 49 for $f(x) = x^{4/3}$.

51. Repeat Exercise 49 for $f(x) = x|x|$.

52. Repeat Exercise 49 for $f(x) = x^2 \sin(1/x)$.

2.10 The Mean Value Theorem for Derivatives

This section presents a fundamental result about the derivative that underlies many of the applications of differentiation to be discussed in the next chapter. It seems to have first been given by Lagrange in the latter part of the eighteenth century.

Suppose that f is continuous on a closed interval $[a, b]$ and differentiable on the open interval (a, b). Consider the secant S joining the points $A(a, f(a))$ and $B(b, f(b))$ in **Figure 10.1.** The mean value theorem states that for some point c in the open interval (a, b), the tangent line T to the graph of $y = f(x)$ at $(c, f(c))$ is parallel to the secant. In Figure 10.1, we have labeled a point $C(c, f(c))$, where the tangent seems to have this property. Since the tangent line at $(c, f(c))$ has

FIGURE 10.1

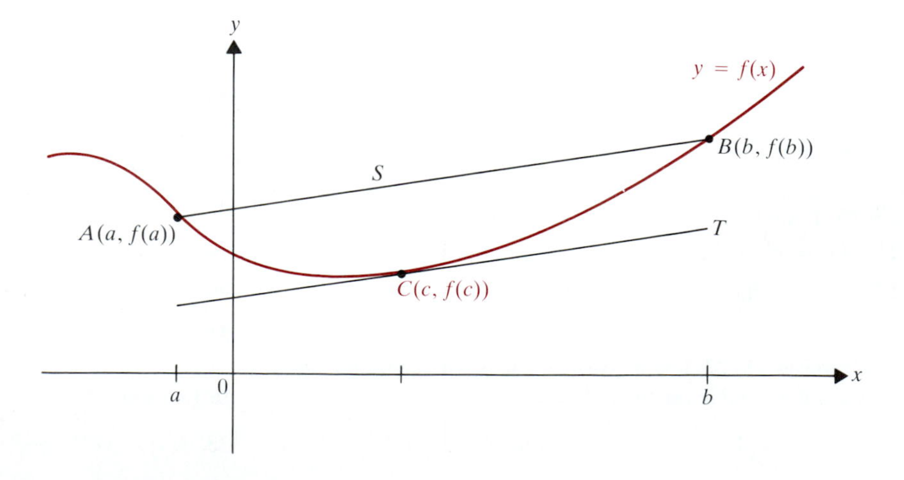

slope $f'(c)$ and the secant has slope

$$\frac{f(b)-f(a)}{b-a},$$

the mean value theorem asserts that for some $c \in (a, b)$,

$$f'(c) = \frac{f(b)-f(a)}{b-a}. \tag{1}$$

Mean is another word for *average*. The mean value theorem says that for a function continuous on $[a, b]$ and differentiable on (a, b), its *average* rate of change over the interval $[a, b]$ is equal to its *instantaneous* rate of change $f'(c)$ for at least one number c between a and b. That is, the ***mean*** rate of change $[f(b) - f(a)]/(b - a)$ is the exact ***value*** of the derivative $f'(c)$ for some $c \in (a, b)$.

Before presenting the theorem formally and proving it, we suggest a perhaps startling application, which is easy to put into practice. Prior to 1973, the Kansas Turnpike between Topeka and Wichita had the highest posted speed limit in the United States: 80 miles per hour (about 130 km/h). After the imposition of a national speed limit of 55 miles per hour (about 90 km/h), it became difficult to enforce the drastic curtailment in speed on that road. One simple means of enforcement can be justified by the mean value theorem.

Example 1

Suppose that toll tickets are dispensed to motorists by a machine that prints the time of release on the back. At the end of the 137-mile turnpike, toll collectors are instructed to collect, in addition to the normal toll, a speeding fine from any motorist who completes a trip in less than 2 hours and 17 minutes. Why is it true that any motorist who makes a trip that quickly must at some time have been guilty of speeding?

Solution. Let the number of miles the motorist has traveled be $y = f(t)$, where t is the number of hours since the toll tickets were released by the machine. Then $f(0) = 0$ and $f(b) = 137$, where b is the number of hours needed to reach the end of the turnpike. If a motorist reaches the end in 2 hours and 17 minutes, then $b = 2\frac{17}{60} = 137/60$. Thus the motorists's *average* velocity on the turnpike is

$$m = \frac{f\left(\frac{137}{60}\right) - f(0)}{\frac{137}{60} - 0} = \frac{137}{\frac{137}{60}} = 60$$

miles per hour. At each instant during the trip, the motorist has *some* velocity (possibly even 0). Hence, the function f is differentiable over the interval $(0, 137/60)$. It is also continuous on $[0, 137/60]$. Therefore the mean value theorem guarantees that at some instant of time, say $c \in (0, 137/60)$, the motorist's instantaneous velocity $f'(c)$ was 60 miles per hour. Thus, at least once, the motorist was traveling at least five miles per hour over the legal limit and, hence, was guilty of speeding. ■

FIGURE 10.2

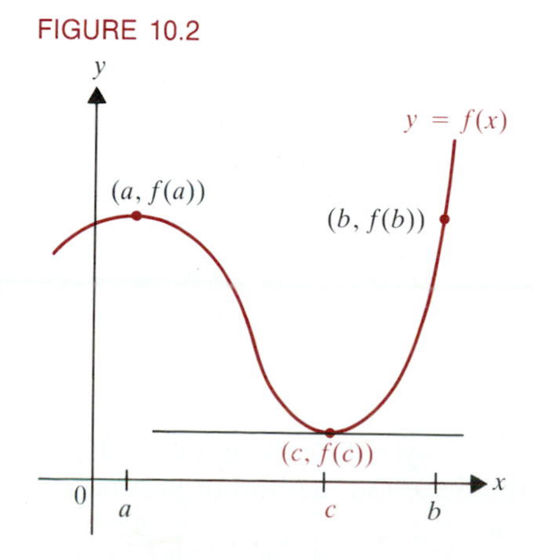

We prove the mean value theorem by first proving the following weaker version (see **Figure 10.2**).

10.1 THEOREM **Rolle's Theorem.** Suppose that f is continuous on $[a, b]$ and differentiable on (a, b). If $f(a) = f(b) = 0$, then there is at least one point $c \in (a, b)$ such that $f'(c) = 0$.

Proof. If $f(x) = 0$ for all x in $[a, b]$, then $f'(x) = 0$ for all $x \in (a, b)$. So $f'(c) = 0$ for *any* c in (a, b). Suppose next that f is not identically 0 on $[a, b]$. Then by Theorem 7.7 of Chapter 1, f has a maximum and a minimum on $[a, b]$. It thus has a positive maximum or a negative minimum (or both) on (a, b). Suppose that f takes on its maximum at the point $c \in (a, b)$. [A similar argument applies if f has its minimum inside (a, b). See Exercise 36.] To show that $f'(c) = 0$, consider

$$f'(c) = \lim_{h \to 0} \frac{f(c + h) - f(c)}{h}. \tag{2}$$

For small values of h, note that

$$f(c + h) \leq f(c),$$

since $f(c)$ is the maximum of f on $[a, b]$. Thus the numerator of (2) is ≤ 0 for small h. Hence, for small positive values of h, the difference quotient in (2) is ≤ 0. By Theorem 5.8(b) of Chapter 1 then,

$$\lim_{h \to 0^+} \frac{f(c + h) - f(c)}{h} \leq 0.$$

Similarly, for small negative values of h, the difference quotient in (2) is ≥ 0. Then Theorem 5.8(a) of Chapter 1 says that

$$\lim_{h \to 0^-} \frac{f(c + h) - f(c)}{h} \geq 0.$$

Since $f'(c)$ exists by hypothesis for every $c \in (a, b)$, the limit in (2) coincides with the left- and right-hand limits (Theorem 6.2 of Chapter 1). Thus, simultaneously, $f'(c) \leq 0$ and $f'(c) \geq 0$. Therefore $f'(c) = 0$. QED

Rolle's theorem asserts the *existence* of the point c, but neither the statement nor the proof provides a means of finding such a point. The next example illustrates how that can be done for some kinds of functions.

Example 2

Verify Rolle's theorem for $f(x) = x^3 - 3x - 2$ on the interval $[-1, 2]$.

Solution. We have $f(-1) = -1 + 3 - 2 = 0$ and $f(2) = 8 - 6 - 2 = 0$. Also, f is differentiable (hence, continuous) on the interval $[-1, 2]$. Thus Rolle's theorem asserts that $f'(c) = 0$ for some $c \in (1, -2)$. To find such a point, we set $f'(x) = 0$ and solve for x. Since $f'(x) = 3x^2 - 3$, we get

$$3(x^2 - 1) = 3(x - 1)(x + 1) = 0,$$

so that $x = 1$ or $x = -1$. We are looking for $c \in (-1, 2)$. Since -1 is not in this open interval, we take $c = 1$. ■

In many cases, trying to solve $f'(x) = 0$ for x may not be as simple as it was in Example 2. In such a case, we could use the bisection method of Section 1.7

to find an approximate solution. But, perhaps fortunately, we are usually less interested in finding the point c in Rolle's theorem than we are in the fact that there *is* some such c. That is how Rolle's theorem is used to prove the mean value theorem, for instance.

10.2 THEOREM

Mean Value Theorem for Derivatives. Suppose that f is continuous on $[a, b]$ and differentiable on (a, b). Then there is some point $c \in (a, b)$ such that

(1) $$f'(c) = \frac{f(b) - f(a)}{b - a}.$$

Proof. To apply Rolle's theorem, we need a function that is continuous on $[a, b]$, differentiable on (a, b), and has value 0 at $x = a$ and at $x = b$. The function we use is one that measures the directed vertical distance from the secant through $A(a, f(a))$ and $B(b, f(b))$ to the curve $y = f(x)$. See **Figure 10.3.** The secant line has slope

$$\frac{f(b) - f(a)}{b - a}$$

FIGURE 10.3

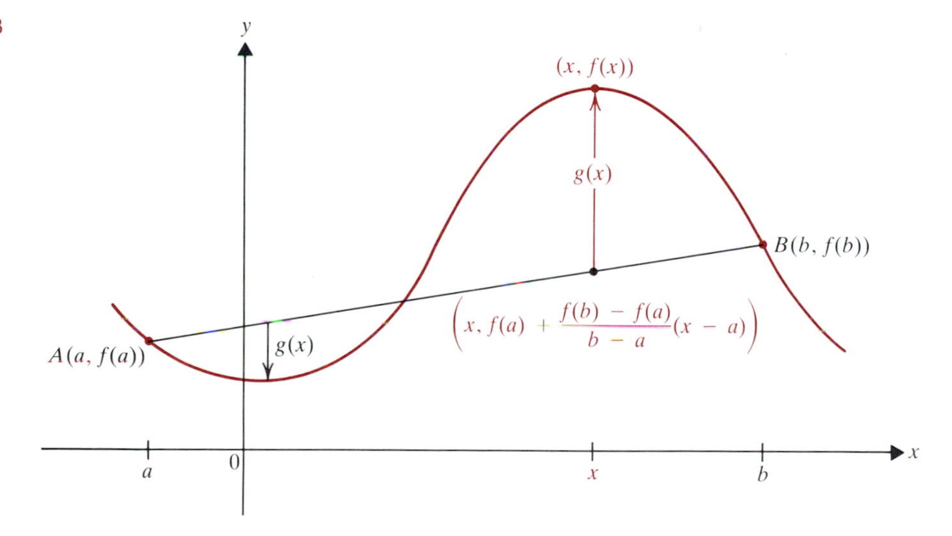

and passes through $A(a, f(a))$. Its equation is therefore

$$y - f(a) = \frac{f(b) - f(a)}{b - a}(x - a),$$

that is,

$$y = f(a) + \frac{f(b) - f(a)}{b - a}(x - a).$$

The vertical directed distance from this line to $y = f(x)$ is thus given by

(3) $$g(x) = f(x) - f(a) - \frac{f(b) - f(a)}{b - a}(x - a).$$

We have $g(a) = 0 = g(b)$, either by direct substitution into (3) or from the fact that at A and B, the distance between the graph of $y = f(x)$ and the secant is 0. By hypothesis, f is differentiable on (a, b) and continuous on $[a, b]$. Hence, so is g because it is the sum of f, the constant $-f(a)$, and a constant multiple of $x - a$. Thus Rolle's theorem guarantees that $g'(c) = 0$ for some $c \in (a, b)$. From (3), we have

$$g'(x) = f'(x) - \frac{f(b) - f(a)}{b - a}.$$

Hence, $g'(c) = 0$ for c in (a, b) tells us that

$$f'(c) = \frac{f(b) - f(a)}{b - a}. \quad \text{QED}$$

FIGURE 10.4

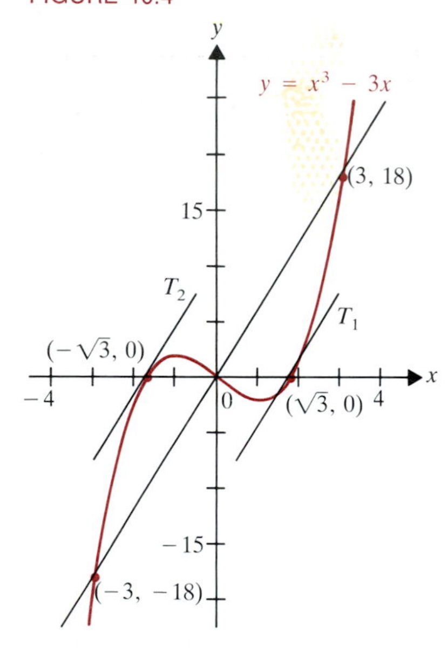

Example 3

Verify the mean value theorem for $f(x) = x^3 - 3x$ on the interval $[-3, 3]$. (See **Figure 10.4.**)

Solution. First, as a polynomial function, f is differentiable on $[-3, 3]$ and so is continuous there. We have $f(-3) = -18$ and $f(3) = 18$. Therefore

$$\frac{f(b) - f(a)}{b - a} = \frac{18 - (-18)}{3 - (-3)} = \frac{36}{6} = 6.$$

We need then to find $c \in (-3, 3)$ such that

$$f'(c) = 3c^2 - 3 = 6 \rightarrow 3c^2 = 9 \rightarrow c^2 = 3.$$

Thus either $c = \sqrt{3}$ or $c = -\sqrt{3}$ will do. ■

The hypothesis that f be continuous on $[a, b]$ and differentiable on (a, b) cannot be relaxed in Theorem 10.2, as the next example illustrates.

Example 4

Determine whether the function

$$f(x) = |x|$$

satisfies the conclusion of the mean value theorem on $[-1, 1]$.

FIGURE 10.5

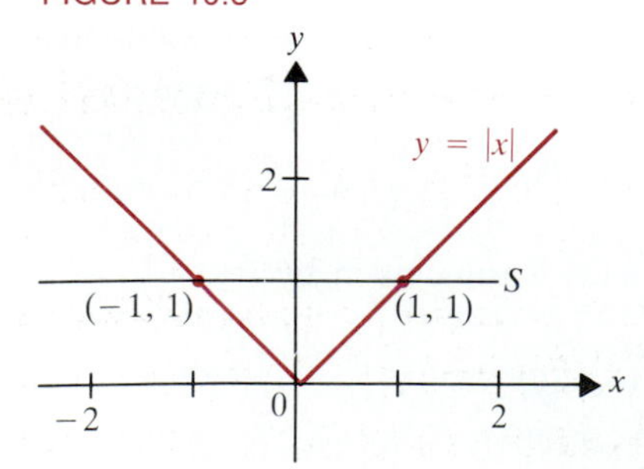

Solution. Here, $a = -1$ and $b = 1$. Thus

$$\frac{f(b) - f(a)}{b - a} = \frac{|1| - |-1|}{1 - (-1)} = \frac{1 - 1}{2} = 0.$$

Since $f(x) = x$ for $x \geq 0$ and $f(x) = -x$ for $x < 0$, we have $f'(x) = 1$ for $x > 0$ and $f'(x) = -1$ for $x < 0$. Hence, there is no point $c \in (-1, 1)$ such that $f'(c) = 0$. Example 6 of Section 1 showed that f fails to be differentiable on $(-1, 1)$, because it is not differentiable at $c = 0$. Thus, even though f is continuous on $[-1, 1]$, it fails to satisfy the hypotheses of the mean value theorem. For this reason, we can't expect the conclusion of the theorem to hold. See **Figure 10.5.** ■

HISTORICAL NOTE

Theorem 10.1 was first stated for polynomial functions in 1690 by the French mathematician Michel Rolle (1652–1719), who must turn over in his grave every time his name appears in a calculus text. He believed that calculus was illogical and full of contradictions and errors, and from his position in the French Academy of Science he loudly denounced it. In this, he set the stage for Bishop Berkeley's more famous criticism. Rolle introduced the notation $\sqrt[n]{x}$ for nth roots and made contributions to number theory and geometry but is, ironically, remembered today for his theorem (for which he offered no proof) in calculus, the subject he so despised.

Theorem 10.2 was first proved by Lagrange in the late 1700s. The proof given by Cauchy in 1821 actually established a stronger result, which we will meet in Chapter 9. The proof we have given, based on Rolle's theorem, seems to have been discovered by the Swiss mathematician P. Ossian Bonnet (1819–1892) in the middle of the last century. Bonnet made fundamental contributions to geometry as well as to analysis, the branch of mathematics that includes calculus and advanced calculus.

Exercises 2.10

1. Suppose in Example 1 that it is desired to stagger speeding fines as follows: \$25 for speed above 55 miles per hour up to 60 miles per hour, \$50 for speed above 60 miles per hour up to 65 miles per hour, and \$100 for speed above 65 miles per hour. What time intervals correspond to each fine level?

2. The Manila North Expressway in the Philippines is an 80-km toll road with a speed limit of 100 km per hour. What is the most rapid legal trip on this road? What time intervals correspond to average speeds of 100 to 110 km/h, 110 to 120 km/h, and above 120 km/h?

In Exercises 3–6, verify the hypotheses and conclusion of Rolle's theorem for the given function on the given interval.

3. $f(x) = x^2 - 3x + 2$ on $[1, 2]$.
4. $h(x) = \sin x$ on $[0, 2\pi]$.
5. $m(x) = x^3 - 12x - 16$ on $[-2, 4]$.
6. $p(x) = x^4 + 3x^2 - 28$ on $[-2, 2]$.

In Exercises 7–14, verify the hypotheses and conclusion of the mean value theorem for the given function on the given interval.

7. $f(x) = 2x^2 - 5x + 6$ on $[-2, 3]$.
8. $h(x) = x^3 - 3x + 2$ on $[1, 3]$.
9. $m(x) = 1/x$ on $[1, 4]$.
10. $k(x) = x^3 - 2x^2 + 1$ on $[-1, 3]$.
11. $n(x) = \sqrt{x}$ on $[4, 9]$.
12. $p(x) = \sqrt{x^2 - 1}$ on $[1, 3]$.
13. $f(x) = (x - 1)^{1/3}$ on $[-1, 1]$.
14. $f(x) = (x - 1)^{1/3}(x + 3)^{5/3}$ on $[-3, 1]$.

In Exercises 15–20, show there is no point c such that (1) holds for the function f on the given interval. Which hypotheses of the mean value theorem fail to hold?

15. $f(x) = 1/x - 1$ on $[-1, 1]$.
16. $f(x) = x + 1/x$ on $[-1, 2]$.
17. $f(x) = |x - 3|$ on $[2, 4]$.
18. $f(x) = |5 - 2x|$ on $[2, 5]$.
19. $f(x) = \sec x$ on $[\pi/3, 2\pi/3]$.
20. $f(x) = \tan x$ on $[-\pi, \pi]$.

21. Does $f(x) = x^{1/3}$ satisfy the conclusion of the mean value theorem on $[-1, 8]$? Does this function satisfy the hypotheses? Are the (sufficient) conditions in Theorem 10.2 necessary for (1) to hold?

22. Repeat Exercise 21 for $f(x) = \sec x$ on the interval $[0, 2\pi]$.

23. Consider $f(x) = ax^2 + bx + c$ on any interval $I = [e_1, e_2]$. Show that the secant joining $(e_1, f(e_1))$ to $(e_2, f(e_2))$ is parallel to the tangent at $(x_0, f(x_0))$ for $x_0 = (e_1 + e_2)/2$.

24. Show that if an object is thrown upward from the ground, then at some point its velocity is 0.

25. Suppose that f and g are continuous on $[a, b]$ and differentiable on (a, b). If $f(a) = g(a)$ and $f(b) = g(b)$, then show that there is some point $c \in (a, b)$ such that $f'(c) = g'(c)$.

26. Use Exercise 25 to show that if two runners in a 100-m dash finish in a dead heat, then for at least one instant during the race, they were running at *exactly* the same speed.

27. Rolle used his theorem to describe the possibilities for roots of polynomials. Consider $f(x) = x^3 + x + 1$.

(a) Show that f has a zero between $x = -1$ and $x = 0$.
(b) Show that $f'(x)$ is never 0.
(c) Use Rolle's theorem to conclude that f has exactly one real zero.

28. Repeat Exercise 27 for $f(x) = x^3 + 2x + 1$.

29. Show that a cubic equation $ax^3 + bx^2 + cx + d = 0$ has exactly one real root if $b^2 - 3ac < 0$. (*Hint:* See Exercise 27.)

30. If $f'(x)$ exists on $[a, b]$, $f(a) \cdot f(b) < 0$, and $f'(x) \neq 0$ on (a, b), then show that there is one and only one $c \in (a, b)$ such that $f(c) = 0$.

31. Show that between every two zeros of a polynomial function

$$f(x) = x^n + a_{n-1}x^{n-1} + \ldots + a_2x^2 + a_1x + a_0,$$

there is a zero of the polynomial function

$$g(x) = nx^{n-1} + (n-1)a_{n-1}x^{n-2} + \ldots + 2a_2x + a_1.$$

32. If f is a polynomial of degree 3, then show that $f(x) = 0$ can have at most 3 real solutions x. (*Hint:* Suppose there were four or more distinct roots, and show this would lead to the quadratic equation $f'(x) = 0$ having three distinct roots, which is impossible.)

33. Use mathematical induction and Rolle's theorem to show that a polynomial equation $f(x) = 0$, where f has degree n, can have at most n roots. (See Exercise 32.)

34. If f is differentiable and $f'(x) \neq 0$ for all x, then show that f is a one-to-one function.

35. Use the mean value theorem to give a proof of (22) in Lemma 3.7.

36. Prove Rolle's theorem in case f has a negative minimum on (a, b).

37. Use the mean value theorem to give a proof of (23) in Lemma 3.7.

38. Use the mean value theorem on $f(x) = \sqrt{1 + x}$ to show that $f(x) < 1 + \frac{1}{2}x$ for all positive real numbers x.

39. Show that the result of Exercise 38 also holds for $x \in [-1, 0]$.

40. If $r \leq 1$ is a positive rational number, then show that $(1 + x)^r \leq 1 + rx$ for all $x \geq -1$.

2.11 Looking Back: What *Is* the Derivative?

This chapter examined the derivative of a function f at a point c and the derivative function f'. In the first section, we saw that $f'(c)$ gives both the instantaneous rate of change of f at $x = c$ and the slope of the tangent line to the graph of $y = f(x)$ at $x = c$. This in turn is the basis for the tangent approximation for $f(x)$ near $x = c$ and the related concept of the differential, which were discussed in Section 5. The chapter ended with the mean value theorem. That relates both instantaneous and average rates of change and the slopes of secants and tangents, which were the basis of our first work with derivatives in Section 1.

The rest of the chapter was devoted mainly to developing rules and methods for calculating derivatives. Section 2 presented the basic rules for differentiating sums, constant multiples, products, reciprocals, quotients, and powers of functions. Section 3 opened with a review of trigonometric functions and then studied some limits needed to derive the formulas for differentiating the trigonometric functions. That was done in Section 4.

The central differentiation rule is the chain rule (Theorem 6.1). The technique of implicit differentiation in Section 8 is based on the chain rule, which was also heavily used in Sections 7 and 9. The main result of Section 7—the inverse function differentiation rule—was used to differentiate rational powers of functions, and so significantly extended the class of functions whose derivatives we can find. Section 9 considered the second and higher derivatives, and use of the first and second derivatives to study straight-line motion.

At this point, you may wonder what the derivative *really* is: a tool for finding instantaneous rates of change and tangent lines to graphs of functions, an approximation tool, a means of producing a function f' closely related to a given function f, or a tool for analyzing motion? It is all of those things, but it is best to think of it as the limit of the difference quotient $\Delta y/\Delta x$ as $\Delta x \to 0$:

$$f'(c) = \lim_{h \to 0} \frac{f(c+h) - f(c)}{h} = \lim_{\Delta x \to 0} \frac{f(c + \Delta x) - f(c)}{\Delta x}.$$

For it is upon this definition that all the ensuing interpretations of the derivative and the various computational rules rest.

CHAPTER CHECKLIST

Section 2.1: instantaneous rate of change; derivative $f'(c)$ of f at c; derivative function; differentiable function; secant, tangent line to graph of $y = f(x)$ at $(c, f(c))$; relation between differentiability and continuity at $x = c$.

Section 2.2: differentiation formulas; constants and linear functions, power rule; constant multiples, sums, differences; product rule, reciprocals, quotient rule, general power rule.

Section 2.3: radian measure, sine and cosine functions; values at multiples of $\pi/6$, $\pi/3$ and $\pi/4$; periods; remaining trigonometric functions; cofunctions; double angle formulas; continuity of sine and cosine; area of a circular sector; limits as $t \to 0$ of $(\sin t)/t$ and $(1 - \cos t)/t$.

Section 2.4: derivatives of sine and cosine; derivatives of the other trigonometric functions; projectile motion.

Section 2.5: tangent approximation; differentials; estimation; error; linear approximation theorem (Theorem 5.3).

Section 2.6: chain rule; u-substitution; growth rate of a composite function.

Section 2.7: inverse functions; one-to-one functions; symmetry properties of graphs of inverse functions; increasing functions, decreasing functions, monotonic functions; inverse function differentiation rule; rational power rule.

Section 2.8: functional equations; method of implicit differentiation; estimation of y near $x = c$ when y is known at $x = c$.

Section 2.9: second derivative; acceleration; straight-line motion; third and higher derivatives.

Section 2.10: Rolle's theorem; mean value theorem for derivatives.

REVIEW EXERCISES 2.11

In Exercises 1–10, differentiate the given functions.

1. $f(x) = 3x^3 - 4x^2 + 5x - 2$
2. $g(x) = \dfrac{1}{x^2 - 7x + 3}$
3. $h(x) = (5x^4 - 3x^2 + x - 1)(3x^5 - 4x^3 + 2x^2 + 6)$
4. $k(x) = \dfrac{6x^2 + x - 3}{x^2 + 3x + 5}$
5. $m(x) = \sin^2 x \cos^3 x - x^2 \cos^2 x$
6. $n(x) = \dfrac{\sin x \tan x - 1}{\sin x \tan x + 2}$
7. $p(x) = (x^2 + 5x - 3)^{3/2}$
8. $q(x) = \sqrt{\sin x^2 + \cos x^2}$
9. $r(x) = (x^3 - 3x^2 + x + 1)^{1/3}$
10. $s(x) = (\sin^2 x \tan x^2 + 1)^{1/2}$

11. Use only the definition to calculate $f'(c)$ for $f(x) = 1/(x + 1)$.
12. Repeat Exercise 11 for $f(x) = 1/x^2$.
13. Explain why the facts that $\dfrac{f(x) + g(x)}{h} = \dfrac{f(x)}{h} + \dfrac{g(x)}{h}$ but $\dfrac{f(x) \cdot g(x)}{h} \neq \dfrac{f(x)}{h} \cdot \dfrac{g(x)}{h}$ (in general) lead to the facts that the derivative of a sum is the sum of the derivatives, but the derivative of a product is *not* in general the product of the derivatives.

In Exercises 14–17, (a) find the equation of the tangent line to the graph of the given function at c, and (b) estimate y for the given x.

14. $y = \sqrt{x^2 + 5}$, $c = 2$; $x = 2.02$
15. $y = x^5 - 5x^4 + x^3 - 2x^2 + 5x - 1$, $c = 1$; $x = 1.01$
16. $y = \sin 2x$, $c = \pi/12$; $x = 15.2°$
17. $y = (x^2 + x + 2)^{1/3}$, $c = 2$; $x = 1.97$

In Exercises 18–20, show the function is *not* differentiable at the given point.

18. $f(x) = \dfrac{1}{x + 2}$ at $c = -2$.

19. $f(x) = \sqrt{x - 2}$ at $c = 2$.

20. $f(x) = |x + 4|$ at $c = -4$.

21. Graph the function $y = \cos 2x$ for $x \in [0, 2\pi]$. What is its period?

22. Find

(a) $\cos \frac{7\pi}{6}$ **(b)** $\cot \frac{7\pi}{12}$ **(c)** $\sin\left(-\frac{2\pi}{3}\right)$ **(d)** $\tan\left(-\frac{3\pi}{4}\right)$

23. Show that

(a) $\cos(\pi - t) = -\cos t$ **(b)** $\sin\left(t + \frac{1}{2}\pi\right) = \sin\left(\frac{1}{2}\pi - t\right)$

24. The ***frequency*** of a trigonometric function is defined to be $1/p$, where p is its period.
(a) What is the period of a function with frequency 10?
(b) What is the frequency of a function with period 2π?

25. In recent years, coffee prices, after adjustment for inflation, have tended to be cyclical with a four-year period. If prices ranged from a high of $4.00 per pound at the start to a low of $2.00 per pound after two years, give a cosine function that models the price. What would be the predicted price at the *end* of the four-year period?

26. Solve for t in the interval $[0, 2\pi]$:

$$2\sin^3 t - 2\sin^2 t - \sin t + 1 = 0.$$

In Exercises 27–30, evaluate the given limit.

27. $\lim_{x\to 0} \frac{\sin 3x}{4x}$ **28.** $\lim_{x\to 0} \frac{1 - \cos 2x}{\sin 2x}$

29. $\lim_{x\to 0} \frac{\sin^3 x}{x^2}$ **30.** $\lim_{x\to 0} \frac{1 - \cos x}{3x}$

31. If $f(x) = \sin x/x$, then what is the domain of f'?

32. For the ideal gas law $PV = nRT$, suppose that

$P = 9$ atm, $V = 10$ liters, and $dV/dt = -0.10$ liters/min.

If $n = 10$ moles, then find dT/dt.

33. Find the rate of change of the volume of a solid cylinder of radius r and height twice r, with respect to r.

34. If $y = f(x) = x^2 - 3x + 4$, find Δy and dy for $c = 1$ and $\Delta x = 0.1$.

35. Estimate the volume of paint needed to paint a room of size $8 \times 16 \times 8$ feet if two coats are needed, each 1/64 inch thick.

36. Estimate the error made in computing the area of a wall of measured width 12 feet and measured height 8 feet if the measurements are accurate to within 1/8 inch.

37. If $V = \pi r^2 h$, then find dh in terms of V, dV, r, and dr.

38. In the 1968 Olympic Games in Mexico City, many athletes were bothered by the altitude. Due to reduced oxygen levels in the air, performance is adversely affected by lower oxygen blood levels. A simple model for the percent of oxygen saturation S in the blood at altitude h meters is

$$S = 0.88 - 1.4 \times 10^{-8} h^2.$$

An approximation for the pressure P of oxygen in red blood cells in terms of S is

$$P = 35 + (10\,S - 6)^3, \qquad \text{where } 0.4 < S \le 0.88$$

in mm of mercury.
(a) Find S and P at sea level and at $h = 2500$ m (the approximate altitude of Mexico City).
(b) Find dP/dt for a marathon runner at $h = 2500$ m if the runner is going up a hill at the rate of 2 m/min.
(c) Repeat (b) for $h = 0$.

In Exercises 39–42, decide whether f is one-to-one. If it is, then find a formula for f^{-1} and $(f^{-1})'$.

39. $f(x) = x^3 + 1$

40. $f(x) = \sqrt{4 - x^2}$, $x \in [-2, 2]$

41. $f(x) = x^2 + 1$ **42.** $f(x) = \frac{x + 3}{x - 3}$

43. How would you restrict the domain of $f(x) = 1 - x^2$ to have an invertible function? In that case, give a formula for $(f^{-1})'$.

44. Repeat Exercise 43 for $f(x) = \sin x$.

In Exercises 45–48, give a formula for dy/dx.

45. $3x^3y^2 = x^2 - y^2 - 3$. **46.** $\sqrt{x^2 + y^2} = x^2y^2 + 1$

47. $y\cos x^2 - x\sin y^2 = 1$ **48.** $x^{2/3} + y^{2/3} = 4$

In Exercises 49 and 50, use the tangent approximation to approximate y for the given value of x.

49. $x^2y + y^2x = 6$ at $x = 2.06$. [Note that (2, 1) satisfies this equation.]

50. $y^3 + 2y^2 + xy = x^3 + 3$ at $x = 0.97$. [Note that (1, 1) satisfies this equation.]

In Exercises 51 and 52, discuss the motion of a body on the x-axis with given directed distance from the origin.

51. $y = \frac{1}{3}t^3 - 2t^2 - 5t + 6$ **52.** $y = t^4 - 2t^2 + 1$

53. Find all derivatives of $p(x) = 6x^4 - 3x^2 + x$.

54. Find all derivatives of $t(x) = 1/x^2$.

55. In Exercise 48, find d^2y/dx^2.

56. If $x^3 - y^3 = 1$, then find d^2y/dx^2.

57. An eighteen year old calculus student weighs 115 pounds. At birth she weighed 7 pounds. Show that at some time during her life, she was gaining weight at the rate of 1/2 pound per month.

58. Show that $f(x) = x^4 + 2x^2 - 8$ satisfies the hypotheses and conclusion of Rolle's theorem on $[-2, 2]$.

59. Does

$$f(x) = \frac{1}{x - 2}$$

satisfy the hypotheses and conclusion of the mean value theorem on $[3, 6]$?

60. Does the function in Exercise 59 satisfy the hypotheses and conclusion of the mean value theorem on $[1, 3]$?

61. If $f(x) = (x - a)^n g(x)$, where $f(x)$ and $g(x)$ are polynomials, then show that $(x - a)^{n-1}$ is a factor of $f'(x)$.

62. If $x = a$ is a double root of any polynomial, that is, if $(x - a)^2$ is a factor, then show that $f(x)$ and $f'(x)$ have a nonconstant common factor.

63. This exercise shows how the tangent approximation is used to derive the fundamental *Michaelis–Menten equation* of biochemistry (see Exercise 39, Section 2). An enzyme reacts with a substance called a *substrate* to form a product. The *reaction velocity* v is given by

$$v = \frac{V_{max}}{2E_T}(A - \sqrt{A^2 - 4E_T S_T}) \tag{1}$$

where E_T is the concentration of the enzyme, S_T is the concentration of the substrate, and $A = K_M + S_T + E_T$, where K_M is a certain constant called the *Michaelis constant*.

(a) Show that the tangent approximation gives

$$\sqrt{A^2 + h} \approx A + \frac{h}{2A} \qquad \text{if } A > 0.$$

(b) With $h = -4E_T S_T$, use the tangent approximation in (a) to approximate v in (1).

(c) Show that the result in (b) simplifies to

$$v \approx \frac{V_{max} S_T}{K_M + S_T + E_T}.$$

(d) If E_T is small compared to S_T, then show that the result in (c) simplifies to a *close* approximation of the equation as given in Exercise 39 of Section 2. (In actual biochemical reactions, E_T usually is negligible compared with S_T.)

CHAPTER 3 Applications of Differentiation

3.0 Introduction

In this chapter we use the formulas and techniques of Chapter 2 to study the behavior of differentiable functions. As we will see, the derivative is very helpful in analyzing and graphing functions.

We begin in Section 1 by using the chain rule to find the rate of change of one quantity with respect to time if we know that of a related quantity. The next six sections concentrate on finding maximum and minimum values of functions and drawing their graphs. Section 2 uses f' to find systematically the maximum and minimum of f on a closed interval $[a, b]$. In the next section, we consider that problem over more general sets. The mean value theorem yields a test for determining when f is increasing and when it is decreasing. This leads to the first-derivative test for classifying candidates for extreme value points. In Section 4, the second derivative of f is used to obtain further information about the shape of the graph of f. This in turn leads to the second-derivative test, another useful tool for classifying candidates for maximum and minimum points. Sections 5 and 7 consider applied optimization problems and accurate graphing of functions. Limits involving infinity, the subject of Section 6, are often helpful in graphing rational functions.

Section 8 presents Newton's method for finding zeros of a differentiable function f. This widely used technique for solving complicated functional equations is based on the tangent approximation.

The last two sections take up the question of finding a function when its rate of change is given. This commonly encountered problem in applications of mathematics is the basis for the branch of mathematics called differential equations. These two sections comprise a brief introduction to that subject.

3.1 Related Rates

In this section, we consider two variables, say x and y, that change over time and are related by an equation $y = f(x)$ or $F(x, y) = 0$. The problem is to find one of the rates of change dy/dt or dx/dt at some time when we know the value of the other one, and the values of x and y as well. To solve such a problem, we use the chain rule of Section 2.6 or the method of implicit differentiation of Section 2.8. For a first illustration of how this is done, we return to Problem 2 of the Prologue.

Example 1

You are challenged to inflate a spherical balloon to a radius of 8 inches in 16 seconds by maintaining a steady rate of radial increase of 1/2 in./s. When the radius is 4 in., how much air per second must your lungs pump into the balloon to maintain the required rate of radial growth?

Solution. The quantities that vary here are

- $t =$ the number of seconds elapsed since you began blowing up the balloon,
- $r =$ the number of inches in the radius of the balloon after t seconds,
- $V =$ the number of cubic inches in the volume of the balloon after t seconds.

The second and third of these are related by the equation

$$V = \frac{4}{3}\pi r^3 \tag{1}$$

for the volume of a ball of radius r. Both V and r are functions of t. We aren't given a formula for either, but dr/dt must be 1/2 for each value of r. We are asked for dV/dt when $r = 4$. We can compute dV/dt from (1) if we use the chain rule, because

$$\frac{dV}{dt} = \frac{dV}{dr} \cdot \frac{dr}{dt} = \frac{4}{3}\pi \cdot 3r^2 \frac{dr}{dt},$$

$$\frac{dV}{dt} = 4\pi r^2 \frac{dr}{dt}. \tag{2}$$

When $r = 4$ and $dr/dt = 1/2$, therefore, we must have

$$\frac{dV}{dt} = 4\pi \cdot 4^2 \cdot \frac{1}{2} = 32\pi \approx 100.53.$$

Thus, when $r = 4$, your lungs must pump more than 100 cubic inches of air into the balloon per second to maintain the desired rate of growth in the radius. ■

Example 1 contains all the elements common to related-rate problems. The volume V and radius r play the roles of x and y in the discussion leading up to the example. Equation (1) gives the relationship between these two variables, each of which is a function of time. Even without explicit formulas for r and

V as functions of time, we can still find dV/dt for given values of r and dr/dt, using (2). When $r = 2$, for instance, we see from (2) that to maintain $dr/dt = 1/2$,

$$\frac{dV}{dt} = 4\pi \cdot 4 \cdot \frac{1}{2} = 8\pi \approx 25.13.$$

Thus doubling the radius from 2 to 4 quadruples the rate of air input needed to maintain the same rate of growth in the radius, 1/2 in./s.

In solving related-rate problems, it is helpful to follow the sequence of steps used in Example 1.

STEPS FOR SOLVING RELATED-RATE PROBLEMS

1. Draw a figure, if appropriate, and label it. Assign a letter to each variable, giving a precise definition in each case.
2. Find an equation that relates the quantity whose rate of change is sought to the quantity whose rate of change is given.
3. Use the chain rule or implicit differentiation to differentiate the equation found in Step 2 with respect to the variable t.
4. Substitute the given data for the variables and one derivative into the result of Step 3 and solve for the desired value of the derivative of the other variable—the rate of change sought.

Although we did not do so in Example 1, drawing a figure is often helpful in finding the equation in Step 2. The next example is one for which a figure is virtually essential.

Example 2

A lighthouse is one mile out to sea from a stretch of straight shore. Its beacon rotates three times per minute. How fast is the spot of light from the beacon traveling as it passes a point three miles down the beach from the shore point nearest the lighthouse?

FIGURE 1.1

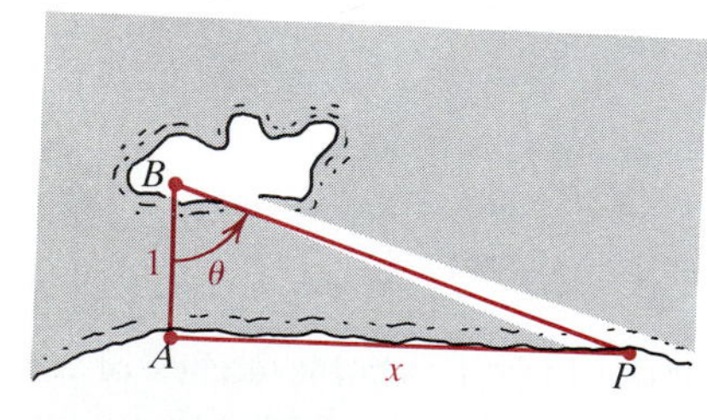

Solution. In **Figure 1.1**, the position of the beacon is labeled B, the shore point nearest the lighthouse is labeled A, and the point where the beacon's spot of light strikes the shore is labeled P. To complete Step 1 above, let

- $x =$ the distance in miles from A to P,
- $\theta =$ the number of radians in angle ABP.

We need to find dx/dt when $x = 3$. Since the beacon makes three revolutions per minute, and since each revolution is 2π radians, we have

$$\frac{d\theta}{dt} = 3 \cdot 2\pi = 6\pi \tag{3}$$

radians per minute. Triangle PAB is a right triangle, so the Pythagorean theorem gives $d(P, B) = \sqrt{1 + x^2}$. Hence, the equation that Step 2 calls for is

$$\cos\theta = \frac{1}{\sqrt{1 + x^2}} = (1 + x^2)^{-1/2}.$$

Step 3 is to differentiate this equation with respect to t. That gives

$$-\sin\theta \frac{d\theta}{dt} = -\frac{1}{2}(1+x^2)^{-3/2} \cdot 2x \cdot \frac{dx}{dt},$$

$$\sin\theta \frac{d\theta}{dt} = \frac{x}{(1+x^2)^{3/2}} \frac{dx}{dt}. \tag{4}$$

When $x = 3$, we find

$$\sin\theta = \frac{x}{\sqrt{1+x^2}} = \frac{3}{\sqrt{10}}.$$

Substituting this and (3) into (4), we complete Step 4:

$$\frac{3}{\sqrt{10}} \cdot 6\pi = \frac{3}{10\sqrt{10}} \frac{dx}{dt} \to \frac{dx}{dt} = 60\pi.$$

Thus the spot of light from the beacon is moving down the shore at a speed of 60π miles per minute. ■

When walking at night, you may have noticed that as you move away from a street light, your shadow lengthens. The next example relates how fast it lengthens to your rate of walking.

Example 3

A girl 5′6″ tall walks along a level sidewalk away from a lamppost. On the post is a street light 16 1/2 ft above the ground. See **Figure 1.2.** At what rate is her shadow lengthening if she is walking at a speed of 3 feet per second? How fast is the tip of her shadow moving?

FIGURE 1.2

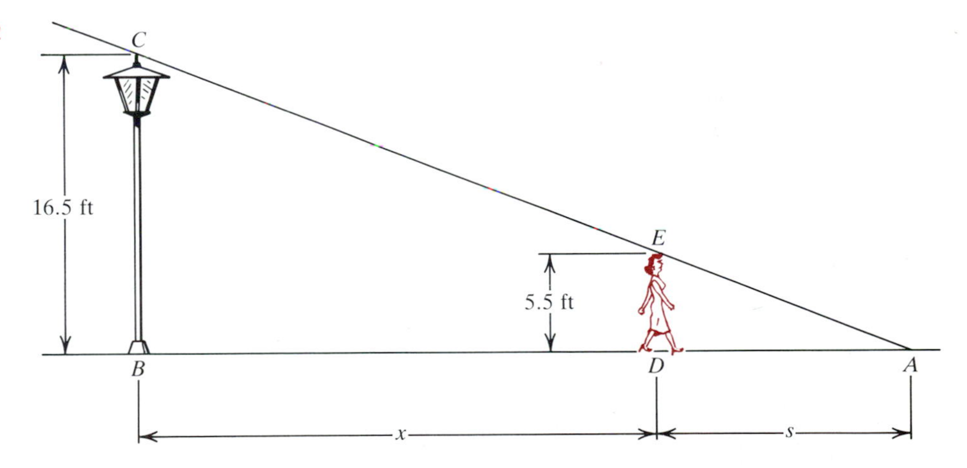

Solution. Figure 1.2 provides the figure for Step 1. To complete that step, let

- t = the number of seconds since the girl passed the street light,
- x = the number of feet the girl has walked since passing the light,
- s = the number of feet in the length of the girl's shadow.

We note that triangle ABC is similar to triangle ADE. That gives the equation needed for Step 2:

$$\frac{x+s}{16.5} = \frac{s}{5.5} \to x + s = 3s \to x = 2s.$$

Differentiating this, we have Step 3:

$$\frac{dx}{dt} = 2\frac{ds}{dt}. \tag{5}$$

To finish the solution, we carry out Step 4. Substituting dx/dt into (5) and solving for ds/dt, we find

$$\frac{ds}{dt} = \frac{1}{2} \cdot 3 = 1.5 \text{ ft/s}.$$

Thus the girl's shadow is lengthening at only *half* the rate she is walking, which may be surprising. The *tip* of her shadow moves at rate

$$\frac{d(x+s)}{dt} = \frac{dx}{dt} + \frac{ds}{dt} = 3 + 1.5 = 4.5 \text{ ft/s}.$$

Thus the *tip* of the shadow (which most people notice moving as they walk) moves *faster* than the rate of walking. ■

FIGURE 1.3

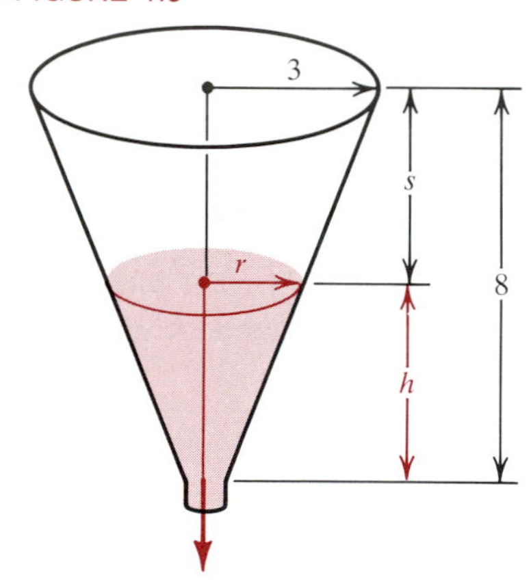

Example 4

Liquid is poured into a conical funnel of top radius 3 in. and height 8 in., as in **Figure 1.3.** How fast is the liquid leaving the funnel if the top of the fluid is 5 in. below the top of the funnel, and the level of the fluid is dropping $\frac{1}{8}$ in. per second?

Solution. In such problems, it is customary to regard the entire funnel as a right circular cone of radius 3 and height 8, ignoring the small discrepancy at the bottom of the funnel. We already have the figure for Step 1, so we identify the variables by letting

- r = the radius of the liquid in inches,
- h = the height of the liquid in inches,
- V = the volume of the liquid in cubic inches.

The equation called for in Step 2 is provided by the formula for the volume of a right circular cone of base radius r and height h:

$$V = \frac{1}{3}\pi r^2 h. \tag{6}$$

We can't perform Step 3 by differentiating (6) with respect to t, because it contains three variables. But the geometry of Figure 1.3 allows us to express r in terms of h. The right triangles formed by the radii, altitude, and edge of the funnel and the liquid are similar. Therefore

$$\frac{r}{3} = \frac{h}{8} \to r = \frac{3}{8}h. \tag{7}$$

If we substitute (7) into (6), we get

$$V = \frac{1}{3}\pi r^2 h = \frac{1}{3}\pi \cdot \frac{9}{64}h^2 \cdot h = \frac{3\pi}{64}h^3. \tag{8}$$

We now do Step 3 by differentiating (8) with respect to t. That gives

$$\frac{dV}{dt} = \frac{9\pi}{64}h^2\frac{dh}{dt}. \tag{9}$$

At the time when we are to find dV/dt, $s = 5$, so that $h = 8 - 5 = 3$. We are also given that at this time $dh/dt = 1/8$. To perform Step 4 and finish the solution, we substitute this information into (9):

$$\frac{dV}{dt} = \frac{9\pi}{64} \cdot 3^2 \cdot \frac{1}{8} = \frac{81\pi}{512}.$$

Therefore the fluid is leaving the funnel at the rate of $81\pi/512 \approx 0.497$ cubic inch per second. ■

Our final example illustrates still another way in which the geometry of a figure can help relate the variables in Step 2 of our suggested solution procedure.

Example 5

An air traffic controller observes two planes at the same altitude on a radar screen. The first is on a northbound course at 200 miles per hour and the second is on an eastbound course at 160 miles per hour. If the first plane is 8 miles south of the control tower, and the second is 6 miles west of the tower, then how fast are they approaching each other?

FIGURE 1.4

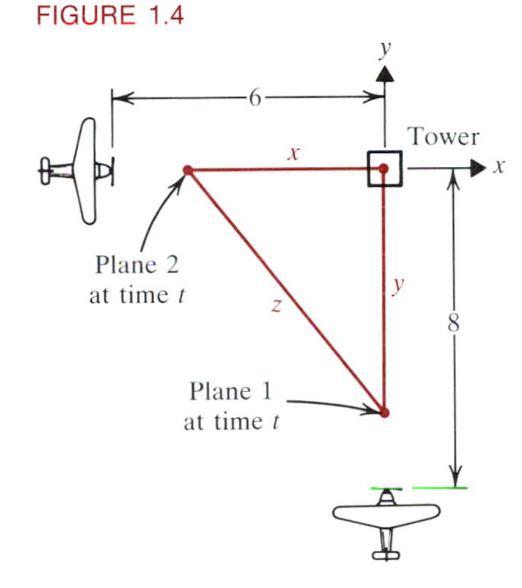

Solution. We diagram the situation in **Figure 1.4,** where

- z = the distance in miles between the planes at time t,
- y = the distance in miles between the tower and the first plane at time t,
- x = the distance in miles between the tower and the second plane at time t.

This completes Step 1. The figure and the Pythagorean theorem provide the equation called for in Step 2:

$$z^2 = x^2 + y^2. \tag{10}$$

Step 3 consists of differentiating (10) implicitly with respect to time, to obtain

$$2z\frac{dz}{dt} = 2x\frac{dx}{dt} + 2y\frac{dy}{dt}. \tag{11}$$

At the time of observation, each plane is approaching the control tower. This means that its distance from the tower is decreasing, so the rate of change of that distance is *negative*. Thus $dx/dt = -160$, $dy/dt = -200$, $x = 6$, and $y = 8$. From (10) we therefore get

$$z^2 = 36 + 64 = 100 \rightarrow z = 10.$$

Step 4 is done by substituting the values of x, y, z, dx/dt, and dy/dt into (11):

$$10\frac{dz}{dt} = 6(-160) + 8(-200) \rightarrow \frac{dz}{dt} = -96 - 160 = -256.$$

The two planes are therefore approaching each other at 256 miles/h, a *greater* speed than either of their individual speeds. ■

Exercises 3.1

1. A 10-foot ladder leans against a wall. The bottom starts slipping away from the wall at the rate of 1 foot per second. How fast is the top of the ladder sliding down the wall when the top is 6 ft above the ground?

2. In Exercise 1, suppose the top of the ladder is dropping at 2 ft per second. How fast is the bottom of the ladder slipping away from the wall when its foot is 3 ft from the wall?

3. When oil spills from a tanker into the ocean, it spreads away from the site in a circular pattern. If a helicopter observing the spill reports that the radius of the oil is increasing at 10 ft per minute, then how fast is the area of the oil increasing when its radius is 100 ft?

4. A rock is thrown into a pond. If the area of the ripple is 4π square feet and increasing at the rate of 3 square feet per second, then how fast is the radius increasing?

5. A watering trough is 10 ft long. Its ends are equilateral triangles with side 3 ft. Water is flowing into the trough at the rate of 1 cubic foot per minute. How fast is the level of water rising when it is 1.5 ft deep?

6. In Exercise 5, suppose that the level of water is rising 1/2 ft/min when the depth of the water is 1 ft. How fast is water being pumped into the trough?

7. According to Poiseuille's law, the velocity of blood flow at a distance r from the center of an artery is

$$v = C(R^2 - r^2),$$

where R is the radius of the artery and C is a constant that depends on blood pressure, the length of the artery, and other factors such as blood cholesterol. Suppose a person takes two aspirin, which causes R to increase at the rate of 0.0003 cm/min. If $R = 0.03$ cm and $C = 1$, find dv/dt.

8. In Exercise 7, suppose the person is given epinephrine to constrict the blood vessels. If R is decreasing at 0.0002 cm/min and $R = 0.03$ cm, then what is dv/dt?

9. After inflating the balloon in Example 1, you let go of its neck and the balloon flies off. If its radius decreases $2/\pi$ in. per second, then how much air is coming out of the balloon when its radius is 3 in.?

10. How fast is the radius of a mothball shrinking if its volume is decreasing at a rate of π cm^3 per week and its radius is 2 cm?

11. A volume of gas governed by Boyle's law $PV = C$ is increasing at the rate of 10 cm^3/min. At what rate is the pressure changing at a time when $V = 2$ liters and $P = 2$ atm?

12. In Exercise 11, suppose that the pressure is increasing at the rate of 1 atm/min. Find how fast the volume is changing when V is 3 liters and P is 2 atm.

13. A 6-ft tall man walks away from a light that is 18 ft above the ground. If he is walking 3 ft/s, then how fast is his shadow lengthening? How fast is the tip of the shadow moving?

14. A 5-ft tall girl walks toward a light that is 18 ft above the ground, at a speed of 2 ft/s. How fast does her shadow shorten? How fast does the tip of her shadow move?

15. Suppose that the funnel in Example 4 is being filled with liquid. If the liquid is 3 in. from the top and rising 1/4 in./s, then how fast is the volume of liquid increasing?

16. A funnel has radius 4 in. and height 10 in. If fluid is leaving the funnel at the rate of $3\pi^3$ in./s, then how fast is the height of the fluid falling when it is 4 in. high?

17. Two trains leave the same station one hour apart. The first travels north at 60 km/h. The second travels west at 80 km/h. Find how fast the trains are separating from each other one hour after the second train has left.

18. A ship steaming south at 20 knots passes a certain point at 1:00 PM. An hour later a second ship steaming west at 24 knots passes the same point. How fast is the distance between the ships increasing at 4:00 PM?

19. A rocket 2000 ft above its launch platform is rising at a speed of 600 ft/s. How fast is the rocket moving away from an observer located on the ground 2500 ft from the launch site?

20. An airplane takes off from an airport along a path inclined 30° to the ground. If its airspeed is 300 km/h, then how fast is its altitude above the level of the runway increasing when it is 100 m off the ground?

21. A kite is 30 m above its control reel and is moving in a path away from the control reel parallel to the ground at 3 m/s. How fast is its string being played out when 50 m of string are out?

22. A plane is flying east at 2000 km/h toward an antiaircraft radar station and is descending at the rate of 100 m/min. How fast is the plane approaching the station when it is 3 km above the ground and is over a point that is exactly 4 km from the emplacement?

23. A man stands on a dock 7 ft above water level and pulls in a boat by means of a line hooked to the boat 1 ft above the water line. If he pulls in the line at the rate of 1 ft/s, then how fast is the boat approaching its berth at the dock when it is 8 ft away from the berth?

24. One end of a rope is attached to a coal car on a track. The rope passes through a pulley mounted above the track. The other end of the rope is hooked to a machine 8 ft directly below the pulley, at the same height as the coal car. See **Figure 1.5.** The machine pulls in the rope at the rate of 3 ft/s. How fast is the coal car moving toward the machine when it is 15 ft from it?

FIGURE 1.5

25. In baseball, the base path surrounding the infield is a square of side 90 ft. Suppose a player rounds second base and tries to advance to third on a throw to the catcher. The runner runs 30 ft/s. The catcher throws the ball toward third at a speed of 90 ft/s. How fast is the distance between the runner and the ball changing when the catcher releases the ball, if the runner is 30 ft from third base when the ball is released?

26. A basketball player releases a free throw (15 ft from the basket) at a height of 6 ft above the floor. Taking coordinate system with origin at the point of release, the ball follows a parabolic path

$$y = -\frac{16}{150}x^2 + \frac{28}{15}x.$$

[The basket is at (15, 4).] How fast is the distance from the ball to the basket changing when $x = 5$ if $dx/dt = 1$?

27. A gas expands *adiabatically* if no heat is transferred into or out of the system. If oxygen expands adiabatically at a constant temperature, then it follows the law

$$PV^{1.40} = C,$$

where P is the pressure, V is volume, and C is constant. If $P = 2$ atm, $dP/dt = -0.1$ atm/min, and $V = 3$ liters, then what is dV/dt?

28. When carbon dioxide expands adiabatically (see Exercise 27) at constant temperature, it follows the law

$$PV^{1.29} = C.$$

If $P = 1.5$ atm, $dP/dt = -0.2$ atm/min, and $V = 2.5$ liters, then find dV/dt.

29. A particle moves along the circle $x^2 + y^2 = 25$. Find the rate of change of its distance from the x-axis when it is at (3, 4) if its x-coordinate is increasing at the rate of 3 units per second.

30. A particle moves on the graph of $xy^2 + yx^3 = 6$. Find the rate of change of the distance from the particle to the point $(-2, 1)$ when it passes the point (1, 2) if its y coordinate is increasing at the rate of 1 unit per second.

31. The illumination E (in lumens per square meter) from a light source is given by $E = kI/r^2$, where I is the intensity of the source in candelas, and r is the distance from the source in meters. Suppose that E is 100 when $r = 4$ and the illuminated object is moving directly toward the light source at the rate of 1 m/min. Find the rate of change of E assuming that I is fixed.

32. The gravitational force of the earth on a rocket traveling toward the moon is

$$F = G\frac{mM}{r^2},$$

where G is the gravitational constant, m is the mass of the rocket, M is the mass of the earth, and r is distance of the rocket from the center of the earth. Suppose $m = 2000$ kg and when $r = 100{,}000$ km, $dr/dt = 5$ km/s. Find dF/dt. (You may leave your answer in terms of G and M, or use the values given for them in Exercise 37 of Section 2.2 if you prefer.)

33. A ferris wheel of radius 10 ft rotates at one revolution per minute. Find how fast a rider is rising when she is 18 ft above ground level.

34. In Exercise 33, how fast is the rider's distance from the vertical axis of the ferris wheel changing?

35. If a balloon's volume decreases at a rate proportional to its surface area S, then show that its radius decreases at a constant rate. (*Hint:* $S = 4\pi r^2$, where r is the radius.)

36. If the area enclosed by a circular ripple in a pond (cf. Exercise 4) increases at a rate proportional to the circumference of the circular ripple, then show that the radius is increasing at a constant rate.

37. In Exercise 3, suppose that the thickness of the resulting oil slick is given by $h(t) = 0.3/\sqrt{t}$, t minutes after the spill has occurred. If 90,000 ft^3 of oil were spilled, then find a formula for the rate of growth of the radius r of the oil slick at time t. (*Hint:* The slick is a right circular cylinder of constant volume 90,000.)

38. In Example 1, suppose the balloon has volume 512 in.^3 when a leak starts. If the air is released at a rate of $1/(1 + t^2)$ in.^3/s, then what is the rate of change of the radius of the balloon at the start of the leak? After 10 min has the rate of leakage increased or decreased?

3.2 Extreme Values of a Function Over a Closed Interval

This section and the next three introduce *optimization*, the process of finding extreme (maximum and minimum) values of a function f. This is one of the most important areas to which differential calculus is applied, because of the need to maximize or minimize so many quantities in science and life generally. Much recent technological progress has stemmed from improvements in such diverse areas as minimizing the volumes of products like computer chips and stereo headphones, as well as maximizing things like production and the interior space of an automobile.

The problem to be studied in this section is how to find the extreme values of a continuous function f on a closed interval $[a, b]$. According to the extreme value theorem (Theorem 7.7 of Chapter 1), such a function has a maximum value M and a minimum value m on $[a, b]$. Extreme values of this type are called *absolute*.

2.1 DEFINITION

The function $f(x)$ has ***absolute maximum*** M on $[a, b]$ at the point $x_1 \in [a, b]$ if

(1) $$f(x) \leq M = f(x_1) \qquad \text{for all } x \in [a, b].$$

The function f has ***absolute minimum*** m on $[a, b]$ at $x_2 \in [a, b]$ if

(2) $$f(x) \geq m = f(x_2) \qquad \text{for all } x \in [a, b].$$

Figure 2.1 shows the graph of a function f over an interval $[a, b]$ with maximum M at the point a as well as at the point x_1 inside the interval. This function also takes on its minimum m at two points x_2 and x_3 inside $[a, b]$. Note then that a function may well take on its absolute extreme values on an interval at several points.

FIGURE 2.1

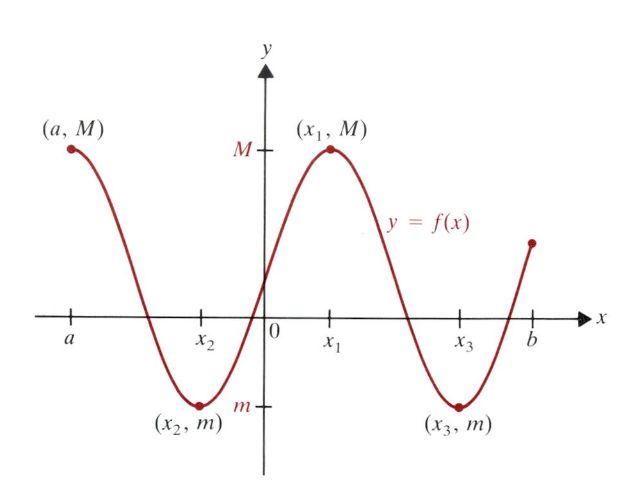

FIGURE 2.2

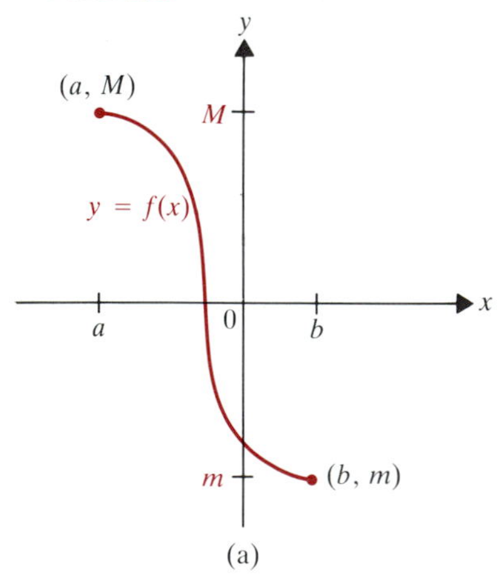

(a)

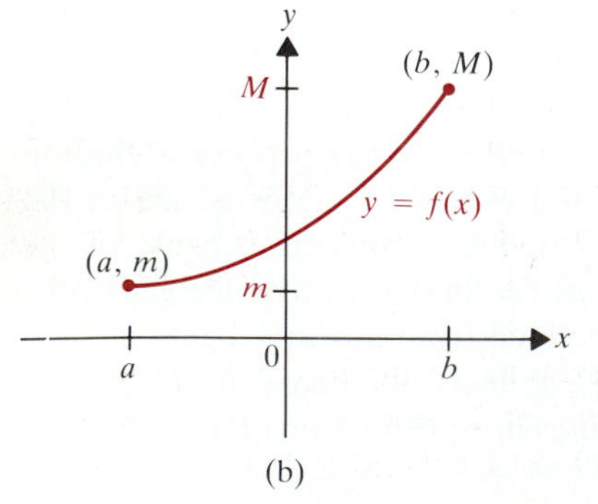

(b)

The extreme value theorem guarantees that a continuous function f has a maximum M and a minimum m on any closed interval $[a, b]$, but it gives no procedure for finding the points x_1 and x_2 where f takes on the values M and m. For increasing or decreasing functions, this is no problem, as **Figure 2.2** illustrates: The maximum and minimum will always occur at the endpoints a and b of the interval. But not all functions are monotonic throughout every interval of interest. To illustrate, we return to Problem 1 of the Prologue.

Example 1

An apartment building has 250 units. At a monthly rental of \$200 per unit, there are no vacancies. In the past, occupancy has dropped an average of one unit in response to each \$5.00 increase in rent. What function must be maximized over what interval in order to maximize revenue?

Solution. We let x be the variable that management can control:

$$x = \text{the number of \$5.00 increases in the rent.}$$

At the corresponding level of rent—$200 + 5x$ dollars—we see that

(3) $$250 - x = \text{the number of units rented,}$$

because each increase of \$5.00 in the rent results in one vacancy. Since the monthly revenue R is the product of the unit rent times the number of units rented, we have

$$R = (200 + 5x)(250 - x) = 50{,}000 + 1050x - 5x^2.$$

The function to be maximized is therefore

$$f(x) = 50{,}000 + 1050x - 5x^2. \tag{4}$$

Theoretically, we can envision an unlimited number of rent increases. But (3) says that after 250 such increases, occupancy would fall to zero, resulting in zero revenue. Hence, it certainly suffices to maximize f over the interval $I = [0, 250]$. ■

FIGURE 2.3

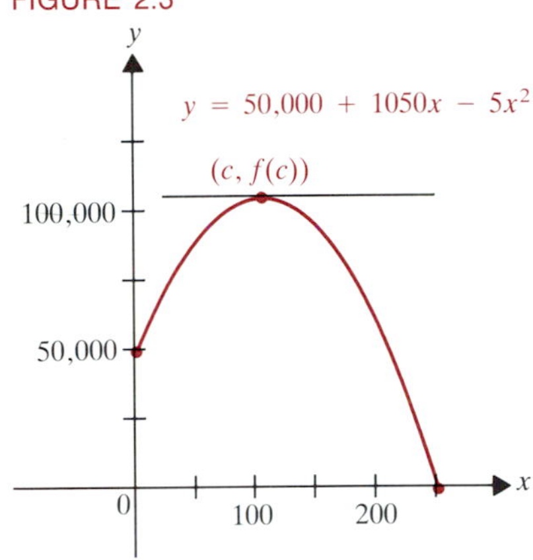

As the following table and **Figure 2.3** suggest, the function f defined by (4) is not monotonic over $[0, 250]$. Rather, f increases up to a maximum value

x	0	20	40	60	80	100	120
$f(x)$	50,000	69,000	84,000	95,000	102,000	105,000	104,000

x	140	160	180	200	220	240
$f(x)$	99,000	90,000	77,000	60,000	39,000	14,000

around $x = 100$ and then decreases. All that is needed to complete the solution of Problem 1 of the Prologue is to determine precisely *which* value c near $x = 100$ gives the largest value of $f(x)$ defined by (4). From Figure 2.3, it appears that at $x = c$, the tangent to $y = f(x)$ is *horizontal* and hence has slope 0. In Figure 2.1, there also seem to be horizontal tangents at the points x_1, x_2, and x_3 inside $[a, b]$ where f takes on extreme values. Points like c in Figure 2.3 and x_1, x_2, and x_3 in Figure 2.1 are very important in optimization.

2.2 DEFINITION A function $f(x)$ has a ***local maximum*** at the point c if there is an open interval I containing c such that

$$f(x) \leq f(c) \qquad \text{for all } x \in I. \tag{5}$$

The function f has a ***local minimum*** at the point d if there is some open interval J containing d such that

$$f(x) \geq f(d) \qquad \text{for all } x \in J. \tag{6}$$

A ***local extreme value*** (or ***local extremum***) of f is a number K which is either a local maximum or a local minimum of f. A ***local extreme point*** is a point where f takes on a local maximum or a local minimum.

Figure 2.4 shows a local maximum of f at c and a local minimum at d. The term *relative* is often substituted for *local* in discussing extreme points of f. This reflects the fact that a local maximum $f(c)$ is relatively large (compared to $f(x)$ for x close to c), and similarly for a local minimum $f(d)$. An absolute maximum or minimum is sometimes called a *global* maximum or minimum, to emphasize the contrast to local extrema. (Note: the plural of *extremum* is *extrema*.) In Figure 2.4, x_1 is a global maximum point on $[a, b]$, while b is a global minimum point.

If a function f achieves its absolute maximum or minimum on $[a, b]$ at a point $c \in (a, b)$, then c is also a local extreme point. Indeed, if c is not an endpoint of $[a, b]$, then there is some open interval I or J around c that is completely

FIGURE 2.4

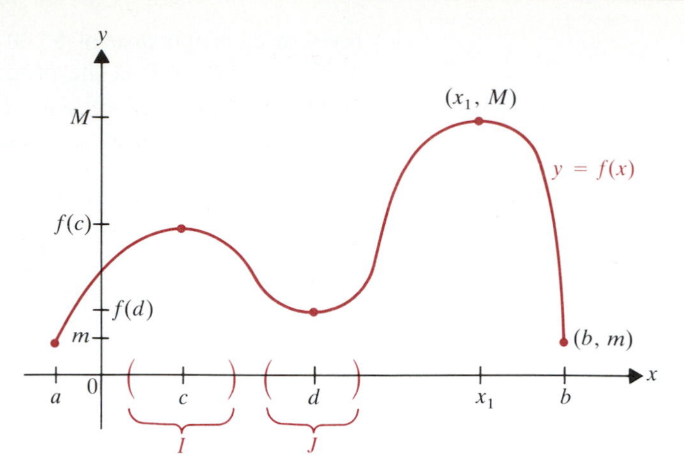

inside $[a, b]$. Since $f(c)$ is an absolute extreme value of f on the whole interval $[a, b]$, the requirement of (5) or (6) for a local extreme value is certainly satisfied. As **Figure 2.5** illustrates, if the graph of f has a tangent line at the point $(c, f(c))$, then the tangent line must be horizontal. This is so because the secants drawn from $(c, f(c))$ to nearby points of the graph have *positive* slope on one side of c and *negative* slope on the other. Consequently, the limits of the slopes of the secants must be simultaneously nonnegative and nonpositive, by Theorem 5.8 of Chapter 1. That is, the slope of the tangent line must be zero. This geometric argument can be formalized exactly as for the proof of Rolle's theorem (Theorem 10.1 of Chapter 2) to rigorously prove the following result. (See Exercise 33.) It was discovered independently by Newton about 1671 and by Leibniz about 1684. It had been used by Fermat around 1635 and stated in noncalculus terms by Kepler (see the historical note to Section 8.2) around 1610.

FIGURE 2.5

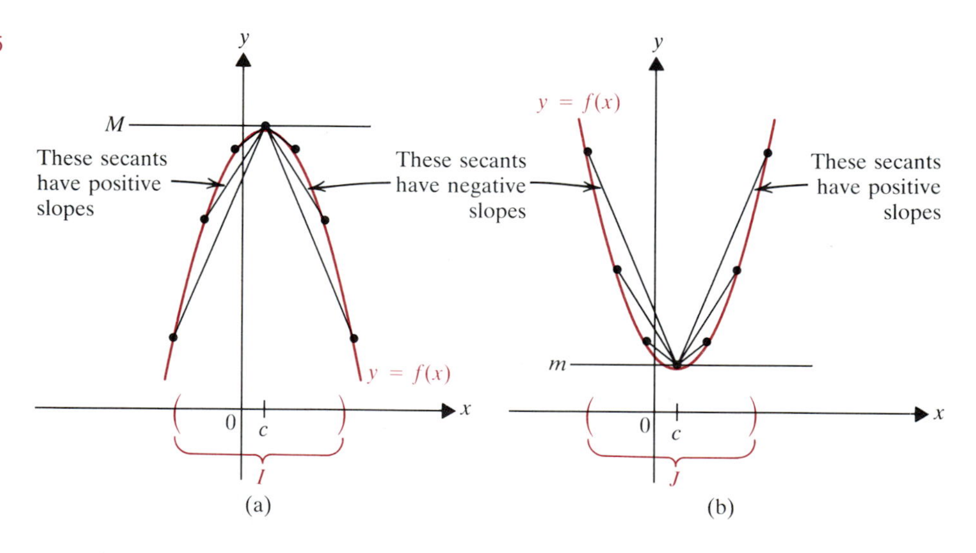

2.3 THEOREM

> If f has a local extreme value at a point c where f is differentiable, then $f'(c) = 0$.

In Example 1, it is clear from the graph in Figure 2.3 that the absolute maximum of f on $[0, 250]$ occurs at a local maximum point c. Since f is differentiable for all x, we can use Theorem 2.3 to find c.

Example 2

What rent should be charged for each apartment in Example 1 in order to maximize revenue, if we assume that (4) models the revenue accurately?

Solution. As noted above, the maximum revenue is realized at a local extreme point c. Since f is differentiable for all x in $[0, 250]$, all such points can be found by solving $f'(x) = 0$ for $x \in [0, 250]$. We have

$$f(x) = 50{,}000 + 1050x - 5x^2,$$

so

$$f'(x) = 1050 - 10x = 0$$

when

$$10x = 1050 \rightarrow x = 105.$$

Thus the maximum revenue will result from 105 increases of \$5.00 in the rent. That produces a per-unit rent of $200 + 5 \cdot 105 = 725$ dollars per month. ■

At this high level of rent (more than 3 1/2 times the initial level of \$200), the occupancy data from the past five years would predict 105 vacancies: more than 40% of the building! In itself, that could easily reduce the profitability of such a huge increase in rent. (We will consider profit later.)

Theorem 2.3 says that if f is differentiable at a local extreme point c, then $f'(c) = 0$. We can use it to search for local extreme points c in the following way.

2.4 COROLLARY

> If f has a local extreme value at the point c in its domain, then either $f'(c) = 0$ or else $f'(c)$ fails to exist.

This follows at once from Theorem 2.3, because if $f'(c)$ exists at a local extreme point c, then $f'(c) = 0$.

Example 3

Find all local extreme values of
(a) $f(x) = x^2$ (b) $g(x) = \sin x$ (c) $h(x) = |x|$.

Solution. (a) This function is differentiable everywhere, so its local extreme points c satisfy $f'(c) = 0$. Since $f'(x) = 2x$, the only such point is $c = 0$, which is clearly a local and global minimum point. The absolute minimum of f is 0 at $x = 0$.

(b) This function also has a derivative everywhere: $g'(x) = \cos x$. So again local extreme values occur at solutions of $g'(x) = 0$, that is, at odd multiples $(2n - 1)\pi/2$ of $\pi/2$. As **Figure 2.6** shows, $x = (2n - 1)\pi/2$ for $n = \pm 1, \pm 3, \pm 5, \ldots$ is a local maximum point, and $x = (2n - 1)\pi/2$ for $n = 0, \pm 2, \pm 4, \ldots$ is a local minimum point. The corresponding local maximum values (1) and minimum values (-1) are also global.

FIGURE 2.6

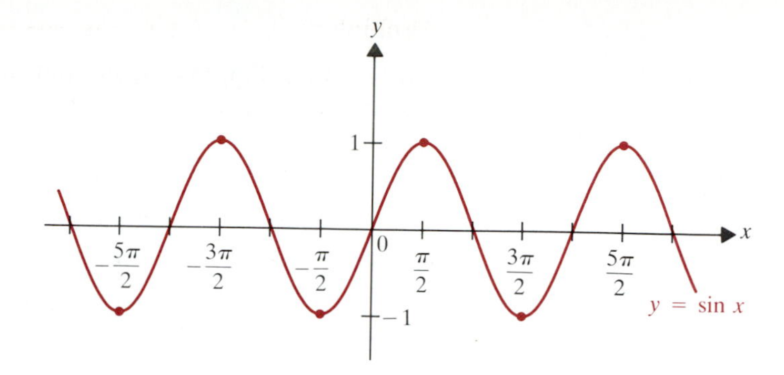

FIGURE 2.7

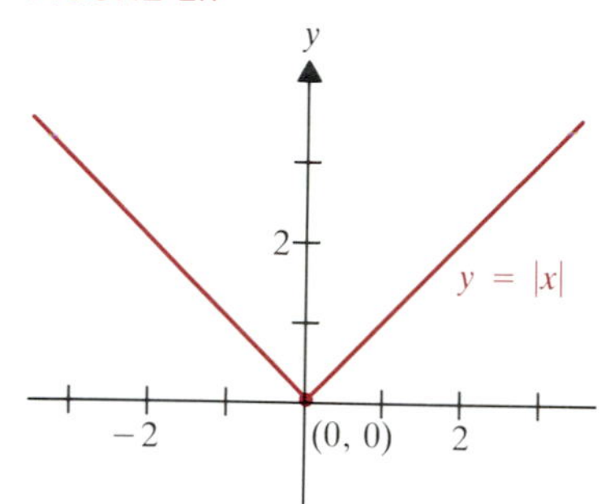

(c) Since

$$h(x) = \begin{cases} x & \text{for } x > 0 \\ -x & \text{for } x < 0, \end{cases}$$

we have

$$h'(x) = \begin{cases} 1 & \text{for } x > 0 \\ -1 & \text{for } x < 0. \end{cases}$$

Example 6 of Section 2.1 showed that h is not differentiable at 0, that is, $h'(0)$ fails to exist. This is a local (and global) minimum point, because the absolute minimum of h is 0 at $x = 0$ as **Figure 2.7** shows. ■

The two alternatives in Corollary 2.4 lead to the following terminology.

2.5 DEFINITION A point c in the domain of f is a ***critical point*** of a function f if either

$$f'(c) = 0 \qquad \text{or} \qquad f'(c) \text{ fails to exist.}$$

Corollary 2.4 leads to the following systematic procedure for finding the maximum and minimum values of a continuous function f on a closed interval $[a, b]$.

OPTIMIZATION OF A CONTINUOUS FUNCTION f ON A CLOSED INTERVAL $[a,b]$

1. Find all critical points c by solving $f'(c) = 0$ for c in (a, b), and determine where f' fails to exist on (a, b).
2. Evaluate f at all critical points and also at the endpoints a and b of $[a, b]$.
3. The largest $f(d)$ found in Step 2 is the absolute maximum of f on $[a, b]$. The smallest $f(e)$ found is the absolute minimum of f on $[a, b]$.

This program works because Step 1 finds *all* candidates for local extreme points, by Corollary 2.4. In Step 2 we compute $f(a)$ and $f(b)$ and f at all the points found in Step 1. If f has its absolute extreme values at *local* extreme points, then we find them among the values of f at its critical points. Otherwise, f has at least one of its absolute extreme values at an *endpoint* a or b of $[a, b]$, as in Figures 2.1 and 2.2. In that case, we discover this by computing $f(a)$ and $f(b)$.

FIGURE 2.8

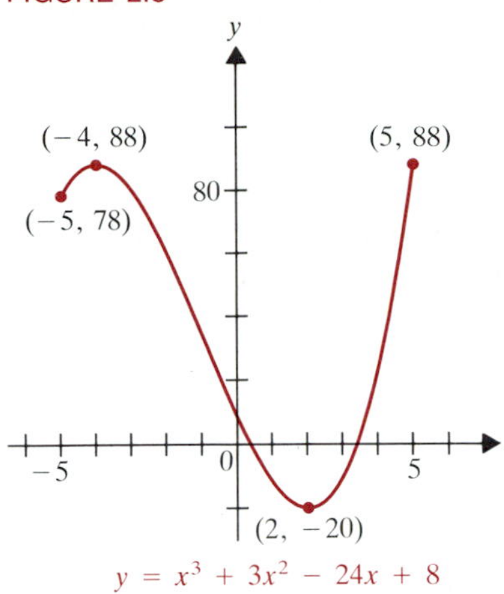

Example 4

Find the absolute maximum and minimum of $f(x) = x^3 + 3x^2 - 24x + 8$ on the interval $[-5, 5]$.

Solution. Following our program, we first find all critical points.

$$f'(x) = 3x^2 + 6x - 24 = 3(x^2 + 2x - 8) = 3(x - 2)(x + 4).$$

Thus $f'(x) = 0$ when $x = 2$ or $x = -4$. Since f is a polynomial function, it is differentiable everywhere, so there are no further critical points. Evaluating f at the critical points 2 and -4 and at the endpoints -5 and 5 of the given interval, we obtain

$$f(2) = -20, \qquad f(-4) = 88, \qquad f(-5) = 78, \qquad f(5) = 88.$$

Hence, the absolute minimum occurs at $x = 2$ and is -20. The absolute maximum occurs at $x = -4$ and at the endpoint $x = 5$ and is 88. The graph of f is sketched in **Figure 2.8.** ■

In Example 4, both critical points are local (and absolute) extreme points. But not all critical points need produce local extreme values.

FIGURE 2.9

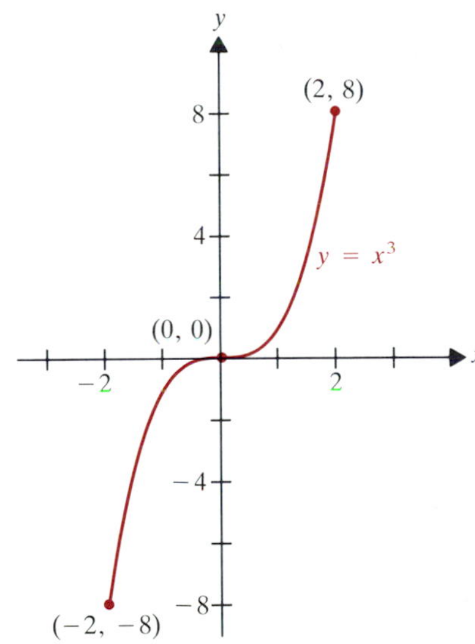

Example 5

Find all local extreme points of $f(x) = x^3$ on $[-2, 2]$.

Solution. Here $f'(x) = 3x^2$ is 0 only when $x = 0$. Since f is a polynomial function, there are no other critical points. Thus $x = 0$ is the only candidate for a local extreme point. But it is *not* one, because $f(0) = 0$ and $f(x) < 0$ for x negative, while $f(x) > 0$ for x positive. Thus f has no local extreme points. As **Figure 2.9** shows, f is an increasing function on $[-2, 2]$. Thus its absolute minimum occurs at the left endpoint $x = -2$, and its absolute maximum occurs at $x = 2$. ■

As Example 4 suggests, we need some tests to determine whether critical points are actually local extreme points. The next two sections are devoted to developing such tests. If we restrict attention to a closed interval, however, we can locate absolute extreme values very well by using the boxed program above.

Example 6

Find the maximum and minimum values of $f(x) = (2x + 4)\sqrt{4 - x^2}$ on the interval $[-2, 2]$.

Solution. We have for $x \neq \pm 2$

$$f'(x) = 2\sqrt{4 - x^2} + (2x + 4) \cdot \frac{1}{2}(4 - x^2)^{-1/2}(-2x)$$

$$= 2\sqrt{4 - x^2} - \frac{(2x^2 + 4x)}{\sqrt{4 - x^2}}$$

$$= \frac{8 - 2x^2 - 2x^2 - 4x}{\sqrt{4 - x^2}} = -4 \cdot \frac{x^2 + x - 2}{\sqrt{2 - x}\sqrt{2 + x}},$$

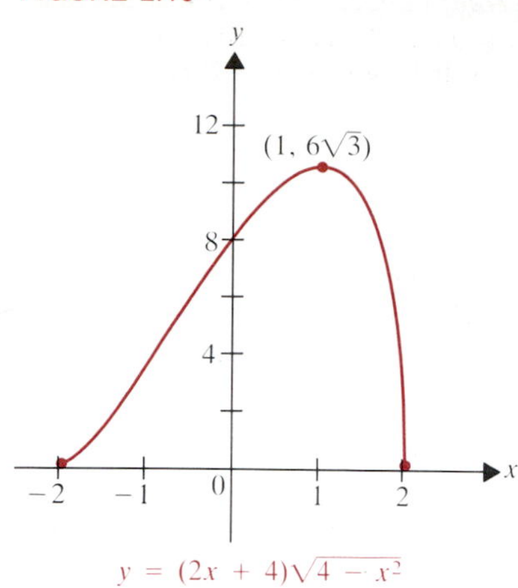

FIGURE 2.10

that is,

$$f'(x) = \frac{-4(x+2)(x-1)}{\sqrt{2-x}\sqrt{x+2}} \quad \text{if } x \neq \pm 2. \tag{7}$$

Just as $D_x\sqrt{x}$ fails to exist at $x = 0$, so $f'(x)$ fails to exist at $x = \pm 2$, where (7) is undefined. Besides these critical points, which also happen to be endpoints of the given interval, we see from (7) that $x = 1$ is a critical point, since $f'(1) = 0$. Computing the values of f at these points, we get

$$f(1) = 6\sqrt{3}, \qquad f(2) = f(-2) = 0.$$

Thus f has its maximum at $x = 1$ and its minimum at $x = \pm 2$ over the interval $[-2, 2]$. The graph is sketched in **Figure 2.10.** ■

Exercises 3.2

1. Suppose in Example 1 that a management survey of other buildings suggests that for every ten dollar increase in rent, five units become vacant. Management decides to use this model in place of the one used in Example 1.
   (a) Give the revenue function.
   (b) Over what interval must it be maximized?
   (c) What rent will maximize revenue, according to this model?
   (d) Do you find your result in (c) more or less reasonable than the result of Example 2?
2. An apple grower harvests an average of 100 apples per tree when the trees are planted 200 to the acre. For every additional 20 trees per acre planted, the yield per tree drops by 5 apples.
   (a) Set up the function that models yield per acre as a function of the number of additional 20-tree per acre plantings.
   (b) Over what interval should the function in (a) be maximized?
   (c) What density of trees per acre results in the highest yield per acre?

In Exercises 3–24, find the absolute minimum and maximum of the given function over the given interval.

3. $f(x) = x^2 - 4x + 1$ on $[0, 3]$.
4. $g(x) = x^2 - 6x - 2$ on $[2, 4]$.
5. $h(x) = x^3 - 6x^2 + 9x + 1$ on $[0, 4]$.
6. $k(x) = x^3 - 9x^2 + 24x - 2$ on $[1, 5]$.
7. $m(x) = x^{1/3}$ on $[-8, 8]$.
8. $n(x) = x^{1/5} + 1$ on $[-32, 32]$.
9. $f(x) = x^4 - 2x^2 + 3$ on $[-1, 1]$.
10. $g(x) = \sin x + 1$ on $[0, 2\pi/3]$.
11. $k(x) = |2x - 3|$ on $[-1, 1]$.
12. $m(x) = |5 - 3x|$ on $[-2, 2]$.
13. $n(x) = \dfrac{x}{x+1}$ on $[1, 3]$.
14. $r(x) = \dfrac{x-1}{2x+1}$ on $[1, 4]$.
15. $t(x) = \sqrt{1 - x^2}$ on $[-1, 1]$.
16. $f(x) = x^{2/3} + 1$ on $[-1, 1]$.
17. $g(x) = (x + 1)\sqrt{1 - x^2}$ on $[-1, 1]$.
18. $h(x) = (x + 3)\sqrt{9 - x^2}$ on $[-3, 3]$.
19. $f(x) = \cos^2 x + \sin x$ on $[0, \pi]$.
20. $f(x) = \sin^2 x + \cos x$ on $[0, \pi]$.
21. $f(x) = \sec x$ on $[-\pi/3, \pi/3]$.
22. $f(x) = \csc x$ on $[3\pi/4, 4\pi/3]$.
23. $g(x) = -\cot x + \csc x$ on $[-\pi/6, \pi/4]$.
24. $g(x) = \tan x - \sec x$ on $[\pi/6, \pi/3]$.
25. Let $f(x) = \sin x - x$. What is the maximum value of f on $[0, 2\pi]$?
26. **(a)** From Exercise 25, show that $\sin x \leq x$ on $[0, 2\pi]$.
    (b) From (a), conclude that $\sin x \leq x$ on $[0, +\infty]$.
27. Describe the extreme values of $f(x) = mx + b$ on any interval $[c, d]$.
28. **(a)** Show that $f(x) = (ax + b)/(cx + d)$ has no critical points unless $ad - bc = 0$.
    (b) If $ad - bc = 0$, then show that f is a constant function k on its domain.
29. If f is an odd function $[f(-x) = -f(x)]$ on $[-a, a]$, show that f has a local maximum at c if and only if it has a local minimum at $-c$.
30. Show that f has a local maximum at c if and only if $-f$ has a local minimum at c.
31. Show that every quadratic function $f(x) = ax^2 + bx + c$ has exactly one critical point.

32. If $f(x) = x^n$ for $n > 1$ a positive integer, then show that f has exactly one critical point, which is an absolute minimum point if n is even but is not an extreme value point if n is odd.

33. Prove Theorem 2.3. [*Hint:* Suppose that c is a maximum point. Show that $f'(c) \leq 0$ and $f'(c) \geq 0$, using the reasoning in the proof of Rolle's theorem (Theorem 10.1 of Chapter 2).]

3.3 Increasing and Decreasing Functions / The First-Derivative Test

Increasing and decreasing functions were introduced in Definition 7.4 of Chapter 2. Recall that increasing functions are just those whose graphs rise as x increases from left to right along the x-axis. Similarly, decreasing functions are those whose graphs fall as x increases. Thus far, the only way for us to tell whether a given function f is increasing or decreasing on some interval is to plot its graph over the interval. In this section, we will use the mean value theorem to derive a test for finding the intervals on which a differentiable function f increases and the intervals on which it decreases. That leads to a test for classifying critical points of f as local maximum points, local minimum points, or neither. These tests enable us to study the behavior of f over its entire domain, rather than just over a closed interval $[a, b]$ as in the last section.

Figure 3.1 shows the graphs of two differentiable functions, one of which is increasing and the other of which is decreasing. It also shows tangent lines to each graph at several points. The pictures suggest that when the tangent lines have positive slope, then f is increasing, and when the tangents have negative slope, f is decreasing. The slope of the tangent line to the graph of f at each point $(c, f(c))$ is given by $f'(c)$. Thus it appears that functions with positive derivatives are increasing, and those with negative derivatives are decreasing.

FIGURE 3.1

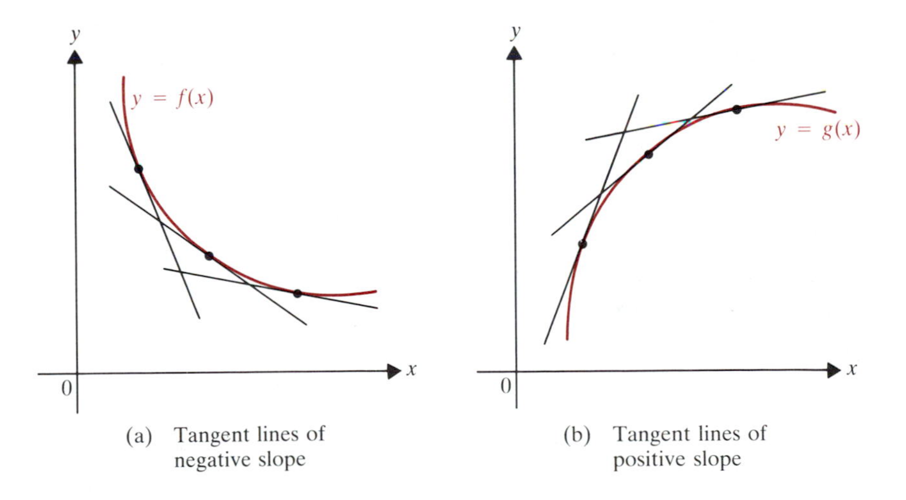

(a) Tangent lines of negative slope

(b) Tangent lines of positive slope

3.1 THEOREM

Let f be a continuous function on an interval I.

(a) If $f'(x) > 0$ for all $x \in I$, then f is increasing on I.

(b) If $f'(x) < 0$ for all $x \in I$, then f is decreasing on I.

(c) If $f'(x) = 0$ for all $x \in I$, then f is constant on I.

Proof. Suppose that x_1 and x_2 belong to I, with $x_1 < x_2$. Then on $[x_1, x_2]$, f satisfies the hypotheses of the mean value theorem. Therefore for some $c \in (x_1, x_2)$,

$$\frac{f(x_2) - f(x_1)}{x_2 - x_1} = f'(c). \tag{1}$$

(a) Here $f'(c) > 0$. Multiplying both sides of (1) by the positive number $x_2 - x_1$, we have $f(x_2) - f(x_1) > 0$, that is, $f(x_2) > f(x_1)$. Therefore f is increasing on I.

(b) Here $f'(c) < 0$. Thus multiplication of (1) through by the positive number $x_2 - x_1$ gives $f(x_2) - f(x_1) < 0 \rightarrow f(x_2) < f(x_1)$. Hence, f is decreasing on I.

(c) This time, if we multiply (1) through by $x_2 - x_1$, we get $f(x_2) - f(x_1) = 0$; that is, $f(x_2) = f(x_1)$. Taking x_1 to be any fixed point $a \in I$, we conclude that $f(x_2) = f(a)$ for all $x_2 \in I$. That is, f is constant on I. QED

Part (c) is an extra dividend, which we included to make the theorem complete. It is an interesting converse to the differentiation rule

$$\frac{d}{dx}(c) = 0,$$

for any constant function $f(x) = c$. It says that the *only* functions f that have zero as derivative throughout an entire interval are constant functions $f(x) = c$. This will be used to good advantage in Section 9. For now, parts (a) and (b) provide an immediate means of determining the intervals where a given function increases and where it decreases. Note that Thoerem 3.1 applies to any type of interval: closed, open, half-lines, and so on.

Example 1

If $f(x) = x^3 + 6x^2 - 15x + 2$, then determine where f is increasing and where it is decreasing.

Solution. We have

$$f'(x) = 3x^2 + 12x - 15 = 3(x^2 + 4x - 5) = 3(x + 5)(x - 1).$$

As in Section 1 of Chapter 1, we construct a table to determine where f' is positive and where it is negative. Then we apply Theorem 3.1 to fill in the final row.

x:	$(-\infty, -5)$	$(-5, 1)$	$(1, +\infty)$
$x + 5$:	$-$	$+$	$+$
$x - 1$:	$-$	$-$	$+$
$f'(x) = (x + 5)(x - 1)$:	$+$	$-$	$+$
Conclusions:	f increases	f decreases	f increases

Thus f is increasing on $(-\infty, -5)$ and $(1, +\infty)$ and decreasing on $(-5, 1)$. Its graph is sketched in **Figure 3.2.** ■

FIGURE 3.2

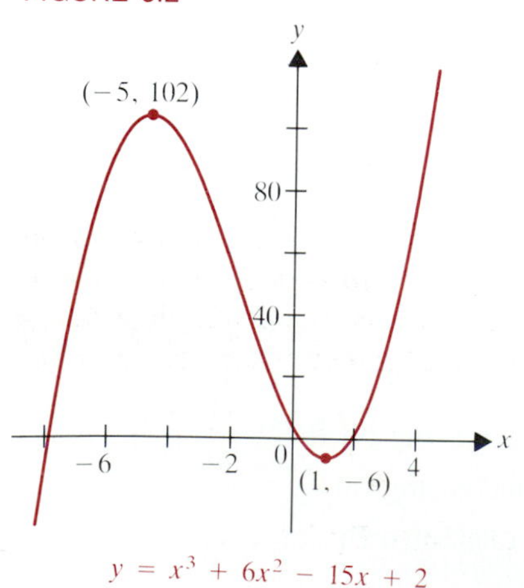

$y = x^3 + 6x^2 - 15x + 2$

The solution of Example 1 contains the idea behind the promised test for classifying critical points. The function f in Example 1 increases up to the critical point $x = -5$ and then decreases. So, near $x = -5$ the graph of f must look like that in **Figure 3.3**(a). Similarly, since f decreases until the critical point $x = 1$ and then increases, the graph of f near $x = 1$ must look like that in Figure 3.3(b). Therefore f must have a local maximum at $x = -5$ and a local minimum at $x = 1$. The same kind of reasoning works in general to give the following test for local extreme values.

3.2 THEOREM

First-Derivative Test. Let c be a critical point of f. Let I be an open interval containing c such that f is continuous on I and $f'(x)$ exists for all $x \neq c$ in I.

(a) If $f'(x) < 0$ for all $x < c$ in I and if $f'(x) > 0$ for all $x > c$ in I, then f has a local minimum at $x = c$.

(b) If $f'(x) > 0$ for all $x < c$ in I and $f'(x) < 0$ for all $x > c$ in I, then f has a local maximum at $x = c$.

(c) If $f'(x)$ has the same sign on both sides of c in I, then c is not a local extreme point for f.

FIGURE 3.3

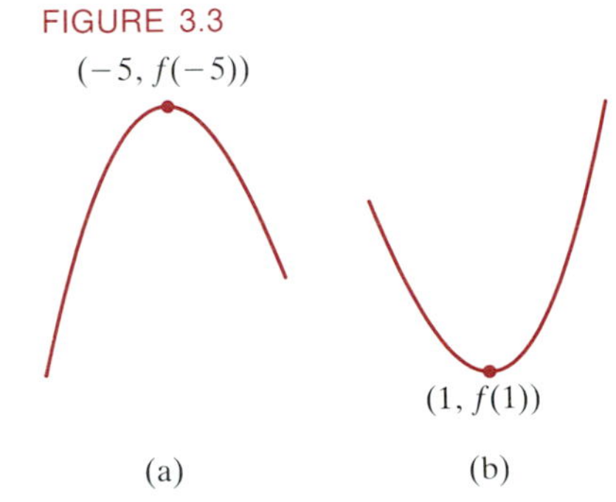

FIGURE 3.4

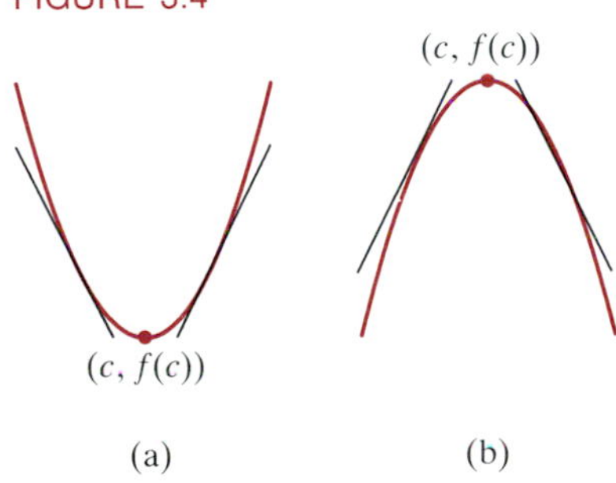

Proof. (a) Since $f'(x) < 0$ for $x < c$ in I, we have from Theorem 3.1(b) that f is decreasing on the part of I to the left of c. Thus

(1) $$f(x) > f(c) \qquad \text{for } x < c \text{ in } I.$$

Also, since $f'(x) > 0$ for $x > c$ in I, Theorem 3.1(a) says that f is increasing on the part of I to the right of c. Consequently,

(2) $$f(c) < f(x) \qquad \text{for } c < x \text{ on } I.$$

Together, (2) and (3) give

$$f(x) > f(c) \qquad \text{for all } x \neq c \text{ in } I.$$

That is, $f(c)$ is a local minimum of f. The proofs of (b) and (c) proceed in the same way. (See Exercises 35 and 36.) QED

FIGURE 3.5

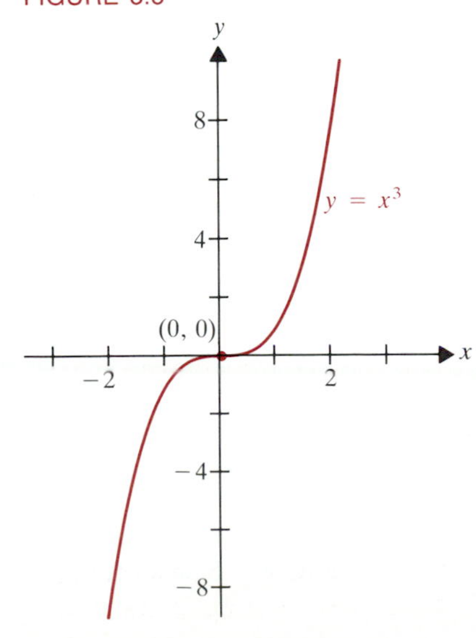

To apply the first-derivative test, it is helpful to make a table like the one in the solution to Example 1. Rather than memorizing (a) and (b) of Theorem 3.2, make a rough drawing of the tangent lines to the graph of $y = f(x)$ on either side of each critical point $x = c$ as in **Figure 3.4.** If $f'(x) < 0$ for $x < c$ and $f'(x) > 0$ for $x > c$, then the graph looks like the one in Figure 3.4(a). The tangent line falls for $x < c$ and rises for $x > c$. By filling in a curve tangent to the lines, a picture like Figure 3.3(b) is obtained, so f must have a local minimum at c. Similarly, $f'(c) > 0$ for $x < c$ and $f'(x) < 0$ for $x > c$ give a situation like the one in Figure 3.4(b).

Similar pictures can be drawn for Theorem 3.3(c), but it is probably simpler just to remember that *when the first derivative fails to change sign at $x = c$, then c is not a local extreme point.* For instance, we saw in Example 4 of the last section that the critical point $x = 0$ is not a local extreme point of $f(x) = x^3$, as **Figure 3.5** shows. For this function,

$$f'(x) = 3x^2 > 0 \qquad \text{for } x \neq 0,$$

so the first derivative fails to change sign at $x = 0$. By Theorem 3.1(a), f is increasing on $(-\infty, 0)$ and $(0, +\infty)$.

FIGURE 3.6

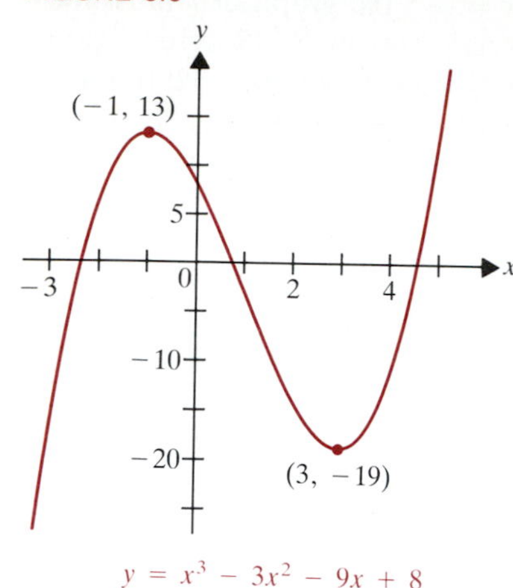

$y = x^3 - 3x^2 - 9x + 8$

Example 2

Identify all local extreme points of $f(x) = x^3 - 3x^2 - 9x + 8$. Draw a rough sketch of the graph.

Solution. We have

$$f'(x) = 3x^2 - 6x - 9 = 3(x^2 - 2x - 3) = 3(x - 3)(x + 1).$$

Thus $f'(x) = 0$ when $x = -1$ and $x = 3$. A table including the critical points makes it easy to apply both Theorem 3.1 and Theorem 3.2. The conclusions at the bottom were filled in from Theorem 3.2 after the rest of the table was completed by using Theorem 3.1. The short table of values together with the information obtained above enables us to draw the sketch of the graph shown in **Figure 3.6.** ■

x:	$(-\infty, -1)$	-1	$(-1, 3)$	3	$(3, +\infty)$ → **R**
$x + 1$:	−		+		+
$x - 3$:	−		−		+
$f'(x) = (x + 1)(x - 3)$:	+	0	−	0	+
Conclusions:	Function increases		Function decreases		Function increases
		Local maximum point		Local minimum point	

x	-3	-1	0	1	3	5
$f(x)$	-19	13	8	-3	-19	13

We next consider an example in which not all critical points are obtained by solving $f'(x) = 0$ for x.

Example 3

Identify all local extreme points of $g(x) = x^{1/3}(x - 3)^{2/3}$. Sketch the graph.

Solution. Using the product and power rules, we obtain

$$g'(x) = \frac{1}{3}x^{-2/3}(x - 3)^{2/3} + \frac{2}{3}x^{1/3}(x - 3)^{-1/3}$$

$$= \frac{1}{3}x^{-2/3}[(x - 3)^{2/3} + 2x(x - 3)^{-1/3}]$$

$$= \frac{1}{3}x^{-2/3}(x - 3)^{-1/3}[x - 3 + 2x]$$

$$= \frac{3x - 3}{3x^{2/3}(x - 3)^{1/3}} = \frac{x - 1}{x^{2/3}(x - 3)^{1/3}}.$$

Thus $g'(x) = 0$ at $x = 1$, and $g'(x)$ fails to exist at $x = 0$ and $x = 3$. We tabulate the behavior of $g'(x)$ as follows.

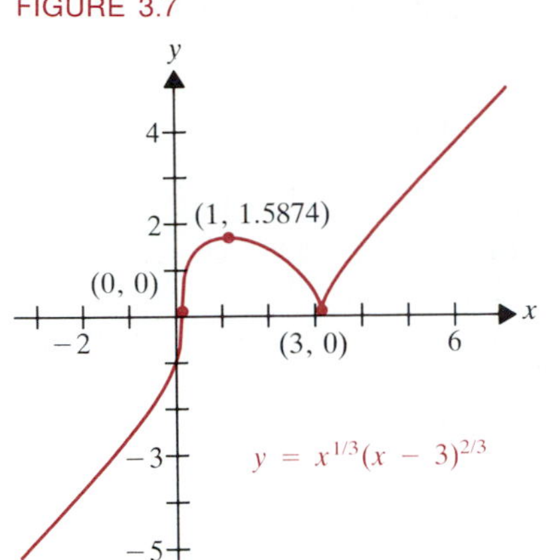

FIGURE 3.7

x:	$(-\infty, 0)$	0	$(0, 1)$	1	$(1, 3)$	3	$(3, +\infty)$
$x - 1$:	$-$		$-$		$+$		$+$
$x^{2/3}$:	$+$		$+$		$+$		$+$
$(x - 3)^{1/3}$:	$-$		$-$		$-$		$+$
$g'(x)$:	$+$	Does not exist	$+$	0	$-$	Does not exist	$+$
Conclusions:	Function increases		Function increases		Function decreases		Function increases
		Not an extreme point		Local maximum point		Local minimum point	

With the aid of a calculator, we can compile the following table of values (most approximate) to draw the graph shown in **Figure 3.7.**

x	0	1	2	3	4	5	-1	-2	-3	-4	-5
$g(x)$	0	1.59	1.26	0	1.59	2.71	-2.52	-3.68	-4.76	-5.81	-6.84

The sharp point at $x = 3$ is called a ***cusp.*** ■

Our final example considers a fourth-degree polynomial function.

Example 4

Find all local extreme value points for $h(x) = x^4 - 2x^2 + 1 = (x^2 - 1)^2$.

Solution. Here we have

$$h'(x) = 4x^3 - 4x = 4x(x^2 - 1) = 4x(x - 1)(x + 1).$$

Thus $h'(x) = 0$ when $x = 0$, 1, or -1. We obtain the following table by using Theorems 3.1 and 3.2.

x:	$(-\infty, -1)$	-1	$(-1, 0)$	0	$(0, 1)$	1	$(1, +\infty)$
x:	$-$		$-$		$+$		$+$
$x - 1$:	$-$		$-$		$-$		$+$
$x + 1$:	$-$		$+$		$+$		$+$
$h'(x)$:	$-$	0	$+$	0	$-$	0	$+$
Conclusions:	Function decreases		Function increases		Function decreases		Function increases
		Local minimum point		Local maximum point		Local minimum point	

The table of values

x	-3	-2	-1	0	1	2	3
$h(x)$	64	9	0	1	0	9	64

enables us to draw a sketch of the curve (**Figure 3.8**). ■

FIGURE 3.8

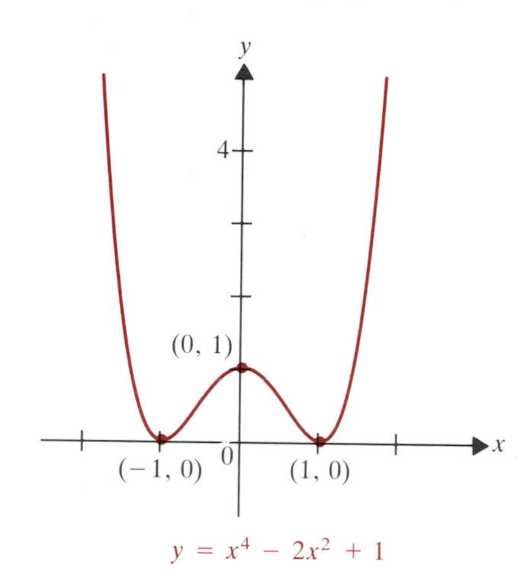

Exercise 3.3

In Exercises 1–10, find where f is increasing and where it is decreasing.

1. $f(x) = x^2 - 3x + 2$
2. $f(x) = 2 + 7x - 2x^2$
3. $f(x) = x^3 - \frac{3}{2}x^2 + 2x - 3$
4. $f(x) = x^4 - 8x^2 + 5$
5. $f(x) = x^4 - 4x + 8$
6. $f(x) = x^5 - 5x^2 + 1$
7. $f(x) = x + \sin x$
8. $f(x) = \sin x + \cos x$
9. $f(x) = \sin^2 x$
10. $f(x) = \cos^2 x$

In Exercises 11–34, find all critical points, classify them using the first-derivative test, and draw a sketch of the graph.

11. $f(x) = x^2 - 5x - 3$
12. $f(x) = 2 + 4x - 3x^2$
13. $f(x)$ of Exercise 3.
14. $f(x) = x^3 - 12x + 7$
15. $f(x)$ of Exercise 5.
16. $f(x)$ of Exercise 4.
17. $f(x)$ of Exercise 7.
18. $f(x)$ of Exercise 8.
19. $f(x) = \sin^2 x$
20. $f(x) = \cos^2 x$
21. $f(x) = x^{1/3}(x - 4)^{2/3}$
22. $f(x) = x^{2/3}(x^2 - 16)$
23. $f(x) = x^{4/3} + 8x^{1/3}$
24. $f(x) = x^{5/3} - (5/2)x^{2/3}$
25. $f(x) = x^3 + 48/x$
26. $f(x) = x^2 - 2/x$
27. $f(x) = x^3 + x^2 - x + 1$
28. $f(x) = x^3 - 3x^2 - 2x + 1$
29. $f(x) = 1/(1 + x^2)$
30. $f(x) = \sqrt{x^2 + 2}$
31. $f(x) = (x + 1)^2(x - 1)^2$
32. $f(x) = (x - 2)^2(x + 1)^3$
33. Prove Theorem 3.2(b).
34. Prove Theorem 3.2(c).
35. Use Theorem 3.2(c) to show that $f(x) = x^3 + 1$ has no extreme value at $x = 0$.
36. Repeat Exercise 35 for $f(x) = x^{2k+1} + c$ where k is a positive integer and c is any real constant.
37. Suppose $f(0) \geq g(0)$ and $f'(x) > g'(x)$ for all $x \in (0, +\infty)$. Then use Theorem 3.1 to show that $f(x) > g(x)$ on $(0, +\infty)$.
38. **(a)** Use Exercise 37 to show that $\sin x < x$ on $(0, \pi/2)$.
    (b) Conclude that $\sin x < x$ for $x \in (0, +\infty)$.
39. If $f(x) = x^3 + ax^2 + bx + c$ where $a^2 \leq 3b$, then show that f is increasing on $(-\infty, +\infty)$.
40. Suppose that f' exists on $[a, b]$. If $f'(a) < 0$ and $f'(b) > 0$, then show that there is some $c \in (a, b)$ such that $f'(c) = 0$. (*Hint:* Show that f is decreasing near a and increasing near b, so there must be a minimum value between a and b. Then use Theorem 1.3.)
41. Use the result of Exercise 40 to prove the *intermediate value theorem for derivatives:* If f' exists on $[a, b]$ and $f'(x_1) < f'(x_2)$ for $x_1, x_2 \in [a, b]$, then for any k between $f'(x_1)$ and $f'(x_2)$, there exists $c \in (a, b)$ such that $f'(c) = k$. (*Hint:* Consider $g(x) = f(x) - kx$.)
42. Suppose that f and g are increasing and differentiable on $[a, b]$. Show that
    (a) $f + g$ is increasing.
    (b) $f \cdot g$ may or may not be increasing on $[a, b]$.
43. Show that if the differentiable function f is increasing on $[a, b]$, then $f'(x) \geq 0$ for all $x \in (a, b)$.
44. Show that if the differentiable function f is decreasing on $[a, b]$, then $f'(x) \leq 0$ for all $x \in (a, b)$.

3.4 Concavity / The Second-Derivative Test

The last section showed how to use the derivative f' to determine where the graph of a function f rises or falls. This led to the first-derivative test for classifying critical points. It is called the *first-derivative* test because there is also a *second-derivative* test for local extreme values. Just as the first-derivative test grew out of the question of finding where the graph of f rises and where it falls, so to the second-derivative test arises from a graphical concept.

Consider a function f on an open interval where its second derivative f'' is positive. On such an interval, f' is an increasing function, by Theorem 3.1. Hence, the slopes of the tangent lines to the graph of f grow larger as x increases. In **Figure 4.1,** we show some tangents at various points. To fit a graph (see the dashed outline in Figure 4.1) to the tangent lines, near each point $(x_i, f(x_i))$ of tangency, the curve has to lie *above* its tangent line at the corresponding point. That is, for x near x_i, we must have

(1) $$f(x) > T(x),$$

where the tangent line has the equation $y = T(x) = f(x_i) + f'(x_i)(x - x_i)$. Inequality (1) is the algebraic requirement for the tangent line to lie below the graph of f near x_i. We single out that geometric property for further study.

FIGURE 4.1

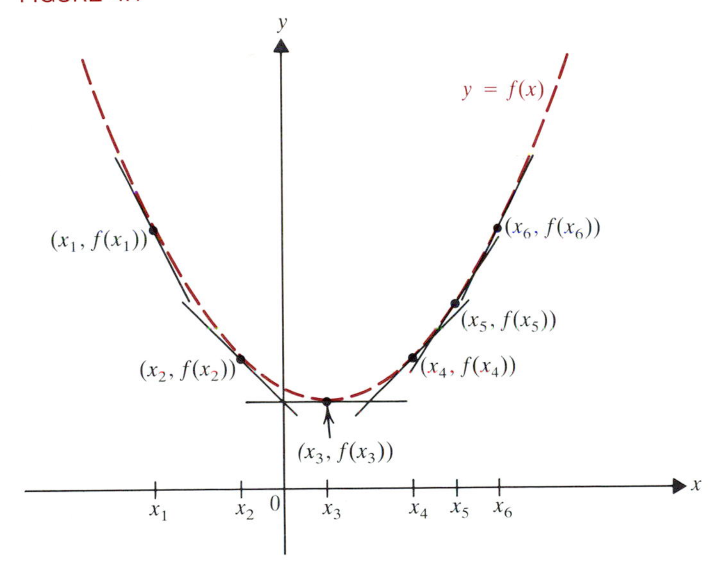

FIGURE 4.2

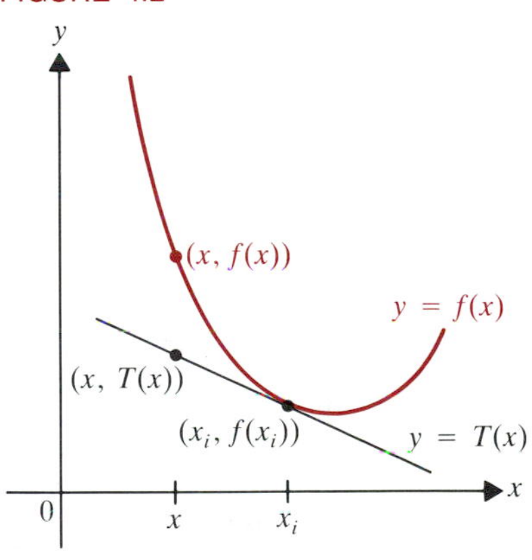

4.1 DEFINITION Let c be a point where f is differentiable. The function f is ***concave up*** (or ***upward***) at c if there is some open interval I containing c such that for all $x \neq c$ in I,

(1) $$f(x) > f(c) + f'(c)(x - c) = T(x).$$

Thus a function is concave up at c if for all values of x near c, the point $(x, f(x))$ on the graph of $y = f(x)$ lies above the tangent line

$$y = T(x) = f(c) + f'(c)(x - c)$$

to the graph at the point $(c, f(c))$. See **Figure 4.2.** As Figure 4.1 shows, if f is concave up over an entire interval, then its graph appears concave when viewed from above. This accounts for the term *concave up*. If we imagine water being poured down into the graph of a function that is concave up, then the graph will tend to "hold" the water.

There is a symmetric notion to concave up.

4.2. DEFINITION Let c be a point where f is differentiable. The function f is ***concave down*** (or ***downward***) at $x = c$ if there is some open interval I containing c such that for all $x \neq c$ in I,

$$f(x) < f(c) + f'(c)(x - c). \tag{2}$$

Thus f is concave down at c if for all x near c, the point $(x, f(x))$ on the graph of f lies *below* the tangent line $y = T(x) = f(c) + f'(c)(x - c)$ to the graph at $(c, f(c))$. See **Figure 4.3.** As **Figure 4.4** illustrates, if f is concave down over an entire interval, then the graph of f appears concave when viewed from below. This time, if we imagine water being poured onto the graph from above, it will run off.

FIGURE 4.3

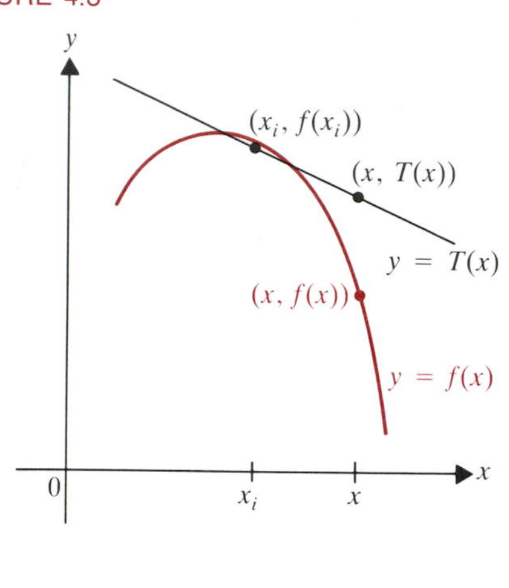

FIGURE 4.4

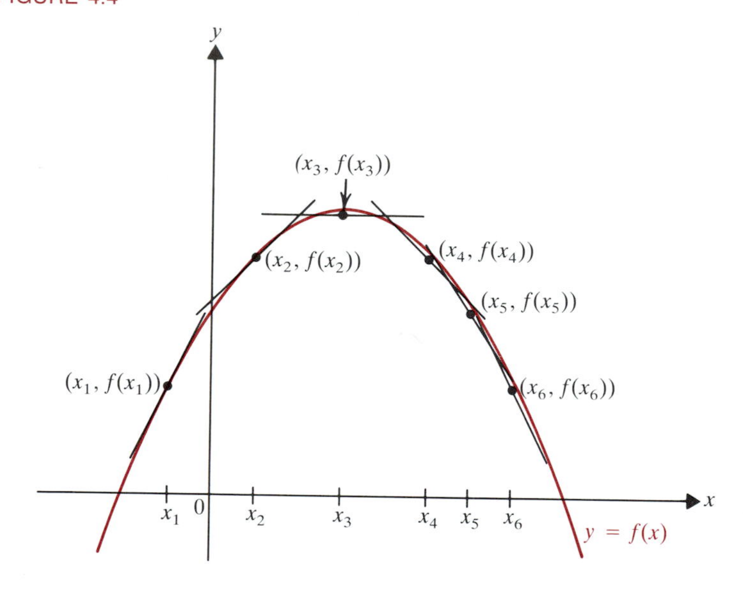

To apply Definitions 4.1 and 4.2 directly, we would have to draw both the graph and several tangent lines. But the discussion leading up to Definition 4.1 suggests an easier way to determine the concavity of f. The following result formalizes the remarks we made there about the sign of f''. It may be thought of as an analogue of Theorem 3.1, since it gives a test for determining a geometric feature (concavity) of f by means of differentiation. Its proof, which like that of Theorem 3.1 is based on the mean value theorem, can be found at the end of the section.

4.3 THEOREM

Suppose that f'' exists on some open interval I.

(a) If $f''(x) > 0$ for all $x \in I$, then f is concave up on I.

(b) If $f''(x) < 0$ for all $x \in I$, then f is concave down on I.

If we use Theorem 4.3 to determine the concavity of f on intervals in its domain and identify the local maximum and local minimum points of f, then we can sketch a pretty accurate graph of f. As the next examples illustrate, it is helpful to make a table to analyze the behavior of f'', just as we did in the last section for f'.

Example 1

For the function f defined by $f(x) = x^5 - 10x^3 - 35x + 4$, determine where f is concave up and where it is concave down. Sketch its graph.

Solution. The first step is to find $f'(x)$ and $f''(x)$. We have

$$f'(x) = 5x^4 - 30x^2 - 35 = 5(x^4 - 6x^2 - 7) = 5(x^2 - 7)(x^2 + 1),$$

$$f''(x) = 20x^3 - 60x = 20x(x^2 - 3) = 20x(x - \sqrt{3})(x + \sqrt{3}).$$

Since $x^2 + 1 > 0$ for every x, the following table is sufficient to identify local extreme points and intervals where f is monotonic.

x:	$(-\infty, -\sqrt{7})$	$-\sqrt{7}$	$(-\sqrt{7}, \sqrt{7})$	$\sqrt{7}$	$(\sqrt{7}, +\infty)$ → **R**
$x - \sqrt{7}$:	$-$		$-$		$+$
$x + \sqrt{7}$:	$-$		$+$		$+$
$f'(x)$:	$+$	0	$-$	0	$+$
Conclusions:	Increasing		Decreasing		Increasing
		Local maximum point		Local minimum point	

We use the factors of $f''(x)$ to construct the table

x:	$(-\infty, \sqrt{3})$	$-\sqrt{3}$	$(-\sqrt{3}, 0)$	0	$(0, \sqrt{3})$	$\sqrt{3}$	$(\sqrt{3}, +\infty)$ → **R**
$x - \sqrt{3}$:	$-$		$-$		$-$		$+$
x:	$-$		$-$		$+$		$+$
$x + \sqrt{3}$:	$-$		$+$		$+$		$+$
$f''(x)$:	$-$	0	$+$	0	$-$	0	$+$
Conclusions:	Concave down		Concave up		Concave down		Concave up

With the help of these tables, we can draw an accurate graph without plotting many points. Using the table of values

x	-4	$-\sqrt{7}$	$-\sqrt{3}$	0	$\sqrt{3}$	$\sqrt{7}$	4
y	-240	152.16	100.99	4	-92.99	-144.16	248

in conjunction with the information on extreme values and concavity, we get the sketch shown in **Figure 4.5.** ■

FIGURE 4.5

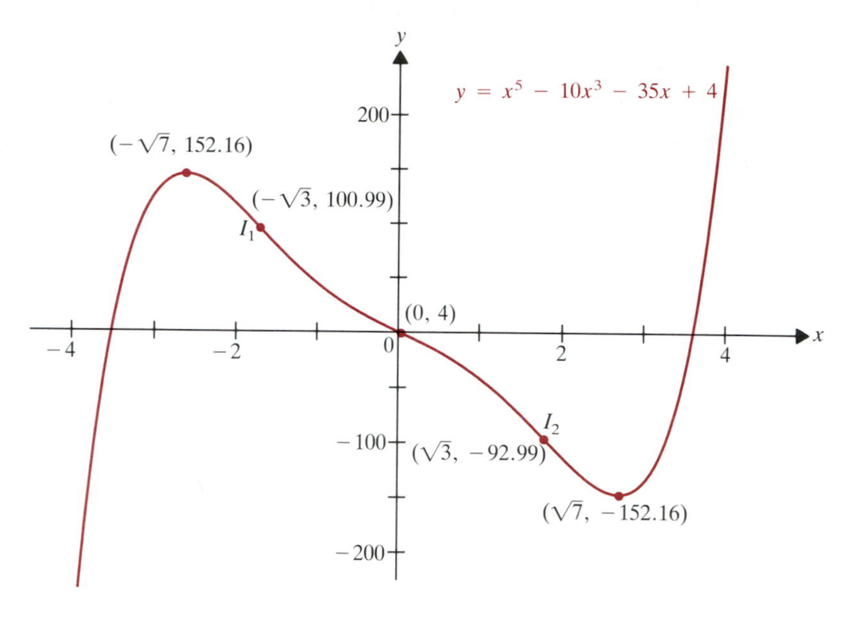

Example 2

For the function g of Example 3 of the last section,

$$g(x) = x^{1/3}(x-3)^{2/3},$$

determine where g is concave up and where it is concave down.

Solution. In the last section, we found that

$$g'(x) = \frac{x-1}{x^{2/3}(x-3)^{1/3}}.$$

This is analyzed in the table in Example 3, p. 159. The quotient rule gives

$$g''(x) = \frac{x^{2/3}(x-3)^{1/3} - (x-1)\left[\frac{2}{3}x^{-1/3}(x-3)^{1/3} + \frac{1}{3}x^{2/3}(x-3)^{-2/3}\right]}{x^{4/3}(x-3)^{2/3}}$$

$$= \frac{x(x-3) - (x-1)\left[\frac{2}{3}(x-3) + \frac{1}{3}x\right]}{x^{5/3}(x-3)^{4/3}}$$

multiplying numerator and denominator by $x^{1/3}(x-3)^{2/3}$

$$= \frac{x^2 - 3x - (x-1)(x-2)}{x^{5/3}(x-3)^{4/3}}$$

$$= \frac{x^2 - 3x - x^2 + 3x - 2}{x^{5/3}(x-3)^{4/3}} = \frac{-2}{x^{5/3}(x-3)^{4/3}}.$$

Notice that $g''(x)$ is never 0, but is undefined at $x = 0$ and $x = 3$. The following table analyzes the sign of $g''(x)$.

FIGURE 4.6

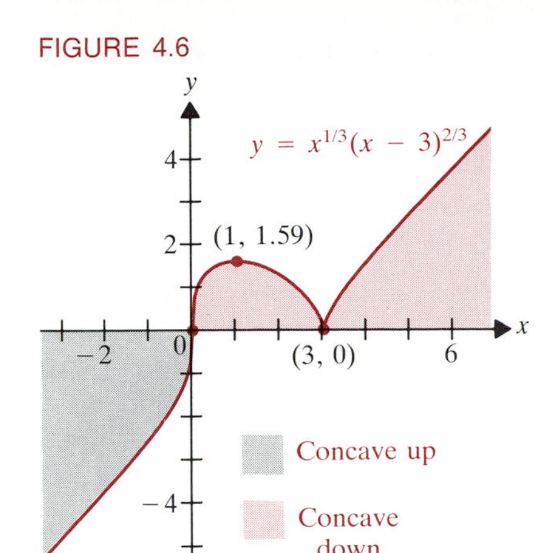

x:	$(-\infty, 0)$	0	$(0, 3)$	3	$(3, +\infty)$
$x^{5/3}$:	−		+		+
$(x-3)^{4/3}$:	+		+		+
$x^{5/3}(x-3)^{4/3}$:	−		+		+
$g''(x)$:	+	Does not exist	−	Does not exist	−
Conclusions:	Concave up		Concave down		Concave down

Again, the conclusions in the final row seem compatible with the graph shown in **Figure 4.6,** which is a reproduction of Figure 3.7. Points like $x = 0$ in Example 2 and $x = -\sqrt{3}, 0$, and $\sqrt{3}$ in Example 1, where f changes from upward to downward concavity (or vice versa), are given a special name.

4.4 DEFINITION Suppose that f'' exists on an open interval containing c except possibly at c itself. The point $P(c, f(c))$ is an ***inflection point*** (or ***point of inflection***) of the graph of f if f changes concavity at c. That is, there is some interval I containing c such that either

(3) $f''(x) < 0$ for $x < c$ in I and $f''(x) > 0$ for $x > c$ in I

or

(4) $f''(x) > 0$ for $x < c$ in I and $f''(x) < 0$ for $x > c$ in I.

Notice particularly that an inflection point is a point of the graph of f. Thus there can be an inflection point at $x = c$ *only* if f is defined at c. A second necessary condition for an inflection point to exist at $x = c$ can be gleaned from (3) and (4). If (3) holds, then by Theorem 3.1 the first derivative f' of f is decreasing for $x < c$ in I and is increasing for x beyond c in I. Similarly, if (4) holds, then f' is increasing for $x < c$ in I and is decreasing for $x > c$. In either case, f' has a local extreme value at c. Theorem 1.3 then gives the following necessary condition for inflection points.

> If $P(c, f(c))$ is an inflection point of the graph of the function f, then either $f''(c) = 0$ or else $f''(c)$ fails to exist.

To find inflection points, therefore, we need to find all points c such that $f''(c) = 0$ or $f''(c)$ fails to exist. We then have to test all such candidates c to determine whether there is really an inflection point at $x = c$. Just as not every critical point of f is a local extreme point, so it can easily happen that a point where $f''(c) = 0$ or $f''(c)$ fails to exist may not be a point of inflection.

Example 3

Find all inflection points of the function
(a) f in Example 1 (b) g in Example 2

Solution. (a) In Example 1, we saw that

$$f''(x) = 20x(x - \sqrt{3})(x + \sqrt{3})$$

is zero at the points $x = 0$, $x = \sqrt{3}$, and $x = -\sqrt{3}$. From the table for f'' in Example 1, we see that f'' changes sign at $x = -\sqrt{3}$, $x = 0$, and at $x = \sqrt{3}$. Thus each of these points is a point of inflection. In Figure 4.5 we have labeled these three inflection points. The points where $x = -\sqrt{3}$ and $x = \sqrt{3}$ are labeled I_1 and I_2. The point where $x = 0$ is labeled by its coordinates (0, 4).

(b) From the solution of Example 2,

$$g''(x) = \frac{-2}{x^{5/3}(x - 3)^{4/3}}.$$

This is never 0, but it does fail to exist at $x = 0$ and $x = 3$. The table in the solution of Example 2 shows that at $x = 0$, f'' changes from positive to negative. Thus (0, 0) is an inflection point. However, at $x = 3$, f'' fails to change sign. On both sides of $x = 3$, the function is concave down. Hence, (3, 0) is *not* an inflection point of the graph. As we saw in Example 3 of the last section, it is a local low point of the graph. See Figure 4.6. ■

The foregoing leads to a test for deciding whether critical points of f give a local maximum or a local minimum. This test goes back to the English mathematician Colin Maclaurin (1698–1746) in 1742.

4.5 THEOREM

> ***Second-Derivative Test.*** Suppose that $f'(c) = 0$ and $f''(x)$ exists on some interval I surrounding the point c.
>
> (a) If $f''(c) > 0$, then c is a local minimum point.
>
> (b) If $f''(c) < 0$, then c is a local maximum point.

FIGURE 4.7

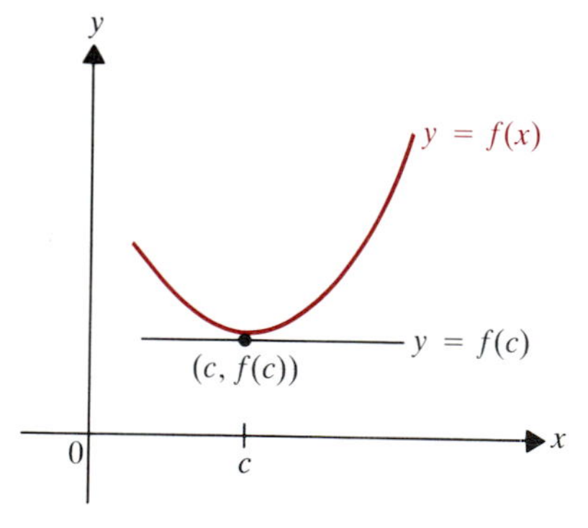

Proof. We prove (a) and leave (b) for Exercise 35. Since $f'(c) = 0$ and $f''(c) > 0$, the graph of f has a horizontal tangent at $(c, f(c))$, and f is concave upward at c. Thus the graph of f lies above its tangent $y = f(c)$ at $(c, f(c))$, as shown in **Figure 4.7.** For $x \neq c$ near c, (1) and the fact that $f'(c) = 0$ give

$$f(x) > f(c).$$

This means that c is a local minimum point of f. QED

It is important to note that *the second-derivative test gives no information at all if $f''(c) = 0$ or if $f''(c)$ fails to exist.* Thus it is not as decisive a test as the first-derivative test. In particular, it cannot be applied to a function such as $g(x)$ in Example 2 at the critical point $c = 3$. Nevertheless, it is useful for functions whose second derivatives are easy to calculate.

Example 4

Find all local extreme points of $f(x) = \frac{1}{4}x^4 - x^3 - \frac{1}{2}x^2 + 3x + 1$.

Solution. We have

(5)
$$\begin{aligned} f'(x) &= x^3 - 3x^2 - x + 3 = x^2(x - 3) - (x - 3) \\ &= (x^2 - 1)(x - 3) = (x - 1)(x + 1)(x - 3), \end{aligned}$$

FIGURE 4.8

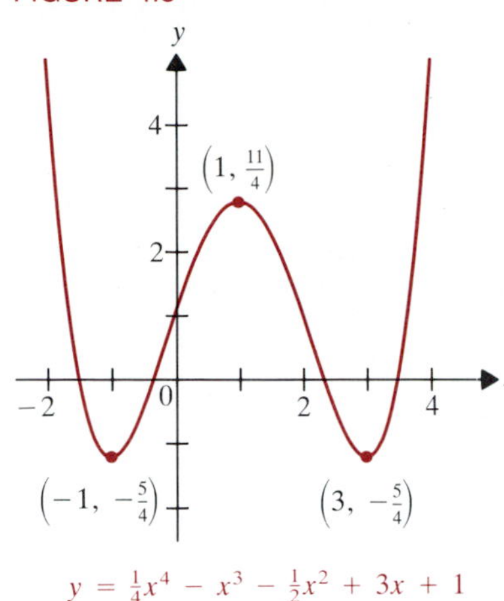

$y = \frac{1}{4}x^4 - x^3 - \frac{1}{2}x^2 + 3x + 1$

so the critical points are $x = -1$, $x = 1$, and $x = 3$. Since a table analyzing the sign of f' would be rather large, it is easier to compute $f''(x)$ and use the second-derivative test. We get from (5)

$$f''(x) = 3x^2 - 6x - 1.$$

We thus find

$$f''(-1) = 3 + 6 - 1 > 0, \qquad f''(1) = 3 - 6 - 1 < 0,$$

and

$$f''(3) = 27 - 18 - 1 > 0.$$

By the second-derivative test, f has a local minimum at $x = -1$ and at $x = 3$, and a local maximum at $x = 1$. The table of values

x	-2	-1	0	1	2	3	4
y	5	$-5/4$	1	$11/4$	1	$-5/4$	5

was used to make the sketch shown in **Figure 4.8.** You might wish to check that there are inflection points at $x = 1 + 2/\sqrt{3}$ and $x = 1 - 2/\sqrt{3}$. ■

Our final example shows that when $f''(c) = 0$ for a critical point c, virtually *anything* can happen.

Example 5

What is the nature of the critical point $c = 0$ of f if
(a) $f(x) = x^3$? (b) $f(x) = x^4$? (c) $f(x) = 1 - x^4$?

Solution. (a) We have $f'(x) = 3x^2$ and $f''(x) = 6x$, so that $f'(c) = f''(c) = 0$ at $c = 0$. Since $f'(x) > 0$ for all $x \neq 0$, we see that $c = 0$ is *not* a local extreme value point, by Theorem 3.2. In fact, $(0, 0)$ is an inflection point since $f''(x) < 0$ for $x < 0$, and $f''(x) > 0$ for $x > 0$. See **Figure 4.9.**

(b) Here $f'(x) = 4x^3$ and $f''(x) = -12x^2$, so that $f'(c) = f''(c) = 0$ at $c = 0$. Since $f'(x) < 0$ for $x < 0$ and $f'(x) > 0$ for $x > 0$, we see that $c = 0$ is a local minimum point, by Theorem 3.2. See **Figure 4.10.**

(c) This time, $f'(x) = -4x^3$ and $f''(x) = -12x^2$, so that $f'(c) = 0$ at $c = 0$. Since $f'(x) > 0$ for $x < 0$ and $f'(x) < 0$ for $x > 0$, we see that $c = 0$ is a local maximum point, by Theorem 3.2. See **Figure 4.11.** ■

FIGURE 4.9

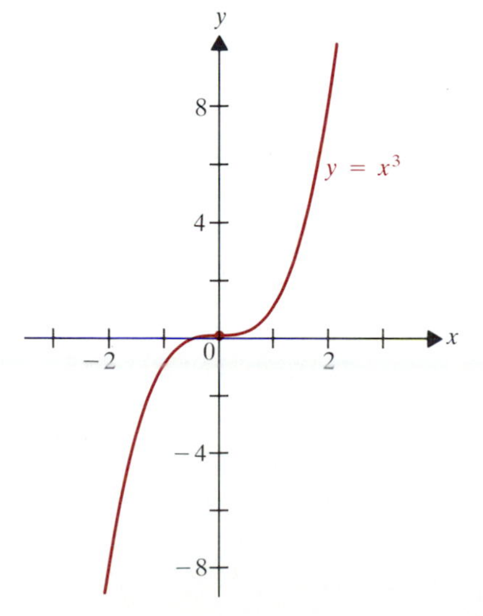

FIGURE 4.10

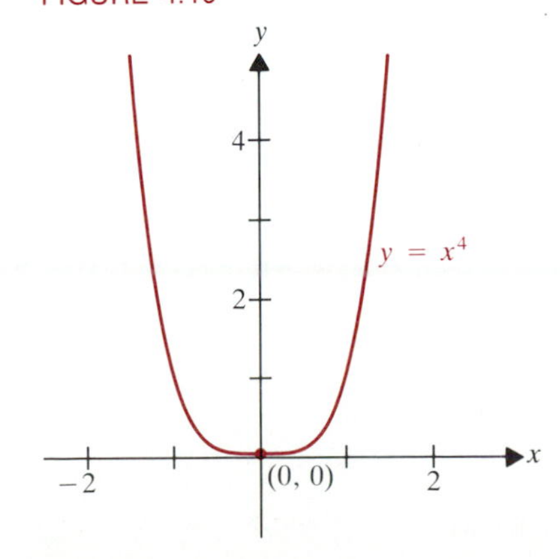

FIGURE 4.11

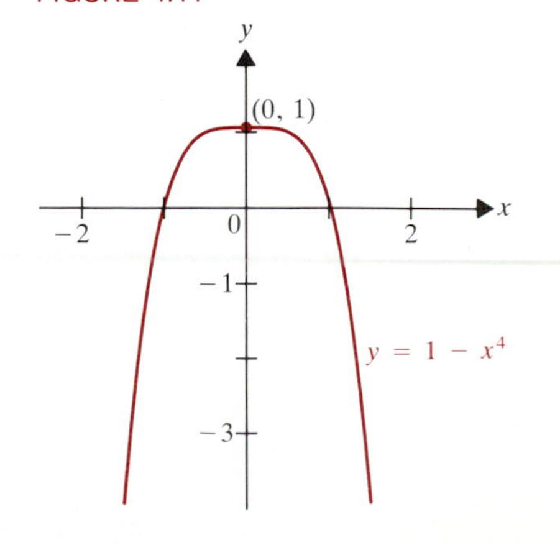

Proof of Theorem 4.3*

4.3 THEOREM Suppose that f'' exists on some interval I.

(a) If $f''(x) > 0$ for all $x \in I$, then f is concave up on I.

(b) If $f''(x) < 0$ for all $x \in I$, then f is concave down on I.

Proof. We leave (b) for Exercise 33. To prove (a), pick any $x_i \in I$. Then the equation of the tangent line T to the graph of f at $(x_i, f(x_i))$ is

$$y = f(x_i) + f'(x_i)(x - x_i) = T(x).$$

For any $x \neq x_i$ in I, we thus have

$$f(x) - T(x) = f(x) - f(x_i) - f'(x_i)(x - x_i). \tag{6}$$

Since f'' exists throughout I, f is differentiable on I. It thus satisfies the hypotheses of the mean value theorem on the subinterval of I with endpoints x and x_i. Therefore, for some c in the open interval between x and x_i, we have

$$f(x) - f(x_i) = f'(c)(x - x_i). \tag{7}$$

Substituting (7) into (6), we obtain

$$f(x) - T(x) = f'(c)(x - x_i) - f'(x_i)(x - x_i),$$

$$f(x) - T(x) = [f'(c) - f'(x_i)](x - x_i). \tag{8}$$

By hypothesis, $f''(x) > 0$ for all x in I. It thus follows from Theorem 3.1 that f' is an increasing function on I. Hence, the two factors on the right side of (8) have the same sign. For instance, if $x - x_i > 0$, then

$$x > c > x_i \rightarrow f'(c) > f'(x_i) \rightarrow f'(c) - f'(x_i) > 0 \rightarrow f(x) - T(x) > 0,$$

so that (1) holds for any $x \neq x_i$ in I. By Definition 4.1 then, f is concave upward. QED

Exercises 3.4

In Exercises 1–10, for the corresponding exercise in Exercises 3.3 determine where f is concave upward, where it is concave downward, and where there are inflection points.

In Exercises 11–32, find all local extreme values of the given function. Use the second-derivative test when possible.

11. $f(x) = 5x^2 - 3x - 8$

12. $f(x) = 8 + 3x - 2x^2$

13. $f(x) = \frac{1}{3}x^3 - 4x^2 + 12x - 3$

14. $f(x) = \frac{2}{3}x^3 - 2x^2 - 6x + 2$

15. $f(x) = \frac{1}{4}x^4 - \frac{2}{3}x^3 - \frac{1}{2}x^2 + 2x - 1$

16. $f(x) = \frac{1}{4}x^4 - \frac{4}{3}x^3 - \frac{1}{2}x^2 + 4x - 1$

17. $f(x) = 1 + 2x^2 - x^4$

18. $f(x) = 3 + 8x^2 - x^4$

19. $f(x) = x^4 - 4x^3 + 4x^2 - 3$

20. $f(x) = x^3(x - 4)^4 + 1$

21. $f(x) = x^5 - 5x + 2$

22. $f(x) = x^5 - \frac{20}{3}x^3 + 15x + 6$

23. $f(x) = x^5 - x^4 + 3$

24. $f(x) = 10x^6 - 24x^5 + 15x^4 - 2$

25. $f(x) = \dfrac{x^2 - 8}{x - 2}$

26. $f(x) = \dfrac{x^2 + 3}{x + 2}$

27. $f(x) = 1 + \sin^2 x$

28. $f(x) = \cos^2 x + 4$

29. $f(x) = \csc x$

30. $f(x) = \sec x$

31. $f(x) = x + \sin 2x$

32. $f(x) = \sin 2x + \cos 2x$

33. Prove Theorem 4.3(b).

34. In the proof of Theorem 4.3(a), show that in (8) if $x - x_i < 0$, then $f'(c) - f'(x_i) < 0$, and hence $f(x) - T(x) > 0$.

* Optional

35. Prove Theorem 4.5(b).

36. If f'' exists on an open interval I containing x and f is concave upward on I, then show that $f''(x) \geq 0$ for all $x \in I$.

37. Show that $f(x) = ax^2 + bx + c$ has a local minimum if $a > 0$ and a local maximum if $a < 0$.

38. Show that $f(x) = x^2 + 1/x$ has a local minimum but no local maximum.

39. Show that $(0, 0)$ is an inflection point of the graph $f(x) = x|x|$.

40. Show that $f''(0)$ fails to exist if $f(x) = x^2 + x|x|$. Is $(0, 0)$ an inflection point?

41. Give an example of a function f that is concave upward at a point $x = c$ but for which $f''(c)$ fails to exist.

42. Give an example of a function f that is concave downward at a point $x = c$ but for which $f''(c)$ fails to exist.

43. In optimization theory, a function f is said to be *convex* on an open interval I if for a and $b \in I$ and $t \in [0, 1]$

(9) $$f(ta + (1 - t)b) \leq tf(a) + (1 - t)f(b).$$

If a function f is concave upward in the sense of Definition 4.1, then show that (9) holds.

44. If a function f is concave downward in the sense of Definition 4.2 then show that for a and b in an open interval I and $t \in (0, 1)$,

(10) $$f(ta + (1 - t)b) \geq tf(a) + (1 - t)f(b).$$

(This is the definition of a *concave* function f on I in optimization theory.)

3.5 Applications Involving Extreme Values

We began our discussion of optimization in Section 2 with an applied problem: What per-unit rent in a large apartment building produces maximum revenue? In this section, we use the techniques developed in the past three sections to analyze several more applied optimization problems. The first of these is geometric.

Example 1

A discount department store manager wants to fence in a portion of the parking lot to make a rectangular garden shop with 800 square feet of display space. If the fence is to hook onto the wall of the store as in **Figure 5.1,** then find the dimensions that require the least amount of fencing material.

FIGURE 5.1

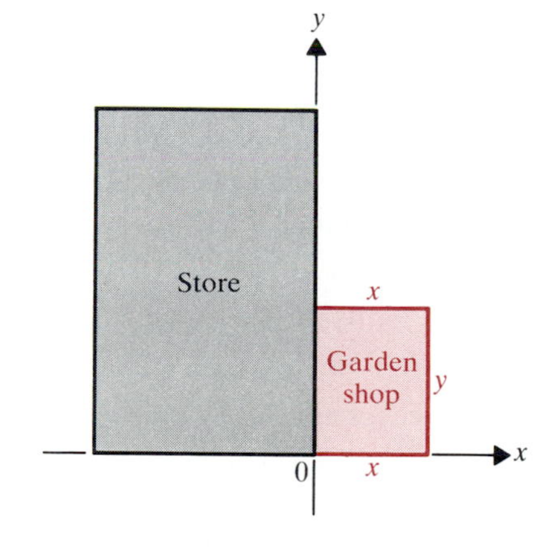

Solution. In Figure 5.1, the dimensions of the garden shop are x and y, where x is the length of fence perpendicular to the wall, in feet, and y is the length of fence parallel to the wall, in feet. Here x and y must be positive. The quantity to be minimized is the number of feet of fencing used to make the rectangle:

(1) $$F = 2x + y.$$

It appears that we have to minimize a function of *two* variables, x and y. But they are related by the condition that the garden shop must have area 800 ft^2. This means that

(2) $$xy = 800 \to y = \frac{800}{x} = 800x^{-1}.$$

Substitution of (2) into (1) gives

(3) $$F = 2x + 800x^{-1}.$$

Since x must be positive, we solve the problem by minimizing (3) over the open interval $(0, +\infty)$. The solution must be a local minimum, since the endpoint $x = 0$ is not in the domain of the function defined by (3). Differentiation of (3) gives

$$\frac{dF}{dx} = 2 - 800x^{-2}. \tag{4}$$

This is undefined at $x = 0$, but that is not a critical point since it is not a point of the domain $(0, +\infty)$ of F. The critical points of F are therefore found by solving $dF/dx = 0$ for x:

$$2 = \frac{800}{x^2} \rightarrow 2x^2 = 800 \rightarrow x^2 = 400 \rightarrow x = 20. \tag{5}$$

(The other root of (5), $x = -20$, is not in the interval $(0, +\infty)$, so we ignore it.) Since

$$\frac{d^2F}{dx^2} = 1600x^{-3} > 0 \qquad \text{when } x = 20,$$

we conclude that F is minimized when $x = 20$. From (2), the corresponding value of y is $800/20 = 40$ ft. The most economical dimensions for the garden shop, therefore, are 40 by 20 ft. ■

The steps followed in Example 1 suggest the following general procedure for solving applied optimization problems. (A sequence of steps designed to solve a problem or class of problems is called an *algorithm.*)

OPTIMIZATION ALGORITHM

1. Identify the central quantities in the problem and introduce variables $x, y, \ldots$ to represent them. Draw a figure, if appropriate.
2. Express the quantity z to be maximized or minimized in terms of the other variable(s).
3. If z is a function of two (or more) variables, then translate verbal relationships among the variables into mathematical equations. Use those to express z as a function of just one variable: $z = f(x)$.
4. Find the critical points of the function f in Step 3.
5. Test the critical points in Step 4 and any endpoints to find the required extreme point of f.
6. Check that the solution found in Step 5 is reasonable.

In Example 1, the central quantities referred to in Step 1 were the dimensions of the garden shop, its required area of 800 ft^2, and the amount of fencing used. It is important to write out precisely what each variable represents, as we did in Example 1. If a problem asks *what value*, *how much*, *when*, or a similar question, then it is usually helpful to let x stand for the quantity referred to in the question. For instance, in Example 1 we let x and y stand for the dimensions of the rectangle. We also noted that the allowable domain for each was $(0, +\infty)$. It is always helpful, and often essential, to determine the domain of the variables in Step 1.

We carried out Step 2 in Equation (1), which expresses the perimeter F of the enclosing fence in terms of the variables x and y. In Example 1, we had to express the area condition as

(2) $$xy = 800,$$

and solve this for y in terms of x for the perimeter F to become a function of the single variable x [Equation (3)]. As is often the case, a figure (Figure 5.1) was helpful in seeing the relationship (2) between the variables x and y that was needed to express y in terms of x.

Steps 4 and 5 simply say to apply the techniques developed earlier in the chapter to find the extreme points of f. We determine where $f'(x) = 0$ and locate any points in the domain of f where f is not differentiable. If the domain of f is a closed interval I, then we have to compare the values of f at its critical points to its values at the endpoints of I. (See the boxed program on p. 152, and the first- and second-derivative tests.)

In applied problems, the final step is often the most important of all. For example, if your work finds that a negative price maximizes profit, then you know that there is an error to look for!

In Example 2 of Section 2, we found that rental revenue in a 250-unit apartment building would be maximized by raising the rent on each unit from \$200 to \$725. Management policy is to maximize *profit*, rather than just gross revenue from rent. The next example proceeds from this point, using the model of Exercise 1 of Section 2: Each \$10.00 increase in rent results in five vacancies. As we work through the example, notice how we follow the six steps suggested above.

Example 2

Suppose that the apartment building of Example 2 of Section 2 costs \$100 per unit per month to operate when each unit is rented. Because of increased unit costs for insurance and other items that do not decline when units become vacant (taxes, personnel, etc.), for each 5 units that become vacant, average operating costs increase \$7.00 per unit. What rent should be charged for maximum monthly profit?

Solution. As in Example 2 of Section 1, the controllable variable for management is the rent. We thus let

$$x = \text{the number of \$10 increases in rent.}$$

Then

(7) $200 + 10x =$ the number of dollars rent per unit,

(8) $250 - 5x =$ the number of units occupied,

(9) $100 + 7x =$ the number of dollars in operating costs per unit.

The total monthly revenue at this level of rent is the product of the number of units rented times the monthly rent per unit. Therefore, from (7) and (8),

(10) $$R(x) = (250 - 5x)(200 + 10x).$$

In the same way, we see from (8) and (9) that the total monthly operating cost is

(11) $$C(x) = (250 - 5x)(100 + 7x).$$

Therefore, from (10) and (11), the monthly profit is

$$\begin{aligned} P(x) &= R(x) - C(x) \\ &= (250 - 5x)(200 + 10x) - (250 - 5x)(100 + 7x) \\ &= (250 - 5x)[200 + 10x - (100 + 7x)] = (250 - 5x)(100 + 3x), \\ &= 25{,}000 + 250x - 15x^2. \end{aligned}$$

Since $P(50) = 0$, we can maximize P over the interval $[0, 50]$. We have

$$P'(x) = 250 - 30x.$$

Hence, $P'(x) = 0$ when $x = 25/3$. Since $P''(x) = -36 < 0$, we see that $x = 25/3$ produces a local maximum. Since $P(0) = 25{,}000$, $P(50) = 0$, and $P(25/3) = 26041.67$, for maximum profit we conclude that per unit rent should be raised $(25/3)(10) = 83.33$ dollars, to about \$283.33. (Since rents are usually in whole dollar amounts, management might decide to implement a series of eight \$10.00 increases, ending at \$280.00 per month rent. For $x = 8$, profit is $P(8) = 210 \cdot 124 = 26{,}040$ dollars.) ■

Our solution to Example 2 measures up better on the reasonableness test than the one found in Section 2. It stands to reason that management would find a series of eight increases in rent acceptable, especially if they were phased in over a period of time. The resulting 40 vacancies would be offset by a boost of more than \$1000 per month in profits and would still leave an occupancy rate of 84%, usually regarded as an acceptable level in an apartment building.

One way to make cardboard boxes is to cut the corners out of a rectangular piece of cardboard and fold up the edges to make the sides. See **Figure 5.2,** where we show a pattern for making such a box.

FIGURE 5.2

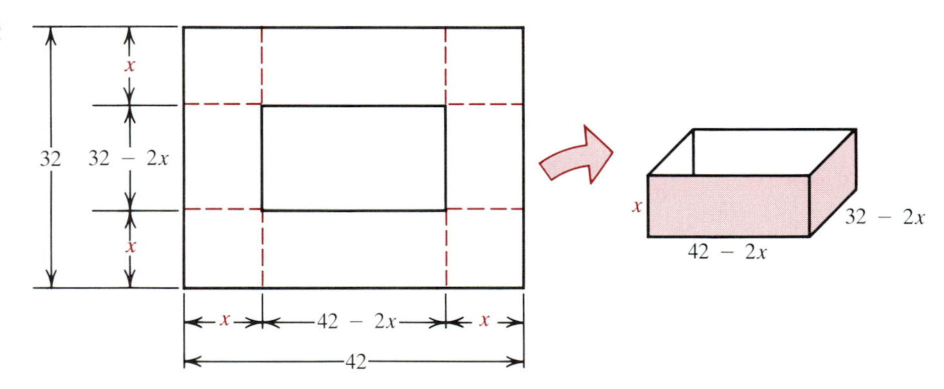

Example 3

Squares of cardboard are cut out from a 32 by 42 in. piece of cardboard, and the sides are folded up as in Figure 5.2 to make a box. What side length should the cut-out squares have to produce a box of maximum volume?

Solution. We let x be the number of inches in the side of the piece cut out. Figure 5.2 shows that the resulting box has height x, length $42 - 2x$, and width $32 - 2x$. Thus the volume V of the box is given by

$$\begin{aligned} V(x) &= x(42 - 2x)(32 - 2x) = 4x(21 - x) \cdot (16 - x) \\ &= 4x(336 - 37x + x^2) = 4(336x - 37x^2 + x^3). \end{aligned}$$

Before proceeding, we point out that we can confine x to the closed interval [0, 16]. This is so because the length of material cut out must not exceed either dimension of the cardboard. That is,

$$2x \leq 32 \qquad \text{and} \qquad 2x \leq 42.$$

Hence, $x \leq 16$. This lets us use the technique of Section 2 to maximize V. Differentiating, we obtain

$$V'(x) = 4(336 - 74x + 3x^2) = 4(3x - 56)(x - 6). \tag{12}$$

Thus $V'(x) = 0$ when

$$3x - 56 = 0 \quad \text{or} \quad x - 6 = 0 \quad \rightarrow \quad x = \frac{56}{3} \quad \text{or} \quad x = 6.$$

The point $x = 56/3$ is not a *feasible* solution, because it is not in [0, 16]. So we need test only $x = 6$. From (12),

$$V''(x) = 4(-74 + 6x),$$

so $V''(6) = 4(-74 + 36) < 0$. Hence, $x = 6$ gives a local maximum. Since $V = 0$ when $x = 0$ or $x = 16$, the absolute maximum of V doesn't occur at an endpoint. Therefore $x = 6$ produces the maximum value of V. In that case, we have a box of length 30, width 20, and height 6. The volume is therefore 3600 cubic in., a little over 2 cubic ft. ■

Example 3 asked how big a container could be made from a given amount of material. We can also ask for the most economical container that will hold a given volume. For example, consider cylindrical cans that hold certain prescribed volumes of liquid. The company packaging the liquid products wants its container costs to be as small as possible.

Example 4

Some motor oil cans are cylindrical with cardboard sides and metal top and bottom. Outside the United States, the standard volume for an oil can is one liter. If the metal costs twice as much per square centimeter as the cardboard, what dimensions should the can have to minimize the cost?

FIGURE 5.3

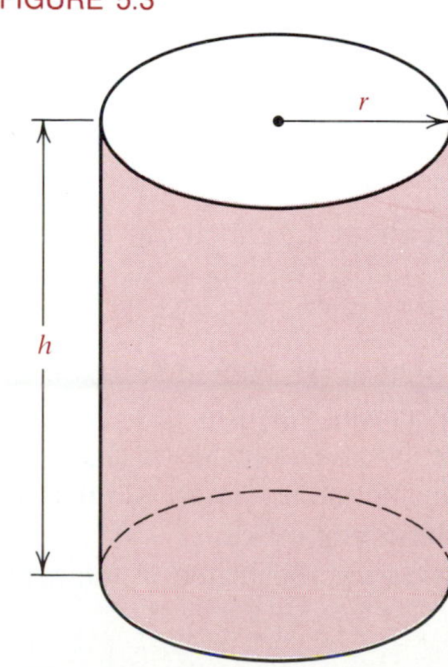

Solution. Referring to **Figure 5.3,** we let

- r = the radius of the can in centimeters,
- h = the height of the can in centimeters,
- V = the volume of the can in cubic centimeters.

Then

$$V = \pi r^2 h = 1000 \rightarrow h = 1000/\pi r^2.$$

since 1 liter = 1000 cm^3. Here r must be a positive real number. Recall that the top and bottom of the can have total area

$$\pi r^2 + \pi r^2 = 2\pi r^2. \tag{13}$$

The cardboard side has area

$$2\pi r h = 2\pi r \cdot \frac{1000}{\pi r^2} = \frac{2000}{r}. \tag{14}$$

(To see that the lateral area is $2\pi rh$, imagine slicing the side vertically and opening it. The result is a rectangle of height h and length $2\pi r$, the circumference of the top and bottom of the can. See **Figure 5.4.**) Suppose that the cardboard costs c dollars per square centimeter. Then the metal costs $2c$ dollars/cm^2. Thus, from (13) and (14), the total cost of the can is

$$C(r) = \frac{2000}{r} \cdot c + 2 \cdot 2\pi r^2 c = 2c(1000r^{-1} + 2\pi r^2).$$

FIGURE 5.4

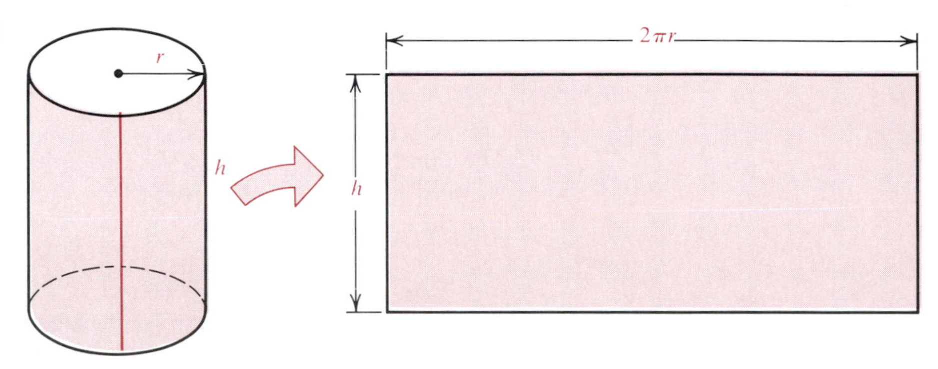

Since c is constant, we have

$$C'(r) = 2c(-1000r^{-2} + 4\pi r) = 2c\left(4\pi r - \frac{1000}{r^2}\right).$$

Thus

$$C'(r) = 8c\left(\frac{\pi r^3 - 250}{r^2}\right) \tag{15}$$

is zero when

$$\pi r^3 = 250 \rightarrow r = \sqrt[3]{250/\pi} \approx 4.30127.$$

$C'(r)$ fails to exist when $r = 0$, which is not a *feasible* value for r (since then $V = 0$ rather than 1000). From (15), we see that $C'(r)$ changes from negative to positive at $r = \sqrt[3]{250/\pi}$, so this value gives an absolute minimum: C decreases up to this point and then increases. The corresponding value of h is

$$\frac{1000}{\pi r^2} \approx 17.21.$$

Thus the most economical can should have radius about 4.3 cm (about 1.69 in.) and height about 17.2 cm (about 6.77 in.). ■

Actual quart (slightly less than a liter) oil cans sold in the United States have radius about 5 cm and height about 14 cm, indicating the top and bottom probably cost slightly less than double the sides. This also suggests that our solution to Example 4 is not only reasonable but actually is quite similar to the analysis made by the can manufacturers in designing oil cans.

Another common problem encountered in business is illustrated in the next example.

Example 5

A cable television company has to lay cable in an area with underground utilities. Two subdivisions are located on opposite sides of a river, one a bit upstream from the other as in **Figure 5.5.** The company must connect points A and B, where A is 1000 ft north and 1000 ft west of B. If it costs \$16 per foot to lay cable underground and \$20 per foot to lay it underwater, then what is the most economical way to lay the cable?

FIGURE 5.5

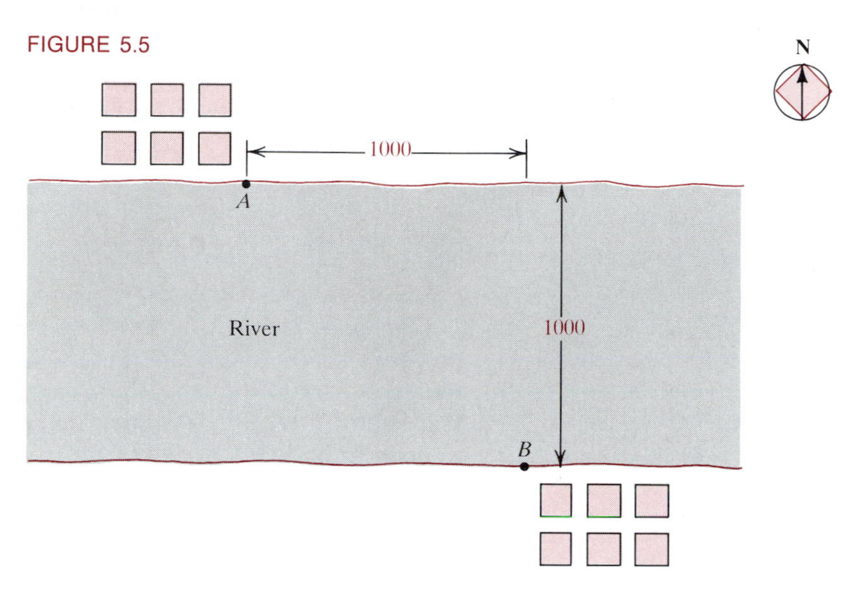

FIGURE 5.6

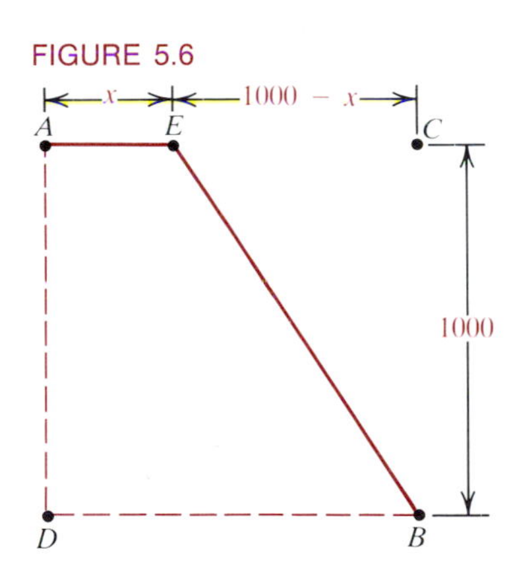

Solution. To analyze this problem, we suppose that the company proceeds east from A for x feet to E and then lays submerged cable across the river to B, as suggested by **Figure 5.6.** To include the full range of feasible possibilities, we restrict x to the closed interval $[0, 1000]$. If $x = 0$, the company puts *all* the cable underwater. If $x = 1000$, it lays cable first to C and then crosses under water. (That would cost the same as going first to D and then to B. In general, then, all *costs* are represented by this assumption even though for each path, there is a symmetric one that goes under water *first* from A.) The underground portion costs

$$16x \tag{16}$$

dollars to lay. The underwater portion costs

$$20 \cdot d(E, B) \tag{17}$$

dollars. Since triangle ECB is a right triangle,

$$d(E, B) = \sqrt{1000^2 + (1000 - x)^2}. \tag{18}$$

Hence, from (16), (17), and (18), the total cost of laying cable along path AEB is

$$C(x) = 16x + 20[1000^2 + (1000 - x)^2]^{1/2}. \tag{19}$$

Thus

$$C'(x) = 16 + 10[10^6 + (10^3 - x)^2]^{-1/2} \cdot 2(10^3 - x)(-1)$$

$$= 16 - \frac{20(10^3 - x)}{[10^6 + (10^3 - x)^2]^{1/2}}. \tag{20}$$

Hence, $C'(x) = 0$ when

$$16[10^6 + (10^3 - x)^2]^{1/2} = 20(10^3 - x) \to 4[10^6 + (10^3 - x)^2]^{1/2} = 5(10^3 - x),$$

$$16[10^6 + (10^3 - x)^2] = 25(10^3 - x)^2 \to 16 \cdot 10^6 = (25 - 16)(10^3 - x)^2,$$

$$\frac{16}{9} 10^6 = (10^3 - x)^2 \to \pm\frac{4}{3} 10^3 = 10^3 - x,$$

$$\pm\frac{4000}{3} = 1000 - x \to x = 2333.33 \quad \text{or} \quad -333.33.$$

But neither of these values of x is in the interval $[0, 1000]$! Thus the *only* candidates for an answer are the endpoints $x = 0$ and $x = 1000$. We find that

$$C(0) \approx 28284.27 \quad \text{and} \quad C(1000) = 36{,}000.$$

So the cheapest path is one that proceeds east from A to the point C and then under the river to B. ■

Our final examples illustrate that the inherent efficiency of many natural processes makes optimization useful in science. In the first case, the efficiency referred to relates to the path taken by light. *Fermat's principle* in optics says that light will take the path that minimizes the time of travel. This enables us to derive a basic fact about reflection of light from a surface.

Example 6

Use Fermat's principle to show that if light strikes a surface from which it is reflected, then the angle θ_1 of incidence in **Figure 5.7** equals the angle θ_2 of reflection.

FIGURE 5.7

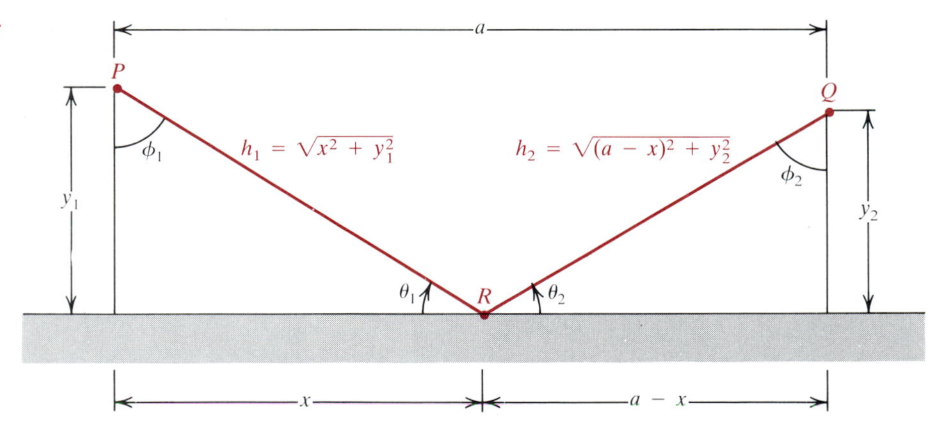

Solution. In Figure 5.7, the light proceeds from P to Q via some intermediate point R at the speed c of light, where $c = 2.998 \times 10^8$ m/s is a constant. The time t required for the light to travel that path is

$$(21) \qquad t = \frac{d(P, R) + d(R, Q)}{c} = \frac{1}{c}\{[x^2 + y_1{}^2]^{1/2} + [(a - x)^2 + y_2{}^2]^{1/2}\}$$

where x varies over the interval $[0, +\infty)$. Since the elevations y_1 and y_2 of P and Q above the surface are fixed, (21) expresses t as a function of x. To find the value of x that minimizes t, we differentiate (21), getting

$$\frac{dt}{dx} = \frac{1}{c} \cdot \frac{1}{2}(x^2 + y_1^2)^{-1/2} \cdot (2x) + \frac{1}{2}[(a-x)^2 + y_2^2]^{-1/2} \cdot 2(a-x)(-1)$$

$$= \frac{1}{c}\left[\frac{x}{\sqrt{x^2 + y_1^2}} - \frac{a-x}{\sqrt{(a-x)^2 + y_2^2}}\right].$$

Thus $dt/dx = 0$ when

$$\frac{x}{\sqrt{x^2 + y_1^2}} - \frac{a-x}{\sqrt{(a-x)^2 + y_2^2}} = 0 \rightarrow \frac{x}{\sqrt{x^2 + y_1^2}} = \frac{a-x}{\sqrt{(a-x)^2 + y_2^2}}.$$

From Figure 5.7 this holds when

$$\cos\theta_1 = \cos\theta_2. \tag{22}$$

Since θ_1 and θ_2 are two angles between 0 and π, (22) holds precisely when $\theta_1 = \theta_2$. Noting that if $x = 0$, then $a = 0$ and so $\theta_1 = \theta_2 = \pi/2$, we see that the minimum time of path, whether it occurs for $x > 0$ or at the endpoint $x = 0$, entails $\theta_1 = \theta_2$. ■

In optics, angle ϕ_1 (rather than θ_1) is called the *angle of incidence* and ϕ_2 is called the *angle of reflection*. Since $\theta_1 = \theta_2$, it follows that $\phi_1 = \phi_2$ also, since $\phi_1 + \theta_1 + \pi/2 = \pi = \phi_2 + \theta_2 + \pi/2$.

FIGURE 5.8

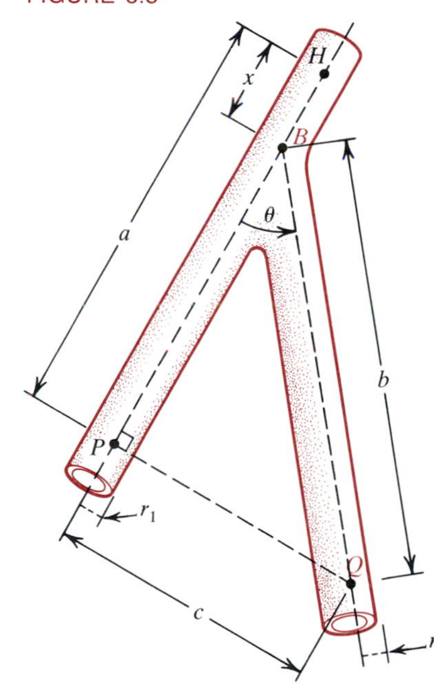

We close the section by considering a mathematical model for arterial networks in man and other mammals. Blood flows from the heart to various vital organs in the body. In **Figure 5.8,** suppose that a main artery flows from H to P, and an organ Q is located at a perpendicular distance c from HP as shown. A branch artery is formed to carry blood to Q. Where along HP is the branch point B? This is equivalent to asking what is the angle θ between the main artery HP and the branch artery BQ. The principle of natural efficiency says that B will be located so as to minimize the energy loss E due to the functioning of the blood pumping system. That energy loss is given by a law of Jean L. M. Poiseuille (1799–1869), a French physiologist who pioneered the study of blood flow. Poiseuille's law says that in an artery of length l and radius r,

$$E = \frac{kl}{r^4}, \tag{23}$$

where k is a fixed constant of proportionality. In Figure 5.8, consider the total energy loss E in the system of arteries HB and BQ. By (23), this is

$$E = \frac{kx}{r_1^4} + \frac{kb}{r_2^4}, \tag{24}$$

where x is the length HB, b is the length BQ, and r_1 and r_2 are the radii of the main branch arteries. We want to express E in terms of the angle $\theta = PBQ$. From the figure,

$$\cos\theta = \frac{a-x}{b} \quad \text{and} \quad \sin\theta = \frac{c}{b}. \tag{25}$$

Thus

$$a - x = b\cos\theta \rightarrow x = a - b\cos\theta. \tag{26}$$

We also get from (25)

$$b = \frac{c}{\sin\theta} = c\csc\theta.$$

Substituting this into (26), we have

$$x = a - \frac{c}{\sin\theta}\cos\theta \to x = a - c\cot\theta. \tag{27}$$

Substituting (27) into (24), we obtain

$$E = k\left(\frac{a - c\cot\theta}{r_1^{\,4}} + \frac{c\csc\theta}{r_2^{\,4}}\right). \tag{28}$$

This gives E as a function of the branching angle θ. We can now find θ for minimum E.

Example 7

What value of θ in (28) minimizes E?

Solution. Differentiating with respect to θ and recalling that r_1 and r_2 are constants, we get

$$\frac{dE}{d\theta} = k\left(\frac{c\csc^2\theta}{r_1^{\,4}} - \frac{c\csc\theta\cot\theta}{r_2^{\,4}}\right) = kc\csc^2\theta\left(\frac{1}{r_1^{\,4}} - \frac{\cot\theta}{r_2^{\,4}\csc\theta}\right).$$

Thus $dE/d\theta = 0$ when

$$\frac{1}{r_1^{\,4}} = \frac{1}{r_2^{\,4}}\,\frac{\cos\theta/\sin\theta}{1/\sin\theta}, \tag{29}$$

since $\csc^2\theta$ is never 0. Solving (29) for $\cos\theta$ we get

$$\cos\theta = \frac{r_2^{\,4}}{r_1^{\,4}} = \left(\frac{r_2}{r_1}\right)^4. \tag{30}$$

Thus θ should be a first-quadrant angle whose cosine is $(r_2/r_1)^4$. It is not hard to show this results in a local minimum, which is in fact an absolute minimum (Exercise 45). ■

In many species of animals, the angle θ between a main and a branch artery and the arterial radii r_1 and r_2 very nearly satisfy (30).

Exercises 3.5

1. In 1979, the Organization of Petroleum Exporting Countries (OPEC) set its selling price for crude oil at \$28.00 per barrel. Sales were approximately 24 million barrels per day at that price. Prices were then raised, eventually reaching \$36.00 per barrel briefly. It seemed that for each \$1.00 increase in the cost of a barrel, demand fell by about one million barrels per day. Assuming this, what price per barrel would maximize revenue for OPEC?

2. Suppose in Example 2 that for each \$10 increase in rent, six units become vacant. Find the most profitable rent. How many units will be rented and what will be the monthly profit?

3. An orange grove containing 60 trees per acre yields an average of 500 oranges per tree. For each additional tree planted per acre, the yield per tree drops by 5 oranges. How many trees should be planted for maximal yield per acre?

4. A rapid transit system carries 80,000 passengers per day at a fare of 50¢ per ride. For each 10¢ increase in fare, surveys predict ridership will drop by 1000 passengers. What fare maximizes revenue?

5. A box with a lid is made from the pattern shown in **Figure 5.9(a).** Cuts are made along the colored lines, and folds are made along the dashed lines to produce the box pictured in Figure 5.9(b). Find the dimensions of the box of largest volume that can be made from the cardboard in this way.

FIGURE 5.9

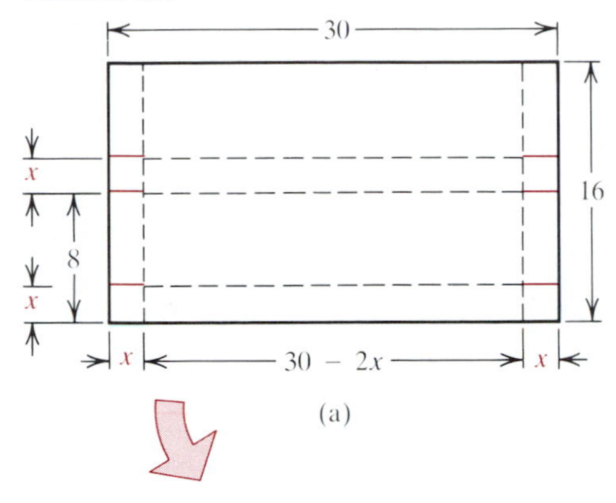

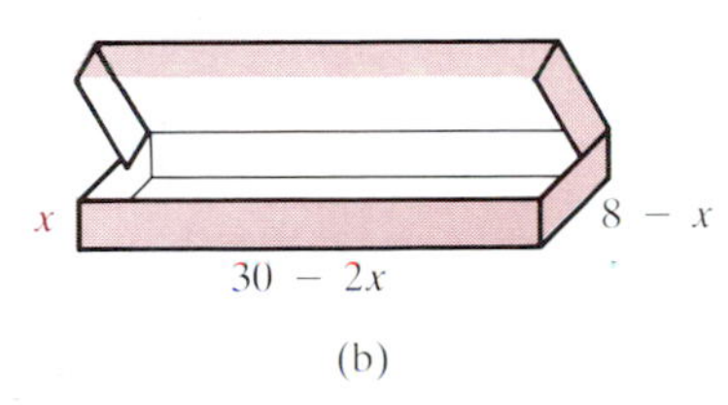

6. Suppose in Exercise 5 that a box is to be made from 144 square inches of cardboard with fixed height $x = h$. Your job is to determine the width w and length l ($= 144/w$) of the pieces of cardboard that the box maker should order if boxes of maximum volume are desired.

(a) Express the volume V of the box as a function of w.

(b) Recalling that h is constant, show that V is maximized when $w = 12$. Thus what shape will the pieces of cardboard have?

(c) Using the value $w = 12$ found in (b), find the value of the height h that maximizes the volume of the box.

(d) What is the maximum volume of the box, and what are its dimensions?

7. A manufacturer wants to make wooden crates with square bases and no top, of capacity 4 cubic feet. Find the dimensions that minimize the amount of wood used.

8. A box with square base and no top is to be made from 12 square feet of cardboard. What dimensions give the largest volume?

9. A standard cylindrical soup can has capacity of about 316 cm^3, about 10 3/4 fluid ounces. Find the dimensions that require the minimum amount of material.

10. A cylindrical can of soy sauce has capacity 1.77 liters (60 fl. oz.). What dimensions require the minimum amount of material?

11. In Example 5, if it costs \$12 per foot to lay the cable underground and \$20 per foot to lay it underwater, then find the most economical path.

12. If in Example 5 it costs twice as much to lay the cable underwater as to lay it underground, then how should it be laid to minimize cost?

13. A man in a rowboat B on a lake is 1 mile from the nearest shore point A. See **Figure 5.10.** He hears a fire alarm and must report to the fire station, which is 5 miles down shore from A. If he can row 4 miles per hour and run 5 miles per hour, how should he travel to reach the fire station most quickly?

FIGURE 5.10

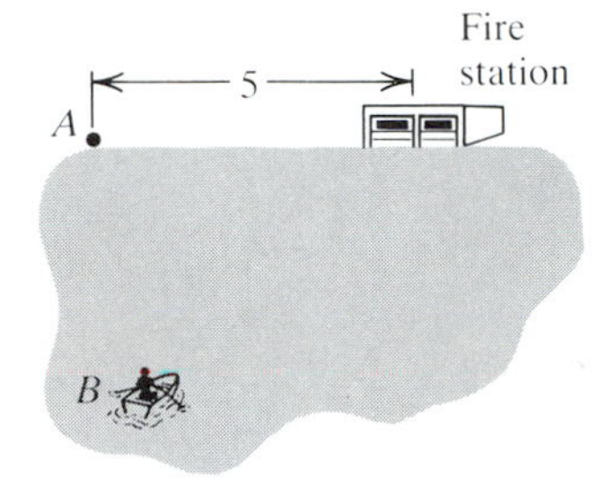

14. **(a)** If the man in Exercise 13 can row only 3 miles per hour, then how should he proceed?

(b) If the fire station in Exercise 13 is only two miles down shore from A, and the man can row 4 miles per hour, then how should he proceed?

15. A woman wishes to fence a rectangular garden plot bordering a canal, with no fencing along the canal. If she has 40 ft of fencing, how large a garden can be made?

16. Show that among all rectangles of fixed perimeter p, the one with largest area is a square.

17. Show that among all rectangles of fixed area A, the one with minimal perimeter is a square.

18. A trapezoid has three sides of length l. What is the largest possible area it can enclose? (*Hint:* The area of a trapezoid is half the sum of the bases times the height. See **Figure 5.11.**)

FIGURE 5.11

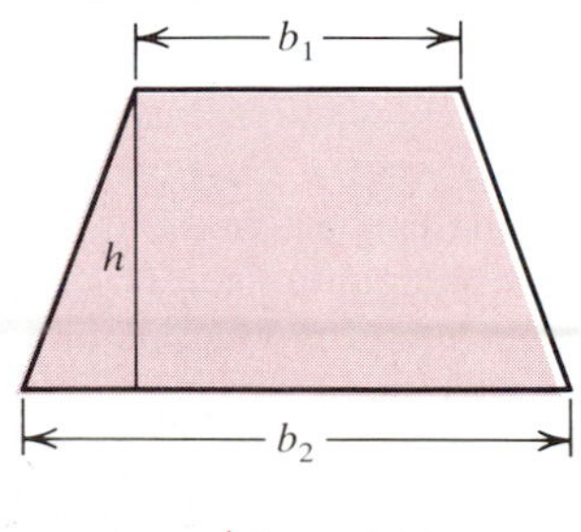

$A = \frac{1}{2}(b_1 + b_2)h$

19. A rain gutter is made by bending up the edges of a piece of aluminum as in Figure 5.12(a). The aluminum is 25 in.

wide. If both sides of the gutter have the same height and are perpendicular to the bottom of the gutter, then find h so that the flow capacity of the gutter will be as large as possible. (The flow capacity is proportional to the cross-sectional area. See **Figure 5.12(a)**.)

FIGURE 5.12

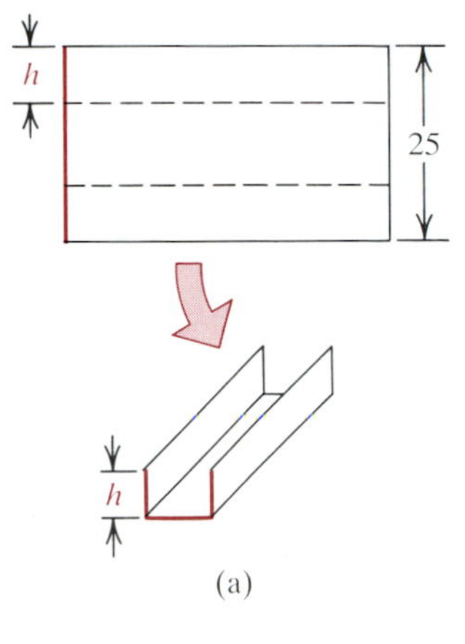

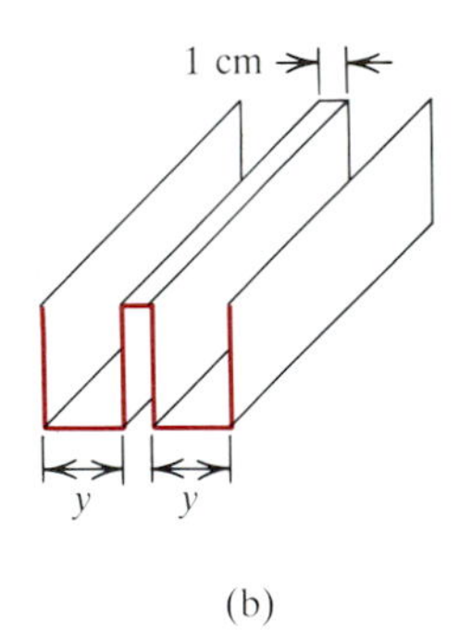

20. A certain manufacturing process produces two types of waste liquid. One must be treated before being discharged from the plant. The other can be discharged directly. You are asked to design a double gutter to carry the two products away from the production area. You decide to fold a 25-in. piece of aluminum as in Figure 5.12(b). How high should the gutter be for maximum flow rate?

21. In chemistry, a reaction whose product acts as a catalyst is called *autocatalytic*. (A catalyst is a substance that alters the rate of a reaction without being consumed by it.) Suppose that an autocatalytic reaction begins with an amount a of substance A. Suppose that the rate v of reaction is jointly proportional to the amount of product and the amount of substance A left. Suppose further that x moles of product require x moles of A for their production.

(a) Write a formula for v.

(b) Find the value of x that corresponds to $v_{\max}$, the maximum rate of reaction.

22. When running, a four-legged animal at times has all four legs off the ground (floating phase of a gallop) and at other times two legs on the ground as its legs propel it forward (stepping phase). Suppose that during each stepping phase the animal covers a units and during each floating phase it covers b units. The quantity $j = b/a$ is called the jumpiness of the animal's gait. It can be shown* that the work done per unit time in running is

$$W = Aj\frac{L^4}{s} + B\frac{s^3L^2}{1+j}, \tag{31}$$

where L is the length of the animal, s is its speed, and A and B are constants.

(a) Find dW/dj.

(b) Find j corresponding to $dW/dj = 0$.

(c) Show that for the optimal value of j, corresponding to minimum W, $j + 1$ is proportional to s^2 and inversely proportional to L.

(d) Explain why j increases with s. Interpret in terms of a horse walking, trotting, and galloping.

(e) Explain why the larger the animal, the lower j tends to be. (It is in fact almost 0 for an elephant compared with 0.3 for a horse.)

23. *Snell's law.* When light travels from P in one medium to Q in another, it bends (refracts). See **Figure 5.13.** Snell's law states that

$$\frac{\sin\phi_1}{\sin\phi_2} = \frac{c_1}{c_2},$$

where c_1 is the speed of light in the first medium, and c_2 is the speed of light in the second medium. (Both c_1 and c_2 are constant.)

(a) Express the time t of travel as a function of x in Figure 5.13.

FIGURE 5.13

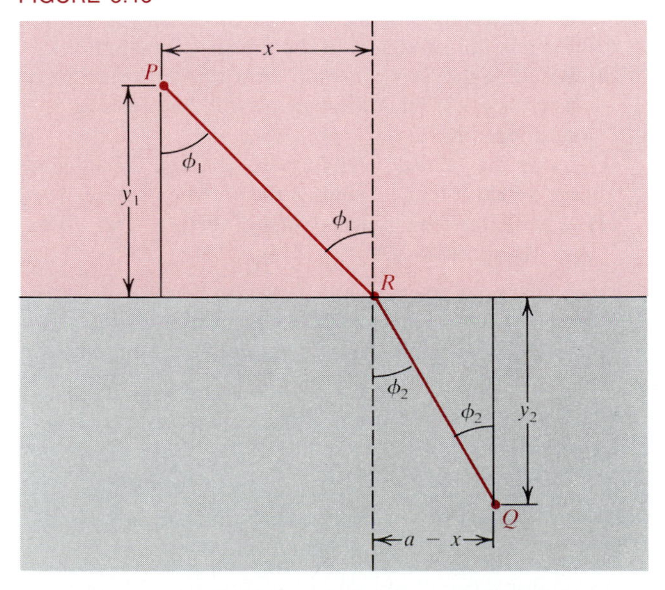

* See, for instance, J. M. Smith, *Mathematical Ideas in Biology* (Cambridge: Cambridge University Press, 1968), p. 15.

(b) Show that t minimal (Fermat's principle) gives Snell's law.

24. Suppose that we assume Snell's law as a basic law of optics.
(a) Show that Fermat's principle then holds for the path followed by the light in Figure 5.13.
(b) Show that Fermat's principle also holds for the path followed by the light in Figure 5.7.

25. A sawmill foreman wishes to make a rectangular wooden beam of largest possible cross-sectional area from a log of uniform diameter two feet. See **Figure 5.14.**
(a) Show that the width x and height y of the beam satisfy $x^2 + y^2 = 4$.
(b) Show that the beam of maximal area is square of edge $\sqrt{2}$ feet.

FIGURE 5.14

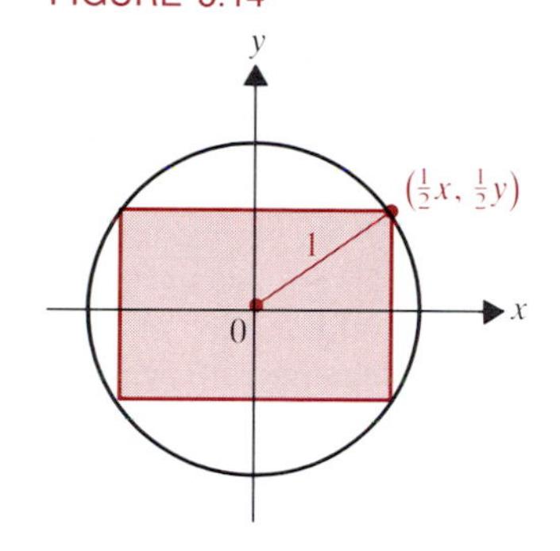

26. Show that the rectangle of largest area which can be inscribed in a circle of radius a is a square of edge $\sqrt{2}a$.

27. In Exercise 25, the strength of the beam is proportional to xy^2. What are the dimensions of the *strongest* beam that can be cut from the log?

28. If the *stiffness* of the beam in Exercise 25 is proportional to xy^3, then what are the dimensions of the stiffest beam that can be cut from the log?

29. A *Norman window* consists of a rectangle surmounted by a semicircular region, as in **Figure 5.15.** The amount of light passing through the window is proportional to the area of the glass. Find the dimensions of the window admitting maximal light if the perimeter is 10 feet.

FIGURE 5.15

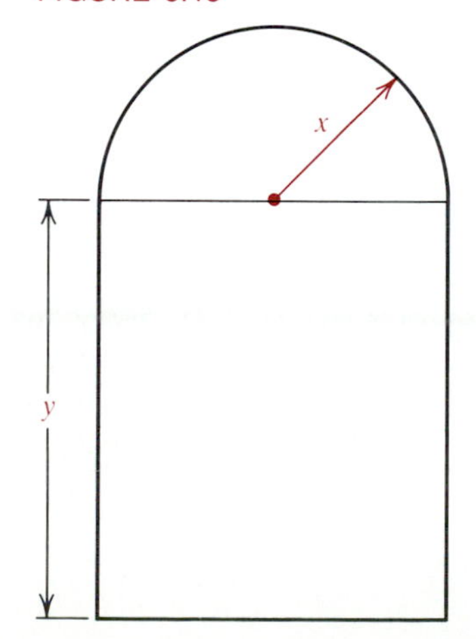

30. Find the trapezoid of largest area that can be inscribed in the region bounded by the graph of $y = 4 - x^2$ and the x-axis. See **Figure 5.16.**

FIGURE 5.16

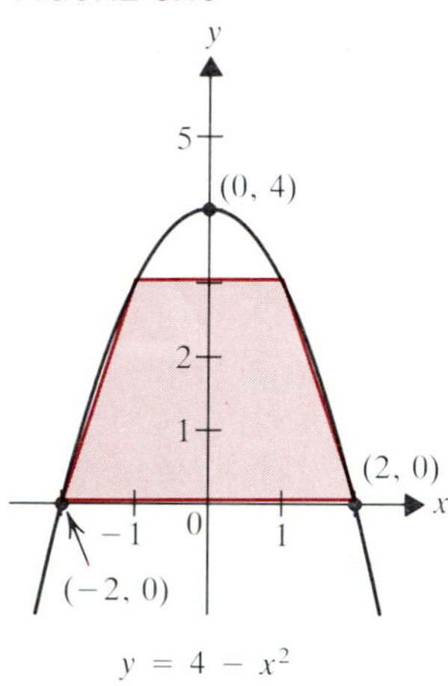

31. The illumination at P from a light source at O is proportional to the strength of the source and inversely proportional to the square of the distance from O to P. Suppose two sources of strengths y_1 and y_2 are a units apart at O_1 and O_2. At what point on the segment joining the two points O_1 and O_2 is the illumination least?

32. A woman sits in a park between two highways 100 yards apart. One highway noise level is 64 decibels; the other is 27 decibels. If noise level follows the same law as that for illumination in Exercise 31, then how far between the two highways should the woman sit if she wants the quietest spot?

33. Find the volume of the largest right circular cylinder that can be inscribed in a sphere of radius a. See **Figure 5.17.**

FIGURE 5.17

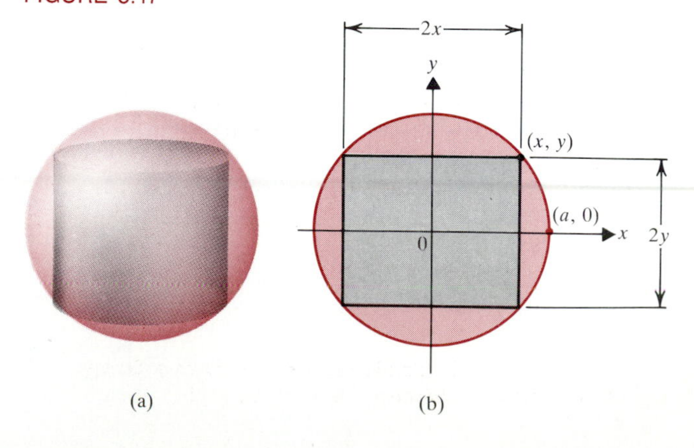

34. Find the volume of the largest right circular cylinder that can be inscribed in a right circular cone of radius r and height h. See **Figure 5.18.**

FIGURE 5.18

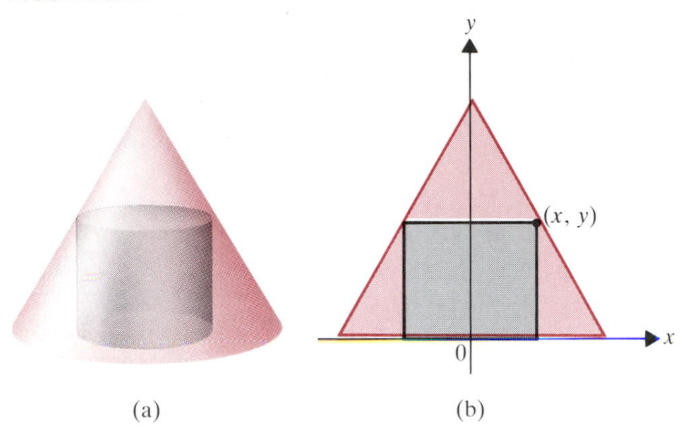

35. Two roads, Main (north–south) and Broadway (east–west) intersect at P. At noon, a westbound car passes P at a constant speed of 25 miles per hour. At that time, another car is 1 mile south of P, going north at 30 miles per hour. When are the two cars closest, and what is the distance between them then?

36. Ship A steams north past a point P at 20 knots. At that moment, ship B is 5 nautical miles west of P steaming east at 25 knots. When are the two ships closest, and what is their distance then?

37. Find which point on the line $x + y = 2$ is nearest (5, 2).

38. Find which point on the line $ax + by + c = 0$ is nearest the point (x_1, y_1). What is the minimum distance?

39. A calculus text is to have pages with area 72 square inches. There are to be one-inch margins on the right and left of each page, with half-inch margins on the top and bottom. What dimensions should be chosen for maximum printed area?

40. A tricolor flag is being designed for a new nation. It is to have red and white rectangles with a blue stripe between them. If the red and white parts are to have area 9 square ft and the entire perimeter is to be 14 ft, then what design maximizes the width of the blue stripe?

41. A 4-ft wide corridor meets a hall at right angles. A 16-ft pole is being pushed along the floor of the corridor. How wide must the hall be for the pole to make it around the corner? See **Figure 5.19.**

FIGURE 5.19

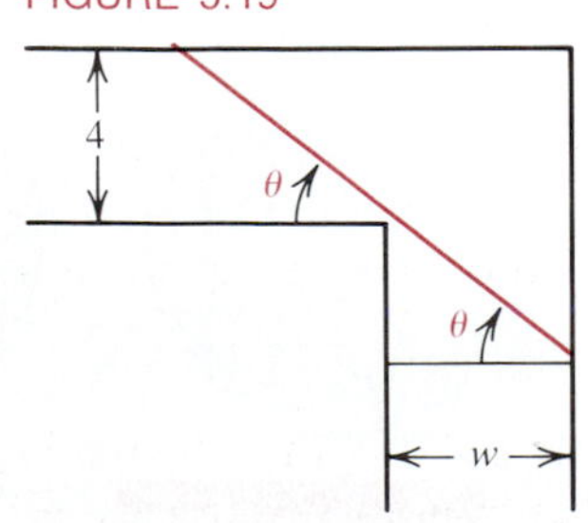

42. Find the trapezoid of largest area that can be inscribed in a semicircle of radius a. See **Figure 5.20.**

FIGURE 5.20

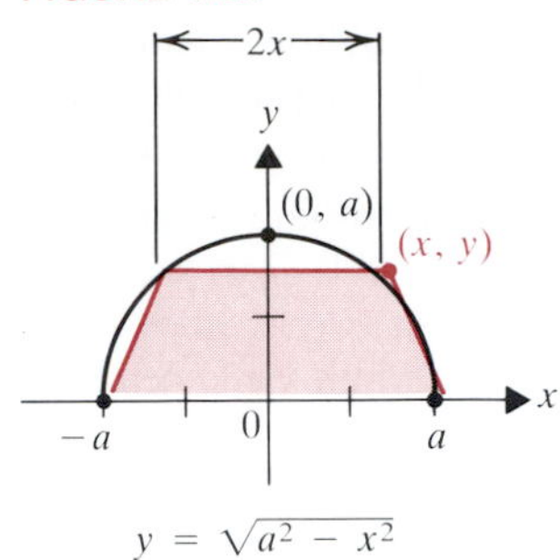

43. The reaction of a patient to a dosage level x of an allergy shot is modeled by

$$R = x^2\left(\frac{1}{2}M - \frac{1}{3}x\right),$$

where M is the maximum permissible dose of the drug, and R is a measure of redness and swelling at the site of the shot. The *sensitivity* of the body to the drug is defined to be the rate R' of change of R with respect to x. Find x so that R' is maximized. At this dosage, explain why the body is *really* most sensitive to the shot.

44. In coughing, the windpipe contracts in such a way as to impart maximal velocity to the air being expelled. A model for v is

$$v = cP\pi r^2,$$

where r is the radius of the windpipe during a cough, and P is the air pressure in the windpipe during a cough. P is modeled by

$$P = \frac{1}{k}(r_0 - r),$$

where r_0 is the normal radius of the trachea. Find r for maximum v. (Experiments have shown that the windpipe contracts to about this radius when a person coughs. Here c and k are constants.)

45. Show that the value of θ in (30) gives the absolute minimum of E.

46. The surface area of a single cell of a beehive is given by

$$S(x) = 6ck + \frac{3}{2}k^2(-\cot x + \sqrt{3}\csc x),$$

where $x \in (0, \pi/2)$, and k and c are constants. Show that S is minimized if $\cos x = 1/\sqrt{3}$. (The number x is a certain angle formed by two edges of the cell. Actual beehives have been found to have almost exactly this value of $\cos x$, another instance of the efficiency principle.)

HISTORICAL NOTE

Willebrord Snell van Royen (1591–1626) was professor of mathematics and Hebrew at the University of Leyden, Holland during the last 13 years of his short life. The law that carries his name was formulated by him during this period, but he never published it, or any of his work. His work is known through its citation by his countryman, Christian Huygens (1629–1695), who is regarded as the founder of the modern theory of light. Using a telescope, Huygens discovered Saturn's rings but is better known for the principle he formulated that bears his name and is basic to the study of wave motion. The first publication of Snell's law seems to have been by Descartes, who independently arrived at it in 1638.

3.6 Limits Involving Infinity

We turn now to the graphical use of calculus. This section is devoted to some new types of limits that are used to discover when the graph of a function f approaches a straight line. The first type is the limit of a function f as x increases without bound, or decreases without bound.

Such limits are of more than just graphical interest. For instance, you may some day go to your bookstore to sell back one of your textbooks, only to be told that a new edition is being published, so your text has no resale value. Why are new editions published so frequently? Publishers receive no revenue at all from the sale of used texts by bookstores. The longer a book is in print, the larger the number of used books for sale. Thus, even if a book is adopted by more colleges, total sales of new copies dwindle, and so do the profits realized by the publishing company. The following example suggests how sales decline.

Example 1

The sales of a calculus text during its tth year of publication are approximated by the function S, where

$$S(t) = \frac{20{,}000}{1 + \frac{1}{4}(t-1)^2}. \tag{1}$$

What is the long-term trend for the sales?

Solution. Using (1), we tabulate the sales in each year after publication.

t	1	2	3	4	5	6	7	8	9	10
$S(t)$	20,000	16,000	10,000	6153	4000	2758	2000	1509	1176	941

This shows a rapid fall-off in sales, as might be accounted for by resale of used copies. If the publisher needs to sell at least 5000 new copies in any year to realize a profit on the text, then after the fourth year the text is no longer profitable. ■

The table in Example 1 shows that the tenth-year sales are less than 5% of the first year total. What eventually happens to $S(t)$ if no new edition appears?

As t becomes large, $(t-1)^2$ becomes huge. Hence, the denominator of (1) becomes arbitrarily large as t increases without bound. Division of 20,000 by that enormous amount gives a tiny answer. As t gets larger and larger, $S(t)$ thus approaches 0. The statement "t becomes arbitrarily large" is written symbolically as $t \to +\infty$. This brings us to the following definition.

6.1 DEFINITION Suppose that for some real number b, f is defined for all $x > b$. Then ***the limit of $f(x)$ as $x \to +\infty$ is L***, written

$$\lim_{x \to +\infty} f(x) = L,$$

means that $f(x)$ is arbitrarily close to L for every sufficiently large positive number x. If f is defined for all $x < a$, then ***the limit of $f(x)$ as $x \to -\infty$ is M***, written

$$\lim_{x \to -\infty} f(x) = M,$$

means that $f(x)$ is arbitrarily close to M for every sufficiently large negative number x.

Definition 6.1 can be expressed in the language of the definition of ordinary limits (Definition 4.1 of Chapter 1) as follows:

- $\lim_{x \to +\infty} f(x) = L$ means that for any $\varepsilon > 0$, there is some real number $N > 0$ such that

$$|f(x) - L| < \varepsilon \qquad \text{whenever } x > N; \tag{2}$$

- $\lim_{x \to -\infty} f(x) = M$ means that for any $\varepsilon > 0$, there is some real number $K < 0$ such that

$$|f(x) - M| < \varepsilon \qquad \text{whenever } x < K. \tag{2'}$$

In Example 1, we have

$$\lim_{t \to +\infty} s(t) = \lim_{t \to +\infty} \frac{20{,}000}{1 + \frac{1}{4}(t-1)^2} = 0.$$

Similarly, for any real numbers a and c and any positive rational number m, it is easy to see that

$$\lim_{x \to +\infty} \frac{c}{(x-a)^m} = \lim_{x \to -\infty} \frac{c}{(x-a)^m} = 0. \tag{3}$$

Example 2

Evaluate

(a) $\displaystyle\lim_{x \to +\infty} \frac{2}{(x+1)^5}$ (b) $\displaystyle\lim_{x \to -\infty} \frac{2}{x^3 - 8}$ (c) $\displaystyle\lim_{x \to +\infty} \frac{1}{\sqrt{x+2}}$

Solution. (a) By (3) with $a = 1$ and $c = 2$,

$$\lim_{x \to +\infty} \frac{2}{(x+1)^5} = 0.$$

(b) As x tends to $-\infty$, so does x^3 since the cube of a large negative number is a huge negative number. Division of 2 by the huge negative number $x^3 - 8$

produces a tiny negative quotient. Thus

$$\lim_{x \to -\infty} \frac{2}{x^3 - 8} = 0.$$

(c) Here $m = 1/2$, $a = -2$, and $c = 1$. So (3) gives $\lim_{x \to +\infty} 1/\sqrt{x+2} = 0$. ■

Equation (3) is helpful in evaluating limits of rational functions

$$r(x) = \frac{p(x)}{q(x)}, \qquad \text{where } p(x) \text{ and } q(x) \text{ are polynomials,}$$

as $x \to +\infty$ or $x \to -\infty$. We first **divide the numerator and denominator of $r(x)$ by the highest power of x present in $p(x)$ or $q(x)$.** We then use the fact (see Exercises 49–54) that the basic limit theorems of Section 1.5 continue to hold as $x \to +\infty$ or $x \to -\infty$.

FIGURE 6.1

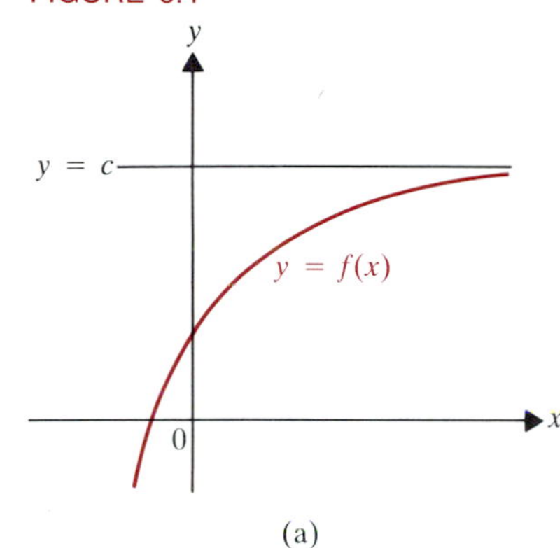

(a)

(b)

Example 3

Evaluate the following limits.

$$\text{(a)} \lim_{x \to +\infty} \frac{5x^3 - 8x^2 + x - 8}{3x^3 - 2x^2 - 5x + 1} \qquad \text{(b)} \lim_{x \to -\infty} \frac{7x^2 - x - 1}{x^3 + x + 5}$$

Solution. We use the above strategy. Here x^3 is the highest power of x, so we get

$$\text{(a)} \lim_{x \to +\infty} \frac{5x^3 - 8x^2 + x - 8}{3x^3 - 2x^2 - 5x + 1} = \lim_{x \to +\infty} \frac{(5x^3 - 8x^2 + x - 8) \div x^3}{(3x^3 - 2x^2 - 5x + 1) \div x^3}$$

$$= \lim_{x \to +\infty} \frac{5 - 8/x + 1/x^2 - 8/x^3}{3 - 2/x - 5/x^2 + 1/x^3}$$

$$= \frac{5 - 0 + 0 - 0}{3 - 0 - 0 + 0} = \frac{5}{3},$$

where we used the analog of Theorems 5.4 and 5.1 of Chapter 1.

$$\text{(b)} \lim_{x \to -\infty} \frac{7x^2 - x - 1}{x^3 + x + 5} = \lim_{x \to -\infty} \frac{(7x^2 - x - 1) \div x^3}{(x^3 + x + 5) \div x^3}$$

$$= \lim_{x \to -\infty} \frac{7/x - 1/x^2 - 1/x^3}{1 + 1/x^2 + 5/x^3} = \frac{0 - 0 - 0}{1 + 0 + 0} = \frac{0}{1} = 0. \quad ■$$

There is an important geometric interpretation of Definition 6.1. In general, an ***asymptote*** of a curve is a line that is approached arbitrarily closely by a curve. This somewhat imprecise statement is easy to clarify for horizontal lines if we use Definition 6.1.

6.2 DEFINITION

The line $y = c$ is a ***horizontal asymptote*** of the graph of $y = f(x)$ if $\lim_{x \to +\infty} f(x) = c$ or $\lim_{x \to -\infty} f(x) = c$. See **Figure 6.1.**

Knowing that the graph of $y = f(x)$ approaches a horizontal line as $x \to +\infty$ or $x \to -\infty$ can make it possible to graph the function rather accurately without plotting many points.

Example 4

Draw a sketch of the graph of $y = f(x) = \dfrac{2x + 4}{\sqrt{x^2 + 4}}$.

Solution. Using a calculator, we compile a table of values of the function.

x	0	2	4	10	100	1000	-2	-4	-10	-100	-1000
y	2	2.8284	2.6833	2.3534	2.0396	2.0040	0	-0.8944	-1.5689	-1.9596	-1.9960

Next, to compute $\lim_{x\to+\infty} f(x)$, note first that since $x \to +\infty$, we can restrict attention to positive x. We thus divide numerator and denominator by $\sqrt{x^2} = |x|$. That gives

FIGURE 6.2

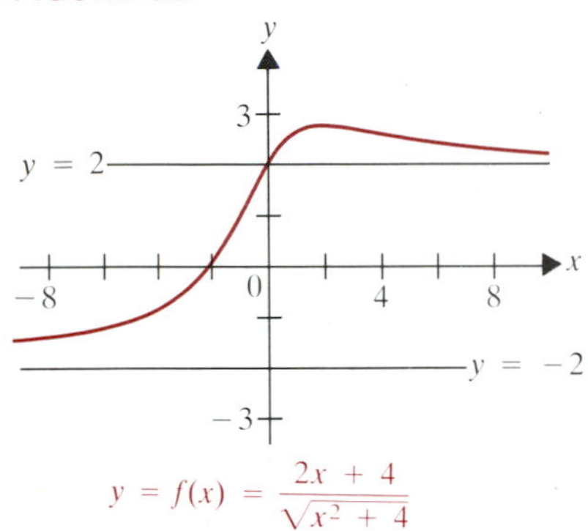

$y = f(x) = \dfrac{2x + 4}{\sqrt{x^2 + 4}}$

$$\lim_{x\to+\infty} f(x) = \lim_{x\to+\infty} \frac{2 + 4/x}{\sqrt{1 + 4/x^2}} = \frac{2 + 0}{\sqrt{1 + 0}} = 2.$$

To find $\lim_{x\to-\infty} f(x)$, note that $\sqrt{x^2} = |x| = -x$ as $x \to -\infty$. So we divide numerator and denominator by $\sqrt{x^2} = -x$, getting

$$\lim_{x\to-\infty} \frac{-2 - 4/x}{\sqrt{1 + 4/x^2}} = \frac{-2 - 0}{\sqrt{1 + 0}} = -2.$$

Thus the graph has horizontal asymptotes $y = 2$ and $y = -2$. A sketch is shown in **Figure 6.2.** Note that the curve does intersect the asymptote $y = 2$, but at $x = 0$, on its way to what appears to be a high point at $x = 2$, *before* the curve starts to approach the asymptote later on. ■

We can certainly think of functions f that do *not* have limits as $x \to +\infty$ or $x \to -\infty$. One such function is $f(x) = x^3$. Moreover, since x^3 becomes huge (and is positive) as $x \to +\infty$, while it becomes huge in size (but is negative) as $x \to -\infty$, it is natural to write

$$\lim_{x\to+\infty} x^3 = +\infty \qquad \text{and} \qquad \lim_{x\to-\infty} x^3 = -\infty.$$

We define the same idea for more general functions, as follows.

6.3 DEFINITION Suppose that for some real number b, the function f is defined for all $x > b$. We say that f has an *infinite limit* as $x \to +\infty$ in the following cases:

(a) $\lim_{x\to+\infty} f(x) = +\infty$ if $f(x)$ is arbitrarily large and positive for every sufficiently large positive number x.

(b) $\lim_{x\to+\infty} f(x) = -\infty$ if $f(x)$ is arbitrarily large and negative for every sufficiently large negative number x.

There is a parallel definition for $\lim_{x\to-\infty} f(x) = +\infty$ and $\lim_{x\to-\infty} f(x) = -\infty$. [See Exercise 42(a).] It is worth emphasizing that *a function with an infinite limit does not have a limit.* This seeming paradox results from the fact that **for a function to have a limit, it must approach a *real number*.** As we mentioned in Chapter 1, $+\infty$ and $-\infty$ are *not* numbers. The statement that $\lim_{x\to+\infty} f(x) = +\infty$ or $-\infty$ is, however, more informative than merely saying that f fails to have a limit as $x \to +\infty$. It conveys the *reason* that f has no limit as $x \to +\infty$: The function is *unbounded* and either positive or negative as x increases without bound. By contrast, $\lim_{x\to+\infty} \sin x$ fails to have a limit. But the reason in that case is that the sine function takes on every value between -1 and $+1$ as x travels through intervals of length 2π. In giving answers to problems that ask you to compute limits, it is good practice to say $+\infty$ (no limit) or $-\infty$ (no limit) rather than to simply say that there is no limit.

A common situation in which infinite limits arise as $x \to +\infty$ or $x \to -\infty$ involves rational functions $r(x)$ with numerator of higher degree than the denominator.

Example 5

Determine the following limits.

$$\text{(a)}\ \lim_{x \to +\infty} \frac{-3x^3 + 2x - 5}{2x^2 + 8x - 3} \qquad \text{(b)}\ \lim_{x \to -\infty} \frac{-2x^2 + 3x + 1}{5x - 7}$$

Solution. (a) We could again divide numerator and denominator by the highest power of x. That would give

$$\lim_{x \to +\infty} \frac{-3x^3 + 2x - 5}{2x^2 + 8x - 3} = \lim_{x \to +\infty} \frac{-3 + 2/x^2 - 5/x^3}{2/x + 8/x^2 - 3/x^2}.$$

Here we *cannot* simply use (3), because all three terms in the denominator approach 0. This difficulty can be overcome by carefully analyzing how each term in the denominator goes to zero. But it is easier to use the following rule-of-thumb. **In evaluating $\lim_{x \to \pm\infty} r(x)$ for a rational function $r(x)$ whose numerator has higher degree than the denominator, divide numerator and denominator by the highest power of x *in the denominator*.** Here this gives

$$\lim_{x \to +\infty} \frac{-3x^3 + 2x - 5}{2x^2 + 8x - 3} = \lim_{x \to +\infty} \frac{-3x + 2/x - 5/x^3}{2 + 8/x - 3/x^3} = -\infty,$$

since the numerator becomes huge in size but negative as $x \to +\infty$, and the denominator approaches 2.

(b) Applying the rule of thumb in (a), we get

$$\lim_{x \to -\infty} \frac{-2x^2 + 3x + 1}{5x - 7} = \lim_{x \to -\infty} \frac{-2x + 3 + 1/x}{5 - 7/x} = +\infty,$$

since the numerator again becomes huge in size, but positive this time, while the denominator approaches 5. ■

There is one final type of infinite limit that is especially useful. It also arises naturally in many applied contexts. Consider, for instance, the ***ideal gas law*** for a gas of volume V (in liters) subject to pressure P (in atmospheres) and temperature T (in Kelvin):

$$PV = nRT, \tag{4}$$

where n is the number of moles of the gas, and R is the *gas constant* 0.0821 liter·atm/mol·K. To *compress* the gas (that is, decrease its volume) while keeping the temperature T and the number of moles n constant, we must increase the pressure P. Mathematically, we might ask what happens to P in (4) as $V \to 0^+$. We have

$$P = \frac{nRT}{V}, \tag{5}$$

where n, R, and T are all positive. Thus P is the ratio of a fixed positive number nRT to the shrinking positive quantity V. As we divide nRT by less and less, P grows larger and larger, as the table shows.

V	10	1	0.1	0.01	10^{-6}
P	$\frac{1}{10}nRT$	nRT	$10nRT$	$100nRT$	10^6nRT

It seems that as V approaches 0, P becomes arbitrarily large. We might express this by saying, $\lim_{V\to 0^+} P = +\infty$. In practical terms, it means that we can't compress a volume of gas to 0—that would require *infinite* pressure! This is an instance of the following definition.

6.4 DEFINITION Suppose that f is defined on some open interval having left endpoint a. Then

(a) $\lim\limits_{x\to a^+} f(x) = +\infty$ means that $f(x)$ is arbitrarily large and positive whenever $x > a$ is sufficiently close to a.

(b) $\lim\limits_{x\to a^+} f(x) = -\infty$ means that $f(x)$ is arbitrarily large and negative whenever $x > a$ is sufficiently close to a.

We similarly can define $\lim_{x\to b^-} f(x) = +\infty$ and $\lim_{x\to b^-} f(x) = -\infty$ [see Exercise 42(b)]. These kinds of limits have an important geometric interpretation in terms of asymptotic behavior of the graph of f.

6.5 DEFINITION The line $x = a$ is a ***vertical asymptote*** of the graph of f if either the left- or right-hand limit of f as $x \to a$ is $+\infty$ or $-\infty$. See **Figure 6.3.**

FIGURE 6.3

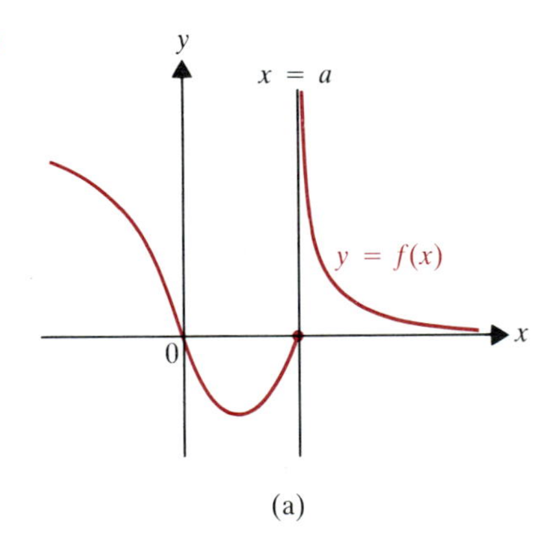

(a)

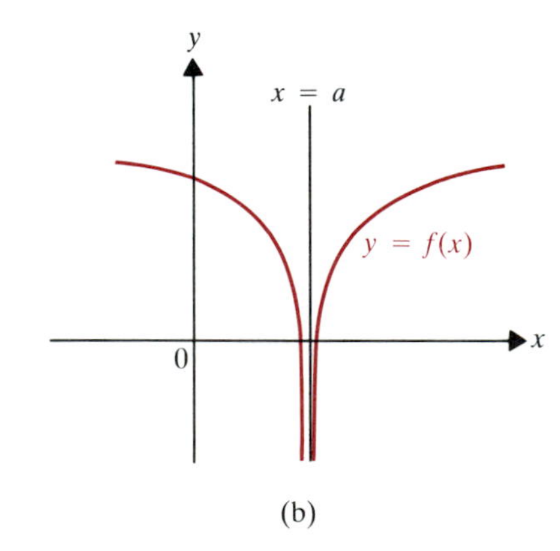

(b)

FIGURE 6.4

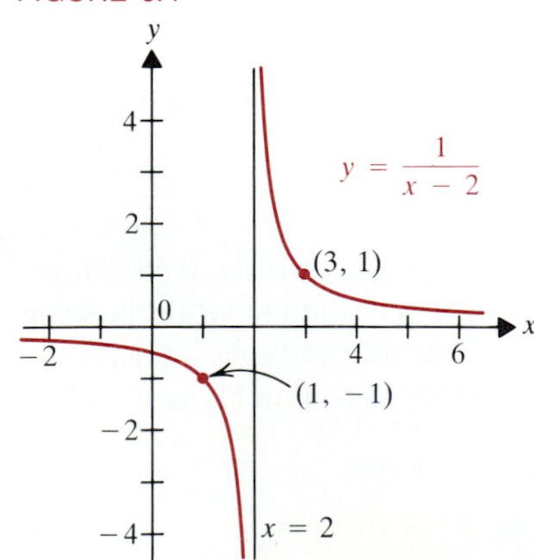

Example 6

Sketch the graph of $y = f(x) = \dfrac{1}{x-2}$.

Solution. First, we compile a table of values.

x	0	1	3/2	7/4	15/8	31/16	33/16	17/8	9/4	5/2	3	4	12	102
y	−1/2	−1	−2	−4	−8	−16	16	8	4	2	1	1/2	1/10	1/100

The x-axis $(y = 0)$ is a horizontal asymptote since $\lim_{x\to+\infty} f(x) = 0 = \lim_{x\to-\infty} f(x)$. As $x \to 2^+$, we divide 1 by tiny positive numbers, obtaining huge quotients. Thus $\lim_{x\to 2^+} f(x) = +\infty$. As $x \to 2^-$, we divide 1 by tiny negative numbers, obtaining quotients that are huge in size but negative. Hence, $\lim_{x\to 2^-} f(x) = -\infty$. Consequently, $x = 2$ is a vertical asymptote. This information enables us to sketch the graph. See **Figure 6.4.** ■

FIGURE 6.5

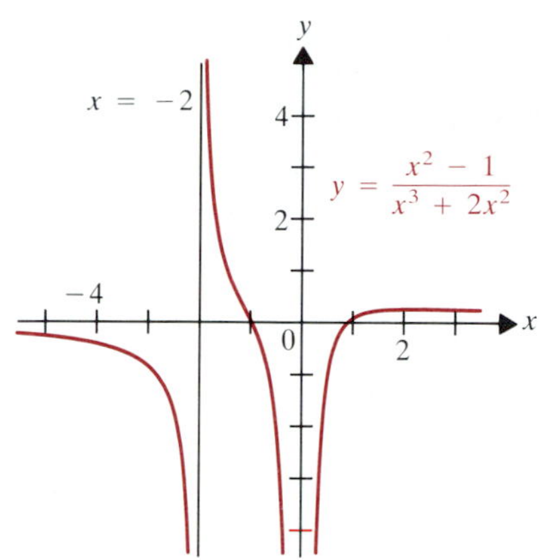

To find the vertical asymptotes of the graph of a rational function

$$y = r(x) = \frac{p(x)}{q(x)},$$

factor the denominator completely. After canceling any common factors that may occur in the numerator also, let x approach each remaining root a of the denominator. Then each vertical line $x = a$ should be a vertical asymptote. For example, the function

$$f(x) = \frac{x^2 - 1}{x^3 + 2x^2} = \frac{(x-1)(x+1)}{x^2(x+2)}$$

has vertical asymptotes $x = 0$ and $x = -2$. Its graph is shown in **Figure 6.5.**

Exercises 3.6

In Exercises 1–32, find the limits.

1. $\lim_{x\to+\infty} \frac{x^2 - 4x + 1}{2x^2 + 5x - 8}$

2. $\lim_{x\to+\infty} \frac{x^3 - 4x^2 + 6}{-7x^3 - 6x + 2}$

3. $\lim_{x\to-\infty} \frac{-5x^3 + 3x - 8}{2x^3 - x^2 + 6}$

4. $\lim_{x\to-\infty} \frac{7x^2 - 3x - 8}{2x^2 - 4x + 6}$

5. $\lim_{x\to+\infty} \frac{5x^2 - 6x + 11}{x^3 - 2x + 8}$

6. $\lim_{x\to+\infty} \frac{-x^2 - x + 9}{3x^3 + 6x + 1}$

7. $\lim_{x\to-\infty} \frac{7x^2 + 5x + 1}{-2x + 8}$

8. $\lim_{x\to-\infty} \frac{x^2 - 4x + 9}{-x + 10}$

9. $\lim_{x\to+\infty} \frac{-2x^3 + 8x}{x^2 - 4}$

10. $\lim_{x\to+\infty} \frac{x^4 - 16}{-x^2 + 8}$

11. $\lim_{x\to+\infty} \frac{\sqrt{9x^2 - 3x + 1}}{5x - 8}$

12. $\lim_{x\to+\infty} \frac{\sqrt{x^2 - 8x - 1}}{-x + 1}$

13. $\lim_{x\to-\infty} \frac{\sqrt[3]{8x^3 - 16x^2 + 8}}{5x - 1}$

14. $\lim_{x\to-\infty} \frac{11x + 17}{\sqrt[3]{27x^3 - 16}}$

15. $\lim_{x\to+\infty} x \sin\frac{1}{x}$

16. $\lim_{x\to-\infty} x \cos\frac{1}{x}$

17. $\lim_{x\to+\infty} \cos\frac{1}{x}$

18. $\lim_{x\to-\infty} \sin\frac{1}{x}$

19. $\lim_{x\to+\infty} \frac{\sin x}{x}$

20. $\lim_{x\to-\infty} \frac{\cos x}{x}$

21. $\lim_{x\to4^+} \frac{5 - 2x}{(x-4)^{3/2}}$

22. $\lim_{x\to3^+} \frac{5x - 2}{(x-3)^3}$

23. $\lim_{x\to2^-} \frac{x^2 - 5}{x^2 - x - 2}$

24. $\lim_{x\to3^-} \frac{5x - 6}{x^2 - 2x - 3}$

25. $\lim_{x\to1^-} \frac{x^2 - 2x + 5}{x^2 - 4x + 3}$

26. $\lim_{x\to-1^-} \frac{2x + 3}{x^2 + 3x + 2}$

27. $\lim_{x\to0^+} \frac{\cos x}{x}$

28. $\lim_{x\to0^-} \frac{\cos x}{x}$

29. $\lim_{x\to+\infty} (\sqrt{x^2 + 4} - x)$. (*Hint:* Consider the limit of the equivalent expression with denominator 1 and use the approach of Example 4, Section 2.1.)

30. $\lim_{x\to-\infty} (\sqrt{x^2 + 9} - x)$

31. $\lim_{x\to-\infty} (\sqrt[3]{x^3 + 8} - x)$

32. $\lim_{x\to+\infty} (\sqrt[3]{x^3 + 27} - x)$

In Exercises 33–42, determine horizontal and vertical asymptotes and sketch the graph.

33. $y = \frac{1}{x}$

34. $y = \frac{1}{x + 1}$

35. $y = \frac{x - 1}{x + 1}$

36. $y = \frac{x + 2}{x + 1}$

37. $y = \frac{x}{x^2 + 1}$

38. $y = \frac{x - 1}{x^2 + 1}$

39. $y = \frac{x^2}{x^2 + 1}$

40. $y = \frac{1 - x^2}{x^2 + 1}$

41. $y = \dfrac{x^2 - 1}{x^2 + 2x - 3}$

42. $y = \dfrac{x^2 - 1}{x^2 + 4x + 3}$

43. We introduced the *Michaelis–Menten equation* in Exercise 39 of Section 2.2. It describes the rate of reaction between an enzyme and a *substrate* (a chemical compound acted on by the enzyme). The equation is

$$v = \frac{S \cdot V_{\max}}{K_m + S}, \tag{6}$$

where K_m is a constant (called the *Michaelis constant*), S is the concentration of the substrate, v is the rate of the reaction, and $V_{\max}$ is the *limiting* (or *maximal*) *rate of reaction at "infinite" substrate concentration.* Compute the limit of v in (6) as $S \to +\infty$, and use your answer to explain the italicized description given for $V_{\max}$.

44. With reference to the ideal gas law (4), is it physically possible to achieve a temperature of 0 K (called *absolute zero*)? Explain.

45. In the ideal gas law (4), what happens to V as $P \to 0^+$?

46. **(a)** Give the definition of $\lim_{x \to -\infty} f(x) = +\infty$ and $\lim_{x \to -\infty} f(x) = -\infty$.

(b) Give the definition of $\lim_{x \to b^-} f(x) = +\infty$ and $\lim_{x \to b^-} f(x) = -\infty$.

(c) The condition in Definition 6.3(a) that $\lim_{x \to +\infty} f(x) = +\infty$ is formulated more precisely by saying that for any positive real number M, there is some positive real number K such that $f(x) > M$ whenever $x > K$. Give a similar formulation for Definition 6.3(b).

(d) Formulate Definition 6.4 using the language of (c) and Definition 6.1 of Chapter 1.

47. If $r(x) = \dfrac{p(x)}{q(x)}$ is a rational function, then show that $\lim_{x \to \pm\infty} r(x) = 0$ if $p(x)$ has smaller degree than $q(x)$.

48. If $r(x)$ is as in Exercise 47 but $p(x)$ has larger degree than $q(x)$, then show that $r(x)$ has infinite limits as $x \to +\infty$ and $x \to -\infty$.

In Exercises 49–54, prove the limit theorems for infinite limits.

49. If $\lim_{x \to c} f(x) = \lim_{x \to c} g(x) = +\infty$, then $\lim_{x \to c} (f + g)(x) = +\infty$.

50. If $\lim_{x \to c} f(x) = +\infty$ and $\lim_{x \to c} g(x) = d$, then $\lim_{x \to c} (f + g)(x) = +\infty$.

51. If $\lim_{x \to c} f(x) = +\infty$ and $\lim_{x \to c} g(x) = d > 0$, then $\lim_{x \to c} (f \cdot g)(x) = +\infty$.

52. If $\lim_{x \to c} f(x) = +\infty$ and $\lim_{x \to c} g(x) = d < 0$, then $\lim_{x \to c} (f \cdot g)(x) = -\infty$.

53. If $\lim_{x \to c} f(x) = d \neq 0$ and $\lim_{x \to c} g(x) = 0$, then $\lim_{x \to c} \left|\dfrac{f(x)}{g(x)}\right| = +\infty$.

54. If $\lim_{x \to c} f(x) = +\infty$ or $\lim_{x \to c} f(x) = -\infty$, then $\lim_{x \to c} \dfrac{1}{f(x)} = 0$.

55. Consider the third-degree polynomial function $p(x) = ax^3 + bx^2 + cx + d$ where $a \neq 0$.

(a) Show that $\lim_{x \to +\infty} p(x) = +\infty$ if $a > 0$ and $\lim_{x \to +\infty} p(x) = -\infty$ if $a < 0$.

(b) What is $\lim_{x \to -\infty} p(x)$?

(c) Use (a), (b), and the intermediate theorem for continuous functions (Theorem 7.6 of Chapter 1) to show that the equation $p(x) = 0$ has at least one real root.

56. Use the approach of Exercise 55 to show that *any* odd-degree polynomial equation $p(x) = 0$ has at least one real root. Is this true of every *even*-degree polynomial equation also? Explain.

3.7 Curve Sketching

This section puts together the graphical tools developed earlier. The result is a program for constructing accurate graphs of given functions f. Since little in the way of new theory is required, most of this section consists of working through examples that bring out some of the possibilities that arise. Before starting, we give a brief discussion of symmetry criteria, which can often save considerable work in graphing.

In Theorem 7.3 of Chapter 2, we saw that the graph of $y = f^{-1}(x)$ is the reflection of the graph of $y = f(x)$ in the line $y = x$. One says that the graphs of $y = f(x)$ and $y = f^{-1}(x)$ are *symmetric* in the line $y = x$. This is an instance of the following formal definition.

7.1 DEFINITION A curve in the xy-plane is ***symmetric in*** (or ***relative to***) ***a line*** l if for every point $P_1(x_1, y_1)$ on the curve, there is another point $P_2(x_2, y_2)$ on the curve such that l is the perpendicular bisector of the segment P_1P_2.

Figure 7.1 illustrates Definition 7.1. The line l is referred to as a *line* (or *axis*) *of symmetry* of the curve. Symmetry is also defined relative to a point.

FIGURE 7.1

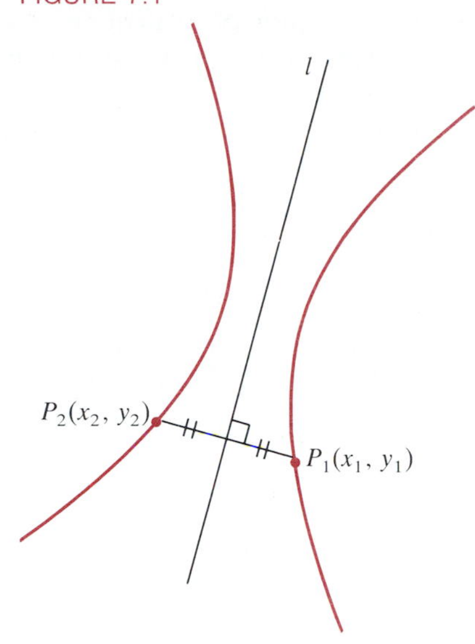

FIGURE 7.2

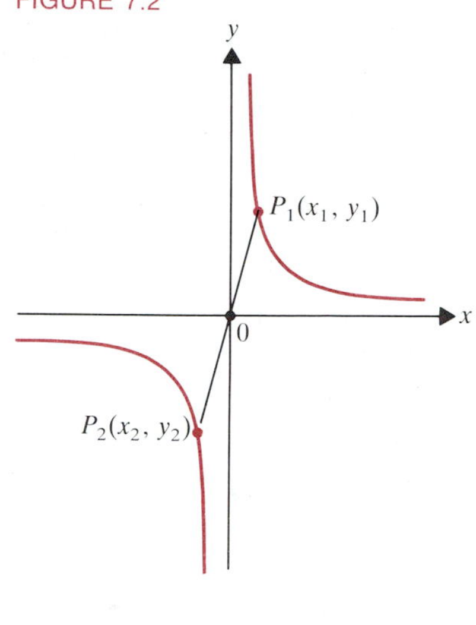

7.2 DEFINITION A curve is ***symmetric in*** (or ***relative to***) ***a point*** P_0 if for every point $P_1(x_1, y_1)$ on the curve, there is another point $P_2(x_2, y_2)$ on the curve such that P_0 is the midpoint of the segment P_1P_2.

The curve shown in **Figure 7.2** is symmetric in the origin 0. This is usually the only point of symmetry considered for plane curves. The following theorem gives a simple test for the symmetry of the graph of a functional equation $F(x, y) = 0$ in a coordinate axis or in the origin. A partial proof is given at the end of the section.

7.3 THEOREM

(a) The graph of an equation $F(x, y) = 0$ is symmetric in the x-axis if and only if $F(x, -y) = 0$ is equivalent to $F(x, y) = 0$.

(b) The graph of $F(x, y) = 0$ is symmetric in the y-axis if and only if $F(-x, y) = 0$ is equivalent to $F(x, y) = 0$.

(c) The graph of $F(x, y) = 0$ is symmetric in the origin if and only if $F(-x, -y) = 0$ is equivalent to $F(x, y) = 0$.

Theorem 7.3 says that the graph of a function f is symmetric in the y-axis if and only if f is an *even* function, that is, for every x in its domain, $-x$ is also in the domain and

$$f(-x) = f(x).$$

Similarly, f is symmetric in the origin if and only if f is an *odd* function, that is,

$$f(-x) = -f(x),$$

for every x in the domain of f.

FIGURE 7.3

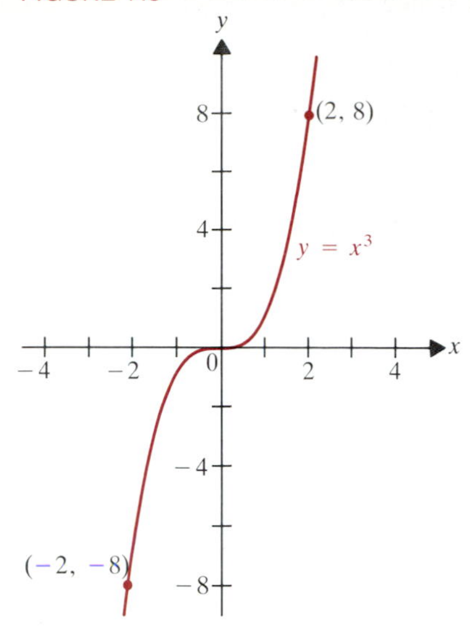

FIGURE 7.4

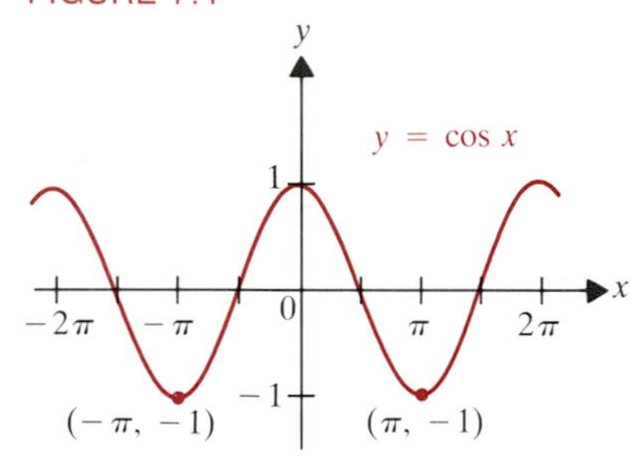

FIGURE 7.5

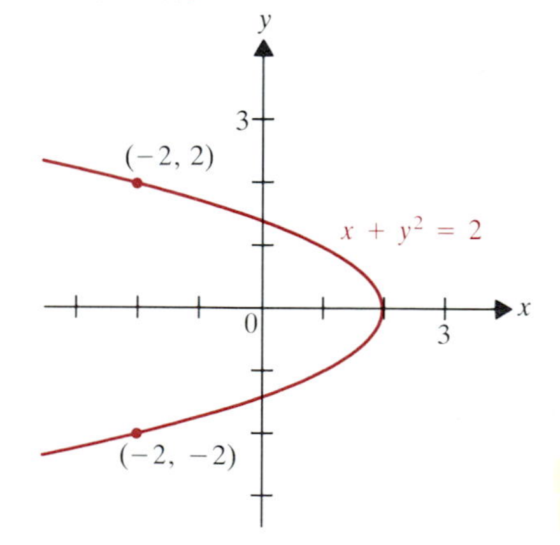

Example 1

Determine whether the following graphs are symmetric in the x- or y-axis or the origin.

(a) $y = x^3$ (b) $y = \cos x$ (c) $x + y^2 = 2$

Solution. (a) The equation $y = x^3$ is *not* equivalent to $y = (-x)^3$. For example, (2, 8) is on the graph of $y = x^3$ but not on the graph of $y = (-x)^3 = -x^3$. Similarly, $y = x^3$ is not equivalent to $-y = x^3$, that is, to $y = -x^3$. So the graph is not symmetric in either coordinate axis. The equation $y = x^3$ *is* equivalent to $-y = (-x)^3$, that is, to $-y = -x^3$. So the graph is symmetric in the origin. See **Figure 7.3.**

(b) The cosine function is even, so the equation $y = \cos x$ is equivalent to $y = \cos(-x) = \cos x$. Hence, the graph is symmetric in the y-axis. Since $y = \cos x$ is *not* equivalent to $-y = \cos x$ or to $-y = \cos(-x)$, the graph is not symmetric in either the x-axis or the origin. See **Figure 7.4.**

(c) The equation is unchanged if y is replaced by $-y$, so the graph is symmetric in the x-axis. The equations $-x + y^2 = 2$ and $-x + (-y)^2 = 2$ are *not* equivalent to the original equation, so the graph is not symmetric in either the y-axis or the origin. See **Figure 7.5.** ■

As we did in Section 1 for related rates, we offer a list of suggested steps to follow in constructing an accurate graph of an equation $y = f(x)$.

ALGORITHM FOR CURVE SKETCHING

1. Determine the domain and, if convenient, the range of f.
2. Use Theorem 7.3 to determine whether the graph is symmetric relative to one or both coordinate axes or the origin.
3. Find f' and f''. Use them to determine where f is increasing, decreasing, concave upward, and concave downward.
4. Find all local extreme points and points of inflection.
5. Find any horizontal or vertical asymptotes of the graph.
6. Plot enough points to show the behavior found in Steps 1–5 and draw a smooth curve through those points.

Before illustrating the use of the algorithm, we make some comments about it. First, some of the steps can be used in graphing a functional equation $F(x, y)$ that cannot be conveniently put into the form $y = f(x)$. In graphing a function, determining the domain helps you restrict attention at the outset to the portion of the x-axis where there is a graph. Finding the range of f may be more work than it is worth, but in several of the following examples we can find it fairly easily. Example 1 illustrates how Step 2 is carried out.

Step 3 is done using the techniques of Sections 3 and 4. Recall that f is increasing where $f'(x) > 0$, decreasing where $f'(x) < 0$, concave up where $f''(x) > 0$, and concave down where $f''(x) < 0$. To do Step 4, use the first- or second-derivative tests and find where f'' changes sign.

To find horizontal asymptotes, evaluate $\lim_{x \to +\infty} f(x)$ and $\lim_{x \to -\infty} f(x)$. The line $y = d$ is a horizontal asymptote if either of these limits is d. To identify vertical asymptotes, we must find all points c for which $\lim_{x \to c+} f(x)$ or $\lim_{x \to c-} f(x)$ is infinite. If f is a rational function,

$$f(x) = \frac{p(x)}{q(x)}, \qquad \text{where } p(x) \text{ and } q(x) \text{ are polynomials,}$$

then zeros of the denominator are obvious candidates for such points c. If f does have an infinite limit at c, then $x = c$ is a vertical asymptote.

Now for some examples!

Example 2

Draw an accurate graph of f, if $f(x) = x^3 - 9x^2 + 24x - 18$.

Solution. The domain of f is the entire real line $\mathbf{R} = (-\infty, +\infty)$. Since

$$\lim_{x \to \pm\infty} f(x) = \lim_{x \to \pm\infty} x^3\left(1 - \frac{9}{x} + \frac{24}{x^2} - \frac{18}{x^3}\right) = \pm\infty,$$

f takes on arbitrarily large positive and negative values. By the intermediate value theorem (Theorem 7.6 of Chapter 1), f takes on all values between such large numbers P and N. Its range is therefore $(-\infty, +\infty)$. The graph fails to be symmetric in either coordinate axis or the origin, because

$$f(-x) = (-x)^3 - 9(-x)^2 + 24(-x) - 18 = -x^3 - 9x^2 - 24x - 18 \neq \pm f(x),$$

so f is neither an even nor an odd function. Similarly,

$$-y = f(x) \text{ is not equivalent to } y = f(x).$$

We have

$$f'(x) = 3x^2 - 18x + 24 = 3(x - 4)(x - 2),$$

so

$$f'(x) = 0 \qquad \text{when } x = 2 \text{ or } x = 4.$$

Since

$$f''(x) = 6x - 18 = 6(x - 3),$$

we see that $f''(x) = 0$ when $x = 3$. We tabulate the behavior of f as follows.

x:	$(-\infty, 2)$	2	$(2, 3)$	3	$(3, 4)$	4	$(4, +\infty)$ → $\mathbf{R}$
$x - 2$:	$-$		$+$		$+$		$+$
$x - 4$:	$-$		$-$		$-$		$+$
$f'(x)$:	$+$	0	$-$	-3	$-$	0	$+$
$x - 3$:	$-$		$-$		$+$		$+$
$f''(x)$:	$-$	$-$	$-$	0	$+$	$+$	$+$

Conclusions:
$(-\infty, 2)$: Increasing, concave downward
$(2, 3)$: Decreasing, concave downward
$(3, 4)$: Decreasing, concave upward
$(4, +\infty)$: Increasing, concave upward

$x = 2$ is a local maximum
$x = 3$ is an inflection point
$x = 4$ is a local minimum

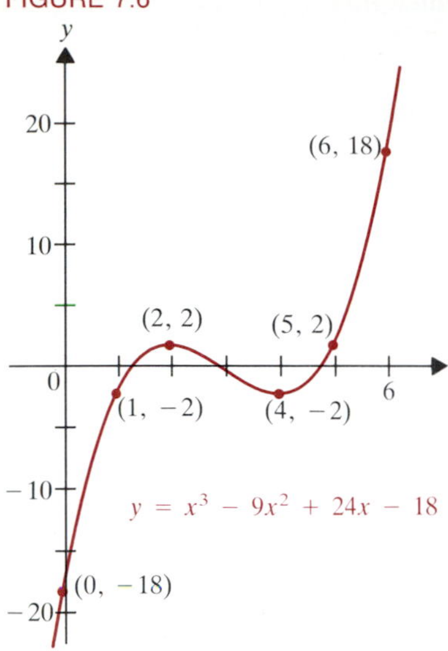

FIGURE 7.6

There are no vertical asymptotes, because f is continuous everywhere: There are no holes in the graph where an asymptote could pass through. Since $\lim_{x \to +\infty} f(x)$ is infinite, and the same holds true for $\lim_{x \to -\infty} f(x)$, there are no horizontal asymptotes. The following table serves for Step 6.

x	0	1	2	3	4	5	6
y	-18	-2	2	0	-2	2	18

The graph is shown in **Figure 7.6.** ■

We next consider a graph of a rational function. Vertical and horizontal asymptotes are apt to arise for such functions.

Example 3

Draw an accurate graph of the function $r(x) = \dfrac{x}{x^2 - 1}$.

Solution. Since r is an odd function, the graph is symmetric in the origin. But the graph is not symmetric in either coordinate axis. The domain of r consists of all real numbers except 1 and -1, where the denominator becomes 0. Thus $x = 1$ and $x = -1$ are candidates for vertical asymptotes. The behavior of r as x nears these points confirms that they are vertical asymptotes, and suggests that the range of r is $(-\infty, +\infty)$ (see Exercise 48):

$$\lim_{x \to -1^-} r(x) = \lim_{x \to -1^-} \frac{x}{x^2 - 1} = -\infty = \lim_{x \to 1^-} r(x),$$

$$\lim_{x \to -1^+} r(x) = \lim_{x \to -1^+} \frac{x}{x^2 - 1} = +\infty = \lim_{x \to 1^+} r(x).$$

From the fact that $\lim_{x \to \pm\infty} r(x) = 0$, we see that the x-axis is a horizontal asymptote. The quotient rule gives

$$r'(x) = \frac{(x^2 - 1) \cdot 1 - x \cdot 2x}{(x^2 - 1)^2} = \frac{-(x^2 + 1)}{(x^2 - 1)^2}.$$

Notice that $r'(x) < 0$ for all $x \neq \pm 1$, and so there are no critical points. We also have

$$\begin{aligned} r''(x) &= \frac{-(x^2 - 1)^2 \cdot 2x + (x^2 + 1) \cdot 2(x^2 - 1) \cdot 2x}{(x^2 - 1)^4} \\ &= \frac{-(x^2 - 1)2x + 4x(x^2 + 1)}{(x^2 - 1)^3} \\ &= \frac{-2x^3 + 2x + 4x^3 + 4x}{(x^2 - 1)^3} = \frac{2x(x^2 + 3)}{(x^2 - 1)^3}. \end{aligned}$$

Thus $r''(x) = 0$ when $x = 0$. Both $r'(x)$ and $r''(x)$ fail to exist at $x = \pm 1$, since r is not defined there. Also, $-(x^2 + 1)$ is *always* negative, and $2(x^2 + 3)$ is *always* positive. Hence, these factors do not need to be tabulated in determining the signs of $r'(x)$ and $r''(x)$.

x:	$(-\infty, -1)$	-1	$(-1, 0)$	0	$(0,1)$	1	$(1, +\infty)$
$(x^2-1)^2$:	$+$		$+$		$+$		$+$
$r'(x)$:	$-$	does not exist	$-$	-1	$-$	does not exist	$-$
x:	$-$		$-$		$+$		$+$
$(x^2-1)^3$:	$+$		$-$		$-$		$+$
$r''(x)$:	$-$	does not exist	$+$	0	$-$	does not exist	$+$

Conclusions: $(-\infty, -1)$: Decreasing, concave downward
$(-1, 0)$: Decreasing, concave upward
$(0, 1)$: Decreasing, concave downward
$(1, +\infty)$: Decreasing, concave upward

$x = -1$ is a vertical asymptote
$x = 0$ is an inflection point
$x = 1$ is a vertical asymptote

Using the table of values

x	0	1/2	4/5	6/5	2	5
$r(x)$	0	$-2/3$	$-20/9$	30/11	2/3	5/24

and the fact that r is an odd function, we draw the graph in **Figure 7.7.** ■

FIGURE 7.7

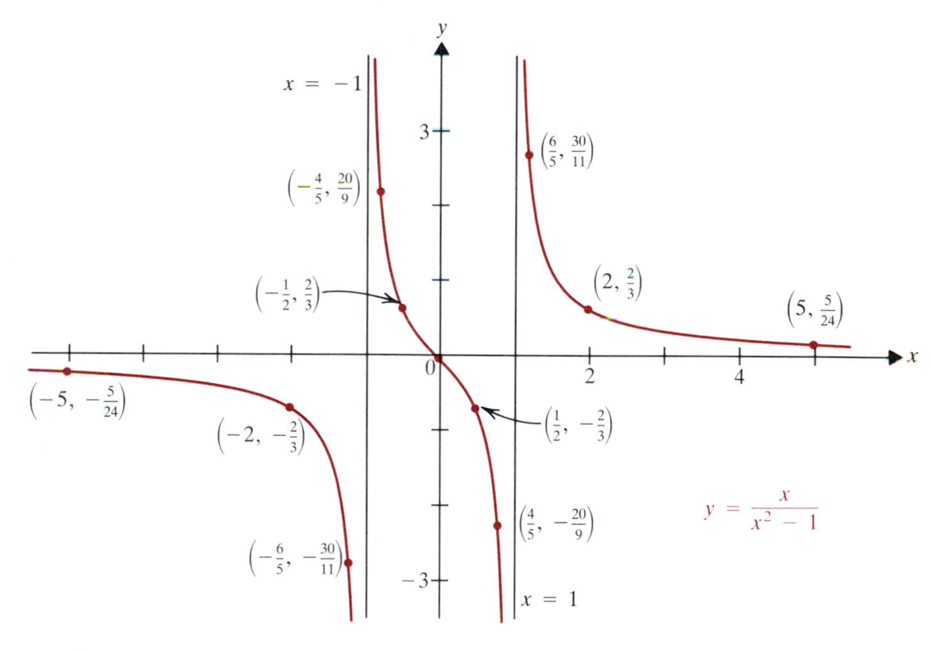

With our collection of tools, we can draw accurate graphs of trigonometric functions. We begin with a simple example.

Example 4

Draw an accurate graph of $h(x) = \cos 2x$.

Solution. The domain of h is $(-\infty, +\infty)$ and its range is $[-1, 1]$. Since cosine is an even function, the graph is symmetric in the y-axis. We have

$$h'(x) = -(\sin 2x) \cdot 2 = -2 \sin 2x.$$

Thus

$$h'(x) = 0 \qquad \text{when} \qquad x = 0 \pm \tfrac{1}{2}n\pi,$$

for any positive integer n. Also,

$$h''(x) = -4\cos 2x,$$

which is zero when $x = \frac{1}{4}\pi \pm n\pi$ for any positive integer n. Since cosine is periodic with period 2π, $\cos 2x$ is periodic with period π. It is therefore enough to draw the graph over the interval $[0, \pi]$.

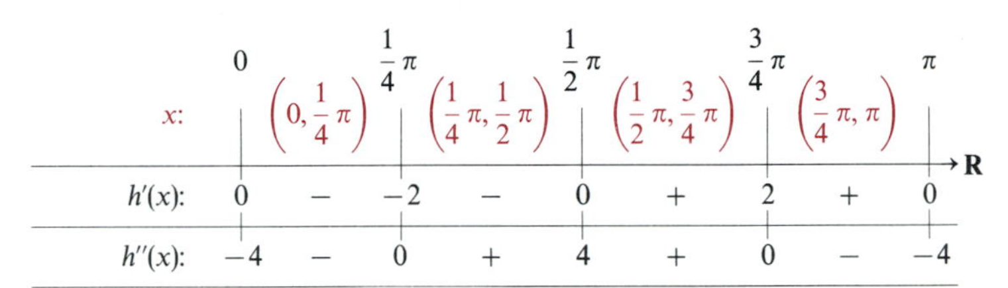

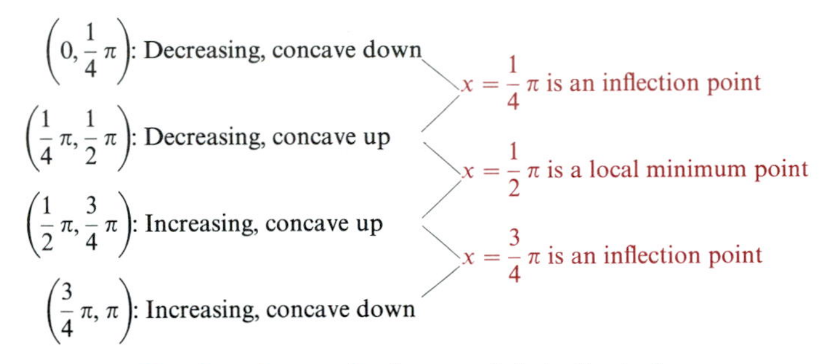

Using the data from above and the following short table of values, the graph shown in **Figure 7.8** was plotted. You might compare it to the graph of $y = \cos x$ in Figure 7.4.

x	0	$1/4\pi$	$1/2\pi$	$3\pi/4$	π
$h(x)$	1	0	-1	0	1

■

FIGURE 7.8

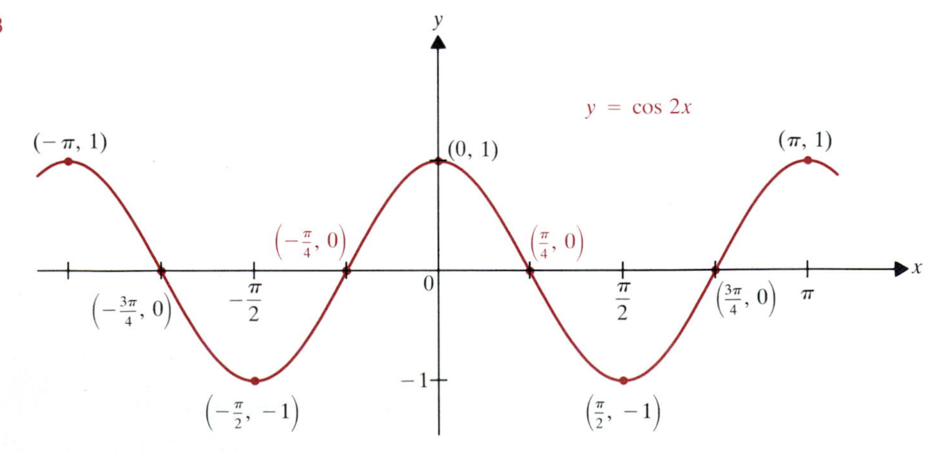

We did not draw the graphs of the tangent, cotangent, secant, or cosecant functions in Section 2.3, because we said they are better drawn using calculus. We conclude this section by discussing the graphs of the tangent and secant functions.

Example 5

Draw an accurate graph of $y = \tan x$.

Solution. In Definition 3.4 of Chapter 2, we excluded all odd multiples $(2n + 1)\pi/2$ of $\pi/2$ from the domain of the tangent function, because $\cos x = 0$ at such points. Moreover, we have

$$(1) \qquad \lim_{x\to\pi/2^-} \tan x = \lim_{x\to\pi/2^-} \frac{\sin x}{\cos x} = +\infty,$$

and

$$(2) \qquad \lim_{x\to\pi/2^+} \tan x = \lim_{x\to\pi/2^+} \frac{\sin x}{\cos x} = -\infty.$$

Since the period of the tangent function is π, we see that the lines

$$x = \pi/2 \pm n\pi, \qquad n = 0, \pm 1, \pm 2, \ldots$$

are all vertical asymptotes. It can be shown (Exercise 49) from (1) and (2) that the tangent function has range $(-\infty, +\infty)$. We have for all x in the domain of the tangent function

$$D_x \tan x = \sec^2 x = \frac{1}{\cos^2 x} > 0,$$

so the tangent function is increasing on any interval of its domain. There are thus no local extreme points. Since

$$(3) \qquad \frac{d^2}{dx^2}(\tan x) = D_x[(\cos x)^{-2}] = \frac{2\sin x}{\cos^3 x},$$

the second derivative is 0 when $\sin x = 0$, that is, at $x = 0, \pm\pi, \pm 2\pi, \ldots$. We see from (3) that the concavity changes at $x = 0$:

$$\text{For } x < 0,\ \sin x < 0 \text{ and } \cos x > 0, \text{ so } \frac{d^2}{dx^2}(\tan x) < 0.$$

$$\text{For } x > 0,\ \sin x > 0 \text{ and } \cos x > 0, \text{ so } \frac{d^2}{dx^2}(\tan x) > 0.$$

Hence, (0, 0) is an inflection point: *tan* changes from being concave down on $(-\pi/2, 0)$ to being concave up on $(0, \pi/2)$. Since the tangent function has period π, the concavity is preserved if any integer multiple of π is added to these intervals. With the aid of the following table of values and the above information, we can draw an accurate graph. In **Figure 7.9** we carefully plotted the tabulated points, taking account of concavity, and drew the vertical asymptote $x = \pi/2$. The fact that $\tan x$ has period π enabled us to draw the other vertical asymptotes $x = \pi/2 \pm \pi, \pi/2 \pm 2\pi, \ldots$ and trace curves parallel to those drawn through the plotted points. ■

FIGURE 7.9

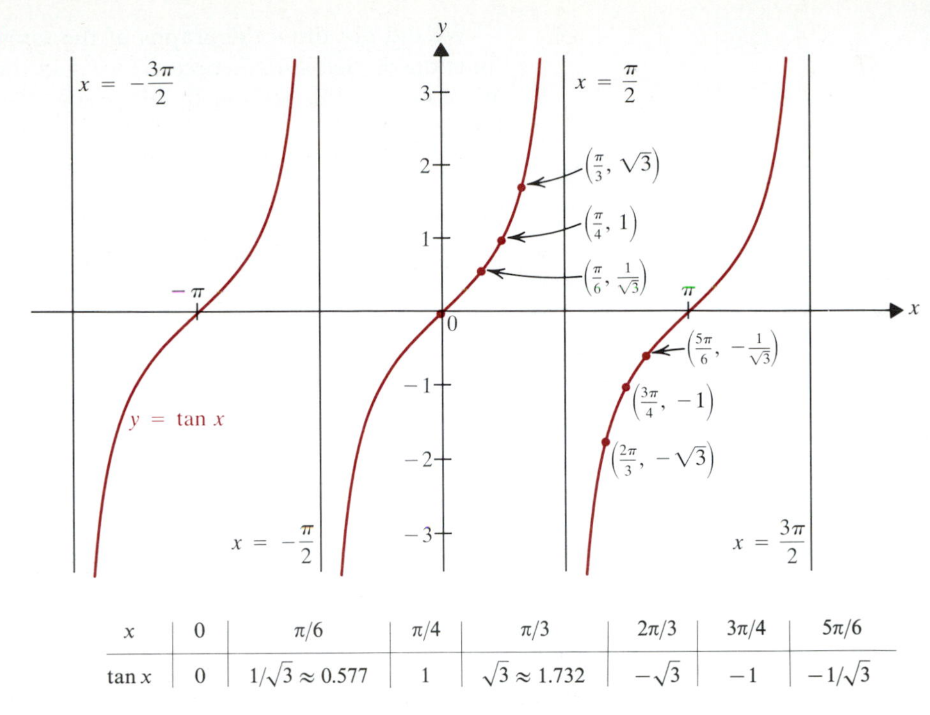

x	0	$\pi/6$	$\pi/4$	$\pi/3$	$2\pi/3$	$3\pi/4$	$5\pi/6$
$\tan x$	0	$1/\sqrt{3} \approx 0.577$	1	$\sqrt{3} \approx 1.732$	$-\sqrt{3}$	-1	$-1/\sqrt{3}$

A similar discussion for the cotangent function (Exercise 43) produces **Figure 7.10.**

FIGURE 7.10

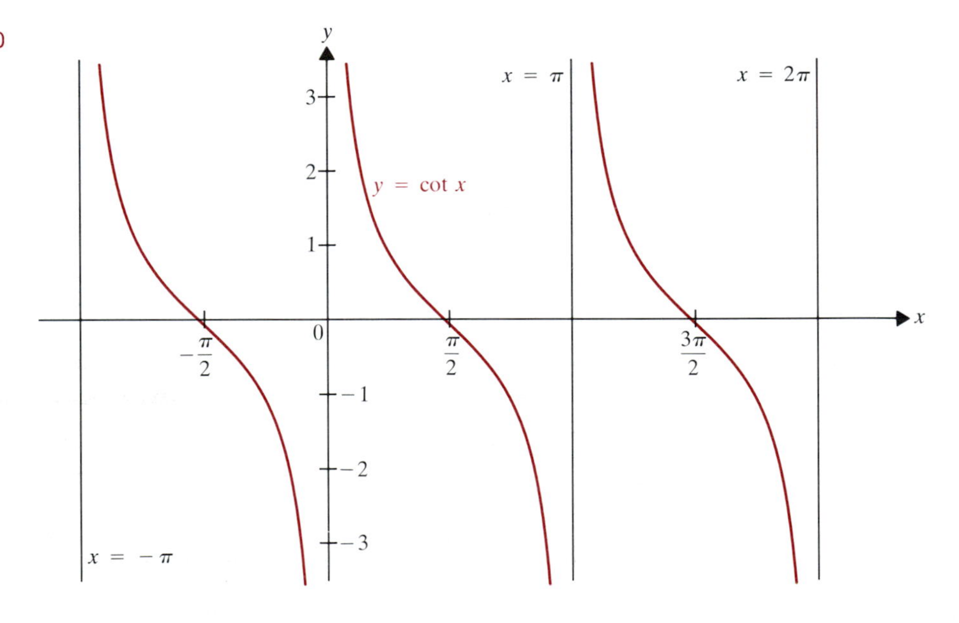

Example 6

Draw an accurate graph of $y = \sec x$.

Solution. The domain of the secant function is the same as that for the tangent function, and we see as in Example 5 that

$$\lim_{x \to \pi/2^-} \sec x = \lim_{x \to \pi/2^-} \frac{1}{\cos x} = +\infty,$$

$$\lim_{x \to \pi/2^+} \sec x = \lim_{x \to \pi/2^+} \frac{1}{\cos x} = -\infty,$$

so that the lines $x = \pm\pi/2, \pm 3\pi/2, \pm 5\pi/2, \ldots$ are vertical asymptotes. Noting that $|\cos x| \le 1$, we see that $|\sec x| \ge 1$ for all x in the domain of the secant function. We have

$$D_x \sec x = \sec x \tan x = \frac{1}{\cos x} \cdot \frac{\sin x}{\cos x} = \frac{\sin x}{\cos^2 x},$$

so every integer multiple of π is a critical point. (The secant is not differentiable where $\cos x = 0$, but such points are outside the domain, so the only critical points are those where $\sec x \tan x = 0$.) We use the second-derivative test to classify these critical points:

$$\frac{d^2}{dx^2}(\sec x) = D_x(\sec x \tan x) = \sec x \tan^2 x + \sec x \sec^2 x$$

$$= \sec x(\tan^2 x + \sec^2 x).$$

This has the sign of $\sec x$, because $\tan^2 x + \sec^2 x$ is never negative. Since $\sec x$ has the sign of $\cos x$, the secant function is concave down on $(\pi/2, 3\pi/2)$ and concave up on $(0, \pi/2) \cup (3\pi/2, 2\pi)$. In particular, there is a local minimum at $x = 0$ and a local maximum at $x = \pi$. If we combine this information with the above facts about $\sec x$ and plot the points in the table, then we get the graph shown in **Figure 7.11.**

FIGURE 7.11

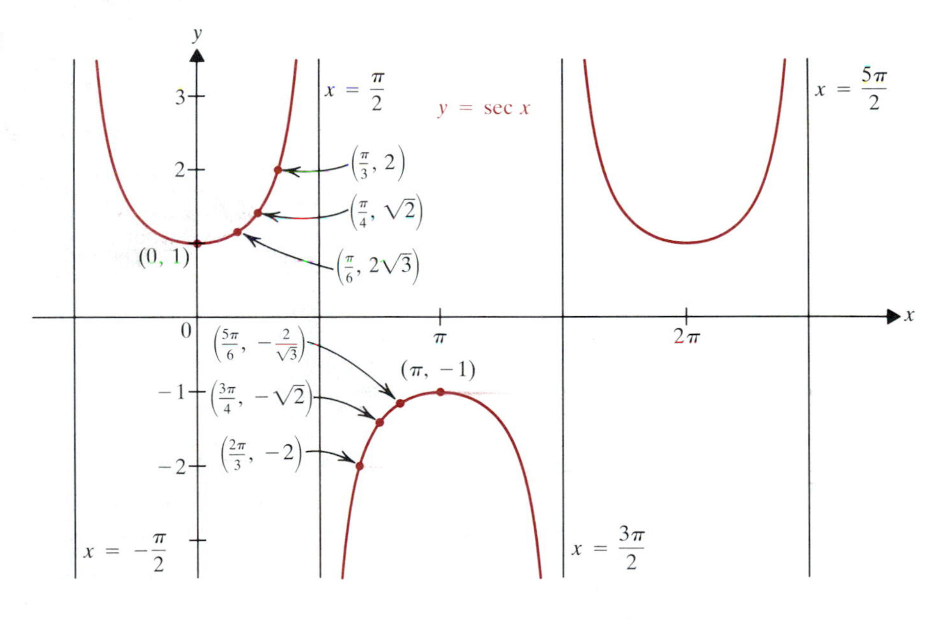

x	0	$\pi/6$	$\pi/4$	$\pi/3$	$2\pi/3$	$3\pi/4$	$5\pi/6$
$\sec x$	1	$2/\sqrt{3} \approx 1.155$	$\sqrt{2}$	2	-2	$-\sqrt{2}$	$-2/\sqrt{3} \approx -1.155$

x	π	$7\pi/6$	$5\pi/4$	$4\pi/3$	$5\pi/3$	$7\pi/4$	$11\pi/6$	2π
$\sec x$	-1	$-2/\sqrt{3}$	$-\sqrt{2}$	-2	2	$\sqrt{2}$	$2/\sqrt{3}$	1

As in the case of tangent and cotangent, a discussion similar to the one we have just given (Exercise 44) yields the graph of $y = \csc x$ shown in **Figure 7.12.**

FIGURE 7.12

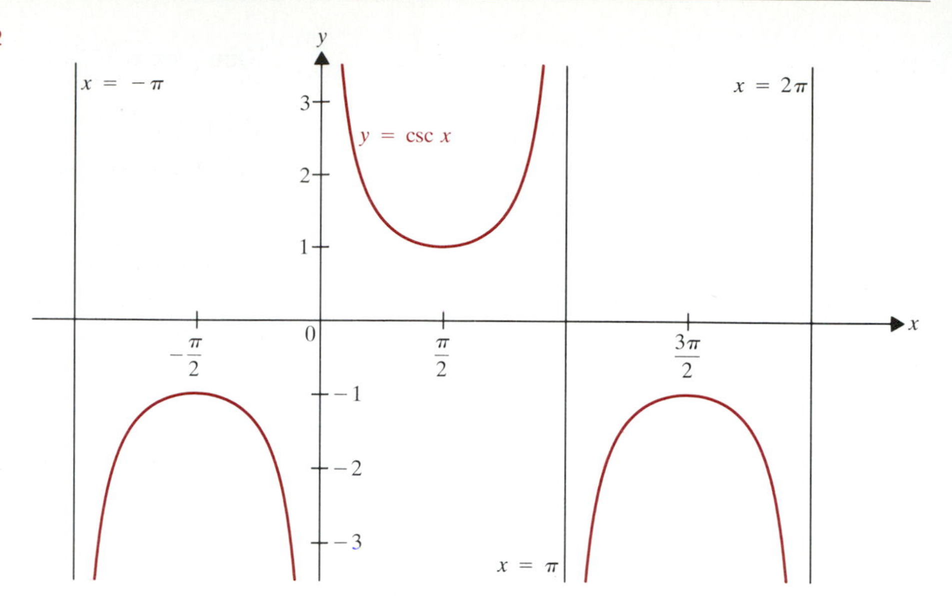

Proof of Theorem 7.3*

Since the proofs of all three parts are similar, we prove the first part and leave the rest for Exercises 45 and 46.

7.3(a) THEOREM

The graph of an equation $F(x, y) = 0$ is symmetric in the x-axis if and only if

$$F(x, -y) = 0 \text{ is equivalent to } F(x, y) = 0.$$

FIGURE 7.13

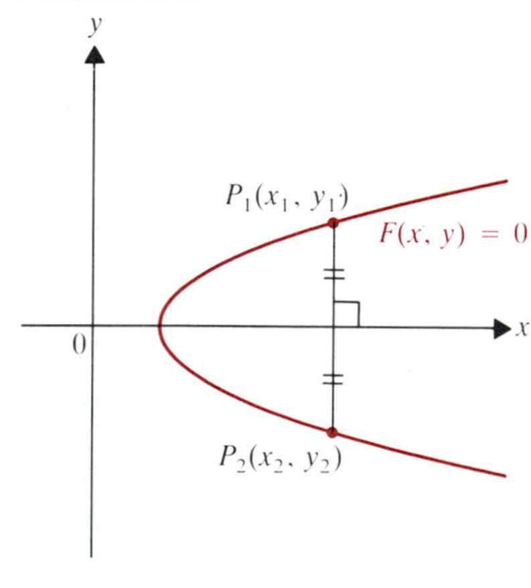

Proof. (i) Suppose that the graph is symmetric in the x-axis. Then for each point $P_1(x_1, y_1)$ on the graph, there is a point $P_2(x_2, y_2)$ on the graph such that the x-axis is the perpendicular bisector of P_1P_2. From **Figure 7.13,** we see that $x_2 = x_1$ since P_1P_2 is vertical. Also, $y_2 = -y_1$ since P_1 and P_2 are equidistant from the x-axis. Thus, if $P_1(x_1, y_1)$ is on the graph, then so is $P_2(x_1, -y_1)$. This means that if $F(x_1, y_1) = 0$, then $F(x_1, -y_1) = 0$. Similarly, if $P_2(x_1, -y_1)$ is on the graph, then so is $P_1(x_1, y_1)$. Thus if $F(x_1, -y_1) = 0$, then $F(x_1, y_1) = 0$. This shows that if the graph is symmetric in the x-axis, then $F(x, y) = 0$ is equivalent to $F(x, -y) = 0$.

(ii) Conversely, suppose that $F(x, y) = 0$ is equivalent to $F(x, -y) = 0$. Then if $P_1(x_1, y_1)$ is on the graph, so is $P_1(x_1, -y_1)$. Hence, the graph is symmetric in the x-axis. This shows that if $F(x, y) = 0$ is equivalent to $F(x, -y) = 0$, then the graph is symmetric in the x-axis.

From (i) and (ii), the graph is symmetric in the x-axis if and only if $F(x, -y) = 0$ is equivalent to $F(x, y) = 0$. QED

Exercises 3.7

In Exercises 1–10, draw accurate graphs of the functions in Exercises 1–10 of Exercises 3.3

In Exercises 11–44, draw accurate graphs of the functions.

11. $f(x) = x^3 - 12x + 5$

12. $f(x) = x^3 - 27x + 1$

13. $f(x) = \sin 2x$

14. $f(x) = \cos 3x$

15. $f(x) = \sin^2 x$

16. $f(x) = \cos^2 x$

17. $f(x) = x^{1/3}(x - 4)^{2/3}$

18. $f(x) = x^{2/3}(x^2 - 16)$

19. $f(x) = x^{4/3} + 8x^{1/3}$

20. $f(x) = x^{5/3} - \frac{5}{2}x^{2/3}$

* Optional

21. $f(x) = \dfrac{x}{x - 3}$

22. $f(x) = \dfrac{x^2}{x + 3}$

23. $f(x) = \dfrac{1}{x^2 + 1}$

24. $f(x) = \dfrac{x^2 + 1}{x^2 + x}$

25. $f(x) = \dfrac{x}{x^2 + 4}$

26. $f(x) = \dfrac{x + 1}{(x + 9)^2}$

27. $f(x) = \dfrac{x}{x^2 - 4}$

28. $f(x) = \dfrac{x^2 - 1}{x^2 - 4}$

29. $f(x) = \dfrac{x^2}{x^2 - 4}$

30. $f(x) = \dfrac{x^2 - 4}{x^2 + 3}$

31. $f(x) = \dfrac{x^2}{1 + x}$

32. $f(x) = \dfrac{x^2 + 2}{x - 3}$

33. $f(x) = \dfrac{x^2 + 1}{x}$

34. $f(x) = \dfrac{x^3}{x^2 - 4}$

35. $g(x) = \dfrac{1}{x^3 - 6x}$

36. $g(x) = \dfrac{1}{x^3 + 4x^2}$

37. $h(x) = \dfrac{x - 5}{(x - 1)^3}$

38. $h(x) = \dfrac{(x + 1)^3}{x - 2}$

39. $f(x) = \sin x \cos x$

40. $f(x) = \sin x + \cos^2 x$

41. $f(x) = \dfrac{\sin x}{2 + \cos x}$

42. $f(x) = \dfrac{\cos x}{1 + 2 \sin x}$

43. $f(x) = \cot x$

44. $f(x) = \csc x$

45. Prove Theorem 7.3(b).

46. Prove Theorem 7.3(c).

47. If $r(x) = \dfrac{p(x)}{q(x)}$ is a rational function such that $\deg(p(x)) = 1 + \deg(q(x))$, then show that the graph of r has an asymptote that is neither vertical nor horizontal. [*Hint:* Divide $p(x)$ by $q(x)$. What is the degree of the quotient $Q(x)$? Since the remainder has smaller degree than $q(x)$, you can show that as $x \to \pm\infty$, $r(x) \to$ the line $y = Q(x)$.]

48. Use the intermediate value theorem for continuous functions (Theorem 7.6 of Chapter 1) to show that the range of the function r in Example 3 is $(-\infty, +\infty)$. (*Hint:* Consider r on $(-1, 1)$.)

49. Use the approach of Exercise 48 to show that the tangent function has range $(-\infty, +\infty)$.

50. Show that the secant function has range $(-\infty, -1] \cup [1, +\infty)$. (See Example 6 and Exercise 48.)

3.8 Newton's Method

There is one weak link in the optimization theory presented in the past several sections. It requires us to solve the equation $f'(x) = 0$ for x in order to find critical points of a function f. Most of the optimization examples and exercises in this chapter have involved functions for which this was relatively easy. However, problems arising in real life need not be so considerate of our ability to solve equations.

Suppose for instance that we wished to find the critical points of

$$f(x) = \frac{1}{4}x^4 - x^2 - 5x + 3.$$

Since

$$f'(x) = x^3 - 2x - 5,$$

we would be faced with solving the equation

$$x^3 - 2x - 5 = 0. \tag{1}$$

Noting that

$$f'(2) = 8 - 4 - 5 = -1 < 0$$

and

$$f'(3) = 27 - 6 - 5 = 16 > 0,$$

the intermediate value theorem (Theorem 7.6 of Chapter 1) says that there is a root of (1) between 2 and 3. How can we find it? There *are* algebraic formulas, resembling the quadratic formula, for the roots of third- and fourth-degree polynomials. But they are very complicated and cumbersome. Moreover, there are no general formulas at all for roots of polynomial equations of degree 5 or higher. Thus we lack any procedure for finding roots of an equation like

$$x^5 - 2x - 1 = 0. \tag{2}$$

Observe that (2) has a root between 1 and 2, because

$$1^5 - 2 \cdot 1 - 1 = -2 < 0$$

and

$$2^5 - 2 \cdot 2 - 1 = 27 > 0.$$

But how can we find it? This is Problem 4 of the Prologue, and the method of solution we develop will enable you to solve it (see Exercise 11).

The solution scheme applies to any equation $f(x) = 0$, where f is differentiable. By calculating $f(x)$ for a few choices of x, we can locate an interval $[a, b]$ where f changes sign. Since f is differentiable, it is continuous by Theorem 1.3 of Chapter 2. Thus there is a zero x^* of f between a and b by the intermediate value theorem. We begin to search for it by choosing a point x_1 such that $f(x_1)$ is small. For x_1 we might choose one of the endpoints or midpoint of $[a, b]$, for instance. Since we can't solve $f(x) = 0$ for x, we do the next best thing. We solve

$$T(x) = 0 \tag{3}$$

for x, where

$$y = T(x) = f(x_1) + f'(x_1)(x - x_1) \tag{4}$$

is the equation of the tangent line to the graph of f at $(x_1, f(x_1))$. So instead of finding x^*, where the graph of $y = f(x)$ intersects the x-axis, we find the point x_2, where the *tangent line* to the graph at $(x_1, f(x_1))$ intersects the x-axis. Since the tangent approximation $T(x)$ is close to $f(x)$ for x near x_1, we expect x_2 to be close to x^*. **Figure 8.1** suggests that this expectation is a reasonable one. Moreover, it is easy to find x_2. From (3) and (4) we have

FIGURE 8.1

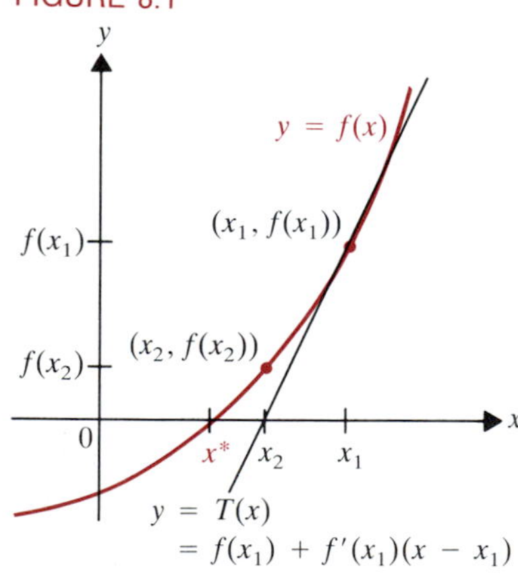

$$f(x_1) + f'(x_1)(x_2 - x_1) = 0 \rightarrow f'(x_1)x_2 - f'(x_1)x_1 = -f(x_1),$$
$$f'(x_1)x_2 = f'(x_1)x_1 - f(x_1).$$

Thus if $f'(x_1) \neq 0$, we obtain

$$x_2 = x_1 - \frac{f(x_1)}{f'(x_1)}. \tag{5}$$

[If we were unlucky enough to choose as x_1 a point where $f'(x_1) = 0$, we would choose a different x_1 so that we could use (5).]

What (5) provides is a rough approximation to x^*, that is, to a solution of $f(x) = 0$. We next try to improve on it by using the tangent approximation for $f(x)$ near $f(x_2)$. **Figure 8.2,** which shows part of Figure 8.1 under a microscope, illustrates that process. A better approximation to x^* than x_2 is the point x_3 where the tangent line to the graph of f at $(x_2, f(x_2))$ intersects the x-axis. Reasoning identical to that preceding Equation (5) leads to the formula

FIGURE 8.2

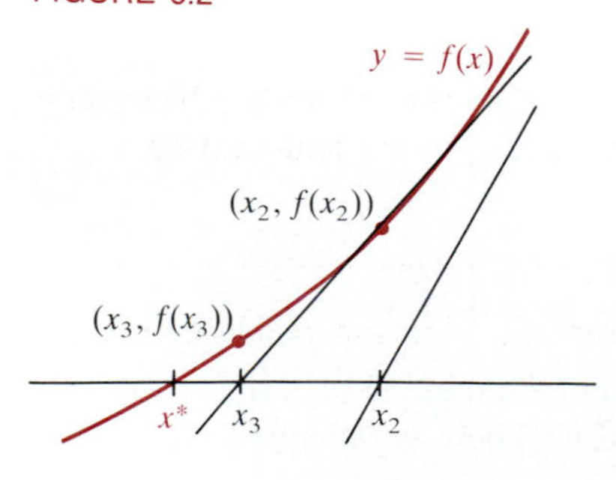

$$x_3 = x_2 - \frac{f(x_2)}{f'(x_2)}$$

if $f'(x_2) \neq 0$. If we continue this scheme, then after n *iterations* (i.e., repetitions) of the process, we obtain the approximation

(6) $$x_{n+1} = x_n - \frac{f(x_n)}{f'(x_n)}$$

for x^*, if $f'(x_n) \neq 0$.

It is reasonable to expect $x_1, x_2, \ldots, x_n, x_{n+1}, \ldots$ to be a sequence of successively closer approximations to x^*. And in fact, iteration of (5) and (6) usually produces a sequence of approximations that rapidly close in on x^*. By this we mean that after a few iterations, x_{n+1} will agree with x_n to as many decimal places as your calculations involve—say as many as your calculator displays. Use of (5) and (6) to approximate a root x^* of $f(x) = 0$ is referred to as ***Newton's method*** because it was introduced by Newton in a book he wrote in 1669, which was published in 1711. To illustrate the process, we return to Equation (1), which was the very example Newton used to introduce the method.

Example 1

Find a root of $x^3 - 2x - 5 = 0$.

Solution. Letting $f(x) = x^3 - 2x - 5$, we have $f'(x) = 3x^2 - 2$. Since $f(2) = -1$ is pretty close to 0, we use $x_1 = 2$. Then (5) gives

$$x_2 = 2 - \frac{f(2)}{f'(2)} = 2 - \frac{-1}{10} = 2.1.$$

From (6) we obtain

$$x_3 = 2.1 - \frac{f(2.1)}{f'(2.1)} = 2.1 - \frac{0.061}{11.23} \approx 2.09456812,$$

$$x_4 = x_3 - \frac{f(x_3)}{f'(x_3)} \approx 2.09456812 - \frac{0.00018572}{11.1616468} \approx 2.09455148,$$

$$x_5 = x_4 - \frac{f(x_4)}{f'(x_4)} \approx 2.09455148 - \frac{0.00000006}{11.16143773} \approx 2.09455148.$$

Since the calculator used for the calculations could not distinguish between x_5 and x_4, we conclude that

(7) $$x^* \approx 2.09455148$$

is an accurate approximation. (Indeed, the calculator shows $f(x_5) \approx 5 \times 10^{-9} \approx f(x^*) = 0$.) Since f is continuous and increasing on $(\sqrt{2/3}, +\infty)$, it has a continuous inverse f^{-1} near x^* (p. 117). Thus $f(x_5)$ very close to $f(x^*)$ means that

$$f^{-1}(f(x_5)) \text{ is very close to } f^{-1}(f(x^*)) \quad \rightarrow \quad x_5 \text{ is very close to } x^*.$$

See **Figure 8.3.** ■

FIGURE 8.3

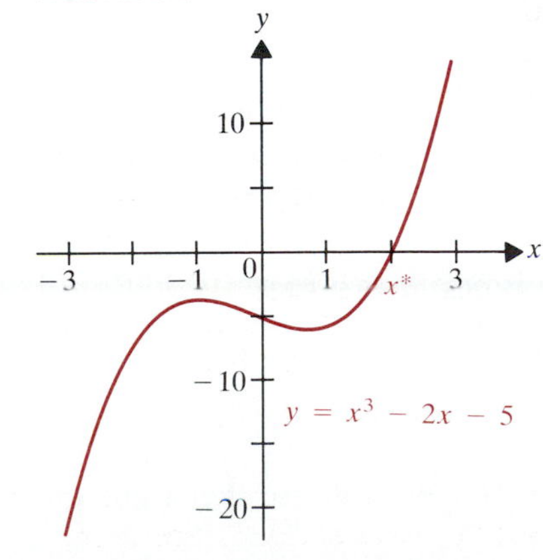

The fact that x_{n-1} is obtained directly from x_n using (6) makes Newton's method ideal for computer or programmable calculator implementation.

Example 2

(a) Find the Newton's method formula for x_{n+1} for the positive root of $x^2 - k = 0$, $k \geq 0$.

(b) Use the result of (a) to approximate $\sqrt{2}$.

Solution. (a) We let $f(x) = x^2 - k$. Then $f'(x) = 2x$. Hence (6) takes the form

$$x_{n+1} = x_n - \frac{x_n^2 - k}{2x_n}. \tag{8}$$

Equation (8) is the one programmed onto the chip of most calculators that have a square-root key. The calculator iterates (8) internally until x_{n+1} and x_n agree to the number of decimal places available.

(b) Noting that $x^2 - 2$ is negative for $x = 1$ and positive for $x = 2$, we can use $x_1 = 1$. Then from (8), with $k = 2$, a calculator gives

$$x_2 = x_1 - \frac{1 - 2}{2} = 1 - \frac{-1}{2} = 1.5$$

$$x_3 = x_2 - \frac{x_1^2 - 2}{2x_1} \approx 1.4166667$$

$$x_4 = x_3 - \frac{x_2^2 - 2}{2x_2} \approx 1.4142157$$

$$x_5 = x_4 - \frac{x_3^2 - 2}{2x_3} \approx 1.4142136 = x_6.$$

Substitution of this value into $f(x) = x^2 - 2$ gives 0 on the calculator used to do these calculations. So we conclude that, to seven decimal places, $\sqrt{2} = 1.4142136$. (In fact, to nine decimal places, $\sqrt{2} = 1.414213562$.) ■

It may be of interest that (8) in the form

$$x_{n+1} = \frac{x_n^2 + k}{2x_n} = \frac{1}{2}\left(x_n + \frac{k}{x_n}\right)$$

is an ancient algorithm for calculating $\sqrt{k}$, called the *Babylonian square root* method. It approximates $\sqrt{k}$ by *averaging* the previous approximation x_n with the quotient k/x_n.

So far, we have used Newton's method only for polynomial functions. Much of its usefulness stems from the fact that it works just as well for nonalgebraic functions.

Example 3

Solve $3 \sin x - x = 0$ on $[0, 2\pi]$.

Solution. One obvious root is $x = 0$. To search for others, we let $f(x) = 3 \sin x - x$ and compile a table of values for f. (Be sure that your calculator is in *radian* mode to check the entries in the table.)

x	0	$\pi/6$	$\pi/3$	$\pi/2$	$2\pi/3$
$f(x)$	0	0.9764012	1.5508787	1.4292037	0.5036811

x	$3\pi/4$	$5\pi/6$	π	$3\pi/2$	2π
$f(x)$	-0.2348741	-1.1179939	$-\pi$	-7.712389	-2π

We see that there is a root near $3\pi/4 \approx 2.3561945$. We will use that value for

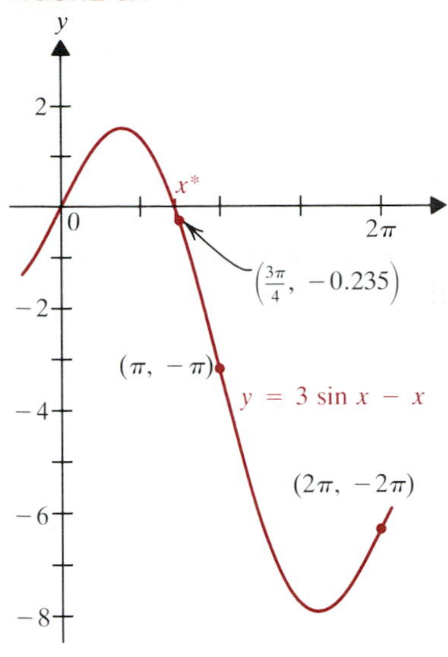

FIGURE 8.4

x_1. Since $f'(x) = 3\cos x - 1$, we have from (6)

$$x_{n+1} = x_n - \frac{3\sin x_n - x_n}{3\cos x_n - 1}. \tag{9}$$

From (9) we obtain

$$x_2 = 2.2809462,$$
$$x_3 = 2.2788643,$$
$$x_4 = 2.2788627 = x_5.$$

We conclude then that $x^* \approx 2.2788627$ is another root of $3\sin x - x = 0$ in $[0, 2\pi]$. As a check, a calculator gives $f(x_5) = -6 \times 10^{-10}$, so as in Example 1, x_5 is indeed a very good approximation to x^*. From **Figure 8.4,** there appear to be no other roots of the given equation in $[0, 2\pi]$. ■

Further Remarks on Newton's Method*

The examples we have worked using Newton's method are representative of how well the method usually works. The Newton approximations $x_1, x_2, \ldots, x_n, \ldots$ usually close in on the root x^* quite rapidly—so rapidly that if x_n is an approximation to x^* correct in the first k digits, then x_{n+1} will often approximate x^* correctly to $2k$ digits. We leave a detailed discussion of how fast the x_n's close in on x^* for a course in numerical analysis. There are, though, situations in which the x_n's don't close in on x^* at all, or take so long to do so that one doesn't want to get caught in such a sequence. Consider **Figure 8.5,** where we again show the graph of $y = 3\sin x - x$, this time between $x = -\pi$ and $x = \pi$. In Example 3, we used $x_1 = 3\pi/4$, which was close to the root we were seeking. Suppose, however, that we started with a point near the local maximum, which occurs when $\cos x = 1/3$. If you have a calculator with the inverse cosine function programmed, then you will find the corresponding x to be approximately 1.23096. If we start with the slightly smaller value 1.23 for x, then the figure suggests that we will shoot off almost horizontally to a point x_2 very far from any of the three roots of the equation $3\sin x - x = 0$. As Table 1 shows, this is indeed what happens. Table 1 gives the results of a computer program that generates Newton approximations but stops after 50 iterations if two successive x_n fail to be within 0.000001 of each other. Hopeless as the entries of the table make it seem, this sequence of approximations actually closes in on the root found in Example 3, but only after more than 225 iterations! That certainly is not how you would want to locate this root. Thus it is wise to build into computer programs for implementing Newton's method a counter to stop the program after a certain number of iterations.

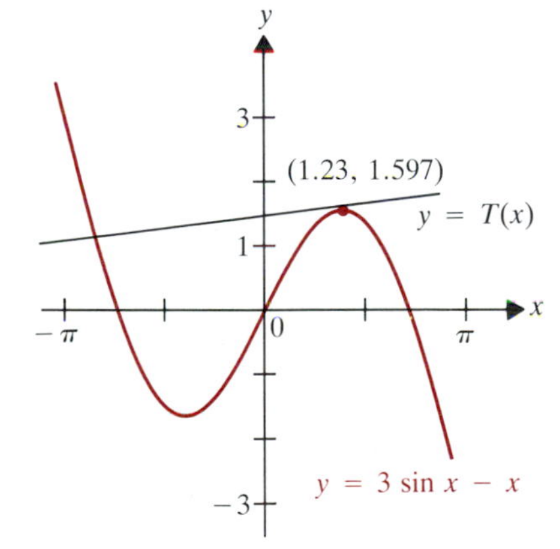

FIGURE 8.5

Newton's method approximations can actually become arbitrarily large as n keeps increasing. (See Exercise 45.) They can also cycle back and forth between two values and so never close in on the desired root.

Example 4

Use Newton's method to look for the root of the equation $2x^{1/3} + x = 0$, starting with $x_1 = 1$.

* Optional

TABLE 1 Newton's Method for $f(x) = 3\sin x - x$, starting from $x_1 = 1.23$

n	x_n	$f(x_n)$	n	x_n	$f(x_n)$
2	−587.550275	587.767433	26	18484.4084	−18486.3929
3	−440.318744	438.891611	27	33275.5957	−33276.0549
4	−708.130746	710.998172	28	50213.0241	−50215.462
5	−330.335534	331.689658	29	31941.3017	−31943.3005
6	−240.129023	237.190513	30	22073.4708	−22071.6772
7	359.26307	−356.560801	31	37785.6227	−37788.6022
8	1536.12939	−1535.8018	32	−20385.0715	20383.0859
9	1150.44919	−1148.69079	33	−14111.3154	14113.2908
10	1953.36891	−1955.30551	34	−25331.7418	25334.3881
11	3467.71018	−3469.42413	35	−14833.6564	14836.0863
12	5840.47951	−5841.24718	36	−34370.1562	34367.4397
13	4342.7687	−4340.11208	37	−160241.162	160238.506
14	15366.5401	−15369.0813	38	−564862.932	564865.903
15	9442.82379	−9444.98363	39	−166345.424	166348.255
16	18171.295	−18170.3425	40	−82801.5074	82798.5292
17	28021.0269	−28023.7798	41	−21962.7841	21962.5141
18	15238.6527	−15235.8423	42	−16455.3931	16456.1911
19	7805.82434	−7803.24721	43	−25153.4835	25150.6376
20	4728.47483	−4729.58169	44	−12250.0907	12252.6486
21	3480.01817	−3482.30435	45	−7478.11795	7475.40801
22	7174.67568	−7176.65821	46	−33529.7679	33528.4617
23	12908.7912	−12908.752	47	−24469.749	24469.4012
24	9681.39684	−9683.90774	48	−18321.3085	18322.6399
25	24771.2579	−24770.6613	49	−29173.5254	29171.6018
			50	−51576.4865	51578.9437

The number of iterations specified failed to find a root with the required accuracy. The method failed.

TABLE 2 Newton's Method for $f(x) = 2x^{1/3} + x = 0$, starting with $x_1 = 1$

n	x_n	$f(x_n)$	n	x_n	$f(x_n)$
2	−.8	−2.65663553	27	.544338614	2.17733934
3	.697879071	2.47189175	28	.544336094	−2.1773343
4	−.640205994	−2.36393864	29	.544334414	2.17733094
5	.60561899	2.29773379	30	−.544333294	−2.1773287
6	−.584089541	−2.25591066	31	.544332547	2.1773272
7	.57036661	2.22899092	32	−.54433205	−2.17732621
8	−.561484493	−2.211454	33	.544331718	2.17732554
9	.555677726	2.19993962	34	−.544331497	−2.1773251
10	−.551856413	−2.19234052	35	.544331349	2.17732481
11	.549330718	2.18730831	36	−.544331251	−2.17732461
12	−.547656538	−2.18396844	37	.544331185	2.17732448
13	.546544666	2.18174845	38	−.544331141	−2.17732439
14	−.545805298	−2.18027138	39	.544331112	2.17732433
15	.545313219	2.17928796	40	−.544331093	−2.1773243
16	−.544985535	−2.17863292	41	.54433108	2.17732427
17	.544767243	2.17819648	42	−.544331071	−2.17732425
18	−.544621789	−2.17790563	43	−.544331066	2.17732424
19	.544524851	2.17771179	44	−.544331062	−2.17732423
20	−.544460241	−2.17758258	45	.544331059	2.17732423
21	.544417174	2.17749645	46	−.544331058	−2.17732423
22	−.544388465	−2.17743904	47	.544331057	2.17732422
23	.544369327	2.17740076	48	−.544331056	−2.17732422
24	−.544356569	−2.17737525	49	.544331055	2.17732422
25	.544348064	2.17735824	50	−.544331055	−2.17732422
26	−.544342394	−2.1773469	51	.544331054	2.17732422

The number of iterations specified failed to find a root with the required accuracy. The method failed.

Solution. Letting $f(x) = 2x^{1/3} + x$, we have

$$f'(x) = \frac{2}{3}x^{-2/3} + 1.$$

Using (6), a computer program with a limit of 51 iterations produced Table 2. ■

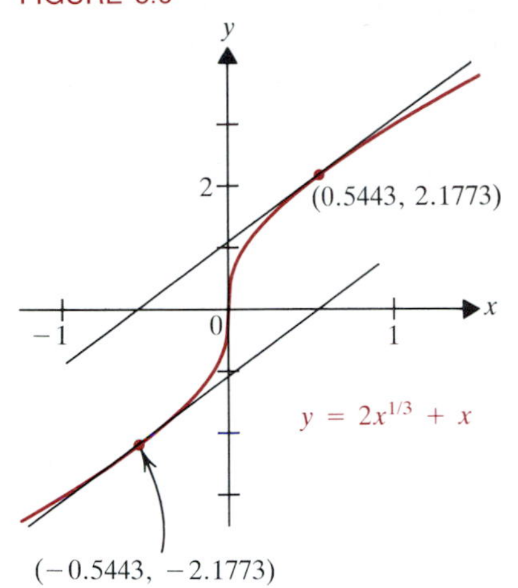

FIGURE 8.6

The behavior in Table 2 is rather different from that in Table 1. Here the x_n's seem to cycle back and forth between values around 0.5443 and -0.5443. Indeed, if we continue Table 2 out to any number of iterations, we find that the values of successive x_n jump back and forth between x_{51} and $-x_{51}$. This time, the problem is again shown by the graph (**Figure 8.6**). The shape of the graph is such that the tangent line drawn near $x = x_{51}$ seems to hit the x-axis very near $x = x_{52}$. This sets up a cycle in which we get caught and can never escape.

It is not easy to produce examples in which Newton's method fails, so the preceding discussion shouldn't discourage you. It does, however, suggest that some care is helpful to avoid problems. Notice in Example 4 that $x^* = 0$ is such that f is not differentiable at x^*. The following theorem, whose proof is given in numerical analysis, says that for most functions actually dealt with in practice, Newton's method will *not* fail if we start with an initial guess x_1 fairly close to x^*.

8.1 THEOREM

Suppose that f'' exists and is continuous on an interval I containing x^* where $f'(x^*) \neq 0$. Then there exists a $\delta > 0$ such that *any* choice of x_1 in $(x^* - \delta, x^* + \delta)$ will produce a sequence of Newton approximations x_n that close in on the root x^*.

In view of Theorem 8.1, the best practice is simply to follow the procedure in the first three examples. As long as your initial guess x_1 is close to x^*, Newton's method should find a very close approximation to x^* quite rapidly. If it doesn't, then either your initial value for x_1 was chosen too far from x^* (so try a different x_1) or the function f is badly behaved near x^*.

Exercises 3.8

In Exercises 1–20, find the root of the given equation in the given interval. Express your answers to at least five decimal places.

1. $x^2 - 3 = 0$ on $[1, 2]$. That is, find $\sqrt{3}$ by Newton's method.
2. $x^2 - 5 = 0$ on $[2, 3]$. That is, find $\sqrt{5}$ by Newton's method.
3. $x^3 - 2 = 0$ on $[1, 2]$. That is, find $\sqrt[3]{2}$ by Newton's method.
4. $x^3 - 10 = 0$ on $[2, 3]$. That is, find $\sqrt[3]{10}$ by Newton's method.
5. $x^4 - 3 = 0$ on $[1, 2]$. That is, find $3^{1/4}$ by Newton's method.
6. $x^4 - 18 = 0$ on $[2, 3]$. That is, find $18^{1/4}$ by Newton's method.
7. $x^5 - 10 = 0$ on $[1, 2]$. That is, find $10^{1/5}$ by Newton's method.
8. $x^5 - 45 = 0$ on $[2, 3]$. That is, find $45^{1/5}$ by Newton's method.
9. $x^6 - 100 = 0$ on $[2, 3]$. That is, find $100^{1/6}$ by Newton's method.
10. $x^{10} - 100 = 0$ on $[1, 2]$. That is, find $\sqrt[10]{100}$ by Newton's method.
11. $x^5 - 2x - 1 = 0$ on $[1, 2]$.
12. $x^5 - 2x - 1 = 0$ on $[-1, 0]$.
13. $x^5 - 2x - 5 = 0$ on $[1, 2]$.
14. $x^3 - x - 1 = 0$ on $[1, 2]$.
15. $x^4 - 4x^2 + x + 1 = 0$ on $[0, 1]$.
16. $x^4 - x^3 - 2x - 3 = 0$ on $[1, 2]$.
17. $\cos x - \sin x = 0$ on $[0, 1]$.
18. $\cos x - x = 0$ on $[0, 1]$.

19. $2 \sin x - x = 0$ on $[1, 2]$.

20. $2 \sin x - x^2 = 0$ on $[1, 2]$.

21. Since calculators divide more slowly than they add or multiply, the $1/x$ key on scientific calculators is programmed using Newton's method. Give the formula for x_{n+1} in terms of x_n.

22. Most calculators don't have a cube-root key. Use Newton's method to construct a program that such a key could use to evaluate $\sqrt[3]{a}$. (See Example 2.)

23. Give an iteration formula for $\sqrt[m]{a}$ for m a positive integer.

24. Repeat Exercise 23 for a^2.

A slightly modified version of (6) that is easier to use on a nonprogrammable calculator is

$$x_{n+1} = x_n - \frac{f(x_n)}{f'(x_1)}, \tag{10}$$

if $f'(x_1) \neq 0$. Use (10) to solve the indicated earlier exercises.

25. Exercise 1 **26.** Exercise 3

27. Exercise 5 **28.** Exercise 7

29. Exercise 9 **30.** Exercise 11

31. Exercise 13 **32.** Exercise 15

33. Exercise 17 **34.** Exercise 19

PC **Exercises 35–44 are intended for those who have access to a computer program for Newton's method. It should first evaluate the function at points between −20 and 20, searching for intervals where $f(x)$ changes sign. Find to at least five decimal places *all* roots of the equation in the interval [−20, 20].**

35. Exercise 11 **36.** $x^5 - 3x - 2 = 0$

37. Exercise 13 **38.** Exercise 14

39. Exercise 15 **40.** Exercise 16

41. Exercise 17 **42.** Exercise 18

43. Exercise 19 **44.** Exercise 20

45. What happens if you try to apply Newton's method to $f(x) = x^{1/3}$ starting at $x_1 = 1/2$?

46. In Example 4, what happens if you start from $x_1 = 0.5$?

47. In Example 4, what happens if you start from $x_1 = 0.25$?

48. On the basis of Exercises 46 and 47, what do you conclude about the differentiability hypothesis in Theorem 8.1?

3.9 Antidifferentiation

Many scientific laws involve rates of change. A familiar example is Newton's second law of motion,

$$F = ma = m\frac{dv}{dt} = m\frac{d^2y}{dt^2}, \tag{1}$$

where F is the force acting on a body of mass m moving along a line with velocity v whose position at time t is given by $y = y(t)$. If the body is falling freely from rest, the acceleration a is just that produced by gravity, that is, $a = -g$. (The minus sign indicates that the force of gravity is directed downward.) In that case, (1) becomes

$$m\frac{dv}{dt} = -mg. \tag{2}$$

Cancellation of the constant m from both sides of (2) gives

$$\frac{dv}{dt} = -g. \tag{3}$$

Falling *from rest* means that $v = 0$ when $t = 0$. Using this information in conjunction with (3), we can derive a formula for the velocity v of the body at time t. Before doing so, we remark that this is a problem of the following general type:

Given the derivative $f'(x)$ of a function $f(x)$, determine the function f.

This leads to a basic definition:

9.1 DEFINITION A function F is an ***antiderivative*** of a function f on the interval I if $F'(x) = f(x)$ for all x in I.

For instance, $F(x) = x^2$ is an antiderivative of $f(x) = 2x$ on any interval I, since $F'(x) = 2x$. Another antiderivative of f is $G(x) = x^2 + 1$. In fact, for any constant C, $F(x) + C$ is an antiderivative of $f(x)$. That these are the only antiderivatives of $f(x) = 2x$ follows from Theorem 3.1(c).

9.2 THEOREM If F and G are both antiderivatives of f on an interval I, then F and G differ by a constant on I. That is, for all x in I,

$$F(x) = G(x) + C \tag{4}$$

for some constant C.

Proof. For all x in I, we have $F'(x) = G'(x) = f(x)$. Thus

$$(F - G)'(x) = F'(x) - G'(x) = f(x) - f(x) = 0.$$

By Theorem 3.1(c), $F - G$ is therefore constant on I. That is,

$$(F - G)(x) = F(x) - G(x) = C. \quad \boxed{\text{QED}}$$

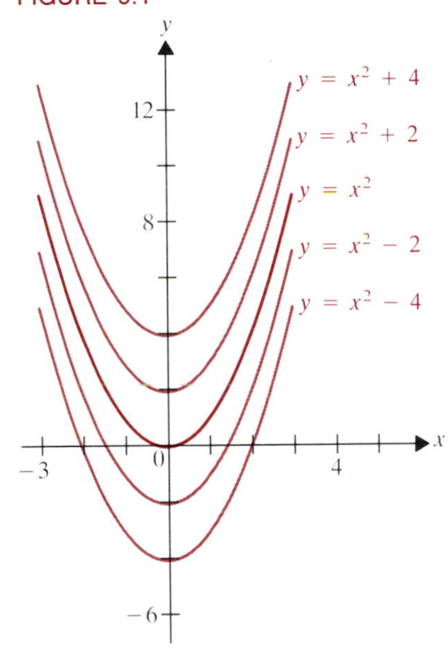

FIGURE 9.1

Figure 9.1 shows the graphical interpretation of Theorem 9.2 for the function $f(x) = 2x$. As noted above, $F(x) = x^2$ is one antiderivative of $f(x)$. By Theorem 9.2, *any* antiderivative G of f is of the form $x^2 + C$ for some constant C. So its graph is a vertical translate of the graph of $y = x^2$. Figure 9.1 shows the graph of $y = x^2 + C$ for several values of C. (If one graph is a vertical translate of another, then the slopes at each value of x are the same, and one graph is a fixed distance above the other.)

Returning to the discussion of the falling body, we see that finding a formula for v from (3) is the same as finding an antiderivative G of the constant function $-g$. Since v is 0 when $t = 0$, the function G must also satisfy the condition $G(0) = 0$. Note that

$$\frac{d}{dt}(-gt) = -g.$$

Thus one possible formula for $G(t)$ is $-gt$. This function also satisfies the condition $G(0) = 0$. To see that it is the *only* such function, let F be any antiderivative of the constant function $-g$ such that $F(0) = 0$. By Theorem 9.2, $F(t)$ can differ from the antiderivative $-gt$ by only a constant. Thus for some number C, we have

$$F(t) - (-gt) = C. \tag{5}$$

In particular, (5) must hold for the value $t = 0$. Since $F(0) = 0$, we find

$$0 - 0 = C \rightarrow C = 0.$$

Therefore (5) says that

$$F(t) = -gt = G(t). \tag{6}$$

Since $v = dy/dt$, we can also antidifferentiate (6) to get a formula for y. Noting that $(d/dt)(\frac{1}{2}t^2) = t$, we have from (6) and Theorem 9.2

$$y = -\frac{1}{2}gt^2 + K, \tag{7}$$

for some constant K. If we measure the body's position from its starting point, then $y = 0$ when $t = 0$. Hence, (7) gives

$$0 = -\frac{1}{2} g \cdot 0^2 + K \to K = 0.$$

We therefore have the formula

$$y = -\frac{1}{2} g t^2, \tag{8}$$

which you may recall from physics.

The foregoing discussion suggests the importance of learning to antidifferentiate given functions. Before getting to that, we introduce some notation for antiderivatives. Since antidifferentiation is the process that *undoes* differentiation, we will sometimes use the notation

$$D_x^{-1}(f(x)) \qquad \text{or} \qquad D^{-1}(f) \tag{9}$$

for it. (Read as *D sub x inverse of f*(*x*) or *D inverse of f*.) A more common notation due to Leibniz is better suited to one of the important techniques for finding antiderivatives. This is the *indefinite integral* notation

$$\int f(x)\, dx \tag{10}$$

[read (*indefinite*) *integral of f*(*x*) *dx*]. We won't be in a position to explain the origin of the integral sign until the next chapter, so for now we simply treat it as a formal symbol. Theorem 9.2 says that if F is one antiderivative of f on an interval I, then *any* antiderivative G of f on I has the form

$$G(x) = F(x) + C,$$

for some constant C. We write

$$\int f(x)\, dx = D_x^{-1}(f(x)) = F(x) + C \tag{11}$$

to express this fact. Here C is a constant that can be *any* real number. For this reason, this ***constant of antidifferentiation*** or ***constant of integration*** is referred to as an *arbitrary constant*. For each value of C, we get one particular antiderivative. As was the case above, many problems come with conditions that determine C, so that *one* of the infinitely many antiderivatives (11) is the unique solution of the problem.

The differentiation rules of the last chapter provide formulas for antiderivatives. For example, the power rule for derivatives,

$$D_x(x^{r+1}) = (r + 1)x^r,$$

gives the ***power rule for antiderivatives:***

$$\int x^r\, dr = \frac{x^{r+1}}{r+1} + C, \qquad \text{if } r \neq -1. \tag{12}$$

To verify (12), simply differentiate the right side:

$$D_x\left(\frac{x^{r+1}}{r+1}\right) = \frac{1}{r+1} D_x(x^{r+1}) = \frac{1}{\cancel{(r+1)}} \cancel{(r+1)} x^r = x^r.$$

(The excluded case $r = -1$ in (12) will be considered in Chapter 6.)

The rules for differentiating the sum (or difference) of two functions, and a constant times a function, similarly give antidifferentiation formulas, namely,

$$\int [f(x) \pm g(x)]\,dx = \int f(x)\,dx \pm \int g(x)\,dx \tag{13}$$

and

$$\int kf(x)\,dx = k\int f(x)\,dx. \tag{14}$$

To verify these, let F and G be antiderivatives of f and g respectively on I. Then $D_xF(x) = f(x)$ and $D_xG(x) = g(x)$. Hence, for any x in I,

$$D_x[F(x) \pm G(x)] = D_xF(x) \pm D_xG(x) = f(x) \pm g(x),$$

which gives (13). Similarly,

$$D_x[kF(x)] = kD_xF(x) = kf(x),$$

for any x in I, so that (14) holds.

Example 1

Find

(a) $D_x^{-1}(x^3 - 3x^2 + x - 1)$ (b) $\int\left(\sqrt{x} - \frac{1}{\sqrt[3]{x}}\right)dx$

Solution.

(a) Using (12), (13), and (14), we find

$$\begin{aligned} D_x(x^3 - 3x^2 + x - 1) &= D_x^{-1}(x^3) - D_x^{-1}(3x^2) + D_x^{-1}(x) - D_x^{-1}(-1) \\ &= \frac{1}{4}x^4 - x^3 + \frac{1}{2}x^2 - x + C. \end{aligned}$$

(b)
$$\begin{aligned} \int\left(\sqrt{x} - \frac{1}{\sqrt[3]{x}}\right)dx &= \int (x^{1/2} - x^{-1/3})\,dx = \int x^{1/2}\,dx - \int x^{-1/3}\,dx && \text{by (13)} \\ &= \frac{x^{3/2}}{3/2} - \frac{x^{2/3}}{2/3} + C && \text{by (12)} \\ &= \frac{2}{3}x^{3/2} - \frac{3}{2}x^{2/3} + C. \quad ■ \end{aligned}$$

The chain rule provides the following result, which is just as important to antidifferentiation as the chain rule itself is to differentiation.

9.3 THEOREM

Chain Rule for Antiderivatives. Suppose that F is an antiderivative of f on an interval I, and g is a differentiable function whose range is in I. Then

$$\int f(g(x))g'(x)\,dx = F(g(x)) + C. \tag{15}$$

If we let $u = g(x)$, then (15) becomes

$$\int f(u)\,du = F(u) + C. \tag{16}$$

Proof. By the chain rule,

$$\frac{d}{dx}(F(g(x)) = (F \circ g)'(x) = F'(g(x))g'(x) = f(g(x))g'(x).$$

Therefore

$$\int f(g(x))g'(x)\,dx = F(g(x)) + C.$$

This proves (15). To get (16), we make the substitution $u = g(x)$. Then $du = g'(x)\,dx$, so (15) becomes

$$\int f(u)\,du = F(u) + C. \quad \boxed{\text{QED}} \tag{16}$$

Example 2

Find $\int x^2\sqrt{x^3+1}\,dx$ on the interval $[-1, +\infty)$.

Solution. Observe that, except for a constant factor of 3, x^2 is the derivative of $x^3 + 1$. Because (14) makes it possible to antidifferentiate constant multiples of any function whose antiderivative we can find, the absence of the factor 3 is not a problem. To apply (16), we let

$$u = g(x) = x^3 + 1. \tag{17}$$

We next calculate the factor du that appears in (16):

$$du = g'(x)\,dx = 3x^2\,dx. \tag{18}$$

To transform the given integral into a u-integral, we express $x^2\,dx$ in terms of u:

$$\frac{1}{3}\,du = \frac{1}{3}\,g'(x)\,dx = x^2\,dx. \tag{19}$$

Theorem 9.3 then gives

$$\int x^2\sqrt{x^3+1}\,dx = \int \frac{1}{3}\sqrt{u}\,du \tag{20}$$

$$= \frac{1}{3}\int u^{1/2}\,du \qquad \text{by (14)} \tag{21}$$

$$= \frac{1}{3}\frac{u^{3/2}}{3/2} + C \qquad \text{by (12)} \tag{22}$$

$$= \frac{2}{9}(x^3+1)^{3/2} + C, \tag{23}$$

since $u = x^3 + 1$. ■

Antidifferentiation plays a prominent role in subsequent chapters. It is therefore important to develop skill in finding antiderivatives. In our solution of Example 2, we followed a sequence of steps aimed at making Theorem 9.3 easy to apply. We summarize the system for you in the following algorithm for antidifferentiating a function to which (12), (13), and (14) don't apply immediately.

ANTIDIFFERENTIATION BY SUBSTITUTION

1. In the function to be antidifferentiated, try to find a factor $h(x)$ that is, up to a constant multiple, the derivative of *some other part* $g(x)$ of the given function.
2. Set u equal to the $g(x)$ identified in Step 1. Then $du = g'(x)\,dx$ will be a constant multiple of the factor $h(x)$ in Step 1.
3. In the function to be antidifferentiated, substitute u and du for their respective expressions in Step 2.
4. Use Theorem 9.3 (in conjunction with (12), (13), (14), or other known formulas) to antidifferentiate the resulting function of u.
5. Substitute $g(x)$ back for u, to obtain a formula for the antiderivative as a function of x.

In Example 2, Step 1 was carried out first. Step 2 was done in (17) and (18). In (19) we solved *explicitly* for the factor $x^2\,dx$ identified in Step 1. That made Step 3—Equation (20)—easy. Equations (21) and (22) correspond to Step 4. Finally, Step 5 simply consisted of using (17) to make the transition from (22) to (23). Carefully following the suggested sequence of steps should help you develop a systematic approach to antidifferentiation. In Section 2.6, u-substitution introduced to help apply the chain rule for derivatives was quickly discarded. By contrast, you are likely to find Step 2 difficult to short cut until you have had considerable practice using it.

It is often worthwhile to check your work by differentiating your answer to see that you get the function you were given to antidifferentiate. (Try this in Example 2: Differentiation of (23) should give $x^2\sqrt{x^3+1}$.)

The rules in Section 2.4 for differentiating trigonometric functions give the following antidifferentiation formulas.

9.4 THEOREM

(a) $\int \sin x\,dx = -\cos x + C$ (b) $\int \cos x\,dx = \sin x + C$

(c) $\int \sec^2 x\,dx = \tan x + C$ (d) $\int \csc^2 dx = -\cot x + C$

(e) $\int \sec x \tan x\,dx = \sec x + C$ (f) $\int \csc x \cot x\,dx = -\csc x + C$

Proof. Since each part is proved in the same way, we derive only the first formula and leave the rest for Exercise 43. Theorem 4.1 of Chapter 2 says that $D_x(\cos x) = -\sin x$. Hence,

$$\frac{d}{dx}(-\cos x) = -(-\sin x) = \sin x \rightarrow \int \sin x\,dx = -\cos x + C. \quad \boxed{\text{QED}}$$

Example 3

Find (a) $\int \sin 3x\,dx$ (b) $\int \frac{\cos\sqrt{x}}{\sqrt{x}}\,dx$.

Solution. (a) While no factor of $\sin 3x$ is, up to a constant multiple, the derivative of another factor, our early experience with the chain rule for derivatives (see Example 3 of Section 2.6) suggests the substitution

$$u = 3x.$$

Then

$$du = 3\,dx,$$

so that

$$\frac{1}{3}\,du = dx.$$

Thus Theorem 9.3 gives

$$\int \sin 3x\,dx = \frac{1}{3}\int \sin u\,du = \frac{1}{3}(-\cos u) + C$$
$$= -\frac{1}{3}\cos 3x + C.$$

(b) Here $1/\sqrt{x} = x^{-1/2}$ is, except for a constant factor, the derivative of $\sqrt{x} = x^{1/2}$. We therefore let

$$u = \sqrt{x} = x^{1/2}.$$

Then

$$du = \frac{1}{2}x^{-1/2}\,dx = \frac{1}{2\sqrt{x}}\,dx,$$

so that

$$2\,du = \frac{1}{\sqrt{x}}\,dx.$$

Thus by Theorem 9.3,

$$\int \frac{\cos\sqrt{x}}{\sqrt{x}}\,dx = 2\int \cos u\,du = 2\sin u + C = 2\sin\sqrt{x} + C. \quad ■$$

If Example 3(a) strikes you as too much work to antidifferentiate such a simple function, then you may wish to memorize at least the first two of the following formulas. The proofs involve simply using the idea of Example 3(a). (See Exercises 44–47.)

$$\int \sin kx\,dx = -\frac{1}{k}\cos kx + C \qquad \int \cos kx\,dx = \frac{1}{k}\sin kx + C$$

$$\int \sec^2 kx\,dx = \frac{1}{k}\tan kx + C \qquad \int \csc^2 kx\,dx = -\frac{1}{k}\cot kx + C$$

$$\int \sec kx \tan kx\,dx = \frac{1}{k}\sec kx + C \qquad \int \csc kx \cot kx\,dx = -\frac{1}{k}\csc kx + C$$

Example 4

Find (a) $\int \sec^2 3x\,dx$ (b) $\int \frac{\csc x \cot x}{(1 + \csc x)^3}\,dx$

Solution. (a) The third boxed formula above, with $k = 3$, gives

$$\int \sec^2 3x\, dx = \frac{1}{3} \tan 3x + C.$$

(b) Here the factor $\csc x \cot x$ is, except for a minus sign, the derivative of the factor $1 + \csc x$ in the denominator. So we let

$$u = 1 + \csc x.$$

Then we have

$$du = -\csc x \cot x\, dx \rightarrow \csc x \cot x\, dx = -du.$$

Thus Theorem 9.3 says that

$$\begin{aligned}\int \frac{\csc x \cot x}{(1 + \csc x)^3}\, dx &= \int \frac{-du}{u^3} = -\int u^{-3}\, du \\ &= -\frac{u^{-2}}{-2} + C = \frac{1}{2u^2} + C \\ &= \frac{1}{2(1 + \csc x)^2} + C. \quad \blacksquare\end{aligned}$$

As our final example illustrates, sometimes basic trigonometric identities can be helpful in working out antiderivatives of trigonometric functions.

Example 5

Find (a) $\int \sin^3 x\, dx$ (b) $\int \tan^2 x\, dx$

Solution. (a) Using the trigonometric identity $\sin^2 x + \cos^2 x = 1$ in the form $\sin^2 x = 1 - \cos^2 x$, we get

$$\begin{aligned}\int \sin^3 x\, dx &= \int \sin x \sin^2 x\, dx = \int \sin x(1 - \cos^2 x)\, dx \\ &= \int \sin x\, dx - \int \cos^2 x \sin x\, dx.\end{aligned}$$

In the second integral, since $\sin x$ is—except for a minus sign—the derivative of $\cos x$, we let $u = \cos x$. Then $du = -\sin x\, dx$, so

$$\sin x\, dx = -du,$$

and we get

$$\begin{aligned}\int \sin^3 x\, dx &= \int \sin x\, dx - \int u^2(-du) = \int \sin x\, dx + \int u^2\, du \\ &= -\cos x + \frac{u^3}{3} + C = -\cos x + \frac{\cos^3 x}{3} + C.\end{aligned}$$

(b) Since $\int \sec^2 x\, dx = \tan x + C$ [Theorem 9.4(c)], we use the basic trigonometric identity $1 + \tan^2 x = \sec^2 x$ in the form $\tan^2 x = \sec^2 x - 1$ to get

$$\int \tan^2 x\, dx = \int (\sec^2 x - 1)\, dx = \int \sec^2 x\, dx - \int 1\, dx = \tan x - x + C. \quad \blacksquare$$

HISTORICAL NOTE

The power rule for antiderivatives seems to have first appeared in the work of the Italian priest Bonaventura Cavalieri (1598–1647). His book *Geometria Indivisibilis*, published in 1635, contained a derivation that fell short of the standards of mathematical rigor even of that era. More rigorously based work of Fermat and Blaise Pascal* (1623–1662), two of the prominent seventeenth-century French mathematicians, was published a few years after Cavalieri's book and contained derivations of Theorem 9.3.

The chain rule for antiderivatives was developed by Newton in a form very close to equation (16) in Theorem 9.3. His original work in this direction was done about 1665 but was not published until 1704. Leibniz, in 1673, also independently derived Theorem 9.3. That work was published in 1686.

Exercises 3.9

In Exercises 1–14, find a formula for the antiderivative.

1. $\int \frac{1}{\sqrt{x}}\,dx$
2. $\int (\sqrt[3]{x} + 2)\,dx$
3. $\int (3x^4 - 5x)\,dx$
4. $\int \left(\frac{1}{x^2} + 3\sin x\right)dx$
5. $\int (x^2 - 3x + 1)\,dx$
6. $\int (2x^3 - 3x^2 + 5x - 4)\,dx$
7. $\int \left(\frac{1}{x^3} - \frac{1}{x^2} + 3\right)dx$
8. $\int \left(2\sqrt[3]{x} - \frac{3}{\sqrt{x}} + 1\right)dx$
9. $\int \left(x^2\sqrt{x} - \frac{1}{x\sqrt{x}} - \frac{1}{\sqrt{x}}\right)dx$
10. $\int \frac{x-2}{\sqrt{x}}\,dx$
11. $\int \frac{x^2 - 3x + 1}{x\sqrt{x}}\,dx$
12. $\int \frac{x^3 - 5x^2 + x - 1}{x^2\sqrt{x}}\,dx$
13. $\int \sin 2x \cos 2x\,dx$
14. $\int \frac{\sin x}{\cos^3 x}\,dx$

In Exercises 15–40, use Theorem 9.3 to find a formula for the most general antiderivative of the given function.

15. $f(x) = 6x(x^2 - 3)^5$
16. $f(x) = (x^3 - 2x + 1)^3(6x^2 - 4)$
17. $f(x) = (x^2 - x + 1)^{-3}(6x - 3)$
18. $f(x) = (3x - 2)^8$
19. $f(x) = \sqrt{x^4 - 3x^2 + 5}(2x^3 - 3x)$
20. $f(x) = \frac{1}{(6x + 1)^4}$
21. $g(t) = \frac{t}{\sqrt{1 + 2t^2}}$
22. $g(t) = \frac{t - 3}{\sqrt{1 + 6t - t^2}}$
23. $h(x) = \sqrt{2x - 1}$
24. $g(t) = \sqrt{t^3 - 2t^2 + 5}(4t - 3t^2)$
25. $h(x) = \sin^2 x \cos x$
26. $k(x) = \frac{\cos x}{\sqrt{1 - \sin x}}$
27. $k(x) = 3\sin 2x - 2\sin 3x$
28. $m(x) = (1 + \sin^2 x)\sin 2x$
29. $n(x) = \frac{(\sqrt{x} + 1)^{2/3}}{\sqrt{x}}$
30. $p(x) = \frac{1}{\sqrt{x}(\sqrt{x} + 2)^3}$
31. $r(x) = \sin 2x \cos^5 2x$
32. $s(x) = \frac{\cos x}{\sin^4 x}$
33. $t(x) = \sec^2 5x\,dx$
34. $u(x) = \sec^2 x \tan x + \sec^4 x$
35. $v(x) = \csc^2 3x \cot^2 3x$
36. $w(x) = \frac{\csc^2 x \cot x}{(1 - \cot^2 x)^{3/2}}$
37. $f(x) = \sqrt{\cot 3x}\csc^2 3x$
38. $g(x) = \frac{\cot\sqrt{x}\csc\sqrt{x}}{\sqrt{x}}$
39. $h(x) = x^2 \sec x^3 \tan x^3$
40. $j(x) = \sec^{3/2} x \tan x$
41. Find $D_x^{-1}(\sin 2x)$ using Theorem 9.3 and the identity $\sin 2x = 2\sin x \cos x$. Reconcile your answer with that given by the first boxed formula on p. 214.
42. (a) Find $\int x(x^2 + 1)^2\,dx$ by multiplying out $x(x^2 + 1)^2$.
    (b) Find $\int x(x^2 + 1)^2\,dx$ by letting $u = x^2 + 1$ and applying Theorem 9.3.
    (c) Reconcile your answers in (a) and (b).
43. Prove the second and third parts of Theorem 9.4.
44. Prove the last three formulas in Theorem 9.4.
45. Prove the first two boxed formulas on p. 214.
46. Prove the third and fourth boxed formulas on p. 214.
47. Prove the last two boxed formulas on p. 214.

* See the historical note in Section 5.6 for more about Pascal.

48. Show that $D_x^{-1}(\sqrt{ax+b}) = \frac{2}{3a}(ax+b)^{3/2} + C$, if $a \neq 0$.

49. Let

$$F(x) = \begin{cases} x^2 & \text{for } x > 0 \\ -x^2 & \text{for } x < 0 \end{cases}$$

and

$$G(x) = \begin{cases} x^2 + 3 & \text{for } x > 0 \\ -x^2 - 1 & \text{for } x < 0. \end{cases}$$

(a) Show that F and G are antiderivatives of the function f defined by

$$f(x) = \begin{cases} 2x & \text{for } x > 0 \\ -2x & \text{for } x < 0. \end{cases}$$

(b) Do $F(x)$ and $G(x)$ differ by a constant?

(c) Why do parts (a) and (b) not constitute a violation of Theorem 9.2?

50. Repeat Exercise 49 for $F(x) = -\frac{1}{x^2}$,

$$G(x) = \begin{cases} \dfrac{x^2 - 1}{x^2} & \text{for } x < 0 \\ -\dfrac{x^2 + 1}{x^2} & \text{for } x > 0 \end{cases}$$

and $f(x) = 2x^{-3}$.

51. Find the most general antiderivative of the function $f(x) = |x|$.

52. Repeat Exercise 51 for the function $g(x) = x|x|$.

3.10 Simple Differential Equations

The last section introduced the antiderivative by investigating the velocity and position of a falling body with constant acceleration $-g$. Since the acceleration is the derivative of the velocity v and the second derivative of the position y, we have the equations

$$\frac{dv}{dt} = -g \tag{1}$$

and

$$\frac{d^2y}{dt^2} = -g. \tag{2}$$

These equations made it possible to find formulas for v by antidifferentiating (2) and for y by antidifferentiating (1). They are two simple instances of *differential equations.*

10.1 DEFINITION A ***differential equation*** is an equation of the form

$$F\left(x, y, \frac{dy}{dx}, \frac{d^2y}{dx^2}, \ldots, \frac{d^ny}{dx^n}\right) = 0, \tag{3}$$

where y is a function $f(x)$ of x. The ***order*** of the differential equation is n if the highest order derivative in (3) is the nth derivative d^ny/dx^n.

Thus (1) is a *first-order* differential equation and (2) is a *second-order* differential equation.

Differential equations are used to model a number of phenomena outside mathematics. As an illustration, our first example solves Problem 5 of the Prologue.

Example 1

A driver in a car traveling 50 miles per hour applies the brakes hard. The car's braking system produces a maximum deceleration of 20 feet per second each second. If the car maintains that deceleration, how long will the car take to stop, and how many feet will it travel before coming to a stop? If a second car is stopped at a red light 130 feet ahead of the point where the brakes of the first car were applied, then will the first car strike the second? If so at what speed?

Solution. Let y be the number of feet the first car has traveled since braking began. Let v be its speed in feet per second t seconds after braking began. Since the speed decreases 20 ft/s each second, we have

$$\frac{d^2y}{dt^2} = \frac{dv}{dt} = -20. \tag{4}$$

Antidifferentiating (4), we get

$$v = \frac{dy}{dt} = -20t + C. \tag{5}$$

Since $v = 50$ miles/hour when $t = 0$, and since

$$50\,\frac{\text{mile}}{\text{hour}} = 50\,\frac{\cancel{\text{mile}}}{\cancel{\text{hour}}} \cdot \frac{5280\text{ ft}}{\cancel{\text{mile}}} \cdot \frac{1\,\cancel{\text{hour}}}{3600\text{ s}} = \frac{220}{3}\text{ ft/s},$$

we have

$$\frac{220}{3} = -(20)\cdot 0 + C = C.$$

Therefore (5) becomes

$$v = \frac{dy}{dt} = -20t + \frac{220}{3}. \tag{6}$$

Antidifferentiation of (6) gives

$$y = -20\cdot\frac{t^2}{2} + \frac{220}{3}t + K. \tag{7}$$

Since $y = 0$ when $t = 0$, we find that $K = 0$. Thus (7) becomes

$$y = -10t^2 + \frac{220}{3}t. \tag{8}$$

To find how long it takes the car to stop, we set $v = 0$ in (6) and solve for t. That gives

$$20t = \frac{220}{3} \rightarrow t = \frac{11}{3} = 3\frac{2}{3}. \tag{9}$$

Substituting this into (8), we find that the distance traveled by the car before it stops is

$$y = -10\left(\frac{11}{3}\right)^2 + \frac{220}{3}\left(\frac{11}{3}\right) = \frac{-1210 + 2420}{9} = \frac{1210}{9} \approx 134.44.$$

The car will therefore stop after $3\frac{2}{3}$ seconds and $134\frac{4}{9}$ feet of travel. Hence, it *will* strike the second car. The final question asks for v when $y = 130$. Since (6) gives v as a function of t, we need to find t when $y = 130$. This we do from

(8), getting

$$130 = -10t^2 + \frac{220}{3}t \rightarrow 390 = -30t^2 + 220t,$$

$$39 = -3t^2 + 22t \rightarrow 3t^2 - 22t + 39 = 0,$$

(10) $$(3t - 13)(t - 3) = 0.$$

From (10) we find that $t = 3$ or $t = 13/3$. But we can disregard the second value, because (8) is valid *only* while the car is moving. From (9) then, (8) is valid only for $t \leq 11/3$. Putting $t = 3$ into (6), we find

$$v = -60 + \frac{220}{3} = \frac{40}{3} \text{ ft/s.}$$

Thus the second car is hit by the first one at a speed of

$$\frac{40}{3}\frac{\text{ft}}{\text{s}} \cdot \frac{3600 \text{ s}}{\text{hour}} \cdot \frac{1 \text{ mile}}{5280 \text{ ft}} \approx 9.09 \text{ miles/h.} \quad ■$$

Simple differential equations can be used to model many kinds of phenomena besides motion problems. The next example comes from medicine.

Example 2

A patient takes a 5-mg dose of cortisone on waking up in the morning. The rate at which the body uses up the dose decreases over the ensuing 24 hours. After t hours, the rate of use is approximately

$$\frac{1}{2\sqrt{t}} \text{ mg/h.}$$

Find how long it takes for the dose to be used up. If the patient goes to sleep 16 hours after waking, how much cortisone remains at that time?

Solution. Let $c(t)$ be the number of milligrams (mg) of cortisone still in the patient's system t hours after taking the dose. Then

$$\frac{dc(t)}{dt} = -\frac{1}{2\sqrt{t}} = -\frac{1}{2}t^{-1/2}.$$

Therefore antidifferentiation gives

(11) $$c(t) = -\frac{1}{2} \cdot \frac{t^{1/2}}{1/2} + K \rightarrow c(t) = K - t^{1/2}.$$

When $t = 0$, $c(t) = 5$, so from (11) we have $K = 5$. Hence,

(12) $$c(t) = 5 - t^{1/2}.$$

All the dose has been used up when

$$c(t) = 0 = 5 - t^{1/2} \rightarrow t^{1/2} = 5 \rightarrow t = 25.$$

So it takes slightly more than one day for the dose to be used up. When $t = 16$, we get from (12)

$$c(t) = 5 - 16^{1/2} = 5 - 4 = 1.$$

Thus 80% of the dose is used up during waking hours. ■

In Examples 1 and 2, we were able to reduce to a problem of the form

$$\frac{dy}{dx} = f(x),$$

which can be solved by simple antidifferentiation. Many differential equations not initially in such a form can be reduced to it by some algebraic manipulation that isolates the two variables on separate sides of the equation. Such equations are called *separable.*

10.2 DEFINITION A first-order differential equation is ***separable*** if it can be put in the form

$$g(y)\frac{dy}{dx} = h(x). \tag{13}$$

To solve a separable differential equation, we need only take the antiderivative of each side of (13). If G is an antiderivative of g and H is an antiderivative of h, then we obtain the ***general solution***

$$G(y) = H(x) + C. \tag{14}$$

In (14), y is only *implicitly* given as a function of x. But in view of Section 2.8, such functions can be studied without undue effort.

Example 3

Find all differentiable functions whose graphs have the property that at each point $P(x, y)$ not on the x-axis, the slope of the tangent line is the ratio of the square of the x-coordinate to the y-coordinate.

Solution. Since the slope of the tangent line at $P(x, y)$ is dy/dx, we are told that when $y \neq 0$,

$$\frac{dy}{dx} = \frac{x^2}{y}. \tag{15}$$

It is easy to separate the variables in (15). Multiplying each side by y, we get

$$y\frac{dy}{dx} = x^2. \tag{16}$$

By the chain rule for antiderivatives, $\frac{1}{2}y^2$ is an antiderivative of the left side of (16). Hence,

$$\frac{1}{2}y^2 = \frac{1}{3}x^3 + C. \tag{17}$$

If we want an *explicit* function $y = f(x)$, we can solve (17) for y, getting

$$y^2 = \frac{2}{3}x^3 + 2C' = \frac{2}{3}x^3 + K. \tag{18}$$

So

$$y = \sqrt{\frac{2}{3}x^3 + K} \quad \text{or} \quad y = -\sqrt{\frac{2}{3}x^3 + K}.$$

(Without further information, we cannot tell which formula for y to favor. See below.) ■

We have seen several examples in which a condition of the form $y = y_0$ when $x = x_0$ determines a *unique* solution to a given differential equation. Such a condition is called an ***initial condition,*** and a first-order differential equation

$$F\left(x, y, \frac{dy}{dx}\right) = 0$$

with an initial condition

$$y = y_0 \quad \text{when} \quad x = x_0$$

is called a ***first-order initial value problem.*** The following theorem, whose proof we leave for a course in differential equations, guarantees that most initial value problems have unique solutions.

10.3 THEOREM Let g and h be continuous functions, where $g(y_0) \neq 0$. Then a separable first-order initial value problem

$$g(y)\frac{dy}{dx} = h(x), \qquad \text{where } y = y_0 \text{ when } x = x_0,$$

has a unique solution $y = y(x)$ on some interval containing x_0.

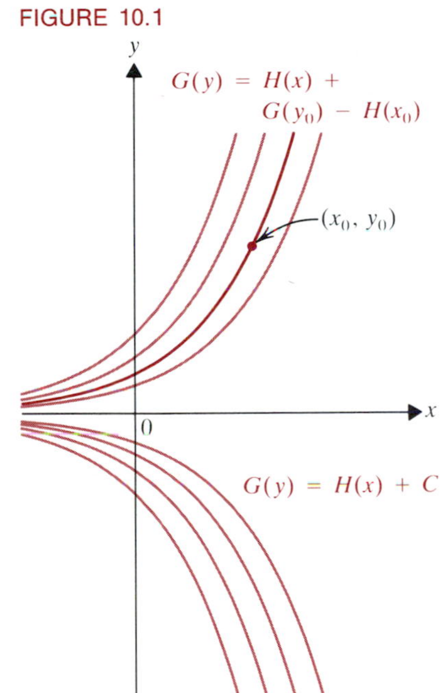

FIGURE 10.1

The idea behind Theorem 10.3 is the following. If G and H are antiderivatives of g and h respectively, then as noted before Example 3, the differential equation

$$g(y)\frac{dy}{dx} = h(x) \tag{13}$$

has the general solution

$$G(y) = H(x) + C, \tag{14}$$

where C is an arbitrary constant. As C varies through all possible real values, infinitely many solutions of the form (14) are generated. **Figure 10.1** shows the graphs of some of those functions. The effect of an initial condition such as

$$y = y_0 \quad \text{when} \quad x = x_0 \tag{19}$$

is to select from the infinitely many functions of the form (14), the one whose graph passes through (x_0, y_0). Algebraically, we do this by finding the value that C must have in (14) in order for (19) to hold:

$$G(y_0) = H(x_0) + C \rightarrow C = G(y_0) - H(x_0).$$

Example 4

In Example 3, find the function satisfying the given conditions whose graph passes through the point $(3, -5)$.

Solution. By the solution of Example 3, all functions satisfying the given condition must also satisfy

$$y^2 = \frac{2}{3}x^3 + K. \tag{18}$$

The function we are looking for has $y = -5$ when $x = 3$. Putting those values into (18), we get

$$25 = \frac{2}{3} \cdot 27 + K = 18 + K \rightarrow K = 7.$$

FIGURE 10.2

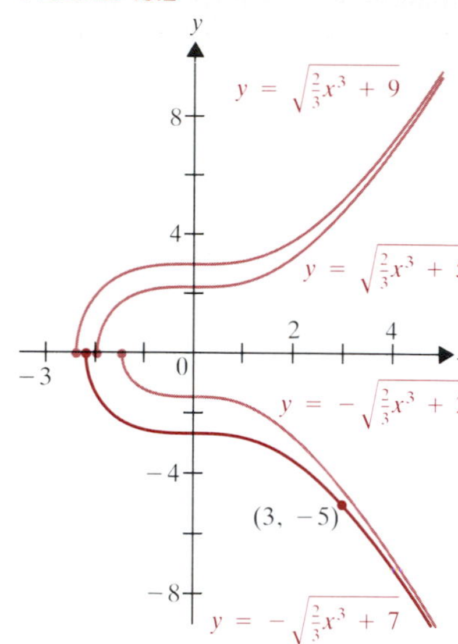

We thus have

$$y^2 = \frac{2}{3}x^3 + 7.$$

Moreover, the function we seek must be the *negative* square root of $\frac{2}{3}x^3 + 7$, because $y = -5$ is negative when $x = 3$. Hence,

$$y = -\sqrt{\frac{2}{3}x^3 + 7}. \tag{20}$$

The initial condition thus determines *exactly one* of the infinitely many functions $y = f(x)$ that satisfy (18). **Figure 10.2** shows the graphs of (20) and three other functions of the form (18). ■

Exercises 3.10

In Exercises 1–18, set up and solve a differential equation that models the given situation.

1. In Example 1, suppose that the second car is 109.9 ft ahead of the point where the brakes are applied. At what speed will the first car strike it? How long after the first car's brakes are applied will the collision occur?
2. In Example 1, suppose that the first car is traveling 30 miles/h and decelerates at 12 ft/s². How long will it take to stop, and how far will it travel before stopping? If a car is stopped 80 ft ahead of the point where the brakes are applied, will it be hit? At what speed and after how many seconds? If the second car's bumper can withstand a 5 mile-per-hour impact without damage, will it sustain any damage?
3. A drag racer accelerates at a constant rate from 0 to 120 miles/h over a 1/4-mile track. What is the acceleration, and how long does it take for the car to cover the 1/4 mile?
4. What constant rate of deceleration is required to bring a car traveling 60 miles/h to a stop in 200 ft? How long will the brakes have to be applied to do this?
5. When an object is thrown vertically upward from ground level with an initial velocity of 64 ft/s, find a formula for its height above the ground after t seconds. When does it reach its highest point? How high is that? How long does it take to fall back to the ground? With what velocity does it strike the ground? (Use $g = -32$ ft/s².)
6. What initial velocity needs to be imparted to a ball thrown vertically upward in order for it to reach a height of 256 feet?
7. In the 1930s, a baseball was dropped from the top of the Washington Monument for a player on the ground to catch. If the player caught the ball 500 feet below the point of release, then how fast was the ball traveling when it was caught? How long did it take to reach the glove of the catcher? (For comparison, the fastest recorded speed of a pitched baseball is about 104 miles per hour.)
8. A ball is thrown upward from a point 256 ft above ground with an initial speed of 64 ft/s. How long will it take to hit the ground, and how fast will it be traveling when it hits the ground?
9. A patient takes a 1/8-milligram dose of synthetic thyroid hormone in the morning. The rate of use by the body after t hours is about $1/(80\sqrt{t})$ mg/h. How long does it take for the entire dose to be used? How much is left 16 hours after it is taken?
10. A person with a cold takes two time-released 10-mg capsules. At that time 4 mg of the previous dose are still in the person's bloodstream. If the capsules are metabolized (used up) by the body at a constant rate of 2 mg/h, find a formula for the amount left after t hours. When will all the drug be metabolized (assuming the patient takes no more capsules)?
11. When a diabetic (with insulin output 0) takes insulin, the rate at which glucose is broken down increases at a constant rate. If the rate changes from 0 mg/min to 1.2 mg/min in 3 min, then how much glucose is broken down during that time?
12. A flu epidemic is detected in a community when 100 people are infected and 96 new cases per day are being reported. If the rate of spread at time t is approximately

$100t - 4t^2$ people per day, then how many people have become infected 5 days after the disease was detected?

13. When a speck of dust is in a region of the atmosphere that is saturated with moisture, water vapor condenses on the particle at a constant rate of k g per cm of travel. This forms a water droplet, which falls as its mass grows. What is the acceleration in terms of its velocity v and length x of fall? (Assume that the only force on the droplet is gravity, and use Newton's law for *variable* mass,

$$F = \frac{d}{dt}(m(t)v(t)),$$

and the chain rule to express dm/dt in terms of dm/dx and v.)

14. In Exercise 13, assume that the initial mass of the dust is negligible.

(a) Give a formula for $m(t)$ in terms of x.

(b) Use the result of part (a) and the formula for the acceleration in Exercise 13 to obtain a second-order differential equation for the motion of the water droplet (don't try to solve the equation).

(c) Verify that $x = \frac{1}{6}gt^2$ is a solution to your equation in part (b). What then are v and acceleration?

15. Data from a glacier often can be used to model temperature variation over a long period of time. Suppose that the rate of change in the annual temperature in northern Canada is estimated from such data to have been approximately

$$0.02 \sin \frac{\pi t}{100}, \qquad \text{where } t \text{ is in years.}$$

Find a formula for the temperature in year t if when $t = -5000$, T is estimated to have been $-10°$C. What was the period of this temperature function?

16. According to *Hooke's law* [formulated in 1676 by the English physicist Robert Hooke (1635–1703)], when a spring is stretched or compressed x units from its natural length, the restoring force is

$$F = -kx,$$

where k is a constant (called the *spring constant*). The *work* done in moving the spring from its natural length to displacement x is the antiderivative of the force $-F$ needed to overcome the restoring force. Give a formula for that work.

17. In Exercise 16, suppose that a force of 8 newtons is required to displace a spring 1/4 meter from its natural length. Find the work done in stretching the spring over a distance of 1/2 meter.

18. In Exercise 16, suppose that a force of 4 pounds is required to compress a spring 3 inches from its natural length. Find the work done in compressing it 4 inches.

In Exercises 19–26, solve the given separable differential equation.

19. $\dfrac{dy}{dx} = \dfrac{x}{y}$

20. $\dfrac{dy}{dx} = \dfrac{x^3}{y}$

21. $\dfrac{dy}{dx} = \dfrac{\cos 2x}{y + 2}$

22. $\dfrac{dy}{dx} = \dfrac{\sin 2x}{y - 1}$

23. $2(y - 1)\dfrac{dy}{dx} - 3x^2 - 34x = 2$

24. $2(y^2 + 1)\dfrac{dy}{dx} + x^2 + 5x = 3$

25. $\dfrac{dy}{dx} = \dfrac{1}{(\cos^2 x)(3y^2 + y^{1/3})}$

26. $x \sec y \tan y \dfrac{dy}{dx} + x^2 \cos y = 0$

In Exercises 27–34, solve the given initial value problem.

27. Exercise 19 if $y = -3$ when $x = 3$.

28. Exercise 20 if $y = 2$ when $x = 2$.

29. Exercise 21 if $y = 1$ when $x = 0$.

30. Exercise 22 if $y = 2$ when $x = \pi$.

31. Exercise 23 if $y = -1$ when $x = 0$.

32. Exercise 24 if $y = 2$ when $x = 1$.

33. Exercise 25 if $y = 0$ when $x = \pi/4$.

34. Exercise 26 if $y = \pi/4$ when $x = 1$.

3.11 Looking Back

This chapter has presented a sequence of applications of differentiation. Sections 2 through 4 discussed methods for finding local and absolute extreme values of functions. For a continuous function f over a closed interval $[a, b]$, this is done by following the boxed scheme on p. 152. To find the intervals of increase and decrease of a differentiable function, Theorem 3.1 is used. To classify critical points, either the first-derivative test (Theorem 3.2) or the second-derivative test (Theorem 4.5) can be used. Theorem 4.3 is used to find intervals of constant concavity. To draw an accurate graph of a function f, we use these procedures, plus those introduced in Section 6 for locating horizontal and vertical asymptotes. Refer to the boxed algorithm (p. 192) for a step-by-step summary.

A major theme of the chapter is the usefulness of the derivative in analyzing applied problems. Sections 1 and 5 were completely devoted to this, as were portions of Sections 9 and 10. On pages 142 and 170 there are boxed algorithms to follow in solving related-rate and applied optimization problems.

The remaining topics of the chapter include Newton's method, antidifferentiation, and separable differential equations. Newton's method is based on Formula (6) on p. 203 for x_{n+1} in terms of x_n. The main results on antidifferentiation are the power rule (p. 210) and the chain rule (Theorem 9.3). The boxed suggestions on p. 213 provide a systematic approach to antidifferentiation problems. Section 10 on separable differential equations introduces an important application of antidifferentiation and indicates how it is used to analyze problems in other fields.

CHAPTER CHECKLIST

Section 3.1: four-step procedure to set up related-rate problems.

Section 3.2: absolute maximum and minimum; local maximum and minimum; extreme value (extremum), extreme value point; critical point; three-step procedure to optimize a continuous function on a closed interval.

Section 3.3: increasing and decreasing differentiable functions (Theorem 3.1); first-derivative test; tabular analysis of f and its critical points.

Section 3.4: functions concave up and concave down; use of f'' to find intervals of constant concavity (Theorem 4.3); inflection points; second-derivative test.

Section 3.5: six-step procedure for setting up applied optimization problems; constraints.

Section 3.6: limit of a function as $x \to +\infty$ or $x \to -\infty$; limits of rational functions as $x \to \pm\infty$; horizontal asymptotes of curves; infinite limits; vertical asymptotes.

Section 3.7: symmetry conditions relative to x-axis, y-axis, and the origin (Theorem 7.3;) six-step procedure for making an accurate graph of a function.

Section 3.8: Newton's method of approximating a root x^* of $f(x) = 0$ for a differentiable function f; formula for x_{n+1} in terms of x_n.

Section 3.9: antiderivatives of a function; indefinite integrals; constant of integration; power rule; chain rule; antiderivatives of trigonometric functions (Theorem 9.4); five-step procedure for finding antiderivatives by change of variable (u-substitution).

Section 3.10: differential equations; separable equations; initial conditions; initial value problems; uniqueness of solution.

REVIEW EXERCISES 3.11

In Exercises 1–6, find the absolute maximum and minimum of the given function on the given interval.

1. $f(x) = \dfrac{1}{3}x^3 - 3x^2 + 8x - 4$ on $[0, 5]$.

2. $f(x) = x^4 - 2x^2 + 1$ on $[-2, 2]$.

3. $f(x) = (x-3)^2(x+3)^3$ on $[-3, 3]$.

4. $f(x) = (16 - x^2)^{3/2}$ on $[-4, 4]$.

5. $g(x) = x - \cos x$ on $[0, \pi]$.

6. $g(x) = \dfrac{\cos x}{1 + 2\cos x}$ on $[0, \pi/2]$.

In Exercises 7–10, find where f is increasing, where it is decreasing, and all local extreme value points.

7. The function of Exercise 2.

8. The function of Exercise 4.

9. $f(x) = \dfrac{1}{3}x^3 - 5x^2 + 24x - 8$

10. $f(x) = \dfrac{1}{2}x + \cos x, \qquad x \geq 0$

In Exercises 11 and 12, find where f is concave up and where it is concave down and locate all inflection points.

11. $f(x) = \frac{1}{3}x^3 - 3x^2 + 8x - 4$

12. $f(x) = \sin^2 x$ on $[0, 2\pi]$

13. A jet plane has 200 seats, of which the airline can sell on an average only 30% at the present full fare. But for each $10 reduction in fare, an average of five more seats can be sold. Find the fare that maximizes revenue on a route with a current full fare of $400. How many seats will be sold at that fare?

14. A playing field is to be built in the shape of a rectangle with two semicircles of the same radius at each end, as in **Figure 11.1.** If the perimeter is to be a 400-meter-long track, then what dimensions x and y will make the area of the rectangular part as large as possible? What is the resulting shape of the field?

FIGURE 11.1

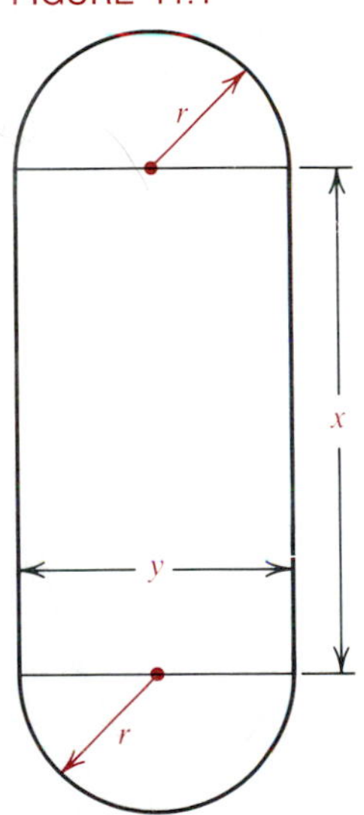

15. A feedlot operator wants to build a rectangular feeding pen with an enclosed area of 400 square feet. The pen is to be divided into three sections—one each for cattle, sheep, and pigs—by running two fences parallel to one of the outer fences. What is the least number of feet of fence needed? What dimensions will the pen have?

16. Design a rectangular box (with no lid), having square ends and a volume of 64 cubic inches, so that the amount of cardboard in the box is as small as possible.

In Exercises 17–26, evaluate the limits.

17. $\lim_{x \to +\infty} \frac{5x^3 - 3x^2 + x - 1}{2x^3 + 6x - 5}$

18. $\lim_{x \to -\infty} \frac{3x^4 - 2x + 5}{5x^4 + 2x^3 - 1}$

19. $\lim_{x \to +\infty} \frac{2x^2 - 3x + 1}{x^3 + 5}$

20. $\lim_{x \to -\infty} \frac{x^2 - 4x + 1}{x^4 - 1}$

21. $\lim_{x \to +\infty} \frac{5x^3 - 6}{89x^2 - 3x + 1}$

22. $\lim_{x \to -\infty} \frac{x^3 - 3x}{2x^2 + x + 1}$

23. $\lim_{x \to 1^+} \frac{4x - 3}{x^2 + x - 2}$

24. $\lim_{x \to 2^+} \frac{5x - 2}{x^2 + x - 2}$

25. $\lim_{x \to 1^-} \frac{4x - 3}{x^2 + x - 2}$

26. $\lim_{x \to 1^-} \frac{5x - 2}{x^2 + x - 2}$

In Exercises 27–34, draw an accurate graph of the given function.

27. $f(x) = \frac{1}{3}x^3 - 3x^2 + 8x - 4$

28. $h(x) = \sin x \cos^2 x$

29. $m(x) = \frac{1}{3}x^3 + x + 2$

30. $p(x) = \frac{2x + 4}{\sqrt{x^2 + 4}}$

31. $f(x) = \frac{4}{(x - 2)^2(x + 1)}$

32. $g(x) = \frac{x^2 + 1}{x^2 - 1}$

33. $h(x) = \frac{3 - x}{3 + x}$

34. $h(x) = \frac{x^2 - 3x + 2}{x^2 - 4}$

35. A man six feet tall walks toward a street light ten feet above street level at a rate of four feet per second. At what rate is the length of his shadow decreasing when he is fifteen feet from the base of the light? At what rate is the tip of the shadow moving?

36. At noon, train A is 30 miles west of a certain station and is heading east at 40 miles per hour. At the same time, train B is 40 miles north of the station, heading north at 60 miles per hour. At that time, how fast is the distance between the two trains changing?

37. Water is running out of a conical funnel at 2 cm^3/s. The funnel has radius 10 cm and height 10 cm. How fast is the water level dropping when the water is 5 cm high?

38. A patient receives glucose from a cylindrical bottle of radius 5 cm and height 20 cm. If the fluid drops at the rate of 0.02 cm/min, how fast is the solution entering the patient's body?

In Exercises 39 and 40, use Newton's method to find a root of each given equation.

39. $x^3 - x^2 + x + 1 = 0$

40. $\cos x = x$

In Exercises 41–46, find $D_x^{-1}(f(x))$ for the given function f.

41. $f(x) = \frac{3}{\sqrt{x}} - \frac{\sqrt{x}}{3} + \sqrt{x + 1}$

42. $f(x) = \frac{x^2 - 2x + 3}{x^{3/2}}$

43. $f(x) = x\sqrt{x^2 - 1}$

44. $f(x) = \frac{x - 2}{\sqrt{x^2 - 4x + 12}}$

45. $f(x) = x \sin x^2 - \sin^2 x \cos x$

46. $f(x) = \dfrac{\sin x}{\sqrt{\cos x + 1}}$

47. A rock is dropped from a height of 144 feet. How long does it take to hit the ground? If a second rock is thrown down from the same spot one second later, what initial velocity does it need to hit the ground at the same time as the first rock?

48. A curve passes through the point (3, 4), and its tangent line at every point (x, y) of the curve not on the x-axis has slope $-x/y$. Find the equation of the curve.

In Exercises 49 and 50, solve the given differential equation.

49. $\dfrac{dy}{dx} = \dfrac{x-2}{y+2}$

50. $(y-1)\dfrac{dy}{dx} = \cos x - x$ if $y = 0$ when $x = 0$.

51. (*Economic Equilibrium*). Consider two groups of n people who mine gold. At present, each A person has 4 gold bars and expects to have 3 gold bars next year. Each B person has 2 gold bars now but expects to have 5 next year. Each group uses the *utility function* $u(x, y) = xy$ to value holding x gold bars now and y gold bars next year. A gold market forms at which B people try to borrow gold lent by A people, for one year. Borrowers and lenders try to *maximize* their individual utility.

(a) Write the utility function for an A person who lends x gold bars at interest rate r. (*Hint:* How much gold will such a person have now? How much next year when his borrower repays the loan with interest?) What value of x maximizes his utility?

(b) Write the utility function for a B person who borrows x gold bars at interest rate r. What value of x maximizes her utility?

(c) The gold market will be in *equilibrium* when the total gold lent at interest rate r exactly balances the demand at interest rate r. Assuming equilibrium evolves, what will be the corresponding free market interest rate r?

(d) How much gold will each A person lend at the rate in (c)? How much gold will each B person borrow at this rate?

52. Repeat Exercise 51 if each A person has 2 gold bars now and expects to have 2 next year, while each B person has 1 bar now and expects to have 3 next year.

CHAPTER 4 Integrals

4.0 Introduction

The last two chapters were devoted to the first of two main topics of calculus, the derivative. This chapter and the next discuss the other central concept of calculus, the definite integral of a function f over a closed interval $[a, b]$. As we will see, the integral is a powerful tool for solving a wide range of interesting problems.

FIGURE 0.1

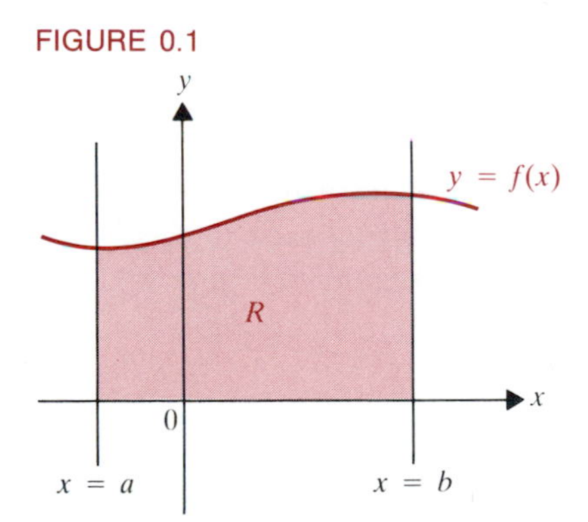

We begin with the question of how to calculate the area of a region R lying above the x-axis and under the graph of a positive-valued function f over $[a, b]$. See **Figure 0.1.** First we define that area to be the limit of sums of the areas of approximating rectangles. That leads to the definition of the definite integral of a (not necessarily positive-valued) function f over $[a, b]$. Among the basic properties of the definite integral is the mean value theorem for integrals. Like its namesake for derivatives, it is helpful in deriving more computationally useful facts. In Section 4, we use it to show that differentiation and integration are essentially *inverse* operations. This important fact constitutes the *fundamental theorem of calculus.* For functions whose antiderivatives can be calculated, it provides a direct means of evaluating definite integrals without any limit computations. It also serves to explain the indefinite integral notation for antiderivatives introduced in Section 3.9.

In Section 5, we apply the change-of-variable technique of Section 3.9 to definite integrals. That permits us to work out complicated definite integrals by substitution without having to separately work out the indefinite integrals.

The final section of this chapter considers the question of how to evaluate the definite integral of functions that we *can't* antidifferentiate. Three increasingly accurate schemes for numerically approximating the definite integral by appropriate sums are presented. These are rather a chore for hand computation, but are well suited to computers or programmable calculators and are widely used in practice when the fundamental theorem fails, or is difficult, to apply.

4.1 Area Under a Curve

More than 2000 years ago, Archimedes was able to find the areas of many plane regions bounded by curves, including the area πr^2 enclosed by a circle of radius r. You long ago learned the area formulas for triangles and rectangles (see **Figure 1.1**). The area of a polygonal figure in the plane—that is, a figure bounded by straight line segments—can be calculated by triangulating the polygon as in **Figure 1.2.** Its area is the sum of the areas of the constituent triangles.

FIGURE 1.1

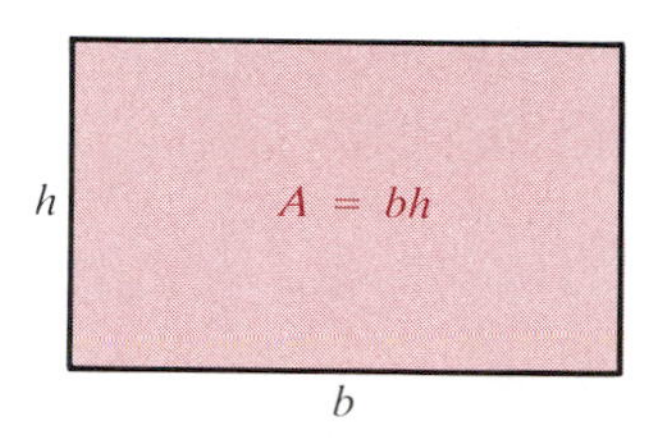

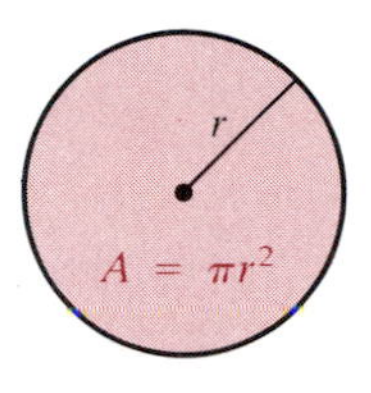

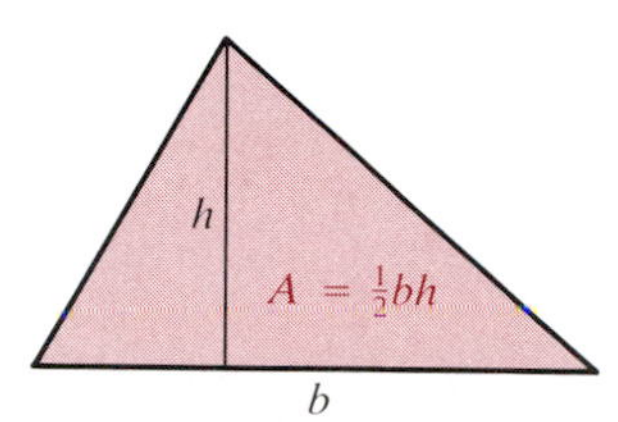

FIGURE 1.2

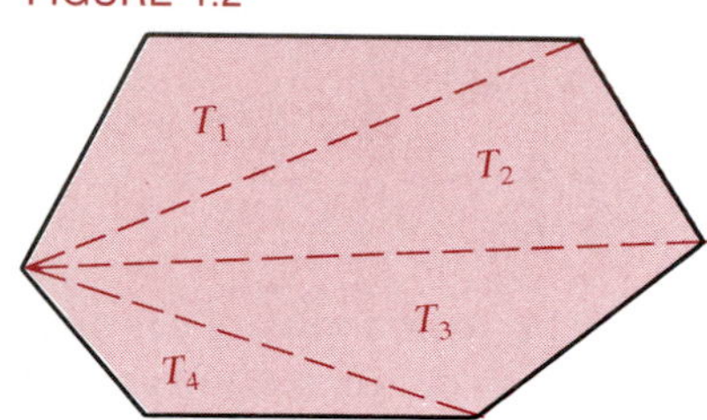

We need calculus to find the area of regions with curved boundaries. **Figure 1.3** suggests why such areas are of interest. It shows the graph of an airplane's velocity v in miles per hour during a one-hour flight. The plane becomes airborne at a velocity of 150 miles per hour, and its speed climbs rapidly until a cruising velocity of 540 miles per hour is reached. That takes about 20 minutes (1/3 hour). The plane maintains its cruising speed until 20 minutes before landing, when it starts to slow and begin its descent for landing, which occurs at 150 miles per hour. During the cruising phase, the distance traveled by the plane is simply the product of its velocity and the length of time it is cruising:

$$540 \frac{\text{miles}}{\cancel{\text{hour}}} \times \frac{1}{3} \cancel{\text{hour}} = 180 \text{ miles}.$$

Notice that this distance is exactly the area of the rectangle R in Figure 1.3, since R has base 1/3 and height 540.

During the intervals [0, 1/3] and [2/3, 1], the plane's velocity varies. But, by analogy to the cruising phase, the distance traveled during each of these time intervals should be the area of the respective shaded region of Figure 1.3. The

FIGURE 1.3

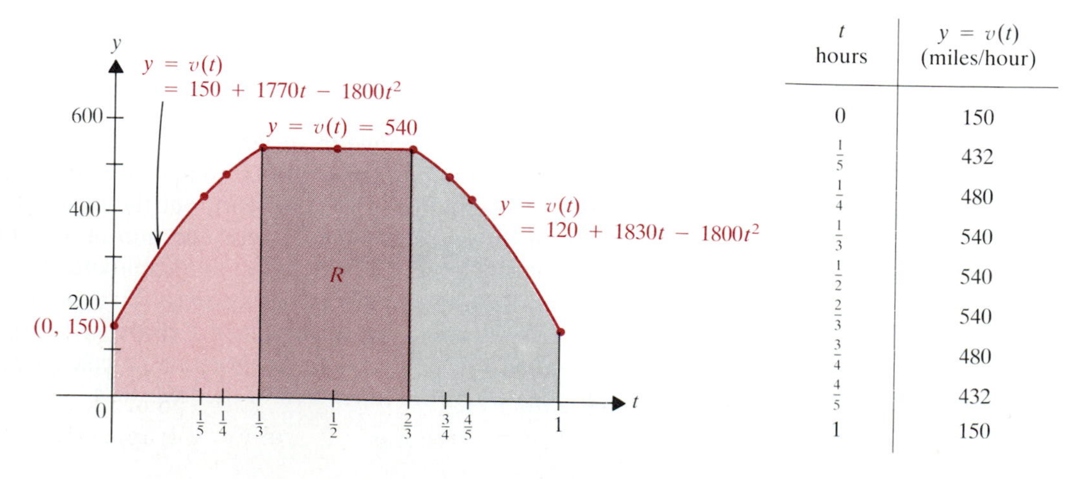

t hours	$y = v(t)$ (miles/hour)
0	150
$\frac{1}{5}$	432
$\frac{1}{4}$	480
$\frac{1}{3}$	540
$\frac{1}{2}$	540
$\frac{2}{3}$	540
$\frac{3}{4}$	480
$\frac{4}{5}$	432
1	150

velocity function v has formula

$$v(t) = \begin{cases} 150 + 1770t - 1800t^2 & \text{for } t \in [0, 1/3], \\ 540 & \text{for } t \in (1/3, 2/3), \\ 120 + 1830t - 1800t^2 & \text{for } t \in [2/3, 1]. \end{cases}$$

The region shaded in color lies under the graph of $y = 150 + 1770t - 1800t^2$ between $t = 0$ and $t = 1/3$. The gray-shaded region lies under the graph of $y = 120 + 1830t - 1800t^2$ between $t = 2/3$ and $t = 1$. To compute these areas, we need the new mathematical tool that this chapter presents. Later (Example 1, Section 5.1), we will return to the question of determining the distance the airplane flies during its journey. For now, we formulate the general problem of finding the area of regions like the shaded portions of Figure 1.3.

> ***Area Problem.*** Suppose that $f(x) \geq 0$ for $x \in [a, b]$. Find the area lying above the x-axis and under the graph of f between $x = a$ and $x = b$.

Figure 1.4 shows a representation of the problem for a continuous positive-valued function f on $[a, b]$. We approach the area problem by first chopping up $[a, b]$ into a collection of subintervals. Such a subdivision is determined by specifying a ***partition*** $P = \{x_0, x_1, \ldots, x_n\}$ of $[a, b]$, which is a set of points satisfying

$$a = x_0 < x_1 < \cdots < x_{i-1} < x_i < \ldots < x_{n-1} < x_n = b.$$

FIGURE 1.4

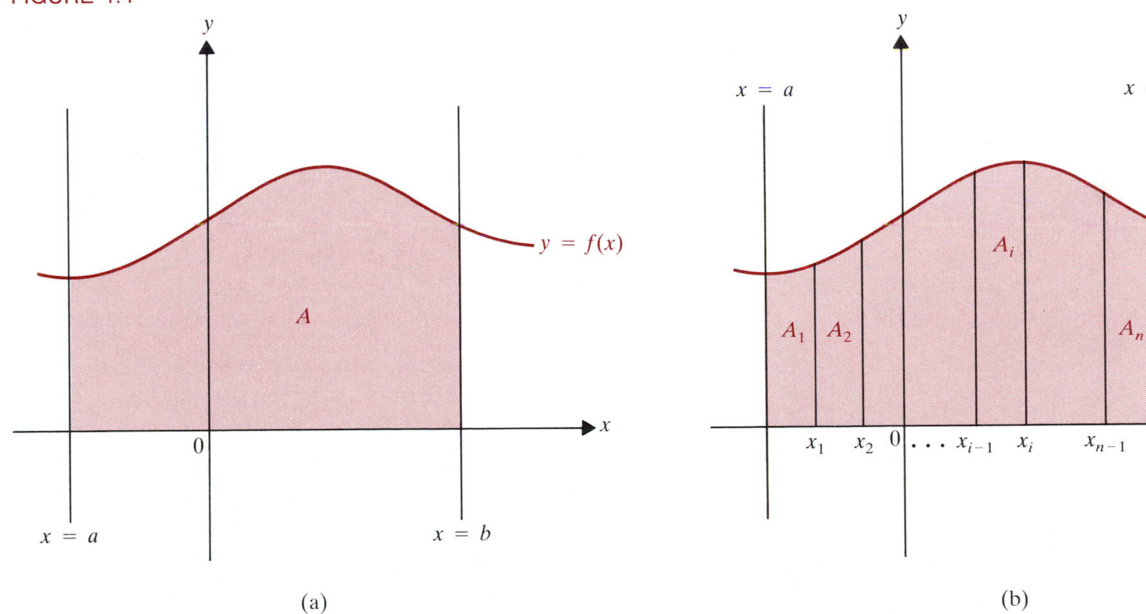

A partition determines the n subintervals

$$[x_0, x_1], [x_1, x_2], \ldots, [x_{i-1}, x_i], \ldots, [x_{n-1}, x_n],$$

whose union is $[a, b]$. The entire area A between $x = a$ and $x = b$ in Figure 1.4(b) is the sum of the areas $A_1, A_2, \ldots, A_n$ of the regions lying above those intervals. To approximate each area A_i, first choose a point z_i in $[x_{i-1}, x_i]$. Then the area under $y = f(x)$ between $x = x_{i-1}$ and $x = x_i$ is approximately the area of the rectangle with base $[x_{i-1}, x_i]$ and height $f(z_i)$. See **Figure 1.5.** The area of that rectangle, which we denote by ΔA_i, is

$$f(z_i)(x_i - x_{i-1}). \tag{1}$$

If we let Δx_i stand for the base length $x_i - x_{i-1}$, then (1) says that

$$\Delta A_i = f(z_i)\,\Delta x_i. \tag{2}$$

Adding the approximations (2) over all the subintervals $[x_{i-1}, x_i]$, we get an approximation to the area of the entire region under $y = f(x)$ between $x = a$ and $x = b$. See **Figure 1.6.**

FIGURE 1.5

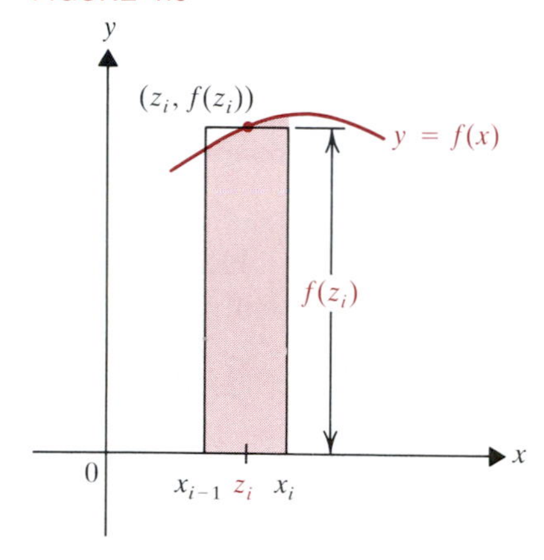

FIGURE 1.6

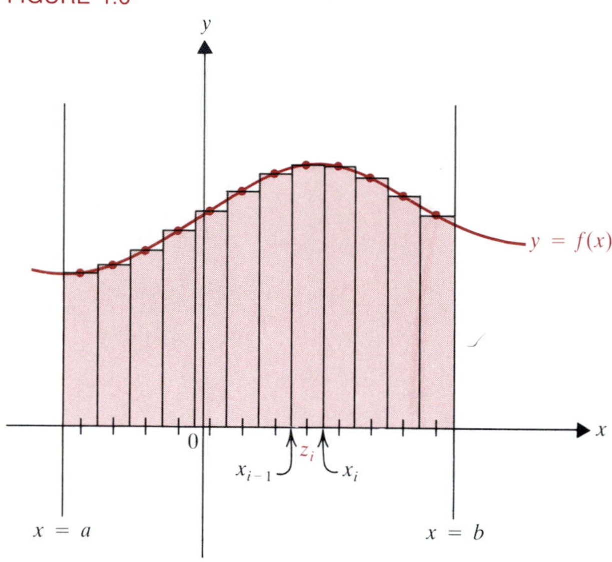

1.1 DEFINITION Let $f(x) \geq 0$ on $[a, b]$. Let $P = \{x_0, x_1, \ldots, x_n\}$ be a partition of $[a, b]$. In each subinterval $[x_{i-1}, x_i]$, choose a point z_i. The corresponding ***rectangular approximation*** to the area A of the region lying under $y = f(x)$ between $x = a$ and $x = b$ is

$$R_P = f(z_1)\,\Delta x_1 + f(z_2)\,\Delta x_2 + \ldots + f(z_n)\,\Delta x_n. \tag{3}$$

The idea behind the rectangular approximation scheme is reminiscent of the tangent approximation. But this time, instead of using the *tangent* line $y = T(x)$ to approximate $f(x)$, we approximate $f(x)$ over $[x_{i-1}, x_i]$ by the *horizontal* line $y = f(z_i)$, where z_i is an arbitrary point in $[x_{i-1}, x_i]$. The notation R_P reflects the fact that the rectangular approximation depends on P and the choice of the points z_i in the subintervals $[x_{i-1}, x_i]$. **Figure 1.7** suggests that the accuracy of the approximation of A by R_P should improve if we increase the number of subintervals and let their lengths Δx_i approach 0. The next example supports that impression. We use a function f for which we can calculate the area A exactly, so that we can measure the accuracy of the approximation of A by rectangular approximations R_P for different partitions P.

Example 1

For the function $f(x) = 4x$ over the interval $[0, 2]$, find R_P if

(a) $P = \{0, 1/2, 3/4, 1, 3/2, 2\}$ and $z_i = x_i$ for each i.

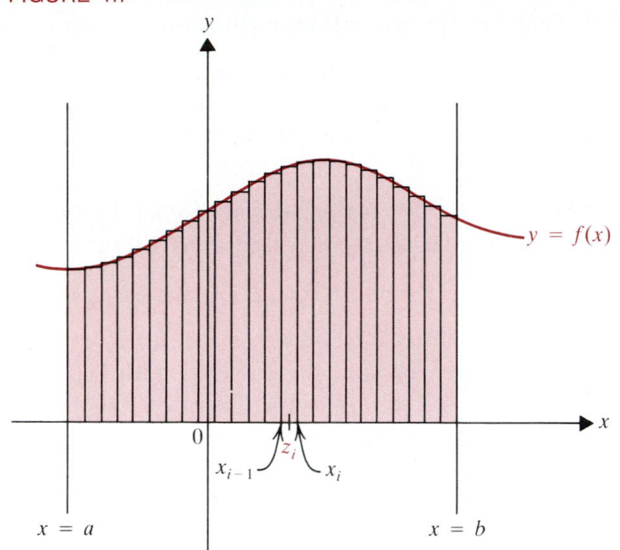

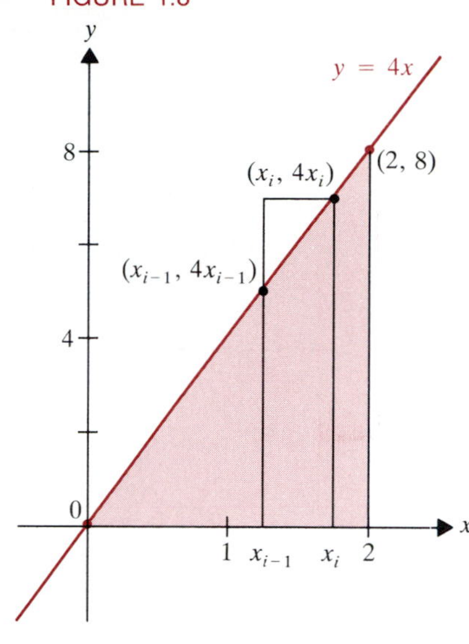

(b) $P = \{0, 1/4, 1/2, 3/4, 1, 5/4, 3/2, 7/4, 2\}$ and $z_i = x_{i-1}$ for each i.

(c) Compare the accuracy of the approximations in (a) and (b).

See **Figure 1.8.**

Solution. (a) Here $z_1 = x_1 = 1/2$, $z_2 = x_2 = 3/4$, $z_3 = x_3 = 1$, $z_4 = x_4 = 3/2$, and $z_5 = x_5 = 2$. The subintervals of $[0, 2]$ determined by P are $[0, 1/2]$, $[1/2, 3/4]$, $[3/4, 1]$, $[1, 3/2]$, and $[3/2, 2]$. Thus

$$\Delta x_1 = \frac{1}{2} - 0 = \frac{1}{2}, \qquad \Delta x_2 = \frac{3}{4} - \frac{1}{2} = \frac{1}{4},$$

$$\Delta x_3 = 1 - \frac{3}{4} = \frac{1}{4}, \qquad \Delta x_4 = \frac{3}{2} - 1 = \frac{1}{2},$$

$$\Delta x_5 = 2 - \frac{3}{2} = \frac{1}{2}.$$

Hence, from (3) we get

$$R_P = f\left(\frac{1}{2}\right)\cdot\frac{1}{2} + f\left(\frac{3}{4}\right)\cdot\frac{1}{4} + f(1)\cdot\left(\frac{1}{4}\right) + f\left(\frac{3}{2}\right)\cdot\left(\frac{1}{2}\right) + f(2)\cdot\frac{1}{2}$$

$$= 2\cdot\frac{1}{2} + 3\cdot\frac{1}{4} + 4\cdot\frac{1}{4} + 6\cdot\frac{1}{2} + 8\cdot\frac{1}{2}$$

$$= 1 + \frac{3}{4} + 1 + 3 + 4 = 9.75.$$

(b) Here z_i is the left endpoint of the subinterval $[x_{i-1}, x_i]$. Since every subinterval has length 1/4, we get from (3)

$$R_P = \frac{1}{4}\left[f(0) + f\left(\frac{1}{4}\right) + f\left(\frac{1}{2}\right) + f\left(\frac{3}{4}\right) + f(1) + f\left(\frac{5}{4}\right) + f\left(\frac{3}{2}\right) + f\left(\frac{7}{4}\right)\right]$$

$$= \frac{1}{4}[0 + 1 + 2 + 3 + 4 + 5 + 6 + 7]$$

$$= \frac{28}{4} = 7.$$

(c) The area A is the area of a triangle with base 2 and height 8. So $A = \frac{1}{2} \cdot 2 \cdot 8 = 8$. Thus the approximation in (b) is more accurate. ■

To define the area of a region lying under the graph of a positive-valued function f and above a closed interval $[a, b]$, we need a measure of how finely a partition chops up $[a, b]$.

1.2 DEFINITION Let $P = \{x_0, x_1, x_2, \ldots, x_n\}$ be a partition of $[a, b]$. The ***norm*** of P is the length of the longest subinterval determined by the partition:

$$|P| = \max\{\Delta x_1, \Delta x_2, \ldots, \Delta x_n\}.$$

In Example 1(a), $|P| = 1/2$, because the longest subinterval—$[0, 1/2]$, $[1, 3/2]$, and $[3/2, 2]$ are all equally long—has length 1/2. In Example 1(b), each subinterval has length 1/4, so $|P| = 1/4$. Partitions of this sort have a special name.

1.3 DEFINITION A partition $P = \{x_0, x_1, x_2, \ldots, x_n\}$ of $[a, b]$ is ***uniform*** if every subinterval has length

$$\frac{b - a}{n}.$$

In this case, we write Δx instead of Δx_i for the length of the ith subinterval.

Thus, in the case of a uniform partition, $|P| = \Delta x$. Uniform partitions are the easiest to use with computers and calculators, so we tend to favor them.

We can now define the area A under a curve $y = f(x)$ between $x = a$ and $x = b$ if $f(x) \geq 0$ on $[a, b]$. As in Example 1, we expect the rectangular approximations R_P to approximate the area A more closely as $|P|$ approaches 0. As $|P| \to 0$, the number n of subintervals becomes larger and larger without bound. Thus R_P is the sum of more and more terms, each of which seems to be getting smaller and smaller since each $\Delta x_i \to 0$. While we aren't sure that the approximations R_P will in fact approach a fixed number, when they *do*, we call that limit the area A.

1.4 DEFINITION Suppose $f(x) \geq 0$ on $[a, b]$. The ***area*** of the region lying under the graph of $y = f(x)$, above the x-axis, and between $x = a$ and $x = b$ is A if

$$\lim_{|P| \to 0} R_P = A. \tag{4}$$

Here the limit statement means the following. Given any $\varepsilon > 0$, there is a $\delta > 0$ such that if $P = \{x_0, x_1, \ldots, x_n\}$ is *any* partition of $[a, b]$ with $|P| < \delta$, and if z_i is *any* point of $[x_{i-1}, x_i]$ for $i = 1, 2, \ldots, n$, then

$$|A - R_P| < \varepsilon. \tag{5}$$

The area A is a limit of the rectangular approximations R_P. Note that $f(x) \geq 0$ for all $x \in [a, b]$, and $\Delta x_i = x_i - x_{i-1} > 0$ since it is the *length* of the subinterval $[x_{i-1}, x_i]$. Thus every $R_P \geq 0$, because it is the sum of products of $f(z_i) \geq 0$ with $\Delta x_i \geq 0$. So the limit in (4) is nonnegative. **Area is always a nonnegative quantity.** If your calculations produce a negative value for A, then there is a computational error.

The limit in (4) is conceptually different from those in Chapter 1, because R_P is *not* simply a function of $|P|$. Rather, it also depends on the choice of points z_i from the subintervals determined by P. The requirement in (5) is particularly demanding: It must hold for *any* sufficiently fine partition, *no matter how* the points z_i in $[x_{i-1}, x_i]$ are chosen. To avoid the intricate arguments needed to establish that, we use the following theorem of advanced calculus. It guarantees that for most cases of interest to us, the limit in Definition 1.4 does exist.

1.5 THEOREM

If f is continuous on $[a, b]$, then $\lim_{|P| \to 0} R_P$ exists.

This theorem says that for any continuous function there *is* an area A for the region between $y = f(x)$, the x-axis, $x = a$, and $x = b$. To calculate it, we can use *any* partition scheme and *any* choice of points z_i, because the stringent requirement in Definition 1.4 is *known* to be met. We will then almost always use uniform partitions and choose either the left endpoint, the right endpoint, or the midpoint of $[x_{i-1}, x_i]$ for z_i.

Example 2

Compute R_P for $f(x) = x^2$ over $[0, 1]$ using uniform partitions and midpoints and (a) $n = 5$, (b) $n = 10$. What limit does R_P seem to be approaching?

Solution. In each case, $z_i = \frac{1}{2}(x_{i-1} + x_i)$ is the midpoint of $[x_{i-1}, x_i]$.

(a) Here

$$\Delta x = \frac{1-0}{5} = 0.2.$$

Since $x_1 = 0.2$, $x_2 = 0.4$, $x_3 = 0.6$, $x_4 = 0.8$, and $x_5 = 1$, we have

$$z_1 = 0.1,\ z_2 = 0.3,\ z_3 = 0.5,\ z_4 = 0.7, \text{ and } z_5 = 0.9.$$

Therefore

$$\begin{aligned} R_P &= 0.2[f(z_1) + f(z_2) + f(z_3) + f(z_4) + f(z_5)] \\ &= 0.2[0.01 + 0.09 + 0.25 + 0.49 + 0.81] = (0.2)(1.65) = 0.33. \end{aligned}$$

(b) Here

$$\Delta x = \frac{1-0}{10} = 0.1.$$

The set of points then is $\{z_1, z_2, \ldots, z_{10}\} = \{0.05, 0.15, \ldots, 0.95\}$. With the aid of a calculator, we find

$$\begin{aligned} R_P = (0.1)[&(0.05)^2 + (0.15)^2 + (0.25)^2 + (0.35)^2 + (0.45)^2 \\ &+ (0.55)^2 + (0.65)^2 + (0.75)^2 + (0.85)^2 + (0.95)^2] \\ = (0.1)&(3.325) = 0.3325. \end{aligned}$$

It seems that R_P may be approaching $1/3 = 0.333\ldots$. ■

Indeed, a computer gives the following results if we start doubling n in Example 2 and continue to use $z_i = \frac{1}{2}(x_{i-1} + x_i)$.

n	20	40	80	160	320	640
$R(f, P, z)$	0.333125	0.3332813	0.3333203	0.3333301	0.3333325	0.3333331

So it seems safe to conclude that the area under $y = x^2$ between $x = 0$ and $x = 1$ is indeed $1/3$. To see *why* R_P approaches $1/3$ as $P \to 0$, we need a formula for calculating the sums of squares.

1.6 THEOREM

Let n be a positive integer. Then

(a) $1 + 2 + 3 + \ldots + n = \frac{1}{2} n(n+1)$

(b) $1^2 + 2^2 + 3^2 + \ldots + n^2 = \frac{n(n+1)(2n+1)}{6}$

(c) $1^3 + 2^3 + 3^3 + \ldots + n^3 = \left[\frac{1}{2} n(n+1)\right]^2$

Proof. We prove (a) and (b) and leave (c) for Exercise 31.

(a) Let

$$S_n = 1 + 2 + 3 + \ldots + (n-1) + n. \tag{6}$$

This can also be written as

$$S_n = n + (n-1) + (n-2) + \ldots + 2 + 1. \tag{7}$$

Adding (6) and (7) gives

$$2S_n = (n+1) + (n+1) + \ldots + (n+1) + (n+1) = n(n+1).$$

Thus

$$S_n = \frac{1}{2} n(n+1).$$

(b) The easiest way to prove this is by mathematical induction (see Appendix III), which says that a theorem about positive integers n is true for all n if it is true for $n = 1$ and is true for $n = k + 1$ whenever it holds for $n = k$. Formula (b) holds for $n = 1$:

$$1^2 = \frac{1(1+1)(2+1)}{6} = \frac{1 \cdot 2 \cdot 3}{6}.$$

Moreover, it holds for $n = k + 1$ whenever it holds for $n = k$. For if

$$1^2 + 2^2 + \ldots + k^2 = \frac{k(k+1)(2k+1)}{6},$$

then

$$\begin{aligned}
1^2 + 2^2 + \ldots + k^2 + (k+1)^2 &= \frac{k(k+1)(2k+1)}{6} + (k+1)^2 \\
&= (k+1)\left[\frac{k(2k+1)}{6} + k + 1\right] \\
&= (k+1)\left[\frac{2k^2 + k}{6} + \frac{6k+6}{6}\right] \\
&= \frac{1}{6}(k+1)(2k^2 + 7k + 6) \\
&= \frac{1}{6}(k+1)(k+2)(2k+3).
\end{aligned}$$

Thus (b) holds for every positive integer n. QED

FIGURE 1.9

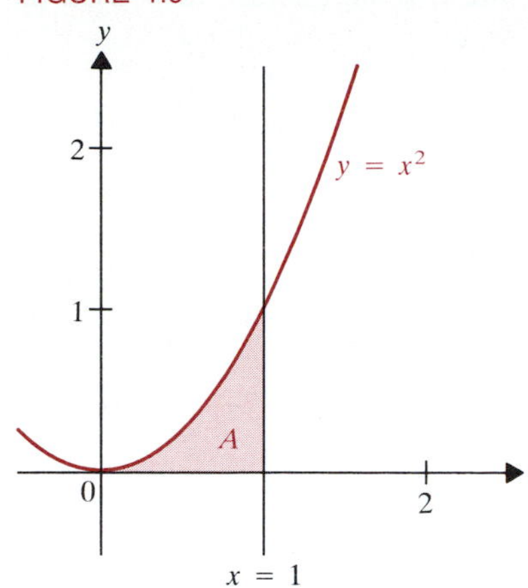

Now we can show why R_P approaches 1/3 in Example 2.

Example 3

Show that the area under $y = x^2$ between $x = 0$ and $x = 1$ is 1/3. See **Figure 1.9.**

Solution. We know the area exists by Theorem 1.5. So to compute it, we can use a uniform partition $P = \{x_0, x_1, \ldots, x_n\}$ with $|P| = \Delta x = (1 - 0)/n = 1/n$ and $z_i = x_i$, the right endpoint of $[x_{i-1}, x_i]$. We have

$$z_1 = x_1 = \frac{1}{n}, \ldots, z_2 = x_2 = \frac{2}{n}, \ldots, z_i = x_i = \frac{i}{n}, \ldots, z_n = x_n = \frac{n}{n} = 1.$$

Thus

$$\begin{aligned} R_P &= \frac{1}{n}[z_1{}^2 + z_2{}^2 + \ldots + z_n{}^2] = \frac{1}{n}\left[\frac{1}{n^2} + \frac{4}{n^2} + \ldots + \frac{n^2}{n^2}\right] \\ &= \frac{1}{n^3}[1 + 4 + \ldots + n^2] \\ &= \frac{1}{n^3} \cdot \frac{n(n+1)(2n+1)}{6} \qquad \textit{by Theorem 1.6(b)} \\ &= \frac{(n+1)(2n+1)}{6n^2} = \frac{2n^2 + 3n + 1}{6n^2} \\ &= \frac{1}{3} + \frac{1}{2n} + \frac{1}{6n^2} \end{aligned}$$

As $|P| \to 0$, $n \to +\infty$, so

$$\begin{aligned} \lim_{|P| \to 0} R_P &= \lim_{n \to +\infty} \left(\frac{1}{3} + \frac{1}{2n} + \frac{1}{6n^2}\right) \\ &= \frac{1}{3} + 0 + 0 = \frac{1}{3}. \quad \blacksquare \end{aligned}$$

Even though n is *not* a real variable (because it takes on *only* integer values), we nevertheless can see that as n becomes arbitrarily large, the terms $1/n$, $1/n^2, \ldots$ become arbitrarily close to 0. We express that by writing

$$\lim_{n \to +\infty} \frac{1}{n} = 0, \qquad \lim_{n \to +\infty} \frac{1}{n^2} = 0, \quad , \quad \text{and so forth.}$$

(Section 9.1 discusses such limits in more detail.)

Because writing rectangular approximations term-by-term as in Example 3 is time consuming, there is a short-hand ***summation notation,*** which provides a much more compact way of writing sums. Its key feature is the Greek capital sigma: $\sum$.

1.7 DEFINITION

Let $f(n)$ be an expression involving the nonnegative integer n. Then if $k < m$,

$$\sum_{i=k}^{m} f(i) = f(k) + f(k+1) + \ldots + f(m-1) + f(m).$$

Thus $\Sigma_{i=k}^{m} f(i)$ stands for the sum of all terms starting with $f(k)$ and ending with $f(m)$ and including $f(i)$ for every integer between k and m. The letter i is called the ***index of summation.*** Other letters like j, k, m, etc. are also used as summation indices:

$$2^2 + 3^2 + 4^2 + 5^2 + 6^2 = \sum_{i=2}^{6} i^2, \qquad 1 + \frac{1}{2} + \frac{1}{4} + \frac{1}{8} = \sum_{k=0}^{3} \frac{1}{2^k},$$

$$1 + 3 + 5 + 7 + 9 = \sum_{m=1}^{5} (2m - 1)..$$

Theorem 1.6 asserts that

$$\sum_{i=1}^{n} i = \frac{1}{2} n(n+1),$$

$$(8) \qquad \sum_{i=1}^{n} i^2 = \frac{n(n+1)(2n+1)}{6},$$

$$\sum_{i=1}^{n} i^3 = \frac{1}{4} n^2(n+1)^2.$$

In summation notation, the rectangular approximation (3) becomes

$$R_P = \sum_{i=1}^{n} f(z_i)\,\Delta x_i.$$

Summation notation often saves time and space. For instance, in summation notation, the first four lines of the calculation of R_P in Example 3 are

$$R_P = \frac{1}{n} \sum_{i=1}^{n} z_i{}^2 = \frac{1}{n} \sum_{i=1}^{n} \frac{i^2}{n^2} = \frac{1}{n^3} \sum_{i=1}^{n} i^2$$

$$= \frac{1}{n^3} \cdot \frac{n(n+1)(2n+1)}{6}. \qquad \textit{by (8)}$$

The following result summarizes the computational rules for summation notation.

1.8 THEOREM

(a) $\sum_{i=1}^{n} c = nc,$ for any real number c.

(b) $\sum_{i=1}^{n} [f(i) + g(i)] = \sum_{i=1}^{n} f(i) + \sum_{i=1}^{n} g(i)$

(c) $\sum_{i=1}^{n} cf(i) = c \sum_{i=1}^{n} f(i),$ for any real number c.

(d) **Telescoping Property.**

$$\sum_{i=1}^{n} [f(i+1) - f(i)] = f(n+1) - f(1).$$

Partial Proof. We prove (a) and (d), leaving (b) and (c) for Exercises 33 and 34.

(a)
$$\sum_{i=1}^{n} c = c + c + \ldots + c \qquad (n\,c\text{'s})$$
$$= nc.$$

(d)
$$\sum_{i=1}^{\infty} [f(i+1) - f(i)] = [f(2) - f(1)] + [f(3) - f(2)] + \ldots$$
$$+ [f(n) - f(n-1)] + [f(n+1) - f(n)]$$
$$= f(n+1) - f(1). \quad \text{QED}$$

Our concluding example illustrates how summation notation is used and that Definition 1.4 is consistent with earlier area formulas.

Example 4

Use Definition 1.4 to find the area under the graph of $y = 2x$ between $x = 0$ and $x = 2$.

Solution. Theorem 1.5 guarantees that the area exists since $f(x) = 2x$ defines a continuous function. Partition $[0, 2]$ into n subintervals of uniform length:

$$\Delta x = \frac{2-0}{n} = \frac{2}{n}.$$

As in Example 3, let z_i be the right endpoint $x_i = i \cdot 2/n$ of the subinterval $[x_{i-1}, x_i]$. Then

$$R_P = \sum_{i=1}^{n} f(z_i)\,\Delta x = \frac{2}{n}\sum_{i=1}^{n} f(x_i) \qquad \textit{by Theorem 1.8(c)}$$

$$= \frac{2}{n}\sum_{i=1}^{n} 2x_i = \frac{4}{n}\sum_{i=1}^{n}\frac{2i}{n} \qquad \textit{by Theorem 1.8(c)}$$

$$= \frac{8}{n^2}\sum_{i=1}^{n} i = \frac{8}{n^2}\cdot\frac{n(n+1)}{2} \qquad \textit{by Theorem 1.6(a)}$$

$$= \frac{4}{n}(n+1) = 4\left(1+\frac{1}{n}\right).$$

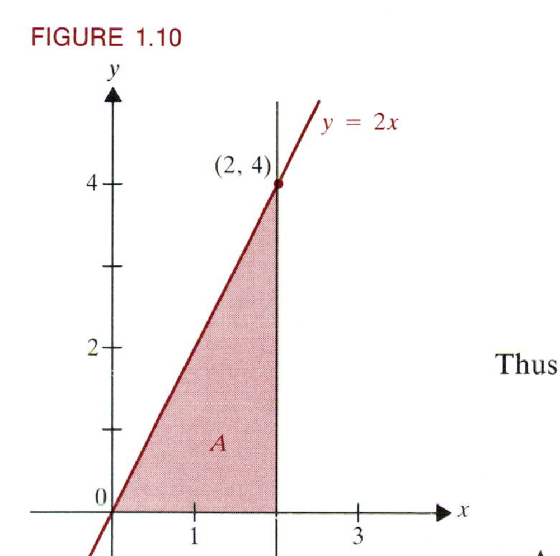

FIGURE 1.10

Thus

$$A = \lim_{|P|\to 0} R_P = \lim_{n\to\infty} 4\left(1+\frac{1}{n}\right) = 4. \quad \blacksquare$$

As **Figure 1.10** shows, the region in Example 4 is a right triangle of height 4 and base 2, so indeed $A = (1/2)\cdot 4 \cdot 2 = 4$.

HISTORICAL NOTE

We could have proved Theorem 1.6(a) by mathematical induction also, but have instead given a proof discovered by the great German mathematician Karl Friedrich Gauss (1777–1855) in 1787. Gauss's teacher in elementary school was irked by his class's behavior and, as punishment, told the students they had to add the first 100 positive integers. Little Karl astonished the teacher by immediately walking up with the answer 5050. He had added $1 + 2 + \ldots + 100$ and $100 + 99 + \ldots + 1$ as we did in (6) and (7), noting that the answer was $100 \cdot 101$. He then divided by 2 to get $50 \cdot 101$ ($= 5050$)! Thus began the mathematical career of one of the giants of mathematical history, who also made many important contributions to physics and astronomy.

The idea of defining area as in Definition 1.4 goes back to Cauchy in 1823. Cauchy used the left endpoint x_{i-1} as his z_i, and was able to prove Theorem 1.5. He derived many of the basic facts about integration and area from Definition 1.4.

Exercises 4.1

In Exercises 1–12, calculate the indicated rectangular approximation R_P.

1. $f(x) = x + 1$ on $[0, 2]$, $P = \{0, 1/2, 1, 5/4, 2\}$, $z_1 = 0$, $z_2 = z_3 = 1$, $z_4 = 3/2$

2. $f(x) = 2x + 1$ on $[0, 2]$, P and z_i as in Exercise 1

3. $f(x) = \cos x$ on $[-\pi/2, \pi/2]$, $P = \{-\pi/2, -\pi/4, 0, \pi/4, \pi/2\}$, $z_1 = -\pi/2$, $z_2 = -\pi/4$, $z_3 = \pi/4$, $z_4 = \pi/2$

4. $f(x) = \cos x$ on $[-\pi/2, \pi/2]$, P as in Exercise 3, $z_1 = -\pi/4$, $z_2 = z_3 = 0$, $z_4 = \pi/4$

5. $f(x) = 4x + 2$ on $[0, 2]$, P and z_i as in Example 2(a)

6. $f(x) = 4x + 1$ on $[0, 2]$, P and z_i as in Example 2(b)

7. $f(x) = x^2$ over $[0, 2]$, P a uniform partition with $z_i = (1/2)(x_{i-1} + x_i)$ and (a) $n = 5$; (b) $n = 10$

8. $f(x) = 1 + x^2$ over $[0, 3]$, P a uniform partition with $z_i = (1/2)(x_{i-1} + x_i)$ and (a) $n = 5$; (b) $n = 10$

9. Repeat Exercise 7 using $z_i = x_{i-1}$.

10. Repeat Exercise 8 using $z_i = x_{i-1}$.

11. Repeat Exercise 7 using $z_i = x_i$.

12. Repeat Exercise 8 using $z_i = x_i$.

In Exercises 13–22, use Definition 1.4 to find the indicated area above the x-axis.

13. Under the graph of $y = 2x + 1$ between $x = 0$ and $x = 3$

14. Under the graph of $y = 3x + 2$ between $x = 0$ and $x = 2$

15. Under the graph of $y = x^2$ between $x = 0$ and $x = 2$ **(This exercise is referred to in Section 4.2.)**

16. Under the graph of $y = 1 + x^2$ between $x = 0$ and $x = 3$

17. Under the graph of $y = x^3$ between $x = 0$ and $x = 2$

18. Under the graph of $y = 1 + x^3$ between $x = 0$ and $x = 1$

19. Under the graph of $y = x + x^2$ between $x = 0$ and $x = 2$

20. Under the graph of $y = 2x + 3x^2$ between $x = 0$ and $x =$

21. Under the graph of $y = 9 - x^2$ between $x = 0$ and $x = 3$

22. Under the graph of $y = 4 - x^2$ between $x = 0$ and $x = 2$

23. Evaluate the sums

(a) $\sum_{k=1}^{4} (k^2 + 2k - 1)$ **(b)** $\sum_{j=0}^{3} (j^3 - 3j^2 + 2)$

(c) $\sum_{i=1}^{5} i \sin \frac{i\pi}{2}$ **(d)** $\sum_{m=1}^{4} \frac{1 - m}{1 + m}$

24. Write in summation notation:

(a) $2 + 4 + 6 + 8 + \ldots + 18$

(b) $1 + \frac{3}{2} + \frac{9}{4} + \ldots + \frac{531441}{4096}$

(c) $1 - \frac{1}{2} + \frac{1}{4} - \frac{1}{8} + \frac{1}{16} - \frac{1}{32}$

(d) $1 \cdot 2 + 3 \cdot 4 + 5 \cdot 6 + 7 \cdot 8 + 9 \cdot 10$

25. Evaluate

(a) $\sum_{n=1}^{50} \left(\frac{1}{n} - \frac{1}{n+1}\right)$ **(b)** $\sum_{n=1}^{100} (2^{n+1} - 2^n)$

26. Evaluate

(a) $\sum_{i=1}^{n} (\sqrt{i+1} - \sqrt{i-1})$ **(b)** $\sum_{k=1}^{25} \cos \pi k$

27. Use Definition 1.4 to derive the formula for the area of
(a) a rectangle of base b and height h,
(b) a triangle of base b and height h.

28. Use Definition 1.4 to derive the formula for the area under the parabola $y = x^2$ between $x = 0$ and $x = a$.

29. If $y = ax^2 + bx + c \geq 0$ on $[0, d]$, then show that the area under the graph between $x = 0$ and $x = d$ is

$$\frac{1}{3}ad^3 + \frac{1}{2}bd^2 + cd.$$

30. Find a formula for the area under $y = x^3$ from $x = 0$ to $x = a$.

31. Prove Theorem 1.6(c) by mathematical induction.

32. Prove Theorem 1.6(b) by using the fact that $(i + 1)^3 - i^3 = 3i^2 + 3i + 1$ and applying Theorem 1.8 to $\Sigma_{i=1}^{n} [(i + 1)^3 - i^3] = \Sigma_{i=1}^{n} (3i^2 + 3i + 1)$.

33. Prove Theorem 1.8(b).

34. Prove Theorem 1.8(c).

35. If g is the Dirichlet function of Exercise 37, Section 1.4, then show that the region lying under the graph of $y = g(x)$ between $x = 0$ and $x = 1$ has no area. [*Hint:* What happens for any partition P if every z_i is rational (respectively, irrational)?]

36. Repeat Exercise 35 for the schizophrenic function f defined by

$$f(x) = \begin{cases} x & \text{if } x \text{ is rational,} \\ 1 & \text{if } x \text{ is irrational.} \end{cases}$$

PC **Exercises 37–48 are intended for solution by computer or programmable calculator. Write or use an available program for computing the rectangular approximation (3) to the area under the graph of the given function above the specified interval. Start with $n = 10$, use midpoints, and let n double after each calculation, stopping at $n = 320$. On the basis of your results, what do you think the exact area is? Express all answers to at least five decimal places.**

37. $f(x) = x^2$ over $[0, 3]$

38. $f(x) = 2 - x^2$ over $[-1, 1]$

39. $f(x) = \sin x$ over $[0, \pi]$

40. $f(x) = \cos x$ over $[0, \pi/2]$

41. $f(x) = x^4$ over $[0, 1]$

42. $f(x) = x^5 - 3x + 3$ over $[0, 2]$

43. $f(x) = x^2 - 4x + 5$ over $[1, 3]$

44. $f(x) = 2x^2 - 3x + 2$ over $[1, 2]$

45. $f(x) = 1/x$ over $[1, 2]$

46. $f(x) = 1/x$ over $[1, 3]$

47. $f(x) = \sqrt{1 + x^2}$ over $[0, 1]$

48. $f(x) = \sqrt{1 + x^3}$ over $[0, 1]$

4.2 The Definite Integral

FIGURE 2.1

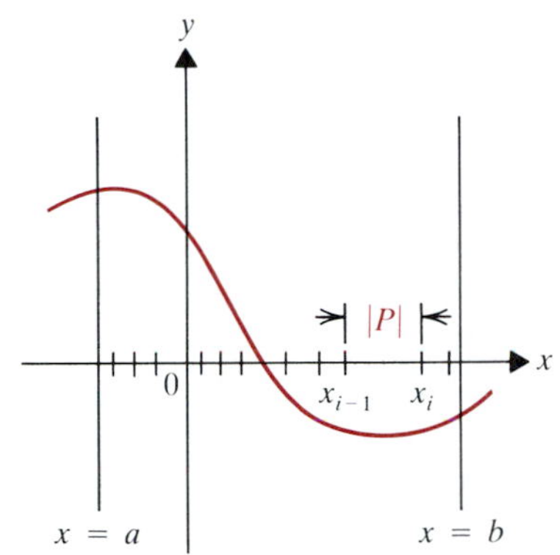

In this section we generalize Definition 1.4 by making it apply to *any* function whose domain includes $[a, b]$. We start by examining Definition 1.4, which was stated for *nonnegative* functions f. It defined the area A under the graph of f over $[a, b]$ to be the limit of the rectangular approximations

(1) $$R_P = \sum_{i=1}^{n} f(z_i)\,\Delta x_i$$

as $|P| \to 0$. Since a partition P is defined without reference to the function f, the definition of the norm of P needs no generalizing. It is the length Δx_i of the longest subinterval $[x_{i-1}, x_i]$. Formula (1) for the rectangular approximation also extends to any function f defined on $[a, b]$, since nothing in (1) depends on f having positive values. See **Figure 2.1.** When f is an arbitrary function, the quantity R_P defined by (1) has a special name, which honors the German mathematician G. F. Bernhard Riemann (1826–1866), who first singled it out for study in 1854.

2.1 DEFINITION Suppose that f is defined on $[a, b]$, and $P = \{x_0, x_1, \ldots x_n\}$ is a partition of $[a, b]$. A ***Riemann sum R_P*** of f relative to P over $[a, b]$ is

(2) $$R_P = \sum_{i=1}^{n} f(z_i)\,\Delta x_i = f(z_1)\,\Delta x_1 + \ldots + f(z_n)\,\Delta x_n,$$

where $z_i \in [x_{i-1}, x_i]$ for $i = 1, 2, \ldots, n$.

No new skills are needed to evaluate Riemann sums. In case the function f is positive valued on $[a, b]$, the Riemann sum R_P represents an approximation to the area under the graph of $y = f(x)$ over the interval $[a, b]$. But if the function takes on any negative values over $[a, b]$, then the Riemann sum does *not* give an approximation to the area of any region.

Example 1

FIGURE 2.2

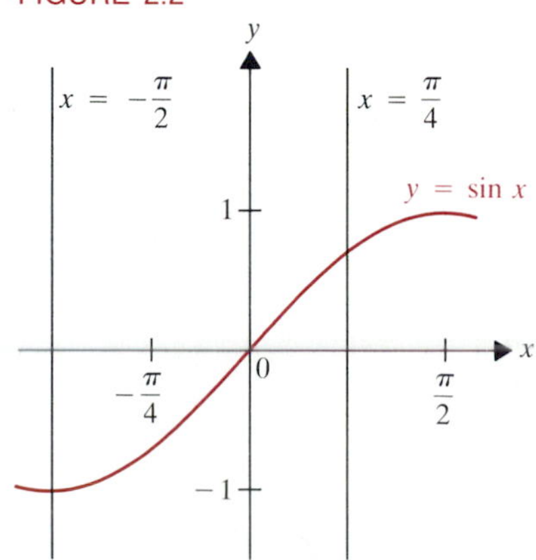

Find R_P over the interval $[-\pi/2, \pi/4]$ if $f(x) = \sin x$, $P = \{-\pi/2, -\pi/3, -\pi/4, -\pi/6, 0, \pi/6, \pi/4\}$ and $z_i = x_i$, the right endpoint of $[x_{i-1}, x_i]$. See **Figure 2.2.**

Solution. We have

$$\begin{aligned}
R_P &= \sum_{i=1}^{6} f(z_i)\,\Delta x_i \\
&= [\sin(-\pi/3)](\pi/6) + [\sin(-\pi/4)](\pi/12) + [\sin(-\pi/6)](\pi/12) \\
&\quad + [\sin 0](\pi/6) + [\sin(\pi/6)](\pi/6) + [\sin(\pi/4)](\pi/12) \\
&= \left[-\frac{1}{2}\sqrt{3}\cdot 2 - \frac{1}{2}\sqrt{2} - \frac{1}{2} + 0\cdot 2 + \frac{1}{2}\cdot 2 + \frac{1}{2}\sqrt{2}\right](\pi/12) \\
&= \left[-\sqrt{3} + \frac{1}{2}\right](\pi/12) \approx -0.32255. \blacksquare
\end{aligned}$$

We can now give Riemann's definition of the definite integral of f over $[a, b]$.

2.2 DEFINITION

Let f be defined on $[a, b]$. The ***definite integral of f over [a, b]*** is

$$\int_a^b f(x)\,dx = \lim_{|P|\to 0} R_P \tag{3}$$

if this limit exists. If the limit does exist, then f is said to be ***integrable*** over $[a, b]$. The function f is called the ***integrand,*** and the numbers a and b are called the ***limits of integration.***

The limit statement (3) has the same meaning as in Definition 1.4. The only difference between Definition 1.4 and Definition 2.2 is that here R_P is a Riemann sum, rather than a rectangular approximation of an area. The symbol $\int_a^b f(x)\,dx$ is read *(definite) integral from a to b of f* (or *of* $f(x)\,dx$). The symbol $\int$ is called the *integral sign*, and it is really the seventeenth-century printing symbol for the lower-case letter *s* (standing for *sum*). It was introduced by Leibniz in work done in 1675 and published in 1686. By 1676 Leibniz included the differential dx in his integral notation. In Section 4, we will explain how the definite integral and the indefinite integral (antiderivative) of Section 3.9 are related. For now, think of $\int_a^b f(x)\,dx$ as a limiting version of the symbol $\Sigma_{i=1}^n f(z_i)\,\Delta x_i$.

Note that if it exists, *the definite integral is a real number*, the limit of the Riemann sums R_P as $|P| \to 0$. The variable x in (3) is called a *dummy variable*, like the index i of summation in $\Sigma_{i=1}^n f(i)$. Just as the latter can be replaced by j, k, m, or other letters, so in (3) we can replace x by any other letter without changing the definite integral at all. Thus

$$\int_a^b f(x)\,dx = \int_a^b f(t)\,dt = \int_a^b f(u)\,du, \qquad \text{and so on.}$$

The following extension of Theorem 1.5 is proved in advanced calculus.

2.3 THEOREM

> If f is continuous on $[a, b]$, then f is integrable over $[a, b]$.

The definite integral of a function f is defined over a closed interval $[a, b]$. Thus, in the notation $\int_a^b f(x)\,dx$, a is restricted to be *less than* b. It is useful to give meaning to the definite integral when $a = b$ or $a > b$ also. What is a reasonable definition of $\int_a^a f(x)\,dx$? Since the interval from a to a has length 0, any attempt to partition it would produce subintervals of length 0 too! Thus any Riemann sum $R_P = \Sigma_{i=1}^n f(a) \cdot 0 = 0$. For any real number a in the domain D_f of f, it thus seems appropriate to define

$$\int_a^a f(x)\,dx = 0. \tag{4}$$

As for $\int_b^a f(x)\,dx$, think of how, for a uniform partition, the Riemann sum R_P would relate to a Riemann sum for $f(x)$ over the closed interval $[a, b]$. If we use n subintervals, with $\overline{\Delta x} = \dfrac{a-b}{n}$, the *negative* of the Δx corresponding to a uniform partition of $[a, b]$, then every Riemann sum for $\int_b^a f(x)\,dx$ is the *negative* of a corresponding Riemann sum for $\int_a^b f(x)\,dx$. Hence, we define

$$\int_b^a f(x)\,dx = -\int_a^b f(x)\,dx. \tag{5}$$

This means that *reversing the order of the limits of integration changes the sign of the definite integral.*

To avoid evaluating limits of complicated Riemann sums, we want to develop formulas for the definite integral of elementary functions. We make a start in this direction with the following result, which is illustrated in **Figure 2.3** for the case of a positive constant c.

2.4 THEOREM If $f(x) = c$ for all $x \in [a, b]$, then

$$\int_a^b f(x)\,dx = \int_a^b c\,dx = c(b-a).$$

Proof. Let P be *any* partition of $[a, b]$. In each subinterval $[x_{i-1}, x_i]$ select any point z_i. Then $f(z_i) = c$. Hence

FIGURE 2.3

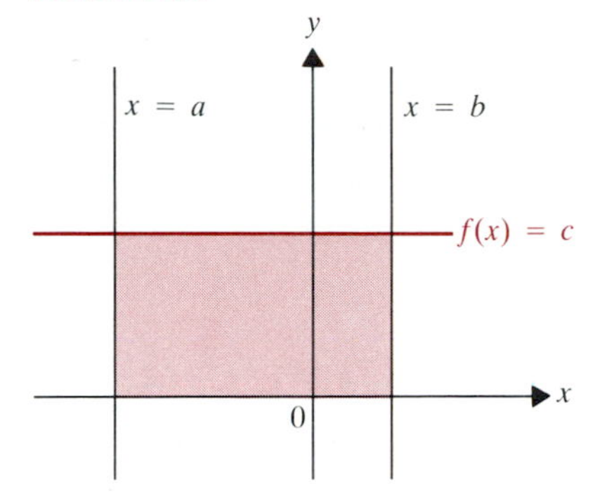

$$R_P = \sum_{i=1}^{n} f(z_i)\Delta x_i = \sum_{i=1}^{n} c\,\Delta x_i = c\sum_{i=1}^{n}\Delta x_i = c(b-a).$$

Thus

$$\lim_{|P|\to 0} R_P = c(b-a). \quad \boxed{\text{QED}}$$

Example 2

Evaluate

$$\text{(a)} \int_a^b 0\,dx \qquad \text{(b)} \int_a^b 1\,dx \qquad \text{(c)} \int_{-1}^{3} (-5)\,dx$$

Solution. (a) Using Theorem 2.4 with $c = 0$, we find

$$\int_a^b 0\,dx = 0\cdot(b-a) = 0.$$

(b) In this case, $c = 1$ in Theorem 2.4. Therefore

$$\int_a^b 1\,dx = 1(b-a) = b-a = \text{length of } [a, b].$$

(c) Here, $c = -5$, $b = 3$, and $a = -1$. Hence

$$\int_{-1}^{3} (-5)\,dx = (-5)[3-(-1)] = -5\cdot 4 = -20. \quad \blacksquare$$

Theorem 2.2 of Chapter 2 said that the derivative of cf is cf' and the derivative of $f+g$ (or $f-g$) is $f'+g'$ (respectively, $f'-g'$) if f and g are differentiable functions. Definite integrals have analogous properties. The proof is given at the end of the section.

2.5 THEOREM

If f and g are integrable over $[a, b]$, then

(a) $\int_a^b cf(x)\,dx = c\int_a^b f(x)\,dx$;

(b) $\int_a^b (f+g)(x)\,dx = \int_a^b f(x)\,dx + \int_a^b g(x)\,dx$;

(c) $\int_a^b (f-g)(x)\,dx = \int_a^b f(x)\,dx - \int_a^b g(x)\,dx$.

Repeated application of Theorem 2.5(b) shows that for any collection f_1, $f_2, \ldots, f_k$ of integrable functions,

$$(6) \quad \int_a^b (f_1+f_2+\ldots+f_k)(x)\,dx = \int_a^b f_1(x)\,dx + \int_a^b f_2(x)\,dx + \ldots + \int_a^b f_k(x)\,dx.$$

FIGURE 2.4

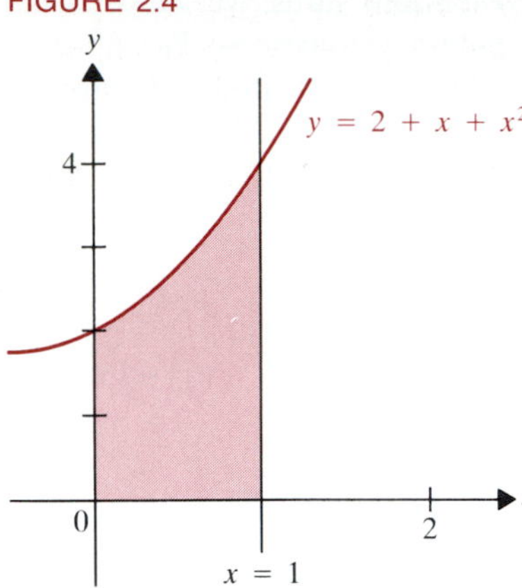

Example 3

Find $\int_0^1 (2 + x + x^2)\, dx$. (See **Figure 2.4.**)

Solution. Using (6) and Theorem 2.4, we have

$$\begin{aligned}\int_0^1 (2 + x + x^2)\, dx &= \int_0^1 2\, dx + \int_0^1 x\, dx + \int_0^1 x^2\, dx \\ &= 2(1 - 0) + \int_0^1 x\, dx + \int_0^1 x^2\, dx.\end{aligned}$$

As **Figure 2.5**(a) shows, $\int_0^1 x\, dx$ is the area of the triangle bounded by the x-axis, the line $y = x$, and the line $x = 1$. So

$$\int_0^1 x\, dx = \frac{1}{2} \cdot 1 \cdot 1 = \frac{1}{2}.$$

Similarly, Figure 2.5(b) shows that $\int_0^1 x^2\, dx$ is the area under $y = x^2$ between $x = 0$ and $x = 1$. In Example 3 of the last section, we found $\int_0^1 x^2\, dx = 1/3$. We thus have

$$\int_0^1 (2 + x + x^2)\, dx = 2 + \frac{1}{2} + \frac{1}{3} = \frac{17}{6}. \quad \blacksquare$$

FIGURE 2.5

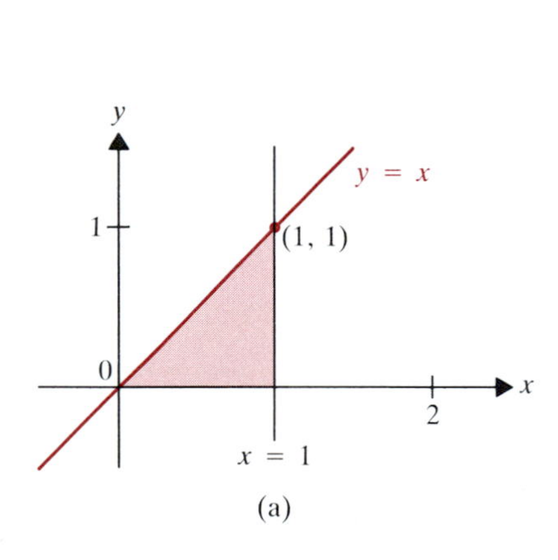

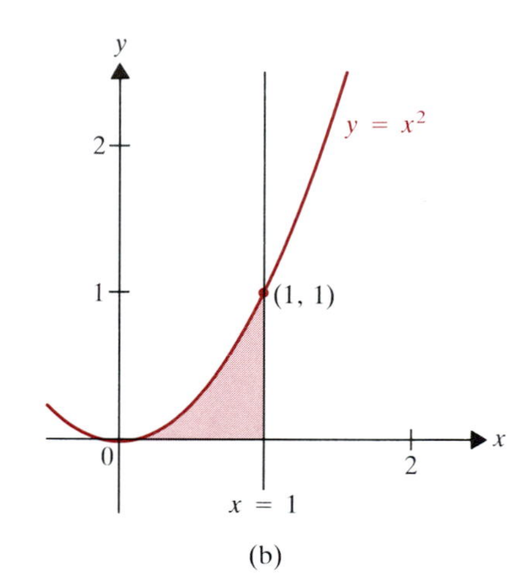

Take careful note that Theorem 2.5(a) permits *constants* to be factored out of definite integrals but says nothing about trying to integrate products of *functions.* In fact, generally,

$$\int_a^b f(x) g(x)\, dx \neq \left(\int_a^b f(x)\, dx\right)\left(\int_a^b g(x)\, dx\right).$$

Example 3 shows that **the integral of a product is not, in general, the product of the integrals.** As noted in the course of working the example,

$$\int_0^1 x\, dx = \frac{1}{2} \quad \text{and} \quad \int_0^1 x^2\, dx = \frac{1}{3}.$$

Thus

$$\int_0^1 x^2\, dx \neq \left(\int_0^1 x\, dx\right) \cdot \left(\int_0^1 x\, dx\right), \quad \text{since} \quad \frac{1}{3} \neq \frac{1}{2} \cdot \frac{1}{2}.$$

FIGURE 2.6

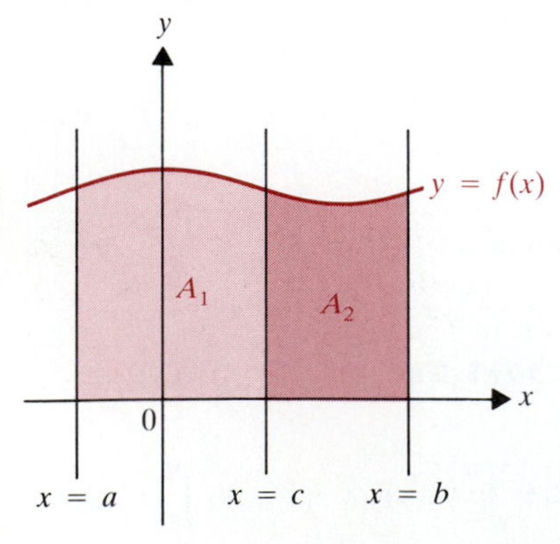

As **Figure 2.6** illustrates, the area under $y = f(x)$ between $x = a$ and $x = b$ is the sum of the areas A_1 (of the region under $y = f(x)$ between $x = a$ and $x = c$)

and A_2 (of the region under $y = f(x)$ between $x = c$ and $x = b$), if c is any point between a and b. This observation makes the following result, which is proved in advanced calculus, easy to accept.

2.6 THEOREM If f is integrable over $[a, c]$ and $[c, b]$ where c is any number in (a, b), then f is integrable over $[a, b]$, and

$$\int_a^b f(x)\,dx = \int_a^c f(x)\,dx + \int_c^b f(x)\,dx. \tag{7}$$

Example 4

Evaluate $\int_{-1}^{2} |x|\,dx$.

Solution. Since

$$|x| = \begin{cases} -x & \text{for } x < 0 \\ x & \text{for } x \geq 0, \end{cases}$$

it is natural to break the interval $[-1, 2]$ of integration into the subintervals $[-1, 0]$ and $[0, 2]$. Theorem 2.6 then gives

$$\int_{-1}^{2} |x|\,dx = \int_{-1}^{0} |x|\,dx + \int_0^2 |x|\,dx = \int_{-1}^{0} -x\,dx + \int_0^2 x\,dx.$$

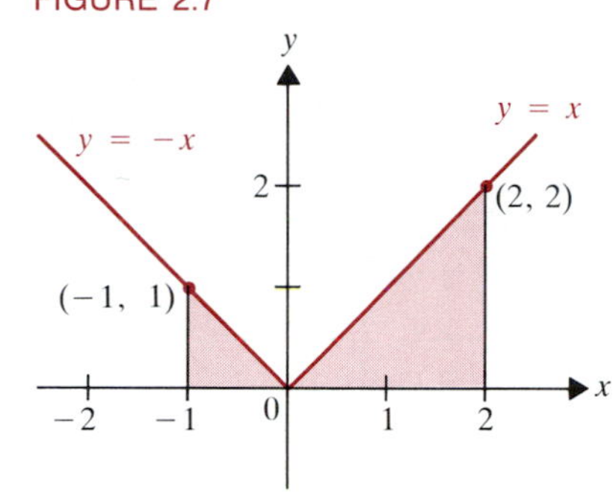

FIGURE 2.7

As **Figure 2.7** shows, each integral on the right is the area of a triangle. Therefore

$$\int_{-1}^{2} |x|\,dx = 1 \cdot \frac{1}{2} + 2 \cdot \frac{2}{2} = \frac{1}{2} + 2 = \frac{5}{2}. \quad ■$$

The next result says that (7) holds even if we don't have $a < c < b$.

2.7 THEOREM If f is integrable on an interval containing a, b, and c, then

$$\int_a^c f(x)\,dx = \int_a^b f(x)\,dx + \int_b^c f(x)\,dx, \tag{8}$$

regardless of the order of the real numbers a, b, and c.

To illustrate the idea in Theorem 2.7, suppose that the order is $a < c < b$. Suppose also that f is integrable on $[a, c]$ and $[c, b]$. Then Theorem 2.6 says that

$$\int_a^b f(x)\,dx = \int_a^c f(x)\,dx + \int_c^b f(x)\,dx.$$

Therefore we have

$$\begin{aligned}\int_a^c f(x)\,dx &= \int_a^b f(x)\,dx - \int_c^b f(x)\,dx \\ &= \int_a^b f(x)\,dx + \int_b^c f(x)\,dx. && \text{by (5)}\end{aligned}$$

Thus (8) follows in this case. The proofs in the other cases are similar.

Figure 2.8 shows the region under the graph of a continuous positive-valued *even* function f over a symmetric interval $[-a, a]$ around the origin. (Recall that this means $f(-x) = f(x)$ for all x in the domain of f.) It is clear from the figure that the area lying under the graph of $y = f(x)$ between $x = -a$ and $x = a$ is twice the area between $x = 0$ and $x = a$, so

$$\int_{-a}^{a} f(x)\,dx = 2\int_0^a f(x)\,dx.$$

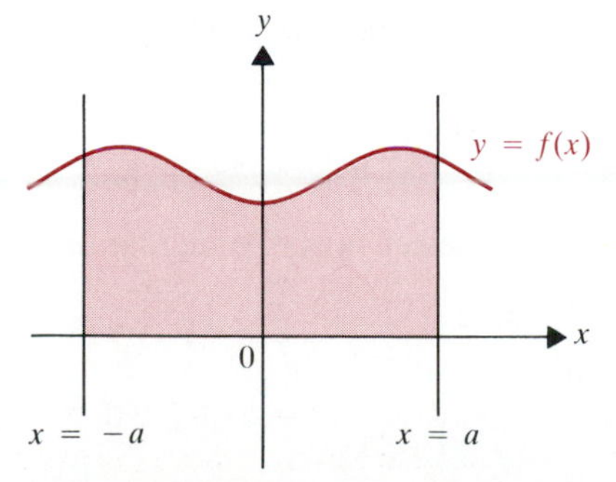

FIGURE 2.8

The same holds true if f is negative valued over the entire interval. If f is sometimes positive and sometimes negative, then dividing the interval into subintervals of constant sign, we can prove the first part of the following useful result. The second part is similar (see Exercises 37 and 38).

2.8 THEOREM

Suppose that f is integrable over $[-a, a]$.

(a) If f is an even function, then

$$\int_{-a}^{a} f(x)\,dx = 2\int_{0}^{a} f(x)\,dx.$$

(b) If f is an odd function (that is, $f(-x) = -f(x)$), then

$$\int_{-a}^{a} f(x)\,dx = 0.$$

Example 5

Use Theorem 2.8 to evaluate (a) $\int_{-1}^{1} x^2\,dx$ (b) $\int_{-\pi}^{\pi} \sin x\,dx$.

Solution. (a) We saw in Example 3 of the last section that

$$\int_{0}^{1} x^2\,dx = 1/3.$$

Theorem 2.8(a) therefore says that

$$\int_{-1}^{1} x^2\,dx = 2\int_{0}^{1} x^2\,dx = 2\cdot\frac{1}{3} = \frac{2}{3}.$$

(b) The sine function is odd. Consequently,

$$\int_{-\pi}^{\pi} \sin x\,dx = 0,$$

by Theorem 2.8(b). ■

Proof of Theorem 2.5.*

We assume that f and g are integrable over $[a, b]$. We prove (b) and leave the other parts for Exercise 37. Part (b) asserts that

$$\int_{a}^{b} (f+g)(x)\,dx = \int_{a}^{b} f(x)\,dx + \int_{a}^{b} g(x)\,dx.$$

To prove this, let $I_1 = \int_a^b f(x)\,dx$ and $I_2 = \int_a^b g(x)\,dx$. Then for any $\varepsilon > 0$, there exist $\delta_1 > 0$ and $\delta_2 > 0$ such that whenever $|P| < \delta = \min(\delta_1, \delta_2)$, we have

$$|I_1 - R_P(f)| < \frac{1}{2}\varepsilon \quad \text{and} \quad |I_2 - R_P(g)| < \frac{1}{2}\varepsilon \tag{9}$$

for any choice of points $z_i \in [x_{i-1}, x_i]$. If we select any particular set $\{z_1, z_2, \ldots, z_n\}$, then

$$R_P(f+g) = R_P(f) + R_P(g), \tag{10}$$

* Optional

since

$$\begin{aligned} R_P(f+g) &= \sum_{i=1}^{n} (f+g)(z_i)\,\Delta x_i \\ &= \sum_{i=1}^{n} [f(z_i) + g(z_i)]\,\Delta x_i \\ &= \sum_{i=1}^{n} f(z_i)\,\Delta x_i + \sum_{i=1}^{n} g(z_i)\,\Delta x_i \qquad \textit{by Theorem 1.8(b)} \\ &= R_P(f) + R_P(g). \end{aligned}$$

From (9) and (10), we have for any choice of points $z_1, z_2, \ldots, z_n$,

$$\begin{aligned} |I_1 + I_2 - R_P(f+g)| &= |I_1 - R_P(f) + I_2 - R_P(g)| \\ &\le |I_1 - R_P(f)| + |I_2 - R_P(g)| \\ &< \frac{1}{2}\varepsilon + \frac{1}{2}\varepsilon = \varepsilon, \end{aligned}$$

whenever $|P| < \delta$. Thus

$$I_1 + I_2 = \lim_{|P| \to 0} R_P(f+g).$$ QED

> HISTORICAL NOTE
>
> Bernhard Riemann was a Ph.D. student of Gauss at Goettingen University in Germany. He received his degree in 1851, and three years later formulated essentially the same definition we have given of $\int_a^b f(x)\,dx$. In doing so, he generalized Cauchy's approach of three decades earlier. Later Riemann developed a kind of noneuclidean geometry, now called *Riemannian geometry*, which Einstein used as the mathematical basis for much of his theory of relativity a half century later. Riemann also made fundamental and enduring contributions to the theory of functions of a complex variable, to differential equations, and to mathematical physics. A famous conjecture of his, now called the *Riemann hypothesis*, is still a major topic of study in higher mathematics. His achievements are all the more remarkable in view of his brief career, which was cut short by tuberculosis.

Exercises 4.2

In Exercises 1–4, find Δx_i for each subinterval determined by the partition P. What is $|P|$?

1. $P = \{0, 0.26, 0.50, 0.75, 1, 5/4\}$
2. $P = \{-1, -3/8, 0, 2/5, 1, 5/3\}$
3. $P = \{0, 1/6, 1/5, 1/4, 1/3, 1/2, 1\}$
4. $P = \{-\pi/2, -\pi/4, 0, \pi/6, \pi/4, \pi/3, \pi/2\}$

In Exercises 5–12, find the indicated Riemann sum.

5. $f(x) = 1 - x$ over $[0, 2]$, $P = \{0, 1/2, 1, 5/4, 2\}$, $z_1 = 0$, $z_2 = z_3 = 1$, $z_4 = 3/2$
6. $f(x) = \sin x$, P as in Example 1, $z_i = x_{i-1}$, the left endpoint of $[x_{i-1}, x_i]$
7. $f(x) = 1 - x^2$ on $[0, 2]$, P a uniform partition with $z_i = x_{i-1}$ and
   (a) $n = 5$ **(b)** $n = 10$
8. $f(x) = 1 - 2x^2$ on $[0, 1]$, P a uniform partition with $z_i = x_{i-1}$ and
   (a) $n = 5$ **(b)** $n = 10$
9. Repeat Exercise 7 with $z_i = x_i$.
10. Repeat Exercise 8 with $z_i = x_i$.
11. Repeat Exercise 7 with $z_i = \frac{1}{2}(x_{i-1} + x_i)$.
12. Repeat Exercise 8 with $z_i = \frac{1}{2}(x_{i-1} + x_i)$.

In Exercises 13–16, express each limit as a definite integral over the given interval.

13. $\lim_{|P|\to 0}\left(\sum_{i=1}^{n}\sqrt{1-z_i^2}\,\Delta x_i\right)$, $[0, 1]$

14. $\lim_{|P|\to 0}\left(\sum_{i=1}^{n}\left(\frac{2}{z_i}+z_i^2\right)\Delta x_i\right)$, $[1, 2]$

15. $\lim_{|P|\to 0}\left(\sum_{i=1}^{n}(z_i^2+\cos z_i \sin z_i)\Delta x_i\right)$, $[0, 2]$

16. $\lim_{|P|\to 0}\left(\sum_{i=1}^{n}\frac{2}{\sqrt{z_i^2+1}}\Delta x_i\right)$, $[0, 4]$

In Exercises 17–20, use facts about areas to find the value of the definite integral.

17. $\int_{-2}^{1}(2x+4)\,dx$ **18.** $\int_{-3}^{3}(x+3)\,dx$

19. $\int_{0}^{3}\sqrt{9-x^2}\,dx$ **20.** $\int_{0}^{4}\sqrt{16-x^2}\,dx$

In Exercises 21–26, use the results of Example 3, Exercises 15–28 of Section 1, and information from this section to evaluate the given integral.

21. $\int_{-1}^{1}(2x^2+1)\,dx$ **22.** $\int_{-2}^{2}(4x^3-3)\,dx$

23. $\int_{0}^{2}(2x+1)\,dx-\int_{0}^{2}x^2\,dx + \int_{2}^{3}(2x+1)\,dx+\int_{-2}^{2}(x^3-x)\,dx$

24. $\int_{0}^{4}(9-x^2)\,dx+\int_{0}^{1}x^2\,dx - \int_{3}^{4}(9-x^2)\,dx+\int_{-1}^{1}(x^2+4)\,dx$

25. $\int_{0}^{3}(x+x^2)\,dx+\int_{0}^{2}(9-x^2)\,dx - \int_{2}^{3}(x+x^2)\,dx+\int_{2}^{3}(9-x^2)\,dx$

26. $\int_{0}^{3}x^2\,dx+\int_{0}^{3}(x+x^2)\,dx-\int_{0}^{1}x^2\,dx-\int_{2}^{3}(x+x^2)\,dx$

In Exercises 27–32, write each expression as a single definite integral.

27. $\int_{1}^{3}f(x)\,dx+\int_{-2}^{1}f(x)\,dx$

28. $\int_{-1}^{3}f(x)\,dx+\int_{-3}^{-1}f(x)\,dx$

29. $\int_{1}^{4}f(x)\,dx-\int_{5}^{4}f(x)\,dx$

30. $\int_{-1}^{3}f(x)\,dx-\int_{6}^{3}f(x)\,dx$

31. $\int_{a}^{x_0+h}f(x)\,dx-\int_{a}^{x_0}f(x)\,dx$

32. $\int_{x_0}^{b}f(x)\,dx+\int_{b}^{x_0}f(x)\,dx$

In Exercises 33–36, evaluate the definite integral.

33. $\int_{0}^{2}|x-1|\,dx$ **34.** $\int_{-3}^{2}|x+1|\,dx$ **35.** $\int_{-1}^{2}|x|\,dx$

36. $\int_{-1}^{2}[2x-1]\,dx$, where $[x]$ is the greatest integer $\le x$

37. Use Definition 2.2 to prove Theorem 2.8(a).

38. Use Definition 2.2 to prove Theorem 2.8(b).

39. **(a)** Prove Theorem 2.5(a). **(b)** Prove Theorem 2.5(c).

40. If f and g are integrable over $[a, b]$ and c_1 and c_2 are real numbers, then show that

$$\int_a^b [c_1 f(x)+c_2 g(x)]\,dx = c_1\int_a^b f(x)\,dx + c_2\int_a^b g(x)\,dx.$$

41. If f is integrable on an interval containing $a_1, a_2, \ldots, a_k$, then show that

$$\int_{a_1}^{a_k} f(x)\,dx = \sum_{i=1}^{k-1}\left[\int_{a_i}^{a_{i+1}} f(x)\,dx\right].$$

42. Show that the Dirichlet function of Exercise 37, Section 1.4, is not integrable over any interval $[a, b]$.

43. Show that if $f(x)=1/x$ for $x\neq 0$ and $f(0)=0$, then f is not integrable over $[0, 1]$. (Use uniform partitions with z_i the midpoint of $[x_{i-1}, x_i]$.)

44. Suppose that both I_1 and I_2 satisfy the following condition: For any $\varepsilon>0$, there is a $\delta>0$ such that for all partitions P of $[a, b]$ with $|P|<\delta$, $|I_1-R_P|<\varepsilon$ and $|I_2-R_P|<\varepsilon$, for every choice of points $z_1, z_2, \ldots, z_n$, where z_i belongs to $[x_{i-1}, x_i]$. Then prove that $I_1=I_2$. (*Hint:* If $I_1\neq I_2$, then use (5) of Section 1 with $\varepsilon=|I_1-I_2|$ to make $|I_1-R_P|<\frac{1}{2}\varepsilon$ and $|I_2-R_P|<\frac{1}{2}\varepsilon$. From this and the triangle inequality, obtain the contradiction $|I_1-I_2|<|I_1-I_2|$.) **This shows that the definite integral as defined by (2) is unique if it exists.**

4.3 The Mean Value Theorem for Integrals

One of the most important results about differentiation is the mean value theorem for derivatives (Theorem 10.2 of Chapter 2). There is an analogous result for definite integrals, whose usefulness is suggested by the next section, where it plays a key role in deriving the fundamental theorem of calculus.

What is meant by the *average* (or *mean*) *value* of a function f over an interval $[a, b]$? If f is defined at only a finite number of points $x_1, x_2, \ldots, x_n$ in $[a, b]$, then the arithmetic mean of the values $f(x_1), f(x_2), \ldots, f(x_n)$ is

$$\bar{f}(x) = \frac{f(x_1) + f(x_2) + \ldots + f(x_n)}{n}. \tag{1}$$

That is the starting point for defining the average value of an integrable function f over $[a, b]$.

If f is integrable over $[a, b]$, then let $P = \{x_0, x_1, \ldots, x_n\}$ be a uniform partition of $[a, b]$ with associated subintervals $[x_{i-1}, x_i]$, $i = 1, \ldots, n$, each of length

$$\Delta x = \frac{b - a}{n}. \tag{2}$$

Pick $z_i \in [x_{i-1}, x_i]$ for $i = 1, 2, \ldots, n$. Then (1) says that the arithmetic mean of the values $f(z_1), f(z_2), \ldots, f(z_n)$ is

$$\bar{f}(z) = \frac{f(z_1) + f(z_2) + \ldots + f(z_n)}{n}. \tag{3}$$

We can interpret (3) as a Riemann sum. From (2),

$$\frac{1}{n} = \frac{\Delta x}{b - a},$$

and so (3) can be put in the form

$$\bar{f}(z) = \sum_{i=1}^{n} \frac{1}{b - a} f(z_i)\,\Delta x.$$

Thus $\bar{f}(z)$ is a Riemann sum of the function g defined by

$$g(x) = \frac{1}{b - a} f(x).$$

As the number of subdivision points increases, the arithmetic mean of the function values $f(z_i)$, $i = 1, 2, \ldots, n$, should more and more closely approximate the average value of f over $[a, b]$. We are thus led to the following definition.

3.1 DEFINITION Suppose that f is an integrable function over a closed interval $[a, b]$. Then the *average value* of f over $[a, b]$ is

$$\bar{f} = \int_a^b \frac{1}{b - a} f(x)\,dx = \frac{1}{b - a} \int_a^b f(x)\,dx. \tag{4}$$

Example 1

What is the average value of the function f defined by $f(x) = x^2$ over the interval $[0, 1]$?

Solution. In Example 3 of Section 1, we found that

$$\int_0^1 x^2\,dx = \frac{1}{3}.$$

Since $b - a = 1$, the average value of f over $[0, 1]$ is 1/3. ■

That 1/3 is a reasonable value of $f(x) = x^2$ over $[0, 1]$ can be seen with the help of the following table of function values.

x	0.05	0.15	0.25	0.35	0.45	0.55	0.65	0.75	0.85	0.95
x^2	0.0025	0.0225	0.0625	0.1225	0.2025	0.3025	0.4225	0.5625	0.7225	0.9025

The arithmetic mean of the values shown in the table is exactly the Riemann sum R_P calculated in Example 2(b) of Section 1, as you can check. Thus, that mean is 0.3325, which is quite close to 1/3.

The mean value theorem for integrals is a result about the average value of f defined by (4). To derive it, we use two preliminary results about the size of $\int_a^b f(x)\,dx$. The formal proof of the first one is given at the end of the section, but intuitively it just says that if all Riemann sums R_P lie between two numbers l and u, then so does their limit.

3.2 LEMMA Let f be integrable over $[a, b]$, and let u and l be real numbers. If for every partition $P = \{x_0, x_1, \ldots, x_n\}$ of $[a, b]$, and every choice of points z_i in $[x_{i-1}, x_i]$, $i = 1, 2, \ldots, n$, we have

$$l \leq R_P \leq u,$$

then

$$l \leq \int_a^b f(x)\,dx \leq u.$$

The other preliminary result used to prove the mean value theorem for integrals is a sandwich result on the size of $\int_a^b f(x)\,dx$.

3.3 THEOREM Suppose that the integrable function f satisfies

$$m \leq f(x) \leq M$$

for all $x \in [a, b]$. Then

$$m(b-a) \leq \int_a^b f(x)\,dx \leq M(b-a). \tag{5}$$

Proof. Let $P = \{x_0, x_1, \ldots, x_n\}$ be any partition of $[a, b]$. For $i = 1, 2, \ldots, n$ and any z_i in $[x_{i-1}, x_i]$, we have

$$m \leq f(z_i) \leq M.$$

Every $\Delta x_i \geq 0$, so that

$$m\,\Delta x_i \leq f(z_i)\,\Delta x_i \leq M\,\Delta x_i, \tag{6}$$

for $i = 1, 2, \ldots, n$. Summing (6) for $i = 1$ to n, we thus get

$$\sum_{i=1}^{n} m\,\Delta x_i \leq \sum_{i=1}^{n} f(z_i)\,\Delta x_i \leq \sum_{i=1}^{n} M\,\Delta x_i. \tag{7}$$

From Theorem 1.8c and the fact that $[a, b]$ is the union of the nonoverlapping subintervals $[x_{i-1}, x_i]$, we obtain

$$\sum_{i=1}^{n} m\,\Delta x_i = m \sum_{i=1}^{n} \Delta x_i = m(b-a)$$

and

$$\sum_{i=1}^{n} M\,\Delta x_i = M \sum_{i=1}^{n} \Delta x_i = M(b-a).$$

Substituting these facts into (7), we find

$$m(b-a) \le R_P \le M(b-a), \tag{8}$$

for any partition P and any choice of points $z_i \in [x_{i-1}, x_i]$. From (8) and Lemma 3.2, we conclude that

$$m(b-a) \le \int_a^b f(x)\,dx \le M(b-a). \quad \text{QED}$$

Theorem 3.3 can be used to obtain rough bounds for definite integrals.

Example 2

Estimate $\int_0^1 x^3\,dx$.

Solution. We have $x^3 \ge 0$ for $x \in [0, 1]$. Also $x^3 \le 1$ for $x \in [0, 1]$. Using $m = 0$ and $M = 1$, we get from Theorem 3.3

$$0 \cdot (1-0) \le \int_0^1 x^3\,dx \le 1(1-0) \to 0 \le \int_0^1 x^3\,dx \le 1. \quad ■$$

FIGURE 3.1

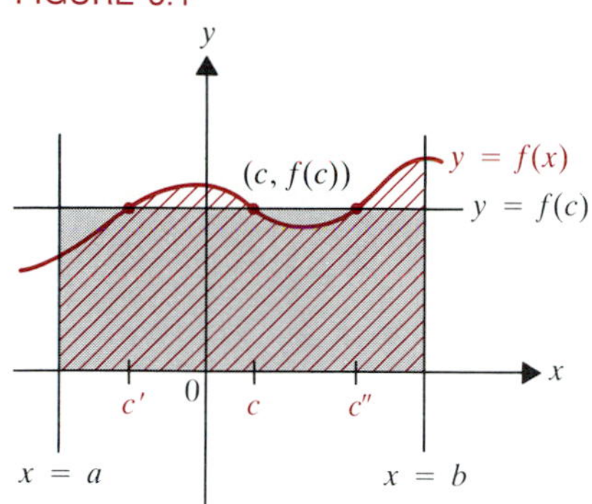

We are now ready for the mean value theorem for integrals, first formulated and proved by Cauchy in 1823. **Figure 3.1** illustrates its geometric content for a continuous, positive-valued function f on $[a, b]$. The theorem asserts that for at least one c in $[a, b]$, the area under the graph of $y = f(x)$ between $x = a$ and $x = b$ is the same as the area of the rectangle of length $b - a$ and height $f(c)$.

3.4 THEOREM

Mean Value Theorem for Integrals. If f is continuous on $[a, b]$, then there is a number c in $[a, b]$ such that

$$\int_a^b f(x)\,dx = f(c)(b-a). \tag{9}$$

Proof. If f is continuous on $[a, b]$, then f has an absolute maximum M and an absolute minimum m on $[a, b]$, by Theorem 7.7 of Chapter 1. Thus for all x in $[a, b]$ we have

$$m \le f(x) \le M.$$

It then follows from Theorem 3.3 that

$$m(b-a) \le \int_a^b f(x)\,dx \le M(b-a). \tag{10}$$

If we divide (10) through by the positive number $b - a$, we obtain

$$m \le \frac{1}{b-a}\int_a^b f(x)\,dx \le M. \tag{11}$$

Since f is continuous on $[a, b]$, it takes on every value between its maximum M and minimum m on $[a, b]$, by Theorem 7.6 of Chapter 1. Thus there is some c in $[a, b]$ such that

$$f(c) = \frac{1}{b-a}\int_a^b f(x)\,dx.$$

Multiplying through by $b - a$, we obtain

$$\int_a^b f(x)\,dx = f(c)(b - a). \quad \boxed{\text{QED}}$$

The name of Theorem 3.4 is explained by the following corollary.

3.5 COROLLARY If f is continuous on $[a, b]$, then there is a number c in $[a, b]$ such that $f(c)$ is the average value of f over $[a, b]$. That is,

$$f(c) = \frac{1}{b - a}\int_a^b f(x)\,dx = \bar{f},$$

for some c in $[a, b]$.

Proof. This follows immediately from Theorem 3.4: Just divide both sides of (9) by $b - a$. $\boxed{\text{QED}}$

Example 3

Find c in $[0, 1]$ such that c^2 is the average value of $f(x) = x^2$ over $[0, 1]$.

Solution. The c we are looking for satisfies

$$f(c) = c^2 = \frac{1}{1 - 0}\int_0^1 x^2\,dx = \frac{1}{3},$$

by Example 1. Thus the only candidates are $c = \sqrt{1/3}$ and $c = -1/\sqrt{3}$. Since $-1/\sqrt{3}$ does not belong to $[0, 1]$, the desired c is $1/\sqrt{3}$. ■

The concluding result of this section gives some additional useful properties of definite integrals that are easy to prove from the mean value theorem for integrals.

3.6 THEOREM Suppose that f and g are continuous on $[a, b]$.

(a) If $0 \le f(x)$, for all $x \in [a, b]$, then

$$0 \le \int_a^b f(x)\,dx.$$

(b) If $f(x) \le g(x)$ for all $x \in [a, b]$, then

$$\int_a^b f(x)\,dx \le \int_a^b g(x)\,dx.$$

(c)
$$\left|\int_a^b f(x)\,dx\right| \le \int_a^b |f(x)|\,dx.$$

Proof. We prove (a) and leave the other two parts for Exercises 25 and 26. The mean value theorem for integrals applies to give

$$\int_a^b f(x)\,dx = f(c)(b - a) \tag{9}$$

for some c in $[a, b]$. Since $f(x) \ge 0$ for every x in $[a, b]$, we have $f(c) \ge 0$. Since $[a, b]$ is an interval, $b - a \ge 0$. Therefore it follows from (9) that

$$\int_a^b f(x)\,dx \ge 0. \quad \boxed{\text{QED}}$$

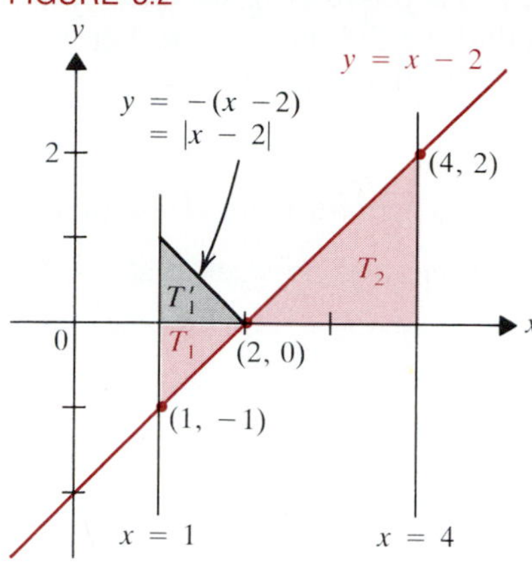

FIGURE 3.2

Example 4

Verify Theorem 3.6(c) for the function $f(x) = x - 2$ over $[1, 4]$.

Solution. In **Figure 3.2,**

$$\int_1^2 [-f(x)]\,dx = -\int_1^2 f(x)\,dx = A(T'_1) = A(T_1).$$

Thus

$$\int_1^2 f(x)\,dx = -A(T_1) = -\frac{1}{2}\cdot 1\cdot 1 = -\frac{1}{2}.$$

We therefore have

$$\int_1^4 f(x)\,dx = \int_1^2 f(x)\,dx + \int_2^4 f(x)\,dx = -A(T_1) + A(T_2)$$

$$= -\frac{1}{2} + \frac{1}{2}\cdot 2\cdot 2 = \frac{3}{2}.$$

On the other hand,

$$\int_1^4 |f(x)|\,dx = \int_1^2 -(x-2)\,dx + \int_2^4 (x-2)\,dx = A(T'_1) + A(T_2)$$

$$= 1/2 + 2 = 5/2.$$

So in this case, we do have

$$\left|\int_1^4 f(x)\,dx\right| = \frac{3}{2} \le \int_1^4 |f(x)|\,dx = \frac{5}{2}. \quad \blacksquare$$

We can use Theorem 3.6 to get a more accurate bound for the definite integral in Example 2.

Example 5

Improve the estimate of $\int_0^1 x^3\,dx$ obtained in Example 2.

Solution. For x in $[0, 1]$, we have

$$0 \le x^3 \le x^2.$$

Hence, by Theorem 3.6(b) and Example 3 of Section 1

$$0 \le \int_0^1 x^3\,dx \le \int_0^1 x^2\,dx = \frac{1}{3}. \quad \blacksquare$$

Proof of Lemma 3.2.*

Let f be integrable over $[a, b]$, and let u and l be real numbers. If for every partition $P = \{x_0, x_1, \ldots, x_n\}$ of $[a, b]$, and every choice of points z_i in $[x_{i-1}, x_i]$, $i = 1, 2, \ldots, n$, we have $l \le R_P \le u$, then

$$l \le \int_a^b f(x)\,dx \le u. \tag{12}$$

Proof. Let $I = \int_a^b f(x)\,dx$. We must prove two inequalities in (12). We show

$$I \le u \tag{13}$$

* Optional

and leave the other inequality for Exercise 27. To prove (13), we show that $I > u$ is impossible. Suppose to the contrary that we *did* have $I > u$. Let $\varepsilon = I - u > 0$. Then by Definition 2.2, there is a $\delta > 0$ such that

$$|I - R_P| < \varepsilon$$

for all choices of P and z_i such that $|P| < \delta$. Thus, by Theorem 1.2(a) of Chapter 1,

$$-\varepsilon < I - R_P < \varepsilon = I - u. \tag{14}$$

Subtracting I in (14), we get

$$-R_P < -u,$$

that is,

$$R_P > u. \tag{15}$$

But (15) contradicts the hypothesis that $R_P \le u$ for every partition P. Since the assumption that $I > u$ has led to a contradiction, we conclude that $I \le u$ must be true. QED

Exercises 4.3

In Exercises 1–12, find (a) the average value of f over $[a, b]$ and (b) the point(s) c satisfying (9) in Theorem 3.4. Use results of earlier examples and exercises as needed.

1. $f(x) = 2x + 1$ on $[0, 3]$

2. $f(x) = 1 - x$ on $[0, 3]$

3. $f(x) = 2x^2 + 1$ on $[-1, 1]$

4. $f(x) = 3x^2 - 1$ on $[0, 2]$

5. $f(x) = x^3$ on $[0, 2]$

6. $f(x) = 4x^3 - 3$ on $[0, 2]$

7. $f(x) = 9 - x^2$ on $[0, 3]$

8. $f(x) = x + x^2$ on $[0, 2]$

9. $f(x) = x^3 - x$ on $[-2, 2]$

10. $f(x) = x^2 + 1$ on $[-2, 2]$

11. $f(x) = \sqrt{4 - x^2}$ on $[0, 2]$

12. $f(x) = \sqrt{9 - x^2}$ on $[-3, 3]$

In Exercises 13–20, use Theorem 3.3 to find upper and lower bounds for $\int_a^b f(x)\,dx$.

13. $f(x) = \sin x$, $a = 0$, $b = \pi/4$

14. $f(x) = \cos x$, $a = \pi/3$, $b = 2\pi/3$

15. $f(x) = \sin x + \cos x$, $a = 0$, $b = \pi/2$

16. $f(x) = \sin x - \cos x$, $a = 0$, $b = \pi$

17. $f(x) = x^3 - 3x + 4$, $a = 0$, $b = 2$

18. $f(x) = x^3 - 6x + 2$, $a = -3$, $b = 3$

19. $f(x) = x^4 - \dfrac{16}{3}x^3 + 1$, $a = -1$, $b = 5$

20. $f(x) = x^4 + \dfrac{8}{3}x^3 + 3$, $a = -5$, $b = 5$

21. A body falls freely from rest under the influence of gravity subject to the constant acceleration $a = -g$. Recall that its velocity after t seconds is $v = -gt$, and the distance it has fallen after t seconds is $y = -\frac{1}{2}gt^2$. Find the average velocity over $[0, 2]$.

22. In Exercise 21, find the average position over $[0, 2]$.

23. If f is continuous on $[a, b]$ and positive valued there, then show that $\int_a^b f(x)\,dx$ is positive.

24. If f is continuous and negative valued on $[a, b]$, then show that $\int_a^b f(x)\,dx$ is negative.

25. Prove Theorem 3.6(a).

26. Prove Theorem 3.6(b).

27. In Lemma 3.2, show that $l \le \int_a^b f(x)\,dx$.

28. Suppose that $f(x) = k > 0$ for all x in $[a, b]$. Then what does Theorem 3.4 reduce to?

29. If f is continuous on $[a, b]$ and $\int_a^b f(x)\,dx = 0$, then show that for at least one $c \in [a, b]$, $f(c) = 0$.

30. Show that the result of Exercise 29 need no longer be true if f is just integrable on $[a, b]$. That is, give an example of a function f on an interval $[a, b]$ such that $f(x) \ne 0$ for all $x \in [a, b]$, even though $\int_a^b f(x)\,dx = 0$.

31. If f is continuous and $f(x) \ge 0$ on $[0, \pi]$, then prove that

$$\int_0^\pi f(x)\sin x\,dx \le \int_0^\pi f(x)\,dx.$$

32. If f is continuous and $f(x) \ge 0$ on $[-\pi/2, \pi/2]$, then prove that

$$\int_{-\pi/2}^{\pi/2} f(x)\cos x\,dx \le \int_{-\pi/2}^{\pi/2} f(x)\,dx.$$

33. Weighted Mean Value Theorem.* Suppose that f and g are continuous on $[a, b]$ and g does not change sign in $[a, b]$.

* This result is used to give an optional proof in Section 9.6.

(a) If $\int_a^b g(x)\,dx = 0$, then show that $\int_a^b f(x)g(x)\,dx = 0$. (*Hint:* Show that $g(x) = 0$ for all x in $[a, b]$.)

(b) Show that

$$\int_a^b f(x)g(x)\,dx = f(c)\int_a^b g(x)\,dx$$

for some c in $[a, b]$. (*Hint:* Suppose that $g(x) \geq 0$. Apply Theorem 7.7 of Chapter 1 to obtain x_1 and x_2 in $[a, b]$ such that

$$f(x_1)g(x) \leq f(x)g(x) \leq f(x_2)g(x).$$

Then use Theorem 3.6(b) and Theorem 7.6 of Chapter 1.)

34. Show that Exercise 33 is a generalization of Theorem 3.4 by obtaining Theorem 3.4 from Exercise 33 for a particular function g.

In Exercises 35–38, use Exercise 33 to obtain the given inequalities.

35. $\int_0^2 \frac{x+1}{x^2+1}\,dx \leq \int_0^2 (x+1)\,dx$

36. $\int_{-1}^1 \frac{1+x^2}{\sqrt{x^2+2}}\,dx \leq \frac{1}{\sqrt{2}}\int_{-1}^1 (1+x^2)\,dx$

37. $\frac{1}{2\sqrt{2}} \leq \int_0^1 \frac{x}{\sqrt{1+x}}\,dx \leq \frac{1}{2}$

38. $\frac{1}{3\sqrt{2}} \leq \int_0^1 \frac{x^2}{\sqrt{1+x}}\,dx \leq \frac{1}{3}$

39. Suppose that f is continuous on $[a, b]$ and $\int_a^b f(x)g(x)\,dx = 0$ for every continuous function g on $[a, b]$. Show that $f(x) = 0$ for all $x \in [a, b]$.

40. Given that

$$\int_{2\pi}^{4\pi} \sin x\,dx = 0,$$

show that

$$\int_{2\pi}^{4\pi} \frac{\sin x}{x}\,dx = 0.$$

4.4 The Fundamental Theorem of Calculus

FIGURE 4.1

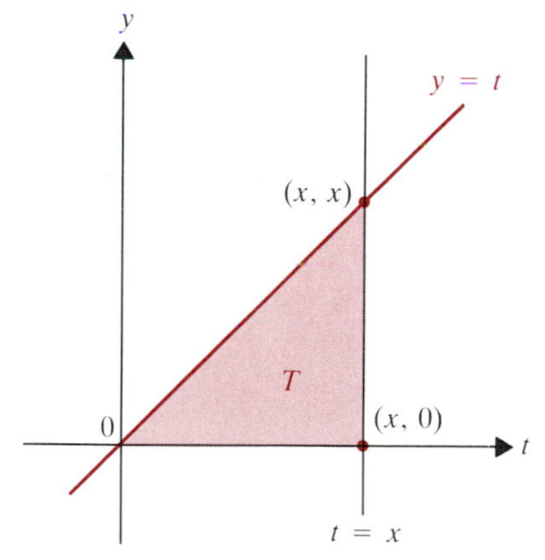

As its title suggests, this is a climactic section. It presents a two-part theorem which says that differentiation and integration are essentially *inverse* operations. These two basic concepts of calculus are thus intimately linked.

To indicate how this basic result came to be recognized, we first calculate the area $A(x)$ of the triangular region T in **Figure 4.1.** It is the region in the ty-plane lying under the graph of the line $y = t$ between $t = 0$ and $t = x$, where $x > 0$. Since the area of a triangle is half the base times the height, we have

$$A(x) = \frac{1}{2}\cdot x \cdot x = \frac{1}{2}x^2 = \int_0^x t\,dt.$$

Noting that

$$\frac{d}{dx}\left(\frac{1}{2}x^2\right) = x,$$

we see that the function F defined by

$$F(x) = \int_0^x t\,dt$$

is an *antiderivative* of the function f given by $f(t) = t$. Newton used ideas of his colleague at Cambridge, Isaac Barrow (1630–1677), to show more generally that the function G defined by

$$G(x) = \int_0^x t^k\,dt, \qquad k > 0,$$

is an antiderivative of $f(t) = t^k$ on any interval where f is defined. With the help of the mean value theorem for integrals, we will establish a still more general result of this type.

4.1 THEOREM

Fundamental Theorem of Calculus, Part 1. Suppose that f is continuous on an open interval I containing the closed interval $[a, b]$. For $x \in I$, define G by

$$G(x) = \int_a^x f(t)\,dt. \tag{1}$$

Then G is an antiderivative of f on $[a, b]$. That is, for any x in $[a, b]$,

$$G'(x) = \frac{d}{dx}\left[\int_a^x f(t)\,dt\right] = f(x). \tag{2}$$

Proof. We need to show that

$$\lim_{h \to 0} \frac{G(x+h) - G(x)}{h} = f(x) \tag{3}$$

for any x in $[a, b]$. Since $h \to 0$ in (3), we consider only values of h small enough that $x + h$ is in I. Then

$$G(x+h) - G(x) = \int_a^{x+h} f(t)\,dt - \int_a^x f(t)\,dt = \int_x^{x+h} f(t)\,dt, \tag{4}$$

by Theorem 2.7 in the form

$$\int_a^{x+h} f(t)\,dt = \int_a^x f(t)\,dt + \int_x^{x+h} f(t)\,dt.$$

[Here, x plays the role of b and $x + h$ plays the role of c in Equation (8), p. 243.] Since x and $x + h$ both belong to I, where f is continuous, we can apply Theorem 3.4 to get

$$\int_x^{x+h} f(t)\,dt = f(c)(x + h - x) = f(c)h \tag{5}$$

for some c between x and $x + h$, inclusive. Substituting (5) into (4), we have

$$G(x+h) - G(x) = f(c)h. \tag{6}$$

Division of (6) through by h gives

$$\frac{G(x+h) - G(x)}{h} = f(c). \tag{7}$$

Now let $h \to 0$ in (7). Since c is between x and $x + h$, by the sandwich theorem for limits (Theorem 5.7 of Chapter 1) we have $c \to x$. Also $f(c) \to f(x)$ as $c \to x$, because f is continuous on $[a, b]$. Consequently,

$$G'(x) = \lim_{h \to 0} \frac{G(x+h) - G(x)}{h} = f(x). \quad \text{QED}$$

Formula (1) says that if we *first integrate* a continuous function f and *next differentiate* the result, then we arrive back at f. That is, differentiation *undoes* integration. The next example illustrates how Theorem 4.1 can make seemingly difficult differentiation problems quite simple.

Example 1

If $G(x) = \int_0^x [t \sin t^2 + t^5 \cos(t^2 + 1)]\,dt$, then find $G'(x)$.

Solution. We don't yet know any way to evaluate the integral for $G(x)$. Even if we use later methods, it is tedious to find such a formula. But Theorem 4.1 makes a formula unnecessary, because it says that

$$G'(x) = x \sin x^2 + x^5 \cos(x^2 + 1). \quad ■$$

If we combine Theorem 4.1 with the general chain rule for derivatives in Section 2.6, then we get a formula for differentiating still more complicated integrals. Suppose that g is differentiable on $[a, b]$ and f is continuous on the range of g. If

$$G(x) = \int_a^{g(x)} f(t)\,dt, \tag{8}$$

then to differentiate G, we let $u = g(x)$ and $y = G(x)$. Then (8) becomes

$$y = \int_a^u f(t)\,dt.$$

The general chain rule for derivatives and Theorem 4.1 then give

$$\frac{dy}{dx} = \frac{dy}{du}\frac{du}{dx} = \frac{d}{du}\left[\int_a^u f(t)\,dt\right]\frac{du}{dx} = f(u)\frac{du}{dx}.$$

That is,

$$G'(x) = f(g(x))\,g'(x). \tag{9}$$

Example 2

Find $\dfrac{d}{dx}\left[\displaystyle\int_{-2}^{\sqrt{1+x^2}} \frac{\sin^2(3t^3 + 1)}{(1 + t^2)^{3/2}}\,dt\right]$.

Solution. Here $g(x) = \sqrt{1 + x^2} = (1 + x^2)^{1/2}$ so that

$$g'(x) = \frac{1}{2}(1 + x^2)^{-1/2}(2x) = \frac{x}{\sqrt{1 + x^2}}.$$

Then (9) gives

$$\frac{d}{dx}\left[\int_{-2}^{\sqrt{1+x^2}} \frac{\sin^2(3t^3 + 1)}{(1 + t^2)^{3/2}}\,dt\right] = \frac{\sin^2[3(1 + x^2)^{3/2} + 1]}{\{1 + [(1 + x^2)^{1/2}]^2\}^{3/2}} \cdot \frac{x}{(1 + x^2)^{1/2}}$$

$$= \frac{x \sin^2[3(1 + x^2)^{3/2} + 1]}{(1 + x^2)^{1/2}(2 + x^2)^{3/2}}. \quad ■$$

The idea behind (9) can also be used to differentiate functions defined by integrals with two variable limits.

Example 3

Find $\dfrac{d}{dx}\left[\displaystyle\int_{\sin 2x}^{\cos 2x} \sqrt{1 + t^2}\,dt\right]$.

Solution. We first express the integral as a sum of integrals of the form (8). Theorem 2.7 gives

$$\int_{\sin 2x}^{\cos 2x} \sqrt{1 + t^2}\,dt = \int_0^{\cos 2x} \sqrt{1 + t^2}\,dt - \int_0^{\sin 2x} \sqrt{1 + t^2}\,dt.$$

Each integral on the right has the form (8), with $g_1(x) = \cos 2x$ and $g_2(x) = \sin 2x$. Thus

$$g_1'(x) = -2\sin 2x \qquad \text{and} \qquad g_2'(x) = 2\cos 2x.$$

Then we get from (9)

$$\begin{aligned}\frac{d}{dx}\left[\int_{\sin 2x}^{\cos 2x}\sqrt{1+t^2}\,dt\right] &= \frac{d}{dx}\left[\int_0^{\cos 2x}\sqrt{1+t^2}\,dt\right] - \frac{d}{dx}\left[\int_0^{\sin 2x}\sqrt{1+t^2}\,dt\right]\\ &= \sqrt{1+\cos^2 2x}\cdot(-2\sin 2x) - \sqrt{1+\sin^2 2x}\cdot 2\cos 2x\\ &= -2\sin 2x\sqrt{1+\cos^2 2x} - 2\cos 2x\sqrt{1+\sin^2 2x}. \quad ■\end{aligned}$$

We are ready to derive the second part of the fundamental theorem, which makes it possible to evaluate many definite integrals by antidifferentiation.

4.2 THEOREM

Fundamental Theorem of Calculus, Part 2. Suppose that f is continuous on an open I containing $[a, b]$. If F is any antiderivative of f on $[a, b]$, then

(10) $$\int_a^b f(x)\,dx = F(b) - F(a).$$

Proof. Let G be defined by Equation (1):

(1) $$G(x) = \int_a^x f(t)\,dt.$$

Then Theorem 4.1 says that G is an antiderivative of f on $[a, b]$. By Theorem 9.2 of Chapter 3, if F is *any* antiderivative of f on $[a, b]$, then

$$F(x) = G(x) + C$$

for some constant C. Therefore

$$\begin{aligned}F(b) - F(a) &= G(b) + C - [G(a) + C] = G(b) - G(a)\\ &= \int_a^b f(x)\,dx - \int_a^a f(x)\,dx = \int_a^b f(x)\,dx - 0\\ &= \int_a^b f(x)\,dx. \quad \text{QED}\end{aligned}$$

Theorem 4.2 provides an easy way to work out definite integrals of any function f that we can antidifferentiate. Since we can use *any* antiderivative in (10), we use the simplest one—the one with constant of integration $C = 0$. Because (10) is so widely used, special notation has been developed for its right side, namely, $F(x)|_a^b$ or $F(x)]_a^b$ or $[F(x)]_a^b$. In these notations, (10) becomes

(11) $$\int_a^b f(x)\,dx = F(x)\Big|_a^b = F(x)\Big]_a^b = \Big[F(x)\Big]_a^b.$$

The next example shows how Theorem 4.2 is used to work out definite integrals.

Example 4

Evaluate

(a) $\int_1^3 (x^2 - 4x + 5)\,dx$ (b) $\int_0^8 (2x^{2/3} - x^{1/3} + 1)\,dx$

(c) $\int_1^4 \left(x\sqrt{x} - x - 2\sqrt{x} + \frac{2}{x^2}\right) dx$ (d) $\int_0^{\pi/2} \cos x \, dx$

(e) $\int_{\pi/4}^{\pi/3} \tan^2 x \, dx$

Solution. (a) The power rule for antiderivatives (p. 210) gives

$$\begin{aligned}\int_1^3 (x^2 - 4x + 5)\, dx &= \frac{x^3}{3} - 4\cdot\frac{x^2}{2} + 5x\bigg|_1^3 \\ &= \left[\frac{3^3}{3} - 4\cdot\frac{3^2}{2} + 5\cdot 3\right] - \left[\frac{1^3}{3} - 4\cdot\frac{1^2}{2} + 5\cdot 1\right] \\ &= 9 - 18 + 15 - \left(\frac{1}{3} - 2 + 5\right) \\ &= 6 - \frac{10}{3} = \frac{8}{3}.\end{aligned}$$

(b)
$$\begin{aligned}\int_0^8 (2x^{2/3} - x^{1/3} + 1)\, dx &= 2\frac{x^{5/3}}{5/3} - \frac{x^{4/3}}{4/3} + x\bigg]_0^8 \\ &= \frac{6}{5}8^{5/3} - \frac{3}{4}8^{4/3} + 8 - 0 \\ &= \frac{6}{5}\cdot 32 - \frac{3}{4}\cdot 16 + 8 \\ &= \frac{192}{5} - 12 + 8 = \frac{172}{5}.\end{aligned}$$

(c)
$$\begin{aligned}\int_1^4 \left(x\sqrt{x} - x - 2\sqrt{x} + \frac{2}{x^2}\right) dx &= \int_1^4 (x^{3/2} - x - 2x^{1/2} + 2x^{-2})\, dx \\ &= \left[\frac{x^{5/2}}{5/2} - \frac{x^2}{2} - 2\frac{x^{3/2}}{3/2} + 2\frac{x^{-1}}{-1}\right]_1^4 \\ &= \left[\frac{2}{5}4^{5/2} - \frac{16}{2} - \frac{4}{3}4^{3/2} - 2\cdot\frac{1}{4}\right] \\ &\quad - \left[\frac{2}{5} - \frac{1}{2} - \frac{4}{3} - 2\right] \\ &= \left[\frac{64}{5} - 8 - \frac{32}{3} - \frac{1}{2}\right] \\ &\quad - \left[\frac{2}{5} - \frac{1}{2} - \frac{4}{3} - 2\right] \\ &= \frac{62}{5} - \frac{28}{3} - 6 = -\frac{44}{15}.\end{aligned}$$

(d) $\int_0^{\pi/2} \cos x\, dx = \sin x\Big|_0^{\pi/2} = \sin\frac{\pi}{2} - \sin 0 = 1 - 0 = 1.$

(e)
$$\begin{aligned}\int_{\pi/4}^{\pi/3} \tan^2 x\, dx &= \int_{\pi/4}^{\pi/3} (\sec^2 x - 1)\, dx = \tan x - x\bigg]_{\pi/4}^{\pi/3} \\ &= \sqrt{3} - \frac{\pi}{3} - \left(1 - \frac{\pi}{4}\right) = \sqrt{3} - 1 - \frac{\pi}{12}.\end{aligned}$$ ■

We are now equipped to solve a wider class of area problems than those treated in Section 1 and a wider class of average value problems than those considered in Section 3.

Example 5

(a) Find the area of the region bounded by $y = x^4$, $y = 1$, and the $y-$axis.

(b) Find the average value of $\sin x$ over $[0, \pi/2]$.

FIGURE 4.2

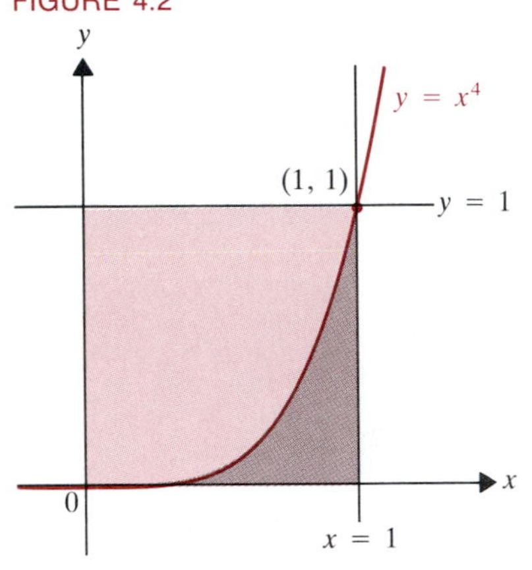

Solution. (a) The region is shown in **Figure 4.2** shaded in color. Its area A is $A_1 - A_2$, where A_1 is the area of the square formed by the coordinate axes and the lines $x = 1$ and $y = 1$ and where A_2 is the area of the darker shaded region. Since A_2 is the area under $y = x^4$ between $x = 0$ and $x = 1$, we have from Definitions 2.2 and 1.4 and Theorem 4.2,

$$A_2 = \int_0^1 x^4\,dx = \frac{1}{5}x^5\Big|_0^1 = \frac{1}{5} - 0 = \frac{1}{5}.$$

Hence,

$$A = A_1 - A_2 = 1 - \frac{1}{5} = \frac{4}{5}.$$

(b) By Definition 3.1, the average value of $f(x) = \sin x$ over the closed interval $[0, \pi/2]$ is

$$\bar{f} = \frac{1}{\pi/2 - 0}\int_0^{\pi/2} \sin x\,dx = \frac{2}{\pi}\Big[-\cos x\Big]_0^{\pi/2}$$

$$= \frac{2}{\pi}[-0 + 1] = \frac{2}{\pi} \approx 0.63662. \blacksquare$$

Our final example illustrates how Theorem 2.8 can be used in conjunction with Theorem 4.2 to simplify evaluation of definite integrals over intervals that contain 0.

Example 6

Evaluate

(a) $\int_{-2}^{2} (x^4 + x^2)\,dx$ (b) $\int_{-\pi/2}^{\pi/2} \sin x\,dx$

(c) $\int_{-1}^{2} (x^3 + x)\,dx$ (d) $\int_{-2}^{2} (x^2 + x)\,dx$

Solution. (a) This function is even. So by Theorem 2.8(a),

$$\int_{-2}^{2} (x^4 + x^2)\,dx = 2\int_0^2 (x^4 + x^2)\,dx = 2\left[\frac{x^5}{5} + \frac{x^3}{3}\right]_0^2$$

$$= 2\left[\frac{32}{5} + \frac{8}{3}\right] - 0 = \frac{272}{15}.$$

Note that use of Theorem 2.8(a) avoids evaluation of the antiderivative at $x = -2$.

(b) The function is odd, since $\sin(-x) = -\sin x$. Then Theorem 2.8(b) tells us immediately that

$$\int_{-\pi/2}^{\pi/2} \sin x \, dx = 0.$$

(c) Note that the function $g(x) = x^3 + x$ is odd. Then Theorems 2.6 and 2.8 give

$$\begin{aligned}\int_{-1}^{2} (x^3 + x)\, dx &= \int_{-1}^{1} (x^3 + x)\, dx + \int_{1}^{2} (x^3 + x)\, dx \\ &= 0 + \left[\frac{x^4}{4} + \frac{x^2}{2}\right]_1^2 = 0 + [4 + 2] - \left[\frac{1}{4} + \frac{1}{2}\right] \\ &= 6 - \frac{3}{4} = \frac{21}{4}.\end{aligned}$$

(d) $$\begin{aligned}\int_{-2}^{2} (x^2 + x)\, dx &= \int_{-2}^{2} x^2\, dx + \int_{-2}^{2} x\, dx \\ &= 2\int_0^2 x^2\, dx + 0 \qquad \textit{by Theorem 2.8(b)} \\ &= 2\left[\frac{x^3}{3}\right]_0^2 = \frac{16}{3}. \quad \blacksquare\end{aligned}$$

Differentiation and Integration as Inverse Operations*

At the start of the section, we mentioned that differentiation and integration are *essentially* inverse operations. This seems to have first been observed geometrically by Barrow in a series of lectures given around 1663 at Cambridge. Newton and Leibniz showed it analytically shortly thereafter. To describe the inverse relationship more precisely, we introduce the indefinite integration operator I_x by defining

$$I_x(f) = \int_a^x f(t)\, dt,$$

if f is a continuous function on an open interval J containing $[a, b]$. Then Theorem 4.1 says that

$$D_x[I_x(f)] = D_x \int_a^x f(t)\, dt = f(x),$$

that is,

(12) $$(D_x \circ I_x) f(x) = f(x),$$

where D_x is the differentiation operator. Equation (12) says that D_x undoes the effect of I_x. For D_x and I_x to be inverse operations, I_x should also undo the effect of D_x. Suppose that f is a differentiable function on J, so that $D_x[f(x)] = f'(x)$ is defined for x in $[a, b]$. Suppose also that f' is continuous on J. Then we have

$$\begin{aligned}(I_x \circ D_x)(f(x)) &= I_x[D_x(f(x))] = I_x(f'(x)) \\ &= \int_a^x f'(t)\, dt = f(x)\Big]_a^x \qquad \textit{by Theorem 4.2}\end{aligned}$$

(13) $$(I_x \circ D_x)(f(x)) = f(x) - f(a).$$

Thus $(I_x \circ D_x)(f(x))$ is $f(x)$ only if $f(a)$ happens to be 0. So I_x doesn't *quite* undo D_x in all cases. Consequently, we can't say that I_x and D_x are actually

* Optional

inverse operations. Nevertheless, since $(I_x \circ D_x)(f(x))$ differs from $f(x)$ *only* by the constant $-f(a)$, we say that differentiation and integration are *essentially* inverse operations.

> **HISTORICAL NOTE**
>
> The development of the fundamental theorem given here closely parallels Cauchy's presentation in his 1823 text, *Cours d' analyse.* While the fact that integrals could be evaluated by antidifferentiation was known (and used) by both Newton and Leibniz a century and a half before Cauchy, it was not until the work of the great French mathematician that this was formulated precisely and proved rigorously.

Exercises 4.4

In Exercises 1–20, use Theorem 4.2 to evaluate the definite integrals.

1. $\int_1^2 (x^2 - x + 1)\,dx$
2. $\int_{-1}^2 (x^2 - 3x + 2)\,dx$
3. $\int_0^2 (x^3 - x^2 + 2)\,dx$
4. $\int_1^2 \frac{2x^2 - 1}{x^2}\,dx$
5. $\int_{-2}^2 (x^2 + 1)\,dx$
6. $\int_{-1}^1 (\tan^3 x - x)\,dx$
7. $\int_{-\pi/4}^{\pi/4} (\tan^2 x + 1)\,dx$
8. $\int_{-1}^2 (x^3 + 2x)\,dx$
9. $\int_{-1}^2 (2x^2 - 8)\,dx$
10. $\int_{-\pi/2}^{\pi/2} (\cos x + 1)\,dx$
11. $\int_{-\pi/2}^{\pi/2} (x + \sin x)\,dx$
12. $\int_1^4 \sqrt{x}(1 - x)\,dx$
13. $\int_1^4 \frac{x - 1}{\sqrt{x}}\,dx$
14. $\int_1^3 \frac{2x + 3}{\sqrt{x}}\,dx$
15. $\int_0^1 \frac{x^2 - 1}{x + 1}\,dx$
16. $\int_0^2 \frac{4 - x^2}{2 + x}\,dx$
17. $\int_0^{\pi/4} \sec x \tan x\,dx$
18. $\int_{\pi/6}^{\pi/3} \csc^2 x\,dx$
19. $\int_{\pi/6}^{\pi/3} \cot^2 x\,dx$
20. $\int_{\pi/4}^{\pi/3} \tan x \sec^2 x\,dx$

In Exercises 21–32, find the derivative of the function defined by the given integral.

21. $F(x) = \int_0^x \sqrt{t^2 - t + 1}\,dt$
22. $F(x) = \int_2^x \sqrt[3]{t^3 - 3t^2 + 1}\,dt$
23. $G(x) = \int_1^x \frac{1}{\sin^2 t + t^{3/2}}\,dt$
24. $G(x) = \int_0^x \frac{1}{\cos^2 t + t^3}\,dt$
25. $F(x) = \int_0^{x^2 + x} \sqrt{1 + t^3}\,dt$
26. $G(x) = \int_0^{\sqrt{1 + x^2}} (t^2 + 1)\,dt$
27. $H(x) = \int_x^3 (\sin^3 t + t^2)\,dt$
28. $H(x) = \int_x^1 (\tan^2 t + t^3)\,dt$
29. $H(x) = \int_{x^2}^{1+x} \sqrt{1 + t^2}\,dt$
30. $H(x) = \int_{2x}^{x^2} \sqrt{4 + t^2}\,dt$
31. $H(x) = \int_{\sqrt{1-x^2}}^{\sec 2x} (t^2 - 3)\,dt$
32. $H(x) = \int_{\cot^2 x}^{\sqrt{4+x^2}} (t^2 - 4)\,dt$

In Exercises 33–36, find the area under the graph of the given $y = f(x)$ between $x = a$ and $x = b$. Make a sketch

33. $f(x) = x^2 + x + 1,\ a = 0,\ b = 2$
34. $f(x) = x^3 + 1,\ a = 0,\ b = 1$
35. $f(x) = \cos x,\ a = 0,\ b = \pi/2$
36. $f(x) = \sec^2 x,\ a = 0,\ b = \pi/4$

In Exercises 37–40, find the average value of the given function between $x = a$ and $x = b$.

37. $f(x) = x^2 - 3x - 2,\ a = 0,\ b = 2$
38. $f(x) = x^3 + x - 1,\ a = -1,\ b = 2$
39. $f(x) = x + \cos x,\ a = 0,\ b = \pi/2$
40. $f(x) = 2x - \sin x,\ a = -\pi/4,\ b = \pi/6$
41. Let $y = y(t)$ give the position at time t of a body moving along the t-axis. Recall that the average velocity of the body between $t = t_0$ and $t = t_1$ is
$$\frac{\Delta y}{\Delta t} = \frac{y(t_1) - y(t_0)}{t_1 - t_0}.$$
Show that the average value of the velocity function $v(t) = dy/dt$ over $[t_0, t_1]$, in the sense of Section 3, is indeed $\Delta y/\Delta t$. (Assume that v and y are continuous functions of t.)

42. Repeat Exercise 41 for the average acceleration.

43. Suppose in Exercise 41 that $v(t) = gt$. Find the average velocity over $[0, 4]$.

44. In Exercises 41 and 42, show that

$$v(t_1) = v(t_0) + \int_{t_0}^{t_1} a(t)\,dt.$$

45. We obtained Theorem 4.2 from Theorem 4.1. It can be proved without using Theorem 4.1 as follows. Let P_n be a uniform partition of $[a, b]$ into n subintervals.

(a) For each subinterval $[x_{i-1}, x_i]$ determined by P_n, show that

(14) $$F(x_i) - F(x_{i-1}) = f(c_i)\,\Delta x$$

for some $c_i \in (x_{i-1}, x_i)$.

(b) Sum Equations (14) over $i = 1, 2, \ldots, n$ to show that

$$\sum_{i=1}^{n} f(c_i)\,\Delta x = F(b) - F(a).$$

(c) Take the limits in (b) as $n \to \infty$ to obtain (10).

46. Find an f and an a satisfying

$$\int_a^x tf(t)\,dt = \cos x + x \sin x - \frac{x^2}{2} - 1.$$

47. Find an f and an a satisfying

$$\int_a^x f(t)\,dt = \sin x - \frac{1}{2}.$$

4.5 Change of Variable

The fundamental theorem of calculus shows why the indefinite integral notation $\int f(x)\,dx$ introduced in Section 3.9 is appropriate. Indeed, since Theorems 4.1 and 4.2 say that differentiation and integration are essentially inverse operations, we will henceforth use the indefinite integral notation

$$\int f(x)\,dx$$

exclusively to represent the most general antiderivative of f over an interval I. Thus we write

(1) $$\int f(x)\,dx = F(x) + C,$$

where F is an antiderivative of f on I and C is an arbitrary constant.

In using indefinite integral notation, it is important to keep firmly in mind the the distinction between the *real number*

(2) $$\int_a^b f(x)\,dx,$$

and the indefinite integral

(3) $$\int f(x)\,dx.$$

Recall from Section 3.9 that (3) stands for a family of *functions*

(4) $$F(x) + C,$$

where $F'(x) = f(x)$. One remarkable thing about the fundamental theorem is that it *relates* (2) and (3): we evaluate the real number (2) by calculating the difference between the values of *any* one of the functions (4) at b and at a.

To evaluate a definite integral

$$\int_a^b f(x)\,dx$$

by the fundamental theorem, we need to find an antiderivative F of f. We therefore need to work out the indefinite integral

$$\int f(x)\,dx.$$

In the last section, we considered functions for which this could be done directly. However, we often need to use the method of substitution based on the chain rule for antidifferentiation. (Before proceeding, you may find it helpful to review Section 3.9, especially p. 213.) To illustrate, we consider the definite integral of a function we antidifferentiated in Example 2 of Section 3.9.

Example 1

Evaluate $\int_0^2 x^2\sqrt{1+x^3}\,dx$.

Solution. The first step in evaluating an indefinite integral like $\int x^2\sqrt{1+x^3}\,dx$ is to identify a factor $h(x)$ that is (up to a constant multiple) the derivative of some other part $g(x)$ of the integrand. We then make the substitution $u = g(x)$. In this integrand, x^2 is (except for the constant multiple 3) the derivative of $1 + x^3$. The next step here then is to let $u = 1 + x^3$. From this, we calculate $du = 3x^2\,dx$, and express $x^2\,dx$ in terms of du:

$$x^2\,dx = \frac{1}{3}\,du.$$

We then substitute into the given integral and antidifferentiate:

$$\int x^2\sqrt{1+x^3}\,dx = \int \frac{1}{3}\sqrt{u}\,du = \frac{1}{3}\int u^{1/2}\,du = \frac{1}{3}\frac{u^{3/2}}{3/2} + C.$$

Finally we express this answer in terms of x:

$$\int x^2\sqrt{1+x^3}\,dx = \frac{2}{9}(1+x^3)^{3/2} + C. \tag{5}$$

We can now apply Theorem 4.2 to evaluate the definite integral:

$$\int_0^2 x^2\sqrt{x^3+1}\,dx = \frac{2}{9}(x^3+1)^{3/2}\Big]_0^2$$

$$= \frac{2}{9}(8+1)^{3/2} - \frac{2}{9}(0+1)^{3/2}$$

$$= \frac{2}{9}[9^{3/2} - 1^{3/2}] = \frac{2}{9}[27-1] = \frac{52}{9}. \tag{6}$$ ■

There is a handy shortcut that can eliminate some of the work done in Example 1. In evaluating $\int f(g(x))\,g'(x)\,dx$, we use a change of variable of the form $u = g(x)$. So it is reasonable to expect the same change of variable to transform the given definite integral to a simpler one:

$$\int_a^b f(g(x))\,g'(x)\,dx \rightarrow \int_c^d f(u)\,du,$$

for some new limits of integration c and d. In Example 1, knowledge of what c and d are would have allowed us to bypass the return to the variable x in (5). We could instead have gone directly to the substitution of the limits of integration $u = d$ and $u = c$ to arrive more quickly at (6). The main result of this section is that c and d are easily calculable from a and b and the substitution formula $u = g(x)$. They are just the values of u that correspond to the given limits of integration $x = a$ and $x = b$.

5.1
THEOREM

Change of Variable in Definite Integrals. Suppose that g' is continuous on the interval $[a, b]$, that $c = g(a)$, and that $d = g(b)$. If f is continuous on $[c, d]$, then

$$\int_a^b f(g(x))\,g'(x)\,dx = \int_c^d f(u)\,du \tag{7}$$

where $u = g(x)$.

Proof. Theorem 4.2 applies to the left side of (7) to give

$$\int_a^b f(g(x))\,g'(x)\,dx = \left[\int f(g(x))\,g'(x)\,dx\right]_a^b. \tag{8}$$

Theorem 9.3 of Chapter 3 says that

$$\int f(g(x))\,g'(x)\,dx = F(g(x)) + C,$$

where F is an antiderivative of f. Thus (8) becomes

$$\int_a^b f(g(x))\,g'(x)\,dx = F(g(b)) - F(g(a)) = F(d) - F(c)$$

$$= F(u)\Big]_c^d = \int_c^d f(u)\,du. \quad \text{QED} \qquad \textit{by Theorem 4.2}$$

To work out a definite integral by substitution, let u equal an appropriate part $g(x)$ of the function to be integrated, as in Steps 1 and 2 of the general procedure on p. 213. In addition, calculate $c = g(a)$ and $d = g(b)$, which are just the values of u corresponding to $x = a$ and $x = b$, and use (7) to transform the given definite integral. Then find an antiderivative F of f and compute $F(d) - F(c)$. To illustrate, we rework Example 1.

Example 2

Evaluate $\int_0^2 x^2\sqrt{x^3 + 1}\,dx$.

Solution. As before, we let

$$u = x^3 + 1 = g(x) \tag{9}$$

and compute

$$du = 3x^2\,dx \rightarrow x^2\,dx = \frac{1}{3}\,du.$$

We also need the u-limits of integration corresponding to $x = 0$ and $x = 2$. From (9), these are

$$g(0) = 0^3 + 1 = 1 = c \qquad \text{and} \qquad g(2) = 2^3 + 1 = 9 = d.$$

Then (7) gives

$$\int_0^2 x^2\sqrt{x^3 + 1}\,dx = \int_1^9 \frac{1}{3}\,u^{1/2}\,du = \frac{1}{3}\,\frac{u^{3/2}}{3/2}\Big|_1^9$$

$$= \frac{2}{9}\,[9^{3/2} - 1^{3/2}] = \frac{52}{9}. \quad ■$$

Comparison of the length of the solutions of Examples 1 and 2 reveals the advantage in using Theorem 5.1. The next example shows how the technique works on a trigonometric integral.

Example 3

Evaluate $\int_0^{\pi/4} \sin^2 x \cos x \, dx$.

Solution. Here $\cos x$ is the derivative of $\sin x$, so we let

$$u = \sin x = g(x).$$

Then

$$du = \cos x \, dx.$$

When $x = 0$, we have $u = \sin 0 = 0$. When $x = \pi/4$, we find $u = \sin \pi/4 = 1/\sqrt{2}$. Therefore

$$\int_0^{\pi/4} \sin^2 x \cos x \, dx = \int_0^{1/\sqrt{2}} u^2 \, du = \frac{u^3}{3}\bigg|_0^{1/\sqrt{2}}$$

$$= \frac{1}{3}\left[\frac{1}{2\sqrt{2}} - 0\right] = \frac{1}{6\sqrt{2}}. \quad ■$$

As you know, your body breaks down the food you eat to produce energy. This process is called *metabolism*, and some of that energy in the form of heat is what causes your body to be at the warm temperature of 37°C (98.6°F). The *basal metabolism* is the amount of energy required for the maintenance of the body's vital functions, such as breathing, digestion, and kidney function.

A person's total basal metabolism over a 24-hour period can be computed from the *basal metabolism rate* $R(t)$, in kilocalories (kcal) per hour. Let $M(t)$ be the number of kcal in total basal metabolism from 0 to t. Then

$$M'(t) = R(t).$$

If we partition $[0, 24]$ into n small subintervals $[t_{i-1}, t_i]$ of length $\Delta t_i = t_i - t_{i-1}$, then for any $z_i \in [t_{i-1}, t_i]$,

$$\frac{M(t_i) - M(t_{i-1})}{\Delta t_i} \approx R(z_i) \to M(t_i) - M(t_{i-1}) \approx R(z_i)\Delta t_i.$$

Summing from $i = 0$ to n and noting that the sum of terms on the left is telescoping [Theorem 1.8(d)], we have

$$M(24) - M(0) = \sum_{i=1}^{n} [M(t_i) - M(t_{i-1})] \approx \sum_{i=1}^{n} R(x_i)\Delta t_i. \tag{10}$$

The right-most sum in (10) is a Riemann sum for the basal metabolism rate function. The approximation in (10) stands to become better as $n \to \infty$. Since $M(0) = 0$, we define the ***total basal metabolism*** over a one-day period to be

$$M = \int_0^{24} R(t) \, dt.$$

Example 4

The basal metabolism rate in kcal/h of a male calculus student is approximated by

$$R(t) = \frac{250}{3} - \frac{1}{6}\cos\frac{\pi t}{12},$$

where t is the time in hours measured from 4:00 AM. What is the student's total basal metabolism over a 24-hour day?

Solution. From the above discussion, we have

$$M = \int_0^{24} \left[\frac{250}{3} - \frac{1}{6}\cos\frac{\pi t}{12}\right] dt$$

$$= \int_0^{24} \frac{250}{3}\, dt - \int_0^{24} \frac{1}{6}\cos\frac{\pi t}{12}\, dt$$

$$= \frac{250}{3}\int_0^{24} dt - \frac{1}{6}\int_0^{24} \cos\frac{\pi t}{12}\, dt$$

Let $u = \frac{\pi t}{12}$, so $du = \frac{\pi}{12} dt$ and $\frac{12}{\pi} du = dt$.
Note that $u = 0$ when $t = 0$, and $u = 2\pi$ when $t = 24$.

$$= \frac{250}{3}\Big[t\Big]_0^{24} - \frac{1}{6}\int_0^{2\pi} (\cos u)\left(\frac{12}{\pi}\, du\right) = \frac{250}{3}[24 - 0] - \frac{2}{\pi}\int_0^{2\pi} \cos u\, du$$

$$= 250 \cdot 8 - \frac{2}{\pi}\Big[\sin u\Big]_0^{2\pi} = 2000 - 0 = 2000. \blacksquare$$

The student therefore needs food supplying about 2000 kcal of energy per day. As you are aware, eating food with a *higher* caloric content leads to storage of unburned energy in the form of fat. Eating food with a *lower* caloric content causes stored energy to be broken down, producing a loss of weight. [*Note:* Nutritionists measure energy content of foods by kcal, which they call *large* calories. The nutrition information found on cereal boxes, peanut butter jars, ice cream cartons, and so forth is given in terms of *these* (large) calories, not the (small) calories of physics or chemistry, which measure the heat needed to raise the temperature of 1 gram of water 1°C.]

Exercises 4.5

In Exercises 1–20, evaluate the given definite integral.

1. $\int_0^1 \sqrt{1 + 3x}\, dx$
2. $\int_{5/3}^{10/3} \sqrt{3x - 1}\, dx$
3. $\int_0^2 x(x^2 + 1)^3\, dx$
4. $\int_0^1 x(x^2 - 1)^3\, dx$
5. $\int_0^{\pi/4} \cos(\pi - x)\, dx$
6. $\int_0^{\pi/6} \sin(\pi - x)\, dx$
7. $\int_0^{\pi/4} \sin^3 2x \cos 2x\, dx$
8. $\int_0^{\pi/3} \cos^3 3x \sin 3x\, dx$
9. $\int_{-1}^1 (t^2 - 1)^2 t\, dt$
10. $\int_{-1}^1 (t^2 - 1)^3 t\, dt$
11. $\int_1^4 \frac{1}{\sqrt{t}(\sqrt{t} + 1)^2}\, dt$
12. $\int_9^{16} \frac{1}{\sqrt{t}(\sqrt{t} - 2)^{3/2}}\, dt$
13. $\int_0^4 \frac{t}{(t^2 + 9)^2}\, dt$
14. $\int_0^3 \frac{t}{\sqrt{t^2 + 16}}\, dt$
15. $\int_0^{\pi/2} \frac{\cos t}{(1 + \sin t)^3}\, dt$
16. $\int_1^4 \frac{(\sqrt{x} - 1)^3}{\sqrt{x}}\, dx$
17. $\int_0^{\pi/4} x^2 \sec x^3 \tan x^3\, dx$
18. $\int_{\pi^2/36}^{\pi^2/4} \frac{\cot\sqrt{x} \csc\sqrt{x}}{\sqrt{x}}\, dx$
19. $\int_0^{\pi/3} \frac{\tan x}{\sqrt{\sec x}}\, dx$
20. $\int_{\pi/6}^{\pi/4} \frac{\sec^2 x}{\cot x}\, dx$

In Exercises 21–24, find the area of the given region. Make a sketch.

21. Above the x-axis, below $y = x/\sqrt{1 + x^2}$ between $x = 0$ and $x = \sqrt{3}$.
22. Above the x-axis, below $y = \sqrt{1 + 1/x}/x^2$, between $x = 1$ and $x = 2$.
23. Above the x-axis, below $y = \sqrt{\cos x} \sin x$ between $x = 0$ and $x = \pi/2$.
24. Above the x-axis, below $y = \sec^2 x/\sqrt{\tan x}$ between $x = \pi/6$ and $x = \pi/4$.

In Exercises 25–28, find the average value of the given function over the indicated interval.

25. $y = \frac{(\sqrt{x} - 1)^3}{\sqrt{x}}$, $[1, 4]$
26. $y = \frac{x}{(1 + x^2)^3}$, $[0, 1]$
27. $y = \frac{\cos t}{(1 + \sin t)^3}$, $[0, \pi/2]$
28. $y = \tan^2 x \sec^2 x$, $[0, \pi/4]$

29. A female calculus student has a basal metabolism rate $R(t) = 125/2 - (1/4)\sin(\pi t/24)$. What is her total daily basal metabolism?

30. Repeat Exercise 29 for a child for whom $R(t) = 125 - (1/3)\cos(\pi t/12)$.

31. Fluid is flowing through a chamber. The rate of inflow at time t is

$$R_i(t) = t^2 + 1.$$

The rate of outflow is

$$R_0(t) = \sqrt{2t + 1}.$$

What is the volume of the fluid in the chamber at $t = 2$ if $V(0) = 1$?

32. Repeat Exercise 31 if $R_i(t) = 3t^2 - 1$ and $R_0(t) = \sqrt{1 + (3/2)t}$.

33. In measuring the cardiac output of heart patients, dye is injected into the bloodstream and its concentration $c(t)$ is measured over a period of 15 seconds at an artery that is "downstream" from the heart and the site of the injection. The average concentration $\bar{c}$ of dye at the artery is computed, and the cardiac output in liters per minute is estimated as $4i/\bar{c}$, where i is the volume of dye injected. Suppose that $c(t)$ is plotted over $t \in [0, 15]$ and is found to be approximated by

$$c(t) = \sqrt{t + 1}$$

in mg/liter.

(a) Find $\bar{c}$.
(b) If $i = 14$ mL, then what is the cardiac output?

34. A rock sample has constant cross-sectional area $A = 1\ \text{cm}^2$. The density of the rock x cm from the top is given by $d(x) = 1/\sqrt{x + 3}$. If the sample is 13 cm long, then find its mass. [*Hint:* The average density between the top and depth x is $M(x)/xA$, where $M(x)$ is the mass to depth x.]

35. What is the average density in Exercise 34?

36. If m and n are positive integers, then show that

$$\int_0^1 x^n(1 - x)^m\, dx = \int_0^1 x^m(1 - x)^n\, dx.$$

37. Show that

$$\int_0^{\pi/2} \sin^n x\, dx = \int_0^{\pi/2} \cos^n x\, dx$$

if n is a positive integer. [*Hint:* Use (16), p.84]

38. Show that

$$\int_0^{\pi/2} \cos^n x \sin^n x\, dx = 2^{-n} \int_0^{\pi/2} \cos^n x\, dx$$

if n is a positive integer. [*Hint:* Use the identity $\cos x \sin x = (1/2)\sin 2x$, the hint for Exercise 37, and Theorem 2.8.]

4.6 Integration by Numerical Methods

At this point, it may seem that the fundamental theorem of calculus makes short work of any definite integral $\int_a^b f(x)\, dx$. Theorem 4.2 says that all we have to do to evaluate such an integral is to find a formula for an antiderivative F of f and then subtract $F(a)$ from $F(b)$.

Unfortunately, finding a formula for F is often not nearly as easy as Sections 4 and 5 may make it appear. Suppose, for instance, that we are confronted with

$$(1) \qquad \int_0^1 \sqrt{1 - x^2}\, dx.$$

No obvious substitution suggests itself, since no part of the integrand looks like the derivative of another part. Later, we will learn a technique for working out (1) by antidifferentiation. For now, we can evaluate it geometrically from our knowledge of area formulas. As **Figure 6.1** shows, (1) gives the area under the graph of $y = \sqrt{1 - x^2}$ between $x = 0$ and $x = 1$. That is one quarter of the area enclosed by the unit circle $x^2 + y^2 = 1$. Since the unit circle encloses an area of exactly $\pi \cdot 1^2 = \pi$ square units, we conclude that

FIGURE 6.1

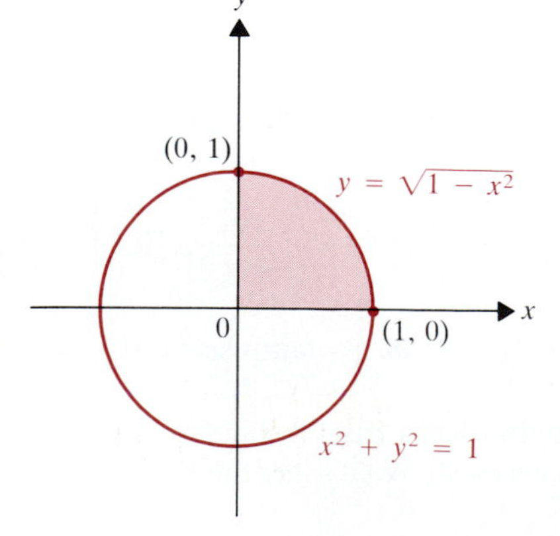

$$(2) \qquad \int_0^1 \sqrt{1 - x^2}\, dx = \frac{\pi}{4} \approx 0.7853982.$$

This section is devoted to a sequence of increasingly accurate numerical methods of approximating $\int_a^b f(x)\, dx$. We will use these methods on (1) to produce answers that we can compare to the decimal approximation in (2), which is correct to seven decimal places.

The methods we present are used primarily to evaluate integrals $\int_a^b f(x)\, dx$ where there is no formula for $\int f(x)\, dx$. It may surprise you, but not only are there many such functions, they frequently come up in applied work. (See

Example 6 below.) Thus, in many cases, the *only* usable techniques to evaluate a given integral are numerical.

Rectangular Approximations

The simplest numerical method for approximating the definite integral of $f(x)$ over $[a, b]$ is the Riemann sum approximation

$$\int_a^b f(x)\,dx \approx R_P = f(z_1)\Delta x_1 + f(z_2)\Delta x_2 + \ldots + f(z_n)\Delta x_n, \tag{3}$$

where $P = \{x_0, x_1, \ldots, x_n\}$ is a partition of $[a, b]$, $\Delta x_i = x_i - x_{i-1}$ is the length of the ith subinterval $[x_{i-1}, x_i]$, and $z_i \in [x_{i-1}, x_i]$. To make the calculations as simple as possible, we will always use a *uniform* partition P, for which each subinterval is of length

$$\Delta x = h = \frac{b-a}{n}. \tag{4}$$

Moreover, we will use a standard method to choose $z_i \in [x_{i-1}, x_i]$. There are three common schemes for doing so.

The first scheme is the ***left endpoint approximation***, for which $z_i = x_{i-1}$ for $i = 1, 2, \ldots, n$. That is,

$$\int_a^b f(x)\,dx \approx [f(x_0) + f(x_1) + \ldots + f(x_{n-1})]h = L_n(f), \tag{5}$$

where h is given by (4) and

$$\begin{aligned} x_0 = a, \qquad x_1 &= a + h, \ldots, \\ x_j &= a + j \cdot h, \ldots \\ x_{n-1} &= a + (n-1)h. \end{aligned} \tag{6}$$

The ***right endpoint approximation*** is the one with $z_i = x_i$ for $i = 1, 2, \ldots, n$. That is,

$$\int_a^b f(x)\,dx \approx [f(x_1) + f(x_2) + \ldots + f(x_n)]h = R_n(f), \tag{7}$$

where the x_j are again given by (6). (Thus $x_n = a + nh = b$.)

The ***midpoint approximation*** is the one for which $z_i = \frac{1}{2}(x_{i-1} + x_i)$ for $i = 1, 2, \ldots, n$. Thus

$$\begin{aligned} z_1 &= \frac{1}{2}(x_0 + x_1) = \frac{1}{2}(a + a + h) = a + \frac{1}{2}h, \\ z_2 &= \frac{1}{2}(x_1 + x_2) = \frac{1}{2}(a + h + a + 2h) = a + \frac{3}{2}h, \\ &\vdots \\ z_j &= \frac{1}{2}(x_{j-1} + x_j) = a + \frac{2j-1}{2}h = a + \left(j - \frac{1}{2}\right)h, \\ &\vdots \\ z_n &= \frac{1}{2}(x_{n-1} + x_n) = a + \frac{2n-1}{2}h = a + \left(n - \frac{1}{2}\right)h. \end{aligned} \tag{8}$$

So the midpoint approximation is

$$\begin{aligned} \int_a^b f(x)\,dx &\approx \left[f\left(a + \frac{1}{2}h\right) + f\left(a + \frac{3}{2}h\right) + \ldots + f\left(a + \left[n - \frac{1}{2}\right]h\right)\right]h \\ &= M_n(f) \end{aligned} \tag{9}$$

These three approximations are quite similar. In each case,

$$z_j = z_{j-1} + h.$$

That is, we get each point of evaluation of f from the preceding one by adding the step size h. This makes the formulas easy to program. Moreover, $R_n(f)$ and $L_n(f)$ differ only in that $R_n(f)$ starts with $f(x_1) = f(a + h)$ and ends with $f(b)$, while $L_n(f)$ starts with $f(x_0) = f(a)$ and ends with $f(x_{n-1})$. We therefore have

$$L_n(f) = R_n(f) + [f(a) - f(b)]h. \tag{10}$$

So we need only compute one of these explicitly. The other can be gotten from (10).

Figure 6.2 illustrates the midpoint approximation $M_n(f)$ geometrically in case the function f is positive-valued over the interval $[a, b]$; $M_n(f)$ approximates the shaded area in Figure 6.2 by the sum of the areas of approximating rectangles based on $[x_{j-1}, x_j]$ with height $f(z_j)$, where $z_j = \frac{1}{2}(x_{j-1} + x_j)$. To get an idea of the accuracy of $M_n(f)$, $R_n(f)$, and $L_n(f)$, we try them on (1).

FIGURE 6.2

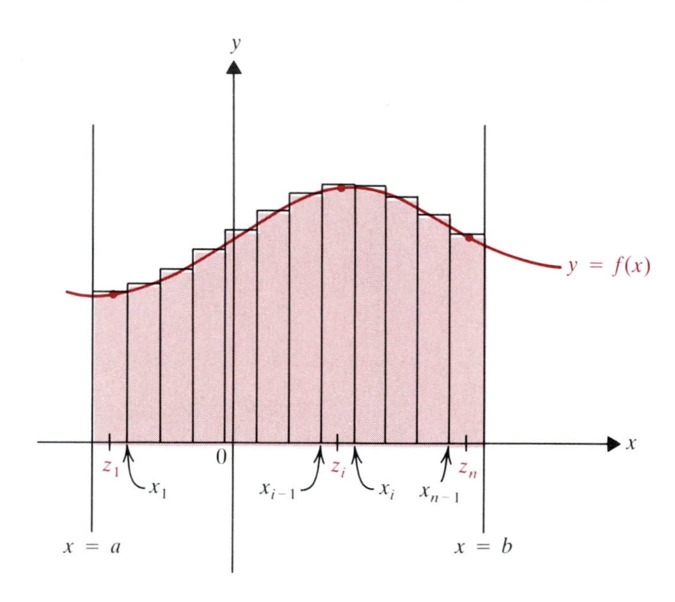

Example 1

Use each type of rectangular approximation with $n = 10$ to approximate $\int_0^1 \sqrt{1 - x^2}\, dx$.

Solution. Since $a = 0$, $b = 1$, and $n = 10$, here $h = (1 - 0)/10 = 0.1$. For each type of approximation, we have

$$\int_a^b f(x)\, dx \approx [f(z_1) + f(z_2) + \ldots + f(z_n)]h.$$

(a) For $L_n(f)$, we start with $z_1 = 0$. Then (5) gives

$$\begin{aligned}\int_0^1 \sqrt{1 - x^2}\, dx \approx L_n &= [f(0) + f(0.1) + f(0.2) + f(0.3) + f(0.4) \\ &\quad + f(0.5) + f(0.6) + f(0.7) + f(0.8) + f(0.9)](0.1) \\ &= [1 + 0.9949874 + 0.9797959 + 0.9539392 \\ &\quad + 0.9165151 + 0.8660254 + 0.8 + 0.7141428 \\ &\quad + 0.6 + 0.4358899](0.1) \\ &= (8.2612958)(0.1) = 0.8261296.\end{aligned}$$

(b) Using (10), we find

$$R_n(f) = L_n(f) + [f(b) - f(a)]h$$
$$= 0.8261296 + (0 - 1)(0.1) = 0.7261296.$$

(c) For the midpoint approximation, we start with 0.05 as z_1. Then (9) gives

$$\begin{aligned}\int_0^1 \sqrt{1 - x^2}\, dx &\approx [f(0.05) + f(0.15) + f(0.25) + f(0.35)\\ &\quad + f(0.45) + f(0.55) + f(0.65)\\ &\quad + f(0.75) + f(0.85) + f(0.95)](0.1)\\ &\approx [0.9987492 + 0.988686 + 0.9682458 + 0.9367497\\ &\quad + 0.8930286 + 0.8351647 + 0.7599342 + 0.6614378\\ &\quad + 0.5267827 + 0.3122499](0.1)\\ &\approx (7.8810286)(0.1) \approx 0.7881029. \quad \blacksquare\end{aligned}$$

In each case, there is considerable error in the rectangular approximation as compared to (2). The midpoint approximation is the most accurate, but even it is correct only to the first two decimal places. As you might expect (and you can check if you have access to a computer or programmable calculator), the accuracy improves if we increase the number of subintervals. Table 1 shows a computer printout of the results of repeatedly doubling the number of subintervals. For 160 subintervals, we get $M_n(f)$ correct to 4 decimal places, and for 640 we get 5-decimal-place accuracy. But for 6-place accuracy, we need more than 5000 subintervals! Accumulation of decimal *round-off error* makes the approximations for $n = 10{,}240$ *less* accurate than those for $n = 5120$, so that those methods *never* can produce 7-decimal place accuracy on the 9-digit device that carried out the calculations in Table 1. As these results suggest, rectangular approximation is not a very accurate technique unless n is *very* large. Thus it can be costly to use. (The last column of Table 1 is discussed below.)

TABLE 1

n	$L_n(f)$	$R_n(f)$	$M_n(f)$	$T_n(f)$
10	.826129582	.726129581	.788102858	.776129581
20	.80711622	.75711622	.786357647	.78211622
40	.796736934	.771736935	.785737966	.784236935
80	.791237451	.778737451	.785518405	.784987451
160	.788377924	.782127924	.785440689	.785252924
320	.786909303	.783784303	.785413198	.785346803
640	.786161269	.784598769	.785403494	.785380019
1280	.785782398	.785001148	.785400072	.785391773
2560	.785591155	.785200529	.785398775	.785395842
5120	.785494897	.785299585	.785398278	.785397241
10240	.785446917	.785349261	.785398459	.785398089

Trapezoidal Rule

In Example 1, the left endpoint approximation $L_n(f)$ is consistently too large and $R_n(f)$ is too small. By using the midpoint (the average of the left and right endpoints) for z_i, we got a more accurate approximation. Another way to improve the accuracy of the endpoint approximation is to *average* the left endpoint approximation and the right endpoint approximation. In case $n = 10$, we would

FIGURE 6.3

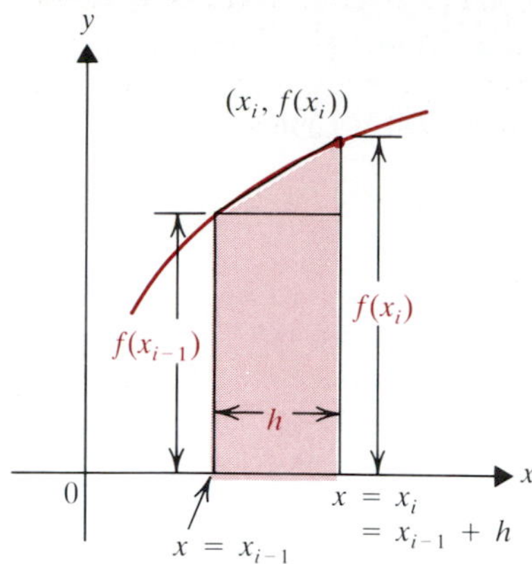

get

$$\int_0^1 \sqrt{1 - x^2}\, dx \approx \frac{1}{2}[0.8261296 + 0.7261296]$$

$$\approx 0.7761296,$$

which is more accurate than either endpoint approximation. Results for larger n are in the $T_n(f)$ column of Table 1. They show accuracy superior to both $L_n(f)$ and $R_n(f)$. Averaging the left and right endpoint approximations has a simple geometric interpretation. The area under the graph of a positive-valued function $y = f(x)$ between $x = x_{i-1}$ and $x = x_i$ can be approximated by the area of the *trapezoid* with parallel sides $f(x_{i-1})$ and $f(x_i)$ and height h, as shown in **Figure 6.3.** (The *height* of a trapezoid is the distance between the parallel sides.) The figure suggests that the trapezoidal approximation is more accurate than either $L_n(f)$ or $R_n(f)$. The colored trapezoid in Figure 6.3 has area

$$\frac{1}{2}h(f(x_{i-1}) + f(x_i)). \tag{11}$$

Summing (11) from $i = 1$ to $i = n$, we get

$$\int_a^b f(x)\, dx \approx \frac{1}{2}h \sum_{i=1}^{n} [f(x_{i-1}) + f(x_i)]$$

$$\approx \frac{1}{2}h[f(x_0) + f(x_1) + f(x_1) + f(x_2) + \ldots + f(x_{n-2})$$

$$+ f(x_{n-1}) + f(x_{n-1}) + f(x_n)].$$

Noting that for $j \neq 0$ and $j \neq n$, $f(x_j)$ occurs *twice* in this sum, we have

$$\int_a^b f(x)\, dx \approx T_n(f) = \frac{1}{2}h[f(x_0) + 2f(x_1) + 2f(x_2) + \ldots$$
$$+ 2f(x_{n-1}) + f(x_n)]. \tag{12}$$

Equation (12) is called the ***trapezoidal rule*** (or ***trapezoidal approximation***) for $\int_a^b f(x)\, dx$. It is usually faster to apply (12) directly than to separately calculate the left and right endpoint approximations and average them. The trapezoidal rule can be a chore to do by hand unless n is fairly small, but it works very well on a programmable calculator or computer.

Table 1 shows $T_n(f)$ for $\int_0^1 \sqrt{1 - x^2}\, dx$. The next example considers a different function.

Example 2

Use the trapezoidal rule to approximate $\int_0^1 \sqrt{1 + t^3}\, dt$ with $n = 10$.

Solution. Here $a = 0$, $b = 1$, so $h = (1 - 0)/10 = 0.1$. Then (12) and a calculator give

$$\int_a^b f(x)\, dx = \frac{1}{2}(0.1)[f(0) + 2f(0.1) + 2f(0.2) + 2f(0.3) + \ldots$$

$$+ 2f(0.9) + f(1)]$$

$$\approx 0.05[22.246648] \approx 1.1123324. \blacksquare$$

Unlike the situation in Example 1, where we had access to a value of $\int_0^1 f(x)\,dx$ to compare our answer with, we have no way to compute $\int_0^1 \sqrt{1+t^3}\,dt$ exactly. The following theorem provides a way of estimating the accuracy of the trapezoidal approximation. The proof is given in numerical analysis.

6.1 THEOREM

Let $T_n(f)$ be the trapezoidal approximation (12) to $\int_a^b f(x)\,dx$. Let

$$E_T = \left| \int_a^b f(x)\,dx - T_n(f) \right|$$

be the error made in approximating $\int_a^b f(x)\,dx$ by $T_n(f)$. If f'' exists on $[a, b]$ and $|f''(x)| \le M$ for all $x \in [a, b]$, then

$$E_T \le M \cdot \frac{(b-a)^3}{12n^2}. \tag{13}$$

Theorem 6.1 fails to apply to Example 1, because $f(x) = \sqrt{1-x^2}$ is not differentiable at $x = 1$, where the graph has a vertical tangent. Thus $f''(x)$ does not exist on all of $[0, 1]$. We can use (13) to estimate the accuracy in Example 2 though.

Example 3

How accurate is the trapezoidal approximation to $\int_0^1 \sqrt{1+t^3}\,dt$ in Example 2?

Solution. Here $f(t) = (1+t^3)^{1/2}$. Thus

$$f'(t) = \frac{1}{2}(1+t^3)^{-1/2}(3t^2) = \frac{3}{2}t^2(1+t^3)^{-1/2}.$$

Therefore

$$\begin{aligned} f''(t) &= 3t(1+t^3)^{-1/2} - \frac{3}{4}t^2(1+t^3)^{-3/2} \cdot 3t^2 \\ &= 3t(1+t^3)^{-1/2} - \frac{9}{4}t^4(1+t^3)^{-3/2} \\ &= 3t(1+t^3)^{-3/2}\left[(1+t^3) - \frac{3}{4}t^3\right] \\ &= \frac{3t\left(1+\frac{1}{4}t^3\right)}{(1+t^3)^{3/2}}. \end{aligned}$$

To find an upper bound for $|f''(x)|$ on $[0, 1]$, note that the numerator is no bigger than

$$3\left(1+\frac{1}{4}\right) = \frac{15}{4},$$

since $t = 1$ makes the numerator its largest. The denominator is no smaller than 1, its value for $t = 0$. So 15/4 is certainly an upper bound for $f''(x)$. In Example 2, we used $n = 10$. Therefore, from (13) in Theorem 6.1,

$$E_T \le \frac{15}{4} \cdot \frac{(1-0)^3}{12 \cdot 10^2} = \frac{15}{4800} = 0.003125.$$

Hence, our answer in Example 2 is correct to *at least* two decimal places. ■

One says that an approximation a_n to a quantity s is ***accurate to k decimal places*** if

$$|a_n - s| < \frac{1}{2}10^{-k}.$$

This means that the true value of s lies between $a_n - (1/2)10^{-k}$ and $a_n + (1/2)10^{-k}$ by Theorem 1.2(c) of Chapter 1. Usually (but because of decimal round-off error, not always), this means that the first k decimal places in a_n are correct.

Simpson's Rule

The third and final scheme for approximating $\int_a^b f(x)\,dx$ is also usually the most accurate. Though it bears the name of the English mathematician Thomas Simpson (1710–1761), who published it in 1737, it was essentially known to and used by Newton more than half a century earlier. His students, Roger Cotes (1682–1716) and James Stirling (1692–1770), both studied and used the method years before Simpson's exposition made it widely known.

The idea behind Simpson's rule is to approximate the graph of $y = f(x)$ between the points $P_1(x_{i-1}, f(x_{i-1}))$, $P_2(x_i, f(x_i))$, and $P_3(x_{i+1}, f(x_{i+1}))$ of a uniform partition by a *parabola*,

$$y = p(x) = Ax^2 + Bx + C,$$

that passes through P_1, P_2, and P_3, as in **Figure 6.4.** From the figure, it appears that the area under the graph of $y = f(x)$ between $x = x_{i-1}$ and $x = x_{i+1}$ will be very close to the area under the parabola. The following theorem gives a formula for the area under that approximating parabola.

FIGURE 6.4

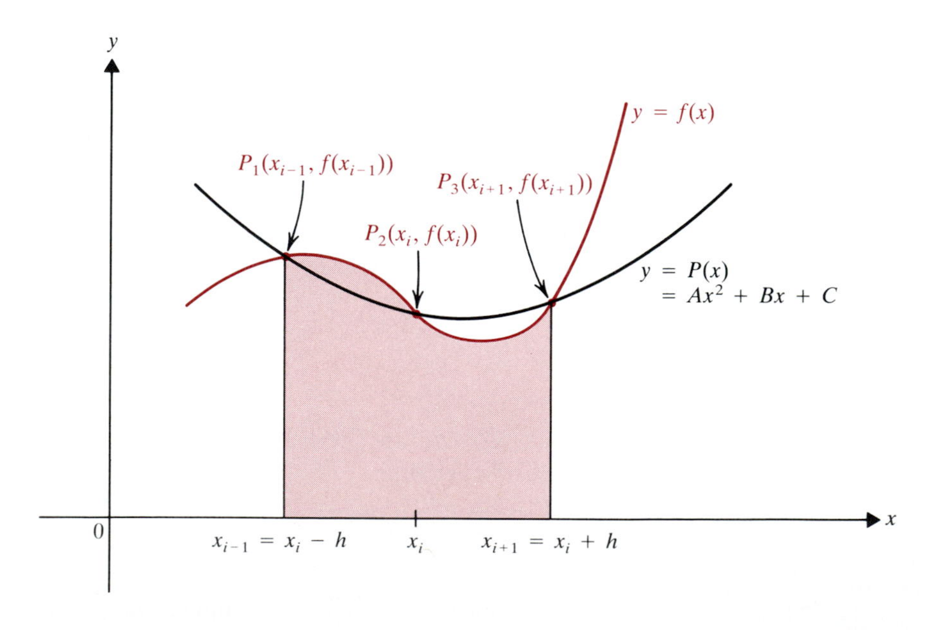

6.2 LEMMA If $P_1(x_i - h, f(x_i - h))$, $P_2(x_i, f(x_i))$, and $P_3(x_i + h, f(x_i + h))$ lie on the parabola $y = p(x) = Ax^2 + Bx + C$, then

$$\int_{x_i-h}^{x_i+h} p(x)\,dx = \frac{h}{3}[f(x_i - h) + 4f(x_i) + f(x_i + h)]. \tag{14}$$

Proof. First we compute the definite integral on the left side of (14). We have

$$\int_{x_i-h}^{x_i+h} p(x)\,dx = \frac{1}{3}Ax^3 + \frac{1}{2}Bx^2 + Cx\Big|_{x_i-h}^{x_i+h}$$

$$= \frac{1}{3}A(x_i+h)^3 + \frac{1}{2}B(x_i+h)^2 + C(x_i+h)$$

$$- \left[\frac{1}{3}A(x_i-h)^3 + \frac{1}{2}B(x_i-h)^2 + C(x_i-h)\right]$$

$$= \frac{1}{3}A(6x_i^2h + 2h^3) + \frac{1}{2}B(4x_ih) + C(2h)$$

$$= \frac{2h}{3}A(3x_i^2 + h^2) + 2Bx_ih + 2Ch.$$

Next we evaluate the right side of (14). The individual terms are

$$f(x_i-h) = p(x_i-h) = A(x_i-h)^2 + B(x_i-h) + C$$
$$= A(x_i^2 - 2x_ih + h^2) + B(x_i-h) + C,$$

$$f(x_i) = p(x_i) = Ax_i^2 + Bx_i + C,$$

$$f(x_i+h) = p(x_i+h) = A(x_i+h)^2 + B(x_i+h) + C$$
$$= A(x_i^2 + 2x_ih + h^2) + B(x_i+h) + C.$$

Therefore the right side of (14) is

$$\frac{h}{3}[f(x_i-h) + 4f(x_i) + f(x_i+h)] = \frac{h}{3}A(6x_i^2 + 2h^2) + \frac{h}{3}B(6x_i) + \frac{h}{3}C(6)$$

$$= \frac{2h}{3}A(3x_i^2 + h^2) + 2Bx_ih + 2Ch$$

$$= \int_{x_i-h}^{x_i+h} p(x)dx. \quad \text{QED}$$

We do not have to find the constants A, B, and C to find the area under the parabola. Thus we can find the area under $y = p(x)$ without even knowing the formula for $p(x)$!

Simpson's rule approximates the area under the graph of f over the *double* subinterval $[x_i - h, x_i + h]$ by using (14). It then sums these approximations for all such subintervals to approximate $\int_a^b f(x)\,dx$. For this scheme to work, we must partition $[a, b]$ so that every subinterval $[x_i - h, x_i]$ has a "mate" $[x_i, x_i + h]$. That is, there must be an *even* number of subintervals, so ***n* must be an even number.** (Since we include $a = x_0$ in the partition, the number of *points* used will be odd.) Using (14) repeatedly, we obtain

$$\int_a^b f(x)\,dx \approx S_n(f) = \int_{x_1-h}^{x_1+h} p(x)\,dx + \int_{x_3-h}^{x_3+h} p(x)\,dx + \ldots + \int_{x_{n-1}-h}^{x_{n-1}+h} p(x)\,dx$$

$$(15) \qquad = \frac{h}{3}\{[f(x_1-h) + 4f(x_1) + f(x_1+h)] + [f(x_3-h) + 4f(x_3) + f(x_3+h)] + \ldots + [f(x_{n-1}-h) + 4f(x_{n-1}) + f(x_{n-1}+h)]\}.$$

Now notice that $x_1 - h = x_0$ and $x_{n-1} + h = x_n$. Also, $x_1 + h = x_3 - h = x_2$; this pattern repeats for all of the intermediate terms, so that $f(x)$ is evaluated twice for each partition point that has an *even* subscript between x_2 and x_{n-2}, inclusive. This observation results in the usual form of Simpson's rule:

$$\int_a^b f(x)\,dx \approx \frac{h}{3}[f(x_0) + 4f(x_1) + 2f(x_2) + 4f(x_3) + 2f(x_4) + \dots + 2f(x_{n-2}) + 4f(x_{n-1}) + f(x_n)]. \tag{16}$$

Formulas (15) and (16) are both well suited for computers or programmable calculators. To use (15), simply sum over double subintervals terms of the form

$$f(x_j - h) + 4f(x_j) + f(x_j + h)$$

from 1 to $n/2$, and multiply the result by $h/3$. To use (16), evaluate f at each partition point from x_1 to x_{n-1}, multiply by alternating coefficients 4 and 2, add on $f(x_0)$ and $f(x_n)$, and then multiply the result by $h/3$. For a nonprogrammable calculator, (16) requires less key punching. The next example enables us to compare the accuracy of Simpson's rule to that of the trapezoidal rule.

Example 4

Approximate $\int_0^1 \sqrt{1 - x^2}\,dx$ using Simpson's rule and $n = 10$.

Solution. Using (16) and a calculator, we obtain

$$\begin{aligned}\int_0^1 \sqrt{1 - x^2}\,dx &\approx \frac{0.1}{3}[f(0) + 4f(0.1) + 2f(0.2) + 4f(0.3) + 2f(0.4) \\ &\quad + 4f(0.5) + 2f(0.6) + 4f(0.7) + 2f(0.8) \\ &\quad + 4f(0.9) + f(1)] \\ &\approx \frac{0.1}{3}[1 + 4(0.9949874) + 2(0.9797959) + 4(0.9539392) \\ &\quad + 2(0.9165151) + 4(0.8660254) + 2(0.8) \\ &\quad + 4(0.7141428) + 2(0.6) + 4(0.4358899) + 0] \\ &\approx \frac{0.1}{3}[23.451561] \approx 0.781752. \ \blacksquare\end{aligned}$$

Simpson's rule in Example 4 gives a better approximation to $\int_0^1 \sqrt{1 - x^2}\,dx$ than did the trapezoidal rule. The improved accuracy of Simpson's rule is brought out by the following theorem of numerical analysis.

6.3 THEOREM Let $S_n(f)$ be the Simpson's rule approximation (16) to $\int_a^b f(x)\,dx$. Let

$$E_S = \left|\int_a^b f(x)\,dx - S_n(f)\right|$$

be the error made in approximating $\int_a^b f(x)\,dx$ by $S_n(f)$. If $f^{(4)}(x)$ exists on $[a, b]$ and $|f^{(4)}(x)| \le M$ for all $x \in [a, b]$, then

$$E_S \le M \cdot \frac{(b-a)^5}{180n^4}. \tag{17}$$

We can't use (17) in Example 4 because f' (hence $f^{(4)}$ also) fails to exist at $x = 1$. In general (17) is not very easy to use because finding $f^{(4)}(x)$ and M often gets quite involved. The next example considers a case in which it is rather straightforward.

Example 5

Approximate

$$\int_1^3 \frac{1}{x}\,dx$$

using Simpson's rule and $n = 10$. Estimate the error.

Solution. In Chapter 6, we will see that the exact value of the given integral is ln 3, which is approximately 1.0986123. Since $n = 10$, we have $h = (3 - 1)/10 = 0.2$. From (16) then,

$$\int_1^3 \frac{1}{x}\,dx \approx \frac{0.2}{3}\left[\frac{1}{1} + 4\frac{1}{1.2} + 2\frac{1}{1.4} + 4\frac{1}{1.6} + 2\frac{1}{1.8} + 4\frac{1}{2} + 2\frac{1}{2.2} + 4\frac{1}{2.4} + 2\frac{1}{2.6} + 4\frac{1}{2.8} + \frac{1}{3}\right]$$

$$\approx \frac{0.2}{3}[16.479909] \approx 1.0986606.$$

This is very close to ln 3. Since

$$f(x) = \frac{1}{x} = x^{-1},$$

we have

$$f'(x) = -x^{-2}, \qquad f''(x) = 2x^{-3},$$
$$f'''(x) = -6x^{-4}, \qquad f^{(4)}(x) = 24x^{-5}.$$

On [1, 3],

$$f^{(4)}(x) = \frac{24}{x^5} \leq 24.$$

So we can use 24 for M. Then Theorem 6.3 gives

$$E_S \leq 24 \cdot \frac{(3-1)^5}{180 \cdot 10^4} = \frac{24 \cdot 32}{180 \cdot 10^4} = \frac{192}{45 \cdot 10^4} \rightarrow E_S \leq 0.00043.$$

Thus Theorem 6.3 guarantees that our answer is accurate to *at least* three decimal places. ■

We have seen several methods of numerically integrating functions f when we can't find formulas for $\int f(x)\,dx$. We tend to think of such functions f as given by complicated formulas like the one in Example 2. But in scientific work, functions are often known only from a table of experimentally obtained values. Even in the absence of a formula for $f(x)$, numerical methods of integration can provide approximations for $\int_a^b f(x)\,dx$. Our final example illustrates this.

Example 6

Suppose that the only information about $y = f(x)$ is contained in the following table of experimental data. Then approximate $\int_0^2 f(x)\,dx$.

x	0	0.4	0.8	1.2	1.6	2.0
y	0.75	1.10	1.25	1.32	1.61	1.75

Solution. Since the number of points x_i is even (6), there are an odd number of subintervals, namely, 5. Thus we *can't* use Simpson's rule. The next most accurate numerical integration method is the trapezoidal rule (12). Here $h = 0.4$, so (12) gives

$$\int_0^2 f(x)\,dx \approx \frac{1}{2}(0.4)[0.75 + 2(1.10) + 2(1.25) + 2(1.32) + 2(1.61) + 1.75]$$

$$\approx 0.2[13.06] \approx 2.612. \blacksquare$$

If y in Example 6 is the speed in meters per second of a moving body at observation times 0.4 seconds apart, then between $t = 0$ and $t = 2$, the body will have moved about 2.6 meters.

Exercises 4.6

In Exercises 1–4, approximate the given definite integral by (a) $L_n(f)$, (b) $R_n(f)$, (c) $M_n(f)$ for the given n.

1. $\int_0^1 \sqrt{1 + x^3}\,dx$, $n = 5$

2. Repeat Exercise 1 with $n = 10$.

3. $\int_0^{\pi/2} \sqrt{\sin x}\,dx$, $n = 4$

4. Repeat Exercise 2 with $n = 8$.

In Exercises 5–10, use the trapezoidal rule to approximate the given definite integral. Express answers to four decimal places.

5. $\int_0^1 \sqrt{1 + t^3}\,dt$, **(a)** $n = 5$ **(b)** $n = 8$

6. $\int_0^{\pi/2} \sqrt{\sin x}\,dx$, **(a)** $n = 5$ **(b)** $n = 8$

7. $\int_0^1 \frac{dx}{1 + x}$, **(a)** $n = 5$ **(b)** $n = 10$

8. $\int_0^2 \frac{dx}{\sqrt{1 + x^2}}$, **(a)** $n = 5$ **(b)** $n = 10$

9. $\int_0^1 \sqrt{1 + x^4}\,dx$, **(a)** $n = 5$ **(b)** $n = 10$

10. $\int_0^{\pi/4} \tan t\,dt$, **(a)** $n = 5$ **(b)** $n = 10$

In Exercises 11–18, use Simpson's rule to approximate the given integral. Express answers to five decimal places.

11. $\int_0^1 \sqrt{1 + t^3}\,dt$, **(a)** $n = 4$ **(b)** $n = 8$

PC **12.** $\int_0^1 \sqrt{1 + t^3}\,dt$, **(a)** $n = 10$ **(b)** $n = 20$

13. $\int_0^1 \frac{dx}{1 + x}$, **(a)** $n = 4$ **(b)** $n = 8$

14. $\int_0^{\pi/2} \sqrt{\sin x}\,dx$, **(a)** $n = 4$ **(b)** $n = 8$

15. $\int_0^1 \sqrt{1 + x^4}\,dx$, **(a)** $n = 4$ **(b)** $n = 10$

16. $\int_0^2 \frac{dx}{\sqrt{1 + x^2}}$, **(a)** $n = 4$ **(b)** $n = 10$

17. $\int_0^{\pi/6} \sec t\,dt$, $n = 4$

18. $\int_0^{\pi/4} \tan t\,dt$, $n = 10$

In Exercises 19–22, estimate $\int_0^2 f(x)\,dx$ from the tabulated data. Use Simpson's rule if possible, otherwise the trapezoidal rule.

19.

x	0	0.4	0.8	1.2	1.6	2.0
$f(x)$	0.34	0.51	0.70	1.03	1.41	1.50

20.

x	0	0.4	0.8	1.2	1.6	2.0
$f(x)$	0.44	0.21	0.35	0.57	0.81	0.77

21.

x	0	0.5	1.0	1.5	2.0
$f(x)$	0.38	1.41	0.63	2.88	1.47

22.

x	0	0.5	1.0	1.5	2.0
$f(x)$	1.81	0.08	1.33	2.84	0.95

23. During one-minute intervals, a car's speedometer readings in miles per hour were observed to be those in the following table.

t	0	1	2	3	4
$f(x)$	53	58	55	56	54

About how far did the car travel during this period? (Use Simpson's rule.)

24. A surveyor takes front-to-back measurements of an irregularly shaped lot of length 250 ft. At each end, the lot is 100 ft deep. At distances 50, 100, and 150 ft from one end, the depths are 115, 95, and 110 ft, respectively. Estimate the area of the lot by Simpson's rule.

In Exercises 25–30, estimate the error using Theorem 6.1 or Theorem 6.3.

25. **(a)** Exercise 5(a) **(b)** Exercise 5(b)

26. **(a)** Exercise 10(a) **(b)** Exercise 10(b)

27. **(a)** Exercise 7(a) **(b)** Exercise 7(b)

28. **(a)** Exercise 13(a) **(b)** Exercise 13(b)

29. **(a)** Exercise 11(a) **(b)** Exercise 11(b)

30. **(a)** Exercise 12(a) **(b)** Exercise 12(b)

31. Show that if $f(x)$ is a polynomial function of degree 3 or less, then Simpson's rule gives the *exact* value of $\int_a^b f(x)\,dx$.

32. If $f(x)$ is a cubic polynomial function, then show that

$$\int_a^b f(x)\,dx = \frac{b-a}{6}\left[f(a) + 4f\left(\frac{1}{2}(a+b)\right) + f(b)\right].$$

This is called the *prismoidal rule.*

33. Use Exercise 32 to evaluate

$$\int_0^2 (5x^3 - 3x^2 + x + 1)\,dx.$$

34. Repeat Exercise 33 for

$$\int_0^4 (x^3 + x^2 - 4x + 2)\,dx.$$

35. How large an n is needed to be sure that the integral in Exercise 7 is approximated with an error less than 0.0005?

36. How large an n is needed to be sure that the integral in Exercise 13 is approximated with an error less than 0.0005?

The following exercises are intended for those with access to a computer or programmable calculator. Use programs for the trapezoidal and Simpson's rule. Find $T_n(f)$ and $S_n(f)$ for the given function for n = 10, 20, 40, 80, 160, and 320.

PC **37.** The function of Exercise 3 over $[0, \pi/2]$

PC **38.** $f(x) = \sqrt{1 + x^2}$ over $[0, 1]$

PC **39.** $f(x) = \dfrac{1}{1 + \cos x}$ over $[0, \pi/2]$

PC **40.** $f(x) = \tan x$ over $[0, \pi/4]$

4.7 Looking Back

In this chapter, we have studied the definite integral $\int_a^b f(x)\,dx$ of a function f over a closed interval $[a, b]$. It arose in the first section as a means of defining—and calculating—the area of a region lying under the graph of a positive-valued function f between $x = a$ and $x = b$. Such an area is the limit (if it exists) of the rectangular approximations $\Sigma_{i=1}^n f(z_i)\,\Delta x_i$ as the norm $|P|$ of the partition $P = \{x_0, x_1, x_2, \ldots, x_n\}$ approaches 0. With the help of some formulas for sums of certain sequences of integers (Theorem 1.6), such limits can be computed for functions such as $f(x) = x$, $g(x) = x^2$, and $h(x) = x^3$.

In Section 2, we extended the rectangular approximation to functions that take on both positive and negative values, obtaining the Riemann sum R_P of f over $[a, b]$ relative to P. The definite integral $\int_a^b f(x)\,dx$ was defined to be the limit of R_P as $|P| \to 0$, if that limit exists. Theorems 2.6 and 2.7 give a basic additivity property for definite integrals. Theorem 2.8 simplifies integration of odd and even functions over intervals $[-a, a]$ that are centered at the origin.

Section 3 discussed the average value of an integrable function over an interval $[a, b]$. The main result was the mean value theorem for integrals (Theorem 3.4), which was the key tool in establishing part 1 of the fundamental theorem of calculus (Theorem 4.1) on differentiation of functions with a variable upper limit. That in turn enabled us to prove part 2 of the fundamental theorem (Theorem 4.2) on evaluation of definite integrals by antidifferentiation. Section 5 presented the technique of evaluating definite integrals by change of variable (Theorem 5.1), which widened the class of functions to which Theorem 4.2 can be applied.

The last section showed how to approximate a definite integral $\int_a^b f(x)\,dx$ numerically when we are unable to find the antiderivative of f. The trapezoidal rule and Simpson's rule are accurate and efficient elementary numerical integration techniques.

CHAPTER CHECKLIST

Section 4.1: area problem; partition, rectangular approximation, norm, uniform partition; area as a limit; summation formulas for sequences of positive integers (Theorem 1.6); summation notation and properties, telescoping sums (Theorem 1.8).

Section 4.2: Riemann sum; definite integral as a limit of Riemann sums; limits of integration; integrability of continuous functions (Theorem 2.3); integration of a constant function (Theorem 2.4); basic properties of definite integrals (Theorems 2.5, 2.6, 2.7, 2.8).

Section 4.3: average value of an integrable function over an interval; upper and lower bounds for the definite integral of a continuous function (Theorem 3.3); mean value theorem for integrals; preservation of inequalities by definite integrals (Theorem 3.6).

Section 4.4: Differentiation of functions defined by integrals with variable upper limit (fundamental theorem of calculus, part 1); evaluation of definite integrals by antidifferentiation (fundamental theorem of calculus, part 2); differentiation and integration as inverse operations.

Section 4.5: Change of variable in definite integrals (Theorem 5.1); integration of rates of change to obtain function values (pp. 264–265).

Section 4.6: left endpoint, midpoint, right endpoint rectangular approximation; trapezoidal rule; approximation accurate to k decimal places; Simpson's rule; error estimates; integration of experimental data.

REVIEW EXERCISES 4.7

In Exercises 1–4, compute the given Riemann sum.

1. $f(x) = 2x - 3$ on $[0, 2]$, $P = \{0, 1/2, 3/4, 5/4, 7/4, 2\}$, left endpoints

2. $f(x) = \sin x$ on $[0, \pi/2]$, $P = \{0, \pi/6, \pi/4, \pi/3, \pi/2\}$, right endpoints

3. $f(x) = 1 + x^3$ on $[0, 2]$, uniform partition, midpoints, $n = 5$

4. $f(x)$ of Exercise 3 on $[0, 2]$, uniform partition, right endpoints, $n = 8$

In Exercises 5–8, compute the indicated area as a limit of rectangular approximations.

5. Under $y = x^2 - 1$ between $x = 1$ and $x = 2$

6. Under $y = 2x - x^2$ between $x = 0$ and $x = 1$

7. Under $y = x^3$ between $x = 0$ and $x = 1$

8. Under $y = 1 - x^3$ between $x = -1$ and $x = 1$

9. **(a)** Evaluate

$$\sum_{i=1}^{5} (i^3 - 2i^2 + 3i + 1).$$

(b) Evaluate

$$\sum_{i=1}^{n} (2i + 1) - (2i - 1).$$

(c) Write in summation notation:

$$1 + 3 + 5 + 7 + \ldots 31.$$

10. **(a)** Evaluate

$$\sum_{i=1}^{4} (2i^3 + 3i^2 - 5i - 3).$$

(b) Evaluate

$$\sum_{i=1}^{n} (3^{i+1} - 3^i).$$

(c) Write in summation notation:

$$1 - 2 + 4 - 8 + \ldots + 256.$$

11. Express

$$\lim_{n\to\infty} \left[\sum_{i=1}^{n} \frac{1 - z_i^2}{1 + z_i^2} \Delta x_i \right]$$

as an integral of a function over $[a, b]$.

12. Repeat Exercise 11 for

$$\lim_{n\to\infty} \left[\sum_{i=1}^{n} \left(1 - \frac{z_i}{1 + z_i^2}\right) \Delta x_i \right].$$

13. Write as a single integral:

(a) $\int_2^5 f(x)\,dx + \int_{-3}^2 f(x)\,dx$

(b) $\int_{-2}^3 f(x)\,dx - \int_1^3 f(x)\,dx$

14. Decide whether the function is integrable over $[0, 1]$.

(a) $f(x) = \begin{cases} 1 & \text{for } x \text{ rational} \\ -1 & \text{for } x \text{ irrational} \end{cases}$

(b) $f(x) = \begin{cases} 1 & \text{for } x \geq 1/2 \\ -1 & \text{for } x < 1/2 \end{cases}$

15. Find an upper and lower bound for

$$\int_0^\pi x \cos x^2\,dx.$$

16. Repeat Exercise 15 for

$$\int_0^3 \sqrt{1 + x^4}\,dx.$$

17. (a) Find the average value of $f(x) = \sin x$ over $[0, 2\pi/3]$.
(b) Find the point or points c satisfying the conclusion of the mean value theorem for integrals.

18. Repeat Exercise 17 for

$$f(x) = x\sqrt{1 - x^2}.$$

19. If $\bar{f}$ is the average value of f over $[a, b]$, then show that

$$\int_a^b [f(x) - \bar{f}]\,dx = 0.$$

20. Suppose that $f'(x)$ is continuous on $[a, b]$. Then show that the average rate of change of f over $[a, b]$ is $\overline{f'}$.

In Exercises 21–28, evaluate the given integral.

21. $\int_{-1}^1 (x^4 + x^2 + 3)\,dx$

22. $\int_1^3 \frac{1 + 3x^2}{x^2}\,dx$

23. $\int_0^1 (t^3 - 3t^2 + 1)(t^2 - 2t)\,dt$

24. $\int_0^{\sqrt{\pi}} x \sin x^2\,dx$

25. $\int_0^{\pi/2} \frac{\sin t}{(1 + \cos t)^2}\,dt$

26. $\int_0^{\pi/6} \sin^3 2t \cos 2t\,dt$

27. $\int_0^9 \frac{t}{\sqrt{t^2 + 16}}\,dt$

28. $\int_0^1 (1 - t)\sqrt{t^2 - 2t + 8}\,dt$

In Exercises 29–32, differentiate the given function.

29. $F(x) = \int_1^x \sqrt{\sin^2 t^2 + 2t^{3/2}}\,dt$

30. $G(x) = \int_x^3 \frac{1}{1 + 8t^2}\,dt$

31. $H(x) = \int_0^{\sqrt{1-x^2}} (t^3 - 2t + 1)\,dt$

32. $K(x) = \int_{2x-1}^{x^2+1} \sqrt{1 + \sec t^2}\,dt$

33. The *marginal revenue* of a product at production level x is defined as $MR(x) = R'(x)$, where $R(x)$ is the revenue realized from the production of x units. If $MR(x) = x^2 - 0.08x + 10$, then find the total revenue generated by producing 100 units.

34. Fluid flows into a chamber at the rate of $\sqrt{8t + 1}$ liter/min. If at time $t = 0$ there was 1 liter in the chamber, how many liters are there at $t = 3$?

In Exercises 35–44, approximate the given integral using the method suggested. Express your answers to four decimal places.

35. $\int_0^1 \sqrt{2 + x^3}\,dx$

(a) Left endpoint, $n = 5$
(b) Right endpoint, $n = 5$
(c) Midpoint, $n = 5$

36. $\int_0^{\pi/2} \sqrt{\cos x}\,dx$

(a) Left endpoint, $n = 4$
(b) Right endpoint, $n = 4$
(c) Midpoint, $n = 4$

37. Repeat Exercise 35 using the trapezoidal rule and $n = 5$.

38. Repeat Exercise 36 with the trapezoidal rule and $n = 5$.

39. Repeat Exercise 37 with $n = 10$.

40. Repeat Exercise 38 with $n = 10$.

41. Repeat Exercise 35 using Simpson's rule and $n = 4$.

42. Repeat Exercise 36 using Simpson's rule and $n = 4$.

43. Repeat Exercise 41 with $n = 8$.

44. Repeat Exercise 42 with $n = 8$.

45. How large an n is needed to estimate

$$\int_1^2 \frac{dx}{x}$$

with an error less than 0.0005 if the trapezoidal rule is used?

46. Repeat Exercise 45 if Simpson's rule is used.

CHAPTER 5

Applications of Integration

5.0 Introduction

Just as we applied the derivative to graphical questions (about extreme values and curve sketching), so in this chapter we apply the definite integral to geometric problems related to the graph of a function: finding areas, volumes, and arc lengths associated with curves.

Section 1 returns to the problem that motivated our initial discussion of the definite integral: calculating the area of a plane region bounded by one or more curves that are the graphs of functions. The next two sections show how the definite integral can be used in two different ways to compute the volume of certain kinds of solids. Section 4 shows how to find the length along the graph of a function $y = f(x)$ between two points $(a, f(a))$ and $(b, f(b))$.

The remaining sections deal with physical applications of integration. Section 5 develops techniques for computing moments and centers of mass, which are studied in physics and engineering in connection with the stability of structures and other objects. Section 6 uses the definite integral to determine the total pressure exerted by a confined fluid (such as water against the side of a dam) and to define and calculate the work done by a variable force in moving a body from one place to another.

The importance of many of these applications, which after three centuries still play prominent roles in physics and engineering, underscores the impact that integral calculus has exerted on the development of those subjects.

FIGURE 1.1

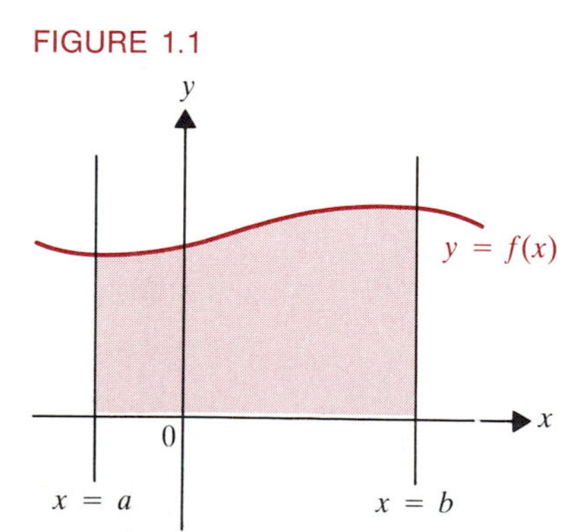

5.1 Area of Plane Regions

In Section 4.1, we considered the *area problem* illustrated in **Figure 1.1:**

Find the area A above the x-axis and under the graph of a positive- valued function f over an interval $[a, b]$.

Recall from Definition 1.4 of the last chapter that

$$A = \int_a^b f(x)\,dx = \lim_{|P|\to 0} R_P, \tag{1}$$

if the limit exists, where R_P is a Riemann sum

$$\sum_{i=1}^{n} f(z_i)\,\Delta x_i$$

corresponding to a partition P of $[a, b]$.

At the beginning of Section 4.1, we gave an interpretation of the area of a region under the graph of the velocity function v of an airplane between $t = a$ and $t = b$. That area represents the distance traveled by the plane between $t = a$ and $t = b$. As promised, we begin this section by calculating this distance, if the velocity function is

$$v(t) = \begin{cases} 150 + 1770t - 1800t^2 & \text{for } t \in [0, 1/3) \\ 540 & \text{for } t \in [1/3, 2/3) \\ 120 + 1830t - 1800t^2 & \text{for } t \in [2/3, 1]. \end{cases} \tag{2}$$

Note first that v is continuous on the entire interval $[0, 1]$. This can be seen from **Figure 1.2.** More formally, v is continuous on $[0, 1/3)$, $(1/3, 2/3)$, and $(2/3, 1]$ since it is a polynomial function on each of those subintervals. It is also continuous at $t = 1/3$ and $t = 2/3$, because

$$\lim_{t\to 1/3} v(t) = 540 = v(1/3) \qquad \text{and} \qquad \lim_{t\to 2/3} v(t) = 540 = v(2/3).$$

FIGURE 1.2

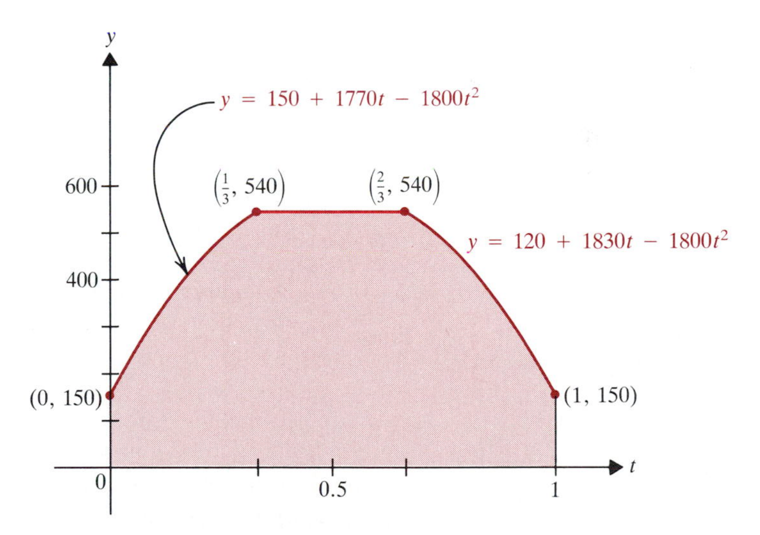

Thus, according to Theorem 1.5 of Chapter 4, the limit in (1) does exist, so we can compute the area under (2) by integration.

Example 1

Compute the distance traveled during a one-hour flight by a plane with velocity function (2).

FIGURE 1.3

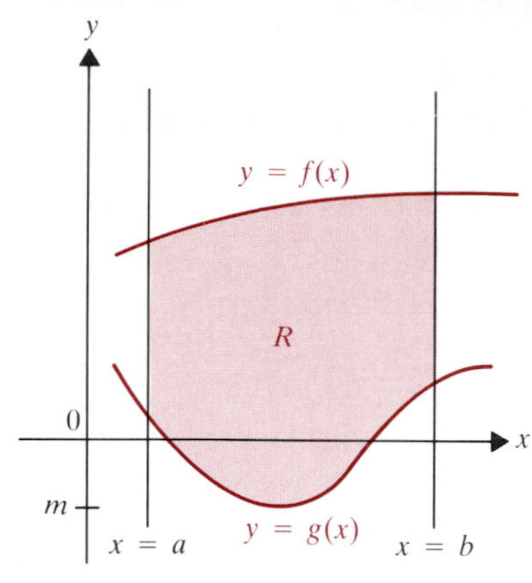

Solution. By the above remarks, the distance is given by $\int_0^1 v(t)\,dt$. In view of (2), the distance is therefore

$$\begin{aligned}
&\int_0^{1/3} (150 + 1770t - 1800t^2)\,dt + \int_{1/3}^{2/3} 540\,dt + \int_{2/3}^{1} (120 + 1830t - 1800t^2)\,dt \\
&\quad = 150t + 885t^2 - 600t^3\Big|_0^{1/3} + 540t\Big|_{1/3}^{2/3} + \Big[120t + 915t^2 - 600t^3\Big]_{2/3}^{1} \\
&\quad = \left(50 + \frac{295}{3} - \frac{200}{9}\right) + (360 - 180) \\
&\qquad + \left(120 + 915 - 600 - 80 - \frac{1220}{3} + \frac{1600}{9}\right) \\
&\quad = \frac{3890}{9} \approx 432.22.
\end{aligned}$$

Thus the flight covered about 430 miles. ■

FIGURE 1.4

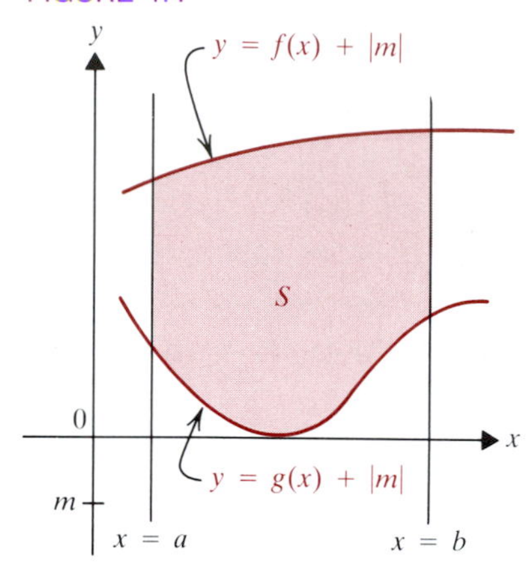

In this section we consider the following more general area problem:

Find the area of the region R lying between the graphs of two continuous functions f and g over an interval $[a, b]$.

Here we do not restrict f and g to be nonnegative over $[a, b]$. But we can solve this area problem by solving an equivalent problem involving nonnegative functions. To do so, suppose, as in **Figure 1.3,** that $g(x) \le f(x)$ on $[a, b]$. Since g is continuous, it has a minimum value m on $[a, b]$, by Theorem 7.7 of Chapter 1. **Figure 1.4** illustrates what happens if we replace $f(x)$ by $f(x) + |m|$ and $g(x)$ by $g(x) + |m|$. The region S between the graphs of $y = f(x) + |m|$ and $y = g(x) + |m|$ is congruent to the original shaded region R in Figure 1.3: We merely translated R upward $|m|$ units to produce S. Since congruent regions have the same area,

$$A(R) = A(S).$$

FIGURE 1.5

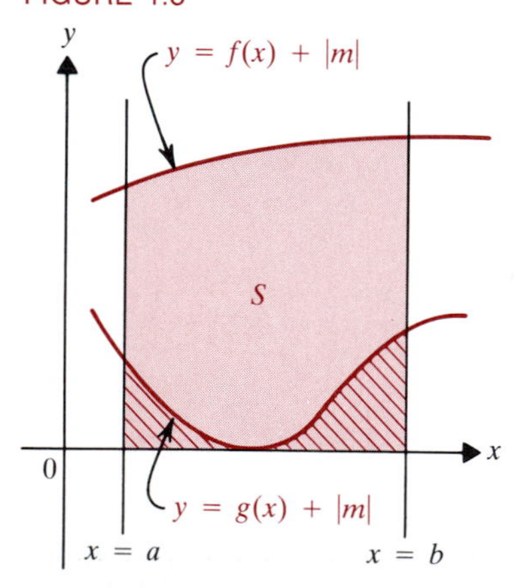

We can calculate $A(S)$ as the difference $A_1 - A_2$, where (see **Figure 1.5**)

$$A_1 = \text{area under } y = f(x) + |m| \text{ between } x = a \text{ and } x = b,$$

and

$$A_2 = \text{area under } y = g(x) + |m| \text{ between } x = a \text{ and } x = b.$$

Thus

$$\begin{aligned}
A(S) &= \int_a^b [f(x) + |m|]\,dx - \int_a^b [g(x) + |m|]\,dx \\
&= \int_a^b [f(x) + |m| - g(x) - |m|]\,dx.
\end{aligned}$$

We therefore have

(3) $$A(S) = \int_a^b [f(x) - g(x)]\,dx,$$

which means we have derived the following result.

1.1 THEOREM Suppose that f and g are continuous functions on $[a, b]$ with $f(x) \ge g(x)$. Then the area of the region R bounded by the graphs of $y = f(x)$ and $y = g(x)$ between $x = a$ and $x = b$ is

(4) $$A(R) = \int_a^b [f(x) - g(x)]\,dx.$$

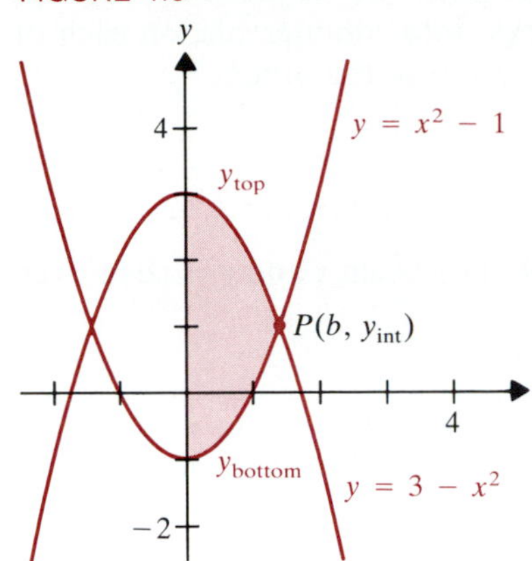

FIGURE 1.6

In words, (4) says that $A(R)$ is the integral over $[a, b]$ of

$$y_{top} - y_{bottom}$$

where y_{top} is the larger of the two functions and y_{bottom} is the smaller of the two functions.

The next example illustrates the use of (4), and also how the limits of integration can be found if they are not initially given.

Example 2

Find the area A of the region between the graphs of $y = x^2 - 1$ and $y = 3 - x^2$.

Solution. The region is shown in **Figure 1.6.** Both f and g are even functions, so the region is symmetric in the y-axis by Theorem 7.3(b) of Chapter 3. Then by Theorem 2.8(a) of Chapter 4, the area is twice the area of the shaded part of Figure 1.6. We therefore have

$$\begin{aligned} A &= 2\int_0^b [y_{top} - y_{bottom}]\,dx \\ &= 2\int_0^b [(3 - x^2) - (x^2 - 1)]\,dx = 2\int_0^b [4 - 2x^2]\,dx, \end{aligned}$$

where $P(b, y_{int})$ is the first-quadrant intersection point of the graphs of $y = 3 - x^2$ and $y = x^2 - 1$. We can find b by equating the formulas for $f(x)$ and $g(x)$, because when $x = b$ each formula gives the value y_{int}. Since $b > 0$, we get

$$x^2 - 1 = 3 - x^2 \rightarrow 2x^2 = 4 \rightarrow x^2 = 2 \rightarrow x = \sqrt{2}.$$

We therefore have

$$\begin{aligned} A &= 2\int_0^{\sqrt{2}} (4 - 2x^2)\,dx = 2\left[4x - \frac{2}{3}x^3\right]_0^{\sqrt{2}} \\ &= 2\left[4\sqrt{2} - \frac{2}{3}\cdot 2\sqrt{2} - 0\right] = 8\sqrt{2}\left(1 - \frac{1}{3}\right) \\ &= \frac{16}{3}\sqrt{2} \approx 7.5425. \ \blacksquare \end{aligned}$$

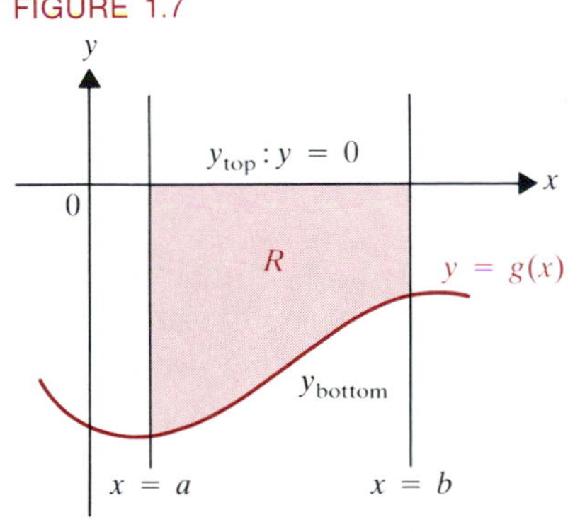

FIGURE 1.7

If $f(x) = 0$ in (4), then we get a formula for the area of a region (like R in **Figure 1.7**) below the x-axis ($y = 0$) and above the graph of $y = g(x)$, where $g(x) \leq 0$ on $[a, b]$. Since $0 - g(x) = -g(x)$, we have

$$A = -\int_a^b g(x)\,dx. \tag{5}$$

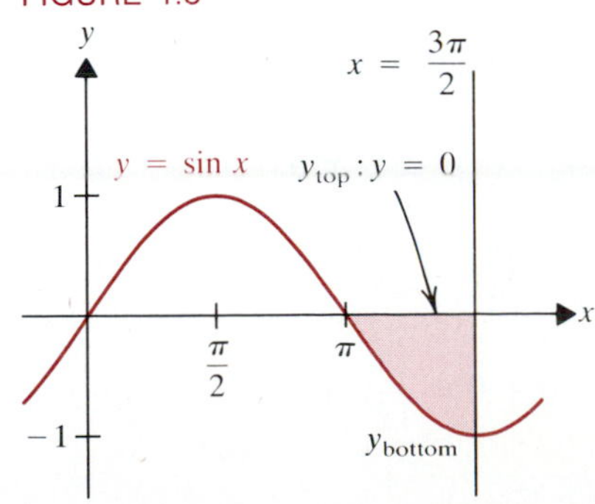

FIGURE 1.8

Example 3

Find the area bounded by the x-axis and the graph of $y = \sin x$ between $x = \pi$ and $x = 3\pi/2$. See **Figure 1.8.**

Solution. Since $\sin x \leq 0$ for $x \in [\pi, 3\pi/2]$, we have from (5)

$$\begin{aligned} A &= -\int_\pi^{3\pi/2} \sin x\,dx = -\Big[-\cos x\Big]_\pi^{3\pi/2} \\ &= \cos x\Big|_\pi^{3\pi/2} = 0 - (-1) = 1. \ \blacksquare \end{aligned}$$

FIGURE 1.9

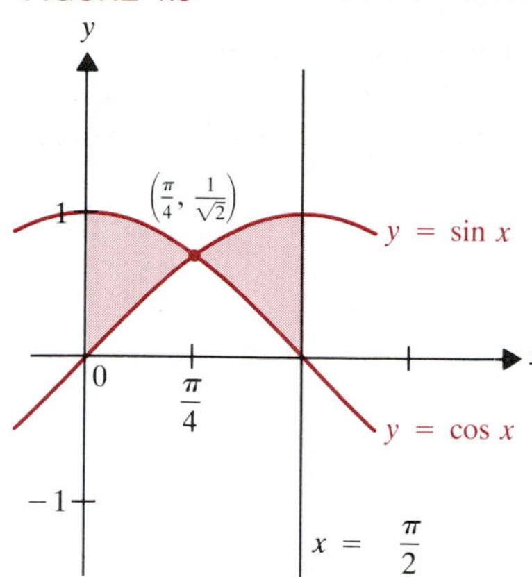

Sometimes, one graph does not lie above the other across the entire interval of interest. In such a case, we break the interval into subintervals, on each of which the graph of one function *is* above the graph of the other.

Example 4

Find the area A of the region between the graphs of $y = \sin x$ and $y = \cos x$ over the interval $[0, \pi/2]$.

Solution. As **Figure 1.9** shows, for $x \in [0, \pi/4]$, the curve $y = \cos x$ lies above $y = \sin x$. On $[\pi/4, \pi/2]$, the graph of $y = \sin x$ is above the graph of $y = \cos x$. The total area is the sum of the two shaded regions lying over the subintervals. We thus have

$$\begin{aligned} A &= \int_0^{\pi/4} (\cos x - \sin x)\,dx + \int_{\pi/4}^{\pi/2} (\sin x - \cos x)\,dx \\ &= \sin x + \cos x\Big|_0^{\pi/4} + \Big[-\cos x - \sin x\Big]_{\pi/4}^{\pi/2} \\ &= \frac{1}{\sqrt{2}} + \frac{1}{\sqrt{2}} - 0 - 1 + \left[0 - 1 + \frac{1}{\sqrt{2}} + \frac{1}{\sqrt{2}}\right] \\ &= 2\sqrt{2} - 2. \quad \blacksquare \end{aligned}$$

FIGURE 1.10

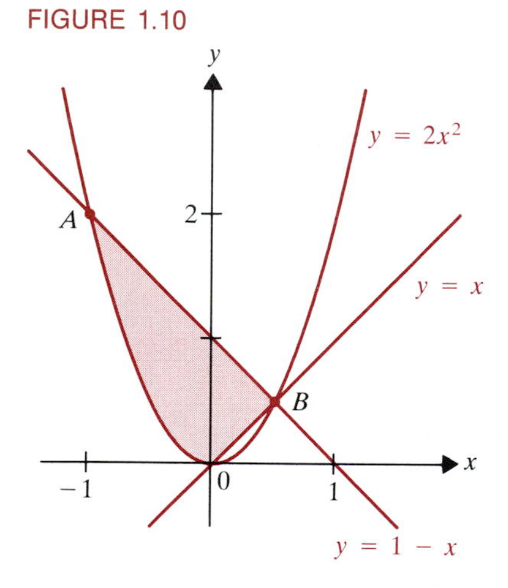

If a region is bounded by the graphs of more than two functions, then we usually must break the interval down into subintervals as in Example 4.

Example 5

Find the area of the region bounded by the graphs of $y = 2x^2$, $y = x$, and $y = 1 - x$.

Solution. The region is shaded in **Figure 1.10.** Throughout, the top boundary is $y = 1 - x$. But the bottom boundary changes at $x = 0$: For $x < 0$ that boundary is $y = 2x^2$, while for positive x it is $y = x$. Before integrating, we find the points A and B in Figure 1.10, which are the intersection points of $y = 1 - x$ and $y = 2x^2$. Equating the two expressions for y, we have

$$2x^2 = 1 - x \to 2x^2 + x - 1 = 0 \to (2x - 1)(x + 1) = 0,$$

so

$$x = 1/2 \qquad \text{or} \qquad x = -1.$$

As the figure shows, $x = -1$ at A and $x = 1/2$ at B. We therefore have

$$\begin{aligned} A &= \int_{-1}^{0} (1 - x - 2x^2)\,dx + \int_0^{1/2} (1 - x - x)\,dx \\ &= x - \frac{x^2}{2} - \frac{2x^3}{3}\Big|_{-1}^{0} + \Big[x - x^2\Big]_0^{1/2} \\ &= 0 + 1 + \frac{1}{2} - \frac{2}{3} + \frac{1}{2} - \frac{1}{4} - 0 + 0 = \frac{13}{12}. \quad \blacksquare \end{aligned}$$

FIGURE 1.11

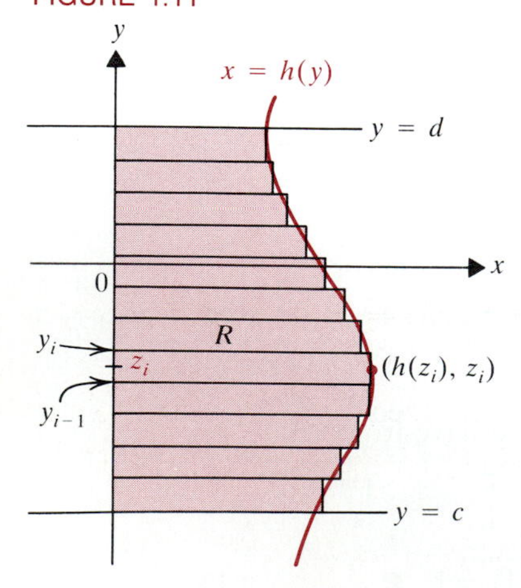

Integrating with Respect to y

As we saw in Section 2.7, it is sometimes convenient to deal with functions $x = h(y)$ in which x is given in terms of y. Everything in Section 4.1 and this section carries over to the calculation of areas involving the graphs of such functions. We need only switch the roles of x and y.

For example, suppose $h(y) \geq 0$ for y in some interval $[c, d]$. To find the area of the region R bounded by the graph of $x = h(y)$, the y-axis, $y = c$, and $y = d$, we partition $[c, d]$ and approximate the area by summing the areas of rectangles like those in **Figure 1.11.** The area is approximated by the Riemann sum

FIGURE 1.12

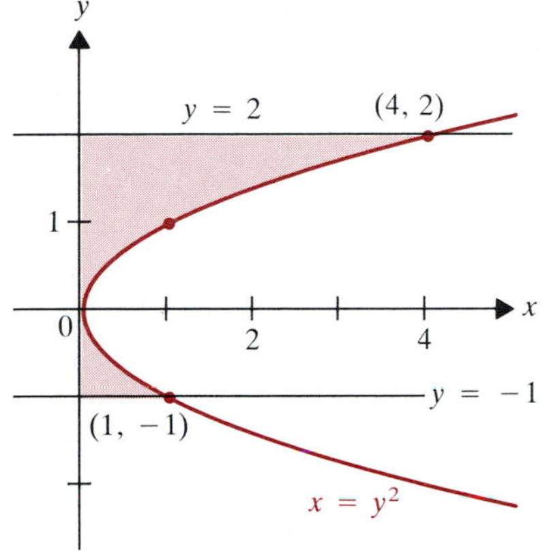

$$R_P = \sum_{i=1}^{n} h(z_i)\Delta y_i.$$

Thus

(6) $$A(R) = \int_c^d h(y)\,dy.$$

Example 6

Find the area of the region R bounded by the graphs of $x = y^2$, the y-axis, $y = -1$, and $y = 2$. See **Figure 1.12.**

Solution. We have from (6),

$$A = \int_{-1}^{2} y^2\,dy = \frac{y^3}{3}\bigg|_{-1}^{2} = \frac{8}{3} - \left(-\frac{1}{3}\right) = 3. \blacksquare$$

For functions $x = h(y)$ and $x = k(y)$, Theorem 1.1 takes the following form.

1.2 THEOREM Suppose that h and k are continuous functions and $h(y) \geq k(y)$ for all $y \in [c, d]$. Then the area of the region R bounded by $x = h(y)$, $x = k(y)$, $y = c$, and $y = d$ is

(7) $$A = \int_c^d [h(y) - k(y)]\,dy.$$

In words, (7) says that **A is the integral from the smallest to the largest y-value** of

$$x_{\text{right}} - x_{\text{left}}$$

where x_{right} is the larger x-value and x_{left} is the smaller one. See **Figure 1.13.**

FIGURE 1.13

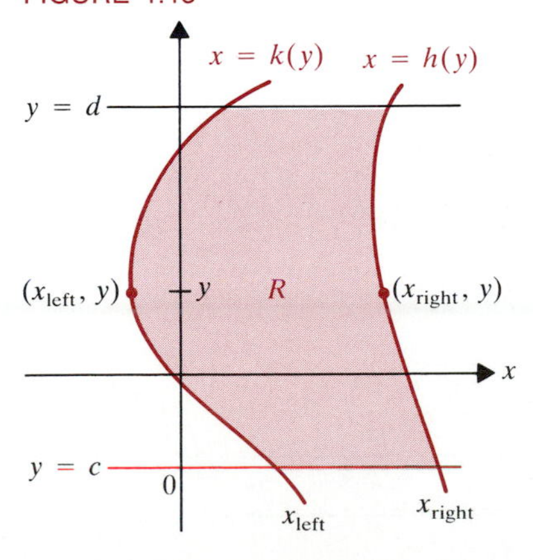

Example 7

Find the area of the region R between the graphs of $x = y^2 - 1$ and $x + y = 1$.

Solution. In **Figure 1.14** the largest x values in the shaded region belong to points on the line $x + y = 1$. The smallest ones belong to points on the curve $x = y^2 - 1$. Thus

$$x_{\text{right}} = 1 - y \quad \text{and} \quad x_{\text{left}} = y^2 - 1.$$

FIGURE 1.14

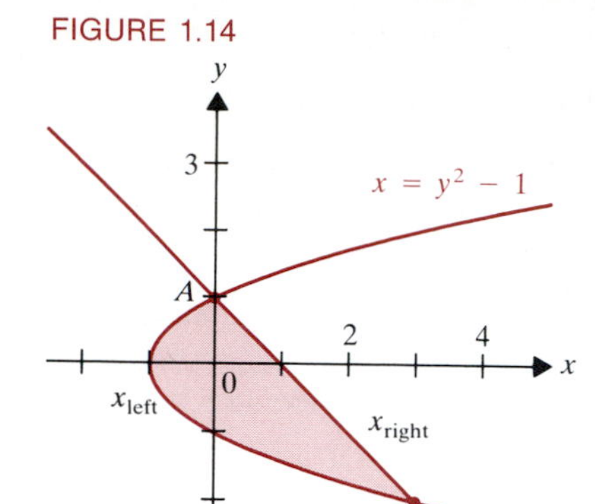

To find the limits of integration, we need to find the y-coordinates of the points A and B in Figure 1.14 where the line intersects the curve. Equating the two expressions for x, we get

$$y^2 - 1 = 1 - y \rightarrow y^2 + y - 2 = 0,$$
$$(y + 2)(y - 1) = 0 \rightarrow y = -2, y = 1.$$

Consequently,

$$A = \int_{-2}^{1} [1 - y - (y^2 - 1)]\,dy = \int_{-2}^{1} (2 - y - y^2)\,dy$$
$$= 2y - \frac{y^2}{2} - \frac{y^3}{3}\bigg|_{-2}^{1} = 2 - \frac{1}{2} - \frac{1}{3} - \left(-4 - 2 + \frac{8}{3}\right)$$
$$= \frac{9}{2}. \quad ■$$

Exercises 5.1

In Exercises 1–32, find the area of the plane region bounded by the graphs of the given functions. Draw a figure.

1. $y = 2 + x$, $y = 4 - x^2$
2. $y = 2 + x$, $y = x^2 - 4$
3. $y = \sqrt{x + 1}$, $y = -\sqrt{x + 1}$, $x = 3$
4. $y = x^2 - 1$, $y = 1 - x^2$
5. $y = x^2$, $x = y^2$
6. $y = 6 - x^2$, $y = 3 - 2x$
7. $y = x^3$, $y = x^2$
8. $y = \sin x$ and $y = 0$ between $x = \pi$ and $x = 2\pi$
9. $y = \cos x$ and $y = 0$ between $x = \pi/2$ and $x = \pi$
10. $y = \csc^2 x$ and $y = 0$ between $x = \pi/6$ and $x = \pi/2$
11. $y = \sec^2 x$ and $y = 0$ between $x = 0$ and $x = \pi/3$
12. $y = \cos x$, $y = \sin x$ between $x = -\pi/2$ and $x = \pi/2$
13. $y = \sin x$, $y = \cos x$ between $x = \pi/4$ and $x = 5\pi/4$
14. $x = y^2$, the y-axis, $y = -2$, $y = 2$
15. $x - y = 5$, $2x - y^2 = 2$
16. $x - 2y = 3$, $x + 2y - y^2 = 0$
17. $x + y = 6$, $x = y^2$, $y = 1$ (smaller region)
18. $x + 4y = y^2$, $y = x$
19. $y^2 = 4x$, $y = 2x - 4$
20. $y^2 = x - 1$, $x + y = 1$
21. $y^2 = x^2(4 - x^2)$
22. $y^2 = x^2 - x^4$
23. $x = y(y - 1)(y - 3)$ between $y = 1$ and $y = 3$, to the left of the y-axis
24. $y = x(x - 1)(x - 2)$ between $x = 1$ and $x = 2$, below the x-axis
25. $y = x + 2$, $x + y = 4$, $y = 0$
26. $x - 2y = 4$, $x + y = 4$, $x = 0$
27. $y = x + 2$, $x + 3y + 2 = 0$, $x + y = 4$
28. $y = 5x - 15$, $x + y = 3$, $2y - x = 6$
29. $y = x + 2$, $x + y = 4$, $x - 3y = 4$, $x + 3y + 2 = 0$
30. $x + y = 0$, $x + y + 2 = 0$, $2x - y = 1$, $2x - y = 3$
31. $y = x^2$, $y = 3x$, $y = x + 2$ (left-most region).
32. $2y = x^2$, $y = 6x$, $3y - 2x = 16$ (left region).
33. Suppose that $p = f(q)$ gives the price of a commodity as a function of demand q. A typical demand function f is decreasing: As supply rises, price falls. See **Figure 1.15.** Suppose s_0 is the amount of supply available at the present selling price $p_0 = f(s_0)$. The *consumer's surplus* resulting from an open economic system (one without price discrimination) is defined to be

$$CS = \int_0^{s_0} (f(q) - p_0)\,dq.$$

[The idea behind this is that if price discrimination existed, manufacturers could sell some of their product at the maximum possible price (nearly p_1 in Figure 1.15). Then they could lower their price a bit to p_2 and sell q_2 units, then lower the price some more, and so on. The total extra cost to consumers under such a discriminating price structure is represented by the shaded area in Figure 1.15.] Suppose that $f(q) = 20 - 0.1\sqrt{q}$. Find the consumer's surplus at price level \$10.00.

FIGURE 1.15

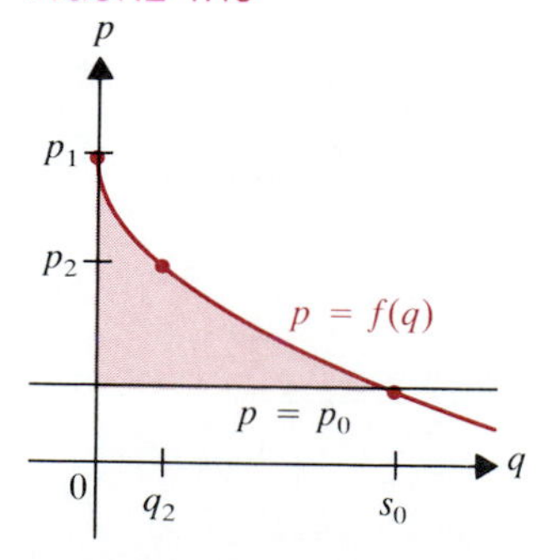

34. Repeat Exercise 33 if $f(q) = 15 - 0.01\sqrt{q}$ and $p_0 = 10$.

35. Let $p = g(q)$ be the supply curve for a commodity, where q is the supply at price p. A typical supply function is increasing: Price rises as demand increases. See **Figure 1.16.** Symmetrically to the consumer's surplus in Exercise 33, economists define the *producer's surplus* in an open economy without price discrimination as

$$PS = \int_0^{s_0} (p_0 - g(q))\,dq,$$

the shaded area in Figure 1.16. (This represents the amount producers gain by selling at price p_0, instead of first offering their product at a bare break-even price p_1, then raising the price to p_2 until supply q_2 is sold, and so on, until price p_0 is finally reached at which the present supply s_0 can be sold.) If $g(q) = 100 + 0.01q^2$ and the present price is \$200, find the producer's surplus.

FIGURE 1.16

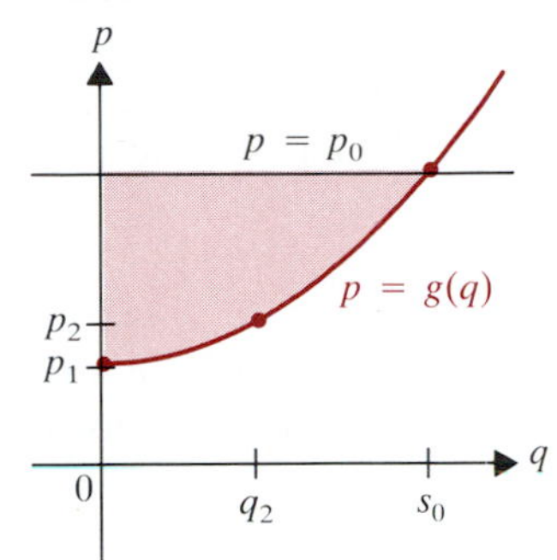

36. Repeat Exercise 35 if $g(q) = 200 + 0.01q^2$ and the present price is \$400.

37. *Perfect competition* exists when supply and demand are in equilibrium, with supply exactly matching demand. If $p = f(q)$ and $\bar{p} = g(q)$ are the demand and supply functions, then explain why under perfect competition both consumers and producers profit. (This gives part of the theoretical economic rationale behind the capitalist system.) Must the surplus of each be the *same* under perfect competition?

5.2 Volumes of Solids / Slicing

Since we approached the definite integral via the area problem (p. 229), it is no surprise that integration is useful in computing areas. In this section and the next, we will see that the definite integral is also a powerful tool for calculating the volume of certain kinds of solids. Elementary mathematics contains formulas for the volumes of rectangular boxes, pyramids, balls, cylindrical solids, and conical solids. Here, we consider the volume of additional solids: those whose cross sections made by planes perpendicular to a coordinate axis are (plane) regions whose area we can calculate. Such volumes are actually defined to be definite integrals.

To see why, consider a solid S and an axis t as in **Figure 2.1.** (In most cases t will be x or y.) A plane perpendicular to the t-axis through the point

FIGURE 2.1

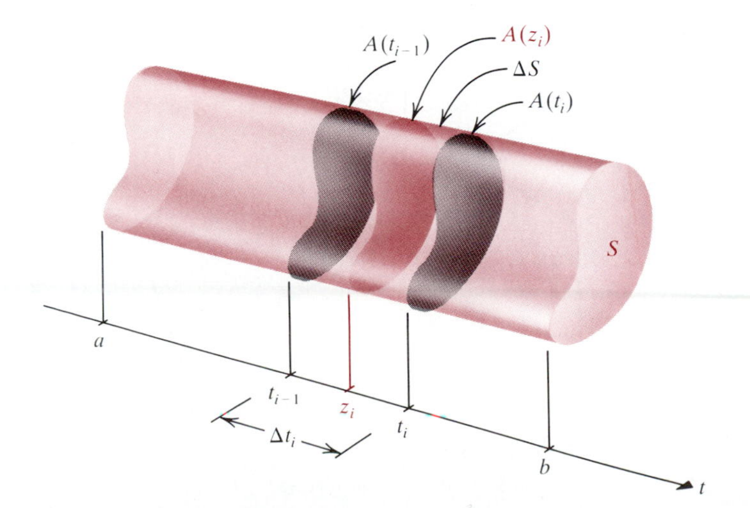

FIGURE 2.2

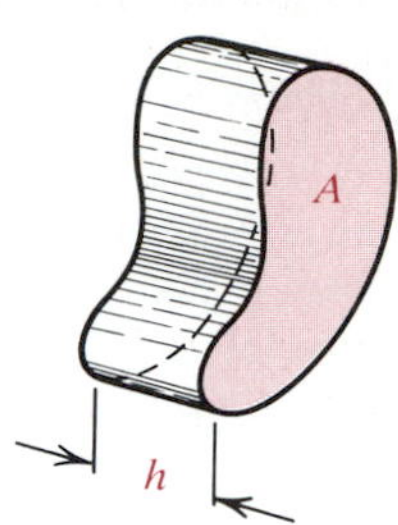

$t = z_i \in [a, b]$ intersects the solid in a region whose area $A(z_i)$ is shown in Figure 2.1. We assume that A is a continuous function as t varies over $[a, b]$. (Thus, the shape of S does not change abruptly as t varies between a and b.) We partition $[a, b]$ into subintervals of length $\Delta t_i = t_i - t_{i-1}$. For a small increment Δt_i, the part of S between $t = t_{i-1}$ and $t = t_i = t_{i-1} + \Delta t_i$ is approximately a solid right cylinder of height Δt_i. Since A is continuous, the base area can be approximated by $A(z_i)$ for any $z_i \in [t_{i-1}, t_i]$. See **Figure 2.2.** By Euclidean geometry, the volume of a solid right cylinder of base area A and height h is

$$V = A \cdot h. \tag{1}$$

Returning to Figure 2.1, the part ΔS of the solid between $t = t_{i-1}$ and $t = t_i$ is a right cylinder of base area $A(z_i)$ and height Δt_i. So ΔS has approximate volume

$$\Delta V_i = A(z_i)\,\Delta t_i. \tag{2}$$

Then the volume of S is approximately the sum of the volumes of approximating solid cylinders lying between $t = t_{i-1}$ and $t = t_i$, for $i = 1, 2, \ldots, n$:

$$V(S) \approx \sum_{i=1}^{n} A(z_i)\,\Delta t_i = A(z_1)\,\Delta t_1 + \ldots + A(z_n)\,\Delta t_n. \tag{3}$$

The right side of (3) is a Riemann sum for the continuous function A over the interval $[a, b]$. The finer the partition we use, the better the approximation should be. It therefore seems compelling to define the volume S to be the limit of the approximating Riemann sums in (3).

2.1 DEFINITION Let S be a solid bounded by planes perpendicular to the t-axis through $t = a$ and $t = b$. Suppose that for each t, the cross-sectional area is $A(t)$, where A is a continuous function on $[a, b]$. Then the ***volume*** of S is

$$V(S) = \int_a^b A(t)\,dt. \tag{4}$$

Our first example illustrates the consistency of Definition 2.1 with the volume formulas of elementary mathematics.

Example 1

Show that (4) gives the correct volume of

(a) the rectangular box of dimensions $l \times w \times h$ in **Figure 2.3;**

(b) the solid right circular cylinder in **Figure 2.4.**

FIGURE 2.3

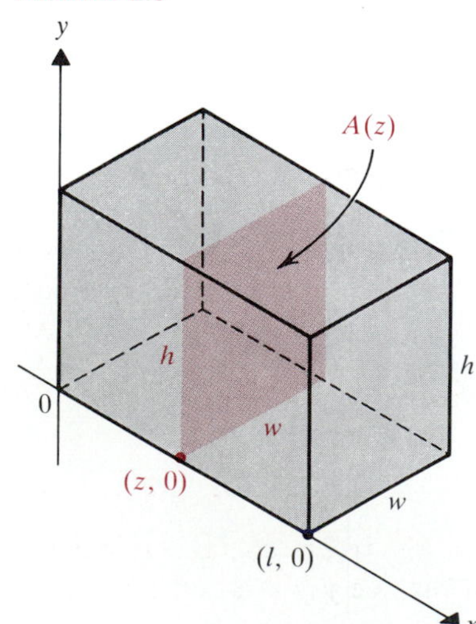

FIGURE 2.4

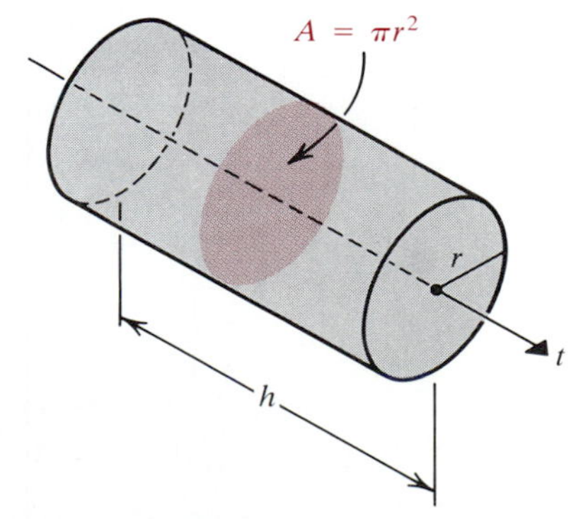

FIGURE 2.5

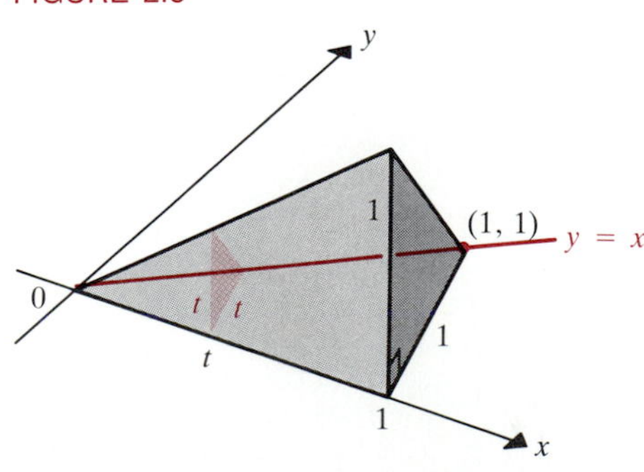

(c) A tetrahedron T lies over the triangle formed by the x-axis and the lines $y = x$ and $x = 1$. Every cross section made by a plane through $x = t$ perpendicular to the x-axis is an isosceles right triangle. (See **Figure 2.5.**) Find the volume of T.

Solution. (a) In Figure 2.3, the box lies between the y-axis and the point $x = l$ on the x-axis. A plane perpendicular to the x-axis through any point $(z, 0)$ makes a rectangular cross section of area

$$A(z) = hw.$$

Hence (4) says that

$$V = \int_0^l hw\,dx = hw\int_0^l dx = lhw,$$

the familiar formula for the volume of a rectangular box.

(b) The solid cylinder in Figure 2.4 can be considered to lie along the t-axis between $t = 0$ and $t = h$. Every circular cross section has area πr^2. Thus from (4) we have

$$V(S) = \int_0^h \pi r^2\,dt = \pi r^2\int_0^h dt = \pi r^2 h,$$

again in agreement with the formula from Euclidean geometry.

(c) By the statement of the problem, the plane through the solid perpendicular to the x-axis at $x = t$ slices out an isosceles right triangle. The color-shaded triangle in Figure 2.5 has base t and height t. So its area is

$$A(t) = \frac{1}{2}t \cdot t = \frac{1}{2}t^2.$$

Thus (4) says that

$$V(T) = \int_0^1 \frac{1}{2}t^2\,dt = \frac{1}{6}t^3\Big|_0^1 = \frac{1}{6}. \quad ■$$

The Euclidean geometry formula for the volume of a tetrahedron like T is $\frac{1}{3}Ah$, where A is the area of the base and h is the height. The tetrahedron's base is the gray-shaded triangle in Figure 2.5, with base 1 and altitude 1. So $A = \frac{1}{2} \cdot 1 \cdot 1 = \frac{1}{2}$. The height of the tetrahedron is $h = 1$ since it extends from $x = 0$ to $x = 1$. Thus the Euclidean geometry formula gives $V(T) = \frac{1}{3} \cdot \frac{1}{2} \cdot 1 = \frac{1}{6}$, again coinciding with the result given by (4).

Example 1 is reassuring, but we have yet to see that Definition 2.1 is truly a *generalization* of the earlier idea of volume. The next example remedies that by using the definition to find a volume that can't be computed by elementary formulas.

Example 2

Find the volume of the solid S of revolution in **Figure 2.6,** which results when the region bounded by $y = x^2$, the y-axis, and the line $y = 4$ is revolved about the y-axis.

FIGURE 2.6

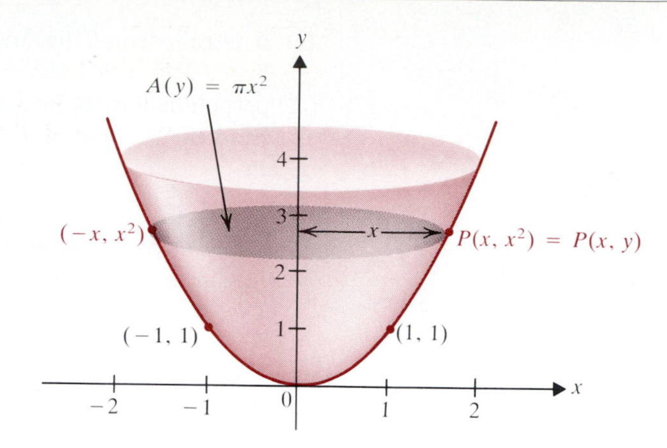

Solution. Each cross section of S by a plane perpendicular to the y-axis is a circular disk of radius x, where $y = x^2$ since $P(x, y)$ lies on the cross section. Thus the disk has area

$$A(y) = \pi x^2 = \pi y.$$

From (4), we therefore obtain

$$V(S) = \int_0^4 \pi y\, dy = \pi \Big[y^2/2\Big]_0^4 = 8\pi. \quad ■$$

Example 2 illustrates how the definite integral can be used to find the volume of a ***solid of revolution*** obtained by revolving the graph of $y = f(x) = x^2$ about the y-axis. The cross sections by planes perpendicular to the axis of revolution are circular disks of area πx^2, where x is the distance from a typical point $P(x, x^2)$ on the graph to the axis of revolution (in this case, the y-axis). If we revolve the graph about the x-axis, we obtain a different solid. However, we can calculate its volume using the *same* method.

FIGURE 2.7

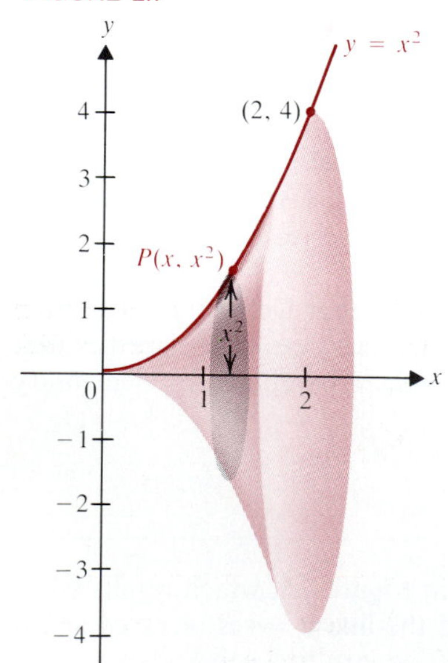

Example 3

Find the volume of the solid R of revolution obtained by revolving the region in the xy-plane bounded by the graph of $y = x^2$, the x-axis, and $x = 2$ about the x-axis. See **Figure 2.7.**

Solution. Any cross section of R by a plane perpendicular to the x-axis is once again a circular disk. This time such a disk has radius $y = x^2$, since $P(x, x^2)$ lies on the curve. Thus (4) gives

$$V(R) = \int_0^2 A(x)\, dx = \int_0^2 \pi(x^2)^2\, dx = \pi \int_0^2 x^4\, dx$$

$$= \pi \left[\frac{x^5}{5}\right]_0^2 = \frac{32}{5}\pi. \quad ■$$

We can summarize the method used in Examples 2 and 3 as follows.

2.2 THEOREM

Circular Disk Method. Suppose that a plane region R is revolved about a line l in the plane to generate a solid of revolution S, as in **Figure 2.8.** Then the volume of S is given by

(4) $$V(S) = \int_a^b A(t)\,dt.$$

In (4), $A(t)$ is the cross-sectional area at t:

(5) $$A(t) = \pi r^2,$$

where $r = r(t)$ is the ***radius of revolution:*** the distance from the bounding curve of R to the axis of revolution l. The variable t measures distances along l.

FIGURE 2.8

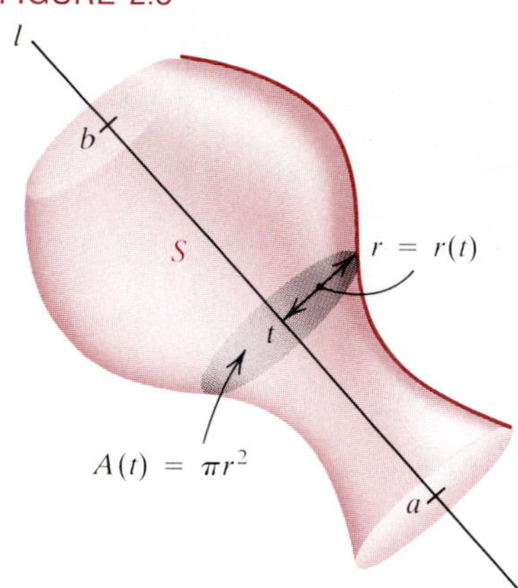

Figure 2.9 shows two common situations in which the circular disk method is used. In Figure 2.9a, the region between the y-axis and the graph of $y = f(x)$ over $[a, b]$ is revolved about the y-axis. In that case, (4) says that

$$V(T) = \pi \int_c^d x^2\,dy = \int_c^d A(y)\,dy.$$

FIGURE 2.9

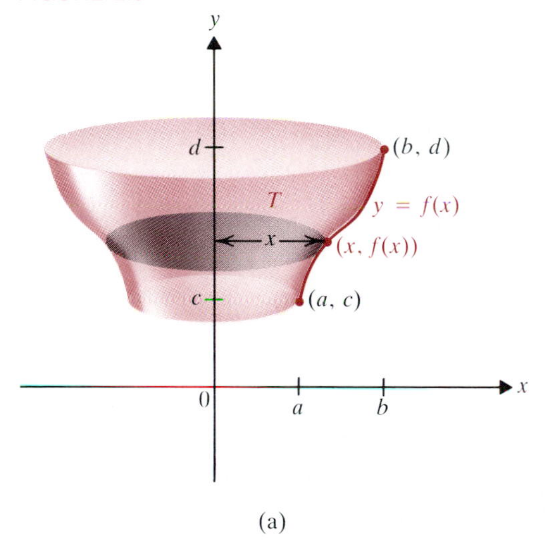

(a)

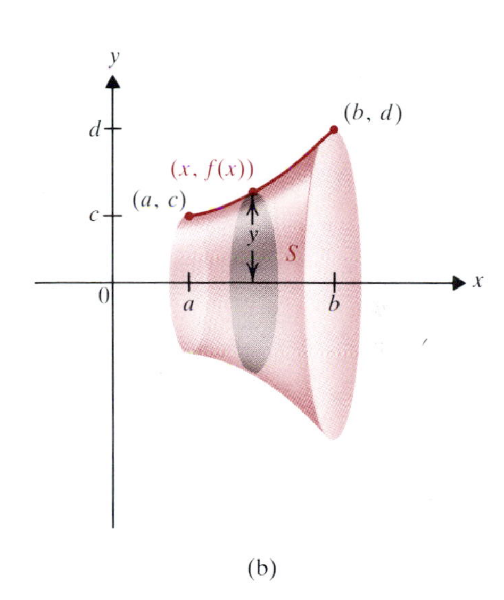

(b)

In Figure 2.9(b), the region between the graph of $y = f(x)$ and the x-axis from $x = a$ to $x = b$ is revolved about the x-axis. In that case, each circular disk has radius $y = f(x)$. Thus its area is $\pi y^2 = \pi[f(x)]^2$. So, by (4),

$$V(S) = \pi \int_a^b [f(x)]^2\,dx.$$

Example 4

The region bounded by the graph of $x = 1 + y^2$ between $y = -1$ and $y = 1$ is revolved about the y-axis. Find the volume of the resulting solid of revolution.

FIGURE 2.10

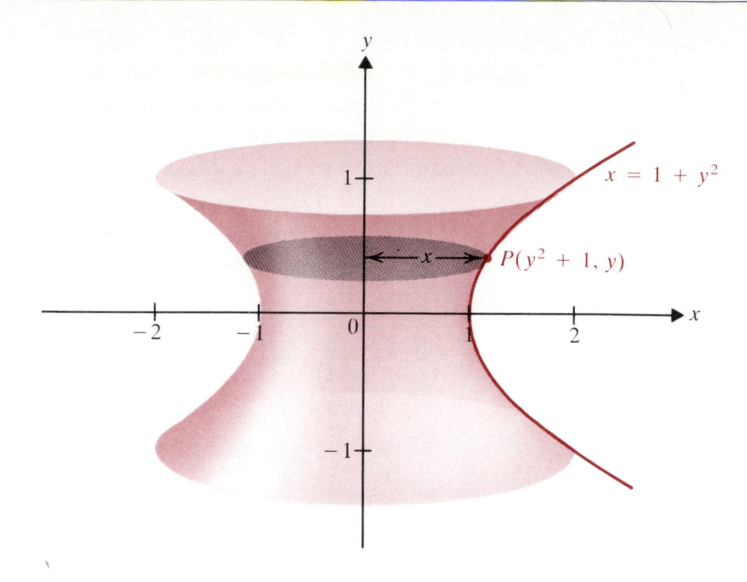

Solution. As **Figure 2.10** shows, the radius of revolution is $x = 1 + y^2$. Thus the circular disk in the figure has area

$$A = \pi x^2 = \pi(y^2 + 1)^2 = \pi(y^4 + 2y^2 + 1).$$

By symmetry, the volume is twice that obtained by revolving the first-quadrant portion around the y-axis. We then get from (4)

$$\begin{aligned} V &= 2\pi \int_0^1 (1 + 2y^2 + y^4)\, dy \\ &= 2\pi\left[y + \frac{2}{3}y^3 + \frac{1}{5}y^5\right]_0^1 = 2\pi\left[1 + \frac{2}{3} + \frac{1}{5}\right] \\ &= 2\pi\left[\frac{15 + 10 + 3}{15}\right] = \frac{56\pi}{15}. \quad \blacksquare \end{aligned}$$

In the last section, we were able to calculate areas of plane regions bounded by two curves. Similarly, we can calculate volumes of solids of revolution obtained by revolving the region between two graphs about an axis. Suppose, for instance, that $0 \le f(x) \le g(x)$ on $[a, b]$ and we revolve about the x-axis the region between the graphs of $y = f(x)$ and $y = g(x)$ from $x = a$ to $x = b$. See **Figure 2.11.** Every cross section of the resulting solid S made by planes perpendicular to the x-axis is an *annular region* (circular ring) between two circles. The inner circle has radius $f(x)$. The outer circle has radius $g(x)$, as shown in Figure 2.11(b). The shaded region in that figure is a disk of radius $g(x)$ from which a disk of radius $f(x)$ has been removed. Its area is therefore

$$A(x) = \pi \cdot (\text{outer radius})^2 - \pi \cdot (\text{inner radius})^2 = \pi[g(x)]^2 - \pi[f(x)]^2.$$

Thus

$$V(S) = \pi \int_a^b ([g(x)]^2 - [f(x)]^2)\, dx.$$

Using this reasoning and Definition 2.1, we get the following general result.

FIGURE 2.11

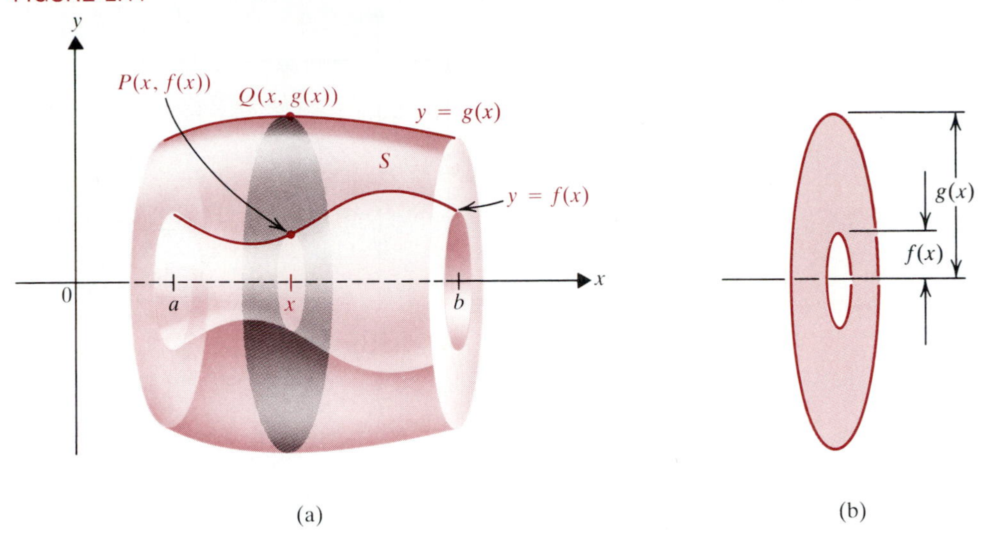

2.3 THEOREM

Circular Ring Method. Suppose that the plane region R lies between the graphs of two functions. Let R be revolved about a line l outside R to generate a solid of revolution S. Then the volume of S is

$$V(S) = \pi \int_a^b (r_{\text{outer}}^2 - r_{\text{inner}}^2)\,dt, \tag{6}$$

where

r_{outer} = distance from axis of revolution l to farther graph,

r_{inner} = distance from axis of revolution l to nearer graph,

and t measures distances along l. See **Figure 2.12.**

FIGURE 2.12

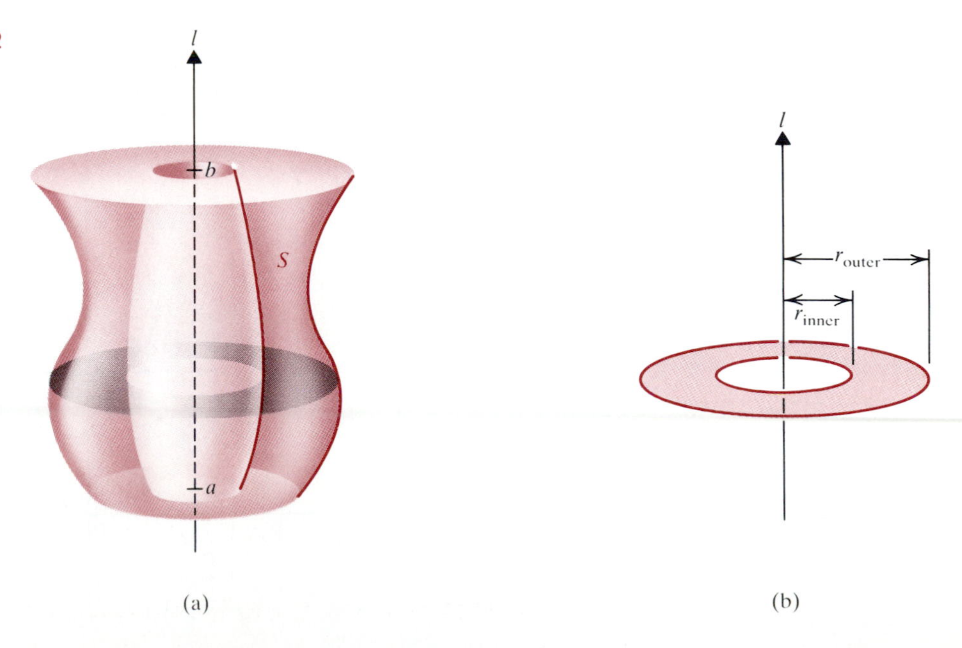

FIGURE 2.13

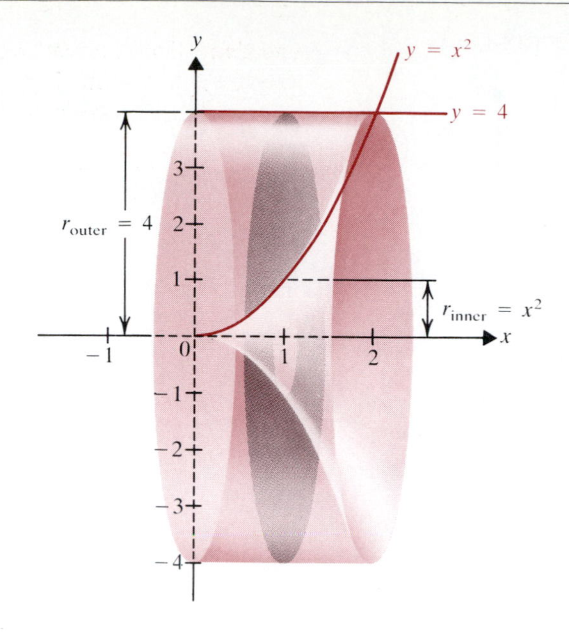

Example 5

Suppose that the region in Example 2 is revolved about the x-axis. Find the volume of the resulting solid.

Solution. The revolved region lies between $y = x^2$ and $y = 4$ above $[0, 2]$. As **Figure 2.13** shows, the inner radius is x^2 and the outer radius is 4. Hence (6) gives

$$\begin{aligned} V(S) &= \pi \int_0^2 [4^2 - (x^2)^2]\,dx = \pi \int_0^2 (16 - x^4)\,dx \\ &= \pi\left[16x - \frac{x^5}{5}\right]_0^2 = \pi\left[32 - \frac{32}{5}\right] = 32\pi\left[1 - \frac{1}{5}\right] \\ &= \frac{128\pi}{5}. \quad \blacksquare \end{aligned}$$

Example 6

The first-quadrant region bounded by $y = x^2$ and $y = x$ is revolved about the y-axis. Find the volume of the resulting solid T in **Figure 2.14.**

FIGURE 2.14

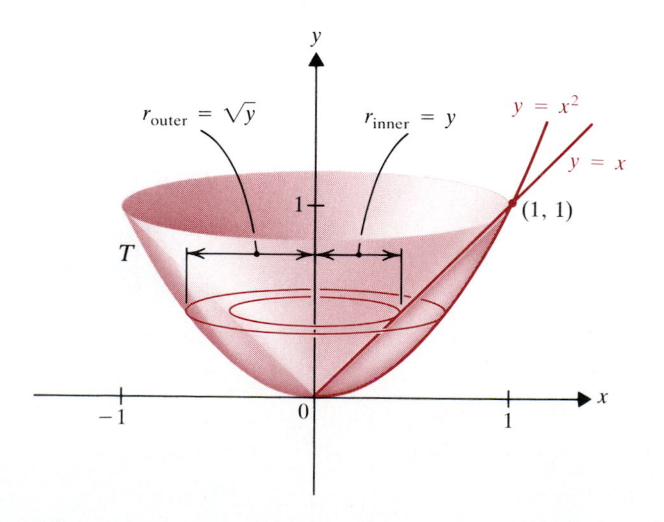

Solution. The revolved region lies between the curves $x = \sqrt{y}$ and $x = y$ for $y \in [0, 1]$. As Figure 2.14 shows, we have $y \leq \sqrt{y}$ for $y \in [0, 1]$. Thus here the outer radius of revolution is $\sqrt{y}$, and the inner radius of revolution is y. Hence

$$V = \pi \int_0^1 (y - y^2)\,dy = \pi\left[\frac{1}{2}y^2 - \frac{1}{3}y^3\right]_0^1 = \pi\left[\frac{1}{2} - \frac{1}{3}\right] = \frac{1}{6}\pi. \quad \blacksquare$$

Exercises 5.2

In Exercises 1–24, find the volume of the solid of revolution formed when the region bounded by the given curves is revolved about the specified axis. Make a sketch.

1. $x = y^2$ and $x = 4$; x-axis
2. $x = 1 + y^2$ and $x = 10$; x-axis
3. $y = \sqrt{\sin x}$, x-axis, $x = 0$, and $x = \pi/2$; x-axis
4. $y = \sqrt{\cos x}$, x-axis, $x = 0$, and $x = \pi/2$; x-axis
5. $y = x^2$ and $y = 4$; x-axis
6. $y = 4 - x^2$ and x-axis; x-axis
7. $y = x^2 + 1$ and x-axis, between $x = 0$ and $x = 2$; x-axis
8. $y = 2 - x^2$ and x-axis between $x = 0$ and $x = 1$; x-axis
9. $y = \sqrt{\sin x}\cos x$ between x-axis, $x = 0$, and $x = \pi/2$; the line $y = 2$
10. $y = \sqrt{\cos x}\sin x$ between x-axis, $x = 0$, and $x = \pi/2$; the line $y = -1$
11. $y = \sec x$, x-axis, $x = 0$ and $x = \pi/4$; x-axis
12. $y = \tan x$, x-axis, $x = 0$ and $x = \pi/3$; x-axis
13. $y = \sec x\sqrt{\tan x}$, x-axis, $x = 0$ and $x = \pi/6$; x-axis
14. $x = \cot y$, y-axis, $y = \pi/6$ and $y = \pi/4$; y-axis
15. $y^3 = x^2$, $y = 1$ and $x = 0$; y-axis
16. $y^3 = x^2$, x-axis, and $x = 1$; x-axis
17. $y = x^2$ and $y = 2x$ (first-quadrant region); x-axis
18. The region in Exercise 17; y-axis
19. $y = x^3$ and $y = x$ (first-quadrant region); y-axis
20. The region in Exercise 19; x-axis
21. The region of Exercise 1; y-axis
22. The region of Exercise 1; the line $x = 4$
23. $y = x^2 + 1$, $y = 5$ and $x = 0$; the line $x = 3$
24. $y = 2 - x^2$ and $y = 1$; the line $x = 2$

In Exercises 25–40, use the methods of this section to find the volume of the given solid.

25. A right circular cone of height h and base radius a
26. A right circular cylinder of height h and base radius a
27. A ball of radius a
28. The *ellipsoid of revolution* that results from revolving the graph of $x^2/a^2 + y^2/b^2 = 1$ about the x-axis. See **Figure 2.15.**

FIGURE 2.15

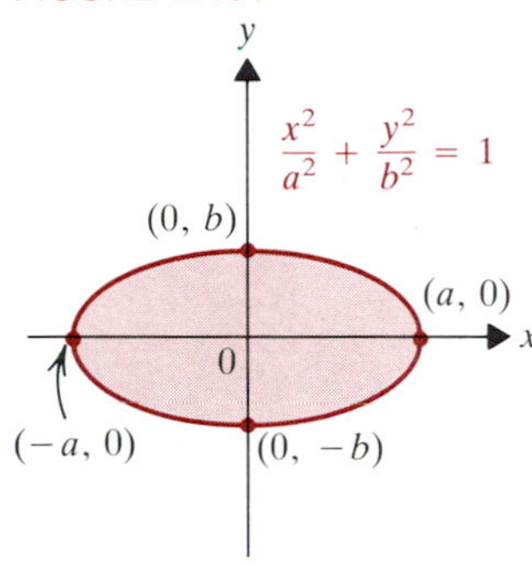

29. The *ellipsoid of revolution* that results from revolving the graph of $x^2/a^2 + y^2/b^2 = 1$ about the y-axis
30. The *hyperboloid of revolution* that results from revolving the graph of $x^2/a^2 - y^2/b^2 = 1$ between $x = a$ and $x = 2a$ about the x-axis
31. The *hyperboloid of revolution* that results from revolving the graph of $x^2 - y^2 = a^2$ between $x = a$ and $x = 2a$ about the y-axis. See **Figure 2.16.**

FIGURE 2.16

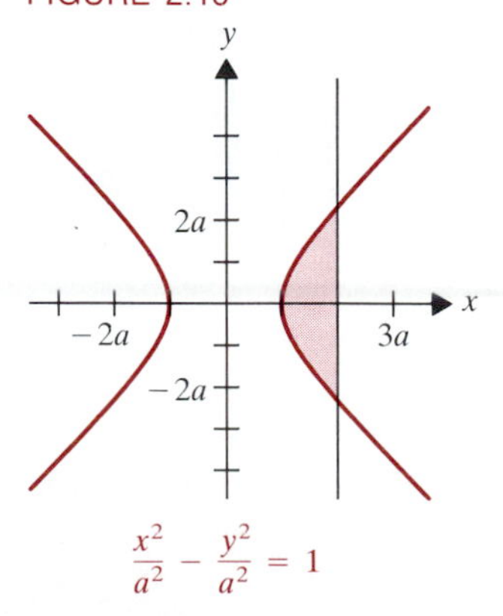

32. The *paraboloid of revolution* that results from revolving the graph of $4py = x^2$ below $y = p > 0$ about the y-axis. See **Figure 2.17.**

FIGURE 2.17

33. The pyramid with square base of side a and height h. See **Figure 2.18.**

FIGURE 2.18

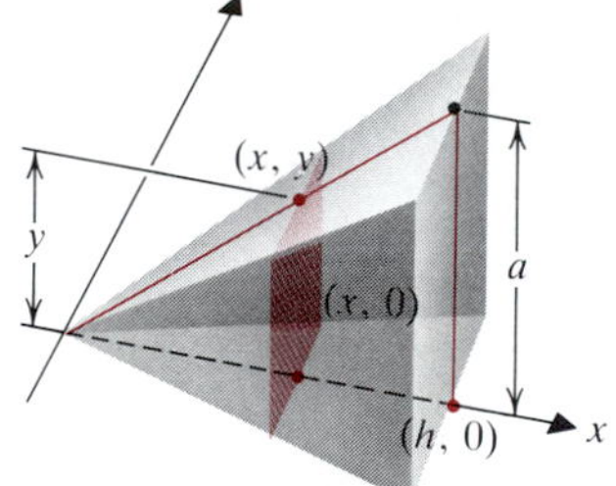

34. A regular octahedron, a solid obtained by gluing together two pyramids all of whose edges have length a. See **Figure 2.19.**

FIGURE 2.19

35. A solid S has base the circular disk $x^2 + y^2 \leq 4$, and every plane section of the solid perpendicular to the x-axis is a square. Find the volume of S.

36. Repeat Exercise 35 if the base is the disk $x^2 + y^2 \leq 9$ and every plane section perpendicular to the y-axis is a square.

37. A solid T has base an equilateral triangle of side 3, one vertex at the origin, and midpoint of the opposite side on the x-axis (**Figure 2.20**). Find the volume, if each plane section perpendicular to the x-axis is a square.

FIGURE 2.20

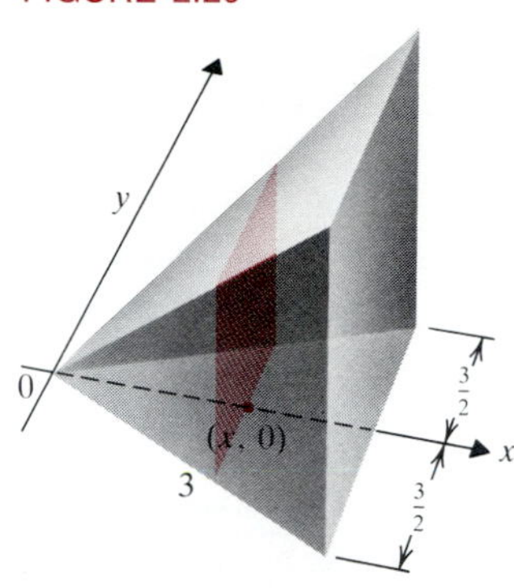

38. Repeat Exercise 37 if the side length is 4.

39. A solid circular cylinder intersects the xy-plane in the disk $x^2 + y^2 \leq 4$. A wedge is cut out of the cylinder by first cutting along the xy-plane and then making a second cut that intersects the xy-plane at the y-axis, making an angle of 30°. See **Figure 2.21.** Find the volume of the wedge, noting that slices perpendicular to the y-axis are right triangles. (Such wedges are cut out of trees before they are felled, to make them fall in a desired direction.)

FIGURE 2.21

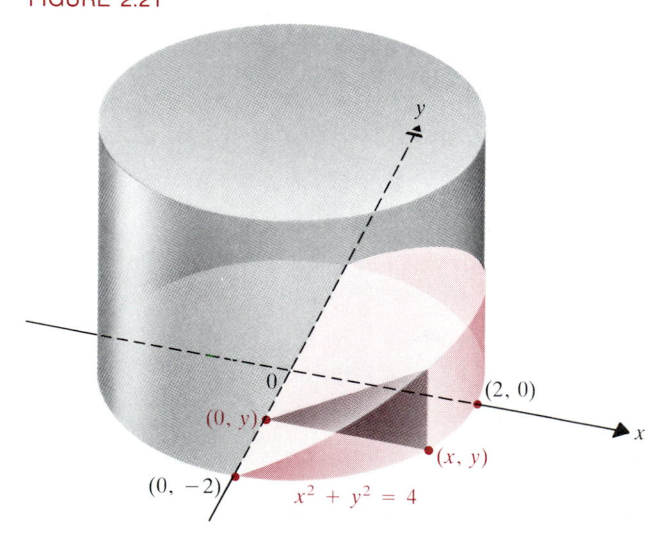

40. Repeat Exercise 39 if the disk has radius 3 and the angle of cutting is 45°.

FIGURE 2.22

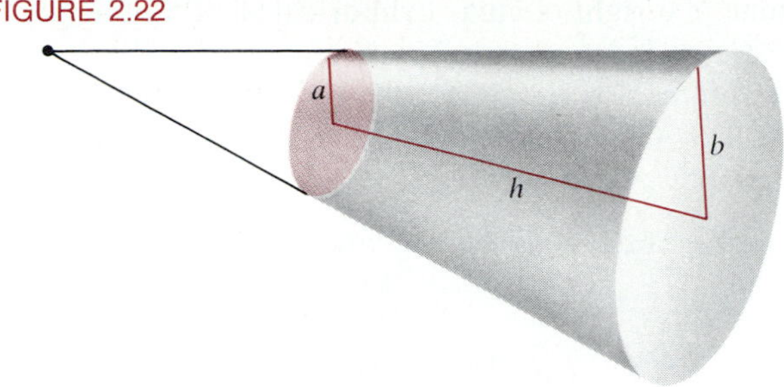

41. Find a formula for the volume of the *frustum* of a cone shown in **Figure 2.22.**

42. The interior of a bowl has the shape of a hemisphere of radius 3 cm. It is filled with soup to a height of 2 cm. What is the volume of soup in the bowl?

43. Derive the formula $V = \frac{1}{3}Ah$ for the volume of a tetrahedron of height h and base area A. [That formula was mentioned in Example 1(c).]

5.3 Volumes of Solids / Cylindrical Shells

The methods of the last section enable us to find the volume of solids with known cross-sectional area. They are particularly effective for finding the volume of a solid of revolution that is obtained by revolving the graph of a function $y = f(x)$ about the x-axis or by revolving the graph of a function $x = g(y)$ about the y-axis.

However, in other cases, these methods may be inconvenient. Consider, for example, the region bounded by the graph of $f(x) = 8x - x^4$, the y-axis, and the line $y = 7$. If we revolve this region about the y-axis, the circular disk method gives the formula

$$V(S) = \pi \int_0^7 x^2\, dy \tag{1}$$

for the volume of the resulting solid S (see **Figure 3.1**). Since $y = f(x) = 8x - x^4$, to evaluate (1) we would have to find the *inverse* function $x = f^{-1}(y)$. But finding such a formula for x would involve solving the fourth-degree equation

$$x^4 - 8x + y = 0$$

for x in terms of y. Such a task is too formidable to contemplate seriously.

The purpose of this section is to develop an alternative method for finding the volume of a solid of revolution for use in problems such as this, in which the circular disk method is inconvenient or inapplicable. The new method is based

FIGURE 3.1

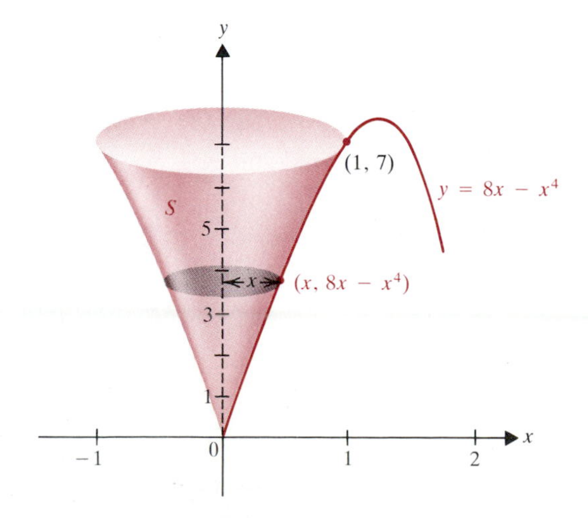

FIGURE 3.2

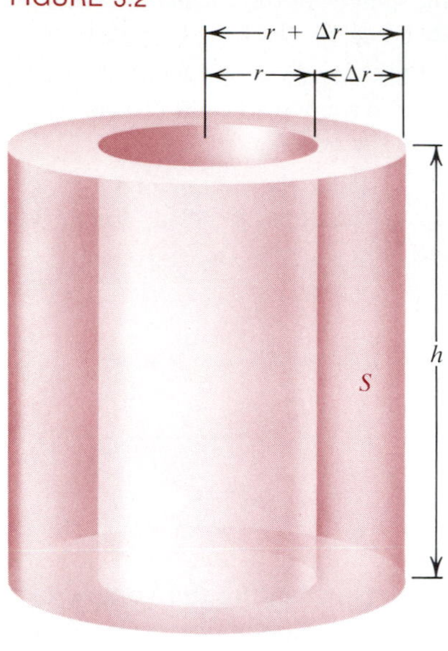

on the formula for the volume of a right circular cylindrical shell S of height h, inner radius r, and outer radius $r + \Delta r$.

As **Figure 3.2** shows, the volume S is the difference $V_1 - V_2$, where V_1 is the volume of a cylinder of radius $r + \Delta r$ and height h, and V_2 is the volume of a cylinder of radius r and height h. Since

$$V_1 = \pi(r + \Delta r)^2 h = \pi(r^2 + 2r\,\Delta r + (\Delta r)^2)h \qquad \text{and} \qquad V_2 = \pi r^2 h,$$

we have

$$\begin{aligned} V(S) = V_1 - V_2 &= \pi r^2 h + 2\pi r h\,\Delta r + \pi(\Delta r)^2 h - \pi r^2 h \\ &= 2\pi r h\,\Delta r + \pi(\Delta r)^2 h = 2\pi h\,\Delta r\left(r + \frac{1}{2}\,\Delta r\right), \end{aligned}$$

that is,

$$V(S) = 2\pi\left(r + \frac{1}{2}\,\Delta r\right)h\,\Delta r.$$

Thus, the volume of the cylindrical shell S is

(2) $$2\pi \cdot (\text{average radius}) \cdot \text{height} \cdot \text{thickness}.$$

Now suppose that we revolve the region bounded by the graph of the continuous function $y = f(x)$ and the lines $y = c$, $x = a$, and $x = b$ about the y-axis (see **Figure 3.3**). As usual, we partition $[a, b]$ into subintervals of equal length $\Delta x = (b - a)/n$. Consider the ith subinterval $[x_{i-1}, x_i]$. When the region above this subinterval between $y = c$ and $y = f(x)$ is revolved about the y-axis, the resulting solid ΔS_i is approximately a right circular cylindrical shell. Referring to **Figure 3.4,** we see that the volume of ΔS_i is approximately the volume of a shell of thickness $\Delta x = x_i - x_{i-1}$, inner radius x_{i-1}, outer radius x_i, and height $f(m_i) - c$, where m_i is the midpoint of $[x_{i-1}, x_i]$. Hence, (2) gives

$$\begin{aligned} V(\Delta S_i) &\approx 2\pi \cdot \text{average radius} \cdot \text{height} \cdot \text{thickness} \\ &\approx 2\pi m_i[f(m_i) - c]\,\Delta x. \end{aligned}$$

The total volume of S is approximated by the sum of the volumes of all such shells (see **Figure 3.5**):

(3) $$V(S) \approx \sum_{i=1}^{n} V(\Delta S_i) \approx \sum_{i=1}^{n} 2\pi m_i[f(m_i) - c]\,\Delta x.$$

FIGURE 3.3

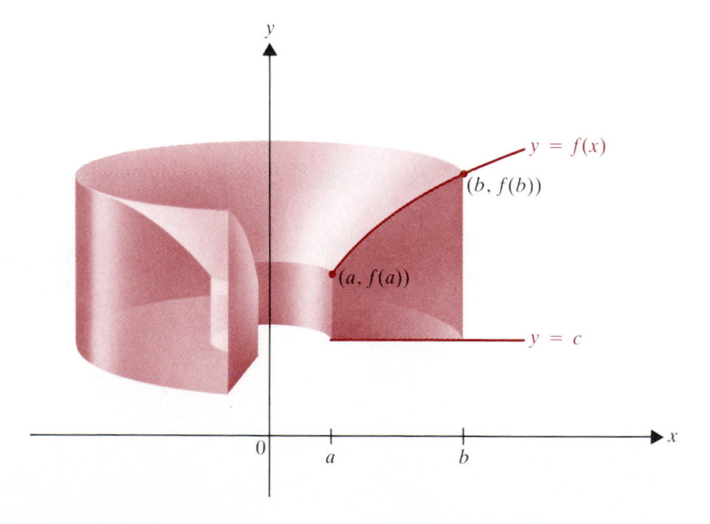

FIGURE 3.4

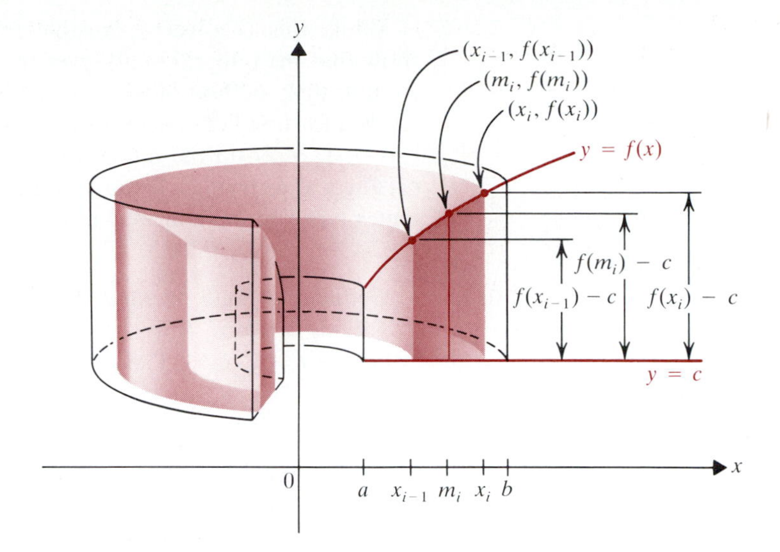

FIGURE 3.5

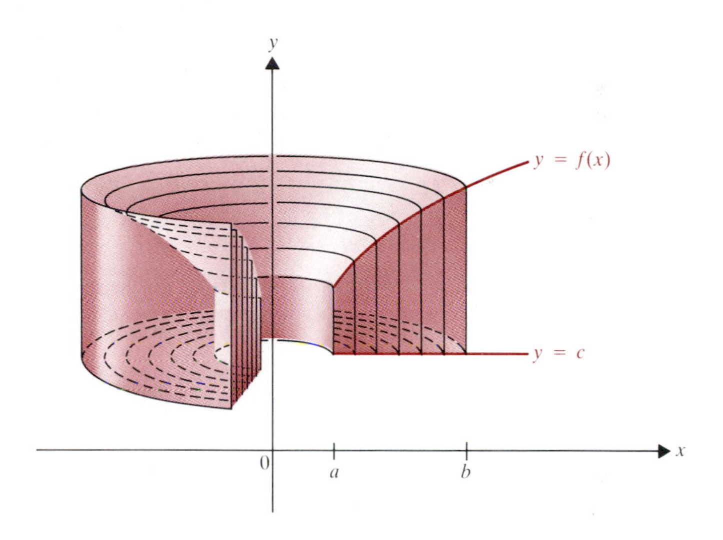

The sum on the right side of (3) is a Riemann sum for the function

$$2\pi x[f(x) - c]$$

over the interval $[a, b]$. Moreover, it appears that as we take finer and finer partitions, the error made in approximating $V(S)$ by the Riemann sum

$$R_P = \sum_{i=1}^{n} 2\pi m_i[f(m_i) - c]\,\Delta x$$

becomes smaller and smaller. The next result therefore seems quite plausible.

3.1 THEOREM Let S be the solid of revolution obtained by revolving the region under the graph of a continuous function f and above $y = c$ between $x = a$ and $x = b$ about the y-axis. Then

$$V(S) = 2\pi \int_a^b x[f(x) - c]\,dx. \tag{4}$$

In particular, if $c = 0$,

$$V(S) = 2\pi \int_a^b xf(x)\,dx. \tag{5}$$

While this is a very plausible result, we do *not* prove it. A proof must show that the integral in (4) gives the same volume as that calculated using the circular disk method of the last section. We will offer a partial proof in Section 7.1, but for now we concentrate on applying this method. As in the last section, it is best to set up the integral for each individual problem, rather than memorizing and applying a formula like (4) or (5). The reasoning leading up to Theorem 3.1 can be summarized in the following general statement.

3.2 THEOREM

Cylindrical Shell Method. Suppose that the solid S is generated by revolving a region R in the plane about a line l perpendicular to the t-axis (where $t = x$ or y). Then

$$V(S) = 2\pi \int_a^b (\text{radius of revolution}) \cdot (\text{height of shell})\, dt \tag{6}$$

Here a and b are the smallest and largest t coordinates in the region R, *radius of revolution* is the distance from the axis of revolution to a typical point (x, y) on the boundary of R, and *height of shell* is the distance between two points on opposite boundaries of R measured parallel to l.

Example 1

Find the volume of the solid that results when the region bounded by $y = x^2$, $y = 1$, and $x = 2$ is revolved about the y-axis.

Solution. We show a typical shell in **Figure 3.6.** The radius of revolution is x, and the height of the shell is $f(x) - 1 = x^2 - 1$. Here $a = 1$ and $b = 2$. We then have from (6)

$$\begin{aligned} V(S) &= 2\pi \int_1^2 x(x^2 - 1)\, dx \\ &= 2\pi \int_1^2 (x^3 - x)\, dx = 2\pi \left[\frac{1}{4}x^4 - \frac{1}{2}x^2\right]_1^2 \\ &= \frac{1}{2}\pi\left[x^4 - 2x^2\right]_1^2 = \frac{1}{2}\pi[16 - 8 - 1 + 2] = \frac{9\pi}{2}. \quad \blacksquare \end{aligned}$$

FIGURE 3.6

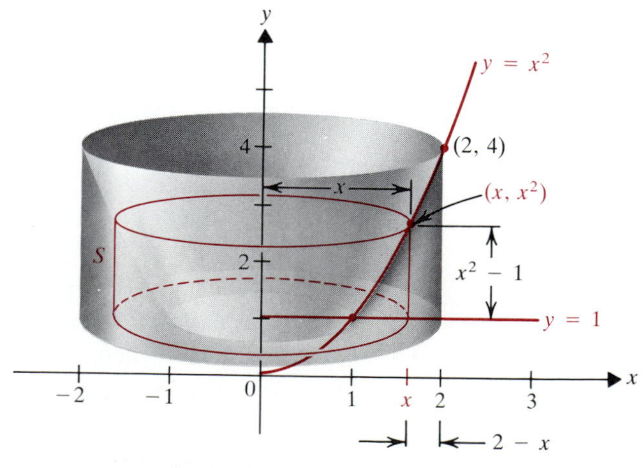

The next example returns to the solid described at the beginning of the section.

Example 2

Find the volume of the solid that results when the region between the y-axis and the graph of $y = 8x - x^4$ from $x = 0$ to $x = 1$ is revolved about the y-axis.

Solution. Referring to **Figure 3.7,** we see that the radius of revolution is x and the height of the shell is $7 - (8x - x^4) = x^4 - 8x + 7$. Since $a = 0$ and $b = 1$, we get from (6)

$$V = 2\pi \int_0^1 x(x^4 - 8x + 7)\,dx = 2\pi \int_0^1 (x^5 - 8x^2 + 7x)\,dx$$

$$= 2\pi\left[\frac{x^6}{6} - \frac{8}{3}x^3 + \frac{7}{2}x^2\right]_0^1 = 2\pi\left[\frac{1}{6} - \frac{8}{3} + \frac{7}{2}\right] = 2\pi. \quad ■$$

FIGURE 3.7

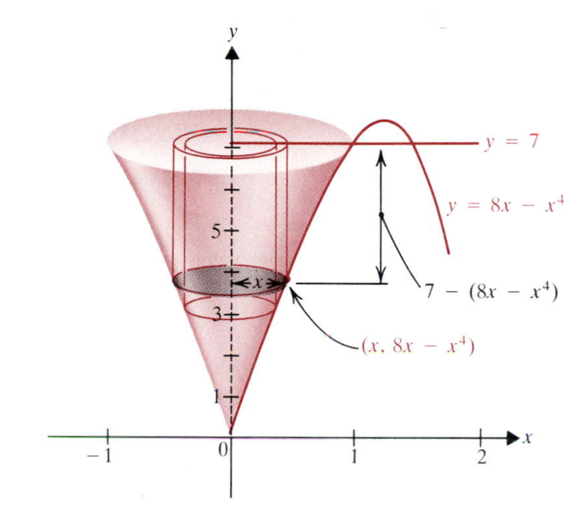

The advantage of using (6) is that it applies to surfaces obtained by revolving regions about *any* line in the plane, not just the y-axis as in (4) and (5).

Example 3

The region in Example 1 is revolved about the line $x = 2$. Find the volume of the resulting solid.

Solution. Referring back to Figure 3.6, we see that the cylindrical shell this time has radius of revolution $2 - x$. Since the height is still $x^2 - 1$ and since a and b are still 1 and 2, respectively, we get from (6)

$$V(S) = 2\pi \int_1^2 (2 - x)(x^2 - 1)\,dx = 2\pi \int_1^2 (2x^2 - x^3 + x - 2)\,dx$$

$$= 2\pi\left[\frac{2}{3}x^3 - \frac{x^4}{4} + \frac{x^2}{2} - 2x\right]_1^2$$

$$= 2\pi\left[\frac{16}{3} - \frac{16}{4} + \not{2} - \not{4} - \frac{2}{3} + \frac{1}{4} - \frac{1}{2} + \not{2}\right] = \frac{5\pi}{6}. \quad ■$$

The case when l is the x-axis in (6) gives the following dual result to Theorem 3.1.

3.2 COROLLARY Let S be the solid of revolution in **Figure 3.8,** which is obtained by revolving the region R between the y-axis, $y = c$, $y = d$, and the graph of a continuous function $x = g(y)$ about the x-axis. Then S has volume

$$V(S) = 2\pi \int_c^d y\,g(y)\,dy. \tag{7}$$

FIGURE 3.8

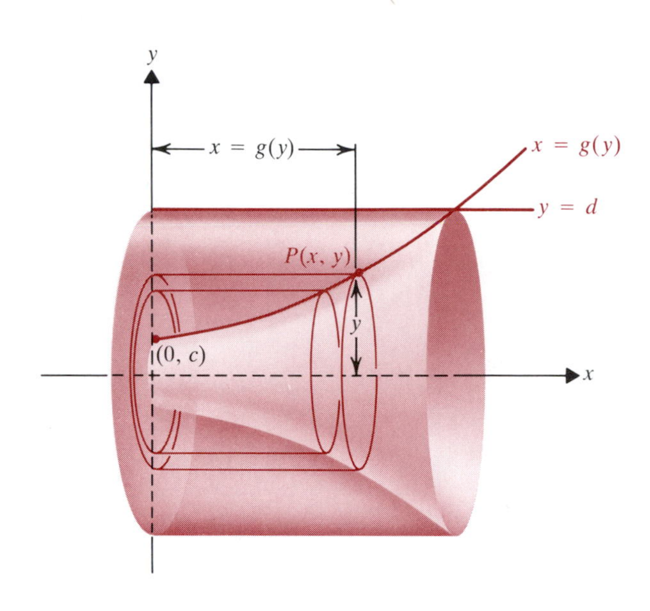

Proof. The cylindrical shell obtained by revolving the segment between $(0, y)$ and $P(g(y), y)$ about the x-axis has height $x = g(y)$ and radius of revolution y. Since y varies from c to d in the region, (6) gives

$$V(S) = 2\pi \int_c^d y\,g(y)dy. \quad \boxed{\text{QED}}$$

To illustrate the consistency between this section's methods and those of the last section, we next rework Example 3 of the last section.

FIGURE 3.9

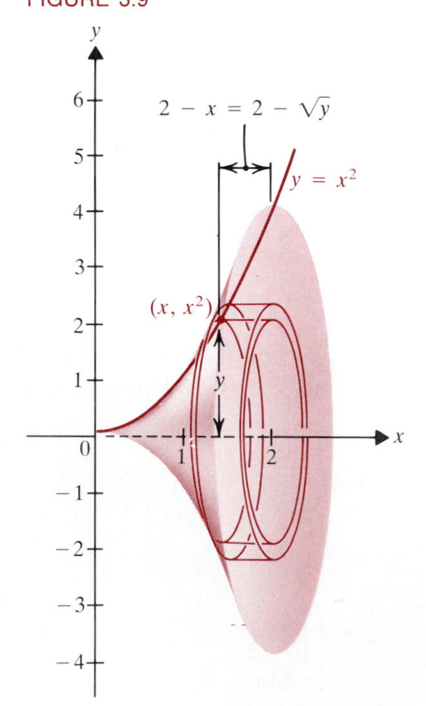

Example 4

Find the volume of the solid of revolution in **Figure 3.9** obtained by revolving the region in the xy-plane between the x-axis, the graph of $y = x^2$, and the line $x = 2$ about the x-axis.

Solution. For the portion of the graph we are considering, $x = g(y) = \sqrt{y}$. The cylindrical shell has radius y and height $2 - x$. Thus from (6),

$$\begin{aligned} V &= 2\pi \int_0^4 y(2 - x)\,dy \\ &= 2\pi \int_0^4 y(2 - \sqrt{y})\,dy = 2\pi \int_0^4 (2y - y^{3/2})\,dy \\ &= 2\pi\left[y^2 - \frac{2}{5}y^{5/2}\right]_0^4 = 2\pi\left[16 - \frac{64}{5}\right] = \frac{32\pi}{5}. \end{aligned}$$ ■

As Example 4 suggests,

> it is frequently better to use the methods of the last section when dealing with solids produced by revolving graphs of functions $y = f(x)$ about the x-axis, or of functions $x = g(y)$ about the y-axis. The cylindrical shell method tends to be easier when dealing with solids obtained by revolving the graph of $y = f(x)$ about the y-axis, or the graph of $x = g(y)$ about the x-axis.

FIGURE 3.10

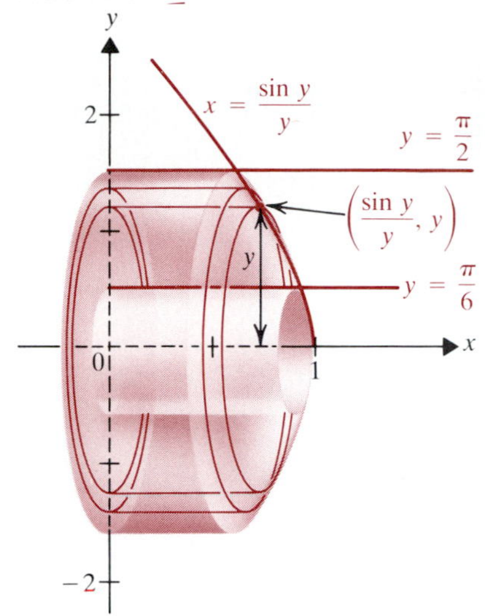

Our final example is one in which the cylindrical shell method is the *only* feasible one.

Example 5

Find the volume of the solid obtained by revolving the region R bounded by the y-axis and the graph of

$$x = \frac{\sin y}{y}$$

between $y = \pi/6$ and $y = \pi/2$ about the x-axis. See **Figure 3.10.**

Solution. The radius of revolution of the cylindrical shell is y, and its height is $x = (\sin y)/y$. Thus (6) gives

$$V = 2\pi \int_{\pi/6}^{\pi/2} y\left[\frac{\sin y}{y}\right] dy = 2\pi \int_{\pi/6}^{\pi/2} \sin y\, dy = 2\pi \Big[-\cos y\Big]_{\pi/6}^{\pi/2}$$

$$= 2\pi\left[0 + \frac{\sqrt{3}}{2}\right] = \pi\sqrt{3} \approx 5.4414. \blacksquare$$

In this example, it is not feasible to use the methods of the last section, because they would require us to solve $x = (\sin y)/y$ for y in terms of x.

Exercises 5.3

In Exercises 1–28, find the volume of the solid of revolution formed when the region bounded by the given curves is revolved about the indicated axis. Make a sketch. Use the cylindrical shell method where possible.

1. $y = x^2$, the x-axis, and $x = 2$; y-axis
2. $y = x^2$, the x-axis, and $x = 1$; y-axis
3. $y = x^2$, the x-axis, and $x = \sqrt{2}$; y-axis
4. $y = x^2$, the x-axis, and $x = 3$; y-axis
5. $y = \sin x^2$, the x-axis, and $x = \sqrt{\pi/2}$; y-axis
6. $y = \cos x^2$, $y = 0$, the x-axis, and $x = \sqrt{\pi/2}$; y-axis
7. $y = x \sec^2 x^3$, the x-axis, $x = 0$, and $x = \sqrt[3]{\pi/4}$; y-axis
8. $y = \tan^2 x^2$, the x-axis, and $x = \sqrt{\pi/3}$; y-axis
9. $x = y^2$, the x-axis, $x = 1$, and $x = 4$ (first-quadrant portion); y-axis
10. $x = y^2$, the x-axis, $x = 1$, and $x = 9$ (first-quadrant portion); y-axis
11. $y = x^2$, $y = 1$, and $x = 3$; x-axis
12. $y = x^2$, $y = 2$, and $x = 2$; x-axis
13. $y = x^2 + 1$, $y = 5$, and $x = 0$ (right portion); the line $x = 3$
14. $y = 2 - x^2$, $y = 1$, and $x = 0$ (right half); the line $x = 2$
15. $y = x^2$, $y = x$, in the first quadrant; y-axis
16. $y = x^2$, $y = 2x$, in the first quadrant; y-axis
17. $y = 2x^2 - x^3$, $y = 0$, between $x = 0$ and $x = 1$; y-axis
18. $y = 2x^2 - x^3$, $y = 0$, between $x = 0$ and $x = 2$; y-axis
19. $x = y^2$, $y = 2$, y-axis; x-axis
20. $x = y^2$, $y = 4$, y-axis; x-axis
21. $x = \cot^2 y^2$, $y = \sqrt{\pi/6}$, y-axis; x-axis

22. $x = y\tan^2 y^3$, $y = \sqrt[3]{\pi/6}$, $y = \sqrt[3]{\pi/3}$, y-axis; x-axis

23. $x = y^2$, $x = y^3$, between $y = 0$ and $y = 1$; x-axis

24. $x = y^2$, $x^2 = 8y$; x-axis

25. $y = 4x^2$, $x = 2$, $y = 0$; $x = 2$

26. $y = 2x^2 + 1$, $x = 0$, $x = 2$, $y = 0$; $x = 2$

27. $x^2 = y^3$, $x = 0$, $y = 4$; y-axis

28. $x^3 = y^2$, $x = 0$, $y = 4$; x-axis.

29. A hole of radius 1 is bored through the center of a ball of radius 4. Find the volume of the resulting solid.

30. Repeat Exercise 29 if the hole has radius 2.

31. If a ball of radius a has a hole bored through it of radius $b < a$, then what is the volume of the resulting solid?

32. A *torus* looks like a doughnut or bagel and is formed by revolving the disk enclosed by $x^2 + y^2 = a^2$ about the line $x = b$, where $b > a$ (see **Figure 3.11**). Find the volume of the torus. (*Hint:* Recall that $\int_0^a \sqrt{a^2 - x^2}\,dx = \frac{1}{4}a^2\pi$ is the area of one quarter of the disk.)

In Exercises 33–38, use the cylindrical shell method to solve the indicated problem from Exercises 5.2.

33. Exercise 25
34. Exercise 26
35. Exericse 27
36. Exercise 28
37. Exercise 31
38. Exercise 32

FIGURE 3.11

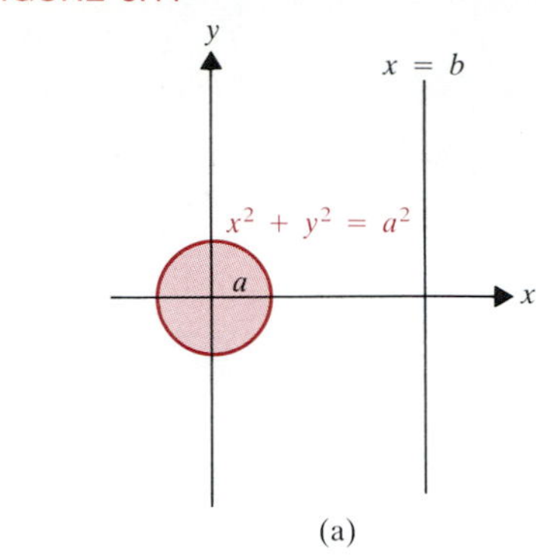

(a)

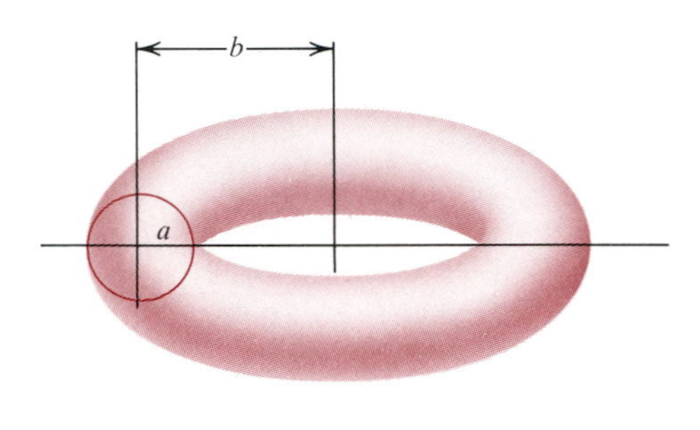

(b)

5.4 Arc Length of a Curve

You learned in elementary school that the circumference of a circle of radius r is $2\pi r$. In this section, we will use integration to calculate the lengths of more general curves. Consider the graph of a function $y = f(x)$ between $x = a$ and $x = b$. Physically, its length L can be measured by superimposing a piece of string on it, and then laying the string along the scaled x-axis to determine its length, as suggested by **Figure 4.1.**

To describe L mathematically, we use the familiar approach of approximating Riemann sums, whose limit is defined to be L. First, partition the interval

FIGURE 4.1

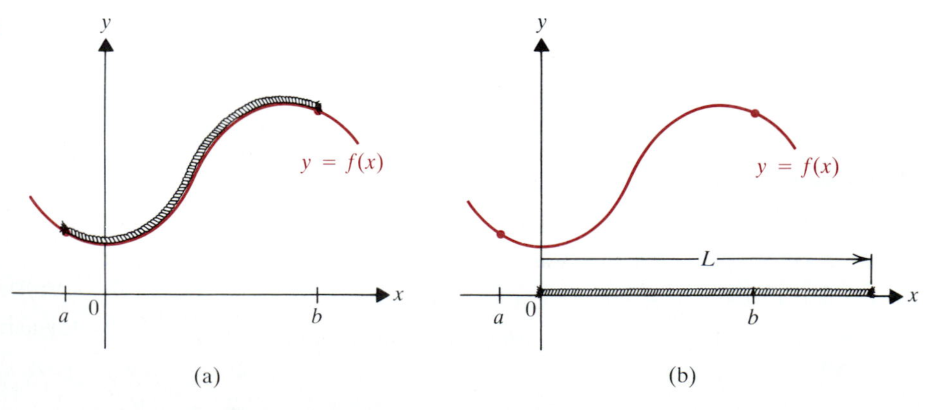

(a) (b)

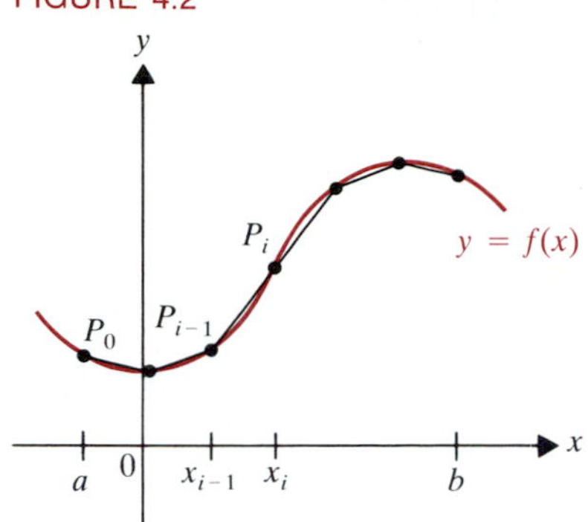

FIGURE 4.2

$[a, b]$ into n subintervals $[x_{i-1}, x_i]$, where $x_0 = a$ and $x_n = b$. Next, join adjacent points $P_{i-1}(x_{i-1}, f(x_{i-1}))$ on the curve by a *secant*, that is, a straight line segment. As **Figure 4.2** suggests, the length of a piece of string laid out along the graph of $y = f(x)$ between $(a, f(a))$ and $(b, f(b))$ will be rather closely approximated by the sum of the lengths of the secants $P_{i-1}P_i$. From the distance formula, that sum is

$$(1) \qquad \sum_{i=1}^{n} d(P_{i-1}, P_i) = \sum_{i=1}^{n} \sqrt{(x_i - x_{i-1})^2 + [f(x_i) - f(x_{i-1})]^2}.$$

Now, it seems natural to take the limit of the approximating sum (1) and define that to be the arc length L. But (1) *isn't* a Riemann sum $\Sigma_{i=1}^{n} g(z_i)\Delta x_i$ for any function g. However, we can use the mean value theorem for derivatives (Theorem 10.2 of Chapter 2) to write it as a Riemann sum. To do so, we must assume that f' exists on (a, b) and that f is continuous on $[a, b]$. Then in each subinterval (x_{i-1}, x_i) there is a number z_i such that

$$(2) \qquad \frac{f(x_i) - f(x_{i-1})}{x_i - x_{i-1}} = f'(z_i).$$

Substituting (2) into (1) and recalling that $x_i - x_{i-1} = \Delta x_i$, we obtain the approximation

$$(3) \qquad \sum_{i=1}^{n} \sqrt{(\Delta x_i)^2 + [f'(z_i)]^2(\Delta x_i)^2} = \sum_{i=1}^{n} \sqrt{1 + [f'(z_i)]^2}\, \Delta x_i$$

for the arc length of the graph of $y = f(x)$ between $x = a$ and $x = b$. This *is* a Riemann sum, for the function $g(x) = \sqrt{1 + [f'(x)]^2}$. Thus we can define L to be its limit as the norm of the partition approaches zero, if that limit exists.

To ensure that the Riemann sum (3) has a limit, we restrict attention to functions that have a continuous derivative, so that $g(x) = \sqrt{1 + [f'(x)]^2}$ is continuous.

4.1 DEFINITION A function f is ***smooth*** over an interval $[a, b]$ if f' is defined and continuous on $[a, b]$.

The condition that a function be smooth is easy to interpret geometrically. Since $f'(x)$ is the slope of the tangent line to the graph of $y = f(x)$, the graph of a smooth function has a *continuously turning tangent*, as in **Figure 4.3.** It is natural then to say that the graph is smooth, because it can have no sharp points. **Figure 4.4** shows such a sharp point on the graph of $y = |x|$. The tangent line to $y = |x|$ is $y = x$ for $x > 0$ and jumps to $y = -x$ for $x < 0$. At $x = 0$, f'

FIGURE 4.3

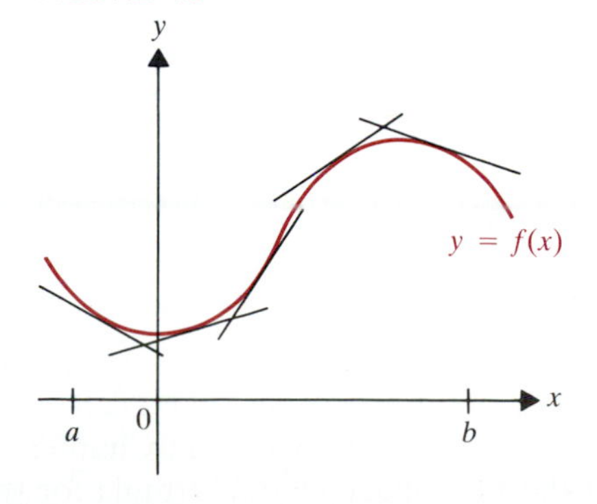

FIGURE 4.4

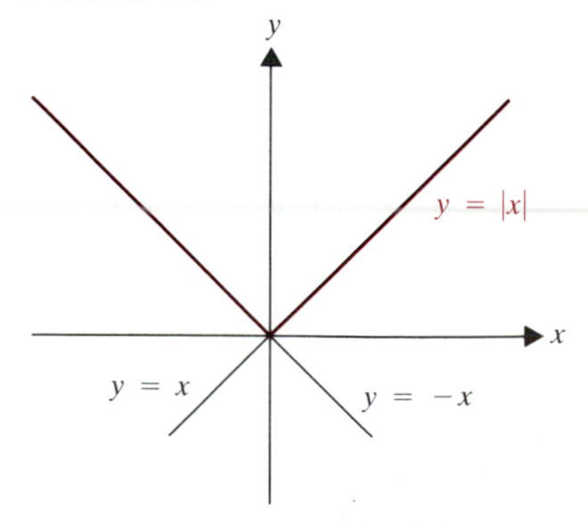

has a jump discontinuity: it is undefined there, and jumps from -1 for $x < 0$ to 1 for $x > 0$.

We can now define the arc length for a smooth function f.

4.2 DEFINITION If f is a smooth function over $[a, b]$, then the ***arc length*** of the graph of $y = f(x)$ between $x = a$ and $x = b$ is

$$L = \int_a^b \sqrt{1 + [f'(x)]^2}\, dx. \tag{4}$$

The integral in (4) exists by Theorem 2.3 of the last chapter, because of the continuity of f'. For since f' is continuous on $[a, b]$, so are the functions $[f']^2$, $1 + [f']^2$, and $\sqrt{1 + [f']^2}$, because $1 + [f'(x)]^2 > 0$ for every x in $[a, b]$.

Definition 4.2 gives the expected results for familiar curves in the plane, such as line segments.

Example 1

Use Definition 4.2 and the distance formula to find the arc length of the line $y = mx + b$ between $x = a$ and $x = c$, if $c > a$ (**Figure 4.5**).

FIGURE 4.5

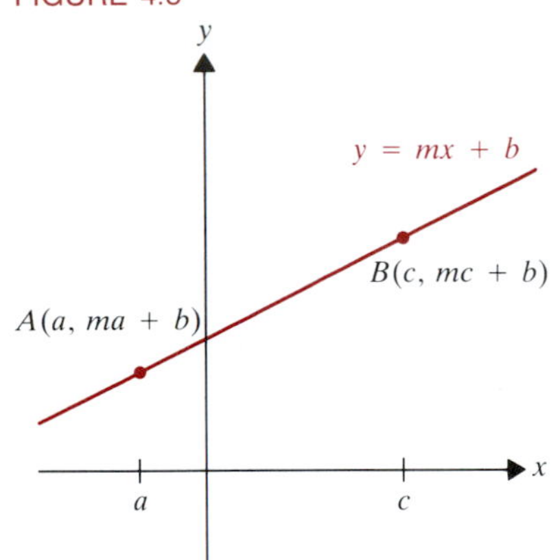

Solution. Here $dy/dx = m$, so (4) says that

$$L = \int_a^c \sqrt{1 + m^2}\, dx = \sqrt{1 + m^2}(c - a).$$

Computing the length of the line segment joining the points $A(a, ma + b)$ and $B(c, mc + b)$ from the distance formula, we get

$$\begin{aligned} d(A, B) &= \sqrt{(c - a)^2 + (mc + b - ma - b)^2} \\ &= \sqrt{(c - a)^2 + m^2(c - a)^2} = (c - a)\sqrt{1 + m^2} \\ &= L. \quad \blacksquare \end{aligned}$$

With this reassuring example under our belts, we proceed to find some more general arc lengths.

Example 2

Find the arc length of the curve

$$y = f(x) = \frac{2}{3}x^{3/2}$$

over $[0, 3]$ (**Figure 4.6**).

FIGURE 4.6

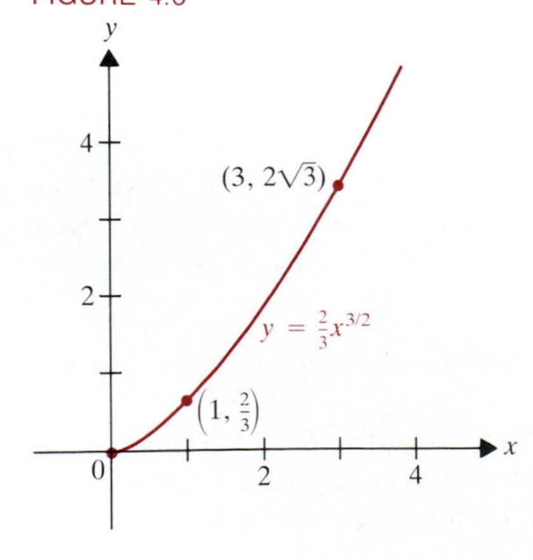

Solution. We have $f'(x) = x^{1/2}$. Since f' is defined and continuous on $[0, 3]$, f is a smooth function. Thus from (4),

$$\begin{aligned} L &= \int_0^3 \sqrt{1 + (x^{1/2})^2}\, dx = \int_0^3 \sqrt{1 + x}\, dx \\ &= \frac{(1 + x)^{3/2}}{3/2}\Bigg]_0^3 = \frac{2}{3}[4^{3/2} - 1^{3/2}] = \frac{14}{3}. \quad \blacksquare \end{aligned}$$

The presence of the radical in (4) can lead to complicated integrals, some of which we will learn to evaluate in Chapter 7. As an example, we can check (4) against the elementary school formula for the circumference of a circle.

FIGURE 4.7

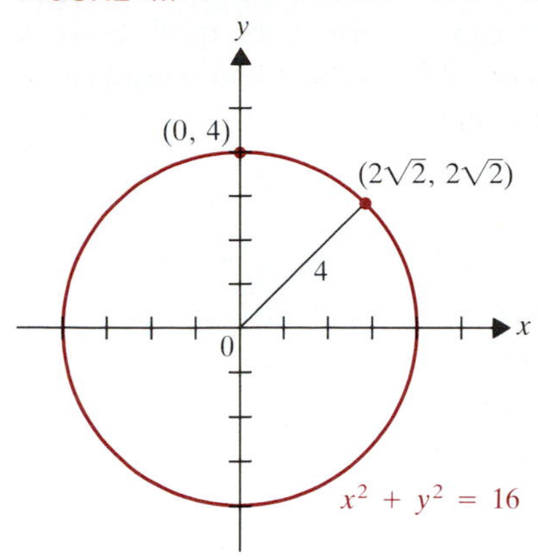

Example 3

Set up the integral for the arc length of the first-quadrant part of the circle $x^2 + y^2 = 16$ between $x = 0$ and $x = 2\sqrt{2}$ (**Figure 4.7**).

Solution. The technique of implicit differentiation gives

$$2x + 2y\frac{dy}{dx} = 0 \rightarrow \frac{dy}{dx} = -\frac{x}{y}.$$

Since $y \neq 0$ for $x \in [0, 2\sqrt{2}]$, dy/dx is continuous on this interval, and we have

$$1 + \left(\frac{dy}{dx}\right)^2 = 1 + \frac{x^2}{y^2} = \frac{y^2 + x^2}{y^2} = \frac{16}{y^2} = \frac{16}{16 - x^2}.$$

Hence (4) says that

$$L = \int_0^{2\sqrt{2}} \frac{4}{\sqrt{16 - x^2}}\, dx. \quad (5)$$

Since the full circle of radius 4 in Example 2 has circumference $2\pi \cdot 4 = 8\pi$, the first-quadrant portion has length 2π. Thus the integral in (5) should be π, since the length of the arc from (0, 4) to $(2\sqrt{2}, 2\sqrt{2})$ is half the first-quadrant arc length. [Note that $(2\sqrt{2}, 2\sqrt{2})$ lies on the line $y = x$ which bisects the first quadrant.] Although we won't learn how to evaluate (5) by antidifferentiation until Chapter 7, we *can* approximate it by Simpson's rule. Taking $n = 10$, we have

$$h = \frac{2\sqrt{2} - 0}{10} \approx 0.2828427.$$

Hence Simpson's rule [Equation (16), p. 274] gives

$$L \approx \frac{h}{3}[g(0) + 4g(h) + 2g(2h) + 4g(3h) + 2g(4h) + 4g(5h)$$
$$+ 2g(6h) + 4g(7h) + 2g(8h) + 4g(9h) + g(2\sqrt{2})]$$
$$\approx 3.1416411 \qquad \text{(from a calculator)},$$

where

$$g(x) = \frac{4}{\sqrt{16 - x^2}}.$$

Thus, to four decimal places, L agrees with the answer given by the familiar formula for the circumference of a circle.

Why was only *half* the first-quadrant arc length of the circle calculated in Example 3? We know that the definite integral (5) exists, because the function being integrated is continuous over the entire interval $[0, 2\sqrt{2}]$ of integration. If we tried to use (4) to compute the entire first-quadrant arc length L', then we would have been confronted by the formula

$$L = \int_0^{4} \frac{4}{\sqrt{16 - x^2}}\, dx. \quad (5')$$

Since its denominator is 0 at $x = 4$, the integrand fails to be continuous on [0, 4]. Thus we don't know whether (5′) is a meaningful formula. [The integral in (5′) is called an *improper integral*. Section 7.6 discusses such integrals.]

The approach of Example 3 can be used to solve Problem 7 of the Prologue (see Exercise 31). The methods of Section 4.5 can sometimes be used to work out integrals by antidifferentiation that at first glance appear too complicated to evaluate directly, as the next example illustrates.

Example 4

Find the arc length of the graph of

$$y = \frac{3}{2}x^{2/3}$$

between $x = 1$ and $x = 8$.

Solution. Here $dy/dx = x^{-1/3}$, which is continuous on $[1, 8]$. We have

$$1 + \left(\frac{dy}{dx}\right)^2 = 1 + \frac{1}{x^{2/3}} = \frac{x^{2/3} + 1}{x^{2/3}}.$$

Thus (4) gives

$$L = \int_1^8 \frac{\sqrt{x^{2/3} + 1}}{x^{1/3}}\, dx = \int_1^8 x^{-1/3}\sqrt{x^{2/3} + 1}\, dx. \tag{6}$$

Let $u = x^{2/3} + 1$. Then

$$du = \frac{2}{3}x^{-1/3}\, dx,$$

so we have

$$x^{-1/3}\, dx = \frac{3}{2}\, du.$$

Note that $u = 2$ when $x = 1$, and $u = 8^{2/3} + 1 = 5$ when $x = 8$. Thus (6) transforms to

$$L = \frac{3}{2}\int_2^5 u^{1/2}\, du = \frac{3}{2}\left[\frac{2}{3}u^{3/2}\right]_2^5 = 5\sqrt{5} - 2\sqrt{2} \approx 8.3519. \quad ■$$

FIGURE 4.8

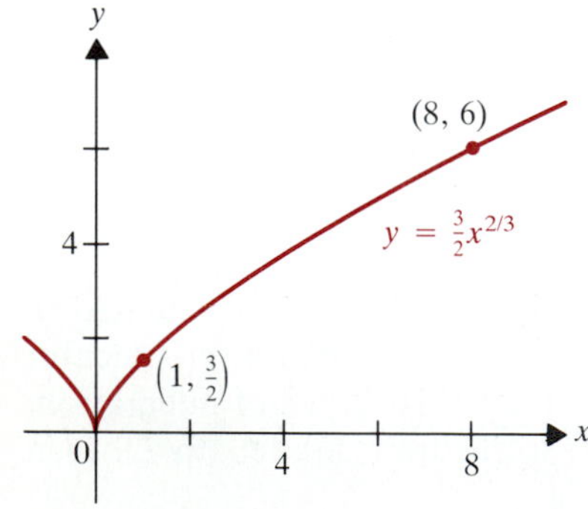

As **Figure 4.8** shows, the graph of $y = \frac{3}{2}x^{2/3}$ has a cusp (sharp point) at the origin. The function $f(x) = x^{2/3}$ can be shown to be nondifferentiable at $x = 0$ (Exercise 35), so is not smooth over any interval containing 0.

To find the arc length of the graph of a function $x = g(y)$ between $y = c$ and $y = d$, we use the analogous formula to (4), namely,

$$L = \int_c^d \sqrt{1 + [g'(y)]^2}\, dy \tag{7}$$

Sometimes it is helpful to use (7) even when we are given a function $y = f(x)$.

Example 5

Use (7) to find the arc length in Example 4.

Solution. For $x \in [1, 8]$ we can solve $y = \frac{3}{2}x^{2/3}$ for x, getting

$$x^{2/3} = \frac{2}{3}y \rightarrow x = (2/3)^{3/2}y^{3/2} = g(y).$$

Then

$$\frac{dx}{dy} = \frac{3}{2} \cdot \frac{2}{3}\left(\frac{2}{3}\right)^{1/2} y^{1/2} = \left(\frac{2}{3}\right)^{1/2} y^{1/2}.$$

When $x = 1$, $y = 3/2$. When $x = 8$, $y = 6$. So from (7),

$$L = \int_{3/2}^{6} \sqrt{1 + \frac{2}{3}y}\, dy = \frac{3}{2}\int_{3/2}^{6}\left(1 + \frac{2}{3}y\right)^{1/2} \cdot \frac{2}{3}\, dy$$

$$= \frac{3}{2} \cdot \frac{[1 + (2/3)y]^{3/2}}{3/2}\Bigg]_{3/2}^{6} = 5^{3/2} - 2^{3/2}. \quad ■$$

The integral in Example 5 was somewhat simpler to evaluate than the integral in Example 4.

Exercises 5.4

In Exercises 1–22, find the arc length of the graph of the given function over the given interval.

1. $f(x) = x^{3/2}$ over $[0, 4/3]$
2. $f(x) = x^{3/2}$ over $[0, 32/9]$
3. $f(x) = \frac{1}{3}(x^2 + 2)^{3/2}$ over $[0, 1]$
4. $f(x) = \frac{1}{3}(x^2 + 2)^{3/2}$ over $[0, 2]$
5. $f(x) = x^3 + \frac{1}{12x}$ over $[1, 2]$
6. $f(x) = \frac{x^3}{2} + \frac{1}{6x}$ over $[1, 2]$
7. $f(x) = \frac{1}{4}x^4 + \frac{1}{8x^2}$ over $[1, 2]$
8. $f(x) = \frac{x^4}{8} + \frac{1}{4x^2}$ over $[1, 3]$
9. $f(x) = \frac{2}{3}x^{3/2} - \frac{1}{2}x^{1/2}$ over $[1, 4]$
10. $f(x) = \frac{2}{3}x^{3/2} - \frac{1}{2}x^{1/2} + 3$ over $[1, 9]$
11. $f(x) = \frac{3}{2}x^{2/3}$ between $x = 1$ and $x = 27$
12. $f(x) = \frac{3}{2}x^{2/3}$ between $x = 1$ and $x = 64$
13. $g(y) = y^{3/2} - \frac{1}{3}y^{1/2}$ over $[1, 9]$
14. $g(y) = y^{3/2} - \frac{1}{3}y^{1/2}$ over $[4, 16]$
15. $g(y) = \frac{1}{3}(y^2 - 2)^{3/2}$ over $[0, 3]$
16. $g(y) = \frac{1}{3}(y^2 - 2)^{3/2}$ over $[0, 1]$
17. $f(x) = \int_1^x \sqrt{t^2 - 1}\, dt$ over $[1, 2]$
18. $f(x) = \int_1^x \sqrt{t^2 + 2t}\, dt$
19. $f(x) = \int_0^x \sqrt{\cos t}\, dt$ over $[0, \pi/2]$ [*Hint:* Use Corollary 3.6(a) of Chapter 2.]
20. $f(x) = \int_0^x \sqrt{\cos t}\, dt$ over $[0, \pi/3]$ (*Hint:* See Exercise 19.)
21. $f(x) = \int_0^x \sqrt{\sec^4 t - 1}\, dt$ over $[0, \pi/4]$
22. $f(x) = \int_0^x \sqrt{\cot^4 t - 1}\, dt$ over $[\pi/6, \pi/4]$
23. The ***hypocycloid of four cusps*** is the graph of $x^{2/3} + y^{2/3} = a^{2/3}$.
    (a) Graph the equation with $a = 1$.
    (b) Find the arc length over $[1/8, 1]$.
24. Repeat Exercise 23 for $x^{2/3} + y^{2/3} = 4$.
25. Can (4) and (7) be used to find the entire arc length of the hypocycloid in Exercise 23? Explain.
26. Use Theorem 7.7 of Chapter 1 to show that if f is a smooth function on a closed interval $[a, b]$, then for some nonnegative real numbers m and M,
$$m \le |f'(x)| \le M.$$
27. Use Exercise 26 and (4) to show that for a smooth function $f(x)$, its arc length L over $[a, b]$ satisfies
$$(b - a)\sqrt{1 + m^2} \le L \le (b - a)\sqrt{1 + M^2}.$$
28. Use Exercise 27 to find upper and lower bounds for L if $f(x) = \cos x$ and $[a, b]$ is the interval $[\pi/3, \pi/2]$.

In Exercises 29–32, use Simpson's rule with $n = 10$ to estimate the given arc length.

PC **29.** $y = x^2$ between $x = 0$ and $x = 1$

PC **30.** $y = x^2$ between $x = -1$ and $x = 2$

PC **31.** $f(x) = \sin x$ between $x = 0$ and $x = \pi/2$

PC **32.** $f(x) = \cos x$ between $x = 0$ and $x = \pi/3$

PC **33.** Use Simpson's rule to estimate the arc length along $y = \tan x$ from $x = 0$ to $x = \pi/4$, correct to six decimal places.

PC **34.** Repeat Exercise 33 for $y = \sec x$ between $x = 0$ and $x = \pi/3$.

35. Show that the function $f(x) = x^{2/3}$ is not differentiable at $x = 0$.

36. Repeat Exercise 35 for the function $f(x) = x^{1/3}$.

37. Is $f(x) = x^{1/2}$ smooth on $[0, 1]$?

38. Is $f(x) = x^{3/2}$ smooth on $[0, 1]$?

5.5 Moments and Center of Mass

FIGURE 5.1

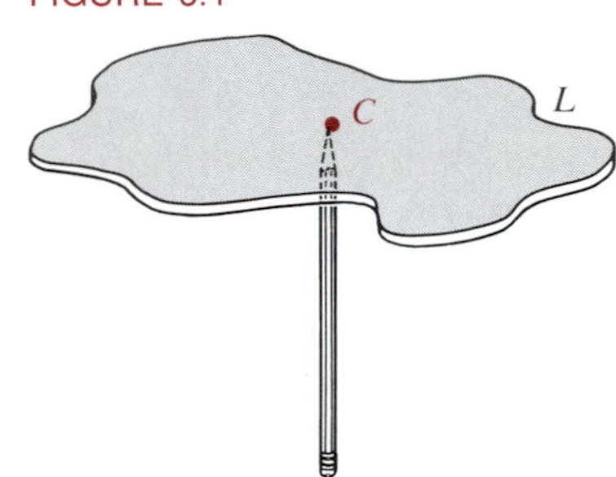

Suppose you try to balance a piece L of cardboard on the point of a pencil, as in **Figure 5.1.** You will find by experimenting that there is exactly one point C at about the middle of L at which the cardboard balances. In this section, we apply the definite integral to calculate the coordinates $(\bar{x}, \bar{y})$ of C, which is called the *center of mass* of L.

Calculation of center of mass involves an important concept in physics called the *moment* of a force. This is easiest to introduce in the context of a simpler one-dimensional problem—namely, what is the center of mass of a discrete system of point masses distributed along a thin rod of negligible mass? In **Figure 5.2,** we show the simplest situation: two masses m_1 and m_2 located at points with coordinates x_1 and x_2. Each point mass exerts a force on the rod. We can think of the rod as a seesaw with a fulcrum at 0, and m_1 and m_2 as children (or corresponding weights w_1 and w_2) seated at the points x_1 and x_2. Experience tells us that among the many possible arrangements, only one produces equilibrium. To describe it, we need the concept of the moments of m_1 and m_2 relative to the point 0. These are defined as

$$M_1 = w_1 x_1 \qquad \text{and} \qquad M_2 = w_2 x_2.$$

FIGURE 5.2

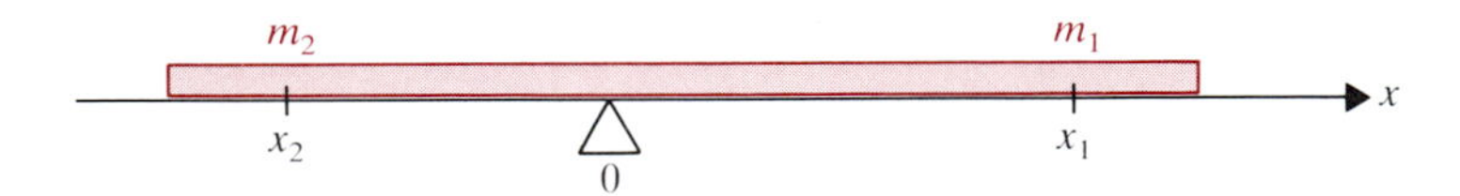

According to a law formulated about 2300 years ago by Archimedes (sometimes called the *law of the lever*), equilibrium is achieved if and only if the *algebraic sum of the moments is zero.* Thus equilibrium occurs precisely in case

(1) $$w_1 x_1 + w_2 x_2 = 0.$$

Since $w_1 = m_1 g$ and $w_2 = m_2 g$, Equation (1) is equivalent to

(2) $$m_1 g x_1 + m_2 g x_2 = 0 \rightarrow m_1 x_1 + m_2 x_2 = 0.$$

The law of the lever is often written as the right side of (2), which involves only mass and distance. The expression

(3) $$m_1 x_1 + m_2 x_2$$

is called the *moment of the system relative to 0.*

If, as in **Figure 5.3,** the system consists of several masses $m_1, m_2, \ldots, m_n$, then, as above, equilibrium exists if and only if the analogue of (2) holds:

$$m_1x_1 + m_2x_2 + \ldots + m_nx_n = \sum_{i=1}^{n} m_ix_i = 0. \tag{4}$$

In analogy with (3), the left side of (4) is called the *moment of the system relative to 0.*

FIGURE 5.3

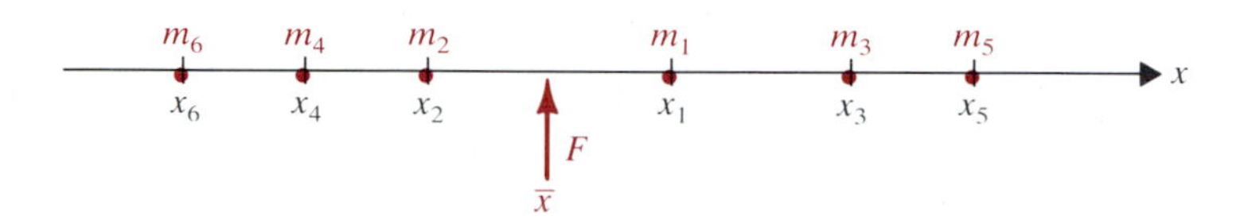

If the system is *not* in equilibrium, we can ask what upward directed force F applied at which point $\bar{x}$ balances the system, that is, produces equilibrium. Since the total downward weight of the system is

$$m_1g + m_2g + \ldots + m_ng = mg,$$

where $m = m_1 + m_2 + \ldots + m_n$, the balancing force F must first satisfy

$$F = -mg.$$

This follows from a basic physical principle that for a body to be in equilibrium, the algebraic sum of the forces acting on it must be 0. In addition, the law of Archimedes says that for equilibrium to hold, the moments must sum to 0. Hence, if we apply $F = -mg$ at point $\bar{x}$, then the system is in equilibrium if

$$-mg\bar{x} + m_1gx_1 + \ldots + m_ngx_n = 0 \rightarrow m\bar{x} = m_1x_1 + m_2x_2 + \ldots + m_nx_n,$$

that is, if

$$\bar{x} = \frac{\sum_{i=1}^{n} m_ix_i}{m}. \tag{5}$$

This expression for $\bar{x}$ represents the *weighted average* of the coordinates $x_1, \ldots, x_n$, weighted by the corresponding masses located at these respective points. The point $\bar{x}$ in (5) is called the ***center of mass*** of the system of n masses.

Example 1

Four particles of mass $m_1 = 1.00$ g, $m_2 = 0.40$ g, $m_3 = 0.50$ g and $m_4 = 1.50$ g are distributed along the x-axis as in **Figure 5.4.** Find the center of mass of the system.

Solution. Here the total mass of the system is

$$m = 1.00 + 0.40 + 0.50 + 1.50 = 3.40 \text{ g}.$$

FIGURE 5.4

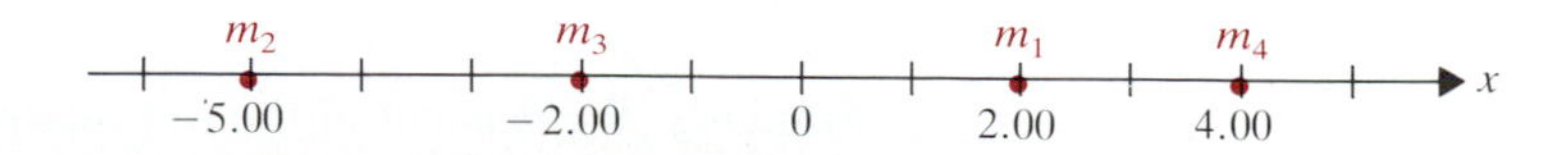

Hence (5) gives

$$\bar{x} = \frac{(1.00)(2.00) + (0.40)(-5.00) + (0.50)(-2.00) + (1.50)(4.00)}{3.40}$$

$$= \frac{2.00 - 2.00 - 1.00 + 6.00}{3.40} = \frac{5.00}{3.40} = 1.47,$$

to three significant figures. ■

If the number of point masses grows larger and larger, and the maximum distance between them declines toward 0, then the simple arithmetic of Example 1 quickly gets out of hand. In such large-scale systems (for example, billions of electrons in a linear accelerator), it is helpful to regard the distribution of mass as *continuous* rather than discrete. We imagine the mass to be distributed along a very thin rod of negligible mass, which is laid along the x-axis over the interval $[a, b]$. See **Figure 5.5.** We associate a continuous density function $\delta(x)$ to the rod, which represents the *mass per unit length* at point x. Thus, if $m(x)$ represents the total mass up to x, then

$$\delta(x) = \lim_{h \to 0} \frac{m(x + h) - m(x)}{h} = m'(x),$$

the rate of change of m with respect to x. If we partition $[a, b]$ into a large number of short subintervals $[x_{i-1}, x_i]$, then for any z_i in $[x_{i-1}, x_i]$, the density at z_i won't differ by much from the average mass per unit length over the subinterval containing z_i. That is,

(6) $$\delta(z_i) \approx \frac{m(x_{i-1} + \Delta x_i) - m(x_{i-1})}{\Delta x_i} \approx \frac{\Delta m_i}{\Delta x_i},$$

where $\Delta x_i = x_i - x_{i-1}$. Isolating Δm_i in (6), we get

(7) $$\Delta m_i \approx \delta(z_i) \Delta x_i.$$

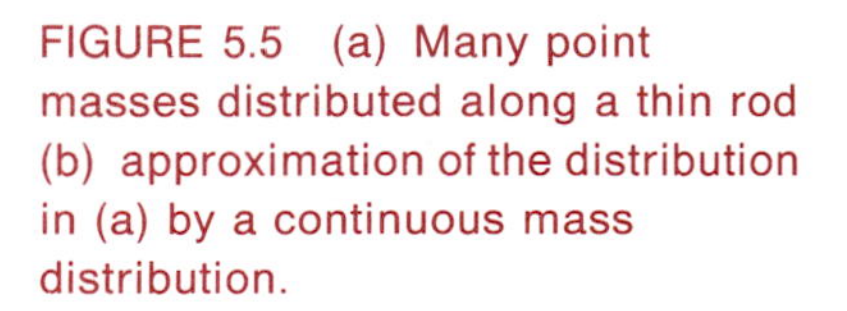

FIGURE 5.5 (a) Many point masses distributed along a thin rod (b) approximation of the distribution in (a) by a continuous mass distribution.

Hence what we would think of as the total mass m of the rod is approximately the sum of all Δm's in (7),

(8) $$m \approx \sum_{i=1}^{n} \delta(z_i) \Delta x_i = R_P,$$

which is a Riemann sum for the density function δ. Hence we define m to be the limit of the Riemann sum (8),

(9) $$m = \int_a^b \delta(x)\, dx.$$

Similarly, the *moment* of the rod relative to the origin is defined by thinking of the mass of the sub-rod lying over the subinterval $[x_{i-1}, x_i]$ as concentrated at z_i. Then the moment of the sub-rod relative to the origin is

$$z_i \Delta m_i \approx z_i \delta(z_i) \Delta x_i, \tag{10}$$

from (7). The total moment of the rod is then the sum of the moments of the sub-rods in (10).

$$\sum_{i=1}^{n} z_i \Delta m_i \approx \sum_{i=1}^{n} z_i \delta(z_i) \Delta x_i. \tag{11}$$

The right side of (11) is a Riemann sum for the function $z\,\delta(z)$, so it is natural to define the ***moment*** M_0 of the rod with respect to the origin to be

$$M_0 = \int_a^b x\,\delta(x)\,dx. \tag{12}$$

The ***center of mass*** $\bar{x}$ of the rod is defined, as in the discrete case, to be the quotient of the moment of the rod and its mass, as given by (9) and (12):

$$\bar{x} = \frac{M_0}{m} = \frac{\int_a^b x\,\delta(x)\,dx}{\int_a^b \delta(x)\,dx}. \tag{13}$$

Example 2

The density of a thin straight wire of length 10 cm is proportional to the distance from the left end of the wire. Where is the center of mass?

Solution. Think of the wire as lying along the x-axis between $x = 0$ and $x = 10$, as in **Figure 5.6.** A point x units from the left end of the rod then has coordinate x. For some constant k,

FIGURE 5.6

$$\delta(x) = kx.$$

Hence from (9) we have

$$m = \int_0^{10} kx\,dx = k \cdot \frac{1}{2}x^2 \Big]_0^{10} = 50k.$$

Equation (12) gives

$$M_0 = \int_0^{10} x \cdot kx\,dx = k \int_0^{10} x^2\,dx = k \cdot \frac{x^3}{3}\Big]_0^{10} = \frac{1000}{3}k.$$

Thus from (13) the center of mass is at

$$\bar{x} = \frac{\frac{1000}{3}k}{50k} = \frac{20}{3}.$$

The wire therefore has center of mass 2/3 of the way toward its right end. ■

If the density is proportional to some function f of x, then the constant of proportionality does not affect the center of mass, since it always cancels just as in Example 2. In fact, (13) gives

$$\bar{x} = \frac{\int_a^b x \cdot kf(x)\,dx}{\int_a^b kf(x)\,dx} = \frac{k\int_a^b xf(x)\,dx}{k\int_a^b f(x)\,dx} = \frac{\int_a^b xf(x)\,dx}{\int_a^b f(x)\,dx}.$$

In particular, if $\delta(x) = k$ is constant, then k again cancels and (13) becomes

$$\bar{x} = \frac{\int_a^b x\,dx}{b-a} = \frac{\frac{1}{2}x^2\Big]_a^b}{b-a} = \frac{\frac{1}{2}(b^2-a^2)}{b-a} = \frac{1}{2}(b+a).$$

Thus for *homogeneous* distributions (those corresponding to constant density), the center of mass is the geometric center of the rod, the point halfway from each end.

Now we return to the two-dimensional problem posed at the beginning of the section. Again we start with the discrete case. Suppose we have a finite number of point masses $m_1, m_2, \ldots, m_n$ located at points $(x_1, y_1), (x_2, y_2), \ldots, (x_n, y_n)$ in the plane, as in **Figure 5.7.** The moment of m_i with respect to the x-axis is $m_i y_i$, and the moment of m_i with respect to the y-axis is $m_i x_i$. The total moment of the system relative to the x-axis is therefore

FIGURE 5.7

$$M_x = \sum_{i=1}^{n} m_i y_i.$$

The total moment relative to the y-axis is

$$M_y = \sum_{i=1}^{n} m_i x_i.$$

The center of the mass of the system is then $(\bar{x}, \bar{y})$, where

$$\bar{x} = \frac{M_y}{m} = \frac{\sum_{i=1}^{n} m_i x_i}{\sum_{i=1}^{n} m_i}, \tag{14}$$

$$\bar{y} = \frac{M_x}{m} = \frac{\sum_{i=1}^{n} m_i y_i}{\sum_{i=1}^{n} m_i}, \tag{15}$$

and, as before, $m = \Sigma_{i=1}^{n} m_i$ is the total mass of the system.

Example 3

Masses $m_1 = 0.7$ g, $m_2 = 1.1$ g, $m_3 = 0.6$ g and $m_4 = 1.2$ g are located at the points shown in **Figure 5.8.** Find the center of mass of the system.

FIGURE 5.8

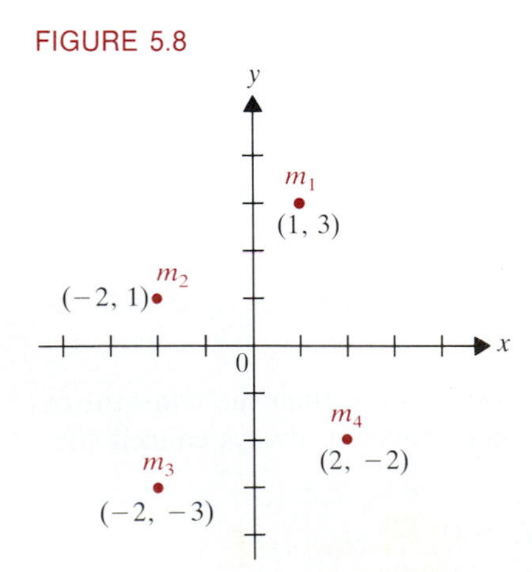

Solution. Note that $m = 0.7 + 1.1 + 0.6 + 1.2 = 3.6$ g. Then (14) and (15) give

$$\bar{x} = \frac{(0.7)(1) + (1.1)(-2) + (0.6)(-2) + (1.2)(2)}{3.6} = \frac{-0.3}{3.6} = -\frac{1}{12}.$$

$$\bar{y} = \frac{(0.7)(3) + (1.1)(1) + (0.6)(-3) + (1.2)(-2)}{3.6} = \frac{-1}{3.6} = -\frac{5}{18}.$$

Hence the center of mass is at $(-1/12, -5/18)$. ■

In (14) and (15), one gets $\bar{x}$ by dividing M_y (**not** M_x) by m, and gets $\bar{y}$ by dividing M_x (**not** M_y) by m. For this reason, some students prefer to write $M\bar{x}$ for M_y and $M\bar{y}$ for M_x. Here we will follow the more common notation in physics and engineering, M_x and M_y.

Next we consider a homogeneous plane ***lamina***, which can be thought of as a thin sheet of material of constant density δ overlying a portion of the xy-plane. (The nonhomogeneous case, in which the density varies, is more easily treated using the calculus of functions of more than one variable.) Suppose that the lamina lies under the graph of $y = f(x)$ between $x = a$ and $x = b$ as in **Figure 5.9**, where f is continuous and nonnegative on $[a, b]$. We partition $[a, b]$ into subintervals $[x_{i-1}, x_i]$. In each subinterval, we select the midpoint z_i and construct the rectangle of height $f(z_i)$ based on $[x_{i-1}, x_i]$. Since the density is constant throughout the rectangle, its center of mass is located at the *centroid* of the rectangle, its geometric center $(z_i, \frac{1}{2}f(z_i))$. If the norm of the partition is small, we can think of the lamina as *approximated* by a discrete system of masses

FIGURE 5.9

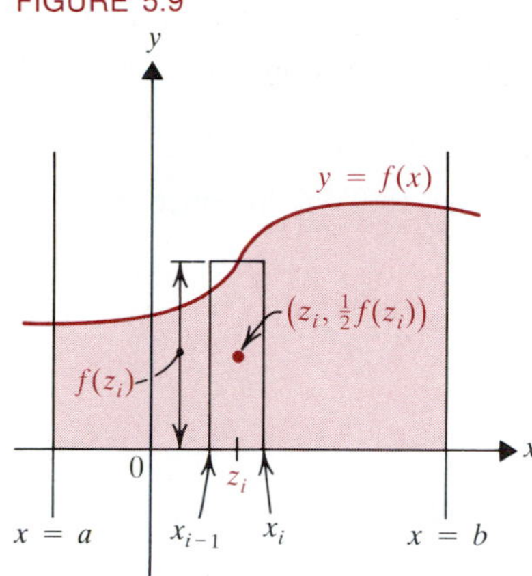

$$m_i = \delta \cdot f(z_i)(x_i - x_{i-1}) = \delta f(z_i)\,\Delta x_i$$

concentrated at the centroid of each rectangle. Note that m_i is the mass of the corresponding rectangle. Thus the approximating discrete system has mass

$$\sum_{i=1}^{n} m_i = \delta \sum_{i=1}^{n} f(z_i)\,\Delta x_i. \tag{16}$$

Its moment with respect to the y-axis is

$$\sum_{i=1}^{n} m_i \cdot z_i = \delta \sum_{i=1}^{n} z_i f(z_i)\,\Delta x_i. \tag{17}$$

Its moment with respect to the x-axis is

$$\sum_{i=1}^{n} m_i \cdot \frac{1}{2} f(z_i) = \sum_{i=1}^{n} \frac{1}{2}\,\delta[f(z_i)]^2\,\Delta x_i. \tag{18}$$

As the partition becomes finer and finer, we expect (16) to more and more closely approximate the mass of the lamina. Also, the center of mass of the lamina should be closely approximated by the point whose coordinates result from division of the expressions in (17) and (18) by the approximate mass (16). This leads to the following definition.

5.1 DEFINITION Let L be a homogeneous lamina of constant density δ lying under the graph of $y = f(x)$ between $x = a$ and $x = b$, where f is continuous and nonnegative on $[a, b]$. Then the ***mass*** of L is

$$m = \delta \int_a^b f(x)\,dx = \delta A, \tag{19}$$

where A is the area of L. The moments M_y and M_x of L relative to the y-axis and x-axis, respectively, are

$$M_y = \delta \int_a^b x f(x)\,dx \qquad \text{and} \qquad M_x = \frac{1}{2}\delta \int_a^b [f(x)]^2\,dx.$$

The ***center of mass*** or ***centroid*** of L is $(\bar{x}, \bar{y})$, where

$$\bar{x} = \frac{M_y}{m} = \frac{\int_a^b x f(x)\,dx}{\int_a^b f(x)\,dx}, \tag{20}$$

and

$$\bar{y} = \frac{M_x}{m} = \frac{\int_a^b \frac{1}{2}[f(x)]^2\,dx}{\int_a^b f(x)\,dx}. \tag{21}$$

It is worthwhile to memorize (19), (20), and (21) to avoid mix-ups in computing $\bar{x}$ and $\bar{y}$. ***Remember***: $\bar{x}$ is the directed distance of the centroid from the y-axis, so we divide M_y (**not** M_x) by m to get $\bar{x}$. Similar remarks apply to $\bar{y}$.

Example 4

Find the centroid of the region lying under $y = (1 - x^2)^{1/2}$ between -1 and 1 (**Figure 5.10**).

FIGURE 5.10

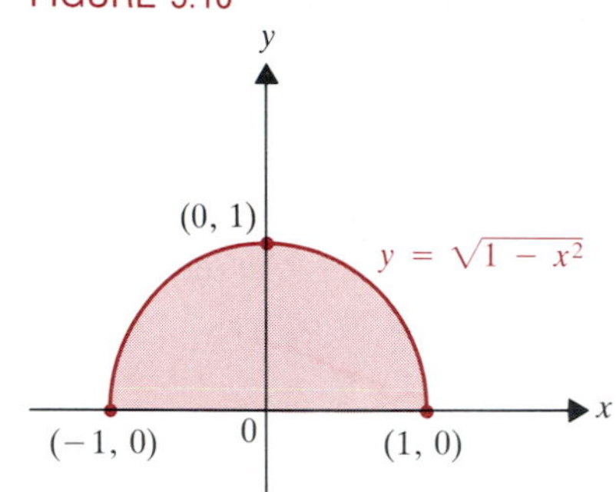

Solution. It follows from the symmetry of the region that $\bar{x} = 0$. Since the area of a circular disk of radius r is πr^2, we have

$$m = \int_{-1}^{1} \delta\sqrt{1 - x^2}\,dx = \delta \int_{-1}^{1} \sqrt{1 - x^2}\,dx = \delta \cdot \frac{1}{2}\pi \cdot 1^2 = \frac{1}{2}\pi\delta.$$

Also,

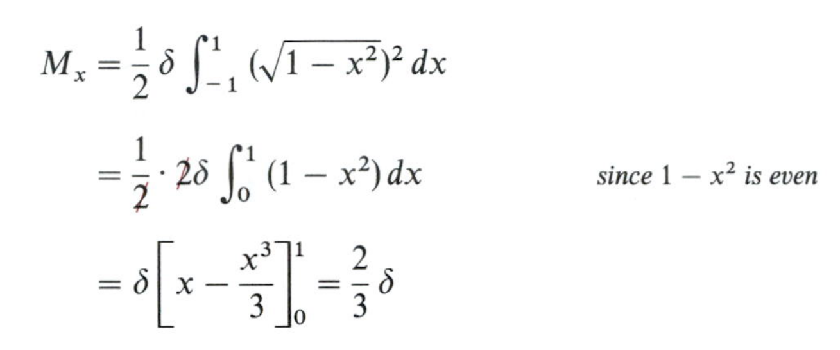

$$\begin{aligned} M_x &= \frac{1}{2}\delta \int_{-1}^{1} (\sqrt{1 - x^2})^2\,dx \\ &= \frac{1}{2} \cdot 2\delta \int_0^1 (1 - x^2)\,dx \qquad \textit{since } 1 - x^2 \textit{ is even} \\ &= \delta\left[x - \frac{x^3}{3}\right]_0^1 = \frac{2}{3}\delta \end{aligned}$$

Hence (21) gives

$$\bar{y} = \frac{\frac{2}{3}\delta}{\frac{1}{2}\pi\delta} = \frac{4}{3\pi}.$$

Thus the center of mass is at $(0, 4/3\pi)$. ■

As in Example 4, where we saw that $\bar{x} = 0$, it is helpful to take advantage of symmetry as fully as possible in working center-of-mass problems to avoid unnecessary integration.

To find the centroids of more general regions, we use (19) in the form

$$m = \delta A$$

and (20) and (21) in the forms

$$\bar{x} = \frac{M_y}{m} \qquad \text{and} \qquad \bar{y} = \frac{M_x}{m}.$$

We can compute A by integration as in Section 1 and we similarly can compute M_y (and M_x) by introducing a factor of x, and a factor $\frac{1}{2}f(x)$, into the integral for A. Rather than stating this formally for the various cases, we illustrate the procedure by example.

Example 5

Find the centroid of the region bounded by $y = x^2$ and the line $y = x + 2$.

FIGURE 5.11

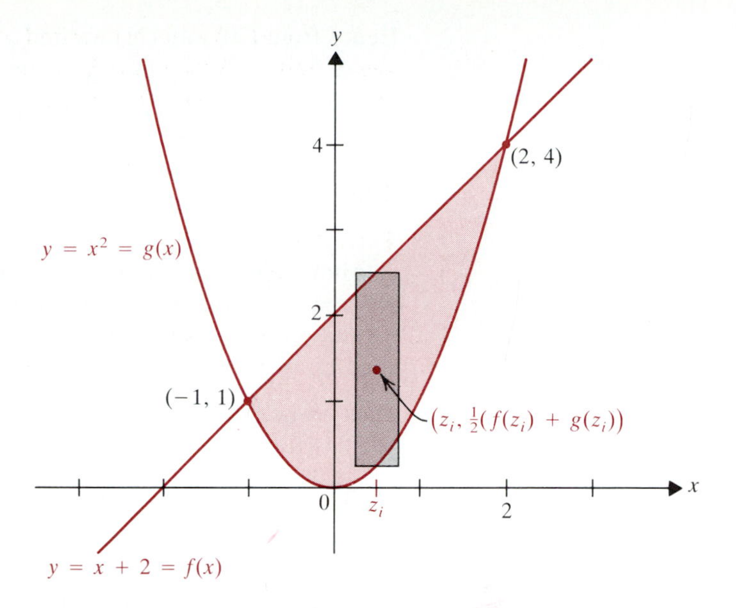

Solution. **Figure 5.11** shows the region. From Section 1, we have

$$\begin{aligned}
m = \delta A &= \delta \int_{-1}^{2} [f(x) - g(x)]\, dx \\
&= \delta \int_{-1}^{2} (x + 2 - x^2)\, dx = \delta \left[\frac{1}{2} x^2 + 2x - \frac{x^3}{3} \right]_{-1}^{2} \\
&= \delta \left[2 + 4 - \frac{8}{3} - \frac{1}{2} + 2 - \frac{1}{3} \right] = \frac{9}{2} \delta.
\end{aligned}$$

The approximating rectangle in Figure 5.11 has centroid $(z_i, \frac{1}{2}[f(z_i) + g(z_i)])$. The reasoning leading up to Definition 5.1 then gives

$$\begin{aligned}
M_y &= \delta \int_{-1}^{2} x[f(x) - g(x)]\, dx \\
&= \delta \int_{-1}^{2} (x^2 + 2x - x^3)\, dx = \delta \left[\frac{x^3}{3} + x^2 - \frac{1}{4} x^4 \right]_{-1}^{2} \\
&= \delta \left[\frac{8}{3} + 4 - 4 + \frac{1}{3} - 1 + \frac{1}{4} \right] = \frac{9}{4} \delta.
\end{aligned}$$

Also, by the same reasoning,

$$\begin{aligned}
M_x &= \delta \int_{-1}^{2} \frac{1}{2} [f(x) + g(x)][f(x) - g(x)]\, dx \\
&= \frac{1}{2} \delta \int_{-1}^{2} ([f(x)]^2 - [g(x)]^2)\, dx = \frac{1}{2} \delta \int_{-1}^{2} (x^2 + 4x + 4 - x^4)\, dx \\
&= \frac{1}{2} \delta \left[\frac{1}{3} x^3 + 2x^2 + 4x - \frac{x^5}{5} \right]_{-1}^{2} \\
&= \frac{1}{2} \delta \left[\frac{8}{3} + 8 + 8 - \frac{32}{5} + \frac{1}{3} - 2 + 4 - \frac{1}{5} \right] \\
&= \frac{36}{5} \delta.
\end{aligned}$$

Hence from (20) and (21) we find

$$\bar{x} = \frac{M_y}{m} = \frac{\frac{9}{4}\delta}{\frac{9}{2}\delta} = \frac{1}{2} \qquad \bar{y} = \frac{M_x}{m} = \frac{\frac{36}{5}\delta}{\frac{9}{2}\delta} = \frac{8}{5}. \quad ■$$

About 1700 years ago, the Greek mathematician Pappus of Alexandria discovered a striking relationship between centroids and the volumes of solids of revolution. We give a derivation that uses the methods of Section 2 of this chapter, which were developed 14 centuries after the time of Pappus.

5.2 THEOREM

Theorem of Pappus. Let R be the region bounded by the graphs of the continuous functions f and g and by the lines $x = a$ and $x = b$, where $f(x) \leq g(x)$ for $x \in [a, b]$. See **Figure 5.12.** Let S be the solid of revolution obtained by revolving R about the x-axis. Then the volume of S is

$$V(S) = 2\pi\bar{y}A(R) \tag{22}$$

where $\bar{y}$ is the y-coordinate of the centroid of R, and $A(R)$ is the area of R.

FIGURE 5.12

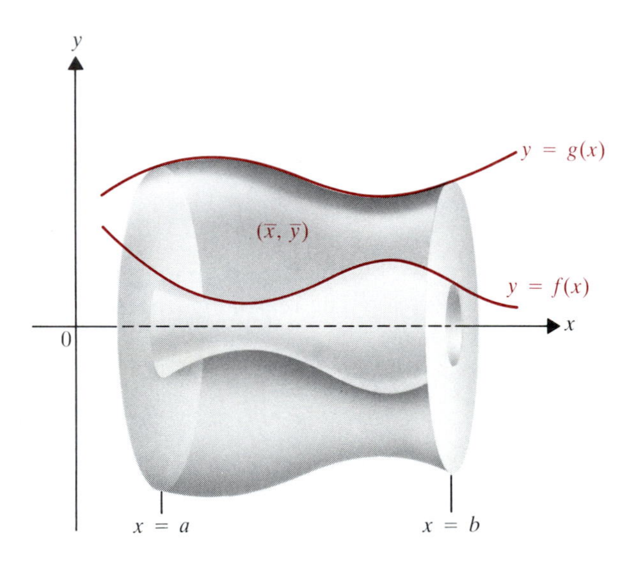

If, instead, R is revolved about the y-axis to generate a solid of revolution T, then

$$V(T) = 2\pi\bar{x}A(R) \tag{23}$$

where $\bar{x}$ is the x-coordinate of the centroid of R.

Proof. We prove (22), leaving (23) for Exercise 49. We calculate $V(S)$ by using the circular ring method of Section 2:

$$V(S) = \pi \int_a^b ([g(x)]^2 - [f(x)]^2)\,dx.$$

We find $\bar{y}$ by using the approach of Example 5. Letting δ stand for the density of R, we have

$$m = \delta A(R) = \delta \int_a^b [g(x) - f(x)]\,dx,$$

and

$$M_x = \delta \int_a^b \frac{1}{2}[g(x) + f(x)][g(x) - f(x)]\,dx$$
$$= \frac{1}{2}\delta \int_a^b ([g(x)]^2 - [f(x)]^2)\,dx.$$

Therefore

$$\bar{y} = \frac{M_x}{m} = \frac{\frac{1}{2}\delta \int_a^b ([g(x)]^2 - [f(x)]^2)\,dx}{\delta \int_a^b [g(x) - f(x)]\,dx},$$
$$= \frac{\frac{1}{2}\int_a^b ([g(x)]^2 - [f(x)]^2)\,dx}{A(R)}.$$

Hence

$$\bar{y}A(R) = \frac{1}{2}\int_a^b ([g(x)]^2 - [f(x)]^2)\,dx,$$
$$2\pi\bar{y}A(R) = \pi \int_a^b ([g(x)]^2 - [f(x)]^2)\,dx = V(S).$$ QED

This classical theorem enables us to easily compute the volumes of doughnuts or tire inner tubes, which have the shape of a *torus*. As shown in **Figure 5.13,** such a solid T is obtained by revolving a circular disk D of radius a about a line l at distance $b > a$ from the center C of D. The centroid of D is, of course, its center $C(k, b)$. So by (22) the volume of T is

$$V(T) = 2\pi\bar{y}A(D) = 2\pi b\pi a^2 = 2\pi^2 a^2 b.$$

If you worked Exercise 32 of Section 3, then you will appreciate the simplification that Pappus's theorem provides: Whenever the centroid of a solid of revolution is easy to calculate, Theorem 5.2 makes its volume just as easy to calculate. In Exercise 41, for instance, you can use Pappus's theorem to give a simple approximate solution to Problem 10 of the Prologue.

FIGURE 5.13

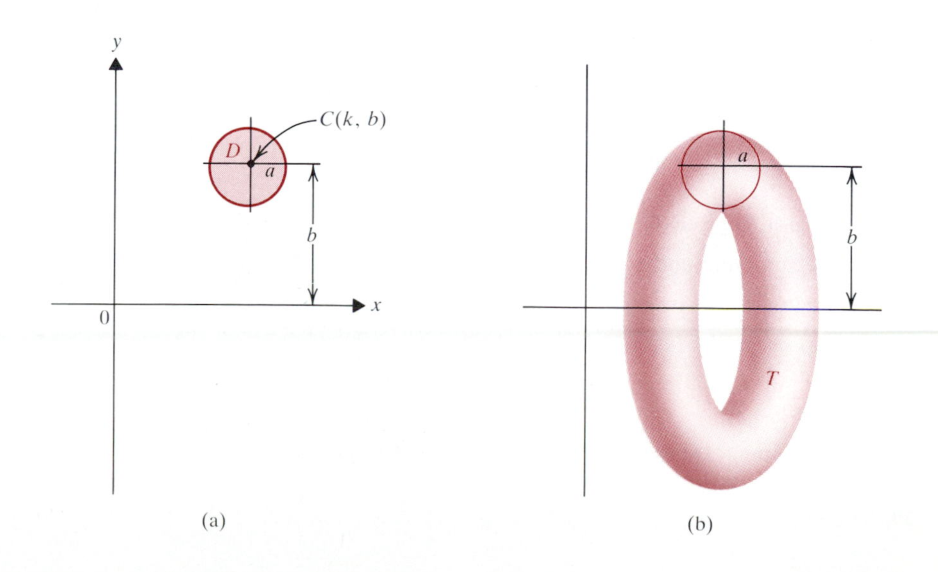

Exercises 5.5

In Exercises 1–8, find the centroid of the given discrete system.

1. Masses $m_1 = 4$, $m_2 = 6$, and $m_3 = 10$ at points $x_1 = -2$, $x_2 = 3$, and $x_3 = 7$ on the x-axis
2. Masses $m_1 = 2$, $m_2 = 3$, and $m_3 = 1$ at points $x_1 = 1.5$, $x_2 = -2/3$, and $x_3 = -2$
3. Masses $m_1 = 4$, $m_2 = 6$, and $m_3 = 10$ at points $(-2, 1)$, $(3, -2)$, and $(7, 2)$
4. Masses $m_1 = 2$, $m_2 = 3$, and $m_3 = 1$ at $(1.5, -2.5)$, $(-2/3, 1/3)$, and $(-2, 1)$
5. Masses $m_1 = 2$, $m_2 = 6$, $m_3 = 4$, and $m_4 = 1$ at $(1, -2)$, $(-4, 1)$, $(-2, 3)$, and $(4, -1)$
6. Masses $m_1 = 2$, $m_2 = 3$, $m_3 = 1$, and $m_4 = 4$ at $(-2, 1)$, $(1, -3)$, $(-2, 5)$, and $(1, 1)$
7. Masses all of the same size m at the points $x_1 = 1$, $x_2 = 2$, $x_3 = 3$, $x_4 = 4$, and $x_5 = 5$
8. Repeat Exercise 7 in the case of n equal masses, located at the points $x_i = i$.

In Exercises 9–18, find the mass and center of mass of the rod with the given density function lying over the given interval. Here x is the directed distance from the origin.

9. $\delta(x) = x^2$ over $[0, 1]$
10. $\delta(x) = x^3$ over $[0, 2]$
11. $\delta(x) = \sqrt{x}$ over $[0, 9]$
12. $\delta(x) = \sqrt[3]{x}$ over $[0, 8]$
13. $\delta(x) = 2 - x$ over $[0, 1]$
14. $\delta(x) = 3 - 2x$ over $[0, 1]$
15. $\delta(x) = (1 - x^2)^{1/2}$ over $[-1, 1]$
16. $\delta(x) = (a^2 - x^2)^{1/2}$ over $[-a, a]$
17. $\delta(x) = 1/x^3$ over $[1, 3]$
18. $\delta(x) = 1/x^3 + 1$ over $[1, 3]$

In Exercises 19–34, find the centroid of the given plane region.

19. The triangle with vertices (0, 0), (1, 0), and (0, 1)
20. The triangle with vertices (0, 0), $(a, 0)$, and (b, c)
21. The region lying under $y = \sqrt{4 - x^2}$ between $x = -2$ and $x = 2$
22. The region lying under $y = \sqrt{a^2 - x^2}$ between $x = -a$ and $x = a$
23. The region lying under $y = x^2$ between $x = 0$ and $x = 1$
24. The region lying under $y = x^3$ between $x = 0$ and $x = 2$
25. The region in the first quadrant under $y = 4 - x^2$
26. The region in the first quadrant under $y = x - x^2$
27. The first-quadrant region below $y = 4x - x^2$ and above the line $y = x$
28. The first-quadrant region below $y = 3x - x^2$ and above the line $y = 2x$
29. The region between $y = x^3$ and $y = \sqrt{x}$
30. The region between $y = 6x - x^2$ and $y = 2x$
31. The first-quadrant region bounded by $y = x^2$ and $y = 2x - x^2$
32. The region between $y = 5x - x^2$ and $y = 2x$
33. The region in the first quadrant bounded by the circles $x^2 + y^2 = 1$ and $x^2 + y^2 = 4$. (Use symmetry.)
34. The region in the first quadrant bounded by the circles $x^2 + y^2 = 1$ and $x^2 + y^2 = a^2$

In Exercises 35–38, find the volume of the given solid.

35. The torus formed when the circular disk of radius 3 centered at (2, 3) is revolved about the line $x = 6$.
36. The torus formed when the circular disk of radius 2 centered at $(-1, 2)$ is revolved about the line $y = 5$.
37. The solid formed when the triangle with vertices (0, 0), $(a, 0)$, and (b, c) is revolved about the line $x = -a$. (Assume that a, b, and $c > 0$.)
38. The solid formed when the top half of the disk $x^2 + y^2 \le a^2$ is revolved about the line $x = -a$.
39. Use Theorem 5.2 and the fact that a ball of radius a has volume $\frac{4}{3}\pi a^3$ to find the centroid of the semidisk $x^2 + y^2 \le a^2$, $y \ge 0$.
40. Use your knowledge of the volume of a cone of height h and radius a (Exercise 25, Section 2) to find the centroid of the triangle with vertices $O(0, 0)$, $A(0, a)$, and $B(h, 0)$.
41. Problem 10 of the Prologue asks for an estimate of the volume of air in a tire on a subcompact car if the inner radius of the tire is 7 inches and the outer radius is

FIGURE 5.14

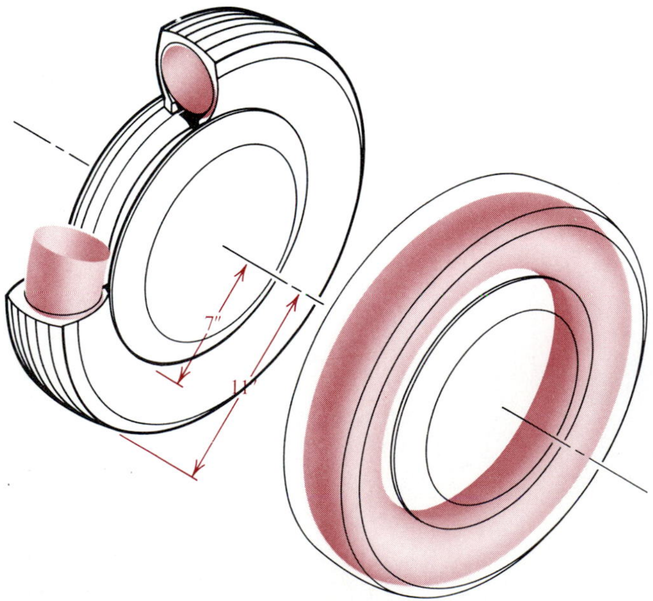

11 inches. Note that this is very close to the volume of air in an *inner tube* (**Figure 5.14**) in the shape of a torus. Use this to approximate the volume of air in the tire.

42. Repeat Exercise 41 for a tire of inner radius 8 inches and outer radius 12 inches.

43. Two suburban towns on a train line are 10 miles apart. One has population 24,000, the other 18,000. Where should a rapid transit station be located if planners want it to be at the population center of the two communities?

44. Use the result of Exercise 20 to show that the centroid of any triangle is at the point of intersection of the *medians.* (The medians of a triangle are the lines drawn from each vertex to the midpoints of the opposite side.)

45. The ***centroid of a curve,*** say the graph of $y = f(x)$ between $x = a$ and $x = b$, is defined to be the point $(\bar{x}, \bar{y})$ such that $\bar{x} = M_y/L$, and $\bar{y} = M_x/L$, where $L = \int_a^b \sqrt{1 + [f'(x)]^2}\, dx$ as in Section 4, $M_x = \int_a^b y\sqrt{1 + [f'(x)]^2}\, dx$, and $M_y = \int_a^b x\sqrt{1 + [f'(x)]^2}\, dx$. Use these formulas and symmetry to find the centroid of the graph of $y = \sqrt{9 - x^2}$ over $[-3, 3]$.

46. Repeat Exercise 45 for the graph of $y = \sqrt{a^2 - x^2}$ over $[-a, a]$.

47. Find a density function δ such that the center of a mass of a line segment of length L is 1/4 of the way from one end to the other.

48. Suppose new coordinate axes are introduced through the line $y = k$ and $x = h$, with (h, k) as the new origin.

(a) Find the moments M_h and M_k of the region R of the xy-plane in terms of M_x, M_y, and the mass m of R.

(b) Express the centroid of R using M_h and M_k in terms of the centroid $(\bar{x}, \bar{y})$ calculated from M_x and M_y.

(c) Explain, using (b), why the centroid is unaffected under translation of axes.

49. Prove (23) in Theorem 5.2.

5.6 Work; Fluid Pressure

The first part of this section discusses the work done by a force moving an object in a straight line. Work is a fundamental idea in physics and is widely applied in engineering. The second part of the section considers fluid pressure exerted against the walls of a container, and the resulting forces that must be calculated in designing structures like dams, indoor and above-ground swimming pools, and oil and water storage tanks.

FIGURE 6.1

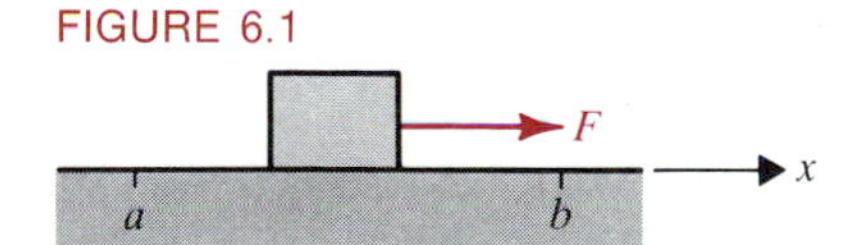

The simplest type of work is done by a constant force F that moves an object along a path parallel to the direction of F from a point a to a point b, as in **Figure 6.1.** In this situation, the work done is defined to be

$$W = F \cdot (b - a). \tag{1}$$

The units for work are *newton-meters* (N-m) or *dyne-centimeters* (dyne-cm) in metric countries and scientific calculations, and foot-pounds in the United States. (A newton-meter is one *joule* (J); a dyne-centimeter is one *erg.*) In **Figure 6.2,** a force of 10 N is exerted by a small package against the floor. The work done by a vertical force F in lifting the package 2 m off the floor is therefore

FIGURE 6.2

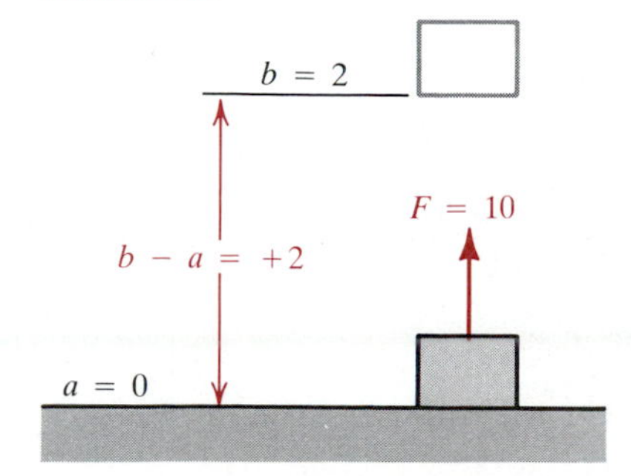

$$W = 10 \cdot 2 = 20 \text{ N-m}.$$

(This is about the same as the work done in lifting a $2\frac{1}{2}$ pound package to a height of 6 feet, i.e., about 15 ft-lb.) The basic idea is that work is done by a force acting over a distance, and the amount of work done is the product of the size of the force and the directed distance through which the object is moved.

Force is a *vector* quantity, having both size and direction. The forces we deal with here are vertical or horizontal. So we indicate their direction in the usual way: Positive signs are attached to forces directed upward or toward the right; negative signs are attached to downward- or leftward-directed forces. We attach the same signs to the directed distances moved in calculating W. Thus work is a *signed* quantity. In Figure 6.2, for instance, F is directed upward, so

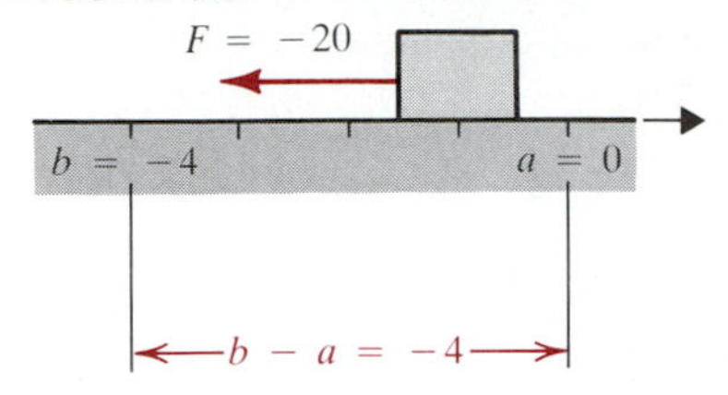

FIGURE 6.3

$W = +20$ N-m. In **Figure 6.3,** F is directed to the left and moves the object 4 cm to the left. If F has magnitude 20 dynes, then the work done is

$$W = (-20)(-4) = 80 \text{ dyne-cm}.$$

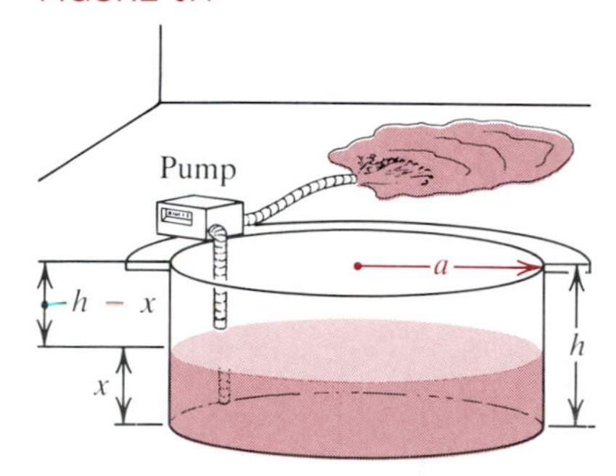

FIGURE 6.4

Formula (1) is inadequate to cover most situations of interest in physics or engineering. First, *variable* forces are more commonly encountered than constant ones. (Think of pushing a car along a level road: The force needed to start the car in motion is greater than that needed to maintain motion.) Second, in pumping liquids (such as water out of the cylindrical swimming pool shown in **Figure 6.4**), even though a constant force may be sufficient, it must act through a *variable* distance. As the water level x in the pool drops, water must be pumped a greater distance $h - x$ to the top of the pool.

Turning to the first problem, consider a continuous variable force F acting on an object over an interval $[a, b]$. Partition $[a, b]$ into a large number n of subintervals $[x_{i-1}, x_i]$ of length $\Delta x_i = x_i - x_{i-1}$. On each small subinterval, the force F varies only slightly, because F is continuous. So for any z_i in $[x_{i-1}, x_i]$, F is approximately $F(z_i)$. According to (1) then, the work done in moving the object through the distance Δx_i is approximately $F(z_i)\Delta x_i$. Hence the total work done in moving the object from a to b is approximated by the Riemann sum

$$R_P = \sum_{i=1}^{n} F(z_i)\Delta x_i.$$

This leads to the following formal definition.

6.1 DEFINITION

The ***work*** done by a continuous force F acting in the direction from a to b in moving an object from $x = a$ to $x = b$ is

$$W = \int_a^b F(x)\,dx. \tag{2}$$

An important problem in physics is the calculation of the work done in compressing or stretching a spring. That calculation uses *Hooke's law*, formulated by the English physicist Robert Hooke (1635–1703) in 1676. When a spring is not acted on by any forces, its length is called its *natural length*. If, as in **Figure 6.5,** the spring is stretched x units beyond its natural length, then Hooke's law states that the spring exerts an oppositely directed *restoring* (or resisting) force R given by

$$R = -kx, \tag{3}$$

where k is a positive constant called the *spring constant*. Thus the force F needed to stretch the spring x units in the positive direction from its natural length is

$$F = kx. \tag{4}$$

(If the spring is compressed, then both x and F are negative.)

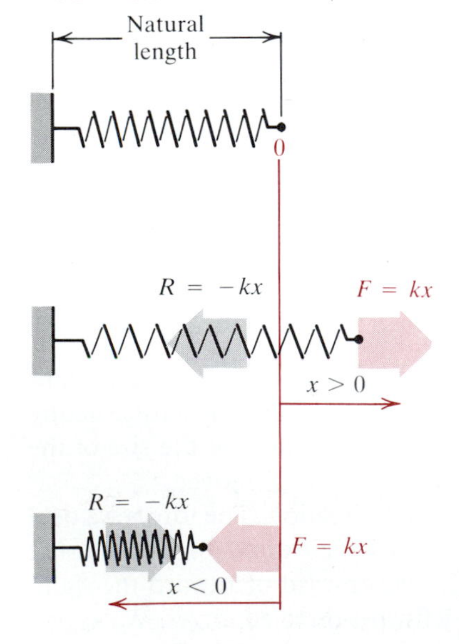

FIGURE 6.5

Example 1

A force of 0.2 N is required to stretch a spring 0.1 m beyond its natural length of 1 m.

(a) How much work is done?

(b) How much work is done compressing the spring from its natural length to 0.7 m?

Solution. (a) To apply (2), we must determine k in (4). We orient the spring as in Figure 6.5, so stretching forces act in the positive direction. Since $F = 0.2$ when

$x = 0.1$, Equation (4) gives

$$0.2 = k(0.1) \rightarrow k = 2.$$

Then we have from (2),

$$W = \int_0^{0.1} 2x\,dx = x^2\Big|_0^{0.1} = 0.01 \text{ J}.$$

(b) In this case, a negative force acts to compress the spring 0.3 m to the left in Figure 6.5. Since $0.7 = 1 + (-0.3)$, we have

$$W = \int_0^{-0.3} -2x\,dx = -x^2\Big|_0^{-0.3} = -0.09 \text{ J}. \quad ■$$

Warning. In problems like Example 1, the upper limit of integration is the *displacement* from the natural length of the spring. Do not succumb to the temptation experienced by many students to use the new length of the spring instead of this displacement.

Recently another kind of work problem involving a variable force has become of practical interest. In launching satellites, booster rockets must do work against the force of gravity. *Newton's law of gravitation* states that the force of gravity on a body of mass m located x kilometers from the center of the earth is given by

$$g(x) = -\frac{GMm}{x^2}, \tag{5}$$

where M is the mass of the earth (about 5.993×10^{24} kg), and G is the *universal gravitational constant* (about 6.674×10^{-17} N-km^2/kg^2). The force F needed to overcome the force $g(x)$ of gravity is then given by the negative of (5),

$$F(x) = \frac{GMm}{x^2}. \tag{6}$$

Example 2

A rocket has mass 2000 kg. Find the work done in launching the rocket from the earth's surface to an altitude of 100 km. (Use 6.400×10^3 km as the radius of the earth and ignore the weight of the fuel burned during the flight.)

Solution. Letting $R_0 = 6400$ and $R_1 = 6500$, we have from (2) and (6) that the work done is

$$\begin{aligned} W &= \int_{R_0}^{R_1} F(x)\,dx = GMm \int_{R_0}^{R_1} x^{-2}\,dx \\ &= GMm\left[-\frac{1}{x}\right]_{R_0}^{R_1} = GMm\left[\frac{1}{R_0} - \frac{1}{R_1}\right] \\ &\approx 6.674 \times 10^{-17} \frac{\text{N-}\cancel{\text{km}^2}}{\cancel{\text{kg}^2}} \cdot 5.993 \times 10^{24}\,\cancel{\text{kg}} \cdot 2000\,\cancel{\text{kg}} \cdot 2.404 \times 10^{-6}\,\cancel{\text{km}^{-1}} \\ &\approx 1.923 \times 10^6 \text{ N-km} = 1.923 \times 10^9 \text{ N-m} = 1.923 \times 10^9 \text{ J}. \end{aligned}$$

The nearly two billion joules of work needed to launch the rocket to a height of 100 km accounts for the massive size of booster engines used for such launchings. ■

FIGURE 6.6

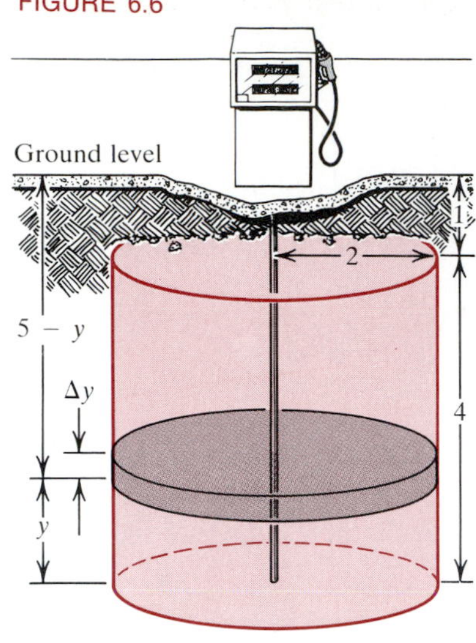

We turn next to the second type of problem mentioned above: one in which a liquid is to be pumped out of a tank. Suppose that the buried cylindrical gasoline tank in **Figure 6.6** of radius $a = 2$ m and height $h = 4$ m has its top 1 m below the surface and is initially full of gasoline of density δ (for unleaded gasoline, $\delta \approx 0.75$ g/cm^3 ≈ 750 kg/m^3).

Example 3

Find the work done in pumping all the gasoline from the tank in Figure 6.6 up to the surface.

Solution. Figure 6.6 shows a small portion of gasoline located between depths y and $y + \Delta y$. This "layer" of gasoline has volume

$$\Delta V = \pi 2^2 h = \pi \cdot 4 \cdot (\Delta y) \text{ m}^3.$$

Gravity exerts on it a downward force

$$-mg = -(\delta \cdot \Delta V)g = -4\pi\,\delta g\,\Delta y \text{ newtons (N)}$$

(where $g \approx 9.80$ m/s^2). Thus the force required to lift this layer to ground level is approximately

$$+4\pi\,\delta g\,\Delta y \text{ N},$$

where the plus sign indicates that we have chosen coordinate axes so that the pumping force is positively directed. If Δy is small, then the work ΔW done in pumping this thin layer of gasoline to the top is approximately

$$4\pi\,\delta g(5 - y)\,\Delta y, \tag{7}$$

since the layer is about $(5 - y)$ m from ground level. Thus the total work done in pumping all the gasoline up to ground level is approximated by the sum of all expressions (7), as y varies from 0 to 4:

$$W = \sum \Delta W \approx \sum 4\pi\,\delta g(5 - y)\,\Delta y. \tag{8}$$

As before, the *exact* work is defined to be the limit of (8) as $\Delta y \to 0$. Thus we have

$$\begin{aligned} W &= \int_0^4 4\pi\,\delta g(5 - y)\,dy = 4\pi\,\delta g \int_0^4 (5 - y)\,dy \\ &= 4\pi\,\delta g\left[5y - \frac{1}{2}y^2\right]_0^4 = 4\pi\,\delta g[20 - 8] \\ &= 48\pi\,\delta g \approx 1108 \text{ kJ.} \quad \blacksquare \end{aligned}$$

A common variation on the preceding example involves an above-ground conical storage tank.

FIGURE 6.7

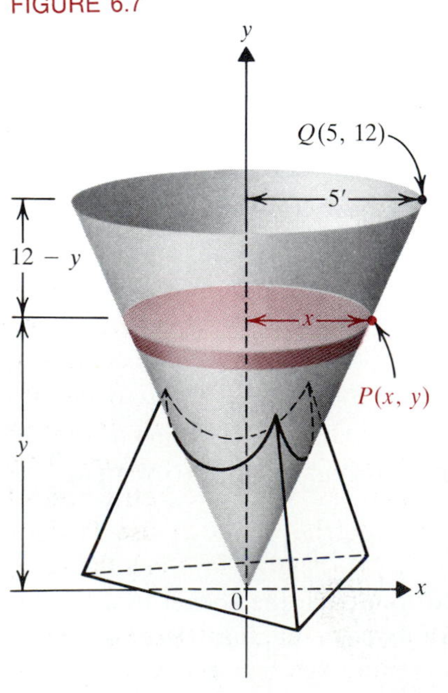

Example 4

The tank shown in **Figure 6.7** is full of water. Find the work done in pumping all the water out over the top of the tank.

Solution. We use the reasoning of Example 3. This time, though, the radius of each "layer" of water is not constant. We have labeled the radius at height y

as x. Noting that $P(x, y)$ lies on the line through $0(0, 0)$ and $Q(5, 12)$, which has slope $12/5$, we find that

$$y = \frac{12}{5}x \rightarrow x = \frac{5}{12}y. \tag{9}$$

As in Example 3, the volume of the layer is approximately $\pi x^2 \, \Delta y$, the volume of a solid cylinder of radius x and height Δy. So the work done in lifting the water "layer" to the top is approximated by

$$\Delta W \approx \pi x^2 \, \Delta y \cdot w \cdot (12 - y), \tag{10}$$

where $w \approx 62.4$ lb is the weight of one cubic foot of water. Hence from (9) and (10),

$$\begin{aligned} W &= \int_0^{12} \pi w \cdot \frac{25}{144} \cdot y^2(12 - y)\,dy = \frac{25\pi w}{144} \int_0^{12} (12y^2 - y^3)\,dy \\ &= \frac{25\pi w}{144}\left[4y^3 - \frac{1}{4}y^4\right]_0^{12} = \frac{25\pi w}{144}[48 \cdot 144 - 36 \cdot 144] \\ &= 25\pi w[48 - 36] = 300\pi w \approx 58{,}800 \text{ ft-lb.} \end{aligned}$$

■

Fluid Force

We now turn to a different kind of problem involving confined liquids: the design of containers strong enough to withstand the force that such liquids exert on the sides of their containers. To analyze that force, we need some facts from physics. First, the ***pressure*** exerted by a liquid at a depth h is defined by

$$p = wh, \tag{11}$$

where w is the weight of one cubic unit (meter, centimeter, or foot) of the liquid. This is interpretable as the weight of a column of water of depth h and cross-sectional area 1, that is, as the force exerted on the base B of the column, which has area 1. (See **Figure 6.8.**) More generally, if the base has area A, then the total force on the base is

FIGURE 6.8

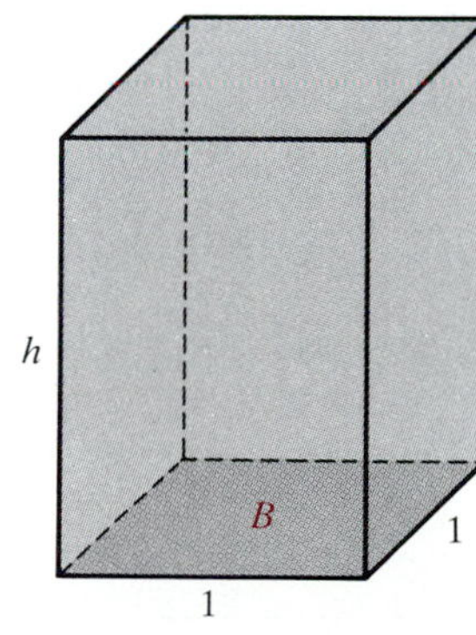

$$F = whA = pA, \tag{12}$$

from which we find

$$p = \frac{F}{A}, \tag{13}$$

which says that *pressure is force per unit area.*

Equation (12) makes it simple to find the total force F on the *bottom* of any container of liquid: We need only measure the area A, the length h, and the weight w of one cubic unit. To find the force on the *sides* of the container, however, we need more than (12), because the depth *varies* from 0 at the top to h at the bottom. Fortunately a second basic principle of physics, known as ***Pascal's principle,*** states that *fluid exerts equal pressure in all directions.* This principle is of such fundamental importance that the standard metric unit of measure for pressure, 1 N/m^2, is called a *pascal* (Pa). We can use Pascal's principle and (12) to derive an integral formula for pressure in much the same way as we did for work.

Consider a flat plate submerged vertically in a liquid as in **Figure 6.9.** Here the surface of the liquid is at $y = c$, and the plate lies between $x = g(y)$ and

FIGURE 6.9

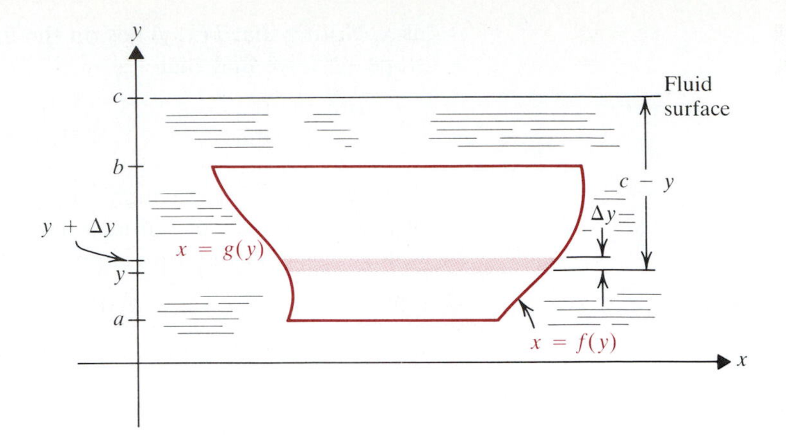

$x = f(y)$, where f and g are continuous functions over $[a, b]$ with

$$g(y) \leq f(y), \qquad y \in [a, b].$$

For a small value of Δy, the shaded region lies about $c - y$ units below the surface of the liquid and has approximate area

$$\Delta A \approx [f(y) - g(y)]\,\Delta y.$$

Thus, by (12) and Pascal's principle, the force on the shaded part of the plate is about

$$\Delta F \approx w(c - y)[f(y) - g(y)]\,\Delta y,$$

where w is the weight of one cubic unit of the liquid. The total force on the lamina is then approximately

$$\sum \Delta F \approx \sum w(c - y)[f(y) - g(y)]\,\Delta y. \tag{14}$$

As $\Delta y \to 0$, we expect (14) to give better and better approximations to this total force. We are thus led to the following definition.

6.2 DEFINITION Suppose that a lamina is bounded by $y = a$, $y = b$, $x = g(y)$, and $x = f(y)$, where $g(y) \leq f(y)$ for $y \in [a, b]$ and where f and g are continuous on $[a, b]$. If this lamina is submerged vertically in a liquid with its top c units below the surface, then the ***force*** exerted on the lamina by the liquid is

$$F = \int_a^b w(c - y)[f(y) - g(y)]\,dy, \tag{15}$$

where w is the weight of one cubic unit of the liquid.

Example 5

A 100-foot-high rectangular dam is built across a river that is 300 feet wide.

(a) What is the total force against the dam when it is full?

(b) What is the total force exerted against the bottom 30 feet of the dam when it is full?

Solution. (a) When the dam is full, $c = 100$. Since the face of the dam is rectangular, we introduce the y-axis along one edge, as in **Figure 6.10.** Then $g(y) = 0$

FIGURE 6.10

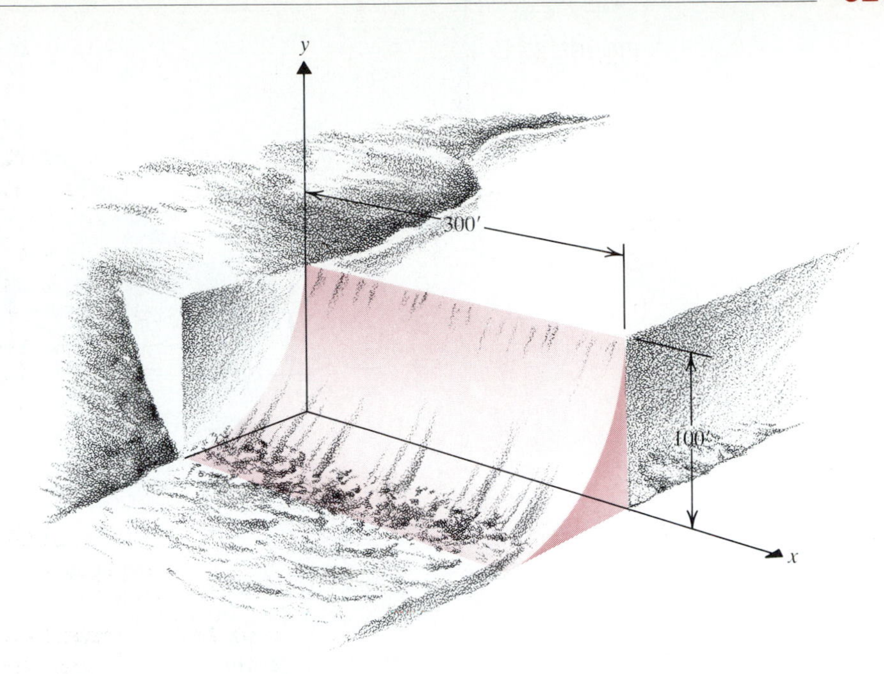

and $f(y) = 300$. Hence (15) gives

$$F = w\int_0^{100}(100 - y)\cdot 300\,dy = 300w\left[100y - \frac{1}{2}y^2\right]_0^{100}$$

$$= 300w\cdot[10{,}000 - 5000] = 1.5\times 10^6\,w \approx 9.36\times 10^7\text{ lb.}$$

(b) Here we need only integrate from 0 to 30, getting

$$F = w\int_0^{30}(100 - y)\cdot 300\,dy = 300w\left[100y - \frac{1}{2}y^2\right]_0^{30}$$

$$= 300w[3000 - 450] = 300w[2550] = 7.65\times 10^5\,w$$

$$\approx 4.77\times 10^7\text{ lb.}$$

Thus *more than half* the total force on the dam is exerted on the lower 30% of the face. This helps explain why dams are thicker at the bottom than at the top. (Notice that the distance behind the dam that the water is backed up has *no effect* on the force exerted against the dam.) ■

The next example illustrates how the strengths of the ends of oil storage tanks can be calculated.

Example 6

A cylindrical oil tank has radius 2 m and length 5 m. It is full of oil that weighs $w = 700\text{ N/m}^3$. Find the force against each end of the tank.

Solution. We show the tank in **Figure 6.11.** The width of the color-shaded strip is

$$2x = 2\sqrt{4 - y^2},$$

so its area is

$$2\sqrt{4 - y^2}\,\Delta y.$$

FIGURE 6.11

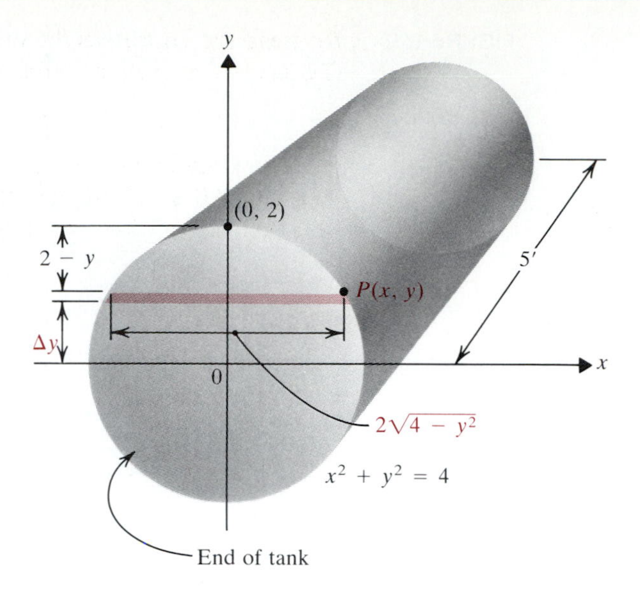

FIGURE 6.12

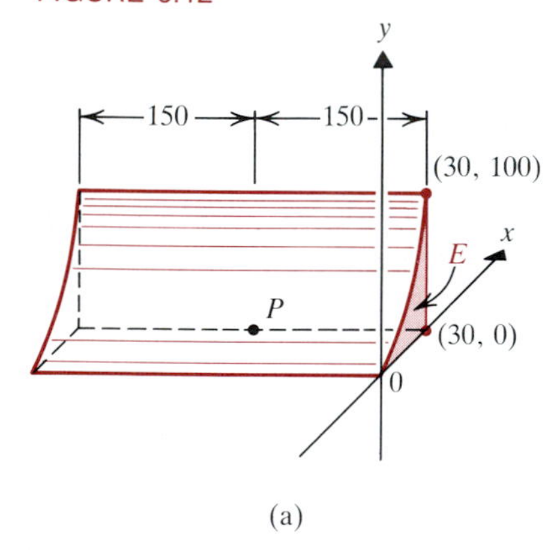

(a)

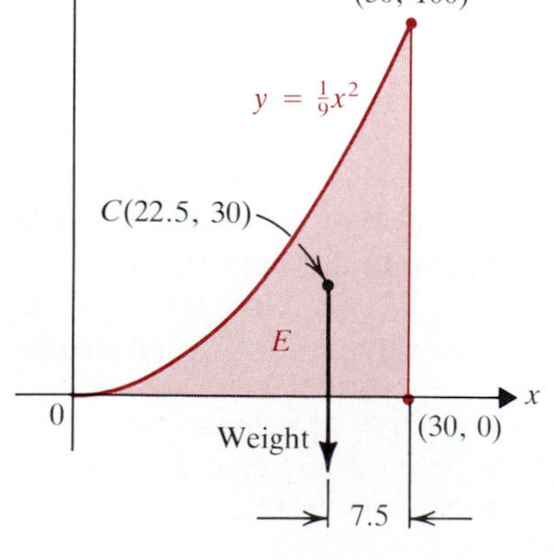

(b)

Hence the force on the strip is

$$\Delta F \approx w(2-y) \cdot 2\sqrt{4-y^2}\,\Delta y.$$

Thus by (15), we have

$$\begin{aligned} F &= 2w \int_{-2}^{2} (2-y)\sqrt{4-y^2}\,dy \\ &= 4w \int_{-2}^{2} \sqrt{4-y^2}\,dy + w \int_{-2}^{2} (-2y)\sqrt{4-y^2}\,dy \\ &= 4w \cdot \frac{1}{2}\pi \cdot 4 + w \cdot \frac{2}{3}(4-y^2)^{3/2}\Big]_{-2}^{2} \qquad \textit{since the first integral is the area of half a disk of radius 2} \\ &= 8\pi w \approx 1.76 \times 10^4 \text{ N}. \end{aligned}$$

Notice that the length of the tank has no effect on the force exerted on the ends. ■

Moments of Force on a Dam*

The rest of this section discusses a refinement of the foregoing ideas that is used in engineering dams and aquariums. For stability, more is needed than just the strength to withstand the force of the water. The *moment of force* on the dam due to its weight must also be great enough to offset the moment of force of the water relative to the point P in **Figure 6.12**(a), which tends to make the dam rotate forward (that is, collapse). In Figure 6.12, we show the dam of Figure 6.10 again, without the water. The coordinate axes are chosen so that the two ends of the dam are the regions E lying under the graph of $y = \frac{1}{9}x^2$ between $x = 0$ and $x = 30$. The methods of the last section show (see Exercise 41) that the centroid of this region is the point $C(\bar{x}, \bar{y})$, where

(16) $$\bar{x} = 22.5 \quad \text{and} \quad \bar{y} = 30.$$

* Optional

We treat the entire weight of the dam as a force acting downward from C. The *moment of the dam relative to the point P* is defined to be

$$M_P = \bar{w} \cdot A(E) \cdot 300 \cdot (7.5),$$

where $\bar{w}$ is the weight of a unit volume of the *dam* material, $A(E) = 1000$ is the area of E, 300 is the length of the dam, and $7.5 = 30 - 22.5$ is the perpendicular distance from (30, 0) to the line of action of the weight. Thus

$$M_P = 2.25 \times 10^6\ \bar{w}.$$

Here assume that $\bar{w} > w$, the weight of a unit volume of water. The *moment of force of the water relative to P* is defined as

$$M_P^* = w \int_0^{100} (100 - y) \cdot 300\, dy = w\Big[3 \times 10^4\ y - 150y^2\Big]_0^{100}$$

$$= w[3 \times 10^6 - 1.5 \times 10^6] = 1.5 \times 10^6\ w.$$

Thus $M_P > M_P^*$, so the water's tendency to make the dam collapse is more than offset by the dam's weight. This (in somewhat simplified form) is the sort of calculation that construction engineers must make in designing a dam. In Exercise 43 you are asked to find the *minimum a* so that M_P is at least as large as M_P^*, if the ends of the dam are represented by the region under the graph of

$$y = \frac{1}{a}x^2 \qquad \text{between } x = 0 \text{ and } x = 10\sqrt{a}.$$

This gives the *minimum* thickness required for safety. (In practice, engineers allow an additional safety margin, but somewhat less than the 50%+ margin by which M_P exceeds M_P^* above.)

HISTORICAL NOTE

Blaise Pascal (1623–1662) was a brilliant, innovative mathematician and physicist. In 1643 he discovered that air pressure decreases with altitude, and shortly afterward he formulated the physical principle that bears his name. He is regarded as one of the cofounders of probability theory (with Fermat); his motivation is said to have stemmed from his love of gambling.

He devoted the last years of his illness-plagued life to philosophy and applied his newly developed principles of probability to religion in "Pascal's wager." It analyzed the question of belief in God as follows. First, if God does *not* exist and one is an atheist, then the person realizes no particular benefits now or after death. But, Pascal reasoned, if there *is* a God and the person failed to believe in God's existence, then he stood to gain nothing now and lose enormously after death. On the other hand, if he believed in God, then he would be rewarded after death if he was right, and lose nothing if he was wrong. He concluded then that the prudent course was to maximize his expected reward by believing in God! (He was, in fact, an intensely religious man.)

The computer language that bears his name honors him as the inventor (at the age of 19) of the first adding machine. That feat, which won him instant fame, constituted the first step along the path leading to the construction of the electronic calculators and computers that have become such a prominent feature of modern life.

Exercises 5.6

In Exercises 1–12, find the work done in carrying out the given task.

1. A 7-pound bag of groceries is carried up a flight of stairs from a garage to a kitchen 10 feet above the garage.
2. A 110-pound student walks up three flights of stairs to a classroom located 30 feet above ground level.
3. A spring of natural length 2.4 cm requires a 5-dyne force to stretch it 1.0 cm. It is stretched from its natural length to a length of 4.2 cm.
4. A spring of natural length 0.1 m requires 1 N of force to stretch it another 0.1 m. It is stretched from its natural length to a length of 0.3 m.
5. A spring that is compressed 2 cm from its natural length exerts a restoring force of 5 dynes. The spring is compressed from that point three more centimeters.
6. A spring compressed 6 in. from its natural length exerts a restoring force of 2 lb. It is compressed six more inches.
7. A large spring has natural length 1 m. A force of 10 N is required to compress it to a length of 1/2 m. It is stretched from its natural length to a length of 2 m.
8. A spring with natural length 18 in. is compressed to a length of 12 in. by an 8-lb force. It is stretched from its natural length to a length of 30 in.
9. A force of 2 N is required to stretch a spring 10 cm from its natural length. It is compressed 50 cm from its natural length.
10. A spring is stretched 8 cm by a force of 3 N. It is compressed 9 cm from its natural length.
11. The rocket in Example 2 is launched to an altitude of 50 km, where the lower stage breaks away and a 500-kg payload is lifted to an altitude of 100 km.
12. In Exercise 11 the first stage is lifted to an altitude of 60 km, and a second stage of 1000 kg is lifted to an altitude of 200 km.

13. When studying the motion of a single body, Newton's law of gravitation can be written as $g(x) = -k/x^2$, for a constant k. Use this to find the work done in launching a 1000-pound rocket to an altitude of 100 miles. (Express your answer in ft-lb.)
14. Repeat Exercise 13 if the vehicle is to be launched to an altitude of 1000 miles, where a stage separation takes place.
15. Two like-charged particles repel each other with force $F(x) = k/x^2$, where x is the distance between them. How much work is done in moving one along a line from distance 2×10^{-6} cm to distance 1.5×10^{-6} cm from the other, if at the first distance F is 10^{-4} dynes?
16. Two oppositely charged particles attract each other with a force $F(x) = k/x^2$, where x is their distance apart. If $F(x) = 2 \times 10^{-7}$ dynes when $x = 1.8 \times 10^{-4}$ cm, then find how much work is done in moving one of them along a line to a point 3×10^{-4} cm from the other.
17. The pressure and volume of a gas in a cylinder are related by $PV^{1.4} = 200$, where P is in pascals, and V is in cubic centimeters. Find the work done by a piston that compresses the gas from 10 cm^3 to 1 cm^3. [*Hint:* Let $V = cx$ for some constant c, where x is the linear distance of compression. What does c represent? Use Equation (13) to express the integral (2) for W as an integral in terms of V between 10 and 1.]
18. In Exercise 17, suppose instead that the gas expands, pushing the piston back, from 9 cm^3 to 16 cm^3. Find the work done by the gas in moving the piston.
19. An above-ground cylindrical oil storage tank of radius 12 feet and height 20 feet is full of oil. Find the work done in pumping all the oil out the top of the tank, if the weight w of one cubic foot of oil is 45 lb.
20. An above-ground cylindrical swimming pool of radius 10 feet and height 8 feet. Find the work done in pumping all the water out of the pool. (Use $w = 62.4$.)
21. A buried storage tank is conical with its circular top at ground level, $r = 4$ m, and $h = 8$ m. It contains oil (weighing 800 N/m^3) whose top is 2 m below ground level. How much work is done in pumping all the oil to the surface?
22. An above-ground conical tank like the one in Figure 6.7 has $r = 5$ ft and $h = 10$ ft and is full of water. Find the work done in pumping all the water to the top.
23. Find the work done in pumping out all the liquid from a filled hemispherical bowl of radius a, if one cubic unit of the liquid has weight w.
24. Repeat Exercise 23 if it is required to pump the liquid to a height h units above the top of the bowl.
25. A cylindrical gasoline storage tank has $r = 10$ m and $h = 8$ m. It contains gasoline to a height of 5 m. Find the work required to pump enough gasoline so that the tank has gasoline only to a height of 1 m. (Express your answer in terms of w.)
26. Repeat Exercise 25 if the tank is a buried conical tank of radius 10 m and height 8 m, and its circular top is at ground level.
27. Find the total force against a 100-foot-high dam across a 200-foot-wide stream if its face is rectangular.
28. Repeat Exercise 27 if the dam is 150 feet high and 250 feet wide.
29. A watering trough for animals has isosceles trapezoids as back and front sides, with bottom 2 m long and top 4 m long. If the trough is 1 m deep, find the force against the front and back when it is full of water. See **Figure 6.13.**

FIGURE 6.13

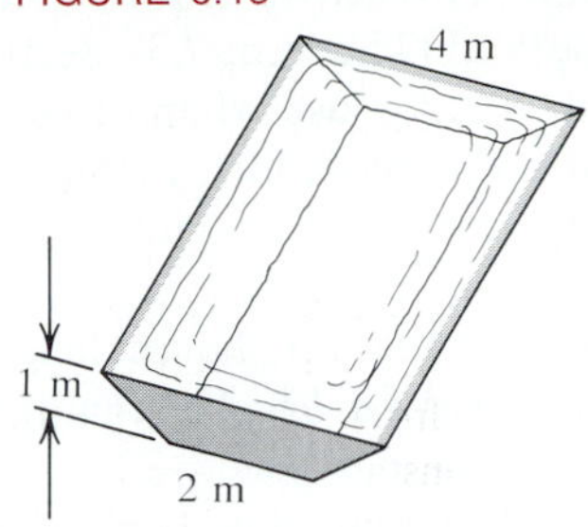

30. Repeat Exercise 29 for a trough 8 feet wide at the top, 6 feet wide at the bottom, and 3 feet deep.

31. Suppose the oil tank in Example 6 has radius 5 feet and height 10 feet and is full of oil weighing 48 pounds per cubic foot. Find the force on the ends of the tank.

32. Repeat Exercise 31 if the tank has radius 3 m and height 5 m, and contains oil weighing 720 N/m^3.

33. An oil tank like that in Example 6 has radius 4 feet and height 8 feet and is half full of oil that weighs 45 lb/ft^3. What is the force on each end of the tank?

34. Repeat Exercise 32 if the tank is half full of oil.

35. An aquarium window is in the shape of an isosceles trapezoid. Its top edge is 6 feet long and is 4 feet below the surface of the water in a tank. Its bottom edge is 8 feet long and is 10 feet below the water surface. What is the force on the window?

36. An aquarium window is circular with radius 3 feet. If the top of the window is 3 feet below water level, find the total force on the window.

37. A basic law of physics is that *work done by a moving body equals the change in its kinetic energy* ($\frac{1}{2}mv^2$). This exercise derives that basic law for forces acting in a straight line.
    (a) Write Newton's second law $F = ma$ in terms of time t as
$$F(t) = m\frac{dv}{dt}.$$
    Then show that
$$F(x) = mv\frac{dv}{dx}.$$
    (b) Integrate $F(x)$ in (a) over $[a, b]$ to show that the work done by $F(x)$ over $[a, b]$ is $\frac{1}{2}mv(b)^2 - \frac{1}{2}mv(a)^2$, i.e., it is the change in kinetic energy over $[a, b]$.

38. If gas expands in a cylinder from volume V_1 to volume V_2, moving a piston, then show that the work done is
$$\int_{V_1}^{V_2} P\,dV,$$
where P is the pressure.

39. *Archimedes' principle* is that a body immersed in a fluid is buoyed up by a force equal to the weight of the liquid displaced by the body. This exercise uses (13) to verify Archimedes' principle in a special case. Suppose that a rectangular beam of horizontal area A and height h is submerged at depth D in water weighing w N/m^3.
    (a) What is the liquid buoyant force F_b on the bottom of the beam?
    (b) What is the liquid force F_t on the top of the beam?
    (c) Find the net force $F = F_b - F_t$ on the beam. How is this related to the volume of the beam?

40. Show that the force F on the face of a dam is the product pA, where A is the area of the face of the dam, and p is the pressure at the centroid $(\bar{x}, \bar{y})$ of the face of the dam.

41. Verify the values of $\bar{x}$ and $\bar{y}$ in (16).

42. Show that $M_P < M_P^*$ if the ends of the dam are the region bounded by $y = \frac{1}{4}x^2$ between $x = 0$ and $x = 20$.

43. Find the minimum a so that $M_P \geq M_P^*$ if the ends of the dam are the region bounded by
$$y = \frac{1}{a}x^2$$
between $x = 0$ and $x = 10\sqrt{a}$.

44. Verify that $a = 16$ works in Exercise 43.

5.7 Looking Back

The goal of this chapter has been to provide practice with definite integrals themselves and in setting up the kinds of problems (in and out of mathematics) in which integrals are frequently put to use.

The first section discussed the area of various plane regions. The principal results are Theorems 1.1 and 1.2 on the area of regions lying between the graphs of two continuous functions and two lines parallel to a coordinate axis.

Sections 2 and 3 presented two different techniques for finding the volume of solids of revolution. In Section 2, the basic Definition 2.1 states that one way to calculate such volumes is to integrate the cross-sectional area made by planes perpendicular to the axis of revolution. This led to the circular disk method (Theorem 2.2) and the circular ring method (Theorem 2.3). Section 3 presented the cylindrical shell method (Theorem 3.2) for use when cross-sectional areas are difficult (or impossible) to compute. The basic idea is that the volume of a cylindrical shell is 2π times the average radius of revolution times the height times the thickness of the shell.

Section 4 provided (Definition 4.2) a formula for the arc length of the graph of a smooth function $f(x)$ over an interval $[a, b]$. An analogous formula—Equation (7), p. 308—holds for functions of y.

Section 5 discussed mass, moments, and centroids, beginning with one dimension and then proceeding to the fundamental Definition 5.1 for plane regions. The theorem of Pappus (Theorem 5.2) showed how centroids are used to find volumes of solids of revolution.

The final section was devoted to two applications to physics—work and fluid force. Definitions 6.1 (for work) and 6.2 (for fluid force) were the foundation for the various kinds of problems treated.

CHAPTER CHECKLIST

Section 1: area of a region between the graphs of continuous functions $f(x)$ and $g(x)$ over an interval $[a, b]$; area of a region between the graphs of continuous functions $h(y)$ and $k(y)$ over an interval $[c, d]$.

Section 2: volume of a solid with cross-sectional area $A(t)$, $t \in [a, b]$; circular disk method; circular ring method for solids of revolution.

Section 3: volume of a right circular cylindrical shell; cylindrical shell method for volumes of solids of revolution.

Section 4: smooth functions; arc length of the graph of a smooth function $y = f(x)$ over an interval $[a, b]$; arc length formula for the graph of $x = g(y)$ over $[c, d]$.

Section 5: moments and centers of mass of discrete systems; mass; moments and centers of mass of continuously distributed masses along an axis; mass, moments, and centers of mass of plane lamina; theorem of Pappus.

Section 6: work; Hooke's law; work done in launching rockets or pumping liquids; fluid force; dams and other containers.

REVIEW EXERCISES 5.7

In Exercises 1–6, find the area of the given region.

1. Bounded by $2y = x^2$ and $y = 2 - x^2$
2. Bounded by $y = x + 3$ and $y = 9 - x^2$
3. Enclosed by $y^2 = x^2(a^2 - x^2)$
4. Bounded by $x + y = 2$ and $x = y^2$
5. Bounded by $x^2 = y - 2$, $x + y = 0$, the y-axis, and $x = 1$
6. Bounded by $y = \sin x$, $y = \cos x$, $x = -\pi/4$, and $x = \pi/4$

In Exercises 7–20, find the volume of the solid of revolution formed when the given region is revolved about the given line.

7. Enclosed by $y^2 = x^2 - x^4$; x-axis
8. Enclosed by $x^2 + 4y^2 = 4$; x-axis
9. Under $y = 4/(x + 1)$ between $x = 2$ and $x = 5$; x-axis
10. Under $y^3 = x^2$ and above $y = 0$ between $x = 0$ and $x = 2\sqrt{2}$; x-axis
11. Under $y = 4x^2$ and above $y = 0$ between $x = 0$ and $x = 2$; $x = 2$
12. Under $y = \sqrt{\sin x}\cos x$ and above the x-axis between $x = 0$ and $x = \pi/2$; $y = -1$
13. Under $y = x^2 - 1$ and above $y = 0$ between $x = 1$ and $x = 2$; y-axis
14. Bounded by $y = x^2$, $y = 1$, and $x = 2$; $y = -2$
15. First-quadrant region bounded by $y = x^2$, y-axis, and $y = 4$; y-axis
16. Bounded by $y = x^3$, y-axis, and $y = 8$; $x = 3$
17. Bounded by y-axis, $x = \dfrac{\cos y}{y}$, $y = \pi/3$, $y = \pi/2$; x-axis

18. Bounded by x-axis, $y = \frac{\sin x}{x}$, $x = \pi/4$, and $x = \pi/3$; y-axis

19. Bounded by $(x-1)^2 + (y-3)^2 = 4$; the line $x = 4$

20. Inside the triangle with vertices (0, 0), (2, 0), and (5, 2); $x = -2$

In Exercises 21–26, find the arc length of the given curve between the given points.

21. $y = (2/3)x^{3/2}$; $x = 0$ to $x = 8$

22. $x = y^{3/2} + 1$; $x = 1$ to $x = 9$

23. $y = (3/2)x^{2/3}$; $x = 1$ to $x = 27$

24. $x = y^{2/3}$; $x = 1$ to $x = 4$

25. $y = \int_1^x \sqrt{t^2 - 1}\,dt$, $x = 1$ to $x = 3$

26. $y = \int_1^x \sqrt{\sec^3 t \tan^2 t - 1}\,dt$, $x = \pi/4$ to $x = \pi/3$

In Exercises 27–36, find the mass and center of mass of the given system.

27. Rod with density $\delta(x) = x^2 - 1$, over $[2, 5]$

28. Rod with density $\delta(x) = \sqrt[3]{x} - 1$, over $[2, 9]$

29. Rod with density $\delta(x) = (4 - x^2)^{1/2}$, over $[-2, 2]$

30. Rod with density $\delta(x) = x^{2/3} + 5$, over $[0, 1]$

31. Lamina with density δ, under $y = 4 - x^2$ and above the x-axis

32. Lamina with constant density δ, under $y = 4x - x^2$ and above $y = x$

33. Lamina with constant density δ, bounded by $y = x^2$ and $y = x^3$

34. Lamina with constant density δ, bounded by $y = 1 - x^2$ and $y = x^2 - 1$

35. Lamina with constant density δ, between $y = x^3$ and $y^2 = x$

36. Lamina with constant density δ, between $y = 4 - x^2$ and $y = x + 2$

In Exercises 37–44, find the work done.

37. A spring that requires a force of $\frac{1}{2}$ N to stretch it 30 cm beyond its natural length is compressed 10 cm from its natural length.

38. A spring that is compressed 10 cm from its natural length by a force of 25 dyne is stretched 15 cm from its natural length.

39. A rocket weighing 1000 pounds is launched to an altitude of 500 miles above the earth's surface. (Use 4000 miles for the radius of the earth.)

40. A rocket that weighs 1200 lb on earth is launched from the surface of the moon to a larger rocket 100 miles above the surface. Gravity on the moon is only 1/6 the strength of gravity on earth. Use 1000 miles for the radius of the moon.

41. An above-ground conical tank filled with water has its pointed end at ground level. It has radius 5 feet and height 10 feet. All the water is pumped out.

42. A buried tank is in the shape of a cone with its round top 1 m below ground level. It has radius 3 m and height 10 m, and has oil up to a point 5 m below ground level. The oil weighs 800 N/m^3. All the contents of the tank are pumped to the surface.

43. Water is pumped out of a hemispherical container of radius 5 m, whose circular top is 6 m below ground level, until the depth is 1 m.

44. A cable is wound around a windlass, with 60 feet hanging loose. The cable weighs 4 pounds per linear foot. Forty feet of the cable are wound up.

45. Find the total force against a 150-foot-high rectangular dam across a 100-foot-wide river, when the reservoir is full of water.

46. A cylindrical oil tánk has radius 4 m and height 7 m, and contains oil weighing 750 N/m^3. If the sides are 3 feet high, find the force against its circular ends if these are vertical.

47. Find the force on a semicircular aquarium window of radius 3 feet whose diameter is at water level.

48. A trough has sides that are isosceles trapezoids, with upper base 3 feet and lower base 2 feet. If the sides are 3 feet long, then find the force on the side when the trough is full of water.

49. *Cavalieri's principle* for two solids S_1 and S_2 with parallel bases says the following: If every plane P parallel to the bases intersects the solids in regions R_1 and R_2 of equal area, then S_1 and S_2 have equal volume. Prove this if the bases are planes perpendicular to the x-axis.

50. A solid has base disk $x^2 + y^2 \leq 25$. Every plane section perpendicular to the x-axis is an isosceles right triangle with a leg in the base (see **Figure 7.1**). Find the volume of the solid.

FIGURE 7.1

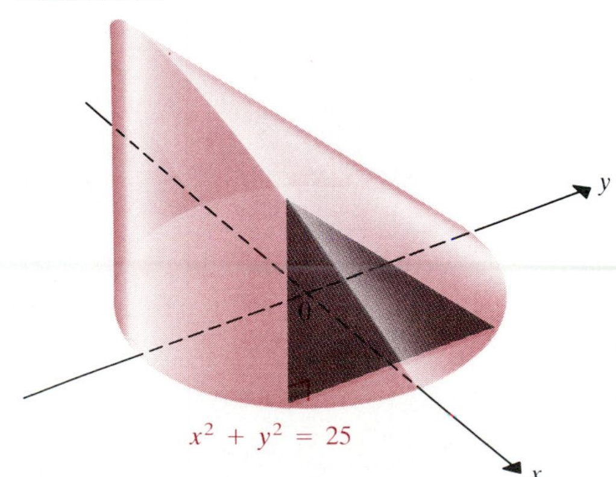

51. The *potential energy* P of a body acted on by a force F over an interval $[a, b]$ is given by

$$P(x) = -\int_a^x F(t)\,dt.$$

Use the fact that

$$W = \int_a^b F(x)\,dx = \frac{1}{2}mv(b)^2 - \frac{1}{2}mv(a)^2$$

is the change in kinetic energy (Exercise 37 of Section 6) to establish the *law of conservation of total energy:*

$$P(b) + \frac{1}{2}mv(b)^2 = P(a) + \frac{1}{2}mv(a)^2.$$

52. ***Power*** is the rate at which work is done, dW/dt. Show that if $s = dx/dt$ is the speed at which a force F moves a body from $x = a$ to $x = b$, then

$$\frac{dW}{dt} = F(x(t))s.$$

Power is measured in *watts* (w) (1 w = 1 J/s).

53. Let $f(x)$ be a smooth increasing function whose graph joins (0, 0) to (1, 1). Show that the arc length of $f(x)$ lies between $\sqrt{2}$ and 2. (*Hint:* Consider any polygonal arc inscribed in the curve. Show that its length must lie between $\sqrt{2}$ and 2.)

54. Show that in Example 4 of Section 6 the same work is done if we regard all the water as concentrated at the centroid of the tank.

CHAPTER 6

Further Transcendental Functions

6.0 Introduction

A function f is *algebraic* if its formula involves only the operations of addition, subtraction, multiplication, division, absolute value, and exponentiation to rational powers. Otherwise the function is called *transcendental*. The six trigonometric functions, which we have been working with since Chapter 2, are familiar transcendental functions. This chapter develops the calculus of additional transcendental functions.

The first five sections are devoted to logarithmic and exponential functions. Following a review of the basic properties of logarithms, Section 1 defines the natural logarithm function *ln* as an antiderivative. Properties of integration are then used to show that this function has all the usual logarithmic properties, and to derive its calculus. The next section defines and studies the natural exponential function *exp* as the inverse function of *ln*. Section 3 uses *ln* and *exp* to analyze several important types of growth and decay. The following section discusses further exponential and logarithmic functions, which arise in several areas outside mathematics.

Section 5 introduces the inverse trigonometric functions and their calculus. These functions are at the heart of one of the techniques of integration presented in the next chapter. Section 6 is devoted to hyperbolic functions, a class of exponential functions with properties similar to those of the trigonometric functions. Section 7 discusses an important application of the hyperbolic functions to physics and engineering, as well as some applications of trigonometric functions to motion problems.

Section 8 presents a technique named for the author of the first calculus textbook, written in 1696. That technique, l'Hôpital's rule, can be used to evaluate a number of limits that, up to now, can only be classified as *indeterminate*, not capable of being determined by earlier methods. Many of those limits involve transcendental functions. The final section gives an optional discussion of some widely used graphical schemes that are well-suited to the mathematical models discussed in Section 3 as well as to power functions.

6.1 Logarithms and the Natural Logarithm Function

In precalculus, logarithmic functions are usually introduced as inverse functions of exponential functions. The logarithmic function $\log_b$ with base b is defined as the function with domain $(0, +\infty)$ such that

(1) $\quad y = \log_b x \quad$ (read: *the logarithm of x to the base b*)

means

(2) $\quad x = b^y$

if $b \neq 1$ is a positive number. The function defined by (2) is called the *exponential function with base b*. Because $b^0 = 1$ for all nonzero numbers b, for any base b we have

(3) $$\log_b 1 = 0.$$

In addition to (3), there are three other basic laws of logarithms:

(4) $$\log_b(uv) = \log_b u + \log_b v, \quad \log_b(u/v) = \log_b u - \log_b v, \quad \log_b u^p = p \log_b u.$$

These are derived from corresponding basic laws of exponents:

(5) $$b^p b^q = b^{p+q}, \quad b^p/b^q = b^{p-q}, \quad (b^p)^q = b^{pq}.$$

For instance, to derive the first law in (4), let $u = b^p$ and $v = b^q$. The equivalence of (1) and (2) says that $p = \log_b u$ and $q = \log_b v$. Then

$$\begin{aligned} \log_b(uv) &= \log_b(b^p \cdot b^q) = \log_b(b^{p+q}) && \text{by (5)} \\ &= p + q && \text{by (1) and (2)} \\ &= \log_b u + \log_b v. \end{aligned}$$

There is a difficulty with this approach, however. It assumes that the definition and properties of exponential functions are well known. This assumption is innocuous enough when the exponents are integers or rational numbers. Indeed, for positive integers n, b^n is simply $b \cdot b \cdot \ldots \cdot b$ (n factors) and b^{-n} is $1/b^n$. For a rational number $r = m/n$ (where m and n are integers with $n > 0$) and a positive base b, one defines

(6) $$b^r = \sqrt[n]{b^m}.$$

But if r is an irrational number, such as $\sqrt{2}$ or π, then what does b^r mean? There is no simple definition like (6) for b to an irrational power. One of the goals of this chapter is to use calculus to develop logarithmic and exponential functions on a solid foundation. That will enable us to give a precise meaning to b^r when r is irrational.

Our approach reverses the path taken in precalculus. The first step is to define the natural logarithm function *ln* as a definite integral; then we develop its algebraic rules and most of its calculus. (We won't learn how to antidifferentiate *ln* until the next chapter.) In the next section, we will define the natural exponential function *exp* as the inverse function of *ln* and show that this definition agrees with (6) when x is a rational number. We will then use our knowledge of inverse functions to develop the algebra and calculus of *exp*.

Recall the power rule for antiderivatives (p. 210):

$$\int x^r\,dx = \frac{x^{r+1}}{r+1} + C$$

for all $r \neq -1$. We have as yet no formula for evaluating

$$\int x^{-1}\,dx = \int \frac{1}{x}\,dx.$$

We are going to introduce a function *ln* to fill in that gap: *ln* will be a function such that $D_x(\ln x) = 1/x$. It will also have domain $(0, +\infty)$ and satisfy laws (3) and (4), and so will deserve the name *logarithm*. To create the function *ln*, we use part 1 of the fundamental theorem of calculus (Theorem 4.1 of Chapter 4). Recall that if f is continuous on an interval containing a and if we let

$$G(x) = \int_a^x f(t)\,dt,$$

then $G'(x) = f(x)$. We thus use the function given by $f(t) = 1/t$ to define *ln*.

1.1 DEFINITION The ***natural logarithm function*** ln is defined by

$$\ln x = \int_1^x \frac{1}{t}\,dt, \tag{7}$$

for $x \in (0, +\infty)$.

It follows immediately from (7) and Theorem 4.1 of Chapter 4 that

$$\frac{d}{dx}(\ln x) = \frac{1}{x} \tag{8}$$

for all positive real numbers x.

Example 1

Find f' for the function f if

(a) $f(x) = \ln(3x^2 - 2x)$ (b) $f(x) = \ln(\sec 2x)$

Solution. We use the chain rule for derivatives and (8), which give

(a) $f'(x) = \dfrac{1}{3x^2 - 2x} \cdot \dfrac{d}{dx}(3x^2 - 2x) = \dfrac{6x - 2}{3x^2 - 2x}$,

(b) $f'(x) = \dfrac{1}{\sec 2x} \cdot \dfrac{d}{dx}(\sec 2x) = \cos 2x \cdot (\sec 2x \tan 2x) \cdot 2 = 2\tan 2x$. ■

For all x in the domain $(0, +\infty)$ of *ln*, $1/x$ is also positive, of course. From this and Theorem 3.1 of Chapter 3, it follows that

the function *ln* is increasing on its entire domain $(0, +\infty)$.

If we let $x = 1$ in (7), then we find

$$\ln 1 = \int_1^1 \frac{1}{t}\,dt = 0, \tag{9}$$

FIGURE 1.1

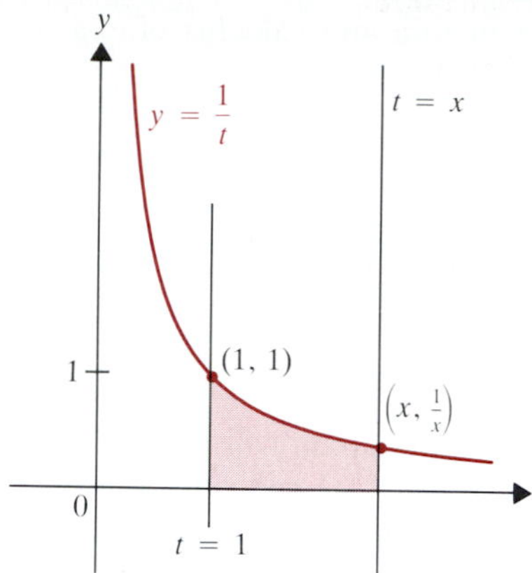

so the function *ln* satisfies Law (3) of logarithms. Since *ln* is an increasing function, we also have

(10) $\ln x > 0$ if $x > 1$ and $\ln x < 0$ if $0 < x < 1$.

Property (10) is easy to see geometrically also. Note in **Figure 1.1** that for $x > 1$, $\ln x$ is the area under the graph of $y = 1/t$ between $t = 1$ and $t = x$. For $x < 1$, on the other hand, $\ln x$ is the *negative* of the shaded area in **Figure 1.2** under $y = 1/t$ between $t = x$ and $t = 1$, because reversing the limits of integration in a definite integral multiplies the integral by -1:

$$\ln x = \int_1^x \frac{1}{t}\, dt = -\int_x^1 \frac{1}{t}\, dt.$$

Having already established Law (3) in (9), we next complete the justification of the name *natural logarithm* for the function *ln*.

1.2 THEOREM

If a and b are positive, then

(a) $\ln(ab) = \ln a + \ln b$;

(b) $\ln a^p = p \ln a$ for any rational number p;

(c) $\ln \frac{a}{b} = \ln a - \ln b$.

FIGURE 1.2

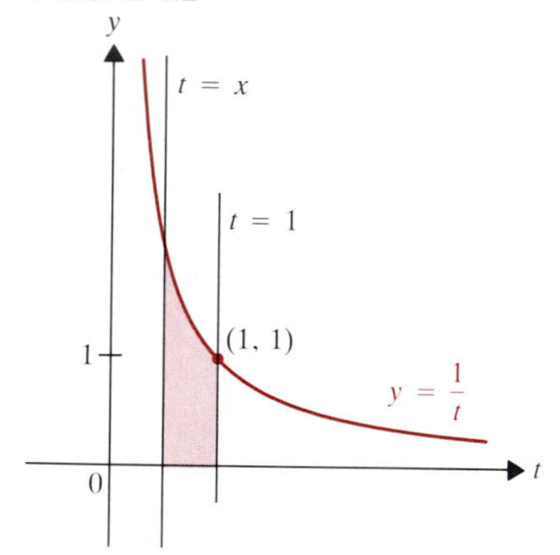

Proof. (a) For any $x > 0$ we have from the chain rule and Theorem 4.1 of Chapter 4

$$\frac{d}{dx}(\ln ax) = \frac{d}{dx}\int_1^{ax} \frac{1}{t}\, dt = \frac{1}{ax} \cdot \frac{d}{dx}(ax)$$

$$= \frac{1}{ax} \cdot a = \frac{1}{x} = \frac{d}{dx}(\ln x).$$

Thus $\ln x$ and $\ln ax$ have the same derivative. So by Theorem 9.2 of Chapter 3, they differ only by a constant C:

(11) $$\ln ax = \ln x + C.$$

To find C, we let $x = 1$ in (11):

$$\ln a = \ln 1 + C = C$$

by (9). Hence (11) becomes

$$\ln ax = \ln x + \ln a$$

for any x. Letting $x = b$, we have (a).

(b) For $x > 0$ consider the function

$$\ln x^p = \int_1^{x^p} \frac{1}{t}\, dt.$$

The chain rule and Theorem 9.2 of Chapter 3 say that

$$\frac{d}{dx}(\ln x^p) = \frac{1}{x^p} \cdot \frac{d}{dx}(x^p) = \frac{1}{x^p} \cdot px^{p-1} = p \cdot \frac{1}{x} = \frac{d}{dx}(p \ln x).$$

Hence $\ln x^p$ and $p \ln x$ can differ by at most a constant:

$$\ln x^p = p \ln x + C. \tag{12}$$

To determine C, we again let $x = 1$:

$$\ln 1^p = p \ln 1 + C.$$

Since $1^p = 1$ and $\ln 1 = 0$, we thus have

$$0 = 0 + C \to C = 0.$$

Hence (12) becomes

$$\ln x^p = p \ln x$$

for any $x > 0$. This proves (b).

(c) Using (a) we have

$$\ln \frac{a}{b} + \ln b = \ln\left(\frac{a}{b} \cdot b\right) = \ln a.$$

Therefore

$$\ln \frac{a}{b} = \ln a - \ln b. \quad \boxed{\text{QED}}$$

Now that we have justified calling *ln* a logarithm, it is natural to ask how to find values of $\ln x$ for given positive numbers x. One way is to use a numerical integration method such as Simpson's rule on $f(x) = 1/x$. Table 1 shows results obtained on a computer when Simpson's rule was used to approximate

$$\ln 2 = \int_1^2 \frac{1}{x}\, dx$$

It suggests that the value of $\ln 2$, to eight decimal places, is 0.69314718. That should agree with your calculator's value for $\ln 2$. (But the same computer gave 0.69314719 for $n = 640$ and 0.69314714 for $n = 5120$. Accumulation of decimal *round-off error* can actually *lessen* the accuracy of the approximation computed from $S_n(f)$ if n is allowed to grow too large.) Theorem 6.3 of Chapter 4 can be used to estimate the accuracy of a Simpson's rule approximation to $\ln x$.

TABLE 1

n	10	20	40	80	160	320
$S_n(f)$	0.69315023	0.69314737	0.69314719	0.69314718	0.69314718	0.69314718

Example 2

Estimate the accuracy of the value for $\ln 2$ in Table 1 when $n = 80$.

Solution. To apply the Simpson-rule error estimate, we need to find an upper bound M for $f^{(4)}(x)$, where $f(x)$ is the integrand $1/x = x^{-1}$. We have

$$f'(x) = -x^{-2}, \qquad f''(x) = 2x^{-3}, \qquad f'''(x) = -6x^{-4}, \qquad \text{and } f^{(4)}(x) = 24x^{-5}.$$

On the interval $[1, 2]$, therefore,

$$f^{(4)}(x) = \frac{24}{x^5} \le \frac{24}{1^5} = 24.$$

So, from Theorem 6.3 of Chapter 4, the error E_S made when ln 2 is approximated by $S_{80}(f)$ satisfies

$$E_S \leq 24\frac{(2-1)^5}{180 \cdot 80^4} \approx 3.255 \times 10^{-9} < \frac{1}{2} \times 10^{-8}.$$

Thus $S_{80}(f)$ should indeed approximate ln 2 accurately to 8 decimal places. ■

FIGURE 1.3

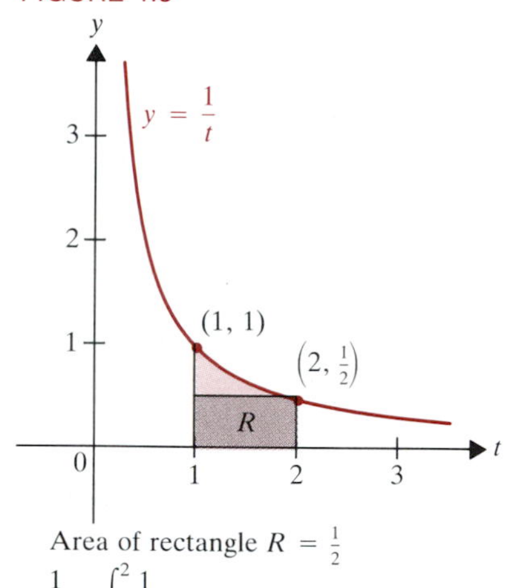

Area of rectangle $R = \frac{1}{2}$

$\frac{1}{2} < \int_1^2 \frac{1}{t}\,dt = \ln 2$

Area under $y = \frac{1}{t}$ from $t = 1$ to $t = 2$

We continue our study of *ln* by next showing that its range is the entire real number system. The first step is to show that $\lim_{x\to\infty} \ln x = +\infty$. For that, note from Example 1 or **Figure 1.3** that

$$\ln 2 > \frac{1}{2}.$$

Then by Theorem 1.2(b), for any positive integer p we have

$$\ln 2^p = p \ln 2 > \frac{1}{2}p.$$

Hence

(13) $$\lim_{p\to+\infty} \ln 2^p = +\infty \rightarrow \lim_{x\to+\infty} \ln x = +\infty.$$

To see that $\lim_{x\to 0^+} \ln x = -\infty$, let $u = 1/x$. Then $x \to 0^+$ is equivalent to $u \to +\infty$. Theorem 1.2(b) with $p = -1$ gives

$$\ln u = \ln\frac{1}{x} = \ln x^{-1} = -\ln x \rightarrow \ln x = -\ln u.$$

Therefore

(14) $$\lim_{x\to 0^+} \ln x = \lim_{x\to 0^+} (-\ln u) = -\lim_{u\to+\infty} \ln u = -\infty$$

by (13). We can now use the intermediate value theorem (Theorem 7.6 of Chapter 1) to show that *ln* has range $(-\infty, +\infty)$. Since by (8) *ln* is differentiable on its entire domain $(0, +\infty)$, it is continuous there by Theorem 1.3 of Chapter 2. Equations (13) and (14) say that *ln* takes on arbitrarily large positive and negative values. Thus, for *any* real number r, there are a and b in $(0, +\infty)$ such that

$$\ln a < r < \ln b.$$

Then by the intermediate value theorem, there is some x in $[a, b]$ such that

$$r = \ln x.$$

This means that *ln* has range **R**.

In particular, the number 1 is in the range of *ln*. There is then some real number e whose natural logarithm is 1. Since *ln* is an increasing function, there is exactly one such real number.

1.3 DEFINITION The number e is the unique real number such that $\ln e = 1$.

Recall from precalculus that $\log_b b = 1$, because $b^1 = b$. We therefore refer to natural logarithms as *logarithms with base e* (that is, $ln = log_e$). The number e is an irrational number, like π, with an infinite nonrepeating decimal representation. Also like π, it has been computed to many decimal places:

$$e = 2.71828\ 18284\ 59045\ 23536\ldots.$$

The approach of Section 3.7 is helpful in sketching the graph of *ln*.

Example 3

Draw an accurate graph of $y = \ln x$.

Solution. As remarked after Example 1, *ln* is increasing on its domain $(0, +\infty)$. Also, since

$$\frac{d^2}{dx^2}(\ln x) = \frac{d}{dx}\left(\frac{1}{x}\right) = -\frac{1}{x^2} < 0,$$

it follows from Theorem 4.3 of Chapter 3 that *ln* is concave downward on its domain. It has x intercept $(1, 0)$ since $\ln 1 = 0$, and it passes through $(e, 1)$ since $\ln e = 1$. It follows from (14) that the y-axis $(x = 0)$ is a vertical asymptote. And from (13), as $x \to +\infty$ we have $\ln x \to +\infty$. **Figure 1.4** was sketched from this information, and the following table of values from Example 2 and a calculator.

FIGURE 1.4

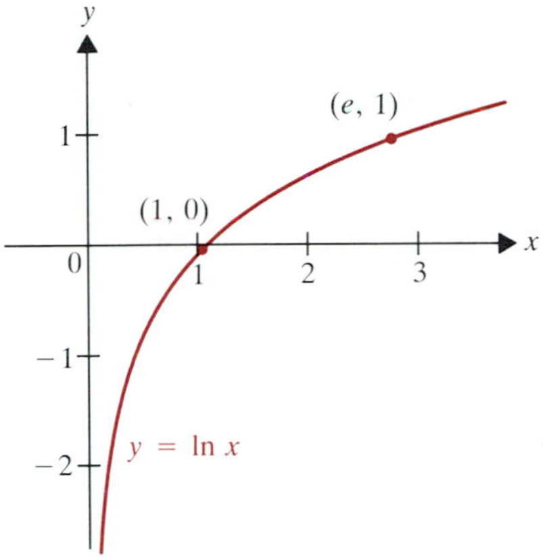

1/4	1/3	1/2	1	2	3	4	5
−1.386	−1.099	−0.693	0	.693	1.099	1.386	1.609

■

Logarithmic Differentiation

The properties in Theorem 1.2 can help simplify certain kinds of formidable differentiation problems, as the following example illustrates.

Example 4

Find $f'(x)$ if $f(x) = \ln\sqrt{\dfrac{x^2+3}{x^2+4}}$.

Solution. At first glance this looks like a messy chain-rule problem. But Theorem 1.2 gives

$$\ln\sqrt{\frac{x^2+3}{x^2+4}} = \ln\left(\frac{x^2+3}{x^2+4}\right)^{1/2} = \frac{1}{2}\ln\left(\frac{x^2+3}{x^2+4}\right)$$
$$= \frac{1}{2}[\ln(x^2+3) - \ln(x^2+4)].$$

Hence

$$f'(x) = \frac{1}{2}\left[\frac{1}{x^2+3}\cdot 2x - \frac{1}{x^2+4}\cdot 2x\right]$$
$$= \frac{x}{x^2+3} - \frac{x}{x^2+4} = \frac{x(x^2+4) - x(x^2+3)}{(x^2+3)(x^2+4)}$$
$$= \frac{x}{(x^2+3)(x^2+4)}.$$

■

This kind of simplification leads to the important technique of ***logarithmic differentiation.*** It is a combination of implicit differentiation and the simplifying power of Theorem 1.2. When faced with a complicated expression $g(x)$ involving powers, roots, products, and quotients, use the following technique.

LOGARITHMIC DIFFERENTIATION

1. Let $y = g(x)$.
2. Take the natural logarithm of each side: $\ln y = \ln g(x)$.
3. Use Theorem 1.2 to reduce $\ln g(x)$ to a simpler expression $h(x)$:

$$\ln y = h(x). \tag{15}$$

4. Use (8) and the chain rule to implicitly differentiate each side of (15) with respect to x:

$$\frac{1}{y}\frac{dy}{dx} = h'(x). \tag{16}$$

5. Solve (16) for $\dfrac{dy}{dx}$.

The next example illustrates the simplification that logarithmic differentiation can produce.

Example 5

Find the derivative of

$$g(x) = \sqrt[3]{\frac{(x^2 + 1)\sin^2 x}{(x^3 + 3)\cos^4 x}}, \qquad \text{where } x \in (0, \pi/2). \tag{17}$$

Solution. Let $y = g(x)$. Then

$$\begin{aligned}
\ln y &= \ln\left(\frac{(x^2 + 1)\sin^2 x}{(x^3 + 3)\cos^4 x}\right)^{1/3} \\
&= \frac{1}{3}(\ln[(x^2 + 1)\sin^2 x] - \ln[(x^3 + 3)\cos^4 x]) \\
&= \frac{1}{3}(\ln(x^2 + 1) + \ln\sin^2 x - \ln(x^3 + 3) - \ln\cos^4 x) \\
&= \frac{1}{3}(\ln(x^2 + 1) + 2\ln\sin x - \ln(x^3 + 3) - 4\ln\cos x).
\end{aligned}$$

Then differentiating with respect to x, we have

$$\begin{aligned}
\frac{1}{y}\frac{dy}{dx} &= \frac{1}{3}\left(\frac{1}{x^2 + 1}\cdot 2x + \frac{2}{\sin x}\cos x - \frac{1}{x^3 + 3}\cdot 3x^2 - \frac{4}{\cos x}(-\sin x)\right) \\
&= \frac{1}{3}\left(\frac{2x}{x^2 + 1} + 2\frac{\cos x}{\sin x} - \frac{3x^2}{x^3 + 3} + 4\frac{\sin x}{\cos x}\right), \\
\frac{dy}{dx} &= \frac{y}{3}\left(\frac{2x}{x^2 + 1} + 2\frac{\cos x}{\sin x} - \frac{3x^2}{x^3 + 3} + 4\frac{\sin x}{\cos x}\right),
\end{aligned}$$

where $y = g(x)$ is given by (17). ■

Although we used several steps to simplify $\ln y$, they were all elementary applications of Theorem 1.2. Likewise, the simplified $\ln g(x)$ was differentiated by simple applications of the chain rule. By contrast, direct differentiation of (17) would have been very complicated.

Logarithmic Integration

If $x > 0$, then it follows immediately from (7) that

$$\int \frac{1}{x}\,dx = \ln x + C. \tag{18}$$

We can extend (18) by considering the function f given by

$$f(x) = \ln|x|,$$

which is defined for all x except $x = 0$. Since $|x| = -x$ if $x < 0$, the chain rule with $x < 0$ and $f(x) = \ln(-x)$ gives

$$D_x \ln(-x) = \frac{1}{-x}\cdot D_x(-x) = -\frac{1}{x}\cdot(-1) = \frac{1}{x}, \qquad \text{for } x < 0. \tag{19}$$

Combining (19) and (8), we have

$$D_x \ln|x| = \frac{1}{x}, \qquad \text{if } x \neq 0. \tag{20}$$

Equation (20) and the chain rule for antidifferentiation (Theorem 9.3 of Chapter 3) give the promised extension of (18). On any interval where a differentiable function g is nonzero, we have

$$\int \frac{g'(x)}{g(x)}\,dx = \ln|g(x)| + C. \tag{21}$$

Many complicated quotients can be integrated by using (21) and substitution.

Example 6

Evaluate the following integrals.

(a) $\displaystyle\int \frac{x-1}{x^2-2x+5}\,dx$ (b) $\displaystyle\int \frac{\sin x}{2+\cos x}\,dx$ (c) $\displaystyle\int \frac{(\ln x)^2}{x}\,dx$

Solution. (a) The numerator looks like the derivative of the denominator except for a constant factor. So we let

$$u = g(x) = x^2 - 2x + 5.$$

Then

$$du = (2x-2)\,dx \Rightarrow \frac{1}{2}\,du = (x-1)\,dx,$$

so we have

$$\int \frac{x-1}{x^2-2x+5}\,dx = \frac{1}{2}\int \frac{du}{u} = \frac{1}{2}\ln|u| + C$$

$$= \frac{1}{2}\ln|x^2-2x+5| + C.$$

Note that $x^2 - 2x + 5 = (x - 1)^2 + 4 > 0$ for all x, so this answer is valid on the interval $(-\infty, +\infty)$.

(b) Here we let

$$u = 2 + \cos x.$$

Then

$$du = -\sin x\, dx \rightarrow -du = \sin x\, dx.$$

We thus have

$$\int \frac{\sin x}{2 + \cos x}\, dx = -\int \frac{du}{u} = -\ln |u| + C$$

$$= \ln |u^{-1}| + C = \ln \frac{1}{2 + \cos x} + C.$$

Since $\cos x \geq -1$ for all x, $2 + \cos x > 0$ for all x, so that $|2 + \cos x| = 2 + \cos x$. Thus our answer is valid on the interval $(-\infty, +\infty)$.

(c) Here we recognize $1/x$ as the derivative of $\ln x$. So we let

$$u = \ln x.$$

Then

$$du = \frac{1}{x}\, dx,$$

so

$$\int \frac{(\ln x)^2}{x}\, dx = \int u^2\, du = \frac{1}{3} u^3 + C = \frac{1}{3} (\ln x)^3 + C.$$

This answer is valid on the interval $(0, +\infty)$, that is, wherever the integrand is defined. ■

HISTORICAL NOTE

The term *natural logarithm* for the function *ln* defined by (7) in Definition 1.1 goes back to the Belgian mathematician Nicolas Mercator (1620–1687), who used the term in 1668 to refer to the area pictured in Figure 1.1. The Scottish mathematician John Napier (1550–1617) is generally recognized as the inventor of logarithms as computational aids, in 1614. Until supplanted by electronic calculators within the last fifteen years, logarithms were still in wide use for calculations. Multiplication and division of large numbers was reduced by properties of logarithms (Theorem 1.2(a) and (c)) to addition and subtraction of logarithms. Textbooks such as this included fairly long tables of logarithms to facilitate that.

The symbol *e* for the base of the natural logarithm function was introduced by Euler in 1748 in the course of giving the first systematic treatment of logarithms and trigonometric functions. A popular legend is that he chose the symbol *e* because of his name! The differentiation rule (8) was derived by Euler in 1755, although not in the way we derived it.

Exercises 6.1

In Exercises 1–26, differentiate the given function.

1. $\ln(2x + 5)$
2. $\ln(1 - 3\sqrt{x})$
3. $\ln\sqrt{2x + 5}$
4. $\ln(1 - 3x)^2$
5. $\ln\sqrt[3]{x^3 - 4x^2 + 6}$
6. $\ln\sqrt[3]{1 - 3x + x^2 - x^3}$
7. $\ln(\sin^2 x \cos^2 x)$
8. $\ln\sqrt{\sec x \tan x}$
9. $\cos(\ln(x + 3))$
10. $\sec(\ln\sqrt{x})$
11. $\ln(\ln x)$
12. $\ln(\sec\sqrt{x})$
13. $\dfrac{\ln x}{x^4}$
14. $\dfrac{x^2 + 1}{\ln x}$
15. $\ln x^2 + (\ln x)^2$
16. $\ln\sqrt{x + 1} - \sqrt{\ln(x + 1)}$
17. $\ln[(2x^2 + 1)^3(x^3 + 1)^4]$
18. $\ln\left[\dfrac{(x^2 - 1)^3}{\sqrt{x^2 + 1}}\right]$
19. $\ln\sqrt[3]{\dfrac{x - 1}{x + 1}}$
20. $\ln\sqrt{\dfrac{x^2 - 1}{x^2 + 1}}$
21. $\ln\dfrac{\sqrt[3]{x^2 - 5}}{(2x^3 + x^2 + 1)^8}$
22. $\ln\dfrac{\sin^3 x \cos^2 x}{(x^2 + 1)^3}$
23. $g(x) = \sqrt{(x^2 - 1)(x^2 + 2)(x^2 + 3)}$
24. $g(x) = \dfrac{\sqrt[3]{1 + x^3}}{\sqrt[4]{1 + x^4}}$
25. $g(x) = \sqrt{\sin^2 x \cos^2 x}$
26. $g(x) = \dfrac{x(x^2 + 1)^3}{(x^2 + 2)^2(x^2 + 3)^3}$

In Exercises 27–42, evaluate the given integral.

27. $\displaystyle\int \frac{dx}{2x + 1}$
28. $\displaystyle\int \frac{x\,dx}{1 + 2x^2}$
29. $\displaystyle\int \frac{(x^2 + 1)\,dx}{x^3 + 3x - 2}$
30. $\displaystyle\int \frac{(2x - 1)\,dx}{2x^2 - 2x + 5}$
31. $\displaystyle\int \frac{\sin x \cos x}{1 + \sin^2 x}\,dx$
32. $\displaystyle\int \frac{\sin x\,dx}{1 + \cos x}$
33. $\displaystyle\int \frac{\cot^2(\ln x)}{x}\,dx$
34. $\displaystyle\int \frac{\sec^2 x\,dx}{5 - 2\tan x}$
35. $\displaystyle\int \frac{(\ln x)^3}{x}\,dx$
36. $\displaystyle\int \frac{\ln\sqrt{x}}{x}\,dx$
37. $\displaystyle\int \frac{dx}{x(\ln x)^3}$
38. $\displaystyle\int \frac{(2 + \ln x)^2}{x}\,dx$
39. $\displaystyle\int_0^1 \frac{x + 2}{x^2 + 4x + 1}\,dx$
40. $\displaystyle\int_0^3 \frac{x - 1}{x^2 - 2x + 5}\,dx$
41. $\displaystyle\int_1^{e^{\pi/2}} \frac{1}{x}\sin(\ln x)\,dx$
42. $\displaystyle\int_1^2 \frac{x^2 + 1}{x^3 + 3x + 1}\,dx$

In Exercises 43 and 44, find the area lying above the x-axis and under the graph of $y = f(x)$ between $x = a$ and $x = b$.

43. $f(x) = \dfrac{2}{x + 1}$, $a = 0$ and $b = 3$
44. $f(x) = \dfrac{\ln x}{x}$, $a = 1$ and $b = 2$

In Exercises 45 and 46, find the volume that results when the region under the graph of $y = f(x)$ and above the x-axis between $x = a$ and $x = b$ is revolved about the x-axis.

45. $f(x) = \dfrac{1}{\sqrt{x}}$, $a = 1$ and $b = 4$
46. $y = \dfrac{\ln x}{\sqrt{x}}$, $a = 1$ and $b = 4$

In Exercises 47 and 48, find the average value of the given function over the given interval.

47. $f(x) = \cot x$ over $[\pi/6, \pi/2]$
48. $f(x) = \tan x$ over $[0, \pi/3]$
49. If $f(x) = \ln x$, then find $f'(x)$, $f''(x)$, and $f'''(x)$. What is $f^{(n)}(x)$?
50. Are the functions $f(x) = \ln x^2$ and $g(x) = 2\ln x$ the same? Explain.

PC **In Exercises 51–54: (a) Use Simpson's rule as in Example 2 to obtain estimates of $\ln x$ for the given x and $n = 10$. (b) Estimate the error in (a). (c) If you can write or have access to a computer program or a programmable calculator program for Simpson's rule, repeat (a) for $n = 20, 40, 60, 80, 160$, and 320 (stop at $n = 80$ if using a programmable calculator). (d) Estimate the error in (c) when $n = 80$.**

51. $x = 3$
52. $x = 5$
53. $x = 7$
54. $x = 11$

PC 55. Use results of Example 2 and Exercises 51–54 to obtain 5-decimal-place approximations for
(a) $\ln 1/2$ (b) $\ln 4$ (c) $\ln 6$
(d) $\ln 10$ (e) $\ln 8$

PC 56. Repeat Exercise 55 for
(a) $\ln 1/4$ (b) $\ln 9$ (c) $\ln 12$
(d) $\ln 14$ (e) $\ln 15$

PC 57. Use Theorem 6.3 of Chapter 4 to estimate the number of subintervals in Simpson's rule needed to find the value of $\ln 2$ accurate to
(a) 6 decimal places (that is, E_S should be less than 0.5×10^{-6}),
(b) 8 decimal places (compare to Example 2).

PC 58. Repeat Exercise 57 for $\ln 3$ and compare to Exercise 51.

59. Show that

$$\int_a^{ab} \frac{1}{t}\,dt = \int_1^b \frac{1}{t}\,dt.$$

60. Use Definition 1.1 to show that ln 3 lies between 2/3 and 2. (*Hint:* See **Figure 1.5.**)

FIGURE 1.5

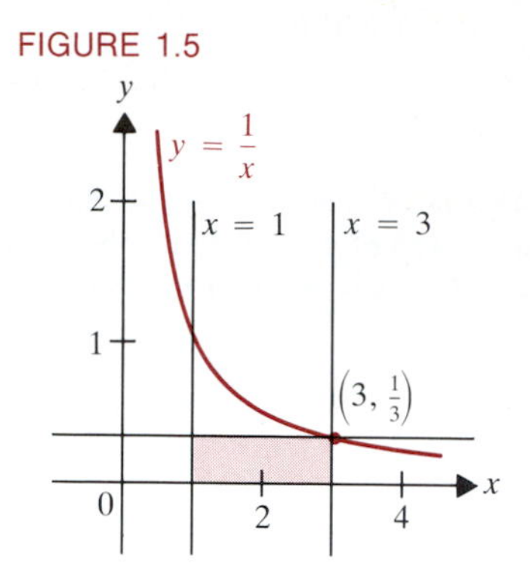

61. A gas is governed by Boyle's law, $PV = C$, where P is the pressure and V is the volume. If the volume changes from V_0 to V_1, find the average pressure during this change.

62. The pressure on a certain type of membrane is given by

$$P(t) = E\int_{t_0}^{t} \frac{1}{x}\,dx,$$

where E is a constant called the elastic modulus of the membrane, t_0 is the natural thickness of the membrane, and t is the thickness under pressure. The *stress energy* is defined by

$$S(t) = -\int_{t_0}^{t} P(u)\,du. \tag{22}$$

Use differentiation to show that

$$S(t) = E\left[t \ln \frac{t_0}{t} + t - t_0\right]$$

satisfies (22).

63. Show that

$$F(x) = \begin{cases} \ln|x| + 1, & \text{for } x < 0 \\ \ln x - 3, & \text{for } x > 0 \end{cases}$$

and

$$G(x) = \begin{cases} \ln|x|, & \text{for } x < 0 \\ \ln x + 5, & \text{for } x > 0 \end{cases}$$

are both antiderivatives of $f(x) = 1/x$ but do not differ by a constant. Why is this not a violation of Theorem 9.2 of Chapter 3?

64. Repeat Exercise 63 for

$$F(x) = \begin{cases} \ln|x - 2|, & \text{for } x < 2 \\ \ln(x - 2), & \text{for } x > 2 \end{cases}$$

and

$$G(x) = \begin{cases} \ln|x - 2| - 3, & \text{for } x < 2 \\ \ln(x - 2) + 1, & \text{for } x > 2 \end{cases}$$

and

$$f(x) = \frac{1}{x - 2}.$$

6.2 The Natural Exponential Function

We saw in the last section that the natural logarithm function *ln* has domain $(0, +\infty)$, the set of positive real numbers, and range $(-\infty, +\infty)$, the entire set of real numbers. In addition, since

$$\frac{d}{dx}(\ln x) = \frac{1}{x} > 0, \qquad \text{for } x \text{ in } (0, +\infty),$$

ln is an increasing function on its entire domain. Thus (see p. 117) the function *ln* has an inverse function, which is the subject of this section.

2.1 DEFINITION The ***natural exponential function exp*** is the function with domain $(-\infty, +\infty)$ and range $(0, +\infty)$ that is the inverse of the natural logarithm function *ln*. That is,

$$y = \exp(x) \qquad \text{if and only if} \qquad x = \ln y. \tag{1}$$

According to Theorem 7.3 of Chapter 2, the graphs of any pair of inverse functions, $y = f^{-1}(x)$ and $y = f(x)$, are symmetric with respect to the line $y = x$. Thus we can obtain the graph of $y = \exp(x)$ by reflecting the graph of $y = \ln x$ in the line $y = x$, as shown in **Figure 2.1.**

Equation (1) gives the basic relationship between the natural logarithm and the natural exponential function. If we substitute $x = \ln y$ into $y = \exp(x)$ in (1),

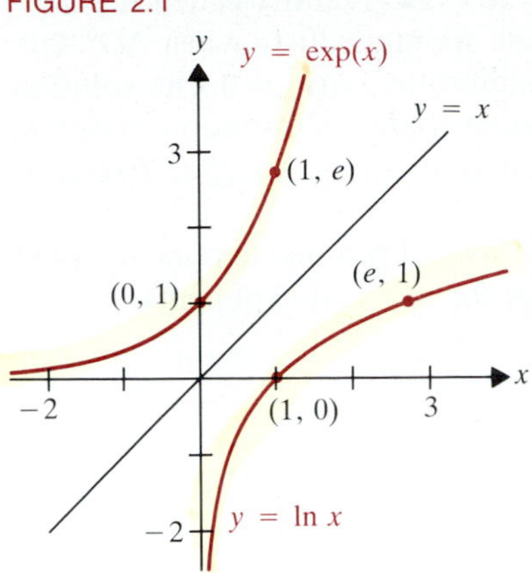

FIGURE 2.1

then we obtain

$$y = \exp(\ln y) \tag{2}$$

for any positive y. On the other hand, if we substitute $y = \exp(x)$ into $x = \ln y$ in (1), then we get for any real number x,

$$x = \ln(\exp(x)). \tag{3}$$

Recall that e is the real number with the property that

$$\ln e = 1. \tag{4}$$

If we apply the function *exp* to both sides of (4) and use (2), then we get

$$\exp(\ln e) = \exp 1 \rightarrow e = \exp(1). \tag{5}$$

Equation (5) says that *e is the value of the natural exponential function at the number 1*. For any rational number r, we have from (2) and Theorem 1.2(b)

$$e^r = \exp(\ln e^r) = \exp(r \ln e) = \exp(r \cdot 1) = \exp(r).$$

Thus *applying exp to any rational number r is the same as raising the number e to the r power*. This helps explain why the function *exp* is called an exponential function and leads to the following definition of e^x for an arbitrary real number x, which extends exponentiation to irrational x.

2.2 DEFINITION For any real number x,

$$e^x = \exp(x). \tag{6}$$

From now on we will write e^u instead of $\exp(u)$, unless u is so complicated that e^u would be hard to read. Our first example illustrates this and also the usefulness of (2) and (3).

Example 1

(a) Solve for x: $e^{1-2x} = 7e^{3x}$.
(b) Solve for K:

$$\Delta G^0 = -RT \ln K. \tag{7}$$

Solution. (a) To solve an exponential equation, take the natural logarithm of each side and use (3). Here that gives

$$\ln e^{1-2x} = \ln(7e^{3x}),$$

$$1 - 2x = \ln 7 + \ln e^{3x}, \qquad \textit{by (3) and Theorem 1.2(a) and (b)}$$

$$1 - 2x = \ln 7 + 3x, \qquad \textit{by (3)}$$

$$1 - \ln 7 = 5x \rightarrow x = \frac{1}{5}(1 - \ln 7).$$

(b) To solve a logarithmic equation, isolate the natural logarithm on one side. Then apply the function *exp* to each side and use (2) to complete the solution. Doing so here, we obtain

$$-\frac{\Delta G^0}{RT} = \ln K \rightarrow \exp(-\Delta G^0/RT) = \exp(\ln K) \rightarrow K = \exp(-\Delta G^0/RT). \quad \blacksquare$$

Equation (7) expresses an important chemical relationship used to determine whether or not a reaction (such as the rusting of iron in moist air) is *spontaneous*. In (7), T is the temperature in Kelvin, R is a constant, K is a quantity called the *equilibrium constant* for the reaction, and ΔG^0 is a quantity called the *standard free energy change*. The reaction proceeds spontaneously when $\Delta G^0 < 0$, is nonspontaneous when $\Delta G^0 > 0$, and is at equilibrium if $\Delta G^0 = 0$. Our solution to Example 1(b) makes it possible to determine useful information about K once it is known whether or not the reaction is spontaneous. (See Exercises 57 and 58.)

We next use Definition 2.2 to show that the laws of rational exponents listed in Section 1 extend to the function $\exp(x) = e^x$ for *any* real exponents.

2.3 THEOREM

If k and l are real numbers, then

(a) $e^0 = 1$ (b) $e^k e^l = e^{k+l}$ (c) $(e^k)^l = e^{kl}$ (d) $\dfrac{e^k}{e^l} = e^{k-l}$

Proof. (a) Since $\ln 1 = 0$, this follows immediately from Definition 2.1.

(b) Let $a = e^k$ and $b = e^l$. Then $k = \ln a$ and $l = \ln b$. Hence

$$k + l = \ln a + \ln b \rightarrow k + l = \ln ab, \tag{8}$$

by Theorem 1.2(a). Now we apply exp to the right portion of (8) and use (6)(2) to get

$$e^{k+l} = e^{\ln ab} = ab = e^k e^l.$$

(c) Using (6) in conjunction with (2), we have

$$\begin{aligned} (e^k)^l &= e^{\ln[(e^k)^l]} = e^{l \ln e^k} && \text{by Theorem 1.2(b)} \\ &= e^{lk \ln e} && \text{by Theorem 1.2(b) again} \\ &= e^{lk} = e^{kl}, \end{aligned}$$

since $\ln e = 1$.

(d) First observe from (c) that $e^{-l} = (e^l)^{-1} = 1/e^l$. Then by (b),

$$e^{k-l} = e^k \cdot e^{-l} = e^k \cdot \frac{1}{e^l} = \frac{e^k}{e^l}. \quad \boxed{\text{QED}}$$

Theorem 2.3 shows that the natural exponential function *exp* satisfies all the usual algebraic laws of exponents given in precalculus. Facts about the calculus of inverse functions from Section 2.7 make it easy to derive the basic calculus properties of *exp*. First, since *ln* is a continuous, increasing function on its domain $(0, +\infty)$, its inverse function *exp* is continuous and increasing on its domain $(-\infty, +\infty)$ (see p. 117).

Next, if $y = e^x$, then

$$x = \ln y.$$

Differentiation of that with respect to y thus gives

$$\frac{dx}{dy} = \frac{1}{y} = \frac{1}{e^x}.$$

Then the inverse function differentiation rule (Theorem 7.5 of Chapter 2) says that

$$\frac{dy}{dx} = \frac{1}{\frac{dx}{dy}} = \quad \rightarrow \frac{d}{dx}(e^x) = e^x. \tag{9}$$

From this it follows immediately that

$$\int e^x\,dx = e^x + C. \tag{10}$$

Combining (9) and (10) with the chain rules for derivatives and antiderivatives, we get the following differentiation and integration rules.

$$\frac{d}{dx}\left[e^{g(x)}\right] = e^{g(x)}g'(x) \tag{11}$$

$$\int e^{g(x)}g'(x)\,dx = e^{g(x)} + C \tag{12}$$

Example 2

If $f(x) = (x+1)e^{x^2+2x}$, then find
(a) $f'(x)$ (b) $\int f(x)\,dx$

Solution. (a) Using the product rule and (11), we get

$$\begin{aligned} f'(x) &= 1 \cdot e^{x^2+2x} + (x+1)e^{x^2+2x}(2x+2) \\ &= e^{x^2+2x} + 2(x+1)^2e^{x^2+2x} = e^{x^2+2x}[1 + 2(x+1)^2]. \end{aligned}$$

(b) To evaluate

$$\int (x+1)e^{x^2+2x}\,dx,$$

we note that $x + 1$ is, except for the constant factor 2, the derivative of $x^2 + 2x$. So we let

$$u = x^2 + 2x.$$

Then $du = (2x+2)\,dx = 2(x+1)\,dx$, so

$$\tfrac{1}{2}\,du = (x+1)\,dx.$$

Therefore (12) gives

$$\begin{aligned} \int (x+1)e^{x^2+2x}\,dx &= \int \frac{1}{2}e^u\,du = \frac{1}{2}e^u + C \\ &= \frac{1}{2}e^{x^2+2x} + C. \quad ■ \end{aligned}$$

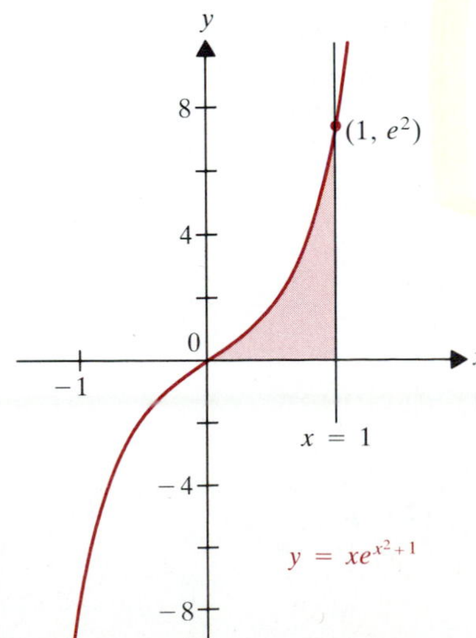

FIGURE 2.2

Example 3

Find the area of the region bounded by $y = xe^{x^2+1}$, the x-axis, and the line $x = 1$.

Solution. The shaded region in **Figure 2.2** has area

$$\begin{aligned} A &= \int_0^1 xe^{x^2+1}\,dx = \frac{1}{2}\int_0^1 2xe^{x^2+1}\,dx \\ &= \frac{1}{2}\left[e^{x^2+1}\right]_0^1 = \frac{1}{2}\left[e^2 - e^1\right] \approx 2.335. \quad ■ \end{aligned}$$

Rather than changing the variable in $\int_0^1 xe^{x^2+1}\,dx$, we noted that only a factor of 2 was needed to make $xe^{x^2+1}\,dx$ exactly of the form $e^{g(x)}g'(x)\,dx$. We then supplied that factor in the second step of the solution and compensated by multiplying outside the integral by $1/2$. Such shortcuts, which avoid explicit substitutions like $u = x^2 + 1$ and corresponding changes of limits of integration, can save time.

Certain exponential functions play a prominent role in probability and statistics. For instance, the ***standard normal probability density*** is the function

$$n(x) = \frac{1}{\sqrt{2\pi}}e^{-x^2/2}. \tag{13}$$

Its graph is referred to as the *standard normal curve*. To draw it, we use the methods of Section 3.7.

Example 4

Sketch the graph of $y = n(x)$.

Solution. First, the domain of n is $(-\infty, +\infty)$. Since *exp* takes on only positive values, its range is confined to $(0, +\infty)$. We have

$$\lim_{x\to\pm\infty} n(x) = \lim_{x\to\pm\infty} \frac{1}{\sqrt{2\pi}e^{x^2/2}} = 0,$$

since $e^{x^2/2} \to +\infty$ as $x \to \pm\infty$. Thus the x-axis is a horizontal asymptote of the graph. Since n is clearly an even function, the graph is symmetric in the y-axis. Differentiating (13), we get

$$n'(x) = \frac{1}{\sqrt{2\pi}}e^{-x^2/2}\cdot(-x) = -\frac{x}{\sqrt{2\pi}e^{x^2/2}}. \tag{14}$$

Hence $n'(x) > 0$ for x negative, $n'(0) = 0$, and $n'(x) < 0$ for x positive. Since n increases up to $x = 0$ and then decreases, there is a local and absolute maximum at $x = 0$. Differentiating the first formula for $n'(x)$ in (14), we get

$$n''(x) = -\frac{1}{\sqrt{2\pi}}e^{-x^2/2} - \frac{x}{\sqrt{2\pi}}e^{-x^2/2}\cdot(-x) = \frac{1}{\sqrt{2\pi}}e^{-x^2/2}(x^2 - 1).$$

Thus $n''(x) = 0$ when $x = \pm 1$. Since $n''(x)$ changes sign at $x = \pm 1$, those are inflection points. Since $\exp(-x^2/2) > 0$ for all x, $n''(x)$ has the sign of $x^2 - 1$. Hence $n''(x) > 0$ for $x > 1$ and $x < -1$, whereas $n''(x) < 0$ for x between -1 and

FIGURE 2.3

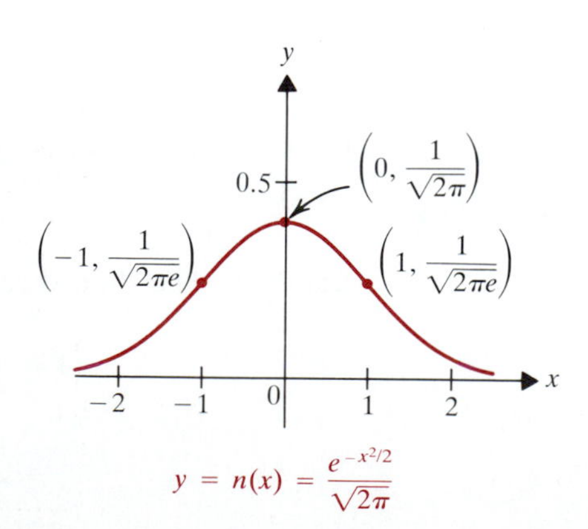

1. Thus the graph is concave upward on $(-\infty, -1) \cup (1, +\infty)$ and concave downward on $(-1, 1)$. With this information we obtain the graph in **Figure 2.3.** ■

In the last section we defined e to be the unique real number whose natural logarithm is 1. There is an interesting and useful alternative definition of e as an intriguing limit. We close the section by deriving it. (Its usefulness will be seen in the next section.)

2.4 THEOREM

$$\lim_{k \to 0} (1 + k)^{1/k} = e$$

Proof. Let $g(x) = \ln x$. Then $g'(1) = 1/1 = 1$. Hence, by the definition of the derivative,

$$1 = g'(1) = \lim_{k \to 0} \frac{g(1 + k) - g(1)}{k} = \lim_{k \to 0} \frac{\ln(1 + k) - \ln 1}{k}$$

$$= \lim_{k \to 0} \frac{1}{k} \ln(1 + k) = \lim_{k \to 0} \ln(1 + k)^{1/k}. \qquad \textit{by Theorem 1.2(b)}$$

Letting $f(k) = \ln(1 + k)^{1/k}$, we therefore have

$$\lim_{k \to 0} f(k) = 1.$$

Noting that

$$\exp(f(k)) = \exp[\ln(1 + k)^{1/k}] = (1 + k)^{1/k},$$

we have by Theorem 5.3 of Chapter 1,

$$\lim_{k \to 0} \exp(f(k)) = \exp\left(\lim_{k \to 0} f(x)\right) = \exp(1) = e.$$

That is,

$$\lim_{k \to 0} (1 + k)^{1/k} = e. \quad \boxed{\text{QED}}$$

Table 1 provides an illustration of Theorem 2.5. It shows the results of a computer program to evaluate $(1 + k)^{1/k}$ near $k = 0$.

TABLE 1

k	$1 + k$	$1/k$	$(1 + k)^{1/k}$
−0.1	0.9	−10	2.867971987
−0.01	0.99	−100	2.731999011
−0.001	0.999	−1,000	2.719641855
−0.0001	0.9999	−10,000	2.718417747
−0.00001	0.99999	−100,000	2.718309013
−0.000001	0.999999	−1,000,000	2.718281830
0.000001	1.000001	1,000,000	2.718281830
0.00001	1.00001	100,000	2.718309013
0.0001	1.0001	10,000	2.718145919
0.001	1.001	1,000	2.716923572
0.01	1.01	100	2.704813814
0.1	1.1	10	2.593742460
0.5	1.5	2	2.25
1	2	1	2

HISTORICAL NOTE

In 1748, Euler was able to obtain a 23-decimal-place approximation to e. He did not use the limit in Theorem 2.4, but that limit was implicit in his discussion. Euler also seems to have been the first mathematician to relate exponentials to logarithms as we have done in Equation (1). In his 1755 text, he derived (9).

Exercises 6.2

In Exercises 1–12, solve for x.

1. $e^{2x} = 5$

2. $e^{1-x} = 5e^{2x}$

3. $e^{x^2-3x-4} = 1$

4. $e^{1-x^2} = 1/2$

5. $e^{2/x} = 5$

6. $e^{2x-1} = 5e^{2x}$

7. $\ln(x^2 - 3x + 1) = \ln 5$

8. $\ln(x + 1)^3 = 0$

9. $\ln x - \ln 3 = 2$

10. $\ln 3x + \ln 2 = 1$

11. $\ln x - \ln x_0 = kt$

12. $\ln x - \ln x_0 = \dfrac{\Delta H(t - t_0)}{Rt_0 t}$

In Exercises 13–34, differentiate the given function.

13. e^{-2x+1}

14. $e^{x/2-2}$

15. e^{x^2+1}

16. e^{5-x^2}

17. $e^{\sin x}$

18. $\tan(e^{x^2})$

19. $\exp[(x + 2)^{1/2}]$

20. $(e^{x^2+1})^{1/2}$

21. $e^x \sin 2x$

22. $2x^2 e^{\cot^2 x}$

23. $\sec e^{x^2}$

24. $e^{\sec x^2}$

25. $x^2 e^{x^2+1}$

26. $[\exp(x^2 + 1)]/x^2$

27. $\dfrac{x + 1}{e^{x^2}}$

28. $\dfrac{e^{x^2}}{x^2 + 1}$

29. $\dfrac{e^x - e^{-x}}{e^{-x} + e^x}$

30. $\exp(e^x)$

31. $\ln(e^x + e^{-x})$

32. $\ln \dfrac{e^{x-1}}{e^{x+1}}$

33. $\exp(\ln(x^2 + 1))$

34. $\ln e^{x^2+1}$

In Exercises 35–50, integrate the given function.

35. e^{2x-1}

36. $\dfrac{e^{3x-1}}{e^{x+1}}$

37. xe^{x^2+5}

38. $\dfrac{x}{e^{x^2+1}}$

39. $\dfrac{\exp[(x + 1)^{1/2}]}{(x + 1)^{1/2}}$

40. $\dfrac{\exp(t^{-2} + 1)}{t^3}$

41. $\dfrac{1 + e^{2x}}{e^x}$

42. $\dfrac{e^x - 1}{e^{2x}}$

43. $\dfrac{e^x - e^{-x}}{e^x + e^{-x}}$

44. $\dfrac{1}{1 + e^x}$

45. $\sec^2 x \tan x \exp(\tan^2 x)$

46. $\csc x \cot x \exp(\csc x)$

47. $x^2 e^{1-x^3/2}$ over $[0, 1]$

48. $\dfrac{x}{e^{x^2}}$ over $[0, 2]$

49. $\dfrac{e^x + e^{-x}}{e^x - e^{-x}}$ over $[1, 2]$

50. $\dfrac{e^x}{1 + e^x}$ over $[0, 1]$

In Exercises 51–56, find the indicated area or volume or arc length.

51. Area above the x-axis and under $y = e^{-x}$ from $x = 0$ to $x = 1$

52. Area between $y = x$ and $y = e^x$ between $x = 1$ and $x = 3$

53. The volume obtained by revolving about the y-axis the region bounded by the x-axis, the y-axis, the graph of $y = \exp(x^2)$, and the line $x = 1$

54. The volume obtained by revolving about the x-axis the region above the x-axis and under $y = xe^{x^3}$, between $x = 0$ and $x = 2$

55. The arc length along $y = \dfrac{e^x + e^{-x}}{2}$ between $x = 0$ and $x = 1$

56. The arc length along $y = \dfrac{e^{2x} + e^{-2x}}{4}$ between $x = 0$ and $x = 1$

57. In Example 1(b), suppose that the reaction is spontaneous, i.e., ΔG^0 is negative. Then show that $K > 1$.

58. In Example 1(b), suppose that ΔG^0 is positive, so the reaction is not spontaneous. Show that $K < 1$.

59. A patient is given a dose of a certain drug. After t hours, the concentration $C(t)$ of the drug in the patient's bloodstream is $C(t) = 500e^{-3t}$. The doses of medicine are scheduled six hours apart. Find the average concentration of drug in the patient's blood over the six hours following a dose.

60. The amount of insulin in a person's blood is affected by the amount of sugar in the bloodstream. When insulin enters the bloodstream, it breaks down sugar and is rapidly used up. In tests for diabetes and hypoglycemia, a large amount of glucose is injected into a patient. Typical insulin readings t minutes later are described by

$$I(t) = \begin{cases} t(10 - t), & \text{for } t \leq 5 \\ 25e^{-k(t-5)}, & \text{for } t > 5. \end{cases}$$

Find the average insulin level over $[0, 60]$ for a person with $k = 0.03$.

PC **61.** On a computer or programmable calculator, evaluate the expression $(1 + k)^{1/k}$ for $k = 1/2, 1/4, 1/8, \ldots$. What number does this sequence appear to approach? Somewhere between the fourth and sixtieth evaluation (depending on your computer or calculator), the values of $(1 + k)^{1/k}$ suddenly change drastically. Explain what happens.

6.3 Growth and Decay

This section presents some of the applications of *ln* and *exp* in modeling phenomena outside mathematics. We will discuss examples from finance, life science, archeology, physics, and chemistry.

We first consider a rather pleasant sort of growth, widely praised in advertisements of financial institutions. Suppose that a person invests an amount P_0 of money in an account that earns an annual interest rate r. The rate r is some percent, commonly between 5% and 15%. We treat percents as decimals. For an annual rate of 10%, then, $r = 0.10$. If the interest is paid once a year (annual compounding), then after a year the initial investment P_0 grows to

$$P_0 + rP_0 = P_0(1 + r). \tag{1}$$

After two years, P_0 grows to

$$P_0(1 + r) + rP_0(1 + r) = P_0(1 + r)^2.$$

Similarly, after n years it grows to $P_0(1 + r)^n$. Savings institutions compound much more frequently than annually, however. Quarterly compounding, in which $\frac{1}{4}rP$ is added to the principal P after each 3 months, has given way to monthly and now daily compounding of interest. Let us consider what happens when interest is compounded m times per year.

After the first compounding period, the amount $(r/m)P_0$ is added to the initial principal P_0. As in (1) then, the new principal is

$$P_1 = P_0 + \frac{r}{m} P_0 = P_0\left(1 + \frac{r}{m}\right).$$

At the next compounding time, $(r/m)P_1$ is added to P_1, giving

$$P_2 = P_1 + \frac{r}{m} P_1 = P_1\left(1 + \frac{r}{m}\right) = P_0\left(1 + \frac{r}{m}\right)^2.$$

Continuing in this way, we see that after m compounding periods have passed (one year), the initial principal P_0 grows to

$$P_m = P_0\left(1 + \frac{r}{m}\right)^m. \tag{2}$$

We use (2) to find out what happens as the frequency m of compounding increases, assuming that the interest rate r is a positive rational number. We use the fact (Exercise 34, Section 1.5) that if $\lim_{x \to a} f(x) = L$, then

$$\lim_{x \to a} (f(x))^r = L^r. \tag{3}$$

Example 1

Find the limit of P_m in (2) as $m \to \infty$, that is, as the frequency of the compounding increases without bound.

Solution. We first rewrite P_m in the form

$$P_m = P_0\left[\left(1 + \frac{r}{m}\right)^{m/r}\right]^r,$$

which is equivalent to (2) by the third law in (5) on p. 336. Next we let

$$k = r/m.$$

Then $m/r = 1/k$. Since $m \to \infty$ is equivalent to $k \to 0$, we get from (3) with $f(k) = (1 + k)^{1/k}$,

$$\begin{aligned}\lim_{m\to\infty} P_m &= \lim_{k\to 0} P_0 \cdot [(1 + k)^{1/k}]^r = P_0 \lim_{k\to 0} [(1 + k)^{1/k}]^r \\ &= P_0\left[\lim_{k\to 0} (1 + k)^{1/k}\right]^r && \textit{by (3)} \\ &= P_0 e^r. && \textit{by Theorem 2.4}\end{aligned}$$

Thus, after one year, P_0 grows to P_0e^r. ■

Interest is said to be *continuously compounded* when the principal amount grows in one year to P_0e^r. Some savings institutions offer continuous compounding, which gives the customer a slight advantage compared to the more common daily compounding.

Example 2

Bank A offers continuous compounding on one-year savings certificates at 10%. Bank B compounds daily. Find the size of a $1000 initial investment at each bank after one year, assuming no further deposits or withdrawals.

Solution. After one 365-day year, the amount on deposit at bank B is, by (2),

$$P_{365} = 1000\left(1 + \frac{0.10}{365}\right)^{365} \approx 1000(1.000273973)^{365} = 1105.16$$

dollars (rounding to the nearest cent). At bank A, the $1000 grows to

$$P_0e^r = 1000e^{0.10} = 1105.17$$

dollars (again, to the nearest cent). Thus a saver at bank A earns only about one cent more on $1000 over this period! ■

Financial advertising quotes the equivalent simple interest rate as the *effective annual yield* of an interest rate r compounded according to the bank's practice. This is the interest rate r_{eff} such that

$$P_0(1 + r_{eff})$$

equals the value of the account *after one full calendar year of 365 days.* In Example 2, for instance, the effective annual yield is 10.516% at bank B and

10.517% at bank A. This is so because

$$1105.17 = 1000(1 + 0.10517) \quad \text{and} \quad 1105.16 = 1000(1 + 0.10516).$$

Warning. While some American banks follow the arithmetic used in Example 2, many do *not*. The Federal Reserve Board permits banks to operate for compounding purposes on 360-day years (12 months of 30 days each) as a bookkeeping simplification. For daily compounding in this system, r is divided by 360 rather than 365. After a calendar year of 365 days, the depositor's account then grows from P_0 to

$$P_{365} = P_0\left(1 + \frac{r}{360}\right)^{365}.$$

Under this system, a \$1000 account at bank B will grow after one year to 1106.69, for an effective annual yield of 10.669%. This compounding method is slightly more favorable to the depositor. For more on this, see Exercises 35–37.

With continuous compounding, after one year an initial deposit P_0 grows to $P(1) = P_0e^r$. After a second year, the account then grows to

$$P(2) = P_1e^r = P_0e^re^r = P_0e^{2r},$$

after three years to

$$P(3) = P_2e^r = P_0e^{2r}e^r = P_0e^{3r},$$

and so on. In general, after t years, the account grows to

$$P(t) = P_0e^{rt}. \tag{4}$$

If we regard t as a real variable, then differentiation of (4) gives

$$\frac{dP}{dt} = P_0e^{rt} \cdot r = rP_0e^{rt} = rP. \tag{5}$$

In general a quantity $P = P(t)$ is said to follow the ***law of natural*** (or ***exponential***) ***growth*** if (5) holds. If $r > 0$, then P is said to have ***growth rate r***. If $r < 0$, then P is said to have ***decay rate r***.

The discussion above shows that any quantity P that is given by a formula of the form (4) obeys the natural growth law (5).

Conversely, suppose that (5) holds. If at any time $P = 0$, then further growth is impossible so that $P = 0$ for all subsequent times t. So suppose that P is never 0. We can then separate the variables in (5), getting

$$\frac{1}{P}\frac{dP}{dt} = r \to \frac{1}{P}\,dP = r\,dt \to \int \frac{1}{P}\,dP = \int r\,dt = r\int dt,$$

$$\ln P = rt + C \tag{6}$$

for some constant C. Since $P = P_0 > 0$ when $t = 0$, (6) gives

$$\ln P_0 = r \cdot 0 + C = C.$$

Hence $C = \ln P_0$, and (6) becomes

$$\ln P = rt + \ln P_0 \to \ln P - \ln P_0 = rt.$$

$$\ln \frac{P}{P_0} = rt \to \frac{P}{P_0} = e^{rt} \to P = P_0e^{rt}.$$

Thus (4) holds. Hence **(4) and (5) are equivalent.**

Among the phenomena governed by (5) is the growth of a colony of bacteria in its early stage. An important strain of bacteria used in research is *Escherichia coli*, commonly abbreviated *E. coli*. It resides naturally in the human digestive system. During the growth phase of a colony of *E. coli*, which typically lasts about 12 hours, the colony grows exponentially. It is impractical to count cells, so the size of the colony is measured by its mass M. (This can be determined from a density reading on a spectrophotometer, for instance.)

Example 3

At the start of an experiment, a bacteria colony has mass $M_0 = 2.0 \times 10^{-9}$ grams. After two hours, $M = 1.5 \times 10^{-8}$ g. Find the rate of growth r of the colony and M_{12}, the size of the colony after 12 hours.

Solution. We use (4) in the form

$$M = M_0 e^{rt},$$

where $M_0 = 2.0 \times 10^{-9}$ g. When $t = 2$, we have $M = 1.5 \times 10^{-8} = 15 \times 10^{-9}$. Thus

$$15 \times 10^{-9} = 2 \times 10^{-9} e^{2r} \rightarrow 7.5 = e^{2r},$$

$$\ln 7.5 = 2r \rightarrow r = \frac{1}{2} \ln 7.5 \approx 1.0.$$

When $t = 12$, we find

$$\begin{aligned} M_{12} = M_0 e^{12r} &= 2.0 \times 10^{-9} \exp\left(12 \cdot \frac{1}{2} \ln 7.5\right) \\ &= 2.0 \times 10^{-9} e^{6 \ln 7.5} = 2.0 \times 10^{-9} e^{\ln(7.5)^6} \\ &= (2.0 \times 10^{-9}) \cdot (7.5)^6 \approx 3.6 \times 10^{-4} \text{ g.} \quad \blacksquare \end{aligned}$$

Substances can decay, as well as grow. For example, radioactive isotopes decay by emitting radioactive particles (alpha, beta, or gamma). This process reduces the mass in any given sample of a radioactive isotope over time. The rate of decay is exponential. That is, if M is the mass of the radioactive isotope at time t, then

$$\frac{dM}{dt} = kM,$$

where the negative constant k is the rate of decay. By (5), the amount present at time t is

$$M = M_0 e^{kt}. \tag{7}$$

An important application of this is the method of ***carbon-14 dating.*** The 1960 Nobel Prize in chemistry was awarded to Willard F. Libby (1908–1980) of the University of California, Los Angeles, for devising a system of measurement of carbon-14 in ancient organic materials. This isotope is produced in the atmosphere from nitrogen as a result of cosmic radiation. Living plants and animals absorb it in minute amounts during their lives. At death, the absorption of carbon-14 stops but the carbon-14 already present in the plant or animal continues to decay. It takes 5730 years for half the amount of carbon-14 present to decay. This period is called the isotope's ***half-life.*** By measuring the amount of carbon-14 left in an ancient relic, one can mathematically estimate its age with a high degree of accuracy. Libby's process of measurement made that practical.

Example 4

(a) Determine the value of k in (7) for carbon-14.

(b) If human hair is found that contains 20% of the concentration present in a living person's hair, then about how long ago did the person live who grew that hair?

Solution. (a) Since the half-life is 5730 years, $M = \frac{1}{2}M_0$ when $t = 5730$. Thus from (7),

$$\frac{1}{2}M_0 = M_0 e^{5730k} \rightarrow \frac{1}{2} = e^{5730k} \rightarrow \ln\frac{1}{2} = 5730k,$$

$$k = \frac{\ln\frac{1}{2}}{5730} = \frac{-\ln 2}{5730} \approx -1.210 \times 10^{-4}.$$

(b) Working from (7), we have

$$\frac{M}{M_0} = e^{kt} \rightarrow \ln\frac{M}{M_0} = kt \rightarrow t = \frac{1}{k}\ln\frac{M}{M_0}.$$

When $M = 0.20\,M_0$, we get

$$t = \frac{1}{k}\ln 0.20 = -\frac{5730}{\ln 2}\ln\frac{1}{5} = 5730\frac{\ln 5}{\ln 2} \approx 13{,}305.$$

Thus the ancient man or woman lived about 133 centuries ago, about 11,300 BC. ■

The functions *ln* and *exp* are used to study processes other than exponential growth and decay. An example from physics is *Newton's law of cooling*, which models the temperature change in an object placed in a surrounding cooler (or hotter) medium. If T is the temperature of the object at time t, and C is the temperature of the surrounding medium (which is kept constant), then the law states that

$$\frac{dT}{dt} = k(T - C). \tag{8}$$

That is, the rate of cooling (or heating) is proportional to the difference between the temperatures of the object and the surrounding medium. This law can be used to solve Problem 8 of the Prologue.

Example 5

An office is maintained at a constant temperature of 21°C. At 2:00 PM, the body of a murdered man is found in the office. The coroner arrives at 3:00 PM and finds the body temperature is 31°C. An hour later, its temperature has dropped to 29°C. If the body temperature was 37°C at the time of death, find at what time the murder took place.

Solution. Using (8), we have

$$\frac{dT}{dt} = k(T - 21) \rightarrow \frac{dT}{T - 21} = k\,dt.$$

Antidifferentiating, we get

$$\ln(T - 21) = kt + C, \tag{9}$$

which is valid as long as $T > 21$. Since $T = 37$ when $t = 0$, we have $\ln 16 = C$. Thus (9) becomes

$$\ln(T - 21) - \ln 16 = kt \rightarrow \ln\frac{T - 21}{16} = kt. \tag{10}$$

Let $t = t_0$ be the number of hours from death until 3:00 PM. Then $T = 31$ when $t = t_0$, so (10) gives

$$\ln\frac{10}{16} = kt_0. \tag{11}$$

At $t = t_0 + 1$, $T = 29$, so that

$$\ln\frac{8}{16} = k(t_0 + 1). \tag{12}$$

Division of (11) by (12) gives

$$\frac{\ln 5/8}{\ln 1/2} = \frac{t_0}{t_0 + 1} \rightarrow t_0 \ln\frac{5}{8} + \ln\frac{5}{8} = t_0 \ln\frac{1}{2},$$

$$\ln\frac{5}{8} = t_0\left(\ln\frac{1}{2} - \ln\frac{5}{8}\right) \rightarrow t_0 = \frac{\ln\frac{5}{8}}{\ln\frac{1}{2} - \ln\frac{5}{8}} \approx 2.106.$$

Thus at 3:00 PM the murder victim had been dead about 2 hours and 6 minutes. The time of death was therefore about 12:54 PM. ■

A chemical reaction involves a reactant A that reacts with itself or other reactants B, C, D, and so on, to produce one or more products. Experiments show that the rate of a reaction is often given by an expression of the form

$$k(\text{conc. } A)^m(\text{conc. } B)^n(\text{conc. } C)^p(\text{conc. } D)^q \ldots, \tag{13}$$

where $k > 0$ is called the *rate constant* and (conc. X) denotes the concentration of substance X at time t. The exponents are usually positive integers. The *order of the reaction* with respect to the reactant A is m, with respect to B is n, with respect to C is p, and so on. The order of the reaction with respect to each reactant is used to calculate the rate at which the individual reactants are being consumed. Our final example illustrates this.

Example 6

If a reaction is second order in a reactant, find the half-life of the reactant if the initial concentration is $(X)_0$.

Solution. By (13), the rate at which the reactant is being used up is proportional to $(X)^2$, where we follow the standard chemical practice of abbreviating (conc. X) to (X). Thus

$$\frac{d(X)}{dt} = -k(X)^2, \tag{14}$$

where the minus sign reflects the fact that as the reaction proceeds, the concentration of the reactant decreases. We can solve (14) as follows.

$$(X)^{-2}\,d(X) = -k\,dt \rightarrow \int (X)^{-2}\,d(X) = -k \int dt,$$

$$-\frac{1}{(X)} = -kt + C. \tag{15}$$

When $t = 0$, $(X) = (X)_0$. Substituting this into (15), we get

$$C = -\frac{1}{(X)_0}.$$

Hence (15) becomes

$$\frac{1}{(X)_0} - \frac{1}{(X)} = -kt.$$

To find the half-life $t_{1/2}$, we let $(X) = \frac{1}{2}(X)_0$. Then we have

$$\frac{1}{(X)_0} - \frac{2}{(X)_0} = -kt_{1/2} \rightarrow t_{1/2} = \frac{1}{k(X)_0}. \quad ■$$

Example 6 shows that half-life for a second-order reaction is completely dependent on the initial concentration. This contrasts with the situation for radioactive decay. In radioactive decay, we have

$$\frac{dX}{dt} = -kX, \tag{16}$$

an equation of the form (4) with $r = -k$. Thus *radioactive decay is a first-order reaction.* From (5),

$$X = X_0 e^{-kt}.$$

To find $t_{1/2}$, we let $X = (1/2)X_0$ in (16), getting

$$\frac{1}{2} = e^{-kt_{1/2}} \rightarrow \ln\frac{1}{2} = -kt_{1/2} \rightarrow t_{1/2} = -\frac{1}{k}\ln\frac{1}{2} = \frac{\ln 2}{k}.$$

Thus, $t_{1/2}$ for radioactive decay is a *constant*, independent of how much of the solution is initially present: We can then speak of *the* half-life of a radioactive isotope. Half of a sample of iodine-131, for instance, decays in 8.07 days, whether the sample is initially 2 g or 0.5 mg.

The above discussion suggests something of the power of mathematics as a tool in other fields. Since the mathematical model (in this case, the law of natural decay) often applies to different fields, it can help provide connections that might otherwise be difficult to notice. Here, for instance, we can see immediately from (16) that radioactive decay is a first-order chemical reaction. This affords a connection between nuclear physics and chemistry that would be difficult to discover experimentally.

Exercises 6.3

1. In October, 1981, banks offered "All Savers Certificates," which initially paid an interest rate of 11.713% compounded daily. What was the effective annual yield? (Assume a uniform 365-day year.)
2. What was the effective annual yield in Exercise 1 if the interest was compounded continuously? (Assume a uniform 365-day year.)
3. In 1985, banks offering negotiable order of withdrawal (NOW) accounts paid 5.25% interest on such accounts. Find the effective annual yield **(a)** with daily compounding over a 365-day year, **(b)** with continuous compounding over a 365-day year.
4. A certain money market mutual fund says its average interest rate over a 30-day period is 9%. What effective annual yield would this produce if maintained for a 365-day year? (The fund uses daily compounding.)
5. "Small saver" certificates are issued by banks for $2\frac{1}{2}$-year terms. During the period August 30–September 14, 1981, the interest rate on these certificates was at a record high of 16.500%. Find the effective annual yield with **(a)** daily compounding, **(b)** continuous compounding. **(c)** How much would a \$1000.00 investment in such a certificate have grown to after $2\frac{1}{2}$ years under continuous compounding? **(d)** Repeat part (c) for daily compounding.
6. Six-month certificates issued by banks during the first week of April 1984 carried an interest rate of 10.00%. Compounding on such certificates was not allowed. A certain money market mutual fund's average yield over the next 6 months was 9.51%. The money market fund compounds daily. How much would a \$10,000 investment of each type have been worth in 6 months?
7. Determine how long is required for \$1000 to double in an investment that grows at a continuous interest rate of **(a)** 10% **(b)** 15%. (Use a 365-day year for compounding and paying.)
8. The "rule of 72" says that an approximation of the time required for an investment to double in value can be obtained by dividing 72 by $100r$, where r is the decimal interest rate r. Check this rule for $r = 0.12$ and 365-day years for both compounding and paying.
9. If a colony of bacteria has initial mass $M_0 = 3.00 \times 10^{-9}$ g and after one hour has mass $M = 15.00 \times 10^{-9}$ g, find the mass after 10 hours. How long is required for the mass to double?
10. Repeat Exercise 9 for a bacteria colony of initial mass $M_0 = 4 \times 10^{-9}$ g that after three hours has mass $M = 2.5 \times 10^{-7}$ g.
11. A piece of charcoal in a cave contains 30% of the amount of carbon-14 found in living trees. How long ago did an ancient caveman cut down the tree and build the fire that produced the charcoal?
12. A fossil is found to contain 10% of the carbon-14 found in living shellfish. About how old is the fossil?
13. A by-product of atmospheric nuclear tests is the isotope strontium-90, which research has linked to bone cancer in humans. The half-life of strontium-90 is 28 years. How much strontium-90 will remain in the year 2000 from the amount created by a nuclear explosion in 1950?
14. In certain thyroid tests, iodine-131 (whose half-life is 8.07 days) is introduced into a patient's body. How long is required for 90% of the iodine to have decayed? How much remains in the patient's body after two weeks?
15. A hospitalized criminal has been running a fever of 40°C. At 11:00 PM a nurse finds him shot to death. At that time his body temperature is 37°C. An hour later, the temperature is down to 35°C. If the room is maintained at 20°C, determine the time of death.
16. A corpse has temperature 32°C and an hour later has cooled to 28°C. The corpse is in a coroner's holding room maintained at 10°C. How long before the first reading was the body temperature 37°C?
17. A cup of black coffee is poured from a pot, whose contents are at 95°C, into a noninsulated cup in a room at 20°C. After a minute, the coffee has cooled to 90°C. How long is required before the coffee reaches a drinkable temperature of 65°C?
18. A metal foundry has a cooling bath maintained at 10°C. Hot metal at 250°C is plunged into the bath. After 30 seconds, it has cooled to 150°C. How long will it take to reach the handling temperature of 60°C?
19. The population of Phoenix, Arizona was about 100,000 in 1950 and about 500,000 in 1970. Assuming an exponential growth rate, what predicted population would have been expected in 1980?
20. The population of the United States was about 92,000,000 in 1910 and about 122,800,000 in 1930. Assuming exponential growth, what population would have been expected for 1950? (The actual 1950 population was slightly less than 151,000,000.)
21. For many drugs, the amount in the bloodstream at time t decays exponentially between doses. A patient is given a 500-mg dose of a certain drug. After 2 hours, 350 mg remain. If at least 250 mg should be in the patient's bloodstream at all times, how far apart should the doses be scheduled?
22. Under the assumptions of Exercise 21, how often should a drug be prescribed if **(a)** its dose is 1.25 mg, **(b)** it is desired to have at least 0.25 mg in the bloodstream at all times, and **(c)** after 3 hours the 1.25 mg dose is down to 1.05 mg?
23. In 1950 the earth's population was about 2.5 billion. By 1975 it was 4 billion. Assuming exponential growth, obtain projections for the population in 2000 and 2050. If the maximum number of people the earth can support with its resources is estimated to be about 25 billion, when will the population reach that level if the 1950–1975 growth pattern persists?

24. Repeat Exercise 23 under the assumption that the growth rate is brought down to half the 1950–1975 rate starting in 1975.

25. Find the half-life of a reactant in a reaction that is third order in the reactant.

26. Repeat Exercise 25 for a fourth-order reaction.

27. Awareness of a company's products often declines exponentially after an advertising campaign ends. A firm finds that one week after its latest campaign ended, awareness of its product was only 93% of its level A_0 a week earlier. If the company decides to put a new campaign in place before the awareness drops below 50% of A_0, when should it launch the next campaign?

28. A calculator company launches a new product with an advertising campaign. Sales rise to a peak of 1000 units per week. A week after the campaign ends, sales are down to 900 units per week. If the company decides to run another campaign when sales fall to 75% of their peak level, when should the next campaign begin? (See Exercise 27.)

29. The Domar growth model, named for the Polish-born economist Evsey D. Domar (1914–) of the Massachusetts Institute of Technology, assumes that a certain fixed percentage k of a country's gross national product $y(t)$ is invested to produce growth of the total capital $x(t)$. The rate of growth of the capital is measured by this investment. It is further assumed that $y(t)$ is a fixed percentage m of the total capital $x(t)$. Show that in this model $x(t)$ grows exponentially.

30. In 1930 the following learning model was proposed by the American psychologist Louis L. Thurstone (1887–1955), a pioneer in learning theory and intelligence measurement. First, a person's knowledge is specified at time t by a function k, where $0 \leq k(t) \leq 1$. [Think of $k(t)$ as the percentage score that the person would achieve at time t on an objective test in the particular field of study.] Next, every act a person performs is either helpful or harmful to learning. Let $s(t)$ be the frequency with which acts contributing to learning are performed, and let $f(t)$ be the frequency with which harmful acts are performed. Then Thurstone asserted that

$$k(t) = \frac{s(t)}{s(t) + f(t)}.$$

He also proposed that the rates of change of $s(t)$ and $f(t)$ should be measured by the formulas

$$\frac{ds}{dt} = Kk(t) \quad \text{and} \quad \frac{df}{dt} = -K(1 - k(t)),$$

where K is a positive constant.

(a) Show that from this the chain rule gives $ds/df = -s/f$.

(b) Solve the equation in (a) to obtain $sf = D$, where the constant D measures the difficulty of the material.

(c) Show that then $k(t) = \dfrac{s^2}{s^2 + D}$.

(d) Show that $\dfrac{dt}{ds} = \dfrac{1}{K} + \dfrac{D}{Ks^2}$.

(e) Separate the variables in (d) and integrate to get an equation defining s implicitly as a function of t, hence by (c) giving $k(t)$ as a function of t.

31. A gas governed by the ideal gas law $PV = nRT$ is confined in a cylinder as in **Figure 3.1.** A piston of cross-sectional area A is used to compress the gas, while T is held constant.

(a) If the piston moves distance Δx against the gas, find a formula for the corresponding change ΔV in volume.

(b) What approximate force ΔF must be applied to cause the compression in part (a)?

(c) Find the total work done in compressing the gas from volume V_1 to volume V_2.

(d) What is the sign of the work?

FIGURE 3.1

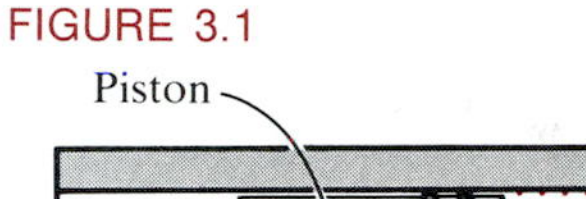

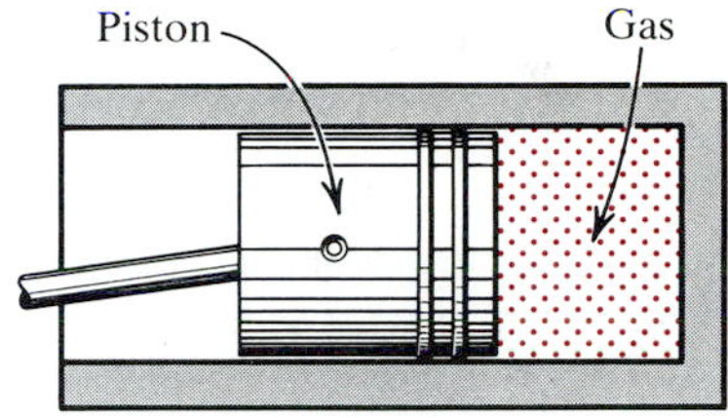

32. The heat absorbed by the system in Exercise 31 is the negative of the work. (This is a consequence of the first law of thermodynamics.) Recall that in the ideal gas law, P is the pressure in atmospheres, n is the number of moles of the gas, $R = 8.314$ joules/mole-K, and T is the temperature in Kelvin (K). Find the amount of heat absorbed by one mole of an ideal gas whose volume triples at a constant temperature $T = 27°C$ (=300 K).

33. Suppose that a bank compounds continuously at rate r. Prove that after t years, an initial deposit P_0 grows to P_0e^{rt}, assuming no deposits and no withdrawals.

34. If in Exercise 33 the bank operates on a 360-day compounding year, find the size of the deposit after t (365-day) years.

35. In Exercise 1 assume that the bank compounds on a 360-day year basis. What is the effective annual rate? That is, find P_{365}.

36. In Exercise 2 assume that the bank compounds on a 360-day year basis. Find the effective annual yield by computing P after a 365-day year elapses.

37. Repeat Exercise 3 for a bank using a 360-day compounding year.

38. By comparing advertised r and effective annual yield for one or more banks in your area, determine whether they use a 360- or 365-day year for compounding interest. (See Exercises 33–36.)

6.4 Bases Other Than e

At the start of Section 1, we mentioned that if $b \neq 1$ is a positive number, then the functions log_b and exp_b are related by the equivalence

$$y = \log_b x \leftrightarrow x = b^y.$$

Thus far we have discussed only the case $b = e$, but outside mathematics other bases are used, particularly $b = 10$. This section discusses such bases. The first step is to define arbitrary powers of a real number b. Up to now, b^x is defined only for rational numbers $r = m/n$:

$$b^r = b^{m/n} = \sqrt[n]{b^m}.$$

To define b^x for an arbitrary real number x, we use a property that holds for rational numbers r. Recall from Theorem 1.2(b) that for any rational number r,

$$\ln b^r = r \ln b.$$

Hence

$$e^{\ln b^r} = e^{r \ln b} \rightarrow b^r = e^{r \ln b}.$$

We use this property to give meaning to real powers of b.

4.1 DEFINITION If x is any real number and $b \neq 1$ is a positive number, then the function $\boldsymbol{exp_b}$ is defined by

$$\text{(1)} \qquad \exp_b(x) = b^x = \exp(x \ln b) = e^{x \ln b}.$$

Definition 4.1 extends exponentiation to the entire real line $(-\infty, +\infty)$, because that is the domain of the natural exponential function, *exp*: $\exp(x \ln b)$ is meaningful for any real number x and any base $b > 0$. And the remarks preceding the definition show that for a rational number $x = m/n$, Equation (1) agrees with the earlier definition of $b^{m/n}$ as $\sqrt[n]{b^m}$.

Scientific calculators and most microcomputers use (1) to compute b^x: First $\ln b$ is approximated, then it is multiplied by x (or a finite decimal approximation), and then *exp* of the result is calculated to some number of decimal places and rounded for display. To be able to compute powers of a *negative* number b, special engineering is required, because if $b < 0$, then the step of approximating $\ln b$ cannot be done directly. [Try $(-2)^3$ on your calculator to see whether the manufacturer built in the extra engineering. If not, an error message or blinking display will occur.]

To complete the justification of Definition 4.1, we use it to extend the basic laws of rational exponents stated at the start of Section 1.

4.2 THEOREM

> If p and q are any real numbers, then
>
> (a) $b^p b^q = b^{p+q}$ (b) $b^p/b^q = b^{p-q}$ (c) $(b^p)^q = b^{pq}$

Proof. We derive these from the corresponding properties of the natural exponential function *exp*.

(a) $b^p b^q = e^{p \ln b} e^{q \ln b} = e^{p \ln b + q \ln b}$ *by Theorem 2.3(b)*

$= e^{(p+q) \ln b} = b^{p+q}.$

(b) $b^p/b^q = e^{p \ln b}/e^{q \ln b} = e^{p \ln b - q \ln b}$ *by Theorem 2.3(d)*

$= e^{(p-q)\ln b} = b^{p-q}.$

(c) $(b^p)^q = (e^{p \ln b})^q = e^{pq \ln b}$ *by Theorem 2.3(c)*

$= b^{pq}.$ QED

The calculus of the function exp_b is also easy to derive from our knowledge of the calculus of e^x. For instance,

(2) $$D_x b^x = D_x e^{x \ln b} = e^{x \ln b} \cdot D_x(x \ln b) = b^x \ln b.$$

Thus

(3) $$\int b^x \, dx = \frac{b^x}{\ln b} + C.$$

Combining (2) and (3) with the chain rules for differentiation and antidifferentiation, we obtain

(4) $$\frac{d}{dx}\left[b^{g(x)}\right] = b^{g(x)} g'(x) \ln b$$

and

(5) $$\int b^{g(x)} g'(x) \, dx = \frac{1}{\ln b} b^{g(x)} + C.$$

Equations (4) and (5) show why the natural exponential function is the most natural base to use in calculus. Since $\ln e = 1$, the choice $b = e$ reduces the above formulas to the simpler ones derived previously for e^x.

Example 1

Evaluate

$$\text{(a) } D_x 10^{x^2+1} \qquad \text{(b) } \int 3^x \, dx \qquad \text{(c) } \int x^2 2^{x^3-3} \, dx$$

Solution. (a) and (b) From (4) and (5) we get

$$D_x(10^{x^2+1}) = 10^{x^2+1} \cdot (\ln 10) \cdot 2x = 2x \cdot 10^{x^2+1} \cdot \ln 10$$

and

$$\int 3^x \, dx = \frac{3^x}{\ln 3} + C.$$

(c) Since x^2 is, except for a constant factor, the derivative of $x^3 - 3$, we let

$$u = x^3 - 3.$$

Then

$$du = 3x^2 \, dx \rightarrow \frac{1}{3} du = x^2 \, dx.$$

Consequently,

$$\int x^2 2^{x^3+3} \, dx = \frac{1}{3} \int 2^u \, du = \frac{1}{3} \frac{2^u}{\ln 2} + C = \frac{2^{x^3-3}}{3 \ln 2} + C. \quad ■$$

Since we can exponentiate x to any base $b > 0$, we can define the function

$$f(x) = x^x \tag{6}$$

when $x > 0$. That is, f has domain $(0, +\infty)$. Based on the form of (2), there are two natural guesses for what $f'(x)$ should be. The first is $x \cdot x^{x-1} = x^x$, in analogy to the formula $D_x(x^r) = rx^{r-1}$ when r is a rational number. The second guess is $x^x \ln x$, in analogy to the formula for $D_x(b^x)$ in (2). Alas, *neither* of these natural conjectures is correct. But adding the two wrong guesses produces the right answer: Two wrongs sometimes *do* make a right!

Example 2

Find $f'(x)$ if $f(x) = x^x$.

Solution. We must use Definition 4.1 on (6):

$$f(x) = x^x = e^{x \ln x}.$$

We find

$$f'(x) = e^{x \ln x} \cdot \frac{d}{dx}(x \ln x) = e^{x \ln x}\left(1 \cdot \ln x + x \cdot \frac{1}{x}\right)$$
$$= x^x(1 + \ln x). \quad ■$$

With the result of Example 2, we can graph the function defined by (6).

Example 3

Sketch the graph of $f(x) = x^x$.

Solution. According to Definition 4.1, the function has domain $(0, +\infty)$. To find where f increases and where it decreases, we examine the sign of $1 + \ln x$, since $x^x = e^{x \ln x}$ is positive for all positive x. We find that f increases when

$$1 + \ln x > 0 \rightarrow \ln x > -1 \rightarrow x > e^{-1},$$

and f decreases when $x < e^{-1}$. The point $x = e^{-1} \approx 0.37$ is the only critical point. Since f changes from decreasing to increasing there, it has an absolute minimum at this point. At $x = e^{-1}$, we have

$$f(x) = (e^{-1})^{e^{-1}} \approx 0.6922.$$

Since

$$\lim_{x \to +\infty} f(x) = \lim_{x \to +\infty} e^{x \ln x} = +\infty,$$

there are no asymptotes. Using Example 2, we can show (Exercise 49) that f is concave up on its domain. That information and the following table give the graph in **Figure 4.1.** It looks rather similar to the graph of $y = e^x$ once x gets beyond 1. From the table, it appears that as $x \to 0^+$, $f(x) \to 1$. We will confirm that in Section 8. ■

FIGURE 4.1

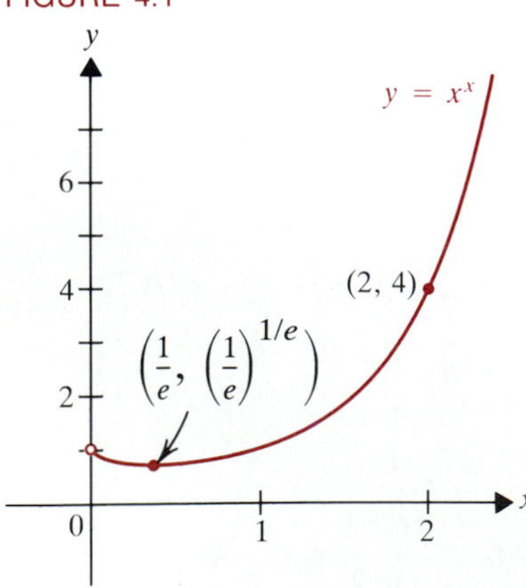

x	0.001	0.01	0.1	1/4	e^{-1}	1	2	2.5	3
$f(x)$	0.993	0.955	0.794	$1/\sqrt{2}$	0.69	1	4	9.88	27

We next define the logarithm function with base $b > 1$, log_b. Since

$$b^x = e^{x \ln b} \qquad \text{and} \qquad D_x b^x = b^x \ln b > 0$$

for all x, the function exp_b is increasing on its domain $\mathbf{R}$, so it has an inverse function.

4.3 DEFINITION The ***logarithm function with base b*** is the inverse of exp_b. That is,

$$y = \log_b x \qquad \text{means} \qquad b^y = x \tag{7}$$

When $b = e$, we get the natural logarithm function. In general we can obtain a formula relating $\log_b x$ and $\ln x$ by combining Definitions 4.1 and 4.3 with a basic property of the natural logarithm function. For $y = \log_b x$ means $b^y = x$, that is,

$$e^{y \ln b} = x. \tag{8}$$

If we take the natural logarithm of each side of (8), we get

$$\ln e^{y \ln b} = \ln x \rightarrow y \ln b = \ln x.$$

Thus

$$y = \log_b x = \frac{\ln x}{\ln b} \tag{9}$$

Differentiating (9), we find

$$D_x(\log_b x) = \frac{1}{x \ln b} \tag{10}$$

Combining (10) with the chain rule, we obtain

$$D_x \log_b g(x) = \frac{1}{g(x) \ln b} g'(x). \tag{11}$$

Before giving an example of these formulas, we remark that the properties of natural logarithms in Theorem 1.2 carry over easily to log_b. You are asked to prove the following result in Exercise 44. Here x and y are positive.

4.4 THEOREM

(a) $\log_b(xy) = \log_b x + \log_b y$

(b) $\log_b\left(\dfrac{x}{y}\right) = \log_b x - \log_b y$

(c) $\log_b x^r = r \log_b x$

Example 4

Differentiate

(a) $f(x) = \log_{10} \sqrt[3]{\dfrac{x-1}{x^2+1}}$ (b) $f(x) = \log_{10}(\sin^2 x)$ on $(0, \pi/2)$

Solution. (a) $f(x) = \log_{10}\left(\dfrac{x-1}{x^2+1}\right)^{1/3} = \dfrac{1}{3}\log_{10}\dfrac{x-1}{x^2+1}$

$$= \frac{1}{3}\log_{10}(x-1) - \frac{1}{3}\log_{10}(x^2+1).$$

Thus

$$f'(x) = \frac{1}{3(x-1)\ln 10} - \frac{2x}{3(x^2+1)(\ln 10)}$$
$$= \frac{1}{3\ln 10}\left(\frac{1}{x-1} - \frac{2x}{x^2+1}\right).$$

(b) By Theorem 4.4(c), we have $f(x) = 2\log_{10}\sin x$. Thus

$$f'(x) = 2\cdot\frac{1}{(\sin x)\ln 10}\cdot\cos x = \frac{2}{\ln 10}\cot x. \quad ■$$

The quantity $\ln 10$, which came up in both parts of Example 4, is approximately 2.30258509. Using (9) and the fact that $\ln e = 1$, we could have expressed the factor $1/\ln 10$ as follows in Example 4:

$$\frac{1}{\ln 10} = \frac{\ln e}{\ln 10} = \log_{10} e.$$

As we mentioned at the beginning of the section, the most common base other than e is $b = 10$. Indeed, logarithms to the base 10 are referred to as *common logarithms.* Scientific calculators usually carry a programmed key for $\log_{10} x$ and for its inverse function 10^x as well. In many areas of science, $\log x$ (without mention of the base) is understood to mean $\log_{10} x$. However, because $\ln x$ is well suited to calculus-based scientific applications, many formulas in science that involve $\log_{10} x$ also involve the factor $\ln 10 \approx 2.3026$. Chemistry texts, for instance, usually write

$$\Delta G^0 = -2.303RT\log_{10} K$$

in place of the formula

$$\Delta G^0 = -RT\ln K$$

in Example 1(b) of Section 2, and perform all associated calculations using base 10. We close the section by mentioning another standard application of log_{10} in chemistry.

The strengths of acids and bases are determined by their concentration of hydrogen ions in moles per liter (mol/L), denoted $[H^+]$. The ***p*H** of such a substance is defined by

$$pH = -\log_{10}[H^+]. \tag{12}$$

(The symbol *p*H stands for *p*otential of *h*ydrogen.) For distilled water the *p*H is about 7. Substances with a lower *p*H are *acids*, while those with higher *p*H are *bases.*

Example 5

(a) What is the hydrogen ion concentration of a solution whose *p*H is 3.15?

(b) If the hydrogen ion concentration is 2.0×10^{-10} mol/L, is the substance acidic or basic?

Solution. (a) $-\log_{10}[H^+] = 3.15 \rightarrow \log_{10}[H^+] = -3.15$

$$[H^+] = 10^{-3.15} \approx 7.08 \times 10^{-4} \text{ mol/L}.$$

(b) Since $[H^+] = 2.0 \times 10^{-10}$,

$$\log_{10}[H^+] = \log_{10} 2 + \log_{10} 10^{-10} = \log_{10} 2 + (-10) \approx -9.7.$$

Thus, $pH \approx 9.7 > 7$, so the substance is basic. ■

Exercises 6.4

In Exercises 1–28, find the derivative of the given function.

1. $f(x) = 2^{3x}$
2. $g(x) = 10^{-2x+5}$
3. $h(x) = 10^{-x^2+x-1}$
4. $k(x) = 3^{\sqrt{x+1}}$
5. $k(x) = \log_{10}(x^2 - 3x + 1)$
6. $m(x) = \log_{10}\sqrt{x^2 - 3}$
7. $n(x) = \log_{10}|x^2 - 3x + 1|$
8. $p(x) = \log_{10}|\tan 2x|$
9. $q(x) = \left(\dfrac{1}{x^2} - 1\right)10^{2/(x-1)}$
10. $r(x) = \dfrac{10^{x^2-2x+1}}{x^2+1}$
11. $r(x) = 10^{\cos 3x}$
12. $g(x) = \log_{10} \sec 2^x$
13. $s(x) = 2^{\tan^2 3x}$
14. $h(x) = \cos(2^{-x^2})$
15. $t(x) = (10^x + 2^{-x})^5$
16. $u(x) = \dfrac{3}{(5^{-x} + 10^{2x})^3}$
17. $v(x) = \log_{10} \ln x$
18. $w(x) = \ln \log_{10} x$
19. $a(x) = (x+1)^{3x}$
20. $b(x) = (x+1)^{x^2-1}$
21. $c(x) = x^{\sqrt{x}}$
22. $e(x) = x^{x^2+1}$
23. $f(x) = \log_{10}\left(\dfrac{1+x^2}{x^2+8}\right)^{3/2}$
24. $s(x) = 2^x \log_{10}\left(\dfrac{x^2+1}{2x+1}\right)$
25. $h(x) = x^{\ln x}$
26. $k(x) = (\ln x)^x$
27. $l(x) = x^{\tan x}$
28. $m(x) = (\tan x)^{\sin x}$

In Exercises 29–42, evaluate the integrals.

29. $\int 10^{2x-1}\,dx$
30. $\int 2^{-3x}\,dx$
31. $\int x\,10^{-3x^2+5}\,dx$
32. $\int \dfrac{10^{\sqrt{x+1}}}{\sqrt{x+1}}\,dx$
33. $\int \dfrac{10^x}{10^x+1}\,dx$
34. $\int \dfrac{2^{\ln x}}{x}\,dx$
35. $\int \dfrac{dx}{x \log_2 x}$
36. $\int \dfrac{[\log_2 x]^5}{x}\,dx$
37. $\int 2^{\tan^2 x} \sec^2 x \tan x\,dx$
38. $\int 20^{\sin^2 x} \sin 2x\,dx$
39. $\int_0^1 \dfrac{2^x - 2^{-x}}{2^x + 2^{-x}}\,dx$
40. $\int_1^3 \dfrac{dx}{10^{-x} + 10^x}$
41. $\int_1^2 x^2 2^{-x^3+1}\,dx$
42. $\int_{-1}^1 x^2 10^{1-x^3}\,dx$

43. Prove that $\log_b 1 = 0$.

44. **(a)** Prove Theorem 4.4(a). **(b)** Prove Theorem 4.4(b). **(c)** Prove Theorem 4.4(c).

45. Show that for any two bases a and b

$$(\log_a b)(\log_b a) = 1.$$

46. Show that $\lim_{n\to\infty} n(x^{1/n} - 1) = \ln x$. $\left[\textit{Hint:}\right.$ Use the fact that $\left.\dfrac{d}{dx}(e^x)\Big|_{x=0} = 1.\right]$

47. The noise level $n(I)$ of a sound wave of intensity I is

$$n(I) = 10 \log_{10} \frac{I}{I_0},$$

where I_0 is the intensity of a sound wave at the point where it becomes audible. The unit for $n(I)$ is the *decibel* (db), named for Alexander Graham Bell. If I doubles, then by how much does $n(I)$ grow?

48. In Exercise 47, what increase in intensity is there between the sound of a whisper at 20 db and the sound of ordinary conversation at 70 db?

49. Earthquakes are measured on the *Richter* scale, named for the American geologist Charles Richter (1900–1985) of the California Institute of Technology. The magnitude m satisfies

$$m = \log \frac{I}{I_0},$$

where I is the intensity of the vibration, and I_0 is the intensity of a "standard" vibration. How much greater was the intensity of the San Francisco earthquake of 1906 (estimated $m = 8.4$) than the San Fernando Valley earthquake of 1971 for which $m = 6.5$?

50. How much greater was the intensity of the Mexico City earthquake of 1985 ($m = 8.1$) than the San Fernando Valley earthquake of 1971 (Exercise 49)?

51. Biological scientists are sometimes interested in studying population in terms of doubling time t_2, the time required for an initial population P_0 to double to $2P_0$.
(a) Show that $t_2 = (\ln 2)/r$, where r is the growth rate.
(b) Show that $P = P_0 \cdot 2^{t/t_2}$.

52. If the United States population almost exactly doubled between 1900 and 1950, use the approach of Exercise 47

to predict the population in 1980 if the 1900 population was 76,000,000. (The actual 1980 population was about 226,500,000.)

53. Show that the graph of $y = x^x$ is concave up on its entire domain.

54. Refer to Exercise 51. Physicists often express the radioactive decay formula $M = M_0 e^{kt}$ in terms of the half-life $t_{1/2}$ instead of the decay constant k.

(a) Show that $t_{1/2} = -\dfrac{\ln 2}{k}$.

(b) Show that $M = M_0 2^{-t/t_{1/2}} = M_0\left(\dfrac{1}{2}\right)^{t/t_{1/2}}$.

6.5 Inverse Trigonometric Functions

In Section 2, we treated the natural exponential function *exp* as the inverse of the natural logarithm *ln* and derived many of its properties from those of *ln*. This section uses a similar approach to define and study the inverse trigonometric functions.

FIGURE 5.1

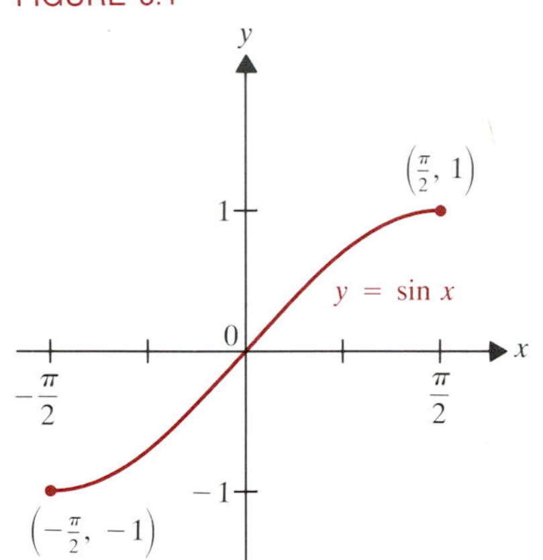

We start with the sine function. We can't simply define an inverse function g for it by saying $y = g(x)$ means that $x = \sin y$. The obstacle is that sine is *not* a one-to-one function. Since, for instance,

$$0 = \sin 0 = \sin \pi = \sin 2\pi = \ldots,$$

there is no unique y to define as $g(0)$. To get around this difficulty, we restrict attention to the interval $[-\pi/2, \pi/2]$, where, as **Figure 5.1** shows, sine takes on all values in its range $[-1, 1]$. Since

$$\frac{d}{dx}(\sin x) = \cos x > 0 \qquad \text{on } (-\pi/2, \pi/2),$$

the sine function is increasing on the interval $[-\pi/2, \pi/2]$, and hence is one-to-one. Thus the sine function with this restricted domain has an inverse function.

5.1 DEFINITION

The ***inverse sine (or arcsine) function $\sin^{-1}$ (or arcsin)*** is the function with domain $[-1, 1]$ and range $[-\pi/2, \pi/2]$ such that

(1) $$y = \sin^{-1} x \qquad \text{if and only if} \qquad x = \sin y,$$

where $x \in [-1, 1]$ and $y \in [-\pi/2, \pi/2]$.

FIGURE 5.2

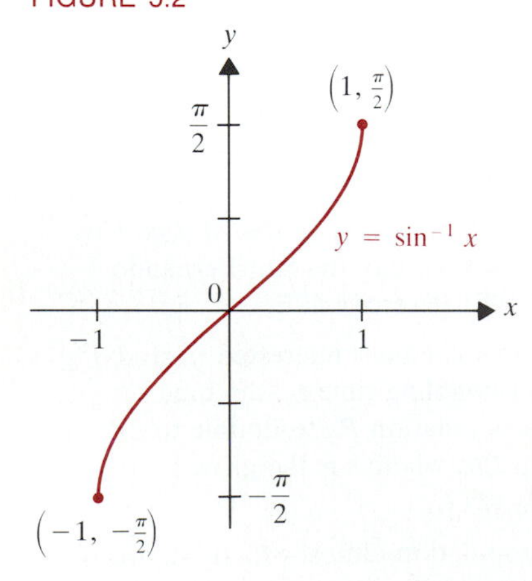

The function $\sin^{-1}$ assigns to each real number x in $[-1, 1]$ the unique angle y in the first or fourth quadrant whose sine is x. The "inverse sine" function programmed into your calculator (in radian mode) closely approximates $\arcsin x$ for real numbers x by displaying a rational approximation to $\arcsin r$, where r is a finite decimal agreeing with x in as many digits as the calculator can manage.

The symbol -1 in $\sin^{-1}$ is *not* an exponent but rather denotes *inverse*, as in the notation f^{-1} in Section 2.7. Be careful not to be confused: $\sin^2 x$ does mean $(\sin x)^2$, but $\sin^{-1} x$ does *not* mean $(\sin x)^{-1} = \csc x$. Similar remarks apply to the other inverse trigonometric functions to be introduced later in this section.

By Theorem 7.3 of Chapter 2, the graph of $y = \sin^{-1} x$ is the reflection of the graph of $y = \sin x$ in the line $y = x$. Thus **Figure 5.2** results from reflecting the curve in Figure 5.1 in the line $y = x$.

If we substitute each equation in (1) into the other, we obtain

(2) $$y = \sin^{-1}(\sin y), \qquad \text{for } y \in [-\pi/2, \pi/2]$$

and

(3) $$x = \sin(\sin^{-1} x), \qquad \text{for } x \in [-1, 1],$$

which are reminiscent of the equations $y = \exp(\ln y)$ and $x = \ln(\exp x)$ in Section 2.

Example 1

Find
(a) $\arcsin(-\sqrt{3}/2)$ (b) $\cos(\sin^{-1} \frac{1}{2})$ (c) $\sin^{-1}(\tan(-\pi/4))$
(d) $\tan(\sin^{-1} x)$ (e) $\arcsin(\sin \pi/3)$ (f) $\arcsin(\sin 2\pi/3)$

Solution. (a) $\arcsin(-\sqrt{3}/2) = \sin^{-1}(-\sqrt{3}/2) = -\pi/3$, since $-\pi/3$ is the angle in $[-\pi/2, \pi/2]$ such that $\sin(-\pi/3) = -\sin \pi/3 = -\sqrt{3}/2$.

(b) Since $\sin^{-1} \frac{1}{2} = \pi/6$, we have $\cos(\sin^{-1} \frac{1}{2}) = \cos \pi/6 = \sqrt{3}/2$.

(c) Since $\tan(-\pi/4) = -1$, we have $\sin^{-1}(\tan(-\pi/4)) = \sin^{-1}(-1) = -\pi/2$.

(d) Let $y = \sin^{-1} x$. Then $\sin y = x$. In **Figure 5.3** we show a representation of the angle y as an acute angle of a right triangle with hypotenuse 1. The opposite side is then x, because $\sin y = x/1 = x$. By the Pythagorean theorem, the third side is $\sqrt{1 - x^2}$. Hence

FIGURE 5.3

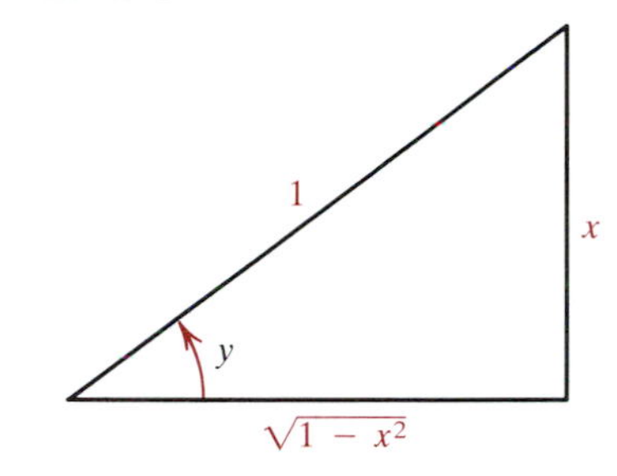

$$\tan y = \tan(\sin^{-1} x) = \frac{x}{\sqrt{1 - x^2}}.$$

(e) Noting that $\pi/3 \in [-\pi/2, \pi/2]$, we get at once from (2) that

$$\arcsin(\sin \pi/3) = \sin^{-1}(\sin \pi/3) = \pi/3.$$

(f) Here $2\pi/3 \notin [-\pi/2, \pi/2]$, so (2) doesn't apply. We have instead

$$\arcsin(\sin 2\pi/3) = \arcsin(\sqrt{3}/2) = \pi/3,$$

because $\pi/3$ is the angle in $[-\pi/2, \pi/2]$ whose sine is $\sqrt{3}/2$. ■

Since the sine function is increasing and differentiable on $[-\pi/2, \pi/2]$, we can use the inverse function differentiation rule (Theorem 7.5 of Chapter 2) to differentiate $\sin^{-1}$. To do so, let $y = \sin^{-1} x$. Then $x = \sin y$, where $y \in [-\pi/2, \pi/2]$. Then the rule gives

(4) $$\frac{dx}{dy} = \cos y \rightarrow \frac{dy}{dx} = \frac{1}{\cos y}.$$

Referring to Figure 5.3 again, we see that

(5) $$\cos y = \frac{\sqrt{1 - x^2}}{1} = \sqrt{1 - x^2}.$$

Substituting (5) into the right side of (4) and writing $D_x \sin^{-1} x$ for dy/dx, we have

(6) $$D_x \sin^{-1} x = \frac{1}{\sqrt{1 - x^2}}$$

From (6), we immediately obtain a new integration formula:

$$\int \frac{dx}{\sqrt{1-x^2}} = \arcsin x + C \tag{7}$$

Example 2

Find (a) $D_x \sin^{-1}(x^2 - 1)$, (b) $\int \frac{dx}{\sqrt{1-3x^2}}$.

Solution. (a) The chain rule and (6) give

$$D_x \sin^{-1}(x^2-1) = \frac{1}{\sqrt{1-(x^2-1)^2}} \cdot 2x = \frac{2x}{\sqrt{-x^4+2x^2}}.$$

(b) Since $1 - 3x^2 = 1 - (\sqrt{3} \cdot x)^2$, we can use (7) if we make the substitution $u = \sqrt{3}x$. Then

$$du = \sqrt{3}\,dx \to dx = \frac{1}{\sqrt{3}}\,du.$$

Hence

$$\int \frac{dx}{\sqrt{1-3x^2}} = \frac{1}{\sqrt{3}} \int \frac{du}{\sqrt{1-u^2}} = \frac{1}{\sqrt{3}} \arcsin u + C$$

$$= \frac{1}{\sqrt{3}} \arcsin(\sqrt{3}x) + C. \blacksquare$$

The inverse functions of the remaining trigonometric functions are defined similarly to the inverse sine function. Noting that cosine is a decreasing function on $[0, \pi]$, we restrict it to $[0, \pi]$ to obtain an inverse function.

5.2 DEFINITION The ***inverse cosine (or arccosine) function*** $\cos^{-1}$ is the function with domain $[-1, 1]$ and range $[0, \pi]$ such that

$$y = \cos^{-1} x \quad \text{if and only if} \quad x = \cos y, \tag{8}$$

where $x \in [-1, 1]$ and $y \in [0, \pi]$.

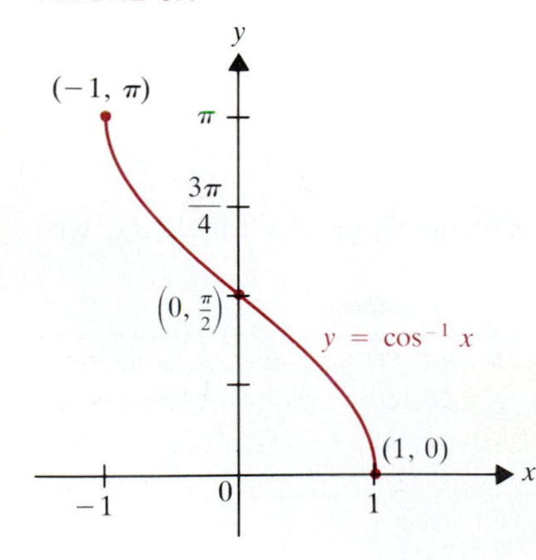

FIGURE 5.4

The inverse cosine function assigns to each real number x in $[-1, 1]$ the unique first- or second-quadrant angle y whose cosine is x. It is approximated by the inverse cosine function programmed into scientific calculators. Note that

$$y = \cos^{-1}(\cos y) \quad \text{and} \quad x = \cos(\cos^{-1} x),$$

for $x \in [-1, 1]$ and $y \in [0, \pi]$. The graph of $y = \cos^{-1} x$ is obtained by reflecting the graph of $y = \cos x$ (for $x \in [0, \pi]$) in the line $y = x$. It is shown in **Figure 5.4.**

Example 3

Find

(a) $\cos^{-1}(-1)$ (b) $\sin(\cos^{-1} x)$

(c) $\cos^{-1}(\cos 2\pi/3)$ (d) $\cos^{-1}(\cos 3\pi/2)$

Solution. (a) Since π is the only x in $[0, \pi]$ for which $\cos x = -1$, we have $\cos^{-1}(-1) = \pi$.

(b) Referring to **Figure 5.5,** which shows $y = \cos^{-1} x$, we see that

$$\sin(\cos^{-1} x) = \sin y = \frac{\sqrt{1-x^2}}{1} = \sqrt{1-x^2}.$$

FIGURE 5.5

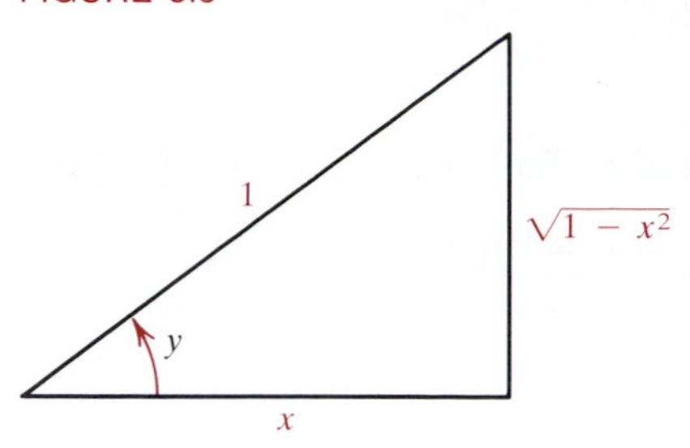

(c) Since $2\pi/3 \in [0, \pi]$, it follows from (8) that $\cos^{-1}(\cos 2\pi/3) = 2\pi/3$.

(d) We can't apply (8) because $3\pi/2 \notin [0, \pi]$. Instead, we have

$$\cos^{-1}(\cos 3\pi/2) = \cos^{-1}(0) = \pi/2. \quad \blacksquare$$

We differentiate $\cos^{-1}$ by using the inverse function differentiation rule in the same way we did for $\sin^{-1}$. Namely, we let $y = \cos^{-1} x$, so that $x = \cos y$, where $y \in [0, \pi]$. Then

$$\frac{dx}{dy} = -\sin y.$$

Hence, by Theorem 7.5 of Chapter 2,

$$\frac{dy}{dx} = -\frac{1}{\sin y} = -\frac{1}{\sin(\cos^{-1} x)} = -\frac{1}{\sqrt{1-x^2}}$$

by Example 3(b). We therefore have

(9)
$$D_x \cos^{-1} x = -\frac{1}{\sqrt{1-x^2}}$$

This gives the integration formula

(10)
$$\int \frac{dx}{\sqrt{1-x^2}} = -\arccos x + C$$

Comparing (10) with (7), we see that $\arcsin x$ must differ from $-\arccos x$ by a constant (see Exercise 51).

Example 4

Find $D_x(\sin^{-1} x - \cos^{-1} x)$.

Solution. Using (6) and (9), we get

$$D_x(\sin^{-1} x - \cos^{-1} x) = \frac{1}{\sqrt{1-x^2}} - \frac{-1}{\sqrt{1-x^2}} = \frac{2}{\sqrt{1-x^2}}. \quad \blacksquare$$

We can define inverse functions for the four remaining trigonometric functions by suitably restricting their domains. The last one that occurs commonly enough to be programmed onto most scientific calculators is $\tan^{-1}$. To define it, we restrict the tangent function to the interval $(-\pi/2, \pi/2)$, where it is increasing with range $(-\infty, +\infty)$.

5.3 DEFINITION The ***inverse tangent*** (*or* ***arctangent***) ***function*** $\tan^{-1}$ (*or* ***arctan***) is the function with domain $(-\infty, +\infty)$ and range $(-\pi/2, \pi/2)$ such that

(11) $$y = \tan^{-1} x \quad \text{if and only if} \quad x = \tan y$$

where $y \in (-\pi/2, \pi/2)$, $x \in (-\infty, +\infty)$.

The graph of $y = \tan^{-1} x$ is shown in **Figure 5.6.** We see from (11) that

$$y = \tan^{-1}(\tan y) \qquad \text{and} \qquad x = \tan(\tan^{-1} x)$$

for any real number x and any $y \in (-\pi/2, \pi/2)$.

FIGURE 5.6

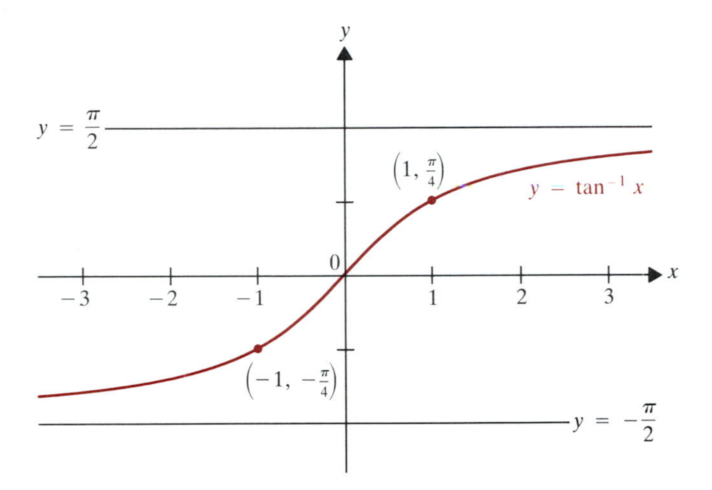

To differentiate arctan, we proceed as above. Let $y = \tan^{-1} x$, so that $x = \tan y$, where $y \in (-\pi/2, \pi/2)$. Then

$$\frac{dx}{dy} = \sec^2 y.$$

Thus Theorem 7.5 of Chapter 2 gives

(12) $$\frac{dy}{dx} = \cos^2 y.$$

FIGURE 5.7

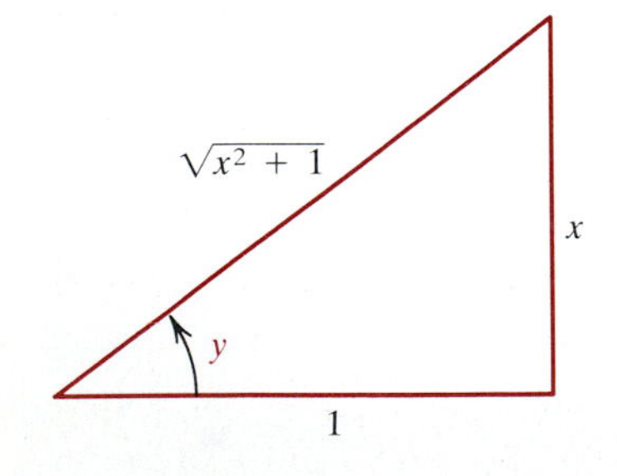

Figure 5.7 shows the angle $y = \tan^{-1} x$. From it, we see that

(13) $$\cos y = \frac{1}{\sqrt{x^2 + 1}}.$$

Substituting (13) into (12), we obtain

(14) $$D_x \tan^{-1} x = \frac{1}{x^2 + 1}$$

This gives the important integration formula

(15) $$\int \frac{dx}{x^2 + 1} = \arctan x + C$$

Example 5

(a) Find $D_x \tan^{-1} \dfrac{x}{x+1}$. (b) Find $\displaystyle\int \frac{dx}{x^2 + a^2}$.

Solution. (a)
$$D_x \tan^{-1} \frac{x}{x+1} = \frac{1}{1 + x^2/(x+1)^2} \cdot \frac{(x+1)\cdot 1 - x \cdot 1}{(x+1)^2}$$
$$= \frac{(x+1)^2}{(x+1)^2 + x^2} \cdot \frac{1}{(x+1)^2}$$
$$= \frac{1}{2x^2 + 2x + 1}.$$

(b) Since

$$\int \frac{dx}{x^2 + a^2} = \int \frac{\frac{1}{a^2}\, dx}{\frac{x^2}{a^2} + 1},$$

it is natural to let $u = x/a$. Then

$$du = \frac{dx}{a} \rightarrow dx = a\, du.$$

Hence

$$\int \frac{dx}{x^2 + a^2} = \int \frac{\frac{1}{a^2} \cdot a\, du}{u^2 + 1} = \frac{1}{a} \int \frac{du}{u^2 + 1} = \frac{1}{a} \tan^{-1} u + C. \blacksquare$$

FIGURE 5.8

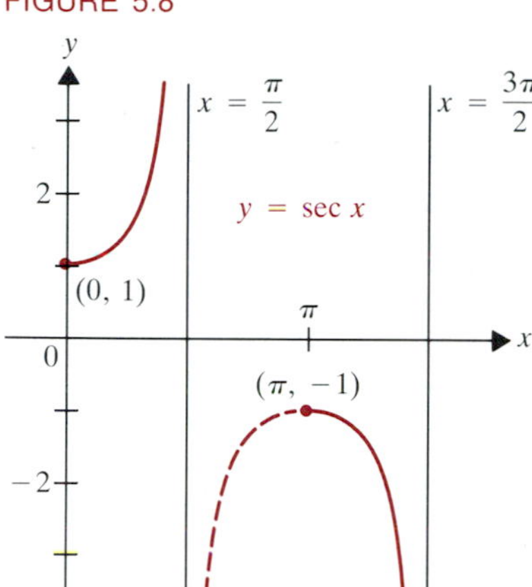

Thus we obtain the following extension of (15).

(16) $$\int \frac{dx}{a^2 + x^2} = \frac{1}{a} \tan^{-1} \frac{x}{a} + C$$

FIGURE 5.9

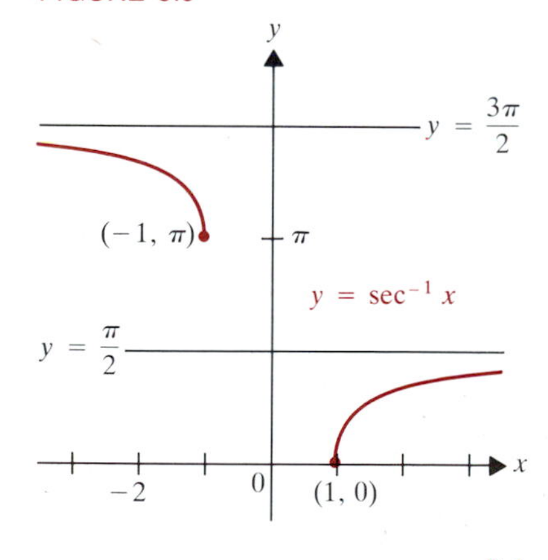

The last inverse trigonometric function we need is the inverse secant function. We restrict attention to the subset $S = [0, \pi/2) \cup [\pi, 3\pi/2)$ of real numbers. While it may seem strange not to use an interval, this choice of S will serve us well in Section 7.3. As **Figure 5.8** shows, the secant is a one-to-one function on S, since it is increasing on $[0, \pi/2)$ and decreasing on $[\pi, 3\pi/2)$. Moreover, every real number in the range of the secant function is taken on as x ranges over S. The following definition is therefore meaningful.

5.4 DEFINITION

The ***inverse secant (or arcsecant) function*** $\boldsymbol{\sec^{-1}}$ **(*or arcsec*)** is the function with domain $(-\infty, -1] \cup [1, +\infty)$ and range $[0, \pi/2) \cup [\pi, 3\pi/2)$ such that

(17) $$y = \sec^{-1} x \quad \text{if and only if} \quad x = \sec y,$$

for $x \in (-\infty, -1] \cup [1, +\infty)$ and $y \in [0, \pi/2) \cup [\pi, 3\pi/2)$.

The graph of the inverse secant function is shown in **Figure 5.9.** We differentiate $\sec^{-1}$ by using the approach followed for the other inverse functions. Letting $y = \sec^{-1} x$, we have $x = \sec y$, where y is a first- or third-quadrant angle. Thus

$$\frac{dx}{dy} = \sec y \tan y. \tag{18}$$

There are two cases, depending on where the angle y lies. First note that since $x \neq \pm 1$, we have $y \neq 0$ and $y \neq \pi$.

Case 1. $y \in (0, \pi/2)$. In this case, *sec* is increasing near y since its derivative, $\sec y \tan y$, is positive in $(0, \pi/2)$. So Theorem 7.5 of Chapter 2 applies to give

$$\frac{dy}{dx} = \frac{1}{\sec y \tan y} = \frac{1}{x\sqrt{x^2 - 1}} \tag{19}$$

from (18) and **Figure 5.10**(a).

FIGURE 5.10

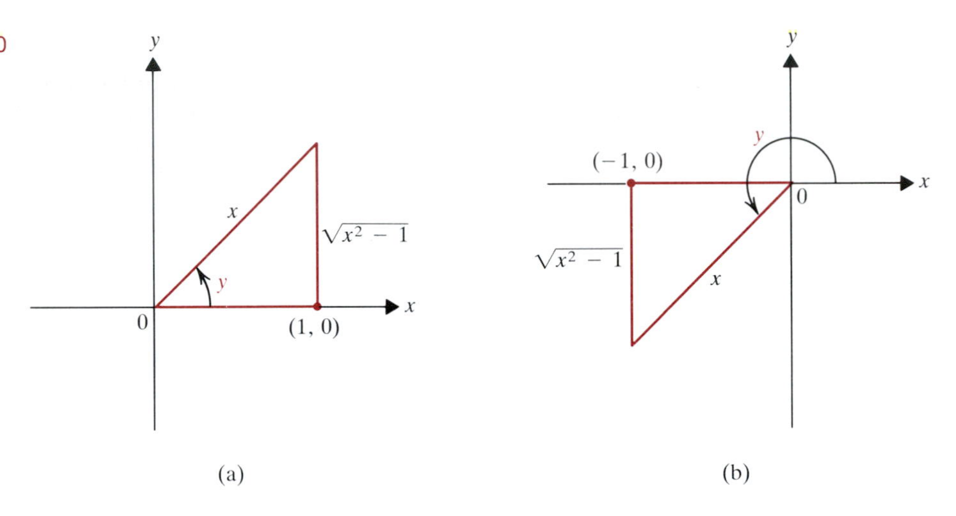

Case 2. $y \in (\pi, 3\pi/2)$. In this case, *sec* is decreasing near y since $\sec y \tan y < 0$ for $y \in (\pi, 3\pi/2)$. Then Theorem 7.5 of Chapter 2 gives

$$\frac{dy}{dx} = \frac{1}{\sec y \tan y} = \frac{1}{x\sqrt{x^2 - 1}},$$

since for a third-quadrant angle y, $\tan y \geq 0$, so

$$\tan y = \sqrt{\tan^2 y} = \sqrt{\sec^2 x - 1} = \sqrt{x^2 - 1}.$$

In both cases then, we have

$$D_x \sec^{-1} x = \frac{1}{x\sqrt{x^2 - 1}}, \qquad \text{if } x \in (-\infty, -1) \cup (1, +\infty) \tag{20}$$

From this we get the integration formula

$$\int \frac{dx}{x\sqrt{x^2 - 1}} = \text{arcsec}\, x + C \tag{21}$$

Warning. Some other texts use the set $[0, \pi/2) \cup (\pi/2, \pi]$ in Definition 5.4 instead of $[0, \pi/2) \cup [\pi, 3\pi/2)$. They obtain a slightly different function, which we will call *Sec*$^{-1}$. For it, formulas (20) and (21) take the more complicated forms,

(22) $$D_x \operatorname{Sec}^{-1} x = \frac{1}{|x|\sqrt{x^2 - 1}}, \qquad x \in (-\infty, -1) \cup (1, +\infty),$$

(23) $$\int \frac{dx}{x\sqrt{x^2 - 1}} = \operatorname{Sec}^{-1}|x| + C.$$

You are asked to derive (22) and (23) in Exercises 49 and 50. (Since the inverse secant function arises so rarely outside calculus courses, this difference should not cause you any difficulty.)

Example 6

Find

(a) $D_x \sec^{-1}\sqrt{x}$ and (b) $\displaystyle\int \frac{dx}{x\sqrt{x^2 - a^2}}$, where $a > 0$

Solution. (a) The chain rule and (20) give

$$D_x \sec^{-1}\sqrt{x} = \frac{1}{\sqrt{x}\sqrt{x - 1}} D_x(\sqrt{x})$$

$$= \frac{1}{\sqrt{x}\sqrt{x - 1}} \frac{1}{2\sqrt{x}} = \frac{1}{2x\sqrt{x - 1}}.$$

(b) We first put the radical into a form to which (21) can be applied:

$$\int \frac{dx}{x\sqrt{x^2 - a^2}} = \int \frac{dx}{ax\sqrt{\left(\dfrac{x}{a}\right)^2 - 1}} = \int \frac{\dfrac{1}{a}\,dx}{x\sqrt{\left(\dfrac{x}{a}\right)^2 - 1}}.$$

Let $u = x/a$. Then

$$du = \frac{1}{a}\,dx, \qquad \text{and} \qquad x = au.$$

We thus obtain

$$\int \frac{dx}{x\sqrt{x^2 - a^2}} = \int \frac{du}{au\sqrt{u^2 - 1}} = \frac{1}{a}\int \frac{du}{u\sqrt{u^2 - 1}}$$

$$= \frac{1}{a}\sec^{-1} u + C. \quad ■$$

We therefore have the following extension of (21).

(24) $$\int \frac{dx}{x\sqrt{x^2 - a^2}} = \frac{1}{a}\sec^{-1}\frac{x}{a} + C$$

We can use (15) and (21) together with the technique of completing the square, to integrate some functions that might appear to be beyond our present methods. In this, the boxed results of Examples 5(b) and 6(b) are helpful.

Example 7

Find (a) $\int \frac{dx}{2x^2 - x + 1}$ and (b) $\int \frac{x\,dx}{2x^2 - x + 1}$.

Solution. (a) We complete the square in the denominator, obtaining

$$\begin{aligned} 2x^2 - x + 1 &= 2\left(x^2 - \frac{1}{2}x \quad\right) + 1 \\ &= 2\left(x^2 - \frac{1}{2}x + \frac{1}{4}\right) + 1 - \frac{1}{2} \\ &= 2\left(x - \frac{1}{2}\right)^2 + \frac{1}{2}. \end{aligned}$$

Then

$$\begin{aligned} \int \frac{dx}{2x^2 - x + 1} &= \int \frac{dx}{2\left(x - \frac{1}{2}\right)^2 + \frac{1}{2}} = \frac{1}{2}\int \frac{dx}{\left(x - \frac{1}{2}\right)^2 + \frac{1}{4}} \\ &= \frac{1}{2} \cdot \frac{1}{\frac{1}{2}} \tan^{-1} \frac{x - \frac{1}{2}}{\frac{1}{2}} + C \qquad \textit{by Example 5(b)} \\ &= \tan^{-1}(2x - 1) + C \end{aligned}$$

(b) We can't use the approach of part (a), because of the factor x in the numerator of the integrand. But we can add and subtract in the numerator to obtain the derivative of the denominator plus or minus a constant. This permits us to work out the integral as a natural logarithm plus or minus an inverse tangent. Since the derivative of the denominator is $4x - 1$, we get

$$\begin{aligned} \int \frac{x\,dx}{2x^2 - x + 1} &= \frac{1}{4}\int \frac{4x\,dx}{2x^2 - x + 1} = \frac{1}{4}\int \frac{4x - 1 + 1}{2x^2 - x + 1}\,dx \\ &= \frac{1}{4}\int \frac{4x - 1}{2x^2 - x + 1}\,dx + \frac{1}{4}\int \frac{dx}{2x^2 - x + 1} \\ &= \frac{1}{4}\ln(2x^2 - x + 1) + \frac{1}{4}\tan^{-1}(2x - 1) + C, \end{aligned}$$

where the second integral was evaluated in part (a). ■

We will see in the next chapter that integrals like the one in Example 7(b) arise often enough to be of significance.

Exercises 6.5

In Exercises 1–6, evaluate the given quantity.

1. (a) $\sin^{-1}(1/\sqrt{2})$ (b) $\arcsin(-1/2)$ (c) $\sin^{-1}(-1)$ (d) $\arcsin 0$ (e) $\sin^{-1}(\sin(3\pi/4))$ (f) $\sin^{-1}(\sin(\pi/6))$
2. (a) $\cos^{-1}(-1/\sqrt{2})$ (b) $\arccos(\sqrt{3}/2)$ (c) $\cos^{-1} 1$ (d) $\arccos 0$ (e) $\cos^{-1}(\cos(-\pi/4))$ (f) $\cos^{-1}(\cos(3\pi/4))$
3. (a) $\tan^{-1}(-\sqrt{3})$ (b) $\arctan(-1)$ (c) $\tan^{-1}(1/\sqrt{3})$ (d) $\arctan 0$ (e) $\tan^{-1}(\tan(-\pi/4))$ (f) $\tan^{-1}(\tan(5\pi/6))$
4. (a) $\sec^{-1}(-\sqrt{2})$ (b) $\operatorname{arcsec} 1$ (c) $\sec^{-1}(2)$ (d) $\operatorname{arcsec}(-1)$
5. (a) $\sec(\sin^{-1} x)$ (b) $\cot(\sin^{-1} x)$ (c) $\tan(\cos^{-1} x)$ (d) $\csc(\cos^{-1} x)$
6. (a) $\sin(\tan^{-1} x)$ (b) $\sec(\tan^{-1} x)$ (c) $\cos(\sec^{-1} x)$ (d) $\sin(\sec^{-1} x)$

In Exercises 7–22, find $f'(x)$ for the given $f(x)$.

7. $\sin^{-1}(\sqrt{x+1})$
8. $\arccos\sqrt{x}$
9. $\arcsin 1/x^2$
10. $(\cos^{-1} e^{-x} + 1)^3$
11. $\tan^{-1}(\ln x)$
12. $\ln(\arctan\sqrt{x})$
13. $\sec^{-1}(\sin x)$
14. $\sec^{-1}(\ln\tan x)$
15. $\sin^{-1}\dfrac{x-1}{x+1}$
16. $\tan^{-1}\dfrac{x-1}{x+1}$
17. $\dfrac{\cos^{-1} x}{\sqrt{1-x^2}}$
18. $\sec^{-1}\sqrt{x^2+1}$
19. $\exp(\sqrt{\sin^{-1} x})$
20. $\left(\dfrac{1}{x} - \arccos\dfrac{1}{x}\right)^3$
21. $(\tan^{-1} x)^{\tan x}$
22. $(\sin x)^{\arcsin x}$

In Exercises 23–46, evaluate the given integral.

23. $\int_0^{1/6} \dfrac{dx}{\sqrt{1-9x^2}}$
24. $\int_0^{1/10} \dfrac{dx}{\sqrt{1-25x^2}}$
25. $\int \dfrac{e^x}{\sqrt{1-e^{2x}}}\,dx$
26. $\int \dfrac{e^{3x}}{\sqrt{1-e^{6x}}}\,dx$
27. $\int \dfrac{e^x}{1+e^{2x}}\,dx$
28. $\int \dfrac{\cos x}{\sin^2 x + 1}\,dx$
29. $\int_0^1 \dfrac{x\,dx}{x^4+1}$
30. $\int_0^1 \dfrac{x^2\,dx}{x^6+1}$
31. $\int \dfrac{dx}{x\sqrt{4x^2-1}}$
32. $\int \dfrac{\cos x\,dx}{\sin x\sqrt{4\sin^2 x - 1}}$
33. $\int \dfrac{dx}{x\sqrt{x^4-1}}$
34. $\int \dfrac{dx}{\sqrt{e^{2x}-1}}$
35. $\int \dfrac{dx}{\sqrt{a^2-x^2}}$, where $a > 0$
36. $\int \dfrac{dx}{\sqrt{9-x^2}}$
37. $\int_0^2 \dfrac{dx}{x^2+4}$
38. $\int_0^4 \dfrac{dx}{x^2+16}$
39. $\int \dfrac{dx}{x^2-4x+13}$
40. $\int \dfrac{dx}{x^2-2x+5}$
41. $\int \dfrac{x\,dx}{x^2-4x+13}$
42. $\int \dfrac{x\,dx}{x^2-2x+5}$
43. $\int_0^{1/3} \dfrac{dx}{\sqrt{4-9x^2}}$
44. $\int_0^{1/5} \dfrac{dx}{\sqrt{9-25x^2}}$
45. $\int \dfrac{dx}{(x+1)\sqrt{x^2+2x}}$
46. $\int \dfrac{dx}{(x-3)\sqrt{x^2-6x+5}}$

47. Give a definition for $\cot^{-1} x$ and find a formula for $D_x \cot^{-1} x$.
48. Repeat Exercise 47 for $\csc^{-1} x$.
49. Derive (22).
50. Derive (23).
51. (a) Show from (6) and (9) that $D_x \sin^{-1} x = -D_x \cos^{-1} x$.
    (b) Use (a) to prove that $\sin^{-1} x + \cos^{-1} x = \pi/2$ if $x \in [0, 1]$.
52. (a) Show that $D_x \cos^{-1}(1/x) = D_x \sec^{-1} x$ if $x > 1$.
    (b) Use (a) to show that $\sec^{-1} x = \cos^{-1}(1/x)$ if $x > 1$.
53. A painting 3 feet high is hung on a wall with its bottom edge 7 feet above the floor. A person whose eyes are 5 feet above the floor observes the painting. How far from the wall on which the painting is hung should the observer stand to maximize angle θ in **Figure 5.11?** (*Hint:* $\theta = \phi + \theta - \phi$. Express $\theta + \phi$ and ϕ as inverse tangents.)

FIGURE 5.11

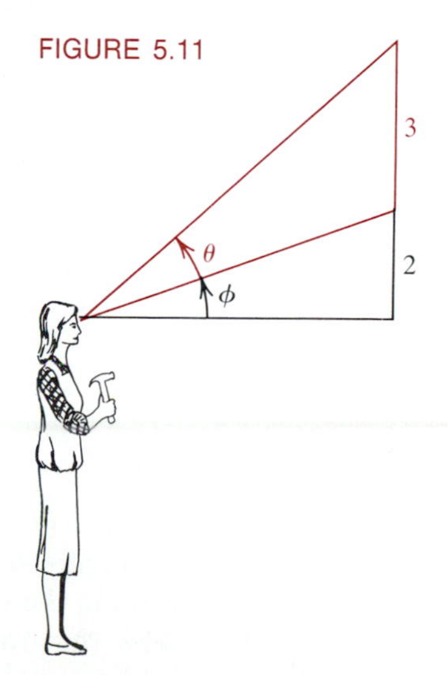

54. Repeat Exercise 53 if the person's eyes are 6 feet above the floor.

55. Suppose you purchase a painting 3 feet high for your living room. If your eyes are 5 1/2 feet above floor level, then where should the center C of the painting be placed so as to maximize angle θ in **Figure 5.12**? (*Hint:* Note $\theta = \phi + \psi$. Suppose that for viewing you stand at a constant k feet from the wall. Express ϕ and ψ as inverse tangents.)

56. During a rocket launch, the angle of elevation of the rocket from a tracking station is monitored. Suppose the tracking device is 5000 meters from the launch site. When the angle of elevation of the base of the rocket is $\pi/4$ radians, it is calculated to be changing at the rate of π radians per minute. Find the velocity of the rocket in meters per minute, when $\theta = \pi/4$. Refer to **Figure 5.13.**

FIGURE 5.13

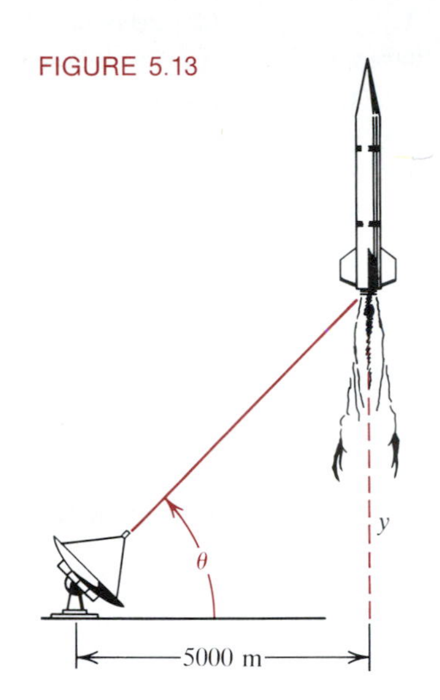

FIGURE 5.12

6.6 Hyperbolic Functions

This section is about a family of transcendental functions that are defined in terms of the exponential functions e^x and e^{-x}. These functions satisfy identities and rules of differentiation and integration remarkably similar to those of the trigonometric functions. They also arise in solving a number of important problems in science and engineering, one of which we discuss in the next section. We begin with the two basic hyperbolic functions.

6.1 DEFINITION

The ***hyperbolic sine function sinh*** (pronounced *cinch*) is

(1) $$\sinh x = \frac{1}{2}(e^x - e^{-x}),$$

where x is any real number. The ***hyperbolic cosine function cosh*** (pronounced *cosh*) is

(2) $$\cosh x = \frac{1}{2}(e^x + e^{-x}),$$

where x is any real number.

FIGURE 6.1

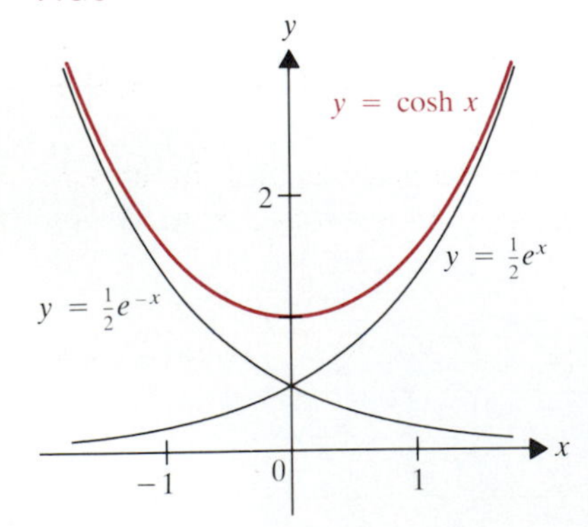

Each of these functions has domain $\mathbf{R} = (-\infty, +\infty)$. The graph of $y = \cosh x$ in **Figure 6.1** can be obtained from the graphs of $y = \frac{1}{2}e^x$ and $y = \frac{1}{2}e^{-x}$ by adding the corresponding y-coordinates for each value of x. **Figure 6.2** shows how the graph of $y = \sinh x$ is similarly obtained from the graphs of $y = \frac{1}{2}e^x$ and $y = -\frac{1}{2}e^{-x}$. Since $e^x \to 0$ as $x \to -\infty$, the graphs of both *cosh* and *sinh*

FIGURE 6.2

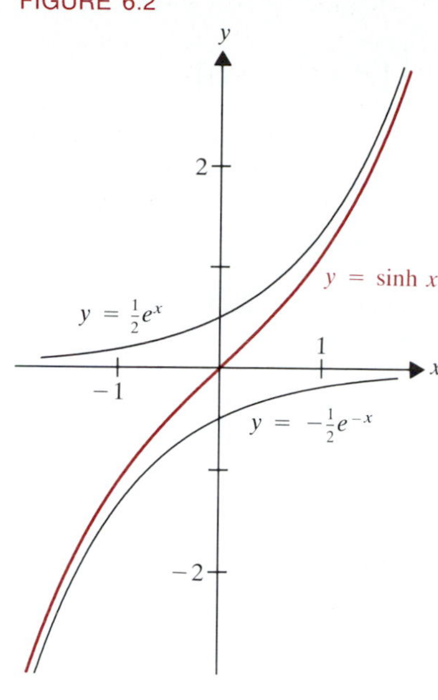

approach the graph of $\frac{1}{2}e^x$ as $x \to +\infty$. (The graphs are called *asymptotic* to the graph of $\frac{1}{2}e^x$.) Thus *cosh* has range $[1, +\infty)$. Similarly, the graph of *sinh* is asymptotic to the graph of $-\frac{1}{2}e^{-x}$ as $x \to -\infty$. Hence *sinh* has range $R = (-\infty, +\infty)$.

If we square $\sinh x$ and $\cosh x$ in (1) and (2), then we get

$$\cosh^2 x = \frac{1}{4}(e^{2x} + 2 + e^{-2x}) \quad \text{and} \quad \sinh^2 x = \frac{1}{4}(e^{2x} - 2 + e^{-2x}).$$

Hence

$$\cosh^2 x - \sinh^2 x = 1. \tag{3}$$

This identity is reminiscent of the basic trigonometric identity $\cos^2 x + \sin^2 x = 1$. It also permits us to explain why the names *hyperbolic sine* and *hyperbolic cosine* are appropriate. Recall that the point $P(\cos t, \sin t)$ lies on the unit circle $x^2 + y^2 = 1$ for every real number t (see **Figure 6.3**). In view of (3), the point $Q(\cosh t, \sinh t)$ lies on the *hyperbola* $x^2 - y^2 = 1$ shown in **Figure 6.4.** (For more on hyperbolas, see Section 8.3.) There is a further analogy between hyperbolic sine and hyperbolic cosine on the one hand, and sine and cosine on the other hand. The shaded region in Figure 6.3 is a *circular sector* with central angle t and radius 1. So by Lemma 3.9 of Chapter 2, its area is

$$A = \frac{1}{2} \cdot 1^2 \cdot t = \frac{1}{2}t.$$

FIGURE 6.3

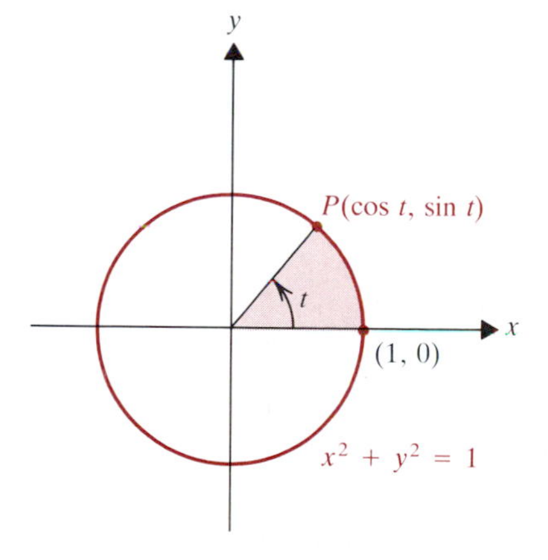

FIGURE 6.4

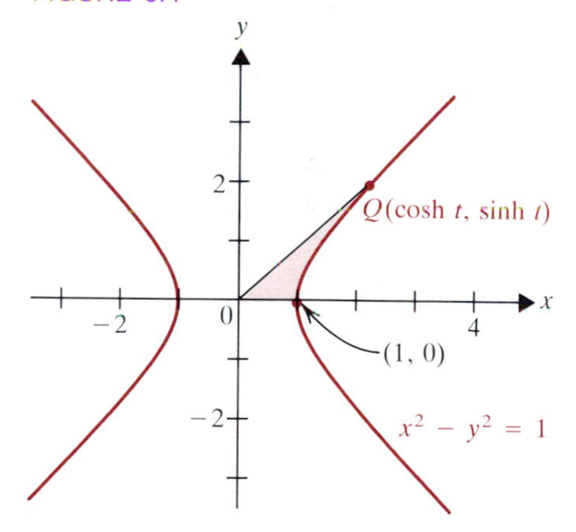

It turns out (Exercise 47) that the analogous shaded region in Figure 7.4, which might be called a *hyperbolic sector*, also has area $\frac{1}{2}t$.

The analogy between the standard trigonometric functions and the hyperbolic ones is even more striking in their calculus. For example, from (1) we have

$$D_x \sinh x = D_x\left[\frac{1}{2}(e^x - e^{-x})\right] = \frac{1}{2}[e^x - e^{-x}(-1)] \qquad \textit{by the chain rule}$$

$$= \frac{1}{2}(e^x + e^{-x}) = \cosh x.$$

Similarly, from (2),

$$D_x \cosh x = D_x\left[\frac{1}{2}(e^x + e^{-x})\right] = \frac{1}{2}[e^x + e^{-x}(-1)]$$

$$= \frac{1}{2}(e^x - e^{-x}) = \sinh x.$$

We therefore have the integration formulas

$$\int \sinh x\, dx = \cosh x + C \qquad \text{and} \qquad \int \cosh x\, dx = \sinh x + C.$$

Example 1

Evaluate (a) $\int \cosh^2 x \sinh x\, dx$ and (b) $D_x\sqrt{1 + \cosh 2x}$.

Solution. (a) If we let $u = \cosh x$, then $du = \sinh x\, dx$. Hence

$$\int \cosh^2 x \sinh x\, dx = \frac{1}{3}\cosh^3 x + C.$$

(b)
$$D_x\sqrt{1 + \cosh 2x} = \frac{1}{2}(1 + \cosh 2x)^{-1/2} \cdot (\sinh 2x) \cdot 2$$

$$= \frac{\sinh 2x}{\sqrt{1 + \cosh 2x}}. \quad \blacksquare$$

Notice that no minus signs occur in the differentiation and integration of either *sinh* or *cosh*. Thus their calculus is not only analogous to that of sine and cosine but also a bit simpler. There are still more analogies between the trigonometric and hyperbolic sine and cosine functions. For instance,

$$\sinh(-x) = \frac{1}{2}(e^{-x} - e^{-(-x)}) = -\frac{1}{2}(e^x - e^{-x}) = -\sinh x \tag{4}$$

and

$$\cosh(-x) = \frac{1}{2}(e^{-x} + e^{-(-x)}) = \cosh x.$$

Thus *sinh* is an *odd* function (like sine), and *cosh* is an *even* function (like cosine). Moreover, these functions also satisfy the following identities, reminiscent of the trigonometric sine and cosine identities (see Exercise 37).

$$\sinh(x + y) = \sinh x \cosh y + \cosh x \sinh y, \tag{5}$$

$$\cosh(x + y) = \cosh x \cosh y + \sinh x \sinh y, \tag{6}$$

$$\sinh 2x = 2 \sinh x \cosh x, \tag{7}$$

$$\cosh 2x = \cosh^2 x + \sinh^2 x, \tag{8}$$

$$\cosh^2 x = \frac{1}{2}(1 + \cosh 2x), \tag{9}$$

$$\sinh^2 x = -\frac{1}{2}(1 - \cosh 2x). \tag{10}$$

The close parallel between cosh and sinh on the one hand, and cosine and sine on the other, makes it natural to define other hyperbolic functions in terms of them. They are analogous to the remaining trigonometric functions.

6.2 DEFINITION

(a) The ***hyperbolic tangent function*** is given by

$$\tanh x = \frac{\sinh x}{\cosh x} = \frac{e^x - e^{-x}}{e^x + e^{-x}},$$

for any real number x.

(b) The ***hyperbolic cotangent function*** is given by

$$\coth x = \frac{\cosh x}{\sinh x} = \frac{e^x + e^{-x}}{e^x - e^{-x}},$$

for any real number $x \neq 0$.

(c) The ***hyperbolic secant function*** is given by

$$\operatorname{sech} x = \frac{1}{\cosh x} = \frac{2}{e^x + e^{-x}},$$

for any real number x.

(d) The ***hyperbolic cosecant function*** is given by

$$\operatorname{csch} x = \frac{1}{\sinh x} = \frac{2}{e^x - e^{-x}},$$

for any real number $x \neq 0$.

If we divide the basic identity

$$\cosh^2 x - \sinh^2 x = 1 \tag{3}$$

through by $\cosh^2 x$ (which, recall, is never 0), then we get

$$1 - \tanh^2 x = \operatorname{sech}^2 x. \tag{11}$$

This is analogous to the trigonometric identity $1 + \tan^2 x = \sec^2 x$; but as in (3), there is a sign difference. If we divide (3) through by $\sinh^2 x$, then we get

$$\coth^2 x - 1 = \operatorname{csch}^2 x, \qquad \text{if } x \neq 0. \tag{12}$$

This is analogous to the trigonometric identity $1 + \cot^2 x = \csc^2 x$. Fortunately, (11) and (12) don't arise frequently enough for the sign differences with their trigonometric analogues to cause confusion.

The graphs of the hyperbolic functions can be accurately drawn by using our knowledge of the exponential function and the techniques of Section 3.7.

Example 2

Construct the graph of $y = \tanh x$.

Solution. First, the domain of tanh is $(-\infty, +\infty)$. Since

$$\tanh x = \frac{e^x - e^{-x}}{e^x + e^{-x}} = \frac{e^{2x} - 1}{e^{2x} + 1} = \frac{1 - e^{-2x}}{1 + e^{-2x}},$$

we have

$$\lim_{x \to +\infty} \tanh x = \lim_{x \to +\infty} \frac{1 - e^{-2x}}{1 + e^{-2x}} = \frac{1 - 0}{1 + 0} = 1,$$

and

$$\lim_{x\to-\infty} \tanh x = \lim_{x\to-\infty} \frac{e^{2x}-1}{e^{2x}+1} = \frac{0-1}{0+1} = -1.$$

Thus the lines $y = \pm 1$ are horizontal asymptotes. There are no critical points, because

$$D_x \tanh x = D_x \frac{\sinh x}{\cosh x} = \frac{\cosh x \cosh x - \sinh x \sinh x}{\cosh^2 x}$$

$$= \frac{\cosh^2 x - \sinh^2 x}{\cosh^2 x} = \frac{1}{\cosh^2 x}.$$

Hence $D_x \tanh x$ is never 0. Moreover, it exists for all x, because $\cosh x$ is never 0. So there are no local extreme value points. If we differentiate $D_x \tanh x = (\cosh x)^{-2}$, then we get

$$D_x^2(\tanh x) = -2(\cosh x)^{-3} \cdot \sinh x = -\frac{2\sinh x}{\cosh^3 x},$$

which is 0 when $\sinh x = 0$, that is, when $x = 0$. Now $\cosh^3 x$ is always positive, while $\sinh x < 0$ for $x < 0$ and $\sinh x > 0$ for $x > 0$. Hence the graph is concave upward for $x < 0$ and concave downward for $x > 0$. Thus (0, 0) is an inflection point. The graph is shown in **Figure 6.5.** ■

FIGURE 6.5

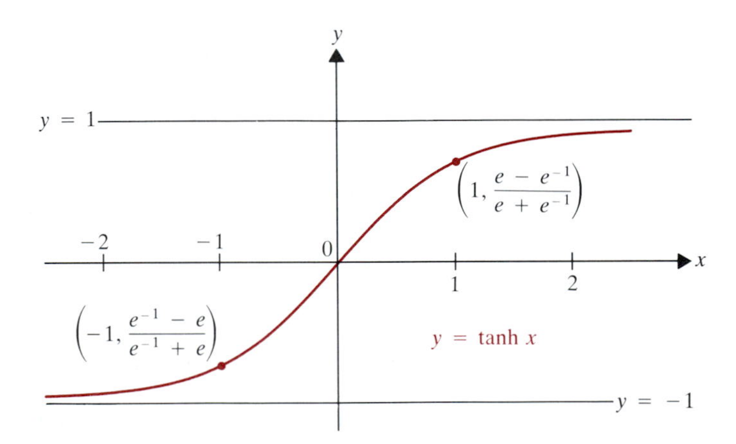

In working out Example 2, we found that $D_x \tanh x = 1/\cosh^2 x = \operatorname{sech}^2 x$. This formula, which is analogous to that for $D_x \tan x$, is the first one listed below. The remaining parts follow easily from the differentiation formulas for sinh and cosh (see Exercise 38).

(13) $D_x \tanh x = \operatorname{sech}^2 x,$ $\qquad D_x \coth x = -\operatorname{csch}^2 x.$

(14) $D_x \operatorname{sech} x = -\operatorname{sech} x \tanh x,$ $\qquad D_x \operatorname{csch} x = -\operatorname{csch} x \coth x.$

(15) $\int \operatorname{sech}^2 x\, dx = \tanh x + C,$ $\qquad \int \operatorname{csch}^2 x\, dx = -\coth x + C.$

(16) $\int \operatorname{sech} x \tanh x = -\operatorname{sech} x + C,$ $\qquad \int \operatorname{csch} x \coth x = -\operatorname{csch} x + C.$

Using (13) and (14), we could proceed as in Example 2 to draw graphs of $y = \coth x$, $y = \operatorname{sech} x$, and $y = \operatorname{csch} x$ (see Exercise 35 and **Figure 6.6**).

FIGURE 6.6

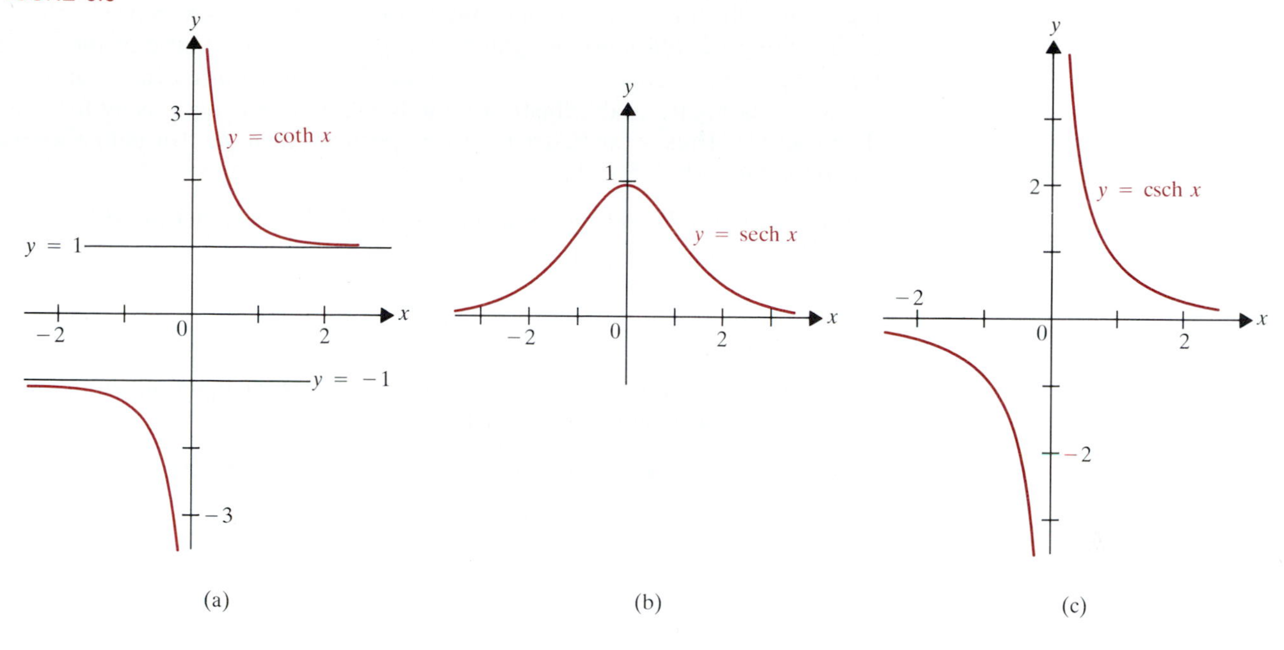

The next example illustrates the use of (15) and (16).

Example 3

Evaluate

(a) $D_x(\exp(\tanh x))$ (b) $\int \frac{\operatorname{sech}^2 x\,dx}{1 + 3\tanh x}$ (c) $\int \sinh^3 x \cosh^2 x\,dx$

Solution. (a) $D_x(\exp(\tanh x)) = [\exp(\tanh x)]\,D_x \tanh x = \operatorname{sech}^2 x \exp(\tanh x)$.

(b) Letting $u = 1 + 3\tanh x$, we have $du = 3\operatorname{sech}^2 x\,dx$, so that the numerator of the integrand is $\frac{1}{3}du$. Hence

$$\int \frac{\operatorname{sech}^2 x\,dx}{1 + 3\tanh x} = \frac{1}{3}\int \frac{du}{u} = \frac{1}{3}\ln|u| + C = \frac{1}{3}\ln|1 + 3\tanh x| + C.$$

(c) Using (3), we have

$$\begin{aligned}\sinh^3 x \cosh^2 x &= \sinh x \sinh^2 x \cosh^2 x = \sinh x(\cosh^2 x - 1)\cosh^2 x \\ &= \sinh x(\cosh^4 x - \cosh^2 x).\end{aligned}$$

Thus,

$$\begin{aligned}\int \sinh^3 x \cosh^2 x\,dx &= \int \cosh^4 x \sinh x\,dx - \int \cosh^2 x \sinh x\,dx \\ &= \frac{1}{5}\cosh^5 x - \frac{1}{3}\cosh^3 x + C. \quad \blacksquare\end{aligned}$$

Inverse Hyperbolic Functions*

The analogy between inverse hyperbolic functions and inverse trigonometric functions is most striking for the hyperbolic cosine and hyperbolic secant. As Figure 6.1 illustrates, *cosh* is *not* one-to-one on its domain. Since $\cosh x = \cosh(-x)$ by Equation (4), the graph of $y = \cosh x$ is symmetric in the y-axis. Since $D_x \cosh x = \sinh x > 0$ for $x > 0$, the function *cosh* increases for $x > 0$. Similarly, as Figure 6.6(b) illustrates, the function *sech* is decreasing for $x > 0$ (Exercise 51). Thus, if we restrict x to be positive, then we can define inverse functions for cosh and sech.

6.3 DEFINITION

(a) The ***inverse hyperbolic cosine function cosh*** $^{-1}$ is the function with domain $[1, +\infty)$ and range $[0, +\infty)$ such that

$$y = \cosh^{-1} x \qquad \text{if and only if} \qquad x = \cosh y,$$

where $x \in [1, +\infty)$ and $y \in [0, +\infty)$.

(b) The ***inverse hyperbolic secant function sech*** $^{-1}$ is the function with domain $(0, 1]$ and range $[0, +\infty)$ such that

$$y = \operatorname{sech}^{-1} x \qquad \text{if and only if} \qquad x = \operatorname{sech} y,$$

where $x \in (0, 1]$ and $y \in [0, +\infty)$.

FIGURE 6.7

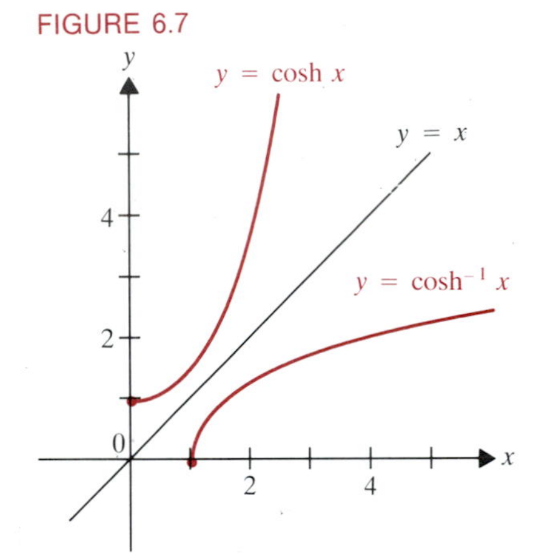

Figure 6.7 shows the graph of $y = \cosh^{-1} x$. It is obtained by reflecting the right half of the graph of $y = \cosh x$ (the half for which $x \geq 0$) in the line $y = x$. (See Exercise 52 for the graph of $y = \operatorname{sech}^{-1} x$.)

The remaining inverse hyperbolic functions are simpler to define. Referring to Figures 6.2, 6.5, and 6.6(a) and (c), we see that sinh and tanh are increasing on their entire domains $(-\infty, +\infty)$, while coth and csch are decreasing on their domains $(-\infty, 0) \cup (0, +\infty)$. Hence every one of these functions is one-to-one and so has an inverse.

6.4 DEFINITION

(a) The ***inverse hyperbolic sine*** and ***inverse hyperbolic tangent functions*** are defined by

$$y = \sinh^{-1} x \qquad \text{if and only if} \qquad x = \sinh y,$$

where x and $y \in (-\infty, +\infty)$;

$$y = \tanh^{-1} x \qquad \text{if and only if} \qquad x = \tanh y,$$

where $x \in (-1, 1)$ and $y \in (-\infty, +\infty)$.

(b) The ***inverse hyperbolic cotangent*** and ***inverse hyperbolic cosecant functions*** are defined by

$$y = \coth^{-1} x \qquad \text{if and only if} \qquad x = \coth y,$$

where $x \in (-\infty, -1) \cup (1, +\infty)$ and $y \in (-\infty, 0) \cup (0, +\infty)$;

$$y = \operatorname{csch}^{-1} x \qquad \text{if and only if} \qquad x = \operatorname{csch} y,$$

where x and $y \in (-\infty, 0) \cup (0, +\infty)$.

Figure 6.8 shows the graph of $\sinh^{-1}$ and $\tanh^{-1}$ (Exercise 53). The inverse hyperbolic functions are of interest primarily for their calculus, rather than for their graphs. So we proceed to their differentiation and integration formulas, which are derived by using the inverse function differentiation rule as in the last section. We begin with $\sinh^{-1}$.

* Optional

FIGURE 6.8

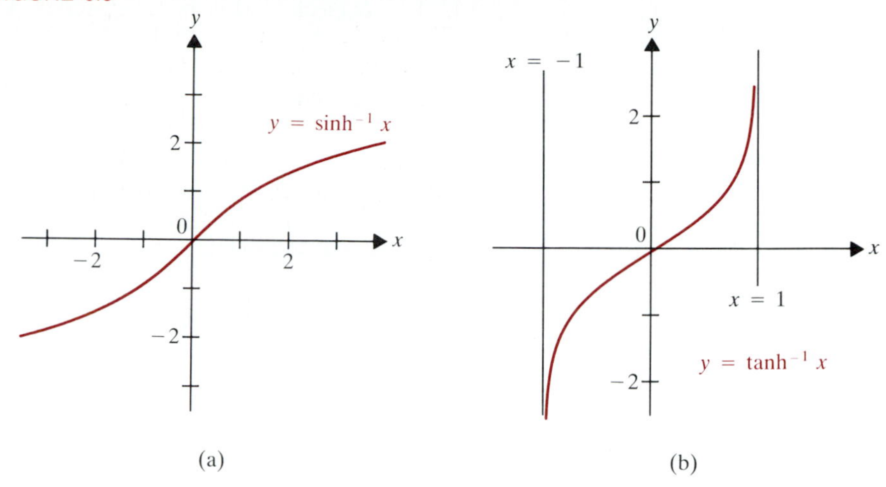

Let $y = \sinh^{-1} x$, so that $x = \sinh y$. Then Theorem 7.5 of Chapter 2 and the formula $D_x(\sinh x) = \cosh x$ give

$$(17) \qquad \frac{dy}{dx} = \frac{1}{\dfrac{dx}{dy}} = \frac{1}{\cosh y}.$$

From Definition 6.1 (or Figure 6.1), we see that $\cosh y \geq 1$ for every y. So (3) in the form

$$\cosh^2 y = 1 + \sinh^2 y$$

gives

$$\cosh y = \sqrt{1 + \sinh^2 y} = \sqrt{1 + x^2}.$$

Substituting this into (17), we find

$$(18) \qquad D_x(\sinh^{-1} x) = \frac{1}{\sqrt{x^2 + 1}} \quad \text{and} \quad \int \frac{dx}{\sqrt{x^2 + 1}} = \sinh^{-1} x + C.$$

If we let $y = \tanh^{-1} x$, then $x = \tanh y$, and from (14), $dy/dx = \operatorname{sech}^2 y$. Then the inverse function differentiation rule and (11) give

$$\frac{dy}{dx} = \frac{1}{\dfrac{dx}{dy}} = \frac{1}{\operatorname{sech}^2 y} = \frac{1}{1 - \tanh^2 y}.$$

Since $x = \tanh y$, we have

$$(19) \qquad D_x(\tanh^{-1} x) = \frac{1}{1 - x^2} \quad \text{and} \quad \int \frac{dx}{1 - x^2} = \tanh^{-1} x + C.$$

In the integration portion of (19), we must make the restriction $|x| < 1$, because $|x| < 1$ if $y = \tanh^{-1} x$, by Definition 6.4.

Similarly (see Exercise 39), we can derive the following additional differentiation and integration formulas.

$$(20) \qquad D_x(\cosh^{-1} x) = \frac{1}{\sqrt{x^2 - 1}}, \quad \int \frac{dx}{\sqrt{x^2 - 1}} = \cosh^{-1} x + C$$

$$(21) \qquad D_x(\coth^{-1} x) = \frac{1}{1 - x^2}, \quad \int \frac{dx}{1 - x^2} = \coth^{-1} x + C, \quad \text{for } |x| > 1$$

(22) $$D_x(\operatorname{sech}^{-1} x) = \frac{-1}{x\sqrt{1-x^2}}, \qquad \int \frac{dx}{x\sqrt{1-x^2}} = -\operatorname{sech}^{-1}|x| + C$$

(23) $$D_x(\operatorname{csch}^{-1} x) = \frac{-1}{|x|\sqrt{x^2+1}}, \qquad \int \frac{dx}{x\sqrt{x^2+1}} = -\operatorname{csch}^{-1}|x| + C$$

Example 4

Evaluate the following.

$$\text{(a) } D_x(\sinh^{-1}\sqrt{x^2-1}) \qquad \text{(b) } \int \frac{dx}{x^2-1}, \text{ if } x > 1$$

Solution. (a) Formula (18) and the chain rule give

$$D_x(\sinh^{-1}\sqrt{x^2-1}) = \frac{1}{\sqrt{x^2-1+1}} \cdot \frac{1}{2}(x^2-1)^{-1/2} \cdot 2x$$
$$= \frac{1}{\sqrt{x^2}} \frac{x}{\sqrt{x^2-1}} = \frac{x}{|x|\sqrt{x^2-1}}$$
$$= \frac{\operatorname{sgn}(x)}{\sqrt{x^2-1}},$$

where $\operatorname{sgn}(x)$ is the *sign* (or *signum*) function defined by

$$\operatorname{sgn}(x) = \begin{cases} 1, & \text{if } x > 0 \\ -1, & \text{if } x < 0. \end{cases}$$

(b) Since $|x| > 1$, we use the integration part of (21):

$$\int \frac{dx}{x^2-1} = -\int \frac{dx}{1-x^2} = -\coth^{-1} x + C. \quad ■$$

The fact that the hyperbolic functions are defined in terms of e^x and e^{-x} makes it possible to express the inverse hyperbolic functions in terms of the natural logarithm function. For example, if $y = \sinh^{-1} x$, then by Definition 6.4 we have

$$x = \sinh y = \frac{e^y - e^{-y}}{2} = \frac{e^{2y}-1}{2e^y}.$$

Multiplying through by $2e^y$, we get

$$2xe^y = e^{2y} - 1 \rightarrow e^{2y} - 2xe^y - 1 = 0.$$

The quadratic formula applied to this quadratic equation in e^y gives

$$e^y = \frac{2x + \sqrt{4x^2+4}}{2} \quad \text{or} \quad e^y = \frac{2x - \sqrt{4x^2+4}}{2}.$$

Only the first alternative is correct, because $e^y > 0$ and $\sqrt{4x^2+4} > 2x$. Thus

$$e^y = x + \sqrt{x^2+1} \rightarrow y = \ln(x + \sqrt{x^2+1}).$$

Thus, we have derived the formula

(24) $$\sinh^{-1} x = \ln(x + \sqrt{x^2+1}).$$

Similar reasoning (see Exercise 40) can be used to obtain the following identities.

(25)	$\cosh^{-1} x = \ln(x + \sqrt{x^2 - 1})$	for $x \geq 1$
(26)	$\tanh^{-1} x = \frac{1}{2} \ln \frac{1 + x}{1 - x}$	for $x \in (-1, 1)$
(27)	$\coth^{-1} x = \frac{1}{2} \ln \frac{x + 1}{x - 1}$	for $\lvert x\rvert > 1$
(28)	$\operatorname{sech}^{-1} x = \ln\left(\frac{1 + \sqrt{1 - x^2}}{x}\right)$	for $x \in (0, 1]$
(29)	$\operatorname{csch}^{-1} x = \ln\left(\frac{1}{x} + \frac{\sqrt{1 + x^2}}{\lvert x\rvert}\right)$	for $x \neq 0$

If we combine Formulas (24)–(29) with the integration portion of Formulas (18)–(23), then we get the following integration formulas. Like the earlier formulas, these are usually not memorized, but are found in tables of integrals like the one in Appendix I.

(30) $$\int \frac{dx}{\sqrt{x^2 + 1}} = \ln(x + \sqrt{x^2 + 1}) + C$$

(31) $$\int \frac{dx}{\sqrt{x^2 - 1}} = \ln\left|x + \sqrt{x^2 - 1}\right| + C$$

(32) $$\int \frac{dx}{1 - x^2} = \frac{1}{2} \ln\left|\frac{1 + x}{1 - x}\right| + C$$

(33) $$\int \frac{dx}{x\sqrt{1 - x^2}} = \ln\left|\frac{1 - \sqrt{1 - x^2}}{x}\right| + C = -\ln\left|\frac{1 + \sqrt{1 - x^2}}{x}\right| + C$$

(34) $$\int \frac{dx}{x\sqrt{x^2 + 1}} = \ln\left|\frac{1 - \sqrt{1 + x^2}}{x}\right| + C$$

It may seem that we have claimed too much in (31)–(34), because the restrictions on x in (25)–(29) have been dropped. But the absolute value signs in (31)–(34) compensate for this. For example, in (32),

$$\begin{aligned} D_x\left[\frac{1}{2} \ln\left|\frac{1 + x}{1 - x}\right| + C\right] &= D_x\left[\frac{1}{2} \ln|1 + x| - \frac{1}{2} \ln|1 - x| + C\right] \\ &= \frac{1}{2} \frac{1}{1 + x} - \frac{1}{2} \frac{1}{1 - x}(-1) \\ &= \frac{1}{2}\left[\frac{1}{1 + x} + \frac{1}{1 - x}\right] = \frac{1}{2}\left[\frac{1 - x + 1 + x}{1 - x^2}\right] \\ &= \frac{1}{1 - x^2} \qquad \text{for all } x \neq \pm 1. \end{aligned}$$

There are more general versions of (30)–(34), in which 1 is replaced by a^2 on the left. They are derived by letting $u = x/a$, so that $dx = a\,du$. See Exercises 49–50 for the derivation of the following formulas.

(35) $\displaystyle\int \frac{dx}{\sqrt{x^2+a^2}} = \sinh^{-1}\frac{x}{a} + C = \ln(x + \sqrt{x^2+a^2}) + C$

(36) $\displaystyle\int \frac{dx}{\sqrt{x^2-a^2}} = \cosh^{-1}\frac{x}{a} + C, \text{ for } x > a = \ln(x + \sqrt{x^2-a^2}) + C$

(37) $$\int \frac{dx}{a^2-x^2} = \begin{cases} \dfrac{1}{2a}\ln\left|\dfrac{a+x}{a-x}\right| + C & \\ \dfrac{1}{a}\tanh^{-1}\dfrac{x}{a} + C & \text{if } a > 0 \text{ and } |x| < a \\ \dfrac{1}{a}\coth^{-1}\dfrac{x}{a} + C & \text{if } a > 0 \text{ and } |x| > a \end{cases}$$

(38) $\displaystyle\int \frac{dx}{x\sqrt{a^2-x^2}} = \frac{1}{a}\ln\left|\frac{a-\sqrt{a^2-x^2}}{x}\right| + C$

(39) $\displaystyle\int \frac{dx}{x\sqrt{x^2+a^2}} = \frac{1}{a}\ln\left|\frac{a-\sqrt{x^2+a^2}}{x}\right| + C$

Our concluding example illustrates the use of this long list of integration formulas.

Example 5

Evaluate

$$\text{(a)} \int \frac{dx}{\sqrt{4x^2-9}} \qquad \text{(b)} \int_0^4 \frac{dx}{\sqrt{x^2+9}} \qquad \text{(c)} \int_0^2 \frac{dx}{9-x^2}$$

Solution. (a) Let $u = 2x$. Then $du = 2\,dx$, so $dx = 1/2\,du$ and $4x^2 = u^2$. So

$$\begin{aligned} \int \frac{dx}{\sqrt{4x^2-9}} &= \frac{1}{2}\int \frac{du}{\sqrt{u^2-3^2}} \\ &= \frac{1}{2}\cosh^{-1}\frac{u}{3} + C = \frac{1}{2}\cosh^{-1}\frac{2x}{3} + C \\ &= \frac{1}{2}\ln(u + \sqrt{u^2-a^2}) + C \qquad \text{by (36)} \\ &= \frac{1}{2}\ln(2x + \sqrt{4x^2-9}) + C. \end{aligned}$$

(b) Here we can use (35) directly, with $a = 3$. That gives

$$\begin{aligned} \int_0^4 \frac{dx}{\sqrt{x^2+9}} &= \ln(x + \sqrt{x^2+9})\Big|_0^4 = \ln(4+\sqrt{25}) - \ln(0+\sqrt{9}) \\ &= \ln 9 - \ln 3 = \ln 3. \end{aligned}$$

(c) Here (37) applies with $a = 3$ to give

$$\int_0^2 \frac{dx}{9-x^2} = \frac{1}{6}\ln\left|\frac{3+x}{3-x}\right|_0^2 = \frac{1}{6}\ln\frac{5}{1} - \frac{1}{6}\ln\frac{3}{3} = \frac{1}{6}\ln 5. \ \blacksquare$$

Exercises 6.6

In Exercises 1–14, find the derivative of the given function.

1. $\sinh(2x + 5)$

2. $\cosh\sqrt{x + 1}$

3. $\tanh\sqrt{x^2 + 4}$

4. $\coth\dfrac{1}{x^2 + 1}$

5. $\operatorname{sech} e^{x^2}$

6. $\operatorname{csch}\dfrac{1}{\sqrt{x + 1}}$

7. $\sinh \ln x$

8. $\sinh^{-1}(\sqrt{x^2 + 4})$

9. $\cosh^3(x^2 - 2)$

10. $\tanh^{-1}(2x + 1)$

11. $\tan^{-1}(\sinh(x^2 + 1))$

12. $\sin^{-1}(\tanh x)$

13. $\tanh^2 x + \operatorname{sech}^2 x$

14. $(\cosh^{-1} 3x)^2$

In Exercises 15–34, evaluate the given integral.

15. $\int x \cosh x^2\, dx$

16. $\int \cosh^3 x\, dx$

17. $\int \dfrac{\cosh x}{1 + \sinh x}\, dx$

18. $\int \dfrac{\sinh x\, dx}{\cosh^4 x}$

19. $\int \sqrt{\sinh x} \cosh x\, dx$

20. $\int \dfrac{\operatorname{sech} x \tanh x}{1 + \operatorname{sech} x}\, dx$

21. $\int \cosh^3 x \sinh^2 x\, dx$

22. $\int \sinh^5 x \cosh^2 x\, dx$

23. $\int \dfrac{\operatorname{csch}\sqrt{x + 1} \coth\sqrt{x + 1}}{\sqrt{x + 1}}\, dx$

24. $\int \tanh 2x\, dx$

25. $\int \dfrac{dx}{\sqrt{9x^2 + 25}}$

26. $\int \dfrac{dx}{\sqrt{16x^2 - 9}}$

27. $\int \dfrac{dx}{4 - x^2}$

28. $\int \dfrac{dx}{x\sqrt{4 - x^2}}$

29. $\int_0^3 \dfrac{dx}{16 - x^2}$

30. $\int_0^{a-1} \dfrac{dx}{a^2 - x^2}$

31. $\int_0^1 \dfrac{dx}{\sqrt{x^2 + 4}}$

32. $\int_0^3 \dfrac{dx}{\sqrt{x^2 + 16}}$

33. $\int_4^5 \dfrac{dx}{\sqrt{x^2 - 9}}$

34. $\int_3^5 \dfrac{dx}{\sqrt{x^2 - 4}}$

35. Sketch the graph of
(a) $y = \coth x$ **(b)** $y = \operatorname{sech} x$ **(c)** $y = \operatorname{csch} x$

36. **(a)** What is $\cosh x + \sinh x$?
(b) What is $\cosh x - \sinh x$?

37. **(a)** Derive (5). **(b)** Derive (6). **(c)** Derive (7).
(d) Derive (8). **(e)** Derive (9). **(f)** Derive (10).

38. **(a)** Derive the second part of (13).
(b) Derive the first part of (14).
(c) Derive the second part of (14)
(d) Derive the first part of (15).
(e) Derive the second part of (15)
(f) Derive the first part of (16).

39. **(a)** Derive (20) **(b)** Derive (21).
(c) Derive (22). **(d)** Derive (23).

40. **(a)** Derive (25). **(b)** Derive (26). **(c)** Derive (27).
(d) Derive (28). **(e)** Derive (29).

41. **(a)** Derive (31).
(b) Derive (32) for the case $x < 1$.
(c) Derive (33), and verify the equality of the two logarithms.
(d) Derive (34).

42. If A and B are constants, then show that $f(x) = A \cosh cx + B \cosh cx$ is a solution of the differential equation

$$f''(x) = c^2 f(x)$$

for any real number c.

43. Find the arc length of the curve $y = \cosh x$ over the interval $[0, 1]$.

44. Repeat Exercise 43 for

$$y = a \cosh \frac{x}{a} \qquad \text{over } [0, b].$$

45. Find the volume of the solid generated when the region under $y = \cosh x$ between $x = 0$ and $x = 1$ is revolved about the x-axis.

46. Repeat Exercise 45 for $y = \sinh x$.

47. **(a)** Express the shaded area A in Figure 6.4 in terms of the area of triangle OQP and the region R under the curve between $x = 1$ and $x = \cosh t$.
(b) Use part 1 of the fundamental theorem of calculus (Theorem 4.1 of Chapter 4) to show that $dA/dt = 1/2$ for all t.
(c) Conclude that $A = \frac{1}{2}t + C$ for some constant C.
(d) Show $C = 0$, so that $A = \frac{1}{2}t$.

48. In Definition 6.1, we defined $\cosh t$ and $\sinh t$ and then noted that $(\cosh t, \sinh t)$ lies on the hyperbola $x^2 - y^2 = 1$. Suppose instead that we start geometrically and define t to be twice the area of the shaded region in Figure 6.4. We can then show that $Q(x, y)$ has coordinates $x = \frac{1}{2}(e^t + e^{-t})$ and $y = \frac{1}{2}(e^t - e^{-t})$ as follows.
(a) Show that $t = x\sqrt{x^2 - 1} - 2\int_1^x \sqrt{u^2 - 1}\, du$.
(b) Use Theorem 4.1 of Chapter 4 to find dt/dx in (a).
(c) Show that $t = \int \dfrac{dx}{\sqrt{x^2 - 1}}$ and apply (31).
(d) Solve for x in (c).
(e) Find y from (d) and the fact that $x^2 - y^2 = 1$.

49. **(a)** Derive (35). **(b)** Derive (36). **(c)** Derive (37).

50. **(a)** Derive (38). **(b)** Derive (39).

51. Show that $\operatorname{sech} x$ is decreasing for $x > 0$.

52. Draw the graph of $y = \operatorname{sech}^{-1} x$.

53. Verify Figure 6.8 by plotting
(a) the graph of $y = \sinh^{-1} x$ and
(b) the graph of $y = \tanh^{-1} x$.

54. Sketch the graph of (a) $y = \coth^{-1} x$ and (b) $y = \operatorname{csch}^{-1} x$.

6.7 Some Applications*

This section is devoted to some applications of functions studied earlier in the chapter. We use them to analyze motion and load distribution, two important topics in physics and engineering. The first kind of motion we consider is *periodic motion.*

A familiar and simple type of periodic motion is that of the sweep second hand of a watch. Each revolution sweeps out a central angle of 2π radians, so the *angular speed* of the second hand is 2π rad/min. In mathematics and science, *counterclockwise* rotation is considered positive. **Figure 7.1** thus shows a clock whose second hand moves "backwards" at a constant angular speed ω radians per minute, starting from $(a, 0)$ at time $t = 0$. Thus at time t, we have

FIGURE 7.1

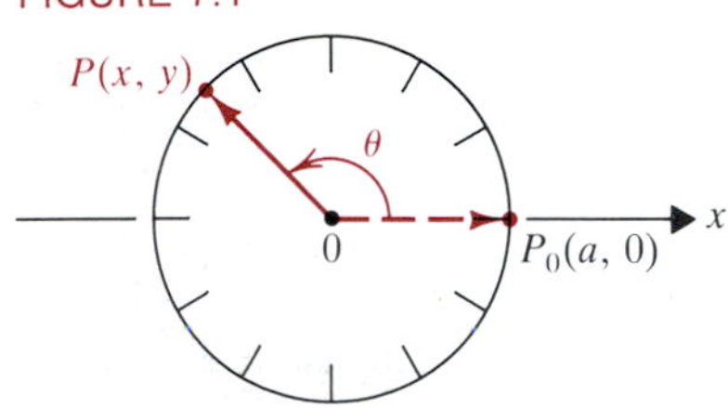

$$\theta = \omega t.$$

Hence the point P has x-coordinate

$$x = a \cos \omega t,$$

where a is the length of the second hand.

More generally, consider the same kind of motion when the second hand starts from *any* point $P_0(x_0, y_0)$ on the circle, as in **Figure 7.2.** If OP_0 makes initial angle ϕ_0 with the positive x-axis, then at time t the angle θ is given by

FIGURE 7.2

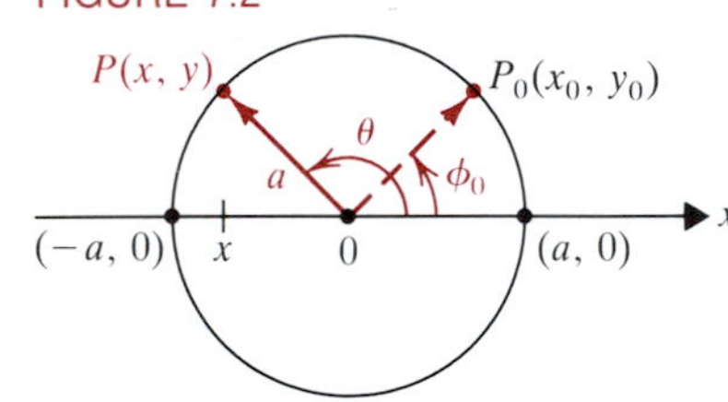

$$\theta = \omega t - \phi_0,$$

because the second hand has moved to make angle ωt with its initial position. The x-coordinate of P is therefore

$$x = a \cos \theta = a \cos(\omega t - \phi_0). \tag{1}$$

The value of x oscillates between a and $-a$ very regularly. Since one revolution of the second hand takes it through all possible positions, x is a periodic quantity with an easily determined period T. The second hand traces out its complete path as θ goes from ϕ_0 to $\phi_0 + 2\pi$, so we have

$$\omega T = 2\pi \to T = \frac{2\pi}{\omega}. \tag{2}$$

The regular oscillatory motion exhibited by P is called ***simple harmonic motion.*** The maximum value of x (in this case, a) is called the ***amplitude.*** The quantity ω is called the ***frequency,*** and as noted in (2), $T = 2\pi/\omega$ is the ***period.*** We measure ω in radians per minute (or other unit of time), although revolutions per minute and hertz (cycles per second, abbreviated Hz) are often used. The angle ϕ_0 is called the ***phase angle.***

If uniform motion around a circle were the only phenomenon in which simple harmonic motion arose, then it would hardly be of much importance. But in reality, this mathematical model applies to a number of different situations. The common mathematical thread linking these phenomena is that the same differential equation provides a mathematical model of each system. That equation is

$$\frac{d^2x}{dt^2} + \omega^2 x = 0, \tag{3}$$

where ω is a constant.

* Optional section

Although we shall not do so here, it is shown in differential equations that the function $x = x(t)$ satisfies (3) if and only if it has the form

$$x = c_1 \cos \omega t + c_2 \sin \omega t \tag{4}$$

for constants c_1 and c_2. It is easy to check that at least every equation of the form (4) does define a solution to (3). Indeed, differentiation of (4) gives

$$\frac{dx}{dt} = -c_1\omega \sin \omega t + c_2\omega \cos \omega t,$$

$$\frac{d^2x}{dt^2} = -c_1\omega^2 \cos \omega t - c_2\omega^2 \sin \omega t = -\omega^2(c_1 \cos \omega t + c_2 \sin \omega t)$$

$$= -\omega^2 x.$$

Thus

$$\frac{d^2x}{dt^2} + \omega^2 x = 0$$

for all functions of the form (4).

Equation (4) looks rather different from (1). But if x is not identically 0, that is, if c_1 and c_2 are not both 0, then we can rewrite (4) as

FIGURE 7.3

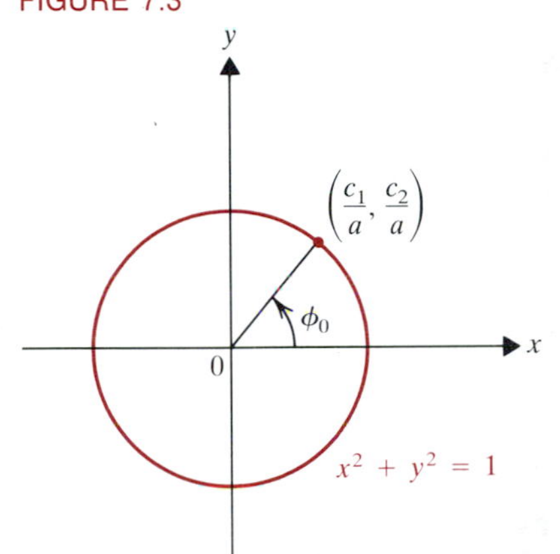

$$x = a\left(\frac{c_1}{a} \cos \omega t + \frac{c_2}{a} \sin \omega t\right), \tag{5}$$

where we let $a = \sqrt{c_1{}^2 + c_2{}^2}$. Then

$$\frac{c_1{}^2}{a^2} + \frac{c_2{}^2}{a^2} = \frac{c_1{}^2 + c_2{}^2}{a^2} = \frac{a^2}{a^2} = 1.$$

Thus $(c_1/a, c_2/a)$ is on the unit circle $x^2 + y^2 = 1$. So there is an angle ϕ_0 such that

$$\cos \phi_0 = \frac{c_1}{a} \quad \text{and} \quad \sin \phi_0 = \frac{c_2}{a},$$

as in **Figure 7.3.** Thus (5) is really

$$x = a(\cos \phi_0 \cos \omega t + \sin \phi_0 \sin \omega t) = a \cos(\omega t - \phi_0). \tag{1}$$

We sum up the foregoing discussion as follows.

> The function $x = x(t)$ is a solution of the differential equation
>
> $$\frac{d^2x}{dt^2} + \omega^2 x = 0 \tag{3}$$
>
> if and only if
>
> $$x = a \cos(\omega t - \phi_0), \tag{1}$$
>
> for constants a, ω, and ϕ_0.

Vibratory Motion

To indicate the scope of the above, we discuss some physical systems modeled by (3). First, consider the simple vibrating system in **Figure 7.4.** If the spring is stretched or compressed, and then released, the mass will oscillate back and

FIGURE 7.4

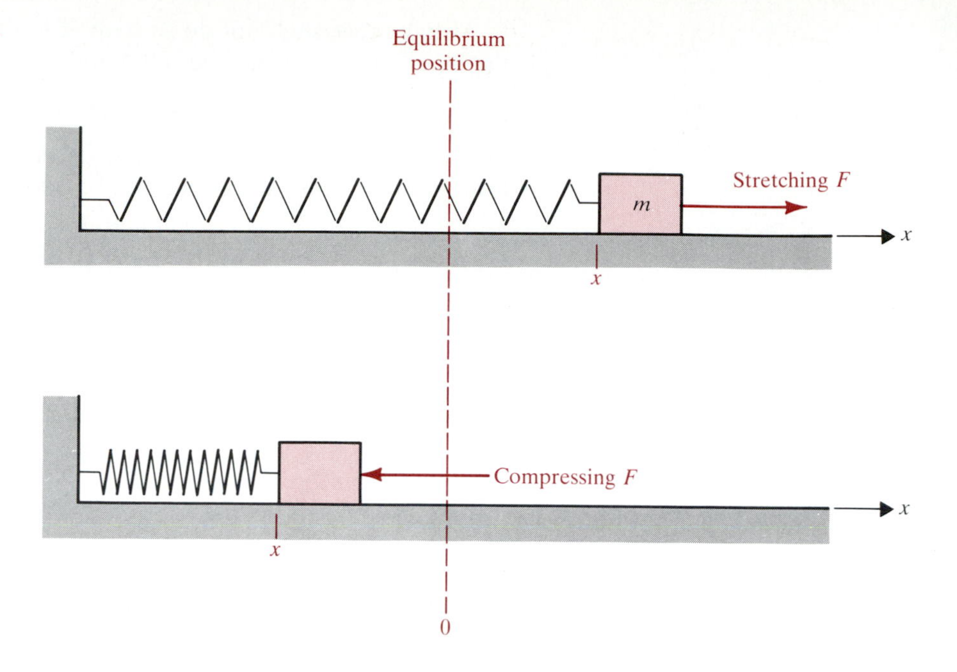

forth about its equilibrium position. Recall that Hooke's law (Section 5.6) states that the spring exerts a resisting force on the mass that is proportional to the displacement x from the equilibrium position. The resisting force acts in the direction opposite to the displacement. We follow the normal convention that forces directed toward the right are positive and those directed toward the left are negative. We then have

$$F = -hx, \tag{6}$$

where x is the directed distance from the equilibrium position to the mass at time t, and h is the spring constant. By Newton's second law of motion, the force is also the product of the mass m and the acceleration (which is d^2x/dt^2). Thus (6) becomes

$$m\frac{d^2x}{dt^2} = -hx \rightarrow \frac{d^2x}{dt^2} + \frac{h}{m}x = 0. \tag{7}$$

If we let $\omega^2 = h/m$, then (7) has the form (3). The displacement x is therefore given by an equation of the form (1).

Example 1

In Figure 7.4, suppose that $m = 3$ kg and a force of 12 newtons is required to move the mass 1 meter to the right. If the spring is compressed 1/2 meter to the left and then released (with initial velocity 0), then find a formula for the displacement x at time t.

Solution. First, we determine the spring constant h. Since a rightward directed force of 12 newtons is needed to overcome the resisting force F of the spring when $x = 1$, F is directed toward the left and

$$-12 = -h \cdot 1 \rightarrow h = 12.$$

Then $x(t)$ is given by

(1) $$x = a\cos(\omega t - \phi_0),$$

where

$$\omega^2 = \frac{h}{m} = \frac{12}{3} = 4,$$

so $\omega = 2$. From (1) we can calculate the speed

(8) $$v = \frac{dx}{dt} = -a\omega\sin(\omega t - \phi_0).$$

We are told that $x = -1/2$ and $v = 0$ when $t = 0$. Putting this information into (1) and (8), we have

(9) $$-\frac{1}{2} = a\cos(-\phi_0) = a\cos\phi_0,$$

$$0 = 2a\sin(-\phi_0) = -2a\sin\phi_0 \rightarrow 0 = \sin\phi_0.$$

Thus $\phi_0 = \sin^{-1} 0 = 0$. Substituting this into (9), we get

$$-\frac{1}{2} = a\cos 0 = a.$$

Thus the equation of the motion (1) is

$$x = -\frac{1}{2}\cos 2t. \blacksquare$$

Electrical Circuits

A phenomenon that appears totally different from oscillatory motion but which is also modeled by (3) is the current flow in the simple electrical circuit shown in **Figure 7.5.** An electromotive force (such as a battery) is connected to an inductor and a capacitor. A law due to the German physicist Gustav R. Kirchhoff (1824–1887) states that the voltage E of the electromotive force (EMF) must equal the sum of the voltage drops across the capacitor and the inductor. Laboratory measurements show that the voltage drop across the inductor is

FIGURE 7.5

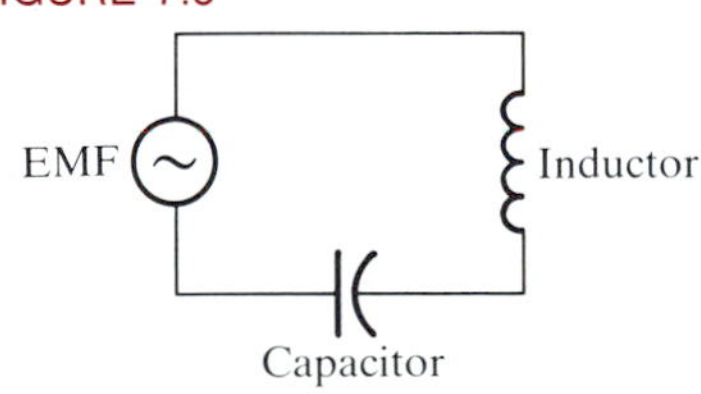

$$L\frac{di}{dt},$$

where i is the current (in amps) and L is the inductance (in henrys) of the inductor. Similar measurements show that the voltage drop across the capacitor is

$$\frac{1}{C}q(t),$$

where C is the capacitance in farads and $q(t)$ is the charge on the capacitor. The current $i = i(t)$ at time t is defined to be dq/dt. Hence

(10) $$L\frac{di}{dt} + \frac{1}{C}q(t) = E.$$

To get an equation involving just i we differentiate (10). Assuming that the voltage E, the inductance L, and the capacitance C are all constant, we find

$$L\frac{d^2i}{dt^2} + \frac{1}{C}i = 0 \rightarrow \frac{d^2i}{dt^2} + \frac{1}{LC}i = 0. \tag{11}$$

Letting $\omega^2 = 1/LC$, we see that (11) has the form (3). Thus the current is given by an equation of the form (1):

$$i(t) = a\cos\left(\frac{1}{\sqrt{LC}}t - \phi_0\right). \tag{12}$$

Therefore the current is periodic.

Example 2

In Figure 8.5, suppose that $E = 120.0$ volts, $L = 2.500 \times 10^{-1}$ henrys, and $C = 4.000 \times 10^{-4}$ farads. Find ω, ϕ_0, and a if $i = 10.00$ amps and $di/dt = 1$ when $t = 0$.

Solution. Here $LC = 10^{-4}$, so $\omega = 1/\sqrt{LC} = 10^2$. When $t = 0$, (12) gives

$$i(0) = a\cos(-\phi_0) = a\cos\phi_0 \rightarrow 10 = a\cos\phi_0. \tag{13}$$

To use the given initial value of di/dt, we differentiate (12), getting

$$\frac{di}{dt} = -\frac{a}{\sqrt{LC}}\sin\left(\frac{1}{\sqrt{LC}}t - \phi_0\right).$$

Thus when $t = 0$, we have

$$1 = -\frac{a}{100}\sin(-\phi_0) = \frac{a}{100}\sin\phi_0 \rightarrow 100 = a\sin\phi_0. \tag{14}$$

If we divide (14) by (13), we find

$$10 = \tan\phi_0 \rightarrow \phi_0 = \tan^{-1} 10 \approx 1.471.$$

Then (13) and a calculator give

$$a = \frac{10}{\cos\phi_0} \approx \frac{10}{0.0995} \approx 100.5.$$

Thus in this circuit, the current is given approximately by

$$i = 100.5\cos(100t - 1.471).$$

The Hanging Cable

A problem of fundamental importance in engineering is determining the shape assumed by a hanging cable. Suppose that the cable is flexible and of uniform density. In **Figure 7.6** it is shown hanging between points B and C located h units above the x-axis. We have denoted the lowest point A and a typical point between A and C by $P(x, y)$. The cable does not move, because the forces on it are in equilibrium. To analyze those forces, engineers consider the part of the cable between A and P, shown in color in Figure 7.6. There is a downward force F_2 acting on this arc, whose magnitude is ws, where w is the weight of one unit of the cable, and s is the arc length from A to P. In addition, there is tension in the cable produced by the supports at B and C. That is due to two tangential

FIGURE 7.6

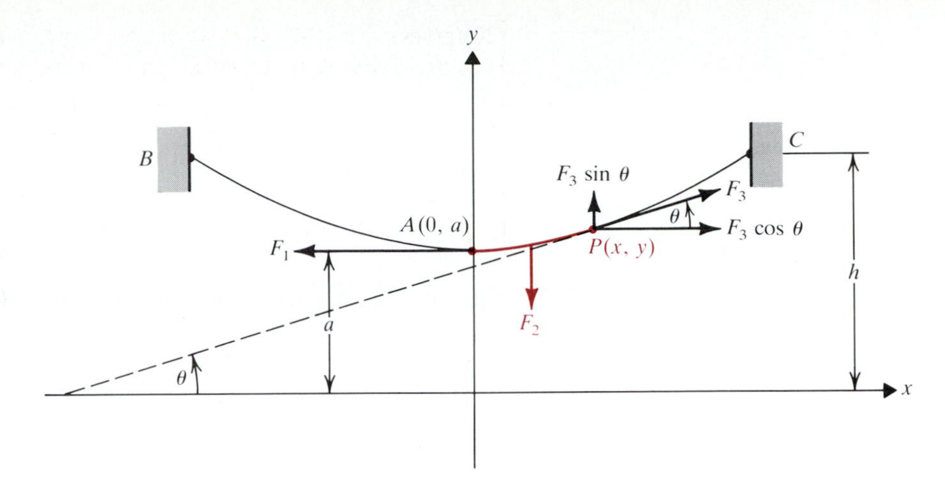

forces F_1 and F_3 acting in the direction of the cable, one at A and the other at P. (The tangential force F_1 is horizontal, since A is the lowest point on the curve.) To be in equilibrium, the sum of all the horizontal components of these forces is 0, and the sum of all the vertical components is 0. The horizontal component of F_3 is $F_3 \cos\theta$, where θ is the angle F_3 makes with the horizontal. Thus

$$-F_1 + F_3 \cos\theta = 0 \rightarrow F_1 = F_3 \cos\theta. \tag{15}$$

The vertical component $F_3 \sin\theta$ of F_3 must similarly balance the downward force $F_2 = -ws$, so

$$-ws + F_3 \sin\theta = 0 \rightarrow ws = F_3 \sin\theta. \tag{16}$$

Dividing (16) by (15), and letting $b = F_1/w$, we obtain

$$\tan\theta = \frac{ws}{F_1} = \frac{s}{b}. \tag{17}$$

Now $\tan\theta$ is the slope of the tangent to the curve at $P(x, y)$. Hence

$$\frac{dy}{dx} = \frac{s}{b} = \frac{1}{b}\int_0^x \sqrt{1 + \left(\frac{dy}{dt}\right)^2}\, dt. \tag{18}$$

Assume that dy/dt is continuous. Then differentiation of (18) gives

$$\frac{d^2y}{dx^2} = \frac{1}{b}\sqrt{1 + \left(\frac{dy}{dx}\right)^2} \tag{19}$$

by part 1 of the fundamental theorem of calculus (Theorem 4.1 of Chapter 5). We find the equation of the curve assumed by the cable by solving (19).

First, let $v = dy/dx$. Then (19) becomes

$$\frac{dv}{dx} = \frac{1}{b}(1 + v^2)^{1/2} \rightarrow \frac{dv}{(1 + v^2)^{1/2}} = \frac{1}{b}\,dx \rightarrow \int \frac{dv}{(1 + v^2)^{1/2}} = \int \frac{1}{b}\,dx,$$

$$\sinh^{-1} v = \frac{x}{b} + C \tag{20}$$

by (18) in Section 6 (p. 385). When $x = 0$, we have $v = dy/dx = 0$, because we are at the lowest point A, where the tangent is horizontal. Since $\sinh^{-1} 0 = 0$,

(20) gives

$$0 = \frac{0}{b} + C \rightarrow C = 0.$$

Thus (20) becomes

$$\sinh^{-1} v = \frac{x}{b}. \tag{21}$$

Applying the hyperbolic sine function to each side of (21), we obtain

$$v = \sinh \frac{x}{b} \rightarrow \frac{dy}{dx} = \sinh \frac{x}{b},$$

$$y = \int \sinh \frac{x}{b}\, dx \rightarrow y = b \cosh \frac{x}{b} + K. \tag{22}$$

When $x = 0$, we have $y = a$. Thus

$$a = b \cosh 0 + K = b + K.$$

Hence $K = a - b$, and (22) becomes

$$y = b \cosh \frac{x}{b} + (b - a). \tag{23}$$

This is therefore the equation of the curve in Figure 7.6.

FIGURE 7.7

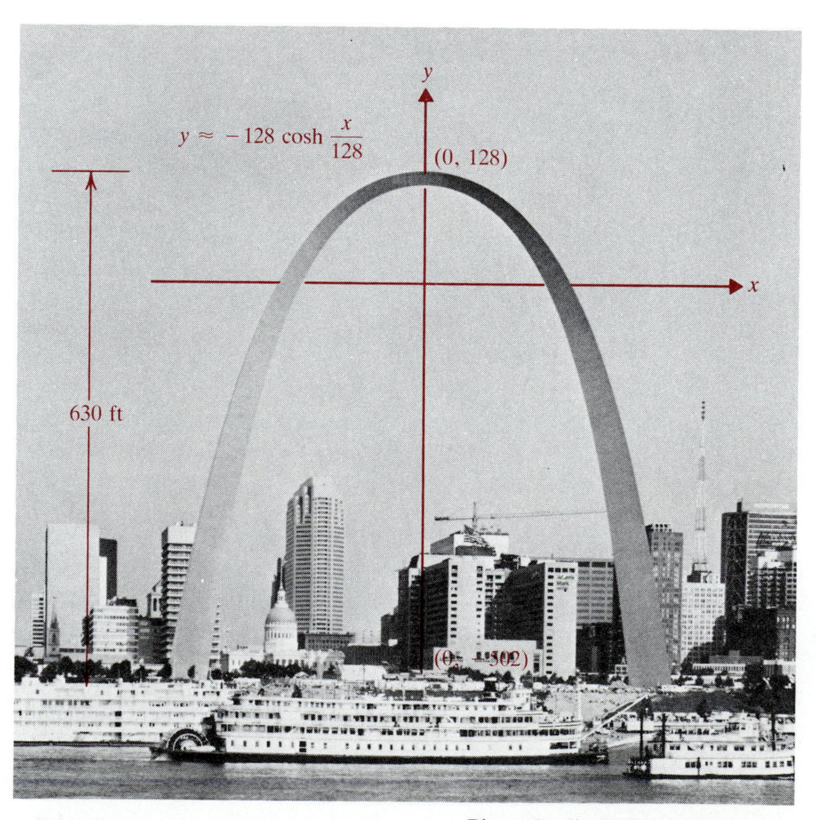

Photo Credit: St. Louis Convention and Visitors Commission.

If we relocate the x-axis, so that A is $b = F_1/w$ units above that axis, then $a = b$ in (23), so the equation becomes

$$y = a\cosh\frac{x}{a},$$

which is the equation of the *catenary*, a word derived from the Latin word for chain. A hanging chain or cable between two posts assumes this shape, as do telephone and electrical power lines. Utility engineers use this to analyze the strength and placement of utility poles. It may be of interest that the Gateway Arch in St. Louis, Missouri, which was erected to commemorate the westward migration in the United States in the nineteenth century, is a catenary with $a \approx -128$ feet. See **Figure 7.7.**

While it might be supposed that the shape of a hanging cable under a load would be described by a more complicated curve, it turns out that the opposite is true. In Exercises 21 and 22, you can work through that analysis. This applies, for instance, to the supporting cables of a suspension bridge that support the roadway. See **Figure 7.8.**

FIGURE 7.8

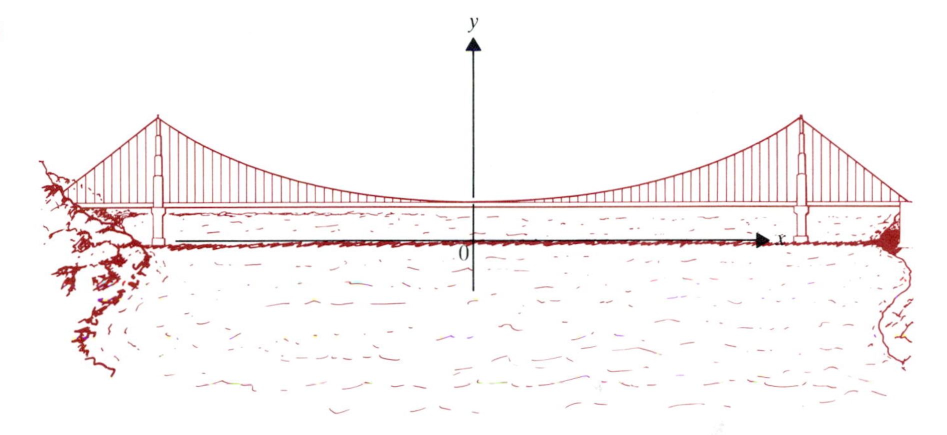

> HISTORICAL NOTE
>
> The catenary was first introduced by Leibniz in 1691. It was studied in detail by James Bernoulli (1654–1705) shortly thereafter. Bernoulli was one of a family of Swiss mathematicians. (We will mention his brother in the next section.) He introduced the word *integral* in place of the term *summatrix*, which had been used by Leibniz.

Exercises 6.7

1. Verify that if x is the x-coordinate of a point moving on a circle of radius a with constant angular speed ω, then x satisfies (3).

2. **(a)** Find an equation for the y-coordinate of a point moving on a circle of radius a with constant angular speed ω.
(b) In part (a), show that y moves in simple harmonic motion. What is the phase angle?

3. **(a)** In uniform circular motion, show that the velocity v_x of the x-coordinate of the moving point satisfies $v_x = -\omega y$.
(b) Find a similar expression for v_y.
(c) Show that $\sqrt{v_x{}^2 + v_y{}^2} = \omega a$, so it is constant.

4. **(a)** In uniform circular motion, show that the acceleration a_x of the x-coordinate of the moving point satisfies $a_x = -\omega^2 x$.

(b) Find a similar expression for a_y.
(c) Show that $\sqrt{a_x^2 + a_y^2} = \omega^2 a$, so it is constant.

5. Suppose that a particle moves counterclockwise on a circle under simple harmonic motion. If at time t (in minutes) $x(t) = \frac{1}{3}\cos 8t$, then
   (a) What is the radius of the circle?
   (b) What is the initial phase angle ϕ_0?
   (c) What is the angular velocity?
   (d) How long is required for the particle to move from its initial position on the x-axis to a point halfway in toward the y-axis?
6. Repeat Exercise 5 if $x(t) = 3\cos 4t$.
7. Suppose that the system in Figure 7.4 has $m = 2$ kg and requires a force of 8 newtons to move the mass 1 meter to the left. If the spring is stretched 1 meter to the right and then released (with initial velocity 0), then find a formula for $x(t)$.
8. Repeat Exercise 7 if $m = 1$ kg, an 18-newton force is needed to move the mass 2 m to the right, and the spring is stretched 1 m to the left and then released with 0 initial velocity.
9. Repeat Exercise 7 if $m = 4$ kg, an 8-newton force is needed to move the mass 1/2 m to the left, and the spring is compressed 1/2 m to the left and set in motion with an initial velocity of $v_0 = -1$.
10. Repeat Exercise 7 if $m = 3$ kg, a 6-newton force is needed to move the mass 1/2 m to the right, and the spring is stretched 1 m to the right and set in motion with an initial velocity $v_0 = \sqrt{3}$.
11. In Figure 7.5, suppose that $E = 115.0$ volts, $L = 2.000 \times 10^{-1}$ henrys, and $C = 5.000 \times 10^{-4}$ farads. Find ω, ϕ_0, and a if at $t = 0$, $i = 12.00$ amps and $di/dt = 2.000$ amps/second.
12. Repeat Exercise 11 if $E = 110.0$ volts, $L = 4.000 \times 10^{-1}$ henrys, and $C = 2.500 \times 10^{-4}$ farads.
13. Repeat Exercise 11 if $E = 120.0$ volts, $L = 4.000 \times 10^{-1}$ henrys, $C = 40.00 \times 10^{-4}$ farads and at $t = 0$, $i = 6.000$ amps and $di/dt = 2.000$ amps/second.
14. Repeat Exercise 11 if $E = 117.0$ volts, $L = 2.000 \times 10^{-1}$ henrys, $C = 20.00 \times 10^{-4}$ farads, and when $t = 0$, $i =$ 8.000 amps and $di/dt = 3.000$ amps/second.
15. A *simple pendulum* is an idealized physical system made up of a point mass hanging from a fixed support by means of a cord of negligible weight. See **Figure 7.9.** When pulled to one side and released, the pendulum swings back and forth periodically, along the arc of the circle of radius L shown in the figure. Let $s(t)$ be the arc length of the circle from 0 to m at time t.
   (a) Show that $s = L\theta$, so $ds/dt = L(d\theta/dt)$.
   (b) The *kinetic energy* of the pendulum at time t is $\frac{1}{2}mv^2$, where v is the speed of the pendulum as it moves along its arc. Show that the kinetic energy is $\frac{1}{2}mL^2(d\theta/dt)^2$.

FIGURE 7.9

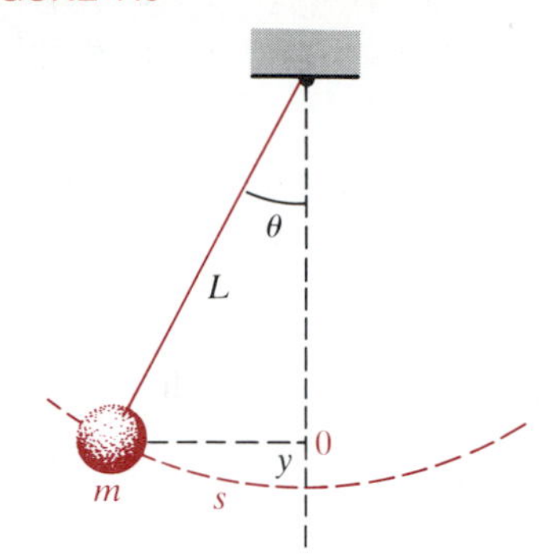

16. In Exercise 15, the *potential energy* of the pendulum is the product of the weight mg of the mass and the vertical displacement y of the mass above its low point 0. Show that the potential energy is $mgL - mgL\cos\theta$.
17. The *law of conservation of energy* in physics states that the sum of the potential energy and kinetic energy is constant. Use the results of Exercises 15(b) and 16 and implicit differentiation to show that this law gives
$$\frac{d^2\theta}{dt^2} + \frac{g}{L}\sin\theta = 0.$$
18. In Exercise 17, argue that for small values of θ, Theorem 3.10 of Chapter 2 says that $\sin\theta$ and θ differ by a negligible amount. Conclude that the motion of a simple pendulum is very close to being simple harmonic with $\omega = \sqrt{g/L}$, if θ is kept small.
19. Use the result of Exercise 18 to show that the approximate period of the pendulum is $T = 2\pi\sqrt{L/g}$.
20. Explain how the result of Exercise 19 provides an experimental method of determining g.
21. Consider the cable in **Figure 7.10,** which supports a load, as in a suspension bridge. We assume that the load is uniformly distributed horizontally between B and C, so that the weight between 0 and p that is supported is $W = kx$ for a constant k.
   (a) Show that $F_1 = F_3\cos\theta$.
   (b) Show that $F_2 = kx = F_3\sin\theta$.

FIGURE 7.10

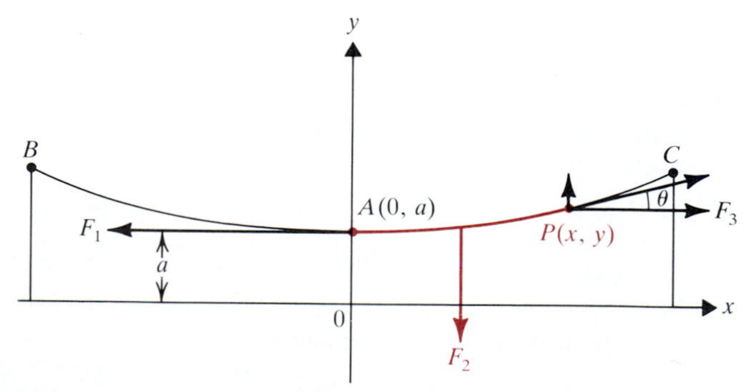

22. **(a)** Proceed as in the analysis preceding (18) to show that the differential equation relating y and x in Exercise 21 is $dy/dx = kx/F_1$.
(b) Letting $b = F_1/k$, solve the equation in part (a).
(c) What shape do the supporting cables of a suspension bridge therefore assume?

23. For a hanging cable in the shape of a catenary

$$y = a\cosh\frac{x}{a},$$

find the length of the portion between $x = 0$ and $x = t$.

24. What is the maximum sag in the catenary

$$y = a\cosh\frac{x}{a}$$

between $x = -c$ and $x = c$?

25. For the catenary

$$y = a\cosh\frac{x}{a},$$

we can calculate the tension produced by F_3 at $P(x, y)$ as follows. Refer to Figure 7.6 and the discussion preceding (17).
(a) Show that $F_3 = F_1 \sec\theta$.
(b) Show that $\tan\theta = \sinh\frac{x}{a}$.
(c) Use the identity $\sec^2\theta = 1 + \tan^2\theta$ and part (b) to show that $\sec\theta = \cosh\frac{x}{a}$.
(d) Show that $F_3 = wa\cosh\frac{x}{a}$.
(e) Conclude that the tension at $P(x, y)$ is the weight wy of a length of cable y units long.

26. Use the result of Exercise 25(e) to show that the cable is in equilibrium if at points B and C having height h above the x-axis, exactly h units of cable are draped over each support and fall straight to the x-axis.

6.8 Indeterminate Forms / L'Hôpital's Rule

A limit that cannot be evaluated by simple substitution is called an *indeterminate form.* The limits

$$\lim_{x\to 2}\frac{x^2-4}{x-2}, \qquad \lim_{x\to 0}\frac{\sin x}{x}, \qquad \text{and} \qquad \lim_{x\to +\infty}\frac{x^2-2x+1}{3x^2+x-3}$$

are all indeterminate forms that we learned how to evaluate in Sections 1.4, 2.3, and 3.6. (The limits are respectively 4, 1, and 1/3.) Probably from considering such examples, around 1691 the Swiss mathematician John Bernoulli (1667–1748) discovered a rule for evaluating

$$\lim_{x\to c}\frac{f(x)}{g(x)}$$

when f and g are differentiable functions that both approach either 0 or $\pm\infty$ as $x \to c$. In the first two examples above, both numerator and denominator approach 0 as $x \to c$ (=2 and 0, respectively). In the third example, both numerator and denominator approach $+\infty$ as $x \to +\infty$. We say the first two limits have the ***indeterminate form 0/0*** and the third limit has the ***indeterminate form ∞/∞.*** (The symbols 0/0 and ∞/∞ are shorthand notation to indicate the limits of the numerator and denominator. *These symbols have no numerical meaning whatever.*)

In the first two cases, the remarkable fact is that

$$\lim_{x\to c}\frac{f(x)}{g(x)} = \lim_{x\to c}\frac{f'(x)}{g'(x)}.$$

That is, the limit of the ratio of $f(x)$ to $g(x)$ is the same as the limit of the ratio of the derivatives! For we have

$$\lim_{x\to 2}\frac{x^2-4}{x-2} = \lim_{x\to 2}\frac{2x}{1} = 4,$$

and

$$\lim_{x\to 0} \frac{\sin x}{x} = \lim_{x\to 0} \frac{\cos x}{1} = 1.$$

In the third case, it turns out that

$$\lim_{x\to +\infty} \frac{f'(x)}{g'(x)} = \lim_{x\to +\infty} \frac{2x - 2}{6x + 1}$$

is also of indeterminate form ∞/∞, but the latter limit is equal to

$$\lim_{x\to +\infty} \frac{f''(x)}{g''(x)} = \lim_{x\to +\infty} \frac{2}{6} = \frac{1}{3}.$$

The fact that three limits as different as these all have answers that can be obtained in the same way suggests the following result.

8.1 THEOREM

L'Hôpital's Rule.

(a) Suppose that f and g are differentiable on (a, b) except possibly at $x = c$. Suppose also that $\lim_{x\to c} f(x) = \lim_{x\to c} g(x) = 0$, but $g(x) \neq 0$ on (a, b) when $x \neq c$. Let L be either a real number or $+\infty$ or $-\infty$.

$$\text{If } \lim_{x\to c} \frac{f'(x)}{g'(x)} = L, \text{ then } \lim_{x\to c} \frac{f(x)}{g(x)} = L = \lim_{x\to c} \frac{f'(x)}{g'(x)}. \tag{1}$$

(b) Statement (a) remains true if $x \to c^+$ or $x \to c^-$.

(c) Statement (a) remains true if $c = +\infty$ [respectively, $-\infty$] and f and g are differentiable on $(a, +\infty)$ [respectively, $(-\infty, b)$].

(d) Statement (a) remains true if $\lim_{x\to c} f(x)$ is $+\infty$ or $-\infty$ and $\lim_{x\to c} g(x)$ is $+\infty$ or $-\infty$.

A formal proof of (a) and (b), based on an extension of the mean value theorem for derivatives, is given later in the section. But it is easy to see the idea behind (1) directly from the definition of derivative when $f(c) = g(c) = 0$. For then we have

$$\lim_{x\to c} \frac{f(x)}{g(x)} = \lim_{x\to c} \frac{f(x) - f(c)}{g(x) - g(c)} = \lim_{x\to c} \frac{\dfrac{f(x) - f(c)}{x - c}}{\dfrac{g(x) - g(c)}{x - c}}.$$

Now let $x = c + h$. Then $x - c = h$, and $x \to c$ is equivalent to $h \to 0$, so we get

$$\lim_{x\to c} \frac{f(x)}{g(x)} = \lim_{h\to 0} \frac{\dfrac{f(c + h) - f(c)}{h}}{\dfrac{g(c + h) - g(c)}{h}} = \frac{f'(c)}{g'(c)}.$$

We now give some examples on the use of l'Hôpital's rule and how it can be applied to other types of indeterminate forms.

Example 1

Evaluate the following limits.

$$\text{(a)}\ \lim_{x\to 1}\frac{x-1}{\ln x} \qquad \text{(b)}\ \lim_{x\to 0+}\frac{\cos x - 1}{\sin x - x} \qquad \text{(c)}\ \lim_{x\to +\infty}\frac{x^2}{e^x}$$

Solution. (a) This limit has indeterminate form 0/0. Theorem 8.1(a) applies to give

$$\lim_{x\to 1}\frac{x-1}{\ln x} = \lim_{x\to 1}\frac{1-0}{1/x} = \frac{1}{1} = 1.$$

(b) This also has the indeterminate form 0/0. Theorem 8.1(b) gives

$$\lim_{x\to 0+}\frac{\cos x - 1}{\sin x - x} = \lim_{x\to 0+}\frac{-\sin x - 0}{\cos x - 1},$$

which is still of the form 0/0. A second application of Theorem 8.1(b) gives

$$\lim_{x\to 0+}\frac{\cos x - 1}{\sin x - x} = \lim_{x\to 0+}\frac{-\cos x}{-\sin x} = +\infty,$$

since $\cos x \to 1$ and $\sin x \to 0^+$ as $x \to 0^+$.

(c) This is of the form ∞/∞, so Theorem 8.1(c) applies. We have

$$\lim_{x\to +\infty}\frac{x^2}{e^x} = \lim_{x\to +\infty}\frac{2x}{e^x}.$$

This is still of the form ∞/∞, so we apply Theorem 8.1(c) again to get

$$\lim_{x\to +\infty}\frac{x^2}{e^x} = \lim_{x\to +\infty}\frac{2}{e^x} = 0. \blacksquare$$

L'Hôpital's rule can apply to further indeterminate forms. If $f(x) \to \pm\infty$ and $g(x) \to 0$ as $x \to c$, then $\lim_{x\to c} f(x)g(x)$ is said to have the ***indeterminate form*** $\boldsymbol{\infty \cdot 0}$. If $f(x) \to +\infty$ and $g(x) \to -\infty$ as $x \to c$, then $\lim_{x\to c}[f(x) + g(x)]$ is said to have the ***indeterminate form*** $\boldsymbol{\infty - \infty}$. (Again, the symbols $0 \cdot \infty$ and $\infty - \infty$ have *no* numerical interpretation.) To apply l'Hôpital's rule to the corresponding limits, we rewrite the limit so that it has the form 0/0 or ∞/∞.

Example 2

Evaluate

$$\text{(a)}\ \lim_{x\to 0+} x^2 \ln x^2 \qquad \text{(b)}\ \lim_{x\to 0+}\left(\csc x - \frac{1}{x}\right).$$

Solution. (a) This has the form $0 \cdot \infty$ but can be transformed to the form ∞/∞ by writing

$$\lim_{x\to 0+} x^2\ln x^2 = \lim_{x\to 0+}\frac{\ln x^2}{1/x^2} = \lim_{x\to 0+}\frac{\left(\frac{1}{x^2}\right)\cdot 2x}{-2/x^3} \quad \textit{by Theorem 8.1(d)}$$

$$= \lim_{x\to 0+}(-x^2) = 0.$$

(b) This has the form $\infty - \infty$, but transforms to the form 0/0 if we write

$$\lim_{x\to 0^+}\left(\csc x - \frac{1}{x}\right) = \lim_{x\to 0^+}\left(\frac{1}{\sin x} - \frac{1}{x}\right) = \lim_{x\to 0^+}\frac{x - \sin x}{x \sin x}$$

$$= \lim_{x\to 0^+}\frac{1 - \cos x}{\sin x + x\cos x} \qquad \textit{by Theorem 8.1(a)}$$

$$= \lim_{x\to 0^+}\frac{\sin x}{\cos x + \cos x - x\sin x} \qquad \textit{by Theorem 8.1(a) again}$$

$$= \frac{0}{2 - 0} = 0. \blacksquare$$

Warning. *Always check that the hypotheses of l'Hôpital's rule are satisfied before trying to apply it.* Some students, in their enthusiasm for it, try to use the rule when it does not apply. Consider, for instance,

$$\lim_{x\to \pi/2^-}\frac{1 + \cos x}{\sin x} = \frac{1 + 0}{1} = 1.$$

If we tried (erroneously) to apply l'Hôpital's rule here, we would get the *incorrect* answer

$$\lim_{x\to \pi/2^-}\frac{-\sin x}{\cos x} = -\infty!$$

There are three more indeterminate forms that l'Hôpital's rule can handle: $\mathbf{0^0}$, $\mathbf{\infty^0}$, and $\mathbf{1^\infty}$, which are again symbols with no numerical meaning. Such forms arise as limits of functions of the form

$$f(x) = g(x)^{h(x)}. \tag{2}$$

To evaluate such a limit, take the natural logarithm of both sides of (2), to get

$$\ln f(x) = h(x)\ln g(x). \tag{3}$$

As Exercise 56 asks you to check, Equation (3) leads to the indeterminate form $\infty \cdot 0$, which we evaluate as in Example 2(a). If we find that for some real number L,

$$\lim_{x\to c}\ln f(x) = L,$$

then we can calculate $\lim_{x\to c} f(x)$ by using the continuity of the natural exponential function exp. Since $f(x) = \exp(\ln f(x))$, we have

$$\lim_{x\to c} f(x) = \lim_{x\to c}\exp(\ln f(x)) = \exp L = e^L, \tag{4}$$

by Theorem 5.3 of Chapter 1. Similarly, if $\lim_{x\to c}\ln f(x) = +\infty$, then

$$f(x) = \exp(\ln f(x)) \to +\infty \qquad \text{as } x \to c. \tag{5}$$

Finally, if $\lim_{x\to c}\ln f(x) = -\infty$, then

$$f(x) = \exp(\ln f(x)) \to 0 \qquad \text{as } x \to c. \tag{6}$$

We can summarize this discussion as follows.

ALGORITHM FOR $\lim_{x \to c} f(x)$ WHEN $f(x) = g(x)^{h(x)}$ TENDS TO THE INDETERMINATE FORM 0^0, ∞^0, OR 1^∞ AS $x \to c$

1. Find $\lim_{x \to c} \ln f(x) = L$ by l'Hôpital's rule.
2. (a) If L is a real number, then $\lim_{x \to c} f(x) = e^L$.
   (b) If $L = +\infty$, then $\lim_{x \to c} f(x) = +\infty$.
   (c) If $L = -\infty$, then $\lim_{x \to c} f(x) = 0$.

This is a *two-step* procedure. After first finding $\lim_{x \to c} \ln f(x) = L$ by l'Hôpital's rule, we have to use (4), (5), or (6) to translate that information into an answer for $\lim_{x \to c} f(x)$.

The next example illustrates how to carry out that procedure. The first limit is one we met in Example 3 of Section 4, where the graph suggested that the limit is 1. The second limit *should* turn out to be e, because Theorem 2.5 says that

$$e = \lim_{t \to 0^+} (1 + t)^{1/t}.$$

If we let $x = 1/t$, then $1/x = t$ and $t \to 0^+$ is equivalent to $x \to +\infty$.

Example 3

Find the following limits.

$$\text{(a)} \lim_{x \to 0^+} x^x \qquad \text{(b)} \lim_{x \to +\infty} \left(1 + \frac{1}{x}\right)^x$$

$$\text{(c)} \lim_{x \to +\infty} (e^{-x})^{1/\ln x} \qquad \text{(d)} \lim_{x \to +\infty} \left(1 + \frac{1}{x}\right)^{x^2}$$

FIGURE 8.1

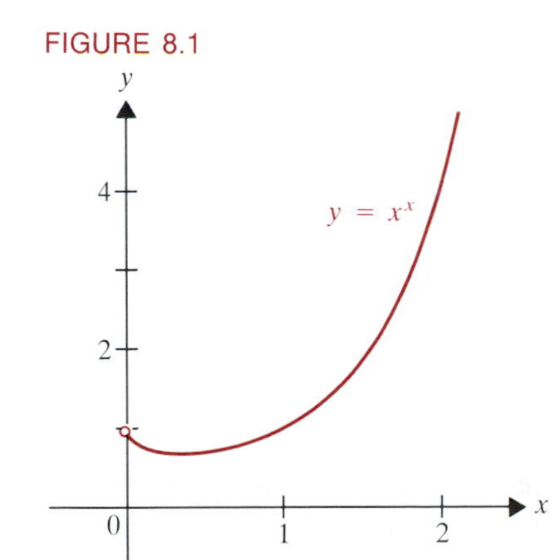

Solution. (a) This limit has the indeterminate form 0^0. Letting $f(x) = x^x$, we have $\ln f(x) = x \ln x$. Then

$$\lim_{x \to 0^+} \ln f(x) = \lim_{x \to 0^+} x \ln x = \lim_{x \to 0^+} \frac{\ln x}{1/x}$$

$$= \lim_{x \to 0^+} \frac{1/x}{-1/x^2} \qquad \textit{by Theorem 8.1(a)}$$

$$= \lim_{x \to 0^+} (-x) = 0.$$

Therefore (4) says that $\lim_{x \to 0^+} f(x) = e^0 = 1$, as suggested by **Figure 8.1.**

(b) This limit has the indeterminate form 1^∞. Let

$$f(x) = \left(1 + \frac{1}{x}\right)^x.$$

Then as $x \to +\infty$,

$$\ln f(x) = x \ln\left(1 + \frac{1}{x}\right) = \frac{\ln\left(1 + \frac{1}{x}\right)}{\frac{1}{x}},$$

assumes the indeterminate form 0/0. Using Theorem 8.1(c), we get

$$\lim_{x\to+\infty} \ln f(x) = \lim_{x\to+\infty} \frac{\dfrac{1}{1+1/x}\cdot(-1/x^2)}{-1/x^2} = \lim_{x\to+\infty} \frac{1}{1+\dfrac{1}{x}} = 1.$$

Thus $\lim_{x\to+\infty} f(x) = e^1 = e$, as we said should be the case.

(c) This has the indeterminate form 0^0 since as $x \to +\infty$, $\ln x \to +\infty$, and so $1/\ln x \to 0$. Also, as $x \to +\infty$, $e^{-x} = 1/e^x$ approaches 0. Letting $f(x) = (e^{-x})^{1/\ln x}$ we have

$$\ln f(x) = \frac{1}{\ln x}\ln e^{-x} = -\frac{x}{\ln x}.$$

Therefore

$$\begin{aligned}\lim_{x\to+\infty} \ln f(x) &= -\lim_{x\to+\infty} \frac{x}{\ln x} \\ &= -\lim_{x\to+\infty}\frac{1}{1/x} && \textit{by Theorem 8.1(d)} \\ &= -\infty.\end{aligned}$$

We conclude from (6) that

$$\lim_{x\to+\infty} f(x) = 0.$$

(d) As in (b), this has the form 1^∞. Letting

$$f(x) = \left(1+\frac{1}{x}\right)^{x^2},$$

we find

$$\ln f(x) = x^2 \ln\left(1+\frac{1}{x}\right).$$

As $x \to +\infty$, this assumes the indeterminate form $\infty \cdot 0$. Then

$$\begin{aligned}\lim_{x\to+\infty} \ln f(x) &= \lim_{x\to+\infty} \frac{\ln\left(1+\dfrac{1}{x}\right)}{x^{-2}} && \text{has indeterminate form } 0/0 \\ &= \lim_{x\to+\infty} \frac{\dfrac{1}{1+\dfrac{1}{x}}\cdot\left(-\dfrac{1}{x^2}\right)}{-2/x^3} && \textit{by Theorem 8.1(c)} \\ &= \lim_{x\to+\infty} \frac{1}{2}\cdot\frac{x}{1+\dfrac{1}{x}} \\ &= \frac{1}{2}\lim_{x\to+\infty}\frac{x^2}{x+1} && \text{of indeterminate form } \infty/\infty \\ &= \frac{1}{2}\lim_{x\to+\infty}\frac{2x}{1} && \textit{by Theorem 8.1(d)} \\ &= +\infty.\end{aligned}$$

Therefore from (5), $\lim_{x\to+\infty} f(x) = +\infty$. ■

The Cauchy Mean Value Theorem and Proof of L'Hôpital's Rule*

To prove the first two parts of Theorem 8.1, we need the following generalization of the mean value theorem for derivatives (Theorem 10.2 of Chapter 2).

8.2 THEOREM **_Cauchy Mean Value Theorem._** Suppose that f and g are continuous on $[a, b]$ and differentiable on (a, b) and that $g'(x) \neq 0$ on (a, b). Then there is a point p in (a, b) such that

$$\frac{f(b) - f(a)}{g(b) - g(a)} = \frac{f'(p)}{g'(p)} \tag{7}$$

Before proving this, we note that in case g is the function given by $g(x) = x$, then Theorem 8.2 becomes the ordinary mean value theorem for derivatives. For in that case, $g'(x) = 1$ for every x, so (7) is the assertion that

$$\frac{f(b) - f(a)}{b - a} = \frac{f'(p)}{1} = f'(p)$$

for some p in (a, b).

Proof. First, from $g'(x) \neq 0$ on (a, b), it follows that $g(b) \neq g(a)$, because if $g(b) - g(a)$ were 0, then by the ordinary mean value theorem, there would be some point $x_0 \in (a, b)$ such that

$$g'(x_0) = \frac{g(b) - g(a)}{b - a} = 0.$$

Thus at least (7) makes sense. Our proof resembles that of the mean value theorem for derivatives: We apply Rolle's theorem (Theorem 10.1 of Chapter 2) to a suitably chosen function. It is not difficult (Exercise 51) to discover that the proper choice for h is given by

$$h(x) = f(x) - f(a) - \frac{f(b) - f(a)}{g(b) - g(a)}[g(x) - g(a)]. \tag{8}$$

This function is clearly differentiable on (a, b) and continuous on $[a, b]$. Moreover, substitution of $x = a$ and $x = b$ into (8) shows that $h(b) = h(a) = 0$. Thus Rolle's theorem guarantees the existence of a point p in (a, b) such that $h'(p) = 0$. Differentiating (8) at $x = p$, we get

$$h'(p) = f'(p) - \frac{f(b) - f(a)}{g(b) - g(a)} g'(p) = 0 \to f'(p) = \frac{f(b) - f(a)}{g(b) - g(a)} g'(p). \tag{9}$$

Division of the second equation in (9) by $g'(p) \neq 0$ gives (7). QED

We next give a proof of the first two parts of l'Hôpital's rule.

Proof of Theorem 8.1(a), (b). We prove part (b) first. In the case $x \to c^+$, we are interested in the behavior of $f(x)/g(x)$ for x in an interval (c, b). By hypothesis, $\lim_{x \to c^+} f(x) = 0 = \lim_{x \to c^+} g(x)$. If the functions aren't initially defined at $x = c$, we can define $f(c) = g(c) = 0$ and thereby make f and g continuous at $x = c$. Also, since f and g are differentiable on (a, b) except possibly at $x = c$, they are continuous there by Theorem 1.3 of Chapter 2. Hence, for any x in (c, b), f and g satisfy the hypotheses of the Cauchy mean value theorem on $[c, x]$. Thus

* Optional

there is some point p in (c, x) such that

$$\frac{f(x) - f(c)}{g(x) - g(c)} = \frac{f'(p)}{g'(p)} = \frac{f(x)}{g(x)},$$

because $f(c) = g(c) = 0$. Then

$$\lim_{x \to c^+} \frac{f(x)}{g(x)} = \lim_{p \to c^+} \frac{f'(p)}{g'(p)} = L.$$

A similar argument applies to the case $x \to c^-$ (Exercise 55). Thus (b) is proved. Now (a) follows since if

$$\lim_{x \to c} \frac{f'(x)}{g'(x)} = L,$$

then

$$\lim_{x \to c^+} \frac{f'(x)}{g'(x)} = \lim_{x \to c^-} \frac{f'(x)}{g'(x)} = L,$$

so (b) gives

$$\lim_{x \to c} \frac{f(x)}{g(x)} = L. \quad \text{QED}$$

Part (c) of Theorem 8.1 can be proved by using the result of part (a) and the chain rule (Exercises 52 and 53), but the proof of Theorem 8.1(d) is more sophisticated, and we leave that for advanced calculus.

Accuracy of the Tangent Approximation*

Lest it appear that the Cauchy mean value theorem is of interest only as a means of proving l'Hôpital's rule, we close the section by using it to obtain an error estimate for the tangent approximation. Recall that if $f'(a)$ exists, then for x near a the tangent approximation to $f(x)$ is $T(x)$, where

(10) $$f(x) \approx T(x) = f(a) + f'(a)(x - a).$$

By Theorem 5.3 of Chapter 2, the error $e(x)$ made in (10) has a smaller order of magnitude than $x - a$, that is,

$$\lim_{x \to a} \frac{e(x)}{x - a} = 0,$$

where

(11) $$e(x) = f(x) - T(x) = f(x) - f(a) - f'(a)(x - a).$$

But we never gave a formula for estimating the size of $e(x)$, that is, for estimating the accuracy of the tangent approximation. Thanks to the Cauchy mean value theorem, we can now do so for most functions f studied in this text.

8.3 THEOREM Suppose that $f''(x)$ exists for all x in an open interval I containing a. If K is an upper bound for $|f''(x)|$ on I, that is, if $|f''(x)| \leq K$ for all x in I, then

(12) $$|e(x)| \leq \frac{1}{2} K(x - a)^2$$

for all x in I.

* Optional

Proof. From (11) we have

$$e'(x) = f'(x) - f'(a), \tag{13}$$

so $e'(a) = 0$. Let t be any point in I. Then the functions $e(x)$ and $g(x) = (x - a)^2$ are both continuous on the closed interval $\bar{J}$ and differentiable on the open interval J with endpoints a and t. Also $g'(x) = 2(x - a) \neq 0$ when $x \neq a$. Hence the Cauchy mean value theorem applies to show that for some c between a and t,

$$\frac{e(x) - e(a)}{g(x) - g(a)} = \frac{e'(c)}{g'(c)} \rightarrow \frac{e(x)}{(x-a)^2} = \frac{e'(c)}{2(c-a)} = \frac{e'(c) - e'(a)}{2(c-a)}.$$

Now the functions $e'(x)$ and $h(x) = 2(x - a)$ are differentiable on $\bar{J}$ and continuous on J. Also $h'(x) = 2 \neq 0$. So we can apply the Cauchy mean value theorem to (15). Noting from (13) that $e''(x) = f''(x)$, we get

$$\frac{e(x)}{(x-a)^2} = \frac{e''(z)}{2} = \frac{1}{2}f''(z)$$

for some z between a and c. Then z is between a and x (since c is), and so

$$e(x) = \frac{1}{2}f''(z)(x-a)^2$$

for some z between a and x. Notice that $z \in I$ since $x \in I$, and

$$|e(x)| = \frac{1}{2}|f''(z)|(x-a)^2. \tag{14}$$

Since $|f''(x)| \leq K$ for all x in I, $|f''(z)| \leq K$. Substituting this into (14), we obtain the desired inequality (12):

$$|e(x)| \leq \frac{1}{2}K(x-a)^2. \quad \boxed{\text{QED}}$$

Theorem 8.3 is a precursor of an important result about polynomial approximation of differentiable functions in Chapter 9. For now we can use it to estimate the accuracy of the tangent approximation.

Example 4

Find the tangent approximation for $f(x) = \sin x$ near $x = \pi/3$. Estimate its accuracy in general and at $x = \pi/3 + \pi/180$ $(=61^\circ)$.

Solution. Since $f'(x) = \cos x$, we have $f'(\pi/3) = 1/2$. Recalling that $\sin \pi/3 = \sqrt{3}/2$, we have

$$f(x) \approx T(x) = f(\pi/3) + f'(\pi/3)(x - \pi/3),$$

$$T(x) = \frac{\sqrt{3}}{2} + \frac{1}{2}\left(x - \frac{\pi}{3}\right).$$

To estimate the accuracy of this approximation, we calculate

$$f''(x) = -\sin x.$$

Certainly $|f''(x)| = |\sin x| \leq 1$, so that by Theorem 8.3

$$|e(x)| \leq \frac{1}{2} \cdot 1\left(x - \frac{\pi}{3}\right)^2 = \frac{1}{2}\left(x - \frac{\pi}{3}\right)^2.$$

At $x = \pi/3 + \pi/180$, we have

$$|e(x)| \le \frac{1}{2}\left(\frac{\pi}{180}\right)^2 \approx 0.0001523 < 0.0005.$$

Thus the tangent approximation is accurate to at least three decimal places when x is within 1° of $a = \pi/3$. ■

As a check, we calculate

$$T\left(\frac{\pi}{3} + \frac{\pi}{180}\right) = \frac{1}{2}\sqrt{3} + \frac{1}{2}\cdot\frac{\pi}{180} \approx 0.87475,$$

which rounds to 0.875 to three decimal places. A scientific calculator gives

$$\sin 61° \approx 0.874619707,$$

which also rounds off to 0.875. The error in $T(\pi/3 + \pi/180)$ is about

$$0.00013 = 0.87475 - 0.87462,$$

which is, indeed, less than the upper bound 0.0001523 calculated in Example 4.

HISTORICAL NOTE

Theorem 8.1 is not called *Bernoulli's rule*, because Bernoulli did not publish his development of it. Instead, the result appeared in print in the first calculus text ever written, *Analysis of the Infinitely Small for Understanding Curves*, published in 1696. As you may have guessed, the author was Guillaume F. A. de l'Hôpital (1661–1704). He was a French nobleman (a marquis) who hired Bernoulli to teach him calculus in the early 1690s. Bernoulli included this result in a letter to de l'Hôpital in 1694. Cauchy used his mean value theorem to provide the first rigorous derivation of l'Hôpital's rule in 1829. Theorem 8.3 was also derived by Cauchy in his text, which was titled *Lessons on Differential Calculus*. That result was, however, derived (somewhat less rigorously) by Joseph L. Lagrange (1736–1813) in 1797 using methods found in Chapter 9.

Exercises 6.8

In Exercises 1–42, evaluate the given limit.

1. $\lim_{x\to 0} \frac{e^x - x - 1}{x^2}$

2. $\lim_{x\to 0} \frac{e^x - e^{-x}}{\sin x}$

3. $\lim_{x\to 1} \frac{\ln x}{x^2 - 1}$

4. $\lim_{x\to \pi/2} \frac{1 - \sin x}{\cos x}$

5. $\lim_{x\to 0} \frac{1 - \cos x}{x^2}$

6. $\lim_{x\to \pi/2} \frac{1 - \sin x}{(1/2) - x}$

7. $\lim_{x\to 0} \frac{e^x}{x^2}$

8. $\lim_{x\to 0} \frac{1 - e^x}{1 - x}$

9. $\lim_{x\to +\infty} \frac{e^x}{x^3 + 1}$

10. $\lim_{x\to +\infty} \frac{x^2 - 3x}{e^x}$

11. $\lim_{x\to 0} \frac{\sin x}{\tan^{-1} x}$

12. $\lim_{x\to \pi/2} \frac{\tan^{-1} x}{\cos x}$

13. $\lim_{x\to 0^+} \frac{\tan x}{x^2}$

14. $\lim_{x\to 0^+} \frac{x^2}{\sin x}$

15. $\lim_{x\to 0^+} x^2 \ln x$

16. $\lim_{x\to +\infty} x^2 e^{-x}$

17. $\lim_{x\to \pi/2} (2x - \pi)\sec x$

18. $\lim_{x\to \pi} (x - \pi)\csc x$

19. $\lim_{x\to 0^+} x\ln(\sin x)$

20. $\lim_{x\to 0^+} x\ln(\tan x)$

21. $\lim_{x\to \infty} x\sin\frac{1}{x}$

22. $\lim_{x\to +\infty} x^2\sin\frac{1}{x}$

23. $\lim_{x\to 0} x\cot x$

24. $\lim_{x\to 0} x^2\sec x$

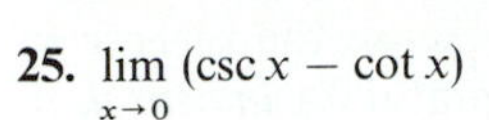

25. $\lim_{x\to 0} (\csc x - \cot x)$

26. $\lim_{x\to 1} \left(\dfrac{1}{\ln x} - \dfrac{1}{x-1}\right)$

27. $\lim_{x\to\pi/2} (\sec x - \tan x)$

28. $\lim_{x\to 0^+} (\cot x - \ln x)$

29. $\lim_{x\to+\infty} (x - \sinh x)$

30. $\lim_{x\to+\infty} (\cosh x - x)$

31. $\lim_{x\to 0^+} (\sin x)^{\tan x}$

32. $\lim_{x\to 0^+} x^{\sin x}$

33. $\lim_{x\to 0} (1 - x)^{1/x}$

34. $\lim_{x\to 0} (2x + 1)^{1/x}$

35. $\lim_{x\to+\infty} x^{1/x}$

36. $\lim_{x\to+\infty} (\ln x)^{1/x}$

37. $\lim_{x\to-\infty} (e^x)^{1/\ln x^2}$

38. $\lim_{x\to+\infty} (e^{-x^2})^{1/\ln x}$

39. $\lim_{x\to+\infty} \left(1 + \dfrac{1}{x}\right)^{x^3}$

40. $\lim_{x\to+\infty} \left(1 - \dfrac{1}{x}\right)^{x^2}$

41. $\lim_{x\to 0^+} (\cot x)^x$

42. $\lim_{x\to\pi/2^-} (\tan x)^{\cos x}$

In Exercises 43–50, obtain an upper bound for the error $e(x)$ in the tangent approximation to the given function near the given point a.

43. $f(x) = x^2$, $a = 2$

44. $f(x) = 1 - x^2$, $a = 3$

45. $f(x) = \cos x$, $a = \pi/4$

46. $f(x) = \tan x$, $a = \pi/4$

47. $f(x) = e^x$, $a = 0$, on $[-1, 1]$

48. $f(x) = \cosh x$, $a = 0$, on $(-1, 1)$

49. $f(x) = \ln x$, $a = 1$, on $(1/2, 3/2)$

50. $f(x) = (1 + x^2)^{1/2}$, $a = 0$, on $(-1/2, 1/2)$

51. In Theorem 8.2, we needed a function h such that $h(a) = h(b) = 0$. Since $h(x) = f(x) - f(a) - m(g(x) - g(a))$ satisfies $h(a) = 0$, we need only choose m so that $h(b) = 0$. Show that m *must* be chosen to be $\dfrac{f(b) - f(a)}{b - a}$.

52. To prove Theorem 8.1c, consider the functions $F(t) = f(1/t)$ and $G(t) = g(1/t)$.
(a) Show that $\lim_{t\to 0^+} F(t) = \lim_{x\to+\infty} f(x)$ and $\lim_{t\to 0^+} G(t) = \lim_{x\to+\infty} g(x)$.
(b) Show that $F'(t) = -\dfrac{1}{t^2} f'\left(\dfrac{1}{t}\right) = f'(x)$ and $G'(t) = -\dfrac{1}{t^2} g'\left(\dfrac{1}{t}\right) = g'(x)$.
(c) Show that if $\lim_{x\to+\infty} \dfrac{f'(x)}{g'(x)} = L$, then $L = \lim_{x\to+\infty} \dfrac{f(x)}{g(x)}$.

53. Repeat Exercise 52 for the case $x \to -\infty$.

54. Consider the function f defined by

$$f(x) = \begin{cases} e^{-1/x^2} & \text{for } x \neq 0 \\ 0 & \text{for } x = 0 \end{cases}$$

(a) Show that $f'(0) = \lim_{h\to 0} \dfrac{e^{-1/h^2}}{h}$.
(b) Show that $f'(0) = 0$. (*Hint:* Make the change of variable $k = 1/t$.)

55. Prove Theorem 8.1(b) for the case $x \to c^-$.

56. Show that if $f(x) = g(x)^{h(x)}$ and $\lim_{x\to c} f(x)$ is one of the indeterminate forms 0^0, ∞^0, or 1^∞, then $\lim_{x\to c} \ln f(x)$ has the indeterminate from $\infty \cdot 0$.

6.9 Log-log and Semi-log Graphs*

Because so many relationships in science and engineering are logarithmic or exponential in nature, special graphing techniques based on logarithmic scaling of one or both of the coordinate axes have been developed. One advantage of these techniques is the simplification they provide for graphs of logarithmic or exponential functions, which turn into *straight lines* when plotted on log-log or semi-log graph paper. Laboratory scientists, in search of functions to model observed data mathematically, frequently plot their data on such graph paper. In that way they can often easily find a logarithmic or exponential modeling function that might otherwise be difficult to discover. This section introduces these graphical tools and explains their mathematical basis.

FIGURE 9.1

First, recall that graphs are drawn to display the relationship between the two variables x and y. A glance at a graph like that of **Figure 9.1** reveals a great deal about the behavior of y over a considerable portion of the x-axis: y increases to a maximum at c and thereafter declines. If y represents a company's sales and x represents time, for instance, then the company has suffered a fairly severe sales loss since time c.

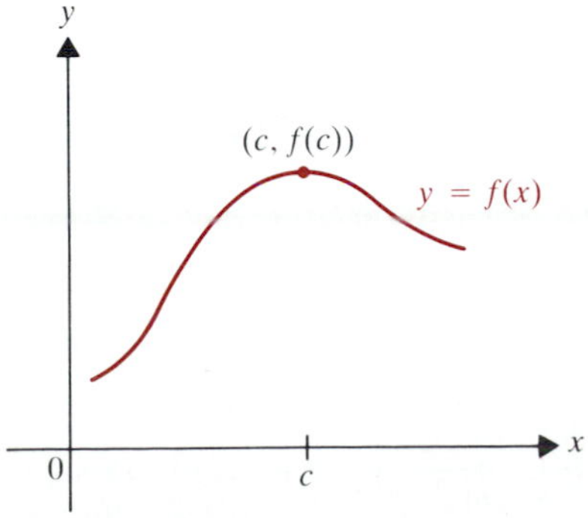

* Optional section

Scientific graphing is not only different but is in some sense the *reverse* of mathematical graphing. Scientists often analyze experimental data *in search of a function* f to serve as a fairly accurate mathematical model. The coordinate axes are scaled so that all data over a meaningful domain *and* range of values can be shown clearly in a reasonably sized graph. In addition, it is most helpful if the plotted points suggest a mathematical relationship between the quantities under study.

There are three common scaling schemes. The first is the one we have already used many times. The units on one axis are a constant multiple k of those on another axis. In this section we introduce the second scheme, *logarithmic* scaling. In that system the point labeled x is really $\log_{10} x$ or $\ln x$ units from the origin. The origin itself, then, is marked 1 rather than 0. In **Figure 9.2** the axis is labeled $\ln x$ rather than x. Thus the point labeled 100 is actually

$$\ln 100 = \ln 10^2 = 2 \ln 10$$

units from the origin. It is therefore twice as far from the origin as is the point labeled 10. (For the third scheme, see Exercises 1–4.)

FIGURE 9.2

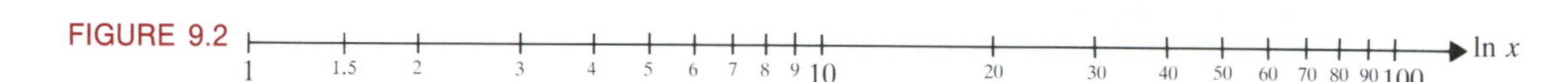

The scaling on a logarithmic axis can be to *any* base $b \neq 1$ in $(0, +\infty)$. Since, for example,

$$\log_{10} x = \frac{\ln x}{\ln 10} = \frac{1}{\ln 10} \ln x,$$

use of base 10 in place of base e merely multiplies each coordinate by the constant $1/\ln 10$. In commercially scaled logarithmic graph paper, the base b is usually not specified, so you can use any convenient base.

Commercial logarithmic graph paper labels the logarithmic axis or axes slightly differently from Figure 9.2. (See **Figure 9.3.**) Since, for example,

$$\log 20 = \log(2 \cdot 10) = \log 2 + \log 10, \tag{1}$$

the second 2 in Figure 9.3 means that the corresponding point P is at distance $\log 2$ to the right of the point labeled 10. Similarly, the 2 to the right of the second 10 means that point Q is at distance $\log 2$ to the right of the second 10. Thus Q represents

$$\log 200 = \log 2 + \log 100. \tag{2}$$

We have omitted the base b in (1) and (2), because any positive base $b \neq 1$ is as good as any other. In science, b is commonly thought of as either 10 or e, but it really doesn't matter because (1) and (2) hold regardless of which base b is used.

Logarithmic scaling is very useful for displaying exponential relationships. For instance, consider the law of exponential growth and decay,

$$P = P_0 e^{kx}. \tag{3}$$

FIGURE 9.3

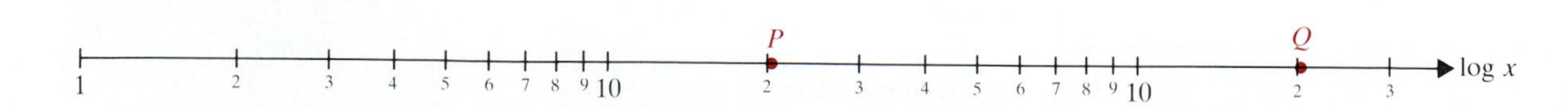

If we take the natural logarithm of each side in (3), we obtain

(4) $$\ln P = \ln(P_0 e^{kx}) = \ln P_0 + \ln e^{kx} = kx + \ln P_0 \to \ln P = kx + B,$$

where $B = \ln P_0$ is a constant. If we scale the y-axis logarithmically and the x-axis as usual, then $y = \ln P$ represents distances along the y-axis. Hence (4) becomes

(5) $$y = kx + B,$$

whose graph is a *straight line of slope k*, the growth rate, and *y-intercept B*, the initial size. Graph paper with one axis scaled uniformly and the other axis scaled logarithmically is called *semi-logarithmic* (or *semi-log*) *paper*.

Example 1

Graph the exponential growth function

(6) $$f(x) = 5e^{0.2x}$$

on semi-log paper.

Solution. We transform (6) to the form (5):

(7) $$y = (0.2)x + B,$$

where $y = \ln f(x)$ and $B = \ln 5$. The graph of (7) is a straight line with slope $m = 0.2$ and y-intercept B. That line passes through the point $(0, \ln 5)$, and when $x = 5$, we have

$$y = 1 + \ln 5 = \ln e + \ln 5 = \ln 5e.$$

We can draw the line simply by connecting $(0, \ln 5)$ and $(5, \ln 5e)$. Note that $5e \approx 13.59$. The graph is shown in **Figure 9.4.** ■

FIGURE 9.4

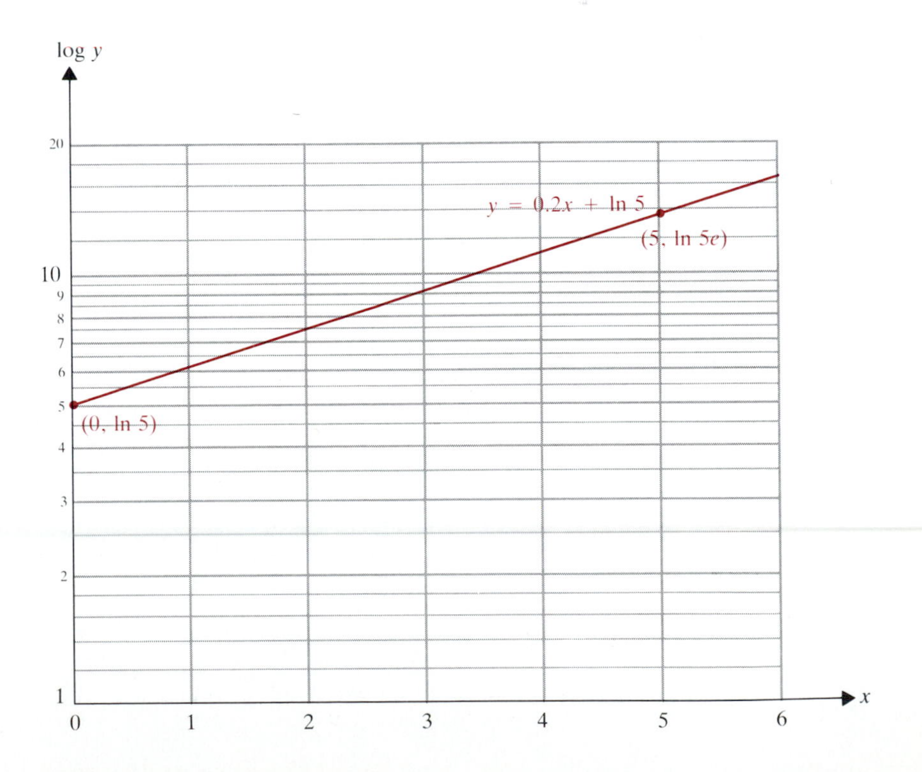

It is standard laboratory practice to plot experimental data on semi-log graph paper, to see whether the observed phenomenon can be modeled as an exponential growth or decay process. If the data points lie approximately on a straight line on semi-log paper, then it is easy to find an approximate modeling function.

Example 2

Graph the data points from the following table on semi-log graph paper. Does the relationship between y and x appear to be exponential? If so, find an approximate modeling function f such that $y \approx f(x)$.

x	0	1	2	3	4	5	6	7	8	9	10
y	15	11	8.0	6.0	4.5	3.3	2.5	1.8	1.4	1	0.75

Solution. **Figure 9.5** shows that the data points appear to lie almost exactly on a straight line when plotted on semi-log paper. That line has an equation of the form (4), with y playing the role of P:

$$\ln y = kx + B. \tag{8}$$

Here k is the slope of the line. Since the line passes through points $(0, \ln 15)$ and $(9, \ln 1)$ on the semi-log paper, the slope k of the line is

$$k = \frac{\ln 1 - \ln 15}{9 - 0} = -\frac{1}{9}\ln 15 \approx -0.30.$$

FIGURE 9.5

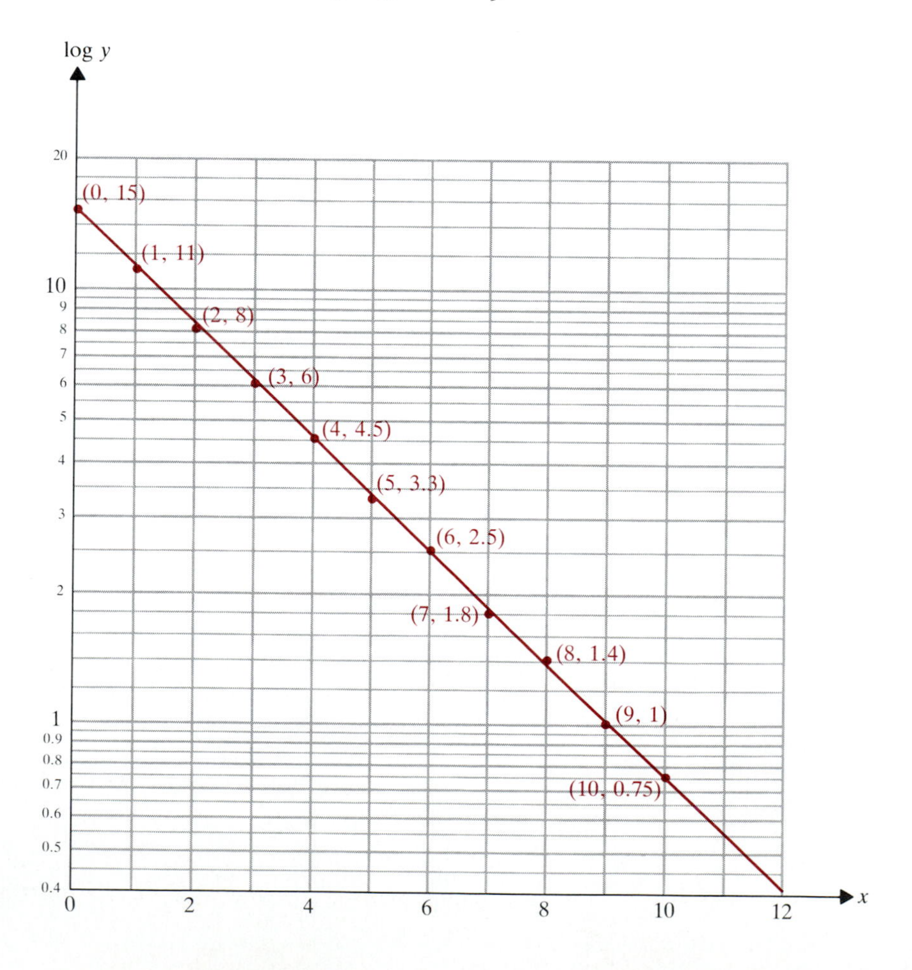

Also in (8), $B = \ln 15$. So (8) says that

$$\ln y = -0.30x + \ln 15 \rightarrow \ln y - \ln 15 = -0.30x,$$

$$\ln \frac{y}{15} = -0.30x \rightarrow \frac{y}{15} = e^{-0.30x} \rightarrow y = 15e^{-0.30x}. \tag{9}$$

If we check (9) against the given table of values, then we find a *very* close correlation:

x	0	1	2	3	4	5	6	7	8	9	10
$y = 15e^{-0.30x}$	15	11.1	8.23	6.10	4.52	3.35	2.48	1.84	1.36	1.01	0.747

Thus we have found a mathematical model for f that is *quite* close to the relationship exhibited by the data. ■

In some circumstances it is helpful to scale *both* axes logarithmically. Graph paper on which that has been done is called *log-log paper* and is also available commercially. Such paper is used in graphing *allometric* functional relations, those of the form

$$y = A(x) = cx^k, \tag{10}$$

where k and c are constants. These functions are difficult to graph on standard Cartesian coordinate graphs unless k is very small. The reason is that, when k is positive, $A(x)$ becomes very large in absolute value very quickly (that is, for relatively small values of x). When k is negative, $A(x)$ quickly approaches 0 as x increases in absolute value, making the graph hard to distinguish from the x-axis—except near the origin, where it is hard to distinguish from the y-axis! If we take logarithms in (10), though, things improve substantially. We get

$$\ln y = \ln(c \cdot x^k) = \ln c + \ln x^k = \ln c + k \ln x.$$

If we introduce logarithmic scaling on each axis by letting $u = \ln x$ and $v = \ln y$, then we have

$$v = ku + C, \tag{11}$$

where $C = \ln c$. The graph of (11) is a straight line with slope k (the exponent) and y-intercept $C = \ln c$.

Example 3

Graph $y = 2x^6$ on log-log paper.

Solution. Taking logarithms, we obtain

$$\ln y = \ln 2 + 6 \ln x \rightarrow v = 6u + \ln 2, \tag{12}$$

where $v = \ln y$ and $u = \ln x$. On log-log paper, the graph of (12) is the line through $(\ln 2, \ln 128)$ with y-intercept $\ln 2$ (See **Figure 9.6**). Note that when $x = 2$, we have

$$\ln y = 6 \ln 2 + \ln 2 = 7 \ln 2 = \ln 2^7 = \ln 128. \quad ■$$

As with semi-log paper, scientists often plot experimental data on log-log paper to see whether an allometric function could be used to model the data.

FIGURE 9.6

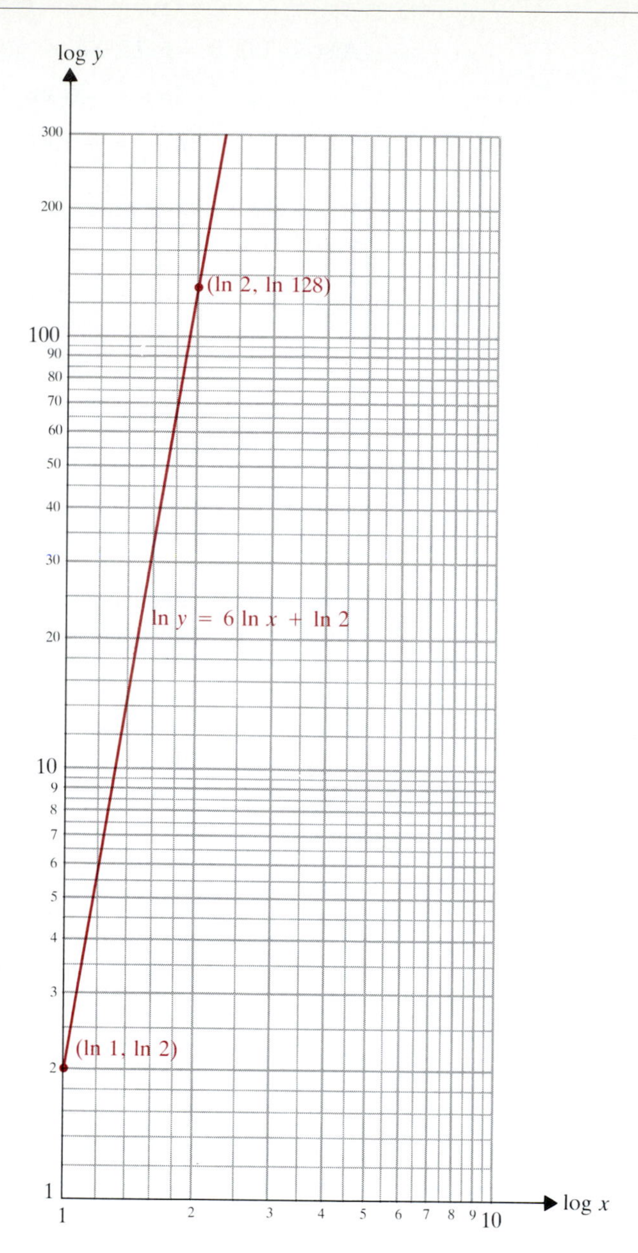

Example 4

An optics laboratory measures the illumination E (in lumens per square meter) produced by a certain light source of constant intensity I (in candelas) at various distances r (in meters) from the light. For a certain light source, $I = 200$. The following table of values is obtained.

r	1	2	3	4	5	6	7	8	9	10
E	200	50	22	12.5	8	5.5	4.1	3.1	2.5	2

Find a formula $E = f(r)$ to model the data.

Solution. We plot the data on the log-log graph in **Figure 9.7.** It appears to lie approximately on the line through $(\ln 1, \ln 200)$ and $(\ln 10, \ln 2)$. That line

FIGURE 9.7

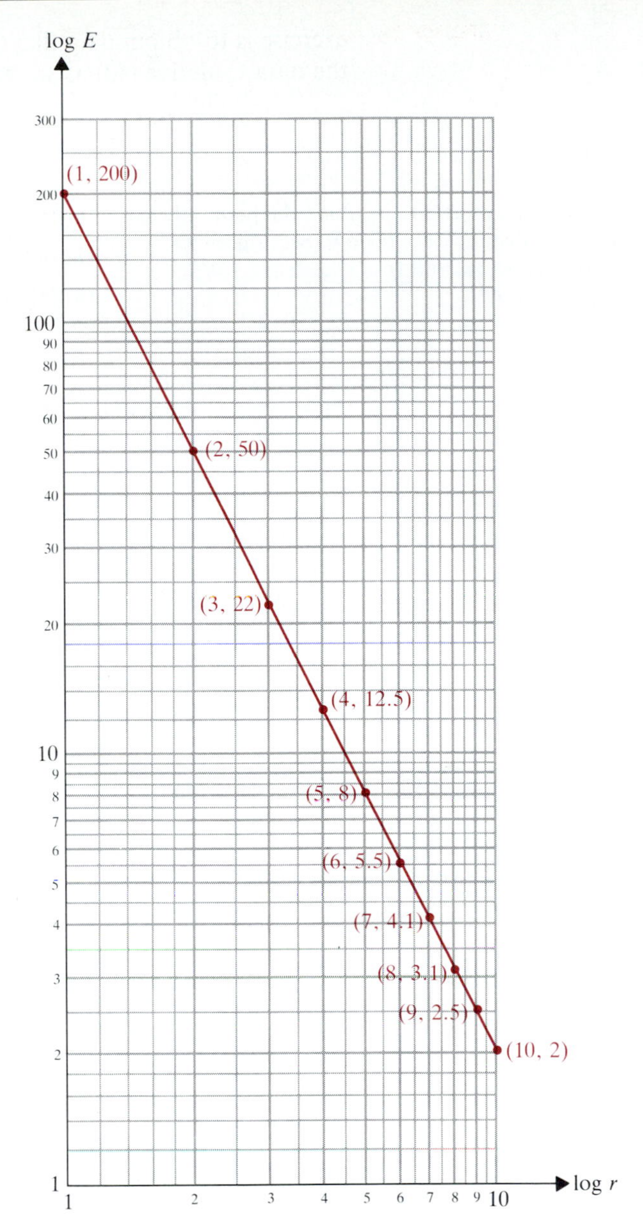

has slope

$$m = \frac{\ln 200 - \ln 2}{\ln 1 - \ln 10} = -\frac{\ln 100}{\ln 10} = -\frac{2 \ln 10}{\ln 10} = -2.$$

Since the line passes through (0, ln 200), its equation is

$$\ln E = -2 \ln r + \ln 200 \rightarrow \ln \frac{E}{200} = \ln r^{-2} \rightarrow \frac{E}{200} = r^{-2},$$

$$E = \frac{200}{r^2}. \tag{13}$$ ■

Equation (13) is a law of physics that relates illumination, intensity, and distance from the light source. It was discovered experimentally by the same procedure used to solve Example 4. A common elementary physics laboratory exercise is to obtain data like that in Example 4 and compare it to (13), or use the data to derive (13) as we have done.

Exercises 6.9

1. A *reciprocally scaled axis* is one in which the point labeled $x \neq 0$ is plotted at distance $1/x$ from 0. See **Figure 9.8.** If $x_1 > x_2 > 0$, then show that $0 < 1/x_1 < 1/x_2$. What happens if $x_1 < x_2 < 0$?

FIGURE 9.8

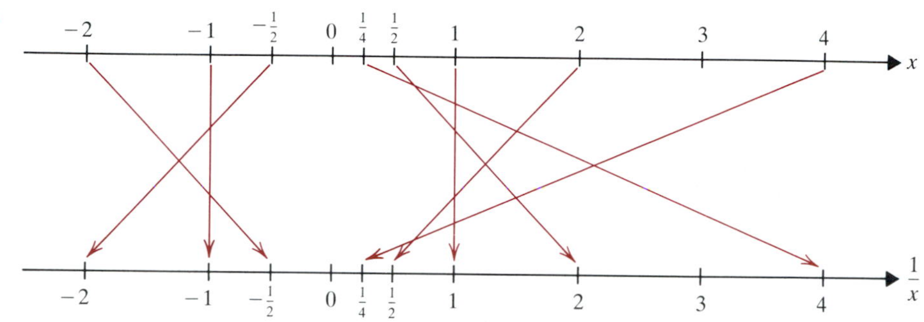

2. In Exercise 1, if $x_1 < 0 < x_2$, then how are $1/x_1$ and $1/x_2$ related?

3. Plot the numbers 2, 3.5, 6, 8, 12, 15, 17, 21, and 25 on
   (a) a uniformly scaled axis;
   (b) an axis with a logarithmic scale;
   (c) a reciprocally scaled axis.

4. Repeat Exercise 3 for the numbers 0.02, 0.003, 0.014, 0.001, 0.03, and 0.82

In Exercises 5–8, plot the given points (x, y) on a semi-log graph, where the y-axis is logarithmically scaled.

5. **(a)** (2, 5) **(b)** (3, 12) **(c)** (4, 30)
6. **(a)** (5, 40) **(b)** (6, 50) **(c)** (7, 60)
7. **(a)** (8, 24) **(b)** (9, 32) **(c)** (10, 43)
8. **(a)** (12, 58) **(b)** (3.2, 4.5) **(c)** (1.8, 1.8)

In Exercises 9–12, plot the given points (x, y) on a log-log graph.

9. **(a)** (1, 5) **(b)** (2, 20) **(c)** (20, 2)
10. **(a)** (3, 17) **(b)** (17, 3) **(c)** (28, 120)
11. **(a)** (200, 300) **(b)** (0.8, 0.2) **(c)** (0.04, 0.06)
12. **(a)** (150, 400) **(b)** (0.75, 0.35) **(c)** (0.02, 0.05)

In Exercises 13–18, plot the given data points (x, y) on a semi-log graph, on which the y-axis is logarithmically scaled. If the points lie on a straight line, find a modeling function $y = f(x)$ for the data.

13.

x	0	1	2	3	4	5	6
y	5	6.1	7.5	9.1	11.1	13.6	16.6

14.

x	0	1	2	3	4	5	6
y	3	3.3	3.7	4	4.5	4.9	5.5

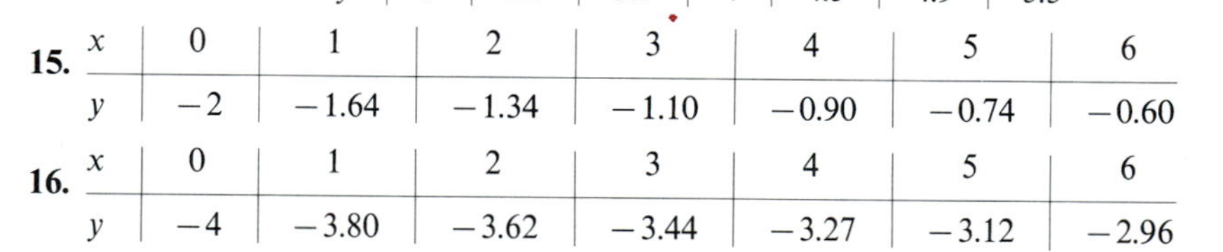

15.

x	0	1	2	3	4	5	6
y	−2	−1.64	−1.34	−1.10	−0.90	−0.74	−0.60

16.

x	0	1	2	3	4	5	6
y	−4	−3.80	−3.62	−3.44	−3.27	−3.12	−2.96

17.

x	0	1	2	3	4	5	6
y	1.6	5.31	17.64	58.56	194.4	645.5	2143

18.

x	0	1	2	3	4	5	6
y	2.3	1.03	0.46	0.21	0.094	0.042	0.019

In Exercises 19–24, plot the given data points (x,y) on a log-log graph. If the points lie on a straight line, find a modeling function $y = f(x)$ for the data.

19.

x	0	1	2	3	4	5	6
y	0	2	2.83	3.46	4	4.47	4.90

20.

x	0	1	2	3	4	5	6
y	0	0.5	1.41	2.60	4	5.59	7.35

21.

x	1	2	3	4	5	6	7	8
y	3	2.12	1.73	1.5	1.34	1.22	1.13	1

22.

x	1	2	3	4	5	6	7	8
y	2	1.26	0.96	0.79	0.68	0.61	0.55	0.5

23.

x	0	1	2	3	4	5	6
y	0	−1.5	−1.98	−2.33	−2.61	−2.86	−3.07

24.

x	1	2	3	4	5	6	9
y	0.2	0.071	0.038	0.025	0.018	0.014	0.0074

In Exercises 25–32, plot the data on semi-log paper and then on log-log paper if the semi-log graph does not appear to be a straight line.

25. Oxygen expands adiabatically in a chamber (no heat transfer takes place). The following data for P and V were obtained experimentally.

V	1	2	3	4	5
P	9.3	3.52	2.00	1.34	0.98

Find an approximate formula for P as a function of V.

26. Carbon dioxide expands adiabatically in a chamber. Find P as a function of V if the following data were obtained.

V	1	2	3	4	5
P	2.80	1.15	0.679	0.468	0.351

27. A signal of strength S is picked up by a receiver at strength R where R and S are related by the following data.

S	1	2	3	5	8
R	2	2.5	2.9	3.4	4

Find an approximate formula for S in terms of R.

28. Repeat Exercise 27 for these data.

S	1	2	3	4	6
R	3	4.55	5.80	6.89	8.79

29. An audio laboratory staff measures the dropoff in the sound level S (in decibels) of a stereophonic cassette player at various frequencies x (in kilohertz). They obtain the following data.

x	10	12	15	18	20
S	0	−5.2	−11.6	−16.8	−20

Find an approximate formula for S as a function of x.

30. Repeat Exercise 29 for a cassette deck for which the following data were obtained.

x	10	12	15	16	19
S	0	−5	−12	−15	−20

31. The following data give the length y of a year (in days) of the first six planets and the maximum distance D (in millions of miles) from the sun to each planet.

y	88	225	365	687	4333	10759
D	36	67	93	142	484	886

Find a formula relating y to D.

32. Illumination E in lumens per square meter is measured at distance r meters from a light. The following data are obtained.

r	1.0	2.0	3.0	4.0	5.0	6.0
E	20.5	5.1	2.5	1.4	1.0	0.7

Find an approximate formula for E as a function of r.

33. Consider the curve $(a - y)(b + x) = c$, where $c = ab$.
(a) Solve for y as a function of x.
(b) Invert the expression found in (a), obtaining an equation for $1/y$ in terms of $1/x$.
(c) What is the graph of the equation found in (b) in a reciprocally scaled coordinate system?

34. The *Michaelis-Menten equation* in biochemistry is

$$(V + v)(K + s) = VK,$$

where V is a constant (the maximum velocity of reaction of an enzyme E with a substrate S), K is the equilibrium constant [cf. Example 1b of Section 2], v is the rate of the reaction, and s is the concentration of S. Proceeding as in Exercise 33, find a formula for $1/v$ as a function of $1/s$. When experimental data are plotted on a reciprocally scaled coordinate system, what kind of graph is obtained?

35. In Exercise 34, explain how K and V can be determined from experimental data plotted on a reciprocally scaled coordinate system.

36. Use the approach of Exercises 34 and 35 to find V and K for the following data, obtained in a biochemistry laboratory for the enzyme sucrase.

s	0.01	0.02	0.04	0.08	0.12	0.21
v	0.0066	0.010	0.015	0.019	0.020	0.022

37. In chemistry, the rate k of reaction is related to activation energy E_a (in joules) by

$$k = ce^{-E_a/RT},$$

where T is the temperature in Kelvin, R is the gas constant (about 8.31 joules/K), and c is a constant. The following measurements are from a freshman laboratory exercise.

T	283	299	313
k	1.5×10^{-5}	2.7×10^{-5}	1.1×10^{-4}

Plot in $\ln k$ versus $1/T$. Then find E_a.

38. Repeat Exercise 37 for the following data.

T	283	293	303
k	1.5×10^{-5}	2.2×10^{-5}	8.9×10^{-5}

HISTORICAL NOTE

In doing Exercise 31, you derive *Kepler's third law of planetary motion*, discovered in 1619 by the German astronomer and mathematician, Johannes Kepler (1571–1630). Kepler arrived at this law from empirical data like that given in the problem. It was later derived by Newton from the general law of gravitation.

The reciprocal plotting scheme described in Exercises 34 and 35 is called the *Lineweaver–Burk* plot of the Michaelis–Menten equation, after the American biochemists Hans Lineweaver (1907–), of the United States Department of Agriculture, and Dean Burk (1904–), of the National Cancer Institute. The graphical method of finding V and K is of fundamental importance and is often used in biochemical research.

6.10 Looking Back

The main focus of this chapter has been on several families of transcendental functions: logarithmic, exponential, inverse trigonometric, and hyperbolic.

The first section defined $\ln x$ as a definite integral, derived its basic properties, and presented the techniques of logarithmic differentiation and integration. The next section defined the natural exponential function *exp* as the inverse of *ln*, and derived the differentiation and integration formulas of *exp* from those of *ln*. Section 4 discussed logarithmic and exponential functions for bases $b \neq e$.

Section 5 defined inverse functions for the sine, cosine, tangent, and secant functions and used the inverse function differentiation rule to find their derivatives. This also produced a number of new integration formulas. Section 6 discussed the hyperbolic functions and their inverses.

Section 8 presented l'Hôpital's rules for evaluating indeterminate limits, many of which involve transcendental functions. Also included was optional material on the Cauchy mean value theorem and its use in estimating the accuracy of the tangent approximation.

The remaining sections of the chapter were devoted to applications of transcendental functions. The law of exponential growth and decay was studied in several different contexts in Section 3, and exponential functions were also applied to physics (Newton's law of cooling) and chemistry (reaction rates). Section 7 discussed applications of trigonometric and hyperbolic functions to oscillatory phenomena and freely hanging cables. The optional Section 9 presented two graphical tools for mathematically modeling scientific data: semi-log and log-log graphing. Plotting data on semi-log or log-log graph paper can lead to an explicit formula for the underlying functional relationship.

Because so many differentiation and integration formulas have been presented in the chapter, we collect the principal ones together here before giving the Chapter Checklist. Ask your instructor how many of them you need to memorize.

$\dfrac{d}{dx}(\ln x) = \dfrac{1}{x}$	$\displaystyle\int \frac{1}{x}\,dx = \ln\lvert x\rvert + C$
$\dfrac{d}{dx}(e^x) = e^x$	$\displaystyle\int e^x\,dx = e^x + C$
$\dfrac{dP}{dt} = rP$	$P = P_0 e^{rt}$
$\dfrac{d}{dx}(b^x) = b^x \ln b$	$\displaystyle\int b^x\,dx = \frac{b^x}{\ln b} + C$
$\dfrac{d}{dx}(\log_b x) = \dfrac{1}{x\ln b}$	$\log_b x = \dfrac{\ln x}{\ln b}$
$D_x(\sin^{-1}x) = \dfrac{1}{\sqrt{1-x^2}}$	$\displaystyle\int \frac{dx}{\sqrt{1-x^2}} = \arcsin x + C$
$D_x(\tan^{-1}x) = \dfrac{1}{x^2+1}$	$\displaystyle\int \frac{dx}{x^2+1} = \arctan x + C$
$\displaystyle\int \frac{dx}{\sqrt{a^2-x^2}} = \arcsin\frac{x}{a} + C$	$\displaystyle\int \frac{dx}{x^2+a^2} = \frac{1}{a}\arctan\frac{x}{a} + C$
$D_x \sinh x = \cosh x$	$\displaystyle\int \cosh x\,dx = \sinh x + C$
$D_x \cosh x = \sinh x$	$\displaystyle\int \sinh x\,dx = \cosh x + C$
$D_x \tanh x = \operatorname{sech}^2 x$	$\displaystyle\int \operatorname{sech}^2 x\,dx = \tanh x + C$
$D_x(\sinh^{-1}x) = \dfrac{1}{\sqrt{x^2+1}}$	$\displaystyle\int \frac{dx}{\sqrt{x^2+1}} = \sinh^{-1}x + C = \ln(x+\sqrt{x^2+1}) + C$
$D_x(\cosh^{-1}x) = \dfrac{1}{\sqrt{x^2-1}}$	$\displaystyle\int \frac{dx}{\sqrt{x^2-1}} = \cosh^{-1}x + C = \ln\left\lvert x+\sqrt{x^2-1}\right\rvert + C$
$D_x(\tanh^{-1}x)\,dx = \dfrac{1}{1-x^2}$	$\displaystyle\int \frac{dx}{1-x^2} = \begin{cases}\tanh^{-1}x + C & \text{if } \lvert x\rvert < 1\\ \coth^{-1}x + C & \text{if } \lvert x\rvert > 1\end{cases} = \frac{1}{2}\ln\left\lvert\frac{1+x}{1-x}\right\rvert + C$

CHAPTER CHECKLIST

Section 6.1: definition of $\ln x$ as a definite integral; domain, range, and algebraic properties of *ln;* the number *e*; logarithmic differentiation; logarithmic integration; integrals giving $\ln\lvert g(x)\rvert + C$.

Section 6.2: the natural exponential function *exp* as the inverse of *ln*; domain, range, and algebraic properties of *exp*; e^x; derivatives and integrals involving e^x; *e* as a limit.

Section 6.3: compound interest; daily and continuous compounding; law of exponential growth; carbon-14 dating, half-life; Newton's law of cooling; order of a chemical reaction.

Section 6.4: definition of b^x and its algebraic properties; differentiation of functions of the form $f(x)^{g(x)}$; definition of $\log_b x$ and its relationship to $\ln x$; pH.

Section 6.5: Inverse sine; inverse cosine, inverse tangent, inverse secant and their derivatives; integrals leading to inverse trigonometric functions.

Section 6.6: hyperbolic sine and cosine; trigonometric-like identities, derivatives, and integrals; hyperbolic tangent and the remaining hyperbolic functions; inverse hyperbolic cosine, sine, and tangent and their derivatives; integrals yielding inverse hyperbolic functions.

Section 6.7: simple harmonic motion, amplitude, phase angle, period, frequency; circular motion; vibrating masses; electrical circuits; hanging cable, catenary.

Section 6.8: indeterminate limits and l'Hôpital's rule; Cauchy mean value theorem; error bound for the tangent approximation.

Section 6.9: logarithmic scaling; semi-log plotting; log-log plotting; fitting functions to data via straight line plots.

REVIEW EXERCISES 6.10

In Exercises 1–20, differentiate the given function.

1. $\ln\sqrt{\sin^2 x + x^3}$

2. $\sqrt{x+1}\ln(3x^2+5)^{2/3}$

3. $\ln\left(\dfrac{1}{\sec 3x + x^2}\right)$

4. $\dfrac{\sqrt[3]{x^5+4x^2+1}\sqrt{x^2-8}}{\sqrt[4]{x^4+8}}$

5. $\sqrt{(x^3-2)^3(x^2+2)^5}$

6. $e^{x^2}\tan(e^{(x^2+3)^{1/2}})$

7. $\ln(e^{x^2}+e^{-x^2})$

8. $\ln\left(\dfrac{e^{x^2}+1}{e^x+1}\right)$

9. $x^{x^2+\ln x}$

10. $(\ln x)^{\sqrt{x}+x^2}$

11. $2^{\cos x^2}$

12. $7^{x\ln x^2}$

13. $\log_2\left(\dfrac{x^2-8}{x^2+8}\right)^{1/3}$

14. $\log_5(\cot x)^{5\ln x}$

15. $\sin^{-1} e^{x^2}$

16. $\tan^{-1}\sqrt{x^2+4}$

17. $\sinh(\sin^{-1} x)$

18. $\tanh^2 e^{\sin x}$

19. $\operatorname{sech}\sqrt{x^2+4}$

20. $\cosh[\ln(x^2+1)]$

In Exercises 21–44, evaluate the given integrals.

21. $\displaystyle\int \frac{5x-1}{10x^2-4x+8}\,dx$

22. $\displaystyle\int \frac{\sin x}{3+\cos x}\,dx$

23. $\displaystyle\int \frac{(\ln x)^2}{x}\,dx$

24. $\displaystyle\int \frac{\sqrt{\ln(x+1)}}{\sqrt{x+1}}\,dx$

25. $\displaystyle\int \frac{t+1}{t}\,dt$

26. $\displaystyle\int \frac{x^2-2x+3}{x^3-3x^2+9x+1}\,dx$

27. $\displaystyle\int \frac{e^{-x}}{1+e^{-x}}\,dx$

28. $\displaystyle\int \csc^2 x \cot x \exp(\cot^2 x)\,dx$

29. $\displaystyle\int \frac{dx}{1-e^{-x}}$

30. $\displaystyle\int \frac{dx}{x\log_{10} x}$

31. $\displaystyle\int x\,10^{-x^2+3}\,dx$

32. $\displaystyle\int \frac{(\log_3 x)^3}{x}\,dx$

33. $\displaystyle\int \frac{e^{-x}}{\sqrt{1-e^{-2x}}}\,dx$

34. $\displaystyle\int \frac{dx}{\sqrt{4-9x^2}}$

35. $\displaystyle\int \frac{dx}{\sqrt{4-x^2}}$

36. $\displaystyle\int_0^{1/4} \frac{dx}{\sqrt{1-4x^2}}$

37. $\displaystyle\int_{-1}^{0} \frac{x\,dx}{1+x^4}$

38. $\displaystyle\int \frac{x\,dx}{x^2-6x+13}$

39. $\displaystyle\int \frac{dx}{x^2-6x+13}$

40. $\displaystyle\int x^2\cosh x^3\,dx$

41. $\displaystyle\int \frac{dx}{\sqrt{x^2+9}}$

42. $\displaystyle\int \frac{dx}{16-x^2}$

43. $\displaystyle\int \frac{dx}{\sqrt{x^2-4}}$

44. $\displaystyle\int \frac{e^x}{9-e^{2x}}\,dx$

In Exercises 45–58, evaluate the given limits.

45. $\displaystyle\lim_{x\to 0} \frac{x}{1-e^x}$

46. $\displaystyle\lim_{x\to 0} \frac{1-x+\ln x}{x^2-1}$

47. $\displaystyle\lim_{x\to +\infty} \frac{e^x}{x^2+1}$

48. $\displaystyle\lim_{x\to +\infty} \frac{\ln x}{x+1}$

49. $\displaystyle\lim_{x\to +\infty} \frac{2x}{3^x+4}$

50. $\displaystyle\lim_{x\to \pi/4} \frac{\tan x}{x-\pi/2}$

51. $\displaystyle\lim_{x\to 0}\left(\frac{1}{e^x-1}-\frac{1}{x}\right)$

52. $\displaystyle\lim_{x\to 0}\left(\frac{1}{x}-\frac{1}{\tan^{-1} x}\right)$

53. $\displaystyle\lim_{x\to 0^+} x^{2/\ln x}$

54. $\displaystyle\lim_{x\to 0^+} (1-x)^{1/x}$

55. $\displaystyle\lim_{x\to +\infty} x^{-x}$

56. $\displaystyle\lim_{x\to 0} (1+x)^{1/2x}$

57. $\displaystyle\lim_{x\to \pi/2}\left(x-\frac{\pi}{2}\right)\sec x$

58. $\displaystyle\lim_{x\to +\infty} x\left(\sqrt{\frac{x+1}{x}}-1\right)$

59. Find the area under the graph of $y = xe^{x^2+1}$ from $x = 0$ to $x = 1$.

60. Find the volume of the solid that results when the graph of $y = e^{-x}$ between $x = 1$ and $x = 2$ is revolved about the x-axis.

61. A gas governed by Boyle's law $PV = c$ (P pressure, V volume) has pressure change from P_1 to P_2. What is the average volume during this period?

63. The magnitude m of an earthquake on the Richter scale is given by $\log(I/I_0)$, where I is the intensity of the earthquake and I_0 is the intensity of normal vibration. Find how much greater was the intensity of an earthquake in Iran in 1980 of magnitude 7.2 on the Richter scale, than an earthquake in Japan the same year of magnitude 6.0.

63. What effective annual yield is advertised by a bank on $2\frac{1}{2}$-year certificates paying a continuously compounded interest rate of 12%?

64. Suppose that the Indians who sold Manhattan for $24 in 1624 had invested their money in a savings account paying interest at the continuously compounded rate of 5%. How much would their account have contained in 1987, assuming no additional deposits and no withdrawals?

65. A bone is found that contains 22% of the carbon-14 present in living bone tissue. About how old is the bone?

66. How long is required for a colony of bacteria to double in size if its initial mass is 1.50×10^{-9} g, and after an hour the mass is 2.08×10^{-9} g?

67. A cup of soup is made by pouring water at 100°C over soup mix. The mixture has initial temperature of 98°C. After two minutes in a room at 20°C, the temperature is 85°C. When will it cool to 59°C?

68. A murderer tries to conceal the time of death of his victim by wrapping the body in an electric blanket. After an hour, the body temperature reached 40°C, and the murderer maintained it at that level for 3 hours. Then he removed the blanket. Three hours later, after the body had been discovered, the coroner found that the body temperature was 35°C. After another hour, it was down to 34°C. The room was at 20°C, and the coroner assumed the body had been at 37°C at the time of death. What time did the coroner report as the time of death, if his first measurement was made at 7:00 PM?

69. A particle moves counterclockwise on a circle in simple harmonic motion with $x(t) = 2\cos 3t$.
(a) What is the radius of the circle?
(b) What is the phase angle?
(c) What is the angular velocity?
(d) How long does it take for the particle to move from the point (2, 0) to a point where $x = 1$?

70. Suppose that a system like the one in Figure 7.4 has $m = 3$ kg and that a force of 6 newtons moves the mass 1/2 meter to the right. If the spring is stretched 1 meter to the right and released with $v_0 = 0$, find a formula for $x(t)$.

71. In Figure 7.5, suppose that $E = 120.0$ volts, $L = 5.000 \times 10^{-1}$ henrys, and $C = 2.000 \times 10^{-4}$ farads. Find ω, ϕ_0, and a if $di/dt = 2.000$ amps/second and $i = 8.000$ amps at $t = 0$.

72. A meter measures illumination E at distance r m from a light source. From the following data, find an approximate formula for E as a function of r.

r	1.00	2.00	3.00	4.00	5.00
E	50.0	11.7	4.98	2.72	1.70

73. Oxygen expands adiabatically in a chamber. The following data lead to what approximate formula for P as a function of V?

V	1	2	3	4	5
P	10	3.79	2.15	1.44	1.05

74. The intriguing monoliths depicted on the cover are thought to have been erected at Stonehenge sometime between 2100 BC and 1500 BC. Part of the evidence for this estimate comes from charcoal discovered near the site, which may have been produced by the builders' fires. If charcoal analyzed in 1967 was found to contain 63.40% of the carbon-14 present in recently cut firewood, then show that the charcoal was formed about 1800 BC.

CHAPTER 7 Techniques of Integration

7.0 Introduction

By now, we have derived most of the basic integration formulas of calculus. This chapter presents techniques for applying those formulas more widely.

The first section begins by completing the list of formulas given at the end of this introduction: Formulas (14)–(17) for integrating the tangent, cotangent, secant, and cosecant functions are derived. Following that, procedures are discussed for integrating powers of trigonometric functions.

Section 2 presents perhaps the most widely used of all integration techniques, which is actually the antidifferentiation version of the product rule for derivatives. Section 3 shows how to simplify integrals that involve $\sqrt{a^2 - x^2}$, $\sqrt{a^2 + x^2}$, or $\sqrt{x^2 - a^2}$ by making certain trigonometric substitutions.

Section 4 discusses integration of rational functions $r(x) = p(x)/q(x)$, where $p(x)$ and $q(x)$ are polynomials. This is done by expressing $r(x)$ as a sum of simpler rational functions, which we can integrate individually. Section 5 considers some special techniques and illustrates the use of integral tables.

The last section is devoted to improper integrals, which are integrals over unbounded intervals or integrals of unbounded functions over ordinary closed intervals. Such integrals are worked out by calculating limits that involve infinity.

Before starting to develop the new integration methods, we list the first 33 integration formulas from Appendix I. These consist mostly of formulas derived in earlier chapters. They are stated in terms of a variable u, to remind you that a change of variable

$$u = g(x) \qquad du = g(x)\,dx$$

may be needed to transform a given integral into one of these forms. Not all of these formulas need to be committed to memory. Ask your instructor which ones *should* be memorized.

BASIC INTEGRATION FORMULAS

1. $\int u^r\,du = \frac{u^{r+1}}{r+1} + C, \qquad \text{if } r \neq -1$

2. $\int \frac{1}{u}\,du = \ln|u| + C$

3. $\int e^u\,du = e^u + C$

4. $\int e^{ku}\,du = \frac{1}{k}e^{ku} + C$

5. $\int b^u\,du = \frac{b^u}{\ln b} + C$

6. $\int \sin u\,du = -\cos u + C$

7. $\int \sin ku\,du = -\frac{1}{k}\cos ku + C$

8. $\int \cos u\,du = \sin u + C$

9. $\int \cos ku\,du = \frac{1}{k}\sin ku + C$

10. $\int \sec^2 u\,du = \tan u + C$

11. $\int \csc^2 u\,du = -\cot u + C$

12. $\int \sec u \tan u\,du = \sec u + C$

13. $\int \csc u \cot u\,du = -\csc u + C$

14. $\int \tan u\,du = -\ln|\cos u| + C = \ln|\sec u| + C$

15. $\int \cot u\,du = \ln|\sin u| + C$

16. $\int \sec u\,du = \ln|\sec u + \tan u| + C$

17. $\int \csc u\,du = \ln|\csc u - \cot u| + C$

18. $\int \frac{du}{\sqrt{1-u^2}} = \sin^{-1} u + C = -\cos^{-1} u + C$

19. $\int \frac{du}{\sqrt{a^2-u^2}} = \sin^{-1}\frac{u}{a} + C$, where $a > 0$

20. $\int \frac{du}{1+u^2} = \tan^{-1} u + C$

21. $\int \frac{du}{a^2+u^2} = \frac{1}{a}\tan^{-1}\frac{u}{a} + C$

22. $\int \frac{du}{u\sqrt{u^2-1}} = \sec^{-1} u + C$

23. $\int \frac{du}{u\sqrt{u^2-a^2}} = \frac{1}{a}\sec^{-1}\frac{u}{a} + C$, where $a > 0$

24. $\displaystyle\int \frac{du}{\sqrt{u^2+1}} = \sinh^{-1} u + C = \ln(u + \sqrt{u^2+1}) + C$

25. $\displaystyle\int \frac{du}{\sqrt{u^2+a^2}} = \sinh^{-1} \frac{u}{a} + C = \ln(u + \sqrt{u^2+a^2}) + C$

26. $\displaystyle\int \frac{du}{\sqrt{u^2-1}} = \cosh^{-1} u + C = \ln|u + \sqrt{u^2-1}|, \qquad \text{if } |u| \geq 1$

27. $\displaystyle\int \frac{du}{\sqrt{u^2-a^2}} = \cosh^{-1} \frac{u}{a} + C = \ln|u + \sqrt{u^2-a^2}|, \qquad \text{if } |u| > a$

28. $\displaystyle\int \frac{du}{1-u^2} = \begin{cases} \tanh^{-1} u + C, & \text{if } |u| < 1 \\ \coth^{-1} u + C, & \text{if } |u| > 1 \end{cases} = \frac{1}{2} \ln\left|\frac{1+u}{1-u}\right| + C$

29. $\displaystyle\int \frac{du}{a^2-u^2} = \begin{cases} \frac{1}{a}\tanh^{-1} \frac{u}{a} + C, & \text{if } |u| < a \\ \frac{1}{a}\coth^{-1} \frac{u}{a} + C, & \text{if } |u| > a \end{cases} = \frac{1}{2a} \ln\left|\frac{a+u}{a-u}\right| + C$

30. $\displaystyle\int \frac{du}{u\sqrt{1-u^2}} = -\operatorname{sech}^{-1}|u| + C = \ln \frac{1-\sqrt{1-u^2}}{|u|} + C, \qquad \text{if } |u| < 1$

31. $\displaystyle\int \frac{du}{u\sqrt{a^2-u^2}} = -\frac{1}{a}\operatorname{sech}^{-1}\left|\frac{u}{a}\right| + C = -\frac{1}{a}\ln\left|\frac{a+\sqrt{a^2-u^2}}{u}\right| + C$

32. $\displaystyle\int \frac{du}{u\sqrt{u^2+1}} = -\operatorname{csch}^{-1}|u| + C = \ln\left|\frac{1-\sqrt{1+u^2}}{u}\right| + C$

33. $\displaystyle\int \frac{du}{u\sqrt{a^2+u^2}} = -\frac{1}{a}\operatorname{csch}^{-1}\left|\frac{u}{a}\right| + C = -\frac{1}{a}\ln\left|\frac{a-\sqrt{a^2+u^2}}{u}\right| + C$

7.1 Trigonometric Integrals

The only formulas from the last section that we have not yet derived are Formulas (14)–(17). The first two are easy: For (14), the fact that $D_x(\cos x) = -\sin x$ gives

$$\begin{aligned}\int \tan x\,dx &= \int \frac{\sin x}{\cos x}\,dx = -\int \frac{-\sin x}{\cos x}\,dx \\ &= -\ln|\cos x| + C = \ln|\cos x|^{-1} + C \\ &= \ln|\sec x| + C.\end{aligned}$$

Thus

$$\int \tan x\,dx = -\ln|\cos x| + C = \ln|\sec x| + C.$$

Similarly,

$$\int \cot x\, dx = \int \frac{\cos x}{\sin x}\, dx = \ln|\sin x| + C,$$

since $D_x(\sin x) = \cos x$.

Example 1

Evaluate $\int x \tan x^2\, dx$.

Solution. Except for a factor of 2, the derivative of x^2 is x, so we let $u = x^2$. Then

$$du = 2x\, dx \rightarrow \frac{1}{2}\, du = x\, dx.$$

We thus obtain

$$\int x \tan x^2\, dx = \frac{1}{2}\int \tan u\, du = \frac{1}{2}\ln|\sec u| + C$$

$$= \frac{1}{2}\ln|\sec x^2| + C. \quad \blacksquare$$

The secant and cosecant functions are not as simple to integrate as tangent and cotangent. To evaluate $\int \sec x\, dx$, for example, a trick is needed: Simultaneously multiply and divide $\sec x$ by $\sec x + \tan x$. We then have

$$\int \sec x\, dx = \int \frac{(\sec x)(\sec x + \tan x)}{\sec x + \tan x}\, dx = \int \frac{\sec^2 x + \sec x \tan x}{\sec x + \tan x}\, dx$$

If we let $u = \sec x + \tan x$, then

$$du = (\sec x \tan x + \sec^2 x)\, dx = (\sec^2 x + \sec x \tan x)\, dx,$$

so we get

$$\int \sec x\, dx = \int \frac{du}{u} = \ln|u| = \ln|\sec x + \tan x| + C,$$

which is Formula (16).

A similar approach works for $\int \csc x\, dx$: Simultaneous multiplication and division by $\csc x - \cot x$ gives

$$\int \csc x\, dx = \int \frac{(\csc x)(\csc x - \cot x)}{\csc x - \cot x}\, dx = \int \frac{\csc^2 x - \csc x \cot x}{\csc x - \cot x}\, dx.$$

Letting $u = \csc x - \cot x$, we have

$$du = (-\csc x \cot x + \csc^2 x)\, dx = (\csc^2 x - \csc x \cot x)\, dx.$$

Therefore

$$\int \csc x\, dx = \int \frac{du}{u} = \ln|u| + C = \ln|\csc x - \cot x| + C,$$

which is Formula (17).

Example 2

Evaluate $\int \cot^3 x\, dx$.

Solution. For such an integral, the identity $1 + \cot^2 x = \csc^2 x$ in the form

$$\cot^2 x = \csc^2 x - 1 \tag{1}$$

is helpful, since, except for sign, $\csc^2 x$ is the derivative of $\cot x$. We thus break off the factor $\cot^2 x$ and make the substitution (1), getting

$$\begin{aligned}\int \cot^3 x\, dx &= \int \cot x \cdot \cot^2 x\, dx = \int \cot x(\csc^2 x - 1)\, dx \\ &= \int \cot x \csc^2 x\, dx - \int \cot x\, dx \\ &= -\int \cot x(-\csc^2 x\, dx) - \int \cot x\, dx \\ &= -\frac{\cot^2 x}{2} - \ln|\sin x| + C. \blacksquare\end{aligned}$$

The rest of the section presents techniques for integrating powers of the various trigonometric functions. As in Example 2, the strategy is to use trigonometric identities like (1) to reduce the integrals to a simpler form.

Powers of sin x and cos x

Consider

$$\int \sin^m x \cos^n x\, dx, \tag{2}$$

where m and n are nonnegative integers. (This includes integrals of the type $\int \sin^m x\, dx$, when $n = 0$, and $\int \cos^n x\, dx$, when $m = 0$.) The strategy for evaluating (2) depends on whether or not m and n are both even. The simpler strategy applies when at least one of them is odd.

Example 3

Evaluate $\int \sin^3 dx$.

Solution. First we write

$$\int \sin^3 x\, dx = \int \sin^2 x \sin x\, dx. \tag{3}$$

Then we use the basic trigonometric identity $\sin^2 x + \cos^2 x = 1$ in the form

$$\sin^2 x = 1 - \cos^2 x. \tag{4}$$

Substituting (4) into (3), we have

$$\int \sin^3 x\, dx = \int (1 - \cos^2 x) \sin x\, dx. \tag{5}$$

Since, except for sign, $\sin x$ is the derivative of $\cos x$, we let

$$u = \cos x. \tag{6}$$

Then $du = -\sin x\,dx$; that is, $\sin x\,dx = -du$. We thus have

$$\int \sin^3 x\,dx = \int (1 - u^2)(-du) = -\int (1 - u^2)\,du \tag{7}$$

$$= -u + \frac{1}{3}u^3 + C = -\cos x + \frac{\cos^3 x}{3} + C. \quad ■ \tag{8}$$

The procedure followed in solving Example 3 is effective for any integral of the form (2) with at least one of m or n odd. It is summarized in the following algorithm.

ALGORITHM FOR $\int \sin^m x \cos^n x\,dx$, WHERE AT LEAST ONE EXPONENT IS ODD

1. Separate out one factor, say $g(x)$, of the odd power.
2. Use the identity

$$\cos^2 x + \sin^2 x = 1 \tag{4}$$

   to express the rest of the integrand as a sum of powers of the cofunction $h(x)$ of $g(x)$.
3. Let $u = h(x)$. Then $du = \pm g(x)\,dx$.
4. Transform the integral (2) to a u-integral, and evaluate using the power rule for antiderivatives.
5. Substitute $h(x)$ for u to express the answer in terms of x.

In the solution of Example 3, Step 1 consisted of separating off the factor $g(x) = \sin x$. Step 2 was done by using (4) in the form $\sin^2 x = 1 - \cos^2 x$ to express the rest of the integrand in terms of $\cos^2 x$ in (5). Step 3 began with the change of variable (6) and concluded with the calculation of $du = -\sin x\,dx$. Step 4 was carried out by (7) and the first equation in (8). Step 5 was done by going back to the variable x in the last step of the solution. The next example shows the use of the algorithm in two more integrals that contain at least one odd power of $\sin x$ or $\cos x$.

Example 4

Evaluate (a) $\int \sin^2 x \cos^3 x\,dx$ and (b) $\int \sin^3 x \cos^3 x\,dx$.

Solution. (a) Following Steps 1 and 2, we separate off one $\cos x$ from the odd power of $\cos^3 x$ and then use (4) in the form

$$\cos^2 x = 1 - \sin^2 x. \tag{4}$$

That gives

$$\begin{aligned}
\int \sin^2 x \cos^3 x\,dx &= \int \sin^2 x \cos^2 x \cos x\,dx \\
&= \int \sin^2 x(1 - \sin^2 x)\cos x\,dx \qquad \textit{by (4)} \\
&= \int \sin^2 x \cos x\,dx - \int \sin^4 x \cos x\,dx \\
&= \frac{\sin^3 x}{3} - \frac{\sin^5 x}{5} + C.
\end{aligned}$$

(b) This time we have a choice. We can separate off a factor $\sin x$ from the odd power $\sin^3 x$ or a factor $\cos x$ from $\cos^3 x$. The first alternative and equation (4) give

$$\begin{aligned}\int \sin^3 x \cos^3 x\,dx &= \int \sin^2 x \cos^3 x \sin x\,dx \\ &= \int (1-\cos^2 x)\cos^3 x \sin x\,dx && \textit{by (4)} \\ &= \int \cos^3 x \sin x\,dx - \int \cos^5 x \sin x\,dx \\ &= -\frac{\cos^4 x}{4} + \frac{\cos^6 x}{6} + C.\end{aligned}$$

In Exercise 58, you are asked to check that separating out the factor $\cos x$ leads to the same answer. ■

If an integral involves only even powers of $\sin x$ and $\cos x$, then we have to use a different method. It is based on two identities from Section 2.3 (see Corollary 3.5 of Chapter 2):

(9) $$\cos^2 t = \frac{1+\cos 2t}{2} \quad \text{and} \quad \sin^2 t = \frac{1-\cos 2t}{2}.$$

The next example shows how (9) is used to integrate $\sin^2 t$ and $\cos^2 t$.

Example 5

Evaluate (a) $\int \cos^2 t\,dt$ and (b) $\int \sin^2 t\,dt$.

Solution. (a)

$$\begin{aligned}\int \cos^2 t\,dt &= \int \frac{1+\cos 2t}{2}\,dt && \textit{by (9)} \\ &= \int \frac{1}{2}\,dt + \frac{1}{2}\int \cos 2t\,dt \\ &= \frac{1}{2}t + \frac{1}{4}\sin 2t + C. && \textit{by Formula 9 of the integral table}\end{aligned}$$

(b)

$$\begin{aligned}\int \sin^2 t\,dt &= \int \frac{1-\cos 2t}{2}\,dt && \textit{by (9)} \\ &= \int \frac{1}{2}\,dt - \frac{1}{2}\int \cos 2t\,dt \\ &= \frac{1}{2}t - \frac{1}{4}\sin 2t + C. && \textit{by Formula 9 of the integral table}\end{aligned}$$

■

To evaluate an integral $\int \sin^m x \cos^n x\,dx$, where m and n are nonnegative even integers, we use the approach of Example 5. The procedure is usually lengthier than the first algorithm.

ALGORITHM FOR $\int \sin^m x \cos^n x\,dx$, WHERE NEITHER m NOR n IS ODD

1. Write $\sin^m x = (\sin^2 x)^{m/2}$ and $\cos^n x = (\cos^2 x)^{n/2}$.
2. Substitute one or both of the formulas in (9) for $\cos^2 x$ and $\sin^2 x$ in the integrand and expand.
3. Repeat Steps 1 and 2 on any even powers of $\cos 2x$ or $\sin 2x$ that remain.
4. Continue until a sum of integrals is obtained in which the terms are constants, $\sin 2kx$, or $\cos 2lx$ for some integers k and l, or terms containing an odd power of $\sin kx$ or $\cos lx$ (to which the last algorithm applies).
5. Evaluate the integrals in Step 4, using Formulas (7) and (9) of the integral table, or the techniques of the last algorithm.

Example 6

Evaluate (a) $\int \cos^4 x\,dx$ and (b) $\int \cos^2 x \sin^2 x\,dx$.

Solution. (a)

$$\int \cos^4 x\,dx = \int (\cos^2 x)^2\,dx = \int \left(\frac{1+\cos 2x}{2}\right)^2 dx \qquad \textit{by (9)}$$

$$= \frac{1}{4}\int [1 + 2\cos 2x + \cos^2 2x]\,dx$$

$$= \frac{1}{4}\int dx + \frac{1}{4}\int 2\cos 2x\,dx + \frac{1}{4}\int \cos^2 2x\,dx$$

$$= \frac{1}{4}x + \frac{1}{4}\sin 2x + \frac{1}{4}\int \frac{1+\cos 4x}{2}\,dx \qquad \textit{by (9)}$$

$$= \frac{1}{4}x + \frac{1}{4}\sin 2x + \frac{1}{8}\int dx + \frac{1}{8}\int \cos 4x\,dx$$

$$= \frac{1}{4}x + \frac{1}{4}\sin 2x + \frac{1}{8}x + \frac{1}{32}\sin 4x + C$$

$$= \frac{3}{8}x + \frac{1}{4}\sin 2x + \frac{1}{32}\sin 4x + C.$$

(b)

$$\int \cos^2 x \sin^2 x\,dx = \int \frac{1+\cos 2x}{2}\cdot\frac{1-\cos 2x}{2}\,dx \qquad \textit{by (9)}$$

$$= \frac{1}{4}\int (1-\cos^2 2x)\,dx = \frac{1}{4}\int \sin^2 2x\,dx$$

$$= \frac{1}{4}\int \frac{1-\cos 4x}{2}\,dx = \frac{1}{8}\int dx - \frac{1}{8}\int \cos 4x\,dx \qquad \textit{by (9)}$$

$$= \frac{1}{8}x - \frac{1}{32}\sin 4x + C. \quad ■$$

Powers of sec x and tan x

Since $D_x(\sec x) = \sec x \tan x$ and $D_x(\tan x) = \sec^2 x$, when $r \neq 1$ the power rule and chain rule for antiderivatives give

(10)
$$\int \sec^r x(\sec x \tan x)\,dx = \frac{\sec^{r+1} x}{r+1} + C,$$

and

(11)
$$\int \tan^r x \sec^2 x\,dx = \frac{\tan^{r+1} x}{r+1} + C.$$

These facts are used in conjunction with the basic identity

(12)
$$1 + \tan^2 x = \sec^2 x,$$

to work out integrals of the form

(13)
$$\int \sec^m x \tan^n x\,dx,$$

where m is even or n is odd. The next algorithm, which is analogous to the first one for $\int \sin^m x \cos^n x\,dx$, applies when m is even.

ALGORITHM FOR $\int \sec^m x \tan^n x\,dx$ WHEN m IS EVEN

1. Separate off a factor $\sec^2 x$.
2. Use Equation (12) to replace the remaining powers of $\sec x$ by a sum of powers of $\tan x$.
3. Use Equation (11) to integrate the result of Step 2.

The next example illustrates this method.

Example 7

Evaluate (a) $\int \sec^4 x\,dx$ and (b) $\int \sec^4 x \tan^3 x\,dx$

Solution. (a)
$$\begin{aligned}\int \sec^4 x\,dx &= \int \sec^2 x \sec^2 x\,dx \\ &= \int (1 + \tan^2 x)\sec^2 x\,dx && \textit{by (12)} \\ &= \int \sec^2 x\,dx + \int \tan^2 x \sec^2 x\,dx \\ &= \tan x + \frac{\tan^3 x}{3} + C && \textit{from (11)}\end{aligned}$$

(b)
$$\begin{aligned}\int \sec^4 x \tan^3 x\,dx &= \int \sec^2 x \sec^2 x \tan^3 x\,dx \\ &= \int \sec^2 x(1 + \tan^2 x)\tan^3 x\,dx && \textit{by (12)} \\ &= \int \tan^3 x \sec^2 x\,dx + \int \tan^5 x \sec^2 x\,dx \\ &= \frac{\tan^4 x}{4} + \frac{\tan^6 x}{6} + C. && \textit{by (11)}\end{aligned}$$

■

When n is odd, we work out (13) by first separating out a factor $\sec x \tan x$.

ALGORITHM FOR $\int \sec^m x \tan^n x\,dx$ WHEN n IS ODD

1. Separate off a factor $\sec x \tan x$.
2. Use Equation (12) to replace the remaining powers of $\tan x$ by a sum of powers of $\sec x$.
3. Use Equation (10) to integrate the result of Step 2.

This method works similarly to the last one.

Example 8

Evaluate $\int \sec^3 x \tan^5 x\,dx$.

Solution.

$$\int \sec^3 x \tan^5 x\,dx = \int \sec^2 x \tan^4 x \sec x \tan x\,dx$$

$$= \int \sec^2 x(\sec^2 x - 1)^2 \sec x \tan x\,dx \qquad \textit{using (12)}$$

$$= \int \sec^2 x(\sec^4 x - 2\sec^2 x + 1)\sec x \tan x\,dx$$

$$= \int \sec^6 x \sec x \tan x\,dx - 2\int \sec^4 x \sec x \tan x\,dx + \int \sec^2 x \sec x \tan x\,dx$$

$$= \frac{\sec^7 x}{7} - \frac{2\sec^5 x}{5} + \frac{\sec^3 x}{3} + C. \qquad \textit{by (10)}$$

■

We have not given methods to cover *all* integrals of the form $\int \sec^m x \tan^n x\,dx$. If such an integral has *both* m odd *and* n even, then neither of the preceding two algorithms applies. Instead, we can use the identity $1 + \tan^2 x = \sec^2 x$ to reduce to an integral involving only odd powers of $\sec x$ (see Exercises 57 and 58). But to integrate odd powers of $\sec x$, we need an additional method, to be discussed in the next section.

We can evaluate $\int \tan^n x\,dx$, even when n is even, without more sophisticated tools, however.

ALGORITHM FOR $\int \tan^n x\,dx$, $n \geq 2$, n EVEN

1. Separate off $\tan^2 x$, leaving $\tan^{n-2} x$.
2. Use Equation (12) to reduce to
$$\int \tan^{n-2} x \sec^2 x\,dx - \int \tan^{n-2} x\,dx.$$
3. Use Equation (12) and continue using Step 1 until the integration can be completed.

Example 9

Evaluate $\int \tan^4 x\,dx$.

Solution.

$$\int \tan^4 x\,dx = \int \tan^2 x \tan^2 x\,dx$$

$$= \int \tan^2 x(\sec^2 x - 1)\,dx \qquad \textit{by (12)}$$

$$= \int \tan^2 x \sec^2 x\,dx - \int \tan^2 x\,dx$$

$$= \frac{\tan^3 x}{3} - \int (\sec^2 x - 1)\,dx \qquad \textit{using (11) and (12)}$$

$$= \frac{\tan^3 x}{3} - \int \sec^2 x\,dx + \int dx$$

$$= \frac{\tan^3 x}{3} - \tan x + x + C. \quad \blacksquare$$

To handle $\int \csc^m x \cot^n x\,dx$, we proceed as for $\int \sec^m x \tan^n x\,dx$, using the identity

$$(14) \qquad 1 + \cot^2 x = \csc^2 x$$

and the formulas

$$(15) \qquad \int \csc^r x \csc x \cot x\,dx = -\frac{\csc^{r+1} x}{r+1} + C,$$

$$(16) \qquad \int \cot^r x \csc^2 x\,dx = -\frac{\cot^{r+1} x}{r+1} + C,$$

which apply when $r \neq -1$. Since (14) exactly parallels (12), and since (15) and (16) parallel the corresponding formulas (10) and (11), involving $\sec x$ and $\tan x$, the approaches of the last three algorithms will work just as above.

Example 10

Evaluate $\int \csc^3 x \cot^3 x\,dx$.

Solution. Since $\cot x$ appears to an odd power, we use the approach of the algorithm on the top of p. 431. We have

$$\int \csc^3 x \cot^3 x\,dx = \int \csc^2 x \cot^2 x \csc x \cot x\,dx$$

$$= \int \csc^2 x(\csc^2 x - 1)\csc x \cot x\,dx \qquad \textit{by (14)}$$

$$= \int \csc^4 x \csc x \cot x\,dx - \int \csc^2 x \csc x \cot x\,dx$$

$$= -\frac{\csc^5 x}{5} + \frac{\csc^3 x}{3} + C. \quad \blacksquare$$

Exercises 7.1

In Exercises 1–40, evaluate the given integral.

1. $\int \frac{\tan \sqrt[3]{x}}{x^{2/3}}\,dx$
2. $\int \frac{\cot \sqrt{x}}{\sqrt{x}}\,dx$
3. $\int xe^{x^2}\sec(e^{x^2})\,dx$
4. $\int \frac{\csc(\ln x)}{x}\,dx$
5. $\int \cos^3 x\,dx$
6. $\int \sin^5 x\,dx$
7. $\int \cos^7 x\,dx$
8. $\int \sin^7 x\,dx$
9. $\int \sin^3 x\cos^2 x\,dx$
10. $\int \sin^2 x\cos^5 x\,dx$
11. $\int \sin^3 x\cos^4 x\,dx$
12. $\int \sin^4 x\cos^3 x\,dx$
13. $\int \frac{\cos^3 x}{\sqrt{\sin x}}\,dx$
14. $\int \frac{\sin^5 x}{\sqrt{\cos x}}\,dx$
15. $\int \sin^4 x\,dx$
16. $\int \cos^6 x\,dx$
17. $\int \sin^2 x\cos^4 x\,dx$
18. $\int \sin^4 x\cos^2 x\,dx$
19. $\int_0^{\pi} \sin^4 x\cos^4 x\,dx$
20. $\int_0^{2\pi} \sin^6 x\cos^2 x\,dx$
21. $\int \sec^2 x\tan^2 x\,dx$
22. $\int \sec^2 x\sqrt{\tan x}\,dx$
23. $\int \sec^6 x\,dx$
24. $\int \sec^6 3x\,dx$
25. $\int \sec^4 x\tan x\,dx$
26. $\int \sec^4 x\tan^2 x\,dx$
27. $\int \csc^4 x\,dx$
28. $\int \csc^4 x\tan^5 x\,dx$
29. $\int \tan^5 x\,dx$
30. $\int \cot^3 x\,dx$
31. $\int \sec^2 x\tan^3 x\,dx$
32. $\int \csc^2 x\cot^5 x\,dx$
33. $\int_0^{\pi/6} \sec^3 x\tan^3 x\,dx$
34. $\int_{\pi/6}^{\pi/3} \csc^3 x\cot^5 x\,dx$
35. $\int \sqrt{\sec x}\tan^3 x\,dx$
36. $\int \frac{\cot^3 x}{\sqrt{\csc x}}\,dx$
37. $\int \frac{\sec^2 x}{\tan^3 x}\,dx$
38. $\int \frac{\sec^4 x}{\tan^2 x}\,dx$
39. $\int \frac{\cot x}{\csc^2 x}\,dx$
40. $\int \frac{\cot^3 x}{\csc^2 x}\,dx$

41. Find the area under the graph of $y = \tan^6 x$ and above the x-axis between $x = 0$ and $x = \pi/4$.

42. Find the area under the graph of $y = \cos^4 3x$ and above the x-axis between $x = 0$ and $x = \pi$.

43. Find the volume produced by revolving the region above the x-axis and below the graph of $y = \sin^3 x$ between $x = 0$ and $x = \pi/2$ about the x-axis.

44. Find the volume produced by revolving the region above the x-axis and below the graph of $y = \cot^2 x$ between $x = \pi/6$ and $x = \pi/4$ about the x-axis.

45. **(a)** Show that $\sin x \sin y = \frac{1}{2}\cos(x - y) - \frac{1}{2}\cos(x + y)$.
   (b) Show that $\sin x \cos y = \frac{1}{2}\sin(x - y) + \frac{1}{2}\sin(x + y)$.

46. Find an identity similar to those in Exercise 45 for $\cos x \cos y$.

In Exercises 47–50, use the results of the previous two exercises to evaluate the given integral.

47. $\int \sin x\cos 2x\,dx$
48. $\int \cos 2x\cos 3x\,dx$
49. $\int \sin 3x\sin x\,dx$
50. $\int \cos 3x\cos x\,dx$

51. Show that $\int_0^{2\pi} \sin mx \sin nx\,dx = 0$ for any integers m and n, such that $m \neq \pm n$.

52. Show that $\int_0^{2\pi} \cos mx \sin nx\,dx = 0$ for any integers m and n, such that $m \neq \pm n$.

53. Find $\int_0^{2\pi} \cos mx \cos nx\,dx$, if m and n are integers.

54. Find $\int_0^{\pi} \cos mx \cos nx\,dx$, if m and n are integers.

55. Show that $\int \tan^n x\,dx = \tan^{n-1} x/(n - 1) - \int \tan^{n-2} x\,dx$, if n is an integer greater than 1.

56. Reduce $\int \sec^3 x\tan^2 x\,dx$ to an integral involving only odd powers of $\sec x$.

57. Explain why any integral $\int \sec^m x\tan^n x\,dx$ can be reduced to an integral involving only odd powers of $\sec x$, if m is odd and n is even.

58. Verify that if the factor $\cos x$ is separated off in Example 4b, then $\int \sin^3 x\cos^3 x\,dx$ still reduces to $-\frac{1}{4}\cos^4 x + \frac{1}{6}\cos^6 x + C$.

59. Some texts give the formula
$$\int \csc x\,dx = -\ln|\csc x + \cot x| + C.$$
Why is this not in conflict with Formula (17) of the integral table?

7.2 Integration by Parts

In Section 3.9, we derived several antidifferentiation rules from corresponding rules for computing derivatives. For instance, from

$$D_x[f(x) + g(x)] = f'(x) + g'(x),$$

we obtained the rule

$$\int [f(x) + g(x)]\, dx = \int f(x)\, dx + \int g(x)\, dx.$$

In this section, we derive an integration formula from the product rule

$$D_x[f(x)g(x)] = f'(x)g(x) + f(x)g'(x)$$

for derivatives. The first step is to rewrite it in the form

$$f(x)g'(x) = D_x[f(x)g(x)] - g(x)f'(x).$$

Then we antidifferentiate both sides to get

$$\int f(x)g'(x)\, dx = f(x)g(x) - \int g(x)f'(x)\, dx. \tag{1}$$

At first glance, this formula doesn't look very helpful. But it is used in conjunction with a double change of variable to simplify complicated integrals in which we can integrate some *factor* of the integrand. We let

$$u = f(x) \qquad \text{and} \qquad dv = g'(x)\, dx, \tag{2}$$

where we recognize the antiderivative $g(x)$ of $g'(x)$. Then

$$du = f'(x)\, dx \qquad \text{and} \qquad v = g(x).$$

Thus, (1) becomes

$$\int u\, dv = uv - \int v\, du \tag{3}$$

Generally, (1) is applied in the form (3), which is called the ***integration by parts formula.*** The goal is to define u and dv in (2) so that the right side of (3) is simpler to integrate than the left side—the given integral. If we can integrate even *part* of a given integral, then we can try letting $dv = g'(x)\, dx$ be that part and $u = f(x)$ be the remaining factor of the integrand. If $f'(x)$ is simpler than $f(x)$, or $v = g(x)$ is simpler than $g'(x)$, then the integral $\int v\, du$ on the right side of (3) is easier to evaluate than the original integral $\int u\, dv$. (The arbitrary constant of integration C must be supplied when $\int v\, du$ is worked out.)

In choosing u and dv, a good goal to keep in mind is that

du should be no more complicated than u,

and

v should be no more complicated than dv.

Example 1

Evaluate $\int x \sin x\, dx$.

Solution. Here the integrand consists of two factors, x and $\sin x$. We can produce a du that is simpler than u by letting

$$u = x \qquad \text{and} \qquad dv = \sin x\, dx.$$

Then we have

$$du = dx \qquad \text{and} \qquad v = -\cos x,$$

since x "differentiates away." We therefore obtain from (3),

$$\int x\sin x\,dx = \int u\,dv = uv - \int v\,du = x(-\cos x) - \int -\cos x\,dx$$
$$= -x\cos x + \int \cos x\,dx = -x\cos x + \sin x + C.$$

As a check, differentiation of our answer gives

$$D_x[-x\cos x + \sin x + C] = -\cos x + x\sin x + \cos x + 0 = x\sin x. \blacksquare$$

While we could have let u be $\sin x$ and dv be $x\,dx$ in Example 1, that choice would have made $\int v\,du$ more complicated than the original integral $\int u\,dv$. For if $u = \sin x$ and $dv = x\,dx$, then $du = \cos x\,dx$ and $v = x^2/2$, so we would get

$$\int x\sin x\,dx = \frac{1}{2}x^2\sin x - \int \frac{1}{2}x^2\cos x\,dx.$$

Thus, had we started this way, we would have had to go back and try a different choice for u and dv.

The steps used in integration by parts are given in the following algorithm.

INTEGRATION BY PARTS

1. Let dv equal as much of the given integrand as you can integrate directly from known formulas like (1)–(20) in Section 7.0.
2. Let u equal the remaining factor of the integrand.
3. Calculate v from dv, and calculate du from u.
4. Apply Formula (3).
5. If the right side of Formula (3) is *more* complicated than the left, then go back and try another choice for u and dv.
6. If the right side of (3) is *not* more complicated than the left, then proceed. (It may be necessary to apply the method a second time.)

Example 2

Evaluate $\int x^3\sqrt{1+x^2}\,dx$.

Solution. Here we can't integrate $\sqrt{1+x^2}$, but we *can* integrate $x\sqrt{1+x^2}$. So we borrow one factor of x from x^3 and let

$$u = x^2 \qquad \text{and} \qquad dv = x\sqrt{1+x^2}\,dx.$$

Then, since

$$dv = \frac{1}{2}[2x(1+x^2)^{1/2}]\,dx,$$

we have

$$du = 2x\,dx \qquad \text{and} \qquad v = \frac{1}{2}\left[\frac{(1+x^2)^{3/2}}{3/2}\right] = \frac{1}{3}(1+x^2)^{3/2}.$$

Thus

$$\begin{aligned}\int x^3\sqrt{1+x^2}\,dx &= \int u\,dv = uv - \int v\,du \\ &= \frac{1}{3}x^2(1+x^2)^{3/2} - \frac{1}{3}\int 2x(1+x^2)^{3/2}\,dx \\ &= \frac{1}{3}x^2(1+x^2)^{3/2} - \frac{1}{3}\left[\frac{(1+x^2)^{5/2}}{5/2}\right] + C \\ &= \frac{1}{3}x^2(1+x^2)^{3/2} - \frac{2}{15}(1+x^2)^{5/2} + C.\end{aligned}$$

As in Example 1, this result can be checked by differentiation. ■

We can even use integration by parts when the integrand isn't a product, by letting $dv = dx$.

Example 3

Evaluate $\int \cos^{-1} x\,dx$.

Solution. The only part of this that we can integrate is dx. So we let

$$u = \cos^{-1} x \qquad \text{and} \qquad dv = dx.$$

Then

$$du = \frac{-1}{\sqrt{1-x^2}}\,dx \qquad \text{and} \qquad v = x.$$

Even though du is not that much simpler than u, we can proceed since

$$\begin{aligned}\int \cos^{-1} x\,dx &= \int u\,dv = uv - \int v\,du \\ &= x\cos^{-1} x - \int \frac{-x}{\sqrt{1-x^2}}\,dx \\ &= x\cos^{-1} x - \frac{1}{2}\int -2x(1-x^2)^{-1/2}\,dx \\ &= x\cos^{-1} x - \frac{1}{2}\left[\frac{(1-x^2)^{1/2}}{1/2}\right] + C \\ &= x\cos^{-1} x - \sqrt{1-x^2} + C.\end{aligned}$$ ■

The next example illustrates how it is sometimes necessary to apply (5) twice (cf. the last remark in the algorithm on p. 435).

Example 4

Evaluate $\int_0^1 x^2 e^{-x}\,dx$.

Solution. The complexity of the integration can be reduced by letting

$$u = x^2 \qquad \text{and} \qquad dv = e^{-x}\,dx.$$

Then

$$du = 2x\,dx \quad \text{and} \quad v = -e^{-x}.$$

Hence

$$\int x^2 e^{-x}\,dx = \int u\,dv = uv - \int v\,du = -x^2e^{-x} - 2\int -xe^{-x}\,dx$$
$$= -x^2e^{-x} + 2\int xe^{-x}\,dx.$$

The integral on the right is less complicated than the original integral but still cannot be integrated directly. So we use integration by parts a second time, letting

$$u = x \quad \text{and} \quad dv = e^{-x}\,dx.$$

Then

$$du = dx \quad \text{and} \quad v = -e^{-x}.$$

Thus

$$\int x^2e^{-x}\,dx = -x^2e^{-x} + 2\left(-xe^{-x} - \int -e^{-x}\,dx\right)$$
$$= -x^2e^{-x} - 2xe^{-x} - 2e^{-x} + C.$$

Hence

$$\int_0^1 x^2e^{-x}\,dx = -x^2e^{-x} - 2xe^{-x} - 2e^{-x}\Big]_0^1 = -e^{-1}(1 + 2 + 2 + 2) = -\frac{7}{e}. \quad ■$$

Sometimes, especially when working with trigonometric functions, two applications of the integration by parts formula will produce

$$\int u\,dv = f(x) + k\int u\,dv, \tag{4}$$

for some constant k. We can then *solve* (4) for $\int u\,dv$ if $k \neq 1$:

$$(1 - k)\int u\,dv = f(x) \rightarrow \int u\,dv = \frac{1}{1-k}f(x) + C.$$

(Here C must be supplied.) The next example illustrates this situation.

Example 5

Evaluate $\int e^x \cos x\,dx$.

Solution. Let

$$u = e^x \quad \text{and} \quad dv = \cos x\,dx.$$

Then

$$du = e^x\,dx \quad \text{and} \quad v = \sin x.$$

Hence

$$\int e^x\cos x\,dx = \int u\,dv = uv - \int v\,du = e^x\sin x - \int e^x \sin x\,dx. \tag{5}$$

Even though the integral on the right isn't any simpler than the one on the left, at least it's not more complicated. So we use integration by parts on it again, letting

$$u = e^x \qquad \text{and} \qquad dv = \sin x\, dx.$$

Then

$$du = e^x\, dx \qquad \text{and} \qquad v = -\cos x.$$

Substituting into (5) and using (3), we get

$$\int e^x \cos x\, dx = e^x \sin x - \left[e^x(-\cos x) - \int e^x(-\cos x)\, dx \right] \tag{6}$$

$$= e^x \sin x + e^x \cos x - \int e^x \cos x\, dx + \bar{C}.$$

This time we append an arbitrary constant $\bar{C}$ because no more integrations are required. In fact, adding $\int e^x \cos x\, dx$ to both sides of (6), we obtain

$$2 \int e^x \cos x\, dx = e^x \sin x + e^x \cos x + \bar{C},$$

$$\int e^x \cos x\, dx = \frac{1}{2} e^x \sin x + \frac{1}{2} e^x \cos x + C. \quad ■$$

In the last section, we mentioned that integration by parts is needed to evaluate $\int \sec^{2n+1} x\, dx$ for positive integers n. The next example illustrates that. (The same approach works on odd powers of $\csc x$.)

Example 6

Evaluate $\displaystyle\int \sec^3 x\, dx$.

Solution. We know how to integrate both $\sec x$ and $\sec^2 x$. Since $\sec^2 x$ is more complicated, and we recognize it to be $D_x(\tan x)$, we let

$$u = \sec x \qquad \text{and} \qquad dv = \sec^2 x\, dx.$$

Then

$$du = \sec x \tan x\, dx \qquad \text{and} \qquad v = \tan x.$$

Thus

$$\int \sec^3 x\, dx = \int u\, dv = uv - \int v\, du$$

$$= \sec x \tan x - \int \tan^2 x \sec x\, dx$$

$$= \sec x \tan x - \int (\sec^2 x - 1) \sec x\, dx, \qquad \text{since } \tan^2 x = \sec^2 x - 1$$

$$= \sec x \tan x - \int \sec^3 x\, dx + \int \sec x\, dx$$

$$= \sec x \tan x - \int \sec^3 x\, dx + \ln|\sec x + \tan x| + \bar{C}.$$

Since $\int \sec^3 x\, dx$ occurs on both sides, we can solve for it, obtaining

$$2 \int \sec^3 x\, dx = \sec x \tan x + \ln|\sec x + \tan x| + \bar{C},$$

$$\int \sec^3 x\, dx = \frac{1}{2} \sec x \tan x + \frac{1}{2} \ln|\sec x + \tan x| + C. \quad ■$$

An important use of integration by parts is to reduce a complicated integral to simpler form. Our final example illustrates how that can be done for a more complicated version of the integral in Example 4. The result is called a *reduction formula.*

Example 7

Obtain a reduction formula for $\int x^n e^{-x}\,dx$, where n is a positive integer.

Solution. We use the approach of Example 4, letting

$$u = x^n \qquad \text{and} \qquad dv = e^{-x}\,dx.$$

Then

$$du = nx^{n-1}\,dx \qquad \text{and} \qquad v = -e^{-x}.$$

Hence

$$\int x^n e^{-x}\,dx = -x^n e^{-x} + n\int x^{n-1}e^{-x}\,dx. \tag{7}$$

Equation (7) is the desired reduction formula. It can be applied again to the integral on the right, until the power of x has been reduced to 0, so that the integration can be completed. ■

To illustrate the use of (7), consider the case $n = 3$. We have

$$\begin{aligned}
\int x^3 e^{-x}\,dx &= -x^3e^{-x} + 3\int x^2 e^{-x}\,dx \\
&= -x^3e^{-x} + 3\left[-x^2e^{-x} + 2\int xe^{-x}\,dx\right] && \textit{by (7) with } n = 2 \\
&= -x^3e^{-x} - 3x^2e^{-x} + 6\int xe^{-x}\,dx \\
&= -x^3e^{-x} - 3x^2e^{-x} + 6\left[-xe^{-x} + \int e^{-x}\,dx\right] && \textit{by (7) with } n = 1 \\
&= -x^3e^{-x} - 3x^2e^{-x} - 6xe^{-x} - 6e^{-x} + C.
\end{aligned}$$

Consistency of Volume Calculation Methods*

In Sections 5.2 and 5.3, we developed two methods for finding the volume of a solid of revolution. We show here that the two methods give the same result in many cases when both can be used.

Let R be the region lying above the x-axis and below the graph of an increasing differentiable function $f(x)$ between $x = a$ and $x = b$. If, as in **Figure 2.1,** we revolve R about the y-axis to generate a solid of revolution S, then we can compute the volume of S by either the cylindrical shell method or the circular disk method. In the first case,

the radius of revolution is x and the height of the shell is $y = f(x)$.

Thus

$$V = 2\pi\int_a^b xf(x)\,dx = \pi\int_a^b f(x)\cdot 2x\,dx. \tag{8}$$

* Optional

FIGURE 2.1

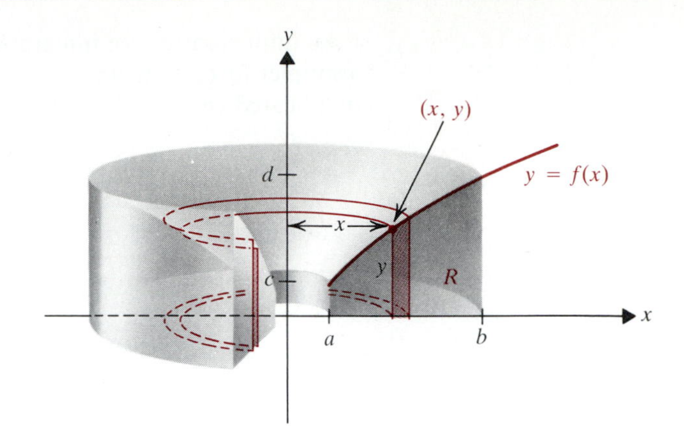

The integral in (8) can be transformed by using the integration by parts substitutions

$$u = f(x) \quad \text{and} \quad dv = 2x\,dx.$$

Then

$$du = f'(x)\,dx \quad \text{and} \quad v = x^2.$$

Therefore

$$\int u\,dv = uv - \int v\,du = x^2 f(x) - \int x^2 f'(x)\,dx.$$

Hence (8) becomes

$$V = \pi\Big[x^2 f(x)\Big]_a^b - \pi\int_a^b x^2 f'(x)\,dx = \pi[b^2 d - a^2 c] - \pi\int_a^b x^2 f'(x)\,dx. \tag{9}$$

Now let us calculate V by the circular disk method. As **Figure 2.2** shows, we need to find the volume $V(S)$ of S in two steps. For the solid S_1 obtained by revolving R_1 about the y-axis,

$$r_{\text{outer}} = b \quad \text{and} \quad r_{\text{inner}} = a.$$

Hence

$$V(S_1) = \pi\int_0^c (b^2 - a^2)\,dy = \pi b^2 c - \pi a^2 c. \tag{10}$$

For the solid S_2 obtained by revolving R_2 about the y-axis,

$$r_{\text{outer}} = b \quad \text{and} \quad r_{\text{inner}} = x,$$

FIGURE 2.2

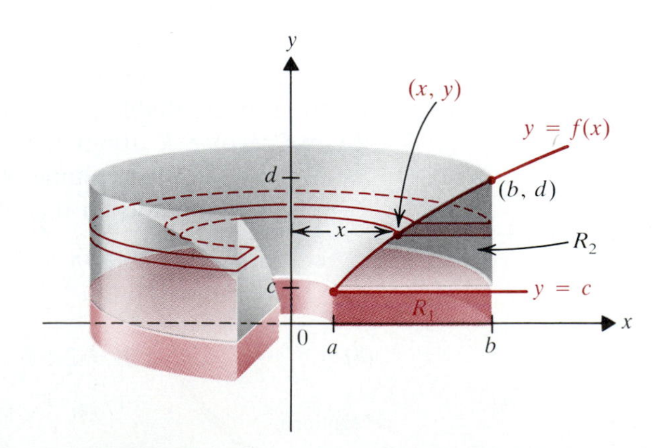

Thus

$$V(S_2) = \pi \int_c^d (b^2 - x^2)\,dy, \tag{11}$$

By assumption, f is an increasing function for $x \in [a, b]$. It therefore has an inverse function g:

$$x = g(y), \qquad \text{for } y \in [c, d].$$

Thus (11) becomes

$$V(S_2) = \pi \int_c^d b^2\,dy - \pi \int_c^d [g(y)]^2\,dy. \tag{12}$$

To evaluate the second integral, we make the change of variable

$$x = g(y).$$

Then the inverse function differentiation rule (Theorem 7.5 of Chapter 2) gives

$$dx = g'(y)\,dy = \frac{1}{f'(x)}\,dy \to dy = f'(x)\,dx.$$

Note that $x = a$ when $y = c$, and $x = b$ when $y = d$. Using Theorem 5.1 of Chapter 4, we then obtain from (12)

$$V(S_2) = \pi \int_c^d b^2\,dy - \pi \int_a^b x^2 f'(x)\,dx,$$

$$= \pi b^2 d - \pi b^2 c - \pi \int_a^b x^2 f'(x)\,dx. \tag{13}$$

Adding (10) and (13), we get

$$V(S) = V(S_1) + V(S_2) = \cancel{\pi b^2 c} - \pi a^2 c + \pi b^2 d - \cancel{\pi b^2 c} - \pi \int_a^b x^2 f'(x)\,dx$$

$$= \pi b^2 d - \pi a^2 c - \pi \int_a^b x^2 f'(x)\,dx,$$

which is the result (9) of calculating V by the cylindrical shell method.

> HISTORICAL NOTE
>
> The technique of integration by parts was developed by Newton around 1671 but was not published until 1736, nine years after his death. He obtained it by generalizing the technique of change-of-variable, much as we did in deriving (1).
>
> The verification that the cylindrical shell and circular disk methods give the same answer for $V(S)$ is based on an article by Charles A. Cable of Allegheny College, published in 1984 in the *American Mathematical Monthly* (volume 91, p. 139).

Exercises 7.2

In Exercises 1–30, evaluate the given integral.

1. $\int xe^x\,dx$

2. $\int x \ln x\,dx$

3. $\int x \cos x\,dx$

4. $\int x \sin 2x\,dx$

5. $\int x^2 \sin x\,dx$

6. $\int x^2 \cos 2x\,dx$

7. $\int_0^1 x^2 e^{2x}\,dx$

8. $\int_{-1}^1 x^3 e^{-x}\,dx$

9. $\int x \sec^2 x\,dx$

10. $\int x \csc^2 x\,dx$

11. $\int \ln x\,dx$

12. $\int (\ln x)^2\,dx$

13. $\int \sin^{-1} x\,dx$

14. $\int \tan^{-1} x\,dx$

15. $\int x^2\sqrt{1+x}\,dx$

16. $\int x^5\sqrt{4-x^3}\,dx$

17. $\int e^x \sin x\,dx$

18. $\int e^{2x}\cos 3x\,dx$

19. $\int \csc^3\theta\,d\theta$

20. $\int \sec^5\theta\,d\theta$

21. $\int_0^1 x\tan^{-1}x\,dx$

22. $\int_0^3 x^2\cot^{-1}x\,dx$

23. $\int x^2\sinh x\,dx$

24. $\int x^3\cosh x\,dx$

25. $\int_1^e \sin(\ln x)\,dx$

26. $\int_e^{e^2} \cos(\ln x)\,dx$

27. $\int e^{ax}\cos bx\,dx$

28. $\int e^{ax}\sin bx\,dx$

29. $\int x^2\tan^{-1}x\,dx$

30. $\int \tan^{-1}\sqrt{x}\,dx$

In Exercises 31–40, derive reduction formulas for the given integrals.

31. $\int \sec^n x\,dx$

32. $\int \csc^n x\,dx$

33. $\int x^n e^x\,dx$

34. $\int x^n e^{ax}\,dx$

35. $\int (\ln x)^n\,dx$

36. $\int x(\ln x)^n\,dx$

37. $\int x^n\sinh x\,dx$

38. $\int x^n\cosh x\,dx$

39. $\int x^{-n}e^x\,dx$

40. $\int x^{-n}e^{ax}\,dx$

41. Find the area bounded by the graph of $y = xe^x$ between $x = 0$ and $x = 1$.

42. Repeat Exercise 41 for $y = xe^{-x}$.

43. The area under $y = \ln x$ between $x = 1$ and $x = 10$ is revolved about the x-axis. Find the volume of the resulting solid.

44. Repeat Exercise 43 for the region in Exercise 41.

45. The region under the graph of $y = \sin x$ from $x = 0$ to $x = \pi/2$ is revolved about the y-axis. Find the volume of the resulting solid.

46. Repeat Exercise 45 for the region under the graph of $y = \ln x$ from $x = 1$ to $x = e$.

47. Show that $\int_0^{\pi/2} \cos^n x\,dx = \frac{n-1}{n}\int_0^{\pi/2} \cos^{n-2} x\,dx.$

48. Show that $\int_0^{\pi/2} \sin^n x\,dx = \frac{n-1}{n}\int_0^{\pi/2} \sin^{n-2} x\,dx.$

49. Criticize the following "application" of integration by parts to $\int x^{-1}\,dx$. Let $u = x$ and $dv = x^{-2}\,dx$. Then $du = dx$ and $v = -1/x$. Hence

$$\int x^{-1}\,dx = -1 + \int x^{-1}\,dx.$$

Subtracting $\int x^{-1}\,dx$ on each side, we have proved that $0 = -1$!

50. The following "application" of integration by parts also "proves" that $0 = -1$. What is wrong? Since $\int \tan x\,dx = \int \sin x \sec x\,dx$, we can let $u = \sec x$ and $dv = \sin x\,dx$. Then $du = \sec x\tan x\,dx$, and $v = -\cos x$. Hence

$$\begin{aligned}\int \tan x\,dx &= -\sec x\cos x + \int \cos x\sec x\tan x\,dx \\ &= -1 + \int \tan x\,dx.\end{aligned}$$

Subtracting $\int \tan x\,dx$ from each side, we have $0 = -1$!

7.3 Trigonometric Substitutions

Problems that involve the Pythagorean theorem, the equation of a circle, or certain scientific laws often lead to integrals that contain $\sqrt{a^2 - x^2}$, $\sqrt{a^2 + x^2}$, or $\sqrt{x^2 - a^2}$, where a is a positive constant. A special integration technique, called ***trigonometric substitution,*** has been devised for such integrals. As the name suggests, it is based on substitution of a trigonometric function for x. It uses the basic identities

$$\sin^2 t + \cos^2 t = 1 \qquad \text{and} \qquad 1 + \tan^2 t = \sec^2 t$$

to eliminate the radical from the integrand. (More generally, this method is useful when the integrand has powers of these radicals.) The technique has three parts, one for each type of radical. But in each case, the underlying idea is the same: We draw a right triangle with the radical as one side, with x and a as the other sides, and with the angle t as one of the acute angles of the triangle.

To begin, suppose that the integrand has a factor $\sqrt{a^2 - x^2}$. Since

$$a^2 - a^2\sin^2 t = a^2(1 - \sin^2 t) = a^2\cos^2 t,$$

letting $x = a\sin t$ will simplify the radical. We thus make the substitution

(1) $$t = \sin^{-1}\frac{x}{a},$$

which by Definition 6.1 of the last chapter means that

(2) $$\sin t = \frac{x}{a} \to x = a\sin t,$$

FIGURE 3.1

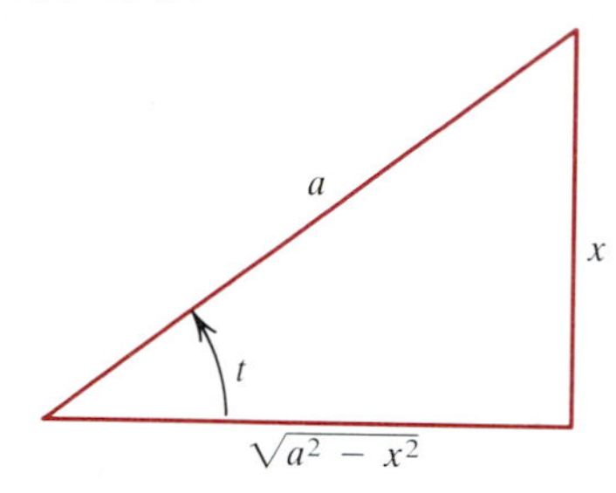

where $t \in [-\pi/2, \pi/2]$. **Figure 3.1** shows the right triangle associated with (1) and (2). We find from (2) that

(3) $$dx = a\cos t\,dt.$$

We also have

(4) $$\sqrt{a^2 - x^2} = \sqrt{a^2 - a^2\sin^2 t} = a\sqrt{1 - \sin^2 t} = a\sqrt{\cos^2 t} = a\cos t,$$

because t lies in the interval $[-\pi/2, \pi/2]$, where $\cos t \geq 0$, so that

$$\sqrt{\cos^2 t} = |\cos t| = \cos t.$$

We summarize this scheme in the following algorithm.

TRIGONOMETRIC SUBSTITUTION FOR INTEGRALS CONTAINING $\sqrt{a^2 - x^2}$, $a > 0$

1. Draw a right triangle with hypotenuse a and legs x and $\sqrt{a^2 - x^2}$.
2. Let $t = \sin^{-1}\frac{x}{a}$, that is, $x = a\sin t$ for $t \in [-\pi/2, \pi/2]$.
3. Compute $dx = a\cos t\,dt$.
4. Substitute Steps 2 and 3 and Equation (4) into the given integral.
5. Integrate the resulting function of t.
6. Use Step 1 and Figure 3.1 to express the answer in Step 5 in terms of the original variable x.

Example 1

Evaluate $\int \sqrt{9 - x^2}\,dx$.

Solution. Here, $a = \sqrt{9} = 3$. Following the boxed algorithm, we first draw **Figure 3.2** and then let

(5) $$x = 3\sin t, \qquad t \in [-\pi/2, \pi/2].$$

FIGURE 3.2

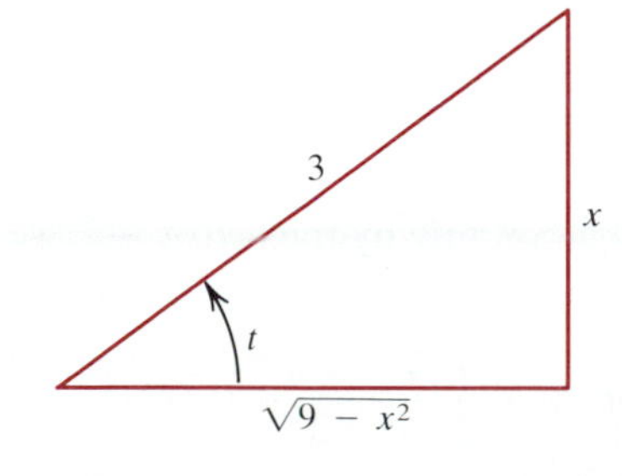

Then Steps 3 and 4 give $dx = 3\cos t\,dt$, $\sqrt{9 - x^2} = 3\cos t$, and

$$\int \sqrt{9 - x^2}\,dx = \int 3\cos t \cdot 3\cos t\,dt = 9\int \cos^2 t\,dt$$

$$= 9\int \frac{1 + \cos 2t}{2}\,dt = \frac{9}{2}\left[t + \frac{1}{2}\sin 2t\right] + C.$$

Having completed Step 5, we have to go back to the variable x. We can't simply use Figure 3.2, because our answer involves $2t$ rather than t. To remedy that, we use the identity $\sin 2t = 2\sin t\cos t$, the fact that $t = \sin^{-1}(x/3)$, and Figure 3.2

to express our answer as

$$\frac{9}{2}[t + \sin t \cos t] + C = \frac{9}{2}\left[\sin^{-1}\frac{x}{3} + \frac{x}{3}\cdot\frac{\sqrt{9-x^2}}{3}\right] + C,$$

$$= \frac{9}{2}\sin^{-1}\frac{x}{3} + \frac{1}{2}x\sqrt{9-x^2} + C. \blacksquare$$

The next example shows how the algorithm is used to evaluate definite integrals, when Step 6 can be avoided by using Theorem 5.1 of Chapter 4. Trigonometric substitution permits us to calculate the value of an integral that previously we could evaluate only numerically or by means of the formula for the area enclosed by a circle (see p. 266).

Example 2

Evaluate $\int_0^a \sqrt{a^2 - x^2}\,dx$.

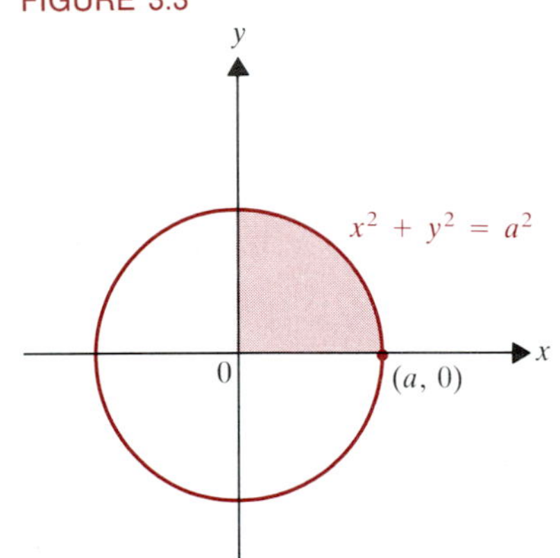

FIGURE 3.3

Solution. Step 1 of the algorithm produces **Figure 3.3.** Steps 2, 3, and 4 give

$$x = a\sin t,\ t \in [-\pi/2, \pi/2] \qquad dx = a\cos t\,dt, \qquad \sqrt{a^2 - x^2} = a\cos t.$$

When $x = 0$, we have $t = \sin^{-1} 0 = 0$; when $x = a$, the corresponding t is

$$t = \sin^{-1}\frac{a}{a} = \sin^{-1} 1 = \pi/2.$$

The given integral then reduces to

$$\int_0^a \sqrt{a^2 - x^2}\,dx = \int_0^{\pi/2} a\cos t \cdot a\cos t\,dt = a^2\int_0^{\pi/2}\cos^2 t\,dt$$

$$= a^2\int_0^{\pi/2}\frac{1 + \cos 2t}{2}\,dt = \frac{1}{2}a^2\left[t + \frac{1}{2}\sin 2t\right]_0^{\pi/2}$$

$$= \frac{1}{2}a^2\left[\frac{1}{2}\pi + 0 - 0 - 0\right] = \frac{1}{4}\pi a^2. \blacksquare$$

As both examples illustrate, the methods of Section 1 may be needed to work out the t-integrals that result from making a trigonometric substitution.

Warning. Don't become so focused on the algorithm that you use it when a simpler solution is available!

Example 3

Evaluate $\int x\sqrt{4 - x^2}\,dx$.

Solution. Since x differs from the derivative of $4 - x^2$ only by the factor -2, we have

$$\int x\sqrt{4 - x^2}\,dx = -\frac{1}{2}\int -2x(4 - x^2)^{1/2}\,dx = -\frac{1}{2}\frac{(4 - x^2)^{3/2}}{3/2} + C$$

$$= -\frac{1}{3}(4 - x^2)^{3/2} + C. \blacksquare$$

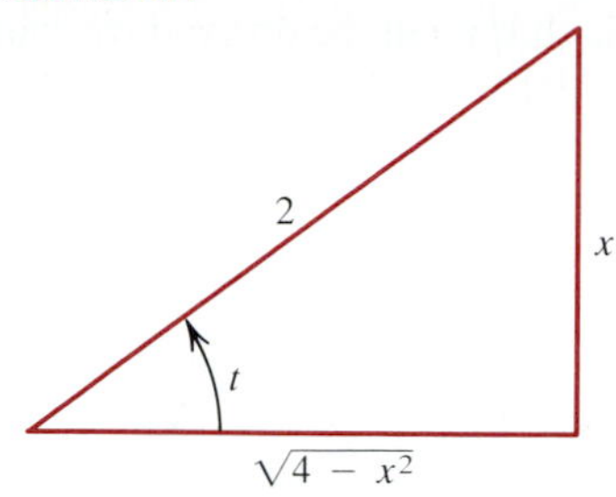

FIGURE 3.4

To use trigonometric substitution in Example 3, we would draw **Figure 3.4** and let

$$x = 2\sin t, \qquad t \in [-\pi/2, \pi/2],$$

so

$$dx = 2\cos t\, dt \qquad \text{and} \qquad \sqrt{4 - x^2} = 2\cos t.$$

The integral would then transform as follows.

$$\begin{aligned}\int x\sqrt{4 - x^2}\, dx &= \int 2\sin t \cdot 2\cos t \cdot 2\cos t\, dt = 8\int \cos^2 t \sin t\, dt \\ &= -8\frac{\cos^3 t}{3} + C = -\frac{8}{3}\left(\frac{\sqrt{4 - x^2}}{2}\right)^3 + C \\ &= -\frac{1}{3}(4 - x^2)^{3/2} + C.\end{aligned}$$

Thus, while the algorithm does eventually produce the correct answer, it requires much more effort than simple substitution.

The next method of trigonometric substitution is designed for integrals that involve $\sqrt{a^2 + x^2}$, where $a > 0$. The idea is to take advantage of the identity $1 + \tan^2 t = \sec^2 t$ to eliminate the radical. That is done by using **Figure 3.5.** We let

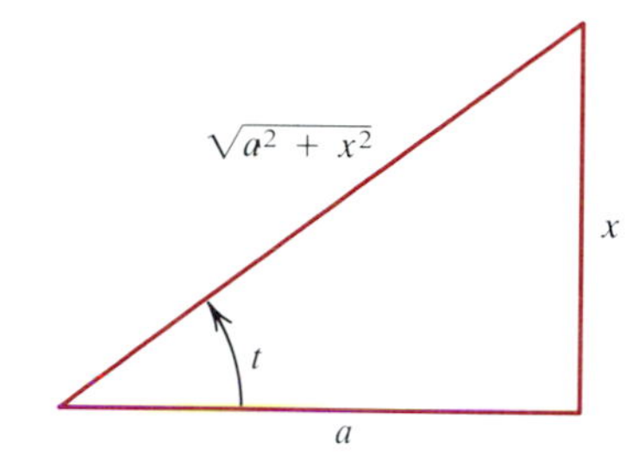

FIGURE 3.5

(6) $$t = \tan^{-1}\frac{x}{a} \rightarrow x = a\tan t.$$

By Definition 6.3 of Chapter 6, this means that t is confined to the interval $(-\pi/2, \pi/2)$. Since $x = a\tan t$, we have

(7) $$dx = a\sec^2 t\, dt,$$

and

(8) $$\sqrt{a^2 + x^2} = \sqrt{a^2 + a^2\tan^2 t} = a\sqrt{1 + \tan^2 t} = a\sqrt{\sec^2 t} = a\sec t,$$

since for $t \in (-\pi/2, \pi/2)$, we have $\sec t = 1/\cos t > 0$. Thus $\sqrt{\sec^2 t} = |\sec t| = \sec t$. We summarize this procedure as follows.

TRIGONOMETRIC SUBSTITUTION FOR INTEGRALS CONTAINING $\sqrt{a^2 + x^2}$, $a > 0$

1. Draw a right triangle with hypotenuse $\sqrt{a^2 + x^2}$ and legs a and x.
2. Let $t = \tan^{-1}(x/a)$, that is, $x = a\tan t$ for $t \in (-\pi/2, \pi/2)$.
3. Compute $dx = a\sec^2 t\, dt$.
4. Substitute Steps 2 and 3 and Equation (8) into the given integral.
5. Integrate the resulting function of t.
6. Use Step 1 and Figure 3.5 to express the answer in Step 5 in terms of x.

The next two examples illustrate how this algorithm is carried out.

Example 4

Find a formula for $\int \frac{dx}{\sqrt{a^2 + x^2}}$, where $a > 0$.

Solution. Such a formula appears as Formula (35) in Section 6.6. We mentioned that it is not usually memorized, and one reason is that it can be derived by using the above algorithm. Referring to Figure 3.5, we let

$$t = \tan^{-1}\frac{x}{a},$$

so that $x = a\tan t$. Using (7) and (8), we obtain

$$\begin{aligned}\int \frac{dx}{\sqrt{x^2+a^2}} &= \int \frac{a\sec^2 t\,dt}{a\sec t} = \int \sec t\,dt \\ &= \ln|\sec t + \tan t| + K = \ln\left|\frac{\sqrt{a^2+x^2}}{a} + \frac{x}{a}\right| + K \\ &= \ln\left|\sqrt{a^2+x^2} + x\right| + C,\end{aligned}$$

where $C = K - \ln a$. ■

The next example illustrates the effectiveness of trigonometric substitution for integrals that involve powers of the radicals $\sqrt{a^2-x^2}$, $\sqrt{a^2+x^2}$, or $\sqrt{x^2-a^2}$.

Example 5

Evaluate $\displaystyle\int \frac{dx}{(x^2+1)^2}$.

FIGURE 3.6

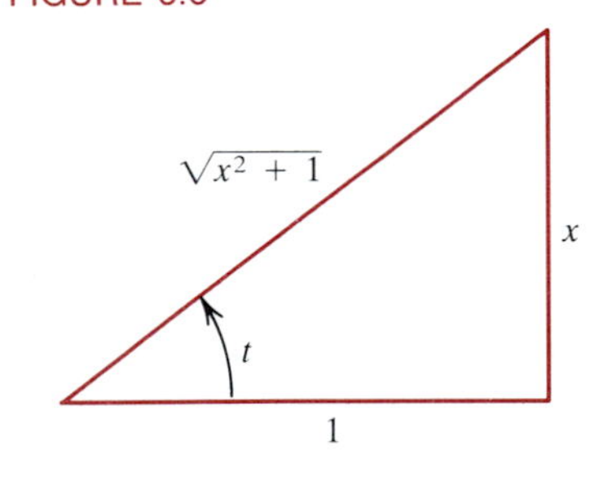

Solution. Even though this integral doesn't involve $\sqrt{x^2+1}$ explicitly, it does involve the fourth power of the radical. So we draw **Figure 3.6** and try letting $x = \tan t$, for $t \in (-\pi/2, \pi/2)$. Then $dx = \sec^2 t\,dt$, and $x^2 + 1 = \tan^2 t + 1 = \sec^2 t$. The integral thus becomes

$$\begin{aligned}\int \frac{dx}{(x^2+1)^2} &= \int \frac{\sec^2 t\,dt}{\sec^4 t} = \int \cos^2 t\,dt = \int \frac{1+\cos 2t}{2}\,dt \\ &= \frac{1}{2}t + \frac{1}{4}\sin 2t + C = \frac{1}{2}t + \frac{1}{2}\sin t\cos t + C \\ &= \frac{1}{2}\tan^{-1}x + \frac{1}{2}\frac{x}{\sqrt{x^2+1}}\cdot\frac{1}{\sqrt{x^2+1}} + C \qquad \textit{from Figure 3.6} \\ &= \frac{1}{2}\tan^{-1}x + \frac{x}{2(x^2+1)} + C. \;■\end{aligned}$$

The last type of radical on which to use a trigonometric substitution is $\sqrt{x^2-a^2}$, where $a > 0$. In this case the substitution is given by **Figure 3.7,** namely,

FIGURE 3.7

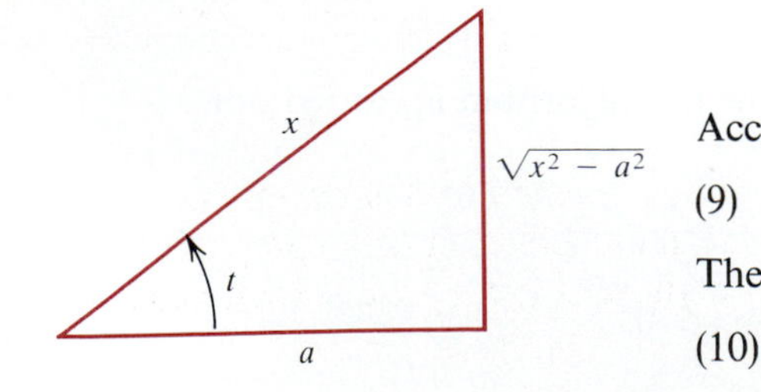

$$t = \sec^{-1}\frac{x}{a}.$$

According to Definition 5.4 of Chapter 6, this means that

(9) $\qquad x = a\sec t, \qquad$ where $t \in [0, \pi/2) \cup [\pi, 3\pi/2)$.

Then

(10) $\qquad dx = a\sec t\tan t\,dt,$

and the apparently strange split domain of $\sec^{-1}$ provides just the simplification we seek:

$$(11)\quad \sqrt{x^2 - a^2} = \sqrt{a^2 \sec^2 t - a^2} = a\sqrt{\sec^2 t - 1} = a\sqrt{\tan^2 t} = a \tan t,$$

because $\tan t \geq 0$ for $t \in [0, \pi/2) \cup [\pi, 3\pi/2)$, so

$$\sqrt{\tan^2 t} = |\tan t| = \tan t.$$

The complete algorithm for evaluating an integral containing $\sqrt{x^2 - a^2}$ is the following.

TRIGONOMETRIC SUBSTITUTION FOR INTEGRALS CONTAINING $\sqrt{x^2 - a^2}$, $a > 0$

1. Draw a right triangle with hypotenuse x and legs a and $\sqrt{x^2 - a^2}$.
2. Let $t = \sec^{-1} \dfrac{x}{a}$, that is, $x = a \sec t$ for $t \in [0, \pi/2) \cup [\pi, 3\pi/2)$.
3. Compute $dx = a \sec t \tan t\, dt$.
4. Substitute Steps 2 and 3 and Equation (11) into the given integral.
5. Integrate the resulting function of t.
6. Use Step 2 and Figure 3.7 to express the answer in Step 5 in terms of x.

Example 6

Evaluate $\displaystyle\int \frac{dx}{(x^2 - 9)^{3/2}}$.

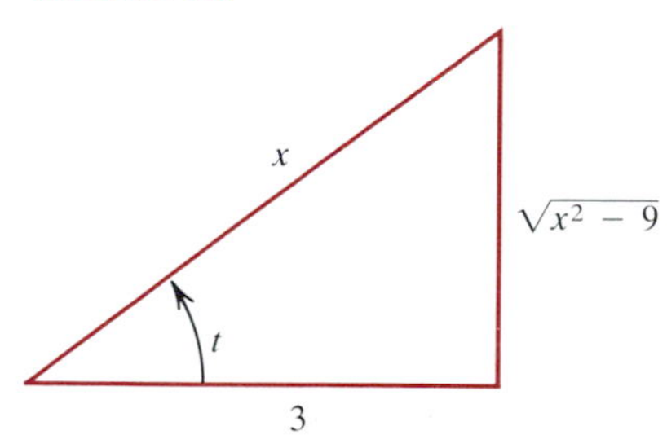

FIGURE 3.8

Solution. From the right triangle in **Figure 3.8,** we let $x = 3 \sec t$, where t is a first- or third-quadrant angle. Then $dx = 3 \sec t \tan t\, dt$, and

$$(x^2 - 9)^{3/2} = (\sqrt{x^2 - 9})^3 = (3 \tan t)^3 = 27 \tan^3 t,$$

where we used (11) with $a = 3$. Then

$$\int \frac{dx}{(x^2 - 9)^{3/2}} = \int \frac{3 \sec t \tan t\, dt}{27 \tan^3 t} = \frac{1}{9} \int \frac{\sec t}{\tan^2 t}\, dt$$

$$= \frac{1}{9} \int \frac{1}{\cos t} \cdot \frac{\cos^2 t}{\sin^2 t}\, dt = \frac{1}{9} \int \frac{\cos t}{\sin^2 t}\, dt.$$

Letting $u = \sin t$, we have $du = \cos t\, dt$, so the integral becomes

$$\frac{1}{9} \int u^{-2}\, du = \frac{1}{9} \frac{u^{-1}}{-1} + C = -\frac{1}{9} \frac{1}{\sin t} + C = -\frac{x}{9\sqrt{x^2 - 9}} + C. \quad ■$$

Table 1 summarizes the three algorithms presented in this section.

Trigonometric substitution can be used in combination with completing the square to integrate powers of quadratic polynomials, $ax^2 + bx + c$. Our closing example demonstrates how that is done.

TABLE 1

Integrand contains	$\sqrt{a^2 - x^2}$	$\sqrt{x^2 - a^2}$	$\sqrt{a^2 + x^2}$
Triangle	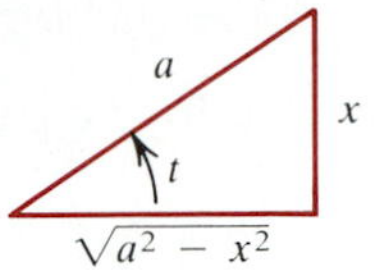	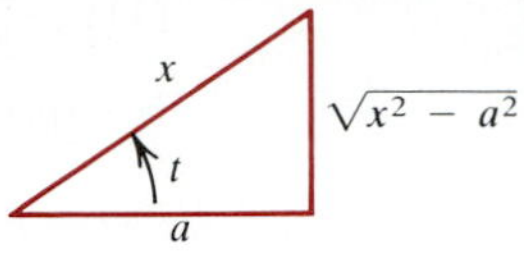	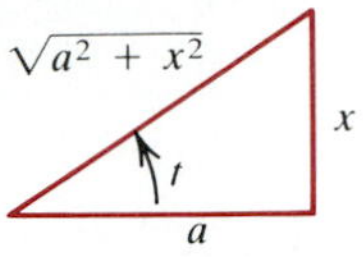
Substitution	$t = \sin^{-1} x/a$ $x = a \sin t$, $t \in [-\pi/2, \pi/2]$ $dx = a \cos t\, dt$	$t = \sec^{-1} x/a$ $x = a \sec t$, $t \in [0, \pi/2) \cup [\pi, 3\pi/2)$ $dx = a \sec t \tan t\, dt$	$t = \tan^{-1} x/a$ $x = a \tan t$, $t \in (-\pi/2, \pi/2)$ $dx = a \sec^2 t\, dt$
Radical becomes	$a \cos t$	$a \tan t$	$a \sec t$

Example 7

Evaluate $\displaystyle\int \frac{dx}{\sqrt{4x^2 - 8x - 5}}$.

Solution. First we complete the square in the radical, getting

$$\begin{aligned} 4x^2 - 8x - 5 &= 4(x^2 - 2x + 1) - 4 - 5 = 4(x-1)^2 - 9 \\ &= [2(x-1)]^2 - 9 = u^2 - a^2, \end{aligned}$$

where $u = 2(x - 1) = 2x - 2$ and $a = 3$. The radical thus is $\sqrt{u^2 - a^2}$. So we use the middle column of Table 1. In **Figure 3.9** we have

FIGURE 3.9

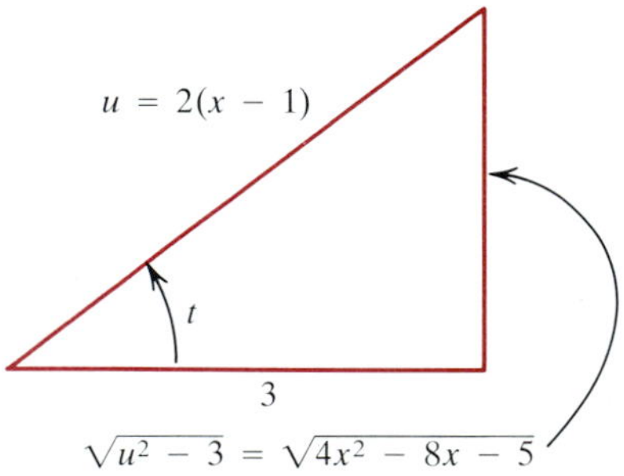

$$t = \sec^{-1}(u/3),$$

so $u = 3 \sec t$, and t is a first- or third-quadrant angle. Then

$$du = 2\,dx = 3 \sec t \tan t\, dt \rightarrow dx = \frac{3}{2} \sec t \tan t\, dt.$$

The denominator of the given integral is then

$$\sqrt{u^2 - 9} = \sqrt{[2(x-1)]^2 - 9} = \sqrt{9 \sec^2 t - 9} = 3 \tan t.$$

Hence

$$\begin{aligned} \int \frac{dx}{\sqrt{4x^2 - 8x - 5}} &= \int \frac{\frac{3}{2} \sec t \cancel{\tan t}}{3 \cancel{\tan t}}\, dt = \frac{1}{2} \int \sec t\, dt \\ &= \frac{1}{2} \ln|\sec t + \tan t| + K \\ &= \frac{1}{2} \ln\left|\frac{2x-2}{3} + \frac{\sqrt{4x^2 - 8x - 5}}{3}\right| + K \qquad \textit{from Figure 3.9} \\ &= \frac{1}{2} \ln|2x - 2 + \sqrt{4x^2 - 8x - 5}| + C \end{aligned}$$

where $C = K - \frac{1}{2} \ln 3$. ■

In Example 7 we introduced u as a *temporary* crutch. Although we could have worked with it, this would only have added another layer of substitution. Whenever possible, it is best to change variables just once.

Exercises 7.3

In Exercises 1–26, evaluate the given integral.

1. $\int \frac{dx}{\sqrt{1-x^2}}$
2. $\int \frac{\sqrt{4-x^2}}{x^2}\,dx$
3. $\int \frac{dx}{(1-x^2)^{3/2}}$
4. $\int \frac{dx}{x\sqrt{9-x^2}}$
5. $\int x^3\sqrt{1-x^2}\,dx$
6. $\int x^5\sqrt{4-x^2}\,dx$
7. $\int_0^{\sqrt{3}/2} \frac{dx}{(1-x^2)^3}$
8. $\int_0^{1/2} \frac{x^2}{(1-x^2)^3}\,dx$
9. $\int \frac{dx}{x\sqrt{4+x^2}}$
10. $\int \sqrt{x^2+1}\,dx$
11. $\int \frac{dx}{x^2\sqrt{x^2+1}}$
12. $\int \frac{dx}{(x^2+4)^2}$
13. $\int_0^1 \frac{x^2\,dx}{\sqrt{x^2+4}}$
14. $\int_0^2 \frac{x^2\,dx}{(x^2+1)^2}$
15. $\int \frac{\sqrt{x^2-9}}{x}\,dx$
16. $\int \frac{dx}{x^2\sqrt{x^2-1}}$
17. $\int \frac{dx}{x\sqrt{x^2-1}}$
18. $\int \sqrt{x^2-1}\,dx$
19. $\int_5^6 \frac{dx}{\sqrt{x^2-16}}$
20. $\int_1^{\sqrt{10}} \frac{\sqrt{x^2-1}}{x^2}\,dx$
21. $\int \frac{dx}{\sqrt{9-16x-4x^2}}$
22. $\int \frac{dx}{\sqrt{8+2x-x^2}}$
23. $\int_0^1 \frac{dx}{\sqrt{x^2+4x+13}}$
24. $\int_0^3 \frac{dx}{\sqrt{4x^2+8x+8}}$
25. $\int \frac{dx}{\sqrt{x^2-6x-7}}$
26. $\int \frac{dx}{\sqrt{4x^2-16x-9}}$

27. The result of Example 4 gives only part of Formula (35) of Section 6.6. To get the part involving $\sinh^{-1}(x/a)$, we can use the *hyperbolic* substitution $t = \sinh^{-1}(x/a)$, i.e., $x = a\sinh t$. Show that
    (a) $\sqrt{a^2+x^2} = a\cosh t$
    (b) The integral equals $\sinh^{-1}(x/a) + C$.

In Exercises 28–30, use a hyperbolic substitution (cf. Exercise 27) to evaluate the integral.

28. $\int \frac{dx}{\sqrt{x^2-a^2}}$
29. $\int \frac{dx}{x\sqrt{1-x^2}}$
30. $\int \frac{dx}{x\sqrt{x^2+1}}$, where $x > 0$
31. Find the arc length along $y = x^2$ between $x = 0$ and $x = 1$. Compare your answer with that for Exercise 29 of Section 5.4.
32. Repeat Exercise 31 between $x = -1$ and $x = 2$. Compare your answer with that for Exercise 30 of Section 5.4.
33. Find the arc length along $y = \ln x$ between $x = 1$ and $x = 2$.
34. Repeat Exercise 33 over the interval from $x = 1$ to $x = 3$.
35. Show that the area of a circular sector of radius r and central angle θ (**Figure 3.10**) is $\frac{1}{2}r^2\theta$. (*Hint:* Show that
$$A = \tfrac{1}{2}r^2\sin\theta\cos\theta + \int_{r\cos\theta}^{r} (r^2-x^2)^{1/2}\,dx,$$
where r and θ are treated as constants.)

FIGURE 3.10

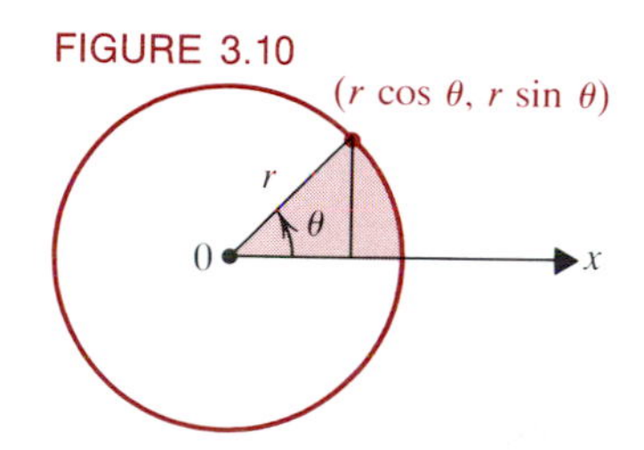

36. Can we validly assert that Exercise 35 proves Lemma 3.9 of Chapter 2?
37. Obtain the "reduction formula"
$$\int \frac{dx}{(x^2+a^2)^n} = \frac{1}{a^{2n-1}}\int \cos^{2n-2} t\,dt.$$
38. Obtain the "reduction formula"
$$\int \frac{dx}{(a^2-x^2)^n} = \frac{1}{a^{2n-1}}\int \sec^{2n-1} t\,dt.$$

7.4 Integration of Rational Functions

Recall that a *rational function* has the form $r(x) = p(x)/q(x)$, where $p(x)$ and $q(x)$ are polynomials. We have already seen how to integrate some kinds of rational functions. For example,

$$\int \frac{dx}{x-a} = \ln|x-a| + C, \tag{1}$$

and if $n > 1$,

$$\text{(2)} \qquad \int \frac{dx}{(x-a)^n} = \frac{(x-a)^{-n+1}}{-n+1} + C = -\frac{1}{(n-1)(x-a)^{n-1}} + C.$$

We also have seen how the technique of completing the square can reduce integrals of the type

$$\text{(3)} \qquad \int \frac{Ax+B}{Cx^2+Dx+E}\,dx$$

to a form we can integrate. (See Example 7 of Section 6.5.) In Example 7 of the last section, we saw that trigonometric substitution in combination with completing the square could be used to evaluate integrals of the form

$$\text{(4)} \qquad \int \frac{Ax+B}{(Cx^2+Dx+E)^m}\,dx,$$

where m is an integer or rational number.

This section is devoted to an important algebraic technique for reducing integrals of more complicated rational functions $r(x)$ to a sum of integrals of the forms (1), (2), (3), and (4). That technique is called ***partial fraction decomposition.*** It is based on results from abstract algebra, so we will just state the needed results without proof. First, we remind you of how the denominator of any rational function can be factored.

4.1 THEOREM ***Polynomial Factorization Theorem.*** Any polynomial $q(x)$ with real coefficients can be factored into the product of terms of the form $(ex-d)^k$ and $(ax^2+bx+c)^m$, where the factors ax^2+bx+c are *irreducible* (that is, $ax^2+bx+c=0$ has no real roots).

Recall that by the quadratic formula

$$\frac{-b \pm \sqrt{b^2-4ac}}{2a}$$

for the roots of $ax^2+bx+c=0$, the quadratic polynomial ax^2+bx+c is irreducible if and only if $b^2-4ac<0$. For instance, x^2+1 is irreducible, because for this polynomial $a=c=1$ and $b=0$, so that $b^2-4ac=-4<0$.

Example 1

Factor $q(x) = x^4 - 2x^3 + 2x^2 - 2x + 1$ into irreducible factors.

Solution. Recall that **$q(x)$ has a linear factor $x-d$ if and only if $q(d)=0$.** To factor $q(x)$, therefore, we look for roots of the equation $q(x)=0$. The easiest roots to look for are integer roots d. A theorem of algebra states that **the integer d is a root of a polynomial equation with integer coefficients**

$$x^n + a_{n-1}x^{n-1} + \ldots + a_2x^2 + a_1x + a_0 = 0$$

if and only if d divides a_0. For the given polynomial $q(x)$, we have $a_0 = 1$. The only possible integer roots are therefore -1 and 1. We find that $q(1)=0$. Division of $q(x)$ by $x-1$ gives

$$q(x) = (x-1)(x^3-x^2+x-1) = (x-1)q_1(x).$$

The only possible integer roots of $q_1(x)$ are again 1 and -1. We find that $x = 1$

is a root. Division of $x^3 - x^2 + x - 1$ by $x - 1$ gives

$$q(x) = (x - 1)^2(x^2 + 1),$$

where, as noted above, $x^2 + 1$ is irreducible. ■

To integrate a rational function $r(x) = p(x)/q(x)$, we first factor $q(x)$ as in Theorem 4.1. We then express $r(x)$ as a sum of *partial fractions* corresponding to the factors of $q(x)$. For example, we can integrate

$$\int \frac{dx}{x^2 - 1}$$

by first noting that $x^2 - 1 = (x - 1)(x + 1)$. It also turns out that

$$\frac{1}{x^2 - 1} = \frac{-1/2}{x + 1} + \frac{1/2}{x - 1}, \tag{5}$$

as you can check by adding the terms on the right. Therefore

$$\int \frac{dx}{x^2 - 1} = -\frac{1}{2}\int \frac{dx}{x + 1} + \frac{1}{2}\int \frac{dx}{x - 1} = -\frac{1}{2}\ln|x + 1| + \frac{1}{2}\ln|x - 1| + C$$

$$= \frac{1}{2}\ln\left|\frac{x - 1}{x + 1}\right| + C.$$

This section develops a technique for finding decompositions like (5). The technique is based on the following result of abstract algebra.

4.2 THEOREM

Partial Fraction Decomposition. Every rational function $r(x) = p(x)/q(x)$, *where* $p(x)$ *has smaller degree than* $q(x)$, can be written as a sum of fractions as follows.

(a) Factor $q(x)$ into the product of factors $(ex - d)^k$ and $(ax^2 + bx + c)^m$ as in Theorem 4.1.

(b) To each factor $(ex - d)^k$ there corresponds a sum

$$\frac{A_1}{ex - d} + \frac{A_2}{(ex - d)^2} + \ldots + \frac{A_k}{(ex - d)^k}, \tag{6}$$

where the A_i are constants.

(c) To each factor $(ax^2 + bx + c)^m$ there corresponds a sum

$$\frac{B_1x + C_1}{ax^2 + bx + c} + \frac{B_2x + C_2}{(ax^2 + bx + c)^2} + \ldots + \frac{B_mx + C_m}{(ax^2 + bx + c)^m}, \tag{7}$$

where the B_i and C_i are constants.

(d) The function $r(x)$ is the sum of all of the partial fractions found in (b) and (c).

Warning. In applying Theorem 4.2, *be sure that the rational function* $r(x)$ *to be integrated is written in the form stipulated*, that is, $\boldsymbol{r(x) = p(x)/q(x)}$**, where** $\boldsymbol{p(x)}$ **has *smaller* degree than** $\boldsymbol{q(x)}$**.** If this is *not* initially the case, then divide $p(x)$ by $q(x)$ to obtain a quotient polynomial $s(x)$ and a remainder $t(x)$:

$$p(x) = q(x)s(x) + t(x), \tag{8}$$

where $t(x)$ has smaller degree than $q(x)$. Then divide (8) through by $q(x)$ to get

(9) $$r(x) = \frac{p(x)}{q(x)} = s(x) + \frac{t(x)}{q(x)}.$$

We can integrate $s(x)$ in (9) because it is just a polynomial, and we can apply Theorem 4.2 to the rational function $t(x)/q(x)$ since $\deg t(x) < \deg q(x)$ for that function.

To determine the coefficients A_i, B_i, and C_i in (6) and (7), we first use Theorem 4.2 to write $p(x)/q(x)$ or $t(x)/q(x)$ as a sum of partial fractions, and then multiply through by the common denominator $q(x)$. This gives an equation between two polynomials, $p(x)$ or $t(x)$ on the left and a polynomial with coefficients A_i, B_i, and C_i on the right. Those coefficients are found by the *method of judicious substitution*, which is illustrated in the next two examples, and the *method of equating coefficients*, which we introduce in Example 4.

Example 2

Evaluate $\displaystyle\int \frac{x^2 + x - 1}{x^3 - x^2 - 6x}\,dx.$

Solution. Since the numerator has smaller degree than the denominator, the integrand is in proper form to apply Theorem 4.2. The denominator factors as

$$x^3 - x^2 - 6x = x(x^2 - x - 6) = x(x - 3)(x + 2).$$

Since each factor is first degree, parts (b) and (d) of Theorem 3.2 say that

(10) $$\frac{x^2 + x - 1}{x^3 - x^2 - 6x} = \frac{A}{x} + \frac{B}{x - 3} + \frac{C}{x + 2},$$

for some real constants A, B, and C. To find their values, we first multiply both sides by the common denominator $x^3 - x^2 - 6x$. That gives

(11) $$x^2 + x - 1 = A(x - 3)(x + 2) + Bx(x + 2) + Cx(x - 3).$$

The two polynomials in (11) are equal, so they have the same value at any real number c. By choosing values of c judiciously, we can eliminate two of the constants A, B, and C in (11) and determine the remaining one. For instance, if we evaluate (11) at $x = 0$, then we get

$$-1 = A(-3)(2) + 0 + 0 \rightarrow A = 1/6.$$

Similarly, evaluation of (8) successively at $x = 3$ and $x = -2$ gives

$$9 + 3 - 1 = 0 + 15B + 0 \rightarrow B = 11/15,$$

$$4 - 2 - 1 = 0 + 0 + 10C \rightarrow C = 1/10.$$

Thus from (10),

$$\int \frac{x^2 + x - 1}{x^3 - x^2 - 6x}\,dx = \frac{1}{6}\int \frac{dx}{x} + \frac{11}{15}\int \frac{dx}{x - 3} + \frac{1}{10}\int \frac{dx}{x + 2}$$

$$= \frac{1}{6}\ln|x| + \frac{11}{15}\ln|x - 3| + \frac{1}{10}\ln|x + 2| + C.$$

(This can be checked by differentiating the answer and adding the resulting fractions.) ■

The next example shows how to proceed when the numerator $p(x)$ of the rational function is *not* of lower degree than the denominator $q(x)$.

Example 3

Evaluate $\int \frac{x^3 + x^2 - 3}{x^2 - x - 2}\,dx$.

Solution. Here the numerator has higher degree (3) than the denominator, so the rational function is *not* in proper form. As suggested in the Warning following Theorem 4.2, we divide the numerator by the denominator:

$$\begin{array}{r|l} & x + 2 \\ \hline x^2 - x - 2 & x^3 + x^2 \qquad -3 \\ & x^3 - x^2 - 2x \\ \hline & \quad 2x^2 + 2x - 3 \\ & \quad 2x^2 - 2x - 4 \\ \hline & \qquad\quad 4x + 1 \end{array}.$$

According to (9), we then have

$$\frac{x^3 + x^2 - 3}{x^2 - x - 2} = x + 2 + \frac{4x + 1}{x^2 - x - 2}. \tag{12}$$

Since $x^2 - x - 2 = (x - 2)(x + 1)$, Theorem 4.2 says that the second fraction on the right in (12) decomposes as

$$\frac{4x + 1}{x^2 - x - 2} = \frac{A}{x - 2} + \frac{B}{x + 1}. \tag{13}$$

Multiplying through by $(x - 2)(x + 1)$, we get

$$4x + 1 = A(x + 1) + B(x - 2). \tag{14}$$

As in Example 1, we successively evaluate both sides of (14) at $x = -1$ and $x = 2$ to obtain

$$-4 + 1 = 0 - 3B \rightarrow B = 1,$$
$$8 + 1 = 3A + 0 \rightarrow A = 3.$$

Then, from (12) and (13), we find

$$\int \frac{x^3 + x^2 - 3}{x^2 - x - 2}\,dx = \int (x + 2)\,dx + 3\int \frac{dx}{x - 2} + 1\int \frac{dx}{x + 1}$$
$$= \frac{1}{2}x^2 + 2x + 3\ln|x - 2| + \ln|x + 1| + C. \quad ■$$

The *method of judicious substitution* works best when there are no repeated factors in the denominator. When a repeated factor is present, the *method of equating coefficients* is helpful.

Example 4

Evaluate $\int \frac{x^2 - 3}{x^3 - 4x^2 + 4x}\,dx$.

Solution. The integrand is in proper form, so we factor the denominator, getting

$$x^3 - 4x^2 + 4x = x(x^2 - 4x + 4) = x(x-2)^2.$$

Then (6) of Theorem 4.2 gives

$$\frac{x^2 - 3}{x^3 - 4x^2 + 4x} = \frac{A}{x} + \frac{B}{x-2} + \frac{C}{(x-2)^2},$$

since the first-degree factor $x - 2$ appears twice. We have

$$x^2 - 3 = A(x-2)^2 + Bx(x-2) + Cx. \tag{15}$$

Letting $x = 2$ and $x = 0$, we find

$$1 = 2C \rightarrow C = 1/2,$$

$$-3 = 4A \rightarrow A = -3/4.$$

To find B, we could let x be some other number, such as 1, but we can also reason as follows. **Since the two polynomials in (15) are equal, the coefficients of corresponding powers of x are the same, and the constant terms are the same.** We can therefore equate the coefficients of the x^2 terms:

$$1 = A + B.$$

Since we already know $A = -3/4$, we have $B = 1 + 3/4 = 7/4$. Hence

$$\begin{aligned}\int \frac{x^2 - 3}{x^3 - 4x^2 + 4x}\,dx &= -\frac{3}{4}\int \frac{dx}{x} + \frac{7}{4}\int \frac{dx}{x-2} + \frac{1}{2}\int \frac{dx}{(x-2)^2}\\ &= -\frac{3}{4}\ln|x| + \frac{7}{4}\ln|x-2| + \frac{1}{2}\cdot\frac{(x-2)^{-1}}{-1} + C\\ &= -\frac{3}{4}\ln|x| + \frac{7}{4}\ln|x-2| - \frac{1}{2(x-2)} + C. \quad ■\end{aligned}$$

Note that equating constant terms is the *same* as substituting $x = 0$. In (15) for instance, that would give $-3 = 4A$. It is usually best to find as many coefficients as possible by judicious substitution and then to equate coefficients to find the others. This typically gives a system of linear equations that can be solved without too much work by using the values already found.

Thus far we have had only first-degree factors in the denominator. The next example has a quadratic factor in the denominator, so both (6) and (7) of Theorem 4.2 are used.

Example 5

Evaluate $\displaystyle\int \frac{x^2 + x + 2}{x^3 - 2x^2 + 4x - 8}\,dx.$

Solution. Again the degree of the numerator is less than the degree of the denominator, so the integrand is in proper form. The denominator can be factored by grouping the first two terms and the second two, since

$$\begin{aligned}x^3 - 2x^2 + 4x - 8 &= x^2(x-2) + 4(x-2)\\ &= (x^2 + 4)(x-2).\end{aligned}$$

The term $x^2 + 4$ is an irreducible quadratic factor, so Theorem 4.2 says that

$$\frac{x^2 + x + 2}{x^3 - 2x^2 + 4x - 8} = \frac{Ax + B}{x^2 + 4} + \frac{C}{x - 2}.$$

Clearing of fractions, we get

$$x^2 + x + 2 = (Ax + B)(x - 2) + C(x^2 + 4). \tag{16}$$

We first substitute $x = 2$ into (16), which gives

$$4 + 2 + 2 = 8C \rightarrow C = 1.$$

No substitution can make $x^2 + 4$ equal to zero, so we equate the coefficients of x^2 and the constants, obtaining

$$1 = A + C = A + 1 \rightarrow A = 0,$$

$$2 = -2B + 4C = -2B + 4 \rightarrow -2 = -2B \rightarrow B = 1.$$

Hence

$$\int \frac{x^2 + x + 2}{x^3 - 2x^2 + 4x - 8}\,dx = 1\int \frac{dx}{x^2 + 4} + 1\int \frac{dx}{x - 2}$$

$$= \frac{1}{2}\tan^{-1}\frac{x}{2} + \ln|x - 2| + C. \quad ■$$

As you might imagine, one is seldom fortunate enough to have one of the constants turn out to be 0, as happened in Example 5. The next example shows what happens when all the constants are nonzero and when we need to complete the square in a quadratic factor to finish the integration.

Example 6

Evaluate $\displaystyle\int \frac{x^2 + 2}{x^3 + x^2 + x}\,dx.$

Solution. Again the integrand is in proper form, so we factor the denominator.

$$x^3 + x^2 + x = x(x^2 + x + 1).$$

The factor $x^2 + x + 1$ is an irreducible quadratic factor since $x^2 + x + 1 = 0$ has only complex roots: $b^2 - 4ac = 1 - 4 < 0$. Thus

$$\frac{x^2 + 2}{x^3 + x^2 + x} = \frac{A}{x} + \frac{Bx + C}{x^2 + x + 1},$$

$$x^2 + 2 = A(x^2 + x + 1) + x(Bx + C).$$

We let $x = 0$ (that is, we equate constant terms) to get $A = 2$. Equating coefficients of x^2 and of x, we get

$$1 = A + B = 2 + B \rightarrow B = -1,$$

$$0 = A + C = 2 + C \rightarrow C = -2.$$

Then

$$\int \frac{x^2+2}{x^3+x^2+x}\,dx = 2\int \frac{dx}{x} - 1\int \frac{x+2}{x^2+x+1}\,dx$$

$$= 2\ln|x| - \frac{1}{2}\int \frac{2x+4}{x^2+x+1}\,dx$$

$$= 2\ln|x| - \frac{1}{2}\int \frac{2x+1+3}{x^2+x+1}\,dx$$

$$= 2\ln|x| - \frac{1}{2}\int \frac{2x+1}{x^2+x+1}\,dx - \frac{3}{2}\int \frac{dx}{x^2+x+1}$$

where we recognize $2x+1 = D_x(x^2+x+1)$

$$= 2\ln|x| - \frac{1}{2}\ln|x^2+x+1| - \frac{3}{2}\int \frac{dx}{(x^2+x+1/4)+3/4}$$

completing the square

$$= 2\ln|x| - \frac{1}{2}\ln(x^2+x+1) - \frac{3}{2}\int \frac{dx}{(x+1/2)^2+(\sqrt{3}/2)^2}$$

$$= 2\ln|x| - \frac{1}{2}\ln(x^2+x+1) - \frac{3}{2}\cdot\frac{2}{\sqrt{3}}\tan^{-1}\frac{x+1/2}{\sqrt{3}/2} + C$$

$$= 2\ln|x| - \frac{1}{2}\ln(x^2+x+1) - \sqrt{3}\tan^{-1}\frac{2x+1}{\sqrt{3}} + C. \quad \blacksquare$$

An Application to Population Growth

As an application of the technique of integration by partial fractions, we consider Problem 6 of the Prologue:

The rate of population increase in the United States has been slackening recently. How can census data be used to estimate the long-term trend for this population?

In Section 6.3 we used the exponential growth equation

$$\frac{dP}{dt} = kP, \tag{17}$$

where $k > 0$ is the growth constant, to study bacteria populations. As we mentioned, (17) is an accurate model for the growth of laboratory cultures of bacteria during their initial growth phase, which typically lasts 12 to 18 hours. Since (17) leads to

$$P = P_0e^{kt}, \tag{18}$$

the growth rate k is *not sustainable*. For as $t \to +\infty$, so does e^{kt} (since $k > 0$), and hence $P_0e^{kt} \to +\infty$. But no population can grow without bound, due to the constraints of finite living space and finite food supply. Observation of bacteria and insect colonies, and even colonies of rats, suggests that once P reaches a certain level, reproduction rates start to decline. This has led to the *logistic growth* equation

$$\frac{dP}{dt} = kP - mP^2 = kP\left(1 - \frac{m}{k}P\right), \tag{19}$$

where k and m are positive constants. The idea behind (19) is that the very size P of the population acts as a brake on its growth rate. The rate of growth is then jointly proportional to P and the quantity $1 - (m/k)P$, which measures how close P is to the maximum sustainable population. The first thing to do in studying this model is to solve the differential equation (19) to obtain a formula for P.

Example 7

Find $P = P(t)$ from (19).

Solution. It is helpful to write (19) as

$$\frac{dP}{dt} = P(k - mP).$$

Assuming that $P \neq 0$ and $P \neq k/m$, we obtain

$$\int \frac{1}{P(k - mP)}\, dP = \int dt. \tag{20}$$

To integrate (20), we find the partial fraction decomposition of the left side. We have

$$\frac{1}{P(k - mP)} = \frac{A}{P} + \frac{B}{k - mP} \rightarrow 1 = A(k - mP) + BP.$$

Letting $P = 0$, and then letting $P = k/m$, we find

$$1 = Ak \rightarrow A = 1/k,$$
$$1 = B\, k/m \rightarrow B = m/k.$$

Thus (20) gives

$$\frac{1}{k}\int \frac{dP}{P} + \frac{m}{k}\int \frac{dP}{k - mP} = \int dt \rightarrow \frac{1}{k}\ln P - \frac{1}{k}\int \frac{-m\, dP}{k - mP} = t + C,$$

$$\frac{1}{k}\ln P - \frac{1}{k}\ln|k - mP| = t + C \rightarrow \ln\left|\frac{P}{k - mP}\right| = k(t + C),$$

$$\left|\frac{P}{k - mP}\right| = e^{kt}e^{kC} \rightarrow \pm\frac{P}{k - mP} = e^{kC}e^{kt},$$

$$\frac{P}{k - mP} = Ke^{kt}, \tag{21}$$

where $K = \pm e^{kC}$. We can solve for P in (21) by first clearing it of fractions:

$$P = kKe^{kt} - mKe^{kt}P \rightarrow P(1 + mKe^{kt}) = kKe^{kt} \rightarrow P = \frac{kKe^{kt}}{1 + mKe^{kt}}. \tag{22}$$

Rewriting the last equation in (22), we obtain

$$P = \frac{kK}{e^{-kt} + mK},$$

so that as $t \rightarrow +\infty$, $P \rightarrow (kK/mK) = k/m$. Thus the logistic growth model predicts that eventually the population will tend toward a *stable* level k/m, which is the ratio of the growth constant k in (19) to the braking constant m. This certainly is a more reasonable long-term prediction than that given by (18). Successive refinements of this model have been used over the last half century by the United States Bureau of the Census to predict population growth in the United States. Recently the census bureau has predicted a long-term stable population of about 265,000,000 for the United States. See Exercises 45 and 46 for more about this. ■

HISTORICAL NOTE

The method of judicious substitution was developed by the English mathematician and physicist Oliver Heaviside (1850–1925) about a century ago. The logistic growth model was introduced in 1846 by the Belgian sociologist Pierre-François Verhulst (1804–1849) to model Belgium's population. He obtained a long-range stable population estimate of about 9,400,000. In 1970 (a century and a quarter later), the actual population of Belgium was 9,650,000. The use of this model in demographic studies is widespread because of such accuracy.

Verhulst's pioneering work attracted little attention at first and lay virtually unknown for many years. The logistic growth model was rediscovered in 1920 by the American biologists Raymond Pearl (1879–1940) and Louis J. Reed (1886–1966). They studied United States census data from 1790 through 1910 and on this basis proposed the logistic growth equation (19) with $k = 3.13395 \times 10^{-2}$ and $m = 1.58864 \times 10^{-10}$ as a model for the growth of the population of the United States. This led to a predicted stable population $k/m = 197{,}272{,}500$, whereas the 1970 census showed a population of 203,184,772. The present model for the United States population used by the census bureau is a descendant of the Pearl-Reed model.

Exercises 7.4

In Exercises 1–40, evaluate the given integrals.

1. $\int \frac{dx}{x^2 - 1}$

2. $\int \frac{3\,dx}{x^2 - 4}$

3. $\int \frac{2x}{x^2 - x - 2}\,dx$

4. $\int \frac{4x - 1}{x^2 + 2x - 3}\,dx$

5. $\int_3^4 \frac{x - 1}{x^3 - x^2 - 2x}\,dx$

6. $\int_{-4}^{-3} \frac{3 - x}{x^3 - 2x^2 - 8x}\,dx$

7. $\int \frac{4x^2 - 3x - 4}{x^3 + x^2 - 2x}\,dx$

8. $\int \frac{x^2 + 2x - 3}{x^3 + 2x^2 - 8x}\,dx$

9. $\int_2^3 \frac{x\,dx}{x^2 - 2x + 1}$

10. $\int_0^1 \frac{2 - 3x}{x^2 - 6x + 9}\,dx$

11. $\int \frac{(x^2 + 1)\,dx}{(x - 1)^2(x + 3)^2}$

12. $\int \frac{x^2 - x + 2}{(x + 1)^2(x - 2)^2}\,dx$

13. $\int \frac{x + 5}{x^3 - 2x^2 + x}\,dx$

14. $\int \frac{x^2 - x + 3}{x^3 - 4x^2 + 4x}\,dx$

15. $\int_2^3 \frac{x + 1}{x^3 - x^2}\,dx$

16. $\int_1^2 \frac{2x^2 - x + 1}{x^3 - 9x^2}\,dx$

17. $\int \frac{x^2 + 1}{x^2 - 1}\,dx$

18. $\int \frac{x^2 + x + 2}{x^2 - 4}\,dx$

19. $\int \frac{x^3 + x^2}{x^2 + x - 2}\,dx$

20. $\int \frac{x^3 + 2x^2 - x + 1}{x^2 - x - 6}\,dx$

21. $\int_{-1}^{0} \frac{3x + 2}{(x - 1)(x^2 + 1)}\,dx$

22. $\int_0^1 \frac{2x - 3}{(x - 3)(x^2 + 4)}\,dx$

23. $\int \frac{x^2\,dx}{x^4 - 16}$

24. $\int \frac{x^3 + x^2 + 1}{x^4 - 81}\,dx$

25. $\int \frac{x^2 + x + 1}{(2x + 1)(x^2 + 1)}\,dx$

26. $\int \frac{x^2 + 5x + 2}{(2x - 3)(x^2 + 4)}\,dx$

27. $\int \frac{3x^2 + 1}{x^3 + x}\,dx$

28. $\int \frac{3x^2 - 10x + 6}{x^3 - 5x^2 + 6x}\,dx$

29. $\int \frac{3x^2\,dx}{(x^2 - x + 1)(x + 1)}$

30. $\int \frac{3x^2\,dx}{(x^2 - 4x + 4)(x + 2)}$

31. $\int \frac{x^2 - 3x - 2}{x^3 + x^2 + x}\,dx$

32. $\int \frac{x^2 - 2x + 1}{x^3 + 3x^2 + x}\,dx$

33. $\int_0^1 \frac{x^3\,dx}{(x^2 + 1)^2}$

34. $\int_0^2 \frac{x^3 + 2}{(x^2 + 4)^2}\,dx$

35. $\int \frac{2x^2 + x + 7}{(x^2 + 4)^2}\, dx$

36. $\int \frac{x^2 - 3x + 5}{(x^2 + 9)^2}\, dx$

37. $\int \frac{x^3 - 3x^2 + 2x - 3}{(x^2 + 1)^2}\, dx$

38. $\int \frac{x^3 + 4x^2 - x - 3}{(x^2 + 1)^2}\, dx$

39. $\int \frac{x^3 - 2x}{(x^2 + 2x + 2)^2}\, dx$

40. $\int \frac{x^3 - x^2 + 4}{(x^2 + x + 1)^2}\, dx$

In Exercises 41–44, change variables to transform the given integral to a rational function of *u*. Then use partial fractions to complete the integration.

41. $\int \frac{\cos\theta\, d\theta}{\sin^2\theta - \sin\theta - 6}$

42. $\int \frac{\sin\theta\, d\theta}{\cos^2\theta + 5\cos\theta + 4}$

43. $\int \frac{e^x\, dx}{e^x + e^{2x}}$

44. $\int \frac{e^{2t}\, dt}{(1 + e^t)^2}$

45. In Example 7, denote the limiting population k/m by L.
   (a) Show that $\frac{P}{L - P} = mKe^{kt}$.
   (b) If we measure time from the instant when $P = L/2$, show then that $P = \frac{L}{1 + e^{-kt}}$. (In the Pearl–Reed model, time was measured from April 1, 1914.)

46. The United States population in 1940 was about 131,700,000 and in 1960 was about 179,300,000. Use the formula in Exercise 45 to model the population, measuring t since the time of the 1940 census.
   (a) Find k so that $P(20) = 179{,}300{,}000$.
   (b) Predict the population in 1970, 1980, and 2000. If the 1970 population was 204,000,000, and the 1980 population was 226,500,000, then what is the percent error in the model for $t = 30$ and 40?
   (c) What is the limiting population of this model?

47. Consider a chemical reaction in which single molecules of S and T react to form one molecule of a product U. The *law of mass action* says that the rate of production of U at any instant is proportional to the concentrations of S and T present at that instant. Suppose that the starting concentrations are $S_0 = T_0 = a$ and $U_0 = 0$ (moles per liter). If $U = a/2$ at $t = 10$, find a formula for the concentration of U at time t. What is the long-term trend?

48. In Exercise 47, suppose that 2 molecules of S combine with one molecule of T to form each molecule of U. If $S_0 = a$, $T_0 = a/2$, and $U_0 = 0$ (mol/L), find a formula for the concentration of U present at time t. What is the long-term trend?

49. In Exercise 47, suppose that $S_0 = a \neq b = T_0$ and $U_0 = 0$ (mol/L). Then find a formula for the concentration of U present at time t.

50. A simple model for the spread of epidemics is based on the differential equation $dy/dt = ky(1 - y)$, where $y \in [0, 1]$ is the fraction of the population infected at time t. Assuming that $y = y_0 > 0$ when $t = 0$, find y at time t. What happens as $t \to \infty$?

7.5 Further Techniques / Use of Integral Tables

In the first two parts of this section, we present some special methods that apply to narrower classes of integrals than the techniques presented earlier in this chapter. They are, however, quite effective. The last part of the section discusses how to take advantage of tables of integrals such as the one found in Appendix I.

We first give an algorithm for integrating expressions that contain fractional powers of x.

INTEGRATION OF EXPRESSIONS CONTAINING FRACTIONAL POWERS OF x

1. Let $u = x^{1/n}$, where n is the least common denominator of all the powers of x that occur.
2. Express $x = u^n$ in terms of u, and find $dx = nu^{n-1}\, du$.
3. Transform the given integral to an integral involving a rational function of u.
4. Integrate (use partial fractions if necessary).
5. Express the answer in terms of x.

Example 1

Evaluate $\int \frac{2\sqrt[6]{x} + \sqrt{x}}{1 + \sqrt[3]{x}}\,dx$.

Solution. Here the fractional powers of x are $\sqrt[6]{x} = x^{1/6}$, $\sqrt{x} = x^{1/2}$, and $\sqrt[3]{x} = x^{1/3}$. The least common denominator of the exponents is 6. We therefore let $u = x^{1/6}$. Since $u^6 = x$, Step 2 is completed by calculating $dx = 6u^5\,du$. Also, we have

$$\sqrt{x} = x^{1/2} = (x^{1/6})^3 = u^3, \qquad \text{and} \qquad \sqrt[3]{x} = x^{1/3} = (x^{1/6})^2 = u^2.$$

Thus Step 3 gives

$$\int \frac{2\sqrt[6]{x} + \sqrt{x}}{1 + \sqrt[3]{x}}\,dx = \int \frac{(2u + u^3)\cdot 6u^5\,du}{u^2 + 1} = 6\int \frac{u^8 + 2u^6}{u^2 + 1}\,du$$

$$= 6\int \left(u^6 + u^4 - u^2 + 1 - \frac{1}{u^2 + 1}\right)du \qquad \textit{by long division}$$

$$= 6\left[\frac{u^7}{7} + \frac{u^5}{5} - \frac{u^3}{3} + u - \tan^{-1} u\right] \qquad \textit{which completes Step 4}$$

$$= \frac{6}{7}x^{7/6} + \frac{6}{5}x^{5/6} - 2x^{1/2} + 6x^{1/6} - 6\tan^{-1} x^{1/6} + C. \quad \blacksquare$$

Example 2

Evaluate $\int \frac{2 + \sqrt{x}}{x + \sqrt[4]{x}}\,dx$.

Solution. We let $u = \sqrt[4]{x}$, so $u^2 = \sqrt{x}$ and $u^4 = x$. Then Step 2 gives $dx = 4u^3\,du$. Hence for Step 3 we obtain

$$\int \frac{2 + \sqrt{x}}{x + \sqrt[4]{x}}\,dx = \int \frac{2 + u^2}{u^4 + u}\cdot 4u^3\,du = 4\int \frac{u^5 + 2u^3}{u(u^3 + 1)}\,du$$

$$= 4\int \frac{u^4 + 2u^2}{u^3 + 1}\,du = 4\int \left(u + \frac{2u^2 - u}{u^3 + 1}\right) \qquad \textit{by division}$$

$$= 2u^2 + 4\int \frac{2u^2 - u}{(u + 1)(u^2 - u + 1)}\,du.$$

To evaluate the last integral, we use partial fractions:

$$\frac{2u^2 - u}{(u + 1)(u^2 - u + 1)} = \frac{A}{u + 1} + \frac{Bu + C}{u^2 - u + 1},$$

$$2u^2 - u = A(u^2 - u + 1) + (Bu + C)(u + 1).$$

Letting $u = -1$, we obtain

$$3 = 3A \rightarrow A = 1.$$

Equating coefficients of u^2 and constants, we have

$$2 = A + B = 1 + B \rightarrow B = 1$$

and

$$0 = A + C = 1 + C \rightarrow C = -1.$$

Now we are ready to complete Step 4. We get

$$\int \frac{2u^2 - u}{(u+1)(u^2 - u + 1)} = \int \frac{du}{u+1} + \int \frac{u-1}{u^2 - u + 1}\, du$$

$$= \ln|u+1| + \frac{1}{2}\int \frac{2u-2}{u^2 - u + 1}\, du$$

$$= \ln|u+1| + \frac{1}{2}\int \frac{2u-1-1}{u^2 - u + 1}\, du$$

$$= \ln|u+1| + \frac{1}{2}\int \frac{2u-1}{u^2 - u + 1}\, du - \frac{1}{2}\int \frac{du}{u^2 - u + 1}$$

$$= \ln|u+1| + \frac{1}{2}\ln|u^2 - u + 1| - \frac{1}{2}\int \frac{du}{(u - 1/2)^2 + 3/4}$$

$$= \ln|u+1| + \frac{1}{2}\ln|u^2 - u + 1|$$

$$-\frac{1}{2}\frac{2}{\sqrt{3}}\tan^{-1}\frac{u - 1/2}{\sqrt{3}/2} + C$$

$$= \ln|u+1| + \frac{1}{2}\ln|u^2 - u + 1| - \frac{1}{\sqrt{3}}\tan^{-1}\frac{2u-1}{\sqrt{3}} + C.$$

We then finish the integration via Step 5, getting

$$\int \frac{2 + \sqrt{x}}{x + \sqrt[4]{x}}\, dx = 2u^2 + 4\ln|u+1| + 2\ln|u^2 - u + 1| - \frac{4}{\sqrt{3}}\tan^{-1}\frac{2u-1}{\sqrt{3}} + C$$

$$= 2x^{1/2} + 4\ln(x^{1/4} + 1) + 2\ln(\sqrt{x} - x^{1/4} + 1)$$

$$-\frac{4}{\sqrt{3}}\tan^{-1}\frac{2x^{1/4} - 1}{\sqrt{3}} + C. \quad ■$$

A similar strategy can be used on integrals that contain $\sqrt[n]{f(x)}$ for a function f. Rather than restating the above algorithm for such more general integrals, we illustrate the procedure to follow by reworking Example 2 of Section 2.

Example 3

Evaluate $\int x^3\sqrt{1 + x^2}\, dx$.

Solution. We let $u = \sqrt{1 + x^2}$, so $u^2 = 1 + x^2$. Then

$$2u\, du = 2x\, dx \rightarrow x\, dx = u\, du.$$

Since $x^2 = u^2 - 1$, we have

$$\int x^3\sqrt{1 + x^2}\, dx = \int x^2\sqrt{1 + x^2}\, x\, dx = \int (u^2 - 1)u \cdot u\, du$$

$$= \int (u^4 - u^2)\, du = \frac{1}{5}u^5 - \frac{1}{3}u^3 + C$$

$$= \frac{1}{5}(1 + x^2)^{5/2} - \frac{1}{3}(1 + x^2)^{3/2} + C. \quad ■$$

At first glance we seem to have gotten a different answer from that found in Section 2 using integration by parts, but a little algebra shows that the answers are in fact the same (Exercise 39).

Rational Functions of Sine and Cosine*

Rational functions of sine and cosine can sometimes be integrated by first changing variable and then using partial fractions, as Exercises 41 and 42 of the preceding section suggest. A special technique for this uses the change of variable

$$u = \tan \frac{1}{2} x. \tag{1}$$

Then

$$\tan^{-1} u = \frac{1}{2} x \rightarrow x = 2 \tan^{-1} u.$$

Hence

$$dx = \frac{2}{1+u^2} du. \tag{2}$$

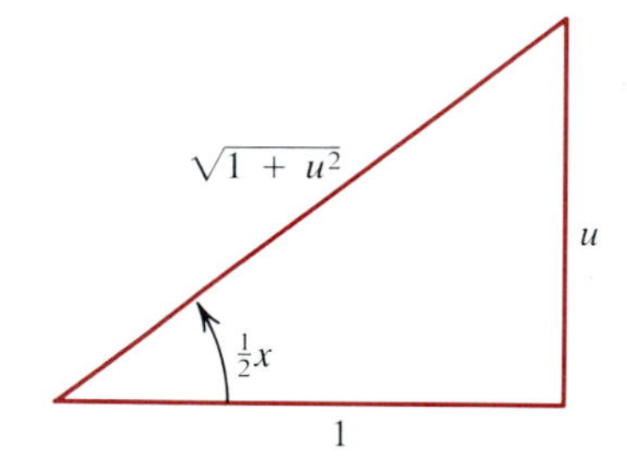

FIGURE 5.1

We can get $\sin x$ and $\cos x$ in terms of u from (1), **Figure 5.1,** and the trigonometric identities

$$\sin x = 2 \sin \frac{1}{2} x \cos \frac{1}{2} x \quad \text{and} \quad \cos x = \cos^2 \frac{1}{2} x - \sin^2 \frac{1}{2} x. \tag{3}$$

These give

$$\sin x = 2 \frac{u}{\sqrt{1+u^2}} \cdot \frac{1}{\sqrt{1+u^2}} = \frac{2u}{1+u^2} \tag{4}$$

and

$$\cos x = \frac{1}{1+u^2} - \frac{u^2}{1+u^2} = \frac{1-u^2}{1+u^2} \tag{5}$$

Using (2), (4), and (5) we can transform the rational function of $\sin x$ and $\cos x$ to a rational function of u. This leads to the following algorithm.

INTEGRATION OF RATIONAL FUNCTIONS OF $\sin x$ AND $\cos x$

1. Let $u = \tan \frac{1}{2} x$.
2. Substitute the value of dx given by Equation (2).
3. Use Equations (4) and (5) to express $\sin x$ and $\cos x$ in terms of u.
4. Integrate the resulting rational function of u (use partial fractions if necessary).
5. Use Equation (1) and, if necessary, the identities (3) to express the answer in terms of x.

* Optional

Example 4

Evaluate $\int \frac{dx}{1 + \cos x}$.

Solution. Following the suggested approach, we let $u = \tan \frac{1}{2}x$ and get from (2) and (5)

$$\int \frac{dx}{1 + \cos x} = \int \frac{2}{1 + u^2} \cdot \frac{1}{1 + (1 - u^2)/(1 + u^2)}\, du = \int \frac{2}{(1 + u^2) + (1 - u^2)}\, du$$

$$= \int du = u + C = \tan \frac{1}{2} x + C. \quad \blacksquare$$

The next example illustrates how the method of partial fractions can be combined with the algorithm for rational functions of $\sin x$ and $\cos x$.

Example 5

Evaluate $\int \frac{\sin x}{1 + \sin x}\, dx$.

Solution. Here $u = \tan \frac{1}{2}x$ and (2) and (4) give

$$\int \frac{\sin x}{1 + \sin x}\, dx = \int \frac{2u}{1 + u^2} \cdot \frac{1}{1 + 2u/(1 + u^2)} \cdot \frac{2\, du}{1 + u^2}$$

$$= \int \frac{4u}{(1 + u^2 + 2u)(1 + u^2)}\, du$$

$$= \int \frac{4u}{(1 + u)^2(1 + u^2)}\, du.$$

We do the last integral by partial fractions:

$$\frac{4u}{(1 + u)^2(1 + u^2)} = \frac{A}{1 + u} + \frac{B}{(1 + u)^2} + \frac{Cu + D}{u^2 + 1},$$

$$4u = A(1 + u)(u^2 + 1) + B(u^2 + 1) + (Cu + D)(1 + u)^2.$$

Letting $u = -1$, and then letting $u = 0$, we get

$$-4 = 2B \rightarrow B = -2.$$

$$0 = A + B + D \rightarrow 2 = A + D. \tag{6}$$

Equating coefficients of u^3 and u^2 gives

$$0 = A + C, \tag{7}$$

$$0 = A + B + 2C + D \rightarrow 2 = A + 2C + D. \tag{8}$$

From (6) we have $D = 2 - A$; and it follows from (8) that $C = -A$. Substituting these facts into (8), we find

$$2 = A - 2A + 2 - A \rightarrow 0 = -2A \rightarrow A = 0.$$

Then (7) gives $C = 0$ and (6) gives $D = 2$. We therefore have

$$\int \frac{4u}{(1+u)^2(1+u^2)}\,du = -2\int \frac{du}{(1+u)^2} + 2\int \frac{du}{u^2+1}$$

$$= -2\frac{(1+u)^{-1}}{-1} + 2\tan^{-1}u + C$$

$$= \frac{2}{1+\tan(x/2)} + 2\cdot\frac{1}{2}x + C$$

$$= x + \frac{2}{1+\tan(x/2)} + C. \ \blacksquare$$

Use of Tables

Before closing this final section on techniques of indefinite integration, we give some suggestions for using tables of integrals like the one in Appendix I. While other, more complete, tables are available, it is usually necessary to do some algebraic manipulation to convert a given integral to a form found in a table. Sometimes a change of variable or integration by parts or partial fractions may be needed to reduce the integral to a form in an integral table. Reduction formulas often must be used repeatedly to complete an integration.

Example 6

Use the table of integrals to evaluate $\int \frac{dx}{\sqrt{2x^2 - 3x}}$.

Solution. The closest formula we find is Formula 27, which involves $1/\sqrt{u^2 - a^2}$. To apply that, we first complete the square in the radical:

$$\sqrt{2x^2 - 3x} = \sqrt{2\left(x^2 - \frac{3}{2}x + \frac{9}{16}\right) - \frac{9}{8}} = \sqrt{2\left(x - \frac{3}{4}\right)^2 - \left(\frac{3}{2\sqrt{2}}\right)^2}.$$

It is natural to let $u = \sqrt{2}(x - \frac{3}{4})$ and $a = 3/(2\sqrt{2})$. We then have $du = \sqrt{2}\,dx$, so that $dx = du/\sqrt{2}$. Thus we obtain

$$\int \frac{dx}{\sqrt{2x^2 - 3x}} = \frac{1}{\sqrt{2}}\int \frac{du}{\sqrt{u^2 - a^2}} = \frac{1}{\sqrt{2}}\ln(u + \sqrt{u^2 - a^2}) + C$$

$$= \frac{1}{\sqrt{2}}\ln\left(\sqrt{2}\left(x - \frac{3}{4}\right) + \sqrt{2x^2 - 3x}\right) + C,$$

bearing in mind that $\sqrt{u^2 - a^2}$ is the original denominator $\sqrt{2x^2 - 3x}$. $\blacksquare$

Example 7

Use the table of integrals to evaluate $\int x^2 \sin 2x\,dx$.

Solution. We use Formula 35 for $\int u^n \sin u\, du$, letting $u = 2x$. Then $du = 2\,dx$, so $dx = du/2$. Also, $x = u/2$, so $x^2 = u^2/4$. We therefore have

$$\int x^2 \sin 2x\, dx = \int \frac{1}{4}u^2 \sin u \cdot \frac{1}{2}\,du = \int \frac{1}{8}u^2 \sin u\, du$$

$$= \frac{1}{8}\left[-u^2 \cos u + 2\int u \cos u\, du\right] \qquad \textit{by Formula 35, with } n = 2$$

$$= -\frac{1}{8}u^2 \cos u + \frac{1}{4}\int u \cos u\, du$$

$$= -\frac{1}{8}u^2 \cos u + \frac{1}{4}\left[u \sin u - \int \sin u\, du\right] \qquad \textit{by Formula 34, with } n = 1$$

$$= -\frac{1}{8}(4x^2)\cos 2x + \frac{1}{4}\cdot 2x \sin 2x + \frac{1}{4}\cos u + C$$

$$= -\frac{1}{2}x^2 \cos 2x + \frac{1}{2}x \sin 2x + \frac{1}{4}\cos 2x + C. \quad \blacksquare$$

Exercise 7.5

In Exercises 1–28, use the algorithms of this section to evaluate the given integral.

1. $\int \frac{dx}{1 + \sqrt{x}}$
2. $\int \frac{dx}{1 - \sqrt{x}}$
3. $\int_0^1 \frac{\sqrt{x}\, dx}{1 + \sqrt[3]{x}}$
4. $\int_0^8 \frac{\sqrt[6]{x}\, dx}{1 + \sqrt[3]{x}}$
5. $\int \frac{1 - \sqrt{x}}{x + \sqrt[4]{x}}\, dx$
6. $\int \frac{\sqrt{x}}{\sqrt[3]{x} + \sqrt{x}}\, dx$
7. $\int \frac{\sqrt{x}}{x^2 - 1}\, dx$
8. $\int \frac{\sqrt{x}}{x^{3/2} - 8}\, dx$
9. $\int x^2\sqrt{x + 1}\, dx$
10. $\int x^3\sqrt{x^2 + 3}\, dx$
11. $\int \frac{x\, dx}{\sqrt{1 + 2x}}$
12. $\int \frac{x\, dx}{\sqrt{1 + kx}}$
13. $\int_0^2 \frac{x^2\, dx}{\sqrt{2 + x}}$
14. $\int_{-4}^0 \frac{x^2\, dx}{\sqrt{1 - 2x}}$
15. $\int \frac{x^3\, dx}{\sqrt[3]{x^2 + 4}}$
16. $\int x^3 \sqrt[3]{x^2 + 1}\, dx$
17. $\int x^3\sqrt{x^2 - 1}\, dx$
18. $\int x^3\sqrt{x^2 - 4}\, dx$
19. $\int_0^4 x^3\sqrt{x^2 + 9}\, dx$
20. $\int_0^3 x^3\sqrt{x^2 + 16}\, dx$
21. $\int \frac{dx}{1 + \sin x}$
22. $\int \frac{dx}{1 - \cos x}$
23. $\int \frac{dx}{1 - \sin x + \cos x}$
24. $\int \frac{dx}{1 + \sin x + \cos x}$
25. $\int_0^{\pi/2} \frac{dx}{2 + \sin x}$
26. $\int_0^{\pi/2} \frac{dx}{2 + \cos x}$
27. $\int \frac{dx}{\sin x(1 + \cos x)}$
28. $\int \frac{dx}{(\cos x)(1 - \cos x)}$

In Exercises 29–38, use Appendix I to evaluate the given integral.

29. $\int \frac{dx}{\sqrt{x^2 + 3x}}$
30. $\int \frac{dx}{\sqrt{x^2 - x}}$
31. $\int \frac{dx}{\sqrt{2x^2 - 5}}$
32. $\int \frac{dx}{\sqrt{3 - 2x^2}}$
33. $\int_0^1 x^3 e^{-2x}\, dx$
34. $\int_0^{\sqrt{\pi}} x e^{-2x^2} \cos 3x^2\, dx$
35. $\int x^3 \cos 3x\, dx$
36. $\int x^2 \sin \frac{1}{3}x\, dx$
37. $\int_0^{\sqrt{5}} (9 - x^2)^{3/2}\, dx$
38. $\int_2^{\sqrt{13}} (x^2 - 4)^{3/2}\, dx$
39. Show that the answers to Example 3 of this section and Example 2 of Section 2 are equivalent.
40. Use the methods of this section to derive the following alternative formula to Formula 16 of the table of integrals.

$$\int \sec x\, dx = \ln\left|\frac{1 + \tan\frac{1}{2}x}{1 - \tan\frac{1}{2}x}\right| + C.$$

41. Multiply numerator and denominator of $1/\cos x$ and use partial fractions to derive the alternative formula

$$\int \sec x\, dx = \ln\left|\frac{1 + \sin x}{1 - \sin x}\right|^{1/2} + C.$$

42. Use the approach of Exercise 40 to derive an alternative to Formula 17 of the table of integrals.

43. Evaluate $\int \sqrt{1 + e^x}\, dx$.

44. Repeat Exercise 43 for $\int \sqrt{1 + e^{4x}}\, dx$.

45. If $a > b > 0$, then show that

$$\int \frac{dx}{a + b\cos x} = \frac{2}{\sqrt{a^2 - b^2}} \tan^{-1}\left(\sqrt{\frac{a - b}{a + b}} \tan \frac{1}{2} x\right) + C.$$

46. What happens in Exercise 45 if $b > a > 0$?

7.6 Improper Integrals

Thus far in this chapter we have concentrated on indefinite integrals. In this section, we extend the definite integral $\int_a^b f(x)\, dx$ to two new contexts: (a) unbounded intervals $(-\infty, b)$, $(a, +\infty)$, and $(-\infty, +\infty)$ and (b) unbounded functions f on closed intervals $[a, b]$. The resulting concept is important in statistics and probability, physics, engineering, physical chemistry, and economics.

We begin by returning to a question first considered in Section 5.6. In Example 2 of that section, we calculated the work done against the force of gravity in lifting a rocket of mass 2000 kg from the earth's surface to an altitude of 100 km. We used Newton's law of gravitation,

$$F(x) = \frac{GMm}{x^2},$$

where x is the distance in kilometers from the center of the earth to the center of mass of the rocket, $m = 2000$ is the mass of the rocket in kilograms, M is the mass of the earth (about 5.993×10^{24} kg), and G is the universal gravitational constant, approximately 6.674×10^{-17} N-km^2/kg^2. We saw that if the mass of burned fuel is neglected, then

$$W = \int_{R_0}^{R_1} F(x)\, dx = GMm \int_{R_0}^{R_1} x^{-2}\, dx = GMm\left(\frac{1}{R_0} - \frac{1}{R_1}\right), \tag{1}$$

where $R_0 \approx 6400$ is the radius of the earth in kilometers and $R_1 = R_0 + 100 \approx 6500$. Multistage rockets are used to launch deep space probes in addition to putting satellites in orbit.

Example 1

At altitude R_1, the first stage drops off, and a second stage of remaining mass 500 kg continues on. What is the total work done against gravity by the time the probe has completely left the influence of the earth's gravity? (Neglect the weight of the fuel burned.)

Solution. As the probe travels farther from the earth, less and less residual gravitational force acts on it. The work done in reaching R_1 is given by (1). The work done in going from R_1 to a point at distance R from the earth's center is

$$W_1 = GMm_1 \int_{R_1}^{R} x^{-2}\, dx = GMm_1\left(\frac{1}{R_1} - \frac{1}{R}\right), \tag{2}$$

where $m_1 = 500$ is the mass of the remaining pay load. The total work done in sending the probe from the earth's surface to R is therefore $W + W_1$. As R

gets larger and larger, the term $1/R$ in (2) becomes more and more negligible. Eventually W_1 becomes indistinguishable from GMm_1/R_1. The total work done in sending the probe completely out of the influence of the earth's gravity is thus

$$\bar{W} = \frac{GMm}{R_0} - \frac{GMm}{R_1} + \frac{GMm_1}{R_1} = GM\left(\frac{m}{R_0} - \frac{m - m_1}{R_1}\right). \blacksquare$$

In finding $\bar{W}$, we calculated $\lim_{R \to +\infty} W_1$. It is natural to call this the work W_∞ done in moving the probe from R_1 to $+\infty$ and to write

$$W_\infty = \int_{R_1}^{+\infty} \frac{GMm_1}{x^2}\, dx. \tag{3}$$

This indeed is what is done in physics and engineering. The following mathematical definition is implicit in our solution to Example 1.

6.1 DEFINITION

Suppose that f is a continuous function on $[a, +\infty)$. Then

$$\int_a^{+\infty} f(x)\, dx = \lim_{b \to +\infty} \int_a^b f(x)\, dx, \tag{4}$$

if this limit exists as a real number.

If f is continuous on $(-\infty, b]$, then

$$\int_{-\infty}^{b} f(x)\, dx = \lim_{a \to -\infty} \int_a^b f(x)\, dx, \tag{5}$$

if this limit exists as a real number.

If f is continuous on $\mathbf{R} = (-\infty, +\infty)$, then

$$\int_{-\infty}^{+\infty} f(x)\, dx = \int_{-\infty}^{0} f(x)\, dx + \int_0^{+\infty} f(x)\, dx, \tag{6}$$

provided *both* integrals on the right exist as real numbers in the sense of (4) and (5).

The three integrals defined by (4), (5), and (6) are called ***improper integrals.*** The term *improper* reflects the fact that until now definite integrals $\int_a^b f(x)\, dx$ made sense only when a and b were both real numbers. It was not proper to allow $a = -\infty$ or $b = +\infty$. If a limit in (4), (5), or (6) fails to exist, then we say that the corresponding improper integral ***diverges*** or ***is divergent*** or ***does not exist.*** When the limits *do* exist, we say the improper integral ***converges*** or ***is convergent*** to its limit. Sometimes two improper integrals that *seem* quite similar can have totally different convergence characteristics.

Example 2

Evaluate if possible (a) $\int_1^{+\infty} \frac{1}{x}\, dx$ and (b) $\int_1^{+\infty} \frac{\pi}{x^2}\, dx$.

Solution. (a) $\int_1^{+\infty} \frac{1}{x}\, dx = \lim_{b \to +\infty} \int_1^b \frac{1}{x}\, dx = \lim_{b \to +\infty} \ln x \Big]_1^b$

$$= \lim_{b \to +\infty} [\ln b - 0] = +\infty.$$

Thus the improper integral $\int_1^{+\infty} (1/x)\, dx$ is divergent.

$$\text{(b)} \int_1^{+\infty} \frac{\pi}{x^2}\,dx = \pi \lim_{b\to+\infty} \int_1^b x^{-2}\,dx = \pi \lim_{b\to+\infty} \left[\frac{x^{-1}}{-1}\right]_1^b$$

$$= \pi \lim_{b\to+\infty} \left[-\frac{1}{b} + 1\right] = \pi.$$

Hence this improper integral converges to the value π. ■

There is a curious geometrical interpretation of Example 2. In **Figure 6.1**(a) we show the region lying under the graph of $y = 1/x$ to the right of $x = 1$. Example 2(a) can be looked upon as answering (negatively) the question *Is the shaded area in Figure 6.1(a) finite?* Just as we used (3) to define the work done in transporting an object from R_1 out of the earth's gravitational influence, so it is natural to define the area under $y = 1/x$ to the right of $x = 1$ to be

$$A = \int_1^{+\infty} \frac{1}{x}\,dx,$$

if this improper integral exists. Since the integral diverges to $+\infty$, the area is not finite.

FIGURE 6.1

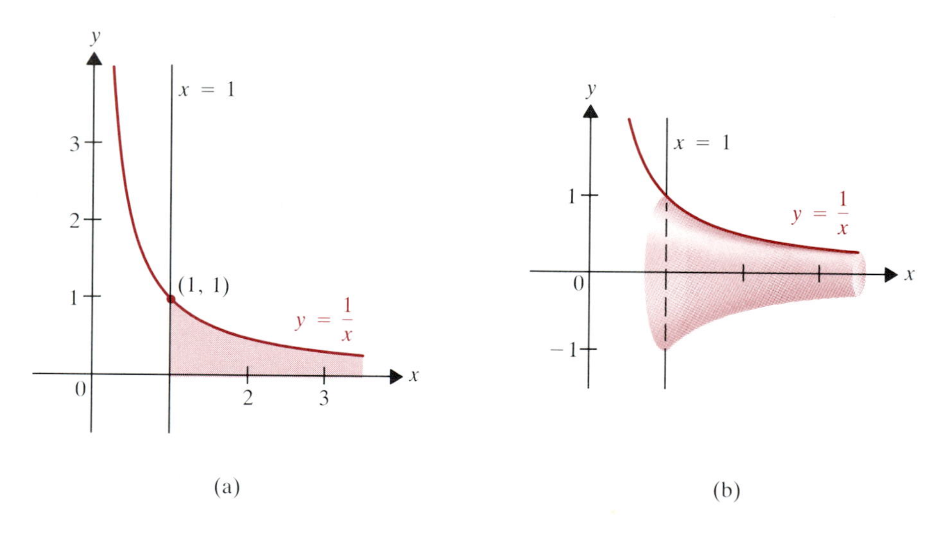

Figure 6.1(b) shows the solid of revolution that results from revolving the shaded region in Figure 6.1(a) about the x-axis. We can ask if that solid has finite volume. It is natural to try to calculate the volume by the circular disk method. That is, we say that the volume is

$$V = \int_1^{+\infty} \pi\left(\frac{1}{x}\right)^2 dx = \int_1^{+\infty} \frac{\pi}{x^2}\,dx,$$

if this improper integral exists. Example 2(b) says that this volume is indeed finite and in fact is π. Example 2 thus shows that when a region of infinite area is revolved about an axis, the resulting solid may have finite volume! This may seem paradoxical, but it reflects the fact that planar areas and three-dimensional volumes, while both measures of extent, do measure *different* things. Plane regions are quite different from solids, as exemplified by our present conception of the earth versus that prevailing before Columbus!

L'Hôpital's rule can be helpful in determining whether improper integrals converge or diverge, as the next example illustrates.

Example 3

Evaluate $\int_{-\infty}^{+\infty} xe^{-x}\,dx$ if possible.

Solution. We have to see whether $\int_0^{+\infty} xe^{-x}\,dx$ and $\int_{-\infty}^0 xe^{-x}\,dx$ exist. If they do, then $\int_{-\infty}^{+\infty} xe^{-x}\,dx$ converges to their sum. We use integration by parts with $u = x$ and $dv = e^{-x}\,dx$. Then $du = dx$ and $v = -e^{-x}$, so

$$\int xe^{-x}\,dx = uv - \int v\,du = -xe^{-x} - \int -e^{-x}\,dx = -xe^{-x} + \int e^{-x}\,dx$$

$$= -\frac{x}{e^x} - \frac{1}{e^x} + C = -\frac{x+1}{e^x} + C.$$

We thus obtain

$$\int_0^{+\infty} xe^{-x}\,dx = -\lim_{b\to+\infty}\left[\frac{x+1}{e^x}\right]_0^b = -\lim_{b\to+\infty}\left[\frac{b+1}{e^b} - 1\right]$$

$$= 1 - \lim_{b\to+\infty}\frac{b+1}{e^b}$$

$$= 1 - \lim_{b\to+\infty}\frac{1}{e^b} \qquad \textit{by l'Hôpital's rule}$$

$$= 1.$$

Similarly,

$$\int_{-\infty}^{0} xe^{-x}\,dx = -\lim_{a\to-\infty}\left[\frac{x+1}{e^x}\right]_a^0 = -\lim_{a\to-\infty}\left[1 - (a+1)e^{-a}\right]$$

$$= +\infty,$$

since as $a \to -\infty$, $e^{-a} \to +\infty$ and $a + 1 \to -\infty$. Thus, although $\int_0^{+\infty} xe^{-x}\,dx$ converges, $\int_{-\infty}^{+\infty} xe^{-x}\,dx$ diverges because $\int_{-\infty}^0 xe^{-x}\,dx$ fails to exist. ■

Pay careful attention to (6) in Definition 6.1. Notice, in particular, that

$$\int_{-\infty}^{+\infty} f(x)\,dx \qquad \text{is } \textbf{not} \qquad \lim_{a\to+\infty}\int_{-a}^{a} f(x)\,dx.$$

This is easy to see by considering the function $f(x) = x$. We have

$$\lim_{a\to+\infty}\int_{-a}^{a} f(x)\,dx = \lim_{a\to+\infty}\int_{-a}^{a} x\,dx$$

$$= \lim_{a\to+\infty} 0 \qquad \textit{by Theorem 2.8(b) of Chapter 4}$$

$$= 0.$$

But the improper integral $\int_{-\infty}^{+\infty} f(x)\,dx$ *diverges*, because

$$\int_0^{+\infty} f(x)\,dx = \lim_{b\to+\infty}\int_0^b x\,dx = \lim_{b\to+\infty}\left[\frac{1}{2}x^2\right]_0^b = \lim_{b\to+\infty}\left[\frac{1}{2}b^2\right] = +\infty.$$

Integrals of Unbounded Functions

There is a second type of improper integral, namely, the integral of an unbounded function over a closed interval $[a, b]$. Although this appears to be quite different from the integrals we have been considering, the next example shows that geometrically it is closely related to the first type of improper integral.

Example 4

Is the area under the curve $y = 1/\sqrt{x}$ between $x = 0$ and $x = 1$ finite?

FIGURE 6.2

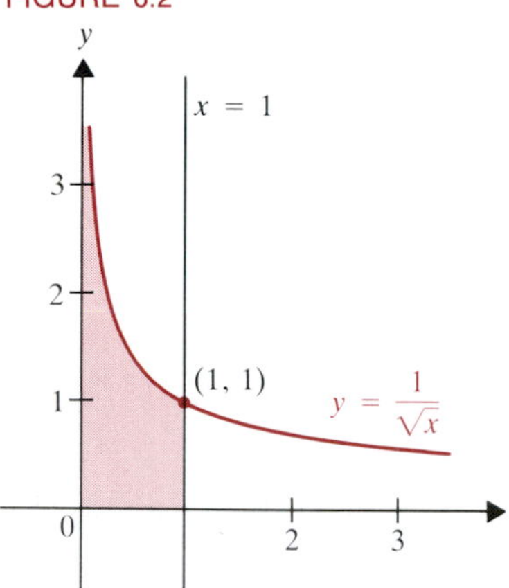

Solution. The area is shaded in **Figure 6.2.** The problem posed here is very much like the one in Example 2(a): Is the area between the curve $y = 1/\sqrt{x}$, the y-axis (rather than the x-axis), and the line $x = 1$ finite? As before, we consider the areas of the *bounded* regions between the lines $x = r$ and $x = 1$, and let $r \to 0^+$. Since

$$\int_r^1 \frac{1}{\sqrt{x}}\,dx = \int_r^1 x^{-1/2}\,dx = 2x^{1/2}\Big]_r^1 = 2 - 2\sqrt{r},$$

we have

$$\lim_{r\to 0^+} \int_r^1 \frac{1}{\sqrt{x}}\,dx = \lim_{r\to 0^+} (2 - 2\sqrt{r}) = 2 - 0 = 2.$$

Hence the area is finite—equal to 2 square units. ■

Example 4 suggests the next definition.

6.2 DEFINITION

Suppose that f is continuous on $(a, b]$ and $\lim_{x\to a^+} f(x)$ fails to exist. Then

$$\int_a^b f(x)\,dx = \lim_{r\to a^+} \int_r^b f(x)\,dx, \tag{7}$$

if this limit exists.

If f is continuous on $[a, b)$ and $\lim_{x\to b^-} f(x)$ fails to exist, then

$$\int_a^b f(x)\,dx = \lim_{r\to b^-} \int_a^r f(x)\,dx, \tag{8}$$

if this limit exists.

If f is continuous on $[a, c)$ and $(c, b]$, but $\lim_{x\to c^-} f(x)$ or $\lim_{x\to c^+} f(x)$ (or both) fails to exist, then

$$\int_a^b f(x)\,dx = \int_a^c f(x)\,dx + \int_c^b f(x)\,dx, \tag{9}$$

if *both* integrals on the right exist as real numbers.

We apply the same terminology about convergence and divergence that we used before. Thus Example 4 shows that $\int_0^1 x^{-1/2}\,dx$ converges to 2. A classic test of whether a calculus student looks before leaping is given by

$$\int_{-1}^1 1/x^2\,dx.$$

Failure to notice the infinite discontinuity at $x = 0$ would lead to an "evaluation" of this integral as

$$-\frac{1}{x}\bigg|_{-1}^1 = -\frac{1}{1} + \frac{1}{-1} = -2.$$

FIGURE 6.3

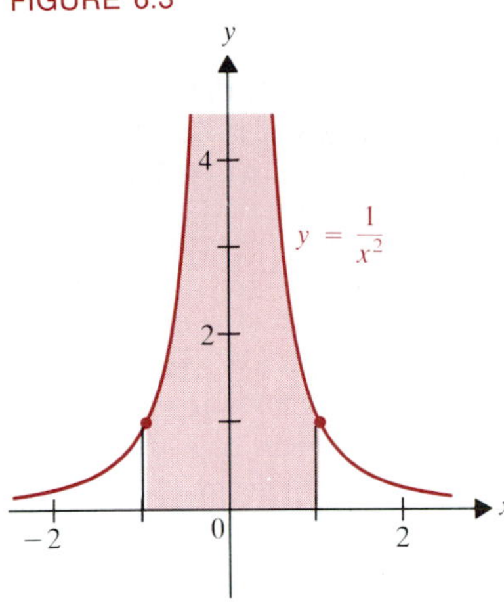

However, since the integrand is positive for all x where it is defined, we would expect (cf. Theorem 3.6 of Chapter 4) the value of the integral to be *positive* if it exists. The next example uses Definition 6.2 to determine whether this integral (equivalently, the shaded area in **Figure 6.3**) is finite.

Example 5

Does $\int_{-1}^{1} \frac{1}{x^2}\,dx$ exist?

Solution. We have to investigate $\int_{-1}^{0} x^{-2}\,dx$ and $\int_{0}^{1} x^{-2}\,dx$. If both of these improper integrals converge, then $\int_{-1}^{1} x^{-2}\,dx$ converges to their sum, by (9). Starting with the first integral, we have from (8)

$$\int_{-1}^{0} x^{-2}\,dx = \lim_{r\to 0^-} \int_{-1}^{r} x^{-2}\,dx = \lim_{r\to 0^-} \left[-\frac{1}{x}\right]_{-1}^{r}$$

$$= \lim_{r\to 0^-} \left[-\frac{1}{r} + \frac{1}{1}\right] = +\infty.$$

Since the integral over $[-1, 0]$ diverges, there is no need to investigate the other integral: both would have to converge for $\int_{-1}^{1} x^{-2}\,dx$ to exist. We conclude then that the given improper integral diverges. ■

In Section 5.4 (p. 307) we saw that an attempt to use the formula

$$L = \int_a^b \sqrt{1 + [f'(x)]^2}\,dx \tag{10}$$

to compute the arc length of the first-quadrant portion of the circle $x^2 + y^2 = 16$ would lead to an improper integral. Here

$$y = f(x) = \sqrt{16 - x^2} \to f'(x) = -\frac{x}{\sqrt{16 - x^2}},$$

so (10) would give

$$L = \int_0^4 \sqrt{1 + \frac{x^2}{16 - x^2}}\,dx = \int_0^4 \frac{4}{\sqrt{16 - x^2}}\,dx.$$

Example 6

Does $\int_0^4 \frac{4}{\sqrt{16 - x^2}}\,dx$ exist?

Solution. Using Formula (19) of the table of integrals in conjunction with Definition 6.2, we get

$$\int_0^4 \frac{4}{\sqrt{16 - x^2}}\,dx = 4\int_0^4 \frac{dx}{\sqrt{16 - x^2}} = 4 \lim_{b\to 4^-} \int_0^b \frac{dx}{\sqrt{4^2 - x^2}}$$

$$= 4 \lim_{b\to 4^-} \left[\sin^{-1}\frac{x}{4}\right]_0^b = 4[\sin^{-1} 1 - \sin^{-1} 0]$$

$$= 4 \cdot \frac{\pi}{2} = 2\pi.$$

FIGURE 6.4

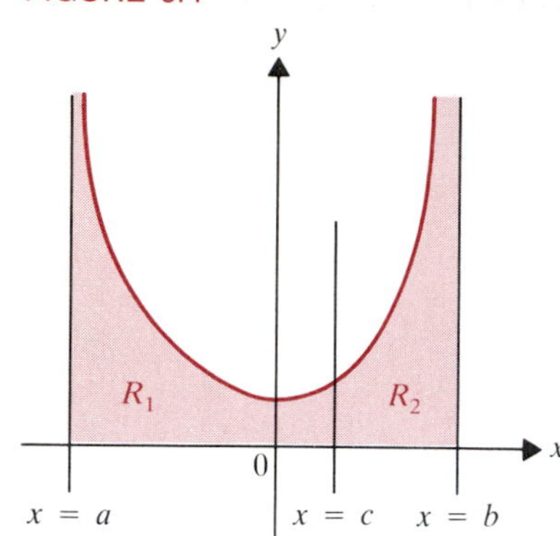

Therefore (10) gives an improper integral that converges to the correct arc length

$$\frac{1}{4}(2\pi r) = \frac{1}{4}(2\pi \cdot 4) = 2\pi. \quad ■$$

It is possible for an integral $\int_a^b f(x)\,dx$ to be improper with *both* $\lim_{x\to a^+} f(x)$ and $\lim_{x\to b^-} f(x)$ failing to exist. In that case, we pick any c in (a, b) and investigate $\int_a^c f(x)\,dx$ and $\int_c^b f(x)\,dx$. If both integrals exist, then we define

$$\text{(11)} \qquad \int_a^b f(x)\,dx = \int_a^c f(x)\,dx + \int_c^b f(x)\,dx.$$

In case $f(x) \geq 0$ on $[a, b]$, (11) amounts to computing the shaded area in **Figure 6.4** as the sum of the areas of the regions R_1 and R_2. It is shown in advanced calculus that if both integrals on the right side of (11) exist for one choice of c in (a, b), then they exist and have the same sum for all choices of c. (See Exercises 57–59.) In applying (11) then, we use any convenient c from the interval (a, b).

Example 7

Evaluate $\displaystyle\int_{-1}^{1} \frac{dx}{\sqrt{1 - x^2}}$.

FIGURE 6.5

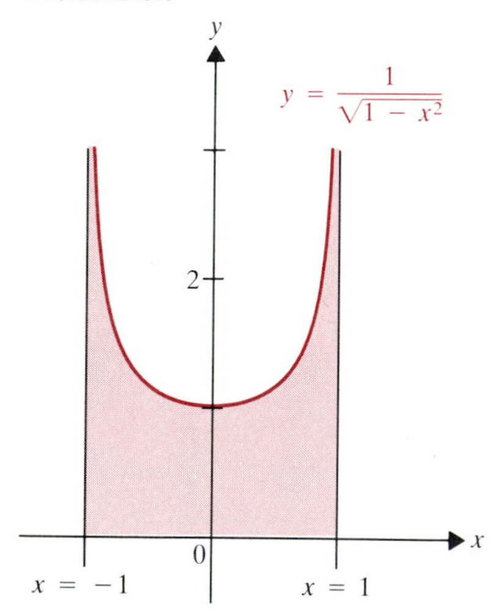

Solution. This improper integral corresponds to the area of the shaded region in **Figure 6.5,** which by symmetry is twice the area of the region in the first quadrant. Using $c = 0$ in (11), and Formula (18) of the table of integrals, we get

$$\int_{-1}^{1} \frac{dx}{\sqrt{1 - x^2}} = \int_{-1}^{0} \frac{dx}{\sqrt{1 - x^2}} + \int_{0}^{1} \frac{dx}{\sqrt{1 - x^2}} = 2\int_{0}^{1} \frac{dx}{\sqrt{1 - x^2}}$$

$$= 2 \lim_{b\to 1^-} \int_0^b \frac{dx}{\sqrt{1 - x^2}} = 2 \lim_{b\to 1^-} \left[\sin^{-1} x\right]_0^b$$

$$= 2 \lim_{b\to 1^-} [\sin^{-1} b - 0] = 2 \cdot \frac{\pi}{2} = \pi. \quad ■$$

Present Value and Stock Prices*

In Section 6.3 we saw that a sum P_0 of money growing at a continuously compounded interest rate r grows after t years to

$$\text{(12)} \qquad P = P_0 e^{rt}.$$

This leads to the important concept of *present value.* To have $P = P_0 e^{rt}$ dollars available t years from now, P_0 dollars must be invested now at interest rate r. We say that P_0 is the ***present value*** of the amount P available in t years at interest rate r. Solving (12) for P_0, we obtain the formula

$$\text{(13)} \qquad P_0 = Pe^{-rt}.$$

The relationship expressed in (13) does more than allow you to figure how much P_0 to set aside now to meet an expense P that will arise t years from now. As

* Optional

the next example shows, it also lets you compare the value of a currently available sum of money and one available later.

Example 8

An elderly person passes away. Part of the estate is a $50,000 life insurance policy. The company offers the beneficiary two options:

(a) $50,000 lump sum payment now;

(b) $76,000 lump sum payment in six years, when the beneficiary's first child will enter college.

If the beneficiary expects to be able to realize an average return of 8% (compounded continuously) on investments, then which option is better?

Solution. We can calculate the present value of option (b) from (13). It is

$$P_0 = 76{,}000e^{(-0.08)\cdot 6} = 47{,}027.54$$

Thus option (a) is better. ■

In practice, insurance companies provide several choices. In Example 8, another payment option might be monthly payments of $600 for ten years. To compare that option to the others, we need to compute the present value of that series of payments. By (13), the present value of a payment of $600 to be made after t_i years is $600e^{-rt_i}$, where r is the interest rate obtainable. Thus the total present value of the series of payments made at intervals of 1/12 year is

$$(14) \qquad P_0 = 600[e^{-r/12} + e^{-2r/12} + \ldots + e^{-120r/12}] = \sum_{i=1}^{120} 7200e^{-rt_i}\Delta t,$$

where $\Delta t = (10 - 0)/120 = 1/12$. This sum can be evaluated by using the formula for the sum of a geometric sequence (see Section 9.2), but instead we note that (14) is a Riemann sum for the function $f(t) = 7200e^{-rt}$ over the interval $[0, 10]$. For a large value of n (like 120), its value is therefore approximately

$$\int_0^{10} 7200e^{-rt}\,dt = 7200\int_0^{10} e^{-rt}\,dt.$$

Thus, using the value $r = 0.08$ from Example 8, we find

$$P_0 \approx 7200\int_0^{10} e^{-0.08t}\,dt = 7200\left[\frac{e^{-0.08t}}{-0.08}\right]_0^{10}$$

$$\approx 90000[-e^{-0.8} + e^0] \approx \$49560.39.$$

Hence the installment payment is better than the option (b) but still not as advantageous as a lump sum distribution of $50,000 paid immediately.

The foregoing involved an integral, but not an improper one. It sets the scene, however, for our concluding example. A large corporation receives income from its operations almost continuously. For this reason, financial analysts treat its earnings per share of common stock as a continuous function f. This is used to estimate the value of such a corporation's stock, which is regarded as wholly determined by the firm's earnings. If the long-term inflation rate is estimated to be r, investors assume that they can obtain interest rate r from banks or money market funds. Over the next N years, the company's earnings per share at time t will be $f(t)$. As in (14), the present value P_0 of those earnings is approximated by a Riemann sum

$$\sum_{i=1}^{N} f(t_i)e^{-rt_i}\,\Delta t, \tag{15}$$

for $f(t)e^{-rt}$ over the integral $[0, N]$. Economists define the ***exact present value*** to be the limit of (15) as $N \to \infty$. That is,

$$P_0 = \int_0^{+\infty} f(t)e^{-rt}\,dt, \tag{16}$$

an improper integral of the type covered by Definition 6.1. For regulated companies, $f(t)$ is relatively stable.

Example 9

A large telephone utility is projected to earn \$5.00 per share of common stock indefinitely. How much is a share of its stock worth if long-term interest rates are projected to average (a) 5% or (b) 10%?

Solution. Since here $f(t) = 5$, we get from (16)

$$\begin{aligned} P_0 &= \int_0^{+\infty} 5e^{-rt}\,dt = 5\int_0^{+\infty} e^{-rt}\,dt \\ &= 5 \lim_{b \to +\infty} \int_0^b e^{-rt}\,dt = 5 \lim_{b\to+\infty}\left[\frac{e^{-rt}}{-r}\right]_0^b \\ &= 5 \lim_{b\to+\infty} \frac{1}{r}[1 - e^{-rb}] = \frac{5}{r}. \end{aligned}$$

In case (a),

$$P_0 = \frac{5}{0.05} = 100.$$

In case (b),

$$P_0 = \frac{5}{0.10} = 50. \quad ■$$

Example 9 suggests why high inflation (and concurrent high interest rates) are bad for stock prices. As the inflation rate rises, the present value of future earnings is *severely* discounted, in this case cut in *half*. Intelligent investors are therefore willing to pay much less for a share in those future profits. (The assumption that future earnings will not grow is, of course, an oversimplification. For more true-to-life examples, see Exercises 45 and 46.)

Lest it appear that improper integrals like (16) are of interest only in finance, we mention that

$$F(r) = \int_0^{+\infty} f(t)e^{-rt}\,dt$$

is the *Laplace transform* $\mathscr{L}(f)$ of the function f. The Laplace transform is an important tool in solving certain kinds of differential equations with initial conditions, including important equations that arise in physics and electrical engineering. The integral in (16) is the Laplace transform of the constant function $f(t) = 5$ for all t. Thus $F(r) = 5/r$.

Exercise 7.6

In Exercises 1–40, evaluate the given integral, if possible.

1. $\int_1^{+\infty} \frac{1}{x^2+1}\,dx$

2. $\int_1^{+\infty} \frac{dx}{x^2+4}$

3. $\int_{-\infty}^{+\infty} xe^{-x^2}\,dx$

4. $\int_{-\infty}^{+\infty} \frac{dx}{x^2+1}$

5. $\int_2^{+\infty} \frac{dx}{\sqrt{x-1}}$

6. $\int_3^{+\infty} \frac{dx}{(x-1)^{1/3}}$

7. $\int_{-\infty}^{+\infty} \frac{dx}{x^2+1}$

8. $\int_{-\infty}^{+\infty} \frac{x\,dx}{x^4+1}$

9. $\int_0^{+\infty} \frac{x\,dx}{x^2+1}$

10. $\int_0^{+\infty} \frac{dx}{x+1}$

11. $\int_2^{+\infty} \frac{dx}{x\ln x}$

12. $\int_1^{+\infty} \ln x\,dx$

13. $\int_0^{+\infty} xe^{-2x}\,dx$

14. $\int_0^{+\infty} xe^{-x/2}\,dx$

15. $\int_0^{+\infty} \sin 2x\,dx$

16. $\int_1^{+\infty} \cos 2x\,dx$

17. $\int_0^{+\infty} e^{-x}\cos x\,dx$

18. $\int_0^{+\infty} e^{-2x}\sin x\,dx$

19. $\int_0^1 \frac{dx}{x^{1/3}}$

20. $\int_1^2 \frac{dx}{\sqrt{x-1}}$

21. $\int_3^4 \frac{dx}{(x-3)^2}$

22. $\int_{-1}^0 \frac{dx}{(1+x)^2}$

23. $\int_0^2 \frac{x\,dx}{x^2-4}$

24. $\int_0^2 \frac{dx}{\sqrt{4-x^2}}$

25. $\int_0^3 \frac{dx}{(x-1)^{2/3}}$

26. $\int_0^4 \frac{dx}{(x-2)^{1/3}}$

27. $\int_0^2 \frac{3\,dx}{x-1}$

28. $\int_0^3 \frac{dx}{(3-x)^{3/2}}$

29. $\int_0^{\pi} \sec^2 x\,dx$

30. $\int_0^{\pi} \sec x\tan x\,dx$

31. $\int_0^1 x\ln x\,dx$

32. $\int_{-1}^2 x\ln(x+1)\,dx$

33. $\int_1^2 \frac{dx}{x\sqrt{x^2-1}}$

34. $\int_0^1 \frac{dx}{\sqrt{1-x^2}}$

35. $\int_0^{\pi} \frac{\cos x}{\sqrt{1-\sin x}}\,dx$

36. $\int_{-\pi/2}^{\pi/2} \frac{\sin x}{\sqrt{1-\cos x}}\,dx$

37. $\int_{-2}^2 \frac{dx}{\sqrt{4-x^2}}$

38. $\int_{-3}^3 \frac{dx}{\sqrt{9-x^2}}$

39. $\int_1^{+\infty} \frac{dx}{x\sqrt{x^2-1}}$

40. $\int_2^{+\infty} \frac{dx}{x\sqrt{x^2-4}}$

41. Find the *escape velocity* v_0 needed to send a rocket of mass m out of the earth's gravitational force, by using the fact that the initial kinetic energy $mv_0{}^2/2$ must supply the needed work.

42. With reference to Example 1, note that $F(x) = k/x^2$ for a constant k. Find the work needed to send a load weighing 1000 pounds out of the earth's gravity, using 4000 miles for the radius of the earth.

43. Suppose that a house can be rented for a constant yearly rental of \$5000.
(a) How much is a fair market price for the house if an inflation rate of 10% is assumed?
(b) What if the inflation rate is only 6%?
(c) Explain how high interest rates will depress demand for houses, if buyers regard a house as an investment.

44. To be *actuarially sound*, a retirement fund should collect sufficient contributions to pay each retiree the agreed-upon pension. Funds with many members can effectively deal with this by using mortality tables. If a firm with *only one* employee wants to pay that employee a \$10,000 initial annual pension, with a 5% (continuously compounded) annual cost-of-living increase, then how much in premiums should it have at the time the person retires, if it forecasts inflation to be 8%? (It should have an amount equal to an indefinitely long pay-out, since it cannot project an individual's life span.)

45. In Example 9, suppose that earnings of a growing company are 50¢ per share and are projected to increase at a constant rate to \$1.00 per share in five years. How much is the stock worth under inflation rate r? If r is 10%, how much is the stock worth?

46. Repeat Exercise 45 if earnings are now \$1.00 per share and are forecast to increase at a constant rate to \$2.00 per share in three years.

47. Show that $\int_1^{+\infty} e^{-x^2}\,dx$ is convergent, by comparing it to $\int_1^{+\infty} e^{-x}\,dx$.

48. Generalize Exercise 47 as follows. **(a)** If $0 \le f(x) \le g(x)$ are continuous functions and $\int_1^{+\infty} g(x)\,dx$ converges, then show that $\int_1^{+\infty} f(x)\,dx$ must also converge. **(b)** If $0 \le f(x) \le g(x)$ and $\int_1^{+\infty} f(x)\,dx$ diverges, then so must $\int_1^{+\infty} g(x)\,dx$. **(c)** State results analogous to (a) and (b) for improper integrals $\int_a^b f(x)\,dx$ and $\int_a^b g(x)\,dx$.

49. Show that $\int_1^{+\infty} x^{-r}\,dx$ converges if $r > 1$ and diverges otherwise.

50. Show that $\int_0^1 x^{-r}\,dx$ converges if $r < 1$ and diverges otherwise.

In Exercises 51–56, use Exercise 48 to determine whether the given integral converges.

51. $\int_0^{\pi/2} \frac{\sin^2 x}{\sqrt{x}}\,dx$

52. $\int_2^{+\infty} \frac{dx}{\sqrt{x}-1}$

53. $\int_2^{+\infty} \frac{dx}{\ln x}$

54. $\int_0^{+\infty} \frac{x^2\,dx}{x^5+1}$

55. $\int_2^{+\infty} \frac{dx}{\sqrt{x^2-1}}$

56. $\int_1^{+\infty} \frac{dx}{\sqrt{x^3+1}}$

57. Suppose that f is continuous on $[a, b]$ and that $\lim_{x\to a^+} f(x)$ fails to exist. Suppose that $\int_a^b f(x)\,dx$ exists. Then show that for any $c \in (a, b)$, $\int_a^c f(x)\,dx$ exists and

$$(17) \qquad \int_a^b f(x)\,dx = \int_a^c f(x)\,dx + \int_c^b f(x)\,dx.$$

58. Suppose that f is continuous on $[a, b)$ and $\lim_{x\to b^-} f(x)$ fails to exist. Suppose that $\int_a^b f(x)\,dx$ exists. Then show that for any $d \in (a, b)$, $\int_d^b f(x)\,dx$ exists and

$$(18) \qquad \int_a^b f(x)\,dx = \int_a^d f(x)\,dx + \int_d^b f(x)\,dx.$$

59. Suppose that both $\lim_{x\to a^+} f(x)$ and $\lim_{x\to b^-} f(x)$ fail to exist, and $\int_a^b f(x)\,dx$ exists for two choices $c_1 < c_2$ of c in (11). Then use (17) and (18) to show that

$$(19) \quad \int_a^{c_1} f(x)\,dx + \int_{c_1}^b f(x)\,dx = \int_a^{c_2} f(x)\,dx + \int_{c_2}^b f(x)\,dx.$$

This is part of what is needed to show that the choice of c in (11) does not matter. [*Hint:* Show that both sides of (19) equal

$$\int_a^{c_1} f(x)\,dx + \int_{c_1}^{c_2} f(x)\,dx + \int_{c_2}^b f(x)\,dx.]$$

PC **60.** On p. 350, we introduced the standard normal probability density function

$$n(x) = \frac{1}{\sqrt{2\pi}} e^{-x^2/2}.$$

An important fact about n is that $\int_{-\infty}^{+\infty} n(x)\,dx = 1$. Use Simpson's rule to assemble numerical evidence for this as follows.

(a) Show that n is an even function.

(b) Assuming that $\int_0^{+\infty} n(x)\,dx$ exists, show that $\int_{-\infty}^{+\infty} n(x)\,dx = 2\int_0^{+\infty} n(x)\,dx$.

(c) Use Simpson's rule and $n = 320$ to estimate $\int_0^b n(x)\,dx$ for $b = 1, 2, 3, 4, 5$, and 10. What does $\int_0^{+\infty} n(x)\,dx$ appear to be? (Use 3.141593 for π, and round off your approximations to 6 decimal places.)

7.7 Looking Back

This chapter discussed several major techniques of integration. The most important of the methods is integration by parts, due to its applicability to so many types of integration problems and its use in obtaining reduction formulas. These allow many complicated integrals to be evaluated by a sequence of repeated application of the formula.

Section 1 presented the integration formulas for the trigonometric functions besides sine and cosine and gave methods for integrating powers of trigonometric functions.

Section 3 discussed the three trigonometric substitution schemes for formulas containing $\sqrt{a^2 - x^2}$, $\sqrt{x^2 - a^2}$, or $\sqrt{a^2 + x^2}$. Section 4 presented the technique of partial fraction decomposition (Theorem 4.2) for rational functions $p(x)/q(x)$, *where* $p(x)$ *has smaller degree than* $q(x)$. (Remember to **first divide $p(x)$ by $q(x)$ if the degree of $q(x)$ is not bigger than the degree of $p(x)$:** Example 3 of Section 4.) We saw how to use this in conjunction with judicious substitution and equating of coefficients to work out

$$\int \frac{p(x)}{q(x)}\,dx.$$

By considering population modeling, we also saw how this technique can be helpful in applied work.

Section 5 discussed some methods for integrals that involve nth roots or rational functions of sine and cosine, as well as the efficient use of tables of integrals (Examples 6 and 7).

The focus shifted in Section 6 from indefinite to definite integrals. In Definition 6.1, we saw how to extend the meaning of $\int_a^b f(x)\,dx$ to the cases where a is $-\infty$ and/or b is $+\infty$. In Definition 6.2, we extended the meaning of the definite integral to include some functions f that have infinite discontinuity points at a, or b, or an interior point of $[a, b]$.

CHAPTER CHECKLIST

Section 1: integrals of tangent, cotangent, secant, and cosecant functions; $\int \sin^m x \cos^n x\,dx$ when at least one exponent is odd or when both are even; $\int \sec^m x \tan^n x\,dx$ when m is even, or n is odd, or $m = 0$ and n is even.

Section 2: integration by parts formula; strategy for using integration by parts (p. 435); recurrence of the given integral (cf. Examples 5 and 6); reduction formulas (cf. Example 7).

Section 3: substitutions for integrals containing $\sqrt{a^2 - x^2}$, $\sqrt{a^2 + x^2}$, or $\sqrt{x^2 - a^2}$; use of right triangles to express answers in terms of x; completing the square in a radical before substituting (cf. Example 7).

Section 4: polynomial factorization over **R** (Theorem 4.1); decomposition of $p(x)/q(x)$ into partial fractions when $p(x)$ has smaller degree than $q(x)$ (Theorem 4.2); judicious substitution; equating coefficients; completing the square in quadratic factors (cf. Example 5); logistic growth and population modeling.

Section 5: integration of expressions containing fractional powers of x or $f(x)$; integration of rational functions of $\sin x$ and $\cos x$; use of tables of integrals.

Section 6: improper integrals over infinite integrals; improper integrals of functions with infinite discontinuity points; areas and volumes of unbounded regions; present value of money; stock pricing.

REVIEW EXERCISES 7.7

In Exercises 1–84, evaluate the given integral.

1. $\int \sin^5 x \cos^2 x\,dx$

2. $\int \sin^4 x \cos^5 x\,dx$

3. $\int_0^{\pi/2} \sin^2 x \cos^2 x\,dx$

4. $\int_0^{\pi} \cos^4 x\,dx$

5. $\int \sec^4 x \tan^2 x\,dx$

6. $\int \sec^6 x\,dx$

7. $\int \sec^3 x \tan^3 x\,dx$

8. $\int \sec^4 x \tan^3 x\,dx$

9. $\int_0^{\pi/3} \tan^6 x\,dx$

10. $\int_0^{\pi/4} \tan^4 x\,dx$

11. $\int x^2 \cos x\,dx$

12. $\int x^2 \ln x\,dx$

13. $\int \frac{x^3}{\sqrt{1 + x^2}}\,dx$

14. $\int \frac{x^3}{\sqrt{x^2 + 5}}\,dx$

15. $\int_0^{\pi/3} e^{3x} \sin 2x\,dx$

16. $\int_0^{\pi/2} e^{2x} \cos 3x\,dx$

17. $\int \sec^{-1} x\,ax$

18. $\int (\sin^{-1} x)^2\,dx$

19. $\int x \sin^{-1} x\,dx$

20. $\int x(\ln x)^2\,dx$

21. $\int x^2 e^{2x}\,dx$

22. $\int x^2 e^{-3x}\,dx$

23. $\int x^2 \sin^{-1} x\,dx$

24. $\int x^2 \tan^{-1} 2x\,dx$

25. $\int \frac{x^2}{\sqrt{4 - x^2}}\,dx$

26. $\int \frac{\sqrt{9 - x^2}}{x^2}\,dx$

27. $\int_0^1 (1 - x^2)^{3/2}\,dx$

28. $\int_1^2 x^2\sqrt{4 - x^2}\,dx$

29. $\int \sqrt{a^2 + x^2}\,dx$

30. $\int \frac{\sqrt{a^2 + x^2}}{x}\,dx$

31. $\int_1^{2\sqrt{2}} \frac{\sqrt{1 + x^2}}{x^2}\,dx$

32. $\int_0^{\sqrt{5}} \frac{x^2}{\sqrt{4 + x^2}}\,dx$

33. $\int \sqrt{x^2 - a^2}\,dx$

34. $\int \frac{x^2}{\sqrt{x^2 - 9}}\,dx$

35. $\int_{-3}^{-2} \frac{\sqrt{x^2 - 4}}{x}\,dx$

36. $\int_4^5 \frac{\sqrt{x^2 - 16}}{x^2}\,dx$

37. $\int \frac{dx}{\sqrt{x^2 - 4x + 5}}$

38. $\int \frac{dx}{\sqrt{x^2 + 6x - 12}}$

39. $\int \frac{x\,dx}{\sqrt{5 + 4x - x^2}}$

40. $\int \frac{dx}{\sqrt{4x - x^2}}$

41. $\int \frac{x + 1}{x^2 - x}\,dx$

42. $\int \frac{3x^2 + x - 1}{x^3 - 5x^2 + 6x}\,dx$

43. $\int_2^3 \frac{x\,dx}{(x - 1)^2}$

44. $\int_0^1 \frac{(2x + 1)\,dx}{x^3 + 2x^2 + x}$

45. $\int \frac{x^2 + 4x - 1}{x^3 - x}\,dx$

46. $\int \frac{x^2 - 3x + 2}{x^3 + 8x}\,dx$

47. $\int \frac{x^2 + 1}{x^3 + x^2 - 2x}\,dx$

48. $\int \frac{2 - 4x - x^2}{x^3 - 3x^2 + 2x}\,dx$

49. $\int_0^1 \frac{x^2 + x + 1}{(2x + 1)(x^2 + 1)}\,dx$

50. $\int_1^2 \frac{x^2 - 3x + 3}{(3x - 2)(x^2 + 4)}\,dx$

51. $\int \frac{2x^2 + 1}{(x - 2)^3}\,dx$

52. $\int \frac{x^2 - 3x + 5}{x(x + 1)^2}\,dx$

53. $\int \frac{x^3}{x^2 - 2x - 3}\,dx$

54. $\int \frac{x^3 + 1}{x^3 - 4x}\,dx$

55. $\int \frac{dx}{x^5 + 2x^3 + x}$

56. $\int \frac{x^4 + 9x^2 + 15}{(x - 1)(x^2 + 4)^2}\,dx$

57. $\int \frac{dx}{\sqrt{x} + \sqrt[3]{x}}$

58. $\int \frac{dx}{\sqrt[4]{x + 1} + \sqrt{x + 1}}$

59. $\int_0^1 \frac{dx}{1 + x^{1/4}}$

60. $\int_0^{16} \frac{x^{3/4}\,dx}{1 + \sqrt{x}}$

61. $\int x\sqrt{x - 1}\,dx$

62. $\int x\sqrt{2x + 5}\,dx$

63. $\int xe^{\sqrt{2-x}}\,dx$

64. $\int x \sin\sqrt{1 - x}\,dx$

65. $\int_0^{\pi/6} \frac{dx}{1 - \sin x}$

66. $\int_{1/4\pi}^{1/2\pi} \frac{dx}{1 + \sin x - \cos x}$

67. $\int \frac{dx}{4\sin x - 3\cos x}$

68. $\int \frac{dx}{\sin x - \tan x}$

69. $\int_1^{+\infty} x^{-3/2}\,dx$

70. $\int_{-\infty}^{2} (3 - x)^{-2}\,dx$

71. $\int_1^{+\infty} \frac{dx}{2x^2 + 2}$

72. $\int_1^{+\infty} \frac{dx}{2x + 2}$

73. $\int_{-\infty}^{0} \frac{dx}{(1 - 2x)^{3/2}}$

74. $\int_{-\infty}^{0} \frac{dx}{\sqrt{1 - 2x}}$

75. $\int_{-\infty}^{+\infty} \frac{e^x}{1 + e^{2x}}\,dx$

76. $\int_{-\infty}^{+\infty} e^{-x}\,dx$

77. $\int_0^1 \frac{dx}{(x - 1)^{2/3}}$

78. $\int_0^1 \frac{dx}{(x - 1)^{4/3}}$

79. $\int_1^3 \frac{dx}{(x - 1)^{2/3}}$

80. $\int_1^3 \frac{dx}{(x - 1)^{4/3}}$

81. $\int_1^2 \frac{dx}{x^2 - 2x}$

82. $\int_0^{\pi/2} \sec x\,dx$

83. $\int_0^{+\infty} \frac{dx}{\sqrt{x}(1 + x)}$

84. $\int_0^{+\infty} \frac{dx}{x^2 + 2x}$

In Exercises 85 and 86, obtain a reduction formula for the given integrals.

85. $\int \frac{dx}{(1 + x^2)^n}$

86. $\int \frac{dx}{x(1 + x^2)^n}$

In Exercises 87 and 88, derive the given reduction formula.

87. $\int x^n\sqrt{a + bx}\,dx$

$$= \frac{2}{b(2n + 3)}\left[x^n(a + bx)^{3/2} - an\int x^{n-1}\sqrt{a + bx}\,dx\right]$$

88. $\int \frac{x^n\,dx}{\sqrt{a + bx}} = \frac{2}{(2n + 1)b}\left[x^n\sqrt{a + bx} - na\int \frac{x^{n-1}\,dx}{\sqrt{a + bx}}\right]$

89. Show that the area enclosed by the ellipse $x^2/a^2 + y^2/b^2 = 1$ is πab.

90. A magnet of length $2l$ generates a magnetic field, whose strength S on a particle P at perpendicular distance x units from the center of the magnet (see **Figure 7.1**) is

$$S = \frac{2pl}{(x^2 + l^2)^{3/2}}.$$

Here p is the strength of the magnetic field at the poles of the magnet. Find the average strength of the field over the interval $[0, D]$, where $D > 0$ is some distance from the center of the magnet.

FIGURE 7.1

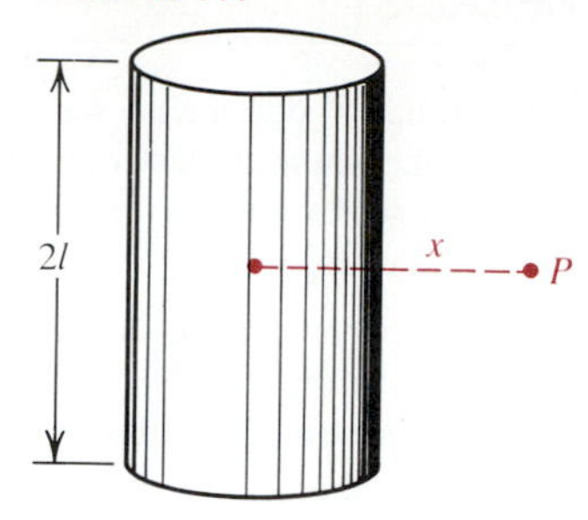

91. Find the arc length of the curve $y = x^2$ between $x = 0$ and $x = 2$.

92. Find the arc length of the curve $y = \ln x$ from $x = 1$ to $x = 3$.

93. Suppose that an amount of substance S is placed in a reacting chamber with an amount of a substance T. If the rate of production of the substance U (formed by the reaction of one molecule of S with one molecule of T to produce each molecule of U) is proportional to the product of concentrations of S and T present, then find a formula for U at time t.

94. Solve the differential equation

$$\frac{dy}{dx} = \frac{k}{(x + 1)(c - x)}$$

for $y = f(x)$.

95. The *gamma function* is defined by

$$\Gamma(x) = \int_0^{+\infty} t^{x-1}e^{-t}\,dt, \qquad \text{if } x > 0.$$

(a) Show that for any fixed real number r, $\Gamma(r + 1) = r\Gamma(r)$.
(b) Show that if r is a positive integer, then $\Gamma(r + 1) = r!$.

96. **(a)** Obtain the reduction formula (for $n > 1$):

$$\int_0^{+\infty} \frac{x^{2n-1}}{(x^2 + 1)^{n+3}}\,dx = \frac{n - 1}{n + 2}\int_0^{+\infty} \frac{x^{2n-3}}{(x^2 + 1)^{n+2}}\,dx.$$

(b) Find $\int_0^{+\infty} \frac{x^3}{(x^2 + 1)^5}\,dx$.

97. A large international oil company earns \$6.00 per share of common stock. The earnings are expected to remain at this level indefinitely. How much is the firm's common stock worth, assuming long-term inflation will average 12%?

98. A book earns yearly royalties of \$10,000. The author grows dissatisfied with his publisher and offers to buy the publication rights from the publisher. If the royalties are 10% of net revenue to the publisher, and inflation is expected to average 10%, how much should the publisher ask for the rights to the book, assuming that the book will remain profitable indefinitely?

CHAPTER 8 Some Plane Analytic Geometry

8.0 Introduction

The first four sections of this chapter are devoted to an important class of plane curves called *conic sections*. They can be obtained by cutting a right circular cone with a plane. **Figure 0.1** shows such a cone, obtained by revolving the line l: $y = mx$ in the xy-plane about the axis. The cone consists of an upper and lower portion, which are called *nappes*. Cutting the cone by a plane Π parallel

FIGURE 0.1

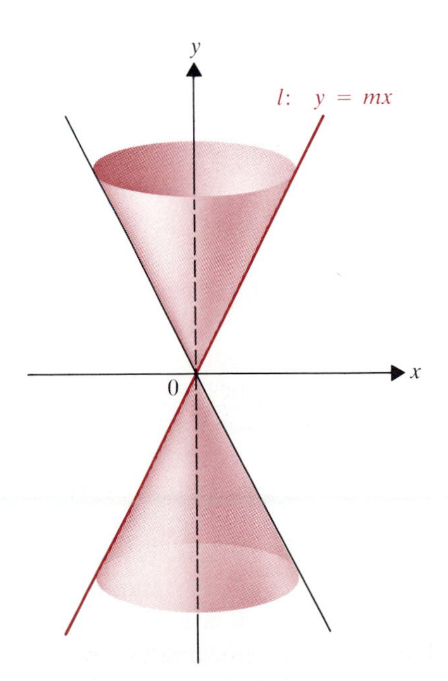

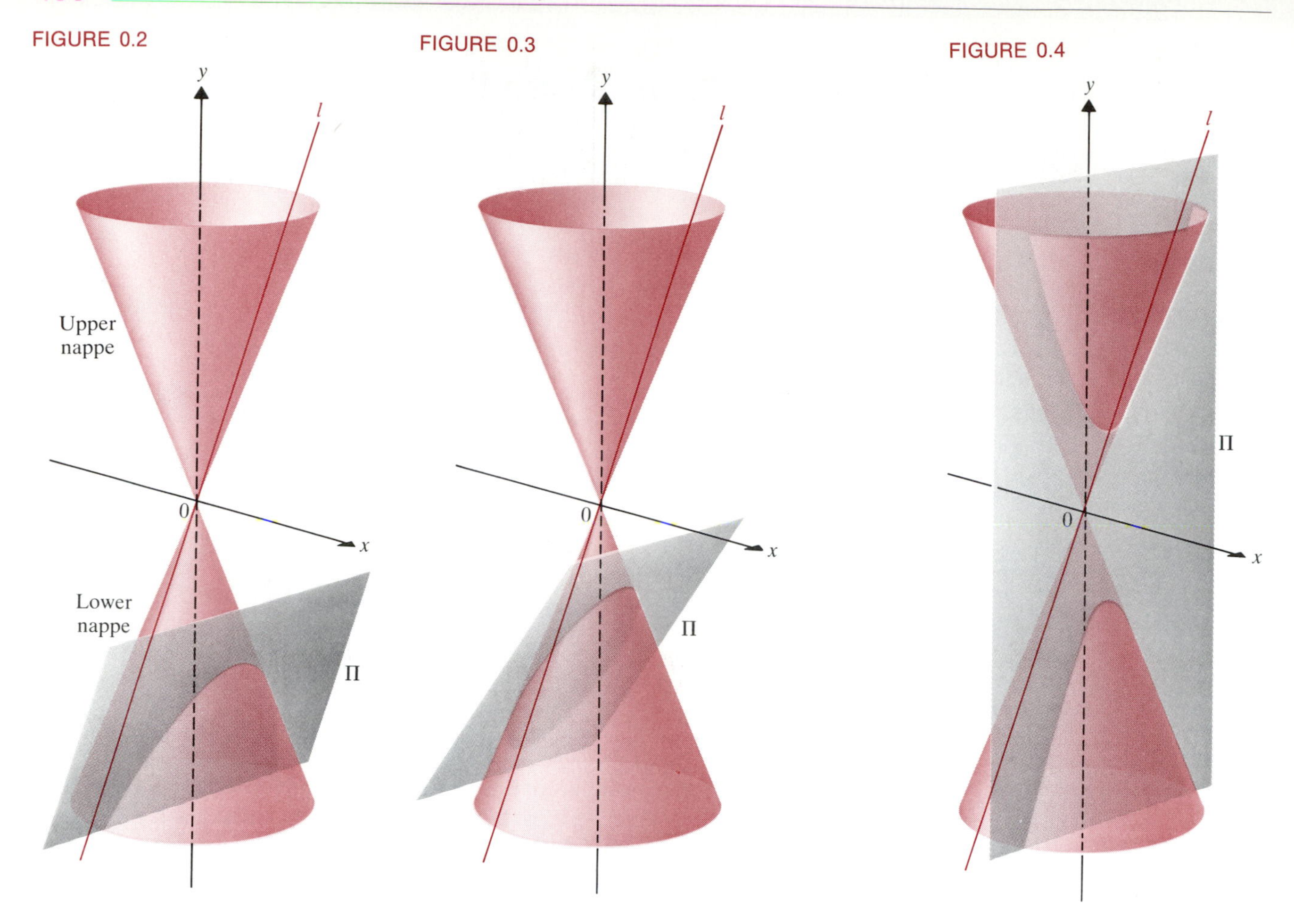

to l produces the *parabola* shown in **Figure 0.2.** (Note that only one nappe is cut by Π.) If we incline Π more horizontally, then it cuts completely across one nappe, producing the *ellipse* shown in **Figure 0.3.** (If Π is exactly horizontal, then we get a *circle* as the intersection. Thus a circle is a special case of an ellipse.) If Π is tilted more vertically, then it cuts both nappes, producing the *hyperbola* shown in **Figure 0.4.**

In Sections 1–3, we use Cartesian coordinates to study the parabola, ellipse, and hyperbola as plane curves. In Section 4, we present the technique of rotation of axes, which makes it possible to determine algebraically the nature of the graph of any second-degree polynomial equation $ax^2 + bxy + cy^2 + ex + fy + g = 0$. It turns out that the graph must be one of the conic sections mentioned above or else a *degenerate* conic section: that is, a point, one line, two lines, or an empty graph (no points). As **Figure 0.5** illustrates, the first three degenerate conic sections are also intersections of a plane with a cone.

The remainder of the chapter considers new coordinate schemes in the plane. In Section 5, we introduce parametric equations for plane curves. They give the coordinates (x, y) of points on a curve in terms of a variable t, often time. The last two sections present the polar coordinate system, a useful alternative to the standard Cartesian coordinate system in the plane. In the final section, we use calculus to find areas and arc lengths of curves described by polar coordinates.

FIGURE 0.5

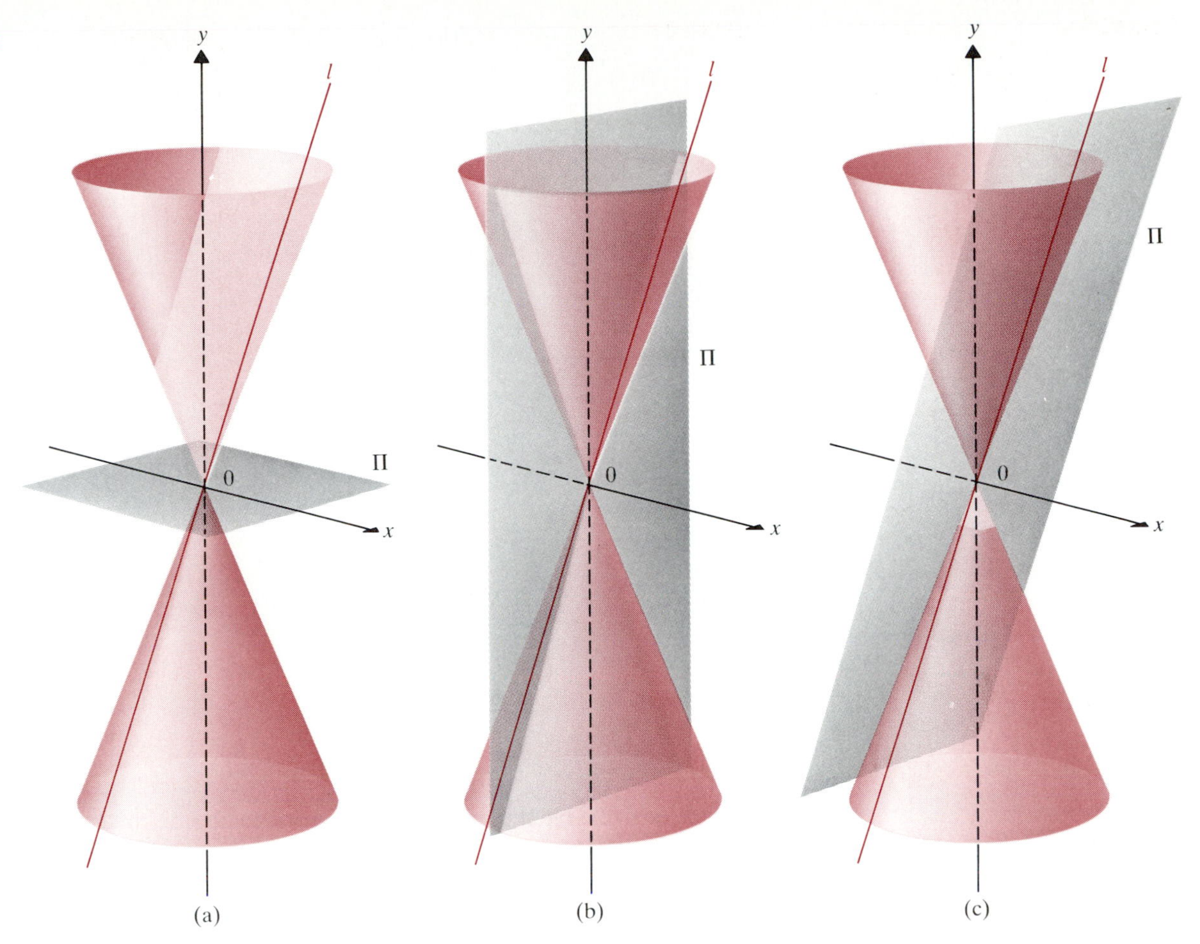

8.1 Parabolas

The following two-dimensional definition makes it possible to study parabolas as plane curves.

1.1 DEFINITION A ***parabola*** is the set of all points $P(x, y)$ in the xy-plane that are equidistant from a given line l (called the ***directrix***) and a given point F not on l (called the ***focus***).

To derive an equation for the parabola, we first select a focus F and a directrix l as in **Figure 1.1.** The resulting equation is simplest if the directrix is parallel to one coordinate axis, the focus is on the other axis, and the origin is midway between F and l, as in **Figure 1.2.** If $P(x, y)$ is any point on the parabola, then Definition 1.1 applied to Figure 1.2 gives

(1) $d(F, P) = d(P, Q) \rightarrow \sqrt{(x-0)^2 + (y-p)^2} = \sqrt{(x-x)^2 + (y-(-p))^2}.$

Squaring both sides of the second part of (1), we get

(2) $$x^2 + y^2 - 2py + p^2 = y^2 + 2py + p^2 \rightarrow x^2 = 4py.$$

Thus every point on the parabola satisfies (2). Conversely, every point $P(x, y)$ whose coordinates satisfy (2) is on the parabola. To see that, suppose that x

FIGURE 1.1

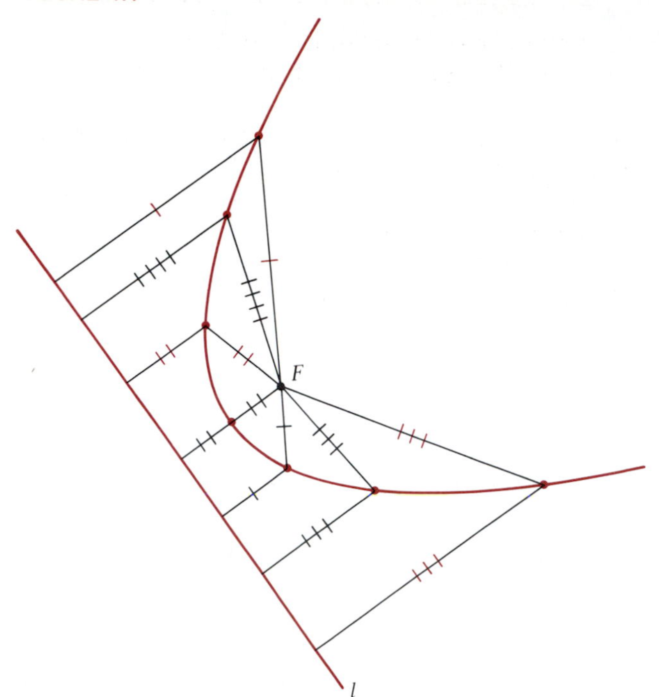

FIGURE 1.2

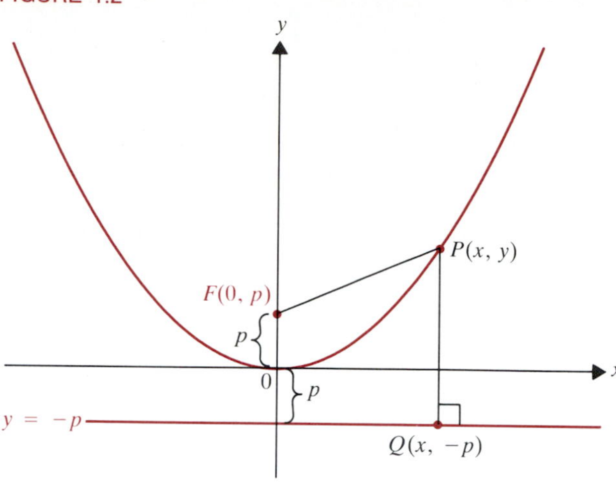

and y satisfy (2). Then subtracting $2py$ from each side, we have

$$x^2 - 2py = 2py.$$

Adding $y^2 + p^2$ to each side and taking positive square roots gives

$$x^2 + y^2 - 2py + p^2 = y^2 + 2py + p^2,$$

$$\sqrt{(x-0)^2 + (y-p)^2} = \sqrt{(x-x)^2 + (y+p)^2} \rightarrow d(F, P) = d(P, Q).$$

FIGURE 1.3

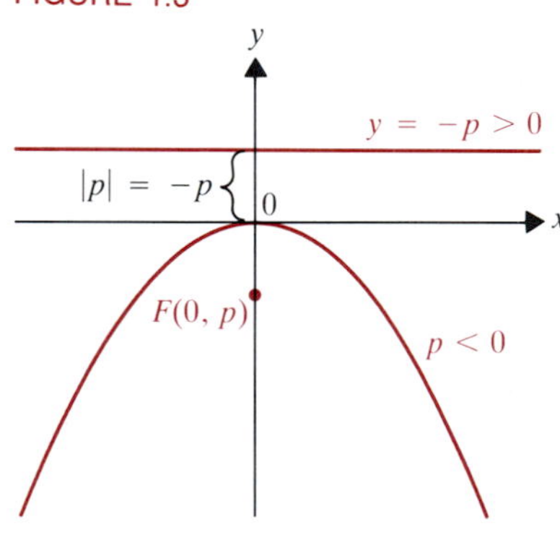

Thus, by Definition 1.1, $P(x, y)$ is on the parabola, and so (2) is its equation.

In Figure 1.1 the parabola curves around the focus F. Thus we see at once from (2) that

(3) if $p > 0$, the parabola opens upward;
if $p < 0$, the parabola opens downward.

Figure 1.3 illustrates the second alternative in (3). Equation (2) is the ***standard form equation*** of a parabola with focus $(0, p)$ and directrix $y = -p$. The origin is called the ***vertex*** of the parabola.

In **Figure 1.4,** the parabola has focus $F(p, 0)$ on the x-axis and directrix $x = -p$ parallel to the y-axis. Its equation is obtained by interchanging x and y in (2):

FIGURE 1.4

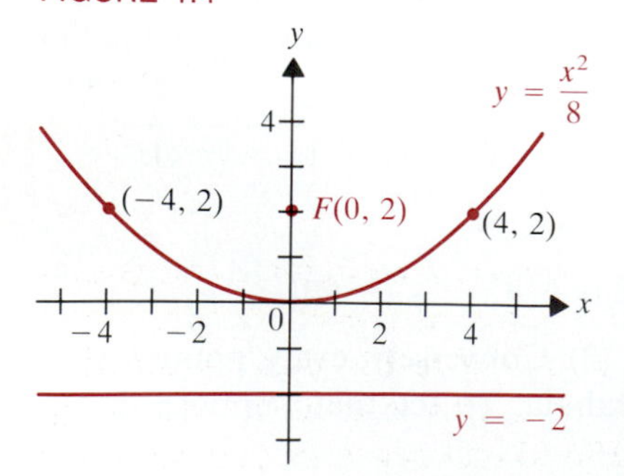

(4) $$y^2 = 4px.$$

Again, since the parabola curves around its focus, we see that

(5) if $p > 0$, the parabola opens to the right;
if $p < 0$, the parabola opens to the left.

Equation (4) is the *standard form equation* of a parabola with focus $(p, 0)$ and directrix $x = -p$. Using (3) and (5), we can easily sketch the graph of a parabola with vertex at the origin and focus on either coordinate axis.

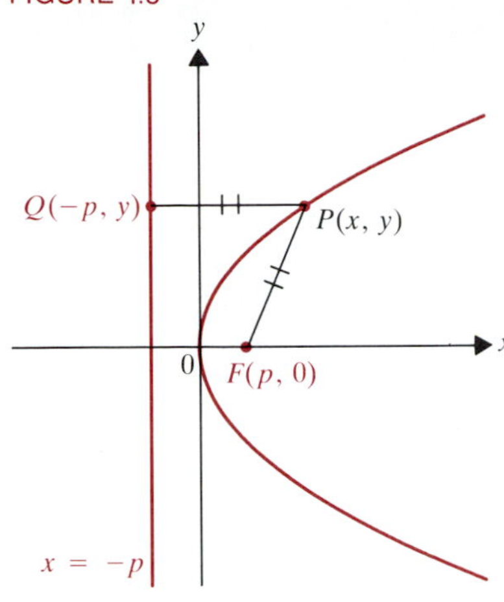

FIGURE 1.5

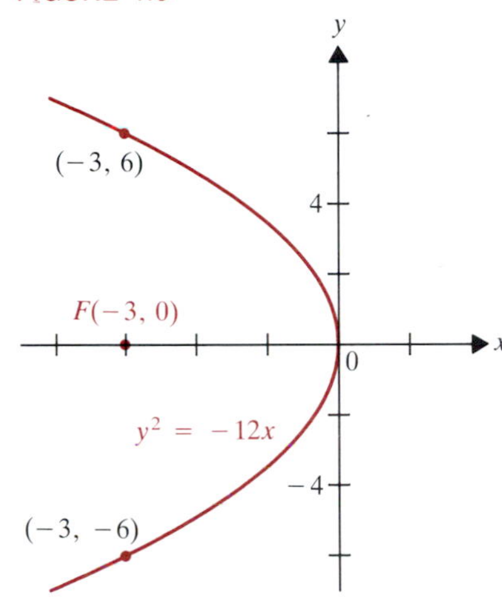

FIGURE 1.6

Example 1

Find the equation and make a sketch of the following parabolas.
(a) Focus (0, 2), directrix $y = -2$. (b) Focus $(-3, 0)$, directrix $x = 3$.

Solution. (a) The equation is $x^2 = 4py$, where $p = 2$, so $x^2 = 8y$. It is sketched in **Figure 1.5.** Since we know the vertex and have (3) available, we need plot only two other points to construct a rough graph. It is often convenient to use the points on the graph with y-coordinate p, as in Figure 1.5.

(b) This time, the equation has the form $y^2 = 4px$, with $p = -3$. So the equation is $y^2 = -12x$. It is sketched in **Figure 1.6,** where we used (5). ■

Example 2

(a) Find the focus and directrix of the parabola $3x^2 + 4y = 0$.
(b) Find the equation of the parabola that is symmetric relative to the x-axis, has vertex at (0, 0), and passes through the point (3, 2).

Solution. (a) We first put the equation in standard form:

$$3x^2 = -4y \to x^2 = -\frac{4}{3}y = -4 \cdot \frac{1}{3}y.$$

This is of the form (2): Thus $p = -1/3$. Hence the focus is at $(0, -1/3)$, and the directrix is $y = 1/3$. The parabola is sketched in **Figure 1.7.**

(b) Since the axis of symmetry is the x-axis, and the vertex is at the origin, the parabola has standard form equation (4): $y^2 = 4px$. Since $P(3, 2)$ is on the curve, we have

$$4 = 12p \to p = 1/3.$$

Hence the standard form equation is $y^2 = (4/3)x$. The curve is sketched in **Figure 1.8.** ■

FIGURE 1.7
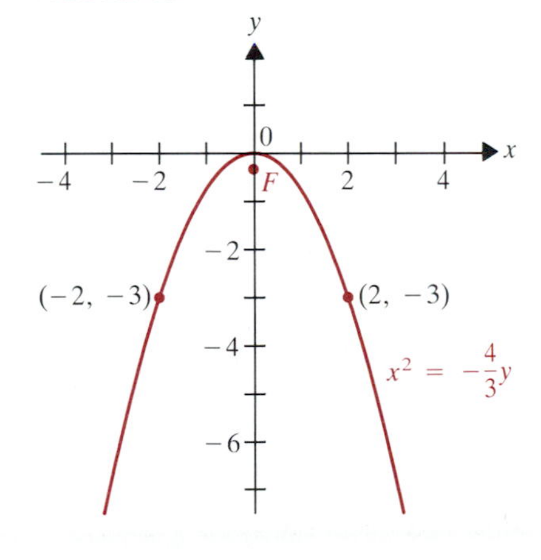

FIGURE 1.8
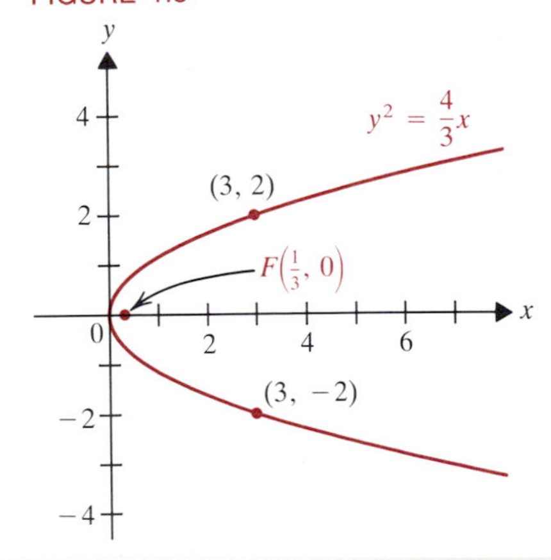

A parabola with vertex at the origin and focus on a coordinate axis is said to be in *standard position.* If a parabola has vertex at the point $H(h, k)$ and focus on one of the lines $x = h$ or $y = k$, as in **Figure 1.9,** then its equation can be found by ***translation of coordinate axes.*** This is done by setting up a new coordinate system with origin 0′ at $H(h, k)$ and new axes parallel to the

FIGURE 1.9

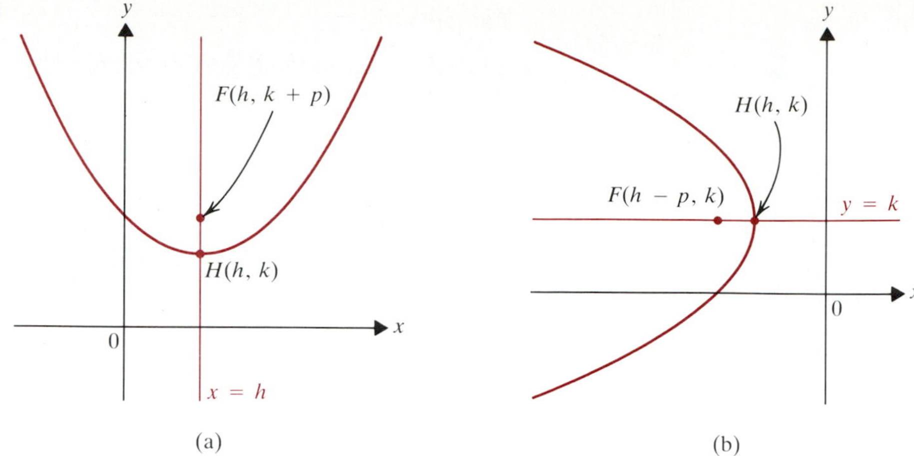

original ones, as in **Figure 1.10.** A point P with coordinates (x, y) in the old system has new coordinates (x', y') in the new system, where Figure 1.10 shows that

$$x' = x - h \quad \text{and} \quad y' = y - k. \tag{6}$$

The equation of the parabola with vertex at $H(h, k)$ and focus on $x = h$ or $y = k$ can be found with respect to the translated $x'y'$-coordinate system from (2) or (4). If the focus is $F(h, k + p)$ in the standard coordinate system, then from (6) the translated coordinates of F are $(x', y') = (0, p)$. Thus by (2) the equation in the new coordinate system is

$$x'^2 = 4py'. \tag{7}$$

Then (6) gives the equation in the original coordinate system:

$$(x - h)^2 = 4p(y - k). \tag{8}$$

Similarly, if a parabola has vertex $H(h, k)$ and focus $F(h + p, k)$, then its equation is

$$(y - k)^2 = 4p(x - h). \tag{9}$$

Note that in both (8) and (9), the distance from the focus to the vertex is $|p|$.

FIGURE 1.10

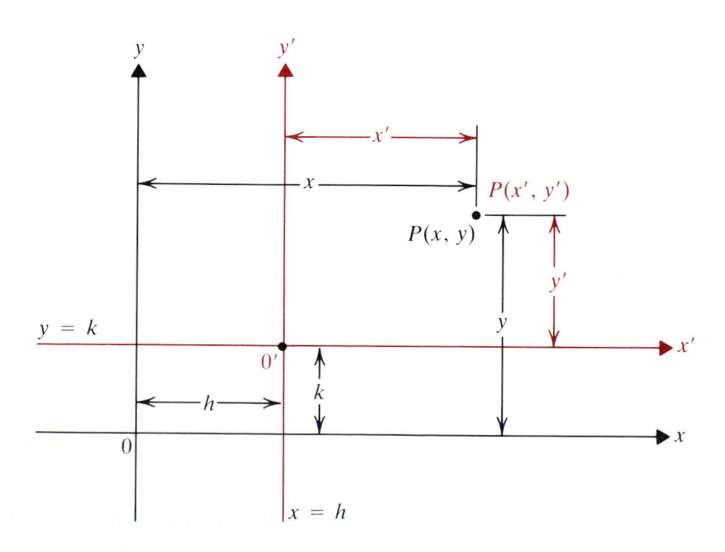

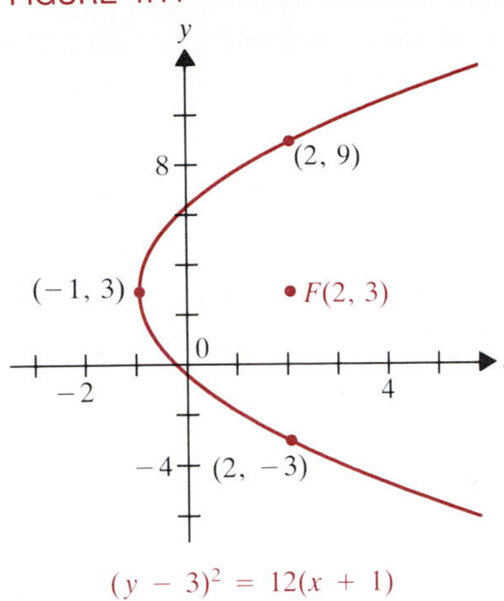

FIGURE 1.11

$(y-3)^2 = 12(x+1)$

Example 3

A parabola has vertex at $(-1, 3)$ and focus at $(2, 3)$. Find its equation and draw a sketch.

Solution. Since the distance from the focus to the vertex is $|p|$, we have

$$|p| = 2 - (-1) = 3. \tag{10}$$

The focus is to the right of the vertex, so the parabola opens to the right. Hence p is positive. Thus (10) gives $p = 3$. Because the vertex and the focus have the same y-coordinate, the parabola has equation (9), with $h = -1$ and $k = 3$: $(y-3)^2 = 4p(x+1) = 12(x+1)$. See **Figure 1.11** for the graph. ■

The graph of any second-degree equation in x and y of either of the forms

$$ax^2 + ex + fy + g = 0, \qquad \text{where } a \neq 0, \tag{11}$$

or

$$cy^2 + ex + fy + g = 0, \qquad \text{where } c \neq 0, \tag{12}$$

is a parabola. To draw it, we complete the square to obtain an equation of the form (8) or (9).

Example 4

Find the vertex, focus, and directrix of the parabola with equation $2y^2 - x - 6y + 3 = 0$. Draw the graph.

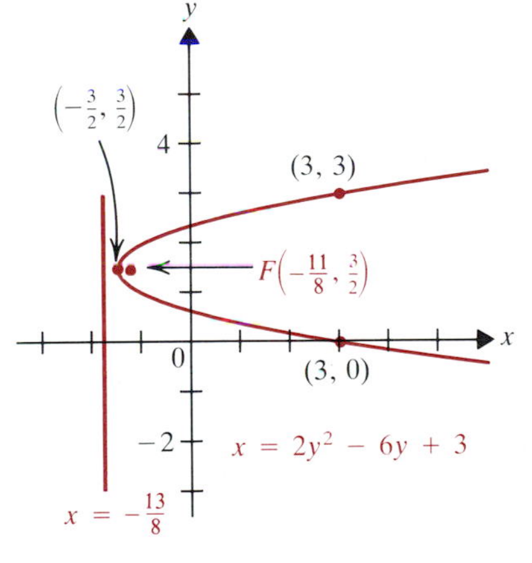

FIGURE 1.12

Solution. We separate the variables and complete the square, getting

$$x - 3 = 2(y^2 - 3y) \rightarrow x - 3 + \frac{9}{2} = 2\left(y^2 - 3y + \frac{9}{4}\right),$$

$$x + \frac{3}{2} = 2\left(y - \frac{3}{2}\right)^2 \rightarrow \frac{1}{2}\left(x + \frac{3}{2}\right) = \left(y - \frac{3}{2}\right)^2. \tag{13}$$

This is an equation of the form (9) with $4p = 1/2$ (so $p = 1/8$), $h = -3/2$, and $k = 3/2$. Therefore the vertex is at $(-3/2, 3/2)$. Since $p > 0$, the parabola opens to the right. The focus is at the point $(-3/2 + 1/8, 3/2) = (-11/8, 3/2)$. The directrix is $x = -3/2 - 1/8 = -13/8$. The graph is easiest to plot from (13) and is shown in **Figure 1.12**, which shows the points that are listed in the following table.

y	$3/2$	3	0
x	$-3/2$	3	3

■

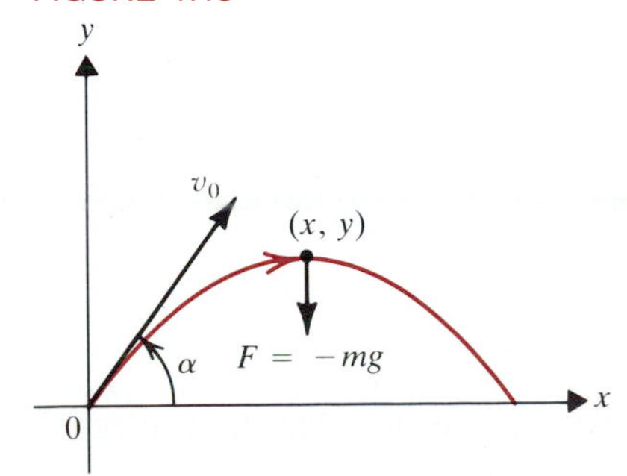

FIGURE 1.13

Projectile Motion*

The paths followed by a basketball shot toward a goal, a fly ball in baseball, and a shell fired from a howitzer are all parabolic. To see why, we consider a projectile of mass m fired at angle of elevation α (as in **Figure 1.13**) over level ground. At time t seconds after firing, let (x, y) be the coordinates of the tip of the projectile, where we choose axes so that the projectile is fired from the origin.

* Optional

The firing imparts an initial velocity v_0, where $dx/dt = v_0 \cos\alpha$ and $dy/dt = v_0 \sin\alpha$ at $t = 0$. We neglect all forces except gravity. In particular then, the horizontal force on the projectile is 0. Thus Newton's second law of motion says that

$$(14) \qquad 0 = m\frac{d^2x}{dt^2} \to \frac{d^2x}{dt^2} = 0,$$

where m is the mass of the projectile. Since the projectile is subject to a downward force $F = -mg$, where g is the acceleration due to gravity, we have

$$(15) \qquad -mg = m\frac{d^2y}{dt^2} \to \frac{d^2y}{dt^2} = -g.$$

Integrating (14) and (15), we get

$$(16) \qquad \frac{dx}{dt} = C_1 \quad \text{and} \quad \frac{dy}{dt} = -gt + C_2.$$

Using the values $dx/dt = v_0 \cos\alpha$ and $dy/dt = v_0 \sin\alpha$, when $t = 0$, we obtain from (16)

$$C_1 = v_0 \cos\alpha \quad \text{and} \quad C_2 = v_0 \sin\alpha.$$

Thus

$$\frac{dx}{dt} = v_0 \cos\alpha \quad \text{and} \quad \frac{dy}{dt} = -gt + v_0 \sin\alpha.$$

Integration of these gives

$$(17) \qquad x = (v_0 \cos\alpha)t + k_1 \quad \text{and} \quad y = -\frac{1}{2}gt^2 + (v_0 \sin\alpha)t + k_2.$$

Since the projectile was fired from the origin, we have $x = y = 0$ when $t = 0$. Therefore (17) says that

$$k_1 = 0 \quad \text{and} \quad k_2 = 0.$$

Hence

$$(18) \qquad x = (v_0 \cos\alpha)t \quad \text{and} \quad y = -\frac{1}{2}gt^2 + (v_0 \sin\alpha)t.$$

To find the path followed by the projectile, we solve the first equation in (18) for t and substitute that value into the second equation. We get

$$t = \frac{x}{v_0 \cos\alpha} \to y = -\frac{1}{2}g\frac{x^2}{{v_0}^2 \cos^2\alpha} + \frac{\cancel{v_0} \sin\alpha}{\cancel{v_0} \cos\alpha}x,$$

$$(19) \qquad y = -\frac{g}{2{v_0}^2 \cos^2\alpha}x^2 + (\tan\alpha)x.$$

This is the equation of a downward-opening parabola. We can use (19) to find information about the flight of the projectile.

Example 5

If $\alpha = \pi/6$ radians and $v_0 = 42$ meters per second, find how far the projectile will travel before hitting the ground.

FIGURE 1.14

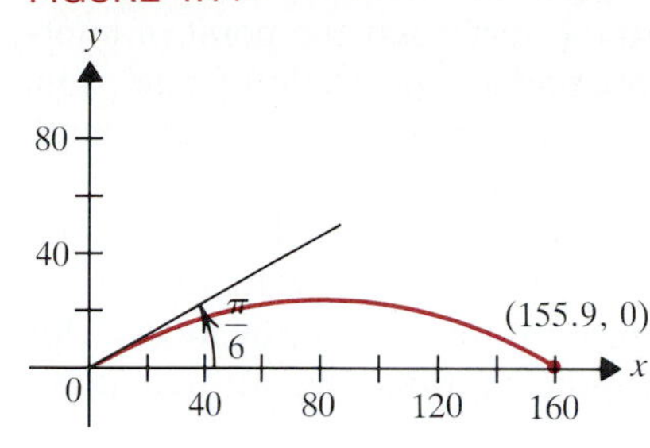

Solution. The projectile hits the ground when $y = 0$ in (19). That occurs when

$$0 = -\frac{g}{2v_0^2 \cos^2 \alpha} x^2 + (\tan \alpha)x \rightarrow 0 = x(-gx + 2v_0^2 \sin \alpha \cos \alpha).$$

Since $x = 0$ occurs at the start, we equate $-gx + 2v_0^2 \sin \alpha \cos \alpha$ to 0, to find x at the end. That gives

$$x = \frac{2v_0^2 \sin \alpha \cos \alpha}{g} = \frac{2 \cdot 42^2 \cdot 1/2 \cdot \sqrt{3}/2}{9.80} \approx 155.9 \text{ meters.}$$

The projectile's path is graphed in **Figure 1.14.** ■

Reflection Property of Parabolas*

The following property of parabolas is extensively used in applications.

1.2 THEOREM Let F be the focus of a parabola, $P_0(x_0, y_0)$ a point on the parabola, and l the tangent line to the parabola at P_0. Then l makes equal angles with FP_0 and the line through P_0 parallel to the axis of symmetry of the parabola. That is, in **Figure 1.15,** the angles α and β are equal.

FIGURE 1.15

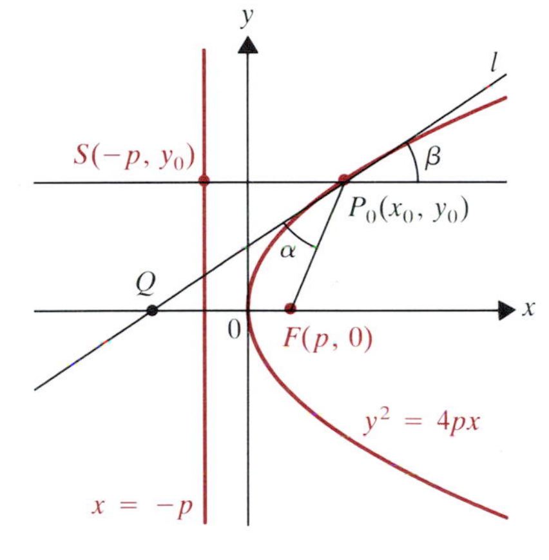

Proof. Without loss of generality, we can suppose that the parabola has vertex at (0, 0) and equation $y^2 = 4px$, as in Figure 1.15. Then the directrix is the line $x = -p$. We first find the equation of the tangent line l to the parabola at $P_0(x_0, y_0)$. Differentiating $y^2 = 4px$, we get

$$2yy' = 4p,$$

so the slope of l is

$$\left.\frac{dy}{dx}\right|_{(x_0, y_0)} = \frac{2p}{y_0}.$$

Thus l has equation

$$y - y_0 = \frac{2p}{y_0}(x - x_0). \tag{20}$$

At the point $Q(x, 0)$ where l meets the x-axis, we have $y = 0$ in (20). Therefore

$$-y_0^2 = 2p(x - x_0) \rightarrow -4px_0 = 2px - 2px_0 \rightarrow -2px_0 = 2px,$$

$$x = -x_0. \tag{21}$$

Angle $FQP_0 = \beta$, since it is the angle that l makes with the horizontal. Therefore $\alpha = \beta$ if FQP_0 is an isosceles triangle. To show that FQP_0 *is* isosceles, we prove that $d(Q, F) = d(F, P_0)$. From Definition 1.1 we have

$$\begin{aligned} d(F, P_0) &= d(P_0, S) = x_0 + p = d(Q, 0) + d(0, F) \qquad \text{by (21)} \\ &= d(Q, F). \end{aligned}$$

Thus the triangle is isosceles, so $\alpha = \beta$. QED

FIGURE 1.16

The reflection property in Theorem 1.2 is at the heart of several important technological applications. For example, the headlamp reflectors on automobiles have the shape of a *paraboloid,* the surface obtained when a parabola $y^2 = 4px$ is revolved about the x-axis. See **Figure 1.16.** When a tiny bulb is

* Optional

placed at the focus, then every light beam coming from F has angle of incidence α with the surface (that is, with the tangent to the surface at the point of intersection). By a basic law of physics, α is also equal to the angle of reflection. Theorem 1.2 says that $\alpha = \beta$, the angle between the surface and the horizontal. Therefore the beam is reflected outward horizontally. The same principle is used, in reverse, to make reflecting and radio telescopes, as well as radar and microwave antennas. Distant light or signals come to the receiving dish in essentially parallel lines. They then are all reflected off the dish to the focus F for viewing or reception. The dish thus magnifies the captured signal by concentrating it all at F.

Exercises 8.1

In Exercises 1–22, find the equation of the parabola and sketch its graph. Here V stands for vertex, F for focus, and l for directrix.

1. $F(0, -3)$; l: $y = 3$
2. $F(0, 4)$; l: $y = -4$
3. $F(2, 0)$; l: $x = -2$
4. $F(-1, 0)$; l: $x = 1$
5. $F(0, -3)$; l: $y = 1$
6. $F(0, 4)$; l: $y = -2$
7. $F(2, 0)$; l: $x = -1$
8. $F(-1, 0)$; l: $x = 2$
9. $V(1, 2)$; $F(4, 2)$
10. $V(2, -3)$; $F(0, -3)$
11. $V(3, 1)$; $F(3, 4)$
12. $V(1, -2)$; $F(1, -4)$
13. $F(-3, 7)$; l: $y = 1$
14. $F(2, 1)$; l: $y = 4$
15. $F(-3, 7)$; l: $x = 1$
16. $F(2, 1)$; l: $x = -3$
17. $V(1, 1)$; symmetric in the line $x = 1$; passes through $(6, 3)$
18. $V(-1, 3)$; symmetric in the line $y = 3$; passes through $(4, 7)$
19. $V(-3, 5)$; symmetric in the line $y = 5$; passes through $(5, 9)$
20. $V(2, -3)$; symmetric in the line $y = -3$; passes through $(-1, 7)$
21. l: $y = 4$; symmetric in the line $x = 4$; passes through $(7, 9)$
22. l: $x = 2$; symmetric in the line $y = 2$; passes through $(7, 5)$

In Exercises 23–36, find the vertex, focus, and directrix of the parabola whose equation is given, and draw a sketch.

23. $y^2 = 7x$
24. $y^2 = -8x$
25. $x^2 = -6y$
26. $x^2 = 3y$
27. $(y - 3)^2 = -4(x + 1)$
28. $(y + 2)^2 = 6(x - 2)$
29. $(x + 1)^2 = 2(y + 2)$
30. $(x - 3)^2 = -3(y - 2)$
31. $8x = 4y^2 - 12y + 1$
32. $2x = -3y^2 + 6y - 1$
33. $x^2 + y = 2x$
34. $x - 4y^2 = 6y$
35. $6y = x^2 + 2x - 17$
36. $3y = 2x^2 - 12x + 7$

37. **(a)** Show that the tangent line to the parabola $x^2 = 4py$ at the point (x_0, y_0) has equation

$$y = \frac{x_0}{2p} x - y_0.$$

(b) Show that the tangent line in (a) intersects the y-axis at the point $(0, -y_0)$.

38. The *latus rectum* of a parabola is the line segment, with endpoints on the parabola, that is perpendicular to the axis of symmetry and passes through the focus. See **Figure 1.17.** Show that its length is $4|p|$.

FIGURE 1.17

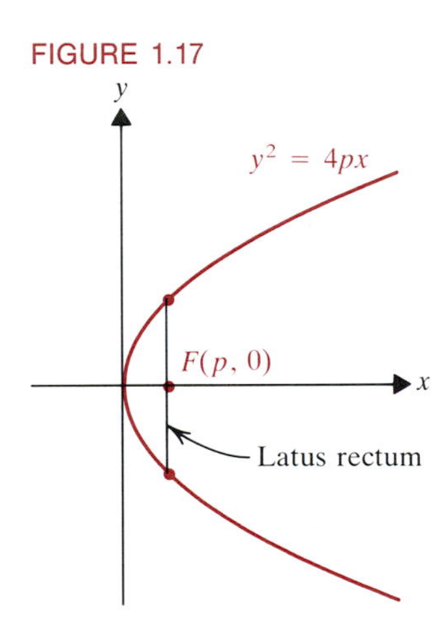

39. Show that the point on any parabola that is nearest the focus is the vertex.

The remaining exercises deal with projectile motion and reflection.

40. **(a)** Use (19) to find how long the projectile is in flight.
(b) Use (a) to show that the range of the projectile is

$$\frac{v_0^2}{g} \sin 2\alpha.$$ (This generalizes the result of Example 5.)

41. **(a)** Find t corresponding to the maximum height of the projectile in Exercise 40.
(b) What is the maximum height?

42. To hit a target 1000 meters from a gun for which $v_0 = 149$ meters per second, what angle of elevation should be used? (Take $g = 9.8$ m/s^2.)

43. If $\alpha = \pi/4$, and a projectile fired from a howitzer travels exactly 2 kilometers over level ground before striking its target, what is v_0 for the howitzer?

44. What angle of elevation produces maximum range for a shell fired from a gun for which v_0 is fixed?

45. A headlamp reflector is 5 inches high and $2\frac{1}{2}$ inches deep. Where should the bulb be placed?

46. The telescope at Mt. Palomar observatory in California is 200 inches across its surface (edge to edge). Its shape is a paraboloid of revolution. The vertex is $3\frac{3}{4}$ inches below the level of the edges of the mirror. Where is the focus of this reflecting telescope?

47. Assume that the cannon in Example 5 is located at the base of a hillside that has an angle of elevation of 30° (**Figure 1.18**). Show that the shell hits the hillside at point P whose x-coordinate is

$$\frac{2v_0^2}{g}\left(\sin\alpha\cos\alpha - \frac{1}{\sqrt{3}}\cos^2\alpha\right).$$

(*Hint:* The shell hits the hillside at the point where $y/x = \tan 30° = 1/\sqrt{3}$.)

FIGURE 1.18

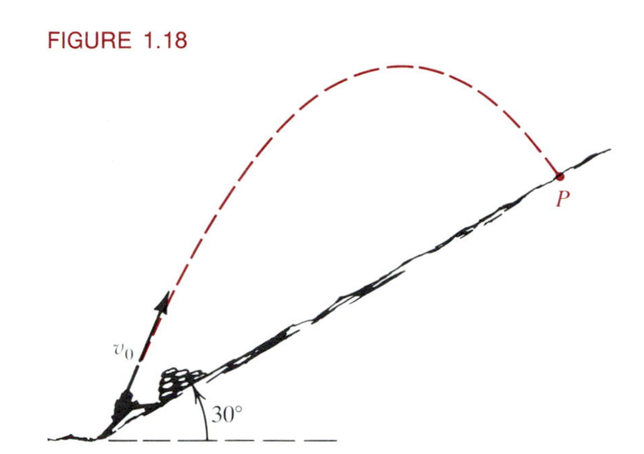

48. How long is the shell in Exercise 47 in flight?

8.2 Ellipses

The defining condition

$$d(P, F) = d(P, l)$$

for a parabola with focus F and directrix l is equivalent to

$$(1) \qquad \frac{d(P, F)}{d(P, l)} = 1.$$

Thus a parabola consists of all points for which the ratio of $d(P, F)$ to $d(P, l)$ is constantly 1. For an ellipse, the ratio of $d(P, F)$ to $d(P, l)$ is a constant *less* than 1.

2.1 DEFINITION Let l be a line, F a point not on l, and $E < 1$ a fixed positive number. An ***ellipse*** is the set of all points $P(x, y)$ such that

$$(2) \qquad \frac{d(P, F)}{d(P, l)} = E.$$

That is, an ellipse consists of all points P the ratio of whose distances from a fixed point F (the ***focus***) and a fixed line l (the ***directrix***) is a constant E (the ***eccentricity***) less than 1.

In **Figure 2.1,** we show part of an ellipse and some points on it. The ratio E of $d(P, F)$ to $d(P, l)$ appears to be about 1/2.

To derive the equation of an ellipse, we select a focus F on a coordinate axis and a directrix l so that the resulting equation will be as simple as possible: in Figure 2.1, F is the point $(c, 0)$ and l is the line $x = c/E^2$. For any point $P(x, y)$ on the ellipse, (2) then becomes

$$(3) \quad d(P, F) = Ed(P, l) \rightarrow \sqrt{(x-c)^2 + (y-0)^2} = E\sqrt{\left(\frac{c}{E^2} - x\right)^2 + (y-y)^2}.$$

FIGURE 2.1

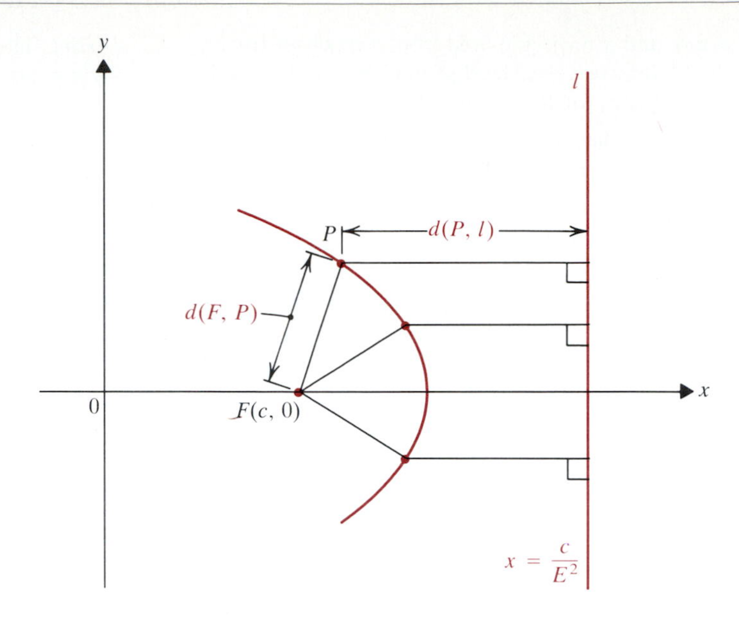

Squaring, we get

$$x^2 - 2cx + c^2 + y^2 = E^2\left(\frac{c^2}{E^4} - \frac{2c}{E^2}x + x^2\right),$$

$$x^2 - 2cx + c^2 + y^2 = \frac{c^2}{E^2} - 2cx + E^2x^2,$$

$$x^2(1 - E^2) + y^2 = \frac{c^2}{E^2} - c^2 = c^2\left(\frac{1}{E^2} - 1\right) = \frac{c^2}{E^2}(1 - E^2). \tag{4}$$

To simplify this equation, we let $a = c/E$. We then get

$$x^2(1 - E^2) + y^2 = a^2(1 - E^2),$$

which when divided by $a^2(1 - E^2)$ becomes

$$\frac{x^2}{a^2} + \frac{y^2}{a^2(1 - E^2)} = 1. \tag{5}$$

Since $E < 1$, both $1 - E^2$ and $a^2(1 - E^2)$ are positive. We let $b^2 = a^2(1 - E^2)$. Then

$$b^2 = a^2 - a^2E^2 = a^2 - c^2 \rightarrow a^2 = b^2 + c^2. \tag{6}$$

We can then rewrite (5) as

$$\frac{x^2}{a^2} + \frac{y^2}{b^2} = 1, \qquad a^2 > b^2. \tag{7}$$

The substitution $a = c/E$ made in deriving (7) transforms the equation $x = c/E^2$ of the directrix to the simpler form $x = a/E$. (Note that dividing $a = c/E$ by E gives $a/E = c/E^2$.)

The derivation of (7) shows that the coordinates of every point $P(x, y)$ on the ellipse satisfy (7). In Exercise 45 you are asked to check that every point $P(x, y)$ whose coordinates satisfy (7) also lies on the ellipse. Equation (7) is the ***standard form equation*** of the ellipse with focus $F(c, 0)$, eccentricity $E = c/a$, and directrix $x = a/E$, where $c^2 = a^2 - b^2$. **Figure 2.2** shows the graph of (7).

FIGURE 2.2

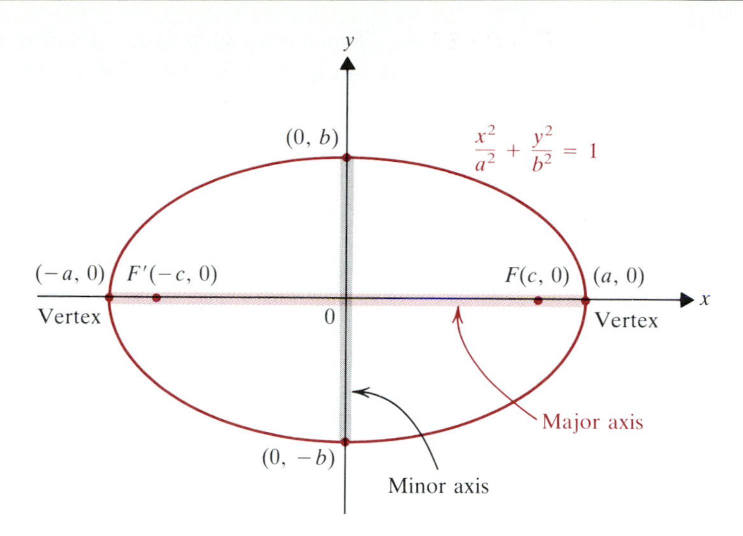

The ellipse crosses the x-axis at $(a, 0)$ and $(-a, 0)$, which are called the ***vertices*** of the ellipse. The part of the x-axis joining the vertices is called the ***major axis*** of the ellipse. The ellipse crosses the y-axis at $(0, b)$ and $(0, -b)$. The segment joining those points is called the ***minor axis.*** The point $(0, 0)$ halfway between the vertices is called the ***center*** of the ellipse.

The symmetric nature of the vertices $(a, 0)$ and $(-a, 0)$ suggests that $(-c, 0)$ might be a second focus. Indeed, if we use $(-c, 0)$ as focus F' and $x = -c/E^2$ as directrix l', then from

$$\frac{d(P, F')}{d(P, l')} = E \tag{2'}$$

and **Figure 2.3** we obtain

$$\sqrt{(x + c)^2 + (y - 0)^2} = E\sqrt{\left(-\frac{c}{E^2} - x\right)^2 + (y - y)^2}, \tag{3'}$$

instead of (3). Squaring (3′) will once again lead to (7). Thus the ellipse with focus $(-c, 0)$ and directrix l': $x = -c/E^2 = -a/E$ *coincides* with the original ellipse, since it has the same equation. Therefore the ellipse has two foci $F(c, 0)$ and $F'(-c, 0)$ and two directrices, $x = c/E^2$ and $x = -c/E^2$.

FIGURE 2.3

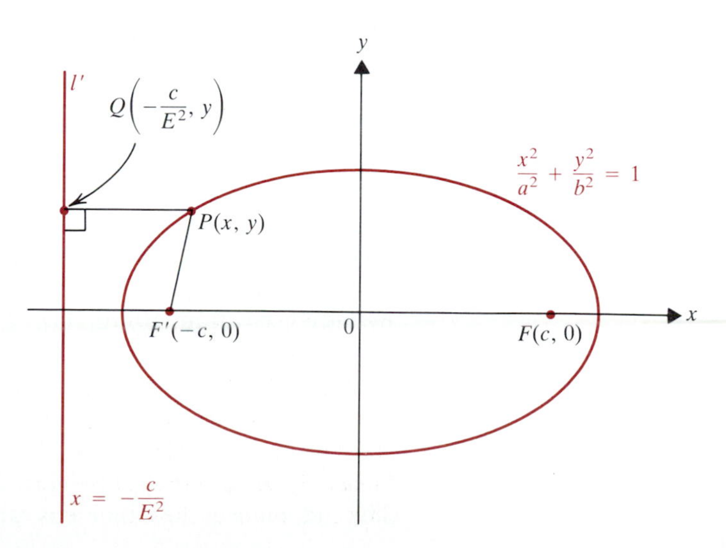

There is an alternative definition of the ellipse, using the two foci F and F' as the basic data. The ellipse can be defined as the set of all points $P(x, y)$ in the plane such that

$$d(F, P) + d(F', P) = 2a.$$

(See Exercise 48.)

If the foci are $F(0, c)$ and $F'(0, -c)$ on the y-axis and the corresponding directrices are $y = c/E^2$ and $y = -c/E^2$, then we obtain (Exercise 46) the standard form equation

$$\frac{x^2}{b^2} + \frac{y^2}{a^2} = 1, \qquad a^2 > b^2, \tag{8}$$

where still

$$a^2 = b^2 + c^2. \tag{6}$$

FIGURE 2.4

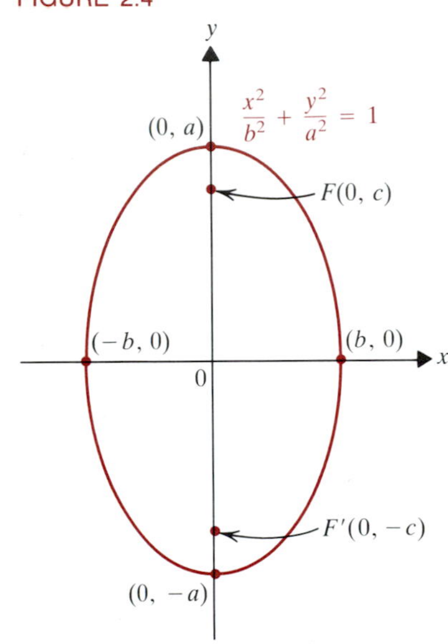

The graph of (8) is shown in **Figure 2.4.**

The symmetry apparent in the graphs of the ellipses we have drawn follows from Equations (7) and (8) and Theorem 7.3 of Chapter 3.

A circle can be thought of as an ellipse with eccentricity 0, since if $E = c/a = 0$, then $c = 0$, so $a = b$. If $a = b$ in (7) or (8), then we get the circle

$$\frac{x^2}{a^2} + \frac{y^2}{a^2} = 1 \to x^2 + y^2 = a^2$$

of radius a centered at the origin.

The graph of any quadratic equation in x and y of the form

$$kx^2 + ly^2 = m, \tag{9}$$

where k, l, and m are positive numbers, is always an ellipse (or circle). Dividing (9) through by m will produce an equation of the form (7) or (8), depending on whether x^2 or y^2 is divided by the larger constant a^2.

Example 1

Find the vertices, foci, and endpoints of the minor axis of the ellipse with equation

$$2x^2 + 9y^2 = 8. \tag{10}$$

Draw a sketch of the curve.

Solution. Dividing (10) through by 8, we get

$$\frac{x^2}{4} + \frac{y^2}{8/9} = 1. \tag{11}$$

Since $4 > 8/9$, this is of the form (7), so the major axis lies along the x-axis. The vertices are the points of the curve on the x-axis, which can be obtained by letting $y = 0$ in (10) or (11): $(2, 0)$ and $(-2, 0)$. Since $a^2 = 4$ and $b^2 = 8/9$, (6) gives

$$c^2 = a^2 - b^2 = 4 - \frac{8}{9} = \frac{28}{9}.$$

Hence $c = \sqrt{28/9} = 2\sqrt{7}/3$. Thus the foci are $F'(-2\sqrt{7}/3, 0)$ and $F(2\sqrt{7}/3, 0)$. The endpoints of the minor axis can be found by letting $x = 0$ in (11): $(0, 2\sqrt{2}/3)$

and $(0, -2\sqrt{2}/3)$. Without plotting any more points, we can sketch the graph as in **Figure 2.5.** ■

FIGURE 2.5

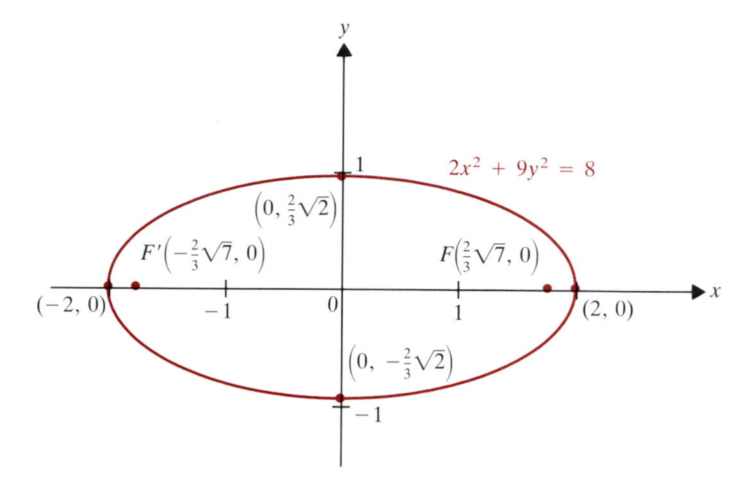

The nearer the eccentricity $E = c/a$ is to 1, the more elongated the ellipse, since the foci will be almost as far from the center as the vertices. In Example 1, for instance, we have

$$E = \frac{c}{a} = \frac{2\sqrt{7}/3}{2} = \frac{\sqrt{7}}{3} \approx 0.882.$$

The nearer to 0 the eccentricity is, the more nearly circular the ellipse, since the foci are relatively close to each other. (As we saw above, when $E = 0$ we get a circle.) One of Kepler's laws of planetary motion is that celestial bodies have elliptical orbits. As Table 1 shows, the eccentricity of each planet except Mercury and Pluto is very close to 0, so the orbits are very nearly circular.

Using (6), (7), and (8), we can find the equation of an ellipse, given its vertices and foci or equivalent data, as in the next example.

TABLE 1

Planetary Data

Name	*Value of a (in millions of km)*	*E*	*Orbital period (in days)*
Mercury	57.9	0.2056	87.96
Venus	108.2	0.0068	224.68
Earth	149.6	0.0167	365.26
Mars	227.9	0.0934	686.95
Jupiter	778.3	0.0483	4 337
Saturn	1427.0	0.0560	10 760
Uranus	2871.0	0.0461	30 700
Neptune	4497.1	0.0100	60 200
Pluto	5913.5	0.2484	90 780

Example 2

Find the standard form equation of the ellipse centered at the origin if its minor axis has length 8 and its foci are $(0, \pm 3)$.

Solution. The major axis is along the y-axis, and since the foci are $(0, \pm 3)$, we have $c = 3$. We are also given $2b = 8$, so $b = 4$. Then (6) gives

$$a^2 = b^2 + c^2 = 16 + 9 = 25.$$

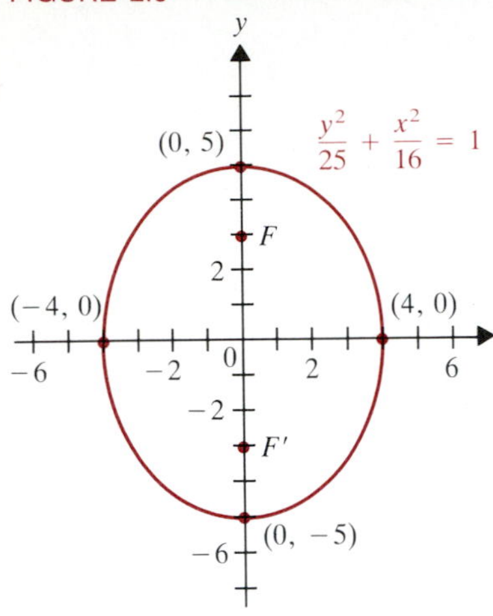

FIGURE 2.6

Thus $a = 5$. Because the foci are on the y-axis, the ellipse has an equation of the form (8):

$$\frac{y^2}{25} + \frac{x^2}{16} = 1.$$

The ellipse is sketched in **Figure 2.6.** ■

Thus far we have restricted attention to ellipses centered at (0, 0). If, instead, the ellipse is centered at $C(h, k)$, then we can use the method of translation of axes from the last section. If we translate axes so that the new origin $0'$ is at $C(h, k)$, then the ellipse has one of the equations

$$\frac{x'^2}{a^2} + \frac{y'^2}{b^2} = 1 \quad \text{or} \quad \frac{x'^2}{b^2} + \frac{y'^2}{a^2} = 1.$$

Recalling from the last section [Equation (6), p. 484] that $x' = x - h$ and $y' = y - k$, we see that the ellipse has the equation

$$\frac{(x-h)^2}{a^2} + \frac{(y-k)^2}{b^2} = 1 \tag{12}$$

or

$$\frac{(x-h)^2}{b^2} + \frac{(y-k)^2}{a^2} = 1. \tag{13}$$

Example 3

Find the vertices, foci, and endpoints of the minor axis of the ellipse

$$\frac{(x+1)^2}{8} + \frac{(y-3)^2}{4} = 1. \tag{14}$$

Draw a sketch.

Solution. This ellipse is centered at $(-1, 3)$, because its equation has the form (12). Here $a^2 = 8$ and $b^2 = 4$. Hence

$$c^2 = a^2 - b^2 = 8 - 4 = 4.$$

The vertices are obtained by letting $y' = 0$, that is, $y = 3$. From (14) then,

$$\frac{(x+1)^2}{8} + 0 = 1 \rightarrow (x+1)^2 = 8,$$

$$x + 1 = \pm 2\sqrt{2} \rightarrow x = -1 + 2\sqrt{2} \quad \text{or} \quad x = -1 - 2\sqrt{2}.$$

Therefore the vertices are $V_1(-1 - 2\sqrt{2}, 3)$ and $V_2(-1 + 2\sqrt{2}, 3)$. The foci are also on the line $y = 3$; since $c^2 = 4$, they have $x'y'$-coordinates $(\pm 2, 0)$. So their coordinates are $y = 3$ and

$$x = -1 \pm 2 \rightarrow x = 1 \text{ and } -3.$$

Thus the foci are $F_1(-3, 3)$ and $F_2(1, 3)$. Since $b^2 = 4$, $b = 2$. So the minor axis, which lies along the line $x = -1$, has length $2b = 4$. Its endpoints are b units above and below the center $(-1, 3)$ and so are $(-1, 5)$ and $(-1, 1)$. The graph is sketched in **Figure 2.7.** ■

FIGURE 2.7

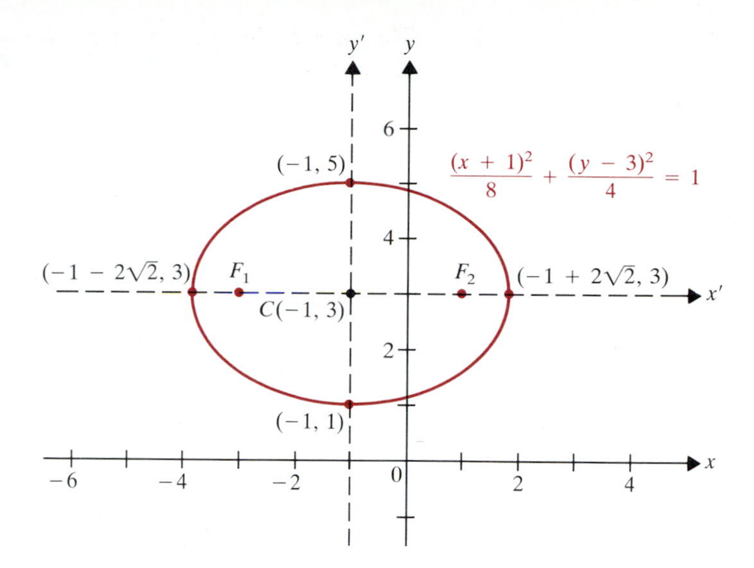

FIGURE 2.8

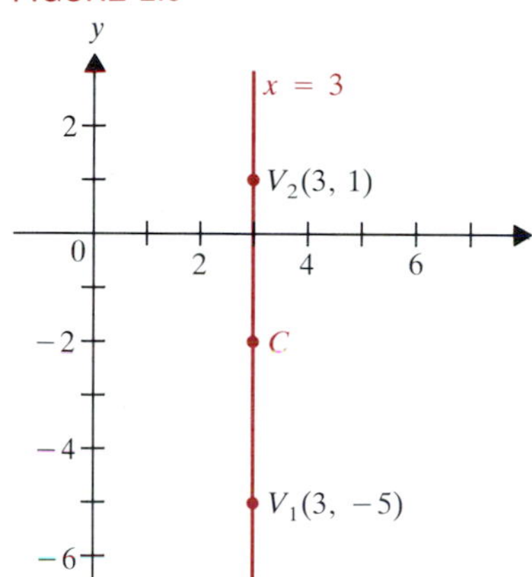

Translation of axes also can be used to find the equation of an ellipse centered at some point C other than the origin.

Example 4

An ellipse has vertices $V_1(3, -5)$ and $V_2(3, 1)$. Its minor axis has length 3. Find its equation. Where are the foci? What is the eccentricity?

Solution. We show some of the given data in **Figure 2.8.** The major axis lies in the line $x = 3$, since the two vertices are on that line. We are also told that $2b = 3$, so $b = 3/2$. Since the center is halfway between the vertices, it is the point $C(3, -2)$. Now $2a$ is the distance between the vertices, so $2a = 6 \rightarrow a = 3$. Thus from (8) the equation is

$$\frac{y'^2}{9} + \frac{x'^2}{9/4} = 1 \rightarrow \frac{(y+2)^2}{9} + \frac{(x-3)^2}{9/4} = 1.$$

FIGURE 2.9

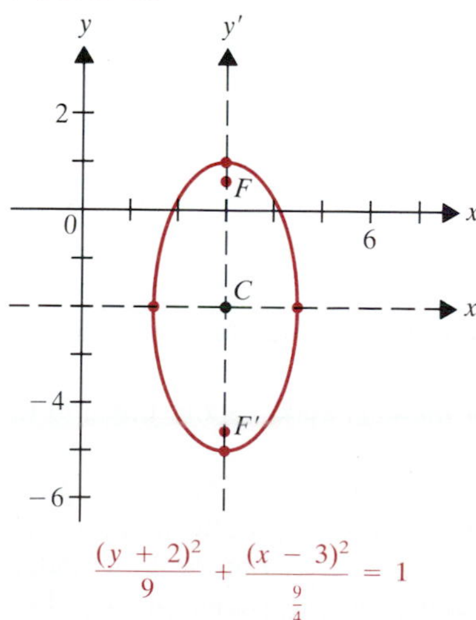

Since $c^2 = a^2 - b^2 = 9 - 9/4 = 27/4$, we have $c = 3\sqrt{3}/2$. So

$$E = c/a = \sqrt{3}/2.$$

The foci are then at $(3, -2 - 3\sqrt{3}/2)$ and $(3, -2 + 3\sqrt{3}/2)$. The graph is shown in **Figure 2.9.** ■

The graph of an equation of the form

$$ax^2 + cy^2 + ex + fy + g = 0,$$

where a and c have the same sign, is either an ellipse, a circle, a single point, or the empty set. To find out which, we complete the square in x and y as in the last section.

Example 5

Identify and sketch the graph of the equation $3x^2 + 4y^2 - 9x + 8y + 27/4 = 0$.

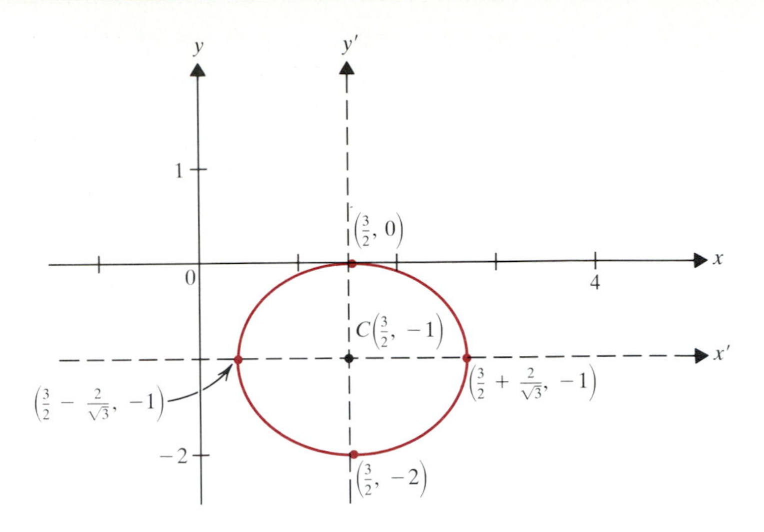

FIGURE 2.10

Solution. Completing the square, we get

$$3(x^2 - 3x) + 4(y^2 + 2y) = -\frac{27}{4},$$

$$3\left(x - \frac{3}{2}\right)^2 + 4(y + 1)^2 = -\frac{27}{4} + 3 \cdot \frac{9}{4} + 4 = 4,$$

$$\frac{\left(x - \frac{3}{2}\right)^2}{\frac{4}{3}} + (y + 1)^2 = 1.$$

Thus the graph of the equation is an ellipse centered at $(3/2, -1)$ with $a^2 = 4/3$ and $b^2 = 1$. The vertices are on the line $y = -1$, with x-coordinate satisfying

$$\left(x - \frac{3}{2}\right)^2 = \frac{4}{3} \rightarrow x - \frac{3}{2} = \pm\frac{2}{\sqrt{3}} \rightarrow x = \frac{3}{2} \pm \frac{2}{\sqrt{3}}.$$

Since $b^2 = 1$, the minor axis has endpoints $(3/2, 0)$ and $(3/2, -2)$. The graph is sketched in **Figure 2.10.** ■

FIGURE 2.11

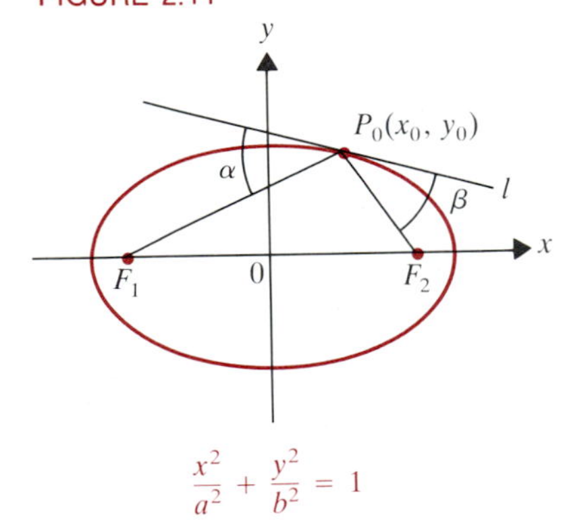

$$\frac{x^2}{a^2} + \frac{y^2}{b^2} = 1$$

There is a reflecting property of ellipses somewhat akin to that of Theorem 1.2 for parabolas. In Exercises 47 and 49 you are asked to prove the following result.

2.2 THEOREM Let l be the tangent to an ellipse at a point $P_0(x_0, y_0)$. Then l makes equal angles with the segments F_1P_0 and F_2P_0. That is, in **Figure 2.11** the angles α and β are equal.

This theorem implies that light or sound waves originating from either focus and hitting any point P on the ellipse will be reflected to the other focus. This accounts for the phenomenon of a *whispering gallery*, such as the one found in the statuary hall (formerly the House of Representatives chamber) in the United States Capitol building. The ceiling of that room has the shape of an *ellipsoid of revolution*, obtained by revolving the graph of an ellipse about its major axis. A representative seated at one focus, say F_2 in Figure 2.11, under this ceiling could hear whispered conversation of a colleague seated at the other focus F_1, even though no one else in the room could hear it.

HISTORICAL NOTE

Johannes Kepler (1571–1630) taught mathematics at the University of Graz in Austria from 1594 to 1601, before becoming the imperial astronomer and mathematician at the court of the Holy Roman Emperor Rudolph II. Besides his discovery that planetary orbits are elliptical, he formulated two other basic astronomical laws. The first states that a line drawn from the sun to a planet sweeps out equal areas during equal time periods, irrespective of the planet's position in its orbit. The third law says that the cube of a in Table 1 is proportional to the square of the orbital period. (The constant of proportionality is 25.1, as you can check from the data in Table 1.) Kepler's work led Newton to formulate his law of gravitation. As mentioned in Section 3.2, Kepler also discovered Theorem 2.3 of Chapter 3: $f'(c) = 0$ at a local extreme point c of a differentiable function f.

The whispering gallery phenomenon in the old House of Representatives was the subject of much gossip. Former President John Quincy Adams (1767–1848) served in the House from 1831 to 1848. He was rumored to have discovered that his desk was located directly under F_2 and to have feigned sleep while in fact listening to political opponents' whispered plans. Visitors to the Capitol's statuary hall can see a bronze disk embedded in the floor where Mr. Adams' desk was located. He suffered a fatal seizure at that spot in 1848 while delivering an impassioned attack on the Treaty of Guadalupe-Hidalgo, which ended the Mexican War. **Figure 2.12** is a painting by the American artist Samuel F.B. Morse (1791–1872), more famous as the inventor of the telegraph. It shows the House in session in 1822. The thick drapes were hung in a futile attempt to "deaden unwanted echoes."

FIGURE 2.12 Samuel Finley Breese Morse: *The Old House of Representatives.* In the collection of The Corcoran Gallery of Art, Museum Puchase, 1911.

Exercises 8.2

In Exercises 1–24, find the center, vertices, foci, and endpoints of the minor axes of the ellipse with the given equation. Make a sketch.

1. $\frac{x^2}{16} + \frac{y^2}{25} = 1$

2. $\frac{x^2}{4} + \frac{y^2}{9} = 1$

3. $\frac{x^2}{16} + \frac{y^2}{9} = 1$

4. $\frac{x^2}{25} + \frac{y^2}{9} = 1$

5. $4x^2 + 9y^2 = 36$

6. $25x^2 + 9y^2 = 225$

7. $4x^2 + y^2 = 16$

8. $16x^2 + 9y^2 = 144$

9. $\frac{x^2}{9} + \frac{(y-3)^2}{4} = 1$

10. $\frac{(x+3)^2}{16} + \frac{y^2}{4} = 1$

11. $\frac{(x+3)^2}{9} + \frac{(y-1)^2}{4} = 1$

12. $\frac{(x-2)^2}{16} + \frac{(y+3)^2}{36} = 1$

13. $9(x-3)^2 + 4(y+2)^2 = 36$

14. $4(x+1)^2 + (y-3)^2 = 4$

15. $25(x+1)^2 + 9(y+2)^2 = 225$

16. $16(x+4)^2 + 9(y+3)^2 = 144$

17. $4x^2 + 9y^2 - 8x = 32$

18. $9x^2 + 25y^2 - 100y - 125 = 0$

19. $9x^2 + 36x + 4y^2 - 8y + 4 = 0$

20. $25x^2 + 150x + 16y^2 - 96y - 31 = 0$

21. $x^2 + y^2 + 4x - 4y + 8 = 0$

22. $9x^2 + 6y^2 - 54x - 24y + 105 = 0$

23. $16x^2 + 25y^2 - 128x + 150y + 545 = 0$

24. $4x^2 + 3y^2 + 16x - 30y + 100 = 0$

In Exercises 25–44, find the equation of the ellipse satisfying the given conditions.

25. Foci $(\pm 3, 0)$, major axis of length 10

26. Foci $(\pm 12, 0)$, major axis of length 26

27. Foci $(0, \pm 8)$, major axis of length 20

28. Foci $(0, \pm 7)$, major axis of length 16

29. $C(0, 0)$, $F_2(4, 0)$, $V_2(5, 0)$

30. $C(0, 0)$, $F_1(-3, 0)$, $V_1(-5, 0)$

31. Vertices $(\pm 4, 0)$, foci $(\pm 2, 0)$

32. Vertices $(\pm 6, 0)$, foci $(\pm 4, 0)$

33. Vertices $(0, \pm 5)$, foci $(0, \pm 3)$

34. Vertices $(0, \pm 4)$, foci $(0, \pm 3)$

35. Foci $(\pm 5, 0)$, minor axis endpoints $(0, \pm 3)$

36. Foci $(0, \pm 4)$, minor axis endpoints $(\pm 3, 0)$

37. Foci $(2, -1)$ and $(2, 7)$, major axis of length 12

38. Foci $(-2, 2)$ and $(-2, 6)$, major axis of length 10

39. Foci $(0, 1)$ and $(4, 1)$, major axis of length 6

40. Foci $(3, -2)$ and $(-1, -2)$, major axis of length 6

41. Foci $(1, -2)$, $(7, -2)$, minor axis of length 8

42. Foci $(-3, 2)$, $(-3, -4)$, minor axis of length 8

43. Vertices $(-1, 0)$ and $(-1, 4)$, $E = \sqrt{3}/2$

44. Vertices $(3, 2)$ and $(-1, 2)$, $E = \sqrt{3}/2$

45. Show that every point $P(x, y)$ whose coordinates satisfy (7) lies on the ellipse determined by (3). [*Hint:* Let $c^2 = a^2 - b^2$ and let $E = c/a$. Show that $E < 1$. Next show that (7) implies (5). Multiply (5) through by $a^2(1 - E^2)$ and rewrite to get (4). Then reverse the steps in the text to get (3).]

46. Derive Equation (8) for the ellipse with foci $F(0, c)$ and $F'(0, -c)$ and directrices $y = \pm c/E^2$.

47. **(a)** Show that the equation of the tangent line to the ellipse (7) at $P_0(x_0, y_0)$ is

$$\frac{xx_0}{a^2} + \frac{yy_0}{b^2} = 1.$$

(b) What is the equation of the tangent line to the ellipse (8) at $P_0(x_0, y_0)$?

48. Verify the alternative definition of the ellipse mentioned on p. 492 as follows. Given the foci $F(c, 0)$ and $F'(-c, 0)$, show that the set S of all points such that $d(F, P) + d(F', P) = 2a$ is exactly the set of all points whose coordinates satisfy (7). (Here let $b^2 = a^2 - c^2$.)

49. Use the trigonometric identity

$$\tan(\theta + \beta) = \frac{\tan\theta + \tan\beta}{1 - \tan\theta\tan\beta}$$

and Exercise 47 to prove Theorem 2.2. [*Hint:* Let θ be the angle between l and the horizontal in Figure 2.11. Then use (6) and (7) to show that $\tan\alpha = \tan\beta = b^2/cy_0$.]

50. Under what conditions on a, b, and c is the graph of $x^2 + ax + y^2 + by = c$ **(a)** an ellipse, **(b)** a single point, and **(c)** empty?

51. The moon's orbit about the earth, which is at a focus, has major axis of length 478,400 miles and minor axis of length 477,600 miles. Find the **(a)** greatest, **(b)** least, and **(c)** mean distance of the moon from the earth.

52. An equation for the earth's orbit can be found by measuring the greatest and least distances from the earth to the sun throughout a year and using the fact that the sun is at a focus of the earth's orbit. If those distances are 94,560,000 and 91,450,000 miles (to the nearest 10,000 miles), find an equation for the earth's orbit. Find the eccentricity and compare it with the value shown in Table 1.

53. The *apogee* A of a satellite's orbit is its greatest distance from the center of the earth. The *perigee* P is the least distance from the center.

(a) Show that the eccentricity is given by

$$E = \frac{A - P}{A + P}.$$

(b) If a satellite has $A = 7000$ and $P = 6600$, find the equation of its orbit. What is E?

54. Show that the arc length around the ellipse $x^2/a^2 + y^2/b^2 = 1$ is given by $L = 4a \int_0^{\pi/2} \sqrt{1 - E^2 \sin^2 \theta}\, d\theta$, where E is the eccentricity. (The integral is called an *elliptic integral*. It cannot be worked out by antidifferentiation.)

PC **55.** If you have access to a computer or programmable calculator program for Simpson's rule, find an approximate value for L in Exercise 54 if $a = 4$ and $b = 2$. Iterate until successive results agree in the first six decimal places.

PC **56.** Repeat Exercise 54 for $a = 5$ and $b = 3$.

57. There is a mechanical scheme based on Exercise 48 for drawing an ellipse. Cut a string of length $2a$ and pin its ends to F and F'. Stretch the string taut as in **Figure 2.13.** As the pencil in Figure 2.13 moves around the midpoint 0 of FF', it traces out an ellipse. Why?

FIGURE 2.13

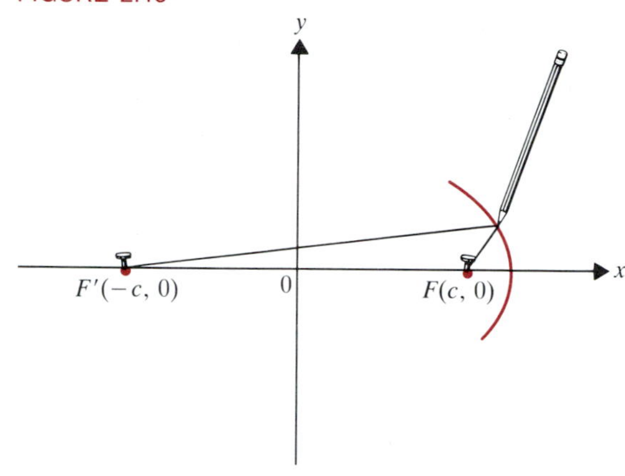

8.3 Hyperbolas

The third and last conic section is the hyperbola. Its definition looks just like that of an ellipse, except that for a hyperbola the eccentricity E is *greater* than one. This small change produces a curve with two branches.

3.1 DEFINITION

Let l be a line, F a point not on l, and $E > 1$ a fixed positive number. A ***hyperbola*** is the set of all points $P(x, y)$ such that

$$(1) \qquad \frac{d(P, F)}{d(P, l)} = E.$$

That is, a hyperbola consists of all points P the ratio of whose distances from a fixed point F (the ***focus***) and a fixed line l (the ***directrix***) is a constant E (the ***eccentricity***) greater than 1 (**Figure 3.1**).

FIGURE 3.1

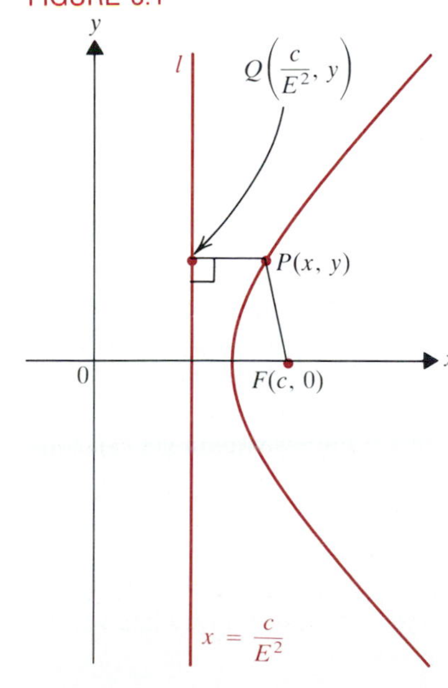

To derive the equation of a hyperbola, we select the same focus $F(c, 0)$ and directrix $x = c/E^2$ used in the last section. For any point $P(x, y)$ on the hyperbola, we have from (1),

$$d(P, F) = Ed(P, l),$$

$$\sqrt{(x - c)^2 + (y - 0)^2} = E\sqrt{(x - c/E^2)^2 + (y - y)^2},$$

$$x^2 - 2cx + c^2 + y^2 = E^2x^2 - 2cx + \frac{c^2}{E^2},$$

$$c^2 - \frac{c^2}{E^2} = (E^2 - 1)x^2 - y^2 \rightarrow \frac{c^2}{E^2}(E^2 - 1) = (E^2 - 1)x^2 - y^2.$$

Now let $a = c/E$, just as for the ellipse. Then the equation becomes

$$(2) \qquad (E^2 - 1)x^2 - y^2 = a^2(E^2 - 1).$$

Since $E > 1$, it follows that $E^2 - 1 > 0$. Thus we can let

$$b^2 = a^2(E^2 - 1) = c^2 - a^2.$$

If we substitute this into (2) and then divide by b^2, we get

$$\frac{x^2}{a^2} - \frac{y^2}{b^2} = 1. \tag{3}$$

As for ellipses, the fact that $a = c/E$ leads to the simpler equation $x = a/E$ for the directrix. We leave it to you (Exercise 45) to check that any point $P(x, y)$ whose coordinates satisfy (3) must lie on the hyperbola. Equation (3) is called the ***standard form equation*** of the hyperbola with focus $F(c, 0)$, eccentricity E, and directrix $x = a/E$, where

$$c^2 = a^2 + b^2. \tag{4}$$

Equation (3) differs from the equation of an ellipse only in the sign between the terms on the left side and the relation between a, b, and c. As with ellipses, the hyperbola is symmetric in both coordinate axes. But **Figure 3.2** shows that hyperbolas look markedly different from ellipses. To start with, the hyperbola consists of two branches: There are no points on the curve between $x = -a$ and $x = a$:

$$\frac{y^2}{b^2} = \frac{x^2}{a^2} - 1 = \frac{x^2 - a^2}{a^2} \rightarrow y = \pm\frac{b}{a}\sqrt{x^2 - a^2}. \tag{5}$$

Thus y is defined *only* for $x^2 - a^2 \geq 0$, that is, for $x \leq -a$ or $x \geq a$.

FIGURE 3.2

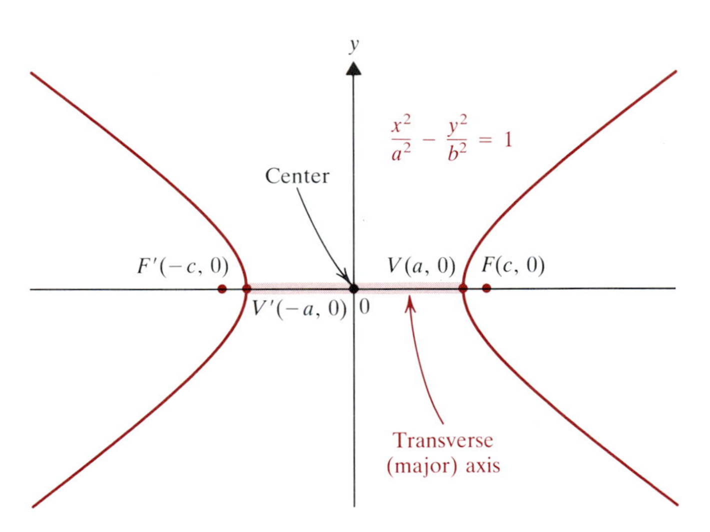

The points $(a, 0)$ and $(-a, 0)$ are called the ***vertices*** of the hyperbola. The origin is located halfway between the vertices and is referred to as the *center* of the hyperbola. As for ellipses, the segment joining the vertices has a special name. Traditionally, it is called the *transverse axis*, although there is little harm in calling it the *major axis*, in analogy to ellipses. Again, symmetry suggests that $F'(-c, 0)$ is another focus. It is not hard to check (Exercise 46) that if we replace F by F' and the directrix l: $x = c/E^2$ by l': $x = -c/E^2$, then we again get (3) from (1). Thus the hyperbola has two foci and two directrices.

Example 1

Find the equation of the hyperbola that is centered at the origin and has vertex $(5, 0)$ and focus $(13, 0)$.

Solution. Here $a = 5$ and $c = 13$. Thus from (4),

$$b^2 = c^2 - a^2 = 169 - 25 = 144 \rightarrow b = 12.$$

By (3) the equation is therefore

$$\frac{x^2}{25} - \frac{y^2}{144} = 1. \blacksquare$$

FIGURE 3.3

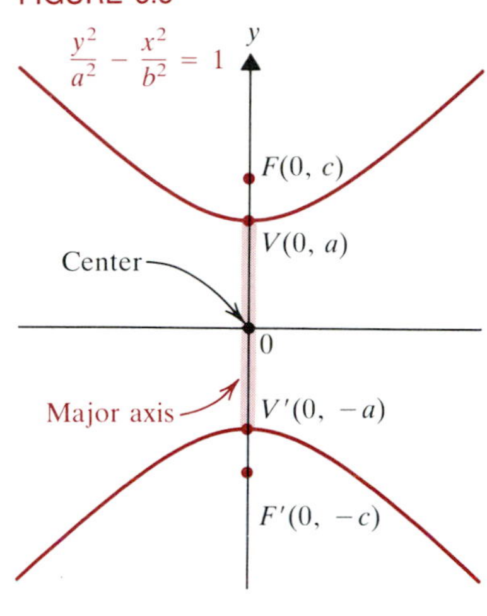

There is an alternative definition for the hyperbola in terms of its foci F and F', just as there is for the ellipse. As Exercise 54 asks you to show, the hyperbola with equation (3) can also be defined as the set of all points $P(x, y)$ in the xy-plane such that

$$|d(F, P) - d(F', P)| = 2a,$$

where F is $(c, 0)$ and F' is $(-c, 0)$.

If we start with focus $F(0, c)$ (or $F'(0, -c)$) and directrix $y = c/E^2$ (or $y = -c/E^2$), then we obtain (Exercise 47) the standard form equation

$$\frac{y^2}{a^2} - \frac{x^2}{b^2} = 1, \tag{6}$$

where (4) still holds. The graph is shown in **Figure 3.3.** The center is still (0, 0), but this time the major axis lies on the y-axis, and there are no points on the curve for $-a < y < a$.

For an ellipse we identify the major axis by finding the *larger* term dividing x^2 or y^2. (That term is then a^2.) But for a hyperbola the major axis is determined by the *sign* of the terms:

$$\frac{u^2}{a^2} - \frac{v^2}{b^2} = 1$$

has major axis on the u-axis, *regardless* of whether a^2 or b^2 is larger.

Another major difference between hyperbolas and ellipses is that *each hyperbola has a pair of asymptotes*, which are very helpful in drawing the graph of the hyperbola. In Section 3.6, we introduced vertical and horizontal asymptotes. We extend the concept as follows.

3.2 DEFINITION

The line $y = mx + b$ is an ***asymptote*** for the graph of $y = f(x)$ if

$$\lim_{x \to +\infty} [mx + b - f(x)] = 0, \tag{7}$$

or

$$\lim_{x \to -\infty} [mx + b - f(x)] = 0. \tag{8}$$

FIGURE 3.4

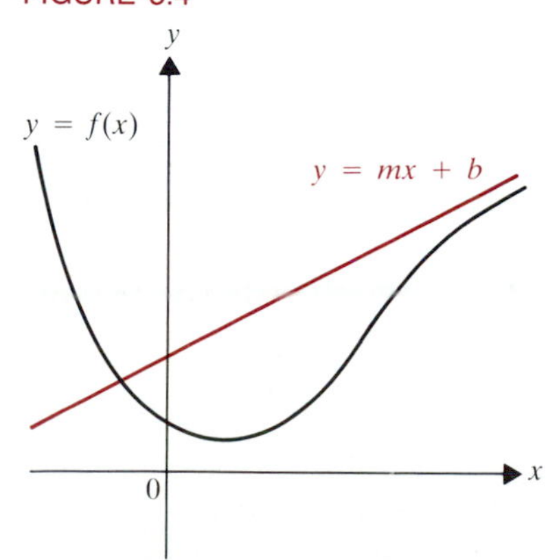

We say that the graph is ***asymptotic*** to the graph of $y = mx + b$ in case (7) or (8) holds. [**Figure 3.4** illustrates (7).]

The hyperbola

$$\frac{x^2}{a^2} - \frac{y^2}{b^2} = 1 \tag{3}$$

is the graph of *two* functions h_1 and h_2 given by (5):

$$y = \begin{cases} h_1(x) = \dfrac{b}{a}\sqrt{x^2 - a^2} \\[2ex] h_2(x) = -\dfrac{b}{a}\sqrt{x^2 - a^2}. \end{cases}$$

For large positive x, notice that since $x = \sqrt{x^2}$ we have

$$\frac{b}{a}\frac{\sqrt{x^2-a^2}}{x} = \frac{b}{a}\sqrt{1-\frac{a^2}{x^2}} \approx \frac{b}{a}.$$

Thus for large positive x it follows that

$$\frac{h_1(x)}{x} \approx \frac{b}{a} \quad \text{and} \quad \frac{h_2(x)}{x} \approx -\frac{b}{a}.$$

This suggests that the lines

$$\frac{y}{x} = \frac{b}{a} \rightarrow y = \frac{b}{a}x \quad \text{and} \quad \frac{y}{x} = -\frac{b}{a} \rightarrow y = -\frac{b}{a}x$$

are asymptotes to the hyperbola (3). To confirm this we compute

$$\begin{aligned}
\lim_{x\to+\infty}\left(\frac{b}{a}x - h_1(x)\right) &= \frac{b}{a}\lim_{x\to+\infty}(x-\sqrt{x^2-a^2}) \\
&= \frac{b}{a}\lim_{x\to+\infty}(x-\sqrt{x^2-a^2})\cdot\frac{x+\sqrt{x^2-a^2}}{x+\sqrt{x^2-a^2}} \\
&= \frac{b}{a}\lim_{x\to+\infty}\frac{x^2-(x^2-a^2)}{x+\sqrt{x^2-a^2}} = \frac{b}{a}\lim_{x\to+\infty}\frac{a^2}{x+\sqrt{x^2-a^2}} = 0.
\end{aligned}$$

According to Definition 3.2,

$$y = \frac{b}{a}x$$

is thus an asymptote to the graph of h_1, the top half of the hyperbola (3). Similarly,

$$\lim_{x\to-\infty}\left(-\frac{b}{a}x - h_1(x)\right) = -\frac{b}{a}\lim_{x\to-\infty}(x+\sqrt{x^2-a^2}) = 0,$$

so

$$y = -\frac{b}{a}x$$

is also an asymptote of the graph of h_1. (Exercise 48 asks you to check that these lines are also asymptotic to the graph of h_2, the bottom half of the hyperbola.)

There is a simple way to calculate the asymptotes of (3), namely, just replace the 1 by 0 and solve for y:

$$\frac{x^2}{a^2} - \frac{y^2}{b^2} = 0 \rightarrow \frac{y^2}{b^2} = \frac{x^2}{a^2} \rightarrow \frac{y}{b} = \pm\frac{x}{a} \rightarrow y = \pm\frac{b}{a}x.$$

The same procedure produces the asymptotes of

$$\frac{y^2}{a^2} - \frac{x^2}{b^2} = 1 \tag{6}$$

also. As Exercises 49 and 50 ask you to check, the lines

$$y = \pm\frac{a}{b}x$$

FIGURE 3.5

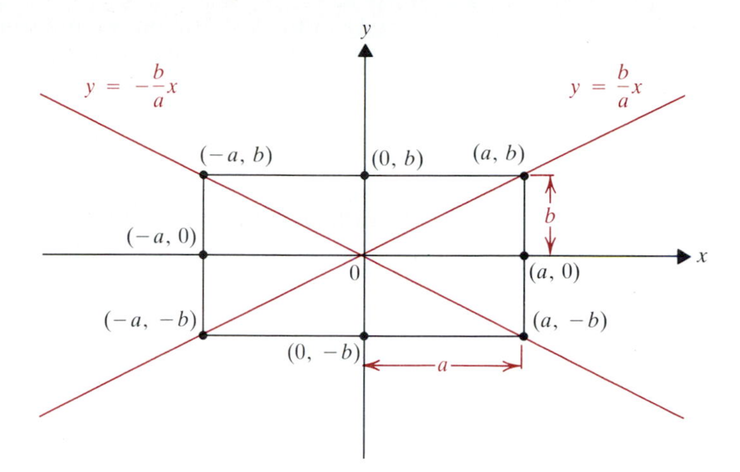

are asymptotic to (6) and are obtainable by replacing the 1 in (6) by 0 and solving for y. To summarize,

(9) The asymptotes of $\frac{x^2}{a^2} - \frac{y^2}{b^2} = 1$ are $y = \pm\frac{b}{a}x$.

(10) The asymptotes of $\frac{y^2}{a^2} - \frac{x^2}{b^2} = 1$ are $y = \pm\frac{a}{b}x$.

FIGURE 3.6

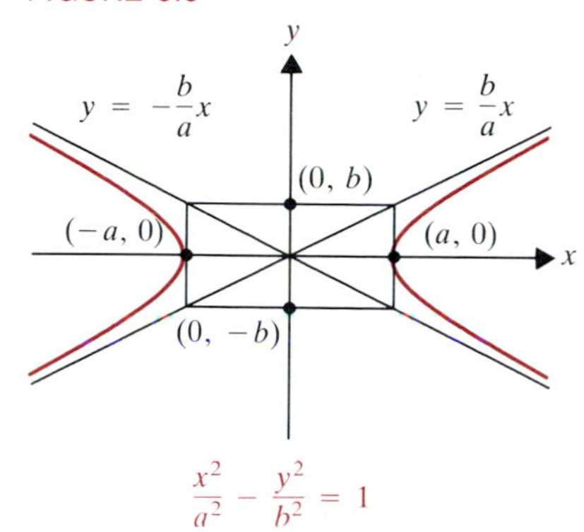

$\frac{x^2}{a^2} - \frac{y^2}{b^2} = 1$

These facts are helpful in sketching the graphs of hyperbolas. Consider (3), for instance. The vertices of the graph are at $(a, 0)$ and $(-a, 0)$, and by (9) the graph is asymptotic to the lines $y = (b/a)x$ and $y = -(b/a)x$. These are lines through the origin of slopes b/a and $-b/a$ and so are easily plotted. In fact, the quickest way to plot them is to construct the ***asymptotic*** (or *auxiliary*) ***rectangle*** shown in **Figure 3.5.** That is the rectangle with vertices (a, b), $(-a, b)$, $(-a, -b)$, and $(a, -b)$. It intersects the major axis at the vertices of the hyperbola. (The segment of the other coordinate axis that it cuts off is called the *conjugate axis* or *minor axis* of the hyperbola.) Since the hyperbola is asymptotic to lines through the diagonals of this rectangle, we can sketch the hyperbola as in **Figure 3.6** from just this information: We sketch in a curve through the vertices that approaches the asymptotes as $x \to +\infty$ and $x \to -\infty$.

FIGURE 3.7

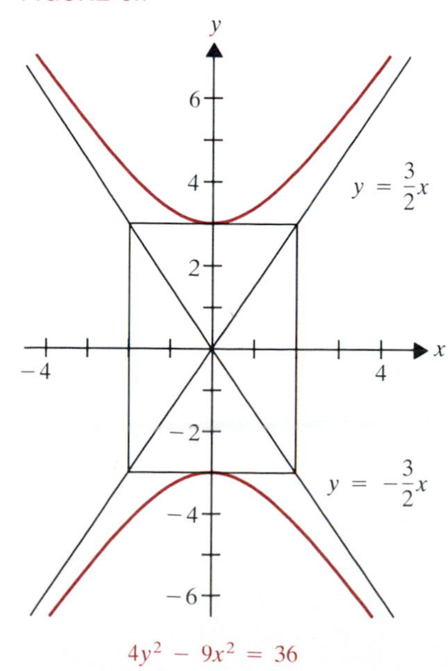

$4y^2 - 9x^2 = 36$

Example 2

Find the asymptotes and sketch the graph of the hyperbola $4y^2 - 9x^2 = 36$.

Solution. The standard form equation can be obtained by dividing the given equation through by 36. That gives

$$\frac{y^2}{9} - \frac{x^2}{4} = 1,$$

which is of the form (6). So the vertices are on the y-axis. The asymptotes are found by changing the 1 in the last equation to 0:

$$\frac{y^2}{9} - \frac{x^2}{4} = 0 \to y^2 = \frac{9}{4}x^2 \to y = \pm\frac{3}{2}x.$$

The graph is shown in **Figure 3.7.** ■

As in the last section, we use translation of axes for a hyperbola centered at $C(h, k)$ with horizontal or vertical major axis. Such a hyperbola has equation

$$\frac{x'^2}{a^2} - \frac{y'^2}{b^2} = 1 \quad \text{or} \quad \frac{y'^2}{a^2} - \frac{x'^2}{b^2} = 1,$$

where $x' = x - h$ and $y' = y - k$. Equations (9) and (10) applied in the $x'y'$-coordinate systems give the asymptotes

$$y' = \pm\frac{b}{a}x' \quad \text{or} \quad y' = \pm\frac{a}{b}x',$$

where x' and y' are again given by $x - h$ and $y - k$, respectively.

The hyperbola with center $C(h, k)$ and horizontal major axis has equation

$$\frac{(x-h)^2}{a^2} - \frac{(y-k)^2}{b^2} = 1$$

and asymptotes $y = k \pm \frac{b}{a}(x - h)$. See **Figure 3.8.**

The hyperbola with center $C(h, k)$ and vertical major axis has equation

$$\frac{(y-k)^2}{a^2} - \frac{(x-h)^2}{b^2} = 1$$

and asymptotes $y = k \pm \frac{a}{b}(x - h)$. See **Figure 3.9.**

Example 3

Find the equation of the hyperbola with vertices $(-1, 3)$ and $(5, 3)$ and foci $(-2, 3)$ and $(6, 3)$.

FIGURE 3.8

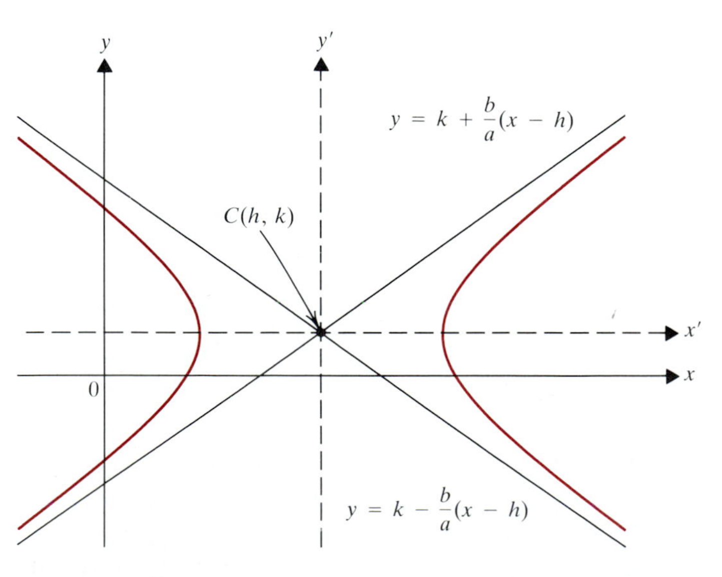

FIGURE 3.9

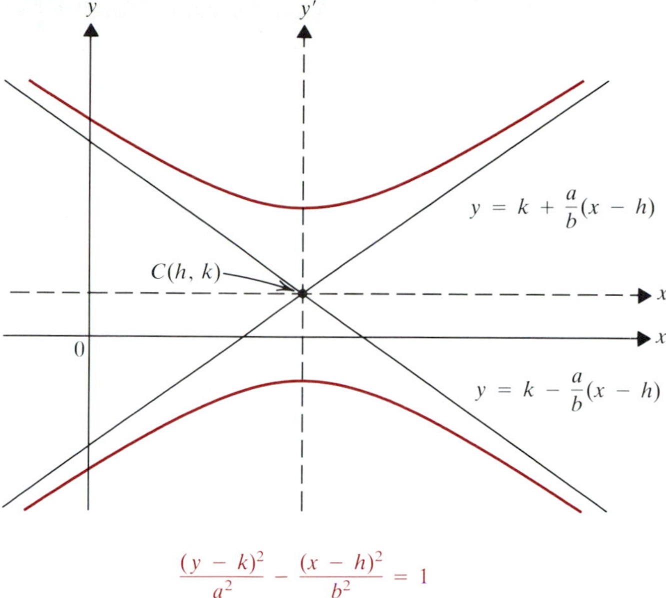

$$\frac{(y-k)^2}{a^2} - \frac{(x-h)^2}{b^2} = 1$$

Solution. This hyperbola has major axis on the line $y = 3$ and center $C(h, k)$ midway between its vertices. Thus $k = 3$ and $h = \frac{1}{2}(-1 + 5) = 2$. The distance between the vertices is

$$2a = 5 - (-1) = 6,$$

so $a = 3$. The distance between the foci is

$$2c = 6 - (-2) = 8,$$

so $c = 4$. We thus get from (2),

$$b^2 = c^2 - a^2 = 16 - 9 = 7.$$

Relative to an $x'y'$-coordinate system centered at $C(2, 3)$, the hyperbola therefore has equation

$$\frac{x'^2}{9} - \frac{y'^2}{7} = 1.$$

So its xy-equation is

$$\frac{(x-2)^2}{9} - \frac{(y-3)^2}{7} = 1.$$

The asymptotes are

$$\frac{y'^2}{7} = \frac{x'^2}{9} \to y' = \pm\frac{\sqrt{7}}{3}x' \to y - 3 = \pm\frac{\sqrt{7}}{3}(x-2) \to y = 3 \pm \frac{\sqrt{7}}{3}(x-2).$$

See **Figure 3.10.** ■

Any second-degree equation in x and y of the form

$$ax^2 + cy^2 + ex + fy + g = 0, \tag{11}$$

where a and c have *opposite* signs, is either a hyperbola or a degenerate hyperbola consisting of two intersecting lines. (See Exercise 51.) We complete the square to identify the graph of (11).

FIGURE 3.10

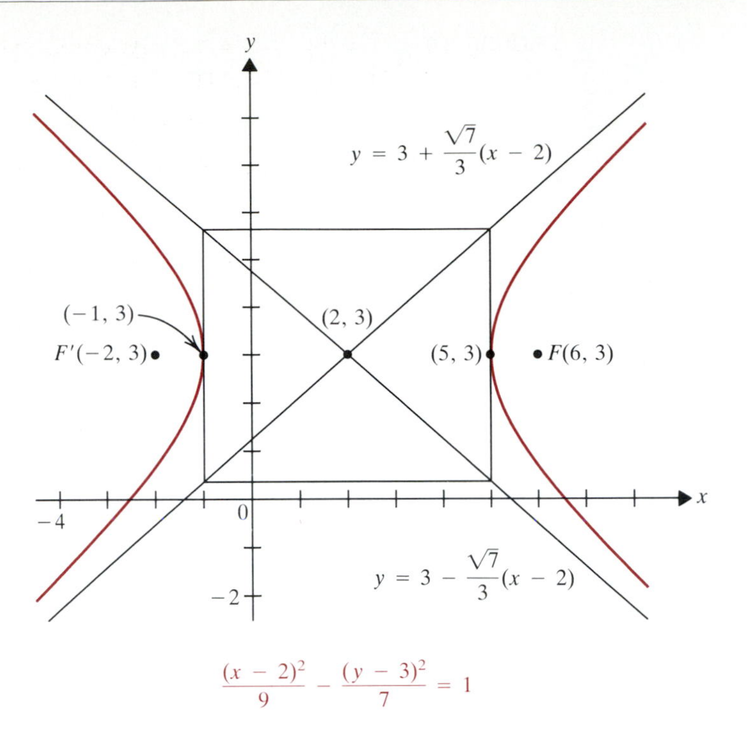

$$\frac{(x-2)^2}{9} - \frac{(y-3)^2}{7} = 1$$

Example 4

Show that the graph of $x^2 - 4y^2 + 6x - 16y + 9 = 0$ is a hyperbola, and find its center, vertices, foci, and asymptotes.

Solution. Completing the square, we get

$$x^2 + 6x + 9 - 4(y^2 + 4y) = 0 \rightarrow (x + 3)^2 - 4(y^2 + 4y + 4) = -16,$$
$$(x + 3)^2 - 4(y + 2)^2 = -16,$$

which, when divided by -16, becomes

$$\frac{(y + 2)^2}{4} - \frac{(x + 3)^2}{16} = 1.$$

FIGURE 3.11

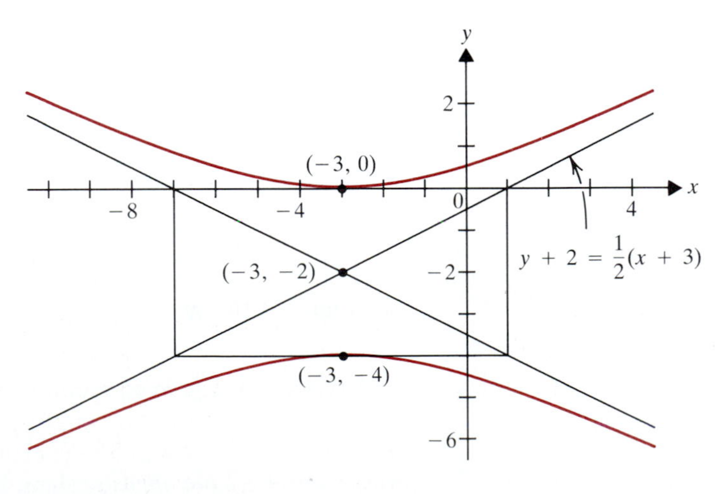

$$\frac{(y+2)^2}{4} - \frac{(x+3)^2}{16} = 1$$

Thus the graph is a hyperbola centered at $(-3, -2)$ with vertical major axis $x = -3$. The vertices are on $x = -3$, where we have

$$(y + 2)^2 = 4 \rightarrow y + 2 = \pm 2 \rightarrow y = 0 \text{ or } -4.$$

FIGURE 3.12

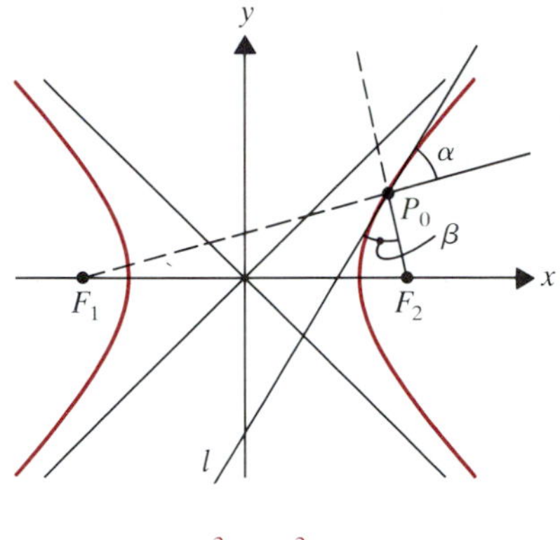

$$\frac{x^2}{a^2} - \frac{y^2}{b^2} = 1$$

Thus the vertices are $(-3, 0)$ and $(-3, -4)$. From (4),

$$c^2 = a^2 + b^2 = 4 + 16 = 20,$$

so $c = 2\sqrt{5}$. The foci are therefore at $(-3, -2 \pm 2\sqrt{5})$. The asymptotes are

$$\frac{(y + 2)^2}{4} - \frac{(x + 3)^3}{16} = 0 \rightarrow (y + 2)^2 = \frac{1}{4}(x + 3)^2,$$

$$y + 2 = \pm\frac{1}{2}(x + 3).$$

The graph is sketched in **Figure 3.11.** ■

Reflecting Property of Hyperbolas*

Hyperbolas have a reflecting property similar to the one for ellipses. The proof of the following result is asked for in Exercise 53.

3.3 THEOREM Let l be the tangent line to a hyperbola at a point $P_0(x_0, y_0)$. Then l makes equal angles with F_1P_0 and F_2P_0. That is, the angles α and β in **Figure 3.12** are equal.

FIGURE 3.13

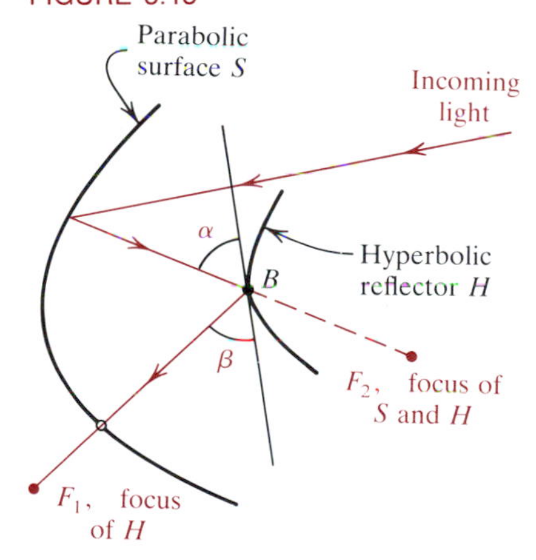

This leads to an important optical application. A small hyperbolic reflector is often inserted between the focus F_2 and surface S of a reflecting telescope to direct incoming light rays to an eyepiece *behind* the parabolic surface for viewing, as indicated schematically in **Figure 3.13.** The incoming light is reflected by the parabolic mirror toward F_2, because of Theorem 1.2. The angle of incidence of this light against the hyperbolic reflector H is $\alpha = \beta$, so the light is then reflected toward the other focus F_1 of H, where the eyepiece is located behind a small hole in the parabolic mirror.

Hyperbolas are also used to determine the source of a radio signal broadcast by a transmitter at T (see **Figure 3.14**) and received by monitors at F_1 and F_2. Since radio waves travel at a constant speed, the difference Δt between the times the signals reach F_1 and F_2 is proportional to

$$|d(T, F_1) - d(T, F_2)|.$$

FIGURE 3.14

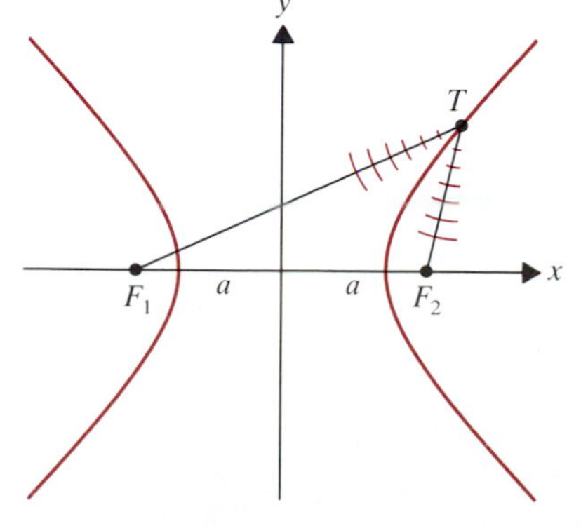

According to Exercise 54, T must then lie somewhere on the hyperbola determined by

$$|d(T, F_1) - d(T, F_2)| = 2a,$$

where $2a$ is Δt times the speed of the waves. If a second pair of monitors is used, then a second hyperbola is determined, and the transmitter must be at one of their points of intersection. This approach is used to pinpoint the location of illegal or hostile radio broadcasts. Ships at sea can also use it to determine their position. Exercises 54–56 explore the details, which form the basis for the Loran C navigation system.

* Optional

Exercises 8.3

In Exercises 1–24, find the vertices, foci, and asymptotes of the given hyperbola and make a sketch.

1. $\frac{x^2}{16} - \frac{y^2}{9} = 1$
2. $\frac{x^2}{36} - \frac{y^2}{16} = 1$
3. $\frac{y^2}{25} - \frac{x^2}{9} = 1$
4. $\frac{y^2}{36} - \frac{x^2}{4} = 1$
5. $16x^2 - 9y^2 = 144$
6. $25y^2 - 9x^2 = 225$
7. $144y^2 - 25x^2 = 3600$
8. $36x^2 - 64y^2 = 2304$
9. $x^2 - y^2 = 1$
10. $y^2 - x^2 = 9$
11. $4x^2 - y^2 = 4$
12. $9y^2 - x^2 = 9$
13. $x^2 - 4y^2 = 9$
14. $x^2 - 9y^2 = 16$
15. $y^2 - 16x^2 = 25$
16. $y^2 - 9x^2 = 25$
17. $\frac{(x-3)^2}{16} - \frac{(y-2)^2}{9} = 1$
18. $\frac{(x-2)^2}{25} - \frac{(y+3)^2}{16} = 1$
19. $\frac{(y+3)^2}{36} - \frac{(x-2)^2}{64} = 1$
20. $\frac{(y-2)^2}{9} - \frac{(x+5)^2}{16} = 1$
21. $4x^2 + 8x - 3y^2 + 16 = 0$
22. $9y^2 - 4x^2 - 36y + 8x = 4$
23. $x^2 - 4y^2 - 4x - 8y - 9 = 0$
24. $16y^2 - x^2 - 32y - 6x = 57$

In Exercises 25–44, find an equation for the hyperbola.

25. $V_2(3, 0)$, $F_2(5, 0)$, $C(0, 0)$
26. $V_1(-4, 0)$, $F_1(-5, 0)$, $C(0, 0)$
27. $V_1(0, -4)$, $F_1(0, -6)$, $C(0, 0)$
28. $V_2(0, 3)$, $F_2(0, 4)$, $C(0, 0)$
29. $V(\pm 6, 0)$, asymptotes $y = \pm\frac{4}{3}x$
30. $V(\pm 4, 0)$, asymptotes $y = \pm\frac{3}{2}x$
31. $V(0, \pm 8)$, asymptotes $y = \pm\frac{4}{3}x$
32. $V(0, \pm 6)$, asymptotes $y = \pm\frac{3}{2}x$
33. $F(\pm 5, 0)$, asymptotes $y = \pm\frac{3}{4}x$
34. $F(\pm 4, 0)$, asymptotes $y = \pm\frac{2}{3}x$
35. $F(0, \pm\sqrt{8})$, asymptotes $y = \pm x$
36. $F(0, \pm\sqrt{5})$, asymptotes $y = \pm\frac{1}{2}x$
37. $V(\pm 3, 0)$, through $P(5, 2)$
38. $V(\pm 4, 0)$, through $P(8, 2)$
39. $F_1(-1, 2)$, $F_2(5, 2)$, $V_1(0, 2)$, $V_2(4, 2)$
40. $F_1(3, -6)$, $F_2(3, 4)$, $V_1(3, -4)$, $V_2(3, 2)$
41. $F_1(-2, -1)$, $F_2(-2, 9)$, major axis of length 6
42. $V_1(3, 2)$, $V_2(7, 2)$, minor axis of length 8
43. $V_1(3, -5)$, $V_2(3, 1)$, asymptotes $y + 2 = \pm 2(x - 3)$
44. $V_1(1, -3)$, $V_2(5, -3)$, asymptotes $y + 3 = \pm\frac{1}{2}(x - 3)$
45. Show that if $P(x, y)$ is any point whose coordinates satisfy (3), then (1) holds for $F(c, 0)$ and l: $x = c/E^2$.
46. Show that if (1) is replaced by $d(P, F')/d(P, l') = E$, where F' is the point $(-c, 0)$ and l' is the line $x = -c/E^2$, then the coordinates of P satisfy (3).
47. Show that if in (1) F is $(0, c)$ and l is $y = c/E^2$, then the coordinates of P satisfy (6).
48. Show that the graph of $h_2(x) = -\frac{b}{a}\sqrt{x^2 - a^2}$ is asymptotic to $y = -\frac{b}{a}x$ as $x \to +\infty$ and to $y = \frac{b}{a}x$ as $x \to -\infty$.
49. Show that $y = \frac{a}{b}x$ is an asymptote of $y = \frac{a}{b}\sqrt{x^2 + b^2}$.
50. Show that $y = -\frac{a}{b}x$ is an asymptote of $y = -\frac{a}{b}\sqrt{x^2 + b^2}$.
51. Show that the graph of (11) is either a hyperbola or two intersecting straight lines.
52. Show that the tangent line to the hyperbola $x^2/a^2 - y^2/b^2 = 1$ at $P_0(x_0, y_0)$ has equation $xx_0/a^2 - yy_0/b^2 = 1$.
53. Use the result of Exercise 52 to prove Theorem 3.3.
54. An alternative definition of the hyperbola is the set of all points $P(x, y)$ such that
$$|d(F, P) - d(F', P)| = 2a,$$
where F and F' are the foci. If F is $(c, 0)$ and F' is $(-c, 0)$, then show that S consists of all points whose coordinates satisfy (3), so S is indeed a hyperbola. [Here b^2 is defined by (4).]

FIGURE 3.15

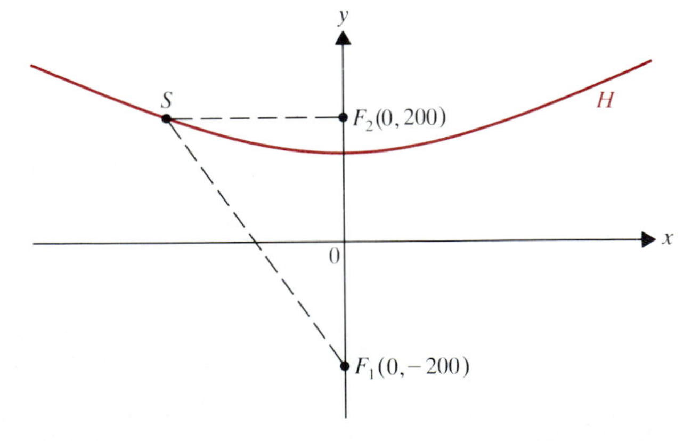

55. A ship is due west of an oil-drilling platform at a point F_2 off the coast of California. A signal is broadcast simultaneously from F_2 and also from another platform at F_1, which is 400 km south of F_2. The ship receives the signal from F_2 2000 microseconds (μs) before it receives the signal from F_1. The signal travels 0.16 km/μs. Find how far from F_2 the ship is. (See **Figure 3.15.** Find the equation of hyperbola H.)

56. (a) If in Exercise 55 it is known only that the ship is *somewhere* off the coast (not necessarily due west of F_2), what can be concluded about its position?

(b) Explain how its *exact* position can be determined if it also monitors signals from a second pair of transmitters located on floating platforms that are respectively 50 and 150 km west of F_2, and both 400 km north of F_2.

8.4 Rotation of Axes*

The discussion of conic sections in the preceding three sections makes it possible to identify the graph of any second-degree equation in x and y of the form

$$ax^2 + cy^2 + ex + fy + g = 0. \tag{1}$$

We begin by restating the three remarks made just before Example 4 of Section 1, Example 5 of Section 2, and Example 4 of Section 3.

> The graph of any second-degree equation of the form (1) is
>
> (a) a parabola if $ac = 0$ but a or c is nonzero.
>
> (b) an ellipse (or a circle, a single point, or empty) if $ac > 0$.
>
> (c) a hyperbola (or pair of intersecting lines) if $ac < 0$.

FIGURE 4.1

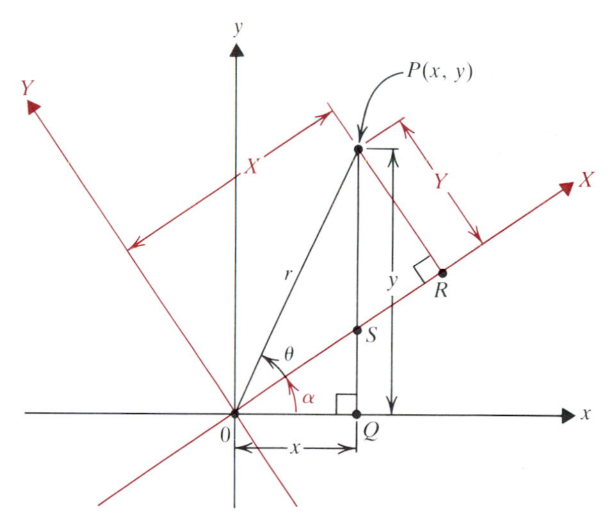

To sketch the graph of (1), we complete the square and apply the techniques of the first three sections. Unfortunately, though, (1) is not the most general second-degree polynomial equation in x and y, because it has no xy-term. The ***general second degree polynomial equation in x and y*** is

$$ax^2 + bxy + cy^2 + ex + fy + g = 0. \tag{2}$$

Thus (1) is a special case of (2), corresponding to $b = 0$. To graph (2) when $b \neq 0$, we change coordinates by rotating the coordinate axes, as in **Figure 4.1,**

* Optional

through a certain angle α. That produces a new pair of coordinate axes, labeled X and Y in the figure. The angle α is chosen so that when (2) is translated into an equation in the new coordinate variables X and Y, there is no XY-term. That is, (2) transforms to

$$AX^2 + CY^2 + EX + FY + G = 0.$$

The first step in finding such an angle α is to use Figure 4.1 to express the new coordinates (X, Y) of any point P in terms of its original coordinates (x, y). We have

$$\text{(3)} \qquad \begin{cases} X = |OP|\cos\theta = r\cos\theta, \\ Y = |OP|\sin\theta = r\sin\theta. \end{cases}$$

Next, Equations (12) and (17) of Section 2.3 (p. 83 and p. 84) and Figure 4.1 give

$$\text{(4)} \qquad \begin{cases} x = r\cos(\theta + \alpha) = r\cos\theta\cos\alpha - r\sin\theta\sin\alpha \\ y = r\sin(\theta + \alpha) = r\sin\theta\cos\alpha + r\cos\theta\sin\alpha. \end{cases}$$

Substitution of (3) into (4) gives the ***change-of-coordinate equations***

$$\text{(5)} \qquad \begin{aligned} x &= X\cos\alpha - Y\sin\alpha \\ y &= X\sin\alpha + Y\cos\alpha \end{aligned}$$

To illustrate the use of (5) we investigate the graph of the simple equation $xy = 4$.

Example 1

Use (5) with $\alpha = \pi/4$ to identify the graph of $xy = 4$.

Solution. Since $\sin\pi/4 = \cos\pi/4 = 1/\sqrt{2}$, Equations (5) become

$$x = \frac{1}{\sqrt{2}}X - \frac{1}{\sqrt{2}}Y = \frac{1}{\sqrt{2}}(X - Y),$$

$$y = \frac{1}{\sqrt{2}}X + \frac{1}{\sqrt{2}}Y = \frac{1}{\sqrt{2}}(X + Y).$$

Thus $xy = 4$ transforms to

$$\left[\frac{1}{\sqrt{2}}(X - Y)\right]\left[\frac{1}{\sqrt{2}}(X + Y)\right] = 4 \rightarrow \frac{1}{2}(X^2 - Y^2) = 4,$$

$$X^2 - Y^2 = 8 \rightarrow \frac{X^2}{8} - \frac{Y^2}{8} = 1.$$

This is the equation of a hyperbola with vertices at $(X, Y) = (\pm 2\sqrt{2}, 0)$, and foci at $(X, Y) = (\pm c, 0)$, where

$$c^2 = a^2 + b^2 = 8 + 8 = 16.$$

So the foci are at $(X, Y) = (\pm 4, 0)$. Using (5), we find that the original coordinates of the vertices $(X, Y) = (\pm 2\sqrt{2}, 0)$ are (x, y), where

$$x = \frac{1}{\sqrt{2}}(\pm 2\sqrt{2}) - 0 = \pm 2 \quad \text{and} \quad y = \frac{1}{\sqrt{2}}(\pm 2\sqrt{2}) + 0 = \pm 2.$$

So the vertices are $(x, y) = (2, 2)$ and $(x, y) = (-2, -2)$. Similarly, the foci are $(x, y) = (2\sqrt{2}, 2\sqrt{2})$ and $(-2\sqrt{2}, -2\sqrt{2})$, since

$$x = \frac{1}{\sqrt{2}}(\pm 4) - 0 = \pm 2\sqrt{2} \quad \text{and} \quad y = \frac{1}{\sqrt{2}}(\pm 4) - 0 = \pm 2\sqrt{2}.$$

FIGURE 4.2

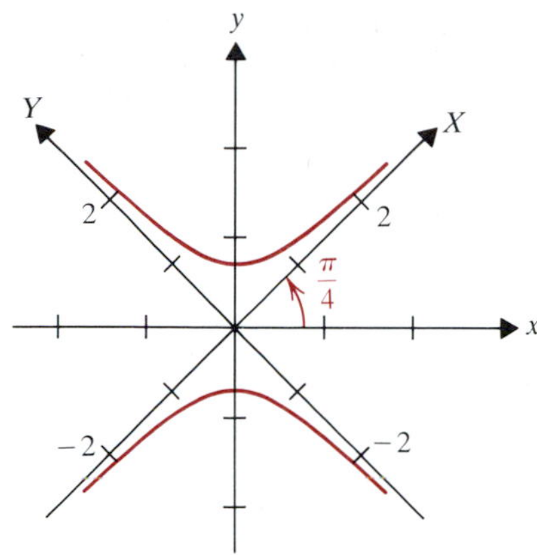

The graph is shown in **Figure 4.2.** ■

Equations (5) give the old coordinates x and y in terms of the new coordinates X and Y. To obtain equations for X and Y in terms of x and y, we solve (5) as follows. If we multiply the first equation in (5) by $\cos\alpha$, the second by $\sin\alpha$, and then add, we get

$$\begin{aligned} x\cos\alpha &= X\cos^2\alpha - Y\sin\alpha\cos\alpha \\ y\sin\alpha &= X\sin^2\alpha + Y\sin\alpha\cos\alpha \quad (+) \\ \hline x\cos\alpha + y\sin\alpha &= X(\cos^2\alpha + \sin^2\alpha) = X. \end{aligned}$$

Similarly, multiplying the first equation in (5) by $-\sin\alpha$, the second by $\cos\alpha$, and adding, we obtain

$$\begin{aligned} -x\sin\alpha &= -X\sin\alpha\cos\alpha + Y\sin^2\alpha \\ y\cos\alpha &= X\sin\alpha\cos\alpha + Y\cos^2\alpha \quad (+) \\ \hline -x\sin\alpha + y\cos\alpha &= Y(\sin^2\alpha + \cos^2\alpha) = Y. \end{aligned}$$

Thus we have

$$(6) \qquad \begin{aligned} X &= x\cos\alpha + y\sin\alpha \\ Y &= -x\sin\alpha + y\cos\alpha \end{aligned}$$

With these equations, we can confirm that in Example 1 the asymptotes of the hyperbola are the x- and y-axes, as Figure 4.2 makes it appear.

Example 2

Find the asymptotes to the hyperbola in Example 1.

Solution. The asymptotes are

$$Y = \pm\frac{b}{a}X = \pm X,$$

since $a = b = 2\sqrt{2}$. We can use (6) with $\alpha = \pi/4$ to transform these equations to xy-equations. The equation $Y = X$ becomes

$$-\frac{1}{\sqrt{2}}x + \frac{1}{\sqrt{2}}y = \frac{1}{\sqrt{2}}x + \frac{1}{\sqrt{2}}y \rightarrow \frac{2}{\sqrt{2}}x = 0 \rightarrow x = 0,$$

the y-axis. The equation $Y = -X$ becomes

$$-\frac{1}{\sqrt{2}}x + \frac{1}{\sqrt{2}}y = -\frac{1}{\sqrt{2}}x - \frac{1}{\sqrt{2}}y \rightarrow \frac{2}{\sqrt{2}}y = 0 \rightarrow y = 0.$$

Thus the asymptotes are the original coordinate axes. ■

The angle $\alpha = \pi/4$ was a good choice in Example 1, because it transformed the given equation to a standard-form equation. We now want to see how to choose α in order to transform *any* given quadratic equation

$$ax^2 + bxy + cy^2 + ex + fy + g = 0, \tag{2}$$

where $b \neq 0$, to an equation of the form (1):

$$AX^2 + BXY + CY^2 + EX + FY + G = 0,$$

where $B = 0$. In general, substitution of the change-of-coordinate equations (5) into (2) produces

$$\begin{aligned} a(X\cos\alpha - Y\sin\alpha)^2 &+ b(X\cos\alpha - Y\sin\alpha)(X\sin\alpha + Y\cos\alpha) \\ &+ c(X\sin\alpha + Y\cos\alpha)^2 + e(X\cos\alpha - Y\sin\alpha) \\ &+ f(X\sin\alpha + Y\cos\alpha) + g = 0. \end{aligned}$$

To express B in terms of the coefficients in (2) and the angle α, we collect the XY-terms in the last equation:

$$\begin{aligned} XY[-2a\sin\alpha\cos\alpha &+ b(\cos^2\alpha - \sin^2\alpha) + 2c\sin\alpha\cos\alpha] \\ &= XY[-a\sin 2\alpha + b\cos 2\alpha + c\sin 2\alpha]. \end{aligned}$$

Hence we have

$$B = b\cos 2\alpha - (a - c)\sin 2\alpha.$$

Thus $B = 0$ if

$$b\cos 2\alpha - (a - c)\sin 2\alpha = 0 \rightarrow b\cos 2\alpha = (a - c)\sin 2\alpha,$$

that is, if

$$\cot 2\alpha = \frac{a - c}{b}. \tag{7}$$

As a simple illustration of (7), in the equation $xy = 4$ of Example 1, $a = c = 0$. Then the angle of rotation α in (7) is given by

$$\cot 2\alpha = \frac{0}{1} = 0 \rightarrow 2\alpha = \cot^{-1} 0 = \frac{\pi}{2} \rightarrow \alpha = \frac{\pi}{4}.$$

As Example 1 confirmed, rotation of axes through the angle $\alpha = \pi/4$ did lead to an equation in X and Y that contained no XY-term.

Once α is calculated from (7) in the form

$$2\alpha = \cot^{-1}\frac{a - c}{b} \rightarrow \alpha = \frac{1}{2}\cot^{-1}\frac{a - c}{b},$$

change-of-coordinate equations (5) can be used to transform (2) to an equation in X and Y. In doing so, the identity

$$\sin^2\alpha = \frac{1 - \cos 2\alpha}{2}$$

is again helpful, as it was in integrating powers of sine and cosine in Section 7.1. The next example illustrates how (7) is used to identify and graph a given equation (2).

Example 3

Identify and sketch the graph of the equation

$$13x^2 + 5xy + y^2 = 4. \tag{8}$$

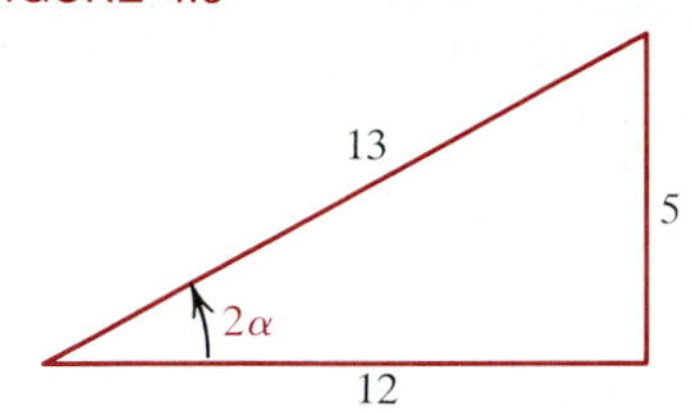

FIGURE 4.3

Solution. Here $a = 13$, $b = 5$, and $c = 1$. So we need to rotate axes through angle α, where

$$\cot 2\alpha = \frac{a - c}{b} = \frac{12}{5} \rightarrow \tan 2\alpha = \frac{5}{12}.$$

Thus $\alpha = \frac{1}{2}\tan^{-1}(5/12)$ is a first-quadrant angle, about 0.197 rad (11.3°). We see from **Figure 4.3** that $\cos 2\alpha = 12/13$. Hence

$$\sin^2 \alpha = \frac{1 - \cos 2\alpha}{2} = \frac{1 - (12/13)}{2} = \frac{1}{26}$$

and

$$\cos^2 \alpha = 1 - \sin^2 \alpha = 1 - \frac{1}{26} = \frac{25}{26}.$$

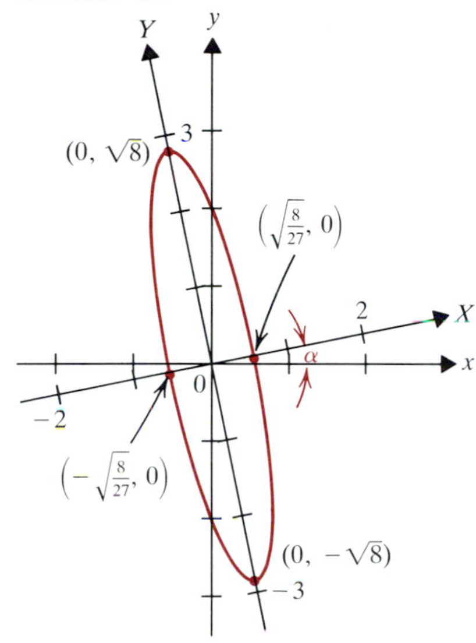

FIGURE 4.4

Thus $\sin \alpha = 1/\sqrt{26}$ and $\cos \alpha = 5/\sqrt{26}$. Equations (5) then are

$$x = \frac{1}{\sqrt{26}}(5X - Y), \qquad y = \frac{1}{\sqrt{26}}(X + 5Y).$$

Hence (8) transforms to

$$\frac{13}{26}(5X - Y)^2 + \frac{5}{26}(5X - Y)(X + 5Y) + \frac{1}{26}(X + 5Y)^2 = 4,$$

$$13(25X^2 - 10XY + Y^2) + 5(5X^2 + 24XY - 5Y^2) + X^2 + 10XY + 25Y^2 = 104,$$

$$351X^2 + 13Y^2 = 104 \rightarrow 27X^2 + Y^2 = 8,$$

$$\frac{X^2}{8/27} + \frac{Y^2}{8} = 1.$$

Thus the graph is an ellipse with $a^2 = 8$ and $b^2 = 8/27$, and major axis rotated about 11° from the vertical. See **Figure 4.4.** ■

You may have guessed that because a and c were both positive in Example 3, the graph was necessarily an ellipse. The next example shows that such reasoning is, unfortunately, *invalid*.

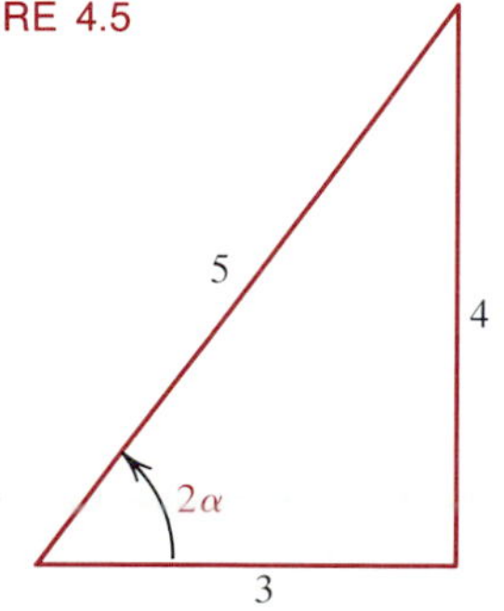

FIGURE 4.5

Example 4

Identify and sketch the graph of the equation

(9) $$4x^2 + 4xy + y^2 + 20x - 10y = \sqrt{5}.$$

Solution. Here $a = 4$, $b = 4$, and $c = 1$, so we rotate the axes through the angle α such that

$$\cot 2\alpha = \frac{a - c}{b} = \frac{4 - 1}{4} = \frac{3}{4} \rightarrow \tan 2\alpha = \frac{4}{3}.$$

Hence $2\alpha = \tan^{-1}\frac{4}{3}$, $\alpha = \frac{1}{2}\tan^{-1}\frac{4}{3} \approx 0.464$ rad (26.6°). From **Figure 4.5,** $\cos 2\alpha = 3/5$. Hence

$$\sin^2 \alpha = \frac{1 - \cos 2\alpha}{2} = \frac{1 - \frac{3}{5}}{2} = \frac{1}{5},$$

and

$$\cos^2 \alpha = 1 - \sin^2 \alpha = 1 - \frac{1}{5} = \frac{4}{5}.$$

Thus $\sin \alpha = 1/\sqrt{5}$, $\cos \alpha = 2/\sqrt{5}$, and Equations (5) are

$$x = \frac{1}{\sqrt{5}}(2X - Y) \qquad \text{and} \qquad y = \frac{1}{\sqrt{5}}(X + 2Y).$$

Then (9) transforms to

$$\frac{4}{5}(2X - Y)^2 + \frac{4}{5}(2X - Y)(X + 2Y) + \frac{1}{5}(X + 2Y)^2$$

$$+ \frac{20}{\sqrt{5}}(2X - Y) - \frac{10}{\sqrt{5}}(X + 2Y) = \sqrt{5},$$

$$4(4X^2 - 4XY + Y^2) + 4(2X^2 + 3XY - 2Y^2) + X^2 + 4XY + 4Y^2$$

$$+ 40\sqrt{5}X - 20\sqrt{5}Y - 10\sqrt{5}X - 20\sqrt{5}Y = 5\sqrt{5},$$

$$25X^2 + 30\sqrt{5}X - 40\sqrt{5}Y = 5\sqrt{5},$$

$$\sqrt{5}X^2 + 6X - 8Y = 1.$$

This curve is therefore a *parabola*,

$$8Y + 1 + \frac{9\sqrt{5}}{5} = \sqrt{5}\left(X^2 + \frac{6}{\sqrt{5}}X + \frac{9}{5}\right) \to 8Y + \frac{5 + 9\sqrt{5}}{5} = \sqrt{5}\left(X + \frac{3}{\sqrt{5}}\right)^2,$$

$$8\left(Y + \frac{5 + 9\sqrt{5}}{40}\right) = \sqrt{5}\left(X + \frac{3}{\sqrt{5}}\right)^2,$$

with vertex at $X = -3/\sqrt{5} \approx -1.34$, where $Y = -(5 + 9\sqrt{5})/40 \approx -0.628$.

The following table of approximate values was used to draw **Figure 4.6.**

X	-8	-6	-4	-2	-1.34	0	2	4	6
Y	11.764	5.437	1.347	-0.507	-0.628	-0.125	2.493	7.347	14.437

FIGURE 4.6

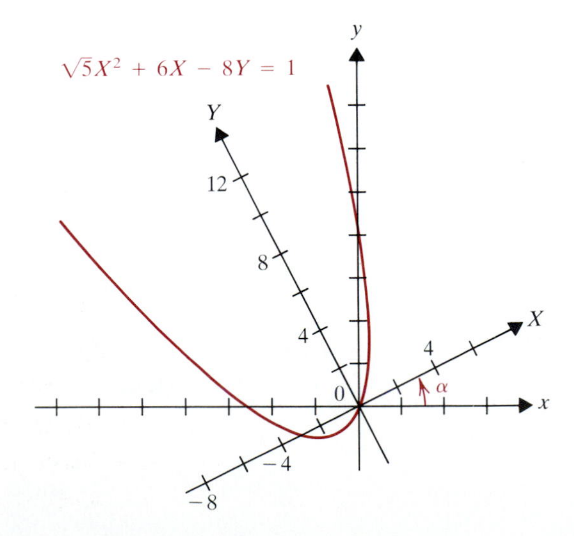

Instead of transforming (2) each time we use (7) and calculating the resulting coefficients of the transformed equation

$$AX^2 + BXY + CY^2 + EX + FY + G = 0, \tag{10}$$

as in Example 4, you may prefer to use Table 1, which gives the coefficients of (10) in terms of those of (2). We verified the formula for B in the course of deriving (7). Similar calculations will verify the other entries (see Exercise 40).

TABLE 1

Coefficients of $AX^2 + BXY + CY^2 + EX + FY + G = 0$ in Terms of the Coefficients of $ax^2 + bxy + cy^2 + ex + fy + g = 0$ and the Angle α of Rotation
$A = a\cos^2\alpha + b\cos\alpha\sin\alpha + c\sin^2\alpha$
$B = b\cos 2\alpha - (a - c)\sin 2\alpha = b(\cos^2\alpha - \sin^2\alpha) - 2(a - c)\sin\alpha\cos\alpha$
$C = a\sin^2\alpha - b\cos\alpha\sin\alpha + c\cos^2\alpha$
$E = e\cos\alpha + f\sin\alpha$
$F = -e\sin\alpha + f\cos\alpha$
$G = g$

While the simple conjecture made before Example 4 is false, it *is* possible to determine quickly which type of conic section is the graph of a given second-degree polynomial equation in x and y. This is done by using the boxed facts on p. 509 and the following result about the transformed version of (2) produced by rotating the coordinate axes.

4.1 LEMMA If the equation

$$ax^2 + bxy + cy^2 + ex + fy + g = 0 \tag{2}$$

transforms to

$$AX^2 + BXY + CY^2 + EX + FY + G = 0 \tag{10}$$

under a rotation of axes through the angle α, then

$$B^2 - 4AC = b^2 - 4ac. \tag{11}$$

The expression $b^2 - 4ac$ is called the ***discriminant*** of (2).

Proof. To verify (11) we use the first three formulas in Table 1, which give A, B, and C in terms of a, b, c, and the angle α. These give

$$\begin{aligned} B^2 - 4AC &= [b(\cos^2\alpha - \sin^2\alpha) + 2(c - a)\cos\alpha\sin\alpha]^2 - 4a^2\cos^2\alpha\sin^2\alpha \\ &\quad - 4ab\cos\alpha\sin^3\alpha - 4ac\sin^4\alpha + 4ab\cos^3\alpha\sin\alpha \\ &\quad + 4b^2\cos^2\alpha\sin^2\alpha + 4bc\cos\alpha\sin^3\alpha - 4ac\cos^4\alpha \\ &\quad - 4bc\cos^3\alpha\sin\alpha - 4c^2\cos^2\alpha\sin^2\alpha \\ &= b^2(\cos^4\alpha - 2\cos^2\alpha\sin^2\alpha + \sin^4\alpha) + 4b(c - a)(\cos^3\alpha\sin\alpha \\ &\quad - \cos\alpha\sin^3\alpha) + 4(c - a)^2\cos^2\alpha\sin^2\alpha + 4b^2\cos^2\alpha\sin^2\alpha \\ &\quad - 4bc\cos^3\alpha\sin\alpha + 4ab\cos^3\alpha\sin\alpha + 4bc\cos\alpha\sin^3\alpha \\ &\quad - 4ab\cos\alpha\sin^3\alpha - 4a^2\cos^2\alpha\sin^2\alpha - 4c^2\cos^2\alpha\sin^2\alpha \\ &\quad - 4ac(\sin^4\alpha + \cos^4\alpha) \end{aligned}$$

$$= b^2(\cos^4\alpha + 2\cos^2\alpha\sin^2\alpha + \sin^4\alpha) + 4c^2\cos^2\alpha\sin^2\alpha$$
$$- 8ac\cos^2\alpha\sin^2\alpha + \cancel{4a^2\cos^2\alpha\sin^2\alpha} - 4c^2\cos^2\alpha\sin^2\alpha$$
$$- \cancel{4a^2\cos^2\alpha\sin^2\alpha} - 4ac(\sin^4\alpha + \cos^4\alpha)$$
$$= b^2(\cos^2\alpha + \sin^2\alpha)^2 - 4ac(\sin^2\alpha + \cos^2\alpha)^2$$
$$= b^2 - 4ac. \quad \boxed{\text{QED}}$$

Now, if in Lemma 4.1 we take α so that $\cot 2\alpha = (a - c)/b$, then $B = 0$. From the boxed facts on p. 509, it thus follows that the graph of the transformed Equation (10) is

(a) a parabola (or degenerate parabola) if $AC = 0$ but $A \neq 0$ or $C \neq 0$,

(b) an ellipse (or degenerate ellipse) if $AC > 0$,

(c) a hyperbola (or degenerate hyperbola) if $AC < 0$.

But since $B = 0$, we have

$$B^2 - 4AC = -4AC = b^2 - 4ac.$$

The fact that $-4AC < 0$ if and only if $AC > 0$ gives the following result.

4.2 THEOREM

> The graph of
>
> (2) $$ax^2 + bxy + cy^2 + ex + fy + g = 0$$
>
> is
>
> (a) a parabola (or degenerate parabola) if $b^2 - 4ac = 0$,
>
> (b) an ellipse (or degenerate ellipse) if $b^2 - 4ac < 0$,
>
> (c) a hyperbola (or degenerate hyperbola) if $b^2 - 4ac > 0$.

This makes it possible to identify the graph of a given second-degree equation without rotating axes.

FIGURE 4.7

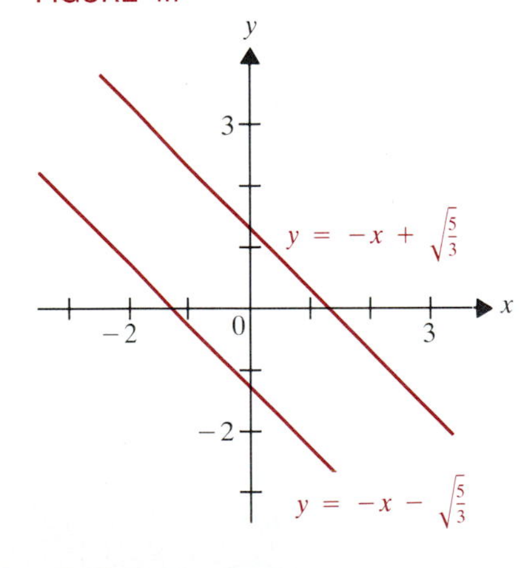

Example 5

Identify the graphs of

(a) $3x^2 + 6xy + 3y^2 = 5$

(b) $2x^2 - 8xy + 3y^2 - 8x - 6y = 3$

(c) $5x^2 + 3xy + y^2 + 6x + 5y = 1$

Solution. (a) $b^2 - 4ac = 36 - 4 \cdot 9 = 0$, so the graph is a parabola. (It is in fact a degenerate parabola, since the graph of $3(x + y)^2 = 5$ consists of two parallel lines. (See **Figure 4.7.**)

(b) $b^2 - 4ac = 64 - 24 = 40 > 0$, so the graph is a hyperbola.

(c) $b^2 - 4ac = 9 - 20 = -11 < 0$, so the graph is an ellipse. ■

Exercises 8.4

In Exercises 1–6, find the vertices and foci of the graph of the given equation.

1. $xy = -4$

2. $xy = -9$

3. $xy = 1$

4. $xy = -1$

5. $xy = 2$

6. $xy = -1/2$

In Exercises 7–16, find the equation of the curve if the given rotation of coordinate axes is carried out. (Express in terms of X and Y.)

7. $x - y = 2$; $\alpha = -\pi/6$

8. $x + 3y = 5$; $\alpha = \pi/3$

9. $4x^2 + 9y^2 = 36$; $\alpha = -\pi/2$

10. $x^2 + 4y^2 = 9$; $\alpha = \pi/2$

11. $4x^2 - 25y^2 = 100$; $\alpha = \pi/2$

12. $x^2 - 9y^2 = 16$; $\alpha = -\pi/2$

13. $3x^2 + 10xy + 3y^2 = -32$; $\alpha = \pi/4$

14. $x^2 - xy + y^2 = 10$; $\alpha = \pi/4$

15. $6x^2 + 4\sqrt{3}xy + 2y^2 + x - \sqrt{3}y = 0$; $\alpha = \pi/6$

16. $x^2 + 2xy + y^2 - \sqrt{2}x + \sqrt{2}y = 0$; $\alpha = -\pi/4$

In Exercises 17–28, rotate the coordinate axes through an angle so that the transformed equation has no *xy*-term. Identify and sketch the graph.

17. $x^2 + xy + y^2 = 3$

18. $x^2 + xy + y^2 = 1$

19. $73x^2 + 72xy + 52y^2 + 30x - 40y - 75 = 0$

20. $5x^2 + 8xy + 5y^2 + 16\sqrt{2}x + 20\sqrt{2}y + 31 = 0$

21. $x^2 - 3xy + y^2 = 5$

22. $2x^2 - 5xy + 2y^2 = 3$

23. $11x^2 - 10\sqrt{3}xy + y^2 + 32\sqrt{3}x - 32y + 80 = 0$

24. $23x^2 - 72xy + 2y^2 + 140x + 20y - 75 = 0$

25. $x^2 + 2xy + y^2 - 3\sqrt{2}x - 5\sqrt{2}y + 12 = 0$

26. $18x^2 - 48xy + 32y^2 + 80x - 65y = 0$

27. $x^2 + 2xy + y^2 = 8$

28. $x^2 + 4xy + y^2 = 16$

In Exercises 29–34, use the discriminant and Theorem 4.2 to determine the general form of the graph of the given equation.

29. $x^2 - 2xy + y^2 - \sqrt{2}x + 3\sqrt{3}y + 16 = 0$

30. $5x^2 - 8xy + 5y^2 - 8x - 9y + 10 = 0$

31. $2x^2 - 3xy + y^2 + 5\sqrt{3}x - 8\sqrt{2}y + 1 = 0$

32. $x^2 - 4xy + 3y^2 - 3x - 15x + 6 = 0$

33. $5x^2 - 2xy + y^2 + 3x - 2y + 4 = 0$

34. $2x^2 + xy + y^2 - 8x - 10y - 10 = 0$

35. Show that when (5) is substituted into (2), the resulting equation $AX^2 + BXY + CY^2 + EX + FY + G = 0$ satisfies $A + C = a + c$.

36. Can (2) be the equation of a circle if $b \neq 0$?

37. Use (5) to show that the equation of a circle centered at (0, 0) is not changed by any rotation.

38. Show that a rotation through $\alpha = -\pi/4$ also eliminates the xy-term in (2) if $a = c$.

39. Derive (6) from (5) by exchanging the roles of the xy and XY coordinates and letting the angle of rotation be $-\alpha$.

40. Verify the entries for A, C, E, and F in Table 1.

8.5 Parametric Equations

Our study of geometry in the plane has thus far dealt with curves as graphs of functions $y = f(x)$ or of equations $F(x, y) = 0$ relating x and y directly. There is a more general way of dealing with curves, which is especially useful in studying motion. Suppose that a particle moves in the plane, and we keep track of its coordinates over time. Then as t varies over some interval I, its x-coordinate is some function of t, say

$$x = f(t), \qquad t \in I, \tag{1}$$

and its y-coordinate is also some function of t, say

$$y = g(t), \qquad t \in I. \tag{2}$$

As t varies, the path of the particle traces out a curve in the plane.

Example 1

Plot the curve determined by the equations

$$x = \cos t, \qquad y = \sin t, \tag{3}$$

as t varies between 0 and 2π.

Solution. The following table gives the values of x and y for several values of t between 0 and $\frac{1}{2}\pi$.

t	0	$\pi/6$	$\pi/4$	$\pi/3$	$\pi/2$
$x(t) = \cos t$	1	$\sqrt{3}/2$	$\sqrt{2}/2$	1/2	0
$y(t) = \sin t$	0	1/2	$\sqrt{2}/2$	$\sqrt{3}/2$	1

FIGURE 5.1

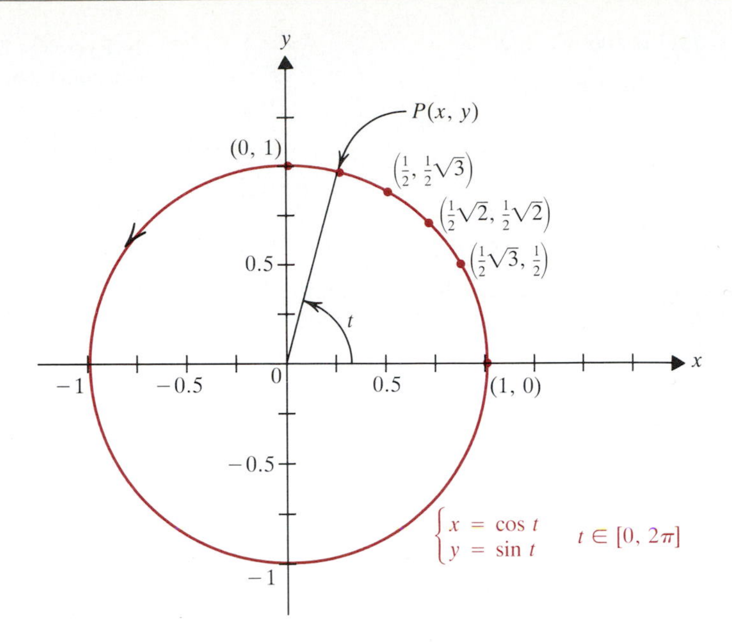

The corresponding points (x, y) in **Figure 5.1** appear to lie on the unit circle $x^2 + y^2 = 1$, which we have traced in through the points. To confirm this, note that

$$x^2 = \cos^2 t \qquad \text{and} \qquad y^2 = \sin^2 t,$$

so

(4)
$$x^2 + y^2 = \cos^2 t + \sin^2 t = 1.$$

FIGURE 5.2

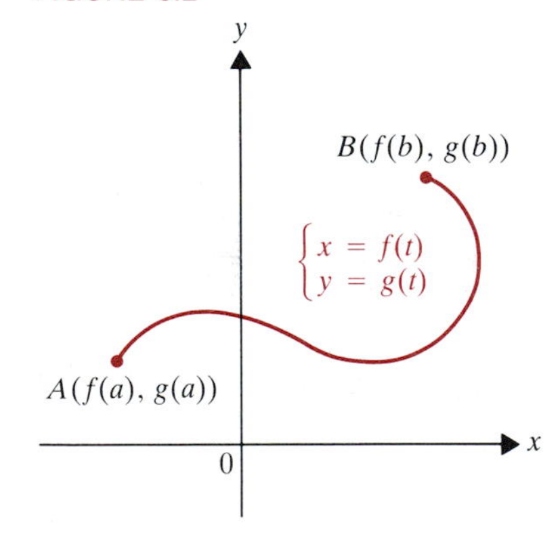

Thus every point satisfying (3) lies on the unit circle. Recall from Section 2.3 that the coordinates of any point $P(x, y)$ on the unit circle satisfy (3) for the angle $t \in [0, 2\pi]$ made by $\overline{0P}$ and the positive x-axis, measured counterclockwise from the x-axis. Thus the graph of the equations (3) is indeed the unit circle. ■

If in Example 1, t varies over $[0, \pi]$ instead of $[0, 2\pi]$, then (4) still holds for every point satisfying (3). So every such point lies on the unit circle. But the graph is only the *top half* of the unit circle; since $\sin t \geq 0$ on $[0, \pi]$, there are no points below the y-axis.

The ideas in Example 1 lead to the following definition.

5.1 DEFINITION

A ***parametrized curve*** in the plane is the set of all points $P(x, y)$ such that

(5)
$$x = f(t), \; y = g(t), \qquad t \in I$$

for some continuous functions f and g and some interval I of real numbers. Equations (5) are called ***parametric equations*** of the curve. The variable t is called a ***parameter***. If I is a closed interval $[a, b]$, then the points $A(f(a), g(a))$ and $B(f(b), g(b))$ are called ***endpoints*** of the curve (see **Figure 5.2**).

It is sometimes helpful to eliminate the parameter t in (5) and find a Cartesian coordinate equation $F(x, y) = 0$ satisfied by $x = f(t)$ and $y = g(t)$. This can show, as above, that the parametric curve is part or all of a familiar curve.

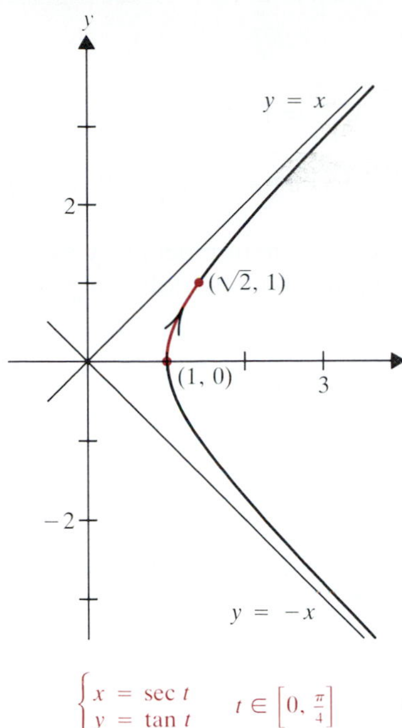

FIGURE 5.3

Example 2

Identify the curve

$$x = \sec t,\ y = \tan t, \qquad t \in [0, \pi/4].$$

Solution. Here the trigonometric identity $1 + \tan^2 x = \sec^2$ gives

$$1 + y^2 = x^2 \to x^2 - y^2 = 1.$$

So the curve lies on the hyperbola with vertices $(\pm 1, 0)$ and asymptotes $y = \pm x$. When $t = 0$ we get the point $(1, 0)$. When $t = \pi/4$, we get the point $(\sqrt{2}, 1)$. Thus the parametric curve is the colored arc shown in **Figure 5.3.** ■

An advantage of studying curves parametrically is that as t varies, we can see *how* the curve is traced out, not just its shape. For instance, in Example 1 the unit circle is traced out counterclockwise from the starting point $(1, 0)$ back to the same point. The arrowheads in Figures 5.1 and 5.3 indicate the direction of tracing in Examples 1 and 2.

A parametric curve whose two endpoints coincide is called a ***closed curve.*** The unit circle is one such curve. **Figure 5.4** shows two other closed curves. The one in part (a), which does not intersect itself (except at the beginning–endpoint), is said to be ***simple.*** The curve in part (b) is not simple. Parts (c) and (d) show respectively a simple and nonsimple curve, neither of which is closed.

It is easy to find parametric equations for a curve that is the graph of a function $y = f(x)$ or $x = g(y)$. In the first case we simply use x as a parameter, by letting

(6) $\qquad x = t \quad$ and $\quad y = f(t).$

In the second case we let

(7) $\qquad y = t \quad$ and $\quad x = g(t),$

so that y is the parameter.

FIGURE 5.4

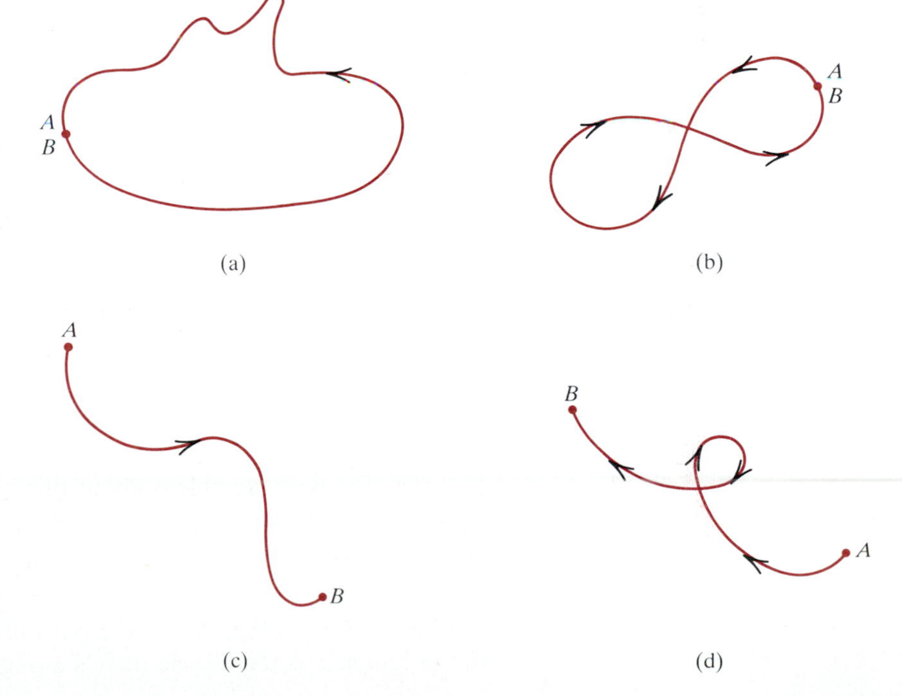

FIGURE 5.5

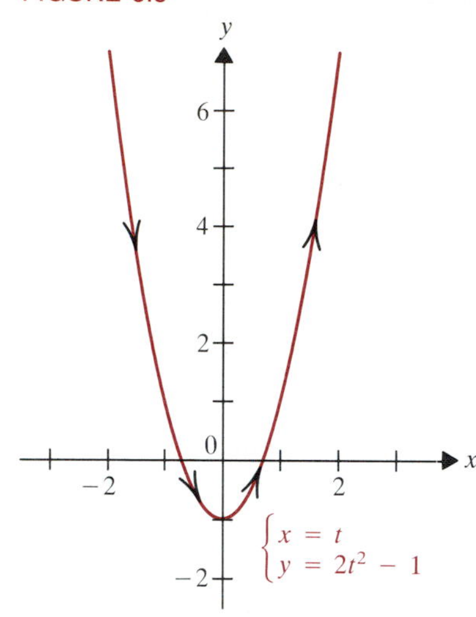

Example 3

Parametrize the curves that are the graphs of

$$\text{(a) } y = 2x^2 - 1 \qquad \text{(b) } \frac{x^2}{9} + \frac{y^2}{16} = 1.$$

Solution. (a) Using (6) we let $x = t$, $y = 2t^2 - 1$, $t \in (-\infty, +\infty)$. (See **Figure 5.5.**)

(b) We can't solve for $y = f(x)$, because the given equation defines y as *two* functions of x. Instead we use the idea in Example 1 in the following way. Recalling that $\cos^2 t + \sin^2 t = 1$, we parametrize so that

$$\frac{x^2}{9} = \cos^2 t \qquad \text{and} \qquad \frac{y^2}{16} = \sin^2 t.$$

A simple way to do this is to let

$$\text{(8)} \qquad x = 3\cos t \qquad \text{and} \qquad y = 4\sin t, \qquad t \in [0, 2\pi].$$

As t increases from 0 to 2π, the ellipse is traversed counterclockwise, starting and ending at the point (3, 0). (See **Figure 5.6.**) ■

FIGURE 5.6

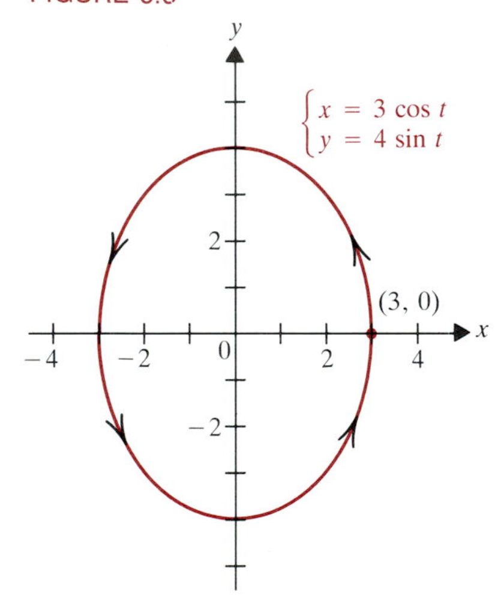

There are two points to note in Example 3. First, whenever a curve to be parametrized has a Cartesian coordinate equation of the form $u^2 + v^2 = 1$, it can be parametrized by letting $u = \cos t$ and $v = \sin t$, $t \in [0, 2\pi]$. Second, there are *many* different ways to parametrize a given curve. In Example 3(a), for instance, the parametrization

$$x = t - 1 \qquad \text{and} \qquad y = 2t^2 - 4t + 1, \qquad t \in (-\infty, +\infty)$$

also works, because we have

$$2x^2 - 1 = 2(t-1)^2 - 1 = 2(t^2 - 2t + 1) - 1 = 2t^2 - 4t + 1 = y.$$

Similarly, in Example 3(b) we could change t to $2\pi - t$ without changing the *curve*, although we *would* change the way the curve is traced. As t goes from 0 to 2π, the parametrization (8) traces the curve *counterclockwise* from (3, 0). However, as t goes from 0 to 2π in the parametrization

$$\text{(9)} \qquad x = 3\cos(2\pi - t) = 3\cos t \qquad \text{and} \qquad y = 4\sin(2\pi - t) = -4\sin t,$$

the curve is traced in the *clockwise* sense from (3, 0). See **Figure 5.7** and the following table, which compares (8) and (9).

FIGURE 5.7

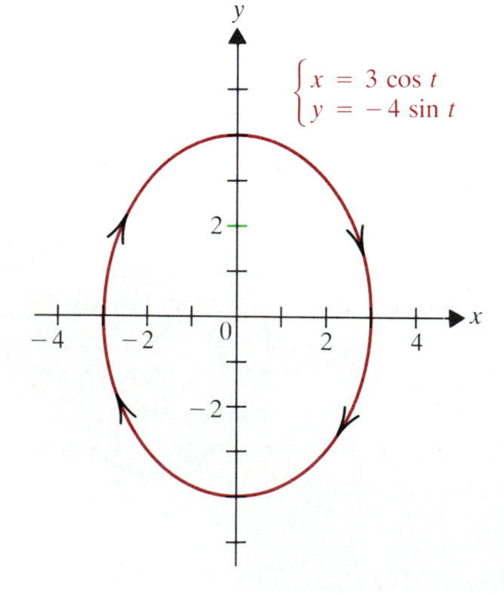

t	0	$\pi/2$	π	$3\pi/2$	2π
$3\cos t$	3	0	−3	0	3
$4\sin t$	0	4	0	−4	0
$-4\sin t$	0	−4	0	4	0

Parametric equations are most useful in describing curves traced by moving particles. The next example is about one such curve, for which it is best to use something *other* than time for parameter.

Example 4

Find parametric equations for the path of a reflector on the edge of a bicycle tire of radius a as the bicycle travels along a straight, level path.

FIGURE 5.8

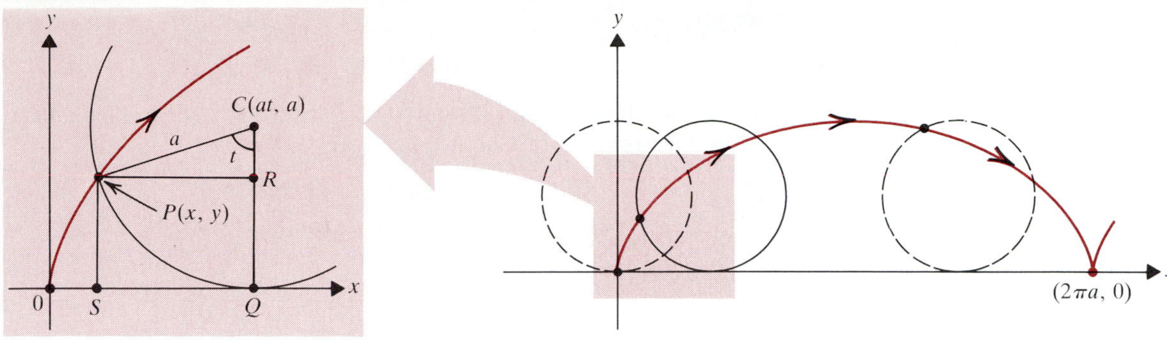

Solution. In **Figure 5.8** the path of the bicycle tire is along the x-axis, with the initial ($t = 0$) position of the reflector at the origin. As a parameter, we use the number t of radians in the angle through which the tire has turned. Then the distance $|0Q|$ that the wheel has rolled along the ground is equal to the arc length $\widehat{PQ}$. Thus $|0Q| = at$. The center C then has coordinates (at, a). Referring to right triangle PCR, we see that

$$\cos t = \frac{|RC|}{|PC|} = \frac{|RC|}{a} \rightarrow |RC| = a\cos t,$$

and

$$\sin t = \frac{|PR|}{|PC|} = \frac{|PR|}{a} \rightarrow |PR| = a\sin t.$$

Since $|0Q| = at$ and $|QR| = y$, we have

$$\begin{aligned} x &= |0S| = |0Q| - |SQ| = |0Q| - |PR| \\ &= at - a\sin t = a(t - \sin t), \\ y &= |QR| = |QC| - |RC| = a - a\cos t = a(1 - \cos t). \end{aligned}$$

Therefore the parametric equations for the path are

$$x = a(t - \sin t),\ y = a(1 - \cos t), \qquad t \in [0, +\infty). \ \blacksquare$$

The path in Example 4 is called a *cycloid.* (See the historical note at the end of this section for more about this curve.) If the reflector is located on a spoke of the tire, then its path is similar to the cycloid, but it does not hit the x-axis or have sharp points like the one at $(2\pi a, 0)$. That curve is called a *trochoid,* and you are asked to derive its equation in Exercise 39.

To discuss the tangent line to a parametric curve at a point $P(x_0, y_0)$, we restrict attention to *smooth curves,* those parametric curves

$$x = f(t),\ y = g(t), \qquad t \in I \tag{10}$$

for which dx/dt and dy/dt are both continuous and never simultaneously 0. This guarantees that near any point $P_0(x_0, y_0)$ Equation (10) defines either y as a differentiable function h of x or x as a differentiable function k of y. To see why, note first that if dx/dt is continuous and nonzero near $x = x_0$, then x is a monotonic function of t by Theorem 3.1 of Chapter 3. Thus this function has an inverse $t = f^{-1}(x)$ on an interval I containing x_0 (see p. 117). Substitution into (10) gives

$$y = g(t) = g(f^{-1}(x)) = h(x). \tag{11}$$

If instead, $dy/dt \neq 0$, then near $y = y_0$ we have $t = g^{-1}(y)$, and substituting into (10), we get

$$x = f(t) = f(g^{-1}(y)) = k(y). \tag{12}$$

We can use the chain rule on (11) and (12) because f^{-1} (or g^{-1}) is differentiable in view of Theorem 7.5 of Chapter 2. From (11) we get

$$\frac{dy}{dt} = \frac{dh}{dx}\frac{dx}{dt} = \frac{dy}{dx}\frac{dx}{dt}.$$

From (12) we get

$$\frac{dx}{dt} = \frac{dk}{dy}\frac{dy}{dt} = \frac{dx}{dy}\frac{dy}{dt}.$$

In either case, since

$$\frac{dy}{dx} = \frac{1}{dx/dy}$$

by the inverse function differentiation rule, we have

$$\frac{dy}{dx} = \frac{dy/dt}{dx/dt} \tag{13}$$

if $dx/dt \neq 0$. If $dx/dt = 0$, then we can be sure that $dy/dt \neq 0$, since we are working only with smooth curves. In that case, the curve has a vertical tangent.

We illustrate (13) by returning to the curves in Example 4 and Example 3(b).

Example 5

Find the equation of the tangent line to the given parametric curve at the given point.

(a) $x = a(t - \sin t)$, $y = a(1 - \cos t)$, at $t = \frac{1}{2}\pi$

(b) $x = 3\cos t$, $y = 4\sin t$, at $t = 0$

Solution. (a) Here $x = a(t - \sin t)$ and $y = a(1 - \cos t)$. Therefore

$$\frac{dx}{dt} = a(1 - \cos t) \quad \text{and} \quad \frac{dy}{dt} = a\sin t.$$

Hence from (13),

$$\frac{dy}{dx} = \frac{a\sin t}{a(1 - \cos t)} = \frac{\sin t}{1 - \cos t}$$

if $dx/dt \neq 0$. At $t = \frac{1}{2}\pi$, $\sin t = 1$ and $\cos t = 0$, so

$$\frac{dy}{dx} = \frac{1}{1 - 0} = 1.$$

At $t = \frac{1}{2}\pi$, $x = a\left(\frac{\pi}{2} - 1\right)$ and $y = a$. Thus the equation of the tangent line is

$$y - a = 1\left(x - \frac{a\pi}{2} + a\right) \rightarrow y = x + a\left(2 - \frac{\pi}{2}\right).$$

(See **Figure 5.9.**)

FIGURE 5.9

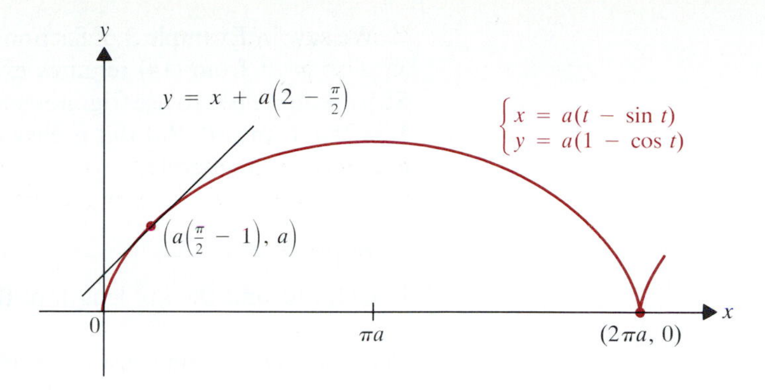

(b) From (8) we have $x = 3$ and $y = 0$ when $t = 0$. Also,

$$\frac{dx}{dt} = -3 \sin t = 0$$

at $t = 0$. The tangent is therefore vertical, the line $x = 3$, as Figure 5.6 suggests. ■

Arc Length of a Parametric Curve

The arc length of a parametric curve can be calculated by using the formula

$$\int_a^b \sqrt{1 + (dy/dx)^2}\, dx \tag{14}$$

from Section 5.4 in conjunction with (13). Recall in (14) that $a < b$, that is, we integrate in the direction of increasing x. We can change the variable x in (14) by letting $x = x(t)$. Then

$$dx = \frac{dx}{dt}\, dt,$$

where we first assume that x is increasing: $(dx/dt) > 0$. From (13) we then have

$$1 + \left(\frac{dy}{dx}\right)^2 = 1 + \frac{(dy/dt)^2}{(dx/dt)^2} = \frac{(dx/dt)^2 + (dy/dt)^2}{(dx/dt)^2}.$$

Thus, since $(dx/dt) > 0$,

$$\sqrt{1 + (dy/dx)^2} = \frac{1}{|dx/dt|} \sqrt{(dx/dt)^2 + (dy/dt)^2} = \frac{1}{dx/dt} \sqrt{(dx/dt)^2 + (dy/dt)^2}.$$

If $x(t_0) = a$ and $x(t_1) = b$, then by the change-of-variable theorem for definite integrals (Theorem 5.1 of Chapter 4), (14) becomes

$$L = \int_{t_0}^{t_1} \frac{1}{\cancel{dx/dt}} \sqrt{(dx/dt)^2 + (dy/dt)^2}\, \cancel{dx/dt}\, dt = \int_{t_0}^{t_1} \sqrt{(dx/dt)^2 + (dy/dt)^2}\, dt.$$

We thus have proved the following theorem when $dx/dt > 0$. [For the case $dx/dt < 0$, see Exercise 51(a).]

5.2 THEOREM

If dx/dt and dy/dt are continuous and never simultaneously zero on $[t_0, t_1]$, then the parametric curve

$$x = x(t),\ y = y(t), \qquad t \in [t_0, t_1]$$

has arc length

$$L = \int_{t_0}^{t_1} \sqrt{(dx/dt)^2 + (dy/dt)^2}\, dt. \tag{15}$$

We saw in Example 3 of Section 5.4 that computing the arc length of a circle $x^2 + y^2 = a^2$ from (14) requires evaluation of an integral involving $\sqrt{a^2 - x^2}$. So we would need to use trigonometric substitution to derive the familiar formula $L = 2\pi a$ from (14). But this is easy to derive from (15) and the usual parametric equations of the circle.

Example 6

Use (15) to find the arc length of the circle $x^2 + y^2 = a^2$.

Solution. We parametrize the circle as in Example 1 by

$$x = a\cos t,\ y = a\sin t, \qquad t \in [0, 2\pi].$$

Then

$$\frac{dx}{dt} = -a\sin t \qquad \text{and} \qquad \frac{dy}{dt} = a\cos t.$$

Since

$$\left(\frac{dx}{dt}\right)^2 + \left(\frac{dy}{dt}\right)^2 = a^2\sin^2 t + a^2\cos^2 t = a^2,$$

(15) gives

$$L = \int_0^{2\pi} a\,dt = a\int_0^{2\pi} dt = 2\pi a. \quad ■$$

With some more effort, we can use (15) to find the length of one arch of the cycloid of Example 4.

Example 7

Find the arc length of the cycloid

$$x = a(t - \sin t),\ y = a(1 - \cos t), \qquad t \in [0, 2\pi]. \tag{16}$$

Solution. From (16) we find

$$\frac{dx}{dt} = a(1 - \cos t) \qquad \text{and} \qquad \frac{dy}{dt} = a\sin t.$$

Thus

$$\sqrt{(dx/dt)^2 + (dy/dt)^2} = a\sqrt{1 - 2\cos t + \cos^2 t + \sin^2 t} = a\sqrt{2}\sqrt{1 - \cos t}.$$

As t varies from 0 to 2π, $\frac{1}{2}t$ goes from 0 to π. Hence, taking radicals in the identity

$$\sin^2\frac{1}{2}t = \frac{1 - \cos t}{2},$$

we obtain

$$\sin\frac{1}{2}t = \sqrt{\frac{1 - \cos t}{2}},$$

because $\sin\frac{1}{2}t \geq 0$ for $\frac{1}{2}t \in [0, \pi]$. Therefore

$$\sqrt{(dx/dt)^2 + (dy/dt)^2} = a\sqrt{2}\sqrt{2}\sin\frac{1}{2}t = 2a\sin\frac{1}{2}t.$$

Hence from (15),

$$L = 2a \int_0^{2\pi} \sin \frac{1}{2} t \, dt = -4a \cos \frac{1}{2} t \Big|_0^{2\pi} = 8a. \quad ■$$

HISTORICAL NOTE

The cycloid is a curve with a rich mathematical history. In 1643, a generation before the invention of calculus, the French mathematician Gilles P. de Roberval (1602–1675) was able to find the tangent to the cycloid by using vector methods. The famous architect of St. Paul's Cathedral in London, Christopher Wren (1632–1723), who also was a mathematician and astronomer, obtained the result of Example 7 by purely geometric methods in 1658. That year Pascal had issued a challenge to the mathematical world to solve several such problems about the cycloid. In 1669, Newton found the Cartesian coordinate equation

$$y = \sqrt{x - x^2} + \sin \sqrt{x}, \qquad x \in [-1, 1]$$

for the cycloid. In 1696, John Bernoulli posed the problem of finding the shape of a wire joining a point P to a point Q at a lower level so that a drop of water rolling down the wire would reach Q in the least time. Bernoulli and his brother James both solved it, as did Newton, Leibniz, and l'Hôpital. The answer is an arc of an *inverted* cycloid. See **Figure 5.10.** This problem, called the *brachistochrone problem* (from the two Greek words for *shortest time*), gave impetus to the branch of higher mathematics called *calculus of variations.*

FIGURE 5.10

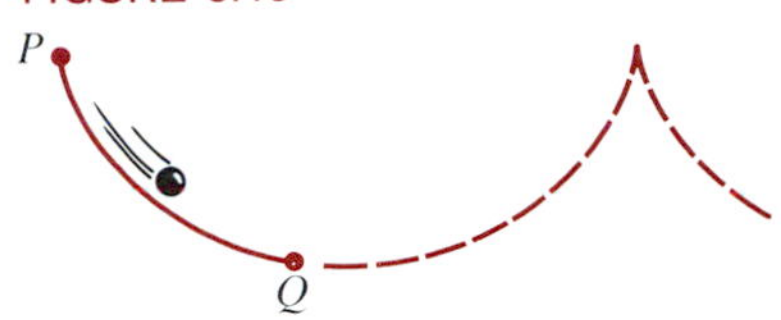

Exercises 8.5

In Exercises 1–20, sketch the graph of the given curve and label the direction of increasing t. Identify the curve and find a Cartesian equation for it.

1. $x = t - 1,\ y = 2t + 1,\ t \in [0, 4]$
2. $x = \frac{1}{2}t + 1,\ y = 3t - 4,\ t \in [0, 5]$
3. $x = 1 - 3t,\ y = 2 + \frac{5}{2}t,\ t \in [-3, 3]$
4. $x = -2 - 4t,\ y = -5 + 3t,\ t \in [1, 8]$
5. $x = t^2 - 3,\ y = t^2 + 2,\ t \in [0, 1]$
6. $x = 1 - t^3,\ y = 1 + t^3,\ t \in [-3, 3]$
7. $x = 3t^2,\ y = 2t,\ t \in [-2, 2]$
8. $x = 2t,\ y = t^2 - 1,\ t \in [-2, 2]$
9. $x = 2t,\ y = \sqrt{4 - t},\ t \le 4$
10. $x = -3t,\ y = \sqrt{4 + 2t},\ t \ge -2$
11. $x = 3\cos t,\ y = 2\sin t,\ t \in [0, 2\pi]$
12. $x = -3\cos t,\ y = 4\sin t,\ t \in [0, 2\pi]$
13. $x = -2 + \cos t,\ y = 4 - \sin t,\ t \in [0, 2\pi]$
14. $x = 1 - 3\cos t,\ y = 2 + 3\sin t,\ t \in [0, 2\pi]$
15. $x = 3 - 2\cos t,\ y = 5 + 5\sin t,\ t \in [0, 2\pi]$
16. $x = 2\sin t - \frac{3}{2},\ y = -3\cos t + 2,\ t \in [0, 2\pi]$
17. $x = e^t,\ y = e^{-t},\ t \in [0, +\infty)$
18. $x = -2e^{-t},\ y = 3e^t,\ t \in [0, +\infty)$
19. $x = 2\cosh t,\ y = 3\sinh t$
20. $x = -5\sinh t,\ y = 2\cosh t$

In Exercises 21–28, find the slope of the tangent line to the given curve at the given point.

21. Curve of Exercise 7, $t = 1$
22. Curve of Exercise 8, $t = -1$
23. Curve of Exercise 9, $t = 4$
24. Curve of Exercise 10, $t = -2$
25. Curve of Exercise 11, $t = \pi/4$
26. Curve of Exercise 13, $t = 2\pi/3$
27. Curve of Exercise 17, $t = 2$
28. Curve of Exercise 18, $t = -2$

In Exercises 29–38, find the indicated arc length.

29. $x = 2t - 1,\ y = -t + 3,\ t \in [1, 5]$

30. $x = 1 - 3t,\ y = 2t - 2,\ t \in [0, 4]$

31. $x = \dfrac{1}{2}(t-1)^2,\ y = \dfrac{4}{3}t^{3/2},\ t \in [0, 4]$

32. $x = \dfrac{1}{3}t^3,\ y = \dfrac{1}{2}t^2,\ t \in [0, 2]$

33. $x = \sin t + \cos t,\ y = \sin t - \cos t,\ t \in [0, \pi/2]$

34. $x = e^t \cos t,\ y = e^t \sin t,\ t \in [0, \pi/3]$

35. $x = t^2,\ y = 2t^3,\ t \in [-1, 1]$

36. $x = t^2 \cos t,\ y = t^2 \sin t,\ t \in [0, 1]$

37. $x = \cos^3 t,\ y = \sin^3 t,\ t \in [0, 2\pi]$ (*Hint:* Be careful in working with the radical.)

38. $x = 1 - 2\sin^2 t,\ y = 2 \sin t \cos t,\ t \in [0, 2\pi]$

39. Suppose the reflector in Example 4 is located on a spoke of the bicycle at distance $b < a$ from the center. Find parametric equations for its path (the *trochoid*).

40. A circle of radius b rolls on the inside of a circle of radius $a > b$. The curve traced by a point P on the smaller circle is called an *astroid* or *hypocycloid*. Taking the initial position of P as $(a, 0)$, show that

$$x = (a - b)\cos t + b \cos \frac{a-b}{b} t$$

$$y = (a - b)\sin t - b \sin \frac{a-b}{b} t$$

are parametric equations of the astroid. (*Hint:* How are arcs $\overset{\frown}{TP}$ and $\overset{\frown}{AT}$ in **Figure 5.11** related?)

FIGURE 5.11

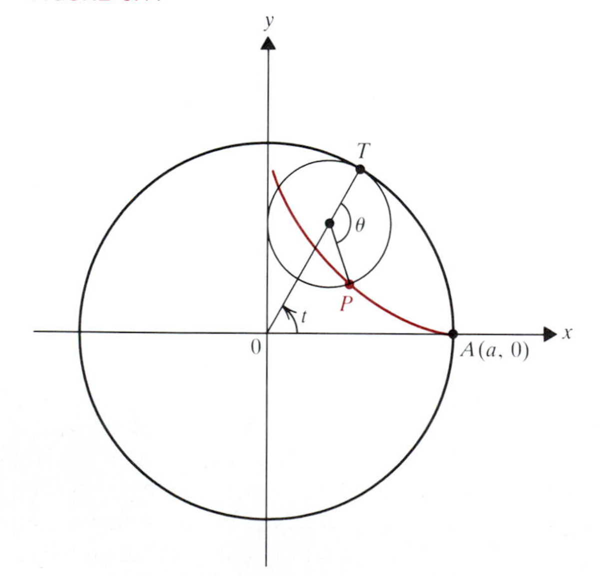

41. If in Exercise 40, $b = \frac{1}{4}a$, the curve is called a *hypocycloid of four cusps.*

(a) Show that the curve has parametric equations

$$x = a\cos^3 t,\ y = a \sin^3 t.$$

(*Hint:* Use Exercises 15 and 16 of Section 2.3.)

(b) Show that the curve has Cartesian coordinate equation

$$x^{2/3} + y^{2/3} = a^{2/3}.$$

(c) Is this curve smooth? Explain.

(d) Find the arc length of this curve by first finding the length of its first-quadrant portion.

42. Suppose that a particle moves on the curve $x = f(t)$, $y = g(t)$, $t \in [0, a]$, where t is the number of seconds elapsed since motion started.

(a) What is the distance $s(u)$ the particle has moved in u seconds?

(b) Explain how

$$\left.\frac{ds}{du}\right|_{u=t_0}$$

gives the speed of the particle at time t_0.

43. Show that if $x = f(\theta)\cos\theta$ and $y = f(\theta)\sin\theta$, then

$$\left(\frac{dx}{d\theta}\right)^2 + \left(\frac{dy}{d\theta}\right)^2 = [f(\theta)]^2 + [f'(\theta)]^2.$$

44. Show that the formula

$$L = \int_c^d \sqrt{1 + \left(\frac{dx}{dy}\right)^2}\, dy$$

for the arc length along the graph of $x = g(y)$ between $y = c$ and $y = d$ is a special case of (15) in Theorem 5.2.

45. Show that the area under one arch of the cycloid in Example 4 is $3\pi a^2$. (*Hint:* Change $A = \int_0^{2\pi a} y\, dx$ into a t-integral.)

46. Use the approach of Exercise 45 to show that the area enclosed by the ellipse $x = a\cos t,\ y = b \sin t,\ t \in [0, 2\pi]$ is πab.

47. Use the approach of Exercise 45 to show that the area enclosed by the hypocycloid of four cusps (Exercise 41) is $3\pi a^2/8$.

48. The *spiral of Archimedes* can be given parametrically by

$$x = t\cos t,\ y = t \sin t, \qquad t \geq 0.$$

(a) Draw the curve for $t \in [0, 4\pi]$.

(b) Find the arc length of this curve between $t = 0$ and $t = 2\pi$.

49. Find a formula for d^2y/dx^2 from (13).

50. Use the results of Example 5(a) and Exercise 49 to show that the graph of the cycloid in Example 4 is concave downward for all t (except even multiples of π).

51. **(a)** Prove Theorem 5.2 in the case $dx/dt < 0$.

(b) Why are the discussion in the text and part (a) enough to prove Theorem 5.2?

52. Explain the relation between the discussion of projectile motion in Section 1 and this section.

PC **53.** If various values are assigned to a and b in Exercise 40, then interesting curves result. Do computer plots for
(a) $a = 7, b = 2, t \in [0, 4\pi]$
(b) $a = 9, b = 2, t \in [0, 4\pi]$
(c) $a = 7, b = 3, t \in [0, 6\pi]$
(d) $a = 11, b = 3, t \in [0, 6\pi]$
(e) $a = 11, b = 4, t \in [0, 8\pi]$
(f) $a = 6, b = 2, t \in [0, 4\pi]$
(g) $a = 12, b = 4, t \in [0, 8\pi]$
What do you conjecture about the curve for general positive integers a and b?

PC **54.** Continue Exercise 53 for
(a) $a = 33, b = 13, t \in [0, 26\pi]$
(b) $a = 17, b = 19, t \in [0, 38\pi]$
(c) $a = 59, b = 17, t \in [0, 34\pi]$
(d) $a = 5000, b = 1250, t \in [0, 2500\pi]$
(e) Experiment with your choice of values for a and b.

8.6 Polar Coordinates

Up to now we have exclusively used the rectangular Cartesian coordinate system. In this section, we introduce a second coordinate system for the plane, which is well suited to curves and regions that are symmetric about a point. The key ingredients of the alternative coordinate system are a point and a ray (half-line) emanating from the point. The point is called the ***pole*** and the ray is the ***polar axis.*** It is convenient in many cases to superimpose the new system of coordinates on the usual Cartesian coordinate system. This is done by taking the origin 0 as the pole and the positive x-axis as the polar axis, as in **Figure 6.1.** Then, given a point P in the plane, we specify its location by giving the pair of numbers $[r, \theta]$ where

FIGURE 6.1

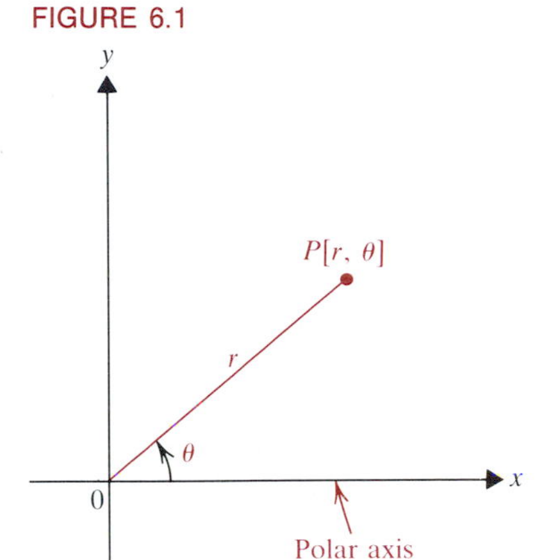

$$r = |0P| = \text{distance of } P \text{ from the origin}$$

and

$$\theta = \text{angle between } 0P \text{ and the positive } x\text{-axis},$$

measured counterclockwise from the axis.

6.1 DEFINITION

Given a point P in the real plane with rectangular coordinates (x, y), it has ***polar coordinates*** $[r, \theta]$, where

$r = \pm|0P|$ and θ is an angle that $0P$ makes with the positive x-axis.

Positive angles are measured *counterclockwise* from the axis. If θ is negative, then we interpret the angle as being measured *clockwise* from the positive axis. If r is negative, then $P[r, \theta]$ stands for the same point as $P[-r, \theta \pm \pi]$ (for which $-r > 0$).

The origin 0 is called the ***pole.***

The angle θ is usually restricted to the interval $[0, 2\pi)$, and r is often restricted to nonnegative real values, so that every point P other than the pole has only one pair $[r, \theta]$ of polar coordinates. (Note that the pole can be represented as $[0, \theta]$ for any θ whatsoever.) But Definition 6.1 allows the possibility of using θ outside $[0, 2\pi)$ and using $r < 0$. Thus there is *not* a one-to-one correspondence between points P in the plane and polar coordinates $[r, \theta]$. In fact, *each point has infinitely many polar coordinates.* For example, the point R in **Figure 6.2** has polar coordinates $[2, 5\pi/4]$ as well as $[2, -3\pi/4]$. Also, P has coordinates $[2, \pi/4]$ and $[2, 9\pi/4]$. In fact, P has coordinates $[2, \pi/4 \pm 2n\pi]$ for *any* integer n.

FIGURE 6.2

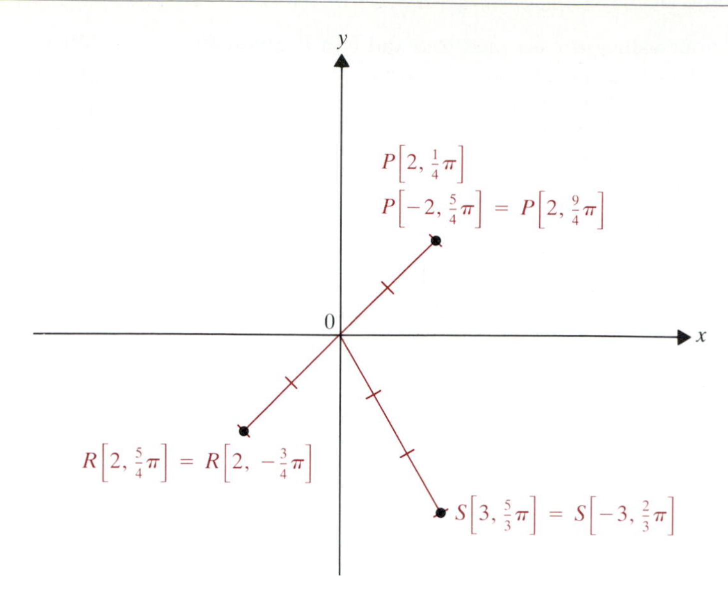

Similarly, any point $P[r, \theta]$ also has polar coordinates $[-r, \theta \pm \pi]$. In Figure 6.2, for example, $P[2, \pi/4]$ is also representable as $[-2, \pi/4 + \pi] = [-2, 5\pi/4]$, and $S[3, 5\pi/3]$ has polar coordinates $[-3, 5\pi/3 - \pi] = [-3, 2\pi/3]$ as well. You may wish to view r as the *directed* distance from the pole to the point P along the terminal side of angle θ. In Figure 6.2, for instance, we reach $P[-2, 5\pi/4]$ by measuring the counterclockwise angle $5\pi/4$ from the positive x-axis, and then proceeding from 0 along the ray $\theta = 5\pi/4 + \pi = 9\pi/4$ for 2 units. This can also be done directly by proceeding from the origin along the ray $\theta = 5\pi/4$ in the *negative* direction for 2 units.

FIGURE 6.3

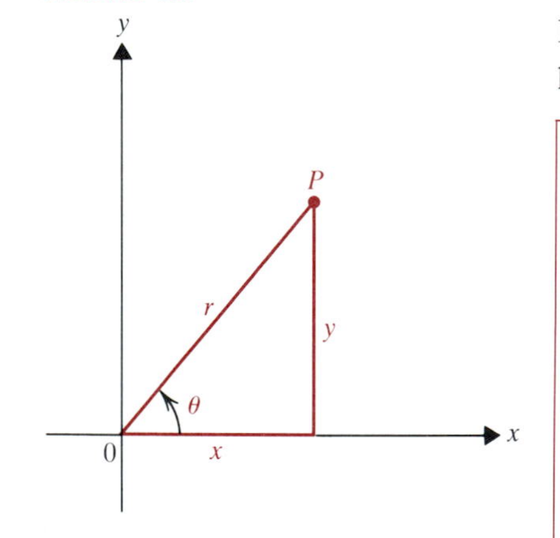

The following change of coordinate equations, which can be read from Definition 6.1 and **Figure 6.3,** make it easy to change back and forth between polar and rectangular coordinates.

CHANGE OF COORDINATE EQUATIONS

If P has rectangular coordinates (x, y) and polar coordinates $[r, \theta]$, then

(1) $$x = r\cos\theta$$

(2) $$y = r\sin\theta$$

(3) $$r^2 = x^2 + y^2$$

(4) $$\tan\theta = y/x \text{ if } x \neq 0$$

(5) $$\sin\theta = y/r \text{ if } (x, y) \neq (0, 0)$$

(6) $$\cos\theta = x/r \text{ if } (x, y) \neq (0, 0).$$

Example 1

Find (a) the rectangular coordinates of $[-2, \pi]$ and $[3, \pi/6]$ and (b) polar coordinates for $(5, 0)$ and $(-2\sqrt{3}, -2)$.

Solution. (a) For $[-2, \pi]$ we have from (1) and (2),

$$x = r\cos\theta = -2\cos\pi = (-2)(-1) = 2,$$

$$y = r\sin\theta = -2\sin\pi = 0.$$

FIGURE 6.4

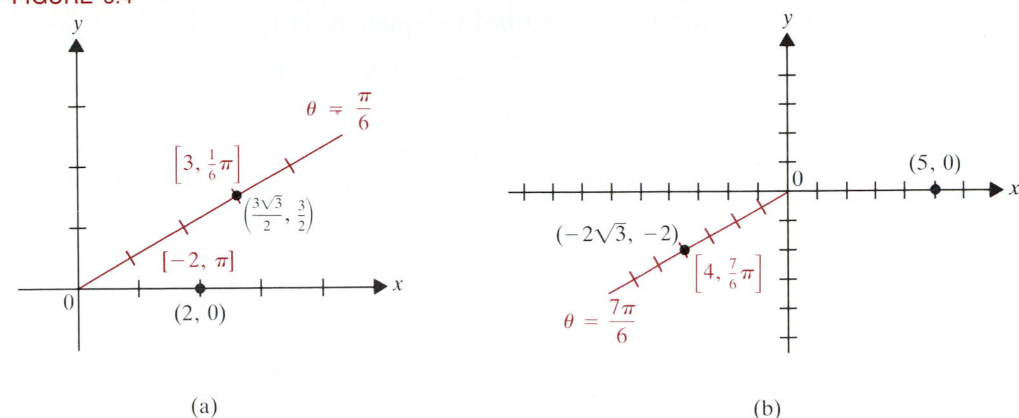

So $[-2, \pi] = (2, 0)$. For $[3, \pi/6]$ we have

$$x = r\cos\theta = 3\cos\frac{\pi}{6} = \frac{3\sqrt{3}}{2},$$

$$y = r\sin\theta = 3\sin\frac{\pi}{6} = \frac{3}{2}.$$

So $[3, \pi/6] = (3\sqrt{3}/2, 3/2)$.

(b) For (5, 0), in view of (3) we can use

$$r = \sqrt{x^2 + y^2} = \sqrt{25 + 0} = 5.$$

We also have from (4),

$$\tan\theta = \frac{y}{x} = 0,$$

so 0 is one value we can assign to θ. Then $(5, 0) = [5, 0]$, that is, the polar coordinates in this case coincide with the rectangular coordinates. (You should verify that this *always* happens for points on the x-axis if we take $\theta = 0$.) For $(-2\sqrt{3}, -2)$, we have

$$r = \sqrt{4 \cdot 3 + 4} = \sqrt{16} = 4,$$

and

$$\tan\theta = \frac{-2}{-2\sqrt{3}} = \frac{1}{\sqrt{3}}.$$

From this, θ is either $\pi/6$ or $7\pi/6$ (if we restrict θ to the range 0 to 2π). But since x and y are both negative, θ must be a third-quadrant angle, so $\theta = 7\pi/6$. Thus $(-2\sqrt{3}, -2) = [4, 7\pi/6]$. See **Figure 6.4.** ■

FIGURE 6.5

FIGURE 6.6

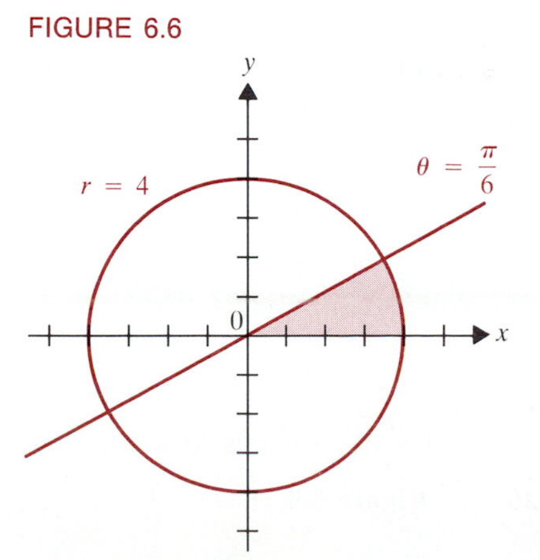

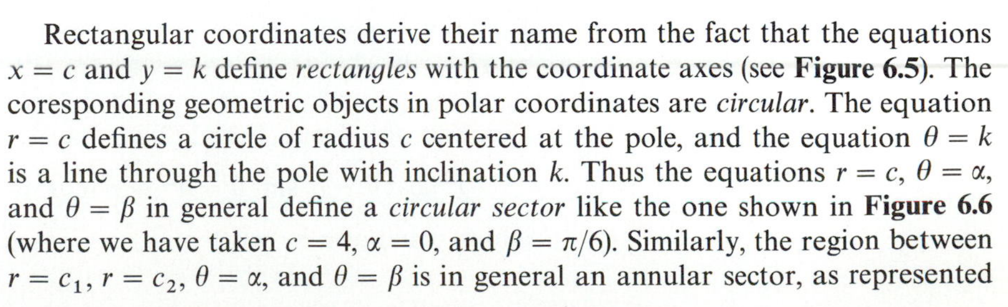

Rectangular coordinates derive their name from the fact that the equations $x = c$ and $y = k$ define *rectangles* with the coordinate axes (see **Figure 6.5**). The coresponding geometric objects in polar coordinates are *circular*. The equation $r = c$ defines a circle of radius c centered at the pole, and the equation $\theta = k$ is a line through the pole with inclination k. Thus the equations $r = c$, $\theta = \alpha$, and $\theta = \beta$ in general define a *circular sector* like the one shown in **Figure 6.6** (where we have taken $c = 4$, $\alpha = 0$, and $\beta = \pi/6$). Similarly, the region between $r = c_1$, $r = c_2$, $\theta = \alpha$, and $\theta = \beta$ is in general an annular sector, as represented

FIGURE 6.7

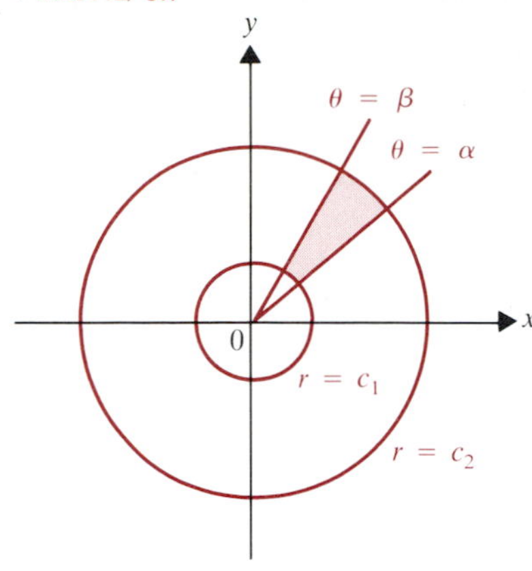

in **Figure 6.7.** These sectors correspond to rectangles in Cartesian coordinates, since they are described by equations of the form

$$\text{coordinate variable} = \text{constant}.$$

Besides circles centered at the pole, circles centered on the x-axis or y-axis and tangent to the other axis have simple equations. For example, consider the circle in **Figure 6.8**(a) centered at $(a/2, 0)$ with radius $a/2$. Its rectangular equation is

$$\left(x - \frac{1}{2}a\right)^2 + y^2 = \left(\frac{1}{2}a\right)^2 \to x^2 - ax + \frac{1}{4}a^2 + y^2 = \frac{1}{4}a^2 \to x^2 + y^2 = ax.$$

By (3) and (1), this transforms to the polar coordinate equation

(7) $$r^2 = ar\cos\theta.$$

For every point except the origin $r \neq 0$, so we can divide (7) through by r to get

(8) $$r = a\cos\theta.$$

When $\theta = \pi/2$, we also get the origin in (8) as the point $[0, \pi/2]$, so (8) is the equation of the given circle. The same approach shows that the polar coordinate equation for the circle in Figure 6.8(b) with center $(0, a/2)$ and radius $a/2$ is

(9) $$r = a\sin\theta.$$

FIGURE 6.8

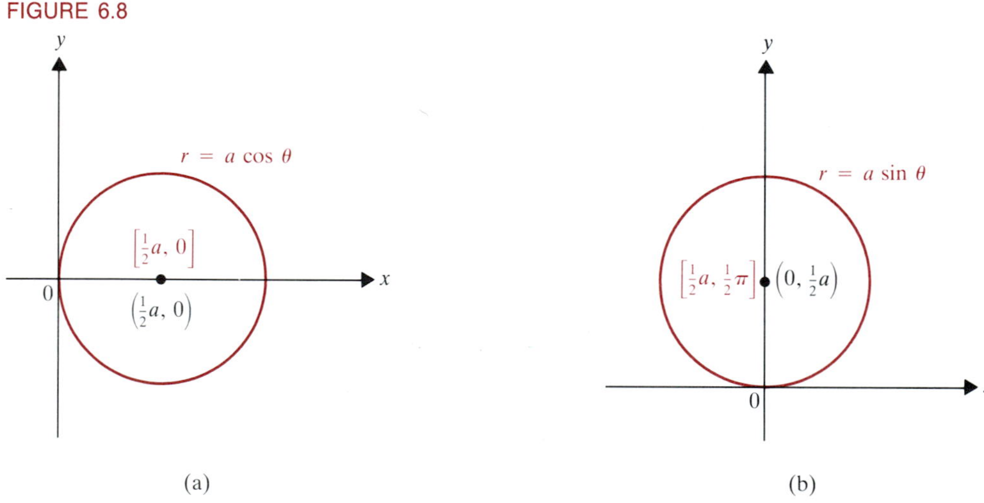

There are a few other types of rectangular equations that are easily converted to polar coordinates. The next example considers two of them.

Example 2

Find polar coordinates for (a) the hyperbola $x^2 - y^2 = a^2$ and (b) the lines $x = c$ and $y = d$.

Solution. (a) From (1) and (2) we have

$$r^2\cos^2\theta - r^2\sin^2\theta = a^2 \to r^2(\cos^2\theta - \sin^2\theta) = a^2 \to r^2\cos 2\theta = a^2,$$

which could also be written as $r^2 = a^2\sec 2\theta$. See **Figure 6.9**(a).

FIGURE 6.9

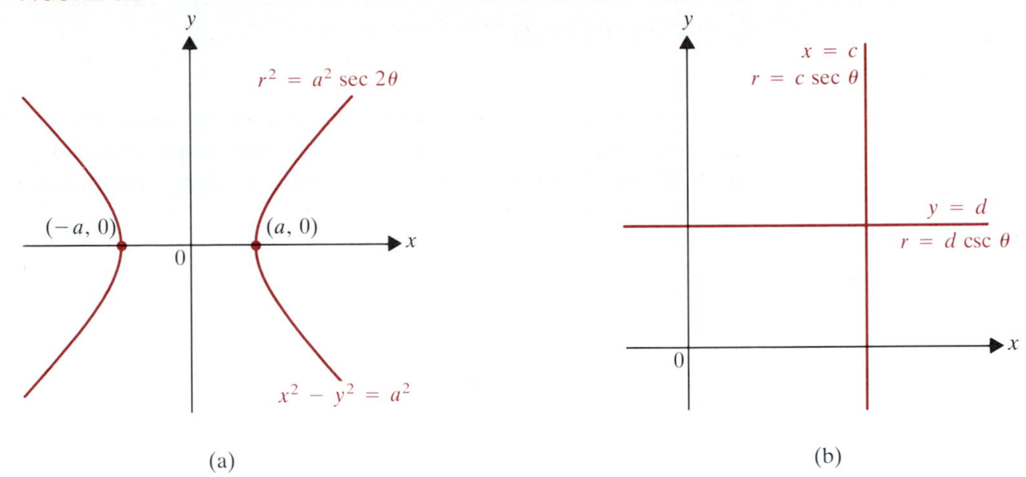

(b) Again, (1) and (2) can be used to transform the equations $x = c$ and $y = d$ to

$$r\cos\theta = c \rightarrow r = c\sec\theta \qquad \text{and} \qquad r\sin\theta = d \rightarrow r = d\csc\theta.$$

See Figure 6.9(b). ■

Before studying the graphs of polar equations, we give conditions for such graphs to be symmetric in an axis or relative to the pole. The proof is left for Exercises 59–61.

6.2 THEOREM

If an equation $f(r, \theta) = 0$ is unchanged when

(a) $[r, \theta]$ is replaced by $[r, -\theta]$ (or $[-r, -\theta + \pi]$), then the graph is symmetric in the polar axis ($\theta = 0$);

(b) $[r, \theta]$ is replaced by $[r, \pi - \theta]$ (or $[-r, -\theta]$), then the graph is symmetric in the y-axis ($\theta = \pi/2$);

(c) $[r, \theta]$ is replaced by $[-r, \theta]$ (or by $[r, \theta + \pi]$), then the graph is symmetric relative to the pole.

To illustrate Theorem 6.2, the graph of

$$r = \frac{Ed}{1 + E\cos\theta}$$

is symmetric in the x-axis by part (a), since $\cos(-\theta) = \cos\theta$, and the graph of

$$r = \frac{Ed}{1 + E\sin\theta}$$

is symmetric in the y-axis by part (b), since $\sin(\pi - \theta) = \sin\theta$. (These graphs will be discussed in more detail shortly.) Also by part (c), the graph of $r^2 = 4\cos 2\theta$ is symmetric in the pole, since the equation is unchanged when r is replaced by $-r$.

Example 3

Graph $r = 1 + \cos\theta$.

Solution. Since $\cos(-\theta) = \cos\theta$, Theorem 6.2(a) says that the graph is symmetric in the polar axis. It is only necessary then to plot points $[r, \theta]$ on the graph for θ between 0 and π. The following table of values is enough to draw the graph shown in **Figure 6.10.**

FIGURE 6.10

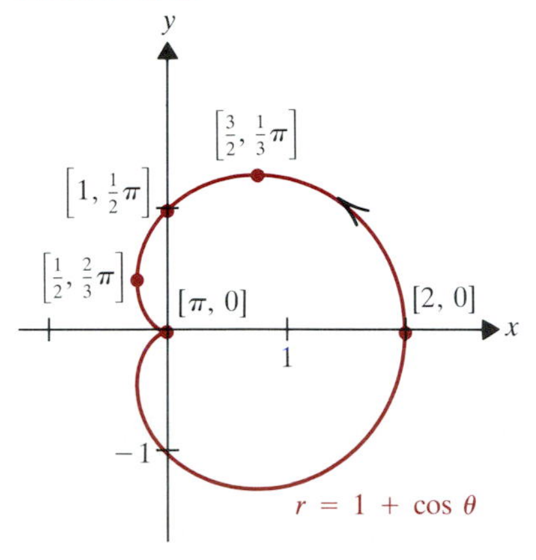

θ	0	$\pi/3$	$\pi/2$	$2\pi/3$	π
r	2	3/2	1	1/2	0

The graph is called a ***cardioid*** since it is roughly heart-shaped. The arrowhead in the figure indicates the direction in which the curve is traced as θ increases. ■

Another pleasing graph, called a ***four-petal rose***, is plotted in the next example, which shows that the parenthetical possibilities in Theorem 6.2 are not frivolous.

Example 4

Graph $r = 3\sin 2\theta$.

Solution. If we replace $[r, \theta]$ by $[r, -\theta]$, we do *not* get an equivalent equation, since

$$r = 3\sin(-2\theta) = -3\sin 2\theta$$

is not equivalent to $r = 3\sin 2\theta$. However, if we replace $[r, \theta]$ by $[-r, -\theta + \pi]$, then we get

$$-r = 3\sin 2(\pi - \theta) = 3\sin(2\pi - 2\theta) = -3\sin 2\theta,$$

which *is* equivalent to $r = 3\sin 2\theta$. Thus the equation is symmetric in the polar axis. It is also symmetric in the y-axis, because replacement of $[r, \theta]$ by $[-r, -\theta]$ gives

$$-r = 3\sin(-2\theta) = -3\sin 2\theta,$$

which is again equivalent to $r = 3\sin 2\theta$. Finally, the graph is also symmetric in the pole, since if we replace $[r, \theta]$ by $[r, \theta + \pi]$, we get

$$r = 3\sin 2(\theta + \pi) = 3\sin(2\theta + 2\pi) = 3\sin 2\theta.$$

In view of the high degree of symmetry present, it is enough to plot just the first-quadrant points in the following table.

θ	0	$\pi/12$	$\pi/6$	$\pi/4$	$\pi/3$	$5\pi/12$	$\pi/2$
r	0	3/2	$3\sqrt{3}/2$	3	$3\sqrt{3}/2$	3/2	0

The first-quadrant part of the graph has a symmetric image in the fourth quadrant, by symmetry in the polar axis. The first- and fourth-quadrant portions have symmetric images in the second and third quadrants, because of symmetry in the y-axis. The curve is shown in **Figure 6.11.** ■

FIGURE 6.11

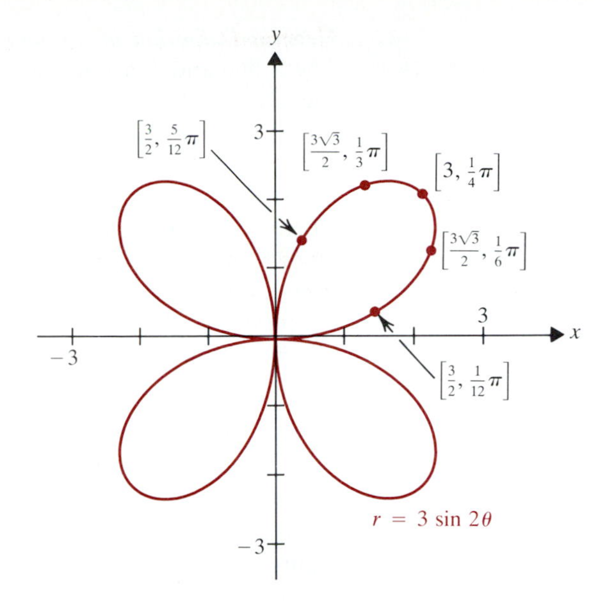

For any integer n, the graph of $r = a \sin n\theta$ resembles the graph in Example 4. When n is even, there are $2n$ petals. When n is odd, there are n petals. The graph of $r = a \cos n\theta$ behaves the same way. (See Exercises 35–38, 62, and 63.)

Commercially available polar coordinate paper, or a protractor, is helpful in plotting polar coordinate graphs like the one in Figure 6.11. The next example is one for which it is helpful to analyze the domain of values of θ.

Example 5

Graph $r^2 = 4 \cos 2\theta$.

Solution. We have symmetry in the polar axis, since $r^2 = 4\cos(-2\theta)$ is equivalent to $r^2 = 4\cos 2\theta$. We also have symmetry relative to the pole, since the equation is unchanged if we replace $[r, \theta]$ by $[-r, \theta]$. Thus it is again necessary to compile values only for θ in the first quadrant. The fourth-quadrant portion can be obtained from symmetry in the polar axis, the second- and third-quadrant pieces by reflecting the first and fourth quadrant parts in the origin. Because r^2 is never negative, there are *no points* for θ greater than $\pi/4$ in the first quadrant. For if $\theta \in (\pi/4, \pi/2)$, then

$$\frac{\pi}{2} < 2\theta < \pi,$$

so $\cos 2\theta$ is negative and hence cannot equal r^2. From the table of values

θ	0	$\pi/12$	$\pi/8$	$\pi/6$	$\pi/4$
r	± 2	$\pm 2(\sqrt{3}/2)^{1/2}$	$\pm 2^{3/4}$	$\pm\sqrt{2}$	0

we obtain the graph shown in **Figure 6.12.** The curve is called a ***lemniscate*** and looks like an infinity sign. ■

FIGURE 6.12

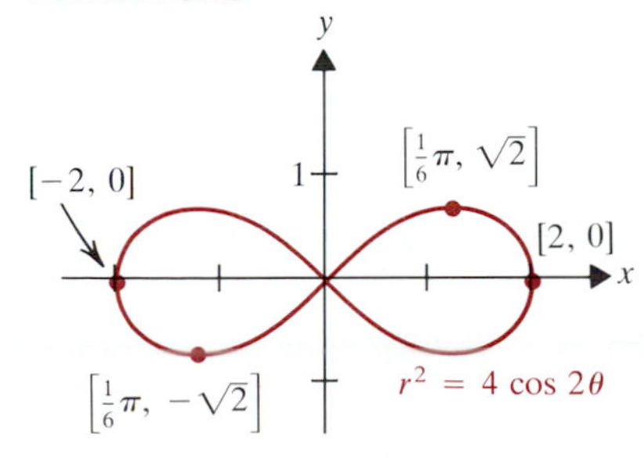

The conic sections discussed in Sections 1, 2, and 3 have a common polar equation. It is easily derived from Definitions 1.1, 2.1, and 3.1, which we combine as follows.

6.3 DEFINITION

General Definition of a Conic Section. Let $E \geq 0$ be a fixed real number. Let l be a line and F a point not on l. A ***conic section*** is the set of all points P in the plane such that

$$\frac{d(P, F)}{d(P, l)} = E,$$

where $d(P, F)$ is the distance from P to F, and $d(P, l)$ is the perpendicular distance from P to the line l (see **Figure 6.13**). The constant E is called the ***eccentricity,*** F is called a ***focus,*** and l is called a ***directrix.*** If $E = 1$, the conic section is a ***parabola.*** If $E < 1$, the conic section is an ***ellipse.*** If $E > 1$, the conic section is a ***hyperbola.***

FIGURE 6.13

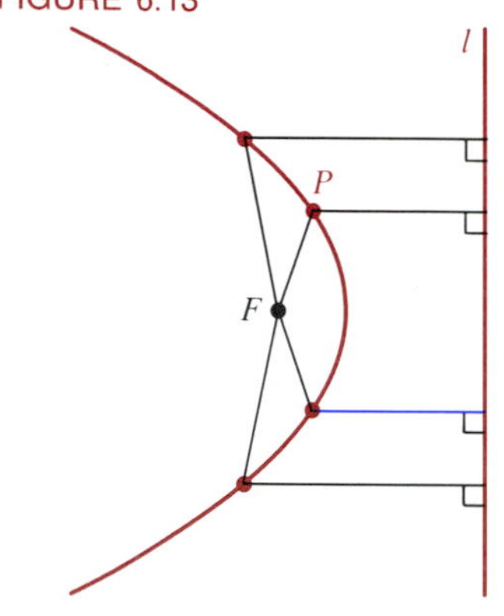

Referring back to the earlier definitions of parabola, ellipse, and hyperbola, we see that Definition 6.3 is simply a synthesis of those earlier definitions. What's more, it makes it possible to derive polar equations for all the conic sections (except the circle) simultaneously.

To do so, we suppose that the directrix is the vertical line $x = d$ and refer to **Figure 6.14.** Let P be any point on C. Let Q be the point of intersection with $x = d$ of the horizontal line drawn from P, which is then perpendicular to the directrix. From Definition 6.3 we have

$$\frac{|0P|}{|PQ|} = E,$$

FIGURE 6.14

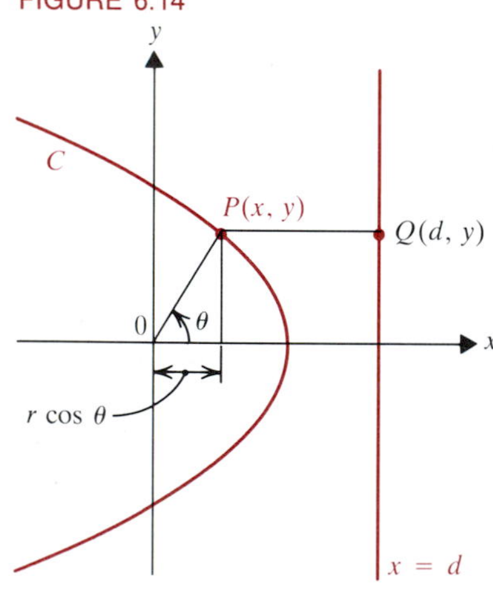

that is,

$$r = E|PQ| = E(d - x) = E(d - r\cos\theta) = Ed - Er\cos\theta.$$

Thus

(10) $$r + rE\cos\theta = Ed \rightarrow r(1 + E\cos\theta) = Ed \rightarrow r = \frac{Ed}{1 + E\cos\theta}.$$

In case the directrix is $x = -d$, a similar argument produces the equation

$$r = \frac{Ed}{1 - E\cos\theta}.$$

Using this approach, we can similarly show (Exercise 64) that if C has eccentricity E, focus F at the pole, and horizontal directrix $y = d$ or $y = -d$, then it has polar coordinate equation

(11) $$r = \frac{Ed}{1 \pm E\sin\theta},$$

where the plus sign applies if the directrix is $y = d$ and the minus sign applies if the directrix is $y = -d$.

Example 6

Identify, find the vertex or vertices, and draw a rough sketch of the graph of

$$r = \frac{18}{3 - 6\cos\theta}.$$

Solution. First divide the numerator and denominator by 3 to put the equation into the standard form (10):

$$r = \frac{6}{1 - 2\cos\theta} = \frac{Ed}{1 - E\cos\theta}.$$

As remarked following Theorem 6.2, the graph is symmetric in the polar axis. Here $E = 2 > 1$, so the conic section is a hyperbola with focus at the pole. The directrix is $x = -d$, where $Ed = 6$, so $d = 6/E = 6/2 = 3$. The vertices lie on the x-axis and so arise when $\theta = 0$ or π. Substituting $\theta = 0$ into the equation, we get $r = \dfrac{6}{1-2} = -6$, so one vertex is $[-6, 0]$. Substituting $\theta = \pi$, we find for the other vertex $r = \dfrac{6}{1+2} = 2$. Hence, that vertex is $P[2, \pi] = P(-2, 0)$. By letting $\theta = \pi/2$ and $3\pi/2$, we can locate the points $(0, \pm 6)$ on the right branch. By symmetry, there are points on the left branch at the same distances from the polar axis, as far to the left of the vertex $(-6, 0)$ as the y-axis is to the right of the vertex $(-2, 0)$. Thus the points $(-8, \pm 6)$ are also on the curve. (We could also have found additional points by assigning values to θ and evaluating r for the equation.) The graph is sketched in **Figure 6.15.** ■

FIGURE 6.15

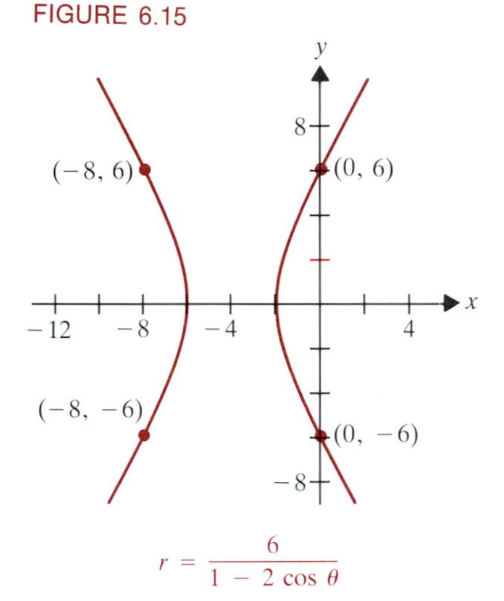

Having to exclude the circle from Definition 6.3 is not a serious problem, because we already have the simple equations $r = c$, $r = a\cos\theta$, and $r = a\sin\theta$ for circles.

Polar coordinates are more than just superficially different from rectangular coordinates. They are designed for use with curves that possess a high degree of central symmetry. They are *not* well suited to the simplest of all curves, straight lines. As Example 2(b) brings out, simple Cartesian equations for lines tend to transform to more complex polar equations. The line $y = mx + b$, for instance, becomes

$$r\sin\theta = mr\cos\theta + b \to r(\sin\theta - m\cos\theta) = b,$$

$$r = \frac{b}{\sin\theta - m\cos\theta} = l(\theta),$$

a *complicated* polar function. For this reason there is little to gain by finding tangent lines to the graphs of polar functions $r = f(\theta)$, because complexity rather than simplification would result from approximating a polar curve by its tangent line. Thus, while formulas exist for the slope of a polar curve, we do not discuss them here.

HISTORICAL NOTE

Polar coordinates grew out of the work of Abraham DeMoivre (1667–1754), who was born in Vitry, France. At an early age his family emigrated to England, where he pursued a career in mathematics. If you studied complex numbers $z = x + iy$ in high school, then you may recall a formula for z^n that bears his name. That is a consequence of the basic formulas $x = r\cos\theta$ and $y = r\sin\theta$ of polar coordinates.

Exercises 8.6

In Exercises 1 and 2, plot the points having the given polar coordinates.

1. **(a)** $[2, \pi/3]$ **(b)** $[-2, \pi/2]$ **(c)** $[-3, \pi]$ **(d)** $[1, 2\pi/3]$ **(e)** $[0, 7\pi/20]$

2. **(a)** $[-1, \pi/6]$ **(b)** $[3, -5\pi/4]$ **(c)** $[2, \pi]$ **(d)** $[-2, -\pi/3]$ **(e)** $[0, -8\pi/5]$

3. Find the Cartesian coordinates of the points in Exercise 1.

4. Find the Cartesian coordinates of the points in Exercise 2.

In Exercises 5 and 6, find polar coordinates for the points having the given rectangular coordinates.

5. **(a)** $(1, 2)$ **(b)** $(0, 2)$ **(c)** $(2, 3)$ **(d)** $(-2, -3)$ **(e)** $(2, -3)$

6. **(a)** $(-1, 1)$ **(b)** $(2, -1)$ **(c)** $(2, 0)$ **(d)** $(-2, 3)$ **(e)** $(1, 2)$

In Exercises 7–16, find a polar equation for the given rectangular equation.

7. $x^2 + y^2 = 9$

8. $x^2 + y^2 = 2$

9. $x^2 - y^2 = 4$

10. $y^2 - x^2 = 9$

11. $(x - 2)^2 + y^2 = 4$

12. $(x + 6)^2 + y^2 = 36$

13. $x^2 + (y + 4)^2 = 16$

14. $x^2 + (y - 10)^2 = 100$

15. $4x^2 + y^2 = 16$

16. $x^2 + 9y^2 = 36$

In Exercises 17–32, find a rectangular equation for the given polar equation.

17. $\theta = \pi/6$

18. $\theta = 2\pi/3$

19. $r = 5 \sec \theta$

20. $r + 2 \csc \theta = 0$

21. $r = 6$

22. $r^2 - 9 = 0$

23. $r - 2 \sin \theta = 0$

24. $r + 4 \cos \theta = 0$

25. $r = 1 + \cos \theta$

26. $r = 3 - \cos \theta$

27. $r^2 = \sec 2\theta$

28. $r^2 = 4 \sec 2\theta$

29. $r = \dfrac{1}{\sin \theta + 2 \cos \theta}$

30. $r = \dfrac{-3}{5 \sin \theta - 2 \cos \theta}$

31. $r = \dfrac{1}{1 + \sin \theta}$

32. $r = \dfrac{4}{2 - \cos \theta}$

In Exercises 33–50, draw the graph of the given polar equation. Identify the graph if it is not named.

33. $r = 1 + 2 \cos \theta$ (limaçon)

34. $r = 1 - 3 \sin \theta$ (limaçon)

35. $r = 3 \cos 2\theta$ (four-petal rose)

36. $r = 3 \cos 3\theta$ (three-petal rose)

37. $r = 3 \sin 5\theta$ (five-petal rose)

38. $r = 3 \sin 4\theta$ (eight-petal rose)

39. $r = \theta$ (spiral of Archimedes)

40. $r = 2\theta$ (spiral of Archimedes)

41. $r = 1 - \sin \theta$

42. $r = 4(1 - \cos \theta)$

43. $r\theta = 1$ (hyperbolic spiral)

44. $r = e^{\theta}$ (logarithmic spiral)

45. $r^2 = 4 \sin 2\theta$

46. $r^2 = 9 \cos 2\theta$

47. $r = \dfrac{2}{1 - \sin \theta}$

48. $r = \dfrac{4}{2 + \cos \theta}$

49. $r = \dfrac{2}{1 + 4 \cos \theta}$

50. $r = \dfrac{2}{1 + 4 \sin \theta}$

In Exercises 51–58, find a polar equation for the given conic section.

51. Focus at the pole, directrix $x = -4$, $E = 1/2$

52. Focus at the pole, directrix $y = 3$, $E = 2/3$

53. Hyperbola, focus at pole, directrix $x = -4$, vertex $(-3, 0)$

54. Hyperbola, focus at pole, directrix $y = 3$, vertex $(0, 2)$

55. Parabola, focus at pole, directrix $x = -3$

56. Parabola, focus at pole, directrix $y = 4$

57. Hyperbola, focus at pole, directrix $y = 3$, vertex $(0, 2)$

58. Hyperbola, focus at pole, directrix $x = 6$, vertex $(4, 0)$

59. Prove Theorem 6.2(a).

60. Prove Theorem 6.2(b).

61. Prove Theorem 6.2(c).

62. Derive Equation (11) for a conic section with focus on the y-axis and horizontal directrix.

PC **63.** If you have access to a computer graphics program for polar coordinates, then use it to graph $r = a \cos n\theta$ for $n = 1, 2, 3, \ldots, 10, 20, 40$, and 80. What happens to the curve?

PC **64.** Repeat Exercise 63 for $r = a \sin n\theta$.

8.7 Area and Arc Length in Polar Coordinates

We saw in the last section that many kinds of curves are best treated via polar coordinates. A case in point is the cardioid $r = 1 + \cos \theta$ of Example 3. If we translate to Cartesian coordinates, we obtain

$$\sqrt{x^2 + y^2} = 1 + \frac{x}{\sqrt{x^2 + y^2}} \rightarrow x^2 + y^2 = \sqrt{x^2 + y^2} + x,$$

$$x^2 + y^2 - x = \sqrt{x^2 + y^2},$$

$$x^4 + y^4 + x^2 + 2x^2y^2 - 2x^3 - 2xy^2 = x^2 + y^2, \tag{1}$$

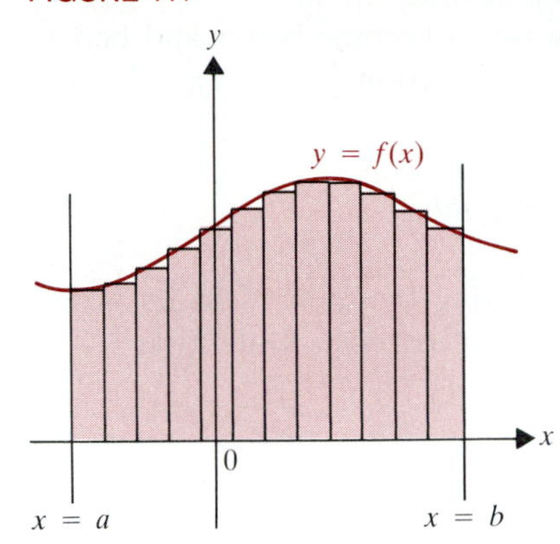

FIGURE 7.1

a *very* complicated equation indeed. If we wish to find the area enclosed by the cardioid, then it certainly seems preferable to work directly with the simple polar equation. In this section we will develop a polar coordinate integration formula to make that possible.

First, recall the Cartesian coordinate formula for area. If f is a continuous function on an interval $[a, b]$ and $f(x) \geq 0$ on $[a, b]$, then the area lying under the graph of $y = f(x)$ between $x = a$ and $x = b$ (see **Figure 7.1**) is given by

$$\int_a^b f(x)\,dx.$$

In seeking an analogous formula for curves with polar coordinate equations, we consider a corresponding function,

$$r = g(\theta). \tag{2}$$

We suppose that g is continuous, and we let R be the region bounded by the graph of (2) between the two rays $\theta = \alpha$ and $\theta = \beta$, as shown in **Figure 7.2.** (Here $\alpha < \beta$.)

FIGURE 7.2
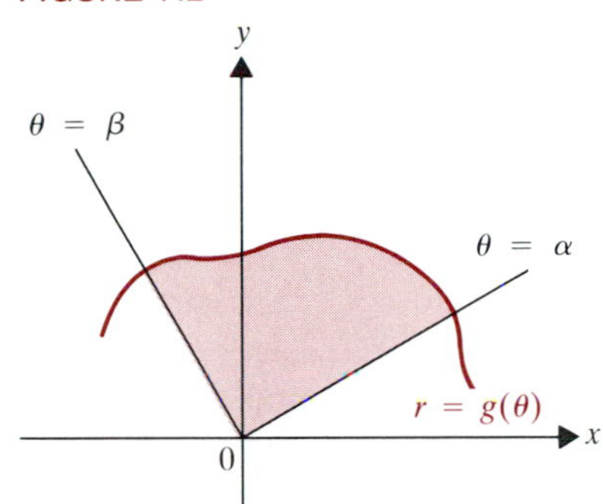

FIGURE 7.3
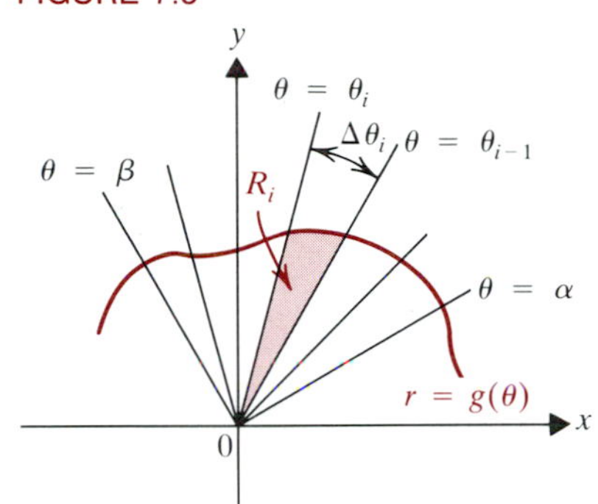

We next partition the interval $[\alpha, \beta]$ by choosing partition points

$$\theta_0 = \alpha < \theta_1 < \theta_2 < \ldots < \theta_{i-1} < \theta_i < \ldots < \theta_n = \beta,$$

and construct the rays $\theta = \theta_i$ for $i = 1, \ldots, n - 1$. Those rays divide R into subregions R_i bounded by the graph of $r = g(\theta)$ (see **Figure 7.3**). Let

$$\Delta\theta_i = \theta_i - \theta_{i-1},$$

and let ΔA_i stand for the area of the subregion R_i. Then

$$A = \Delta A_i + \Delta A_2 + \ldots + \Delta A_n.$$

FIGURE 7.4
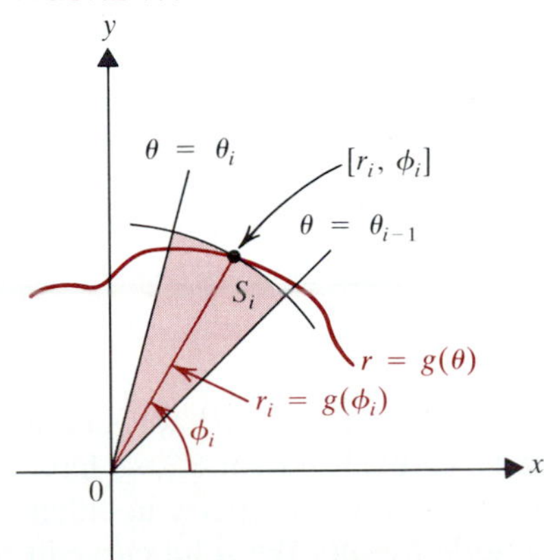

The area ΔA_i is approximated by choosing a value ϕ_i in the interval $[\theta_{i-1}, \theta_i]$ and considering the circular sector S_i of radius $g(\phi_i)$ and central angle $\Delta\theta_i$ shown in **Figure 7.4.** The area of S_i approximates the area of R_i. Since S_i has radius $r_i = g(\phi_i)$, Lemma 3.9 of Chapter 2 says that its area is

$$\frac{1}{2} r_i^2\,\Delta\theta_i = \frac{1}{2}[g(\phi_i)]^2\,\Delta\theta_i. \tag{3}$$

Thus A is approximated by the sum of all such approximations:

$$A \approx \sum_{i=1}^{n} \frac{1}{2} r_i^2\,\Delta\theta_i = \sum_{i=1}^{n} \frac{1}{2}[g(\phi_i)]^2\,\Delta\theta_i. \tag{4}$$

If we take finer and finer partitions of $[\alpha, \beta]$, so that the largest $\Delta\theta_i$ shrinks toward 0, then we expect the approximation (4) to become better and better. The right side of (4) is a Riemann sum for the function $\frac{1}{2}r^2 = \frac{1}{2}[g(\theta)]^2$, so it approaches

$$\int_\alpha^\beta \frac{1}{2} r^2 \, d\theta = \frac{1}{2} \int_\alpha^\beta [g(\theta)]^2 \, d\theta.$$

This suggests the following theorem for calculating the area of R. Its proof, which must show that the formula for A is consistent with Definition 1.4 of Chapter 4, is given in higher mathematics.

7.1 THEOREM

Suppose that g is a continuous function on an interval $[\alpha, \beta]$. Then the area of A of the region R bounded by the graph of $r = g(\theta)$ between $\theta = \alpha$ and $\theta = \beta$ is

$$A = \int_\alpha^\beta \frac{1}{2} r^2 \, d\theta = \frac{1}{2} \int_\alpha^\beta [g(\theta)]^2 \, d\theta. \tag{5}$$

With this result, we can solve the area problem posed at the start of the section.

Example 1

Find the area enclosed by the cardioid $r = 1 + \cos\theta$.

Solution. Recall from Example 3 of the last section that the curve is traced out as θ varies from 0 to 2π and is symmetric in the polar axis. Hence the total area inside the cardioid is twice the area of the shaded region in **Figure 7.5.** We therefore have

FIGURE 7.5

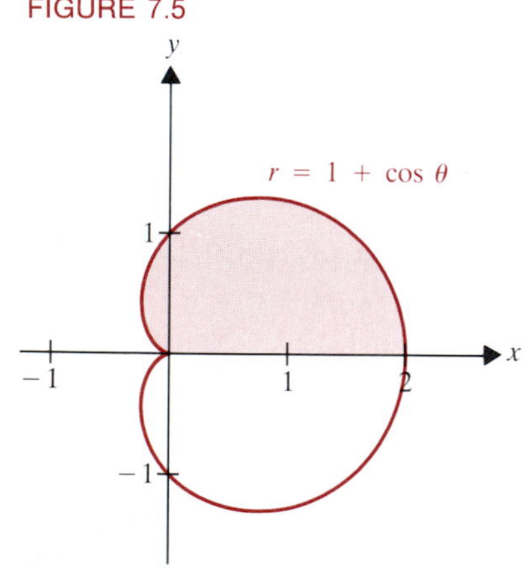

$$\begin{aligned} A &= \int_0^{2\pi} \frac{1}{2}(1 + \cos\theta)^2 \, d\theta = 2 \int_0^{\pi} \frac{1}{2}(1 + \cos\theta)^2 \, d\theta \\ &= \int_0^{\pi} (1 + 2\cos\theta + \cos^2\theta) \, d\theta \\ &= \int_0^{\pi} \left(1 + 2\cos\theta + \frac{1 + \cos 2\theta}{2}\right) d\theta = \int_0^{\pi} \left(\frac{3}{2} + 2\cos\theta + \frac{1}{2}\cos 2\theta\right) d\theta \\ &= \frac{3}{2}\theta + 2\sin\theta + \frac{1}{4}\sin 2\theta \Big]_0^{\pi} = \frac{3\pi}{2}. \quad \blacksquare \end{aligned}$$

In the next example we again make effective use of symmetry.

Example 2

Find the area of the region enclosed by the three-petal rose $r = 2\sin 3\theta$.

Solution. First we graph the curve. If we replace $[r, \theta]$ by $[-r, -\theta]$, we get the equivalent equation $-r = 2\sin(-3\theta) = -2\sin 3\theta$, so by Theorem 6.2 we have symmetry in the y-axis. (You can check that we don't have symmetry in either the polar axis or the pole.) From the following table we plot the solid curve in

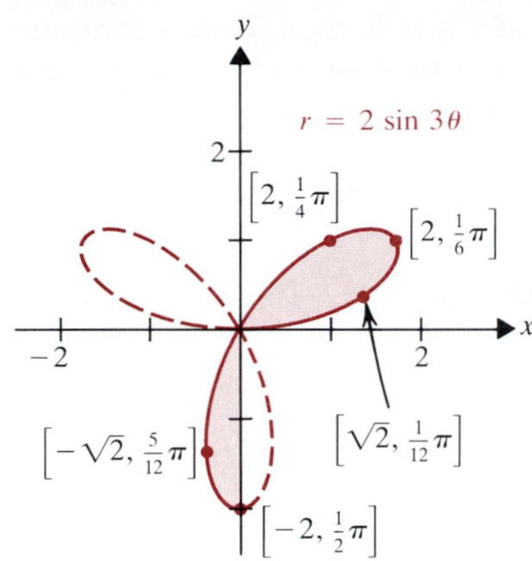

FIGURE 7.6

Figure 7.6, which can be completed by using symmetry in

θ	0	$\pi/12$	$\pi/6$	$\pi/4$	$\pi/3$	$5\pi/12$	$\pi/2$
r	0	$\sqrt{2}$	2	$\sqrt{2}$	0	$-\sqrt{2}$	-2

the y-axis to fill in the dashed portion. The area enclosed by the entire curve is twice the area generated as θ goes from 0 to $\pi/2$. We thus have from (5)

$$A = 2\int_0^{\pi/2} \frac{1}{2} r^2\, d\theta = \int_0^{\pi/2} 4\sin^2 3\theta\, d\theta = 4\int_0^{\pi/2} \frac{1-\cos 6\theta}{2}\, d\theta$$

$$= 2\left[\theta - \frac{1}{6}\sin 6\theta\right]_0^{\pi/2} = 2\left[\frac{1}{2}\pi - \frac{1}{6}\cdot 0 - 0 + 0\right] = \pi. \blacksquare$$

When integrating polar coordinate functions, **do not make the common error of thinking that θ always goes from 0 to 2π.** In Example 2 the curve is traced out completely as θ varies over the interval $[0, \pi]$. We were able to compute the total enclosed area A as twice the area of the shaded region in Figure 7.6. We would have gotten the same answer by computing

$$A = \int_0^{\pi} \frac{1}{2} r^2\, d\theta = \int_0^{\pi} 2\sin^2 3\theta\, d\theta,$$

as you can check. But if we had integrated over $[0, 2\pi]$, we would have gotten *twice* the correct area. The reason is that the entire curve is generated *twice* over that interval: once over $[0, \pi]$ and once more over $[\pi, 2\pi]$. By integrating over $[0, 2\pi]$, we would get the area A generated as θ goes from 0 to π *plus* the area A again, generated as θ goes from π to 2π. The best way to avoid such errors is to *graph the curve carefully in every problem.* The next example illustrates this further.

Example 3

Find the area enclosed by the inner loop of the ***limaçon*** $r = 1 + 2\cos\theta$.

Solution. The graph of $r = 1 + 2\cos\theta$ is symmetric in the polar axis. From the table of values

θ	0	$\pi/4$	$\pi/3$	$\pi/2$	$2\pi/3$	$3\pi/4$	$5\pi/6$	π
r	3	$1+\sqrt{2}$	2	1	0	$1-\sqrt{2}$	$1-\sqrt{3}$	-1

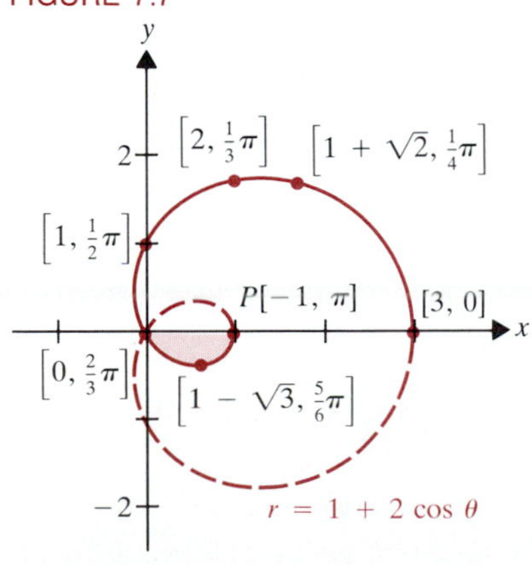

FIGURE 7.7

we plot the solid curve in **Figure 7.7** and use the symmetry to finish it by drawing in the dashed portion. The area inside the inner loop is twice the area of the shaded portion, which is generated as θ varies from $2\pi/3$ to π. Thus by (5),

$$A = 2\int_{2\pi/3}^{\pi} \frac{1}{2}(1 + 2\cos\theta)^2\, d\theta = \int_{2\pi/3}^{\pi} (1 + 4\cos\theta + 4\cos^2\theta)\, d\theta$$

$$= \theta + 4\sin\theta\Big]_{2\pi/3}^{\pi} + 2\int_{2\pi/3}^{\pi} (1 + \cos 2\theta)\, d\theta$$

$$= \theta + 4\sin\theta + 2\theta + \sin 2\theta\Big]_{2\pi/3}^{\pi} = 3\theta + 4\sin\theta + \sin 2\theta\Big]_{2\pi/3}^{\pi}$$

$$= 3\pi + 0 + 0 - 2\pi - 4\cdot\frac{\sqrt{3}}{2} - \left(-\frac{\sqrt{3}}{2}\right) = \pi - \frac{3}{2}\sqrt{3}. \blacksquare$$

FIGURE 7.8

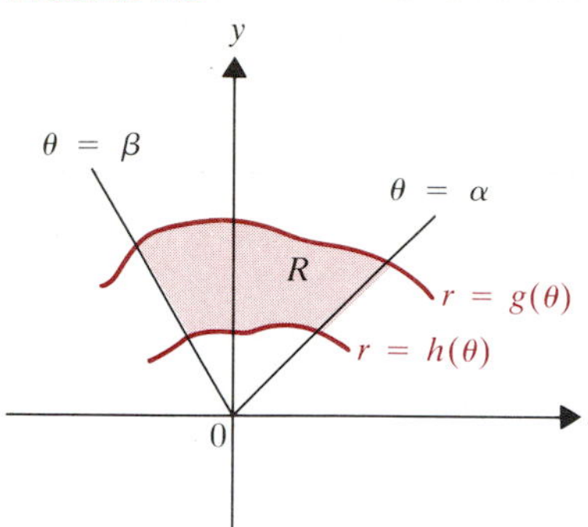

If we have a region R, like the one in **Figure 7.8** that lies between the graphs of $r = g(\theta)$, $r = h(\theta)$, $\theta = \alpha$, and $\theta = \beta$, where $g(\theta) \geq h(\theta)$, then we can compute the area of R as the *difference* between the areas of the regions bounded by $g(\theta)$ and $h(\theta)$ as θ varies from α to β. Theorem 7.1 then gives

$$A = \int_\alpha^\beta \frac{1}{2}[g(\theta)]^2\,d\theta - \int_\alpha^\beta \frac{1}{2}[h(\theta)]^2\,d\theta,$$

$$A = \frac{1}{2}\int_\alpha^\beta ([g(\theta)]^2 - [h(\theta)]^2)\,d\theta. \tag{6}$$

In applying (6) it is often necessary to find the points of intersection of two polar curves to determine α and β. This is not as easy to do algebraically as it is for Cartesian curves, because the same point P may be on two polar curves for *different* values of θ. (If we think of the curves as being traced out parametrically with θ as parameter, an intersection point P may be reached at *different* stages. When you enter a class, your path may intersect a friend's, but you *won't* collide if you pass through the point at different times!) As a simple example, the point $P[1, 0]$ clearly lies on the limaçon $r = 1 + 2\cos\theta$ in Figure 7.7, but those coordinates do *not* satisfy the equation since when $\theta = 0$, $r = 3$. Rather, we get P as the point with polar coordinates $[-1, \pi]$, when $\theta = \pi$. If we tried to find all the points when the limaçon intersects the polar axis by solving the system of equations

$$\begin{cases} \theta = 0 \\ r = 1 + 2\cos\theta, \end{cases} \tag{7}$$

we would find *only* the point $[3, 0]$. We would miss both the point P and the pole (with coordinates $[0, 2\pi/3]$). To avoid this kind of difficulty, **always draw a figure.** It will then be clear, as in Figure 7.7, if any points of intersection of the curves are missed in solving two equations like (7). The next example illustrates this.

Example 4

Find the area common to the two cardioids $r = 1 + \cos\theta$ and $r = 1 - \cos\theta$.

FIGURE 7.9

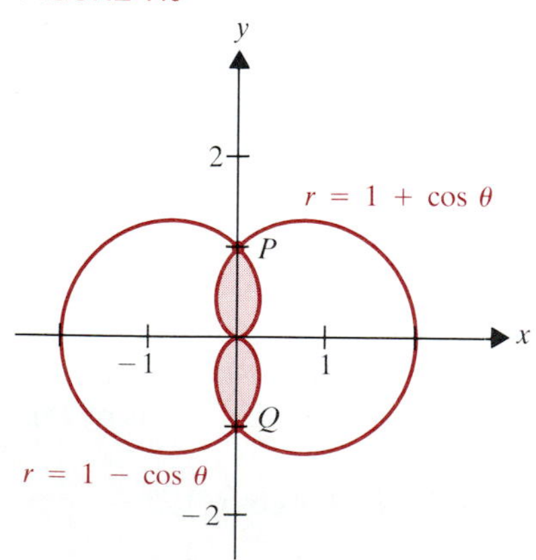

Solution. The area is shaded in **Figure 7.9,** which shows three points of intersection: the pole and points P and Q where $\theta = \pi/2$ and $3\pi/2$. If we were to try to find these points algebraically, we would solve

$$1 + \cos\theta = 1 - \cos\theta \rightarrow 2\cos\theta = 0 \rightarrow \theta = \pi/2 \text{ or } 3\pi/2.$$

So we would find P and Q, but we *wouldn't* get 0. It comes from $\theta = \pi$ on $r = 1 + \cos\theta$, but from $\theta = 0$ on $r = 1 - \cos\theta$.

Having found the intersections, we use symmetry to find the area. It is twice the area of the portion above the polar axis, which in turn is twice that of the first-quadrant region. Thus

$$A = 4\int_0^{\pi/2} \frac{1}{2}(1 - \cos\theta)^2\,d\theta = 2\int_0^{\pi/2} (1 - 2\cos\theta + \cos^2\theta)\,d\theta$$

$$= 2\Big[\theta - 2\sin\theta\Big]_0^{\pi/2} + \int_0^{\pi/2} (1 + \cos 2\theta)\,d\theta$$

$$= 2\left[\frac{\pi}{2} - 2\right] + \left[\theta + \frac{1}{2}\sin 2\theta\right]_0^{\pi/2} = \pi - 4 + \frac{\pi}{2} + 0 - 0 - 0$$

$$= \frac{3}{2}\pi - 4. \quad ■$$

A common type of area problem involving two curves is the subject of the next example.

Example 5

Find the area of the region inside the limaçon $r = \frac{1}{2} + \cos\theta$ and outside the circle $r = 1$.

FIGURE 7.10

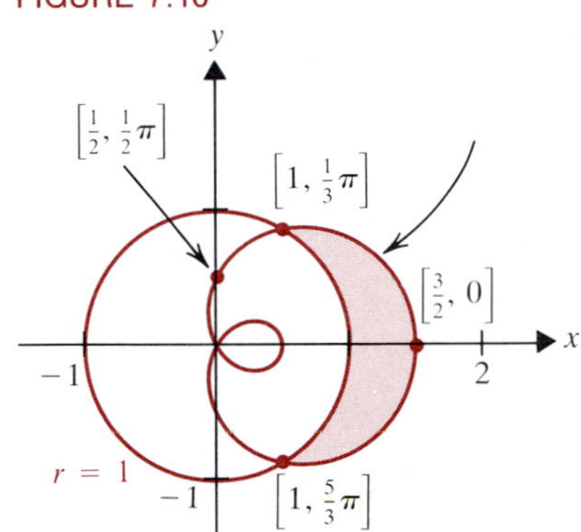

Solution. For each θ, the corresponding r value on the limaçon is half the value found in Example 3. The region is the shaded portion of **Figure 7.10.** To try to find the coordinates of the points of intersection, we equate the two expressions for r. This gives both the intersection points visible in the figure:

$$1 = \frac{1}{2} + \cos\theta \rightarrow \frac{1}{2} = \cos\theta \rightarrow \theta = \pi/3 \text{ or } 5\pi/3.$$

The area of the shaded region is twice the area of its first-quadrant portion, so

$$A = 2\int_0^{\pi/3} \frac{1}{2}\left[\left(\frac{1}{2} + \cos\theta\right)^2 - 1^2\right] d\theta = \int_0^{\pi/3} \left(\frac{1}{4} + \cos\theta + \cos^2\theta - 1\right) d\theta$$

$$= \int_0^{\pi/3} \left[-\frac{3}{4} + \cos\theta + \frac{1}{2}\left(1 + \cos 2\theta\right)\right] d\theta$$

$$= \int_0^{\pi/3} \left(-\frac{1}{4} + \cos\theta + \frac{1}{2}\cos 2\theta\right) d\theta = -\frac{\theta}{4} + \sin\theta + \frac{\sin 2\theta}{4}\bigg|_0^{\pi/3}$$

$$= -\frac{\pi}{12} + \frac{\sqrt{3}}{2} + \frac{\sqrt{3}}{8} = \frac{5}{8}\sqrt{3} - \frac{\pi}{12} \approx 0.8207. \quad \blacksquare$$

We conclude the section by deriving a formula for calculating the arc length of a smooth polar curve. Suppose that the curve is the graph of the function $r = f(\theta)$ for $\alpha \leq \theta \leq \beta$, where we assume that $df/d\theta$ is continuous. We can parametrize the curve by using the change-of-coordinate formulas

$$x = r\cos\theta \quad \text{and} \quad y = r\sin\theta,$$

together with the fact that on the curve $r = f(\theta)$. We then get the parametric equations

$$\begin{cases} x = r\cos\theta = f(\theta)\cos\theta \\ y = r\sin\theta = f(\theta)\sin\theta \end{cases}, \quad \theta \in [\alpha, \beta]. \tag{8}$$

Using the product rule and the chain rule for derivatives, we get from (8)

$$\frac{dx}{d\theta} = \frac{dr}{d\theta}\cos\theta - r\sin\theta \quad \text{and} \quad \frac{dy}{d\theta} = \frac{dr}{d\theta}\sin\theta + r\cos\theta.$$

Then

$$\left(\frac{dx}{d\theta}\right)^2 = \left(\frac{dr}{d\theta}\right)^2 \cos^2\theta - 2r\frac{dr}{d\theta}\sin\theta\cos\theta + r^2\sin^2\theta,$$

$$\left(\frac{dy}{d\theta}\right)^2 = \left(\frac{dr}{d\theta}\right)^2 \sin^2\theta + 2r\frac{dr}{d\theta}\sin\theta\cos\theta + r^2\cos^2\theta.$$

Therefore, since $\sin^2\theta + \cos^2\theta = 1$, we have

$$\left(\frac{dx}{d\theta}\right)^2 + \left(\frac{dy}{d\theta}\right)^2 = \left(\frac{dr}{d\theta}\right)^2 + r^2.$$

FIGURE 7.11

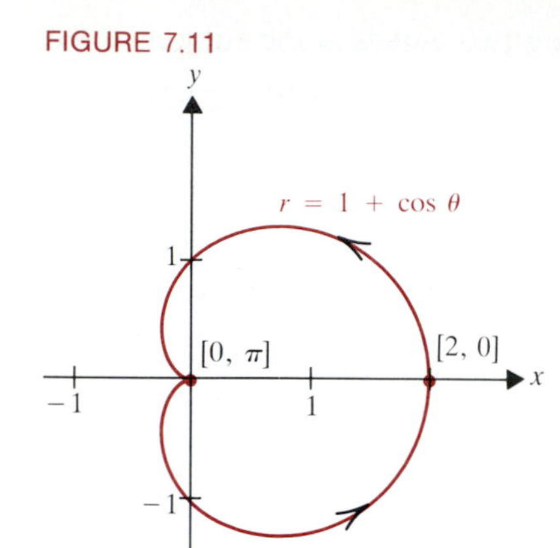

Then Theorem 5.2 gives

(9) $$L = \int_\alpha^\beta \sqrt{\left(\frac{dx}{d\theta}\right)^2 + \left(\frac{dy}{d\theta}\right)^2}\, d\theta = \int_\alpha^\beta \sqrt{\left(\frac{dr}{d\theta}\right)^2 + r^2}\, d\theta.$$

Example 6

Find the arc length of the cardioid $r = 1 + \cos\theta$ (**Figure 7.11**).

Solution. First we find $dr/d\theta = -\sin\theta$. Then we calculate

$$\left(\frac{dr}{d\theta}\right)^2 + r^2 = \sin^2\theta + 1 + 2\cos\theta + \cos^2\theta = 2 + 2\cos\theta.$$

Since the cardioid is traced out as θ goes from 0 to 2π, we get from (9),

(10) $$L = \int_0^{2\pi} \sqrt{2 + 2\cos\theta}\, d\theta = \sqrt{2}\int_0^{2\pi} \sqrt{1 + \cos\theta}\, d\theta.$$

Now

$$\cos^2\frac{1}{2}\theta = \frac{1 + \cos\theta}{2} \rightarrow 2\cos^2\frac{1}{2}\theta = 1 + \cos\theta.$$

Hence $\sqrt{1 + \cos\theta} = \sqrt{2\cos^2(\theta/2)} = \sqrt{2}\cos(\theta/2)$ as long as $0 \le \theta/2 \le \pi/2$, that is for $0 \le \theta \le \pi$. But by symmetry, the length L is twice the length traced as θ goes from 0 to π. So

$$L = 2\sqrt{2}\int_0^{\pi} \sqrt{1 + \cos\theta}\, d\theta = 2\sqrt{2}\sqrt{2}\int_0^{\pi} \cos\frac{1}{2}\theta\, d\theta = 4 \cdot 2\sin\frac{1}{2}\theta\Big]_0^{\pi} = 8. \blacksquare$$

In working with $\sqrt{(dr/d\theta)^2 + r^2}$, be careful that *any square root extraction done is valid over the entire interval of integration* from α to β. In the preceding example, if we had not taken that precaution but had simply put $\sqrt{2}\cos(\theta/2)$ into (10), then we would have gotten $L = 0$, as you should check. This absurdity results from the fact that $\sqrt{2}\cos(\theta/2)$ is *negative* over half the interval of integration, so the positive contribution over $[0, \pi]$ would be canceled by the negative contribution over $[\pi, 2\pi]$.

Exercises 8.7

In Exercises 1–12, find the area enclosed by the given curve.

1. $r = 4\cos\theta$ **2.** $r = 6\sin\theta$

3. $r = 1 - \cos\theta$ **4.** $r = 3 + 3\sin\theta$

5. $r = 3\cos 2\theta$ **6.** $r = \cos 2\theta$

7. $r^2 = 4\sin 2\theta$ **8.** $r^2 = 4\cos 2\theta$

9. $r = 3\cos 3\theta$ **10.** $r = 4\sin 3\theta$

11. $r = 2\sin 2\theta$ **12.** $r = 4\cos 2\theta$

In Exercises 13–18, find the area enclosed by one loop of the graph.

13. $r = 2\cos\theta$ **14.** $r = 3\sin 3\theta$

15. $r = \sin 4\theta$ **16.** $r = 2\sin 5\theta$

17. $r^2 = 9\sin 2\theta$ **18.** $r^2 = 8\cos 2\theta$

In Exercises 19–28, find the area of the region inside the first graph and outside the second.

19. $r = 1 + 2\cos\theta$, $r = 1$ **20.** $r = 1 + \cos\theta$, $r = 1$

21. $r = 1$, $r = 1 - \cos\theta$ **22.** $r = 2$, $r = 2 - 2\sin\theta$

23. $r = 3\cos\theta$, $r = 2 - \cos\theta$ **24.** $r = 4\sin\theta$, $r = 3 - \sin\theta$

25. $r = 5\cos\theta$, $r = 2 + \cos\theta$ **26.** $r = 3\sin\theta$, $r = 2 - \sin\theta$

27. $r = 1 + 2\sin\theta$, $r = 2$ **28.** $r = 2$, $r = 1 + \sin\theta$

In Exercises 29–38, find the area of the given region.

29. The inner loop of the limaçon $r = 1 - 2\sin\theta$

30. The inner loop of the limaçon $r = 2 - 3\cos\theta$

31. Inside the spiral $r = \theta/2$ between $\theta = 0$ and $\theta = \pi$

32. Inside the spiral $r = 2\theta$ between $\theta = \pi/6$ and $\theta = \pi/2$

33. Inside the cardioid $r = 1 + \cos\theta$ and the circle $r = 1$

34. Inside the cardioid $r = 2 - 2\sin\theta$ and the circle $r = 2$

35. Common to the two disks bounded by $r = 4\cos\theta$ and $r = 4\sin\theta$

36. Common to the two disks bounded by $r = \cos\theta$ and $r = \sin\theta$

37. The smaller loop of $r = 1 + 2\sin 2\theta$

38. Common to $r^2 = \cos 2\theta$ and $r^2 = \sin 2\theta$

In Exercises 39–48, find the arc length.

39. $r = 1 + \cos\theta$, $\theta \in [0, \pi/3]$

40. $r = 1 - \sin\theta$, $\theta \in [0, \pi/6]$

41. $r = \theta^2$, $0 \le \theta \le 2\pi$

42. $r = \theta$, $0 \le \theta \le 2\pi$

43. $r = \sin^3(\theta/3)$, entire curve

44. $r = e^{-\theta}$, $0 \le \theta \le \pi$

45. $r = e^{\theta/2}$, $\theta \in [0, 2\pi]$

46. $r = e^{-\theta}$, $\theta \in [0, +\infty]$

47. $r = 2\sin^2(\theta/2)$, $\theta \in [0, \pi]$. (*Hint:* First find a familiar equation for the curve.)

48. $r = \cos^2(\theta/2)$, $\theta \in [0, 2\pi]$

In Exercises 49–52, use Simpson's rule with $n = 10, 20, 40, \ldots$ to approximate the given arc length correct to six decimal places.

PC **49.** $r = \sin\theta$, $\theta \in [0, \pi]$

PC **50.** $r = \cos 2\theta$, $\theta \in [0, 2\pi]$

PC **51.** $r = \dfrac{2}{2 + \cos\theta}$, $\theta \in [0, 2\pi]$

PC **52.** $r = \dfrac{2}{1 - \sin\theta}$, $\theta \in [-\pi/4, \pi/4]$

8.8 Looking Back

This chapter separates naturally into two pieces. The first four sections studied the conic sections (and degenerate special cases), which arise as the graphs of second-degree polynomial equations in x and y. The second part discussed alternative approaches for graphing plane curves.

Besides defining and deriving the standard equations of a parabola, in Definition 1.1 and Equations (2) and (4), Section 1 also introduced the important technique of translation of coordinate axes (p. 484). This led to Equations (8) and (9) for parabolas with vertices at points (h, k) other than the origin. We saw that projectiles follow parabolic paths and gave the reflection property of parabolas in Theorem 1.2.

Section 2 was devoted to ellipses (Definition 2.1) and their standard equations (7) and (8). We also used translation of axes to find equations of ellipses centered at points $C(h, k)$ other than $(0, 0)$ (see p. 494).

Section 3 discussed hyperbolas (Definition 3.1). In addition to deriving their standard equations (3) and (6), we gave a simple procedure for finding their asymptotes and sketching them. Translation of axes again produced equations of hyperbolas centered at points $C(h, k)$ other than the origin (see p. 504).

Section 4 developed machinery to classify the graph of the general polynomial equation

$$(1) \qquad ax^2 + bxy + cy^2 + ex + fy + g = 0.$$

First, Equation (7) on p. 512 gives the angle α of rotation that transforms (1) to an equation with no xy-term. Table 1 (p. 515) can then be used to find an equation of the graph of (1) in the rotated coordinate system. Perhaps the most important outgrowth of these ideas (Theorem 4.2) is the use of the discriminant $b^2 - 4ac$ of (1) to identify the graph of any given equation (1), without having to carry out a rotation.

Section 5 discussed parametric representation of curves in the xy-plane. Following Definition 5.1, we saw the connection between parametric equations and graphs of functions on p. 519. Equation (13) on p. 522 gives the formula for the slope of a parametric curve, and Theorem 5.2 contains the formula for the arc length of such a curve.

The last two sections of the chapter presented polar coordinates and some calculus associated with polar coordinate equations of curves. Besides the basic Definition 6.1, Section 6 gave the important change-of-coordinate equations (1)–(6) on p. 528 and Equations (8) and (9) on p. 530 for circles centered on one coordinate axis and tangent to the other. The symmetry criteria in Theorem 6.2 proved helpful in drawing polar curves, often making it sufficient to plot only first-quadrant points. The general Definition 6.3 of a conic section led to the uniform polar equations for these curves given on p. 534. Theorem 7.1 gives the basic area formula for regions defined by polar coordinate equations. Equation (9) on p. 542, derived from Theorem 5.2, is the formula for the arc length along the graph of a polar equation $r = r(\theta)$ between $\theta = \alpha$ and $\theta = \beta$, if $dr/d\theta$ is continuous.

CHAPTER CHECKLIST

Section 1: parabola, directrix, focus; standard form equations; vertex; translation of axes, change-of-coordinate formulas; projectile motion; reflection property.

Section 2: ellipse, eccentricity, directrices; foci; vertices, major axis, minor axis; relation among a^2, b^2, and c^2; standard form equations; translation to a center other than (0, 0).

Section 3: hyperbola, eccentricity, foci, vertices; standard form equations; center; transverse (major) axis; asymptotes; asymptotic (auxiliary) rectangle; conjugate (minor) axis; relation among a^2, b^2, and c^2; translation to a center other than (0, 0).

Section 4: general second-degree polynomial equation in x and y; rotational change-of-coordinate equations; choice of angle of rotation to eliminate xy-term; discriminant; classification of graphs of second-degree polynomial equations.

Section 5: parametrized curve, parametric equations, endpoints; closed curve, simple curve; orientation; cycloid, trochoid; smooth curve; slope formula for parametric curve; arc length formula.

Section 6: polar coordinates, pole, polar axis; change-of-coordinate formulas; symmetry criteria; equations of circles centered on x- or y-axis tangent to the other axis; cardioids, roses, lemniscates; general definition of conic sections; polar coordinate equations of conic sections with focus at the pole.

Section 7: area formula for a region enclosed by $r = f(\theta)$ between $\theta = \alpha$ and $\theta = \beta$; limaçon; area of a region between $r = g(\theta)$ and $r = h(\theta)$; arc length formula for a polar curve.

REVIEW EXERCISES 8.8

In Exercises 1–16, find the equation of the given conic section with vertex V, focus F, center C, and directrix l. Sketch.

1. Parabola; $F(-3, 0)$; l: $x = 3$
2. Parabola; $F(4, 0)$; l: $x = -2$
3. Parabola; $V(2, 1)$; $F(2, 0)$
4. Parabola; $V(-3, 2)$; $F(0, 2)$
5. Parabola; $V(2, 2)$; symmetric in the line $x = 2$; passes through (0, 6)
6. Parabola; $V(-1, 3)$; symmetric in the line $y = 3$; passes through (2, 0)
7. Ellipse; $F(0, \pm 4)$; major axis of length 10
8. Ellipse; $V(\pm 5, 0)$; minor axis of length 6
9. Ellipse; $C(2, -1)$; $F_2(2, 3)$; $V_2(2, 4)$
10. Ellipse; $F_1(2, -3)$; $F_2(2, 1)$; major axis of length 5
11. Ellipse; $V(\pm 13, 0)$; $E = 12/13$
12. Ellipse; $V(0, \pm 10)$; $F(0, \pm 8)$
13. Hyperbola; $V_2(0, 3)$; $F_2(0, 5)$
14. Hyperbola; $V(\pm 8, 0)$; asymptotes $y = \pm \frac{3}{4}x$
15. Hyperbola; $V(0, \pm 3)$; through (2, 5)
16. Hyperbola; $V_1(-5, 3)$; $V_2(1, 3)$; asymptotes $y - 3 = \pm 2(x + 2)$

In Exercises 17–24, identify the conic section with the given equation. Find its vertices, foci, and asymptotes, if any. Sketch the graph.

17. $3x = 4y^2 + 8y - 2$

18. $3x^2 + 12x = 8y - 2$

19. $25x^2 - 100x + 9y^2 + 36 = 0$

20. $6x^2 - 24x + 9y^2 - 54y + 80 = 0$

21. $4y^2 + 8y - 3x^2 + 16 = 0$

22. $4x^2 - 8x + y^2 - 4y - 9 = 0$

23. $xy = 9$

24. $2xy - 1 = 0$

In Exercises 25–28, rotate the coordinate axes through an angle α to eliminate the xy-term. Identify and sketch the graph

25. $7x^2 - 6\sqrt{3}xy + 13y^2 - 16 = 0$

26. $17x^2 - 12xy + 8y^2 - 80 = 0$

27. $x^2 + 2\sqrt{3}xy - y^2 = 4$

28. $8x^2 + 6xy = 9$

In Exercises 29–34, use the discriminant to identify the general type of graph of the given equation.

29. $5x^2 - 3xy - y^2 + 3x - 8y = 4$

30. $x^2 - 2xy + y^2 - 8x + 11y = 62$

31. $x^2 - 2xy + 2y^2 - 8x + 11 = 62$

32. $x^2 + 2xy + y^2 - 3y = 8$

33. $x^2 - 7xy + 8x - 3y = 16$

34. $3x^2 - 4xy + 2y^2 - 8x - 4y + 1 = 0$

In Exercises 35–40, graph the parametric curve, identify it by finding a Cartesian equation, and find the slope of the tangent line at the given point.

35. $x = t^2,\ y = 2\ln t,\ t \in (0, +\infty);\ t = 1$

36. $x = 2\cos t,\ y = 1 + \cos 2t,\ t \in [0, \pi/2];\ t = \pi/2$

37. $x = \sqrt{t},\ y = t + 1,\ t \in [0, +\infty);\ t = 4$

38. $x = 1 - \cos t,\ y = 2 + \sin t,\ t \in [0, 2\pi];\ t = \pi/2$

39. $x = 2 - \frac{1}{2}\cos t,\ y = 3 + \frac{3}{2}\sin t,\ t \in [0, 2\pi];\ t = 2\pi/3$

40. $x = 2\cosh t,\ y = 3\sinh t,\ t \in (0, +\infty)$

In Exercises 41–46, find the indicated arc length.

41. $x = \frac{4}{3}t^{3/2},\ y = \frac{1}{2}t^2 - t,\ t \in [0, a]$

42. $x = t + \ln t,\ y = t - \ln t,\ t \in [1, 4]$

43. $x = t\cos t,\ y = t\sin t,\ t \in [0, \pi]$

44. $x = 3\sin t\cos t,\ y = 3\sin^2 t,\ t \in [0, \pi]$

45. $r = \sin^2 \frac{1}{2}\theta,\ \theta \in [0, \pi]$

46. $r = 1 + \sin\theta,\ \theta \in [0, \pi]$

47. Find a polar equation for $y^2 - x^2 = 4$.

48. Find a polar equation for $(x - 3)^2 + y^2 = 9/4$.

49. Find a Cartesian equation for $r = 2\sin\theta + 2\cos\theta$.

50. Find a Cartesian equation for $r^2 = \csc 2\theta$.

In Exercises 51–56, draw the graph. Identify the graph if it is not named.

51. $x = \dfrac{3t}{1 + t^3},\ y = \dfrac{3t^2}{1 + t^3},\ t \in (-\infty, +\infty)$; the *folium* of Descartes

52. $x = 2\cot t,\ y = 2\sin^2 t$; the *witch of Agnesi*

53. $r = 1 + 2\sin\theta$

54. $r = 3 - 3\cos\theta$

55. $r = 2\sin 2\theta$

56. $r = 2\sin 3\theta$

In Exercises 57–62, find the area of the given region.

57. Enclosed by $r = 2\cos 2\theta$

58. Enclosed by the curve of Exercise 54

59. Inside $r = 3\cos\theta$ and outside $r = 1 + \cos\theta$

60. Inside both curves in Exercise 59

61. Inside the right petal of $r = 2\cos 2\theta$ and outside the circle $r = 1$

62. Inside the loop of $r = 1 + 2\cos\theta$

63. **(a)** Identify the curve $x = 2\sin(t + \pi/3),\ y = 2\cos(t - \pi/3),\ t \in [0, 2\pi]$.
(b) Find where the curve in part (a) has horizontal tangents.

64. Let Q be the point of intersection of the directrix of the parabola $y^2 = 4px$ and its tangent line at $P(x_0, y_0)$. If $F(p, 0)$ is the focus, then show that FQ is perpendicular to FP.

CHAPTER 9 Infinite Sequences and Series

9.0 Introduction

When we discussed Newton's method of approximating a root x^* of an equation $f(x) = 0$, we spoke of the approximations $x_1, x_2, \ldots, x_n$ *closing in* on x^*. We treated this as intuitively understandable but did not give it a precise mathematical meaning. The basic theme of this chapter is an outgrowth of the idea behind Newton's method, namely, the approximation of unknown irrational quantities by easily calculable rational quantities. We begin by defining what it means for a sequence of numbers to *converge* to (close in on) a limit.

In Section 4.2 we defined $\int_a^b f(x)\,dx$ to be the limit of the Riemann sums

$$R_P = \sum_{i=1}^{n} f(z_i)\,\Delta x_i \tag{1}$$

as $|P| \to 0$. As the norm $|P|$ of P approaches 0, the number n of subintervals increases without bound. Thus the limit in (1) seems to involve an infinite sum, called an *infinite series*. We begin to study such sums in Section 2. The origins of this topic go back to Greek philosophy and mathematics, but it came to be understood precisely only after the advent of calculus. In the ensuing three sections, we present various tests for convergence of infinite series. We start with series of positive terms, then consider series whose terms are alternately positive and negative, and finally deal with series of arbitrary terms.

The rest of the chapter is devoted to applying the ideas of convergence of infinite series to fundamental questions about differentiable transcendental functions f. First, how can the values of such a function be computed? How, for example, were the values of $\sin x$ on your calculator found? If the transcendental function f can be differentiated repeatedly, then $f(x)$ can be accurately approximated by the values $p_n(x)$ of certain *polynomial functions*. Moreover, the approximating polynomial functions p_n can also be used to accurately compute derivatives and integrals of f.

The last section presents an interesting extension of the binomial theorem of elementary algebra. It is perhaps fitting that the final section of the last chapter of single-variable calculus presents ideas that originated with Newton. In fact, Newton's discussion of binomial series in 1665 constituted the first step in his development of calculus. That most modern texts *end* their study of single-variable calculus where Newton *began* his stands as a tribute to his remarkable genius and suggests the gigantic leap forward he made into uncharted new areas.

9.1 Infinite Sequences

A shopping list, a telephone number, and a ZIP code are all simple examples of *finite* sequences. In Section 3.8 we considered the finite sequence $(x_1, x_2, \ldots, x_n, x_{n+1})$ of Newton approximations to the solution x^* of an equation $f(x) = 0$, where f is a differentiable function. To have confidence in the accuracy of Newton's method, we need to know that the terms become arbitrarily close to x^* as we go farther and farther into the sequence. We are thus led to consider *infinite* sequences.

A clue to what is meant by such a thing is provided by finite sequences, which are just lists with a first entry, a second entry, a third entry, and so on. To capture the order of occurrence of the entries, we view the sequence

$$\mathbf{a} = (a_1, a_2, \ldots, a_n)$$

as a *function* f with domain $\{1, 2, \ldots, n\}$, where

$$f(i) = a_i \qquad \text{for } i = 1, 2, \ldots, n.$$

We are interested mainly in sequences whose entries

$$f(1) = a_1, f(2) = a_2, \ldots, f(n) = a_n$$

are real numbers. The first entry is the value of f at 1, the second entry is the value of f at 2, and so on. With this point of view, the finite sequence

$$(-2, \pi, 3, 1/2, -1)$$

is a real-valued function f whose domain is $\{1, 2, 3, 4, 5\}$ and whose range is $\{-2, \pi, 3, 1/2, -1\}$ since $f(1) = -2$, $f(2) = \pi$, $f(3) = 3$, $f(4) = 1/2$, and $f(5) = -1$.

This notion of finite sequences extends easily to infinite sequences.

1.1 DEFINITION An ***infinite sequence a*** $= (a_1, a_2, a_3, \ldots, a_n, \ldots)$ is a function $f\colon \mathbf{N} \to \mathbf{R}$, where $\mathbf{N}$ is the set $\{1, 2, 3, \ldots, n, \ldots\}$ of positive integers. The numbers $a_1 = f(1)$, $a_2 = f(2)$, $a_3 = f(3), \ldots$ are called the ***entries*** of the sequence. The *first entry* is a_1, the *second entry* is a_2, and so on.

Thus the set of entries of a given sequence is just the range of the associated function $f\colon \mathbf{N} \to \mathbf{R}$. In writing a sequence, we usually list its first few entries and then its nth entry $a_n = f(n)$ to indicate the formula for the function f. Standardized tests often ask you to *guess* such formulas, and in this chapter occasionally you will be expected to come up with the formula for f from knowledge of the first few terms. This is always a bit risky, because from a *few* examples you can only guess a *reasonable* formula for $f(n)$ (see Exercise 52).

Infinite sequences are determined by their entries *and* the order of occurrence of the entries. As the following example illustrates, a sequence is *not* simply its set of entries.

Example 1

Show that the *sets* of entries of the sequences

$$\boldsymbol{a} = (1, 1/2, 1/4, 1/8, \ldots, 1/2^{n-1}, \ldots)$$

and

$$\boldsymbol{b} = (1, 1/2, 1, 1/4, 1, 1/8, \ldots, g(n), \ldots),$$

FIGURE 1.1

where

$$g(n) = \frac{1}{2^{n/2}} \text{ for } n \text{ even}, \qquad g(n) = 1 \text{ for } n \text{ odd},$$

are the same. Are the sequences the same?

Solution. The range of the first sequence consists of all rational numbers of the form $a_n = 1/2^{n-1}$ for n a positive integer. Clearly the range of the second sequence is contained in this set, since only $1 = 1/2^0$ and $1/2^m$ for $m \geq 1$ occur as entries. Conversely, all entries of the first sequence occur in the second: Any $a_n = 1/2^{n-1}$ is $b_{2(n-1)}$, because

FIGURE 1.2

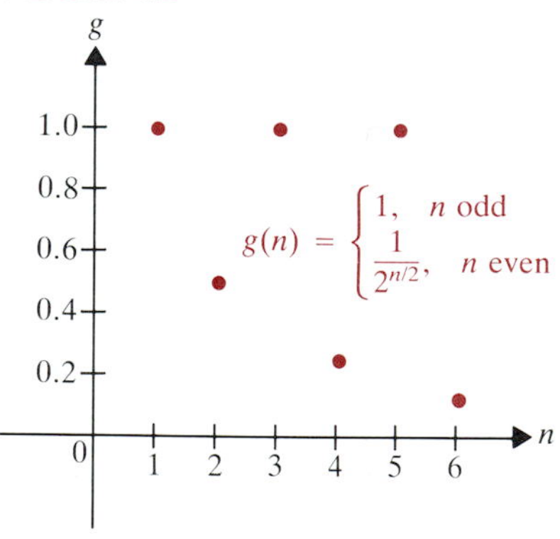

$$b_{2(n-1)} = \frac{1}{2^{2(n-1)/2}} = \frac{1}{2^{n-1}}.$$

So the respective sets of entries are identical. However, the sequences are *not* the same, since the defining functions f and g differ. For example, $f(3) = a_3 = 1/4$, but $g(3) = b_3 = 1$. ■

The sequence $\boldsymbol{a}$ in Example 1 appears to be quite orderly: Its entries seem to concentrate near 0. In Figures 1.1 and 1.2 we graph the sequences of Example 1 by plotting points $(n, f(n))$. **Figure 1.1** shows the orderly behavior of the terms of $\boldsymbol{a}$. **Figure 1.2** shows disorderly behavior of $\boldsymbol{b}$, whose terms don't close in on any real number. Rather, they oscillate back and forth between 1 and smaller and smaller intervals around 0. Even more disorderly conduct is exhibited by the sequence

FIGURE 1.3

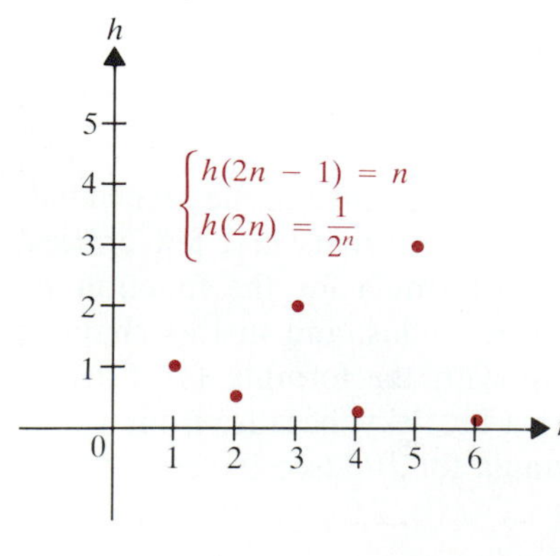

$$\boldsymbol{c} = (1, 1/2, 2, 1/4, 3, 1/8, 4, 1/16, \ldots),$$

whose defining function h is given by

$$h(2n-1) = n, \qquad h(2n) = \frac{1}{2^n}.$$

In this case, some of the entries again concentrate near 0, but the rest march off toward $+\infty$. See **Figure 1.3.**

As you would expect, the sequences studied in calculus are the well-behaved ones, those whose entries a_n approach a real number a as $n \to \infty$. This means that a_n is arbitrarily close to a for large enough n. The formal statement of this resembles the formal definition of limit (Definition 4.1 of Chapter 1).

1.2 DEFINITION A sequence $\boldsymbol{a} = (a_1, a_2, \ldots, a_n, \ldots)$ is said to ***converge*** (or be ***convergent***) to the real number a if $\lim_{n\to\infty} a_n = a$ in the following sense: For every $\varepsilon > 0$, there is some natural number N such that $|a_n - a| < \varepsilon$ for all $n > N$. A sequence that does not converge is called ***divergent*** (or is said to ***diverge***).

Considering the sequences above,

$$\boldsymbol{a} = (1, 1/2, 1/4, \ldots, 1/2^{n-1}, \ldots),$$

$$\boldsymbol{b} = (1, 1/2, 1, 1/4, \ldots, g(n), \ldots),$$

$$\boldsymbol{c} = (1, 1/2, 2, 1/4, 3, 1/8, \ldots, h(n), \ldots),$$

we see that $\boldsymbol{a}$ converges to

$$0 = \lim_{n\to\infty} \frac{1}{2^{n-1}},$$

whereas $\boldsymbol{b}$ and $\boldsymbol{c}$ diverge since $\lim_{n\to\infty} g(n)$ and $\lim_{n\to\infty} h(n)$ fail to exist.

Example 2

If

$$s_n = \frac{2n^2 - 5n}{3n^2 - 2},$$

determine whether $\boldsymbol{s} = (s_1, s_2, \ldots, s_n, \ldots)$ is convergent or divergent. If it converges, find its limit.

Solution. Dividing numerator and denominator by n^2, the highest power of n in the fraction s_n, we have

$$\lim_{n\to\infty} s_n = \lim_{n\to\infty} \frac{2n^2 - 5n}{3n^2 - 2} = \lim_{n\to\infty} \frac{2 - \frac{5}{n}}{3 - \frac{2}{n^2}} = \frac{2}{3}.$$

Thus $\boldsymbol{s}$ converges to 2/3. ■

In Example 2 we calculated $\lim_{n\to\infty} s_n$ just as we would have calculated

$$\lim_{x\to+\infty} \frac{2x^2 - 5x}{3x^2 - 2}.$$

This is an illustration of a useful general principle.

1.3 THEOREM

> Suppose that
>
> $$\lim_{x\to+\infty} f(x) = a.$$
>
> Then the sequence
>
> $$\boldsymbol{a} = (a_1, a_2, a_3, \ldots, a_n, \ldots), \text{ where } a_n = f(n),$$
>
> converges to a. If $\lim_{x\to+\infty} f(x) = \pm\infty$, then $\boldsymbol{a}$ diverges.

Proof. If $\lim_{x\to+\infty} f(x) = a$, then $f(x)$ is arbitrarily close to a for all sufficiently large real numbers x. Thus, for natural numbers n that are sufficiently large, that is, greater than some N, we also have $f(n)$ arbitrarily close to a. Hence $\boldsymbol{a}$ converges to a.

If, on the other hand $\lim_{x\to+\infty} f(x) = \pm\infty$, then for all sufficiently large positive integers n, $|f(n)| = |a_n|$ is arbitrarily large. Thus $\boldsymbol{a}$ cannot converge to any real number. QED

Example 3

Decide whether the sequences $\boldsymbol{a}$, $\boldsymbol{b}$, and $\boldsymbol{c}$ converge if

$$\text{(a)}\ a_n = n\sin\frac{1}{n}, \qquad \text{(b)}\ b_n = ne^{-n}, \qquad \text{(c)}\ c_n = \frac{e^n}{n^2}.$$

Solution. (a) Let $f(x) = x\sin\frac{1}{x}$. Then

$$\lim_{x\to+\infty} f(x) = \lim_{x\to+\infty} \frac{\sin\frac{1}{x}}{\frac{1}{x}} = \lim_{t\to0+} \frac{\sin t}{t}, \qquad \text{where } t = \frac{1}{x}.$$

Therefore $\lim_{x\to+\infty} f(x) = 1$, by Theorem 3.10 of Chapter 2. Thus by Theorem 1.3 the given sequence $\boldsymbol{a}$ converges to 1.

(b) Let $g(x) = xe^{-x} = x/e^x$. Then l'Hôpital's rule gives

$$\lim_{x\to+\infty} g(x) = \lim_{x\to+\infty} \frac{x}{e^x} = \lim_{x\to+\infty} \frac{(x)'}{(e^x)'} = \lim_{x\to+\infty} \frac{1}{e^x} = 0.$$

So the given sequence $\boldsymbol{b}$ converges to 0.

(c) Here let $h(x) = e^x/x^2$. We can use l'Hôpital's rule twice to get

$$\lim_{x\to+\infty} h(x) = \lim_{x\to+\infty} \frac{e^x}{2x} = \lim_{x\to+\infty} \frac{e^x}{2} = +\infty,$$

so the given sequence $\boldsymbol{c}$ diverges to $+\infty$. ■

Unfortunately, Theorem 1.3 is not universally applicable. For example, if $f(n) = (-1)^n$, then the corresponding sequence

$$(-1, 1, -1, 1, -1, 1, \ldots, (-1)^n, \ldots)$$

clearly diverges. But we can't apply Theorem 1.3 to it, because the function

$$\bar{f}(x) = (-1)^x$$

is not defined for irrational numbers x:

$$(-1)^x = \exp(x\ln(-1))$$

is meaningless, since -1 is not in the domain $(0, +\infty)$ of the natural logarithm function. Neither can we apply Theorem 1.3 to the sequences

$$\boldsymbol{b} = (1, 1/2, 1, 1/4, \ldots, g(n), \ldots) \qquad \text{and} \qquad \boldsymbol{c} = (1, 1/2, 2, 1/4, \ldots, h(n), \ldots)$$

considered above. Since $\lim_{n\to\infty} g(n)$ and $\lim_{n\to\infty} h(n)$ do not exist, neither do $\lim_{x\to+\infty} g(x)$ and $\lim_{x\to+\infty} h(x)$.

Limits of sequences have many properties in common with limits of functions. Although we shall not prove the following analogue of Theorem 5.1 and 5.4 of Chapter 1, we will feel free to apply it.

1.4 THEOREM If $\boldsymbol{a}$ and $\boldsymbol{b}$ are convergent sequences with limits a and b, respectively, and c is a constant, then

(a) $c\boldsymbol{a} \equiv (ca_1, ca_2, \ldots, ca_n, \ldots)$ converges to ca.

(b) $\boldsymbol{a} + \boldsymbol{b} \equiv (a_1 + b_1, a_2 + b_2, \ldots, a_n + b_n, \ldots)$ converges to $a + b$, and $\boldsymbol{a} - \boldsymbol{b}$ converges to $a - b$.

(c) $\boldsymbol{ab} \equiv (a_1b_1, a_2b_2, \ldots, a_nb_n, \ldots)$ converges to ab.

(d) If $b \neq 0$ and $b_i \neq 0$ for all i, then

$$\boldsymbol{a}/\boldsymbol{b} \equiv (a_1/b_1, a_2/b_2, \ldots, a_n/b_n, \ldots) \text{ converges to } a/b.$$

There is an important criterion for convergence of *monotonic sequences.* It is related to the basic nature of the real-number system.

1.5 DEFINITION A sequence $\boldsymbol{a} = (a_1, a_2, \ldots, a_n, \ldots)$ is called ***monotonic*** if one of the following properties is true.

(i) $a_n \leq a_{n+1}$, for all n ($\boldsymbol{a}$ is ***increasing***), or

(ii) $a_n \geq a_{n+1}$, for all n ($\boldsymbol{a}$ is ***decreasing***).

If $a_n < a_{n+1}$ (respectively, $a_n > a_{n+1}$) for all n, then we say that the monotonic sequence a is *strictly increasing* (respectively, *strictly decreasing*). See **Figure 1.4.**

If f is a monotonic real-valued function (Definition 7.4 of Chapter 2), then the sequence $\boldsymbol{a} = (a_1, a_2, \ldots, a_n, \ldots)$, where $a_n = f(n)$, is a monotonic sequence.

FIGURE 1.4

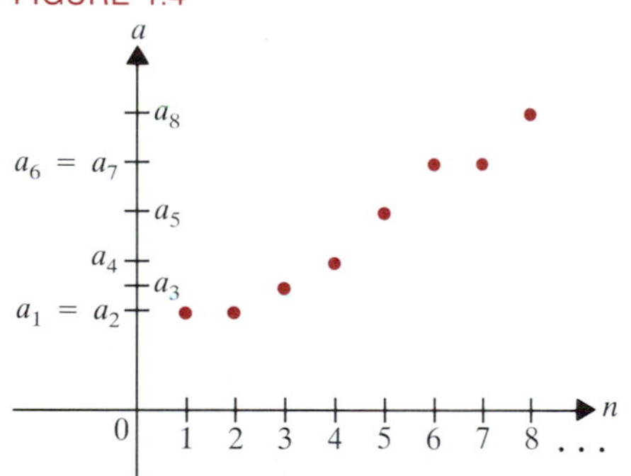

(a) Monotonic increasing sequence

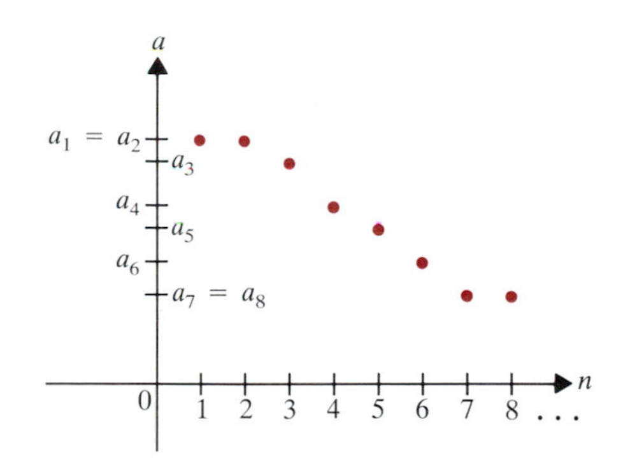

(b) Monotonic decreasing sequence

Example 4

Determine whether $\boldsymbol{a}$ and $\boldsymbol{b}$ are monotonic sequences if

$$\text{(a) } a_n = \frac{n^2 + 1}{2n + 3}, \qquad \text{(b) } b_n = \frac{(-1)^n}{n}.$$

Solution. (a) The first few entries of $\boldsymbol{a}$ are 2/5, 5/7, 10/9, and 17/11. So we suspect that the series is monotonic increasing. To confirm this we investigate the corresponding function f given by

$$f(x) = \frac{x^2 + 1}{2x + 3}$$

on $(1, +\infty)$. We have

$$f'(x) = \frac{(2x+3)(2x) - (x^2+1)\cdot 2}{(2x+3)^2} = \frac{4x^2 + 6x - 2x^2 - 2}{(2x+3)^2}$$

$$= \frac{2(x^2+3x-1)}{(2x+3)^2} = \frac{2\left(x+\dfrac{3}{2}\right)^2 - \dfrac{13}{2}}{(2x+3)^2} > 0 \text{ if } x \geq 1.$$

By Theorem 3.1 of Chapter 3, f is therefore increasing on $[1, +\infty)$. Hence the sequence $\boldsymbol{a}$ is monotonic increasing.

(b) The first few entries of $\boldsymbol{b}$ are -1, $1/2$, $-1/3$, $1/4$, $-1/5$, and $1/6$. It is clear that this sequence is *not* monotonic: Each even-numbered entry is positive and each odd-numbered entry is negative. ■

To give the criterion we mentioned for convergence of monotonic sequences, we need the notion of a *bounded* sequence.

1.6 DEFINITION Let $\boldsymbol{s}$ be a sequence of real numbers. Then:

(a) A number l is a ***lower bound*** of $\boldsymbol{s}$ if $l \leq s_n$ for every n.
A number u is an ***upper bound*** of $\boldsymbol{s}$ if $u \geq s_n$ for every n.
The sequence $\boldsymbol{s}$ is ***bounded*** if it has both an upper bound and a lower bound.

(b) A ***greatest lower bound*** **(glb)** of $\boldsymbol{s}$ is a number g such that
(1) g is a lower bound of $\boldsymbol{s}$, and
(2) $g \geq l$ for every lower bound l of $\boldsymbol{s}$.
A ***least upper bound*** **(lub)** of $\boldsymbol{s}$ is a number b such that
(3) b is an upper bound of $\boldsymbol{s}$, and
(4) $b \leq u$ for every upper bound of $\boldsymbol{s}$.

Example 5

(a) Let $\boldsymbol{a} = (1, 1/2, 1, 1/4, 1, 1/8, \ldots)$. Is $\boldsymbol{a}$ bounded? Does it have a least upper bound? A greatest lower bound?

(b) Answer the same questions for $\boldsymbol{b} = \left(2/5, 5/7, 10/9, \ldots, \dfrac{n^2+1}{2n+3}, \ldots\right)$.

Solution. (a) Here *any* number $l \leq 0$ will do for a lower bound. The *greatest* lower bound is 0, since any larger number fails to be a lower bound: $1/2^k$ becomes arbitrarily close to zero as k becomes larger and larger. Also, any number $u \geq 1$ will do as an upper bound. The least upper bound is then 1. The sequence is bounded, because its set of entries has both an upper bound and a lower bound. See **Figure 1.5.**

(b) We saw in Example 4(a) that this sequence is strictly increasing. Because

$$\lim_{n\to\infty} \frac{n^2+1}{2n+3} = \lim_{n\to\infty} \frac{1 + \dfrac{1}{n^2}}{\dfrac{2}{n} + \dfrac{3}{n^2}} = +\infty,$$

it cannot have any upper bounds. Since it is strictly increasing, any number $l \leq 2/5 = b_1$ will do for a lower bound. The greatest lower bound is then $2/5$.

FIGURE 1.5

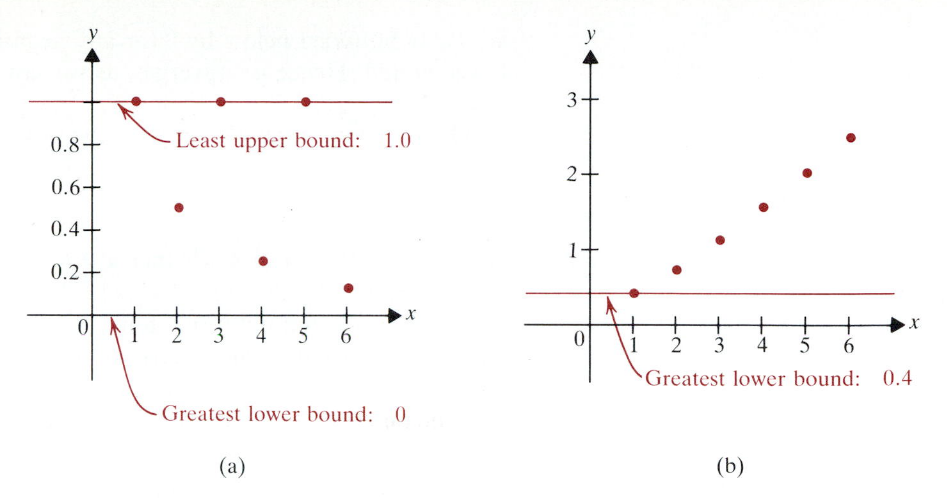

Since the set of entries is just bounded below (and not both above and below), it is *not* a bounded sequence. ■

As you can see from Example 5, the least upper bound b of a sequence $\boldsymbol{s}$ is the *most efficient* upper bound u that can be used. Geometrically, $y = b$ is the lowest horizontal line above and never crossed by the graph of the sequence. Similarly, if g is the greatest lower bound of $\boldsymbol{s}$, then $y = g$ is the highest horizontal line below and never crossed by the graph. See **Figure 1.6.**

FIGURE 1.6

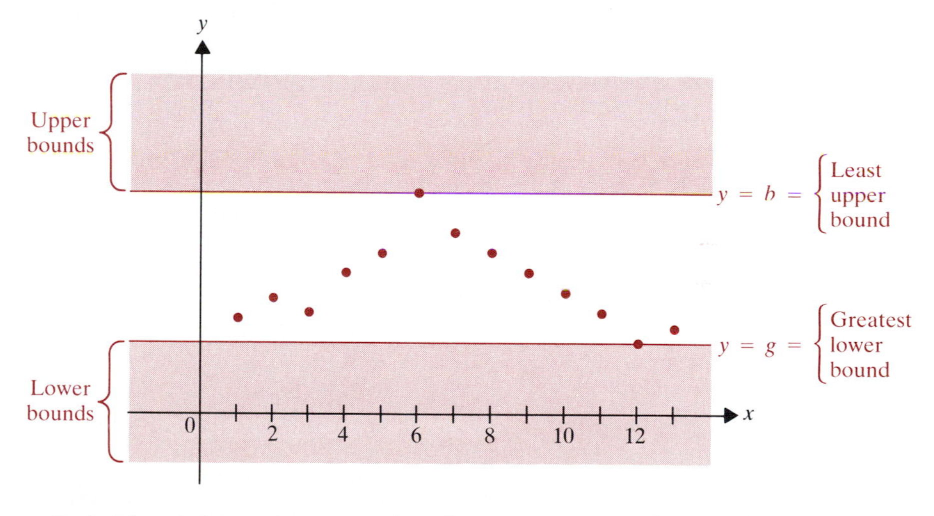

Definition 1.6 permits us to give the convergence criterion mentioned just before Definition 1.5. The proof of the following basic result is given below in the optional subsection on the axiom of completeness.

1.7 THEOREM

> Every bounded monotonic sequence $\boldsymbol{a}$ of real numbers converges to a limit a in $\mathbf{R}$. In fact, a bounded increasing sequence converges to its least upper bound, and a bounded decreasing sequence converges to its greatest lower bound.

Looking back at the sequence $\boldsymbol{a} = (1, 1/2, 1/4, \ldots, 1/2^{n-1}, \ldots)$ in Example 1, we see that it is monotonic decreasing, since for all n we have $1/2^n < 1/2^{n-1}$. It is clearly bounded below by 0 (or any negative number) and so has a greatest lower bound. Hence $\boldsymbol{a}$ converges, as we saw by other means following Definition 1.2.

The following example suggests the power of Theorem 1.7.

Example 6

If $a_n = n!/n^n$, then decide whether $\boldsymbol{a} = (a_1, a_2, a_3, \ldots, a_n, \ldots)$ converges.

Solution. Since computation of $\lim_{n\to\infty} a_n$ appears to be difficult, we try to determine whether the sequence is monotonic. We don't know how to define $x!$ for a real number x, so we can't use Theorem 1.3. Instead we study the ratio of successive terms. We have

$$a_{n+1} = \frac{(n+1)!}{(n+1)^{n+1}},$$

so

$$\frac{a_{n+1}}{a_n} = \frac{\dfrac{(n+1)!}{(n+1)^{n+1}}}{\dfrac{n!}{n^n}} = \frac{(n+1)!}{(n+1)^{n+1}} \cdot \frac{n^n}{n!}$$

$$= \frac{(n+1)!}{n!} \frac{n^n}{(n+1)^n(n+1)} = (n+1)\left(\frac{n}{n+1}\right)^n \frac{1}{n+1}$$

$$= \left(\frac{n}{n+1}\right)^n < 1, \qquad \text{for all } n.$$

Thus $a_{n+1} < a_n$, so the sequence is monotonic *decreasing*. It is clearly bounded: above by the first term, below by any negative number. Hence by Theorem 1.7 it converges to its greatest lower bound. ■

Theorem 1.7 permits us to conclude that $\boldsymbol{a}$ converges without finding $\lim_{n\to\infty} a_n$. (It is a bit more work to show that $\lim_{n\to\infty} a_n = 0$. See Exercise 48.)

Axiom of Completeness*

We do calculus using *real* variables x, not *rational* variables, even though, algebraically speaking, the real and rational numbers seem to have identical properties. What distinguishes the real number system $\mathbf{R}$ from the rationals is that $\mathbf{R}$ is a *continuum*—a number line with no holes in it, such as those that occur in the rational number system $\mathbf{Q}$ at the places where irrational numbers like $\sqrt{2}$, π, $\sqrt{3}$, and e "should" be. The rational number system must be expanded to $\mathbf{R}$ in order for every bounded sequence of numbers to have a greatest lower bound and a least upper bound in the set. For example, the sequence $\boldsymbol{l} = (1, 1.4, 1.41, 1.414, 1.4142, \ldots)$ of rational numbers is bounded (4, 2, and 3/2 are upper bounds, and -10, -1, and $-3/2$ are lower bounds). But it has no least upper bound *in the set of rational numbers:* The obvious candidate for the least upper bound, $\sqrt{2}$, is not in $\mathbf{Q}$! This reflects the fact that the sequence $(r_1, r_2, r_3, \ldots)$ of rational approximations to $\sqrt{2}$ which comes from the square

* Optional

root algorithm does not have a limit in **Q**. The limit of $\boldsymbol{l}$ "should be" $\sqrt{2}$, but $\sqrt{2}$ is not a member of **Q**. So we don't do calculus in **Q**, because *we can't take limits satisfactorily if we restrict ourselves to rational numbers only*. That this sort of problem can't arise in **R** is guaranteed by the fundamental *axiom of continuity* (or *completeness*) for **R**.

1.8 AXIOM

Axiom of Completeness. Every infinite sequence $\boldsymbol{s}$ of real numbers that has an upper bound has a real least upper bound.

This is a fundamental assumption about the real numbers, not a theorem to prove or a definition. It is rather a description of the inherent property of **R** that distinguishes it from **Q**. The following consequence of the axiom is left as Exercise 35.

1.9 COROLLARY

Every infinite sequence $\boldsymbol{t}$ of real numbers that has a lower bound has a greatest lower bound in **R**.

The sequence $\boldsymbol{l}$ above is a nonempty sequence of real numbers that is bounded above, and it does have a least upper bound, namely, $\sqrt{2}$. See **Figure 1.7.**

FIGURE 1.7

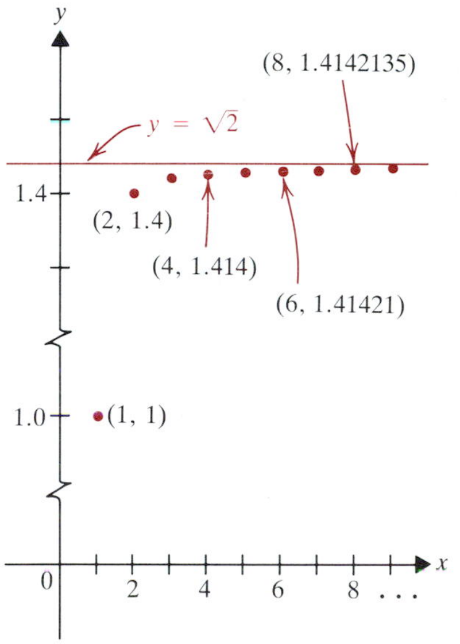

While the set **R** has the completeness property, it pays a price for this in terms of decimal representation of its elements. Recall that every rational number $r = m/n$ (where m and $n \neq 0$ are integers) has a decimal representation that is either *finite* (for example, $3/20 = 0.15$) or *infinite repeating* (for example, $3/11 = 0.272727\ldots$). In either case, it is easy to perform arithmetic with these decimals. But irrational real numbers (such as $\sqrt{2}$, e, π, and $\sqrt{3}$) have *infinite nonrepeating* decimal representations. To do arithmetic with $e = 2.718281828459\ldots$, for example, we must approximate it by a rational number by rounding off to a finite decimal. For instance, $e \approx 2.7183$ is a 4-decimal-place approximation, which is adequate for many calculations. Values of functions like $\ln x$, e^x, and $\sin x$ that are programmed into calculators consist of rational approximations to usually irrational quantities. This almost gives us the best of both worlds: *We have completeness*, hence limits of convergent sequences in **R** are also in **R**, and *we have a finitely calculable arithmetic* of real numbers using *rational approximations.* (Decimal round-off error, however, can cause inaccuracy, as we saw on p. 269.) The idea of approximating infinite quantities (in this case, nonterminating, nonrepeating decimals) by finite quantities (in this case, finite decimals) is the basic theme of this chapter.

We can use the axiom of completeness to prove Theorem 1.7.

1.7 THEOREM

A bounded increasing sequence of real numbers converges to its least upper bound. A bounded decreasing sequence converges to its greatest lower bound.

Proof. We give the proof for the case in which $\boldsymbol{a}$ is a bounded *increasing* sequence. The bounded decreasing case is similar: see Exercise 37. Since $\boldsymbol{a}$ is bounded, the axiom of completeness says that it has a least upper bound a. We claim that $a = \lim_{n\to\infty} a_n$.

To see this, let $\varepsilon > 0$ be given. Since $a - \varepsilon < a$, the number $a - \varepsilon$ can't be an upper bound for $\boldsymbol{a}$. So some entry, say a_N, of $\boldsymbol{a}$ satisfies

$$a - \varepsilon < a_N \leq a. \tag{1}$$

But since $\boldsymbol{a}$ is an increasing sequence, for all $n > N$, we then have

$$a_N \leq a_n. \tag{2}$$

Combining (1) and (2), we see that for all $n > N$,

$$a - \varepsilon < a_n \leq a < a + \varepsilon.$$

FIGURE 1.8

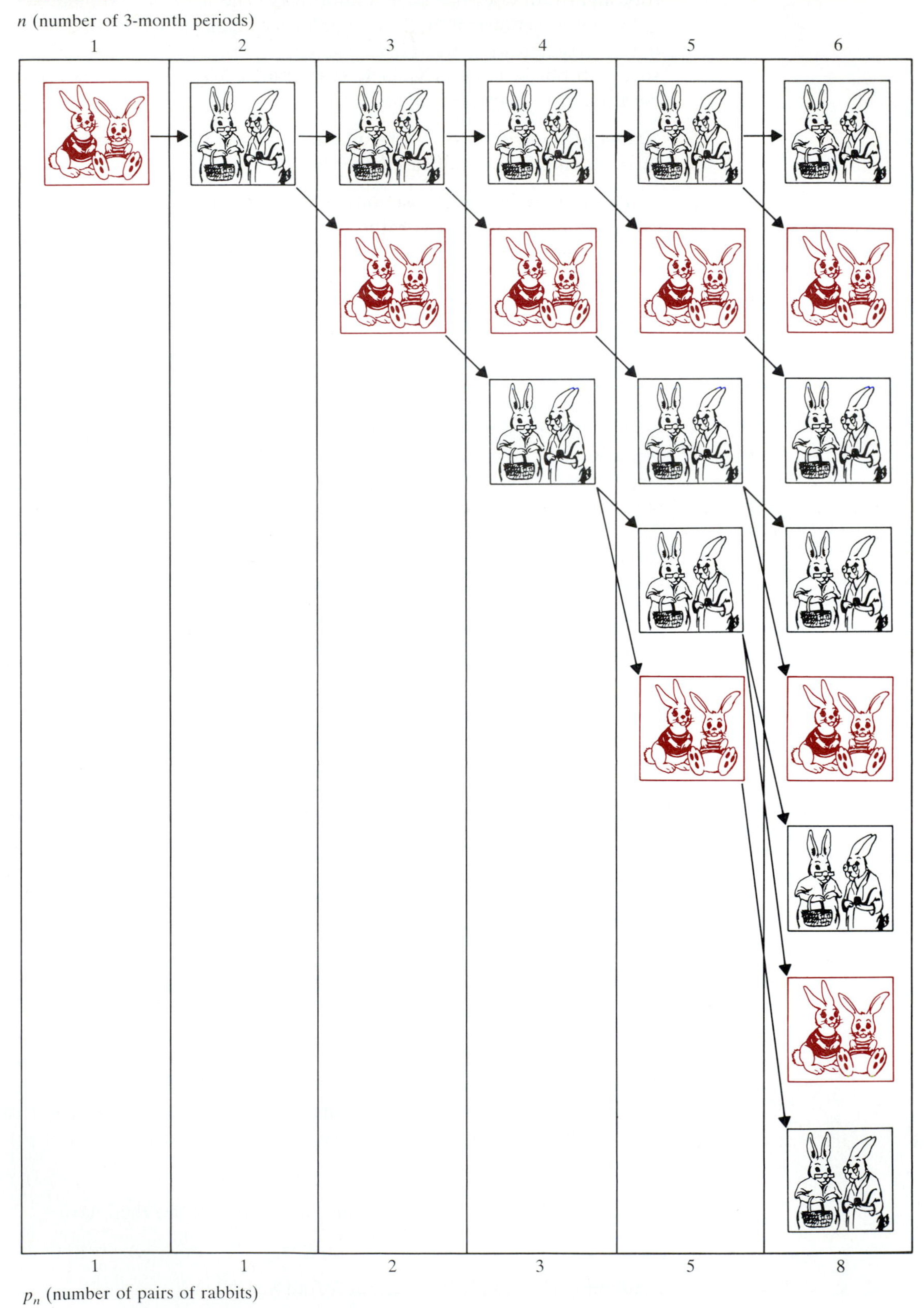
n (number of 3-month periods)
1
2
3
4
5
6
1
1
2
3
5
8
p_n (number of pairs of rabbits)

Hence

$$|a_n - a| < \varepsilon, \qquad \text{for } n > N.$$

By Definition 1.2 then, $\lim_{n\to\infty} a_n = a$, so $\boldsymbol{a}$ converges to its least upper bound a. QED

The Fibonacci Sequence*

Although divergent sequences are usually not of much interest, there is one famous exception. The ***Fibonacci sequence*** is

$$f = (1, 1, 2, 3, 5, 8, 13, 21, 34, 55, 89, 144, 233, 377, \ldots).$$

Beginning with f_3, each term of this sequence is the sum of the two terms just before it. That is,

$$f_n = f_{n-1} + f_{n-2}, \qquad \text{if } n \geq 3.$$

The sequence was discovered by Leonardo Fibonacci (c. 1170–1230) in studying rabbit reproduction. It is not hard to show (Exercise 53) that the sequence diverges. This corresponds to the explosive population growth of rabbit colonies. Fibonacci observed that a pair of mature rabbits tended to reproduce an average of two offspring every three months. Newborn rabbits produced their first offspring after 6 months.

Example 7

Suppose that each litter of rabbits is made up of one male and one female. After n 3-month periods how many pairs of rabbits are there in a colony that started with one newly born couple (one male, one female)? (Assume that no rabbits die during the experiment.)

Solution. Let p_n be the number of pairs of rabbits after n periods. When $n = 1$, there is a single pair. Thus $p_1 = 1$. After three months this pair has not yet reproduced, so $p_2 = 1$. But after another three months the first pair of offspring are born, so $p_3 = 2$. Three months later the original parents give birth to another couple, so that $p_4 = 1 + 2 = 3$. After three more months the entire population of six months earlier (the original parents and their first offspring) produce new litters, increasing the population by two couples. Thus $p_5 = 2 + 3 = 5$. After n periods the population six months earlier (p_{n-2}) gives birth to an equal number of new rabbits, who are added to the population p_{n-1}. Since no deaths occur, $p_n = p_{n-1} + p_{n-2}$. Hence the sequence of populations is the Fibonacci sequence (see **Figure 1.8**). ■

Exercises 9.1

In Exercises 1–6, give as simple a formula for $f(n) = a_n$ as you can.

1. $\left(2, 2, \frac{8}{6}, \frac{16}{24}, \frac{32}{120}, \frac{64}{720}, \ldots\right)$

2. $\left(-1, \frac{3}{4}, \frac{4}{-8}, \frac{5}{16}, \frac{6}{-32}, \frac{7}{64}, \ldots\right)$

3. $(-1, 1, -1, 1, -1, 1, \ldots)$

4. $(1, -1, 1, -1, 1, -1, \ldots)$

5. $\left(1, \frac{1}{4}, \frac{1}{9}, \frac{1}{16}, \frac{1}{25}, \frac{1}{36}, \ldots\right)$

6. $\left(1, \frac{\sqrt{2}}{4}, \frac{\sqrt{3}}{9}, \frac{1}{8}, \frac{\sqrt{5}}{25}, \frac{\sqrt{6}}{36}, \ldots\right)$

* Optional

In Exercises 7–30, determine whether the given sequence converges or diverges by investigating $\lim_{n\to\infty} a_n$.

7. $a_n = \dfrac{2n^2 + 5n + 2}{3n^2 - 7}$

8. $a_n = \dfrac{2^n(n-1)}{n\,2^{n-1} + 7}$

9. $a_n = \dfrac{(n^2+1)3^n}{n\,2^{n+100}}$

10. $\mathbf{a} = \left(2, \dfrac{1}{2}, \dfrac{3}{2}, \dfrac{1}{4}, \dfrac{7}{4}, \dfrac{1}{8}, \ldots\right)$

11. $a_n = \dfrac{\ln n}{n}$

12. $a_n = \dfrac{3n^2 - 5n + 6}{e^n}$

13. $a_n = \dfrac{\sin \frac{1}{2} n\pi}{n}$

14. $a_n = \dfrac{\tan\left[\left(\frac{n+1}{n}\right)\frac{1}{2}\pi\right]}{\left(\frac{n+1}{n}\right)\frac{1}{2}\pi}$

15. $a_n = \dfrac{n}{n+1}\sin\dfrac{1}{2}n\pi$

16. $a_n = \dfrac{n}{n+1}\cos n\pi$

17. $a_n = n \sin \dfrac{1}{n}$

18. $a_n = \dfrac{\sin n}{n}$

19. $a_n = \dfrac{e^n + 1}{2^n}$

20. $a_n = \dfrac{2^n + 1}{e^n}$

21. $a_n = \tan^{-1} n$

22. $a_n = \dfrac{\tan^{-1} n}{n}$

23. $a_n = \left(1 + \dfrac{a}{n}\right)^n$

24. $a_n = \dfrac{n^p}{e^n}$, where $p > 0$

25. $a_n = \sqrt[n]{n}$

26. $a_n = \sqrt[n]{3^{n+1}}$

27. $a_n = \left(\dfrac{1 - n^2}{2 + n^2}\right)^n$

28. $a_n = (\cos n\pi)^n$

29. $a_n = \sqrt{n} - \sqrt{n-1}$

30. $a_n = \sqrt{n+2} - \sqrt{n}$

In Exercises 31–38, show that each sequence is monotonic and determine whether it converges.

31. $a_n = \dfrac{2^n}{n!}$

32. $a_n = \dfrac{3^{n+1}}{(n+1)!}$

33. $a_n = \dfrac{n!}{3n}$

34. $a_n = \dfrac{(n+1)!}{3^n}$

35. $a_n = \dfrac{n!}{3 \cdot 5 \cdot 7 \ldots (2n+1)}$

36. $a_n = \dfrac{1 \cdot 4 \cdot 7 \cdot \ldots \cdot (3n-2)}{n!}$

37. $a_n = \dfrac{n^p}{n!}$, where $0 < p < 1$

38. $a_n = \dfrac{p^n}{n!}$, where $0 < p < 2$

39. Prove Corollary 1.9.

40. **(a)** Prove that if an infinite sequence of real numbers has a least upper bound l, then l is unique.

(b) Show that if an infinite sequence of real numbers has a greatest lower bound g, then g is unique.

41. Prove Theorem 1.7 if $\mathbf{a}$ is a bounded decreasing sequence.

42. Prove that if $|r| < 1$, then $\mathbf{r} = (r, r^2, r^3, \ldots, r^n, \ldots)$ converges to 0. (***This result is needed in the next section.***)

43. Prove that if $|r| > 1$, then the sequence $(r, r^2, r^3, \ldots, r^n, \ldots)$ diverges by showing $\lim_{n\to\infty} |r|^n = +\infty$. What happens if $|r| = 1$? (***This result is also used in the next section.***)

44. Give an example to show that if the hypothesis "monotonic" is removed from Theorem 1.7, the sequence need no longer be convergent.

45. Prove parts (a) and (b) of Theorem 1.4.

46. Give an example of a sequence that is not monotonic but diverges to $+\infty$.

47. Give an example of a sequence that is not monotonic but converges to 0.

48. Show that the sequence in Example 6 actually converges to 0.

49. In Example 7, suppose that the rabbit colony starts with *two* newly born couples. Write down the sequence of populations of the colony over the first twelve 3-month periods.

50. Repeat Exercise 49 if the colony starts with three pairs of newly born rabbits, but at six months, four more pairs of mature rabbits are added.

51. Let $r_n = f_{n+1}/f_n$ be the ratio of two consecutive Fibonacci numbers. Show that if $\mathbf{r} = (r_1, r_2, \ldots, r_n, \ldots)$ converges to a limit L, then

(a) $L^2 = L + 1$. (*Hint:* Recall that $f_{n+1} = f_n + f_{n-1}$. What happens if we assume that $L = \lim_{n\to\infty} f_{n+1}/f_n$?)

(b) Conclude that if L exists, then $L = \dfrac{1 + \sqrt{5}}{2}$.

52. Here is an example of how an inductive guess of a_n from the first terms of a sequence $\mathbf{a}$ may be risky. If

$$f(n) = 2n + (n-1)(n-2)(n-3)(n-4),$$

then calculate $f(1)$, $f(2)$, $f(3)$, and $f(4)$. What would a reasonable person guess the formula for a_n to be?

53. Show that the Fibonacci sequence $f_1, f_2, \ldots, f_n, \ldots$ diverges. (*Hint:* Show that for $n \geq 5$, $f_n \geq n$.)

54. A sequence (a_n) is called ***eventually monotonic*** if for some $k \geq 1$, $(a_k, a_{k+1}, a_{k+2}, \ldots)$ is monotonic.

(a) Show that a bounded, eventually increasing sequence converges to its least upper bound.

(b) Show that a bounded, eventually decreasing sequence converges to its greatest lower bound.

55. If $\lim_{x\to+\infty} f(x)$ fails to exist in Theorem 1.3, then show by example that the associated sequence $\mathbf{a} = (f(1), f(2), f(3), \ldots)$ need not be divergent.

9.2 Infinite Series

Zeno of Elea (c. 495–430 BC) formulated a number of tantalizing paradoxes, which have been troublesome to philosphers from Socrates (who in Plato's *Dialogues* gives a wholly unsatisfactory "refutation") down to modern times. One of his paradoxes leads naturally to the subject of this section, infinite series.

According to Zeno, you cannot leave the room where you are now. For, to reach the door at distance d from you, you first must cover the distance $d/2$. From that point, you must move half ($d/4$) the *remaining* distance to the door. Then you must traverse half ($d/8$) the still-remaining distance. Zeno says that continuing this process of halving makes it clear that you will never reach the door, because you will never reach the end of the sequence of points at distance $\frac{1}{2}d$, $\frac{1}{4}d$, $\frac{1}{8}d$, $\frac{1}{16}d$, etc. from the door. Since you have only a finite life span, you will die of old age before reaching the door! See **Figure 2.1.**

FIGURE 2.1

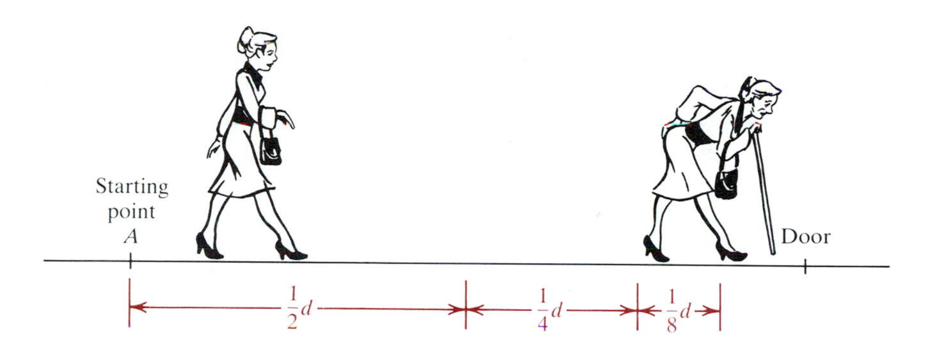

How can we answer Zeno's contention? First, if T is the time used up in covering the distance $\frac{1}{2}d$, and you maintain a constant speed, then it will take you $\frac{1}{2}T$ to cover the distance $\frac{1}{4}d$, and $\frac{1}{4}T$ to cover the distance $\frac{1}{8}d$, etc. Thus, Zeno's contention seems to be that

$$T + \frac{1}{2}T + \frac{1}{4}T + \ldots + \frac{1}{2^{n-1}}T$$

becomes arbitrarily large as n increases without bound. We should then investigate this sum of a finite *geometric sequence*. Each term of such a sequence (after the first term) is a constant r times the preceding term.

2.1 DEFINITION A ***geometric sequence*** is a sequence

$$\boldsymbol{g} = (a, ar, ar^2, ar^3, \ldots, ar^{n-1}, \ldots),$$

where r is a fixed real number called the ***common ratio*** of the sequence.

In Zeno's case, the sequence $\boldsymbol{g} = (T, T/2, T/4, \ldots, T/2^{n-1}, \ldots)$ is geometric, with $r = 1/2$. The ancient Greeks derived the formula for the sum of a finite geometric sequence $(a, ar, ar^2, \ldots, ar^{n-1})$ with $r \neq 1$. Let

$$S = a + ar + ar^2 + \ldots + ar^{n-2} + ar^{n-1}.$$

Multiplying through by r, we get

$$rS = ar + ar^2 + \ldots + ar^{n-2} + ar^{n-1} + ar^n.$$

If we subtract the second equation from the first, most of the terms cancel, giving

$$(1 - r)S = a - ar^n = a(1 - r^n).$$

Since $1 - r \neq 0$, we can divide this equation through by it, obtaining

$$(1) \qquad S = a + ar + ar^2 + \dots + ar^{n-1} = \frac{a(1 - r^n)}{1 - r}.$$

Applying (1) to Zeno's argument, we get

$$T + \frac{1}{2}T + \frac{1}{4}T + \dots + \frac{1}{2^{n-1}}T = \frac{T\left(1 - \left(\frac{1}{2}\right)^n\right)}{1 - \frac{1}{2}} = 2T\left(1 - \frac{1}{2^n}\right).$$

As Zeno keeps dividing your distance from the door into halves, the index n grows without bound, that is, $n \to \infty$. The time required to reach the door then approaches

$$\lim_{n \to \infty} 2T\left(1 - \frac{1}{2^n}\right) = 2T,$$

which is *finite!* Hence you *can* get to the door even if your lifespan is finite! To avoid further pitfalls like Zeno's, we should carefully investigate what happens when infinitely many terms are added together.

2.2 DEFINITION

An ***infinite series*** $\sum_{n=1}^{\infty} a_n$ is a sequence $\mathbf{s} = (s_1, s_2, s_3, \dots, s_n, \dots)$ where

$$\begin{aligned} s_1 &= a_1 \\ s_2 &= a_1 + a_2 &&= s_1 + a_2 \\ s_3 &= a_1 + a_2 + a_3 &&= s_2 + a_3 \\ &\vdots &&\vdots \\ s_n &= a_1 + \dots + a_n &&= s_{n-1} + a_n. \end{aligned}$$

The numbers a_n are called the ***terms*** of the series. The numbers s_n are called the ***partial sums*** of the series. The series is said to ***converge*** (or be ***convergent***) ***to the sum*** S if the sequence of partial sums converges in the sense of Definition 1.2, that is, $\lim_{n \to \infty} s_n = S$. Otherwise the series is said to ***diverge*** (to be ***divergent***).

Definition 2.2 reduces the question of convergence of an infinite series $\Sigma_{n=1}^{\infty} a_n$ to the convergence of the *sequence* $\mathbf{s}$ *of partial sums*, where

$$s_n = a_1 + a_2 + \dots + a_n.$$

Note that the convergence of $\Sigma_{n=1}^{\infty} a_n$ is **not** defined to be equivalent to the convergence of the sequence $\mathbf{a} = (a_1, a_2, \dots, a_n, \dots)$. In particular, the sum S of a convergent infinite series is given by

$$S = \lim_{n \to \infty} s_n, \qquad \textbf{not} \qquad \lim_{n \to \infty} a_n.$$

In Zeno's example, for instance, we saw that

$$\lim_{n \to \infty} s_n = 2T \neq \lim_{n \to \infty} a_n = \lim_{n \to \infty} \frac{T}{2^{n-1}} = 0.$$

This generalizes as follows.

2.3 THEOREM

Geometric Series Test. If $a \neq 0$ and $|r| < 1$, then the geometric series

$$\sum_{n=1}^{\infty} ar^{n-1} = a + ar + ar^2 + \ldots + ar^{n-1} + \ldots$$

converges to

$$S = \frac{a}{1-r}.$$

For all other r, the series diverges.

Proof. If $|r| < 1$, then (1) and Exercise 42 of the last section give

$$\lim_{n\to\infty} s_n = \lim_{n\to\infty} \frac{a(1-r^n)}{1-r} = \frac{a(1-0)}{1-r} = \frac{a}{1-r}.$$

If $r = 1$, then the series is

$$\sum_{n=1}^{\infty} ar^{n-1} = a + a + a + \ldots + a + \ldots,$$

which clearly diverges to $+\infty$ or $-\infty$ since $a \neq 0$. If $r = -1$, then we have

$$\sum_{n=1}^{\infty} ar^{n-1} = a - a + a - a + - \ldots,$$

whose partial sums s_n are alternately 0 (for n even) and a (for n odd). So $\lim_{n\to\infty} s_n$ fails to exist. Finally, if $|r| > 1$, then

$$\lim_{n\to\infty} s_n = \lim_{n\to\infty} \frac{a(1-r^n)}{1-r} = \pm\infty,$$

since if $|r| > 1$, then $\lim_{n\to\infty} |r|^n = +\infty$, by Exercise 43 of Section 1. QED

Theorem 2.3 provides examples of convergent series and of divergent series: Any geometric series converges if its common ratio is strictly between -1 and 1, and diverges otherwise. If $-1 < r < 1$, then by Exercise 42 of Section 1 the terms ar^{n-1} of the geometric series tend to zero as $n \to \infty$. The next theorem says that **any given series $\Sigma_{n=1}^{\infty} a_n$ will diverge unless $a_n \to 0$ as $n \to \infty$.**

2.4 THEOREM

Divergence Test. If $\lim_{n\to\infty} a_n \neq 0$, then $\sum_{n=1}^{\infty} a_n$ is divergent. Thus $\sum_{n=1}^{\infty} a_n$ can converge *only if* $\lim_{n\to\infty} a_n = 0$.

Proof. Suppose that $\lim_{n\to\infty} a_n \neq 0$. If $\Sigma_{n=1}^{\infty} a_n$ converged, then since

$$s_n = s_{n-1} + a_n,$$

we would have by Theorem 1.4(b) that

$$\begin{aligned} S = \lim_{n\to\infty} s_n &= \lim_{n\to\infty} (s_{n-1} + a_n) = \lim_{n\to\infty} s_{n-1} + \lim_{n\to\infty} a_n \\ &= S + \lim_{n\to\infty} a_n, \end{aligned}$$

since as $n \to \infty$, s_{n-1} and s_n would both approach S. Then subtracting S from both sides, we would have

$$0 = \lim_{n\to\infty} a_n,$$

contrary to hypothesis. Thus $\Sigma_{n=1}^{\infty} a_n$ *must* diverge if $\lim_{n\to\infty} a_n \neq 0$. QED

Once we know that the nth term of a series must approach 0 if there is to be *any chance* for it to converge, it is natural to ask whether every series whose nth term approaches 0 is in fact convergent. The answer is a resounding ***NO!***

Example 1

Show that the ***harmonic series*** $\sum_{n=1}^{\infty} \frac{1}{n}$ is divergent.

Solution. Our reasoning is a preview of the integral test of the next section. Let s_n be the nth partial sum of the harmonic series:

$$s_n = 1 + \frac{1}{2} + \frac{1}{3} + \ldots + \frac{1}{n}.$$

Now $1/n$ is the area of the rectangle R_n in **Figure 2.2,** with base the interval $[n, n+1]$ and height $1/n$. So s_n is the sum of the areas of all the rectangles $R_1, R_2, \ldots, R_n$. That sum is a Riemann sum for the function $f(x) = 1/x$ between $x = 1$ and $x = n + 1$. As Figure 2.2 makes clear, this Riemann sum is bigger than

$$\int_1^{n+1} \frac{1}{x}\, dx = \ln x \Big]_1^{n+1} = \ln(n+1),$$

which is the area under $y = 1/x$ from $x = 1$ to $x = n + 1$. Thus, $s_n > \ln(n + 1)$.

FIGURE 2.2

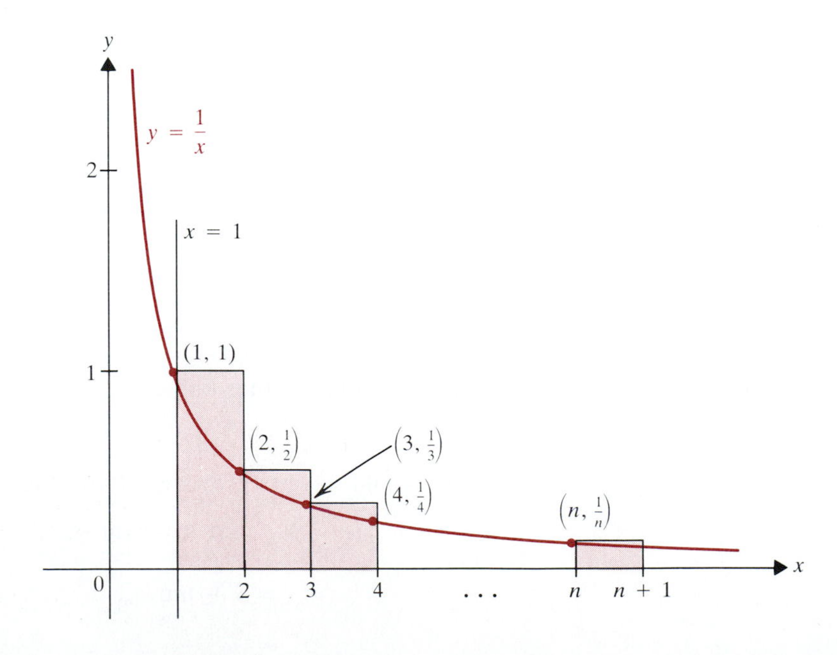

It therefore follows that

$$\lim_{n\to\infty} s_n \geq \lim_{n\to\infty} \ln(n+1) = +\infty,$$

so the sequence of partial sums $(s_1, s_2, \ldots, s_n, \ldots)$ diverges to $+\infty$. Thus

$$\sum_{n=1}^{\infty} \frac{1}{n}$$

is divergent. ■

The preceding example is classical and means that the theory of convergence of infinite series is not trivial, as it would have been if convergence *were* equivalent to the nth term of the series approaching 0 as $n \to \infty$. So ***in testing a given series $\Sigma_{n=1}^{\infty} a_n$ for convergence, first see if $\lim_{n\to\infty} a_n = 0$. If not, then the divergence test*** (*Theorem 2.4*) ***guarantees that the series diverges. If $\lim_{n\to\infty} a_n = 0$, then try to use some other tool*** (such as Theorem 2.3) ***to decide whether the series converges or diverges.*** We will consider some applications of this rule as soon as we have a few more tools to work with. The next result states that convergence depends on the *ultimate* behavior of the series—the first few terms have no effect. The proof is given at the end of the section.

2.5 THEOREM

The series $\sum_{n=1}^{\infty} a_n$ converges if and only if the series

$$\sum_{n=k}^{\infty} a_n = a_k + a_{k+1} + a_{k+2} + \ldots$$

converges for any $k > 1$.

Example 2

If $a_n = 1000$ for $n \leq 100{,}000$ and $a_n = 1/2^n$ for $n > 100{,}000$, then does $\Sigma_{n=1}^{\infty} a_n$ converge?

Solution. Yes, it does converge, since $\Sigma_{n=k}^{\infty} a_n$ converges for $k = 100{,}001$ as a geometric series with common ratio 1/2. ■

The next result says that convergent series possess a property we have seen before, when we studied differentiation and integration. Its proof is asked for in Exercises 37 and 38.

2.6 THEOREM

If $\sum_{n=1}^{\infty} a_n$ and $\sum_{n=1}^{\infty} b_n$ are convergent series with respective sums S and T, and c is a real number, then

$$\sum_{n=1}^{\infty} (a_n + b_n)$$

converges to $S + T$, and $\sum_{n=1}^{\infty} ca_n$ converges to cS.

As a consequence of Theorem 2.6, it follows that **if $\Sigma_{n=1}^{\infty} a_n$ diverges, then so does any series $\Sigma_{n=1}^{\infty} ca_n$ for $c \neq 0$.** (For if $\Sigma_{n=1}^{\infty} ca_n$ converged, then so would

$$\sum_{n=1}^{\infty} \frac{1}{c}(ca_n) = \sum_{n=1}^{\infty} (a_n).$$

Example 3

Decide whether the following series converge.

$$\text{(a)} \sum_{n=1}^{\infty} \frac{3}{2n}, \qquad \text{(b)} \sum_{n=1}^{\infty} \left(\frac{1}{2^n} - \frac{1}{4^n}\right).$$

Solution. (a) First,

$$\lim_{n \to \infty} \frac{3}{2n} = 0,$$

so the divergence test does not apply. However, we can use the remark above:

$$\sum_{n=1}^{\infty} \frac{3}{2n} = \sum_{n=1}^{\infty} \frac{3}{2}\left(\frac{1}{n}\right).$$

Thus the given series must diverge, since its terms are multiples of the terms of the divergent harmonic series of Example 1.

(b) Here

$$\sum_{n=1}^{\infty} \left(\frac{1}{2^n} - \frac{1}{4^n}\right) = \sum_{n=1}^{\infty} \left[\frac{1}{2^n} + \left(-\frac{1}{4^n}\right)\right]$$

converges by Theorem 2.6, since

$$\sum_{n=1}^{\infty} \frac{1}{2^n} \quad \text{and} \quad \sum_{n=1}^{\infty} -\frac{1}{4^n}$$

are convergent geometric series ($r = 1/2$ and $r = 1/4$, respectively). The sum is

$$\frac{\frac{1}{2}}{1 - \frac{1}{2}} - \frac{\frac{1}{4}}{1 - \frac{1}{4}} = 1 - \frac{1}{3} = \frac{2}{3}. \quad ■$$

Sometimes we can find the sum of a convergent series even if it is not geometric. One such class of series consists of ***telescoping series*** of the form $\sum_{n=1}^{\infty} (a_n - a_{n+1})$. The partial sums of such a series are

$$\begin{aligned}
s_1 &= a_1 - a_2, \\
s_2 &= a_1 - a_2 + a_2 - a_3 = a_1 - a_3, \\
s_3 &= s_2 + a_3 - a_4 = a_1 - a_4, \\
&\vdots \\
s_n &= a_1 - a_{n+1}.
\end{aligned}$$

Thus, if we can calculate $\lim_{n \to \infty} a_n$, then we can find $\lim_{n \to \infty} s_n$.

Example 4

Find the sum of the series $\displaystyle\sum_{n=2}^{\infty} \frac{1}{n^2 - 1}$.

Solution. The term $1/(n^2 - 1)$ decomposes into partial fractions:

$$\frac{1}{n^2 - 1} = \frac{1}{(n-1)(n+1)} = \frac{1/2}{n-1} - \frac{1/2}{n+1} = \frac{1}{2}\left(\frac{1}{n-1} - \frac{1}{n+1}\right).$$

So the first few partial sums are

$$s_2 = \frac{1}{2}\left(1 - \frac{1}{3}\right),$$

$$s_3 = s_2 + \frac{1}{2}\left(\frac{1}{2} - \frac{1}{4}\right) = \frac{1}{2}\left(1 - \frac{1}{3} + \frac{1}{2} - \frac{1}{4}\right) = \frac{1}{2}\left(1 - \frac{1}{3} + \frac{1}{4}\right),$$

$$s_4 = s_3 + \frac{1}{2}\left(\frac{1}{3} - \frac{1}{5}\right) = \frac{1}{2}\left(1 - \frac{1}{3} + \frac{1}{4} + \frac{1}{3} - \frac{1}{5}\right) = \frac{1}{2}\left(1 + \frac{1}{4} - \frac{1}{5}\right),$$

$$s_5 = s_4 + \frac{1}{2}\left(\frac{1}{4} - \frac{1}{6}\right) = \frac{1}{2}\left(1 + \frac{1}{4} - \frac{1}{5} + \frac{1}{4} - \frac{1}{6}\right) + \frac{1}{2}\left(\frac{3}{2} - \frac{1}{5} - \frac{1}{6}\right),$$

$$s_6 = s_5 + \frac{1}{2}\left(\frac{1}{5} - \frac{1}{7}\right) = \frac{1}{2}\left(\frac{3}{2} - \frac{1}{5} - \frac{1}{6} + \frac{1}{5} - \frac{1}{7}\right) = \frac{1}{2}\left(\frac{3}{2} - \frac{1}{6} - \frac{1}{7}\right),$$

$$s_7 = s_6 + \frac{1}{2}\left(\frac{1}{6} - \frac{1}{8}\right) = \frac{1}{2}\left(\frac{3}{2} - \frac{1}{6} - \frac{1}{7} + \frac{1}{6} - \frac{1}{8}\right) = \frac{1}{2}\left(\frac{3}{2} - \frac{1}{7} - \frac{1}{8}\right).$$

Thus we see that as $n \to \infty$,

$$s_n = \frac{1}{2}\left(\frac{3}{2} - \frac{1}{n} - \frac{1}{n+1}\right) = \frac{1}{2}\left(\frac{3}{2} - \frac{n+1+n}{n(n+1)}\right)$$

$$= \frac{1}{2}\left(\frac{3}{2} - \frac{2n+1}{n^2+n}\right) \text{ approaches } \frac{3}{4} - 0 = \frac{3}{4}.$$

Therefore the given series converges to 3/4. ■

Proof of Theorem 2.5*

2.5 THEOREM The series $\sum_{n=1}^{\infty} a_n$ converges if and only if the series

$$\sum_{n=k}^{\infty} a_n = a_k + a_{k+1} + a_{k+2} + \ldots$$

converges for any $k > 1$.

Proof. Let $s_n = \Sigma_{i=1}^{n} a_i$ and $t_n = \Sigma_{i=k}^{n} a_i$, where for $n < k$ we understand t_n to be 0. Then for $n \geq k$, we have

$$s_n = t_n + \sum_{i=1}^{k-1} a_i = t_n + c,$$

where $c = \Sigma_{i=1}^{k-1} a_1$ is fixed. It follows that the sequences of partial sums of the two series in the statement of the theorem are

$$\boldsymbol{s} = (s_1, s_2, \ldots, s_k, s_{k+1}, s_{k+2}, \ldots, s_n, \ldots)$$

and

$$\boldsymbol{t} = (0, 0, \ldots, 0, t_k, t_{k+2}, \ldots, t_n, \ldots)$$
$$= (0, 0, \ldots, 0, s_k - c, s_{k+1} - c, \ldots, s_n - c, \ldots).$$

Note that $t_n = s_n - c$ for $n \geq k$. Thus, by Theorem 1.4(b), $\boldsymbol{t}$ converges if and only if

$$\lim_{n\to\infty} t_n = \lim_{n\to\infty} (s_n - c) = \lim_{n\to\infty} s_n - c,$$

that is, if and only if $\boldsymbol{s}$ converges. QED

* Optional

Exercises 9.2

In Exercises 1–26, determine whether or not the given series is convergent. If it is convergent, then find its sum.

1. $\sum_{n=1}^{\infty} \frac{1}{3n}$

2. $\sum_{n=1}^{\infty} \frac{1}{7n}$

3. $\sum_{n=1}^{\infty} \left(\frac{3}{2}\right)^n$

4. $\sum_{n=1}^{\infty} \left(\frac{4}{3}\right)^n$

5. $\sum_{n=1}^{\infty} 5\left(\frac{3}{4}\right)^{n-1}$

6. $\sum_{n=1}^{\infty} \frac{7}{3^{n+1}}$

7. $1 - \frac{1}{2} + \frac{1}{4} - \frac{1}{8} + - \ldots$

8. $3 - 2 + \frac{4}{3} - \frac{8}{9} + - \ldots$

9. $\frac{1}{4} + \frac{1}{3} + \frac{4}{9} + \frac{16}{27} + \ldots$

10. $-\frac{1}{8} + \frac{5}{32} - \frac{25}{128} + \frac{125}{512} - + \ldots$

11. $\frac{1}{2}\sin\theta - \frac{1}{4}\sin^2\theta + \frac{1}{8}\sin^3\theta - \frac{1}{16}\sin^4\theta + - \ldots, \theta \in [0, 2\pi]$

12. $\frac{2}{3}\cos\theta - \frac{4}{9}\cos^2\theta + \frac{8}{27}\cos^3\theta - \frac{16}{81}\cos^4\theta + - \ldots, \theta \in [0, 2\pi]$

13. $\sum_{n=1}^{\infty} \left(\frac{4}{3^n} + \frac{(-2)^n}{3^n}\right)$

14. $\sum_{n=1}^{\infty} \left(\frac{3}{2n} - \frac{1}{2^n}\right)$

15. $\sum_{n=1}^{\infty} \frac{3^{n-2}}{4^n}$

16. $\sum_{n=1}^{\infty} \frac{3^{n-1}}{4^{n+1}}$

17. $\sum_{n=1}^{\infty} \left(\frac{1}{3n} - \frac{1}{4n}\right)$

18. $\sum_{n=1}^{\infty} e^{-n}$

19. $\sum_{n=1}^{\infty} \frac{n^2}{10n - 5}$

20. $\sum_{n=1}^{\infty} \frac{2n^3 - 5n - 100}{100n^3 + 100n + 1}$

21. $-1 - 2 - \ldots - 1000 + \sum_{n=1}^{\infty} \frac{3}{4n}$

22. $100 + 100^2 + \ldots + 100^{100} + \sum_{n=1}^{\infty} \frac{2^n}{3^{n-1}}$

23. $\sum_{n=1}^{\infty} \frac{1}{n(n+1)}$ (*Hint:* See Example 4.)

24. $\sum_{n=3}^{\infty} \frac{1}{n^2 - 4}$

25. $\sum_{n=1}^{\infty} \ln \frac{n}{n+1}$

26. $\sum_{n=3}^{\infty} \ln\left(\frac{1}{n^2 + n}\right)$

27. Prove that if $\Sigma_{n=1}^{\infty} a_n$ converges and $\Sigma_{n=1}^{\infty} b_n$ diverges, then $\Sigma_{n=1}^{\infty} (a_n + b_n)$ diverges. (*Hint:* Use Theorem 2.6.)

28. Show that if $\Sigma_{n=1}^{\infty} a_n$ and $\Sigma_{n=1}^{\infty} b_n$ both diverge, then we can't conclude that $\Sigma_{n=1}^{\infty} (a_n + b_n)$ diverges.

29. Show that the telescoping series $\Sigma_{n=1}^{\infty} (a_n - a_{n+1})$ converges if and only if the sequence $\boldsymbol{a} = (a_1, a_2, \ldots, a_n, \ldots)$ converges.

30. A rubber ball is dropped from a height of six feet. Each time it hits the floor, it rebounds 5/8 of the distance it fell. What is the total distance traveled by the ball?

31. ***Cantor's middle-third set*** (named for the nineteenth-century German mathematician Georg Cantor [1835–1918], the founder of modern set theory) is constructed as follows. The open interval (1/3, 2/3) is erased from the closed interval [0, 1]. Next, the two open intervals (1/9, 2/9) and (7/9, 8/9) are erased from the remnant. One continues in this way, at each stage erasing the open middle third of each remaining subinterval of [0, 1].
(a) Find the total length of all the subintervals erased.
(b) Give an example of a point that is never erased.

32. Zeno claimed that the tortoise would always outrace the swiftest hare, if only the tortoise could have a head start. For suppose the tortoise starts at P, at some positive distance from the starting point of the hare. Then by the time the hare has arrived at P, the tortoise has proceeded on to Q, at some positive distance from P. Again, by the time the hare has reached Q, the tortoise has lumbered on to a new point R, still a positive distance from Q. "Clearly" we can continue this analysis and thus prove that the hare can never catch the tortoise! Can you explain this paradox?

33. Show that $1 - 1 + 1 - 1 + - \ldots + (-1)^{n-1} + \ldots$ diverges.

34. Refer to Exercise 33. Explain the fallacy in the following "proof" that the series converges to 1/2. Let

$$S = 1 - 1 + 1 - 1 + - \ldots.$$

Then

$$S - 1 = -1 + 1 - 1 + - \ldots = -S.$$

Hence adding S to both sides, we obtain $2S = 1$, that is, $S = 1/2$.

35. Following Example 8 in Section 7.6, we approximated the present value of a ten-year series of \$600 monthly payments. Using a projected interest rate of 8%, we obtained

$$600[e^{-r/12} + e^{-2r/12} + \ldots + e^{-120r/12}] \approx 7200 \int_0^{10} e^{-0.08t}\,dt.$$

The value of the integral was computed as \$49,560.39. Use (1) to find the *exact* present value of the series of payments.

36. A certain state lottery pays the winner a "million dollar jackpot" in 20 annual payments of \$50,000. Assuming an interest rate of 10% over that period, find the present value of the jackpot.

37. Prove the first assertion in Theorem 2.6.

38. Prove the second assertion in Theorem 2.6.

39. Prove the following part of ***Cauchy's criterion***, which extends Theorem 2.4. If $\Sigma_{n=1}^{\infty} a_n$ converges, then for any $\varepsilon > 0$, there is an N such that $|s_m - s_n| < \varepsilon$ for all $m, n > N$. Then show that the choice $m = n + 1$ yields Theorem 2.4 as a special case.

40. Use the Cauchy criterion of the preceding exercise to give another proof that the harmonic series

$$\sum_{n=1}^{\infty} \frac{1}{n}$$

diverges. (*Hint:* Note that

$$\frac{1}{n+k} \geq \frac{1}{2n} \qquad \text{for } 0 \leq k \leq n.$$

Use this to show that $|s_{2n} - s_n| \geq 1/2$, i.e., that the criterion in Exercise 39 fails for the particular case $m = 2n$.)

9.3 Series of Positive Terms

The last section's convergence tests often fail to apply to given series $\Sigma_{n=1}^{\infty} a_n$. For example, consider

$$\sum_{n=1}^{\infty} \frac{1}{3n+2}.$$

It seems that for large values of n the terms of this series are nearly the same as the corresponding terms of

$$\sum_{n=1}^{\infty} \frac{1}{3n} = \sum_{n=1}^{\infty} \left(\frac{1}{3}\right)\left(\frac{1}{n}\right),$$

which diverges by the remark following Theorem 2.6 (since the harmonic series diverges). So a reasonable guess is that

$$\sum_{n=1}^{\infty} \frac{1}{3n+2}$$

diverges. But at present we can't verify that guess. This section presents some further convergence tests for series of positive terms, that is, series $\Sigma_{n=1}^{\infty} a_n$ where $a_n > 0$. (All the results have symmetric versions for series of negative terms, but we won't state those explicitly.) First we give two important *comparison tests.* The first one says that *a series whose terms are smaller than those of a convergent series must also converge,* and *a series whose terms are larger than those of a divergent series must diverge.*

3.1 THEOREM

Term-Size Comparison Test. Let $\sum_{n=1}^{\infty} a_n$ and $\sum_{n=1}^{\infty} b_n$ be series of positive terms.

(a) If $\sum_{n=1}^{\infty} b_n$ converges and if for some $k \geq 1$, we have $a_n \leq b_n$ for all $n \geq k$, then $\sum_{n=1}^{\infty} a_n$ also converges.

(b) If $\sum_{n=1}^{\infty} b_n$ diverges and if for some $k \geq 1$, we have $a_n \geq b_n$ for all $n \geq k$, then $\sum_{n=1}^{\infty} a_n$ also diverges.

Proof. In view of Theorem 2.5 we may as well assume that $k = 1$.

(a) We denote the nth partial sum of $\Sigma_{n=1}^{\infty} a_n$ by s_n and the nth partial sum of $\Sigma_{n=1}^{\infty} b_n$ by t_n. Then $\boldsymbol{s} = (s_1, s_2, \ldots, s_n, \ldots)$ and $\boldsymbol{t} = (t_1, t_2, \ldots, t_n, \ldots)$ are strictly increasing sequences, because $s_n - s_{n-1} = a_n > 0$ and $t_n - t_{n-1} = b_n > 0$. Moreover, the hypothesis that $\Sigma_{n=1}^{\infty} b_n$ converges means that $\boldsymbol{t}$ converges to $T = \lim_{n\to\infty} t_n$. That is, T is the limit of a strictly increasing sequence. We also have for any n,

$$s_n = \sum_{j=1}^{n} a_j \leq \sum_{j=1}^{n} b_n = t_n,$$

so $s_n \leq t_n \leq T$. Hence T is an upper bound for the strictly increasing sequence $\boldsymbol{s}$. Then by Theorem 1.7 the sequence $\boldsymbol{s}$ converges. Thus by Definition 2.2 $\Sigma_{n=1}^{\infty} a_n$ converges.

(b) If $\Sigma_{n=1}^{\infty} a_n$ converged, then we could apply (a) to conclude that $\Sigma_{n=1}^{\infty} b_n$ also converged, since $b_n \leq a_n$ for all sufficiently large n. But by hypothesis $\Sigma_{n=1}^{\infty} b_n$ diverges. Therefore $\Sigma_{n=1}^{\infty} a_n$ must also diverge. QED

To apply Theorem 3.1 you must first make a conjecture that a given series $\Sigma_{n=1}^{\infty} a_n$ converges or diverges. The best way to do so is by spotting a resemblance to some known series $\Sigma_{n=1}^{\infty} b_n$. **If you think that $\Sigma_{n=1}^{\infty} a_n$ converges, then you need to find a convergent series $\Sigma_{n=1}^{\infty} b_n$ such that**

$$a_n \leq b_n \tag{1}$$

for all n beyond some point. Then Theorem 3.1(a) says that $\Sigma_{n=1}^{\infty} a_n$ also converges. **If you think that $\Sigma_{n=1}^{\infty} a_n$ diverges, then you need to find a divergent series $\Sigma_{n=1}^{\infty} b_n$ for which**

$$a_n \geq b_n \tag{2}$$

for all n beyond some point. Then Theorem 3.1(b) says that $\Sigma_{n=1}^{\infty} a_n$ diverges.

Example 1

Decide whether or not $\displaystyle\sum_{n=1}^{\infty} \frac{3}{4^n + 1}$ is convergent.

Solution. The terms of the given series resemble those of

$$\sum_{n=1}^{\infty} \frac{3}{4^n} = \sum_{n=1}^{\infty} 3\left(\frac{1}{4}\right)^n,$$

a convergent geometric series by Theorems 2.6 and 2.3 ($r = 1/4 < 1$). Since

$$\frac{3}{4^n + 1} < \frac{3}{4^n}$$

for all $n \geq 1$, we have (1). Therefore Theorem 3.1(a) says that the given series converges. ■

In Example 1 the nth term $3/4^n$ of the given series approaches 0 as $n \to \infty$, but *that is not enough to imply convergence.* Be careful not to try to make the divergence test (Theorem 2.4) into more than it is!

Example 2

Decide whether or not $\sum_{n=1}^{\infty} \frac{1}{3n+2}$ converges.

Solution. In the discussion at the start of the section we guessed that it diverges, because of its resemblance to $\Sigma_{n=1}^{\infty} (1/3n)$. Theorem 3.1(b) will confirm this if we can find a divergent series $\Sigma_{n=1}^{\infty} b_n$ such that (2) holds:

$$\frac{1}{3n+2} > b_n$$

for all sufficiently large n. Since

$$\frac{1}{3n+2} < \frac{1}{3n}$$

for all n, we can't use $\Sigma_{n=1}^{\infty} (1/3n)$ as the comparison series $\Sigma_{n=1}^{\infty} b_n$. Instead, we use $\Sigma_{n=1}^{\infty} (1/4n)$, because

$$\frac{1}{3n+2} \geq \frac{1}{3n+n} = \frac{1}{4n} = b_n$$

for all $n \geq 2$. Since

$$\sum_{n=1}^{\infty} \frac{1}{4n} = \sum_{n=1}^{\infty} \frac{1}{4}\left(\frac{1}{n}\right)$$

diverges by Theorems 2.3 and 2.6, we conclude from Theorem 3.1(b) that

$$\sum_{n=1}^{\infty} \frac{1}{3n+2}$$

also diverges. ■

Note again that the fact that the nth term of the series in Example 2 approaches 0 as $n \to \infty$ does *not* imply that the series is convergent!

There is a second comparison test, which is often easier to use than the term-size comparison test.

3.2 THEOREM

Limit Comparison Test. Suppose that $\sum_{n=1}^{\infty} a_n$ and $\sum_{n=1}^{\infty} b_n$ are series of positive terms. If

$$\lim_{n \to \infty} \frac{a_n}{b_n} = l,$$

where l is a positive real number, then either both series converge or both diverge.

Proof. Suppose that

$$\lim_{n \to \infty} \frac{a_n}{b_n} = l > 0.$$

Then $\frac{1}{2}l > 0$. Thus by Definition 1.2 there is an N such that

$$\left|\frac{a_n}{b_n} - l\right| < \frac{1}{2}l \qquad \text{for all } n > N.$$

Then for all $n > N$, Theorem 1.2 of Chapter 1 says that

$$(3) \qquad -\frac{1}{2}l < \frac{a_n}{b_n} - l < \frac{1}{2}l \rightarrow \frac{1}{2}l < \frac{a_n}{b_n} < \frac{3}{2}l.$$

Hence we have

$$(4) \qquad \frac{1}{2}lb_n < a_n \rightarrow b_n < \frac{2}{l}a_n.$$

Now if $\sum_{n=1}^{\infty} a_n$ converges, then so does

$$\sum_{n=1}^{\infty} \frac{2}{l}a_n$$

by Theorem 2.6. Therefore $\sum_{n=1}^{\infty} b_n$ converges by (4) and Theorem 3.1(a).

On the other hand, if $\sum_{n=1}^{\infty} a_n$ diverges, then so does

$$\sum_{n=1}^{\infty} \frac{2}{3l}a_n$$

by the remark following Theorem 2.6. For $n > N$, it follows from the last part of (3) that

$$a_n < \frac{3}{2}lb_n \rightarrow \frac{2}{3l}a_n < b_n.$$

Then by Theorem 3.1(b) $\sum_{n=1}^{\infty} b_n$ also diverges. Thus $\sum_{n=1}^{\infty} b_n$ converges (respectively, diverges) if $\sum_{n=1}^{\infty} a_n$ does. The same reasoning will also show (Exercise 33) that $\sum_{n=1}^{\infty} a_n$ converges (respectively, diverges) if $\sum_{n=1}^{\infty} b_n$ does. Thus either both series converge or both diverge. QED

If the limit l in Theorem 3.2 is 0, then the limit comparison test fails (Exercise 31). It also does not apply if $l = +\infty$. (But see Exercises 35 and 37 for a partial extension of the test to the cases $l = 0$ and $l = +\infty$.)

Theorem 3.2 makes it possible to use the resemblance of

$$\sum_{n=1}^{\infty} \frac{1}{3n+2} \qquad \text{to} \qquad \sum_{n=1}^{\infty} \frac{1}{3n}$$

to obtain a quick solution to Example 2. We have

$$\lim_{n\to\infty} \frac{\dfrac{1}{3n+2}}{\dfrac{1}{3n}} = \lim_{n\to\infty} \frac{3n}{3n+2} = \lim_{n\to\infty} \frac{3}{3+2/n} = 1,$$

a positive real number. Theorem 3.2 thus says that since $\sum_{n=1}^{\infty} (1/3n)$ is divergent, so is

$$\sum_{n=1}^{\infty} \frac{1}{3n+2}.$$

The following example further illustrates the value of the limit comparison test.

Example 3

Determine whether or not $\sum_{n=1}^{\infty} \frac{3n+1}{5n^2-2}$ converges.

Solution. Since the numerator of each term has degree one less than the denominator, we can reason that for large n,

$$a_n = \frac{3n+1}{5n^2-2} \approx \frac{3}{5n},$$

so the given series probably diverges. To confirm this we let $b_n = 1/n$ and compute $\lim_{n\to\infty} a_n/b_n$. We get

$$\frac{a_n}{b_n} = \frac{\frac{3n+1}{5n^2-2}}{\frac{1}{n}} = \frac{3n^2+n}{5n^2-2} = \frac{3+\frac{1}{n}}{5-\frac{2}{n^2}} \to \frac{3}{5} > 0,$$

as $n \to \infty$. Thus, since $\sum_{n=1}^{\infty} (1/n)$ diverges, so does

$$\sum_{n=1}^{\infty} \frac{3n+1}{5n^2-2}. \blacksquare$$

The following more general version of the reasoning used in Example 1 of the last section goes back at least as far as the Scottish mathematician Colin Maclaurin (1698–1746), but the proof found at the end of the section is due to Cauchy.

3.3 THEOREM

Integral Test. Suppose that f is a continuous, positive, decreasing function on $(k, +\infty)$. If $f(n) = a_n$ for all positive integers $n \geq k$, then $\sum_{n=1}^{\infty} a_n$ is convergent if and only if $\int_k^{+\infty} f(x)\,dx$ is convergent.

Note that the **integral test does not say that $\Sigma_{n=1}^{\infty} a_n$ and $\int_1^{+\infty} f(x)\,dx$ converge to the same limit if the improper integral is convergent.** In fact, that almost never happens. The integral test can only show that a given series is convergent; it does *not* find its sum. The next example illustrates its use.

Example 4

Determine whether or not $\sum_{n=2}^{\infty} \frac{1}{n \ln n}$ converges.

Solution. The integral test is worth trying, since we can calculate

$$\int \frac{1}{x \ln x}\,dx = \ln|\ln x| + C.$$

To apply the test we must first check that its hypotheses are satisfied by the function $f(x) = 1/(x \ln x)$. First, for $n \geq 2$ this function is continuous and has positive values. To check that it is decreasing we calculate

$$f'(x) = \frac{(x \ln x) \cdot 0 - 1 \cdot \left(\ln x + x \cdot \dfrac{1}{x}\right)}{x^2(\ln x)^2} = -\frac{1 + \ln x}{x^2(\ln x)^2},$$

which is negative for $x \geq 2$. Thus, by Theorem 3.1 of Chapter 3, f is decreasing for $x \geq 2$. Now we calculate

$$\int_2^{+\infty} \frac{1}{x \ln x}\, dx = \lim_{k \to +\infty} \int_2^k \frac{1}{x \ln x}\, dx = \lim_{k \to +\infty} \ln(\ln x)\Big|_2^k = +\infty.$$

Since the improper integral is divergent, so is the given series, by Theorem 3.3. ■

We can also use the integral test to derive a useful generalization of the result of Example 1 of the last section that $\Sigma_{n=1}^{\infty} (1/n)$ is divergent. Consider the ***hyperharmonic series*** $\Sigma_{n=1}^{\infty} (1/n^p)$, where p is a fixed real number. (This is also sometimes called the ***p-series.***) If $p < 0$, then it diverges by the divergence test (Theorem 2.4), because

$$\lim_{n \to \infty} \frac{1}{n^p} = \lim_{n \to \infty} n^{-p} = +\infty,$$

since $-p > 0$. Suppose next that $p > 0$. Define the function f by $f(x) = 1/x^p = x^{-p}$. This is positive and continuous on $(1, +\infty)$. It is also decreasing since $p > 0$:

$$f'(x) = -px^{-p-1} < 0 \qquad \text{for } x > 1.$$

For each integer n we have $f(n) = 1/n^p$, so we can use the integral test: The given series converges if and only if the improper integral $\int_1^{+\infty} f(x)\, dx$ converges. If $p = 1$, the series is the divergent harmonic series. Suppose then that $p \neq 1$. We find

$$\int_1^{+\infty} x^{-p}\, dx = \lim_{b \to +\infty} \int_1^p x^{-p}\, dx = \lim_{b \to +\infty} \left[\frac{x^{-p+1}}{-p+1}\right]_1^b$$

$$= \lim_{b \to +\infty} \frac{1}{1-p}[b^{-p+1} - 1] = \frac{1}{1-p} \lim_{b \to +\infty} \left[\frac{1}{b^{p-1}} - 1\right].$$

If $p - 1 > 0$ (that is, if $p > 1$) then the limit in brackets is -1, since $1/b^{p-1} \to 0$ as $b \to +\infty$. If $p - 1 < 0$ (that is, if $p < 1$) then the limit in brackets fails to exist, since $1/b^{p-1} = b^{1-p} \to +\infty$ as $b \to +\infty$. The integral test therefore says that

(5) $\displaystyle\sum_{n=1}^{\infty} \frac{1}{n^p}$ converges if and only if $p > 1$.

Example 5

Determine whether or not $\displaystyle\sum_{n=1}^{\infty} \frac{1}{n^{3/2} + 5}$ converges.

Solution. The series $\Sigma_{n=1}^{\infty} (1/n^{3/2})$ converges by (5), since $p = 3/2 > 1$. We can then use the limit comparison test (Theorem 3.2). Letting $a_n = 1/n^{3/2}$ and $b_n = 1/(n^{3/2} + 5)$, we have

$$\frac{a_n}{b_n} = \frac{n^{3/2} - 5}{n^{3/2}}.$$

Then

$$\lim_{n\to\infty} \frac{a_n}{b_n} = \lim_{n\to\infty} \frac{n^{3/2} + 5}{n^{3/2}} = \lim_{n\to\infty} \left(1 + \frac{5}{n^{3/2}}\right) = 1,$$

a positive real number. Since $\Sigma_{n=1}^{\infty} (1/n^{3/2})$ converges, so does

$$\sum_{n=1}^{\infty} \frac{1}{n^{3/2} + 5}$$

by Theorem 3.2. ■

We now have a number of tests for convergence of an infinite series for which all the terms have the same sign. Theorem 2.3 completely classifies geometric series and (5) does the same for hyperharmonic series. The divergence test (Theorem 2.4) disposes of any series whose nth term doesn't approach zero. The integral test (Theorem 3.3) gives a criterion for convergence of series whose defining function is readily integrable. And the two comparison tests (Theorems 3.1 and 3.2) allow us to determine convergence or divergence of many series that closely resemble standard series. With several tools now available, some thought is needed in choosing the one best suited to a given problem.

Example 6

Decide whether or not $\sum_{n=1}^{\infty} ne^{-n^2}$ converges.

Solution. This does not resemble any of the standard series, so no obvious comparison suggests itself. We might try taking the limit of its nth term as $n \to \infty$. But l'Hôpital's rule gives

$$\lim_{x\to+\infty} xe^{-x^2} = \lim_{x\to+\infty} \frac{x}{e^{x^2}} = \lim_{x\to+\infty} \frac{1}{2xe^{x^2}} = 0,$$

so by Theorem 1.3 the nth term tends to 0. Thus the divergence test doesn't apply. Having eliminated virtually every other tool, we try the integral test. Let $f(x) = xe^{-x^2}$. Then f is positive and continuous on $[1, +\infty)$ and

$$f'(x) = e^{-x^2} - 2x^2e^{-x^2} = e^{-x^2}(1 - 2x^2) < 0 \qquad \text{for } x > 1,$$

so f is decreasing on $(1, +\infty)$. We have

$$\int_1^{+\infty} xe^{-x^2}\,dx = \lim_{b\to+\infty} \int_1^b xe^{-x^2}\,dx = \lim_{b\to+\infty} \left[-\frac{1}{2}\int_1^b -2xe^{-x^2}\,dx\right]$$

$$= \left[-\frac{1}{2}\lim_{b\to+\infty} e^{-x^2}\right]_1^b = -\frac{1}{2}\lim_{b\to+\infty} \left[e^{-b^2} - e^{-1}\right] = \frac{1}{2e}.$$

Since $\int_1^{+\infty} xe^{-x^2}\,dx$ converges, the integral test says that $\Sigma_{n=1}^{\infty} ne^{-n^2}$ also converges. (Remember though, that the integral test does *not* say that the series converges to $1/2e$.) ■

Proof of the Integral Test (Theorem 3.3)*

Proof. Suppose that f is continuous, decreasing, and positive-valued on $[k, +\infty)$. If $f(n) = a_n$ for $n \geq k$, then we must show that $\Sigma_{n=1}^{\infty} a_n$ converges if and only if the improper integral $\int_k^{+\infty} f(x)\,dx$ is convergent.

First, in view of Theorem 2.5 we may as well assume that $k = 1$. For any n, partition the interval $[1, n]$ into subintervals of length 1 and form the Riemann sums

$$R_1 = \qquad f(2) + f(3) + \ldots + f(n-1) + f(n)$$

and

$$R_2 = f(1) + f(2) + \qquad \ldots + f(n-1).$$

See **Figure 3.1.** Since f is decreasing, its maximum on $[i-1, i]$ is $f(i-1)$, and its minimum is $f(i)$. So R_2 is the sum of the areas of the larger rectangles, and R_1 is the sum of the areas of the smaller rectangles. We then have

$$(6) \qquad R_1 \leq \int_1^n f(x)\,dx \leq R_2,$$

since $\int_1^n f(x)\,dx$ is the area under the curve between $x = 1$ and $x = n$. If, as usual, we write s_n for the nth partial sum of the series, then (6) becomes

$$(7) \qquad s_n - a_1 \leq \int_1^n f(x)\,dx \leq s_{n-1}.$$

Now we are ready to complete the proof.

FIGURE 3.1

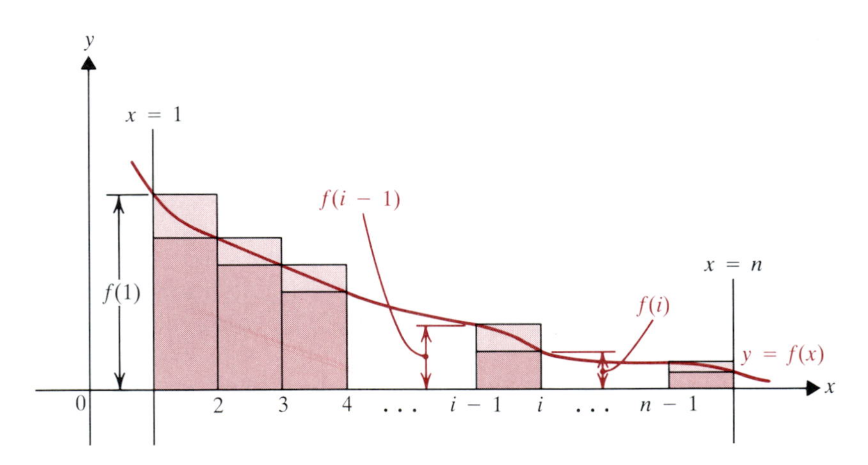

(a) Suppose that the given series converges. Since each term a_i is positive, the sequence of partial sums s_{n-1} is strictly increasing. Thus for each positive integer n,

$$s_{n-1} \leq s_n \leq \lim_{n\to\infty} s_n.$$

We therefore get from (7)

$$\int_1^n f(x)\,dx \leq \lim_{n\to\infty} s_n.$$

Now the sequence $(I_1, I_2, \ldots, I_n, \ldots)$ given by $I_n = \int_1^n f(x)\,dx$ is an increasing sequence since $f(x) \geq 0$ on $[1, +\infty)$. This sequence is also

* Optional

bounded above by $\lim_{n\to\infty} s_n$. So by Theorem 1.7,

$$\lim_{n\to\infty} I_n = \lim_{n\to\infty} \int_1^n f(x)\,dx$$

exists. Therefore the improper integral $\int_1^{+\infty} f(x)\,dx$ converges.

(b) Conversely, suppose that the improper integral $\int_1^{+\infty} f(x)\,dx$ converges. That is, $\lim_{b\to\infty} \int_1^b f(x)\,dx = I$, some real number. Then for all n,

$$\int_1^n f(x)\,dx \le I \xrightarrow{(7)} s_n - a_1 \le I \longrightarrow s_n \le a_1 + I.$$

Thus the sequence $s = (s_1, s_2, \ldots, s_n, \ldots)$ of partial sums is bounded above. Since each $a_n > 0$, this is an increasing sequence. Hence by Theorem 1.7, s is convergent. Therefore $\Sigma_{n=1}^{\infty} a_n$ converges. QED

Exercises 9.3

In Exercises 1–22, determine whether or not the given series converges.

1. $\sum_{n=1}^{\infty} \frac{1}{n^2+n}$

2. $\sum_{n=1}^{\infty} \frac{1}{2n^2+3n}$

3. $\sum_{n=2}^{\infty} \frac{1}{n^2-n}$

4. $\sum_{n=1}^{\infty} \frac{1}{3n^2-n}$

5. $\sum_{n=1}^{\infty} \frac{1}{\sqrt{2n+1}}$

6. $\sum_{n=1}^{\infty} \frac{1}{\sqrt{3n-1}}$

7. $\sum_{n=2}^{\infty} \frac{\ln n}{2n}$

8. $\sum_{n=1}^{\infty} \frac{\ln \sqrt{n}}{n}$

9. $\sum_{n=2}^{\infty} \frac{\tan^{-1} n}{3n^2+3}$

10. $\sum_{n=2}^{\infty} \frac{1}{n(\ln n)^2}$

11. $\sum_{n=1}^{\infty} \frac{3n^2-4}{n^3+2n^2+1}$

12. $\sum_{n=1}^{\infty} \frac{n^2-2}{n^3+5n+1}$

13. $\sum_{n=1}^{\infty} \frac{n-1}{n^3+4n^2+1}$

14. $\sum_{n=1}^{\infty} \frac{2n-3}{n^3+2n+1}$

15. $\sum_{n=1}^{\infty} \frac{1}{\sqrt{n^3+2n^2-1}}$

16. $\sum_{n=1}^{\infty} \frac{n+1}{\sqrt[3]{n^4+2n+1}}$

17. $\sum_{n=1}^{\infty} \frac{1}{(n^2+5n+1)^{1/3}}$

18. $\sum_{n=1}^{\infty} \frac{1}{(n^3+2n+3)^{1/3}}$

19. $\sum_{n=1}^{\infty} \frac{\ln n}{e^n}$

20. $\sum_{n=1}^{\infty} \frac{\ln n}{n^2}$

21. $\sum_{n=1}^{\infty} \frac{1}{\sqrt{4n^2-3}}$

22. $\sum_{n=1}^{\infty} \frac{1}{\sqrt{4n^2+8}}$

Recall that for every positive integer n, n-factorial is defined by

$$n! = n(n-1)\ldots 3\cdot 2\cdot 1,$$

and $0! = 1$.

23. (a) Show that $\frac{1}{n!} \le \frac{1}{n^2-n}$ for $n \ge 2$.

(b) Show that $\sum_{n=1}^{\infty} \frac{1}{n!}$ converges.

24. Decide whether $\sum_{n=1}^{\infty} \frac{n^3}{n!}$ converges. (See Exercise 23(b).)

25. Decide whether $\sum_{n=1}^{\infty} \frac{n^2}{n!}$ converges. (See Exercise 23(b).)

26. (a) Show that $\ln n < n$ for any positive integer n.

(b) Decide whether $\sum_{n=1}^{\infty} \frac{\ln n}{n^3}$ converges.

27. Decide whether $\sum_{n=1}^{\infty} e^{-n^2}$ converges. (See Example 6.)

28. Show that $\sum_{n=1}^{\infty} \frac{1}{n^3} \sin\left(\frac{1}{n^2}\right)$ converges.

29. Prove that if $\Sigma_{n=1}^{\infty} a_n$ and $\Sigma_{n=1}^{\infty} b_n$ are both convergent series of positive terms, then so is $\Sigma_{n=1}^{\infty} a_n b_n$. Hence, in particular, if $\Sigma_{n=1}^{\infty} a_n$ converges, then so does $\Sigma_{n=1}^{\infty} a_n^2$.

30. If $a_n > 0$ for all n and $\Sigma_{n=1}^{\infty} a_n^2$ converges, must $\Sigma_{n=1}^{\infty} a_n$ converge? If so, prove it. If not, give a counterexample.

31. If in Theorem 3.2, $\lim_{n\to\infty} \frac{a_n}{b_n} = 0$, then show by example that it no longer follows that either both series converge or both diverge. (But see Exercises 35 and 37.)

32. Suppose for a certain series that $\lim_{n\to\infty} na_n$ is a nonzero real number. Then does $\Sigma_{n=1}^{\infty} a_n$ converge? Explain.

33. In Theorem 3.2 show that if $\Sigma_{n=1}^{\infty} b_n$ converges, then so does $\Sigma_{n=1}^{\infty} a_n$.

34. Repeat Exercise 33 for the case of divergence of $\Sigma_{n=1}^{\infty} b_n$.

35. Suppose in Theorem 3.2 that $\lim_{n\to\infty} (a_n/b_n) = 0$. If $\Sigma_{n=1}^{\infty} b_n$ converges, then prove that $\Sigma_{n=1}^{\infty} a_n$ must also converge.

36. Use Exercise 35 and a suitable choice of b_n to give a quicker solution to Example 6.

37. Suppose in Theorem 3.2 that $\lim_{n\to\infty} (a_n/b_n) = +\infty$, and $\Sigma_{n=1}^{\infty} b_n$ is divergent. Then show that $\Sigma_{n=1}^{\infty} a_n$ must also diverge.

9.4 Alternating Series

We turn now to series with *both* positive and negative terms. In this section we discuss series whose terms are alternately positive and negative, and in the next section we consider arbitrary series of positive and negative terms. Alternating series are relatively easy to handle. The main result of this section goes back to Leibniz in the late 17th century, whereas those of the next section are usually credited to Cauchy and appear to go back no further than the mid-18th century at the earliest.

4.1 DEFINITION

An ***alternating series*** $\sum_{n=1}^{\infty} a_n$ is a series for which $a_k a_{k+1} < 0$ for all $k = 1, 2, 3, \ldots$.

This just says that the signs of every pair of successive terms are $+-$ or $-+$. Thus an alternating series has one of the two forms

$$(1) \qquad \sum_{n=1}^{\infty} (-1)^n b_n = -b_1 + b_2 - b_3 + b_4 - + \ldots,$$

or

$$(2) \qquad \sum_{n=1}^{\infty} (-1)^{n+1} b_n = b_1 - b_2 + b_3 - b_4 + - \ldots,$$

where $b_n > 0$ for all n. It is sufficient to consider only series of type (2), since (1) can be obtained from (2) simply by multiplying by -1 (cf. Theorem 2.6 and the remark following it, with $c = -1$).

An alternating series whose terms decrease in size and approach 0 as $n \to \infty$ is convergent. The proof of the following formal statement of this is given at the end of the section.

4.2 THEOREM

Alternating Series Test. Suppose that for some $k \geq 1$ we have $0 < b_{n+1} \leq b_n$ for all $n \geq k$ and that $\lim_{n\to\infty} b_n = 0$. Then $\sum_{n=1}^{\infty} (-1)^{n+1} b_n$ converges.

Our first example of this test suggests that it is easier for an alternating series to converge than it is for a series of positive terms to do so.

Example 1

Determine whether the ***alternating harmonic series***

$$\sum_{n=1}^{\infty} (-1)^n \frac{1}{n} \qquad \text{and} \qquad \sum_{n=1}^{\infty} (-1)^{n+1} \frac{1}{n}$$

converge.

Solution. In each series $b_n = 1/n$, so $b_{n+1} \leq b_n$ holds, since $1/(n + 1) < 1/n$ for all n. Also, $\lim_{n\to\infty} b_n = \lim_{n\to\infty} 1/n = 0$. Theorem 4.2 then says that both the alternating harmonic series converge. ■

While Example 1 shows that it is fairly easy for alternating series to converge, not *all* alternating series are convergent.

Example 2

Determine whether the alternating series $\sum_{n=1}^{\infty} (-1)^{n+1} \dfrac{n}{(2n+1)}$ converges.

Solution. It might be tempting to first see whether the terms decrease in size, but it is better to check whether the limit of the nth term is zero. If it *isn't*, then the series diverges, by the divergence test (Theorem 2.4). Here

$$\lim_{n\to\infty} \frac{n}{2n+1} = \lim_{n\to\infty} \frac{1}{2+\dfrac{1}{n}} = \frac{1}{2} \neq 0,$$

so the given series is divergent. ■

In Section 1 we talked about the importance of finitely approximating infinite decimals (see p. 555). We now consider this process in more detail. Thus far we have almost no tools for finding the sum S of a convergent series. Indeed, only in the case of geometric series (Theorem 2.3) do we have a formula for S. For more general series, the best we can usually do is approximate S by s_n for as large a value of n as necessary to get the degree of accuracy desired. To determine how large an n *is* required, we need to be able to calculate the ***error*** made in using this approximation.

4.3 DEFINITION

The ***error*** in approximating the sum S of a convergent series $\sum_{n=1}^{\infty} a_n$ by s_n is

$$e_n = |S - s_n|.$$

Usually, we can't apply Definition 4.3 directly, because we don't have the exact value of S available. However, if we can find an upper bound for the error e_n, then we can estimate how close s_n is to S for a given n. In the case of an alternating series that satisfies the hypotheses of Theorem 4.2, there is a particularly simple bound on the size of e_n. The following result is proved at the end of the section.

4.4 THEOREM

Suppose that $0 \leq b_{n+1} \leq b_n$ for all $n \geq$ some k and that $\lim_{n\to\infty} b_n = 0$. Then the error e_n made in approximating the sum S of

$$\sum_{n=1}^{\infty} (-1)^n b_n \qquad \text{or} \qquad \sum_{n=1}^{\infty} (-1)^{n+1} b_n$$

by s_n *is at most* b_{n+1}, the absolute value of the first term omitted from s_n.

In applying this theorem, the terminology introduced on p. 272 is helpful. Recall that an approximation s_n to a quantity S is called ***accurate to k decimal places*** if

$$e_n = |S - s_n| < \frac{1}{2} \cdot 10^{-k}.$$

This means that the true value of S lies between $s_n - \frac{1}{2} \cdot 10^{-k}$ and $s_n + \frac{1}{2} \cdot 10^{-k}$. As an example, 1.41 is an approximation to $\sqrt{2}$ that is accurate to 2 decimal places, because

$$|\sqrt{2} - 1.41| < 0.0043 < 0.005.$$

The assertion that 1.41 is accurate to two decimal places means that

$$1.41 - 0.005 < \sqrt{2} < 1.41 + 0.005 \to 1.405 < \sqrt{2} < 1.415.$$

Example 3

Find an approximation to the sum of the series $\sum_{n=1}^{\infty} \frac{(-1)^n}{n^4}$ that is accurate to three decimal places.

Solution. First, we check that this series satisfies the hypotheses of Theorem 4.4. We have

$$b_n = \frac{1}{n^4} \to \lim_{n\to\infty} b_n = 0.$$

Also,

$$b_{n+1} = \frac{1}{(n+1)^4} < \frac{1}{n^4} = b_n.$$

So the series converges by Theorem 4.2, and $e_n = |S - s_n| < b_{n+1}$ by Theorem 4.4. So it suffices to take n large enough that

$$b_{n+1} = \frac{1}{(n+1)^4} < 0.0005.$$

This is a good job for a calculator: We successively try $n = 2, 3, 4$, and so on, until we find the first n such that $1/(n+1)^4 < 0.0005$. For $n = 6$ we get

$$e_n < b_{n+1} = \frac{1}{7^4} = \frac{1}{2401} < 0.00042.$$

So s_6 gives the sum, accurate to three decimal places. We have

$$\begin{aligned} s_6 &= -1 + \frac{1}{16} - \frac{1}{81} + \frac{1}{256} - \frac{1}{625} + \frac{1}{1296} \\ &\approx -1.00000 + 0.06250 - 0.01235 + 0.00391 - 0.00160 + 0.00077 \\ &\approx -0.94677. \end{aligned}$$

Since s_6 is accurate to three decimal places, we express our answer as $S \approx -0.947$. ■

Warning. Be careful to take into account the *accuracy* of your calculations. Since in the preceding example our approximation to S is accurate only to *three* decimal places, a reasonable procedure is to calculate s_6 using 5-decimal-place approximations and then round off to 3 places to get the approximation to S. Be careful not to let your calculator display lure you into expressing your approximation for S as an 8-, 10-, or 12-place decimal. Since the approximation is accurate to only 3 decimal places, any digits after that are *not* significant.

Proof of Theorems 4.2 and 4.4*

4.2 THEOREM Suppose that for some $k \geq 1$, we have $0 < b_{n+1} \leq b_n$ for all $n > k$ and that $\lim_{n\to\infty} b_n = 0$. Then $\sum_{n=1}^{\infty} (-1)^{n+1} b_n$ is convergent.

* Optional

Proof. As usual, in view of Theorem 2.6 we may assume that $k = 1$. Let s_n be the nth partial sum of the series,

$$s_n = \sum_{i=1}^{n} (-1)^{i+1} b_i.$$

To show that the sequence s of partial sums converges, we show that the two sequences $(s_2, s_4, s_6, \ldots, s_{2n}, \ldots)$ and $(s_1, s_3, s_5, \ldots, s_{2n-1}, \ldots)$ of even and odd partial sums both converge to the same limit. First, the hypothesis that $b_{n+1} \le b_n$ tells us that every odd partial sum

$$\begin{aligned} s_{2n-1} &= (b_1 - b_2) + (b_3 - b_4) + (b_5 - b_6) + \ldots + (b_{2n-3} - b_{2n-2}) + b_{2n-1} \\ &\ge 0, \end{aligned}$$

since it is a sum of nonnegative terms. It is also bounded below by 0 and is a decreasing sequence, because

$$s_{2n+1} = s_{2n-1} - (b_{2n} - b_{2n+1}) \le s_{2n-1}.$$

Thus Theorem 1.7 guarantees that it converges. Let $S = \lim_{n\to\infty} s_{2n-1}$. Then note that

$$s_{2n} = s_{2n-1} + (-1)^{2n+1} b_{2n} = s_{2n-1} - b_{2n}.$$

Hence

$$\begin{aligned} \lim_{n\to\infty} s_{2n} &= \lim_{n\to\infty} (s_{2n-1} - b_{2n}) = \lim_{n\to\infty} s_{2n-1} - \lim_{n\to\infty} b_{2n}, \qquad \textit{by Theorem 1.4(b)} \\ &= S - 0 = S, \end{aligned}$$

since $b_{2n} \to 0$ as $n \to \infty$ by hypothesis. Now since

$$\lim_{n\to\infty} s_{2n-1} = \lim_{n\to\infty} s_{2n} = S,$$

we can show that $\lim_{n\to\infty} s_n = S$. To do so, let $\varepsilon > 0$ be given. Then we can find N_1 and N_2 such that

$$|s_{2n} - S| < \varepsilon \qquad \text{for } n > N_1$$

and

$$|s_{2n-1} - S| < \varepsilon \qquad \text{for } n < N_2.$$

Hence if n is greater than the larger of N_1 and N_2, then

$$|s_{2n} - S| < \varepsilon \qquad \text{and} \qquad |s_{2n-1} - S| < \varepsilon.$$

So by Definition 1.2, $\lim_{n\to\infty} s_n = S$. QED

We next prove Theorem 4.4.

4.4 THEOREM Suppose that an alternating series satisfies the hypotheses of Theorem 4.2. Then the error made in approximating its sum S by s_n is at most b_{n+1}.

Proof. We can assume that the series is $\Sigma_{n=1}^{\infty} (-1)^{n+1} b_n$. In the proof of Theorem 4.2 we saw that the sequence $(s_1, s_3, s_5, \ldots, s_{2n-1}, \ldots)$ is decreasing and converges to $S = \Sigma_{n=1}^{\infty} (-1)^{n+1} b_n$. In a similar way we can see that $(s_2, s_4, s_6, \ldots, s_{2n}, \ldots)$ is an increasing sequence:

$$s_{2n+2} = s_{2n} + (-1)^{2n+2} b_{2n+1} + (-1)^{2n+3} b_{2n+2} = s_{2n} + b_{2n+1} - b_{2n+2} \ge s_{2n},$$

since the terms b_n decrease in size. Thus $(s_2, s_4, s_6, \ldots, s_{2n}, \ldots)$ is an increasing sequence converging to S. So for any n we have

$$s_{2n} \le S \le s_{2n-1}.$$

Then

$$0 \le S - s_{2n} \le s_{2n+1} - s_{2n} = (-1)^{2n+2} b_{2n+1} = b_{2n+1},$$

so

$$|S - s_{2n}| \le b_{2n+1}.$$

Also, since the odd terms s_{2n-1} decrease to limit $S \ge s_{2n}$,

$$0 \le s_{2n-1} - S \le s_{2n-1} - s_{2n} = b_{2n}.$$

Thus

$$|S - s_{2n-1}| \le b_{2n}.$$

Therefore, for any n, even or odd,

$$e_n = |S - s_n| \le b_{n+1}. \quad \boxed{\text{QED}}$$

Exercises 9.4

In Exercises 1–14, determine whether the given alternating series converges.

1. $\sum_{n=1}^{\infty} (-1)^n \dfrac{n^2 - 1}{3n^2 + 1}$
2. $\sum_{n=1}^{\infty} (-1)^{n+1} \dfrac{n^3 - 10}{2n^3 + 5n + 1}$
3. $\sum_{n=1}^{\infty} \dfrac{(-1)^n}{\sqrt{n}}$
4. $\sum_{n=2}^{\infty} \dfrac{(-1)^{n+1}}{\ln n}$
5. $\sum_{n=1}^{\infty} (-1)^n n e^{-n}$
6. $\sum_{n=1}^{\infty} (-1)^{n+1} n^2 e^{-n^3}$
7. $\sum_{n=1}^{\infty} \cos n\pi$
8. $\sum_{n=1}^{\infty} \sin \dfrac{1}{2}(2n - 1)\pi$
9. $\sum_{n=1}^{\infty} \dfrac{\cos n\pi}{n}$
10. $\sum_{n=1}^{\infty} \dfrac{\sin \frac{1}{2}(2n - 1)\pi}{n}$
11. $\sum_{n=1}^{\infty} \dfrac{(-1)^n n}{3n + 1}$
12. $\sum_{n=1}^{\infty} \dfrac{(-1)^n \ln n}{n}$
13. $\sum_{n=1}^{\infty} \dfrac{(-1)^n n!}{3 \cdot 5 \cdot \ldots \cdot (2n + 1)}$
14. $\sum_{n=2}^{\infty} \dfrac{(-1)^n n}{\ln n}$

In Exercises 15–24, find an approximation to the sum of the given series that is accurate to three decimal places.

15. $\sum_{n=1}^{\infty} \dfrac{(-1)^n}{n!}$
16. $\sum_{n=1}^{\infty} \dfrac{(-1)^{n+1}}{n^2}$
17. $\sum_{n=1}^{\infty} \dfrac{(-1)^n}{n^3}$
18. $\sum_{n=1}^{\infty} \dfrac{(-1)^{n+1}}{n^5}$
19. $\sum_{n=1}^{\infty} \dfrac{\sin \frac{1}{2}(2n - 1)\pi}{2^n}$
20. $\sum_{n=1}^{\infty} \dfrac{\cos n\pi}{n!}$
21. $\sum_{n=1}^{\infty} \dfrac{(-1)^n}{n2^n}$
22. $\sum_{n=1}^{\infty} \dfrac{(-1)^{n+1}}{n3^n}$
23. $\sum_{n=1}^{\infty} \dfrac{(-1)^{n+1}}{n!2^n}$
24. $\sum_{n=1}^{\infty} \dfrac{(-1)^n}{(n + 1)!3^n}$

In Exercises 25–28, determine how large n must be for s_n to approximate the sum S of the given series accurately to the specified number of decimal places.

25. $\sum_{n=1}^{\infty} \dfrac{(-1)^n}{n!}$; 6 places
26. $\sum_{n=1}^{\infty} \dfrac{(-1)^{n+1}}{n^3}$; 6 places
27. $\sum_{n=1}^{\infty} \dfrac{(-1)^{n+1}}{n}$; 4 places
28. $\sum_{n=1}^{\infty} \dfrac{(-1)^n}{n^2}$; 4 places

Exercise 29 is designed to obtain an upper bound for e_n in case $\Sigma_{n=1}^{\infty} a_n$ is a series of positive terms.

29. Suppose that $f\colon [k, +\infty) \to \mathbf{R}$ is a continuous, decreasing, positive-valued function. If $\Sigma_{n=1}^{\infty} a_n$ is a series for which $a_n = f(n)$ for $n \ge k$, then show that

$$e_k \le \int_k^{+\infty} f(x)\,dx.$$

(*Hint:* Use the technique in the proof of Theorem 3.3 on the interval $[k, n]$.)

30. Use Exercise 29 to find an upper bound on e_{20} for $\Sigma_{n=1}^{\infty} (1/n^3)$.

31. What is an upper bound on e_{20} for

$$\sum_{n=1}^{\infty} \frac{(-1)^n}{n^3}?$$

Compare the usefulness of Exercise 29 with Theorem 4.4.

32. Use the estimate in Exercise 29 to find how large n must be so that s_n is an approximation to

$$\sum_{n=1}^{\infty} \frac{n}{2^{n^2}}$$

accurate to three decimal places.

33. Use Exercises 29 and 32 to find an approximation to $\Sigma_{n=1}^{\infty} (n/2^{n^2})$ accurate to three decimal places.

34. Use Exercise 29 to find an approximation to $\Sigma_{n=1}^{\infty} (n/e^{n^2})$ accurate to four decimal places.

PC **35.** If you have a computer program to evaluate sums, compute the sum of the series $\Sigma_{n=1}^{\infty} (-1)^{n+1}/n$ accurate to two decimal places.

PC **36.** Repeat Exercise 35 to three decimal places.

37. By considering

$$b_n = \frac{(-1)^n}{\sqrt{n}},$$

show that the condition in Exercise 29 of the last section that $\Sigma_{n=1}^{\infty} a_n$ and $\Sigma_{n=1}^{\infty} b_n$ be series of *positive* terms cannot be dropped.

9.5 Ratio Test; Root Test

The two preceding sections contained tests for determining whether series of positive terms (or negative terms) or alternating series converge. This present section takes its name from two tests that can give useful information about *arbitrary* series. The first test plays a major role later in the chapter. But before we can state it, we need a new idea.

In Example 1 of the last section, we saw that the alternating harmonic series $\Sigma_{n=1}^{\infty} (-1)^n/n$ and $\Sigma_{n=1}^{\infty} (-1)^{n+1}/n$ converge, even though the corresponding series of absolute values is the divergent harmonic series $\Sigma_{n=1}^{\infty} (1/n)$. There is special terminology for such behavior.

5.1 DEFINITION An infinite series $\sum_{n=1}^{\infty} a_n$ is called ***absolutely convergent*** if $\sum_{n=1}^{\infty} |a_n|$ converges. If $\sum_{n=1}^{\infty} a_n$ converges but $\sum_{n=1}^{\infty} |a_n|$ diverges, then $\sum_{n=1}^{\infty} a_n$ is said to be ***conditionally convergent.***

The following example illustrates these two types of convergence.

Example 1

Classify the following series as absolutely convergent, conditionally convergent, or divergent.

$$\text{(a)}\ \sum_{n=1}^{\infty} \frac{(-1)^n}{n^4}, \qquad \text{(b)}\ \sum_{n=1}^{\infty} \frac{(-1)^n}{n}, \qquad \text{(c)}\ \sum_{n=1}^{\infty} \frac{(-1)^n n}{n+3}.$$

Solution. (a) The series of absolute values is the hyperharmonic series $\Sigma_{n=1}^{\infty} (1/n^4)$, which converges by (5) of Section 3, since $4 > 1$. Thus the given series is absolutely convergent.

(b) This is the alternating harmonic series, which we saw in Example 1 of the last section is convergent. Since the series of absolute values is the divergent harmonic series $\Sigma_{n=1}^{\infty} (1/n)$, the alternating harmonic series is conditionally convergent.

(c) Since

$$\lim_{n\to\infty} \frac{n}{n+3} = \lim_{n\to\infty} \frac{1}{1 + 3/n} = \frac{1}{1+0} = 1 \neq 0,$$

it is clear that the nth term of this series does not tend to 0 as $n \to \infty$. Therefore the series diverges by Theorem 2.4 (the divergence test). ■

When we say that a series is absolutely convergent, we mean that another series, the series of absolute values of its terms, converges. The term *conditional* was chosen because it is commonly used as a contrasting term to *absolute*. The following theorem says that absolute convergence is a stronger property than ordinary convergence.

5.2 THEOREM If $\sum_{n=1}^{\infty} a_n$ is absolutely convergent, then it is convergent.

Proof. Suppose that $\Sigma_{n=1}^{\infty} a_n$ is absolutely convergent. Then $\Sigma_{n=1}^{\infty} |a_n|$ converges. Hence so does $\Sigma_{n=1}^{\infty} 2|a_n|$, by Theorem 2.6. By adding $|a_n|$ to

$$-|a_n| \leq a_n \leq |a_n|,$$

which holds for all n, we get

$$0 \leq a_n + |a_n| \leq 2|a_n|.$$

Thus by the term-size comparison test (Theorem 3.1), $\sum_{n=1}^{\infty} (a_n + |a_n|)$ converges. Then by Theorem 2.6 again,

$$\sum_{n=1}^{\infty} a_n = \sum_{n=1}^{\infty} [(a_n + |a_n|) - |a_n|]$$

also converges. QED

The next result about absolutely convergent series states that the absolute value of the sum of an absolutely convergent series is no greater than the sum of the series of absolute values. It may be thought of as an extension of the triangle inequality [Theorem 1.2(d) of Chapter 1] to infinite sums. The proof is given at the end of the section.

5.3 THEOREM If $\sum_{n=1}^{\infty} a_n$ converges, then

$$\left| \sum_{n=1}^{\infty} a_n \right| \leq \sum_{n=1}^{\infty} |a_n|,$$

where we consider any real number to be less than $+\infty$.

Example 2

Verify Theorem 5.3 for $\displaystyle\sum_{n=0}^{\infty} \frac{(-1)^n 2^{n+1}}{3^n}$.

Solution. The given series is an alternating geometric series whose first term is

$$a = \frac{(-1)^0 2^{0+1}}{3^0} = 2$$

and whose common ratio r is $-2/3$. Thus, by the geometric series test (Theorem 2.3), it converges to

$$S = \frac{a}{1-r} = \frac{2}{1+2/3} = \frac{6}{3+2} = \frac{6}{5}.$$

The series of absolute values is a geometric series with first term $|a| = 2$ and ratio $|r| = 2/3$. So it converges to

$$\frac{|a|}{1 - |r|} = \frac{2}{1 - 2/3} = \frac{2}{1/3} = 6.$$

Thus we indeed have

$$\left|\sum_{n=0}^{\infty} a_n\right| = \frac{6}{5} \leq \sum_{n=0}^{\infty} |a_n| = 6. \quad \blacksquare$$

The next result is the main tool for investigating absolute convergence.

5.4
THEOREM

Ratio Test. Let $\sum_{n=1}^{\infty} a_n$ be an infinite series such that $a_n \neq 0$ for all sufficiently large n. Suppose that $\lim_{n \to \infty} \left|\frac{a_{n+1}}{a_n}\right| = l$.

(a) If $l < 1$, then the series is absolutely convergent.

(b) If $l > 1$ or $l = +\infty$, then the series diverges.

(c) If $l = 1$, then no conclusion follows from this information alone.

Proof. (a) Suppose that

$$\lim_{n \to \infty} \left|\frac{a_{n+1}}{a_n}\right| = \lim_{n \to \infty} \frac{|a_{n+1}|}{|a_n|} = l,$$

where $l < 1$. Then $|a_{n+1}|/|a_n|$ will be as close to l as we please if n is sufficiently large. So if we fix a number r such that $l < r < 1$, then there is some N such that

$$\text{(1)} \qquad \frac{|a_{n+1}|}{|a_n|} \leq r \qquad \text{for all } n \geq N.$$

For $n = N$, this says that

$$\frac{|a_{N+1}|}{|a_N|} \leq r \to |a_{N+1}| \leq r|a_N|.$$

Also, letting $n = N + 1$, we get from (1)

$$\frac{|a_{N+2}|}{|a_{N+1}|} \leq r \to |a_{N+2}| \leq r|a_{N+1}| \leq r^2|a_N|.$$

In general, if we assume that

$$|a_{N+k}| \leq r^k|a_N|,$$

then we have

$$\frac{|a_{N+k+1}|}{|a_{N+k}|} \leq r \to |a_{N+k+1}| \leq r|a_{N+k}| \leq r^{k+1}|a_N|.$$

Hence by mathematical induction it follows that

$$\text{(2)} \qquad |a_{N+m}| \leq r^m|a_N| \qquad \text{for all } m \geq 1.$$

Now the geometric series $\Sigma_{m=1}^{\infty} |a_N| r^m$ converges, because $|r| < 1$. Thus by (2) and the term-size comparison test (Theorem 3.1) with $k = N$, $\Sigma_{n=1}^{\infty} |a_n|$ converges. Therefore $\Sigma_{n=1}^{\infty} a_n$ is absolutely convergent.

(b) If

$$\lim_{n\to\infty} \left|\frac{a_{n+1}}{a_n}\right| = l > 1,$$

then let s be a number such that $l > s > 1$. Then there is an M such that

$$\left|\frac{a_{n+1}}{a_n}\right| \geq s \qquad \text{for all } n \geq M.$$

Hence

$$\frac{|a_{M+1}|}{|a_M|} \geq s \to |a_{M+1}| \geq s|a_M|.$$

Then, as in part (a), $|a_{M+m}| > s^m$. Since $s > 1$, it follows that $s^m \to +\infty$ as $m \to +\infty$. Therefore $\lim_{n\to\infty} a_n \neq 0$, so by Theorem 2.4 $\Sigma_{n=1}^{\infty} a_n$ diverges.

(c) Consider the hyperharmonic series $\Sigma_{n=1}^{\infty} (1/n^p)$. We have

$$\frac{|a_{n+1}|}{|a_n|} = \frac{n^p}{(n+1)^p} = \left(\frac{n}{n+1}\right)^p.$$

So

$$\lim_{n\to\infty} \frac{|a_{n+1}|}{|a_n|} = \lim_{n\to\infty} \left(\frac{n}{n+1}\right)^p = \lim_{n\to\infty} \left(\frac{1}{1+\frac{1}{n}}\right)^p = 1.$$

We know from Section 3 that sometimes ($p \leq 1$) this series diverges and sometimes ($p > 1$) it converges. So no conclusion can be drawn simply from the fact that $\lim_{n\to\infty} |a_{n+1}|/|a_n| = 1$. QED

Example 3

Determine whether $\sum_{n=1}^{\infty} (-1)^n \frac{2^n}{n!}$ is absolutely convergent, conditionally convergent, or divergent.

Solution. Here we have

$$\frac{|a_{n+1}|}{|a_n|} = \frac{2^{n+1}/(n+1)!}{2^n/n!} = \frac{2}{n+1}.$$

Hence

$$\lim_{n\to\infty} \left|\frac{a_{n+1}}{a_n}\right| = \lim_{n\to\infty} \frac{2}{n+1} = 0 < 1.$$

Thus by Theorem 5.4(a) the given series is absolutely convergent. ■

Example 4

Determine whether $\sum_{n=1}^{\infty} (-1)^n \frac{3^n}{5 \cdot 2^{(3/2)n}}$ is absolutely convergent, conditionally convergent, or divergent.

Solution. Here

$$\left|\frac{a_{n+1}}{a_n}\right| = \frac{3^{n+1}/5 \cdot 2^{(3/2)(n+1)}}{3^n/5 \cdot 2^{(3/2)n}} = \frac{3}{2^{3/2}} = \frac{3}{2\sqrt{2}}.$$

Therefore

$$\lim_{n\to\infty}\left|\frac{a_{n+1}}{a_n}\right| = \frac{3}{2\sqrt{2}} > 1.$$

Thus the given series diverges, by Theorem 5.4(b). ■

The ratio test can also be used on series of positive terms.

Example 5

Determine whether $\displaystyle\sum_{n=1}^{\infty} \frac{1 \cdot 3 \cdot 5 \dots (2n+1)}{1 \cdot 4 \cdot 9 \dots n^2}$ converges.

Solution. Since every term is positive, it equals its absolute value. Thus

$$\left|\frac{a_{n+1}}{a_n}\right| = \frac{\cancel{1 \cdot 3 \cdot 5 \dots (2n-1)}(2(n+1)+1)/\cancel{1 \cdot 4 \cdot 9 \dots n^2}(n+1)^2}{\cancel{1 \cdot 3 \cdot 5 \dots (2n-1)}/\cancel{1 \cdot 4 \cdot 9 \dots n^2}}$$

$$= \frac{2n+3}{(n+1)^2} = \frac{2n+3}{n^2+2n+1} \to 0 < 1,$$

as $n \to \infty$. Hence the given series is convergent by Theorem 5.4(a). ■

The series in the following example can be thought of as an analogue to geometric series and will be of use in Section 8.

Example 6

Show that $\Sigma_{n=1}^{\infty} nr^{n-1}$ converges absolutely if $|r| < 1$ and diverges if $|r| \geq 1$.

Solution. We have

$$\left|\frac{a_{n+1}}{a_n}\right| = \frac{n+1}{n}|r| = \left(1 + \frac{1}{n}\right)|r|.$$

Thus

$$\lim_{n\to\infty}\left|\frac{a_{n+1}}{a_n}\right| = \lim_{n\to\infty}\left(1 + \frac{1}{n}\right)|r| = |r|.$$

So by Theorem 5.4 the given series diverges if $|r| > 1$ and converges absolutely if $|r| < 1$. What about $|r| = 1$? In this case $\lim_{n\to\infty} |a_n| = \lim_{n\to\infty} n = \infty$, so by Theorem 2.4 the given series diverges. ■

The last few examples might give the impression that the ratio test is universally applicable. Alas, *no* convergence test is that good. Keep Theorem 5.4(c) in mind: The ratio test *fails* if

$$\lim_{n\to\infty}\left|\frac{a_{n+1}}{a_n}\right| = 1,$$

which happens for infinitely many series, including all hyperharmonic series. Moreover, Theorem 5.4 is of no use for conditionally convergent series: It only distinguishes between absolutely convergent and divergent series. Thus don't discard the tests given in earlier sections!

The ratio test *is* often effective for series whose terms involve repeated products (like those in Examples 3, 4, and 5). Taking the ratio of a_{n+1} to a_n reduces complexity, because many factors can be canceled. There is another test that is well suited to series whose terms involve powers. This time, complexity is reduced by taking roots.

5.5 THEOREM

Root Test. Let $\sum_{n=1}^{\infty} a_n$ be an infinite series. Suppose that $\lim_{n\to\infty} \sqrt[n]{|a_n|} = l$.

(a) If $0 \le l < 1$, then the series is absolutely convergent.

(b) If $l > 1$ or $l = +\infty$, then the series diverges.

(c) If $l = 1$, then no conclusion follows from this information alone.

Proof. The proof parallels that of Theorem 5.4. We prove (a) and leave the other parts as Exercises 40 and 41. Suppose then that $0 \le l < 1$. Choose a number r such that $l < r < 1$. Then there is some N such that

$$\sqrt[n]{|a_n|} < r \qquad \text{for all } n \ge N.$$

Thus

$$|a_n| < r^n \qquad \text{for all } n \ge N. \tag{3}$$

Since $r < 1$, the geometric series $\Sigma_{n=1}^{\infty} r^n$ converges by Theorem 2.3. From (3) and the term-size comparison test (Theorem 3.1), we see that $\Sigma_{n=1}^{\infty} |a_n|$ converges. That is, $\Sigma_{n=1}^{\infty} a_n$ is absolutely convergent. QED

Example 7

Determine whether the series

$$\text{(a)} \sum_{n=1}^{\infty} \frac{(2n^3-1)^n}{n^{3n}} \qquad \text{and} \qquad \text{(b)} \sum_{n=1}^{\infty} \frac{n^n}{3^{n^2}}$$

converge.

Solution. (a) We have

$$\sqrt[n]{|a_n|} = \frac{(2n^3-1)^{n \cdot 1/n}}{n^{3n \cdot 1/n}} = \frac{2n^3-1}{n^3} = 2 - \frac{1}{n^3}.$$

Thus $\lim_{n\to\infty} \sqrt[n]{|a_n|} = 2 > 1$. By Theorem 5.5(b), the series is then divergent.

(b) In this case

$$\sqrt[n]{|a_n|} = \frac{n}{3^n}.$$

Thus $\lim_{n\to\infty} \sqrt[n]{|a_n|}$ is indeterminate of the form ∞/∞. So we investigate

$$\lim_{x\to+\infty} \frac{x}{3^x}.$$

Using l'Hôpital's rule, we get

$$\lim_{x\to+\infty} \frac{x}{3^x} = \lim_{x\to+\infty} \frac{1}{3^x \ln 3} = 0.$$

Therefore, by Theorem 1.3, $\lim_{n\to\infty} \sqrt[n]{|a_n|} = 0$. Then Theorem 5.5(a) says that the given series converges. ■

By now we have presented several convergence tests. We close the section with a suggested order in which to try them on a given series.

ALGORITHM FOR TRYING CONVERGENCE TESTS ON $\sum_{n=1}^{\infty} a_n$

1. *If the series is geometric, use the* ***geometric*** *series test* (Theorem 2.3). *Otherwise, go on.*
2. *Check whether* $\lim_{n\to\infty} a_n = 0$. If not, then the series diverges by Theorem 2.5. If $\lim_{n\to\infty} a_n = 0$, then go on.
3. *Try the ratio test* (Theorem 5.3). If
$$\lim_{n\to\infty} \left|\frac{a_{n+1}}{a_n}\right| < 1,$$
then the series is absolutely convergent. If this limit is $+\infty$ or a real number $l > 1$, then the series diverges. If the limit is 1 or fails to exist, then go on.
4. *If the series is an alternating series, see if the alternating series test* (Theorem 4.2) *applies.* If not, go on.
5. *If it is not alternating, then try one of the comparison tests* (Theorems 3.1 and 3.2). If you can't find a useful series to compare the given series to, then go on.
6. *If you can integrate $f(x)$ where $f(n) = a_n$, then see if the integral test* (Theorem 3.3) *can be applied.* If not, go on.
7. *If the series involves nth powers*, then try the root test.

Proof of Theorem 5.3*

5.3 THEOREM If $\sum_{n=1}^{\infty} a_n$ converges, then

$$\left|\sum_{n=1}^{\infty} a_n\right| \le \sum_{n=1}^{\infty} |a_n|,$$

where we consider any real number to be less than $+\infty$.

Proof. Let s_n be the nth partial sum of $\Sigma_{n=1}^{\infty} a_n$. Then the sequence $\mathbf{s} = (s_1, s_2, s_3, \ldots, s_n, \ldots)$ converges to some limit S by hypothesis. We also have for each n

$$\left|\sum_{i=1}^{n} a_i\right| = |s_n| \le \sum_{i=1}^{n} |a_i|, \tag{4}$$

* Optional

by the triangle inequality. Since $|a_i| \geq 0$, the sequence $\boldsymbol{t} = (t_1, t_2, \ldots, t_n, \ldots)$ of partial sums of $\Sigma_{n=1}^{\infty} |a_n|$ is increasing, so either converges or diverges to $+\infty$. In either case, for each n we have

$$t_n = \sum_{i=1}^{n} |a_i| \leq \lim_{n\to\infty} t_n.$$

Then from (4), $|s_n| \leq \lim_{n\to\infty} t_n$ for each n. Thus $\lim_{n\to\infty} |s_n| \leq \lim_{n\to\infty} t_n$. Since $\lim_{n\to\infty} |s_n| = |\lim_{n\to\infty} s_n|$, it follows that

$$\left|\sum_{n=1}^{\infty} a_n\right| \leq \sum_{n=1}^{\infty} |a_n|,$$

where we understand the right side to be $+\infty$ if the sequence $\boldsymbol{t}$ diverges. QED

> HISTORICAL NOTE
>
> Both the ratio test and the root test were proved rigorously by Cauchy in 1837. But both had been used nearly a century earlier by the French mathematician Jean d'Alembert (1717–1783). D'Alembert was also one of the first who tried to treat the derivative as the limit of the difference quotient, to avoid the logical flaws in the work of Newton and Leibniz.

Exercises 9.5

In Exercises 1–38, classify the given series as absolutely convergent, conditionally convergent, or divergent.

1. $\sum_{n=1}^{\infty} \frac{(-1)^n n}{2^n}$

2. $\sum_{n=1}^{\infty} \frac{(-1)^{n+1}(n+1)}{3^n}$

3. $\sum_{n=1}^{\infty} \frac{(-1)^n n^2}{3^n}$

4. $\sum_{n=1}^{\infty} \frac{(-1)^{n+1}(n^2+n+1)}{2^n}$

5. $\sum_{n=1}^{\infty} \frac{(-1)^n 4^n}{3^{4n/3}}$

6. $\sum_{n=1}^{\infty} \frac{(-1)^{n+1} n! 2^n}{4^{(1/2)n}}$

7. $\sum_{n=1}^{\infty} \frac{(-1)^n 1000^n}{n!}$

8. $\sum_{n=1}^{\infty} \frac{(-1)^n n!}{100^{n+1}}$

9. $\sum_{n=2}^{\infty} \frac{(-1)^n n}{n^2-1}$

10. $\sum_{n=1}^{\infty} \frac{(-1)^n 5n}{n^2+2n+3}$

11. $\sum_{n=1}^{\infty} \frac{(-1)^n 1 \cdot 3 \cdot 5 \ldots (2n-1)}{n!}$

12. $\sum_{n=1}^{\infty} \frac{(-1)^n 2 \cdot 4 \cdot 6 \ldots 2n}{n!}$

13. $\sum_{n=1}^{\infty} \frac{(n!)^2}{(3n)!}$

14. $\sum_{n=1}^{\infty} \frac{(-1)^n \tan^{-1} n}{n^3}$

15. $\sum_{n=1}^{\infty} \frac{(-1)^n}{e^n - n}$

16. $\sum_{n=1}^{\infty} \frac{(-1)^n}{\ln n - n}$

17. $\sum_{n=1}^{\infty} \frac{(-1)n^n}{n!}$

18. $\sum_{n=1}^{\infty} \frac{(-1)^n 2^n n^3}{n!}$

19. $\sum_{n=1}^{\infty} \frac{(-1)^n}{n\sqrt{n+1}}$

20. $\sum_{n=1}^{\infty} \frac{(-1)^n(n^2+1)}{\sqrt{n^2+1}}$

21. $\sum_{n=1}^{\infty} \frac{\sin n\pi/4}{n^2}$

22. $\sum_{n=1}^{\infty} \frac{\cos n\pi/3}{n}$

23. $\sum_{n=1}^{\infty} (-1)^n \frac{2^{n^2}}{n!}$

24. $\sum_{n=1}^{\infty} \frac{(-1)^n 2^{2n}}{n!}$

25. $\sum_{n=1}^{\infty} \frac{(-1)^n \ln n}{n}$

26. $\sum_{n=2}^{\infty} \frac{(-1)^n n^2}{2 \ln n}$

27. $\sum_{n=1}^{\infty} \frac{(-1)^n \ln n}{n^3}$

28. $\sum_{n=1}^{\infty} \frac{(-1)^{n+1} \ln n}{n^2}$

29. $\sum_{n=1}^{\infty} \frac{(-1)^{n+1} n^{2n}}{(3n^2+1)^n}$

30. $\sum_{n=1}^{\infty} \frac{(-1)^n 5n^{3n}}{(n^2+4)^n}$

31. $\sum_{n=1}^{\infty} \frac{(n+1)^{2n}}{(n^2+1)^{4n}}$

32. $\sum_{n=1}^{\infty} \frac{(-1)^{n+1} 2^{3n+2}}{n^n}$

33. $\sum_{n=1}^{\infty} \frac{n^{n+1}}{10^{10n-1}}$

34. $\sum_{n=1}^{\infty} \frac{n^n}{3^{4n+1}}$

35. $\sum_{n=1}^{\infty} \frac{(-1)^n 2^{n^2}}{n^{3n}}$

36. $\sum_{n=1}^{\infty} \frac{(-1)^{n+1} n^{2n}}{2^{n^2}}$

37. $\sum_{n=2}^{\infty} (\ln n)^{-n}$

38. $\sum_{n=1}^{\infty} \frac{n^n}{e^n}$

39. In the proof of Theorem 5.4(b), prove in detail that $|a_{M+m}| \geq s^m |a_M|$.

40. Prove Theorem 5.5(b).

41. Show that if $\lim_{n\to\infty} \sqrt[n]{|a_n|} = 1$, then $\Sigma_{n=1}^{\infty} a_n$ may converge or diverge. (*Hint:* Consider the hyperharmonic series.)

42. Use the ratio test on the series in Example 6 of Section 3. Which solution seems simpler?

43. Investigate the convergence of $\Sigma_{n=1}^{\infty} n^n/3n^2$ using Theorems 5.4 and 5.5. Which method works better on this series?

9.6 Taylor Polynomials

We now return to a question raised in the Introduction: How can one calculate the value $f(x)$ of a transcendental function f for given x? Where, for example, do values like

$$e^2 = 7.3890561, \quad \sin(0.5) = 0.4794255, \quad \text{and} \quad \tan^{-1}(-0.3) = -0.2914568,$$

which are given by a scientific calculator, come from? They can be found from accurate polynomial approximations to $f(x) = e^x$, $g(x) = \sin x$, and $h(x) = \tan^{-1} x$, respectively. This section presents a method for finding an approximating nth degree polynomial $p_n(x)$ to any function f that is differentiable at least n times. This important method is named for the English mathematician Brook Taylor (1685–1731).

In contrast to transcendental functions, polynomial functions are easy to evaluate. At any point x, $p_n(x)$ can be computed by *finitely many additions* and *multiplications:*

$$(1) \qquad p_n(x) = a_n x^n + a_{n-1}x^{n-1} + a_{n-2}x^{n-2} + \ldots + a_1 x + a_0.$$

Mathematicians have found that the most efficient way to compute the values of a polynomial function is to use *Horner's method*, given by the English mathematician William G. Horner (1773–1827). It evaluates $p_n(x)$ as

$$(2) \qquad p_n(x) = (\ldots((a_n x + a_{n-1})x + a_{n-2})x + \ldots + a_1)x + a_0$$

by first multiplying x by a_n, adding the next coefficient, multiplying by x again, and continuing until the coefficient a_0 is added. On a microcomputer, for instance, (2) is noticeably faster than (1) for a polynomial with more than a few terms. Once we have a method for approximating a transcendental function f by polynomials p_n over an interval I, it is easy to construct a table of approximate rational values of $f(x)$ for x in I.

Taylor's scheme for producing such polynomials $p_n(x)$ is an extension of the differential (or tangent) approximation $T(x)$ discussed in Section 2.5. Recall that if f is differentiable at $x = a$, then for x near a, the graph of $y = f(x)$ is closely approximated by its tangent line drawn to the point $(a, f(a))$. See **Figure 6.1.** The equation of the tangent line is

FIGURE 6.1

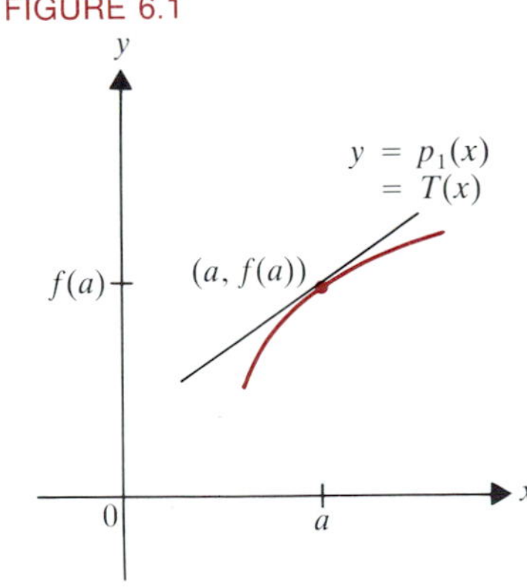

$$(3) \qquad y = T(x) = f(a) + f'(a)(x - a).$$

Thus (3) is a *first-degree* polynomial approximation to $f(x)$ near $x = a$. We then define the *first-degree Taylor polynomial* $p_1(x)$ *of* f *(expanded) about* $x = a$ to be the tangent approximation $T(x)$. That is,

$$(4) \qquad p_1(x) = f(a) + f'(a)(x - a).$$

Both

$$(5) \qquad p_1(a) = f(a) \quad \text{and} \quad p_1'(a) = f'(a),$$

are clear from (4). Thus $p_1(x)$ agrees with $f(x)$ at $x = a$, and $p_1'(x)$ agrees with $f'(x)$ at $x = a$. Moreover, as $x \to a$, Theorem 5.2 of Chapter 2 says that $p(x)$ approaches $f(x)$ at a rate faster than the rate of approach of x to a. These are the calculus properties that correspond to the geometric closeness of the graphs of $y = f(x)$ and $y = p_1(x)$ near the point $(a, f(a))$.

For $n > 1$ we want to produce a polynomial $p_n(x)$ that will approximate $f(x)$ more closely than $p_1(x)$ does near $x = a$. To do so, we assume that f and its first n derivatives $f', f'', f''', \ldots, f^{(n)}$ exist at $x = a$. The next step is to impose

FIGURE 6.2

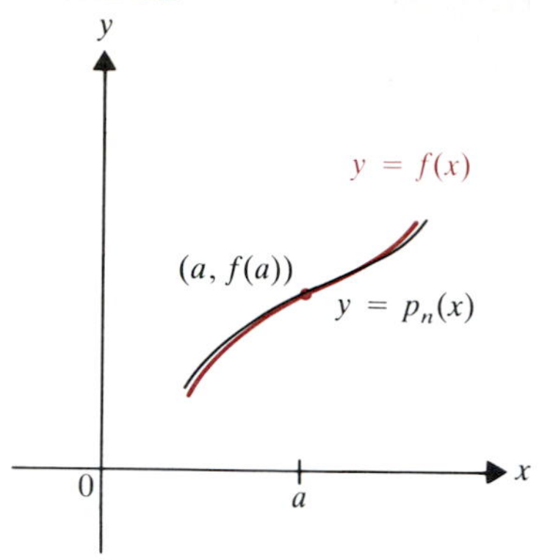

a condition to make $p_n(x)$ closely approximate $f(x)$ near $x = a$. The condition we use is the nth-degree version of (5):

$$\text{(6)} \qquad p_n(a) = f(a),\ p_n'(a) = f'(a),\ p_n''(a) = f''(a),$$
$$p_n'''(a) = f'''(a), \ldots, p_n^{(n)}(a) = f^{(n)}(a).$$

Geometrically, we are requiring that the graphs of $y = f(x)$ and $y = p_n(x)$ go through the point $(a, f(a))$ and that the two graphs have the same slope, the same concavity, the same rate of change of the concavity, and so on, at $x = a$. Thus near $x = a$ the two graphs are almost indistinguishable. Under a microscope, the graphs should appear as in **Figure 6.2.**

It turns out that the conditions (6) are enough to *determine* the polynomial $p_n(x)$. This is a consequence of the following basic fact about polynomials.

6.1 LEMMA For any real number a, every polynomial

$$q(x) = a_n x^n + a_{n-1}x^{n-1} + \ldots + a_1 x + a_0$$

of degree n can be written as a polynomial in powers of $x - a$:

$$\text{(7)} \qquad q(x) = b_n(x - a)^n + b_{n-1}(x - a)^{n-1} + \ldots + b_1(x - a) + b_0,$$

where $b_0, b_1, \ldots, b_{n-1}, b_n$ are real constants.

Proof. The lemma simply says that we can change the variable in $q(x)$ from x to $x - a$. We thus let $u = x - a$. Then $x = u + a$, and so

$$q(x) = a_n(u + a)^n + a_{n-1}(u + a)^{n-1} + \ldots + a_1(u + a) + a_0.$$

If we expand the powers $(u + a)^i$, multiply the results by the coefficients a_i, and regroup, we will obtain an expression for $q(x)$ as a sum of constants times powers of u. Since $u = x - a$, $q(x)$ will then be expressed in the form (7). QED

Example 1

Express $q(x) = 2x^2 - x + 1$ as a polynomial in $x - 3$.

Solution. As in the proof of the lemma, we let $u = x - 3$. Then $x = u + 3$. So

$$\begin{aligned} q(x) &= 2(u + 3)^2 - (u + 3) + 1 = 2(u^2 + 6u + 9) - u - 3 + 1 \\ &= 2u^2 + 11u + 16 \\ &= 2(x - 3)^2 + 11(x - 3) + 16. \end{aligned}$$

Thus in this case, $b_2 = 2$, $b_1 = 11$, and $b_0 = 16$ in (7). ■

To show that the conditions (6) determine a *unique* polynomial $p_n(x)$, we use Lemma 6.1 to express $p_n(x)$ in powers of $x - a$:

$$\text{(8)} \qquad p_n(x) = b_0 + b_1(x - a) + b_2(x - a)^2 + b_3(x - a)^3 + b_4(x - a)^4 + \ldots + b_n(x - a)^n.$$

Then the requirement in (6) that $p_n(a) = f(a)$ says that

$$p_n(a) = b_0,$$

because substitution of a for x in (8) makes every term on the right equal 0, except for the constant term. To apply the remaining conditions in (6), we first calculate $p_n'(x)$, $p_n''(x)$, $p_n'''(x)$, and so on, from (8). We get

$$\begin{aligned}
p_n'(x) &= b_1 + 2b_2(x-a) + 3b_3(x-a)^2 + 4b_4(x-a)^3 + \ldots + nb_n(x-a)^{n-1}\\
p_n''(x) &= 2b_2 + 3\cdot 2b_3(x-a) + 4\cdot 3b_4(x-a)^2 + \ldots + n(n-1)b_n(x-a)^{n-2}\\
p_n'''(x) &= 3\cdot 2b_3 + 4\cdot 3\cdot 2b_4(x-a) + \ldots + n(n-1)(n-2)b_n(x-a)^{n-3}\\
&\vdots\\
p_n^{(n-1)}(x) &= (n-1)(n-2)\ldots 1b_{n-1} + n(n-1)(n-2)\ldots 1b_n(x-a)\\
p^{(n)}(x) &= n(n-1)(n-2)\ldots 1b_n = n!b_n.
\end{aligned}$$

The conditions in (6) that $p_n'(a) = f'(a)$, $p_n''(a) = f''(a)$, $p_n'''(a) = f_n'''(a), \ldots,$ $p_n^{(n)}(a) = f^{(n)}(a)$ then give

$$\begin{aligned}
f'(a) &= b_1 & &\rightarrow & b_1 &= f'(a)\\
f''(a) &= 2b_2 & &\rightarrow & b_2 &= \frac{1}{2!}f''(a)\\
f'''(a) &= 3\cdot 2b_3 & &\rightarrow & b_3 &= \frac{1}{3!}f'''(a)\\
&\vdots & & & &\vdots\\
f^{(n-1)}(a) &= (n-1)(n-2)\ldots 1b_{n-1} & &\rightarrow & b_{n-1} &= \frac{1}{(n-1)!}f^{(n-1)}(a)\\
f^{(n)}(a) &= n!b_n & &\rightarrow & b_n &= \frac{1}{n!}f^{(n)}(a).
\end{aligned}$$

Thus every coefficient $b_0, b_1, b_2, \ldots, b_{n-1}, b_n$ in (8) is determined by (6). Therefore, there is only one polynomial $p_n(x)$ that satisfies the conditions (6).

6.2 DEFINITION

Let f be at least n times differentiable at $x = a$. Then the ***nth-degree Taylor polynomial of f (expanded) about x = a*** is

$$p_n(x) = f(a) + f'(a)(x-a) + \frac{1}{2!}f''(a)(x-a)^2 + \frac{1}{3!}f'''(a)(x-a)^3 + \ldots + \frac{1}{n!}f^{(n)}(a)(x-a)^n. \tag{9}$$

When $a = 0$, the Taylor polynomial is sometimes called the *Maclaurin polynomial of f expanded in powers of x.*

Example 2

Find $p_5(x)$ expanded about $x = 0$ if $f(x) = \sin x$.

Solution. To fill in the right side of (9), we have to evaluate f and its first five derivatives at $x = a = 0$:

$$\begin{aligned}
f(x) &= \sin x & &\rightarrow & f(0) &= 0,\\
f'(x) &= \cos x & &\rightarrow & f'(0) &= 1,\\
f''(x) &= -\sin x & &\rightarrow & f''(0) &= 0,\\
f'''(x) &= -\cos x & &\rightarrow & f'''(0) &= -1,\\
f^{(4)}(x) &= \sin x & &\rightarrow & f^{(4)}(0) &= 0,\\
f^{(5)}(x) &= \cos x & &\rightarrow & f^{(5)}(0) &= 1.
\end{aligned}$$

Putting this information into (9), we get

$$p_5(x) = f(0) + f'(0)x + \frac{1}{2!}f''(0)x^2 + \frac{1}{3!}f'''(0)x^3 + \frac{1}{4!}f^{(4)}(0)x^4 + \frac{1}{5!}f^{(5)}(0)x^5$$

$$= 0 + x + 0 - \frac{1}{3!}x^3 + 0 + \frac{1}{5!}x^5 = x - \frac{1}{6}x^3 + \frac{1}{120}x^5.$$

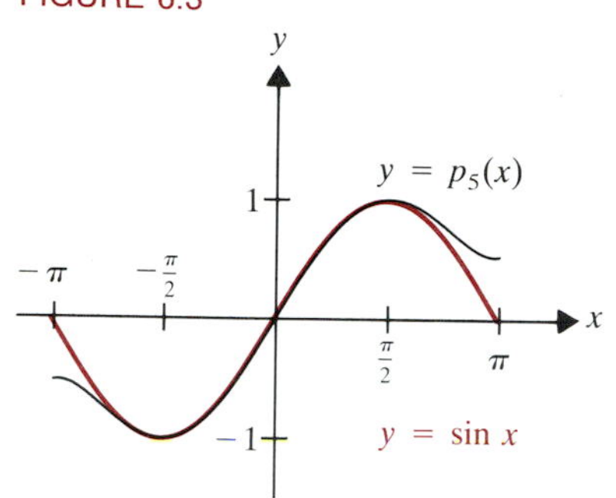

FIGURE 6.3

Figure 6.3 shows the graphs of $y = p_5(x)$ and $y = \sin x$ over the interval $[-\pi, \pi]$. These graphs, which were drawn with the aid of a computer, are virtually indistinguishable between $x = -\pi/2$ and $x = \pi/2$. ■

Example 3

Find the fourth-degree Taylor polynomial of $f(x) = \ln x$ about $x = 1$.

Solution. Computing $f(x)$ and its first four derivatives at $x = 1$, we find

$$f(x) = \ln x \qquad \rightarrow \quad f(1) = 0$$

$$f'(x) = \quad \frac{1}{x} = x^{-1} \rightarrow f'(1) = 1$$

$$f''(x) = \quad -x^{-2} = -\frac{1}{x^2} \rightarrow f''(1) = -1$$

$$f'''(x) = \quad 2x^{-3} = \frac{2}{x^3} \rightarrow f'''(1) = 2$$

$$f^{(4)}(x) = -2 \cdot 3x^{-4} = -\frac{3!}{x^4} \rightarrow f^{(4)}(1) = -3!.$$

Then from (9) we get

$$p_4(x) = f(1) + f'(1)(x-1) + \frac{1}{2!}f''(1)(x-1)^2 + \frac{1}{3!}f'''(1)(x-1)^3 + \frac{1}{4!}f^{(4)}(1)(x-1)^4$$

$$= 0 + (x-1) - \frac{1}{2}(x-1)^2 + \frac{2}{3!}(x-1)^3 - \frac{3!}{4!}(x-1)^4$$

$$= (x-1) - \frac{1}{2}(x-1)^2 + \frac{1}{3}(x-1)^3 - \frac{1}{4}(x-1)^4.$$

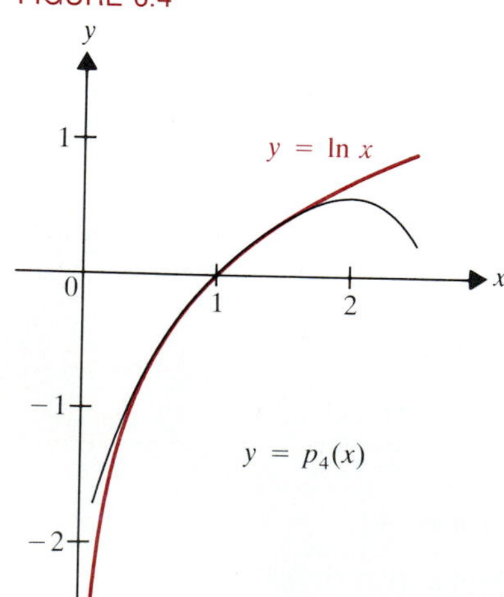

FIGURE 6.4

Figure 6.4 shows the graphs of $y = p_4(x)$ and $y = \ln x$ near $x = 1$. Again, they virtually coincide for $x \in [1/2, 3/2]$. ■

Earlier we recalled that Theorem 5.2 of Chapter 2 says that

$$\lim_{x \to a} \frac{f(x) - p_1(x)}{x - a} = 0,$$

that is, $p_1(x) \rightarrow f(x)$ at a rate faster than the rate of approach of x to a. For $n > 1$, $f(x)$ seems to be approximated more closely by $p_n(x)$ than by $p_1(x)$. It is not surprising then that as x approaches a, $p_n(x) \rightarrow f(x)$ even more rapidly than $p_1(x) \rightarrow f(x)$. To describe this precisely, we give a name to the difference $f(x) - p_n(x)$, which measures the error in the approximation of $f(x)$ by $p_n(x)$.

6.3 DEFINITION

The ***remainder when $f(x)$ is approximated by $p_n(x)$*** is

(10) $$r_n(x) = f(x) - p_n(x).$$

The following formula for $r_n(x)$ is derived at the end of the section.

6.4 THEOREM

Cauchy Remainder Formula. Suppose that f is at least $(n + 1)$ times differentiable and that $f^{(n+1)}$ is continuous on an interval I containing a. Then for any x in I,

(11) $$r_n(x) = \frac{1}{n!}\int_a^x (x - t)^n f^{(n+1)}(t)\,dt.$$

The complicated integral in (11) makes it hard to use, but fortunately we will be primarily interested in estimating the *size* of $r_n(x)$ rather than in computing its value. The next result gives an upper bound for the error made when $f(x)$ is approximated by $p_n(x)$ over an interval I (such as $[-\pi, \pi]$ in Example 1). It is a generalization of Theorem 8.3 of Chapter 6, which gave an upper bound for the error $|r_2(x)| = |e(x)|$ made in approximating $f(x)$ by the tangent approximation $T(x) = p_1(x)$.

6.5 COROLLARY

Remainder Estimate. Let I be an interval containing a. Suppose that B is an upper bound for $|f^{(n+1)}(x)|$ on I, that is, B is a constant such that

(12) $$|f^{(n+1)}(x)| \le B \qquad \text{for all } x \text{ in } I.$$

Then for any x in I,

(13) $$|r_n(x)| \le \frac{B}{(n+1)!}|x - a|^{n+1}.$$

Proof. We give the proof assuming that $x > a$. (A symmetric argument applies in case $x < a$. See Exercise 42.) From (11) we have

$$|r_n(x)| = \frac{1}{n!}\left|\int_a^x (x - t)^n f^{(n+1)}(t)\,dt\right|$$

$$\le \frac{1}{n!}\int_a^x |(x - t)^n f^{(n+1)}(t)|\,dt \qquad \textit{by Theorem 3.6 of Chapter 4}$$

$$\le \frac{B}{n!}\int_a^x (x - t)^n\,dt \qquad \textit{by (12) and the fact that } |x - t| = x - t \textit{ for } t \in [a, x]$$

$$\le -\frac{B}{n!}\frac{(x - t)^{n+1}}{n + 1}\Bigg|_{t=a}^{t=x} = \frac{B}{(n+1)!}|x - a|^{n+1}. \quad \boxed{\text{QED}}$$

Inequality (13) says that $r_n(x)$ is *no larger* than $B|x - a|^{n+1}/(n + 1)!$. Often it will be significantly smaller than this upper bound, as the next example illustrates.

Example 4

Find an upper bound for

(a) $|r_5(x)|$ in Example 2 over $I = [-\pi, \pi]$

(b) $|r_4(x)|$ in Example 3 over $I = [1/2, 3/2]$

Solution. (a) In Example 2 we computed the first five derivatives of $f(x) = \sin x$. To find an upper bound for $|r_5(x)|$ from (13), we find an upper bound B for $|f^{(6)}(x)|$. Since $f^{(5)}(x) = \cos x$, we have

$$|f^{(6)}(x)| = |-\sin x| \leq 1$$

for *all* x, so certainly for $x \in [-\pi, \pi]$. We can therefore use $B = 1$ in (13), getting

$$|r_5(x)| \leq \frac{1}{6!}|x - 0|^6 = \frac{|x|^6}{6!}.$$

The largest value for $|x|^6$ on $I = [-\pi, \pi]$ is π^6, which occurs at $x = \pm\pi$. Thus on I,

$$|r_5(x)| \leq \frac{\pi^6}{6!} \approx 1.335263.$$

This upper bound is *very* generous. A calculator gives $p_5(\pi) - \sin \pi \approx 0.524044$. (Look back at Figure 6.3.)

(b) In Example 3 we found for $f(x) = \ln x$ that $f^{(4)}(x) = -3!x^{-4}$. Thus

$$f^{(5)}(x) = 4!x^{-5} = \frac{24}{x^5}.$$

This function has its maximum on $I = [1/2, 3/2]$ at the left endpoint $x = 1/2$, where 24 is divided by the least amount. Thus for B in (12) we can use

$$4!/(1/2)^5 = 2^5 \cdot 4!.$$

Then (13) says that

$$|r_4(x)| \leq \frac{2^5 \cdot 4!}{5!}|x - 1|^5 = \frac{2^5}{5}|x - 1|^5$$

for $x \in I = [1/2, 3/2]$. The largest value of $|x - 1|^5$ on this interval is $1/2^5$, which occurs at each endpoint. Thus on $[1/2, 3/2]$ we have

$$|r_4(x)| \leq \frac{2^5}{5} \cdot \frac{1}{2^5} = 0.2.$$

Glancing back at Figure 6.3, we get the impression that this bound is also very generous. Indeed, a calculator confirms that at $x = 3/2$, $\ln x - p_4(x) \approx 0.00442344$. ■

We next generalize Theorem 5.2 of Chapter 2 by showing that $p_n(x)$ approaches $f(x)$ faster than $(x - a)^n$ approaches 0, if the hypotheses of Corollary 6.5 hold.

6.6 COROLLARY Let I be an interval containing a. Suppose that $|f^{(n+1)}(x)|$ is bounded by some real number B on the interval I. Then

$$\lim_{x \to a} \frac{r_n(x)}{(x - a)^n} = 0. \tag{14}$$

Proof. From (13) we have

$$0 \leq \left|\frac{r_n(x)}{(x - a)^n}\right| \leq \frac{B}{(n + 1)!}|x - a|$$

for $x \in I$. As $x \to a$, the term on the right approaches 0. By the sandwich theorem for limits (Theorem 5.7 of Chapter 1), the middle term also approaches 0. Therefore (14) holds. QED

There is another formula for $r_n(x)$, due to the French mathematician Joseph L. Lagrange (1736–1813). Its proof is also given at the end of the section.

6.7 THEOREM

Lagrange Remainder Formula. Suppose that $f^{(n+1)}$ is continuous on an open interval I containing a. Then

$$r_n(x) = \frac{f^{(n+1)}(c)}{(n+1)!}(x-a)^{n+1} \tag{15}$$

for some c between a and x.

This result provides an "almost" formula for $r_n(x)$. In Example 2, for instance, since $f^{(6)}x = -\sin x$, Theorem 6.7 says that

$$r_5(x) = \frac{-\sin c}{6!}x^6 \tag{16}$$

for some c between 0 and x. Since we don't know c exactly—only that it lies somewhere between 0 and x—(16) is not quite an exact formula for $r_5(x)$.

The next two examples show how Taylor polynomials and the remainder estimate (13) of Corollary 6.5 can be used to calculate values of transcendental functions as accurately as desired.

Example 5

Use Taylor polynomials to find an approximation to e that is accurate to 5 decimal places.

Solution. The only value of $f(x) = e^x$ that we know exactly is $f(0) = 1$. So we find the Taylor polynomial about $x = 0$. For every positive integer k, we have $f^{(k)}(x) = e^x$ since $f'(x) = e^x = f(x)$. Therefore

$$p_n(x) = f(0) + f'(0) + \frac{1}{2!}f''(0)x^2 + \frac{1}{3!}f'''(0)x^3 + \dots + \frac{1}{n!}f^{(n)}(0)x^n$$

$$= 1 + x + \frac{1}{2!}x^2 + \frac{1}{3!}x^3 + \dots + \frac{1}{n!}x^n.$$

We are asked to find n so that

$$|r_n(1)| = |e^1 - p_n(1)| \le \frac{1}{2}\cdot 10^{-5} = 5\cdot 10^{-6}.$$

For x in the interval $[0, 1]$, we have from (13) that

$$|r_n(1)| \le \frac{B}{(n+1)!}x^{n+1} \le \frac{B}{(n+1)!}1^{n+1} = \frac{B}{(n+1)!}, \tag{17}$$

where B is an upper bound for $f^{(n+1)}(x) = e^x$. Since e^x has its largest value on $[0, 1]$ at $x = 1$, we might try $B = e$. That is not helpful for (17), however, since we are trying to *compute* e here. But $\ln e = 1$, and in the absence of a table, $\ln 3$ can be computed from Simpson's rule to be about 1.0986123 (Exercise 51 of Section 6.1). Thus, since e^x is an increasing function, $e = e^1 < e^{\ln 3} = 3$. So we can use $B = 3$ in (17). Then

$$|r_n(1)| < \frac{3}{(n+1)!}.$$

Evaluating the right side at successive integers, we find

$$|r_8(1)| < \frac{3}{9!} = \frac{1}{3 \cdot 8!} \approx 8 \cdot 10^{-6},$$

$$|r_9(1)| < \frac{3}{10!} = \frac{1}{30 \cdot 8!} \approx 8 \cdot 10^{-7} < 5 \cdot 10^{-6}.$$

So we can use $p_9(1)$ to get the desired approximation. We have

$$\begin{aligned} p_9(1) &= 1 + 1 + \frac{1}{2} + \frac{1}{3!} + \frac{1}{4!} + \frac{1}{5!} + \frac{1}{6!} + \frac{1}{7!} + \frac{1}{8!} + \frac{1}{9!} \\ &\approx 2.500000 + 0.166667 + 0.041667 + 0.008333 + 0.001389 \\ &\quad + 0.000198 + 0.000025 + 0.000003 \\ &\approx 2.718282, \end{aligned}$$

which rounds off to 2.71828. ■

Taylor polynomials can be used to generate tables of values for e^x in the same way we used $p_n(1)$ to approximate e. This in fact is how such tables were first made, by *hand calculation* in the 1600s.

To generate a 5-place table of values for $\sin x$ and $\cos x$, we could proceed in the same way. All magnitudes (absolute values) assumed by $\sin x$ and $\cos x$ occur for $x \in [0, \pi/2]$. Also,

$$\cos\left(\frac{1}{2}\pi - x\right) = \sin x \qquad \text{and} \qquad \sin\left(\frac{1}{2}\pi - x\right) = \cos x,$$

so we need to tabulate the functions only for $x \in [0, \pi/4]$.

Example 6

Let $p_n(x)$ be the nth-degree Taylor polynomial for $f(x) = \sin x$ about $x = 0$. Find n so that $|r_n(x)| < 5 \cdot 10^{-6}$ for $x \in [0, \pi/4]$.

Solution. From the work in Example 2, we have

$$p_3(x) = p_4(x) = x - \frac{x^3}{6}.$$

Thus for $x \in [0, \pi/4]$ it follows that

$$|r_3(x)| = |r_4(x)| \leq \frac{1}{5!}x^5,$$

since we can use $B = 1$ in Corollary 6.5. Hence $|r_3(x)| < 5 \cdot 10^{-6}$ when

$$\frac{1}{5!}x^5 < 5 \cdot 10^{-6} \rightarrow x^5 < 120 \cdot 5 \cdot 10^{-6} = 6 \times 10^{-4},$$

that is (from a calculator), when $x < 0.2268$. So for $x < \pi/15 \approx 0.2094$, we can use $p_3(x)$ to generate 5-place tables. Similar analysis (Exercises 35 and 36) shows that

$$p_5(x) = x - \frac{x^3}{3!} + \frac{x^5}{5!}$$

gives 5-place accuracy for $x < 0.59$ ($\approx 33.8°$), and

$$p_7(x) = x - \frac{x^3}{3!} + \frac{x^5}{5!} - \frac{x^7}{7!}$$

gives 5-place accuracy for the remaining x up to $\pi/4$. **Figure 6.5** shows the graphs of $y = p_3(x)$, $y = p_5(x)$, and $y = p_7(x)$ over the interval $[0, 2\pi]$. Notice how closely $p_7(x)$ approximates $\sin x$ over $[0, \pi]$. ■

FIGURE 6.5

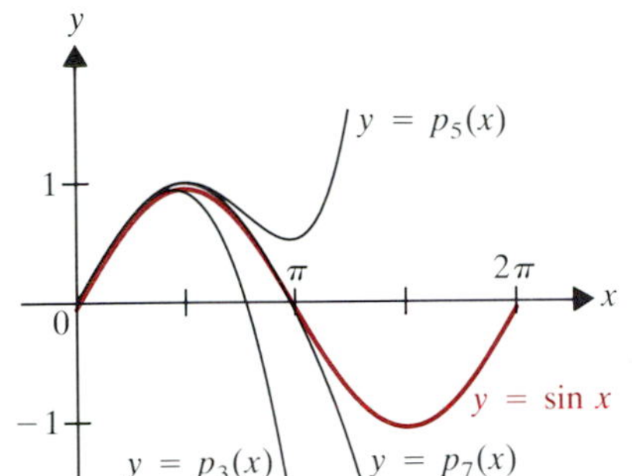

Example 6 shows that *low-degree Taylor polynomials* (no more than degree 7) are sufficient to generate 5-place tables of $\sin x$ (and $\cos x$). This explains how it was possible to compile such tables by hand calculation.

Our final example shows the advantage in choosing $a \neq 0$ in some cases.

Example 7

Obtain an approximation of $\sin 31°$ accurate to 5 decimal places.

Solution. Although Example 6 shows that we could use $p_5(x)$ already computed about $x = 0$, if we were approaching this problem from scratch, it would be better to expand about $a = \pi/6$. We get

$$\begin{aligned} f(x) &= \sin x \rightarrow f(\pi/6) = 1/2 \\ f'(x) &= \cos x \rightarrow f'(\pi/6) = \sqrt{3}/2 \\ f''(x) &= -\sin x \rightarrow f''(\pi/6) = -1/2 \\ f'''(x) &= -\cos x \rightarrow f'''(\pi/6) = -\sqrt{3}/2. \end{aligned}$$

We are interested in computing $\sin 31°$, where

$$31° = \frac{\pi}{6} + \frac{\pi}{180}$$

radians. Thus from (13) in Corollary 6.5 (with $B = 1$),

$$(18) \qquad |r_2(x)| \leq \frac{1}{3!}\left(x - \frac{\pi}{6}\right)^3$$

for x in an interval I containing $\pi/6$. Letting I be the interval $(\pi/6 - \pi/180, \pi/6 + \pi/180)$, we see that the right side of (18) is its largest at $x = \pi/6 + \pi/180$. This means that

$$|r_3(x)| \leq \frac{1}{3!}\left(\frac{\pi}{180}\right)^3 \approx 8.86 \cdot 10^{-7} < 5 \cdot 10^{-6}.$$

We will then have an approximation of $\sin 31°$ accurate to 5 decimal places if we use

$$p_2(x) = f\left(\frac{\pi}{6}\right) + f'\left(\frac{\pi}{6}\right)\left(x - \frac{\pi}{6}\right) + \frac{1}{2}f''(x)\left(\frac{\pi}{6}\right)\left(x - \frac{\pi}{6}\right)^2$$

$$= \frac{1}{2} + \frac{1}{2}\sqrt{3}(\pi/180) - \frac{1}{4}(\pi/180)^2 \approx 0.500000 + 0.015115 - 0.000076$$

$$\approx 0.515039 \approx 0.51504. \blacksquare$$

Proof of the Remainder Theorems*

We conclude the section by deriving the remainder formulas (11) and (15). The first is the Cauchy remainder formula.

6.4 THEOREM Suppose that f is at least $(n + 1)$ times differentiable on I and that $f^{(n+1)}$ is continuous on an interval I containing a. Then for any x in I,

$$r_n(x) = \frac{1}{n!}\int_a^x (x - t)^n f^{(n+1)}(t)\,dt. \tag{11}$$

Proof. We use mathematical induction. For $n = 1$ the integral in (11) is

$$\int_a^x (x - t)f''(t)\,dt = \int_a^x xf''(t)\,dt - \int_a^x tf''(t)\,dt.$$

We integrate by parts in the second integral on the right, letting

$$u = t \qquad \text{and} \qquad dv = f''(t)\,dt.$$

Then

$$du = dt \qquad \text{and} \qquad v = f'(t).$$

We therefore obtain

$$\int_a^x (x - t)f''(t)\,dt = x\int_a^x f''(t)\,dt - \Big[tf'(t)\Big]_{t=a}^{t=x} + \int_a^x f'(t)\,dt$$

$$= xf'(x) - xf'(a) - xf'(x) + af'(a) + f(x) - f(a)$$

$$= f(x) - f(a) - f'(a)(x - a) = r_1(x).$$

Thus (11) holds for $n = 1$. Next suppose that (11) holds for $n = k$:

$$r_k(x) = \frac{1}{k!}\int_a^x (x - t)^k f^{(k+1)}(t)\,dt.$$

Then we need to show that it holds for $n = k + 1$, that is, that $r_{k+1}(x)$ is given by

$$\frac{1}{(k+1)!}\int_a^x (x - t)^{k+1} f^{(k+2)}(t)\,dt.$$

We again use integration by parts, letting

$$u = (x - t)^{k+1} \qquad \text{and} \qquad dv = f^{(k+2)}(t)\,dt.$$

Then

$$du = -(k + 1)(x - t)^k\,dt \qquad \text{and} \qquad v = f^{(k+1)}(t).$$

* Optional

We therefore have

$$\frac{1}{(k+1)!}\int_a^x (x-t)^{k+1} f^{(k+2)}(t)\,dt = \frac{1}{(k+1)!}(x-t)^{k+1} f^{(k+1)}(t)\Big|_{t=a}^{t=x}$$

$$+\frac{k+1}{(k+1)!}\int_a^x (x-t)^k f^{(k+1)}(t)\,dt$$

$$= -\frac{1}{(k+1)!}(x-a)^{k+1} f^{(k+1)}(a)$$

$$+\frac{1}{k!}\int_a^x (x-t)^k f^{(k+1)}(t)\,dt$$

$$= -\frac{1}{(k+1)!}(x-a)^{k+1} f^{(k+1)}(a)$$

$$+ r_k(x) \qquad \textit{by the induction hypothesis for } n = k$$

$$= f(x) - p_k(x) - \frac{f^{(k+1)}(a)}{(k+1)!}(x-a)^{k+1}$$

$$= f(x) - p_{k+1}(x) = r_{k+1}(x).$$

Thus (11) holds for $n = k + 1$ whenever it holds for $n = k$. It therefore holds for all n. QED

Finally, we prove the Lagrange remainder formula.

6.7 THEOREM If $f^{(n+1)}$ is continuous on an open interval I containing a, then

$$r_n(x) = \frac{f^{(n+1)}(c)}{(n+1)!}(x-a)^{n+1} \tag{15}$$

for some c between x and a.

Proof. We give the proof for the case $x \geq a$; the case $x < a$ is Exercise 43. If $x = a$, then $r_n(a) = f(a) - p_n(a) = 0$, and the right side of (15) is 0. Thus (15) holds in this case. The case $a < x$ remains. It follows from Theorem 6.4 by the weighted mean value theorem for integrals (see Exercise 33, Section 4.3). For $t \neq x$ in the interval $[a, x]$, the function

$$g(t) = (x-t)^n$$

is positive for $t < x$. Let

$$b = \frac{x-a}{2} \qquad \text{and} \qquad k = \frac{x+a}{2}.$$

Then $b > 0$, $k > a$, and on the interval $[a, k]$ we have

$$g(t) \geq \left(x - \frac{a+x}{2}\right)^n = \left(\frac{x-a}{2}\right)^n = b^n > 0.$$

Therefore

$$\int_a^x g(t)\,dt = \int_a^k g(t)\,dt + \int_k^x g(t)\,dt \qquad \textit{by Theorem 2.6 of Chapter 4}$$

$$\geq \int_a^k g(t)\,dt \qquad \textit{by Theorem 3.6(a) of Chapter 4}$$

$$\geq \int_a^k b^n\,dt = b^n(k-a) > 0. \qquad \textit{by Theorem 3.6(b) of Chapter 4}$$

Thus by Exercise 33(b) of Section 4.3, for some c between a and x, we have

$$\int_a^x f^{(n+1)}(t)(x-t)^n\,dt = \int_a^x f^{(n+1)}(t)g(t)\,dt$$

$$= f^{(n+1)}(c)\int_a^x g(t)\,dt = f^{(n+1)}(c)\int_a^x (x-t)^n\,dt$$

$$= f^{(n+1)}(c)\left[-\frac{(x-t)^{n+1}}{n+1}\right]_a^x$$

$$= \frac{f^{(n+1)}(c)}{(n+1)}(x-a)^{n+1}.$$

Then from (11) we have

$$r_n(x) = \frac{1}{n!}\int_a^x f^{(n+1)}(t)(x-t)^n\,dt = \frac{f^{(n+1)}(c)}{(n+1)!}(x-a)^{n+1},$$

for some c between a and x. QED

HISTORICAL NOTE

Brook Taylor, a student of Newton, first published the main ideas of this section in 1715. However, in 1671 the Scottish mathematician James Gregory (1638–1675) had worked out a number of particular examples, including Exercises 7 and 18. Colin Maclaurin (1698–1746) of Scotland was another student of Newton. He used Taylor polynomials expanded about $x = 0$ as a major tool in a text he published in 1742, which also contained the integral test.

Lagrange, one of the great mathematicians of the 18th century, used the ideas of this section to help develop calculus without any use of infinitesimals. Many of his methods led to important numerical techniques that came into wide use with the advent of computers.

Exercises 9.6

In Exercises 1–20, obtain the indicated Taylor polynomial of $f(x)$ about the given a. Give a formula for $r_n(x)$.

1. $f(x) = \sin x$; $p_9(x)$; $a = 0$
2. $f(x) = \cos x$; $p_8(x)$; $a = 0$
3. $f(x) = \ln(1+x)$; $p_6(x)$; $a = 0$
4. $f(x) = (1+x)^{1/2}$; $p_5(x)$; $a = 4$
5. $f(x) = e^{-x/2}$; $p_6(x)$; $a = 0$
6. $f(x) = e^{2x}$; $p_5(x)$; $a = 0$
7. $f(x) = \tan x$; $p_4(x)$; $a = 0$
8. $f(x) = x^2 \ln x$; $p_4(x)$; $a = 5$
9. $f(x) = \sin(x/2)$; $p_5(x)$; $a = \pi/4$
10. $f(x) = \sin x + \cos x$; $p_5(x)$; $a = 0$
11. $f(x) = 3\sin x - \cos x$; $p_4(x)$; $a = \pi/4$
12. $f(x) = \sin x$; $p_5(x)$; $a = \pi/3$
13. $f(x) = \cos x$; $p_6(x)$; $a = \pi/3$
14. $f(x) = \sqrt{x}$; $p_4(x)$; $a = 4$
15. $f(x) = \sqrt[3]{x}$; $p_4(x)$; $a = 8$
16. $f(x) = 4x^3 + 5x^2 - 2x + 1$; $p_3(x)$; $a = 2$
17. $f(x) = x^3 - 2x^2 + 3x - 1$; $p_3(x)$; $a = 1$
18. $f(x) = \tan^{-1} x$; $p_3(x)$; $a = 0$
19. $f(x) = \ln\cos x$; $p_3(x)$; $a = \pi/3$
20. $f(x) = \ln\sin x$; $p_4(x)$; $a = \pi/6$

In Exercises 21–30, approximate the given expression accurately to 4 decimal places. Use results of Exercises 1–20 as needed.

21. $\sin 12^\circ$
22. $\cos 12^\circ$
23. $\ln 1.02$
24. $\tan 3^\circ$
25. $e^{1/2}$
26. $\sin 23^\circ$
27. $\sin 62^\circ$
28. $\cos 58^\circ$
29. $\sqrt{4.2000}$
30. $\sqrt[3]{7.9000}$

31. Show that $p_4(x)$ with $a = 0$ approximates $f(x) = e^x$ to 3 decimal places on the interval $(-1/2, 1/2)$.

32. Show that $p_{10}(x)$ with $a = 0$ satisfies $|e^x - p_{10}(x)| < 0.001$ on the interval $[-2, 2]$.

33. Show that $n = 12$ gives a Taylor polynomial $p_n(x)$ about $a = 0$ such that on the interval $[-1, 1]$, we have
$$|e^{-x^2} - p_n(x)| < 0.001.$$

34. What degree Taylor polynomial $p_n(x)$ about $x = 0$ should be used to approximate $\cos 40°$ if we want the approximation to be accurate to at least 5 decimal places?

35. In Example 6 show that $p_5(x)$ gives values for $\sin x$ accurate to at least 5 decimal places if $|x| < 0.59$.

36. In Example 6 show that $p_7(x)$ gives values of $\sin x$ accurate to at least 6 decimal places if $|x| < \pi/4$.

37. Suppose that $f(x)$ is a polynomial of degree n, and $p_n(x)$ is its nth-degree Taylor polynomial. Then show that $f(x) = p_n(x)$. What can you say about $p_k(x)$ for $k > n$?

The next two exercises consider the Newton approximations $x_1, x_2, \ldots, x_n, \ldots$ to a root x^* of $f(x) = 0$, assuming that $f''(x)$ exists.

38. **(a)** Expand $f(x)$ about $x = x_k$ and use Theorem 6.7 to give a formula for $f(x)$ in terms of $p_1(x)$ and $r_1(x)$.

(b) Let $x = x^*$ in part (a). From the resulting equation, subtract the Newton iteration relation
$$0 = f(x_k) + f'(x_k)(x_{k+1} - x_k)$$
and solve the equation you obtain for $x^* - x_{k+1}$. (Assume that $f'(x_k) \neq 0$.)

39. In Exercise 38, suppose that for x near x_k we have
$$|f''(x)| \leq 2|f'(x_k)|.$$
If x_k is an approximation to x^* that is accurate to r decimal places, then show that x_{k+1} is an approximation to x^* accurate to $2r$ decimal places. (This is one case in which the number of accurate digits in Newton's method doubles with each iteration.)

The next two exercises provide an alternative proof of the second-derivative test (Theorem 4.5 of Chapter 3).

40. Suppose that f satisfies the hypotheses of Theorem 6.7 for $n = 2$. Show that if $f'(a) = 0$ and $f''(a) < 0$, then f has a local maximum at a.

41. Suppose that f is as in Exercise 40 except that $f''(a) > 0$. Then show that f has a local minimum at a.

42. Prove Corollary 6.5 for the case $x < a$.

43. Prove Theorem 6.7 if $x < a$.

9.7 Power Series

We saw in the last section how to approximate any sufficiently differentiable function f near a point a by finding its Taylor polynomial $p_n(x)$ about $x = a$. We saw in Example 7 that the closeness of the approximation of $\sin x$ by $p_n(x)$ improved as n increased. For instance, for x near $\pi/4$, $p_7(x)$ gave a smaller error than $p_5(x)$, which in turn was a better approximation than $p_3(x)$. As n increases, the Maclaurin polynomial
$$p_n(x) = x - \frac{x^3}{3!} + \frac{x^5}{5!} - \frac{x^7}{7!} + - \ldots + \frac{(-1)^{n-1}x^{2n-1}}{(2n-1)!}$$
thus approximates $\sin x$ with greater accuracy. This suggests that the infinite series
$$\sum_{n=1}^{\infty} (-1)^{n-1}\frac{x^{2n-1}}{(2n-1)!} = x - \frac{x^3}{3!} + \frac{x^5}{5!} - \frac{x^7}{7!} + - \ldots$$
should converge to $\sin x$. That turns out to be true, and we pursue it in the next section. Before we can do so, we need to study the convergence properties of power series.

7.1 DEFINITION Let a be a fixed real number. A ***power series in $x - a$*** is a series

$$(1) \quad \sum_{n=0}^{\infty} b_n(x-a)^n = b_0 + b_1(x-a) + b_2(x-a)^2 + \ldots + b_n(x-a)^n + \ldots$$

If $a = 0$, then we get a *power series in x*,

(2) $$\sum_{n=0}^{\infty} b_n x^n = b_0 + b_1 x + b_2 x^2 + \ldots + b_n x^n + \ldots.$$

Here x is a real variable.

Comparison of Definitions 7.1 and 6.2 reveals that every Taylor polynomial is a partial sum of a power series $\Sigma_{n=0}^{\infty} b_n(x-a)^n$, where

$$b_0 = f(a),\ b_1 = f'(a),\ b_2 = \frac{1}{2!} f''(a), \ldots, b_n = \frac{1}{n!} f^{(n)}(a), \ldots.$$

When x is replaced by a specific real number in (1), we get an ordinary infinite series whose convergence can be investigated by the methods of Sections 2–5. Different values of x may produce series with different convergence properties. For instance, the power series

(3) $$\sum_{n=0}^{\infty} (x-1)^n = 1 + (x-1) + (x-1)^2 + (x-1)^3 + \ldots$$

gives the convergent geometric series

$$\sum_{n=0}^{\infty} \left(\frac{1}{2}\right)^n = 1 + \frac{1}{2} + \frac{1}{4} + \frac{1}{8} + \ldots$$

for $x = 3/2$. When $x = 5/2$, however, we get the divergent geometric series

$$\sum_{n=0}^{\infty} (3/2)^n = 1 + \frac{3}{2} + \frac{9}{4} + \frac{27}{8} + \ldots.$$

We can think of (1) as defining a function whose domain consists of all real numbers x for which the series is convergent.

Rather than testing individual values of x to see whether (1), or (2), converges for that value, we prefer to find all numbers for which the series converges. The next example illustrates how that is done for the power series (3) as well as for two other rather extreme examples.

Example 1

Determine for which values of x the following power series converge.

$$\text{(a)} \sum_{n=0}^{\infty} n!x^n, \qquad \text{(b)} \sum_{n=1}^{\infty} \frac{x^n}{n!}, \qquad \text{(c)} \sum_{n=0}^{\infty} (x-1)^n.$$

Solution. (a) Let $a_n = n!x^n$. We apply the ratio test, getting

$$\lim_{n\to\infty} \left|\frac{a_n + 1}{a_n}\right| = \lim_{n\to\infty} \left|\frac{(n+1)!x^{n+1}}{n!x^n}\right| = \lim_{n\to\infty} (n+1)|x|.$$

This limit is $+\infty$ unless $x = 0$, when it is 0. Thus by Theorem 5.4 the given series converges *only* for $x = 0$.

(b) In this case we let $a_n = x^n/n!$. Then we have

$$\lim_{n\to\infty} \left|\frac{a_{n+1}}{a_n}\right| = \lim_{n\to\infty} \frac{|x^{n+1}|}{(n+1)!} \frac{n!}{|x^n|} = \lim_{n\to\infty} \frac{|x|}{n+1} = 0 < 1,$$

for *all* x. So this time the ratio test says that the series is absolutely convergent for every real number x.

(c) This time we let $a_n = (x-1)^n$. We get

$$\lim_{n\to\infty}\left|\frac{a_{n+1}}{a_n}\right| = \lim_{n\to\infty}\left|\frac{(x-1)^{n+1}}{(x-1)^n}\right| = \lim_{n\to\infty}|x-1| = |x-1|.$$

The ratio test says the series is absolutely convergent when

$$|x-1| < 1 \rightarrow -1 < x-1 < 1 \rightarrow 0 < x < 2,$$

that is, for $x \in (0, 2)$. It also says the series diverges for $|x-1| > 1$, that is, for $x < 0$ or $x > 2$. The only points x not covered by the ratio test are then $x = 0$ and $x = 2$. Substitution of $x = 0$ and $x = 2$ into (3) produces the series

$$\sum_{n=0}^{\infty} (-1)^n \quad \text{and} \quad \sum_{n=0}^{\infty} 1^n,$$

which diverge by the divergence test (Theorem 2.4). Thus the given series is absolutely convergent on the open interval (0, 2) and is divergent otherwise. ■

The approach followed in Example 1 is used to prove the following general result about convergence of power series. The details are given at the end of the section.

7.2 THEOREM

For the power series $\sum_{n=0}^{\infty} b_n(x-a)^n$, suppose that

$$\lim_{n\to\infty}\left|\frac{b_{n+1}}{b_n}\right| = q.$$

(a) If $q = +\infty$, then the series converges only for $x = a$.

(b) If $q = 0$, then the series is absolutely convergent for all real numbers x.

(c) If $q \in (0, +\infty)$, then let $r = 1/q$. The series is absolutely convergent on $(a-r, a+r)$ and divergent on $(-\infty, a-r) \cup (a+r, +\infty)$. Convergence of the series at $x = a-r$ and at $x = a+r$ must be checked separately.

The number r introduced in Theorem 7.2(c) is called the ***radius of convergence*** of the series $\Sigma_{n=0}^{\infty} b_n(x-a)^n$. We extend this term to cases (a) and (b) of Theorem 7.2 by defining r to be 0 in case (a) and to be ∞ in case (b). The ***interval of convergence*** of $\Sigma_{n=0}^{\infty} (x-a)^n$ is the set of all real numbers x for which the power series converges. **Figure 7.1** shows the three cases in Theorem 7.2. In part (a), the interval of convergence is just $[a, a] = \{a\}$, the single point a. In part (b), the interval of convergence is $(-\infty, +\infty)$, the set of all real numbers. In part (c), the theorem says only that there is absolute convergence on $(a-r, a+r)$ and divergence on $(-\infty, a-r) \cup (a-r, +\infty)$. It says *nothing* about what happens at $x = a-r$ or $x = a+r$. The reason is that at those points, we have

$$l = q|x-a| = \frac{1}{r}\cdot r = 1,$$

so the test fails ratio. What happens at the endpoints of the interval of convergence must be investigated separately in every example. In some cases the

FIGURE 7.1

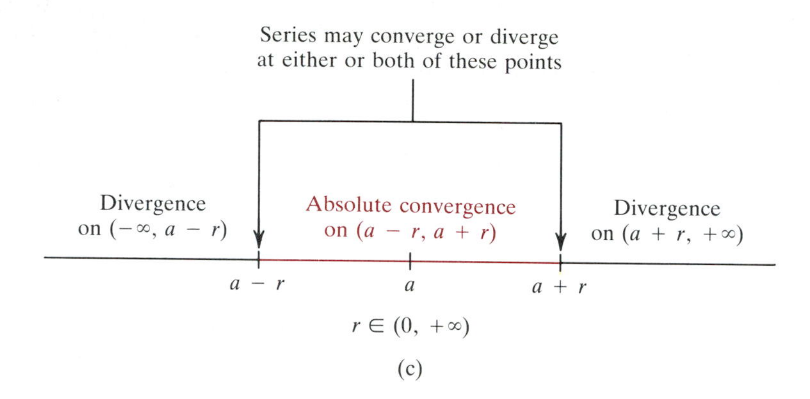

series will converge at both $a - r$ and $a + r$. In other cases it will converge at one endpoint but not at the other. In still other cases it will converge at neither endpoint. Thus the interval of convergence can be any one of the following:

(a) $\{a\}$ (b) $(-\infty, +\infty) = \mathbf{R}$

(c) $(a - r, a + r)$, $[a - r, a + r)$, $(a - r, a + r]$, or $[a - r, a + r]$.

We saw illustations of (a) and (b) in Example 1, parts (a) and (b). Another series for which the radius of convergence is infinite is the power series

(4) $$\sum_{n=0}^{\infty} \frac{(-1)^{n-1}x^{2n-1}}{(2n-1)!} = x - \frac{x^3}{3!} + \frac{x^5}{5!} - \frac{x^7}{7!} + - \cdots.$$

This is not of the form specified in Theorem 7.2, because it does not involve every positive integral power of x (there are no even powers). While we shall not do so, it can be shown that Theorem 7.2 remains true for power series of the more general form $\Sigma_{n=0}^{\infty} b_n(x - a)^{kn+l}$ where k and l are fixed integers. The series (4) is such a series, where $k = 2$ and $l = 1$, for example. So we will use the approach of Theorem 7.2 to investigate its convergence.

Example 2

Find all real numbers x such that (4) converges. Is the convergence always absolute?

Solution. We have

$$b_{n+1} = \frac{(-1)^{n+1}}{(2n+1)!} \quad \text{and} \quad b_n = \frac{(-1)^n}{(2n-1)!}.$$

Thus

$$\lim_{n\to\infty} \left|\frac{a_{n+1}}{a_n}\right| = \lim_{n\to\infty} \frac{(2n-1)!}{(2n+1)!}|x| = \lim_{n\to\infty} \frac{|x|}{2n(2n+1)} = 0.$$

Hence by Theorem 7.2(b), $r = \infty$, and so the series is absolutely convergent for all real numbers x. ■

The series in Example 1(c) had $r = 1$, and its interval of convergence was of the form $(a - r, a + r)$, namely, $(0, 2) = (1 - 1, 1 + 1)$. The next series has interval of convergence of the form $[a - r, a + r)$.

Example 3

Find all real numbers x such that

$$\sum_{n=1}^{\infty} \frac{(x-2)^n}{n} = (x-2) + \frac{(x-2)^2}{2} - \frac{(x-2)^3}{3} + \dots \tag{5}$$

converges. Is the series absolutely convergent for all x where it converges?

Solution. Here $b_n = 1/n$, so we have

$$\lim_{n\to\infty} \left|\frac{a_{n+1}}{a_n}\right| = \lim_{n\to\infty} \left|\frac{(x-2)^{n+1}}{(n+1)} \cdot \frac{n}{(x-2)^n}\right| = \lim_{n\to\infty} \frac{n}{n+1} |x-2| = 1 \cdot |x-2|.$$

Thus $r = 1/1 = 1$. By Theorem 7.2(c), we have absolute convergence for x such that

$$|x-2| < 1 \rightarrow -1 < x - 2 < 1 \rightarrow 1 < x < 3,$$

and we have divergence on $(-\infty, 1) \cup (3, +\infty)$. We have to examine (5) at the endpoints $x = 1$ and $x = 3$ of the interval of convergence. When $x = 1$, we get the alternating harmonic series

$$\sum_{n=0}^{\infty} \frac{(-1)^n}{n} = -1 + \frac{1}{2} - \frac{1}{3} + \frac{1}{4} - + \dots,$$

which converges, but not absolutely by Example 1(b), Section 5. When $x = 3$, we get

$$\sum_{n=1}^{\infty} \frac{1^n}{n} = \sum_{n=1}^{\infty} \frac{1}{n},$$

the divergent harmonic series. Thus the interval of convergence is $[1, 3)$, with conditional convergence at $x = 1$ and absolute convergence on $(1, 3)$. ■

It is easy to give an example of power series with interval of convergence $(a - r, a + r]$. In fact, consider

$$\sum_{n=1}^{\infty} \frac{(-1)^n(x-2)^n}{n} = -(x-2) + \frac{(x-2)^2}{2} - \frac{(x-3)^3}{3} + - \dots. \tag{6}$$

The calculations of Example 3 show that $r = 1$, so there is again absolute convergence on $(1, 3)$. At $x = 1$ we get

$$\sum_{n=1}^{\infty} \frac{(-1)^n(-1)^n}{n} = \sum_{n=1}^{\infty} \frac{(-1)^{2n}}{n} = \sum_{n=1}^{\infty} \frac{1}{n},$$

the divergent harmonic series. At $x = 3$ we have

$$\sum_{n=1}^{\infty} \frac{(-1)^n 1^n}{n} = \sum_{n=1}^{\infty} \frac{(-1)^n}{n},$$

the convergent alternating harmonic series. Thus the interval of convergence of (6) is $(1, 3]$.

The next example is where the interval of convergence is a closed interval $[a - r, a + r]$.

Example 4

Find all real numbers x for which

$$\sum_{n=1}^{\infty} \frac{x^n}{n^3} \tag{7}$$

converges. Is the series absolutely convergent for all x where it converges?

Solution. We have

$$\lim_{n\to\infty} \left|\frac{a_{n+1}}{a_n}\right| = \lim_{n\to\infty} \left|\frac{x^{n+1}}{(n+1)^3} \cdot \frac{n^3}{x^n}\right| = \lim_{n\to\infty} \left(\frac{n}{n+1}\right)^3 |x| = 1 \cdot |x|.$$

Thus $r = 1$, so there is absolute convergence for $|x| < r$, that is, on $(-1, 1)$. At $x = 1$ we get the convergent hyperharmonic series

$$\sum_{n=1}^{\infty} \frac{1}{n^3}. \tag{8}$$

At $x = -1$ we get the convergent alternating series

$$\sum_{n=1}^{\infty} \frac{(-1)^n}{n^3},$$

which is absolutely convergent, since (8) converges. Thus (7) has interval of convergence $[-1, 1]$, where it converges absolutely. ■

Our final example deals with a more complicated series.

Example 5

Find all real numbers x for which the series

$$\sum_{n=0}^{\infty} \frac{x^n}{3^{2n} + n} = 1 + \frac{x}{3^2 + 1} + \frac{x^2}{3^4 + 2} + \frac{x^3}{3^6 + 3} + \dots$$

converges. Is the series absolutely convergent where it converges?

Solution.

$$\lim_{n\to\infty} \left|\frac{a_{n+1}}{a_n}\right| = \lim_{n\to\infty} \left|\frac{x^{n+1}}{3^{2n+2} + n + 1} \cdot \frac{3^{2n} + n}{x^n}\right|$$

$$= \lim_{n\to\infty} \frac{1 + \dfrac{n}{3^{2n}}}{3^2 + \dfrac{n+1}{3^{2n}}} |x| = \frac{1}{9} |x|,$$

since l'Hôpital's rule shows that

$$\lim_{x\to+\infty}\frac{x}{3^{2x}}=\lim_{x\to+\infty}\frac{1}{3^{2x}\cdot 2\ln 3}=0.$$

Hence $r=9$, and so there is absolute convergence for $|x|<9$ i.e., for $x\in(-9,9)$. When $x=9$ we get

$$1+\frac{9}{3^2+1}+\frac{9^2}{3^4+2}+\ldots+\frac{9^n}{3^{2n}+n}+\ldots$$
$$=1+\frac{3^2}{3^2+1}+\frac{3^4}{3^4+2}+\ldots+\frac{3^{2n}}{3^{2n}+n}+\ldots.$$

Note that

$$\lim_{n\to\infty}\frac{3^{2n}}{3^{2n}+n}=\lim_{n\to\infty}\frac{1}{1+\dfrac{n}{3^{2n}}}=1\neq 0,$$

so the series diverges by Theorem 2.4. Similarly, at $x=-9$ we get the alternating version of this series, which also diverges by the divergence test. The interval of convergence is therefore $(-9,9)$, where the convergence is absolute. ■

Proof of Theorem 7.2*

7.2 THEOREM Given the series $\sum_{n=0}^{\infty} b_n(x-a)^n$, suppose that

$$\lim_{n\to\infty}\left|\frac{b_{n+1}}{b_n}\right|=q.$$

(a) If $q=+\infty$, then the series converges only for $x=a$.

(b) If $q=0$, then the series is absolutely convergent for all real numbers x.

(c) If $q\in(0,+\infty)$, then let $r=1/q$. The series is absolutely convergent on $(a-r,a+r)$ and divergent on $(-\infty,a-r)\cup(a+r,+\infty)$.

Proof. We use the ratio test as in Example 1, letting $a_n=b_n(x-a)^n$. Then

$$\lim_{n\to\infty}\left|\frac{a_{n+1}}{a_n}\right|=\lim_{n\to\infty}\left|\frac{b_{n+1}(x-a)^{n+1}}{b_n(x-a)^n}\right|=\lim_{n\to\infty}\left|\frac{b_{n+1}}{b_n}\right||x-a|=q|x-a|=l.$$

(a) If $q=+\infty$, then $l=+\infty$ also, unless $x=a$. So the series diverges unless $x=a$, by Theorem 5.4(b). When $x=a$, we get $b_0+0+0+\ldots$, which converges to b_0.

(b) If $q=0$, then $l=0<1$, so the series converges for every value of x by Theorem 5.4(a).

(c) Suppose that $q\in(0,+\infty)$. Then by Theorem 5.4 the power series is absolutely convergent if $l=q|x-a|<1$ and divergent if $l=q|x-a|>1$. Thus we have absolute convergence for those values of x satisfying

$$q|x-a|<1\rightarrow|x-a|<1/q=r\rightarrow -r<x-a<r\rightarrow a-r<x<a+r,$$

* Optional

that is, for $x \in (a + r, a - r)$. Likewise, we have divergence for those x such that

$$q|x - a| > 1 \rightarrow |x - a| > \frac{1}{q} = r,$$

that is, for $x \in (-\infty, a - r) \cup (a + r, +\infty)$. QED

Exercises 9.7

In Exercises 1–26, find the interval of convergence of the given series and determine where the convergence is absolute.

1. $\sum_{n=0}^{\infty} \frac{(x-a)^{n+1}}{(n+1)!}$ (This exercise is used in the next section.)

2. $\sum_{n=0}^{\infty} \frac{(-1)^n x^{2n}}{(2n)!}$

3. $\sum_{n=1}^{\infty} \frac{n! x^{2n}}{2n - 1}$

4. $\sum_{n=1}^{\infty} \frac{n! x^{2n-1}}{n + 1}$

5. $\sum_{n=0}^{\infty} \frac{(-1)^{n-1} 2^n x^n}{n 3^n}$

6. $\sum_{n=0}^{\infty} \frac{(-1)^n 3^n x^n}{n 5^n}$

7. $\sum_{n=0}^{\infty} n(x-1)^n$

8. $\sum_{n=0}^{\infty} \frac{n^2}{n+1}(x-2)^n$

9. $\sum_{n=0}^{\infty} \frac{x^n}{n^2+1}$

10. $\sum_{n=0}^{\infty} \frac{(-1)^n e^n x^n}{(n+1)^2}$

11. $\sum_{n=0}^{\infty} \frac{(-1)^n (x+1)^n}{n+1}$

12. $\sum_{n=0}^{\infty} \frac{\cos \pi n}{n+1} x^n$

13. $\sum_{n=0}^{\infty} \frac{n^2 (x-1)^n}{2^n}$

14. $\sum_{n=1}^{\infty} \frac{2^n (x+1)^n}{n^2}$

15. $\sum_{n=1}^{\infty} \frac{2^n}{n}(2x-1)^n$

16. $\sum_{n=1}^{\infty} \frac{n}{3^n}(3x+1)^n$

17. $\sum_{n=1}^{\infty} \frac{(\ln n)(x-2)^n}{n}$

18. $\sum_{n=1}^{\infty} \frac{\ln(n+1)(x-3)^n}{n+2}$

19. $\sum_{n=1}^{\infty} \frac{(x-3)^n}{n(n+1)}$

20. $\sum_{n=1}^{\infty} \frac{(x+1)^n}{n^2+n+1}$

21. $\sum_{n=0}^{\infty} \frac{x^{3n}}{n+5}$

22. $\sum_{n=1}^{\infty} \frac{(n+1)x^n}{n^2+3n+2}$

23. $1 + 3x + x^2 + 3x^3 + x^4 + 3x^5 + \ldots$

24. $1 + 2x + x^2 + 2x^3 + x^4 + 2x^5 + \ldots$

25. $\sum_{n=0}^{\infty} \frac{x^n}{2^{3n} + n}$

26. $\sum_{n=0}^{\infty} \frac{(-1)^n x^n}{3^n + n + 1}$

In Exercises 27–30, find the radius of convergence.

27. $\sum_{n=1}^{\infty} (-1)^{n-1} \frac{1 \cdot 3 \cdot 5 \ldots (2n-1)}{2^n n!} x^{2n-1}$

28. $\sum_{n=1}^{\infty} \frac{1 \cdot 3 \cdot 5 \ldots (2n-1)}{2^n n!(2n+1)} x^{2n+1}$

29. $\sum_{n=2}^{\infty} (-1)^{n-1} \frac{2 \cdot 5 \ldots (3n-4)}{n! 3^n} x^n$

30. $\sum_{n=1}^{\infty} (-1)^{2n-1} \frac{2 \cdot 5 \ldots (3n-4)}{n! 3^n} (-2x^2)^n$

31. Show that if $\Sigma_{n=0}^{\infty} b_n(x-a)^n$ converges absolutely at either end of its interval of convergence, then it converges absolutely on the entire closed interval $[a - r, a + r]$.

32. Show that if $\Sigma_{n=0}^{\infty} b_n(x-a)^n$ converges at only one endpoint of its interval of convergence, then it is conditionally convergent at that endpoint.

33. Let k be a positive integer. Show that if $\Sigma_{n=0}^{\infty} b_n(x-a)^n$ has radius of convergence r, then the series $\Sigma_{n=0}^{\infty} b_n(x-a)^{kn}$ has radius of convergence $\sqrt[k]{r}$.

34. Suppose that $\Sigma_{n=0}^{\infty} b_n(x-a)^n$ has radius of convergence r, and $\Sigma_{n=0}^{\infty} c_n(x-a)^n$ has radius of convergence s. Then show that the series $\Sigma_{n=0}^{\infty} (b_n + c_n)(x-a)^n$ is absolutely convergent on $(a - \delta, a + \delta)$, where δ is the smaller of r and s. (Thus two convergent power series in $x - a$ can be added on the interval $(a - \delta, a + \delta)$, where δ is the smaller radius of convergence.)

9.8 Taylor Series

In Section 6 we saw how to approximate an arbitrary sufficiently differentiable function f by its Taylor polynomial p_n about a point $x = a$. In the last section, we learned how to find the values of x for which a given power series $\Sigma_{n=0}^{\infty} b_n(x-a)^n$ converges. We now put these two ideas together in the notion of a *power series representation for a function f about a point a.*

Suppose that we are given a function f, such as the sine function, that has derivatives of every order at the point a. (Such a function is called *infinitely differentiable* at $x = a$.) Then we can form the power series whose nth partial sum is the nth-degree Taylor polynomial $p_n(x)$.

8.1 DEFINITION If f is infinitely differentiable at $x = a$, then the ***Taylor series of f (expanded) about x = a*** is

$$(1) \quad \sum_{n=0}^{\infty} \frac{f^{(n)}(a)}{n!}(x-a)^n = f(a) + f'(a)(x-a) + \frac{1}{2}f''(a)(x-a)^2 + \frac{1}{3!}f'''(a)(x-a)^3 + \ldots + \frac{1}{n!}f^{(n)}(a)(x-a)^n + \ldots.$$

If $a = 0$, then the Taylor series is also called the ***Maclaurin series*** of f.

From the last section, (1) has an interval of convergence, which may be just a single point, an interval of finite length, or the entire real line. We are interested in whether, as seemed to be the case in Example 1 of Section 6, (1) converges to $f(x)$ when it converges. Our first theorem says that happens if and only if the remainder

$$r_n(x) = f(x) - p_n(x)$$

approaches 0 as $n \to \infty$.

8.2 THEOREM

> Let f be infinitely differentiable on an interval $I = (a - r, a + r)$, where if $r = \infty$, then I is understood to be $(-\infty, +\infty)$. Suppose that the Taylor series (1) is convergent on I. Then it converges to $f(x)$ on I if and only if $\lim_{n\to\infty} r_n(x) = 0$ for all $x \in I$.

Proof. We have

$$\lim_{n\to\infty} p_n(x) = f(x) \quad \text{if and only if } f(x) - \lim_{n\to\infty} p_n(x) = 0,$$

$$\text{if and only if } \lim_{n\to\infty} f(x) - \lim_{n\to\infty} p_n(x) = 0,$$

$$\text{if and only if } \lim_{n\to\infty} [f(x) - p_n(x)] = 0,$$

$$\text{if and only if } \lim_{n\to\infty} r_n(x) = 0. \quad \boxed{\text{QED}}$$

When $r_n(x)$ does approach 0 for all x in some interval I, that is, when the Taylor series converges to $f(x)$ on I, we say that ***the Taylor series represents f(x) on that interval, or that it is a power series representation of f(x) on I.***

The next example shows how Theorem 8.2 can be used in conjunction with the Lagrange remainder theorem to show that the Taylor series for a given f converges to $f(x)$.

Example 1

Show that the Maclaurin series for $f(x) = e^x$ converges to $f(x)$ for all x.

Solution. Since $f^{(i)}(x) = e^x$ for every positive integer i, we have $f^{(i)}(0) = 1$. So the Maclaurin series is

$$\sum_{n=0}^{\infty} \frac{x^n}{n!} = 1 + x + \frac{x^2}{2!} + \frac{x^3}{3!} + \ldots + \frac{x^n}{n!} + \ldots.$$

From Theorem 6.7, $r_n(x)$ is given by

$$r_n(x) = \frac{f^{(n+1)}(c)}{(n+1)!} x^{n+1} = \frac{e^c}{(n+1)!} x^{n+1},$$

where c is between 0 and x. If x is positive then $e^c < e^x$. If x is negative, then $e^c < e^0 = 1$. Let B be the larger of the two numbers 1 and e^x. Then B does not depend on n, and we have

$$0 \leq |r_n(x)| \leq \frac{B|x|^{n+1}}{(n+1)!}.$$

Recall from Example 1(b) of the last section that

$$\sum_{n=1}^{\infty} \frac{x^n}{n!} = \sum_{n=0}^{\infty} \frac{x^{n+1}}{(n+1)!}$$

converges for every real number x. Hence by Theorem 2.4

$$\lim_{n \to \infty} \frac{x^{n+1}}{(n+1)!} = 0$$

for every real number x. We thus have

$$0 \leq \lim_{n \to \infty} |r_n(x)| \leq B \lim_{n \to \infty} \frac{|x|^{n+1}}{(n+1)!} = 0.$$

So by the sandwich theorem for limits (Theorem 5.7 of Chapter 1), we have $\lim_{n \to \infty} r_n(x) = 0$. Hence the Maclaurin series for e^x converges to e^x for all real numbers x, as required. ■

We next apply Corollary 6.5 to obtain a test of whether $r_n(x)$ does tend to 0 as $n \to \infty$.

8.3 COROLLARY ***Taylor Series Convergence Test.*** Suppose that f is infinitely differentiable on an interval I as in Theorem 8.2. If there is some constant B such that for every n

$$|f^{(n+1)}(x)| \leq B$$

on I, then the Taylor series converges to $f(x)$ for all $x \in I$.

Proof. From Corollary 6.5 we have

$$|r_n(x)| \leq \frac{B}{(n+1)!} |x - a|^{n+1}.$$

Then

$$\lim_{n \to \infty} |r_n(x)| \leq B \lim_{n \to \infty} \frac{|x - a|^{n+1}}{(n+1)!}. \tag{2}$$

But $\dfrac{(x-a)^{n+1}}{(n+1)!}$ is the nth term of the power series

$$\sum_{n=0}^{\infty} \frac{(x-a)^{n+1}}{(n+1)!} = (x-a) + \frac{(x-a)^2}{2!} + \frac{(x-a)^3}{3!} + \dots. \tag{3}$$

By Exercise 1 of the last section, (3) converges for all x. Hence by Theorem 2.4 its nth term must approach 0 as $n \to \infty$. That is,

$$\lim_{n\to\infty} \frac{(x-a)^{n+1}}{(n+1)!} = 0.$$

It therefore follows from (2) that

$$0 \le \lim_{n\to\infty} |r_n(x)| \le B \lim_{n\to\infty} \frac{|x-a|^{n+1}}{(n+1)!} = 0.$$

Thus by the sandwich theorem for limits, $\lim_{n\to\infty} |r_n(x)| = 0$. Hence $\lim_{n\to\infty} r_n(x) = 0$. Then by Theorem 8.2 the Taylor series converges to $f(x)$ on I. QED

The following example shows how this convergence test is applied.

Example 2

Show that the Taylor series for $f(x) = \cos x$ about $\pi/3$ converges to $\cos x$ for all x.

Solution. We have

$$\begin{aligned}
f(x) &= \cos x & &\to & f(\pi/3) &= \frac{1}{2},\\
f'(x) &= -\sin x & &\to & f'(\pi/3) &= -\frac{\sqrt{3}}{2},\\
f''(x) &= -\cos x & &\to & f''(\pi/3) &= -\frac{1}{2},\\
f'''(x) &= \sin x & &\to & f'''(\pi/3) &= \sqrt{3}/2,\\
f^{(4)}(x) &= \cos x = f(x) & &\to & f^{(4)}(\pi/3) &= \frac{1}{2}.
\end{aligned}$$

Since $f^{(4)}(x) = f(x)$, the derivatives repeat in blocks of four: $f^{(5)}(x) = f'(x)$, $f^{(6)} = f''(x), f^{(7)} = f'''(x), f^{(8)}(x) = f^{(4)}(x) = f(x)$, and so on. So the Taylor series is

$$\frac{1}{2} - \frac{\sqrt{3}}{2}\left(x - \frac{\pi}{3}\right) + \frac{1}{2!}\left(-\frac{1}{2}\right)\left(x - \frac{\pi}{3}\right)^2 + \frac{1}{3!}\left(\frac{\sqrt{3}}{2}\right)\left(x - \frac{\pi}{3}\right)^3 + \frac{1}{4!}\left(\frac{1}{2}\right)\left(x - \frac{\pi}{3}\right)^4 + \dots.$$

Here $|f^{(n)}(x)| \le 1$ for all x, since $f^{(n)}(x) = \pm\sin x$ or $\pm\cos x$. Hence with $B = 1$ in Corollary 8.3, we conclude that the Taylor series converges to $\cos x$ for every real number x. ■

The next example shows that the work done to determine whether $p_n(x)$ converges to $f(x)$ is really necessary.

Example 3

Let

$$f(x) = \begin{cases} e^{-1/x^2} & \text{for } x \neq 0 \\ 0 & \text{for } x = 0. \end{cases}$$

(a) Find the Maclaurin series of $f(x)$.

(b) Show that the Maclaurin series converges for every real number x.

(c) Show that the Maclaurin series converges to $f(x)$ *only* at the single value $x = 0$.

Solution. (a, b) First we need to compute the derivatives of f at 0. We have

$$\lim_{h\to 0+} \frac{e^{-1/h^2} - 0}{h} = \lim_{t\to +\infty} \frac{1}{\frac{1}{t} e^{t^2}} \qquad \textit{letting } t = 1/h$$

$$= \lim_{t\to +\infty} \frac{t}{e^{t^2}} = \lim_{t\to +\infty} \frac{1}{2te^{t^2}} = 0. \qquad \textit{by l'Hôpital's rule}$$

Similarly,

$$\lim_{h\to 0-} \frac{e^{-1/h^2} - 0}{h} = 0.$$

Thus $f'(0) = 0$. Similar calculations (Exercise 42) show that

$$f''(0) = f'''(0) = \ldots = f^{(n)}(0) = 0.$$

Hence the Maclaurin series for $f(x)$ about $x = 0$ is

$$0 + 0x + \frac{0}{2!}x^2 + \frac{0}{3!}x^3 + \ldots + \frac{0}{n!}x^n, \tag{4}$$

which obviously converges to 0 for all real numbers x.

(c) $f(x) = e^{-1/x^2} \neq 0$ for $x \neq 0$. Thus the Maclaurin series (4) converges to $f(x)$ *only* when $x = 0$! ■

Example 3 underscores the importance of checking that $\lim_{n\to\infty} r_n(x) = 0$ before working with the Taylor series for f. Even if the Taylor series converges, it may not converge to $f(x)$!

The methods of Section 6 make it possible to compute values of $f(x)$ when the Taylor series does converge to it. While this is of conceptual as well as computational interest, the importance of a Taylor series representation for a function f goes beyond computing values of $f(x)$. When f is represented by its Taylor series, then inside the interval of convergence we can differentiate or integrate f by term-by-term differentiation or integration of its Taylor series. Thus $\int_c^d f(x)\,dx$ can be approximated by $\int_c^d p_n(x)\,dx$, which is easy to evaluate. The next result states this formally. It is easier to understand and use than to prove, so we leave the proof for advanced calculus.

8.4 THEOREM

Suppose that the power series $\sum_{n=0}^{\infty} b_n(x-a)^n$ converges to $f(x)$ on $I = (a-r, a+r)$, where $[c, d] \subseteq I$. Then

$$\int_c^d f(x)\,dx = \lim_{n\to\infty} \int_c^d s_n(x)\,dx,$$

where

$$s_n(x) = \sum_{k=0}^{n} b_k(x-a)^k.$$

In other words,

$$\int_c^d \left[\sum_{n=0}^{\infty} b_n(x-a)^n\right] dx = \sum_{n=0}^{\infty} \left[\int_c^d b_n(x-a)^n\,dx\right],$$

that is,

$$\int_c^d \lim_{n\to\infty} s_n(x)\,dx = \lim_{n\to\infty} \int_c^d s_n(x)\,dx,$$

so the integral of the limit is the limit of the integral.

The next example illustrates the technique of approximating $\int_c^d f(x)\,dx$ by $\int_c^d p_n(x)\,dx$ when f is representable by its Taylor series.

Example 4

Find an approximation of $\int_0^{1/3} e^{-t^2}\,dt$ accurate to 5 decimal places.

Solution. The function $f(x) = e^{-t^2}$ does not have an antiderivative expressible in terms of elementary functions. But this function is used to define the normal probability distribution in statistics, and so it is necessary to work with it. If we replace x by $-t^2$ in the Taylor series for e^x in Example 1, then for every real number t we have

$$e^{-t^2} = \sum_{n=0}^{\infty} \frac{(-1)^n t^{2n}}{n!} \tag{5}$$

This series converges on any interval I. So we let $I = (-1, 1)$, which contains $[0, 1/3]$. Then by Theorem 8.4 we have

$$\begin{aligned}
\int_0^{1/3} e^{-t^2}\,dt &= \lim_{n\to\infty} \int_0^{1/3} \sum_{k=0}^{n} \frac{(-1)^k t^{2k}}{k!}\,dt \\
&= \lim_{n\to\infty} \sum_{n=0}^{n} \left.\frac{(-1)^k t^{2k+1}}{(2k+1)k!}\right]_0^{1/3} \\
&= \sum_{k=0}^{\infty} \frac{(-1)^k (1/3)^{2k+1}}{(2k+1)\cdot k!} \\
&= \frac{1}{3^1\cdot 1\cdot 0!} - \frac{1}{3^3\cdot 3\cdot 1} + \frac{1}{3^5\cdot 5\cdot 2!} - \frac{1}{3^7\cdot 7\cdot 3!} \\
&\quad + \frac{1}{3^9\cdot 9\cdot 4!} - + \cdots.
\end{aligned}$$

The last series converges by the alternating series test, so by Theorem 4.4 the error made in approximating it by the partial sum s_n is at most $|a_{n+1}|$. Here we find

$$|a_3| = \frac{1}{3^7 \cdot 7 \cdot 3!} \approx 0.000011,$$

$$|a_4| = \frac{1}{3^9 \cdot 9 \cdot 4!} \approx 0.0000001 < 0.000005.$$

So s_4 gives S accurate to 5 decimal places. From a calculator we get

$$\int_0^{1/3} e^{-t^2}\,dt \approx 0.333333 - 0.012345 + 0.000412 - 0.000011$$

$$\approx 0.321388 \approx 0.32139. \blacksquare$$

In Example 4 we approximated $\int_0^{1/3} e^{-t^2}\,dt$ by $\int_0^{1/3} p_3(t)\,dt$, where $p_3(t)$ is the third-degree Taylor polynomial of e^{-t^2} about 0. Thus the hitherto *intractable* integral $\int_a^b f(x)\,dx$ was approximated by integrating a *polynomial approximation* to f over $[a, b]$. This technique is advantageous when Taylor polynomials can be found without undue work, and when they converge rapidly to $f(x)$.

Our next result is also of far-reaching importance. Its proof is again left for advanced calculus.

8.5 THEOREM Suppose that the power series $\sum_{n=0}^{\infty} b_n(x-a)^n$ converges to $f(x)$ on $I = (a - r, a + r)$. Then f is differentiable on I and

$$f'(x) = \sum_{n=1}^{\infty} nb_n(x-a)^{n-1} \qquad \text{for } x \in I.$$

In other words, the power series for $f'(x)$ can be obtained by term-by-term differentiation of the power series for $f(x)$.

Example 5

Differentiate the power series for e^{-x^2} term-by-term.

Solution. From (5) in Example 4,

$$e^{-x^2} = \sum_{n=0}^{\infty} \frac{(-1)^n x^{2n}}{n!} = 1 - x^2 + \frac{x^4}{2!} - \frac{x^6}{3!} + \frac{x^8}{4!} - \frac{x^{10}}{5!} + - \cdots.$$

Term-by-term differentiation gives

$$-2x + 2x^3 - x^5 + \frac{1}{3}x^7 - + \ldots = -2x\left(1 - x^2 + \frac{x^4}{2!} + \frac{x^6}{3!} + - \ldots\right)$$

$$= -2x \cdot \sum_{n=0}^{\infty} \frac{(-x^2)^n}{n!},$$

which is the series for $-2x \cdot e^{-x^2}$, as Theorem 8.5 asserts. $\blacksquare$

Whereas Theorem 8.4 had obvious practical implications for evaluating integrals, at first glance Theorem 8.5 seems less impressive. After all, it is usually easier to evaluate derivatives than integrals, so no work seems to be saved by using Taylor series in a problem like Example 5. But Theorem 8.5 may be an even *bigger* labor saver than Theorem 8.4, for it leads to the following very useful result.

8.6 THEOREM

If $\sum_{n=0}^{\infty} b_n(x-a)^n$ converges to $f(x)$ for all x in $I = (a-r, a+r)$, then

$$b_n = \frac{f^{(n)}(a)}{n!}$$

for all n. That is,

$$\sum_{n=0}^{\infty} b_n(x-a)^n$$

is the Taylor series for $f(x)$ about $x = a$.

Proof. We have $f(a) = b_0$, and by Theorem 8.5

$$f'(x) = \sum_{n=1}^{\infty} nb_n(x-a)^{n-1}.$$

Thus $f'(a) = b_1$. But we can apply Theorem 8.5 again to the series for f'. We get

$$f''(x) = \sum_{n=2}^{\infty} n(n-1)b_n(x-a)^{n-2}.$$

Hence

$$f''(a) = 2 \cdot 1b_2 \rightarrow b_2 = \frac{f''(a)}{2!}.$$

We can continue differentiating. After k applications of Theorem 8.5 we get

$$f^{(k)}(x) = \sum_{n=k}^{\infty} n(n-1)\ldots(n-k+1)b_n(x-a)^{n-k}.$$

Thus

$$f^{(k)}(a) = k(k-1)\ldots(k-k+1) = k! \rightarrow b_k = \frac{f^{(k)}(a)}{k!}.$$ QED

Theorem 8.6 says that once you find, *by any method*, a power series that converges to $f(x)$ on $(a-r, a+r)$, you have found the *Taylor series*, because any power series that converges to $f(x)$ near a must coincide with the Taylor series.

Example 6

Find the Taylor series for $\ln(1+x)$ near $a = 0$. [This series was first obtained by Nicholas Mercator (1620–1687) in 1668.]

Solution. We use the facts that

$$\int_0^x \frac{1}{1+t}\,dt = \ln(1+x) \qquad \text{for } x \in (-1, 1)$$

and

$$\frac{1}{1+t} = 1 - t + t^2 - t^3 + - \ldots + (-1)^n t^n + - \ldots \tag{6}$$

for $t \in (-1, 1)$, which follows because the power series in (6) is geometric with common ratio $-t$. Then we have from Theorem 8.4

$$\ln(1+x) = \int_0^x \frac{1}{1+t}\,dt = \int_0^x (1 - t + t^2 - t^3 +- \ldots + (-1)^n t^n +- \ldots)\,dt$$

$$= t - \frac{t^2}{2} + \frac{t^3}{3} - \frac{t^4}{4} +- \ldots + \frac{(-1)^n t^{n+1}}{n+1} +- \ldots \Big]_0^x,$$

$$\ln(1+x) = x - \frac{1}{2}x^2 + \frac{x^3}{3} - \frac{1}{4}x^4 +- \ldots + \frac{(-1)^n x^{n+1}}{n+1} +- \ldots,$$

for $x \in (-1, 1)$. By Theorem 8.6 the last series must be the Taylor series for $\ln(1+x)$ near $x = 0$. ■

Example 7

Find the Taylor series for $\ln\left[\dfrac{1+x}{1-x}\right]$ near $x = 0$.

Solution. From the preceding example,

$$\ln(1+x) = x - \frac{1}{2}x^2 + \frac{x^3}{3} - \frac{1}{4}x^4 +- \ldots + \frac{(-1)^n x^{n+1}}{n+1} +- \ldots,$$

for $x \in (-1, 1)$. Replacement of x by $-x$ gives

$$\ln(1-x) = -x - \frac{1}{2}x^2 - \frac{x^3}{3} - \frac{1}{4}x^4 - \ldots - \frac{x^{n+1}}{n+1} - \ldots,$$

for $x \in (-1, 1)$. If we subtract the two series, then by Theorem 2.6 the resulting series converges to

$$\ln(1+x) - \ln(1-x) = \ln\left(\frac{1+x}{1-x}\right)$$

for $x \in (-1, 1)$. We thus have

$$\ln\frac{1+x}{1-x} = 2x + \frac{2x^3}{3} + \ldots + \frac{2x^{2n-1}}{2n-1} + \ldots$$

$$= 2\left(x + \frac{x^3}{3} + \ldots + \frac{x^{2n-1}}{2n-1} + \ldots\right).$$

Again, this series must be the Taylor series for $\ln\left(\dfrac{1+x}{1-x}\right)$. ■

The series in Example 7 can be used to compute a rational approximation for $\ln y$ for any positive real number y. One lets

$$x = \frac{y-1}{y+1}. \tag{7}$$

Then $-1 < x < 1$, and multiplication of both sides of (7) by $y + 1$ gives

$$xy + x = y - 1 \rightarrow x + 1 = y - xy = y(1 - x) \rightarrow y = \frac{1 + x}{1 - x}, \quad (8)$$

for $x \in (-1, 1)$. So $\ln y$ is given by the series for

$$f(x) = \ln \frac{1 + x}{1 - x}.$$

Exercises 34–36 explore this further. Exercises 21 and 37 explore how similar ideas can be used to approximate π.

We close by giving Table 1, which lists important Maclaurin series, and the intervals I on which they represent (that is, converge to) their function f. Familiarity with these series can prove helpful in working homework and examination problems.

TABLE 1

Some Important Maclaurin Series

$f(x)$	*Maclaurin Series*	*Interval I*
e^x	$\sum_{n=0}^{\infty} \frac{x^n}{n!}$	$(-\infty, +\infty)$
$\sin x$	$\sum_{n=1}^{\infty} (-1)^{n-1} \frac{x^{2n-1}}{(2n-1)!}$	$(-\infty, +\infty)$
$\cos x$	$\sum_{n=0}^{\infty} (-1)^n \frac{x^{2n}}{(2n)!}$	$(-\infty, +\infty)$
$\ln(1 + x)$	$\sum_{n=1}^{\infty} (-1)^{n-1} \frac{x^n}{n}$	$(-1, 1)$
$\frac{1}{1 + x}$	$\sum_{n=0}^{\infty} (-1)^n x^n$	$(-1, 1)$

Exercises 9.8

In Exercises 1–12, show that the Taylor series for the given function f about the given point a converges to $f(x)$ for x in the interval I.

1. $f(x) = \sin x$; $a = 0$; $I = (-\infty, +\infty)$
2. $f(x) = \cosh x = \frac{1}{2}(e^x + e^{-x})$; $a = 0$; $I = (-\infty, +\infty)$
3. $f(x) = \sin x$; $a = \pi/4$; $I = (-\infty, +\infty)$
4. $f(x) = \cos x$; $a = \pi/4$; $I = (-\infty\ +\infty)$
5. $f(x) = \ln(1 - x)$; $a = 0$; $I = (0, 1)$
6. $f(x) = \ln x$; $a = 5$; $I = [4, 6]$
7. $f(x) = \sqrt{x}$; $a = 4$; $I = (1, 7)$
8. $f(x) = \sin^2 x$; $a = 0$; $I = (-\infty, +\infty)$ (*Hint:* $\cos 2x = 1 - 2\sin^2 x$)
9. $f(x) = \cos^2 x$; $a = 0$; $I = (-\infty, +\infty)$ (See Exercise 8.)
10. $f(x) = e^{-x^2}$; $a = 0$; $I = (-\infty, +\infty)$
11. $f(x) = \dfrac{1}{1 - x^2}$; $a = 0$; $I = (-1, 1)$ (*Hint:* To save work, see Equation (6).)
12. $f(x) = 1/(1 + x)$; $a = 0$; $I = (-1, 1)$

In Exercises 13–20, find an approximation of the given integral accurate to four decimal places.

13. $\int_0^{1/2} e^{-t^2}\,dt$
14. $\int_0^{1/2} \frac{dt}{1 + t^3}$ [*Hint:* See Equation (6).]
15. $\int_0^1 \sin t^2\,dt$
16. $\int_0^1 \cos t^2\,dt$
17. $\int_0^{0.1} e^{x^3}\,dx$
18. $\int_{1/2}^1 \frac{\sin x}{x}\,dx$
19. $\int_0^{0.1} \ln(1 + \sin x)\,dx$ (*Hint:* Use Example 6 with $\sin x$ in place of x.)
20. $\int_0^{0.1} \ln(1 - \sin x)\,dx$
21. **(a)** Show that

$$1 - x^2 + x^4 - x^6 + - \ldots + (-1)^n x^{2n} + \ldots = \frac{1}{1 + x^2}, \quad \text{for } x \text{ in } (-1, 1).$$

(b) Obtain the Taylor series for $\tan^{-1} x$ for x in $(-1, 1)$.
(c) If $f(x) = \tan^{-1} x$, then find $f^{(5)}(0)$ and $f^{(6)}(0)$.

22. **(a)** Find the first three terms of the Taylor series for $\tan x$ about 0.
(b) Find the first three terms of the Taylor series for $\sec^2 x$ about 0.
(c) Find the first three terms of the Taylor series for $\ln|\cos x|$ about 0.

23. **(a)** Show that

$$\frac{1}{1-x} = 1 + x + x^2 + \ldots + x^n + \ldots,$$

for x in $(-1, 1)$.
(b) Obtain the Taylor series for

$$\frac{1}{1-2x+x^2}, \qquad \text{for } x \text{ in } (-1, 1).$$

24. Use Exercise 21(a) to find a power series for

$$\frac{x}{(1+x^2)^2} \qquad \text{if } x \text{ belongs to } (-1, 1).$$

25. **(a)** Find the interval of convergence for

$$\sum_{n=1}^{\infty} (-1)^{n-1} \frac{x^n}{n}.$$

(b) Find the interval of convergence of $\Sigma_{n=1}^{\infty} (-1)^{n-1} x^{n-1}$. (Thus the series obtained by differentiating a given power series term-by-term need not have the same interval of convergence as the original series.)

26. Show that term-by-term differentiation of the Taylor series for $\cosh x = \frac{1}{2}(e^x + e^{-x})$ about 0 gives the Taylor series for $\sinh x = \frac{1}{2}(e^x - e^{-x})$ about 0. What does term-by-term integration give?

27. Find the Taylor series for f about 0 if $f(0) = 0$ and $f(x) = (e^x - 1)/x$ for $x \neq 0$. Differentiate term-by-term to show that $\Sigma_{n=1}^{\infty} n/(n+1)!$ converges to sum 1.

28. Find the Taylor series for $x^2 e^{-x}$ about $x = 0$. Differentiate term-by-term to show that

$$\sum_{n=1}^{\infty} (-2)^{n-1} \frac{n+2}{n!}$$

converges to the sum 1.

29. Find the Taylor series for xe^x about $x = 0$. Integrate term-by-term to show that

$$\sum_{n=0}^{\infty} \frac{1}{n!(n+2)}$$

converges to sum 1. (*Hint:* Use integration by parts on xe^x.)

30. Find the Taylor series for $x^2 e^x$ about $x = 0$. Integrate term-by-term to show that

$$2 + \left(\frac{1}{3} + \frac{1}{4} + \frac{1}{5 \cdot 2!} + \frac{1}{6 \cdot 3!} + \ldots + \frac{1}{(n+3)n!} + \ldots\right)$$

converges to e.

31. The ***error function*** $\operatorname{erf}(x)$ is defined by

$$\operatorname{erf}(x) = \frac{2}{\sqrt{\pi}} \int_0^x e^{-t^2}\, dt.$$

Find the Taylor series for $\operatorname{erf}(x)$ near 0. (The error function is widely applied in scientific and statistical work.)

32. **(a)** Show that $\int \ln(1-x)\,dx = -x - (1-x)\ln(1-x) + C$.
(b) Integrate the Taylor series for $\ln(1-t)$ to show that

$$\sum_{n=2}^{\infty} \frac{x^n}{(n-1)n} = x + (1-x)\ln(1-x).$$

33. **(a)** Show that $\int x \tan^{-1} x\, dx = \frac{1}{2}(x^2+1)\tan^{-1} x - \frac{1}{2}x + C$.
(b) Integrate the Taylor series for $t \tan^{-1} t$ to show that

$$\sum_{n=1}^{\infty} (-1)^{n+1} \frac{x^{2n+1}}{(2n-1)(2n+1)} = \frac{1}{2}(x^2+1)\tan^{-1} x - \frac{1}{2}x.$$

34. If $t \in (-1, 1)$, then show that

$$\frac{1}{1-t} = 1 + t + t^2 + \ldots + t^n + \frac{t^{n+1}}{1-t}$$

and

$$\frac{1}{1+t} = 1 - t + t^2 - + \ldots + (-1)^n t^n + \frac{(-1)^{n+1} t^{n+1}}{1+t}.$$

35. Integrate the formulas in Exercise 34 and add to show that

$$\ln \frac{1+x}{1-x} = 2x + \frac{2}{3}x^3 + \ldots + \frac{[1+(-1)^n]x^{n+1}}{n+1} + r_n(x),$$

where

$$r_n(x) = \int_0^x \left[t^{n+1}\left(\frac{1}{1-t} + \frac{(-1)^{n+1}}{1+t}\right)\right] dt,$$

for $x \neq 1$ in the interval $(-1, 1)$. (See Exercise 36.)

36. To find $\ln 2$ accurate to four decimal places, we let $y = 2$ and use the last equation in (8) to find

$$x = \frac{2-1}{2+1} = \frac{1}{3}.$$

(a) Use the formula from Exercise 35 to show that

$$|r_n(x)| \leq \int_0^{1/3} \frac{5}{2} t^{n+1}\, dt = \frac{5}{2(n+2)3^{n+2}}.$$

(b) Show that $n = 6$ gives $|r_n(x)| < 0.00005$.
(c) Find an approximation of $\ln 2$ that is accurate to at least four decimal places.

37. **(a)** The series in Exercise 21(b) converges to $\tan^{-1} x$ on $[-1, 1]$. Show from this that

$$\frac{\pi}{4} = 1 - \frac{1}{3} + \frac{1}{5} - \frac{1}{7} + \frac{1}{9} - \frac{1}{11} + - \ldots.$$

How many terms should be added to approximate π to an accuracy of four decimal places?

(b) Use the formula

$$\tan(\alpha + \beta) = \frac{\tan\alpha + \tan\beta}{1 - \tan\alpha\tan\beta}$$

with $\alpha = \tan^{-1}(1/2)$ and $\beta = \tan^{-1}(1/3)$ to show that

$$\pi = 4\tan^{-1}(1/2) + 4\tan^{-1}(1/3). \tag{9}$$

Compare your calculator's value of π with the value for the right side of (9).

38. **(a)** Use the basic identities

$$\sin^2 x = \frac{1 - \cos 2x}{2}, \qquad \cos^2 x = \frac{1 + \cos 2x}{2},$$

and $1 - \cos^2 x = \sin^2 x$ to show that for $y \in (-\pi/2, \pi/2)$, we have

$$\tan \frac{1}{2} y = \frac{\sin y}{1 + \cos y}.$$

(b) Deduce that

$$\tan \frac{1}{2} y = \frac{\tan y}{1 + \sqrt{1 + \tan^2 y}}. \tag{10}$$

39. To accurately compute $\tan^{-1} x$, let x be any real number and $y = \tan^{-1} x$.

(a) Show by applying $\tan^{-1}$ to both sides of (10) that

$$\tan^{-1} x = 2\tan^{-1}\left(\frac{x}{1 + \sqrt{1 + x^2}}\right),$$

where $\quad x/(1 + \sqrt{1 + x^2}) \in (-1, 1)$.

(b) Let

$$z = \frac{y}{1 + \sqrt{1 + y^2}}, \qquad \text{where} \qquad y = \frac{x}{1 + \sqrt{1 + x^2}}.$$

Then show that $\tan^{-1} x = 4\tan^{-1} z$.

PC **40.** **(a)** Write or use a computer program to compute partial sums of the series in Exercise 21(b) to within an input accuracy E for $x \in (0, 1)$. (Note that the series is alternating.)

(b) Modify your program in (a) to compute $\tan^{-1} x$ as $4\tan^{-1} z$ for z in Exercise 39(b).

(c) Use the programs in (a) and (b) to calculate $\tan^{-1}(1/3)$ and $\tan^{-1}(1/2)$ to an accuracy of at least 6 decimal places. Which method is faster?

PC **41.** Use your program in Exercise 40(b) to compile a 5-place table of values of the function $\tan^{-1} x$ for $x \in [0, 1/2]$, with increments of 1/64. Compare with the inverse tangent function on your computer (which was probably programmed to use this procedure).

42. **(a)** Show that for the function of Example 3,

$$f'(x) = -\frac{2}{x^3} e^{-1/x^2} \qquad \text{for } x \neq 0.$$

(b) Use l'Hôpital's rule to show that $f''(0) = 0$.

9.9 Binomial Series*

The main result of this section is an extension of the binomial theorem of elementary algebra. That theorem states that if p is a positive integer, then $(1 + x)^p$ is a polynomial in x with special integer coefficients. Namely,

$$(1 + x)^p = \sum_{n=0}^{p} \binom{p}{n} x^n, \tag{1}$$

where

$$\binom{p}{n} = \frac{p(p-1)\ldots(p-n+1)}{n!} = \frac{p!}{n!(p-n)!} \tag{2}$$

is an integer called the nth ***binomial coefficient.*** Binomial coefficients are important in combinatorial mathematics generally and in probability and statistics in particular. (For instance, $\binom{p}{n}$ is the number of n-element subsets that can be formed from a set of p elements.) Since $0! = 1$, when $n = 0$ or $n = p$ we have

$$\binom{p}{n} = \frac{p!}{0!p!} = 1.$$

In 1665, Newton discovered a generalization of (1) called the ***binomial series.*** He let p be any *rational number* and computed coefficients

$$\frac{p(p-1)(p-2)\ldots(p-n+1)}{n!}$$

* Optional

for $n = 1, 2, 3, \ldots$. Unlike the case when p is a positive integer, the factor $p - n + 1$ in the numerator never becomes 0. So instead of the finite sum (1) we get the infinite series

$$(3) \qquad 1 + px + \frac{p(p-1)}{2!}x^2 + \frac{p(p-1)(p-2)}{3!}x^3 + \ldots + \frac{p(p-1)\ldots(p-n+1)}{n!}x^n + \ldots .$$

For example, if $p = 1/2$, then we get

$$1 + \frac{1}{2}x + \frac{\frac{1}{2}\cdot\left(-\frac{1}{2}\right)}{2}x^2 + \frac{\frac{1}{2}\cdot\left(-\frac{1}{2}\right)\left(-\frac{3}{2}\right)}{3!}x^3 + \ldots + \frac{\frac{1}{2}\left(-\frac{1}{2}\right)\ldots\left(-\frac{2n-1}{2}\right)}{n!}x^n + \ldots$$

$$= 1 + \frac{1}{2}x + \frac{(-1)}{2^2\cdot 2!}x^2 + \frac{(-1)^2 1\cdot 3}{2^3\cdot 3!}x^3 + \ldots + \frac{(-1)^{n-1}\frac{1}{2}\left(\frac{1}{2}\right)\ldots\left(\frac{2n-1}{2}\right)}{2^n n!}x^n + \ldots$$

$$= \sum_{n=0}^{\infty} \frac{(-1)^{n-1} 1\cdot 3\ldots(2n-1)}{2^n n!}x^n.$$

As you can readily verify, this is the Maclaurin series of the function $f(x) = (1 + x)^p$. In the 18th century, Euler, a master of formal manipulation, studied this series, and one of his formal calculations produced a paradox that became famous.

Euler *assumed* that (3) converged to $(1 + x)^p$ for all x and p and performed calculations on that basis. In one of them, he let $p = -1$ and $x = -2$ and noted that

$$(1 + x)^p = (1 - 2)^{-1} = (-1)^{-1} = \frac{1}{-1} = -1.$$

And yet (3) gives

$$(1-2)^{-1} = 1 + (-1)(-2) + \frac{(-1)(-2)}{2}\cdot(-2)^2 + \frac{(-1)(-2)(-3)}{3\cdot 2}(-2)^3 + \ldots$$

$$= 1 + 2 + 4 + 8 + \ldots + 2^n + \ldots,$$

which obviously diverges to $+\infty$. Euler shrugged this off with a remark about the wonders of the infinite: Where else but in that realm could we get -1 by adding together a sequence of positive integers! Having begun our study of infinite series with a paradox of Zeno, it is perhaps fitting that we conclude that study by straightening out this situation.

9.1 THEOREM The binomial series (3) is absolutely convergent for $|x| < 1$ and divergent for $|x| > 1$.

Proof. We use the ratio test (Theorem 5.4). Here

$$a_{n+1} = \frac{p(p-1)\ldots(p-n+1)(p-n)}{(n+1)!}x^{n+1}$$

and

$$a_n = \frac{p(p-1)\ldots(p-n+1)}{n!}x^n.$$

Then

$$\left|\frac{a_{n+1}}{a_n}\right| = \frac{|p-n|}{n+1}|x|,$$

so

$$\lim_{n\to\infty}\left|\frac{a_{n+1}}{a_n}\right| = \lim_{n\to\infty}\frac{|p/n-1|}{1+1/n}|x| = |x|.$$

Thus the series is absolutely convergent for $|x| < 1$ and divergent for $|x| > 1$. QED

When $p = -1$, it therefore *makes no sense to try computing the binomial series* for $x = -2$, or any other number of absolute value greater than 1. What about $x = 1$ and $x = -1$? It can be shown, but requires more work than we can spare here, that when $x = 1$, the series converges to $(1+1)^p = 2^p$ provided $p > -1$, and when $x = -1$, the series converges to $0 = (1+(-1))^p$ provided $p > 0$. We *will* show that (3) converges to $(1+x)^p$ for any p, if $|x| < 1$.

9.2 THEOREM

The binomial series

(3) $$1 + \sum_{n=1}^{\infty}\frac{p(p-1)\ldots(p-n+1)}{n!}x^n$$

converges to $(1+x)^p$ if $|x| < 1$.

Proof. We could try to use Theorem 8.2, but it turns out to be much easier to let $f(x)$ be the value to which (3) converges on $(-1, 1)$ and then show that $f(x) = (1+x)^p$ for $x \in (-1, 1)$. From (3) and Theorem 8.5 we have

(4) $$f'(x) = \sum_{n=1}^{\infty}\frac{p(p-1)\ldots(p-n+1)}{n!}nx^{n-1}$$

$$= p + \sum_{n=2}^{\infty}\frac{p(p-1)\ldots(p-n+1)}{(n-1)!}x^{n-1},$$

(5) $$f'(x) = p + \sum_{m=1}^{\infty}\frac{p(p-1)\ldots(p-m+1)(p-m)}{m!}x^m,$$

where in (5) we have changed to a new summation index, $m = n - 1$, which will be more convenient for upcoming calculations. If we multiply each side of (4) by x, we get

$$xf'(x) = \sum_{n=1}^{\infty}\frac{p(p-1)\ldots(p-n+1)}{n!}nx^n,$$

(6) $$xf'(x) = \sum_{m=1}^{\infty}\frac{p(p-1)\ldots(p-m+1)}{m!}mx^m,$$

where we have replaced the dummy index n by m. Adding (5) and (6) we find

$$(7)\qquad f'(x) + xf'(x) = p + \sum_{m=1}^{\infty} \frac{p(p-1)\ldots(p-m+1)}{m!}(p-m+m)x^m$$

$$= p + \sum_{m=1}^{\infty} \frac{p(p-1)\ldots(p-m+1)}{m!}px^m$$

$$= p\left(1 + \sum_{m=1}^{\infty} \frac{p(p-1)\ldots(p-m+1)}{m!}x^m\right) = pf(x).$$

We thus have

$$(8)\qquad (1+x)f'(x) = pf(x).$$

By hypothesis $x \neq -1$, so $1 + x \neq 0$, and $(1+x)^p \neq 0$. Thus (8) gives

$$(9)\qquad \frac{f'(x)}{f(x)} = \frac{p}{1+x}.$$

Since

$$\frac{d}{dx}[\ln f(x)] = \frac{f'(x)}{f(x)} \qquad \text{and} \qquad \frac{d}{dx}[\ln(1+x)^p] = \frac{p}{1+x},$$

we see from (9) that $\ln f(x)$ and $\ln(1+x)^p$ have the same derivative, so they differ by some constant c. That is,

$$(10)\qquad \ln f(x) = \ln(1+x)^p + c.$$

When $x = 0$, we have $f(x) = 1$ by (3). Then (10) gives

$$\ln 1 = \ln 1^p + c \rightarrow 0 = 0 + c \rightarrow c = 0.$$

Hence (10) becomes

$$(11)\qquad \ln f(x) = \ln(1+x)^p.$$

Then exponentiating each side of (11), we get

$$f(x) = (1+x)^p. \quad \boxed{\text{QED}}$$

Theorem 9.2 helps produce useful Taylor series for functions that involve fractional powers.

Example 1

Find the Taylor series for $f(x) = (1-x)^{-1/2}$, $x \in (-1, 1)$.

Solution. For $x \in (-1, 1)$, the binomial series for $f(x) = [1 + (-x)]^{-1/2}$ converges to $(1-x)^{-1/2}$. So by Theorem 8.6 this is the Taylor series for f near 0. We can then use Theorem 9.2 with $-x$ in place of x. Setting $p = -1/2$ in (3), we get

$$(1-x)^{-1/2} = 1 - \frac{1}{2}(-x) + \frac{\left(-\frac{1}{2}\right)\left(-\frac{1}{2}-1\right)}{2!}(-x)^2$$

$$+ \frac{\left(-\frac{1}{2}\right)\left(-\frac{1}{2}-1\right)\left(-\frac{1}{2}-2\right)}{3!}(-x)^3 + \ldots$$

$$+ \frac{\left(-\frac{1}{2}\right)\left(-\frac{3}{2}\right)\left(-\frac{5}{2}\right)\ldots\left(-\frac{1}{2}-n+1\right)}{n!}(-x)^n + \ldots.$$

We can simplify the general term as follows:

$$\frac{\left(-\frac{1}{2}\right)\left(-\frac{3}{2}\right)\left(-\frac{5}{2}\right)\cdots\left(\frac{-1-2n+2}{2}\right)}{n!}(-x)^n$$

$$= \frac{\left(-\frac{1}{2}\right)\left(-\frac{3}{2}\right)\left(-\frac{5}{2}\right)\cdots\left(\frac{-2n+1}{2}\right)}{n!}(-1)^n x^n$$

$$= \frac{(-1)^n\left(\frac{1}{2}\right)\left(\frac{3}{2}\right)\left(\frac{5}{2}\right)\cdots\left(\frac{2n-1}{2}\right)}{n!}(-1)^n x^n$$

$$= (-1)^{2n}\frac{1\cdot 3\cdot 5\ldots(2n-1)}{2^n n!}x^n$$

$$= \frac{1\cdot 3\cdot 5\ldots(2n-1)}{2^n n!}x^n.$$

Thus for $x \in (-1, 1)$, we have

$$(1-x)^{-1/2} = 1 + \sum_{n=1}^{\infty}\frac{1\cdot 3\cdot 5\ldots(2n-1)}{2^n n!}x^n. \blacksquare$$

Example 2

Find a power series for $\sin^{-1} x$.

Solution. Recall that

$$\sin^{-1} x = \int_0^x \frac{dt}{\sqrt{1-t^2}} = \int_0^x (1-t^2)^{-1/2}\,dt.$$

We can thus use Theorem 8.4 to get the Taylor series for $\sin^{-1} x$. First we can get the Taylor series for $(1-t^2)^{-1/2}$, $t \in (-1, 1)$, by replacing x in the preceding example by t^2. That gives

$$(1-t^2)^{-1/2} = 1 + \sum_{n=1}^{\infty}\frac{1\cdot 3\cdot 5\ldots(2n-1)}{2^n n!}t^{2n}.$$

Then if $x \in (-1, 1)$, we can use Theorem 8.4 to integrate this series term-by-term:

$$\int_0^x (1-t^2)^{-1/2}\,dt = \sin^{-1} x = \int_0^x\left(1 + \sum_{n=1}^{\infty}\frac{1\cdot 3\cdot 5\ldots(2n-1)}{2^n n!}t^{2n}\right)dt$$

$$= x + \sum_{n=1}^{\infty}\frac{1\cdot 3\cdot 5\ldots(2n-1)}{2^n n!}\left.\frac{t^{2n+1}}{2n+1}\right]_0^x$$

$$= x + \sum_{n=1}^{\infty}\frac{1\cdot 3\cdot 5\ldots(2n-1)}{2^n n!(2n+1)}x^{2n+1}$$

for $x \in (-1, 1)$. $\blacksquare$

Computational applications of binomial series involve the approach of the previous section. The binomial series gives a new and often simple way to generate the Taylor series of functions. We close with an indication of how they can be used to compute nth roots.

Example 3

Compute $\sqrt[3]{5/4}$ to an accuracy of at least four decimal places.

Solution. Since

$$\frac{5}{4} = 1 + \frac{1}{4},$$

the binomial series $(1 + x)^{1/3}$ with $x = 1/4$ gives

$$(1+x)^{1/3} = 1 + \frac{1}{3}x + \frac{\frac{1}{3}\left(\frac{1}{3}-1\right)}{2!}x^2 + \frac{\frac{1}{3}\left(\frac{1}{3}-1\right)\left(\frac{1}{3}-2\right)}{3!}x^3 + \ldots$$

$$+ \frac{\frac{1}{3}\left(\frac{1}{3}-1\right)\left(\frac{1}{3}-2\right)\cdots\left(\frac{1}{3}-n+1\right)}{n!}x^n + \ldots$$

$$= 1 + \frac{1}{3}x + \frac{\left(\frac{1}{3}\right)\left(\frac{-2}{3}\right)}{2!}x^2 + \frac{\left(\frac{1}{3}\right)\left(-\frac{2}{3}\right)\left(-\frac{5}{3}\right)}{3!}x^3$$

$$+ \ldots + \frac{\left(\frac{1}{3}\right)\left(-\frac{2}{3}\right)\left(-\frac{5}{3}\right)\cdots}{n!}x^n + \ldots.$$

Consider the nth term,

$$\frac{\left(\frac{1}{3}\right)\left(-\frac{2}{3}\right)\left(-\frac{5}{3}\right)\cdots\left(\frac{4-3n}{3}\right)}{n!}x^n = \frac{(-1)^{n-1}2\cdot 5\ldots(3n-4)}{n!3^n}x^n.$$

For $x = 1/4$ we have

$$\left(1+\frac{1}{4}\right)^{1/3} = 1 + \frac{1}{3}x + \sum_{n=2}^{\infty}(-1)^{n-1}\frac{2\cdot 5\ldots(3n-4)}{n!3^n4^n}.$$

By Theorem 4.5 the error made in approximating $(1 + 1/4)^{1/3}$ as the nth partial sum of this alternating series is at most the absolute value of the first term omitted. By trial and error we find that the fifth term is the first one that is less than 0.00005:

$$(-1)^4\cdot\frac{2\cdot 5\cdot 8\cdot 11}{5!3^54^5} = \frac{2\cdot 5\cdot 8\cdot 11}{5\cdot 24\cdot 3^5\cdot 4\cdot 4^4} = \frac{\not{5}\cdot\not{8}\cdot\not{2}\cdot 11}{\not{5}\cdot\not{8}\cdot 3\cdot 3^5\cdot\not{2}\cdot 2\cdot 4^4}$$

$$= \frac{11}{2\cdot 3^6\cdot 4^4} = \frac{11}{373248} < 0.00003.$$

So to get $\sqrt[3]{5/4}$ with an accuracy of at least four decimal places, we can use the terms of the infinite series from $n = 0$ through $n = 4$:

$$\sqrt[3]{\frac{5}{4}} = \left(1+\frac{1}{4}\right)^{1/3} \approx 1 + \frac{1}{12} - \frac{1}{144} + \frac{5}{3^44^3} - \frac{5}{2\cdot 3^54^3}$$

$$\approx 1.00000 + 0.08333 - 0.00694 + 0.00096 - 0.00016$$

$$\approx 1.07719 \approx 1.0772. \blacksquare$$

Exercises 9.9

In Exercises 1–10, find a power series for the given function. Give the largest open interval on which you know the power series converges to the given function.

1. $\sqrt{1+x}$
2. $\dfrac{1}{x^2+1}$
3. $\dfrac{1}{\sqrt[3]{1+x}}$
4. $(1+x^2)^{1/5}$
5. $(1-x)^{1/3}$
6. $\dfrac{1}{(1-x^2)^{1/3}}$
7. $\sqrt{2+x} = \sqrt{2}\sqrt{1+x/2}$
8. $(8-x)^{1/3}$
9. $(1+2x^2)^{1/3}$
10. $\dfrac{1}{\sqrt{1-\frac{1}{4}x^2}}$

In Exercises 11–14, compute the first four terms of the Taylor series for the function.

11. $\sqrt{1+xe^x}$
12. $\sqrt{1-x^2e^{-2x}}$
13. $\sqrt{1-k^2\sin^2 t}$
14. $\sinh x = \displaystyle\int_0^x \frac{1}{\sqrt{1+t^2}}\,dt$

In Exercises 15–22, use the binomial series to find an approximation of the given root with an accuracy of at least four decimal places.

15. $\sqrt{26}$
16. $\sqrt{35}$
17. $(6)^{1/3}$ (*Hint:* $(6)^{1/3} = (8)^{1/3}(3/4)^{1/3} = 2\sqrt[3]{1-\frac{1}{4}}$.)
18. $(62)^{1/3}$
19. $(630)^{1/4}$
20. $(14)^{1/4}$
21. $(1001)^{1/3}$
22. $(30)^{1/5}$

In Exercises 23–30, compute the value of the given definite integral to an accuracy of at least four decimal places. (Use the results of earlier exercises and examples as needed.)

23. $\displaystyle\int_0^{1/3} \frac{dx}{(1+x^2)^{1/3}}$
24. $\displaystyle\int_0^{1/2} \frac{dx}{(1-x^2)^{1/3}}$
25. $\displaystyle\int_0^{1/2} \frac{dx}{\sqrt{1-x^3}}$
26. $\displaystyle\int_0^{1/4} \frac{dx}{\sqrt{1+x^3}}$
27. $\displaystyle\int_0^{1/4} \sqrt{1+x^3}\,dx$
28. $\displaystyle\int_0^{1/2} \sqrt{1+x^3}\,dx$
29. $\displaystyle\int_0^{1/2} \frac{dx}{1+x^3}$
30. $\displaystyle\int_0^{1/3} \frac{dx}{1+x^4}$

The following exercises involve *elliptic integrals* $\int_0^{\pi/2} \sqrt{1-k^2\sin^2 t}\,dt$, which arise in applied work.

31. Find accurately to four decimal places the length of the first-quadrant arc of the ellipse $x(t) = 2\cos t$, $y(t) = \sqrt{5}\sin t$. Express your answer as a decimal multiple of $\sqrt{5}\pi$. (See Exercise 13.)
32. Find accurately to four decimal places the length of the first-quadrant arc of the ellipse $x(t) = 2\sqrt{2}\cos t$, $y(t) = 3\sin t$. (Express your answer as a decimal multiple of π.)
33. PC Rework Exercise 55 of Section 8.2 using a binomial series. Which method is better suited to this problem?
34. PC Rework Exercise 56 of Section 8.2 using a binomial series. Which method is better suited to this problem?

9.10 Looking Back

This chapter has dealt mainly with convergence and divergence of series and sequences. Section 1 introduced infinite sequences, defined convergence and divergence, and derived basic facts about convergent sequences. A basic tool is Theorem 1.3, which permits us to use l'Hôpital's rule to determine whether a sequence converges. Another useful criterion is given in Theorem 1.7: Bounded monotonic sequences always converge.

Section 2 defined convergence for series and derived two basic tests: Theorem 2.3 for geometric series and the divergence test (Theorem 2.4). We also met the harmonic series in Example 1. Section 3 presented the two comparison tests (Theorems 3.1 and 3.2) and the integral test (Theorem 3.3) for determining convergence of series of positive terms. The latter allowed us to find for which p the hyperharmonic series $\Sigma_{n=1}^{\infty} 1/n^p$ converges. Section 4 was devoted to the alternating series test (Theorem 4.2) and a useful approximation result, Theorem 4.4. Section 5 introduced the notions of absolute and conditional convergence and presented the most useful test for convergence of series, the ratio test (Theorem 5.4), as well as the root test (Theorem 5.5).

Section 6 discussed approximation of repeatedly differentiable functions by Taylor polynomials. The key result was the remainder estimate (Corollary 6.5), which makes it possible to determine the accuracy of the approximation of $f(x)$ by the Taylor polynomial $p_n(x)$. Section 7 introduced power series and the important notion of the interval of convergence (Theorem 7.2). Section 8 "took the lid" off Taylor polynomials by introducing Taylor series and their convergence to $f(x)$ (Theorem 8.2). We saw that we could approximate derivatives and integrals of $f(x)$ by differentiating and integrating $p_n(x)$ (Theorems 8.4 and 8.5), and that Taylor series could often be produced more simply than by calculating successive Taylor polynomials (Theorem 8.6). The final section pursued this idea through the example of the binomial series.

CHAPTER CHECKLIST

Section 1: infinite sequence, entries; convergence, divergence; monotonic sequence; bounded sequence, greatest lower bound, least upper bound; convergence of a bounded monotonic sequence; axiom of completeness; Fibonacci sequence.

Section 2: geometric sequence; infinite series, terms, partial sums, convergence, divergence; geometric series; divergence test; harmonic series; telescoping series.

Section 3: term-size comparison test; limit comparison test; integral test; hyperharmonic series.

Section 4: alternating series test; alternating harmonic series; error in approximation by s_n; approximation accurate to k decimal places.

Section 5: absolutely convergent series; conditionally convergent series; ratio test; root test.

Section 6: nth-degree Taylor polynomial; Maclaurin polynomial; Cauchy remainder formula; remainder estimate; Lagrange remainder formula.

Section 7: power series in $x - a$; radius of convergence; interval of convergence; investigation of convergence at endpoints of interval of convergence.

Section 8: Taylor series; Maclaurin series; power series expansion of $f(x)$; integration and differentiation of power series; uniqueness of Taylor series.

Section 9: binomial series; radius of convergence; approximation using binomial series.

REVIEW EXERCISES 9.10

1. Give a formula for a_n if the sequence $\boldsymbol{a}$ is

(a) $\left(-1, \frac{4}{9}, -\frac{5}{27}, \frac{6}{81}, -\frac{7}{243}, \ldots\right)$

(b) $\left(3, \frac{6}{5}, \frac{12}{25}, \frac{24}{125}, \frac{48}{625}, \ldots\right)$

2. Determine whether $\boldsymbol{a}$ converges by computing $\lim_{n\to\infty} a_n$ if

(a) $a_n = \dfrac{n^3 + 5n^2 + 5n + 4}{3n^3 + n^2 + 1}$

(b) $a_n = \dfrac{n \cdot 5^{n-1}}{4^{n+10}}$

(c) $a_n = \left(0, 1, \frac{1}{2}, \frac{1}{2}, \frac{1}{4}, \frac{3}{4}, \frac{1}{8}, \frac{7}{8}, \ldots\right)$

(d) $a_n = \dfrac{1}{n^2} \cos n\pi$

3. Show that the sequence $\boldsymbol{a}$ is monotonic and determine whether it converges if

(a) $a_n = \dfrac{(n+1)!}{4n}$

(b) $a_n = \dfrac{n!}{2 \cdot 4 \cdot 6 \cdot \ldots \cdot 2n}$

4. Determine whether the given series converges and, if it does, find its sum.

(a) $-\frac{1}{3} + \frac{1}{9} - \frac{1}{27} + \frac{1}{81} - + \ldots$

(b) $\displaystyle\sum_{n=1}^{\infty} \left(\frac{3}{2^n} - \frac{(-1)^n 5}{2^n}\right)$

(c) $\displaystyle\sum_{n=1}^{\infty} \frac{5^{n-1}}{4^{n+3}}$

(d) $\displaystyle\sum_{n=1}^{\infty} \left(\frac{2}{5n} - \frac{1}{n}\right)$

(e) $\displaystyle\sum_{n=1}^{\infty} \frac{3n^2 - 20n - 400}{1000n^2 + 1000n + 595}$

5. Explain the fallacy in the following. From Theorem 2.3 with $r = 2$, we have

$$1 + 2 + 4 + \ldots + 2^n + \ldots = \frac{1}{1-2} = -1.$$

Thus we can obtain -1 by adding infinitely many powers of 2.

In Exercises 6–19, determine whether the given series converges. If it converges, is the convergence absolute?

6. $\sum_{n=1}^{\infty} \frac{1}{n^3+n}$

7. $\sum_{n=1}^{\infty} \frac{1}{\sqrt{2n-1}}$

8. $\sum_{n=1}^{\infty} \frac{2n^2+3}{n^3+n^2+4}$

9. $\sum_{n=1}^{\infty} \frac{2n-5}{n^3-2n+5}$

10. $\sum_{n=1}^{\infty} \frac{1}{\sqrt[3]{2n^2+3n+1}}$

11. $\sum_{n=1}^{\infty} \frac{(-1)^{n+1}}{\sqrt[3]{n}}$

12. $\sum_{n=1}^{\infty} (-1)^{n+1}n^2e^{-n}$

13. $\sum_{n=1}^{\infty} \frac{(-1)^{n+1}n!}{2\cdot 4 \ldots (2n)}$

14. $\sum_{n=1}^{\infty} \frac{(-1)^{n+1}}{1+n!}$

15. $\sum_{n=1}^{\infty} \frac{(-5)^{n+1}}{4^{4n/3}}$

16. $\sum_{n=1}^{\infty} \frac{(-1)^{n+1}100^{n+100}}{2n!}$

17. $\sum_{n=1}^{\infty} (-1)^{n+1}\frac{2\cdot 4\cdot 6 \ldots (2n)}{n!}$

18. $\sum_{n=1}^{\infty} \frac{(-1)^{n+1}}{(n+1)\sqrt{n+2}}$

19. $\sum_{n=1}^{\infty} \frac{(-1)^{n+1}\ln(2n+1)}{3n+5}$

20. Obtain the Taylor polynomial of f near the given point a and give a formula for $r_n(x)$.
(a) $f(x) = \ln x$, $a = 5$. Find $p_6(x)$.
(b) $f(x) = e^x \sin x$, $a = 0$. Find $p_5(x)$.

21. Use Taylor polynomials to approximate to an accuracy of at least four decimal places:
(a) $\ln 5.02$ (b) $e^{1/3}$

22. Determine the interval of convergence of the given series and find where the convergence is absolute.

(a) $\sum_{n=0}^{\infty} \frac{(2x)^n}{n!}$ (b) $\sum_{n=2}^{\infty} \frac{x^n}{n^2-1}$

(c) $\sum_{n=1}^{\infty} \frac{(-1)^{n-1}(x-3)^n}{n+2}$ (d) $\sum_{n=1}^{\infty} \frac{(-2)^n n! x^n}{1\cdot 3\cdot 5 \ldots (2n-1)}$

(e) $\sum_{n=1}^{\infty} \frac{x^{5n}}{2n+1}$

23. Show that the Taylor series for f about a converges to $f(x)$ on the interval I.
(a) $f(x) = \cos x$, $a = 0$, $I = (-\infty, +\infty)$
(b) $f(x) = \sinh x = \frac{1}{2}(e^x - e^{-x})$, $a = 0$, $I = (-\infty, +\infty)$
(c) $f(x) = \sqrt{x}$, $a = 9$, $I = (1, 17)$
(d) $f(x) = e^{-x^2/2}$, $a = 0$, $I = (-\infty, +\infty)$

24. Show that the given power series converges to the given function f on the given interval.
(a) The Maclaurin series for e^{-x^2} to $f(x) = e^{-x^2}$ on $[0, 1]$.
(b) $1 + x^2 + x^4 + \ldots + x^{2n} + - \ldots$ to $f(x) = \frac{1}{1-x^2}$ on $(-1, 1)$.

25. Show that the following power series converge on $[-1/2, 1/2]$.

(a) $\sum_{n=0}^{\infty} nx^n$ (b) $\sum_{n=0}^{\infty} \frac{(-1)^{n+1}x^n}{n^2+e^n}$

26. Approximate to an accuracy of at least four decimal places:

(a) $\int_0^{2/5} e^{-t^2}\,dt$ (b) $\int_0^{1/3} \frac{dt}{1+t^3}$

(c) $\int_0^{1/5} e^{x^3}\,dx$ (d) $\int_{1/2}^{1} \frac{\cos x}{x}\,dx$

27. Show that term-by-term differentiation of the Taylor series for $\sinh x = \frac{1}{2}(e^x - e^{-x})$ near 0 gives the Taylor series for $\cosh x = \frac{1}{2}(e^x + e^{-x})$. What does term-by-term integration give?

28. Find a power series for the given $f(x)$ and give the largest open interval on which you know the power series converges to $f(x)$.

(a) $\sqrt{1-x}$ (b) $\frac{1}{\sqrt[3]{1-x^2}}$ (c) $\sqrt[4]{1-x}$

29. Compute the first four terms of the Taylor series for the function

$$f(x) = \sqrt{1 - x^2e^{2x}}.$$

30. Compute to an accuracy of at least four decimal places:
(a) $\sqrt{24}$ (b) $\sqrt[3]{7}$

(c) $\int_0^{1/3} \frac{dx}{\sqrt[3]{1-x^2}}$ (d) $\int_0^{1/2} \frac{dx}{\sqrt{1+x^3}}$

CHAPTER 10 Vector Geometry and Algebra

10.0 Introduction

The earlier chapters have presented the calculus of real-valued functions f of a single variable—which we have usually labeled x or t. Both the domain and range of such a function are subsets of the real-number system $\mathbf{R}$, and its graph can be drawn in the Cartesian coordinate plane $\mathbf{R}^2 = \{(x, y) | x, y \in \mathbf{R}\}$.

Although we have seen many applications of such functions, a host of important situations are modeled by functions whose domain or range (or both) involves more than one real variable. For example, we need three variables—length l, width w, and height h—to describe a three-dimensional rectangular box, and its volume is given by the function

$$f: \mathbf{R}^3 \to \mathbf{R}, \qquad \text{where } f(l, w, h) = lwh.$$

Another example arises from the parametric equations

$$x = g(t), \qquad y = h(t), \quad t \in [a, b]$$

for the path of a particle moving in the plane. Those equations determine a function

$$\mathbf{f}: \mathbf{R} \to \mathbf{R}^2, \qquad \text{where } \boldsymbol{f}(t) = (g(t), h(t)).$$

The next five chapters are concerned with the calculus of functions

$$\mathbf{f}: \mathbf{R}^n \to \mathbf{R}^m,$$

where m and n are 1, 2, or 3 (or occasionally, larger positive integers). Before we can begin to study such functions, we need a working familiarity with the algebra and geometry of $\mathbf{R}^2$ and $\mathbf{R}^3$. Developing that is the business of this chapter.

What links the algebra of these *Cartesian spaces* to their geometry is the concept of *vector*. Geometrically, vectors are objects that have both length and direction. Algebraically, they can be represented as ordered pairs (or ordered triples) of real numbers which are combined by performing ordinary algebraic operations on their real coordinates. This is explored in the first two sections, which introduce one-, two-, and three-dimensional vectors and relate them to plane and solid analytic geometry.

Sections 3 and 6 develop additional algebraic tools for studying the geometry of $\mathbf{R}^3$, which is the main subject of Sections 4 and 5. Those sections treat lines and planes in $\mathbf{R}^3$, which are the simplest curves and surfaces in three dimensions.

10.1 One- and Two-Dimensional Vectors

FIGURE 1.1

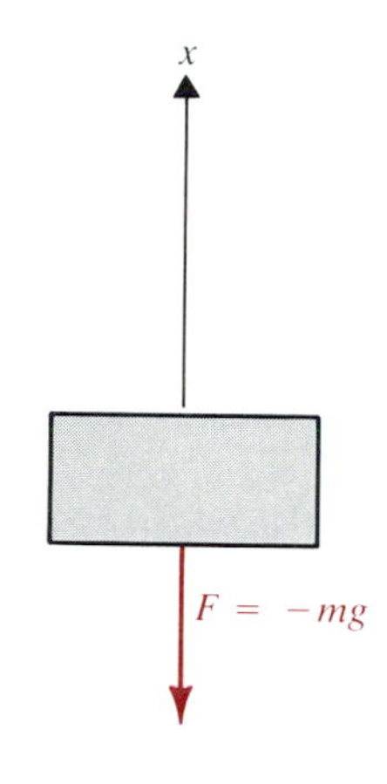

In physics and engineering, vectors are defined as geometric objects having both *length* and *direction.* They are represented as line segments with arrowheads (to indicate their direction). For example, **Figure 1.1** shows the weight of an object of mass m as a downward-directed force of magnitude mg. (The minus sign indicates the downward orientation.)

A major conceptual advance of the nineteenth century was the development of an *algebraic representation* for vectors. (See the historical note at the end of the section.) This section and the next introduce that algebra in dimensions one, two, and three.

We start with the algebraic representation of a vector $\mathbf{v}$ in one dimension. (Vectors are usually typeset in bold face but written with an arrow over them, since $\vec{v}$ is easier to write, but $\mathbf{v}$ is easier to typeset.) To represent an object having length and direction in $\mathbf{R}^1 = \mathbf{R}$, the familiar real line, note first that there are only two directions: positive (upward in Figure 1.1) and negative (downward in Figure 1.1). The usual assignment of coordinates to $\mathbf{R}^1$ gives a natural measurement of length. For example, a positively directed vector $\mathbf{v}$ that is three units long can be represented as $(+3)$, while a negatively directed vector $\mathbf{w}$ of the same length is representable as (-3). This scheme provides a one-to-one correspondence between one-dimensional vectors in $\mathbf{R}^1$ and real numbers x in $\mathbf{R}$ if we add one special vector: $\mathbf{0}$, the *zero vector*, which has length 0 and *no direction.* The real number x corresponding to a one-dimensional vector $\mathbf{v}$ is called its *coordinate.* The *length* of any vector $\mathbf{R}^1$ is the absolute value of its coordinate. For example, the vectors $\mathbf{v} = (+3)$ and $\mathbf{w} = (-3)$ both have length 3, because

$$|(+3)| = 3 \qquad \text{and} \qquad |(-3)| = 3.$$

In general, a vector $\mathbf{x} = (x)$ has length

$$|(x)| = \sqrt{x^2}. \tag{1}$$

This simple observation is the basis for our later notation for the length of higher dimensional vectors.

The above description of vectors by real numbers contains no information about where a vector begins (or ends). In **Figure 1.2,** for instance, both $\mathbf{v}_1$ and $\mathbf{v}_2$ are represented by $(+2)$. The reason for this is that *geometric vectors of the same length and direction are considered identical*, regardless of where their initial points may be.

FIGURE 1.2

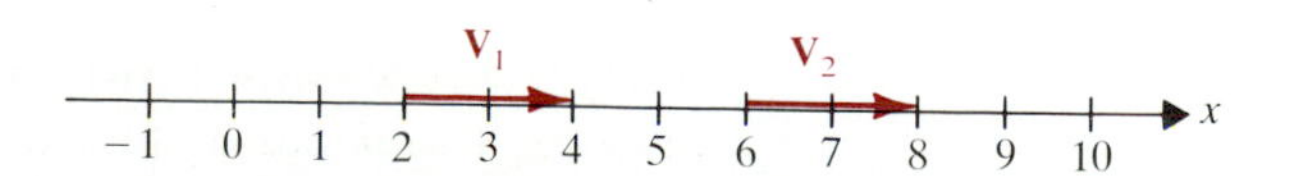

FIGURE 1.3

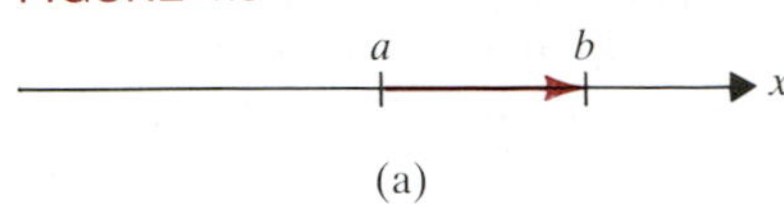

(a)

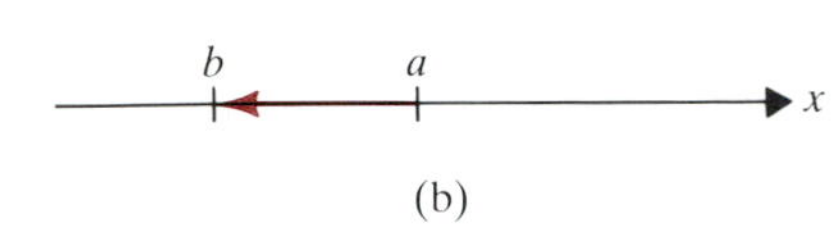

(b)

The vector drawn from the point a on the real line to the line point b has coordinate representation $(b - a)$, since $b - a$ is the directed distance from a to b: positive if $b > a$, negative if $b < a$, zero if $b = a$. See **Figure 1.3.**

The *resultant* of two geometric vectors $\mathbf{v}_1$ and $\mathbf{v}_2$ is defined as follows. The initial point of $\mathbf{v}_2$ is placed at the endpoint of $\mathbf{v}_1$, and the resultant is the vector drawn from the initial point of $\mathbf{v}_1$ to the endpoint of $\mathbf{v}_2$. (See **Figure 1.4.**) This corresponds to simply *adding* the coordinates of $\mathbf{v}_1$ and $\mathbf{v}_2$. For example, the resultant of positively directed vector $\mathbf{u}$ of length 2 and a negatively directed vector $\mathbf{v}$ of length 3 is a negatively directed vector $\mathbf{w}$ of length 1. This is easiest to see from the coordinate equation

$$(+2) + (-3) = (-1).$$

(See also **Figure 1.5.**) Comparison of the complexity of this equation with the verbal description illustrates the simplification provided by the algebraic representation of vectors.

Finally, there is a natural multiplication of vectors $\mathbf{v}$ by real numbers a. The vector $a\mathbf{v}$ is $\mathbf{0}$ if $a = 0$ or $\mathbf{v} = \mathbf{0}$. Otherwise, it points in the same direction as $\mathbf{v}$ if $a > 0$, in the opposite direction from $\mathbf{v}$ if $a < 0$, and has length $|a|$ times the length of $\mathbf{v}$. See **Figure 1.6.**

FIGURE 1.4

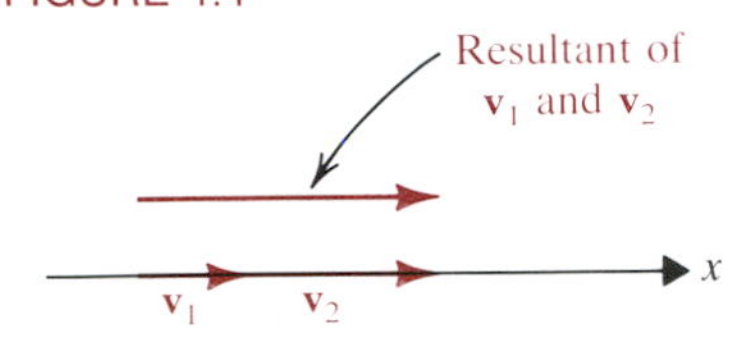

FIGURE 1.5

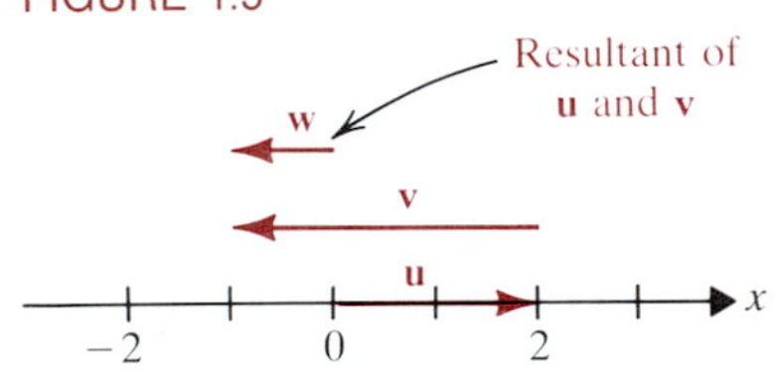

FIGURE 1.6

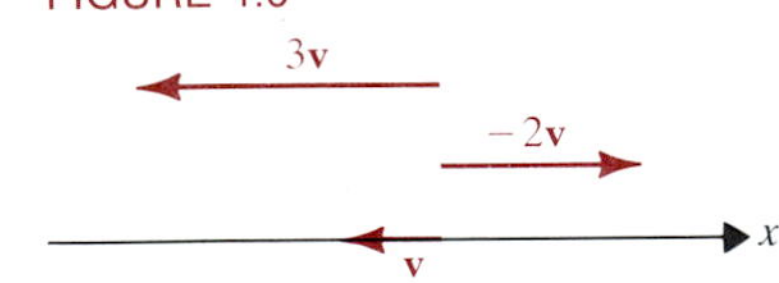

Plane Vectors

Plane vectors are two-dimensional objects with length and direction, which are *identified* (that is, regarded as identical) if their lengths and directions are the same. Since only length and direction matter, we can translate any given vector $\mathbf{v}$ parallel to itself until its initial point is at the origin (see **Figure 1.7**). We then use the coordinates (a, b) of its endpoint to represent $\mathbf{v}$ and call a and b the ***coordinates*** of $\mathbf{v}$.

Recall that there is a one-to-one correspondence between points in the plane $\mathbf{R}^2$ and ordered pairs (a, b) of real numbers. If we again add the zero vector $\mathbf{0}$ (the vector of length 0 and no direction), then we have a one-to-one correspondence between plane vectors and coordinate pairs (a, b). That is, two plane vectors $\mathbf{v} = (a, b)$ and $\mathbf{w} = (c, d)$ are equal if and only if both $a = c$ and $b = d$.

By the distance formula, the length of a vector $\mathbf{v} = (a, b)$ is

$$\sqrt{(a - 0)^2 + (b - 0)^2} = \sqrt{a^2 + b^2}.$$

Note the analogy with (1): The length of a one-dimensional vector (a) is $|(a)| = \sqrt{a^2}$. Because of that analogy, we use the notation

$$|\mathbf{v}| = |(a, b)| = \sqrt{a^2 + b^2} \tag{2}$$

for the length of a plane vector $\mathbf{v} = (a, b)$. The remaining algebraic operations on plane vectors are also defined by analogy to those in $\mathbf{R}^1$.

1.1 DEFINITION

Let $\mathbf{v} = (a, b)$ and $\mathbf{w} = (c, d)$ be plane vectors, and let r be a real number. Then the ***vector sum*** of $\mathbf{v}$ and $\mathbf{w}$ is

$$\mathbf{v} + \mathbf{w} = (a, b) + (c, d) = (a + c, b + d). \tag{3}$$

The ***product*** of $\mathbf{v}$ by r is

$$r\mathbf{v} = r(a, b) = (ra, rb). \tag{4}$$

The ***vector difference*** of $\mathbf{v}$ and $\mathbf{w}$ is

$$\mathbf{v} - \mathbf{w} = \mathbf{v} + (-\mathbf{w}) = (a - c, b - d), \qquad \text{where } -\mathbf{w} = (-1)\mathbf{w}. \tag{5}$$

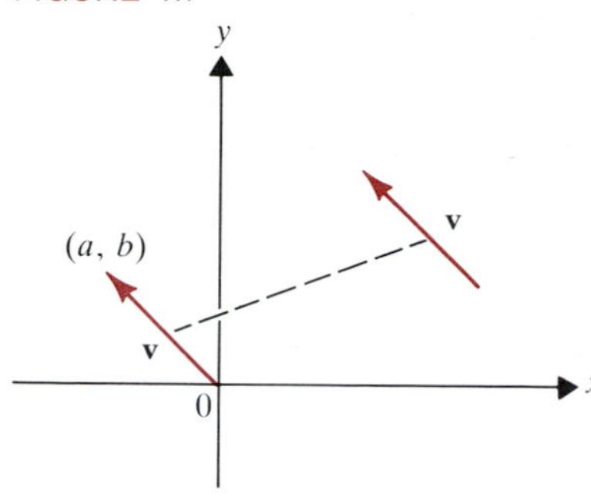

FIGURE 1.7

We add one-dimensional vectors by adding their coordinates. In two dimensions, we add the *corresponding* coordinates. Similarly, to multiply a plane vector by a real number r, we multiply each of its coordinates by r. In working with vectors, real numbers are often referred to as ***scalars*** (because they are used to scale the coordinate axes when we draw vectors). The quantity $r\mathbf{v}$ is called the ***scalar multiple*** of $\mathbf{v}$ by r, or the ***scalar product*** of r and $\mathbf{v}$. If $\mathbf{v} = (a, b)$, then

$$|r\mathbf{v}| = |(ra, rb)| = \sqrt{r^2a^2 + r^2b^2} = |r|\sqrt{a^2 + b^2} = |r|\,|\mathbf{v}|. \tag{6}$$

Thus $r\mathbf{v}$ has length $|r|$ times the length of $\mathbf{v}$, just as for one-dimensional vectors. It is also easy to see (**Figure 1.8**) that just as in $\mathbf{R}^1$, $r\mathbf{v}$ points in the same direction as $\mathbf{v}$ if $r > 0$ but in the opposite direction if $r < 0$.

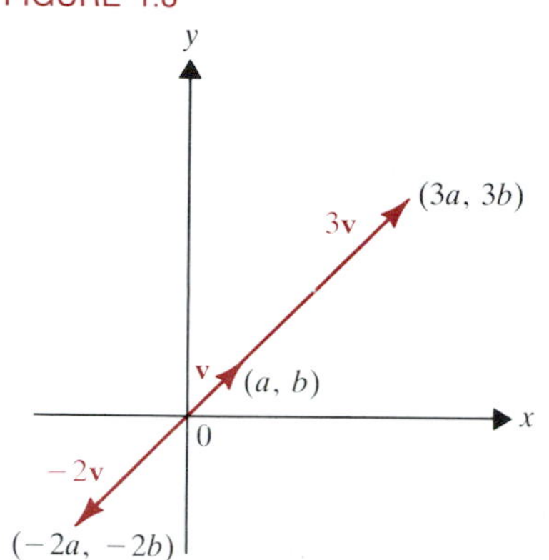

FIGURE 1.8

Example 1

If $\mathbf{v} = (-2, 5)$ and $\mathbf{w} = (3, -1)$, then compute

(a) $\mathbf{v} + \mathbf{w}$ (b) $\mathbf{v} - \mathbf{w}$ (c) $-3\mathbf{v} + 2\mathbf{w}$
(d) $2\mathbf{v} - 3\mathbf{w}$ (e) Find $|2\mathbf{v} - 3\mathbf{w}|$

Solution.

(a) $\mathbf{v} + \mathbf{w} = (-2, 5) + (3, -1) = (1, 4)$

(b) $\mathbf{v} - \mathbf{w} = (-2, 5) - (3, -1) = (-2, 5) + (-3, 1) = (-5, 6)$

(c) $-3\mathbf{v} + 2\mathbf{w} = -3(-2, 5) + 2(3, -1) = (6, -15) + (6, -2) = (12, -17)$

(d) $2\mathbf{v} - 3\mathbf{w} = 2(-2, 5) - 3(3, -1) = (-4, 10) + (-9, 3)$
$= (-13, 13) = 13(-1, 1)$

(e) By (3) and (6), $|2\mathbf{v} - 3\mathbf{w}| = 13|(-1, 1)| = 13\sqrt{1 + 1} = 13\sqrt{2}$ ■

Vector addition defined by (3) has an important geometric interpretation called the ***parallelogram law.*** That law states that if $\mathbf{v}$ is not a scalar multiple of $\mathbf{w}$, then $\mathbf{v} + \mathbf{w}$ is the diagonal of the parallelogram formed when $\mathbf{v}$ and $\mathbf{w}$ are drawn from the same initial point (**Figure 1.9**). That is, the quadrilateral $0ADB$ is a parallelogram. To see this, note that if $c = 0$, then AD and $0B$ are vertical, hence parallel. If $c \neq 0$, then $0B$ has slope

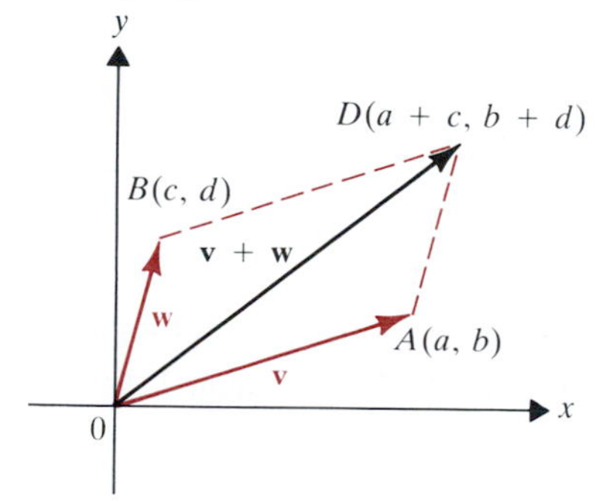

FIGURE 1.9

$$\frac{d}{c} = \frac{d - 0}{c - 0} = \frac{(b + d) - b}{(a + c) - a} = \text{slope } AD.$$

Hence AD and $0B$ are parallel. Next, if $a = 0$, then $0A$ and BD are both vertical. Otherwise, both have slope

$$\frac{b}{a} = \frac{(b + d) - d}{(a + c) - c}$$

and so are parallel. Thus $0ABD$ is a parallelogram.

It can actually be further shown (Exercise 40) that for the parallelogram law to hold, vector addition in $\mathbf{R}^2$ *must* be defined by (3). We can use the parallelogram law to easily find the algebraic representation of the vector $\mathbf{AB}$ in **Figure 1.10** drawn from an initial point $A(a_1, a_2)$ to the endpoint $B(b_1, b_2)$. Since $\mathbf{0B} = (b_1, b_2)$ is the sum of $\mathbf{0A} = (a_1, a_2)$ and $\mathbf{AB}$, we have

$$(b_1, b_2) = (a_1, a_2) + \mathbf{AB}.$$

FIGURE 1.10

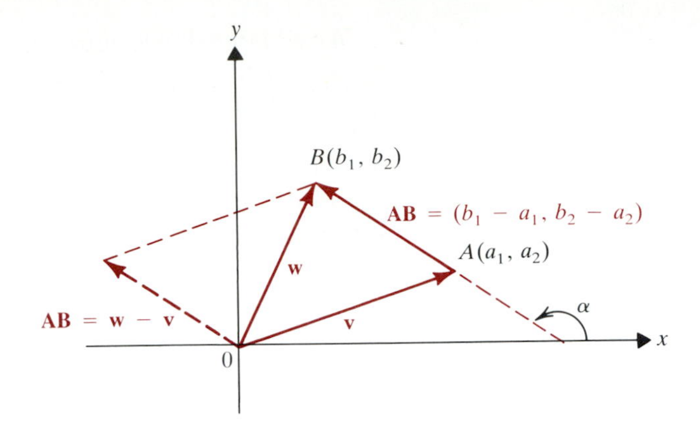

Thus, subtracting (a_1, a_2) from each side, we get

$$-(a_1, a_2) + (b_1, b_2) = -(a_1, a_2) + (a_1, a_2) + \mathbf{AB},$$
$$(b_1 - a_1, b_2 - a_2) = (0, 0) + \mathbf{AB} = \mathbf{AB}.$$

Thus the vector **AB** with beginning point $A(a_1, a_2)$ and endpoint $B(b_1, b_2)$ has algebraic representation

$$\mathbf{AB} = (b_1 - a_1, b_2 - a_2), \tag{7}$$

whose coordinates are obtained by subtracting the respective coordinates of the initial point from those of the endpoint.

In Figure 1.10, we also have labeled the vector **0A** as **v** and the vector **0B** as **w**. Since $\mathbf{v} = (a_1, a_2)$ and $\mathbf{w} = (b_1, b_2)$, we see that

$$\mathbf{AB} = (b_1 - a_1, b_2 - a_2) = \mathbf{w} - \mathbf{v} = \mathbf{0B} - \mathbf{0A}. \tag{8}$$

In the figure, we drew **AB** with its initial point at A and also with its initial point at 0, and labeled as $\mathbf{w} - \mathbf{v}$. We can interpret (8) as saying that

> $\mathbf{w} - \mathbf{v}$ is represented by the vector drawn from the endpoint of **v** to the endpoint of **w**.

This is another instance of the parallelogram law, for Figure 1.10 also shows the parallelogram for constructing the resultant of **v** and $\mathbf{w} - \mathbf{v}$.

If **AB** is not vertical in Figure 1.10, then the quotient

$$\frac{b_2 - a_2}{b_1 - a_1}$$

is the slope m of the line through **AB**, and $m = \tan \alpha$, where α is the counterclockwise angle from the positive x-axis to the line. If **AB** is vertical, then B is the point (a_1, b_2) and so

$$\mathbf{AB} = (0, b_2 - a_2).$$

In this case, the representation (7) for **AB** *shows* that **AB** is vertical, because there is a 0 in the first coordinate. The algebraic representation of **AB** thus contains enough information about the vector from A to B to specify *both* its length $(= \sqrt{(b_1 - a_1)^2 + (b_2 - a_2)^2})$ *and* its direction (unless $A = B$, so that $\mathbf{AB} = (0, 0)$ is the zero vector, which has no direction).

Example 2

Find the vectors from $A(-2, 1)$ to $B(1, -3/2)$; $C(1, -2/3)$ to $D(1, 1/3)$; and $E(2, 5)$ to $F(5, 5/2)$. Sketch and compute the length of each vector.

Solution. We have

$$\mathbf{AB} = (1 - (-2), -3/2 - 1) = (3, -5/2)$$
$$\mathbf{CD} = (1 - 1, 1/3 - (-2/3)) = (0, 1)$$
$$\mathbf{EF} = (5 - 2, 5/2 - 5) = (3, -5/2)$$
$$|\mathbf{AB}| = \sqrt{3^2 + (-5/2)^2} = \sqrt{9 + 25/4} = \sqrt{61/4} = \sqrt{61}/2$$
$$|\mathbf{CD}| = \sqrt{0^2 + 1^2} = 1$$
$$|\mathbf{EF}| = \sqrt{3^2 + (-5/2)^2} = \sqrt{61}/2$$

FIGURE 1.11

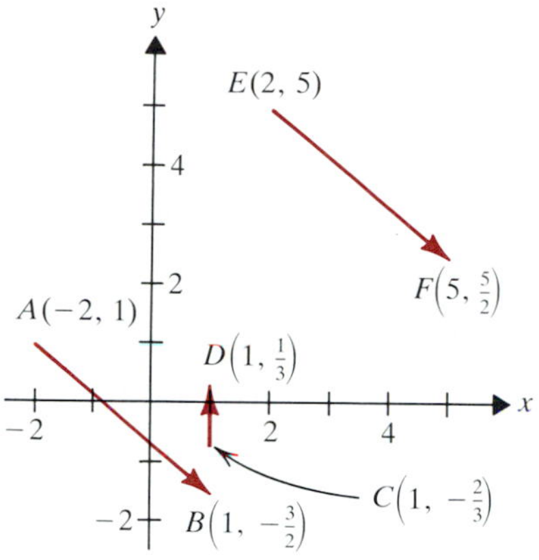

We sketch these results in **Figure 1.11.** Note that $|\mathbf{AB}| = |\mathbf{EF}|$, and the vectors are parallel because

$$\text{slope } \mathbf{AB} = \frac{-\frac{3}{2} - 1}{1 - (-2)} = \frac{-\frac{5}{2}}{3} = -\frac{5}{6} = \frac{\frac{5}{2} - 5}{5 - 2} = \text{slope } \mathbf{EF}.$$

Thus $\mathbf{AB} = \mathbf{EF}$ geometrically. Algebraically, $\mathbf{AB} = \mathbf{EF} = (3, -5/2)$. ■

Definition 1.1 is given in terms of the usual operations of addition, subtraction, and multiplication of real numbers, so it is not surprising that familiar algebraic properties of those operations also hold for the corresponding vector operations.

1.2 THEOREM

For all vectors $\mathbf{u}$, $\mathbf{v}$, and $\mathbf{w}$ in $\mathbf{R}^2$ and all scalars s and t in $\mathbf{R}$,

(a) $\mathbf{v} + \mathbf{w} = \mathbf{w} + \mathbf{v}$. (Addition is *commutative.*)

(b) $\mathbf{u} + (\mathbf{v} + \mathbf{w}) = (\mathbf{u} + \mathbf{v}) + \mathbf{w}$. (Addition is *associative.*)

(c) $\mathbf{v} + \mathbf{0} = \mathbf{v}$, where $\mathbf{0} = (0, 0) \in \mathbf{R}^2$ is the *additive identity.*

(d) $\mathbf{v} + (-\mathbf{v}) = \mathbf{0}$, where $-\mathbf{v} = (-a, -b)$ is the *negative* of $\mathbf{v} = (a, b)$.

(e) $s(\mathbf{v} + \mathbf{w}) = s\mathbf{v} + s\mathbf{w}$. (A kind of *distributive law.*)

(f) $(s + t)\mathbf{v} = s\mathbf{v} + t\mathbf{v}$. (Another kind of distributive law.)

(g) $s(t\mathbf{v}) = (st)\mathbf{v}$. (A kind of *associative law for scalar multiplication.*)

(h) $1\mathbf{v} = \mathbf{v}$. (Thus 1 acts as an *identity for scalar multiplication.*)

(i) $0\mathbf{v} = \mathbf{0}$.

Partial Proof. These properties are proved from the corresponding properties of the coordinates. We prove parts (b) and (e). (See Exercises 37–39 for proofs of other parts.) In (b), let

$$\mathbf{u} = (x, y), \qquad \mathbf{v} = (a, b), \qquad \text{and} \qquad \mathbf{w} = (p, q).$$

Then

$$\mathbf{u} + (\mathbf{v} + \mathbf{w}) = (x, y) + (a + p, b + a) = (x + (a + p), y + (b + q)),$$

and

$$(\mathbf{u} + \mathbf{v}) + \mathbf{w} = (x + a, y + b) + (p, q) = ((x + a) + p, (y + b) + q).$$

Each coordinate of $\mathbf{u} + (\mathbf{v} + \mathbf{w})$ equals the corresponding coordinate of $(\mathbf{u} + \mathbf{v}) + \mathbf{w}$ by the associative law of addition of real numbers. Hence (b) follows. For (e) we have

$$\begin{aligned} s(\mathbf{v} + \mathbf{w}) &= s(a + p, b + q) = (s(a + p), (b + q)) \\ &= (sa + sp, sb + sq) \qquad \textit{by the distributive law for real numbers} \\ &= (sa, sb) + (sp, sq) = s(a, b) + s(p, q) = s\mathbf{v} + s\mathbf{w}. \end{aligned}$$ QED

We have seen that the length of a vector $\mathbf{a} = (a, b)$ is

$$|\mathbf{a}| = \sqrt{a^2 + b^2}.$$

A vector $\mathbf{u}$ is called a ***unit vector*** if $|\mathbf{u}| = 1$. Two important unit vectors are (1, 0) and (0, 1).

1.3 DEFINITION

The ***standard basis*** for $\mathbf{R}^2$ is the *ordered* set $\{\mathbf{i}, \mathbf{j}\}$ (that is, $\mathbf{i}$ is the *first* vector, and $\mathbf{j}$ is the *second*), where

$$\mathbf{i} = (1, 0) \qquad \text{and} \qquad \mathbf{j} = (0, 1).$$

Thus $\mathbf{i}$ points in the direction of the positive x-axis, and $\mathbf{j}$ points in the direction of the positive y-axis. See **Figure 1.12.**

FIGURE 1.12

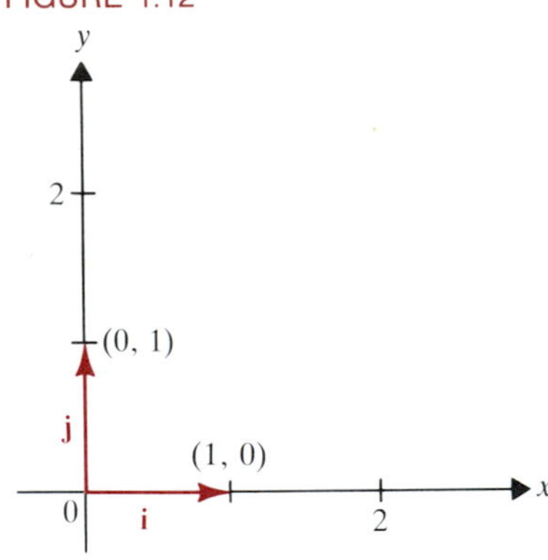

The term *basis* derives from the fact that every vector in $\mathbf{R}^2$ can be uniquely expressed as a simple combination of the basic vectors $\mathbf{i}$ and $\mathbf{j}$. Indeed, for any vector $\mathbf{a} = (a, b)$, we have

$$\mathbf{a} = a\mathbf{i} + b\mathbf{j}, \tag{9}$$

because (3) and (4) give

$$a\mathbf{i} + b\mathbf{j} = a(1, 0) + b(0, 1) = (a, 0) + (0, b) = (a, b).$$

Figure 1.13 illustrates this graphically.

A ***linear combination*** of vectors $\mathbf{v}$ and $\mathbf{w}$ is an expression of the form $c_1\mathbf{v} + c_2\mathbf{w}$, where c_1 and c_2 are real numbers. Thus (9) expresses $\mathbf{a}$ as a linear combination of $\mathbf{i}$ and $\mathbf{j}$.

FIGURE 1.13

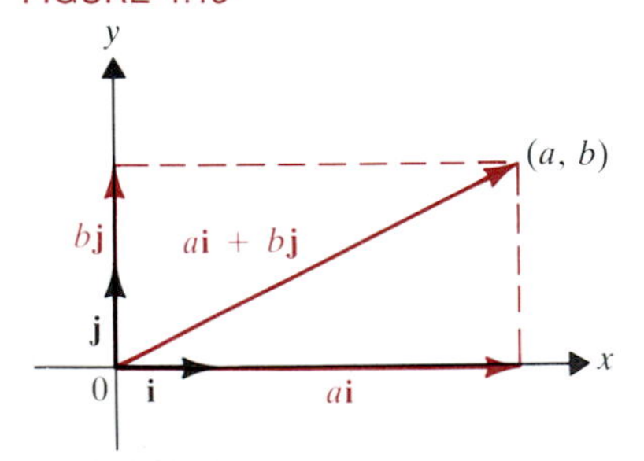

Example 3

(a) Express the vectors $\mathbf{v} = (-3, 1)$ and $\mathbf{w} = (2, 0)$ in terms of $\mathbf{i}$ and $\mathbf{j}$.

(b) Express the vectors $\mathbf{z} = 4\mathbf{i} - 5\mathbf{j}$ and $\mathbf{r} = -3\mathbf{j}$ in coordinate form.

Solution. (a) Equation (9) gives

$$\mathbf{v} = -3\mathbf{i} + \mathbf{j} \qquad \text{and} \qquad \mathbf{w} = 2\mathbf{i}.$$

(b) We have from (9)

$$\mathbf{z} = 4\mathbf{i} - 5\mathbf{j} = (4, -5) \qquad \text{and} \qquad \mathbf{r} = 0\mathbf{i} - 3\mathbf{j} = (0, -3).$$ ■

Every nonzero vector $\mathbf{v}$ has a naturally associated unit vector pointing in the same direction as $\mathbf{v}$.

1.4 DEFINITION

If $\mathbf{v} \neq \mathbf{0}$ is a plane vector, then the ***unit vector in the direction of*** $\mathbf{v}$ is

$$\mathbf{u} = \frac{1}{|\mathbf{v}|}\mathbf{v}.$$

FIGURE 1.14

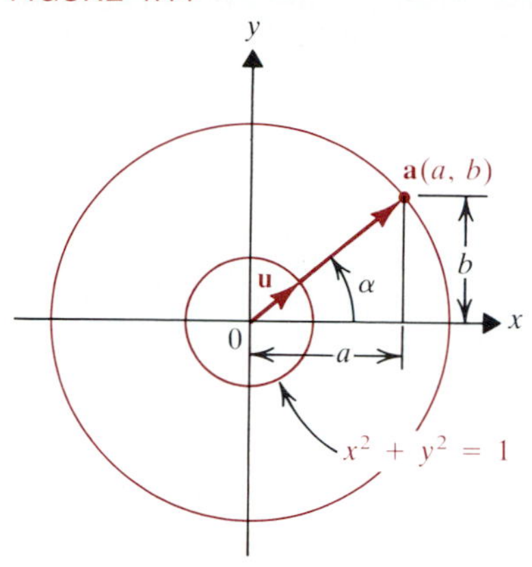

For the vector $\mathbf{v} = (-3, 1)$ in Example 3(a), the unit vector in the direction of $\mathbf{v}$ is thus

$$\mathbf{u} = \frac{1}{|(-3, 1)|}(-3, 1) = \frac{1}{\sqrt{9+1}}(-3, 1) = \left(-\frac{3}{\sqrt{10}}, \frac{1}{\sqrt{10}}\right).$$

Notice that $\mathbf{u}$ is indeed a unit vector, because

$$|\mathbf{u}| = \sqrt{\frac{9}{10} + \frac{1}{10}} = \sqrt{1} = 1.$$

The unit vector in the direction of $\mathbf{v}$ is also called the ***direction vector*** of $\mathbf{v}$. The reason is that $\mathbf{u}$ carries enough information to find the angle α that $\mathbf{v}$ makes with the positive x-axis. We have

$$\mathbf{u} = \frac{1}{\sqrt{a^2 + b^2}}(a, b) = \left(\frac{a}{\sqrt{a^2 + b^2}}, \frac{b}{\sqrt{a^2 + b^2}}\right) = (\cos\alpha, \sin\alpha).$$

Since the tips of unit vectors lie on the unit circle $x^2 + y^2 = 1$, each one determines a unique angle α. See **Figure 1.14.**

Definition 1.4 leads to the following definition of parallel vectors.

1.5 DEFINITION The nonzero vectors $\mathbf{v}$ and $\mathbf{w}$ are ***parallel*** if they have the same direction vector or if their direction vectors differ by a factor of -1.

There is a simple algebraic criterion for two vectors to be parallel.

1.6 THEOREM The nonzero vectors $\mathbf{v}$ and $\mathbf{w}$ are parallel if and only if

$$\mathbf{w} = k\mathbf{v} \tag{10}$$

for some nonzero real number k.

Proof. Let $\mathbf{u}_w$ be the direction vector of $\mathbf{w}$ and $\mathbf{u}_v$ the direction vector of $\mathbf{v}$. Then $\mathbf{v}$ and $\mathbf{w}$ are parallel if and only if

$$\mathbf{u}_w = \frac{1}{|\mathbf{w}|}\mathbf{w} = \pm\mathbf{u}_v = \pm\frac{1}{|\mathbf{v}|}\mathbf{v}.$$

That is, $\mathbf{v}$ and $\mathbf{w}$ are parallel if and only if

$$\mathbf{w} = \pm\frac{|\mathbf{w}|}{|\mathbf{v}|}\mathbf{v} = k\mathbf{v},$$

where $k \neq 0$. QED

Example 4

Determine whether the given vectors are parallel.

(a) $\mathbf{v} = -6\mathbf{i} - \mathbf{j}$, $\mathbf{w} = 2\mathbf{i} + (1/3)\mathbf{j}$

(b) $\mathbf{AB}$ and $\mathbf{CD}$, where $A(2, 1)$, $B(-1, 3)$, $C(-3, 5)$, $D(6, 11)$.

Solution. (a) Since $\mathbf{v} = -3\mathbf{w}$, the vectors are parallel by Theorem 1.6.
(b) Here we have

$$\mathbf{AB} = (-1 - 2, 3 - 1) = (-3, 2),$$

$$\mathbf{CD} = (6 + 3, 11 - 5) = (9, 6).$$

Since $9 = (-3) \cdot (-3)$ but $6 = (+3) \cdot 2$, we see that **CD** is not a multiple of **AB**. Hence the given vectors are *not* parallel. ■

HISTORICAL NOTE

Vectors were dealt with as far back as Newton's time. Many of his laws of physics involve vector quantities such as force, velocity, and acceleration. The algebraic treatment of vectors is the contribution of two nineteenth-century mathematicians, William Rowan Hamilton (1805–1865) of Ireland (who also made important contributions to physics) and Hermann G. Grassmann (1809–1877) of Germany (who also made contributions to philosophy). Their work profoundly influenced the evolution of modern mathematics, physics, and engineering. Contemporary mathematics studies many concepts built on these men's work. The branch of mathematics called *linear algebra* grew from the idea of studying vectors algebraically.

Exercises 10.1

In Exercises 1–10, find (a) $\mathbf{v} + \mathbf{w}$; (b) $\mathbf{v} - \mathbf{w}$; (c) $-2\mathbf{v} + 3\mathbf{w}$; (d) $3\mathbf{v} - 2\mathbf{w}$, (e) $|\mathbf{v}|$, $|\mathbf{w}|$; (f) $\mathbf{u}_v$, $\mathbf{u}_w$.

1. $\mathbf{v} = (2, 5)$, $\mathbf{w} = (3, 2)$
2. $\mathbf{v} = (-2, 3)$, $\mathbf{w} = (3, -1)$
3. $\mathbf{v} = (1, -2)$, $\mathbf{w} = (-2, 3)$
4. $\mathbf{v} = (-2, -5)$, $\mathbf{w} = (-1, -3)$
5. $\mathbf{v} = 3\mathbf{i} + \mathbf{j}$, $\mathbf{w} = \mathbf{i} + 4\mathbf{j}$
6. $\mathbf{v} = -3\mathbf{i} + 2\mathbf{j}$, $\mathbf{w} = \mathbf{i} - 3\mathbf{j}$
7. $\mathbf{v} = 2\mathbf{i} - \mathbf{j}$, $\mathbf{w} = -3\mathbf{i} + 4\mathbf{j}$
8. $\mathbf{v} = -5\mathbf{i} - 12\mathbf{j}$, $\mathbf{w} = -3\mathbf{i} - 2\mathbf{j}$
9. $\mathbf{v} = -3\mathbf{i}$, $\mathbf{w} = 2\mathbf{j}$
10. $\mathbf{v} = -\mathbf{j}$, $\mathbf{w} = \mathbf{i} + \mathbf{j}$

In Exercises 11–20, find the vector from A **to** B **and its length. Make a sketch.**

11. $A(0, 0)$, $B(3, 2)$
12. $A(0, 0)$, $B(-3, 5)$
13. $A(2, 3)$, $B(5, 7)$
14. $A(2, 3)$, $B(1, 4)$
15. $A(2, 3)$, $B(-1, -1)$
16. $A(-1, 3)$, $B(-3, 2)$
17. $A(3, -2)$, $B(8, 10)$
18. $A(-1, -5)$, $B(4, -3)$
19. $A(-2, -7)$, $B(-1, -8)$
20. $A(-3, -5)$, $B(-6, -2)$

In Exercises 21–32, find which pairs of vectors are parallel.

21. $\mathbf{v} = (-3, 2)$, $\mathbf{w} = (1, -2/3)$
22. $\mathbf{v} = 5\mathbf{i} + \sqrt{2}\mathbf{j}$, $\mathbf{w} = \sqrt{50}\mathbf{i} + 2\mathbf{j}$
23. $\mathbf{v} = (-2\sqrt{5}, -5)$, $\mathbf{w} = (\frac{1}{2}, 1/\sqrt{5})$
24. $\mathbf{v} = -\sqrt{3}\mathbf{i} + \frac{1}{4}\mathbf{j}$, $\mathbf{w} = \frac{1}{2}\mathbf{i} + \sqrt{12}\mathbf{j}$
25. $\mathbf{v} = (-2\sqrt{3}, -3)$, $\mathbf{w} = (\frac{1}{2}, 1/\sqrt{3})$
26. $\mathbf{v} = \sqrt{2}\mathbf{i} - \mathbf{j}$, $\mathbf{w} = \frac{1}{5}\mathbf{i} - \sqrt{2}\mathbf{j}$
27. **AB** and **CD**: $A(2, 1)$, $B(-3, 3)$, $C(4, -1)$, $D(8, 11)$
28. **AB** and **CD**: $A(5, -1)$, $B(2, 2)$, $C(-1, -3)$, $D(5, -9)$
29. **AB** and **CD**: $A(-1, 3)$, $B(4, 0)$, $C(-3, 1)$, $D(3, 11)$
30. **AB** and **CD**: $A(-2, -3)$, $B(-1, -3)$, $C(5, -8)$, $D(5, 17)$
31. **AB** and **CD**: $A(4, 0)$, $B(0, -3)$, $C(2, -1)$, $D(6, 2)$
32. **AB** and **CD**: $A(-3, 1)$, $B(0, -1)$, $C(4, -1)$, $D(2, -4)$

33. Find the resultant of the two forces $\mathbf{F}_1 = 3\mathbf{i} - 2\mathbf{j}$ and $\mathbf{F}_2 = -4\mathbf{i} + \mathbf{j}$.
34. Repeat Exercise 33 for $\mathbf{F}_1 = -2\mathbf{i} - 3\mathbf{j}$ and $\mathbf{F}_2 = 5\mathbf{i} + \mathbf{j}$.
35. Find the mass of an object if at some time the force $\mathbf{F}$ on it is $6\mathbf{i} - 2\mathbf{j}$ and its acceleration is $\mathbf{a} = 9\mathbf{i} - 3\mathbf{j}$. (*Hint:* Use Newton's second law $\mathbf{F} = m\mathbf{a}$.)
36. If an object of mass 1.5 kg is acted on by a force $\mathbf{F} = 5\mathbf{i} - 2\mathbf{j}$, then what is the resulting acceleration?
37. Prove parts (a) and (c) of Theorem 1.2.
38. Prove parts (d) and (f) of Theorem 1.2.
39. Prove parts (g), (h), and (i) of Theorem 1.2.
40. Suppose we represent $\mathbf{v}$ in $\mathbf{R}^2$ by (a, b) and $\mathbf{w}$ in $\mathbf{R}^2$ by (c, d). If addition is defined by the parallelogram law, then show that $\mathbf{v} + \mathbf{w}$ *must* be the vector $(a + c, b + d)$.
41. Use vectors to show that the line segment connecting the midpoints of two sides of a triangle is parallel to the third side and has length half that of the third side. (*Hint:* Take the third side through $\mathbf{i}$.)
42. Use vectors to prove that the diagonals of a parallelogram bisect each other.
43. If A, B, and C are three distinct points in the plane, then show that $\mathbf{AB} + \mathbf{BC} + \mathbf{CA} = \mathbf{0}$.

10.2 The Space R^3

The last section discussed the spaces $\mathbf{R}^1(=\mathbf{R})$ and $\mathbf{R}^2$. We now move up to the case $n = 3$. The space $\mathbf{R}^3$ is called ***three-dimensional real Euclidean*** (or ***Cartesian***) ***space,*** and it plays a central role in applications of calculus to science and engineering. Recalling that the real numbers supply coordinates for the points on a line and that ordered pairs of real numbers coordinatize the points in a plane, we would expect that the set of all ordered *triples* (x, y, z) of real numbers can be used to assign coordinates to three-dimensional space. We begin our discussion of $\mathbf{R}^3$ by describing how that is done.

FIGURE 2.1

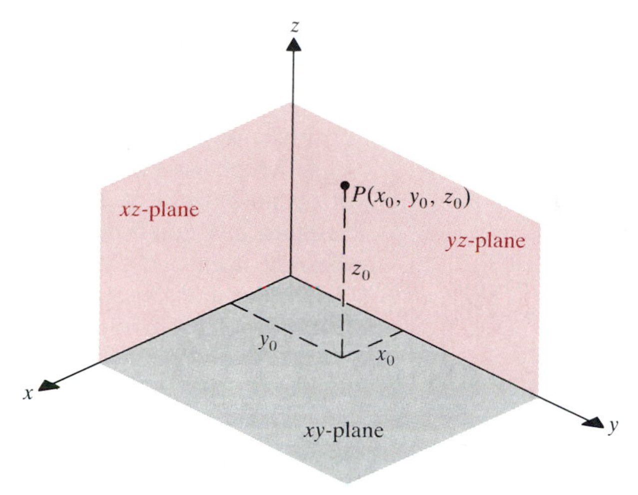

Three mutually perpendicular lines (called the ***x-***, ***y-***, and ***z-axes*** and labeled as in **Figure 2.1**) are drawn from a common point 0 selected as the origin. Since any two lines determine a plane, this procedure determines three ***coordinate planes*** called the *xy-plane*, *yz-plane*, and *xz-plane*. The coordinates of a point P in space are then (x_0, y_0, z_0), where x_0 is the directed distance from the yz-plane to P, y_0 is the directed distance from the xz-plane to P, and z_0 is the directed distance from the xy-plane to P. Directed distances are positive when measured to the front, right, or upward (that is, in the direction of the positive x-axis, positive y-axis, or positive z-axis). To locate P, we measure x_0 and y_0 as usual

FIGURE 2.2

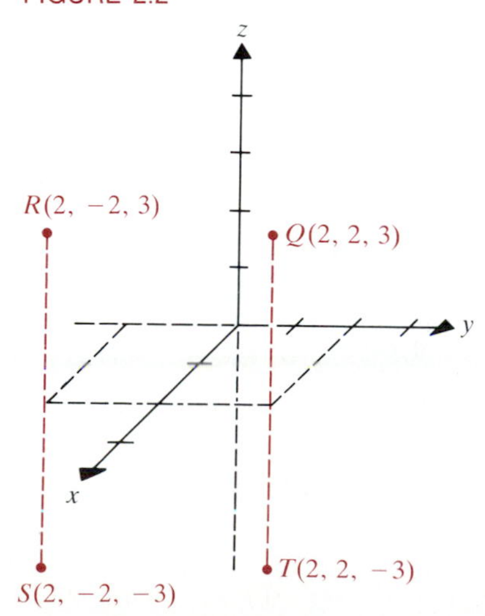

FIGURE 2.3

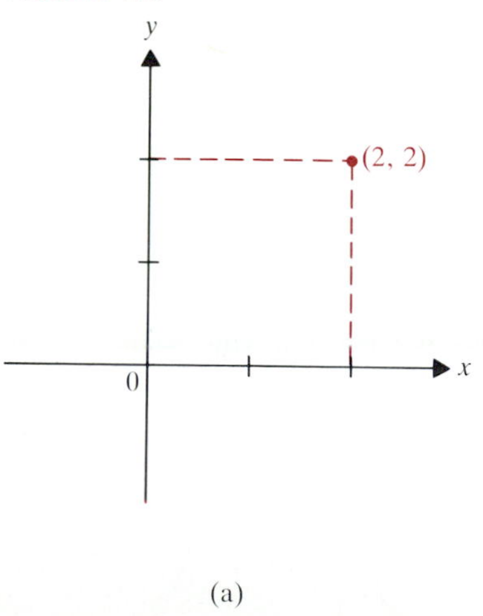

(a)

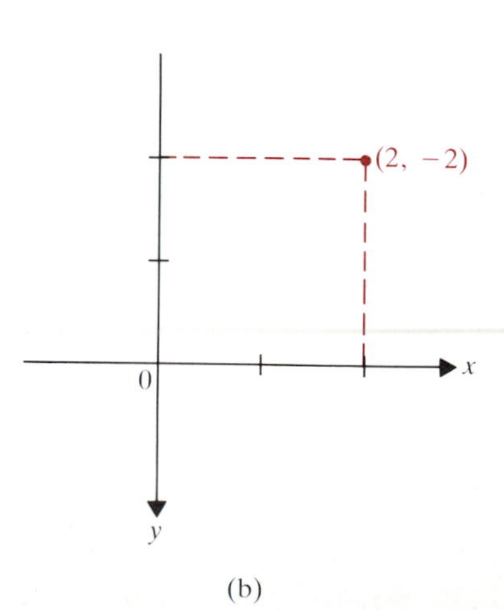

(b)

FIGURE 2.4

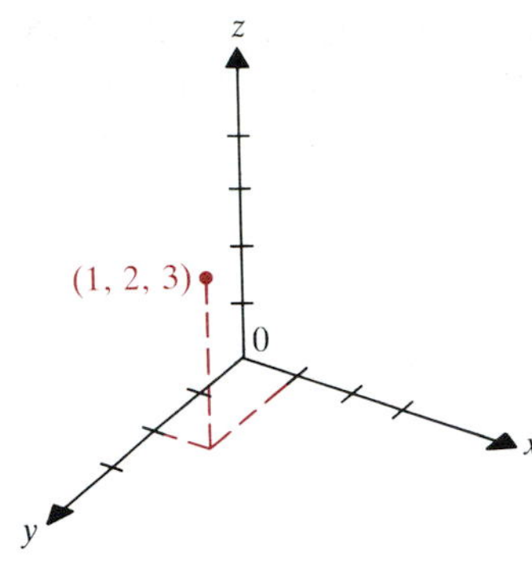

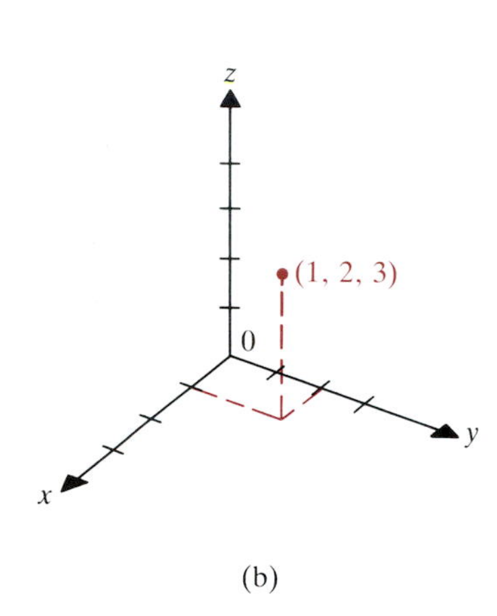

(b)

in the xy-plane and then go up or down to P, according to whether z_0 is positive or negative. See **Figure 2.2.**

Just as the coordinate axes in $\mathbf{R}^2$ divide the plane into four quadrants, so the three coordinate planes divide $\mathbf{R}^3$ into eight ***octants.*** Customarily, the only octant assigned a number is the *first* octant, all of whose points have all three coordinates nonnegative. In drawing three-dimensional pictures, clutter is often minimized by showing only the first octant portion completely, either omitting the rest or drawing it with dotted lines.

Just as for $\mathbf{R}^2$, the *orientation* of the axes in $\mathbf{R}^3$ has to be prescribed. The orientations shown in **Figures 2.3** and **2.4** give different coordinatizations of $\mathbf{R}^2$ and $\mathbf{R}^3$. The orientation used throughout this book (and most other mathematics texts) is the ***right-handed*** one of Figures 2.1 and 2.4(b). The terminology results from the following model. If you orient your right index finger in the direction of the positive x-axis and your right middle finger in the direction of the positive y-axis, then your thumb will point upward in the direction of the positive z-axis. See **Figure 2.5.**

Although we will not study planes in detail until Section 5, the equations of planes parallel to a coordinate plane are easy to derive. For example, the xz-plane has equation $y = 0$, since a point is on the xz-plane if and only if its distance from that plane is 0. Similarly, $z = -3$ is the equation of the plane parallel to and three units below the xy-plane. We can also identify a plane whose equation involves only two coordinate variables.

FIGURE 2.5

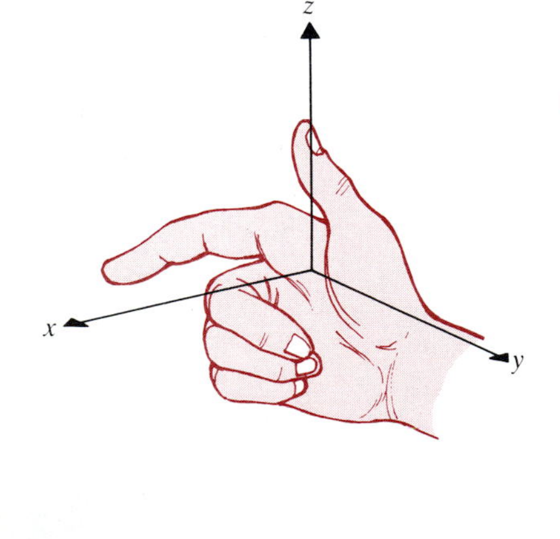

Example 1

Identify the plane in $\mathbf{R}^3$ whose equation is $x + y = 1$.

Solution. Since z is not restricted by the equation, it ranges over the entire interval $(-\infty, +\infty)$. When $z = 0$, we get the line l in **Figure 2.6**(a), the intersection of the plane with the xy-plane. Letting z vary, we get lines above and below l. Therefore, the given plane is the plane through l perpendicular to the xy-plane, shown in Figure 2.6(b). ■

FIGURE 2.6

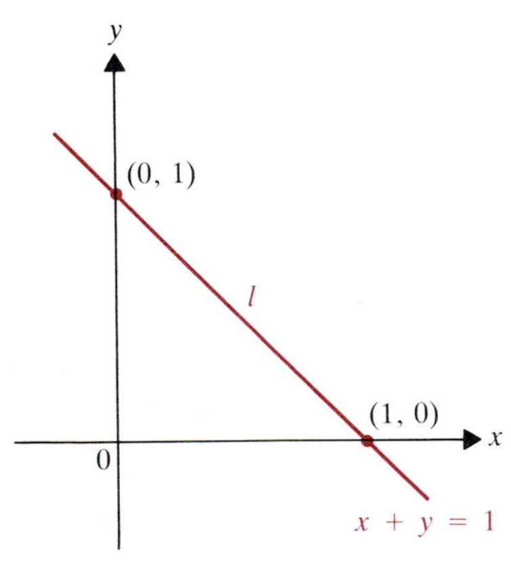

(a)

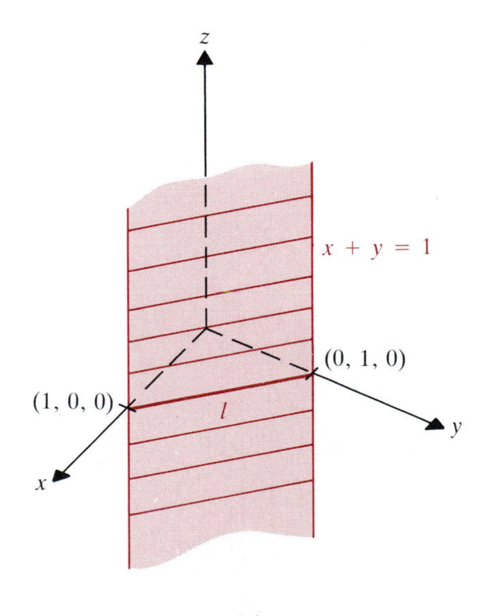

(b)

The algebraic representation of vectors in $\mathbf{R}^3$ is obtained just as in $\mathbf{R}^2$. Thus we again translate a given vector $\mathbf{v}$ in $\mathbf{R}^3$ parallel to itself until its initial point is at the origin and use the coordinates (a, b, c) of the endpoint P of the translated $\mathbf{v}$ as its algebraic representation. See **Figure 2.7.** Addition, scalar multiplication, and length are then defined as in Definition 1.1.

2.1 DEFINITION If $\mathbf{v} = (a, b, c)$ and $\mathbf{w} = (d, e, f)$ are vectors in $\mathbf{R}^3$, then

(1) $$\mathbf{v} + \mathbf{w} = (a + d, b + e, c + f)$$

and

(2) $$r\mathbf{v} = (ra, rb, rc)$$

for any real number r. The ***vector difference*** $\mathbf{v} - \mathbf{w}$ is $\mathbf{v} + (-\mathbf{w})$, where $-\mathbf{w} = (-1)\mathbf{w}$. The ***length*** of $\mathbf{v}$ is $|\mathbf{v}| = \sqrt{a^2 + b^2 + c^2}$.

FIGURE 2.7

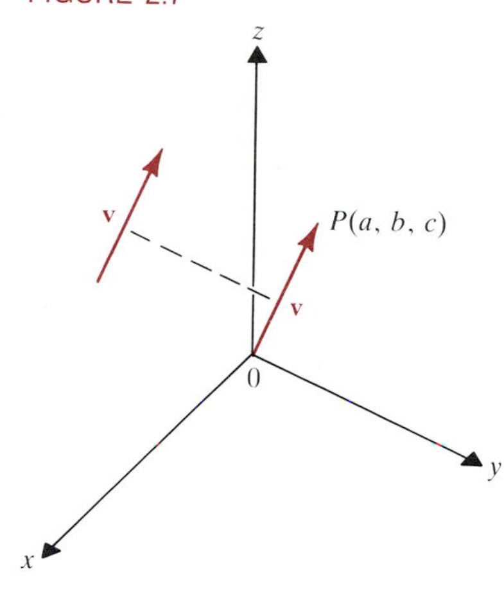

As in $\mathbf{R}^1$ and $\mathbf{R}^2$, it is easy to see that if r is a nonzero scalar, then $r\mathbf{v}$ is a vector of length $|r||\mathbf{v}|$, pointing in the same direction as $\mathbf{v}$ if $r > 0$ but in the opposite direction if $r < 0$ (see Exercise 33).

Example 2

If $\mathbf{v} = (-4, 2, 0)$ and $\mathbf{w} = (6, -3, 4)$, then find $\mathbf{v} + \mathbf{w}$, $\mathbf{v} - \mathbf{w}$, and $2\mathbf{v} + \mathbf{w}$. Plot these vectors.

Solution. From Definition 2.1, we have

$$\mathbf{v} + \mathbf{w} = (-4 + 6, 2 + (-3), 0 + 4) = (2, -1, 4),$$

$$\mathbf{v} - \mathbf{w} = \mathbf{v} + (-1)\mathbf{w} = (-4, 2, 0) + (-6, 3, -4) = (-10, 5, -4),$$

and

$$2\mathbf{v} + \mathbf{w} = (-8, 4, 0) + (6, -3, 4) = (-2, 1, 4).$$

These vectors are shown in **Figure 2.8,** where $\mathbf{v} - \mathbf{w}$ is drawn both with its initial point at the tip of $\mathbf{w}$ and with its initial point at 0. (Compare Figure 1.10.) ■

As on p. 631, it can be shown that the parallelogram law for addition follows from (1). Indeed, in Exercise 37 you are asked to show that (1) is unavoidable if we want the parallelogram law to hold for vector addition. As in

FIGURE 2.8

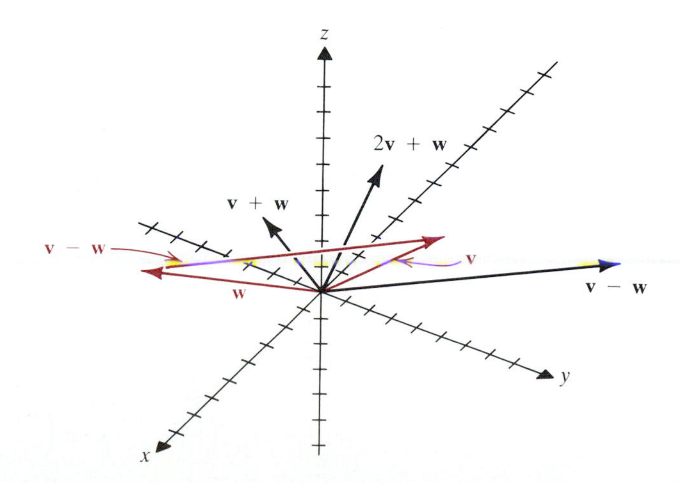

FIGURE 2.9

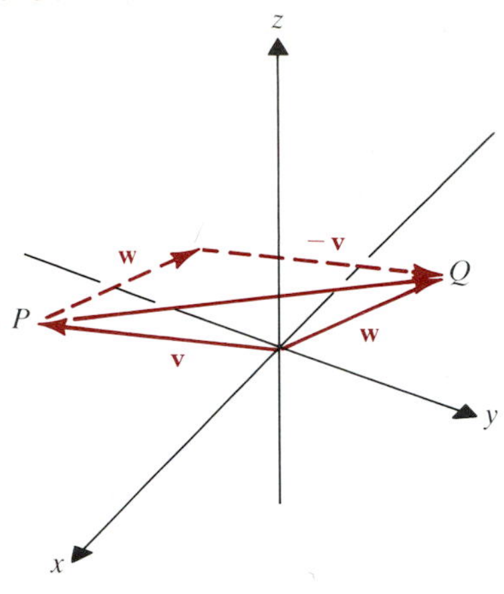

the last section, the parallelogram law gives the algebraic representation of the vector **PQ** joining any point $P(x_1, y_1, z_1)$ in $\mathbf{R}^3$ to another point $Q(x_2, y_2, z_2)$. (Compare the following with the discussion on p. 632.) Let $\mathbf{v} = \mathbf{0P}$ and $\mathbf{w} = \mathbf{0Q}$. Referring to **Figure 2.9,** we see that **PQ** is the diagonal of the parallelogram formed by **w** and $-\mathbf{v}$. Since $\mathbf{w} = (x_2, y_2, z_2)$ and $-\mathbf{v} = (-x_1, -y_1, -z_1)$, we have

$$\mathbf{PQ} = \mathbf{w} + (-\mathbf{v}) = (x_2 - x_1, y_2 - y_1, z_2 - z_1) = \mathbf{w} - \mathbf{v}. \tag{3}$$

To complete the analogy with $\mathbf{R}^2$, we need a distance formula for $\mathbf{R}^3$. Let $P(x_1, y_1, z_1)$ and $Q(x_2, y_2, z_2)$ be two points in $\mathbf{R}^3$. The distance between them is then $|\mathbf{PQ}|$. To calculate it, we construct the rectangular box shown in **Figure 2.10** with faces parallel to the coordinate planes and vertices P and Q. The vertex directly above Q is labeled B and has coordinates (x_2, y_2, z_1). We have

$$|\mathbf{BQ}| = |z_2 - z_1| = \sqrt{(z_2 - z_1)^2}$$

and

$$|\mathbf{PB}| = \sqrt{(x_2 - x_1)^2 + (y_2 - y_1)^2},$$

FIGURE 2.10

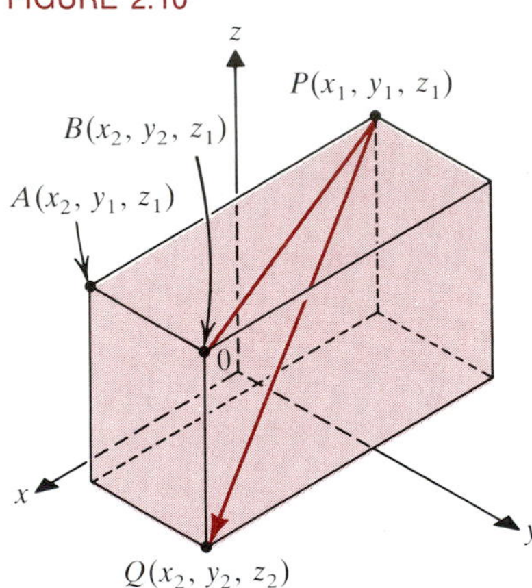

by the distance formulas for $\mathbf{R}^1$ and $\mathbf{R}^2$. (**BQ** can be viewed as a vector in the z-axis, which is a copy of $\mathbf{R}^1$, and **PB** can be viewed as a vector in the plane through P, A, and B, which is a copy of the xy-plane.) Triangle PBQ is a right triangle with hypotenuse $|\mathbf{PQ}|$. Hence the Pythagorean theorem gives

$$|\mathbf{PQ}| = \sqrt{|\mathbf{PB}|^2 + |\mathbf{BQ}|^2} = \sqrt{(x_2 - x_1)^2 + (y_2 - y_1)^2 + (z_2 - z_1)^2}. \tag{4}$$

From (4) it follows immediately that the vector **0P** drawn from the origin to any point $P(x, y, z)$ in $\mathbf{R}^3$ has length

$$|\mathbf{0P}| = \sqrt{x^2 + y^2 + z^2}. \tag{5}$$

Example 3

(a) Find the length of the vector from $P(1, -1/2, 3)$ to $Q(-2, 3/2, 5)$.

(b) **Midpoint Formula.** Show that

$$M\left(\frac{a_1 + b_1}{2}, \frac{a_2 + b_2}{2}, \frac{a_3 + b_3}{2}\right)$$

is the midpoint of the segment from $A(a_1, a_2, a_3)$ to $B(b_1, b_2, b_3)$.

Solution. (a) From (4) we get

$$|\mathbf{PQ}| = \sqrt{(-2 - 1)^2 + \left(\frac{3}{2} - \left(-\frac{1}{2}\right)\right)^2 + (5 - 3)^2} = \sqrt{(-3)^2 + 2^2 + 2^2}$$
$$= \sqrt{9 + 4 + 4} = \sqrt{17}.$$

(b) We have

$$|\mathbf{AM}| = \sqrt{\left(\frac{a_1 + b_1}{2} - a_1\right)^2 + \left(\frac{a_2 + b_2}{2} - a_2\right)^2 + \left(\frac{a_3 + b_3}{2} - a_3\right)^2}$$
$$= \sqrt{\left(\frac{b_1 - a_1}{2}\right)^2 + \left(\frac{b_2 - a_2}{2}\right)^2 + \left(\frac{b_3 - a_3}{2}\right)^2}$$

Also,

$$|\mathbf{MB}| = \sqrt{\left(\frac{a_1 + b_1}{2} - b_1\right)^2 + \left(\frac{a_2 + b_2}{2} - b_2\right)^2 + \left(\frac{a_3 + b_3}{2} - b_3\right)^2}$$
$$= \sqrt{\left(\frac{a_1 - b_1}{2}\right)^2 + \left(\frac{a_2 - b_2}{2}\right)^2 + \left(\frac{a_3 - b_3}{2}\right)^2}.$$

Note that

$$|\mathbf{AM}| = |\mathbf{MB}| = \frac{1}{2}\sqrt{(b_1 - a_1)^2 + (b_2 - a_2)^2 + (b_3 - a_3)^2} = \frac{1}{2}|\mathbf{AB}|.$$

Since a line segment is the shortest distance between two points, and since $|\mathbf{AM}| + |\mathbf{MB}| = |\mathbf{AB}|$, M is on the segment $\mathbf{AB}$. Thus M is the midpoint of the segment joining A to B. ■

In $\mathbf{R}^3$ the ***sphere*** of radius r and center C is defined to be the set of all points P whose distance from C is r.

Example 4

Find an equation for the sphere of radius 5 centered at $C(3, -1, 2)$. (See **Figure 2.11.**)

FIGURE 2.11

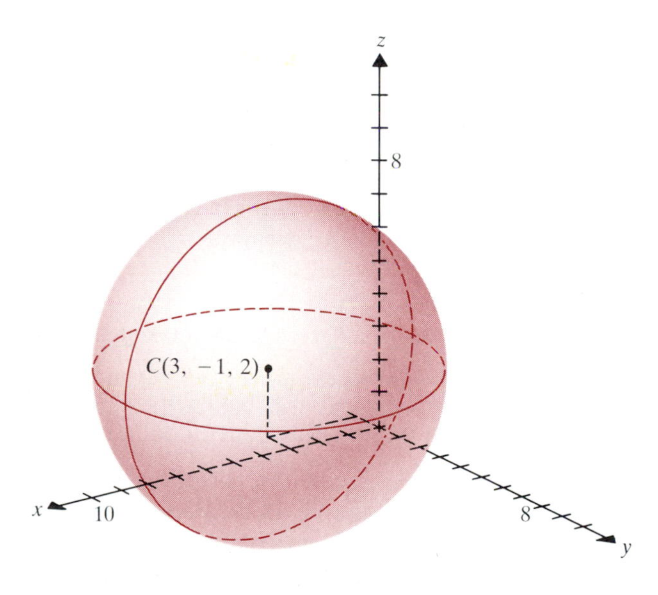

FIGURE 2.12

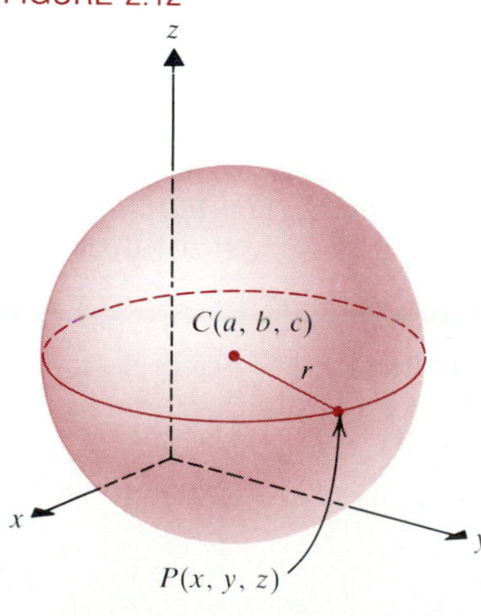

Solution. If $P(x, y, z)$ is any point on the sphere, then $|\mathbf{PC}| = 5$. Hence, since the sphere has radius 5,

$$\sqrt{(x - 3)^2 + (y + 1)^2 + (z - 2)^2} = 5.$$

Squaring both sides, we get

$$(x - 3)^2 + (y + 1)^2 + (z - 2)^2 = 25. \quad ■$$

Example 4 generalizes immediately: The equation of the sphere in **Figure 2.12** with center $C(a, b, c)$ and radius r is

$$(x - a)^2 + (y - b)^2 + (z - c)^2 = r^2. \tag{6}$$

Example 5

Find the center and radius of the sphere whose equation is $x^2 + y^2 + z^2 - 3x + 6y = 1$.

Solution. Grouping terms of the same variable together and completing the square, we have

$$x^2 - 3x + y^2 + 6y + z^2 = 1,$$

$$x^2 - 3x + \frac{9}{4} + y^2 + 6y + 9 + z^2 = \frac{9}{4} + 9 + 1,$$

$$\left(x - \frac{3}{2}\right)^2 + (y + 3)^2 + z^2 = \frac{9}{4} + \frac{36}{4} + \frac{4}{4} = \frac{49}{4}.$$

So the center is $(3/2, -3, 0)$, and the radius is $7/2$. ■

As in $\mathbf{R}^2$, a vector $\mathbf{u}$ in $\mathbf{R}^3$ is called a ***unit vector*** if $|\mathbf{u}| = 1$. The most natural unit vectors in $\mathbf{R}^3$ are $(1, 0, 0)$, $(0, 1, 0)$, and $(0, 0, 1)$.

2.2 DEFINITION

The ***standard basis*** for $\mathbf{R}^3$ is the ordered set $\{\mathbf{i}, \mathbf{j}, \mathbf{k}\}$, where

$$\mathbf{i} = (1, 0, 0), \qquad \mathbf{j} = (0, 1, 0), \qquad \mathbf{k} = (0, 0, 1).$$

See **Figure 2.13.**

FIGURE 2.13

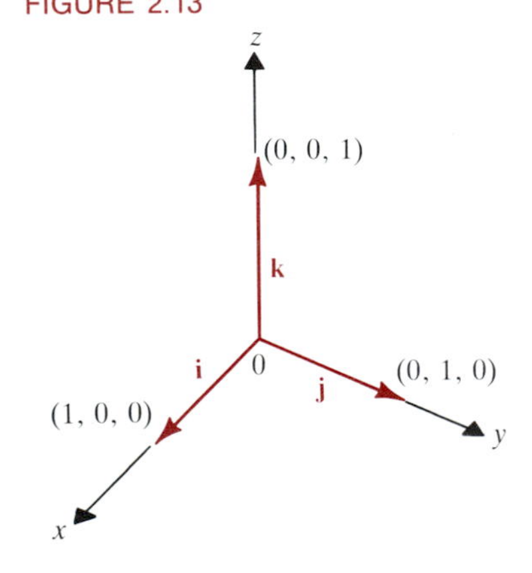

These three unit vectors point in the directions of the positive x-, y-, and z-coordinate axes. (Note that $\mathbf{i}$ and $\mathbf{j}$ in $\mathbf{R}^2$ have the same length and direction as their counterparts in $\mathbf{R}^3$, if we identify $\mathbf{R}^2$ with the xy-plane in $\mathbf{R}^3$.) Every vector in $\mathbf{R}^3$ can be expressed as a linear combination of $\mathbf{i}$, $\mathbf{j}$, and $\mathbf{k}$, just as every vector in $\mathbf{R}^2$ is a linear combination of $\mathbf{i}$ and $\mathbf{j}$. In fact, for any $\mathbf{v} = (a, b, c)$ in $\mathbf{R}^3$, we have

$$\mathbf{v} = a\mathbf{i} + b\mathbf{j} + c\mathbf{k}.$$

Exercise 34 considers the standard basis further.

Higher-Dimensional Cartesian Spaces*

We close the section by giving a general description of $\mathbf{R}^1$, $\mathbf{R}^2$, and $\mathbf{R}^3$, which can be used to define $\mathbf{R}^n$ for $n > 3$. There is no intuitive notion of a vector as a geometric object with length and direction in more than three dimensions. Instead, vectors are defined in an entirely algebraic way as follows.

2.3 DEFINITION

The set $\mathbf{R}^n = \{(x_1, x_2, \ldots, x_n) | x_i \in \mathbf{R}\}$ of all real n-tuples (that is, all sequences of n real numbers) is called the ***n-dimensional Cartesian*** (or ***Euclidean***) ***space.*** A ***vector*** in $\mathbf{R}^n$ is a real n-tuple. The numbers $x_1, x_2, \ldots, x_n$ are called the ***coordinates*** of the vector $\mathbf{x} = (x_1, x_2, \ldots, x_n)$.

Thus, when $n = 1$, 2, or 3, Definition 2.3 just reduces to the algebraic representation given earlier for vectors in $\mathbf{R}^1$, $\mathbf{R}^2$, or $\mathbf{R}^3$. We add, subtract, and multiply vectors in $\mathbf{R}^n$ by scalars coordinate-wise, exactly as we did for $\mathbf{R}^1$, $\mathbf{R}^2$, and $\mathbf{R}^3$. Namely, if $\mathbf{x} = (x_1, x_2, \ldots, x_n)$ and $\mathbf{y} = (y_1, y_2, \ldots, y_n)$ are vectors in $\mathbf{R}^n$,

* Optional

and a is a real number, then we define

$$\mathbf{x} + \mathbf{y} = (x_1 + y_1, x_2 + y_2, \ldots, x_n + y_n),$$
$$a\mathbf{x} = (ax_1, ax_2, \ldots, ax_n),$$
$$\mathbf{x} - \mathbf{y} = \mathbf{x} + (-1)\mathbf{y},$$

and

$$|\mathbf{x}| = \sqrt{x_1^2 + x_2^2 + \ldots + x_n^2}.$$

If $|\mathbf{x}| = 1$, then $\mathbf{x}$ is called a ***unit vector.***

For each value of n, there is a standard basis for $\mathbf{R}^n$. It consists of the n different unit vectors having all zero coordinates except for a single 1.

2.4 DEFINITION The ***standard basis*** for $\mathbf{R}^n$ is the ordered set $\{\mathbf{e}_1, \mathbf{e}_2, \ldots, \mathbf{e}_n\}$, where $\mathbf{e}_1 = (1, 0, 0, \ldots, 0)$, $\mathbf{e}_2 = (0, 1, 0, \ldots, 0)$, $\ldots$, $\mathbf{e}_n = (0, 0, 0, \ldots, 1)$. If $n = 2$ (or 3), one writes $\mathbf{i}$ in place of $\mathbf{e}_1$ and $\mathbf{j}$ in place of $\mathbf{e}_2$ (and $\mathbf{k}$ in place of $\mathbf{e}_3$).

Since we cannot draw pictures of vectors in $\mathbf{R}^4$ and higher-dimensional spaces, we will work almost exclusively with $\mathbf{R}^2$ and $\mathbf{R}^3$. However, most of the results that we will derive are valid for all dimensions. For instance, it is easy to show that all of the properties of vector arithmetic in Theorem 1.2 are shared by $\mathbf{R}^n$ for all n (Exercises 29–32).

Although all vectors in this section have been written algebraically as row vectors $(x_1, x_2, \ldots, x_n)$, there is no compelling reason (aside from making the typesetting simpler) for preferring $(x_1, x_2, \ldots, x_n)$ to the column

$$\begin{pmatrix} x_1 \\ x_2 \\ \vdots \\ x_n \end{pmatrix}$$

as a representation for a vector in $\mathbf{R}^n$. Later we shall write vectors in column form as well as in row form, whichever is more appropriate to a given context. (See Exercise 35.)

Exercises 10.2

In Exercises 1 and 2, carefully plot the given points in $\mathbf{R}^3$.

1. **(a)** $(2, 6, 4)$ **(b)** $(-2, 6, -4)$
(c) $(2, -6, 4)$ **(d)** $(2, 6, -4)$
(e) $(-2, -6, 4)$ **(f)** $(2, -6, -4)$

2. **(a)** $(3, 2, 5)$ **(b)** $(-3, 2, 5)$
(c) $(3, -2, 5)$ **(d)** $(3, -2, -5)$
(e) $(-3, -2, 5)$ **(f)** $(3, 2, -5)$

In Exercises 3 and 4, draw the rectangular box with faces parallel to the coordinate planes and the given points as vertices.

3. **(a)** $(2, -3, 4)$ and $(-1, 2, 1)$
(b) $(3, 1, -2)$ and $(5, -1, 3)$

4. **(a)** $(3, 2, 5)$ and $(-2, 0, 1)$
(b) $(-2, 1, 3)$ and $(1, -1, 0)$

In Exercises 5 and 6, find the length and midpoint of the line segment joining the given points.

5. **(a)** $(1, -2, 3)$ and $(-2, 4, 5)$
(b) $(0, 3/2, \sqrt{2})$ and $(2, 7/2, -\sqrt{2})$

6. **(a)** $(-1, 3, 2)$ and $(1, 1, 3)$
(b) $(1/4, 1/2, 1)$ and $(-1/2, 3/4, 1/4)$

In Exercises 7–10, give the equation of the plane.

7. **(a)** The xy-plane
(b) The plane parallel to and two units to the left of the xz-plane

8. **(a)** The yz-plane
(b) The plane parallel to the xy-plane and four units above it

9. **(a)** The xz-plane
(b) The plane parallel to and three units in front of the yz-plane

10. (a) The plane through the origin, (2, 0, 0), and (0, 3, 0)
(b) The plane through the origin, (0, 1, 1) and (0, 0, 1)

In Exercises 11–14, compute (a) $\mathbf{v} + \mathbf{w}$, (b) $\mathbf{v} - \mathbf{w}$, (c) $\mathbf{w} - \mathbf{v}$, (d) $4\mathbf{v} - 2\mathbf{w}$, (e) $-2\mathbf{v} + 4\mathbf{w}$, (f) $|\mathbf{v}|$, (g) $|\mathbf{w}|$. (h) **Find the unit vectors in the directions of v and w.**

11. $\mathbf{v} = (-1, 3, 2)$, $\mathbf{w} = (2, -1, 1)$

12. $\mathbf{v} = (1, 2, -2)$ and $\mathbf{w} = (3, -1, 1)$

13. $\mathbf{v} = (1, -2, 0, 3)$ and $\mathbf{w} = (-1, 1, 2, 1)$ in $\mathbf{R}^4$

14. $\mathbf{v} = (1, 1, 2, 1)$ and $\mathbf{w} = (0, -1, 1, 4)$

In Exercises 15–18, find the coordinate representation of the vector AB and compute its length. Draw a figure if the vector belongs to $\mathbf{R}^3$.

15. (a) $A = (1, -1, 3)$ and $B = (-2, -1, 1)$
(b) $A = (0, 2, -1)$ and $B = (1, -1, 1)$

16. (a) $A = (3, 1, 1)$, $B = (1, -3, 4)$
(b) $A = (-1, 1, 2)$, $B = (2, -3, 1)$

17. (a) $A = (1, 2, -3, 1)$, $B = (-1, 2, 1, 3)$
(b) $A = (0, 2, 1, -1)$, $B = (3, -2, 4, 1)$

18. (a) $A = (1, 1, 2, 1)$, $B = (0, -1, 1, 4)$
(b) $A = (-1, 2, -1, 2)$, $B = (1, -3, 0, 1)$

In Exercises 19 and 20, describe the plane in $\mathbf{R}^3$ with the given equation.

19. (a) $3x - 2y = 6$ **(b)** $-2y + 3z = 0$

20. (a) $y = x$ **(b)** $z = x$

In Exercises 21 and 22, find the equation of the sphere with given center C and radius r.

21. (a) $C(1, -2, -2)$, $r = 2$ **(b)** $C(2, 0, 1)$, $r = 3$

22. (a) $C(1, -1, 4)$, $r = 1/2$ **(b)** $C(0, -3, 2)$, $r = 1$

In Exercises 23 and 24, find the center and radius of the sphere with the given equation.

23. $x^2 + y^2 + z^2 - 3x + 2y - 4z = 7/4$

24. $x^2 + y^2 + z^2 + 4x - 6y + 2z = 13$

In Exercises 25 and 26, determine whether the given points lie on a line.

25. (a) $(-1/2, -3/2, 5)$, $(5/2, -7/2, 0)$, and $(-7/2, 13/2, 5)$
(b) $(2, -2, -3)$, $(-1, 1, 1)$, and $(5, -5, -7)$

26. (a) $(1/3, -2/3, 1)$, $(2/3, -1, 3)$, $(2, 1/3, -5)$
(b) $(-1, 3, 4)$, $(3, 0, -1)$, $(1, 3/2, -7/2)$

27. Show that any vector $\mathbf{v}$ in $\mathbf{R}^3$ can be written as $\mathbf{v} = |\mathbf{v}|\mathbf{u}$, where $\mathbf{u}$ is a unit vector.

28. Repeat Exercise 27 for $\mathbf{R}^n$.

29. (a) Prove (a) of Theorem 1.2 for $\mathbf{R}^3$.
(b) Prove (d) of Theorem 1.2 for $\mathbf{R}^3$.

30. (a) Prove (f) of Theorem 1.2 for $\mathbf{R}^3$.
(b) Prove (g), (h), and (i) of Theorem 1.2 for $\mathbf{R}^3$.

31. For the analogue of Theorem 1.2 in $\mathbf{R}^n$, write out the proof of parts (b) and (e). State parts (c) and (d).

32. State and prove the following parts of the analogue for $\mathbf{R}^n$ of Theorem 1.2. **(a)** Parts (a) and (g); **(b)** Parts (f) and (h).

33. If $\mathbf{v} \in \mathbf{R}^3$ and $r \in \mathbf{R}$, then show that $|r\mathbf{v}| = |r|\,|\mathbf{v}|$ and $r\mathbf{v}$ points in the same direction as $\mathbf{v}$ if $r > 0$, or in the opposite direction if $r < 0$.

34. Show that any vector in $\mathbf{R}^n$ can be expressed in exactly one way in the form $\Sigma_{i=1}^n r_i\mathbf{e}_i$ for suitable real numbers $r_1, r_2, \ldots, r_n$.

35. In applications, vectors are often written as columns. Suppose that a manufacturing company produces four products, which are shipped to four regional distributors located in the northeast, southeast, northwest, and southwest. Each distributor submits his order to the parent company on a form with Product A listed first, Product B second, Product C third, and Product D fourth. We can represent such an order by a column vector

$$\mathbf{x} = \begin{pmatrix} x_1 \\ x_2 \\ x_3 \\ x_4 \end{pmatrix}$$

for the northeast distributor, and vectors

$$\mathbf{y} = \begin{pmatrix} y_1 \\ y_2 \\ y_3 \\ y_4 \end{pmatrix}, \qquad \mathbf{z} = \begin{pmatrix} z_1 \\ z_2 \\ z_3 \\ z_4 \end{pmatrix}, \qquad \text{and} \qquad \mathbf{w} = \begin{pmatrix} w_1 \\ w_2 \\ w_3 \\ w_4 \end{pmatrix}$$

for the southeast, northwest, and southwest distributors, respectively.

(a) What do the vectors $\mathbf{x} + \mathbf{y}$, $\mathbf{x} + \mathbf{z}$, $\mathbf{y} + \mathbf{w}$, and $\mathbf{z} + \mathbf{w}$ represent?

(b) If next month the distributors' orders change according to the following table, then compute for each the vector giving his new order.

Distributor	Change in order
NE	+2%
SE	−2.2%
NW	+1.1%
SW	−0.2%

36. If a force $\mathbf{F}$ acts in space, then at any point $\mathbf{x}$ we have

$$\mathbf{F}(\mathbf{x}) = (F_1(\mathbf{x}), F_2(\mathbf{x}), F_3(\mathbf{x})).$$

(a) Express $\mathbf{F}$ as the resultant of three forces acting parallel to the three coordinate axes.

(b) If a particle of mass m has acceleration $\mathbf{a}(\mathbf{x})$ at $\mathbf{x}$, then Newtonian physics says that $\mathbf{F}(\mathbf{x}) = m\mathbf{a}(\mathbf{x})$. Show that this *single* vector equation is equivalent to *three* scalar equations, each having the same general form. (This suggests some of the simplification that the use of vectors produces in physics and engineering.)

37. Suppose that $\mathbf{v} = (a_1, a_2, a_3)$ and $\mathbf{w} = (b_1, b_2, b_3)$ in $\mathbf{R}^3$. If addition is defined by the parallelogram law, then show that $\mathbf{v} + \mathbf{w}$ *must* be the vector $(a_1 + b_1, a_2 + b_2, a_3 + b_3)$. (*Hint:* See Exercise 40 of the last section.)

FIGURE 2.14

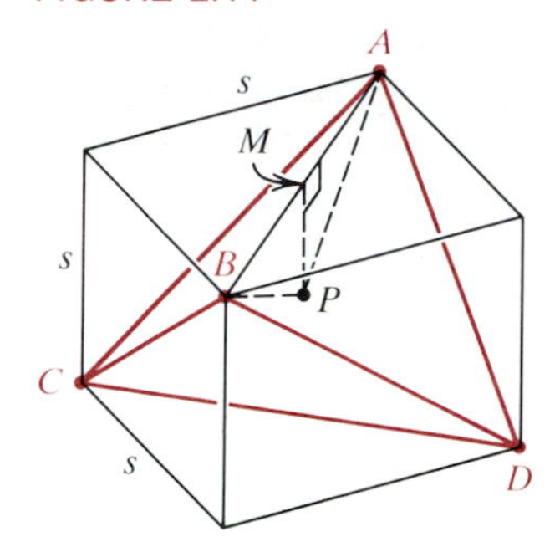

38. A tetrahedron is formed by connecting the vertices A, B, C, and D of the cube in **Figure 2.14.** The center of the tetrahedron is the same as the center P of the cube. If $|AB| = d$, then find $|AP|$ in terms of d. (*Hint:* Triangle APM in Figure 2.14 is a right triangle, where M is the midpoint of AB.)

39. In Exercise 38, find angle APB. (This is an important *bond angle* in chemistry, the angle made by an atom bonded to two nearby atoms in a molecule. This angle, called the *tetrahedral bond angle*, occurs in the diamond crystal and the methane molecule, among others.)

40. Suppose that four congruent spherical atoms of radius a are mutually tangent and have their centers at the corners of a tetrahedron. Suppose also that there is a fifth spherical atom of radius b located at the center of the tetrahedron and tangent to the other four atoms. Find b in terms of a. (*Hint:* First find b in terms of the edge length d of the tetrahedron.)

10.3 Dot Product and Angles

In the preceding two sections, we have seen how to perform arithmetic with two- and three-dimensional vectors and how to compute their lengths. This section develops a means of algebraically measuring the direction of vectors. We begin by recalling the observation following Definition 1.4, that the direction of any plane vector $\mathbf{x} = x_1\mathbf{i} + x_2\mathbf{j} = (x_1, x_2)$ is described by the angle α between $\mathbf{x}$ and $\mathbf{i} = (1, 0)$, the unit vector in the direction of the positive x-axis. As **Figure 3.1** illustrates, we lose no generality by restricting α to be a first- or second-quadrant angle. We can calculate α from the simple formula

$$(1) \qquad \cos\alpha = \frac{x_1}{\sqrt{x_1{}^2 + x_2{}^2}} = \frac{x_1 \cdot 1 + x_2 \cdot 0}{|\mathbf{x}|\,|\mathbf{i}|},$$

where the reason for the second way of writing the numerator will be seen shortly.

FIGURE 3.1

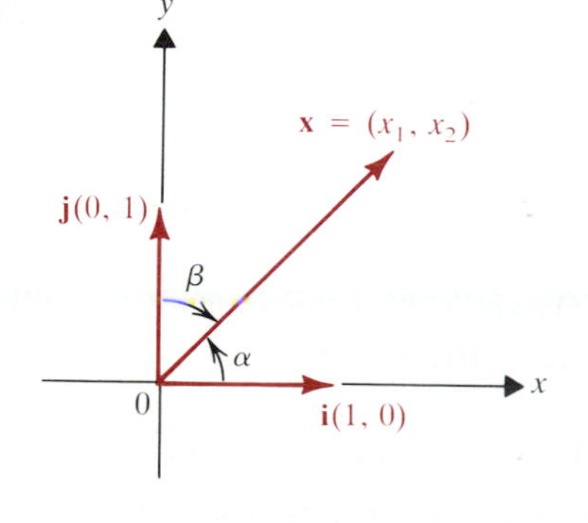

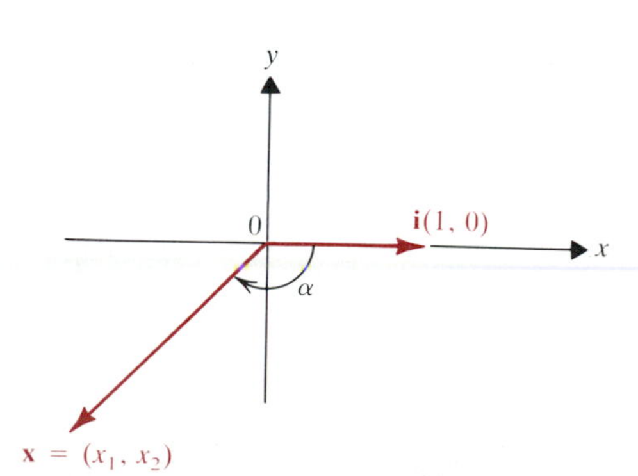

We could just as well describe the direction of **x** in terms of the angle β between **x** and **j**, which is easily calculated from the formula

$$\cos\beta = \frac{x_2}{\sqrt{x_1^2 + x_2^2}} = \frac{x_1 \cdot 0 + x_2 \cdot 1}{|\mathbf{x}|\,|\mathbf{j}|}. \tag{2}$$

In both (1) and (2), the cosine of the angle between **x** and the unit vector is given by a quotient. The denominator in each case is the product of the lengths of **x** and the unit vector (which has length 1). The numerator is the sum of the products of the corresponding coordinates of **x** and the unit vector. The geometric importance of that numerator is indicated by the following example.

Example 1

Let $\mathbf{x} = x_1\mathbf{i} + x_2\mathbf{j} = (x_1, x_2)$ and $\mathbf{y} = y_1\mathbf{i} + y_2\mathbf{j} = (y_1, y_2)$ be any two nonzero plane vectors. Find a formula for $\cos\theta$, where θ is the angle between **x** and **y**.

FIGURE 3.2

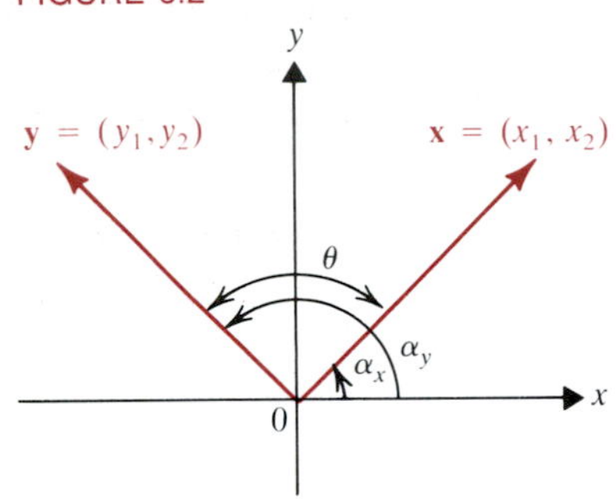

Solution. From **Figure 3.2** we see that $\theta = \alpha_y - \alpha_x$. (Since $\cos\theta = \cos(-\theta)$, it doesn't matter whether we measure θ from **x** to **y** or from **y** to **x**.) We have

$$\cos\theta = \cos(\alpha_y - \alpha_x) = (\cos\alpha_y)(\cos\alpha_x) + (\sin\alpha_y)(\sin\alpha_x)$$ *y Equation (13) of Section 2.3*

$$= \frac{y_1}{\sqrt{y_1^2 + y_2^2}} \cdot \frac{x_1}{\sqrt{x_1^2 + x_2^2}} + \frac{y_2}{\sqrt{y_1^2 + y_2^2}} \cdot \frac{x_2}{\sqrt{x_1^2 + x_2^2}}$$

$$= \frac{x_1y_1 + x_2y_2}{\sqrt{x_1^2 + x_2^2}\sqrt{y_1^2 + y_2^2}} = \frac{x_1y_1 + x_2y_2}{|\mathbf{x}|\,|\mathbf{y}|}. \blacksquare$$

In Example 1, the numerator of the formula for $\cos\theta$ is again the sum of the products of the corresponding coordinates of **x** and **y**. Such quantities are the main objects of study in this section.

3.1 DEFINITION If $\mathbf{x} = x_1\mathbf{i} + x_2\mathbf{j} = (x_1, x_2)$ and $\mathbf{y} = y_1\mathbf{i} + y_2\mathbf{j} = (y_1, y_2)$ are plane vectors, then their ***dot product*** (or ***inner product***) is the scalar

$$\mathbf{x} \cdot \mathbf{y} = x_1y_1 + x_2y_2. \tag{3}$$

Similarly, if $\mathbf{x} = x_1\mathbf{i} + x_2\mathbf{j} + x_3\mathbf{k} = (x_1, x_2, x_3)$ and $\mathbf{y} = y_1\mathbf{i} + y_2\mathbf{j} + y_3\mathbf{k} = (y_1, y_2, y_3)$ are three-dimensional vectors, then their ***dot product*** is the scalar

$$\mathbf{x} \cdot \mathbf{y} = x_1y_1 + x_2y_2 + x_3y_3. \tag{4}$$

For any $n \geq 2$, the dot product of the vectors $\mathbf{x} = (x_1, x_2, \ldots, x_n)$ and $\mathbf{y} = (y_1, y_2, \ldots, y_n)$ in $\mathbf{R}^n$ is the scalar

$$\mathbf{x} \cdot \mathbf{y} = x_1y_1 + x_2y_2 + \ldots + x_ny_n. \tag{5}$$

The result of Example 1 thus takes the form

$$\cos\theta = \frac{\mathbf{x} \cdot \mathbf{y}}{|\mathbf{x}|\,|\mathbf{y}|}. \tag{6}$$

Formulas (1) and (2) for the angles α and β between **x** and the coordinate axes are instances of (6), with $\mathbf{y} = \mathbf{i}$ and $\mathbf{y} = \mathbf{j}$, respectively.

Example 2

(a) Find the angle between $\mathbf{x} = (1, 2)$ and $\mathbf{y} = (3, 1)$.

(b) Find $\mathbf{x} \cdot \mathbf{y}$ if $\mathbf{x} = \mathbf{i} - \mathbf{j}$ and $\mathbf{y} = -\mathbf{i} + \mathbf{j} + \sqrt{2}\mathbf{k}$.

Solution. (a) From (6) we get

$$\cos\theta = \frac{\mathbf{x}\cdot\mathbf{y}}{|\mathbf{x}|\,|\mathbf{y}|} = \frac{1\cdot 3 + 2\cdot 1}{\sqrt{5}\sqrt{10}} = \frac{5}{5\sqrt{2}} = \frac{1}{\sqrt{2}}.$$

So $\theta = \pi/4$ radians (45°).

(b) Since $\mathbf{x} = \mathbf{i} - \mathbf{j} = (1, -1, 0)$ and $\mathbf{y} = -\mathbf{i} + \mathbf{j} + \sqrt{2}\mathbf{k} = (-1, 1, \sqrt{2})$, (4) gives

$$\mathbf{x}\cdot\mathbf{y} = 1(-1) + (-1)(1) + 0(\sqrt{2}) = -2. \quad ■$$

The following theorem, which holds for vectors in $\mathbf{R}^n$, $n \geq 2$, summarizes the main algebraic properties of the dot product.

3.2 THEOREM For all vectors $\mathbf{x}$, $\mathbf{y}$, and $\mathbf{z}$ and all real numbers a,

(a) $\mathbf{x}\cdot\mathbf{y} = \mathbf{y}\cdot\mathbf{x}$.

(b) $\mathbf{x}\cdot(\mathbf{y}+\mathbf{z}) = \mathbf{x}\cdot\mathbf{y} + \mathbf{x}\cdot\mathbf{z}$ and $(a\mathbf{x})\cdot\mathbf{y} = \mathbf{x}\cdot(a\mathbf{y}) = a(\mathbf{x}\cdot\mathbf{y})$.

(c) $\mathbf{0}\cdot\mathbf{x} = 0$ for all $\mathbf{x}$. If $\mathbf{v}\cdot\mathbf{x} = 0$ for all $\mathbf{x}$, then $\mathbf{v} = \mathbf{0}$.

(d) $\mathbf{x}\cdot\mathbf{x} \geq 0$ for all $\mathbf{x}$, and $\mathbf{x}\cdot\mathbf{x} = 0$ only when $\mathbf{x} = \mathbf{0}$.

(e) $\mathbf{x}\cdot\mathbf{x} = |\mathbf{x}|^2$.

Proof. We prove parts (c), (d), and (e), leaving the rest for Exercise 32.

(c) Substitution of $\mathbf{0} = (0, 0, \ldots, 0)$ in Definition 3.1 gives $\mathbf{0}\cdot\mathbf{x} = 0$, proving the first assertion. Suppose next that for some vector $\mathbf{v} = (v_1, v_2, \ldots, v_n)$ we have $\mathbf{v}\cdot\mathbf{x} = 0$ for *all* $\mathbf{x}$ in $\mathbf{R}^n$. To show that $\mathbf{v} = \mathbf{0}$, we prove that each coordinate $v_i = 0$. Since $\mathbf{v}\cdot\mathbf{e}_i = 0$ for every basis vector $\mathbf{e}_i = (0, 0, \ldots, 0, 1, 0, \ldots, 0)$, we have

$$0 = \mathbf{v}\cdot\mathbf{e}_i = v_1 0 + v_2 0 + \ldots + v_i 1 + \ldots + v_n 0 = v_i.$$

(d and e) For $\mathbf{x} = (x_1, x_2, \ldots, x_n)$ in $\mathbf{R}^n$, we get from (5) and the definition of $|\mathbf{x}|$,

$$\mathbf{x}\cdot\mathbf{x} = x_1{}^2 + x_2{}^2 + \ldots + x_n{}^2 = |\mathbf{x}|^2 \geq 0,$$

since it is a sum of squares. Moreover, the only time a sum of squares $x_i{}^2$ can be zero is when each summand $x_i{}^2 = 0$, that is, when each $x_i = 0$. Thus $\mathbf{x}\cdot\mathbf{x} = 0$ only when $\mathbf{x} = \mathbf{0}$. QED

Example 3

Verify part (a) and the first assertions in parts (b) and (d) of Theorem 3.2, if $\mathbf{x} = -\mathbf{i} + 2\mathbf{j} + 3\mathbf{k}$, $\mathbf{y} = 3\mathbf{i} - \mathbf{j} + 2\mathbf{k}$, and $\mathbf{z} = 2\mathbf{i} - \mathbf{k}$.

Solution. (a) We have

$$\mathbf{x}\cdot\mathbf{y} = -1\cdot 3 + 2\cdot(-1) + 3\cdot 2 = 1 = 3\cdot(-1) + (-1)\cdot 2 + 2\cdot 3 = \mathbf{y}\cdot\mathbf{x}.$$

(b) We get from (3),

$$\begin{aligned}\mathbf{x}\cdot(\mathbf{y}+\mathbf{z}) &= (-1, 2, 3)\cdot[(3, -1, 2) + (2, 0, -1)]\\ &= (-1, 2, 3)\cdot(5, -1, 1) = -5 - 2 + 3 = -4.\end{aligned}$$

We also have

$$\begin{aligned}\mathbf{x}\cdot\mathbf{y} + \mathbf{x}\cdot\mathbf{z} &= (-1, 2, 3)\cdot(3, -1, 2) + (-1, 2, 3)\cdot(2, 0, -1)\\ &= -3 - 2 + 6 - 2 + 0 - 3 = -4.\end{aligned}$$

Thus $\mathbf{x}\cdot(\mathbf{y}+\mathbf{z}) = \mathbf{x}\cdot\mathbf{y} + \mathbf{x}\cdot\mathbf{z}$.

(d) $\mathbf{x}\cdot\mathbf{x} = (-1, 2, 3)\cdot(-1, 2, 3) = (-1)^2 + 2^2 + 3^2 = 14 > 0$. ■

We come now to an important result that carries the names of Cauchy, and Hermann A. Schwarz (1843–1921) of Poland, who made many important contributions to advanced calculus. The result appears to have been discovered independently by the Russian mathematician Victor Bunyakovsky (1804–1889). A proof is given at the end of the section.

3.3 THEOREM

Cauchy–Schwarz Inequality. For any two vectors $\mathbf{x}$ and $\mathbf{y}$ in $\mathbf{R}^n$,

$$|\mathbf{x}\cdot\mathbf{y}| \le |\mathbf{x}|\,|\mathbf{y}|. \tag{7}$$

We can use the Cauchy–Schwarz inequality to define the angle θ between any two nonzero vectors $\mathbf{x}$ and $\mathbf{y}$ in a Euclidean space $\mathbf{R}^n$, where $n > 2$. Theorem 1.2(a) of Chapter 1 says that if $\mathbf{x}$ and $\mathbf{y}$ are nonzero, then (7) is equivalent to

$$-|\mathbf{x}|\,|\mathbf{y}| \le \mathbf{x}\cdot\mathbf{y} \le |\mathbf{x}|\,|\mathbf{y}| \to -1 \le \frac{\mathbf{x}\cdot\mathbf{y}}{|\mathbf{x}|\,|\mathbf{y}|} \le 1.$$

Thus there is a unique angle θ between 0 and π (inclusive) such that

$$\cos\theta = \frac{\mathbf{x}\cdot\mathbf{y}}{|\mathbf{x}|\,|\mathbf{y}|}. \tag{8}$$

3.4 DEFINITION

If $\mathbf{x}$ and $\mathbf{y}$ are nonzero vectors in $\mathbf{R}^n$, where $n \ge 2$, then the ***angle θ between* x *and* y** is

$$\theta = \cos^{-1}\frac{\mathbf{x}\cdot\mathbf{y}}{|\mathbf{x}|\,|\mathbf{y}|} \in [0, \pi]. \tag{9}$$

Note that we do *not* define an angle between any vector $\mathbf{x}$ and the (directionless) zero vector $\mathbf{0}$. A useful formula for $\mathbf{x}\cdot\mathbf{y}$ can be obtained by multiplying both sides of (8) by $|\mathbf{x}|\,|\mathbf{y}|$:

$$\mathbf{x}\cdot\mathbf{y} = |\mathbf{x}|\,|\mathbf{y}|\cos\theta. \tag{10}$$

FIGURE 3.3

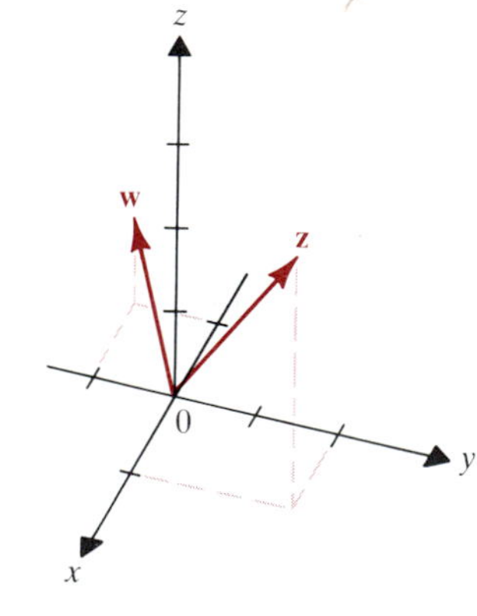

Example 4

(a) Find the angle between $\mathbf{x} = (1, -1, 0)$ and $\mathbf{y} = (-1, 1, \sqrt{2})$.

(b) Show that the vectors $\mathbf{z} = (1, 2, 3)$ and $\mathbf{w} = (-1, -1, 1)$ are perpendicular.

Solution. (a) We have from (8),

$$\cos\theta = \frac{\mathbf{x}\cdot\mathbf{y}}{|\mathbf{x}|\,|\mathbf{y}|} = \frac{(1)(-1) + (-1)(1) + (0)(\sqrt{2})}{\sqrt{1^2 + (-1)^2 + 0^2}\sqrt{(-1)^2 + 1^2 + 2}} = \frac{-2}{\sqrt{2}\cdot 2} = -\frac{1}{\sqrt{2}}.$$

Thus $\theta = 3\pi/4$ (135°).

(b) The dot product of the given vectors is $\mathbf{z}\cdot\mathbf{w} = -1 - 2 + 3 = 0$. Hence $\cos\theta = 0$, so $\theta = \pi/2$. Thus the vectors are perpendicular. See **Figure 3.3.** ■

Example 4(b) generalizes to any Euclidean space $\mathbf{R}^n$, because two vectors are perpendicular if and only if the angle θ between them is $\pi/2$, that is, if and only if $\cos\theta = 0$. By Definition 3.4, that is equivalent to

$$\cos\theta = \frac{\mathbf{x}\cdot\mathbf{y}}{|\mathbf{x}|\,|\mathbf{y}|} = 0,$$

which in turn is equivalent to $\mathbf{x}\cdot\mathbf{y} = 0$. We therefore see that

(11) two nonzero vectors $\mathbf{x}$ and $\mathbf{y}$ are perpendicular if and only if $\mathbf{x}\cdot\mathbf{y} = 0$.

The term *orthogonal* is often used as a synonym for perpendicular, and we will do so occasionally. As an illustration of (11), for the standard basis vectors **i**, **j**, and **k**, which we know are mutually perpendicular, we have

$$\mathbf{i} \cdot \mathbf{j} = (1, 0, 0) \cdot (0, 1, 0) = 0,$$

$$\mathbf{j} \cdot \mathbf{k} = (0, 1, 0) \cdot (0, 0, 1) = 0,$$

$$\mathbf{i} \cdot \mathbf{k} = (1, 0, 0) \cdot (0, 0, 1) = 0.$$

FIGURE 3.4

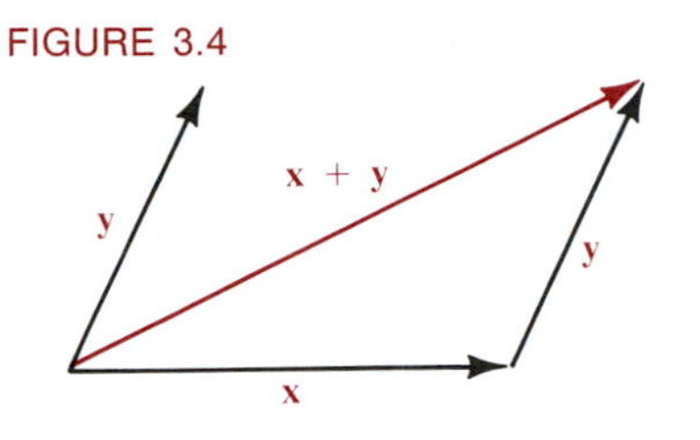

In the plane it is clear geometrically that the length of $\mathbf{x} + \mathbf{y}$ is no greater than $|\mathbf{x}| + |\mathbf{y}|$, because $\mathbf{x}$, $\mathbf{y}$, and $\mathbf{x} + \mathbf{y}$ usually form a triangle. See **Figure 3.4.** With the help of the Cauchy–Schwarz inequality, we can prove this for any Euclidean space $\mathbf{R}^n$ and thereby generalize the triangle inequality of Theorem 1.2(d) of Chapter 1.

3.5 THEOREM

Triangle Inequality. Let $\mathbf{x}$ and $\mathbf{y}$ be any two vectors in $\mathbf{R}^n$. Then

(12) $$|\mathbf{x} + \mathbf{y}| \le |\mathbf{x}| + |\mathbf{y}|.$$

Proof. We start by applying part (e) of Theorem 3.2 to $|\mathbf{x} + \mathbf{y}|$. That gives

$$|\mathbf{x} + \mathbf{y}|^2 = (\mathbf{x} + \mathbf{y}) \cdot (\mathbf{x} + \mathbf{y}) = \mathbf{x} \cdot \mathbf{x} + \mathbf{y} \cdot \mathbf{x} + \mathbf{x} \cdot \mathbf{y} + \mathbf{y} \cdot \mathbf{y} \qquad \text{by Theorem 3.2(b)}$$

$$= \mathbf{x} \cdot \mathbf{x} + 2\mathbf{x} \cdot \mathbf{y} + \mathbf{y} \cdot \mathbf{y} \qquad \text{by Theorem 3.2(a)}$$

$$\le |\mathbf{x}||\mathbf{x}| + 2|\mathbf{x}||\mathbf{y}| + |\mathbf{y}||\mathbf{y}| = (|\mathbf{x}| + |\mathbf{y}|)^2. \qquad \text{by the Cauchy–Schwarz inequality}$$

Now extract the positive square roots to get (12). QED

Example 5

Verify Theorem 3.5 for the vectors

(a) $\mathbf{x} = 2\mathbf{i} - \mathbf{j}$, $\mathbf{y} = \mathbf{i} + 2\mathbf{j}$, (b) $\mathbf{x} = (1, -2, 2)$, $\mathbf{y} = (1/2, -1, 1)$.

Solution. (a) Here $\mathbf{x} + \mathbf{y} = 3\mathbf{i} + \mathbf{j}$. We thus have

$$|\mathbf{x} + \mathbf{y}| = \sqrt{9 + 1} = \sqrt{10},$$

$$|\mathbf{x}| = \sqrt{4 + 1} = \sqrt{5}, \quad \text{and} \quad |\mathbf{y}| = \sqrt{1 + 4} = \sqrt{5}.$$

Since

$$|\mathbf{x} + \mathbf{y}| = \sqrt{10} = \sqrt{2}\sqrt{5} < 2\sqrt{5} = \sqrt{5} + \sqrt{5} = |\mathbf{x}| + |\mathbf{y}|,$$

we see that (12) holds.

(b) In this case $\mathbf{x} + \mathbf{y} = (3/2, -3, 3)$, so

$$|\mathbf{x} + \mathbf{y}| = 3\sqrt{1/4 + 1 + 1} = 3\sqrt{9/4} = 9/2,$$

$$|\mathbf{x}| = \sqrt{1 + 4 + 4} = 3, \quad \text{and} \quad |\mathbf{y}| = \sqrt{1/4 + 1 + 1} = 3/2.$$

We then have

$$|\mathbf{x} + \mathbf{y}| = \frac{9}{2} = 3 + \frac{3}{2} = |\mathbf{x}| + |\mathbf{y}|,$$

which again satisfies (12). ■

In the second line of the proof of Theorem 3.5, notice that the term $\mathbf{x} \cdot \mathbf{y}$ is 0 if $\mathbf{x}$ and $\mathbf{y}$ are perpendicular. In that case we get the following version of the Pythagorean theorem, which holds in any dimension.

3.6 THEOREM If $\mathbf{x}$ and $\mathbf{y}$ are perpendicular, then

$$|\mathbf{x} + \mathbf{y}|^2 = |\mathbf{x}|^2 + |\mathbf{y}|^2.$$

For instance, in Example 5(a), $\mathbf{x}$ and $\mathbf{y}$ are perpendicular, since

$$\mathbf{x} \cdot \mathbf{y} = (2, -1) \cdot (1, 2) = 2 - 2 = 0.$$

FIGURE 3.5

And we see in **Figure 3.5** that

$$|\mathbf{x}|^2 + |\mathbf{y}|^2 = 5 + 5 = 10 = |\mathbf{x} + \mathbf{y}|^2.$$

Recall from Definition 1.4 that the unit vector in the direction of a nonzero vector $\mathbf{x}$ is $\mathbf{u} = \dfrac{1}{|\mathbf{x}|}\mathbf{x}$. This extends to n-dimensional vectors $\mathbf{x} \neq \mathbf{0}$ as well. In the three-dimensional case, for example,

$$\mathbf{u} = (u_1, u_2, u_3) = \frac{1}{\sqrt{x_1{}^2 + x_2{}^2 + x_3{}^2}}(x_1, x_2, x_3)$$

is called the ***unit vector in the direction of* x**, or the ***direction vector of* x**. As for plane vectors, $\mathbf{u}$ carries enough information to find the angles $\mathbf{x}$ makes with the standard basis vectors $\mathbf{i}, \mathbf{j}$, and $\mathbf{k}$, that is, with the positive x-, y-, and z-coordinate axes. Letting α_1, α_2, and α_3 be the respective angles (see **Figure 3.6**), we have

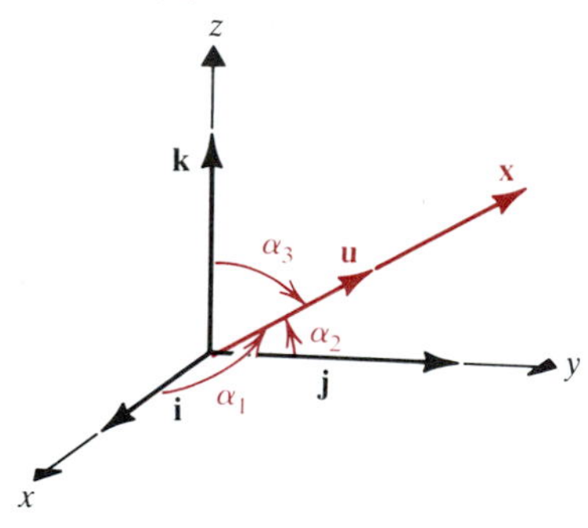

FIGURE 3.6

$$\cos\alpha_1 = \frac{\mathbf{x} \cdot \mathbf{i}}{|\mathbf{x}|\,|\mathbf{i}|} = \frac{x_1}{\sqrt{x_1{}^2 + x_2{}^2 + x_3{}^2}} = u_1,$$

$$\cos\alpha_2 = \frac{\mathbf{x} \cdot \mathbf{j}}{|\mathbf{x}|\,|\mathbf{j}|} = \frac{x_2}{\sqrt{x_1{}^2 + x_2{}^2 + x_3{}^2}} = u_2,$$

$$\cos\alpha_3 = \frac{\mathbf{x} \cdot \mathbf{k}}{|\mathbf{x}|\,|\mathbf{k}|} = \frac{x_3}{\sqrt{x_1{}^2 + x_2{}^2 + x_3{}^2}} = u_3.$$

Thus the direction $\mathbf{u}$ is given by

$$\mathbf{u} = (\cos\alpha_1, \cos\alpha_2, \cos\alpha_3).$$

The coordinates of $\mathbf{u}$ are called the ***direction cosines*** of the vector $\mathbf{x}$. The angles α_1, α_2, and α_3 are called the ***direction angles*** of $\mathbf{x}$. (See Exercise 31.)

We next extend Definition 1.5 to higher dimensions.

3.7 DEFINITION Let $\mathbf{x}$ and $\mathbf{y}$ be nonzero vectors in $\mathbf{R}^3$ (or $\mathbf{R}^n$ for any $n > 2$). Then $\mathbf{x}$ is ***parallel*** to $\mathbf{y}$ if $\mathbf{u_x} = \pm\mathbf{u_y}$, where $\mathbf{u_x}$ and $\mathbf{u_y}$ are the direction vectors of $\mathbf{x}$ and $\mathbf{y}$ respectively.

In view of this, it is easy to see that Theorem 1.6 holds more generally:

$\mathbf{x}$ is parallel to $\mathbf{y}$ if and only if $\mathbf{x} = a\mathbf{y}$ for some scalar $a \neq 0$.

Example 6

Find the direction vector and direction angles of the vector drawn from $P(1, -1, 2)$ to $Q(-1, 0, 4)$.

Solution. Here $\mathbf{x} = \mathbf{PQ} = (-2, 1, 2)$. Thus

$$|\mathbf{x}| = \sqrt{(-2)^2 + 1^2 + 2^2} = \sqrt{9} = 3,$$

so the direction vector of $\mathbf{x}$ is $\mathbf{u} = \frac{1}{3}(-2, 1, 2) = (-2/3, 1/3, 2/3)$. The direction angles are $\alpha = \cos^{-1}(-2/3)$, $\beta = \cos^{-1}(1/3)$, and $\gamma = \cos^{-1}(2/3)$. From a calculator, $\alpha \approx 2.30$ radians (about 132°), $\beta \approx 1.23$ radians (about 71°), and $\gamma \approx 0.84$ radians (about 48°). ■

FIGURE 3.7

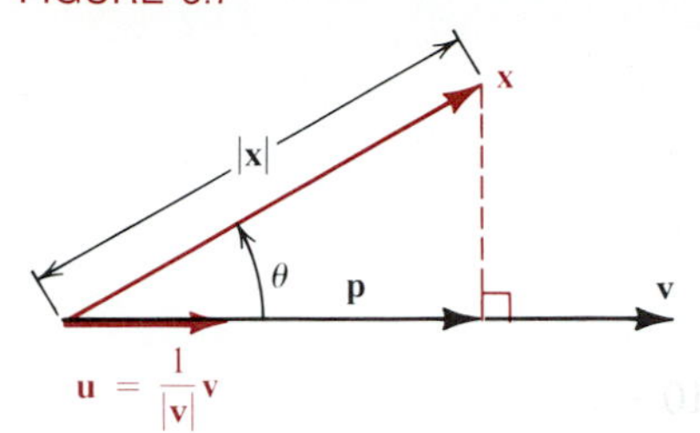

It is easy to express a three-dimensional vector $\mathbf{x} = (x_1, x_2, x_3)$ in terms of the standard basis vectors **i**, **j**, and **k**: $\mathbf{x} = x_1\mathbf{i} + x_2\mathbf{j} + x_3\mathbf{k}$. In physics and engineering, it is often necessary to express a vector **x** in terms of some other known vector **v**. (For instance, it is sometimes required to express the force on a body in terms of the direction vector of its velocity.) Figure 3.7 illustrates this type of geometric problem. Suppose we project **x** perpendicularly onto **v** to get the vector **p** in **Figure 3.7.** Since **p** points in the same direction as **v**, its direction vector is

$$\mathbf{u} = \frac{1}{|\mathbf{v}|}\mathbf{v} = \frac{1}{|\mathbf{p}|}\mathbf{p}. \tag{13}$$

From the figure and equation (8) we have

$$|\mathbf{p}| = |\mathbf{x}|\cos\theta = |\mathbf{x}|\frac{\mathbf{x}\cdot\mathbf{v}}{|\mathbf{x}||\mathbf{v}|} = \frac{\mathbf{x}\cdot\mathbf{v}}{|\mathbf{v}|}.$$

Thus

$$\mathbf{p} = |\mathbf{p}|\mathbf{u} = \frac{\mathbf{x}\cdot\mathbf{v}}{|\mathbf{v}|}\mathbf{u}.$$

We give to **p** and its length the names suggested by this discussion.

3.8 DEFINITION Let **x** and $\mathbf{v} \neq \mathbf{0}$ be vectors in $\mathbf{R}^n$ for any $n > 2$. The ***coordinate of* x *in the direction of* v** is the scalar

$$x_{\mathbf{v}} = \frac{\mathbf{x}\cdot\mathbf{v}}{|\mathbf{v}|}. \tag{14}$$

The ***component of* x *in the direction of* v** (or the ***projection of* x *onto* v**) is the vector

$$x_{\mathbf{v}}\mathbf{u} = \frac{\mathbf{x}\cdot\mathbf{v}}{|\mathbf{v}|}\frac{1}{|\mathbf{v}|}\mathbf{v}, \tag{15}$$

where **u** is the direction vector of **v**.

Equation (14) simply says that *the coordinate of* **x** *in the direction of a nonzero vector* **v** *is the dot product of* **x** *with the unit vector* $\frac{1}{|\mathbf{v}|}\mathbf{v}$ *in the direction of* **v**.

Example 7

Find the vector obtained when $\mathbf{x} = -2\mathbf{i} + 4\mathbf{j} + 10\mathbf{k}$ is projected onto $\mathbf{v} = 12\mathbf{i} + 9\mathbf{k}$.

Solution. Refer to **Figure 3.8.** The vector sought is the component of **x** in the direction of **v**. Here the direction vector of **v** is

$$\mathbf{u} = \frac{1}{|\mathbf{v}|}\mathbf{v} = \frac{1}{\sqrt{144+0+81}}(12, 0, 9) = \frac{1}{15}(12, 0, 9) = \left(\frac{4}{5}, 0, \frac{3}{5}\right).$$

Then the projection of **x** onto **v** is

$$x_{\mathbf{v}}\mathbf{u} = (\mathbf{x}\cdot\mathbf{u})\mathbf{u} = \left(-\frac{8}{5} + 0 + \frac{30}{5}\right)\left(\frac{4}{5}, 0, \frac{3}{5}\right) = \frac{22}{5}\left(\frac{4}{5}, 0, \frac{3}{5}\right) = \left(\frac{88}{25}, 0, \frac{66}{25}\right). \quad ■$$

FIGURE 3.8

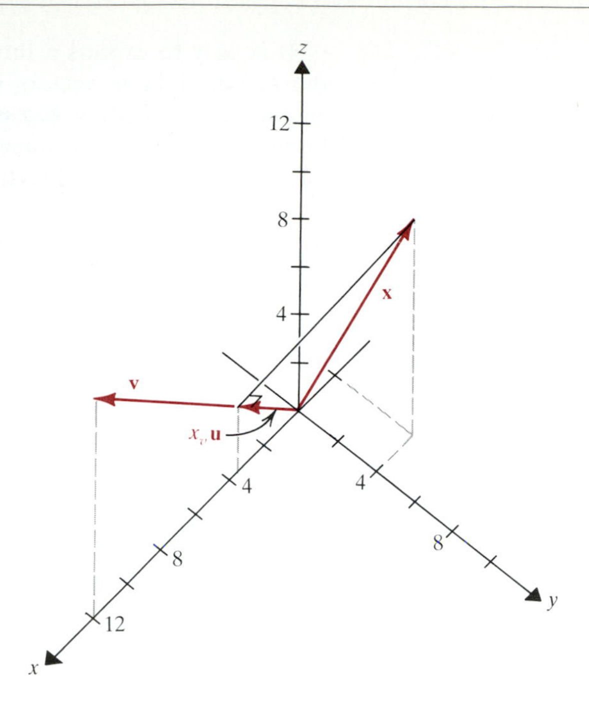

Tangential and Normal Components*

A common problem in physics and engineering is to find the components of a vector force **F** in the direction of a given vector **v** and in the direction of a vector **w** perpendicular to **v** (see Exercise 38). This kind of procedure is called the ***resolution*** of **x** relative to **v** and **w**. We close the section with an example of such a resolution.

FIGURE 3.9

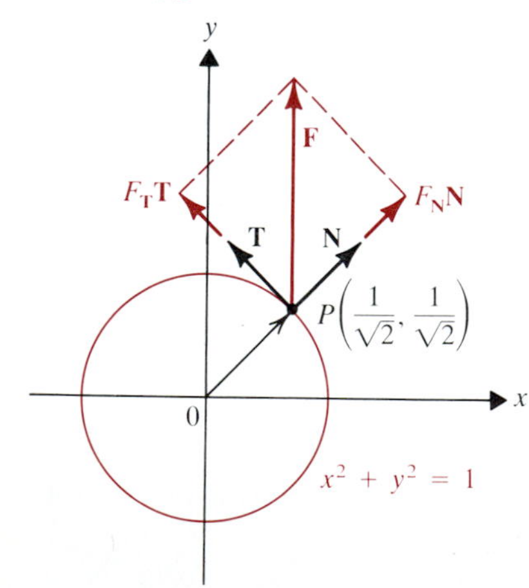

Example 8

A particle moving counterclockwise along the unit circle $x^2 + y^2 = 1$ as in **Figure 3.9** is acted on by a force **F**. When the particle is at the point $P(1/\sqrt{2}, 1/\sqrt{2})$, the force is given by $\mathbf{F} = 2\mathbf{j}$. Find the components of **F** in the directions of **T** and **N** in Figure 3.9.

Solution. Since the tangent to a circle is perpendicular to its radius, **N** is in the direction of the radius vector $\mathbf{0P} = (1/\sqrt{2}, 1/\sqrt{2})$. Thus

$$\mathbf{N} = \frac{1}{|\mathbf{0P}|}\,\mathbf{0P} = \frac{1}{1}\left(\frac{1}{\sqrt{2}}, \frac{1}{\sqrt{2}}\right) = \frac{1}{\sqrt{2}}\mathbf{i} + \frac{1}{\sqrt{2}}\mathbf{j}.$$

Then for **T** we can use

$$-\frac{1}{\sqrt{2}}\mathbf{i} + \frac{1}{\sqrt{2}}\mathbf{j},$$

since T is a unit vector pointing upward and to the left and satisfying $\mathbf{T} \cdot \mathbf{N} = 0$. The coordinates of **F** in the directions of **T** and **N** are calculated from (14). Keeping in mind that $|\mathbf{T}| = |\mathbf{N}| = 1$, we find

$$F_{\mathbf{T}} = \frac{\mathbf{F} \cdot \mathbf{T}}{|\mathbf{T}|} = \mathbf{F} \cdot \mathbf{T} = 2\mathbf{j} \cdot \frac{1}{\sqrt{2}}(-\mathbf{i} + \mathbf{j}) = \frac{2}{\sqrt{2}}[0 \cdot (-1) + 1 \cdot 1] = \sqrt{2},$$

* Optional

and

$$F_{\mathbf{N}} = \frac{\mathbf{F} \cdot \mathbf{N}}{|\mathbf{N}|} = \mathbf{F} \cdot \mathbf{N} = 2\mathbf{j} \cdot \frac{1}{\sqrt{2}}(\mathbf{i} + \mathbf{j}) = \sqrt{2}.$$

Then from (15), the tangential and normal components of $\mathbf{F}$ are

$$F_{\mathbf{T}}\mathbf{T} = (\mathbf{F} \cdot \mathbf{T})\mathbf{T} = \sqrt{2}\frac{1}{\sqrt{2}}(-\mathbf{i} + \mathbf{j}) = -\mathbf{i} + \mathbf{j},$$

and

$$F_{\mathbf{N}}\mathbf{N} = (\mathbf{F} \cdot \mathbf{N})\mathbf{N} = \sqrt{2}\frac{1}{\sqrt{2}}(\mathbf{i} + \mathbf{j}) = \mathbf{i} + \mathbf{j}.$$

Note that $\mathbf{F}$ is the sum of its tangential and normal components, since

$$F_{\mathbf{T}}\mathbf{T} + F_{\mathbf{N}}\mathbf{N} = (-\mathbf{i} + \mathbf{j}) + (\mathbf{i} + \mathbf{j}) = 2\mathbf{j} = \mathbf{F}. \quad ■$$

In the next chapter, we will see that the resolution equation

$$\mathbf{F} = (\mathbf{F} \cdot \mathbf{T})\mathbf{T} + (\mathbf{F} \cdot \mathbf{N})\mathbf{N} = F_{\mathbf{T}}\mathbf{T} + F_{\mathbf{N}}\mathbf{N},$$

obtained in Example 8 for circular motion, has an important general counterpart for motion along a smooth curve.

Proof of the Cauchy–Schwarz Inequality*

3.3 THEOREM For any two vectors $\mathbf{x}$ and $\mathbf{y}$ in $\mathbf{R}^n$,

$$(7) \qquad |\mathbf{x} \cdot \mathbf{y}| \leq |\mathbf{x}|\,|\mathbf{y}|$$

Proof. Clearly (7) holds (as an equality) if either $\mathbf{x} = \mathbf{0}$ or $\mathbf{y} = \mathbf{0}$. So we can confine attention to the case $\mathbf{x} \neq \mathbf{0}$ and $\mathbf{y} \neq \mathbf{0}$. We first show that for the unit vectors

$$(16) \qquad \mathbf{u} = \frac{1}{|\mathbf{x}|}\mathbf{x} \quad \text{and} \quad \mathbf{v} = \frac{1}{|\mathbf{y}|}\mathbf{y}$$

in the respective directions of $\mathbf{x}$ and $\mathbf{y}$, we have

$$(17) \qquad \mathbf{u} \cdot \mathbf{v} \leq |\mathbf{u}|\,|\mathbf{v}|.$$

We then derive (7) from (17). From part (d) of Theorem 3.2, with $\mathbf{x} = \mathbf{u} - \mathbf{v}$, we have

$$\begin{aligned} 0 \leq (\mathbf{u} - \mathbf{v}) \cdot (\mathbf{u} - \mathbf{v}) &= (\mathbf{u} - \mathbf{v}) \cdot \mathbf{u} + (\mathbf{u} - \mathbf{v}) \cdot (-\mathbf{v}) && \textit{by Theorem 3.2(b)} \\ &= \mathbf{u} \cdot \mathbf{u} - \mathbf{v} \cdot \mathbf{u} - \mathbf{u} \cdot \mathbf{v} + \mathbf{v} \cdot \mathbf{v} && \textit{by Theorem 3.2(b)} \\ &= 2 - 2(\mathbf{u} \cdot \mathbf{v}). && \textit{by Theorem 3.2(a)} \end{aligned}$$

Hence

$$2(\mathbf{u} \cdot \mathbf{v}) \leq 2 \rightarrow \mathbf{u} \cdot \mathbf{v} \leq 1 = |\mathbf{u}|\,|\mathbf{v}|.$$

From (16) and Theorem 3.2(b) again, we therefore have

$$(18) \qquad \frac{1}{|\mathbf{x}|}\mathbf{x} \cdot \frac{1}{|\mathbf{y}|}\mathbf{y} \leq 1 \rightarrow \frac{1}{|\mathbf{x}|}\frac{1}{|\mathbf{y}|}\mathbf{x} \cdot \mathbf{y} \leq 1 \rightarrow \mathbf{x} \cdot \mathbf{y} \leq |\mathbf{x}|\,|\mathbf{y}|.$$

* Optional

The last inequality holds for any nonzero vectors $\mathbf{x}$ and $\mathbf{y}$ and so remains true if we replace $\mathbf{x}$ by $-\mathbf{x}$. That is, we also have

(19) $$(-\mathbf{x}) \cdot \mathbf{y} \le |-\mathbf{x}|\,|\mathbf{y}| \to -(\mathbf{x} \cdot \mathbf{y}) \le |\mathbf{x}|\,|\mathbf{y}| \to \mathbf{x} \cdot \mathbf{y} \ge -|\mathbf{x}|\,|\mathbf{y}|.$$

Together, the final inequalities in (18) and (19) say that

$$-|\mathbf{x}|\,|\mathbf{y}| \le \mathbf{x} \cdot \mathbf{y} \le |\mathbf{x}|\,|\mathbf{y}|,$$

which by Theorem 1.2(a) of Chapter 1 is equivalent to (7). QED

Exercises 10.3

In Exercises 1–6, find the angle between each pair of vectors.

1. (a) $(3, -1)$ and $(2, 1)$
   (b) $\sqrt{3}\mathbf{i} + \frac{\sqrt{3}}{2}\mathbf{j}$ and $\mathbf{i} + 2\mathbf{j}$
2. (a) $(-1, 3)$ and $(2, -1)$
   (b) $\frac{1}{7}\mathbf{i} - 3\mathbf{j}$ and $28\mathbf{i} + \frac{4}{3}\mathbf{j}$
3. (a) $(1, -1, 0)$ and $(-1, 2, 1)$
   (b) $-\mathbf{i} + \mathbf{j}$ and $\mathbf{j} - \mathbf{k}$
4. (a) $(\sqrt{2}, 1, 1)$ and $(0, 1, 1)$
   (b) $-\frac{3}{2}\mathbf{i} + \mathbf{k}$ and $2\mathbf{i} - 8\mathbf{j} + 3\mathbf{k}$
5. (a) $(1, -\sqrt{5}, 2)$ and $(3, 0, 1)$
   (b) $2\mathbf{i} + \mathbf{j} + 2\mathbf{k}$ and $\mathbf{i} - 2\mathbf{j} + 2\mathbf{k}$
6. (a) $(1, -1, 1, 1)$ and $(\sqrt{3}, -\sqrt{3}, \sqrt{5}, -\sqrt{5})$ in $\mathbf{R}^4$
   (b) $(-1, 2, 1, 1)$ and $(2, 0, 1, 1)$ in $\mathbf{R}^4$

In Exercises 7–10, find the direction vector and direction angles of the given vectors.

7. (a) $(1, -1, 0)$ (b) $\sqrt{3}\mathbf{i} + \mathbf{k}$
8. (a) $(-4, 3, 5)$ (b) $\mathbf{i} - 2\mathbf{j} + 2\mathbf{k}$
9. (a) $(\sqrt{2}, 1, 1)$ (b) $4\mathbf{i} + 2\mathbf{j} - \sqrt{5}\mathbf{k}$
10. (a) $(1, 2, -1, \sqrt{3})$ (b) $(-3, 2, 1, -\sqrt{2})$

In Exercises 11–16, find a unit vector orthogonal to the given vector v. Is there only one such unit vector? If not, then how many are there?

11. $\mathbf{v} = 2\mathbf{i} - 3\mathbf{j}$ (in $\mathbf{R}^2$)
12. $\mathbf{v} = \mathbf{i} + \mathbf{j}$ (in $\mathbf{R}^2$)
13. $\mathbf{v} = (-1, 3, 2)$
14. $\mathbf{v} = (2, 5, -1)$
15. $\mathbf{v} = (1, 2, 1, 3)$
16. $\mathbf{v} = (-1, 3, -1, 1)$

In Exercises 17–20, find *x* so that the two given vectors are orthogonal. Is there a value of *x* so that the two vectors are parallel?

17. $2\mathbf{i} + x\mathbf{j} - 3\mathbf{k}$, $-\mathbf{i} + 3\mathbf{j} - 2\mathbf{k}$
18. $x\mathbf{i} - \mathbf{j} + 2\mathbf{k}$, $-2\mathbf{i} + 3\mathbf{j} + \mathbf{k}$
19. $(-2, 1, x)$, $(6/5, -3/5, 1)$
20. $(x, 6, -4)$, $(\frac{1}{4}, -3/2, 1)$

In Exercises 21–26, find the coordinate and component of the first vector x in the direction of the second vector v.

21. $\mathbf{x} = (3, 1)$, $\mathbf{v} = (1, 2)$
22. $\mathbf{x} = (3, 1)$, $\mathbf{v} = (-2, 1)$
23. $\mathbf{x} = 2\mathbf{i} - 3\mathbf{j} + 4\mathbf{k}$, $\mathbf{v} = \mathbf{i} + \mathbf{j} + 2\mathbf{k}$
24. $\mathbf{x} = 3\mathbf{i} + \mathbf{j} - 2\mathbf{k}$, $\mathbf{v} = -3\mathbf{i} + 6\mathbf{j} + 2\mathbf{k}$
25. $\mathbf{x} = (1, 2, 1, 1)$, $\mathbf{v} = (-1, 2, 3, \sqrt{2})$
26. $\mathbf{x} = (-1, 0, 1, 2)$, $\mathbf{v} = (1, -5, 2, \sqrt{6})$
27. In physics, the ***work*** done by a constant force $\mathbf{F}$ in moving a particle along a straight line segment is defined as the product of the distance moved times the coordinate of the force in the direction of the motion. A constant force $\mathbf{F} = \mathbf{i} + 4\mathbf{j}$ acts on a particle as it moves from the origin to the point (3, 3). Find the work done. See **Figure 3.10.**

FIGURE 3.10

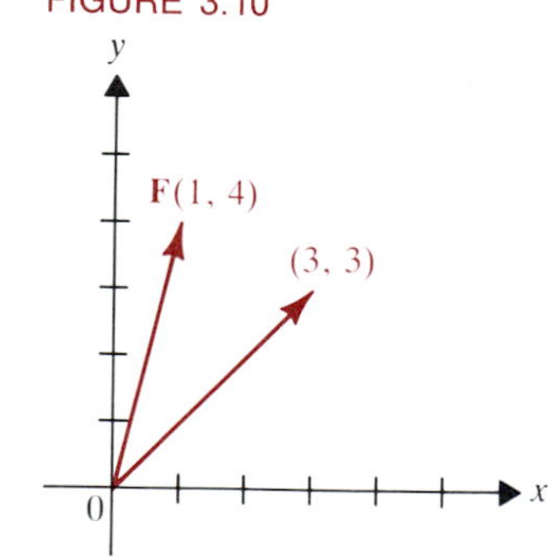

28. Find the work done by $\mathbf{F} = 3\mathbf{i} - \mathbf{j}$ in moving a particle from the origin to $(1, -2)$.
29. Under what circumstances does equality hold in the Cauchy–Schwarz inequality? Explain your answer thoroughly.
30. Under what circumstances can equality hold in the triangle inequality? Explain your answer.
31. (a) If $\cos\alpha_1, \cos\alpha_2, \ldots, \cos\alpha_n$ are the direction cosines of $\mathbf{v}$, then show that
    $\cos^2\alpha_1 + \cos^2\alpha_2 + \cos^2\alpha_3 + \ldots + \cos^2\alpha_n = 1.$
    (b) Specialize to $\mathbf{R}^2$ and show how part (a) gives the trigonometric identity $\sin^2\theta + \cos^2\theta = 1$.

32. **(a)** If $\{\mathbf{e}_1, \mathbf{e}_2, \ldots, \mathbf{e}_n\}$ is the standard basis of $\mathbf{R}^n$, then show that $\mathbf{e}_i \cdot \mathbf{e}_j = 0$, when $i \neq j$ and $\mathbf{e}_i \cdot \mathbf{e}_i = 1$, for all $i = 1, 2, \ldots, n$.
(b) Prove parts (a) and (b) of Theorem 3.2.
(c) Show that for any nonzero vector $\mathbf{v}$ in $\mathbf{R}^n$, $\mathbf{v} = |\mathbf{v}|(\cos\alpha_1, \cos\alpha_2, \ldots, \cos\alpha_n)$.

33. Prove the *Law of Cosines* (a generalization of the Pythagorean theorem):
$$|\mathbf{x} - \mathbf{y}|^2 = |\mathbf{x}|^2 + |\mathbf{y}|^2 - 2|\mathbf{x}|\,|\mathbf{y}|\cos\theta,$$
where θ is the angle between $\mathbf{x}$ and $\mathbf{y}$. Refer to **Figure 3.11.**

FIGURE 3.11

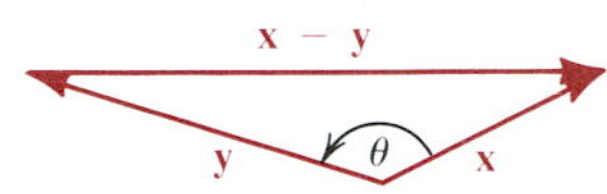

34. Use vectors to prove the *parallelogram identity:* The sum of the squares of the lengths of the diagonals of a parallelogram is the same as the sum of the squares of the lengths of the sides. (*Hint:* Consider the parallelogram formed by $\mathbf{x}$ and $\mathbf{y}$.)

35. Prove the *polarization identity:*
$$|\mathbf{x} + \mathbf{y}|^2 - |\mathbf{x} - \mathbf{y}|^2 = 4\mathbf{x} \cdot \mathbf{y}.$$

36. **(a)** If $\mathbf{x}$ and $\mathbf{y}$ are orthogonal, then show that $\mathbf{x} + \mathbf{y}$ and $\mathbf{x} - \mathbf{y}$ have the same length.
(b) If $\mathbf{x}$ and $\mathbf{y}$ have the same length, then show that $\mathbf{x} + \mathbf{y}$ and $\mathbf{x} - \mathbf{y}$ are orthogonal.
(c) Use part (b) to prove that an angle inscribed in a semicircle is a right angle. See **Figure 3.12.**

FIGURE 3.12

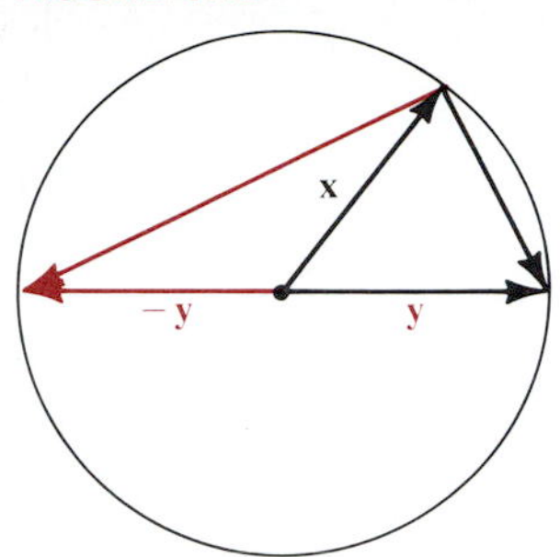

37. Use vectors to show that the line segment connecting the midpoints of two sides of a triangle is parallel to the third side and has length half the length of the third side.

38. If $\mathbf{x}$ is any vector and $\mathbf{u}$ is a unit vector (any direction), then show that $\mathbf{x} - (\mathbf{x} \cdot \mathbf{u})\mathbf{u}$ is perpendicular to $\mathbf{u}$. Use this and the component of $\mathbf{x}$ in the direction of $\mathbf{u}$ to resolve $\mathbf{x}$ into two vectors, one of which is perpendicular to $\mathbf{u}$ and the other of which is parallel to $\mathbf{u}$.

39. In physics, the *power* P (measured in watts) expended by a force $\mathbf{F}$ to maintain a velocity $\mathbf{v}$ is given by $P = \mathbf{F} \cdot \mathbf{v}$. Find the power expended by a force $\mathbf{F} = -2\mathbf{i} + 3\mathbf{j} + \mathbf{k}$ to maintain the velocity $\mathbf{v} = 2\mathbf{i} + 3\mathbf{j} + 2\mathbf{k}$.

40. Find the power at $t = \pi/2$ if $\mathbf{F} = 2\mathbf{i} - \mathbf{j} + 3\mathbf{k}$ and $\mathbf{v} = (\cos t)\mathbf{i} - (\sin t)\mathbf{j} + t\mathbf{k}$ at time t.

41. Show that the plane vector $\mathbf{v} = a\mathbf{i} + b\mathbf{j}$ is perpendicular to the line $ax + by + c = 0$ in $\mathbf{R}^2$. (*Hint:* Consider the vector $\mathbf{AB}$ for any two points A and B on the line and use (11).)

10.4 Lines in Space

The simplest curves in the plane are straight lines. Their equations are linear and can be put in the form $ax + by = c$. In three dimensions we still have a clear picture of what a line is, and in this section we give vector and scalar equations for lines in $\mathbf{R}^3$.

We begin by looking at lines in the plane from the vector point of view. Let l be a line in the plane through two points $P_0(x_0, y_0)$ and $P_1(x_1, y_1)$ (**Figure 4.1**). Let $P(x, y)$ be an arbitrary point on l. If we let $\mathbf{x} = \mathbf{0P}$, then
$$\mathbf{P}_0\mathbf{P}_1 = (x_1 - x_0, y_1 - y_0)$$
is a vector in the direction of the line. We can get to P by first going from 0 to P_0 and then adding a suitable multiple of $\mathbf{P}_0\mathbf{P}_1$ to $\mathbf{0P}_0$, for we will then be moving along l from P_0. In Figure 4.1 it appears that

$$\mathbf{x} = \mathbf{0P} = \mathbf{0P}_0 + t\mathbf{P}_0\mathbf{P}_1, \tag{1}$$

where t is approximately 1/2. Letting $v = \mathbf{P}_0\mathbf{P}_1$ and $x_0 = \mathbf{0P}_0$, we then have

$$\mathbf{x} = \mathbf{x}_0 + t\mathbf{v}. \tag{2}$$

FIGURE 4.1

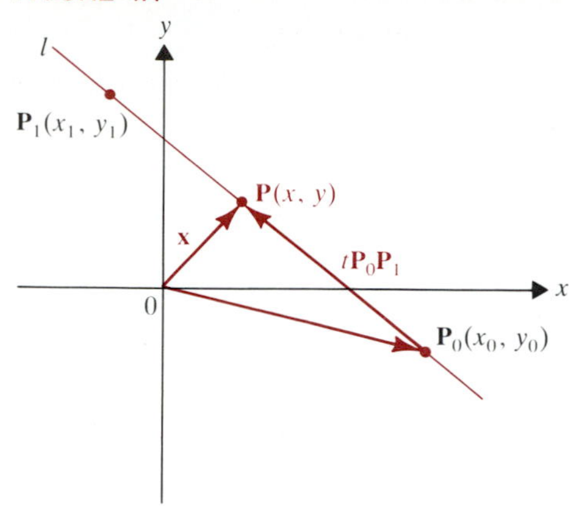

FIGURE 4.2

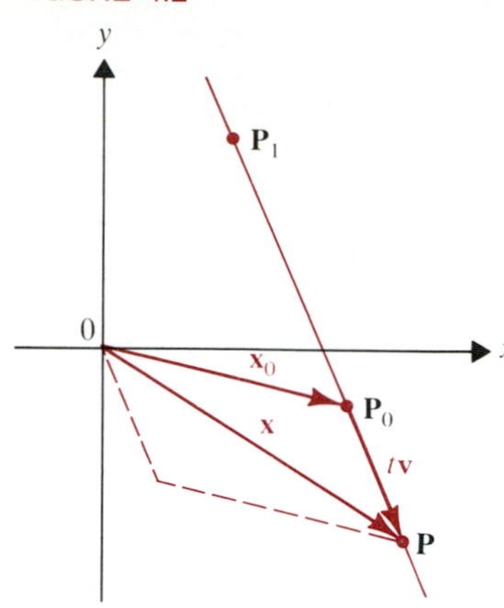

Thus, for any point P on l, $\mathbf{x} = \mathbf{0P}$ is given by (2) for some real number t. Conversely, if P is the endpoint of any vector $\mathbf{x}$ of the form (2), then P is on l. For, as **Figure 4.2** shows, P is reached by proceeding from the origin to the given point P_0 on the line and then traveling along the line to P. That is, $\mathbf{x}$ is the resultant of $\mathbf{x}_0$ and $t\mathbf{v}$. Thus (2) holds *if and only if* the endpoint P of $\mathbf{x}$ lies on the line l. It is then natural to call (2) a vector equation of l.

4.1 DEFINITION

A ***vector equation of the line l through two points P_0 and P_1*** is

$$\mathbf{x} = t\mathbf{v} + \mathbf{x}_0, \tag{2}$$

where $\mathbf{v} = \mathbf{P}_0\mathbf{P}_1$, $\mathbf{x}_0 = \mathbf{0P}_0$, and t is a real variable called a ***parameter.*** The vector $\mathbf{v}$ is called a ***vector in the direction of l*** (or a ***direction vector of l***).

Note that in (2), t is the *only* variable. Both $\mathbf{x}_0$ and $\mathbf{v}$ are *fixed* vectors. Equation (2) can be thought of as a vector analogue of the scalar equation

$$y = mx + b,$$

in which m and b are fixed in $\mathbf{R}^1$, and x is a real variable.

If we put (2) in coordinate form, then we have

$$(x, y) = t(x_1 - x_0, y_1 - y_0) + (x_0, y_0) = (t(x_1 - x_0) + x_0, t(y_1 - y_0) + y_0). \tag{3}$$

From (3) we can obtain scalar equations of l.

4.2 DEFINITION

If $P_0(x_0, y_0)$ and $P_1(x_1, y_1)$ are two points on l, then l has ***parametric (scalar) equations***

$$x = t(x_1 - x_0) + x_0 \quad \text{and} \quad y = t(y_1 - y_0) + y_0. \tag{4}$$

In neither Definition 4.1 nor Definition 4.2 have we suggested that a line has *unique* vector or parametric scalar equations. The reason is that Equations (2) and (4) are *not* uniquely determined by l, as the following example illustrates.

FIGURE 4.3

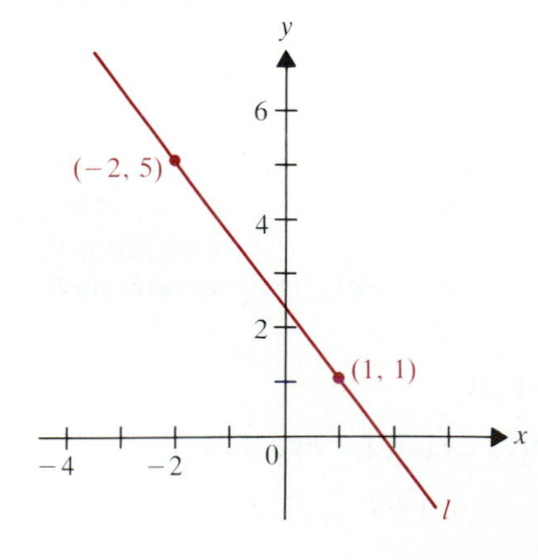

Example 1

Find vector and parametric equations of the line l through $(-2, 5)$ and $(1, 1)$, using (a) $P_0 = (1, 1)$ and (b) $P_0 = (-2, 5)$. Reconcile the two answers.

Solution. Refer to **Figure 4.3.** A vector in the direction of l is

$$\mathbf{v} = (-2 - 1, 5 - 1) = (-3, 4) = -3\mathbf{i} + 4\mathbf{j}.$$

(a) $\mathbf{x} = t(-3, 4) + (1, 1)$ is a vector equation for l, since $\mathbf{0P}_0 = (1, 1)$. The parametric equations are

$$(x, y) = (-3t, 4t) + (1, 1) = (-3t + 1, 4t + 1),$$

$$x = -3t + 1, \qquad y = 4t + 1. \tag{5}$$

(b) This time $\mathbf{x} = s(-3, 4) + (-2, 5)$ is a vector equation for l. Combining the vectors, we get

$$(x, y) = (-3s, 4s) + (-2, 5) = (-3s - 2, 4s + 5),$$

$$x = -3s - 2, \qquad y = 4s + 5. \tag{6}$$

Equations (5) and (6) *appear* different. But what has *really* changed is the parameter. For if we eliminate t and s, we get in each case the familiar scalar equation of l. In (5), solving $x = -3t + 1$ for t, we have

$$3t = -x + 1 \rightarrow t = -\frac{1}{3}x + \frac{1}{3}.$$

Substituting this into $y = 4t + 1$, we get

$$y = 4\left(-\frac{1}{3}x + \frac{1}{3}\right) + 1 = -\frac{4}{3}x + \frac{7}{3}.$$

In (6), solving for s in $x = -3s - 2$, we find

$$3s = -x - 2 \rightarrow s = -\frac{1}{3}x - \frac{2}{3}.$$

Substituting this into $y = 4s + 5$, we again obtain

$$y = 4\left(-\frac{1}{3}x - \frac{2}{3}\right) + 5 = -\frac{4}{3}x + \frac{7}{3}.$$

We can express s in terms of t by equating the two expressions for x (or y):

$$x = -3t + 1 = -3s - 2, \qquad y = 4t + 1 = 4s + 5,$$

$$-3s = -3t + 3, \qquad 4s = 4t - 4,$$

$$s = t - 1. \qquad s = t - 1. \quad ■$$

Example 1 suggests why lines in the plane are often studied via the scalar equation $y = mx + b$. That *is* uniquely determined for each nonvertical line l. In checking answers, it is often simplest just to verify that the coordinates of the given points P_0 and P satisfy the vector (or parametric) equation(s).

A close look at Definition 4.1 reveals that there is no explicit mention of the number of coordinates of the points. We can therefore use Definition 4.1 to define a vector equation of the line through two points P_0 and P in $\mathbf{R}^3$ or in $\mathbf{R}^n$ for any $n \geq 2$.

Given two points $P_0(x_0, y_0, z_0)$ and $P_1(x_1, y_1, z_1)$ in $\mathbf{R}^3$, then just as in the plane,

$$v = \mathbf{P}_0\mathbf{P}_1 = (x_1 - x_0, y_1 - y_0, z_1 - z_0)$$

is a vector in this line. Then Definition 4.1 gives the vector equation

$$\mathbf{x} = t\mathbf{v} + \mathbf{x}_0 = t(x_1 - x_0, y_1 - y_0, z_1 - z_0) + (x_0, y_0, z_0).$$

So the parametric equations are

(7) $x = t(x_1 - x_0) + x_0, \quad y = t(y_1 - y_0) + y_0, \quad z = t(z_1 - z_0) + z_0.$

In Example 1, elimination of the parameter t from the two parametric scalar equations of a line l in $\mathbf{R}^2$ gave a single scalar equation for l. In $\mathbf{R}^3$, elimination of the parameter t leads to 2 (= 3 − 1) scalar equations. If $x_1 - x_0$, $y_1 - y_0$, and $z_1 - z_0$ are all nonzero, then from (7) we obtain

$$x - x_0 = t(x_1 - x_0), \qquad y - y_0 = t(y_1 - y_0), \qquad z - z_0 = t(z_1 - z_0),$$

$$\frac{x - x_0}{x_1 - x_0} = \frac{y - y_0}{y_1 - y_0} = \frac{z - z_0}{z_1 - z_0} = t.$$

4.3 DEFINITION The ***symmetric scalar equations*** of a line in $\mathbf{R}^3$ through two points $P_0(x_0, y_0, z_0)$ and $P_1(x_1, y_1, z_1)$ are

(8) $$\frac{x - x_0}{x_1 - x_0} = \frac{y - y_0}{y_1 - y_0} = \frac{z - z_0}{z_1 - z_0}$$

if $x_1 - x_0$, $y_1 - y_0$, and $z_1 - z_0$ are all nonzero.

If one or two of the numbers $x_1 - x_0$, $y_1 - y_0$, or $z_1 - z_0$ in (8) are zero, then the line is parallel to one or two of the coordinate planes (see **Figure 4.4**). (If all three are zero, then P_0 and P_1 are the same point.) There is no standard set of symmetric scalar equations in those cases; instead, the parametric scalar equations (for instance, $x = x_0$, $y = y_0$, and $z = t(z_1 - z_0) + z_0$) are used.

FIGURE 4.4

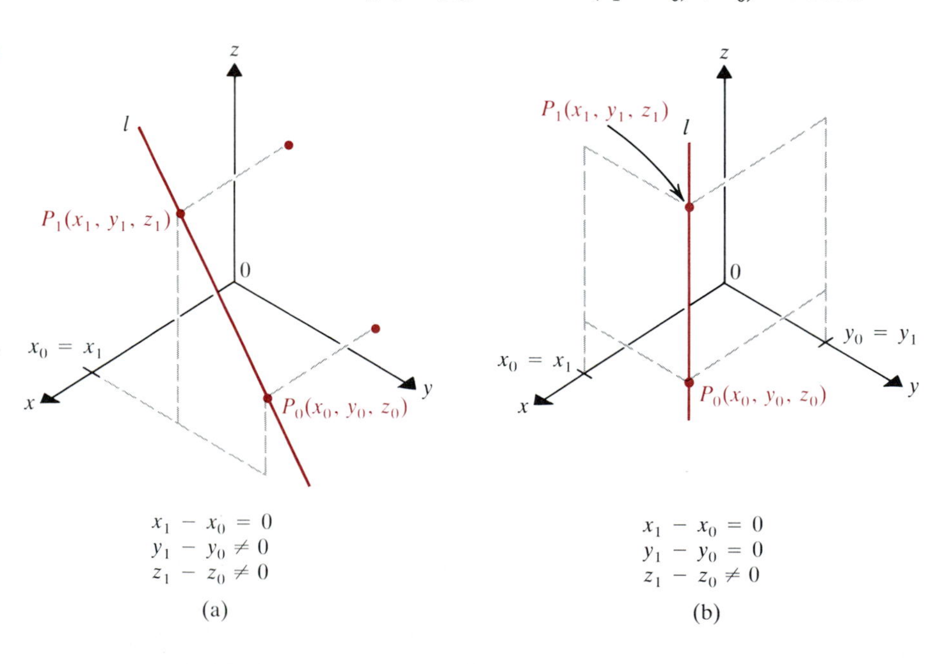

In $\mathbf{R}^n$, if the parameter t is eliminated from the vector equation $\mathbf{x} = t\mathbf{v} + \mathbf{x_0}$, then we are left with $n - 1$ scalar equations, which assert the equality of

$$\frac{x - x_0}{x_1 - x_0}, \quad \frac{y - y_0}{y_1 - y_0}, \quad \frac{z - z_0}{z_1 - z_0}, \ldots.$$

In particular in $\mathbf{R}^2$, *and only there*, we get exactly one scalar equation

$$\frac{y - y_0}{y_1 - y_0} = \frac{x - x_0}{x_1 - x_0},$$

FIGURE 4.5

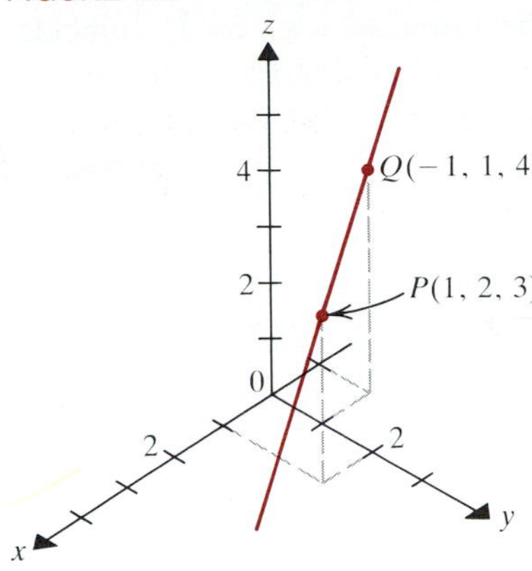

which is equivalent to the *point–slope* equation

$$y - y_0 = \frac{y_1 - y_0}{x_1 - x_0}(x - x_0) = m(x - x_0).$$

Take another look now at Example 1 in light of these ideas.

Example 2

Find vector and scalar equations of the line through the two points $P(1, 2, 3)$ and $Q(-1, 1, 4)$. See **Figure 4.5.**

Solution. Here $\mathbf{v} = \mathbf{PQ} = (-1 - 1, 1 - 2, 4 - 3) = (-2, -1, 1)$. So a vector equation for the line is

$$\mathbf{x} = t(-2, -1, 1) + (1, 2, 3).$$

To get the scalar equations, we simply combine the terms on the right side of the vector equation. That gives

$$\mathbf{x} = (x, y, z) = (-2t, -t, t) + (1, 2, 3) = (-2t + 1, -t + 2, t + 3).$$

Thus the parametric scalar equations are

$$x = -2t + 1, \qquad y = -t + 2, \qquad z = t + 3.$$

The symmetric scalar equations are obtained from these:

$$x - 1 = -2t, \quad y - 2 = -t, \quad z - 3 = t \rightarrow \frac{(x-1)}{-2} = t, \quad \frac{(y-2)}{-1} = t, \quad \frac{(z-3)}{1} = t.$$

So

$$\frac{x-1}{-2} = \frac{y-2}{-1} = \frac{z-3}{1}. \quad ■$$

FIGURE 4.6

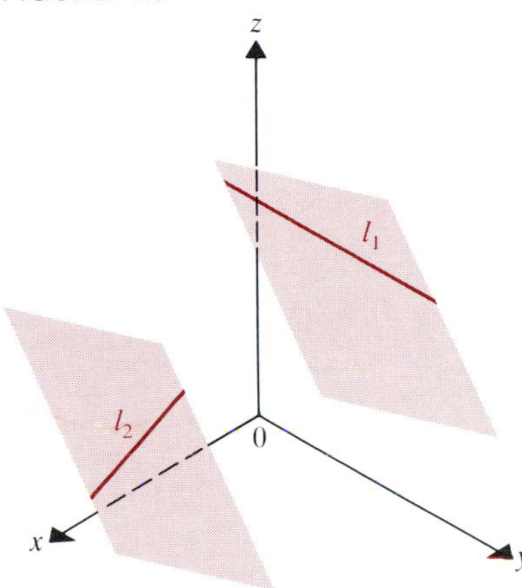

l_1: $\mathbf{x} = s\mathbf{v} + \mathbf{x}_1$
l_2: $\mathbf{x} = t\mathbf{w} + \mathbf{x}_2$

In the plane, any two given lines are either parallel or intersecting. In three-dimensional space, there is a third possibility. **Figure 4.6** shows two lines l_1 and l_2 that lie in parallel planes, but are not parallel because their direction vectors $\mathbf{v}$ and $\mathbf{w}$ are not parallel. The next definition formally defines parallel and skew lines, as well as perpendicular lines.

4.4 DEFINITION

Let two lines l_1 and l_2 in $\mathbf{R}^3$ have vector equations $\mathbf{x} = s\mathbf{v} + \mathbf{x}_1$ and $\mathbf{x} = t\mathbf{w} + \mathbf{x}_2$.

(a) The lines are ***parallel*** if and only if the direction vectors $\mathbf{v}$ and $\mathbf{w}$ are parallel, that is, if and only if $\mathbf{v} = a\mathbf{w}$ for some real number a.

(b) The lines are ***perpendicular*** if and only if they intersect *and* $\mathbf{v}$ and $\mathbf{w}$ are perpendicular vectors, that is, $\mathbf{v} \cdot \mathbf{w} = 0$.

(c) The lines are ***skew*** if and only if they do not intersect and the vectors $\mathbf{v}$ and $\mathbf{w}$ are not parallel.

To apply Definition 4.4(b) or (c), it is necessary to determine whether two given lines in $\mathbf{R}^3$,

$$l_1: \mathbf{x} = s\mathbf{v} + \mathbf{x}_1 \qquad \text{and} \qquad l_2: \mathbf{x} = t\mathbf{w} + \mathbf{x}_2$$

intersect. This can be done algebraically by equating the expressions

$$s\mathbf{v} + \mathbf{x}_1 = t\mathbf{w} + \mathbf{x}_2 \tag{9}$$

for $\mathbf{x}$ and trying to solve (9) for s and t. The lines intersect if and only if there are particular values of s and t such that the point $s\mathbf{v} + \mathbf{x}_1$ on l_1 coincides with the point $t\mathbf{w} + \mathbf{x}_2$ on l_2. If $\mathbf{v} = (v_1, v_2, v_3)$, $\mathbf{w} = (w_1, w_2, w_3)$, $\mathbf{x}_1 = (a_1, b_1, c_1)$, and $\mathbf{x}_2 = (a_2, b_2, c_2)$, then equating coordinates in (9) leads to the system of equations

$$\begin{cases} sv_1 + a_1 = tw_1 + a_2 \\ sv_2 + b_1 = tw_2 + b_2 \\ sv_3 + c_1 = tw_3 + c_2, \end{cases}$$

in the unknowns s and t. Usually, two of these equations can be solved to find a value for s and a value for t. If those values also satisfy the third equation, then the point $\mathbf{x} = s\mathbf{v} + \mathbf{x}_1 = t\mathbf{w} + \mathbf{x}_2$ is the point of intersection (and we call the system of equations *consistent*). However, if those values of s and t do not satisfy the third equation, then there is no point of intersection (and we call the system of equations *inconsistent*).

Example 3

Find the point of intersection, if any, of each pair of lines.

(a) l_1: $\mathbf{x} = s(1, 1, 0) + (2, 0, 4)$ and l_2: $\mathbf{x} = t(1, -1, 3) + (0, 0, 1)$

(b) l_1: $\mathbf{x} = s(\mathbf{i} + \mathbf{j})$ and l_2: $\mathbf{x} = t(\mathbf{i} - \mathbf{j} + 3\mathbf{k}) + \mathbf{k}$

Solution. (a) If there is an intersection point, then there must be values s and t of the parameters such that

$$s(1, 1, 0) + (2, 0, 4) = t(1, -1, 3) + (0, 0, 1),$$

that is, so that

$$(s + 2, s, 4) = (t, -t, 3t + 1).$$

Equating the corresponding coordinates, we get the system

$$\begin{cases} s + 2 = t \\ s \quad\;\; = -t \\ \quad\;\; 4 = 3t + 1. \end{cases}$$

The third equation says $t = 1$, and substitution in the second equation then gives $s = -1$. These values do satisfy the first equation, so the system is consistent. From either vector equation we find the point of intersection $\mathbf{x} = (1, -1, 4)$. (See **Figure 4.7**(a).)

(b) This time the vector equation is

$$s(\mathbf{i} + \mathbf{j}) = t(\mathbf{i} - \mathbf{j} + 3\mathbf{k}) + \mathbf{k},$$

that is, $s\mathbf{i} + s\mathbf{j} = t\mathbf{i} - t\mathbf{j} + (3t + 1)\mathbf{k}$. Thus, equating coefficients of $\mathbf{i}$, $\mathbf{j}$, and $\mathbf{k}$ gives

$$\begin{cases} s = t \\ s = -t \\ 0 = 3t + 1. \end{cases}$$

Adding the first two equations, we get $s = 0$. Thus $t = 0$ also. However, substitution into the third equation gives the absurdity $0 = 1$, so the system is inconsistent (it has no solution). Therefore there is no point of intersection. The lines are skew, since the direction vectors $\mathbf{i} + \mathbf{j}$ and $\mathbf{i} - \mathbf{j} + 3\mathbf{k}$ are not parallel: For no real number a is $\mathbf{i} + \mathbf{j} = a(\mathbf{i} - \mathbf{j} + 3\mathbf{k})$. ■

FIGURE 4.7

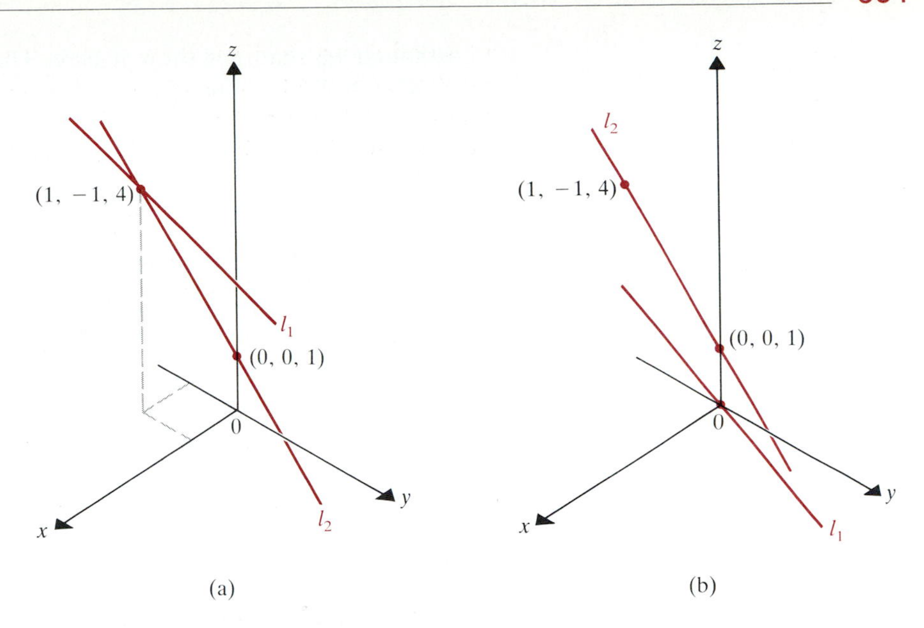

Comparison of the graphs of the lines in Figure 4.7(b) with those in Figure 4.7(a) shows that changing the constant vector in l_1 from $(2, 0, 4)$ to $\mathbf{0}$ in effect translated l_1 parallel to itself so that it passed through the origin. In that position, it cannot intersect l_2.

Since the vectors $\mathbf{v} = (1, 1, 0)$ and $\mathbf{w} = (1, -1, 3)$ are perpendicular (note that $\mathbf{v} \cdot \mathbf{w} = 0$), *and* the lines $\mathbf{x} = s(1, 1, 0) + (2, 0, 4)$ and $\mathbf{x} = t(1, -1, 3) + (0, 0, 1)$ intersect [see Example 3(a)], the two lines are perpendicular. The fact that the lines in Example 3(b) have perpendicular direction vectors $\mathbf{v} = \mathbf{i} + \mathbf{j}$ and $\mathbf{w} = \mathbf{i} - \mathbf{j} + 3\mathbf{k}$ *isn't enough* to make the lines perpendicular, because the lines fail to intersect.

We close the section with an often useful vector representation of the line segment joining two points.

4.5 THEOREM If $P(x_1, x_2, x_3)$ and $Q(y_1, y_2, y_3)$ are two points in $\mathbf{R}^3$, then the line segment joining P to Q consists of the endpoints of all the vectors $(1 - t)\mathbf{x}_0 + t\mathbf{y}_0$ for $0 \leq t \leq 1$, where $\mathbf{x}_0 = \mathbf{0P}$ and $\mathbf{y}_0 = \mathbf{0Q}$.

Proof. The line through P and Q has equation

$$\mathbf{x} = \mathbf{x}_0 + t(\mathbf{y}_0 - \mathbf{x}_0) = \mathbf{x}_0 + t\mathbf{y}_0 - t\mathbf{x}_0 = (1 - t)\mathbf{x}_0 + t\mathbf{y}_0.$$

When $t = 0$, we get $\mathbf{x}_0$; when $t = 1$, we get $\mathbf{y}_0$. For $0 < t < 1$, we get $x_0 + t(\mathbf{y}_0 - \mathbf{x}_0)$, a vector whose endpoint is on the segment from P to Q, since we are adding to $\mathbf{x}_0$ a fractional part of the vector $\mathbf{y}_0 - \mathbf{x}_0$, which joins P to Q. See **Figure 4.8.** QED

FIGURE 4.8

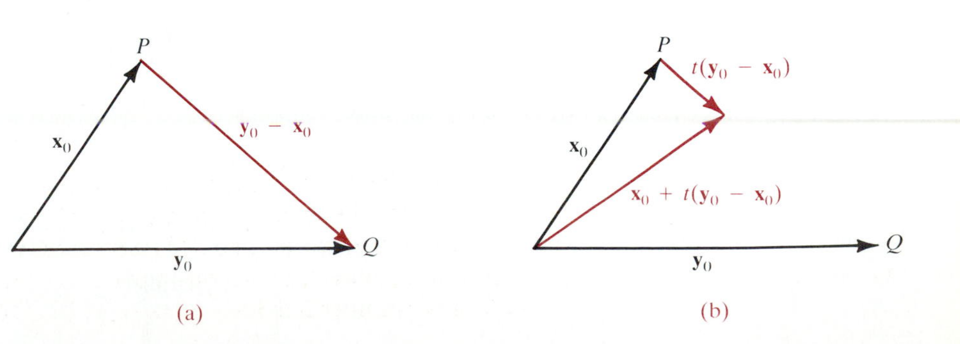

Although we shall not show it here, Theorem 4.5 holds for *any* Euclidean space $\mathbf{R}^n$, $n \geq 2$.

Example 4

Find the vector equation for the segment joining $P(1, -1, 2)$ to $Q(0, 1, -3)$.

Solution. Since $\mathbf{x}_0 = (1 - 1, 2)$ and $\mathbf{y}_0 = (0, 1, -3)$, by Theorem 4.5 the segment has vector equation

$$\begin{aligned}\mathbf{x} &= (1 - t)(1, -1, 2) + t(0, 1, -3)\\ &= (1 - t, -1 + 2t, 2 - 5t), \qquad \text{where } t \in [0, 1]. \quad \blacksquare\end{aligned}$$

Exercises 10.4

In Exercises 1–10, find a vector equation for the given line.

1. **(a)** The line through (2, 1) and (−1, 5)
   (b) The y-axis in $\mathbf{R}^2$
   (c) The line through (−7, −2) and (4, 4)
2. **(a)** The line through (1, −2) in the direction of (1, −2)
   (b) The line through (0, 0) in the direction of $\mathbf{i} + \mathbf{j}$
   (c) The x-axis in $\mathbf{R}^2$
3. **(a)** The line through (1, 2, 3) and (−1, −2, −3)
   (b) The line through (0, 1, −2) and (3, 1, −2)
   (c) The y-axis in $\mathbf{R}^3$
4. **(a)** The line through (1, −1, 4) and (7, −1, −1)
   (b) The line through (2, 0, −1) and (−1, 2, 0)
   (c) The x-axis in $\mathbf{R}^3$
5. **(a)** The line through (0, 0, 0) in the direction of (1, 2, 0)
   (b) The line through (1, 2, −1) in the direction of $\mathbf{i} + 2\mathbf{j} + \sqrt{2}\mathbf{k}$
6. **(a)** The line through (1, −3, 2) in the direction of (1/3, −2/3, 2/3)
   (b) The line through (0, 1, −1) in the direction of $(1/4, 3/4, \sqrt{6}/4)$
7. **(a)** The line through (2, −1, 3) and (−1, 1, 2)
   (b) The line through (2, −1, 3) and (−1, 2, 0)
   (c) The line through (1, 0, 0) and (0, 0, 1)
8. **(a)** The line through (−1, 0, −3) and (−3, −2, 1)
   (b) The line through (0, 0, 0) and (−4, 3, 2)
   (c) The line through (0, 1, 0) and (0, 0, −1)
9. **(a)** The line through (1, 2, 1, −1) and (−1, 3, 1, −2) in $\mathbf{R}^4$
   (b) The line through (−2, −1, 0, −1) in the direction of (1/2, −1/2, −1/2, 1/2)
10. **(a)** The line through the axis containing $\mathbf{e}_2$ in $\mathbf{R}^4$
    (b) The line through (0, 0, 1, 0) and (1, 0, 0, 0) in $\mathbf{R}^4$

In Exercises 11–16, find a vector equation, a symmetric scalar equation, and parametric scalar equations for the given line.

11. The line through (1, 2, 3) and (−1, −3, 5)
12. The line through (1, 4, −2) and (3, 4, 5)
13. The line through (−1, 5, 4) in the direction of $\mathbf{v} = \mathbf{i} - \mathbf{j} - 2\mathbf{k}$
14. The line through (2, 5, 4) in the direction of $\mathbf{v} = -\mathbf{i} + \mathbf{j} + 3\mathbf{k}$
15. The line through (−1, 3) and (2, 8)
16. The line through (−1, 3, 0, 2) and (2, −3, −1, 3) in $\mathbf{R}^4$
17. Find parametric equations for the line through (1, 1, 2) that is parallel to the line
$$\frac{x-1}{3} = \frac{y}{-1} = \frac{2z-3}{4}.$$
18. Find parametric scalar equations for the line through (−3, 1, −1) that is parallel to the line
$$\frac{x+1}{-1} = \frac{y-3}{5} = \frac{z}{-5}.$$
19. Find a vector equation for the line through (−1, 3, 2) that is parallel to the line $x = t$, $y = -1$, and $z = 2 + t$.
20. Find symmetric scalar equations for the line through the origin that is parallel to the line through (−1, 3, 1) and (2, −7, 3).

In Exercises 21 and 22, determine whether the two given lines intersect. If they do, then find the point of intersection.

21. The line $x = 1 + 3t$, $y = -2 - t$, $z = 2 + 2t$ and the line $\dfrac{x+1}{-1} = \dfrac{y-3}{2} = \dfrac{z-5}{-1}$.
22. The line $x = (2, -3, 1) + t(1, 2, -3)$ and the line $x = 1 - 2s$, $y = 2 + 3s$, $z = 4 + 6s$.
23. Are the lines $x = -2 + t$, $y = 2t$, $z = -3 - t$ and $x = 5 - \frac{1}{2}s$, $y = -1 - s$, $z = 7 + \frac{1}{2}s$ parallel?
24. Are the lines $x = 2 - t$, $y = 4 + 3t$, $z = \frac{1}{2}t$ and $x = 3s$, $y = -3 - s$, $z = -\frac{1}{6}s$ parallel?
25. Are the lines
$$\frac{x+3}{2} = \frac{y-1}{2} = \frac{z+3}{-3} \quad \text{and} \quad \frac{x-3}{3} = \frac{y}{-3} = \frac{2z-2}{9}$$
parallel?

26. Are the lines

$$\frac{x-1}{-1} = \frac{y}{4} = \frac{z+3}{2}$$

and

$$\frac{x-1}{2} = \frac{y+3}{-8} = \frac{z}{-4}$$

parallel?

27. Show that the lines $\mathbf{x} = (0, -2, 0) + t(1, -2, -3)$ and $\mathbf{x} = (0, 1, 0) + s(-1, -2, 1)$ are not perpendicular, even though their directions are.

28. Repeat Exercise 27 for $\mathbf{x} = (-1, 3, 2) + t(1, 3, -2)$ and $\mathbf{x} = (1, 3, 0) + s(-2, 1, 1/2)$.

29. Is the line

$$x = -1 + s,\ y = 2 + 2s,\ z = \frac{26}{9} + \frac{3}{2}s$$

perpendicular to the line l through $(-1, 2, 1)$ and $(2, -1, 3)$?

30. Is the line

$$x = -1 + 2s,\ y = 2 + s,\ z = -\frac{8}{9} - \frac{3}{2}s$$

perpendicular to the line l in Exercise 29?

31. William Tell was imprisoned for refusing to salute the Duke of Austria's hat. To escape death he promised to shoot an apple from his son's head with a bow and arrow. Suppose you find yourself in a predicament like young Tell's. Knowing that an arrow shot from a nearby bow travels approximately a straight line, and that the top of your head is exactly 6 feet above the floor when you stand perfectly straight, you ask that the coordinates of the tip of Tell's arrow be displayed **(a)** as it leaves the bow and **(b)** as it reaches a point a quarter of the way toward you.
If your feet are at $(0, 0, 0)$, and if you see that $A = (100, 50, 5)$ and that $B = (75, 37.5, 5.2)$, then should you slouch down to make sure the arrow gets the apple and not your head?

32. (True–False) For each part of this question, if the assertion is true, explain why it is true. If it is false, give an example that shows it to be false.

(a) For any three vectors $\mathbf{u}$, $\mathbf{v}$, and $\mathbf{w}$ in $\mathbf{R}^2$, if $\mathbf{u}$ is perpendicular to $\mathbf{v}$ and $\mathbf{v}$ is perpendicular to $\mathbf{w}$, then $\mathbf{u}$ must be perpendicular to $\mathbf{w}$.

(b) For any three vectors $\mathbf{u}$, $\mathbf{v}$, and $\mathbf{w}$ in $\mathbf{R}^3$, if $\mathbf{u}$ is perpendicular to $\mathbf{v}$ and $\mathbf{v}$ is, in turn, perpendicular to $\mathbf{w}$, then $\mathbf{u}$ must be perpendicular to $\mathbf{w}$.

10.5 Planes

FIGURE 5.1

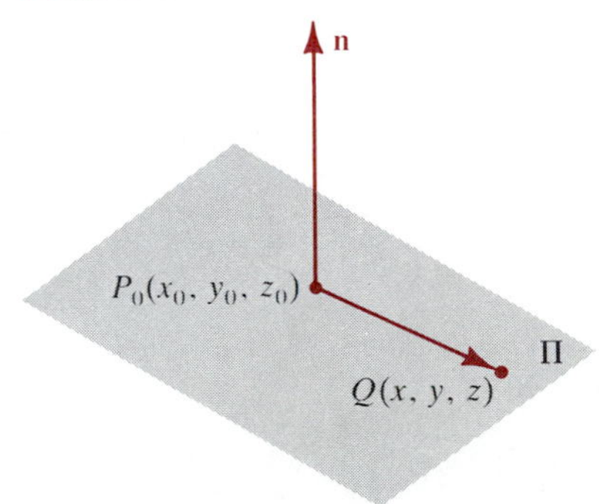

Planes in $\mathbf{R}^3$ play a similar role to that of lines in $\mathbf{R}^2$: Like lines in $\mathbf{R}^2$, their dimension is one less than the dimension of the whole space (3 for $\mathbf{R}^3$). We can think of a plane in $\mathbf{R}^3$ as a copy of $\mathbf{R}^2$ with the coordinate axes erased. If P and Q are two points in a given plane Π, then we say that the vector $\mathbf{PQ}$ is a *vector in the plane* Π. A vector $\mathbf{n}$ in $\mathbf{R}^3$ is called a ***normal vector*** to Π if for any pair of points P and Q in Π, $\mathbf{n}$ is perpendicular to $\mathbf{PQ}$, that is, $\mathbf{n} \cdot \mathbf{PQ} = 0$. Thus a normal vector to a plane is a vector perpendicular to every vector in the plane. See **Figure 5.1.**

Let $\mathbf{n} = (a, b, c) \neq 0$ be a normal vector to a plane Π, and let $P_0(x_0, y_0, z_0)$ be a point in Π. Then for any other point $Q(x, y, z)$ in Π, the vector

$$\mathbf{P_0Q} = (x - x_0, y - y_0, z - z_0)$$

lies in the plane. This leads to a vector equation for Π.

5.1 DEFINITION

The ***vector*** (or ***normal form***) ***equation*** of the plane through $P_0(x_0, y_0, z_0)$ with nonzero normal vector $\mathbf{n} = (a, b, c) = a\mathbf{i} + b\mathbf{j} + c\mathbf{k}$ is

$$\mathbf{n} \cdot (\mathbf{x} - \mathbf{x}_0) = 0, \quad (1)$$

where $\mathbf{x}_0 = \mathbf{0P} = (x_0, y_0, z_0)$ and $\mathbf{x} = \mathbf{0Q} = (x, y, z)$. Here $Q(x, y, z)$ represents an arbitrary point in the plane.

In terms of coordinates, (1) takes the form

$$(a, b, c) \cdot (x - x_0, y - y_0, z - z_0) = 0,$$

$$a(x - x_0) + b(y - y_0) + c(z - z_0) = 0,$$

$$ax + by + cz = d, \tag{2}$$

where $d = ax_0 + by_0 + cz_0$. Thus

> the equation of a plane in $\mathbf{R}^3$ with normal vector $\mathbf{n} = (a, b, c)$ is a linear equation $ax + by + cz = d$, where not all of a, b, and c are 0.

Conversely, any linear equation of the form (2), where not all of a, b, and c are 0, can be written in the form

$$\mathbf{n} \cdot \mathbf{x} = (a, b, c) \cdot (x, y, z) = d. \tag{3}$$

If $P_0(x_0, y_0, z_0)$ is any point of $\mathbf{R}^3$ whose coordinates satisfy this equation, then letting $\mathbf{x}_0 = \mathbf{0P}_0$, we get

$$\mathbf{n} \cdot \mathbf{x}_0 = ax_0 + by_0 + cz_0 = d.$$

Hence (3) becomes

$$(a, b, c) \cdot (x, y, z) = ax_0 + by_0 + cz_0 = (a, b, c) \cdot (x_0, y_0, z_0).$$

Then by Theorem 3.2(b) we have

$$(a, b, c) \cdot [(x, y, z) - (x_0, y_0, z_0)] = 0,$$

$$(a, b, c) \cdot (x - x_0, y - y_0, z - z_0) = 0 \rightarrow \mathbf{n} \cdot (\mathbf{x} - \mathbf{x}_0) = 0.$$

Thus (3) has the form (1), so is the equation of a plane in $\mathbf{R}^3$ with normal (a, b, c). This discussion shows that

> any linear equation $ax + by + cz = d$, where not all of a, b, and c are 0, is the equation of a plane with normal vector (a, b, c).

Example 1

Find vector and scalar equations for the plane through $(-2, 1, 3)$ with normal vector $\mathbf{n} = 3\mathbf{i} + \mathbf{j} + 5\mathbf{k}$. Sketch the plane.

Solution. By Definition 5.1 the vector equation is

$$(3\mathbf{i} + \mathbf{j} + 5\mathbf{k}) \cdot [\mathbf{x} - (-2, 1, 3)] = 0.$$

From this it is easy to get the scalar equation. We have

$$(3, 1, 5) \cdot [(x, y, z) + (2, -1, -3)] = 0 \rightarrow (3, 1, 5) \cdot (x + 2, y - 1, z - 3) = 0,$$

$$3(x + 2) + (y - 1) + 5(z - 3) = 0 \rightarrow 3x + 6 + y - 1 + 5z - 15 = 0,$$

$$3x + y + 5z = 10.$$

To draw the plane, it is helpful to show its ***traces*** in the three coordinate planes. These are the lines in which the plane intersects the xy-, yz-, and xz-planes. To find them, determine the points where the plane intersects each coordinate axis. When $y = 0$ and $z = 0$, we have $x = 10/3$. So $(10/3, 0, 0)$ is on the plane. Similarly, $(0, 10, 0)$ and $(0, 0, 2)$ are on the plane. **Figure 5.2** shows the traces and first octant portion of the plane. ■

FIGURE 5.2

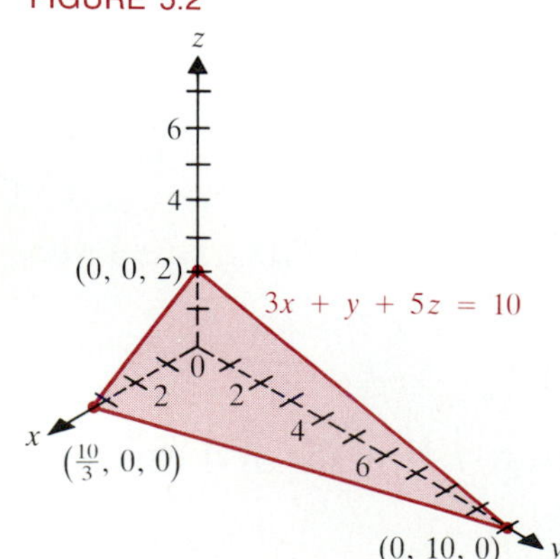

Example 2

Put the equation

$$(4) \qquad 2x - y - z = 5$$

in normal form.

Solution. First, note from (2) that $\mathbf{n} = (2, -1, -1) = 2\mathbf{i} - \mathbf{j} - \mathbf{k}$ is a normal vector to the plane, since the coordinates of $\mathbf{n}$ are the coefficients of x, y, and z in the scalar equation (2). To complete the solution, we need to find a point $P_0(x_0, y_0, z_0)$ whose coordinates satisfy (4). We could proceed by trial and error, but it is more efficient to assign y_0 and z_0 convenient values and then determine a corresponding x_0 by solving (4) for x. Letting $y_0 = 1$ and $z_0 = 0$, for instance, we have

$$2x_0 - 1 - 0 = 5 \rightarrow 2x_0 = 6 \rightarrow x_0 = 3.$$

(We could have chosen $y_0 = 0$ and $z_0 = 0$, or any other convenient values.) Thus (3, 1, 0) is a point on the plane. We can therefore rewrite (4) in the form

$$(2, -1, -1) \cdot (x, y, z) = (2, -1, -1) \cdot (3, 1, 0),$$

$$(2, -1, -1) \cdot [(x, y, z) - (3, 1, 0)] = 0 \rightarrow (2, -1, -1) \cdot [\mathbf{x} - (3, 1, 0)] = 0.$$

This is a normal form equation (1) with $\mathbf{n} = (2, -1, -1)$ and $\mathbf{x}_0 = (3, 1, 0)$. ■

The next example shows how to find the point of intersection of a line and a plane.

Example 3

Find the point of intersection of the line $\mathbf{x} = t(1, -2, 3) + (-1, 1, 3)$ and the plane $2x - y + z = 7$. See **Figure 5.3.**

FIGURE 5.3

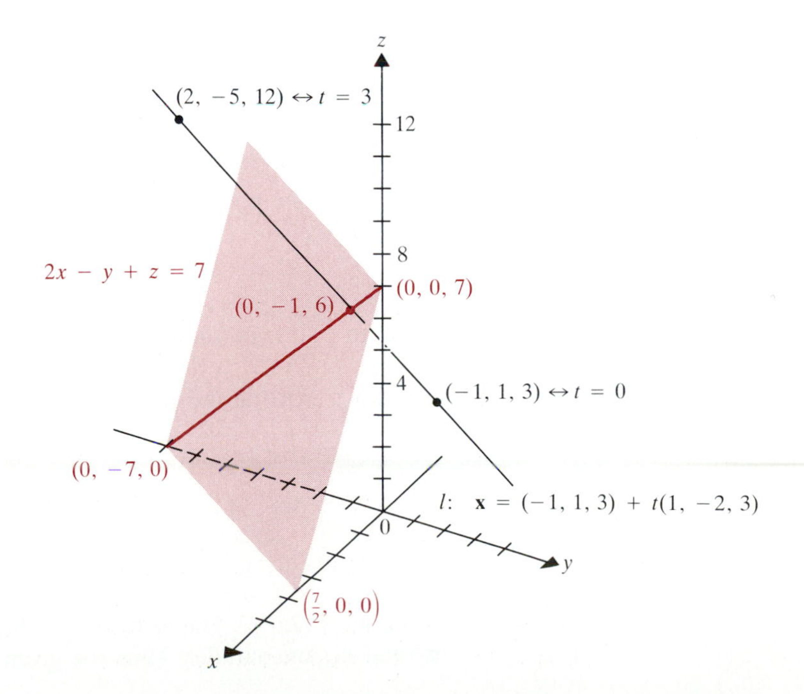

Solution. The equation of the line is

(5) $(x, y, z) = (t - 1, -2t + 1, 3t + 3) \rightarrow x = t - 1, y = -2t + 1, z = 3t + 3.$

At the point of intersection of the line and the plane, the values of x, y, and z in (5) satisfy the equation of the plane:

$$2(t - 1) - (-2t + 1) + 3t + 3 = 7 \rightarrow 2t - 2 + 2t - 1 + 3t + 3 = 7,$$
$$7t = 7 \rightarrow t = 1.$$

From (5) then, the point of intersection is

$$(x, y, z) = (1 - 1, -2 + 1, 3 + 3) = (0, -1, 6). \blacksquare$$

Two planes Π_1 and Π_2 in $\mathbf{R}^3$ are called ***parallel*** if for any normal vectors $\mathbf{n}_1$ of Π_1 and $\mathbf{n}_2$ of Π_2, $\mathbf{n}_1$ is parallel to $\mathbf{n}_2$; see **Figure 5.4**(a). They are ***perpendicular*** planes if $\mathbf{n}_1 \perp \mathbf{n}_2$, that is, if $\mathbf{n}_1 \cdot \mathbf{n}_2 = 0$; see Figure 5.4(b).

FIGURE 5.4

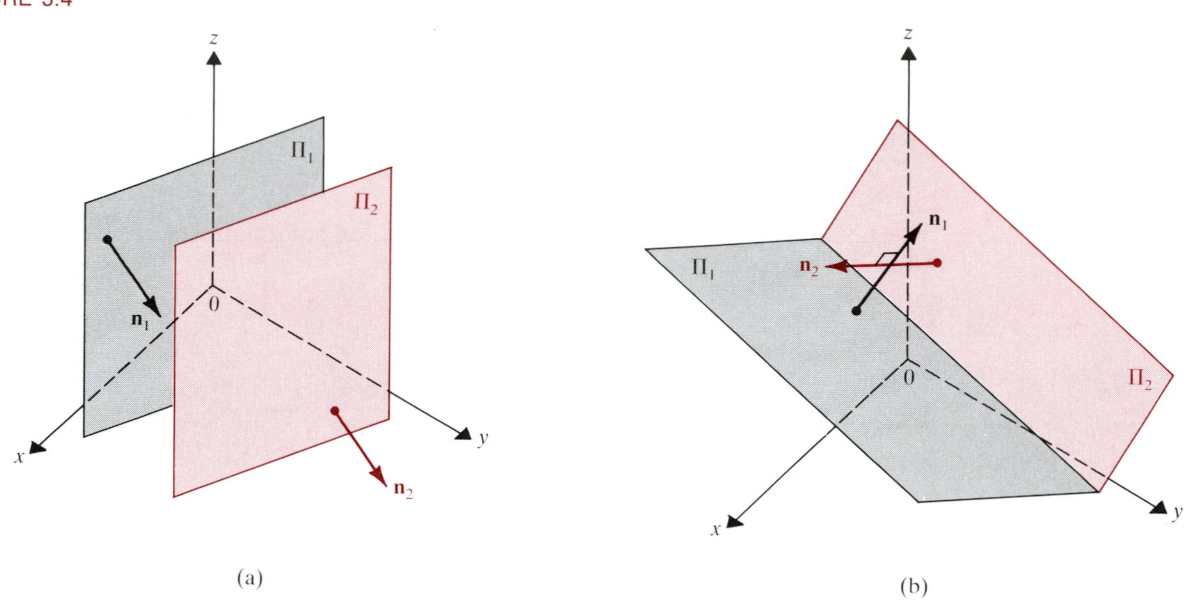

Example 4

Show that the planes

$$2x - 3y + 5z = 7 \qquad \text{and} \qquad -3x + \frac{9}{2}y - \frac{15}{2}z = 5$$

have no points in common.

Solution. Putting the equations in normal form, we get

$$(2, -3, 5) \cdot (x, y, z) = 7 = \mathbf{n}_1 \cdot \mathbf{x}_1$$

and

$$\left(-3, \frac{9}{2}, -\frac{15}{2}\right) \cdot (x, y, z) = 5 = \mathbf{n}_2 \cdot \mathbf{x}_2.$$

Then $\mathbf{n}_1 = (2, -3, 5)$ and $\mathbf{n}_2 = (-3, 9/2, -15/2)$. Since $\mathbf{n}_1 = -\frac{2}{3}\mathbf{n}_2$, the vectors $\mathbf{n}_1$ and $\mathbf{n}_2$ are parallel. Thus the given planes are parallel or coincident. They are not coincident, because the point (7/2, 0, 0) is on the first but not the second plane. Thus the planes are parallel and so have no points in common. ■

When two planes do intersect but don't coincide, their intersection is a line. The following example illustrates how we can find equations for such a line.

Example 5

Find vector and scalar equations for the line of intersection of the two planes

$$3x - 2y + z = 1 \tag{6}$$

and

$$-2x + y + 3z = 2. \tag{7}$$

Solution. The most efficient technique is to eliminate one of the variables. (It is not important which one is eliminated.) Here we can easily eliminate y by multiplying (7) by 2 and adding the result to (6). That gives

$$-x + 7z = 5 \rightarrow x = 7z - 5. \tag{8}$$

We can also express y in terms of z from (7), getting

$$y = 2 + 2x - 3z = 2 + 14z - 10 - 3z \rightarrow y = 11z - 8. \tag{9}$$

If we add the equation $z = t$ to (8) and (9), then we obtain a set of parametric scalar equations for the line of intersection, namely,

$$x = 7t - 5, \qquad y = 11t - 8, \qquad z = t.$$

These equations correspond to the single vector equation

$$\mathbf{x} = (x, y, z) = t(7, 11, 1) + (-5, -8, 0).$$

This is of the form $\mathbf{x} = t\mathbf{v} + \mathbf{x}_0$, where $\mathbf{v} = (7, 11, 1)$ and $\mathbf{x}_0 = (-5, -8, 0)$. See **Figure 5.5.** ■

FIGURE 5.5

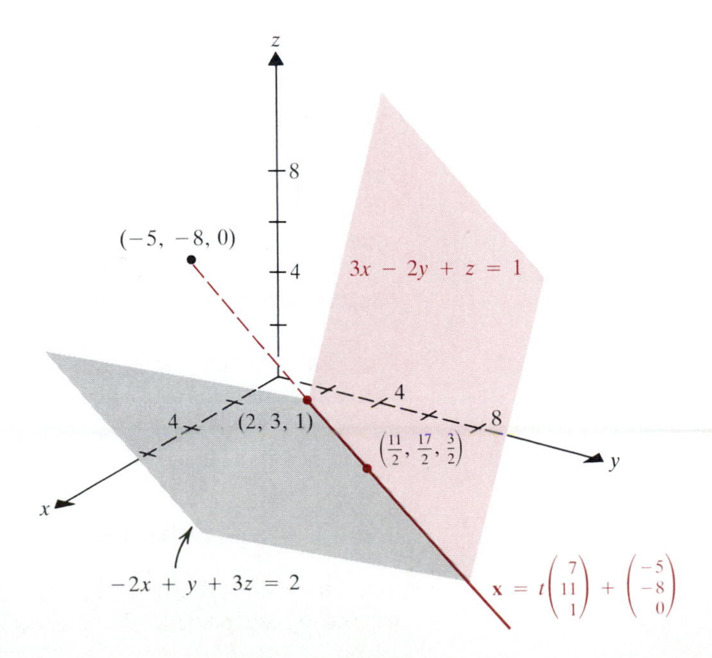

It is often useful to compute the perpendicular distance from a point to a plane. For instance, the intensity of illumination from a point source and the strength of an electron's electric field both depend on such distances. Suppose that a lamp is placed at $P(x_0, y_0, z_0)$, and we want to know the distance to a desktop lying in the plane $ax + by + cz = d$. Let $Q(x, y, z)$ be any point on the plane, and draw both the vector

$$\mathbf{v} = \mathbf{QP} = (x_0 - x, y_0 - y, z_0 - z)$$

FIGURE 5.6

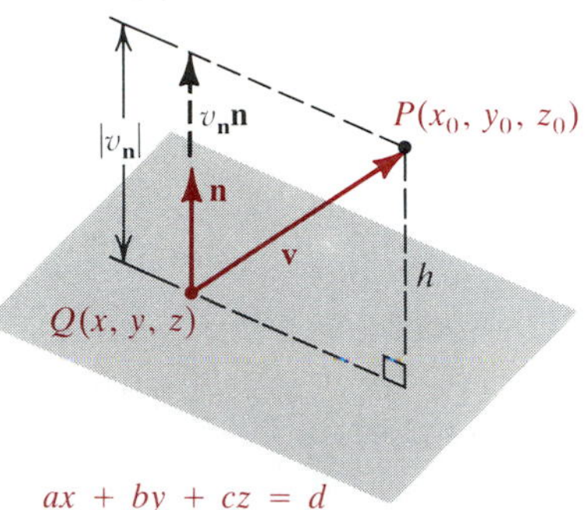

and the normal vector $\mathbf{n} = (a, b, c)$ with their initial points at Q. As **Figure 5.6** shows, the perpendicular distance h from the point P to the plane is just the length of the component of $\mathbf{v}$ in the direction of $\mathbf{n}$ (Definition 3.8), which is the absolute value of the coordinate of $\mathbf{v}$ in the direction of $\mathbf{n}$:

$$h = |v_{\mathbf{n}}| = \frac{|\mathbf{v} \cdot \mathbf{n}|}{|\mathbf{n}|}. \tag{10}$$

(Here $|v_{\mathbf{n}}|$ and $|\mathbf{v} \cdot \mathbf{n}|$ are the *absolute values* of scalars, while $|\mathbf{n}|$ is the *length* of the normal vector.) In terms of the coordinates of the vectors in $\mathbf{R}^3$, (10) says that

$$h = \frac{|(x_0 - x, y_0 - y, z_0 - z) \cdot (a, b, c)|}{\sqrt{a^2 + b^2 + c^2}}$$

$$= \frac{|a(x_0 - x) + b(y_0 - y) + c(z_0 - z)|}{\sqrt{a^2 + b^2 + c^2}},$$

$$h = \frac{|(ax_0 + by_0 + cz_0) - (ax + by + cz)|}{\sqrt{a^2 + b^2 + c^2}}. \tag{11}$$

Now, since Q lies in the plane, its coordinates satisfy the equation $ax + by + cz = d$. Thus (11) simplifies to

$$h = \frac{|ax_0 + by_0 + cz_0 - d|}{\sqrt{a^2 + b^2 + c^2}}. \tag{12}$$

Example 6

Suppose that the plane $3x + y + 5z = 10$ found in Example 1 (Figure 5.2) represents a draftsman's drawing table, and that a lamp is hung at coordinates (x_0, y_0, z_0). A law of physics states that the illumination E (in lumens per square meter) at the table is given by $E = I/h^2$, where I is the intensity of the lamp (in candelas), and h is the perpendicular distance (in meters) from the lamp to the table. Compare the distances and illuminations if the lamp is placed at $P_1(3, 2, 3)$ and then at $P_2(3, 3, 4)$.

Solution. The length of the normal vector $\mathbf{n} = 3\mathbf{i} + \mathbf{j} + 5\mathbf{k}$ is

$$|\mathbf{n}| = \sqrt{9 + 1 + 25} = \sqrt{35}.$$

This does not change when the lamp is moved. From (12), the first distance is

$$h_1 = \frac{|3 \cdot 3 + 1 \cdot 2 + 5 \cdot 3 - 10|}{\sqrt{35}} = \frac{16}{\sqrt{35}} \approx 2.7 \text{ m},$$

and the second distance is

$$h_2 = \frac{|3 \cdot 3 + 1 \cdot 3 + 5 \cdot 4 - 10|}{\sqrt{35}} = \frac{22}{\sqrt{35}} \approx 3.7 \text{ m}.$$

Therefore the corresponding illuminations are

$$E_1 = \frac{I}{\left(\frac{16}{\sqrt{35}}\right)^2} = \frac{35}{256} I \quad \text{and} \quad E_2 = \frac{I}{\left(\frac{22}{\sqrt{35}}\right)^2} = \frac{35}{484} I.$$

The ratio of these illuminations is

$$\frac{E_2}{E_1} = \frac{256}{484} \approx 0.53.$$

Moving the lamp from P_1 to P_2 therefore reduces the illumination by almost half. ■

Exercises 10.5

In Exercises 1–16, find a scalar equation for the plane described.

1. Through $(1, -2, 5)$ perpendicular to $\mathbf{n} = 3\mathbf{i} - 4\mathbf{j} + \mathbf{k}$. Sketch.
2. Through $(-3, 5, 1)$ perpendicular to $\mathbf{n} = \mathbf{i} + 3\mathbf{j} - 4\mathbf{k}$. Sketch.
3. The xy-, xz-, and yz-planes
4. **(a)** Parallel to and 3 units below the xy-plane
   (b) Parallel to and 2 units to the left of the xz-plane
   (c) Parallel to and 1 unit in front of the yz-plane
5. The line through the origin perpendicular to the plane intersects it at the point $(-3, 0, 5)$. Sketch.
6. The line through the origin perpendicular to the plane intersects it at the point $(2, -1, 1)$. Sketch.
7. The plane through $(2, -2, 5)$ perpendicular to the line through $(1, 1, 1)$ and $(-2, 3, 1)$
8. The plane through the point $(-1, 2, 4)$ perpendicular to the line through $(-1, 3, 1)$ and $(2, -1, 3)$
9. The line through $(-1, 3, -2)$ perpendicular to the plane intersects the plane at the point $(2, 3, 1)$.
10. The line through $(2, -1, -3)$ perpendicular to the plane intersects it at the point $(3, -3, 2)$.
11. The plane through $(1, -2, -1)$ and
    (a) perpendicular to the x-axis
    (b) parallel to the xz-plane
12. The plane through $(2, -3, 4)$ and
    (a) perpendicular to the y-axis
    (b) parallel to the yz-plane
13. The plane through $(-1, 3, -5)$ parallel to the plane $3x - 2y - z = 1$
14. The plane through $(1, 2, 3)$ parallel to the plane whose equation is $-x + 3y - 4z = 2$
15. The plane through $(1, -2, 1)$ perpendicular to the y-axis
16. The plane through $(-3, 1, 4)$ perpendicular to the x-axis

In Exercises 17 and 18, determine which of the given planes are coincident, parallel, or perpendicular.

17. **(a)** $-x + 2y - 2z = 3$ **(b)** $\frac{1}{2}x - y + z = -\frac{3}{2}$
    (c) $4x - 2y - 4z = 7$ **(d)** $3x - 6y + 6z = 11$
18. **(a)** $\frac{2}{3}x - y - 5z = 4$ **(b)** $-4x + 6y + 30z = 21$
    (c) $3x - 3y + z = 7$ **(d)** $2x - 3y - 15z = 12$
19. Find the vector and scalar equations for the line of intersection of the planes $x - 2y + 2z = 4$ and $2x - y + 3z = 5$.
20. Find vector and scalar equations for the line of intersection of the two planes $x - 2y + 3z = 5$ and $8x + 7y + z = 2$.
21. Find parametric equations for the traces of the plane $3x - 5y + 2z = 15$ in the three coordinate planes. Draw a picture.
22. Find parametric equations for the traces of the plane $x + 4y + 5z = 10$ in the three coordinate planes. Draw a picture.
23. Find the point of intersection of the plane $2x - 3y + z = 6$ and the line $\mathbf{x} = (2, -1, 3) + t(-1, 3, 1)$.
24. Find the point of intersection of the line
$$\frac{x+1}{-2} = \frac{y-2}{3} = \frac{z-1}{2}$$
    and the plane $-x - 3y + 5z = 4$.
25. Is the line $\mathbf{x} = (2, 1, 1) + t(-1, 3, 2)$ normal to the plane $x - 3y - 2z = 11$? Find the point of intersection.
26. Is the line $\mathbf{x} = (2, 1, 1) + t(-1, 3, 2)$ normal to the plane $x - 3y + 2z = 8$? Find the point of intersection.
27. Show that the line $\mathbf{x} = (3, 1, -2) + t(1, -1, 3)$ does not intersect the plane $2x - y - z = 5$.
28. Show that the line $\mathbf{x} = (2, 0, 3) + t(1, 7, -2)$ does not intersect the plane $-3x + y + 2z = 4$.

29. Prove that the line $\mathbf{x} = \mathbf{x}_0 + t\mathbf{v}$ is parallel to the plane $\mathbf{n} \cdot (\mathbf{x} - \mathbf{x}_1) = 0$ if and only if $\mathbf{n} \cdot \mathbf{v} = 0$.

30. If the line $\mathbf{x} = \mathbf{x}_0 + t\mathbf{v}$ is normal to the plane $ax + by + cz = d$, then what must be true of the vector $\mathbf{v}$?

31. Find the distance from the point $(2, 1, -1)$ to the plane $5x - 2y + 2z = 2$.

32. Find the distance from the point $(2, 3, 1)$ to the plane $2x - y + 2z = 9$.

33. In Example 6, compare the illuminations at $P_3(4, 2, 2)$ and $P_4(4, 3, 3)$.

34. Repeat Exercise 33 if the table lies in the plane $x + 2y + 3z = 8$, P_3 is the point $(2, 3, 2)$, and P_4 is the point $(3, 4, 3)$.

The *dihedral angle* between two planes is defined to be the angle between the normal vectors of the planes. In Exercises 35–38, find the dihedral angle between the given planes.

35. $x - y = 3,\ -x + 2y + z = 3$

36. $x - y = 0,\ y - z = 0$

37. $x - \sqrt{5}y + 2z = 3,\ 3x + z = 0$

38. $x - y = 0,\ -x + y + \sqrt{2}z = 3$

39. A point P_0 lies on a plane, and $\mathbf{0P}_0$ is perpendicular to the plane. Given that $\mathbf{0P}_0$ has direction angles $\alpha = \pi/3$, $\beta = \pi/6$, and $\gamma = \pi/2$, and $d(0, P_0) = 1$, find
(a) the equation of the plane;
(b) a scalar parametric equation of the trace of the plane in the xy-plane.

40. Two planes intersect in a line l. The first plane Π_1 is parallel to the xy-plane; the second plane Π_2 is parallel to the yz-plane. The point $P_0(2, 3, -2)$ is on l.
(a) Find the equation of Π_1.
(b) Find the equation of Π_2.
(c) Find a parametric scalar equation of the line l.
(d) Determine whether l is parallel to or perpendicular to (or neither) each coordinate axis.

10.6 The Vector Cross Product

FIGURE 6.1

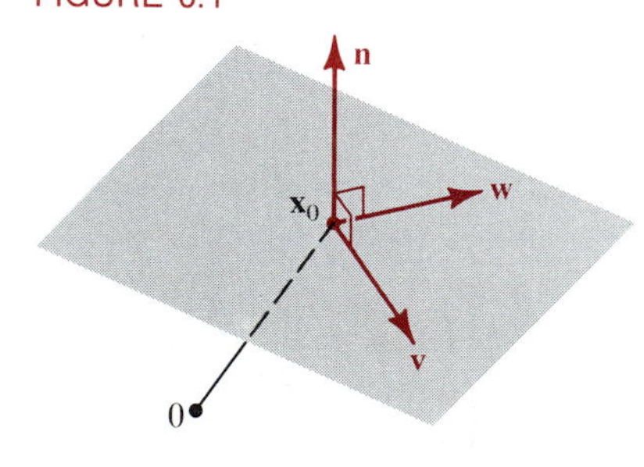

In the last section, we studied planes as objects perpendicular to normal vectors $\mathbf{n}$. But just as a line is determined by two points, so a plane is determined by two (nonskew) lines or three noncollinear points. Accordingly, in this section we present a technique for finding the equation of a plane given (a) two lines l_1 and l_2 in the plane, or (b) three points P, Q, and R on the plane.

First suppose that we are given the two nonskew lines

$$l_1\colon \mathbf{x} = s\mathbf{v} + \mathbf{x}_1 \qquad \text{and} \qquad l_2\colon \mathbf{x} = t\mathbf{w} + \mathbf{x}_2,$$

where $\mathbf{v}$ and $\mathbf{w}$ are direction vectors of l_1 and l_2, respectively. To find the normal-form equation of the plane through l_1 and l_2, we need a vector $\mathbf{n}$ perpendicular to that plane. For then the plane will have equation $\mathbf{n} \cdot (\mathbf{x} - \mathbf{x}_0) = 0$, where the endpoint of $\mathbf{x}_0$ is a point in the plane. See **Figure 6.1.** Such a vector $\mathbf{n}$ must be perpendicular to both $\mathbf{v}$ and $\mathbf{w}$, that is,

$$\mathbf{n} \cdot \mathbf{v} = 0 \qquad \text{and} \qquad \mathbf{n} \cdot \mathbf{w} = 0. \tag{1}$$

Case (b) above brings us to the same problem. For if P, Q, and R are given points in the plane, then let $v = \mathbf{PQ}$ and $w = \mathbf{PR}$. These are two vectors in the plane, so a normal vector $\mathbf{n}$ to the plane must satisfy (1). See **Figure 6.2.**

FIGURE 6.2

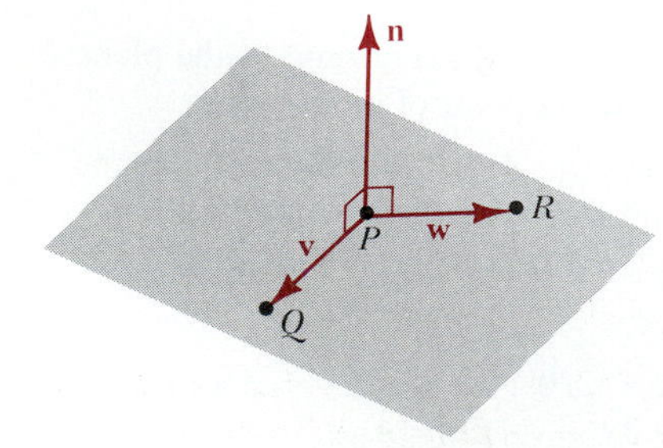

To find a vector $\mathbf{n} = (x, y, z)$ that satisfies (1), let $\mathbf{v} = (v_1, v_2, v_3)$ and $\mathbf{w} = (w_1, w_2, w_3)$. Then the vector equations (1) translate into the scalar equations

$$v_1x + v_2y + v_3z = 0, \tag{2}$$

$$w_1x + w_2y + w_3z = 0. \tag{3}$$

Geometrically, (2) and (3) represent two planes through the origin perpendicular to $\mathbf{v}$ and $\mathbf{w}$. We expect the line of intersection of these planes to be perpendicular to *both* $\mathbf{v}$ and $\mathbf{w}$. We use the technique of elimination from Example 5 of the last section to find a vector lying in that line. To eliminate y, we multiply

(2) by w_2 and (3) by v_2, and subtract:

$$\begin{array}{rl} w_2 \cdot (2): & v_1w_2x + v_2w_2y + v_3w_2z = 0, \\ v_2 \cdot (3): & v_2w_1x + v_2w_2y + v_2w_3z = 0 \quad (-) \\ \hline w_2 \cdot (2) - v_2 \cdot (3): & (v_1w_2 - v_2w_1)x + (v_3w_2 - v_2w_3)z = 0. \end{array}$$

Thus

$$(4) \qquad (v_1w_2 - v_2w_1)x = (v_2w_3 - v_3w_2)z.$$

Similarly, to eliminate x, we multiply (2) by w_1 and (3) by v_1, and subtract:

$$\begin{array}{rl} w_1 \cdot (2): & v_1w_1x + v_2w_1y + v_3w_1z = 0, \\ v_1 \cdot (3): & v_1w_1x + v_1w_2y + v_1w_3z = 0 \quad (-) \\ \hline w_1 \cdot (2) - v_1 \cdot (3): & (v_2w_1 - v_1w_2)y + (v_3w_1 - v_1w_3)z = 0. \end{array}$$

Hence

$$(5) \qquad (v_1w_2 - v_2w_1)y = (v_3w_1 - v_1w_3)z.$$

Now notice that the coefficient of x in (4) is the same as the coefficient of y in (5). If we set z equal to that coefficient, that is, $z = v_1w_2 - v_2w_1$, then $x = v_2w_3 - v_3w_2$ satisfies (4) and $y = v_3w_1 - v_1w_3$ satisfies (5). Hence a vector perpendicular to both $\mathbf{v} = (v_1, v_2, v_3)$ and $\mathbf{w} = (w_1, w_2, w_3)$ is $\mathbf{n} = (v_2w_3 - v_3w_2)\mathbf{i} + (v_3w_1 - v_1w_3)\mathbf{j} + (v_1w_2 - v_2w_1)\mathbf{k}$. This normal vector has a special name.

6.1 DEFINITION The ***cross product*** of the vectors $\mathbf{v} = (v_1, v_2, v_3)$ and $\mathbf{w} = (w_1, w_2, w_3)$ in $\mathbf{R}^3$ is the vector

$$(6) \qquad \begin{aligned} \mathbf{v} \times \mathbf{w} &= (v_2w_3 - v_3w_2, v_3w_1 - v_1w_3, v_1w_2 - v_2w_1) \\ &= (v_2w_3 - v_3w_2)\mathbf{i} + (v_3w_1 - v_1w_3)\mathbf{j} + (v_1w_2 - v_2w_1)\mathbf{k}. \end{aligned}$$

Although the formula for the cross product looks complicated, it is easy to compute $\mathbf{v} \times \mathbf{w}$ using ***determinants.*** The ***2-by-2 determinant function*** **det** is defined on ***2-by-2 matrices,*** that is, on arrays of two rows and two columns of real numbers. Such an array has the form

$$\begin{pmatrix} a & b \\ c & d \end{pmatrix}.$$

It is thus composed of two row vectors

$$(a, b) \text{ and } (c, d) \qquad \text{or two column vectors } \begin{pmatrix} a \\ c \end{pmatrix} \text{ and } \begin{pmatrix} b \\ d \end{pmatrix}.$$

The determinant of this 2-by-2 matrix is defined by

$$\det \begin{pmatrix} a & b \\ c & d \end{pmatrix} = ad - bc.$$

For example,

$$\det \begin{pmatrix} 1 & -2 \\ 3 & 5 \end{pmatrix} = 1 \cdot 5 - 3(-2) = 5 + 6 = 11.$$

We can rewrite (6) in determinant notation as

$$(7) \qquad \mathbf{v} \times \mathbf{w} = \det \begin{pmatrix} v_2 & v_3 \\ w_2 & w_3 \end{pmatrix} \mathbf{i} + \det \begin{pmatrix} v_3 & v_1 \\ w_3 & w_1 \end{pmatrix} \mathbf{j} + \det \begin{pmatrix} v_1 & v_2 \\ w_1 & w_2 \end{pmatrix} \mathbf{k}.$$

In (7) each coordinate of $\mathbf{v} \times \mathbf{w}$ is the determinant formed from the *other two* coordinates of $\mathbf{v}$ and $\mathbf{w}$ in the proper *cyclic* order. The *first* coordinate of $\mathbf{v} \times \mathbf{w}$

FIGURE 6.3

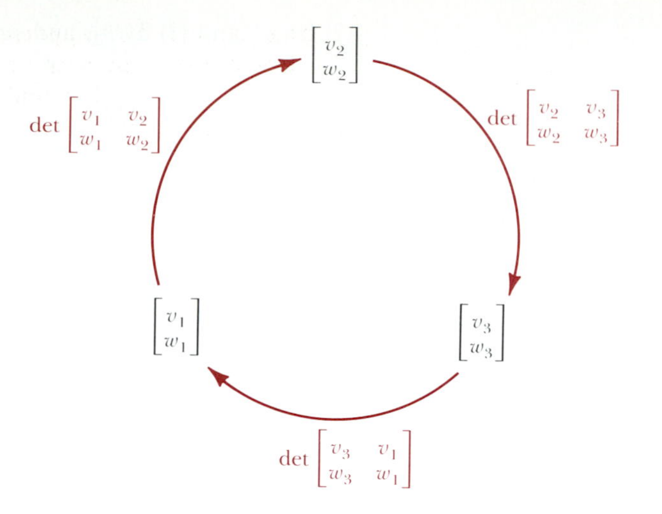

is the determinant of the *second and third coordinates* of **v** and **w**. The *second* coordinate of $\mathbf{v} \times \mathbf{w}$ is the determinant of the *third and first coordinates* (**in *that* order**) of **v** and **w**. The *third* coordinate of $\mathbf{v} \times \mathbf{w}$ is the determinant of the *first and second coordinates* of **v** and **w**. **Figure 6.3** shows this schematically.

Example 1

Find $\mathbf{v} \times \mathbf{w}$ if $\mathbf{v} = (2, -1, 3)$ and $\mathbf{w} = (1, 5, -3)$.

Solution. Applying (7), we get

$$\begin{aligned}
\mathbf{v} \times \mathbf{w} &= \det\begin{pmatrix} -1 & 3 \\ 5 & -3 \end{pmatrix}\mathbf{i} + \det\begin{pmatrix} 3 & 2 \\ -3 & 1 \end{pmatrix}\mathbf{j} + \det\begin{pmatrix} 2 & -1 \\ 1 & 5 \end{pmatrix}\mathbf{k} \\
&= [(-1)(-3) - 5 \cdot 3]\mathbf{i} + [3 \cdot 1 - (-3) \cdot 2]\mathbf{j} + [2 \cdot 5 - 1 \cdot (-1)]\mathbf{k} \\
&= (3 - 15)\mathbf{i} + (3 + 6)\mathbf{j} + (10 + 1)\mathbf{k} \\
&= -12\mathbf{i} + 9\mathbf{j} + 11\mathbf{k} = (-12, 9, 11). \quad \blacksquare
\end{aligned}$$

Although $\mathbf{v} \times \mathbf{w}$ is defined for *three*-dimensional vectors, formula (7) involves *two*-dimensional determinants. There is a three-dimensional version of (7), which uses the **3-by-3 *determinant function* det** defined on **3-by-3 *matrices*.** Such a matrix is an array of real numbers,

$$\begin{pmatrix} a_1 & a_2 & a_3 \\ b_1 & b_2 & b_3 \\ c_1 & c_2 & c_3 \end{pmatrix}$$

It can be thought of as composed of the three row vectors $\mathbf{a} = (a_1, a_2, a_3)$, $\mathbf{b} = (b_1, b_2, b_3)$, and $\mathbf{c} = (c_1, c_2, c_3)$. The determinant of this 3-by-3 matrix is defined by ***expansion by minors of the first row***,

$$\begin{aligned}
\det\begin{pmatrix} a_1 & a_2 & a_3 \\ b_1 & b_2 & b_3 \\ c_1 & c_2 & c_3 \end{pmatrix} &= a_1 \det\begin{pmatrix} b_2 & b_3 \\ c_2 & c_3 \end{pmatrix} - a_2 \det\begin{pmatrix} b_1 & b_3 \\ c_1 & c_3 \end{pmatrix} + a_3 \det\begin{pmatrix} b_1 & b_2 \\ c_1 & c_2 \end{pmatrix} \\
&= a_1(b_2c_3 - b_3c_2) - a_2(b_1c_3 - b_3c_1) + a_3(b_1c_2 - b_2c_1) \\
&= a_1b_2c_3 - a_1b_3c_2 - a_2b_1c_3 + a_2b_3c_1 + a_3b_1c_2 - a_3b_2c_1.
\end{aligned}$$

This can also be computed using the following device, which is **valid *only* for three-dimensional determinants.** Rewrite the first two columns of the matrix next to the final column, and draw an arrow downward along each diagonal of the resulting array.

$$\begin{array}{ccccc} a_1 & a_2 & a_3 & a_1 & a_2 \\ b_1 & b_2 & b_3 & b_1 & b_2 \\ c_1 & c_2 & c_3 & c_1 & c_2 \end{array}$$

Then the determinant is given by the sums of the products of the factors in the diagonals, where a plus sign is attached to products corresponding to arrows pointing toward the right, and a minus sign is attached to products associated with arrows pointing toward the left. That gives

$$a_1b_2c_3 + a_2b_3c_1 + a_3b_1c_2 - a_3b_2c_1 - a_1b_3c_2 - a_2b_1c_3,$$

the same expression obtained above.

This scheme can be used to evaluate $\mathbf{v} \times \mathbf{w}$ as the *symbolic* **three-dimensional determinant**

$$\det\begin{pmatrix} \mathbf{i} & \mathbf{j} & \mathbf{k} \\ v_1 & v_2 & v_3 \\ w_1 & w_2 & w_3 \end{pmatrix}. \tag{8}$$

To illustrate its use, we apply it to the vectors $\mathbf{v}$ and $\mathbf{w}$ in Example 1.

Example 2

Use (8) to compute $\mathbf{v} \times \mathbf{w}$ for $\mathbf{v} = (2, -1, 3)$ and $\mathbf{w} = (1, 5, -3)$.

Solution. From the symbolic array

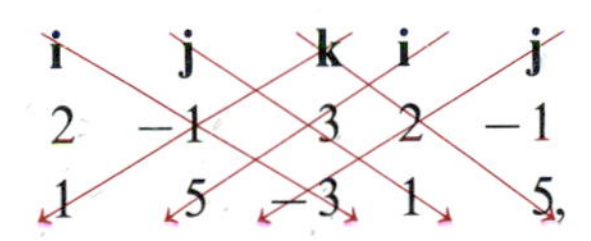

we obtain

$$\begin{aligned} &\mathbf{i}(-1)(-3) + \mathbf{j}(3)(1) + \mathbf{k}(2)(5) - \mathbf{k}(-1)(1) - \mathbf{i}(3)(5) - \mathbf{j}(2)(-3) \\ &\quad = (3 - 15)\mathbf{i} + (3 + 6)\mathbf{j} + (10 + 1)\mathbf{k} \\ &\quad = -12\mathbf{i} + 9\mathbf{j} + 11\mathbf{k} = (-12, 9, 11). \end{aligned}$$

This is the same result we got from (7) in Example 1. ■

Example 3

Find the equation of the plane determined by the two lines $\mathbf{x} = s(2, -1, 3) + (2, 0, 3)$ and $\mathbf{x} = t(1, 5, -3) + (-4, 3, -6)$.

Solution. The vector $\mathbf{v} = (2, -1, 3)$ is a direction vector of the first line, and $\mathbf{w} = (1, 5, -3)$ is one for the second line. So a normal vector to the plane is $\mathbf{v} \times \mathbf{w}$, which from Example 1 or 2 is $(-12, 9, 11)$. Taking $s = 0$, we see that $(2, 0, 3)$ is a point on the plane. So an equation is

$$(-12, 9, 11) \cdot [(x, y, z) - (2, 0, 3)] = 0$$

$$-12x + 24 + 9y + 11z - 33 = 0 \rightarrow -12x + 9y + 11z = 9. \quad ■$$

Example 4

Find the equation of the plane through the three points $A(1, -2, 3)$, $B(-2, 1, 1)$, and $C(1, 3, -2)$.

Solution. The vectors $\mathbf{v} = \mathbf{AB}$ and $\mathbf{w} = \mathbf{AC}$ lie in the plane, so $\mathbf{v} \times \mathbf{w}$ is normal to the plane. We have

$$\mathbf{v} = (-3, 3, -2) \qquad \text{and} \qquad \mathbf{w} = (0, 5, -5).$$

Then

$$\mathbf{v} \times \mathbf{w} = \det \begin{pmatrix} \mathbf{i} & \mathbf{j} & \mathbf{k} \\ -3 & 3 & -2 \\ 0 & 5 & -5 \end{pmatrix} = (-15 + 10)\mathbf{i} + (0 - 15)\mathbf{j} + (-15 - 0)\mathbf{k}$$

$$= -5\mathbf{i} - 15\mathbf{j} - 15\mathbf{k} = -5(\mathbf{i} + 3\mathbf{j} + 3\mathbf{k}) = -5(1, 3, 3).$$

Thus for a normal vector we can take $\mathbf{n} = (1, 3, 3)$. Then an equation for the plane is $\mathbf{n} \cdot (\mathbf{x} - \mathbf{0A}) = 0$, that is,

$$(1, 3, 3) \cdot [(x, y, z) - (1, -2, 3)] = 0 \rightarrow (1, 3, 3) \cdot (x - 1, y + 2, z - 3) = 0,$$

$$x - 1 + 3y + 6 + 3z - 9 = 0 \rightarrow x + 3y + 3z = 4.$$

You should check that the coordinates of all the points A, B, and C satisfy this equation. ■

We turn now to some algebraic properties of the cross product. In the following result, note particularly that the cross product *fails* to be commutative (by part (b)) or associative (see Exercise 46). It is probably the first kind of multiplication you have met that lacks those familiar properties.

6.2 THEOREM

For all vectors $\mathbf{u}$, $\mathbf{v}$, and $\mathbf{w}$ in $\mathbf{R}^3$ and for any real number a,

(a) $(\mathbf{v} \times \mathbf{w}) \cdot \mathbf{v} = (\mathbf{v} \times \mathbf{w}) \cdot \mathbf{w} = 0$

(b) $\mathbf{v} \times \mathbf{w} = -\mathbf{w} \times \mathbf{v}$

(c) $\mathbf{u} \times \mathbf{u} = \mathbf{0}$

(d) $\mathbf{u} \times \mathbf{0} = \mathbf{0} \times \mathbf{u} = \mathbf{0}$

(e) $(\mathbf{u} + \mathbf{v}) \times \mathbf{w} = \mathbf{u} \times \mathbf{w} + \mathbf{v} \times \mathbf{w}$ and $\mathbf{u} \times (\mathbf{v} + \mathbf{w}) = \mathbf{u} \times \mathbf{v} + \mathbf{u} \times \mathbf{w}$

(f) $a(\mathbf{v} \times \mathbf{w}) = (a\mathbf{v}) \times \mathbf{w} = \mathbf{v} \times (a\mathbf{w})$

(g) $\mathbf{u} \cdot (\mathbf{v} \times \mathbf{w}) = \mathbf{v} \cdot (\mathbf{w} \times \mathbf{u}) = \mathbf{w} \cdot (\mathbf{u} \times \mathbf{v})$

(h) $\mathbf{u} \times (\mathbf{v} \times \mathbf{w}) = (\mathbf{u} \cdot \mathbf{w})\mathbf{v} - (\mathbf{u} \cdot \mathbf{v})\mathbf{w}$

Partial Proof. Statement (a) just restates the fact that $\mathbf{v} \times \mathbf{w}$ is perpendicular to both $\mathbf{v}$ and $\mathbf{w}$. Let $\mathbf{u} = (u_1, u_2, u_3)$, $\mathbf{v} = (v_1, v_2, v_3)$, and $\mathbf{w} = (w_1, w_2, w_3)$. Then (b) follows easily: By Definition 6.1 we have

$$\mathbf{v} \times \mathbf{w} = (v_2 w_3 - v_3 w_2, v_3 w_1 - v_1 w_3, v_1 w_2 - v_2 w_1).$$

Also,

$$\begin{aligned} \mathbf{w} \times \mathbf{v} &= (w_2 v_3 - w_3 v_2, w_3 v_1 - w_1 v_3, w_1 v_2 - w_2 v_1) \\ &= (-(v_2 w_3 - v_3 w_2), -(v_3 w_1 - v_1 w_3), -(v_1 w_2 - v_2 w_1)) = -\mathbf{v} \times \mathbf{w}. \end{aligned}$$

We leave (c), (e), and (f) for Exercises 33 and 41. Part (d) is clear. We verify the first equality in (g) and leave the second for you in Exercise 33(d). First,

$$\begin{aligned}\mathbf{u}\cdot(\mathbf{v}\times\mathbf{w}) &= (u_1, u_2, u_3)\cdot(v_2w_3 - v_3w_2, v_3w_1 - v_1w_3, v_1w_2 - v_2w_1)\\ &= u_1v_2w_3 - u_1v_3w_2 + u_2v_3w_1 - u_2v_1w_3 + u_3v_1w_2 - u_3v_2w_1.\end{aligned}$$

Next,

$$\begin{aligned}\mathbf{v}\cdot(\mathbf{w}\times\mathbf{u}) &= (v_1, v_2, v_3)\cdot(w_2u_3 - w_3u_2, w_3u_1 - w_1u_3, w_1u_2 - w_2u_1)\\ &= u_3v_1w_2 - u_2v_1w_3 + u_1v_2w_3 - u_3v_2w_1 + u_2v_3w_1 - u_1v_3w_2\\ &= \mathbf{u}\cdot(\mathbf{v}\times\mathbf{w}).\end{aligned}$$

Finally, we come to (h), which is known as the ***triple vector product*** of **u**, **v**, and **w**. We have $\mathbf{u} = (u_1, u_2, u_3)$ and

$$\mathbf{v}\times\mathbf{w} = (v_2w_3 - v_3w_2, v_3w_1 - v_1w_3, v_1w_2 - v_2w_1),$$

so the first coordinate of $\mathbf{u}\times(\mathbf{v}\times\mathbf{w})$ is

$$u_2(v_1w_2 - v_2w_1) - u_3(v_3w_1 - v_1w_3) = u_2v_1w_2 - u_2v_2w_1 + u_3v_1w_3 - u_3v_3w_1.$$

The first coordinate of $(\mathbf{u}\cdot\mathbf{w})\mathbf{v}$ is

$$(u_1w_1 + u_2w_2 + u_3w_3)v_1 = u_1v_1w_1 + u_2v_1w_2 + u_3v_1w_3.$$

The first coordinate of $(\mathbf{u}\cdot\mathbf{v})\mathbf{w}$ is

$$(u_1v_1 + u_2v_2 + u_3v_3)w_1 = u_1v_1w_1 + u_2v_2w_1 + u_3v_3w_1.$$

Subtracting these last two expressions, we see that the right side of (h) has first coordinate

$$u_2v_1w_2 - u_2v_2w_1 + u_3v_1w_3 - u_3v_3w_1,$$

which is the first coordinate of $\mathbf{u}\times(\mathbf{v}\times\mathbf{w})$ calculated above. Similar computations verify that the second and third coordinates of $\mathbf{u}\times(\mathbf{v}\times\mathbf{w})$ are respectively equal to the second and third coordinates of $(\mathbf{u}\cdot\mathbf{w})\mathbf{v} - (\mathbf{u}\cdot\mathbf{v})\mathbf{w}$. We leave these for you in Exercise 34(a). QED

Theorem 6.2(*g*) says that we can permute **u**, **v**, and **w** *cyclically* without changing the value of $\mathbf{u}\cdot(\mathbf{v}\times\mathbf{w})$, which is called the ***triple scalar product*** of **u**, **v**, and **w**. See **Figure 6.4.**

FIGURE 6.4

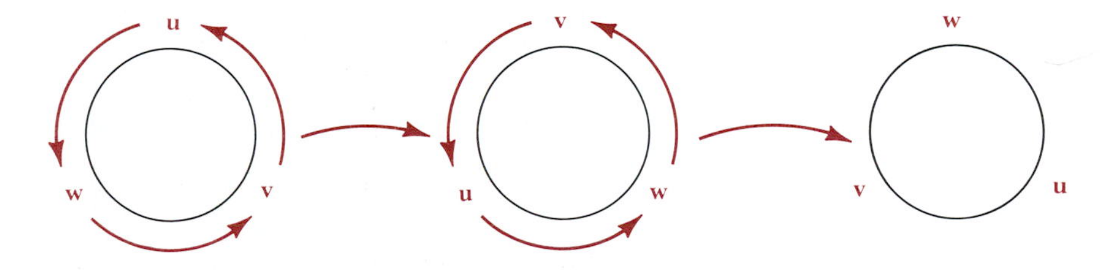

It is a direct consequence of Theorems 6.2(b) and 3.2(b) that an arbitrary permutation of **u**, **v**, and **w** in the triple scalar product changes its value by at most a factor of -1:

$$\mathbf{u}\cdot(\mathbf{w}\times\mathbf{v}) = -\mathbf{u}\cdot(\mathbf{v}\times\mathbf{w}). \tag{9}$$

FIGURE 6.5

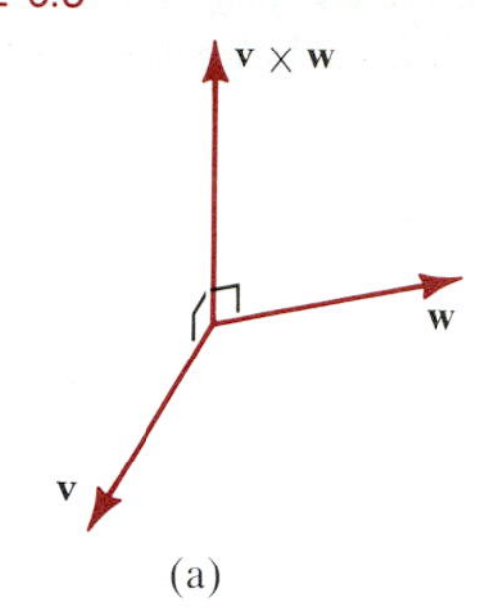

(a)

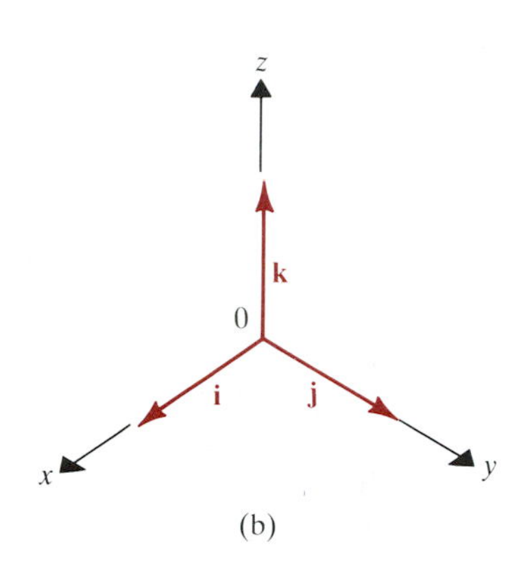

(b)

As an illustration, note that

$$\mathbf{i} \times \mathbf{j} = (1, 0, 0) \times (0, 1, 0) = (0, 0, 1) = \mathbf{k}.$$

Thus

$$\mathbf{k} \cdot (\mathbf{i} \times \mathbf{j}) = \mathbf{k} \cdot \mathbf{k} = (0, 0, 1) \cdot (0, 0, 1) = 1.$$

By contrast, Theorem 6.2(a) says that

$$\mathbf{j} \times \mathbf{i} = -\mathbf{i} \times \mathbf{j} = -\mathbf{k},$$

and so we get the following instance of (9):

$$\mathbf{k} \cdot (\mathbf{j} \times \mathbf{i}) = \mathbf{k} \cdot (-\mathbf{k}) = -1 = -\mathbf{k} \cdot (\mathbf{i} \times \mathbf{j}).$$

The triple $(\mathbf{v}, \mathbf{w}, \mathbf{v} \times \mathbf{w})$ is ***right-handed,*** meaning that it has the same orientation as the triple $(\mathbf{i}, \mathbf{j}, \mathbf{k})$. (See **Figure 6.5.**) This follows from the fact that $\mathbf{k} = \mathbf{i} \times \mathbf{j}$, and it means that *if you point your forefinger in the direction of* $\mathbf{v}$ *and your middle finger in the direction of* $\mathbf{w}$, *then your thumb will point in the direction of* $\mathbf{v} \times \mathbf{w}$.

There is an important geometric property of the cross product that parallels the relation

$$\mathbf{x} \cdot \mathbf{y} = |\mathbf{x}|\,|\mathbf{y}| \cos \theta$$

for the dot product (see p. 648). It is derived from the following technical result, whose proof consists of a verification that the respective sides reduce to the same algebraic expression.

6.3 LEMMA $|\mathbf{v} \times \mathbf{w}|^2 = |\mathbf{v}|^2|\mathbf{w}|^2 - (\mathbf{v} \cdot \mathbf{w})^2.$

Proof.

$$\begin{aligned} |\mathbf{v} \times \mathbf{w}|^2 &= (v_2w_3 - v_3w_2)^2 + (v_3w_1 - v_1w_3)^2 + (v_1w_2 - v_2w_1)^2 \\ &= (v_2w_3)^2 - 2v_2v_3w_2w_3 + (v_3w_2)^2 + (v_3w_1)^2 - 2v_1v_3w_1w_3 \\ &\quad + (v_1w_3)^2 + (v_1w_2)^2 - 2v_1v_2w_1w_2 + (v_2w_1)^2. \end{aligned}$$

$$\begin{aligned} |\mathbf{v}|^2|\mathbf{w}|^2 - |\mathbf{v} \cdot \mathbf{w}|^2 &= (v_1^2 + v_2^2 + v_3^2)(w_1^2 + w_2^2 + w_3^2) - (v_1w_1 + v_2w_2 + v_3w_3)^2 \\ &= (v_1w_1)^2 + (v_1w_2)^2 + (v_1w_3)^2 + (v_2w_1)^2 + (v_2w_2)^2 \\ &\quad + (v_2w_3)^2 + (v_3w_1)^2 + (v_3w_2)^2 + (v_3w_3)^2 - [(v_1w_1)^2 + (v_2w_2)^2 \\ &\quad + (v_3w_3)^2 + 2v_1v_2w_1w_2 + 2v_1v_3w_1w_3 + 2v_2v_3w_2w_3] \\ &= (v_1w_2)^2 + (v_1w_3)^2 + (v_2w_1)^2 + (v_2w_3)^2 + (v_3w_1)^2 \\ &\quad + (v_3w_2)^2 - 2v_1v_2w_1w_2 - 2v_1v_3w_1w_3 - 2v_2v_3w_2w_3 \\ &= |\mathbf{v} \times \mathbf{w}|^2. \end{aligned}$$ QED

From this and Equation (10) of Section 3, we find that

$$\begin{aligned} |\mathbf{v} \times \mathbf{w}| &= \sqrt{|\mathbf{v}|^2|\mathbf{w}|^2 - |\mathbf{v} \cdot \mathbf{w}|^2} = \sqrt{|\mathbf{v}|^2|\mathbf{w}|^2 - |\mathbf{v}|^2|\mathbf{w}|^2 \cos^2 \theta} \\ &= |\mathbf{v}|\,|\mathbf{w}|\sqrt{1 - \cos^2 \theta} = |\mathbf{v}|\,|\mathbf{w}|\sqrt{\sin^2 \theta}. \end{aligned}$$

Since θ is a first- or second-quadrant angle by Definition 3.4, we thus have

$$|\mathbf{v} \times \mathbf{w}| = |\mathbf{v}|\,|\mathbf{w}| \sin \theta. \tag{10}$$

Example 5

Use (10) to find a formula for the area of the parallelogram formed by $\mathbf{v}$ and $\mathbf{w}$.

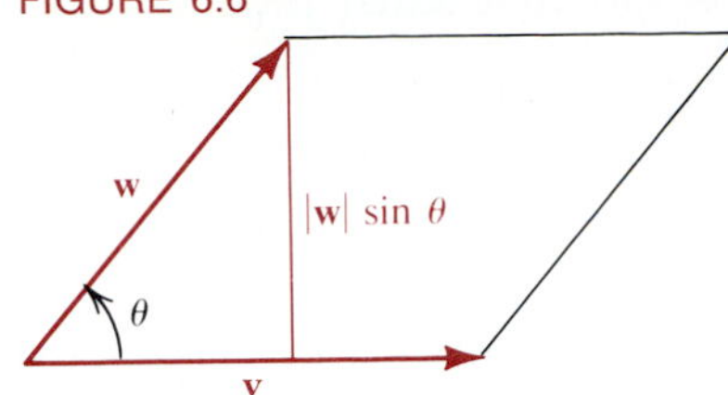

FIGURE 6.6

Solution. The area of a parallelogram is its base times its height. If in **Figure 6.6** we take the base to be $|\mathbf{v}|$, then the height is $|\mathbf{w}|\sin\theta$. So the area of the parallelogram is

$$A = |\mathbf{v}|\,|\mathbf{w}|\sin\theta = |\mathbf{v}\times\mathbf{w}|. \qquad (11)$$ ■

Example 6

Find the area of the parallelogram in the plane determined by the vectors (2, 5) and (−1, 1). See **Figure 6.7.**

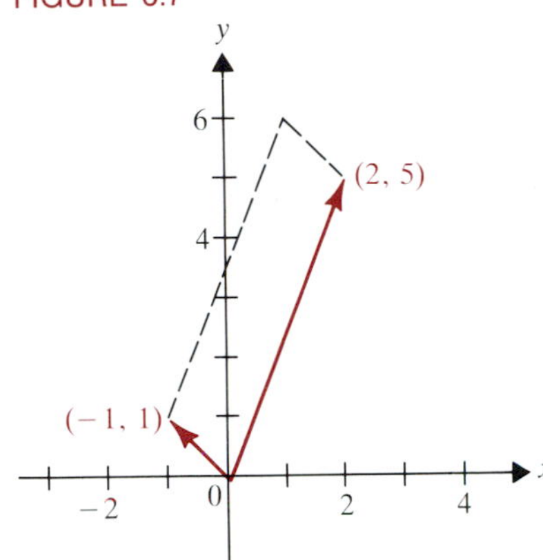

FIGURE 6.7

Solution. We can regard these vectors as vectors in $\mathbf{R}^3$, namely, $\mathbf{v} = (2, 5, 0)$ and $\mathbf{w} = (-1, 1, 0)$. Then by (11) the area of the parallelogram they determine is

$$\begin{aligned}|\mathbf{v}\times\mathbf{w}| &= |(5\cdot 0 - 0\cdot 1, 0\cdot(-1) - 2\cdot 0, 2\cdot 1 - 5\cdot(-1)| \\ &= |(0, 0, 2+5)| = 7.\end{aligned}$$ ■

There is a natural three-dimensional analogue to the above problem of determining the *area of the parallelogram formed by two vectors.* Given *three* vectors $\mathbf{u}$, $\mathbf{v}$, and $\mathbf{w}$ in $\mathbf{R}^3$, we can ask for the *volume of the parallelepiped they form.* Of course, they might fail to form a parallelepiped (if two of the vectors are collinear, for instance). But in most cases (see **Figure 6.8**) we expect that they will form one. In Euclidean solid geometry, the volume of a parallelepiped is defined to be the area of the base times the altitude drawn to that base. If we consider the base formed by $\mathbf{u}$ and $\mathbf{v}$, then the corresponding altitude is the absolute value of the coordinate of $\mathbf{w}$ in the direction of $\mathbf{u}\times\mathbf{v}$ (**Figure 6.9**). Let θ be the angle between $\mathbf{u}\times\mathbf{v}$ and $\mathbf{w}$. Then the altitude is $|\mathbf{w}|\,|\cos\theta|$. The area of the parallelogram formed by $\mathbf{u}$ and $\mathbf{v}$ is $|\mathbf{u}\times\mathbf{v}|$. So the volume V of the parallelepiped is given by

$$V = |\mathbf{u}\times\mathbf{v}|\,|\mathbf{w}|\,|\cos\theta| = |(\mathbf{u}\times\mathbf{v})\cdot\mathbf{w}|. \qquad (12)$$

FIGURE 6.8

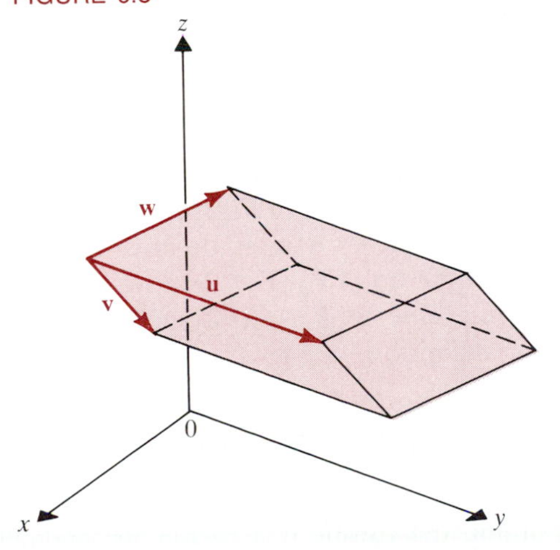

FIGURE 6.9

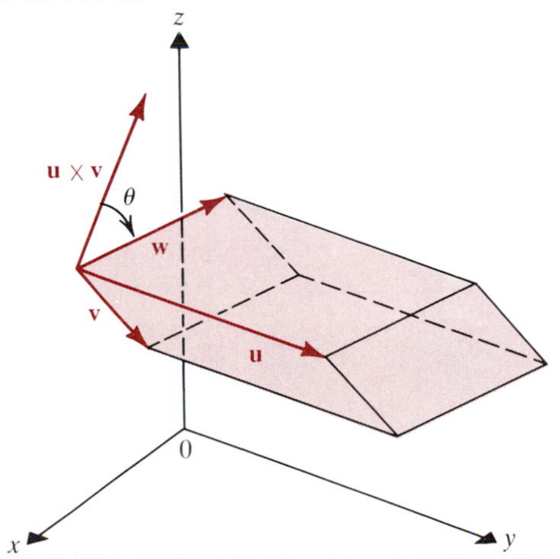

This is absolute value of the triple scalar product of $\mathbf{u}$, $\mathbf{v}$, and $\mathbf{w}$ [Theorem 6.2(d)]. To sum up, this discussion shows that the volume of the parallelepiped formed by vectors $\mathbf{u}$, $\mathbf{v}$, and $\mathbf{w}$ in $\mathbf{R}^3$ is the absolute value of their triple scalar product,

$$V = |(\mathbf{u}\times\mathbf{v})\cdot\mathbf{w}|. \qquad (13)$$

Since we take the absolute value, instead of the triple scalar product

$$(\mathbf{u} \times \mathbf{v}) \cdot \mathbf{w} = \mathbf{u} \cdot (\mathbf{v} \times \mathbf{w}) = \mathbf{v} \cdot (\mathbf{w} \times \mathbf{u}),$$

we could just as well compute its negative, which by (9) and Theorem 6.2(g) is

$$(\mathbf{v} \times \mathbf{u}) \cdot \mathbf{w} = \mathbf{u} \cdot (\mathbf{w} \times \mathbf{v}) = \mathbf{v} \cdot (\mathbf{u} \times \mathbf{w}).$$

FIGURE 6.10

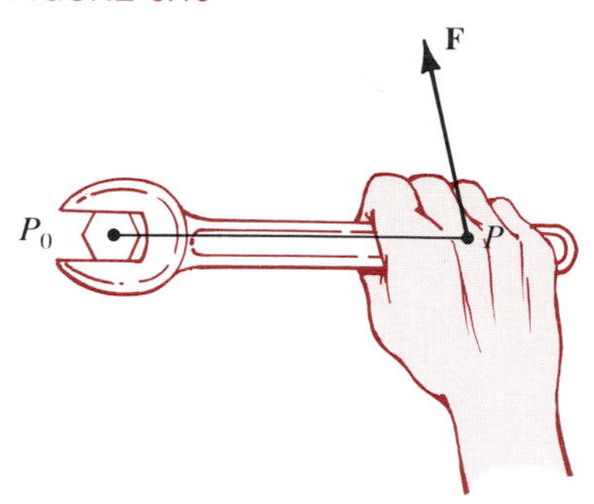

This corresponds to the fact that we could just as well consider the base of the parallelepiped to be the parallelogram formed by **u** and **w** (or **w** and **u**), or the parallelogram formed by **v** and **w** (or **w** and **v**). Even if **u**, **v**, and **w** fail to form a parallelepiped, we can still apply (13) sensibly. For if two of the vectors are collinear, say **u** and **v**, then the angle between them is 0. Then by (10),

$$|\mathbf{u} \times \mathbf{v}| = |\mathbf{u}|\,|\mathbf{v}| \sin 0 = 0.$$

Thus (13) tells us that **u**, **v**, and **w** fail to form a parallelepiped, by giving 0 as its volume.

FIGURE 6.11

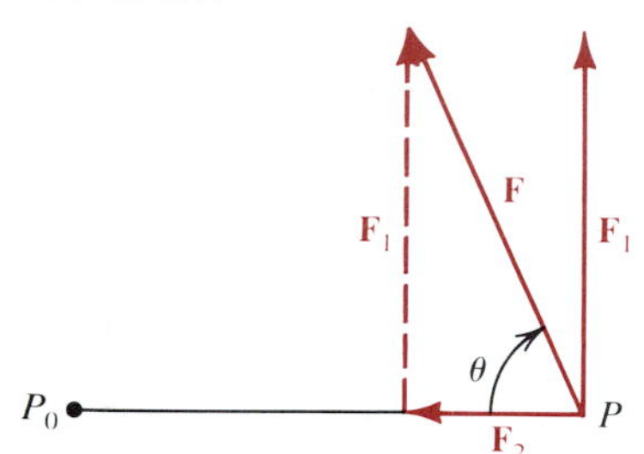

Example 7

Determine whether $\mathbf{u} = (1, 3, -1)$, $\mathbf{v} = (-2, 0, 3)$, and $\mathbf{w} = (2, 1, -2)$ form a parallelepiped in space. If they do, then what is its volume?

Solution. Here $\mathbf{u} \times \mathbf{v} = (9 - 0, 2 - 3, 0 + 6) = (9, -1, 6)$. So

$$(\mathbf{u} \times \mathbf{v}) \cdot \mathbf{w} = (9, -1, 6) \cdot (2, 1, -2) = 18 - 1 - 12 = 5.$$

Therefore, the given vectors do form a parallelepiped, and its volume is 5. ■

Torque*

FIGURE 6.12

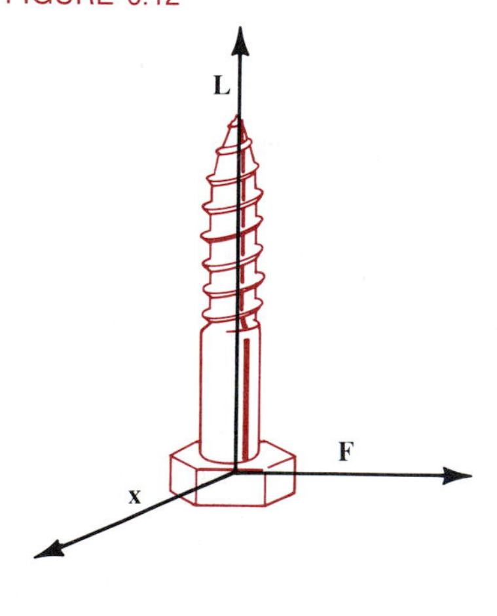

We close this section by examining *torque*, a physical application that helped lead to the definition of the cross product by J. W. Gibbs (see the historical note for Section 1.1). In everyday experience, the application of a force to an object "off-center" tends to cause the object to rotate, as in **Figure 6.10.** Suppose that a force **F** is applied to a wrench at point P, and that a bolt rotates about P_0. The tendency for the force to produce rotation, called torque, is proportional to the distance from P to P_0 and to the coordinate of **F** perpendicular to $\mathbf{P_0P}$.

When **F** is perpendicular to $\mathbf{P_0P}$, there is an ancient definition of Archimedes for the magnitude of the torque: $|\mathbf{x}|\,|\mathbf{F}|$, where $\mathbf{x} = \mathbf{P_0P}$. For a general force **F**, we resolve **F** into components $\mathbf{F}_1$ and $\mathbf{F}_2$ as in **Figure 6.11.** The component $\mathbf{F}_1$ is the one that produces torque, and its magnitude is easily computed if we draw $\mathbf{F}_1$ with its initial point at the tip of $\mathbf{F}_2$. For then we have $|\mathbf{F}_1| = |\mathbf{F}| \sin\theta$, so the magnitude of the torque due to **F** is $|\mathbf{x}|\,|\mathbf{F}| \sin\theta = |\mathbf{x} \times \mathbf{F}|$. It is then natural to give the following definition.

6.4 DEFINITION

The ***torque*** **L** at P_0 produced by a force **F** acting on a mass at point P is $\mathbf{L} = \mathbf{P_0P} \times \mathbf{F}$.

According to this definition, the torque is a *vector* perpendicular to both $\mathbf{P_0P}$ and **F**, that is, a vector in the axis of rotation that the force tends to produce. The direction of this vector is the direction of advance of a screw with a right-handed thread acted on by **F**. See **Figure 6.12.** (This accounts in part for the term *right-handed* as applied to screw threading.) As noted following

* Optional

(9), $\mathbf{k} = \mathbf{i} \times \mathbf{j}$, so the positive z-axis is in the direction of advance of a screw with right-handed thread whose head is at the origin and which is turned in the direction from the positive x-axis to the positive y-axis.

Exercises 10.6

In Exercises 1–12, find the equation of the specified plane.

1. Through the three points $P(1, -1, 3)$, $Q(0, 1, 3)$, and $R(-2, 1, -1)$. Check by verifying that the coordinates of P, Q, and R satisfy your equation.

2. Through the three points $P(2, 1, -1)$, $Q(-3, 1, 2)$, and $R(-1, 3, 3)$. Check by verifying that the coordinates of all three points satisfy your equation.

3. Through the points $P(3, 0, 1)$, $Q(2, -1, 3)$, and $R(-1, 4, 1)$

4. Through the points $(0, 1, 1)$, $(1, 1, 0)$, and $(1, 0, 1)$

5. Through the two lines $\mathbf{x} = s(3, -3, -1) + (2, 0, -1)$, $\mathbf{x} = t(1, -1, 2) + (3, -1, 1)$. Check by verifying that the coordinates of all points on the lines satisfy your equation.

6. Through the lines $\mathbf{x} = s(0, 1, -5) + (3, 1, -1)$ and $\mathbf{x} = t(1, 2, -3) + (0, 3, 4)$. Check as in Exercise 5.

7. Through the lines $\mathbf{x} = s(5, 3, -3) + (1, 1, 2)$ and $\mathbf{x} = t(2, 1, -1) + (5, 3, 0)$

8. Through the lines $\mathbf{x} = s(1, -1, 0) + (-1, 3, 0)$ and $\mathbf{x} = t(-2, 0, 3) + (2, 1, 5)$

9. Containing the point $(2, 5, -1)$ and the line $\mathbf{x} = t(-1, 0, 2) + (-1, 0, 2)$

10. Containing the point $(-3, 1, 1)$ and the line $\mathbf{x} = t(1, -2, 2) + (2, 0, 3)$

11. Containing the point $(0, 1, -1)$ and the line $\mathbf{x} = (2 - t, -3 + 2t, -t)$

12. Containing the point $(-1, 3, -3)$ and the line $\mathbf{x} = (-1 + 2t, -2 + t, 5 - t)$

In Exercises 13–16, find the area of the given parallelogram.

13. Formed by the vectors $\mathbf{u} = \mathbf{i} + 5\mathbf{j} + \mathbf{k}$ and $\mathbf{v} = -2\mathbf{i} + \mathbf{j} + 3\mathbf{k}$

14. With vertices $(2, 1, -1)$, $(-1, 3, 0)$, $(-3, 2, 1)$

15. With vertices $(0, 0)$, $(2, 3)$, $(-1, 2)$, and $(1, 5)$

16. Formed by the vectors $2\mathbf{i} - 3\mathbf{j}$ and $5\mathbf{i} + \mathbf{j}$

17. Show that the area of the triangle whose vertices are the endpoints of the vectors $\mathbf{u}$, $\mathbf{v}$, and $\mathbf{w}$ is

$$A = \frac{1}{2}|(\mathbf{v} - \mathbf{u}) \times (\mathbf{w} - \mathbf{u})|.$$

In Exercises 18–21, find the area of the given triangle. (See Exercise 17.)

18. With vertices $(2, 1)$, $(-1, 1)$, and $(3, 0)$

19. With vertices $(1, -2)$, $(2, 3)$, and $(-1, -3)$

20. With vertices $(-2, 1, 5)$, $(4, 0, 6)$, and $(3, -3, 2)$

21. With vertices $(0, 0, 0)$, $(2, 3, -2)$, and $(-1, 1, 4)$

22. Show that $\mathbf{j} \times \mathbf{k} = \mathbf{i}$ and $\mathbf{k} \times \mathbf{i} = \mathbf{j}$.

In Exercises 23–30, determine whether the given vectors form a parallelepiped in $\mathbf{R}^3$. If they do, find its volume.

23. $(4, 1, -1)$, $(-1, 2, 1)$, $(0, 3, 2)$

24. $(1, -2, 3)$, $(-2, 5, -1)$, $(1, 0, 1)$

25. $(-3, 1, 1)$, $(1, -2, -2)$, $(5, -1, 1)$

26. $(1, 2, 3)$, $(2, 1, 1)$, $(3, 3, 1)$

27. $(1, 3, -1)$, $(2, -1, 2)$, $(-1, 11, -7)$

28. $(2, -2, 0)$, $(-1, 2, -1)$, $(0, -1, 2)$

29. $(2, 0, 1)$, $(-1, 3, 2)$, $(2, 3, 4)$

30. $(3, -1, 1)$, $(2, 0, 3)$, $(4, -2, -1)$

31. A constant force $\mathbf{F} = 3\mathbf{i} - \mathbf{j} + 2\mathbf{k}$ acts on a particle located at $(2, 1, 1)$. Find the resulting torque at **(a)** the origin, **(b)** $(-1, 1, 0)$.

32. A force $\mathbf{F} = 2\mathbf{i} - 5\mathbf{j} + \mathbf{k}$ acts on a particle located at $(1, -1, 3)$. Find the resulting torque at **(a)** the origin, **(b)** $(2, 1, -1)$.

33. **(a)** Prove Theorem 6.2(c). **(b)** Prove Theorem 6.2(e).
(c) Prove Theorem 6.2(f).
(d) Prove that $\mathbf{u} \cdot (\mathbf{v} \times \mathbf{w}) = (\mathbf{u} \times \mathbf{v}) \cdot \mathbf{w}$, the second equality in Theorem 6.2(g).

34. **(a)** Verify the equality of the respective second and third coordinates in Theorem 6.2(h).
(b) Show that $\mathbf{u} \cdot (\mathbf{v} \times \mathbf{w}) = -\mathbf{v} \cdot (\mathbf{u} \times \mathbf{w}) = -\mathbf{w} \cdot (\mathbf{v} \times \mathbf{u})$.

35. If A, B, and C lie on the plane $2x - 3y + 2z = 5$, then what are the possible direction vectors of $\mathbf{AB} \times \mathbf{AC}$?

36. For the plane $-3x + 2y + z = 3$, if A, B, and C lie on the plane, then what are the possible direction vectors of $\mathbf{AB} \times \mathbf{BC}$?

37. If the unit vectors $\mathbf{u}$ and $\mathbf{v}$ make angle $\alpha = \cos^{-1}\frac{3}{5}$, then find the area of the parallelogram formed by $\mathbf{u}$ and $\mathbf{v}$.

38. Repeat Exercise 37 if $\alpha = \cos^{-1}\frac{5}{12}$.

39. Show that

$$(\mathbf{a} \times \mathbf{b}) \cdot (\mathbf{c} \times \mathbf{d}) = \det\begin{pmatrix} \mathbf{a} \cdot \mathbf{c} & \mathbf{b} \cdot \mathbf{c} \\ \mathbf{a} \cdot \mathbf{d} & \mathbf{b} \cdot \mathbf{d} \end{pmatrix}.$$

40. Prove the *Jacobi identity:* For all $\mathbf{u}$, $\mathbf{v}$, and $\mathbf{w}$ in $\mathbf{R}^3$,

$$\mathbf{u} \times (\mathbf{v} \times \mathbf{w}) + \mathbf{v} \times (\mathbf{w} \times \mathbf{u}) + \mathbf{w} \times (\mathbf{u} \times \mathbf{v}) = \mathbf{0}$$

41. A collection L of vectors is called a *Lie algebra* if Theorem 1.2 holds for L and if there is a multiplication of vectors defined on L that satisfies, for all $\mathbf{u}$, $\mathbf{v}$, and $\mathbf{w}$ in L and real numbers a and b,
(a) $\mathbf{vv} = \mathbf{0}$
(b) $(\mathbf{u} + \mathbf{v})\mathbf{w} = \mathbf{uw} + \mathbf{vw}$ and $\mathbf{w}(\mathbf{u} + \mathbf{v}) = \mathbf{wu} + \mathbf{wv}$
(c) $(a\mathbf{v})(b\mathbf{w}) = ab(\mathbf{vw})$
(d) $\mathbf{u}(\mathbf{vw}) + \mathbf{v}(\mathbf{wu}) + \mathbf{w}(\mathbf{uv}) = \mathbf{0}$
Show that $\mathbf{R}^3$ with the vector cross product is a Lie algebra. (Lie algebras have found application in higher mathematics, nuclear particle theory, quantum mechanics, materials science, and control theory.)

42. If $\mathbf{x} = s\mathbf{u} + \mathbf{x}_0$ and $\mathbf{x} = t\mathbf{v} + \mathbf{x}_1$ are two skew lines in $\mathbf{R}^3$, then show that
(a) $\mathbf{N} = \dfrac{\mathbf{u} \times \mathbf{v}}{|\mathbf{u} \times \mathbf{v}|}$ is a unit vector perpendicular to both lines.
(b) the distance between the lines is $|\mathbf{N} \cdot (\mathbf{x}_0 - \mathbf{x}_1)|$. See **Figure 6.13.**

FIGURE 6.13

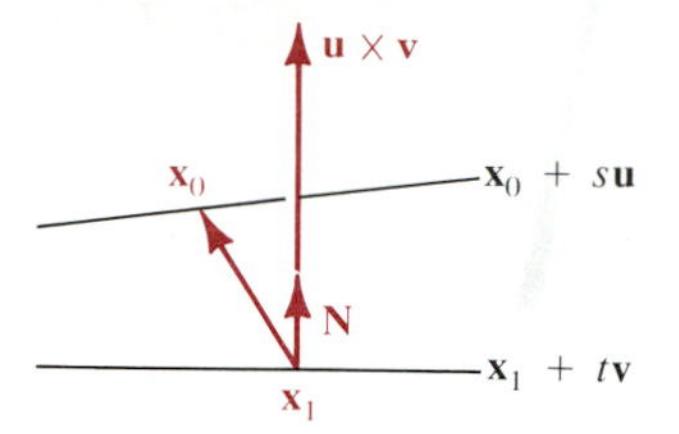

43. Use Exercise 42 to find the distance between the skew lines $\mathbf{x} = s(1, 1, 1)$ and $\mathbf{x} = t(1, 1, 2) + (-1, 0, 2)$.

44. Repeat Exercise 43 for the lines $\mathbf{x} = s(1, -2, 3) + (2, 1, -1)$ and $\mathbf{x} = t(1, -2, 0) + (1, 5, 2)$.

45. For the lines $\mathbf{x} = \mathbf{0} + t(1, 1, 0)$ and $\mathbf{x} = (0, 0, 3) + s(1, -1, 3)$, show that the approach of Example 3 produces the equation $3x - 3y - 2z = 0$. Then show that the points of the second line fail to lie on this plane. (This allows you to conclude that the lines are skew.)

46. Use Theorem 6.2(h) to produce an example showing that the cross product fails to be associative.

> HISTORICAL NOTE
>
> M. Sophus Lie (1842–1899, last name pronounced *Lee*) was professor of mathematics at the University of Christiana in Oslo from 1872 to 1886 and at the University of Leipzig from 1886 to 1898. He helped in the advancement of geometry and differential equations, his greatest contribution being the combination of three of the main branches of mathematics (algebra, analysis, and geometry) in the notion of a Lie group. Those structures are studied through Lie algebras via a generalization of the tangent approximation.

10.7 Looking Back

This chapter has introduced the algebraic and geometric properties of two- and three-dimensional vectors, and indicated how they extend to higher dimensions. We began in Section 1 with plane vectors, and discussed three-dimensional vectors in Section 2 after introducing the Cartesian coordinate system for $\mathbf{R}^3$. The basic arithmetic operations (Definitions 1.1 and 2.1) share the common algebraic properties listed in Theorem 1.2. The distance formula in $\mathbf{R}^3$ [Equation (4), p. 640] parallels the one for $\mathbf{R}^2$. The vector from a point P to another point Q is calculated as $\mathbf{0Q} - \mathbf{0P}$ in both $\mathbf{R}^2$ and $\mathbf{R}^3$. The standard bases in $\mathbf{R}^2$, $\mathbf{R}^3$, and $\mathbf{R}^n$ are all constructed in the same way (Definitions 1.3, 2.2, and 2.4). Section 3 focused on direction and angles (Definitions 3.8 and 3.4) based on the dot product (Definition 3.1). Important facts include the Cauchy–Schwarz inequality (Theorem 3.3), the triangle inequality (Theorem 3.5), and the conditions for vectors to be parallel (Definition 3.7) or perpendicular [Equation (11), p. 648].

In Section 4, the vector, parametric, and symmetric scalar equations of a line are given (Definitions 4.1, 4.2, and 4.3). Parallel and perpendicular lines are described in Definition 4.4. Theorem 4.5 gives an algebraic characterization of the line segment joining two points P and Q.

Sections 5 and 6 treated planes (Definition 5.1) and the vector cross product in $\mathbf{R}^3$ (Definition 6.1). The principal properties of the cross product are listed in Theorem 6.2. Unlike the dot product, which is defined on $\mathbf{R}^n$ for any $n \geq 2$, we defined the cross product *only* in $\mathbf{R}^3$.

CHAPTER CHECKLIST

Section 1: geometric vectors in $\mathbf{R}^1$ and $\mathbf{R}^2$, coordinates; vector sum, scalar product, vector difference, length; parallelogram law; vector from a point A to B; unit vector, standard basis, linear combination; unit vector in the direction of a vector (direction vector); parallel vectors.

Section 2: three-dimensional real Euclidean space; Cartesian coordinates in $\mathbf{R}^3$, coordinate planes, octants; right-handed orientation of axes; vector sum and difference, scalar multiple, length; vector joining two points; distance formula, sphere; standard basis; $\mathbf{R}^n$; coordinates, unit vector, standard basis.

Section 3: dot product; Cauchy–Schwarz inequality, angle between two vectors in $\mathbf{R}^n$, perpendicular vectors; triangle inequality; Pythagorean theorem; direction vector, direction angles, parallel vectors; coordinate and component of one vector in the direction of another; resolution, tangential and normal components.

Section 4: vector, parametric scalar, and symmetric scalar equations of a line in $\mathbf{R}^3$; parallel, perpendicular, and skew lines; line segments.

Section 5: normal-form equation of a plane; traces in coordinate planes; parallel and perpendicular lines; intersections of lines and planes with other planes; distance from a point to a plane.

Section 6: cross product in $\mathbf{R}^3$; determinants of 2-by-2 and 3-by-3 matrices; planes through two lines or three points; triple vector product, triple scalar product; area of parallelogram formed by two vectors, volume of parallelepiped formed by three vectors; torque.

REVIEW EXERCISES 10.7

1. If $\mathbf{v} = -\mathbf{i} + 3\mathbf{j}$ and $\mathbf{w} = 2\mathbf{i} - \mathbf{j}$, then find
(a) $2\mathbf{v} + 3\mathbf{w}$ **(b)** $\mathbf{v} - 4\mathbf{w}$ **(c)** $|\mathbf{v}|$ **(d)** $u_{\mathbf{v}}$

2. If $A(-2, 5)$, $B(1, -2)$, then find $\mathbf{AB}$ and $|\mathbf{AB}|$.

3. **(a)** Is $\mathbf{v} = (2, 1/\sqrt{3})$ parallel or perpendicular to $\mathbf{w} = (1/2, \sqrt{3})$?
(b) Repeat (a) for $\mathbf{v} = \sqrt{3}\mathbf{i} - 2\mathbf{j}$ and $\mathbf{w} = -\dfrac{\sqrt{3}}{2}\mathbf{i} + \mathbf{j}$.
(c) Repeat (a) for $\mathbf{v} = (\sqrt{3}, -2)$ and $\mathbf{w} = -\frac{1}{2}(1, \sqrt{5}/2)$.

4. **(a)** Find a unit vector perpendicular to $\mathbf{v} = -\mathbf{i} + 3\mathbf{j}$.
(b) Find a unit vector parallel to $\mathbf{v} = -\mathbf{i} - 2\mathbf{j}$. What is the vector called?
(c) Find a vector parallel to the line $ax + by + c = 0$.

5. If $\mathbf{v} = (2, -1, 2)$ and $\mathbf{w} = (-3, -1, 1)$, then compute
(a) $-2\mathbf{v} + 3\mathbf{w}$, **(b)** $\frac{1}{3}\mathbf{v} - \frac{4}{5}\mathbf{w}$, **(c)** $|\mathbf{v}|$.
(d) Find the unit vectors in the direction of $\mathbf{v}$ and $\mathbf{w}$.

6. Find the center and radius of the sphere whose equation is

$$x^2 + y^2 + z^2 + 7y - 5z = 7/4.$$

7. Find the angles between
(a) $(1/2, -3/2)$ and $(-9/2, -3/2)$
(b) $\mathbf{i} - \mathbf{j} + 3\mathbf{k}$ and $-4\mathbf{i} + 2\mathbf{j} + 2\mathbf{k}$

8. Find the direction vector and direction angles of $\mathbf{i} + \mathbf{j} + \sqrt{2}\mathbf{k}$.

9. **(a)** Find a unit vector orthogonal to $3\mathbf{i} + 2\mathbf{j} + 6\mathbf{k}$. How many such vectors are there?
(b) Repeat part (a) for $3\mathbf{i} + 5\mathbf{j}$.

10. Find the component of $-2\mathbf{i} + 3\mathbf{j} + \mathbf{k}$ in the direction of $2\mathbf{i} - 3\mathbf{j} + 6\mathbf{k}$.

11. Find vector, parametric scalar, and symmetric scalar equations of the line
(a) through $(-1, 1, 3)$ and $(1, -1, -3)$.
(b) through $(2, -1, 3)$ in the direction of $(1/3, 2/3, 2/3)$.

12. Find parametric equations for the line through $(2, 0, -1)$ which is parallel to the line

$$\frac{x}{-2} = \frac{y - 3}{-2} = \frac{2z + 1}{3}.$$

13. Do the lines $\mathbf{x} = t(-1, 3, 1) + (1, -1, 0)$ and $x = s$, $y = 6 + s$, $z = -4 - 6s$ intersect? If so, find their point of intersection. If not, explain why not.

14. Find the vector equation of the line through the origin that is parallel to the line with symmetric equations

$$\frac{x - 2}{4} = \frac{3y - 2}{6} = \frac{2z - 3}{2}.$$

15. Are the lines

$$\frac{x+2}{-3} = \frac{y}{2} = \frac{z-1}{-1}$$

and

$$x = 1 + t,\ y = -2 - 2t,\ z = 7/2 - 7t$$

perpendicular?

16. Do the lines $\mathbf{x} = s(-1, 2, 2) + (2, 0, 1)$ and $\mathbf{x} = t(-1, 2, 2) + (1, 0, 0)$ intersect? If so, where?

17. The line through $(3, 0, -2)$ perpendicular to a certain plane intersects it at the point $(3, -3, 2)$. Find a scalar equation for the plane.

18. The line through the origin perpendicular to a plane intersects the plane at the point $(1, -3, -1)$. Find the scalar equation of the plane. Sketch.

19. A line l is perpendicular to the plane Π whose equation is $x - 2y + 2z = 11$.

(a) If l intersects the plane at $(1, -2, 3)$, then find a vector equation for l.

(b) Find the equation of the planes parallel to Π and 3 units from Π.

(c) Find the trace of Π in the xy-plane.

20. A plane through $(-1, 5, -2)$ is perpendicular to the line through $(-1, 1, 4)$ and $(2, -1, 3)$. Find a scalar equation of the plane.

21. Find the equation of the plane through $(-1, 4, 1)$ that is parallel to the plane whose equation is $5x - 2y + z = 3$.

22. Find vector and scalar equations for the line of intersection of the planes $-x + 3y - 2z = -4$ and $2x - 5y + z = 3$.

23. Find the distance from the point $(-1, 2, 3)$ to the plane $2x - y + 3z = 5$.

24. Is the line $\mathbf{x} = t(1, 0, 4) + (-1, 3, 1)$ normal to the plane $12x - 7y + 8z = 19$? Find the point of intersection.

25. Find the equation of the plane through $P(2, -1, 2)$, $Q(-1, 3, 2)$, and $R(5, 0, -3)$. Check by verifying that the coordinates of P, Q, and R satisfy your equation.

26. Find the equation of the plane through the two lines $\mathbf{x} = t(1, 1, 3) + (0, 2, 3)$ and $\mathbf{x} = s(1, -5, -1) = (1, -3, 2)$. Check by verifying that the coordinates of all points on the line satisfy your equations.

27. Find the scalar equation of the plane through the points $P(-1, -1, 2)$, $Q(0, 3, -1)$, and $R(2, -3, 5)$. Also find the normal form equation of this plane.

28. A constant force $\mathbf{F} = \mathbf{i} - \mathbf{j} + 5\mathbf{k}$ acts on a particle located at $(-1, 3, -1)$. Find the resulting torque (a) at the origin and (b) at $(2, 3, 0)$.

29. Find the area of the parallelogram in the plane with vertices $(0, 0)$, $(-3, 1)$, $(5, 2)$, and $(2, 3)$.

30. True or false? (If true, give a proof, if false, give an example to show that the statement need not always hold.) If $\mathbf{u} \times \mathbf{v} = \mathbf{0}$ then $\mathbf{u} \cdot \mathbf{v} = a|\mathbf{v}|^2$ for some real number a.

31. Evaluate

$$\det \begin{pmatrix} 2 & -1 & 2 \\ 3 & 2 & -4 \\ 4 & 1 & 0 \end{pmatrix}.$$

32. Do the vectors $-2\mathbf{i} + \mathbf{k}$, $3\mathbf{j} + \mathbf{k}$, and $\mathbf{i} + 5\mathbf{j} - 3\mathbf{k}$ form a parallelepiped in $\mathbf{R}^3$?

33. The atoms of most metallic elements are arranged so that they form *crystals*, which are made up of many unit cells arranged next to each other. One type of unit cell is a cube as in **Figure 7.1,** with one spherical atom centered at each corner and one more atom in the middle, tangent to each of the other atoms. The cell's edge s is measured from the center of the one corner atom to the center of an adjacent corner atom by x-ray diffraction. The diagonal AD in Figure 7.1 is $4r$, where r is the radius of an atom. Find r in terms of s.

FIGURE 7.1

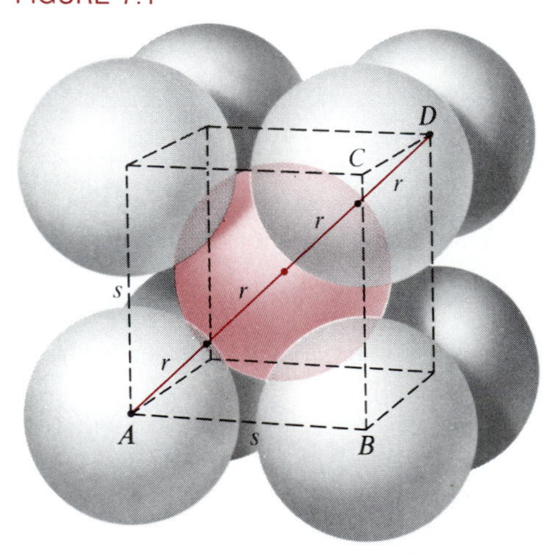

CHAPTER 11 Vector and Scalar Functions

11.0 Introduction

The last chapter provided the working knowledge of the Euclidean spaces $\mathbf{R}^n$ needed to study functions with domain or range (or both) in $\mathbf{R}^2$ or $\mathbf{R}^3$. In this chapter we first consider vector functions $\mathbf{f}: \mathbf{R} \to \mathbf{R}^n$ of one real variable, where almost always $n = 2$ or 3. In the three-dimensional case, we write

$$\mathbf{f}(t) = (f_1(t), f_2(t), f_3(t)) = f_1(t)\,\mathbf{i} + f_2(t)\,\mathbf{j} + f_3(t)\,\mathbf{k},$$

where each f_i is a real-valued function of t. For example, one such function is given by

$$\mathbf{f}(t) = (t^2, e^{-2t}, \tan^3 t) = t^2\,\mathbf{i} + e^{-2t}\mathbf{j} + \tan^3 t\,\mathbf{k}.$$

It is helpful to regard vector functions as resulting from putting the coordinate functions f_1, f_2, and f_3 (if $n = 3$) together to form a vector. The calculus of vector functions is carried out by performing the familiar calculus of earlier chapters on the coordinate functions, and combining the results into a vector.

The graph of a vector function $\mathbf{f}$ is a parametrized curve c in $\mathbf{R}^n$ (**Figure 0.1**). Just as in Chapter 2 we found tangent lines to the graphs of functions $f: \mathbf{R} \to \mathbf{R}$ by differentiation, so we differentiate $\mathbf{f}: \mathbf{R} \to \mathbf{R}^n$ to find tangent vectors and tangent lines to its graph. A formula for arc length analogous to the ones in Sections 5.4 and 8.5 is given in Section 2. Sections 2 and 3 use the calculus of vector functions to study moving particles in the plane or three-space.

In Section 4, we look at functions that are the reverse of vector functions, those whose *domains* are sets of vectors in $\mathbf{R}^n$ and whose values are real numbers. When $n = 2$, there is a natural graph in $\mathbf{R}^3$ of a function $f: \mathbf{R}^2 \to \mathbf{R}$. Such a graph is usually a two-dimensional set, a *surface* in $\mathbf{R}^3$ (see **Figure 0.2**). In Sections 5 and 6, we introduce methods for graphing such surfaces, as well as the slightly more general *quadric surfaces*, which are three-dimensional analogues of conic sections.

FIGURE 0.1

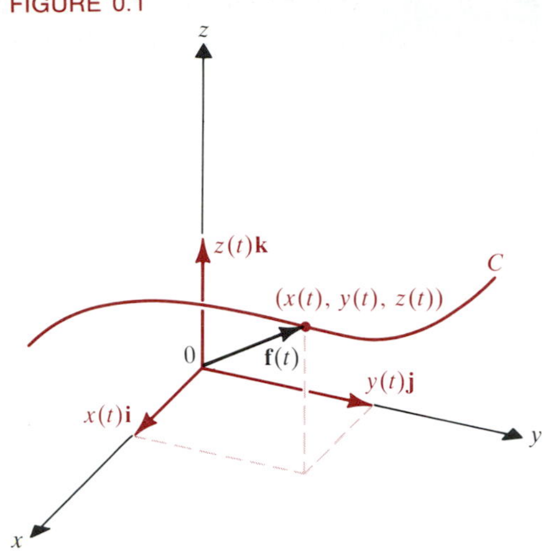

FIGURE 0.2

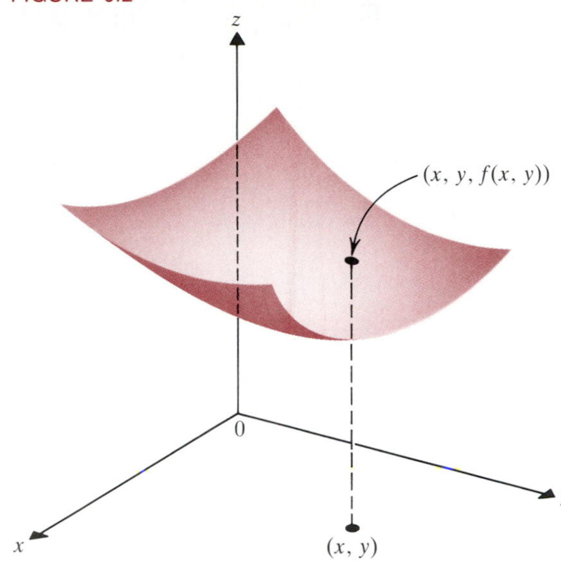

11.1 Vector-Valued Functions

We begin with the formal definition of the functions considered in the first part of the chapter.

1.1 DEFINITION A ***vector*** (or ***vector-valued***) ***function*** on $\mathbf{R}$ is a function $\mathbf{f}\colon \mathbf{R} \to \mathbf{R}^n$, where n is a fixed positive integer (in this text, usually $n = 2$ or 3). We write

$$\mathbf{f}(t) = (x_1(t), \ldots, x_n(t)) = \mathbf{x}(t) \in \mathbf{R}^n.$$

The real-valued functions f_i given by

$$f_i(t) = x_i(t), \qquad i = 1, 2, \ldots, n,$$

are called the ***coordinate functions*** of $\mathbf{f}$.

Vector functions occur frequently in applications. For instance, the path described by a moving particle in $\mathbf{R}^3$ is given by a function such as

$$\mathbf{f}(t) = (x(t), y(t), z(t)) = \left(t + 1, 2t, \frac{1}{2}t^2\right) = (t + 1)\,\mathbf{i} + (2t)\,\mathbf{j} + \left(\frac{1}{2}t^2\right)\mathbf{k}.$$

In studying population growth, demographers in each country try to develop mathematical models to describe the population at time t. The English economist Thomas R. Malthus (1766–1834) constructed the *exponential growth* model described by $dP/dt = kP$, where the constant k is called the *continuous growth rate*. Then (see Section 6.3), $P(t) = P_0 e^{kt}$. A world population growth model is a function $\mathbf{f}\colon \mathbf{R} \to \mathbf{R}^n$, where n is the number of countries in the world (about 160), and where $f_i(t)$ is the population in country i at time t. The total world population at time t, which is the sum of $f_i(t)$ over all i, might be less interesting than the nature of certain f_i corresponding to countries where the growth rate is high.

In both the above situations, it seems natural to study the vector function f by studying its coordinate functions f_i. That is how we will develop the calculus of vector functions f.

1.2 DEFINITION

Let $\mathbf{v} = (v_1, v_2, \ldots, v_n) \in \mathbf{R}^n$. Let $\mathbf{f}: \mathbf{R} \to \mathbf{R}^n$ be a vector function. Suppose for each $j = 1, 2, \ldots, n$ that $f_j(t)$ is defined on some open interval I_j containing a, except perhaps at $t = a$. Then

$$\lim_{t \to a} \mathbf{f}(t) = \mathbf{v}$$

means that $\lim_{t \to a} f_j(t) = v_j$ for each $j = 1, 2, \ldots, n$.

Example 1

If $\mathbf{f}(t) = \cos t\,\mathbf{i} + \sin t\,\mathbf{j} + t\,\mathbf{k}$, then find $\lim_{t \to \pi/2} \mathbf{f}(t)$.

Solution. $\displaystyle \lim_{t \to \pi/2} \mathbf{f}(t) = \lim_{t \to \pi/2} (\cos t\,\mathbf{i} + \sin t\,\mathbf{j} + t\,\mathbf{k}) = \mathbf{j} + \frac{\pi}{2}\mathbf{k}.$ ■

Definition 1.2 says that $\mathbf{f}$ approaches $\mathbf{v}$ as $t \to a$ precisely when each of the coordinate functions f_i approaches the corresponding coordinate of $\mathbf{v}$. It can be shown [Exercise 28(a)] that Definition 1.2 is equivalent to the condition that the distance between the endpoints of $\mathbf{f}(t)$ and $\mathbf{v}$ approach 0 as $t \to a$, that is,

$$|\mathbf{f}(t) - \mathbf{v}| \to 0 \qquad \text{as } t \to a.$$

This in turn is equivalent (Exercise 28) to the following analogue of the epsilon–delta definition of limit for functions $f: \mathbf{R} \to \mathbf{R}$ (Definition 4.1 of Chapter 1).

$$\lim_{t \to a} \mathbf{f}(t) = \mathbf{v}$$

if and only if for any $\varepsilon > 0$ there is some $\delta > 0$ such that

$$|\mathbf{f}(t) - \mathbf{v}| < \varepsilon \qquad \text{for all } 0 < |t - a| < \delta.$$

Continuity and differentiability for a vector function $\mathbf{f}$ are also defined coordinate-wise.

1.3 DEFINITION

A vector function $\mathbf{f}: \mathbf{R} \to \mathbf{R}^n$ is ***continuous*** at the point $t = a$ if and only if each of its coordinate functions f_i is continuous there. The function $\mathbf{f}$ is ***differentiable*** at $t = a$ if and only if each of its coordinate functions is differentiable at a. In this case the ***derivative*** of $\mathbf{f}$ at a is a vector in $\mathbf{R}^n$:

(1) $$\frac{d\mathbf{f}}{dt}(a) = \left(\frac{df_1}{dt}(a), \frac{df_2}{dt}(a), \ldots, \frac{df_n}{dt}(a)\right) = \mathbf{f}'(a).$$

Example 2

Is $\mathbf{f}$ in Example 1 continuous at $t = \pi/2$? Is it differentiable there? If so, find $\mathbf{f}'(\pi/2)$.

Solution. Each coordinate function of $\mathbf{f}$ is not only continuous but also differentiable at every point $t \in \mathbf{R}$, hence at $t = \pi/2$. We have

$$\mathbf{f}'(t) = -\sin t\,\mathbf{i} + \cos t\,\mathbf{j} + \mathbf{k} \to \mathbf{f}'(\pi/2) = -\mathbf{i} + \mathbf{k}. \quad ■$$

In view of Definition 1.2 we could write the condition for $\mathbf{f}$ to be continuous at $t = a$ as

(2) $$\lim_{t \to a} \mathbf{f}(t) = \mathbf{f}(a),$$

since our definition requires for each i that

$$\lim_{t \to a} f_i(t) = f_i(a).$$

This is again equivalent [by Exercise 29(a)] to the condition

$$\lim_{t \to a} |\mathbf{f}(t) - \mathbf{f}(a)| = 0. \tag{3}$$

Similarly, (1) is equivalent [by Exercise 29(b)] to the condition that

$$\lim_{h \to 0} \frac{1}{h} [\mathbf{f}(a + h) - \mathbf{f}(a)] \tag{4}$$

should exist. (When it does exist, its value is $\mathbf{f}'(a)$.)

Conditions (2) and (4) look exactly like Definitions 7.1 of Chapter 1 and 1.1 of Chapter 2. In each case the calculus concept for vector functions results from putting together the corresponding notion for the coordinate functions, just as we create vector functions by putting together the real-valued coordinate functions. If we wanted to take the time, we could prove that the sum of continuous functions is continuous, and other results akin to those found in Section 1.7. Some of these are in the exercises, but we will concentrate on the following collection of properties of differentiation.

1.4 THEOREM

Let $\mathbf{f}$ and $\mathbf{g}$ be differentiable vector functions and h a differentiable scalar function. Then

(a) $\dfrac{d}{dt}[\mathbf{f}(t) + \mathbf{g}(t)] = \mathbf{f}'(t) + \mathbf{g}'(t)$.

(b) $\dfrac{d}{dt}[\mathbf{f}(t) \cdot \mathbf{g}(t)] = \mathbf{f}'(t) \cdot \mathbf{g}(t) + \mathbf{f}(t) \cdot \mathbf{g}'(t)$.

(c) $\dfrac{d}{dt}[|\mathbf{f}(t)|] = \dfrac{1}{|\mathbf{f}(t)|}[\mathbf{f}(t) \cdot \mathbf{f}'(t)]$, if $\mathbf{f}(t) \neq \mathbf{0}$.

(d) $\dfrac{d}{dt}[h(t)\mathbf{f}(t)] = h'(t)\mathbf{f}(t) + h(t)\mathbf{f}'(t)$.

(e) If $\mathbf{f}, \mathbf{g}\colon \mathbf{R} \to \mathbf{R}^3$, then

$$\frac{d}{dt}[\mathbf{f}(t) \times \mathbf{g}(t)] = \mathbf{f}'(t) \times \mathbf{g}(t) + \mathbf{f}(t) \times \mathbf{g}'(t).$$

(f) **Chain Rule.** $\dfrac{d}{ds}\mathbf{f}(t(s)) = \dfrac{d\mathbf{f}}{dt}(t(s))\dfrac{dt}{ds}$, where $t = t(s)$ is a scalar function.

Partial Proof. We leave (b) and (d) for Exercise 27.

(a)
$$\begin{aligned}
\frac{d}{dt}[\mathbf{f}(t) + \mathbf{g}(t)] &= \frac{d}{dt}[(f_1(t), f_2(t), \ldots, f_n(t)) + (g_1(t), g_2(t), \ldots, g_n(t))] \\
&= \frac{d}{dt}[(f_1(t) + g_1(t), f_2(t) + g_2(t), \ldots, f_n(t) + g_n(t))] \\
&= (f'_1(t) + g'_1(t), f'_2(t) + g'_2(t), \ldots, f'_n(t) + g'_n(t)) \\
&= (f'_1(t), f'_2(t), \ldots, f'_n(t)) + (g'_1(t), g'_2(t), \ldots, g'_n(t)) \\
&= \mathbf{f}'(t) + \mathbf{g}'(t).
\end{aligned}$$

(c) If $\mathbf{f}(t) = (f_1(t), f_2(t), f_3(t))$ is a three-dimensional vector function, then

$$|\mathbf{f}(t)| = [(f_1(t))^2 + (f_2(t))^2 + (f_3(t))^2]^{1/2}.$$

This is just a real-valued function of the real variable t, so we can use the chain rule (Theorem 6.1 of Chapter 2) to differentiate it, getting

$$\begin{aligned}\frac{d}{dt}|\mathbf{f}(t)| &= \frac{1}{2}[(f_1(t))^2 + (f_2(t))^2 + (f_3(t))^2]^{-1/2} \\ &\quad \times \left[2f_1(t)\frac{df_1}{dt}(t) + 2f_2(t)\frac{df_2}{dt}(t) + 2f_3(t)\frac{df_3}{dt}(t)\right] \\ &= \frac{1}{|\mathbf{f}(t)|}[\mathbf{f}(t)\cdot\mathbf{f}'(t)],\end{aligned}$$

because the 1/2 cancels all the 2s. Similar reasoning applies in any dimension.

(e) Here we can write $\mathbf{f}(t) = f_1(t)\mathbf{i} + f_2(t)\mathbf{j} + f_3(t)\mathbf{k}$ and $\mathbf{g}(t) = g_1(t)\mathbf{i} + g_2(t)\mathbf{j} + g_3(t)\mathbf{k}$. Then

$$\begin{aligned}\mathbf{f}(t)\times\mathbf{g}(t) = {} & [f_2(t)g_3(t) - f_3(t)g_2(t)]\mathbf{i} + [f_3(t)g_1(t) - f_1(t)g_3(t)]\mathbf{j} \\ & + [f_1(t)g_2(t) - f_2(t)g_1(t)]\mathbf{k}.\end{aligned}$$

We use the product rule (Theorem 2.3 of Chapter 2) to differentiate $\mathbf{f}(t)\times\mathbf{g}(t)$ term-by-term, getting

$$\begin{aligned}\frac{d}{dt}[\mathbf{f}(t)\times\mathbf{g}(t)] &= [f'_2(t)g_3(t) - f'_3(t)g_2(t) + f_2(t)g'_3(t) - f_3(t)g'_2(t)]\mathbf{i} \\ &\quad + [f'_3(t)g_1(t) - f'_1(t)g_3(t) + f_3(t)g'_1(t) - f_1(t)g'_3(t)]\mathbf{j} \\ &\quad + [f'_1(t)g_2(t) - f'_2(t)g_1(t) + f_1(t)g'_2(t) - f_2(t)g'_1(t)]\mathbf{k} \\ &= [f'_2(t)g_3(t) - f'_3(t)g_2(t)]\mathbf{i} + [f'_3(t)g_1(t) - f'_1(t)g_3(t)]\mathbf{j} \\ &\quad + [f'_1(t)g_2(t) - f'_2(t)g_1(t)]\mathbf{k} + [f_2(t)g'_3(t) - f_3(t)g'_2(t)]\mathbf{i} \\ &\quad + [f_3(t)g'_1(t) - f_1(t)g'_3(t)]\mathbf{j} + [f_1(t)g'_2(t) - f_2(t)g'_1(t)]\mathbf{k} \\ &= \mathbf{f}'(t)\times\mathbf{g}(t) + \mathbf{f}(t)\times\mathbf{g}'(t).\end{aligned}$$

(f) The chain rule applied to each coordinate function gives

$$\begin{aligned}\frac{d}{ds}[\mathbf{f}(t(s))] &= \left(\frac{df_1}{ds}(t(s)), \frac{df_2}{ds}(t(s)), \ldots, \frac{df_n}{ds}t(s))\right) \\ &= \left(\frac{df_1}{dt}\frac{dt}{ds}, \frac{df_2}{dt}\frac{dt}{ds}, \ldots, \frac{df_n}{dt}\frac{dt}{ds}\right) \\ &= \left(\frac{df_1}{dt}, \frac{df_2}{dt}, \ldots, \frac{df_n}{dt}\right)\frac{dt}{ds} = \frac{d\mathbf{f}(t)}{dt}\frac{dt}{ds}.\end{aligned}$$ QED

In Definition 5.1 of Chapter 8 we defined a parametrized curve in the plane as the set of points $P(x, y)$ satisfying parametric equations on an interval I:

$$x = f(t),\ y = g(t), \qquad t \in I,$$

where f and g are continuous functions. Each such point is the endpoint of a two-dimensional vector $\mathbf{f}(t) = f(t)\mathbf{i} + g(t)\mathbf{j}$, where $t \in I$. This extends easily to three dimensions: a curve in $\mathbf{R}^3$ can be thought of as the set of endpoints of a three-dimensional vector

$$\mathbf{f}(t) = (f(t), g(t), h(t)) = f(t)\mathbf{i} + g(t)\mathbf{j} + h(t)\mathbf{k}$$

as t varies over some interval I of real numbers. This leads to the following formal definition.

1.5 DEFINITION A ***parametrized curve*** in $\mathbf{R}^n$ is a continuous vector function $\mathbf{f}: \mathbf{R} \to \mathbf{R}^n$.

We define a *parametrized* curve as the parametrizing function $\mathbf{f}$, rather than the range of f, to avoid some technical problems too subtle to go into here. To draw a parametrized curve in $\mathbf{R}^3$, simply plot some points $\mathbf{f}(t) = (x(t), y(t), z(t))$ in $\mathbf{R}^3$ and connect them smoothly, as suggested by **Figure 1.1.**

FIGURE 1.1

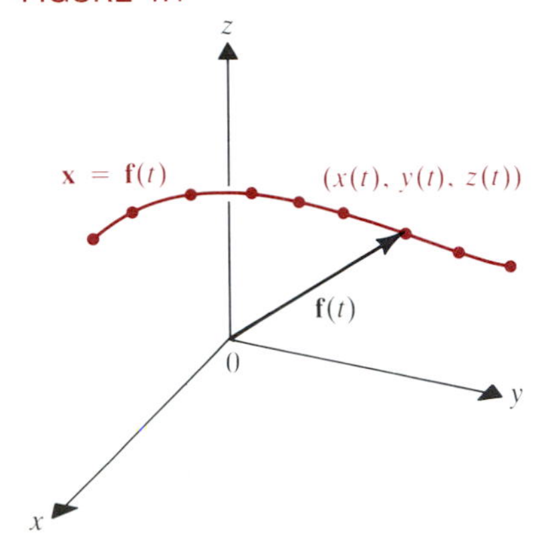

Example 3

Sketch the parametrized curve in Example 1,

$$\mathbf{f}(t) = \cos t\,\mathbf{i} + \sin t\,\mathbf{j} + t\,\mathbf{k}, \qquad t \in (-\infty, +\infty).$$

Solution. For each point $\mathbf{f}(t) = (x(t), y(t), z(t))$, we have $x(t)^2 + y(t)^2 = \cos^2 t + \sin^2 t = 1$. Thus each point on the curve is one unit from the z-axis. The z- coordinate of the endpoint of $\mathbf{f}(t)$ is t. Hence, as t increases, the point $\mathbf{x}(t) = (x(t), y(t), z(t))$ winds around the z-axis as it ascends. The resulting curve in $\mathbf{R}^3$ is called a ***circular helix***. It is sketched in **Figure 1.2.** ■

Recall from Section 8.5 that a *simple curve*, like the curve in Figure 1.2, does not intersect itself. To avoid ambiguities that might arise in finding tangent lines to curves, we restrict attention almost entirely to simple curves. This is not a severe limitation, because in most cases of interest, nonsimple curves like those in **Figure 1.3** can be broken up into two or more simple curves.

FIGURE 1.2

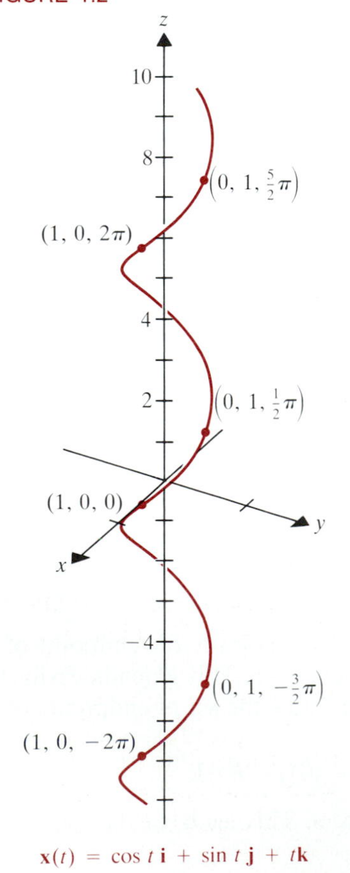

$\mathbf{x}(t) = \cos t\,\mathbf{i} + \sin t\,\mathbf{j} + t\mathbf{k}$

FIGURE 1.3

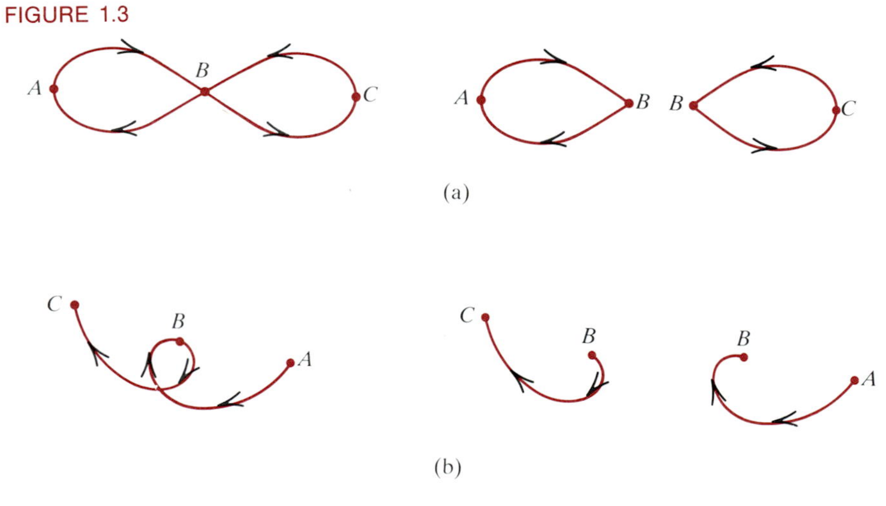

We imagine a simple curve

$$\mathbf{f}(t) = (f_1(t), f_2(t), \ldots, f_n(t)) = (x_1, x_2, \ldots, x_n) = \mathbf{x}(t),$$

being traced out as t varies over an interval I. **Figure 1.4** suggests a natural orientation for the curve, obtained by proceeding along it in the direction of increasing t. That is, $\mathbf{x}(t_1)$ ***precedes*** $\mathbf{x}(t_2)$ if $t_1 < t_2$. This orientation is particularly convenient when discussing particle motion.

As in Section 8.5 the real variable t in Definition 1.5 is called a ***parameter***, and the equations $x_i = f_i(t)$ are called ***parametric equations*** for the curve. The process of describing a curve by giving a vector function of the form (2) is called ***parametrization*** of the curve. As you may recall, recognizing common curves given in parametric form often involves eliminating the parameter t.

FIGURE 1.4

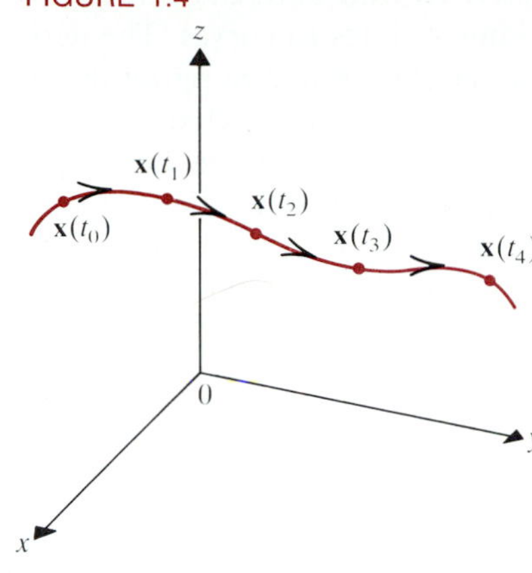

Example 4

Identify the plane curve defined by

$$\text{(a) } \mathbf{x}(t) = \left(\frac{t^2}{4p}, t\right), \quad p > 0, \qquad \text{(b) } \mathbf{y}(s) = \left(\frac{s^2 - 2s + 1}{4p}, 1 - s\right), \quad p > 0,$$

and show its orientation.

Solution. (a) We can calculate as many points as we like on the curve by assigning values to t. We show some such points in tabular form.

t	$-2\sqrt{p}$	-4	-2	-1	0	1	2	4	$2\sqrt{p}$
x	1	$4/p$	$1/p$	$1/4p$	0	$1/4p$	$1/p$	$4/p$	1
y	$-2\sqrt{p}$	-4	-2	-1	0	1	2	4	$2\sqrt{p}$

If we plot these points in order of increasing t, then we get a picture like **Figure 1.5**(a), where we have taken $p = 1/2$. The curve looks parabolic. To confirm this, we eliminate t from the equations. Since $y = t$ and $x = t^2/4p$, we have

$$x = \frac{y^2}{4p} \rightarrow y^2 = 4px.$$

We recognize this as the equation of a parabola with vertex at (0, 0). Thus we can fill in Figure 1.5(a) to get Figure 1.5(b).

FIGURE 1.5

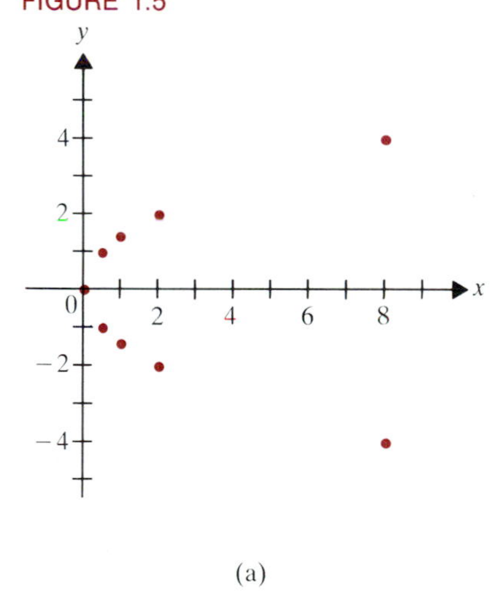

(a)

$\mathbf{x}(t) = \dfrac{t^2}{4p}\mathbf{i} + t\mathbf{j}$

(b)

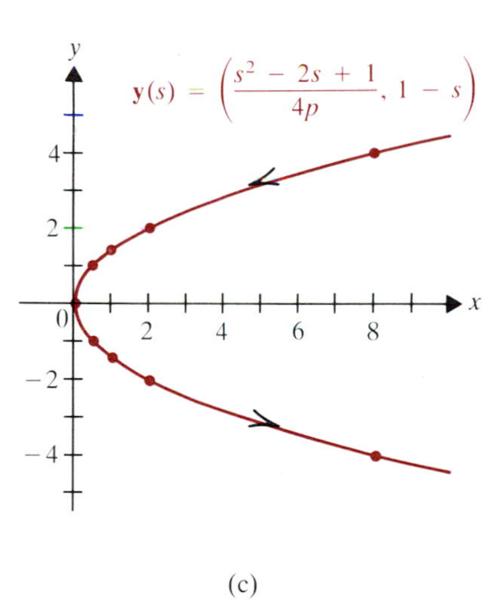

(c)

(b) We could tabulate points as above, but it is more instructive to first eliminate the parameter. Since $y = 1 - s$, we have $s = 1 - y$. Therefore

$$x = \frac{(1-y)^2 - 2(1-y) + 1}{4p} = \frac{1 - 2y + y^2 - 2 + 2y + 1}{4p} = \frac{y^2}{4p},$$

the *same* relation we got in (a). Thus the curve is the parabola of part (a). When $s = 0$, we get the point $(1/4p, 1)$. When $s = 2$, we obtain the point $(1/4p, -1)$. We therefore have the curve of Figure 1.5(c), the same parabola as in Figure 1.5(b) but with the *opposite* orientation. ■

FIGURE 1.6

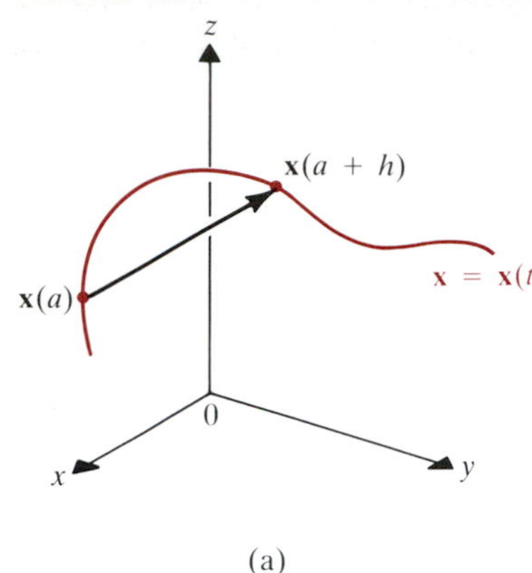

(a)

In Sections 2.1 and 2.9 we used the derivative to find both the velocity of particles moving along straight lines and the tangent lines to curves. The derivative of a vector function can be put to the same kind of use. Suppose that **f**: $\mathbf{R} \to \mathbf{R}^3$ is differentiable at $t = a$. Then it is continuous (Exercise 31), so for small values of h the endpoints of $\mathbf{x}(a)$ and $\mathbf{x}(a + h)$ are close. See **Figure 1.6**(a). The vector $\mathbf{x}(a + h) - \mathbf{x}(a)$ is a *secant* vector from the endpoint of $\mathbf{x}(a)$ to the endpoint of $\mathbf{x}(a + h)$. Then for $h > 0$, the secant vector

$$\begin{aligned} \mathbf{s}_h &= \frac{1}{h}[\mathbf{x}(a+h) - \mathbf{x}(a)] \\ &= \left(\frac{x_1(a+h) - x_1(a)}{h}, \frac{x_2(a+h) - x_2(a)}{h}, \frac{x_3(a+h) - x_3(a)}{h}\right) \end{aligned} \tag{5}$$

points in the same direction, approximately the direction of motion of a particle moving along the curve. See **Figure 1.6**(b). Moreover, the magnitude of $\mathbf{s}_h$,

$$\frac{1}{|h|}|\mathbf{x}(a+h) - \mathbf{x}(a)|,$$

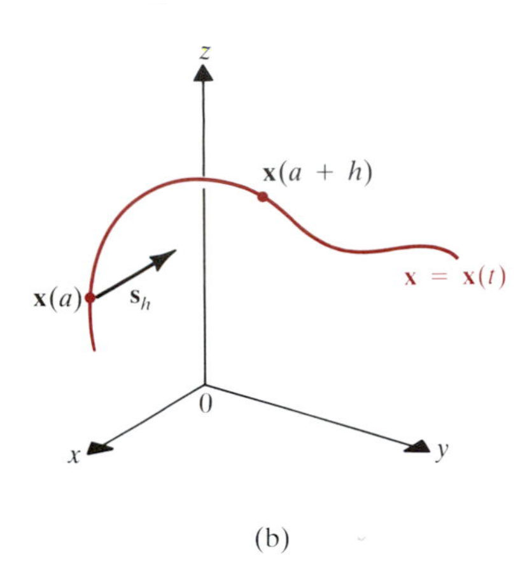

(b)

approximates the speed of motion along the curve, because it measures the average displacement from $\mathbf{x}(a)$ per unit time. Now as $h \to 0$, $\mathbf{s}_h$ seems to approach a tangent vector, and its magnitude seems to approach what we would think of as the instantaneous speed of the particle at $t = a$. Since **f** is differentiable, we see from (5) that the ith coordinate function approaches $(dx_i/dt)(a)$ for $i = 1, 2$, and 3. Thus $\mathbf{s}_h$ approaches

$$\mathbf{x}'(a) = \left(\frac{dx_1}{dt}(a), \frac{dx_2}{dt}(a), \frac{dx_3}{dt}(a)\right) = \frac{dx_1}{dt}(a)\,\mathbf{i} + \frac{dx_2}{dt}(a)\,\mathbf{j} + \frac{dx_3}{dt}(a)\,\mathbf{k}.$$

This leads to the following definition.

1.6 DEFINITION Let a particle move on the curve $\mathbf{x} = \mathbf{x}(t)$ in $\mathbf{R}^n$. If $\mathbf{x}'(a)$ exists, then the ***velocity vector*** along the curve at $\mathbf{x}(a)$ (or at $t = a$) is

$$\mathbf{v}(a) = \mathbf{x}'(a) = \frac{d\mathbf{x}}{dt}(a).$$

The ***speed*** of motion at $\mathbf{x}(a)$ is the magnitude $|\mathbf{v}(a)|$ of the velocity at $t = a$. If $\mathbf{x}'(a) \neq 0$, then the ***unit tangent vector*** at $\mathbf{x}(a)$ is

$$\mathbf{T}(a) = \frac{1}{|\mathbf{x}'(a)|}\mathbf{x}'(a).$$

See **Figure 1.7.**

FIGURE 1.7

$\mathbf{x} = f(t) = (x(t), y(t), z(t))$

In Definition 1.6 the velocity vector $\mathbf{v}(a) = \mathbf{x}'(a)$ is itself a tangent vector to the curve at each point $\mathbf{x}(a)$ where $\mathbf{x}(t)$ is differentiable. In fact, our development gives precisely the notion of tangent in Section 2.1 in case $n = 2$. A parametric representation of the graph of $y = g(x)$ can be obtained by letting $x = t$ and $y = g(t)$. That gives

$$\mathbf{x}(t) = (t, g(t)) = t\,\mathbf{i} + g(t)\,\mathbf{j}.$$

A tangent vector at $(a, g(a))$ is then

$$\mathbf{v}(a) = \mathbf{i} + g'(a)\,\mathbf{j} = \left(1, \frac{dg}{dx}(a)\right),$$

since $t = x$. Because the tangent line l is determined by the direction of $\mathbf{v}(a)$, we see that l has slope $(dg/dx)(a)$, in agreement with the definition of tangent line in Section 2.1.

FIGURE 1.8

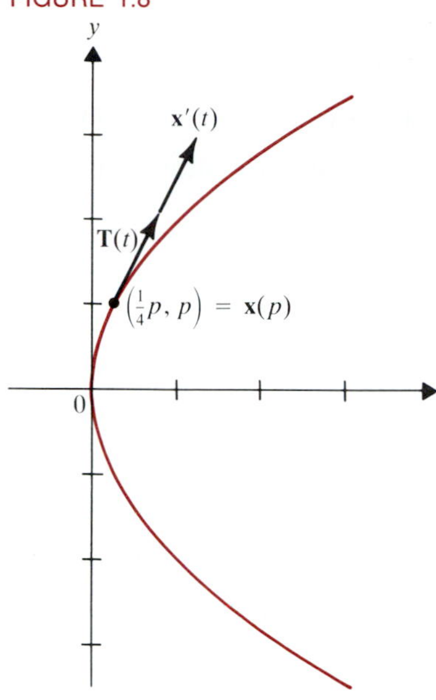

Example 5

Find the velocity of a particle moving on the curve

$$\mathbf{x}(t) = \frac{t^2}{4p}\mathbf{i} + t\mathbf{j}$$

when $t = p$. What is the unit tangent at that point?

Solution. Here $\mathbf{x}'(t) = (t/2p)\mathbf{i} + \mathbf{j}$. When $t = p$, the velocity is $\mathbf{x}'(p) = (1/2)\mathbf{i} + \mathbf{j}$, and so the speed is

$$|\mathbf{x}'(p)| = \sqrt{1/4 + 1} = \frac{1}{2}\sqrt{5}.$$

The unit tangent vector is

$$\frac{1}{|\mathbf{x}'(p)|}\mathbf{x}'(p) = \frac{2}{\sqrt{5}}\left(\frac{1}{2}\mathbf{i} + \mathbf{j}\right) = \frac{1}{\sqrt{5}}\mathbf{i} + \frac{2}{\sqrt{5}}\mathbf{j}.$$

We show these in **Figure 1.8,** where we have taken $p = 1$. ■

To complete the extension of earlier calculus to vector functions, we introduce the notion of the integral of a vector function $\mathbf{f}$ over a closed interval of $\mathbf{R}$.

1.7 DEFINITION If $\mathbf{f}: \mathbf{R} \to \mathbf{R}^n$ is a vector function such that each coordinate function $f_i(t)$ is integrable over $[a, b]$, then

$$\int_a^b \mathbf{f}(t)\,dt = \left(\int_a^b f_1(t)\,dt, \int_a^b f_2(t)\,dt, \ldots, \int_a^b f_n(t)\,dt\right).$$

Example 6

If $\mathbf{f}(t) = \dfrac{t^2}{4p}\mathbf{i} + t\mathbf{j}$, then find $\int_0^1 f(t)\,dt$.

Solution. Definition 1.7 gives

$$\int_0^1 \mathbf{f}(t)\,dt = \int_0^1 \frac{t^2}{4p}\,dt\,\mathbf{i} + \int_0^1 t\,dt\,\mathbf{j} = \left[\frac{t^3}{12p}\right]_0^1\mathbf{i} + \left[\frac{t^2}{2}\right]_0^1\mathbf{j} = \frac{1}{12p}\mathbf{i} + \frac{1}{2}\mathbf{j}. \quad ■$$

Just as the distance moved by a particle can be found by integrating its velocity with respect to time, so we can integrate the velocity vector $\mathbf{v}(t)$ with respect to t to find the vector equation of the path of a moving particle.

Example 7

A particle in $\mathbf{R}^3$ has constant velocity $\mathbf{v} = c_1\mathbf{i} + c_2\mathbf{j} + c_3\mathbf{k}$. What is its path?

Solution. We have

$$\mathbf{x}'(t) = \frac{dx}{dt}\mathbf{i} + \frac{dy}{dt}\mathbf{j} + \frac{dz}{dt}\mathbf{k} = c_1\mathbf{i} + c_2\mathbf{j} + c_3\mathbf{k}.$$

Integration gives

$$\mathbf{x}(t) = \int \mathbf{x}'(t)\,dt = \int \frac{dx}{dt}\,dt\,\mathbf{i} + \int \frac{dy}{dt}\,dt\,\mathbf{j} + \int \frac{dz}{dt}\,dt\,\mathbf{k}$$
$$= \int c_1\,dt\,\mathbf{i} + \int c_2\,dt\,\mathbf{j} + \int c_3\,dt\,\mathbf{k}$$
$$= (c_1 t + m_1)\mathbf{i} + (c_2 t + m_2)\mathbf{j} + (c_3 t + m_3)\mathbf{k} = t\mathbf{v} + \mathbf{m},$$

where $\mathbf{m} = m_1\mathbf{i} + m_2\mathbf{j} + m_3\mathbf{k}$. Thus the particle moves in a straight line. ■

Exercises 11.1

1. (a) Show that $\mathbf{f}(t) = (\cos t, \sin t)$ is continuous and differentiable for all t.
   (b) What is the curve $\mathbf{x}(t) = (\cos t, \sin t)$, $t \in [0, 2\pi]$?
   (c) Find a tangent vector at $t = 0$.
   (d) Show that $\mathbf{x}'(t)$ is orthogonal to $\mathbf{x}(t)$ for all t.
2. (a) Show that $\mathbf{f}(t) = (2\cos t, 3\sin t)$ is continuous and differentiable for all t.
   (b) Identify the curve $\mathbf{x}(t) = (2\cos t, 3\sin t)$, $t \in [0, 2\pi]$.
   (c) Find a tangent vector at $t = \pi/4$.

In Exercises 3–10, (a) find the equation of the tangent line to the curve $x = x(t)$ at $t = a$; (b) find the speed at $t = a$.

3. $\mathbf{x}(t) = (t - 1, t^2 + 1, t^3 - 1)$, $a = 2$
4. $\mathbf{x}(t) = (\cos t, \sin t, t)$, $a = \pi/2$
5. $\mathbf{x}(t) = (2t^2, t, t^3)$, $a = 1$
6. $\mathbf{x}(t) = (t\sin t, 3t, \cos t)$, $a = \pi$
7. $\mathbf{x}(t) = (t, \sin 4t, \cos 4t)$, $a = \pi/8$
8. $\mathbf{x}(t) = (2e^t, 2e^{-t}, 5\cos t)$, $a = 0$
9. $\mathbf{x}(t) = (e^{t^2}, \ln t^2, 1/t)$, $a = 1$
10. $\mathbf{x}(t) = (te^t, -t^2e^{2t}, t)$, $a = 0$
11. If $\mathbf{f}(t) = [(\sin t)/t]\mathbf{i} + [(1 - \cos t)/t]\mathbf{j} + t^2\mathbf{k}$, then find $\lim_{t\to 0} \mathbf{f}(t)$. Is $\mathbf{f}$ continuous at $t = 0$?
12. If $\mathbf{f}(t) = (t^2/\sin t)\mathbf{i} + [t/(1 - \cos t)]\mathbf{j} + t^2\mathbf{k}$, then find $\lim_{t\to 0} \mathbf{f}(t)$. Is $\mathbf{f}$ continuous at $t = 0$?

In Exercises 13–16, find (a) $\dfrac{d}{dx}[x(t) \times y(t)]$ and (b) $\dfrac{d}{dt}[x(t) \cdot y(t)]$.

13. $\mathbf{x}(t) = \cos t\,\mathbf{i} + \sin t\,\mathbf{j} + t\,\mathbf{k}$, $\mathbf{y}(t) = 2\mathbf{i} + t^2\mathbf{j} - 4t\,\mathbf{k}$
14. $\mathbf{x}(t) = e^t\mathbf{i} + \ln t\,\mathbf{j} + t\,\mathbf{k}$, $\mathbf{y}(t) = \cos t\,\mathbf{i} + \sin t\,\mathbf{j} + t\,\mathbf{k}$
15. $\mathbf{x}(t) = t\,\mathbf{i} + t^2\mathbf{j} + t^3\mathbf{k}$, $\mathbf{y}(t) = 4t\,\mathbf{i} + t\sin t\,\mathbf{j} + t\cos t\,\mathbf{k}$
16. $\mathbf{x}(t) = te^t\mathbf{i} + t^2e^{2t}\mathbf{j} + \dfrac{1}{t}\mathbf{k}$, $\mathbf{y}(t) = (8 - t)\mathbf{i} + (2 + t^2)\mathbf{j} + \dfrac{1}{t}\mathbf{k}$

In Exercises 17–22, find $\int_a^b f(t)\,dt$.

17. $\mathbf{f}(t) = (t^2 + 1)\mathbf{i} + (1 - t^2)\mathbf{j}$, $a = 0$, $b = 2$
18. $\mathbf{f}(t) = \cos t\,\mathbf{i} + \sin t\,\mathbf{j}$, $a = 0$, $b = \pi$
19. $\mathbf{f}(t) = \cos t\,\mathbf{i} + \sin t\,\mathbf{j} + t\,\mathbf{k}$, $a = \pi/2$, $b = \pi$
20. $\mathbf{f}(t) = (3\cos 4t)\mathbf{i} + (2\cos 3t)\mathbf{j} + 5t\,\mathbf{k}$, $a = \pi/6$, $b = 2\pi$
21. $\mathbf{f}(t) = (t\cos t)\mathbf{i} + (t\sin t)\mathbf{j} + (1 - t)\mathbf{k}$, $a = 0$, $b = \pi$
22. $\mathbf{f}(t) = \ln t\,\mathbf{i} + \sec^3 t\,\mathbf{j} + \sin^2 t\,\mathbf{k}$, $a = \pi/6$, $b = \pi/2$
23. A particle moving in $\mathbf{R}^3$ has velocity $\mathbf{v}(t) = \mathbf{i} + t\,\mathbf{j} + t^2\mathbf{k}$. Find the path if, when $t = 0$, the particle is at the point $(-1, 2, 2)$.
24. Repeat Exercise 23 if $\mathbf{v}(t) = \sin t\,\mathbf{i} + \cos t\,\mathbf{j} + \mathbf{k}$ and $\mathbf{x}(0) = (1, 0, 0)$.
25. If a particle moves on a curve so that $|\mathbf{x}(t)|$ is constant, then show that $\mathbf{x}(t)$ is perpendicular to $\mathbf{v}(t)$ for all t. (*Hint:* Express $|\mathbf{x}(t)|$ in terms of the dot product and differentiate.)
26. If $\mathbf{x} = \mathbf{x}(t)$ is a curve whose tangent vector is always perpendicular to $\mathbf{x}(t)$, then show that $|\mathbf{x}(t)|$ is constant. (*Hint:* Differentiate $|\mathbf{x}(t)|^2$.)
27. (a) Prove Theorem 1.4(b). (b) Prove Theorem 1.4(d).
28. (a) Prove that Definition 1.2 is equivalent to the condition
$$|\mathbf{f}(t) - \mathbf{v}| \to 0 \qquad \text{as } t \to a.$$
    (b) Prove that the condition in (a) is equivalent to the epsilon–delta condition on p. 685.
29. (a) Prove that (2) is equivalent to (3).
    (b) Prove that (1) is equivalent to (4).
30. If $\mathbf{f}$ and $\mathbf{g}$ are continuous at $t = c$, then prove that any linear combination $a\mathbf{f} + b\mathbf{g}$ is continuous at $t = c$, where a and b are real constants.
31. Prove that if $\mathbf{f}: \mathbf{R} \to \mathbf{R}^n$ is differentiable at $t = a$, then it is continuous there.
32. Prove that if $\mathbf{f}$ and $\mathbf{g}$ are differentiable at $t = c$, then so is any linear combination $a\mathbf{f} + b\mathbf{g}$ for real numbers a and b.
33. If $\mathbf{x} = \mathbf{x}(t)$ is the path of a particle with variable mass $m(t)$, then the *linear momentum* of the particle is the vector $\mathbf{p}(t) = m(t)\mathbf{x}'(t)$. Show that if
$$\frac{d}{dt}(m(t)\mathbf{x}'(t)) = \mathbf{0}$$
for all t, then $\mathbf{p}(t)$ is constant. (This is one form of the *law of conservation of linear momentum.*)

34. If $\mathbf{x} = \mathbf{x}(t)$ is the path of a particle with variable mass $m(t)$, then the torque about 0 is defined to be

$$\mathbf{L}(t) = \mathbf{x}(t) \times \frac{d}{dt}(m(t)\mathbf{x}'(t)).$$

The *angular momentum* about 0 is defined to be $\mathbf{M}(t) = \mathbf{x}(t) \times m(t)\mathbf{x}'(t)$. Show that

$$\frac{d\mathbf{M}(t)}{dt} = \mathbf{L}(t).$$

35. Refer to Exercise 34. Show that if the torque is identically zero, then the angular momentum is constant. (This is one form of the *law of conservation of angular momentum.*)

36. If $\mathbf{f}$ and $\mathbf{g}$ are integrable over $[a, b]$ and $c, k \in \mathbf{R}$, then show that

$$\int_a^b [c\mathbf{f}(t) + k\mathbf{g}(t)]\,dt = c\int_a^b \mathbf{f}(t)\,dt + k\int_a^b \mathbf{g}(t)\,dt.$$

Exercises 37–39 construct vector calculus proofs of the reflecting properties for conic sections.

37. **(a)** Show that the parabola in **Figure 1.9** has vector equation

$$\mathbf{x}(t) = \frac{t^2}{4p}\mathbf{i} + t\mathbf{j}.$$

FIGURE 1.9

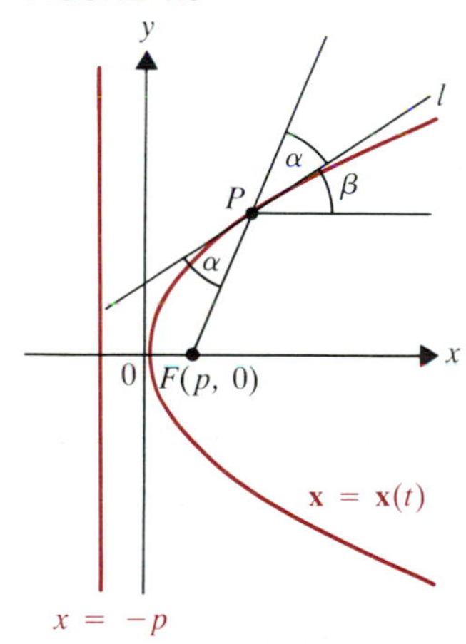

(b) In **Figure 1.10**, show that $\mathbf{0Q} = -p\,\mathbf{i} + t\,\mathbf{j}$, so $\mathbf{0Q}' = \mathbf{j}$.

(c) Show that differentiation of the defining condition $|\mathbf{x}(t) - \mathbf{0Q}| = |\mathbf{x}(t) - \mathbf{0F}|$ gives

$$\frac{\mathbf{x}(t) - \mathbf{0Q}}{|\mathbf{x}(t) - \mathbf{0Q}|} \cdot (\mathbf{x}'(t) - \mathbf{j}) = \frac{(\mathbf{x}(t) - \mathbf{0F})}{|\mathbf{x}(t) - \mathbf{0F}|} \cdot \mathbf{x}'(t).$$

(d) Show that the result of part (c) leads to $\cos\alpha = \cos\beta$, and hence $\alpha = \beta$ in Figure 1.9, thereby reproving Theorem 1.2 of Chapter 8. (*Hint:* Use the fact that $\mathbf{x}(t) - \mathbf{0Q}$ is perpendicular to the y-axis.)

FIGURE 1.10

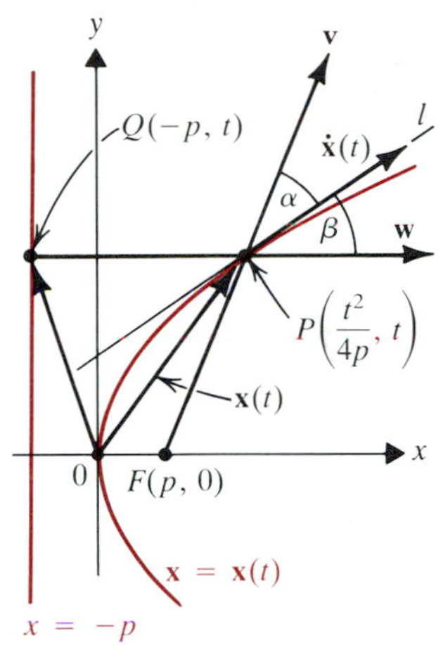

38. Establish the *reflecting property for ellipses.* If F_1 and F_2 are the foci of an ellipse, and P is any point on the ellipse, then the line segments F_1P and F_2P make equal angles with the tangent line at P. (*Hint:* As defining equation of the ellipse, use $|\mathbf{x}(t) - \mathbf{0F}_1| + |\mathbf{x}(t) - \mathbf{0F}_2| = 2a$.)

39. Establish the *reflecting property for hyperbolas:* If F_1 and F_2 are the foci of a hyperbola, and P is any point on the hyperbola, then the tangent vector $\mathbf{x}'(t)$ at P makes equal angles with the vectors $\mathbf{F}_1\mathbf{P}$ and $\mathbf{F}_2\mathbf{P}$. (*Hint:* As defining equation of the hyperbola, use $|\mathbf{x}(t) - \mathbf{0F}_1| - |\mathbf{x}(t) - \mathbf{0F}_2| = -2a$, if $\mathbf{x}(t)$ is on the branch nearer F_1.)

11.2 Velocity, Acceleration, and Arc Length

According to Definition 1.6, if a particle moves along the path $\mathbf{x} = \mathbf{x}(t)$ in $\mathbf{R}^n$, then its velocity at time $t = c$ is

$$\mathbf{v}(c) = \mathbf{x}'(c) = \lim_{t\to c} \frac{\mathbf{x}(t) - \mathbf{x}(c)}{t - c}.$$

Since the velocity vector $\mathbf{v}(c)$ is tangent to the curve $\mathbf{x} = \mathbf{x}(t)$ at $t = c$, at any instant the velocity vector points in the direction of motion. To complete the Newtonian model for the motion, we need two additional definitions.

2.1 DEFINITION If a particle of mass m moves on the curve $\mathbf{x} = \mathbf{x}(t)$ in $\mathbf{R}^n$, then its ***acceleration*** at any time $t = c$ is

$$\mathbf{a}(c) = \lim_{t \to c} \frac{\mathbf{v}(t) - \mathbf{v}(c)}{t - c} = \mathbf{v}'(c) = \mathbf{x}''(c)$$

$$= \left(\frac{dv_1}{dt}, \frac{dv_2}{dt}, \ldots, \frac{dv_n}{dt}\right)\bigg|_{t=c} = \left(\frac{d^2x_1}{dt^2}, \frac{d^2x_2}{dt^2}, \ldots, \frac{d^2x_n}{dt^2}\right)\bigg|_{t=c}.$$

The ***force*** acting on the particle at time $t = c$ is

$$\mathbf{F}(c) = m\mathbf{a}(c)$$

This definition of force is actually ***Newton's second law of motion.*** From it, our first example derives his ***first law of motion.***

Example 1

Describe the path in $\mathbf{R}^3$ of a particle of mass $m \neq 0$ on which there is an equilibrium of forces, that is, the resultant of all forces on the particle is $\mathbf{0}$.

Solution. Since $m \neq 0$, Newton's second law implies that $\mathbf{a}(t) = \mathbf{0} = (0, 0, 0)$. Thus we have

$$\left(\frac{d^2x}{dt^2}, \frac{d^2y}{dt^2}, \frac{d^2z}{dt^2}\right) = (0, 0, 0).$$

It then follows from Definition 1.7 that

$$\mathbf{v}(t) = \int \left(\frac{d^2x}{dt^2}, \frac{d^2y}{dt^2}, \frac{d^2z}{dt^2}\right) dt = \left(\int 0\,dt, \int 0\,dt, \int 0\,dt\right) = (c_1, c_2, c_3),$$

for constants c_1, c_2, and c_3. This means that

$$\mathbf{v} = \left(\frac{dx}{dt}, \frac{dy}{dt}, \frac{dz}{dt}\right) = (c_1, c_2, c_3) = c_1\,\mathbf{i} + c_2\,\mathbf{j} + c_3\,\mathbf{k}$$

is constant. Then, as in Example 7 of the last section, integration gives

$$\mathbf{x} = t(c_1, c_2, c_3) + (d_1, d_2, d_3) = t\mathbf{v} + \mathbf{d},$$

the vector equation of a straight line. Hence the particle moves along a straight line with constant velocity $\mathbf{v} = c_1\,\mathbf{i} + c_2\,\mathbf{j} + c_3\,\mathbf{k}$. (This is Newton's first law of motion.) ■

Example 2

A particle moves with constant speed. Show that when its acceleration vector is not zero, it is perpendicular to the velocity vector of the particle.

Solution. We have $\sqrt{\mathbf{v}(t) \cdot \mathbf{v}(t)} = c$. Thus $\mathbf{v}(t) \cdot \mathbf{v}(t) = c^2$. If we differentiate each side of the latter equation, and use Theorem 1.4(b), then we get $2\mathbf{v}(t) \cdot \mathbf{v}'(t) = 0$. That is, $2\mathbf{v}(t) \cdot \mathbf{a}(t) = 0$. Hence $\mathbf{v}(t) \cdot \mathbf{a}(t) = 0$. So either $\mathbf{a}(t) = \mathbf{0}$ or else $\mathbf{a}(t) \perp \mathbf{v}(t)$. ■

Example 3

A particle moves on the ellipse

$$\frac{x^2}{a^2} + \frac{y^2}{b^2} = 1.$$

Parametrize and find the velocity, speed, and acceleration, in general and at $t = \pi/4$.

FIGURE 2.1

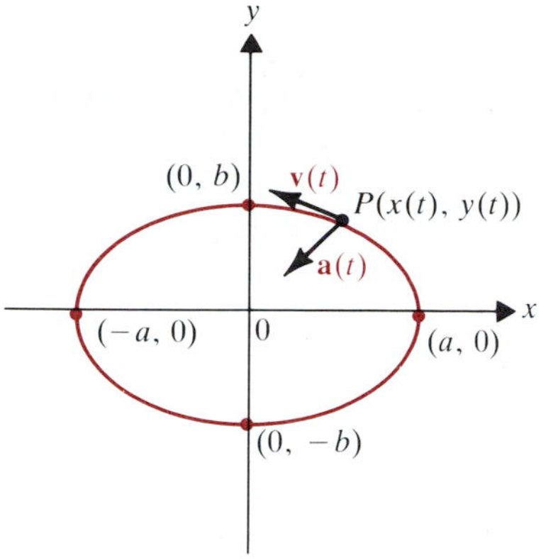

Solution. The ellipse is shown in **Figure 2.1.** Since we have a sum of squares equaling 1, we let

$$x = a\cos t, \qquad y = b\sin t, \qquad t \in [0, 2\pi].$$

The corresponding vector equation is then

$$\mathbf{x}(t) = a\cos t\,\mathbf{i} + b\sin t\,\mathbf{j}.$$

Differentiation gives

$$\mathbf{v}(t) = -a\sin t\,\mathbf{i} + b\cos t\,\mathbf{j} \qquad \text{and} \qquad \mathbf{a}(t) = -a\cos t\,\mathbf{i} - b\sin t\,\mathbf{j} = -\mathbf{x}(t).$$

Thus the speed is

$$|\mathbf{v}(t)| = \sqrt{a^2\sin^2 t + b^2\cos^2 t}.$$

At $t = \pi/4$, we therefore have

$$\mathbf{v}\left(\frac{\pi}{4}\right) = \left(-\frac{a}{\sqrt{2}}\right)\mathbf{i} - \left(\frac{b}{\sqrt{2}}\right)\mathbf{j}, \qquad \left|\mathbf{v}\left(\frac{\pi}{4}\right)\right| = \frac{\sqrt{a^2+b^2}}{\sqrt{2}},$$

and

$$\mathbf{a}\left(\frac{\pi}{4}\right) = \left(-\frac{a}{\sqrt{2}}\right)\mathbf{i} - \left(\frac{b}{\sqrt{2}}\right)\mathbf{j} = -\mathbf{x}\left(\frac{\pi}{4}\right). \quad ■$$

Now we turn to the problem of calculating the arc length of a parametrized curve $\mathbf{x} = \mathbf{x}(t)$ between $t = a$ and $t = b$. There is no problem if the speed is a constant r. For then the formula $d = r(b - a)$ applies, because the total length covered in a time interval of length $b - a$ is the product of the constant speed r and the elapsed time.

In the general case, we assume that the path is *smooth.* By this we mean that $\mathbf{x} = \mathbf{x}(t)$ has a continuous derivative $\mathbf{x}'(t)$ that is nonzero for $t \in [a, b]$. As in Section 5.4, we partition the interval into a large number of subintervals $[t_{i-1}, t_i]$, each of very short length. Then, since $\mathbf{x}$ is smooth, the speed $|\mathbf{x}'|$ is continuous on each short interval $[t_{i-1}, t_i]$. It therefore will not vary much over each interval. Hence we won't err by much if we approximate $|\mathbf{x}'(t)|$ by $|\mathbf{x}'(u_i)|$ for any point $u_i \in [t_{i-1}, t_i]$. Then the arc length over $[t_{i-1}, t_i]$ is approximately

FIGURE 2.2

$$|\mathbf{x}'(u_i)|(t_i - t_{i-1}) = |\mathbf{x}'(u_i)|\,\Delta t_i.$$

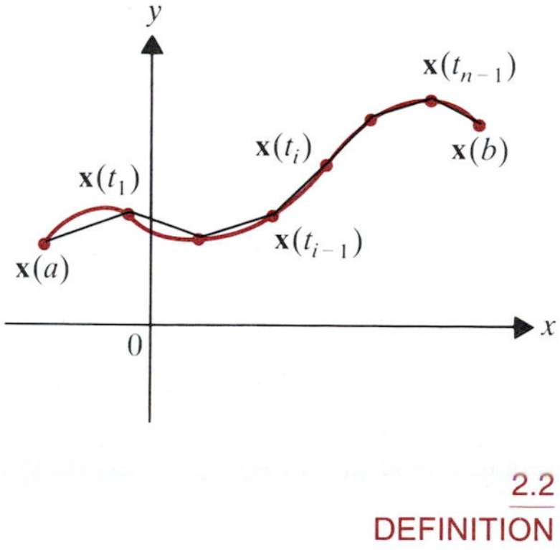

The arc length over the whole interval $[a, b]$ is the sum of the arc lengths over the subintervals $[t_{i-1}, t_i]$ (see **Figure 2.2**). Hence the arc length over $[a, b]$ is approximated by $\sum_i |\mathbf{x}'(u_i)|\,\Delta t_i$. The latter sum is a Riemann sum for the speed function over the interval $[a, b]$. As the subintervals become shorter and shorter we should get a closer and closer approximation to the actual arc length over $[a, b]$. We therefore *define* the arc length over $[a, b]$ to be the definite integral of the speed.

2.2 DEFINITION

If $\mathbf{x} = \mathbf{x}(t)$ defines a smooth curve in $\mathbf{R}^n$, then the ***arc length*** of the curve between $t = a$ and $t = b$ is

$$L = \int_a^b |\mathbf{x}'(t)|\,dt = \int_a^b \sqrt{x_1'(t)^2 + x_2'(t)^2 + \ldots + x_n'(t)^2}\,dt. \tag{1}$$

The ***arc-length function*** is the real-valued function $s\colon [a, b] \to \mathbf{R}$ defined by

$$s(t) = \int_a^t |\mathbf{x}'(u)|\,du. \tag{2}$$

Definition 2.2 simply extends Theorem 5.2 of Chapter 8 to $\mathbf{R}^n$. (Compare the next example with Example 6 of Section 8.5.)

Example 4

FIGURE 2.3

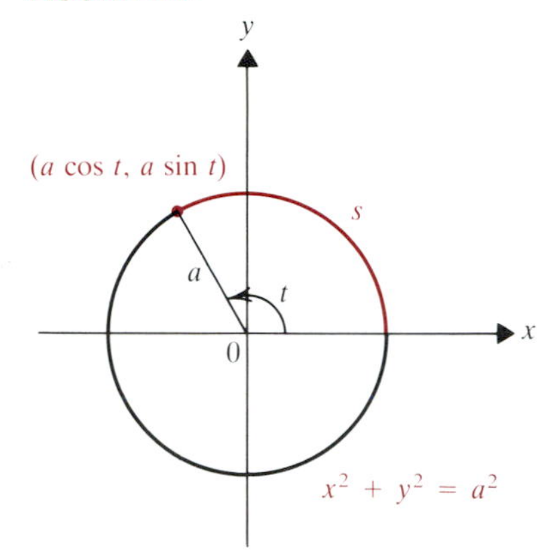

Find the arc length of one quarter of the circle $x^2 + y^2 = a^2$ (see **Figure 2.3**). Also find the arc-length function.

Solution. We parametrize the curve as $\mathbf{x} = \mathbf{x}(t) = a \cos t\,\mathbf{i} + a \sin t\,\mathbf{j}$, $0 \le t \le 2\pi$. Then $\mathbf{x}'(t) = -a \sin t\,\mathbf{i} + a \cos t\,\mathbf{j}$, so the speed is

$$|\mathbf{x}'(t)| = \sqrt{a^2 \sin^2 t + a^2 \cos^2 t} = a.$$

Therefore, by (1), the required arc length is

$$\int_0^{\pi/2} a\,dt = \frac{\pi}{2} a.$$

From (2) the arc-length function s is given by

$$s = \int_0^t a\,dt = at,$$

which is in accord with the relation between arc length along a circle and the central angle measured in radians. ■

Example 5

FIGURE 2.4

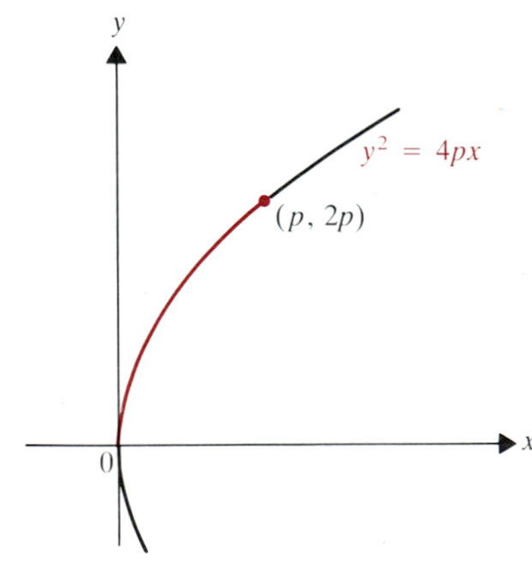

Find the arc length of the parabola $y^2 = 4px$ between $x = 0$ and $x = p$. See **Figure 2.4.**

Solution. As in Example 4 of the last section, $\mathbf{x}(t) = (t^2/4p)\,\mathbf{i} + t\,\mathbf{j}$ is a vector equation. For $x = 0$ we have $t = 0$. For $x = p$ we need $(t^2/4p) = p$, that is, $t^2 = 4p^2$, so $t = 2p$ will do. Since $\mathbf{x}'(t) = (t/2p)\,\mathbf{i} + \mathbf{j}$, we have

$$|\mathbf{x}'(t)| = \sqrt{\frac{t^2}{4p^2} + 1} = \frac{\sqrt{t^2 + 4p^2}}{2p}.$$

From (1) the arc length is therefore

$$\int_0^{2p} \frac{1}{2p} \sqrt{t^2 + 4p^2}\,dt = \frac{1}{2p} \int_0^{2p} \sqrt{t^2 + 4p^2}\,dt.$$

To evaluate this integral, we use Formula 53 in the Table of Integrals, which gives

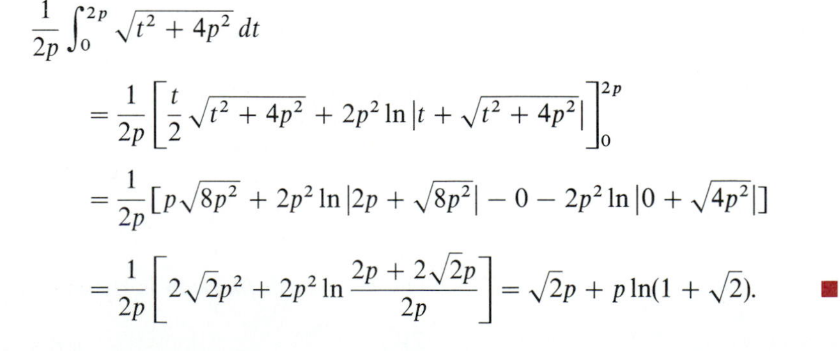

$$\begin{aligned}
&\frac{1}{2p} \int_0^{2p} \sqrt{t^2 + 4p^2}\,dt \\
&= \frac{1}{2p} \left[\frac{t}{2} \sqrt{t^2 + 4p^2} + 2p^2 \ln \left| t + \sqrt{t^2 + 4p^2} \right| \right]_0^{2p} \\
&= \frac{1}{2p} [p\sqrt{8p^2} + 2p^2 \ln |2p + \sqrt{8p^2}| - 0 - 2p^2 \ln |0 + \sqrt{4p^2}|] \\
&= \frac{1}{2p} \left[2\sqrt{2}p^2 + 2p^2 \ln \frac{2p + 2\sqrt{2}p}{2p} \right] = \sqrt{2}p + p \ln(1 + \sqrt{2}).
\end{aligned}$$

■

Definition 2.2 generalizes Definition 4.2 of Chapter 5 for the arc length of a smooth curve that is the graph of an equation $y = f(x)$ between $x = a$ and $x = b$. A natural parametrization of the latter is

$$\mathbf{x}(t) = t\,\mathbf{i} + f(t)\,\mathbf{j}.$$

Then $x'(t) = \mathbf{i} + f'(t)\mathbf{j}$, so the speed is $|\mathbf{x}(t)| = \sqrt{1 + f'(t)^2}$. Thus Definition 2.2 says that the arc length is

$$L = \int_a^b \sqrt{1 + f'(t)^2}\, dt = \int_a^b \sqrt{1 + f'(x)^2}\, dx,$$

which agrees with Definition 4.2 of Chapter 5.

In Definition 2.2 we assume that $\mathbf{x}$ is smooth. Thus $\mathbf{x}'$ is nonzero and continuous. So then is its length $|\mathbf{x}'|$. Thus Theorem 1.5 of Chapter 4 guarantees that every smooth curve has an arc length over a closed interval $[a, b]$. In addition, Theorem 4.1 of Chapter 4 guarantees that s is a differentiable function on $[a, b]$. Since $\mathbf{x}$ is smooth, for every t in $[a, b]$ it also follows from that theorem and (2) that

$$\frac{ds}{dt} = |\mathbf{x}'(t)| > 0. \tag{3}$$

Thus, the rate of change of arc length with respect to time is the *positive* speed of travel along the curve $\mathbf{x} = \mathbf{x}(t)$. (So in traversing the curve, we never "double back.")

It appears from Definition 2.2 that the arc length of a curve might depend on how the curve is parametrized. Fortunately, it does not. To see that, we consider two parametrizations,

$$\mathbf{x} = \mathbf{f}(t), \quad t \in [a, b], \qquad \text{and} \qquad \mathbf{x} = \mathbf{g}(u), \quad u \in [c, d], \tag{4}$$

where the parameters t and u are related by a differentiable function

$$t = h(u), \qquad \text{with } a = h(c) \text{ and } h = h(d), \tag{5}$$

and h' is continuous and positive-valued on the interval $[c, d]$. This may seem to be a highly restrictive set of conditions, but (5) just says that t is an increasing, smooth function of u. Thus the parameters correspond to two consistent "time scales." The next result states that we get the same arc length in Definition 2.2 whether we use t or u as parameter.

2.3 THEOREM Suppose that the curve C has the two parametrizations (4), where t and u are related by (5). Then

$$\int_a^b |\mathbf{f}'(t)|\, dt = \int_c^d |\mathbf{g}'(u)|\, du. \tag{6}$$

That is, the arc length of C calculated from Definition 2.2 is the same whether C is parametrized using t or u as parameter.

Proof. We use the change-of-variable theorem (Theorem 5.1 of Chapter 4). From (4) and (5) we have

$$\mathbf{x} = \mathbf{f}(t) = \mathbf{f}(h(u)) = \mathbf{g}(u).$$

Theorem 1.4(f) then gives

$$\mathbf{g}'(u) = \frac{d\mathbf{x}}{du} = \frac{d\mathbf{f}}{dt}\frac{dt}{du}.$$

Therefore

$$\int_c^d |\mathbf{g}'(u)|\,du = \int_c^d \left|\frac{d\mathbf{x}}{du}\right| du = \int_c^d \left|\frac{d\mathbf{f}}{dt}\frac{dt}{du}\right| du$$

$$= \int_c^d \left|\frac{d\mathbf{f}}{dt}\right| \frac{dt}{du}\,du \qquad \text{since } \frac{dt}{du} > 0$$

$$= \int_a^b |\mathbf{f}'(t)|\,dt. \qquad \textit{by the change-of-variable theorem}$$

QED

The assumption that the curve C is smooth, so that (3) holds, means that s is an increasing function of t, by Theorem 3.1 of Chapter 3. So there is a differentiable inverse function

$$t = t(s) \tag{7}$$

(see p. 117). When we can find a formula like (7) for t in terms of s, we can ***parametrize the curve C by arc length:***

$$\mathbf{x} = \mathbf{x}(t) = \mathbf{x}(t(s)) = \mathbf{y}(s),$$

where $\mathbf{y}$ is a smooth function of s. In this case the chain rule of Theorem 1.4(f) says that

$$\mathbf{y}'(s) = \frac{d}{ds}(\mathbf{x}(t(s))) = \frac{d\mathbf{x}}{dt}\cdot\frac{dt}{ds}. \tag{8}$$

Thus

$$|\mathbf{y}'(s)| = \left|\frac{d\mathbf{x}}{dt}\right|\cdot\frac{dt}{ds} = |\mathbf{x}'(t)|\,\frac{dt}{ds}$$

$$= \frac{ds}{dt}\cdot\frac{1}{\frac{ds}{dt}} \qquad \textit{by (3) and Theorem 7.5 of Chapter 2}$$

$$= 1.$$

Conversely, if $\mathbf{x} = \mathbf{x}(t)$ has speed identically 1, then for each value of t,

$$|\mathbf{x}'(t)| = \frac{ds}{dt} = 1.$$

Hence, starting from $t = 0$, we find from (2) that

$$s = \int_0^t |\mathbf{x}'(u)|\,du = \int_0^t du = t,$$

so the curve is parametrized by arc length. To summarize,

> a curve C is parametrized by arc length if and only if the speed is identically 1.

This makes it easy to check whether a given parametrization parametrizes a curve by arc length.

Example 6

Show that the curve C is parametrized by arc length if

$$\mathbf{x}(t) = \left(\frac{t}{3}, 2\sqrt{2}\cos\frac{1}{3}t, 2\sqrt{2}\sin\frac{1}{3}t\right), \qquad 0 \le t \le 2\pi.$$

Solution. We have

$$\mathbf{x}'(t) = \left(\frac{1}{3}, -\frac{2\sqrt{2}}{3}\sin\frac{1}{3}t, \frac{2\sqrt{2}}{3}\cos\frac{1}{3}t\right),$$

FIGURE 2.5

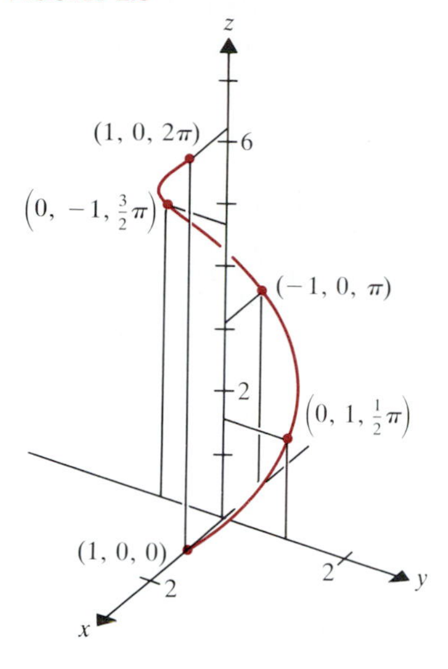

$\mathbf{x}(t) = \cos t\,\mathbf{i} + \sin t\,\mathbf{j} + t\mathbf{k}$
$0 \leq t \leq 2\pi$

so

$$|\mathbf{x}'(t)| = \sqrt{\frac{1}{9} + \frac{8}{9}\sin^2\frac{1}{3}t + \frac{8}{9}\cos^2\frac{1}{3}t} = \sqrt{\frac{1}{9} + \frac{8}{9}} = 1.$$

Hence the curve *is* parametrized by arc length. ■

The next example illustrates how we can sometimes reparametrize a given curve by arc length.

Example 7

For the curve $\mathbf{x} = \mathbf{x}(t) = \cos t\,\mathbf{i} + \sin t\,\mathbf{j} + t\,\mathbf{k}$, $0 \leq t \leq 2\pi$, find a parametrization by arc length.

Solution. We first find the arc-length function $s = s(t)$ for the given parametrization. We have $\mathbf{x}'(t) = -\sin t\,\mathbf{i} + \cos t\,\mathbf{j} + \mathbf{k}$, so the speed is $|\mathbf{x}'(t)| = \sqrt{\sin^2 t + \cos^2 t + 1} = \sqrt{2}$. Hence $s(t) = \int_0^t |\mathbf{x}'(u)|\,du = \int_0^t \sqrt{2}\,dt = \sqrt{2}t$. Thus $t = s/\sqrt{2}$ gives t as a function of s. So if we parametrize the curve in terms of the new parameter $t(s) = s/\sqrt{2}$, then we should obtain a parametrization by arc length. To check that, from

$$\mathbf{x}(t(s)) = \left(\cos\frac{s}{\sqrt{2}}, \sin\frac{s}{\sqrt{2}}, \frac{s}{\sqrt{2}}\right) = \mathbf{y}(s),$$

we find

$$\left|\frac{d\mathbf{y}}{ds}\right| = \left|\left(-\frac{1}{\sqrt{2}}\sin\frac{s}{\sqrt{2}}, \frac{1}{\sqrt{2}}\cos\frac{s}{\sqrt{2}}, \frac{1}{\sqrt{2}}\right)\right| = \sqrt{\frac{1}{2}\sin^2\frac{s}{\sqrt{2}} + \frac{1}{2}\cos^2\frac{s}{\sqrt{2}} + \frac{1}{2}}$$
$$= \sqrt{\frac{1}{2} + \frac{1}{2}} = 1.$$

The curve, shown in **Figure 2.5,** is the same circular helix discussed in Example 3 of Section 1. ■

FIGURE 2.6
The DNA Molecule

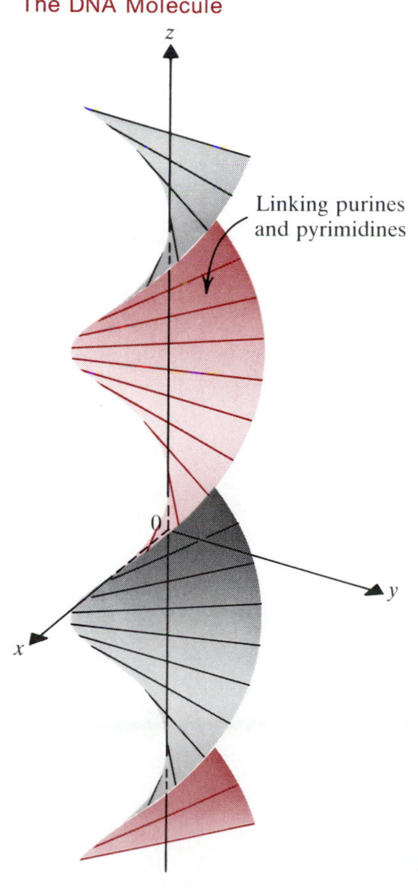

$\mathbf{x} = (\cos t, \sin t, t)$

One of the greatest achievements of modern biochemistry was the characterization of the molecular structure of deoxyribonucleic acid (DNA, the genetic material of living cells). This was achieved in 1953 at Cambridge University by the American biologist James D. Watson (1928–) working with the English biochemist Francis H.C. Crick (1916–). They shared the 1962 Nobel Prize for physiology and medicine for this work. They first proposed and then verified experimentally that the DNA molecule consists of two parallel almost circular helices of sugar and phosphate linked by certain organic bases called purines and pyrimidines. The two helices can be approximately parametrized as

$$\mathbf{x}(t) = \cos t\,\mathbf{i} + \sin t\,\mathbf{j} + t\,\mathbf{k}$$

and

$$\mathbf{y}(t) = \cos(t - 2\pi)\,\mathbf{i} + \sin(t - 2\pi)\,\mathbf{j} + (t - 2\pi)\,\mathbf{k} = \cos t\,\mathbf{i} - \sin t\,\mathbf{j} + (t - 2\pi)\,\mathbf{k},$$

where t varies over an interval $[0, 2k\pi]$ for some k, a positive integer that varies from molecule to molecule. As t increases, the helices ascend and wind around the z-axis in a counterclockwise sense, as viewed from above. See **Figure 2.6,** where some of the linking bases are shown. As Example 7 suggests, the calculus of the circular helix is relatively simple.

Exercises 11.2

In Exercises 1–12, a particle moves on the given path in $\mathbf{R}^3$. Find the velocity, acceleration, and speed. In Exercises 1–6, also sketch the path between $t = a$ and $t = b$.

1. The helix $\mathbf{x} = \mathbf{x}(t) = \cos\frac{1}{2}t\,\mathbf{i} + \sin\frac{1}{2}t\,\mathbf{j} + t\,\mathbf{k}$; $a = 0$, $b = 2\pi$
2. The elliptical spiral $\mathbf{x} = \mathbf{x}(t) = \cos t\,\mathbf{i} + 2\sin t\,\mathbf{j} + 3t\,\mathbf{k}$; $a = 0$, $b = 2\pi$
3. The expanding spiral $\mathbf{x} = \mathbf{x}(t) = t\cos t\,\mathbf{i} + t\sin t\,\mathbf{j} + t\,\mathbf{k}$; $a = 0$, $b = 2\pi$
4. $\mathbf{x}(t) = 2t\,\mathbf{i} + 3t\,\mathbf{j} + t^2\,\mathbf{k}$; $a = 0$, $b = 2$
5. $\mathbf{x}(t) = e^t\,\mathbf{i} + 2e^t\,\mathbf{j} + e^{2t}\,\mathbf{k}$; $a = 0$, $b = 1$
6. $\mathbf{x}(t) = (e^t\sin t, e^t\cos t, 1)$; $a = 0$, $b = 1$
7. $\mathbf{x}(t) = (t, \sin 2\pi t, \cos 2\pi t)$
8. $\mathbf{x}(t) = (\cos\pi t, \sin\pi t, t)$
9. $\mathbf{x}(t) = (\sin t - t\cos t)\,\mathbf{i} + (\cos t + t\sin t)\,\mathbf{j} + t^2\,\mathbf{k}$
10. $\mathbf{x}(t) = t\ln t\,\mathbf{i} + e^{t^2+1}\,\mathbf{k}$
11. $\mathbf{x}(t) = t\sin t\,\mathbf{i} + t\cos t\,\mathbf{j} + (t^2 - 1)\,\mathbf{k}$
12. $\mathbf{x}(t) = (t - \sin t)\,\mathbf{i} + (1 - \cos t)\,\mathbf{j} + \sqrt{t^2 + 1}\,\mathbf{k}$

In Exercises 13–20, find the arc length of the given curve over the given interval.

13. The curve of Exercise 1, between a and b
14. $\mathbf{x}(t) = t\,\mathbf{i} + \sin 4t\,\mathbf{j} + \cos 4t\,\mathbf{k}$ between 0 and π
15. The curve of Exercise 7, between 0 and 2
16. The curve of Exercise 8, between 1 and 4
17. The curve of Exercise 5, between 0 and 1
18. The curve of Exercise 9, between 2 and 5
19. The curve of Exercise 3, between 0 and 1
20. The curve of Exercise 4, between a and b

In Exercises 21–24, parametrize the given curve by arc length.

21. The curve of Exercise 14.
22. $\mathbf{x}(t) = 2\cos 3t\,\mathbf{i} + 2\sin 3t\,\mathbf{j} + 3t\,\mathbf{k}$, $t \in [0, 2\pi]$
23. $\mathbf{x}(t) = 2t\,\mathbf{i} + t\,\mathbf{j} + (1 - t)\,\mathbf{k}$, $t \in [0, 1]$
24. $\mathbf{x}(t) = (\frac{1}{2}t^2, \frac{1}{3}t^3, 0)$, $t \in [0, 2]$

In Exercises 25 and 26, show that the given curve is parametrized by arc length.

25. $\mathbf{x}(t) = \left(\dfrac{1}{3\sqrt{2}}\cos 3t, \dfrac{1}{3\sqrt{2}}\sin 3t, \dfrac{1}{\sqrt{2}}t\right)$, $t \in [0, 1]$
26. $\mathbf{x}(s) = \left(\sin\dfrac{s}{\sqrt{5}}, -\cos\dfrac{s}{\sqrt{5}}, \dfrac{2s}{\sqrt{5}}\right)$, $s \in [0, 1]$
27. A curve in the plane given in polar coordinates as the graph of $r = f(\theta)$, $a \le \theta \le b$, can be parametrized by
$$\mathbf{x} = \mathbf{x}(\theta) = r\cos\theta\,\mathbf{i} + r\sin\theta\,\mathbf{j} = f(\theta)\cos\theta\,\mathbf{i} + f(\theta)\sin\theta\,\mathbf{j}$$
$$= f(\theta)(\cos\theta\,\mathbf{i} + \sin\theta\,\mathbf{j}).$$
(a) Find the velocity (as a function of θ).
(b) Find the speed.
(c) Show that the arc length is given by
$$L = \int_a^b \sqrt{r^2 + \left(\frac{dr}{d\theta}\right)^2}\,d\theta.$$
(d) What happens in (a), (b), and (c) if we regard $\theta = \theta(t)$ as a function of t?
28. Find the arc length of the logarithmic spiral $r = e^\theta$ between 0 and π. [Use Exercise 27(c).]
29. A shell of mass m is fired from a cannon whose angle of elevation with the ground is α (see **Figure 2.7**). The initial velocity is $\mathbf{v}_0 = (v_0\cos\alpha, v_0\sin\alpha)$, where v_0 is the muzzle speed. Assume that the only force acting on the shell is $\mathbf{F} = -mg\,\mathbf{j}$, where g is the acceleration due to gravity.
(a) Describe the path $\mathbf{x} = \mathbf{x}(t)$ of the shell. Identify the curve. (*Hint:* Use $\mathbf{F} = m\mathbf{a} = m\mathbf{x}''(t)$. Take the origin at the gun.)
(b) Find the linear distance from the cannon to the point of impact of the shell, assuming level ground.

FIGURE 2.7

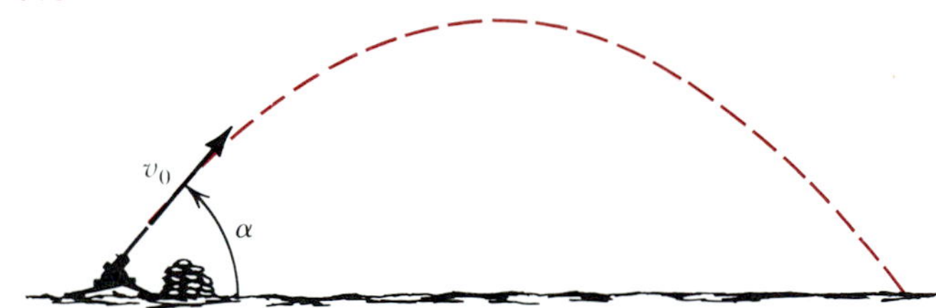

30. Assume that the cannon in Exercise 29 is located at the base of a hillside that has an angle of elevation of 30° (**Figure 2.8**). Show that the shell hits the hillside at point P whose x-coordinate is
$$\frac{2v_0{}^2}{g}\left(\sin\alpha\cos\alpha - \frac{1}{\sqrt{3}}\cos^2\alpha\right).$$
Coordinatize as in Exercise 29.

FIGURE 2.8

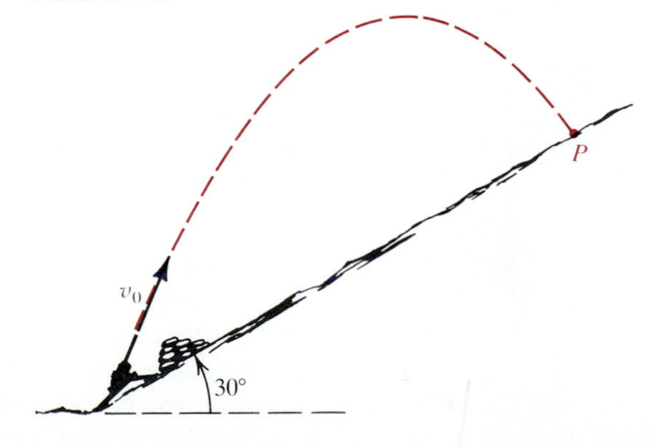

(*Hint:* The shell hits the hillside at the point where $y/x = \tan 30° = 1/\sqrt{3}$.)

31. Describe the path in $\mathbf{R}^3$ of a particle subjected to a constant force $\mathbf{F} = c_1\mathbf{i} + c_2\mathbf{j} + c_3\mathbf{k}$.

32. Describe the path in $\mathbf{R}^3$ of a particle acted on by a force $\mathbf{F} = c\mathbf{i}$ for some constant c.

33. If a wire in space lies along the curve $\mathbf{x} = (\mathbf{x}(t), \mathbf{y}(t), \mathbf{z}(t))$ between $t = a$ and $t = b$, and the mass per unit length at $\mathbf{x}$ is $\delta(\mathbf{x}(t))$, then the total ***mass*** of the wire is defined to be $M = \int_a^b \delta(\mathbf{x}(t))|\mathbf{x}'(t)|\,dt$. Find this mass if $\mathbf{x}(t) = (\cos t, t, \sin t)$, $0 \leq t \leq 2\pi$ and $\delta(\mathbf{x}(t)) = y(t)^2$.

34. Use the definition in Exercise 33 to find the mass of a wire lying along the graph of $\mathbf{x}(t) = (t, t^2, t^2)$ between $t = 0$ and $t = 1$ if the density at $(x(t), y(t), z(t))$ is $6t$.

35. When a body rotates about the z-axis through the origin, its *angular velocity* is defined as $\boldsymbol{\omega} = \theta'\mathbf{k}$, where θ' is the number of radians per unit time in the rotation. See **Figure 2.9.** The *linear velocity* is defined to be $\mathbf{v} = \boldsymbol{\omega} \times \mathbf{x}$, where $\mathbf{x}(t)$ is the position of the body at time t. Show that the linear speed is the product of the angular speed, the displacement $|\mathbf{x}(t)|$ from the origin, and $\cos\phi$, where ϕ is the angle formed by $\mathbf{x}(t)$ and the xy-plane.

FIGURE 2.9

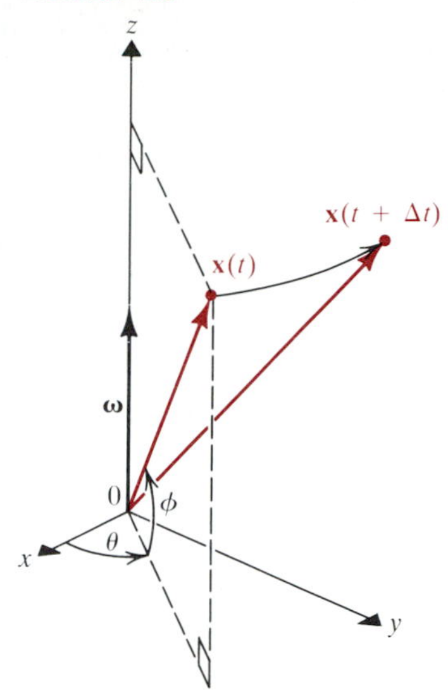

36. The earth rotates about its axis at a speed of 2π radians per day. Assume that the earth is a ball of radius 4000 miles, and put the origin at its center. Find the linear speed in miles per hour of a point on the surface of the earth at **(a)** the equator; **(b)** Manila, Philippines (located at 15° north $= \pi/12$ radians north of the equator); **(c)** Bordeaux, France (located at 45° north); **(d)** the North Pole. (See Exercise 35.)

37. Suppose that the parametrized curve $\mathbf{x} = \mathbf{x}(t)$ has $|\mathbf{x}'(t)| = c$. Then show that $t(s) = s/c$ gives a parametrization by arc length. (This generalizes Example 7.)

11.3 Curvature; Unit Tangent and Normal

In this section we capitalize on the arc length function introduced in Definition 2.2 to study some properties of curves that are helpful in describing the behavior of moving particles in $\mathbf{R}^2$ or $\mathbf{R}^3$.

Recall from Definition 1.6 that the unit tangent vector to a curve C parametrized by $\mathbf{x} = \mathbf{x}(t)$ is

$$\text{(1)} \qquad \mathbf{T}(t) = \frac{1}{|\mathbf{x}'(t)|}\,\mathbf{x}'(t), \qquad \text{if } \mathbf{x}'(t) \neq \mathbf{0}.$$

The unit tangent vector to a curve should be determined by the curve itself and thus be independent of the particular parametrization used. To describe $\mathbf{T}$ in a more intrinsic way, we use the arc-length function $s = s(t)$, which by Theorem 2.3 is independent of the parametrization and is given by

$$s(t) = \int_a^t |\mathbf{x}'(u)|\,du.$$

If the curve is smooth, then $\mathbf{x}'(t)$ is never zero. Hence, as we saw on p. 697, we have

$$\text{(2)} \qquad \frac{ds}{dt} = |\mathbf{x}'(t)| > 0,$$

and so there is an inverse function $t = t(s)$ that leads to the parametrization of C by arc length:

$$\mathbf{x} = \mathbf{x}(t) = \mathbf{x}(t(s)) = \mathbf{y}(s).$$

According to equation (8) of the last section,

$$\frac{d\mathbf{y}}{ds} = \frac{d\mathbf{x}}{dt}\frac{dt}{ds} \rightarrow \frac{d\mathbf{x}}{dt} = \frac{d\mathbf{y}}{ds}\frac{ds}{dt}. \tag{3}$$

Substituting (2) and the second part of (3) into (1), we obtain

$$\mathbf{T} = \frac{\mathbf{x}'(t)}{\dfrac{ds}{dt}} = \frac{\dfrac{d\mathbf{y}}{ds}\dfrac{ds}{dt}}{\dfrac{ds}{dt}} = \frac{d\mathbf{y}}{ds}. \tag{4}$$

This description of the unit tangent vector is more intrinsic to the curve than is (1), because it says that $\mathbf{T}$ is the rate of change of the curve with respect to its arc length. It also helps quantify the notion of the *curvature* of a curve.

We have an intuitive idea of how sharply a curve turns. For example, speed-limit signs are often seen on highways, warning of how fast a car can safely travel around a curve. The *sharper* the curve (i.e., the greater its curvature), the *lower* the speed limit. This seems to reflect an increasing tendency for the car to leave the road as the curvature becomes greater. To measure the intuitive idea of curvature mathematically, we need to first characterize what distinguishes a curve from a straight patch of road. **Figure 3.1** shows a particularly formidable curve. But between t_0 and t_1, as well as between t_4 and t_5, the graph of $\mathbf{x} = \mathbf{x}(t)$ is a straight line segment. So in each of those intervals, $\mathbf{x}(t) = t\mathbf{a} + \mathbf{b}$ for appropriate $\mathbf{a}$ and $\mathbf{b}$. Thus $\mathbf{x}'(t) = \mathbf{a}$ is constant, and so $\mathbf{T}(t)$ is constant with length one. Between t_1 and t_2, however, $\mathbf{x}(t)$ is given by a nonlinear expression. Hence $\mathbf{x}'(t)$ is no longer constant, and neither is $\mathbf{T}(t)$. While the length of $\mathbf{T}(t)$ is always one, it turns quite sharply near the points $\mathbf{x}(t_2)$ and $\mathbf{x}(t_3)$. The curvature can thus be measured by how much the unit tangent vector turns per unit length of arc as we move along the curve.

FIGURE 3.1

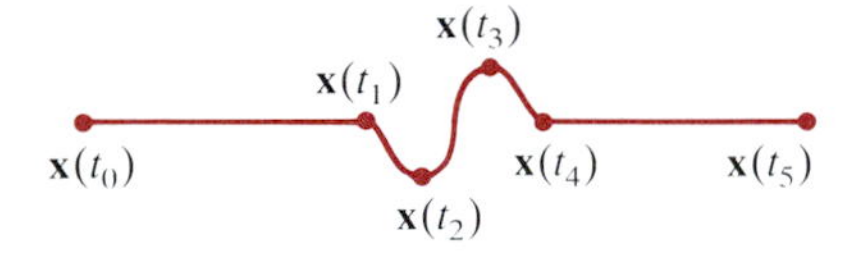

3.1 DEFINITION Let C: $\mathbf{x} = \mathbf{x}(t)$, $t \in I$, be a smooth parametrized curve. The ***curvature*** $\mathbf{K}$ at a point $\mathbf{x}(a)$ is the magnitude of the rate of change of the direction of the curve with respect to arc length. That is,

$$K = \left|\frac{d\mathbf{T}(t(s))}{ds}\right|_{t=a}, \tag{5}$$

where $\mathbf{T}(t(s))$ is the unit tangent vector.

Definition 3.1 has the conceptual advantage of being phrased in terms of the arc length of the curve, which we know is independent of parametrization. However, since it defines the curvature in terms of a derivative with respect to s rather than with respect to t, we are faced with the problem of computing K for a given parametrized curve C: $\mathbf{x} = \mathbf{x}(t)$. Rather than trying to parametrize C by arc length as in Example 7 of the last section, we can derive a direct formula for K in terms of the parameter t. To do so, we use the chain rule, Theorem 1.4(f), to write

$$\frac{d\mathbf{T}}{ds} = \frac{d\mathbf{T}}{dt}\frac{dt}{ds}. \tag{6}$$

By the inverse function differentiation rule (Theorem 7.5 of Chapter 2),

$$\frac{dt}{ds} = \frac{1}{\frac{ds}{dt}}.$$

Thus (6) becomes

$$\frac{d\mathbf{T}}{ds} = \frac{d\mathbf{T}}{dt}\frac{1}{\frac{ds}{dt}} = \frac{\mathbf{T}'(t)}{|\mathbf{x}'(t)|}, \tag{7}$$

which together with (5) gives the formula

$$K = \frac{|\mathbf{T}'(t)|}{|\mathbf{x}'(t)|}. \tag{8}$$

Example 1

Show that the curvature of a line in $\mathbf{R}^n$ is 0.

Solution. If we parametrize the line by $\mathbf{x}(t) = t\mathbf{a} + \mathbf{b}$, then $\mathbf{x}'(t) = \mathbf{a}$. Thus $\mathbf{T} = \mathbf{a}/|\mathbf{a}|$. Hence $\mathbf{T}'(t) = \mathbf{0}$, since $\mathbf{a}$ is constant. Therefore, from (8),

$$\mathbf{K} = \frac{|\mathbf{T}'(t)|}{|\mathbf{x}'(t)|} = \frac{0}{|\mathbf{a}|} = 0,$$

in conformity with our intuitive notion of curvature derived from highway driving. ■

Example 2

Find the curvature of a circle of radius a.

Solution. We might as well view the circle as being in $\mathbf{R}^2$, where we can parametrize it by $x(t) = a\cos t\,\mathbf{i} + a\sin t\,\mathbf{j}$. See **Figure 3.2.** Then $\mathbf{x}'(t) = -a\sin t\,\mathbf{i} + a\cos t\,\mathbf{j}$, so

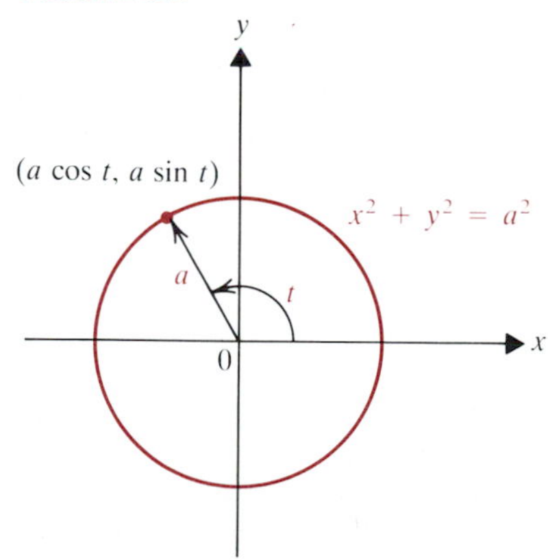

FIGURE 3.2

$$|\mathbf{x}'(t)| = \sqrt{a^2\sin^2 t + a^2\cos^2 t} = a.$$

We therefore have

$$\mathbf{T} = \frac{\mathbf{x}'(t)}{|\mathbf{x}'(t)|} = \frac{1}{a}\mathbf{x}'(t) = -\sin t\,\mathbf{i} + \cos t\,\mathbf{j}.$$

Then $\mathbf{T}'(t) = -\cos t\,\mathbf{i} + \sin t\,\mathbf{j}$. Hence by (8),

$$K = \frac{|\mathbf{T}'(t)|}{|\mathbf{x}'(t)|} = \frac{\sqrt{\cos^2 t + \sin^2 t}}{a} = \frac{1}{a}.$$

Thus the curvature of a circle is the reciprocal of its radius. This is consistent with our intuitive picture of circles with small radii curving more sharply than circles with large radii. ■

Since the unit tangent vector $\mathbf{T}(t)$ to any smooth curve has constant length one, we have $\mathbf{T}\cdot\mathbf{T} = 1$. If we differentiate with respect to t and use Theorem 1.4(b), we get $\mathbf{T}'\cdot\mathbf{T} + \mathbf{T}\cdot\mathbf{T}' = 0$, that is, $2\mathbf{T}\cdot\mathbf{T}' = 0$; hence $\mathbf{T}\cdot\mathbf{T}' = 0$. Thus $\mathbf{T}'$ either is $\mathbf{0}$ or is a *normal* vector to the curve $\mathbf{x} = \mathbf{x}(t)$.

3.2 DEFINITION

If $\mathbf{T}'(t) \neq \mathbf{0}$, then the ***unit normal vector*** to the parametrized curve C: $\mathbf{x} = \mathbf{x}(t)$ is

(9) $$\mathbf{N} = \frac{1}{|\mathbf{T}'(t)|}\,\mathbf{T}'(t).$$

You might expect that $|\mathbf{T}'| = 1$, since $|\mathbf{T}| = 1$. That happened in Example 2, where $\mathbf{T}' = -\cos t\,\mathbf{i} + \sin t\,\mathbf{j}$, but it does *not* happen in general, as shown by the following example, which also shows that the curvature is not always constant (as it was in Examples 1 and 2).

Example 3

Find the unit tangent and normal vectors and the curvature for C: $\mathbf{x}(t) = \frac{1}{2}t^2\,\mathbf{i} + t\,\mathbf{j}$ in general and at $t = 1$.

Solution. This curve is the parabola $y^2 = 2x$ (cf. Example 5 of the last section with $p = 1/2$). Differentiation gives $\mathbf{x}'(t) = t\,\mathbf{i} + \mathbf{j}$. Thus the speed is

(10) $$|\mathbf{x}'(t)| = \sqrt{t^2 + 1}.$$

From (1) then, we have

$$\mathbf{T}(t) = \frac{t}{\sqrt{t^2+1}}\,\mathbf{i} + \frac{1}{\sqrt{t^2+1}}\,\mathbf{j} \to \mathbf{T}(1) = \frac{1}{\sqrt{2}}\,\mathbf{i} + \frac{1}{\sqrt{2}}\,\mathbf{j}.$$

Differentiating the formula for $\mathbf{T}(t)$, we find

$$\mathbf{T}'(t) = \left[\frac{\sqrt{t^2+1} - t \cdot \frac{1}{2}(t^2+1)^{-1/2} \cdot 2t}{t^2+1}\right]\mathbf{i} - \frac{1}{2}(t^2+1)^{-3/2} \cdot 2t\,\mathbf{j}$$

$$= \frac{1}{(t^2+1)^{3/2}}\,\mathbf{i} - \frac{t}{(t^2+1)^{3/2}}\,\mathbf{j},$$

which is never $\mathbf{0}$. Thus

$$\mathbf{T}(1) = \frac{1}{2\sqrt{2}}\,\mathbf{i} - \frac{1}{2\sqrt{2}}\,\mathbf{j}.$$

In general, we have

(11) $$|\mathbf{T}'(t)| = \frac{1}{(t^2+1)^{3/2}}\sqrt{1+t^2} = \frac{1}{t^2+1}.$$

Therefore, from (8),

$$\mathbf{N}(t) = \frac{\mathbf{T}'(t)}{|\mathbf{T}'(t)|} = (t^2+1)\left(\frac{1}{(t^2+1)^{3/2}}\,\mathbf{i} - \frac{t}{(t^2+1)^{3/2}}\,\mathbf{j}\right) = \frac{1}{\sqrt{t^2+1}}\,\mathbf{i} - \frac{1}{\sqrt{t^2+1}}\,\mathbf{j}.$$

When $t = 1$, we find

$$|\mathbf{N}(1)| = \frac{1}{\sqrt{2}}\,\mathbf{i} - \frac{1}{\sqrt{2}}\,\mathbf{j}.$$

Finally, from (8), (10), and (11), we get

$$K(t) = \frac{|\mathbf{T}'(t)|}{|\mathbf{x}'(t)|} = \frac{1}{\sqrt{t^2+1}} \cdot \frac{1}{t^2+1} = \frac{1}{(t^2+1)^{3/2}},$$

FIGURE 3.3

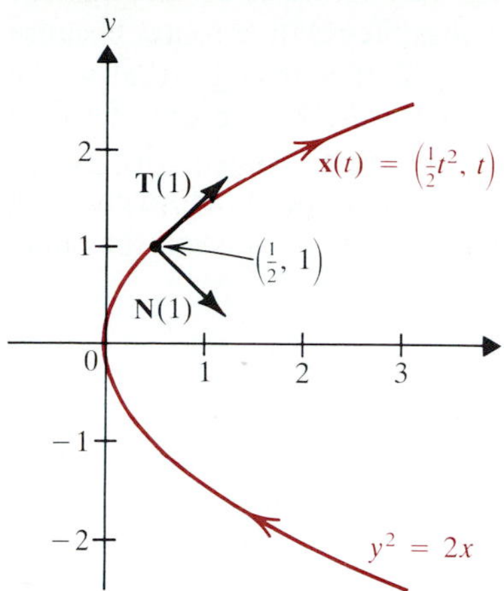

so

$$K(1) = \frac{1}{2\sqrt{2}}.$$

In **Figure 3.3** we show the curve, $\mathbf{T}(1)$, and $\mathbf{N}(1)$. ■

Like the unit tangent, the unit normal is intrinsic to the geometry of a given curve, but (8) seems to depend on the parametrization. To derive a formula for $\mathbf{N}$ in terms of the arc-length variable s [like the parameter-independent formula (4) for $\mathbf{T}$], we use

(7) $$\frac{d\mathbf{T}}{ds} = \frac{\mathbf{T}'(t)}{|\mathbf{x}'(t)|},$$

from p. 703. In conjunction with (8), it gives

$$\frac{1}{K}\frac{d\mathbf{T}}{ds} = \frac{|\mathbf{x}'(t)|}{|\mathbf{T}'(t)|}\frac{\mathbf{T}'(t)}{|\mathbf{x}'(t)|} = \frac{\mathbf{T}'(t)}{|\mathbf{T}'(t)|} = \mathbf{N}.$$

We thus have

(12) $$\mathbf{N} = \frac{1}{K}\frac{d\mathbf{T}}{ds}.$$

This formula puts us in a position to give a mathematical explanation of the tendency of an automobile to leave the road on a sharp curve. We have from (1) and (2)

$$\mathbf{x}'(t) = |\mathbf{x}'(t)|\mathbf{T}(t) = \frac{ds}{dt}\mathbf{T}.$$

Differentiating with respect to t and using Theorem 1.4(d), we get

$$\mathbf{a}(t) = \mathbf{x}''(t) = \frac{d}{dt}(\mathbf{x}'(t)) = \frac{d^2s}{dt^2}\mathbf{T}(t) + \frac{ds}{dt}\frac{d\mathbf{T}(t)}{dt}$$

$$= \frac{d^2s}{dt^2}\mathbf{T} + \frac{ds}{dt}\frac{d\mathbf{T}}{ds}\frac{ds}{dt} \qquad \textit{by the chain rule (Theorem 1.6f)}$$

$$= \frac{d^2s}{dt^2}\mathbf{T} + \left(\frac{ds}{dt}\right)^2\frac{d\mathbf{T}}{ds},$$

which by (12) gives

(13) $$\mathbf{a}(t) = \frac{d^2s}{dt^2}\mathbf{T} + \left(\frac{ds}{dt}\right)^2 K\mathbf{N}.$$

Equation (13) is a resolution of the acceleration vector $\mathbf{a}(t)$ into two components,

$$\mathbf{a_T} = \frac{d^2s}{dt^2}\mathbf{T}$$

tangent to the curve and

$$\mathbf{a_N} = \left(\frac{ds}{dt}\right)^2 K\mathbf{N}$$

FIGURE 3.4

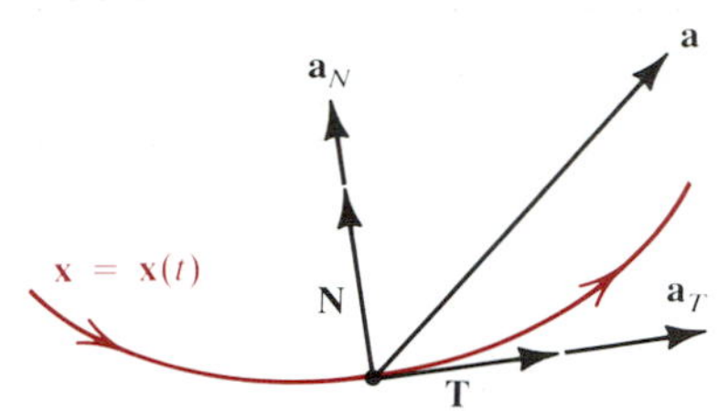

normal to the curve (**Figure 3.4**). To help your car stay on the road through the curve C: $\mathbf{x} = \mathbf{x}(t)$, you want $\mathbf{a_T}$ to be as large as possible relative to $\mathbf{a_N}$, because the direction of the curve at each point is given by **T**. If K is large (that is, the curve is sharp), then since the magnitude of $\mathbf{a_N}$ is $(ds/dt)^2K$, the only feasible way to make $\mathbf{a_N}$ small is to make ds/dt small relative to K. That is the reason for the speed-limit signs posted to warn drivers to reduce speed before entering a curve. In fact, if you wait until entering the curve to brake, then in the curve you will be decreasing your speed. That is,

$$\frac{d^2s}{dt^2} = \frac{d}{dt}\left(\frac{ds}{dt}\right) < 0.$$

FIGURE 3.5

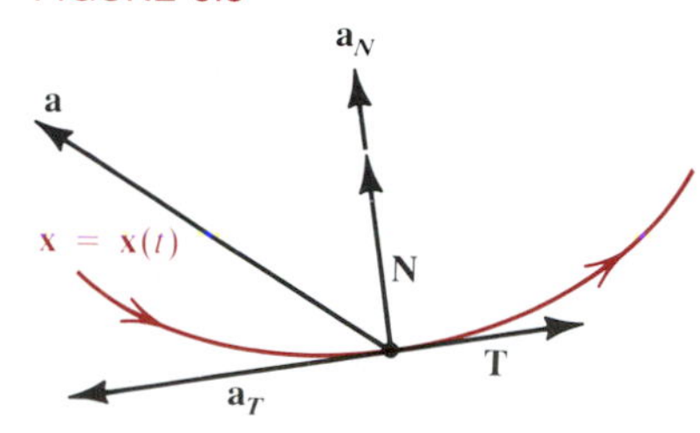

As you can see from **Figure 3.5,** $\mathbf{a_T}$ will no longer be pointing in the direction of motion, but in the *opposite* direction. Then **a** points in the direction of the resultant force $\mathbf{F} = m\mathbf{a}$ acting on the car. That force *hinders* rather than helps the car in negotiating the curve, since it is far removed from the direction of motion. On the other hand, if you slow below the speed limit *before* entering the curve and then increase your speed slightly (i.e., engage the accelerator moderately) in the curve, then $d^2s/dt^2 > 0$ in the curve. Hence as in Figure 3.4, the tangential component $\mathbf{a_T}$ is a positive factor in keeping your car on the road.

This discussion is just one example of the importance of (13) in dynamics. Because of their importance, the two summands in (13) are given special names.

3.3 DEFINITION

The ***tangential*** and ***normal components of acceleration*** of a particle moving along the path $\mathbf{x} = \mathbf{x}(t)$ are

$$\mathbf{a_T} = \frac{d^2s}{dt^2}\mathbf{T} \quad \text{and} \quad \mathbf{a_N} = \left(\frac{ds}{dt}\right)^2 K\mathbf{N}. \tag{14}$$

An alternative formula for $\mathbf{a_N}$ results from substituting the value

$$\frac{|\mathbf{T}'(t)|}{\dfrac{ds}{dt}}$$

of K from (8) into the second formula in (14). That gives

$$\mathbf{a_N} = \left(\frac{ds}{dt}\right)^2 \frac{1}{\dfrac{ds}{dt}}|\mathbf{T}'|\mathbf{N} = \frac{ds}{dt}|\mathbf{T}'|\mathbf{N}. \tag{15}$$

Although the formulas in (14) are very useful in contexts such as a car rounding a curve in the road, the fact that they do not involve t can make them a chore to use for a curve given parametrically as $\mathbf{x} = \mathbf{x}(t)$. When dealing with such curves, it is often easier to compute $\mathbf{a_T}$ directly from **x** and **v**, as the component of **a** in the direction of **v** (Definition 3.8 of Chapter 10). That gives

$$\mathbf{a_T} = (\mathbf{a} \cdot \mathbf{T})\mathbf{T} = \frac{\mathbf{a} \cdot \mathbf{v}}{|\mathbf{v}|}\frac{\mathbf{v}}{|\mathbf{v}|}. \tag{16}$$

Since $\mathbf{a} = \mathbf{a_T} + \mathbf{a_N}$, we also have

$$\mathbf{a_N} = \mathbf{a} - \mathbf{a_T}. \tag{17}$$

The advantage of (17) is that it permits us to compute **N** without using either Definition 3.2 or Equation (8). We can thereby save considerable labor in differentiation, as the next example illustrates.

Example 4

Find the tangential and normal components of acceleration of a particle moving on the curve C of Example 3 at $t = 1$. Use this information to recalculate $\mathbf{N}(1)$.

Solution. We have $\mathbf{x}(t) = \frac{1}{2}t^2\,\mathbf{i} + t\,\mathbf{j}$. Thus

$$\mathbf{v}(t) = \mathbf{x}'(t) = t\,\mathbf{i} + \mathbf{j} \rightarrow \mathbf{v}(1) = \mathbf{i} + \mathbf{j}$$

and

$$\mathbf{a}(t) = \mathbf{x}''(t) = \mathbf{i} \rightarrow \mathbf{a}(1) = \mathbf{i}.$$

In Example 3 we found that

$$\mathbf{T}(1) = \frac{1}{\sqrt{2}}\mathbf{i} + \frac{1}{\sqrt{2}}\mathbf{j}.$$

Hence at $t = 1$ we have

$$\mathbf{a_T} = (\mathbf{a} \cdot \mathbf{T})\mathbf{T} = \left(\frac{1}{\sqrt{2}} + 0\right)\mathbf{T} = \frac{1}{2}\mathbf{i} + \frac{1}{2}\mathbf{j}.$$

Therefore, when $t = 1$, Equation (17) gives

$$\mathbf{a_N} = \mathbf{a} - \mathbf{a_T} = \mathbf{i} - \left(\frac{1}{2}\mathbf{i} + \frac{1}{2}\mathbf{j}\right) = \frac{1}{2}\mathbf{i} - \frac{1}{2}\mathbf{j}.$$

Since $\mathbf{N}$ is the direction vector of $\mathbf{a_N}$, we have

$$\mathbf{N} = \frac{1}{|\mathbf{a_N}|}\mathbf{a_N} = \frac{1}{\sqrt{1/4 + 1/4}}\left(\frac{1}{2}\mathbf{i} - \frac{1}{2}\mathbf{j}\right) = \sqrt{2}\left(\frac{1}{2}\mathbf{i} - \frac{1}{2}\mathbf{j}\right) = \frac{1}{\sqrt{2}}\mathbf{i} - \frac{1}{\sqrt{2}}\mathbf{j},$$

which agrees with the result of Example 3 and can be computed *without* differentiating the complicated function $\mathbf{T}(t)$. ■

The next example further illustrates the use of (16) and (17) as computational tools.

Example 5

FIGURE 3.6

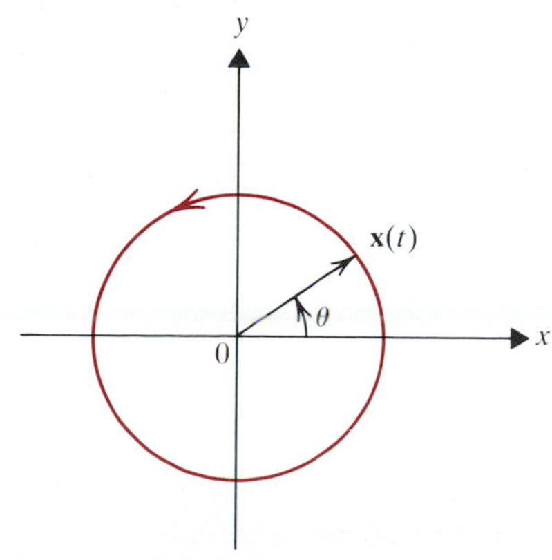

A particle moves on a circle of radius a with angular speed $\omega(t) = \theta'(t)$, where $\theta(t)$ is the central angle shown in **Figure 3.6.** Find the tangential and normal components of the acceleration. Also find $\mathbf{N}$.

Solution. We have $\mathbf{x}(t) = a\cos\theta(t)\,\mathbf{i} + a\sin\theta(t)\,\mathbf{j}$. So by the chain rule,

$$\mathbf{v}(t) = a(-\omega\sin\theta(t)\,\mathbf{i} + \omega\cos\theta(t)\,\mathbf{j}) = a\omega(-\sin\theta(t)\,\mathbf{i} + \cos\theta(t)\,\mathbf{j}). \tag{18}$$

Also,

$$\begin{aligned}\mathbf{a}(t) &= a\omega'(-\sin\theta(t)\,\mathbf{i} + \cos\theta(t)\,\mathbf{j}) + a\omega(-\omega\cos\theta(t)\,\mathbf{i} - \omega\sin\theta(t)\,\mathbf{j})\\ &= a(-\omega'\sin\theta(t) - \omega^2\cos\theta(t))\,\mathbf{i} + a(\omega'\cos\theta(t) - \omega^2\sin\theta(t))\,\mathbf{j}.\end{aligned} \tag{19}$$

Further,

$$\mathbf{T} = \frac{\mathbf{x}'(t)}{|\mathbf{x}'(t)|} = \frac{\mathbf{v}(t)}{|\mathbf{v}(t)|} = -\sin\theta(t)\,\mathbf{i} + \cos\theta(t)\,\mathbf{j}. \tag{20}$$

So (16), (19), and (20) give

$$(21) \quad \mathbf{a_T} = a(\omega' \sin^2 \theta(t) + \cancel{\omega^2 \sin \theta(t) \cos \theta(t)} + \omega' \cos^2 \theta(t) - \cancel{\omega^2 \sin \theta(t) \cos \theta(t)})\mathbf{T}$$
$$= a\omega'\mathbf{T} = a\omega'(-\sin \theta(t)\,\mathbf{i} + \cos \theta(t)\,\mathbf{j}).$$

Since from (18) we have

$$\frac{ds}{dt} = |\mathbf{v}(t)| = a\omega,$$

we can compute $\mathbf{a_N}$ from (14) and the fact that $K = 1/a$:

$$\mathbf{a_N} = \left(\frac{ds}{dt}\right)^2 K\mathbf{N} = a^2\omega^2 \frac{1}{a} \mathbf{N} = a\omega^2\mathbf{N}.$$

To obtain an explicit formula for $\mathbf{a_N}$ in terms of $\theta(t)$, we can use (17), (19), and (21):

$$\mathbf{a_N} = \mathbf{a} - \mathbf{a_T}$$
$$= a(-\cancel{\omega' \sin \theta} - \omega^2 \cos \theta)\,\mathbf{i} + a(\cancel{\omega' \cos \theta} - \omega^2 \sin \theta)\,\mathbf{j}$$
$$- a(-\cancel{\omega' \sin \theta}\,\mathbf{i} + \cancel{\omega' \cos \theta}\,\mathbf{j})$$
$$(22) \quad = -a\omega^2(\cos \theta\,\mathbf{i} + \sin \theta\,\mathbf{j}) = a\omega^2(-\cos \theta\,\mathbf{i} - \sin \theta\,\mathbf{j}).$$

From (22), we can see immediately that $\mathbf{N} = -\cos \theta\,\mathbf{i} - \sin \theta\,\mathbf{j}$, without having to compute $\mathbf{T}'$. ■

In $\mathbf{R}^3$ there is a formula for K purely in terms of the velocity and acceleration along a parametrized curve $\mathbf{x} = \mathbf{x}(t)$. It is derived from (13) and (1), which say that

$$\mathbf{v} = \frac{ds}{dt}\mathbf{T} \qquad \text{and} \qquad \mathbf{a} = \frac{d^2s}{dt^2}\mathbf{T} + \left(\frac{ds}{dt}\right)^2 K\mathbf{N}.$$

Hence

$$\mathbf{v} \times \mathbf{a} = \frac{ds}{dt}\frac{d^2s}{dt^2}\mathbf{T} \times \mathbf{T} + \left(\frac{ds}{dt}\right)^3 K(\mathbf{T} \times \mathbf{N})$$
$$= \left(\frac{ds}{dt}\right)^3 K(\mathbf{T} \times \mathbf{N}). \qquad \textit{since } T \times T = 0$$

Then by Equation (10) of Section 10.6 and the fact that $\mathbf{T} \perp \mathbf{N}$,

$$|\mathbf{v} \times \mathbf{a}| = \left(\frac{ds}{dt}\right)^3 K|\mathbf{T}|\,|\mathbf{N}| \sin \frac{\pi}{2} = \left(\frac{ds}{dt}\right)^3 K.$$

Therefore we get

$$(23) \qquad K = \frac{|\mathbf{v} \times \mathbf{a}|}{|\mathbf{v}(t)|^3}.$$

Example 6

Find the curvature of $\mathbf{x}(t) = te^t\,\mathbf{i} - t^2e^{2t}\,\mathbf{j} + t\,\mathbf{k}$ at $t = 0$.

Solution. We have

$$\mathbf{v}(t) = \mathbf{x}'(t) = (e^t + te^t)\,\mathbf{i} - (2te^{2t} + 2t^2e^{2t})\,\mathbf{j} + \mathbf{k},$$

and

$$\mathbf{a}(t) = \mathbf{x}''(t) = (2e^t + te^t)\mathbf{i} - (2e^{2t} + 8te^{2t} + 4t^2e^{2t})\mathbf{j}.$$

When $t = 0$, we therefore obtain

$$\mathbf{v}(0) = \mathbf{i} + \mathbf{k} \quad \text{and} \quad \mathbf{a}(0) = 2\mathbf{i} - 2\mathbf{j}.$$

We have then $|\mathbf{v}(0)| = \sqrt{2}$, and

$$\mathbf{v}(0) \times \mathbf{a}(0) = \det \begin{vmatrix} \mathbf{i} & \mathbf{j} & \mathbf{k} \\ 1 & 0 & 1 \\ 2 & -2 & 0 \end{vmatrix} = 2\mathbf{i} + 2\mathbf{j} - 2\mathbf{k},$$

which gives $|\mathbf{v}(0) \times \mathbf{a}(0)| = 2\sqrt{3}$. According to (23) the curvature is thus

$$K = \frac{2\sqrt{3}}{(\sqrt{2})^3} = \sqrt{\frac{3}{2}}. \quad ■$$

Exercises 11.3

In Exercises 1–12, find K, T, and N.

1. $\mathbf{x}(t) = \cos t\,\mathbf{i} + \sin t\,\mathbf{j} + t\,\mathbf{k}$
2. $\mathbf{x}(t) = e^t \cos t\,\mathbf{i} + e^t \sin t\,\mathbf{j}$
3. $\mathbf{x}(t) = (3 \sin t)\mathbf{i} + (3 \cos t)\mathbf{j} + 4t\,\mathbf{k}$
4. $\mathbf{x}(t) = (e^t \cos t)\mathbf{i} + (e^t \sin t)\mathbf{j} + e^t\,\mathbf{k}$
5. $\mathbf{x}(t) = (\sin t - t\cos t, \cos t + t \sin t, t^2)$
6. $\mathbf{x}(t) = (\sin t + \cos t, \sin t - \cos t, t/2)$
7. $\mathbf{x}(t) = (\sqrt{2}\cos 3t)\mathbf{i} + (\sqrt{2}\cos 3t)\mathbf{j} + (2\sin 3t)\mathbf{k}$. What is the curve?
8. $\mathbf{x}(t) = (\sqrt{2}\sin 3t)\mathbf{i} + (2\cos 3t)\mathbf{j} + (\sqrt{2}\sin 3t)\mathbf{k}$. What is the curve?
9. $\mathbf{x}(t) = t\,\mathbf{i} + t^2\mathbf{j} + t^3\mathbf{k}$ at $t_0 = 1$
10. $\mathbf{x}(t) = (t^2 - 1)\mathbf{i} + (1 + t)\mathbf{j} + t^3\mathbf{k}$ at $t_0 = 1$
11. $\mathbf{x}(t) = (t \sin t)\mathbf{i} + (t \cos t)\mathbf{j} + (t^2 - 1)\mathbf{k}$ at $t_0 = \pi$
12. $\mathbf{x}(t) = te^t\mathbf{i} - t^2e^{2t}\mathbf{j} + t\,\mathbf{k}$, at $t_0 = 0$

In Exercises 13–22, find the tangential and normal components of acceleration of the parametrized curve in the indicated exercise above.

13. Exercise 1
14. Exercise 2
15. Exercise 3
16. Exercise 4
17. Exercise 5
18. Exercise 6
19. Exercise 9
20. Exercise 10
21. Exercise 11
22. Exercise 12

23. For a certain curve $\mathbf{x} = \mathbf{x}(t)$, it is known that $\mathbf{a}_\mathbf{T} = (\sin^2 t)\mathbf{j} + t\,\mathbf{k}$ and $\mathbf{a}_\mathbf{N} = \mathbf{i} + (\cos^2 t)\mathbf{j} + (2 - t)\mathbf{k}$. If $\mathbf{v}(0) = \mathbf{i}$, then find K at $t = 1$.
24. For a certain curve $\mathbf{x} = \mathbf{x}(t)$, it is known that $\mathbf{x}(0) = \mathbf{i} + \mathbf{j}$ and
$$\mathbf{v}(t) = (1 + 2te^{t^2})\mathbf{i} + (t(t^2 + 1)^{-1/2})\mathbf{j} + (3t^2 + 2)\mathbf{k}.$$
Find the equation of the plane determined by the tangent and normal lines to the curve at $t_0 = 0$.
25. For the curve $\mathbf{x}(t) = (a\cos t, a\sin t, bt)$, show that the tangential component of acceleration is $\mathbf{0}$ and the normal component is $|\mathbf{a}|\mathbf{N}$.
26. If $\mathbf{x} = \mathbf{x}(t)$, then show that
$$K = \frac{\sqrt{|\mathbf{x}'|^2|\mathbf{x}''|^2 - (\mathbf{x}' \cdot \mathbf{x}'')^2}}{|\mathbf{x}'|^3} \quad \text{if } \mathbf{x}' \neq \mathbf{0}.$$
27. If $\mathbf{x} = (x(t), y(t))$ is a smooth curve in $\mathbf{R}^2$, then show that
$$K = \frac{|x'y'' - y'x''|}{[(x')^2 + (y')^2]^{3/2}}, \quad \text{where } x' = \frac{dx}{dt} \text{ and } y' = \frac{dy}{dt}.$$
28. If a plane curve is the graph of $y = f(x)$, then show that
$$K = \frac{|f''(x)|}{[1 + f'(x)^2]^{3/2}}.$$
[*Hint:* Parametrize the curve in the form $\mathbf{x}(t) = (t, f(t))$.]

In Exercises 29–32, use the formula in Exercise 28 to find the curvature of the graph of the given function at the given point.

29. $y = \sin x$ at $x = \pi/2$
30. $y = \cos x$ at $x = 0$
31. $y = x^2$ at $x = 1$
32. $y = \ln x$ at $x = 1$

33. What is the curvature of a plane curve $y = f(x)$ at a point of inflection where f'' exists?
34. Prove that if a curve in $\mathbf{R}^m$ has curvature identically 0, then the curve is a straight line. (This is the converse of Example 1.)
35. The ***binormal vector*** to the curve $\mathbf{x} = \mathbf{x}(t)$ is $\mathbf{B}(t) = \mathbf{T} \times \mathbf{N}$.

    (a) Show that $\mathbf{B}$ is a unit vector and $\mathbf{B} = \dfrac{1}{|\mathbf{v} \times \mathbf{a}|}\mathbf{v} \times \mathbf{a}$.

(b) Show $\dfrac{d\mathbf{B}}{ds} \cdot \mathbf{B} = 0$.

(c) Show that $\dfrac{d\mathbf{B}}{ds} = \mathbf{T} \times \dfrac{d\mathbf{N}}{ds}$.

(d) Conclude that $\dfrac{d\mathbf{B}}{ds} \cdot \mathbf{T} = 0$.

36. Use Exercise 35 to show that

$$\frac{d\mathbf{B}}{ds} = \tau\mathbf{N}, \qquad \text{where } \tau = \frac{d\mathbf{B}}{ds} \cdot \mathbf{N}.$$

The quantity τ is called the ***torsion*** of the curve.

37. **(a)** Show that $\mathbf{N} = \mathbf{B} \times \mathbf{T}$.

(b) Show that $d\mathbf{N}/ds = -K\mathbf{T} - \tau\mathbf{B}$.

HISTORICAL NOTE

The formulas $d\mathbf{T}/ds = K\mathbf{N}$, $d\mathbf{B}/ds = \tau\mathbf{N}$, and the formula in Exercise 37(b) are basic tools in a part of mathematics—called *metric differential geometry*—that studies curves in space and has many applications to physics and astronomy. Those three formulas are part of a set of nine fundamental formulas in differential geometry, called the *Frenet–Serret* formulas. They are named for the nineteenth-century French mathematicians Jean-Frédéric Frenet (1816–1900) and Joseph A. Serret (1819–1885). Frenet's basic theory of space curves was part of his Ph.D. dissertation from the University of Toulouse in 1847. He was later professor of mathematics at the University of Lyons and director of its astronomical observatory. He wrote a very successful calculus text in 1856, which remained in use for a half-century. In it, he pointed the way for modern calculus authors by including many more fully worked out problems than had previously been the custom in such books. He also pioneered the inclusion of historical notes, such as this one, a practice that later fell out of favor until the past few years.

Serret was professor of celestial mechanics at the College de France and later held the title of professor of differential and integral calculus at the Sorbonne. He was one of the inventors of differential geometry and independently arrived at the formulas mentioned above. The triple of vectors $(\mathbf{T}, \mathbf{N}, \mathbf{B})$ is a right-handed system, since $\mathbf{B} = \mathbf{T} \times \mathbf{N}$. It provides a *local coordinate* system at each point $\mathbf{x} = \mathbf{x}(t)$ on the curve. The system $(\mathbf{T}, \mathbf{N}, \mathbf{B})$, which is called the *moving trihedral* of the curve, was the first example of the important idea of local coordinate systems.

11.4 Scalar Functions of Several Variables

Scalar functions act in the reverse direction from the vector functions considered in the last three sections. They assign to every vector $\mathbf{x}$ in some subset D of $\mathbf{R}^n$ a unique real number $y = f(\mathbf{x})$. We use the notation $f\colon \mathbf{R}^n \to \mathbf{R}$ for such functions and write

$$y = f(\mathbf{x}) = f(x_1, x_2, \ldots, x_n)$$

for the real value that the function assigns to the vector $\mathbf{x} = (x_1, x_2, \ldots, x_n)$ in $\mathbf{R}^n$.

We have already worked with one scalar function, namely, the length function for vectors. It assigns to each vector $\mathbf{x} = x\mathbf{i} + y\mathbf{j} + z\mathbf{k}$ in $\mathbf{R}^3$ the nonnegative real number $|\mathbf{x}| = \sqrt{x^2 + y^2 + z^2}$. Another example is the *ideal gas law*, which says that the volume V of a gas is a function of two variables P (pressure) and

T (temperature):

$$V = nR\frac{T}{P},$$

where n is the number of moles of the gas, and R is the gas constant. For each first-quadrant vector $P\mathbf{i} + T\mathbf{j} = (P, T)$, we get a positive volume V. (In working with scalar functions, it is often helpful to identify a vector $\mathbf{x}$ in $\mathbf{R}^n$ with its endpoint, as we have done in **Figure 4.1** with (P, T).)

FIGURE 4.1

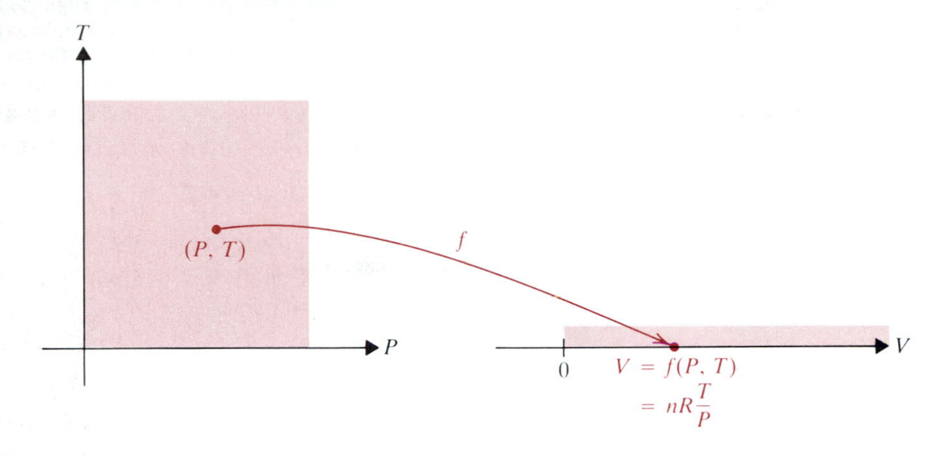

An important quantity in economics that is prominently reported every month is the *cost-of-living index*. The prices p_i of certain key commodities (goods and services) are monitored each month. A fixed weight w_i is assigned to each commodity to reflect its relative importance in a people's lives. For example, housing costs are weighted more heavily than new car prices. The cost-of-living index C is then the weighted average of the p_i:

$$C = f(p_1, p_2, \ldots, p_n) = w_1p_1 + \ldots + w_np_n.$$

The weights are chosen so that C averages 100 for some period, such as 1967–69, to allow easy comparison of today's cost of living with that of several years ago.

The first step in developing the calculus of scalar functions is to define the limit L of f as $\mathbf{x}$ approaches $\mathbf{a}$ in $\mathbf{R}^n$. We do so by requiring that $f(\mathbf{x})$ can be made as close to L as we please by taking $\mathbf{x}$ close enough to $\mathbf{a}$. To parallel Definition 4.1 of Chapter 1, we need the analogue in $\mathbf{R}^n$ of an open interval around a real number c. Recall that

$$(c - \delta, c + \delta) = \{x \in \mathbf{R} \mid d(x, c) < \delta\},$$

where $d(x, c) = |x - c|$ is the distance between x and c on the real-number line. The parallel concept in $\mathbf{R}^n$ is called an *open ball*.

4.1 DEFINITION Let $\mathbf{a} = (a_1, a_2, \ldots, a_n) \in \mathbf{R}^n$. The ***open ball*** $B(\mathbf{a}, \delta)$ about $\mathbf{a}$ of radius δ is

$$B(\mathbf{a}, \delta) = \{\mathbf{x} \in \mathbf{R}^n \mid d(\mathbf{x}, \mathbf{a}) < \delta\},$$

where

$$d(\mathbf{x}, \mathbf{a}) = \sqrt{(x_1 - a_1)^2 + (x_2 - a_2)^2 + \ldots + (x_n - a_n)^2} = |\mathbf{x} - \mathbf{a}|$$

is the distance between $\mathbf{x}$ and $\mathbf{a}$ in $\mathbf{R}^n$. The ***closed ball*** $\bar{B}(\mathbf{a}, \delta)$ about $\mathbf{a}$ of radius δ is

$$\bar{B}(\mathbf{a}, \delta) = \{\mathbf{x} \in \mathbf{R}^n \mid d(\mathbf{x}, \mathbf{a}) \le \delta\}.$$

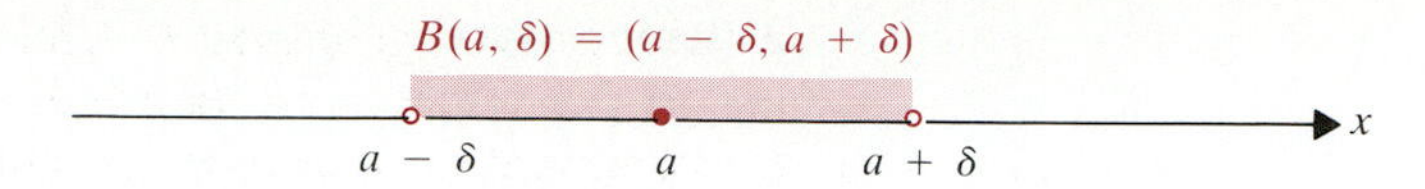

FIGURE 4.2

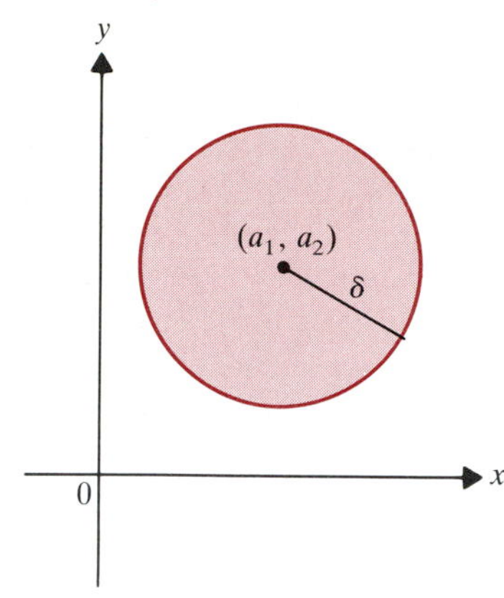

FIGURE 4.3

In one dimension, the open ball $B(a, \delta) = \{x \in \mathbf{R} \,|\, |x - a| < \delta\}$ is the *open interval* $(a - \delta, a + \delta)$ shown in **Figure 4.2.** The closed ball $\bar{B}(a, \delta)$ is the closed interval $[a - \delta, a + \delta]$. When $n = 2$, the vector $\mathbf{a} = a_1\mathbf{i} + a_2\mathbf{j} = (a_1, a_2)$, and so the open ball

$$B(\mathbf{a}, \delta) = \{(x, y) \in \mathbf{R}^2 \,|\, \sqrt{(x - a_1)^2 + (y - a_2)^2} < \delta\}$$

is the interior of the circular disk centered at the endpoint (a_1, a_2) of $\mathbf{a}$ with radius δ. The closed ball $\bar{B}(\mathbf{a}, \delta)$ is just the open ball $B(\mathbf{a}, \delta)$ with the bounding circle $(x - a_1)^2 + (y - a_2)^2 = \delta$ added. See **Figure 4.3.** When $n = 3$, the vector $\mathbf{a} = a_1\mathbf{i} + a_2\mathbf{j} + a_3\mathbf{k} = (a_1, a_2, a_3)$. Thus, in three dimensions, the open ball

$$B(\mathbf{a}, \delta) = \{(x, y, z) \in \mathbf{R}^3 \,|\, \sqrt{(x - a_1)^2 + (y - a_2)^2 + (z - a_3)^2} < \delta\}$$

is the region of $\mathbf{R}^3$ inside the sphere of radius δ centered at the point (a_1, a_2, a_3). The closed ball $\bar{B}(\mathbf{a}, \delta)$ is the open ball $B(\mathbf{a}, \delta)$ plus the bounding sphere $(x - a_1)^2 + (y - a_2)^2 + (z - a_3)^2 = \delta$. See **Figure 4.4.**

FIGURE 4.4

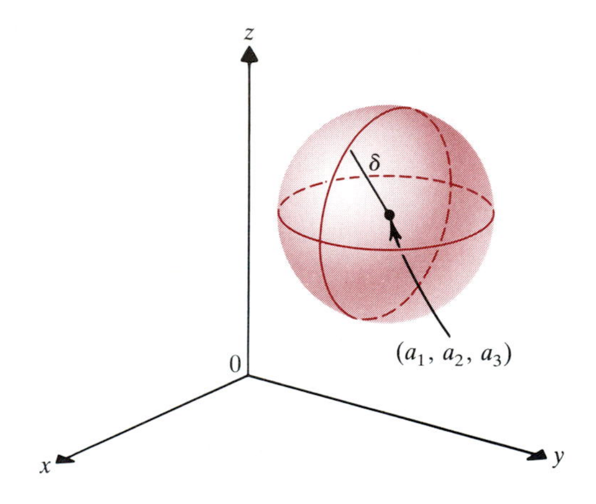

We are now ready to define the limit of a scalar function of several real variables.

4.2 DEFINITION Let $f: \mathbf{R}^n \to \mathbf{R}$ be defined on some open ball $B(\mathbf{a}, \delta)$, except possibly at $\mathbf{a}$ itself. Then

$$\lim_{\mathbf{x} \to \mathbf{a}} f(\mathbf{x}) = L$$

means that $f(\mathbf{x})$ is arbitrarily close to L when $\mathbf{x}$ is sufficiently close to $\mathbf{a}$. That is, for every $\varepsilon > 0$, there is some $\delta > 0$ such that

$$(1) \qquad |f(\mathbf{x}) - L| < \varepsilon \qquad \text{whenever } 0 < |\mathbf{x} - \mathbf{a}| < \delta.$$

This means that, given any positive real number ε, for some real number $\delta > 0$, we have $f(\mathbf{x})$ in the open interval $(L - \varepsilon, L + \varepsilon) = B(L, \varepsilon)$ in $\mathbf{R}^1$ for all $\mathbf{x} \neq \mathbf{a}$ in the open ball $B(\mathbf{a}, \delta)$ in $\mathbf{R}^n$.

Because f must be defined throughout an open ball $B(\mathbf{a}, \delta)$ about $\mathbf{a}$, except possibly at $\mathbf{x} = \mathbf{a}$, the condition $|f(\mathbf{x}) - L| < \varepsilon$ is meaningful for all $\mathbf{x}$ near $\mathbf{a}$. The set $B(\mathbf{a}, \delta) - \{\mathbf{a}\}$ is sometimes called the *punctured* (or *deleted*) *open ball* about $\mathbf{a}$ of radius δ. Condition (1) generalizes Definition 4.1 of Chapter 1 by requiring that $f(\mathbf{x})$ be close to L for *all* $\mathbf{x}$ near $\mathbf{a}$. Thus *no matter how* $\mathbf{x}$ *approaches* $\mathbf{a}$, $f(\mathbf{x})$ *must approach* L. In one dimension there are only two common ways for a real variable x to approach a number a: from the left or the right. But in two or more dimensions there are infinitely many paths of approach. In fact $\mathbf{x}$ can approach $\mathbf{a}$ along any line

$$\mathbf{x} = t\mathbf{v} + \mathbf{a},$$

or more generally, along any curve $\mathbf{x} = \mathbf{x}(t)$ through the point $\mathbf{a}$. **Figure 4.5** illustrates five ways that $\mathbf{x}$ in $\mathbf{R}^2$ can approach $\mathbf{a} = (0, 0)$, for instance: along the x-axis or y-axis, along the line $y = 2x$, along the parabola $y = x^2$, and along the sine curve $y = \sin x$. So there is no analogue of Theorem 6.2 of Chapter 1 asserting that $\lim_{\mathbf{x}\to\mathbf{a}} f(\mathbf{x})$ exists if $f(\mathbf{x}) \to f(\mathbf{a})$ as $\mathbf{x} \to \mathbf{a}$ along two (or any number of) paths. This means that it is often harder to determine whether the limit of a scalar function of several variables exists than whether a function $f: \mathbf{R} \to \mathbf{R}$ of one variable has a limit.

FIGURE 4.5

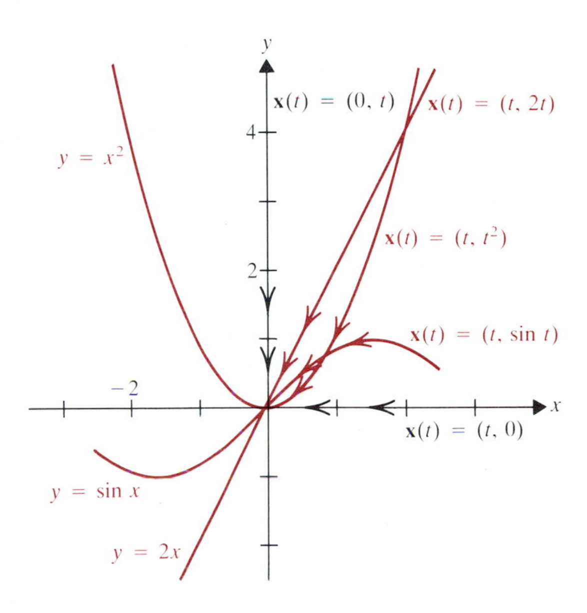

Example 1

Does f defined by

$$f(x, y) = \frac{x^4 - y^4}{x^2 + y^2}$$

have a limit at $(0, 0)$?

Solution. As long as $(x, y) \neq (0, 0)$, we have

$$f(x, y) = \frac{(x^2 + y^2)(x^2 - y^2)}{x^2 + y^2} = x^2 - y^2.$$

So f is defined on any punctured ball about $(0, 0)$. Moreover, no matter how (x, y) may approach $(0, 0)$, we have $x \to 0$ and $y \to 0$. Therefore $x^2 - y^2 \to 0$. Hence

$$\lim_{(x,y)\to(0,0)} f(x, y) = \lim_{(x,y)\to(0,0)} (x^2 - y^2) = 0,$$

so the limit exists and equals 0. ■

Example 1 illustrates why we required f to be defined only *near* (rather than near and also at) $\mathbf{x} = \mathbf{a}$ in Definition 4.2. We won't emphasize the technical aspects of the definition, but instead will rely on intuition and previous experience with limits. The following theorem, whose proof we largely omit, is proved by using (1). The reasoning is the same as that needed to prove the corresponding properties of limits for functions of a single real variable from Definition 4.1 of Chapter 1.

4.3 THEOREM Let f and g be scalar functions defined on a deleted open ball about $\mathbf{x} = \mathbf{a}$. If $\lim_{\mathbf{x}\to\mathbf{a}} f(\mathbf{x}) = L$ and $\lim_{\mathbf{x}\to\mathbf{a}} g(\mathbf{x}) = M$, then

(a) $\lim_{\mathbf{x}\to\mathbf{a}} [f(\mathbf{x}) \pm g(\mathbf{x})] = L \pm M$;

(b) $\lim_{\mathbf{x}\to\mathbf{a}} kf(\mathbf{x}) = kL$, for any real number k;

(c) $\lim_{\mathbf{x}\to\mathbf{a}} [f(\mathbf{x})g(\mathbf{x})] = LM$;

(d) $\lim_{\mathbf{x}\to\mathbf{a}} \dfrac{f(\mathbf{x})}{g(\mathbf{x})} = \dfrac{L}{M}$, if $M \neq 0$.

Partial Proof. We prove part (b), leave part (a) for Exercises 21 and 22, and leave the remaining parts for advanced calculus. First note that if $k = 0$, then clearly $\lim_{\mathbf{x}\to\mathbf{a}} kf(\mathbf{x}) = kL = 0$. If $k \neq 0$, then given $\varepsilon > 0$, choose $\delta > 0$ such that $|f(\mathbf{x}) - L| < \varepsilon/k$ when $\mathbf{x} \neq \mathbf{a}$ is in $B(\mathbf{a}, \delta)$. Then for such $\mathbf{x}$, we have

$$|kf(\mathbf{x}) - kL| = |k|\,|f(\mathbf{x}) - L| < |k|\frac{|\varepsilon|}{|k|} = \varepsilon.$$

So by Definition 4.2, $\lim_{\mathbf{x}\to\mathbf{a}} kf(\mathbf{x}) = kL$. QED

Continuity for scalar functions is defined exactly as for functions of a single real variable.

4.4 DEFINITION A scalar function $f\colon \mathbf{R}^n \to \mathbf{R}$ is ***continuous*** at $\mathbf{x} = \mathbf{a}$ if $\lim_{\mathbf{x}\to\mathbf{a}} f(\mathbf{x}) = f(\mathbf{a})$.

Just as in Section 1.7, the next theorem follows at once from Theorem 4.3 and the definition of continuity (see Theorem 7.3 of Chapter 1).

4.5 THEOREM If f and g are continuous at $\mathbf{x} = \mathbf{a}$, then so are

(a) $f \pm g$;

(b) kf, for any real number k;

(c) $f \cdot g$;

(d) f/g, provided that $g(\mathbf{a}) \neq 0$.

This theorem guarantees that many common scalar functions of two and three variables are continuous. For example, consider the *projection functions* P_x and P_y from $\mathbf{R}^2$ to $\mathbf{R}$, where

$$P_x(x, y) = x \qquad \text{and} \qquad P_y(x, y) = y.$$

Since it is clear for every $\mathbf{a} = (a, b)$ in $\mathbf{R}^2$ that

$$\lim_{\mathbf{x}\to\mathbf{a}} P_x(x, y) = a = P_x(a, b) \qquad \text{and} \qquad \lim_{\mathbf{x}\to\mathbf{a}} P_y(x, y) = b = P_y(a, b),$$

we see that P_x and P_y are continuous at every point in the plane. Then by Theorem 4.5 (a, b, and c), any polynomial function in x and y is continuous on the whole plane, since any such function is a sum of products of P_x and/or

P_y times constants. It then follows from Theorem 4.5(d) that every rational function in x and y is continuous wherever the denominator is nonzero. This leads to a wealth of problems involving the continuity of schizophrenic functions

$$r(x, y) = \begin{cases} \dfrac{f(x, y)}{g(x, y)} & \text{if } g(x, y) \neq 0, \\ k & \text{if } g(x, y) = 0, \end{cases}$$

where $f(x, y)$ and $g(x, y)$ are polynomials, and k is some real number.

Example 2

Determine all points of continuity of the function f if

$$f(x, y) = \begin{cases} \dfrac{x^2 - y^2}{x^2 + 2y^2} & \text{if } (x, y) \neq (0, 0) \\ 0 & \text{at } (0, 0). \end{cases}$$

Solution. By Theorem 4.5(d) this function is continuous at all points (x, y) such that $x^2 + 2y^2 \neq 0$, that is, everywhere except possibly at $(0, 0)$. To determine whether f is continuous at $(0, 0)$, we must first guess an answer and then try to verify our guess. The following rule of thumb is helpful in making the initial guess.

Guess continuity if we can divide out of the numerator and denominator all factors that are 0 at the point of schizophrenia (as in Example 1) *or if the degree of every term of the numerator exceeds the degree of the denominator. Otherwise, guess discontinuity.*

So here we guess discontinuity. To verify this we look for a path of approach to $(0, 0)$ along which $f(x, y)$ either has no limit or tends to a limit *other* than $f(0, 0) = 0$. If we try the x-axis ($y = 0$), then as $(x, 0) \to (0, 0)$, we have

$$f(x, y) = \frac{x^2 - 0}{x^2 + 0} = 1 \to 1 \neq 0 = f(0, 0).$$

As noted following Definition 4.2, for $\lim_{\mathbf{x} \to \mathbf{0}} f(\mathbf{x})$ to be $f(\mathbf{0})$, $f(x, y)$ must approach $f(0, 0)$ no matter how $\mathbf{x}$ approaches $(0, 0)$. Since $f(x, y)$ does *not* approach $f(0, 0)$ as $(x, y) \to (0, 0)$ along the line $y = 0$, we conclude that $f(x, y)$ is *discontinuous* at $(0, 0)$. ■

In Example 2, if we had tried letting $(x, y) \to (0, 0)$ along the line $y = x$, then we would have found that

$$f(x, y) = \frac{x^2 - x^2}{x^2 + 2x^2} = 0 \to 0 = f(0, 0),$$

which would *not* have established our guess. Sometimes it is necessary to try more than once to settle a continuity problem.

Example 3

Decide whether the function f is continuous at $(0, 0)$, if

$$f(x, y) = \begin{cases} \dfrac{x^3y - xy^3}{x^2 + y^2} & \text{if } (x, y) \neq (0, 0), \\ 0 & \text{if } (x, y) = (0, 0). \end{cases}$$

Solution. Each term of the numerator has *higher* degree than the denominator, so we expect that f is continuous at $(0, 0)$. To verify this we must show that $f(x, y) \to 0 = f(0, 0)$ as $(x, y) \to (0, 0)$. The best way to do so is to find an expression whose absolute value is *larger* than $|f(x, y)|$ and which clearly approaches zero as $(x, y) \to (0, 0)$. That is done using the facts that

$$|x| \le \sqrt{x^2 + y^2} = |\mathbf{x}| \qquad \text{and} \qquad |y| \le \sqrt{x^2 + y^2} = |\mathbf{x}|.$$

If $(x, y) \neq (0, 0)$, then

$$\begin{aligned} |f(x, y)| = \frac{|xy(x^2 - y^2)|}{x^2 + y^2} &= \frac{|x|\,|y|\,|x + y|\,|x - y|}{x^2 + y^2} \\ &\le \frac{|x|\,|y|(|x| + |y|)(|x| + |y|)}{x^2 + y^2} \qquad \textit{by the triangle inequality, Theorem 1.2(d) of Chapter 1} \\ &\le \frac{|\mathbf{x}|\,|\mathbf{x}|(|\mathbf{x}| + |\mathbf{x}|)(|\mathbf{x}| + |\mathbf{x}|)}{|\mathbf{x}|^2} \\ &= 4|\mathbf{x}|^2 = 4(x^2 + y^2). \end{aligned}$$

As $(x, y) \to (0, 0)$, we have $4(x^2 + y^2) \to 0$, and so $|f(x, y)| \to 0$ also. We conclude that $\lim_{\mathbf{x} \to \mathbf{0}} f(\mathbf{x}) = 0 = f(\mathbf{0})$, and hence f is continuous at $\mathbf{0} = (0, 0)$. ■

It is usually more challenging to verify a guess that a function is continuous at $\mathbf{x} = \mathbf{a}$ than to show that a function is not continuous there. In the latter case we just need to find a path of approach of $\mathbf{x}$ to $\mathbf{a}$ along which $f(\mathbf{x})$ does not approach $f(\mathbf{a})$. But to prove continuity, we have to come up with some reasoning to show that $f(\mathbf{x}) \to f(\mathbf{a})$ no matter how $\mathbf{x}$ approaches $\mathbf{a}$. As long-time Speaker of the United States House of Representatives Sam Rayburn (1882–1961; Congressman from Texas, 1913–1961) used to say, "Any dadburn fool can tear down a barn, but it takes a lot of know-how to build one!"

The composite of a continuous scalar function and a continuous function of a single real variable is a continuous scalar function, according to the next result. (Compare Theorem 7.5 of Chapter 1.)

4.6 THEOREM Suppose that $f\colon \mathbf{R} \to \mathbf{R}$ is continuous at $t = t_0$ and that $g\colon \mathbf{R}^n \to \mathbf{R}$ is continuous at $\mathbf{x} = \mathbf{x}_0$, where $g(\mathbf{x}_0) = t_0$. Then the composite function $f \circ g\colon \mathbf{R}^n \to \mathbf{R}$ is continuous at $\mathbf{x} = \mathbf{x}_0$. See **Figure 4.6.**

FIGURE 4.6

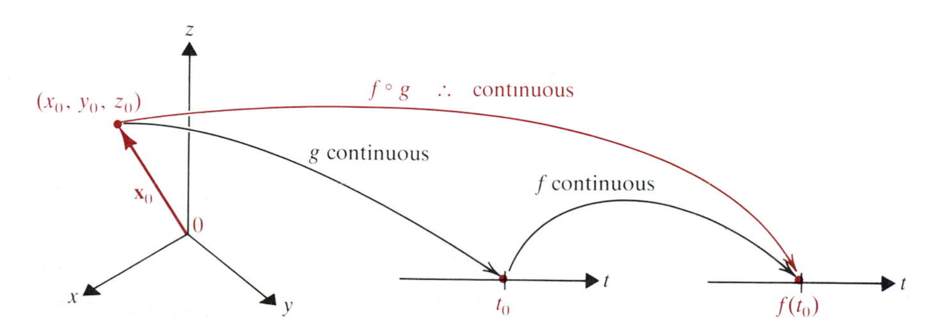

Proof. To show that $\lim_{\mathbf{x} \to \mathbf{x}_0} (f \circ g)(\mathbf{x}) = (f \circ g)(\mathbf{x}_0) = f(g(\mathbf{x}_0)) = f(t_0)$, we first note that

$$g(\mathbf{x}) = t \to g(\mathbf{x}_0) = t_0 \qquad \text{as } \mathbf{x} \to \mathbf{x}_0,$$

because g is continuous at $\mathbf{x} = \mathbf{x}_0$. Also, as $t \to t_0$, we have $f(t) \to f(t_0)$, because f is continuous at $t = t_0$. This means that as $\mathbf{x} \to \mathbf{x}_0$, we have $f(g(\mathbf{x})) \to f(g(\mathbf{x}_0))$. [For a more formal proof based on (1), refer to Exercise 36.] QED

As a consequence of Theorem 4.6, we can be sure that a host of scalar functions are continuous, including, for example, the function h defined by

$$h(x, y) = \sin xy.$$

The polynomial function $g(x, y) = xy$ is continuous on all of $\mathbf{R}^2$, and the sine function is continuous on all of $\mathbf{R}$. Hence the composite function $h = f \circ g$ is continuous on all of $\mathbf{R}^2$, where $f = $ sine.

The final result of this section is the analogue for scalar functions of Theorem 7.7 of Chapter 1 (the extreme value theorem); it guarantees that most scalar functions dealt with in this text have a maximum and a minimum on any closed ball.

4.7 THEOREM If $f: \mathbf{R}^n \to \mathbf{R}$ is continuous on some closed ball B, then f has an absolute maximum and an absolute minimum on B. That is, there are points $\mathbf{x}_0$ and $\mathbf{x}_1$ in B such that for all $\mathbf{x}$ in B, we have

$$f(\mathbf{x}_0) \le f(\mathbf{x}) \le f(\mathbf{x}_1).$$

Like its analogue for functions of one variable, this result is proved in advanced calculus. The theorem assures us that functions such as

$$f(x, y) = x^4 + y^2 + 3$$

have an absolute maximum and an absolute minimum on the closed unit disk

$$B(\mathbf{0}, 1) = \{(x, y) \mid x^2 + y^2 \le 1\}$$

in $\mathbf{R}^2$. However, again like its one-variable counterpart, the theorem provides no method for finding the points $\mathbf{x}_0$ and $\mathbf{x}_1$ where the maximum and the minimum occur. As our experience in the one-variable case suggests, efficient techniques for locating $\mathbf{x}_0$ and $\mathbf{x}_1$ must await a study of differentiation of scalar functions, which is the subject of the next chapter.

Exercises 11.4

In Exercises 1–10, evaluate the limit if it exists. If it does not, then explain why not.

1. $\displaystyle\lim_{(x,y)\to(2,0)} \ln \frac{x^2 - y^2}{x - y}$

2. $\displaystyle\lim_{(x,y)\to(2,1)} \frac{x^3 - y^3}{x^2 - y^2}$

3. $\displaystyle\lim_{(x,y)\to(2,0)} \frac{\ln xy}{x^2 + y^2}$

4. $\displaystyle\lim_{(x,y)\to(2,1)} \frac{x^2 + y^2}{xy^2 - 2}$

5. $\displaystyle\lim_{(x,y)\to(1,1)} \frac{x^2 - y^2}{x - y}$

6. $\displaystyle\lim_{(x,y)\to(2,1)} \frac{x^2 - 4y^2}{x - 2y}$

7. $\displaystyle\lim_{(x,y,z)\to(1,1,3)} \frac{x^2 + y^2 - z + 1}{(x - 1)^2 + (y - 1)^2 + z^2}$

8. $\displaystyle\lim_{(x,y,z)\to(0,0,1)} \frac{x^2y^2 + 3z}{z^2 - 1}$

9. $\displaystyle\lim_{(x,y,z)\to(1,0,3)} \frac{x^2 + 3y^2 + 2z}{xyz}$

10. $\displaystyle\lim_{(x,y,z)\to(1,0,3)} \frac{xyz}{x^2y + z^2}$

In Exercises 11–20, find all points at which the given function is continuous.

11. $f(x, y) = \begin{cases} \dfrac{xy}{x^2 + y^2} & \text{if } (x, y) \ne (0, 0) \\ 0 & \text{if } (x, y) = (0, 0) \end{cases}$

12. $g(x, y) = \begin{cases} \dfrac{x^2y}{x^2 + y^2} & \text{if } (x, y) \ne (0, 0) \\ 0 & \text{if } (x, y) = (0, 0) \end{cases}$

13. $h(x, y) = \begin{cases} \dfrac{x^2y + xy^2}{x^2 + y^2} & \text{if } (x, y) \ne (0, 0) \\ 0 & \text{if } (x, y) = (0, 0) \end{cases}$

14. $f(x, y) = \begin{cases} \dfrac{\sin xy}{xy} & \text{if } xy \neq 0 \\ 1 & \text{if } xy = 0 \end{cases}$

15. $g(x, y) = \begin{cases} \dfrac{x^2y + 3x^2}{x^2 + y^2} & \text{if } (x, y) \neq (0, 0) \\ 0 & \text{if } (x, y) = (0, 0) \end{cases}$

16. $h(x, y) = \begin{cases} \dfrac{xy^2 - y^2}{x^2 + y^2} & \text{if } (x, y) \neq (0, 0) \\ 0 & \text{if } (x, y) = (0, 0) \end{cases}$

17. $f(x, y) = \begin{cases} \dfrac{x^2}{x^2 + y^2} & \text{if } (x, y) \neq (0, 0) \\ \frac{1}{2} & \text{if } (x, y) = (0, 0) \end{cases}$

18. $g(x, y) = \begin{cases} \dfrac{x^2 - y^2}{x^2 + y^2} & \text{if } (x, y) \neq (0, 0) \\ 0 & \text{if } (x, y) = (0, 0) \end{cases}$

19. $f(x, y) = \begin{cases} x^2/y & \text{when } y \neq 0 \\ 1 & \text{when } y = 0 \end{cases}$

20. $f(x, y) = \begin{cases} \dfrac{y^2}{x - 1} & \text{when } x \neq 1 \\ 2 & \text{when } x = 1 \end{cases}$

21. Prove Theorem 4.3(a) for the case of a plus sign.

22. Prove Theorem 4.3(a) for the case of a minus sign.

23. Prove Theorem 4.5(a) and Theorem 4.5(c).

24. Prove Theorem 4.5(b) and Theorem 4.5(d).

25. Prove that the length function on $\mathbf{R}^n$ is continuous everywhere.

26. Discuss the continuity of V in the ideal gas equation $PV = nRT$, taking into account the physical possibility of $P = 0$.

27. Economists regard the prices $p_1, p_2, \ldots, p_n$ of the commodities in the cost-of-living index as nonnegative real variables. Is then the cost-of-living index a continuous function of $p_1, p_2, \ldots, p_n$?

28. Prove that the function f is continuous on $\mathbf{R}^3$ if

$$f(x, y, z) = \frac{x \sin yz - e^x \cos yz}{1 + x^2 + y^2 + z^2}.$$

29. If $f\colon \mathbf{R}^2 \to \mathbf{R}$ is continuous at (x_0, y_0), then show that f_{x_0} defined by $f_{x_0}(y) = f(x_0, y)$ is a continuous function at $y = y_0$. Also show that f_{y_0} defined by $f_{y_0}(x) = f(x, y_0)$ is continuous at $x = x_0$.

30. The continuity of f_{x_0} at $y = y_0$ and f_{y_0} at $x = x_0$ is *not* enough to conclude that f is continuous at (x_0, y_0). Show this by considering the function in Exercise 11 again, with $(x_0, y_0) = (0, 0)$.

In Exercises 31–34, decide whether the given limit exists. If it does, then evaluate.

31. $\displaystyle\lim_{(x,y)\to(0,1)} \frac{x^2 - y^2}{\sqrt{x^2 + y^2}}$

32. $\displaystyle\lim_{(x,y)\to(0,0)} \frac{\sqrt{x^2 + y^2}}{xy}$

33. $\displaystyle\lim_{(x,y,z)\to(0,0,0)} \frac{xy + yz}{x^2 + y^2 + z^2}$

34. $\displaystyle\lim_{(x,y,z)\to(0,0,0)} \frac{x^2 + y^2 + z^2}{xy + xz + yz}$

35. Give an example to show that the "rule of thumb" in Example 2 need not always produce a correct guess. (*Hint:* Consider an earlier exercise.)

36. Give a formal proof of Theorem 4.6 by showing that for any $\varepsilon > 0$, there is some $\delta > 0$ such that

$$|f(g(\mathbf{x})) - f(g(\mathbf{x}_0))| < \varepsilon \qquad \text{whenever } |\mathbf{x} - \mathbf{x}_0| < \delta.$$

(*Hint:* First, show that there is a δ_1 such that $|f(t) - f(t_0)| < \varepsilon$ whenever $|t - t_0| < \delta_1$. Then apply (1) to the continuous function $g\colon \mathbf{R}^n \to \mathbf{R}$.)

11.5 Three-Dimensional Graphing

According to the definition given on p. 23; the graph of a function $g\colon \mathbf{R} \to \mathbf{R}$ consists of all ordered pairs $(x, y) \in \mathbf{R}^2$ for which x is in the domain of g and $y = g(x)$. For a function $f\colon \mathbf{R}^n \to \mathbf{R}$, we analogously define its graph to be the subset of $\mathbf{R}^{n+1}$

$$\{(\mathbf{x}, y) = (x_1, x_2, \ldots, x_n, y) \,|\, \mathbf{x} \in D \text{ and } y = f(\mathbf{x})\},$$

where the subset D of $\mathbf{R}^n$ is the domain of f. This means that the graph of $f\colon \mathbf{R}^n \to \mathbf{R}$ can be drawn *only* when $n = 1$ or 2, because we have pictorial representations only for $\mathbf{R}^2$ and $\mathbf{R}^3$ among the Euclidean spaces $\mathbf{R}^n$.

5.1 DEFINITION The ***graph of a function*** $f: \mathbf{R}^2 \to \mathbf{R}$ is the set of all points (x, y, z) in $\mathbf{R}^3$ such that $z = f(x, y)$. The ***graph of a functional equation*** $g(x, y, z) = 0$ is the set of all points $P(x, y, z)$ in $\mathbf{R}^3$ whose coordinates satisfy the equation.

The graph of a function $f: \mathbf{R} \to \mathbf{R}$ of one variable is a *curve* in $\mathbf{R}^2$, a *one-dimensional* set that pictures how the function transforms real numbers in its one-dimensional domain to give values in its range. The domain of a function $f: \mathbf{R}^2 \to \mathbf{R}$ is generally a *two-dimensional* subset of $\mathbf{R}^2$, and its graph is usually a two-dimensional subset of $\mathbf{R}^3$, called a *surface*. In drawing such a surface, it is often helpful to show the curves of intersection of the surface with planes parallel to the xz- and yz-planes, as in **Figure 5.1.** (Most surface-plotting computer routines do this.)

FIGURE 5.1

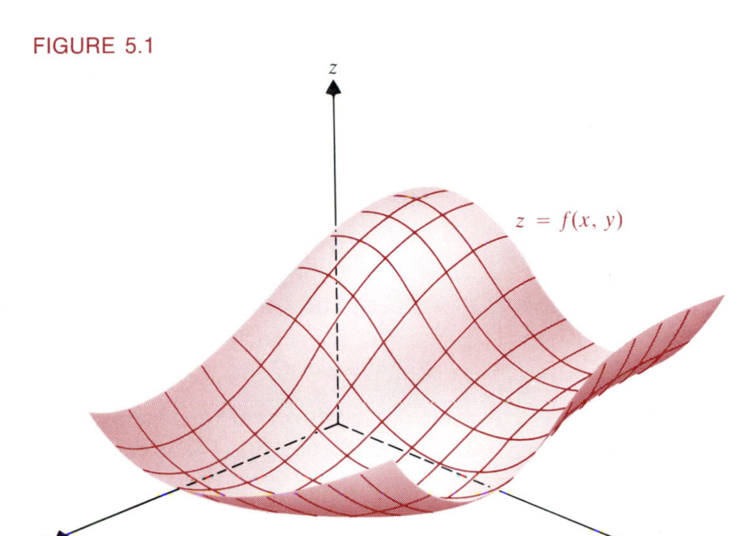

FIGURE 5.2

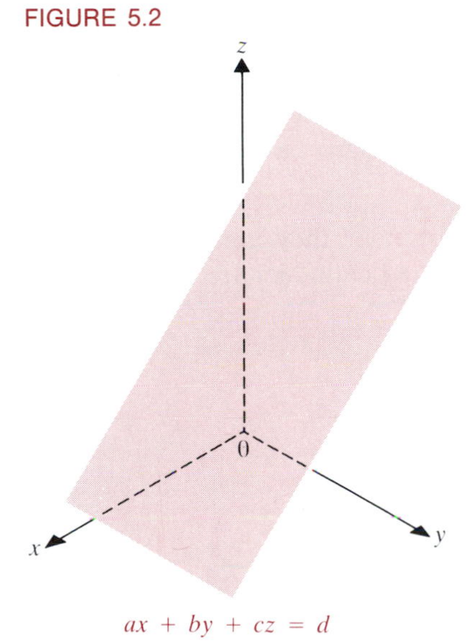

$ax + by + cz = d$

FIGURE 5.3

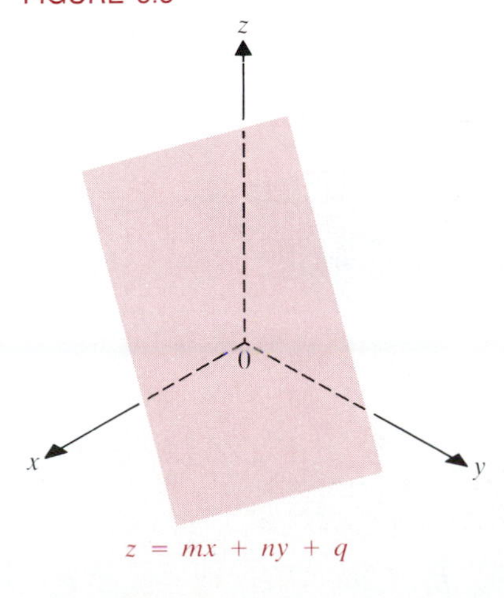

$z = mx + ny + q$

To develop facility in drawing the graphs of functions and equations in $\mathbf{R}^3$ and to become familiar with the relationships between equations and their graphs, there is no substitute for practice. Much of the rest of this chapter is intended to provide such practice. Familiarity with graphs in $\mathbf{R}^3$ will be helpful in the next chapter, where the differential calculus of scalar functions is developed, and it is *essential* in Chapter 13, where the integral calculus of scalar functions is presented.

We are already familiar with some graphs in $\mathbf{R}^3$. For example, in Section 10.5 we saw that a functional equation $ax + by + cz - d = 0$ has a plane as its graph. See **Figure 5.2.** Equivalently, the graph of a linear function $f: \mathbf{R}^2 \to \mathbf{R}$, say $f(x, y) = mx + ny + q$, is a plane. This is so because a point (x, y, z) is on that graph if and only if $z = mx + ny + q$, that is, if and only if

$$mx + ny + (-1)z = -q.$$

That is the equation of a plane with normal vector $m\mathbf{i} + n\mathbf{j} - \mathbf{k}$. See **Figure 5.3.**

We also already know the graph of one type of quadratic polynomial equation: The graph of $x^2 + y^2 + z^2 = a^2$ is a sphere of radius a and center $\mathbf{0}$ (see **Figure 5.4**), and the graph of $(x - h)^2 + (y - k)^2 + (z - l)^2 = a^2$ is a sphere of radius a and center (h, k, l).

Cylinders comprise another important class of graphs of functional equations.

FIGURE 5.4

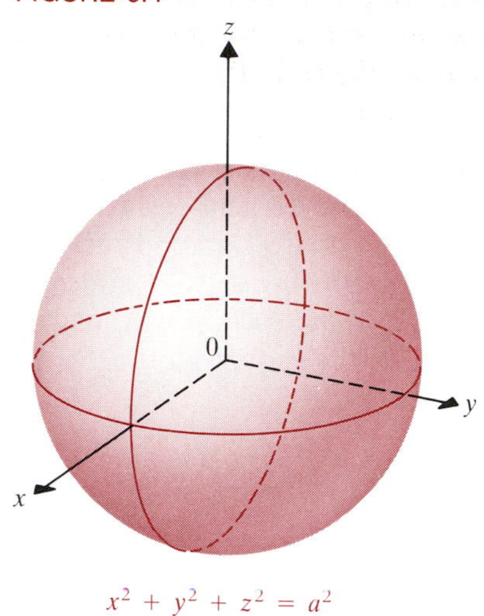

FIGURE 5.5

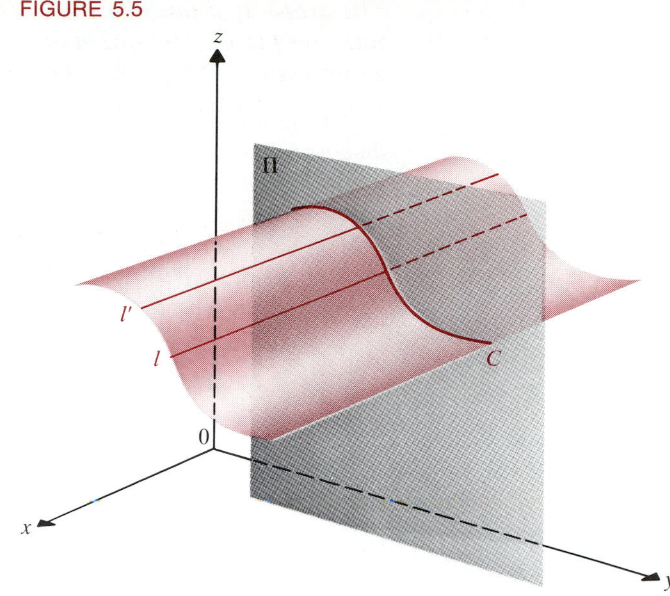

5.2 DEFINITION Let C be a plane curve in $\mathbf{R}^3$, that is, a curve lying entirely in some plane Π. Let l be a line that intersects C and is perpendicular to Π. A ***right cylinder*** is the set of all points $P(x, y, z)$ in $\mathbf{R}^3$ on lines l' that are parallel to l and also intersect C. See **Figure 5.5.**

Cylinders are *unbounded* in $\mathbf{R}^3$, because, in particular, they contain their entire generating line l. The most commonly encountered cylinders are right *circular*

FIGURE 5.6

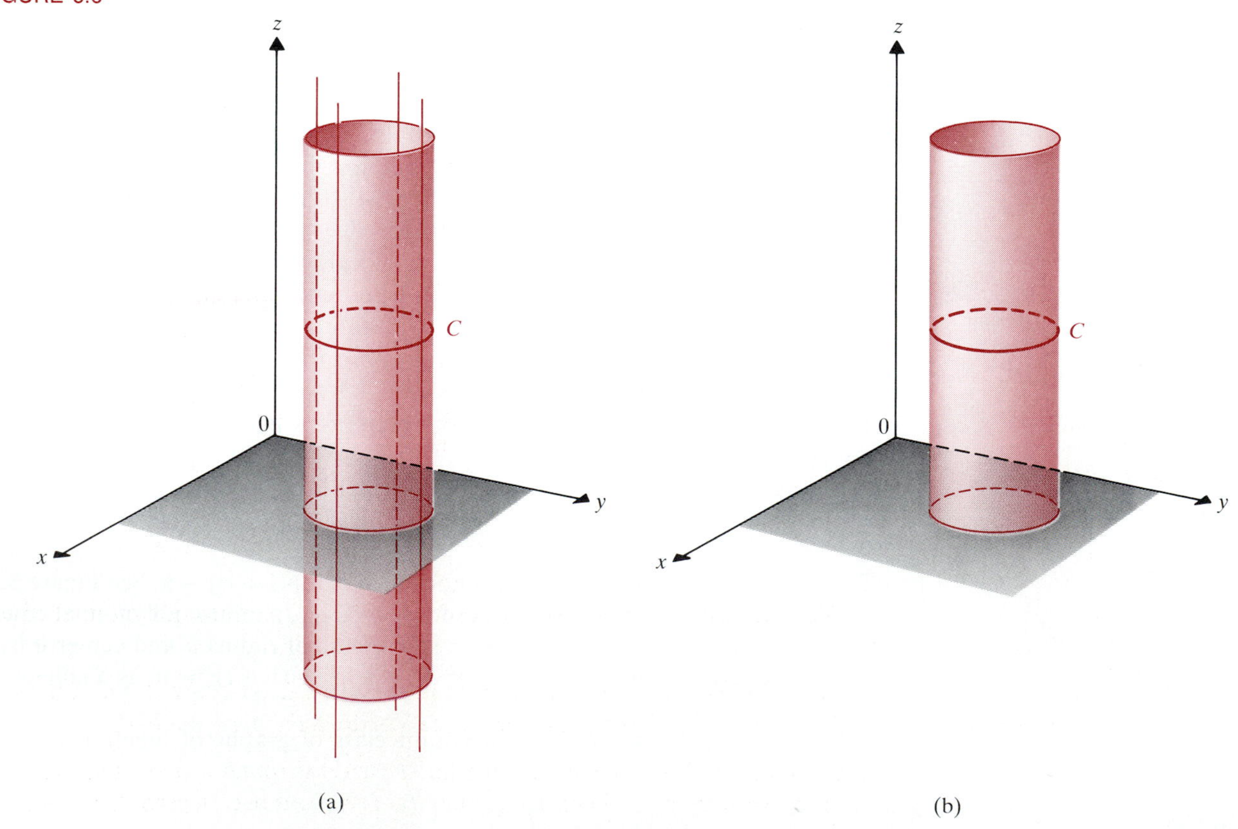

(a) (b)

cylinders. For these, C is a circle, as in **Figure 5.6**(a). Cylindrical cans that contain such products as soup, fruit juices, and oil are *truncated* right circular cylinders, cut off by two planes parallel to C, as in **Figure 5.6**(b).

In general, the graph of a functional equation of the form

(1) $$f(x, y) = 0, \qquad g(y, z) = 0, \qquad \text{or} \qquad h(x, z) = 0,$$

which involves just two of the coordinate variables in $\mathbf{R}^3$, is a right cylinder. (There are some degenerate cases in which the graphs can be empty, a point, a line, or one or more planes.) The key to drawing the graph of equations such as (1) is to realize that the variable not present in the equation can take on *all* values in $(-\infty, +\infty)$. A good picture of the graph can be obtained by drawing the curve(s) of intersection of the cylinder with the coordinate plane Π of the two variables that do occur in the equation, and with one or more planes parallel to Π.

FIGURE 5.7

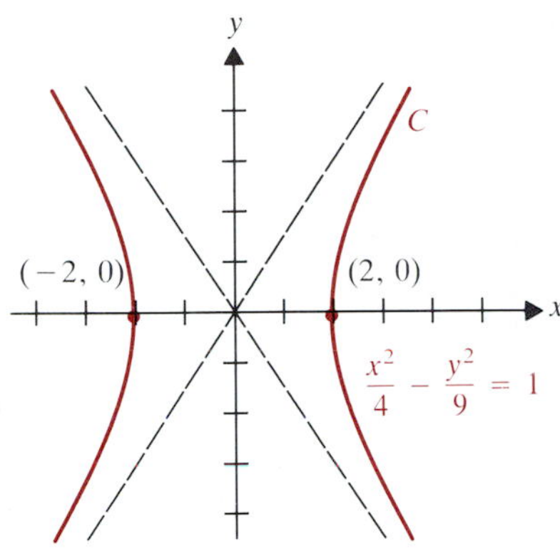

Example 1

Draw the graph of $x^2/4 - y^2/9 = 1$ in $\mathbf{R}^3$.

Solution. Here the variable z is missing. The intersection of the surface with the plane $z = 0$ (the xy-plane) is the hyperbola $x^2/4 - y^2/9 = 1$ with vertices $(\pm 2, 0)$ and asymptotes $y = \pm\frac{3}{2}x$ (**Figure 5.7**). The intersection with any plane $z = k$, where $k \neq 0$, is another hyperbola lying directly above (if $k > 0$) or below (if $k < 0$) the hyperbola in Figure 5.7. The cylinder, pictured in **Figure 5.8,** is called a *hyperbolic* cylinder. It looks like two curved sheets of paper opening away from the yz-plane, one in front and the other behind that plane. ■

FIGURE 5.8

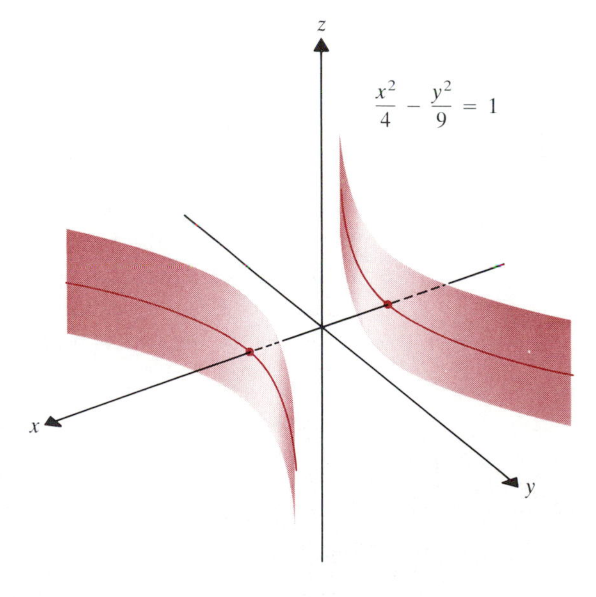

We next discuss the type of surface whose volume was considered in Sections 5.2 and 5.3.

5.3 DEFINITION A ***surface of revolution*** results when a plane curve C is revolved about a line l in the plane of the curve. (The line is called the *axis of revolution.*)

If the curve C is a circle, and the line l passes through a diameter, then the resulting surface is a sphere. If the curve C is a parabola, hyperbola, or ellipse,

FIGURE 5.9
Some Solids of Revolution

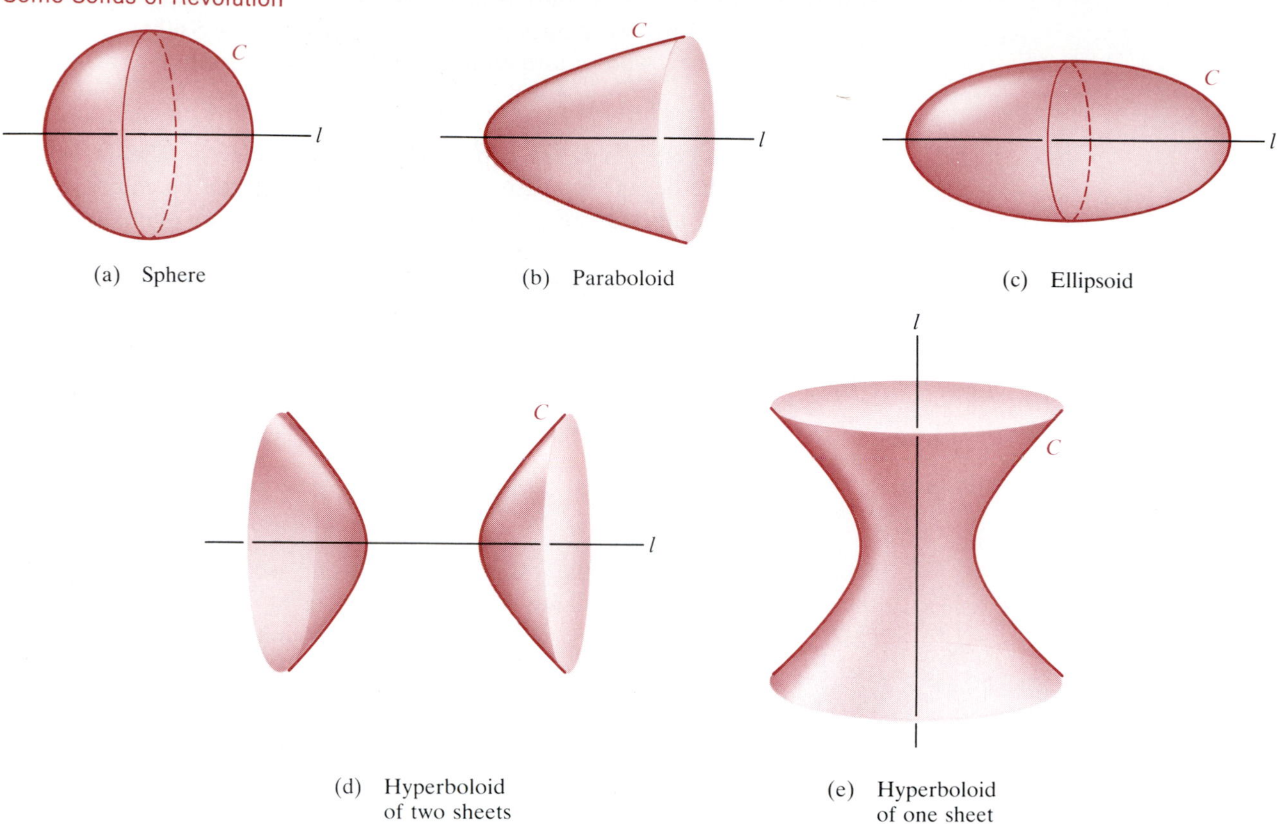

and the line l is the principal axis or a minor axis, then the resulting surface is called a *paraboloid, hyperboloid*, or *ellipsoid of revolution*, respectively. See **Figure 5.9.**

Suppose that we revolve the graph of $y = f(x)$ in the xy-plane about the x-axis. See **Figures 5.10** and **5.11,** where we picture the rotation as taking place in $\mathbf{R}^3$. To find the equation of the resulting surface, let $P(x, y, z)$ be a point on the surface. Then $Q(x, y_1, 0)$ is on the curve in the xy-plane, as shown in Figure 5.11. Let P_0 be the point $(x, 0, 0)$ on the x-axis. Since we revolve the curve about the x-axis, the radius $|\mathbf{P}_0\mathbf{P}|$ is the same as $|\mathbf{P}_0\mathbf{Q}|$. Hence

$$\sqrt{y^2 + z^2} = |y_1|.$$

FIGURE 5.10

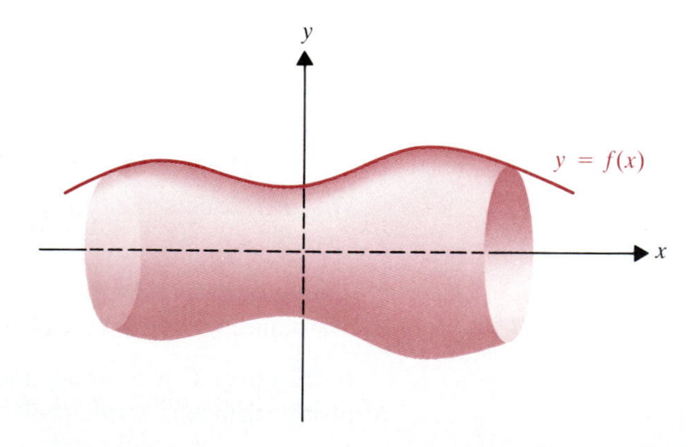

FIGURE 5.11

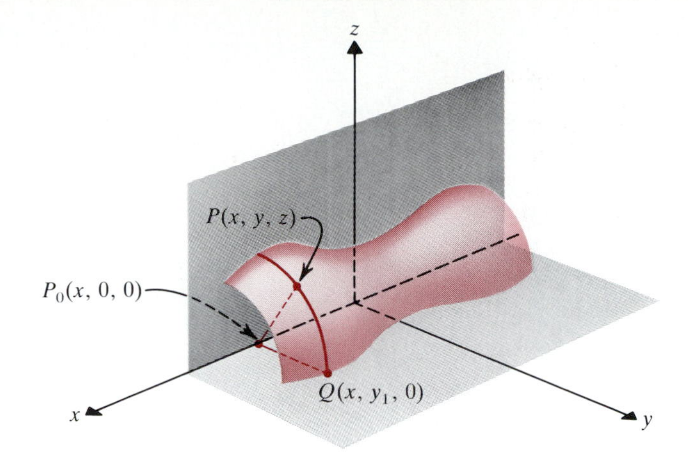

Squaring this, we get

$$y^2 + z^2 = y_1{}^2.$$

But $y_1 = f(x)$ because (x, y_1) is on the curve that we revolved. So we have

$$y^2 + z^2 = f(x)^2.$$

A similar derivation could be given if we revolve any curve in any coordinate plane about one of the plane's coordinate axes. The following theorem applies to any situation in which a curve in any coordinate plane is revolved about a coordinate axis in that plane. Suppose that the curve is the graph of $s = f(r)$ in the rs coordinate plane, where r and s are x, y, or z. Let t denote the third coordinate variable.

5.4 THEOREM

If the graph of $s = f(r)$ in the rs-plane is revolved about the r-axis, then the surface of revolution generated has the equation

$$s^2 + t^2 = f(r)^2. \qquad (2)$$

Example 2

Find the equation of the surface of revolution formed by revolving the given curve about the given coordinate axis.

(a) $\frac{1}{4}x^2 + y^2 = 1$ about the x-axis in the xy-plane;

(b) $\frac{1}{4}x^2 + y^2 = 1$ about the y-axis in the xy-plane;

(c) $x^2 = 8z$ about the z-axis in the xz-plane.

Solution. (a) The axis of revolution is the x-axis, so here $r = x$. Also $s = y$, since y is the other variable in the equation of the curve to be revolved. Hence $t = z$. We don't need to solve $y^2 = 1 - \frac{1}{4}x^2$ for $y = f(x)$, because $s^2[= y^2]$ and $f(x)^2[= f(r)^2]$ occur in (2). We can simply write down the equation of the surface from (2):

$$s^2 + t^2 = f(r)^2 \rightarrow y^2 + z^2 = 1 - \frac{1}{4}x^2 \rightarrow x^2 + 4y^2 + 4z^2 = 4.$$

This surface is called an *ellipsoid of revolution* and is shown in **Figure 5.12.**

FIGURE 5.12

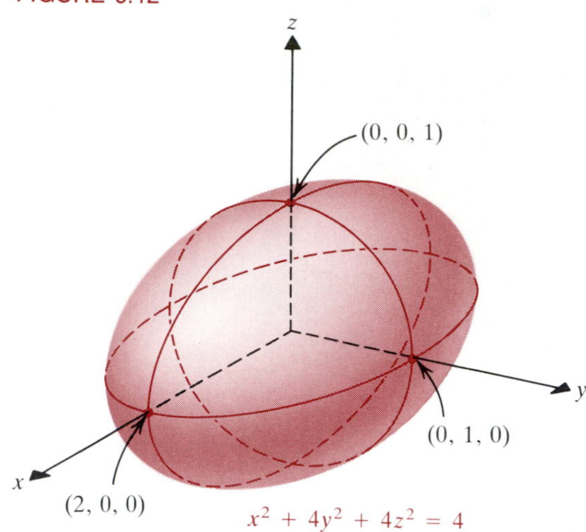

FIGURE 5.13

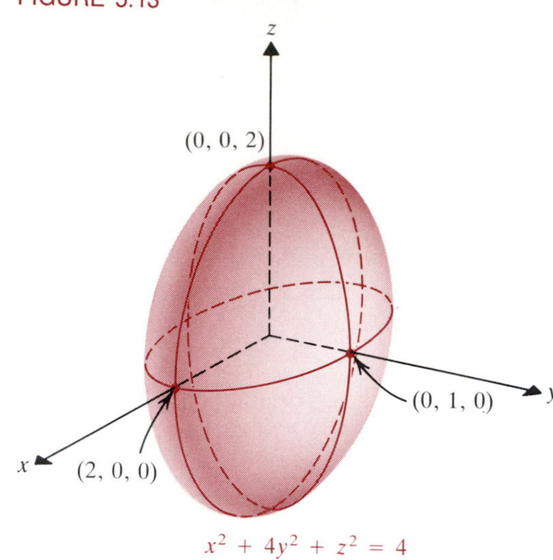

(b) Here $r = y$, so $s = x$. Then $t = z$ again. Isolating $x\,(= s)$, we get $x^2 = 4 - 4y^2$. Thus, from (2), the surface of revolution has equation

$$x^2 + z^2 = 4 - 4y^2 \rightarrow x^2 + 4y^2 + z^2 = 4,$$

another type of ellipsoid of revolution shown in **Figure 5.13.**

(c) This time the axis of revolution is the z-axis, so $r = z$, and thus $s = x$. Hence $t = y$. Therefore (2) says that the equation of the surface of revolution is

$$x^2 + y^2 = 8z.$$

The surface, called a *paraboloid of revolution*, is shown in **Figure 5.14.** ■

FIGURE 5.14

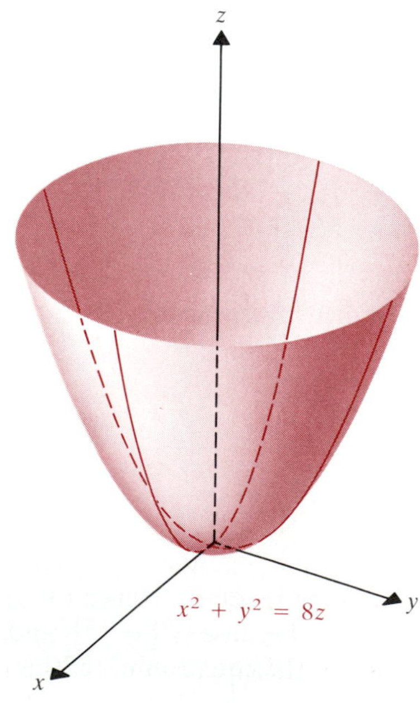

Example 3

Identify the surface of revolution whose equation is $x^2 - 9y^2 + z^2 = 36$.

Solution. We have $x^2 + z^2 = 36 + 9y^2$. So in the theorem, $x^2 + z^2$ corresponds to $s^2 + t^2$, and $36 + 9y^2$ corresponds to $f(r)^2$. Thus $f(r) = \sqrt{36 + 9y^2}$. We have then the curve $x = \pm\sqrt{36 + 9y^2}$ revolved about the y-axis. The generating curve C is thus $x^2 - 9y^2 = 36$, a hyperbola. The given surface is a *hyperboloid of revolution*, shown in **Figure 5.15.** ■

Level Curves and Surfaces

There is another important graphical representation of functions $f\colon \mathbf{R}^2 \rightarrow \mathbf{R}$ that is used in making weather maps and contour maps of a geographical region. In the *two*-dimensional real plane, the curves $f(x, y) = C$ are drawn for various values of the constant C. By looking at these ***level curves*** (called *contour lines* in geography and *isothermal* or *isobaric* lines in meteorology), you can visualize the behavior of the function being mapped. Refer to **Figure 5.16,** which is a contour map of a region with two mountain peaks at A and B and a pass between them at C. The elevations shown are in feet.

On a weather map, isobaric lines are used to locate centers of high pressure or low pressure. In **Figure 5.17** we show a weather map with low-pressure centers off Virginia and near Seattle and a high-pressure center over the Rocky

FIGURE 5.15

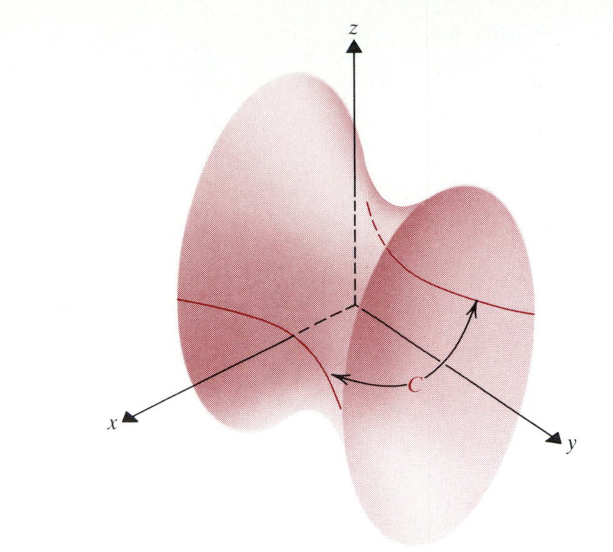

FIGURE 5.16

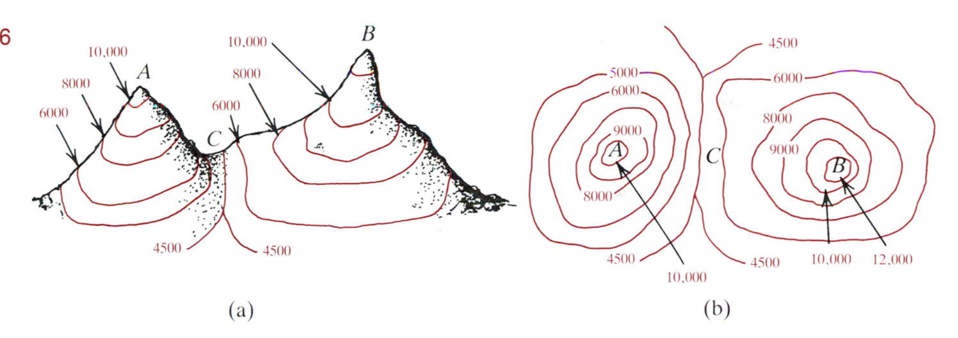

FIGURE 5.17

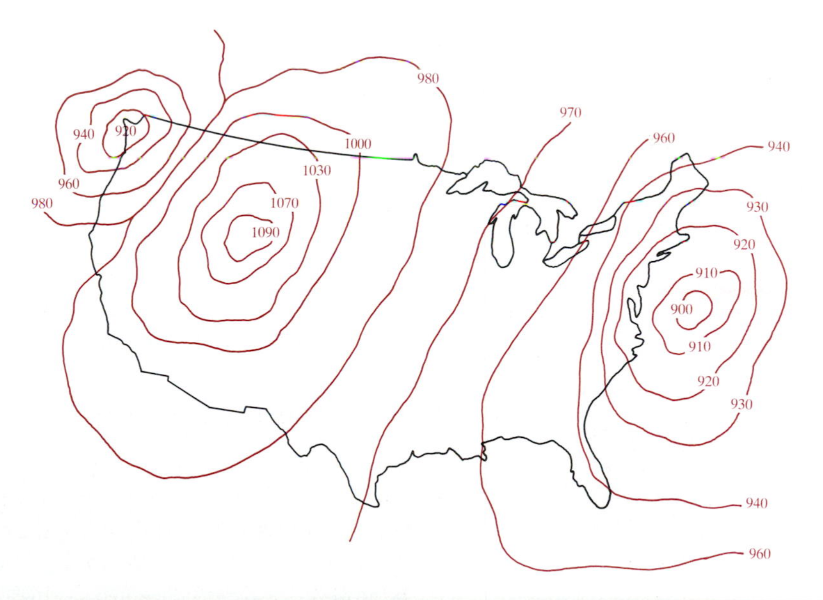

Mountains. (The common meteorological unit is the *millibar:* 1 mb = 100 Pa [Pascals] = 1000 dynes/cm^2.)

Level curves are also associated with functional equations $g(x, y, z) = 0$. Consider the curves $g(x, y, c) = 0$ obtained by setting the variable z equal to some constant c. These are called the *level curves of the surface* that is the graph of $g(x, y, z) = 0$.

FIGURE 5.18

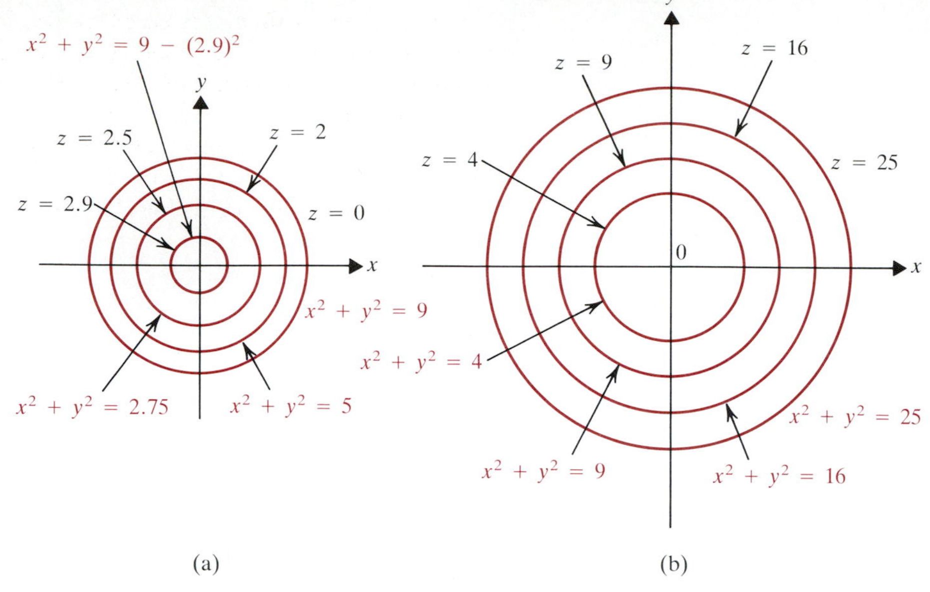

Example 4

Plot level curves of the sphere $x^2 + y^2 + z^2 = 9$ and of the paraboloid $z = x^2 + y^2$. What conclusion can you draw?

Solution. In **Figure 5.18**(a) we show the level curves $z = C$ for the sphere. In Figure 5.18(b) we show level curves $z = C$ for the paraboloid.

In the first case, the level curves are of the form $x^2 + y^2 = 9 - C^2$. Here C can vary between -3 and $+3$. In the case of the paraboloid, the level curves are of the form $x^2 + y^2 = C$, and so C can be any nonnegative number.

FIGURE 5.19

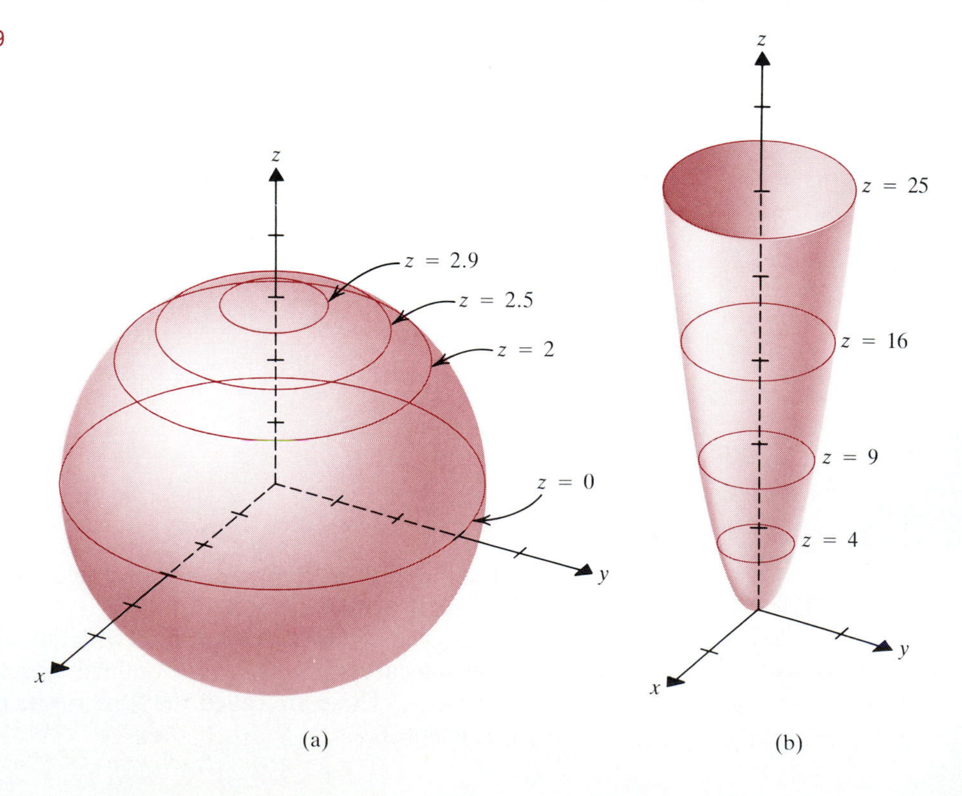

The level curves for the sphere *contract* from the circle $x^2 + y^2 = 9$ $(C = 0)$ to the point $(0, 0)(C = 3)$, whereas those of the paraboloid *expand*. So even though the level curves are circles in both cases, we can see that the surfaces they come from are quite different. The sphere has no points outside the circle of radius 3. As the elevation increases toward 3, there are smaller and smaller circles of intersection. See **Figure 5.19**(a). The paraboloid, however, has points for *all* values of $z > 0$, and as the elevation increases, there are expanding circles of intersection. Thus the level curves give a picture of a surface with a circular cross section parallel to the xy-plane that becomes larger as z grows. See Figure 5.19(b). ■

Although level curves afford an alternative graphical representation of functions $f\colon \mathbf{R}^2 \to \mathbf{R}$, the idea behind them is used to produce the *only* kind of graphical representation for functions $f\colon \mathbf{R}^3 \to \mathbf{R}$. The graph of such a function is the set of points $\{(x, y, z, w) \mid w = f(x, y, z)\}$ in $\mathbf{R}^4$, for which no pictorial representation on a two-dimensional surface such as a page or a blackboard is available. But we can draw the ***level surfaces*** $f(x, y, z) = C$ for such a function.

FIGURE 5.20

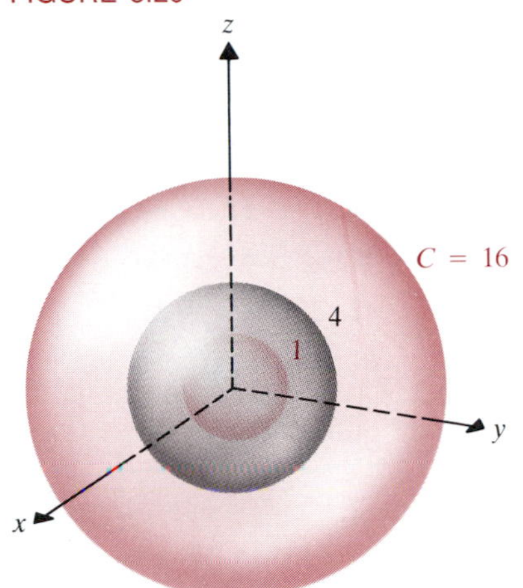

Example 5

Describe the level surfaces of the function $f(x, y, z) = x^2 + y^2 + z^2$.

Solution. If C is a positive constant, then the level surface $f(x, y, z) = C$ is the sphere $x^2 + y^2 + z^2 = C$ of radius $\sqrt{C}$ centered at the origin. We get a sequence of expanding spheres (shown in **Figure 5.20** for $C = 1$, 4, and 16) as C increases. So we can think of the graph of $w = x^2 + y^2 + z^2$ as a hypersurface in $\mathbf{R}^4$ analogous to the paraboloid in $\mathbf{R}^3$ that is the graph of $z = x^2 + y^2$. (Take another look at Example 4 to bring the analogy into sharper focus.) ■

Exercises 11.5

In Exercises 1–8, draw the graphs of the given equations in $\mathbf{R}^3$. Name the surface.

1. $z = 1 - x^2$
2. $z = y^2 + 4$
3. $x^2 + y^2 = 4$
4. $3x^2 + 4y^2 = 12$
5. $4x^2 - 9y^2 = 36$
6. $x^2 + y^2 = 4z$
7. $4x^2 + 9y^2 + 4z^2 = 36$
8. $x^2 + y^2 - 4z^2 = 4$

In Exercises 9–20, find the equation of the surface of revolution.

9. The graph of $y = x^2$ in the xy-plane revolved about the x-axis
10. The graph of $y = x^2$ in the xy-plane revolved about the y-axis
11. The graph of $y^2 = 4z$ in the yz-plane revolved about the y-axis
12. The graph of $y^2 = 4z$ in the yz-plane revolved about the z-axis
13. The graph of $x^2/4 - y^2/9 = 1$ in the xy-plane revolved about the x-axis
14. The graph of $x^2/4 - y^2/9 = 1$ in the xy-plane revolved about the y-axis
15. The graph of $y^2/16 - z^2/9 = 1$ in the yz-plane revolved about the y-axis
16. The graph of $x^2/4 - z^2/16 = 1$ in the xz-plane revolved about the z-axis
17. The graph of $x^2 + y^2 = 4$ in the xy-plane revolved about the y-axis
18. The graph of $y = 3x$ in the xy-plane revolved about the y-axis
19. The graph of $y = 3x$ in the xy-plane revolved about the x-axis
20. The graph of $y = \sin x$ in the xy-plane revolved about the x-axis
21. Is the graph of $9x^2 - 4y^2 + z^2 = 36$ a surface of revolution? Of what curve revolved about which axis?

22. Is the graph of $x^2 + 4y^2 + 16z^2 = 64$ a surface of revolution? Of what curve revolved about which axis?

23. Is the graph of $4x^2 - 3y^2 + 4z^2 = 9$ a surface of revolution? Of what curve revolved about which axis?

24. Is the graph of $9x^2 + 4y^2 + 9z^2 = 36$ a surface of revolution? Of what curve revolved about which axis?

In Exercises 25–30, plot the level curves of the given surface for the values suggested.

25. $z = y^2 - x^2$ for $z = 1, 4, 9, 16$

26. $z = y^2 + 4x$ for $z = 1, 4, 9, 16$

27. $z = x^2 + 9y^2$ for $z = 1, 9, 81$

28. $z = 4x^2 + y^2$ for $z = 1, 4, 16$

29. $z = ye^x$ for $z = 1, 5, 10$

30. $z = xe^{-y}$ for $z = 1, 5, 10$

31. Suppose that the electrical potential at a point in the plane (other than the origin) is given by

$$P(x, y) = \frac{1}{\sqrt{x^2 + y^2}}.$$

Draw the equipotential curves (that is, the level curves) for $P = 1/2, 1, 2,$ and 4.

32. Suppose that the potential at a point in the plane (other than $(1, 2)$) is given by

$$P(x, y) = \frac{1}{\sqrt{(x - 1)^2 + (y - 2)^2}}.$$

Draw the equipotential curves for $P = 1/2, 1, 2,$ and 4.

In Exercises 33–36 describe the level surfaces of the function and draw those for the indicated values of w.

33. $x^2 + y^2 + z^2 + w^2 = 9 \quad w = 0, 2, 3$

34. $x^2 + y^2 + z^2 - w^2 = 9 \quad w = 0, 1, 3$

35. $x^2 + y^2/9 + z^2/9 + w^2 = 4 \quad w = 0, 1, 2$

36. $x^2 - y + z^2 - w = 9 \quad w = 0, 1, 3$

In Exercises 37–44, degenerate right cylinders are given. In each case, identify the graph.

37. $x^2 - y^2 = 0$

38. $x^2 - 5xz + 6z^2 = 0$

39. $y^2 + 2yz + z^2 = 0$

40. $x^2 - 6xy + 9y^2 = 0$

41. $x^2 + y^2 + 2y + 4 = 0$

42. $2y^2 + 6y + 3z^2 + 25 = 0$

43. $y^2 + z^2 - 4y + 4 = 0$

44. $x^2 + 2x + y^2 = 0$

11.6 Quadric Surfaces

Since the graph in $\mathbf{R}^2$ of a second-degree polynomial in two variables is a (possibly degenerate) conic section, it is natural to inquire about the graphs in $\mathbf{R}^3$ of second-degree polynomial equations in x, y, and z. In this section we describe a family of surfaces (and degenerate special cases) that arise as such graphs. They are called ***quadric surfaces***, and their high degree of symmetry gives the graphs a pleasing appearance. Their equations are built up by adding or subtracting terms of the form

$$\frac{x^2}{a^2}, \frac{y^2}{b^2}, \frac{z^2}{c^2}, \frac{x}{d}, \frac{y}{e}, \text{ and } \frac{z}{f},$$

and equating to a constant.

Before discussing these surfaces, we give criteria for the graph of an equation $f(x, y, z) = 0$ to be symmetric with respect to a coordinate plane. As in $\mathbf{R}^2$, taking advantage of such symmetry helps reduce plotting. First, the graph is symmetric in the xy-plane if $(x, y, -z)$ is on the graph whenever (x, y, z) is. See **Figure 6.1.**

(i) If the equation is unchanged when z is replaced by $-z$, then the graph is symmetric relative to the xy-plane.

The criteria for symmetry in the other two coordinate planes are similar.

(ii) If the equation is unchanged when x is replaced by $-x$, then the graph is symmetric relative to the yz-plane. See **Figure 6.2**(a).

FIGURE 6.1

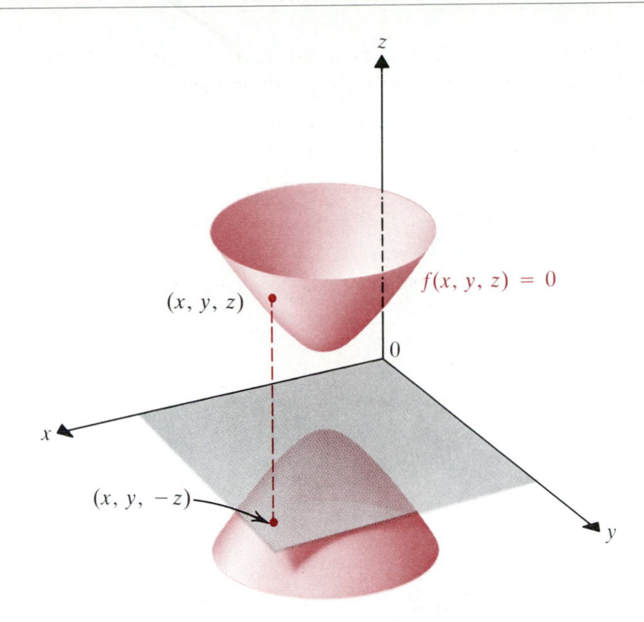

(iii) If the equation is unchanged when y is replaced by $-y$, then the graph is symmetric relative to the xz-plane. See Figure 6.2(b).

We now list the common types of quadric surfaces. Reference to this catalog should help you name and draw a rough sketch of most quadric surfaces.

Right Cylinders

We have already seen (following Definition 5.2) that the graph of a quadratic polynomial function involving only two of the variables x, y, z is generally a ***right cylinder*** perpendicular to the coordinate plane of the two variables which occur. Figure 6.2(a) shows the graph of a cylinder $x^2/a^2 - y^2/b^2 = 1$.

FIGURE 6.2

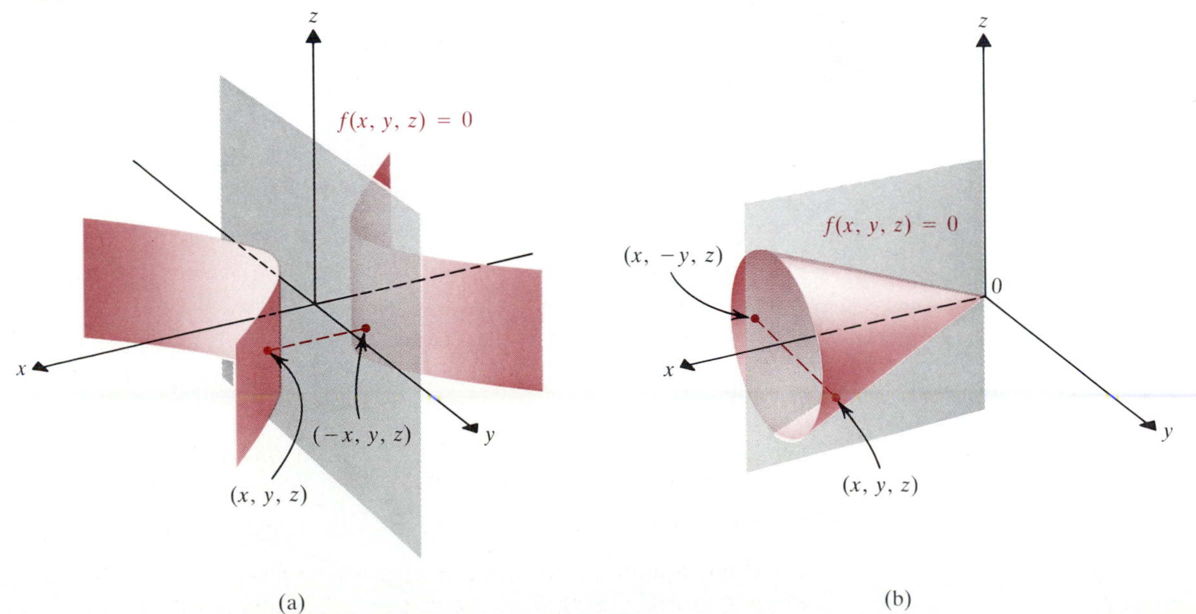

FIGURE 6.3

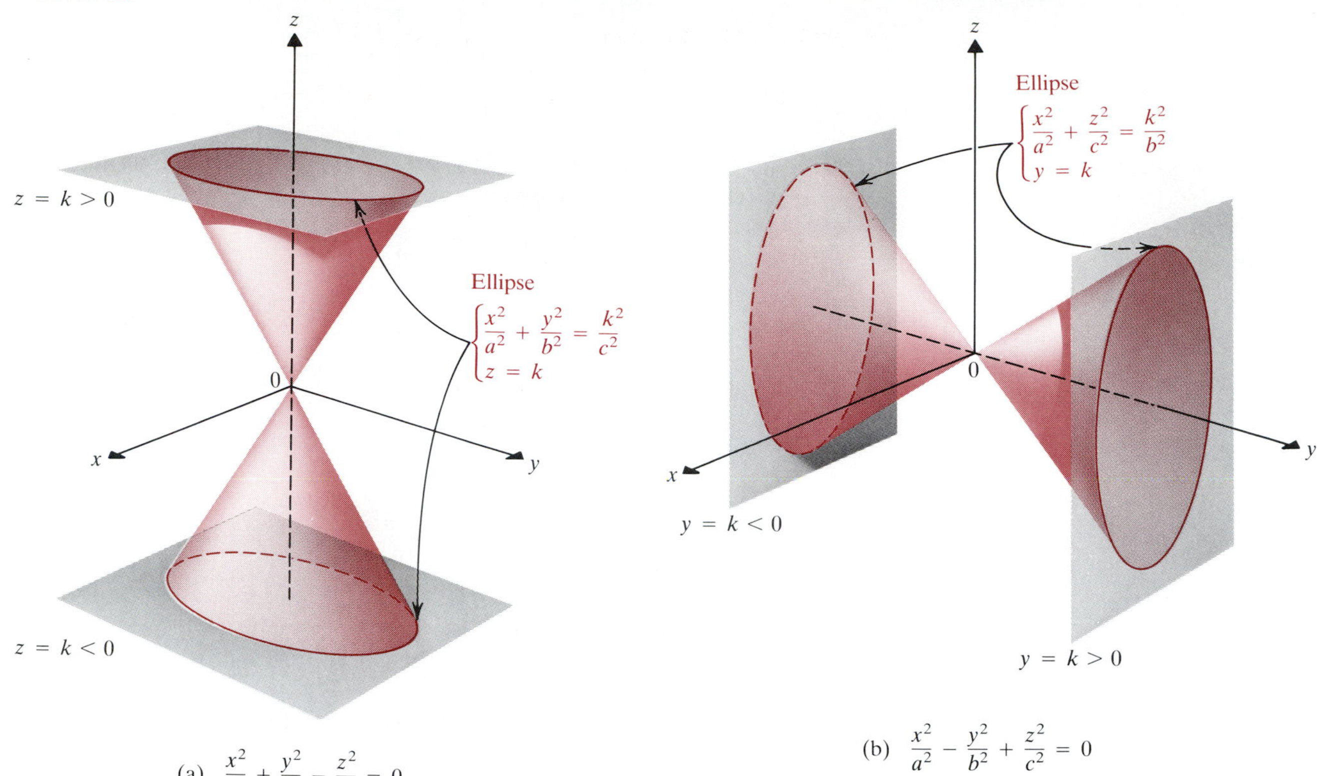

(a) $\frac{x^2}{a^2} + \frac{y^2}{b^2} - \frac{z^2}{c^2} = 0$

(b) $\frac{x^2}{a^2} - \frac{y^2}{b^2} + \frac{z^2}{c^2} = 0$

z

$x = k < 0$

0

y

Ellipse

$\begin{cases} \frac{y^2}{b^2} + \frac{z^2}{c^2} = \frac{k^2}{a^2} \\ x = k \end{cases}$

x

$x = k > 0$

(c) $-\frac{x^2}{a^2} + \frac{y^2}{b^2} + \frac{z^2}{c^2} = 0$

Right Cones

The graph of the equation

$$\frac{x^2}{a^2} + \frac{y^2}{b^2} - \frac{z^2}{c^2} = 0$$

is a ***right (elliptical) cone.*** Its axis is the z-axis. The cross sections made by planes $z = k$ parallel to the xy-plane (that is, the level curves) are ellipses

$$\frac{x^2}{a^2} + \frac{y^2}{b^2} = \frac{k^2}{c^2}, \qquad z = k,$$

centered on the z-axis. See **Figure 6.3**(a). The graph is symmetric in all three coordinate planes. If $a = b$, then the cross sections parallel to the xy-plane are circular, so we have a ***right circular cone.***

Variations. The graph of

$$\frac{x^2}{a^2} - \frac{y^2}{b^2} + \frac{z^2}{c^2} = 0$$

is a right elliptical cone whose axis is the y-axis. See Figure 6.3(b). The graph of

$$-\frac{x^2}{a^2} + \frac{y^2}{b^2} + \frac{z^2}{c^2} = 0$$

is a right elliptical cone whose axis is the x-axis. See Figure 6.3(c).

Elliptical Paraboloids

The graph of

$$\frac{x^2}{a^2} + \frac{y^2}{b^2} = \frac{\pm z}{c^2} \tag{1}$$

is an ***elliptical paraboloid*** with the z-axis as the axis. The cross sections made by planes $z = \pm k$ are again ellipses, which expand as k increases. When $x = 0$, we have

$$z = \pm\frac{c^2}{b^2}y^2,$$

and when $y = 0$ we have

$$z = \pm\frac{c^2}{a^2}x^2.$$

More generally, the cross sections made by planes $x = k$ are parabolas

$$z = \pm c^2\left(\frac{k^2}{a^2} + \frac{y^2}{b^2}\right), \qquad x = k,$$

and those made by planes $y = k$ are parabolas

$$z = \pm c^2\left(\frac{x^2}{a^2} + \frac{k^2}{b^2}\right), \qquad y = k.$$

Those parabolas all open upward if the plus sign is present, downward if the minus sign is present, so the same is true of the surface. See **Figure 6.4,** where we have taken the plus sign for z/c^2. There is symmetry relative to the yz-

FIGURE 6.4
Paraboloid $\frac{x^2}{a^2} + \frac{y^2}{b^2} = \frac{z}{c^2}$

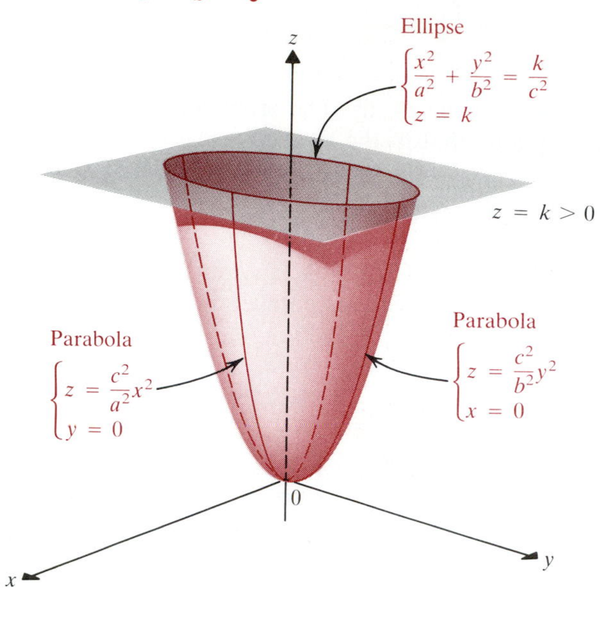

FIGURE 6.5
Computer Plot of $\frac{z}{c^2} = \frac{x^2}{a^2} + \frac{y^2}{b^2}$
(where $a = 2$, $b = 3$, and $c = 1$)

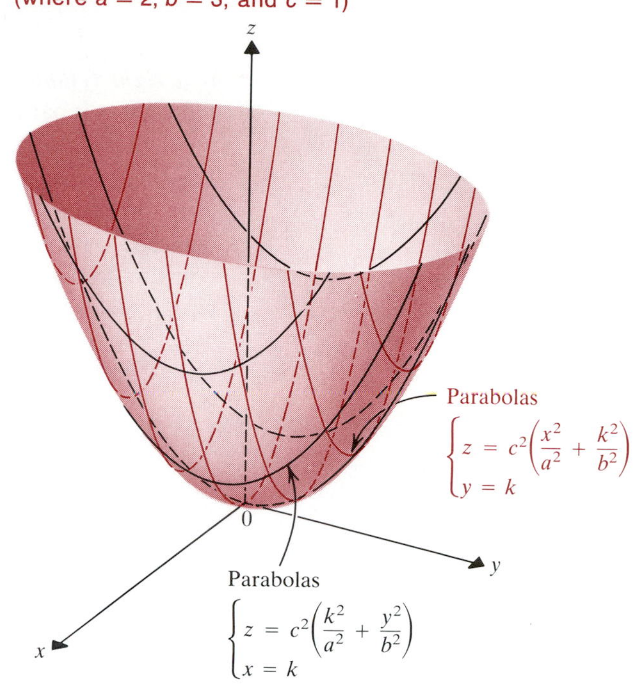

and xz-planes. If $a = b$, then we have a paraboloid of revolution. **Figure 6.5** is a computer-generated sketch of a paraboloid showing the parabolic cross-sectional "wires" made by planes $x = k$ and $y = k$.

Example 1

Sketch the graph of $x^2 + 4y^2 + 4z = 0$.

Solution. Division by 4 and transposition of the z-term give

$$\frac{x^2}{4} + y^2 = -z,$$

which has the standard form (1). Since the coefficient of z is negative, the paraboloid opens downward. The cross section made by the plane $z = -4$ is

$$\frac{x^2}{16} + \frac{y^2}{4} = 1, \qquad z = -4,$$

an ellipse with major axis below the x-axis and minor axis below the y-axis. Drawing in this ellipse, we obtain the sketch shown in **Figure 6.6.** ■

Variations. The graph of

$$\frac{x^2}{a^2} + \frac{z^2}{c^2} = \pm\frac{y}{b^2}$$

is an elliptical paraboloid with the y-axis as its axis. It opens to the right if the positive sign is associated with y/b^2, to the left if the negative sign is associated

FIGURE 6.6

Paraboloid $z = -\frac{x^2}{4} - y^2$

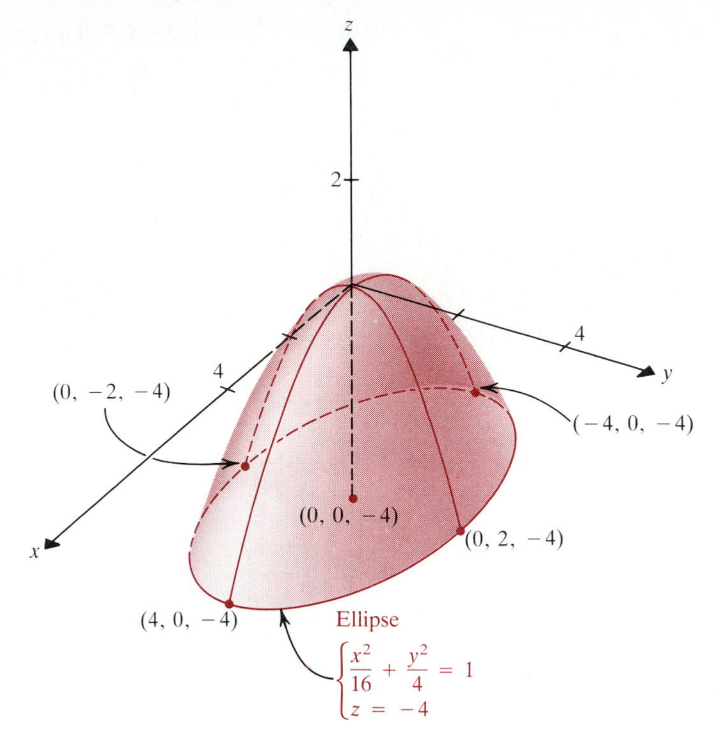

with y/b^2. See **Figure 6.7.** The graph of

$$\frac{y^2}{b^2} + \frac{z^2}{c^2} = \pm\frac{x}{a^2}$$

is an elliptical paraboloid with the x-axis as its axis. It opens to the front if the positive sign is associated with x/a^2, to the rear if the negative sign is associated

FIGURE 6.7

Paraboloid

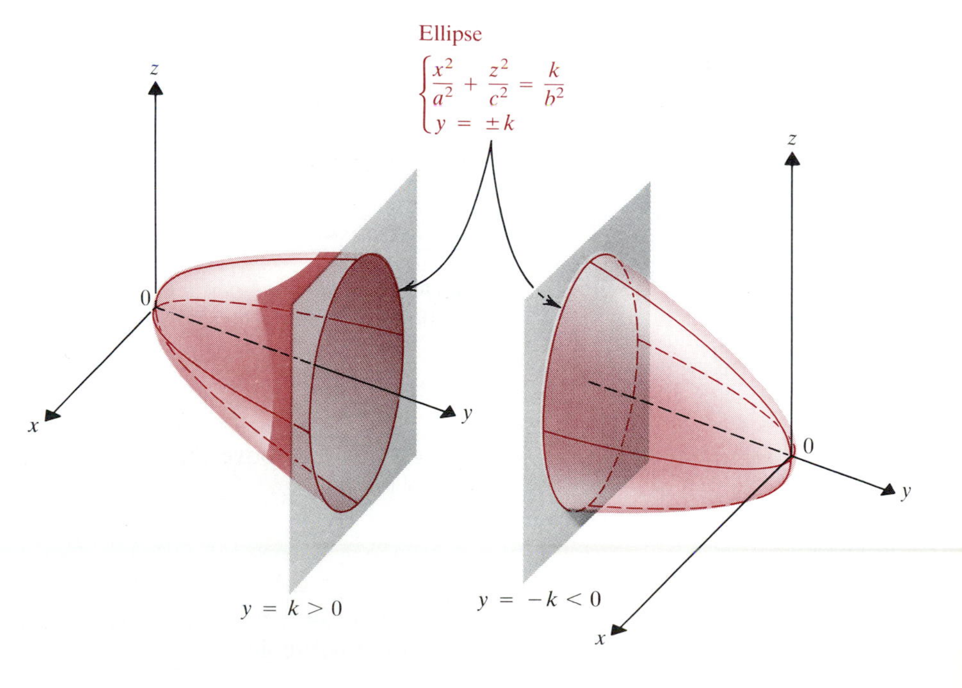

(a) $\frac{y^2}{b^2} = \frac{x^2}{a^2} + \frac{z^2}{c^2}$ (b) $\frac{y^2}{b^2} = -\frac{x^2}{a^2} - \frac{z^2}{c^2}$

with x/a^2. See **Figure 6.8.** The graph of

$$\frac{x^2}{a^2} + \frac{y^2}{b^2} + d = \frac{\pm z}{c^2}$$

has the same shape as the surface (1), but its vertex is at $(0, 0, \pm c^2 d)$ instead of (0, 0, 0).

FIGURE 6.8
Paraboloid

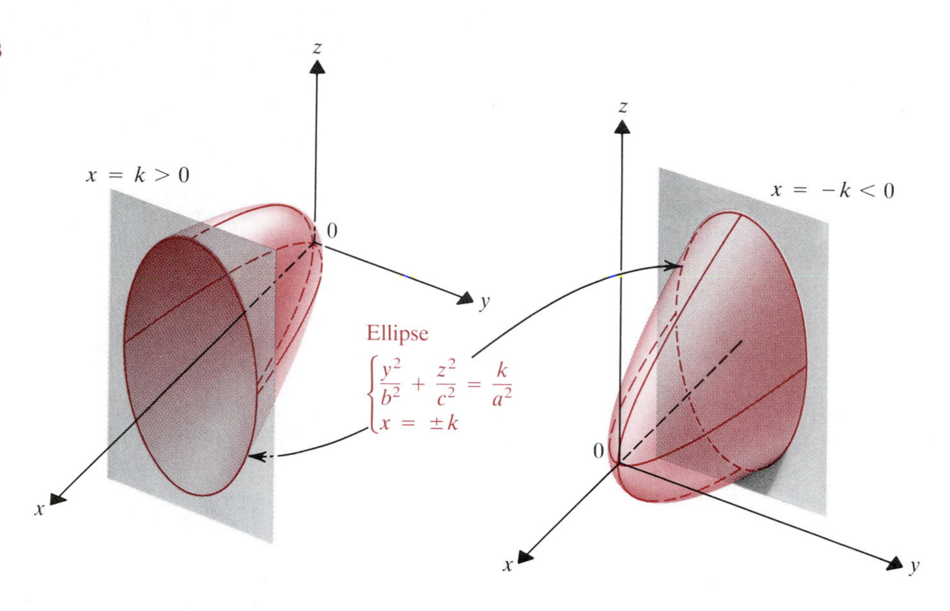

Hyperbolic Paraboloids

The graph of

(2)
$$-\frac{x^2}{a^2} + \frac{y^2}{b^2} = \frac{z}{c^2}$$

is a ***hyperbolic paraboloid*** (or ***saddle***). The horizontal cross sections $z = k > 0$ are hyperbolas

$$\frac{y^2}{b^2} - \frac{x^2}{a^2} = \frac{k}{c^2}$$

whose vertices lie above the y-axis. Planes $z = k < 0$ intersect the surface in hyperbolas

$$\frac{x^2}{a^2} - \frac{y^2}{b^2} = -\frac{k}{c^2}$$

whose vertices lie below the x-axis. The cross section made by the yz-plane $(x = 0)$ is the parabola

$$z = \frac{c^2}{b^2} y^2,$$

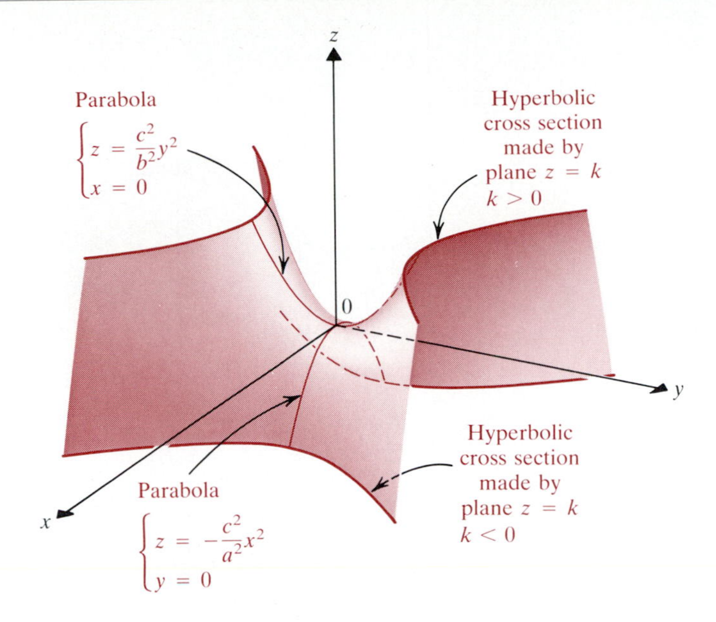

FIGURE 6.9
Hyperbolic Paraboloid $\frac{z}{c^2} = \frac{y^2}{b^2} - \frac{x^2}{a^2}$

which opens upward. The cross section made by the xz-plane ($y = 0$) is the parabola

$$z = -\frac{c^2}{a^2} x^2,$$

which opens downward. There is symmetry relative to the xz- and yz-planes but not relative to the xy-plane. **Figure 6.9** is a sketch of the surface made by this information. **Figure 6.10** shows a computer-generated plot consisting of cross sections made by planes $x = k$ and $y = k$. Those cross sections are, respectively, upward-opening parabolas,

$$z = c^2\left(\frac{y^2}{b^2} - \frac{k^2}{a^2}\right), \qquad x = k,$$

and downward-opening parabolas,

$$z = \left(-\frac{x^2}{a^2} + \frac{k^2}{b^2}\right), \qquad y = k.$$

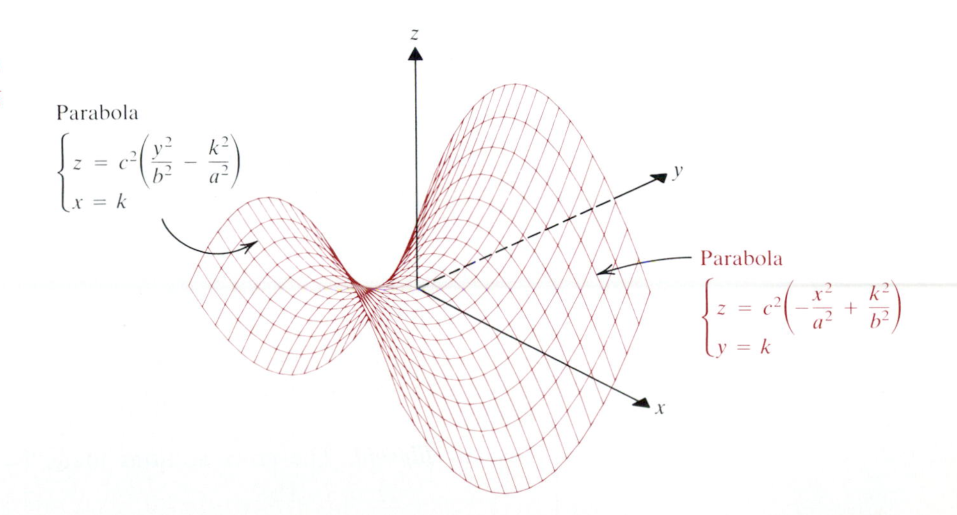

FIGURE 6.10
Computer Plot of $\frac{z}{c^2} = \frac{y^2}{b^2} - \frac{x^2}{a^2}$
(where $a = b = c = 1$)

FIGURE 6.11

Hyperbolic Paraboloid $\frac{z}{c^2} = \frac{x^2}{a^2} - \frac{y^2}{b^2}$

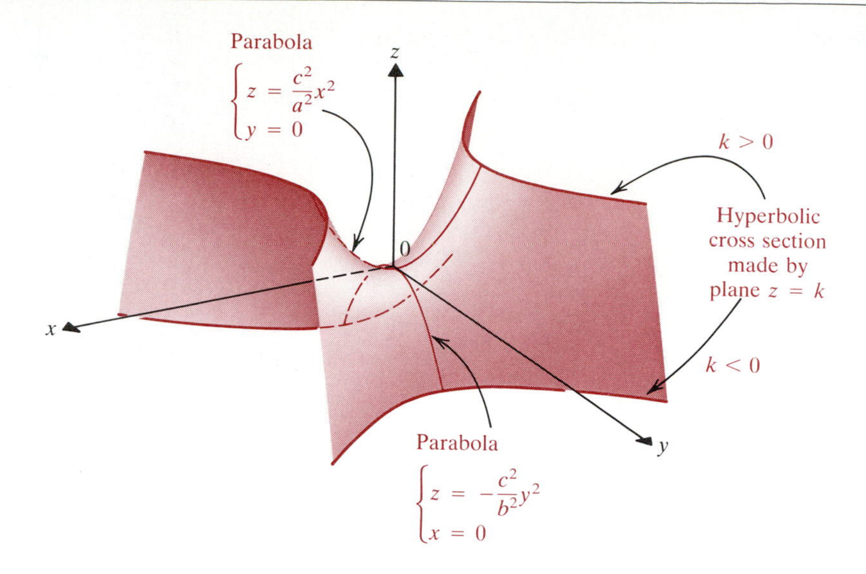

FIGURE 6.12

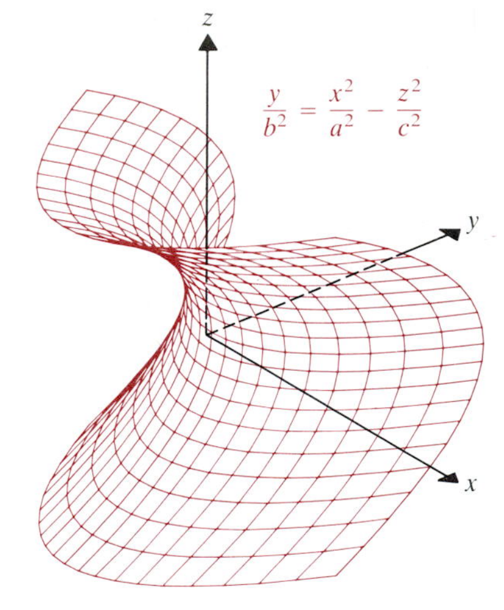

FIGURE 6.13

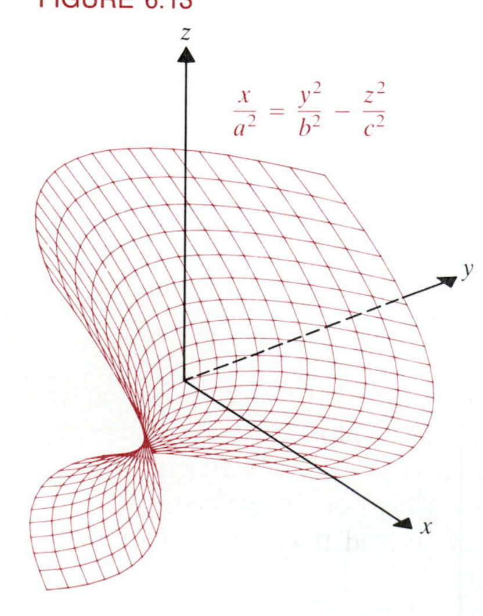

Variations. The graph of

$$\frac{x^2}{a^2} - \frac{y^2}{b^2} = \frac{z}{c^2}$$

is a hyperbolic paraboloid rotated 90° about the z-axis from the one shown in Figure 6.9. [Alternatively, you can turn Figure 6.9 upside down, since z has been replaced by $-z$ in (2).] If we invert the saddle in Figure 6.9, then we still have a saddle, but one rotated 90° laterally: Compare **Figure 6.11** with what you see when you turn Figure 6.9 upside down. The graph of

$$\pm\frac{x^2}{a^2} \mp \frac{z^2}{c^2} = \frac{y}{b^2} \tag{3}$$

is a hyperbolic paraboloid rotated 90° about the x-axis from the one shown in Figure 6.9. [If you imagine a horse under the saddle S in Figure 6.9, then the saddle (3) is the position of S if the horse reared up so that its body became perpendicular to the ground.] See **Figure 6.12.** The graph of

$$\pm\frac{y^2}{b^2} \mp \frac{z^2}{c^2} = \frac{x}{a^2}$$

is a hyperbolic paraboloid rotated 90° about the y-axis from its position in Figure 6.9 (see **Figure 6.13**).

Ellipsoids

The graph of

$$\frac{x^2}{a^2} + \frac{y^2}{b^2} + \frac{z^2}{c^2} = 1$$

is an ***ellipsoid.*** The cross sections made by planes $z = k$ are curves having equations

$$\frac{x^2}{a^2} + \frac{y^2}{b^2} = 1 - \frac{k^2}{c^2}, \qquad z = k,$$

which are ellipses as long as $k^2/c^2 < 1$. When $k = \pm c$, we get a single point. When $k^2 > c^2$, there is no intersection. So the graph lies between $z = -c$ and $z = +c$. Cross sections made by planes $x = k$ and $y = k$ are of the same sort as those made by planes $z = k$. If any two of a^2, b^2, and c^2 are equal, then we get (cf. Theorem 5.4) an ellipsoid of revolution (which looks something like a football). In all cases the graph is symmetric relative to all three coordinate planes. See **Figure 6.14.** (Note: if the 1 in the defining equation is replaced by a 0, we get the single point (0, 0, 0) as a degenerate ellipsoid.) An ellipsoid looks like a distorted sphere, which it is: When $a = b = c$, we get a sphere of radius a.

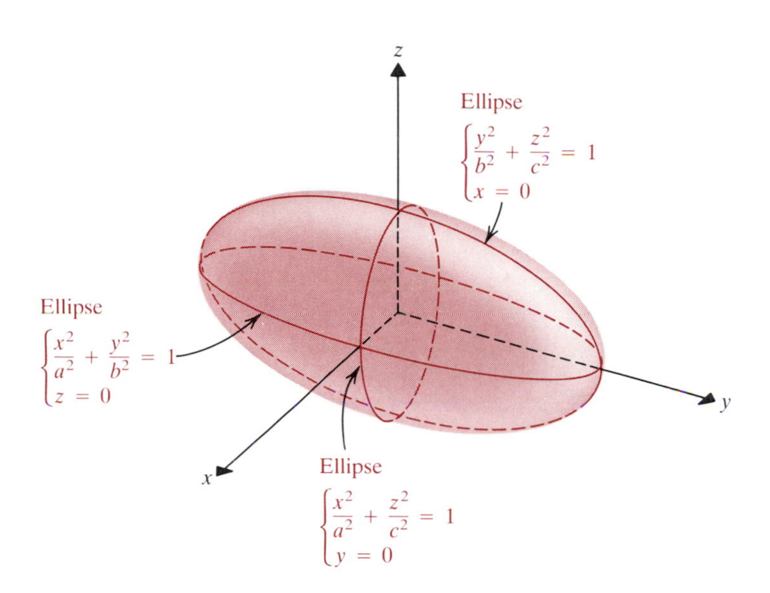

FIGURE 6.14
Ellipsoid $\frac{x^2}{a^2} + \frac{y^2}{b^2} + \frac{z^2}{c^2} = 1$

Example 2

Sketch the graph of $9x^2 + y^2 + 4z^2 = 36$.

Solution. Division by 36 produces the standard form equation

$$\frac{x^2}{4} + \frac{y^2}{36} + \frac{z^2}{9} = 1.$$

To find the points of this ellipsoid that lie on the coordinate axes, we set two coordinate variables equal to 0. Setting $x = y = 0$, we find the points $(0, 0, \pm 3)$. Setting $x = z = 0$, we get $(0, \pm 6, 0)$. Finally, letting $y = z = 0$, we obtain $(\pm 2, 0, 0)$. The surface is sketched in **Figure 6.15.** ■

The 1975 Nobel Prize for physics was awarded to Aage Bohr (1922–) [son of Niels Bohr (1885–1962), the 1922 Nobel laureate in physics for his pioneering work in nuclear physics] and Benjamin Mottelson (1926–) of the Niels Bohr Institute in Denmark, and James Rainwater (1917–) of Columbia University for work done in the 1940s and 1950s that determined that the nucleus of the atom has the shape of an ellipsoid.

FIGURE 6.15

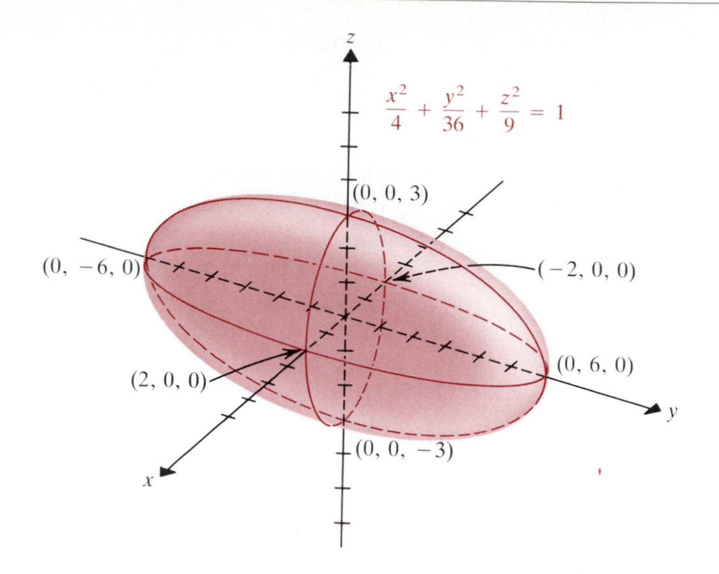

Hyperboloid of One Sheet

The graph of

$$\frac{x^2}{a^2} + \frac{y^2}{b^2} - \frac{z^2}{c^2} = 1$$

is a ***hyperboloid of one sheet*** with the z-axis as its axis. The cross sections made by the planes $z = k$ have equations

$$\frac{x^2}{a^2} + \frac{y^2}{b^2} = 1 + \frac{k^2}{c^2}, \qquad z = k,$$

and so are ellipses for all values of k. The cross section with the yz-plane is the hyperbola

$$\frac{y^2}{b^2} - \frac{z^2}{c^2} = 1, \qquad x = 0.$$

The cross section with the xz-plane is the hyperbola

$$\frac{x^2}{a^2} - \frac{z^2}{c^2} = 1, \qquad y = 0.$$

If $a^2 = b^2$, then we have a hyperboloid of revolution about the z-axis. (See Theorem 5.4.) The graph is symmetric relative to all three coordinate planes. It is sketched in **Figure 6.16**(a).

Variations. The graph of

$$\frac{x^2}{a^2} - \frac{y^2}{b^2} + \frac{z^2}{c^2} = 1 \tag{4}$$

is a hyperboloid of one sheet having the y-axis as axis. See Figure 6.16(b). The graph of

$$-\frac{x^2}{a^2} + \frac{y^2}{b^2} + \frac{z^2}{c^2} = 1$$

is a hyperboloid of one sheet having the x-axis as axis. See Figure 6.16(c), in which we sketch a hyperboloid of revolution.

FIGURE 6.16

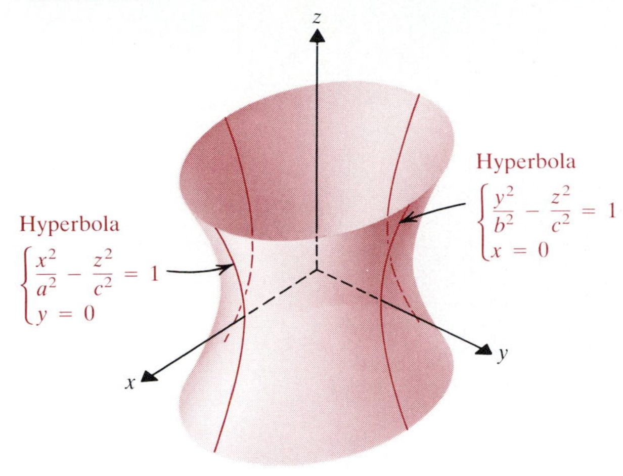

(a) Hyperboloid $\frac{x^2}{a^2} + \frac{y^2}{b^2} - \frac{z^2}{c^2} = 1$

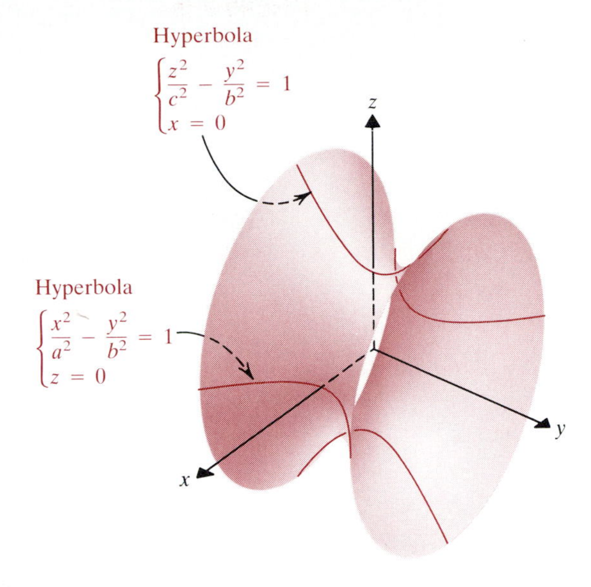

(b) Hyperboloid $\frac{x^2}{a^2} - \frac{y^2}{b^2} + \frac{z^2}{c^2} = 1$

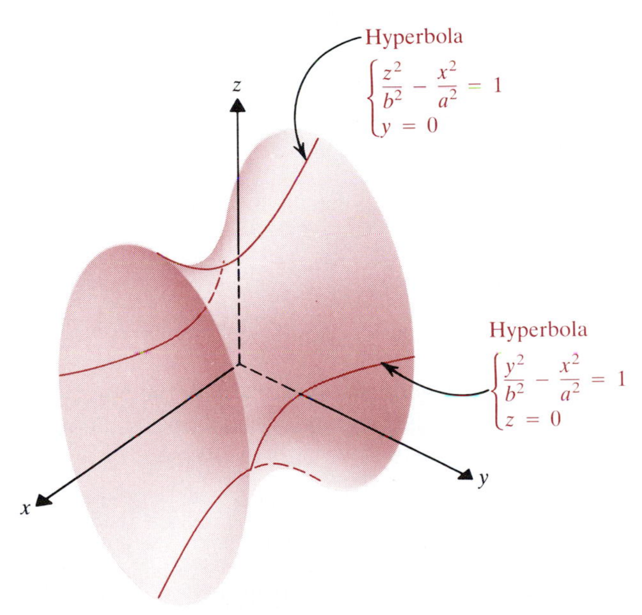

(c) Hyperboloid of revolution
$-\frac{x^2}{a^2} + \frac{y^2}{b^2} + \frac{z^2}{c^2} = 1$

Example 3

Name and draw a rough sketch of the quadric surface $4x^2 - 9y^2 + z^2 = 36$.

Solution. First, divide through by 36 to put the equation in standard form:

$$\frac{x^2}{9} - \frac{y^2}{4} + \frac{z^2}{36} = 1.$$

FIGURE 6.17

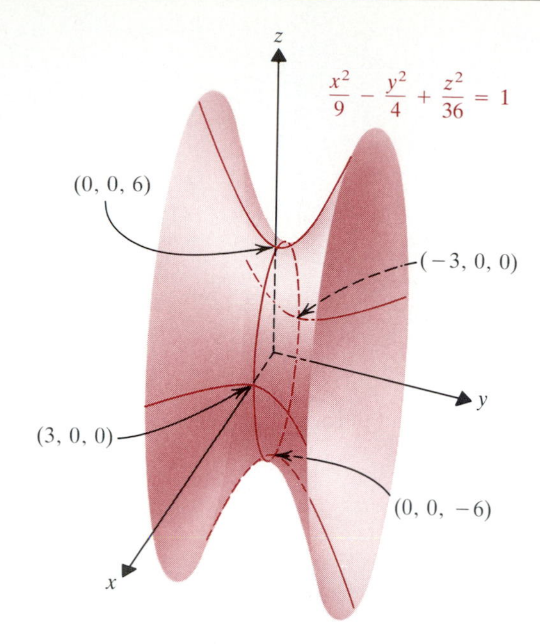

This is an equation of the form (4), so its graph is a hyperboloid of one sheet with the y-axis as axis. To make a rough sketch we find the intersection points of the surface with the coordinate axes. There is no intersection with the y-axis, as we can see in two ways: (a) the y-axis is the axis of the hyperboloid; (b) when $x = z = 0$, we get $y^2 = -4$, so no such y exists. When x and y are zero, we have $z = \pm 6$. When y and z are zero, we find $x = \pm 3$. The surface is sketched in **Figure 6.17.** ■

Hyperboloid of Two Sheets

The graph of

$$-\frac{x^2}{a^2} + \frac{y^2}{b^2} - \frac{z^2}{c^2} = 1$$

is a ***hyperboloid of two sheets*** with the y-axis as axis. That is the only coordinate axis on which the graph has any points. The cross sections made by planes $z = k$ are hyperbolas.

$$\frac{y^2}{b^2} - \frac{x^2}{a^2} = 1 + \frac{k^2}{c^2}, \qquad z = k,$$

with vertices on the y-axis in the $z = k$ copy of the xy-plane. Similarly, the cross sections made by planes $x = k$ are hyperbolas

$$\frac{y^2}{b^2} - \frac{z^2}{c^2} = 1 + \frac{k^2}{a^2}, \qquad x = k,$$

with vertices on the y-axis in the $x = k$ copy of the yz-plane. Cross sections made by the planes $y = k$ have equations

$$\frac{x^2}{a^2} + \frac{z^2}{c^2} = \frac{k^2}{b^2} - 1, \qquad y = k.$$

So there is no cross section if $k^2 < b^2$, a single point if $k^2 = b^2$, and an ellipse if $k^2 > b^2$. If $a^2 = c^2$, we have a hyperboloid of revolution about the y-axis. The graph is symmetric relative to all three coordinate planes. See **Figure 6.18.**

FIGURE 6.18

Hyperboloid $-\frac{x^2}{a^2}+\frac{y^2}{b^2}-\frac{z^2}{c^2}=1$

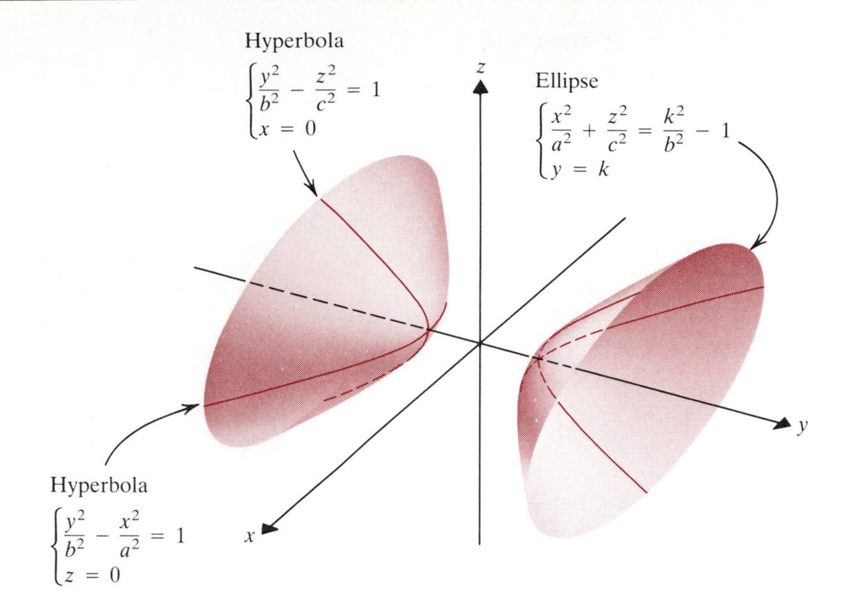

Variations. The graph of

$$-\frac{x^2}{a^2}-\frac{y^2}{b^2}+\frac{z^2}{c^2}=1$$

is a hyperboloid of two sheets with the z-axis as axis. (Rotate Figure 6.18 by 90° about the x-axis.)

$$\frac{x^2}{a^2}-\frac{y^2}{b^2}-\frac{z^2}{c^2}=1$$

is a hyperboloid of two sheets with the x-axis as axis. (Rotate Figure 6.18 by 90° about the z-axis.)

Thus far, all the quadric surfaces have been centered at the origin (0, 0, 0). The equation of a quadric surface centered at (h, k, l) is obtained by replacing x, y, and z in the standard-form equation by translated coordinate variables $x' = x - h$, $y' = y - k$, and $z' = z - l$, respectively.

Example 4

Identify and sketch the quadric surface

$$x^2 - 2y^2 - 4z^2 - 4x + 12y + 32z = 94.$$

Solution. We complete the squares in x, y, and z, getting

$$x^2 - 4x + 4 - 2(y^2 - 6y + 9) - 4(z^2 - 8z + 16) = 94 + 4 - 18 - 64,$$

$$(x-2)^2 - 2(y-3)^2 - 4(z-4)^2 = 16,$$

$$\frac{(x-2)^2}{16} - \frac{(y-3)^2}{8} - \frac{(z-4)^2}{4} = 1. \tag{5}$$

This is a hyperboloid of two sheets with the point $0'(2, 3, 4)$ as center and axis parallel to the x-axis. The vertex has y-coordinate 3 and z-coordinate 4. The x-coordinate can be found by substituting $y = 3$ and $z = 4$ into (5):

$$\frac{(x-2)^2}{16} = 1 \rightarrow \frac{x-2}{4} = \pm 1 \rightarrow x = 2 \pm 4 = 6 \text{ or } -2.$$

FIGURE 6.19

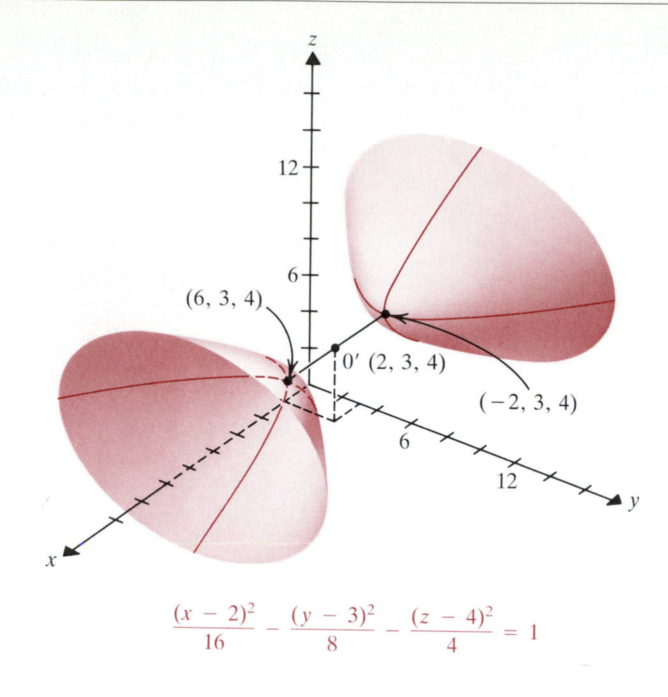

$$\frac{(x-2)^2}{16} - \frac{(y-3)^2}{8} - \frac{(z-4)^2}{4} = 1$$

The vertices are therefore (6, 3, 4) and (−2, 3, 4). The surface is sketched in **Figure 6.19.** ■

Exercises 11.6

In Exercises 1–36, identify and sketch the graph of the given equation.

1. $y^2 + 3z^2 = 9$
2. $y^2 - x^2 = 4$
3. $y^2 + 4z = 4$
4. $x^2 - 2y^2 + 4z^2 = 0$
5. $z^2 = x^2 + y^2$
6. $4x^2 + y^2 - z = 0$
7. $x^2 + 2y + 2z^2 = 0$
8. $z = x^2 + y^2$
9. $x^2 - 2y^2 - 4z = 0$
10. $y^2 = z + x^2$
11. $y^2 - 2x^2 - z = 0$
12. $x^2 + 3y^2 + 2z^2 = 6$
13. $3x^2 + y^2 + 3z^2 = 9$
14. $x^2 + y^2 - 2z^2 = 4$
15. $2x^2 - y^2 + 4z^2 = 4$
16. $y^2 = 1 + x^2 + z^2$
17. $2x^2 - y^2 - 4z^2 = 4$
18. $z^2 = 2 + 2x^2 + y^2$
19. $x^2 - 2z^2 = 8$
20. $x^2 - 9z = 16$
21. $y^2 - 3x^2 + z^2 = 0$
22. $x^2 + 2y^2 - 3z^2 = 0$
23. $x^2 + 2y^2 - 3z = 0$
24. $x^2 - 6y + 3z^2 = 0$
25. $4x - y^2 + z^2 = 0$
26. $x^2 - 3y - 4z^2 = 0$
27. $3x^2 + 2y^2 + 8z^2 = 0$
28. $x^2 + 5y^2 + 25z^2 = 125$
29. $x^2 + y^2 - 3z^2 = 12$
30. $y^2 - z^2 = x^2 + 4$
31. $x^2 - y^2 + 3z^2 - 6x - 4y + 6z = 1$
32. $-x^2 + 2y^2 + 3z^2 - 6x - 4y + 6z = 0$
33. $3x^2 + 4y^2 + 2z^2 - 12x + 8y - 12z = 0$
34. $3x^2 + 5y^2 + z^2 - 18x + 20y + 4z = 4$
35. $3x^2 + 4y^2 - 6x - 16y - 12z + 31 = 0$
36. $3x^2 + 4y^2 + z^2 - 6x - 16y + 12z + 55 = 0$

11.7 Looking Back

The first three sections of this chapter were devoted to the calculus of vector functions $f\colon \mathbf{R} \to \mathbf{R}^n$. Section 1 gave the basic definitions of coordinate function, limits, continuity, derivative, and integral. The principal properties of differentiation were given in Theorem 1.4. Section 2 discussed curvilinear motion, that is, motion by a particle along a parametrized curve. Besides velocity and acceleration, the section also considered arc-length calculation and the associated arc-length function (Definition 2.2) as well as parametrization by arc length (p. 698). Section 3 introduced curvature [Equations (8) and (23)], the unit normal (Definition 3.2), and the tangential and normal components of acceleration [Definition 3.3, Formulas (13), (14), (16), and (17).]

In Section 4, attention shifted to scalar functions $f: \mathbf{R}^n \to \mathbf{R}$ of several real variables. The notions of limit and continuity were introduced for those functions in Definitions 4.2 and 4.4, and the main emphasis in the section was on determining the set of points where a given schizophrenic function is continuous.

The final two sections were devoted to graphing. The main algebraic result was Theorem 5.4 on the equation of a surface of revolution. The last section catalogued the various quadric surfaces arising as the graphs of quadratic equations in three variables.

CHAPTER CHECKLIST

Section 1: vector function, coordinate functions; limits, continuity, derivatives; chain rule; parametrized curve, simple curve, orientation, parameter; velocity, unit tangent vector; integral.

Section 2: acceleration, force; Newton's first and second laws of motion; arc length, arc-length function; parametrization by arc length; circular helix.

Section 3: curvature; unit normal vector; tangential and normal components of acceleration.

Section 4: scalar functions; open ball, closed ball, punctured open ball; limits, continuity, schizophrenic functions.

Section 5: graph of a function $f: \mathbf{R}^2 \to \mathbf{R}$, graph of a functional equation; right cylinder; surface of revolution, equation of a surface of revolution; level curves, level surfaces.

Section 6: right cylinder, right cone, elliptical paraboloid, hyperbolic paraboloid, ellipsoid, hyperboloid of one sheet, hyperboloid of two sheets.

REVIEW EXERCISES 11.7

1. If $\mathbf{x}(t) = \cos 2t\,\mathbf{i} + \sin 2t\,\mathbf{j} + t\,\mathbf{k}$, then find an equation of the tangent line at $t = \pi/4$. What is the speed there?

2. If
$$\mathbf{f}(t) = \left(\frac{t^2 - 4}{t - 2}\right)\mathbf{i} + 4t\,\mathbf{j} + 5\,\mathbf{k},$$
then is $\mathbf{f}$ continuous at $t = 2$? Can you define $\mathbf{f}(2)$ so that $\mathbf{f}$ becomes continuous at $t = 2$?

3. If $\mathbf{f}(t) = \cos 2t\,\mathbf{i} + \sin 2t\,\mathbf{j} + t\,\mathbf{k}$, then find $\int_0^{\pi/4} \mathbf{f}(t)\,dt$.

4. A particle moves on the parabola $y^2 = 4x$. Parametrize and find the velocity, acceleration, and speed at any time t.

5. Describe the path in $\mathbf{R}^2$ of a particle subjected to a force $\mathbf{F}(t) = -\cos t\,\mathbf{i} - \sin t\,\mathbf{j}$.

6. Find the length of arc along $\mathbf{x}(t) = (2t, t^2, t)$ between $t = 0$ and $t = 1$.

7. Show that the curve
$$\mathbf{x}(t) = \left(\frac{1}{3}t, 4 - \frac{2}{3}t, \frac{2}{3}t\right)$$
is parametrized by arc length.

8. If $\mathbf{x}(t) = (1, e^t \cos t, e^t \sin t)$, then find K, $\mathbf{T}$, and $\mathbf{N}$.

9. Repeat Exercise 8 for $\mathbf{x}(t) = (\sin 2t, t, \cos 2t)$.

10. If $\mathbf{x}(t) = (t, \sin 4t, \cos 4t)$, $0 \le t \le \pi$, then find:

(a) the velocity, speed, and acceleration at any time t.
(b) the arc length of the curve.
(c) a parametrization of the curve by arc length.
(d) $\mathbf{T}$, $\mathbf{N}$, and K at $t = \pi/2$.
(e) the tangential and normal components of acceleration.

11. A particle moves on the curve $\mathbf{x} = \mathbf{x}(t) = (t \sin t, t \cos t, t^2 - 1)$, $t \in [0, 2\pi]$. Find:

(a) the velocity, speed, and acceleration in general and at $t = \pi$.

(b) the arc length.

(c) the tangential and normal components of acceleration at $t = \pi$.

12. Find the limit if it exists, or explain why there is no limit.

(a) $\lim_{(x,y)\to(2,2)} \dfrac{x^3 - y^3}{x - y}$ **(b)** $\lim_{(x,y)\to(0,0)} \dfrac{x^6 + y^6}{x^2 + y^2}$

In Exercises 13–16, find all points where the given function is continuous.

13. $f(x, y) = \begin{cases} \dfrac{x^2 y}{x^3 + y^3}, & \text{if } x + y \neq 0 \\ 0, & \text{if } x + y = 0 \end{cases}$

14. $g(x, y) = \begin{cases} \dfrac{x^2 y^2}{x^2 + y^2}, & \text{if } (x, y) \neq (0, 0) \\ 0, & \text{if } (x, y) = (0, 0) \end{cases}$

15. $f(x, y) = \begin{cases} (x^2 + y^2)\sin\dfrac{1}{x}, & \text{if } x \neq 0 \\ 0, & \text{if } x = 0 \end{cases}$

16. $g(x, y) = \begin{cases} \dfrac{x^3}{y^2 - 1}, & \text{if } y \neq 1 \\ 1, & \text{if } y = 1 \end{cases}$

In Exercises 17–20, name and sketch the graph of the given functional equation $g(x, y, z) = 0$.

17. $z = x^2 + y^2$

18. $9z^2 - 16y^2 = 144$

19. $x^2 + 4y^2 + z^2 = 4$

20. $4x^2 - 9y^2 + 4z^2 = 36$

In Exercises 21 and 22, find the equation of the given surface of revolution.

21. The graph of $y = x^2 + 4$ in the xy-plane is revolved about **(a)** the x-axis and **(b)** the y-axis.

22. The graph of $y^2/4 - z^2/9 = 1$ in the yz-plane is revolved about **(a)** the y-axis and **(b)** the z-axis.

23. Plot the level curves of $z = x^2 + 4y$ for $z = 1, 4, 9, 16$.

24. Plot the level surfaces of $x^2/4 + y^2 + z^2/4 + w^2 = 4$ for $w = 0, 1, 2$.

In Exercises 25–36, identify and sketch the graph of the given equation.

25. $2z^2 - y^2 = 4$

26. $x^2 - 4y = 8$

27. $2y^2 - x^2 + 3z^2 = 0$

28. $y^2 = x^2 - z^2$

29. $x - 2y^2 - 3z^2 = 0$

30. $y^2 = 4x^2 + 2z$

31. $x^2 + 2y^2 + z^2 = 4$

32. $x^2 - 3y^2 + 6z^2 = 6$

33. $3y^2 - 5x^2 - 6z^2 = 30$

34. $y^2 - z^2 = x^2 + 4$

35. $x^2 - 2y^2 + 4z^2 + 4x - 8y = 8$

36. $2x^2 - 3y^2 + 4z^2 + 12x - 12y + 8z + 26 = 0$

CHAPTER 12

Differentiation of Scalar Functions

12.0 Introduction

This chapter continues the development of the calculus of scalar functions $f\colon \mathbf{R}^n \to \mathbf{R}$ that was begun in Section 11.4. Differentiation of such functions is carried out in two stages. The first freezes all but one of the variables and allows only that one to vary. This brings us back to the situation of Chapter 2, because it essentially reduces f to a real-valued function of a single real variable. This first stage is presented in Section 1, where the *partial* derivatives of f are defined. The second stage involves giving meaning to the *total* derivative of f. To do so, we use the characterization of differentiation of $f\colon \mathbf{R} \to \mathbf{R}$ given in Theorem 5.3 of Chapter 2. That result states that f is differentiable at $x = a$ if and only if f can be accurately approximated near a by the tangent approximation

$$T(x) = f(a) + m \cdot (x - a),$$

where $m = f'(a)$. In Section 2, we use this property to define the total derivative of $f\colon \mathbf{R}^n \to \mathbf{R}$ for any $n > 1$.

The rest of the chapter is concerned with developing differentiation techniques and applications analogous to those in Chapters 2 and 3 for functions of one variable. For instance, in Sections 1 and 4 you will learn how to find equations of tangent planes to surfaces. Chapters 2 and 3 discussed the chain rule, implicit differentiation, higher derivatives and the use of the first and second derivatives to find maxima and minima. As the titles of Sections 4 (*The Chain Rule*), 5 (*Implicit Differentiation*), 6 (*Higher Partial Derivatives*), and 7 (*Extreme Values*) suggest, those topics all have several-variable counterparts. The final section presents the technique of *Lagrange multipliers*, a tool for finding extreme values of a function that is subject to some restriction. Such problems arise frequently in applications where physical, economic, or other factors restrict the domains of certain functions.

12.1 Partial Derivatives

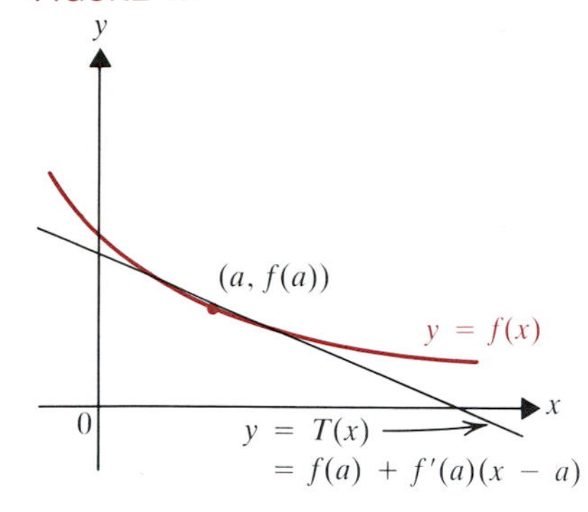

FIGURE 1.1

One of the central facts about the derivative $f'(a)$ of a function of one variable is that it gives the slope of the tangent line to the graph of f at $x = a$. The tangent line has equation

$$y = T(x) = f(a) + f'(a)(x - a), \tag{1}$$

and provides a close approximation [the *tangent approximation* $T(x)$] to $f(x)$ near $x = a$. See **Figure 1.1.**

The graph of a scalar function $f\colon \mathbf{R}^2 \to \mathbf{R}$ of two real variables is a surface in $\mathbf{R}^3$. As we saw in Section 10.5, planes in $\mathbf{R}^3$ are analogous in many ways to lines in $\mathbf{R}^2$. **Figure 1.2** shows how the *tangent plane* Π to the graph of $z = f(x, y)$ at a point $P(a, b, f(a, b))$ might look. What precisely is that plane? It would be reasonable to expect it to be determined by the tangent *vectors* to the two curves

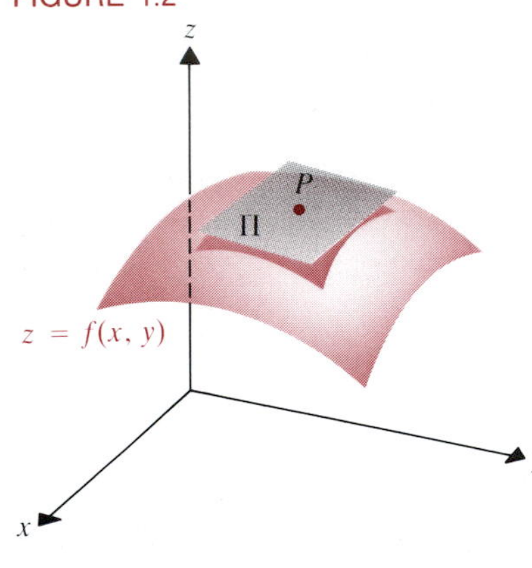

FIGURE 1.2

$$\mathbf{x} = \mathbf{x}_1(t) = (a, t, f(a, t)) = a\,\mathbf{i} + t\,\mathbf{j} + f(a, t)\,\mathbf{k}$$

and

$$\mathbf{x} = \mathbf{x}_2(t) = (t, b, f(t, b)) = t\,\mathbf{i} + b\,\mathbf{j} + f(t, b)\,\mathbf{k}.$$

Those curves are the intersections of the surface with the planes $x = a$ and $y = b$ parallel respectively to the yz- and xz-coordinate planes. See **Figure 1.3.** Tangent vectors to the curves at P are

$$\mathbf{x}_1'(b) = \left(0, 1, \left.\frac{df(a, t)}{dt}\right|_{t=b}\right) \quad \text{and} \quad \mathbf{x}_2'(t) = \left(1, 0, \left.\frac{df(t, b)}{dt}\right|_{t=a}\right).$$

The third coordinates are the derivatives of *single*-variable functions obtained from $f(x, y)$ by freezing one of the variables to its value at $P(a, b, f(a, b))$ and letting the other variable vary. Those derivatives play a prominent role in this chapter.

1.1 DEFINITION Let $f\colon \mathbf{R}^2 \to \mathbf{R}$ be a scalar function and $P(a, b, f(a, b))$ a point on its graph. The ***partial derivative of f with respect to x*** at (a, b) is

$$\frac{\partial f}{\partial x}(a, b) = \left.\frac{df(x, b)}{dx}\right|_{x=a} = \lim_{h \to 0} \frac{f(a + h, b) - f(a, b)}{h},$$

FIGURE 1.3

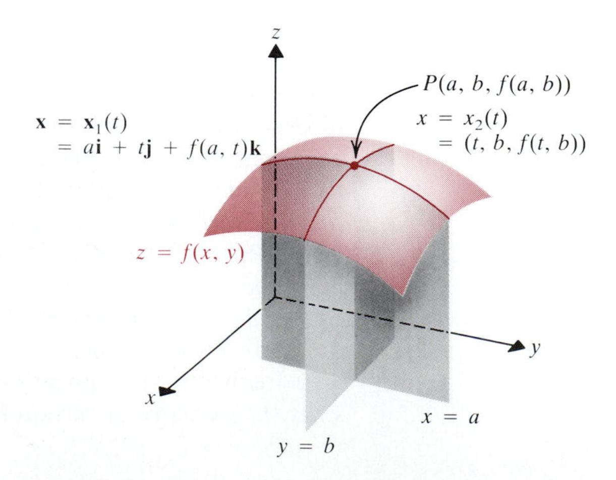

if this limit exists. The ***partial derivative of f with respect to y*** at (a, b) is

$$\frac{\partial f}{\partial y}(a, b) = \left.\frac{df(a, y)}{dy}\right|_{y=b} = \lim_{h \to 0} \frac{f(a, b + h) - f(a, b)}{h},$$

if this limit exists.

The symbols

$$\frac{\partial f}{\partial x}(a, b) \quad \text{and} \quad \frac{\partial f}{\partial y}(a, b)$$

are read "the partial (derivative) of f with respect to x (respectively, y) at (a, b)". Other notations in use for $(\partial f/\partial x)(a, b)$ include

$$f_x(a, b), \frac{\partial z}{\partial x}(a, b), f_1(a, b), z_x(a, b), D_x f(a, b), \text{ and } D_1 f(a, b).$$

Similarly, $(\partial f/\partial y)(a, b)$ is also denoted

$$f_y(a, b), \frac{\partial z}{\partial y}(a, b), f_2(a, b), z_y(a, b), D_y f(a, b), \text{ and } D_2 f(a, b).$$

We shall use several of these notations.

The partial derivatives $(\partial f/\partial x)(a, b)$ and $(\partial f/\partial y)(a, b)$ measure how fast the function f is increasing or decreasing in the x- and y-directions at $(x, y) = (a, b)$. They can be thought of as measuring the rate of rise or fall of land whose surface has equation $z = f(x, y)$:

$$\frac{\partial f}{\partial x}(a, b)$$

is the rate of change in some standard direction (such as east–west) that might be measured by a surveyor. Similarly, $(\partial f/\partial y)(a, b)$ measures the rate of change along a perpendicular direction (for example, north–south).

Example 1

(a) Find $\partial f/\partial x$ and $\partial f/\partial y$ at the point $P(1, 2, 5)$ if $f(x, y) = x^2 + y^2$.

(b) Find formulas for $\partial z/\partial x$ and $\partial z/\partial y$ if $z = \sin x^2 + 3x^2y^3 + e^{-2y}$.

Solution. (a) First we check that $P(1, 2, 5)$ is on the surface. Since

$$1^2 + 2^2 = 5,$$

P does lie on the graph. If we freeze y to its value 2 at P and differentiate the resulting function of x alone, we get

$$\frac{\partial f}{\partial x}(1, 2) = \left.\frac{d}{dx}(x^2 + 4)\right|_{x=1} = \left.2x\right|_{x=1} = 2.$$

Similarly, holding x constant to its value (1) at P, we find

$$\frac{\partial f}{\partial y}(1, 2) = \left.\frac{d}{dy}(1 + y^2)\right|_{y=2} = \left.2y\right|_{y=2} = 4.$$

(b) If we hold x constant, we obtain

$$\frac{\partial z}{\partial y} = 0 + 3x^2 \cdot 3y^2 + e^{-2y}(-2) = 9x^2y^2 - 2e^{2y}.$$

Holding y constant, we get

$$\frac{\partial z}{\partial x} = (\cos x^2) \cdot 2x + 6xy^3 + 0 = 2x \cos x^2 + 6xy^3. \blacksquare$$

We could also have used the approach of Example 1(b) in working Example 1(a). Holding y constant and differentiating with respect to x, we would have found

$$f_x = \frac{\partial}{\partial x}(x^2 + y^2) = 2x. \tag{2}$$

Similarly, holding x constant and differentiating with respect to y we would have found

$$f_y = \frac{\partial}{\partial y}(x^2 + y^2) = 2y. \tag{3}$$

Then to evaluate $f_x(1, 2)$ and $f_y(1, 2)$, we could have substituted 1 for x and 2 for y in (2) and (3).

We return to the discussion of the tangent plane to $z = f(x, y)$ at $P(a, b, f(a, b))$, by using Definition 1.1 to rewrite the tangent vector formulas:

$$\mathbf{x}_1'(b) = \left(0, 1, \frac{\partial f}{\partial y}(a, b)\right) = \mathbf{j} + f_y(a, b)\,\mathbf{k}$$

and

$$\mathbf{x}_2'(a) = \left(1, 0, \frac{\partial f}{\partial x}(a, b)\right) = \mathbf{i} + f_x(a, b)\,\mathbf{k}.$$

The tangent plane to the graph of $z = f(x, y)$ at $P(a, b, f(a, b))$ is defined to be the plane determined by $\mathbf{x}_1'(b)$ and $\mathbf{x}_2'(a)$. As in Section 10.6, that plane has normal vector

$$\mathbf{n} = \mathbf{x}_1'(b) \times \mathbf{x}_2'(a) = \det\begin{pmatrix} \mathbf{i} & \mathbf{j} & \mathbf{k} \\ 0 & 1 & f_y(a, b) \\ 1 & 0 & f_x(a, b) \end{pmatrix}$$

$$= \det\begin{pmatrix} 1 & f_y(a, b) \\ 0 & f_x(a, b) \end{pmatrix}\mathbf{i} + \det\begin{pmatrix} f_y(a, b) & 0 \\ f_x(a, b) & 1 \end{pmatrix}\mathbf{j} + \det\begin{pmatrix} 0 & 1 \\ 1 & 0 \end{pmatrix}\mathbf{k},$$

$$\mathbf{n} = f_x(a, b)\,\mathbf{i} + f_y(a, b)\,\mathbf{j} - \mathbf{k} = (f_x(a, b), f_y(a, b), -1). \tag{4}$$

Thus a vector equation of the plane determined by $\mathbf{x}_1'(b)$ and $\mathbf{x}_2'(a)$ is

$$\mathbf{n} \cdot (\mathbf{x} - \mathbf{x}_0) = 0, \tag{5}$$

where $\mathbf{x} = (x, y, z)$ and $\mathbf{x}_0 = (a, b, f(a, b))$. Expanding (5), we get

$$(f_x(a, b), f_y(a, b), -1) \cdot (x - a, y - b, z - f(a, b)) = 0,$$

$$f_x(a, b)(x - a) + f_y(a, b)(y - b) - (z - f(a, b)) = 0.$$

This leads to the following definition.

1.2 DEFINITION Suppose that $\partial f/\partial x$ and $\partial f/\partial y$ exist on an open disk containing (a, b) and that they are continuous at (a, b). Then the ***tangent plane*** to the graph of $z = f(x, y)$ at $P(a, b, f(a, b))$ is the plane with normal-form vector equation

$$(f_x(a, b), f_y(a, b), -1) \cdot (x - a, y - b, z - c) = 0, \tag{6}$$

where $c = f(a, b)$, and scalar equation

$$z - f(a, b) = \frac{\partial f}{\partial x}(a, b)(x - a) + \frac{\partial f}{\partial y}(a, b)(y - b). \tag{7}$$

The reason for the continuity assumptions for the partial derivatives $\partial f/\partial x$ and $\partial f/\partial y$ at (a, b) will be seen in the next section. For now, notice the striking similarity between (7) and Equation (1) for the *tangent line* to the graph of $y = f(x)$ at the point $(a, f(a))$. If we write (1) in the form

$$y - f(a) = \frac{df}{dx}(a)(x - a), \tag{8}$$

and write (7) as

$$z - f(a,b) = \left(\frac{\partial f}{\partial x}(a, b), \frac{\partial f}{\partial y}(a, b)\right) \cdot (x - a, y - b) \tag{9}$$

then (9) looks like an exact two-dimensional analogue of (8). In the next section we will see that the vector of partial derivatives of a scalar function is indeed the two-dimensional analogue of the derivative of a function of one variable.

Example 2

Find the tangent plane to the paraboloid $z = x^2 + y^2$ at $P(1, 2, 5)$.

Solution. In Example 1(a), we found that $f_x(1, 2) = 2$ and $f_y(1, 2) = 4$. Substituting this information into (7), we get

$$z - 5 = 2(x - 1) + 4(y - 2) \rightarrow z - 5 = 2x - 2 + 4y - 8,$$

$$2x + 4y - z = 5. \tag{10}$$ ■

By a ***normal vector to a surface*** $z = f(x, y)$ at a point $P(a, b, f(a, b))$, we mean a normal vector to the tangent plane at P.

Example 3

Find a unit normal vector to the paraboloid in Example 2 at the point $P(1, 2, 5)$.

Solution. We found in (4) that

$$\mathbf{n} = f_x(a, b)\,\mathbf{i} + f_y(a, b)\,\mathbf{j} - \mathbf{k}$$

is a normal vector to the tangent plane at P. Since $f_x(1, 2) = 2$ and $f_y(1, 2) = 4$, the vector $\mathbf{n} = 2\mathbf{i} + 4\mathbf{j} - \mathbf{k}$ is normal to the tangent plane. [Alternatively, we can see this immediately from (10).] Thus a unit normal vector to the surface is

$$\mathbf{N} = \frac{1}{|\mathbf{n}|}\mathbf{n} = \frac{1}{\sqrt{21}}(2\mathbf{i} + 4\mathbf{j} - \mathbf{k}) = \left(\frac{2}{\sqrt{21}}, \frac{4}{\sqrt{21}}, -\frac{1}{\sqrt{21}}\right).$$

The vector $\mathbf{N}$ is shown in **Figure 1.4.** ■

The idea of partial differentiation extends easily to scalar functions of three (or more) variables. In the three-dimensional case we get three partial derivatives by freezing two of the variables and differentiating the resulting function of the remaining variable.

FIGURE 1.4

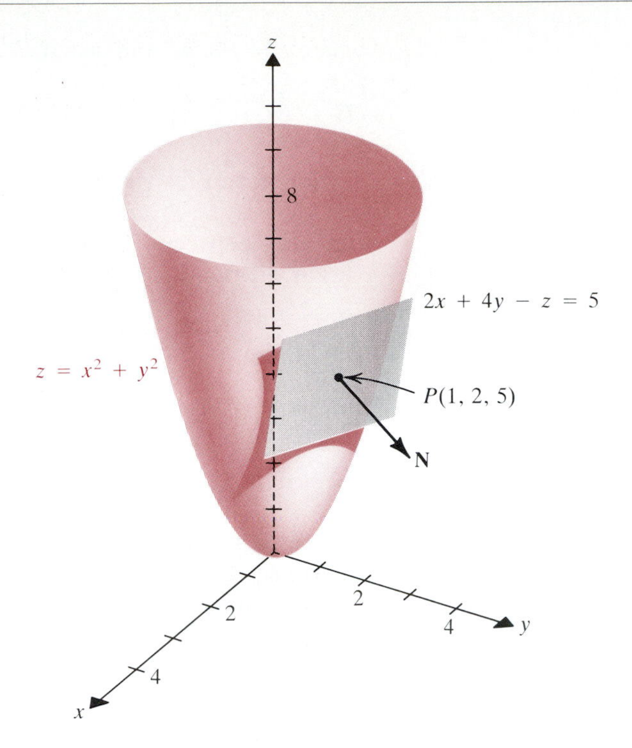

1.3 DEFINITION Let f: $\mathbf{R}^3 \to \mathbf{R}$ be a scalar function, and (a, b, c) a point in the domain of f. Then the ***partial derivatives*** of f at (a, b, c) are

$$f_x(a, b, c) = \frac{\partial f}{\partial x}(a, b, c) = \left.\frac{df(x, b, c)}{dx}\right|_{x=a} = \lim_{h \to 0} \frac{f(a + h, b, c) - f(a, b, c)}{h},$$

$$f_y(a, b, c) = \frac{\partial f}{\partial y}(a, b, c) = \left.\frac{df(a, y, c)}{dy}\right|_{y=b} = \lim_{h \to 0} \frac{f(a, b + h, c) - f(a, b, c)}{h}.$$

$$f_z(a, b, c) = \frac{\partial f}{\partial z}(a, b, c) = \left.\frac{df(a, b, z)}{dz}\right|_{z=c} = \lim_{h \to 0} \frac{f(a, b, c + h) - f(a, b, c)}{h}$$

if these limits exist.

Example 4

Find the partial derivatives of f: $\mathbf{R}^3 \to \mathbf{R}$ if $f(x, y, z) = e^{xy} \sin^2 yz + y^2 \ln xyz$.

Solution. Since $\ln xyz = \ln x + \ln y + \ln z$, we can simplify:

$$f(x, y, z) = e^{xy} \sin^2 yz + y^2(\ln x + \ln y + \ln z).$$

Holding y and z constant and differentiating with respect to x, we obtain

$$\frac{\partial f}{\partial x} = ye^{xy} \sin^2 yz + y^2\left(\frac{1}{x} + 0 + 0\right) = ye^{xy} \sin^2 yz + \frac{y^2}{x}.$$

Similarly, holding x and z constant and differentiating with respect to y, we get

$$\begin{aligned}\frac{\partial f}{\partial y} &= (xe^{xy}) \sin^2 yz + e^{xy}(2z \sin yz \cos yz) + 2y(\ln x + \ln y + \ln z) + y^2 \frac{1}{y} \\ &= xe^{xy} \sin^2 yz + 2ze^{xy} \sin yz \cos yz + 2y(\ln x + \ln y + \ln z) + y.\end{aligned}$$

Finally, holding x and y constant and differentiating with respect to z, we find

$$\frac{\partial f}{\partial z} = 2ye^{xy} \sin yz \cos yz + y^2 \frac{1}{z} = 2ye^{xy} \sin yz \cos yz + \frac{y^2}{z}. \quad \blacksquare$$

Partial differentiation is of considerable importance in economics, where many quantities are functions of several variables. For instance, the output of an industry is a function of the number x of workers employed in the industry and of the number y of dollars of capital invested in the industry (and perhaps of other variables, too). One attempt to describe this relationship mathematically uses the modeling function $p(x, y) = ax^{1-e}y^e$, where a and e are constants and e is between 0 and 1. As another example, the cost of producing one unit of a given commodity can be modeled by a function $C(x, y) = b + cx + fy$, where b is a constant (which can be interpreted as "plant overhead"—utility costs, upkeep, taxes on the property, and so on), x is the number of dollars in labor costs needed to produce one unit of the commodity, y is the number of dollars in the cost of raw materials needed to produce one unit, and c and f are constants. In economics, an important tool in studying a variable such as production or cost is the *marginal production* or *cost* relative to each independent variable.

1.4 DEFINITION If z is a function of x and y, then the ***marginal value of z relative to x*** is $\partial z/\partial x$ and the ***marginal value of z relative to y*** is $\partial z/\partial y$.

Example 5

Compute the marginal costs for the model $C(x, y) = b + cx + fy$ mentioned above. Interpret the results.

Solution. Partial differentiation of the cost function gives

$$\frac{\partial C(x, y)}{\partial x} = c \quad \text{and} \quad \frac{\partial C(x, y)}{\partial y} = f.$$

These are the marginal costs relative to labor and relative to raw materials, respectively (the rate of increase of the cost with respect to labor and raw materials). Roughly speaking, if labor costs rise one dollar per unit, the cost of the product will increase c dollars per unit. If the raw materials cost goes up one dollar per unit, then the cost of the product will increase f dollars per unit. While this is an extremely simple model in which only two variables contribute to cost, there is nothing to prevent calculation of marginal cost for cost functions of more variables by taking partial derivatives. Using these ideas, economists can calculate such quantities as the effect on the price of a new automobile resulting from a collective bargaining agreement increasing salaries, or from an increase in the price of steel used to make cars, or from an increase in the price of petroleum used in transporting steel to the automobile factory. ■

Exercises 12.1

In Exercises 1–10, find all the partial derivatives at a general point, and evaluate them at the given point.

1. $f(x, y) = \sqrt{x^2 + y^2}$; (1, 2)

2. $f(x, y) = \sqrt{4 - x^2 - y^2}$; (1,1)

3. $f(x, y) = 6e^y \cos x - 5e^{xy} \ln x^2y^2$; $(\pi/2, 1)$

4. $f(x, y) = (\sin xy)\sqrt{x^2 + y^2}$; $(1, \pi/2)$

5. $f(x, y) = \sqrt{x^2 + y^2 + z^2}$; (1, 1, 1)

6. $f(x, y) = \ln \cos x^2y - e^{xy^2} \tan xy$; $(1, 2\pi)$

7. $f(x, y, z) = x^2 y \sin xyz - e^{z^2 y} \cos xyz$; $(1, 1, \pi)$

8. $g(x, y, z) = \tan^{-1} xyz$; $(1, 1/2, 2)$

9. $g(x, y, z) = \tan^2 xy^2 z$; $(\pi/4, 1, 1)$

10. $g(x, y, z) = e^{\sin xy} z^2 - \ln \cos x^2 yz$; $(1, \pi, 1)$

In Exercises 11–16, find a scalar equation of the tangent plane to the surface at the given point.

11. $z = 3x^2 + 5y^2$ at $(1, 1, 8)$

12. $z = \sqrt{9 - x^2 - y^2}$ at $(2, 1, 2)$

13. $z = -x^2/4 + y^2/9$ at $(2, 3, 0)$

14. $z = \sqrt{x^2 + y^2}$ at $(4, 3, 5)$

15. $z = x^2 \ln yx + e^{xy}$ at $(1, 1, e)$

16. $z = \sin \sqrt{xy + 3y^2} + \tan \sqrt{x^2 + x - y}$ at $(\pi/4, \pi/4, 2)$

In Exercises 17–20, find a unit normal vector to the surface at the given point.

17. The surface of Exercise 11

18. The surface of Exercise 12

19. The surface of Exercise 13

20. The surface of Exercise 14

21. Find the distance from the origin to the tangent plane to the surface $z = x^2 + y$ at $(1, 1, 2)$. (*Hint:* Use Formula (12), p. 668.)

22. Find the distance from the origin to the tangent plane to the surface $z = x^2 + y^2$ at $(1, 1, 2)$. (See Exercise 21.)

23. For the ideal gas law

$$P = nR\frac{T}{V},$$

where n is the number of moles, R is a proportionality constant, T is the temperature in Kelvin, V is the volume in cubic centimeters, and P is the pressure in atmospheres, find $\partial P/\partial T$ and $\partial P/\partial V$. Interpret in terms of the effect on the pressure of a unit increase or decrease in V and T.

24. Find the marginal production relative to labor cost and capital investment in the model $P(x, y) = ax^{1-e}y^e$ mentioned after Example 4. Interpret.

25. Suppose that a beer company tries to model the demand for its product by $d(p, x, y) = a - 10p + 5x + y$, where a is a constant (representing those unshakeably loyal to the firm's beer), p is the number of cents in the price of a bottle of beer, x is the number of cents in the average price per bottle of competing beer, and y is the number of cents per bottle spent in aging the beer. Discuss the marginal demand for the beer and the sensitivity of demand to changes in p, x, and y.

26. Suppose in Exercise 25 that the beer company becomes dissatisfied with the simple linear model of demand and changes to a nonlinear model $d(p, x, y) = ap^{-10}x^5 y$, where p, x, and y are as in Exercise 25. Discuss the marginal demand and the sensitivity of the demand to change in p, x, and y.

27. In a certain market situation, the demands for two goods are given by $d_1 = f(x, y)$ and $d_2 = g(x, y)$, where x is the price of the first commodity and y is the price of the second. Assume $\partial d_1/\partial x < 0$ and $\partial d_2/\partial y < 0$. If the goods compete, then what do you expect of the signs of $\partial d_1/\partial y$ and $\partial d_2/\partial x$? Explain.

28. Refer to Exercise 27. Suppose that the two goods complement each other (such as new housing construction and furniture). Again assuming $\partial d_1/\partial x < 0$ and $\partial d_2/\partial y < 0$, what do you expect of the signs of $\partial d_1/\partial y$ and $\partial d_2/\partial x$? Explain.

29. In the economic theory of consumption, the preferences of a household are measured by a *utility function* u: $\mathbf{R}^n \to \mathbf{R}$. The domain of u is the set $D = \{\mathbf{x} = (x_1, \ldots, x_n) | x_i \geq 0$ for $i = 1, 2, \ldots, n\}$ where each x_i represents the quantity of some commodity i (good or service) that can be purchased by the household. A vector $\mathbf{x}$ in D is called a *commodity bundle*, and u is supposed to assign some numerical measure to the attractiveness of commodity bundles to the household—the higher the utility, the more attractive the bundle is. The "Axiom of Greed" assumes that $\partial u/\partial x_i > 0$ for each i. Show that under this assumption, u is an increasing function in each individual variable x_i.

30. Refer to Exercise 29. If $\mathbf{x} = (x_1, \ldots, x_n)$ and $\mathbf{y} = (y_1, \ldots, y_n)$ satisfy $x_i \geq y_i$ for each i and if $\mathbf{x} \neq \mathbf{y}$, then show that $u(\mathbf{x}) > u(\mathbf{y})$. Thus the bigger the commodity bundle, the greater its appeal.

31. Discuss the accuracy of the economic model in Exercises 29 and 30 as a reflection of real life. Is it reasonable to assume that "bigger is always better"? Are there *limits* to greed? Should an axiom be added that u is a *bounded* function?

32. Let $p(x_1, x_2, \ldots, x_n)$ be the production of a certain plant as a function of the number x_1 of workers and other variables. The average **productivity** of a worker is defined to be $p(\mathbf{x})/x_1$. Show that as x_1 increases, the average productivity also increases if the marginal production $\partial p/\partial x_1$ exceeds the average productivity $p(\mathbf{x})/x_1$. In such a situation the plant hires new workers.

33. Refer to Exercise 32. Show that if the marginal productivity $\partial p/\partial x_1 < p(\mathbf{x})/x_1$, then productivity increases as x_1 decreases. In such a situation a plant lays off workers.

34. In determining dosage levels of cortisone prescribed for cortisone-deficient children, endocrinologists use the *body surface area* as a guide. For a child of weight x kilograms and height y centimeters, the approximation $A = 0.007x^{0.425}y^{0.725}$ is used for body surface area in square meters. Compute $\partial A/\partial x$ and $\partial A/\partial y$ when $x = 30$ and $y = 125$. What is the physical significance of these partial derivatives? (If you don't have a calculator with

an x^y key, use the approximations $(30)^{-0.575} \approx 0.141$, $(125)^{-0.275} \approx 0.265$, $(125)^{0.725} \approx 33.133$, and $(30)^{0.425} \approx 4.244$.)

35. Suppose that the model $Y = AK^{0.2}L^{0.8}$ is used for the output Y in billions of dollars per year in the United States, where A is a constant that measures productivity, K is total capital investment in billions of dollars, and L is total workforce in billions of workers. If in 1982, Y was about 3.000×10^3, $K \approx 6.000 \times 10^3$ and $L \approx 0.1000$, then find **(a)** A in 1982 and **(b)** $\partial Y/\partial K$ and $\partial Y/\partial L$ in 1982.

36. Suppose that the total output Y of the American economy in billions of dollars a year is modeled by $Y = A(1 + r)^t K^a L^{1-a}$, where A, K, and L are as in Exercise 35, and r is the annual rate of growth of productivity. If marginal output $\partial Y/\partial t$ is $0.03Y$, $\partial K/\partial t = 0.04K$, $\partial L/\partial t = 0.01L$, and $A = 0.2$, then find a formula for $\partial r/\partial t$, the marginal productivity growth rate.

12.2 Differentiable Functions

As suggested after Definition 1.2, the multidimensional analogue of the derivative $f'(a)$ of a function of one variable is obtained by putting the partial derivatives of a scalar function into a vector. This section describes how that is done. The first step is to define the *total derivative* of a function $f: \mathbf{R}^n \to \mathbf{R}$, where $n = 2$ or 3, at a point $\mathbf{x} = \mathbf{a} = (a_1, a_2, a_3)$ in $\mathbf{R}^n$. (If $n = 2$, then a_3 is not present.) After defining the total derivative, we will see that it is calculated from the partial derivatives by arraying them in a vector. (The same is true for $n > 3$, but we leave the more general discussion for later courses.)

It is not possible to extend Definition 1.1 of Chapter 2 to a scalar function of more than one variable. In the two-dimensional case, suppose we try to define the derivative of $f(x, y)$ at $(x, y) = (a, b)$ as

$$\lim_{\mathbf{h} \to \mathbf{0}} \frac{f(\mathbf{a} + \mathbf{h}) - f(\mathbf{a})}{|\mathbf{h}|}, \tag{1}$$

where $\mathbf{a} = (a, b)$ and $\mathbf{h} = (h, k)$. This looks like the closest possible analogue of the limit of the ordinary difference quotient: Since we can't divide the numerator $f(\mathbf{a} + \mathbf{h}) - f(\mathbf{a})$—a real number—by the vector $\mathbf{h}$, we divide by the *length* of the vector instead. The problem is that the limit in (1) is not unique: It usually depends on the *direction* of approach of $\mathbf{h}$ to $\mathbf{0}$. Consider, for instance, the simple function $f(x, y) = x$. As $\mathbf{h} \to \mathbf{0}$ along the positive x-axis, that is, for $\mathbf{h} = (h, 0)$ where $h > 0$, we have

$$\frac{f(\mathbf{0} + \mathbf{h}) - f(\mathbf{0})}{|h|} = \frac{f(0 + h, 0 + 0) - f(0, 0)}{h} = \frac{h}{h} \to 1.$$

But as $h \to 0$ along the negative x-axis, that is, for $h = (h, 0)$ where $h < 0$, we have

$$\frac{f(\mathbf{0} + \mathbf{h}) - f(\mathbf{0})}{|h|} = \frac{f(0 + h, 0 + 0)}{-h} = \frac{h}{-h} \to -1.$$

This is enough to show that the limit in (1) fails to exist, since we don't get the same limit L for all paths of approach of $\mathbf{h}$ to $\mathbf{0}$ (cf. the remarks following Definition 4.2 of the last chapter).

Since we can't extend Definition 1.1 of Chapter 2 to functions of several variables, we instead focus on the *equivalent* condition of Theorem 5.3 of Chapter 2. That result states that $f: \mathbf{R} \to \mathbf{R}$ is differentiable at $x = a$ if and only if there is a real number m such that

$$\lim_{x \to a} \frac{f(x) - f(a) - m \cdot (x - a)}{x - a} = 0. \tag{2}$$

It further says that m is really $f'(a)$. We *can* extend (2) to functions of several variables, by requiring that the analogous limit be 0 as $\mathbf{x} \to \mathbf{a}$.

2.1 DEFINITION Suppose that the scalar function f of n (usually 2 or 3) real variables is defined on an open ball containing $\mathbf{a} \in \mathbf{R}^n$. Then f is ***differentiable*** at $\mathbf{x} = \mathbf{a}$ if and only if there is a vector $\mathbf{m}$ in $\mathbf{R}^n$ such that

$$\lim_{\mathbf{x}\to\mathbf{a}} \frac{f(\mathbf{x}) - f(\mathbf{a}) - \mathbf{m}\cdot(\mathbf{x}-\mathbf{a})}{|\mathbf{x}-\mathbf{a}|} = 0. \tag{3}$$

If f is differentiable at $\mathbf{x} = \mathbf{a}$, then the vector $\mathbf{m}$ is called the (***total***) ***derivative*** of f at $\mathbf{x} = \mathbf{a}$.

Condition (3) is the closest possible n-dimensional analogue of (2). In the numerator we take the *dot product* of the vectors $\mathbf{m}$ and $\mathbf{x} - \mathbf{a}$ to get a real number that can be subtracted from $f(\mathbf{x}) - f(\mathbf{a})$. In the denominator we use the length of the vector $\mathbf{x} - \mathbf{a}$ to make division possible.

As mentioned on p. 749, in the two-dimensional case we expect the derivative $\mathbf{m}$ of $f(x, y)$ at (a, b) to be the vector

$$\left(\frac{\partial f}{\partial x}(a, b), \frac{\partial f}{\partial y}(a, b)\right)$$

of partial derivatives. If it is, then the equation of the tangent plane to $z = f(x, y)$ at the point $(a, b, f(a, b))$,

$$z = f(a, b) = \left(\frac{\partial f}{\partial x}(a, b), \frac{\partial f}{\partial y}(a, b)\right)\cdot(x - a, y - b),$$

exactly parallels the equation

$$y - f(a) = f'(a)(x - a)$$

for the tangent line to the graph of $y = f(x)$ at the point $(a, f(a))$. The first fact we prove about a differentiable function f of 2 or 3 variables is that $\mathbf{m}$ in Definition 2.1 *is* the vector of partial derivatives of f at $\mathbf{x} = \mathbf{a}$. (In particular, then, when $\mathbf{m}$ exists, it is *unique*.)

2.2 THEOREM Suppose that $f\colon \mathbf{R}^n \to \mathbf{R}$ is differentiable at $\mathbf{x} = \mathbf{a}$ in $\mathbf{R}^n$, where $n = 2$ or 3. Then the vector $\mathbf{m}$ in Definition 2.1 is

$$\mathbf{m} = \frac{\partial f}{\partial x}(\mathbf{a})\,\mathbf{i} + \frac{\partial f}{\partial y}(\mathbf{a})\,\mathbf{j} + \frac{\partial f}{\partial z}(\mathbf{a})\,\mathbf{k}, \tag{4}$$

where the $\mathbf{k}$-component is not present if $n = 2$.

Proof. We give the proof in case $n = 2$. (The case $n = 3$ is Exercise 38.) Then we have $\mathbf{m} = m_1\mathbf{i} + m_2\mathbf{j}$ for some real numbers m_1 and m_2. We need to show that

$$m_1 = \frac{\partial f}{\partial x}(a, b) \qquad \text{and} \qquad m_2 = \frac{\partial f}{\partial y}(a, b).$$

To do so, let $\mathbf{x} = (x, b)$. That is, we freeze y to its value at $\mathbf{a} = (a, b)$ and let x vary. Then $\mathbf{x} - \mathbf{a} = (x - a, 0) = (x - a)\mathbf{i}$ and $\mathbf{x} \to \mathbf{a}$ is equivalent to $x \to a$.

Therefore (3) says that

$$0 = \lim_{x \to a} \frac{f(x, b) - f(a, b) - (m_1 \mathbf{i} + m_2 \mathbf{j}) \cdot (x - a)\mathbf{i}}{|(x - a)\mathbf{i}|}$$

$$= \lim_{x \to a} \frac{f(x, b) - f(a, b) - m_1(x - a)}{|x - a|}$$

$$= \lim_{x \to a} \frac{f(x, b) - f(a, b) - m_1(x - a)}{\pm(x - a)} = \pm \lim_{x \to a} \left[\frac{f(x, b) - f(a, b)}{x - a} - m_1 \right].$$

Since the limit is 0, the $\pm$ can be dropped. That gives

$$(5) \qquad 0 = \lim_{x \to a} \frac{f(x, b) - f(a, b)}{x - a} - m_1 = \frac{\partial f}{\partial x}(a, b) - m_1,$$

because writing h for $x - a$, we have $a + h = x$, so

$$\frac{\partial f}{\partial x}(a, b) = \lim_{h \to 0} \frac{f(a + h, b) - f(a, b)}{h} = \lim_{x \to a} \frac{f(x, b) - f(a, b)}{x - a}.$$

Addition of m_1 to each side of (5) gives

$$m_1 = \frac{\partial f}{\partial x}(a, b),$$

as required. The same reasoning on the second coordinate shows (Exercise 39) that $m_2 = \dfrac{\partial f}{\partial y}(a, b)$. QED

Because of Theorem 2.2, we introduce the notation

$$(6) \qquad \mathbf{f}'(\mathbf{a}) = \left(\frac{\partial f}{\partial x}(\mathbf{a}), \frac{\partial f}{\partial y}(\mathbf{a}), \frac{\partial f}{\partial z}(\mathbf{a}) \right) = \frac{\partial f}{\partial x}(\mathbf{a})\,\mathbf{i} + \frac{\partial f}{\partial y}(\mathbf{a})\,\mathbf{j} + \frac{\partial f}{\partial z}(\mathbf{a})\,\mathbf{k}$$

for the total derivative of $f\colon \mathbf{R}^n \to \mathbf{R}$ at $\mathbf{x} = \mathbf{a}$. That is, the total derivative of a differentiable scalar function of n variables is the n-dimensional vector whose coordinates are the respective partial derivatives of f. Note that *although* $f(\mathbf{a})$ *is a scalar, the derivative* $\mathbf{f}'(\mathbf{a})$ *is a vector* and so is printed in boldface. (If $n = 2$, the last coordinate, the $\mathbf{k}$ component, is not present.)

Example 1

Show that the function f defined by $f(x, y) = x^2 + y^2$ is differentiable at $\mathbf{a} = (2, 3)$ and find $\mathbf{f}'(\mathbf{a})$.

Solution. Because of Theorem 2.2, there is only one candidate for $\mathbf{f}'(\mathbf{a})$, namely,

$$\frac{\partial f}{\partial x}(\mathbf{a})\,\mathbf{i} + \frac{\partial f}{\partial y}(\mathbf{a})\,\mathbf{j} = 2x\,\mathbf{i} + 2y\,\mathbf{j}\Big|_{(2,3)} = 4\,\mathbf{i} + 6\,\mathbf{j}.$$

We have to show then that $\mathbf{m} = 4\,\mathbf{i} + 6\,\mathbf{j}$ satisfies Definition 2.1. We have

$$f(\mathbf{a}) = f(2, 3) = 2^2 + 3^2 = 13$$

and

$$\mathbf{x} - \mathbf{a} = (x, y) - (2, 3) = (x - 2, y - 3) = (x - 2)\,\mathbf{i} + (y - 3)\,\mathbf{j}.$$

So, computing the limit in (3), we get

$$\begin{aligned}
\lim_{(x,y)\to(2,3)} &= \frac{x^2 + y^2 - 13 - (4\mathbf{i} + 6\mathbf{j}) \cdot [(x-2)\mathbf{i} + (y-3)\mathbf{j}]}{|(x, y) - (2, 3)|} \\
&= \lim_{(x,y)\to(2,3)} \frac{x^2 + y^2 - 13 - 4x + 8 - 6y + 18}{\sqrt{(x-2)^2 + (y-3)^2}} \\
&= \lim_{(x,y)\to(2,3)} \frac{x^2 - 4x + y^2 - 6y + 13}{\sqrt{(x-2)^2 + (y-3)^2}} \\
&= \lim_{(x,y)\to(2,3)} \frac{x^2 - 4x + 4 + y^2 - 6y + 9}{\sqrt{(x-2)^2 + (y-3)^2}} \\
&= \lim_{(x,y)\to(2,3)} \frac{(x-2)^2 + (y-3)^2}{\sqrt{(x-2)^2 + (y-3)^2}} \\
&= \lim_{(x,y)\to(2,3)} \sqrt{(x-2)^2 + (y-3)^2} = 0.
\end{aligned}$$

This shows that f is differentiable at $\mathbf{a} = (2, 3)$, and so $\mathbf{f}'(2, 3) = 4\mathbf{i} + 6\mathbf{j}$. ■

We were fortunate in Example 1 that the limit in (3) of Definition 2.1 worked out easily. In general, Definition 2.1 is more useful in theoretical rather than computational work. A good illustration of its theoretical value is provided by the following n-dimensional version of Theorem 1.3 of Chapter 2.

2.3 THEOREM If the scalar function f is differentiable at $\mathbf{x} = \mathbf{a}$, then it is continuous at $\mathbf{x} = \mathbf{a}$.

Proof. We have to show that $\lim_{\mathbf{x}\to\mathbf{a}} f(\mathbf{x}) = f(\mathbf{a})$, which is equivalent to

$$\lim_{\mathbf{x}\to\mathbf{a}} [f(\mathbf{x}) - f(\mathbf{a})] = 0.$$

Since f is differentiable at $\mathbf{x} = \mathbf{a}$, Definition 2.1 says that

$$\lim_{\mathbf{x}\to\mathbf{a}} \frac{f(\mathbf{x}) - f(\mathbf{a}) - \mathbf{m} \cdot (\mathbf{x} - \mathbf{a})}{|\mathbf{x} - \mathbf{a}|} = 0.$$

The only way that the limit of a quotient of two functions can be 0 is for the numerator to have limit 0. We therefore have

$$\lim_{\mathbf{x}\to\mathbf{a}} [f(\mathbf{x}) - f(\mathbf{a}) - \mathbf{m} \cdot (\mathbf{x} - \mathbf{a})] = 0. \tag{7}$$

Now,

$$g(\mathbf{x}) = \mathbf{m} \cdot (\mathbf{x} - \mathbf{a}) = m_1(x - a_1) + m_2(y - a_2) + m_3(z - a_3)$$

is a first-degree polynomial where if $n = 2$, the last term does not appear. The function g is continuous at $\mathbf{x} = \mathbf{a}$ (see the remark following Theorem 4.5 of the last chapter), so we have

$$\lim_{\mathbf{x}\to\mathbf{a}} \mathbf{m} \cdot (\mathbf{x} - \mathbf{a}) = \mathbf{m} \cdot (\mathbf{a} - \mathbf{a}) = \mathbf{m} \cdot \mathbf{0} = 0.$$

We therefore get from (7)

$$\begin{aligned}
0 &= \lim_{\mathbf{x}\to\mathbf{a}} [f(\mathbf{x}) - f(\mathbf{a}) - \mathbf{m} \cdot (\mathbf{x} - \mathbf{a})] = \lim_{\mathbf{x}\to\mathbf{a}} [f(\mathbf{x}) - f(\mathbf{a})] + \lim_{\mathbf{x}\to\mathbf{a}} [\mathbf{m} \cdot (\mathbf{x} - \mathbf{a})] \\
&= \lim_{\mathbf{x}\to\mathbf{a}} [f(\mathbf{x}) - f(\mathbf{a})].
\end{aligned}$$

QED

Since it can be a chore to use Definition 2.1, it would be nice if all we had to do to find $\mathbf{f}'(\mathbf{a})$ would be to compute the vector of partial derivatives

$$\left(\frac{\partial f}{\partial x}(\mathbf{a}), \frac{\partial f}{\partial y}(\mathbf{a}), \frac{\partial f}{\partial z}(\mathbf{a})\right) = \frac{\partial f}{\partial x}(\mathbf{a})\,\mathbf{i} + \frac{\partial f}{\partial y}(\mathbf{a})\,\mathbf{j} + \frac{\partial f}{\partial z}(\mathbf{a})\,\mathbf{k}. \tag{8}$$

Unfortunately, even though (8) is the only *candidate* for $\mathbf{f}'(\mathbf{a})$, it can happen that (8) may exist but still not be $\mathbf{f}'(\mathbf{a})$. The problem is that a scalar function f can be nondifferentiable at $\mathbf{x} = \mathbf{a}$ and still have a partial derivative at $\mathbf{x} = \mathbf{a}$ with respect to each of its variables. This is somewhat akin to continuity in each separate variable not being enough to ensure continuity of a scalar function (see Exercises 29 and 30 of Exercises 11.4.) The next example is one in which (8) exists but f is not differentiable.

Example 2

Determine whether the function

$$f(x, y) = \begin{cases} \dfrac{xy}{x^2 + y^2}, & \text{if } (x, y) \neq (0, 0), \\ 0, & \text{if } (x, y) = (0, 0), \end{cases}$$

is differentiable at $(x, y) = (0, 0)$.

Solution. To determine

$$\frac{\partial f}{\partial x}(0, 0) \qquad \text{and} \qquad \frac{\partial f}{\partial y}(0, 0),$$

we freeze either x or y to its value at $(0, 0)$. Then $f(x, y)$ is 0 for all values of the other variable and so in each case f has partial derivative 0. That is, $f_x(0, 0) = f_y(0, 0) = 0$. Thus there *is* a candidate for $\mathbf{f}'(0, 0)$, namely, $\mathbf{0} = 0\,\mathbf{i} + 0\,\mathbf{j}$. But f is *not* differentiable at $(x, y) = (0, 0)$, because it is not continuous there. This can be seen as in Example 2 of Section 11.4: If $(x, y) \to (0, 0)$ along the line $y = x$, then

$$f(x, y) = \frac{x^2}{x^2 + y^2} \to \frac{1}{2} \neq 0 = f(0, 0).$$

Theorem 2.3 says that if f were differentiable at $(0, 0)$, then it would have to be continuous there. Therefore f is not differentiable at $(0, 0)$. (It can also be shown directly that Definition 2.1 fails for $\mathbf{m} = \mathbf{0}$. See Exercise 40.) ■

Because of the kind of behavior exhibited by the function in Example 2, there is a special name and notation given to the vector quantity (8).

2.4 DEFINITION Suppose that $n = 2$ or 3 and the scalar function $f\colon \mathbf{R}^n \to \mathbf{R}$ has a partial derivative with respect to each of the variables x, y, and (if present) z at $\mathbf{x} = \mathbf{a}$. Then the ***gradient*** of f at $\mathbf{x} = \mathbf{a}$ is

$$\nabla f(\mathbf{a}) = \left(\frac{\partial f}{\partial x}(\mathbf{a}), \frac{\partial f}{\partial y}(\mathbf{a}), \frac{\partial f}{\partial z}(\mathbf{a})\right) = \frac{\partial f}{\partial x}(\mathbf{a})\,\mathbf{i} + \frac{\partial f}{\partial y}(\mathbf{a})\,\mathbf{j} + \frac{\partial f}{\partial z}(\mathbf{a})\,\mathbf{k}.$$

(The symbol ∇ is read *del*. If $n = 2$, the last coordinate is not present.)

Example 3

Find the gradient of the function f of Example 2 at any point (x, y) in $\mathbf{R}^2$.

Solution. The work in Example 2 shows that $\nabla f(0, 0) = 0\mathbf{i} + 0\mathbf{j} = \mathbf{0}$. For any point other than (0, 0), we can calculate

$$\frac{\partial f}{\partial x} = \frac{(x^2 + y^2)y - xy(2x)}{(x^2 + y^2)^2} = \frac{y^3 - x^2y}{(x^2 + y^2)^2}, \qquad \frac{\partial f}{\partial y} = \frac{x^3 - xy^2}{(x^2 + y^2)^2}.$$

[We can get $\partial f/\partial y$ by simply exchanging x and y in the formula for $\partial f/\partial x$, since x and y occur symmetrically in the formula for $f(x, y)$.] Thus for $(x, y) \neq (0, 0)$ we have

$$\nabla f(x, y) = \frac{y^3 - x^2y}{(x^2 + y^2)^2}\mathbf{i} + \frac{x^3 - xy^2}{(x^2 + y^2)^2}\mathbf{j} = \frac{1}{(x^2 + y^2)^2}[(y^3 - x^2y)\mathbf{i} + (x^3 - xy^2)\mathbf{j}].$$ ■

Notice in Example 3 that both $\partial f/\partial x$ and $\partial f/\partial y$ are discontinuous at (0, 0), since neither has a limit at (0, 0). As $(x, y) \to (0, 0)$ along the line $y = 2x$, for instance, we see that

$$f_x(x, y) = \frac{8x^3 - 2x^3}{(x^2 + 4x^2)^2} = \frac{6x^3}{25x^4} = \frac{6}{25x}$$

has no limit. Similar remarks apply to $f_y(x, y)$. The next result says that when all the entries of $\nabla f(x, y)$ *are* continuous at $\mathbf{x} = \mathbf{a}$, then f is differentiable at $\mathbf{x} = \mathbf{a}$, and so by Theorem 2.2, $\nabla f(\mathbf{a})$ is $\mathbf{f}'(\mathbf{a})$. (This serves to explain why we required $\partial f/\partial x$ and $\partial f/\partial y$ to be continuous in Definition 1.2.) The proof can be found at the end of the section.

2.5 THEOREM

> Suppose that the scalar function f: $\mathbf{R}^n \to \mathbf{R}$ is defined on an open ball B containing $\mathbf{x} = \mathbf{a}$. If all the partial derivatives of f exist on B and are continuous at $\mathbf{x} = \mathbf{a}$, then f is differentiable at $\mathbf{x} = \mathbf{a}$, and so $\nabla f(\mathbf{a})$ is the derivative $\mathbf{f}'(\mathbf{a})$ of f at $\mathbf{x} = \mathbf{a}$.

Theorem 2.5 makes it possible to bypass Definition 2.1 in determining whether most functions f of several variables are differentiable at a point $\mathbf{x} = \mathbf{a}$. We just calculate $\nabla f(x, y, z)$ and check that all its entries are continuous at $\mathbf{x} = \mathbf{a}$. If, as usually happens, they *are* all continuous there, then f is differentiable at $\mathbf{x} = \mathbf{a}$, and so $\mathbf{f}'(\mathbf{a}) = \nabla f(\mathbf{a})$. If any of the partial derivatives $\partial f/\partial x_i$ is discontinuous at $\mathbf{x} = \mathbf{a}$, then we have to take a close look at Definition 2.1 or Theorem 2.2, as in Example 2, to determine whether f is differentiable at $\mathbf{x} = \mathbf{a}$.

Example 4

Determine whether $f(x, y, z) = \ln(x + y + z) + e^{x^2+y^2+z^2}$ is differentiable at (1, 2, 2). If it is, find $\mathbf{f}'(1, 2, 2)$.

Solution. To find ∇f we calculate

$$\frac{\partial f}{\partial x} = \frac{1}{x + y + z} + 2xe^{x^2+y^2+z^2}, \qquad \frac{\partial f}{\partial y} = \frac{1}{x + y + z} + 2ye^{x^2+y^2+z^2},$$

$$\frac{\partial f}{\partial z} = \frac{1}{x + y + z} + 2ze^{x^2+y^2+z^2}.$$

Each of these functions is continuous on the entire first octant set of points (x, y, z) with $x, y, z > 0$. So, in particular, each one is continuous at (1, 2, 2). Therefore

$$\mathbf{f}'(1, 2, 2) = \nabla f(1, 2, 2) = \left(\frac{1}{5} + 2e^9\right)\mathbf{i} + \left(\frac{1}{5} + 4e^9\right)\mathbf{j} + \left(\frac{1}{5} + 4e^9\right)\mathbf{k}.$$ ■

FIGURE 2.1

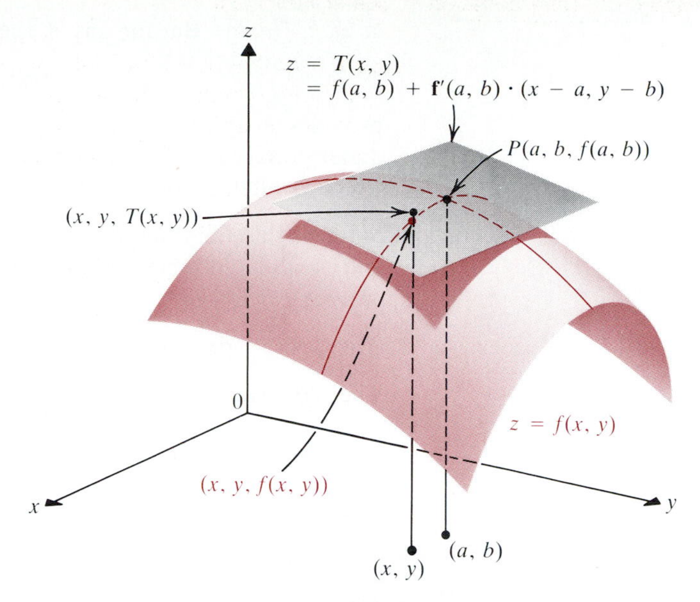

When a scalar function f is differentiable at $\mathbf{x} = \mathbf{a}$, we can use this fact to approximate $f(\mathbf{x})$ near $\mathbf{x} = \mathbf{a}$ by the ***tangent approximation*** $T(\mathbf{x})$, which is given by

$$T(\mathbf{x}) = f(\mathbf{a}) + \mathbf{f}'(\mathbf{a}) \cdot (\mathbf{x} - \mathbf{a}), \tag{9}$$

the multidimensional analogue of the tangent approximation of a function of one variable. In the case of a differentiable function of two variables, we can draw the tangent plane

$$\begin{aligned} z &= f(a, b) + \left(\frac{\partial f}{\partial x}(a, b), \frac{\partial f}{\partial y}(a, b)\right) \cdot (x - a, y - b) \\ &= f(\mathbf{a}) + \mathbf{f}'(\mathbf{a}) \cdot (\mathbf{x} - \mathbf{a}) = T(x, y) \end{aligned}$$

to the graph of $z = f(x, y)$ at the point $P(a, b, f(a, b))$. As **Figure 2.1** suggests, for (x, y) near (a, b), $T(x, y)$ is a close approximation to $f(x, y)$.

Example 5

For the function $f(x, y) = x^2 + y^2$, compare $f(x, y)$ at the point $(x, y) = (2.01, 2.98)$ to the tangent approximation (9), where $\mathbf{a} = (2, 3)$.

Solution. From Example 1, $\mathbf{f}'(2, 3) = 2x\,\mathbf{i} + 2y\,\mathbf{j}|_{(2,3)} = 4\,\mathbf{i} + 6\,\mathbf{j}$. Thus

$$\begin{aligned} T(x, y) &= f(2, 3) + \mathbf{f}'(2, 3) \cdot [(x - 2)\,\mathbf{i} + (y - 3)\,\mathbf{j}] \\ &= 13 + (4\,\mathbf{i} + 6\,\mathbf{j}) \cdot [(x - 2)\,\mathbf{i} + (y - 3)\,\mathbf{j}] = 13 + 4(x - 2) + 6(y - 3). \end{aligned}$$

Therefore

$$T(2.01, 2.98) = 13 + 4(0.01) + 6(-0.02) = 13 + 0.04 - 0.12 = 12.92.$$

Direct calculation gives

$$f(2.01, 2.98) = (2.01)^2 + (2.98)^2 = 12.9205.$$

Thus

$$f(x, y) - T(x, y) = 0.0005,$$

so the two numbers differ by about 0.00387% of $f(x, y)$. ■

We can draw the graph of the tangent approximation only in the case of $f\colon \mathbf{R}^2 \to \mathbf{R}$, when it coincides with the tangent plane to the graph of $z = f(x, y)$

at $(a, b, f(a, b))$. But for any n, the linear function

$$f(\mathbf{a}) + \mathbf{f}'(\mathbf{a}) \cdot (\mathbf{x} - \mathbf{a})$$

provides an approximation to $f(\mathbf{x}) = f(x_1, x_2, \ldots, x_n)$ near $\mathbf{x} = \mathbf{a}$. As in the one-dimensional case (Definition 5.1 of Chapter 2), we introduce *differential* notation, which is often used in applications involving the tangent approximation.

2.6 DEFINITION

Suppose that the scalar function $f: \mathbf{R}^n \to \mathbf{R}$ is defined on an open ball $B = B(\mathbf{a}, \delta)$ and is differentiable at $\mathbf{x} = \mathbf{a} = (a_1, a_2, a_3)$. The **(*total*) *differential of*** $\mathbf{x} = (x, y, z) \in B$ is

(10) $$\mathbf{dx} = \mathbf{x} - \mathbf{a} = (x - a_1, y - a_2, z - a_3) = (dx, dy, dz).$$

The ***differential of f*** **near** $\mathbf{x} = \mathbf{a}$ is

$$df = \nabla f(\mathbf{a}) \cdot \mathbf{dx} = \left(\frac{\partial f}{\partial x}(\mathbf{a})\,\mathbf{i} + \frac{\partial f}{\partial y}(\mathbf{a})\,\mathbf{j} + \frac{\partial f}{\partial z}(\mathbf{a})\,\mathbf{k}\right) \cdot (dx\,\mathbf{i} + dy\,\mathbf{j} + dz\,\mathbf{k}),$$

(11) $$df = \frac{\partial f}{\partial x}(\mathbf{a})\,dx + \frac{\partial f}{\partial y}(\mathbf{a})\,dy + \frac{\partial f}{\partial z}(\mathbf{a})\,dz.$$

(If $n = 2$, then the third coordinates are not present.)

With this notation, Formula (9) for the tangent approximation becomes

(12) $$T(\mathbf{x}) = f(\mathbf{a}) + \mathbf{f}'(\mathbf{a}) \cdot \mathbf{dx} = f(\mathbf{a}) + df.$$

If we use (10) and (11) in Example 5, where $\mathbf{a} = (2, 3)$, the calculation of $T(x, y)$ takes the form

$$\begin{aligned} T(x, y) &= f(2, 3) + f'(2, 3) \cdot (dx\,\mathbf{i} + dy\,\mathbf{j}) \\ &= 13 + (4\,\mathbf{i} + 6\,\mathbf{j}) \cdot (dx\,\mathbf{i} + dy\,\mathbf{j}) \\ &= 13 + 4\,dx + 6\,dy, \end{aligned}$$

where $dx = x - 2$ and $dy = y - 3$.

Example 5 and our experience with functions of one variable suggest that the tangent approximation $T(\mathbf{x})$ should closely approximate $f(\mathbf{x})$ near $\mathbf{x} = \mathbf{a}$. As yet, however, we have no upper bound on the error

(13) $$|e(\mathbf{x})| = |f(\mathbf{x}) - T(\mathbf{x})| = |f(\mathbf{x}) - f(\mathbf{a}) - \mathbf{f}'(\mathbf{a}) \cdot (\mathbf{x} - \mathbf{a})| = |f(\mathbf{x}) - f(\mathbf{a}) - df|$$

made in approximating $f(\mathbf{x})$ by $T(\mathbf{x})$. As in the case of a function of one variable, such a bound is given in terms of higher derivatives, so we will come back to it in Section 6. In the meantime, we assume that (3) ensures that $T(\mathbf{x})$ is very close to $f(\mathbf{x})$ as long as $\mathbf{x}$ is close to $\mathbf{a}$. One says that

$$e(\mathbf{x}) = f(\mathbf{x}) - T(\mathbf{x})$$

has a *smaller order of magnitude* than $d\mathbf{x} = \mathbf{x} - \mathbf{a}$ because of (3).

Our final example illustrates how the tangent approximation (9) is used in applied work. Recall that Boyle's law for an ideal gas of constant mass maintained at a constant temperature is

$$PV = c$$

where P is the pressure (in atmospheres, say), and V is the volume (in cubic centimeters). The constant c can't be measured directly but is computed from measured values of P and V, which are subject to error that is in turn propagated to c when the values are multiplied. The tangent approximation provides a convenient way to estimate the error in c if the accuracy of the measurements of P and V is known.

Example 6

In a certain gas cylinder, the pressure is measured to be $P_0 = 10.0$ atmospheres, where the measurement is accurate to within 2%. The cylinder is manufactured to have interior volume $V_0 = 400 \text{ cm}^3$, subject to an error of 0.5%. Use the tangent approximation to estimate the accuracy of the value of c calculated from this information.

Solution. The calculated value of c is

$$c(P_0, V_0) = P_0 V_0 = (10.0)(400) = 4000 \text{ atm}\cdot\text{cm}^3$$

The maximum possible error in $P_0 = 10.0$ is 2%. That is, if P is the actual pressure, then the largest possible value of $dP = P - P_0$ is

$$dP = (0.02)(10.0) = 0.200 \text{ atm}.$$

Similarly, the maximum possible error in $V_0 = 400$ is 0.5%. Thus, if V is the actual volume, then the largest possible value of $dV = V - V_0$ is

$$dV = (0.005)(400) = 2.00 \text{ cm}^3.$$

Since $c = PV$, we regard c as a function of P and V. Then we have

$$\frac{\partial c}{\partial P} = V \quad \text{and} \quad \frac{\partial c}{\partial V} = P.$$

Hence, at $(P_0, V_0) = (10.0, 400)$,

$$\nabla c = V_0\mathbf{i} + P_0\mathbf{j} = 400\,\mathbf{i} + 10.0\,\mathbf{j}.$$

The tangent approximation for c is

$$\begin{aligned} c(P, V) \approx T(P, V) &= c(P_0, V_0) + \left(\frac{\partial c}{\partial P}(P_0, V_0)\mathbf{i} + \frac{\partial c}{\partial V}(P_0, V_0)\mathbf{j}\right)\cdot(dP\,\mathbf{i} + dV\,\mathbf{j}) \\ &= 4000 + (400\,\mathbf{i} + 10.0\,\mathbf{j})\cdot(dP\,\mathbf{i} + dV\,\mathbf{j}) \\ &= 4000 + 400\,dP + 10.0\,dV. \end{aligned}$$

The maximum error $|c(P, V) - c(P_0, V_0)|$ in the calculated value of c from P_0 and V_0 can then be approximated as follows.

$$\begin{aligned} |c(P, V) - c(P_0, V_0)| &\approx |T(P, V) - 4000| = |400\,dP + 10.0\,dV| \\ &\le |(400)(0.200)| + |(10.0)(2.00)| = 80.0 + 20.0 = 100.0. \end{aligned}$$

Consequently, an upper bound for the error in c is 100. So we estimate that c lies in the interval [3900, 4100]. The maximum *percent error* in c is therefore about $100/4000 = 2.5\%$, the *sum* of the individual maximum errors in P and V. ■

Proof of Theorem 2.5*

2.5 THEOREM If all the partial derivatives $\partial f/\partial x_i$, $i = 1, 2, \ldots, n$ of a scalar function f are defined on an open ball B containing $\mathbf{x} = \mathbf{a}$ and are continuous at $\mathbf{x} = \mathbf{a}$, then f is differentiable at $\mathbf{x} = \mathbf{a}$.

Proof. We give the proof for a scalar function f of two variables. The same reasoning works for $n = 3$ (see Exercise 41). Let $\mathbf{x} = (x, y)$ and $\mathbf{a} = (a, b)$ both be in the open ball B. To show that f is differentiable at (a, b), we must show that

* Optional

(3) in Definition 2.1 holds for

$$\mathbf{m} = \nabla f(a, b) = \left(\frac{\partial f}{\partial x}(a, b), \frac{\partial f}{\partial y}(a, b)\right) = \frac{\partial f}{\partial x}(a, b)\mathbf{i} + \frac{\partial f}{\partial y}(a, b)\mathbf{j}.$$

Before proceeding to (3), we rewrite the first two terms of the numerator

$$e(\mathbf{x}) = f(\mathbf{x}) - T(\mathbf{x}) = f(\mathbf{x}) - f(\mathbf{a}) - \nabla f(\mathbf{a}) \cdot (\mathbf{x} - \mathbf{a}) \tag{14}$$

as

$$f(\mathbf{x}) - f(\mathbf{a}) = [f(x, y) - f(x, b)] + [f(x, b) - f(a, b)],$$

by subtracting and adding back $f(x, b)$. Each term in brackets is the difference between two real-valued functions of a *single* real variable: y in the first bracket, x in the second. We can then apply the mean value theorem (Theorem 10.2 of Chapter 2) to each of these functions of one variable, because each one has a partial derivative on B. There is then some point y_1 between b and y such that

$$f(x, y) - f(x, b) = \frac{\partial f}{\partial y}(x, y_1)(y - b),$$

and also some point x_1 between a and x such that

$$f(x, b) - f(a, b) = \frac{\partial f}{\partial x}(x_1, b)(x - a).$$

Substitution of these two equations into (14) makes the numerator of (3)

$$\begin{aligned} e(\mathbf{x}) &= f(x, y) - f(a, b) - \left[\frac{\partial f}{\partial x}(a, b)\mathbf{i} + \frac{\partial f}{\partial y}(a, b)\mathbf{j}\right] \cdot [(x - a)\mathbf{i} + (y - b)\mathbf{j}] \\ &= \left[\frac{\partial f}{\partial y}(x, y_1)(y - b) + \frac{\partial f}{\partial x}(x_1, b)(x - a)\right] \\ &\quad - \left[\frac{\partial f}{\partial x}(a, b)(x - a) + \frac{\partial f}{\partial y}(a, b)(y - b)\right] \\ &= (x - a)\left[\frac{\partial f}{\partial x}(x_1, b) - \frac{\partial f}{\partial x}(a, b)\right] + (y - b)\left[\frac{\partial f}{\partial y}(x, y_1) - \frac{\partial f}{\partial y}(a, b)\right], \end{aligned}$$

The triangle inequality (Theorem 3.5 of Chapter 10) then gives

$$\begin{aligned} |e(\mathbf{x})| &\le |x - a|\left|\frac{\partial f}{\partial x}(x_1, b) - \frac{\partial f}{\partial x}(a, b)\right| + |y - b|\left|\frac{\partial f}{\partial y}(x, y_1) - \frac{\partial f}{\partial y}(a, b)\right| \\ &\le |\mathbf{x} - \mathbf{a}|\left[\left|\frac{\partial f}{\partial x}(x_1, b) - \frac{\partial f}{\partial x}(a, b)\right| + \left|\frac{\partial f}{\partial y}(x, y_1) - \frac{\partial f}{\partial y}(a, b)\right|\right], \end{aligned}$$

since $|\mathbf{x} - \mathbf{a}| = \sqrt{(x - a)^2 + (y - b)^2}$ is greater than or equal to $\sqrt{(x - a)^2} = |x - a|$ and $\sqrt{(y - b)^2} = |y - b|$. Thus, dividing by $|\mathbf{x} - \mathbf{a}|$, we have

$$\frac{|e(\mathbf{x})|}{|\mathbf{x} - \mathbf{a}|} \le \left|\frac{\partial f}{\partial x}(x_1, b) - \frac{\partial f}{\partial x}(a, b)\right| + \left|\frac{\partial f}{\partial y}(x, y_1) - \frac{\partial f}{\partial y}(a, b)\right|. \tag{15}$$

As $\mathbf{x} \to \mathbf{a}$, each absolute value on the right side of (15) approaches 0, because $\partial f/\partial x$ and $\partial f/\partial y$ are continuous at $\mathbf{x} = \mathbf{a}$. Thus we have

$$\lim_{\mathbf{x} \to \mathbf{a}} \frac{|e(\mathbf{x})|}{|\mathbf{x} - \mathbf{a}|} = 0 \to \lim_{\mathbf{x} \to \mathbf{a}} \frac{e(\mathbf{x})}{|\mathbf{x} - \mathbf{a}|} = 0.$$

Therefore (3) holds, and so f is differentiable at $\mathbf{x} = \mathbf{a}$. QED

> **HISTORICAL NOTE**
>
> The idea of partial differentiation goes back to the origins of calculus three centuries ago. However, the notions of differentiability, the total derivative, and the total differential belong to this century. The total derivative introduced in Definition 2.1 was defined in 1908 by the English mathematician William H. Young (1863–1942), head of a family of prominent English and American mathematicians of this century. He applied the theory of vectors of Hamilton and Grassmann and the linear algebra developed in the last half of the nineteenth century by the English mathematicians Arthur Cayley (1821–1895) and James J. Sylvester (1814–1897). The notion of differentiability given in Definition 2.1 has been used for more general functions than those from $\mathbf{R}^n$ to $\mathbf{R}$ and is one of the fundamental concepts of modern analysis and differential geometry. It has thus turned out that the real essence of differentiation lies in property (3) rather than in the three-century-old limit of difference quotients (which means that Definition 1.1 of Chapter 2 is less basic than Theorem 5.3 of Chapter 2). The understanding of the real meaning of differentiation, which took nearly 250 years to come about, has played a major role in recent mathematical progress.

Exercises 12.2

In Exercises 1–14, find the total derivative of the given function at the given point.

1. $f(x, y) = x^2 + 3y$ at $(2, 1)$
2. $g(x, y) = \cos xy$ at $(1, \pi/3)$
3. $f(x, y) = x^2 + 3xy^2 + x^2y - y^3$ at $(1, 1)$
4. $h(x, y) = x^2 \cos y + y^2 e^{x^2}$ at $(1, \pi/3)$
5. $f(x, y) = e^{x^2+y^2}$ at $(0, 1)$
6. $k(x, y) = \ln(x^2 + y^2)$ at $(1, 2)$
7. $f(x, y) = \sqrt{x^2 + y^2}$ at $(1, 1)$
8. $g(x, y) = \dfrac{x + y}{x^2 + y^2}$ at $(1, 1)$
9. $f(x, y) = \tan^{-1} \dfrac{y}{x}$ at $(3, 4)$
10. $k(x, y, z) = 3x^2 + y^2 - 4z^2$ at $(1, 0, 1)$
11. $f(x, y, z) = e^{x^2+y^2} \sin z$ at $(1, 1, \pi/3)$
12. $f(x, y, a) = \ln(x^2 + xy) + x^2y^2z$ at $(2, 1, 1)$
13. $f(x, y, z) = \sin(x^2 + y^2 + z^2)$ at $(0, 0, \sqrt{\pi})$
14. $h(x, y, z) = \tan^{-1} \sqrt{x^2 + y^2 + z^2}$ at $(1, 2, 2)$

In Exercises 15–20, find $\nabla f(\mathbf{x})$ and state for which values of x and y you can be sure that f is differentiable, so that $f'(\mathbf{x}) = \nabla f(\mathbf{x})$.

15. $f(x, y) = \sqrt{x^2 + y^2}$
16. $f(x, y) = \dfrac{x + y}{x - y}$
17. $f(x, y) = \ln(1 + x^2 - y)$
18. $f(x, y) = \tan^{-1} \dfrac{y}{x}$
19. $f(x, y, z) = \sqrt{16 - x^2 - 4y^2 - 4z^2}$
20. $f(x, y, z) = \sin(x + y \ln z)$, $z > 0$

In Exercises 21 and 22, use Definition 2.1 to show that the given function is differentiable at the given point.

21. $f(x, y) = x^2 + y^2$ at $(2, 1)$
22. $f(x, y) = 3x + 2y$ at $(2, 1)$

In Exercises 23–28, use the tangent approximation to approximate the given quantity. Compare with a calculator value.

23. $\sqrt{(3.01)^2 + (3.98)^2}$
24. $\sqrt{(1.1)^3 + (1.98)^3}$
25. $(1.99)(8.03)^{4/3}$
26. $\dfrac{1}{2}\left(3 + \dfrac{1}{12}\right)^2 \sin\left(\dfrac{\pi}{6} + 0.04\right)$
27. $\sqrt{(2.01)^2 + (1.94)^2 + (0.98)^2}$
28. $\sqrt{(1.99)^2 + (2.97)^3 + (1.01)^2}$
29. The dimensions of a rectangular room are measured as 9.00, 12.00, and 8.00 feet respectively, with possible errors of ± 0.01, ± 0.01, and ± 0.04 feet respectively. Calculate the length of the diagonal across the room and estimate the maximum error possible in this. (The *diagonal* is the line drawn from the floor in one corner to the ceiling in the opposite corner.)
30. The dimensions of a box are measured as 12.00, 4.00, and 3.00 inches respectively, with possible errors of ± 0.03, ± 0.01, and ± 0.01 inches respectively. Calculate the length of the diagonal of the box and find the maximum possible error in this.

31. An aluminum can in the shape of a right circular cylinder has inner radius 2 inches and interior height 6 inches. The aluminum sides, top, and bottom are 0.02-inch thick. Use the tangent approximation to compute the approximate amount of aluminum in each can. (*Hint:* $V = \pi x^2 y$, where x is the radius and y is the height.)

32. A cylindrical oil storage tank has inner radius of 20 feet and inner height of 30 feet. The sides, floor, and top are a metallic alloy 2 inches thick. Find the approximate amount of alloy needed to make one such oil tank.

33. If $f: \mathbf{R}^n \to \mathbf{R}$ is a linear function, $f(\mathbf{x}) = \mathbf{m} \cdot \mathbf{x} + \mathbf{b}$ for some constant vector $\mathbf{m}$ in $\mathbf{R}^n$ and some real constant $\mathbf{b}$, then find ∇f and show that f is differentiable at any $\mathbf{x} = \mathbf{a}$ in $\mathbf{R}^n$.

34. At which points $\mathbf{x} = \mathbf{a}$ in $\mathbf{R}^n$ is the length function $f(\mathbf{x}) = |\mathbf{x}|$ differentiable? Give the formula for $\mathbf{f}'(\mathbf{x})$.

35. Show that the function

$$f(x, y) = \begin{cases} \dfrac{3x^2 y}{x^2 + y^2}, & \text{for } (x, y) \neq (0, 0), \\ 0, & \text{for } (x, y) = (0, 0), \end{cases}$$

is not differentiable at (0, 0).

36. Show that

$$f(x, y) = \begin{cases} \dfrac{x^2 + y^2}{xy^2 - 2}, & \text{for } xy^2 \neq 2, \\ 0, & \text{for } xy^2 = 2, \end{cases}$$

is not differentiable at (2, 1).

37. Show that the function

$$f(x, y) = \begin{cases} \dfrac{x^2 y + xy^2}{x^2 + y^2}, & \text{for } (x, y) \neq (0, 0), \\ 0, & \text{for } (x, y) = (0, 0), \end{cases}$$

is not differentiable at (0, 0) even though it is continuous there.

38. Prove Theorem 2.2 for $n = 3$.

39. Prove that $m_2 = \dfrac{\partial f}{\partial y}(a, b)$ in Theorem 2.2.

40. Show that (3) in Definition 2.1 fails for $\mathbf{m} = \mathbf{0}$ in Example 2.

41. Prove Theorem 2.5 when $n = 3$.

12.3 Directional Derivatives

We saw in Section 1 that the partial derivatives

$$\frac{\partial f}{\partial x}(a, b) \qquad \text{and} \qquad \frac{\partial f}{\partial y}(a, b)$$

measure the rate of change of a scalar function $f(x, y)$ in the respective directions of the x- and y-axes. Recall that

$$\frac{\partial f}{\partial x}(a, b)$$

is the slope of the curve

$$\mathbf{x} = \mathbf{x}_2(t) = t\,\mathbf{i} + b\,\mathbf{j} + f(t, b)\,\mathbf{k} = (t, b, f(t, b))$$

of intersection of the surface $z = f(x, y)$ with the plane $y = b$ at the point $P(a, b, (a, b))$. (See **Figure 3.1.**) Similarly,

$$\frac{\partial f}{\partial y}(a, b)$$

is the slope of the curve

$$\mathbf{x} = \mathbf{x}_1(t) = a\,\mathbf{i} + t\,\mathbf{j} + f(a, t)\,\mathbf{k} = (a, t, f(a, t))$$

of intersection of the surface with the plane $x = a$ at P. As mentioned following Definition 1.1, if a piece of land has surface equation $z = f(x, y)$, then $(\partial f/\partial x)(a, b)$ and $(\partial f/\partial y)(a, b)$ measure the rates of rise or fall of the land at P in two perpendicular standard directions such as east–west and north–south. To describe

FIGURE 3.1

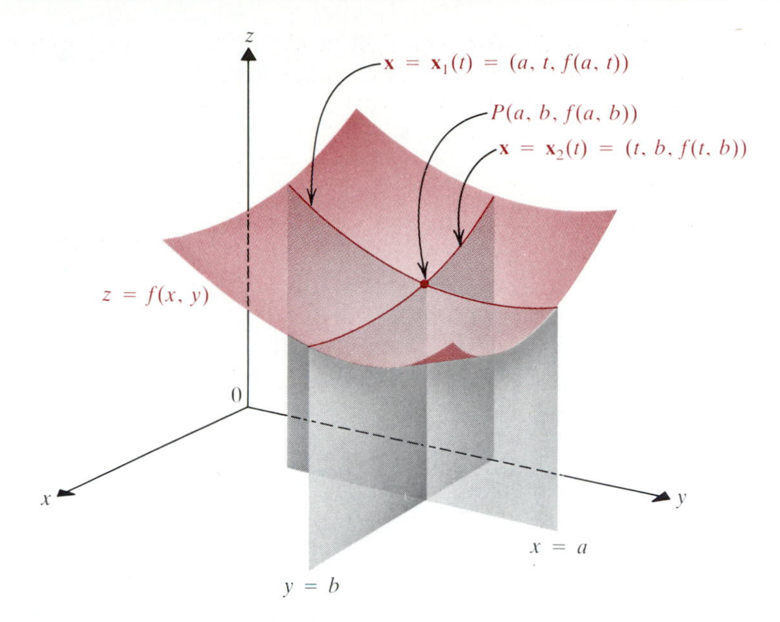

the rate of rise of the land in other directions, such as southwest–northeast, we need to investigate more closely the directions associated with $(\partial f/\partial x)(a, b)$ and $(\partial f/\partial y)(a, b)$.

In Definition 1.1 we turned $f(x, y)$ into a function of one variable by freezing the value of the other variable. In terms of the standard basis vectors **i** and **j** of $\mathbf{R}^2$, we were actually considering the two functions

$$f(\mathbf{a} + t\,\mathbf{i}) = f((a, b) + t(1, 0)) = f(a + t, b)$$

and

$$f(\mathbf{a} + t\,\mathbf{j}) = f((a, b) + t(0, 1)) = f(a, b + t).$$

As t varies, only one coordinate changes in each of those functions. With this notation,

$$\frac{\partial f}{\partial x}(\mathbf{a}) \quad \text{and} \quad \frac{\partial f}{\partial y}(\mathbf{a})$$

can be written as

$$\frac{\partial f}{\partial x}(\mathbf{a}) = \lim_{t \to 0} \frac{f(\mathbf{a} + t\,\mathbf{i}) - f(\mathbf{a})}{t} \quad \text{and} \quad \frac{\partial f}{\partial y}(\mathbf{a}) = \lim_{t \to 0} \frac{f(\mathbf{a} + t\,\mathbf{j}) - f(\mathbf{a})}{t}.$$

To measure the rate of change of f in the direction of an *arbitrary* unit vector $\mathbf{u} = u_1\mathbf{i} + u_2\mathbf{j}$ in $\mathbf{R}^2$, we substitute **u** for **i** or **j** in these limits. That corresponds geometrically to finding the slope of the curve

$$\mathbf{x} = \mathbf{x}(t) = (a + tu_1, b + tu_2, f(a + tu_1, b + tu_2))$$

of intersection of the graph of $z = f(x, y)$ and the plane Π in **Figure 3.2.** That plane is perpendicular to the xy-plane and cuts through the vector **u** drawn from the point $(a, b, 0)$. We are thus led to the next definition.

3.1 DEFINITION Let $\mathbf{a} = (a_1, a_2, \ldots, a_n) \in \mathbf{R}^n$. Let $f\colon \mathbf{R}^n \to \mathbf{R}$ be defined in an open ball $B(\mathbf{a}, \delta)$ in $\mathbf{R}^n$. Let **v** be a nonzero vector in $\mathbf{R}^n$ and

$$\mathbf{u} = \frac{1}{|\mathbf{v}|}\,\mathbf{v}$$

FIGURE 3.2

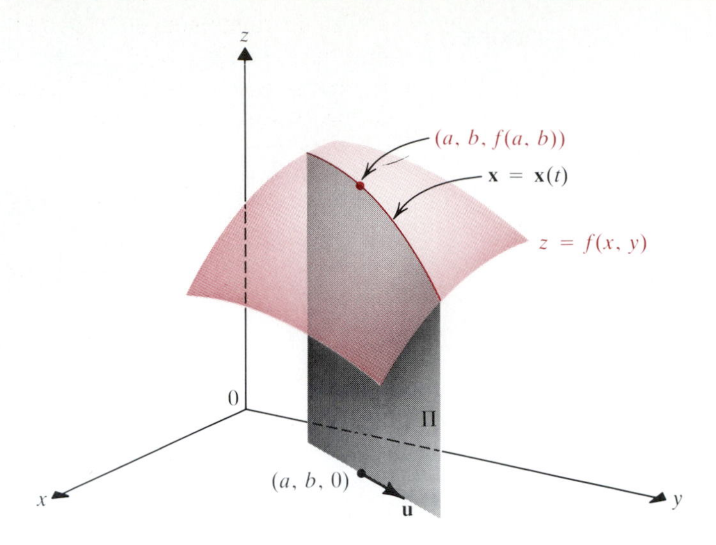

its direction vector. The ***directional derivative of f in the direction of* v** at **x** = **a** is

$$D_{\mathbf{u}}f(\mathbf{a}) = \lim_{t\to 0} \frac{f(\mathbf{a} + t\mathbf{u}) - f(\mathbf{a})}{t}, \tag{1}$$

if this limit exists.

Whereas it is easy to compute the partial derivative of f with respect to any coordinate variable x_i, there is no obvious way around (1), which even for simple functions involves some work, as the following example shows.

Example 1

Find the directional derivative of $f(x, y) = x^2 + y^2$ in the direction of $3\mathbf{i} + 4\mathbf{j}$ at the point (2, 1).

Solution. Here $\mathbf{v} = 3\mathbf{i} + 4\mathbf{j}$ has length $|\mathbf{v}| = \sqrt{9 + 16} = 5$, so

$$\mathbf{u} = \frac{3}{5}\mathbf{i} + \frac{4}{5}\mathbf{j}.$$

Then (1) gives

$$\begin{aligned} D_{\mathbf{u}}f(2, 1) &= \lim_{t\to 0} \frac{f\left((2, 1) + t\left(\frac{3}{5}, \frac{4}{5}\right)\right) - f(2, 1)}{t} \\ &= \lim_{t\to 0} \frac{f\left(2 + \frac{3t}{5}, 1 + \frac{4t}{5}\right) - f(2, 1)}{t} \\ &= \lim_{t\to 0} \frac{4 + \frac{12t}{5} + \frac{9t^2}{25} + 1 + \frac{8t}{5} + \frac{16t^2}{25} - 4 - 1}{t} \\ &= \lim_{t\to 0} \frac{4t + t^2}{t} = \lim_{t\to 0} (4 + t) = 4. \quad \blacksquare \end{aligned}$$

Our instincts suggest that the right differentiation rule could substantially shorten the solution of Example 1. The next result provides that rule, which can be used whenever f is differentiable at $\mathbf{x} = \mathbf{a}$. In its statement, n is 2 or 3, and we again understand that in the two-dimensional case, the third coordinates are not present.

3.2 THEOREM If $f: \mathbf{R}^n \to \mathbf{R}$ is differentiable at $\mathbf{x} = \mathbf{a} = (a_1, a_2, a_3)$ and $\mathbf{v} \neq \mathbf{0}$ is a vector in $\mathbf{R}^n$, then

$$(2) \qquad D_{\mathbf{u}}f(\mathbf{a}) = \mathbf{f}'(\mathbf{a}) \cdot \mathbf{u} = \nabla f(\mathbf{a}) \cdot \mathbf{u}$$
$$= \left(\frac{\partial f}{\partial x}(\mathbf{a})\,\mathbf{i} + \frac{\partial f}{\partial y}(\mathbf{a})\,\mathbf{j} + \frac{\partial f}{\partial z}(\mathbf{a})\,\mathbf{k}\right) \cdot (u_1\mathbf{i} + u_2\mathbf{j} + u_3\mathbf{k}),$$

where $\mathbf{u} = \dfrac{1}{|\mathbf{v}|}\mathbf{v} = (u_1, u_2, u_3)$. That is, *the directional derivative of f in the direction of $\mathbf{v}$ at $\mathbf{x} = \mathbf{a}$ is the coordinate of the total derivative of f at $\mathbf{x} = \mathbf{a}$ in the direction of $\mathbf{u}$.*

Proof. We need to show that the limit in (1) equals $\mathbf{f}'(\mathbf{a}) \cdot \mathbf{u}$, which is equivalent to showing that

$$\lim_{t \to 0}\left[\frac{f(\mathbf{a} + t\mathbf{u}) - f(\mathbf{a})}{t} - \mathbf{f}'(\mathbf{a}) \cdot \mathbf{u}\right] = 0.$$

That limit is

$$\lim_{t \to 0}\frac{f(\mathbf{a} + t\mathbf{u}) - f(\mathbf{a}) - t\mathbf{f}'(\mathbf{a}) \cdot \mathbf{u}}{t} = \lim_{t \to 0}\frac{f(\mathbf{a} + t\mathbf{u}) - f(\mathbf{a}) - \mathbf{f}'(\mathbf{a}) \cdot t\mathbf{u}}{t}.$$

We can put it into the form of Definition 2.1 by letting

$$\mathbf{x} = \mathbf{a} + t\mathbf{u}.$$

Then $t\mathbf{u} = \mathbf{x} - \mathbf{a}$. Equating the lengths of these two vectors and using the fact that $|\mathbf{u}| = 1$, we get

$$|t| = |\mathbf{x} - \mathbf{a}| \to t = \pm|\mathbf{x} - \mathbf{a}|.$$

Thus $t \to 0$ is equivalent to $\mathbf{x} \to \mathbf{a}$. Hence

$$\lim_{t \to 0}\frac{f(\mathbf{a} + t\mathbf{u}) - f(\mathbf{a}) - \mathbf{f}'(\mathbf{a}) \cdot t\mathbf{u}}{t} = \lim_{\mathbf{x} \to \mathbf{a}}\frac{f(\mathbf{x}) - f(\mathbf{a}) - \mathbf{f}'(\mathbf{a}) \cdot (\mathbf{x} - \mathbf{a})}{\pm|\mathbf{x} - \mathbf{a}|}$$
$$= \pm\lim_{\mathbf{x} \to \mathbf{a}}\frac{f(\mathbf{x}) - f(\mathbf{a}) - \mathbf{f}'(\mathbf{a}) \cdot (\mathbf{x} - \mathbf{a})}{|\mathbf{x} - \mathbf{a}|}$$
$$= \pm 0 = 0,$$

by Definition 2.1, since f is differentiable at $\mathbf{x} = \mathbf{a}$. QED

Figure 3.3 illustrates the interpretation of $D_{\mathbf{u}}f(\mathbf{a})\,\mathbf{u}$ as the component of the total derivative $\mathbf{f}'(\mathbf{a}) = \nabla f(\mathbf{a})$ in the direction $\mathbf{u}$ for a differentiable function f. Comparison of the next example to Example 1 suggests the simplification Theorem 3.2 provides.

Example 2

Use Theorem 3.2 to find the directional derivative of $f(x, y) = x^2 + y^2$ in the direction of $3\mathbf{i} + 4\mathbf{j}$ at the point (2, 1).

FIGURE 3.3

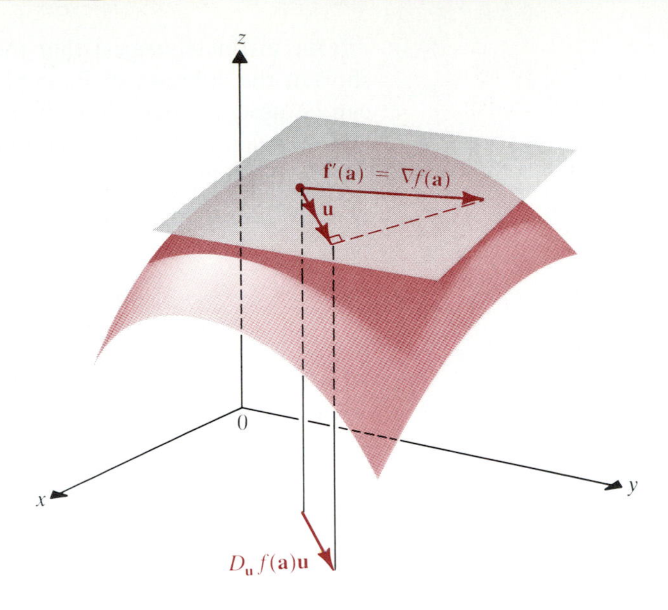

Solution. As in Example 1 we have

$$\mathbf{u} = \frac{3}{5}\mathbf{i} + \frac{4}{5}\mathbf{j}.$$

The total derivative of f at (2, 1) is

$$\nabla f(\mathbf{a}) = 2x\,\mathbf{i} + 2y\,\mathbf{j}\Big|_{(2,1)} = 4\,\mathbf{i} + 2\,\mathbf{j}.$$

Thus (2) gives

$$D_{\mathbf{u}}f(\mathbf{a}) = \nabla f(\mathbf{a}) \cdot \mathbf{u} = (4\,\mathbf{i} + 2\,\mathbf{j}) \cdot \left(\frac{3}{5}\mathbf{i} + \frac{4}{5}\mathbf{j}\right) = \frac{12}{5} + \frac{8}{5} = 4. \quad \blacksquare$$

Theorem 3.2 works equally well in three dimensions, as the next example illustrates.

Example 3

Find the directional derivative of $f(x, y, z) = x^2 + y^2 - x\cos \pi y - y \sin \pi x + \sin z^2$ in the direction of $\mathbf{v} = 2\,\mathbf{i} + \mathbf{j} - 2\,\mathbf{k}$ at $\mathbf{a} = (-1, 2, \sqrt{\pi})$.

Solution. The total derivative of f is

$$f'(x, y, z) = \nabla f(x, y, z) = (2x - \cos \pi y - \pi y \cos \pi x)\,\mathbf{i} + (2y + \pi x \sin \pi y - \sin \pi x)\,\mathbf{j} + 2z \cos z^2\,\mathbf{k}.$$

At $\mathbf{a} = (-1, 2, \sqrt{\pi})$ then, we have

$$\begin{aligned}\mathbf{f}'(\mathbf{a}) &= (-2 - 1 + 2\pi)\,\mathbf{i} + (4 - 0 - 0)\,\mathbf{j} + 2\sqrt{\pi}(-1)\,\mathbf{k} \\ &= (-3 + 2\pi)\,\mathbf{i} + 4\,\mathbf{j} - 2\sqrt{\pi}\,\mathbf{k}.\end{aligned}$$

Since $|\mathbf{v}| = \sqrt{4 + 1 + 4} = 3$, the direction vector of $\mathbf{v}$ is

$$\mathbf{u} = \frac{2}{3}\mathbf{i} + \frac{1}{3}\mathbf{j} - \frac{2}{3}\mathbf{k}.$$

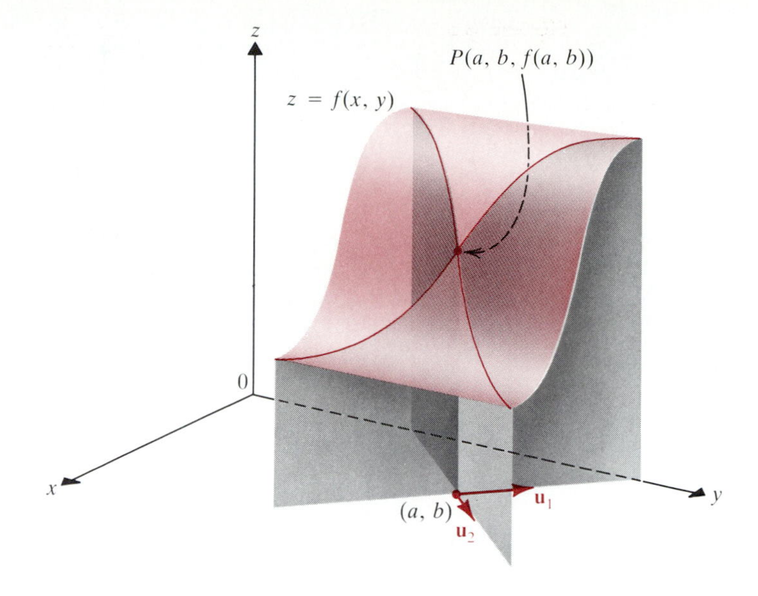

FIGURE 3.4
f increases in the direction of u_1
f decreases in the direction of u_2

We therefore get from (2)

$$D_{\mathbf{u}}f(\mathbf{a}) = [(-3 + 2\pi)\mathbf{i} + 4\mathbf{j} - 2\sqrt{\pi}\,\mathbf{k}] \cdot \left[\frac{2}{3}\mathbf{i} + \frac{1}{3}\mathbf{j} - \frac{2}{3}\mathbf{k}\right]$$

$$= -2 + \frac{4\pi}{3} + \frac{4}{3} + \frac{4\sqrt{\pi}}{3} = -\frac{2}{3} + \frac{4}{3}(\pi + \sqrt{\pi}). \quad \blacksquare$$

In the case of a differentiable function f of one variable, we can tell from computing $f'(a)$ whether f is increasing, decreasing, or neither at $x = a$ (Theorem 3.1 of Chapter 3). For a function of several variables, the situation is more complicated: The rate of change of f may be positive in one direction but negative in another, as **Figure 3.4** illustrates. However, at any point $\mathbf{a}$ where $\nabla f(\mathbf{a}) \neq \mathbf{0}$, there is one direction in which the rate of increase of f is largest, and one direction in which that rate is smallest. We can use Theorem 3.2 to find those directions:

$$\text{(3)} \qquad \begin{aligned} D_{\mathbf{u}}f(\mathbf{a}) &= \mathbf{f}'(\mathbf{a}) \cdot \mathbf{u} = \nabla f(\mathbf{a}) \cdot \mathbf{u} \\ &= |\nabla f(\mathbf{a})|\,|\mathbf{u}| \cos\theta \qquad \textit{by (10), p. 648} \\ &= |\nabla f(\mathbf{a})| \cos\theta, \end{aligned}$$

where θ is the angle between $\nabla f(\mathbf{a})$ and $\mathbf{u}$. The maximum value of $\cos\theta$ is 1, when $\theta = 0$. In that case, $\mathbf{u}$ is the direction of $\nabla f(\mathbf{a})$. The minimum value of $\cos\theta$ is -1, when $\theta = \pi$. In this case, $\mathbf{u}$ is the direction of $-\nabla f(\mathbf{a})$. Hence we have the following result, which is illustrated geometrically in **Figure 3.5.**

3.3 THEOREM

If f is differentiable at $\mathbf{x} = \mathbf{a}$ and $\nabla f(\mathbf{a}) \neq \mathbf{0}$, then the maximum rate of increase of f is in the direction of the gradient $\nabla f(\mathbf{a})$, and the maximum rate of decrease of f is in the direction of $-\nabla f(\mathbf{a})$. The maximum rate of increase is $|\nabla f(\mathbf{a})|$, and the maximum rate of decrease is $-|\nabla f(\mathbf{a})|$.

Proof. Our discussion establishes everything except the maximum rates of increase and decrease. The maximum rate of increase occurs when $\theta = 0$, that is, when u is the unit vector in the direction of $\nabla f(\mathbf{a})$. Then $\cos\theta = 1$, so the rate

FIGURE 3.5

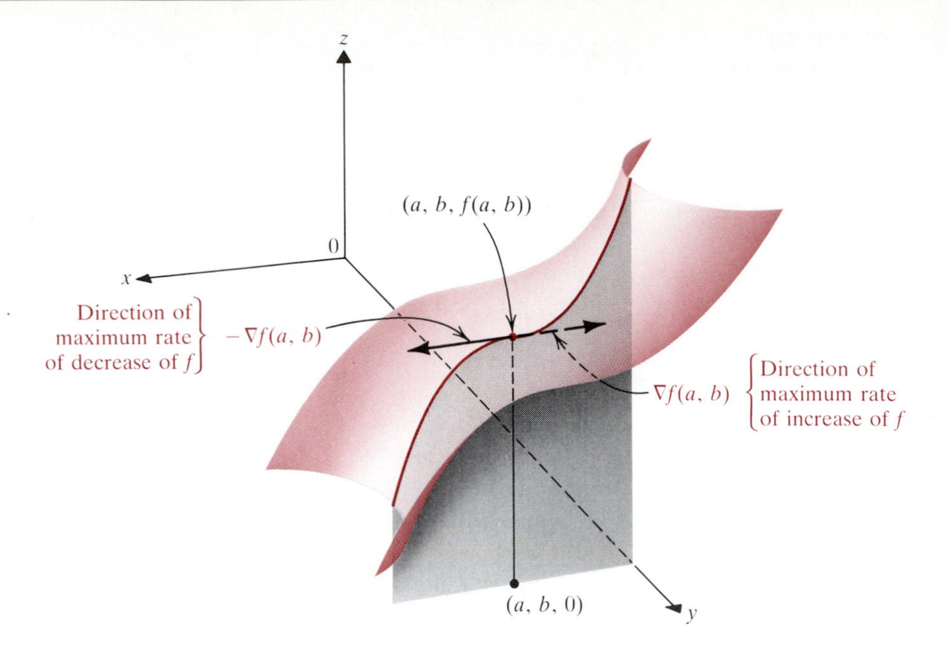

of increase of f in that direction is

$$|\mathbf{f}'(\mathbf{a})| = |\nabla f(\mathbf{a})|.$$

The maximum rate of decrease occurs when $\theta = \pi$, that is, when $\mathbf{u}$ is the unit vector in the direction of $-\nabla f(\mathbf{a})$. Since $\cos \pi = -1$, the rate of increase of f in that direction is

$$-|\mathbf{f}'(\mathbf{a})| = -|\nabla f(\mathbf{a})|.$$ QED

Example 4

Find the magnitude and direction of the maximum rates of increase and decrease of f at $\mathbf{a} = (-3, \sqrt{3}, 2)$ if $f(x, y, z) = \sqrt{x^2 + y^2 + z^2}$.

Solution. The function $f(x, y, z)$ is differentiable at every point (x, y, z) of $\mathbf{R}^3$ except the origin $(0, 0, 0)$ and so is differentiable at $\mathbf{a}$. We have

$$\nabla f(x, y, z) = \frac{x}{\sqrt{x^2 + y^2 + z^2}}\mathbf{i} + \frac{y}{\sqrt{x^2 + y^2 + z^2}}\mathbf{j} + \frac{z}{\sqrt{x^2 + y^2 + z^2}}\mathbf{k}.$$

Thus, since $\sqrt{(-3)^2 + (\sqrt{3})^2 + 2^2} = \sqrt{16} = 4$, we find

$$\nabla f(-3, \sqrt{3}, 2) = \frac{-3}{4}\mathbf{i} + \frac{\sqrt{3}}{4}\mathbf{j} + \frac{1}{2}\mathbf{k}.$$

This is the direction of maximum increase; since

$$|\nabla f(-3, \sqrt{3}, 2)| = \sqrt{\frac{9}{16} + \frac{3}{16} + \frac{4}{16}} = 1,$$

the gradient $\nabla f(-3, \sqrt{3}, 2)$ is its own direction vector. Thus the maximum rate of increase at $(-3, \sqrt{3}, 2)$ is 1, which occurs in the direction of

$$-\frac{3}{4}\mathbf{i} + \frac{\sqrt{3}}{4}\mathbf{j} + \frac{1}{2}\mathbf{k}.$$

FIGURE 3.6

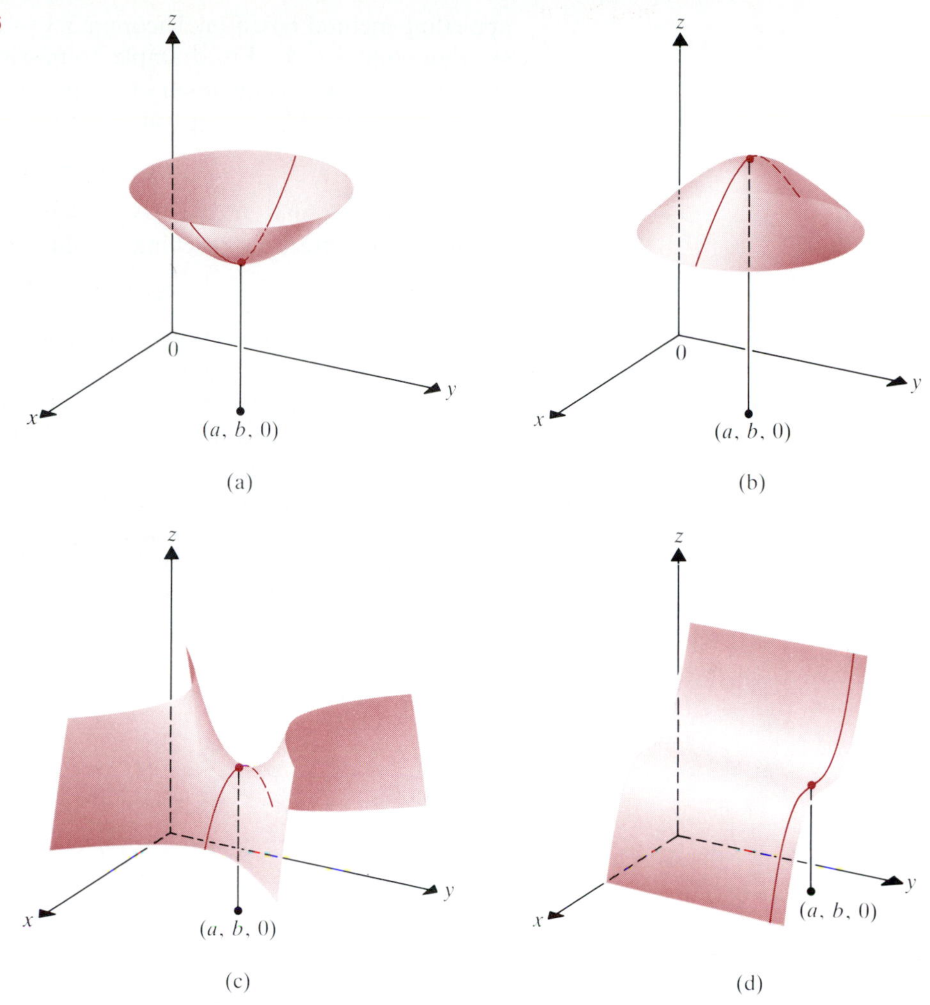

The maximum rate of decrease is -1, which occurs in the direction of

$$\frac{3}{4}\mathbf{i} - \frac{\sqrt{3}}{4}\mathbf{j} - \frac{1}{2}\mathbf{k}. \quad \blacksquare$$

What happens in the excluded case in Theorem 3.3, when f is differentiable at $\mathbf{x} = \mathbf{a}$, but $\nabla f(\mathbf{a}) = 0$? By Theorem 3.2, $D_{\mathbf{u}}f(\mathbf{a}) = 0$ for every unit vector $\mathbf{u}$. So at $\mathbf{x} = \mathbf{a}$, there is zero rate of change in f in every direction. By analogy to the one-variable theory (Theorem 2.3 of Chapter 3), we expect $\mathbf{a}$ to be a candidate for an extreme point of f. We will pursue this further in Section 7, but for now you can see from **Figure 3.6** that, as in the one-dimensional case, such candidates must be tested to see whether they are extreme points.

Numerical Optimization of Scalar Functions: The Gradient Method*

As the above remarks suggest, one way to find extreme points of a scalar function is to solve the equation $\nabla f(\mathbf{x}) = \mathbf{0}$ to find points $\mathbf{x} = \mathbf{a}$ and then to test those points. If it is not possible to solve $\nabla f(\mathbf{x}) = \mathbf{0}$ for $\mathbf{x}$, then we can use an intuitively

* Optional

appealing method based on Theorem 3.3 to numerically optimize f near some starting point $\mathbf{x} = \mathbf{x}_1$. For example, to maximize f, first compute $\nabla f(\mathbf{x}_1)$. Since the gradient gives the direction of steepest increase of f at $\mathbf{x} = \mathbf{x}_1$, travel in the direction of $\nabla f(\mathbf{x}_1)$ for a short distance brings us to a second point,

$$\mathbf{x}_2 = \mathbf{x}_1 + t\,\nabla f(\mathbf{x}_1),$$

where we expect that $f(\mathbf{x}_2) > f(\mathbf{x}_1)$. Next, compute $\nabla f(\mathbf{x}_2)$ and travel a short distance in its direction, reaching a point

$$\mathbf{x}_3 = \mathbf{x}_2 + t\,\nabla f(\mathbf{x}_2)$$

where we expect that $f(\mathbf{x}_3) > f(\mathbf{x}_2)$. In this way, a sequence of points is generated that should approach a point $\mathbf{x}_m$ where f has a local maximum. To minimize f take each step in the direction of $-\nabla f(\mathbf{x}_i)$. This scheme is the basis of a numerical optimization technique called the ***gradient method*** or ***method of steepest ascent/descent.*** Although it is effective, the sequence of points it generates may approach the maximum or minimum point in a roundabout way, leading to slow convergence. The idea is similar to hiking up a mountain by following a stream: Although at each point the water has taken the path of swiftest descent, in the process it may have meandered around quite a bit, thus making the hiker's climb a long and slow one.

In practice, sophisticated refinements of the gradient method are used, but the following simple example illustrates the underlying ideas.

Example 5

Let $f(x, y) = e^{-x^2-y^2}$. Starting from $\mathbf{x}_1 = (1, 1)$, use the gradient method with $t = 1/4$ to generate a sequence of ten points $\mathbf{x}_i$ to approximate a local maximum point of $f(x, y)$ near $\mathbf{x}_1$. Do the points seem to be approaching a limit?

Solution. This function is differentiable over the entire plane $\mathbf{R}^2$. Its total derivative is

$$\nabla f(x, y) = \left(\frac{\partial f}{\partial x}, \frac{\partial f}{\partial y}\right) = (e^{-x^2-y^2}(-2x), e^{-x^2-y^2}(-2y))$$
$$= (-2xe^{-x^2-y^2}, -2ye^{-x^2-y^2}).$$

Using $(x_1, y_1) = (1, 1)$ and $t = 1/4$, we get

$$(x_2, y_2) = (x_1, y_1) + t\,\nabla f(1, 1) = (1, 1) + \frac{1}{4}(-2e^{-2}, -2e^{-2})$$
$$= \left(1 - \frac{1}{2}e^{-2}, 1 - \frac{1}{2}e^{-2}\right) \approx (0.932332, 0.932332).$$

Then

$$(x_3, y_3) = (x_2, y_2) + \nabla f(0.932332, 0.932332)$$
$$\approx (0.932332, 0.932332)$$
$$+ \frac{1}{4}(-1.864665e^{-1.738487}, -1.864665e^{-1.738487})$$
$$\approx (0.850387, 0.850387).$$

TABLE 1

N	XN	YN	F(XN, YN)
0	1	1	.135335283
1	.932332358	.932332358	.17578612
2	.850386815	.850386815	.235436164
3	.75028091	.75028091	.324378937
4	.628593248	.628593248	.453727291
5	.485988292	.485988292	.623523241
6	.334475795	.334475795	.799516496
7	.200766337	.200766337	.922549502
8	.108157895	.108157895	.976875310
9	.0553295064	.0553295064	.993895997
10	.0278336189	.0278336189	.998451779

Continuing in this way, we can obtain Table 1, which is the output of a computer program for the gradient method. Table 1 suggests that the points (x_n, y_n) are converging to a local maximum at (0, 0). This suggestion is consistent with the computer-generated graph of the surface in **Figure 3.7** for $x \in [-3, 3]$ and $y \in [-3, 3]$. And if we write the formula for $f(x, y)$ in the form

$$f(x, y) = \frac{1}{e^{x^2 + y^2}},$$

then it is clear that f has its absolute maximum at (0, 0). The points $(x_n, y_n, f(x_n, y_n))$ shown in Figure 3.7 seem to be climbing upward along the surface toward (0, 0, 1). ■

FIGURE 3.7

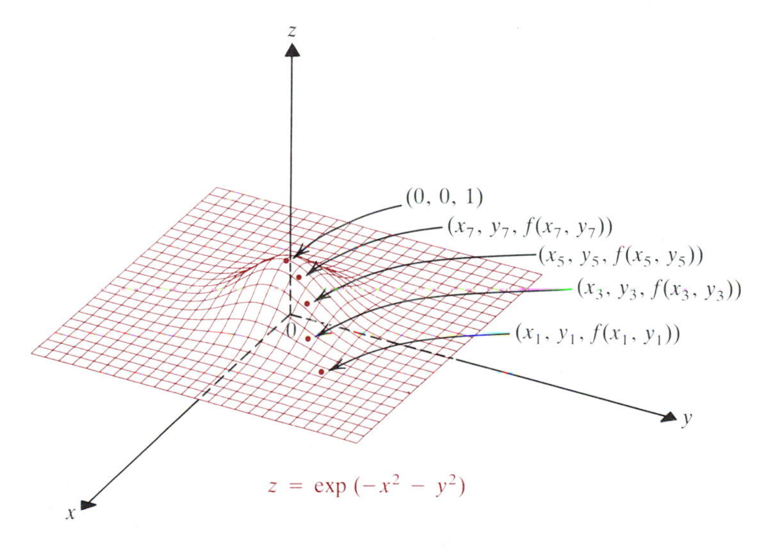

Exercises 29–35 explore the gradient method further; they are best done with the aid of a computer or programmable calculator because many repetitive calculations are involved.

Exercises 12.3

In Exercises 1–10, find the directional derivative of f at the point a in the given direction.

1. $f(x, y) = x^2 + y^2$, $\mathbf{a} = (1, 2)$, $\mathbf{u} = (1/\sqrt{2})\mathbf{i} + (1/\sqrt{2})\mathbf{j}$
2. $f(x, y) = x^2 - y^2$, $\mathbf{a} = (2, 1)$, $\mathbf{u} = (1/\sqrt{2})\mathbf{i} + (1/\sqrt{2})\mathbf{j}$
3. $f(x, y) = e^x \cos y$, $\mathbf{a} = (0, \pi/2)$, $\mathbf{u} = (4/5)\mathbf{i} + (3/5)\mathbf{j}$
4. $f(x, y) = e^y \sin x$, $\mathbf{a} = (\pi, 1)$, $\mathbf{u} = (3/5)\mathbf{i} + (4/5)\mathbf{j}$
5. $f(x, y, z) = x^2 + y^2 - z^2$, $\mathbf{a} = (1, 2, 2)$, $\mathbf{u} = (1/3)\mathbf{i} + (2/3)\mathbf{j} + (2/3)\mathbf{k}$

6. $f(x, y, z) = 2x^2 - 4y^2 + z^2$, $\mathbf{a} = (2, 1, 3)$, $\mathbf{u} = (2/3)\mathbf{i} + (2/3)\mathbf{j} + (1/3)\mathbf{k}$

7. $f(x, y, z) = xyz$, $\mathbf{a} = (1, -1, 2)$, in the direction of $\mathbf{v} = \mathbf{i} + 2\mathbf{j} + 2\mathbf{k}$

8. $f(x, y, z) = xye^{xyz} + yze^x$, $\mathbf{a} = (1, 1, 1)$, in the direction of $\mathbf{v} = 2\mathbf{i} + 2\mathbf{j} + \mathbf{k}$

9. $f(x, y, z) = x^2 - 2yz$, $\mathbf{a} = (1, 2, 3)$, in the direction of $\mathbf{v} = 2\mathbf{i} + 2\mathbf{j} + \mathbf{k}$

10. $f(x, y, z) = \ln(1 - x^2 + y^2 + z^2)$, $\mathbf{a} = (1, -1, 1)$, in the direction of $\mathbf{v} = 3\mathbf{i} - 2\mathbf{j} + 2\mathbf{k}$

In Exercises 11–18, find the direction and the rate of maximum increase and maximum decrease for f at the given point.

11. $f(x, y) = x^2 + y^2$, $\mathbf{a} = (1, 2)$

12. $f(x, y) = x^2 - 2y^2$, $\mathbf{a} = (3, 1)$

13. $f(x, y) = x^2y - xy^2$, $\mathbf{a} = (3, 4)$

14. $f(x, y) = \tan^{-1}\dfrac{x}{y}$, $\mathbf{a} = (-2, 1)$

15. $f(x, y, z) = x^2 + y^2 + z^2$, $\mathbf{a} = (1, 2, 1)$

16. $f(x, y, z) = 2x^2 - y^2 - 3z^2$, $\mathbf{a} = (2, 1, 0)$

17. $f(x, y, z) = e^{-(x^2+y^2+z^2)}$, $\mathbf{a} = (1, 2, -1)$

18. $f(x, y, z) = x^3y^2 + \sqrt{yz}$, $\mathbf{a} = (-2, 1, 3)$

19. The *Newtonian potential* is

$$f(x, y, z) = \frac{1}{|\mathbf{x}|} = \frac{1}{\sqrt{x^2 + y^2 + z^2}}, \text{ for } \mathbf{x} \neq 0.$$

Find the direction of maximum increase in the Newtonian potential function at the point (1, 2, 2).

20. The *logarithmic potential* is

$$g(x, y, z) = -\frac{1}{2}\ln(x^2 + y^2 + z^2) = \ln f(\mathbf{x}) \qquad \text{for } \mathbf{x} \neq \mathbf{0},$$

where $f(\mathbf{x})$ is the Newtonian potential of Exercise 19. Find the direction of maximum increase in the logarithmic potential at the point (1, 2, 2).

21. Suppose that the temperature at (x, y, z) is given by the Newtonian potential function in Exercise 19. Assuming that heat always flows in the direction of maximum decrease in temperature, describe the resulting heat flow across a sphere of radius 3.

22. Repeat Exercise 21, assuming that the temperature is given by the logarithmic potential function of Exercise 20.

23. A skier starts his run at the point (2, 1, 8) of a mountain whose equation is $z = 14 - x^2 - 2y^2$, where x, y, and z are measured in thousands of feet. If he wants to take the path of steepest descent, then along what vector in the xy-plane does he travel?

24. In Exercise 23, suppose the skier starts at (9/4, 5/4, 93/16). If he wants to follow the path of steepest descent, what path should he follow? (Note that if the skier stayed on his course of steepest descent in Exercise 23, then he *would* reach the point (9/4, 5/4, 93/16).)

25. On the top hemisphere of the unit sphere $x^2 + y^2 + z^2 = 1$, the equation defines z as a function of x and y, $z = f(x, y)$. Find $\nabla f(x, y)$ for this function.

26. In Exercise 25 show that the direction of ∇f is the direction of the vector pointing from $(x, y, 0)$ to $(0, 0, 0)$. Then describe the path of steepest ascent from any point (x, y, z) on the hemisphere.

27. Let $f(x, y) = 1 - x^2 + 2y^2$. A particle starts moving from the point (1, 1) in the direction of maximum increase of f.
(a) Find the unit tangent vector $\mathbf{T}$ to the path at (x, y) as a function of x and y.
(b) Equate your answer in part (a) to

$$\frac{\left(\dfrac{dx}{dt}, \dfrac{dy}{dt}\right)}{\sqrt{\left(\dfrac{dx}{dt}\right)^2 + \left(\dfrac{dy}{dt}\right)^2}}$$

to obtain two differential equations in dx/dt, dy/dt, x, and y.
(c) Divide the two equations obtained in (b) to obtain an equation for dy/dx in terms of x and y.
(d) Separate the variables in part (c) and integrate to get an equation $y = y(x)$ for the path of motion.
(e) Use the fact that (1, 1) is on the path of motion to solve for the constant of integration in part (d), thereby finding the equation for the unique path of the particle.

28. Repeat Exercise 27 for $f(x, y) = 1 - x^2 - 3y^2$.

29. It is clear that the function $f(x, y) = 1 - x^2 - y^2$ has its maximum value at $(x^*, y^*) = (0, 0)$. (Why?) Starting from the initial guess $(x_1, y_1) = (1, 1)$ as an approximation of (x^*, y^*), apply the gradient method described in the text, with $t = 0.1$ for each step. Find (x_i, y_i) for $i = 2, 3, 4,$ and 5.

PC **30.** Continue the iteration process in Exercise 29. **(a)** Find (x_i, y_i) for $i = 6, 7, 8, 9,$ and 10. **(b)** How large an n is required to reach a point (x_n, y_n) such that both coordinates are smaller than 0.05?

31. Refer to Exercise 29. Use the gradient method with $t = 0.2$ to approximate the minimum value of the function $f(x, y) = x^2 + y^2 - 2x + 4y$ starting from the point $(x_1, y_1) = (1, 1)$. Find (x_i, y_i) for $i = 2, 3, 4,$ and 5.

PC **32.** **(a)** Repeat Exercise 30(a) for the function in Exercise 31.
(b) What do the points (x_i, y_i) seem to be converging to?
(c) What appears to be the minimum value of $f(x, y)$ near (1, 1)?

PC **33.** In Example 5, use $t = 1/10$ instead of 1/4.

PC **34.** Repeat Exercise 33 with $t = 1/3$.

PC **35.** From Example 5 and Exercises 33 and 34, what relation seems to exist between t and the rate of convergence of the points (x_i, y_i) to (0, 0)?

12.4 The Chain Rule

In single-variable calculus the most important differentiation rule is the chain rule of Section 2.6, which applies to composite function $f \circ g$. If f and g are differentiable functions, then the chain rule says that $f \circ g$ is differentiable and

$$(f \circ g)'(x) = f'(g(x)) \cdot g'(x).$$

This section gives multivariable versions of the chain rule. They cover situations such as the composite of two differentiable functions

$$f\colon \mathbf{R}^n \to \mathbf{R} \qquad \text{and} \qquad \mathbf{g}\colon \mathbf{R} \to \mathbf{R}^n,$$

FIGURE 4.1

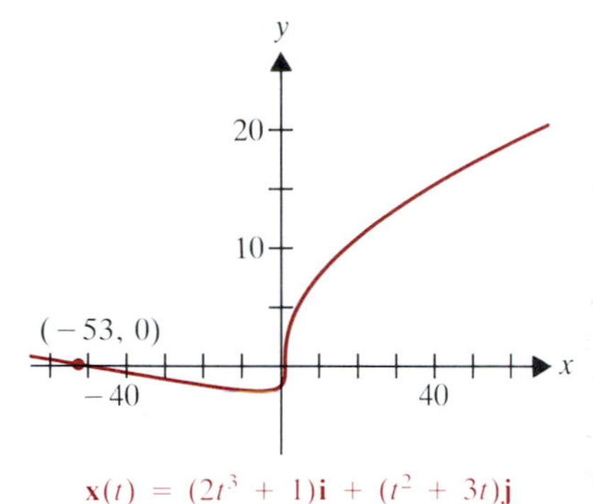

$\mathbf{x}(t) = (2t^3 + 1)\mathbf{i} + (t^2 + 3t)\mathbf{j}$

where n is 2 or 3. For example, suppose that we want to study the behavior of the function f given by $f(x, y) = e^{xy}$ along the curve with parametric equations

$$x = 2t^3 + 1 \qquad \text{and} \qquad y = t^2 + 3t.$$

Then the equations for x and y specify the coordinate functions of the vector function $\mathbf{g}\colon \mathbf{R} \to \mathbf{R}^2$ given by

$$\mathbf{g}(t) = (2t^3 + 1)\,\mathbf{i} + (t^2 + 3t)\,\mathbf{j}.$$

(See **Figure 4.1.**) The chain rule makes it possible to find df/dt without substituting the formulas for x and y in terms of t into the formula e^{xy} for f.

Although we could give a proof like the one in Section 2.6, based on Definition 2.1 of a differentiable scalar function, we can avoid some complexity by assuming throughout that the partial derivatives of f are continuous. This not only guarantees that f is differentiable (Theorem 2.5) but also permits us to derive the chain rule by using the multidimensional version of the mean value theorem for derivatives (Theorem 10.2 of Chapter 2). The first step is to state this version of the mean value theorem. The proof, whose details are not directly related to the chain rule, is given at the end of the section.

4.1 THEOREM

Mean Value Theorem. Let f be a differentiable scalar function defined on an open ball B containing the line segment joining the distinct points $\mathbf{a}$ and $\mathbf{b}$ in $\mathbf{R}^n$. Then there is a point c on this segment strictly between $\mathbf{a}$ and $\mathbf{b}$ such that

$$f(\mathbf{b}) - f(\mathbf{a}) = \mathbf{f}'(\mathbf{c}) \cdot (\mathbf{b} - \mathbf{a}). \tag{1}$$

Theorem 4.5 of Chapter 10 says that the segment joining the endpoints of $\mathbf{a}$ and $\mathbf{b}$ in $\mathbf{R}^n$ has vector equation

$$\mathbf{x}(t) = (1 - t)\,\mathbf{a} + t\,\mathbf{b}, \qquad 0 \le t \le 1.$$

Thus the mean value theorem asserts that there is some t_1 strictly between 0 and 1 such that $\mathbf{c} = (1 - t_1)\,\mathbf{a} + t_1\,\mathbf{b}$ satisfies (1). This is illustrated for three dimensions in **Figure 4.2.** When $n = 1$, Theorem 4.1 reduces to the ordinary mean value theorem, because the segment joining a to b is just the closed interval $I = [a, b]$. Since f is assumed to be differentiable on an open ball (= open interval when $n = 1$) containing I, it is continuous on I and certainly differentiable on the open interval (a, b). Therefore the hypotheses of Theorem 4.1 become those of the one-variable mean value theorem, and (1) is exactly the conclusion of that result. So Theorem 4.2 is a generalization of that theorem to higher dimensions.

The mean value theorem is of theoretical rather than computational importance. The following example is thus intended primarily to illustrate what the theorem says in a concrete situation.

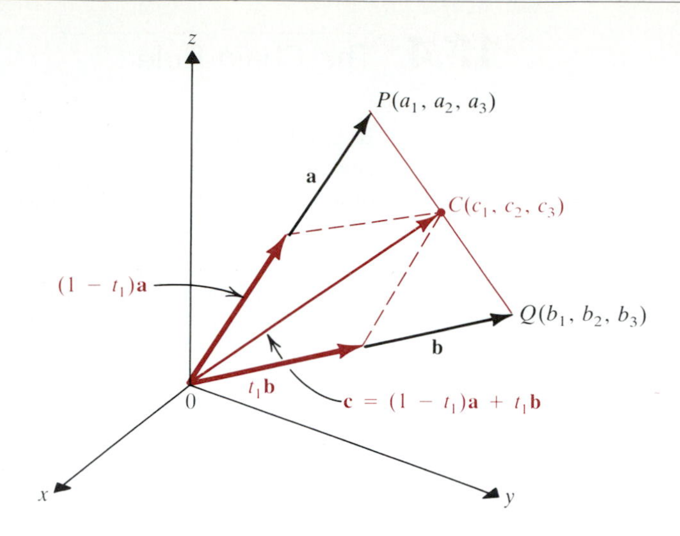

FIGURE 4.2
At $\mathbf{c}$, $f(\mathbf{b}) - f(\mathbf{a}) = \nabla f(\mathbf{c}) \cdot (\mathbf{b} - \mathbf{a})$

Example 1

If $f\colon \mathbf{R}^2 \to \mathbf{R}$ is given by

$$f(x, y) = \frac{x^2}{4} + \frac{y^2}{9},$$

then find $\mathbf{c}$ on the line segment joining (0, 0) to (2, 3) such that

$$\mathbf{f}'(\mathbf{c}) \cdot (2\mathbf{i} + 3\mathbf{j}) = f(2, 3) - f(0, 0).$$

Solution. We have

$$f(2, 3) = 2, \qquad f(0, 0) = 0, \qquad \text{and} \qquad \mathbf{f}'(x, y) = \frac{x}{2}\mathbf{i} + \frac{2y}{9}\mathbf{j}.$$

We must find $\mathbf{c} = (c_1, c_2)$ such that

$$\left(\frac{c_1}{2}\mathbf{i} + \frac{2c_2}{9}\mathbf{j}\right) \cdot (2\mathbf{i} + 3\mathbf{j}) = 2 - 0 \to c_1 + \frac{2}{3}c_2 = 2. \tag{2}$$

Since (c_1, c_2) is to be on the line segment strictly between (0, 0) and (2, 3), we also have

$$(c_1, c_2) = (0, 0) + t(2, 3) = (2t, 3t), \qquad t \in (0, 1).$$

Thus

$$2t + \frac{2}{3}(3t) = 2 \to 4t = 2 \to t = \tfrac{1}{2}.$$

Hence $\mathbf{c} = (1, 3/2)$ works. (You can readily check this if you substitute in (2).) ■

We are now ready to derive the version of the chain rule for the composite of a scalar function f of n variables following an n-dimensional vector function $\mathbf{g}$. **Figure 4.3** shows a schematic representation of f and $\mathbf{g}$ for $n = 3$.

4.2 THEOREM ***Chain Rule.*** Suppose that the vector function $\mathbf{g}\colon \mathbf{R} \to \mathbf{R}^n$ is differentiable at $t = t_0$. Suppose also that every partial derivative of the scalar function f exists and is continuous on an open ball B containing $\mathbf{x}_0 = \mathbf{g}(t_0)$. Then $f \circ \mathbf{g}\colon \mathbf{R} \to \mathbf{R}$ is differentiable at t_0 and

$$(f \circ \mathbf{g})'(t_0) = \mathbf{f}'(\mathbf{x}_0) \cdot \mathbf{g}'(t_0) = \nabla f(\mathbf{x}_0) \cdot \frac{d\mathbf{g}}{dt}(t_0) \tag{3}$$

FIGURE 4.3

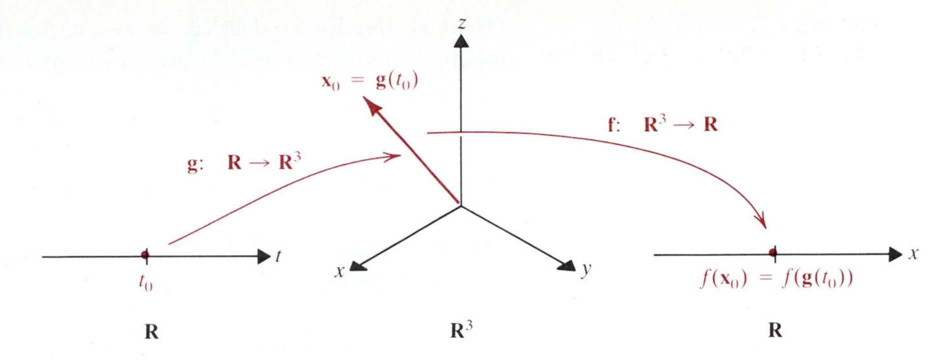

Proof. First, Theorem 2.5 assures us that f is differentiable at $\mathbf{x}_0$. Since $\mathbf{g}$ is differentiable at t_0, it is continuous there. Thus when t is sufficiently close to t_0, $\mathbf{g}(t)$ is in B. Then Theorem 4.1 applied to the function $\mathbf{g}$ gives

$$(f \circ \mathbf{g})'(t_0) = \lim_{t \to t_0} \frac{f(\mathbf{g}(t)) - f(\mathbf{g}(t_0))}{t - t_0} = \lim_{t \to t_0} \frac{\mathbf{f}'(\mathbf{c}(t)) \cdot (\mathbf{g}(t) - \mathbf{g}(t_0))}{t - t_0}$$

for some $\mathbf{c}(t) = (1 - t_1)\mathbf{g}(t_0) + t_1\mathbf{g}(t)$ on the line segment joining $\mathbf{g}(t_0)$ and $\mathbf{g}(t)$. Note that $\mathbf{g}(t) \to \mathbf{g}(t_0)$ as $t \to t_0$, because $\mathbf{g}$ is continuous at t_0. Thus $\mathbf{c}(t) \to \mathbf{g}(t_0)$ as $t \to t_0$, since $\mathbf{c}(t)$ lies on the segment between $\mathbf{g}(t_0)$ and $\mathbf{g}(t)$. See **Figure 4.4.** Now, $\mathbf{f}'$ is continuous on that segment, so as $t \to t_0$,

$$\mathbf{f}'(\mathbf{c}(t)) \to \mathbf{f}'(\mathbf{g}(t_0)) = \mathbf{f}'(\mathbf{x}_0).$$

Hence

$$(f \circ \mathbf{g})'(t_0) = \mathbf{f}'(\mathbf{x}_0) \cdot \lim_{t \to t_0} \frac{\mathbf{g}(t) - \mathbf{g}(t_0)}{t - t_0} = \mathbf{f}'(\mathbf{x}_0) \cdot \mathbf{g}'(t_0). \quad \boxed{\text{QED}}$$

FIGURE 4.4

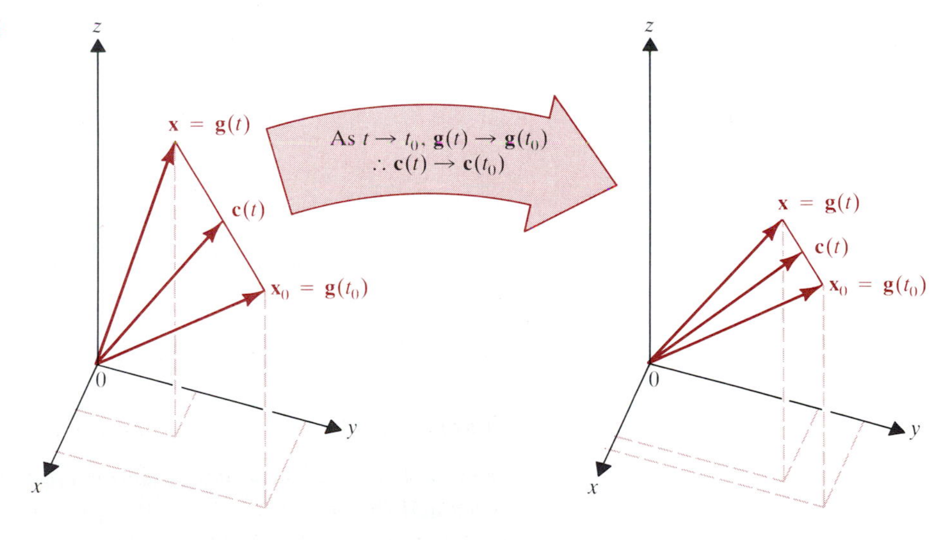

Instead of $(f \circ \mathbf{g})'(t_0)$ in (3), we sometimes write

$$\frac{d(f \circ \mathbf{g})}{dt}(t_0) \quad \text{or} \quad \left.\frac{dz}{dt}\right|_{t_0}, \qquad \text{where } z = f(\mathbf{g}(t)) = f(\mathbf{x}).$$

Note the analogy between Equation (3) and the chain rule of Section 2.6: *To compute $(f \circ \mathbf{g})'(t_0)$, take the total derivative of f with respect to the intermediate variable $\mathbf{x}$, take the derivative of $\mathbf{g}$ with respect to t, and then multiply* *by taking the* dot *product of the two vectors.* We illustrate by returning to the function mentioned at the start of the section.

Example 2

If $z = e^{xy}$, where $x = 2t^3 + 1$ and $y = t^2 + 3t$, then find dz/dt.

Solution. Here $z = e^{xy} = f(\mathbf{x})$, where $\mathbf{x} = \mathbf{g}(t) = (2t^3 + 1)\mathbf{i} + (t^2 + 3t)\mathbf{j}$. Applying (3), we obtain

$$\frac{dz}{dt} = \mathbf{f}'(\mathbf{x}) \cdot \mathbf{g}'(t) = \nabla f(\mathbf{x}) \cdot \frac{d\mathbf{g}(t)}{dt} = [ye^{xy}\mathbf{i} + xe^{xy}\mathbf{j}] \cdot [6t^2\mathbf{i} + (2t + 3)\mathbf{j}]$$

$$= 6t^2 ye^{xy} + (2t + 3)xe^{xy}.$$

We can express dz/dt as a function of t alone by substituting $x = 2t^3 + 1$ and $y = t^2 + 3t$. That gives

$$\frac{dz}{dt} = [6t^2(t^2 + 3t) + (2t + 3)(2t^3 + 1)]e^{(2t^3+1)(t^2+3t)}. \quad ■$$

We could have found dz/dt by substituting $x = 2t^3 + 1$ and $y = t^2 + 3t$ into the formula for z and then differentiating with respect to t, but that is more work than using the chain rule.

We can extend Chain Rule 4.2 to differentiate a composite function $f \circ \mathbf{g}$, where f is a scalar function of n variables, each of which is in turn a function of m variables $t_1, t_2, \ldots, t_m$. The schematic situation is shown in **Figure 4.5** for the case $m = 2$ and $n = 3$. The next example shows how such composite functions are differentiated.

FIGURE 4.5

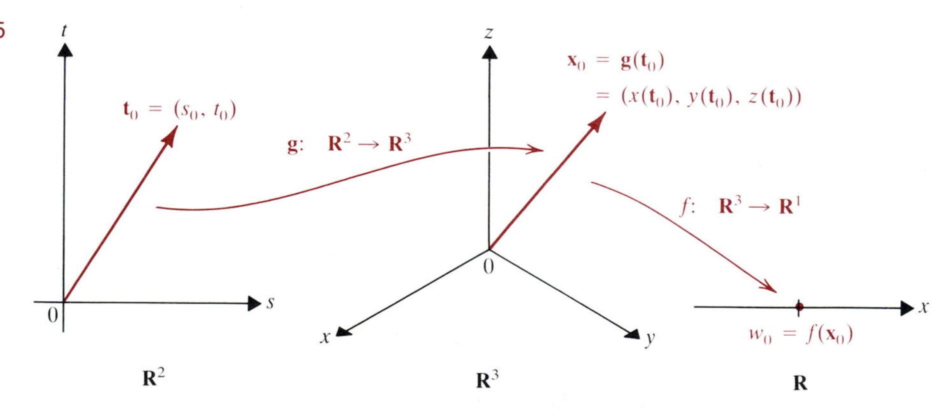

Example 3

Let $w = e^{x^2+y^2+z^2}$, where $x = \sin s + \cos t$, $y = \tan s^2 t$, and $z = s^2 - t^2$. Find $\partial w/\partial s$ and $\partial w/\partial t$.

Solution. Here $w = f(x, y, z) = e^{x^2+y^2+z^2}$, where each intermediate variable x, y, and z is a function of two variables s and t:

$$\begin{cases} x = \sin s + \cos t = g_1(s, t), \\ y = \tan s^2 t \qquad\quad = g_2(s, t), \\ z = s^2 - t^2 \qquad\quad = g_3(s, t). \end{cases} \tag{4}$$

These three equations define a vector function $\mathbf{g}:\mathbf{R}^2 \to \mathbf{R}^3$ of *two* variables s and t, which we represent by turning (4) into the vector equation

$$\mathbf{x} = \begin{pmatrix} x \\ y \\ z \end{pmatrix} = \mathbf{g}(t) = \begin{pmatrix} g_1(s, t) \\ g_2(s, t) \\ g_3(s, t) \end{pmatrix} = \begin{pmatrix} \sin s + \cos t \\ \tan s^2 t \\ s^2 - t^2 \end{pmatrix}. \tag{5}$$

Here it is most natural to translate the vertically displayed Equations (4) into a *column* vector equation, because all we need to do is draw in large parentheses to obtain (5) from (4). To find $\partial w/\partial s$, we freeze the variable t and regard (5) as a vector function of the single variable s, so that we can use (3). In this way we obtain from Chain Rule 4.2

$$\frac{\partial w}{\partial s} = \nabla f(x, y, z) \cdot \left(\frac{\partial \mathbf{g}}{\partial s}\right) = \left(\frac{\partial f}{\partial x}, \frac{\partial f}{\partial y}, \frac{\partial f}{\partial z}\right) \cdot \begin{pmatrix} \partial x/\partial s \\ \partial y/\partial s \\ \partial z/\partial s \end{pmatrix}$$

$$= (2xe^{x^2+y^2+z^2}, 2ye^{x^2+y^2+z^2}, 2ze^{x^2+y^2+z^2}) \cdot \begin{pmatrix} \partial x/\partial s \\ \partial y/\partial s \\ \partial z/\partial s \end{pmatrix}$$

$$= 2e^{x^2+y^2+z^2}(x, y, z) \cdot \begin{pmatrix} \cos s \\ 2st \sec^2 s^2 t \\ 2s \end{pmatrix}$$

$$= 2e^{x^2+y^2+z^2}[x \cos s + 2st\, y \sec^2 s^2 t + 2sz].$$

Similarly, we find $\partial w/\partial t$ by freezing s and treating (5) as a vector function of t alone. Then (3) gives

$$\frac{\partial w}{\partial t} = \nabla f(x, y, z) \cdot \frac{\partial \mathbf{g}}{\partial t} = 2e^{x^2+y^2+z^2}(x, y, z) \cdot \begin{pmatrix} -\sin t \\ s^2 \sec^2 s^2 t \\ -2t \end{pmatrix}$$

$$= 2e^{x^2+y^2+z^2}(-x \sin t + s^2 y \sec^2 s^2 t - 2tz). \blacksquare$$

The reasoning used in Example 3 can be used (Exercise 35) to prove the following general rule. As usual, if m or n is 2, then the third coordinate(s) do(es) not appear.

4.3 THEOREM

General Chain Rule for Partial Derivatives. Suppose that the vector function $\mathbf{g}:\mathbf{R}^m \to \mathbf{R}^n$, where m and n are 2 or 3, is given by

$$\mathbf{x} = \begin{pmatrix} x \\ y \\ z \end{pmatrix} = \mathbf{g}(t) = \mathbf{g}(s, t, u) = \begin{pmatrix} g_1(s, t, u) \\ g_2(s, t, u) \\ g_3(s, t, u) \end{pmatrix} \in \mathbf{R}^n.$$

Suppose that each partial derivative of g_1, g_2, and g_3 if present, exists with respect to each coordinate variable t_j of $\mathbf{t} = (t_1, t_2, t_3)$. Suppose also that the scalar function $f: \mathbf{R}^n \to \mathbf{R}$ has continuous partial derivatives for all $\mathbf{x}$ near $\mathbf{x}_0 = \mathbf{g}(\mathbf{t}_0)$. Then

$$\frac{\partial(f \circ \mathbf{g})}{\partial t_j}(\mathbf{t}_0) = \mathbf{f}'(\mathbf{x}_0) \cdot \frac{\partial \mathbf{g}}{\partial t_j}(\mathbf{t}_0) = \nabla f(\mathbf{x}_0) \cdot \frac{\partial \mathbf{g}}{\partial t_j}(\mathbf{t}_0), \tag{6}$$

where

$$\frac{\partial \mathbf{g}}{\partial t_j} = \begin{pmatrix} \dfrac{\partial g_1}{\partial t_j} \\ \dfrac{\partial g_2}{\partial t_j} \\ \dfrac{\partial g_3}{\partial t_j} \end{pmatrix} = \begin{pmatrix} \dfrac{\partial x}{\partial t_j} \\ \dfrac{\partial y}{\partial t_j} \\ \dfrac{\partial z}{\partial t_j} \end{pmatrix} = \frac{\partial x}{\partial t_j}\mathbf{i} + \frac{\partial y}{\partial t_j}\mathbf{j} + \frac{\partial z}{\partial t_j}\mathbf{k}.$$

In other words, *to compute the partial derivative of* $f \circ \mathbf{g}:\mathbf{R}^n \to \mathbf{R}$ *with respect to its jth ultimate variable* t_j, *take the dot product of the total derivative of f with respect to the intermediate variable* $\mathbf{x}$ *and the partial derivative of* $\mathbf{x}$ *with respect to* t_j.

In Example 3 the total derivative of w with respect to its ultimate variables s and t is obtained by putting the continuous partial derivatives $\partial w/\partial s$ and $\partial w/\partial t$ into the vector

$$\nabla w = \left(\frac{\partial w}{\partial s}, \frac{\partial w}{\partial t}\right) = \frac{\partial w}{\partial s}\mathbf{i} + \frac{\partial w}{\partial t}\mathbf{j}$$

$$= 2e^{x^2+y^2+z^2}[(x\cos z + 4sty\sec^2 s^2t + 2sz)\mathbf{i} - (x\sin t - s^2y\sec^2 s^2t + 2tz)\mathbf{j}].$$

More generally, if in Theorem 4.3 each partial derivative $\partial g_i/\partial t_j$ is continuous at $\mathbf{t}_0$, then so is $\partial(f \circ \mathbf{g})/\partial t_j$, because it is the sum of the products of the continuous functions $\partial f/\partial x_i$ and $\partial g_i/\partial t_j$, where $x_i = x$, y, or z. For we have

$$\mathbf{f}'(\mathbf{x}) = \left(\frac{\partial f}{\partial x}, \frac{\partial f}{\partial y}, \frac{\partial f}{\partial z}\right) = \frac{\partial f}{\partial x}\mathbf{i} + \frac{\partial f}{\partial y}\mathbf{j} + \frac{\partial f}{\partial z}\mathbf{k}$$

and

$$\frac{\partial \mathbf{g}}{\partial t_j} = \begin{pmatrix} \partial x/\partial t_j \\ \partial y/\partial t_j \\ \partial z/\partial t_j \end{pmatrix} = \frac{\partial x}{\partial t_j}\mathbf{i} + \frac{\partial y}{\partial t_j}\mathbf{j} + \frac{\partial z}{\partial t_j}\mathbf{k},$$

so that

$$\mathbf{f}'(\mathbf{x}) \cdot \frac{\partial \mathbf{g}}{\partial t_j} = \frac{\partial f}{\partial x}\frac{\partial x}{\partial t_j} + \frac{\partial f}{\partial y}\frac{\partial y}{\partial t_j} + \frac{\partial f}{\partial z}\frac{\partial z}{\partial t_j}.$$

Since each $\partial(f \circ \mathbf{g})/\partial t_j$ is continuous, $f \circ \mathbf{g}$ is differentiable by Theorem 2.5, so its total derivative is just the result of putting together the $\partial(f \circ \mathbf{g})/\partial t_j$ for $j = 1, 2, \ldots, m$. Thus we have proved the following theorem.

4.4 THEOREM ***General Chain Rule for Total Derivatives.*** If in Theorem 4.3 all the $\partial g_i/\partial t_j$ are continuous at $\mathbf{t}_0$, then $f \circ \mathbf{g}:\mathbf{R}^m \to \mathbf{R}$ defined by $w = f(\mathbf{g}(t))$ is differentiable at $\mathbf{t}_0$ and its total derivative is given by

$$(\mathbf{f} \circ \mathbf{g})'(\mathbf{t}_0) = \left(\frac{\partial(f \circ \mathbf{g})}{\partial t_1}(\mathbf{t}_0), \frac{\partial(f \circ \mathbf{g})}{\partial t_2}(\mathbf{t}_0), \frac{\partial(f \circ \mathbf{g})}{\partial t_3}(\mathbf{t}_0)\right)$$

$$= \left(\nabla f(\mathbf{x}_0) \cdot \frac{\partial \mathbf{g}}{\partial t_1}(\mathbf{t}_0), \nabla f(\mathbf{x}_0) \cdot \frac{\partial \mathbf{g}}{\partial t_2}(\mathbf{t}_0), \nabla f(\mathbf{x}_0) \cdot \frac{\partial \mathbf{g}}{\partial t_3}(\mathbf{t}_0)\right).$$

Example 4

Let $w = x^2 + y^2 + z^2$, where $x = s^2 + t^2 + u^2$, $y = stu$, and $z = s^2 + t^2u$. Find the total derivative of w as a function of s, t, and u.

Solution. Here $w = f(x, y, z) = x^2 + y^2 + z^2$, where

$$\text{(7)} \qquad \begin{pmatrix} x \\ y \\ z \end{pmatrix} = \begin{pmatrix} s^2 + t^2 + u^2 \\ stu \\ s^2 + t^2 u \end{pmatrix} = \begin{pmatrix} g_1(s, t, u) \\ g_2(s, t, u) \\ g_3(s, t, u) \end{pmatrix} = \mathbf{g}(s, t, u).$$

Proceeding as in Example 3, we get from (6)

$$\frac{\partial w}{\partial s} = \nabla f(x) \cdot \frac{\partial \mathbf{g}}{\partial s} = (2x, 2y, 2z) \cdot \begin{pmatrix} 2s \\ tu \\ 2s \end{pmatrix} = 4sx + 4tuy + 4sz,$$

$$\frac{\partial w}{\partial t} = \nabla f(x) \cdot \frac{\partial \mathbf{g}}{\partial t} = (2x, 2y, 2z) \cdot \begin{pmatrix} 2t \\ su \\ 2tu \end{pmatrix} = 4tx + 2suy + 4tuz,$$

$$\frac{\partial w}{\partial u} = \nabla f(x) \cdot \frac{\partial \mathbf{g}}{\partial u} = (2x, 2y, 2z) \cdot \begin{pmatrix} 2u \\ st \\ t^2 \end{pmatrix} = 4ux + 2sty + 2t^2 z.$$

Thus

$$\mathbf{w}'(t) = (4sz + 4tuy + 4sz)\,\mathbf{i} + (4tx + 2suy + 4tuz)\,\mathbf{j} + (4ux + 2sty + 2t^2 z)\,\mathbf{k}.$$

If we wished, we could use (7) to express $\mathbf{w}'(t)$ in terms of s, t, and u. ■

Tangent Planes to Level Surfaces

The chain rule provides an easy means of finding the tangent plane to a *level surface* $f(x, y, z) = c$. We once again assume that $\partial f/\partial x$, $\partial f/\partial y$, and $\partial f/\partial z$ are continuous. Let $\mathbf{x} = \mathbf{g}(t) = (x(t), y(t), z(t))$ be a curve on the level surface, passing through the point $\mathbf{x}_0 = \mathbf{g}(t_0) = (x(t_0), y(t_0), z(t_0))$. Then, since the curve lies on the level surface, we have

$$f(\mathbf{g}(t)) = c \rightarrow (f \circ \mathbf{g})(t) = c.$$

If we differentiate this equation with respect to t and use Theorem 4.2, then we get

$$(f \circ \mathbf{g})'(t_0) = \nabla f(\mathbf{x}(t_0)) \cdot \mathbf{g}'(t_0) = \frac{d}{dt}(c) = 0.$$

This says that the total derivative $\nabla f(\mathbf{x}_0)$ is either $\mathbf{0}$ or perpendicular to the tangent vector $\mathbf{g}'(t_0)$ to the curve $\mathbf{x} = \mathbf{g}(t)$ at $\mathbf{g}(t_0) = \mathbf{x}_0$. See **Figure 4.6.** Thus, if $\nabla f(\mathbf{x}_0) \neq \mathbf{0}$, then it is perpendicular to the level surface at $\mathbf{x}_0 = (x(t_0), y(t_0), z(t_0))$, because it is perpendicular to *any* curve $\mathbf{x} = \mathbf{g}(t)$ on the surface that passes through $\mathbf{x}_0$. Recall that the normal-form equation of a plane through $\mathbf{x}_0$ with normal vector $\mathbf{n}$ is $\mathbf{n} \cdot (\mathbf{x} - \mathbf{x}_0) = 0$ (Definition 5.1 of Chapter 10). We thus make the following definition.

4.5 DEFINITION Suppose that the scalar function $f\colon \mathbf{R}^3 \rightarrow \mathbf{R}$ has continuous partial derivatives $\partial f/\partial x$, $\partial f/\partial y$, and $\partial f/\partial z$ near $\mathbf{x} = \mathbf{x}_0$. If $\nabla f(\mathbf{x}_0) \neq \mathbf{0}$, then the ***tangent plane to the level surface*** $f(x, y, z) = c$ at $\mathbf{x}_0$ is the plane with normal-form equation

$$\nabla f(\mathbf{x}_0) \cdot (\mathbf{x} - \mathbf{x}_0) = 0.$$

This means that $\nabla f(\mathbf{x}_0)$ is a normal vector to the level surface $f(x, y, z) = c$.

Example 5

Find the equation of the tangent plane to the level surface $xyz = 2$ at the point $\mathbf{x}_0 = (1, 1, 2)$.

FIGURE 4.6

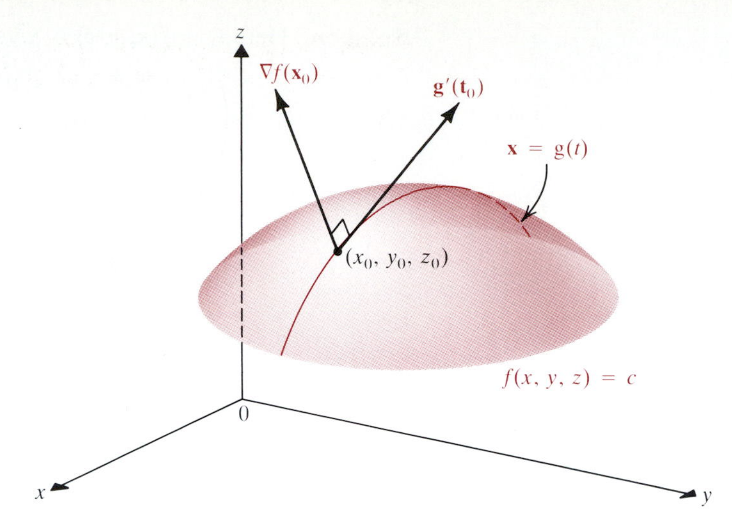

Solution. Here $f(x, y, z) = xyz$. We find $\nabla f(\mathbf{x}) = yz\,\mathbf{i} + xz\,\mathbf{j} + xy\,\mathbf{k}$, so $\nabla f(1, 1, 2) = 2\,\mathbf{i} + 2\,\mathbf{j} + \mathbf{k}$. The tangent plane then has the equation

$$(2\,\mathbf{i} + 2\,\mathbf{j} + \mathbf{k}) \cdot (\mathbf{x} - \mathbf{x}_0) = 0 \rightarrow (2\,\mathbf{i} + 2\,\mathbf{j} + \mathbf{k}) \cdot [(x-1)\,\mathbf{i} + (y-1)\,\mathbf{j} + (z-2)\,\mathbf{k}] = 0,$$

$$2(x-1) + 2(y-1) + z - 2 = 0 \rightarrow 2x + 2y + z = 6.$$

Figure 4.7 shows the surface and the tangent plane, which lies below the surface. ■

FIGURE 4.7

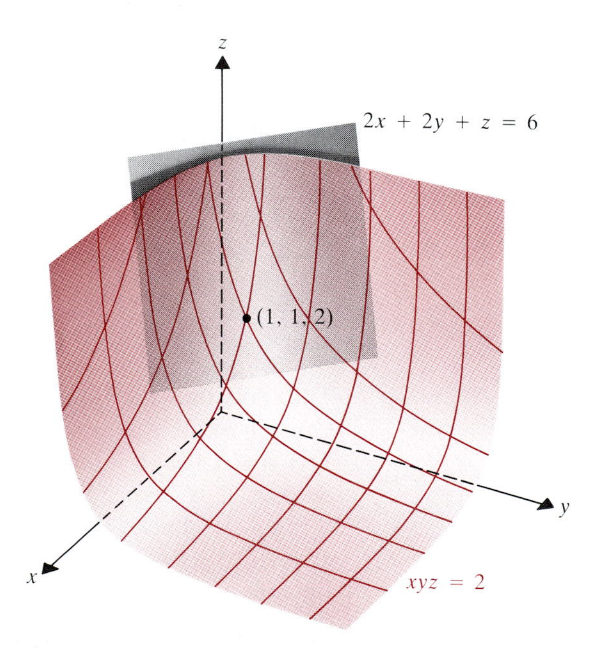

Proof of the Mean Value Theorem*

4.1 THEOREM Let f be a differentiable scalar function defined on an open ball B containing the line segment joining the points $\mathbf{a} \neq \mathbf{b}$ in $\mathbf{R}^n$. Then there is a point $\mathbf{c}$ on this segment strictly between $\mathbf{a}$ and $\mathbf{b}$ such that

$$f(\mathbf{b}) - f(\mathbf{a}) = f'(\mathbf{c}) \cdot (\mathbf{b} - \mathbf{a}). \tag{1}$$

* Optional

Proof. As in the proof of Theorem 10.2 of Chapter 2, we introduce an auxiliary function $g\colon \mathbf{R} \to \mathbf{R}$, which is defined for $t \in [0, 1]$ by

$$g(t) = f(\mathbf{a} + t(\mathbf{b} - \mathbf{a})).$$

Notice that

$$g(1) = f(\mathbf{b}) \qquad \text{and} \qquad g(0) = f(\mathbf{a}).$$

Consequently,

$$f(\mathbf{b}) - f(\mathbf{a}) = g(1) - g(0) = \frac{g(1) - g(0)}{1 - 0}. \tag{8}$$

To prove that (1) holds for some $\mathbf{c}$ between $\mathbf{a}$ and $\mathbf{b}$, we apply Theorem 10.2 of Chapter 2 to g on the interval $[0, 1]$, and use (8). First we show that g satisfies the hypotheses of the ordinary mean value theorem. That is, we must show that g is continuous on $[0, 1]$ and differentiable on $(0, 1)$. The vector function $\mathbf{x}$ given by

$$\mathbf{x}(t) = \mathbf{a} + t(\mathbf{b} - \mathbf{a}) \tag{9}$$

is continuous on $[0, 1]$, and by hypothesis f is differentiable and hence continuous by Theorem 2.3. Therefore (Exercise 36),

$$g = f \circ \mathbf{x}$$

is continuous on $[0, 1]$. To see that g is differentiable on $(0, 1)$, let t_0 be any real number strictly between 0 and 1. Let $\mathbf{x} = \mathbf{x}(t)$ and let $\mathbf{x}_0 = \mathbf{x}(t_0)$ in (9). That is, $\mathbf{x} = \mathbf{a} + t(\mathbf{b} - \mathbf{a})$ and

$$\mathbf{x}_0 = \mathbf{a} + t_0(\mathbf{b} - \mathbf{a}). \tag{10}$$

Then from (9) and (10) we see that

$$\mathbf{x} - \mathbf{x}_0 = (t - t_0)(\mathbf{b} - \mathbf{a}). \tag{11}$$

Therefore, writing $e(\mathbf{x})$ for $f(\mathbf{x}) - f(\mathbf{x}_0) - \mathbf{f}'(\mathbf{x}_0) \cdot (\mathbf{x} - \mathbf{x}_0)$, we have

$$\begin{aligned}\frac{g(t) - g(t_0)}{t - t_0} = \frac{f(\mathbf{x}) - f(\mathbf{x}_0)}{t - t_0} &= \frac{\mathbf{f}'(\mathbf{x}_0) \cdot (\mathbf{x} - \mathbf{x}_0) + e(\mathbf{x})}{t - t_0} \\ &= \frac{\mathbf{f}'(\mathbf{x}_0) \cdot (t - t_0)(\mathbf{b} - \mathbf{a})}{t - t_0} + \frac{e(\mathbf{x})}{t - t_0}.\end{aligned}$$

Thus

$$\begin{aligned}g'(t_0) &= \lim_{t \to t_0} \frac{g(t) - g(t_0)}{t - t_0} \\ &= \lim_{t \to t_0} \left[\mathbf{f}'(\mathbf{x}_0) \cdot (\mathbf{b} - \mathbf{a}) + \frac{e(\mathbf{x})}{t - t_0} \right]\end{aligned}$$

$$g'(t_0) = \mathbf{f}'(\mathbf{x}_0) \cdot (\mathbf{b} - \mathbf{a}), \tag{12}$$

because f is differentiable at $\mathbf{x}_0$: Definition 2.1 says that

$$\begin{aligned}0 = \lim_{t \to t_0} \frac{f(\mathbf{x}) - f(\mathbf{x}_0) - \mathbf{f}'(\mathbf{x}_0) \cdot (\mathbf{x} - \mathbf{x}_0)}{|\mathbf{x} - \mathbf{x}_0|} &= \lim_{t \to t_0} \frac{e(\mathbf{x})}{|t - t_0|\,|\mathbf{b} - \mathbf{a}|} \qquad \textit{by (11)} \\ = \frac{1}{\pm|\mathbf{b} - \mathbf{a}|} \lim_{t \to t_0} \frac{e(\mathbf{x})}{t - t_0}.\end{aligned}$$

That is, we have

$$\lim_{t \to t_0} \frac{e(\mathbf{x})}{t - t_0} = 0.$$

Thus g is differentiable on $(0, 1)$.

We can now apply Theorem 10.2 of Chapter 2 to g. We get from (8)

$$f(\mathbf{b}) - f(\mathbf{a}) = \frac{g(1) - g(0)}{1 - 0} = g'(t_1) \cdot (1 - 0)$$

$$= g'(t_1), \qquad \text{for some } t_1 \in (0, 1),$$

$$= f'(\mathbf{a} + t_1(\mathbf{b} - \mathbf{a})) \cdot (\mathbf{b} - \mathbf{a}), \qquad \textit{by (12) and (10)}$$

$$f(\mathbf{b}) - f(\mathbf{a}) = \mathbf{f}'(\mathbf{c}) \cdot (\mathbf{b} - \mathbf{a}),$$

where $\mathbf{c} = \mathbf{a} + t_1(\mathbf{b} - \mathbf{a})$ lies on the segment joining **a** and **b**, by Theorem 4.5 of Chapter 10. Since t is strictly between 0 and 1, **c** is strictly between **a** and **b**.

QED

Exercises 12.4

In Exercises 1–6, use Chain Rule 4.2 to find a formula for dw/dt. Evaluate at the given point.

1. $w = f(x, y) = xy^2$, where $\mathbf{x} = \cos t\,\mathbf{i} + \sin t\,\mathbf{j}$, $\mathbf{x}_0 = (0, 1)$
2. $w = \sqrt{x^2 + y^2}$, where $\mathbf{x} = t\,\mathbf{i} + t^2\mathbf{j}$, $\mathbf{x}_0 = (-2, 4)$
3. $w = \tan^{-1}(x^2 + y^2 + z^2)$, where $\mathbf{x} = t\,\mathbf{i} + t^2\,\mathbf{j} + t^3\,\mathbf{k}$, $\mathbf{x}_0 = (1, 1, 1)$
4. $w = e^{x^2 - y^2 - z^2}$, where $\mathbf{x} = t\,\mathbf{i} + 2t\,\mathbf{j} + t^2\,\mathbf{k}$, $\mathbf{x}_0(1, 2, 1)$
5. $w = \ln\sqrt{x^2 + y^2 + z^2}$, where $\mathbf{x} = \cos t\,\mathbf{i} + \sin t\,\mathbf{j} + t\,\mathbf{k}$, $\mathbf{x}_0 = (0, 1, \pi/2)$
6. $w = \sqrt{x^2 + y^2 + z^2}$, where $\mathbf{x} = t\,\mathbf{i} + t^2\,\mathbf{j} + t^3\,\mathbf{k}$, $\mathbf{x}_0 = (-1, 1, -1)$

The next two exercises illustrate Theorem 4.1.

7. Suppose that $f\colon \mathbf{R}^2 \to \mathbf{R}$ is given by $f(x, y) = x^2 + y^2$. Find a point **c** on the segment joining (0, 0) to (2, 1) such that $\mathbf{f}'(\mathbf{c}) \cdot (2\,\mathbf{i} + \mathbf{j}) = 5$.
8. Suppose that $f\colon \mathbf{R}^3 \to \mathbf{R}$ is given by $f(x, y, z) = x^2 + y^2 + z^2$. Find a point **c** on the segment joining (2, 1, 1) to (3, 2, −1) such that $f(3, 2, -1) - f(2, 1, 1) = \mathbf{f}'(\mathbf{c}) \cdot (\mathbf{i} + \mathbf{j} - 2\,\mathbf{k})$.

In Exercises 9–14 use Theorem 4.3 to find $\partial w/\partial s$ and $\partial w/\partial t$.

9. $w = x^2 + y^2$, $x = s\cos t$, $y = s\sin t$
10. $w = x^2y^2$, $x = s^2 + st + 2$, $y = 3t^2 + 2st - 1$
11. $w = x^2yz^2$, $x = ts$, $y = t/s$, $z = s/t$
12. $w = x^2e^{yz} + z^2e^{xy}$, $x = s + t$, $y = s^2 - t^2$, $z = s^2 + t^2$
13. $w = \ln\sqrt{x^2 + y^2 + z^2}$, $x = \cos st$, $y = \sin st$, $z = st$
14. $w = x^2 + y^2 + z^2$, $x = s^2 + st^2$, $y = 2st$, $z = s^3t^2$

In Exercises 15–22, use Theorem 4.4 to find the formula for the total derivative of w and evaluate it at the indicated point.

15. $w = x^2 - y^2$, $(x, y) = (s^2t + u, s^2u + t)$, at $(s_0, t_0, u_0) = (1, 2, -2)$
16. $w = \tan^{-1}\dfrac{y}{x}$, $(x, y) = (s^2 + t^2 + u^2, stu)$, at $(s_0, t_0, u_0) = (1, 1, 1)$
17. $w = xy$, $(x, y) = (r\cos\theta, r\sin\theta)$, at $(r_0, \theta_0) = (1, \pi/2)$
18. $w = x^2y + xy$, $(x, y) = (e^{st}, s^2 + t^2)$, at $(s_0, t_0) = (2, 1)$
19. $w = xyz$, $(x, y, z) = (te^s, se^t, st)$, at $(s_0, t_0) = (1, 1)$
20. $w = x^2 + y^2 - z^2$, $(x, y, z) = (r\cos\theta, r\sin\theta, r\theta)$, at $(r_0, \theta_0) = (1, \pi)$
21. $w = \ln\sqrt{x^2 + y^2 + z^2}$, $(x, y, z) = (s + t + u, stu, se^{tu})$, at $(s_0, t_0, u_0) = (1, 2, 0)$
22. $w = x^2e^{yz}$, $(x, y, z) = (st + u^2, s^2ut, s + tu)$, at $(s_0, t_0, u_0) = (-1, 1, 2)$

In Exercises 23–27, find the tangent plane to the given level surface at the given point.

23. $x^2y - 2yz + 4 = 0$ at $(-2, 2, 3)$
24. $xy^2 + yz^2 + z^3 + x^3 = 64$ at $(0, 0, 4)$
25. $x^2 - 3y^2 + z^2 = 5$ at $(2, 1, -2)$
26. $x^2 + y^2 + z^2 = 9$ at $(-2, 1, 2)$
27. $x^2 - y^2 - z^2 = 1$ at $(3, 2, -2)$
28. A plane is flying due east with a speed of 400 miles per hour and is climbing at the rate of 528 feet per minute. It is being tracked by an airport tower. Find how fast the plane is approaching the tower when it is 3 miles above the ground over a point exactly 4 miles due west of the tower.

29. A particle of mass m moves on the surface $z = f(x, y)$ along the curve $\mathbf{x} = (x(t), y(t), f(x(t), y(t)))$. Its *kinetic energy* k at time t is $\frac{1}{2}m\,|\mathbf{v}(t)|^2$, where $\mathbf{v}(t) = \mathbf{x}'(t)$. Show that k is given by

$$k = \frac{1}{2}m\left[(1 + f_x^2)\left(\frac{dx}{dt}\right)^2 + 2f_x f_y \frac{dx}{dt}\frac{dy}{dt} + (1 + f_y^2)\left(\frac{dy}{dt}\right)^2\right].$$

30. A quantity of gas is governed by the *ideal gas law* $PV = nRT$ (n, R constant). (Here recall that P is the pressure, V is the volume, and T is the temperature in Kelvin (K).) The constant factor nR is determined to be 1 liter·atmosphere/K. At a certain time, $T = 300$ K and is being raised by 5 K per minute. The pressure at this time is 2 atmospheres and is decreasing at the rate of 0.01 atmospheres/second. Find dV/dt at this time.

31. The temperature at a point (x, y, z) is given by $T = \sqrt{x^2 + y^2 + z^2}$. A particle traces the path $\mathbf{x}(t) = (\cos t, \sin t, t)$, $0 \le t \le 2\pi$. At what point of the path is the lowest temperature experienced? (*Think* before calculating any derivatives!)

32. If $z = f(x, y)$ where $x = r\cos\theta$ and $y = r\sin\theta$, then

(a) find $\partial z/\partial r$ and $\partial z/\partial\theta$, and **(b)** show that

$$\left(\frac{\partial z}{\partial x}\right)^2 + \left(\frac{\partial z}{\partial y}\right)^2 = \left(\frac{\partial z}{\partial r}\right)^2 + \frac{1}{r^2}\left(\frac{\partial z}{\partial\theta}\right)^2.$$

33. A region U is called *convex* if whenever $\mathbf{x}_1$ and $\mathbf{x}_2$ are in U, the entire line segment joining $\mathbf{x}_1$ and $\mathbf{x}_2$ is contained in $\mathbf{U}$. Suppose that $f\colon \mathbf{R}^3 \to \mathbf{R}$ is differentiable on a convex set U and $\mathbf{f}'(\mathbf{x}) \equiv \mathbf{0}$ on U. Suppose also that for every point P of U, there is an open ball $B \subseteq U$ containing P. Then show that f must be constant on U.

34. Use Exercise 33 to show that two functions that have the same total derivative on the set U must differ by a constant on U.

35. Prove Theorem 4.3.

36. In the proof of Theorem 4.1, show that $g = f \circ \mathbf{x}$ is continuous on $[0, 1]$.

12.5 Implicit Differentiation

In Section 2.8 we presented the technique of implicit differentiation of a functional equation

$$(1) \qquad F(x, y) = 0$$

to find dy/dx when (1) defines y as one or more functions of x near a point (c, d) on its graph. That technique makes it possible to easily approximate y for values of x near a by means of the tangent approximation

$$f(x) \approx f(a) + f'(a)(x - a).$$

The procedure followed in Section 2.8 was to use the chain rule to differentiate both sides of (1) with respect to x. The variable y was treated as a function of x defined by (1), so that, for example, differentiation of x^5y^6 gave

$$\frac{d}{dx}[x^5y^6] = 5x^4y^6 + 6x^5y^5\frac{dy}{dx}.$$

The differential calculus of scalar functions provides a simpler, more direct way to compute dy/dx from a given functional equation (1). In place of the chain rule of Section 2.8, we can use Theorem 4.2 to derive a formula for dy/dx in terms of the partial derivatives F_x and F_y of $F(x, y)$. That reduces the calculation of implicit derivatives to a simple exercise in partial differentiation.

To carry this out, we suppose that (1) defines y as one or more functions f of x near a point (a, b) on the graph of (1). Suppose also that the partial derivatives F_x and F_y are continuous on some open ball B around (a, b) and that $F_y(x, y) \neq 0$ for any point (x, y) in B. We let $F(x, f(x))$ play the role of $f(x, y)$ in Theorem 4.2, and we let the vector function $\mathbf{g}$ be

$$\mathbf{g}(t) = (x, y) = (t, f(t)) = t\,\mathbf{i} + f(t)\,\mathbf{j}.$$

Then Theorem 4.2 says that the derivative of $F(x, y)$ is

$$\begin{aligned}\frac{d}{dx}[F(x, f(x))] &= \frac{d}{dt}[F(x(t), y(t))] = \frac{d}{dt}[(F \circ \mathbf{g})(t)] \\ &= \nabla F(x, y) \cdot \left[\frac{dx}{dt}\mathbf{i} + \frac{dy}{dt}\mathbf{j}\right] \\ &= \left[\frac{\partial F}{\partial x}\mathbf{i} + \frac{\partial F}{\partial y}\mathbf{j}\right] \cdot \left[\mathbf{i} + \frac{dy}{dt}\mathbf{j}\right] \qquad \text{since } x = t, \\ &= \frac{\partial F}{\partial x} + \frac{\partial F}{\partial y}\frac{dy}{dt}.\end{aligned}$$

Since $F(x, y) = 0$, and the derivative of 0 is 0, we have

$$\frac{\partial F}{\partial x} + \frac{\partial F}{\partial y}\frac{dy}{dt} = 0 \rightarrow \frac{\partial F}{\partial y}\frac{dy}{dt} = -\frac{\partial F}{\partial x}.$$

We have assumed that $\partial F/\partial y \neq 0$, so we can divide through by it to get

$$\frac{dy}{dt} = -\frac{\dfrac{\partial F}{\partial x}}{\dfrac{\partial F}{\partial y}} = -\frac{F_x}{F_y}.$$

Since $x = t$, we thus have derived the promised formula for dy/dx on B:

$$\frac{dy}{dx} = -\frac{\dfrac{\partial F}{\partial x}}{\dfrac{\partial F}{\partial y}} = -\frac{F_x}{F_y} \tag{2}$$

To illustrate (2), we use it to rework Examples 3 and 4 of Section 2.8 (Problem 3 of the Prologue).

Example 1

If $y^7 - 11x^5y^6 - x^{15} + 11 = 0$ defines y as a function of x, then find dy/dx. Evaluate

$$\left.\frac{dy}{dx}\right|_{(1,1)}.$$

Use this to find an approximate value of y when $x = 1.002$.

Solution. Letting $F(x, y) = y^7 - 11x^5y^6 - x^{15} + 11$, we find

$$\frac{\partial F}{\partial x} = -55x^4y^6 - 15x^{14} \qquad \text{and} \qquad \frac{\partial F}{\partial y} = 7y^6 - 66x^5y^5.$$

So when $\partial F/\partial y \neq 0$, we have from (2)

$$\frac{dy}{dx} = -\frac{-55x^4y^6 - 15x^{14}}{7y^6 - 66x^5y^5} = \frac{55x^4y^6 + 15x^{14}}{7y^6 - 66x^5y^5}. \tag{3}$$

We thus have

$$\left.\frac{dy}{dx}\right|_{(1,1)} = \frac{70}{-59} \approx -1.186.$$

For $x = 1.002$, we therefore get

$$f(x) \approx f(1) + f'(1)(x - 1) \approx 1 + (-1.186)(1.002 - 1.000)$$
$$\approx 1 + (-1.186)(0.002) \approx 0.998. \quad \blacksquare$$

It is instructive to compare the work required to obtain (3) here with that used to solve Example 3 of Section 2.8.

The assumption that $\partial F/\partial y \neq 0$ is by no means incidental in the derivation of (2). Consider, for example, the simple case of the circle $x^2 + y^2 = 16$. Setting $F(x, y) = x^2 + y^2 - 16$, we have $F_x = 2x$ and $F_y = 2y$. Thus $F_y = 0$ when $y = 0$, that is, at the two points $(4, 0)$ and $(-4, 0)$. There the whole technique of implicit differentiation breaks down. First, the graph of $x^2 + y^2 = 16$ has vertical tangents at those points, so there is no derivative dy/dx. See **Figure 5.1.** Moreover, in any open ball (= disk) in $\mathbf{R}^2$ that contains $x = 4$ or $x = -4$, there is no single function f defined by $F(x, y) = 0$. Above the x-axis, we have $y = \sqrt{16 - x^2}$, whereas below the x-axis the formula for y is $y = -\sqrt{16 - x^2}$. Finally, if we tried to use (2), we would be stymied because that would involve division by zero at $x = \pm 4$.

To avoid all the chaos enountered in the last paragraph, it is enough to assume that

$$\frac{\partial F}{\partial y}(c, d) \neq 0.$$

The following theorem of advanced calculus makes this precise.

5.1 THEOREM **_Implicit Function Theorem._** Let F be defined on an open disk D containing the point (a, b) in $\mathbf{R}^2$. Suppose that $F(a, b) = 0$, that $F_y(a, b) \neq 0$, and that F_x and F_y are continuous on D. Then

$$F(x, y) = 0 \tag{1}$$

defines y as a function of x near the point (a, b). That is, there is an open interval I containing a, and there is a function $f: I \to \mathbf{R}$ such that for every real number x in I,

(a) the corresponding function value $y = f(x)$ is such that the point (x, y) belongs to D,

(b) $F(x, f(x)) = 0$, and

(c) $f'(x) = \dfrac{dy}{dx} = -\dfrac{F_x}{F_y}$.

FIGURE 5.1

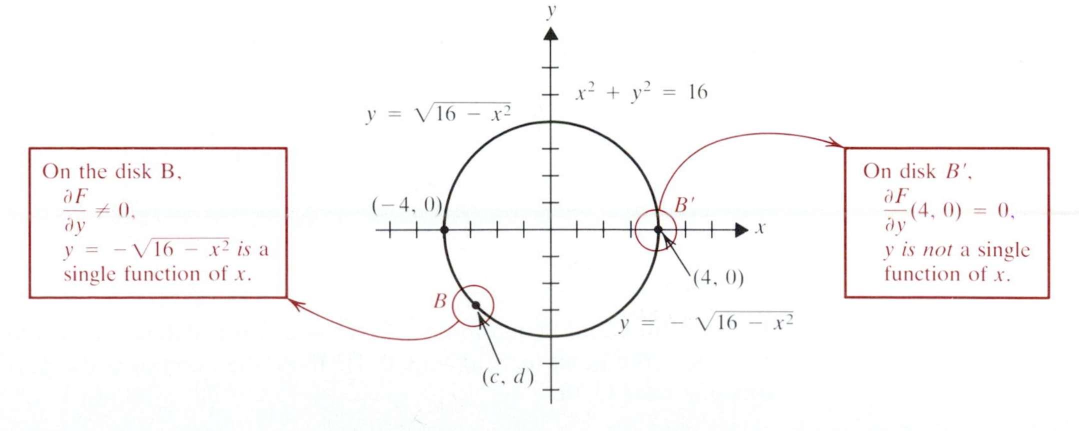

Returning to the circle $x^2 + y^2 = 16$, we see that for $(a, b) \neq (\pm 4, 0)$ we can safely assert that the equation does define y as a function of x near (a, b) and that for this function we have

$$\frac{dy}{dx} = -\frac{F_x}{F_y} = -\frac{x}{y},$$

for all (x, y) in some disk D containing (a, b) [that is, for all (x, y) near (a, b)].

Although we stated Theorem 5.1 just for functions $F\colon \mathbf{R}^2 \to \mathbf{R}$, it generalizes to functions $F\colon \mathbf{R}^n \to \mathbf{R}$ for any $n \geq 2$. To see how it applies to the case $n = 3$, consider

$$x^2 + 4y^2 - 3z^2 = 6 \tag{4}$$

near the point $(a, b, c) = (3, 0, 1)$. Note that F_x, F_y, and F_z are all continuous near this point. Assuming that (4) defines z as a differentiable function of x and y near $(a, b) = (3, 0)$, we might wish to find formulas like (2) for $\partial z/\partial x$ and $\partial z/\partial y$. Here z, as the dependent variable, plays the role of y in Theorem 5.1. Hence it is natural to conjecture that $\partial z/\partial x$ should be given by

$$\frac{\partial z}{\partial x} = -\frac{F_x}{F_z} = -\frac{2x}{-6z} = \frac{x}{3z}.$$

if $F_z \neq 0$. Similarly, we would expect that

$$\frac{\partial z}{\partial y} = -\frac{F_y}{F_z} = -\frac{8y}{-6z} = \frac{4y}{3z},$$

again provided that $F_z \neq 0$. These expectations are confirmed by the following extension of Theorem 5.1 to three dimensions.

5.2 THEOREM **_Implicit Function Theorem._** Suppose that $F\colon \mathbf{R}^3 \to \mathbf{R}$ is defined on an open ball B containing (a, b, c). Suppose further that $F(a, b, c) = 0$ and that F_x, F_y, and F_z are continuous on B. Finally, suppose that $F_z(a, b, c) \neq 0$. Then

$$F(x, y, z) = 0 \tag{5}$$

defines z as function $f(x, y)$ of the other two variables x and y near the point (a, b, c). That is, there is some open disk D in $\mathbf{R}^2$ containing (a, b) and such that for any (x, y) in D,

(a) $(x, y, f(x, y))$ is in B,

(b) $F(x, y, f(x, y)) = 0$,

(c) the function f is differentiable on D, and

$$\frac{\partial z}{\partial x} = -\frac{\dfrac{\partial F}{\partial x}}{\dfrac{\partial F}{\partial z}} = -\frac{F_x}{F_z}, \qquad \frac{\partial z}{\partial y} = -\frac{\dfrac{\partial F}{\partial y}}{\dfrac{\partial F}{\partial z}} = -\frac{F_y}{F_z}. \tag{6}$$

As with Theorem 5.1, we leave the proof of this result to advanced calculus. The next example illustrates how it is used to analyze an equation like (4) above.

Example 2

Does the functional equation $x^2 + 4y^2 - 3z^2 = 6$ define z as a function of x and y near the point $(a, b, c) = (3, 0, 1)$? If so, then compute the partial derivatives of z near $(3, 0)$.

FIGURE 5.2

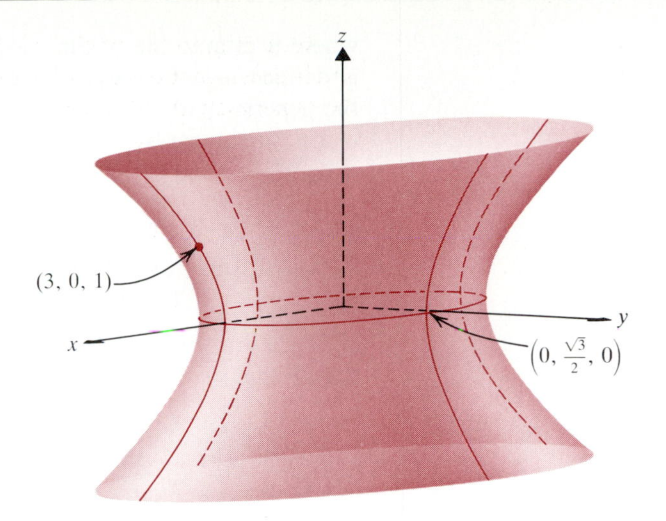

Solution. **Figure 5.2** shows the graph of the given functional equation, a hyperboloid of one sheet. Let $F(x, y, z) = x^2 + 4y^2 - 3z^2 - 6$. Then we have

$$F_z = \frac{\partial F}{\partial z} = -6z = -6 \neq 0 \qquad \text{at } (3, 0, 1).$$

Hence Theorem 5.2 says that the given equation does define z as a function $f(x, y)$ of x and y on an open disk D containing $(a, b) = (3, 0)$. Moreover, on D we get from (6) precisely the formulas conjectured before stating Theorem 5.2:

$$\frac{\partial z}{\partial x} = -\frac{F_x}{F_z} = -\frac{2x}{-6z} = \frac{x}{3z}, \qquad \frac{\partial z}{\partial y} = -\frac{F_y}{F_z} = -\frac{8y}{-6z} = \frac{4y}{3z}. \quad ■$$

Theorem 5.2 speaks just of z being defined as a function of the other two variables in a functional equation $F(x, y, z) = 0$. However, as suggested by the form of (4), in which the variables x, y, and z play symmetric roles, we can use the theorem to investigate whether any variable is defined as a function of the others. In fact, if we assume that F_x, F_y, and F_z are continuous near (a, b, c) and that $F_y(a, b, c) \neq 0$, then $F(x, y, z) = 0$ defines y as a differentiable function of x and z, and we have the following analogue of (6):

$$\text{(7)} \qquad \frac{\partial y}{\partial x} = -\frac{F_x}{F_y} = -\frac{\dfrac{\partial F}{\partial x}}{\dfrac{\partial F}{\partial y}}, \qquad \frac{\partial y}{\partial z} = -\frac{F_z}{F_y} = -\frac{\dfrac{\partial F}{\partial z}}{\dfrac{\partial F}{\partial y}}.$$

Similarly, if we assume that $F_x(a, b, c) \neq 0$, and the partial derivatives F_x, F_y, F_z are continuous near the point (a, b, c), then $F(x, y, z) = 0$ defines x as a differentiable function of y and z near (a, b, c), and

$$\text{(8)} \qquad \frac{\partial x}{\partial y} = -\frac{F_y}{F_x} = -\frac{\dfrac{\partial F}{\partial y}}{\dfrac{\partial F}{\partial x}}, \qquad \frac{\partial x}{\partial z} = -\frac{F_z}{F_x} = -\frac{\dfrac{\partial F}{\partial z}}{\dfrac{\partial F}{\partial x}}.$$

Formulas (6), (7), and (8) at first glance may seem difficult to keep straight, but happily there is a simple mnemonic for them. In each case, if $F_w(a, b, c) \neq 0$, where w is any one of the variables x, y, or z, then near the point (a, b, c), w is defined as a function of the other two variables, which we will denote by the generic symbols u and v. The partial derivatives of the dependent variable w are then given by

$$\text{(9)} \qquad \frac{\partial w}{\partial u} = -\frac{F_u}{F_w} = -\frac{\dfrac{\partial F}{\partial u}}{\dfrac{\partial F}{\partial w}}, \qquad \frac{\partial w}{\partial v} = -\frac{F_v}{F_w} = -\frac{\dfrac{\partial F}{\partial v}}{\dfrac{\partial F}{\partial w}}.$$

If $\partial F/\partial u$ and $\partial F/\partial w$ had an interpretation as quotients, then we could cancel the ∂F factors on the right side of the formula for $\partial w/\partial u$:

$$\frac{\dfrac{\partial F}{\partial u}}{\dfrac{\partial F}{\partial w}} = \frac{\dfrac{1}{\partial u}}{\dfrac{1}{\partial w}} = \frac{\partial w}{\partial u},$$

as desired! This partly explains the historical origins of the notation $\partial F/\partial x$ for F_x. However, since $\partial F/\partial x$ *isn't* interpretable as a quotient, the rationale we have given for the mnemonic must be classified as fortuitous nonsense. There is also a final caveat in using it: *don't forget the minus sign in Formula (6)*, which the cancellation scheme does not remind you of!

The next example illustrates the widened Theorem 5.2.

Example 3

Does the equation $x^2 + 4y^2 - 3z^2 = 6$ in Example 2 define y as a function of x and z near the point $(3, 0, 1)$? If so, then use (7) to find its partial derivatives.

Solution. Here y plays the role of the variable w in the discussion leading up to (9). According to Theorem 5.2, y is a function of x and z if F_x, F_y, and F_z are all continuous near $(a, b, c) = (3, 0, 1)$ and if F_y is not zero at this point. Clearly, all the partial derivatives are continuous everywhere, so that requirement is not a problem. However,

$$\frac{\partial F}{\partial y} = F_y = 8y = 0 \text{ at } (3, 0, 1).$$

Thus we cannot conclude from the theorem that y is defined as a function of x and z near $(3, 0, 1)$. Although the theorem does not provide any conclusion when $F_w(a, b, c)$ is zero, the discussion above for the equation $x^2 + y^2 = 16$ suggests that the given equation fails to define y as a unique function of x and z near $(3, 0, 1)$. Confirming such a suspicion requires a careful analysis of the given functional equation. For (4) we can do so as follows.

If we move from the point $(3, 0, 1)$ by increasing x slightly, then there is no y value defined by the given equation. Geometrically, we have moved toward the right and therefore off the surface in Figure 5.2. Algebraically, any y such that $F(x, y, z) = 0$ would have to satisfy

$$y^2 = 6 + 3z^2 - x^2 < 6 + 3(1)^2 - 3^2 = 6 + 3 - 9 = 0,$$

which is impossible, because y^2 cannot be negative. Hence, for $x > 3$ near $(3, 0, 1)$ there is no corresponding y defined by the equation. Since any open disk containing $(3, 0)$ would contain points (x, y) with $x > 3$, there can be no such open disk D around $(3, 0)$ where y is defined as a function of x and z. ■

Implicit partial differentiation can be used as in Example 1 to numerically approximate the value of a dependent variable w near a point (a, b, c) that is known to satisfy $F(x, y, z) = 0$. Exercises 17–20, for instance, use the tangent approximation

$$w = w_0 + \left[\frac{\partial w}{\partial u}\mathbf{i} + \frac{\partial w}{\partial v}\mathbf{j}\right]_{(u_0, v_0)} \cdot [(u - u_0)\mathbf{i} + (v - v_0)\mathbf{j}] \qquad (10)$$

to approximate $w = f(u, v)$ near a known point (u_0, v_0, w_0) on the graph of the functional equation (5). The next example illustrates the procedure.

Example 4

Note that $(1, -1, 2)$ satisfies the equation $x^2y - y^3z + x^4z^3 = 9$. Find an approximate value of z when $x = 1.0013$ and $y = -0.987$.

Solution. We first check that the given equation defines z as a function of x and y near the given point $(1, -1, 2)$. At that point we have

$$F_z = -y^3 + 3x^4z^2 = -(-1) + 3 \cdot 1 \cdot 4 = 13 \neq 0.$$

Hence Theorem 5.2 guarantees that z is a differentiable function of x and y near $(1, -1, 2)$. To use (10) we need to compute

$$\frac{\partial z}{\partial x} = -\frac{F_x}{F_z} = -\frac{2xy + 4x^3z^3}{-y^3 + 3x^4z^2} = -\frac{-2 + 32}{13} = -\frac{30}{13}$$

and

$$\frac{\partial z}{\partial y} = -\frac{F_y}{F_z} = -\frac{x^2 - 3y^2z}{-y^3 + 3x^4z^2} = -\frac{1 - 3 \cdot 2}{13} = \frac{5}{13},$$

at $(a, b, c) = (1, -1, 2)$. Then at the point $(1.0013, -0.987)$, (10) gives

$$\begin{aligned} z &\approx 2 + \left[-\frac{30}{13}\mathbf{i} + \frac{5}{13}\mathbf{j}\right] \cdot [(1.0013 - 1)\mathbf{i} + (-0.987 - (-1))\mathbf{j}] \\ &\approx 2 + \left[-\frac{30}{13}\mathbf{i} + \frac{5}{13}\mathbf{j}\right] \cdot [0.0013\,\mathbf{i} + 0.013\,\mathbf{j}] \\ &\approx 2 - 0.003 + 0.005 \approx 2.002. \quad \blacksquare \end{aligned}$$

Theorem 5.2 provides formulas for $\partial z/\partial x$ and $\partial z/\partial y$ when z is defined as a function of x and y near a point (a, b, c) that satisfies $F(a, b, c) = 0$. We can therefore use implicit differentiation to find the equation of the tangent plane to a level surface $f(x, y, z) = c$ by considering the functional equation

$$F(x, y, z) = f(x, y, z) - c = 0.$$

That gives an alternative to use of the chain rule (Definition 4.5). The following example lets you compare the two approaches directly.

Example 5

Find in two ways the equation of the tangent plane to the hyperboloid $x^2 + 4y^2 - 3z^2 = 6$ at the point $(3, 0, 1)$.

Solution. (a) We use the technique of Example 5 of the last section. Letting $G(x, y, z) = x^2 + 4y^2 - 3z^2$, we have

$$\nabla G(x, y, z) = 2x\,\mathbf{i} + 8y\,\mathbf{j} - 6z\,\mathbf{k} = 6\,\mathbf{i} - 6\,\mathbf{k}$$

at the given point (3, 0, 1). So an equation of the tangent plane is

$$[6\,\mathbf{i} - 6\,\mathbf{k}] \cdot [(x-3)\,\mathbf{i} + (y-0)\,\mathbf{j} + (z-1)\,\mathbf{k}] = 0$$

$$6x - 18 - 6z + 6 = 0 \rightarrow x - z = 2.$$

(b) We use the result of Example 2:

$$\frac{\partial z}{\partial x} = \frac{x}{3z} = \frac{3}{3} = 1, \qquad \frac{\partial z}{\partial y} = \frac{4y}{3z} = 0.$$

So by Definition 1.2, the tangent plane at (3, 0, 1) has equation

$$z - 1 = 1 \cdot (x - 3) + 0 \cdot (y - 0) \rightarrow z = x - 2.$$

Figure 5.3 repeats the graph of the hyperboloid of one sheet and shows the tangent plane at (3, 0, 1). ■

FIGURE 5.3

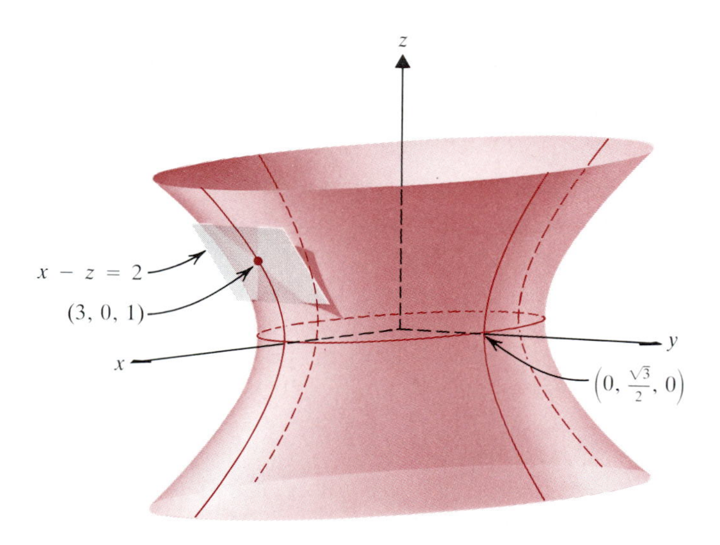

Just as there is a three-dimensional version (Theorem 5.2) of the implicit function theorem, so there is an n-dimensional version. We leave the formal statement and proof for advanced calculus, but the following example indicates that, computationally, things work just as in the two- and three-dimensional cases.

Example 6

Find a formula for $\partial w/\partial y$ valid near the point $(a, b, c, d) = (1, -2, 1, 0)$ if

$$x^3 - 3yz^4w^3 - x^2z^3 \cos w + xzw = 0.$$

Solution. In the ordered quadruple $(1, -2, 1, 0)$, the order of the variables is understood to be (x, y, z, w). The given point satisfies the functional equation, because

$$1^3 - 3(-2)1^4 0^3 - 1^2 \cdot 1^3 \cos 0 = 1 - 0 - 1 = 0.$$

Also, at $(1, -2, 1, 0)$ we have

$$F_w = -9yz^4w^2 + x^2z^3\sin w + xz = -9(-2)1^40^2 + 1^21^3\sin 0 + 1\cdot 1$$
$$= 1 \neq 0.$$

Therefore we conclude that w is a function of the variables x, y, and z near the given point and

$$\frac{\partial w}{\partial y} = -\frac{F_y}{F_w} = -\frac{-3z^4w^3}{-9yz^4w^2 + x^2z^3\sin w + xz} = \frac{-3z^4w^3}{-9yz^4w^2 + x^2z^3\sin w + xz}$$

near that point. ■

Exercises 12.5

In Exercises 1–8, find a formula for dy/dx and state where it is valid. Then check that the given point $(x, y) = (a, b)$ satisfies the given functional equation, and use the tangent approximation to estimate y for the given value of x.

1. $x^2y^2 - 3x - y^3 = 19$; $(a, b) = (3, 2)$; $x = 2.984$
2. $x^2y^3 - x^2 + y^2x = 11$; $(a, b) = (1, 2)$; $x = 0.992$
3. $xy + x^3 + y^3 = 13$; $(a, b) = (-2, 3)$; $x = -2.005$
4. $x^2 + 2xy + 3y^2 = 6$; $(a, b) = (1, 1)$; $x = 1.04$
5. $7x^2y^6 - 5xy^5 + 3y^3 - 2 = 3$; $(a, b) = (1, 1)$; $x = 1.05$
6. $xy^3 - x^2y + y^2 - x^5 - 19 = 0$; $(a, b) = (2, 3)$; $x = 2.10$
7. $x\cos y - x^2e^{xy} + y^2\ln x - 3x^5 + 3 = 0$; $(a, b) = (1, 0)$; $x = 0.95$
8. $e^{x^2y^4} - x^4 + y^5 - \sin y^3 + 5x^2 - 5 = 0$; $(a, b) = (2, 0)$; $x = 1.98$

In Exercises 9 and 10, estimate the value of y at each of the given values of x.

9. $x = 2.968, 2.952, 2.936, 2.920$, for the functional equation in Exercise 1
10. $x = 0.984, 0.976, 0.968, 0.960$, for the functional equation in Exercise 2

In Exercises 11–16, find formulas for $\partial z/\partial x$ and $\partial z/\partial y$ and state where they are valid.

11. $x^2\cos xyz + y^2\sin xyz = z^2$
12. $x + y + z - \sin xyz = 0$
13. $ye^z + xz - x^2 - y^2 = 0$
14. $z^2e^{xy} - 5x^2\dfrac{y}{z} = 7$
15. $x^3e^{y+z} - y\sin(x - z) = 0$
16. $y^2ze^{xy} - x\cos(y^2 - zx) = 0$

In Exercises 17–20, check that $(x, y, z) = (a, b, c)$ satisfies the given functional equation. Then use the tangent approximation to approximate z for the given x and y.

17. $x^2 + y^2 + z^2 = 9$; $(a, b, c) = (2, 1, 2)$; $x = 2.02$, $y = 0.97$
18. $z^3 - (x^2 + y^2)z - 3 = 0$; $(a, b, c) = (2, 2, 3)$; $x = 1.98$, $y = 2.03$
19. $z^3 - x^2z - y^2z = 3$; $(a, b, c) = (-2, 2, 3)$; $x = -2.29$, $y = 1.90$
20. $x^3y - x^2z + y^2z = 52$; $(a, b, c) = (-2, 4, 7)$; $x = -1.88$, $y = 4.12$

In Exercises 21–24, check whether $(x, y, z, w) = (a, b, c, d)$ satisfies the given equation, and find a formula for the indicated partial derivatives near (a, b, c, d) provided they exist.

21. $x^2 - 2y^2z^2 + z^3w^3 - x^3yw^4 + 7 = 0$; $(a, b, c, d) = (1, 0, -1, 2)$; $\partial x/\partial y$, $\partial z/\partial w$
22. $x^2y^3 - 3z^3 + z^2w^3 + 3x^2yw^2 + 47 = 0$; $(a, b, c, d) = (2, -1, 1, 2)$; $\partial w/\partial x$, $\partial z/\partial y$
23. $x^2y^3\cos z + x^3y^3\cos wz + 12 = 0$; $(a, b, c, d) = (2, -1, 0, 1)$; $\partial y/\partial w$, $\partial x/\partial w$
24. $x^2y^2e^{y^2z^3} + \ln(x^2 + z^3w) = 1$; $(a, b, c, d) = (1, 1, 0, 2)$; $\partial y/\partial x$, $\partial z/\partial x$

In Exercises 25–28, find the equation of the tangent plane to the given surface at the given point. Sketch.

25. $x^2 + y^2 - z^2 = 1$ at $(3, 1, -3)$
26. $x^2 - y^2 - z^2 = 3$ at $(4, 3, 2)$
27. $4x^2 + 9y^2 + 4z^2 = 36$ at $(1, 1, \sqrt{23}/2)$
28. $x^2 + y^2 + z^2 = 49$ at $(2, 3, 6)$
29. Show that the surfaces $x^2 + y^2 + 4z = 0$ and $x^2 + y^2 + z^2 - 6y + 7 = 0$ are tangent at the point $(0, 2, -1)$.
30. Show that the surfaces $x^2 + y^2 + z^2 = 8$ and $xz = 4$ are tangent at the point $(2, 0, 2)$.
31. Suppose that two surfaces $F(x, y, z) = 0$ and $G(x, y, z) = 0$ intersect in a curve. Show that if $\nabla F \times \nabla G \neq \mathbf{0}$, then it is a tangent vector to the curve of intersection at each point of intersection.
32. Find a tangent vector to the curve of intersection of $x^2 + y^2 - z = 8$ and $x - y^2 + z^2 = -2$ at $(2, -2, 0)$.
33. Find a tangent vector to the curve of intersection of $3x^2 + 2y^2 + z^2 = 49$ and $x^2 + y^2 - 2z^2 = 10$ at $(3, -3, 2)$.

In numerical analysis, our technique of using the tangent line to approximate a function $y = f(x)$ defined by an equation $F(x, y) = 0$ is used to obtain a *piecewise linear* approximation. We illustrate this in Exercises 34–36. (For simplicity we do only two steps, but on a computer one can readily generate a more complete approximate graph of $y = f(x)$ in this way.)

34. Let $F(x, y) = x^2 + 2xy + 3y^2 - 6$. The graph of $F(x, y) = 0$ contains (1, 1). Show that an approximate value of $y = f(x)$ for $x = 1.50$ is $y \approx 0.75$ by using the tangent approximation.

35. Refer to Exercise 34. Show that the approximate slope of the curve $y = f(x)$ when $x = 1.50$ is -0.60.

36. Refer to Exercise 34. Use the line through (1.50, 0.75) with slope -0.60 to find an approximate value of $y = f(x)$ when $x = 2$.

In Exercises 37–40, show that the given equation *fails* to define the indicated variable as a function of the other variables near the given point.

37. $x^2 + y^2 + z^2 = 9$; $(a, b, c) = (0, 0, 3)$; y

38. $9x^2 + 4y^2 + 4z^2 = 36$; $(a, b, c) = (2, 0, 0)$; z

39. $x^2y^2 - 2y^2z^2 - x^3yw^4 = 0$; $(a, b, c, d) = (1, 0, -1, 2)$; w

40. $x^2 + e^{y^2} + w^2 + z^2 = 4$; $(a, b, c, d) = (\sqrt{2}, 0, 1, 0)$; y

12.6 Higher Partial Derivatives

The *second derivative* f'' of a function $f: \mathbf{R} \to \mathbf{R}$ is often useful in studying the behavior of f. For instance, f'' is essential to the study of the concavity of the graph of f, and the second-derivative test is often the best way to decide whether a critical point c where $f'(c) = 0$ is a local extreme point for scalar functions. We could define a second total derivative for $f: \mathbf{R}^n \to \mathbf{R}$, but we need only the second partial derivatives for the multivariable version of the second-derivative test in the next section. As in Section 1 we start with the two-dimensional case.

6.1 DEFINITION Let $f: \mathbf{R}^2 \to \mathbf{R}$ be a differentiable scalar function. Then the ***second partial derivatives of f*** are

$$\frac{\partial^2 f}{\partial x^2} = \frac{\partial}{\partial x}\left(\frac{\partial f}{\partial x}\right), \qquad \frac{\partial^2 f}{\partial x\,\partial y} = \frac{\partial}{\partial x}\left(\frac{\partial f}{\partial y}\right), \frac{\partial^2 f}{\partial y\,\partial x} = \frac{\partial}{\partial y}\left(\frac{\partial f}{\partial x}\right), \qquad \frac{\partial^2 f}{\partial y^2} = \frac{\partial}{\partial y}\left(\frac{\partial f}{\partial y}\right).$$

The symbols $\partial^2 f/\partial x^2$ and $\partial^2 f/\partial y^2$ are read "second partial (derivative) of f with respect to x" and "second partial (derivative) of f with respect to y". The symbol $\partial^2 f/\partial y\,\partial x$ is read "second partial of f with respect to x and y". This may seem backwards, but it reflects the fact that we get $\partial^2 f/\partial y\,\partial x$ by differentiating *first* with respect to x and *then* with respect to y. Another common notation for second partial derivatives is

$$f_{xx} = \frac{\partial^2 f}{\partial x^2}, \qquad f_{yx} = \frac{\partial^2 f}{\partial x\,\partial y}, \qquad f_{xy} = \frac{\partial^2 f}{\partial y\,\partial x}, \qquad f_{yy} = \frac{\partial^2 f}{\partial y^2}.$$

Note the *reversed* order of writing x and y. In both the ∂- and the f-subscript notations, the partial derivative of the variable *nearest* f is found first. Thus, in the ∂-notation, the second variable with respect to which we differentiate goes on the *left*; but in the f-subscript notation, that second variable goes on the *right*! This appears at first glance to be intolerably confusing, and might well be so except for one saving fact. As we shall see below, we generally get the *same* function whether we compute f_{xy} or f_{yx}. Variants of the notation in color above are

$$D_xD_yf = D_{xy}f = D_{12}f \qquad \text{for } f_{yx} = \frac{\partial^2 f}{\partial x \partial y}.$$

The next example illustrates the ∂- and f-subscript notations.

Example 1

Find all the second partial derivatives of the scalar function f, where $f(x, y) = x^2 \sin xy$.

Solution. First, we compute the two first partial derivatives, getting

$$f_x = \frac{\partial f}{\partial x} = 2x \sin xy + x^2 y \cos xy \qquad \text{and} \qquad f_y = \frac{\partial f}{\partial y} = x^3 \cos xy.$$

Next we differentiate f_x with respect to x and y. That gives

$$f_{xx} = \frac{\partial^2 f}{\partial x^2} = 2 \sin xy + 2xy \cos xy + 2xy \cos xy - x^2 y^2 \sin xy$$
$$= 2 \sin xy + 4xy \cos xy - x^2 y^2 \sin xy$$

and

$$f_{xy} = \frac{\partial^2 f}{\partial y\, \partial x} = 2x^2 \cos xy + x^2 \cos xy - x^3 y \sin xy = 3x^2 \cos xy - x^3 y \sin xy.$$

Finally, differentiation of f_y with respect to x and y produces

$$f_{yx} = \frac{\partial^2 f}{\partial x\, \partial y} = 3x^2 \cos xy - x^3 y \sin xy$$

and

$$f_{yy} = \frac{\partial^2 f}{\partial y^2} = -x^4 \sin xy.$$

Notice that $f_{xy} = f_{yx}$, as we promised would usually happen. ∎

Second partial derivatives of functions of three (or more) variables are computed in the same way as for functions of two variables. For a scalar function f of three variables, the second partial derivatives f_{xx}, f_{yx}, f_{xy}, and f_{yy} are defined exactly as in Definition 6.1. In addition, there are five more second partial derivatives:

$$f_{zx} = \frac{\partial^2 f}{\partial x\, \partial z}, \qquad f_{zy} = \frac{\partial^2 f}{\partial y\, \partial z}, \qquad f_{xz} = \frac{\partial^2 f}{\partial z\, \partial x}, \qquad f_{yz} = \frac{\partial^2 f}{\partial z\, \partial y}, \qquad f_{zz} = \frac{\partial^2 f}{\partial z^2}.$$

Again, for most common functions we have $f_{zx} = f_{xz}$ and $f_{yz} = f_{zy}$.

Example 2

Find all the second partial derivatives of the scalar function f given by $f(x, y, z) = x^2 yz^2 - 2xy^3 + x \sin z$.

Solution. As in Example 1 we first compute the first partial derivatives,

$$f_x = 2xyz^2 - 2y^3 + \sin z, \qquad f_y = x^2 z^2 - 6xy^2, \qquad f_z = 2x^2 yz + x \cos z.$$

We then differentiate each of these with respect to x, getting

$$f_{xx} = 2yz^2, \qquad f_{yx} = 2xz^2 - 6y^2, \qquad f_{zx} = 4xyz + \cos z.$$

Differentiation of the first partials with respect to y gives

$$f_{xy} = 2xz^2 - 6y^2, \qquad f_{yy} = -12xy, \qquad f_{zy} = 2x^2 z.$$

Finally, differentiating the first partials with respect to z, we find

$$f_{xz} = 4xyz + \cos z, \qquad f_{yz} = 2x^2z, \qquad f_{zz} = 2x^2y - x\sin z.$$

Notice that $f_{xy} = f_{yx}$, $f_{yz} = f_{zy}$, and $f_{xz} = f_{zx}$. ■

The results of Examples 1 and 2 support the claim made above that mixed second partial derivatives usually are equal, regardless of the order in which they are computed. The following theorem guarantees that this always happens when the mixed second partials are continuous. Since in every case only two variables appear in the mixed second partials (any others are held constant), it is enough to state the theorem for a function of two variables. The proof is left for advanced calculus.

6.2 THEOREM Suppose that the scalar function $f\colon \mathbf{R}^2 \to \mathbf{R}$ is differentiable on an open disk D containing the point (a, b). If the mixed second partial derivative functions $f_{xy}(x, y)$ and $f_{yx}(x, y)$ both exist on D and are continuous at (a, b), then

$$\frac{\partial^2 f}{\partial y\, \partial x}(a, b) = f_{xy}(a, b) = f_{yx}(a, b) = \frac{\partial^2 f}{\partial x\, \partial y}(a, b)$$

As in the one-variable case, we can use implicit differentiation to find second partial derivatives of implicitly defined functions. The next example illustrates the procedure (compare Example 2 of Section 2.9).

Example 3

If $x^2 + y^2 + z^2 = 9$, find a formula for $\partial^2 z/\partial x^2$ valid near $(1, 2, 2)$.

Solution. Note first that $F(x, y, z) = x^2 + y^2 + z^2 - 9 = 0$ defines z as a differentiable function of x and y near the point $(1, 2, 2)$, because

$$F_z(1, 2, 2) = 2z\Big|_{(1,2,2)} = 4 \neq 0.$$

Theorem 5.2 therefore gives

$$\frac{\partial z}{\partial x} = -\frac{F_x}{F_z} = -\frac{2x}{2z} = -\frac{x}{z}.$$

If we differentiate this with respect to x once more, we get

$$\frac{\partial^2 z}{\partial x^2} = \frac{\partial}{\partial x}\left[\frac{\partial z}{\partial x}\right] = \frac{\partial}{\partial x}\left[-\frac{x}{z}\right] = \frac{z \cdot (-1) + x \cdot \dfrac{\partial z}{\partial x}}{z^2}$$

$$= \frac{-z + x \cdot \left[-\dfrac{x}{z}\right]}{z^2} = \frac{-z^2 - x^2}{z^3} = \frac{y^2 - 9}{z^3},$$

because from $x^2 + y^2 + z^2 = 9$ it follows that $y^2 - 9 = -x^2 - z^2$.

It is sometimes necessary to use the chain rule of Section 4 to find second partial derivatives. The next example illustrates this with a calculation that has important uses in physics and physical chemistry.

Example 4

If $w = f(x, y, z)$, where $x = r\cos\theta$, $y = r\sin\theta$, and $z = c$ is constant, then find a formula for $\partial^2 w/\partial r^2$ in terms of f_{xx}, f_{xy}, f_{yy}, f_{xz}, f_{yz}, f_{zz}, r, and θ.

Solution. First we use Theorem 4.3 to find

$$\begin{aligned}\frac{\partial w}{\partial r} &= \nabla f \cdot \left[\frac{\partial x}{\partial r}\mathbf{i} + \frac{\partial y}{\partial r}\mathbf{j} + \frac{\partial z}{\partial r}\mathbf{k}\right] \\ &= [f_x\mathbf{i} + f_y\mathbf{j} + f_z\mathbf{k}] \cdot [(\cos\theta)\mathbf{i} + (\sin\theta)\mathbf{j} + 0\,\mathbf{k}] \\ &= f_x\cos\theta + f_y\sin\theta.\end{aligned}$$

Then, differentiating this with respect to r once again and keeping in mind that θ is still being held constant, we obtain

$$\begin{aligned}\frac{\partial^2 w}{\partial r^2} &= \frac{\partial}{\partial r}\left[\frac{\partial w}{\partial r}\right] = \frac{\partial}{\partial r}[f_x\cos\theta + f_y\sin\theta] \\ &= \frac{\partial}{\partial r}[f_x\cos\theta] + \frac{\partial}{\partial r}[f_y\sin\theta] \\ &= \left[\frac{\partial}{\partial r}(f_x)\right]\cos\theta + f_x\left[\frac{\partial}{\partial r}(\cos\theta)\right] + \left[\frac{\partial}{\partial r}(f_y)\right]\sin\theta \\ &\quad + f_y\left[\frac{\partial}{\partial r}(\sin\theta)\right] && \textit{by the product rule} \\ &= \nabla(f_x)\cdot\left[\frac{\partial x}{\partial r}\mathbf{i} + \frac{\partial y}{\partial r}\mathbf{j} + \frac{\partial z}{\partial r}\mathbf{k}\right]\cos\theta + f_x\cdot 0 \\ &\quad + \nabla(f_y)\cdot\left[\frac{\partial x}{\partial r}\mathbf{i} + \frac{\partial y}{\partial r}\mathbf{j} + \frac{\partial z}{\partial r}\mathbf{k}\right]\sin\theta + f_y\cdot 0 && \textit{by Theorem 4.3 and the fact that } \theta \textit{ is constant} \\ &= \left[\left[\frac{\partial}{\partial x}(f_x)\mathbf{i} + \frac{\partial}{\partial y}(f_x)\mathbf{j} + \frac{\partial}{\partial z}(f_x)\mathbf{k}\right]\cdot[(\cos\theta)\mathbf{i} + (\sin\theta)\mathbf{j}]\right]\cos\theta \\ &\quad + \left[\left[\frac{\partial}{\partial x}(f_y)\mathbf{i} + \frac{\partial}{\partial y}(f_y)\mathbf{j} + \frac{\partial}{\partial z}(f_y)\mathbf{k}\right]\cdot[(\cos\theta)\mathbf{i} + (\sin\theta)\mathbf{j}]\right]\sin\theta \\ &= [f_{xx}\mathbf{i} + f_{xy}\mathbf{j} + f_{xz}\mathbf{k}]\cdot[(\cos^2\theta)\mathbf{i} + (\sin\theta\cos\theta)\mathbf{j}] \\ &\quad + [f_{yx}\mathbf{i} + f_{yy}\mathbf{j} + f_{yz}\mathbf{k}]\cdot[(\cos\theta\sin\theta)\,\mathbf{i} + (\sin^2\theta)\mathbf{j}] \\ &= f_{xx}\cos^2\theta + f_{xy}\sin\theta\cos\theta + f_{yx}\cos\theta\sin\theta + f_{yy}\sin^2\theta \\ &= f_{xx}\cos^2\theta + 2f_{xy}\sin\theta\cos\theta + f_{yy}\sin^2\theta,\end{aligned}$$

where we used Theorem 6.2 in the last step. ■

Second-Degree Taylor Polynomials*

Although we could define third (and still higher) partial derivatives of scalar functions, second-order partial derivatives suffice for many applications, including the optional material on thermodynamics at the end of this section. We need second partial derivatives for the multivariable second-derivative test. To

* Optional

see how the second-derivative test comes about, and to prove it, we use the second-degree Taylor polynomial $p_2(x, y)$.

Before defining $p_2(x, y)$, we review the one-variable Taylor polynomial $p_2(x)$ and mention its connection to optimization in the one-variable case. For a function $y = f(x)$ with a continuous second derivative $f''(x)$ in an open interval I containing $x = a$, recall that

$$p_2(x) = f(a) + f'(a)(x - a) + \frac{1}{2!} f''(a)(x - a)^2. \tag{1}$$

In general, Corollary 6.6 of Chapter 9 says that this is a closer approximation to $f(x)$ near $x = a$ than the tangent approximation

$$T(x) = p_1(x) = f(a) + f'(a)(x - a), \tag{2}$$

since $p_2(x) \to f(a)$ faster than $(x - a)^2 \to 0$ as $x \to a$.

If $f'(a) = 0$, then the second-derivative test (Theorem 4.5 of Chapter 3) is often used to determine whether $x = a$ is a local extreme point. Recall that if $f''(a) < 0$, then $x = a$ is a local maximum point, and if $f''(a) > 0$, then $x = a$ is a local minimum point. This can be seen from (1). For if $f'(a) = 0$, then the approximation of f by $p_2(x)$ near $x = a$ becomes

$$f(x) \approx f(a) + \frac{1}{2} f''(a)(x - a)^2. \tag{3}$$

The second-degree term $\frac{1}{2}f''(a)(x - a)^2$ has the same sign as $f''(a)$. If $f''(a) < 0$, then for x near a,

$$f(x) \approx f(a) + \frac{1}{2} f''(a)(x - a)^2 < f(a),$$

so $x = a$ is a local maximum point. Similarly, if $f''(a) > 0$, then for x near a,

$$f(x) \approx f(a) + \frac{1}{2} f''(a)(x - a)^2 > f(a),$$

so $x = a$ is a local minimum point. See **Figure 6.1.**

We have already extended the first-degree Taylor polynomial $p_1(x) = T(x)$ to more than one variable, namely,

$$p_1(\mathbf{x}) = f(\mathbf{a}) + \mathbf{f}'(\mathbf{a}) \cdot (\mathbf{x} - \mathbf{a})$$

FIGURE 6.1

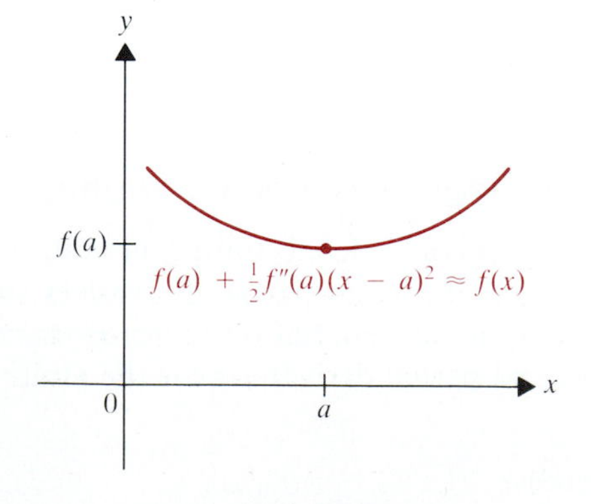

(a)

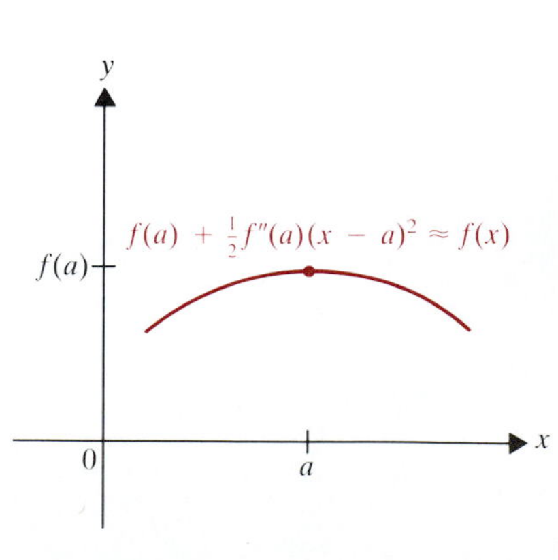

(b)

which in the two-variable case assumes the form

$$p_1(x, y) = f(a, b) + \left[\frac{\partial f}{\partial x}(a, b)\,\mathbf{i} + \frac{\partial f}{\partial y}(a, b)\,\mathbf{j}\right] \cdot \left[(x - a)\,\mathbf{i} + (y - b)\,\mathbf{j}\right]$$

$$= f(a, b) + \frac{\partial f}{\partial x}(a, b)(x - a) + \frac{\partial f}{\partial y}(a, b)(y - b).$$

This polynomial is formed by adding to $f(a, b)$ the products of the first partial derivatives $(\partial f/\partial x)(a, b)$ and $(\partial f/\partial y)(a, b)$ with the respective differentials $dx = x - a$ and $dy = y - b$. The second-degree Taylor polynomial $p_2(x, y)$ is formed analogously.

6.3 DEFINITION Let $f\colon \mathbf{R}^2 \to \mathbf{R}$ be a scalar function such that f_{xx}, f_{xy}, f_{yx}, and f_{yy} are continuous on an open disk D containing the point $p(x, y) = (a, b)$. Then the ***second-degree Taylor polynomial of f (expanded) about (x, y) = (a, b)*** is

$$p_2(x, y) = f(a, b) + \frac{\partial f}{\partial x}(a, b)(x - a) + \frac{\partial f}{\partial y}(a, b)(y - b) + \frac{1}{2!}\left[\frac{\partial^2 f}{\partial x^2}(a, b)(x - a)^2 + 2\frac{\partial^2 f}{\partial x\,\partial y}(a, b)(x - a)(y - b) + \frac{\partial^2 f}{\partial y^2}(a, b)(y - b)^2\right]. \tag{4}$$

By Theorem 6.2 the two separate terms $f_{xy}(x - a)(y - b)$ and $f_{yx}(x - a)(y - b)$ add to give $2f_{xy}(x - a)(y - b)$ in (4), because we have assumed that f_{xy} and f_{yx} are continuous.

Example 5

Find the second-degree Taylor polynomial of f about $(2, 0)$ if $f(x, y) = xe^y$.

Solution. First we compute the first and second partial derivatives in (4) at the point $(a, b) = (2, 0)$. We have

$$f_x = e^y, \qquad f_y = xe^y.$$

Thus

$$f_{xx} = 0, \qquad f_{xy} = e^y, \qquad f_{yy} = xe^y.$$

Therefore

$$f_x(2, 0) = e^0 = 1, \qquad f_y(2, 0) = 2e^0 = 2,$$

$$f_{xx}(2, 0) = 0, \qquad f_{xy}(2, 0) = e^0 = 1, \qquad f_{yy}(2, 0) = 2e^0 = 2.$$

Since $f(2, 0) = 2e^0 = 2$, we obtain from (4)

$$p_2(x, y) = 2 + 1 \cdot (x - 2) + 2(y - 0) + \frac{1}{2}[0 \cdot (x - 2)^2 + 2 \cdot 1(x - 2)(y - 0) + 2(y - 0)^2]$$

$$= 2 + (x - 2) + 2y + (x - 2)y + y^2. \quad ■$$

We could have simplified $p_2(x, y)$ to get $x + xy + y^2$, but keeping it in powers of $x - a(= x - 2)$ and $y - b(= y)$ is usually best. Sometimes the single-variable Taylor polynomial can be used to save work in calculating $p_2(x, y)$.

Example 6

Find the second-degree Taylor polynomial of f about $(0, 0)$ if $f(x, y) = e^{xy}$.

Solution. From Example 5 of Section 9.6 the second-degree Taylor polynomial of $f(t) = e^t$ near 0 is $p_2(t) = 1 + t + t^2/2$. This holds, in particular, if the variable t is itself the product of two variables x and y. If we replace t by xy in the second-degree Taylor polynomial for $f(t)$, then we get

$$p_2(t) = 1 + xy + \frac{1}{2}x^2y^2.$$

The only terms of degree 2 or less in this polynomial in x and y are the first two summands. Hence the second-degree Taylor polynomial for f about (0, 0) is $p_2(x, y) = 1 + xy$. (You can verify this by using (4).) ■

From our experience with the second-degree Taylor polynomial of a function of one variable, we expect $p_2(x, y)$ to be an approximation to $f(x, y)$ near $(x, y) = (a, b)$ that is more accurate than the tangent approximation (2). The next example illustrates this.

Example 7

The pressure P inside a gas cylinder is measured as $P_0 = 10.0$ atmospheres (accurate to within 2%). To within 0.5%, its volume is 400 cm^3. Use $p_2(P, V)$ to find approximate bounds on the true value of the constant c in Boyle's law.

Solution. As in Example 6 of Section 2,

$$c_0 = P_0V_0 = 4000 \text{ atm cm}^3.$$

The maximum error in P is

$$dP = P - P_0 = (0.02)(10.0) = 0.200,$$

and the maximum error in V is

$$dV = V - V_0 = (0.005)(400) = 2.00 \text{ cm}^3.$$

Since $c = PV$, we have

$$\frac{\partial c}{\partial P} = V, \qquad \frac{\partial c}{\partial V} = P, \qquad \frac{\partial^2 c}{\partial V \partial P} = 1, \qquad \frac{\partial^2 c}{\partial P^2} = \frac{\partial^2 c}{\partial V^2} = 0.$$

Hence

$$\begin{aligned} p_2(P, V) &= P_0V_0 + \frac{\partial c}{\partial P}(P_0, V_0)(P - P_0) + \frac{\partial c}{\partial V}(P_0, V_0)(V - V_0) \\ &\quad + \frac{1}{2}2(P - P_0)(V - V_0) \\ &= 4000 + V_0(P - P_0) + P_0(V - V_0) + (P - P_0)(V - V_0) \\ &= 4000 + 400(P - 10.0) + 10.0(V - 400) + (P - 10.0)(V - 400). \end{aligned}$$

We can therefore approximate the maximum error $|c(P, V) - c_0(P_0, V_0)|$ in the calculated value of c as

$$\begin{aligned} |c(P, V) - c(P_0, V_0)| &\approx |p_2(P, V) - c(P_0, V_0)| = |p_2(P, V) - 4000| \\ &= |400(P - 10.0) + 10.0(V - 400) + (P - 10.0)(V - 400)| \\ &\le 400|P - 10.0| + 10.0|V - 400| + |P - 10.0|\,|V - 400| \\ &\le 400(0.200) + (10.0)(2.00) + (0.200)(2.00) = 100.4. \end{aligned}$$

We conclude that the actual value of c lies in the interval $[4000 - 100.4, 4000 + 100.4] = [3899.6, 4100.4]$. The increase in the precision of the error estimate can be seen by referring back to Example 6 of Section 2, where we estimated c to be in the interval $[3900, 4100]$. ■

By multiplying out $p_2(P, V)$ in Example 7, you can verify that it reduces to PV, so the bounds obtained are *exact*, not just approximate. (Exercise 46 asks you to verify that $p_2(x, y) = f(x, y)$ for any second-degree polynomial function $f(x, y)$.) When $f(x, y) \neq p_2(x, y)$, the procedure of Example 7 can be used to approximate a value $f(x, y)$ calculated from measured data (Exercises 39 and 40).

Error in the Tangent Approximation *

In Section 2 we mentioned the desirability of having an upper bound for the error $e(x, y)$ made when a scalar function $f(x, y)$ is approximated by the tangent approximation

$$T(x, y) = f(a, b) + \frac{\partial f}{\partial x}(x - a) + \frac{\partial f}{\partial y}(y - b). \tag{5}$$

An upper bound for $e(x, y)$ analogous to the one for the single-variable case (Theorem 8.3 of Chapter 6) is provided by the next theorem, which is proved in advanced calculus.

6.4 THEOREM Suppose that f_{xx}, $f_{xy}(=f_{yx})$, and f_{yy} are continuous on a closed disk D containing $(x, y) = (a, b)$. Let

$$e(x, y) = f(x, y) - T(x, y) = f(x, y) - p_1(x, y).$$

Suppose that $|f_{xx}(x, y)|$, $|f_{xy}(x, y)|$ and $|f_{yy}(x, y)|$ are all $\leq M$, some fixed real number. Then for all $(x, y) \in D$,

$$|e(x, y)| \leq M|\mathbf{x} - \mathbf{a}|^2 = M\,[(x - a)^2 + (y - b)]^2$$

Example 8

Estimate the accuracy of the tangent approximation $T(x, y)$ in the closed disk $(x - 2)^2 + y^2 \leq 0.01$ about $(2, 0)$ of radius 0.1, if $f(x, y) = x^3 + y^3$.

Solution. The first partial derivatives of f are

$$f_x = 3x^2, \qquad f_y = 3y^2.$$

Thus

$$f_{xx} = 6x, \qquad f_{xy} = 0, \qquad f_{yy} = 6y.$$

From **Figure 6.2** the maximum value for $|f_{xx}|$ on D is 12.6 at the point (2.1, 0). Similarly, the maximum value for $|f_{yy}|$ on D is 0.6 at the point (2, 0.1). Thus we can use 12.6 for the bound M in Theorem 6.4. We then have

$$|e(x, y)| \leq (12.6)(0.1)^2 = 0.126. \quad ■$$

* Optional

FIGURE 6.2

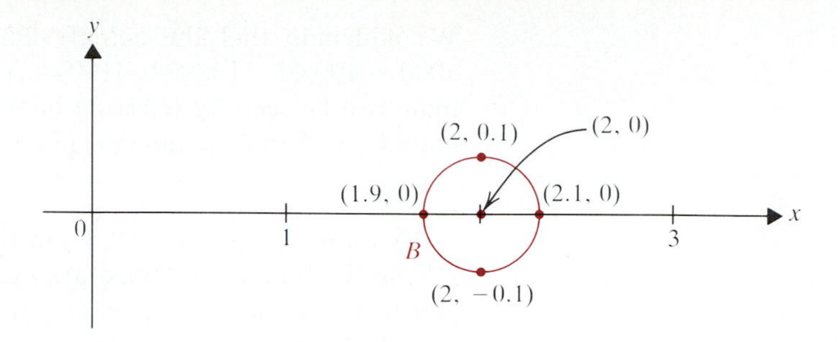

All the results given in this section hold equally true for scalar functions f of three or more variables. In the three-variable case, when all the mixed second partials are continuous, we define $p_2(x, y, z)$ about a point (a, b, c) by using the approach of Definition 6.3:

$$\begin{aligned} p_2(x, y, z) &= f(a, b, c) + \frac{\partial f}{\partial x}(a, b, c)(x-a) + \frac{\partial f}{\partial y}(a, b, c)(y-b) \\ &\quad + \frac{\partial f}{\partial z}(a, b, c)(z-c) + \frac{1}{2!}\left[\frac{\partial^2 f}{\partial x^2}(a, b, c)(x-a)^2\right. \\ &\quad + 2\frac{\partial^2 f}{\partial x\,\partial y}(a, b, c)(x-a)(y-b) \\ &\quad + 2\frac{\partial^2 f}{\partial x\,\partial z}(a, b, c)(x-a)(z-c) + 2\frac{\partial^2 f}{\partial y\,\partial z}(a, b, c)(y-b)(z-c) \\ &\quad \left. + \frac{\partial^2 f}{\partial y^2}(a, b, c)(y-b)^2 + \frac{\partial^2 f}{\partial z^2}(a, b, c)(z-c)^2\right]. \end{aligned}$$

See Exercises 33–36 for some computations involving three variables.

An Application to Thermodynamics*

In thermodynamics there are five basic variables: T (temperature), V (volume), P (pressure), E (energy), and S (entropy). Physicists allow *any* two of these to be regarded as independent, with the remaining three variables viewed as functions of those two. All partial derivatives of all orders are assumed to be continuous. In differentiating, the other independent variable is indicated by notation such as $(\partial S/\partial T)_V$, which means that T and V are the independent variables. The *first and second laws of thermodynamics* can be expressed as

$$\text{(I)}\quad T\left(\frac{\partial S}{\partial T}\right)_V = \left(\frac{\partial E}{\partial T}\right)_V \qquad \text{and} \qquad \text{(II)}\quad T\left(\frac{\partial S}{\partial V}\right)_T = \left(\frac{\partial E}{\partial V}\right)_T + P.$$

From these laws many relationships can be deduced.

Example 9

Show that $\left(\dfrac{\partial P}{\partial T}\right)_V = \left(\dfrac{\partial S}{\partial V}\right)_T.$

* Optional

Solution. We differentiate (I) with respect to V and (II) with respect to T. That gives

$$T\frac{\partial^2 S}{\partial V\,\partial T} = \frac{\partial^2 E}{\partial V\,\partial T}, \tag{6}$$

and

$$\left(\frac{\partial S}{\partial V}\right)_T + T\frac{\partial^2 S}{\partial T\,\partial V} = \frac{\partial^2 E}{\partial T\,\partial V} + \left(\frac{\partial P}{\partial T}\right)_V. \tag{7}$$

We now substitute (6) into (7), getting

$$\left(\frac{\partial S}{\partial V}\right)_T + T\frac{\partial^2 S}{\partial T\,\partial V} = T\frac{\partial^2 S}{\partial V\,\partial T} + \left(\frac{\partial P}{\partial T}\right)_V.$$

Since all partials of every order are assumed continuous, we can cancel

$$T\frac{\partial^2 S}{\partial T\,\partial V} = T\frac{\partial^2 S}{\partial V\,\partial T}$$

by Theorem 6.2. That leaves

$$\left(\frac{\partial S}{\partial V}\right)_T = \left(\frac{\partial P}{\partial T}\right)_V.$$

Exercises 43–45 pursue these ideas further. ■

Exercises 12.6

In Exercises 1–8, find all first and second partial derivatives.

1. $f(x, y) = 3x^2 - 2y^2 - 5xy + 7$
2. $f(x, y) = x\cos(x + y)$
3. $f(x, y) = x^2 \sin xy$
4. $f(x,y) = y\ln(x^2 + y^2)$
5. $f(x, y) = \ln(x^2 + y^2)$
6. $f(x, y) = y^2 e^{xy}$
7. $f(x, y, z) = x^2 - y^2 + 3z^2 - 5$
8. $f(x, y, z) = x^2 + 2y^2 + 5z^2 - 8xy - 4yz - 5xz + 1$

In Exercises 9–12 find a formula for the indicated second partial derivative near the given point.

9. **(a)** $\partial^2 z/\partial y^2$ near $(1, 2, 3)$ if $x^2 + 2y^2 - z^2 = 0$
   (b) $\partial^2 z/\partial x^2$ in part (a)
10. **(a)** $\partial^2 z/\partial x\,\partial y$ in Exercise 9 **(b)** $\partial^2 z/\partial y\,\partial x$ in Exercise 9
11. **(a)** $\partial^2 z/\partial x^2$ near $(-1, 1, 2)$ if $x^2 - y^2 + 3z^2 - 2xy + yz = 16$
    (b) $\partial^2 z/\partial y\,\partial x$ in part (a)
12. **(a)** $\partial^2 z/\partial y^2$ in Exercise 11 **(b)** $\partial^2 z/\partial x\,\partial y$ in Exercise 11

Exercises 13–16 deal with *Laplace's equation*, which is of importance in electricity, optics, thermodynamics, and other applied areas. (Assume that all *second partial derivatives are continuous.*)

13. Show that if $f(x, y) = \ln(x^2 + y^2)$, then

$$\frac{\partial^2 f}{\partial x^2} + \frac{\partial^2 f}{\partial y^2} = 0 \quad \textbf{(the two-dimensional Laplace equation).}$$

14. Show that if

$$f(x, y) = \tan^{-1}\frac{2xy}{x^2 - y^2},$$

then f satisfies the two-dimensional Laplace equation.

15. Show that if

$$f(x, y, z) = \frac{1}{|(x, y, z)|} = (x^2 + y^2 + z^2)^{-1/2},$$

then f satisfies

$$\frac{\partial^2 f}{\partial x^2} + \frac{\partial^2 f}{\partial y^2} + \frac{\partial^2 f}{\partial z^2} = 0$$

(the **three-dimensional Laplace equation**).

16. Repeat Exercise 15 for $f(x, y, z) = 3x^2 + 2y^2 - 5z^2$.

In Exercises 17 and 18, $z = f(x, y)$, where $x = r\cos\theta$ and $y = r\sin\theta$.

17. **(a)** Find $\partial^2 z/\partial r^2$ in terms of f_{xx}, f_{xy}, f_{yy}, r, and θ.
    (b) Repeat (a) for $\partial^2 z/\partial\theta^2$.
18. Show that

$$\frac{\partial^2 z}{\partial x^2} + \frac{\partial^2 z}{\partial y^2} = \frac{\partial^2 z}{\partial r^2} + \frac{1}{r^2}\frac{\partial^2 z}{\partial\theta^2} + \frac{1}{r}\frac{\partial z}{\partial r}.$$

In Exercises 19 and 20, w is as in Example 4.

19. Find $\partial^2 w/\partial\theta^2$ in terms of f_{xx}, f_{xy}, f_{yy}, r, and θ.

20. Show that

$$\frac{\partial^2 w}{\partial x^2} + \frac{\partial^2 w}{\partial y^2} + \frac{\partial^2 w}{\partial z^2} = \frac{\partial^2 w}{\partial r^2} + \frac{1}{r^2}\frac{\partial^2 w}{\partial \theta^2} + \frac{1}{r}\frac{\partial w}{\partial r} + \frac{\partial^2 w}{\partial z^2}.$$

In Exercises 21–36, compute when possible the second-degree Taylor polynomials of f about the given point.

21. $\sin(xy)$; $(0, 0)$

22. $\cos(xy)$; $(0, 0)$

23. $\tan(xy)$; $(0, 0)$

24. $\ln(2x + y)$; $(0, 1)$

25. $\sqrt{1 - x^2 - y^2}$; $(1, 0)$

26. $\sqrt{25 - x^2 - y^2}$; $(3, 4)$

27. $xy + y^2$; $(1, 1)$

28. $2xy + x + 2$; $(1, 1)$

29. $\sqrt{1 + x^2 + y^2}$; $(0, 0)$

30. $\sqrt{1 - x^2 - y^2}$; $(0, 0)$

31. $y^2 e^x$; $(0, 0)$

32. $x \tan y$; $(0, 0)$

33. $\sin(x + y + z)$; $(\pi/6, \pi/6, \pi/6)$

34. $\cos(x + y + z)$; $(\pi/6, \pi/6, \pi/6)$

35. $\sqrt{x^2 + y^2 + z^2}$; $(0, 0, 0)$

36. $\sqrt{1 - x^2 - 2y + z}$; $(1, 0, 0)$

In Exercises 37–40, use the second-degree Taylor polynomial to approximate the quantity in the given problem. Compare with your earlier answers using the tangent approximation.

37. Exercise 23, Section 2

38. Exercise 25, Section 2

39. Exercise 29, Section 2

40. Exercise 31, Section 2

In Exercises 41 and 42, estimate the accuracy of the tangent approximation to the given function in the given closed ball.

41. $f(x, y) = x^2 + y^3$ on $\bar{B}((1, 0), 0.1)$

42. $f(x, y) = e^{x^2} + y^2$ on $\bar{B}((0, 0), 0.1)$

43. Refer to Example 9. Suppose that T and P are chosen as independent variables. Then in Example 9, P is a function of T and V. Also, V is now a function of T and P. So P is ultimately a function of P and T. Show that

(a) $1 = \left(\frac{\partial P}{\partial T}\right)_V \left(\frac{\partial T}{\partial P}\right)_T + \left(\frac{\partial P}{\partial V}\right)_T \left(\frac{\partial V}{\partial P}\right)_T$

(b) $1 = \left(\frac{\partial P}{\partial V}\right)_T \left(\frac{\partial V}{\partial P}\right)_T$

44. Refer to Example 9. Suppose that T and P are chosen as independent variables. Then in Example 9, P is a function of T and V, where T and V are functions of T and P. Then P is ultimately a function of T and P. Show that

$$\left(\frac{\partial P}{\partial T}\right)_V = -\left(\frac{\partial P}{\partial V}\right)_T \left(\frac{\partial V}{\partial T}\right)_P$$

by computing

$$\left(\frac{\partial P}{\partial T}\right)_P \ (=0).$$

45. In Exercise 44, regard S as a function first of T and V (Example 9) and then ultimately as a function of T and P. Use Exercise 44 and the results of Example 9 and Exercise 43(b) to show that

$$\left(\frac{\partial S}{\partial P}\right)_T = -\left(\frac{\partial V}{\partial T}\right)_P.$$

46. **(a)** Suppose $f(x, y) = a_0 + a_1 x + a_2 y + a_3 x^2 + a_4 xy + a_5 y^2$ is a second-degree polynomial and $(a, b) = (0, 0)$. Show that $p_2(x, y) = f(x, y)$.

(b) Repeat (a) in case (a, b) is an arbitrary point in the plane.

47. Suppose that a thin, flexible membrane is stretched over a portion of the xy-plane and is made to vibrate. Let $z = f(x, y, t)$ be the distance from the xy-plane to the membrane at time t. The two-dimensional normalized ***wave equation*** is

$$\frac{\partial^2 z}{\partial x^2} + \frac{\partial^2 z}{\partial y^2} - \frac{\partial^2 z}{\partial t^2} = 0.$$

Show that

$$z(x, y, t) = e^{x-t} - 3e^{y+t} + 7xy - t + 5$$

satisfies the wave equation.

48. Refer to Exercise 47. Show that $z(x, y, t) = (3x - 4y + 5t)^{1/2}$ satisfies the wave equation.

12.7 Extreme Values

This section is the multivariable analogue of Sections 3.3 and 3.4. We develop means for finding local maxima and minima of scalar functions of more than one variable. A detailed development is given only for the two-variable case, but we include a brief discussion of optimization of functions of three variables. First we need to define our terms precisely. The following is the multivariable version of Definitions 2.1 and 2.2 of Chapter 3.

7.1 DEFINITION

Let $f: \mathbf{R}^2 \to \mathbf{R}$ be a scalar function. Then

(a) f has its ***absolute maximum*** at $(x, y) = (a_1, a_2)$ if

$$f(x, y) \leq f(a_1, a_2),$$

for all (x, y) in the domain of f. The function f has its ***absolute minimum*** at $(x, y) = (b_1, b_2)$ if

$$f(x, y) \geq f(b_1, b_2),$$

for all (x, y) in the domain of f.

(b) The point $\mathbf{x} = \mathbf{c} = (c_1, c_2)$ is a ***local maximum*** point for f if

$$f(x, y) \leq f(c_1, c_2),$$

for all (x, y) in some open disk $B(\mathbf{c}, \delta)$ surrounding the point $\mathbf{c}$ in $\mathbf{R}^2$. The point $\mathbf{x} = \mathbf{d} = (d_1, d_2)$ is a ***local minimum*** point for f if

$$f(x, y) \geq f(d_1, d_2),$$

for all $\mathbf{x}$ in some open disk $B(\mathbf{d}, \delta)$ about the point $\mathbf{d}$.

As in the one-variable case, the terms ***extremum*** and ***extreme value*** are sometimes used instead of *maximum or minimum (value)*. The terms ***global maximum*** and ***global minimum*** are synonyms for *absolute maximum* and *absolute minimum.* **Figure 7.1** illustrates Definition 7.1(a) in the case of the function $f(x, y) = x^2 + y^2$. Since the graph is an upward-opening paraboloid of revolution, the point $(x, y) = (0, 0)$ is the absolute minimum point for f. The corresponding absolute minimum value is 0. **Figure 7.2** illustrates a local maximum point (c_1, c_2) of a function $z = f(x, y)$ and a local minimum point (d_1, d_2).

Example 1

Discuss the extreme values of $f(x, y) = 6 - 3x^2 - y^2$.

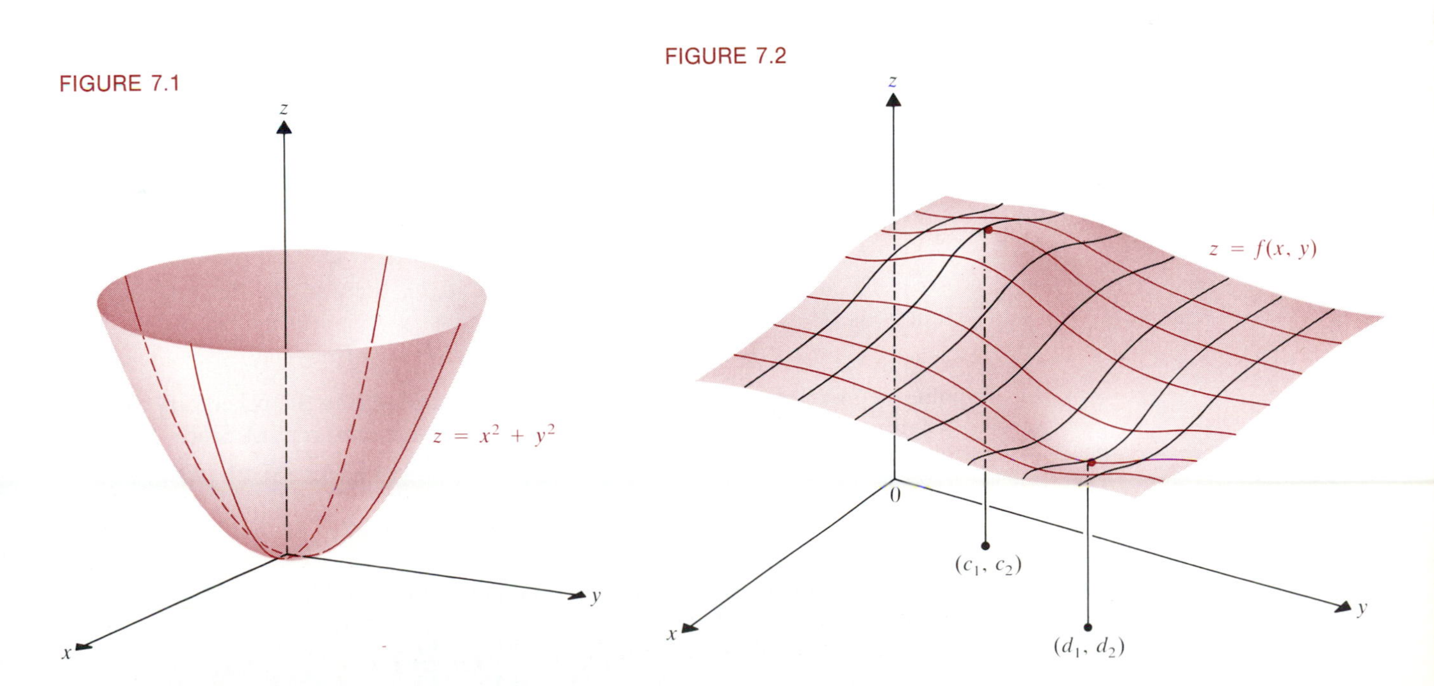

FIGURE 7.1

FIGURE 7.2

FIGURE 7.3

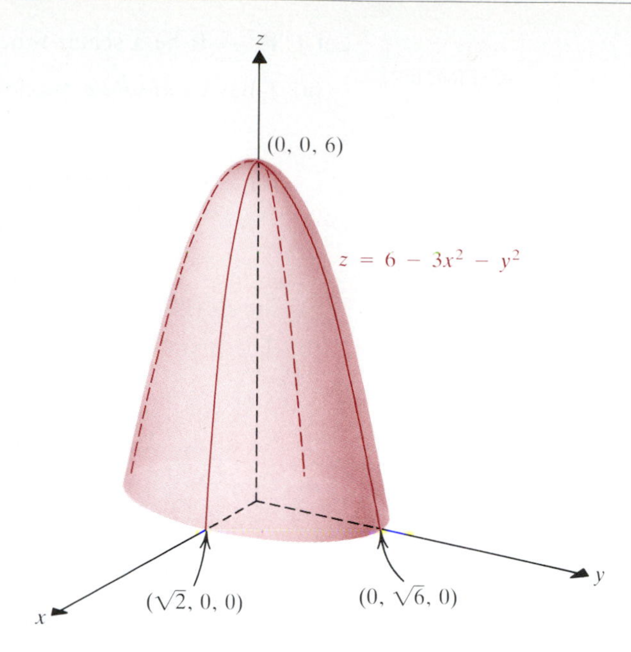

Solution. Since $f(x, y) = 6 - (3x^2 + y^2)$ and $3x^2 + y^2 \geq 0$ for all (x, y), the absolute maximum of f is 6. It occurs when as little as possible is subtracted from 6, that is, $(x, y) = (0, 0)$. As x increases (or y increases) in absolute value, f decreases, so there is no local or absolute minimum. The sole extreme value then is 6. See **Figure 7.3.** ■

In Example 1, $f_x = -6x$ and $f_y = -2y$ are both 0 at the extreme point (0, 0). That is, $f'(0, 0) = \nabla f(0, 0) = \mathbf{0}$. This is what we expect at a local extreme point of a differentiable function f, by analogy to Theorem 2.3 of Chapter 2. It is easy to confirm this expectation.

7.2 THEOREM

> Suppose that the scalar function $f\colon \mathbf{R}^2 \to \mathbf{R}$ is differentiable on an open disk D containing $(x, y) = (a, b)$. If f has a local extreme value at (a, b) then
>
> $$\mathbf{f}'(a, b) = \nabla f(a, b) = \mathbf{0}.$$

Proof. If f has a local extreme value at (a, b), then so do the one-variable functions

$$g(x) = f(x, b) \qquad \text{and} \qquad h(y) = f(a, y),$$

obtained by freezing one coordinate variable and letting the other vary. (**Figure 7.4** illustrates this for the case of a local maximum point.) Since $f(x, y)$ is differentiable near (a, b), so are $g(x)$ and $h(y)$: The partial derivatives

$$\frac{\partial f}{\partial x} = g'(x) \qquad \text{and} \qquad \frac{\partial f}{\partial y} = h'(y)$$

must exist near (a, b), because the total derivative

$$\mathbf{f}'(x, y) = \frac{\partial f}{\partial x}\mathbf{i} + \frac{\partial f}{\partial y}\mathbf{j}$$

FIGURE 7.4

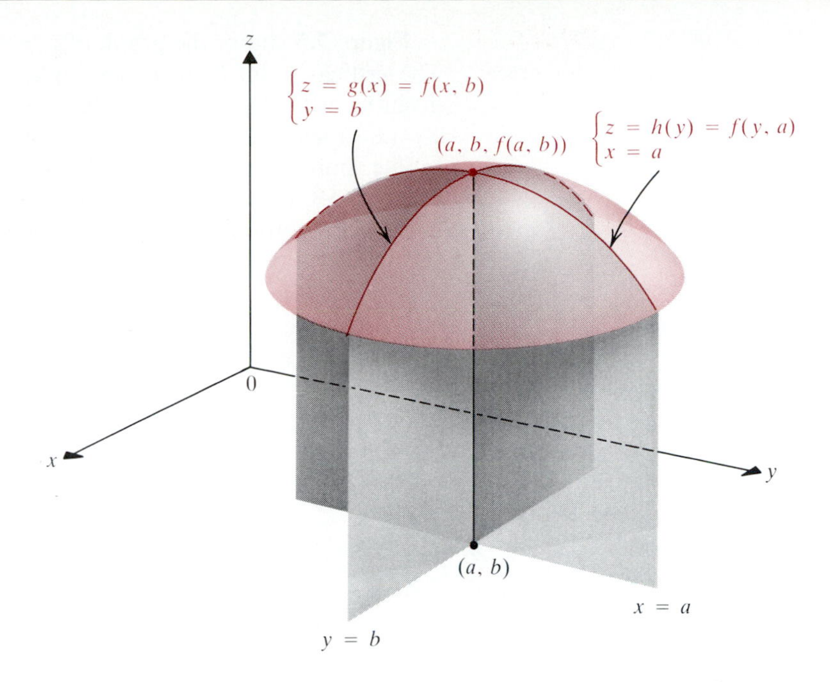

does. Then by Theorem 2.3 of Chapter 3,

$$g'(a) = \frac{\partial f}{\partial x}(a, b) = 0 \qquad \text{and} \qquad h'(b) = \frac{\partial f}{\partial y}(a, b) = 0.$$

Therefore $\nabla f(a, b) = \mathbf{f}'(a, b) = \mathbf{0}$. QED

As in the one-variable case, then, a necessary condition for a differentiable function $f(x, y)$ to have a local extreme value at $(x, y) = (a, b)$ is that $\mathbf{f}'(a, b) = \mathbf{0}$. The following definition is thus a natural extension of Definition 2.5 of Chapter 3.

7.3 DEFINITION

A point $\mathbf{c} = (c_1, c_2)$ of $\mathbf{R}^2$ is called a ***critical point*** of $f(x, y)$ if either $\mathbf{f}'(\mathbf{c}) = \mathbf{0}$ or $\mathbf{f}'(\mathbf{c})$ fails to exist.

As in Chapter 3, critical points of a scalar function of two variables are only *candidates* for local extreme-value points. The next example illustrates this.

Example 2

Discuss the behavior of the function $f(x, y) = xy$ at its critical points.

Solution. The function is differentiable for all points (x, y) in $\mathbf{R}^2$. Thus to find the critical points, we solve $\mathbf{f}'(x, y) = \mathbf{0}$ for x and y. We have

$$\mathbf{f}'(x, y) = \nabla f(x, y) = y\mathbf{i} + x\mathbf{j}.$$

If we set this equal to $\mathbf{0}$ and solve for x and y, then we get

$$y\mathbf{i} + x\mathbf{j} = 0\mathbf{i} + 0\mathbf{j} \rightarrow x = 0,\ y = 0.$$

Thus the only critical point is (0, 0). This is not a local extreme point. For if x and y are very small and have the same sign, then $f(x, y) > 0$. But if x and y are small and of opposite sign, then $f(x, y) < 0$. ■

Figure 7.5 shows the graph of $z = xy$, which is saddle-shaped near (0, 0, 0). The critical point (0, 0) in Example 2 is called a ***saddle point.*** This name is given to critical points (a, b) that are not extreme value points. The graph of $z = (x, y)$ need not, however, resemble a saddle near $(a, b, f(a, b))$ if (a, b) is a saddle point. For example, **Figure 7.6** shows the graph of the function $f(x, y) = x^3$, which has a critical point but not a local extreme value at (0, 0, 0) and whose graph is not saddle-shaped near the origin.

FIGURE 7.5

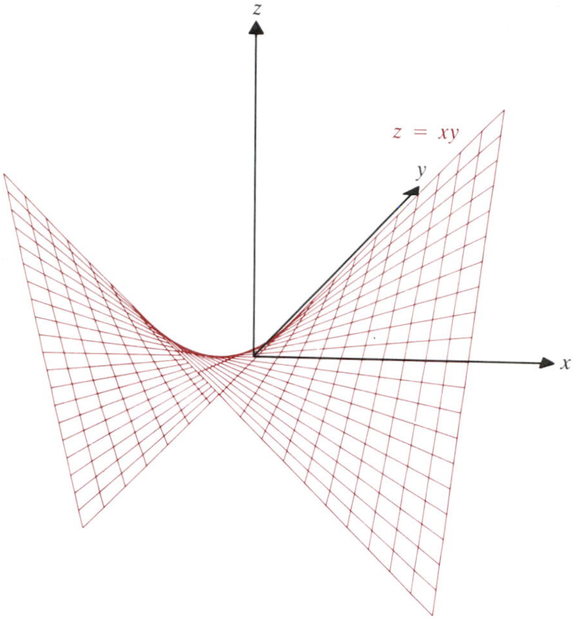

FIGURE 7.6

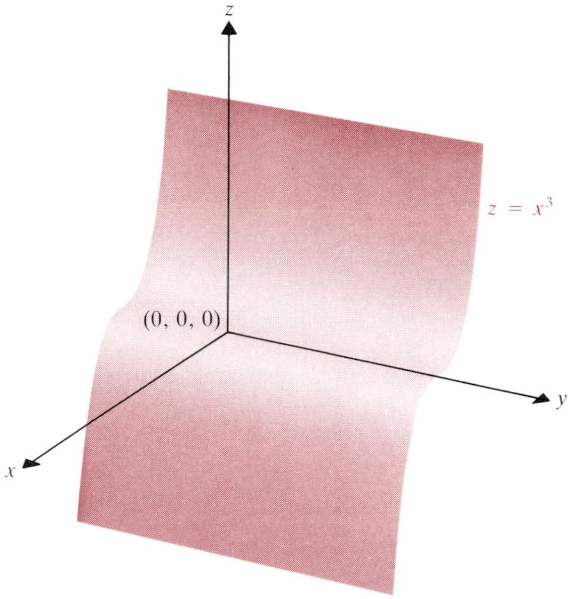

Recall from Definition 1.2 that for a differentiable function $z = f(x, y)$, the tangent plane to the graph at $(a, b, f(a, b))$ is

$$z - f(a, b) = \frac{\partial f}{\partial x}(a, b)(x - a) + \frac{\partial f}{\partial y}(a, b)(y - b).$$

Hence, at a critical point (a, b) of a differentiable function f, the tangent plane is

$$z - f(a, b) = 0(x - a) + 0(y - b) = 0 \rightarrow z = f(a, b)$$

and so is *horizontal.* Once again, this is analogous to the fact that the tangent line to the graph of a differentiable function $y = f(x)$ at a critical point $x = a$ is also horizontal. **Figure 7.7** shows the tangent planes $z = 6$ and $z = 0$ (the xy-plane) to the graphs of the respective functions $f(x, y) = 6 - 3x^2 - y^2$ and $f(x, y) = xy$ considered in Examples 1 and 2.

Examples 1 and 2 suggest the desirability of a test for determining whether a critical point $x = a$ is an extreme value point. It is natural to expect there to be a test corresponding to the first-derivative test (Theorem 3.2 of Chapter 3): If $\mathbf{f}'(a, b) = \mathbf{0}$ and if $\mathbf{f}'(a, b)$ changes sign at (x, y), then we expect (a, b) to be an extreme value point. There is an obstacle, however: What does it *mean* for the *vector*

$$\mathbf{f}'(x, y) = \frac{\partial f}{\partial x}\mathbf{i} + \frac{\partial f}{\partial y}\mathbf{j}$$

FIGURE 7.7

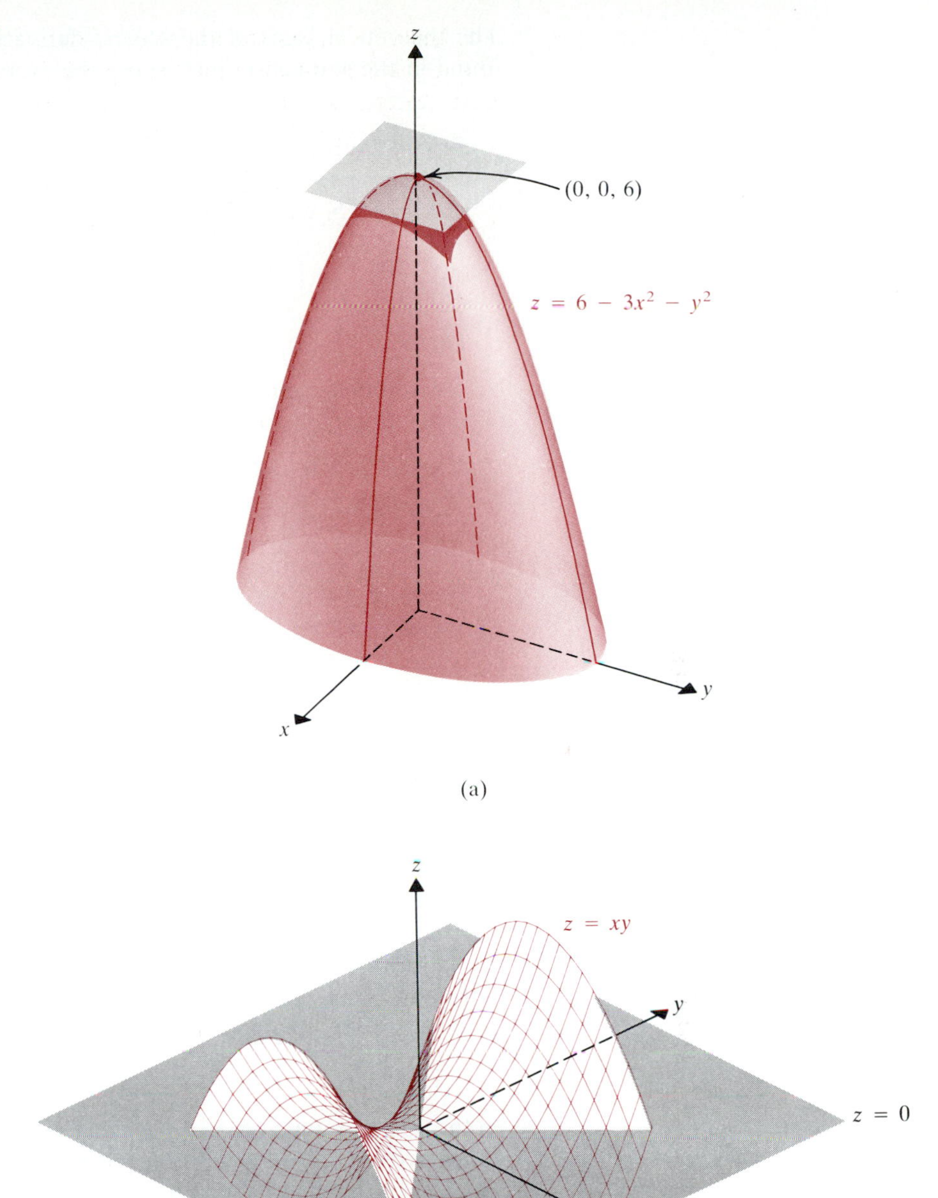

to change sign at (a, b)? There are a number of schemes for assigning a meaning to positive and negative vectors, but none leads to a feasible first-derivative test. (See Exercise 39, for example.)

That brings us to the second-derivative test of Section 3.4. That test does have a two-variable version, which involves the 2-by-2 determinant function *det* introduced on p. 671:

$$\det\begin{pmatrix} a & b \\ c & d \end{pmatrix} = ad - bc.$$

The theoretical basis of the second-derivative test, including a proof, can be found in the subsection just before the exercises.

7.4 THEOREM

Second-Derivative Test. Suppose that the scalar function $f: \mathbf{R}^2 \to \mathbf{R}$ has continuous second partial derivatives on an open disk containing the point $(x, y) = (a, b)$ where $\mathbf{f'(x, y)} = (0, 0)$. Let

$$D = \det\begin{pmatrix} f_{xx}(a, b) & f_{xy}(a, b) \\ f_{xy}(a, b) & f_{yy}(a, b) \end{pmatrix} = f_{xx}(a, b)f_{yy}(a, b) - [f_{xy}(a, b)]^2.$$

Then

(a) If $D > 0$ and $f_{xx}(a, b) > 0$, then (a, b) is a local minimum point for f.

(b) If $D > 0$ and $f_{xx}(a, b) < 0$, then (a, b) is a local maximum point for f.

(c) If $D < 0$, then (a, b) is a saddle point for f.

(d) If $D = 0$, then nothing can be concluded.

Theorem 7.4 is used just like Theorem 4.5 of Chapter 3, as the next two examples illustrate.

Example 3

Find and classify all candidates for extreme points of f given by $f(x, y) = 3x + 12y - x^3 - y^3$.

Solution. The function f is differentiable everywhere, and so all its critical points can be found by solving $\mathbf{f'(x)} = \mathbf{0}$. Since

$$f_x = 3 - 3x^2 \qquad \text{and} \qquad f_y = 12 - 3y^2,$$

that gives

$$\begin{cases} 3 - 3x^2 = 0 \\ 12 - 3y^2 = 0 \end{cases} \to \begin{cases} x^2 = 1 \\ y^2 = 4 \end{cases} \to \begin{cases} x = \pm 1 \\ y = \pm 2. \end{cases}$$

So the critical points are $(1, -2)$, $(1, 2)$, $(-1, 2)$, and $(-1, -2)$. (Note that these points are all of the form (a, b), where a is a critical point of $g(x) = 3x - x^3$, and b is a critical point of $h(y) = 12y - y^3$. This is always the case for functions $f(x, y) = g(x) + h(y)$ that are the sum of separate functions of x alone and y alone. For more on this, see Exercises 40–42.) To classify these, we find the second partial derivatives:

$$f_{xx} = \frac{\partial}{\partial x}(3 - 3x^2) = -6x, \qquad f_{xy} = 0, \qquad f_{yy} = \frac{\partial}{\partial y}(12 - 3y^2) = -6y.$$

Therefore

$$D = \det\begin{pmatrix} f_{xx} & f_{xy} \\ f_{xy} & f_{yy} \end{pmatrix} = \det\begin{pmatrix} -6x & 0 \\ 0 & -6y \end{pmatrix} = 36xy.$$

At the critical point $(1, 2)$ we then have

$$D = 36 \cdot 2 = 72 > 0 \qquad \text{and} \qquad f_{xx}(1, 2) = -6 \cdot 1 = -6 < 0.$$

Thus Theorem 7.4(b) says that $(1, 2)$ is a local maximum point. Here

$$f(1, 2) = 3 \cdot 1 + 12 \cdot 2 - 1^3 - 2^3 = 3 + 24 - 1 - 8 = 18.$$

FIGURE 7.8

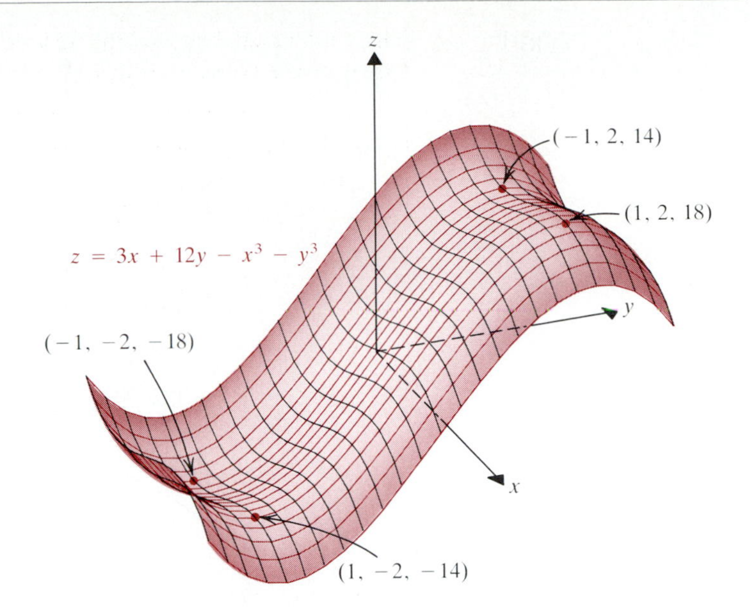

At the critical points $(1, -2)$ and $(-1, 2)$,

$$D = -36 \cdot 1 \cdot 2 = -72 < 0,$$

so these are saddle points by Theorem 7.4(c). At the critical point $(-1, -2)$,

$$D = 36(-1)(-2) = 72 > 0 \quad \text{and} \quad f_{xx}(-1, -2) = -6(-1) > 0.$$

So $(-1, -2)$ is a local minimum point, by Theorem 7.4(a), and

$$f(-1, -2) = 3(-1) + 12(-2) - (-1)^3 - (-2)^3 = -3 - 24 + 1 + 8 = -18.$$

Figure 7.8 shows a computer-generated graph of $f(x, y)$ for $-2 \leq x \leq 2$ and $-3 \leq y \leq 3$. ■

As in the one-variable case, the second-derivative test fails when $D = 0$—this is the content of the last statement (d) in the theorem. At first glance it may appear that we have also excluded $f_{xx}(a, b) = 0$, but we really haven't. For if $f_{xx}(a, b) = 0$, then

$$D = -[f_{xy}(a, b)]^2 \leq 0.$$

If $f_{xy}(a, b) \neq 0$, then $D < 0$, and so Theorem 7.4(c) applies. If $f_{xy}(a, b) = 0$, then $D = 0$, a case already excluded in (d). As the next example suggests, when $D = 0$, anything can happen!

Example 4

Find and classify all critical points of

(a) $f(x, y) = x^2 - 2xy + y^2$ and (b) $f(x, y) = x^3 - 3xy^2 + y^2$.

Solution. (a) Here we have

$$f'(x, y) = \nabla f(x, y) = (2x - 2y)\,\mathbf{i} + (-2x + 2y)\,\mathbf{j},$$

which is $\mathbf{0}$ when $y = x$. So we have the entire line of critical points $y = x$ in the xy-plane shown in **Figure 7.9.** Since $f_{xx} = 2$, $f_{xy} = -2$, and $f_{yy} = 2$, we have

FIGURE 7.9

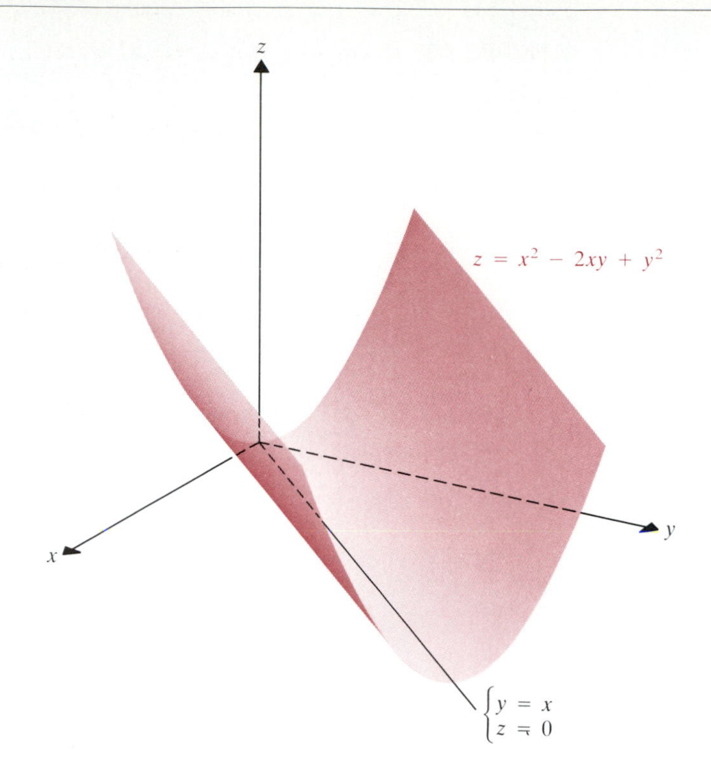

at each critical point

$$D = \det\begin{pmatrix} f_{xx} & f_{xy} \\ f_{xy} & f_{yy} \end{pmatrix} = \det\begin{pmatrix} 2 & -2 \\ -2 & 2 \end{pmatrix} = 4 - 4 = 0.$$

Thus we get no conclusion from Theorem 7.4. However, we can write the function as

$$f(x, y) = (x - y)^2,$$

which shows that $f(x, y) \geq 0$ for all (x, y) in $\mathbf{R}^2$. Thus whenever $y = x$, we get the absolute minimum value 0 of $f(x, y)$. The graph shown in Figure 7.9 reflects this.

(b) We have

$$\mathbf{f}'(x, y) = \nabla f(x, y) = (3x^2 - 3y^2)\,\mathbf{i} + (-6xy + 2y)\,\mathbf{j},$$

which is $\mathbf{0}$ when

$$\begin{cases} 3(x^2 - y^2) = 3(x - y)(x + y) = 0, \\ -6xy + 2y = 2y(-3x + 1) = 0. \end{cases}$$

From the first equation, we find that

$$y = x \qquad \text{or} \qquad y = -x.$$

The second equation says that

$$y = 0 \qquad \text{or} \qquad x = \frac{1}{3}.$$

Putting these together, we get the critical points (0, 0), (1/3, 1/3), and (1/3, −1/3). Differentiating f_x and f_y, we obtain

$$f_{xx} = 6x, \qquad f_{yy} = -6x + 2, \qquad f_{xy} = -6y.$$

Evaluating these at $(1/3, \pm 1/3)$, we get

$$f_{xx} = 2, \qquad f_{yy} = 0, \qquad f_{xy} = -6\left(\pm\frac{1}{3}\right) = \mp 2.$$

Therefore

$$D = \det\begin{pmatrix} 2 & \mp 2 \\ \mp 2 & 0 \end{pmatrix} = 0 - 4 = -4 < 0.$$

Hence by Theorem 7.4(c), $(1/3, 1/3)$ and $(1/3, -1/3)$ are saddle points. At these points,

$$f\left(\frac{1}{3}, \pm\frac{1}{3}\right) = \frac{1}{27} - 3\cdot\frac{1}{3}\cdot\frac{1}{9} + \frac{1}{9} = \frac{1}{27}.$$

Evaluating the second partials at $(0, 0)$, we obtain

$$f_{xx} = 0, \qquad f_{yy} = 2, \qquad f_{xy} = 0.$$

Thus

$$D = \det\begin{pmatrix} 0 & 0 \\ 0 & 2 \end{pmatrix} = 0,$$

and no conclusion follows from Theorem 7.4. But if $y = 0$, the function $f(x, y) = x^3$ has no local extremum at $x = 0$. Therefore $(0, 0)$ is also a saddle point. A computer-generated graph is shown in **Figure 7.10** for $-1 \leq x \leq 1$ and $-1 \leq y \leq 1$. ■

FIGURE 7.10

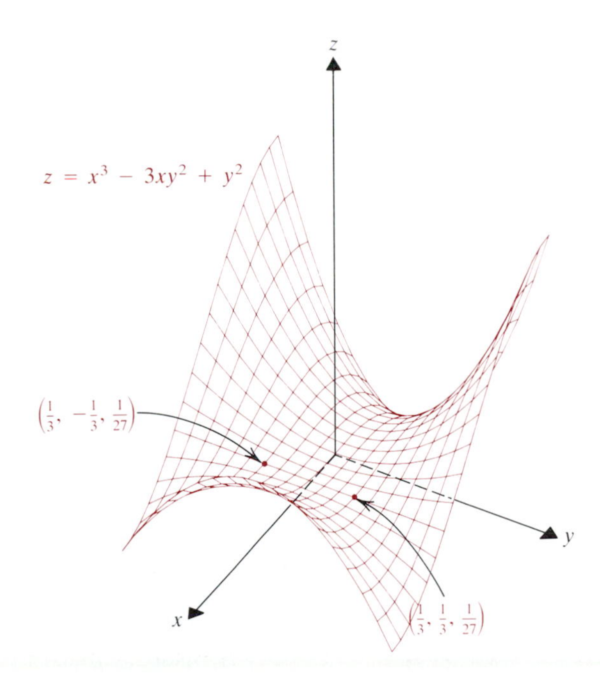

Optimization of Functions of Three Variables*

There is a second-derivative test for scalar functions $f: \mathbf{R}^n \to \mathbf{R}$ for $n > 2$, but we leave it for advanced calculus. It is best stated using concepts of linear algebra that we cannot present here. We do state without proof the form this general

* Optional

second-derivative test assumes for scalar functions of three variables. The test involves 3-by-3 determinants, which you may wish to review if you need to apply the test. (See pp. 672 and 673.)

7.5 THEOREM Suppose that $f\colon \mathbf{R}^3 \to \mathbf{R}$ is differentiable on an open ball B containing the critical point $\mathbf{x} = \mathbf{a} = (a_1, a_2, a_3)$. Suppose also that all second partial derivatives are continuous on B. Let

$$H = \begin{pmatrix} f_{xx} & f_{xy} & f_{xz} \\ f_{yx} & f_{yy} & f_{yz} \\ f_{zx} & f_{zy} & f_{zz} \end{pmatrix}.$$

Then

(a) if $f_{xx} > 0$,

$$\det\begin{pmatrix} f_{xx} & f_{xy} \\ f_{yx} & f_{yy} \end{pmatrix} > 0,$$

and $\det H > 0$ at $\mathbf{x} = \mathbf{a}$, then $f(\mathbf{a})$ is a local minimum;

(b) if $f_{xx} < 0$,

$$\det\begin{pmatrix} f_{xx} & f_{xy} \\ f_{yx} & f_{yy} \end{pmatrix} > 0,$$

and $\det H < 0$ at $x = a$, then $f(\mathbf{a})$ is a local maximum.

Example 5

Find and classify the critical points of $f(x, y, z) = x^2 + 8y^2 + 6z^2 + 4xy - 2xz + 4yz - 9$.

Solution. The function f is differentiable on all $\mathbf{R}^3$, so to find its critical points we solve $\nabla f(x, y, z) = \mathbf{0}$. Since

$$\nabla f(x, y, z) = (2x + 4y - 2z)\,\mathbf{i} + (16y + 4x + 4z)\,\mathbf{j} + (12z - 2x + 4y)\,\mathbf{k},$$

we need to solve the system of linear equations

$$\begin{cases} 2x + 4y - 2z = 0 \\ 4x + 16y + 4z = 0 \\ -2x + 4y + 12z = 0 \end{cases} \to \begin{cases} x + 2y - z = 0 \\ x + 4y + z = 0. \\ -x + 2y + 6z = 0 \end{cases}$$

Such a system can be solved by elimination (as on p. 671). Subtracting the first equation from the second, we get

$$(1) \qquad 2y + 2z = 0 \to y = -z.$$

Addition of the first and third equations gives

$$4y + 5z = 0.$$

Substituting (1) into this, we find

$$-4z + 5z = 0 \to z = 0.$$

Then from (1), $y = 0$, and so $x = 0$ also. The only critical point is therefore $(0, 0, 0)$. Calculation of the second partial derivatives gives the 3-by-3 matrix H:

$$H = \begin{pmatrix} 2 & 4 & -2 \\ 4 & 16 & 4 \\ -2 & 4 & 12 \end{pmatrix}.$$

We find then that $f_{xx} = 2 > 0$,

$$\begin{aligned}\det H &= 2 \cdot 16 \cdot 12 + 4 \cdot 4 \cdot (-2) + (-2) \cdot 4 \cdot 4 \\ &\quad - (-2)^2 \cdot 16 - 4 \cdot 4 \cdot 2 - 4 \cdot 4 \cdot 12 \\ &= 384 - 32 - 32 - 64 - 32 - 192 = 32 > 0,\end{aligned}$$

$$\det\begin{pmatrix} f_{xx} & f_{xy} \\ f_{yx} & f_{yy} \end{pmatrix} = \det\begin{pmatrix} 2 & 4 \\ 4 & 16 \end{pmatrix} = 32 - 16 = 16 > 0.$$

Theorem 7.5(a) then says that (0, 0, 0) is a local minimum point. At this point $f(x, y, z) = -9$. ■

FIGURE 7.11

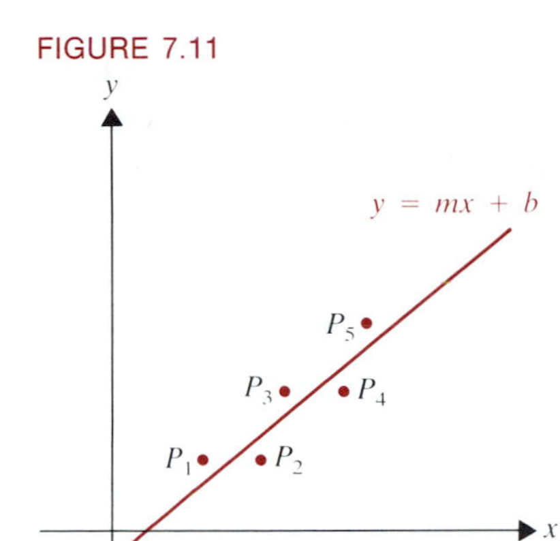

Application: Least-Squares Fits*

Experimental scientists often look for *linear* models to fit data. That is, given a set of n data points $P_1(x_1, y_1), P_2(x_2, y_2), \ldots, P_n(x_n, y_n)$, the straight line that *best fits* the data is sought. **Figure 7.11** shows a line $l: y = mx + b$ that appears to reasonably approximate five such data points, for example. Before discussing how to calculate m and b—that is, how to find the line l of best fit—we discuss what it means for a line l to "best fit" a set of data points.

FIGURE 7.12

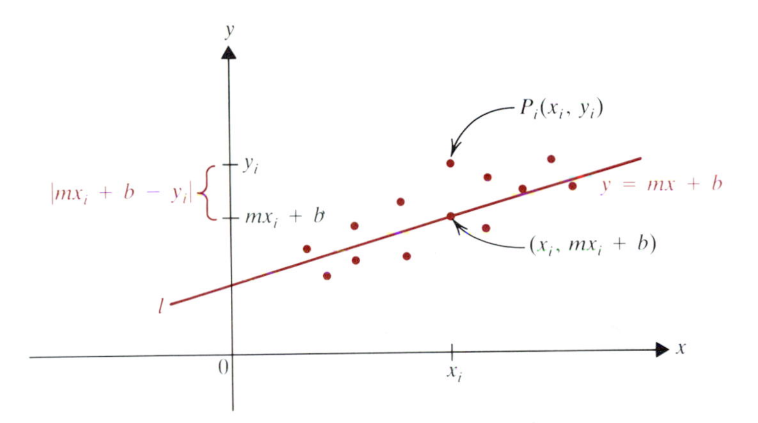

A number of definitions have been suggested, but the most widely accepted was given by Gauss. (See **Figure 7.12.**) He said that the line l: $y = mx + b$ best fits the points $P_1(x_1, y_1), P_2(x_2, y_2), \ldots, P_n(x_n, y_n)$ if the sum of the squares of the errors,

$$(mx_1 + b - y_1)^2 + (mx_2 + b - y_2)^2 + \ldots + (mx_n + b - y_n)^2$$

is as small as possible. When $x = x_i$, the corresponding y on the line is $mx_i + b$, and so $(mx_i + b - y_i)$ measures the vertical directed distance from the approximating point $(x_i, mx_i + b)$ on the line l to the data point $P_i(x_i, y_i)$. Gauss noted that by squaring $(mx_i + b - y_i)$, we eliminate the sign of the error. This means that no attention is paid to whether a point $P_i(x_i, y_i)$ lies above or below l. Gauss's approach is called the ***method of least squares*** or ***linear regression.***

To apply the method of least squares, we must find the real numbers m and b that minimize the function

$$z = f(m, b) = (mx_1 + b - y_1)^2 + (mx_2 + b - y_2)^2 + \ldots + (mx_n + b - y_n)^2. \tag{2}$$

* Optional

To find those values of m and b, we first find the critical points of f, which is differentiable everywhere. So we proceed to solve $\nabla f(m, b) = \mathbf{0}$. From (2) we have

$$\begin{aligned}\frac{\partial z}{\partial m} &= 2(mx_1 + b - y_1)\cdot x_1 + 2(mx_2 + b - y_2)\cdot x_2 + \ldots + 2(mx_n + b - y_n)\cdot x_n \\ &= 2\sum_{i=1}^{n} x_i(mx_i + b - y_i) = 2\sum_{i=1}^{n}(mx_i^2 + bx_i - x_iy_i),\end{aligned}$$

and

$$\begin{aligned}\frac{\partial z}{\partial b} &= 2(mx_1 + b - y_1) + 2(mx_2 + b - y_2) + \ldots + 2(mx_n + b - y_n) \\ &= 2\sum_{i=1}^{n}(mx_i + b - y_i) = 2\sum_{i=1}^{n} mx_i + nb - 2\sum_{i=1}^{n} y_i.\end{aligned}$$

Setting these partial derivatives equal to 0, we obtain

$$(3)\qquad m\sum_{i-1}^{n} x_i^2 + b\sum_{i=1}^{n} x_i = \sum_{i=1}^{n} x_iy_i$$

and

$$(4)\qquad m\sum_{i=1}^{n} x_i + nb = \sum_{i=1}^{n} y_i.$$

From (4) we get

$$(5)\qquad b = \frac{1}{n}\left(\sum_{i=1}^{n} y_i - m\sum_{i=1}^{n} x_i\right).$$

Substitution of (5) into (3) gives

$$m\sum_{i=1}^{n} x_i^2 + \frac{1}{n}\left(\sum_{i=1}^{n} x_i\right)\left(\sum_{i=1}^{n} y_i\right) - \frac{m}{n}\left(\sum_{i=1}^{n} x_i\right)^2 = \sum_{i=1}^{n} x_iy_i,$$

$$mn\sum_{i=1}^{n} x_i^2 - m\left(\sum_{i=1}^{n} x_i\right)^2 = n\sum_{i=1}^{n} x_iy_i - \left(\sum_{i=1}^{n} x_i\right)\left(\sum_{i=1}^{n} y_i\right),$$

$$m\left[n\sum_{i=1}^{n} x_i^2 - \left(\sum_{i=1}^{n} x_1\right)^2\right] = n\sum_{i=1}^{n} x_iy_i - \sum_{i=1}^{n} x_i \cdot \sum_{i=1}^{n} y_i,$$

$$(6)\qquad m = \frac{n\,\Sigma_{i=1}^{n}\, x_iy_i - \Sigma_{i=1}^{n}\, x_i \cdot \Sigma_{i=1}^{n}\, y_i}{n\,\Sigma_{i=1}^{n}\, x_i^2 - (\Sigma_{i=1}^{n}\, x_i)^2}.$$

Substitution of (6) into (5) then gives

$$\begin{aligned}b &= \frac{1}{n}\sum_{i=1}^{n} y_i - \frac{m}{n}\sum_{i=1}^{n} x_i \\ &= \frac{1}{n}\sum_{i=1}^{n} y_i - \frac{1}{n}\sum_{i=1}^{n} x_i \cdot \frac{n\,\Sigma_{i=1}^{n}\, x_iy_i - \Sigma_{i=1}^{n}\, x_i \cdot \Sigma_{i=1}^{n}\, y_i}{n\,\Sigma_{i=1}^{n}\, x_i^2 - (\Sigma_{i=1}^{n}\, x_i)^2} \\ &= \frac{\Sigma_{i=1}^{n}\, y_i\,\Sigma_{i=1}^{n}\, x_i^2 - \cancel{\frac{1}{n}\Sigma_{i=1}^{n}\, y_i(\Sigma_{i=1}^{n}\, x_i)^2} - \Sigma_{i=1}^{n}\, x_i\,\Sigma_{i=1}^{n}\, x_iy_i + \cancel{\frac{1}{n}(\Sigma_{i=1}^{n}\, x_i)^2\,\Sigma_{i=1}^{n}\, y_i}}{n\,\Sigma_{i=1}^{n}\, x_i^2 - (\Sigma_{i=1}^{n}\, x_i)^2},\end{aligned}$$

$$(7)\qquad b = \frac{\Sigma_{i=1}^{n}\, y_i\,\Sigma_{i=1}^{n}\, x_i^2 - \Sigma_{i=1}^{n}\, x_i\,\Sigma_{i=1}^{n}\, x_iy_i}{n\,\Sigma_{i=1}^{n}\, x_i^2 - (\Sigma_{i=1}^{n}\, x_i)^2}.$$

Formulas (6) and (7) are programmed onto some calculators. They are relatively easy to evaluate on a computer or programmable calculator but tedious to use otherwise. The next example illustrates their use in a small-scale problem.

Example 6

Find the line that best fits the following data.

x_i	0	1	2	3	4	5
y_i	3	4.1	4.4	4.3	5.1	5.2

Solution. Here $n = 6$. We have

$$\sum_{i=1}^{6} x_i y_i = 0 + 4.1 + 8.8 + 12.9 + 20.4 + 26 = 72.2,$$

$$\sum_{i=1}^{6} x_i = 0 + 1 + 2 + 3 + 4 + 5 = 15 \rightarrow \left(\sum_{i=1}^{6} x_i\right)^2 = 225,$$

$$\sum_{i=1}^{6} y_i = 3 + 4.1 + 4.4 + 4.3 + 5.1 + 5.2 = 26.1,$$

$$\sum_{i=1}^{6} x_i^2 = 0 + 1 + 4 + 9 + 16 + 25 = 55.$$

FIGURE 7.13

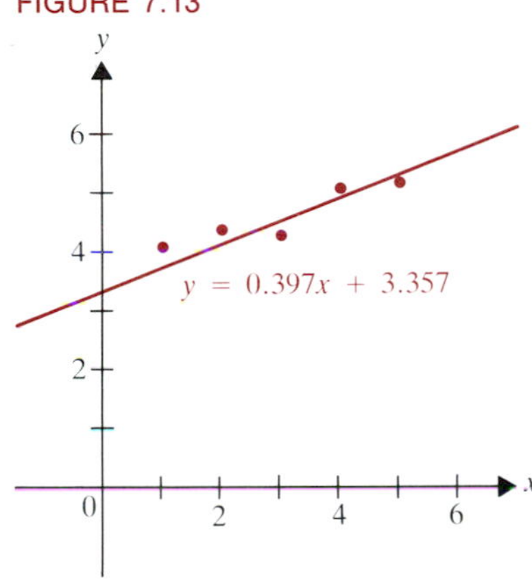

Then (6) gives

$$m = \frac{6 \cdot 72.2 - 15 \cdot 26.1}{6 \cdot 55 - 225} = \frac{433.2 - 391.5}{105} \approx 0.397,$$

and from (7) we get

$$b = \frac{26.1 \cdot 55 - 15 \cdot 72.2}{105} = \frac{1435.5 - 1083}{105} \approx 3.357.$$

Thus the line of best fit is the line

$$y = 0.397x + 3.357.$$

It is plotted in **Figure 7.13,** along with the given data points. ■

Derivation of the Second-Derivative Test*

In the last section we presented the second-degree Taylor polynomial

$$\begin{aligned} p_2(x, y) &= f(a, b) + f_x(a, b)(x - a) + f_y(a, b)(y - b) \\ &\quad + \frac{1}{2!}[f_{xx}(a, b)(x - a)^2 + 2f_{xy}(a, b)(x - a)(y - b) \\ &\quad + f_{yy}(a, b)(y - b)^2], \end{aligned} \tag{8}$$

which closely approximates $f(x, y)$ near $(x, y) = (a, b)$, under the assumption that f_{xx}, f_{xy}, f_{yx}, and f_{yy} are continuous near (a, b). At a critical point (a, b) where $\nabla f(a, b) = \mathbf{0}$, (8) simplifies to

$$\begin{aligned} p_2(x, y) &= f(a, b) + \frac{1}{2!}[f_{xx}(a, b)(x - a)^2 + 2f_{xy}(a, b)(x - a)(y - b) \\ &\quad + f_{yy}(a, b)(y - b)^2] \end{aligned} \tag{9}$$

* Optional

Now $p_2(x, y)$ is a close approximation to $f(x, y)$ near $(x, y) = (a, b)$. Thus, if the bracketed expression in (9) is positive for all $(x, y) \neq (a, b)$ close to (a, b), then $f(a, b)$ is a local minimum. Similarly, if the bracketed expression is negative for all $(x, y) \neq (a, b)$ near (a, b), then $f(a, b)$ is a local maximum. Finally, if the bracketed expression is sometimes positive and sometimes negative, near (a, b), then (a, b) is a saddle point. The proof of Theorem 7.4 just consists in relating these observations to $f_{xx}(a, b)$ and

$$D = \det\begin{pmatrix} f_{xx}(a, b) & f_{xy}(a, b) \\ f_{xy}(a, b) & f_{yy}(a, b) \end{pmatrix} = f_{xx}f_{yy} - f_{xy}{}^2\Big|_{(a, b)}.$$

To carry out the proof, we complete the square in the bracketed part of (9),

$$f_{xx}(a, b)(x - a)^2 + 2f_{xy}(a, b)(x - a)(y - b) + f_{yy}(a, b)(y - b)^2. \tag{10}$$

Supposing that $f_{xx}(a, b) \neq 0$, we factor it from the first two terms, and writing $\mathbf{a} = (a, b)$, we get

$$f_{xx}(\mathbf{a})\left[(x - a)^2 + \frac{2f_{xy}(\mathbf{a})}{f_{xx}(\mathbf{a})}(x - a)(y - b)\right] + f_{yy}(\mathbf{a})(y - b)^2.$$

Now we take half the coefficient of $x - a$, that is,

$$\frac{f_{xy}(\mathbf{a})}{f_{xx}(\mathbf{a})}(y - b),$$

square it, and add the result to the bracket. To maintain an expression equal to (10), we compensate by subtracting the same quantity outside the bracket. That gives

$$f_{xx}(\mathbf{a})\left[(x - a)^2 + \frac{2f_{xy}(\mathbf{a})}{f_{xx}(\mathbf{a})}(x - a)(y - b) + \frac{f_{xy}{}^2(\mathbf{a})}{f_{xx}{}^2(\mathbf{a})}(y - b)^2\right]$$
$$+ f_{yy}(\mathbf{a})(y - b)^2 - \frac{f_{xy}{}^2(\mathbf{a})}{f_{xx}(\mathbf{a})}(y - b)^2.$$

We can rewrite this in the form

$$f_{xx}(\mathbf{a})\left[(x - a) + \frac{f_{xy}(\mathbf{a})}{f_{xx}(\mathbf{a})}(y - b)\right]^2 + \frac{1}{f_{xx}(\mathbf{a})}[f_{xx}(\mathbf{a})f_{yy}(\mathbf{a}) - f_{xy}{}^2(\mathbf{a})](y - b)^2,$$

which is

$$f_{xx}(\mathbf{a})\left[(x - a) + \frac{f_{xy}(\mathbf{a})}{f_{xx}(\mathbf{a})}(y - b)\right]^2 + \frac{1}{f_{xx}(\mathbf{a})}(y - b)^2 D. \tag{11}$$

Now if $f_{xx}(\mathbf{a}) > 0$, the first term in (11) is positive for $\mathbf{x} \neq \mathbf{a}$. If also $D > 0$, then the second term is also positive, so (10) is positive. Thus Theorem 7.4(a) follows from the observations made after (9).

If $f_{xx}(\mathbf{a}) < 0$ and $D > 0$, then both terms in (11) are negative for $\mathbf{x} \neq \mathbf{a}$, so Theorem 7.4(b) follows since the bracketed expression in (9) is negative.

If $D < 0$, then there are several cases. If $f_{xx}(\mathbf{a}) > 0$, then we find that (11) is positive if $y = b$ and x is close to a, because (11) reduces to

$$f_{xx}(\mathbf{a})(x - a)^2.$$

Suppose, however, that we take y close to b and

$$x = a - \frac{f_{xy}(\mathbf{a})}{f_{xx}(\mathbf{a})}(y - b).$$

The first term in (11) is then 0, and the second is negative since $D < 0$. In this case then, (11) is negative. Hence (a, b) is a saddle point. A similar argument applies if $f_{xx}(\mathbf{a}) < 0$. If $f_{xx}(\mathbf{a}) = 0$ but $f_{yy}(\mathbf{a}) \neq 0$, then we can factor out $f_{yy}(\mathbf{a})$ in (10) and complete the square as above; we can then reason, as we just have, to conclude that **a** is a saddle point. (See Exercise 36). If both $f_{xx}(\mathbf{a})$ and $f_{yy}(\mathbf{a})$ are 0, then (10) reduces to $2f_{xy}(\mathbf{a})(x - a)(y - b)$. Whatever sign $f_{xy}(\mathbf{a})$ has, we can make the expression sometimes positive and sometimes negative by taking x slightly greater or smaller than **a**, and y slightly greater than b. Then (a, b) is a saddle point. Hence Theorem 7.4(c) is proved. Example 4 is enough to demonstrate Theorem 7.4(d). QED

Exercises 12.7

In Exercises 1–18, find and classify all critical points of the given functions.

1. $f(x, y) = 4x^2 - xy + y^2$
2. $f(x, y) = 6x - 4y - x^2 - 2y^2$
3. $f(x, y) = x^2 - xy - y^2 + 5y - 1$
4. $f(x, y) = x^3 + y^3 + 3x^2 - 3y^2 - 8$
5. $f(x, y) = x^4 + y^2 - 8x^2 - 6y + 16$
6. $f(x, y) = 2x^4 + y^4 - 2x^2 - 2y^2 + 3$
7. $f(x, y) = 2x + 12y - x^2 - y^3$
8. $f(x, y) = x^3 + 3xy^2 - 3x^2 - 3y^2 + 7$
9. $f(x, y) = 2x^3 + 2y^3 - 3x^2 - 6y^2 - 36x + 5$
10. $f(x, y) = x^4 + 3x^2y^2 + y^4 + 3$
11. $f(x, y) = e^{-x^4 - y^4 + 1}$
12. $f(x, y) = e^{x^3 + y^3}$
13. $f(x, y) = y \cos x$
14. $f(x, y) = e^x \cos y$
15. $f(x, y, z) = x^2 + 3y^2 + 4z^2 - 2xy - 2yz + 2xz + 3$
16. $f(x, y, z) = -x^2 - 2y^2 - z^2 + yz + 5$
17. $f(x, y, z) = 2x^2 + y^2 + 2z^2 + 2xy + 2yz + 2xz + x - 3z - 5$
18. $f(x, y, z) = -2x^2 - y^2 - 3z^2 + 2xy - 2xz + 1$

19. A box for oranges is to be made in the shape of a rectangular parallelepiped with no lid, from 12 square feet of cardboard. Find the dimensions and volume of the largest such box.

20. A tool box is to be constructed from 24 square feet of lumber. It will be in the shape of a rectangular parallelepiped (including a lid). Find the dimensions for maximum volume.

21. A manufacturer wants to make boxes (without lids) of 4 cubic feet volume so as to minimize the amount of cardboard used. What dimensions should he make his boxes?

22. In Exercise 21 suppose that the boxes are made of two kinds of cardboard: The bottom is to be three times as thick (and three times as expensive per unit area) as the sides. Find the dimensions of the most economical box of 4 cubic feet capacity that can be made. (Assume that the sides cost \$1.00 a square foot.)

23. In the 1930s a demand function z for beef in the United States was developed. It was of the form $z = a + bx + cy + dw$, where x is the price of beef, y is a measure of the price of pork, lamb, and poultry, and w is a measure of the income of the consuming public. Here a, c, and d are positive constants, and b is a negative constant. Show that there are no points (x_0, y_0, w_0) where z achieves a local extreme value.

24. Refer to Exercise 23. Another attempt to model the demand for beef is the function $z = ax^by^cw^d$, where a, c, and d are positive constants, and b is a negative constant. Show that in this model, z has local minima but no local maxima. Are the local minima of interest? What would happen if the price of beef fell toward zero but the price of other meats did not? What would happen if the price of other meats increased without bound?

25. Heat loss through each wall of a structure is proportional to the surface area of the wall. The effect of insulation is to reduce the proportionality constant. Suppose that a house is being designed with a volume of 20,000 cubic feet. The house, of length x, width y, and height z feet, is calculated to have daily heat loss of
$$f(x, y, z) = 20xy + 10yz + 10xz$$
at 20°F outside temperature and 68°F inside temperature. Find the dimensions of the house for minimum daily heat loss at these temperatures. Would you find a house of those dimensions attractive?

26. Consider the linear manufacturing cost function (from p. 751) $C(x, y) = b + cx + fy$, where x represents labor costs and y fixed costs. Find all local extrema. Do these furnish useful information?

27. A firm produces two products, A and B. The daily demand function for A is $x = 14 - 2p + 2q$, and the daily demand function for B is $y = 16 + 2p - 6q$, where p and q are the respective unit prices of A and B. It costs \$3.00 to produce a unit of A and \$2.00 to produce a unit of B.
    (a) Write a formula for the company's profit from producing A and B.
    (b) Find the prices it should charge to maximize its profits. How many units should be produced per day for maximum profit?

28. Refer to Exercise 27. Suppose that the demand functions are $x = 16 - 2p + 2q$, $y = 8 + 2p - 4q$, and the production costs are to be the same as in Exercise 27. Find the prices to be charged for maximum profits.

29. Find the line best fitting the points (0, 1), (2, 3), (3, 6), and (4, 8).

30. Repeat Exercise 29 for the points (−2, 1), (0, 3), (1, 6), and (2, 8).

PC **31.** Repeat Exercise 29 for

x	0	1	2	3	4	5	6	7	8	9	10
y	−2	3	1	3	2	6	8	5	7	12	8

PC **32.** Repeat Exercise 29 for

x	0	1	2	3	4	5	6	7	8	9	10
y	5	2	6	2	3	1	0	2	−3	−1	−2

33. In mechanics, a *conservative force field* is one given by a formula $\mathbf{F} = \nabla p(x, y)$, where $-p(x, y)$ is the potential energy at the point (x, y). A *stable equilibrium point* is a point (a, b) where p has a local minimum. An *unstable equilibrium point* is a point (a, b) such that either $\mathbf{p}'(a, b)$ fails to exist or $\mathbf{p}'(a, b) = \mathbf{0}$, but p does not have a local minimum at (a, b). Suppose that a particle moves in the xy-plane with potential energy $-\sqrt{x^2 + y^2}$. Find all equilibrium points and determine whether there is any stable equilibrium point.

34. Refer to Exercise 33. Suppose that a particle moves in space with potential energy $-\sqrt{x^2 + y^2 + z^2}$. Find all equilibrium points and determine whether there is a stable equilibrium point.

35. If x and y are nonnegative, then show that the minimum value of

$$f(x, y) = x + y - 2\sqrt{xy}$$

is 0. That is, show that the geometric mean of x and y is less than or equal to the arithmetic mean of x and y.

36. Prove Theorem 7.4(c) for the case $f_{xx}(\mathbf{a}) = 0$, $f_{yy}(\mathbf{a}) \neq 0$.

We can actually draw slightly more information from the second-derivative test than is stated in Theorem 7.4 itself. This is the point of Exercises 37 and 38.

37. Suppose that f and $\mathbf{a}$ are as in Theorem 7.4. If $D = 0$ and $f_{xx}(\mathbf{a}) > 0$, then show that $f(\mathbf{a})$ is a local minimum of the function $z = f(x, b)$ of x alone. What is the analogous result for $f(a, y)$?

38. Suppose that f and $\mathbf{a}$ are as in Theorem 7.4. If $D = 0$ and $f_{xx}(\mathbf{a}) < 0$, then show that $f(\mathbf{a})$ is a local maximum of the function $z = f(x, b)$ of x alone. What is the analogous result for $f(a, y)$?

39. Suppose that we decide to define *positivity* for a vector $\mathbf{x} = x\mathbf{i} + y\mathbf{j}$ in $\mathbf{R}^2$ by saying that

$$\mathbf{x} > \mathbf{0} \tag{12}$$

means that $x > 0$ and $y > 0$. Use (12) to formulate an analogue for $f: \mathbf{R}^2 \to \mathbf{R}$ of the first-derivative test of elementary calculus, but then show by an example that your analogue fails to be correct.

40. Show that every critical point of a function $f(x, y) = g(x) + h(y)$ is of the form (a, b), where a is a critical point of g and b is a critical point of h.

41. In Example 3, verify that (a, b) is a local maximum point of f if and only if a is a local maximum point for g and b is a local maximum point for h, where $g(x) = 3x - x^3$ and $h(y) = 12y - y^3$. Verify also that the corresponding statement is true for local minimum points.

42. Show that Exercise 41 generalizes to any function $f(x, y) = g(x) + h(y)$ for which g'' and h'' exist and are nonzero at each critical point.

12.8 Lagrange Multipliers

The methods of the last section apply to many optimization problems but are not directly applicable to *constrained* extreme value problems, in which the variables x and y (and possibly z) are restricted by some side condition (called a *constraint*). For example, parcel post regulations stipulate that the *height* plus *girth* of a package cannot exceed 108 inches (see **Figure 8.1**), where the height is defined as the length of the longest edge, and the girth is twice the sum of the lengths of the other edges. Suppose that we want to find the dimensions of the mailable box of largest possible volume. Denoting the length, width, and height (in inches) by x, y, and z respectively, we need to maximize

$$V = xyz,$$

FIGURE 8.1

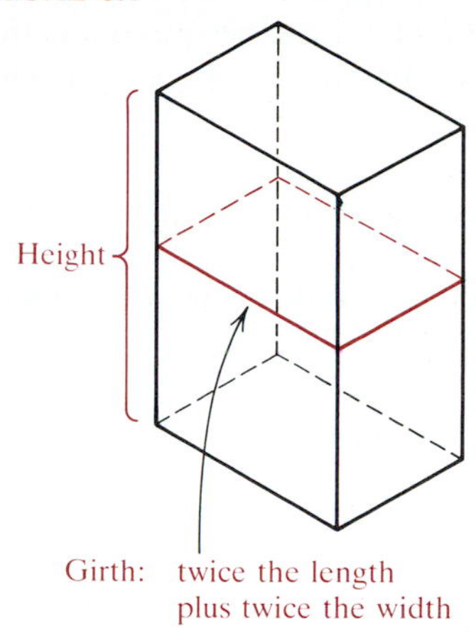

subject to the constraint that

$$(1) \qquad 2x + 2y + z = 108.$$

We could eliminate z by using the constraint equation in the form $z = 108 - 2x - 2y$ and then substitute this into the formula for V. (This method was intended for Exercises 19–22 in the last section.) In this section we present an alternative method originated by the French mathematician Joseph L. Lagrange (1736–1813).

In the parcel post example, the Lagrange method may not seem so advantageous, since z is so easily eliminated from (1). But consider the problem of maximizing or minimizing the electrical potential $f(x, y) = 1/\sqrt{x^2 + y^2}$ over the edge of a television screen with equation $2x^4 + 3y^4 = 32$ (see **Figure 8.2**). In this case the complexity of the equation will increase markedly if we try to eliminate y from the constraint equation.

FIGURE 8.2

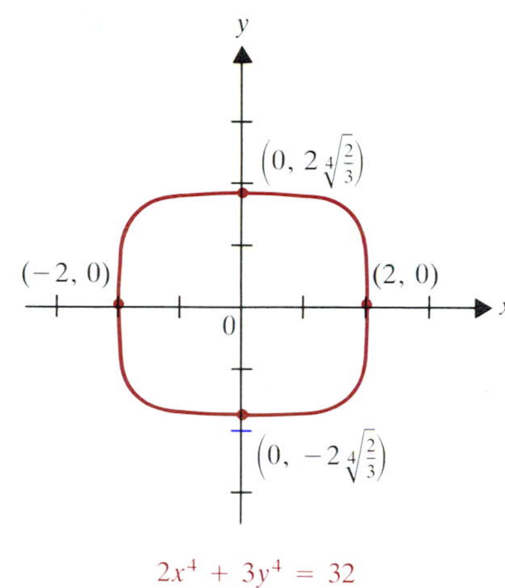

So consider the general problem of minimizing a differentiable function $f\colon \mathbf{R}^2 \to \mathbf{R}$ subject to the constraint $g(x, y) = 0$. For example, $f(x, y)$ might be $1/\sqrt{x^2 + y^2}$ and $g(x, y)$ might be $2x^4 + 3y^4 - 32$. Geometrically, we seek the point (or points) on $g(x, y) = 0$ where f assumes its least value on the curve. For *any* value k, f assumes the value k precisely on the level curve $f(x, y) = k$. In the television example, those level curves are circles $x^2 + y^2 = 1/k^2$ that radiate inward toward the origin as k increases. Eventually they reach the curve $g(x, y) = 0$, as in **Figure 8.3.** The first points of intersection $P_0(a, b)$ are the points on $g(x, y) = 0$ where $f(x, y) = k_0$. No other points on $g(x, y) = 0$ give so small a value to f, because any other point P_1 on $g(x, y) = 0$ is on a level curve of smaller radius $1/k_1$ corresponding to a *larger* value $k_1 > k_0$.

To find the points $P_0(a, b)$ where $f(x, y)$ achieves an extreme value k_0 on the curve $g(x, y) = 0$, note first that the coordinates of such points satisfy both $f(x, y) = k_0$ and $g(x, y) = 0$. One way to find their coordinates would be to solve $g(x, y) = 0$ for y or x in terms of the other variable and then substitute into $f(x, y) = k_0$. The implicit function theorem (Theorem 5.1) says that one of y or x is a function of the other provided that

$$\nabla g(a, b) = \frac{\partial g}{\partial x}(a, b)\,\mathbf{i} + \frac{\partial g}{\partial y}(a, b)\,\mathbf{j} \neq \mathbf{0},$$

FIGURE 8.3

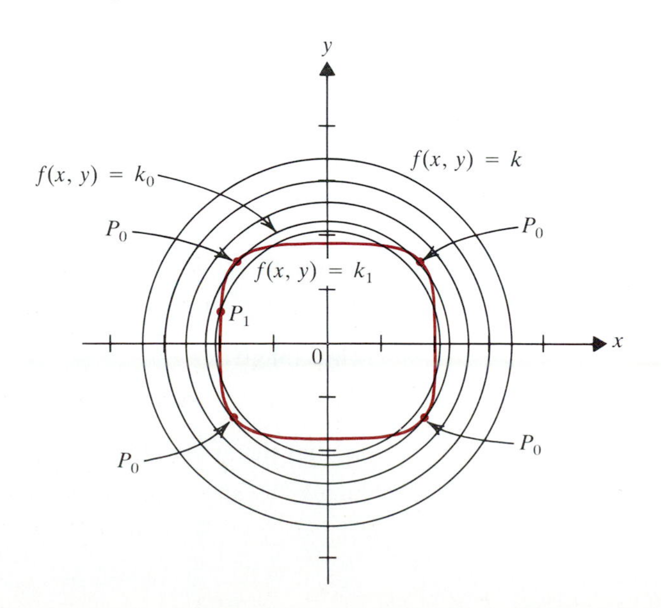

and that $\partial g/\partial x$ and $\partial g/\partial y$ are continuous near (a, b). Suppose then that g satisfies those conditions. For definiteness, assume that $g(x, y) = 0$ defines y as a function $y = y(x)$ of x. Then we can use Theorem 4.2 to differentiate $g(x, y) = 0$. That gives

$$\frac{dg}{dx} = \nabla g(x, y) \cdot \left[\frac{\partial x}{\partial x}\mathbf{i} + \frac{\partial \mathbf{y}}{\partial \mathbf{x}}\mathbf{j}\right] = \nabla g(x, y) \cdot [\mathbf{i} + y'(x)\mathbf{j}] = 0.$$

Hence at a point $P_0(a, b)$ of intersection of $f(x, y) = k_0$ and $g(x, y) = 0$, we have

$$\nabla g(a, b) \cdot [\mathbf{i} + y'(a)\mathbf{j}] = 0. \tag{2}$$

By assumption, the gradient vector $\nabla g(a, b)$ is nonzero. Thus Equation (2) says that it is perpendicular to $\mathbf{i} + y'(a)\mathbf{j}$. Since f has a local extreme value at (a, b), Theorems 4.2 and 7.2 give

$$\frac{df}{dx}(a, b) = \nabla f(a, b) \cdot [\mathbf{i} + y'(a)\mathbf{j}] = 0.$$

Thus if $\nabla f(a, b) \neq \mathbf{0}$, then it is also perpendicular to $\mathbf{i} + y'(a)\mathbf{j}$. Therefore $\nabla f(a, b)$ is parallel to $\nabla g(a, b)$. Recall from Theorem 1.6 of Chapter 10 that two plane vectors are parallel if and only if one is a scalar multiple of the other. We thus have

$$\nabla f(a, b) = \lambda \nabla g(a, b) \rightarrow \frac{\partial f}{\partial x}(a, b)\mathbf{i} + \frac{\partial f}{\partial y}(a, b)\mathbf{j} = \lambda\frac{\partial g}{\partial x}(a, b)\mathbf{i} + \lambda\frac{\partial g}{\partial y}(a, b)\mathbf{j}, \tag{3}$$

for some real number λ.

If $\nabla f(a, b) = 0$, then (3) holds with $\lambda = 0$. Thus (3) is a necessary condition for f to have a local extreme value on the curve $g(x, y) = 0$. Consequently, to find candidates (a, b) for local extreme points, we need to solve the system of scalar equations obtained by equating the respective $\mathbf{i}$ and $\mathbf{j}$ coordinates in (3) on the curve $g(x, y) = 0$:

$$\begin{cases} \dfrac{\partial f}{\partial x}(x, y) = \lambda\dfrac{\partial g}{\partial x}(x, y) \\ \dfrac{\partial f}{\partial y}(x, y) = \lambda\dfrac{\partial g}{\partial x}(x, y) \\ g(x, y) = 0. \end{cases} \tag{4}$$

This is a system of three equations in the variables x, y, and λ, which often has several solutions, as the following example illustrates.

Example 1

Find the extreme values of $f(x, y) = 2x - y$ on the circle $x^2 + y^2 = 16$.

Solution. We have

$$\nabla f(x, y) = 2\mathbf{i} - \mathbf{j} \quad \text{and} \quad \nabla g(x, y) = 2x\mathbf{i} + 2y\mathbf{j}.$$

Thus (3) gives

$$2\mathbf{i} - \mathbf{j} = 2x\lambda\mathbf{i} + 2y\lambda\mathbf{j} \quad \rightarrow \quad 2 = 2x\lambda, \ -1 = 2y\lambda,$$

so that the system (4) is

$$\begin{cases} 1 = \lambda x \\ -1 = 2\lambda y \\ x^2 + y^2 = 16. \end{cases}$$

The first two equations say that $x = 1/\lambda$, and $y = -1/2\lambda$. Substitution of that in the third equation gives

$$\frac{1}{\lambda^2} + \frac{1}{4\lambda^2} = 16 \rightarrow 4 + 1 = 64\lambda^2 \rightarrow \lambda^2 = \frac{5}{64}.$$

Therefore $\lambda = \pm\sqrt{5}/8$. This means that $x = 1/\lambda = 8/\sqrt{5}$ and $y = -1/2\lambda = -x/2 = -4/\sqrt{5}$ or $x = -8/\sqrt{5}$ and $y = 4/\sqrt{5}$. Thus the only candidates for extreme points are $(8/\sqrt{5}, -4/\sqrt{5})$ and $(-8/\sqrt{5}, 4/\sqrt{5})$. We have

$$f(8/\sqrt{5}, -4/\sqrt{5}) = 16/\sqrt{5} + 4/\sqrt{5} = 20/\sqrt{5} = 4\sqrt{5}$$

and

$$f(-8/\sqrt{5}, 4/\sqrt{5}) = -16/\sqrt{5} - 4/\sqrt{5} = -4\sqrt{5}.$$

According to a theorem of advanced calculus, a continuous function such as f has a maximum and a minimum value on the circle $x^2 + y^2 = 16$. We can thus conclude that the maximum value of f on the circle is $4\sqrt{5}$ and its minimum value point is $-4\sqrt{5}$. ■

A similar analysis applies to the case of three variables. To find extreme values of $f(x, y, z) = 0$ on a surface

$$g(x, y, z) = 0, \tag{5}$$

we assume that ∇f and ∇g are continuous and that $\nabla g(\mathbf{a}) \neq 0$ at an extreme point $\mathbf{a} = (a, b, c)$. For instance, if the third coordinate $g_z(a, b, c)$ is nonzero, then Theorem 5.2 guarantees that $g(x, y, z) = 0$ defines $z = z(x, y)$, so that $f(x, y, z) = f(x, y, z(x, y))$. At a local extreme point $\mathbf{a} = (a, b, c)$, we have from Theorem 7.2

$$\frac{\partial f}{\partial x} = 0 = \frac{\partial f}{\partial y},$$

so by Theorem 4.3,

$$\nabla f(\mathbf{a}) \cdot \left[1\,\mathbf{i} + 0\,\mathbf{j} + \frac{\partial f}{\partial x}(\mathbf{a})\,\mathbf{k} \right] = 0 = \nabla f(\mathbf{a}) \cdot \left[0\,\mathbf{i} + 1\,\mathbf{j} + \frac{\partial z}{\partial y}(\mathbf{a})\,\mathbf{k} \right].$$

Also, since $g(x, y, z(x, y)) = 0$, we similarly obtain

$$\frac{\partial g}{\partial x}(\mathbf{a}) = \nabla g(\mathbf{a}) \cdot \left[\mathbf{i} + \frac{\partial z}{\partial x}(\mathbf{a})\,\mathbf{k} \right] = 0 = \nabla g(\mathbf{a}) \cdot \left[\mathbf{j} + \frac{\partial z}{\partial y}(\mathbf{a})\,\mathbf{k} \right] = \frac{\partial g}{\partial y}(\mathbf{a}).$$

Hence if $\nabla f(\mathbf{a}) \neq \mathbf{0}$, then $\nabla f(\mathbf{a})$ and $\nabla g(\mathbf{a})$ are perpendicular to both

$$\mathbf{i} + \frac{\partial z}{\partial x}(\mathbf{a})\,\mathbf{k} \qquad \text{and} \qquad \mathbf{j} + \frac{\partial z}{\partial y}(\mathbf{a})\,\mathbf{k},$$

which by Definition 1.2 determine the tangent plane to $g(x, y, z) = 0$ at (a, b, c). Thus $\nabla f(\mathbf{a})$ and $\nabla g(\mathbf{a})$ are both perpendicular to that tangent plane and hence are parallel. We therefore have

$$\nabla f(\mathbf{a}) = \lambda\,\nabla g(\mathbf{a}), \tag{3}$$

that is

$$\frac{\partial f}{\partial x}(\mathbf{a})\,\mathbf{i} + \frac{\partial f}{\partial y}(\mathbf{a})\,\mathbf{j} + \frac{\partial f}{\partial z}(\mathbf{a})\,\mathbf{k} = \lambda \frac{\partial g}{\partial x}(\mathbf{a})\,\mathbf{i} + \lambda \frac{\partial g}{\partial y}(\mathbf{a})\,\mathbf{j} + \lambda \frac{\partial g}{\partial z}(\mathbf{a})\,\mathbf{k},$$

for some real number λ. Equation (3) continues to hold even if $\nabla f(\mathbf{a}) = \mathbf{0}$, since $\lambda = 0$ is permissible. Similar reasoning applies if $g_x(\mathbf{a})$ or $g_y(\mathbf{a})$ is a nonzero coordinate of $\nabla g(\mathbf{a})$. Thus (3) is once again a necessary condition for $\mathbf{a} = (a, b, c)$ to be a local extreme point. So to find candidates for extreme points of f subject to the constraint $g(x, y, z) = 0$, we solve the system obtained by equating the respective $\mathbf{i}$, $\mathbf{j}$, and $\mathbf{k}$ coordinates in (3) on the surface (5):

$$(6) \qquad \begin{cases} \dfrac{\partial f}{\partial x}(x, y, z) = \lambda \dfrac{\partial g}{\partial x}(x, y, z) \\ \dfrac{\partial f}{\partial y}(x, y, z) = \lambda \dfrac{\partial g}{\partial y}(x, y, z) \\ \dfrac{\partial f}{\partial z}(x, y, z) = \lambda \dfrac{\partial g}{\partial z}(x, y, z) \\ \quad g(x, y, z) = 0, \end{cases}$$

a system of four equations in the four variables x, y, z, and λ.

We summarize the foregoing discussion as follows.

8.1
ALGORITHM

Lagrange Multiplier Method. Let f and g be differentiable scalar functions of two or three variables. To find the extreme values of f subject to the constraint $g(\mathbf{x}) = 0$,

(a) solve the system of equations

$$(7) \qquad \begin{cases} \nabla f(\mathbf{x}) = \lambda \nabla g(\mathbf{x}) \\ \quad g(\mathbf{x}) = 0 \end{cases}$$

for $\mathbf{x}$ and λ;

(b) evaluate f at each candidate $\mathbf{x} = \mathbf{a}$ found in (a). The largest value obtained is the maximum of f subject to the given constraint; the smallest value obtained is the minimum of f subject to $g(\mathbf{x}) = 0$.

The scalar λ is called a ***Lagrange multiplier.*** To illustrate 8.1, we use a Lagrange multiplier to solve the parcel post problem posed at the beginning of the section.

Example 2

Find the dimensions of the box of maximum volume whose height plus girth is 108 inches.

Solution. As on p. 820, we have to maximize $V = xyz$ subject to the constraint $g(x, y, z) = 2x + 2y + z - 108 = 0$. Here $\nabla V = yz\,\mathbf{i} + xz\,\mathbf{j} + xy\,\mathbf{k}$ and $\nabla g = 2\,\mathbf{i} + 2\,\mathbf{j} + \mathbf{k}$. To find candidates for a maximum point $\mathbf{a} = (a, b, c)$ we have to solve the system (7), which in this case is

$$\begin{cases} yz = 2\lambda \\ xz = 2\lambda \\ xy = \lambda \\ 2x + 2y + z - 108 = 0. \end{cases}$$

Subtraction of the second equation from the first gives

$$(y - x)z = 0.$$

Since $z \neq 0$ for a box, we get $y = x$. The third equation then gives $x^2 = \lambda$, which, when substituted into the second equation, yields

$$z = \frac{2\lambda}{x} = \frac{2x^2}{x} = 2x.$$

We can now substitute for y and z in the final equation to obtain

$$2x + 2x + 2x = 108 \rightarrow 6x = 108 \rightarrow x = 18.$$

Hence $y = 18$ and $z = 36$. Therefore the dimensions of the mailable box of largest volume are 18 by 18 by 36 inches. ■

The electrical potential problem mentioned on p. 12-8-1 leads to a more involved Lagrange multiplier problem.

Example 3

Find the extreme values of $f(x, y) = 1/\sqrt{x^2 + y^2}$ on the curve $2x^4 + 3y^4 = 32$.

Solution. Here $g(x, y) = 2x^4 + 3y^4 - 32$. Thus

$$\nabla g = 8x^3\mathbf{i} + 12y^3\mathbf{j}.$$

We also have

$$\nabla f = \frac{-x}{(x^2 + y^2)^{3/2}}\mathbf{i} - \frac{y}{(x^2 + y^2)^{3/2}}\mathbf{j}.$$

We therefore have to solve the system

$$(8) \qquad \begin{cases} \dfrac{-x}{(x^2 + y^2)^{3/2}} = 8\lambda x^3 \\[2ex] \dfrac{-y}{(x^2 + y^2)^{3/2}} = 12\lambda y^3 \\[2ex] 2x^4 + 3y^4 = 32. \end{cases}$$

In the top equation, $x = 0$ is a possible root. By the bottom equation, the corresponding y satisfies $y^4 = 32/3$, so $y = \pm\sqrt[4]{32/3} = \pm 2\sqrt[4]{2/3}$. In the middle equation, $y = 0$ is a possible root, and from the bottom equation the corresponding x is ± 2. For other candidates, $x \neq 0$ and $y \neq 0$. Then we can cancel x in the first equation and y in the second equation, getting

$$\begin{cases} \dfrac{-1}{(x^2 + y^2)^{3/2}} = 8\lambda x^2, \\[2ex] \dfrac{-1}{(x^2 + y^2)^{3/2}} = 12\lambda y^2. \end{cases}$$

So $8\lambda x^2 = 12\lambda y^2$. Now, λ isn't zero here: If it were zero, then x and y would also be zero by the first equation in (8). Hence we can cancel λ and get $x^2 = \frac{12}{8}y^2 = \frac{3}{2}y^2$. Substituting this into $2x^4 + 3y^4 = 32$, we find

$$2\left(\frac{9}{4}y^4\right) + 3y^4 = 32 \rightarrow 9y^4 + 6y^4 = 64 \rightarrow 15y^4 = 64 \rightarrow y^4 = \frac{64}{15},$$

$$y = \pm\sqrt[4]{\frac{64}{15}} = \pm\frac{2\sqrt{2}}{\sqrt[4]{15}}.$$

Then

$$x^2 = \frac{3}{2}y^2 = \frac{3}{2}\cdot\frac{8}{\sqrt{15}} = \frac{12}{\sqrt{15}}, \qquad \text{so} \qquad x = \pm\frac{2\sqrt{3}}{\sqrt[4]{15}}.$$

The candidates for extreme values of f are then

$$(\pm 2, 0), (0, \pm 2\sqrt[4]{\tfrac{2}{3}}), \qquad \text{and} \qquad \frac{1}{\sqrt[4]{15}}(\pm 2\sqrt{3}, \pm 2\sqrt{2}).$$

To test these points we compute f at each of them

$$f(\pm 2, 0) = \frac{1}{\sqrt{4+0}} = \frac{1}{2}, \qquad f(0, \pm 2\sqrt[4]{\tfrac{2}{3}}) = \frac{1}{\sqrt{0 + 4\sqrt{2/3}}} = \frac{1}{2}\sqrt[4]{\tfrac{3}{2}},$$

$$f\left(\frac{1}{\sqrt[4]{15}}(\pm 2\sqrt{3}, \pm 2\sqrt{2})\right) = \frac{1}{\frac{1}{\sqrt[4]{15}}\sqrt{12+8}} = \sqrt{\frac{\sqrt{15}}{20}} = \frac{1}{2}\left(\frac{\sqrt{15}}{5}\right)^{1/2}$$

$$= \frac{1}{2}\left(\frac{15}{25}\right)^{1/4} = \frac{1}{2}\left(\frac{3}{5}\right)^{1/4}$$

Since

$$\frac{1}{2}\sqrt[4]{\tfrac{3}{2}} > \frac{1}{2}\sqrt[4]{1} = \frac{1}{2},$$

we see that

$$f(0, \pm 2\sqrt[4]{\tfrac{2}{3}}) > f(\pm 2, 0).$$

Also, since

$$\frac{1}{2}\left(\frac{3}{5}\right)^{1/4} < \frac{1}{2}\sqrt[4]{1} = \frac{1}{2},$$

we have

$$f\left(\frac{1}{\sqrt[4]{15}}(\pm 2\sqrt{3}, \pm 2\sqrt{2})\right) < f(\pm 2, 0).$$

Thus the minimum value of f on the curve $2x^4 + 3y^4 = 32$ is $\frac{1}{2}\sqrt[4]{3/5}$, occurring at the four points

$$\frac{1}{\sqrt[4]{15}}(\pm 2\sqrt{3}, \pm 2\sqrt{2}).$$

The maximum value of f on the same curve is $\frac{1}{2}\sqrt[4]{3/2}$, occurring at the two points $(0, \pm 2\sqrt[4]{2/3})$. (The points $(\pm 2, 0)$ are not extreme points.) ■

We can combine Lagrange multipliers with the methods of the last section to find the extreme values of a function over a region consisting of a simple closed curve plus its interior. The next example uses this approach to find the extreme values that Theorem 4.7 of Chapter 11 told us must exist.

Example 4

Find the extreme values of $f(x, y) = x^4 + y^2 + 3$ over the closed unit disk $D = \{(x, y) \in \mathbf{R}^2 | x^2 + y^2 \leq 1\}$.

Solution. We have $\nabla f(x, y) = 4x^3\mathbf{i} + 2y\mathbf{j}$, which is $\mathbf{0}$ only at the origin. That point clearly gives the absolute minimum value, 3, for f on D (in fact on all of $\mathbf{R}^2$), so we don't need to test this candidate further. By Theorem 4.7 of Chapter 11, f has a maximum value on D. Since (0, 0) is the only critical point, the maximum is not a local maximum, so it *must* occur on the boundary of D, that is, on the unit circle $x^2 + y^2 = 1$. To find the point or points where the maximum value is taken on, we let $g(x, y) = x^2 + y^2 - 1$ and apply Method 8.1. We then have to solve $\nabla f = \lambda \nabla g$ on the bounding curve $g(x, y) = 0$. Since

$$\nabla g = 2x\,\mathbf{i} + 2y\,\mathbf{j},$$

we have the system of equations

$$\begin{cases} 4x^3 = 2\lambda x \\ 2y = 2\lambda y \\ x^2 + y^2 = 1. \end{cases}$$

In the first equation, $x = 0$ is a possible root. From the last equation, it corresponds to $y = \pm 1$. In the second equation, $y = 0$ is likewise a possible root, and it corresponds to $x = \pm 1$. The other candidates must satisfy $x \neq 0$ and $y \neq 0$, so x and y can be cancelled from the first two equations. That gives

$$2x^2 = \lambda \qquad \text{and} \qquad \lambda = 1.$$

Then $x^2 = 1/2$, so $x = \pm\sqrt{1/2}$. From the third equation $y^2 = 1/2$ also; hence $y = \pm\sqrt{1/2}$. Therefore the candidates for extreme points of f on $x^2 + y^2 = 1$ are $(\pm 1, 0)$, $(0, \pm 1)$, $(\pm\sqrt{1/2}, \sqrt{1/2})$, and $(\pm\sqrt{1/2}, -\sqrt{1/2})$. Evaluating f at each of these, we find that

$$f(\pm 1, 0) = f(0, \pm 1) = 4 \qquad \text{and} \qquad f\left(\pm\sqrt{\frac{1}{2}}, \pm\sqrt{\frac{1}{2}}\right) = \frac{1}{4} + \frac{1}{2} + 3 = 3.75.$$

Thus the maximum of f on D is 4, which occurs at the points (1, 0), (−1, 0), (0, 1), and (0, −1). Our work has also shown that the minimum value of $f(x, y)$ on the bounding circle $x^2 + y^2 = 1$ is 3.75, but that is not the minimum on D, which we already found occurs at the origin. **Figure 8.4** shows the graph of f

FIGURE 8.4

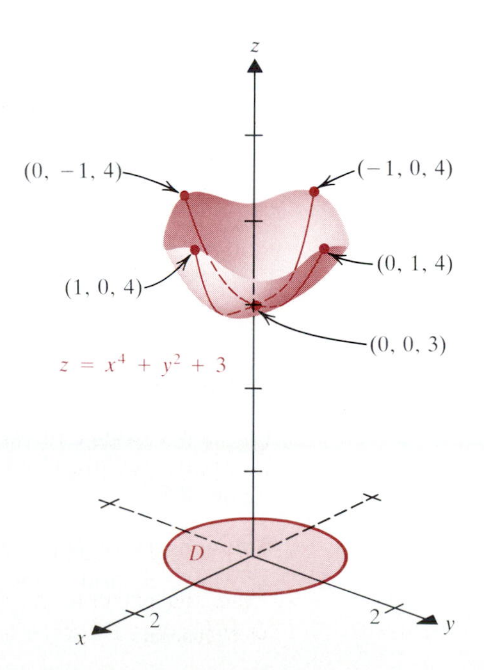

over the disk D and the various extreme value points on D and its bounding circle. ■

Multiple Constraints*

Suppose that we wish to find the extreme values of a function f subject to *several* constraints, say,

$$g_1(\mathbf{x}) = 0,\ g_2(\mathbf{x}) = 0, \ldots, g_m(\mathbf{x}) = 0. \tag{9}$$

There is also a Lagrange multiplier method for this type of problem. Notice in Method 8.1 that we introduce a single Lagrange multiplier λ that multiplies the gradient vector $\nabla g(x)$ in (7). In the case of m constraints (9), the analogous product is

$$\lambda_1 \nabla g_1(\mathbf{x}) + \lambda_2 \nabla g_2(\mathbf{x}) + \ldots + \lambda_m \nabla g_m(\mathbf{x}).$$

In (7) the vector equation

$$\nabla f(\mathbf{x}) = \lambda \nabla g(\mathbf{x})$$

can be written

$$\nabla(f(\mathbf{x}) - \lambda g(\mathbf{x})) = \mathbf{0}.$$

Thus the candidates for extreme value points found by solving (7) are critical points of the function $f - \lambda g$. In the case of m constraints, we look for critical points of

$$f - \boldsymbol{\lambda} \cdot \mathbf{g} = f - (\lambda_1 g_1 + \lambda_2 g_2 + \ldots + \lambda_m g_m),$$

where $\boldsymbol{\lambda} = (\lambda_1, \ldots, \lambda_m)$ and $\mathbf{g}\colon \mathbf{R}^n \to \mathbf{R}^m$ is given by $\mathbf{g}(x) = (g_1(x), g_2(x), \ldots, g_m(x))$. The Lagrange multiplier method for several constraints thus can be described by the following, for which justification is given in advanced calculus.

8.2 ALGORITHM

Lagrange Multiplier Method for m Constraints. Let f and $g_1, g_2, \ldots, g_m$ be differentiable functions of two or three variables. Suppose that (9) defines a nonempty region R in $\mathbf{R}^2$ or $\mathbf{R}^3$. To find the extreme values of f on $\mathbf{R}$,

(a) solve the system of equations

$$\begin{cases} \nabla f(\mathbf{x}) = \lambda_1 \nabla g_1(\mathbf{x}) + \lambda_2 \nabla g_2(\mathbf{x}) + \ldots + \lambda_m \nabla g_m(\mathbf{x}) \\ g_1(\mathbf{x}) = 0,\ g_2(\mathbf{x}) = 0, \ldots, g_m(\mathbf{x}) = 0, \end{cases} \tag{10}$$

for $\mathbf{x}$ and $\boldsymbol{\lambda} = (\lambda_1, \ldots, \lambda_m)$;

(b) evaluate f at each candidate $\mathbf{x} = \mathbf{a}$ found in (a). The largest value obtained is the maximum of f subject to the given constraints; the smallest value obtained is the minimum of f subject to (9).

Example 5

Find the extreme values of $f(x, y, z) = xy + xz$ on the curve of intersection of the right circular cylinder $x^2 + y^2 = 1$ and the hyperbolic cylinder $xz = 1$. See **Figure 8.5.**

Solution. Here $f(x, y, z) = xy + xz$ and the two constraints are $g_1(x, y, z) = x^2 + y^2 - 1$ and $g_2(x, y, z) = xz - 1$. Since $\nabla f(x, y, z) = (y + z)\mathbf{i} + x\mathbf{j} + x\mathbf{k}$,

* Optional

FIGURE 8.5

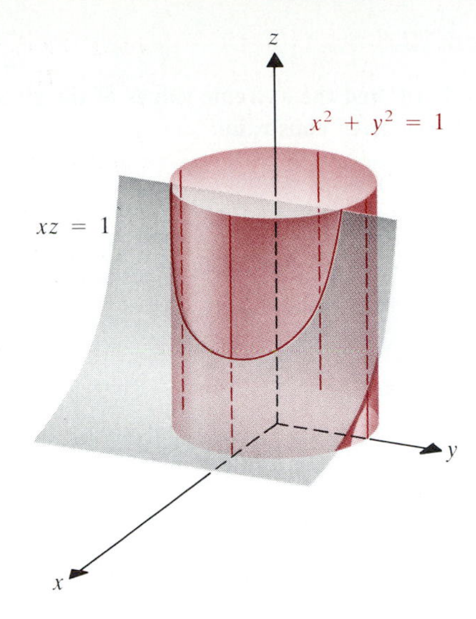

$\nabla g_1 = 2x\mathbf{i} + 2y\mathbf{j}$ and $\nabla g_2 = z\mathbf{i} + x\mathbf{k}$, the system (10) to be solved is

$$\begin{cases} (y+z)\,\mathbf{i} + x\,\mathbf{j} + x\,\mathbf{k} = \lambda_1(2x\,\mathbf{i} + 2y\,\mathbf{j}) + \lambda_2(z\,\mathbf{i} + x\,\mathbf{k}), \\ xz - 1 = 0, \qquad x^2 + y^2 - 1 = 0. \end{cases}$$

We thus have

$$\begin{cases} y + z = 2\lambda_1 x + \lambda_2 z & (11) \\ x = 2\lambda_1 y & (12) \\ x = \lambda_2 x & (13) \\ xz = 1 & (14) \\ x^2 + y^2 = 1 & (15) \end{cases}$$

From (14), x cannot be 0. Then (13) gives $\lambda_2 = 1$, and we have $z = 1/x$ from (14). If we put this information together with (12) into (11), then we obtain

$$y + \frac{1}{x} = 2\lambda_1 x + \frac{1}{x} \rightarrow y - 4\lambda_1{}^2 y = 0 \rightarrow y(1 - 4\lambda_1{}^2) = 0.$$

Hence either $y = 0$ or

$$\lambda_1{}^2 = \frac{1}{4} \rightarrow \lambda_1 = \pm\frac{1}{2}.$$

But from (12), $y = 0$ would give $x = 0$, which we saw is impossible by (14). Hence $\lambda_1 = \pm 1/2$, and so from (12) $x = \pm y$. Then from (15),

$$2x^2 = 1 \rightarrow x = \pm\sqrt{\frac{1}{2}}.$$

Equation (14) then gives $z = 1/x = \pm\sqrt{2}$. Since $y = \pm x$, the candidates for extreme points are $(\sqrt{1/2}, \pm\sqrt{1/2}, \sqrt{2})$ and $(-\sqrt{1/2}, \pm\sqrt{1/2}, -\sqrt{2})$. We calculate f at these points:

$$f\left(\sqrt{\frac{1}{2}}, \sqrt{\frac{1}{2}}, \sqrt{2}\right) = \frac{1}{2} + 1 = \frac{3}{2}, \qquad f\left(-\sqrt{\frac{1}{2}}, -\sqrt{\frac{1}{2}}, -\sqrt{2}\right) = \frac{1}{2} + 1 = \frac{3}{2},$$

$$f\left(\sqrt{\frac{1}{2}}, -\sqrt{\frac{1}{2}}, \sqrt{2}\right) = -\frac{1}{2} + 1 = \frac{1}{2}, \qquad f\left(-\sqrt{\frac{1}{2}}, \sqrt{\frac{1}{2}}, -\sqrt{2}\right) = -\frac{1}{2} + 1 = \frac{1}{2}.$$

So the maximum value is 3/2 and the minimum value is 1/2. ■

Exercises 12.8

In Exercises 1–10, find the extreme values of the given function subject to the given constraint.

1. $f(x, y) = x + 2y$; $x^2 + y^2 = 1$
2. $f(x, y) = 3x - y$; $x^2 + y^2 = 4$
3. $f(x, y) = x + y$; $9x^2 + 4y^2 = 36$
4. $f(x, y) = x - 3y$; $4x^2 + y^2 = 16$
5. $f(x, y) = xy$; $x^2 + y^2 = 1$
6. $f(x, y) = xy$; $x^2 + y^2 = 4$
7. $f(x, y, z) = x - y + z$; $x^2 + y^2 + z^2 = 1$
8. $f(x, y, z) = 2x - y + z$; $x^2 + y^2 + z^2 = 4$
9. $f(x, y, z) = x - 2y + 2z$; $x^2 + y^2 + z^2 = 1$
10. $f(x, y, z) = x + y + z$; $x^2 + y^2 + z^2 = 9$

In Exercises 11–14, solve the indicated problem from the last section by the method of Lagrange multipliers.

11. Exercise 19
12. Exercise 20
13. Exercise 21
14. Exercise 22

In Exercises 15–18, use Method 8.2.

15. Find the extreme values of $f(x, y, z) = 2x + y + 2z$ on the curve of intersection of $x^2 + y^2 - 4 = 0$ and $x + z = 2$.
16. Find the extreme values of $f(x, y, z) = xz + yz$ on the curve of intersection of $x^2 + z^2 = 2$ and $yz = 2$.
17. Find the distance from (1, 1, 1) to the line of intersection of the planes $2x + y - z = 1$ and $x - y + z = 2$. [*Hint:* Let (x, y, z) be a point on the line and minimize the square of its distance from (1, 1, 1).]
18. Find the distance from (1, 1, 0) to the line of intersection of the planes $x - 2y + z = 1$ and $2x - y - z = 2$.
19. What point on the sphere $x^2 + y^2 + z^2 = 4$ is farthest from the point $(1, -1, 1)$?
20. What point on the sphere $x^2 + y^2 + z^2 = 4$ is farthest from the point (0, 1, 1)?
21. Find the volume of the largest rectangular parallelepiped, with sides parallel to the coordinate planes, that can be inscribed in the ellipsoid $x^2 + y^2/9 + z^2/4 = 1$.
22. Find the volume of the largest rectangular parallelepiped, with sides parallel to the coordinate planes, that can be inscribed in the ellipsoid $x^2/4 + y^2/9 + z^2/4 = 1$.
23. Find the maximum and minimum distances from the origin to the ellipsoid $x^2 + y^2/9 + z^2/4 = 1$. (*Hint:* Find the extreme values of the square of the distance.)
24. Find the maximum and minimum distances from the origin to the ellipsoid $4x^2 + 9y^2 + 36z^2 = 1$.
25. What is the rectangle of minimum perimeter and area 1?
26. What is the rectangular parallelepiped of largest volume that can be inscribed in a sphere of radius a?

In Exercises 27–32, find the extreme values of the given function over the given set.

27. $f(x, y) = x^2 + y^2 - 5$; the set $E = \{(x, y) \in \mathbf{R}^2 \mid 4x^2 + 9y^2 \leq 36\}$
28. $f(x, y) = x^2 + 2y^4 + 1$; the set $E = \{(x, y) \in \mathbf{R}^2 \mid x^2 + 4y^2 \leq 4\}$
29. $f(x, y) = x^6 + 2y^2$; the disk $D = \{(x, y) \in \mathbf{R}^2 \mid x^2 + y^2 \leq 1\}$
30. $f(x, y) = 3x^2 + y^4$; the disk $D = \{(x, y) \in \mathbf{R}^2 \mid x^2 + y^2 \leq 4\}$
31. $f(x, y, z) = x^2 + y^4 + z^6 - 3$; the closed ball $\bar{B}((0, 0, 0), 2)$ in $\mathbf{R}^3$ of radius 2 about the origin
32. $f(x, y, z) = x^4 + y^4 + z^2 + 5$; the region $E = \{(x, y, z) \in \mathbf{R}^3 \mid x^2 + 4y^2 + 4z^2 \leq 4\}$
33. In the production model $p(x, y) = ax^{1-e}y^e$ mentioned after Example 4 of Section 1, suppose that $e = 1/4$, $a = 6$, labor costs \$10.00 per employee, and capital costs \$20.00 per unit. Assume that \$3000.00 is budgeted for production costs. How many employees and how many units of capital should there be for maximum production?
34. In Exercise 33 verify that for x and y at their optimal levels x_0 and y_0, the ratio of marginal productivity of labor to marginal productivity of capital is the same as the ratio of unit labor cost to unit capital costs. (This is a basic law of economics.)
35. In Exercise 33 show that, at the optimal levels x_0 and y_0 of x and y, the expenditure of one additional dollar allows the production of λ_0 additional units of production, where λ_0 is the optimal value of λ, i.e., the value corresponding to $x = x_0$ and $y = y_0$.

12.9 Looking Back

This chapter has presented the differential calculus of scalar functions f of several real variables. The first step was the introduction of the partial derivatives of f and the tangent plane to its graph at a point $(a, b, f(a, b))$ in Definitions 1.1, 1.2, and 1.3. In Section 2 we put the partial derivatives together to form the gradient vector (Definition 2.4), which is usually the total derivative (Definition

2.1, Theorems 2.2 and 2.5). We also used the total differential of f (Definition 2.6) to construct the multidimensional tangent approximation.

The notion of directional derivative (Definition 3.1) extended partial differentiation to arbitrary directions. According to Theorem 3.2, the directional derivative of a differentiable function is the coordinate of the gradient (total derivative) in the given direction. Theorem 3.3 says that a differentiable function increases most rapidly in the direction of its gradient.

The next several sections contained several-variable versions of the mean value theorem for derivatives (Theorem 4.1), chain rule (Theorems 4.2, 4.3, and 4.4), method of implicit differentiation (Theorems 5.1 and 5.2), second derivative (Definition 6.1), second-degree Taylor polynomial (Definition 6.3), and second-derivative test for local extreme values (Theorem 7.4). We also found (Theorem 6.2) that for most familiar functions $f\colon \mathbf{R}^n \to \mathbf{R}$ the order of partial differentiation is immaterial in computing second derivatives. Finally, the last section presented the method of Lagrange multipliers (8.1 and 8.2), which facilitates the finding of extreme values of a function subject to one or more constraints. An important class of such problems involves maximizing and minimizing a function f over a region such as a closed disk in $\mathbf{R}^2$ or a closed ball in $\mathbf{R}^3$ (see Example 4 of Section 8).

CHAPTER CHECKLIST

Section 1: partial derivatives; tangent plane to the graph of $z = f(x, y)$ at a point $(a, b, f(a, b))$; marginal values of a quantity.

Section 2: differentiable functions, total derivative, gradient; tangent approximation, total differential.

Section 3: directional derivative, calculation for a differentiable function (Theorem 3.2); maximum rate of increase and decrease of a function at a point, gradient method of numerical optimization.

Section 4: mean value theorem for derivatives for $f\colon \mathbf{R}^n \to \mathbf{R}$; chain rules for partial and total derivatives; tangent planes to level surfaces.

Section 5: implicit differentiation of $F(x, y) = 0$ and $F(x, y, z) = 0$; estimation of an implicitly defined function near a known point; definition of w as a function of u and v when $F(u, v, w) = 0$.

Section 6: second partial derivatives; equality of mixed partial derivatives when they are continuous; second-degree Taylor polynomial of $f(x, y)$ about (a, b), accuracy of tangent approximation of a differentiable function; Laplace's equation.

Section 7: absolute and local extreme values; critical points; saddle points; second-derivative test; linear regression.

Section 8: Lagrange multipliers; condition $\nabla f(\mathbf{a}) = \lambda \nabla g(\mathbf{a})$ for extreme values of $f(\mathbf{x})$ under the constraint $g(\mathbf{x}) = 0$; condition $\nabla f(\mathbf{a}) = \boldsymbol{\lambda} \cdot \nabla \mathbf{g}(\mathbf{a})$ for a function f constrained by $\mathbf{g}(\mathbf{x}) = (g_1(\mathbf{x}), g_2(\mathbf{x}), \ldots, g_m(\mathbf{x})) = 0$.

REVIEW EXERCISES 12.9

1. Find a unit normal vector and the scalar equation of the tangent plane to the surface $z = \sqrt{25 - x^2 - y^2}$ at the point $(2, \sqrt{5}, 4)$.

2. A firm that markets lunch meat uses the demand function

$$d(p, x, y, z, v, w) = a - 10p + 3x + 2y + 2z + 3w - v,$$

where a is a constant, p is the firm's price, x, y, and z are prices of competing firms' lunch meats, w is the amount of meat products used, and v is the amount of cereal filler used (all measured in appropriate units). Discuss the marginal demand for the lunch meat and its sensitivity to price and ingredient fluctuation.

3. Find the total derivative of f at $(1, \pi/2, 3)$ if $f(x, y, z) = x^2yz - z\cos xy$. Why can you be sure that f is differentiable at $(1, \pi/2, 3)$?

4. A lead chamber used to contain radioactive wastes is a box with inner dimensions 3 meters by 4 meters by 5 meters. Its sides are all 5 centimeters thick. Use the tangent approximation to compute the approximate volume of lead in the container.

5. Suppose the pressure P of $n = 1.000$ mol of gas in a chamber is measured to be 2.000 atm, with a possible error of ± 0.005 atm. Suppose also that the volume of the chamber is 250.0 ml (accurate to within 0.1 ml). Use the ideal gas law $PV = nRT$ to compute the Kelvin temperature T of the gas Kelvin if the gas constant R is 82.05 ml·atm/mol·K when expressed to four significant figures. Estimate the maximum possible error in the computed volume of T.

6. Use Definition 2.1 to show that the function $f(x, y) = x^2 + y^2$ is differentiable at $(-1, 2)$.

7. Find the directional derivative of f at $\mathbf{a} = (1, 2, 3)$ in the direction of $\mathbf{v} = \mathbf{i} + 2\mathbf{j} + 2\mathbf{k}$ if $f(x, y, z) = 8xyz - 2x^2yz - xy^2z + 2xyz^2$. What are the direction and rate of maximum increase of f at $\mathbf{a}$?

8. If $\mathbf{w} = \sqrt{x^2 + y^2}$, where $\mathbf{x}(t) = (x(t), y(t)) = (t, 2t^2)$, then find dw/dt at $\mathbf{a} = (-1, 2)$.

9. If $f(x, y, z) = xyz$, then find a point $\mathbf{c}$ on the segment joining $(1, 1, 1)$ to $(-1, 3, 2)$ such that $f(-1, 3, 2) - f(1, 1, 1) = \mathbf{f}'(\mathbf{c}) \cdot [-2\mathbf{i} + 2\mathbf{j} + \mathbf{k}]$.

10. If $w = x^2 - y^2 + 2z^2$, where $x = s^2 - t^2$, $y = 2st$, and $z = s/t$, then find $\partial w/\partial s$ and $\partial w/\partial t$. What is the total derivative of w at the point $(s_0, t_0) = (1, 1)$?

11. Find the equation of the tangent plane to the surface $2yz^3 - x^2y = 4z$ at the point $(1, 4, 1)$.

12. Suppose that $T(x, y, z) = \sqrt{x^2 + y^2 + z^2}$, where $x = r\cos\theta$, $y = r\sin\theta$ and $z = 4$. Find $\partial T/\partial r$ and $\partial T/\partial\theta$.

13. If $x^2y^3 - 3x^4 + 2y^3 = 0$, then find a formula for dy/dx and state where it is valid. Use the tangent approximation to approximate y when $x = 0.98$. (Note that $(1, 1)$ satisfies the given equation.)

14. If $x^2 - 4y^2 + z^2 = 1$, then find formulas for $\partial z/\partial x$ and $\partial z/\partial y$ and state where they are valid. Then use the tangent approximation to approximate z when $x = 2.03$ and $y = 1.01$. (Note that $(2, 1, 1)$ satisfies the given equation.)

15. If $f(x, y) = x^2y^3 + 4x^2y^2 + xy - x^2$, then find the second partial derivatives f_{xx}, f_{xy}, and f_{yy}.

16. Find a formula for $\partial^2 z/\partial x^2$ valid near $(1, 2, -1)$ if $x^2 + 2y^2 + 3z^2 = 12$.

17. Estimate the accuracy of the tangent approximation to $f(x, y) = x^3 + y^2$ on the closed ball of radius 0.1 about the point $(1, 0)$.

18. Find the second-degree Taylor polynomial of f given by $f(x, y) = ye^x$ about the point $(x_0, y_0) = (1, 1)$.

In Exercises 19–21, find and classify all critical points of the given function.

19. $f(x, y) = x^3 + 3xy^2 - 3x^2 - 3y^2 + 7$

20. $f(x, y) = x^4 + 3x^2y^2 + y^4$

21. $f(x, y) = 2x^2 - xy - 3y^2 - 3x + 7y$

22. A crate with a lid is to be made from 18 square feet of lumber. It is to have the shape of a rectangular parallelepiped. Find the dimensions and volume of the largest such crate.

23. Find the extreme values of $f(x, y) = x - 3y$ on the ellipse $4x^2 + y^2 = 16$.

24. Find the extreme values of $f(x, y, z) = xy + yz$ on the curve of intersection $x^2 + y^2 = 4$ and $yz = 4$.

25. Find the maximum and minimum of $f(x, y) = 2x^4 + y^2 + 3$ on the closed unit disk $\{(x, y) | x^2 + y^2 \leq 1\}$.

CHAPTER 13 Multiple Integration

13.0 Introduction

This chapter is devoted to the integral calculus of scalar functions of two and three real variables. As in Chapter 12, the approach is to extend the single-variable theory by analogy. We begin in Section 1 by introducing the *double integral* of a scalar function of two variables over a rectangle in $\mathbf{R}^2$, which is the two-dimensional analogue of a closed interval on the real line. In Section 2, we see that the two-variable version of the fundamental theorem of calculus is powerful enough to also handle double integrals over plane regions more general than rectangles. The next section discusses double integrals in polar coordinates, which, as for one-variable functions, are often more appropriate than rectangular coordinates.

In Section 4, we consider integration of scalar functions of three variables, defining the *triple integral* first over rectangular boxes in $\mathbf{R}^3$ and then over more general regions of three-space. Sections 5 and 6 discuss triple integrals in alternative coordinate systems for $\mathbf{R}^3$, which resemble the polar coordinate system for $\mathbf{R}^2$.

The final section discusses some geometric and physical applications of double and triple integrals.

13.1 The Double Integral over a Rectangle

To see how to define the two-variable analogue of $\int_a^b f(x)\,dx$, we first review the definition of the definite integral over a closed interval $[a, b]$ in $\mathbf{R} = \mathbf{R}^1$. The two-dimensional analogue of a closed interval is obtained by pairing a closed interval $I_1 = [a, b]$ along the x-axis copy of $\mathbf{R}$ in $\mathbf{R}^2$ with a closed interval $I_2 = [c, d]$ along the y-axis copy of $\mathbf{R}$. The result is a *closed rectangle*

$$\mathbf{R} = \{(x, y) | x \in I_1, y \in I_2\} = I_1 \times I_2 = [a, b] \times [c, d]$$

FIGURE 1.1

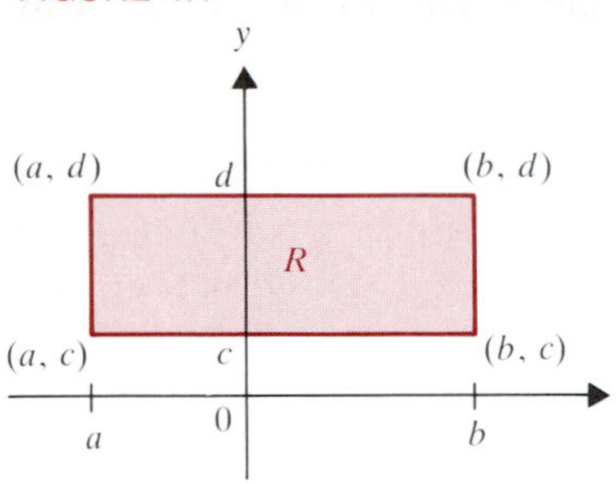

in $\mathbf{R}^2$, made up of all points (x, y) such that $a \le x \le b$ and $c \le y \le d$. See **Figure 1.1.**

To define $\int_a^b f(x)\,dx$, we partitioned $[a, b]$ into subintervals:

$$P\colon a = x_0 < x_1 < x_2 < \ldots < x_{i-1} < x_i < \ldots < x_{m-1} < x_m = b.$$

We then formed the *Riemann sum*

$$R_P = \sum_{i=1}^{m} f(z_i)\,\Delta x_i \tag{1}$$

by selecting a point z_i from the ith subinterval $[x_{i-1}, x_i]$, whose length is $\Delta x_i = x_1 - x_{i-1}$. The definite integral of f over $[a, b]$ was defined by

$$I = \int_a^b f(x)\,dx = \lim_{|P|\to 0} R_P,$$

if the limit exists. (Recall that $|P|$ is the largest Δx_i for $i = 1, 2, \ldots, m$ and that as $|P| \to 0$, $m \to \infty$.)

The analogue for a rectangle $R = [a, b] \times [c, d]$ in $\mathbf{R}^2$ of a partition of $[a, b]$ is obtained by putting together two partitions, P_1 for $I_1 = [a, b]$ and P_2 for $I_2 = [c, d]$. If P_1 is like P above, and the other partition is

$$P_2\colon c = y_0 < y_1 < y_2 < \ldots < y_{j-1} < y_j < \ldots < y_n = d,$$

then a ***grid*** $G = P_1 \times P_2$ of $R = I_1 \times I_2$ is a collection of *subrectangles* $R_{ij} = [x_{i-1}, x_i] \times [y_{j-1}, y_j]$ defined by vertical lines $x = x_i$, $i = 1, 2, \ldots, m$, and horizontal lines $y = y_j$, $j = 1, 2, \ldots, n$. See **Figure 1.2,** where a typical R_{ij} is shown in (a), and an entire grid is shown in (b) for $m = 8$ and $n = 5$. Note that

$$R_{ij} = [x_{i-1}, x_i] \times [y_{j-1}, y_j] = \{(x, y) \mid x \in [x_{i-1}, x_i],\ y \in [y_{j-1}, y_j]\}.$$

For each i between 1 and m, and each j between 1 and n, let $\mathbf{z}_{ij} = (\bar{x}_i, \bar{y}_j)$ be a point in R_{ij}. Then we form a double Riemann sum by analogy with (1), which is a sum of products of $f(z_i)$ times Δx_i, where Δx_i is the length (that is, *one*-dimensional measure of extent) of the subinterval containing z_i. The $\mathbf{R}^2$-analogue of that is a sum of products of $f(\mathbf{z}_{ij})$ times ΔA_{ij}, where ΔA_{ij} is the area (that is, *two*-dimensional measure of extent) of the subrectangle R_{ij} containing $\mathbf{z}_{ij}$. A ***double Riemann sum*** for $f\colon \mathbf{R}^2 \to \mathbf{R}$ corresponding to the grid G is

$$R_G = \sum_{i=1}^{m} \sum_{j=1}^{n} f(\mathbf{z}_{ij})\,\Delta A_{ij} = \sum_{i=1}^{m} \sum_{j=1}^{n} f(\mathbf{z}_{ij})\,\Delta x_i\,\Delta y_j,$$

where $\Delta x_i = x_i - x_{i-1}$ and $\Delta y = y_j - y_{j-1}$. As the notation is meant to indicate, a double sum consists of mn terms, one for each value of i between 1 and m and each value of j between 1 and n. Such a sum can be evaluated by adding all the terms

$$f(\mathbf{z}_{i1})\,\Delta A_{i1} + f(\mathbf{z}_{i2})\,\Delta A_{i2} + \ldots + f(\mathbf{z}_{in})\,\Delta A_{in} \tag{2}$$

FIGURE 1.2

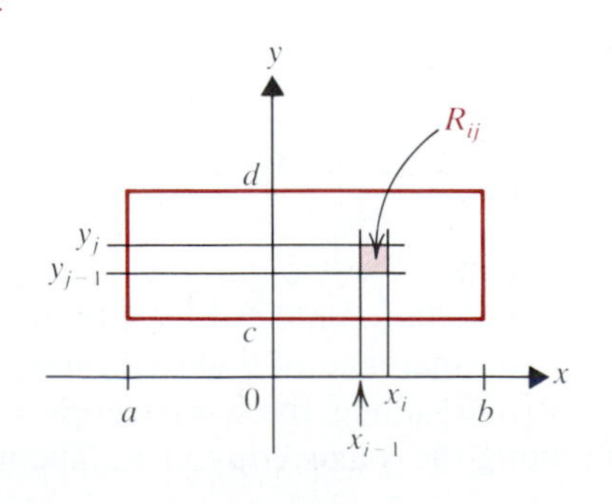

(a)

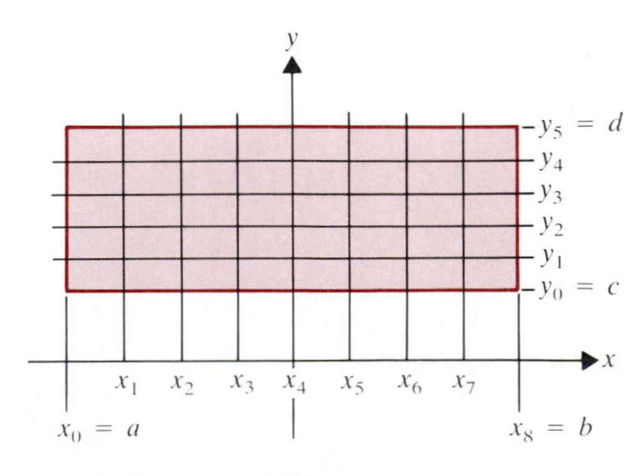

(b)

for each $i = 1, 2, \ldots, m$ and then summing the results, as in the following example.

Example 1

Let $f(x, y) = x^2y^2$ and $R = [0, 1] \times [0, 1/2]$. Evaluate R_G for the grid of R made by the lines $x = 0, 0.2, 0.4, 0.6, 0.8$, and 1 and $y = 0, 1/4$, and $1/2$, where $\mathbf{z}_{ij}$ is the center of the subrectangle R_{ij}. See **Figure 1.3.**

FIGURE 1.3

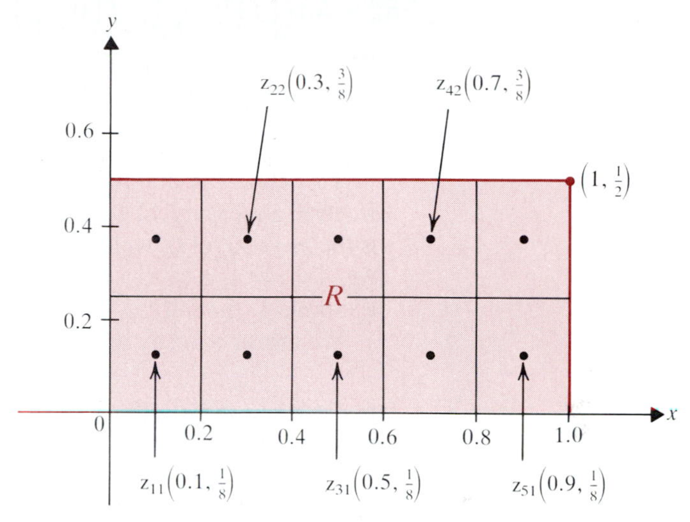

Solution. There are 10 subrectangles R_{ij}, each of base width $\Delta x_i = 0.2$ and height $\Delta y_j = 1/4 = 0.25$. Thus for each i and j, we have

$$\Delta A_{ij} = (0.2)(0.25) = 0.05 = \Delta A.$$

The centers of the rectangles lie on the lines

$$y = \frac{1}{2}\left(\frac{1}{4} - 0\right) = \frac{1}{8} \quad \text{and} \quad y = \frac{1}{2}\left(\frac{1}{2} - \frac{1}{4}\right) = \frac{3}{8},$$

halfway between the lines $y = 0$ and $y = 1/4$, and $y = 1/4$ and $y = 1/2$. Similarly, they lie on the lines $x = 0.1, 0.3, 0.5, 0.7$, and 0.9. Equation (2) then gives

$$\begin{aligned}
R_G &= \sum_{i=1}^{5} \sum_{j=1}^{2} f(\mathbf{z}_{ij})\Delta A = \Delta A \sum_{i=1}^{5} \left(\sum_{j=1}^{2} f(\mathbf{z}_{ij}) \right) \\
&= \Delta A \sum_{i=1}^{5} (f(\mathbf{z}_{i1}) + f(\mathbf{z}_{i2})) \\
&= \Delta A([f(\mathbf{z}_{11}) + f(\mathbf{z}_{12})] + [f(\mathbf{z}_{21}) + f(\mathbf{z}_{22})] + [f(\mathbf{z}_{31}) + f(\mathbf{z}_{32})] \\
&\qquad + [f(\mathbf{z}_{41}) + f(\mathbf{z}_{42})] + [f(\mathbf{z}_{51}) + f(\mathbf{z}_{52})]) \\
&= 0.05\left(\left[f\left(0.1, \frac{1}{8}\right) + f\left(0.1, \frac{3}{8}\right)\right] + \left[f\left(0.3, \frac{1}{8}\right) + f\left(0.3, \frac{3}{8}\right)\right]\right. \\
&\qquad + \left[f\left(0.5, \frac{1}{8}\right) + f\left(0.5, \frac{3}{8}\right)\right] + \left[f\left(0.7, \frac{1}{8}\right) + f\left(0.7, \frac{3}{8}\right)\right] \\
&\qquad \left. + \left[f\left(0.9, \frac{1}{8}\right) + f\left(0.9, \frac{3}{8}\right)\right]\right) \\
&\approx 0.05([0.00016 + 0.00141] + [0.00141 + 0.01266] + [0.00391 + 0.03516] \\
&\qquad + [0.00766 + 0.06891] + [0.01266 + 0.11391]) \\
&\approx 0.01289.
\end{aligned}$$

If instead of using (2), we evaluate the double sum by first summing the terms

$$f(\mathbf{z}_{1i})\,\Delta A_{1j} + f(\mathbf{z}_{2j})\,\Delta A_{2j} + \ldots + f(\mathbf{z}_{mj})\,\Delta A_{mj}$$

for $j = 1, 2, \ldots, n$ and then adding, we just get a rearrangement of the above calculation:

$$R_G = \Delta A\left(\left[f\left(0.1, \frac{1}{8}\right) + f\left(0.3, \frac{1}{8}\right) + f\left(0.5, \frac{1}{8}\right) + f\left(0.7, \frac{1}{8}\right) + f\left(0.9, \frac{1}{8}\right)\right]\right.$$
$$\left. + \left[f\left(0.1, \frac{3}{8}\right) + f\left(0.3, \frac{3}{8}\right) + f\left(0.5, \frac{3}{8}\right) + f\left(0.7, \frac{3}{8}\right) + f\left(0.9, \frac{3}{8}\right)\right]\right). \blacksquare$$

To define the double integral of $f(x, y)$ as the limit of Riemann sums R_G, we need a measure, like $|P|$, for the *fineness* of the grid G. A simple such measure is the ***mesh*** $|G|$ of G, which is defined as the largest of all the dimensions Δx_i and Δy_j, for $i = 1, 2, \ldots, m$ and $j = 1, 2, \ldots, n$, of any rectangle R_{ij} in the grid. (In Example 1, then, $|G| = 0.25$.)

We are now ready to define the double integral over a rectangle.

1.1 DEFINITION Let $f\colon \mathbf{R}^2 \to \mathbf{R}$ be defined on the rectangle $R = [a, b] \times [c, d]$. Then the ***double integral of f over R*** is the limit as $|G| \to 0$ of the double Riemann sums of f over R (if this limit exists). That is,

$$I = \iint_R f(\mathbf{x})\,dA = \iint_R f(x, y)\,dA = \lim_{|G|\to 0} R_G \tag{3}$$

in the sense that given any $\varepsilon > 0$, there is a $\delta > 0$ such that $|I - R_G| < \varepsilon$ for all Riemann sums of f corresponding to any grid G with $|G| < \delta$. If the double integral exists, then f is said to be ***integrable*** over R.

Instead of $\iint_R f(x, y)\,dA$, we could write

$$\iint_R f(x, y)\,dx\,dy, \tag{4}$$

which amounts to the same thing, if we think of dx and dy as representing limiting forms of Δx_i and Δy_j. Since the subregion R_{ij} is a *rectangle*, its area ΔA_{ij} is $\Delta x_i \Delta y_j$. So in the limiting forms, we can think of dA as given by $dx\,dy$. Later on (Section 3), we will see that (3) is *more basic* notation, because it still applies even when we have *nonrectangular* subregions R_{ij}. The form (4) applies *only* when $dA = dx\,dy$, that is, only when subregions are rectangular.

Properties of double integrals are derived from (4) in the same way that properties of the definite integral are derived from Definition 2.2 of Chapter 4.

Example 2

If $f\colon \mathbf{R}^2 \to \mathbf{R}$ is a constant function, $f(x, y) = c$ for all (x, y), then show that

$$\iint_R f(x, y)\,dA = cA(R), \tag{5}$$

where $A(R)$ is the area of the rectangle R.

Solution. Let $G = P_1 \times P_2$ be a grid of R. Then we have

$$R_G = \sum_{i=1}^{m} \sum_{j=1}^{n} f(\mathbf{z}_{ij}) \Delta A_{ij} = \sum_{i=1}^{m} \sum_{j=1}^{n} cA(R_{ij}) \qquad \textit{because } f(\mathbf{z}_{ij}) = c \textit{ and } \Delta A_{ij} = A(R_{ij})$$

$$= c \sum_{i=1}^{m} \sum_{j=1}^{n} A(R_{ij}) \qquad \textit{since the constant } c \textit{ can be factored from each term}$$

$$= cA(R). \qquad \textit{since the sum of all the areas of the subregions is } A(R)$$

Hence $\iint_R f(x, y)\, dA = \lim_{|G| \to 0} R_G = cA(R)$, as required. ■

As an illustration of (5), if $R = [0, 2] \times [-1, 2]$, and if $f(x, y) = 1/2$, then

$$\iint_R f(x, y)\, dA = \frac{1}{2} A(R) = \frac{1}{2} \cdot 2 \cdot 3 = 3.$$

The following simple but important area formula follows immediately from Example 2 by letting $c = 1$ in (5).

1.2 COROLLARY

$$\iint_R 1\, dA = A(R)$$

This result has the notational appeal of asserting that the double integral of dA over a rectangle R is simply the area A of R. This can be thought of as generalizing the single-variable formula

$$\int_a^b dx = b - a,$$

where $b - a$ is the length of the interval $[a, b]$.

There is also a two-variable version of Theorem 2.5 of Chapter 4. The proof is given at the end of the section.

1.3 THEOREM If f and g are integrable over R, then so are $f + g$ and kf for any real number k, and

$$\iint_R (f + g)(\mathbf{x})\, dA = \iint_R f(\mathbf{x})\, dA + \iint_R g(\mathbf{x})\, dA,$$

$$\iint_R kf(\mathbf{x})\, dA = k \iint_R f(\mathbf{x})\, dA.$$

The next result, whose proof is left for Exercise 33, says that the analogues of Theorems 3.6 and 3.3 of Chapter 4 hold for functions of two variables.

1.4 THEOREM Let f and g be integrable on a rectangle R.

(a) If $f(x, y) \leq g(x, y)$ on R, then

$$\iint_R f(x, y)\, dA \leq \iint_R g(x, y)\, dA.$$

(b) If $m \leq f(x, y) \leq M$ for all (x, y) in R, then

$$mA(R) \leq \iint_R f(x, y)\, dA \leq MA(R).$$

The two-variable version of Theorem 2.6 of Chapter 4 also holds for integrable functions in R^2. Its proof is also deferred to the end of the section.

1.5 THEOREM Suppose that the rectangle R is the union of two rectangles R_1 and R_2 that intersect only along a common edge. If f is integrable over R_1 and R_2, then it is integrable over R and

$$\iint_R f(x)\,dA = \iint_{R_1} f(x)\,dA + \iint_{R_2} f(x)\,dA.$$

As for single-variable functions, every continuous function f of two variables is integrable over any rectangle, according to the following theorem of advanced calculus.

1.6 THEOREM If $f\colon \mathbf{R}^2 \to \mathbf{R}$ is continuous on a rectangle R, then f is integrable over R.

This is reassuring, but it doesn't tell us how to evaluate any double integrals. What we need is a tool like the fundamental theorem of calculus (Theorem 4.2 of Chapter 4), which makes it possible to evaluate many definite integrals by antidifferentiation. The analogue for scalar functions f of the antiderivative of a one-variable function is a type of inverse partial derivative. To compute f_x and f_y, we freeze the value of one variable and differentiate f as a function of the remaining variable. For example, if we freeze x, then f becomes a function of y alone. If we *integrate* that function over an interval $[c, d]$, then we get a function of x,

$$g(x) = \int_c^d f(x, y)\,dy,$$

which is shown in advanced calculus to be continuous if f is continuous. Then g can be integrated over the interval $[a, b]$. The result is the tool we are seeking for evaluating double integrals.

1.7 DEFINITION Let $f\colon \mathbf{R}^2 \to \mathbf{R}$ be integrable over $R = [a, b] \times [c, d]$. Then the ***iterated integrals of f over R*** are

$$\int_a^b \int_c^d f(x, y)\,dy\,dx = \int_a^b \left(\int_c^d f(x, y)\,dy \right) dx \tag{6}$$

and

$$\int_c^d \int_a^b f(x, y)\,dx\,dy = \int_c^d \left(\int_a^b f(x, y)\,dx \right) dy, \tag{7}$$

where in $\int_c^d f(x, y)\,dy$ the variable x is frozen, and in $\int_a^b f(x, y)\,dx$ the variable y is frozen.

In an iterated integral the *order* of writing the differentials dy and dx is *crucial.* **The innermost differential gives the first variable of integration.** In (6) we integrate first with respect to y (freezing x). In (7) the first integration is done with respect to x (y is frozen). The next example illustrates the computational procedure.

Example 3

If $f(x, y) = x^2y - x + y$, then find both iterated integrals of f over $R = [1, 2] \times [-1, 1]$.

Solution. To use (6) we initially freeze x and integrate with respect to y, getting

$$\int_1^2 \int_{-1}^1 (x^2y - x + y)\,dy\,dx = \int_1^2 \left[\frac{1}{2}x^2y^2 - xy + \frac{1}{2}y^2\right]_{y=-1}^{y=1} dx$$

$$= \int_1^2 \left[\frac{1}{2}x^2 - x + \frac{1}{2} - \left(\frac{1}{2}x^2 + x + \frac{1}{2}\right)\right] dx$$

$$= \int_1^2 -2x\,dx = -x^2\Big]_1^2 = -(4-1) = -3.$$

Alternatively, we can freeze y and integrate first relative to x. Then (7) gives

$$\int_{-1}^1 \int_1^2 (x^2y - x + y)\,dx\,dy = \int_{-1}^1 \left[\frac{x^3y}{3} - \frac{x^2}{2} + xy\right]_{x=1}^{x=2} dy$$

$$= \int_{-1}^1 \left[\frac{8}{3}y - 2 + 2y - \left(\frac{y}{3} - \frac{1}{2} + y\right)\right] dy$$

$$= \int_{-1}^1 \left(\frac{10}{3}y - \frac{3}{2}\right) dy = \frac{5}{3}y^2 - \frac{3}{2}y\Big]_{-1}^1$$

$$= \frac{5}{3} - \frac{3}{2} - \frac{5}{3} - \frac{3}{2} = -3. \quad ■$$

We got the same answer in both cases, and we will see in Theorem 1.9 that this always happens for continuous functions f.

The definite integral grew out of the problem of calculating areas in $\mathbf{R}^2$. By analogy the double integral is used to calculate volumes in $\mathbf{R}^3$, and the procedure involved provides some informal justification for evaluating double integrals as iterated integrals. Consider the volume of the solid S under a surface $z = f(x, y)$ and above a rectangle $R = [a, b] \times [c, d]$ in the xy-plane, where f is continuous and $f(x, y) \geq 0$ on R. See **Figure 1.4.** If we make a grid $G = P_1 \times P_2$ of R, then the volume S is the sum of the volumes of the solids ΔS_{ij} lying above the subrectangles R_{ij}, one of which is shown in Figure 1.4. If G has small mesh,

FIGURE 1.4

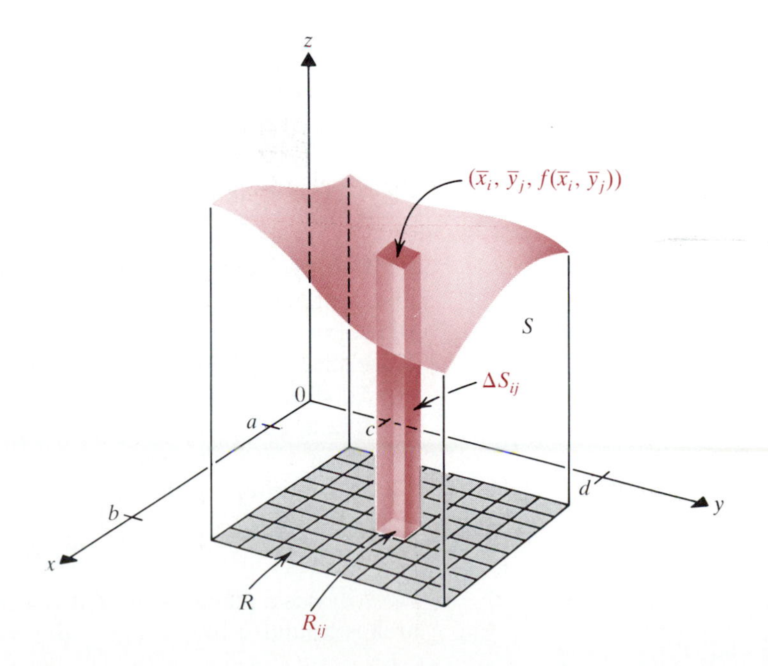

FIGURE 1.5

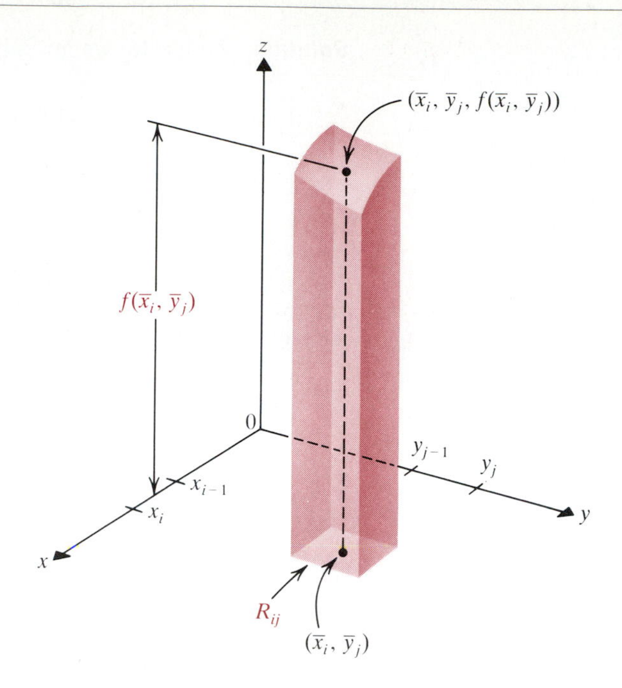

then each such solid ΔS_{ij} is approximately a rectangular box (parallelepiped) with base R_{ij} and height $f(\bar{x}_i, \bar{y}_j)$, for any point $(\bar{x}_i, \bar{y}_j)$ in R_{ij} (see **Figure 1.5**). We can thus approximate the volume ΔV_{ij} of ΔS_{ij} as

$$f(\bar{x}_i, \bar{y}_j)\,\Delta x_i\,\Delta y_j,$$

where ΔA_{ij} is the area $\Delta x_i \Delta y_j$ of R_{ij}. Therefore

$$V = \sum_{i=1}^{m}\sum_{j=1}^{n} \Delta V_{ij} \approx \sum_{i=1}^{m}\sum_{j=1}^{n} f(\mathbf{z}_{ij})\,\Delta A_{ij},$$

which is a Riemann sum R_G for f over the rectangle R corresponding to the grid G. As the grid becomes finer and finer, this approximation appears to become better and better. We are thus led to the following definition.

1.8 DEFINITION Suppose that f is continuous over the rectangle $R = [a, b] \times [c, d]$ and that $f(x, y) \geq 0$ for all (x, y) in R. Then the ***volume V of the solid S under the graph of z = f(x, y) above R*** is

$$(8) \qquad V = \lim_{|G| \to 0} R_G = \lim_{|G| \to 0} \sum_{i=1}^{m}\sum_{j=1}^{n} f(\bar{x}_i, \bar{y}_j)\,\Delta x_i\,\Delta y_j = \iint_R f(x, y)\,dA.$$

With the help of (8), we can see the connection between iterated integrals and double integrals. To do so, we use the method of Section 5.2 for calculating the volume of a solid like S by slicing and integrating the cross-sectional area. If we slice S by planes $x = t$ parallel to the yz-plane, as in **Figure 1.6,** then a typical slice produces a plane cross-sectional area $A(t)$. It is the area of the regions bounded by the lines $y = c$, $y = d$, and $z = 0$ and the curve $z = f(t, y)$ in the plane $x = t$. Then since f is continuous, $A(t)$ is continuous for $t \in [a, b]$. Thus Definition 2.1 of Chapter 5 says that

$$(9) \qquad V = \int_a^b A(t)\,dt = \int_a^b A(x)\,dx.$$

Now $A(x)$ is a planar area, so it too can be calculated as an integral. The planar area is bounded by $y = c$, $y = d$, $z = 0$, and $z = f(x, y)$. Therefore by Theorem

FIGURE 1.6

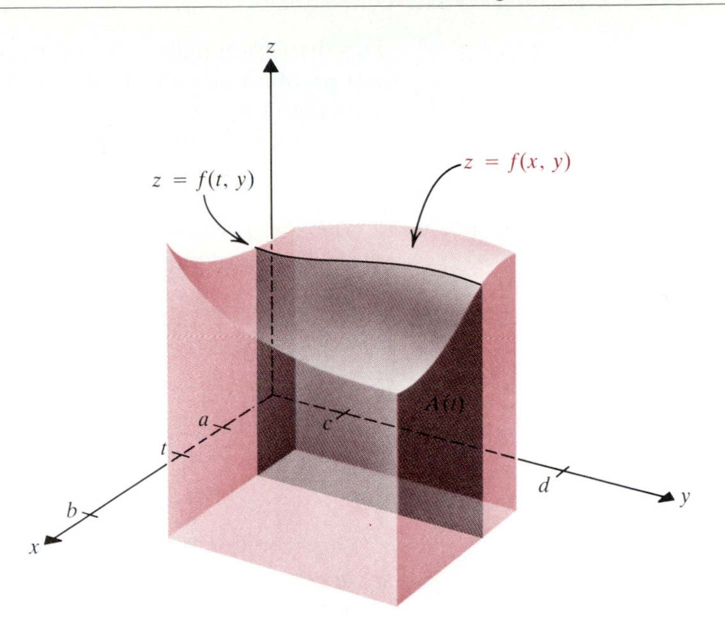

1.1 of Chapter 5 we have

$$A(x) = \int_c^d f(x, y)\, dy. \tag{10}$$

Combining (9) and (10), we get

$$V = \int_a^b A(x)\, dx = \int_a^b \left(\int_c^d f(x, y)\, dy \right) dx = \int_a^b \int_c^d f(x, y)\, dy\, dx.$$

Now there's nothing special about plane sections $x = t$ parallel to the yz-plane. We could just as well slice S by planes $y = u$ parallel to the xz-plane, as in **Figure 1.7.** That would bring us to the formula

$$V = \int_c^d \int_a^b f(x, y)\, dx\, dy.$$

FIGURE 1.7

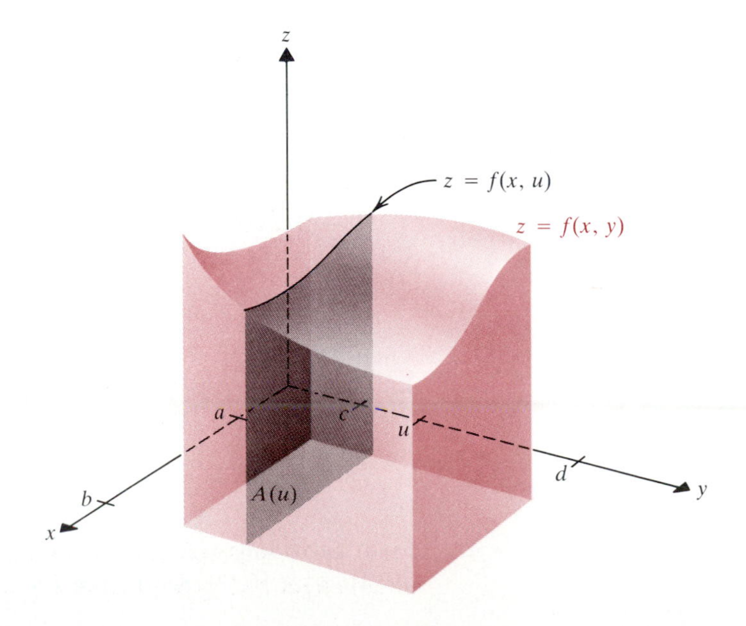

This discussion makes the following theorem quite believable. We leave the formal proof for advanced calculus. [The theorem is named for the Italian mathematician Guido G. Fubini (1879–1943), who made a number of fundamental contributions to the theory of integration.]

1.9 THEOREM

Fubini's Theorem. If $f\colon \mathbf{R}^2 \to \mathbf{R}$ is continuous over the rectangle $R = [a, b] \times [c, d]$, then

$$\iint_R f(x, y)\, dA = \int_a^b \int_c^d f(x, y)\, dy\, dx = \int_c^d \int_a^b f(x, y)\, dx\, dy. \tag{11}$$

Moreover, if $f(x, y) \geq 0$ for all (x, y) in R and if S is the solid region in R^3 lying above R and below the graph of $z = f(x, y)$, then the volume V of S is given by either iterated integral in (11).

This theorem at one fell swoop shows how to evaluate a double integral as an iterated integral and further assures us that the order of iteration used is immaterial. Thus this result is a sort of inverse to Theorem 6.2 of the last chapter, which says that we can interchange the order of partial differentiation for most functions f. In applying (11), we are free to use whichever order of iteration is easier.

Example 4

Evaluate $\iint_R e^y \sin \frac{x}{y}\, dA$, where $R = \left[-\frac{\pi}{2}, \frac{\pi}{2}\right] \times [1, 2]$.

Solution. We can work out either

$$\int_{-\pi/2}^{\pi/2} \left(\int_1^2 e^y \sin \frac{x}{y}\, dy \right) dx \qquad \text{or} \qquad \int_1^2 \left(\int_{-\pi/2}^{\pi/2} e^y \sin \frac{x}{y}\, dx \right) dy.$$

The first integral looks very difficult. We might try integration by parts, but even if that method worked, it would take considerable effort. In the second integral the factor e^y does not involve x, so it can be treated as a constant when we integrate with respect to x. We then obtain

$$\begin{aligned}
\int_1^2 \left(\int_{-\pi/2}^{\pi/2} e^y \sin \frac{x}{y}\, dx \right) dy &= \int_1^2 e^y \left(\int_{-\pi/2}^{\pi/2} \sin \frac{x}{y}\, dx \right) dy \\
&= \int_1^2 e^y \left[-y \cos \frac{x}{y} \right]_{x=-\pi/2}^{x=\pi/2} dy \\
&= \int_1^2 e^y \left[-y \cos \frac{\pi/2}{y} + y \cos \left(\frac{-\pi/2}{y} \right) \right] dy.
\end{aligned}$$

But $\cos(-\theta) = \cos \theta$ for all θ, so the expression in brackets is 0. Hence the given integral is 0. ■

FIGURE 1.8

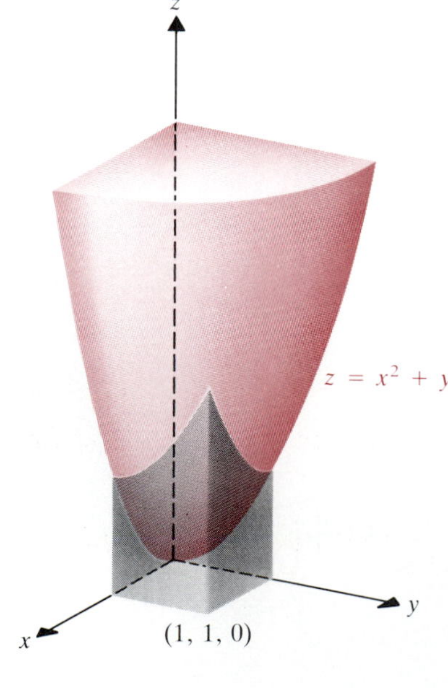

Example 5

Find the volume below the paraboloid $z = x^2 + y^2$ and above the unit square $S = [0, 1] \times [0, 1]$. See **Figure 1.8.**

Solution. By Theorem 1.9,

$$V = \iint_R (x^2 + y^2)\, dA = \int_0^1 \int_0^1 (x^2 + y^2)\, dy\, dx$$

$$= \int_0^1 \left[x^2 y + \frac{y^3}{3} \right]_{y=0}^{y=1} dx = \int_0^1 \left[x^2 + \frac{1}{3} - 0 - 0 \right] dx$$

$$= \int_0^1 \left(x^2 + \frac{1}{3} \right) dx = \frac{x^3}{3} + \frac{x}{3} \Big]_0^1 = \frac{1}{3} + \frac{1}{3} = \frac{2}{3}. \quad ■$$

In this case the order of iteration doesn't affect the complexity of the work.

We close the section by giving a special case of Fubini's theorem, one that can save work when it applies. It deals with functions that can be expressed as the product of a function of x alone and a function of y alone.

1.10 COROLLARY

> If $h: \mathbf{R}^2 \to \mathbf{R}$ satisfies $h(x, y) = f(x) \cdot g(y)$, where f is continuous on $[a, b]$ and g is continuous on $[c, d]$, then
>
> $$\iint_R h(x, y)\, dA = \int_a^b f(x)\, dx \cdot \int_c^d g(y)\, dy,$$
>
> where $R = [a, b] \times [c, d]$. That is, the double integral of the product of a function $f(x)$ of x alone and a function $g(y)$ of y alone is the product of the respective definite integrals.

Proof. If we apply (11) to the function h, then we get

$$\iint_R h(x, y)\, dA = \int_a^b \int_c^d f(x) g(y)\, dy\, dx = \int_a^b f(x)\, dx \cdot \int_c^d g(y)\, dy,$$

because the factors $f(x)$ and dx are constant relative to the y-integration. QED

Example 6

Evaluate $\iint_R e^x \cos y\, dA$ if $R = [0, 2] \times [\pi/6, \pi/2]$.

Solution. Corollary 1.10 gives

$$\iint_R e^x \cos y\, dA = \left(\int_0^2 e^x\, dx \right) \left(\int_{\pi/6}^{\pi/2} \cos y\, dy \right) = \left[e^x \right]_0^2 \left[\sin y \right]_{\pi/6}^{\pi/2}$$

$$= (e^2 - e^0) \left(\sin \frac{\pi}{2} - \sin \frac{\pi}{6} \right) = (e^2 - 1) \left(1 - \frac{1}{2} \right) = \frac{e^2 - 1}{2}. \quad ■$$

Proof of Theorems 1.3 and 1.5*

1.3 THEOREM If f and g are integrable over $R = [a, b] \times [c, d]$, then so are $f + g$ and kf for any real number k, and

$$\iint_R (f + g)(\mathbf{x})\, dA = \iint_R f(\mathbf{x})\, dA + \iint_R g(\mathbf{x})\, dA, \qquad \iint_R kf(\mathbf{x})\, dA = k \iint_R f(\mathbf{x})\, dA.$$

* Optional

Proof. First, any Riemann sum for $f + g$ has the form

$$\sum_{i=1}^{m}\sum_{j=1}^{n}(f+g)(\mathbf{z}_{ij})\,\Delta A_{ij} = \sum_{i=1}^{m}\sum_{j=1}^{n}[f(\mathbf{z}_{ij}) + g(\mathbf{z}_{ij})]\,\Delta A_{ij}$$

$$(12) \qquad = \sum_{i=1}^{m}\sum_{j=1}^{n} f(\mathbf{z}_{ij})\,\Delta A_{ij} + \sum_{i=1}^{m}\sum_{j=1}^{n} g(\mathbf{z}_{ij})\,\Delta A_{ij}.$$

Similarly, any Riemann sum for kf has the form

$$(13) \qquad \sum_{i=1}^{m}\sum_{j=1}^{n}(kf)(\mathbf{z}_{ij})\,\Delta A_{ij} = \sum_{i=1}^{m}\sum_{j=1}^{n} kf(\mathbf{z}_{ij})\,\Delta A_{ij} = k\sum_{i=1}^{m}\sum_{j=1}^{n} f(\mathbf{z}_{ij})\,\Delta A_{ij}.$$

Now take limits in (12) and (13) as $|G| \to 0$ and use Theorem 4.3(b) of Chapter 11, which still holds here [Exercise 39(b)] and says that the limit of a constant times a function is the constant times the limit. We get

$$\iint_R (f+g)(x, y)\,dA = \iint_R f(x, y)\,dA + \iint_R g(x, y)\,dA$$

and

$$\iint_R kf(x, y)\,dA = k\iint_R f(x, y)\,dA.$$

Hence $\iint_R (f + g)(x, y)\,dA$ and $\iint_R kf(x, y)\,dA$ exist and have the required values. QED

1.5 THEOREM Suppose that the rectangle R is the union of rectangles R_1 and R_2 that intersect only along a common edge. If f is integrable over R_1 and R_2, then it is integrable over R, and

$$\iint_R f(\mathbf{x})\,dA = \iint_{R_1} f(\mathbf{x})\,dA + \iint_{R_2} f(\mathbf{x})\,dA.$$

Proof. Suppose that R is as in **Figure 1.9.** Then for any grid G of R there are corresponding grids G_1 of R_1 and G_2 of R_2 obtained by partitioning $[a, e]$ and $[e, b]$ by lines $x = x_i$ and $y = y_j$ (and $x = e$ if it is not already present in G). As $|G| \to 0$,

FIGURE 1.9

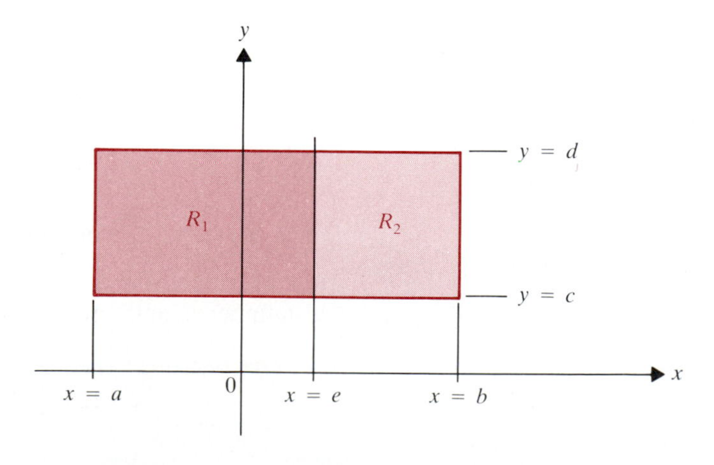

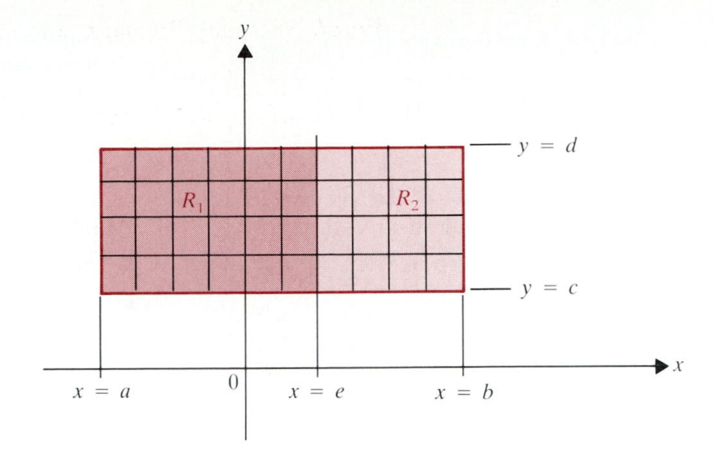

FIGURE 1.10

it is clear that $|G_1| \to 0$ and $|G_2| \to 0$. See **Figure 1.10.** Therefore

$$\iint_{R_1} f(x, y)\, dA + \iint_{R_2} f(x, y)\, dA = \lim_{|G_1|\to 0} R_{G_1} + \lim_{|G_2|\to 0} R_{G_2}$$

$$= \lim_{|G|\to 0} R_G$$

by the analogue of Theorem 4.3(a) of Chapter 11 [(Exercise 39(a)]

$$= \iint_R f(\mathbf{x})\, dA. \quad \boxed{\text{QED}}$$

Exercises 13.1

In Exercises 1–6, evaluate the Riemann sum of the given function over the rectangle $R = [0, 1] \times [0, 1]$. Use the center of R_{ij} for $\mathbf{z}_{ij}$ and use $G = P_1 \times P_2$, where $P_1 = \{0, 1/4, 1/2, 3/4, 1\}$ and $P_2 = \{0, 1/3, 2/3, 1\}$.

1. $f(x, y) = x^2 + y^2$

2. $f(x, y) = x^2 - 3y + 1$

3. $f(x, y) = \sqrt{x^2 + y^2}$

4. $f(x, y) = \dfrac{1}{1 + x^2 + y^2}$

5. $f(x, y) = e^{-x^2 - y^2}$

6. $f(x, y) = \ln(1 + x^2 y^2)$

In Exercises 7–24, evaluate the double integral as an iterated integral.

7. $\iint_R (x^2 - 2xy)\, dA$, $R = [-1, 2] \times [1, 4]$

8. $\iint_R (x^2 - 3x + 2y)\, dA$, $R = [-1, 2] \times [1, 4]$

9. $\iint_R (x + y)^{-2}\, dA$, $R = [0, 1] \times [1, 2]$

10. $\iint_R 3x(x^2 + y)^{-1}\, dA$, $R = [0, 2] \times [1, 2]$

11. $\iint_R (x^2 y + xy^2)\, dA$, $R = [-1, 3] \times [1, 4]$

12. $\iint_R (x^3 y + 2x^2 y^2)\, dA$, $R = [-1, 2] \times [0, 2]$

13. $\iint_R \sin(x + y)\, dA$, $R = [0, \pi/2] \times [0, \pi/2]$

14. $\iint_R \dfrac{x}{1 + y^2}\, dA$, $R = [0, 2] \times [0, 1]$

15. $\iint_R y \cos xy\, dA$, $R = [0, 1] \times [0, \pi]$

16. $\iint_R x \ln xy\, dA$, $R = [1, 2] \times [2, 3]$

17. $\iint_R \dfrac{1}{x + y}\, dA$, $R = [0, 1] \times [1, 2]$

18. $\iint_R \dfrac{y^2}{1 + x^2}\, dA$, $R = [0, 1] \times [0, 1]$

19. $\iint_R e^{x+y}\, dA$, $R = [0, 1] \times [0, 1]$

20. $\iint_R e^x \cos y\, dA$, $R = [0, 1] \times [\pi/2, \pi]$

21. $\iint_R x^2 \sin \dfrac{y}{x}\, dA$, $R = [1, 3] \times [-\pi/4, \pi/4]$

22. $\iint_R y^3 \sin \dfrac{x}{y}\, dA$, $R = [-\pi/6, \pi/6] \times [-4, 2]$

23. $\iint_R (2x + 3)\tan y\, dA$, $R = [0, \pi/4] \times [-\pi/6, \pi/6]$

24. $\iint_R e^{y^2 - 3} \tan x\, dA$, $R = [-\pi/4, \pi/4] \times [0, 3]$

In Exercises 25–32, find the volume of the region in R^3 below the given surface and above the indicated rectangle.

25. The paraboloid $z = 9 - x^2 - y^2$; $R = [0, 1] \times [0, 2]$

26. The paraboloid $z = 16 - x^2 - y^2$; $R = [0, 2] \times [0, 1]$

27. The paraboloid $z = x^2 + y^2$; $R = [-1, 1] \times [0, 2]$

28. The paraboloid $z = x^2 + y^2$; $R = [-1, 2] \times [-1, 1]$

29. The surface $z = xe^{xy}$; $R = [0, 1] \times [1, 2]$

30. The surface $z = ye^{xy}$; $R = [0, 1] \times [1, 2]$

31. The saddle $z = x^2 - y^2$; $R = [3, 4] \times [0, 1]$

32. The saddle $z = y^2 - x^2$; $R = [0, 1] \times [2, 4]$

33. **(a)** If $f(x, y) \geq 0$ on R and if f is integrable over R, then show that $\iint_R f(x, y)\, dA \geq 0$.
(b) Prove Theorem 1.4(a). **(c)** Prove Theorem 1.4(b).

34. If f and $|f|$ are integrable over R, then show that $|\iint_R f(x, y)\, dA| \leq \iint_R |f(x, y)|\, dA$.

35. If f is continuous on R, then show that $mA(R) \leq \iint_R f(x, y)\, dA \leq MA(R)$, where M is an upper bound and m is a lower bound for $f(x, y)$ on R.

36. **(a)** If

$$f(x, y) = \begin{cases} 1 & \text{if } y \text{ is rational} \\ x & \text{if } y \text{ is irrational} \end{cases}$$

then show that f is not integrable over $R = [0, 1] \times [0, 1]$ by considering Riemann sums.

(b) If

$$f(x, y) = \begin{cases} 2 & \text{if } x \text{ is rational} \\ y & \text{if } x \text{ is irrational} \end{cases}$$

then show that f is not integrable over $R = [0, 1] \times [0, 1]$.

37. Use Exercise 35 to show that if $f(x, y)$ is continuous, then

$$\lim_{h \to 0} \frac{1}{A(S_h)} \iint_{S_h} f(\mathbf{x})\, dA = f(\mathbf{x}_0)$$

if S_h is the square of side h and center $\mathbf{x}_0$.

38. Use Exercise 35 to prove the *mean value theorem for double integrals:* If f is continuous on $\mathbf{R}^2$, then

$$\iint_R f(x, y)\, dA = f(x_0, y_0) A(R)$$

for some point (x_0, y_0) in R. (*Hint:* Use, but don't attempt to prove, the *intermediate value theorem for continuous functions* $f: \mathbf{R}^2 \to \mathbf{R}$, which states that f takes on all values between its maximum and minimum on a rectangle.)

39. Suppose that $\lim_{|G| \to 0} R_G(f)$ and $\lim_{|G| \to 0} R_G(g)$ both exist. Then show that
(a) $\lim_{|G| \to 0} R_G(f + g) = \lim_{|G| \to 0} R_G(f) + \lim_{|G| \to 0} R_G(g)$, where the limits are taken in the sense of Definition 1.1;
(b) $\lim_{|G| \to 0} R_G(cf) = c \lim_{|G| \to 0} R_G(f)$, for any constant c.

Exercises 40–43, which comprise the analogue of Theorem 2.8 of Chapter 4, can often save labor.

40. A function $f: \mathbf{R}^2 \to \mathbf{R}$ is called an ***odd function in x*** if $f(-x, y) = -f(x, y)$ whenever (x, y) and $(-x, y)$ both belong to the domain of f. Suppose that f is continuous and odd in x and that the rectangle R is symmetric in the y-axis. Then show that $\iint_R f(x, y)\, dA = 0$.

41. Suppose that f is a continuous ***odd function in y***, that is, $f(x, -y) = -f(x, y)$ whenever both (x, y) and $(x, -y)$ are in the domain of f. If the rectangle R is symmetric in the x-axis, then show that $\iint_R f(x, y)\, dA = 0$.

42. Suppose that f is a continuous ***even function in x***, that is $f(-x, y) = f(x, y)$ whenever (x, y) and $(-x, y)$ both belong to the domain of f. If the rectangle R is symmetric in the y-axis, then show that

$$\iint_R f(x, y)\, dA = 2 \iint_{R'} f(x, y)\, dA$$

where R' is the portion of R lying to the right of the y-axis.

43. Formulate and prove the analogue of Exercise 42 for a continuous function $f(x, y)$ that is *even in y*.

44. *The trapezoidal rule for double integrals.*
(a) Show that

$$\iint_R f(x, y) dA = \int_b^{b + \Delta y} \int_a^{a + \Delta x} f(x, y)\, dx\, dy$$

if $R = [a, a + \Delta x] \times [b, b + \Delta y]$.
(b) Let

$$I = \int_b^{b + \Delta y} g(y)\, dy, \qquad \text{where} \qquad g(y) = \int_a^{a + \Delta x} f(x, y)\, dx.$$

Approximate $g(y)$ by the trapezoidal rule of Section 4.6

$$g(y) \approx \frac{\Delta x}{2} [f(a, y) + f(a + \Delta x, y)].$$

Then do the same for I, to obtain

$$I \approx \frac{\Delta x\, \Delta y}{4} [f(a, b) + f(a + \Delta x, b) + f(a, b + \Delta y) + f(a + \Delta x, b + \Delta y)].$$

This is the trapezoidal approximation rule for a subrectangle $[a, a + \Delta x] \times [b, b + \Delta y]$. Over an entire rectangle R, one approximates $\iint_R f(x, y)\, dA$ by adding these subrectangular approximations.

PC **45.** Use the rule of Exercise 44 to approximate $\iint_R e^{-(x^2 + y^2)}\, dA$, where $R = [-1, 1] \times [-1, 1]$. Take $\Delta x = \Delta y = 1/4$.

PC **46.** Use the rule of Exercise 44 with $\Delta x = \Delta y = 1/4$ to approximate $\iint_R (x+y)^2\,dA$, where $S = [0, 1] \times [0, 1]$. Compare the result with the exact value.

47. *Simpson's rule for double integrals.* Let $R = [a, a + 2\,\Delta x] \times [b, b + 2\,\Delta y]$.

(a) Show that

$$\iint_R f(x, y)\,dA = \int_b^{b+2\Delta y} h(y)\,dy,$$

where

$$h(y) = \int_a^{a+2\Delta x} f(x, y)\,dx.$$

(b) Let

$$I' = \int_b^{b+2\Delta y} h(y)\,dy.$$

Approximate $h(y)$ by Simpson's rule from Section 4.6

$$h(y) \approx \frac{\Delta x}{3}\,[f(a, y) + 4f(a + \Delta x, y) + f(a + 2\,\Delta x, y)].$$

Then do the same for I' to obtain

$$\begin{aligned} I' \approx \frac{\Delta x\,\Delta y}{9}\,[&f(a, b) + f(a + 2\,\Delta x, b) + f(a, b + 2\,\Delta y) \\ &+ f(a + 2\,\Delta x, b + 2\,\Delta y) + 4f(a + \Delta x, b) \\ &+ 4f(a + \Delta x, b + 2\,\Delta y) + 4f(a, b + \Delta y) \\ &+ 4f(a + 2\,\Delta x, b + \Delta y) + 16f(a + \Delta x, b + \Delta y)]. \end{aligned}$$

This is Simpson's approximation rule for a subrectangle. Over an entire rectangle, one approximates $\iint_R f(x, y)\,dA$ by adding these subrectangular approximations. A computer or programmable calculator is needed to manage the arithmetic.

PC **48.** Apply Simpson's rule to the integral of Exercise 45, using $\Delta x = \Delta y = 1/4$.

PC **49.** Apply Simpson's rule to the integral of Exercise 46, using $\Delta x = \Delta y = 1/4$.

13.2 Integrals over General Regions

As you may already be thinking, not everything of interest in $\mathbf{R}^3$ lies over a rectangle! But at this point, we can perform double integration only over rectangles $R = [a, b] \times [c, d]$ in $\mathbf{R}^2$, which seems to limit unduly the usefulness of double integration. For instance, just as in Chapter 5 we could find the area of the region D lying above the x-axis and below the parabola $y = 4 - x^2$ in $\mathbf{R}^2$ (**Figure 2.1**), so we ought to be able to find the volume of the solid S in $\mathbf{R}^3$ that lies above the xy-plane and below the paraboloid $z = 4 - x^2 - y^2$ (**Figure 2.2**). But at the moment we cannot, because S does not lie over a rectangle R in the xy-plane.

To remedy this, we extend the double integral to arbitrary bounded regions. A plane region D is said to be *bounded* if for some $r > 0$, $D \subseteq B(\mathbf{0}, r)$. That is, a bounded region is one that can be enclosed in a disk of sufficiently large radius

FIGURE 2.1

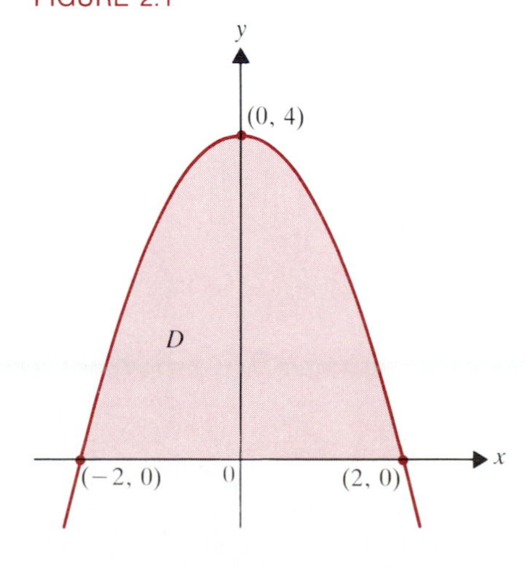

FIGURE 2.2

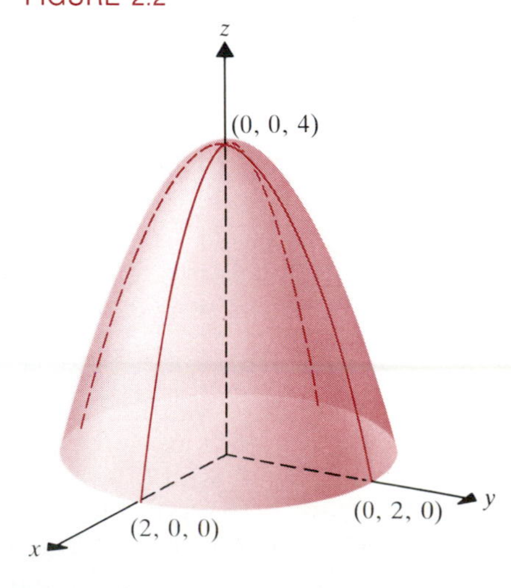

centered at the origin. For example, any rectangle R or disk S is bounded, but the first quadrant is unbounded. See **Figure 2.3.**

2.1 DEFINITION Suppose that $f: \mathbf{R}^2 \to \mathbf{R}$ is defined on a bounded region D. Let R be a rectangle that completely contains D. Define $\bar{f}$ on R by

$$\bar{f}(x, y) = \begin{cases} f(x, y) & \text{if } (x, y) \in D \\ 0 & \text{if } (x, y) \notin D \end{cases} \tag{1}$$

for $x \in R$. Then f is ***integrable over D*** if and only if $\bar{f}$ is integrable over R. If $\bar{f}$ is integrable over R, then the ***double integral of f over D*** is

$$\iint_D f(x, y)\, dA = \iint_R \bar{f}(x, y)\, dA. \tag{2}$$

FIGURE 2.3

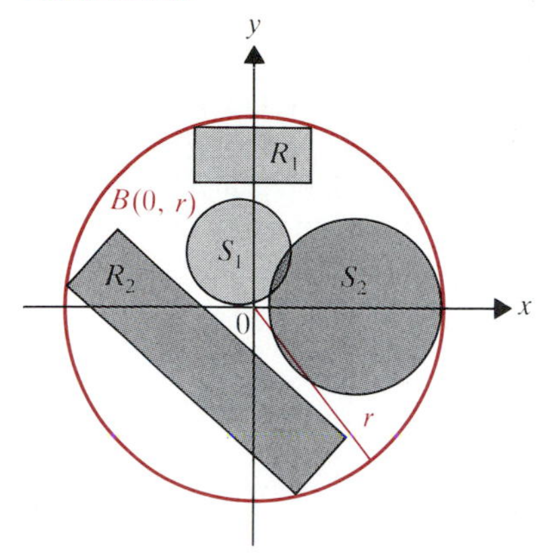

The idea is that we put the region D into some rectangle R and extend the function f to that entire rectangle by making $f(x, y) = 0$ for points (x, y) of the rectangle that are not already in D. We calculate the integral of f over D by evaluating the integral of the extended version of f over R. The rationale for this is that the extended function $\bar{f}$ contributes only zero to any Riemann sum, except at points in D. See **Figure 2.4.**

Although we won't prove them, the results of Section 1 (in particular, Fubini's theorem) hold verbatim for double integrals over bounded regions D. Suppose, for example, that D lies between the lines $x = a$ and $x = b$ and between the curves $y = g_1(x)$ and $y = g_2(x)$ (**Figure 2.5**). If we apply Fubini's theorem to evaluate $\iint_D f(x, y)\, dA$, then referring to Figure 2.5 we have

$$\iint_D f(x, y)\, dA = \iint_R \bar{f}(x, y)\, dA = \int_a^b \int_c^d \bar{f}(x, y)\, dy\, dx$$

$$= \int_a^b \left[\int_c^{g_1(x)} \bar{f}(x, y)\, dy + \int_{g_1(x)}^{g_2(x)} \bar{f}(x, y)\, dy + \int_{g_2(x)}^{d} \bar{f}(x, y) \right] dx$$

$$= \int_a^b \left[\int_c^{g_1(x)} 0\, dy + \int_{g_1(x)}^{g_2(x)} f(x, y)\, dy + \int_{g_2(x)}^{d} 0\, dy \right] dx$$

$$= \int_a^b \left[0 + \int_{g_1(x)}^{g_2(x)} f(x, y)\, dy + 0 \right] dx,$$

FIGURE 2.4

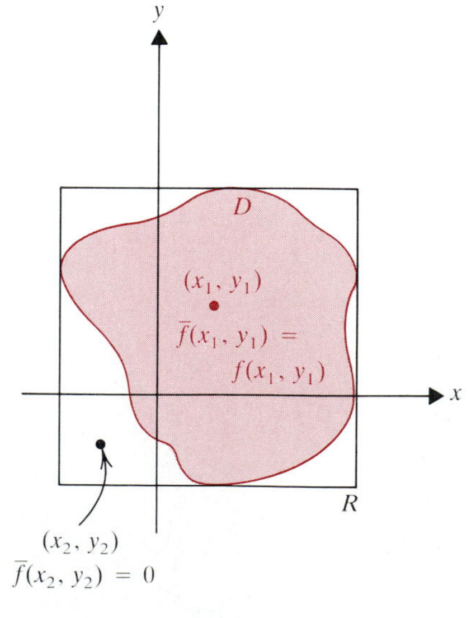

(a)

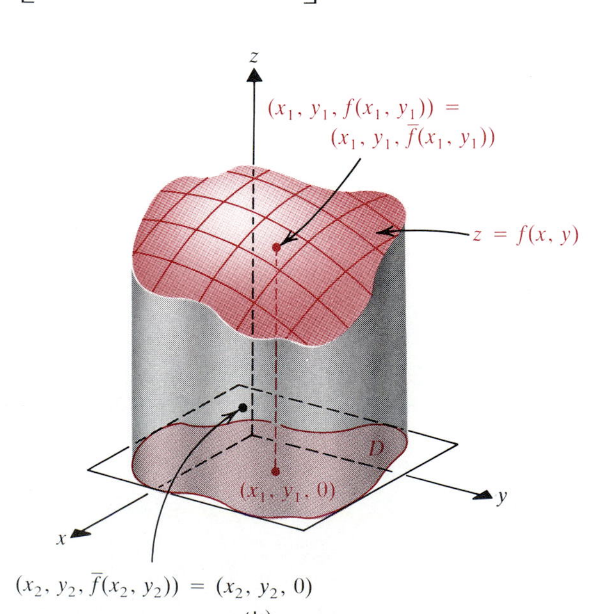

(b)

FIGURE 2.5

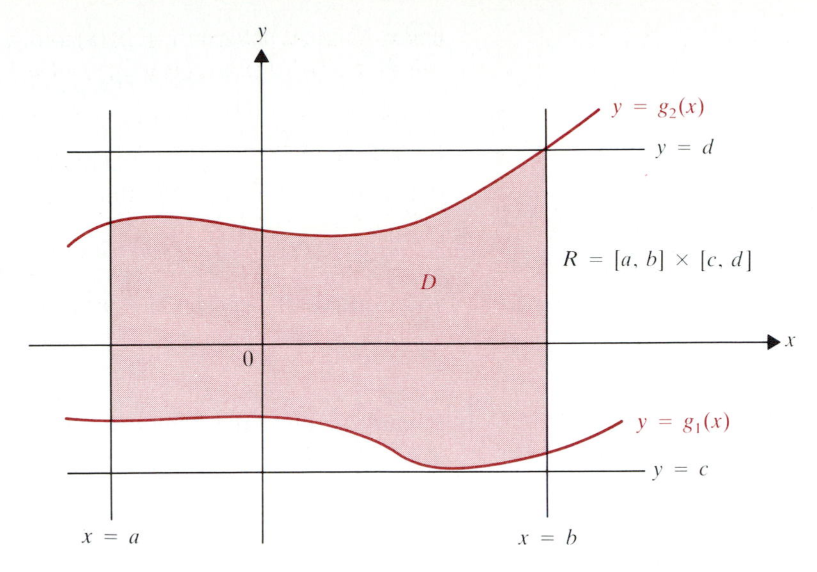

that is,

$$\iint_D f(x, y)\,dA = \int_a^b \int_{g_1(x)}^{g_2(x)} f(x, y)\,dy\,dx. \tag{3}$$

Similarly, if D is a region bounded by curves $x = h_1(y)$ and $x = h_2(y)$ and lines $y = c$ and $y = d$ (**Figure 2.6**), then

$$\iint_D f(x, y)\,dA = \iint_R \bar{f}(x, y)\,dA = \int_c^d \int_a^b \bar{f}(x, y)\,dx\,dy,$$

which as above reduces to

$$\iint_D f(x, y)\,dA = \int_c^d \int_{h_1(y)}^{h_2(y)} f(x, y)\,dx\,dy. \tag{4}$$

In (3) the variable x is held constant in the first integration. We find a partial antiderivative function

$$F(x, y) = \int f(x, y)\,dy$$

FIGURE 2.6

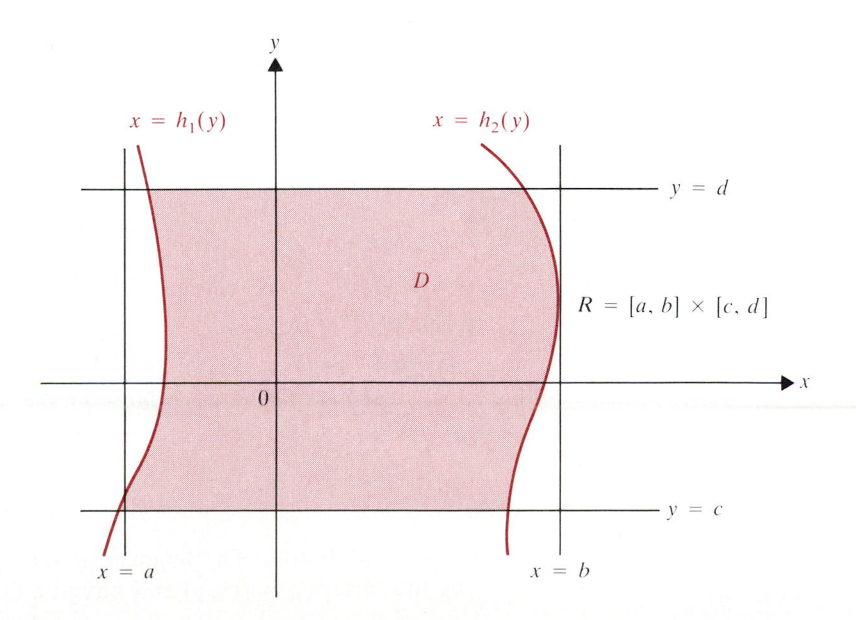

and evaluate it between $y = g_1(x)$ and $y = g_2(x)$. That gives a function of x alone,

$$F(x, g_2(x)) - F(x, g_1(x)),$$

to be integrated between a and b, which yields the number that is the value of $\iint_D f(x, y)\,dA$. If instead we first integrate with respect to x, then y is held constant in the first integration. We find the partial antiderivative

$$G(x, y) = \int f(x, y)\,dx$$

and evaluate it between $x = h_1(y)$ and $x = h_2(y)$. That produces a function of y alone,

$$G(h_2(y), y) - G(h_1(y), y).$$

Integration of this between $y = c$ and $y = d$ then gives the value of $\iint_D f(x, y)\,dA$.

FIGURE 2.7

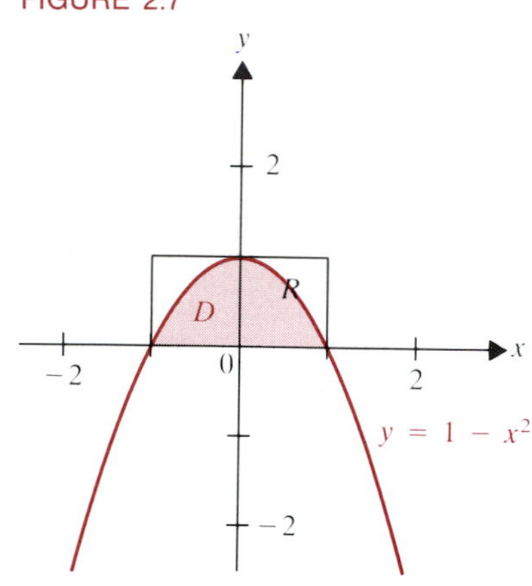

Example 1

Evaluate $\iint_D (xy + x^2)\,dA$, where D is the region in **Figure 2.7** between the x-axis and the parabola $y = 1 - x^2$.

Solution. The region lies between $y = g_1(x) = 0$ and $y = g_2(x) = 1 - x^2$. We can then use (3), with $a = -1$ and $b = 1$, because the rectangle $R = [-1, 1] \times [0, 1]$ in Figure 2.7 completely contains D. We get from (3)

$$\iint_D (xy + x^2)\,dA = \int_{-1}^{1} \int_0^{1-x^2} (xy + x^2)\,dy\,dx = \int_{-1}^{1} \left[\frac{1}{2}x^2y^2 + x^2y\right]_{y=0}^{y=1-x^2} dx$$

$$= \int_{-1}^{1} x^2\left[\frac{1}{2}(1-x^2)^2 + 1 - x^2 - 0 - 0\right] dx$$

$$= \int_{-1}^{1} x^2\left[\frac{1}{2} - x^2 + \frac{1}{2}x^4 + 1 - x^2\right] dx$$

$$= \int_{-1}^{1} \left[\frac{3}{2}x^2 - 2x^4 + \frac{1}{2}x^6\right] dx$$

$$= 2\int_0^1 \left[\frac{3}{2}x^2 - 2x^4 + \frac{1}{2}x^6\right] dx \qquad \textit{by Theorem 2.8(a) of Chapter 4}$$

$$= 2\left[\frac{1}{2}x^3 - \frac{2}{5}x^5 + \frac{1}{14}x^7\right]_0^1 = 2\left[\frac{1}{2} - \frac{2}{5} + \frac{1}{14}\right]$$

$$= 1 - \frac{4}{5} + \frac{1}{7} = \frac{12}{35}. \quad ■$$

We next extend Definition 1.8 to more general regions of $\mathbf{R}^3$ than those lying over rectangles.

2.2 DEFINITION Suppose that $f(x, y)$ is integrable over a bounded set $D \subseteq \mathbf{R}^2$ and $f(x, y) \geq 0$ for $(x, y) \in D$. Then the ***volume of the solid S lying above D and below the graph of z = f(x, y) is***

$$V(S) = \iint_D f(x, y)\,dA. \tag{5}$$

As in Definition 1.8, the integral in (5) gives the volume of a region bounded by the surface $z = f(x, y)$ and a region D of the xy-plane *only* if $f(x, y)$ is nonnegative on D. We can use (5) to find the volume that we said in the opening paragraph of the section should be calculable by double integration.

Example 2

Find the volume lying below the paraboloid $z = 4 - x^2 - y^2$ and above the xy-plane.

Solution. The region is shown in **Figure 2.8.** When $z = 0$, the equation reduces to $x^2 + y^2 = 4$, which means that the paraboloid intersects the xy-plane in that circle. The volume we seek is therefore

$$V = \iint_D (4 - x^2 - y^2)\,dA,$$

where D is the set of points (x, y) in $\mathbf{R}^2$ satisfying $x^2 + y^2 \leq 4$. By symmetry,

$$V = 4 \iint_{D'} (4 - x^2 - y^2)\,dA,$$

FIGURE 2.8

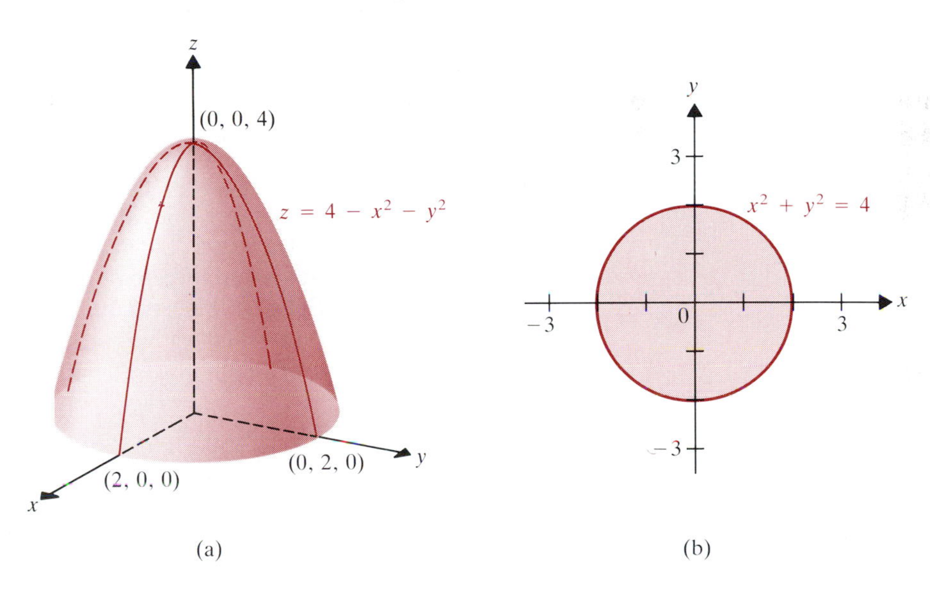

where D' is the first-quadrant portion of D (**Figure 2.9**). The region D' is bounded by the curves $y = 0$ and $y = \sqrt{4 - x^2}$ and by the lines $x = 0$ and $x = 2$ (even though only one point of the latter line is actually needed to bound D' on the right). So by Formula (3),

FIGURE 2.9

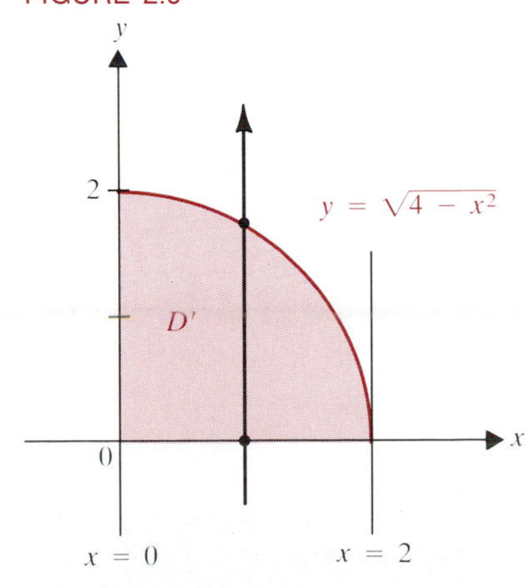

$$\begin{aligned}
V &= 4 \int_0^2 \int_0^{\sqrt{4-x^2}} (4 - x^2 - y^2)\,dy\,dx \\
&= 4 \int_0^2 \left[4y - x^2 y - \frac{y^3}{3} \right]_{y=0}^{y=\sqrt{4-x^2}} dx \\
&= 4 \int_0^2 \left[(4 - x^2)y - \frac{y^3}{3} \right]_{y=0}^{y=\sqrt{4-x^2}} dx \\
&= 4 \int_0^2 \left[(4 - x^2)(4 - x^2)^{1/2} - \frac{1}{3}\,(4 - x^2)^{3/2} \right] dx \\
&= 4 \cdot \frac{2}{3} \int_0^2 (4 - x^2)^{3/2}\,dx.
\end{aligned}$$

This integral can be found in the table of integrals (Formula 56) or can be worked out by using the trigonometric substitution $x = 2\sin\theta$ for $-\pi/2 \le \theta \le \pi/2$. In any event, we get

$$V = \frac{8}{3}\left[\frac{x}{4}(4 - x^2)^{3/2} + \frac{3}{8}\cdot 4x(4 - x^2)^{1/2} + \frac{3}{8}\cdot 16\sin^{-1}\frac{x}{2}\right]_0^2$$

$$= \frac{8}{3}\left[0 + 0 + 6\frac{\pi}{2} - 0 - 0 - 0\right] = 8\pi. \blacksquare$$

In Example 2 we integrated first with respect to y, where y varied between $y = g_1(x) = 0$ and $y = g_2(x) = \sqrt{4 - x^2}$. Then we integrated the resulting $F(x) = (4 - x^2)^{3/2}$ with respect to the single remaining variable x between $x = 0$ and $x = 2$. The directed arrow in Figure 2.9 is intended to help make the limits $g_1(x)$ and $g_2(x)$ in (3) easier to find. The equation of the *lower boundary* of D' (where the arrow first enters the region) gives the *lower limit* $y = g_1(x)$. The equation of the *upper boundary* (where the arrow emerges from the region) gives the *upper limit* $y = g_2(x)$. Once the first integration with respect to y is completed, we are back to a one-variable situation, just as in Chapters 4 and 5. So there is no need to draw any arrows to determine the x limits—the only arrow required is the x-axis. The first integration "integrates out" the variable y. That brings us down to the x-axis for the second integration, where the limits are respectively the smallest and largest values of x that lie under the region of integration.

FIGURE 2.10

There is a parallel strategy for use with (4), when we integrate first with respect to x. The limits $x = h_1(y)$ and $x = h_2(y)$ are the *left* and *right boundaries* of the region of integration. These can be determined by seeing where an arrow parallel to the x-axis and directed toward the right *enters* and *leaves* the region of integration. Refer to **Figure 2.10.** Once x is integrated out, the limits on y are, respectively, the smallest and largest numerical values of y over the entire region. The next example illustrates the use of (4).

Example 3

Find the volume of the solid S in the first octant bounded by the cylinder $y^2 + z^2 = 16$, the plane $3y - 2x = 0$, and the plane $x = 0$.

Solution. The solid S is shown in **Figure 2.11.** It lies below $z = \sqrt{16 - y^2}$ and above the region D of the xy-plane bounded by $x = 0$ (the y-axis), $y = 4$, and

FIGURE 2.11

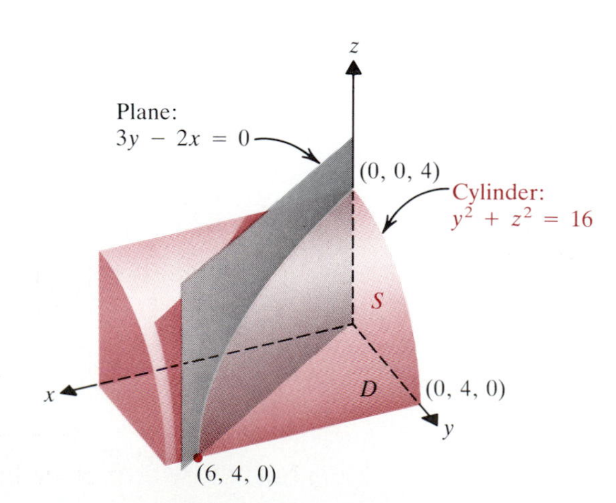

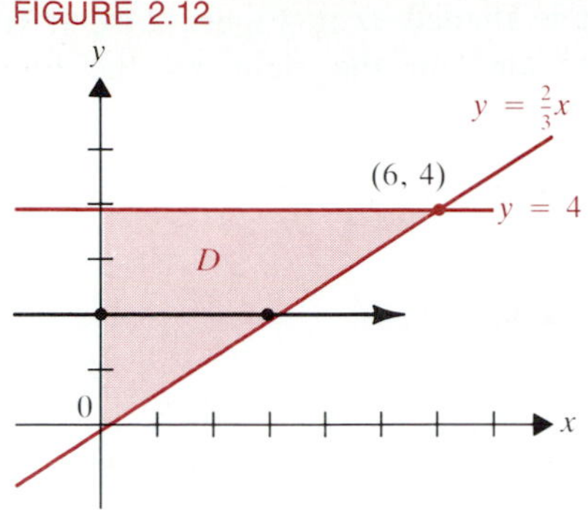

FIGURE 2.12

$x = \frac{3}{2}y$ (the line $3y - 2x = 0$). See **Figure 2.12.** Thus

$$V(S) = \iint_D \sqrt{16 - y^2}\, dA = \int_0^4 \int_0^{3y/2} \sqrt{16 - y^2}\, dx\, dy$$

$$= \int_0^4 \sqrt{16 - y^2} \Big[x\Big]_0^{3y/2} dy = \frac{3}{2}\int_0^4 y\sqrt{16 - y^2}\, dy$$

$$= \frac{3}{2}\left(-\frac{1}{2}\right)\int_0^4 -2y\sqrt{16 - y^2}\, dy = -\frac{3}{4}\frac{(16 - y)^{3/2}}{3/2}\Bigg]_0^4 = \frac{3}{4}\cdot\frac{2}{3}\cdot 16^{3/2}$$

$$= \frac{1}{2}\cdot 64 = 32. \blacksquare$$

As Example 3 illustrates, it is often helpful to draw both two- and three-dimensional figures when finding the limits of integration. In this example we integrated first with respect to x to avoid working with the radical until we had a factor of y to go with it, which let the integration proceed easily.

We next extend Corollary 1.2, so that we can use double integration to find the area of plane regions more general than rectangles.

2.3 THEOREM

If D is a bounded region in the xy-plane between the curves $y = g_1(x)$ and $y = g_2(x)$ and the lines $x = a$ and $x = b$, then the area of D is

(6) $$A(D) = \iint_D 1\, dA.$$

The same formula holds if D is bounded by curves $x = h_1(y)$ and $x = h_2(y)$ and by lines $y = c$ and $y = d$.

Proof. If we put D inside a rectangle R, then

$$\iint_D 1\, dA = \iint_R \bar{f}(x, y)\, dA = \int_a^b \int_c^d f(x, y)\, dy\, dx,$$

where

$$\bar{f}(x, y) = \begin{cases} 1 & \text{if } (x, y) \in D \\ 0 & \text{if } (x, y) \notin D. \end{cases}$$

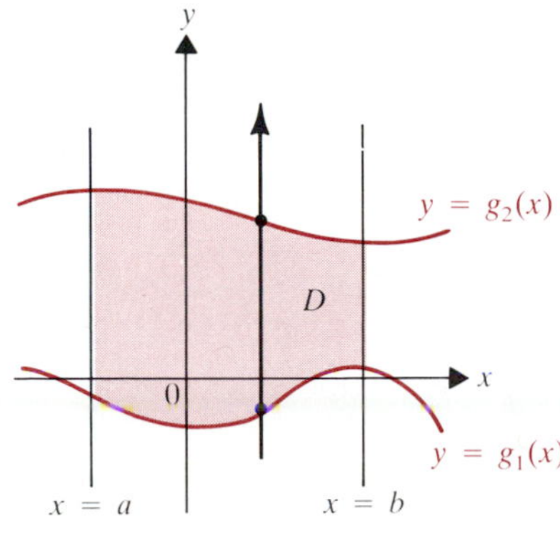

FIGURE 2.13

We thus have

$$\iint_D 1\, dA = \int_a^b \left[\int_c^{g_1(x)} \bar{f}(x, y)\, dy + \int_{g_1(x)}^{g_2(x)} \bar{f}(x, y)\, dy + \int_{g_2(x)}^d \bar{f}(x, y)\, dy\right] dx$$

$$= \int_a^b \left[0 + \int_{g_1(x)}^{g_2(x)} 1\, dy + 0\right] dx = \int_a^b [g_2(x) - g_1(x)]\, dx = A(D),$$

by Theorem 1.1 of Chapter 5 (see **Figure 2.13**). The same reasoning applies if D is bounded by curves $x = h_1(y)$ and $x = h_2(y)$ and by lines $y = c$ and $y = d$. 

Example 4

Find the area in the plane bounded by $x = y^2$ and $3x + 2y = 8$.

FIGURE 2.14

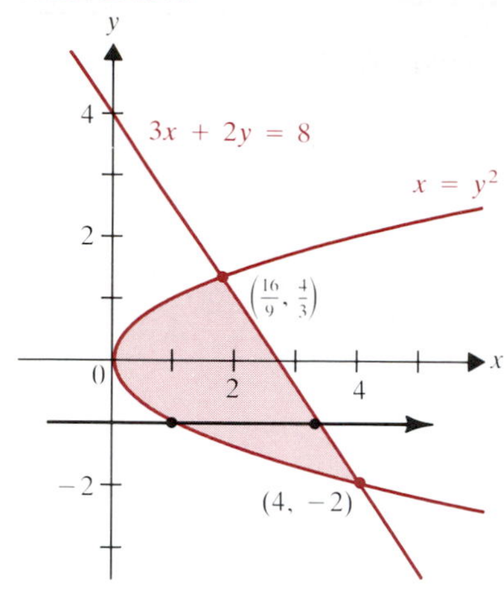

Solution. The region whose area is sought is shaded D in **Figure 2.14.** It is bounded on the left by the curve $x = y^2$ and on the right by the line $3x + 2y = 8$, that is,

$$x = \frac{8 - 2y}{3}.$$

It lies between $y = -2$ and $y = 4/3$. So (6) gives

$$A(D) = \iint_D 1\, dA = \int_{-2}^{4/3} \int_{y^2}^{(8-2y)/3} dx\, dy = \int_{-2}^{4/3} \left(\frac{8}{3} - \frac{2}{3}y - y^2\right) dy$$

$$= \frac{8}{3}y - \frac{1}{3}y^2 - \frac{1}{3}y^3 \Big]_{-2}^{4/3} = \frac{32}{9} - \frac{16}{27} - \frac{64}{81} + \frac{16}{3} + \frac{4}{3} - \frac{8}{3} = \frac{500}{81}. \blacksquare$$

In Figure 2.14 the directed arrow entered the region D from the left by crossing the boundary $x = y^2$ and emerged from the region on the right by crossing the boundary $x = (8 - 2y)/3$. The problem is *much* simpler to do if we use (4) rather than (3). To use (3) we would have to break D into *two* subregions D_1 and D_2 by the line $x = 16/9$, then integrate f over each subregion, and add the results (by Theorem 1.5) to get $A(D)$. See **Figure 2.15.** We would have

FIGURE 2.15

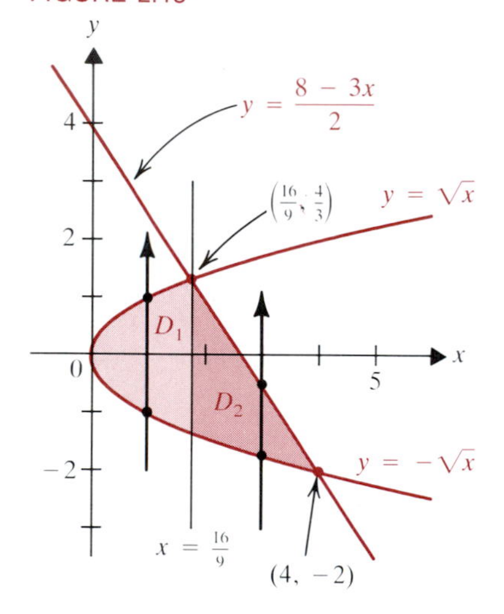

$$A = \int_0^{16/9} dx \int_{-\sqrt{x}}^{\sqrt{x}} dy + \int_{16/9}^{4} dx \int_{-\sqrt{x}}^{(8-3x)/2} dy.$$

Sometimes the integrand itself can be dramatically easier to deal with if one order of integration is used rather than the other. The next example is such a case.

Example 5

Evaluate $\iint_D \sqrt{1 + x^2}\, dA$ over the first-quadrant region T bounded by the lines $y = 0$, $y = x$, and $x = 3$.

Solution. The triangular region T is shown in **Figure 2.16.** We have

FIGURE 2.16

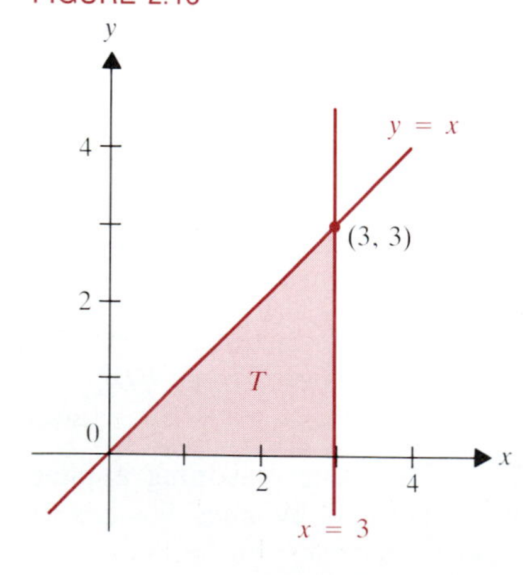

$$\iint_D \sqrt{1 + x^2}\, dA = \int_0^3 \int_0^x \sqrt{1 + x^2}\, dy\, dx = \int_0^3 \int_y^3 \sqrt{1 + x^2}\, dx\, dy.$$

To work out the second iterated integral, a table or trigonometric substitution would be required. The result would be a y-integral that would have to be evaluated numerically. But the first iterated integral is easy to evaluate. We have

$$\int_0^3 \int_0^x \sqrt{1 + x^2}\, dy\, dx = \int_0^3 \sqrt{1 + x^2} \Big[y\Big]_{y=0}^{y=x} dx = \int_0^3 x\sqrt{1 + x^2}\, dx$$

$$= \frac{1}{2}\int_0^3 2x(1 + x^2)^{1/2}\, dx = \frac{1}{2}\frac{(1 + x^2)^{3/2}}{3/2}\Big]_0^3$$

$$= \frac{1}{3}[10\sqrt{10} - 1]. \blacksquare$$

Sometimes the first integration in (3) or (4) inescapably leads to a definite integral that cannot be evaluated by antidifferentiation. In such a case a numerical method must be used.

FIGURE 2.17

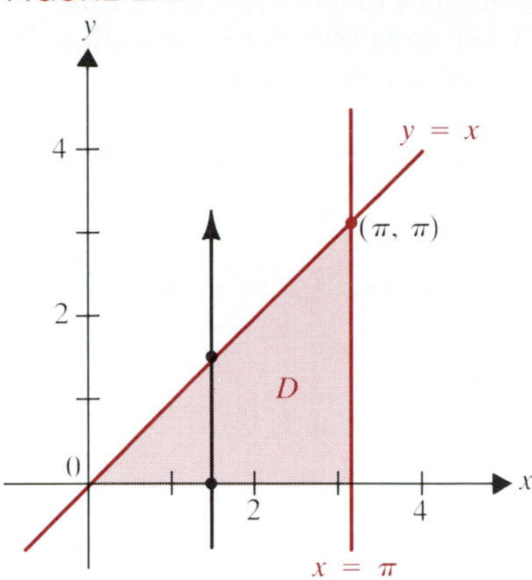

Example 6

Evaluate $\iint_D \sqrt{y} \sin x \, dA$ over the region D in **Figure 2.17.**

Solution. The region D lies above the x-axis and under the line $y = x$ between $x = 0$ and $x = \pi$. Thus (3) gives

$$\iint_D \sqrt{y} \sin x \, dA = \int_0^\pi \int_0^x y^{1/2} \sin x \, dy \, dx = \int_0^\pi \sin x \left[\frac{2}{3} y^{3/2}\right]_{y=0}^{y=x} dx$$

$$= \int_0^\pi \frac{2}{3} x^{3/2} \sin x \, dx.$$

The last definite integral can be approximated by Simpson's rule (p. 274). Table 1 shows the output of a computer program to compute Simpson's approximation

$$S_n(f) \approx \frac{h}{3} [f(x_0) + 4f(x_1) + 2f(x_2) + \ldots + 2f(x_{n-2}) + 4f(x_{n-1}) + f(x_n)]$$

TABLE 1

n	10	20	40	80	160	320
$S_n(f)$	2.817294	2.817375	2.817386	2.817387	2.817387	2.817387

for $f(x) = (2/3)x^{3/2} \sin x$, $x_0 = 0$, $x_n = \pi$, and $h = \pi/n$ for various values of n. From this it appears that the value of the given double integral is approximately 2.81739. ■

Exercises 13.2

In Exercises 1–12, evaluate the given double integral. Draw a figure showing D in each case.

1. $\iint_D xy \, dA$, D the region bounded by the x-axis and the lines $x = 2$ and $y - 2x = 0$
2. $\iint_D (2 - 3x + xy) \, dA$, D the region bounded by the x-axis and the lines $x = 1$ and $y - 3x = 0$
3. $\iint_D (4 - x^2 - y) \, dA$, D the first-quadrant region bounded by the coordinate axes and the parabola $y = 4 - x^2$
4. $\iint_D (x - y^2) \, dA$, D the region bounded by the parabola $x = y^2$ and the line $x = 1$
5. $\iint_D (3x + 2y) \, dA$, D the region bounded by the lines $x + y = 1$, $y - x = 1$, and $x = 1$
6. $\iint_D (2x - 5y) \, dA$, D the region bounded by the lines $x + y = 2$, $y - x = 2$, and $x = 2$
7. $\iint_D e^{-x-y} \, dA$, D the region bounded by the lines $y = 0$, $y = x$, $x = 1/2$, and $x = 1$
8. $\iint_D e^{x+y} \, dA$, D the region of Exercise 7
9. $\iint_D (1 - x^2 - y^2) \, dA$, D the square with vertices $(1, 0)$, $(-1, 0)$, $(0, 1)$, and $(0, -1)$
10. $\iint_D (1 + x^2 + y^2) \, dA$, D the square of Exercise 9
11. $\iint_D \sqrt{9 - x^2} \, dA$, D the region bounded by $y = 0$, $y = 2x$, and $x = 2$
12. $\iint_D \sqrt{y^2 - 4} \, dA$, D the region bounded by $y = 4$, $x = 2$, and $y = x$

In Exercises 13–20 find the area of the given region in the xy-plane.

13. D the region bounded by the parabolas $y^2 = 12x$ and $y = \frac{2}{3}x^2$
14. D the region bounded by the parabolas $y^2 = 4x$ and $y = 2x^2$
15. D the region in the first quadrant bounded by the curves $y = x^2$ and $y = x^4$
16. D the region in the first quadrant bounded by $x = y^2$ and $x = y^4$
17. D the region bounded by $y = e^x$, $y = 0$, $x = 0$, and $x = 1$
18. The region in the first quadrant enclosed by both the circles $x^2 + y^2 = 1$ and $x^2 + (y - 1)^2 = 1$
19. The region bounded by $y = x$ and the curve $x + y^2 = 1$
20. The region bounded by $y = x$ and the curve $x + y^2 = 4$

In Exercises 21–30, find the volume of the given solid.

21. Under the plane $2x + y + z = 5$ and above the rectangle $R = [0, 2] \times [0, 1]$
22. Under the plane $x + y + 2z = 4$ and above the rectangle $R = [0, 1] \times [0, 2]$
23. Under the paraboloid $z = x^2 + y^2$ and above the domain D bounded by the x-axis and the lines $y = 2x$ and $x = 1$
24. Under the paraboloid $z = x^2 + 2y^2$ and above the domain D bounded by the coordinate axes and the line $x + y = 1$
25. Under the surface $z = xy$ and above the domain D lying between $y = 2x$ and $y = x^2$
26. Under the surface $z = xy$ and above the domain lying between $y = x$ and $y = x^3$
27. The first octant region bounded by the two cylinders $x^2 + y^2 = 9$ and $x^2 + z^2 = 9$
28. The first octant region bounded by the two cylinders $x^2 + y^2 = 1$ and $x^2 + z^2 = 1$
29. The first octant region bounded by the cylinder $y^2 + z^2 = 9$, the plane $y = x$, and the yz-plane
30. The first octant region bounded by the cylinder $y^2 + z^2 = 4$, the plane $y = 2x$, and the yz-plane

In Exercises 31–36, draw a sketch of the region D over which the iterated integral is being evaluated. Express the given integral as an iterated integral or (sum of iterated integrals) in which the order of integration is reversed.

31. $\int_0^2 \int_0^{x^2} f(x, y)\,dy\,dx$
32. $\int_0^2 \int_0^{\sqrt{x}} f(x, y)\,dy\,dx$
33. $\int_0^1 \int_{e^y}^{e} f(x, y)\,dx\,dy$
34. $\int_1^e \int_0^{\ln y} f(x, y)\,dx\,dy$
35. $\int_{-1}^2 \int_{x^2}^{x+2} f(x, y)\,dy\,dx$
36. $\int_{-1}^3 \int_{y^2}^{2y+3} f(x, y)\,dx\,dy$

PC **In Exercises 37–40, use Simpson's rule to approximate the definite integral obtained from using (3) or (4) on the given double integral. Use $n = 10, 20, 40, 80, 160$, and 320.**

37. $\iint_D y \sin\sqrt{x}\,dA$, D the region bounded by the x-axis, $y = x$, and $x = \pi$
38. $\iint_D ye^{\sqrt{x^2+y^2}}\,dA$, D the region bounded by the x-axis, $y = x^2$, and the line $x = 1$
39. $\iint_D x\sqrt{x^2 + y^2}\,dA$, D the region bounded by the y-axis, $y^2 = x$, and the lines $y = 0$ and $y = 1$
40. $\iint_D \frac{\cos(\sqrt{x} + y^2)}{\sqrt{x}}\,dy$, D the region of Exercise 37

Exercises 41 and 42 will be used in Section 7. They extend Exercises 40–43 of the last section.

41. Let D be the region between $x = a$, $x = b$, $y = -g(x)$, and $y = g(x)$.
    (a) If $f(x, y)$ is a continuous odd function in y, then show that $\iint_D f(x, y)\,dA = 0$.
    (b) If $f(x, y)$ is a continuous even function in y, and D' is the part of D lying above the x-axis, then show that $\iint_D f(x, y)\,dA = 2\iint_{D'} f(x, y)\,dA$.
42. Let D be the region between $y = c$, $y = d$, $x = -h(y)$, and $x = h(y)$.
    (a) If $f(x, y)$ is a continuous odd function in x, then show that $\iint_D f(x, y)\,dA = 0$.
    (b) If $f(x, y)$ is a continuous even function in x, and D' is the part of D lying to the right of the y-axis, then show that $\iint_D f(x, y)\,dA = 2\iint_{D'} f(x, y)\,dA$.

13.3 Polar Double Integrals

The last two sections of Chapter 8 introduced the polar coordinate system for the plane and discussed definite integrals in polar coordinates. In this section we consider double integrals over plane regions described by polar coordinates. The techniques for evaluating such integrals are quite similar to those given in the last two sections for rectangular double integrals. As the following example suggests, a complicated double integral can often be considerably simplified by changing to polar coordinates.

Example 1

Set up an iterated integral to evaluate $\iint_D f(x, y)\,dA$, where $f(x, y) = x^2 + y^2 - xy$, and D is the closed unit disk $x^2 + y^2 \le 1$.

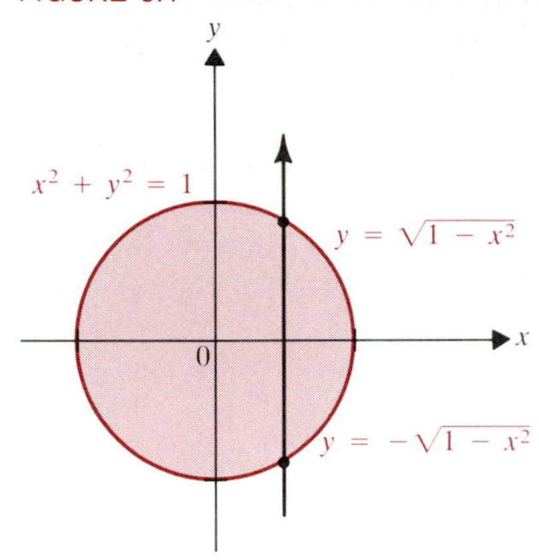

FIGURE 3.1

Solution. As **Figure 3.1** shows, an upward-directed arrow enters the region D by crossing the lower boundary $y = -\sqrt{1 - x^2}$ and leaves D by crossing the upper boundary $y = \sqrt{1 - x^2}$. Therefore

$$\iint_D f(x, y)\,dA = \int_{-1}^{1} \int_{-\sqrt{1-x^2}}^{\sqrt{1-x^2}} (x^2 + y^2 - xy)\,dy\,dx. \blacksquare$$

The iterated integral looks rather formidable, and as you can check, it leads to the complicated definite integral

$$\iint_D f(x, y)\,dA = 2\int_{-1}^{1} \left[x^2\sqrt{1 - x^2} + \frac{1}{3}(1 - x^2)^{3/2}\right]dx.$$

Changing from rectangular to polar coordinates can reduce some of the complexity. Since $x = r\cos\theta$ and $y = r\sin\theta$, for example, we have

$$\begin{aligned} f(x, y) = f(r\cos\theta, r\sin\theta) &= r^2\cos^2\theta + r^2\sin^2\theta - r^2\sin\theta\cos\theta \\ &= r^2(1 - \sin\theta\cos\theta) = g(r, \theta). \end{aligned}$$

The polar coordinate formula $g(r, \theta)$ is much simpler than the Cartesian coordinate formula $f(x, y)$: It is a product of the function r^2 of r alone and the function $1 - \sin\theta\cos\theta$ of θ alone. Moreover, the polar equation $r = 1$ of the boundary of D is so much simpler than its Cartesian equations, $y = -\sqrt{1 - x^2}$ and $y = \sqrt{1 - x^2}$, that we would expect to be able to evaluate $\iint_D g(r, \theta)\,dA$ much more easily than $\iint_D f(x, y)\,dA$. To realize such simplification, we proceed to define double and iterated integrals for a continuous polar coordinate function $g(r, \theta)$ over a plane region T described by polar coordinates.

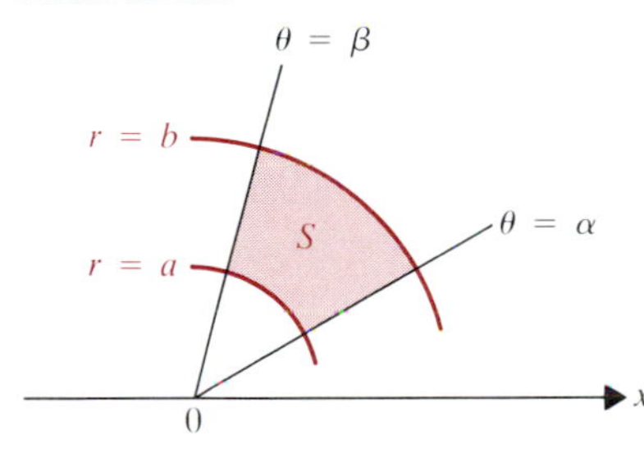

FIGURE 3.2

We begin with the polar coordinate analogue of a rectangle: a region S in the polar coordinate plane bounded by lines $\theta = \alpha$ and $\theta = \beta$ and by circular arcs $r = a$ and $r = b$. See **Figure 3.2.** To define $\iint_S g(r, \theta)\,dA$, we first make a grid G of S by partitioning the two intervals $[\alpha, \beta]$ and $[a, b]$. Accordingly, let

$$P_1\colon a = r_0 < r_1 < \ldots < r_{i-1} < r_i < \ldots < r_{m-1} < r_m = b$$

and

$$P_2\colon \alpha = \theta_0 < \theta_1 < \theta_2 < \ldots < \theta_{j-1} < \theta_j < \ldots < \theta_{n-1} < \theta_n = \beta$$

be partitions of $[a, b]$ and $[\alpha, \beta]$. Putting them together, we get a polar grid of S into subregions S_{ij}, pictured in **Figure 3.3.** To form a Riemann sum of g over

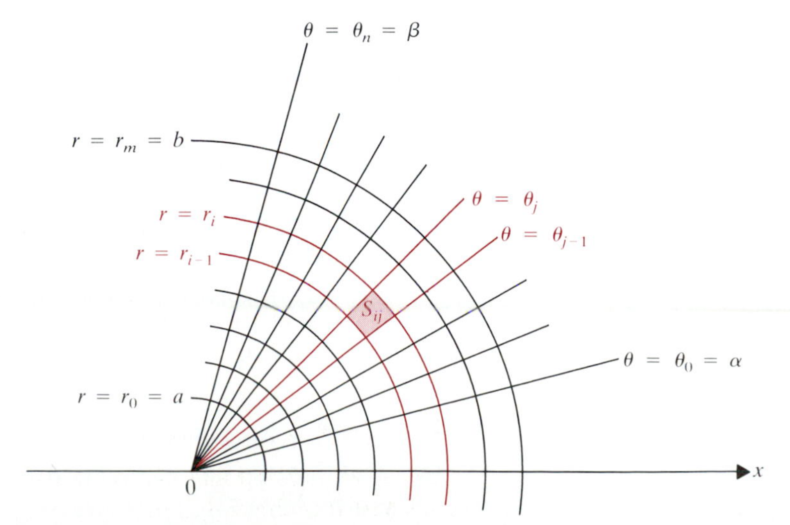

FIGURE 3.3

S, we pick a point $[\bar{r}_i, \bar{\theta}_j]$ in each S_{ij}, evaluate g at this point, multiply by ΔA_{ij} (the area of S_{ij}), and sum over all i and j. There is no difficulty computing $g(\bar{r}_i, \bar{\theta}_j)$, but S_{ij} is *not* a rectangle, so there is no obvious formula (like $\Delta x_i \Delta y_j$ in Section 1) for ΔA_{ij}.

The region S_{ij} lies between two circular sectors of the same central angle $\Delta\theta_j = \theta_j - \theta_{j-1}$. The larger sector has radius r_i, and the smaller has radius r_{i-1}. Therefore ΔA_{ij} is the difference between the areas of these two sectors. According to Lemma 3.9 of Chapter 2, the area of a circular sector of radius r and central angle θ is $(1/2)r^2\theta$. We thus have

$$\begin{aligned}\Delta A_{ij} &= \frac{1}{2} r_i^2 \,\Delta\theta_j - \frac{1}{2} r_{i-1}^2 \,\Delta\theta_j = \frac{1}{2}(r_i^2 - r_{i-1}^2)\,\Delta\theta_j \\ &= \frac{1}{2}(r_i + r_{i-1})(r_i - r_{i-1})\,\Delta\theta_j = \frac{1}{2}(r_i + r_{i-1})\,\Delta r_i \,\Delta\theta_j \\ &= r_i' \,\Delta r_i \,\Delta\theta_j,\end{aligned}$$

where $r_i' = (r_i + r_{i-1})/2$ is in the interval $[r_{i-1}, r_i]$. If the number m of subdivisions of $[a, b]$ is large, then for any $\bar{r}_i$ in the short interval $[r_{i-1}, r_i]$, $\bar{r}_i$ will not differ much from the midpoint r_i'. Thus

$$\Delta A_{ij} \approx \bar{r}_i \,\Delta r_i \,\Delta\theta_j, \qquad \text{for any } \bar{r}_i \text{ in } [r_{i-1}, r_i].$$

A polar Riemann sum of g over S is defined as

$$(1) \qquad R_G = \sum_{i=1}^{m} \sum_{j=1}^{n} g(\bar{r}_i, \bar{\theta}_j)\,\Delta A_{ij} = \sum_{i=1}^{m} \sum_{j=1}^{n} g(\bar{r}_i, \bar{\theta}_i)\bar{r}_i \,\Delta r_i \,\Delta\theta_j.$$

As the mesh $|G|$ of G approaches 0, the limit of the Riemann sum (1) can be shown to exist as a consequence of the continuity of g. The details, which resemble those needed to prove Theorem 1.6, are left for advanced calculus. The double integral of g over S is defined to be the limit of (1) as $|G| \to 0$.

3.1 DEFINITION Let S be a region in the polar coordinate plane lying between $\theta = \alpha$, $\theta = \beta$, $r = a$, and $r = b$. Let g be continuous on S and its boundary. Then the ***polar double integral*** of g over S is

$$\iint_S g(r, \theta)\,dA = \iint_S g(r, \theta) r\,dr\,d\theta = \lim_{|G| \to 0} R_G,$$

where the limit is taken in the same sense as in Definition 1.1.

At this point there is a consistency question about double integrals. Does Definition 3.1 give the same number as Definition 2.1 when S is a region of $\mathbf{R}^2$ described *both* by rectangular and by polar coordinates? It does, but a discussion of how the expressions from these two definitions can be reconciled is left for more advanced courses. The important point is that for polar double integrals to be consistent with rectangular double integrals, we must have

$$dA = r\,dr\,d\theta \qquad \textbf{\textit{NOT}} \qquad dr\,d\theta.$$

There is a version of Fubini's theorem for polar double integrals, which says that they too can be evaluated by iteration. The proof is omitted.

3.2 THEOREM

Let S be the polar region bounded by the lines $\theta = \alpha$ and $\theta = \beta$ and the circles $r = a$ and $r = b$. Let $g(r, \theta)$ be continuous on S and its boundary. Then

(2) $$\iint_S g(r, \theta)\, dA = \int_\alpha^\beta \left[\int_a^b g(r, \theta) r\, dr\right] d\theta = \int_a^b \left[\int_\alpha^\beta g(r, \theta)\, d\theta\right] r\, dr.$$

The following example illustrates how this theorem is used.

FIGURE 3.4

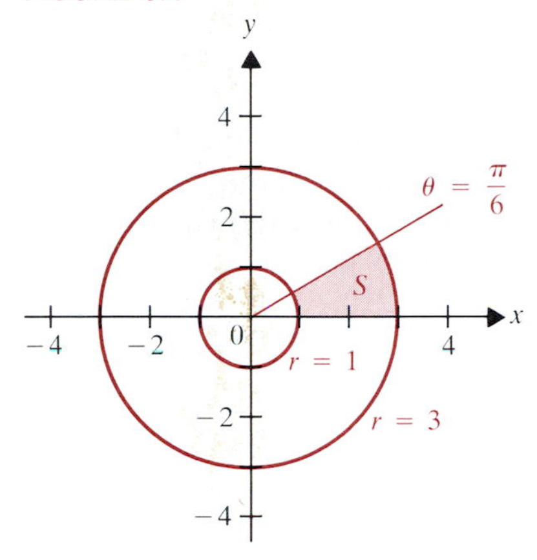

Example 2

Evaluate $\iint_S r \tan 2\theta\, dA$, where S is bounded by the circles $r = 1$ and $r = 3$ and the lines $\theta = 0$, $\theta = \pi/6$. See **Figure 3.4.**

Solution. Equation (2) with $\alpha = 0$, $\beta = \pi/6$, $a = 1$, and $r = 3$ gives

$$\iint_S r \tan 2\theta\, dA = \iint_S (r \tan 2\theta) r\, dr\, d\theta = \int_0^{\pi/6} \left[\int_1^3 r^2\, dr\right] \tan 2\theta\, d\theta$$

$$= \int_0^{\pi/6} \tan 2\theta\, d\theta \int_1^3 r^2\, dr \qquad \textit{by Corollary 1.10}$$

$$= \left[\frac{1}{2} \ln |\sec 2\theta|\right]_0^{\pi/6} \left[\frac{1}{3} r^3\right]_1^3 = \frac{1}{6}[\ln 2 - \ln 1][27 - 1]$$

$$= \frac{13}{3} \ln 2. \quad \blacksquare$$

As suggested at the start of the section, one of the main advantages of polar integration is that it can reduce the complexity of a given rectangular double integral. Suppose that S is a region of the plane given in rectangular coordinates and that T is the same region described by polar coordinates. Then it can be shown that

$$\iint_S f(x, y)\, dA = \iint_T g(r, \theta)\, dA = \iint_T g(r, \theta) r\, dr\, d\theta,$$

where $g(r, \theta) = f(x(r, \theta), y(r, \theta)) = f(r \cos \theta, r \sin \theta)$. That is,

(3) $$\iint_S f(x, y)\, dA = \iint_T f(r \cos \theta, r \sin \theta) r\, dr\, d\theta.$$

Although the formal derivation of (3) is given in more advanced courses, we can see that it is a *reasonable* formula. It says that if we make the substitutions $x = r \cos \theta$ and $y = r \sin \theta$ in $f(x, y)$ and replace the xy-limits of integration by the corresponding $r\theta$-limits, then the value of the double integral is unchanged. This is the analogue for double integrals of the change-of-variable procedure for definite integrals (Theorem 5.1 of Chapter 4). We can use it to evaluate the integral in Example 1.

Example 3

Use polar integration to evaluate $\iint_D f(x, y)\, dA$, where $f(x, y) = x^2 + y^2 - xy$, and D is the closed unit disk $x^2 + y^2 \leq 1$.

Solution. Letting $x = r\cos\theta$ and $y = r\sin\theta$, we find as on p. 857 that

$$f(x, y) = r^2(1 - \sin\theta\cos\theta).$$

The disk D is described in polar coordinates by $T\colon 0 \le r \le 1,\ 0 \le \theta \le 2\pi$. So (3) gives

$$\begin{aligned}\iint_D f(x, y)\,dA &= \iint_T r^2(1 - \sin\theta\cos\theta) r\,dr\,d\theta \\ &= \int_0^1 r^3\,dr \int_0^{2\pi} (1 - \sin\theta\cos\theta)\,d\theta \qquad \text{by Corollary 1.10} \\ &= \frac{1}{4}\left[\theta - \frac{1}{2}\sin^2\theta\right]_0^{2\pi} = \frac{\pi}{2}. \ \blacksquare\end{aligned}$$

Change of variable from rectangular to polar coordinates as in Example 3 is often helpful in volume problems based on Definition 2.2. Recall that if $f(x, y)$ is continuous and nonnegative over a plane region D, then the volume lying below the graph of $z = f(x, y)$ and above the region D is given by

$$V = \iint_D f(x, y)\,dA. \tag{4}$$

The next example is an illustration of how polar coordinates can be used with (4).

Example 4

Find the volume of the region S lying under the graph of $z = f(x, y) = 1/\sqrt{x^2 + y^2}$ and above the first-quadrant region D between the circles $x^2 + y^2 = 1$ and $x^2 + y^2 = 4$ in the xy-plane. See **Figure 3.5.**

FIGURE 3.5

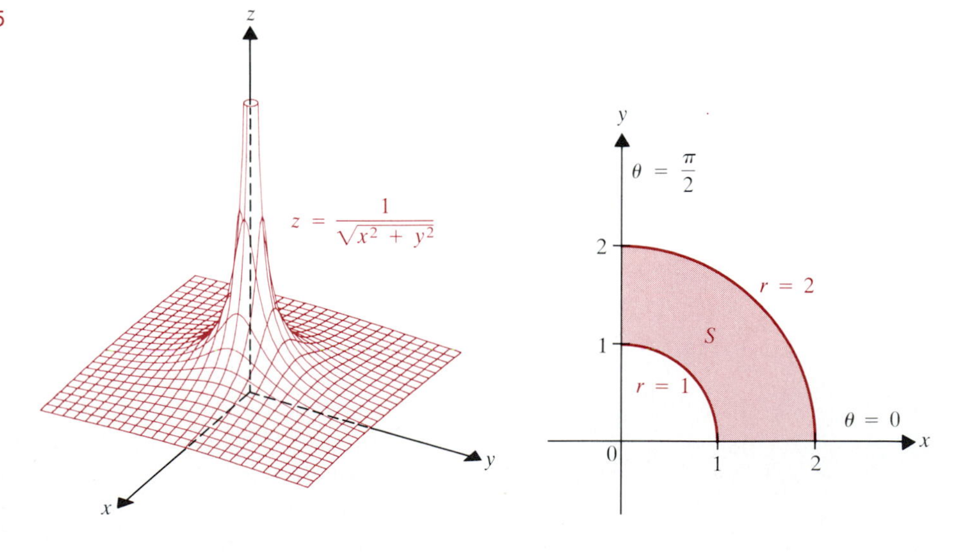

Solution. The presence of the factor $\sqrt{x^2 + y^2} = r$ in the formula for $f(x, y)$ suggests that a change to polar coordinates may be helpful. This is underscored by the fact that in polar coordinates D is the region E: $1 \le r \le 2,\ 0 \le \theta \le \pi/2$.

Since $f(r\cos\theta, y\sin\theta) = 1/r$, we have from (4) and (3)

$$V = \iint_D f(x, y)\,dA = \iint_E \frac{1}{r}\,r\,dr\,d\theta$$

$$= \int_0^{\pi/2} d\theta \int_1^2 dr = \frac{1}{2}\pi(2-1) = \frac{1}{2}\pi. \quad ■$$

The power of polar coordinates is brought out strikingly by using them in Example 2 of the last section.

Example 5

Use polar coordinates to find the volume of the region S lying below the paraboloid $z = 4 - x^2 - y^2$ and above the xy-plane.

Solution. As **Figure 3.6** shows, the region lies over the disk D: $x^2 + y^2 \leq 4$. In polar coordinates, this is the region E: $0 \leq r \leq 2, 0 \leq \theta \leq 2\pi$. From (4) we thus have

$$V = \iint_D (4 - x^2 - y^2)\,dA = \iint_E (4 - r^2)r\,dr\,d\theta$$

$$= \int_0^{2\pi}\int_0^1 (4r - r^3)\,dr\,d\theta = \int_0^{2\pi} d\theta \int_0^2 (4r - r^3)\,dr$$

$$= 2\pi\left[2r^2 - \frac{1}{4}r^4\right]_0^2 = 2\pi[8 - 4] = 8\pi. \quad ■$$

As in the rectangular case, we can extend the polar double integral of g to a more general bounded region D lying between two polar curves $r = h_1(\theta)$ and $r = h_2(\theta)$ and the lines $\theta = \alpha$ and $\theta = \beta$. See **Figure 3.7.** We proceed as in Definition 2.1. Since D is bounded, we can find a basic polar region S containing D, where S lies between $\theta = \alpha$, $\theta = \beta$, $r = a$, and $r = b$ as in Figure 3.7.

FIGURE 3.6

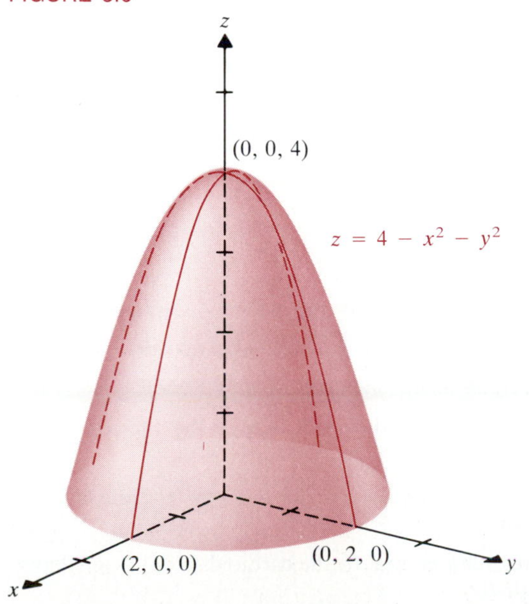

FIGURE 3.7

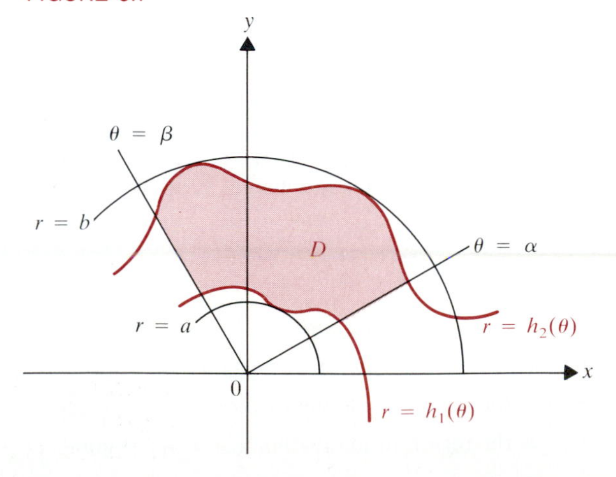

We define $\bar{g}$ on S by

$$\bar{g}(r, \theta) = \begin{cases} g(r, \theta) & \text{if } [r, \theta] \in D \\ 0 & \text{if } [r, \theta] \notin D. \end{cases} \tag{5}$$

Then we define $\iint_D g(r, \theta)\, dA$ to be $\iint_S \bar{g}(r, \theta)\, dA$. As in the rectangular case, Fubini's theorem allows us to evaluate this polar integral by iterated integration. In this case it takes the form

$$\iint_D g(r, \theta)\, dA = \int_\alpha^\beta \left[\int_{h_1(\theta)}^{h_2(\theta)} g(r, \theta) r\, dr \right] d\theta = \int_\alpha^\beta \int_{h_1(\theta)}^{h_2(\theta)} g(r, \theta) r\, dr\, d\theta. \tag{6}$$

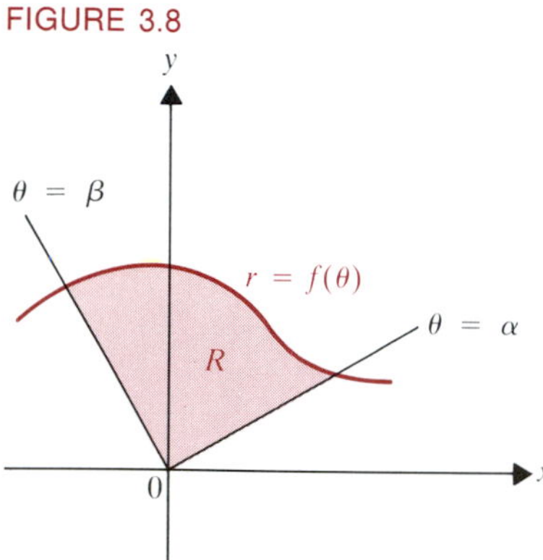

FIGURE 3.8

With (6) available, we can give an alternative derivation for the formula in Theorem 7.1 of Chapter 8 for the area of the region R in **Figure 3.8** bounded by $r = f(\theta)$ and the lines $\theta = \alpha$ and $\theta = \beta$:

$$A(R) = \int_\alpha^\beta \frac{1}{2} f(\theta)^2\, d\theta.$$

We have from Theorem 2.3

$$A = \iint_R 1\, dA = \iint_R 1 r\, dr\, d\theta = \int_\alpha^\beta \int_0^{f(\theta)} r\, dr\, d\theta$$

$$= \int_\alpha^\beta \left[\frac{1}{2} r^2 \right]_0^{f(\theta)} d\theta = \int_\alpha^\beta \frac{1}{2} f(\theta)^2\, d\theta.$$

We next give a more concrete illustration of (6).

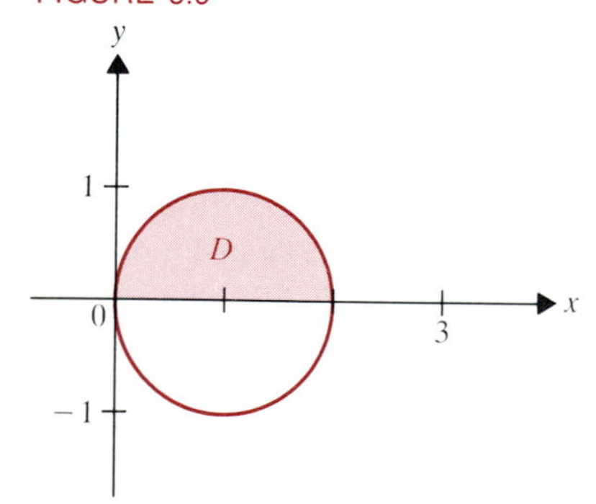

FIGURE 3.9

Example 6

If $g(r, \theta) = \sec \theta$, then evaluate $\iint_D g(r, \theta)\, dA$, where D is the upper half of the disk bounded by the circle $r = 2 \cos \theta$, shown in **Figure 3.9.**

Solution. The region D is swept out as θ goes from 0 to $\pi/2$. So we have

$$\iint_D g(r, \theta)\, dA = \int_0^{\pi/2} \int_0^{2 \cos \theta} (\sec \theta) r\, dr\, d\theta$$

$$= \int_0^{\pi/2} \sec \theta \left[\frac{1}{2} r^2 \right]_0^{2 \cos \theta} d\theta = \frac{4}{2} \int_0^{\pi/2} \sec \theta \cos^2 \theta\, d\theta$$

$$= 2 \int_0^{\pi/2} \cos \theta\, d\theta = 2 \sin \theta \Big]_0^{\pi/2} = 2. \quad \blacksquare$$

Exercises 13.3

In Exercises 1–12, use $A(S) = \iint_S 1\, dA$ to find the area of the given region S. Use polar coordinates in evaluating all integrals.

1. $S = \{(x, y) | x^2 + y^2 \leq 5\}$.
2. $S = \{(x, y) | x^2 + y^2 \leq 9\}$.
3. S is one petal of the rose $r = 2 \sin 3\theta$.
4. S is one petal of the rose $r = 3 \sin 3\theta$.
5. S is the region inside the cardioid $r = 1 + \cos \theta$.
6. S is the region inside the limaçon $r = 2 + \cos \theta$.
7. S is the region common to the circles $r = 2$ and $r = 4 \sin \theta$.
8. S is the region inside $r = \frac{1}{2} + \sin \theta$ and outside $r = 1$.
9. S is the region inside one loop of $r^2 = 8 \cos \theta$.
10. S is the region inside the small loop of the limaçon $r = 1 - 2 \sin \theta$.
11. S is the region inside the circle $r = 3 \sin \theta$ and outside $r = 2 - \sin \theta$.
12. S is the region common to the cardioids $r = 1 + \sin \theta$ and $r = 1 - \sin \theta$.

In Exercises 13–22, use polar integration to find the volume of the given region.

13. Under the cone $z = \sqrt{x^2 + y^2}$ and above the disk $x^2 + y^2 \leq 1$ in the xy-plane

14. Under the cone $z = 4 - \sqrt{x^2 + y^2}$ and above the xy-plane

15. Under the paraboloid $z = x^2 + y^2$ and above the annular region between $x^2 + y^2 = 1$ and $x^2 + y^2 = 4$ in the xy-plane

16. Under the paraboloid $z = 4 - x^2 - y^2$ and above the disk $x^2 + y^2 \leq 1$ in the xy-plane

17. Between the paraboloids $z = 4 - 3x^2 - 3y^2$ and $z = x^2 + y^2$

18. Between the paraboloids $z = 9 - 5x^2 - 5y^2$ and $z = 4x^2 + 4y^2$

19. The hemisphere of radius a

20. The right circular cone of base radius a and height h

21. A cylindrical drill of radius b drills through a sphere of radius a, passing straight through the center. Find the volume of the hole created.

22. A cylindrical drill of radius 1 drills through a sphere of radius 2, passing straight through the center. Find the volume of the resulting solid.

In Exercises 23–30, evaluate the given double integral using polar coordinates.

23. $\iint_D (1/\sqrt{x^2 + y^2})\, dA$, where D is the region bounded by the x-axis, the circle $x^2 + y^2 = 1$, and the lines $y = x$ and $x = 2$

24. $\iint_D (1/\sqrt{x^2 + y^2})\, dA$, where D is the first-quadrant region bounded by the lines $y = 1$ and $y = x$ and the circle $x^2 + y^2 = 4$

25. $\iint_D \sqrt{x^2 + y^2}\, dA$, where D is the upper half of the disk $r \leq 2\cos\theta$

26. $\iint_D \sqrt{x^2 + y^2}\, dA$, where D is the lower half of the region $r \leq 2\cos\theta$

27. $\iint_D x^2 y\, dA$, where D is the disk $x^2 + y^2 \leq 1$

28. $\iint_D xy^2\, dA$, where D is the disk $x^2 + y^2 \leq 1$

29. $\iint_D (xy - x^2 - y^2)\, dA$, where D is the annular region $1 \leq x^2 + y^2 \leq 9$

30. $\iint_D (x^2 + y^2 - 3xy)\, dA$, where D is the annular region $4 \leq x^2 + y^2 \leq 16$

13.4 Triple Integrals

We turn now to the integral calculus of scalar functions of three variables, such as $g(x, y, z) = x^2 z - y^2 z$. The three-variable analogue of the double integral over a rectangle is the triple integral over a *rectangular box* (parallelepiped),

$$B = [a, b] \times [c, d] \times [e, f] = \{(x, y, z) \mid a \leq x \leq b, c \leq y \leq d, e \leq z \leq f\}.$$

We define this by the same procedure used in Section 1. First we chop B up into a grid G of sub-boxes B_{ijk} (see **Figure 4.1**) by partitioning each of the intervals $[a, b]$, $[c, d]$, and $[e, f]$ by partitions P_1, P_2, and P_3. A typical sub-box

FIGURE 4.1

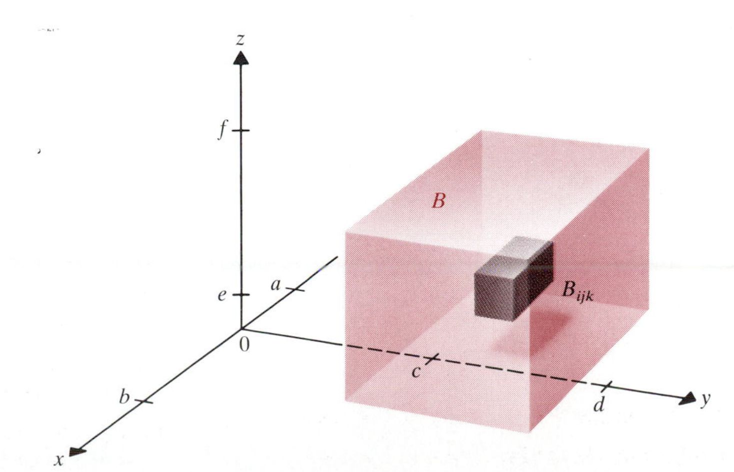

B_{ijk} is $[x_{i-1}, x_i] \times [y_{j-1}, y_j] \times [z_{k-1}, z_k]$. The volume of B_{ijk} is

$$\Delta V_{ijk} = \Delta x_i \Delta y_j \Delta z_k = (x_i - x_{i-1})(y_j - y_{j-1})(z_k - z_{k-1}).$$

A Riemann sum of g corresponding to G is then a triple sum

$$R_G = \sum_{i=1}^{m} \sum_{j=1}^{n} \sum_{k=1}^{p} g(\bar{x}_i, \bar{y}_j, \bar{z}_k) \Delta V_{ijk}$$

where $\bar{\mathbf{x}}_{ijk} = (\bar{x}_i, \bar{y}_j, \bar{z}_k) \in B_{ijk}$. As $|P_1|$, $|P_2|$, and $|P_3| \to 0$, the size of the sub-boxes diminishes toward 0, so R_G ought to approach the triple integral we wish to define.

4.1 DEFINITION The ***triple integral*** of $g\colon \mathbf{R}^3 \to \mathbf{R}$ over a rectangular parallelepiped B in $\mathbf{R}^3$ is defined as

$$\iiint_B g(x, y, z)\, dV = \lim_{|G| \to 0} \sum_{i=1}^{m} \sum_{j=1}^{n} \sum_{k=1}^{p} g(\bar{x}_i, \bar{y}_j, \bar{z}_k) \Delta V_{ijk}.$$

This means that given $\varepsilon > 0$, there is a $\delta > 0$ such that for any grid G with $|G| < \delta$ (where $|G| = \max_{i,j,k}(\Delta x_i, \Delta y_j, \Delta z_k)$),

$$\left| \iiint_B g(x, y, z)\, dV - R_G \right| < \varepsilon$$

for all Riemann sums R_G of g over B.

There is a version of Fubini's theorem for triple integrals over rectangular boxes B. It says that triple integrals can be evaluated by iteration. For example, if $B = [a, b] \times [c, d] \times [e, f]$, then

$$\iiint_B g(x, y, z)\, dV = \int_a^b \left(\int_c^d \left(\int_e^f g(x, y, z)\, dz \right) dy \right) dx$$

$$(1) \qquad = \int_a^b \int_c^d \int_e^f g(x, y, z)\, dz\, dy\, dx.$$

As for double integrals, the order of writing differentials indicates the order of integration, so that parentheses are not needed. Any of the five other possible orders ($dz\, dx\, dy$, $dy\, dz\, dx$, $dy\, dx\, dz$, $dx\, dz\, dy$, or $dx\, dy\, dz$) produces the same value as the one calculated for $\iiint_B g(x, y, z)\, dV$ from (1).

Example 1

Evaluate $\iiint_B g(x, y, z)\, dV$ if B is the box $[1, 2] \times [-1, 1] \times [2, 4]$ and $g(x, y, z) = x^2 z - y^2 z$.

Solution. Here (1) gives

$$\iiint_B g(x, y, z)\, dV = \int_1^2 \int_{-1}^1 \int_2^4 (x^2 - y^2) z\, dz\, dy\, dx$$

$$= \int_1^2 \int_{-1}^1 (x^2 - y^2) \frac{1}{2} z^2 \Big]_{z=2}^{z=4} dy\, dx$$

$$= \int_1^2 \int_{-1}^1 (x^2 - y^2)(8 - 2)\, dy\, dx$$

$$= 6 \int_1^2 x^2 y - \frac{1}{3} y^3 \Big]_{y=-1}^{y=1} = 6 \int_1^2 \left(2x^2 - \frac{2}{3} \right) dx$$

$$= 12 \int_1^2 \left(x^2 - \frac{1}{3} \right) dx = 12 \left[\frac{1}{3} x^3 - \frac{1}{3} x \right]_1^2$$

$$= 4[8 - 2 - 1 + 1] = 24. \quad \blacksquare$$

In solving Example 1, we could have used any other order of iteration we pleased (see Exercises 25 and 26). Notice that once the integration with respect to z was performed in Example 1, we were left with

$$\iint_R 6(x^2 - y^2)\,dA = \int_1^2 \int_{-1}^1 6(x^2 - y^2)\,dy\,dx,$$

the *double* integral over $R = [1, 2] \times [-1, 1]$ of $\int_2^4 (x^2 - y^2)z\,dz$. Thus

$$\iiint_B g(x, y, z)\,dV = \iint_R \left(\int_2^4 g(x, y, z)\,dz \right) dA.$$

This is an instance of a general fact: The triple integral of $g(x, y, z)$ over a box $B = [a, b] \times [c, d] \times [e, f]$ is equal to the double integral over $R = [a, b] \times [c, d]$ of the partial integral $\int_e^f g(x, y, z)\,dz$. The idea is similar to the one we saw in evaluating a double integral: The integration with respect to z reduces the evaluation of the triple integral $\iiint_B g(x, y, z)\,dV$ to a calculation of an integral in one less dimension. That gives the *double* integral over the rectangle R that results from projecting B onto the xy-plane. See **Figure 4.2.**

FIGURE 4.2

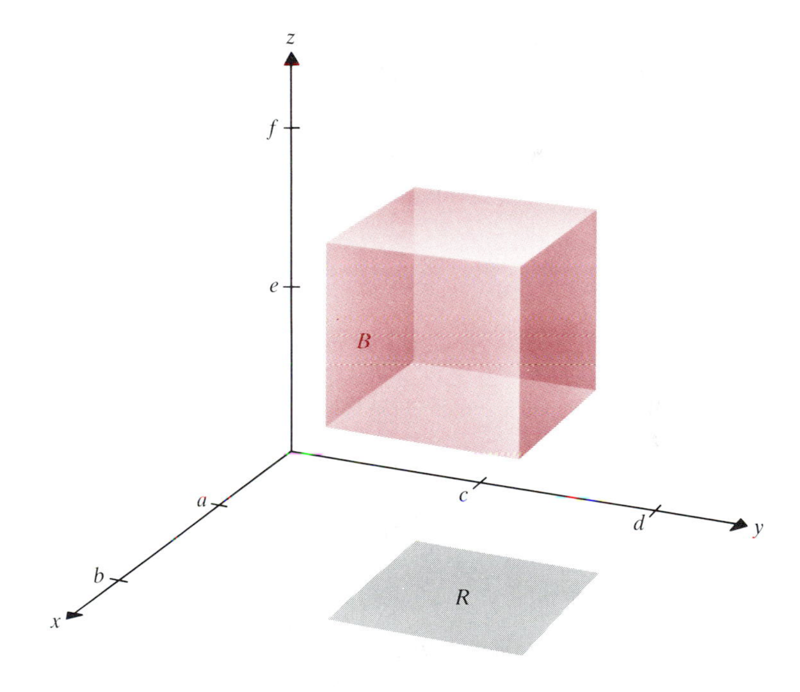

To define $\iiint_E f(x, y, z)\,dV$ for more general regions E in $\mathbf{R}^3$, we use the approach of Definition 2.1. If E is a bounded region of $\mathbf{R}^3$, then we enclose E in a box B and define $\bar{f}$ on B to coincide with f for points x in E and to be 0 for points x outside E. Then the triple integral of f over E is defined by

(2)
$$\iiint_E f(x, y, z)\,dV = \iiint_B \bar{f}(x, y, z)\,dV.$$

A version of Fubini's theorem asserts that the triple integral over the region E can be evaluated as an iterated integral and that we can evaluate the iterated integrals in any order we please. Again, the proof is beyond our scope here, but we do give a precise statement for the case of a region E bounded below by a surface $z = g_1(x, y)$ and above by a surface $z = g_2(x, y)$.

4.2 THEOREM

Suppose that $E \subseteq \mathbf{R}^3$ is bounded below by the surface $z = g_1(x, y)$, and above by the surface $z = g_2(x, y)$. If f is continuous on E and its boundary, then

$$(3) \qquad \iiint_E f(x, y, z)\, dV = \iint_D \left(\int_{g_1(x, y)}^{g_2(x, y)} f(x, y, z)\, dz \right) dA,$$

where D is the perpendicular projection of E onto the xy-plane. (See **Figure 4.3.**)

FIGURE 4.3

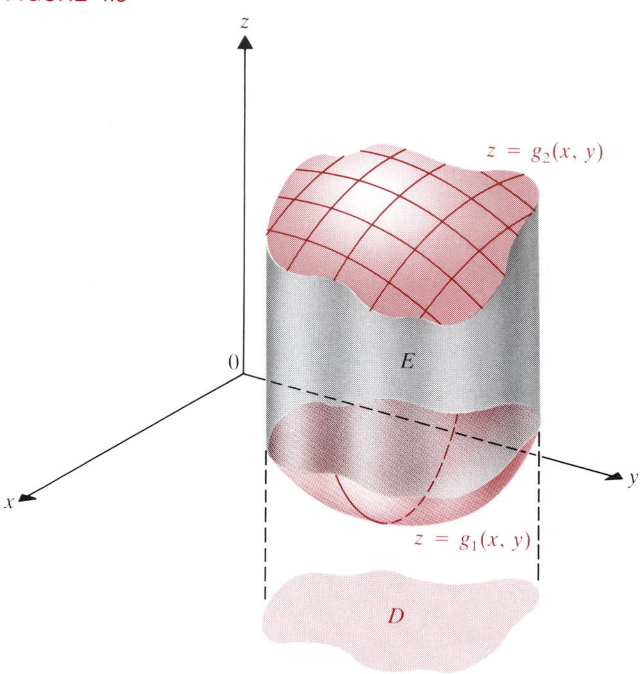

The right side of (3) is an ordinary double integral over D of the function

$$F(x, y) = \int_{g_1(x, y)}^{g_2(x, y)} f(x, y)\, dz,$$

and so it can be evaluated by the methods of Section 2. (The variable z disappears when we calculate

$$\int_{g_1(x, y)}^{g_2(x, y)} f(x, y, z)\, dz,$$

just as when we integrate over a box B.)

Example 2

Evaluate $\iiint_E f(x, y, z)\, dV$ if $f(x, y, z) = 2x + yz$, and E is bounded by the xy-plane, the cylinder $z = 1 - x^2$, and the planes $y = 0$ and $y = 4$.

Solution. The most important step in evaluating a triple integral is drawing a careful sketch from which the limits of integration can be determined. From Section 11.5, the graph of $z = 1 - x^2$ in $\mathbf{R}^3$ is a parabolic cylinder perpendicular

FIGURE 4.4

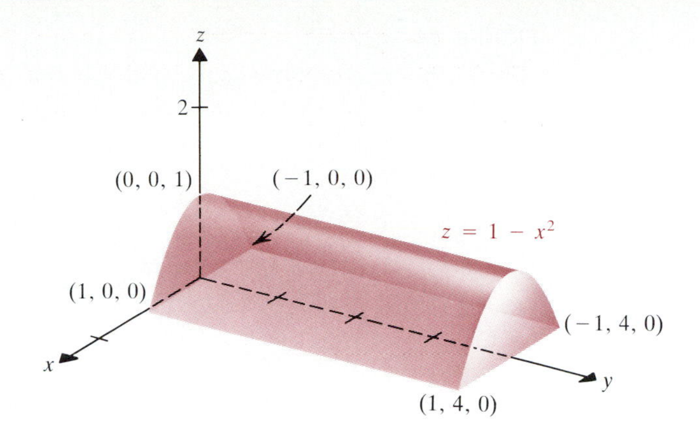

to the xz-plane and opening downward. The region E is shown in **Figure 4.4.** It lies under the cylinder and above the xy-plane between the xz-plane and the plane $y = 4$, so has the shape of a quonset hut plus its interior. We have from (3)

$$\iiint_E f(x, y, z)\,dV = \iint_D \left[\int_0^{1-x^2} (2x + yz)\,dz\right] dA,$$

FIGURE 4.5

where D is the rectangle $[-1, 1] \times [0, 4]$ shown in **Figure 4.5.** It is bounded by the lines $x = -1$, $x = 1$, $y = 0$, and $y = 4$ in the xy-plane. We thus obtain

$$\iiint_E f(x, y, z)\,dV = \int_{-1}^{1}\int_0^4 2xz + \frac{1}{2}yz^2\Big]_{z=0}^{z=1-x^2} dy\,dx$$

$$= \int_{-1}^{1}\int_0^4 \left[2x(1 - x^2) + \frac{1}{2}y(1 - x^2)^2 - 0 - 0\right] dy\,dx$$

$$= \int_{-1}^{1}\int_0^4 \left[2x - 2x^3 + \frac{1}{2}y - yx^2 + \frac{1}{2}yx^4\right] dy\,dx$$

$$= \int_{-1}^{1}\int_0^4 \left[2x - 2x^3 + \left(\frac{1}{2} - x^2 + \frac{1}{2}x^4\right)y\right] dy\,dx$$

$$= \int_{-1}^{1} \left[(2x - 2x^3)y + \left(\frac{1}{2} - x^2 + \frac{1}{2}x^4\right)\frac{1}{2}y^2\right]_{y=0}^{y=4} dx$$

$$= 8\int_{-1}^{1} \left[x - x^3 + \frac{1}{2} - x^2 + \frac{1}{2}x^4\right] dx$$

$$= 8\int_{-1}^{1} \left[\frac{1}{2} - x^2 + \frac{1}{2}x^4\right] dx \qquad \textit{by Theorem 2.8(b) of Chapter 4}$$

$$= 16\int_{0}^{1} \left[\frac{1}{2} - x^2 + \frac{1}{2}x^4\right] dx \qquad \textit{by Theorem 2.8(a) of Chapter 4}$$

$$= 16\left[\frac{1}{2}x - \frac{1}{3}x^3 + \frac{1}{10}x^5\right]_0^1 = 16\left[\frac{1}{2} - \frac{1}{3} + \frac{1}{10}\right]$$

$$= 16\left[\frac{15 - 10 + 3}{30}\right] = \frac{64}{15}. \quad \blacksquare$$

To evaluate $\iiint_E f(x, y, z)\,dV$, we can integrate first with respect to y (or x) instead of with respect to z, if that is easier to do. If, for example, E lies to the

FIGURE 4.6

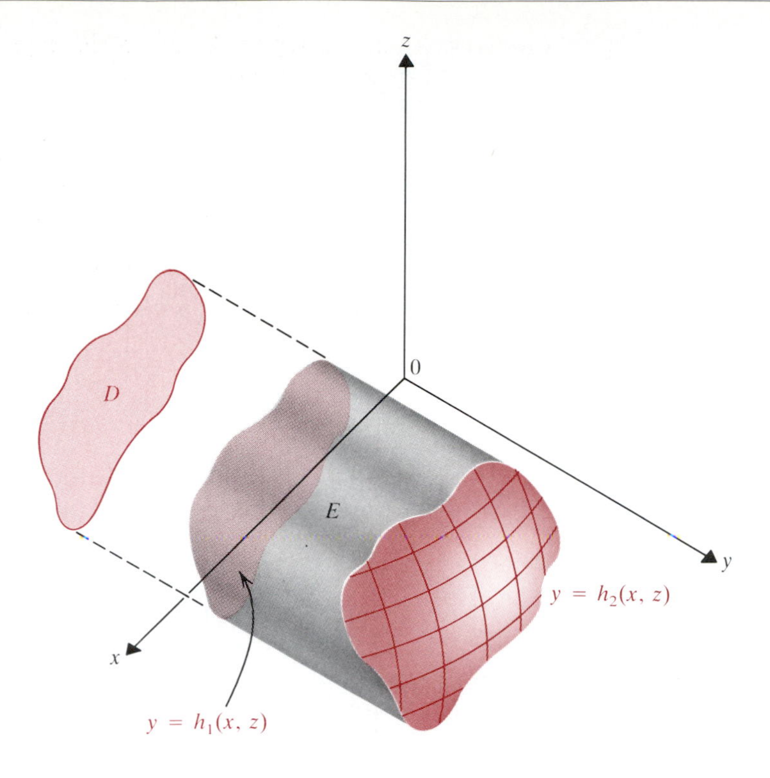

right of $y = h_1(x, z)$ and to the left of $y = h_2(x, z)$, then

$$\iiint_E f(x, y, z)\, dV = \iint_D \left[\int_{h_1(x,z)}^{h_2(x,z)} f(x, y, z)\, dy \right] dA, \tag{4}$$

where D is the perpendicular projection of E onto the xz-plane. See **Figure 4.6.**

Example 3

Evaluate $\iiint_E g(x, y, z)\, dV$ if $g(x, y, z) = 2xyz^2$, and E is bounded by the cylinders $y = x^2$ and $y = 2 - x^2$ and the planes $x = 0$, $z = 0$, and $z = 2$.

Solution. The two cylinders are perpendicular to the xy-plane. The cylinder $y = x^2$ opens to the right, while $y = 2 - x^2$ opens to the left. **Figure 4.7** shows

FIGURE 4.7

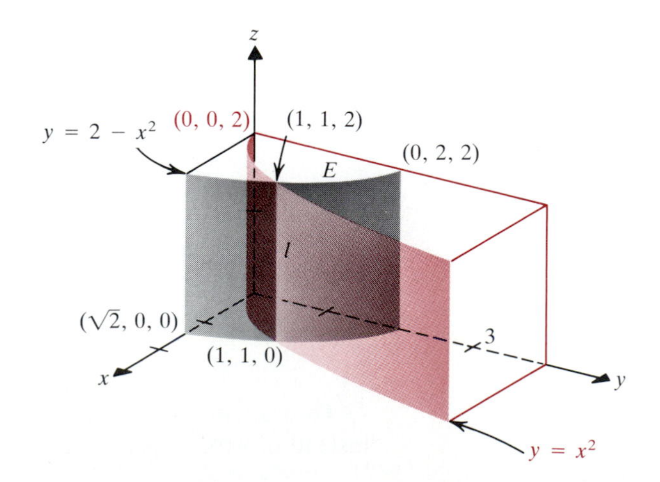

that the region E lies to the right of $y = x^2$ and to the left of $y = 2 - x^2$. Therefore (4) gives

$$\text{(5)} \qquad \iiint_E g(x, y, z)\,dV = \iint_D \left[\int_{x^2}^{2-x^2} 2xyz^2\,dy\right] dA,$$

where D is the perpendicular projection of E onto the xz-plane. From the statement of the problem, z varies between 0 and 2 in E. To determine the range of values of x in E, we first find the line l of intersection of $y = x^2$ and $y = 2 - x^2$. Equating the expressions for y, we obtain

$$x^2 = 2 - x^2 \rightarrow 2x^2 = 2 \rightarrow x = 1,$$

since the region E lies in the first octant. Then $y = 1$ also, whereas z can take on any real value. Thus l is the line with parametric equations $x = 1$, $y = 1$, and $z = t$. From this and the fact that z is between 0 and 2 in E, we see that D is the rectangle shown in **Figure 4.8.** We thus have from (5)

FIGURE 4.8

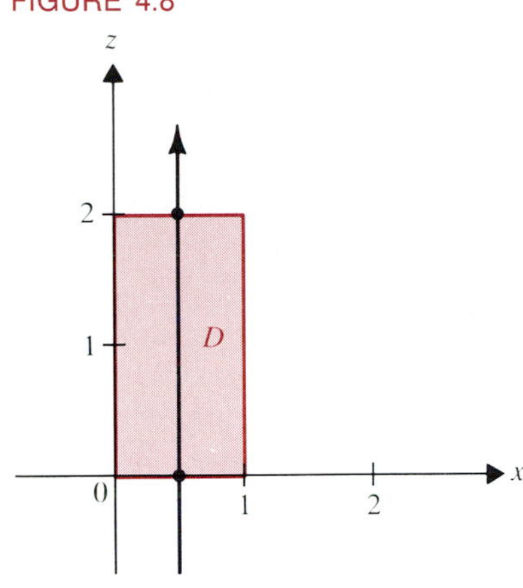

$$\iiint_E g(x, y, z)\,dV = \int_0^1 \int_0^2 \left[xy^2z^2\right]_{y=x^2}^{y=2-x^2} dz\,dx$$

$$= \int_0^1 \int_0^2 [x(2 - x^2)^2 - x^5]z^2\,dz\,dx$$

$$= \int_0^1 [x(4 - 4x^2 + x^4) - x^5]\,\frac{1}{3}z^3\Big|_{z=0}^{z=2} dx$$

$$= \frac{8}{3}\int_0^1 (4x - 4x^3)\,dx = \frac{8}{3}\left[2x^2 - x^4\right]_0^1 = \frac{8}{3}. \quad ■$$

In case E lies in front of $x = k_1(y, z)$ and behind $x = k_2(y, z)$, we evaluate the triple integral of $f(x, y, z)$ over E by using the formula

$$\text{(6)} \qquad \iiint_E (x, y, z)\,dV = \iint_D \left[\int_{k_1(y,z)}^{k_2(y,z)} f(x, y, z)\,dx\right] dA,$$

where D is the perpendicular projection of E onto the yz-plane. See **Figure 4.9.**

In practice we use whichever of Equations (3), (4), or (6) involves the easiest calculations. If there is no difference (for example, if E is a rectangular box),

FIGURE 4.9

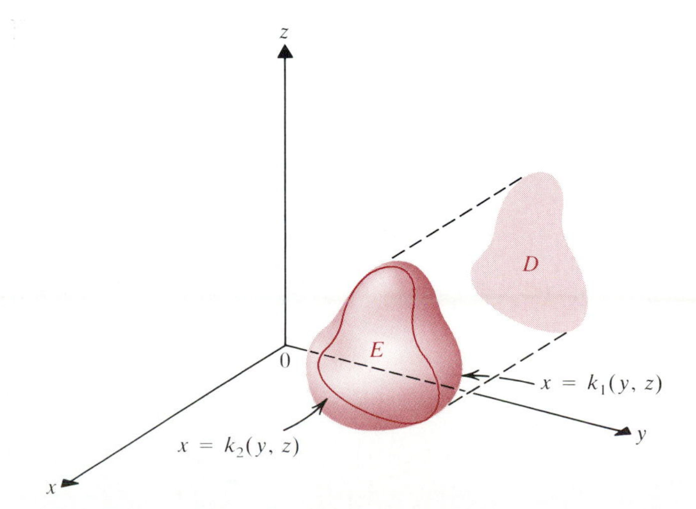

FIGURE 4.10

FIGURE 4.11

then (3) is possibly the best path to follow, because we obtain an xy-double integral, which we have been dealing with in the earlier sections.

The implementation of (3), (4), and (6) requires you to correctly determine the limits of integration. This can be done via the approach of Section 3. For instance, in (3) imagine yourself *under* the region E and shooting an arrow up through it. The equation of the first surface, say $z = g_1(x, y)$, that your arrow would puncture provides the *lower limit* in (3). Similarly, the equation of the surface from which the arrow emerges at the top provides the *upper limit* $g_2(x, y)$. See **Figure 4.10.** In using (4), imagine yourself shooting an arrow from the *left* of the region E through it parallel to the y-axis. The *left bounding surface* gives the lower limit $h_1(x, z)$, and the *right bounding surface* gives the upper limit $h_2(x, z)$, as in **Figure 4.11.** Finally, to apply (6) imagine yourself behind the region E shooting an arrow through it parallel to the x-axis. The *rear bounding surface* is hit first, so it gives the lower limit $k_1(y, z)$. The *front bounding surface* then gives the upper limit $k_2(y, z)$. See **Figure 4.12.**

In all three schemes, once the first integration is complete, the variable with respect to which it was performed has been "integrated out." The result is a

FIGURE 4.12

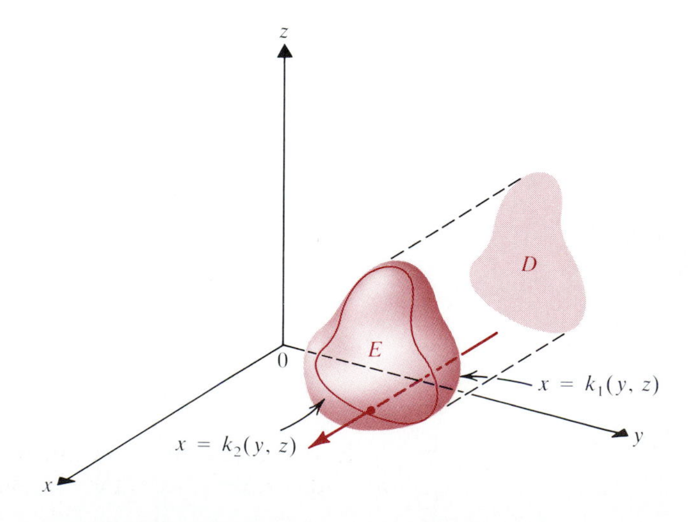

double integral over a two-dimensional region, which you evaluate as in Section 2. The next example shows the details in applying (6) to a complicated region.

Example 4

Evaluate $\iiint_E f(x, y, z)\,dV$, where $f(x, y, z) = 1/x$, and E is the first-octant region bounded behind by the plane $x = 1$, below by the cylinder $x = e^z$, and above by the plane $y = z$ and cylinder $y^2 + z^2 = 4$. See **Figure 4.13.**

FIGURE 4.13

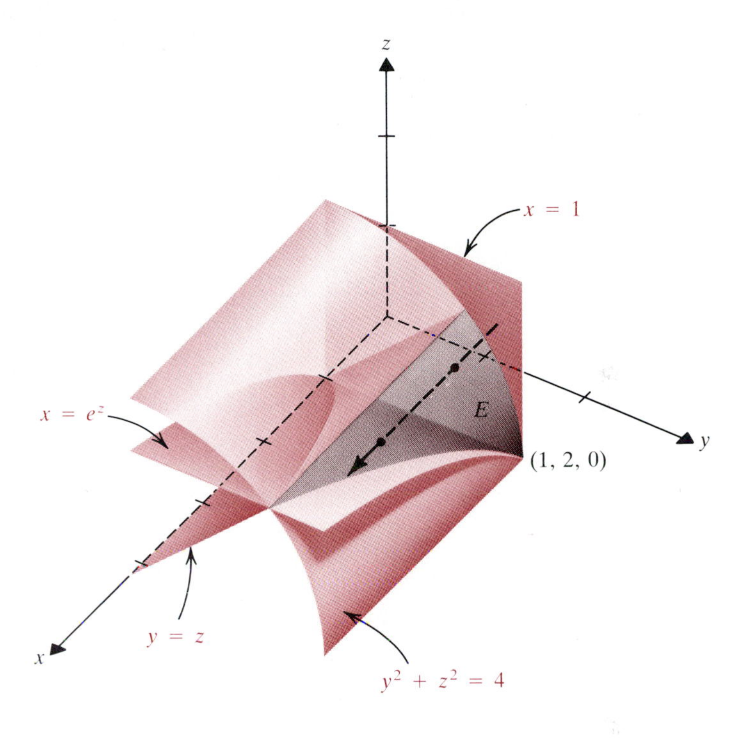

FIGURE 4.14

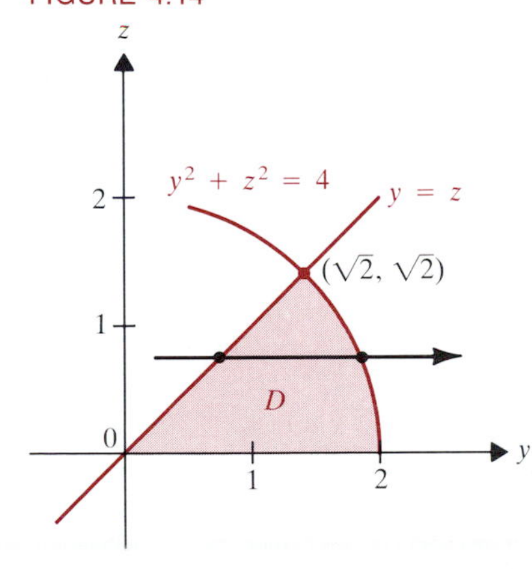

Solution. The region lies in front of the plane $x = 1$ and behind the cylinder $x = e^z$, so we apply (6). Imagining an arrow shot parallel to the x-axis from the yz-plane, we get

$$\iiint_E f(x, y) = \iint_D \left[\int_1^{e^z} \frac{1}{x}\,dx\right] dA = \iint_D \Big[\ln x\Big]_{x=1}^{x=e^z} dA = \iint_D z\,dA,$$

where D is the region of the yz-plane shown in **Figure 4.14.** Evaluating the double integral over D, we obtain

$$\begin{aligned}\iint_D z\,dA &= \int_0^{\sqrt{2}} \int_z^{\sqrt{4-z^2}} z\,dy\,dz = \int_0^{\sqrt{2}} z(\sqrt{4-z^2} - z)\,dz \\ &= -\frac{1}{2}\int_0^{\sqrt{2}} (-2z)\sqrt{4-z^2}\,dz - \int_0^{\sqrt{2}} z^2\,dz \\ &= -\frac{1}{2}\cdot\frac{2}{3}(4-z^2)^{3/2} - \frac{z^3}{3}\Bigg]_0^{\sqrt{2}} = -\frac{2\sqrt{2}}{3} + \frac{8}{3} - \frac{2\sqrt{2}}{3} = \frac{8-4\sqrt{2}}{3}. \quad ■\end{aligned}$$

If we compute the triple integral of the function f given by $f(x, y, z) = 1$ over a solid region $E \subseteq \mathbf{R}^3$, then we simply get the *volume* $V(E)$ of E. For we have from (2),

$$\iiint_E f(x, y, z)\, dV = \iiint_B \bar{f}(x, y, z)\, dV,$$

where $E \subseteq B$, $\bar{f}(x, y, z) = 1$ on E, and $\bar{f}(x, y, z) = 0$ on the part of B that is not in E. If E is a region like the one in **Figure 4.15,** then

$$\iiint_B \bar{f}(x, y, z)\, dV = \iint_D \left[\int_e^{g_1(x, y)} 0\, dz \right] dA + \iint_D \left[\int_{g_1(x, y)}^{g_2(x, y)} 1\, dz \right] dA$$
$$+ \iint_D \left[\int_{g_2(x, y)}^{f} 0\, dz \right] dA$$
$$= \iint_D [g_2(x, y) - g_1(x, y)]\, dA = V(E).$$

FIGURE 4.15

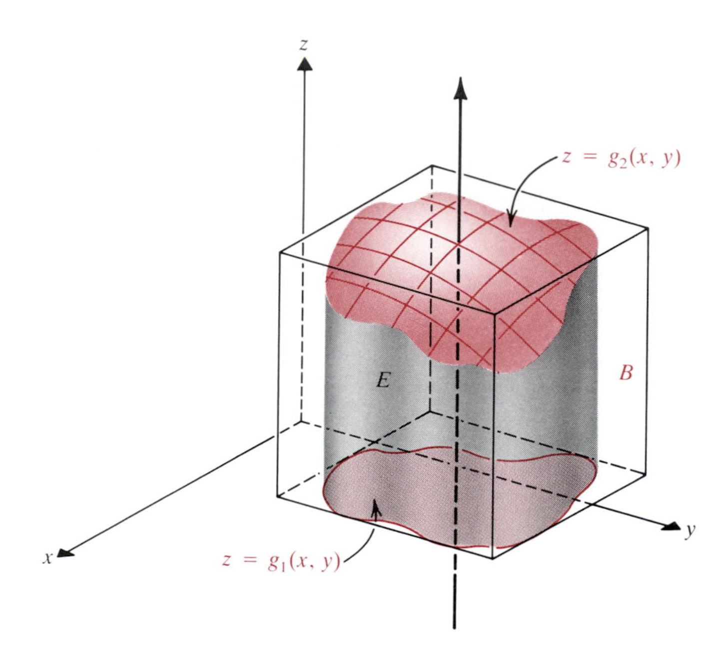

Hence we have proved the following analogue of Theorem 2.3.

4.3 THEOREM

> If E is a bounded region in $\mathbf{R}^3$, then the volume V of E is given by
>
> $$V = \iiint_E 1\, dV,$$
>
> where 1 denotes the function f given by $f(x, y, z) = 1$ for all (x, y, z) in $\mathbf{R}^3$.

Example 5

Find the volume of the region $E \subseteq \mathbf{R}^3$ bounded below by the xy-plane, above by the plane $z = x$, and by the parabolic cylinder $y^2 = 4 - 2x$.

Solution. The region E is sketched in **Figure 4.16.** It lies below the plane $z = x$ and above the region D of the xy-plane shown in **Figure 4.17.** So using (3) of

FIGURE 4.16

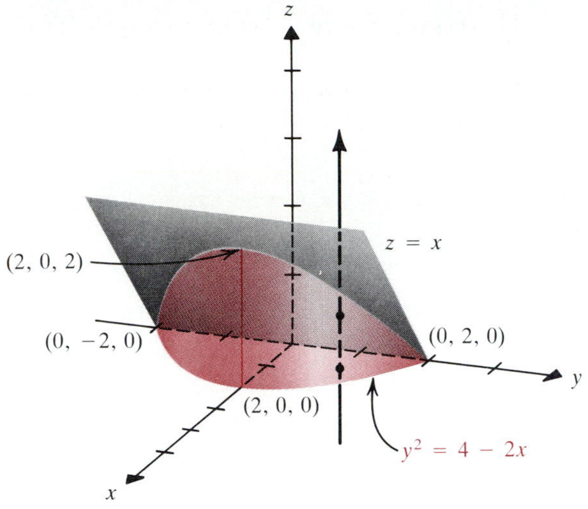

FIGURE 4.17

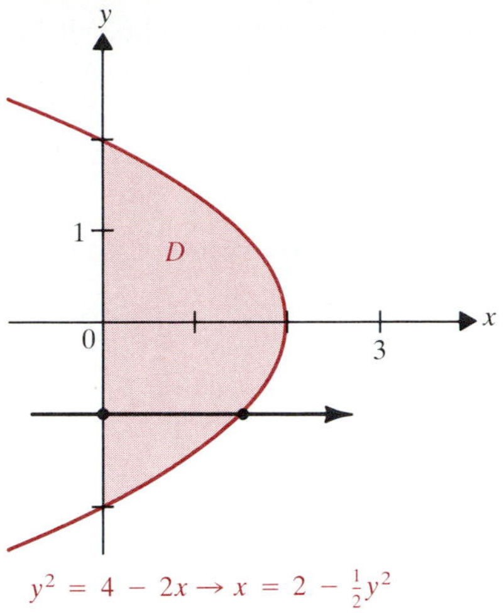

Theorem 4.2 in conjunction with Theorem 4.3, we shoot our arrow upward. It pierces the floor of E at $z = 0$ and leaves through the ceiling $z = x$. We therefore get

$$V = \iiint_E 1\,dV = \iint_D \left[\int_0^x 1\,dz\right] dA = \iint_D \Big[z\Big]_0^x\,dA = \iint_D x\,dA$$

$$= \int_{-2}^{2}\int_0^{2-y^2/2} x\,dx\,dy = \int_{-2}^{2}\left[\frac{1}{2}x^2\right]_0^{2-y^2/2} dy$$

$$= \frac{1}{2}\int_{-2}^{2}\left(4 - 2y^2 + \frac{1}{4}y^4\right)dy = \frac{1}{2}\left[4y - \frac{2y^3}{3} + \frac{1}{20}y^5\right]_{-2}^{2}$$

$$= \frac{1}{2}\left(8 - \frac{2\cdot 8}{3} + \frac{32}{20} + 8 - \frac{2\cdot 8}{3} + \frac{32}{20}\right) = \frac{1}{2}\left(16 - \frac{32}{3} + \frac{16}{5}\right) = 8 - \frac{16}{3} + \frac{8}{5}$$

$$= \frac{120 - 80 + 24}{15} = \frac{64}{15}. \quad \blacksquare$$

Exercises 13.4

In Exercises 1–14, use (3), (4), or (6) to evaluate the given triple integral. Sketch the region of integration.

1. $\iiint_B x\,dV$, where $B = [-1, 1] \times [1, 2] \times [0, 3]$
2. $\iiint_B y\,dV$, where $B = [0, 1] \times [-1, 1] \times [0, 2]$
3. $\iiint_E (1 - z)\,dV$, where E lies above the triangle with vertices (0, 0, 0), (1, 0, 0), and (0, 1, 0) and below the cylinder $z = 1 - x^2$
4. $\iiint_E z\,dV$, where E is the pyramid with apex (0, 0, 1) and with the square base having vertices (1, 0, 0), (0, 1, 0), (1, 1, 0), and (0, 0, 0)
5. $\iiint_E 24yz\,dV$, where E is the region in the first octant bounded by the cylinders $y = x^2$ and $z = 1 - y^2$
6. $\iiint_E (x - yz)\,dV$, where E is the solid tetrahedron formed by the three coordinate planes and the plane $x + y + z = 1$
7. $\iiint_E (xy - z)\,dV$, where E is the region of Exercise 6
8. $\iiint_E (4x + 8y)\,dV$ where E is the solid tetrahedron formed by the three coordinate planes and the plane $6x + 3y + 2z = 6$
9. $\iiint_E (x + y)\,dV$, where E is the region bounded by the planes $z = 0$, $x = z$, and $y = x$ and by the cylinder $y = x^2$
10. $\iiint_E (2x - y)\,dV$, where E is the region of Exercise 9
11. $\iiint_E x^2yz\,dV$, where E is the region bounded by the planes $y = 0$, $z = 0$, $x = 1$, and $x = 2$ and by the cylinders $y = x^2$ and $z = 1/x$

12. $\iiint_E x^3yz\,dV$, where E is the region of Exercise 11

13. $\iiint_E z^2\,dV$, where E is the region bounded below by the xy-plane and above by the cylinders $x^2 + z = 1$ and $y^2 + z = 1$

14. $\iiint_E (z^2 - 1)\,dV$, where E is the region of Exercise 13

In Exercises 15–24, find the volume of the given region E.

15. E is a tetrahedron formed by the three coordinate planes and the plane $x/a + y/b + z/c = 1$.

16. E is the tetrahedron formed by the three coordinate planes and $x + y + z = 1$.

17. E is the first-octant region bounded by $y = 4 - x^2$ and the planes $z = x$, $y = 0$, and $z = 0$.

18. E is the first-octant region bounded by $z = 4 - x^2$ and the planes $y = x$, $y = 0$, and $z = 0$.

19. E is the region below $z = 4 - x^2 - y^2$ and above the xy-plane.

20. E is the first-octant region bounded by the planes $x = 0$, $y = 0$, and $z = 2$ and by the surface $z = x^2 + y^2$.

21. E is the region bounded below by the xy-plane, above by $z = y$, and on the right by $y = 1 - x^2$.

22. E is bounded below by the plane $z = 3$ and above by the sphere $x^2 + y^2 + z^2 = 25$.

23. E is the region bounded by the paraboloids $z = 8 - x^2 - y^2$ and $z = x^2 + 3y^2$.

24. E is the region between the paraboloids $z = 12 - x^2 - y^2$ and $z = 2x^2 + 5y^2$.

25. Rework Example 1 by integrating in the order **(a)** $dz\,dx\,dy$ **(b)** $dy\,dz\,dx$

26. Repeat Example 1, integrating in the order **(a)** $dx\,dy\,dz$ **(b)** $dy\,dx\,dz$

27. Suppose that $\lim_{|G|\to 0} R_G(f)$ and $\lim_{|G|\to 0} R_G(g)$ both exist. Then show that

$$\lim_{|G|\to 0} R_G(f + g) = \lim_{|G|\to 0} R_G(f) + \lim_{|G|\to 0} R_G(g) \quad \text{and}$$

$$\lim_{|G|\to 0} R_G(af) = a \lim_{|G|\to 0} R_G(f).$$

28. Use Exercise 27 to show that if f and g are integrable over E, and a is a real number, then

$$\iiint_E [f(x, y, z) + g(x, y, z)]\,dV = \iiint_E f(x, y, z)\,dV + \iiint_E g(x, y, z)\,dV,$$

$$\iiint_E (af)(x, y, z)\,dV = a \iiint_E f(x, y, z)\,dV.$$

In Exercises 29–34, assume that all functions are continuous.

29. If $B = B_1 \cup B_2$ where B_1 and B_2 are nonoverlapping boxes, or overlap only to the extent of having a common face, then show that

$$\iiint_B f(x, y, z)\,dV = \iiint_{B_1} f(x, y, z)\,dV + \iiint_{B_2} f(x, y, z)\,dV.$$

30. If $f(x, y, z) \geq 0$ on E, then show that

$$\iiint_E f(x, y, z)\,dV \geq 0.$$

31. If $f(x, y, z) \leq g(x, y, z)$ on E, then show that

$$\iiint_E f(x, y, z)\,dV \leq \iiint_E g(x, y, z)\,dV.$$

32. If $B \subseteq \mathbf{R}^3$ is a rectangular box, then show that

$$mV(B) \leq \iiint_B f(x, y, z)\,dV \leq MV(B),$$

where M is the maximum of f on B and m is the minimum of f on B.

33. Use Exercise 32 to show that if f is continuous on an open ball containing $\mathbf{x}_0$, then

$$\lim_{\delta\to 0} \frac{1}{V(B(\mathbf{x}_0, \delta))} \iiint_{B(\mathbf{x}_0,\delta)} f(x, y, z)\,dV = f(x_0, y_0, z_0),$$

where $B(\mathbf{x}_0, \delta)$ is the ball about $\mathbf{x}_0 = (x_0, y_0, z_0)$ of radius δ.

34. Use Exercise 32 to prove the *mean value theorem for triple integrals:* If f is continuous on a box B, then

$$\iiint_B f(x, y, z)\,dV = f(x_0, y_0, z_0)V(B),$$

for some (x_0, y_0, z_0) in B. (See the hint for Exercise 38 of Exercises 13.1.)

The next exercises will be of use in Section 7.

35. A function f is said to be *odd* relative to x if $f(x, y, z) = -f(-x, y, z)$. If f is odd relative to x, and E is symmetric in the yz-plane, then show that

$$\iiint_E f(x, y, z)\,dV = 0.$$

(*Hint:* Use the analogue of Exercise 29.)

36. If f is odd relative to y (see Exercise 35), and E is symmetric in the xz-plane, then show that

$$\iiint_E f(x, y, z)\,dV = 0.$$

(*Hint:* Use the analogue of Exercise 29.)

37. State and prove the analogue of Exercises 35 and 36 for an odd function in z.

38. Formulate analogues of Exercises 42 and 43 of Section 13.1 for triple integrals of even functions. Then show that your analogues are true.

13.5 Cylindrical Coordinates

This section and the next are in many ways the three-dimensional analogues of Section 3 on polar double integrals. Triple integrals that are difficult to evaluate in Cartesian coordinates can often be reduced to more manageable forms by changing to one of two alternative coordinate systems for $\mathbf{R}^3$. In this section we introduce the first of these, the cylindrical coordinate system. It is especially well suited to surfaces whose cross sections by planes $z = c$ are circles, that is, to surfaces that are symmetric in the z-axis. The cylindrical coordinate system is a hybrid produced by crossing polar coordinates (in the xy-plane) with rectangular coordinates (along the z-axis.)

5.1 DEFINITION The point $P(x, y, z) \in \mathbf{R}^3$ has ***cylindrical coordinates*** $\mathbf{r}$, $\boldsymbol{\theta}$, and $\mathbf{z}$, where $Q[r, \theta]$ is a polar-coordinate representation of the perpendicular projection $Q(x, y, 0)$ of P onto the xy-plane. See **Figure 5.1.** We write $P = P[r, \theta, z]$.

The change-of-coordinate equations on p. 528 give

(1) $$x = r\cos\theta, \qquad y = r\sin\theta, \qquad \text{and} \qquad z = z,$$

where for points not on the z-axis,

$$r = \sqrt{x^2 + y^2} > 0, \qquad \frac{x}{r} = \cos\theta, \qquad \frac{y}{r} = \sin\theta,$$

and

$$\frac{y}{x} = \tan\theta \qquad \text{if } x \neq 0.$$

As with polar coordinates for $\mathbf{R}^2$, there is *not* a one-to-one correspondence between points in $\mathbf{R}^3$ and sets of cylindrical coordinates $[r, \theta, z]$, even with the restriction $r \geq 0$. In particular, the origin can be coordinatized as $[0, \theta, 0]$ for any choice of θ.

Which surfaces have simple cylindrical coordinate equations $r = c$, $\theta = k$, and $z = a$? The third is still a plane $|a|$ units from the xy-plane—above it if $a > 0$, below it if $a < 0$. The first is a right circular cylinder ($x^2 + y^2 = c^2$ in rectangular coordinates) perpendicular to the circle in the xy-plane with radius c and center at the origin. See **Figure 5.2.** The origin of the term *cylindrical*

FIGURE 5.1

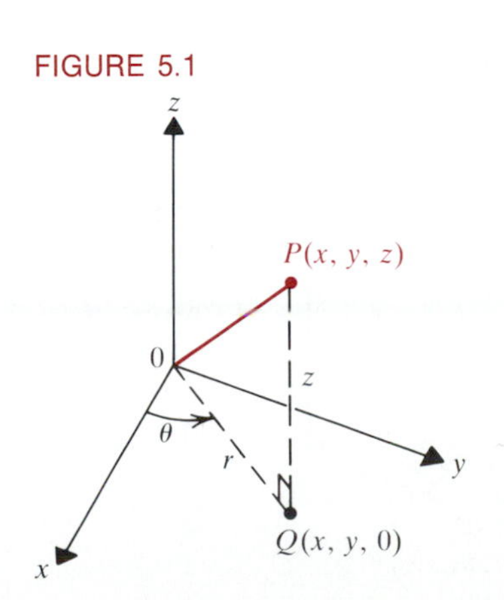

FIGURE 5.2

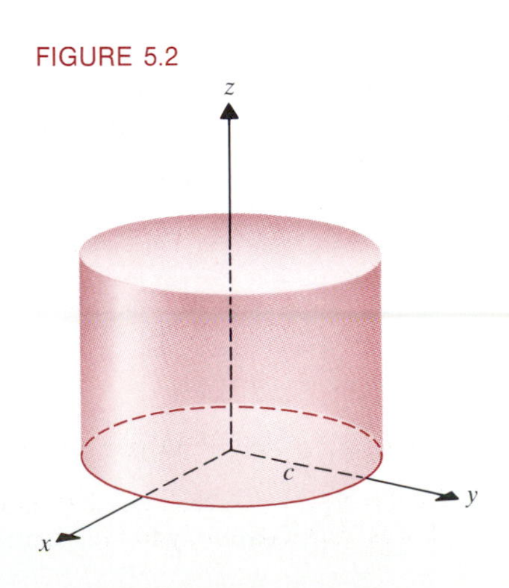

FIGURE 5.3

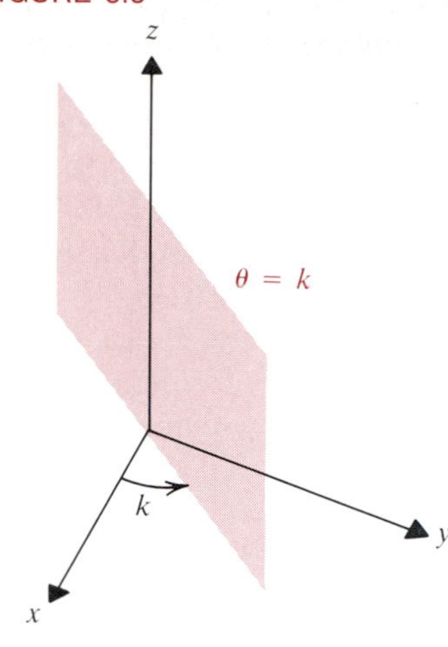

coordinates is now clear. These coordinates give the simple equation $r = c$ for right circular cylinders centered on the z-axis (which are thus analogous to the planes $x = c$ and $y = c$ in rectangular coordinates). The surface $\theta = k$ is a plane through the z-axis perpendicular to the xy-plane and making angle k with the xz-plane. See **Figure 5.3.** In particular, if we restrict r to be nonnegative, then $\theta = 0$ is the front half of the xz-plane and $\theta = \pi$ is its back half; $\theta = \frac{1}{2}\pi$ is the right half of the yz-plane, and $\theta = \frac{3}{2}\pi$ is its left half.

The next example illustrates the simplification that can result from the fact that $x^2 + y^2 = r^2$.

Example 1

Find cylindrical coordinate equations of (a) the paraboloid $z = x^2 + y^2$ and (b) the cone $z^2 = x^2 + y^2$.

Solution. (a) From (1) we have $z = r^2$. See **Figure 5.4**(a).

(b) Similarly, $z^2 = r^2$; the top nappe has the equation $z = r$ and the bottom nappe has the equation $z = -r$, if we restrict r to be nonnegative so that $r = \sqrt{x^2 + y^2}$. See **Figure 5.4**(b). ■

FIGURE 5.4

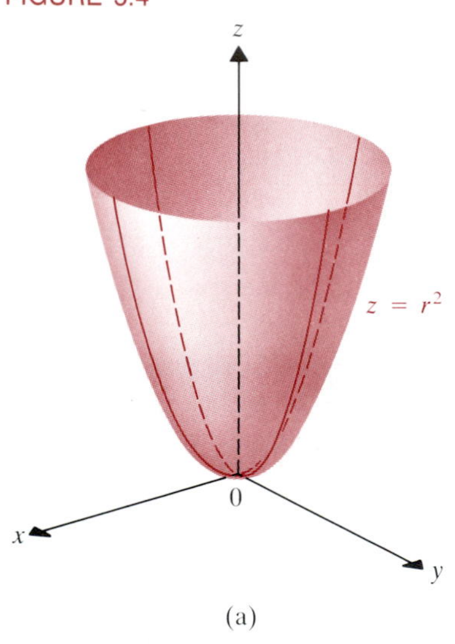

(a)

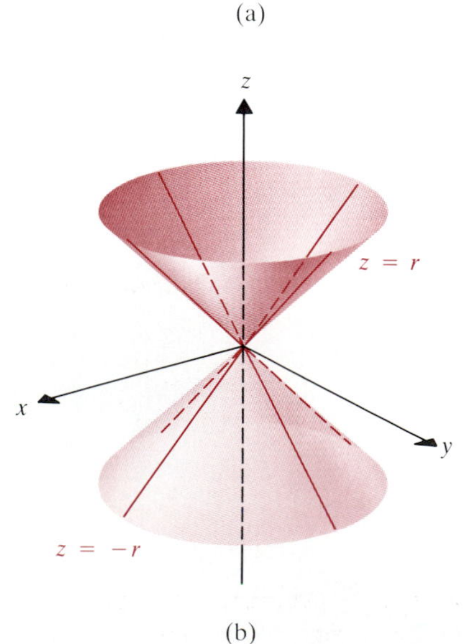

(b)

Our principal use for cylindrical coordinates will be in evaluating triple integrals. We begin with the triple integral of a function $g(r, \theta, z)$ defined over the cylindrical coordinate analogue of a rectangular box. That is a region T bounded by the planes $\theta = \alpha$ and $\theta = \beta$, $z = e$ and $z = f$, and right circular cylinders $r = a$ and $r = b$. See **Figure 5.5,** which looks like a piece of pie with a bite out of it. A cylindrical grid G of the region produces small subregions like the one in **Figure 5.6.** These are cylinders of height Δz_k and base area $\Delta A_{ij} \approx \bar{r}_i \Delta r_i \Delta\theta_j$ as on p. 858. So the subregion has approximate volume

$$\Delta V_{ijk} \approx \bar{r}_i \Delta r_i \Delta\theta_j \Delta z_k.$$

We are thus led to the following definition.

FIGURE 5.5

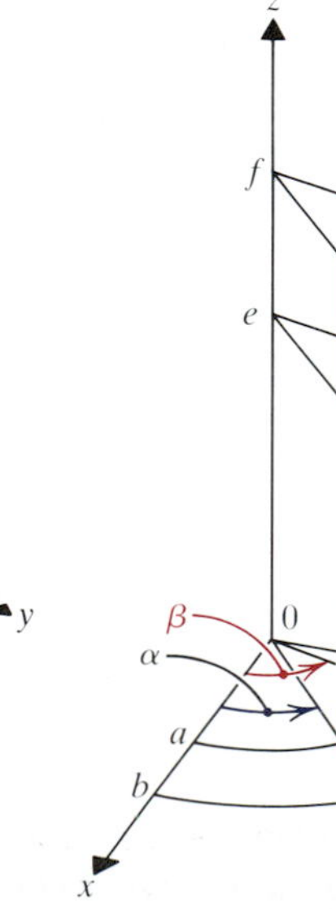

FIGURE 5.6

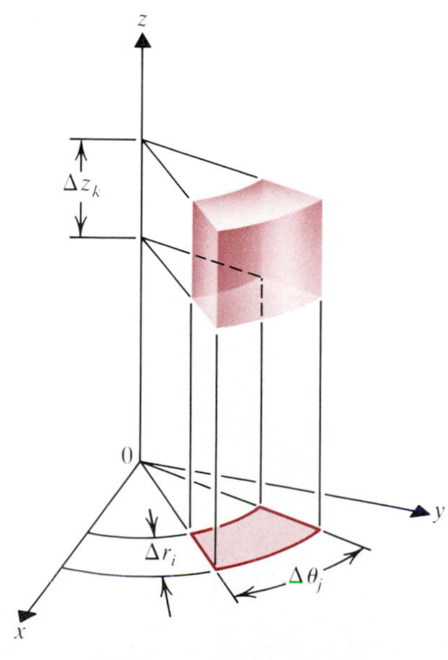

5.2 DEFINITION Let $T \subseteq \mathbf{R}^3$ be the region bounded by $\theta = \alpha$, $\theta = \beta$, $z = e$, $z = f$, $r = a$, and $r = b$. Let $g(r, \theta, z)$ be defined on T. Then the ***cylindrical triple integral*** of g over T is

$$\iiint_T g(r, \theta, z)\,dV = \iiint_T g(r, \theta, z) r\,dr\,d\theta\,dz$$

$$= \lim_{|G| \to 0} \sum_{i=1}^{m} \sum_{j=1}^{n} \sum_{k=1}^{p} g(\bar{r}_i, \bar{\theta}_j, \bar{z}_k)\bar{r}_i\,\Delta r_i\,\Delta\theta_j\,\Delta z_k,$$

where the limit is taken in the same sense as in Definition 4.1.

We again leave a discussion of the consistency of Definition 5.2 and Equation (2) on p. 865 for advanced calculus. That consistency once more requires that

$$dV = r\,dr\,d\theta\,dz \qquad \textit{NOT} \qquad dr\,d\theta\,dz.$$

As in the rectangular case, a cylindrical integral is evaluated by iteration. We state the appropriate version of Fubini's theorem, but its justification is left to more advanced texts.

5.3 THEOREM If T is the region in $\mathbf{R}^3$ bounded by $\theta = \alpha$, $\theta = \beta$ (where $\alpha < \beta$), $z = e$, $z = f$ (where $e < f$), $r = a$, and $r = b$ (where $a < b$), and if g is continuous on T and its boundary, then

$$\iiint_T g(r, \theta, z)\,dV = \int_\alpha^\beta \int_a^b \int_e^f g(r, \theta, z)\,dz\,r\,dr\,d\theta. \tag{2}$$

Moreover, the triple integral can also be calculated by changing to any of the five other orders of iteration.

The last sentence means that we also have

$$\iiint_T g(r, \theta, z)\,dV = \int_a^b \int_\alpha^\beta \int_e^f g(r, \theta, z)\,dz\,d\theta\,r\,dr = \int_e^f \int_\alpha^\beta \int_a^b g(r, \theta, z) r\,dr\,d\theta\,dz,$$

and so forth. The following example illustrates the use of (2).

Example 2

If $g(r, \theta, z) = zr^2 \cos\theta$, then evaluate $\iiint_T g(r, \theta, z)\,dV$ if T is bounded by $\theta = 0$, $\theta = \frac{1}{2}\pi$, $z = 0$, $z = 1$, $r = 1$, and $r = 3$.

Solution. From (2) we have

$$\iiint_T zr^2 \cos\theta\,dV = \int_0^{\pi/2} \int_1^3 \int_0^1 r^2 \cos\theta\, z\,dz\,r\,dr\,d\theta$$

$$= \int_0^{\pi/2} \cos\theta\,d\theta \int_1^3 r^3\,dr \int_0^1 z\,dz \qquad \text{by the analogue of Corollary 1.10}$$

$$= \int_0^{\pi/2} \cos\theta\,d\theta \int_1^3 r^3\,dr \left[\frac{1}{2}\right] = \frac{1}{2}\int_0^{\pi/2} \cos\theta \left[\frac{1}{4}r^4\right]_1^3 d\theta$$

$$= \frac{1}{2}\cdot\frac{80}{4}\sin\theta\Big]_0^{\pi/2} = 10. \blacksquare$$

If U is a more general bounded region described by cylindrical coordinates, then we extend Definition 5.2 as follows. Let T be a region containing U and bounded by $\theta = \alpha$, $\theta = \beta$, $z = e$, $z = f$, $r = a$, and $r = b$. Define $\bar{g}$ on T by

$$\bar{g}(r, \theta, z) = \begin{cases} g(r, \theta, z) & \text{if } [r, \theta, z] \in U \\ 0 & \text{if } [r, \theta, z] \notin U. \end{cases}$$

Then we define

$$\iiint_U g(r, \theta, z)\, dV = \iiint_T \bar{g}(r, \theta, z)\, dV.$$

Fubini's theorem extends to such regions U. Rather than state all possible forms, we content ourselves with the following representative version. (Compare Theorem 4.2.)

5.4 THEOREM

If U is bounded by $\theta = \alpha$, $\theta = \beta$ (where $\alpha < \beta$), $z = e$, $z = f$ (where $e < f$), $r = h_1(\theta)$, and $r = h_2(\theta)$ (where $h_1(\theta) < h_2(\theta)$ for $\theta \in [\alpha, \beta]$), and if g is continuous on U and its boundary, then

$$\iiint_U g(r, \theta, z)\, dV = \int_\alpha^\beta \int_{h_1(\theta)}^{h_2(\theta)} \int_e^f g(r, \theta, z)\, dz\, r\, dr\, d\theta. \tag{3}$$

As in the case of double integrals, we can change an intractable rectangular triple integral to cylindrical form. If U is a subset of $\mathbf{R}^3$ described by rectangular coordinates, and W is the same region described in cylindrical coordinates, then

$$\iiint_U f(x, y, z)\, dV = \iiint_W f(r\cos\theta, r\sin\theta, z)\, dV = \iiint_W g(r, \theta, z)\, dV$$

where $g(r, \theta, z) = f(x(r, \theta), y(r, \theta), z) = f(r\cos\theta, r\sin\theta, z)$, and on the right $dV = r\, dr\, d\theta\, dz$. The next example shows how to use this result, whose justification again involves more subtlety than we can go into here.

Example 3

Evaluate $\iiint_U \sqrt{x^2 + y^2}\, dV$, where U is the region lying above the xy-plane and below the cone $z = 4 - \sqrt{x^2 + y^2}$ (see **Figure 5.7**).

FIGURE 5.7

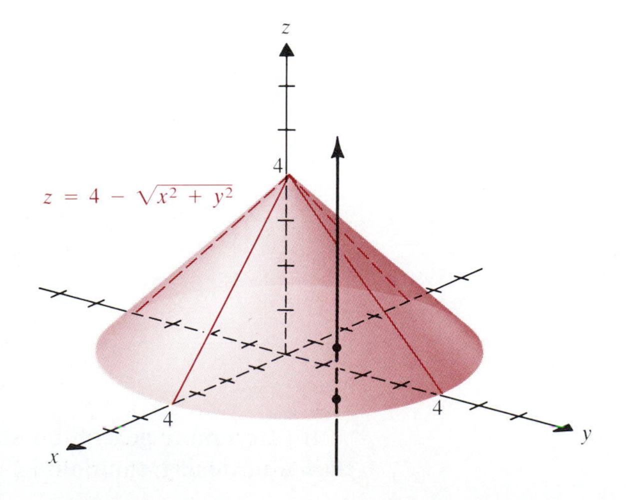

Solution. In cylindrical coordinates, $\sqrt{x^2 + y^2}$ is just r. An arrow shot upward from the xy-plane exits from U by crossing the cone, whose cylindrical coordinate equation is $z = 4 - r$. Since U lies over the disk $x^2 + y^2 \leq 16$ in the xy-plane, we have

$$\iiint_U \sqrt{x^2 + y^2}\, dV = \iiint_U rr\, dr\, d\theta\, dz = \int_0^{2\pi} \int_0^4 \int_0^{4-r} r^2\, dz\, dr\, d\theta$$

$$= \int_0^{2\pi} \int_0^4 r^2(4 - r)\, dr\, d\theta = \int_0^{2\pi} d\theta \int_0^4 (4r^2 - r^3)\, dr$$

$$= \int_0^{2\pi} d\theta \left[\frac{4r^3}{3} - \frac{1}{4}r^4\right]_0^4$$

$$= 2\pi\left[\frac{256}{3} - 64\right] = 2\pi\left[\frac{64}{3}\right] = \frac{128\pi}{3}. \blacksquare$$

Example 4

Find the volume of the region U in Example 3.

Solution. Theorem 4.3 gives

$$V = \iiint_U 1\, dV = \iiint_U r\, dr\, d\theta\, dz = \int_0^{2\pi} \int_0^4 \int_0^{4-r} r\, dz\, dr\, d\theta$$

$$= \int_0^{2\pi} \int_0^4 r(4 - r)\, dr\, d\theta = \int_0^{2\pi} \int_0^4 (4r - r^2)\, dr\, d\theta$$

$$= \int_0^{2\pi} \left[2r^2 - \frac{r^3}{3}\right]_0^4 d\theta = 2\pi\left[32 - \frac{64}{3}\right] = \frac{64\pi}{3}. \blacksquare$$

The formula $V = \frac{1}{3}\pi a^2 h$ for the volume of a right circular cone of base radius a and height h, where $a = 4$ and $h = 4$, gives the same result. (See Exercise 13, where that formula for V is derived.)

Our final example involves the version of (3) needed when the region U lies between $\theta = \alpha$ and $\theta = \beta$ (where $\alpha < \beta$), $r = a$ and $r = b$ (where $a < b$), and surfaces $z = k_1(r, \theta)$ and $z = k_2(r, \theta)$ (where $k_1(r, \theta) < k_2(r, \theta)$):

$$\text{(4)} \qquad \iiint_U g(r, \theta, z)\, dV = \int_\alpha^\beta \int_a^b \int_{k_1(r,\theta)}^{k_2(r,\theta)} g(r, \theta, z)\, dz\, r\, dr\, d\theta.$$

Example 5

Find the volume of the region U between the paraboloids $z = x^2 + y^2$ and $z = 8 - x^2 - y^2$.

Solution. The region U lies inside the shaded surfaces in **Figure 5.8.** An arrow shot upward through U enters at the boundary $z = x^2 + y^2 = r^2$ and exits at the boundary $z = 8 - x^2 - y^2 = 8 - r^2$. Theorem 4.3 thus says that the volume of U is

$$\text{(5)} \qquad V = \iiint_U 1\, dV = \iiint_U dz\, r\, dr\, d\theta = \iint_D \left[\int_{r^2}^{8-r^2} dz\right] r\, dr\, d\theta,$$

where D is the region of the xy-plane obtained by projecting the curve C and its interior in Figure 5.8 downward onto the xy-plane. The bounding equation

FIGURE 5.8

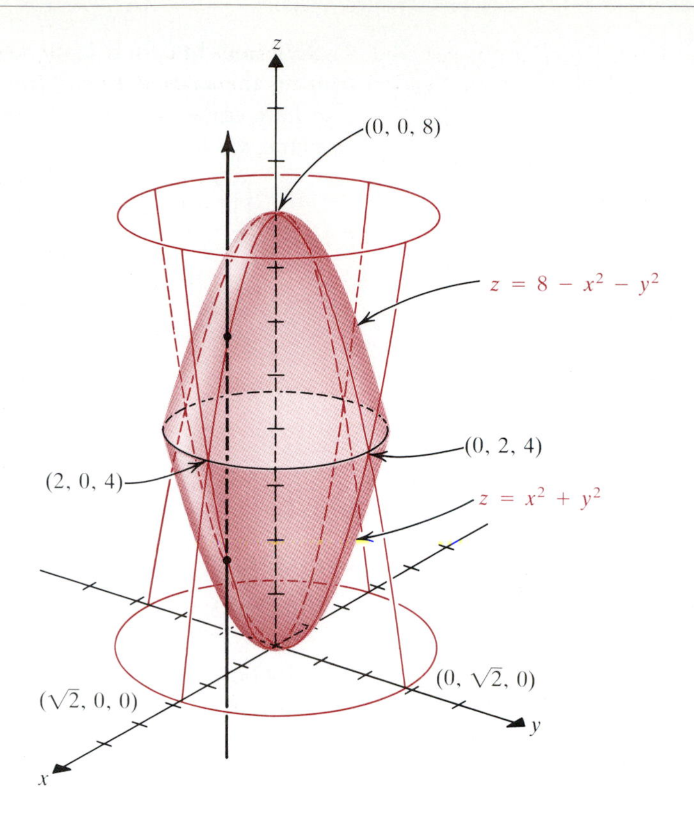

of D can be found by first determining the equation of C. Since C is the curve of intersection of the two paraboloids, its equation is found by equating the respective expressions for z. That gives

$$z = x^2 + y^2 = 8 - x^2 - y^2 \rightarrow 2x^2 + 2y^2 = 8 \rightarrow x^2 + y^2 = 4.$$

Thus C has cylindrical coordinate equations

$$\begin{cases} z = 4 \\ r = 2. \end{cases}$$

Therefore the bounding curve of D in the xy-plane has polar equation $r = 2$. We thus get from (5)

$$V = \int_0^{2\pi} \int_0^2 \int_{r^2}^{8-r^2} r\,dz\,dr\,d\theta = \int_0^{2\pi} \int_0^2 (8 - r^2 - r^2) r\,dr\,d\theta$$

$$= 2 \int_0^{2\pi} d\theta \int_0^2 (4r - r^2)\,dr \qquad \textit{by Corollary 1.10}$$

$$= 4\pi \left[2r^2 - \frac{1}{3} r^3 \right]_0^2 = 4\pi \left[8 - \frac{8}{3} \right] = \frac{64}{3}\pi. \quad \blacksquare$$

Exercises 13.5

In Exercises 1 and 2, find cylindrical coordinates for each of the points with given rectangular coordinates.

1. **(a)** $(1, 2, 3)$ **(b)** $(0, 2, -2)$ **(c)** $(2, 3, -1)$ **(d)** $(-2, -3, 1)$ **(e)** $(2, -3, -1)$

2. **(a)** $(-1, 1, 1)$ **(b)** $(2, -1, 3)$ **(c)** $(2, 0, -3)$ **(d)** $(-2, 3, -1)$ **(e)** $(1, 2, -3)$

In Exercises 3 and 4, find rectangular coordinates for each of the points with given cylindrical coordinates.

3. **(a)** $[2, 2\pi/3, 1]$ **(b)** $[2, \pi/6, 2]$ **(c)** $[1, \pi, 2]$ **(d)** $[2, 7\pi/6, -1]$ **(e)** $[2, 5\pi/3, -1]$

4. **(a)** $[-2, 5\pi/6, 1]$ **(b)** $[-2, \pi/3, 1]$ **(c)** $[-1, \pi, 2]$ **(d)** $[-2, 4\pi/3, -1]$ **(e)** $[-3, 11\pi/6, -2]$

In Exercises 5–8, a rectangular coordinate equation of a surface is given. Obtain a cylindrical coordinate equation of the same surface.

5. **(a)** $x^2 + y^2 = 16$ **(b)** $x^2 - y^2 = 9$
6. **(a)** $4x^2 + y^2 = 16$ **(b)** $x^2 + y^2 = 4z$
7. **(a)** $x^2 + y^2 + 9z^2 = 9$ **(b)** $2x^2 - 4y^2 + 4z^2 = 4$
8. **(a)** $3x^2 - 4y^2 - 12z^2 = 12$ **(b)** $3x^2 + 3y^2 - 5z^2 = 0$

In Exercises 9–12, a cylindrical coordinate equation of a surface is given. Sketch the surface and obtain a rectangular coordinate equation for it.

9. **(a)** $r = 3$ **(b)** $r = 4\cos\theta$ **(c)** $r = 2\csc\theta$
10. **(a)** $r = 4\sin\theta$ **(b)** $r = 2\sec\theta$ **(c)** $r = 2/(1 - \cos\theta)$
11. **(a)** $r = 6/(2 - \cos\theta)$ **(b)** $z = 2r$ **(c)** $r^2 + z^2 = 9$
12. **(a)** $r = 3/(2 + 4\cos\theta)$ **(b)** $z^2 = 4r^2$ **(c)** $r^2 + z^2 = 4$

In Exercises 13–22, find the volume of the given region.

13. The region enclosed by a right circular cone of base radius a and height h
14. The ball enclosed by a sphere of radius a
15. The region bounded by $x^2 + y^2 + z = 4$ and the xy-plane
16. The region bounded by $x^2 + y^2 + z = 9$ and the xy-plane
17. The region inside the sphere $x^2 + y^2 + z^2 = 9$ and outside the cone $z = 3 - \sqrt{x^2 + y^2}$
18. The region lying under the graph of $z = \sqrt{x^2 + y^2}$ and above the unit disk $\{(x, y) | x^2 + y^2 \le 1\}$
19. The solid generated by revolving the region in the xz-plane bounded by $z = 2x^2$, the x-axis, and the line $x = 1$ about the z-axis
20. The solid generated by revolving the region in the xz-plane bounded by $z = x^2$, the x-axis, and the line $x = 2$ about the z-axis
21. The solid generated by revolving the region in the yz-plane bounded by $z = y^2/2$ and the line $y = 2$ about the z-axis
22. The solid generated by revolving the region in the yz-plane bounded by $z = 4y^2$ and the line $y = 1$ about the z-axis

In Exercises 23–30, evaluate the triple integral.

23. $\iiint_U \sqrt{x^2 + y^2}\, z\, dV$, where U is bounded on the left by $y = 0$, below by $z = 0$, and above by $z = 1 - \frac{1}{2}\sqrt{x^2 + y^2}$
24. $\iiint_U \sqrt{x^2 + y^2}\, z\, dV$, where U is bounded on the left by $y = 0$, below by $z = 0$, and above by $z = 1 - \sqrt{x^2 + y^2}$
25. $\iiint_U xyz\, dV$, where U is the region in the first octant bounded by $x^2 + y^2 = a^2$, the xy-plane, and $z = 4$
26. $\iiint_U yz\, dV$, where U is bounded by the planes $z = 0$, and $z = y$, and the cylinder $x^2 + y^2 = 1$
27. $\iiint_U z^4\, dV$, where U is the region bounded by the cones $z = \sqrt{x^2 + y^2}$ and $z = -\sqrt{x^2 + y^2}$, the cylinder $x^2 + y^2 = 1$, in back by $x = 0$, and to the left by $y = 0$
28. $\iiint_U \sqrt{x^2 + y^2}\, dV$, where U is the first-octant region bounded by $x^2 + y^2 = 2x$ and $z^2 = x^2 + y^2$
29. $\iiint_U \sqrt{x^2 + y^2}\, dV$, where U is the region under $z = 4$ and above $z = x^2 + y^2$
30. $\iiint_U z^3\, dV$, where U is the region above the cone $z = \sqrt{x^2 + y^2}$ and below the sphere $x^2 + y^2 + z^2 = 1$

13.6 Spherical Coordinates

The second alternative coordinate system for $\mathbf{R}^3$ is the result of applying the idea behind polar coordinates to three dimensions. We will have more to say in that direction after we have defined the *spherical coordinates* of a point P in $\mathbf{R}^3$.

6.1 DEFINITION The ***spherical coordinates*** of a point $P(x, y, z)$ in $\mathbf{R}^3$ are $\boldsymbol{\rho}$, $\boldsymbol{\phi}$, and $\boldsymbol{\theta}$, where

(1) $$\rho = |\mathbf{0P}| = \sqrt{x^2 + y^2 + z^2},$$

ϕ is the angle between $\mathbf{k}$ and $\mathbf{0P}$, and θ is the polar (or cylindrical) angle between $\mathbf{i}$ and the perpendicular projection $\mathbf{0Q}$ of $\mathbf{0P}$ on the xy-plane, measured counterclockwise. We write $P = P\{\rho, \phi, \theta\}$.

To obtain equations connecting the rectangular and spherical coordinates of P, it is helpful to note that $|\mathbf{0Q}|$ is just the cylindrical coordinate r of P (see

FIGURE 6.1

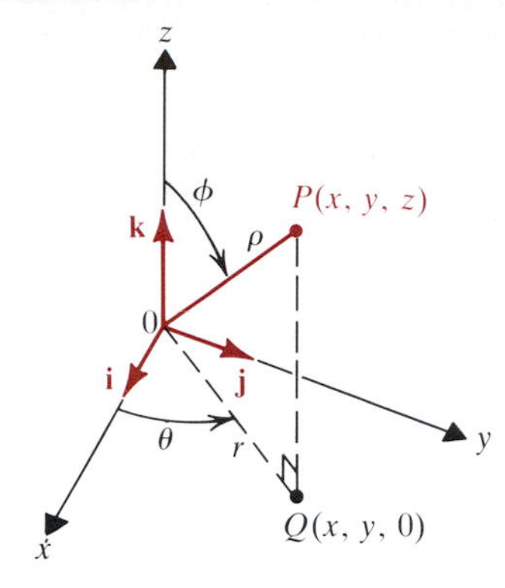

FIGURE 6.2

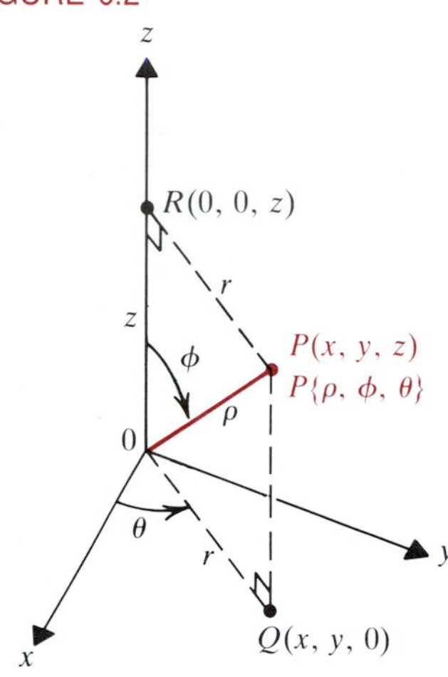

Figure 6.1). Thus Equation (1) on p. 875 gives

$$x = r\cos\theta \quad \text{and} \quad y = r\sin\theta, \tag{2}$$

where $r = \sqrt{x^2 + y^2}$. In **Figure 6.2** triangle $PR0$ has a right angle at R, so

$$\sin\phi = \frac{r}{\rho} \to r = \rho\sin\phi. \tag{3}$$

Substituting this into the Formulas (2) for x and y in terms of θ, we get

$$x = r\cos\theta = \rho\sin\phi\cos\theta \quad \text{and} \quad y = r\sin\theta = \rho\sin\phi\sin\theta.$$

Figure 6.2 also gives

$$\cos\phi = \frac{z}{\rho} \to z = \rho\cos\phi. \tag{4}$$

If $x^2 + y^2 \neq 0$, then we can rewrite (2) as

$$\cos\theta = \frac{x}{r} = \frac{x}{\sqrt{x^2 + y^2}} \quad \text{and} \quad \sin\theta = \frac{y}{r} = \frac{y}{\sqrt{x^2 + y^2}}.$$

Finally, if $\rho \neq 0$, then (3) and (4) say that

$$\cos\phi = \frac{z}{\rho} \quad \text{and} \quad \sin\phi = \frac{\sqrt{x^2 + y^2}}{\rho}.$$

We therefore have the following collection of equations for translating between Cartesian and spherical coordinates.

CHANGE OF COORDINATE EQUATIONS

If P in $\mathbf{R}^3$ has rectangular coordinates (x, y, z) and spherical coordinates $\{\rho, \phi, \theta\}$, then

(1) $\rho = \sqrt{x^2 + y^2 + z^2}$, (5) $x = \rho\sin\phi\cos\theta$,

(6) $y = \rho\sin\phi\sin\theta$, (7) $z = \rho\cos\phi$.

If $\sqrt{x^2 + y^2} \neq 0$ (so that $\rho \neq 0$ also), then

(8) $\cos\theta = \dfrac{x}{\sqrt{x^2 + y^2}}$, (9) $\sin\theta = \dfrac{y}{\sqrt{x^2 + y^2}}$,

(10) $\cos\phi = \dfrac{z}{\rho}$, (11) $\sin\phi = \dfrac{\sqrt{x^2 + y^2}}{\rho}$.

The set of all points in $\mathbf{R}^3$ is not in one-to-one correspondence with the set of all triples $\{\rho, \phi, \theta\}$ of spherical coordinates, even if we limit θ to the interval $[0, 2\pi)$ and ϕ to the interval $[0, \pi]$. As with cylindrical coordinates, there are infinitely many coordinatizations of the origin: $\{0, 0, \theta\}$ or $\{0, \pi, \theta\}$ for any θ.

Spherical coordinates are very nearly those used in navigation, which considers the surface of the earth to be approximately a sphere of radius $\rho_0 \approx$ 4000 miles (6400 km). If the origin is taken to be the center of the earth, then any point P on the surface has spherical coordinates $\{\rho, \phi, \theta\}$, where ρ is nearly

FIGURE 6.3

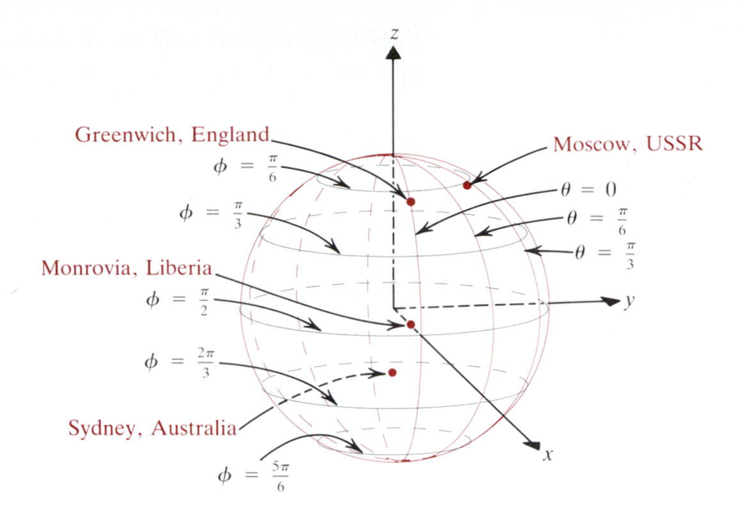

the same for all points. It is thus sufficient to use just $\{\phi, \theta\}$ to locate P. The curves $\theta = c$ are *meridians* (great semicircles passing through the poles), and the curves $\phi = k$ are *parallels* of latitude. See **Figure 6.3.**

In this scheme the *prime meridian* $\theta = 0$ passes through Greenwich, England. Instead of measuring θ from 0 to 2π eastward around the earth, navigators measure θ from 0 to π east from Greenwich and from 0 to π west from Greenwich. Moreover, latitude is not measured simply by ϕ but rather by $\pi/2 - \phi$ for localities north of the equator and by $\phi - \pi/2$ for localities south of the equator. For this reason, ϕ is called the *co-latitude.* Navigators usually measure in degrees rather than radians. Thus Moscow is located roughly at latitude 56°N and longitude 37°E, meaning $\phi \approx 90° - 56° = 34°$, $\theta \approx 37°$. Sydney, Australia is at approximate latitude 33°S and longitude 152°E, meaning $\phi \approx 90° + 33° = 123°$, $\theta \approx 152°$. Monrovia, Liberia is at about longitude 10°W and latitude 8°N, meaning $\theta \approx 350°$, $\phi \approx 82°$. In weather reports the positions of winter storms, hurricanes, and typhoons are usually given in terms of longitude and latitude.

We next examine the simple equations of the form

$$\text{spherical coordinate} = \text{constant.}$$

First, $\rho = a$ is a sphere of radius a centered at the origin. See **Figure 6.4.** Its rectangular coordinate equation is $x^2 + y^2 + z^2 = a^2$. Thus we have the exact

FIGURE 6.4

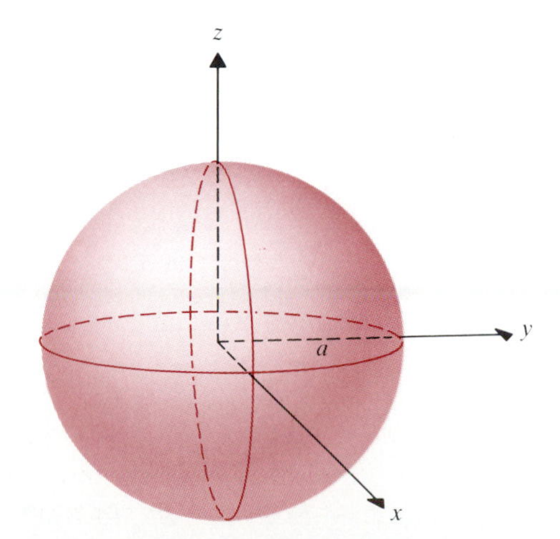

FIGURE 6.5

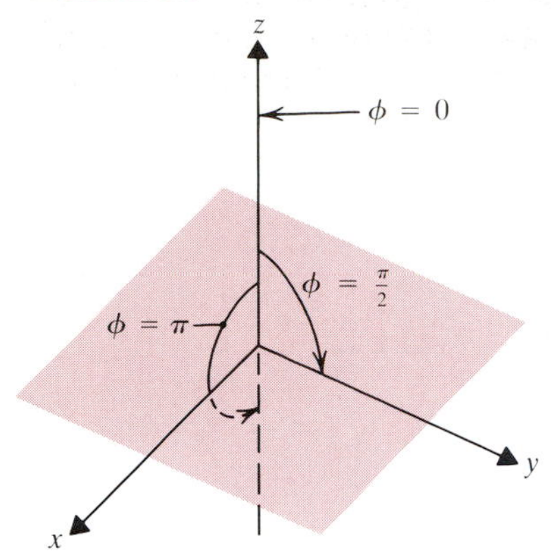

three-dimensional analogue of a polar coordinate equation $r = a$, explaining in part why we said that the spherical coordinate system is the three-dimensional version of the polar coordinate system for $\mathbf{R}^2$. Since the angle θ in spherical coordinates is the same as the angle θ in cylindrical coordinates, and since $\rho \geq 0$, the graph of $\theta = c$ is a half-plane through the z-axis perpendicular to the xy-plane. See Figure 5.3. Finally, we come to $\phi = k$. In this case there is some degeneracy. If $k = 0$, we get the positive z-axis. If $k = \pi$, then we get the negative z-axis. If $k = \pi/2$, we get the xy-plane. (See **Figure 6.5.**) In all other cases we get one nappe of a cone, as shown in **Figures 6.6**(a) and (b). Hence, just as cylindrical coordinates were well suited to both right circular cones and right circular cylinders, so spherical coordinates are appropriate for both right circular cones and spheres.

FIGURE 6.6

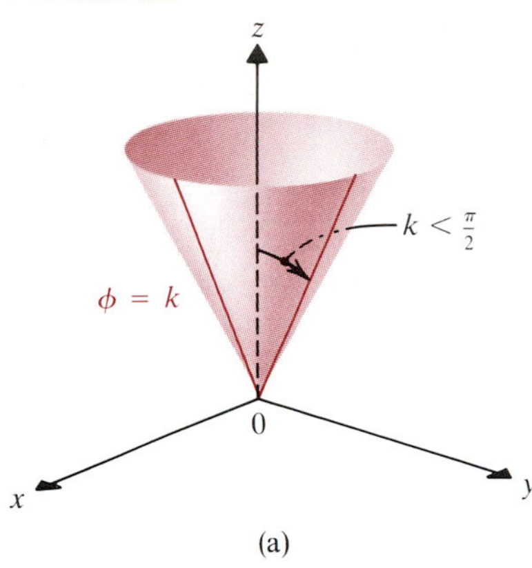

(a)

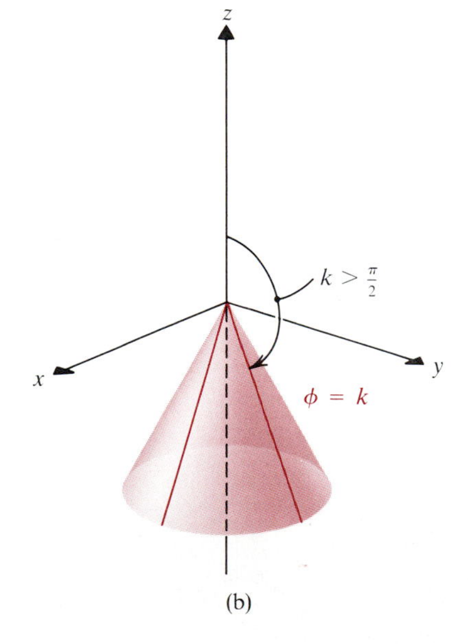

(b)

Example 1

Identify the surface whose spherical coordinate equation is $\rho = 4\cos\phi$.

Solution. This looks reminiscent of the polar equation $r = 4\cos\theta$, which is a circle centered at (2, 0) tangent to the y-axis. We might guess then that this is a sphere of radius 2. A good way to check that guess is to obtain the rectangular equation of the surface. Equation (10) gives

$$\rho = 4\frac{z}{\rho} \to \rho^2 = 4z.$$

We therefore have

$$x^2 + y^2 + z^2 = 4z \to x^2 + y^2 + z^2 - 4z = 0 \to x^2 + y^2 + (z - 2)^2 = 4,$$

which indeed is a sphere of radius 2, centered at (0, 0, 2) and therefore tangent to the xy-plane. It is shown in **Figure 6.7.** (Our reasoning-by-analogy was exactly right in this case. Analogy does have its limits, of course. See Exercise 30.) ■

FIGURE 6.7

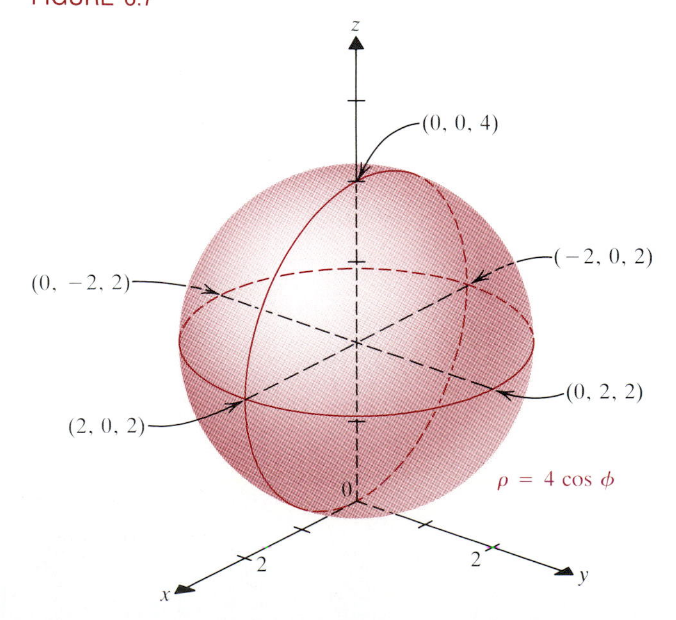

As was the case with cylindrical coordinates, our principal application for spherical coordinates will be to triple integration. Given a region T of $\mathbf{R}^3$ bounded by spheres $\rho = a$ and $\rho = b$, cones $\phi = \alpha$ and $\phi = \beta$, and planes $\theta = \gamma$ and $\theta = \delta$, we thus want to define

$$\iiint_T g(\rho, \phi, \theta)\, dV.$$

This is the limit of Riemann sums

$$\sum_{i=1}^{m} \sum_{j=1}^{n} \sum_{k=1}^{p} g(\bar{\rho}_i, \bar{\phi}_j, \bar{\theta}_k)\, \Delta V_{ijk},$$

where ΔV_{ijk} is the volume of a small region T_{ijk} formed by making a spherical coordinate grid of U. In **Figure 6.8** we try to picture T_{ijk}. It lies between spherical arcs $\rho = \rho_i$ and $\rho = \rho_{i+1} = \rho_i + \Delta\rho_i$, between cones $\phi = \phi_j$ and $\phi = \phi_{j+1} = \phi_j + \Delta\phi_j$, and between planes $\theta = \theta_k$ and $\theta = \theta_{k+1} = \theta_k + \Delta\theta_k$, so it is not easy to draw. You can, however, visualize T_{ijk} as a deformed rectangular box. The base "length" is $\Delta\rho_i$. The base "width" is the length of a circular arc of approximate radius $r = \rho_i \sin\phi_j$ and central angle $\Delta\theta_k$, so it is approximately $\rho_i \sin\phi_j \Delta\theta_k$. The "height" of the battered box is a circular arc of approximate radius ρ_i and central angle $\Delta\phi_j$, so it is approximately $\rho_i \Delta\phi_j$. Thus we can approximate the volume of T_{ijk} by

$$\Delta V_{ijk} \approx (\Delta\rho_i)(\rho_i \sin\phi_j \Delta\theta_k)\rho_i \Delta\phi_j \approx \rho_i^2 \sin\phi_j \Delta\rho_i \Delta\phi_j \Delta\theta_k.$$

This won't change much if we vary ρ_i and ϕ_j slightly. That is,

$$\Delta V_{ijk} \approx \bar{\rho}_i^2 \sin\bar{\phi}_j \Delta\rho_i \Delta\phi_j \Delta\theta_k.$$

Thus the following definition seems reasonable.

FIGURE 6.8

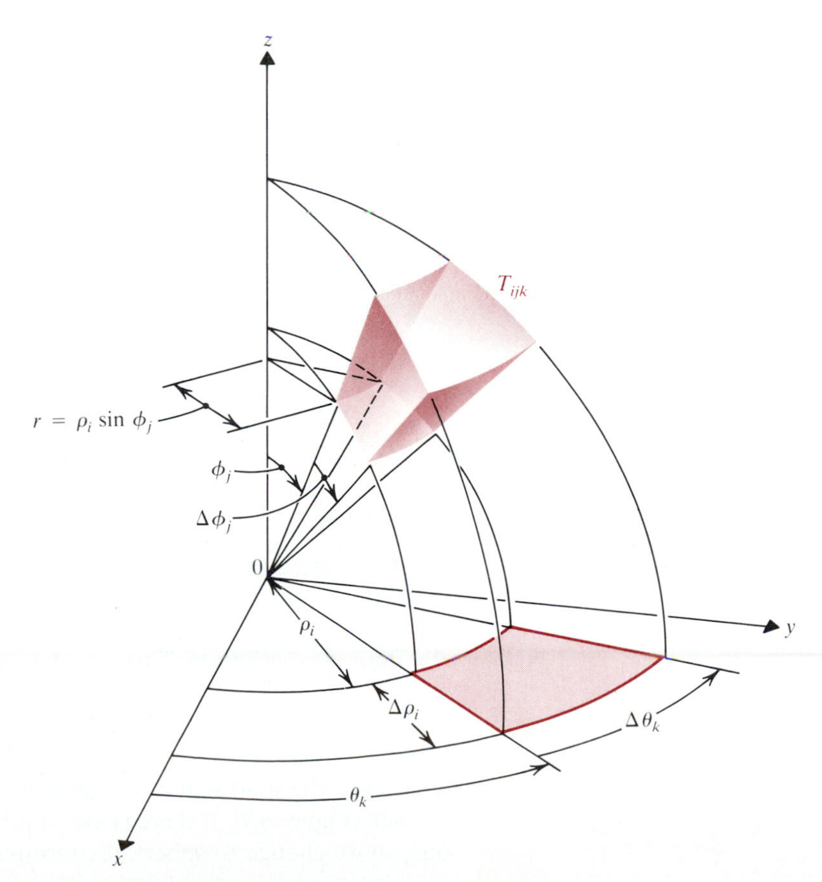

6.2 DEFINITION Let $T = \{\{\rho, \phi, \theta\} | a \le \rho \le b,\ \alpha \le \phi \le \beta,\ \gamma \le \theta \le \delta\} \subseteq \mathbf{R}^3$ be a region on which $g(\rho, \phi, \theta)$ is defined. Then the ***spherical triple integral*** of g over T is

$$\iiint_T g(\rho, \phi, \theta)\, dV = \iiint_T g(\rho, \phi, \theta)\rho^2 \sin\phi\, d\rho\, d\phi\, d\theta$$

$$= \lim_{|G| \to 0} \sum_{i=1}^{m} \sum_{j=1}^{n} \sum_{k=1}^{p} g(\bar{\rho}_i, \bar{\phi}_j, \bar{\theta}_k)\bar{\rho}_i{}^2 \sin\phi_j \Delta\rho_i \Delta\phi_j \Delta\theta_k,$$

where the limit is taken in the same sense as in Definition 4.1.

As before, we can extend the definition to an arbitrary bounded region U by enclosing U in a basic spherical coordinate region T of Definition 6.2. Let $\bar{g}$ be defined on T by

$$\bar{g}(\rho, \phi, \theta) = \begin{cases} 0 & \text{if } \{\rho, \phi, \theta\} \notin U \\ g(\rho, \phi, \theta) & \text{if } \{\rho, \phi, \theta\} \in U. \end{cases}$$

Then we define

$$\iiint_U g(\rho, \phi, \theta)\, dV = \iiint_T \bar{g}(\rho, \phi, \theta)\, dV.$$

Again, Definition 6.3 can be shown to be consistent with Definitions 4.1 and 5.2. Note that in *spherical coordinates*,

$$dV = \rho^2 \sin\phi\, d\rho\, d\phi\, d\theta, \qquad \textbf{NOT} \qquad d\rho\, d\phi\, d\theta.$$

There is a version of Fubini's theorem that permits spherical triple integrals to be evaluated by iteration, without regard to the order of iteration. The following version covers most cases of interest.

6.3 THEOREM

If T is as in Definition 6.2, and g is continuous on T and its boundary, then

$$\iiint_T g(\rho, \phi, \theta)\, dV = \int_\gamma^\delta \int_a^b \int_\alpha^\beta g(\rho, \phi, \theta)\rho^2 \sin\phi\, d\phi\, d\rho\, d\theta.$$

If U is bounded by cones $\phi = \alpha$ and $\phi = \beta$ (where $\alpha \le \beta$), planes $\theta = \gamma$ and $\theta = \delta$ (where $\gamma \le \delta$), and surfaces $\rho = h_1(\phi)$ and $\rho = h_2(\phi)$ where $h_1(\phi) \le h_2(\phi)$, then

$$\iiint_U g(\rho, \phi, \theta)\, dV = \int_\gamma^\delta \int_\alpha^\beta \int_{h_1(\phi)}^{h_2(\phi)} g(\rho, \phi, \theta)\rho^2 \sin\phi\, d\rho\, d\phi\, d\theta.$$

FIGURE 6.9

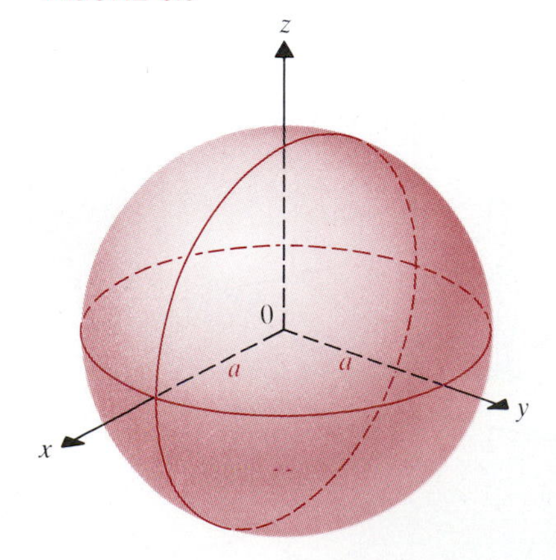

Example 2

Use triple integration to derive the formula for the volume enclosed by a sphere of radius a.

Solution. According to Theorem 4.3 and Definition 6.2, we have

$$V = \iiint_B 1\, dV = \iiint_B \rho^2 \sin\phi\, d\rho\, d\phi\, d\theta,$$

where B is the solid ball of radius a. We can set up coordinates at the center of the ball (**Figure 6.9**). If there is ever a job for spherical coordinates, this must be one, so we change to spherical coordinates to evaluate the integral. The ball is described in spherical coordinates by the conditions $0 \le \rho \le a$, $0 \le \phi \le \pi$, and $0 \le \theta \le 2\pi$. Hence

$$V = \int_0^{2\pi} \int_0^{\pi} \int_0^{a} \rho^2 \sin\phi \, d\rho \, d\phi \, d\theta$$

$$= \int_0^{2\pi} \int_0^{\pi} \sin\phi \left[\frac{\rho^3}{3}\right]_0^a d\phi \, d\theta = \frac{a^3}{3} \int_0^{2\pi} \int_0^{\pi} \sin\phi \, d\phi \, d\theta$$

$$= \frac{a^3}{3} \cdot 2\pi \cdot \Big[-\cos\phi\Big]_0^{\pi} = \frac{2\pi a^3}{3}[1+1] = \frac{4}{3}\pi a^3. \quad \blacksquare$$

As in the case of cylindrical coordinates, we can change an intractable rectangular triple integral $\iiint_U f(x, y, z)\, dV$ to a spherical integral. If U is described by rectangular coordinates and W is the same region described in spherical coordinates, then

$$\text{(12)} \qquad \iiint_U f(x, y, z)\, dV = \iiint_W f(x(\rho, \phi, \theta), y(\rho, \phi, \theta), z(\rho, \phi, \theta))\, dV$$

$$= \iiint_W f(\rho \sin\phi \cos\theta, \rho \sin\phi \sin\theta, \rho\cos\phi)\, dV,$$

where in the last two integrals $dV = \rho^2 \sin\phi \, d\rho \, d\phi \, d\theta$. The equality of the integrals in (12) is proved in advanced calculus. The next example illustrates how (12) is used to simplify stubborn rectangular triple integrals.

Example 3

Evaluate $\iiint_C \sqrt{x^2 + y^2 + z^2}\, dV$, where C is the "ice cream cone," $C = \{(x, y, z) \mid x^2 + y^2 + z^2 \le 1,\ x^2 + y^2 \le z^2/3,\ z \ge 0\}$.

Solution. To use spherical coordinates on the given triple integral, we have to give a spherical coordinate description of C. First, $z \ge 0$ corresponds to $\phi \le \pi/2$, and $x^2 + y^2 + z^2 \le 1$ corresponds to $\rho \le 1$. The bounding surface is

$$x^2 + y^2 = \frac{z^2}{3} \to r^2 = \frac{\rho^2 \cos^2\phi}{3},$$

where $r = \rho \sin\phi$, so $\rho^2 \sin^2\phi = (\rho^2 \cos^2\phi)/3$. Since $\phi \le \pi/2$, this simplifies to

$$\tan^2\phi = \frac{1}{3} \to \tan\phi = \frac{1}{\sqrt{3}}.$$

FIGURE 6.10

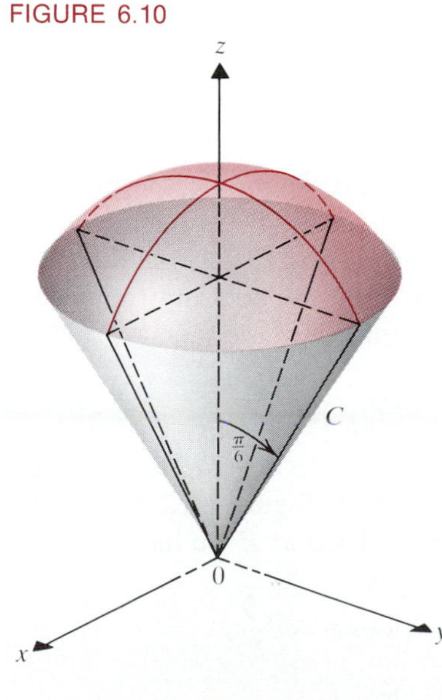

Thus $\phi = \pi/6$ (see **Figure 6.10**). Since $\sqrt{x^2 + y^2 + z^2} = \rho$, the given integral is $\iiint_C \rho\, dV$. The conditions in color say that on C, the angle ϕ varies over $[0, \pi/6]$, and ρ varies over $[0, 1]$. Since θ varies over $[0, 2\pi]$, we have

$$\iiint_C \rho\, dV = \iiint_C \rho^3 \sin\phi \, d\rho\, d\phi\, d\theta = \int_0^{2\pi} \int_0^{\pi/6} \int_0^1 \rho^3 \sin\phi \, d\rho \, d\phi \, d\theta$$

$$= \int_0^{2\pi} d\theta \int_0^{\pi/6} \sin\phi \, d\phi \int_0^1 \rho^3 \, d\rho \qquad \textit{by the analogue of Corollary 1.10}$$

$$= 2\pi \Big[-\cos\phi\Big]_0^{\pi/6} \cdot \left[\frac{1}{4}\rho^4\right]_0^1$$

$$= 2\pi \cdot \frac{1}{4}\left[-\frac{1}{2}\sqrt{3} + 1\right] = \frac{1}{2}\pi\left[1 - \frac{1}{2}\sqrt{3}\right]. \quad \blacksquare$$

We were fortunate in Example 3 to wind up with constant limits of integration. Our concluding example involves a variable limit in one of the iterated integrals.

Example 4

Find the volume of the region E inside the sphere $\rho = 4\cos\phi$ of Example 1 and above the plane $z = 3$.

Solution. The region E is shown in **Figure 6.11.** Theorem 4.3 and Definition 6.2 give

$$V = \iiint_E 1\,dV = \iiint_E \rho^2 \sin\phi\,d\rho\,d\phi\,d\theta.$$

FIGURE 6.11

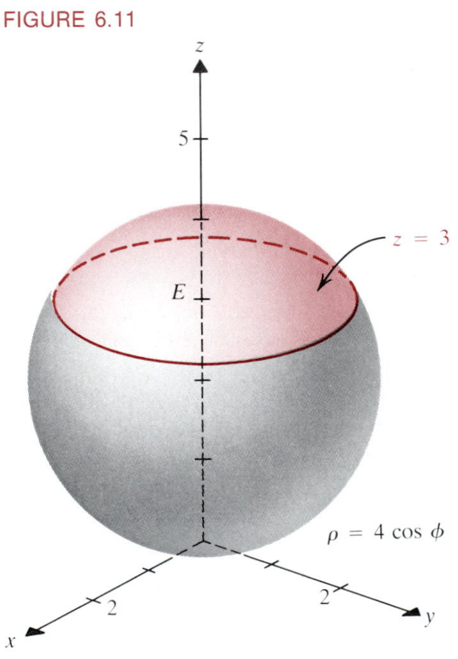

FIGURE 6.12

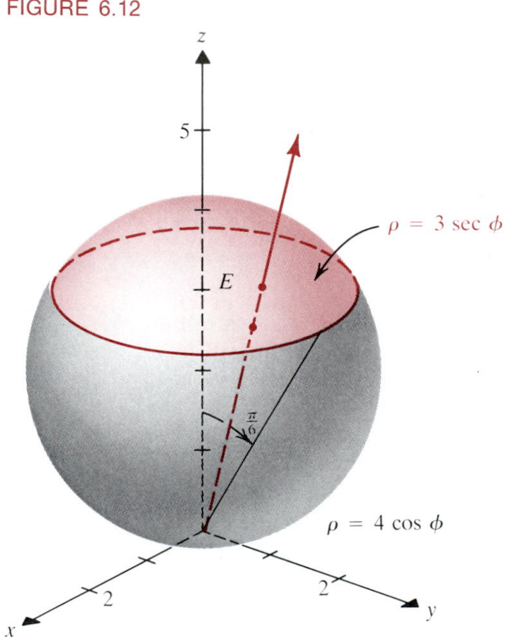

To evaluate this triple integral by iteration, we must determine the limits of integration. Clearly, θ ranges from 0 to 2π. To find the limits on ρ, imagine an arrow shot outward from the origin toward E. As **Figure 6.12** shows, the arrow enters E by crossing the plane $z = 3$ and exits from E by crossing the sphere $\rho = 4\cos\phi$. From (10), $z = 3$ has spherical coordinate equation

$$\rho\cos\phi = 3 \rightarrow \rho = 3\sec\phi.$$

Thus ρ ranges from $3\sec\phi$ to $4\cos\phi$. Lastly, we need to determine the limits on ϕ. It ranges from 0 to its value at the point P in Figure 6.12, where the plane $\rho = 3\sec\phi$ intersects the sphere $\rho = 4\cos\phi$. Thus at P we have

$$3\sec\phi = 4\cos\phi \rightarrow \frac{3}{4} = \cos^2\phi \rightarrow \cos\phi = \frac{\sqrt{3}}{2},$$

since ϕ is a first-quadrant angle. Hence $\phi = \cos^{-1}(\sqrt{3}/2) = \pi/6$. Therefore

$$V = \iiint_E \rho^2 \sin\phi \, d\rho \, d\phi \, d\theta = \int_0^{\pi/6} \int_{3\sec\phi}^{4\cos\phi} \int_0^{2\pi} \rho^2 \sin\phi \, d\theta \, d\rho \, d\phi$$

$$= 2\pi \int_0^{\pi/6} \int_{3\sec\phi}^{4\cos\phi} \rho^2 \sin\phi \, d\rho \, d\phi = \frac{2\pi}{3} \int_0^{\pi/6} \sin\phi \left[\rho^3\right]_{3\sec\phi}^{4\cos\phi} d\phi$$

$$= \frac{2\pi}{3} \int_0^{\pi/6} (64\cos^3\phi \sin\phi - 27\sec^3\phi \sin\phi) \, d\phi$$

$$= \frac{2\pi}{3} \int_0^{\pi/6} (64\cos^3\phi \sin\phi - 27\tan\phi \sec^2\phi) \, d\phi$$

$$= \frac{2\pi}{3}\left[-16\cos^4\phi - 27 \cdot \frac{1}{2}\tan^2\phi\right]_0^{\pi/6}$$

$$= \frac{2\pi}{3}\left[-16 \cdot \frac{9}{16} - \frac{27}{2} \cdot \frac{1}{3} + 16 + 0\right] = \frac{2\pi}{3}\left[7 - \frac{9}{2}\right]$$

$$= \frac{5\pi}{3}. \quad \blacksquare$$

Exercises 13.6

In Exercises 1 and 2, find spherical coordinates for each of the points with the given rectangular coordinates.

1. **(a)** $(1, 2, 3)$ **(b)** $(-1, 1, 1)$ **(c)** $(0, 2, -2)$ **(d)** $(2, 0, -3)$ **(e)** $(-2, 3, -1)$

2. **(a)** $(2, -1, 3)$ **(b)** $(2, 3, -1)$ **(c)** $(-2, -3, 1)$ **(d)** $(2, -3, -1)$ **(e)** $(3, -1, -2)$

In Exercises 3 and 4, find rectangular coordinates for each of the points with the given spherical coordinates.

3. **(a)** $\{3, \pi/6, \pi/4\}$ **(b)** $\{2, \pi/2, \pi/3\}$ **(c)** $\{2, 3\pi/4, \pi/6\}$ **(d)** $\{2, 3\pi/4, 7\pi/6\}$ **(e)** $\{2, \pi, 3\pi/2\}$

4. **(a)** $\{2, \pi/3, \pi/2\}$ **(b)** $\{3, \pi/2, \pi/2\}$ **(c)** $\{2, \pi/6, 3\pi/4\}$ **(d)** $\{2, 3\pi/4, 5\pi/3\}$ **(e)** $\{2, \pi/4, \pi/6\}$

In Exercises 5 and 6, a rectangular coordinate equation of a surface is given. Obtain a spherical coordinate equation for the surface.

5. **(a)** $x^2 + y^2 + z^2 = 9$ **(b)** $x^2 + y^2 + z^2 - 6z = 0$ **(c)** $x^2 + y^2 = 16$

6. **(a)** $2x^2 + 2y^2 + z^2 - 6z = 0$ **(b)** $x^2 + y^2 - 3z^2 = 0$ **(c)** $x^2 + y^2 = z$

In Exercises 7 and 8, a spherical coordinate equation of a surface is given. Sketch the surface and obtain a rectangular coordinate equation for it.

7. **(a)** $\rho = 5$ **(b)** $\theta = \pi/6$ **(c)** $\rho = 8\cos\phi$

8. **(a)** $\phi = \pi/4$ **(b)** $\rho = 8\sec\phi$ **(c)** $\rho\sin\phi = 8$

In Exercises 9–16, use spherical coordinates to find the volume of the given region.

9. A solid right circular cylinder of radius a and height h. (*Hint:* See Exercises 5(c) and 8(c).)

10. A solid right circular cone of base radius a and height h

11. The "ice cream cone" in Example 3

12. The region described by $x^2 + y^2 + z^2 \leq 1$, $x^2 + y^2 \leq 3z^2$, and $z \geq 0$

13. The region lying between the spheres $x^2 + y^2 + z^2 = 1$ and $x^2 + y^2 + z^2 = 4$

14. The region bounded above by the sphere $\rho = 2$, below by the xy-plane, and on the sides by $x^2 + y^2 = 1$. (*Hint:* You can save some work by using the formula from Exercise 9.)

15. The region enclosed by the surface $\rho = 1 - \cos\phi$. Sketch it. (The region can be obtained by revolving the region bounded by the cardioid $r = 1 - \cos\theta$ about the polar (z) axis in the zx-plane.)

16. The region enclosed by the surface $\rho = a(1 - \cos\phi)$

In Exercises 17–24, evaluate the triple integral.

17. $\iiint_B \sqrt{x^2 + y^2 + z^2} \, dV$, where B is the ball of radius 1 centered at the origin

18. $\iiint_B z \, dV$, where B is the first-octant part of the ball in Exercise 17

19. $\displaystyle\iiint_A \frac{1}{x^2 + y^2 + z^2} \, dV$, where A is the region between the spheres $x^2 + y^2 + z^2 = 1$ and $x^2 + y^2 + z^2 = 4$

20. $\displaystyle\iiint_A \frac{1}{x^2 + y^2 + z^2}\,dV$, where A is the region between the spheres $x^2 + y^2 + z^2 = 4$ and $x^2 + y^2 + z^2 = 9$

21. $\iiint_B (x^2 + y^2)\,dV$, where B is the region of Exercise 18

22. $\iiint_U \sqrt{x^2 + y^2 + z^2}\,dV$, where U is the region enclosed by $\rho = 1 - \cos\phi$. (See Exercise 15.)

23. $\displaystyle\iiint_E \frac{1}{\sqrt{x^2 + y^2 + z^2}}\,dV$, where E is the region of Example 4

24. $\displaystyle\iiint_E \frac{1}{\sqrt{x^2 + y^2 + z^2}}\,dV$, where E is the region above the plane $z = 2$ and below the sphere $\rho = 4\cos\phi$

25. If a curve in $\mathbf{R}^3$ is parametrized in spherical coordinates by $\rho = \rho(t)$, $\phi = \phi(t)$, and $\theta = \theta(t)$, then use the chain rule to show that its arc length between $t = a$ and $t = b$ is

$$L = \int_a^b \sqrt{\dot{\rho}^2 + (\rho^2 \sin^2 \phi)\dot{\theta}^2 + \rho^2\dot{\phi}^2}\,dt,$$

where the raised dots indicate differentiation with respect to t: $\dot{\rho} = d\rho/dt$, $\dot{\theta} = d\theta/dt$, and $\dot{\phi} = d\phi/dt$. (*Hint:* Use the arc length formula in rectangular coordinates.)

26. A submarine sails due north from the equator in the mid-Pacific to the north pole. Find the length of its journey. (THINK before resorting to Exercise 25.)

27. A submarine sails due south from a point on the equator in the Indian Ocean. Find the length it must travel to reach the south pole. (See Exercise 26.)

28. A *conical helix* is given parametrically by $\rho = t$, $\phi = \pi/4$, and $\theta = t$. (This curve winds around the right circular cone $\phi = \pi/4$ as it ascends.) Find its arc length from $t = 0$ to $t = \pi$. (See Exercise 25.)

29. Find the arc length of the curve on the surface of a right circular cone given by $\rho = t^2$, $\phi = \pi/6$, and $\theta = t$ between $t = 0$ and $t = 1$. (See Exercise 25.)

30. The reasoning-by-analogy used in Example 2 would suggest that the graph of $\rho = a\sin\phi$ should *also* be a sphere of radius $a/2$. Is it? Draw the graph. Is this related in *any* way to the graph of $r = a\sin\theta$ in polar coordinates?

31. Compute $\iiint_E z\,dV$, where E is the region enclosed by the graph of $\rho = a\sin\phi$.

32. Compute $\iiint_E \sqrt{x^2 + y^2 + z^2}\,dV$, where E is as in Exercise 31.

13.7 Applications of Multiple Integration

In earlier chapters we saw a number of geometrical and physical applications of the ordinary definite integral $\int_a^b f(x)\,dx$ of a function $f\colon \mathbf{R} \to \mathbf{R}$. We have already used double and triple integrals to calculate areas and volumes, and in this section we show how multiple integration is used to calculate additional quantities of interest in mathematics and physics.

First we consider the *mass* of a nonhomogeneous plane lamina, a sheet of material so thin that we neglect its thickness and regard its (variable) density $\delta(x, y)$ as giving its mass per unit area. Suppose that the lamina occupies a bounded region D in $\mathbf{R}^2$. To define its mass, we subdivide the region D by means of a grid (as in **Figure 7.1**) of a rectangle $R = [a, b] \times [c, d]$ containing D. If the grid is fine, and the density function δ is continuous, then on a small rectangle R_{ij} of the grid, $\delta(x, y)$ is nearly constant. Hence the mass of material in this subrectangle will be approximately

$$\delta(\bar{x}_i, \bar{y}_j)\,\Delta x_i\,\Delta y_j,$$

where $(\bar{x}_i, \bar{y}_j)$ is any point in R_{ij}, and as usual,

$$\Delta x_i = x_i - x_{i-1} \qquad \text{and} \qquad \Delta y_j = y_j - y_{j-1}$$

are the dimensions of R_{ij}. Consequently the mass of the lamina is approximated by the Riemann sum

$$R_G(f) = \sum_{i=1}^{m} \sum_{j=1}^{n} \delta(\bar{x}_i, \bar{y}_j)\,\Delta x_i\,\Delta y_j.$$

FIGURE 7.1

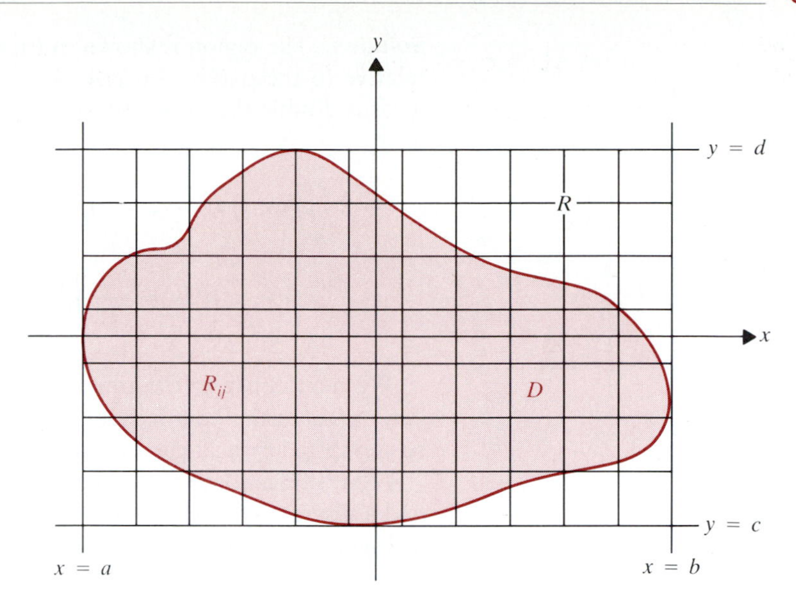

This approximation stands to get better as we make the grid G finer, and so we make the following definition.

7.1 DEFINITION If a lamina with a continuous density function δ occupies the region $D \subseteq \mathbf{R}^2$, then the ***mass*** of the lamina is

$$m = \iint_D \delta(x, y)\, dA.$$

FIGURE 7.2

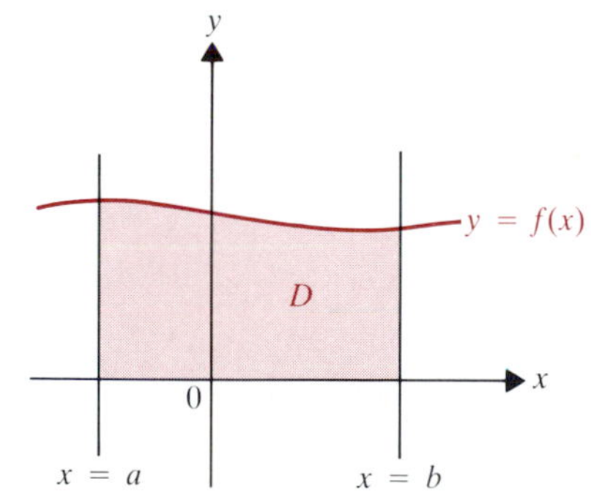

This definition is consistent with Definition 5.1 of Chapter 5, which defined the mass of a plane lamina of constant density covering the region D pictured in **Figure 7.2.** That region lies under the graph of $y = f(x)$ and above the x-axis between $x = a$ and $x = b$.

Example 1

Show that the mass of the lamina of constant density $\delta(x, y) = k$ in Figure 7.2 as computed from Definition 7.1 is the same as the mass computed from Definition 5.1 of Chapter 5.

Solution. Definition 7.1 gives

$$m = \iint_D \delta(x, y)\, dA = k \iint_D dA = k \int_a^b \int_0^{f(x)} dy\, dx = k \int_a^b f(x)\, dx,$$

which is the same as the expression for m from Definition 5.1 of Chapter 5 [see Equation (19), p. 315]. ■

The definition of mass in Chapter 5 was given only for laminas of constant density. The greater generality of Definition 7.1 is indicated by the following example.

Example 2

A lamina occupies the region bounded by $y = 1 - x^2$ and the x-axis. Its density at each point (x, y) is $\delta(x, y) = |xy|$. Find the mass.

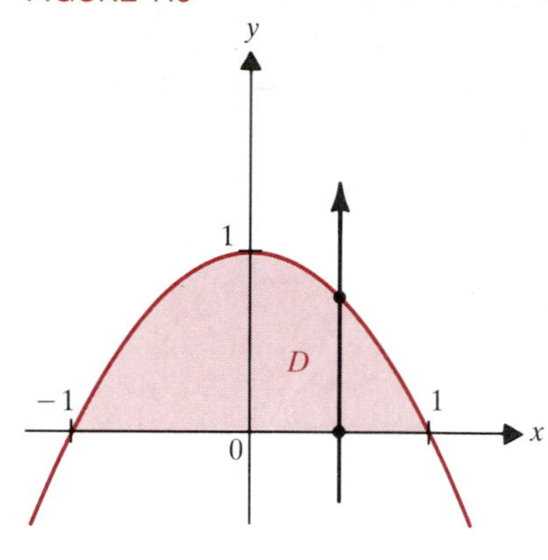

FIGURE 7.3

Solution. The region is shown in **Figure 7.3.** The density function is symmetric relative to the y-axis: $\delta(x, y) = \delta(-x, y) = |x|y$ for any point $(x, y) \in D$. So the mass is double the mass of the first-quadrant portion of the lamina. Hence we have

$$m = 2\iint_D \delta(x, y)\,dA = 2\int_0^1 \int_0^{1-x^2} xy\,dy\,dx = 2\int_0^1 x\left[\frac{1}{2}y^2\right]_{y=0}^{y=1-x^2} dx$$

$$= \int_0^1 x(1-x^2)^2\,dx = -\frac{1}{2}\int_0^1 -2x(1-x^2)^2\,dx = -\frac{1}{2}\frac{(1-x^2)^3}{3}\Bigg]_0^1 = \frac{1}{6}. \quad ■$$

We next consider the center of mass $C(\bar{x}, \bar{y})$ of a nonhomogeneous plane lamina. As in the homogeneous case, this is the point at which the lamina can be supported by a single pin placed under C. The approach used to define $\bar{x}$ and $\bar{y}$ is that of Section 5.5. Suppose for the moment that, instead of a lamina, we had a discrete system of point masses at the points $(x_1, y_1), (x_2, y_2), \ldots, (x_k, y_k)$ in **Figure 7.4,** where at the point (x_i, y_i) there is a mass m_i. Recall from p. 314 that the center of mass of such a system is simply $(\bar{x}, \bar{y})$, where $\bar{x}$ is the weighted average of the x-coordinates of the point masses and $\bar{y}$ is the weighted average of their y-coordinates. Thus

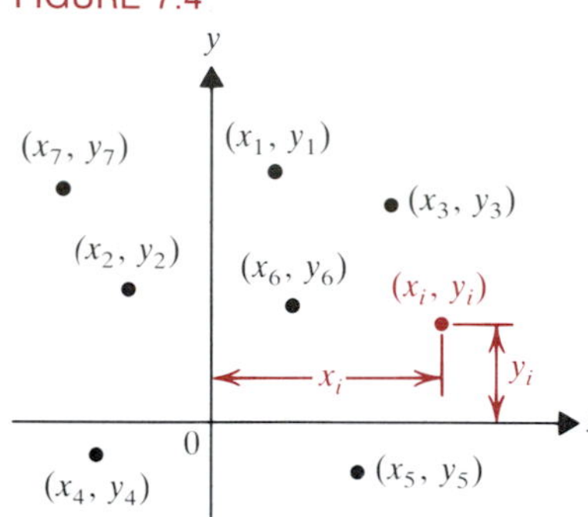

FIGURE 7.4

$$\bar{x} = \frac{\sum_{i=1}^k m_i x_i}{\sum_{i=1}^k m_i} \quad \text{and} \quad \bar{y} = \frac{\sum_{i=1}^k m_i y_i}{\sum_{i=1}^k m_i}.$$

While we don't have such a system, we can *approximate* our lamina in this way; namely, if we finely partition the domain D, as in Figure 7.1, then we regard the entire mass of each subrectangle R_{ij} as concentrated at the point $(\bar{x}_i, \bar{y}_j)$. This gives a discrete system close to the situation in the lamina. Then an approximation to the coordinates $(\bar{x}, \bar{y})$ of the center of mass of the lamina is given by the weighted averages of the $\bar{x}_i$ and $\bar{y}_j$, respectively. Now, we saw above that the mass of each subrectangle is approximately $\delta(\bar{x}_i, \bar{y}_j)\Delta A_{ij}$ where $\Delta A_{ij} = A(R_{ij})$. We thus have

$$\bar{x} \approx \frac{\sum_{i=1}^m \sum_{j=1}^n \bar{x}_i \delta(\bar{x}_i, \bar{y}_j)\Delta A_{ij}}{\sum_{i=1}^m \sum_{j=1}^n \delta(\bar{x}_i, \bar{y}_j)\Delta A_{ij}} \quad \text{and} \quad \bar{y} \approx \frac{\sum_{i=1}^m \sum_{j=1}^n \bar{y}_i \delta(\bar{x}_i, \bar{y}_j)\Delta A_{ij}}{\sum_{i=1}^m \sum_{j=1}^n \delta(\bar{x}_i, \bar{y}_j)\Delta A_{ij}}.$$

As the grid gets finer and finer, these approximations should approach the exact coordinates of the center of mass. Thus the following definition is natural.

7.2 DEFINITION

If a lamina occupies the region $D \subseteq \mathbf{R}^2$ and has a continuous density function $\delta(x, y)$, then the *center of mass* of the lamina is the point $(\bar{x}, \bar{y})$ where

$$\bar{x} = \frac{\iint_D x\,\delta(x, y)\,dA}{m} \quad \text{and} \quad \bar{y} = \frac{\iint_D y\,\delta(x, y)\,dA}{m}.$$

Here the numerators are called the *moments of the lamina with respect to the coordinate axes:*

$$M_y = \iint_D x\,\delta(x, y)\,dA \quad \text{is the } \textit{moment with respect to the y-axis},$$

$$M_x = \iint_D y\,\delta(x, y)\,dA \quad \text{is the } \textit{moment with respect to the x-axis}.$$

Be careful not to be confused by the notation. The formula for M_x has a y in the integral (*not* x), and the formula for M_y has an x in the integral (*not* y). The reason is that x measures the distance from the *y-axis* to (x, y), and y measures the distance from the *x-axis* to (x, y).

Example 3

Find the center of mass of the lamina in Example 2.

Solution. We have already calculated $m = 1/6$. So we need only M_x and M_y. Letting D^+ stand for the first-quadrant part of D, we have

$$M_x = \iint_D y\,\delta(x, y)\,dA = 2\iint_{D^+} xy^2\,dA = 2\int_0^1\int_0^{1-x^2} xy^2\,dy\,dx$$

$$= \frac{2}{3}\int_0^1 x\left[y^3\right]_0^{1-x^2} dx = \frac{2}{3}\int_0^1 x(1-x^2)^3\,dx = -\frac{1}{2}\cdot\frac{2}{3}\int_0^1 -2x(1-x^2)^3\,dx$$

$$= -\frac{1}{3}\frac{(1-x^2)^4}{4}\Bigg]_0^1 = \frac{1}{12}.$$

Thus

$$\bar{y} = \frac{M_x}{m} = \frac{\dfrac{1}{12}}{\dfrac{1}{6}} = \frac{1}{2}.$$

Similarly,

$$M_y = \iint_D x\,\delta(x, y)\,dA = \iint_D x|x|y\,dA.$$

Now

$$x|x|y = x^2 y \quad \text{if } x \geq 0 \qquad \text{and} \qquad x|x|y = -x^2 y \quad \text{if } x < 0,$$

so the integrand is an odd function of x. Since the region D is symmetric in the y-axis, $M_y = \iint_D x|x|y\,dA = 0$ by Exercise 41(a) of Section 2. Thus $(\bar{x}, \bar{y}) = (0, 1/2)$. ■

If the density function is a constant δ, then Definitions 7.1 and 7.2 simplify to

(1) $$M = \iint_D \delta\,dA = \delta A(D),$$

(2) $$M_x = \iint_D \delta y\,dA = \delta\iint_D y\,dA,$$

and

(3) $$M_y = \iint_D \delta x\,dA = \delta\iint_D x\,dA.$$

Recall from Section 5.5 that a lamina with constant density is called *homogeneous*, and its center of mass is referred to as its *centroid.*

Polar integration can often help reduce the complexity of center-of-mass problems. The next example illustrates this for a homogeneous lamina.

FIGURE 7.5

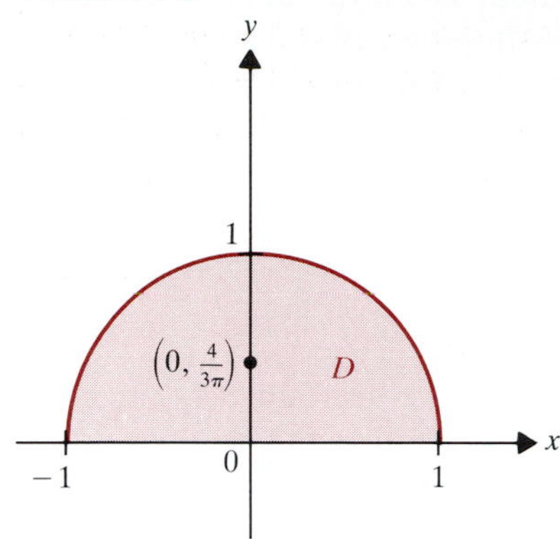

Example 4

Find the centroid of a lamina occupying the upper half of the unit disk $D = \{(x, y) | x^2 + y^2 \le 1\}$. See **Figure 7.5.**

Solution. If δ is the density, then $m = \delta A(D) = \delta\pi/2$. By the symmetry of D in the y-axis, it is clear that $\bar{x} = 0$. From Definition 7.2 and the polar coordinate equations $y = r\sin\theta$ and $dA = r\,dr\,d\theta$, we get

$$M_x = \iint_D y\,\delta\,dA = \delta \iint_D r\sin\theta\, r\,dr\,d\theta$$

$$= \delta \int_0^1 r^2\,dr \int_0^\pi \sin\theta\,d\theta \qquad \textit{by Corollary 1.10}$$

$$= \delta \int_0^1 r^2 \Big[-\cos\theta\Big]_0^\pi dr = 2\delta \frac{1}{3} r^3 \Big]_0^1 = \frac{2\delta}{3}.$$

Hence

$$\bar{y} = \frac{M_x}{m} = \frac{\frac{2\delta}{3}}{\delta\frac{\pi}{2}} = \frac{4}{3\pi}.$$

So the centroid is $(\bar{x}, \bar{y}) = (0, 4/3\pi)$. ■

The concepts of mass and moments carry over easily to regions E in R^3. In fact, they seem more appropriate there, because the density function measures mass per unit volume. Rather than repeat in three dimensions the approximations by discrete mass distributions that led up to Definitions 7.1 and 7.2, we simply state the analogous definition.

7.3 DEFINITION

If a substance S occupies the region $E \subseteq R^3$ and has continuous density function δ, then its ***mass*** is

$$m = \iiint_E \delta(x, y, z)\,dV.$$

The ***center of mass*** of the substance is $(\bar{x}, \bar{y}, \bar{z})$, where

$$\bar{x} = \frac{\iiint_E x\,\delta(x, y, z)\,dV}{m}, \qquad \bar{y} = \frac{\iiint_E y\,\delta(x, y, z)\,dV}{m},$$

and

$$\bar{z} = \frac{\iiint_E z\,\delta(x, y, z)\,dV}{m}.$$

The numerators of $\bar{x}$, $\bar{y}$, and $\bar{z}$ respectively are called the ***moments of S with respect to the yz- , xz- , and xy-planes.*** The notations M_{yz}, M_{xz}, and M_{xy} are used for these moments.

It appears that a considerable amount of calculation is required to find center of mass, but fortunately the symmetry in most commonly encountered solids often helps to reduce the work. We mention two rules that are worth bearing in mind.

1. *If the region E is symmetric in the xy-plane and the density function* δ *is even in* z, that is, satisfies $\delta(x, y, -z) = \delta(x, y, z)$, *then* $\bar{z} = 0$.
2. *If E is symmetric in the z-axis* (that is, whenever (x, y, z) is in E, then so is $(-x, -y, z)$) *and the density function satisfies* $\delta(-x, -y, z) = \delta(x, y, z)$, *then* $\bar{x} = \bar{y} = 0$.

The first rule follows from Exercise 35 of Section 4. We have

$$M_{xy} = \frac{\iiint_E z\,\delta(x, y, z)\,dV}{m} = 0,$$

because $z\,\delta(x, y, z)$ is an odd function in z: At $(x, y, -z)$, its value is $-z\,\delta(x, y, z)$, which is the negative of its value at (x, y, z). The second rule follows similarly. *These rules apply equally well to regions E that are symmetric in the other axes or coordinate planes if the density functions obey the corresponding conditions.*

Applications of these rules, coupled with the judicious use of cylindrical or spherical coordinates, can often keep the work involved in computing centers of mass within reasonable bounds, as the next example suggests.

Example 5

A substance S occupies half a spherical ball of radius 2. Its density at a point P is proportional to the distance of P from the center of the base. Find the center of mass.

FIGURE 7.6

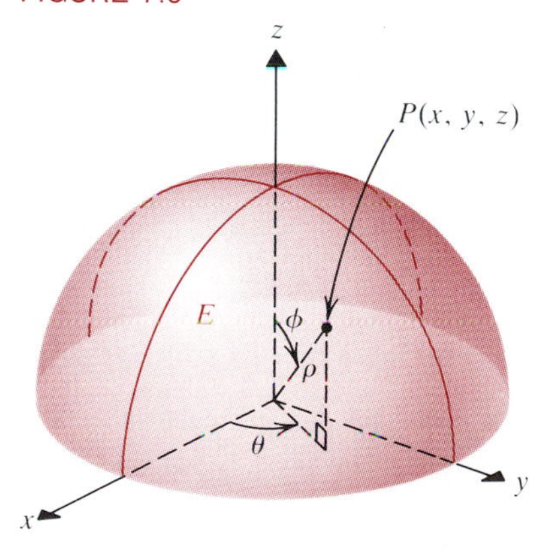

Solution. Here we can represent the solid (see **Figure 7.6**) as occupying the region

$$E = \{(x, y, z) | x^2 + y^2 + z^2 \leq 4, z \geq 0\} = \{\{\rho, \phi, \theta\} | \rho \leq 2, \phi \leq \pi/2\}.$$

The center of the base is then $(0, 0, 0)$, so $\delta(x, y, z) = k\sqrt{x^2 + y^2 + z^2}$. The region E is symmetric in the z-axis, and $\delta(x, y, z) = \delta(-x, -y, z)$, so the second rule above says that $\bar{x} = \bar{y} = 0$. We still need to determine $\bar{z}$. First,

$$m = \iiint_E k\sqrt{x^2 + y^2 + z^2}\,dV = \iiint_E k\rho\,dV = k\iiint_E \rho^3 \sin\phi\,d\rho\,d\phi\,d\theta$$

$$= k\int_0^{2\pi}\int_0^{\pi/2}\int_0^2 \rho^3 \sin\phi\,d\rho\,d\phi\,d\theta$$

$$= k\int_0^{2\pi} d\theta \int_0^{\pi/2} \sin\phi\,d\phi \int_0^2 \rho^3\,d\rho \qquad \textit{by the analogue of Corollary 1.10}$$

$$= k\int_0^{2\pi} d\theta \int_0^{\pi/2} \sin\phi \left[\frac{1}{4}\rho^4\right]_0^2 d\phi = 4k\int_0^{2\pi} d\theta \int_0^{\pi/2} \sin\phi\,d\phi$$

$$= 4k\int_0^{2\pi} d\theta \Big[-\cos\phi\Big]_0^{\pi/2} = 4k\int_0^{2\pi} d\theta = 8\pi k.$$

We also have

$$M_{xy} = k\iiint_E \rho^2\cos\phi\,dV = k\int_0^{2\pi}\int_0^{\pi/2}\int_0^2 \rho^4 \sin\phi\cos\phi\,d\rho\,d\phi\,d\theta$$

$$= k\int_0^{2\pi} d\theta \int_0^{\pi/2} \sin\phi\cos\phi\,d\phi \int_0^2 \rho^4\,d\rho \qquad \textit{by the analogue of Corollary 1.10}$$

$$= k\int_0^{2\pi} d\theta \int_0^{\pi/2} \sin\phi\cos\phi \left[\frac{1}{5}\rho^5\right]_0^2 d\phi = \frac{32}{5}k\int_0^{2\pi} d\theta \left[\frac{1}{2}\sin^2\phi\right]_0^{\pi/2}$$

$$= \frac{16}{5}k\int_0^{2\pi} d\theta = \frac{32}{5}\pi k.$$

Hence

$$\bar{z} = \frac{M_{xy}}{m} = \frac{\frac{32}{5}\pi k}{8\pi k} = \frac{4}{5}.$$

Thus the center of mass is (0, 0, 4/5). ■

The final physical quantities we discuss are of considerable importance in studying dynamical systems. First, recall that the ***kinetic energy*** of a point mass m moving with velocity $\mathbf{v}$ is $K = \frac{1}{2}m|\mathbf{v}|^2$. Suppose next that a solid S rotates with constant angular speed ω about an axis l. (Here ω is a scalar quantity.) See **Figure 7.7,** where we show the path described by a typical point P in the solid. If this point P is at a distance r from the axis l of rotation, then its linear speed is

FIGURE 7.7

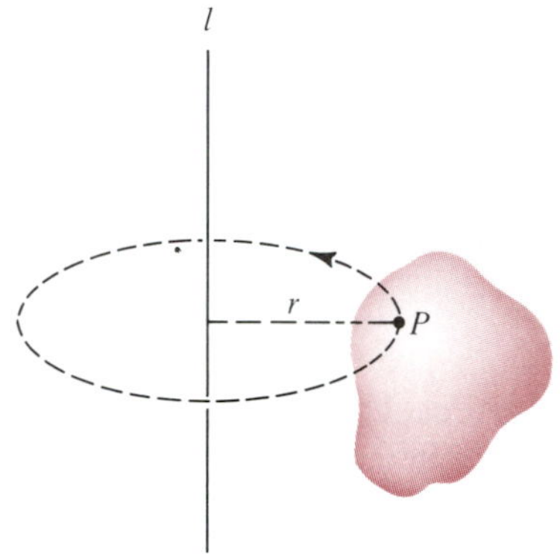

$$|\mathbf{v}| = \omega r.$$

A small part of S surrounding P of mass Δm then has kinetic energy approximately $\frac{1}{2}\Delta m|\mathbf{v}|^2 = \frac{1}{2}\omega^2 r^2\,\Delta m$. Since a grid on S reduces S to many such small parts, the total kinetic energy K of S is approximated by

$$\sum_{i,j,k} \frac{1}{2}\,\omega^2[r(x_i, y_j, z_k)]^2\,\Delta m_{ijk},$$

where

$$\Delta m_{ijk} = \delta(x_i, y_j, z_k)\,\Delta V_{ijk}.$$

Hence we are led to the following definition.

7.4 DEFINITION

If a solid S of density δ occupying a region $E \subseteq \mathbf{R}^3$ rotates about an axis l with constant angular speed ω, then the ***total kinetic energy*** of S is

$$K = \frac{1}{2}\,\omega^2 \iiint_E r^2\,dm = \frac{1}{2}\,\omega^2 \iiint_E r^2\,\delta(x, y, z)\,dV = \frac{1}{2}\,\omega^2 I_l,$$

where $r = r(x, y, z)$ is the distance from $P(x, y, z)$ to the axis l of rotation, and

$$I_l = \iiint_E r^2\,\delta(x, y, z)\,dV$$

is the *moment of inertia* of S about l. The *radius of gyration* R of S about l is defined by

$$R^2 = \frac{I_l}{m},$$

where m is the mass of S. (This represents the distance from l to a point P where the entire mass m of S could be concentrated to yield the same moment of inertia I_l and the same kinetic energy.)

In practice, l is usually the z-axis, x-axis, or y-axis. In these cases we get the following formulas for I_l:

$$I_x = \iiint_E (y^2 + z^2)\,\delta(x, y, z)\,dV, \qquad I_y = \iiint_E (x^2 + z^2)\,\delta(x, y, z)\,dV,$$

$$I_z = \iiint_E (x^2 + y^2)\,\delta(x, y, z)\,dV.$$

Example 6

A bar of gold bullion of constant density δ occupies the box $B = [0, 3] \times [0, 1] \times [0, 1]$. Find its moment of inertia and radius of gyration about the z-axis.

Solution. Here $m = 3\delta$ since $V = 3 \cdot 1 \cdot 1 = 3$. We also have

$$I_z = \iiint_B (x^2 + y^2)\,\delta\, dV = \delta \int_0^3 \int_0^1 \int_0^1 (x^2 + y^2)\, dz\, dy\, dx$$

$$= \delta \int_0^3 \int_0^1 (x^2 + y^2)\, dy\, dx = \delta \int_0^3 \left[x^2 y + \frac{1}{3} y^3 \right]_0^1 dx$$

$$= \delta \int_0^3 \left(x^2 + \frac{1}{3} \right) dx = \delta \left[\frac{27}{3} + \frac{3}{3} \right] = 10\delta.$$

Then $R^2 = 10\not{\delta}/3\not{\delta} = 10/3$, so $R = \sqrt{10/3}$. ■

Exercises 13.7

In Exercises 1–8, use the methods of this section to find the mass and center of mass of a lamina occupying the given region $D \subseteq \mathbf{R}^2$ and having the given density function.

1. The triangle with vertices (0, 0), $(a, 0)$, (b, c), constant density
2. The semidisk $\{(x, y) \mid x^2 + y^2 \le a^2, y \ge 0\}$, constant density
3. The rectangle $[1, 2] \times [1, 3]$, where $\delta(x, y) = xy$
4. The triangular region with vertices (0, 0), (0, 3), (4, 0), where $\delta(x, y) = xy$
5. The region bounded by $y^2 = x$ and $y = x^2$, where $\delta(x, y) = ky$
6. The first-quadrant region enclosed by the circle $x^2 + y^2 = 4$, where $\delta(x, y) = \sqrt{x^2 + y^2}$
7. The first-quadrant region bounded by the circles $x^2 + y^2 = 1$ and $x^2 + y^2 = 4$, constant density δ
8. The region enclosed by the cardioid $r = 2(1 + \cos\theta)$, constant density δ

In Exercises 9–14, find the mass and center of mass of a substance S occupying the given region $E \subseteq \mathbf{R}^3$ and having the given density function.

9. The rectangular box $[0, 1] \times [0, 1] \times [0, 1]$, where $\delta(x, y, z) = z$
10. The rectangular box $[0, 2] \times [0, 1] \times [0, 1]$, where $\delta(x, y, z) = y$
11. The region in Example 5 but of constant density
12. The region described by $x^2 + y^2 + z^2 \le 9$ for $z \ge 0$, where $\delta(x, y, z) = k\sqrt{x^2 + y^2 + z^2}$
13. The solid right circular cone in **Figure 7.8,** constant density

FIGURE 7.8

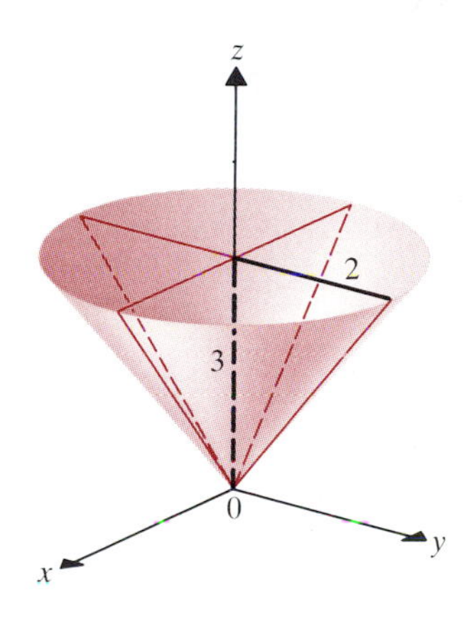

14. The first-octant region enclosed by the sphere $x^2 + y^2 + z^2 = 4$, constant density

In Exercises 15–18, find the moment of inertia and radius of gyration about the axis given.

15. The region of Example 6, x-axis
16. The ball $x^2 + y^2 + z^2 \le a^2$, constant density, z-axis
17. The ball $x^2 + y^2 + z^2 \le 9$, $\delta(x, y, z) = k\sqrt{x^2 + y^2 + z^2}$, z-axis
18. The ball $x^2 + y^2 + z^2 \le a^2$, $\delta(x, y, z) = k\sqrt{x^2 + y^2 + z^2}$, z-axis
19. The ball in Exercise 17 rotates about the z-axis with angular velocity $\omega = 2$. Find the kinetic energy.
20. The ball in Exercise 17 rotates about the x-axis with angular velocity $\omega = 2$. Find the kinetic energy.

21. A 12-inch long-playing record of constant density δ and thickness h revolves at $33\frac{1}{3}$ revolutions per minute. Use Definition 7.4 to express its kinetic energy as a triple integral and evaluate it using cylindrical coordinates.

22. Suppose that a wire is parametrized by arc length (see p. 698): $\mathbf{x} = \mathbf{x}(s)$, $a \le s \le b$, and has density $\delta(s)$. Its mass is defined as $m = \int_a^b \delta(s)\,ds$, and its center of mass by

$$(\bar{x}, \bar{y}) = \frac{1}{m}\left(\int_a^b x(s)\,\delta(s)\,ds, \int_a^b y(s)\,\delta(s)\,ds\right).$$

Find the center of mass of a wire of constant density in the shape $x^2 + y^2 = c^2$, $y \ge 0$.

23. Find the center of mass of the first-quadrant portion of the wire described in Exercise 22.

24. Recall that a theorem of Pappus (Theorem 5.2 of Chapter 5) relates volumes of surfaces of revolution to centroids of plane regions. Another theorem of Pappus says that if a smooth curve $\mathbf{x} = \mathbf{x}(s)$ lying to the right of the y-axis is revolved about the y-axis, then the surface of revolution formed has surface area $A = 2\pi\bar{x}L$, where $\bar{x}$ is the x-coordinate of the center of mass, and L is the length of the curve. Use Exercise 22 to prove this. (*Hint:* At a typical point $\mathbf{x}(s)$, a short portion of the curve ΔS_i, is approximately linear. When revolved it traces an approximate frustum of a cone with area $\Delta A_i \approx 2\pi x\,\Delta S_i$.)

25. Apply Exercise 24 to calculate:
(a) The surface area of a sphere obtained by revolving the graph of $x = \sqrt{a^2 - y^2}$ in the xy-plane about the y-axis.
(b) The surface area of the torus (see p. 319).

26. Apply Exercise 24 to calculate the lateral surface area of the right circular cone $\phi = \pi/4$ from $z = 0$ to $z = h$.

27. Let D be a homogeneous lamina lying under the graph of a continuous positive-valued function $y = f(x)$ between $x = a$ and $x = b$. Show that Formulas (2) and (3) preceding Example 4 reduce to the formulas given in Definition 5.1 of Chapter 5 for M_x and M_y.

28. Use Formulas (1), (2), and (3) to derive definite integral formulas for m, M_x, and M_y for a homogeneous lamina lying between the y-axis, the graph of a continuous, positive-valued function $x = g(y)$, and lines $y = c$ and $y = d$.

29. Suppose that D lies between the graphs of the continuous functions $y = f(x)$ and $y = g(x)$ over the interval $[a, b]$, where $f(x) \le g(x)$ on $[a, b]$. Then use Formulas (1), (2), and (3) to obtain definite integrals for m, M_x, and M_y.

30. Repeat Exercise 29 for D lying between $y = c$, $y = d$, and the graphs of continuous functions $x = h(y)$ and $x = l(y)$, where $h(y) \le l(y)$ on $[c, d]$.

13.8 Looking Back

This chapter has developed the integral calculus of functions $f\colon \mathbf{R}^n \to \mathbf{R}$ for $n = 2$ and 3. We began (Definition 1.1) with the double integral of $f\colon \mathbf{R}^2 \to \mathbf{R}$ over a rectangle $R = [a, b] \times [c, d]$ in $\mathbf{R}^2$. We saw in Theorem 1.9 how iterated integrals (Definition 1.7) are used to evaluate double integrals. Definition 2.1 extended double integration to regions more general than rectangles. The basic connection between area and volume and double integrals was set forth in Corollary 1.2, Definitions 1.8 and 2.2, and Theorem 2.3. Section 3 presented the same ideas within the context of the polar coordinate system. We saw how polar double integrals can often be used to good advantage to evaluate complicated rectangular double integrals, especially when the region of integration has a simple polar coordinate description. Remember the two distinct formulas for dA: $dx\,dy$ for rectangular double integrals and $r\,dr\,d\theta$ for polar double integrals.

The triple integral of a function $f\colon \mathbf{R}^3 \to \mathbf{R}$ was introduced in Section 4, and Equations (3), (4), and (6) of that section are three iteration formulas for evaluating triple integrals. Theorem 4.3 is the three-dimensional version of Theorem 2.3 and the basic formula $\int_a^b 1\,dx = b - a$. It says that the volume (three-dimensional measure of extent) of a region E in $\mathbf{R}^3$ is the triple integral of the constant function $f(x, y, z) = 1$ over E.

Sections 5 and 6 introduced cylindrical (Definition 5.1) and spherical (Definition 6.1) coordinates in $\mathbf{R}^3$. As with polar coordinates in $\mathbf{R}^2$, these can often reduce the complexity of given triple integrals, provided that the respective formulas $dV = r\,dr\,d\theta\,dz$ for *cylindrical* triple integrals, and $dV = \rho^2 \sin\phi\,d\rho\,d\phi\,d\theta$ for *spherical* triple integrals are kept firmly in mind. We saw in Theorems 5.3,

5.4, and 6.3 how to use iteration to evaluate cylindrical and spherical triple integrals.

The last section presented several applications of double and triple integrals. We used them to calculate mass (Definition 7.1), center of mass (Definitions 7.2 and 7.3), kinetic energy, and moments of inertia (Definition 7.4) in two and three dimensions.

CHAPTER CHECKLIST

Section 1: grid, double Riemann sum, mesh; double integral; $dA = dx\,dy$; integrals of constant functions, sums and constant multiples of functions; area; integrability of continuous functions; iterated integrals; volume of a region over a rectangle and under the graph of $z = f(x, y)$; Fubini's theorem, iterated integral of a product of a function of x alone and a function of y alone.

Section 2: double integral over nonrectangular regions of $\mathbf{R}^2$; volumes; directed arrow method for finding limits of integration; area.

Section 3: polar double integral, $dA = r\,dr\,d\theta$; evaluation by iteration; transformation from rectangular to polar double integrals.

Section 4: triple integral over a rectangular box and more general regions, $dV = dx\,dy\,dz$; Fubini's theorem in three dimensions; volumes by triple integration.

Section 5: cylindrical coordinates in $\mathbf{R}^3$; cylindrical triple integral, $dV = r\,dr\,d\theta\,dz$; evaluation by iteration.

Section 6: spherical coordinates in $\mathbf{R}^3$; spherical triple integrals, $dV = \rho^2 \sin\phi\,d\rho\,d\phi\,d\theta$; evaluation by iteration.

Section 7: laminas, mass, moments with respect to coordinate axes; center of mass, centroid; moments of solids with respect to coordinate planes; symmetry rules for triple integration; kinetic energy, moments of inertia about axes, radius of gyration about axes.

REVIEW EXERCISES 13.8

In Exercises 1–12, evaluate the given double or triple integral.

1. $\iint_R e^x \sin y\,dA$, $R = [0, 2] \times [\pi/3, \pi/2]$

2. $\iint_R e^x \sin \frac{y}{x}\,dA$, $R = [1, 2] \times [-\pi/2, \pi/2]$

3. $\iint_D (x + y^2)\,dA$, D the region bounded by the parabola $y = x^2$ and the line $y = 1$

4. $\iint_D (2x - 3y + xy)\,dA$, D the region bounded by the y-axis, the parabola $x = 4 - y^2$, and the line $y = 0$

5. $\iint_D (x^2 + y^2)^{3/2}\,dA$, D the region in the first quadrant bounded by the y-axis, the line $y = \frac{1}{\sqrt{3}}x$, and the circle $x^2 + y^2 = 4$

6. $\iint_D \frac{1}{\sqrt{x^2 + y^2}}\,dA$, where D is the top half of the disk $(x - 1)^2 + y^2 = 1$

7. $\iiint_E xy\,dV$, where E is the first-octant region bounded by the cylinders $y = x^2$ and $z = 1 - y^2$

8. $\iiint_E (x + 2y)\,dV$, where E is enclosed by the tetrahedron formed by the three coordinate planes and the plane $6x + 3y + 2z = 6$

9. $\iiint_U xyz\,dV$, where U is the first-octant region bounded by the xy-plane, the plane $z = 3$, and the cylinder $x^2 + y^2 = a^2$

10. $\iiint_U z^2\,dV$, where U is the region bounded by the cones $z = \sqrt{x^2 + y^2}$ and $z = -\sqrt{x^2 + y^2}$ and the cylinder $x^2 + y^2 = 4$

11. $\iiint_B \sqrt{x^2 + y^2 + z^2}\,dV$, where B is the ball of radius 2 centered at the origin

12. $\iiint_B (x^2 + y^2 + z^2)\,dV$, where B is the region above the cone $z = \sqrt{x^2 + y^2}$ and below the plane $z = 2$

In Exercises 13–19, find the volume of the given region in $\mathbf{R}^3$.

13. In the first octant bounded by the cylinder $y^2 + z^2 = 16$ and the planes $y = x$ and $x = 0$

14. Under the paraboloid $z = x^2 + y^2$ and above the rectangle $[0, 1] \times [0, 2]$

15. Between the paraboloids $z = 2 - 3x^2 - 3y^2$ and $z = -2 + x^2 + y^2$

16. Under the cone $z = \sqrt{x^2 + y^2}$ and above the region between the circles $x^2 + y^2 = 4$ and $x^2 + y^2 = 16$ in the xy-plane

17. Above the xy-plane, below the sphere $x^2 + y^2 + z^2 = 9$, and inside the cylinder $x^2 + y^2 = 4$

18. Inside the sphere $x^2 + y^2 + z^2 = 16$ and outside the cone $z = 4 - \sqrt{x^2 + y^2}$

19. The solid formed by revolving the region in the xz-plane bounded by $z = x^2$, the x-axis, and the line $x = 1$ about the z-axis

20. Reverse the order of integration in $\int_1^e \int_0^{\ln y} f(x, y)\,dx\,dy$.

21. Write $\int_0^1 \int_{e^y}^e f(x, y)\,dx\,dy$ as an iterated integral (or sum of iterated integrals) in which the integration is first performed with respect to y.

22. Find by double integration the area of one petal of the rose $r = \cos 3\theta$.

23. Find by double integration the area bounded by the graphs of $y = e^x$, $x = 0$, $y = 0$, and $x = -1$.

In Exercises 24–26, find the center of mass of the given region.

24. The region in the first quadrant enclosed by $x^2 + y^2 = 9$, if $\delta(x, y) = \sqrt{x^2 + y^2}$

25. The triangular region with vertices (0, 0), (0, 4), and (3, 0), if $\delta(x, y) = xy$

26. The region above the xy-plane and below $x^2 + y^2 + z^2 \leq 16$, if $\delta(x, y, z) = \sqrt{x^2 + y^2 + z^2}$

27. Find the moment of inertia and radius of gyration relative to the z-axis of the region in Exercise 26.

28. The region in Exercise 26 rotates about the z-axis with angular velocity $\omega = 3$. What is the kinetic energy?

CHAPTER 14 Vector Calculus

14.0 Introduction

The last two chapters presented the differential and integral calculus of scalar functions. We turn now to the calculus of vector functions $\mathbf{F}: \mathbf{R}^n \to \mathbf{R}^m$ having domain in $\mathbf{R}^n$ and range in $\mathbf{R}^m$, where m and n are 2 or 3. For example, the formula

$$\mathbf{F}(x, y, z) = (x^2 + y^2 + z^2)\,\mathbf{i} + \frac{x^2}{y^2}\,\mathbf{j} + \cos xyz\,\mathbf{k}$$

gives one such function.

We begin by extending the notions of limit, continuity, and differentiability to vector functions. Their total derivatives are again obtained by putting partial derivatives together. The rest of Section 1 discusses important vector differential operators related to the *gradient* operator. They play a central role in vector integral calculus.

The second section introduces the important notion of the *line integral* of vector functions. That is an extension of the ordinary definite integral over a closed interval $[a, b]$ in $\mathbf{R}$ to more general one-dimensional paths, namely, curves in $\mathbf{R}^2$ or $\mathbf{R}^3$. The next section presents a useful analogue of the fundamental theorem of calculus, called Green's theorem. Section 4 contains a still sharper analogue of the fundamental theorem, this time for line integrals. Sections 5 and 6 introduce parametrized surfaces and surface integrals of vector functions of three variables. In the process we obtain a formula for the area of a parametrized surface in $\mathbf{R}^3$. Parametrized surfaces are two-dimensional analogues in $\mathbf{R}^3$ of curves in $\mathbf{R}^2$. The analogy is so strong that the concluding two sections present analogues for surface integrals of Green's theorem on line integrals over curves. They not only simplify evaluation of surface integrals but also have a number of important applications to physics, chemistry, and engineering, a few of which are discussed in some detail.

14.1 Vector Differentiation and Differential Operators

The first part of Chapter 11 presented the differential calculus of m-dimensional vector functions $\mathbf{g}$ of one real variable (where $m = 2$ or 3), functions such as

$$\mathbf{g}(t) = (t + 1, 2t, \tfrac{1}{2}t^2) = (t + 1)\mathbf{i} + 2t\mathbf{j} + \tfrac{1}{2}t^2\mathbf{k}.$$

In Chapter 12 we developed the differential calculus of scalar functions like

$$f(x, y) = \sqrt{x^2 + y^2} \qquad \text{and} \qquad h(x, y, z) = e^{x^2 + y^2 + z^2}.$$

In this chapter we consider vector functions of several real variables, functions of the form $\mathbf{F}\colon \mathbf{R}^2 \to \mathbf{R}^2$,

$$\mathbf{F}(x, y) = (P(x, y), Q(x, y)) = P(x, y)\mathbf{i} + Q(x, y)\mathbf{j};$$

of the form $\mathbf{F}\colon \mathbf{R}^2 \to \mathbf{R}^3$,

$$\mathbf{F}(x, y) = (P(x, y), Q(x, y), R(x, y)) = P(x, y)\mathbf{i} + Q(x, y)\mathbf{j} + R(x, y)\mathbf{k};$$

of the form $\mathbf{G}\colon \mathbf{R}^3 \to \mathbf{R}^2$,

$$\mathbf{G}(x, y, z) = (P(x, y, z), Q(x, y, z)) = P(x, y, z)\mathbf{i} + Q(x, y, z)\mathbf{j};$$

and of the form $\mathbf{G}\colon \mathbf{R}^3 \to \mathbf{R}^3$,

$$\begin{aligned}\mathbf{G}(x, y, z) &= (P(x, y, z), Q(x, y, z), R(x, y, z)) \\ &= P(x, y, z)\mathbf{i} + Q(x, y, z)\mathbf{j} + R(x, y, z)\mathbf{k}.\end{aligned}$$

Our general notation for such functions is $\mathbf{F}\colon \mathbf{R}^n \to \mathbf{R}^m$ (where $m = 2$ or 3 and $n = 2$ or 3),

$$\mathbf{F}(\mathbf{x}) = \mathbf{F}(x_1, \ldots, x_n) = \mathbf{y} = (y_1, \ldots, y_m) = (f_1(\mathbf{x}), \ldots, f_m(\mathbf{x})).$$

The scalar functions

$$f_i(\mathbf{x}) = f_i(x_1, x_2, \ldots, x_n), \qquad i = 1, 2, \ldots, m,$$

are called the ***coordinate functions*** of $\mathbf{F}$. They are real-valued functions of several real variables, so we know from Section 11.4 how to find their limits and check them for continuity. Following the approach of Section 11.1, we define limits and continuity of vector functions $\mathbf{F}$ by putting together the corresponding concepts for their coordinate functions.

1.1 DEFINITION Let $\mathbf{F}\colon \mathbf{R}^n \to \mathbf{R}^m$ be defined on an open ball $B(\mathbf{a}, \delta)$, except possibly at $\mathbf{x} = \mathbf{a}$. Then

$$\lim_{\mathbf{x} \to \mathbf{a}} \mathbf{F}(\mathbf{x}) = \mathbf{l} = (l_1, l_2, \ldots, l_m)$$

means that for each $i = 1, 2, \ldots, m$,

$$\lim_{\mathbf{x} \to \mathbf{a}} f_i(\mathbf{x}) = l_i.$$

We say that $\mathbf{F}$ is ***continuous*** at $\mathbf{x} = \mathbf{a}$ if f_i is continuous at $\mathbf{x} = \mathbf{a}$ for $i = 1, 2, \ldots, m$, that is, if

$$\lim_{\mathbf{x} \to \mathbf{a}} \mathbf{F}(\mathbf{x}) = \mathbf{F}(\mathbf{a}).$$

Example 1

Determine whether

$$\mathbf{F}(x, y, z) = (x^2 + y^2 + z^2)\,\mathbf{i} + \frac{x^2}{y^2}\,\mathbf{j} + (\cos xyz)\,\mathbf{k}$$

is continuous at (a) $(2, 0, \pi)$ and (b) $(1, 1, \pi)$.

Solution. (a) The second coordinate function $f_2(x, y, z) = x^2/y^2$ is discontinuous at $(2, 0, \pi)$, since it is undefined when $y = 0$. (It also has no limit at this point, because as $(x, y) \to (2, 0)$, we have $x^2/y^2 \to +\infty$.) Thus **F** is discontinuous at $(2, 0, \pi)$.

(b) Writing $\mathbf{a} = (1, 1, \pi)$, we have

$$\lim_{\mathbf{x}\to\mathbf{a}} F(x, y, z) = \left[\lim_{\mathbf{x}\to\mathbf{a}} (x^2 + y^2 + z^2)\right]\mathbf{i} + \left[\lim_{\mathbf{x}\to\mathbf{a}} \frac{x^2}{y^2}\right]\mathbf{j} + \left[\lim_{\mathbf{x}\to\mathbf{a}} \cos xyz\right]\mathbf{k}$$
$$= (2 + \pi^2)\,\mathbf{i} + \mathbf{j} - \mathbf{k} = \mathbf{F}(1, 1, \pi),$$

so **F** is continuous at $(1, 1, \pi)$. ■

If all the coordinate functions of **F** are differentiable at $\mathbf{x} = \mathbf{a}$, then the total derivative of **F** is formed by putting the partial derivatives of the coordinate functions together in a *matrix*. (See pp. 671–72, where matrices were introduced.)

1.2 DEFINITION Let $\mathbf{F}\colon \mathbf{R}^n \to \mathbf{R}^m$ be defined on an open ball $B(\mathbf{a}, \delta)$ in $\mathbf{R}^n$. Then **F** is ***differentiable*** at $\mathbf{x} = \mathbf{a}$ if each of its coordinate functions f_i, $i = 1, 2, \ldots, m$, is differentiable at $\mathbf{x} = \mathbf{a}$. In that case, the ***total derivative*** of **F** at $\mathbf{x} = \mathbf{a}$ is the ***m-by-n Jacobian matrix*** of partial derivatives of the coordinate functions:

$$\mathbf{J}_{\mathbf{F}}(\mathbf{a}) = \begin{pmatrix} \nabla f_1(\mathbf{a}) \\ \vdots \\ \nabla f_i(\mathbf{a}) \\ \vdots \\ \nabla f_m(\mathbf{a}) \end{pmatrix} = \begin{pmatrix} \dfrac{\partial f_1}{\partial x_1}(\mathbf{a}) & \cdots & \dfrac{\partial f_1}{\partial x_j}(\mathbf{a}) & \cdots & \dfrac{\partial f_1}{\partial x_n}(\mathbf{a}) \\ \vdots & & \vdots & & \vdots \\ \dfrac{\partial f_i}{\partial x_1}(\mathbf{a}) & & \dfrac{\partial f_i}{\partial x_j}(\mathbf{a}) & & \dfrac{\partial f_i}{\partial x_n}(\mathbf{a}) \\ \vdots & & \vdots & & \vdots \\ \dfrac{\partial f_m}{\partial x_1}(\mathbf{a}) & & \dfrac{\partial f_m}{\partial x_j}(\mathbf{a}) & & \dfrac{\partial f_m}{\partial x_n}(\mathbf{a}) \end{pmatrix}.$$

The Jacobian matrix is formed by computing the total derivatives (gradients)

$$\nabla f_i(\mathbf{a}) = \left(\frac{\partial f_i}{\partial x_1}(\mathbf{a}), \frac{\partial f_i}{\partial x_2}(\mathbf{a}), \ldots, \frac{\partial f_i}{\partial x_n}(\mathbf{a})\right)$$

at $\mathbf{a} = (a_1, a_2, \ldots, a_n)$ for $i = 1, 2, \ldots, m$ and stacking the results to obtain the m-by-n matrix $\mathbf{J}_{\mathbf{F}}(\mathbf{a})$. Thus $\mathbf{J}_{\mathbf{F}}$ has as many *rows* as the dimension of the *range* of F and as many *columns* as the dimension of the *domain* of F. The next example illustrates the calculation of $\mathbf{J}_{\mathbf{F}}(\mathbf{a})$ for the function **F** of Example 1.

Example 2

Show that the function $\mathbf{F}(x, y, z) = (x^2 + y^2 + z^2)\,\mathbf{i} + (x^2/y^2)\,\mathbf{j} + \cos xyz\,\mathbf{k}$ is differentiable at $\mathbf{a} = (1, 1, \pi)$. Find its total derivative at this point.

Solution. Each coordinate function is differentiable at $\mathbf{x} = \mathbf{a} = (1, 1, \pi)$, where we find

$$\nabla f_1(\mathbf{a}) = 2x\,\mathbf{i} + 2y\,\mathbf{j} + 2z\,\mathbf{k}\big|_{\mathbf{a}} = 2\,\mathbf{i} + 2\,\mathbf{j} + 2\pi\,\mathbf{k},$$

$$\nabla f_2(\mathbf{a}) = \frac{2x}{y^2}\,\mathbf{i} - \frac{2x^2}{y^3}\,\mathbf{j} + 0\,\mathbf{k}\big|_{\mathbf{a}} = 2\,\mathbf{i} - 2\,\mathbf{j},$$

and

$$\nabla f_3(\mathbf{a}) = (-yz \sin xyz)\,\mathbf{i} - (xz \sin xyz)\,\mathbf{j} - (xy \sin xyz)\,\mathbf{k}\big|_{\mathbf{a}} = \mathbf{0}.$$

Thus $\mathbf{F}$ is differentiable at $\mathbf{x} = \mathbf{a}$, and its total derivative is the Jacobian matrix

$$\mathbf{J}_{\mathbf{F}}(1, 1, \pi) = \begin{pmatrix} 2 & 2 & 2\pi \\ 2 & -2 & 0 \\ 0 & 0 & 0 \end{pmatrix}. \quad ■$$

The Jacobian matrix is defined in terms of the gradients ∇f_i of the coordinate functions $f_1, f_2, \ldots, f_m$. Vector integration makes important use of the ***gradient operator*** ∇ (*del*) associated with this process. It is written formally as

$$\nabla = \frac{\partial}{\partial x}\,\mathbf{i} + \frac{\partial}{\partial y}\,\mathbf{j} + \frac{\partial}{\partial z}\,\mathbf{k}, \tag{1}$$

where the colored term is not present if $m = 2$. The gradient operator acts on differentiable scalar functions of two and three variables by the rules

$$\nabla f(x, y) = \left(\frac{\partial}{\partial x}\,\mathbf{i} + \frac{\partial}{\partial y}\,\mathbf{j}\right)(f) = \frac{\partial f}{\partial x}\,\mathbf{i} + \frac{\partial f}{\partial y}\,\mathbf{j},$$

and

$$\nabla g(x, y, z) = \left(\frac{\partial}{\partial x}\,\mathbf{i} + \frac{\partial}{\partial y}\,\mathbf{j} + \frac{\partial}{\partial z}\,\mathbf{k}\right)(g) = \frac{\partial g}{\partial x}\,\mathbf{i} + \frac{\partial g}{\partial y}\,\mathbf{j} + \frac{\partial g}{\partial z}\,\mathbf{k}.$$

Example 3

Apply the gradient operator to
(a) $f(x, y) = x^2y - y^2x$ and (b) $f(x, y, z) = |\mathbf{x}| = \sqrt{x^2 + y^2 + z^2}$.

Solution. (a) We have

$$\nabla f(x, y) = \frac{\partial f}{\partial x}\,\mathbf{i} + \frac{\partial f}{\partial y}\,\mathbf{j} = (2xy - y^2)\,\mathbf{i} + (x^2 - 2xy)\,\mathbf{j}.$$

(b) Here,

$$\nabla f(x, y, z) = \frac{1}{2}(x^2 + y^2 + z^2)^{-1/2} \cdot (2x\,\mathbf{i} + 2y\,\mathbf{j} + 2z\,\mathbf{k})$$

$$= \frac{1}{\sqrt{x^2 + y^2 + z^2}}(x\,\mathbf{i} + y\,\mathbf{j} + z\,\mathbf{k}) = \frac{1}{|\mathbf{x}|}\,\mathbf{x}. \quad ■$$

The calculation in Example 3(b) produces the unit vector in the direction of $\mathbf{x} = (x, y, z)$ at every point of $\mathbf{R}^3$ other than $(0, 0, 0)$. We can thus think of the gradient operator as acting on the length function—a scalar function—to produce unit vectors in $\mathbf{R}^3$.

FIGURE 1.1

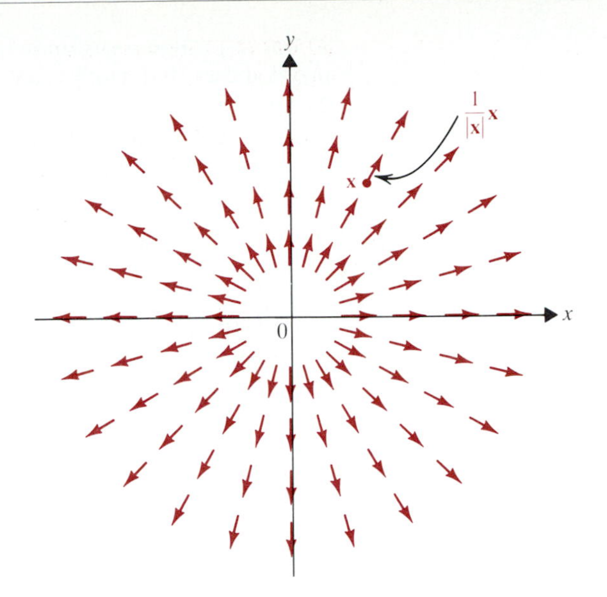

In general the gradient operator ∇ acts on scalar functions of two (or three) variables to produce what is called a ***vector field*** on $\mathbf{R}^2$ (or $\mathbf{R}^3$). That is a function

$$\mathbf{F}: \mathbf{R}^n \to \mathbf{R}^n, \qquad n = 2 \text{ or } 3,$$

which associates to each $\mathbf{x} = (x, y)$ or (x, y, z) in its domain a vector $\mathbf{F}(x, y)$ in $\mathbf{R}^2$ or $\mathbf{F}(x, y, z)$ in $\mathbf{R}^3$. In Example 3(b) the vector field $\mathbf{F} = \nabla f$ has domain $\mathbf{R}^3 - \{\mathbf{0}\}$, and at each point (x, y, z) of this domain, ∇f produces the unit vector in the direction of $\mathbf{x} = x\mathbf{i} + y\mathbf{j} + z\mathbf{k}$. **Figure 1.1** represents this process pictorially. At each point P in the figure, we have drawn a unit vector pointing in the direction of $\mathbf{OP}$. Such diagrams are of interest in physics and engineering, where many of the concepts discussed in this chapter originated (see Exercises 41–44 for additional examples). We will treat vector fields analytically for the most part. The following theorem gives the main analytic properties of the gradient operator.

1.3 THEOREM

Let f and g be differentiable scalar functions of two or three variables. Let h be a real-valued function of one real variable with continuous derivative h', and let a be a real number. Then

(a) $\nabla(f + g) = \nabla f + \nabla g \quad$ and $\quad \nabla(af) = a\nabla f$.

(b) $\nabla(f(\mathbf{x})g(\mathbf{x})) = f(\mathbf{x})\nabla g(\mathbf{x}) + [\nabla f(\mathbf{x})]g(\mathbf{x})$.

(c) $\nabla\left(\dfrac{f(\mathbf{x})}{g(\mathbf{x})}\right) = \dfrac{g(\mathbf{x})\nabla f(\mathbf{x}) - f(\mathbf{x})\nabla g(\mathbf{x})}{g(\mathbf{x})g(\mathbf{x})} \qquad$ if $g(\mathbf{x}) \neq 0$.

(d) $\nabla[h \circ f(\mathbf{x})] = h'(f(\mathbf{x}))\nabla f(\mathbf{x}), \qquad$ that is, $\nabla(h(t)) = \dfrac{dh}{dt}\nabla t$.

Partial Proof. (a) These follow from the corresponding properties of partial differentiation. For instance, for each variable x_i, we have

$$\frac{\partial}{\partial x_i}(f + g) = \frac{\partial f}{\partial x_i} + \frac{\partial g}{\partial x_i},$$

so the respective coordinate functions of $\nabla(f+g)$ and $\nabla f + \nabla g$ agree. A similar proof shows that $\nabla(af) = a\nabla(f)$. We leave the proof of (b) and (c) for Exercises 29 and 30.

(d) This follows from applying the chain rule to the composite function $h \circ f\colon \mathbf{R}^n \to \mathbf{R}^1$, where n is 2 or 3. By Theorem 4.4 of Chapter 12, the total derivative of $h \circ f$ is

$$\nabla(h \circ f)(\mathbf{x}) = \nabla(h(f(\mathbf{x}))) = h'(f(\mathbf{x}))\nabla f(\mathbf{x}). \quad \text{QED}$$

Performing vector operations with the gradient operator leads to the divergence operator, which has many important applications in mechanics, fluid dynamics, electricity, and magnetism.

1.4 DEFINITION Suppose that $\mathbf{F}\colon \mathbf{R}^n \to \mathbf{R}^n$ is a differentiable vector function. Then the ***divergence of*** $\mathbf{F}$ is

$$\text{(2)} \qquad \operatorname{div}\mathbf{F} = \nabla \cdot \mathbf{F} = \frac{\partial f_1}{\partial x_1} + \frac{\partial f_2}{\partial x_2} + \ldots + \frac{\partial f_n}{\partial x_n},$$

where $f_1, f_2, \ldots, f_n$ are the coordinate functions of $\mathbf{F}$.

In the $n = 3$ case we have the formal identity

$$\text{(1)} \qquad \nabla = \left(\frac{\partial}{\partial x}, \frac{\partial}{\partial y}, \frac{\partial}{\partial z}\right) = \frac{\partial}{\partial x}\mathbf{i} + \frac{\partial}{\partial y}\mathbf{j} + \frac{\partial}{\partial z}\mathbf{k},$$

and we can express f in terms of its coordinate functions as

$$\mathbf{F} = (f_1, f_2, f_3) = f_1\,\mathbf{i} + f_2\,\mathbf{j} + f_3\,\mathbf{k}.$$

So if we formally compute the dot product $\nabla \cdot \mathbf{F}$, we obtain

$$\nabla \cdot \mathbf{F} = \frac{\partial}{\partial x}(f_1) + \frac{\partial}{\partial y}(f_2) + \frac{\partial}{\partial z}(f_3) = \frac{\partial f_1}{\partial x} + \frac{\partial f_2}{\partial y} + \frac{\partial f_3}{\partial z},$$

which is the same as (2) when $n = 3$.

We can also use (1) to give meaning to the divergence of a two-dimensional vector field $\mathbf{F}$. Recall that the $\mathbf{k}$-component is not present in (1) in that case. If

$$\mathbf{F}(x, y) = (P(x, y), Q(x, y)) = P(x, y)\,\mathbf{i} + Q(x, y)\,\mathbf{j},$$

then we have from (1)

$$\nabla \cdot \mathbf{F} = \left(\frac{\partial}{\partial x}\mathbf{i} + \frac{\partial}{\partial y}\mathbf{j}\right) \cdot (P\,\mathbf{i} + Q\,\mathbf{j}) = \frac{\partial P}{\partial x} + \frac{\partial Q}{\partial y},$$

which again is (2) when $n = 2$.

The term *divergence* comes from fluid dynamics, where div $\mathbf{F}$ measures the tendency of a fluid to flow outward. In the case of a three-dimensional flow, suppose that flow is described by its differentiable velocity field

$$\mathbf{v}(x, y, z) = v_1(x, y, z)\,\mathbf{i} + v_2(x, y, z)\,\mathbf{j} + v_3(x, y, z)\,\mathbf{k}.$$

Consider a small rectangular box ΔB with edges Δx, Δy, and Δz. See **Figure 1.2.** The component of the flow moving left to right is approximately $v_2(\mathbf{x})\,\mathbf{j}$, where $\mathbf{x}$ is a point on the left wall. At each point $P_1(x, y, z)$ of the left wall, the amount of liquid flowing across that point during time Δt is approximately $v_2(\mathbf{x})\,\Delta t$ linear units. The *total amount* of fluid flowing across the entire wall during time Δt is then approximately

$$\text{(3)} \qquad v_2(\mathbf{x})\,\Delta x\,\Delta z\,\Delta t \text{ cubic units.}$$

FIGURE 1.2

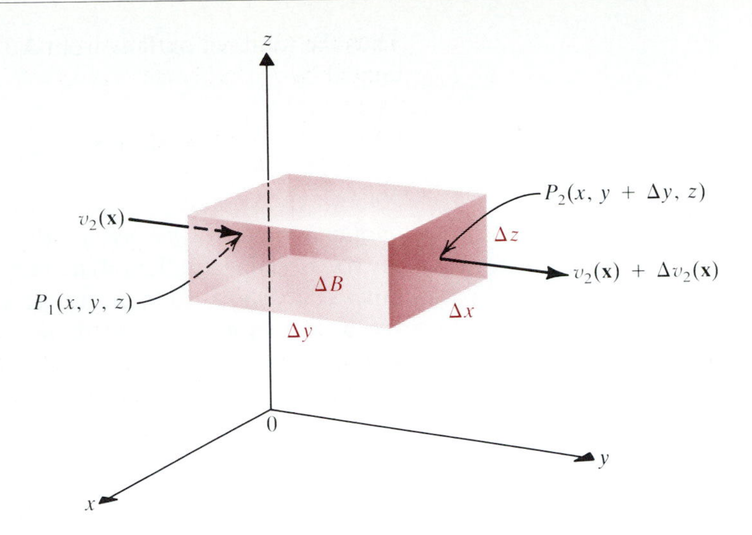

(For example, if $v_2 = 3$ feet per second, then about 3 linear feet per second pass each unit area of the left wall. So $3\,\Delta x\,\Delta z$ cubic feet cross the wall per second.)

At the right wall, the velocity at point $P_2(x, y + \Delta y, z)$ is

$$v_2(x, y + \Delta y, z) \approx v_2(x, y, z) + \frac{\partial v_2(x, y, z)}{\partial y}\,|(x, y + \Delta y, z) - (x, y, z)|,$$

by the tangent approximation for v_2 near (x, y, z). Thus

$$v_2(x, y + \Delta y, z) \approx v_2(x, y, z) + \frac{\partial v_2(\mathbf{x})}{\partial y}\,\Delta y.$$

So the total amount of fluid flowing across the right wall during the time Δt is approximately

$$\left(v_2(\mathbf{x}) + \frac{\partial v_2(\mathbf{x})}{\partial y}\,\Delta y\right)\Delta x\,\Delta z\,\Delta t. \tag{4}$$

Comparing (3) and (4), we see that the net *outflow* of fluid from ΔB in the y-direction, which is the amount leaving the right wall minus the amount entering the left wall, is approximately

$$\frac{\partial v_2(\mathbf{x})}{\partial y}\,\Delta x\,\Delta y\,\Delta z\,\Delta t = \left[v_2(\mathbf{x}) + \frac{\partial v_2(\mathbf{x})}{\partial y}\,\Delta y\right]\Delta x\,\Delta z\,\Delta t - v_2(\mathbf{x})\,\Delta x\,\Delta z\,\Delta t. \tag{5}$$

Identical reasoning on the front and back walls, and on the top and bottom, shows that the net outflow from ΔB in the x-direction is approximately

$$\frac{\partial v_1}{\partial x}(\mathbf{x})\,\Delta x\,\Delta y\,\Delta z\,\Delta t, \tag{6}$$

and that in the z-direction is approximately

$$\frac{\partial v_3}{\partial z}(\mathbf{x})\,\Delta x\,\Delta y\,\Delta z\,\Delta t. \tag{7}$$

Hence the total net outflow from ΔB per unit time is approximated by adding (5), (6), and (7) and dividing by Δt. That gives

$$\left(\frac{\partial v_1}{\partial x} + \frac{\partial v_2}{\partial y} + \frac{\partial v_3}{\partial z}\right)\Delta x\,\Delta y\,\Delta z.$$

Thus the total net outflow from ΔB per unit of volume per unit time is approximated by

$$\text{div } v(\mathbf{x}) = \frac{\partial v_1}{\partial x}(\mathbf{x}) + \frac{\partial v_2}{\partial y}(\mathbf{x}) + \frac{\partial v_3}{\partial z}(\mathbf{x}). \tag{8}$$

As Δx, Δy, Δz, and Δt approach 0, this approximation seems to measure the *outflow per unit volume per unit time* more and more accurately. For this reason, div $\mathbf{v}(\mathbf{x})$ is defined in fluid dynamics as the rate of flow outward from $\mathbf{x}$ per unit time. If div $\mathbf{v}(\mathbf{a}) > 0$, then one says that there is a *source* for the outflow at $\mathbf{x} = \mathbf{a}$. If div $\mathbf{v}(\mathbf{a}) < 0$, then there is a *sink* at $\mathbf{x} = \mathbf{a}$. (There is outflow from a source, inflow to a sink.)

The key properties of the divergence operator are given in the following result.

1.5 THEOREM

Suppose that $\mathbf{F}, \mathbf{G}: \mathbf{R}^n \to \mathbf{R}^n$ and g: $\mathbf{R}^n \to \mathbf{R}$ are differentiable, where $n = 2$ or 3. Let c be a real number. Then

(a) $\text{div}(\mathbf{F} + \mathbf{G}) = \text{div } \mathbf{F} + \text{div } \mathbf{G}$ and $\text{div}(c\mathbf{F}) = c \text{ div } \mathbf{F}$.

(b) $\text{div}(g(\mathbf{x})\mathbf{G}(\mathbf{x})) = g(\mathbf{x}) \text{ div } \mathbf{G}(\mathbf{x}) + \nabla g(\mathbf{x}) \cdot \mathbf{G}(\mathbf{x})$.

Proof. Part (a) is left for Exercise 31. We prove (b) for the case $n = 3$. (The same reasoning applies if $n = 2$.) We have

$$\begin{aligned}
\text{div}(g(\mathbf{x})\mathbf{G}(\mathbf{x})) &= \frac{\partial}{\partial x}(g(\mathbf{x})g_1(\mathbf{x})) + \frac{\partial}{\partial y}(g(\mathbf{x})g_2(\mathbf{x})) + \frac{\partial}{\partial z}(g(\mathbf{x})g_3(\mathbf{x})) \\
&= \frac{\partial g(\mathbf{x})}{\partial x} g_1(\mathbf{x}) + g(\mathbf{x})\frac{\partial g_1(\mathbf{x})}{\partial x} + \frac{\partial g(\mathbf{x})}{\partial y} g_2(\mathbf{x}) + g(\mathbf{x})\frac{\partial g_2(\mathbf{x})}{\partial y} \\
&\quad + \frac{\partial g(\mathbf{x})}{\partial z} g_3(\mathbf{x}) + g(\mathbf{x})\frac{\partial g_3(\mathbf{x})}{\partial z} \\
&= \frac{\partial g(\mathbf{x})}{\partial x} g_1(\mathbf{x}) + \frac{\partial g(\mathbf{x})}{\partial y} g_2(\mathbf{x}) + \frac{\partial g(\mathbf{x})}{\partial z} g_3(\mathbf{x}) + g(\mathbf{x})\frac{\partial g_1(\mathbf{x})}{\partial x} \\
&\quad + g(\mathbf{x})\frac{\partial g_2(\mathbf{x})}{\partial y} + g(\mathbf{x})\frac{\partial g_3(\mathbf{x})}{\partial z} \\
&= \left[\frac{\partial g}{\partial x}\mathbf{i} + \frac{\partial g}{\partial y}\mathbf{j} + \frac{\partial g}{\partial z}\mathbf{k}\right] \cdot [g_1(\mathbf{x})\mathbf{i} + g_2(\mathbf{x})\mathbf{j} + g_3(\mathbf{x})\mathbf{k}] \\
&\quad + g(\mathbf{x}) \text{ div } \mathbf{G}(\mathbf{x}) \\
&= g(\mathbf{x}) \text{ div } \mathbf{G}(\mathbf{x}) + \nabla g(\mathbf{x}) \cdot \mathbf{G}(\mathbf{x}). \quad \boxed{\text{QED}}
\end{aligned}$$

The next example suggests the usefulness of Theorem 1.5 in applied work.

Example 4

Inverse-square Law Fields. Suppose that for nonzero $\mathbf{x} = (x, y, z)$ in $\mathbf{R}^3$,

$$\mathbf{F}(\mathbf{x}) = k\frac{1}{|\mathbf{x}|^2}\mathbf{u},$$

where k is constant and $\mathbf{u}$ is the unit vector giving the direction of $\mathbf{x}$, that is,

$$\mathbf{u} = \frac{1}{|\mathbf{x}|}\,\mathbf{x} = \frac{1}{\sqrt{x^2 + y^2 + z^2}}(x\,\mathbf{i} + y\,\mathbf{j} + z\,\mathbf{k}).$$

Find $\operatorname{div} \mathbf{F}(x, y, z)$.

Solution. We have

$$\mathbf{F}(\mathbf{x}) = k\,\frac{1}{|\mathbf{x}|^3}\,\mathbf{x} = k|\mathbf{x}|^{-3}\mathbf{x} = k(x^2 + y^2 + z^2)^{-3/2}(x\,\mathbf{i} + y\,\mathbf{j} + z\,\mathbf{k}).$$

So if we put

$$g(\mathbf{x}) = k|\mathbf{x}|^{-3} = \frac{k}{(x^2 + y^2 + z^2)^{3/2}} \qquad \text{and} \qquad \mathbf{G}(\mathbf{x}) = \mathbf{x} = x\,\mathbf{i} + y\,\mathbf{j} + z\,\mathbf{k},$$

then we can apply Theorem 1.5(b) to get

$$\begin{aligned} \operatorname{div} \mathbf{F}(\mathbf{x}) &= k|\mathbf{x}|^{-3} \operatorname{div} \mathbf{x} + \nabla(k|\mathbf{x}|^{-3}) \cdot \mathbf{x} \\ &= k|\mathbf{x}|^{-3}(\operatorname{div}[x\,\mathbf{i} + y\,\mathbf{j} + z\,\mathbf{k}]) + k\nabla(|\mathbf{x}|^{-3}) \cdot (x\,\mathbf{i} + y\,\mathbf{j} + z\,\mathbf{k}). \end{aligned}$$

We next apply Theorem 1.3(d) to the second term, with $f(\mathbf{x}) = |\mathbf{x}|$ and $h(t) = t^{-3}$. This, together with the observation that

$$\operatorname{div}[x\,\mathbf{i} + y\,\mathbf{j} + z\,\mathbf{k}] = \frac{\partial x}{\partial x} + \frac{\partial y}{\partial y} + \frac{\partial z}{\partial z} = 1 + 1 + 1 = 3,$$

gives

$$\text{(9)} \quad \operatorname{div} \mathbf{F}(\mathbf{x}) = k|\mathbf{x}|^{-3}(3) + k(-3)|\mathbf{x}|^{-4}\,\nabla(|\mathbf{x}|) \cdot \mathbf{x} = 3k|\mathbf{x}|^{-3} - 3k|\mathbf{x}|^{-4}\,\nabla(|\mathbf{x}|) \cdot \mathbf{x}.$$

Now

$$\begin{aligned} \nabla(|\mathbf{x}|) &= \nabla(\sqrt{x^2 + y^2 + z^2}) \\ &= \frac{x}{\sqrt{x^2 + y^2 + z^2}}\,\mathbf{i} + \frac{y}{\sqrt{x^2 + y^2 + z^2}}\,\mathbf{j} + \frac{z}{\sqrt{x^2 + y^2 + z^2}}\,\mathbf{k} \\ &= \frac{1}{|\mathbf{x}|}\,\mathbf{x}. \end{aligned}$$

Thus

$$\nabla(|\mathbf{x}|) \cdot \mathbf{x} = \frac{1}{|\mathbf{x}|}\,\mathbf{x} \cdot \mathbf{x} = |\mathbf{x}|.$$

So from (9) we have

$$\operatorname{div} \mathbf{F}(\mathbf{x}) = 3k|\mathbf{x}|^{-3} - 3k|\mathbf{x}|^{-4}|\mathbf{x}| = 0. \quad ■$$

Inverse-square law fields play a major role in electromagnetic theory and gravitational theory. For instance, the electric force on a unit charge at $\mathbf{x}$ produced by a unit charge of the same sign at $\mathbf{0}$ is proportional to

$$\frac{1}{|\mathbf{x}|^2}\,\mathbf{u}.$$

The gravitational field of the earth at a point $\mathbf{x}$ is given by

$$G(\mathbf{x}) = -gR^2 \frac{1}{|\mathbf{x}|^2}\mathbf{u},$$

where R is the radius of the earth, and g is the acceleration due to gravity at the surface of the earth. (The origin is taken at the center of the earth.) Thus Example 4 applies to both the electrical and gravitational situations. Inverse-square law fields are part of an important class of vector fields that conserve total (=potential + kinetic) energy. Such fields, which we will study further in Section 4, are defined mathematically as gradients of scalar functions.

1.6 DEFINITION

A vector field $\mathbf{F}: \mathbf{R}^n \to \mathbf{R}^n$ is called ***conservative*** if

$$\mathbf{F}(\mathbf{x}) = \nabla p(\mathbf{x})$$

for some scalar function $p: \mathbf{R}^n \to \mathbf{R}$. In this case, p is called a ***potential function*** for $\mathbf{F}$.

Example 5

Show that the inverse-square law field $\mathbf{F}$ of Example 4 is conservative, with the potential function

$$p(\mathbf{x}) = -k\frac{1}{|\mathbf{x}|}.$$

Solution. We need only verify that $\nabla p(\mathbf{x}) = \mathbf{F}(\mathbf{x})$. Here

$$\nabla p(\mathbf{x}) = -k\nabla\left(\frac{1}{|\mathbf{x}|}\right) = -k\nabla(|\mathbf{x}|^{-1}) = k|\mathbf{x}|^{-2}\nabla(|\mathbf{x}|) \quad \textit{by Theorem 1.3(d)}$$

$$= k\frac{1}{|\mathbf{x}|^2}\frac{1}{|\mathbf{x}|}\mathbf{x} = \mathbf{F}(\mathbf{x}). \blacksquare$$

In Example 4 we showed that $\operatorname{div}\mathbf{F}(\mathbf{x}) = 0$ for the given inverse-square law field F. From Example 5, then, for the potential function

$$p(\mathbf{x}) = -\frac{k}{|\mathbf{x}|} = -\frac{k}{\sqrt{x^2+y^2+z^2}} = -k|\mathbf{x}|^{-1}$$

of the inverse-square law field

$$\mathbf{F}(\mathbf{x}) = \nabla p(\mathbf{x}) = k|\mathbf{x}|^{-3}\mathbf{x},$$

we have

$$\operatorname{div}\nabla p(\mathbf{x}) = 0 = \nabla \cdot \nabla p(\mathbf{x}).$$

Thus

$$\left(\frac{\partial}{\partial x}\mathbf{i} + \frac{\partial}{\partial y}\mathbf{j} + \frac{\partial}{\partial z}\mathbf{k}\right)\cdot\left(\frac{\partial p}{\partial x}\mathbf{i} + \frac{\partial p}{\partial y}\mathbf{j} + \frac{\partial p}{\partial z}\mathbf{k}\right) = 0.$$

This is the same as

$$\frac{\partial^2 p}{\partial x^2} + \frac{\partial^2 p}{\partial y^2} + \frac{\partial^2 p}{\partial z^2} = 0.$$

The latter equation occurs so often that it carries a special name, that of the French mathematician Pierre S. Laplace (1749–1827), who played a major role in the development of mechanics.

1.7 DEFINITION

The operator ∇^2 given by

$$\nabla^2 = \nabla \cdot \nabla = \frac{\partial^2}{\partial x^2} + \frac{\partial^2}{\partial y^2} + \frac{\partial^2}{\partial z^2}$$

is called the ***Laplacian operator.*** The equation

$$\nabla^2 p(\mathbf{x}) = 0$$

is called ***Laplace's equation.*** A function p satisfying Laplace's equation is called a ***harmonic function.***

Many important functions arising in physics, including electrical potential, pressure, and temperature functions, are harmonic functions. Their study forms an important part of modern analysis.

As remarked above, potential functions of inverse-square law fields are always harmonic, so varying k in Example 5 will produce infinitely many harmonic functions. The next example gives a different type of harmonic function.

Example 6

Show that $f(x, y, z) = x^2 - y^2 + e^z \sin y$ is a harmonic function.

Solution. We have

$$\nabla f = 2x\,\mathbf{i} - (2y - e^z \cos y)\,\mathbf{j} + e^z \sin y\,\mathbf{k}.$$

Therefore

$$\nabla^2 f = \nabla \cdot \nabla f = \left(\frac{\partial}{\partial x}\mathbf{i} + \frac{\partial}{\partial y}\mathbf{j} + \frac{\partial}{\partial z}\mathbf{k}\right) \cdot (2x\,\mathbf{i} - [2y - e^z \cos y]\,\mathbf{j} + e^z \sin y\,\mathbf{k})$$
$$= 2 - 2 - e^z \sin y + e^z \sin y = 0. \quad \blacksquare$$

We come now to the final differential operator to be discussed in this section. It also involves the gradient operator ∇. As its name suggests, it is helpful in describing swirling vector fields, such as the velocity field of water in a whirlpool.

1.8 DEFINITION

If $\mathbf{F}\colon \mathbf{R}^3 \to \mathbf{R}^3$ is a differentiable vector field, then its ***curl*** is

$$\textbf{curl}\,\mathbf{F} = \nabla \times \mathbf{F}$$
$$= \left[\frac{\partial}{\partial x}\mathbf{i} + \frac{\partial}{\partial y}\mathbf{j} + \frac{\partial}{\partial z}\mathbf{k}\right] \times [f_1(x, y, z)\,\mathbf{i} + f_2(x, y, z)\,\mathbf{j} + f_3(x, y, z)\,\mathbf{k}]$$
$$= \det\begin{pmatrix} \mathbf{i} & \mathbf{j} & \mathbf{k} \\ \dfrac{\partial}{\partial x} & \dfrac{\partial}{\partial y} & \dfrac{\partial}{\partial z} \\ f_1 & f_2 & f_3 \end{pmatrix} = \left(\frac{\partial f_3}{\partial y} - \frac{\partial f_2}{\partial z}\right)\mathbf{i} + \left(\frac{\partial f_1}{\partial z} - \frac{\partial f_3}{\partial x}\right)\mathbf{j} + \left(\frac{\partial f_2}{\partial x} - \frac{\partial f_1}{\partial y}\right)\mathbf{k}.$$

Note that the curl operator is defined *only* for three-dimensional vector fields.

Example 7

If $\mathbf{F}(x, y, z) = (3x^2yz - 5x^2yz^2)\mathbf{i} + e^{x^2+y^2+z^2}\mathbf{j} + (\sin x + \cos y - \tan z)\mathbf{k}$, then find **curl F**.

Solution. Definition 1.8 gives

$$\begin{aligned}\mathbf{curl\,F} = \nabla \times \mathbf{F} &= \left(\frac{\partial}{\partial x}\mathbf{i} + \frac{\partial}{\partial y}\mathbf{j} + \frac{\partial}{\partial z}\mathbf{k}\right) \times [(3x^2yz - 5x^2yz^2)\mathbf{i} + e^{x^2+y^2+z^2}\mathbf{j} \\ &\qquad + (\sin x + \cos y - \tan z)\mathbf{k}] \\ &= (-\sin y - 2ze^{x^2+y^2+z^2})\mathbf{i} + (3x^2y - 10x^2yz - \cos x)\mathbf{j} \\ &\qquad + (2xe^{x^2+y^2+z^2} - 3x^2z + 5x^2z^2)\mathbf{k}. \quad \blacksquare\end{aligned}$$

Corresponding to Theorems 1.3 and 1.5 is the following collection of properties for the curl operator.

1.9 THEOREM

Let $\mathbf{F}, \mathbf{G}: \mathbf{R}^3 \to \mathbf{R}^3$ and $f: \mathbf{R}^3 \to \mathbf{R}$ be differentiable. Then

(a) $\mathbf{curl}(\mathbf{F} + \mathbf{G}) = \mathbf{curl\,F} + \mathbf{curl\,G}$, $\mathbf{curl}(a\mathbf{F}) = a\,\mathbf{curl\,F}$ for any real number a.

(b) $\mathbf{curl}(f(\mathbf{x})\mathbf{F}(\mathbf{x})) = f(\mathbf{x})\,\mathbf{curl\,F}(\mathbf{x}) + \nabla f(\mathbf{x}) \times \mathbf{F}(\mathbf{x})$.

(c) $\operatorname{div}(\mathbf{F} \times \mathbf{G}) = \mathbf{G} \cdot \mathbf{curl\,F} - \mathbf{F} \cdot \mathbf{curl\,G}$.

(d) If the second partial derivatives of f are continuous, then $\mathbf{curl}\,\nabla f = \mathbf{0}$.

(e) If the second partial derivatives of all the coordinate functions of $\mathbf{F}$ are continuous, then $\operatorname{div}(\mathbf{curl\,F}) = 0$.

Partial Proof. The proof of part (a) is left as Exercise 32. The proofs of (b) and (c) are given by expanding the left sides using Definitions 1.4 and 1.9 and the product rule for derivatives (Exercises 33 and 35). As for (d), we have

$$\begin{aligned}\mathbf{curl}\,\nabla f = \nabla \times \nabla f &= \left(\frac{\partial}{\partial x}\mathbf{i} + \frac{\partial}{\partial y}\mathbf{j} + \frac{\partial}{\partial z}\mathbf{k}\right) \times \left(\frac{\partial f}{\partial x}\mathbf{i} + \frac{\partial f}{\partial y}\mathbf{j} + \frac{\partial f}{\partial z}\mathbf{k}\right) \\ &= \left(\frac{\partial^2 f}{\partial y\,\partial z} - \frac{\partial^2 f}{\partial z\,\partial y}\right)\mathbf{i} + \left(\frac{\partial^2 f}{\partial z\,\partial x} - \frac{\partial^2 f}{\partial x\,\partial z}\right)\mathbf{j} + \left(\frac{\partial^2 f}{\partial x\,\partial y} - \frac{\partial^2 f}{\partial y\,\partial x}\right)\mathbf{k} \\ &= \mathbf{0}. \qquad \textit{by Theorem 6.2 of Chapter 12}\end{aligned}$$

To show (e), we write

$$\begin{aligned}\operatorname{div}\mathbf{curl\,F} &= \frac{\partial}{\partial x}\left(\frac{\partial f_3}{\partial y} - \frac{\partial f_2}{\partial z}\right) + \frac{\partial}{\partial y}\left(\frac{\partial f_1}{\partial z} - \frac{\partial f_3}{\partial x}\right) + \frac{\partial}{\partial z}\left(\frac{\partial f_2}{\partial x} - \frac{\partial f_1}{\partial y}\right) \\ &= \frac{\partial^2 f_3}{\partial x\,\partial y} - \frac{\partial^2 f_2}{\partial x\,\partial z} + \frac{\partial^2 f_1}{\partial y\,\partial z} - \frac{\partial^2 f_3}{\partial y\,\partial x} + \frac{\partial^2 f_2}{\partial z\,\partial x} - \frac{\partial^2 f_1}{\partial z\,\partial y} \\ &= \mathbf{0}. \qquad \textit{again by Theorem 6.2 of Chapter 12} \quad \boxed{\text{QED}}\end{aligned}$$

HISTORICAL NOTE

The Jacobian matrix is named in honor of the German mathematician Carl Gustav J. Jacobi (1804–1851), who made a number of important contributions to nineteenth-century mathematics. He was one of the first to give a systematic theory of determinants and dealt with det $\mathbf{J}_\mathbf{F}$, which is called the *Jacobian determinant,* rather than with the matrix $\mathbf{J}_\mathbf{F}$ per se. He also worked with differential equations.

Pierre-Simon Laplace did extensive work in physics, astronomy, and chemistry, as well as mathematics. He used the equation $\nabla^2 f = 0$ in his work on the gravitational attraction between planets in the solar system. This work helped him explain the regularity and stability of planetary motion.

Exercises 14.1

In Exercises 1–4, find $\lim_{\mathbf{x}\to\mathbf{a}}\mathbf{F}(\mathbf{x})$ for the given function. Is the function continuous at the given point $\mathbf{x} = \mathbf{a}$? Why or why not?

1. $\mathbf{F}(x, y) = \dfrac{x^4 - y^4}{x^2 + y^2}\mathbf{i} + (x^2 + y^2)\mathbf{j} + (3e^{x+y})\mathbf{k}$; $\mathbf{a} = (0, 0)$

2. $\mathbf{F}(x, y) = \tan \pi xy\,\mathbf{i} + e^{x^2 y}\mathbf{j} + \dfrac{x^2 + 1}{y^2 + 1}\mathbf{k}$; $\mathbf{a} = (2, 3)$

3. $\mathbf{F}(x, y, z) = \dfrac{x^2 - y^2}{x + y}\mathbf{i} + yze^{xy}\mathbf{j} + (x^2 - y^2z^2 + e^{xyz})\mathbf{k}$; $\mathbf{a} = (0, 0, 2)$

4. $\mathbf{F}(x, y, z)$ of Exercise 3; $\mathbf{a} = (1, -1, 0)$

In Exercises 5–8, find $\mathbf{J}_\mathbf{F}(\mathbf{a})$ for the given F and a.

5. $\mathbf{F}(x, y, z) = (x^2 + e^y + z^3)\mathbf{i} + (\sin xy + yze^x)\mathbf{j}$; $\mathbf{a} = (1, \pi, -1)$

6. $\mathbf{F}(x, y, z) = \ln(x^2 + y^2 + z^2)\mathbf{i} + 3e^{xyz}\mathbf{j}$; $\mathbf{a} = (1, 0, 1)$

7. $\mathbf{F}(x, y, z) = \begin{pmatrix} x\cos \pi yz - yz\sin \pi x \\ \sqrt{x^2 + 4y^2 + e^{z^2}} \\ \tan \pi xy - 2xyz^2 \end{pmatrix}$; $\mathbf{a} = (1, 1, 0)$

8. $\mathbf{F}(x, y, z) = \begin{pmatrix} x^2 y - z \\ x^2 - z^2 \\ \cos \pi xyz \end{pmatrix}$; $\mathbf{a} = (1, -1, 1/2)$

In Exercises 9–12, compute the divergence of the given vector field F at x and the given a.

9. $\mathbf{F}(\mathbf{x}) = x\,\mathbf{i} + (xz + 3)\mathbf{j} + (yz + x)\mathbf{k}$; $\mathbf{a} = (2, -2, 3)$

10. $\mathbf{F}(\mathbf{x}) = ({x_1}^2 + {x_2}^2 + {x_3}^2 + {x_4}^2, x_1x_2x_3x_4, x_1x_2 - x_3x_4, x_1x_4 - x_2x_3)$; $\mathbf{a} = (1, -1, 2, 3)$

11. $\mathbf{F}(\mathbf{x}) = xy^2z^2\mathbf{i} + z^2\sin y\,\mathbf{j} + x^2e^y\mathbf{k}$; $\mathbf{a} = (1, \pi, -2)$

12. $\mathbf{F}(\mathbf{x}) = \dfrac{x + z}{x^2 + y^2 + z^2}\mathbf{i} + \dfrac{y - x}{x^2 + y^2 + z^2}\mathbf{j} + \dfrac{z - y}{x^2 + y^2 + z^2}\mathbf{k}$; $\mathbf{a} = (1, 0, -2)$

In Exercises 13–16, compute the curl of the given field F at x and at a.

13. $\mathbf{F}(\mathbf{x}) = 3x^2\mathbf{i} + xy\,\mathbf{j} + z\,\mathbf{k}$; $\mathbf{a} = (-3, 4, 2)$

14. The F and a of Exercise 9

15. $\mathbf{F}(\mathbf{x}) = e^x\sin y\cos z\,\mathbf{i} + e^x\cos y\cos z\,\mathbf{j} - e^x\sin y\sin z\,\mathbf{k}$; $\mathbf{a} = (-2, \pi/2, 5\pi/6)$

16. $\mathbf{F}(\mathbf{x}) = (ye^{x^2z} + 2x^2yze^{x^2z}, xe^{x^2z}, x^3ye^{x^2z})$; $\mathbf{a} = (1, 2, 3)$

In Exercises 17–24, show that the given functions are harmonic.

17. $f(x, y) = \ln\sqrt{x^2 + y^2}$; $(x, y) \neq (0, 0)$

18. $f(x, y) = e^x\cos y + 3x^2 - 3y^2$

19. $f(x, y) = \sin x\sinh y$

20. $f(x, y) = \cosh 2x\cos 2y$

21. $f(x, y, z) = x^2 + y^2 - 2z^2$

22. $f(x, y, z) = 3x^2 + 2y^2 - 5z^2$

23. $f(x, y, z) = 1/\sqrt{x^2 + y^2 + z^2}$; $(x, y, z) \neq (0, 0, 0)$

24. $f(x, y, z) = \sin x\sinh y + \cos x\cosh z$

25. Laplacian in Polar Coordinates. Suppose that $z = f(x, y)$, and we introduce polar coordinates $x = r\cos\theta$ and $y = r\sin\theta$. Then $z = f(r\cos\theta, r\sin\theta) = g(r, \theta)$. Show that

$$\nabla^2 z = \frac{\partial^2 g}{\partial r^2} + \frac{1}{r^2}\frac{\partial^2 g}{\partial\theta^2} + \frac{1}{r}\frac{\partial g}{\partial r}.$$

(*Hint:* See Example 4, Section 12.6.)

26. Laplacian in Cylindrical Coordinates. Suppose that $w = f(x, y, z)$, and we introduce cylindrical coordinates $x = r\cos\theta$, $y = r\sin\theta$, and $z = z$. Then $w = f(r\cos\theta, r\sin\theta, z) = g(r, \theta, z)$. Show that

$$\nabla^2 f(x, y, z) = \frac{1}{r}\frac{\partial g}{\partial r} + \frac{\partial^2 g}{\partial r^2} + \frac{1}{r^2}\frac{\partial^2 g}{\partial\theta^2} + \frac{\partial^2 g}{\partial z^2}.$$

(*Hint:* See Example 4, Section 12.6.)

27. Laplacian in Spherical Coordinates. Suppose that $w = f(x, y, z)$, and we introduce spherical coordinates

$x = \rho \sin\phi \cos\theta$, $y = \rho \sin\phi \sin\theta$, and $z = \rho\cos\phi$. Then

$$w = f(\rho \sin\phi\cos\theta, \rho\sin\phi\sin\theta, \rho\cos\phi) = g(\rho, \phi, \theta).$$

Show that

$$\nabla^2 w = \frac{\partial^2 g}{\partial \rho^2} + \frac{2}{\rho}\frac{\partial g}{\partial \rho} + \frac{1}{\rho^2}\frac{\partial^2 g}{\partial \phi^2} + \frac{\cos\phi}{\rho^2 \sin\phi}\frac{\partial g}{\partial \phi} + \frac{1}{\rho^2\sin^2\phi}\frac{\partial^2 g}{\partial \theta^2}.$$

28. **(a)** Show that $\nabla(|\mathbf{x}|) = \frac{1}{|\mathbf{x}|}\mathbf{x}$ if $\mathbf{x} = x\,\mathbf{i} + y\,\mathbf{j} + z\,\mathbf{k} \neq \mathbf{0}$.

(b) Use (a) to find $\mathbf{curl}\left(\frac{1}{|\mathbf{x}|}\mathbf{x}\right)$ without further calculation.

In Exercises 29–33, prove the indicated result.

29. Theorem 1.3(b)

30. Theorem 1.3(c)

31. Theorem 1.5(a)

32. Theorem 1.9(a)

33. Theorem 1.9(b)

34. Prove that if all second partial derivatives of the coordinate functions f_1, f_2, and f_3 of $\mathbf{F}$ are continuous, then

$$\mathbf{curl}(\mathbf{curl}\,\mathbf{F}) = \nabla(\operatorname{div}\mathbf{F}) - \nabla^2\mathbf{F},$$

where $\nabla^2\mathbf{F} = (\nabla^2 f_1)\,\mathbf{i} + (\nabla^2 f_2)\,\mathbf{j} + (\nabla^2 f_3)\,\mathbf{k}$.

35. Prove Theorem 1.9(c).

36. Use Exercise 34 to compute $\mathbf{curl}(\mathbf{curl}\,\mathbf{F})$ if $\mathbf{F}(x, y, z) = (xz + y^2)\,\mathbf{i} + (xy + z^2)\,\mathbf{j} + (x^2 + yz)\,\mathbf{k}$.

37. Verify the formula for $\operatorname{div}(\mathbf{F} \times \mathbf{G})$ in Exercise 35, if $\mathbf{F}(x, y, z) = x^2\,\mathbf{i} + y^2\,\mathbf{j} + z^2\,\mathbf{k}$ and $\mathbf{G}(x, y, z) = y^2\,\mathbf{i} + z^2\,\mathbf{j} + x^2\,\mathbf{k}$.

38. The ***trace*** of a 3-by-3 matrix

$$A = \begin{pmatrix} a_{11} & a_{12} & a_{13} \\ a_{21} & a_{22} & a_{23} \\ a_{31} & a_{32} & a_{33} \end{pmatrix}$$

is defined to be $a_{11} + a_{22} + a_{33}$, the sum of the *main diagonal* entries. If $\mathbf{F}\colon \mathbf{R}^3 \to \mathbf{R}^3$, then what matrix discussed in this section has trace $\operatorname{div}\mathbf{F}$?

39. Refer to Exercise 38. If $f(x, y)$ has continuous second partial derivatives, then express $\nabla^2 f$ as the trace of a matrix we have met.

40. A fluid with velocity field $\mathbf{F}$ is called ***incompressible*** if $\operatorname{div}\mathbf{F}(\mathbf{x}) = 0$. A fluid flow is called ***solenoidal*** if $\mathbf{F}(x) = \mathbf{curl}\,\mathbf{G}(\mathbf{x})$ for some $\mathbf{G}$. What relationship can you conclude exists between incompressible and solenoidal flows? (Assume appropriate differentiability hypotheses.)

In Exercises 41–44, sketch the given vector field as in Figure 1.1.

41. $\mathbf{F}(x, y) = -\frac{1}{2}y\,\mathbf{i} + \frac{1}{2}x\,\mathbf{j}$

42. $\mathbf{F}(x, y) = -\frac{1}{(x^2 + y)^{3/2}}(x\,\mathbf{i} + y\,\mathbf{j})$

43. $\mathbf{F}(x, y, z) = \frac{-1}{\sqrt{x^2 + y^2 + z^2}}(x\,\mathbf{i} + y\,\mathbf{j} + z\,\mathbf{k})$

44. $\mathbf{F}(x, y, z) = -y\,\mathbf{i} + x\,\mathbf{j} + \mathbf{k}$

14.2 Line Integrals

The double and triple integrals of scalar functions $f\colon \mathbf{R}^n \to \mathbf{R}$, where $n = 2$ or 3, were developed in the last chapter as analogues of the ordinary definite integral $\int_a^b f(x)\,dx$ of a function f over an interval $[a, b]$. They are defined over regions D having the *same dimension—n—*as the domain of $f\colon \mathbf{R}^n \to \mathbf{R}$. This corresponds to the fact that a closed interval $[a, b]$ on the real line has the same dimension—1—as the domain of the function f in $\int_a^b f(x)\,dx$.

In this section we define a different kind of generalization of $\int_a^b f(x)\,dx$, this time for vector functions $\mathbf{F}\colon \mathbf{R}^n \to \mathbf{R}^n$ over a *one-dimensional* subset of $\mathbf{R}^n$, namely, a parametrized curve C in $\mathbf{R}^n$, for $n = 2$ or 3. Such a curve is a generalization of a one-dimensional closed interval $[a, b]$ in $\mathbf{R}^1$. We define the new integral in such a way that when $n = 1$ and $C = [a, b]$, we get the ordinary definite integral $\int_a^b f(x)\,dx$.

Before constructing this new integral, we briefly review the main facts about parametrized curves that are needed for its construction. (You may also wish to review Sections 8.5 and 11.1.) A ***parametrized curve*** C in $\mathbf{R}^2$ or $\mathbf{R}^3$ is a continuous vector function $\mathbf{g}\colon \mathbf{R} \to \mathbf{R}^n$ that is defined on some closed interval $[a, b]$ of the real line. We restrict attention to curves that are *piecewise smooth.* This means that $[a, b]$ can be partitioned into a finite number of subintervals $[a_i, b_i]$ on each of which $\mathbf{g}$ is *smooth,* that is, $\mathbf{g}'$ exists and is continuous on

$[a_i, b_i]$. (Recall that the smooth curve C_i thus has an arc length over $[a_i, b_i]$. See Section 11.2.) The curve C might look something like the one in **Figure 2.1.** The parts of the curve between $\mathbf{x}(a)$ and $\mathbf{x}(t_1)$, between $\mathbf{x}(t_i)$ and $\mathbf{x}(t_{i+1})$ for $i = 1$, 2, and 3, and between $\mathbf{x}(t_4)$ and $\mathbf{x}(b)$ are smooth.

FIGURE 2.1

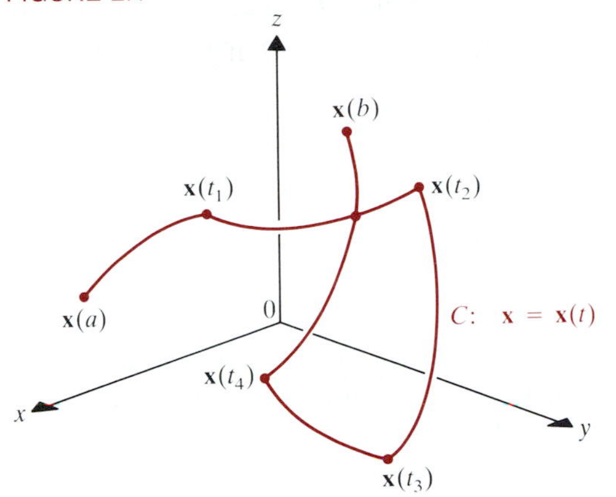

FIGURE 2.2

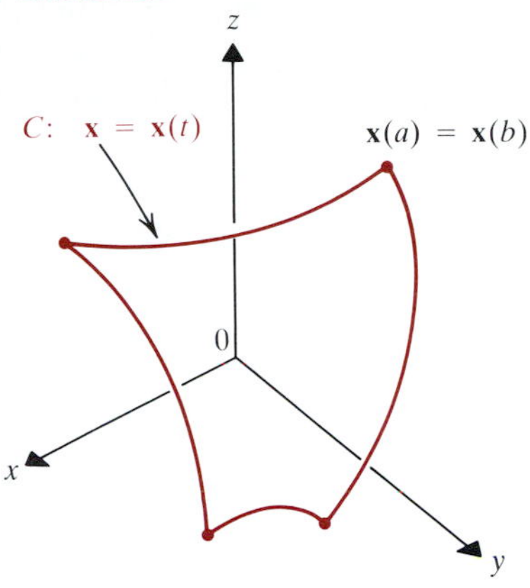

We usually restrict attention to *simple* curves C that do not cross themselves (as C in Figure 2.1 does). Such curves can be parametrized on $[a, b]$ so that $\mathbf{x}(t_1) = \mathbf{x}(t_2)$ can hold only when $t_1 = t_2$ or when $t_1 = a$ and $t_2 = b$. In the latter case, C is a ***simple closed curve*** like the one shown in **Figure 2.2.** Simple curves have a natural *orientation*, namely, $\mathbf{x}(t_1)$ ***precedes*** $\mathbf{x}(t_2)$ if $t_1 < t_2$. As in **Figure 2.3,** arrowheads are often used to indicate the direction of increasing t. Our first example reviews two useful facts about parametrizing curves. *First*, a curve whose Cartesian coordinate equation is reducible to the form $u^2 + v^2 = 1$ can be parametrized by letting $u = \cos t$ and $v = \sin t$. *Second*, if a simple curve C is parametrized by $\mathbf{x} = \mathbf{x}(t)$ for $t \in [a, b]$, then the orientation of C is *reversed* if the parametrization is changed to $\mathbf{x} = \mathbf{x}(a + b - t)$. (See **Figure 2.4,** and Example 3(c) of Section 8.5.) With the latter parametrization, we start at $\mathbf{x}(b)$ when $t = a$ and end at $\mathbf{x}(a)$ when $t = b$. The resulting oriented curve is denoted $-C$ and called the ***negative*** (or ***opposite***) of the original curve C.

FIGURE 2.3

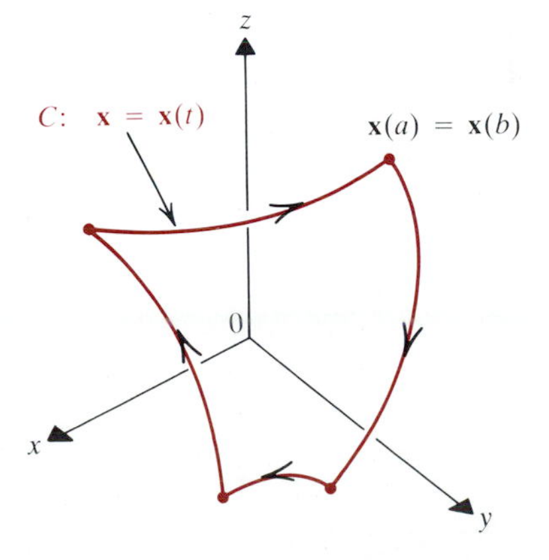

FIGURE 2.4

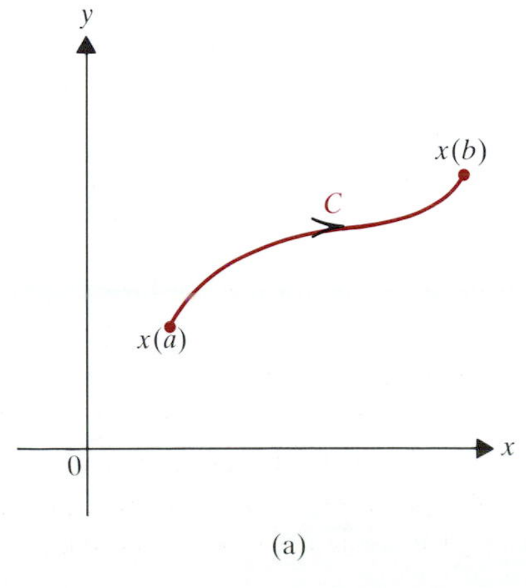

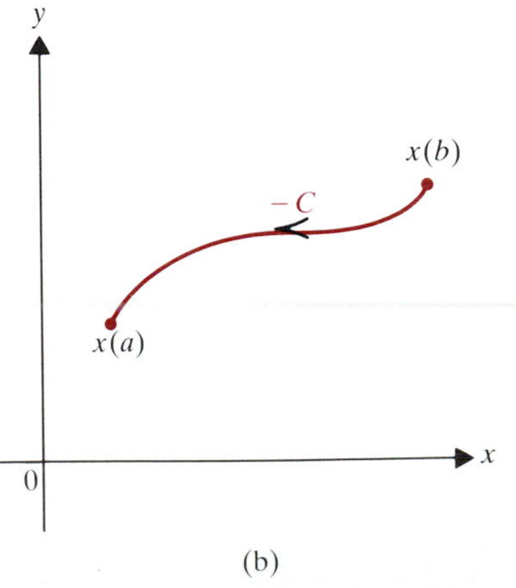

FIGURE 2.5

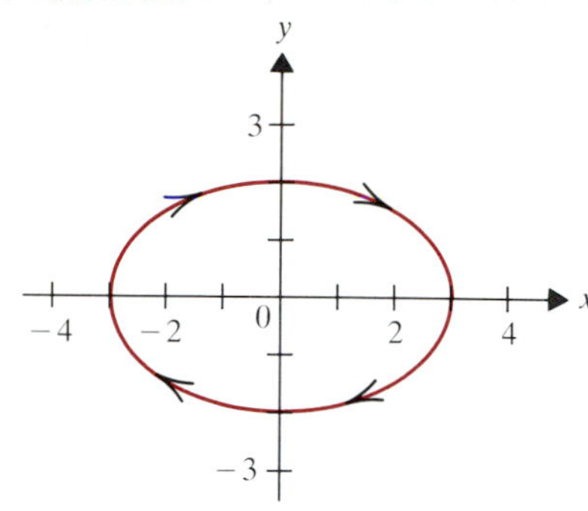

Example 1

Parametrize the simple closed curve shown in **Figure 2.5.**

Solution. The curve is the ellipse $x^2/9 + y^2/4 = 1$ traversed in the *clockwise* sense. The usual parametrization, $\mathbf{x} = \mathbf{x}(t) = (3\cos t, 2\sin t)$ for $0 \leq t \leq 2\pi$, gives this curve, but *not* with this orientation, since as t increases one travels along the curve in the *counterclockwise* sense. As suggested above, to obtain the clockwise orientation we replace t by $2\pi - t$ in the usual parametrization. That gives the parametrization

$$\mathbf{x} = (3\cos(2\pi - t), 2\sin(2\pi - t)) = (3\cos t, -2\sin t),\ t \in [0, 2\pi].$$

Since $\mathbf{x}(0) = (3, 0)$, $\mathbf{x}(\pi/2) = (0, -2)$, $\mathbf{x}(\pi) = (-3, 0)$, and $\mathbf{x}(2\pi) = (3, 0)$, the orientation is indeed clockwise. ■

FIGURE 2.6

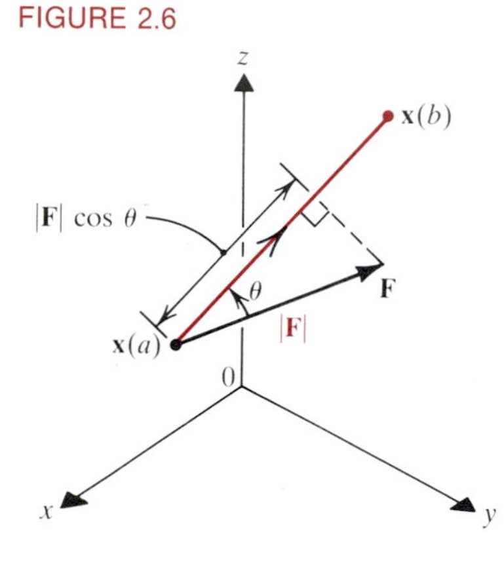

Now we can describe the *line integral* of a continuous vector function $\mathbf{F}: \mathbf{R}^n \to \mathbf{R}^n$ over a curve $C: \mathbf{x} = \mathbf{x}(t)$ between two points $\mathbf{x}(a)$ and $\mathbf{x}(b)$. We approach this via one of the line integral's important physical interpretations, *work*. If a constant force $\mathbf{F}: \mathbf{R}^3 \to \mathbf{R}^3$ acts to move a particle along the line segment joining $\mathbf{x}(a)$ to $\mathbf{x}(b)$, then the ***work*** W done by the force is defined to be the product of the distance $|\mathbf{x}(b) - \mathbf{x}(a)|$ times the coordinate of $\mathbf{F}$ acting in the direction of the motion (see **Figure 2.6**). The latter coordinate is $|\mathbf{F}|\cos\theta$, so we have

$$W = |\mathbf{F}|\,|\mathbf{x}(b) - \mathbf{x}(a)|\cos\theta = \mathbf{F}\cdot(\mathbf{x}(b) - \mathbf{x}(a)). \tag{1}$$

The more typical situation arises when $\mathbf{F}$ is no longer constant but varies over time, and the path of motion is a curve $C: \mathbf{x} = \mathbf{x}(t)$, extending from $\mathbf{x}(a)$ to $\mathbf{x}(b)$, as in **Figure 2.7.**

Suppose then that $\mathbf{F}$ is continuous and that the curve is smooth and simple. If we partition $[a, b]$ by $P: a = t_0 < t_1 < \ldots < t_{i-1} < t_i < \ldots < t_k = b$, then the amount of work ΔW_i done by $\mathbf{F}$ along the arc between $\mathbf{x}(t_{i-1})$ and $\mathbf{x}(t_i)$ is approximately the work done by the constant force $\mathbf{F}(\mathbf{x}(\bar{t}_i))$ for some $\bar{t}_i \in [t_{i-1}, t_i]$ in moving the particle along the straight line segment joining $\mathbf{x}(t_{i-1})$ to $\mathbf{x}(t_i)$ in **Figure 2.8.** Thus from (1),

$$\Delta W_i \approx F(\mathbf{x}(\bar{t}_i))\cdot[\mathbf{x}(t_i) - \mathbf{x}(t_{i-1})]. \tag{2}$$

FIGURE 2.7

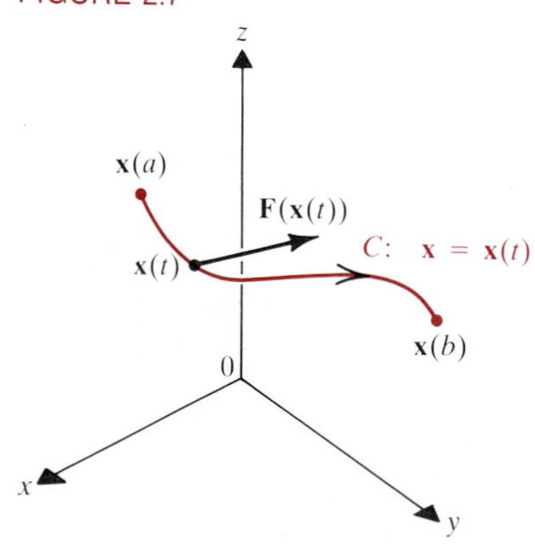

FIGURE 2.8

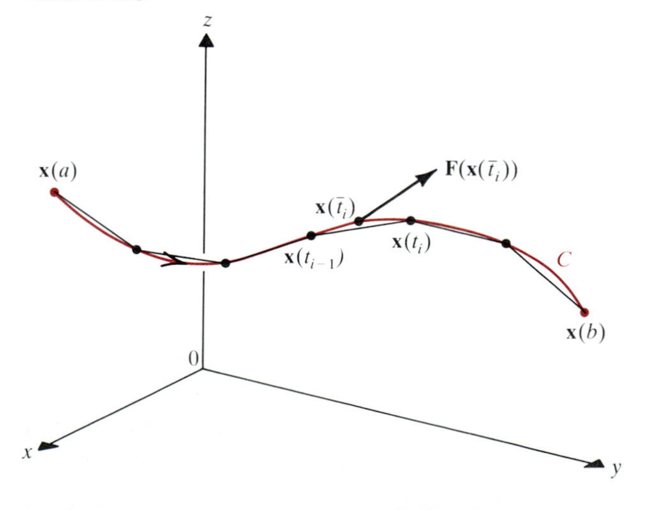

Since C is smooth, $\mathbf{x}$ is differentiable. Hence, for a fine partition P, the tangent approximation gives

(3) $$\mathbf{x}(t_i) - \mathbf{x}(t_{i-1}) \approx \mathbf{x}'(t_{i-1})(t_i - t_{i-1}).$$

The continuity of $\mathbf{x}'(t)$ also says that $\mathbf{x}'(t_{i-1}) \approx \mathbf{x}'(\bar{t}_i)$. Therefore (2) becomes

(4) $$\Delta W_i \approx \mathbf{F}(\mathbf{x}(\bar{t}_i)) \cdot \mathbf{x}'(\bar{t}_i)\,\Delta t_i,$$

where $\Delta t_i = t_i - t_{i-1}$. The *total work* done by $\mathbf{F}$ in moving the particle from $\mathbf{x}(a)$ to $\mathbf{x}(b)$ along C is thus approximately the Riemann sum

$$\sum_{i=1}^{k} \mathbf{F}(\mathbf{x}(\bar{t}_i)) \cdot \mathbf{x}'(\bar{t}_i)\,\Delta t_i$$

for the real-valued function $\mathbf{F}(\mathbf{x}(t)) \cdot \mathbf{x}'(t)$. For this reason, physicists define the work to be the limit of these Riemann sums.

2.1 DEFINITION The ***work*** done by a continuous force $\mathbf{F}: \mathbf{R}^n \to \mathbf{R}^n$ in moving a particle along a smooth simple curve $C: \mathbf{x} = \mathbf{x}(t)$ in $\mathbf{R}^n$ from $\mathbf{x}(a)$ to $\mathbf{x}(b)$ is

(5) $$W = \int_a^b \mathbf{F}(\mathbf{x}(t)) \cdot \mathbf{x}'(t)\,dt.$$

We have assumed that $\mathbf{x}'$ is continuous, so all its coordinate functions are continuous real-valued functions of t. Since $\mathbf{F}$ is also continuous, the coordinate functions of $\mathbf{F} \circ \mathbf{x}$ are continuous real-valued functions of t. Then $(\mathbf{F} \circ \mathbf{x}) \cdot \mathbf{x}'$ is a continuous function of t, so the integral in (5) exists by Theorem 2.3 of Chapter 4.

The integral in (5) is the limit of sums of the terms

$$\mathbf{F}(\mathbf{x}(\bar{t}_i) \cdot \mathbf{x}'(\bar{t}_i))\,\Delta t_i = \mathbf{F}(\mathbf{x}(\bar{t}_i)) \cdot \frac{\mathbf{x}(\bar{t}_i)}{|\mathbf{x}'(\bar{t}_i)|}\,|\mathbf{x}'(\bar{t}_i)|\,\Delta t_i = \mathbf{F}(\mathbf{x}(\bar{t}_i)) \cdot \mathbf{T}(\bar{t}_i))\,\Delta s_i,$$

where $\mathbf{T}(\bar{t}_i)$ is the unit tangent vector at $\mathbf{x}(\bar{t}_i)$ and Δs_i approximates the arc length of C over the subinterval $[t_{i-1}, t_i]$. The factor

(6) $$\mathbf{F}(\mathbf{x}(\bar{t}_i)) \cdot \mathbf{T}(\bar{t}_i)$$

is the coordinate of $\mathbf{F}(\mathbf{x}(\bar{t}_i))$ in the direction of the unit tangent, hence in the direction of the path $\mathbf{x} = \mathbf{x}(t)$. Thus the integral in (5) is the limit of the sum of the values of $\mathbf{F}(\mathbf{x}(t))$ in the direction of the curve multiplied by the numbers Δs_i representing short distances along the curve. We can therefore regard the integral in (5) as the analogue of $\int_a^b f(x)\,dx$, for $\int_a^b f(x)\,dx$ is likewise the limit of the sum of the values $f(\bar{x}_i)$ of f along the interval $[a, b]$ times short distances Δx_i along that interval. We are thus led to the following definition.

2.2 DEFINITION If $\mathbf{F}: \mathbf{R}^n \to \mathbf{R}^n$ is continuous and $C: \mathbf{x} = \mathbf{x}(t)$ for $a \le t \le b$ is a smooth simple curve in $\mathbf{R}^n$, then the ***line integral*** (or *contour integral*) of $\mathbf{F}$ over C is

(7) $$\int_C \mathbf{F} \cdot d\mathbf{x} = \int_a^b \mathbf{F}(\mathbf{x}(t)) \cdot \mathbf{x}'(t)\,dt.$$

Example 2

Let C be the parametrized curve $\mathbf{x}(t) = (3\cos t, -2\sin t)$, $t \in [0, \pi]$, the bottom half of the curve in Example 1. If $\mathbf{F}(x, y) = x^2\,\mathbf{i} + y^2\,\mathbf{j}$, then evaluate the line integral of $\mathbf{F}$ over C.

Solution. To use (7) we need to express the integrand $\mathbf{F}(\mathbf{x}(t)) \cdot \mathbf{x}'(t)$ as a function of t. We have

$$\mathbf{F}(\mathbf{x}(t)) = \mathbf{F}(3\cos t, -2\sin t) = 9\cos^2 t\,\mathbf{i} + 4\sin^2 t\,\mathbf{j},$$

and

$$\mathbf{x}'(t) = (-3\sin t, -2\cos t) = -3\sin t\,\mathbf{i} - 2\cos t\,\mathbf{j}.$$

Thus

$$\mathbf{F}(\mathbf{x}(t)) \cdot \mathbf{x}'(t) = -27\cos^2 t\sin t - 8\sin^2 t\cos t.$$

Then (7) gives

$$\int_C \mathbf{F} \cdot d\mathbf{x} = \int_a^\pi (27\cos^2 t(-\sin t) - 8\sin^2 t\cos t)\,dt$$

$$= 9\cos^3 t - \frac{8}{3}\sin^3 t\Big|_0^\pi = -9 - 9 - 0 + 0 = -18. \blacksquare$$

There are several notations for $\int_C \mathbf{F} \cdot d\mathbf{x}$. The integrand on the right in (7) can be rewritten in terms of the arc length variable s as

$$\int_a^b \mathbf{F}(\mathbf{x}(t)) \cdot \frac{\mathbf{x}'(t)}{|\mathbf{x}'(t)|}\,|\mathbf{x}'(t)|\,dt = \int_{t=a}^{t=b} \mathbf{F}(\mathbf{x}(t)) \cdot \mathbf{T}(t)\,ds$$

$$= \int_{s(a)}^{s(b)} \mathbf{F}(\mathbf{x}(t(s))) \cdot \mathbf{T}(t(s))\,ds,$$

so that an alternate notation to (7) is

$$\int_C \mathbf{F} \cdot \mathbf{T}\,ds = \int_{s(a)}^{s(b)} \mathbf{F}(\mathbf{x}(t(s))) \cdot \mathbf{T}(t(s))\,ds. \tag{8}$$

If C is parametrized by arc length (see p. 698), then $|\mathbf{x}'(t)| = 1$, so $\mathbf{T}(t) = \mathbf{x}'(t)$ and $ds = dt$. In that case then, (7) and (8) coincide.

We can also write (7) in *coordinate form*. Suppose that

$$\mathbf{F}(\mathbf{x}) = P(\mathbf{x})\,\mathbf{i} + Q(\mathbf{x})\,\mathbf{j} + R(\mathbf{x})\,\mathbf{k} = (P(\mathbf{x}), Q(\mathbf{x}), R(\mathbf{x})),$$

where $R(\mathbf{x}) = 0$ if $n = 2$. Since

$$\mathbf{x}'(t) = \left(\frac{dx}{dt}, \frac{dy}{dt}, \frac{dz}{dt}\right) = \frac{dx}{dt}\,\mathbf{i} + \frac{dy}{dt}\,\mathbf{j} + \frac{dz}{dt}\,\mathbf{k},$$

we have in (7)

$$d\mathbf{x} = \mathbf{x}'(t)\,dt = \left(\frac{dx}{dt}, \frac{dy}{dt}, \frac{dz}{dt}\right)dt = (dx, dy, dz) = dx\,\mathbf{i} + dy\,\mathbf{j} + dz\,\mathbf{k}. \tag{9}$$

Then we find

$$\mathbf{F}(\mathbf{x}) \cdot \mathbf{x}'(t) = P(\mathbf{x})\frac{dx}{dt} + Q(\mathbf{x})\frac{dy}{dt} + R(\mathbf{x})\frac{dz}{dt}.$$

So (7) becomes

$$\int_C \mathbf{F} \cdot d\mathbf{x} = \int_a^b \left(P(x, y, z)\frac{dx}{dt} + Q(x, y, z)\frac{dy}{dt} + R(x, y, z)\frac{dz}{dt}\right)dt. \tag{10}$$

Because of (8), equivalent notation for (10) is

$$\int_C \mathbf{F} \cdot d\mathbf{x} = \int_C P(x, y, z)\,dx + Q(x, y, z)\,dy + R(x, y, z)\,dz. \tag{11}$$

When $n = 2$, the variable z and the function R disappear from (8). While the right side of (8) is useful *notation*, it is usually best to stick to (7) when *evaluating* line integrals.

Example 3

Compute $\int_C 2xy\,dx + (x^2 + y^2)\,dy$, where C is the parabola $y^2 = x$ between $(0, 0)$ and $(1, -1)$.

FIGURE 2.9

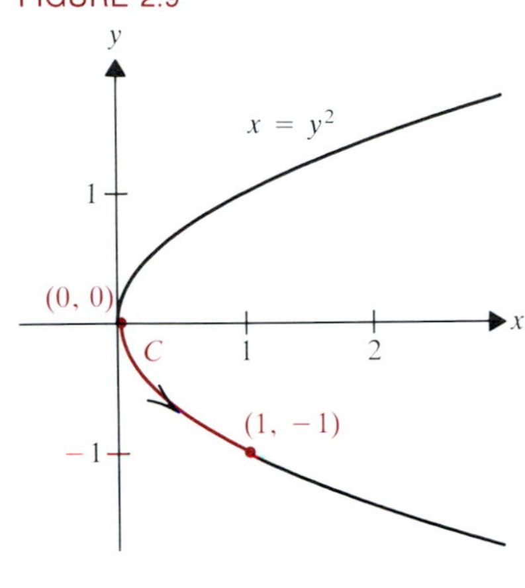

Solution. We show the curve in **Figure 2.9.** To obtain the form (7), we have to parametrize C. Since $y = -\sqrt{x}$ on the lower half of the parabola $x = y^2$, we let $x = t^2$ and $y = -t$ for $t \in [0, 1]$. That is, $\mathbf{x} = \mathbf{x}(t) = (t^2, -t)$. Since $\mathbf{x}(0) = (0, 0)$ and $\mathbf{x}(1) = (1, -1)$, we have

$$d\mathbf{x} = \left(\frac{dx}{dt}, \frac{dy}{dt}\right)dt = (2t, -1)\,dt = (2t\,dt, -dt) = (dx, dy).$$

The integral thus becomes

$$\begin{aligned}\int_C 2xy\,dx + (x^2 + y^2)\,dy &= \int_0^1 2t^2(-t)(2t\,dt) + (t^4 + t^2)(-dt) \\ &= \int_0^1 (-4t^4 - t^4 - t^2)\,dt = \int_0^1 (-5t^4 - t^2)\,dt \\ &= -t^5 - \frac{t^3}{3}\Big]_0^1 = -\frac{4}{3}. \quad ■\end{aligned}$$

We extend line integrals to piecewise smooth paths by adding the line integrals over the smooth pieces.

2.3 DEFINITION If $C = C_1 \cup C_2 \cup \ldots \cup C_k$ is piecewise smooth where each C_i is smooth, then

$$\int_C \mathbf{F} \cdot d\mathbf{x} = \sum_{i=1}^{k} \int_{C_i} \mathbf{F} \cdot d\mathbf{x}. \tag{12}$$

Example 4

Evaluate $\int_C \mathbf{F} \cdot d\mathbf{x}$ if $\mathbf{F}(x, y) = (x + y)\,\mathbf{i} + (x - y)\,\mathbf{j}$, and C is the square shown in **Figure 2.10.**

FIGURE 2.10

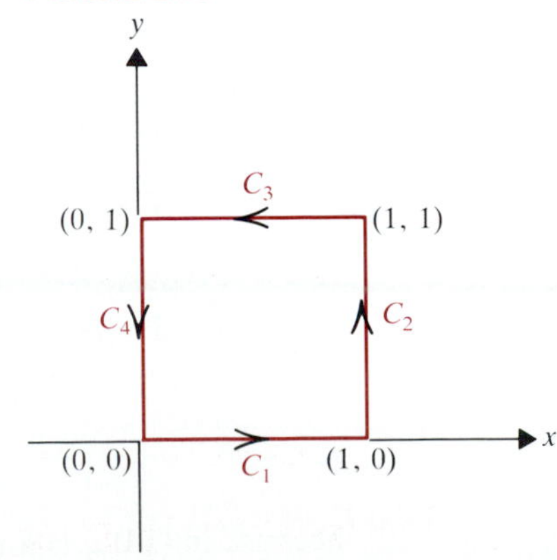

Solution. We have $C = C_1 \cup C_2 \cup C_3 \cup C_4$, where we parametrize each C_i and calculate $\mathbf{F}(\mathbf{x}(t))$ and $d\mathbf{x} = \mathbf{x}'(t)\,dt$ as follows.

C_1: $\mathbf{x}(t) = (t, 0)$, $0 \le t \le 1$, $\mathbf{F}(\mathbf{x}(t)) = t\,\mathbf{i} + t\,\mathbf{j}$, $d\mathbf{x} = (1, 0)\,dt = (dt, 0) = dt\,\mathbf{i}$.

C_2: $\mathbf{x}(t) = (1, t)$, $0 \le t \le 1$, $\mathbf{F}(\mathbf{x}(t)) = (t+1)\,\mathbf{i} + (1-t)\,\mathbf{j}$, $d\mathbf{x} = (0, 1)\,dt = (0, dt) = dt\,\mathbf{j}$.

C_3: $\mathbf{x}(t) = (1-t, 1)$, $0 \le t \le 1$, $\mathbf{F}(\mathbf{x}(t)) = (2-t)\,\mathbf{i} - t\,\mathbf{j}$, $d\mathbf{x} = (-1, 0)\,dt = (-dt, 0) = -dt\,\mathbf{i}$.

C_4: $\mathbf{x}(t) = (0, 1-t)$, $0 \le t \le 1$, $\mathbf{F}(\mathbf{x}(t)) = (1-t)\,\mathbf{i} + (t-1)\,\mathbf{j}$, $d\mathbf{x} = (0, -1)\,dt = (0, -dt) = -dt\,\mathbf{j}$.

From (10), (11), (12), and these calculations, we obtain

$$\begin{aligned}\int_C \mathbf{F}(\mathbf{x}(t)) \cdot d\mathbf{x}(t) &= \int_0^1 [t\,\mathbf{i} + t\,\mathbf{j}] \cdot [dt\,\mathbf{i}] + \int_0^1 [(t+1)\,\mathbf{i} + (1-t)\,\mathbf{j}] \cdot [dt\,\mathbf{j}] \\ &\quad + \int_0^1 [(2-t)\,\mathbf{i} - t\,\mathbf{j}] \cdot [-dt\,\mathbf{i}] \\ &\quad + \int_0^1 [(1-t)\,\mathbf{i} + (t-1)\,\mathbf{j}] \cdot [-dt\,\mathbf{j}] \\ &= \int_0^1 t\,dt + \int_0^1 (1-t)\,dt + \int_0^1 (t-2)\,dt + \int_0^1 (1-t)\,dt \\ &= \int_0^1 0\,dt = 0. \quad \blacksquare\end{aligned}$$

The next result, which will be of great importance throughout the chapter, says that if the orientation of C is *reversed*, then the line integral of $\mathbf{F}$ over C *changes sign*.

2.4 THEOREM

$$\int_{-C} \mathbf{F} \cdot d\mathbf{x} = -\int_C \mathbf{F} \cdot d\mathbf{x}.$$

Proof. Suppose that C is given by $\mathbf{x} = \mathbf{x}(t)$ for $t \in [a, b]$. Then $-C$ is given by $\mathbf{x} = \mathbf{x}(a + b - t)$ for $t \in [a, b]$. So

$$\int_{-C} \mathbf{F} \cdot d\mathbf{x} = \int_a^b \mathbf{F}(\mathbf{x}(a+b-t)) \cdot \mathbf{x}'(a+b-t)\,dt. \tag{13}$$

To work this integral out, we introduce the change of variable

$$u = a + b - t.$$

Then

$$du = -dt, \qquad \text{so} \qquad dt = -du.$$

Also, when $t = a$, then $u = b$, and when $t = b$, we get $u = a$. By the chain rule [Theorem 1.4(f) of Chapter 11], we have

$$\mathbf{x}'(a+b-t) = \frac{d\mathbf{x}(u)}{dt} = \frac{d\mathbf{x}(u)}{du}\frac{du}{dt} = \frac{d\mathbf{x}(u)}{du}(-1) = -\frac{d\mathbf{x}(u)}{du}.$$

Thus (13) becomes

$$\int_{-C} \mathbf{F} \cdot d\mathbf{x} = \int_b^a \mathbf{F}(\mathbf{x}(u)) \cdot \left(-\frac{d\mathbf{x}(u)}{du}\right)(-du) = \int_b^a \mathbf{F}(\mathbf{x}(u)) \cdot \frac{d\mathbf{x}(u)}{du}\,du,$$

$$\int_{-C} \mathbf{F} \cdot d\mathbf{x} = -\int_a^b \mathbf{F}(\mathbf{x}(u)) \cdot \frac{d\mathbf{x}(u)}{du}\,du \tag{14}$$

since reversing the order of limits in a definite integral reverses its sign (p. 240)

$$= -\int_a^b \mathbf{F}(\mathbf{x}(t)) \cdot \frac{d\mathbf{x}(t)}{dt}\,dt = -\int_C \mathbf{F} \cdot d\mathbf{x},$$

because in (14) u is a dummy variable. QED

Example 5

Verify Theorem 2.4 for the line integral $\int_C x^2\,dx + y^2\,dy$ in Example 2.

Solution. The curve C is the bottom half of the ellipse

$$\frac{x^2}{9} + \frac{y^2}{4} = 1$$

traversed clockwise, as in **Figure 2.11**(a). The parametrization of C in Example 2 is

$$\mathbf{x} = \mathbf{x}(t) = (3\cos t,\ -2\sin t), \qquad t \in [0, \pi].$$

FIGURE 2.11

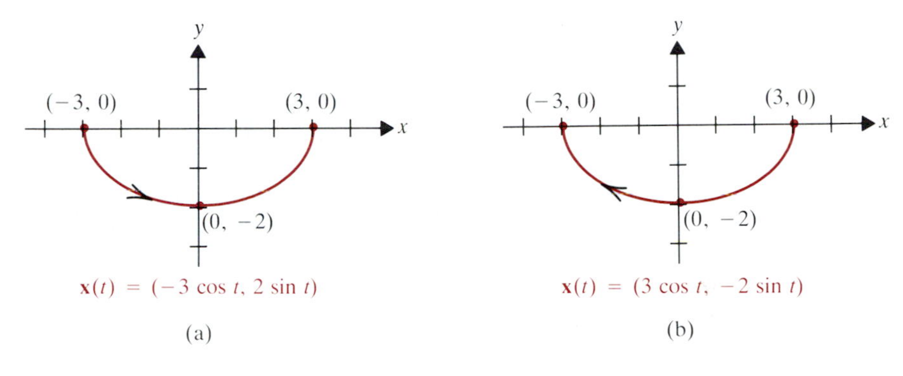

As remarked before Example 1, the curve $-C$ in Figure 2.11(b) can then be parametrized by

$$\mathbf{x} = \mathbf{x}(0 + \pi - t) = (3\cos(\pi - t), 2\sin(\pi - t)) = (-3\cos t, 2\sin t), \qquad t \in [0, \pi].$$

Since $\mathbf{F}(x, y) = (x^2, y^2) = x^2\,\mathbf{i} + y^2\,\mathbf{j}$, we have

$$\mathbf{x}'(t) = (3\sin t, 2\cos t) = 3\sin t\,\mathbf{i} + 2\cos t\,\mathbf{j},$$

and

$$\mathbf{F}(\mathbf{x}(t)) = \mathbf{F}(-3\cos t, 2\sin t) = 9\cos^2 t\,\mathbf{i} + 4\sin^2 t\,\mathbf{j}.$$

Hence

$$\int_{-C} \mathbf{F} \cdot d\mathbf{x} = \int_0^\pi (27\cos^2 t\sin t + 8\sin^2 t\cos t)\,dt = -\int_C \mathbf{F} \cdot d\mathbf{x}. \quad \blacksquare$$

What if we change the parametrization of C but don't change the orientation? That is, we change from $\mathbf{x} = \mathbf{x}(t)$ for $t \in [a, b]$ to $\mathbf{x} = \mathbf{g}(u)$ for $u \in [c, d]$, where $t = h(u)$ has a continuous derivative, and the orientation is preserved. The next result says that the line integral is unchanged. Thus line integrals are *independent of the parametrization of C as long as the orientation is not changed.*

2.5 THEOREM Suppose that C is parametrized by $\mathbf{x} = \mathbf{f}(t)$ for $t \in [a, b]$ and also by $\mathbf{x} = \mathbf{g}(u)$ for $u \in [c, d]$, where $t = h(u)$ with h' continuous. If the orientation of C is preserved (that is, $a = h(c)$ and $b = h(d)$), then the value of $\int_C \mathbf{F} \cdot d\mathbf{x}$ is the same whether computed as

$$\int_a^b \mathbf{F}(\mathbf{f}(t)) \cdot \mathbf{f}'(t)\,dt$$

or as

$$\int_c^d \mathbf{F}(\mathbf{g}(u)) \cdot \mathbf{g}'(u)\,du.$$

Proof. We have $\mathbf{g}(u) = \mathbf{x} = \mathbf{f}(t) = \mathbf{f}(h(u))$. So by the chain rule [Theorem 1.4(f) of Chapter 11]

$$\mathbf{g}'(u) = \mathbf{f}'(t)h'(u) = \frac{d\mathbf{f}}{dt}\frac{dt}{du}.$$

We also have $a = h(c)$ and $b = h(d)$. Hence

$$\int_a^b \mathbf{F}(\mathbf{f}(t)) \cdot \mathbf{f}'(t)\,dt = \int_c^d \mathbf{F}(\mathbf{x}(h(u))) \cdot \frac{d\mathbf{f}}{dt}\frac{dt}{du}\,du = \int_c^d \mathbf{F}(\mathbf{g}(u)) \cdot \mathbf{g}'(u)\,du. \quad \boxed{\text{QED}}$$

Example 6

Verify Theorem 2.5 by computing $\int_C 2xy\,dx + (x^2 + y^2)\,dy$, where C is the parabolic arc of $y^2 = x$ in Example 3 parametrized by

(15) $$\mathbf{x} = \mathbf{g}(u) = (u^2 - 2u + 1, 1 - u), \qquad u \in [1, 2].$$

Solution. When $u = 1$, we have $\mathbf{g}(u) = (0, 0)$, and when $u = 2$, we have $\mathbf{g}(u) = (1, -1)$. For all values of u, we have $x = y^2$. Thus the parametrization (15) is a parametrization of the parabolic arc of $y^2 = x$ from $(0, 0)$ to $(1, -1)$, the curve C of Example 3. We have

$$t = 1 - u = h(u),$$

where the function $h' = -1$ is certainly continuous. The orientation of C in Example 3 is preserved here. So Theorem 2.5 says that we should get $-4/3$ if we calculate $\int_1^2 \mathbf{F}(\mathbf{g}(u)) \cdot \mathbf{g}'(u)\,du$, because in Example 3 we found that $\int_0^1 \mathbf{F}(\mathbf{x}(t)) \cdot \mathbf{x}'(t)\,dt = -4/3$. Here we have

$$\mathbf{g}'(u) = [(2u - 2)\mathbf{i} - \mathbf{j}]\,du = [2(u - 1)\mathbf{i} - \mathbf{j}]\,du$$

and

$$\begin{aligned}\mathbf{F}(g(u)) &= 2(u^2 - 2u + 1)(1 - u)\mathbf{i} + [(u^2 - 2u + 1)^2 + (u - 1)^2]\mathbf{j} \\ &= -2(u - 1)^3\mathbf{i} + [(u - 1)^4 + (u - 1)^2]\mathbf{j}.\end{aligned}$$

Thus

$$\begin{aligned}\int_1^2 \mathbf{F}(g(u)) \cdot \mathbf{g}'(u)\,du &= \int_1^2 [-2(u - 1)^3 \cdot 2(u - 1) - (u - 1)^4 - (u - 1)^2]\,du \\ &= \int_1^2 [-4(u - 1)^4 - (u - 1)^4 - (u - 1)^2]\,du \\ &= \int_1^2 [-5(u - 1)^4 - (u - 1)^2]\,du \\ &= -(u - 1)^5 - \frac{1}{3}(u - 1)^3\Big|_1^2 = -1 - \frac{1}{3} + 0 + 0 = -\frac{4}{3}. \quad ■\end{aligned}$$

The procedure followed in the proof of Theorem 2.5 corresponds to evaluating the integral in Example 6 by means of the change of variable $t = u - 1 = h(u)$, as you can easily check.

Theorem 2.5 says that to evaluate a line integral, we can use any smooth parametrization that does not reverse orientation. Naturally, in practice, to minimize labor the simplest such parametrization is used.

Exercises 14.2

In Exercises 1–14, evaluate the given line integral.

1. $\int_C \mathbf{F} \cdot d\mathbf{x}$, where $\mathbf{F}(x, y) = x^2y\,\mathbf{i} + (x^2 - y)\mathbf{j}$, and C is the curve $\mathbf{x}(t) = (t, 1 - t)$, $0 \le t \le 1$.

2. $\int_C \mathbf{F} \cdot \mathbf{T}\,ds$, where $\mathbf{F}(x, y) = (x + 2y)\mathbf{i} + (x^2 - y^2)\mathbf{j}$, and C is the triangular path from (0, 0) to (1, 0) to (1, 1) to (0, 0).

3. $\int_C \mathbf{F} \cdot d\mathbf{x}$, where $\mathbf{F}(x, y) = \sqrt{x^2 + y^2}\,\mathbf{i} + \sqrt{1 - x^2}\,\mathbf{j}$, and C is the upper half of the unit circle $x^2 + y^2 = 1$, traversed from (1, 0) to (−1, 0).

4. $\int_C \mathbf{F} \cdot d\mathbf{x}$, where $\mathbf{F}(x, y) = \sqrt{1 - y^2}\,\mathbf{i} + \sqrt{1 - x^2}\,\mathbf{j}$, and C is the curve of Exercise 3.

5. $\int_C xy\,dx + (y^2 + 1)\,dy$, where C is the curve $y^2 = x$ from (0, 0) to (1, 1).

6. $\int_C (x^2 - y^2)\,dx + 2xy\,dy$, where C is the curve $y = x^2$ from (−1, 1) to (2, 4).

7. $\int_C \frac{-y}{x^2 + y^2}\,dx + \frac{x}{x^2 + y^2}\,dy$, where C is the circle $x^2 + y^2 = 9$ traversed counterclockwise from (3, 0). **(This exercise is referred to in Section 4.)**

8. $\int_C \frac{x + y}{x^2 + y^2}\,dx - \frac{x - y}{x^2 + y^2}\,dy$, where C is the path in Exercise 7.

9. $\int_C \mathbf{F} \cdot d\mathbf{x}$, where $\mathbf{F}(x, y, z) = y\,\mathbf{i} + z\,\mathbf{j} + x\,\mathbf{k}$, and C is the line segment joining (0, 0, 0) to (2, 4, −1).

10. $\int_C \mathbf{F} \cdot \mathbf{T}\,ds$, where $\mathbf{F}(x, y, z) = x^2\,\mathbf{i} + y^2\,\mathbf{j} + z^2\,\mathbf{k}$, and C is the curve $\mathbf{x}(t) = (t, t^2, t^3)$, $0 \le t \le 1$.

11. $\int_C x\,dx - y\,dy + z\,dz$, where C is the helix $\mathbf{x}(t) = (\cos t, \sin t, t/\pi)$, $0 \le t \le 2\pi$.

12. $\int_C \mathbf{F} \cdot d\mathbf{x}$ where
$$\mathbf{F}(x_1, x_2, x_3, x_4) = (x_1 - x_2, x_2 - x_3, x_4 - x_1, x_1x_2x_3x_4),$$
and C is the curve $\mathbf{x}(t) = (t, t^2, t^3, t^4)$, $0 \le t \le 1$.

13. $\int_C (xz + y^2)\,dx + (yz - x^2)\,dy + (xy - z^2)\,dz$, where C is the line segment from (1, 2, 3) to (4, 3, 5) followed by the segment from (4, 3, 5) to (1, 1, 1).

14. $\int_C y\,dx + z\,dy + x\,dz$, where C is the curve of intersection of the plane $x + y = 2$ and the sphere $(x - 1)^2 + (y - 1)^2 + z^2 = 2$, traversed in the counterclockwise sense viewed from the point (2, 2, 0).

In Exercises 15–20, verify Theorem 2.4 by calculating $\int_{-C} \mathbf{F} \cdot d\mathbf{x}$ for the vector function F and curve C of the indicated exercise.

15. Exercise 1
16. Exercise 3
17. Exercise 5
18. Exercise 7
19. Exercise 9
20. Exercise 11

21. Find the work done by a force $\mathbf{F}(x, y) = (2 - y)\mathbf{i} + x\mathbf{j}$ in moving a particle along one arch of the ***cycloid*** given by $\mathbf{x}(t) = (t - \sin t, 1 - \cos t)$, $0 \le t \le 2\pi$.

22. Find the work done by a force $\mathbf{F}(x, y) = (1 - y)\mathbf{i} + x\mathbf{j}$ in moving a particle along one arch of the cycloid $\mathbf{x}(t) = 2(t - \sin t, 1 - \cos t)$, $0 \le t \le 2\pi$.

23. The repelling force between a charged particle P at the origin and an oppositely charged particle Q at (x, y) is
$$\mathbf{F}(x, y) = \frac{1}{(x^2 + y^2)^{3/2}}[x\,\mathbf{i} + y\,\mathbf{j}].$$
Find the work done by $\mathbf{F}$ as it moves Q along the line segment from (1, 0) to (−1, 2).

24. Refer to Exercise 23. Suppose that $\mathbf{F}$ moves Q along the polygonal path from (1, 0) to (1, 1) to (−1, 2). What is the work done?

25. The ***line integral of a scalar function*** f: $\mathbf{R}^n \to \mathbf{R}$ over a curve C is defined by
$$\int_C f(\mathbf{x})\,ds = \int_C f(\mathbf{x}(t))\,|\mathbf{x}'(t)|\,dt.$$
(Recall that $ds = |\mathbf{x}'(t)|\,dt$.) If $f(\mathbf{x}(t))$ gives the density of a wire in the shape of C at the point $\mathbf{x}(t)$, then the ***mass*** of the wire is defined to be $\int_C f(\mathbf{x})\,ds$. Formulate a definition of the *center of mass* of the wire.

26. Formulate a definition of the *moment of inertia* of a wire in the shape of $\mathbf{x} = \mathbf{x}(t)$ relative to an axis l at distance $d(x, y, z)$ from (x, y, z) on C. (Assume that the wire has density $f(x, y, z)$.)

27. Use Exercise 25 to compute the mass of a helical wire $\mathbf{x}(t) = (\cos t, \sin t, t)$, $0 \le t \le 2\pi$, if its density is given by $\delta(x, y, z) = x^2 + y^2 + z^2$.

28. Find the center of mass of the wire in Exercise 27.

29. Find the moment of inertia of the wire in Exercise 27 about the z-axis.

30. A force $\mathbf{F}$ acts on a particle to move it from $\mathbf{x}(t_0)$ to $\mathbf{x}(t_1)$ along the path $\mathbf{x} = \mathbf{x}(t)$. Show that the work done is the change in kinetic energy, i.e.,
$$\mathbf{W} = \tfrac{1}{2}m|\mathbf{v}(t_1)|^2 - \tfrac{1}{2}m|\mathbf{v}(t_0)|^2.$$
(This is one form of the *law of conservation of energy*.) [*Hint:* Compute $(d/dt)(|\mathbf{v}(t)|^2)$ and use $\mathbf{F} = mx''(t)$, Newton's second law.]

31. If C is a smooth, but not necessarily simple, closed curve that does not pass through the origin, then the *winding number* of C is defined as $(1/2\pi)\int_C \mathbf{F} \cdot d\mathbf{x}$, where
$$\mathbf{F}(x, y) = \frac{-y}{x^2 + y^2}\mathbf{i} + \frac{x}{x^2 + y^2}\mathbf{j}.$$
It is shown in advanced calculus that the winding number is always an integer, which gives the number of times C winds around the origin. If it is positive, then

C encircles the origin in a counterclockwise sense. If it is negative, then C encircles the origin in a clockwise sense. Compute the winding numbers of

$$C_1: \mathbf{x}(t) = (2\cos t, 2\sin t), \qquad 0 \le t \le 6\pi,$$

and

$$C_2: \mathbf{x}(t) = (\cos t, -\sin t), \qquad 0 \le t \le 2\pi.$$

32. If $F(x, y) = -y^{-1/3}\mathbf{i} + x^{-1/3}\mathbf{j}$ then compute $\int_C \mathbf{F} \cdot d\mathbf{x}$ over the hypocycloid of four cusps $x^{2/3} + y^{2/3} = 1$ parametrized in the counterclockwise sense (see Exercise 41, Section 8.5).

14.3 Green's Theorem

This section takes its name from a result that extends the fundamental theorem of calculus to two-dimensional vector fields. To describe the extension precisely, it is helpful to state Theorem 4.2 of Chapter 4 as follows:

If f is a smooth function (that is, f' is continuous) on $[a, b]$, then

$$\int_a^b f'(x)\,dx = f(b) - f(a). \tag{1}$$

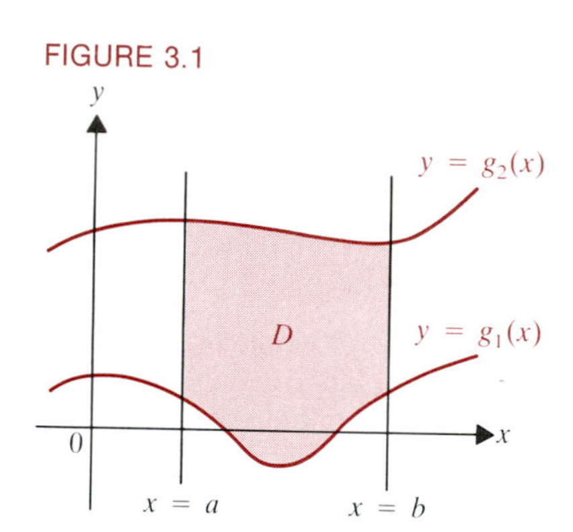

FIGURE 3.1

Qualitatively, this says that *the definite integral of the derivative of f over $[a, b]$ is completely determined by the values of f on the boundary of $[a, b]$*, that is, on the two-point set $\{a, b\}$. The various forms of Fubini's theorem in the last chapter (starting with Theorem 1.9) constitute an extension of this result to scalar functions of two or three variables. In particular, partially antidifferentiated functions, such as $\int f(x, y)\,dy$, are evaluated just on the *boundary* of the region D of integration, such as the curves $y = g_1(x)$ and $y = g_2(x)$ in **Figure 3.1.**

Green's theorem is the first of several analogues of (1) to be presented in this chapter. It says roughly that the double integral of a certain kind of derivative of $\mathbf{F}$ over a standard type of region D is given by the line integral of $\mathbf{F}$ over the *boundary* of D. This lets us evaluate some complicated double integrals as line integrals, and sometimes makes it possible to compute a complicated line integral by working out a less-involved double integral. To state the theorem precisely, we must specify the "standard type" of region and the "certain kind" of derivative mentioned above. We begin with the type of region.

3.1 DEFINITION A plane region D is ***simple*** if it is the part of $\mathbf{R}^2$ enclosed by a piecewise smooth simple closed curve that intersects any vertical or horizontal line either in at most two points A and B or in one line segment AB.

As the name suggests, simple regions are geometrically uncomplicated. **Figure 3.2** shows three examples of simple regions. The region in Figure 3.2(a) is bounded by a *smooth* curve, the circle $x^2 + y^2 = 1$. The regions in Figure 3.2(b) and (c) are bounded by *piecewise smooth* curves C. Points where $\mathbf{x}'$ fails to be continuous are labeled. In all three cases, vertical and horizontal lines intersect C in 0, 1, or 2 points. **Figure 3.3** shows another type of simple region, in this case a square. Notice that some lines parallel to a coordinate axis (such as $x = a$ and $y = c$) intersect C in at most two points, while others (including $x = -1$ and $y = 1$) meet C in entire line segments.

Figure 3.4 shows a nonsimple region. Although some lines (like $x = a$ and $y = c$) intersect C in at most two points, there are others (like $y = d$) that intersect C in four points.

We try to picture a general simple region in **Figure 3.5.** The regions D and R in **Figure 3.6** fail to be simple, because the lines shown in **Figure 3.7** intersect

FIGURE 3.2

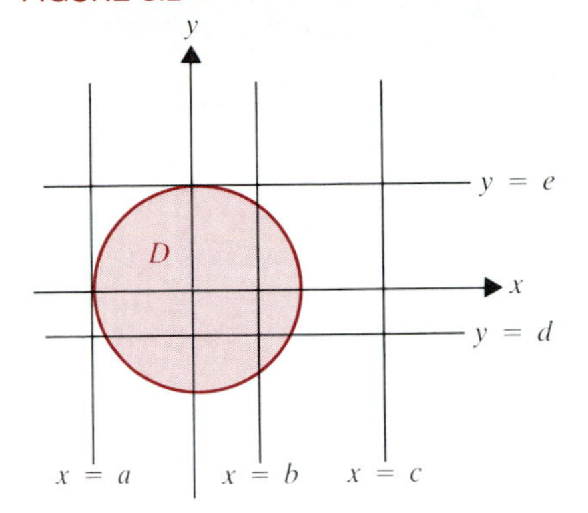

(a)

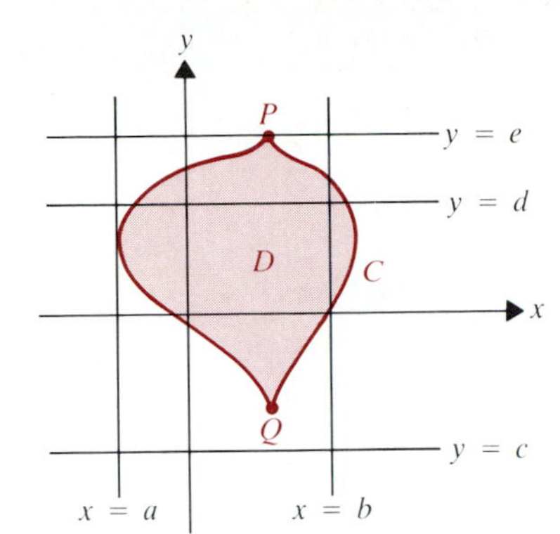

(b)

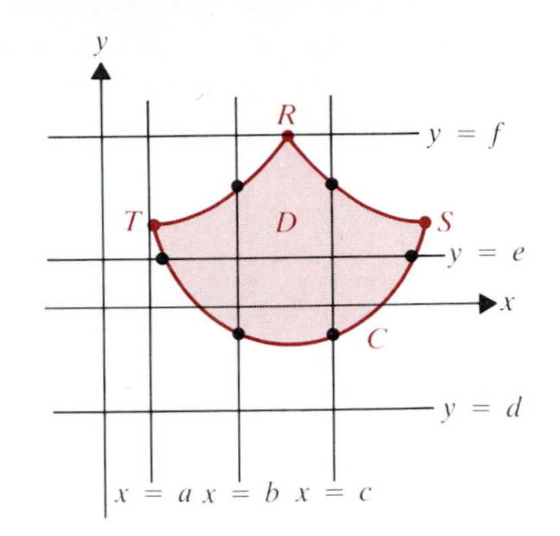

(c)

FIGURE 3.3

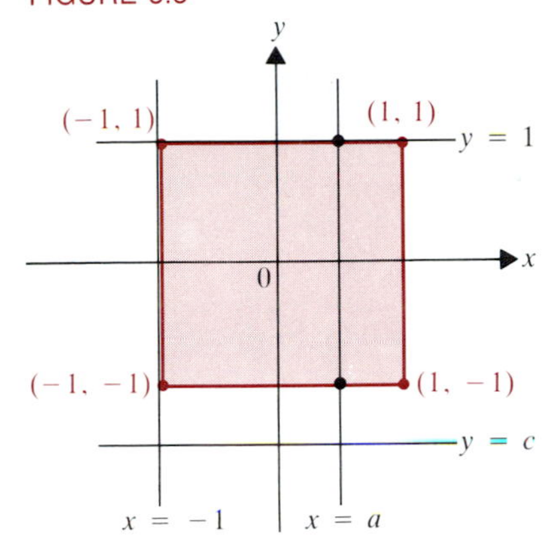

FIGURE 3.4

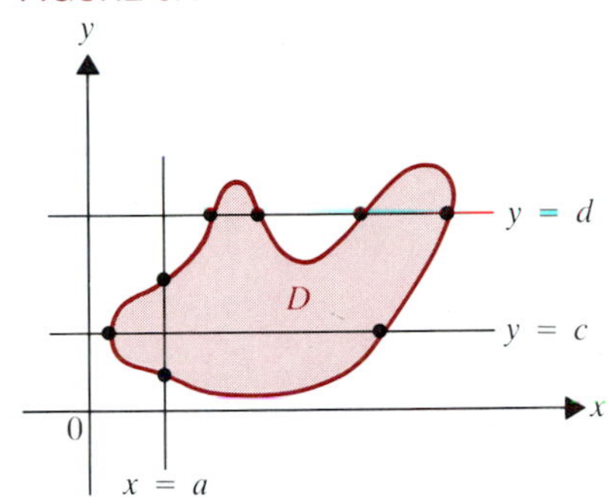

FIGURE 3.5

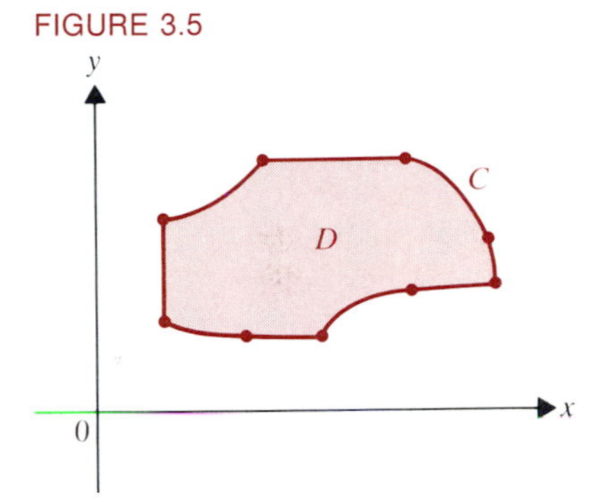

FIGURE 3.6

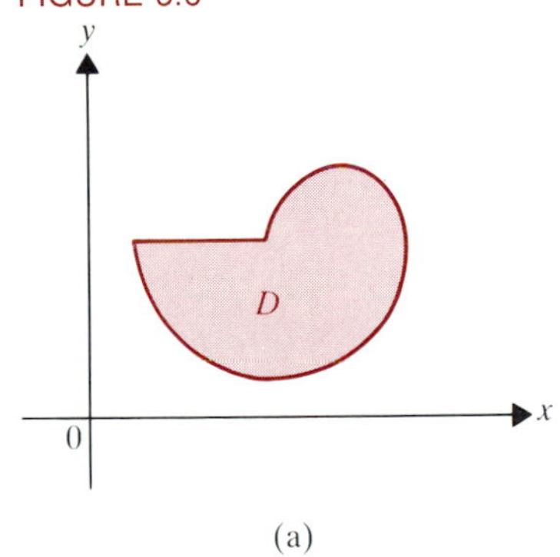

(a)

(b)

FIGURE 3.7

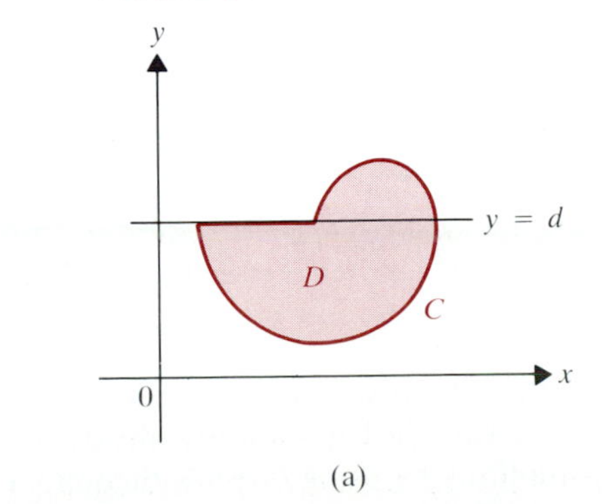

(a)

(b)

FIGURE 3.8

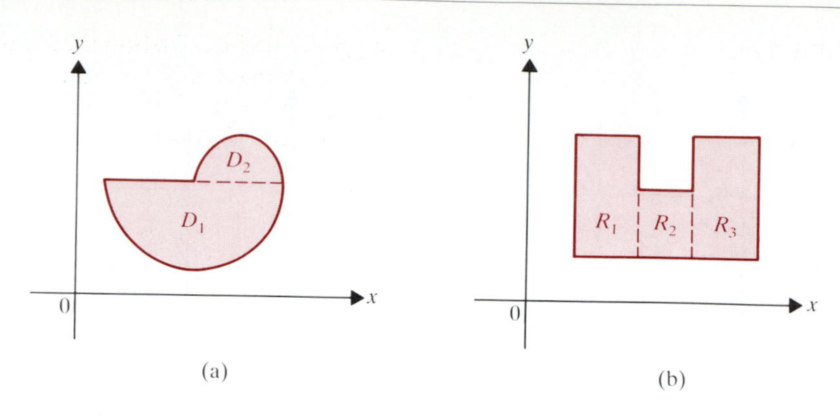

the boundaries in more than two points or a single line segment. However, D and R can be decomposed into subregions that *are* simple, and we show this in **Figure 3.8.**

The first version of Green's theorem applies to simple regions. (We will soon extend it to regions like D in Figure 3.6 that can be decomposed into the union of simple regions.) The proof is given at the end of the section.

3.2 THEOREM

Green's Theorem. Let $\mathbf{F}: \mathbf{R}^2 \to \mathbf{R}^2$ be given by

$$\mathbf{F}(x, y) = P(x, y)\,\mathbf{i} + Q(x, y)\,\mathbf{j},$$

where P and Q have continuous first partial derivatives. Suppose that D is a simple region with boundary C. Then

$$\iint_D \left(\frac{\partial Q}{\partial x} - \frac{\partial P}{\partial y}\right) dA = \oint_C P\,dx + Q\,dy = \oint_C \mathbf{F} \cdot \mathbf{dx} \tag{2}$$

Here $\oint$ means that C is parametrized by $\mathbf{x} = \mathbf{x}(t)$, $a \le t \le b$, in such a way that as t increases from a to b, C is traversed once in a *counterclockwise* sense, that is, so that D is on the left as the boundary is traversed.

The following example suggests how helpful (2) can be.

FIGURE 3.9

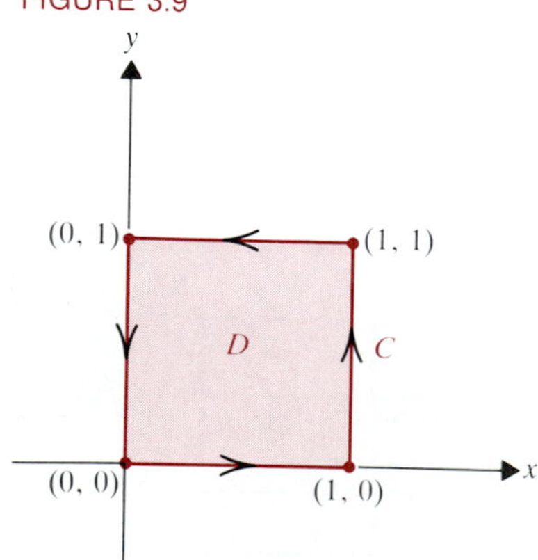

Example 1

If $F(x, y) = (x^2 + y^2)\,\mathbf{i} + (2xy)\,\mathbf{j}$, then compute $\oint_C \mathbf{F} \cdot \mathbf{dx}$, where C is the boundary of the unit square shown in **Figure 3.9.**

Solution. We could compute the line integral as in Example 4 of the last section, but that would require us to parametrize the four sides of D separately, work out each line integral, and add. If we use (2), we have to compute only one double integral:

$$\oint_C \mathbf{F} \cdot \mathbf{dx} = \iint_D \left[\frac{\partial(2xy)}{\partial x} - \frac{\partial}{\partial y}(x^2 + y^2)\right] dA = \iint_D (2y - 2y)\,dA = 0. \quad \blacksquare$$

Although it is unreasonable to expect every application of (2) to turn out *this* simply, Example 1 graphically illustrates how much labor can be saved in certain situations by using Green's theorem.

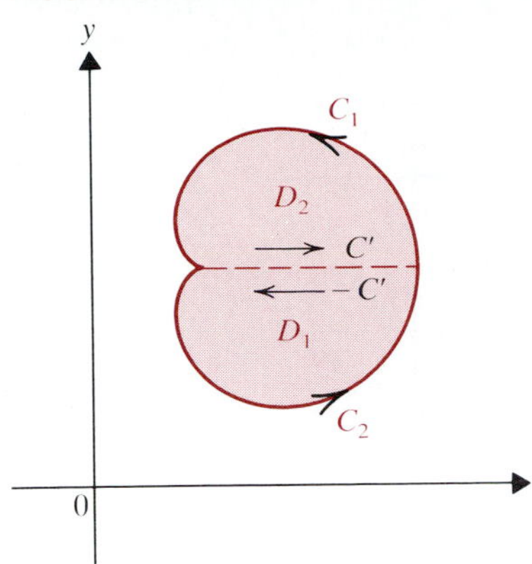

FIGURE 3.10

As suggested before Theorem 3.2, Equation (2) also holds for any region D that is a finite union $D_1 \cup D_2 \cup \ldots \cup D_k$ of nonoverlapping simple regions. Consider, for example, the region D in **Figure 3.10,** which can be subdivided into the two simple subregions D_1 and D_2. On each of these, (2) holds by Theorem 3.2. We thus have

$$\text{(3)} \quad \iint_D \left(\frac{\partial Q}{\partial x} - \frac{\partial P}{\partial y}\right) dA = \iint_{D_1} \left(\frac{\partial Q}{\partial x} - \frac{\partial P}{\partial y}\right) dA + \iint_{D_2} \left(\frac{\partial Q}{\partial x} - \frac{\partial P}{\partial y}\right) dA$$

$$= \oint_{\partial D_1} P\,dx + Q\,dy + \oint_{\partial D_2} P\,dx + Q\,dy,$$

where ∂D_i stands for the boundary of D_i. Notice that each part of ∂D_i that is not part of $C = \partial D$ is traversed twice in *opposite* directions in the evaluation of the line integral in (3) (see Figure 3.10). Hence the net contribution to the line integrals in (3) from the common boundary of D_1 and D_2 is 0:

$$\int_{\partial D_1} P\,dx + Q\,dy + \int_{\partial D_2} P\,dx + Q\,dy$$

$$= \int_{C_1} P\,dx + Q\,dy + \int_{C'} P\,dx + Q\,dy + \int_{C_2} P\,dx + Q\,dy + \int_{-C'} P\,dx + Q\,dy$$

by Definition 2.3

$$= \int_{C_1} P\,dx + Q\,dy + \int_{C'} P\,dx + Q\,dy + \int_{C_2} P\,dx + Q\,dy - \int_{C'} P\,dx + Q\,dy$$

by Theorem 2.4

$$= \int_{C_1} P\,dx + Q\,dy + \int_{C_2} P\,dx + Q\,dy = \int_{\partial D} P\,dx + Q\,dy.$$

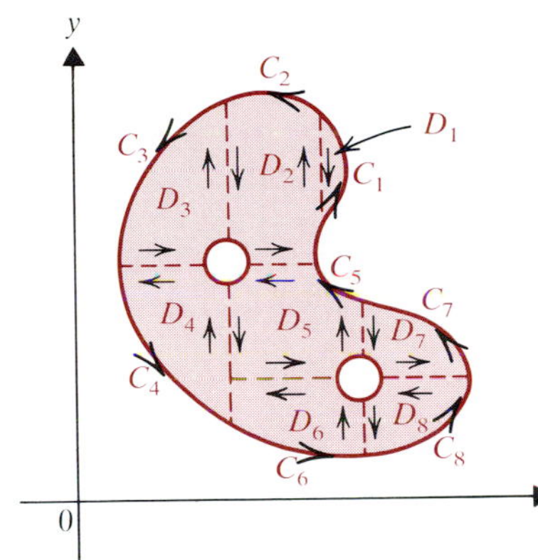

FIGURE 3.11

The same reasoning applies if $D = D_1 \cup D_2 \cup \ldots \cup D_k$, for $k > 2$. Thus (3) becomes

$$\text{(4)} \quad \iint_D \left(\frac{\partial Q}{\partial x} - \frac{\partial P}{\partial y}\right) dA = \sum_{i=1}^{k} \oint_{\partial Di} P\,dx + Q\,dy = \oint_C P\,dx + Q\,dy.$$

We have therefore extended (2) to regions D that can be decomposed into finite unions of simple regions. (See, for example, **Figure 3.11.**) The formal statement is the following.

3.3 THEOREM

Green's Theorem. Let $\mathbf{F}$ be as in Theorem 3.2. Suppose that D is decomposable into the union of a finite number of nonoverlapping simple regions D_i with piecewise smooth counterclockwise oriented simple closed curves C_i as boundaries. Then

$$\text{(2)} \quad \iint_D \left(\frac{\partial Q}{\partial x} - \frac{\partial P}{\partial y}\right) dA = \oint_C P\,dx + Q\,dy,$$

where $C = \partial D = C_1 \cup C_2 \cup \ldots \cup C_k$.

Example 2

Evaluate

$$\oint_C P\,dx + Q\,dy$$

if $P(x, y) = x^2 - 3y$, $Q(x, y) = 2x + y^3 - 3$, and C is the boundary of the shaded region in **Figure 3.12.**

FIGURE 3.12

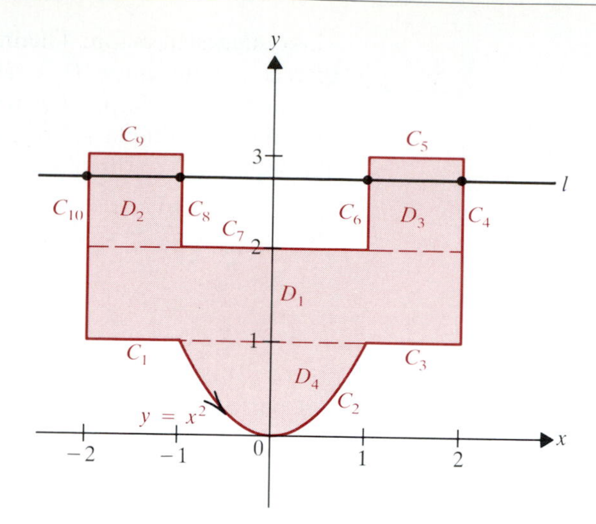

Solution. Here D is bounded by the piecewise smooth simple closed curve

$$C = C_1 \cup C_2 \cup C_3 \cup C_4 \cup C_5 \cup C_6 \cup C_7 \cup C_8 \cup C_9,$$

where the smooth pieces $C_i, i = 1, 2, \ldots, 9$ are labeled in Figure 3.12. The region D is not simple, as can be seen from the four points of intersection of the line l with C in Figure 3.12. However, D can be broken up into four simple regions D_1, D_2, D_3, and D_4, where D_1 is the rectangle $[-2, 2] \times [1, 2]$, D_2 is the square $[-2, -1] \times [2, 3]$, D_3 is the square $[1, 2] \times [2, 3]$, and D_4 is the region lying under $y = 1$ and above $y = x^2$. By Theorem 3.3,

$$\int_C P\,dx + Q\,dy = \iint_D \left(\frac{\partial Q}{\partial x} - \frac{\partial P}{\partial y}\right) dA = \iint_D (2 + 3)\,dA = 5A(D),$$

where $A(D)$ is the area of D. Clearly,

$$A(D_1) = 4 \cdot 1 \qquad \text{and} \qquad A(D_2) = 1 \cdot 1 = 1 = A(D_3).$$

To complete the evaluation of $\int_C P\,dx + Q\,dy$, we need only find $A(D_4)$. We have

$$A(D_4) = \iint_{D_4} 1\,dA = \int_{-1}^{1} \int_{x^2}^{1} dy\,dx = \int_{-1}^{1} (1 - x^2)\,dx$$

$$= 2\int_0^1 (1 - x^2)\,dx = 2\left(1 - \frac{1}{3}\right) = \frac{4}{3}.$$

Therefore

$$\int_C P\,dx + Q\,dy = 5A(D) = 5\left(4 + 1 + 1 + \frac{4}{3}\right) = \frac{110}{3}. \quad ■$$

We have referred to $\partial Q/\partial x - \partial P/\partial y$ as a "kind of derivative" of the vector field

$$\mathbf{F}(x, y) = P(x, y)\,\mathbf{i} + Q(x, y)\,\mathbf{j}.$$

To make this more precise, we regard the two-dimensional vector field $\mathbf{F}: \mathbf{R}^2 \to \mathbf{R}^2$ as a *three*-dimensional vector field $\mathbf{F}: \mathbf{R}^3 \to \mathbf{R}^3$ by writing

$$\mathbf{F}(x, y, z) = P(x, y)\,\mathbf{i} + Q(x, y)\,\mathbf{j} + 0\,\mathbf{k} = (P(x, y), Q(x, y), 0).$$

Then we can apply the differential operator **curl** of Definition 1.8 to $F(x, y, z)$, getting

$$\mathbf{curl}\,\mathbf{F}(x, y, z) = \left(\frac{\partial 0}{\partial y} - \frac{\partial Q}{\partial z}\right)\mathbf{i} + \left(\frac{\partial P}{\partial z} - \frac{\partial 0}{\partial x}\right)\mathbf{j} + \left(\frac{\partial Q}{\partial x} - \frac{\partial P}{\partial y}\right)\mathbf{k} = \left(\frac{\partial Q}{\partial x} - \frac{\partial P}{\partial y}\right)\mathbf{k}.$$

Now, assuming that the partial derivatives of **F** are continuous, we can put Theorem 3.3 into the form

$$(5) \qquad \int_C \mathbf{F} \cdot d\mathbf{x} = \iint_D \left(\frac{\partial Q}{\partial x} - \frac{\partial P}{\partial y}\right) dA = \iint_D \mathbf{curl}\,\mathbf{F} \cdot \mathbf{k}\, dA,$$

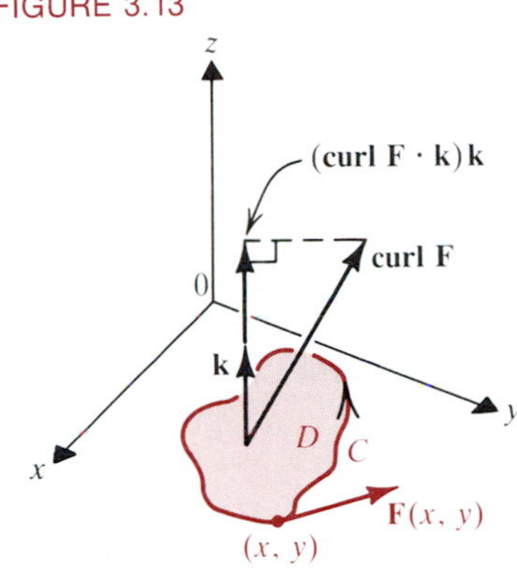

FIGURE 3.13

where $C = \partial D$. This says that the line integral of **F** over C is the coordinate of **curl F** in the direction (in $\mathbf{R}^3$) that is perpendicular to the plane region D (see **Figure 3.13**). Since **curl** is a differential operator, **curl F** (and its coordinate **curl F** · **k** in the direction of **k**) can be thought of as a type of derivative of **F**. The equality (5) says that the double integral of this derivative of **F** over D is completely determined by the value of **F** itself on the *boundary* C of D. Later we shall see that there is a three-dimensional analogue of (5).

An important physical interpretation of (5) helps explain the origin of the word *curl*. In fluid dynamics, the vector field $\mathbf{F}(x, y)$ in Green's theorem is interpreted as the velocity field of a fluid flowing over the plane. The line integral $\int \mathbf{F} \cdot d\mathbf{x}$ is called the ***circulation*** of **F** around C. The terminology is natural because, by (8) on p. 918,

$$\int_C \mathbf{F} \cdot d\mathbf{x} = \int_C \mathbf{F} \cdot \mathbf{T}\, ds$$

measures the strength of **F** in the direction of C. That force tends to produce circulation of the fluid about C. In view of (5) the circulation of **F** around C is the double integral of the coordinate of **curl F** in the direction perpendicular to D. Thus **curl F** itself is a measure of the tendency of the fluid flow to curl around the region enclosed by C.

A vector field **F** is called ***irrotational*** if **curl F** $= \mathbf{0}$. In this case it follows from (5) that the circulation is 0 around every piecewise simple closed curve C. If $\int_C \mathbf{F} \cdot d\mathbf{x}$ is nonzero, then a ***vortex*** is said to exist inside C: The fluid swirls around the region D.

There is another important interpretation of Green's theorem in physics. The intensity of an electrical field in the plane is measured by a vector function $\mathbf{F}\colon \mathbf{R}^2 \to \mathbf{R}^2$ called the ***electric field intensity.*** The ***electromotive force*** (***emf,*** or ***potential difference***) on a simple closed curve C in $\mathbf{R}^2$ is defined to be the line integral of **F** over C.

Example 3

Suppose that the electric field intensity $\mathbf{F}(x, y) = P(x, y)\mathbf{i} + Q(x, y)\mathbf{j}$ is a conservative vector field, that is, $\mathbf{F} = \nabla p$ for some scalar function $p\colon \mathbf{R}^2 \to \mathbf{R}$. Suppose that every entry of the Jacobian matrix of $\mathbf{J_F}$ is continuous on a plane region D. Then show that the electromotive force around any simple closed curve C in D is 0. (In such a case, no current flows along any such curve C in D.)

Solution. Since **F** is conservative, we have

$$\mathbf{F}(x, y) = \nabla p(x, y) = \frac{\partial p}{\partial x}\mathbf{i} + \frac{\partial p}{\partial y}\mathbf{j} = P(x, y)\mathbf{i} + Q(x, y)\mathbf{j}.$$

The entries of $\mathbf{J_F}$ are thus the second partial derivatives of p:

$$\mathbf{J_F} = \begin{bmatrix} \dfrac{\partial P}{\partial x} & \dfrac{\partial P}{\partial y} \\ \dfrac{\partial Q}{\partial x} & \dfrac{\partial Q}{\partial y} \end{bmatrix} = \begin{bmatrix} \dfrac{\partial^2 p}{\partial x^2} & \dfrac{\partial^2 p}{\partial y\,\partial x} \\ \dfrac{\partial^2 p}{\partial x\,\partial y} & \dfrac{\partial^2 p}{\partial y^2} \end{bmatrix}.$$

So by hypothesis all second partial derivatives of f are continuous. Theorem 1.9(d) then says that

$$\textbf{curl}\,\mathbf{F} = \textbf{curl}\,\nabla f = \mathbf{0}.$$

We therefore have from (5) and the definition of electromotive force

$$\text{emf} = \int_C \mathbf{F}\cdot d\mathbf{x} = \iint_{D_C} (\textbf{curl}\,\mathbf{F})\cdot\mathbf{k}\,dA = \iint_{D_C} 0\,dA = 0,$$

where D_C is the region of D enclosed by C. ■

We have seen in Examples 1 and 2 how Green's theorem can save labor in evaluating line integrals. It can also sometimes simplify the evaluation of double integrals. Suppose that the region D satisfies the hypotheses of Theorem 3.3. Then we can apply the theorem to the vector field $\mathbf{F}(x, y) = x\,\mathbf{j}$, getting

$$(6) \qquad \oint_{\partial D} x\,dy = \oint_{\partial D} 0\,dx + x\,dy = \iint_D \left(\frac{\partial x}{\partial x} - \frac{\partial 0}{\partial y}\right) dA = \iint_D 1\,dA = A(D).$$

Similarly, if we apply Green's theorem to the vector field $\mathbf{G}(x, y) = y\,\mathbf{i}$, then we get

$$(7) \qquad -\oint_{\partial D} y\,dx = \oint_{\partial D} (-y)\,dx + 0\,dy = \iint_D \left(\frac{\partial 0}{\partial x} - \frac{\partial(-y)}{\partial y}\right) dA = \iint_D dA = A(D).$$

Combining (6) and (7), we obtain

$$\frac{1}{2}\oint_{\partial D} -y\,dx + x\,dy = \frac{1}{2}\oint_{\partial D} -y\,dx + \frac{1}{2}\oint_{\partial D} x\,dy = \frac{1}{2}A(D) + \frac{1}{2}A(D) = A(D).$$

Summing up, we have three formulas for the area of D:

$$(8) \qquad A(D) = \oint_{\partial D} x\,dy = -\oint_{\partial D} y\,dx = \frac{1}{2}\oint_{\partial D} -y\,dx + x\,dy.$$

Example 4

Find the area enclosed by the ellipse $x^2/16 + y^2/9 = 1$.

FIGURE 3.14

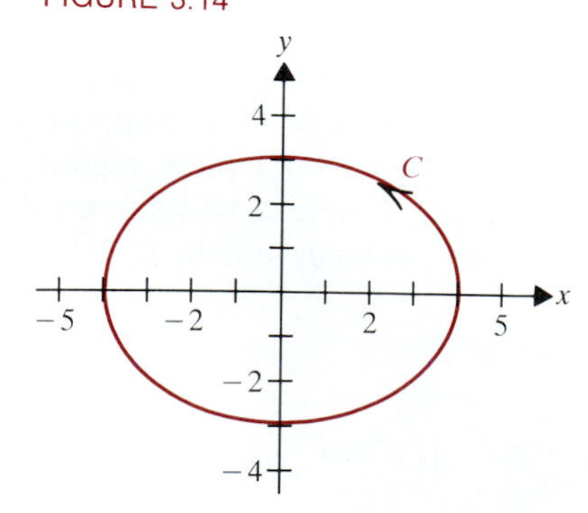

Solution. To do this problem by double integration, we would have to deal with a very complicated definite integral. An alternative is provided by (8), which says that

$$A(D) = \oint_C x\,dy,$$

where C is the ellipse shown in **Figure 3.14.** We parametrize C by $x = 4\cos t$ and $y = 3\sin t$ for $0 \le t \le 2\pi$. Then $dy = 3\cos t\,dt$, so

$$A(D) = \oint_C x\,dy = \int_0^{2\pi} (4\cos t)3\cos t\,dt = 12\int_0^{2\pi} \cos^2 t\,dt$$

$$= 12\left(\frac{1}{2}\right)\int_0^{2\pi} (1 + \cos 2t)\,dt = 6\left[t + \frac{1}{2}\sin 2t\right]_0^{2\pi} = 12\pi. \ ■$$

Sometimes Green's theorem can be used to work out the line integral of a vector function $\mathbf{F}$ over the boundary of a region D, even if $\mathbf{F}$ is not continuous on all D. In the next example we reduce

$$\oint_{\partial D} \mathbf{F} \cdot d\mathbf{x}$$

to the line integral of $\mathbf{F}$ over a simpler curve.

Example 5

If

$$\mathbf{F}(x, y) = P(x, y)\mathbf{i} + Q(x, y)\mathbf{j} = -\frac{y}{x^2 + y^2}\mathbf{i} + \frac{x}{x^2 + y^2}\mathbf{j},$$

then compute $\int_C \mathbf{F} \cdot d\mathbf{x}$ where C is the ellipse of Example 4.

Solution. The given line integral becomes quite complicated if we try to evaluate it directly. Instead we use a trick based on Green's theorem. It takes advantage of the fact that here $\partial Q/\partial x - \partial P/\partial y = 0$, which is seen as follows:

$$\frac{\partial P}{\partial y} = -\frac{(x^2 + y^2) \cdot 1 - y \cdot 2y}{(x^2 + y^2)^2} = \frac{y^2 - x^2}{(x^2 + y^2)^2},$$

and

$$\frac{\partial Q}{\partial x} = \frac{(x^2 + y^2) \cdot 1 - x \cdot 2x}{(x^2 + y^2)^2} = \frac{y^2 - x^2}{(x^2 + y^2)^2}.$$

FIGURE 3.15

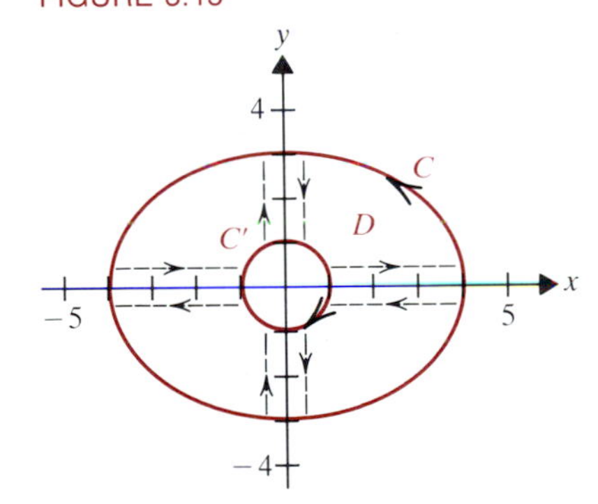

However, we can't apply Green's theorem directly to the interior of the ellipse, since $\mathbf{F}$ is not continuous on the entire interior; in particular, it is not continuous at the origin. To avoid this difficulty, let D be the region shown in **Figure 3.15** between C and the circle C': $x^2 + y^2 = 1$, oriented *clockwise* (so that D is on the *left* as we traverse the circle). Thus C' is parametrized by $x = \cos t$ and $y = -\sin t$ for $t \in [0, 2\pi]$. Now $C \cup C'$ forms the boundary of D, oriented so that D is on the left as $C \cup C'$ is traversed. D can be decomposed into the union of the four simple subregions of D located in the four quadrants. Then (2) and Theorem 3.3 give

$$\oint_{C \cup C'} \mathbf{F} \cdot d\mathbf{x} = \iint_D \left(\frac{\partial Q}{\partial x} - \frac{\partial P}{\partial y}\right) dA = 0.$$

Therefore, since

$$\oint_{C \cup C'} \mathbf{F} \cdot d\mathbf{x} = \oint_C \mathbf{F} \cdot d\mathbf{x} + \oint_{C'} \mathbf{F} \cdot d\mathbf{x} = 0,$$

we obtain

$$\oint_C \mathbf{F} \cdot d\mathbf{x} = -\oint_{C'} \mathbf{F} \cdot d\mathbf{x} = \oint_{-C'} \mathbf{F} \cdot d\mathbf{x},$$

where $-C'$ is parametrized in the *usual* way by

$$\mathbf{x}(t) = (\cos t, \sin t), \; t \in [0, 2\pi].$$

Then

$$\mathbf{x}'(t)\, dt = (-\sin t, \cos t)\, dt.$$

On $-C'$ we have $x^2 + y^2 = 1$, since $x = \cos t$ and $y = \sin t$. Therefore

$$\mathbf{F}(x(t), y(t)) = (-\sin t, \cos t).$$

Thus

$$\oint_C \mathbf{F} \cdot d\mathbf{x} = \oint_{-C'} \mathbf{F}(\mathbf{x}(t)) \cdot \mathbf{x}'(t)\, dt = \oint_{-C'} (-\sin t, \cos t) \cdot (-\sin t, \cos t)\, dt$$
$$= \int_0^{2\pi} (\sin^2 t + \cos^2 t)\, dt = \int_0^{2\pi} 1\, dt = 2\pi. \blacksquare$$

HISTORICAL NOTE

Green's theorem is named for the English mathematician George Green (1793–1841) who first used the identity (2) in 1828. The theorem had actually been discovered a few years earlier by Gauss and Lagrange, but it became well-known through Green's applications of it to electricity and magnetism and to other parts of physics whose modern treatments make heavy use of the result.

The preference for counterclockwise orientation in Green's theorem exemplifies the inclination of mathematicians, and human beings generally, to favor counterclockwise motion over clockwise. Tracks for races of dogs, horses, and athletes, for example, are usually oriented this way.

Proof of Green's Theorem 3.2*

Let D be a simple region, with counterclockwise-oriented boundary C. Suppose that the vector field $\mathbf{F}(x, y) = P(x, y)\mathbf{i} + Q(x, y)\mathbf{j}$ has continuous partial derivatives. To establish

$$\iint_D \left(\frac{\partial Q}{\partial x} - \frac{\partial P}{\partial y}\right) dA = \int_C P\, dx + Q\, dy, \tag{2}$$

we will show separately that

$$\oint_C P(x, y)\, dx = -\iint_D \frac{\partial P}{\partial y}\, dA \tag{9}$$

and

$$\oint_C Q(x, y)\, dy = \iint_D \frac{\partial Q}{\partial x}\, dA. \tag{10}$$

Here by $\oint_C P(x, y)\, dx$ and $\oint_C Q(x, y)\, dy$ we mean, respectively,

$$\int_a^b (P(\mathbf{x}(t)), 0) \cdot \left(\frac{dx}{dt}, \frac{dy}{dt}\right) dt \quad \text{and} \quad \int_a^b (0, Q(\mathbf{x}(t))) \cdot \left(\frac{dx}{dt}, \frac{dy}{dt}\right) dt,$$

if C is parametrized by $\mathbf{x} = \mathbf{x}(t)$ for $t \in [a, b]$. Then (2) follows simply by adding (9) and (10).

First we establish (9). Refer to **Figure 3.16,** which shows a general simple region D. Its boundary C consists of the two vertical segments BA and EF together with the oriented arcs AE and FB. We can parametrize the segment

FIGURE 3.16

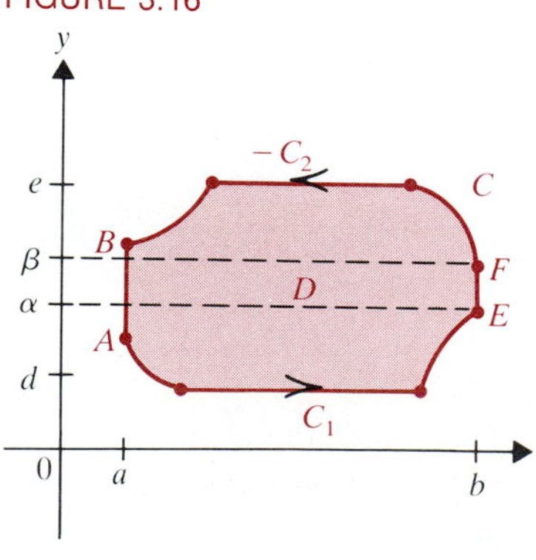

* Optional

EF by

$$\mathbf{x}(t) = (x, y) = (b, t), \qquad \alpha \le t \le \beta.$$

Then $dx/dt = 0$, so $dx = 0$. We therefore get

$$\int_{EF} P(x, y)\, dx = 0.$$

Similarly, $\int_{BA} P(x, y)\, dx = 0$. Thus from Definition 2.3,

$$\int_C P(x, y)\, dx = \int_{AE} P(x, y)\, dx + \int_{FB} P(x, y)\, dx. \tag{11}$$

Since a vertical line crosses each of AE and BF *exactly once*, these arcs are the graphs of *functions* $y = g_1(x)$ and $y = g_2(x)$ for $x \in [a, b]$. So they can be parametrized as curves C_i for $i = 1, 2$, by $x = t$, $y = g_i(t)$. It is natural then to use x as a parameter instead of t. Since C_2 is traced out as x goes from a to b, C_2 joins B to F. Since C traverses the boundary of D in the counterclockwise sense, C proceeds from F to B. Thus we have from (11) and Theorem 2.4

$$\begin{aligned}
\int_C P(x, y)\, dx &= \int_{C_1} P(x, y)\, dx + \int_{-C_2} P(x, y)\, dx \\
&= \int_{C_1} P(x, y)\, dx - \int_{C_2} P(x, y)\, dx \\
&= \int_a^b P(x, g_1(x))\, dx - \int_a^b P(x, g_2(x))\, dx \\
&= -\int_a^b [P(x, g_2(x)) - P(x, g_1(x))]\, dx \\
&= -\int_a^b P(x, y)\Big]_{y=g_1(x)}^{y=g_2(x)} dx = -\int_a^b \int_{g_1(x)}^{g_2(x)} \frac{\partial P(x, y)}{\partial y}\, dy\, dx \\
&= -\iint_D \frac{\partial P}{\partial y}\, dA. \qquad \textit{by Fubini's theorem}
\end{aligned}$$

Thus (9) is proved. The approach to (10) is quite similar and is left as Exercise 37. Addition of (9) and (10) gives (2). QED

Exercises 14.3

In Exercises 1–14, use Green's theorem to evaluate the given integrals.

1. $\oint_C 2y\, dx + 3x\, dy$, where C is the polygonal path from $(-1, 2)$ to $(3, 2)$ to $(3, 5)$ to $(-1, 5)$ to $(-1, 2)$
2. $\oint_C (x^2 - y^3)\, dx + (x^3 + y^2)\, dy$, where C is the unit circle $x^2 + y^2 = 1$ traversed counterclockwise
3. $\oint_C \mathbf{F} \cdot d\mathbf{x}$, where $\mathbf{F}$ is as in Exercise 2, but C is the circle $x^2 + y^2 = 9$ traversed *clockwise*
4. $\oint_C -ye^x\, dx + xe^y\, dy$, where C is the polygonal path from $(1, 1)$ to $(-1, 1)$ to $(-1, -1)$ to $(1, -1)$ to $(1, 1)$
5. $\oint_C y^2 e^x\, dx + 2ye^x\, dy$, where C is the path of Exercise 1
6. $\int_C (e^x + y - 2x^2)\, dx + (7x - \sin y)\, dy$, where C is the triangular path from $(0, 0)$ to $(0, 1)$ to $(1, 0)$ to $(0, 0)$
7. $\int_C (x^2 + y)\, dx + (2x - y)\, dy$, where $C = C_1 \cup C_2$. Here C_1 is the polygonal path of Exercise 1, and C_2 is the unit circle parametrized in the clockwise sense.
8. $\int_C (x^3 + \sin x + y)\, dx + (2x - \sin y \cos y)\, dy$, where C is as in Exercise 7
9. $\oint (\sin y - x^2 y)\, dx + (x \cos y + xy^2)\, dy$, where C is the curve of Exercise 2
10. $\oint_C (x - xy)\, dx + (y^3 + 1)\, dy$, where C is the polygonal path from $(1, 0)$ to $(2, 0)$ to $(2, 1)$ to $(1, 1)$ to $(1, 0)$
11. $\oint_C \frac{-y}{x^2 + y^2}\, dx + \frac{x}{x^2 + y^2}\, dy$, where C is the ellipse $4x^2 + y^2 = 16$, traversed counterclockwise
12. $\oint_C \frac{-y}{x^2 + y^2}\, dx + \frac{x}{x^2 + y^2}\, dy$, where C is the ellipse $5x^2 + y^2 = 25$, traversed *clockwise*
13. $\int_C \frac{x}{x^2 + y^2}\, dx + \frac{y}{x^2 + y^2}\, dy$, where $C = C_1 \cup C_2$. Here C_1 is the parabolic path from $(-1, 0)$ to $(2, 3)$ along $y = x^2 - 1$, and C_2 is the line segment from $(2, 3)$ to $(-1, 0)$.

14. $\int_C \frac{x}{x^2+y^2}\,dx + \frac{y}{x^2+y^2}\,dy$, where $C = C_1 \cup C_2$. Here C_1 is the parabolic path from $(-2, 0)$ to $(3, 5)$ along $y = x^2 - 4$, and C_2 is the line segment $(3, 5)$ to $(-2, 0)$.

In Exercises 15–20, find the area of the given region.

15. The region D enclosed by the ellipse $4x^2 + y^2 = 16$

16. The region D enclosed by the hypocycloid $x^{2/3} + y^{2/3} = 1$. (See Exercise 32, Section 2.)

17. The region D enclosed by the hypocycloid $x^{2/3} + y^{2/3} = 4$

18. The triangle with vertices $(1, 2)$, $(3, 2)$, and $(3, 5)$

19. The annular region between $x^2 + y^2 = 16$ and $x^2 + y^2 = 9$. (*Hint: Think* before resorting to (8)!)

20. The annular region between $x^2 + y^2 = 4$ and $x^2 + y^2 = 16$

21. Show that if $\mathbf{F}(x, y) = y^2e^x\,\mathbf{i} + 2ye^x\,\mathbf{j}$ is the velocity field of a fluid flow, then there is no tendency for a vortex around the origin.

22. Repeat Exercise 21 for

$$\mathbf{F}(x, y) = (x^3 + y^3 + y\sin x)\,\mathbf{i} + (3xy^2 - \cos x)\,\mathbf{j}.$$

23. If D is a region satisfying the hypotheses of Theorem 3.3, then find a formula for $\bar{x}$ and $\bar{y}$, the coordinates of the centroid of D, as a line integral.

24. Generalize Example 5, namely, let

$$\mathbf{F}(\mathbf{x}) = P(\mathbf{x})\,\mathbf{i} + Q(\mathbf{x})\,\mathbf{j}, \text{ where } \frac{\partial P}{\partial y} = \frac{\partial Q}{\partial x} \text{ for } \mathbf{x} \neq \mathbf{0}.$$

Suppose that C_1 and C_2 are smooth simple closed curves with the same orientation, which enclose regions D_i such that $\mathbf{0} \in D_1 \subseteq D_2$. Then show that

$$\oint_{C_1} \mathbf{F} \cdot d\mathbf{x} = \oint_{C_2} \mathbf{F} \cdot d\mathbf{x}.$$

25. Generalize Exercise 11, namely, prove that if

$$\mathbf{F}(\mathbf{x}) = -\frac{y}{x^2+y^2}\,\mathbf{i} + \frac{x}{x^2+y^2}\,\mathbf{j},$$

then for any smooth simple closed curve C

$$\oint_C \mathbf{F} \cdot d\mathbf{x} = \begin{cases} 2\pi & \text{if } \mathbf{0} \text{ is enclosed by } C \\ 0 & \text{if } \mathbf{0} \text{ is not enclosed by } C. \end{cases}$$

26. If C is a smooth curve parametrized by $\mathbf{x} = \mathbf{x}(t) = (x(t), y(t))$ for t in $[a, b]$, then show that

$$\mathbf{N}(t) = \frac{1}{|\mathbf{x}'(t)|}\left(\frac{dy}{dt}\,\mathbf{i} - \frac{dx}{dt}\,\mathbf{j}\right)$$

is a unit normal vector to C at every point on the curve.

27. Use Exercise 26 to show that

$$\oint_C -Q\,dx + P\,dy = \oint_C \mathbf{F} \cdot \mathbf{N}\,ds.$$

If $\mathbf{F} = P(x, y)\,\mathbf{i} + Q(x, y)\,\mathbf{j}$ is the velocity field of a fluid flow in $\mathbf{R}^2$, this integral is called the ***flux*** across C. Explain how it measures the amount of fluid flowing across C. **(This is referred to in Sections 6 and 8.)**

28. Use Exercise 27 and Theorem 3.3 to show that, under suitable hypotheses,

$$\iint_D \operatorname{div} \mathbf{F}\,dA = \int_{\partial D} \mathbf{F} \cdot \mathbf{N}\,ds.$$

(In Section 8 we will see how Green's theorem in this form can be extended to three-dimensional vector fields $\mathbf{F}\colon \mathbf{R}^3 \to \mathbf{R}^3$.)

29. Prove *Green's first identity*, an analogue of integration by parts. If $f\colon \mathbf{R}^2 \to \mathbf{R}$ is differentiable on a simple region D and all second partial derivatives of $g\colon \mathbf{R}^2 \to \mathbf{R}$ are continuous on D, then use Exercise 27 to show

$$\iint_D f(\mathbf{x})\nabla^2 g(\mathbf{x})\,dA = \int_{\partial D} f(\mathbf{x})(\nabla g \cdot \mathbf{N})\,ds - \iint_D \nabla f \cdot \nabla g\,dA$$
$$= \int_{\partial D} f(\mathbf{x}) D_{\mathbf{N}}(g)\,ds - \iint_D \nabla f \cdot \nabla g\,dA,$$

where $D_{\mathbf{N}}(g)$ is the directional derivative of g in the direction of $\mathbf{N}$.

30. In Exercise 29 assume that all second partial derivatives of f and g are continuous on D. Then prove *Green's second identity*,

$$\iint_D [f(\mathbf{x})\nabla^2 g(\mathbf{x}) - g(\mathbf{x})\nabla^2 f(\mathbf{x})]\,dA = \int_{\partial D} f(\mathbf{x})(\nabla g \cdot \mathbf{N})\,ds$$
$$- \int_{\partial D} g(\mathbf{x})(\nabla f \cdot \mathbf{N})\,ds$$
$$= \int_{\partial D} (fD_{\mathbf{N}}(g) - gD_{\mathbf{N}}(f))\,ds.$$

What can you conclude if f and g are harmonic (Definition 1.7)?

31. Suppose that a flowing fluid in the plane has density $\delta(x, y, t)$ at the point (x, y) at time t. Show that $\oint_C \delta(x, y, t)\mathbf{F} \cdot \mathbf{N}\,ds$ is the rate of loss of mass per unit time from the disk D enclosed by a circle C of radius a centered at point (x_0, y_0).

32. Use Exercise 31 to show that

$$\lim_{a \to 0} \frac{1}{\pi a^2} \oint_C \delta(x, y, t)\mathbf{F} \cdot \mathbf{N}\,ds$$

gives the rate of decrease of the density per unit time per unit area at (x_0, y_0). (This is then $-\partial\delta/\partial t$.) Show that this limit is the divergence of the function $\delta\mathbf{F}$ at (x_0, y_0), and hence derive the two-dimensional *continuity equation of hydrodynamics*,

$$\frac{\partial \delta}{\partial t} + \operatorname{div}(\delta\mathbf{F}) = 0.$$

If $\delta(x, y, t)$ is a constant, then conclude that the fluid is incompressible. [*Hint:* Use the mean value theorem for double integrals (Exercise 38, Section 13.1).]

We make the term *curl* even more compelling in the next two exercises, which are set in the *xy*-plane.

33. Suppose that C is a circle enclosing the disk D with center $\mathbf{x}_0$ and radius a. Then use the mean value theorem for double integrals cited in Exercise 32 to show that

$$\oint_C \mathbf{F} \cdot d\mathbf{x} = (\mathbf{curl}\, \mathbf{F}(x_1, y_1) \cdot \mathbf{k})A(D), \text{ for some } (x_1, y_1) \in D.$$

34. Conclude from Exercise 33 that the z-coordinate of **curl F** is the limit of the circulation per unit area at (x_0, y_0) if **curl F** is continuous on an open disk containing (x_0, y_0).

35. Use Exercises 33 and 34 to compute $\mathbf{curl}\, \mathbf{F}(\mathbf{0})$ if $\mathbf{F}(x, y) = -y\mathbf{i} + x\mathbf{j}$. (Find the limit of $1/\pi a^2$ times the circulation around the circle C of radius a and center $(0, 0)$.)

36. A field $F: \mathbf{R}^3 \to \mathbf{R}^3$ is called ***central*** if $F(\mathbf{x}) = f(|\mathbf{x}|)\mathbf{x}$ for some $f: \mathbf{R} \to \mathbf{R}$. Show that at all points $\mathbf{x_0} \neq \mathbf{0}$, the curl of a central field is $\mathbf{0}$.

37. Use the approach of the text in proving (9) to prove (10).

14.4 Independence of Path

If $f': \mathbf{R} \to \mathbf{R}$ is continuous on $[a, b]$, then the fundamental theorem of calculus (Theorem 4.2 of Chapter 4) says that

$$\int_a^b f'(x)\, dx = f(b) - f(a).$$

If $f: \mathbf{R}^n \to \mathbf{R}$ is differentiable, then ∇f is the total derivative of f. The first result of this section is a multivariable version of the fundamental theorem. It says that if ∇f is continuous, then its line integral over a piecewise smooth path C from $\mathbf{a}$ to $\mathbf{b}$ in $\mathbf{R}^n$ is also computable as $f(\mathbf{b}) - f(\mathbf{a})$.

4.1 THEOREM

Suppose that $f: \mathbf{R}^n \to \mathbf{R}$ is differentiable with continuous partial derivatives on a region D containing $\mathbf{a}$ and $\mathbf{b}$. Then for any piecewise smooth path C in D from $\mathbf{a}$ to $\mathbf{b}$,

$$\int_C \nabla f \cdot d\mathbf{x} = f(\mathbf{b}) - f(\mathbf{a}).$$

Proof. Since C is a union of smooth paths C_i, by Definition 2.3 the line integral over C is the sum of the line integrals over those smooth paths. It is enough to show that the result holds on each such smooth path. For suppose that on each smooth portion C_i of C between $t = t_{i-1}$ and $t = t_i$ we have

$$\int_{C_i} \nabla f \cdot d\mathbf{x} = f(\mathbf{x}(t_i)) - f(\mathbf{x}(t_{i-1})).$$

Then, since $\mathbf{x}(t_0) = \mathbf{a}$ and $\mathbf{x}(t_n) = \mathbf{b}$, it follows that

$$\int_C \nabla f \cdot d\mathbf{x} = \sum_{i=1}^{n} \int_{C_i} \nabla f \cdot d\mathbf{x} = \sum_{i=1}^{n} [f(\mathbf{x}(t_i)) - f(\mathbf{x}(t_{i-1}))]$$

$$= f(\mathbf{x}(t_n)) - f(\mathbf{x}(t_0)) = f(\mathbf{b}) - f(\mathbf{a}).$$

So assume that C_i is parametrized by $\mathbf{x} = \mathbf{x}(t)$, $t_{i-1} \leq t \leq t_i$, where $\mathbf{x}'(t)$ is continuous on $[t_{i-1}, t_i]$. Then we have

$$\int_{C_i} \nabla f \cdot d\mathbf{x} = \int_{t_{i-1}}^{t_i} \nabla f(\mathbf{x}) \cdot \mathbf{x}'(t)\, dt$$

$$= \int_{t_{i-1}}^{t_i} \frac{df(\mathbf{x}(t))}{dt} \qquad \textit{by Theorem 4.2 of Chapter 12}$$

$$= f(\mathbf{x}(t))\Big]_{t_{i-1}}^{t_i} \qquad \textit{by Theorem 4.2 of Chapter 4}$$

$$= f(\mathbf{x}(t_i)) - f(\mathbf{x}(t_{i-1})).$$ QED

FIGURE 4.1

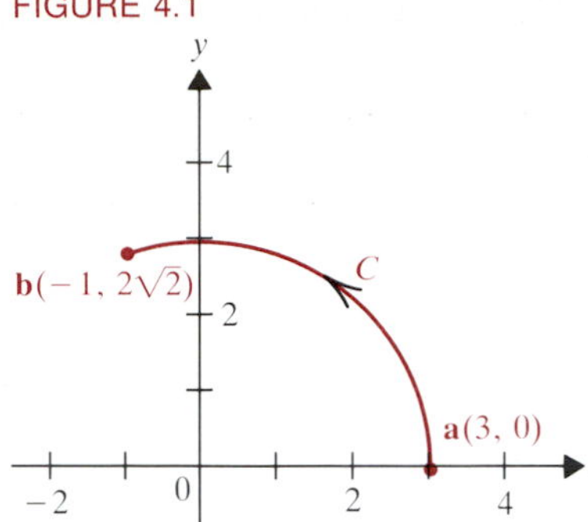

Example 1

If $f(x, y) = x/\sqrt{x^2 + y^2}$, then compute $\int_C \nabla f \cdot d\mathbf{x}$, where C is the arc of the circle $x^2 + y^2 = 9$ joining $\mathbf{a} = (3, 0)$ to $\mathbf{b} = (-1, 2\sqrt{2})$ in the counterclockwise sense (**Figure 4.1**).

Solution. We don't have to compute ∇f or parametrize C! We just use Theorem 4.1 and get

$$\int_C \nabla f \cdot d\mathbf{x} = f(-1, 2\sqrt{2}) - f(3, 0) = -\frac{1}{3} - \frac{3}{3} = -\frac{4}{3}. \quad \blacksquare$$

Suppose that ∇f is continuous on a set $D \subseteq \mathbf{R}^n$. Then the line integral of ∇f over a path C on D depends only on the initial and terminal points of C, as long as C is at least piecewise smooth. Since $\int_C \nabla f \cdot d\mathbf{x} = f(\mathbf{b}) - f(\mathbf{a})$ for every piecewise smooth path C joining $\mathbf{a}$ to $\mathbf{b}$, we say that the line integral of ∇f is ***independent of path*** in D. The impact of this concept is suggested by the following two results.

4.2 THEOREM The line integral $\int_C \mathbf{F} \cdot d\mathbf{x}$ is independent of path in D if and only if $\int_C \mathbf{F} \cdot d\mathbf{x} = 0$ for every piecewise smooth closed curve C in D.

Proof. If the line integral is independent of path, then choose any piecewise smooth closed curve C in D. If $\mathbf{a}$ and $\mathbf{b}$ are any two points on C, then C is decomposed into two piecewise smooth subpaths, C_1 from $\mathbf{a}$ to $\mathbf{b}$ and C_2 from $\mathbf{b}$ to $\mathbf{a}$ (see **Figure 4.2**). Therefore C_1 and $-C_2$ are two paths connecting $\mathbf{a}$ to $\mathbf{b}$. So by independence of path, we have

FIGURE 4.2

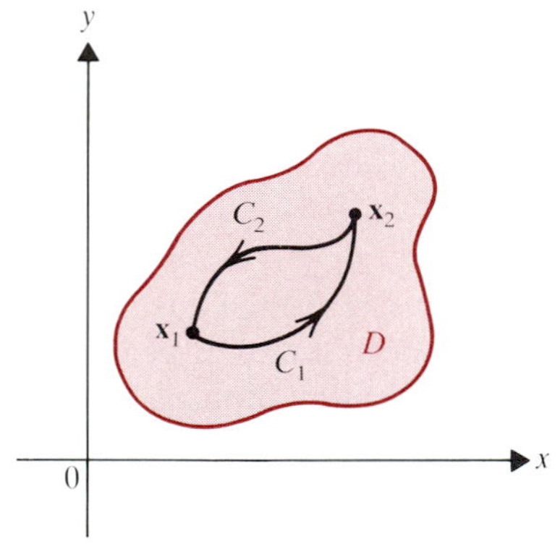

$$\int_{C_1} \mathbf{F} \cdot d\mathbf{x} = \int_{-C_2} \mathbf{F} \cdot d\mathbf{x}.$$

Hence

$$\begin{aligned}\int_C \mathbf{F} \cdot d\mathbf{x} &= \int_{C_1} \mathbf{F} \cdot d\mathbf{x} + \int_{-C_2} \mathbf{F} \cdot d\mathbf{x} && \text{by Definition 2.3}\\ &= \int_{C_1} \mathbf{F} \cdot d\mathbf{x} - \int_{C_2} \mathbf{F} \cdot d\mathbf{x} && \text{by Theorem 2.4}\\ &= 0.\end{aligned}$$

Conversely, if $\int_C \mathbf{F} \cdot d\mathbf{x} = 0$ for any piecewise smooth closed curve C in D, then we claim that $\int_{C_1} \mathbf{F} \cdot d\mathbf{x} = \int_{C_2} \mathbf{F} \cdot d\mathbf{x}$ for any piecewise smooth curves C_1 and C_2 joining $\mathbf{a}$ to $\mathbf{b}$. For $C = C_1 \cup (-C_2)$ is a piecewise smooth closed path (see **Figure 4.3**). Then by hypothesis, $\int_C \mathbf{F} \cdot d\mathbf{x} = 0$. But again by Theorem 2.4, we have

FIGURE 4.3

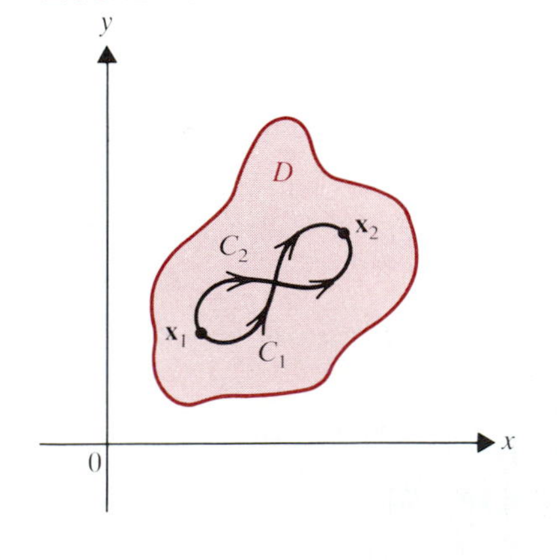

$$0 = \int_C \mathbf{F} \cdot d\mathbf{x} = \int_{C_1} \mathbf{F} \cdot d\mathbf{x} - \int_{C_2} \mathbf{F} \cdot d\mathbf{x},$$

so

$$\int_{C_1} \mathbf{F} \cdot d\mathbf{x} = \int_{C_2} \mathbf{F} \cdot d\mathbf{x}. \quad \boxed{\text{QED}}$$

We next justify the term *conservative vector field* introduced in Definition 1.6 and in the process establish the *law of conservation of energy*, which is of fundamental importance in physics. Recall that $\mathbf{F}: \mathbf{R}^n \to \mathbf{R}^n$ is called *conservative* if $\mathbf{F}(\mathbf{x}) = \nabla f(\mathbf{x})$ for some scalar function f. The function $p = -f$ is called the ***potential energy*** of the field. This definition is made in anticipation of the next example.

Example 2

Conservation of Energy. Suppose that $\mathbf{F}\colon \mathbf{R}^n \to \mathbf{R}^n$ is a conservative vector field on D. Show that if $\mathbf{F}$ moves a particle from $\mathbf{a}$ in D to $\mathbf{b}$ in D along any piecewise smooth path C in D, parametrized by $\mathbf{x} = \mathbf{x}(t)$, then the *total energy* $T(\mathbf{x})$ remains the same, where

$$T(\mathbf{x}) = p(\mathbf{x}) + \frac{1}{2} m|\mathbf{v}(t)|^2$$

is the sum of the potential energy and the kinetic energy.

Solution. We first show that the work done in moving the particle from $\mathbf{a} = \mathbf{x}(t_1)$ to $\mathbf{b} = \mathbf{x}(t_2)$ is the change

$$\frac{1}{2} m|\mathbf{v}(t_2)|^2 - \frac{1}{2} m|\mathbf{v}(t_1)|^2$$

in kinetic energy. For this, recall from Definition 2.1 that the work done in moving the particle from $\mathbf{a}$ to $\mathbf{b}$ is

$$W = \int_{t_1}^{t_2} \mathbf{F}(\mathbf{x}(t)) \cdot \mathbf{x}'(t)\,dt = \int_{t_1}^{t_2} m\mathbf{x}''(t) \cdot \mathbf{x}'(t)\,dt \qquad \text{by Newton's second law of motion}$$

$$= \frac{1}{2} m \int_{t_1}^{t_2} 2\mathbf{x}''(t) \cdot \mathbf{x}'(t)\,dt = \frac{1}{2} m \int_{t_1}^{t_2} \frac{d}{dt}[\mathbf{x}'(t) \cdot \mathbf{x}'(t)]\,dt.$$

Thus

$$W = \frac{1}{2} m|\mathbf{x}'(t)|^2\Big]_{t_1}^{t_2} = \frac{1}{2} m|\mathbf{v}(t_2)|^2 - \frac{1}{2} m|\mathbf{v}(t_1)|^2. \tag{1}$$

We can also use Theorem 4.1 to compute W. That gives

$$W = \int_C \mathbf{F} \cdot d\mathbf{x} = \int_C \nabla f \cdot d\mathbf{x} = -\int_C \nabla p \cdot d\mathbf{x} = -[p(\mathbf{x}_2) - p(\mathbf{x}_1)] = p(\mathbf{x}_1) - p(\mathbf{x}_2). \tag{2}$$

Equating the two expressions for W, we obtain

$$\frac{1}{2} m|\mathbf{v}(t_2)|^2 + p(\mathbf{x}_2) = \frac{1}{2} m|\mathbf{v}(t_1)|^2 + p(\mathbf{x}_1) \to T(\mathbf{x}_2) = T(\mathbf{x}_1). \quad ■$$

Another way of phrasing the conservation of total energy in Example 2 is:

the change in kinetic energy equals the change in potential energy $-p$.

In fact, Equations (1) and (2) say that the total work done from $\mathbf{a}$ to $\mathbf{b}$ is equal to either energy change. (The potential energy is defined to be the *negative* of the potential *function* so that this holds.)

The next result says that the line integral of a continuous vector field F is independent of path on an open ball D *only* if F is conservative, that is, $\mathbf{F} = \nabla f$ for some scalar function f. (In advanced calculus this is shown to hold for more general regions D.) The proof is at the end of the section.

4.3 THEOREM Let D be an open disk in $\mathbf{R}^2$ or an open ball in $\mathbf{R}^3$. If the continuous vector field $\mathbf{F}$ on $\mathbf{R}^2$ or $\mathbf{R}^3$ has a path-independent line integral in D, then $\mathbf{F}$ is a conservative vector field. In fact,

$$\mathbf{F} = \nabla f \qquad \text{for } f(\mathbf{x}) = \int_a^x \mathbf{F} \cdot d\mathbf{x},$$

where the line integral is taken over any piecewise smooth path C in D joining an arbitrarily chosen but fixed point $\mathbf{a}$ to $\mathbf{x}$.

To review the results given thus far:

Theorems 4.1 and 4.3 say that $\mathbf{F}$ is a conservative vector field on an open ball D if and only if $\int_C \mathbf{F} \cdot d\mathbf{x}$ is independent of path in D. This in turn is equivalent, by Theorem 4.2, to $\int_C \mathbf{F} \cdot d\mathbf{x} = 0$ for any piecewise smooth closed curve C in D.

A still-unanswered question is, given a vector field $\mathbf{F}$ on $\mathbf{R}^2$ or $\mathbf{R}^3$, how can we determine whether $\mathbf{F}$ is conservative, that is, whether $\int_C \mathbf{F} \cdot d\mathbf{x}$ is independent of path in reasonable domains D? In the three-dimensional case, $\mathbf{F}$ will be conservative if and only if $\mathbf{F} = \nabla f$ for some scalar function f. If this holds, then

$$\mathbf{F}(x) = \left(\frac{\partial f}{\partial x}, \frac{\partial f}{\partial y}, \frac{\partial f}{\partial z}\right) = (f_1(x, y, z), f_2(x, y, z), f_3(x, y, z)),$$

where f_1, f_2, and f_3 are the coordinate functions of $\mathbf{F}$. If $\mathbf{F}$ is differentiable, then its Jacobian matrix is

$$\mathbf{J_F} = \begin{pmatrix} f_{xx} & f_{xy} & f_{xz} \\ f_{yx} & f_{yy} & f_{yz} \\ f_{zx} & f_{zy} & f_{zz} \end{pmatrix} = \begin{pmatrix} \frac{\partial f_1}{\partial x} & \frac{\partial f_1}{\partial y} & \frac{\partial f_1}{\partial z} \\ \frac{\partial f_2}{\partial x} & \frac{\partial f_2}{\partial y} & \frac{\partial f_2}{\partial z} \\ \frac{\partial f_3}{\partial x} & \frac{\partial f_3}{\partial y} & \frac{\partial f_3}{\partial z} \end{pmatrix}.$$

If the entries of $\mathbf{J_F}$ are continuous, then by Theorem 6.2 of Chapter 12 $\mathbf{J_F}$ is a *symmetric* matrix. This means that $f_{xy} = f_{yx}$, $f_{xz} = f_{zx}$, and $f_{yz} = f_{zy}$, that is,

$$\frac{\partial f_1}{\partial y} = \frac{\partial f_2}{\partial x}, \qquad \frac{\partial f_1}{\partial z} = \frac{\partial f_3}{\partial x}, \qquad \text{and} \qquad \frac{\partial f_2}{\partial z} = \frac{\partial f_3}{\partial y}.$$

Similar reasoning in two dimensions establishes the following result.

4.4 THEOREM If $\mathbf{F}$ is a conservative vector field on $\mathbf{R}^2$ or $\mathbf{R}^3$ whose coordinate functions have continuous partial derivatives on a set D, then $\mathbf{J_F}(\mathbf{x})$ is a ***symmetric matrix,*** that is,

$$\frac{\partial f_i}{\partial x_j} = \frac{\partial f_j}{\partial x_i} \qquad \text{for } i, j = 1, 2, \ldots, n.$$

The condition that $\mathbf{J_F}(\mathbf{x})$ is symmetric is thus *necessary* if there is to be any hope for $\mathbf{F}$ to be a conservative vector field. However, it is not *enough* to assure that $\mathbf{F}$ actually is conservative on an arbitrary domain in $\mathbf{R}^2$ or $\mathbf{R}^3$. This is brought out by the next example.

FIGURE 4.4

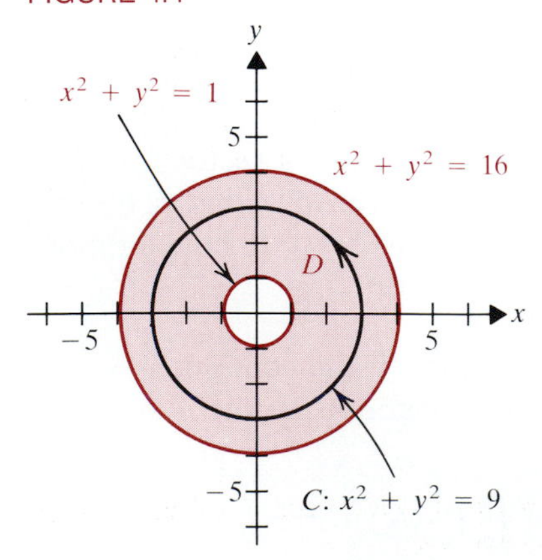

Example 3

Show that the vector field

$$\mathbf{F}(x, y) = -\frac{y}{x^2 + y^2}\mathbf{i} + \frac{x}{x^2 + y^2}\mathbf{j}$$

has a symmetric Jacobian matrix and yet is *not* a conservative vector field on the annular region D lying between the two circles $x^2 + y^2 = 1$ and $x^2 + y^2 = 16$. (See **Figure 4.4.**)

Solution. Here we have

$$f_1(x) = -\frac{y}{x^2 + y^2} \quad \text{and} \quad f_2(x) = \frac{x}{x^2 + y^2}.$$

Thus

$$\frac{\partial f_1}{\partial y} = \frac{-(x^2 + y^2) + 2y^2}{(x^2 + y^2)^2} = \frac{y^2 - x^2}{(x^2 + y^2)^2} \quad \text{and} \quad \frac{\partial f_2}{\partial x} = \frac{x^2 + y^2 - 2x^2}{(x^2 + y^2)^2} = \frac{y^2 - x^2}{(x^2 + y^2)^2}$$

on D. Hence

$$\mathbf{J_F} = \begin{pmatrix} \partial f_1/\partial x & \partial f_1/\partial y \\ \partial f_2/\partial x & \partial f_2/\partial y \end{pmatrix}$$

is symmetric on D. But Example 5 of the last section showed that $\int_C \mathbf{F} \cdot d\mathbf{x} = 2\pi \neq 0$, where C is the circle $x^2 + y^2 = 9$ traversed counterclockwise. So by Theorem 4.2, $\int_C \mathbf{F} \cdot d\mathbf{x}$ is not independent of path on D. Therefore, by Theorem 4.1, F is not a conservative vector field. ■

In the last century it was found that for *some* regions D, the fact that $\mathbf{J_F}(\mathbf{x})$ is symmetric is enough to guarantee that $\mathbf{F}$ is conservative on D. One of the major mathematical problems in the nineteenth century was to characterize the *most general* such regions. It was eventually discovered that the answer is *simply connected* regions. We will use that term for regions D such that every planar simple closed curve C in D bounds a planar surface S completely enclosed in D.* For example, rectangles and disks in $\mathbf{R}^2$ and balls, the interiors of spheres, cubes, tetrahedra, ellipsoids, and similar figures in $\mathbf{R}^3$ are simply connected, as shown in **Figure 4.5.** The region D of Example 3 is *not* simply connected because the

FIGURE 4.5 Some Simply Connected Regions in $\mathbf{R}^2$ and $\mathbf{R}^3$

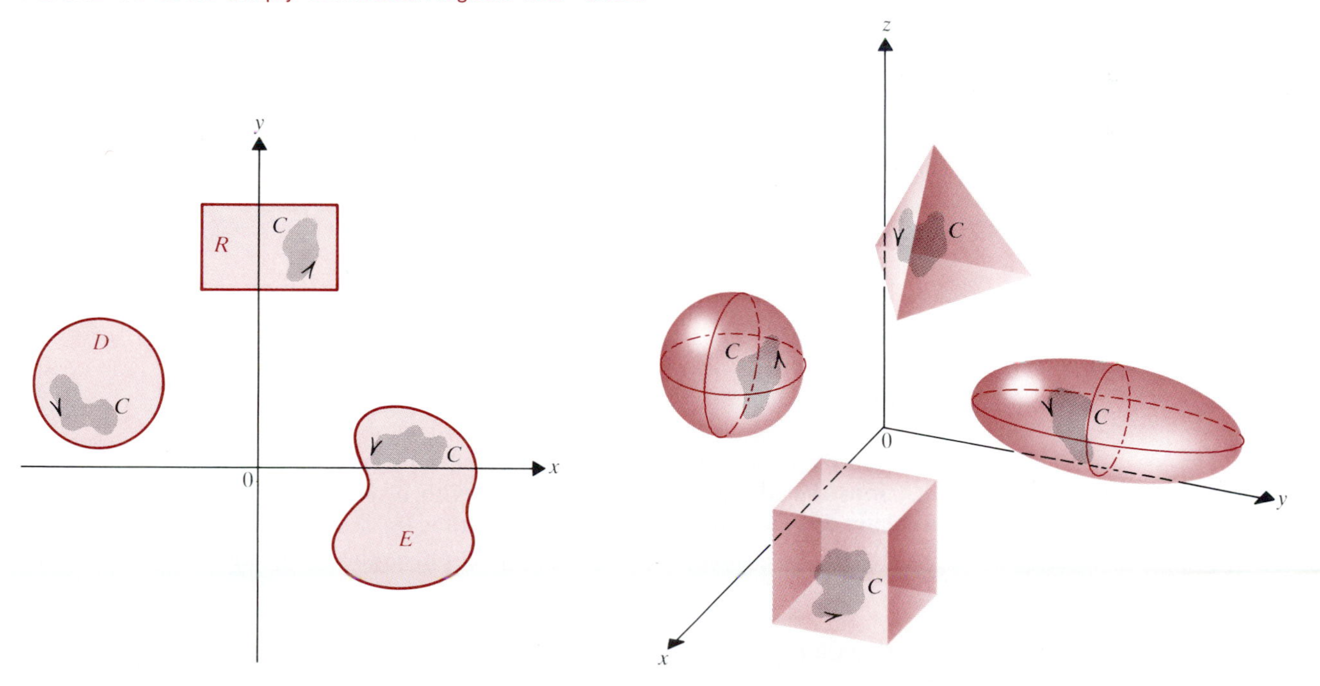

* The term *simply connected* is usually applied to a somewhat wider class of regions in higher mathematics, but we cannot discuss the more general notion here.

simple closed curve C shown in **Figure 4.6** encloses a disk S that is *not* completely contained in D. We can use Green's theorem to prove the following partial converse of Theorem 4.4 for dimension 2. (This result holds in three dimensions, and we will use it, deferring the proof to Section 7. Higher dimensions are left for advanced calculus.)

4.5 THEOREM Suppose that for $\mathbf{F}: \mathbf{R}^2 \to \mathbf{R}^2$, the Jacobian matrix $\mathbf{J_F}$ has continuous entries and is symmetric in a simply connected plane region D. Then $\mathbf{F}$ is a conservative vector field on D.

FIGURE 4.6

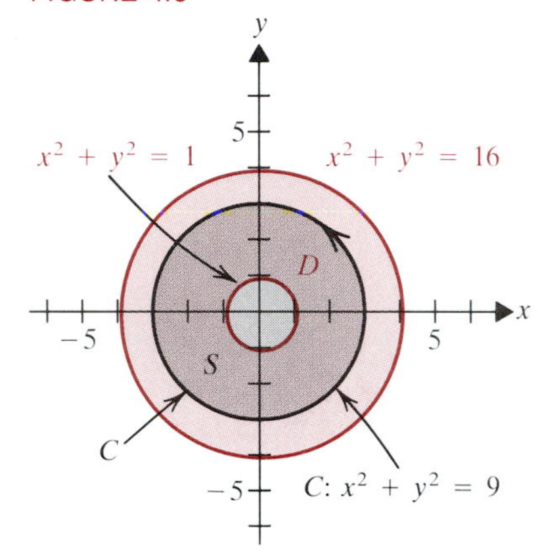

Proof. Let C be any simple closed curve in D. We will show that $\int_C \mathbf{F} \cdot d\mathbf{x} = 0$. Then the result follows from Theorem 4.2. Writing $\mathbf{F} = P\mathbf{i} + Q\mathbf{j}$, Green's theorem says that

$$\int_C \mathbf{F} \cdot d\mathbf{x} = \iint_S \left(\frac{\partial Q}{\partial x} - \frac{\partial P}{\partial y}\right) dA, \tag{3}$$

where S is the region enclosed by C. (S is *completely contained* in D, because D is simply connected.) But since

$$\mathbf{J_F}(\mathbf{x}) = \begin{pmatrix} \partial P/\partial x & \partial P/\partial y \\ \partial Q/\partial x & \partial Q/\partial y \end{pmatrix}$$

is symmetric, we have $\partial P/\partial y = \partial Q/\partial x$. Thus from (3), $\int_C \mathbf{F} \cdot d\mathbf{x} = 0$. QED

Suppose now that we confine ourselves to simply connected regions D. If $\mathbf{J_F}$ is symmetric, then $\mathbf{F}$ is a conservative vector field. So $\mathbf{F} = \nabla f$ for some scalar function f. To find f we solve the partial differential equations

$$\frac{\partial f}{\partial x} = f_1, \qquad \frac{\partial f}{\partial y} = f_2, \qquad \text{and, if present,} \qquad \frac{\partial f}{\partial z} = f_3$$

by partial integration. That is, we freeze y and z in $\partial f/\partial x = f_1$ and integrate with respect to x. Similarly, we integrate $\partial f/\partial y = f_2$ and $\partial f/\partial z = f_3$ with respect to y and z, respectively. We obtain

$$f(x, y, z) = \int f_1(x, y, z)\, dx + c_1(y, z), \tag{4}$$

$$f(x, y, z) = \int f_2(x, y, z)\, dy + c_2(x, z), \tag{5}$$

and

$$f(x, y, z) = \int f_3(x, y, z)\, dz + c_3(x, y), \tag{6}$$

where $c_1(y, z)$, $c_2(x, z)$, and $c_3(x, y)$ are "constants" of integration, that is, functions that don't involve the variable with respect to which we integrated. We have to determine c_1, c_2, and c_3 so that (4), (5), and (6) all give the same function f. This can usually be done by inspection, as the following example illustrates.

Example 4

Determine whether $\mathbf{F}(x, y, z) = 2xy\,\mathbf{i} + (x^2 + z^2)\mathbf{j} + 2yz\,\mathbf{k}$ is conservative on the domain D enclosed by the ellipsoid $x^2 + 2y^2 + 3z^2 = 27$. If it is, find f so that $\mathbf{F} = \nabla f$ on D.

Solution. First note that D is simply connected. We have

$$\mathbf{J_F} = \begin{pmatrix} 2y & 2x & 0 \\ 2x & 0 & 2z \\ 0 & 2z & 2y \end{pmatrix},$$

so $\mathbf{J_F}$ is symmetric. Therefore $\mathbf{F} = \nabla f = (\partial f/\partial x, \partial f/\partial y, \partial f/\partial z)$ for some scalar potential function f. Thus

$$(7) \qquad \frac{\partial f}{\partial x} = 2xy, \qquad \frac{\partial f}{\partial y} = x^2 + z^2, \qquad \frac{\partial f}{\partial z} = 2yz.$$

Integration of (7) with respect to x, y, and z gives

$$(8) \qquad f(x, y, z) = x^2y + c_1(y, z) = x^2y + yz^2 + c_2(x, z) = yz^2 + c_3(x, y).$$

Now, for the equalities in (8) to hold simultaneously, we can take $c_1(y, z) = yz^2$ and $c_3(x, y) = x^2y$. We can also take $c_2(x, z) = 0$, since no term involving x and z turned up in our integrations in (8). Thus

$$f(x, y, z) = x^2y + z^2y$$

will do. (As a check, a simple calculation verifies $\nabla f = \mathbf{F}$.) ■

It turns out that first checking $\mathbf{J_F}$ for symmetry can actually be omitted. We can instead proceed directly to try to find a potential function f by integrating $\nabla f = \mathbf{F}$. If $\mathbf{F}$ is *not* a conservative vector field, then it will be impossible to reconcile (4), (5), and (6) as we did in Example 4. The following example illustrates that situation.

Example 5

Determine whether $\mathbf{F}(x, y) = x^2y\,\mathbf{i} + (x^2 - y^2)\mathbf{j}$ is a conservative vector field on the unit disk enclosed by $x^2 + y^2 = 1$.

Solution. We set $\mathbf{F} = \nabla f$ and try to find f. We have

$$(9) \qquad f_1(x, y) = x^2y = \frac{\partial f}{\partial x} \qquad \text{and} \qquad f_2(x, y) = x^2 - y^2 = \frac{\partial f}{\partial y}.$$

Integrating (9), we obtain

$$(10) \qquad f(x, y) = \frac{x^3y}{3} + c_1(y) \qquad \text{and} \qquad f(x, y) = x^2y - \frac{y^3}{3} + c_2(x).$$

But those two formulas for $f(x, y)$ are irreconcilably different: for no function $c_1(y)$ of y alone and no function $c_2(x)$ of x alone can

$$\frac{x^3y}{3} + c_1(y) = x^2y - \frac{y^3}{3} + c_2(x)$$

hold for all x and y in the unit disk. Therefore F is not conservative. ■

We could have given a quicker solution to Example 5 by computing

$$\mathbf{J_F} = \begin{pmatrix} 2xy & x^2 \\ 2x & -2y \end{pmatrix}$$

and noting that it is not symmetric. Thus **F** can't be a conservative vector field by Theorem 4.4. In practice, computing $\mathbf{J_F}$ saves work when **F** is *not* conservative, and not computing $\mathbf{J_F}$ saves work when **F** is conservative! So no single procedure works best in all cases.

Proof of Theorem 4.3*

4.3 THEOREM

Let D be an open disk in $\mathbf{R}^2$ or an open ball in $\mathbf{R}^3$. If the continuous vector field **F** on $\mathbf{R}^2$ or $\mathbf{R}^3$ has a path-independent line integral in D, then **F** is a conservative vector field. In fact,

$$\mathbf{F} = \nabla f \qquad \text{for } f(\mathbf{x}) = \int_a^x \mathbf{F} \cdot d\mathbf{x},$$

where the line integral is taken over any piecewise smooth path C in D joining an arbitrarily chosen but fixed point **a** to **x**.

Proof. Suppose that D is an open disk in $\mathbf{R}^2$ or an open ball in $\mathbf{R}^3$ and that the continuous vector field F has a path-independent line integral in D. The theorem asserts that in such a case F is the conservative vector field $\mathbf{F} = \nabla f$, where f is the scalar function defined by

$$f(\mathbf{x}) = \int_{\mathbf{a}}^{\mathbf{x}} \mathbf{F} \cdot d\mathbf{x}. \tag{11}$$

The line integral in (11) is taken over any piecewise smooth path in D that joins some point **a** in D to **x** in D. We therefore have to show for each **c** in D that $\nabla f(\mathbf{c}) = \mathbf{F}(\mathbf{c})$ if f is defined by (11). Since D is an open ball, there is some sufficiently small real number h such that $\mathbf{c} + h\mathbf{e}_i \in D$ where $\mathbf{e}_i = (0, \ldots, 0, 1, 0, \ldots, 0)$ is the ith standard basis vector for $\mathbf{R}^n$. We can join **a** to **c** in D by a piecewise smooth path and then join **c** to $\mathbf{c} + h\mathbf{e}_i$ along the line segment $\mathbf{x} = \mathbf{c} + t\mathbf{e}_i$, with t varying between 0 and h. See **Figure 4.7,** where the entire path from **a** to $\mathbf{c} + h\mathbf{e}_i$ is labeled C.

FIGURE 4.7

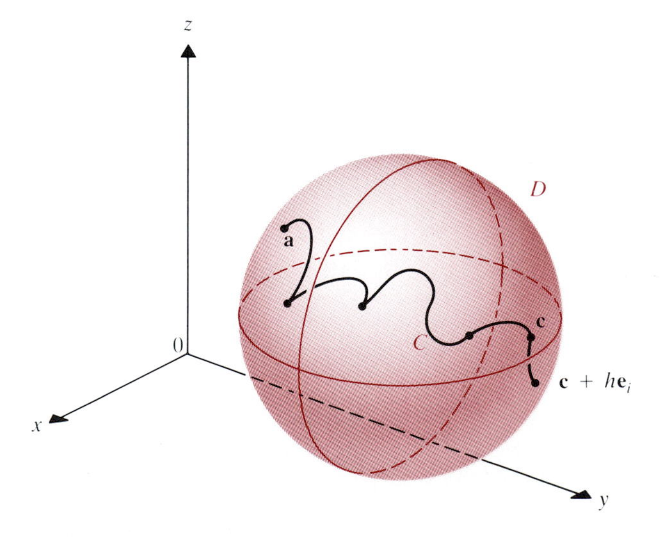

Now, by the independence of path, we have

$$\begin{aligned} f(\mathbf{c} + h\mathbf{e}_i) - f(\mathbf{c}) &= \int_{\mathbf{a}}^{\mathbf{c}+h\mathbf{e}_i} \mathbf{F} \cdot d\mathbf{x} - \int_{\mathbf{a}}^{\mathbf{c}} \mathbf{F} \cdot d\mathbf{x} = \int_{\mathbf{a}}^{\mathbf{c}+h\mathbf{e}_i} \mathbf{F} \cdot d\mathbf{x} + \int_{\mathbf{c}}^{\mathbf{a}} \mathbf{F} \cdot d\mathbf{x} \\ &= \int_{\mathbf{c}}^{\mathbf{c}+h\mathbf{e}_i} \mathbf{F} \cdot d\mathbf{x} = \int_0^h \mathbf{F}(\mathbf{c} + t\mathbf{e}_i) \cdot \mathbf{e}_i \, dt, \end{aligned}$$

* Optional

because on the path $\mathbf{x} = \mathbf{c} + t\mathbf{e}_i$ we have $\mathbf{x}'(t) = \mathbf{e}_i$. We next compute

$$\frac{\partial f}{\partial x_i}(\mathbf{c}) = \lim_{h \to 0} \frac{f(\mathbf{c} + h\mathbf{e}_i) - f(\mathbf{c})}{h} = \lim_{h \to 0} \frac{1}{h} \int_0^h \mathbf{F}(\mathbf{c} + t\mathbf{e}_i) \cdot \mathbf{e}_i \, dt.$$

By the mean value theorem for definite integrals (Theorem 3.4 of Chapter 4), we have

$$\begin{aligned}\int_0^h \mathbf{F}(\mathbf{c} + t\mathbf{e}_i) \cdot \mathbf{e}_i \, dt &= \mathbf{F}(\mathbf{c} + \bar{h}\mathbf{e}_i) \cdot \mathbf{e}_i(h - 0) \\ &= h\mathbf{F}(\mathbf{c} + \bar{h}\mathbf{e}_i) \cdot \mathbf{e}_i, \qquad \text{for some } \bar{h} \text{ between 0 and } h.\end{aligned}$$

Thus

$$\begin{aligned}\frac{\partial f}{\partial x_i}(\mathbf{c}) &= \lim_{h \to 0} \frac{1}{h} \cdot h\mathbf{F}(\mathbf{c} + \bar{h}\mathbf{e}_i) \cdot \mathbf{e}_i \\ &= \mathbf{F}(\mathbf{c}) \cdot \mathbf{e}_i, \qquad \text{since } \mathbf{F} \text{ is continuous and as } h \to 0, \text{ so does } \bar{h} \\ &= f_i(\mathbf{c})\end{aligned}$$

where f_i is the ith coordinate function of $\mathbf{F}$. Therefore

$$\mathbf{F}(\mathbf{c}) = (f_1(\mathbf{c}), \ldots, f_n(\mathbf{c})) = \left(\frac{\partial f}{\partial x_1}(\mathbf{c}), \ldots, \frac{\partial f}{\partial x_n}(\mathbf{c})\right) = \nabla f(\mathbf{c}).$$ QED

Exercises 14.4

In Exercises 1–10, compute the given line integral by finding a potential function for F and using Theorem 4.1.

1. $\int_C 2xy\,dx + (x^2 + 3y)\,dy$, where C is the polygonal path from (0, 0) to (0, 1) to (1, 1)
2. $\int_C y^2\,dx + 2xy\,dy$, where C is the path from (0, 0) to (2, 4) along the parabola $y = x^2$
3. $\int_C \mathbf{F} \cdot d\mathbf{x}$, where
$$\mathbf{F}(x, y) = -\frac{y}{x^2 + y^2}\mathbf{i} + \frac{x}{x^2 + y^2}\mathbf{j},$$
and C is any path from (1, 0) to (1, 1) in the counterclockwise sense that does not enclose (0, 0)
4. $\int_C \mathbf{F} \cdot d\mathbf{x}$, where $\mathbf{F}$ is as in Exercise 3, and C is the straight line segment from (1, 0) to (1, 1)
5. $\oint_C (2xy^3 + 5\cos x)\,dx + (3x^2y^2 - 4e^y)\,dy$, where C is the unit circle $x^2 + y^2 = 1$ traversed counterclockwise
6. $\oint_C (3x^2 \sin y + \cos^2 y)\,dx - x(\sin 2y - x^2 \cos y)\,dy$, where C is the square with vertices $(-1, -1)$, $(1, -1)$, $(1, 1)$, and $(-1, 1)$
7. $\oint_C (6x^2y^2 - 3yz^2)\,dx + (4x^3y - 3xz^2)\,dy - 6xyz\,dz$, where C is the polygonal path from $(-1, 0, 1)$ to $(2, 0, 1)$ to $(2, 1, 1)$ to $(2, 1, -2)$
8. $\int_C \mathbf{F} \cdot d\mathbf{x}$, where $\mathbf{F}(\mathbf{x}) = (x^2 + yz)\mathbf{i} + (y^2 + xz)\mathbf{j} + (z^2 + xy)\mathbf{k}$, and C is the curve $\mathbf{x}(t) = (t, t^2, t^3)$ from (0, 0, 0) to (1, 1, 1)
9. $\int_C ze^{xz} \sin yz\,dx + ze^{xz} \cos yz\,dy + e^{xz}(x \sin yz + y \cos yz)\,dz$, where C is any path from $(-1, 1, \pi/2)$ to $(3, -1, 3\pi/2)$
10. $\int_C \mathbf{F} \cdot d\mathbf{x}$, where
$$\mathbf{F}(\mathbf{x}) = (2xy^2z^2w^2, 2x^2yz^2w^2, 2x^2y^2zw^2, 2x^2y^2z^2w),$$
and C is the curve $\mathbf{x}(t) = (t, t^2, t^3, t^4)$ from (1, 1, 1, 1) to (2, 4, 8, 16) in $\mathbf{R}^4$

In Exercises 11–18, decide whether the given vector field is conservative on the set D. If it is, then find a potential function.

11. $\mathbf{F}(\mathbf{x}) = (3x^2 + y)\mathbf{i} + (e^y + x)\mathbf{j}$, D the unit disk $x^2 + y^2 \le 1$
12. $\mathbf{F}(\mathbf{x}) = (3y^2 + 6xy)\mathbf{i} + (3x^2 + 6y)\mathbf{j}$, D as in Exercise 11
13. $\mathbf{F}(x, y) = \dfrac{x}{\sqrt{x^2 + y^2}}\mathbf{i} + \dfrac{y}{\sqrt{x^2 + y^2}}\mathbf{j}$, D the annular region between the circles $x^2 + y^2 = 1$ and $x^2 + y^2 = 4$
14. $\mathbf{F}(x, y) = \dfrac{y}{x^2 + y^2}\mathbf{i} - \dfrac{x}{x^2 + y^2}\mathbf{j}$, D as in Exercise 13
15. $\mathbf{F}(\mathbf{x}) = (2xyz + z^2 - 2y^2 + 1)\mathbf{i} + (x^2z - 4xy)\mathbf{j} + (x^2y + 2xz - 2)\mathbf{k}$, D the unit ball $x^2 + y^2 + z^2 \le 1$
16. $\mathbf{F}(\mathbf{x}) = 2xy^2\mathbf{i} + x^2z^3\mathbf{j} + 3x^2y^2z^2\mathbf{k}$, D as in Exercise 15
17. $\mathbf{F}(x, y, z) = (3x^2 - 3yz)\mathbf{i} + (3y^2 - 3xz)\mathbf{j} - 3xy\,\mathbf{k}$, D the region between the spheres $x^2 + y^2 + z^2 = 1$ and $x^2 + y^2 + z^2 = 9$
18. $\mathbf{F}(\mathbf{x}) = \dfrac{x}{\sqrt{x^2 + y^2 + z^2}}\mathbf{i} + \dfrac{y}{\sqrt{x^2 + y^2 + z^2}}\mathbf{j} + \dfrac{z}{\sqrt{x^2 + y^2 + z^2}}\mathbf{k}$, D the region between the two cylinders $x^2 + y^2 = 1$ and $x^2 + y^2 = 9$

19. A force $\mathbf{F}$ of friction has constant magnitude and direction always *opposite* to the direction of motion. Show that such a force is not conservative by calculating the line integral of $\mathbf{F}$ over an arbitrary path $\mathbf{x} = \mathbf{x}(t)$ in $\mathbf{R}^n$.

20. Suppose that $\mathbf{F} = \nabla f$ is conservative. Show that at any point $\mathbf{x} = \mathbf{c}$ the force $\mathbf{F}$ is orthogonal to the equipotential surface $f(\mathbf{x}) = f(\mathbf{c})$ through that point.

21. If $\mathbf{F}$ is the inverse-square law field of Example 4 of Section 1, then show that the work done in moving a particle from $\mathbf{a}$ to $\mathbf{b}$ very far from the origin is approximately $k/|\mathbf{a}|$.

22. Using

$$\mathbf{G}(\mathbf{x}) = -gR^2 \frac{1}{|\mathbf{x}|^2} \frac{\mathbf{x}}{|\mathbf{x}|}$$

to represent the gravitational field of the earth at a point $\mathbf{x}$, find the amount of work needed to move a particle from $\mathbf{x} = \mathbf{a}$ on the earth's surface completely out of the gravitational influence of the earth (cf. Exercise 21).

23. Near the surface of the earth the gravitational field $\mathbf{G}$ acts on a particle of mass m approximately as the constant

$$\mathbf{F}(x, y, z) = -mg\,\mathbf{k},$$

where g is the acceleration due to gravity at the surface of the earth. Find the potential function f that is 0 at the origin. (The origin is on the surface of the earth, with z-axis vertical.)

24. Refer to Exercise 23. Find the path followed by a particle of mass m that has velocity $\mathbf{v} = v_1\mathbf{i} + v_2\mathbf{j} + v_3\mathbf{k}$ on the surface of the earth and is acted on only by $\mathbf{F}$.

25. In Exercise 24 show that the total energy is constant at every point of the path $\mathbf{x} = \mathbf{x}(t)$ of the particle.

26. Show that the central field of Exercise 36 of the last section is conservative. (*Hint:* If $p(r) = \int_0^r yf(y)\,dy$, then $\mathbf{F}(\mathbf{x}) = \nabla p(|\mathbf{x}|)$.)

14.5 Surfaces and Surface Area

To state additional analogues of the fundamental theorem of calculus, we need to develop a new type of double integral called a *surface integral.* It bears roughly the same relation to the ordinary double integral that the line integral bears to the ordinary definite integral. Instead of integrating a function f over a region D in $\mathbf{R}^2$, we will integrate vector field $\mathbf{F}$: $\mathbf{R}^3 \to \mathbf{R}^3$ over a surface S in $\mathbf{R}^3$. This is facilitated by using a *vector representation* for surfaces S in $\mathbf{R}^3$ analogous to the vector representation of curves in $\mathbf{R}^2$ presented in Sections 8.5 and 11.2. Recall that a parametrized curve is a continuous function $\mathbf{x} = \mathbf{x}(t)$ that we can think of as stretching and deforming a closed interval $I = [a, b]$ in $\mathbf{R}^1$ to form a plane curve. As t varies over I, the curve is traced out in $\mathbf{R}^2$ (see **Figure 5.1**). We can similarly think of a surface S in $\mathbf{R}^3$ as obtained by *deforming some two-dimensional set* $D \subseteq \mathbf{R}^2$, as in **Figure 5.2.** For example, we can think of a rectangle deformed into a "magic carpet" in $\mathbf{R}^3$ (**Figure 5.3**). Corresponding to the parametrization of a curve in $\mathbf{R}^3$, we parametrize S by a vector function $\mathbf{X}$: $\mathbf{R}^2 \to \mathbf{R}^3$ given by $\mathbf{X} = \mathbf{X}(u, v)$. As $\mathbf{u} = (u, v)$ varies over $D \subseteq \mathbf{R}^2$, we trace out the surface $S \subseteq \mathbf{R}^3$.

FIGURE 5.1

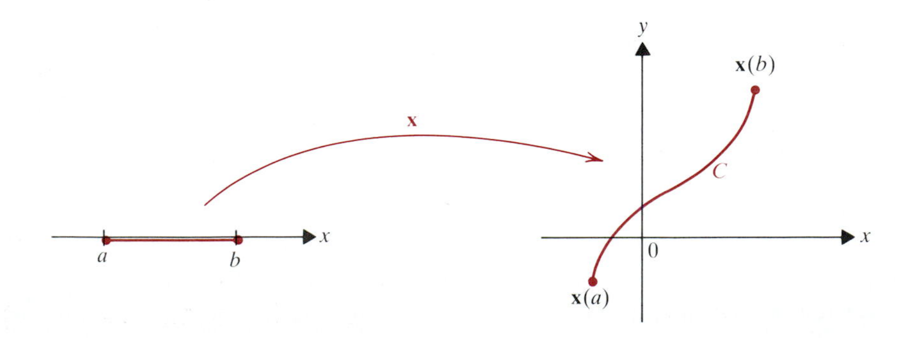

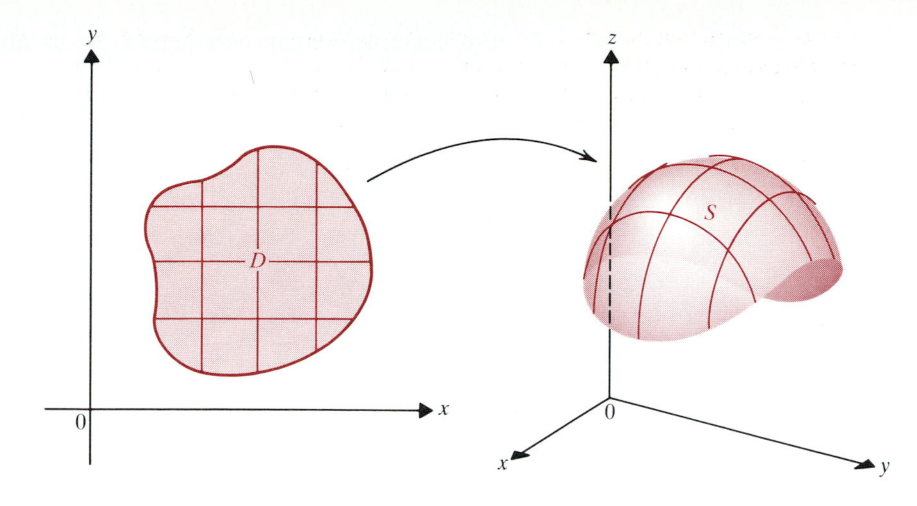

FIGURE 5.2

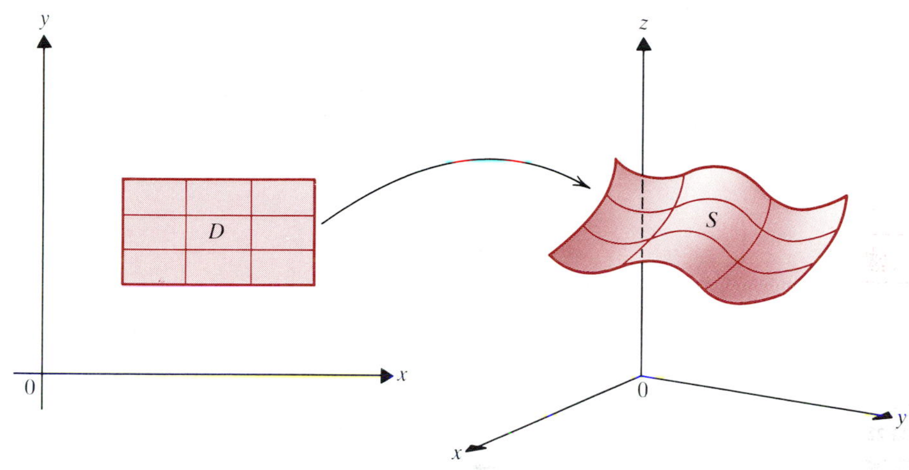

FIGURE 5.3

5.1 DEFINITION A ***parametrization*** of a ***surface patch*** $S \subseteq \mathbf{R}^3$ is a continuous function $\mathbf{X}: \mathbf{R}^2 \to \mathbf{R}^3$ such that for some set $D \subseteq \mathbf{R}^2$,

$$S = \mathbf{X}(D) = \{(x, y, z) \in \mathbf{R}^3 \,|\, (x, y, z) = \mathbf{X}(u, v), \text{ for some } (u, v) \in D\}.$$

A ***parametrized surface*** is a union of parametrized surface patches.

To parametrize a plane $ax + by + cz = d$, we first solve for one of the variables with a nonzero coefficient in terms of the others. If $a \neq 0$, for instance, then

$$x = \frac{d}{a} - \frac{b}{a}y - \frac{c}{a}z.$$

So we can parametrize the plane as a surface patch, by letting $y = u$, $z = v$, and $x = d/a - (b/a)u - (c/a)v$. That gives

$$\mathbf{X}(u, v) = \left(\frac{d}{a} - \frac{b}{a}u - \frac{c}{a}v, u, v\right).$$

A similar approach works for a surface given explicitly as the graph of $z = f(x, y)$. We let

$$x = u, \; y = v, \; z = f(u, v), \qquad \text{for } (u, v) \in D = \text{Domain } f.$$

We therefore have

$$\mathbf{X}(u, v) = (u, v, f(u, v)) = u\,\mathbf{i} + v\,\mathbf{j} + f(u, v)\,\mathbf{k}, \qquad (u, v) \in D.$$

For example, we can parametrize the paraboloid $z = x^2 + y^2$ by letting

$$\mathbf{X}(u, v) = u\,\mathbf{i} + v\,\mathbf{j} + (u^2 + v^2)\,\mathbf{k} = (u, v, u^2 + v^2), \qquad \text{for } (u, v) \in \mathbf{R}^2$$

(see **Figure 5.4**).

FIGURE 5.4

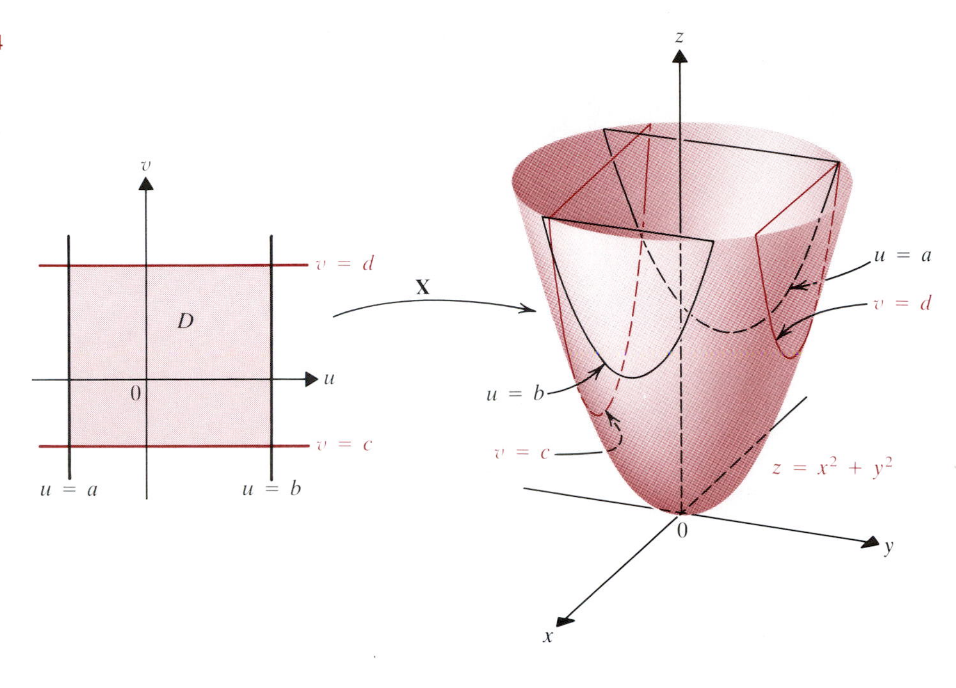

We can handle a wider class of surfaces than those that are graphs of functions, however. The next example provides two illustrations of this. The second part is of additional interest because it shows how some familiar surfaces cannot be parametrized as single patches, but instead must be parametrized as the *union* of several surface patches.

Example 1

Parametrize (a) a sphere of radius a centered at (0, 0, 0); and (b) a right circular cylinder of radius a and height h, with the z-axis as its axis and including a top and bottom (**Figure 5.5**).

FIGURE 5.5

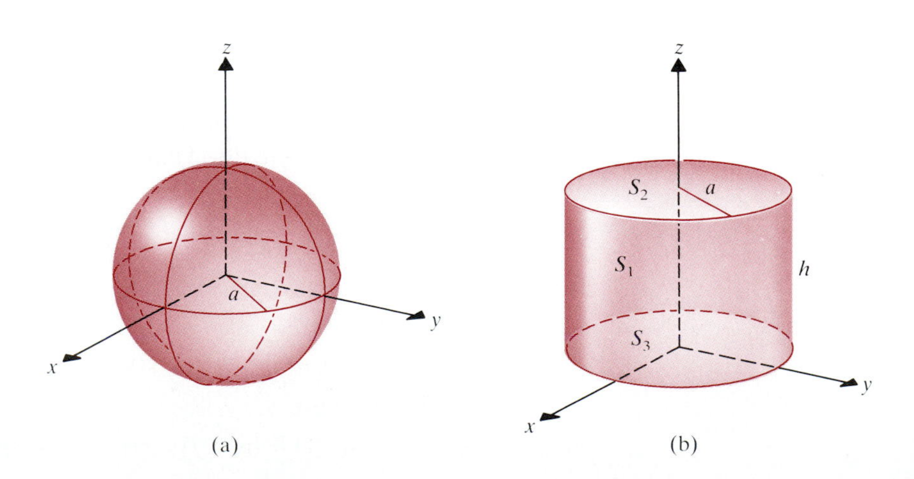

Solution. (a) We parametrize using spherical coordinates on the sphere, which has equation $\rho = a$. We let

$$x = a \sin u \cos v,\ y = a \sin u \sin v,\ z = a \cos u, \qquad (u, v) \in [0, \pi] \times [0, 2\pi].$$

Then we have

$$\begin{aligned}\mathbf{X}(u, v) &= (a \sin u \cos v, a \sin u \sin v, a \cos u)\\ &= a \sin u \cos v\,\mathbf{i} + a \sin u \sin v\,\mathbf{j} + a \cos u\,\mathbf{k},\end{aligned}$$

where $0 \leq u \leq \pi$ and $0 \leq v \leq 2\pi$. Here D is the rectangle $[0, \pi] \times [0, 2\pi]$. **Figure 5.6** shows the meridians corresponding to $u = d_i$ and the great circles corrsponding to $v = c_i$. (Compare the discussion on p. 883.)

FIGURE 5.6

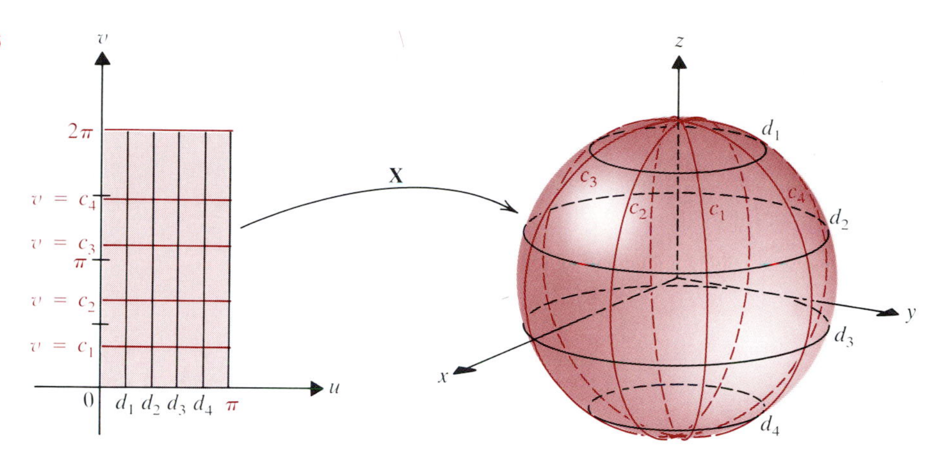

(b) We parametrize the cylinder as three patches: top, curved cylindrical portion, and bottom. For the cylindrical patch S_1, we use cylindrical coordinates with $r = a$. We thus let $\mathbf{X}_1(u, v) = (x, y, z) = x\,\mathbf{i} + y\,\mathbf{j} + z\,\mathbf{k}$, where

$$x = a \cos u, \qquad y = a \sin u, \qquad \text{and} \qquad z = v, \qquad \text{for } u \in [0, 2\pi] \text{ and } v \in [0, h]$$

(see **Figure 5.7**). For the top S_2 and bottom S_3, we use polar coordinates in the planes $z = 0$ and $z = h$. For the top, we let $\mathbf{X}_2(u, v) = (x, y, z) = x\,\mathbf{i} + y\,\mathbf{j} + z\,\mathbf{k}$, where

$$x = u \cos v, \qquad y = u \sin v, \qquad \text{and} \qquad z = h, \qquad \text{for } u \in [0, a] \text{ and } v \in [0, 2\pi].$$

FIGURE 5.7

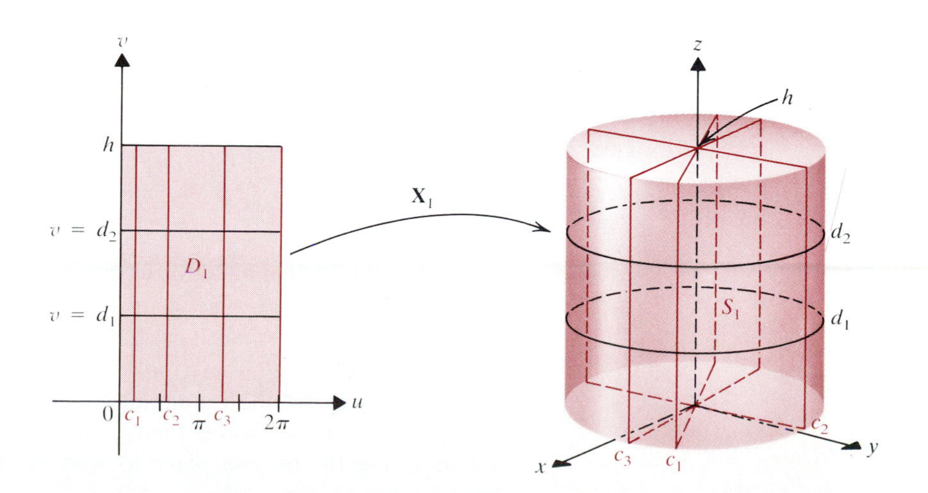

See **Figure 5.8,** which shows that as v increases, the rays $v = c_i$ sweep over S_2 in the counterclockwise sense. For the bottom, we let $\mathbf{X}_3(u, v) = x\,\mathbf{i} + y\,\mathbf{j} + z\,\mathbf{k}$, where

$$x = u\cos(-v), \quad y = u\sin(-v), \quad \text{and} \quad z = 0, \quad \text{for } u \in [0, a] \text{ and } v \in [0, 2\pi].$$

FIGURE 5.8

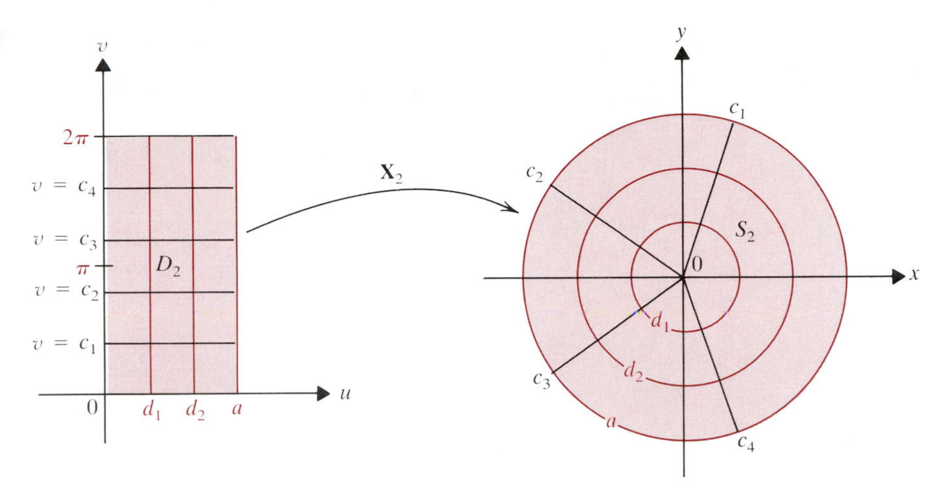

This parametrization is pictured in **Figure 5.9.** Notice that as v increases from 0 to 2π, the rays $v = c_i$ sweep clockwise through S_3. (We *could* have used v instead of $-v$ in parametrizing S_3, of course, but we will see in the next section that this parametrization is a good one.)

FIGURE 5.9

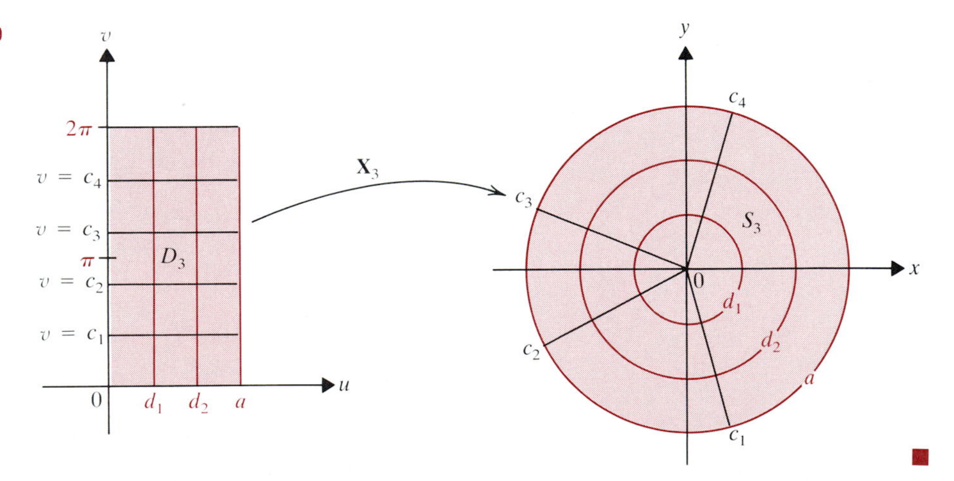

■

In developing line integrals we have generally restricted attention to smooth curves. Likewise, we will concentrate on *smooth surfaces* in defining surface integrals. Intuitively, a smooth surface patch is a magic carpet that is comfortable to ride: there are no sharp points. We were able to describe smooth curves mathematically by the requirement that the tangent vector $\mathbf{x}'(t)$ should vary continuously. It is natural then to describe a smooth surface mathematically by requiring the *tangent plane* to vary continuously. We do so as follows. At

FIGURE 5.10

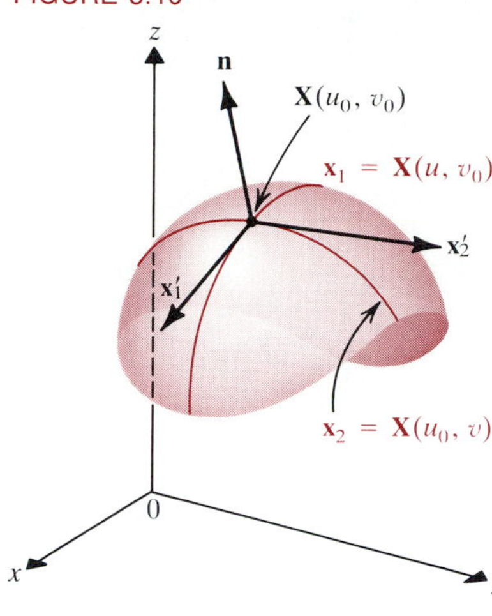

each point $\mathbf{X}(u_0, v_0)$ on a parametrized surface patch S, there are curves (**Figure 5.10**) $\mathbf{x}_1 = \mathbf{X}(u, v_0)$ and $\mathbf{x}_2 = \mathbf{X}(u_0, v)$ passing through that point. Tangent vectors to these curves are given by

$$\mathbf{x}_1' = \mathbf{X}_u(u_0, v_0) = \frac{\partial X}{\partial u}(u_0, v_0) = \frac{\partial x}{\partial u}(u_0, v_0)\,\mathbf{i} + \frac{\partial y}{\partial u}(u_0, v_0)\,\mathbf{j} + \frac{\partial z}{\partial u}(u_0, v_0)\,\mathbf{k}$$

and

$$\mathbf{x}_2' = \mathbf{X}_v(u_0, v_0) = \frac{\partial X}{\partial v}(u_0, v_0) = \frac{\partial x}{\partial v}(u_0, v_0)\,\mathbf{i} + \frac{\partial y}{\partial v}(u_0, v_0)\,\mathbf{j} + \frac{\partial z}{\partial v}(u_0, v_0)\,\mathbf{k}.$$

If these vectors are not collinear, then they determine the plane with normal vector

$$\text{(1)} \qquad \mathbf{n}(u_0, v_0) = (\mathbf{X}_u \times \mathbf{X}_v)(u_0, v_0).$$

If X_u and X_v are continuous, that is, if all entries of the Jacobian matrix

$$\mathbf{J}_{\mathbf{X}} = \begin{pmatrix} \dfrac{\partial x}{\partial u} & \dfrac{\partial x}{\partial v} \\ \dfrac{\partial y}{\partial u} & \dfrac{\partial y}{\partial v} \\ \dfrac{\partial z}{\partial u} & \dfrac{\partial z}{\partial v} \end{pmatrix}$$

are continuous, then $\mathbf{n}$ will vary continuously as u and v vary. So the tangent plane will vary continuously. Hence we give the following definition.

5.2 DEFINITION

A surface patch $S \subseteq \mathbf{R}^3$ is ***smooth*** if it has a parametrization $\mathbf{X}\colon \mathbf{R}^2 \to \mathbf{R}^3$ with

$$\mathbf{X} = \mathbf{X}(u, v) = x(u, v)\,\mathbf{i} + y(u, v)\,\mathbf{j} + z(u, v)\,\mathbf{k},$$

for $(u, v) \in D \subseteq \mathbf{R}^2$ such that all entries of $\mathbf{J}_{\mathbf{X}}$ are continuous on D.

In Example 1 the parametrizations show that the sphere and the top, bottom, and sides of the cylinder are smooth surface patches.

In what follows, we assume that $\mathbf{X}_u$ and $\mathbf{X}_v$ will almost never be collinear at points on a smooth surface patch S, i.e., that $\mathbf{X}_u \times \mathbf{X}_v \neq \mathbf{0}$ is a normal vector to S at almost all points of S.

We turn now to the problem of defining the *area* of a smooth surface patch S parametrized by $\mathbf{X}\colon D \to \mathbf{R}^3$. The idea is familiar. We make a grid of $D \subseteq \mathbf{R}^2$ by rectangles. Consider a subrectangle R_{ij} of D that is mapped by $\mathbf{X}$ onto a small magic subcarpet of S (**Figure 5.11**).

The area of this magic subcarpet should be approximately the same as the area $|\mathbf{w}_1 \times \mathbf{w}_2|$ of the parallelogram determined by

$$\mathbf{w}_1 = \mathbf{X}(u_i + \Delta u_i, v_j) - \mathbf{X}(u_i, v_j) \approx \frac{\partial \mathbf{X}}{\partial u}(u_i, v_j)\,\Delta u_i = \mathbf{X}_u(u_i, v_j)\,\Delta u_i,$$

and

$$\mathbf{w}_2 = \mathbf{X}(u_i, v_j + \Delta v_j) - \mathbf{X}(u_i, v_j) \approx \frac{\partial \mathbf{X}}{\partial v}(u_i, v_j)\,\Delta v_j = \mathbf{X}_v(u_i, v_j)\,\Delta v_j.$$

FIGURE 5.11

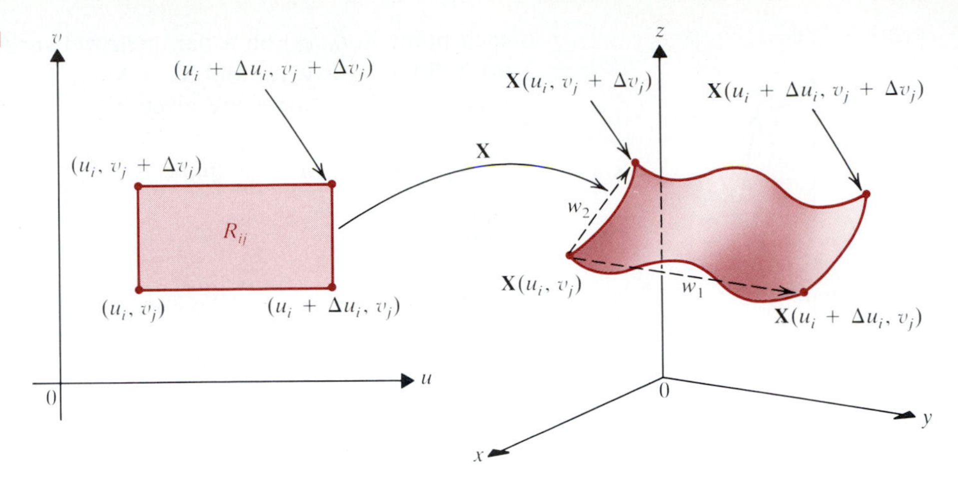

Hence the area of the magic subcarpet is approximately

$$|\mathbf{w}_1 \times \mathbf{w}_2| \approx |\Delta u_i \mathbf{X}_u(u_i, v_j) \times \Delta v_j \mathbf{X}_v(u_i, v_j)| \approx |(\mathbf{X}_u \times \mathbf{X}_v)(u_i, v_j)|\, \Delta u_i\, \Delta v_j.$$

An approximation to the total area $A(S)$ of S is obtained by adding the foregoing approximations over all subrectangles R_{ij}. We get

$$A(S) \approx \sum_i \sum_j |(\mathbf{X}_u \times \mathbf{X}_v)(u_i, v_j)|\, \Delta u_i\, \Delta v_j. \tag{2}$$

We define $A(S)$ to be the limit of the approximation (2) as the grid becomes finer and finer. (It can be shown that this is essentially independent of the parametrization $\mathbf{X}$ used for S. The details are left to advanced calculus.)

5.3 DEFINITION

If S is a smooth surface patch in $\mathbf{R}^3$ parametrized by $\mathbf{X}$: $\mathbf{R}^2 \to \mathbf{R}^3$, where $\mathbf{X}$ maps $D \subseteq \mathbf{R}^2$ onto S, then the ***area*** of S is

$$A(S) = \iint_D |\mathbf{X}_u \times \mathbf{X}_v|\, du\, dv, \tag{3}$$

whenever this integral exists.

Example 2

Use Definition 5.3 to compute the surface area of a sphere of radius a.

Solution. We parametrized the sphere in Example 1(a) as $\mathbf{X}(u, v) = (x, y, z)$ $(a \sin u \cos v,\ a \sin u \sin v,\ a \cos u)$ for $(u, v) \in D = [0, \pi] \times [0, 2\pi]$. We then have

$$\mathbf{X}_u = \left(\frac{\partial x}{\partial u}, \frac{\partial y}{\partial u}, \frac{\partial z}{\partial u}\right) = a \cos u \cos v\, \mathbf{i} + a \cos u \sin v\, \mathbf{j} - a \sin u\, \mathbf{k}$$

$$\mathbf{X}_v = \left(\frac{\partial x}{\partial v}, \frac{\partial y}{\partial v}, \frac{\partial z}{\partial v}\right) = -a \sin u \sin v\, \mathbf{i} + a \sin u \cos v\, \mathbf{j} + 0\, \mathbf{k}.$$

Thus

$$\begin{aligned}\mathbf{X}_u \times \mathbf{X}_v &= a^2[\sin^2 u \cos v\, \mathbf{i} + \sin^2 u \sin v\, \mathbf{j} + (\sin u \cos u \cos^2 v + \sin u \cos u \sin^2 v)\, \mathbf{k}] \\ &= a^2[\sin^2 u \cos v\, \mathbf{i} + \sin^2 u \sin v\, \mathbf{j} + \sin u \cos u\, \mathbf{k}].\end{aligned}$$

Hence

$$\begin{aligned}|\mathbf{X}_u \times \mathbf{X}_v| &= a^2\sqrt{\sin^4 u \cos^2 u + \sin^4 u \sin^2 u + \sin^2 u \cos^2 u} \\ &= a^2\sqrt{\sin^4 u + \sin^2 u(1 - \sin^2 u)} = a^2\sqrt{\sin^2 u} \\ &= a^2 \sin u, \qquad \text{since } u \in [0, \pi].\end{aligned}$$

(Note that $\mathbf{X}_u \times \mathbf{X}_v \neq \mathbf{0}$ except at the two points $u = 0$ and $u = \pi$, the poles.) We therefore get from (4)

$$A(S) = \iint_D a^2 \sin u\, du\, dv = a^2 \int_0^\pi \sin u\, du \int_0^{2\pi} dv = 2\pi a^2\Big[-\cos u\Big]_0^\pi = 4\pi a^2.$$

(You may recall this formula from previous mathematics courses.) ■

When a surface patch $S \subseteq \mathbf{R}^3$ is given explicitly as the graph of a function $f\colon \mathbf{R}^2 \to \mathbf{R}$, then (3) assumes a slightly simpler form, which is reminiscent of the arc length formula of Section 5.4. To obtain it, we use the parametrization mentioned just before Example 1,

$$\mathbf{X}(u, v) = (u, v, f(u, v)) = u\,\mathbf{i} + v\,\mathbf{j} + f(u, v)\,\mathbf{k}, \qquad \text{for } (u, v) \in D.$$

Since

$$\mathbf{X}_u = \left(1, 0, \frac{\partial f}{\partial u}\right) = \mathbf{i} + \frac{\partial f}{\partial u}\mathbf{k} \qquad \text{and} \qquad \mathbf{X}_v = \left(0, 1, \frac{\partial f}{\partial v}\right) = \mathbf{j} + \frac{\partial f}{\partial v}\mathbf{k},$$

we find

$$\mathbf{X}_u \times \mathbf{X}_v = \left(-\frac{\partial f}{\partial u}, -\frac{\partial f}{\partial v}, 1\right) = -\frac{\partial f}{\partial u}\mathbf{i} - \frac{\partial f}{\partial v}\mathbf{j} + \mathbf{k}.$$

Hence

$$|\mathbf{X}_u \times \mathbf{X}_v| = \sqrt{\left(\frac{\partial f}{\partial u}\right)^2 + \left(\frac{\partial f}{\partial v}\right)^2 + 1} = \sqrt{\left(\frac{\partial f}{\partial x}\right)^2 + \left(\frac{\partial f}{\partial y}\right)^2 + 1} = \sqrt{|\nabla f|^2 + 1}.$$

So (3) becomes

$$A(S) = \iint_D \sqrt{1 + |\nabla f(\mathbf{x})|^2}\, dA. \tag{4}$$

If f is differentiable, then ∇f is its total derivative, so that (4) is the exact analogue for $\mathbf{f}\colon \mathbf{R}^2 \to \mathbf{R}$ of the arc length of Definition 4.2 of Chapter 5,

$$L = \int_a^b \sqrt{1 + [f'(x)]^2}\, dx.$$

Example 3

Find the area of the surface $z = x^{3/2} + 2\sqrt{2}y$ that lies above the rectangle $R = [0, 1] \times [1, 2]$. (See **Figure 5.12.**)

Solution. We are given

$$f(x, y) = x^{3/2} + 2\sqrt{2}y.$$

Thus

$$\nabla f(x, y) = \frac{3}{2}x^{1/2}\mathbf{i} + 2\sqrt{2}\mathbf{j} \to |\nabla f(x, y)|^2 = \frac{9}{4}x + 8,$$

$$\sqrt{1 + |\nabla f(x, y)|^2} = \sqrt{\frac{9}{4}x + 9} = \frac{3}{2}\sqrt{x + 4}.$$

FIGURE 5.12

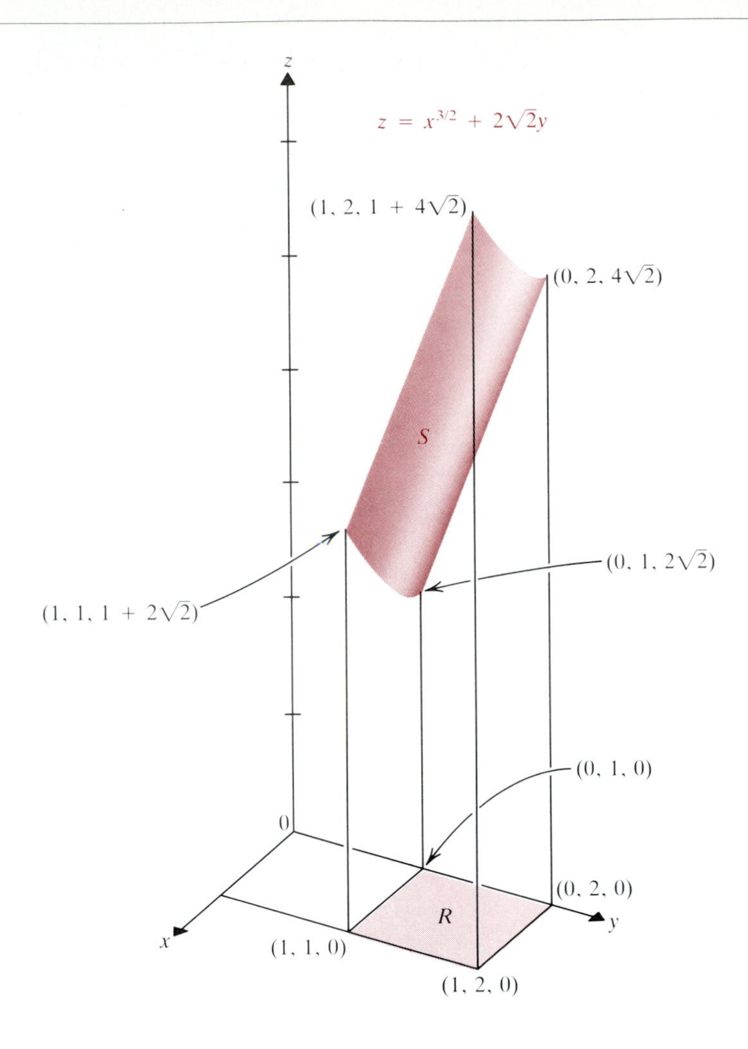

We therefore get from (4)

$$A(S) = \iint_D \frac{3}{2}\sqrt{x+4}\,dx\,dy$$

$$= \frac{3}{2}\int_0^1 \sqrt{x+4}\,dx \int_1^2 dy = \frac{3}{2}\frac{(x+4)^{3/2}}{3/2}\Bigg]_0^1 = 5^{3/2} - 4^{3/2} = 5\sqrt{5} - 8. \quad ■$$

Exercises 14.5

In Exercises 1–12, parametrize the given surface.

1. The paraboloid $x^2 + y^2 - 2z = 0$ above the rectangle $[1, 2] \times [2, 5]$
2. The cone $z = \sqrt{x^2 + y^2}$ lying above the disk $x^2 + y^2 \leq 4$
3. The cone $z = 4 - \sqrt{x^2 + y^2}$ lying above the disk $x^2 + y^2 \leq 16$
4. The part of the plane $z = 2x + 2y$ that lies inside the cylinder $x^2 + y^2 = 1$
5. The part of the plane $z = 3x - 2y$ that lies inside the cylinder $x^2 + y^2 = 4$
6. The part of the cylinder $x^2 + y^2 = 4$ between the planes $z = 0$ and $z = x + 2$
7. The part S of the sphere $x^2 + y^2 + z^2 = 1$ that lies above the cone $z = \sqrt{x^2 + y^2}$
8. The portion of the paraboloid $z = x^2 + y^2$ lying between the planes $z = 1$ and $z = 4$

9. The torus T obtained by revolving a circle in the xz-plane with center $(b, 0, 0)$ and radius $a < b$ about the z-axis (**Figure 5.13**)

FIGURE 5.13

10. The upper part of the sphere $x^2 + y^2 + z^2 = 4$ inside the cylinder $x^2 + y^2 = 2y$. (Recall that the cylinder is $r = 2\sin\theta$ in cylindrical coordinates.)

11. The plane $ax + by + cz = d$, where $c \neq 0$

12. The plane of Exercise 11 if $b \neq 0$

In Exercises 13–22, use (3) or (4) to compute the area of the given surface.

13. The portion of the saddle $z = x^2 - y^2$ that lies inside the cylinder $x^2 + y^2 = 1$

14. The portion of the saddle $z = y^2 - x^2$ that lies inside the cylinder $x^2 + y^2 = 4$

15. The portion of the paraboloid $z = x^2 + y^2$ between the planes $z = 1$ and $z = 4$

16. The portion of the paraboloid $z = x^2 + y^2$ under the plane $z = 4$

17. The ***spiral ramp*** (or ***helicoid***) given by $\mathbf{X}(u, v) = (u\cos v, u\sin v, v)$ for $(u, v) \in [0, 1] \times [0, \pi]$. Sketch the surface.

18. The spiral ramp of Exercise 17, where $D = [0, \sqrt{2}] \times [0, \pi]$

19. The surface S of Exercise 5

20. The surface of intersection of the solid cylinders bounded by $x^2 + y^2 = 1$ and $x^2 + z^2 = 1$

21. The torus T of Exercise 9

22. The surface S of Exercise 10

23. Find the area of the lateral surface S_1 in Example 1(b).

24. Find the area of the lateral surface S of a right circular cone of radius a and height h. Show that your answer reduces to $\pi a l$, where l is the *slant height* $\sqrt{a^2 + h^2}$ shown in **Figure 5.14.**

FIGURE 5.14

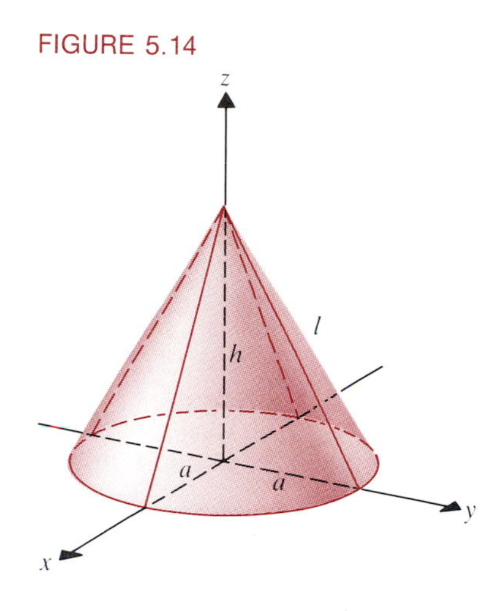

25. Find the area of the triangle T in $\mathbf{R}^3$ with vertices $(a, 0, 0)$, $(0, b, 0)$, and $(0, 0, c)$. Show that this reduces to the familiar formula in $\mathbf{R}^2$ if a or b or c is 0.

26. Set up, but do not attempt to evaluate, the integral for the area of the ellipsoid $x^2/a^2 + y^2/b^2 + z^2/c^2 = 1$.

27. Suppose that a curve C is the graph in the xz-plane of $z = f(x)$, where $0 \leq a \leq x \leq b$. Let S be the surface of revolution obtained by revolving C about the z-axis. Obtain a parametrization of S.

28. Use Exercise 27 to show that $A(S) = 2\pi \int_a^b u\sqrt{1 + f'(u)^2}\, du$.

29. Prove the following *theorem of Pappus*. The area of the surface of revolution in Exercise 27 is $A(S) = 2\pi\bar{x}L$, where $\bar{x}$ is the x-coordinate of the center of mass of the curve C (see Exercise 25 of Section 2), and L is the length of C.

30. Use the theorem of Pappus in Exercise 29 to rework Exercise 21. That is, find the surface area of the torus in Exercise 9.

14.6 Surface Integrals

We are now ready to introduce the last kind of integral that we shall discuss: the *surface integral* of a continuous function over a smooth surface patch $S \subseteq \mathbf{R}^3$. It is defined first for scalar functions and then for vector functions. In the latter case, the surface integral is the natural analogue of the line integral of a two-dimensional vector function.

We define the surface integral of a continuous scalar function over a smooth surface patch S by analogy with the line integral of a scalar function over a curve C in $\mathbf{R}^3$. (See Exercises 25–29 in Section 2.) The latter integral is defined by

$$\int_C f(\mathbf{x})\,ds = \int_a^b f(\mathbf{x}(t))|\mathbf{x}'(t)|\,dt$$

if C is parametrized by $\mathbf{x} = \mathbf{x}(t) = x(t)\mathbf{i} + y(t)\mathbf{j} + z(t)\mathbf{k}$ for $t \in [a, b]$.

Let $\mathbf{X}: D \to \mathbf{R}^3$ be a parametrization of the smooth surface patch S, where $D \subseteq \mathbf{R}^2$. We make a grid of D by rectangles as in the preceding section and define the surface integral of f over S to be the limit of Riemann sums

$$\sum_{i=1}^{m}\sum_{j=1}^{n} f(\mathbf{X}(u_i, v_j))|\mathbf{X}_u \times \mathbf{X}_v(u_i, v_j)|\,\Delta u_i\,\Delta v_j$$

as the mesh of the grid tends to zero. These Riemann sums are obtained by adding the products of the values $f(\mathbf{X}(u_i, v_j)) = f(x(u_i, v_j), y(u_i, v_j), z(u_i, v_j))$ of f at points on the surface times elementary area approximations, for small pieces of S surrounding the points. Refer to **Figure 6.1.** Since f, $\mathbf{X}_u$, and $\mathbf{X}_v$ are all continuous, the Riemann sums approach the ordinary double integral of the real-valued function

$$f(\mathbf{X}(u, v))|\mathbf{X}_u \times \mathbf{X}_v|$$

over the domain D in $\mathbf{R}^2$. We are thus led to the following definition.

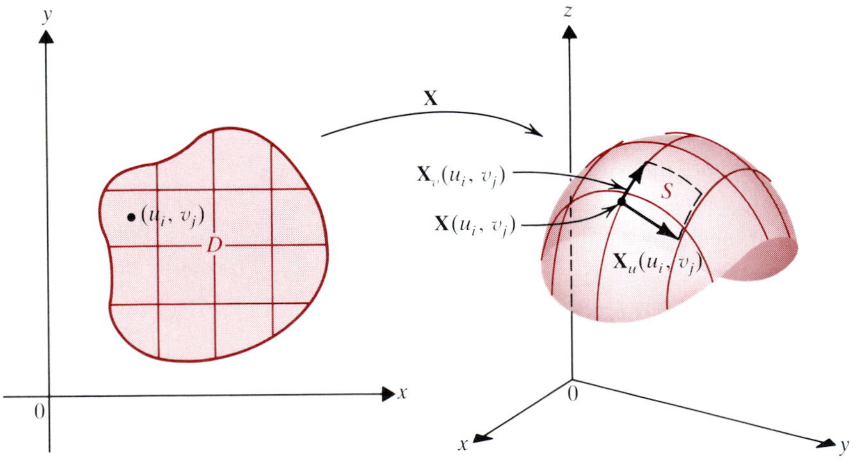

FIGURE 6.1

6.1 DEFINITION Let f be continuous on a smooth surface patch $S \subseteq \mathbf{R}^3$ that is parametrized by $\mathbf{X}: D \to \mathbf{R}^3$ for $D \subseteq \mathbf{R}^2$. Then the ***surface integral*** of f over S is

(1) $$\iint_S f(x, y, z)\,dA = \iint_D f(x(u, v), y(u, v), z(u, v))|\mathbf{X}_u \times \mathbf{X}_v|\,du\,dv.$$

Example 1

Compute the surface integral of $f(x, y, z) = 3z^2 + 4$ over the sphere in Example 2 of the last section.

Solution. In Example 2 of Section 5 we computed $|\mathbf{X}_u \times \mathbf{X}_v| = a^2 \sin u$. Then (1) gives

$$\iint_{\mathbf{X}(D)} f(\mathbf{x})\,dA = \iint_{\mathbf{X}(D)} (3z^2 + 4)\,dA = \iint_D (3a^2\cos^2 u + 4)a^2 \sin u\,du\,dv \qquad \text{since } z = a\cos u$$

$$= a^2 \int_0^{2\pi} dv \int_0^{\pi} 3a^2 \cos^2 u \sin u\,du + 4 \iint_D a^2 \sin u\,du\,dv$$

$$= 2\pi a^2 \left[3a^2\left(-\frac{\cos^3 u}{3}\right)\right]_0^{\pi} + 4(4\pi a^2) = 2\pi a^2[2a^2] + 16\pi a^2$$

$$= 4\pi a^2(a^2 + 4). \quad \blacksquare$$

We next define the surface integral of a continuous vector function $\mathbf{F}: \mathbf{R}^3 \to \mathbf{R}^3$ over a smooth surface patch S. Just as the *line integral* was defined with *work* in mind, so the *surface integral* is defined to describe the physical concept of *flux*. (See Exercise 27 of Section 3.) For this we need the notion of a *standard* unit normal vector to S [cf. Formula (1), p. 949].

6.2 DEFINITION

If $S \subseteq \mathbf{R}^3$ is a smooth surface patch parametrized by $\mathbf{X}: \mathbf{R}^2 \to \mathbf{R}^3$, then the ***standard unit normal*** $\mathbf{N}$ to S at a point $\mathbf{X}(u, v)$ where $\mathbf{X}_u \times \mathbf{X}_v \neq \mathbf{0}$ is

$$\mathbf{N} = \frac{\mathbf{X}_u \times \mathbf{X}_v}{|\mathbf{X}_u \times \mathbf{X}_v|}(u, v). \tag{2}$$

FIGURE 6.2

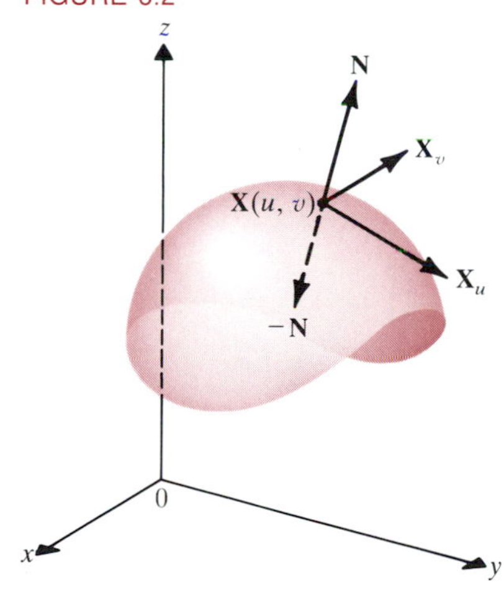

Note that the negative of $\mathbf{N}$ is another unit normal vector pointing in the opposite direction, and that *no matter how S is parametrized*, $(\mathbf{X}_u, \mathbf{X}_v, \mathbf{N})$ is always a *right-handed system*. See **Figure 6.2.**

Example 2

Find the standard unit normal to the lateral portion S_1 of the cylinder of radius a and height h in Example 1(b) of the last section.

Solution. We parametrized S_1 by letting

$$\mathbf{X}(u, v) = a\cos u\,\mathbf{i} + a\sin u\,\mathbf{j} + v\,\mathbf{k}, \qquad \text{for } (u, v) \in [0, 2\pi] \times [0, h].$$

We thus have

$$\mathbf{X}_u = -a\sin u\,\mathbf{i} + a\cos u\,\mathbf{j} \qquad \text{and} \qquad \mathbf{X}_v = \mathbf{k}.$$

Therefore

$$\mathbf{X}_u \times \mathbf{X}_v = \det\begin{pmatrix} \mathbf{i} & \mathbf{j} & \mathbf{k} \\ -a\sin u & a\cos u & 0 \\ 0 & 0 & 1 \end{pmatrix} = a\cos u\,\mathbf{i} + a\sin u\,\mathbf{j}.$$

Since $|\mathbf{X}_u \times \mathbf{X}_v| = \sqrt{a^2\cos^2 u + a^2 \sin^2 u} = a$, we have from (2)

$$\mathbf{N} = \frac{1}{a}[a\cos u\,\mathbf{i} + a\sin u\,\mathbf{j}] = \cos u\,\mathbf{i} + \sin u\,\mathbf{j}.$$

The standard unit normal $\mathbf{N}$ thus points outward from S_1 at each point $P(a\cos u, a\sin u, v)$, as shown in **Figure 6.3.**

FIGURE 6.3

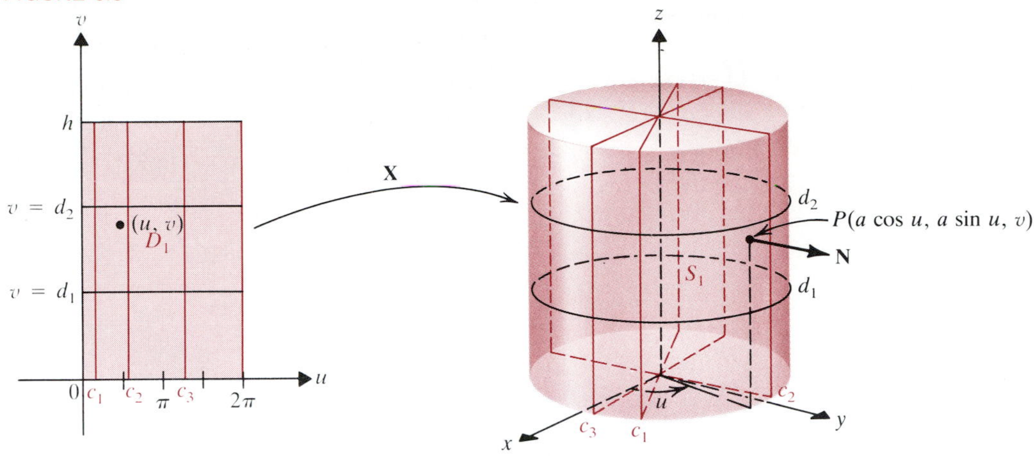

We can now define the surface integral of a continuous vector field.

6.3 DEFINITION Let $\mathbf{F}\colon \mathbf{R}^3 \to \mathbf{R}^3$ be a continuous vector field on a smooth surface patch $S \subseteq \mathbf{R}^3$. Let $\mathbf{X}\colon D \to \mathbf{R}^3$ parametrize S. Then the ***surface integral*** of $\mathbf{F}$ over S is

$$\iint_S \mathbf{F} \cdot d\mathbf{S} = \iint_{\mathbf{X}(D)} \mathbf{F} \cdot \mathbf{N}\, dA = \iint_D \mathbf{F} \cdot \mathbf{N} |\mathbf{X}_u \times \mathbf{X}_v|\, du\, dv. \tag{3}$$

Example 3

Find $\iint_{S_1} \mathbf{F} \cdot d\mathbf{S}$ if $\mathbf{F}(x, y, z) = -y\,\mathbf{i} + (x + y)\,\mathbf{j} + z\,\mathbf{k}$, and S_1 is the cylindrical surface of Example 2 with $a = h = 1$.

Solution: On S_1 we have from Example 2 with $a = h = 1$,

$$x = \cos u,\ y = \sin u, \text{ and } z = v, \qquad \text{for } (u, v) \in [0, 2\pi] \times [0, 1],$$

$$|\mathbf{X}_u \times \mathbf{X}_v| = a = 1,$$

and

$$\mathbf{N} = \cos u\,\mathbf{i} + \sin u\,\mathbf{j}.$$

Thus

$$\mathbf{F}(x(u, v), y(u, v), z(u, v)) = -\sin u\,\mathbf{i} + (\cos u + \sin u)\,\mathbf{j} + v\,\mathbf{k}.$$

We therefore have from (3)

$$\begin{aligned}
\iint_{S_1} \mathbf{F} \cdot d\mathbf{S} &= \iint_{D_1} [-\sin u\,\mathbf{i} + (\cos u + \sin u)\,\mathbf{j} + v\,\mathbf{k}] \cdot [\cos u\,\mathbf{i} + \sin u\,\mathbf{j} + 0\,\mathbf{k}]\, du\, dv \\
&= \iint_{D_1} [-\sin u \cos u + \sin u \cos u + \sin^2 u + 0]\, du\, dv \\
&= \int_0^1 dv \int_0^{2\pi} \sin^2 u\, du = 1 \cdot \int_0^{2\pi} \frac{1 - \cos 2u}{2}\, du \\
&= \frac{1}{2}u - \frac{1}{4}\sin 2u \Big]_0^{2\pi} = \pi.
\end{aligned}$$

Note from (2) that

$$\mathbf{F} \cdot \mathbf{N} |\mathbf{X}_u \times \mathbf{X}_v|\, du\, dv = \mathbf{F} \cdot \frac{\mathbf{X}_u \times \mathbf{X}_v}{\cancel{|\mathbf{X}_u \times \mathbf{X}_v|}} \cancel{|\mathbf{X}_u \times \mathbf{X}_v|}\, du\, dv = \mathbf{F} \cdot (\mathbf{X}_u \times \mathbf{X}_v)\, du\, dv.$$

We can thus rewrite (3) as

(4)
$$\iint_S \mathbf{F} \cdot d\mathbf{S} = \iint_D \mathbf{F} \cdot (\mathbf{X}_u \times \mathbf{X}_v)\, du\, dv.$$

The notation $\iint_S \mathbf{F} \cdot d\mathbf{S}$ parallels the notation $\int_C \mathbf{F} \cdot d\mathbf{x}$ for a line integral. Surface integrals of vector functions over surfaces correspond closely to line integrals of vector functions over curves.

An important illustration of this correspondence concerns the question of uniqueness of the integral. Recall from Section 2 that the line integral of a vector function $\mathbf{F}(x, y, z)$ over a curve C is not unique, but rather depends on the orientation of the curve, that is, on *how* it is parametrized. In fact, Theorem 2.4 says that if we reparametrize C so that its orientation is reversed, then the resulting line integral is the *negative* of the original value. Similarly, if we originally parametrize a surface patch S by

$$x = x(u, v),\ y = y(u, v),\ z = z(u, v), \qquad \text{where} \qquad (u, v) \in D \subseteq \mathbf{R}^2,$$

and compute $\mathbf{N}$ from (2), then we will get a normal vector that points in one of two possible directions (cf. Figure 6.2). One way to reparametrize that will reverse $\mathbf{N}$ is simply to interchange the parameters u and v:

$$x = x(v, u),\ y = y(v, u),\ z = z(v, u), \qquad \text{for} \qquad (v, u) \in \hat{D} = \{(v, u) \mid (u, v) \in D\}.$$

Then from Theorem 6.2(b) of Chapter 10 we have

$$\mathbf{X}_v \times \mathbf{X}_u = -\mathbf{X}_u \times \mathbf{X}_v,$$

so the standard unit normal for the reparametrized S will be $-\mathbf{N}$. Consequently, in (3) the value of the surface integral will be the *negative* of its original value. Keep this in mind when checking your answers. If someone else (such as the answer section) parametrizes a surface so as to obtain $-\mathbf{N}$ when you obtain $\mathbf{N}$, then your answer will be the negative of the other person's.

In Sections 7 and 8 we will tend to prefer whichever parametrization produces an $\mathbf{N}$ that points *outward* from S. In **Figure 6.4**, then, we would prefer $\mathbf{N}_1$ to $\mathbf{N}_2$. Here the surface S is the graph of $z = 4 - x^2 - y^2$. The most natural parametrization of S is

$$\mathbf{X}(u, v) = (u, v, 4 - u^2 - v^2) = u\,\mathbf{i} + v\,\mathbf{j} + (4 - u^2 - v^2)\,\mathbf{k}.$$

FIGURE 6.4

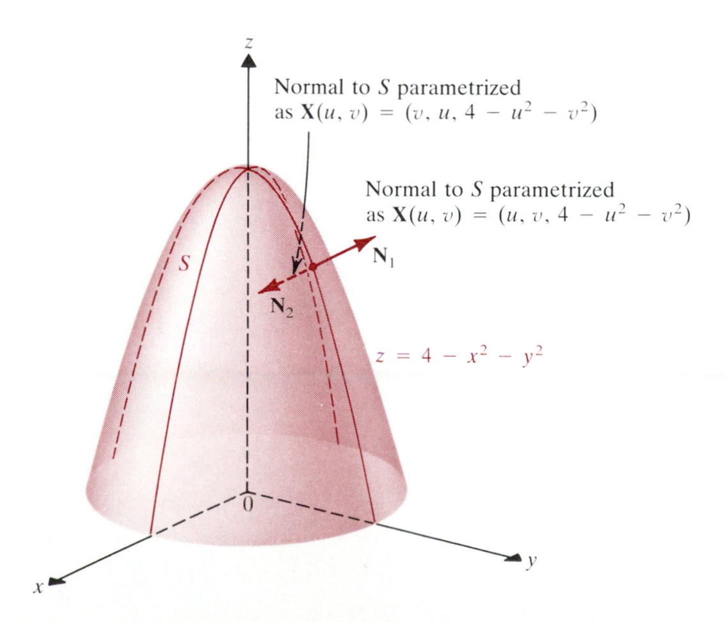

Then

$$\mathbf{X}_u(u, v) = \mathbf{i} - 2u\,\mathbf{k} \qquad \text{and} \qquad \mathbf{X}_v(u, v) = \mathbf{j} - 2v\,\mathbf{k},$$

so

$$\mathbf{X}_u \times \mathbf{X}_v = \det\begin{pmatrix} \mathbf{i} & \mathbf{j} & \mathbf{k} \\ 1 & 0 & -2u \\ 0 & 1 & -2v \end{pmatrix} = 2u\,\mathbf{i} + 2v\,\mathbf{j} + \mathbf{k}$$

points outward from S. Thus

$$\mathbf{N}_1 = \frac{1}{\sqrt{4u^2 + 4v^2 + 1}}(2u\,\mathbf{i} + 2v\,\mathbf{j} + \mathbf{k}).$$

You can check that the alternative parametrization

$$\mathbf{X}(u, v) = (v, u, 4 - u^2 - v^2)$$

gives $\mathbf{X}_u \times \mathbf{X}_v = -2u\,\mathbf{i} - 2v\,\mathbf{j} - \mathbf{k}$, and so produces the *inward*-pointing normal $\mathbf{N}_2$.

In this section we use whichever parametrization seems most natural from our previous experience with surfaces and coordinates in $\mathbf{R}^3$. But different people with the same background may not always agree on what is "most natural" for a given surface. As long as you are not troubled by occasional differences in sign in resulting surface integrals, this need not be a matter of concern. If that *is* bothersome, then you can eliminate the discrepancy by interchanging u and v in your parametrization.

Example 4

If

$$\mathbf{F}(x, y, z) = \frac{x}{\sqrt{x^2 + y^2 + z^2}}\,\mathbf{i} + \frac{y}{\sqrt{x^2 + y^2 + z^2}}\,\mathbf{j} + \frac{z}{\sqrt{x^2 + y^2 + z^2}}\,\mathbf{k} = \frac{1}{|\mathbf{x}|}\,\mathbf{x},$$

then find the surface integral of $\mathbf{F}$ over the sphere of Example 1.

FIGURE 6.5

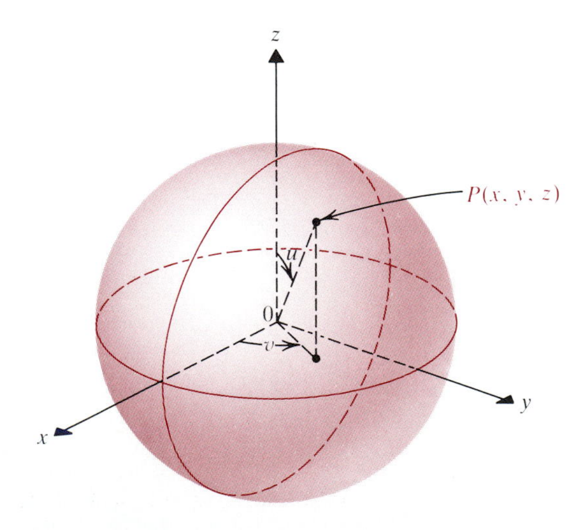

Solution. We have $\sqrt{x^2 + y^2 + z^2} = a$. As in Example 2 of the last section, we parametrize by letting u be the angle ϕ and v the angle θ of spherical coordinates (see **Figure 6.5**):

$$x = a \sin u \cos v,\ y = a \sin u \sin v,\ z = a \cos u, \qquad (u, v) \in [0, \pi] \times [0, 2\pi].$$

Here, on the sphere,

$$\mathbf{F}(x, y, z) = \sin u \cos v\, \mathbf{i} + \sin u \sin v\, \mathbf{j} + \cos u\, \mathbf{k}.$$

From Example 2 of the last section, we have

$$\mathbf{X}_u \times \mathbf{X}_v = a^2[\sin^2 u \cos v\, \mathbf{i} + \sin^2 u \sin v\, \mathbf{j} + \sin u \cos u\, \mathbf{k}].$$

Then from (4) we find

$$\begin{aligned}\iint_S \mathbf{F} \cdot d\mathbf{S} &= \iint_D \mathbf{F} \cdot (\mathbf{X}_u \times \mathbf{X}_v)\, du\, dv \\ &= a^2 \iint_D (\sin^3 u \cos^2 v + \sin^3 u \sin^2 v + \sin u \cos^2 u)\, du\, dv \\ &= a^2 \iint_D (\sin^3 u + \sin u \cos^2 u)\, du\, dv \\ &= a^2 \iint_D \sin u(\sin^2 u + \cos^2 u)\, du\, dv \\ &= a^2 \iint_D \sin u\, du\, dv = 4\pi a^2,\end{aligned}$$

as in Example 2 of Section 5. ■

An important notation for the surface integral of a vector function $\mathbf{F}$ arises by writing $\mathbf{F}$ in coordinate form as $\mathbf{F}(\mathbf{x}) = \mathbf{F}(x, y, z) = (P(x, y, z), Q(x, y, z), R(x, y, z)) = P(x, y, z)\mathbf{i} + Q(x, y, z)\mathbf{j} + R(x, y, z)\mathbf{k}$. Then it is natural to also write $\mathbf{X}\colon \mathbf{R}^2 \to \mathbf{R}^3$ in terms of its coordinate functions as

$$\mathbf{X}(u, v) = (x(u, v), y(u, v), z(u, v)) = x(u, v)\mathbf{i} + y(u, v)\mathbf{j} + z(u, v)\mathbf{k}.$$

From that we obtain

$$\mathbf{X}_u = \frac{\partial x}{\partial u}\mathbf{i} + \frac{\partial y}{\partial u}\mathbf{j} + \frac{\partial z}{\partial u}\mathbf{k} \qquad \text{and} \qquad \mathbf{X}_v = \frac{\partial x}{\partial v}\mathbf{i} + \frac{\partial y}{\partial v}\mathbf{j} + \frac{\partial z}{\partial v}\mathbf{k}.$$

Thus

$$\begin{aligned}\mathbf{X}_u \times \mathbf{X}_v &= \det\begin{pmatrix} \mathbf{i} & \mathbf{j} & \mathbf{k} \\ \partial x/\partial u & \partial y/\partial u & \partial z/\partial u \\ \partial x/\partial v & \partial y/\partial v & \partial z/\partial v \end{pmatrix} \\ &= \left(\frac{\partial y}{\partial u}\frac{\partial z}{\partial v} - \frac{\partial y}{\partial v}\frac{\partial z}{\partial u}, \frac{\partial z}{\partial u}\frac{\partial x}{\partial v} - \frac{\partial z}{\partial v}\frac{\partial x}{\partial u}, \frac{\partial x}{\partial u}\frac{\partial y}{\partial v} - \frac{\partial x}{\partial v}\frac{\partial y}{\partial u}\right) \\ &= \left(\frac{\partial(y, z)}{\partial(u, v)}, \frac{\partial(z, x)}{\partial(u, v)}, \frac{\partial(x, y)}{(u, v)}\right) = \frac{\partial(y, z)}{\partial(u, v)}\mathbf{i} + \frac{\partial(z, x)}{\partial(u, v)}\mathbf{j} + \frac{\partial(x, y)}{\partial(u, v)}\mathbf{k},\end{aligned}$$

where in the last equation we have introduced the *Jacobian determinant* notation. If $\mathbf{F}$ is a vector field on $\mathbf{R}^2$ with coordinate functions x and y,

$$\mathbf{F}(u, v) = (x(u, v), y(u, v)) = x(u, v)\mathbf{i} + y(u, v)\mathbf{j},$$

then the ***Jacobian determinant of x and y with respect to u and v*** is

$$\frac{\partial(x, y)}{\partial(u, v)} = \det \mathbf{J_F} = \det\begin{pmatrix} \dfrac{\partial x}{\partial u} & \dfrac{\partial y}{\partial u} \\ \dfrac{\partial x}{\partial v} & \dfrac{\partial y}{\partial v} \end{pmatrix}$$

$$= \frac{\partial x}{\partial u}\frac{\partial y}{\partial v} - \frac{\partial x}{\partial v}\frac{\partial y}{\partial u}.$$

In this notation, (4) becomes

$$\iint_S \mathbf{F} \cdot d\mathbf{S} = \iint_D \mathbf{F} \cdot (\mathbf{X}_u \times \mathbf{X}_v)\, du\, dv$$

$$= \iint_D P(\mathbf{x}) \frac{\partial(y, z)}{\partial(u, v)}\, du\, dv + Q(\mathbf{x}) \frac{\partial(z, x)}{\partial(u, v)}\, du\, dv + R(\mathbf{x}) \frac{\partial(x, y)}{\partial(u, v)}\, du\, dv.$$

Imagining that symbolic cancellation between $\partial(u, v)$ and $du\, dv$ is possible, we arrive at the notation

$$\iint_S \mathbf{F} \cdot d\mathbf{S} = \iint_D P(\mathbf{x})\, dy \wedge dz + Q(\mathbf{x})\, dz \wedge dx + R(\mathbf{x})\, dx \wedge dy. \tag{5}$$

Recall in double integrals that the order of writing dx and dy indicates which variable is integrated first. Here we introduce the wedge $\wedge$ to remind ourselves that the order of dx and dy also matters in a surface integral: $dx \wedge dy$ really comes via symbolic cancellation from

$$\frac{\partial(x, y)}{\partial(u, v)} = \det\begin{pmatrix} \dfrac{\partial x}{\partial u} & \dfrac{\partial x}{\partial v} \\ \dfrac{\partial y}{\partial u} & \dfrac{\partial y}{\partial v} \end{pmatrix} \to dx \wedge dy = \frac{\partial(x, y)}{\cancel{\partial(u, v)}} \cancel{du\, dv}.$$

Reversing the order of dx and dy in $dx \wedge dy$ then introduces a factor of -1, because

$$\frac{\partial(y, x)}{\partial(u, v)} = -\frac{\partial(x, y)}{\partial(u, v)}.$$

(Recall that the sign of a determinant is *reversed* if we interchange the order of its rows.) The notation (5) leads to the important concept of differential forms, which is studied in advanced calculus.

Example 5

Find $\partial(x, y)/\partial(r, \theta)$ and $\partial(y, x)/\partial(r, \theta)$ if $x = r\cos\theta$ and $y = r\sin\theta$.

Solution. We have

$$\frac{\partial x}{\partial r} = \cos\theta, \qquad \frac{\partial y}{\partial r} = \sin\theta, \qquad \frac{\partial x}{\partial \theta} = -r\sin\theta, \qquad \text{and} \qquad \frac{\partial y}{\partial \theta} = r\cos\theta.$$

Therefore

$$\frac{\partial(x, y)}{\partial(r, \theta)} = \det\begin{pmatrix} \dfrac{\partial x}{\partial r} & \dfrac{\partial x}{\partial \theta} \\ \dfrac{\partial y}{\partial r} & \dfrac{\partial y}{\partial \theta} \end{pmatrix} = \det\begin{pmatrix} \cos\theta & -r\sin\theta \\ \sin\theta & r\cos\theta \end{pmatrix} = r\cos^2\theta + r\sin^2\theta = r,$$

and

$$\frac{\partial(y, x)}{\partial(r, \theta)} = \det\begin{pmatrix} \frac{\partial y}{\partial r} & \frac{\partial y}{\partial \theta} \\ \frac{\partial x}{\partial r} & \frac{\partial x}{\partial \theta} \end{pmatrix} = \det\begin{pmatrix} \sin\theta & r\cos\theta \\ \cos\theta & -r\sin\theta \end{pmatrix} = r\sin^2\theta - r\cos^2\theta = -r. \quad \blacksquare$$

FIGURE 6.6

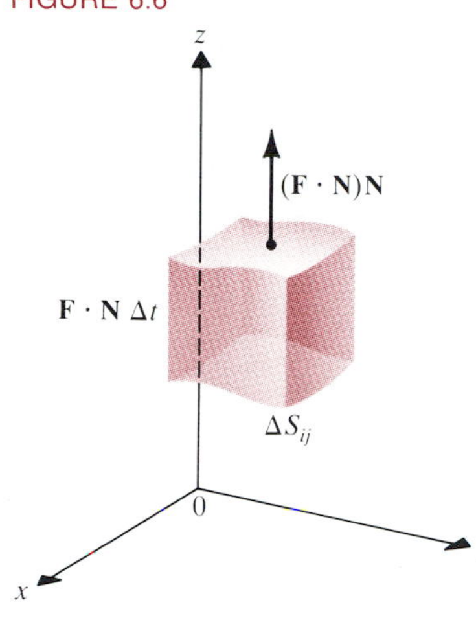

In the study of fluid flows (and also electrical, magnetic, gravitational, and other fields), the *flux* is a measure of the amount of fluid per unit area flowing across a given smooth surface S per unit time. Let $\mathbf{F}: \mathbf{R}^3 \to \mathbf{R}^3$ be the velocity field of the flow. We assume that $\mathbf{F}$ is continuous on S. Consider a small patch of S near a typical point $\mathbf{X}(u_i, v_j)$. Since S is smooth, this patch is nearly planar (flat). By the continuity of $\mathbf{F}$, the velocity at all points of the patch is close to $\mathbf{F}(\mathbf{X}(u_i, v_j))$. Thus the coordinate of the velocity normal to the patch is approximately $\mathbf{F}(\mathbf{X}(u_i, v_j)) \cdot \mathbf{N}(u_i, v_j)$. This measures the rate at which the fluid is flowing across the patch. In a small time period Δt, the mass of the fluid that crosses the patch is therefore approximately

$$\delta \mathbf{F}(\mathbf{X}(u_i, v_j)) \cdot \mathbf{N}(u_i, v_j)\, \Delta S_{ij}\, \Delta t,$$

where δ is the density and ΔS_{ij} is the area of the patch. (Refer to **Figure 6.6,** in which we show the small cylinder filled by the fluid during the period Δt.) Hence the *total mass of the fluid crossing the entire surface* S during time Δt is approximately

$$\delta\, \Delta t \sum_{i,j} \mathbf{F}(\mathbf{X}(u_i, v_j)) \cdot \mathbf{N}(u_i, v_j)\, \Delta S_{ij}.$$

The following definition is thus appropriate.

6.4 DEFINITION

The ***flux*** across S of a fluid flowing with continuous velocity field $\mathbf{F}: \mathbf{R}^3 \to \mathbf{R}^3$ is

$$\Phi(S) = \iint_S \mathbf{F} \cdot d\mathbf{S}.$$

For instance, if the $\mathbf{F}$ of Example 4 is the velocity field of a fluid flow with a source at the origin, then the flux across any sphere of radius a is $4\pi a^2$. In this situation, how the surface is parametrized is of *essential* importance. Since the calculated flux came out positive, we conclude that the parametrization of S is such that the fluid is flowing *outward* from the interior of S. If we were trying to mathematically describe a fluid flow toward a *sink* at the origin, then we would have to reverse the parametrization of S—for example, by interchanging the roles of u and v in the parametrization. Thus, while the orientation of the standard normal may not seem particularly crucial, in applications to physics and engineering problems that have a definite orientation, it is often necessary to check that the parametrization imposed on the problem is faithful to the real-world phenomenon being described.

We were able to define the line integral of a continuous vector function $\mathbf{F}$ over a *piecewise* smooth curve $C = C_1 \cup C_2 \cup \ldots \cup C_k$ (for smooth C_i) to be the sum of the $\int_{C_i} \mathbf{F} \cdot d\mathbf{x}$. By a similar approach we can extend surface integrals to most, *but not all*, surfaces obtained by gluing together smooth surface patches S_i along common boundary curves $\mathbf{X}_i(C_i)$, where $\mathbf{X}_i: D_i \to \mathbf{R}^3$ parametrizes S_i and $C_i = \partial(D_i)$. See **Figure 6.7.**

FIGURE 6.7

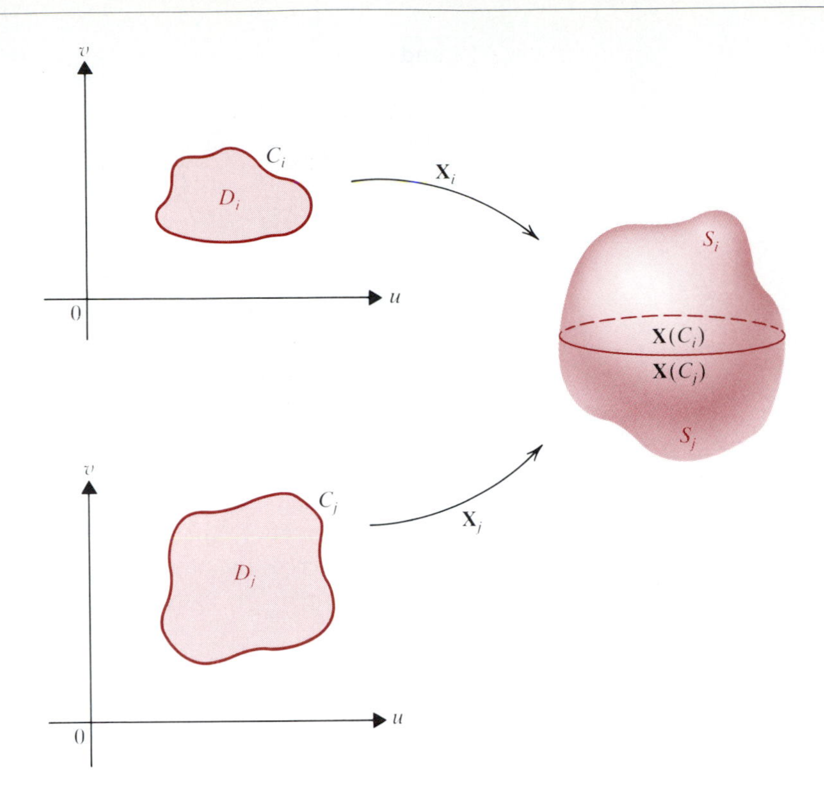

6.5 DEFINITION

A surface $S \subseteq \mathbf{R}^3$ is ***piecewise smooth*** if $S = S_1 \cup S_2 \cup \ldots \cup S_k$ for smooth surface patches S_i such that for $i \neq j$, either $S_i \cap S_j$ is a common boundary curve for both S_i and S_j or else $S_i \cap S_j = \emptyset$. (Here $\emptyset$ is the ***empty set.***)

FIGURE 6.8

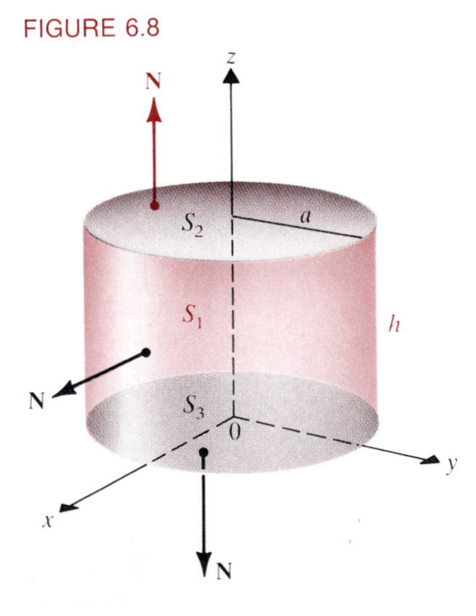

In **Figure 6.8** the cylinder of Example 2 is represented as $S_1 \cup S_2 \cup S_3$, where $S_1 \cap S_2$ is the boundary of the top disk S_2 and part of the curved surface S_1; $S_2 \cap S_3 = \emptyset$; and $S_1 \cap S_3$ is the common boundary of both S_1 and S_3. As you are aware if you have ever stepped on a can of soup, the cylinder is not smooth. But it *is* piecewise smooth.

To satisfactorily extend surface integrals to piecewise smooth surfaces, we require the surfaces to be ***orientable.*** The means that it is possible to parametrize the patches S_i in such a way that the normal vector $\mathbf{X}_u \times \mathbf{X}_v$ does not *reverse* its orientation as we move from one surface patch S_i to an adjacent patch S_j. We would, for example, be able to orient the cylinder in Figure 6.8 by parametrizing it so that $\mathbf{X}_u \times \mathbf{X}_v$ always points outward. The next example shows that not only is that cylinder orientable, but also that the parametrizations in Example 1(b) of the last section orient it.

Example 6

Show that the parametrizations of S_1, S_2, and S_3 given in Example 1 of Section 5 orient the cylinder of **Figure 6.9.**

Solution. In Example 1(b), we parametrized the bottom disk S_3 by

$$\mathbf{X}(u, v) = (u \cos v, -u \sin v, 0), \qquad \text{for } (u, v) \in D_3 = [0, a] \times [0, 2\pi].$$

Therefore

$$\mathbf{X}_u = \cos v\,\mathbf{i} - \sin v\,\mathbf{j}, \qquad \text{and} \qquad \mathbf{X}_v = -u \sin v\,\mathbf{i} - u \cos v\,\mathbf{j},$$

FIGURE 6.9

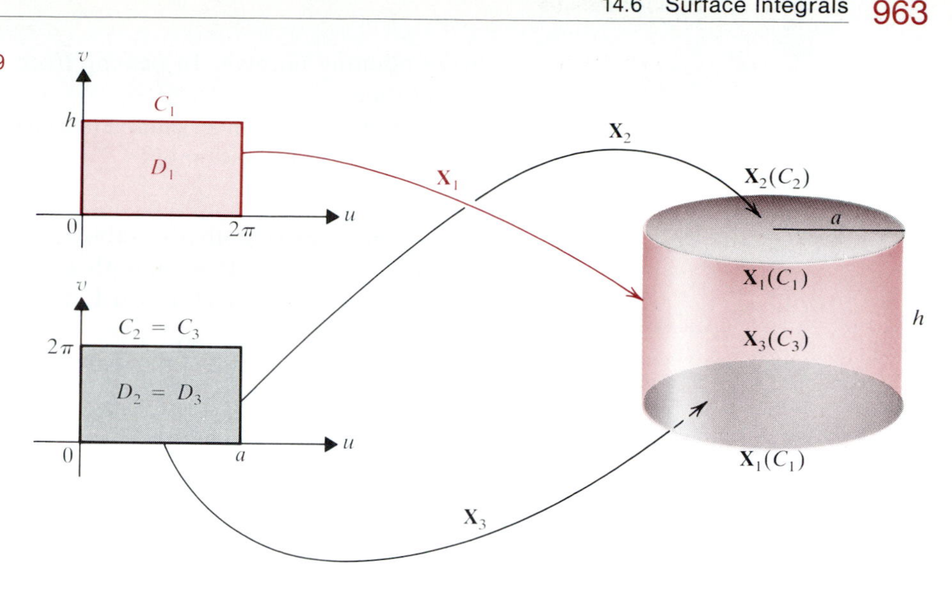

so

$$\mathbf{X}_u \times \mathbf{X}_v = \det\begin{pmatrix} \mathbf{i} & \mathbf{j} & \mathbf{k} \\ \cos v & -\sin v & 0 \\ -u\sin v & -u\cos v & 0 \end{pmatrix} = -u\,\mathbf{k}.$$

The normal vector $\mathbf{X}_u \times \mathbf{X}_v$ thus points *outward* from the surface, because it points downward. On the top we parametrized S_2 by

$$\mathbf{X}(u, v) = (u\cos v, u\sin v, h), \qquad \text{for } (u, v) \in D_2 = [0, a] \times [0, 2\pi].$$

Hence in this case we obtain

$$\mathbf{X}_u = \cos v\,\mathbf{i} + \sin v\,\mathbf{j} \qquad \text{and} \qquad \mathbf{X}_v = -u\sin v\,\mathbf{i} + u\cos v\,\mathbf{j},$$

so

$$\mathbf{X}_u \times \mathbf{X}_v = \det\begin{pmatrix} \mathbf{i} & \mathbf{j} & \mathbf{k} \\ \cos v & \sin v & 0 \\ -u\sin v & u\cos v & 0 \end{pmatrix} = u\,\mathbf{k}.$$

Here $\mathbf{X}_u \times \mathbf{X}_v$ points upward since u is positive. Again, this is *outward* from the surface. Finally, we parametrized the cylindrical middle portion S_1 by

$$\mathbf{X}(u, v) = (a\cos u, a\sin u, v), \qquad \text{for } (u, v) \in D_1 = [0, 2\pi] \times [0, h].$$

In Example 2 we found that

$$\mathbf{X}_u \times \mathbf{X}_v = a\cos u\,\mathbf{i} + a\sin u\,\mathbf{j}.$$

Again, $\mathbf{X}_u \times \mathbf{X}_v$ points *outward* from the surface. As we traverse the surface then, $\mathbf{N}$ always points outward and so never reverses its direction as we pass from one smooth surface patch to another. ■

If we had used the apparently more natural parametrization of S_3 given by

$$\mathbf{X}_4(u, v) = (u\cos v, u\sin v, 0), \text{ for } \qquad (u, v) \in [0, a] \times [0, 2\pi],$$

then we would *not* have had an oriented parametrized surface. The reason is that, as in the case of S_2, the normal vector $\mathbf{X}_u \times \mathbf{X}_v$ would have been $u\,\mathbf{k}$, which points *inward*, that is, *oppositely* relative to the surface from the other two *outward*-pointing normals. In passing from S_1 to S_3, **N** would have reversed its orientation.

For piecewise smooth orientable surfaces S, we can extend the notion of surface integral in Definition 6.3.

6.6 DEFINITION

If S is a piecewise smooth orientable surface in $\mathbf{R}^3$, $S = S_1 \cup S_2 \cup \ldots \cup S_k$ where S_i are parametrized smooth surface patches, then the ***surface integral*** of a continuous vector field $\mathbf{F}: \mathbf{R}^3 \to \mathbf{R}^3$ over S is defined by

$$\iint_S \mathbf{F} \cdot d\mathbf{S} = \sum_{i=1}^{k} \iint_{S_i} \mathbf{F} \cdot d\mathbf{S}. \tag{6}$$

Definition 6.6 bears the same relation to Definition 6.3 that Definition 2.3 bears to Definition 2.2.

FIGURE 6.10

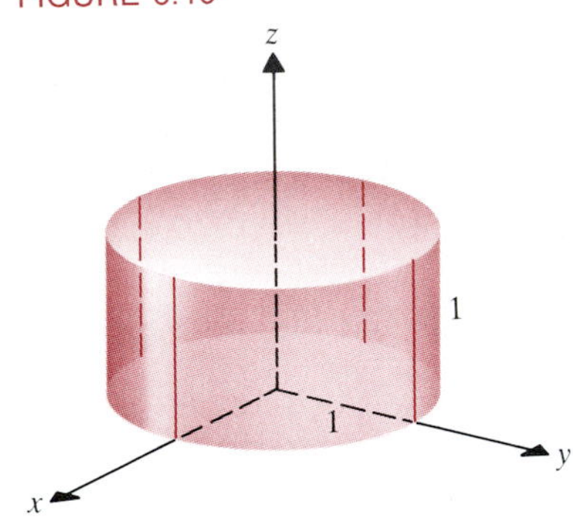

Example 7

Compute the surface integral of $\mathbf{F}(x, y, z) = -y\mathbf{i} + (x + y)\mathbf{j} + z\mathbf{k}$ over the cylinder

$S = \{(x, y, z) | x^2 + y^2 = 1, 0 \le z \le 1\} \cup \{(x, y, z) | 0 \le x^2 + y^2 \le 1, z = 0, z = 1\}$.

See **Figure 6.10.**

Solution. S is a cylinder including top $z = 1$ and bottom $z = 0$. We can use the parametrization of Example 1(b) of the last section with $a = 1$ and $h = 1$. From Example 3 we have

(a) $\displaystyle\iint_{S_1} \mathbf{F} \cdot d\mathbf{S} = \pi$.

(b) On the top S_2, we parametrized by $\mathbf{X}(u, v) = (u\cos v, u\sin v, 1)$ for (u, v) in $[0, 1] \times [0, 2\pi]$. Therefore on S_2,

$$\mathbf{F}(x, y, z) = -u\sin v\,\mathbf{i} + (u\cos v + u\sin v)\mathbf{j} + \mathbf{k}.$$

We also calculated $\mathbf{X}_u \times \mathbf{X}_v = u\mathbf{k}$ in Example 6.

(c) On the bottom S_3, we used the parametrization $\mathbf{X}(u, v) = (u\cos v, -u\sin v, 0)$ for (u, v) in $[0, 1] \times [0, 2\pi]$. Then on S_3,

$$\mathbf{F}(x, y, z) = u\sin v\,\mathbf{i} + (u\cos v - u\sin v)\mathbf{j} + 0\mathbf{k}.$$

We also calculated $\mathbf{X}_u \times \mathbf{X}_v = -u\mathbf{k}$ in Example 6.

From (a), (b), (c), and Definition 6.6, we obtain

$$\begin{aligned}
\iint_S \mathbf{F} \cdot d\mathbf{S} &= \iint_{S_1} \mathbf{F} \cdot d\mathbf{S} + \iint_{S_2} \mathbf{F} \cdot d\mathbf{S} + \iint_{S_3} \mathbf{F} \cdot d\mathbf{S} \\
&= \pi + \iint_{D_2} [-u\sin v\,\mathbf{i} + (u\cos v + u\sin v)\mathbf{j} + \mathbf{k}] \cdot u\mathbf{k}\,du\,dv \\
&\quad + \iint_{D_3} [u\sin v\,\mathbf{i} + (u\cos v - u\sin v)\mathbf{j}] \cdot (-u\mathbf{k})\,du\,dv \\
&= \pi + \iint_{D_2} u\,du\,dv + 0 = \pi + \int_0^{2\pi} dv \int_0^1 u\,du \\
&= \pi + 2\pi\left[\frac{1}{2}u^2\right]_0^1 = \pi + \pi = 2\pi. \quad \blacksquare
\end{aligned}$$

Our remarks about the cylinder of Example 3 may give the impression that any piecewise smooth surface can be oriented if we just take the trouble to parametrize it properly. But that is not true. The celebrated ***Möbius band*** [named for the German mathematician Augustus F. Möbius (1790–1868), who was a student of Gauss] is *not* orientable. The band is made by deforming the rectangle $R = R_1 \cup R_2$ as shown in **Figure 6.11.** One *twists* the rectangle and then glues edge AC to edge DB. It is fun (and easy) to construct the band from a long strip of paper and then study it firsthand. (See Exercise 25.) Figure 6.11 shows the surface and how to construct it. Unlike the oriented cylinder in Example 6, this nonorientable surface has only *one* side. Suppose you begin painting the band on "one side" of AC and never lift your brush from the surface. Then, after passing through the twist, you will arrive at BD on the "reverse side"! So you can continue and thus paint the entire band with one stroke! Have a friend hold a pencil perpendicular to AC while you traverse the band by holding another pencil perpendicular to the band. If you begin with your pencil pointing in the same direction as your friend's, then you will arrive back at BD with your pencil pointing the *opposite* way! This says that $\mathbf{N}$ *reverses* its orientation near the point where you began to traverse the strip. For after a small displacement, $\mathbf{N}$ points the opposite way! See **Figure 6.12.** Such surfaces are studied further in topology, but we will stick to orientable surfaces aside from this curious one-side surface, which is used on many conveyor belts (such as those in supermarket check-out counters) to achieve uniform wear: There is no one side to wear more than the "other"!

FIGURE 6.11

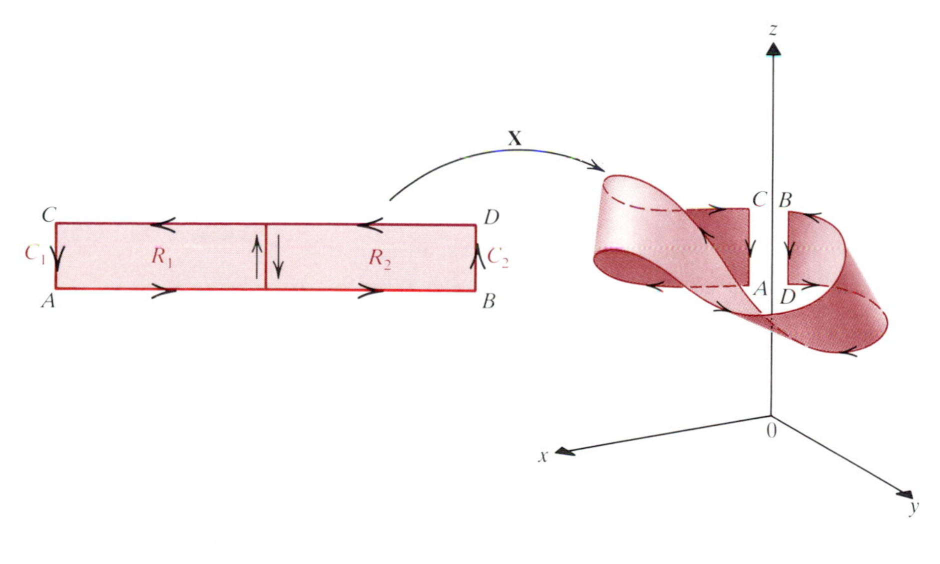

FIGURE 6.12

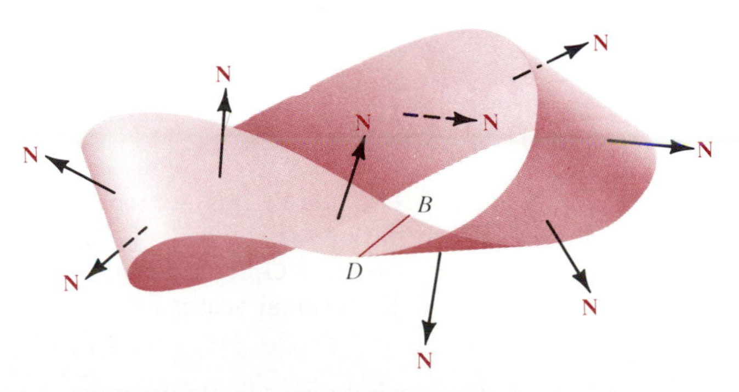

Exercises 14.6

In Exercises 1–18, compute the surface integral of the given function over the given surface.

1. $\iint_S (x^2 + y^2)z\,dA$, where S is the top hemisphere of $x^2 + y^2 + z^2 = 1$
2. $\iint_S (x^2 + y^2)z^2\,dA$, where S is as in Exercise 1
3. $\iint_S z\,dA$, where S is the part of the cylinder $x^2 + y^2 = 4$ between the planes $z = 0$ and $z = x + 2$
4. $\iint_S z^2\,dA$, where S is as in Exercise 3
5. $\iint_S (x^2 + y^2)\,dA$, where S is the part of the sphere $x^2 + y^2 + z^2 = 1$ lying above the cone $z = \sqrt{x^2 + y^2}$.
6. $\iint_S x\,dA$, where S is the triangle with vertices $(1, 0, 0)$, $(0, 1, 0)$, $(0, 0, 1)$
7. $\iint_S x^2\,dA$, where $S = S_1 \cup S_2 \cup S_3$ is the cylinder of Example 6. (Use Definition 6.6.)
8. $\iint_S x\,dA$, where S is the surface of Exercise 3 complete with top and bottom. (Use Definition 6.6.)
9. $\iint_S (x, y, z) \cdot d\mathbf{S}$, where S is the sphere of Example 1 with $a = 1$
10. $\iint_S \mathbf{F} \cdot d\mathbf{S}$, where $\mathbf{F}(x, y, z) = xz\,\mathbf{i} + yz\,\mathbf{j} + x^2\,\mathbf{k}$, and S as in Exercise 9
11. $\iint_S \mathbf{F} \cdot d\mathbf{S}$, where $\mathbf{F}(x, y, z) = y\,\mathbf{i} + (y - x)\,\mathbf{j} + e^x\,\mathbf{k}$, and S is the part of the paraboloid $z = x^2 + y^2$ lying above the rectangle $[0, 1] \times [0, 3]$
12. $\iint_S \mathbf{F} \cdot d\mathbf{S}$, where $\mathbf{F}(x, y, z) = x^2\,\mathbf{i} + y^2\,\mathbf{j} + z^2\,\mathbf{k}$, and S is the portion of $z = \sqrt{x^2 + y^2}$ between the planes $z = 1$ and $z = 2$
13. $\iint_S \mathbf{F} \cdot \mathbf{N}\,dA$, for the $\mathbf{F}$ of Exercise 12, and S the cone $z = \sqrt{x^2 + y^2}$ between $z = 0$ and $z = 3$
14. $\iint_S \mathbf{F} \cdot d\mathbf{S}$, where $\mathbf{F}(x, y, z) = x\,\mathbf{i} + y\,\mathbf{j} + z\,\mathbf{k}$, and S is the cylinder $x^2 + y^2 = 4$ between $z = 0$ and $z = x + 2$
15. $\iint_S \mathbf{F} \cdot d\mathbf{S}$, for the $\mathbf{F}$ of Exercise 14, and S the cylinder $x^2 + y^2 = 1$, $0 \le z \le 1$ with bottom but no top. (Use Definition 6.6.)
16. $\iint_S \mathbf{F} \cdot \mathbf{N}\,dA$, where $\mathbf{F}(x, y, z) = 2x\,\mathbf{i} - 3y\,\mathbf{j} + z\,\mathbf{k}$, and S is the part of the cylinder $x^2 + y^2 = 1$ between $z = 0$ and $z = x + 2$
17. $\iint_S \mathbf{F} \cdot d\mathbf{S}$, where $\mathbf{F}(x, y, z) = (x + y)\,\mathbf{i} + y\,\mathbf{j} + (x + z)\,\mathbf{k}$, and S is the surface of Example 7
18. $\iint_S \mathbf{F} \cdot d\mathbf{S}$, where $\mathbf{F}(x, y, z) = y\,\mathbf{i} + (y - x)\,\mathbf{j} + (y + z)\,\mathbf{k}$, and S is the surface of Example 6
19. The ***electrostatic potential*** V at $\mathbf{0}$ due to a charge distribution of charge density $\delta(x, y, z)$ on S is
$$V(\mathbf{0}) = \iint_S \frac{\delta(\mathbf{x})}{|\mathbf{x}|}\,dA.$$
If $\delta(x, y, z) = \delta$, a constant, then find $V(\mathbf{0})$ for S the sphere $x^2 + y^2 + z^2 = a^2$.
20. Repeat Exercise 19 for the portion of $z = \sqrt{x^2 + y^2}$ lying between $z = 1$ and $z = 3$.
21. If a fluid flows with velocity $\mathbf{F}(x, y, z) = \mathbf{x}/|\mathbf{x}|^3$, then find the flux outward across the sphere $x^2 + y^2 + z^2 = 1$. (Parametrize so that $\mathbf{N}$ points outward. Note that $\mathbf{F}$ is an inverse-square law field. The next exercise says that the answer here does not depend on the radius.)
22. If a fluid flows with velocity
$$\mathbf{F}(x, y, z) = \frac{k}{|\mathbf{x}^2|}\frac{\mathbf{x}}{|\mathbf{x}|},$$
then show that the flux across any sphere of radius a centered at $\mathbf{0}$ is the same as that across any other concentric sphere, if both parametrizations give outward-pointing $\mathbf{N}$.
23. In the study of heat flows, one supposes that the temperature $T(x, y, z)$ has a continuous gradient $\nabla T(\mathbf{x})$. The negative gradient gives the magnitude and direction of the rate of flow of the heat at $\mathbf{x}$. If $T(x, y, z) = x^2 + y^2$, then find the flux across the cylinder $x^2 + y^2 = 1$ between $z = 0$ and $z = 1$. (Parametrize as in Example 6.)
24. Refer to Exercise 23. Find the flux across the sphere $x^2 + y^2 + z^2 = 1$ if $T(x, y, z) = \sqrt{x^2 + y^2 + z^2}$. (Parametrize so that $\mathbf{N}$ points outward.)
25. Construct two paper Möbius bands.
    - **(a)** Using a pair of scissors, pierce one band midway between its "edges," and cut along the entire surface parallel to the "edges." What do you obtain?
    - **(b)** Repeat (a) for the resulting surface. What happens?
    - **(c)** Try making two parallel cuts on the second band, one third of the width of the band from each "edge." What happens?

14.7 Theorem of Stokes

The title of this section refers to an extension of Green's theorem from two to three dimensions. To describe that extension it is helpful to use the formulation of Green's Theorem 3.3 given on p. 929. There we considered a two-dimensional vector field

$$\mathbf{F} = P\,\mathbf{i} + Q\,\mathbf{j}$$

whose coordinate functions P and Q have continuous partial derivatives on a

FIGURE 7.1

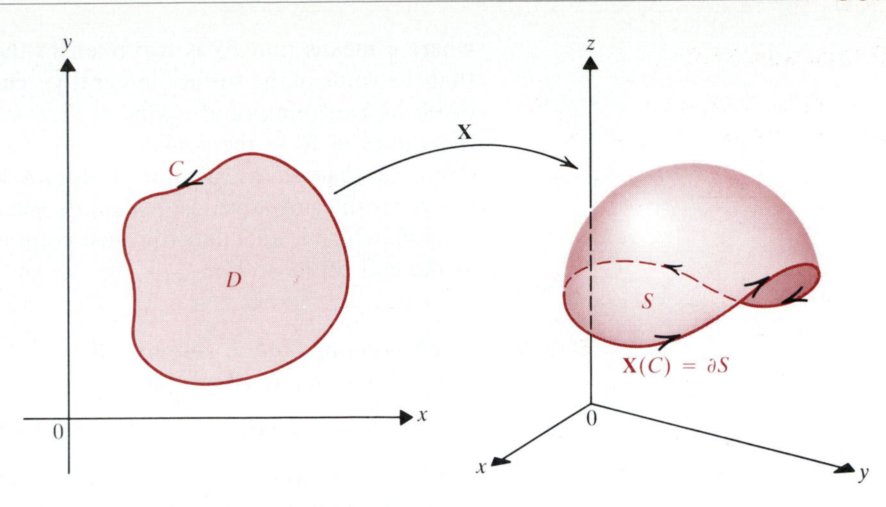

FIGURE 7.2

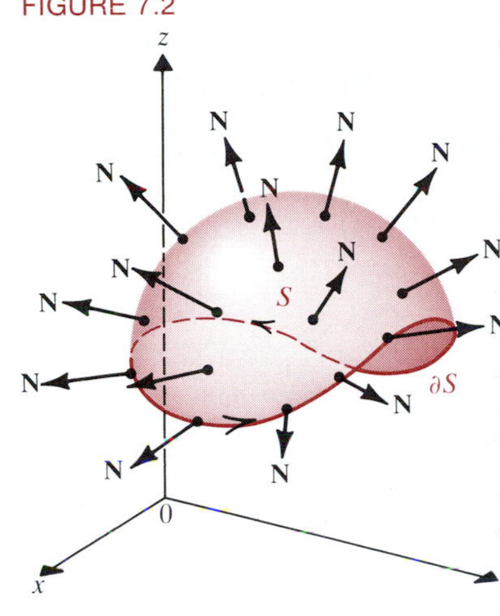

simple region D in the plane. Assuming that the boundary C of D is parametrized in the counterclockwise sense, Green's theorem says that

$$\iint_D (\mathbf{curl\,F} \cdot \mathbf{k})\, dA = \int_C \mathbf{F} \cdot d\mathbf{x}. \tag{1}$$

To extend this to three dimensions we consider a vector field $\mathbf{F}\colon \mathbf{R}^3 \to R^3$ such that $\mathbf{J_F}$ has continuous entries. The analogue of the left side of (1) is the *surface integral* of $\mathbf{curl\,F} \cdot \mathbf{N}$ over an oriented smooth surface patch $S = \mathbf{X}(D)$ (see **Figure 7.1**). S is oriented by choosing a direction for its normal vectors to point relative to S, as in **Figure 7.2**, where S appears to have its normal vectors pointing outward from the surface. The boundary ∂S of the surface S is $\mathbf{X}(C)$, where C is the boundary of D. We orient ∂S by saying that the ***positive direction*** along it is the one induced by the counterclockwise direction along C. That is, as C is traversed in the counterclockwise sense (so that D is on the left), ∂S is traversed in the positive direction. This is equivalent to requiring that if the thumb of your right hand is pointed in the direction of the normal vectors to S, then your right index finger points in the positive direction along ∂S (see **Figure 7.3**). The analogue of (1) is

$$\iint_S \mathbf{curl\,F} \cdot \mathbf{N}\, dA = \oint_{\partial S} \mathbf{F} \cdot d\mathbf{x}, \tag{2}$$

FIGURE 7.3 The right-hand rule for orienting ∂S

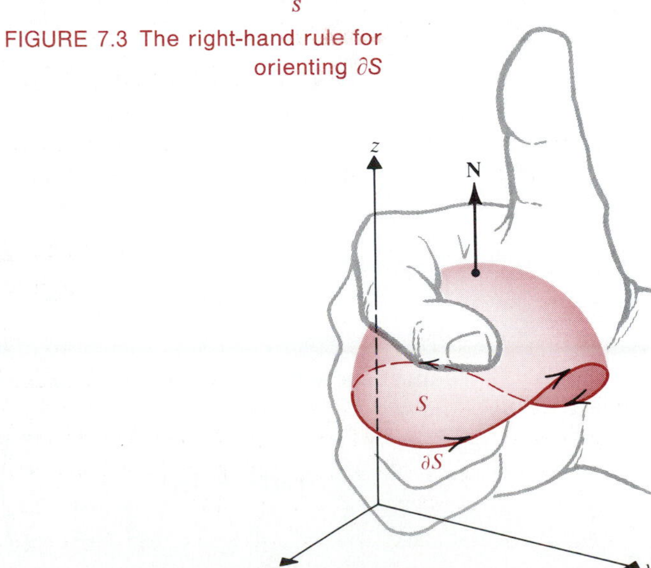

where $\oint$ means that ∂S is traversed in the positive sense. Equation (2) asserts that the value of the surface integral $\iint_S \mathbf{curl\,F} \cdot d\mathbf{S}$ of the quantity $\mathbf{\nabla} \times \mathbf{F}$ (which we again can think of as a *kind of derivative* of $\mathbf{F}$) is completely determined by the values of $\mathbf{F}$ on the *boundary* of S. So (2) is one more analogue of the fundamental theorem of calculus (Theorem 4.2 of Chapter 4). We leave a proof in full generality for advanced calculus. We can, however, establish the following version, which is adequate for most computational purposes. The proof is given at the end of the section.

7.1 THEOREM

Theorem of Stokes. Suppose that $S \subseteq \mathbf{R}^3$ is an oriented smooth surface patch parametrized so that

$$\mathbf{X}(u, v) = (x(u, v), y(u, v), z(u, v)) = x(u, v)\,\mathbf{i} + y(u, v)\,\mathbf{j} + z(u, v)\,\mathbf{k},$$

where x, y, and z have continuous partial derivatives on a simple region $D \subseteq \mathbf{R}^2$ bounded by a piecewise smooth positively oriented simple closed curve ∂S. Suppose that $\mathbf{F}$: $\mathbf{R}^3 \to \mathbf{R}^3$ is given by

$$\mathbf{F}(x, y, z) = (P(x, y, z), Q(x, y, z), R(x, y, z)) = P\,\mathbf{i} + Q\,\mathbf{j} + R\,\mathbf{k},$$

where P, Q, and R have continuous partial derivatives on S. Then

$$\iint_S \mathbf{curl\,F} \cdot d\mathbf{S} = \oint_{\partial S} \mathbf{F} \cdot d\mathbf{x}. \tag{2}$$

Example 1

Verify Theorem 7.1 for the hemisphere S: $x^2 + y^2 + z^2 = 4$, $z \geq 0$ oriented by outward pointing normals, if $F(x, y, z) = y^2\,\mathbf{i} + z^2\,\mathbf{j} + x\,\mathbf{k}$.

FIGURE 7.4

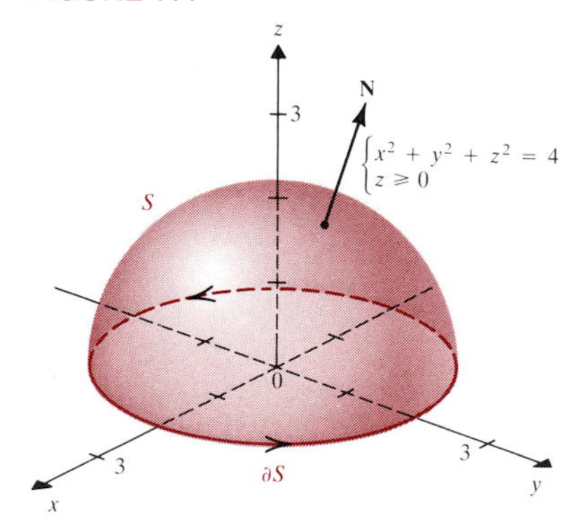

Solution. The right-hand rule says that the positive direction along ∂S is the one indicated in **Figure 7.4.** Here ∂S is the circle $x^2 + y^2 = 4$ in the xy-plane, parametrized in the usual counterclockwise sense:

$$x = 2\cos t, \qquad y = 2\sin t, \qquad z = 0, \qquad t \in [0, 2\pi].$$

Then

$$dx = -2\sin t\,dt, \qquad dy = 2\cos t\,dt, \qquad dz = 0,$$

and on ∂S

$$\mathbf{F}(\mathbf{x}(t)) = 4\sin^2 t\,\mathbf{i} + 0\,\mathbf{j} + 2\cos t\,\mathbf{k} = 2(2\sin^2 t\,\mathbf{i} + \cos t\,\mathbf{j}).$$

Thus

$$\begin{aligned}\int_{\partial S} \mathbf{F} \cdot d\mathbf{x} &= 2\int_0^{2\pi} [2\sin^2 t\,\mathbf{i} + \cos t\,\mathbf{k}] \cdot [-2\sin t\,\mathbf{i} + 2\cos t\,\mathbf{j}]\,dt \\ &= -8\int_0^{2\pi} \sin^3 t\,dt = -8\int_0^{2\pi} (1 - \cos^2 t)\sin t\,dt \\ &= -8\left[-\cos t + \frac{\cos^3 t}{3}\right]_0^{2\pi} = 0.\end{aligned}$$

We must show that this is the surface integral of $\mathbf{curl\,F}$ over S, where

$$\mathbf{curl\,F} = \det\begin{pmatrix} \mathbf{i} & \mathbf{j} & \mathbf{k} \\ \dfrac{\partial}{\partial x} & \dfrac{\partial}{\partial y} & \dfrac{\partial}{\partial z} \\ y^2 & z^2 & x \end{pmatrix} = -2z\,\mathbf{i} - \mathbf{j} - 2y\,\mathbf{k}.$$

We can parametrize S by letting

$$x = 2\sin u\cos v, \quad y = 2\sin u\sin v, \quad z = 2\cos u, \quad u \in [0, \pi/2], v \in [0, 2\pi].$$

Then

$$\mathbf{X}_u = 2\cos u\cos v\,\mathbf{i} + 2\cos u\sin v\,\mathbf{j} - 2\sin u\,\mathbf{k},$$

and

$$\mathbf{X}_v = -2\sin u\sin v\,\mathbf{i} + 2\sin u\cos v\,\mathbf{j} + 0\,\mathbf{k}.$$

Therefore

$$\begin{aligned}\mathbf{X}_u \times \mathbf{X}_v &= \det\begin{pmatrix}\mathbf{i} & \mathbf{j} & \mathbf{k}\\ 2\cos u\cos v & 2\cos u\sin v & -2\sin u\\ -2\sin u\sin v & 2\sin u\cos v & 0\end{pmatrix}\\ &= 4\sin^2 u\cos v\,\mathbf{i} + 4\sin^2 u\sin v\,\mathbf{j}\\ &\quad + [4\sin u\cos u\cos^2 v + 4\sin u\cos u\sin^2 v]\,\mathbf{k}\\ &= 4\sin^2 u\cos v\,\mathbf{i} + 4\sin^2 u\sin v\,\mathbf{j} + 4\sin u\cos u\,\mathbf{k}.\end{aligned}$$

Thus $\mathbf{N} = (\mathbf{X}_u \times \mathbf{X}_v)/|\mathbf{X}_u \times \mathbf{X}_v|$ does point outward since $u \in [0, \pi/2]$, and the outward direction is upward when $u \le \pi/2$. Also,

$$\mathbf{curl}\,\mathbf{F}(\mathbf{X}(u, v)) = -2z\,\mathbf{i} + \mathbf{j} - 2y\,\mathbf{k} = -4\cos u\,\mathbf{i} - \mathbf{j} - 4\sin u\sin v\,\mathbf{k}.$$

Thus

$$\begin{aligned}\mathbf{curl}\,\mathbf{F}(\mathbf{X}(u, v)) \cdot (\mathbf{X}_u \times \mathbf{X}_v) &= -16\sin^2 u\cos u\cos v - 4\sin^2 u\sin v\\ &\quad - 16\sin^2 u\cos u\sin v\\ &= -4\sin^2 u\sin v - 16\sin^2 u\cos u(\cos v + \sin v).\end{aligned}$$

We therefore find

$$\begin{aligned}\iint_S \mathbf{curl}\,\mathbf{F} \cdot d\mathbf{S} &= \iint_D (-4\sin^2 u\sin v - 16\sin^2 u\cos u\,[\cos v + \sin v])\,dA\\ &= -4\int_0^{\pi/2} \sin^2 u\,du \int_0^{2\pi} \sin v\,dv\\ &\quad - 16\int_0^{\pi/2} \sin^2 u\cos u\,du \int_0^{2\pi} (\cos v + \sin v)\,dv\\ &= 0 + 0 = 0.\end{aligned}$$

In this case then, we do have

$$\iint_S \mathbf{curl}\,\mathbf{F} \cdot d\mathbf{S} = \int_{\partial S} \mathbf{F} \cdot d\mathbf{x}. \quad ■$$

As you might expect, Theorem 7.1 can be extended to piecewise smooth orientable surfaces S. We leave the proof for advanced calculus.

7.2 THEOREM

Let $S = S_1 \cup S_2 \cup \ldots \cup S_k$ be a piecewise smooth oriented surface. Let $\mathbf{F} = P\,\mathbf{i} + Q\,\mathbf{j} + R\,\mathbf{k}$ have continuous partial derivatives on S. Let ∂S be positively oriented, piecewise smooth, simple, and closed. Then

$$\iint_S \mathbf{curl}\,\mathbf{F} \cdot \mathbf{N}\,dA = \oint_{\partial S} \mathbf{F} \cdot d\mathbf{x}. \tag{2}$$

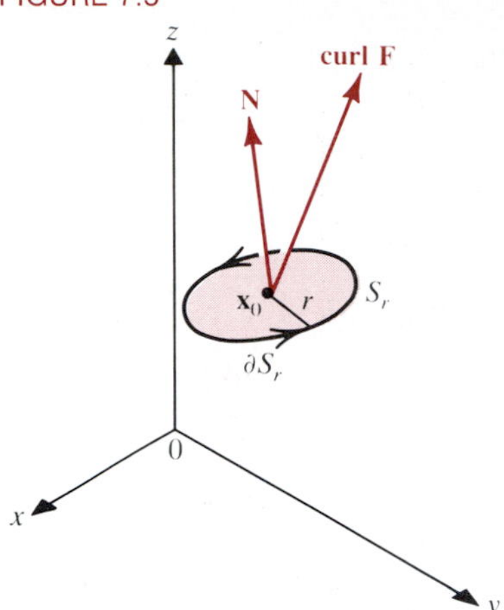

FIGURE 7.5

The remarks on p. 929 about circulation in a fluid flow in $\mathbf{R}^2$ apply equally well to fluid flows in $\mathbf{R}^3$. As before, $\oint_{\partial S} \mathbf{F} \cdot d\mathbf{x}$ is the circulation around ∂S. For a flow in $\mathbf{R}^3$ whose velocity field $\mathbf{F}$ satisfies the hypotheses of Theorems 7.1 or 7.2, the circulation around ∂S is given by $\iint_S \mathbf{curl}\,\mathbf{F} \cdot \mathbf{N}\, dA$. In particular, if S_r is a closed disk of radius r centered at $\mathbf{x}_0$, then the circulation around ∂S_r is $\iint_{S_r} \mathbf{curl}\,\mathbf{F} \cdot \mathbf{N}\, dA$, where $\mathbf{N}$ is perpendicular to the disk. (See **Figure 7.5.**) If r is small, then the surface integral will be approximately $(\mathbf{curl}\,\mathbf{F}(\mathbf{x}_0) \cdot \mathbf{N})A(S_r)$. Hence, among all axes $\mathbf{N}$ of rotation, $\mathbf{curl}\,\mathbf{F}(\mathbf{x}_0)$ is the one that maximizes the circulation around small circles centered at $\mathbf{x}_0$ and perpendicular to $\mathbf{N}$. Physically, a paddle wheel inserted into the flow at $\mathbf{x}_0$ will revolve most rapidly counterclockwise if it is placed with its axle in the direction of $\mathbf{curl}\,\mathbf{F}(\mathbf{x}_0)$. See **Figure 7.6.** If $\mathbf{curl}\,\mathbf{F}(\mathbf{x}_0) = \mathbf{0}$, then the wheel will not revolve at all; this is the origin of the term *irrotational* for flows such that $\mathbf{curl}\,\mathbf{F} = \mathbf{0}$.

Theorems 7.1 and 7.2 can save considerable labor in evaluating surface integrals of curls. The next example illustrates this.

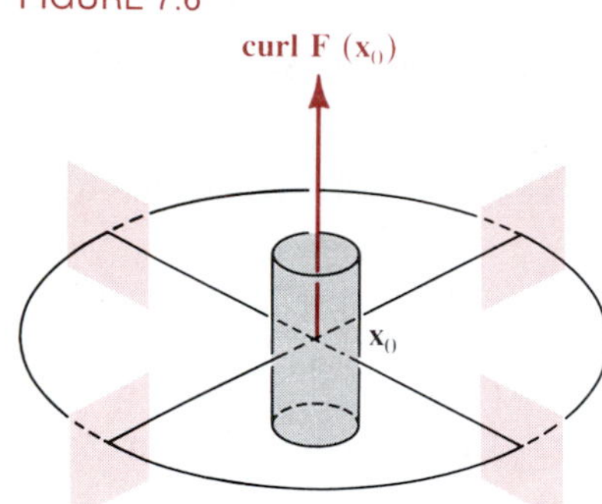

FIGURE 7.6

Example 2

Evaluate $\iint_{S_r} \mathbf{curl}\,\mathbf{F} \cdot \mathbf{N}\, dA$ if $\mathbf{F}(x, y, z) = z\,\mathbf{i} + x\,\mathbf{j} + y\,\mathbf{k}$, and S is the surface $z = x^2 + y^2$ lying below the plane $z = 1$, parametrized so that $\mathbf{N}$ points outward from S.

Solution. The surface can be parametrized in a natural way by letting

$$\mathbf{X}(u, v) = (x, y, z) = (u, v, u^2 + v^2),$$

where D is the disk $0 \le u^2 + v^2 \le 1$. Unfortunately, this natural parametrization *fails* to make $\mathbf{N}$ point outward from S. For we have

$$\mathbf{x}_u = \mathbf{i} + 0\,\mathbf{j} + 2u\,\mathbf{k} \qquad \text{and} \qquad \mathbf{X}_v = 0\,\mathbf{i} + \mathbf{j} + 2v\,\mathbf{k}.$$

Thus

$$\mathbf{X}_u \times \mathbf{X}_v = \det\begin{pmatrix} \mathbf{i} & \mathbf{j} & \mathbf{k} \\ 1 & 0 & 2u \\ 0 & 1 & 2v \end{pmatrix} = -2u\,\mathbf{i} - 2v\,\mathbf{j} + \mathbf{k},$$

which points toward the interior of S, not outward. This can be remedied by interchanging u and v in the parametrization:

$$\mathbf{X}(u, v) = (x, y, z) = (v, u, u^2 + v^2).$$

Then Theorem 7.1 applies to give

$$\iint_S \mathbf{curl}\,\mathbf{F} \cdot \mathbf{N}\, dA = \oint_{\partial S} \mathbf{F} \cdot d\mathbf{x} = \oint_{\partial S} (z, x, y) \cdot d\mathbf{x}, \tag{2}$$

where ∂S is the circle C: $x^2 + y^2 = 1, z = 1$. By the right-hand rule, ∂S is traversed clockwise when viewed from above. (See **Figure 7.7.**)

The circle is parametrized by

$$\mathbf{x}(t) = (x, y, z) = (\cos t, -\sin t, 1) \quad \text{for} \quad 0 \le t \le 2\pi.$$

Then

$$\mathbf{x}'(t) = (-\sin t, -\cos t, 0).$$

FIGURE 7.7

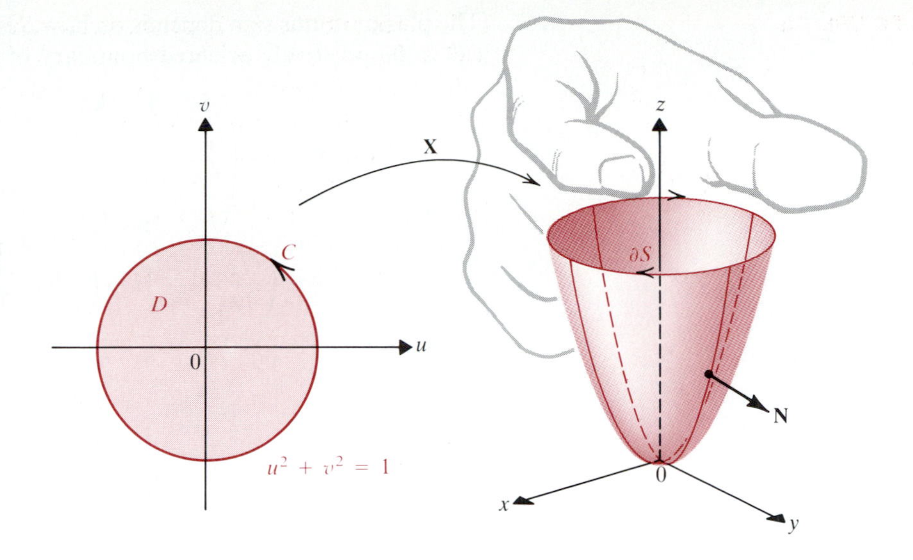

Hence from (2),

$$\iint_S \mathbf{curl}\,\mathbf{F}\cdot\mathbf{N}\,dA = \int_0^{2\pi}(1, -\cos t, \sin t)\cdot(-\sin t, -\cos t, 0)\,dt$$

$$= \int_0^{2\pi}(-\sin t - \cos^2 t)\,dt = \cos t\Big]_0^{2\pi} + \int_0^{2\pi}\frac{1-\cos 2t}{2}\,dt$$

$$= 0 + -\frac{1}{2}\left[t + -\frac{1}{2}\sin 2t\right]_0^{2\pi} = -\frac{1}{2}\cdot 2\pi = -\pi. \quad ■$$

A second look at Example 2 reveals an analogue of independence-of-path for line integrals, namely, the given surface integral is *independent of surface* in the following sense. If S is any surface so parametrized that its positively oriented boundary is ∂S with counterclockwise orientation when viewed from above, then

$$\iint_S \mathbf{curl}\,\mathbf{F}\cdot\mathbf{N}\,dA = -\pi.$$

This is a direct consequence of Stokes's theorem. *The surface integral depends only on ∂S, not on the surface S it bounds, provided that S is parametrized so that ∂S is its positively oriented boundary.*

Speaking of independence of path, we can now establish the three-dimensional analogue of Theorem 4.5, as promised on p. 940.

7.3 THEOREM

Suppose that $\mathbf{F}: \mathbf{R}^3 \to \mathbf{R}^3$ and that $\mathbf{J}_\mathbf{F}(\mathbf{x})$ has continuous entries and is symmetric on a simply connected region $D \subseteq \mathbf{R}^3$. Then $\mathbf{F}$ is a conservative vector field on D.

Proof. Let C be any planar simple closed curve in D. By the definition of a simply connected region, C bounds a surface path $S \subseteq D$. Then by (2) we have

$$\oint_C \mathbf{F}\cdot d\mathbf{x} = \pm\iint_S (\mathbf{curl}\,\mathbf{F}\cdot\mathbf{N})\,dA. \tag{3}$$

(The plus or minus sign depends on how S is parametrized. The plus sign applies if C is the positively oriented boundary of S.) If $\mathbf{F} = f_1\,\mathbf{i} + f_2\,\mathbf{j} + f_3\,\mathbf{k}$, then

$$\mathbf{curl\,F} = \det\begin{pmatrix} \mathbf{i} & \mathbf{j} & \mathbf{k} \\ \dfrac{\partial}{\partial x} & \dfrac{\partial}{\partial y} & \dfrac{\partial}{\partial z} \\ f_1 & f_2 & f_3 \end{pmatrix}$$

$$= \left(\frac{\partial f_3}{\partial y} - \frac{\partial f_2}{\partial z}\right)\mathbf{i} + \left(\frac{\partial f_1}{\partial z} - \frac{\partial f_3}{\partial x}\right)\mathbf{j} + \left(\frac{\partial f_2}{\partial x} - \frac{\partial f_1}{\partial y}\right)\mathbf{k} = \mathbf{0},$$

because

$$\mathbf{J_F} = \begin{pmatrix} \dfrac{\partial f_1}{\partial x} & \dfrac{\partial f_1}{\partial y} & \dfrac{\partial f_1}{\partial z} \\ \dfrac{\partial f_2}{\partial x} & \dfrac{\partial f_2}{\partial y} & \dfrac{\partial f_2}{\partial z} \\ \dfrac{\partial f_3}{\partial x} & \dfrac{\partial f_3}{\partial y} & \dfrac{\partial f_3}{\partial z} \end{pmatrix}$$

is symmetric. Now (3) says that

$$\oint_C \mathbf{F}\cdot d\mathbf{x} = \pm\iint_S \mathbf{0}\cdot\mathbf{N}\,dA = 0.$$

So by Theorem 4.2, $\mathbf{F}$ is conservative. QED

We can often save work by using the approach of the foregoing proof in computational examples. When given a line integral

$$\oint_C \mathbf{F}\cdot d\mathbf{x}$$

to calculate over a closed path C, first compute $\mathbf{curl\,F}$. If $\mathbf{curl\,F} = \mathbf{0}$ (for instance, if $\mathbf{F}$ is conservative and $\mathbf{J_F}$ has continuous entries [cf. Theorem 1.9(d)]), then (2) gives

$$\oint_C \mathbf{F}\cdot d\mathbf{x} = \iint_S \mathbf{0}\cdot d\mathbf{S} = 0$$

for any smooth surface S that C bounds. There is usually no quicker way of computing the answer in this case. Even if $\mathbf{curl\,F}$ *isn't* $\mathbf{0}$, it still might be easier to use (2) than to compute the line integral directly (cf. Example 4 below). So the effort expended in computing $\mathbf{curl\,F}$ is justified by the potential gain.

Theorems 7.1 and 7.2 are important in electrical theory. The ***current density vector*** $\mathbf{J}$ gives the magnitude and direction of a flow of electric charges whose motion constitutes an electrical current. The total current I flowing across a surface $\mathbf{S}$ is defined to be

(4) $$I = \iint_S \mathbf{J}\cdot d\mathbf{S}.$$

The ***magnetic field intensity vector*** $\mathbf{H}$ of a steady magnetic field is related to $\mathbf{J}$ by *Maxwell's second law*

(5) $$\mathbf{J} = \mathbf{curl\,H}.$$

We can use (5) to derive a basic law formulated by André M. Ampère (1775–1836).

Example 3

Establish *Ampère's circuital law,*

$$\oint_C \mathbf{H} \cdot d\mathbf{x} = I, \tag{6}$$

where C is a closed path in the plane traversed counterclockwise.

Solution. Let S be any piecewise smooth, oriented surface for which C is the positively oriented boundary. Then the theorem of Stokes gives

$$\int_C \mathbf{H} \cdot d\mathbf{x} = \iint_S \mathbf{curl}\,\mathbf{H} \cdot d\mathbf{S} = \iint_S \mathbf{J} \cdot d\mathbf{S} \qquad \text{by (5)}$$

$$= I. \qquad \text{by (4)}$$

[Actually (5) and (6) are equivalent. One can also derive (5) from (6).*] ■

The theorem of Stokes is ordinarily a tool for reducing computation of surface integrals to line integrals—which are usually easier than surface integrals to evaluate—but our final example shows that the theorem can also be useful in evaluating line integrals.

FIGURE 7.8

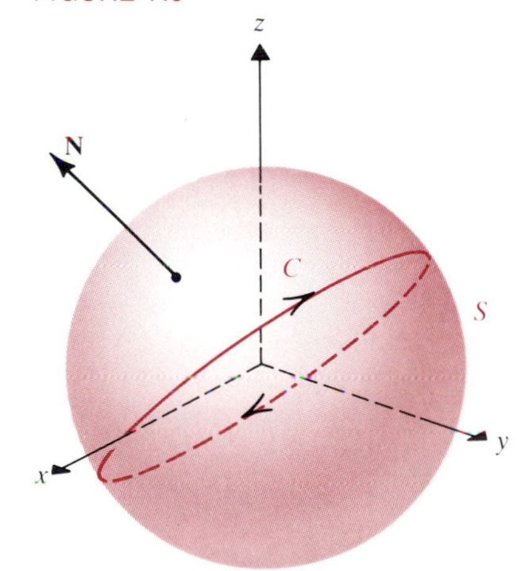

Example 4

Compute the circulation of $\mathbf{F}(x, y, z) = y\,\mathbf{i} + z\,\mathbf{j} + x\,\mathbf{k}$ around the circle C of intersection of the sphere $x^2 + y^2 + z^2 = 1$ with the plane $y = z$ if C is oriented as in **Figure 7.8.**

Solution. Observe that C bounds a disk S, which we can parametrize by

$$X = (x, y, z) = \left(u\cos v, \frac{1}{\sqrt{2}}u\sin v, \frac{1}{\sqrt{2}}u\sin v\right), \qquad \text{for } u \in [0, 1] \text{ and } v \in [0, 2\pi].$$

(We use this parametrization so that $y = z$ and $x^2 + y^2 + z^2 = 1$.) Then we obtain

$$\mathbf{X}_u = \cos v\,\mathbf{i} + \frac{1}{\sqrt{2}}\sin v\,\mathbf{j} + \frac{1}{\sqrt{2}}\sin v\,\mathbf{k}$$

and

$$\mathbf{X}_v = -u\sin v\,\mathbf{i} + \frac{u}{\sqrt{2}}\cos v\,\mathbf{j} + \frac{u}{\sqrt{2}}\cos v\,\mathbf{k}.$$

Hence

$$\mathbf{X}_u \times \mathbf{X}_v = \det\begin{pmatrix} \mathbf{i} & \mathbf{j} & \mathbf{k} \\ \cos v & \frac{1}{\sqrt{2}}\sin v & \frac{1}{\sqrt{2}}\sin v \\ -u\sin v & \frac{u}{\sqrt{2}}\cos v & \frac{u}{\sqrt{2}}\cos v \end{pmatrix} = 0\,\mathbf{i} - \frac{u}{\sqrt{2}}\mathbf{j} + \frac{u}{\sqrt{2}}\mathbf{k}.$$

* See, for example, W. H. Hayt, Jr., *Engineering Electromagnetics*, 3rd Ed. (New York: McGraw-Hill, 1974), p. 255.

Thus $\mathbf{N}$ points outward from S. Since

$$\mathbf{curl\,F} = \det\begin{pmatrix} \mathbf{i} & \mathbf{j} & \mathbf{k} \\ \dfrac{\partial}{\partial x} & \dfrac{\partial}{\partial y} & \dfrac{\partial}{\partial z} \\ y & z & x \end{pmatrix} = -\mathbf{i} - \mathbf{j} - \mathbf{k},$$

Theorem 7.1 says that

$$\oint_C \mathbf{F}\cdot d\mathbf{x} = \iint_S \mathbf{curl\,F}\cdot\mathbf{N}\,dA = \iint_S [-\mathbf{i}-\mathbf{j}-\mathbf{k}]\cdot\left[0\,\mathbf{i} - \frac{1}{\sqrt{2}}\mathbf{j} + \frac{1}{\sqrt{2}}\mathbf{k}\right]u\,du\,dv$$

$$= \iint_S 0\,du\,dv = 0. \quad ■$$

HISTORICAL NOTE

Theorem 7.1 is named for the Irish mathematician and physicist George Gabriel Stokes (1819–1903), who published it in 1854. It was known somewhat earlier to the great English chemist and physicist William Thomson, better known as Lord Kelvin (1824–1907), the designer of the absolute temperature scale. Stokes himself may have learned of the result from Kelvin, his colleague at Cambridge University, where Stokes held a chair in mathematics from 1847 until his retirement. Queen Victoria knighted him in 1889 in recognition of his outstanding scholarship. Stokes played a leading role in the great nineteenth-century advances in physics that came from newly created mathematics. His contributions to hydrodynamics and gravitation were especially noteworthy. In mathematics, Theorem 7.1 pointed the way to a far-reaching generalization of Green's theorem to n dimensions, which is one of the most important results in advanced calculus of several variables.

Lord Kelvin, the son of a mathematics professor at the University of Glasgow, was another major figure in nineteenth-century science. He formulated the second law of thermodynamics: The amount of usable energy in the universe is constantly decreasing; entropy is increasing. He was also an inventor of note, the holder of more than 50 patents. He was raised to the peerage by Queen Victoria in 1892 as Baron Kelvin of Largs, after having been knighted in 1866.

James Clerk Maxwell (1831–1879) was a Scottish mathematician and physicist who published *Treatise on Electricity and Magnetism* in 1873, a work that brought together all the then-known facts about electricity and magnetism. He formulated four basic equations that still carry his name and are fundamental to the theory of electricity and magnetism.

Proof of Theorem 7.1*

The proof is easiest to give in the notation (5) of Section 6 (p. 960). Since

$$\mathbf{curl\,F} = \left(\frac{\partial R}{\partial y} - \frac{\partial Q}{\partial z}, \frac{\partial P}{\partial z} - \frac{\partial R}{\partial x}, \frac{\partial Q}{\partial x} - \frac{\partial P}{\partial y}\right),$$

* Optional

we have in this notation

$$\iint_S \mathbf{curl}\,\mathbf{F}\cdot d\mathbf{S} = \iint_S \left(\frac{\partial R}{\partial y} - \frac{\partial Q}{\partial z}\right) dy \wedge dz + \left(\frac{\partial P}{\partial z} - \frac{\partial R}{\partial x}\right) dz \wedge dx$$
$$+ \left(\frac{\partial Q}{\partial x} - \frac{\partial P}{\partial y}\right) dx \wedge dy.$$

To prove Theorem 7.1, we have to show that

$$\text{(2)} \qquad \oint_{\partial S} \mathbf{F}\cdot d\mathbf{x} = \int_{\partial S} P\,dx + Q\,dy + R\,dz \qquad \text{equals} \qquad \iint_S \mathbf{curl}\,\mathbf{F}\cdot d\mathbf{S}.$$

This will follow if we can prove the following three equalities:

$$\text{(7)} \qquad \int_{\partial S} P\,dx = \iint_S -\frac{\partial P}{\partial y}\,dx \wedge dy + \frac{\partial P}{\partial z}\,dz \wedge dx,$$

$$\text{(8)} \qquad \int_{\partial S} Q\,dy = \iint_S -\frac{\partial Q}{\partial z}\,dy \wedge dz + \frac{\partial Q}{\partial x}\,dx \wedge dy,$$

and

$$\text{(9)} \qquad \int_{\partial S} R\,dz = \iint_S -\frac{\partial R}{\partial x}\,dz \wedge dx + \frac{\partial R}{\partial y}\,dy \wedge dz.$$

Each of these is proved by applying Green's theorem. We prove (7) and leave the similar proofs of (8) and (9) as Exercises 24 and 25. Suppose that C is parametrized by

$$(x, y) = \mathbf{u}(t) = (u(t), v(t)), \qquad \text{for } t \in [a, b].$$

Then ∂S is parametrized by

$$\mathbf{X}(u(t)) = (x(\mathbf{u}(t)),\ y(\mathbf{u}(t)),\ z(\mathbf{u}(t))).$$

We therefore have

$$\begin{aligned}
\oint_{\partial S} P\,dx &= \int_a^b (P(\mathbf{X}(\mathbf{u}(t)), 0, 0)) \cdot \mathbf{X}'(\mathbf{u}(t))\,dt \\
&= \int_a^b P(\mathbf{X}(u(t), v(t)))\,\frac{dx(\mathbf{u}(t))}{dt}\,dt \\
&= \int_a^b P(\mathbf{X}(u(t), v(t)))\left(\frac{\partial x}{\partial u}\frac{du}{dt} + \frac{\partial x}{\partial v}\frac{dv}{dt}\right) dt && \textit{by Chain Rule 4.2 of Chapter 12} \\
&= \oint_C (P \circ \mathbf{X})(u, v)\,\frac{\partial x}{\partial u}\,du + (P \circ \mathbf{X})(u, v)\,\frac{\partial x}{\partial v}\,dv \\
\text{(10)} \qquad &= \iint_D \left[\frac{\partial}{\partial u}\left(P \circ \mathbf{X}(u, v)\,\frac{\partial x}{\partial v}\right) - \frac{\partial}{\partial v}\left(P \circ \mathbf{X}(u, v)\,\frac{\partial x}{\partial u}\right)\right] du\,dv. && \textit{by Green's theorem}
\end{aligned}$$

[Note that $(x, y) = (u(t), v(t))$, so

$$(dx, dy) = \left(\frac{du}{dt}\,dt, \frac{dv}{dt}\,dt\right) = (du, dv)\,.]$$

We are now reduced to calculating the bracket in the double integral. By the product rule for derivatives and the chain rule, we have

$$\text{(11)}\quad \frac{\partial}{\partial u}\left(P \circ \mathbf{X}(u, v)\frac{\partial x}{\partial v}\right) = \frac{\partial}{\partial u}(P \circ \mathbf{X})\frac{\partial x}{\partial v} + (P \circ \mathbf{X})\frac{\partial^2 x}{\partial u\,\partial v}$$

$$= \left(\frac{\partial P}{\partial x}\frac{\partial x}{\partial u} + \frac{\partial P}{\partial y}\frac{\partial y}{\partial u} + \frac{\partial P}{\partial z}\frac{\partial z}{\partial u}\right)\frac{\partial x}{\partial v} + (P \circ \mathbf{X})\frac{\partial^2 x}{\partial u\,\partial v},$$

$$\text{(12)}\quad \frac{\partial}{\partial v}(P \circ \mathbf{X}(u, v))\frac{\partial x}{\partial u} = \frac{\partial}{\partial v}(P \circ \mathbf{X})\frac{\partial x}{\partial u} + (P \circ \mathbf{X})\frac{\partial^2 x}{\partial v\,\partial u}$$

$$= \left(\frac{\partial P}{\partial x}\frac{\partial x}{\partial v} + \frac{\partial P}{\partial y}\frac{\partial y}{\partial v} + \frac{\partial P}{\partial z}\frac{\partial z}{\partial v}\right)\frac{\partial x}{\partial u} + (P \circ \mathbf{X})\frac{\partial^2 x}{\partial v\,\partial u}.$$

The hypotheses say that $\partial^2 x/\partial u\,\partial v$ and $\partial^2 x/\partial v\,\partial u$ are continuous, so they are equal by Theorem 6.2 of Chapter 12. Then subtraction of (12) from (11) gives

$$\frac{\partial}{\partial u}\left(P \circ \mathbf{X}(u, v)\frac{\partial x}{\partial v}\right) - \frac{\partial}{\partial v}\left(P \circ \mathbf{X}(u, v)\frac{\partial x}{\partial u}\right) = \frac{\partial P}{\partial y}\left(\frac{\partial y}{\partial u}\frac{\partial x}{\partial v} - \frac{\partial y}{\partial v}\frac{\partial x}{\partial u}\right)$$

$$+ \frac{\partial P}{\partial z}\left(\frac{\partial z}{\partial u}\frac{\partial x}{\partial v} - \frac{\partial z}{\partial v}\frac{\partial x}{\partial u}\right)$$

$$= -\frac{\partial P}{\partial y}\frac{\partial(x, y)}{\partial(u, v)} + \frac{\partial P}{\partial z}\frac{\partial(z, x)}{\partial(u, v)}.$$

So, in substituting above in (10), we obtain

$$\oint_{\partial S} P\,dx = \iint_D \left[-\frac{\partial P}{\partial y}\frac{\partial(x, y)}{\partial(u, v)} + \frac{\partial P}{\partial z}\frac{\partial(z, x)}{\partial(u, v)}\right] du\,dv$$

$$= \iint_D -\frac{\partial P}{\partial y}\,dx \wedge dy + \frac{\partial P}{\partial z}\,dz \wedge dx.$$

Hence (7) is proved. The same approach establishes (8) and (9). Then (2) follows by adding (7), (8), and (9). QED

Exercises 14.7

In Exercises 1–16, apply the theorem of Stokes to evaluate the given surface or line integral. ***Parametrize so that* N *points outward.***

1. $\oint_C 2xy^2z\,dx + 2x^2yz\,dy + (x^2y^2 - 2z)\,dz$, where C is the curve given by $\mathbf{x}(t) = (\cos t, \sin t, \sin t)$ for $0 \le t \le 2\pi$

2. $\oint_C (6x^2y^2 - 3yz^2)\,dx + (4x^3y - 3xz^2)\,dy - 6xyz\,dz$, where C is the curve of intersection of the surface $z = x^2 + y^2$ with the plane $z = 2 + x + y$

3. $\iint_S \mathbf{curl}\,\mathbf{F} \cdot d\mathbf{S}$, where S is the portion of the paraboloid $z = 4 - x^2 - y^2$ above the xy-plane, and $\mathbf{F}(x, y, z) = (-y + z)\mathbf{i} + (x + z)\mathbf{j} + (x - y)\mathbf{k}$

4. $\iint_S \mathbf{curl}\,\mathbf{F} \cdot d\mathbf{S}$, where $\mathbf{F}(x, y, z) = (y - z)\mathbf{i} + yz\,\mathbf{j} - xz\,\mathbf{k}$, and S is the portion of the unit cube $[0, 1] \times [0, 1] \times [0, 1]$ above the xy-plane, with outward-pointing normal

5. $\oint_C \mathbf{F} \cdot d\mathbf{x}$, where

$$\mathbf{F}(x, y, z) = \frac{x}{\sqrt{x^2 + y^2 + z^2}}\mathbf{i} + \frac{y}{\sqrt{x^2 + y^2 + z^2}}\mathbf{j} + \frac{z}{\sqrt{x^2 + y^2 + z^2}}\mathbf{k},$$

and C is the curve of intersection of the sphere $(x - 5)^2 + (y - 2)^2 + (z - 3)^2 = 1$ with the plane $3x - 5y - z = 2$

6. $\oint_C \mathbf{F} \cdot d\mathbf{x}$, where $\mathbf{F}$ is as in Exercise 5, and C is the curve of intersection of the sphere $(x - 3)^2 + (y - 3)^2 + (z - 2)^2 = 1$ with the plane $2x - y - z = 1$

7. $\iint_S \mathbf{curl}\,\mathbf{F} \cdot \mathbf{N}\,dA$, where $\mathbf{F}(x, y, z) = y\,\mathbf{i} + 2z\,\mathbf{j} + 3x\,\mathbf{k}$, and S is the hemisphere $z = \sqrt{1 - x^2 - y^2}$

8. $\iint_S \textbf{curl}\,\mathbf{F} \cdot \mathbf{N}\,dA$, where $\mathbf{F}(x, y, z) = y^2\,\mathbf{i} + x\,\mathbf{j} - xz\,\mathbf{k}$, and S is the part of $z = 1 - x^2 - y^2$ lying above the xy-plane

9. $\int_C \mathbf{F} \cdot d\mathbf{x}$, where $\mathbf{F}(x, y, z) = z\,\mathbf{i}$, and C is the curve of intersection of the plane $z = x + y$ with the cylinder $x^2 + y^2 = 1$, traversed counterclockwise in the plane $z = x + y$ when viewed from above

10. $\int_C z\,dx + x\,dy + y\,dz$, where C is the circle of intersection of the plane $x + y + z = 0$ and $x^2 + y^2 + z^2 = 1$, traversed counterclockwise as viewed from above the plane $x + y + z = 0$

11. $\iint_S \textbf{curl}\,\mathbf{F} \cdot d\mathbf{S}$, where $\mathbf{F}(x, y, z) = y\,\mathbf{i} + z\,\mathbf{j} + x\,\mathbf{k}$, and S is the portion of the cylinder $x^2 + y^2 = 1$ cut off by the plane $z = x + 3$ and the xy-plane

12. $\iint_S \textbf{curl}\,\mathbf{F} \cdot d\mathbf{S}$, where $\mathbf{F}(x, y, z) = y^2\,\mathbf{i} + xy\,\mathbf{j} - 2xz\,\mathbf{k}$, and S is the hemisphere $z = \sqrt{1 - x^2 - y^2}$

13. $\int_{\partial S} \mathbf{F} \cdot d\mathbf{x}$, where $\mathbf{F}(x, y, z) = e^x \sin y\,\mathbf{i} + (e^x \cos y - z)\,\mathbf{j} + y\,\mathbf{k}$ and $\partial S = C_1 \cup C_2$, where S is the part of $z = \sqrt{x^2 + y^2}$ between $z = 1$ and $z = 2$

14. $\oint_{\partial S} \mathbf{F} \cdot d\mathbf{x}$, where $\mathbf{F}(x, y, z) = z\,\mathbf{i} - x\,\mathbf{k}$, and S is the part of the cylinder $r = 2\cos\theta$ above the xy-plane and below the cone $z = \sqrt{x^2 + y^2}$

15. $\oint_C \mathbf{F} \cdot d\mathbf{x}$, where $\mathbf{F}(x, y, z) = (y - z)\,\mathbf{i} + (z - x)\,\mathbf{j} + (x - y)\,\mathbf{k}$, and C is the intersection of the cylinder $x^2 + y^2 = 1$ and the plane $x + z = 1$, traversed counterclockwise when viewed from above

16. $\oint_C \mathbf{F} \cdot d\mathbf{x}$, where $\mathbf{F}(x, y, z) = (y^2 + z^2)\,\mathbf{i} + (x^2 + z^2)\,\mathbf{j} + (x^2 + y^2)\,\mathbf{k}$, and C is the intersection of the hemisphere $z = \sqrt{4 - (x - 2)^2 - y^2}$ and the cylinder $x^2 + y^2 = 2x$

17. Use Theorem 7.1 to show that for any vector field $\mathbf{F}: \mathbf{R}^3 \to \mathbf{R}^3$ such that $\mathbf{J}_\mathbf{F}$ has continuous entries $\iint_S \textbf{curl}\,\mathbf{F} \cdot d\mathbf{S} = 0$ if S is the sphere $x^2 + y^2 + z^2 = a^2$. (*Hint:* Cut the sphere into two hemispheres along a great circle C. Apply Theorem 7.1 to each hemisphere and note that the two line integrals over C cancel.)

18. Repeat Exercise 17 for the ellipsoid $x^2/a^2 + y^2/b^2 + z^2/c^2 = 1$.

19. Suppose that a fluid flow has velocity vector $\mathbf{F}(x, y, z) = z\,\mathbf{i} + x\,\mathbf{j} + y\,\mathbf{k}$. Let D_r be a disk centered at $\mathbf{x_0} = (x_0, y_0, z_0)$ of radius r with unit normal vector $\mathbf{N} = a\,\mathbf{i} + b\,\mathbf{j} + c\,\mathbf{k}$. Calculate the circulation of $\mathbf{F}$ around ∂D_r, the positively oriented boundary of D_r.

20. In Exercise 19, maximize your answer by maximizing the expression $a + b + c$ subject to the constraint $a^2 + b^2 + c^2 = 1$ (cf. Section 12.8). Is the result

$$\mathbf{N} = \frac{1}{|\textbf{curl}\,\mathbf{F}(\mathbf{x}_0)|}\,\textbf{curl}\,\mathbf{F}(\mathbf{x}_0)?$$

(It *should* be, according to the discussion following Theorem 7.2. See page 970.)

21. Let

$$\mathbf{F}(x, y, z) = -\frac{y}{x^2 + y^2}\,\mathbf{i} + \frac{x}{x^2 + y^2}\,\mathbf{j} + z\,\mathbf{k}.$$

(a) Show that $\textbf{curl}\,\mathbf{F}(\mathbf{x}) = \mathbf{0}$ for all $\mathbf{x}$ not lying in the z-axis.

(b) Let C be the boundary of the disk $x^2 + z^2 = 1$, $y = 1$ traversed counterclockwise as viewed from the right around the y-axis. Show that

$$\oint_C \mathbf{F} \cdot d\mathbf{x} = 0.$$

22. Refer to Exercise 21. Let C be the boundary of the disk $x^2 + y^2 = 1$, $z = 0$, traversed counterclockwise as viewed from above. Can we conclude from Theorem 7.1 that $\oint_C \mathbf{F} \cdot d\mathbf{x} = 0$? Why or why not?

23. Refer to Example 3. In the case of time-varying magnetic fields, Maxwell's second equation has the form

$$\textbf{curl}\,\mathbf{H} = \frac{\partial \mathbf{D}}{\partial t} + \mathbf{J},$$

where $\mathbf{D}$ is the *electric flux density* (or *displacement density*) vector field. Show that in this situation, Ampère's circuital law becomes

$$\oint_C \mathbf{H} \cdot d\mathbf{x} = I + \iint_S \frac{\partial \mathbf{D}}{\partial t} \cdot d\mathbf{S}.$$

24. Derive Equation (8) in the proof of Theorem 7.1.

25. Derive Equation (9) in the proof of Theorem 7.1.

14.8 Theorem of Gauss

There is one more integration theorem relating the integral of a certain kind of derivative of a vector function $\mathbf{F}: \mathbf{R}^n \to \mathbf{R}^n$ over a region D and the values of $\mathbf{F}$ on its boundary ∂D. This theorem was established by Gauss in connection with his work on electricity and, together with the theorems of Green and Stokes, plays a prominent role in that theory.

In the earlier integration theorems of this sort (Theorems 4.1 and 7.1), the "certain kinds" of derivatives referred to were ∇f and $\nabla \times \mathbf{F}$. As you might expect, in the final theorem the "certain kind of derivative" is the divergence $\nabla \cdot \mathbf{F}$ of the vector field $\mathbf{F}$, and the theorem is also frequently called the *divergence theorem.* Exercise 28 in Section 3 was a preview in two dimensions. It showed how Green's theorem in the plane could be formulated as

$$\iint_D \operatorname{div} \mathbf{F}\, dA = \oint_{\partial D} \mathbf{F} \cdot \mathbf{N}\, ds \tag{1}$$

for a two-dimensional vector field $\mathbf{F} = P\mathbf{i} + Q\mathbf{j}$ whose coordinate functions P and Q have continuous partial derivatives on D. Here $\mathbf{N}$ is a unit normal vector to the smooth curve ∂D. To formulate an analogue of (1) for $\mathbf{F}\colon \mathbf{R}^3 \to \mathbf{R}^3$, suppose that $D \subseteq \mathbf{R}^3$ is a solid region whose boundary is a ***positively oriented*** smooth surface patch S. This means that S is parametrized by $\mathbf{X}\colon E \to \mathbf{R}^3$ for some $E \subseteq \mathbf{R}^2$ in such a way that at each point of S, the standard unit normal vector

$$\mathbf{N} = \frac{\mathbf{X}_u \times \mathbf{X}_v}{|\mathbf{X}_u \times \mathbf{X}_v|}$$

points *outward* or, in case D is flat, upward from D. See **Figure 8.1.** If $\mathbf{F} = P\mathbf{i} + Q\mathbf{j} + R\mathbf{k}$, where P, Q, and R have continuous partial derivatives on D, then the analogue of (1) is

$$\iiint_D \operatorname{div} \mathbf{F}\, dV = \iint_S \mathbf{F} \cdot \mathbf{N}\, dA. \tag{2}$$

FIGURE 8.1

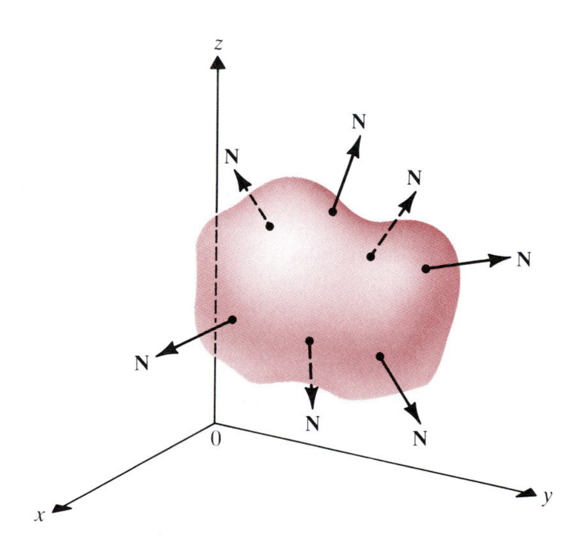

Gauss's theorem says that (2) is correct. It thus may be regarded as a final variation on this chapter's theme that the integral of a derivative of a function $\mathbf{F}$ over a region is determined by the values of $\mathbf{F}$ on the boundary of the region, provided that $\mathbf{F}$, the region, and the boundary satisfy appropriate smoothness and orientation hypotheses. In the present case we will require D to be a *simple region* in $\mathbf{R}^3$ (compare Definition 3.1).

8.1 DEFINITION A region $D \subseteq \mathbf{R}^3$ is ***simple*** if it is the portion of $\mathbf{R}^3$ enclosed by a piecewise smooth orientable surface S that intersects any line parallel to a coordinate axis in either at most two points or one line segment.

Thus balls and solid parallelepipeds are simple regions. The region between two concentric spheres or inside a torus is not simple, however. See **Figure 8.2.** We can now state Gauss's theorem. The proof, which resembles that of Theorem 7.1, is left for advanced calculus.

FIGURE 8.2

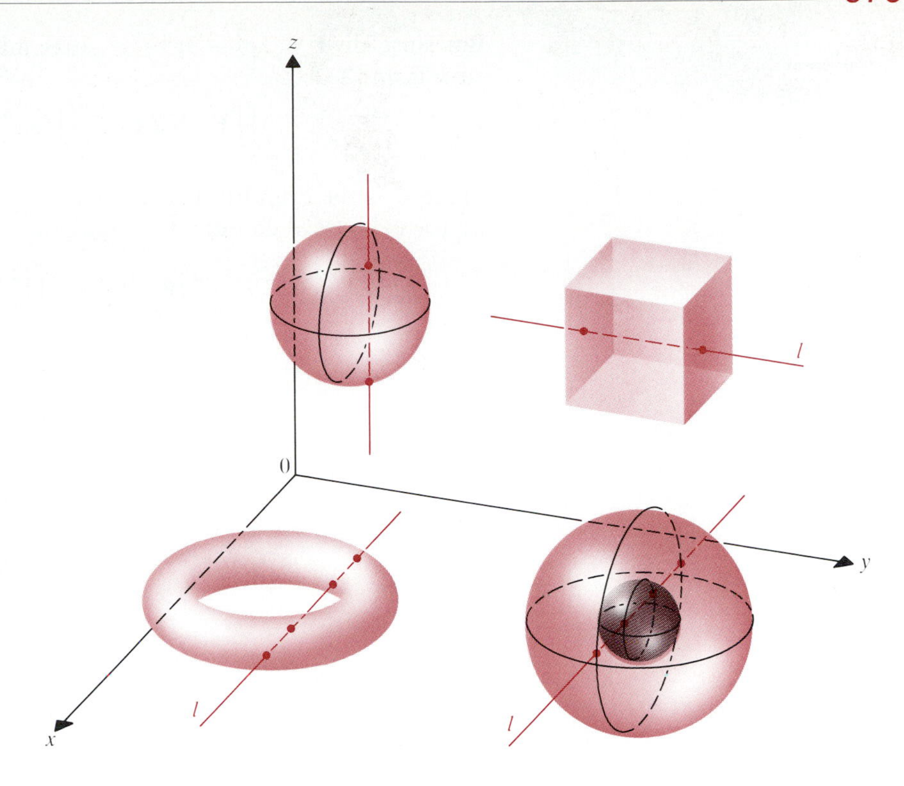

8.2 THEOREM

Theorem of Gauss (Divergence Theorem). Let $D \subseteq \mathbf{R}^3$ be a simple region with positively oriented piecewise smooth boundary S. Suppose that $\mathbf{F} = P\,\mathbf{i} + Q\,\mathbf{j} + R\,\mathbf{k}$, where P, Q, and R have continuous partial derivatives on D and S. Then

$$(2)\qquad \iiint_D \operatorname{div} \mathbf{F}\, dV = \iint_S \mathbf{F} \cdot \mathbf{N}\, dA = \iint_S P\, dy \wedge dz + Q\, dz \wedge dx + R\, dx \wedge dy.$$

Although (2) says that a triple integral over a solid D of the derivative-like quantity $\operatorname{div} \mathbf{F} = \nabla \cdot \mathbf{F}$ is determined by the values of $\mathbf{F}$ over the two-dimensional bounding surface S of D, there is usually no reduction of complexity if we try to evaluate the surface integral $\iint_S \mathbf{F} \cdot \mathbf{N}\, dA$ instead of the triple integral $\iiint_D \operatorname{div} \mathbf{F}\, dV$. Indeed, one of the prime uses of the divergence theorem is to provide a simple means of evaluating surface integrals, as the following example illustrates.

FIGURE 8.3

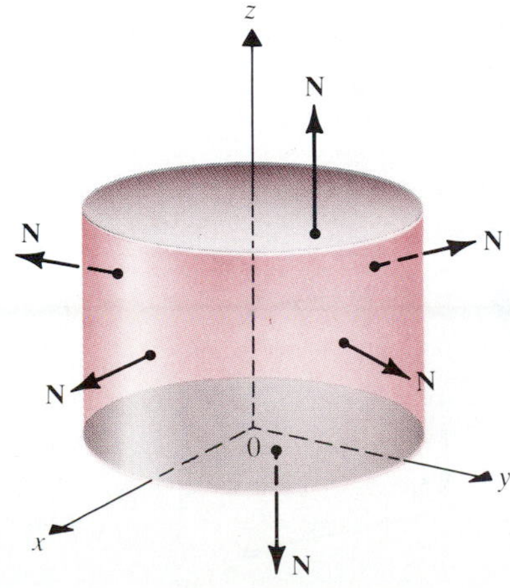

Example 1

If $\mathbf{F}(x, y, z) = x^3\,\mathbf{i} + y^3\,\mathbf{j} - e^z\,\mathbf{k}$, then compute $\iint_S \mathbf{F} \cdot \mathbf{N}\, dA$ if S is the cylinder (including top and bottom) $x^2 + y^2 = 1$, $0 \leq z \leq 1$ with outward-pointing normal. See **Figure 8.3.**

Solution. To do this as a surface integral, we would need to parametrize separately the bottom, top, and cylindrical side of S, evaluate the surface integral over each portion of S, and add the results, as in Example 6 of the last section.

But since $\operatorname{div}\mathbf{F} = 3x^2 + 3y^2 - e^z$, it is much easier to use Theorem 8.1. We have from (2)

$$\iint_S \mathbf{F}\cdot\mathbf{N}\,dA = \iiint_D (3x^2 + 3y^2 - e^z)\,dV,$$

where D is the cylindrical solid enclosed by S. To evaluate the triple integral, we use cylindrical coordinates, obtaining

$$\begin{aligned}
\iiint_D [3(x^2 + y^2) - e^z]\,dx\,dy\,dz &= \iiint_D (3r^2 - e^z)\,dz\,d\theta\,r\,dr \\
&= \int_0^1 \int_0^{2\pi} \int_0^1 r(3r^2 - e^z)\,dz\,d\theta\,dr \\
&= \int_0^1 \int_0^{2\pi} r\Big[3r^2 z - e^z\Big]_{z=0}^{z=1} d\theta\,dr \\
&= \int_0^1 \int_0^{2\pi} r(3r^2 - e + 1)\,d\theta\,dr \\
&= 2\pi \int_0^1 (3r^3 - re + r)\,dr \\
&= 2\pi\left[\frac{3}{4}r^4 - \frac{1}{2}r^2 e + \frac{1}{2}r^2\right]_0^1 \\
&= 2\pi\left[\frac{3}{4} - \frac{1}{2}e + \frac{1}{2}\right] = \pi\left(\frac{5}{2} - e\right). \quad ■
\end{aligned}$$

Just as we were able to extend Green's theorem for simple regions in $\mathbf{R}^2$ to Theorem 3.3 for regions that are decomposable into a finite union of simple regions, so we can extend Theorem 8.2 to more general regions. For this, we consider a region D that is the union $D_1 \cup D_2 \cup \ldots \cup D_k$ of simple regions. On any common boundary surface $S_i \cap S_j$ of two subregions D_i and D_j, the outward-pointing normals will be oppositely directed from each other, as shown in **Figure 8.4.** Then the corresponding surface integrals

$$\iint_{S_i} \mathbf{F}\cdot\mathbf{N}_i\,dS \qquad \text{and} \qquad \iint_{S_j} \mathbf{F}\cdot\mathbf{N}_j\,dS$$

FIGURE 8.4

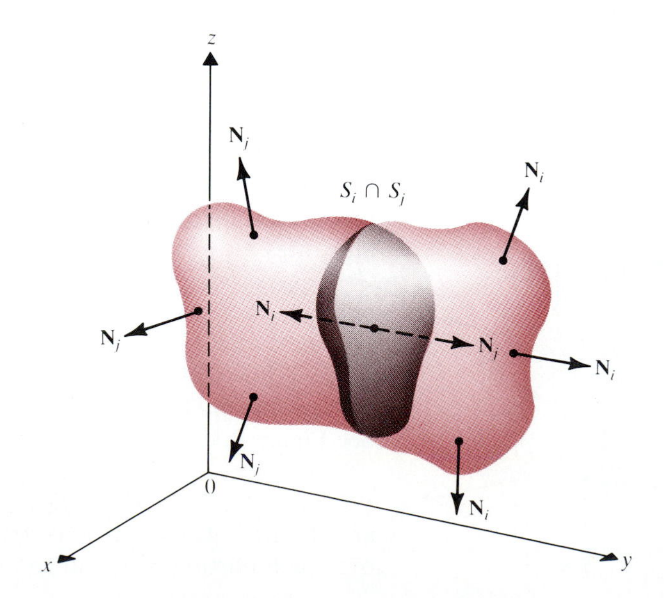

will be negatives of each other. Thus these surface integrals will cancel each other if we compute $\iint_{\cup S_i} \mathbf{F} \cdot \mathbf{N}\, dA$. Hence we have

$$\iiint_D \operatorname{div} \mathbf{F}\, dV = \sum_{i=1}^{k} \iiint_{D_i} \operatorname{div} \mathbf{F}\, dV = \sum_{i=1}^{k} \iint_S \mathbf{F} \cdot \mathbf{N}_i\, dA \qquad \textit{by Theorem 8.2}$$

$$= \iint_S \mathbf{F} \cdot \mathbf{N}\, dA,$$

where S is the union of the *noncoincident* portions of the boundaries of the regions D_i. Thus S is precisely the boundary of $D = D_1 \cup D_2 \cup \ldots \cup D_k$. The following theorem summarizes this discussion.

8.3 THEOREM

Theorem of Gauss. Let D be a finite union of simple regions D_i, each with a positively oriented piecewise smooth boundary S_i. Suppose that $\mathbf{F}: \mathbf{R}^3 \to \mathbf{R}^3$ has coordinate functions with continuous first partial derivatives. Then

$$\iiint_D \operatorname{div} \mathbf{F}\, dV = \iint_S \mathbf{F} \cdot \mathbf{N}\, dA, \tag{2}$$

where $S = \partial D$ is the union of the portions of the S_i not common to any two D_i, oriented consistently with the positive orientation of the S_i.

We illustrate Theorem 8.3 with an example that shows more of the potential in Gauss's theorem for simplification of surface integrals.

Example 2

If $\mathbf{F}(x, y, z) = (y \cos z - 3x)\mathbf{i} + (5y - e^{xz})\mathbf{j} + (z - \sqrt{x^2 + y^2})\mathbf{k}$, then compute $\iint_S \mathbf{F} \cdot \mathbf{N}\, dA$, where S is the boundary of the region D in **Figure 8.5:** D contains

FIGURE 8.5

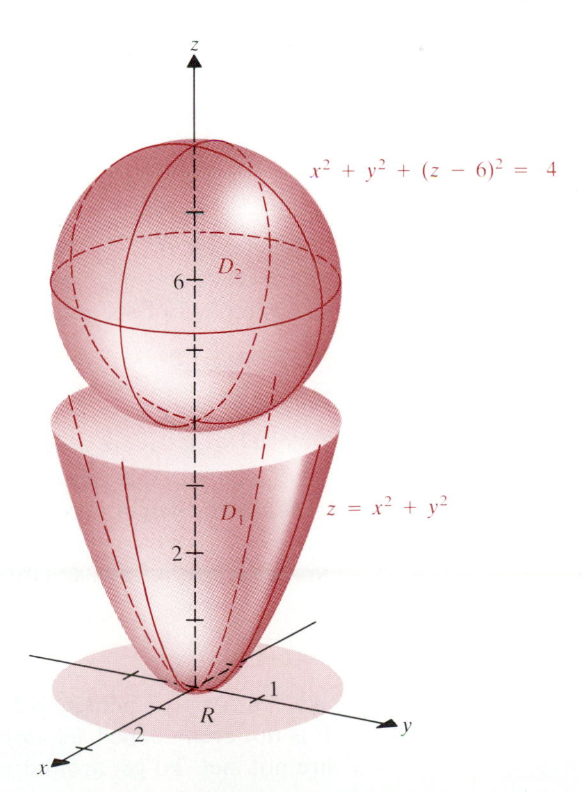

the points inside the sphere $x^2 + y^2 + (z - 6)^2 = 4$, together with those above the paraboloid $z = x^2 + y^2$ and below the plane $z = 4$.

Solution. The nonsimple region D is the union $D_1 \cup D_2$ of two simple regions, where D_1 is the lower portion (lying above $z = x^2 + y^2$ and below $z = 4$) and D_2 is the upper portion (inside the sphere). Since

$$\begin{aligned} \operatorname{div} \mathbf{F} &= \frac{\partial}{\partial x}(y \cos z - 3x) + \frac{\partial}{\partial y}(5y - e^{xz}) + \frac{\partial}{\partial z}(z - \sqrt{x^2 + y^2}) \\ &= -3 + 5 + 1 = 3, \end{aligned}$$

Theorem 8.3 gives

$$\iint_S \mathbf{F} \cdot d\mathbf{S} = \iiint_D 3\,dV = 3V(D) = 3V(D_1) + 3V(D_2).$$

Now the volume of the ball D_2 of radius 2 is

$$V(D_2) = \frac{4}{3}\pi \cdot 2^3 = \frac{32}{3}\pi.$$

Since D_1 lies below $z = 4$ and above $z = x^2 + y^2$, we have

$$V(D_1) = \iiint_{D_1} 1\,dV = \iint_R \left(\int_{x^2+y^2}^{4} dz \right) dA,$$

where R is the region of the xy-plane lying inside the circle $x^2 + y^2 = 4$. Thus, changing to cylindrical coordinates, we obain

$$\begin{aligned} V(D_1) &= \int_0^{2\pi} \int_0^2 \int_{r^2}^4 dz\, r\, dr\, d\theta = \int_0^{2\pi} \int_0^2 (4r - r^3)\, dr\, d\theta \\ &= 2\pi \left[2r^2 - \frac{1}{4} r^4 \right]_0^2 = 2\pi[8 - 4] = 8\pi. \end{aligned}$$

Therefore

$$\iint_S \mathbf{F} \cdot d\mathbf{S} = 3V(D_1) + 3V(D_2) = 3 \cdot 8\pi + 3 \cdot \frac{32}{3}\pi = 56\pi. \quad ■$$

There are many important applications of Gauss's theorem. The next example illustrates how they arise. The technique employed is reminiscent of that in Example 5 of Section 3.

Example 3

If $\mathbf{F}: \mathbf{R}^3 \to \mathbf{R}^3$ is an inverse-square law field (Example 4 of Section 1), then show that the flux across the boundary S of any simple region $D \subseteq \mathbf{R}^3$ that contains the origin is constant.

Solution. An inverse square law field $\mathbf{F}$ is given by

$$\mathbf{F}(\mathbf{x}) = k \frac{1}{|\mathbf{x}|^2} \frac{\mathbf{x}}{|\mathbf{x}|}.$$

We can't apply Theorem 8.2 immediately, because D contains the origin, where $\mathbf{F}$ is not even defined, let alone differentiable, so the hypotheses of Theorem 8.2 are not met. To get around this, we consider the region E inside S and outside

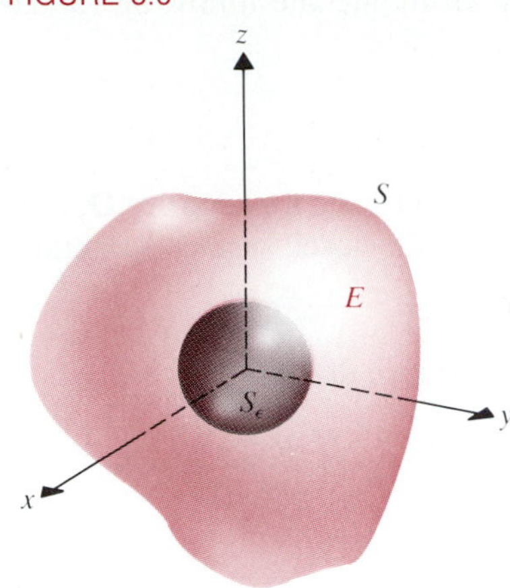

FIGURE 8.6

a sphere S_ε centered at $\mathbf{0}$ and of radius ε so small that $S_\varepsilon \subseteq D$. See **Figure 8.6.** On E all the hypotheses of Theorem 8.3 *are* met, and by Example 4 of Section 1, $\operatorname{div} \mathbf{F} = 0$ on all of E. Now E is bounded by $\partial E = S \cup S_\varepsilon$. If we orient these surfaces so that the normal $\mathbf{N}$ points toward $\mathbf{0}$ on S_ε and away from $\mathbf{0}$ on S, then $\mathbf{N}$ always points *outward* from E. So we can apply Theorem 8.3 and get

$$\iint_{\partial E} \mathbf{F} \cdot \mathbf{N}\, dA = \iiint_E \operatorname{div} \mathbf{F}\, dV = 0.$$

But then

$$\iint_{\partial E} \mathbf{F} \cdot \mathbf{N}\, dA = \iint_{S \cup S_\varepsilon} \mathbf{F} \cdot \mathbf{N}\, dA = \iint_S \mathbf{F} \cdot \mathbf{N}\, dA + \iint_{S_\varepsilon} \mathbf{F} \cdot \mathbf{N}\, dA = 0.$$

Hence the desired flux is

$$\Phi = \iint_S \mathbf{F} \cdot \mathbf{N}\, dA = -\iint_{S_\varepsilon} \mathbf{F} \cdot \mathbf{N}\, dA,$$

where S_ε has inward-pointing normal

$$\mathbf{N} = -\frac{1}{|\mathbf{x}|}\mathbf{x}.$$

We thus have

$$\begin{aligned}\Phi &= -k \iint_{S_\varepsilon} \frac{\mathbf{x}}{|\mathbf{x}|^3} \cdot \left(-\frac{\mathbf{x}}{|\mathbf{x}|}\right) dA = k \iint_{S_\varepsilon} \frac{1}{|\mathbf{x}|^2}\, dS \\ &= \frac{1}{\varepsilon^2} k \iint_{S_\varepsilon} dA = \frac{1}{\varepsilon^2} kA(S_\varepsilon) = \frac{1}{\varepsilon^2} k 4\pi\varepsilon^2 = 4\pi k,\end{aligned}$$

where we used the result of Example 2 of Section 5. ■

We have already mentioned that inverse-square law fields describe a number of important physical phenomena in gravitation, magnetism, electricity, and other areas of science. The result of Example 3 bears on such applications. As an illustration, we return to electricity theory, specifically to *Coulomb's law* (named for the French army engineer and scientist Charles A. Coulomb (1736–1806)). Coulomb determined experimentally that the force between two charged particles in a vacuum is inversely proportional to the square of the distance between them. That is,

$$|\mathbf{F}| = k\frac{q_1 q_2}{r^2},$$

for some constant k, which usually is written as $1/4\pi\varepsilon_0$. Here ε_0 is a constant called the *permittivity* of free space. Example 3 indicates why the factor 4π was introduced. If a charge q_0 is at the origin, then the force on a similarly charged particle at $\mathbf{x}$ carrying a charge q_1 is

$$\mathbf{F} = \frac{1}{4\pi\varepsilon_0} \frac{|\mathbf{x}|}{|\mathbf{x}|^3} q_0 q_1.$$

By Example 2 the flux Φ across any closed surface is then $q_0 q_1$ times the reciprocal of the permittivity. The *electric flux density* is defined as $\mathbf{D} = \varepsilon_0 \mathbf{E}$, where $\mathbf{E} = (1/q_1)\mathbf{F}$, and the *electric flux* across S is then defined as $\Psi = \iint_S \mathbf{D} \cdot d\mathbf{S}$.

Gauss used Theorem 8.2 as in Example 1 to formulate the following fundamental law that bears his name.

Example 4

Obtain *Gauss's law:* If a charge q_0 at $\mathbf{0}$ produces an electric flux density $\mathbf{D}$, then the electric flux passing through any piecewise smooth closed surface S containing $\mathbf{0}$ is q_0.

Solution. Let q_1 be the total charge on S. The electric flux is then

$$\Psi = \iint_S \mathbf{D} \cdot d\mathbf{S} = \iint_S \varepsilon_0 \mathbf{E} \cdot d\mathbf{S} = \varepsilon_0 \iint_S \frac{\mathbf{F}}{q_1} \cdot d\mathbf{S} = \varepsilon_0 \iint_S \frac{q_0}{4\pi\varepsilon_0} \frac{\mathbf{x}}{|\mathbf{x}|^3} \cdot d\mathbf{S}$$

$$= \frac{1}{4\pi} q_0 \iint_S \frac{\mathbf{x}}{|\mathbf{x}|^3} \cdot d\mathbf{S} = \frac{1}{4\pi} q_0 4\pi \qquad \textit{by Example 3}$$

$$= q_0. \quad \blacksquare$$

Electricity is by no means the only important area of application of Gauss's theorem. Further applications occur in Exercises 23–31. Exercises 24–26 are aimed at deriving the *heat equation,* which is fundamental in thermodynamics and describes many important diffusion processes. Those exercises require the following facts from physics about heat flow.

(i) The amount of heat required to raise the temperature of a small body ΔB having constant density δ by T degrees is proportional to T and the mass m of the body. That is, the amount of heat is

$$c\,\delta T V(\Delta B),$$

where $V(\Delta B)$ is the volume of ΔB and c is a constant called the *specific heat* of the material.

(ii) The rate at which heat crosses a small patch of area ΔA on a surface S is

$$-k(\mathbf{N} \cdot \nabla T)\,\Delta A,$$

where $\mathbf{N}$ is the standard unit normal (Definition 6.2), k is a constant called the *coefficient of thermal conductivity,* and $T = T(x, y, z, t)$ is the temperature at $\mathbf{x} = (x, y, z)$ at time t. It is assumed that ∇T has entries whose partial derivatives are continuous on $\mathbf{R}^3$. (The minus sign reflects the fact that heat flows *away* from hot sources toward colder bodies.)

(iii) The total thermal energy (i.e., heat) that enters a body B during time Δt is the amount of heat absorbed by B during this time interval. This is the principle of *conservation of thermal energy.*

Exercises 27–30 derive the celebrated Principle of Archimedes (287–212 BC) for floating bodies, the discovery of which, according to legend, prompted Archimedes to leave his bath and run naked through the streets of ancient Syracuse, shouting "Eureka!"—"I have found it!" Those exercises require the following definition of vector surface integrals and vector triple integrals.

8.4 DEFINITION Suppose that $\mathbf{F} = P\,\mathbf{i} + Q\,\mathbf{j} + R\,\mathbf{k}$ is continuous on a simple region $D \subseteq \mathbf{R}^3$ with positively oriented boundary S. Then the ***vector surface integral*** of $\mathbf{F}$ over S is

$$\iint_S \mathbf{F}(x, y, z)\,dA = \iint_S P(x, y, z)\,dA\,\mathbf{i} + \iint_S Q(x, y, z)\,dA\,\mathbf{j} + \iint_S R(x, y, z)\,dA\,\mathbf{k},$$

where the surface integrals on the right are as in Definition 6.1. The ***vector triple integral*** of $\mathbf{F}$ over D is

$$\iiint_D \mathbf{F}(x, y, z)\,dV = \iiint_D P(x, y, z)\,dV\,\mathbf{i} + \iiint_D Q(x, y, z)\,dV\,\mathbf{j} + \iiint_D R(x, y, z)\,dV\,\mathbf{k}.$$

Example 5

If $\mathbf{F}(x, y, z) = 2\mathbf{i} - \mathbf{j} + z\mathbf{k}$ then find $\iiint_D \mathbf{F}(x, y, z)\,dV$ where D is the unit ball $x^2 + y^2 + z^2 \le 1$.

Solution. Here $P(x, y, z) = 2$ and $Q(x, y, z) = -1$ are constant functions, and $R(x, y, z) = z$. We therefore have

$$\iiint_D \mathbf{F}(x, y, z)\,dV = 2\iiint_D 1\,dV\,\mathbf{i} - 1\iiint_D 1\,dV\,\mathbf{j} + \iiint_D z\,dV\,\mathbf{k}.$$

Since the ball has volume $\frac{4}{3}\pi \cdot 1^3 = \frac{4}{3}\pi$, the first two integrals are $\frac{8}{3}\pi$ and $-\frac{4}{3}\pi$. We can use spherical coordinates to evaluate the last one. We find

$$\begin{aligned}\iiint_D z\,dV &= \int_0^{2\pi}\int_0^{\pi}\int_0^1 \rho\cos\phi\,\rho^2\sin\phi\,d\rho\,d\phi\,d\theta \\ &= \int_0^{2\pi} d\theta \int_0^{\pi} \sin\phi\cos\phi\,d\phi \int_0^1 \rho^3\,d\rho \\ &= 2\pi\left[\frac{1}{2}\sin^2\phi\right]_0^{\pi}\frac{1}{4} = 0. \quad ■\end{aligned}$$

Exercises 14.8

In Exercises 1–14, use Gauss's theorem to evaluate the surface integral $\iint_S \mathbf{F}\cdot\mathbf{N}\,dA$ for the given F and S.

1. $\mathbf{F}(x, y, z) = 2x\,\mathbf{i} + 3y\,\mathbf{j} - 4z\,\mathbf{k}$, S the sphere $x^2 + y^2 + z^2 = 4$ with outward-pointing normal
2. $\mathbf{F}(x, y, z) = 2x\,\mathbf{i} + (x^2 - xz^2)\,\mathbf{j} + (x^2y - y^3)\,\mathbf{k}$, S the sphere $x^2 + y^2 + z^2 = 1$ with inward-pointing normal
3. $\mathbf{F}(x, y, z) = \dfrac{x}{(x^2 + y^2 + z^2)^{3/2}}\,\mathbf{i} + \dfrac{y}{(x^2 + y^2 + z^2)^{3/2}}\,\mathbf{j} + \dfrac{z}{(x^2 + y^2 + z^2)^{3/2}}\,\mathbf{k}$, S the surface that bounds the solid box $[-1, 3] \times [2, 5] \times [-1, 1]$ with outward-pointing normal
4. $\mathbf{F}$ of Exercise 3, S the ellipsoid $x^2/9 + y^2/4 + z^2/16 = 1$ with outward-pointing normal
5. $\mathbf{F}(x, y, z) = (3x^2 + z^2)\,\mathbf{i} + (xy - z^3)\,\mathbf{j} + (z + x^2 - y^2)\,\mathbf{k}$, S the tetrahedral surface formed by the three coordinate planes and the plane $x + y + z = 1$, with outward-pointing normal
6. $\mathbf{F}(x, y, z) = (x^3 + z^2)\,\mathbf{i} + (y^3 + x^2)\,\mathbf{j} + (z^3 + y^2)\,\mathbf{k}$, S the sphere $x^2 + y^2 + z^2 = 1$, with outward-pointing normal
7. $\mathbf{F}(x, y, z) = (x^3 + y^3 + z^3)\,\mathbf{i} + (x^2y + \frac{1}{3}y^3 + z^2)\,\mathbf{j} + (x^2 + y^2)z\,\mathbf{k}$, S the cylindrical surface $x^2 + y^2 = 1$ for $0 \le z \le 1$, including top and bottom, with outward-pointing normal
8. $\mathbf{F}$ as in Exercise 7, S the cone $z = 1 - \sqrt{x^2 + y^2}$ (including the floor in the xy-plane) with outward-pointing normal
9. $\mathbf{F}(x, y, z) = (x^2 - yz)\,\mathbf{i} + (y^2 - xz)\,\mathbf{j} + (z^2 - xy)\,\mathbf{k}$, S the surface of Exercise 3
10. $\mathbf{F}(x, y, z) = (x^2 - \cos yz)\,\mathbf{i} + (y^2 + \sin 2xz)\,\mathbf{j} + (z^2 + \tan 5xy)\,\mathbf{k}$, S the surface of Exercise 3
11. $\mathbf{F}(x, y, z) = (x^2 + ye^z)\,\mathbf{i} + (y^2 + ze^x)\,\mathbf{j} + (z^2 + xe^y)\,\mathbf{k}$, S the part of the cylinder $x^2 + y^2 = 1$ between the xy-plane and $z = x + 1$ (including top and bottom) with outward-pointing normal
12. $\mathbf{F}(x, y, z) = (x^2 + y^2 + z^2)\,\mathbf{i} + (y^2 + 2x^3z^3)\,\mathbf{j} + (z^2 - x^2y^3e^{xy})\,\mathbf{k}$, S as in Exercise 11
13. $\mathbf{F}(x, y, z) = (x^2 + yz)\,\mathbf{i} + (y - xe^z)\,\mathbf{j} + (z - x^2\sqrt{1 - y^2})\,\mathbf{k}$, S the paraboloid $z = x^2 + y^2$ between $z = 0$ and $z = 2x$ (including the top), with outward-pointing normal
14. $\mathbf{F}(x, y, z) = (x^2 - y^3z)\,\mathbf{i} + (y - x^2 + z^2)\,\mathbf{j} + (z - x^2 + y^2)\,\mathbf{k}$, S as in Exercise 11
15. Prove *Green's first identity* in three dimensions: If $f: \mathbf{R}^3 \to \mathbf{R}$ is differentiable on a simple region D and all

second partial derivatives of g are continuous on D, then

$$\iiint_D f(\mathbf{x})\nabla^2 g(\mathbf{x})\,dV = \iint_S [f(\mathbf{x})\nabla g]\cdot \mathbf{N}\,dA - \iiint_D \nabla f \cdot \nabla g\,dV,$$

where S is the positively oriented boundary of D (cf. Exercise 29 of Section 3).

16. Obtain *Green's second identity* from Exercise 15. Assuming the same hypotheses as in Exercise 15, show that

$$\iiint_D [f(\mathbf{x})\nabla^2 g(\mathbf{x}) - g(\mathbf{x})\nabla^2 f(\mathbf{x})]\,dV$$

$$= \iint_S [f(\mathbf{x})(\nabla g \cdot \mathbf{N}) - g(\mathbf{x})(\nabla f \cdot \mathbf{N})]\,dA$$

$$= \iint_S [f(\mathbf{x})D_{\mathbf{N}}(g)(\mathbf{x}) - g(\mathbf{x})D_{\mathbf{N}}(f)(\mathbf{x})]\,dA$$

(cf. Exercise 30 of Section 3).

17. If $u\colon \mathbf{R}^3 \to \mathbf{R}$ and $\mathbf{v}\colon \mathbf{R}^3 \to \mathbf{R}^3$ have Jacobian matrices with continuous entries, and D is a simple region with positively oriented boundary S, then establish the following *integration-by-parts* formula:

$$\iiint_D u \operatorname{div} \mathbf{v}\,dV = \iint_S u\mathbf{v}\cdot \mathbf{N}\,dA - \iiint_D \mathbf{v}\cdot \nabla u\,dV.$$

18. If f is harmonic (Definition 1.7) on D in Exercise 16, then show that $\iint_S D_{\mathbf{N}}(f)(\mathbf{x})\,dA = 0$.

19. Suppose that f is harmonic on D, and $f(\mathbf{x}) = 0$ on $S = \partial D$. Then use Exercise 15 to show that f is identically zero on all of D, if $S \subseteq D$.

20. Let $g, h\colon \mathbf{R}^3 \to \mathbf{R}$. The relations $\nabla^2 f(\mathbf{x}) = g(\mathbf{x})$ for $\mathbf{x} \in D \subseteq \mathbf{R}^3$ and $f(\mathbf{x}) = h(\mathbf{x})$ for $\mathbf{x} \in S = \partial D$ are called *Poisson's equation*, after the French mathematician and physicist Simeon D. Poisson (1781–1840). Show that if g and h are given, then there can be at most one function f that solves Poisson's equation on a simple region D. (Use Exercise 19. The equation $f(\mathbf{x}) = h(\mathbf{x})$ on S is called a *boundary condition*. Assume $S \subseteq D$.)

21. If $\mathbf{F}\colon \mathbf{R}^3 \to \mathbf{R}^3$ is the velocity field of a fluid flow, where the fluid has density $\delta(x, y, z)$, then show that $\iint_S \delta(x, y, z)\mathbf{F}\cdot \mathbf{N}\,dA$ is the net loss of mass per unit time from the ball $B(\mathbf{x}_0, a)$. Here S is the sphere of radius a centered at $\mathbf{x}_0$.

22. Use Exercise 21 to show that

$$\lim_{a\to 0} \frac{1}{\frac{4}{3}\pi a^3} \iint_S \delta(x, y, z, t)\mathbf{F}\cdot \mathbf{N}\,dA$$

is the rate $-\partial\delta/\partial t$ of decrease of the density per unit time per unit volume at $\mathbf{x}_0$.

23. Show that the limit in Exercise 22 is the divergence of $\delta\mathbf{F}$ at $\mathbf{x}_0$ if $\mathbf{F}$ and δ have continuous partial derivatives. Then derive the *continuity equation* of fluid mechanics, $\partial\delta/\partial t + \operatorname{div}(\delta\mathbf{F}) = 0$.

24. **(a)** Show that fact (iii) above (p. 984) can be formulated as

$$\left(\iint_S k(\mathbf{N}\cdot \nabla T)\,dA\right)\Delta t = \iiint_B c\delta\,\Delta T\,dV,$$

where S is the positively oriented boundary of B.

(b) Conclude that

$$\iint_S k(\mathbf{N}\cdot \nabla T)\,dA = \iiint_B c\delta \frac{\partial T(\mathbf{x}, t)}{\partial t}\,dV.$$

25. Use Theorem 8.2 and Exercise 24(b) to show that

$$\iiint_B \left[\nabla\cdot(k\nabla T) - \frac{\partial}{\partial t}(c\,\delta T)\right] dV = 0.$$

26. Use Exercise 25 to show that

$$\text{(iv)}\qquad \frac{\partial T}{\partial t} = \frac{k}{\delta c}\nabla^2 T.$$

The constant $k/\delta c$ is called the *coefficient of thermometric conductivity*. Equation (iv) is called the **heat equation.**

27. Suppose that $f\colon \mathbf{R}^3 \to \mathbf{R}$ has continuous partial derivatives on a simple region D with piecewise smooth positively oriented boundary S.

(a) If $\mathbf{a} \in \mathbf{R}^3$ is fixed, then show that

$$\mathbf{a}\cdot\left(\iint_S f(\mathbf{x})\mathbf{N}\,dA - \iiint_D \nabla f\,dV\right) = 0.$$

[*Hint:* Use Theorems 8.2 and 1.5(b) with $G(x) = a$.]

(b) Since $\mathbf{a}$ is arbitrary, conclude that

$$\iint_S f(\mathbf{x})\mathbf{N}\,dA = \iiint_D \nabla f\,dV.$$

28. Denote by $\mathbf{B}$ the force of buoyancy exerted on a floating body by a surrounding liquid. Let $p(\mathbf{x})$ be the fluid pressure on the body at $\mathbf{x}$. By considering the force on a small area of the surface S of the body near $\mathbf{x}$, justify the definition given in physics,

$$\mathbf{B} = -\iint_S p(\mathbf{x})\mathbf{N}\,dA.$$

29. The *hydrostatic law* states that

$$\nabla p(\mathbf{x}) = \delta(\mathbf{x})\mathbf{g},$$

where $\delta(\mathbf{x})$ is the density of the liquid at $\mathbf{x}$, and $\mathbf{g}$ is the force exerted by gravity. Use this and Theorem 8.2 to show that

$$\mathbf{B} = -\iiint_V \delta(\mathbf{x})\mathbf{g}\,dV.$$

30. Let $\mathbf{W} = m\mathbf{g}$ be the weight vector of the liquid displaced by the body. Show that $\mathbf{W} = \mathbf{g}\iiint_B \delta(x)\,dV$, and hence derive the Principle of Archimedes,

$$\mathbf{W} + \mathbf{B} = \mathbf{0}.$$

31. **(a)** Apply Theorem 8.2 to $\mathbf{F}\times\mathbf{a}$ to show that

$$\mathbf{a}\cdot\left(\iint_S \mathbf{N}\times\mathbf{F}\,dA - \iiint_D \mathbf{curl\,F}\,dV\right) = 0,$$

where $\mathbf{a} \in \mathbf{R}^3$ is fixed. [*Hint:* Use Theorem 1.9(c) with $\mathbf{G}(\mathbf{x}) = \mathbf{a}$.]

(b) Show that $\iint_S \mathbf{N}\times\mathbf{F}\cdot d\mathbf{S} = \iiint_D \mathbf{curl\,F}\,dV$.

14.9 Looking Back

This chapter has two major themes. The first is the extension of integration to vector functions $\mathbf{F}: \mathbf{R}^n \to \mathbf{R}^n$ for $n = 2$ and 3. The second is the variety of analogues of the fundamental theorem of calculus for the different types of vector integrals. Those analogues involve the vector differential operators—gradient, divergence, and curl—which were introduced in Section 1. The inverse-square law fields of Examples 4 and 5 served to introduce conservative vector fields, to which Section 4 is largely devoted.

Section 2 presented the line integral of a vector function over an oriented parametrized smooth curve C in $\mathbf{R}^2$ or $\mathbf{R}^3$. It was introduced through the notion of the work done by a variable force in moving a particle along the curve (Definitions 2.1 and 2.2). Definition 2.3 extended line integrals to piecewise smooth curves. The basic fact that reversing the orientation of C multiplies the line integral by -1 was established in Theorem 2.4. Section 3 was devoted to the first analogue of the fundamental theorem, namely, Green's theorem (Theorems 3.2 and 3.3). Among its applications were computation of area in the plane by line integrals [Equation (8), p. 930], fluid dynamics, and electrical theory (p. 929).

The next section linked path independence with conservative vector fields. Theorem 4.1 is perhaps the sharpest analogue of the fundamental theorem. Path independence is equivalent (Theorem 4.2) to the line integral being 0 over any piecewise smooth closed curve. For an open disk, it is also equivalent to the integrand being a conservative vector field (Theorem 4.3). Theorem 4.4 gives a necessary condition for a vector field $\mathbf{F}$ to be conservative: $\mathbf{J}_\mathbf{F}$ must be a symmetric matrix. If the domain D is simply connected, then the latter condition is enough to guarantee that $\mathbf{F}$ is conservative on D (Theorem 4.5).

Sections 5 and 6 were devoted to parametrizing surface patches, defining the area of a smooth surface patch, and defining the surface integral of a scalar or vector function over a smooth surface patch and more general types of surfaces. This required the notion of standard unit normal and orientable piecewise smooth surface (Definitions 6.2 and 6.5). We also defined the flux (Definition 6.4) across a parametrized surface S associated with a vector function $\mathbf{F}$ to be the surface integral of $\mathbf{F}$ over S.

The final two sections presented the important theorems of Stokes and Gauss (Theorems 7.1, 7.2, 8.2, and 8.3). These results relate surface integrals both to line integrals and to triple integrals and fit the mold of being analogues of the fundamental theorem of calculus. A number of applications to electrical theory and fluid dynamics were discussed in examples in those sections.

CHAPTER CHECKLIST

Section 1: vector functions $\mathbf{F}: \mathbf{R}^n \to \mathbf{R}^n$, coordinate functions; limits, continuity, differentiation; Jacobian matrix; gradient (del); vector field; divergence; inverse-square law fields; conservative vector fields, potential functions; Laplacian operator, Laplace's equation, harmonic functions; curl.

Section 2: parametrized curve; piecewise smooth curve, simple closed curve, orientation; opposite (negative) curve; work, line integral; reversal of orientation of a curve C reverses sign of line integral over C.

Section 3: simple regions D in $\mathbf{R}^2$; Green's theorem, counterclockwise orientation of ∂D; area by line integration; circulation, irrotational fields; electromotive force.

Section 4: path independence; potential energy, conservation of energy; symmetry of Jacobian matrix; simply connected regions.

Section 5: parametrized surface patch, parametrized surface; area of a smooth surface path; area of the graph of a smooth function $f: \mathbf{R}^2 \to \mathbf{R}$.

Section 6: surface integral of a scalar function $f: \mathbf{R}^3 \to \mathbf{R}$; standard unit normal, surface integral of a vector field $\mathbf{F}: \mathbf{R}^3 \to \mathbf{R}^3$; orientation of standard unit normal, Jacobian determinant, flux, piecewise smooth surface, orientable surface, Möbius band.

Section 7: positively oriented boundary surface, right-hand rule; theorem of Stokes; conservative vector field $\mathbf{F}: \mathbf{R}^3 \to \mathbf{R}^3$ and symmetric Jacobian matrix; Ampère's circuital law.

Section 8: positively oriented smooth surface patch; simple regions in $\mathbf{R}^3$, Gauss's theorem, Gauss's law.

REVIEW EXERCISES 14.9

1. Show that

$$\operatorname{div} \frac{\mathbf{x}}{|\mathbf{x}|^3} = 0.$$

2. If $\mathbf{F}(\mathbf{x}) = (x^2 + y^2 - xyz)\mathbf{i} + (xz^2 \cos y)\mathbf{j} + (z^2 e^{x^2+y^2})\mathbf{k}$, then compute div $\mathbf{F}$ and **curl** $\mathbf{F}$ in general and at $\mathbf{x} = (-1, \pi/2, 3)$.

3. Is

$$\mathbf{F}(x, y) = -\frac{x}{(x^2 + y^2)^{3/2}}\mathbf{i} - \frac{y}{(x^2 + y^2)^{3/2}}\mathbf{j}$$

conservative on the annular region between $x^2 + y^2 = 1$ and $x^2 + y^2 = 4$?

4. Is $\mathbf{F}(x, y, z) = (y - 3z + 2x^2)\mathbf{i} + (x + 2z)\mathbf{j} + (2y - 3x + 4xz)\mathbf{k}$ conservative on the unit ball?

5. Is $\mathbf{F}(x, y, z) = (2x^2 + 8xy^2)\mathbf{i} + (3x^3y - 3xy)\mathbf{j} + (4y^2z - 2x^3z)\mathbf{k}$ conservative on the unit ball?

6. Is $F(x, y) = (x^2 - 3xy)\mathbf{i} + (y^2 + 3xy)\mathbf{j}$ conservative on the unit disk?

In Exercises 7–16, evaluate the given integral.

7. $\int_C \mathbf{F} \cdot d\mathbf{x}$, where $\mathbf{F}(x, y) = xy^2\mathbf{i} + (x + y^2)\mathbf{j}$, and C is $\mathbf{x}(t) = (\sin t, 1 - \sin t)$ for $0 \le t \le \frac{1}{2}\pi$

8. $\int_C \mathbf{F} \cdot \mathbf{T}\, ds$, where $\mathbf{F}(x, y) = xy\mathbf{i} + (y^2 + 1)\mathbf{j}$, and C is the triangular path from $(0, 0)$ to $(1, 0)$ to $(1, 1)$

9. $\int_C (2x^2 - y^3)\, dx + (x^3 - 4y^2 + y)\, dy$, where C is the circle $x^2 + y^2 = 4$ traversed clockwise

10. $\int_C -2ye^x\, dx + xe^y\, dy$, where C is the polygonal path from $(1, 1)$, to $(-1, 1)$ to $(-1, -1)$ to $(1, -1)$ to $(1, 1)$

11. $\int_C (-\cos y - y^3)\, dx + (x \sin y - 3xy^2)\, dy$, where C is the ellipse $x^2/4 + y^2/16 = 1$ traversed counterclockwise

12. $\int_{\partial S} \mathbf{F} \cdot d\mathbf{x}$, where $\mathbf{F}(x, y, z) = e^x \sin y\, \mathbf{i} + (e^x \cos y - z)\mathbf{j} + y\mathbf{k}$, and S is the part of the cylinder $x^2 + y^2 = 1$ between $z = 1$ and $z = 2$

13. $\int_{\partial S} \mathbf{F} \cdot d\mathbf{x}$, where $\mathbf{F}(x, y, z) = 2z\mathbf{i} + x\mathbf{k}$, and S is the part of the cylinder $r = 2 \cos \theta$ above the xy-plane and below the cone $z = \sqrt{x^2 + y^2}$

14. $\int_C \mathbf{F} \cdot d\mathbf{x}$, where $\mathbf{F}(x, y, z) = (2y - z)\mathbf{i} + (2z - x)\mathbf{j} + (2x - y)\mathbf{k}$, and C is the intersection of the cylinder $x^2 + y^2 = 4$, and the plane $x + z = 4$ traversed counterclockwise when viewed from above

15. $\int_C 3x^2y^2z\, dx + 2x^3yz\, dy + x^3y^2\, dz$, where C is the polygonal path from $(0, 0, 0)$ to $(2, -1, 0)$ to $(2, 0, 0)$ to $(-1, 2, 1)$

16. $\int_C y \sin z\, dx + x \sin z\, dy + xy \cos z\, dz$, where C is the path $\mathbf{x}(t) = (t, t^2, t^3)$ from $(1, 1, 1)$ to $(2, 4, 8)$

17. **(a)** Parametrize the part of S of the plane $z = x + y$ that lies inside the cylinder $x^2 + y^2 = 1$.
(b) Find the area of S.

18. **(a)** Parametrize the part of S of the sphere $x^2 + y^2 + z^2 = 4$ that lies above the cone $z = \sqrt{x^2 + y^2}$.
(b) Find the area of S.

In Exercises 19–25, compute the given surface integral. Parametrize so that N points outward, or upward, from the surface.

19. $\iint_S (y + z)\, dA$, where S is the surface of Exercise 17

20. $\iint_S y^2\, dA$, where S is the cylinder $x^2 + y^2 = 1$ between $z = 0$ and $z = 1$, including top and bottom

21. $\iint_S \mathbf{F} \cdot d\mathbf{S}$, where $\mathbf{F}(x, y, z) = x\mathbf{i} - 2y\mathbf{j} + \mathbf{k}$, and S is the triangle with vertices $(1, 0, 0)$, $(0, 1, 0)$, and $(0, 0, 1)$

22. $\iint_S \mathbf{F} \cdot \mathbf{N}\, dA$, where $\mathbf{F}(x, y, z) = (x + 2)\mathbf{i} - (2y + 2)\mathbf{j} + 2\mathbf{k}$, and S is the triangle with vertices $(1, 0, 0)$, $(0, 1, 0)$, and $(0, 0, 1)$

23. $\iint_S \mathbf{F} \cdot d\mathbf{S}$, where $\mathbf{F}(x, y, z) = (x^3 + y^3 + e^{z^2})\mathbf{i} + (x^2y + \frac{1}{3}y^3 + z^3)\mathbf{j} + (x^2z/2 + \frac{1}{2}y^2z - x^3y^3)\mathbf{k}$, and S is the cylindrical surface $x^2 + y^2 = 1$ for $0 \le z \le 1$, including top and bottom

24. $\iint_S \mathbf{F} \cdot d\mathbf{S}$, where $\mathbf{F}(x, y, z) = (x^3 + y^2z)\mathbf{i} + (x^2 + y^3 + x^2z^2)\mathbf{j} + (x^2y^3 + z^3)\mathbf{k}$, and S is the sphere $x^2 + y^2 + z^2 = 9$

25. $\iint_S P\, dy \wedge dz + Q\, dz \wedge dx + R\, dx \wedge dy$, where $P(\mathbf{x}) = x + y^2 + z^2$, $Q(\mathbf{x}) = x^2 - 2y + z^3$, $R(\mathbf{x}) = x^2 + y^2 + z$, and S is the ellipsoid $x^2/9 + y^2/36 + z^2/4 = 1$

CHAPTER 15 Differential Equations

15.0 Introduction

Historically, mathematics has developed in response to a number of stimuli. One of these, which we have seen illustrated in multivariable calculus, can be described as the esthetic goal of extending existing mathematical systems, such as single-variable calculus, to newer, more general or abstract settings.

Another strong influence in the development of mathematics has been, and continues to be, the need for mathematics in quantifying other fields. The area of mathematics known as differential equations has received great impetus from such other fields as physics and engineering, and it remains one of the most widely applied parts of mathematics. By a *differential equation* we mean an equation involving one or more variables and their derivatives. This chapter is an introduction to differential equations as a mathematical subject. It includes several illustrations of the role differential equations play in *mathematical modeling:* the process of using mathematics to capture the essential quantitative nature of phenomena in applied fields.

We have previously encountered and studied simple mathematical models involving differential equations. In Section 3.9, for example, we modeled the motion of a body falling freely from rest by the differential equation

$$\frac{d^2y}{dt^2} = g, \tag{1}$$

where y is the position of the body after t seconds, and g is the (downward directed) acceleration due to gravity. [In (1), the positive direction is taken as the direction of motion—downward.] We obtained the formula

$$y = \frac{1}{2}gt^2 \tag{2}$$

by integrating (1) twice and using the initial conditions

$$\left.\frac{dy}{dt}\right|_{t=0} = 0 \qquad \text{and} \qquad y(0) = 0.$$

This formula is one of the laws of motion of Newtonian physics. To test its accuracy we could drop some bodies and measure how far they fall during various time intervals. If we dropped stones or bricks, we would probably be pleased with the model. However, if we dropped paper or feathers, what we would observe would not correlate very well with predictions based on Equation (2). We might then be led to refine the model to take account of air resistance. (See Section 1.)

Before proceeding, you may want to review Section 3.10, which introduced the simplest kinds of differential equations—first-order separable equations. Section 1 of the present chapter considers the next simplest type of differential equations for which a well-developed theory exists—first-order linear differential equations. They play an important role in differential equations just as linear functions are a prominent part of calculus (through the tangent approximation).

Section 2 treats a special type of nonlinear first-order equation, which is closely related to path-independence for line integrals. From then on, the focus of the chapter is almost entirely on linear differential equations. Sections 4 and 5 discuss second-order linear equations, whose solution involves functions with values in the set **C** of complex numbers; the necessary groundwork is laid in Section 3, which reviews complex numbers and introduces complex-valued functions of a real variable.

Applications to mechanics and electricity are discussed in Section 6. The final section takes up power series and numerical approximation of solutions to equations that cannot be solved by the techniques of the earlier sections.

15.1 First-Order Linear Differential Equations

The simplest differential equations are the first-order separable equations

$$g(y)\frac{dy}{dx} = h(x),$$

which can be solved by direct integration (see Section 3.10). The next simplest class of differential equations consists of first-order linear differential equations. *First order* means that only the *first derivative* of y occurs in the equation. *Linear* means that only dy/dx and y, and no higher powers, appear in the equation.

1.1 DEFINITION

The general ***first-order linear differential equation*** is

$$\frac{dy}{dx} + p(x)y(x) = q(x), \tag{1}$$

where $p(x)$ and $q(x)$ are continuous functions, and $y = y(x)$ is a differentiable function to be determined. A function $y = y(x)$ that satisfies (1) is called a ***solution*** of the differential equation.

A simple example of a first-order linear differential equation is Newton's law of cooling [Equation (8), p. 357], which relates the temperature $T(t)$ of a body at time t and the rate of change of T, and which can be put into the form

$$\frac{dT}{dt} - kT = -kC. \tag{2}$$

Equation (2) has the form (1), with t as independent variable in place of x, with $p(t) = -k$ (a constant function), and with $q(t) = -kC$ (another constant function). Equation (2) is also separable, and was solved as such in Section 6.3.

In this section we study and solve (1). The following theoretical characterization of the solution set of (1) also extends to higher-order linear equations and will be helpful in Sections 4 and 5.

1.2 THEOREM

The solution set of (1) is the set of all functions $y(x)$ of the form $y_0(x) + y_p(x)$ where $y_p(x)$ is one fixed particular solution of (1), and $y_0(x)$ is any solution of the *associated homogeneous linear differential equation*

$$y'(x) + p(x)y(x) = 0. \tag{3}$$

Proof. First, every function of the form $y(x) = y_0(x) + y_p(x)$ is a solution of (1), since for any such $y(x)$ we have

$$\begin{aligned}
\frac{dy}{dx} + p(x)y(x) &= \frac{d}{dx}[y_0(x) + y_p(x)] + p(x)[y_0(x) + y_p(x)] \\
&= [y_0'(x) + y_p'(x)] + [p(x)y_0(x) + p(x)y_p(x)] \\
&= [y_0'(x) + p(x)y_0(x)] + [y_p'(x) + p(x)y_p(x)] \\
&= 0 + q(x) \qquad \textit{since } y_0(x) \textit{ is a solution of (3)} \\
&= q(x).
\end{aligned}$$

Next, suppose that $y(x)$ is *any* solution of (1). Then $y(x) - y_p(x)$ is a solution of (3), because

$$\begin{aligned}
&\frac{dy}{dx}[y(x) - y_p(x)] + p(x)[y(x) - y_p(x)] \\
&\quad = [y'(x) - y_p'(x)] + [p(x)y(x) - p(x)y_p(x)] \\
&\quad = [y'(x) + p(x)y(x)] - [y_p'(x) + p(x)y_p(x)] \\
&\quad = q(x) - q(x) = 0.
\end{aligned}$$

Hence $y(x) - y_p(x) = y_0(x)$ for some solution $y_0(x)$ of (3). But then $y(x) = y_0(x) + y_p(x)$. Hence every solution $y(x)$ of (1) must have the required form. So the solution set of (1) consists precisely of the set of all functions of the prescribed form. QED

An expression like $y_0(x) + y_p(x)$, which is a general formula that characterizes the set of all solutions of (1), is called the ***general solution*** of the given differential equation.

Theorem 1.2 gives a nice theoretical characterization of the set of all solutions of (1), but we still face the practical problem of finding a formula that will yield all the solutions of (1). It seems natural to approach this in the two stages suggested by Theorem 1.2, and that approach is used in Sections 4 and 5 to solve second-order linear differential equations. But for first-order equations, we can carry out both steps simultaneously by using an approach Leibniz developed 300 years ago.

He decided to look for a function M such that if

$$\frac{dy}{dx} + p(x)y = q(x) \tag{1}$$

is multiplied through by $M(x)$, then the left side of the resulting equation becomes the derivative of the product of M and the unknown function y. He realized that if he could find such a factor $M(x)$, then (1) would reduce to

$$\frac{d}{dx}(M(x)y) = M(x)q(x),$$

which can be solved by integration:

$$M(x)y = \int M(x)q(x)\,dx + C \rightarrow y = \frac{1}{M(x)}\left[\int M(x)q(x)\,dx + C\right].$$

For this reason, $M(x)$ is called an ***integrating factor*** of (1).

To find an integrating factor $M(x) \neq 0$ for (1), note first that

$$\frac{d}{dx}(M(x)y) = M(x)\frac{dy}{dx} + M'(x)y. \tag{4}$$

Multiplication of the left-hand side of (1) by $M(x)$ gives

$$M(x)\left(\frac{dy}{dx} + p(x)y\right) = M(x)\frac{dy}{dx} + M(x)p(x)y. \tag{5}$$

We want the left sides of (4) and (5) to coincide. To find a suitable $M(x)$, we equate the coefficients of y in (4) and (5) and use the fact that $M(x) \neq 0$:

$$M'(x) = M(x)p(x) \rightarrow \frac{M'(x)}{M(x)} = p(x).$$

Integrating, we get

$$\ln|M(x)| = \int p(x)\,dx \rightarrow |M(x)| = e^{\int p(x)\,dx}.$$

Assuming that $M(x)$ is positive-valued, we obtain the desired formula, namely,

$$M(x) = e^{\int p(x)\,dx}.$$

So to solve (1), we multiply through by $e^{\int p(x)\,dx}$, which gives

$$\frac{dy}{dx}e^{\int p(x)\,dx} + p(x)ye^{\int p(x)\,dx} = q(x)e^{\int p(x)\,dx}. \tag{6}$$

The left-hand side is indeed the derivative of the function $y(x)e^{\int p(x)x\,dx}$. Thus, taking antiderivatives in (6), we obtain

$$ye^{\int p(x)\,dx} = \int q(x)e^{\int p(x)\,dx}\,dx = Q(x) + C, \tag{7}$$

where $Q(x)$ is some fixed antiderivative of $q(x)e^{\int p(x)\,dx}$, and C is an arbitrary constant. Multiplication of (7) through by $e^{-\int p(x)\,dx}$ then gives

$$y = Q(x)e^{-\int p(x)\,dx} + Ce^{-\int p(x)\,dx}. \tag{8}$$

Our work shows that if y is a solution to (1), then y satisfies (8). As written, the formula for y in (8) *looks like* the general solution (3) of (1) in Theorem 1.2, with $Q(x)e^{-\int p(x)\,dx}$ playing the role of $y_p(x)$ and $Ce^{-\int p(x)\,dx}$ playing the role of $y_0(x)$. To check that it really *is*, we must verify that any function $y(x)$ of the form

$$y(x) = e^{-\int p(x)\,dx}\int q(x)e^{\int p(x)\,dx}\,dx + Ce^{-\int p(x)\,dx} \tag{9}$$

is a solution of (1). For that, we work backwards. Multiplication of (9) through by $e^{\int p(x)\,dx}$ gives

$$y(x)e^{\int p(x)\,dx} = \int q(x)e^{\int p(x)\,dx} + C.$$

Differentiating this equation, we get

$$\frac{dy}{dx}e^{\int p(x)\,dx} + p(x)y(x)e^{\int p(x)\,dx} = q(x)e^{\int p(x)\,dx}.$$

Multiplication by the nonzero function $e^{-\int p(x)\,dx}$ then gives (1). Thus every function of the form (9) does solve (1). So (9) gives the general solution of (1).

There is a unique solution of (1) that satisfies the ***initial condition*** $y(x_1) = y_1$. To see that, we write (9) as

$$(10) \qquad y(x) = y_p(x) + Ce^{-P(x)},$$

where $P(x)$ is any antiderivative of $p(x)$. If $y(x_1) = y_1$, then we can solve (10) for the constant C corresponding to this condition:

$$y_1 = y_p(x_1) + Ce^{-P(x_1)} \to C = (y_1 - y_p(x_1))e^{P(x_1)}.$$

Thus there is a *unique function* of the form (9) that satisfies the initial condition $y(x_1) = y_1$

There is a simple geometric interpretation of the fact that the initial-value problem

$$\frac{dy}{dx} + p(x)y = q(x), \qquad y(x_1) = y_1,$$

has a unique solution. The general solution (9) consists of an infinite family of functions, one for each choice of the constant C. As **Figure 1.1** illustrates, the graphs of these functions comprise a family of curves in the plane. The initial condition $y(x_1) = y_1$ singles out the one graph of this family that passes through the point (x_1, y_1).

FIGURE 1.1

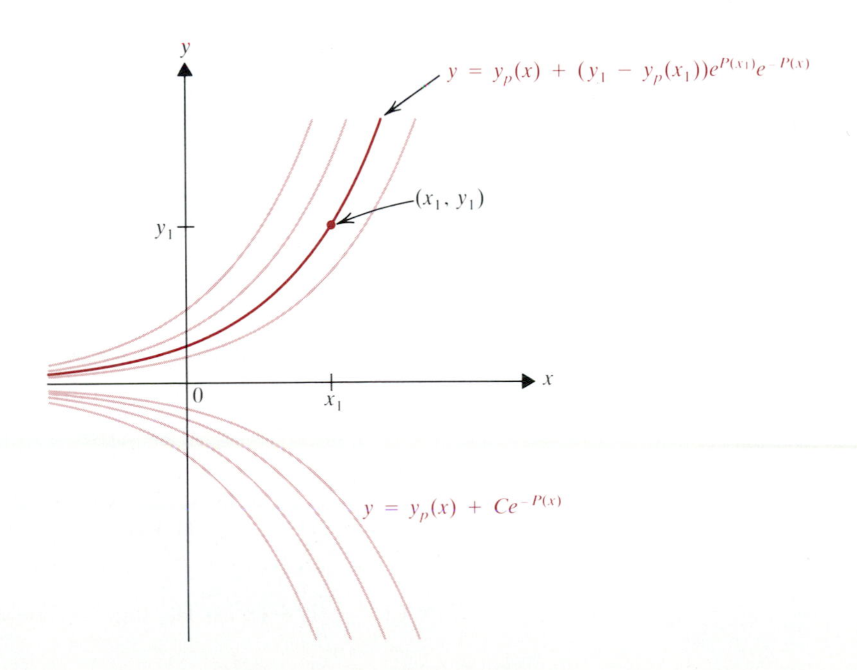

In solving a first-order linear differential equation in the form (1), you can either use Formula (9) or, as many students prefer, multiply through by the integrating factor $e^{\int p(x)\,dx}$ and use the fact that the left-hand side of the resulting equation is the derivative of $ye^{\int p(x)\,dx}$. Then taking antiderivatives will bring you to (9) without having committed that formula to memory. The next example illustrates that either approach leads to the same solution.

Example 1

Find the general solution of

$$(x^2+1)\frac{dy}{dx} - 2xy = x^3 + x = x(x^2+1).$$

Solution. To put this into the form (1), we divide through by the nonzero quantity $x^2 + 1$:

$$\frac{dy}{dx} - \frac{2x}{x^2+1}y = x. \tag{11}$$

We proceed to solve (11) in both of the ways suggested above.

(a) We have

$$p(x) = -\frac{2x}{x^2+1},$$

so

$$\int p(x)\,dx = -\ln|x^2+1| = \ln\frac{1}{x^2+1},$$

since $x^2 + 1$ is always positive. Next, we compute the integrating factor

$$e^{\int p(x)\,dx} = e^{\ln[1/(x^2+1)]} = \frac{1}{x^2+1}.$$

Multiplying both sides of (11) by this factor, we obtain

$$\frac{1}{x^2+1}\frac{dy}{dx} - \frac{2x}{(x^2+1)^2}y = \frac{x}{x^2+1}.$$

Now, recalling (6), we have

$$\frac{d}{dx}\left(\frac{1}{x^2+1}y\right) = \frac{1}{x^2+1}x = \frac{1}{2}\frac{2x}{x^2+1}.$$

Integration of that gives

$$\frac{1}{x^2+1}y = \frac{1}{2}\ln(x^2+1) + C = \ln\sqrt{x^2+1} + C \rightarrow y = (x^2+1)\ln\sqrt{x^2+1} + C(x^2+1).$$

(b) We can use (9). As above, we calculate

$$\int p(x)\,dx = -\ln(x^2+1) \rightarrow e^{-\int p(x)\,dx} = e^{\ln(x^2+1)} = x^2+1.$$

So (9) gives

$$y = C(x^2+1) + (x^2+1)\int xe^{-\ln(x^2+1)}\,dx = C(x^2+1) + (x^2+1)\int\frac{x}{x^2+1}\,dx$$

$$= \bar{C}(x^2+1) + (x^2+1)\frac{1}{2}\ln(x^2+1) = \bar{C}(x^2+1) + (x^2+1)\ln\sqrt{x^2+1}.$$

Note that the constant of integration from

$$\int \frac{x\,dx}{x^2+1}$$

has been consolidated with C in the number $\bar{C}$. ■

When $q(x) = 0$ in (1), the *homogeneous* first-order linear differential equation

$$\frac{dy}{dx} + p(x)y(x) = 0 \tag{3}$$

can, of course, be solved using the approach of this section. But it is also solvable as a *separable* equation

$$\frac{dy}{dx} = -p(x)y \to \frac{dy}{y} = -p(x)\,dx.$$

The two methods can be compared in the following example.

Example 2

Find the general solution of $dy/dx + y\cot x = 0$.

Solution. (a) Separation of the variables gives

$$\frac{dy}{dx} = -y\cot x \to \frac{dy}{y} = -\cot x\,dx \to \int \frac{dy}{y} = -\int \cot x\,dx,$$

$$\ln|y| = -\ln|\sin x| + \ln C = \ln\frac{C}{|\sin x|} \to |y| = \frac{C}{|\sin x|},$$

$$y = \frac{\pm C}{\sin x} \to y = K\csc x, \qquad \text{where } K = \pm C.$$

(b) Treating the problem as a first-order linear differential equation, we have $p(x) = \cot x$, so an integrating factor is

$$e^{\int \cot x\,dx} = e^{\ln|\sin x|} = |\sin x| = \pm\sin x.$$

Multiplying through by $\sin x$, we get

$$\sin x\frac{dy}{dx} + y\cos x = 0 \to \frac{d}{dx}(y\sin x) = 0,$$

$$y\sin x = C \to y = C\csc x. \quad ■$$

We considered the motion of bodies falling freely from rest in Section 0, modeling that situation as follows. Let y be the number of units in the distance the body has fallen after t seconds. Let a be the magnitude of the acceleration of the body, in appropriate units of distance per second per second, at time t. (We are thus measuring both distance and time from the start of the motion.) If we take as positive direction the (downward) direction of motion, then the motion is modeled by

$$a = \frac{d^2s}{dt^2} = g, \tag{12}$$

where g is the magnitude of the acceleration due to gravity (9.80 m/s^2 or 32.2 ft/s^2). This leads to the formula

(13)
$$y = \frac{1}{2} g t^2$$

for the position after t seconds.

As we mentioned in Section 0, the model (12) ignores any air resistance that may be experienced by the falling body. For many objects, physicists have found that such air resistance is a factor in the motion, and is often proportional to the velocity. Since it is exerted in the (upward) direction that is opposite to the direction of motion, the force of air resistance is thus given by

(14)
$$R = -kv$$

for a positive constant k that depends on the shape and mass of the body and possibly other factors. If we incorporate (14) into the model, then we have from $F = ma$,

$$m \frac{d^2 y}{dt^2} = mg - kv.$$

Since $v = dy/dt$, this is equivalent to

(15)
$$m \frac{dv}{dt} = -kv + mg.$$

Example 3

Find formulas for v and y at time t, if (15) is used to model the motion of a body falling from rest. (Assume $k \neq 0$.)

Solution. We have from (15)

$$m \frac{dv}{dt} = -kv + mg \rightarrow \frac{dv}{dt} + \frac{k}{m} v = g.$$

This first-order linear differential equation has integrating factor

$$e^{\int (k/m)\, dt} = e^{(k/m)t}.$$

Multiplying (14) through by this factor, we obtain

$$e^{kt/m} \frac{dv}{dt} + \frac{k}{m} e^{kt/m} v = g e^{kt/m} \rightarrow \frac{d}{dt}(e^{kt/m} v) = g e^{kt/m}.$$

Integration of this gives

$$e^{kt/m} v = \frac{mg}{k} e^{kt/m} + C_1 \rightarrow v = \frac{mg}{k} + C_1 e^{-kt/m}.$$

Since the body is falling from rest, $v = 0$ when $t = 0$. Hence

$$0 = \frac{mg}{k} + C_1 \rightarrow C_1 = -\frac{mg}{k}.$$

We thus have

(16)
$$v = \frac{mg}{k}(1 - e^{-kt/m}).$$

To get y from this, we integrate again, obtaining

$$y + C_2 = \frac{mg}{k}\left(t + \frac{m}{k}e^{-kt/m}\right). \tag{17}$$

Using the fact that $y = 0$ when $t = 0$, we find

$$C_2 = \frac{mg}{k}\left(\frac{m}{k}\right) = \frac{m^2 g}{k^2}.$$

Substituting this into (17), we get

$$y = \frac{mg}{k}t + \frac{m^2 g}{k^2}(e^{-kt/m} - 1). \quad ■ \tag{18}$$

The formulas for y and v derived from (15) are markedly different from those that follow from (12). Most noticeable is the difference in the nature of the speed. Using (12), we get $v = gt$, so the speed is predicted to increase without bound as t increases. But if we let $t \to \infty$ in (16), we find

$$\lim_{t\to\infty} v = \lim_{t\to\infty} \frac{mg}{k}(1 - e^{-kt/m}) = \frac{mg}{k}.$$

Thus in this model, the speed tends to a limiting value that depends upon the mass and coefficient k of resistance.

Which model is "right"? Such a question is really meaningless. Both models give predictions about the behavior of v, and all we can ask is which set of predictions correlates better with the results of experiments. In some cases the model based on (12) is perfectly adequate. In others the model based on (15) is preferable. In still other cases, involving high-speed motion (with an initial velocity), air resistance seems to be better approximated by $-kv^2$. Thus the choice of model to use may be governed not by mathematical factors but by physical considerations. It is to be emphasized that applied mathematics deals with the real world, where the absolute black-or-white, right-or-wrong criteria of pure mathematics are not entirely appropriate. Most often, no model is *absolutely* accurate. Rather, it is a question of which model gives the best *approximation* to the observed behavior.

The preceding example goes back to the beginnings of calculus in the 17th century when the study of physical problems was deeply intertwined with the development of mathematics. We close this section by discussing another application of first-order linear differential equations to a field outside mathematics. This time the application is of recent origin and suggests something of the breadth of use of differential equations as modeling tools.

The model was proposed in 1960 by William K. Estes (1919–) of Indiana University. In it, knowledge is measured by a continuous function $k(t)$, where $0 \le k(t) \le 1$. This function can be thought of as expressing the score an individual would achieve at time t on an examination covering the material to be learned.

To try to capture the observed phenomenon that individuals learn at different rates, Estes modeled the change in $k(t)$ as follows. If a *helpful* act (such as studying, going to class, or doing homework) is performed, then $k(t)$ changes to $k(t) + l(1 - k(t))$, where $l \in [0, 1]$ is the individual's *coefficient of learning* and might be measured by something like I.Q. If a *negative* act is performed (such as cutting class or failing to do homework), then $k(t)$ changes to $k(t) - lk(t)$. This corresponds to the student's falling behind in mastery of the sum total of material covered through time t. If $l = 0$, then the person cannot learn at all. No matter what he or she does, $k(t)$ never changes. On the other hand, if l is close to 1, then positive activity quickly increases $k(t)$ to nearly 1, so the student is a superior learner. Thus the coefficient of learning seems to reflect at least part of the learning process.

Example 4

Suppose that $\lambda \in [0, 1]$ measures the fraction of available time that a student performs helpful acts. Find a formula for her knowledge at time t. Appraise the reasonableness of the result.

Solution. During the fraction λ of available time, the student acts to increase $k(t)$ by the factor $l(1 - k(t))$. The remainder of the time $(1 - \lambda)$, her activity tends to decrease $k(t)$ by the factor $-lk(t)$. Thus the rate of change of $k(t)$ is given by

$$\frac{dk}{dt} = \lambda l(1 - k(t)) + (1 - \lambda)(-lk(t)) = \lambda l - \lambda lk(t) - lk(t) + \lambda lk(t).$$

We therefore have

$$\frac{dk}{dt} + lk(t) = \lambda l.$$

This is a first-order linear differential equation with integrating factor $e^{\int l\,dt} = e^{lt}$. Multiplying through by e^{lt}, we obtain

$$e^{lt}\frac{dk}{dt} + e^{lt}lk(t) = \lambda le^{lt} \rightarrow \frac{d}{dt}(e^{lt}k(t)) = \lambda le^{lt},$$

$$e^{lt}k(t) = \lambda e^{lt} + C \rightarrow k(t) = \lambda + Ce^{-lt}.$$

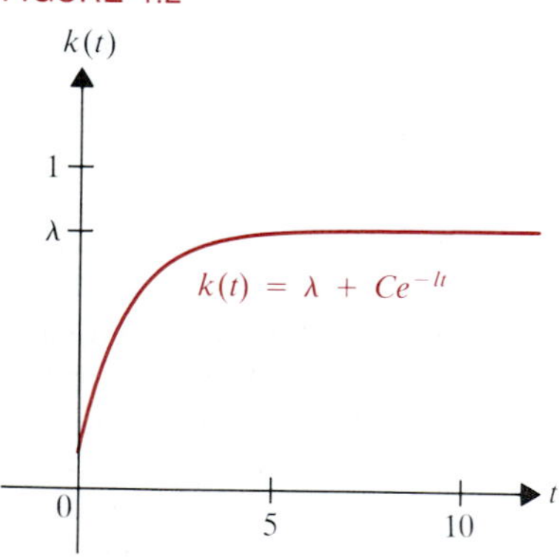

FIGURE 1.2

Thus this model predicts that a person's knowledge is a function not only of his or her native intelligence, but also of how much effort he or she expends in activity related to learning. That seems to correlate rather well with most students' experience. **Figure 1.2** illustrates that as t increases, $k(t)$ tends to be almost entirely determined by how much work the student puts forth. This once more reflects the experience of many students, who often find that increased effort is necessary as they advance into higher levels of education.

The graph was constructed taking $\lambda = l = 0.8$ and $C = -0.7$. It seems to reasonably portray how much might be learned by a student in a 10-week course. The model is far from perfect, however. For example, notice that (for l near 1) it seems to suggest that very bright students not only learn very fast, but also get behind very fast if they don't study hard. For more on this model, see Exercises 30 and 31. ■

Exercises 15.1

In Exercises 1–4, solve the homogeneous linear differential equation.

1. $\dfrac{dy}{dx} + y\sin x = 0$

2. $x\dfrac{dy}{dx} + y = 0$

3. $\cos x\dfrac{dy}{dx} + y\sin x = 0$

4. $\left(\dfrac{1}{x} - x\right)\dfrac{dy}{dx} + y = 0$, where $y = 3$ when $x = 0$

In Exercises 5–15, solve the given first-order linear differential equation.

5. $\dfrac{dy}{dx} - y = e^{3x}$

6. $x\dfrac{dy}{dx} - 2y = x^3 + x$

7. $x\dfrac{dy}{dx} - 3y = x^2 + 1$

8. $\dfrac{dy}{dx} - y = x^2$

9. $x^2\,dy - \sin 3x\,dx = -2xy\,dx$

10. $\cos x \dfrac{dy}{dx} + y \sin x = 1$

11. $(1 - x)\dfrac{dy}{dx} + y = x^2 - 2x + 1$, where $y(0) = 1$

12. $\dfrac{dy}{dx} - 3y = e^{3x} \sin x$, where $y(0) = 1$

13. $x\dfrac{dy}{dx} - 2y = x^5$ where $y(1) = -6$

14. $\dfrac{dy}{dx} - \dfrac{2x}{x^2 + 1} y = 1$ where $y(\pi) = 0$

15. $(x + 1)\dfrac{dy}{dx} - y = x$ where $y(0) = 1$

16. The ***Bernoulli equation*** [first solved by Jakob Bernoulli (1654–1705) in 1695] is

$$\frac{dy}{dx} + p(x)y = q(x)y^n, \tag{19}$$

where $n \neq 0$ and $n \neq 1$. (If $n = 0$ or 1, then we have a first-order linear equation.) Show that if (19) is multiplied through by y^{-n}, and the substitution $z = y^{1-n}$ is made, then a first-order linear equation

$$\frac{dz}{dx} + (1 - n)p(x)z = (1 - n)q(x)$$

results. (This method of solution was published by Leibniz in 1696.)

In Exercises 17–20, use the method of Exercise 16 to solve the given equation.

17. $x\dfrac{dy}{dx} + 3y = x^3y^2$

18. $x\dfrac{dy}{dx} - (3x + 6)y = -9xe^{-x}y^{4/3}$

19. $\dfrac{dy}{dx} + y = xy^3$ where $y(0) = 2$

20. $x^2\dfrac{dy}{dx} + 2xy = y^3$

21. The simple electric circuit shown in **Figure 1.3** is described by a first-order linear differential equation. When the switch at S is closed, current flows. In 1847 the German physicist Gustav R. Kirchhoff (1824–1887) formulated a mathematical model for electric circuits. Now known as *Kirchhoff's voltage law*, it states that the sum of the voltage drops around the circuit at each instant is equal to the electromotive force (EMF) $E(t)$ at that instant. In Figure 1.3 let R be the resistance in ohms, L the inductance in henrys, and E the EMF in volts. From experimentation it is known that the voltage drop across a resistor is $Ri(t)$, where $i(t)$ is the current in amperes at time t. It is also known that the voltage drop across an inductor is

$$L\frac{di}{dt}.$$

(a) Write the differential equation resulting from Kirchhoff's law. **(b)** Find the general solution, assuming that $E(t)$ is constant.

FIGURE 1.3

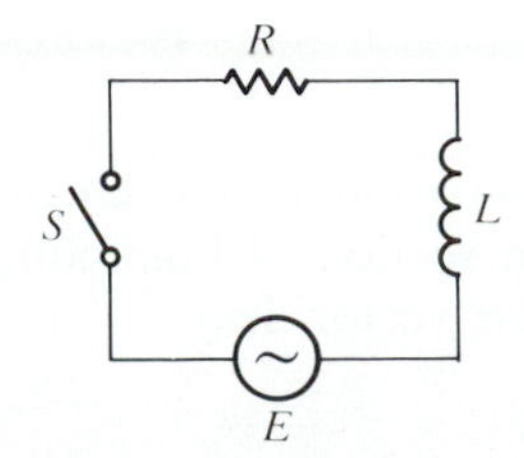

22. In Exercise 21, suppose that $E = 12$, $R = 16$, and $L = 0.02$. Find a formula for $i(t)$ and, by calculating $\lim_{t\to\infty} i(t)$, find the steady-state current flow in the circuit.

23. Brine made up of 25 grams of salt per liter of water is flowing into a tank at the rate of 3 liters per minute. The tank initially held 1 kilogram of salt dissolved in 10 liters of water. As the brine enters, it is mixed thoroughly with the contents of the tank, and the mixture is drawn off at the rate of 3 liters per minute. Find a formula for the amount $s(t)$ of salt in the tank after t minutes. What is the long-term amount of salt in the tank? (*Hint: ds/dt* is the rate at which the salt enters the tank minus the rate at which it leaves the tank.)

24. A tank initially holding 50 gallons of pure water receives a flow of brine containing 2 pounds of salt per gallon at the rate of 3 gallons per minute. The well-stirred mixture is drained off at 3 gallons per minute. Find the long-term concentration of salt in the tank.

25. A room contains 1000 cubic feet of air. Cigarette smokers enter the room and light up, introducing smoke (which contains 4% carbon monoxide) at the rate of 1/10 cubic foot per minute. This diffuses uniformly throughout the room. An air conditioner removes the air and pollutants from the room at the rate of 1/5 cubic foot per minute. Find a formula for the amount of carbon monoxide in the room after t minutes. If a concentration of carbon monoxide of 10 parts per million is harmful to human health, then how long a time is required for the contents of the room to become dangerous?

26. Suppose that a ball is thrown upward with initial speed v_0. If we neglect air resistance, then how far will the ball rise? If air resistance is as in Example 3, how far will the ball rise?

27. Assume in Example 3 that air resistance is proportional to the *square* of the speed. Find formulas for v and y at time t. What is $\lim_{t\to\infty} v$ in this case?

28. A sky diver falls from rest toward the earth. The diver and equipment weigh 192 pounds. Assume that the air resistance in pounds is $v/2$ before the parachute opens

(v measured in feet per second), and $3v^2/4$ after the parachute opens. Find formulas for v before and after the parachute opens. Assume that the parachute opens 12 seconds after free-fall begins. Find the limiting value of v after the parachute opens. What is the limiting value if the parachute fails to open?

29. In launching rockets from the earth's surface, the force due to gravity can no longer be assumed constant. Instead, *Newton's law of gravitational attraction* is used. Assume that the earth is a ball of radius $R \approx 4000$ miles and mass M. If the rocket is x miles from the center of the earth and has mass m, then this law states that $F = kmM/x^2$.

(a) Find k by using the fact that when $x = R$, then $F = -mg$. (Here the positive direction is taken as *upward*.)

(b) Solve the resulting differential equation

$$m\frac{dv}{dt} = \frac{-mgR^2}{x^2}$$

for v by using

$$\frac{dv}{dt} = \frac{dv}{dx}\frac{dx}{dt} = \frac{dv}{dx}v.$$

Assume that the rocket has initial velocity v_0.

(c) How large should v_0 be for the rocket to *escape* the earth's gravity? (That is, for v to always be positive no matter how large x becomes.)

30. Refer to Example 4. Some political scientists have proposed using the Estes learning model for public opinion. Here one interprets "helpful act" as information received that is favorable to a certain product (or governmental policy), and "harmful act" as negative information received. Also, one interprets $k(t)$ as the probability that the person will *respond favorably* to a pollster's question about the product (or policy). The constant l can be interpreted as a stubbornness factor (or a measure of sales resistance or receptiveness). Under this model, what is the long-term tendency for $k(t)$? Explain how this leads to the practice of saturation advertising for a product, or in the case of countries with a controlled press, to reporting of "good news" (favorable to the regime) only.

31. Appraise the accuracy of the Estes model for (a) learning and (b) opinion change. In (b), take into account the issue of *credibility*. Is this factor ignored in the Estes model or is it accounted for? Can you suggest a model that might better weight this factor?

32. The *Solow growth model* was developed by the American economist Robert M. Solow (1924–) of the Massachusetts Institute of Technology during the 1960s. It assumes that the gross national product $y(t)$ is a function $f(x(t), z(t))$ of the total capital $x(t)$ and the size $z(t)$ of the labor force. It is further assumed that for every positive constant c,

$$f(cx(t), cz(t)) = cf(x(t), z(t)).$$

Finally, it is assumed that labor grows exponentially at rate k and capital grows at a rate l proportional to the gross national product. If $w(t) = x(t)/z(t)$, then show that w satisfies the linear differential equation

$$\frac{dw}{dt} + kw = lg(w),$$

where $g(w) = f(w, 1)$.

33. A simple version of the *law of supply and demand* in economics is

$$\frac{dp}{dt} = k(d(t) - s(t)), \tag{20}$$

where $p(t)$ is the price of a given product at time t, and $d(t)$ and $s(t)$ are respectively the demand and supply at time t. If the supply is seasonal (for instance, if the product is agricultural), then a simple model for $s(t)$ is given by

$$s(t) = c(1 - \cos at),$$

where c and a are positive constants. A linear model for demand as a function of price is

$$d(t) = b - lp(t),$$

where again b and l are positive constants, and $0 < p(t) < b/l$. Put these supply and demand expressions into (20) and solve the resulting first-order linear differential equation for $p(t)$. Discuss the long-term behavior of $p(t)$. [*Hint:* For large t, make the substitution $\theta = \tan^{-1}(kl/a)$.]

34. Prove that $y_p(x) = Q(x)e^{-\int p(x)\,dx}$ is a particular solution of (1).

35. Prove that $y_0(x) = Ce^{-\int p(x)\,dx}$ is the general solution of (3).

36. Why do Exercises 34 and 35 provide an alternative to the proof given in the text that (9) is the general solution of (1)?

15.2 Exact Equations*

In this section we consider a reformulation in the language of differential equations of the problem considered in Section 14.4 of finding a potential function for a two-dimensional conservative vector field.

* Section 14.4 is prerequisite to this section.

2.1 DEFINITION A first-order differential equation is ***exact*** if it has the form

(1) $$P(x, y) + Q(x, y)\frac{dy}{dx} = 0,$$

where $P(x, y) = \partial f/\partial x$ and $Q(x, y) = \partial f/\partial y$ for some scalar function $f(x, y)$.

Example 1

Show that

$$\cos x \cos y - \sin x \sin y \frac{dy}{dx} = 0$$

is an exact differential equation.

Solution. We set

$$\frac{\partial f}{\partial x} = P(x, y) = \cos x \cos y \qquad \text{and} \qquad \frac{\partial f}{\partial y} = Q(x, y) = -\sin x \sin y,$$

and try to find the potential function $f(x, y)$ by partial integration, as in Section 14.4. We have

$$f(x, y) = \sin x \cos y + C_1(y) = \sin x \cos y + C_2(x),$$

from which it follows that $f(x, y) = \sin x \cos y$ is a potential function: $\nabla f(x, y) = P(x, y)\mathbf{i} + Q(x, y)\mathbf{j}$. Thus the given equation is exact. ■

Observe that if $y = y(x)$ is a solution to (1), then

$$\frac{\partial f}{\partial x} + \frac{\partial f}{\partial y}\frac{dy}{dx} = 0,$$

so by Theorem 4.2 of Chapter 12, we have

$$\frac{df(x, y)}{dx} = 0.$$

We then can solve (1) for y as an implicit function of x immediately, obtaining

(2) $$f(x, y) = C,$$

where C is an arbitrary constant. Exact equations are thus trivial to solve, *once* they are recognized as exact. [This observation goes back to the French mathematician Alexis C. Clairaut (1713–1765) in 1743.]

Example 2

Solve the exact differential equation

$$\cos x \cos y - \sin x \sin y \frac{dy}{dx} = 0$$

in Example 1.

Solution. From the work above, the function f in (2) is $f(x, y) = \sin x \cos y$. Therefore the general solution is

$$\sin x \cos y = C. \quad ■$$

Examples 1 and 2 are certainly easy, but they are predicated on being *told* that a given equation is exact. What is lacking is a test to determine whether a given first-order differential equation (1) *is* exact. The next result, which follows easily from the work in Sections 12.6 and 14.4, fills this need.

2.2 THEOREM

Suppose that P and Q are continuous and have continuous partial derivatives on a simply connected domain $D \subseteq \mathbf{R}^2$. Then a given first-order differential equation

$$P(x, y) + Q(x, y)\frac{dy}{dx} = 0 \tag{1}$$

is exact if and only if

$$\frac{\partial P}{\partial y} = \frac{\partial Q}{\partial x}.$$

Proof. If (1) is exact, then by Definition 2.1 there is some scalar function f such that

$$P(x, y) = \frac{\partial f}{\partial x} \quad \text{and} \quad Q(x, y) = \frac{\partial f}{\partial y}.$$

Then by Theorem 6.2 of Chapter 12, we find

$$\frac{\partial P}{\partial y} = \frac{\partial^2 f}{\partial y\, \partial x} = \frac{\partial^2 f}{\partial x\, \partial y} = \frac{\partial Q}{\partial x},$$

as desired. On the other hand, if $\partial P/\partial y = \partial Q/\partial x$ holds on D, then the Jacobian matrix

$$\mathbf{J}_{\mathbf{F}} = \begin{pmatrix} \dfrac{\partial P}{\partial x} & \dfrac{\partial P}{\partial y} \\ \dfrac{\partial Q}{\partial x} & \dfrac{\partial Q}{\partial y} \end{pmatrix}$$

of $\mathbf{F} = P\mathbf{i} + Q\mathbf{j}$ is symmetric. Then by Theorem 4.5 of Chapter 14, $P(x, y)\mathbf{i} + Q(x, y)\mathbf{j} = \nabla f(x, y)$ for some scalar function f. Thus $P(x, y) = \partial f/\partial x$ and $Q(x, y) = \partial f/\partial y$, so (1) is exact. QED

Theorem 2.2 essentially reduces the solution of exact differential equations to the determination of the potential function f of the conservative vector field $P\mathbf{i} + Q\mathbf{j}$. Thus no new procedures are needed: just follow the approach of Section 14.4.

Example 3

Solve the differential equation

$$3x^2 + 2y \sin 2x + (2 \sin^2 x + 3y^2)\frac{dy}{dx} = 0.$$

Solution. Here assume that $y = y(x)$ is differentiable for all x. Then

$$P(x, y) = 3x^2 + 2y \sin 2x \quad \text{and} \quad Q(x, y) = 2 \sin^2 x + 3y^2$$

are clearly continuous on *all* of $\mathbf{R}^2$, which is certainly simply connected. Since

$$\frac{\partial P}{\partial y} = 2\sin 2x = 4\sin x\cos x = \frac{\partial Q}{\partial x},$$

the given equation is exact by Theorem 2.2. We can find its solution $f(x, y) = C$ by setting

$$\frac{\partial f}{\partial x} = 3x^2 + 2y\sin 2x \qquad \text{and} \qquad \frac{\partial f}{\partial y} = 2\sin^2 x + 3y^2,$$

and partially integrating. From $\partial f/\partial x$ we obtain

$$f(x, y) = x^3 - y\cos 2x + g(y) \tag{3}$$

for some function $g(y)$ involving only y and constants. From $\partial f/\partial y$ we get

$$f(x, y) = 2y\sin^2 x + y^3 + h(x), \tag{4}$$

where $h(x)$ involves only x and constants. At first glance it appears difficult to reconcile (3) and (4), but we can use the identity $\cos 2x = 1 - 2\sin^2 x$ in (3) to put it in the form

$$f(x, y) = x^3 - y + 2y\sin^2 x + g(y). \tag{5}$$

Comparing (4) and (5), we see that we must have

$$y^3 = -y + g(y) \qquad \text{and} \qquad h(x) = x^3.$$

Thus the solution we seek is given by

$$f(x, y) = x^3 + y^3 + 2y\sin^2 x = C.$$

We could also write the solution in the form (7) as

$$f(x, y) = x^3 - y\cos 2x + y + y^3 = C. \quad \blacksquare$$

If you compare the solution of Example 3 with that of Example 4 of Section 14.4, then you will see that they boil down to the same idea.

Integrating Factors*

Some first-order equations (1) that are not exact can be reduced to exact equations if we multiply through by some expression $M(x, y)$, called an ***integrating factor***. (Compare the discussion leading up to Theorem 1.3.) It is often difficult to find integrating factors, but we will indicate one approach that is sometimes useful. If $M(x, y)$ is an integrating factor for (1), then

$$M(x, y)P(x, y) + M(x, y)Q(x, y)\frac{dy}{dx} = 0 \tag{6}$$

must be exact. Assuming that M, P, and Q are continuous and have continuous partial derivatives on some simply connected domain D, Theorem 2.2 says that

$$\frac{\partial(MP)}{\partial y} = \frac{\partial(MQ)}{\partial x} \to M\frac{\partial P}{\partial y} + \frac{\partial M}{\partial y}P = M\frac{\partial Q}{\partial x} + \frac{\partial M}{\partial x}Q,$$

$$M\left(\frac{\partial Q}{\partial x} - \frac{\partial P}{\partial y}\right) = P\frac{\partial M}{\partial y} - Q\frac{\partial M}{\partial x}. \tag{7}$$

* Optional.

Now (7) is generally *much* harder to try to solve for M than (1) is to solve for y, so we make a simplifying assumption. One such assumption is that $M(x, y)$ *is a function of* x alone. In that case $\partial M/\partial y = 0$, so (7) becomes

(8) $$\frac{dM}{dx} = \frac{-M\left(\frac{\partial Q}{\partial x} - \frac{\partial P}{\partial y}\right)}{Q} \rightarrow \frac{dM}{M} = -\frac{1}{Q}\left(\frac{\partial Q}{\partial x} - \frac{\partial P}{\partial y}\right)dx.$$

If we are lucky enough that the right side of (8) involves only x and we can find an antiderivative $A(x)$ for it, then $\ln M = A(x)$, so that

(9) $$M = e^{A(x)}$$

will be an integrating factor for (1).

Example 4

Solve $2y - e^x + x\frac{dy}{dx} = 0$.

Solution. The equation is not exact, because

$$\frac{\partial Q}{\partial x} = 1 \neq 2 = \frac{\partial P}{\partial y}.$$

Notice that the right side of (8) involves just x:

$$-\frac{1}{Q}\left(\frac{\partial Q}{\partial x} - \frac{\partial P}{\partial y}\right) = -\frac{1}{x}(1 - 2) = \frac{1}{x}.$$

Thus, by (9), an integrating factor is

$$M = e^{\int 1/x\, dx} = e^{\ln|x|} = |x| = \pm x.$$

We need only one integrating factor M. Using $M = x$, we obtain

$$2xy - xe^x + x^2\frac{dy}{dx} = 0.$$

This equation is exact, because

$$\frac{\partial(x^2)}{\partial x} = 2x = \frac{\partial(2xy - xe^x)}{\partial y}.$$

Setting

$$\frac{\partial f}{\partial x} = 2xy - xe^x \qquad \text{and} \qquad \frac{\partial f}{\partial y} = x^2,$$

and computing the partial integrals, we find

$$\begin{aligned} f(x, y) &= x^2y - \int xe^x\, dx = x^2y - xe^x + e^x + C_1(y) \\ &= x^2y + C_2(x). \end{aligned}$$

Hence the general solution is $x^2y - xe^x + e^x = C$. ■

Similar analysis applies if M is assumed to be a function of y alone (Exercise 13) to show that

$$M = e^{B(y)} \tag{10}$$

will be an integrating factor of (1) if $B(y)$ is an antiderivative of

$$\frac{dM}{M} = \frac{1}{P}\left(\frac{\partial Q}{\partial x} - \frac{\partial P}{\partial y}\right) dy, \tag{11}$$

provided (11) involves only y.

Exercises 15.2

In Exercises 1–10, solve the given differential equation.

1. $y + (x - \sin y)\dfrac{dy}{dx} = 0$, where $y(1) = \pi$

2. $2xy + (x^2 + \cos y)\dfrac{dy}{dx} = 0$, where $y(1) = 0$

3. $(3x^2 + 4xy)\,dx + (2x^2 + 2y)\,dy = 0$

4. $(y\cos x + 2xe^y) + (\sin x + x^2e^y + 2)\dfrac{dy}{dx} = 0$

5. $x^2 + y^2 + 2xy\dfrac{dy}{dx} = 0$, where $y(0) = 1$

6. $x^2\dfrac{dy}{dx} + 2xy = 1$

7. $x\dfrac{dy}{dx} + y = y\sin y\dfrac{dy}{dx}$

8. $e^x\sin y + (e^x\cos y + 2y)\dfrac{dy}{dx} = 0$

9. $(3xy + y^2) + (x^2 + xy)\dfrac{dy}{dx} = 0$ [*Hint:* Use Equation (8).]

10. $2x^2y + (x^3 + x\cos y)\dfrac{dy}{dx} = 0$, where $y(1) = \pi$ [*Hint:* Use Equation (8).]

11. Find an integrating factor for $(xy^2 + 4x^2y) + (3x^2y + 4x^3)\dfrac{dy}{dx} = 0$ and solve.

12. Repeat Exercise 11 for $2y^3 - 3xy + x(3x - y^2)\dfrac{dy}{dx} = 0$. (Look for an integrating factor of the form x^my^n.)

13. Show that if $M(y)$ is an integrating factor for (1), then $M = e^{B(y)}$, where $B(y)$ is an antiderivative of

$$\frac{1}{P}\left(\frac{\partial Q}{\partial x} - \frac{\partial P}{\partial y}\right).$$

14. If $a_1b_2 - a_2b_1 \neq 0$, then show that the substitutions $x = X + h$ and $y = Y + k$ reduce the first-order equation

$$(a_1x + b_1y + c_1) + (a_2x + b_2y + c_2)\frac{dy}{dx} = 0 \tag{12}$$

to a homogeneous equation in X and Y, where (h, k) is any solution to

$$\begin{cases} a_1h + b_1k + c_1 = 0 \\ a_2h + b_2k + c_2 = 0. \end{cases}$$

15. If $a_1b_2 - a_2b_1 = 0$, then show that the substitution $z = a_1x + b_1y$ reduces (12) to a separable equation.

16. Solve $2x - y - 4 + (2y - x + 5)\dfrac{dy}{dx} = 0$ by the method of Exercise 14.

17. Solve $x + 2y + 3 + (2x + 4y - 1)\dfrac{dy}{dx} = 0$ by the method of Exercise 15.

18. The *Clairaut equation* is

$$y = px + f(p), \tag{13}$$

where $p = dy/dx$ and f is differentiable. Show that differentiation of (13) gives the equation

$$\frac{dp}{dx}(x + f'(p)) = 0. \tag{14}$$

19. For the particular Clairaut equation $y = px + p^2$, solve (14) by equating each factor to 0. Show that $dp/dx = 0$ leads to the *one-parameter* family of solutions $y = cx + c^2$, where c is the parameter, and $x + f'(p) = 0$ leads to the single solution $y = -\frac{1}{4}x^2$. Describe the geometric relationship between the single solution and the one-parameter family.

20. Repeat Exercise 19 for the equation $y = px - \frac{1}{3}p^3$.

15.3 Complex Numbers and Exponentials

In contrast to the first-order equations, considered so far, second-order linear differential equations are most conveniently solved using the complex exponential function. This section provides the ideas needed, beginning with the system of complex numbers.*

Over time, mathematicians have introduced successively larger number systems in order to solve polynomial equations. For instance, the natural number system **N** is inadequate to solve all equations $ax = b$, where a and $b \in \mathbf{N}$. Thus the system **F** of positive fractions was introduced. But **F** is inadequate for solving equations like $2x + 3 = 0$, so the system **Q** of rational numbers was necessary. But it in turn isn't rich enough to contain the roots of an equation like $x^2 - 2 = 0$. Thus the system **R** of real numbers arose. Still, some quadratic equations like

$$x^2 + 1 = 0 \tag{1}$$

remain unsolvable in **R**. This leads to the introduction of the set **C** of complex numbers. The following scheme, which was devised by Hamilton in the last century, provides the solutions of (1) and also of *any* polynomial equation of any degree.

3.1 DEFINITION The set **C** of ***complex numbers*** consists of all ordered pairs (a, b) of real numbers added, subtracted, and multiplied by the rules

$$(a, b) \pm (c, d) = (a \pm c, b \pm d)$$

$$(a, b)(c, d) = (ac - bd, ad + bc).$$

The complex number $(0, 1)$ is denoted i. A complex number $(a, 0)$ is identified with the real number a, and $(0, b)$ is called a *pure imaginary* number. The number a is called the *real part* of (a, b), and b is called its *imaginary* part. The term *imaginary* arose from the fact that

$$i^2 = i \cdot i = (0, 1)(0, 1) = (0 - 1, 0 + 0) = (-1, 0).$$

Since the complex number $(-1, 0)$ is identified with the real number -1, i is a number with a negative square:

$$i^2 = -1, \tag{2}$$

so we can indeed solve (1): its roots are $x = i$ and $x = -i$. Since the square of every *real* number is nonnegative, mathematicians applied the term *imaginary* to numbers like i. From $i^2 = -1$, it follows that $i^3 = -i$, $i^4 = 1$ and, more generally, for any positive integer n,

$$i^{2n} = (-1)^n, \qquad i^{2n+1} = (-1)^n i. \tag{3}$$

For any real number a we have

$$(a, 0)(c, d) = (ac - 0d, ad + 0c) = (ac, ad).$$

This is the same as the scalar product of (c, d), viewed as a vector in $\mathbf{R}^2$, with the real number a. We can thus identify **C** with the space $\mathbf{R}^2$, since the addition defined on **C** by Definition 3.1 coincides with ordinary addition of vectors. This provides a graphical representation for complex numbers as points (or vectors)

* The first part of this section, up to p. 1011 where the discussion of complex exponentials begins, is for reference and review.

FIGURE 3.1

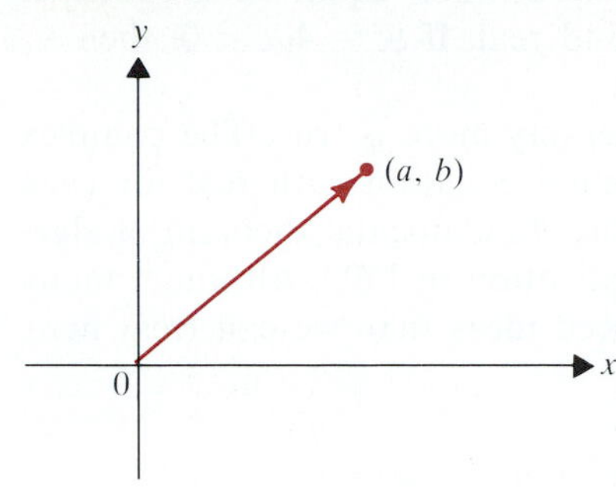

FIGURE 3.2

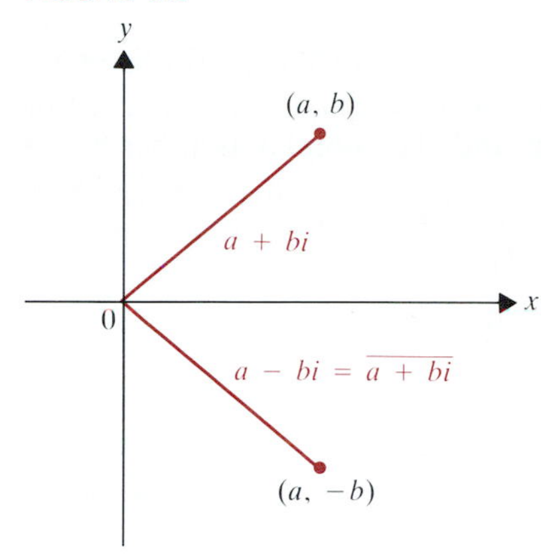

in $\mathbf{R}^2$. That scheme goes back to the Swiss mathematician Jean R. Argand (1768–1822) in 1806 and is illustrated in **Figure 3.1.** Using the standard basis vectors (1, 0) and (0, 1) as usual, we write

$$(a, b) = (a, 0) + (0, b) = a(1, 0) + b(0, 1) = a1 + bi = a + bi.$$

This is the traditional notation for complex numbers, and we shall use it for the most part, rather than (a, b). Since $(a, b) = (c, d)$ if and only if $a = c$ and $b = d$, it follows that $a + bi = c + di$ if and only if $a = c$ and $b = d$.

To divide we need the ***complex conjugate*** $\overline{a + bi} = a - bi$, whose graphical significance is shown in **Figure 3.2:** $\overline{a + bi}$ is the *reflection* of $a + bi$ in the x-axis of $\mathbf{R}^2$. Writing z for an arbitrary complex number $a + bi$, we have the following algebraic properties for the conjugation operation. (Their verification is left for Exercise 1.)

$$\text{(4)} \qquad \overline{z_1 + z_2} = \bar{z}_1 + \bar{z}_2, \qquad \overline{z_1 z_2} = \bar{z}_1 \bar{z}_2, \qquad \text{and} \qquad \overline{az} = a\bar{z},$$

for every real number a. For any complex number z, we also have from Definition 2.1

$$\text{(5)} \qquad z\bar{z} = (a + bi)(a - bi) = a^2 + b^2 + (-ab + ab)i = a^2 + b^2,$$

so $z\bar{z}$ is real and nonnegative.

With the help of (5) we can define the reciprocal of any nonzero complex number $z = a + bi$. Since at least one of a and b is nonzero, $a^2 + b^2 \neq 0$, so we can divide through by it, obtaining

$$z\bar{z} = a^2 + b^2 \to \frac{z\bar{z}}{a^2 + b^2} = 1 \to z\frac{\bar{z}}{a^2 + b^2} = 1.$$

We thus define the *reciprocal* of z to be

$$z^{-1} = \frac{\bar{z}}{z\bar{z}} = \frac{a - bi}{a^2 + b^2}.$$

If $z_1 = a + bi$ and $z_2 = c + di \neq 0$, then we define

$$z_1 \div z_2 = z_1 z_2^{-1} = \frac{(a + bi)(c - di)}{c^2 + d^2}.$$

Example 1

If $z_1 = 2 - 3i$ and $z_2 = -3 + 4i$, then find
(a) $2z_1 - z_2$, (b) $z_1 \div z_2$.

Solution. (a) $2z_1 - z_2 = 4 - 6i - (-3 + 4i) = 4 - 6i + 3 - 4i = 7 - 10i.$

(b) $$z_1 \div z_2 = \frac{(2 - 3i)(-3 - 4i)}{9 + 16} = \frac{-6 - 12 + (9 - 8)i}{25} = -\frac{18}{25} + \frac{1}{25}i. \blacksquare$$

Recall that the quadratic formula provides a complete solution for the general quadratic equation

$$ax^2 + bx + c = 0,$$

where $a \neq 0$, b and c are real numbers. The roots are

$$x_1 = \frac{-b + \sqrt{b^2 - 4ac}}{2a} \qquad \text{and} \qquad x_2 = \frac{-b - \sqrt{b^2 - 4ac}}{2a}.$$

If $b^2 - 4ac = 0$, then $x_1 = x_2$, so the equation has equal real roots. If $b^2 - 4ac > 0$, then x_1 and x_2 are unequal and real. If $b^2 - 4ac < 0$, then x_1 and x_2 are conjugate complex numbers.

The following theorem states that considerably more is true. The complex numbers contain the roots of *every* polynomial equation with real (or even complex) coefficients. This fact, once called the "fundamental theorem of algebra," was proved by Gauss in his Ph.D. dissertation in 1799. Although many proofs are known, they involve more advanced ideas than we can treat here.

3.2 THEOREM Every polynomial equation

$$(6) \qquad a_n x^n + a_{n-1}x^{n-1} + \ldots + a_2 x^2 + a_1 x + a_0 = 0$$

with real or complex coefficients a_i has n roots in **C** (some of which may be repeated roots) if $a_n \neq 0$.

With the aid of this result, it is easy to show that, as for the quadratic equation, complex roots of real polynomial equations occur in conjugate pairs. For if in (6) every coefficient a_i is a real number, and the complex number z is a root, then

$$a_n z^n + \ldots + a_1 z + a_0 = 0.$$

Taking the conjugate of each side and using (4), we get

$$\begin{aligned} \bar{0} = 0 = \overline{a_n z^n + \ldots + a_2 z^2 + a_1 z + a_0} &= \overline{a_n z^n} + \ldots + \overline{a_2 z^2} + \overline{a_1 z} + \overline{a_0} \\ &= a_n \overline{z^n} + \ldots + a_2 \overline{z^2} + a_1 \bar{z} + a_0 && \textit{since the } a_i \textit{ are real} \\ &= a_n \bar{z}^n + \ldots + a_2 \bar{z}^2 + a_1 \bar{z} + a_0. && \textit{by the second equation in (4)} \end{aligned}$$

Thus $\bar{z}$ is also a root of (4).

A special name is given to the length of the complex number $z = (a, b)$ viewed as a vector in $\mathbf{R}^2$.

FIGURE 3.3

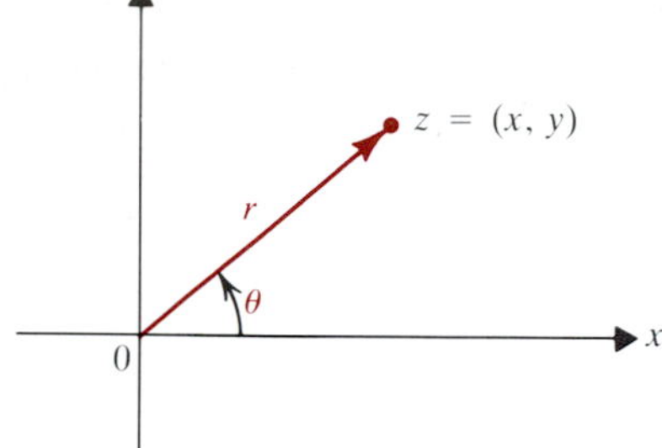

3.3 DEFINITION The ***absolute value*** (or ***modulus***) of the complex number $z = a + bi$ is

$$|z| = \sqrt{z\bar{z}} = \sqrt{a^2 + b^2}.$$

This leads to the *polar form* of $z = a + bi$.

3.4 DEFINITION The *polar form* of the complex number $z = a + bi$ is

$$z = r(\cos\theta + i \sin\theta),$$

where $[r, \theta]$ are polar coordinates for $(x, y) \in \mathbf{R}^2$, with $r = |z|$. The angle θ is called the ***argument*** of z.

Thus, given $z = x + yi$, we take $r = |z|$; and for θ we take the angle from the x-axis to the vector $(x, y) \in \mathbf{R}^2$ measured counterclockwise. See **Figure 3.3.** The polar form is particularly well suited to multiplication of complex numbers, as shown by the following result, whose proof is asked for in Exercises 7 and 8.

3.5 THEOREM

De Moivre's Theorem.

(a) If $z_1 = r_1(\cos\theta_1 + i \sin\theta_1)$ and $z_2 = r_2(\cos\theta_2 + i\sin\theta_2)$, then

$$z_1 z_2 = r_1 r_2[\cos(\theta_1 + \theta_2) + i \sin(\theta_1 + \theta_2)].$$

(b) If $z = r(\cos\theta + i\sin\theta)$, then

$$z^n = r^n(\cos n\theta + i \sin n\theta).$$

FIGURE 3.4

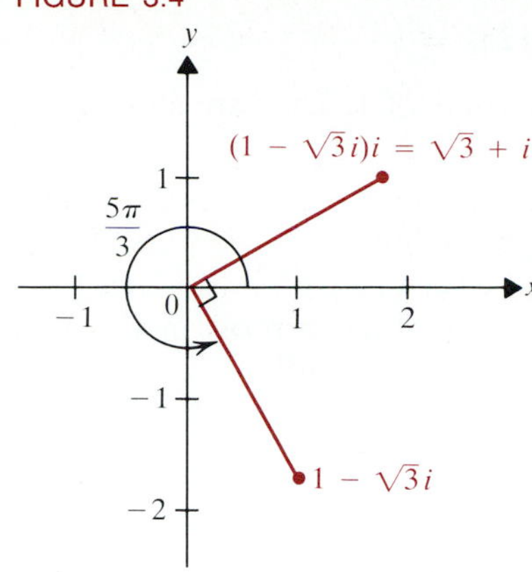

Example 2

(a) Describe the geometric effect of multiplying $1 - \sqrt{3}i$ by i.

(b) Compute $(1 - i)^5$.

Solution. (a) We can express $1 - \sqrt{3}i = (1, -\sqrt{3})$ in polar form as

$$2(\cos 5\pi/3 + i \sin 5\pi/3),$$

and $i = (0, 1)$ as

$$\cos \frac{1}{2}\pi + i \sin \frac{1}{2}\pi.$$

If we multiply $1 - \sqrt{3}i$ by i, then according to Theorem 3.5(a), we obtain a product whose argument is $5\pi/3 + \pi/2 = 13\pi/6$ (or $\pi/6$) and whose absolute value is $2 \cdot 1 = 2$. We can think of $1 - \sqrt{3}i$ as being rotated counterclockwise through the angle $\pi/2$. See **Figure 3.4.**

(b) In polar form we have

$$1 - i = (1, -1) = \sqrt{2}\left(\cos \frac{7\pi}{4} + i \sin \frac{7\pi}{4}\right).$$

Thus by Theorem 3.5(b),

$$(1 - i)^5 = (\sqrt{2})^5\left(\cos \frac{35\pi}{4} + i \sin \frac{35\pi}{4}\right) = 4\sqrt{2}\left(\cos \frac{3\pi}{4} + i \sin \frac{3\pi}{4}\right)$$

$$= 4\sqrt{2}\left(-\frac{1}{\sqrt{2}} + i \frac{1}{\sqrt{2}}\right) = -4 + 4i. \quad ■$$

We can use De Moivre's theorem to find the nth roots of a given complex number, which are determined by the nth roots of 1.

3.6 THEOREM

(a) For each positive integer n, the equation $z^n = 1$ has exactly n distinct complex roots $\omega, \omega^2, \omega^3, \ldots, \omega^{n-1}, \omega^n = 1$, where

$$\omega = \cos \frac{2\pi}{n} + i \sin \frac{2\pi}{n} \tag{7}$$

is called a ***primitive nth root of 1.***

(b) Every nonzero complex number $z = r(\cos \theta + i \sin \theta)$ has exactly n distinct complex nth roots $\sigma, \sigma\omega, \sigma\omega^2, \ldots, \sigma\omega^{n-1}$, where

$$\sigma = r^{1/n}\left(\cos \frac{\theta}{n} + i \sin \frac{\theta}{n}\right) \tag{8}$$

is called a ***primitive nth root of z.***

Proof. (a) It is clear from Theorem 3.5(b) that

$$\omega^n = \cos 2\pi + i \sin 2\pi = 1.$$

Also, for each integer $k (1 \le k \le n)$, we have

$$\omega^k = \cos \frac{2\pi k}{n} + i \sin \frac{2\pi k}{n}.$$

Then

$$(\omega^k)^n = \cos 2\pi k + i \sin 2\pi k = 1.$$

Thus all of the listed powers of ω are the nth roots of 1. They are all distinct, because they have different arguments,

$$\frac{2\pi}{n}, \frac{4\pi}{n}, \frac{6\pi}{n}, \ldots, \frac{(n-1)2\pi}{n}, 2\pi.$$

By Theorem 3.2 the polynomial equation $x^n - 1 = 0$ has n roots in $\mathbf{C}$, so *all* the nth roots of 1 belong to the set $\{\omega, \omega^2, \omega^3, \ldots, \omega^{n-1}, \omega^n\}$.

(b) First, each $\sigma\omega^k$ is an nth root of z, because

$$(\sigma\omega^k)^n = \sigma^n(\omega^k)^n = \sigma^n 1 \qquad \textit{by part (a)}$$

$$= (r^{1/n})^n\left(\cos\frac{n\theta}{n} + i \sin\frac{n\theta}{n}\right) = r(\cos\theta + i\sin\theta) = z.$$

Moreover, we have *all* the nth roots of z. To see this, note that if ϕ is any nth root of z, then $\phi^n = z$. Therefore

$$\left(\frac{\phi}{\sigma}\right)^n = \frac{\phi^n}{\sigma^n} = \frac{z}{z} = 1.$$

Hence ϕ/σ is an nth root of 1. Then, by part (a), $\phi/\sigma = \omega^k$ for some k, $1 \le k \le n$. Thus $\phi = \sigma\omega^k$, as desired. QED

Example 3

(a) Find the fourth roots of 1.

(b) Find the fourth roots of $-8 + 8\sqrt{3}i$.

Solution. (a) We have from (7)

$$\omega = \cos\frac{2\pi}{4} + i\sin\frac{2\pi}{4} = \cos\frac{1}{2}\pi + i\sin\frac{1}{2}\pi = i,$$

$$\omega^2 = \cos\frac{4\pi}{4} + i\sin\frac{4\pi}{4} = i^2 = -1,$$

$$\omega^3 = \omega\omega^2 = (i)(-1) = -i,$$

$$\omega^4 = \omega\omega^3 = i(-i) = -i^2 = 1.$$

So the fourth roots of 1 are i, $-i$, 1, and -1.

(b) Let

$$z = -8 + 8\sqrt{3}i = 16\left(-\frac{1}{2} + \frac{\sqrt{3}}{2}i\right) = 16\left(\cos\frac{2\pi}{3} + i\sin\frac{2\pi}{3}\right).$$

Then from (8) we have

$$\sigma = 16^{1/4}\left(\cos\frac{2\pi}{12} + i\sin\frac{2\pi}{12}\right) = 2\left(\cos\frac{\pi}{6} + i\sin\frac{\pi}{6}\right) = 2\left(\frac{\sqrt{3}}{2} + \frac{1}{2}i\right) = \sqrt{3} + i,$$

$$\sigma\omega = (\sqrt{3} + i)i = -1 + \sqrt{3}i,$$

$$\sigma\omega^2 = (\sigma\omega)\omega = (-1 + \sqrt{3}i)i = -\sqrt{3} - i,$$

$$\sigma\omega^3 = (\sigma\omega^2)\omega = (-\sqrt{3} - i)i = 1 - \sqrt{3}i.$$

So the fourth roots of $-8 + 8\sqrt{3}i$ are $\sqrt{3} + i$, $-\sqrt{3} - i$, $-1 + \sqrt{3}i$, and $1 - \sqrt{3}i$. ■

FIGURE 3.5

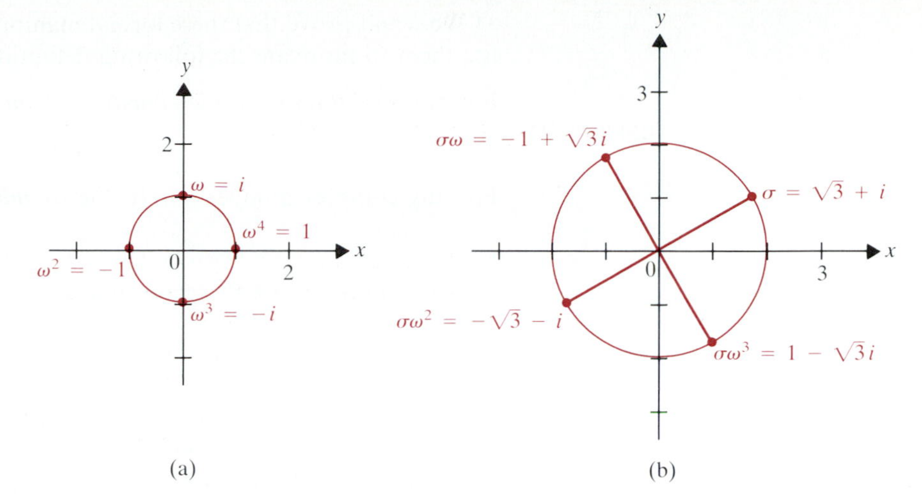

In **Figure 3.5** we show the roots found in parts (a) and (b). Note that in each case the roots are equally spaced around the circle centered at the origin with radius $r^{1/n}$; as the powers of ω increase, we hop around this circle in the counterclockwise sense.

Complex Exponentials

We now come to the main business of this section, the definition of e^z for a complex variable $z = x + iy$. If the usual laws of exponents are to hold, then we should have

$$e^{x+iy} = e^x e^{iy}.$$

The factor e^x presents no difficulty. We already know what that is. So we are faced with defining e^{iy}. Recall from Section 9.8 that for all real numbers y the Taylor series expansion of e^y converges to e^y:

$$e^y = 1 + y + \frac{y^2}{2!} + \frac{y^3}{3!} + \ldots + \frac{y^n}{n!} + \ldots. \tag{9}$$

If we assume that (7) continues to hold for complex numbers and replace y by the *complex* number iy, then we get

$$\begin{aligned}
e^{iy} &= 1 + iy + \frac{(iy)^2}{2!} + \frac{(iy)^3}{3!} + \frac{(iy)^4}{4!} + \frac{(iy)^5}{5!} + \frac{(iy)^6}{6!} + \ldots + \frac{(iy)^{2n}}{(2n)!} \\
&\quad + \frac{(iy)^{2n+1}}{(2n+1)!} + \ldots \\
&= 1 + iy - \frac{1}{2!}y^2 - i\frac{1}{3!}y^3 + \frac{1}{4!}y^4 + i\frac{1}{5!}y^5 - \frac{1}{6!}y^6 + \ldots + \frac{(-1)^n}{(2n)!}y^{2n} \\
&\quad + i\frac{(-1)^n}{(2n+1)!}y^{2n+1} + \ldots && \textit{by (3)} \\
&= \left(1 - \frac{1}{2!}y^2 + \frac{1}{4!}y^4 - \frac{1}{6!}y^6 + \ldots + \frac{(-1)^n}{(2n)!}y^{2n} + \ldots\right) \\
&\quad + i\left(y - \frac{1}{3!}y^3 + \frac{1}{5!}y^5 - \ldots + \frac{(-1)^n}{(2n+1)!}y^{2n+1} + \ldots\right) \\
&= \cos y + i \sin y. && \textit{by Table 1, p. 617}
\end{aligned}$$

We won't prove that these formal manipulations can be justified. Instead we use them to motivate the following definition of complex exponentials.

3.7 DEFINITION For any real number y, the ***complex exponential*** e^{iy} is

$$e^{iy} = \cos y + i \sin y.$$

For any complex number $x + iy$, the ***complex exponential*** e^{x+iy} is

$$e^{x+iy} = e^x(\cos y + i \sin y).$$

Note that e^{iy} is a unit vector and that

$$e^{2\pi i/n} = \cos \frac{2\pi}{n} + i \sin \frac{2\pi}{n} = \omega$$

is a primitive nth root of 1. So *all* nth roots of 1 are of the form $e^{2\pi ri/n}$ for $1 \leq r \leq n$. We next use Definition 3.7 to derive the law of exponents that we assumed should hold. To do so, let $z_1 = x_1 + iy_1$ and $z_2 = x_2 + iy_2$. Then we have

$$\begin{aligned}
e^{z_1}e^{z_2} &= e^{x_1}(\cos y_1 + i \sin y_1)e^{x_2}(\cos y_2 + i \sin y_2) \\
&= e^{x_1}e^{x_2}(\cos y_1 + i \sin y_1)(\cos y_2 + i \sin y_2) \\
&= e^{x_1+x_2}(\cos(y_1 + y_2) + i \sin(y_1 + y_2)) \qquad \textit{by Theorem 3.5(a)} \\
&= e^{z_1+z_2}.
\end{aligned}$$

The real and imaginary parts of e^{iy} are the cosine and sine functions, respectively. Writing $\operatorname{Re} z$ for the real part of z and $\operatorname{Im} z$ for the imaginary part, we thus have

$$\operatorname{Re}(e^{iy}) = \cos y \quad \text{and} \quad \operatorname{Im}(e^{iy}) = \sin y. \tag{10}$$

We now extend the calculus of real-valued functions of a real variable x to complex-valued functions of x in the natural way.

3.8 DEFINITION If $f = g + ih$ is a complex-valued function of the real variable x, where $g(x) = \operatorname{Re} f(x)$ and $h(x) = f(x)$, then

$$\frac{df}{dx} = \frac{dg}{dx} + i \frac{dh}{dx}, \tag{11}$$

and

$$\int f(x)\,dx = \int g(x)\,dx + i \int h(x)\,dx. \tag{12}$$

From this definition we can derive the rules of differentiation and integration of the complex exponential function $e^{\alpha x}$, where α is a complex number $a + bi$. For we have

$$\begin{aligned}
\frac{d}{dx} e^{\alpha x} &= \frac{d}{dx} e^{(a+bi)x} = \frac{d}{dx} e^{ax+bix} \\
&= \frac{d}{dx}[e^{ax}(\cos bx + i \sin bx)] = \frac{d}{dx}(e^{ax}\cos bx) + i \frac{d}{dx}(e^{ax}\sin bx) \qquad \textit{by (11)} \\
&= ae^{ax}\cos bx - be^{ax}\sin bx + i(ae^{ax}\sin bx + be^{ax}\cos bx) \\
&= ae^{ax}(\cos bx + i \sin bx) + ibe^{ax}(\cos bx + i \sin bx) \qquad \textit{since } i^2 = -1 \\
&= (a + bi)e^{ax}(\cos bx + i \sin bx) = \alpha e^{ax+ibx} = \alpha e^{\alpha x}.
\end{aligned}$$

We thus have

$$\frac{d}{dx}(e^{\alpha x}) = \alpha e^{\alpha x}. \tag{13}$$

It then follows just as in the real case that if $\alpha \neq 0$, then

$$\int e^{\alpha x}\,dx = \frac{1}{\alpha}e^{\alpha x} + C. \tag{14}$$

Example 4

Find

(a) $\dfrac{d}{dx}(\tan x - i\sec x)$, (b) $\displaystyle\int(\tan x - i\sec x)\,dx$,

(c) $\dfrac{d}{dx}\left[e^{3x}\left(\cos\frac{1}{2}x + i\sin\frac{1}{2}x\right)\right]$, (d) $\displaystyle\int e^{3x}\left(\cos\frac{1}{2}x + i\sin\frac{1}{2}x\right)dx$.

Solution. (a) From (11) we find

$$\frac{d}{dx}(\tan x - i\sec x) = \sec^2 x - i\sec x\tan x.$$

(b) Using (12) we have

$$\begin{aligned}\int(\tan x - i\sec x)\,dx &= \int\tan x\,dx - i\int\sec x\,dx\\ &= \ln|\sec x| - i\ln|\sec x + \tan x| + C.\end{aligned}$$

(c, d) First, by Definition 3.7 we have

$$e^{3x}\left(\cos\frac{1}{2}x + i\sin\frac{1}{2}x\right) = e^{3x+ix/2} = e^{(3+i/2)x}.$$

Thus (13) with $\alpha = 3 + \frac{1}{2}i$ gives

$$\begin{aligned}\frac{d}{dx}&\left[e^{3x}\left(\cos\frac{1}{2}x + i\sin\frac{1}{2}x\right)\right]\\ &= \left(3 + \frac{1}{2}i\right)e^{(3+i/2)x}\\ &= \left(3 + \frac{1}{2}i\right)e^{3x}\left(\cos\frac{1}{2}x + i\sin\frac{1}{2}x\right)\\ &= e^{3x}\left[3\cos\frac{1}{2}x - \frac{1}{2}\sin\frac{1}{2}x + i\left(\frac{1}{2}\cos x + 3\sin\frac{1}{2}x\right)\right].\end{aligned}$$

Then we get from (14)

$$\begin{aligned}\int &e^{3x}\left(\cos\frac{1}{2}x + i\sin\frac{1}{2}x\right)dx\\ &= \int e^{(3+i/2)x}\,dx = \frac{1}{3+(1/2)i}e^{(3+i/2)x} + C\\ &= \frac{3-(1/2)i}{9+1/4}e^{3x}\left(\cos\frac{1}{2}x + i\sin\frac{1}{2}x\right) + C\\ &= \frac{4}{37}e^{3x}\left[3\cos\frac{1}{2}x + \frac{1}{2}\sin\frac{1}{2}x + i\left(3\sin\frac{1}{2}x - \frac{1}{2}\cos\frac{1}{2}x\right)\right] + C. \quad \blacksquare\end{aligned}$$

Exercises 15.3

1. Prove (4).
2. (a) Show that $z + \bar{z} = 2\,\mathrm{Re}(z)$.
   (b) Show that $z - \bar{z} = 2i\,\mathrm{Im}(z)$.
3. If $z_1 = 1 - i$ and $z_2 = 3 + 2i$, then compute $z_1 + z_2$, $z_1 z_2$, and $z_1 \div z_2$.
4. (a) Prove that multiplication of complex numbers is associative and commutative.
   (b) Show that the distributive law holds for **C**.
5. Find all roots of the polynomial equation $x^4 + 3x^2 + 2 = 0$.
6. Find all roots of the polynomial equation $x^4 + 5x^2 + 6 = 0$.
7. Prove Theorem 3.5(a).
8. Prove Theorem 3.5(b).
9. Describe the geometric effect of multiplying $2 - 3i$ by $1 + i$.
10. Compute $(1 + i)^8$.
11. Compute $(\sqrt{3} - i)^4$.
12. Find all five fifth roots of 1.
13. Find the cube roots of $z = \frac{27}{2}(1 + i\sqrt{3})$.
14. Find the cube roots of $z = 32(\sqrt{3} - i)$.
15. Find the sixth roots of -1.
16. Find the sixth roots of 64.
17. If $\omega \neq 1$ is an nth root of 1, then show that ω is a root of the *cyclotomic polynomial*
$$1 + x + x^2 + \ldots + x^{n-1} = 0.$$
18. A ***primitive nth root of 1*** is an nth root μ of 1 such that every nth root of 1 is some positive integral power of μ. Find all the primitive eighth roots of 1.
19. Establish *Euler's equation*, $e^{i\pi} = -1$. What is e^i? What is $e^{2\pi i}$?
20. Compute $e^{\pi i/3}$ and $e^{\ln 2 + \pi i/4}$.
21. Show from Definition 3.7 that
$$\cos\theta = \frac{e^{i\theta} + e^{-i\theta}}{2} = \mathrm{Re}(e^{i\theta}).$$
22. Show from Definition 3.7 that
$$\sin\theta = \frac{e^{i\theta} - e^{-i\theta}}{2i} = \mathrm{Im}(e^{i\theta}).$$
23. Find the first three derivatives of $e^x(\cos 2x + i\sin 2x)$.
24. Find the first three derivatives of $e^{2x}(\cos x + i\sin x)$.
25. Find the eighth derivative of $e^{x/\sqrt{2}}\left(\cos\frac{x}{\sqrt{2}} + i\sin\frac{x}{\sqrt{2}}\right)$.
26. Find the twelfth derivative of $e^{\sqrt{3}x/2}\left(\cos\frac{x}{2} + i\sin\frac{x}{2}\right)$.
27. By computing $\int e^{\alpha x}\,dx$ for $\alpha = a + ib$, obtain formulas for $\int e^{ax}\cos bx\,dx$ and $\int e^{ax}\sin bx\,dx$. **(This is used in Section 5.)**
28. Using the identities in Exercises 21 and 22, show that $\int_0^{2\pi}\sin mx\cos nx\,dx = 0$ for any positive integers m and n.
29. Assuming that m and n are unequal positive integers, show that $\int_0^{2\pi}\sin mx\sin nx\,dx = 0$. What happens if $m = n$?
30. Compute $\int_0^{2\pi}\cos mx\cos nx\,dx$ for m and n positive integers.

15.4 Second-Order Homogeneous Linear Differential Equations with Constant Coefficients

We turn now to *second-order* differential equations, those that involve the second derivative of a function $y = f(x)$. Apart from some special types considered in Exercises 27 to 33, we will confine ourselves to linear equations. The results we present go back to the Swiss mathematician Johann Bernoulli (1667–1748), the brother of Jakob, and were obtained prior to 1700.

4.1 DEFINITION The ***general second-order linear differential equation*** is

$$\frac{d^2y}{dx^2} + p(x)\frac{dy}{dx} + q(x)y = s(x), \tag{1}$$

where p, q, and s are continuous functions, and $y = y(x)$ is a twice differentiable function to be determined.

Example 1

Classify the following second-order differential equations as linear or nonlinear.

(a) $e^x \dfrac{d^2y}{dx^2} + \sin^2 x \dfrac{dy}{dx} + (\cos x)y = x^2$,

(b) $\left(\dfrac{d^2y}{dx^2}\right)^2 + 6\dfrac{dy}{dx} + 9y = x$.

Solution. Multiplication through by e^{-x} produces

$$\frac{d^2y}{dx^2} + e^{-x}\sin^2 x \frac{dy}{dx} + (e^{-x}\cos x)y = x^2e^{-x},$$

which is of the form (1) with $p(x) = e^{-x}\sin^2 x$, $q(x) = e^{-x}\cos x$, and $s(x) = x^2e^{-x}$. Thus the given equation is linear.

(b) Since d^2y/dx^2 occurs to the second power, the given equation is not of the form (1) or transformable to that form. Thus this second-order differential equation is nonlinear. ■

As the terminology suggests, there is a close analogy between second-order and first-order linear differential equations. Our first result illustrates this. It says that the solution set of (1) is describable in exactly the same way as the solution set of a first-order linear differential equation. The proof is also just like the one given for Theorem 1.2 and so is left for Exercises 23 and 24.

4.2 THEOREM

The solution set of (1) consists of all functions $y = y(x)$ of the form

(2) $$y(x) = y_0(x) + y_p(x),$$

where $y_p(x)$ is one particular solution to (1), and $y_0(x)$ is any solution to the associated homogeneous linear differential equation

(3) $$\frac{d^2y}{dx^2} + p(x)\frac{dy}{dx} + q(x)y = 0.$$

Thus we can solve (1) in two steps. *First*, find the general solution of the associated homogeneous linear differential equation (3); *second*, find one particular solution $y_p(x)$ to (1). In this section we carry out the first step for an important special class of linear differential equations (1) that occur widely in applied work, namely, those for which $p(x) = p$ and $q(x) = q$ are *constants*. That is, we consider linear equations of the form

(3′) $$\frac{d^2y}{dx^2} + p\frac{dy}{dx} + qy = 0,$$

for fixed real numbers p and q. The solution of (3′) is facilitated by introducing some special notation for the fundamental differentiation operator d/dx and the second derivative operator d^2/dx^2.

4.3 DEFINITION

The ***linear differential operators D*** and D^2 are

$$D = \frac{d}{dx} \quad \text{and} \quad D^2 = \frac{d^2}{dx^2}.$$

With this notation (3′) takes the form

$$D^2y + pDy + qy = 0,$$

that is,

$$(D^2 + pD + q)(y) = 0. \tag{4}$$

The expression $D^2 + pD + q$ is called a second-degree ***polynomial differential operator*** in D. We also will work with first-order differential operators of the form $D - r$, where r is a real number. Such an operator acts on the function y by the formula

$$(D - r)y = Dy - ry = \frac{dy}{dx} - ry. \tag{5}$$

Example 2

Express the linear differential equations

(a) $\dfrac{d^2y}{dx^2} - 5\dfrac{dy}{dx} + 6y = 0$ and (b) $\dfrac{dy}{dx} - \dfrac{1}{2}y = x \tan x$

in terms of the operator notation of Definition 4.3.

Solution. (a) This equation is of the form (3′), with $p = -5$ and $q = 6$. Thus from (4) we have

$$(D^2 - 5D + 6)(y) = 0.$$

(b) Here $r = 1/2$, so from (5) the given equation becomes

$$\left(D - \frac{1}{2}\right)(y) = x \tan x. \quad ■$$

Polynomial differential operators in D follow the same rules of algebra as ordinary polynomials, if we "multiply" them by composition. In particular, we have the following useful rule.

4.4 LEMMA If r_1 and r_2 are real numbers, then

$$(D - r_1) \circ (D - r_2) = D^2 - (r_1 + r_2)D + r_1r_2.$$

Proof. We just have to verify that each side has the same effect on any twice differentiable function $y(x)$:

$$\begin{aligned}
(D - r_1) \circ (D - r_2)(y) &= (D - r_1)((D - r_2)(y)) \\
&= (D - r_1)\left(\frac{dy}{dx} - r_2y\right) = D\left(\frac{dy}{dx} - r_2y\right) - r_1\left(\frac{dy}{dx} - r_2y\right) \\
&= \frac{d^2y}{dx^2} - r_2\frac{dy}{dx} - r_1\frac{dy}{dx} + r_1r_2y \\
&= \frac{d^2y}{dx^2} - (r_1 + r_2)\frac{dy}{dx} + r_1r_2y = D^2y - (r_1 + r_2)Dy + r_1r_2y \\
&= (D^2 - (r_1 + r_2)D + r_1r_2)(y). \quad \text{QED}
\end{aligned}$$

In view of this lemma, we can *factor* any second-order polynomial differential operator $D^2 + pD + q$ as

$$D^2 + pD + q = (D - r_1) \circ (D - r_2),$$

where r_1 and r_2 are the roots of the corresponding polynomial $r^2 + pr + q$.

Example 3

Express the second-order polynomial differential operator $D^2 - 5D + 6$ in the form of Lemma 4.4.

Solution. Since we can factor the corresponding polynomial

$$r^2 - 5r + 6 = (r - 3)(r - 2),$$

it follows from Lemma 4.4 that

$$D^2 - 5D + 6 = (D - 3) \circ (D - 2). \blacksquare$$

The polynomial corresponding to a second-order polynomial operator is so important that it has a special name.

4.5 DEFINITION

The ***auxiliary polynomial*** of the operator $D^2 + pD + q$ is the polynomial $r^2 + pr + q$. The ***auxiliary equation*** of the homogeneous linear differential equation

$$(D^2 + pD + q)(y) = 0 \tag{4}$$

is

$$r^2 + pr + q = 0. \tag{6}$$

We are going to use (6) to solve (4). Since the roots of (6) may be complex, we need the following result to find the general real solution of (4).

4.6 LEMMA

The complex function

$$y(x) = f(x) + ig(x)$$

is a solution of (4) if and only if the real and imaginary parts, $\text{Re}(y(x)) = f(x)$ and $\text{Im}(y(x)) = g(x)$, are real solutions of (4).

Proof. If $y(x) = f(x) + ig(x)$ is a solution of (4), then we have

$$Dy = \frac{dy}{dx} = f'(x) + ig'(x) \rightarrow D^2y = \frac{d^2y}{dx^2} = f''(x) + ig''(x).$$

Thus

$$f''(x) + ig''(x) + pf'(x) + ipg'(x) + qf(x) + iqg(x) = 0,$$

$$f''(x) + pf'(x) + qf(x) + i(g''(x) + pg'(x) + qg(x)) = 0 + 0i.$$

Then

$$f''(x) + pf'(x) + qf(x) = 0 \qquad \text{and} \qquad g''(x) + pg'(x) + qg(x) = 0,$$

that is, $f(x)$ and $g(x)$ are both solutions of (4). On the other hand, if $f(x)$ and $g(x)$ are real solutions of (4), then $y(x) = f(x) + ig(x)$ is a complex solution, because

$$\begin{aligned}(D^2 + pD + q)(f(x) + ig(x)) &= f''(x) + ig''(x) + pf'(x) + ipg'(x) + qf(x) + iqg(x)\\ &= f''(x) + pf'(x) + qf(x) + i(g''(x) + pg'(x) + qg(x))\\ &= 0 + i0 = 0. \quad \boxed{\text{QED}}\end{aligned}$$

The method we use to solve (4) is due to the French mathematician Jean d'Alembert (1717–1783) and the Italian mathematician Jacopo Riccati (1676–1754). The idea is to reduce the solution of (4) to the solution of two *first-order linear* differential equations, which we then can easily solve as in Section 1. If r_1 and r_2 are the roots of the auxiliary equation $r^2 + pr + q = 0$, then by Lemma 4.4

$$(D - r_1) \circ (D - r_2)(y) = 0. \tag{7}$$

Now the trick! Let

$$z = (D - r_2)(y). \tag{8}$$

If we substitute (8) into (7), then it becomes

$$(D - r_1)(z) = 0 \rightarrow \frac{dz}{dx} - r_1 z = 0.$$

This is a *first-order homogeneous* linear differential equation whose solution is

$$z(x) = ke^{r_1 x}.$$

Now we can substitute this back into (8), getting

$$ke^{r_1(x)} = (D - r_2)(y).$$

That can be put in the form

$$\frac{dy}{dx} - r_2 y = ke^{r_1 x}, \tag{9}$$

which is a first-order linear differential equation. It thus has integrating factor $e^{-r_2 x}$. If we multiply (9) through by this factor, we obtain, as usual,

$$\frac{d}{dx}(e^{-r_2 x} y) = ke^{(r_1 - r_2)x}. \tag{10}$$

At this point we have to distinguish two cases.

(a) Suppose that $r_1 \neq r_2$. Then we integrate (10) to obtain

$$e^{-r_2 x} y = \frac{k}{r_1 - r_2} e^{(r_1 - r_2)} + c_2 \rightarrow y = \frac{k}{r_1 - r_2} e^{r_1 x} + c_2 e^{r_2 x},$$

$$y = c_1 e^{r_1 x} + c_2 e^{r_2 x}, \tag{11}$$

where

$$c_1 = \frac{k}{r_1 - r_2}.$$

(b) Suppose that $r_1 = r_2$. Then (10) is really

$$\frac{d}{dx}(e^{-r_2 x} y) = k.$$

So integration gives $e^{-r_2 x}y = kx + c_1$, or, equivalently,

(12) $$y = c_1 e^{r_2 x} + c_2 x e^{r_2 x},$$

where we write c_2 in place of k for notational symmetry.

It is easy to verify that any function of the form (11) is a solution to (7) in case $r_1 \neq r_2$ and that any solution of the form (12) is a solution to (7) in case $r_1 = r_2$ (Exercise 25). Thus we have actually found the *general real solution* to (7) provided r_1 and r_2 are real numbers. We state this formally before taking up the case in which r_1 and r_2 are complex numbers.

4.7 THEOREM

If the auxiliary equation for

(4) $$(D^2 + pD + q)(y) = 0$$

has real roots r_1 and r_2, then the general solution of (4) is

(11) $$y = c_1 e^{r_1 x} + c_2 e^{r_2 x}, \qquad \text{if } r_1 \neq r_2,$$

and is

(12) $$y = c_1 e^{rx} + c_2 x e^{rx}, \qquad \text{if } r_1 = r_2 = r.$$

The next two examples illustrate how easy it is to apply Theorem 4.7.

Example 4

Find the general solution of

$$3\frac{d^2y}{dx^2} - 15\frac{dy}{dx} - 42y = 0.$$

Solution. First we put this in the standard form (4) by dividing through by 3 and writing D^2y for d^2y/dx and Dy for dy/dx. This gives $(D^2 - 5D - 14)y = 0$. The auxiliary equation is $r^2 - 5r - 14 = 0$, which factors as $(r - 7)(r + 2) = 0$, so $r_1 = 7$ and $r_2 = -2$. Then from (11) the general solution is

$$y = c_1 e^{7x} + c_2 e^{-2x}. \quad ■$$

Example 5

Find the general solution of

$$5\frac{d^2y}{dx^2} - 40\frac{dy}{dx} + 80y = 0.$$

Solution. Proceeding as in Example 4, we divide through by 5 to get $(D^2 - 8D + 16)y = 0$. The auxiliary equation is $r^2 - 8r + 16 = (r - 4)^2$, which has the single repeated root 4. So by (12) the general solution is

$$y = c_1 e^{4x} + c_2 x e^{4x}. \quad ■$$

What if r_1 and r_2 are complex roots? Then

$$r_1 = \frac{-p + \sqrt{p^2 - 4q}}{2} = a + bi \qquad \text{and} \qquad r_2 = \frac{-p - \sqrt{p^2 - 4q}}{2} = a - bi$$

are conjugate complex numbers. In particular, $r_1 \neq r_2$. Then by (13) of the last section (p. 1012), the integration we performed deriving (11) is valid. Thus

$$y = k_1 e^{(a+bi)x} + k_2 e^{(a-bi)x} \tag{13}$$

is the only possible form for a complex solution to (7). It is easy to check (Exercise 26) that any complex function of that form really *is* a solution, so (13) gives the *general complex solution* to (7). We can use Definition 3.7 to rewrite (13) as

$$\begin{aligned} y &= k_1 e^{ax}(\cos bx + i \sin bx) + k_2 e^{ax}(\cos(-bx) + i \sin(-bx)) \\ &= k_1 e^{ax}(\cos bx + i \sin bx) + k_2 e^{ax}(\cos bx - i \sin bx) \\ &= (k_1 + k_2)e^{ax} \cos bx + i(k_1 - k_2)e^{ax} \sin bx, \end{aligned}$$

or equivalently,

$$y = c_1 e^{ax} \cos bx + i c_2 e^{ax} \sin bx, \tag{14}$$

where $c_1 = k_1 + k_2$ and $c_2 = k_1 - k_2$. From Lemma 4.6 the real and imaginary parts $c_1 e^{ax} \cos bx$ and $c_2 e^{ax} \sin bx$ are real solutions to (7). These in fact *generate* the general real solution, as the following theorem makes precise. The proof is given at the end of the section.

4.8 THEOREM

If the auxiliary equation for

$$(D^2 + pD + q)(y) = 0 \tag{4}$$

has complex roots $r_1 = a + bi$ and $r_2 = a - bi$, then the general real solution of (4) is

$$y = c_1 e^{ax} \cos bx + c_2 e^{ax} \sin bx. \tag{15}$$

It is a purely mechanical exercise to use (11), (12), and (14) to find the general solution of the second-order homogeneous linear differential equation (4) with constant coefficients. *It is thus worthwhile to commit Theorems 4.7 and 4.8 to memory.* The next examples illustrate the use of Theorem 4.8.

Example 6

Find the general real solution of $y'' - 6y' + 10y = 0$.

Solution. In standard form (4) we have

$$(D^2 - 6D + 10)y = 0.$$

The auxiliary equation is $r^2 - 6r + 10 = 0$, whose roots are

$$r_1 = \frac{6 + \sqrt{36 - 40}}{2} = 3 + i \quad \text{and} \quad r_2 = \frac{6 - \sqrt{36 - 40}}{2} = 3 - i.$$

Thus in Theorem 4.8, $a = 3$ and $b = 1$. Then according to (15) the general real solution is

$$y = c_1 e^{3x} \cos x + c_2 e^{3x} \sin x. \quad ■$$

The solutions $e^{r_1 x}$ and $e^{r_2 x}$ in (11) are called *basic solutions.* The same term is applied to e^{rx} and xe^{rx} in (12) and to $e^{ax} \cos bx$ and $e^{ax} \sin bx$ in (15).

Initial-Value and Boundary-Value Problems

In Theorems 4.7 and 4.8 the general solution to (4) has *two* arbitrary constants c_1 and c_2. In Section 1 we saw that for a first-order linear equation

$$\frac{dy}{dx} + p(x)y = q(x), \tag{16}$$

a *single* initial condition $y(x_1) = y_1$ is enough to determine the one arbitrary constant C in the general solution of (16). Thus an initial condition on (4) should similarly determine a pair of values for c_1 and c_2, that is, pick out exactly one of the infinitely many functions that comprise the general solution given in Theorems 4.7 and 4.8. For a second-order linear differential equation like (4), an ***initial condition*** consists of a *pair* of prescribed values for y and y' at a single point $x = x_0$:

$$y(x_0) = y_0, \qquad y'(x_0) = y_1. \tag{17}$$

Although we shall not prove it, the fact is that every second-order linear differential equation (4) subject to an initial condition (17) has a *unique* solution, just as we saw for first-order linear equations on p. 993. The next example shows how to find the unique solution of a given initial-value problem.

Example 7

Solve the initial-value problem

$$3\frac{d^2y}{dx^2} - 15\frac{dy}{dx} - 42y = 0, \qquad y(0) = 1,\ y'(0) = 6.$$

Solution. The general solution was found in Example 4, namely,

$$y = c_1e^{7x} + c_2e^{-2x}.$$

Then

$$y' = 7c_1e^{7x} - 2c_2e^{-2x}.$$

The initial conditions give

$$1 = c_1 + c_2 \tag{18}$$

and

$$6 = 7c_1 - 2c_2. \tag{19}$$

Adding twice (18) to (19), we obtain

$$8 = 9c_1 \to c_1 = \frac{8}{9}.$$

Then from (16), $c_2 = 1/9$. So the unique solution to the given initial-value problem is

$$y = \frac{8}{9}e^{7x} + \frac{1}{9}e^{-2x}. \quad ■$$

There is another type of condition that we can impose on a second-order differential equation to try to determine a unique solution.*

* The remainder of the section is optional.

4.9 DEFINITION

A ***boundary condition*** for a second-order linear differential equation

(4) $$(D^2 + pD + q)(y) = 0$$

consists of a pair of specified values of y or y' at two points $x = x_0$ and $x = x_1$:

$$y(x_0) = y_0 \qquad \text{and} \qquad y(x_1) = y_1$$

or

$$y'(x_0) = y_0 \qquad \text{and} \qquad y'(x_1) = y_1.$$

While a second-order initial-value problem always has a unique solution, a second-order boundary-value problem may or may not have a unique solution. That is illustrated by our concluding example.

Example 9

Determine whether the given boundary-value problems are solvable. If so, find the solution.

(a) $y'' - 6y' + 10y = 0, \quad y(0) = 1, \; y(\pi/2) = 1.$

(b) $y'' - 6y' + 10y = 0, \quad y(0) = 1, \; y(\pi) = 0.$

Solution. (a) From Example 6 the general solution is

$$y = c_1 e^{3x} \cos x + c_2 e^{3x} \sin x.$$

The boundary conditions say that

$$1 = c_1 \qquad \text{and} \qquad 1 = c_2 e^{3\pi/2} \rightarrow c_2 = e^{-3\pi/2}.$$

Thus the desired solution is

$$y = e^{3x} \cos x + e^{3(x - \pi/2)} \sin x.$$

(b) Here the boundary conditions give

$$1 = c_1 \qquad \text{and} \qquad 0 = -c_1 e^{3\pi}.$$

Thus we have the impossible situation $c_1 = 1$ and $c_1 = 0$. Hence this boundary-value problem has no solution. ■

Proof of Theorem 4.8

4.8 THEOREM

If $r_1 = a + bi$ and $r_2 = a - bi$ are the roots of the auxiliary equation for

(4) $$(D^2 + pD + q)(y) = 0,$$

then the general real solution of (4) is

(15) $$y = c_1 e^{ax} \cos bx + c_2 e^{ax} \sin bx.$$

Proof. The discussion leading up to the statement of the theorem on p. 1020 shows that any real function of the form $e^{ax} \cos bx$ or $e^{ax} \sin bx$ is a real solution. Then for any real numbers c_1 and c_2

$$\begin{aligned}(D^2 + pD + q)(c_1 e^{ax} \cos bx + c_2 e^{ax} \sin bx) &= (D^2 + pD + q)(c_1 e^{ax} \cos bx) \\ &\quad + (D^2 + pD + q)(c_2 e^{ax} \sin bx) \\ &= c_1(D^2 + pD + q)(e^{ax} \cos bx) \\ &\quad + c_2(D^2 + pD + q)(e^{ax} \sin bx) \\ &= 0 + 0 = 0.\end{aligned}$$

Thus every real function of the form (5) is a solution of (4). It remains to show that *every* real solution to (4) must have the form (15). To do so, let $y(x) = f(x)$ be any real solution. Then $y_a(x) = f(x) + if(x)$ is a complex solution and so must have the form (14). Hence we have

$$y_a(x) = K_1 e^{ax} \cos bx + K_2 e^{ax} \sin bx,$$

where $K_1 = d_1 + id_2$ and $K_2 = e_1 + ie_2$ are complex constants. Thus

$$\begin{aligned} y_a(x) &= d_1 e^{ax} \cos bx + id_2 e^{ax} \cos bx + ie_1 e^{ax} \sin bx - e_2 e^{ax} \sin bx \\ &= d_1 e^{ax} \cos bx - e_2 e^{ax} \sin bx + i(d_2 e^{ax} \cos bx + e_1 e^{ax} \sin bx) \\ &= f(x) + if(x). \end{aligned}$$

So $f(x)$ must *simultaneously* have the forms

$$d_1 e^{ax} \cos bx - e_2 e^{ax} \sin bx \qquad \text{and} \qquad d_2 e^{ax} \cos bx + e_1 e^{ax} \sin bx.$$

Subtraction of these expressions gives

$$(d_1 - d_2)e^{ax} \cos bx - (e_1 + e_2)e^{ax} \sin bx = 0.$$

Since $e^{ax} \neq 0$, we have

$$(d_1 - d_2) \cos bx - (e_1 + e_2) \sin bx = 0 \tag{20}$$

for all x. When $x = 0$, (20) says that

$$d_1 - d_2 = 0 \rightarrow d_1 = d_2.$$

Thus (20) simplifies to

$$(e_1 + e_2) \sin bx = 0,$$

for all x. Substitution of $x = \pi/2b$ gives

$$e_1 + e_2 = 0 \rightarrow e_1 = -e_2.$$

Hence $f(x)$ has the required form (15), where

$$c_1 = d_1 = d_2 \qquad \text{and} \qquad c_2 = e_1 = -e_2. \quad \boxed{\text{QED}}$$

Exercises 15.4

In Exercises 1–10, find the general solution of the given homogeneous linear differential equation.

1. $y'' - 3y' + 2y = 0$

2. $y'' - 7y' + 6y = 0$

3. $3\dfrac{d^2y}{dx^2} + 15\dfrac{dy}{dx} + 18y = 0$

4. $2\dfrac{d^2y}{dx^2} + 12\dfrac{dy}{dx} + 10y = 0$

5. $3y'' + 18y' + 27y = 0$

6. $-2y'' + 8y' - 8y = 0$

7. $\dfrac{d^2x}{dt^2} + 2\dfrac{dx}{dt} + 5x = 0$

8. $\dfrac{d^2x}{dt^2} + 4\dfrac{dx}{dt} + 13x = 0$

9. $y'' + 9y = 0$

10. $y'' + 4y = 0$

In Exercises 11–18, solve the given initial-value problem. (Use the results of earlier exercises as needed.)

11. $y'' - 3y' + 2y = 0,\ y(0) = 1,\ y'(0) = 3$

12. $y'' - 7y' + 6y = 0,\ y(0) = 1,\ y'(0) = -6$

13. $y'' + 9y = 0,\ y(0) = 1,\ y'(0) = -3$

14. $y'' + 4y = 0,\ y(0) = 1,\ y'(0) = 0$

15. $y'' + 9y = 0,\ y(\pi/6) = 1,\ y'(\pi/6) = 2$

16. $y'' + 4y = 0,\ y(\pi/4) = -1,\ y'(\pi/4) = 0$

17. $3y'' + 18y' + 27y = 0,\ y(0) = 1,\ y'(0) = 1$

18. $-2y'' + 8y' - 8y = 0,\ y(0) = 1,\ y'(0) = 1$

In Exercises 19–22, solve the given boundary-value problems if possible.

19. $y'' + 9y = 0,\ y(0) = 1,\ y(\pi) = 0.$

20. $y'' + 4y = 0,\ y(0) = 1,\ y(\pi) = 0.$

21. $y'' + 9y = 0,\ y(0) = 1,\ y(\pi/2) = 1.$

22. $y'' + 4y = 0,\ y(0) = 1,\ y(\pi/2) = 1.$

23. Prove that every function $y(x)$ of the form (2) solves (1).

24. Prove that every solution of (1) has the form (2).

25. **(a)** Verify that any function of the form (11) is a solution of (4) in case $r_1 \neq r_2$.
(b) Verify that any function of the form (12) is a solution to (4) in case $r_1 = r_2 = r$.

26. Verify that any function of the form (13) is a solution to (4) in case $r_1 = a + bi$ and $r_2 = a - bi$.

Exercises 27–33 are devoted to some special types of nonlinear second-order differential equations for which special techniques exist.

27. **(a)** Show that the substitution $v = y'$ reduces the second-order differential equation $F(x, y', y'') = 0$ to a first-order differential equation.
(b) Show that the substitution $v = dy/dx$ reduces the second-order differential equation $F(y, y', y'') = 0$ to a first-order equation.

$$\left(\textit{Hint: } \frac{d^2y}{dx^2} = \frac{dv}{dx} = v\frac{dv}{dy}\right)$$

28. Solve $x\dfrac{d^2y}{dx^2} = \dfrac{dy}{dx} + 2\sqrt{x^2 + \left(\dfrac{dy}{dx}\right)^2}$.

29. Solve $y\dfrac{d^2y}{dx^2} + \left(\dfrac{dy}{dx}\right)^2 = \dfrac{dy}{dx}$.

30. Solve $y\dfrac{d^2y}{dx^2} - \left(\dfrac{dy}{dx}\right)^2 = 0$.

31. Solve $y\dfrac{d^2y}{dx^2} = \left(\dfrac{dy}{dx}\right)^2 + 2\dfrac{dy}{dx}$.

32. Solve $y'' - 3y^2 = 0$, $y(0) = 2$, $y'(0) = 4$.

33. Solve $y'y'' - x = 0$, $y(1) = 2$, $y'(1) = 1$.

15.5 Nonhomogeneous Second-Order Linear Equations with Constant Coefficients

Returning now to the general second-order linear differential equation with constant coefficients

$$(1) \qquad (D^2 + pD + q)y(x) = s(x),$$

in view of Theorem 4.2, we need to find just one particular solution $y = y_p(x)$ to give the general solution, for Theorems 4.7 and 4.8 provide the general solution to the associated homogeneous linear differential equation

$$(2) \qquad (D^2 + pD + q)y(x) = 0.$$

To find $y_p(x)$ we use the reduction-of-order trick (p. 1018) again, since we can easily solve first-order linear nonhomogeneous differential equations by the methods of Section 1.

Suppose then that r_1 and r_2 are the roots of the auxiliary equation $r^2 + pr + q = 0$. We write (1) as

$$(D - r_1) \circ (D - r_2)(y) = s(x).$$

As in the last section, we let $z = (D - r_2)(y)$. Then (1) becomes

$$(D - r_1)(z) = s(x) \rightarrow \frac{dz}{dx} - r_1 z = s(x).$$

This is first-order linear with integrating factor e^{-r_1x}. Multiplying through by this integrating factor, we obtain

$$\frac{d}{dx}(e^{-r_1x}z) = e^{-r_1x}s(x) \rightarrow e^{-r_1x}z = \int e^{-r_1x}s(x)\,dx + k_1,$$

$$z = e^{r_1x}\int e^{-r_1x}s(x)\,dx + k_1e^{r_1x}.$$

Now substituting

$$(D - r_2)(y) = \frac{dy}{dx} - r_2 y$$

for z, we have

$$\frac{dy}{dx} - r_2 y = e^{r_1 x} \int e^{-r_1 x} s(x)\,dx + k_1 e^{r_1 x}.$$

This again is first-order linear with integrating factor $e^{-r_2 x}$. Multiplying through by that integrating factor, we have

$$\text{(3)} \qquad \frac{d}{dx}(e^{-r_2 x} y) = e^{(r_1 - r_2)x} \int e^{-r_1 x} s(x)\,dx + k_1 e^{(r_1 - r_2)x}.$$

If $r_1 \neq r_2$, then we get

$$e^{-r_2 x} y = \int e^{(r_1 - r_2)x} \left(\int e^{-r_1 x} s(x)\,dx \right) dx + \frac{k_1}{r_1 - r_2} e^{(r_1 - r_2)x} + c_2,$$

$$\text{(4)} \qquad y = e^{r_2 x} \int e^{(r_1 - r_2 x)} \left(\int e^{-r_1 x} s(x)\,dx \right) dx + c_1 e^{r_1 x} + c_2 e^{r_2 x}.$$

This is the general solution of (1), because the last two terms make up the general solution of (2). On the other hand, if $r_1 = r_2 = r$, then (3) has the form

$$\frac{d}{dx}(e^{-rx} y) = \int e^{-rx} s(x)\,dx + c_1.$$

Hence

$$e^{-rx} y = \int \left(\int e^{-rx} s(x)\,dx \right) dx + c_1 x + c_2,$$

$$\text{(5)} \qquad y = e^{rx} \int \left(\int e^{-rx} s(x)\,dx \right) dx + c_1 x e^{rx} + c_2 e^{rx}.$$

The last two terms again constitute the general solution of (2). The parts of (4) and (5) in color are thus particular solutions of (1). In finding them, we can ignore constants of integration like k_1, c_1, and c_2 above, because such constants are associated with the general solution of the corresponding *homogeneous* problem (2), which can be solved more easily using Theorems 4.7 and 4.8.

Example 1

Find the general solution of

$$3\frac{d^2 y}{dx^2} - 15\frac{dy}{dx} - 42y = 3e^{7x} \sin x.$$

Solution. The associated homogeneous problem is the one solved in Example 4 of the last section. Its general solution is $y_0 = c_1 e^{7x} + c_2 e^{-2x}$. To apply the above solution technique we have to divide through by 3, getting

$$(D^2 - 5D - 14)(y) = e^{7x} \sin x \rightarrow (D - 7) \circ (D + 2)(y) = e^{7x} \sin x.$$

Then we let $z = (D + 2)y$, so we have

$$(D - 7)z = e^{7x} \sin x \rightarrow \frac{dz}{dx} - 7z = e^{7x} \sin x.$$

Multiplying through by the integrating factor e^{-7x}, we obtain

$$\frac{d}{dx}(ze^{-7x}) = \sin x.$$

Integrating and suppressing the constant of integration in accord with our remarks above, we get

$$ze^{-7x} = -\cos x \rightarrow z = -e^{7x}\cos x \rightarrow \frac{dy}{dx} + 2y = -e^{7x}\cos x.$$

The latter first-order linear differential equation has integrating factor e^{2x}. Multiplying through by that factor, we get

$$\frac{d}{dx}(e^{2x}y) = -e^{9x}\cos x.$$

Using integration by parts, Formula 52 of the table of integrals, or Exercise 27 of Section 3, we have

$$e^{2x}y = -\frac{1}{82}e^{9x}\sin x - \frac{9}{82}e^{9x}\cos x \rightarrow y = -\frac{1}{82}e^{7x}\sin x - \frac{9}{82}e^{7x}\cos x.$$

The general solution of the given equation is therefore

$$y = \frac{-e^{7x}}{82}(\sin x + 9\cos x) + c_1e^{7x} + c_2e^{-2x}. \quad \blacksquare$$

Method of Undetermined Coefficients

As Example 1 and Formulas (4) and (5) suggest, the integration necessary to find a particular solution via the reduction-of-order trick can be difficult. An additional drawback to using (4) and (5) is their rather complicated nature, which makes them hard to remember. Because of these impediments, an alternative method has been developed. It is called the ***method of undetermined coefficients*** and is basically a catalogue of the results when (4) and (5) are used with various commonly encountered functions $s(x)$ in (1). This method gives the general form of a particular solution $y_p(x)$ to be expected for (1) in the case of several general types of $s(x)$. The exact form of $y_p(x)$ can then be found by substituting the formulas for $y_p'(x)$ and $y_p''(x)$ into (1). We first formulate the method, then illustrate its use.

METHOD OF UNDETERMINED COEFFICIENTS

Given a nonhomogeneous second-order linear differential equation with constant coefficients

(1) $$(D^2 + pD + q)y(x) = s(x),$$

determine whether $s(x)$ has one of the forms listed in Table 1. If it does, then a particular solution $y_p(x)$ exists and has the indicated form, ***unless*** that $y_p(x)$ happens to be a solution of the associated homogeneous problem

(2) $$(D^2 + pD + q)y(x) = 0.$$

In the latter case, there is a particular solution of the form $xy_p(x)$ or $x^2y_p(x)$ that is *not* a solution of (2).

We won't prove that the $y_p(x)$ given in the table (or $xy_p(x)$ or $x^2y_p(x)$ as the case may be) actually *must* be a particular solution of (1). As suggested above, that could be done by a case-by-case study of Formulas (4) and (5). We will

TABLE 1

$s(x)$	$y_p(x)$
$a_0 + a_1x + a_2x^2 + \ldots + a_nx^n$	$A_0 + A_1x + A_2x^2 + \ldots + A_nx^n$
ae^{kx}	Ae^{kx}
$e^{kx}(a_0 + a_1x + \ldots + a_nx^n)$	$e^{kx}(A_0 + A_1x + \ldots + A_nx^n)$
$a\cos cx + b\sin cx$	$A\cos cx + B\sin cx$
$e^{kx}(a\cos cx + b\sin cx)$	$e^{kx}(A\cos cx + B\sin cx)$
$e^{kx}(a_0 + a_1x + \ldots + a_kx^k)\cos cx$ $+ e^{kx}(b_0 + b_1x + \ldots + b_kx^k)\sin cx$	$e^{kx}(A_0 + A_1x + \ldots + A_kx^k)\cos cx$ $+ e^{kx}(B_0 + B_1x \ldots + B_kx^k)\sin cx$

instead illustrate how to use the table. Our first example involves an exponential function $s(x)$ for which the listed $y_p(x)$ has to be modified because it is a homogeneous solution.

Example 2

Solve $\dfrac{d^2y}{dx^2} - 5\dfrac{dy}{dx} - 14y = 5e^{7x}$.

Solution. Here $s(x) = 5e^{7x}$. The table entry $y_p(x)$ for such an $s(x)$ is Ae^{7x}, since $k = 7$. But that $y_p(x)$ is a solution of the associated homogeneous problem

$$\frac{d^2y}{dx^2} - 5\frac{dy}{dx} - 14y = 0$$

(see Example 1), because it has form $c_1e^{7x} + c_2e^{-2x}$ (for $c_1 = A$ and $c_2 = 0$). Instead of the listed $y_p(x)$ then, the method says to use a function of the form

$$y_p(x) = Axe^{7x},$$

since that isn't a homogeneous solution. To complete the problem we have to determine the value of the coefficient A so that $y_p(x)$ is a particular solution. (This suggests how the method was named.) We have

$$y_p'(x) = Ae^{7x} + 7xAe^{7x} \rightarrow y_p''(x) = 7Ae^{7x} + 7Ae^{7x} + 49xAe^{7x}$$
$$= 14Ae^{7x} + 49xAe^{7x}.$$

Then substitution in the given equation gives

$$y_p'' - 5y_p' - 14y = 14Ae^{7x} + 49xAe^{7x} - 5Ae^{7x} - 35xAe^{7x} - 14Axe^{7x} = 9Ae^{7x}$$
$$= 5e^{7x}.$$

Hence $A = 5/9$. Thus $y_p(x) = (5/9)xe^{7x}$. Therefore the general solution is

$$y(x) = \frac{5}{9}xe^{7x} + c_1e^{7x} + c_2e^{-2x}. \blacksquare$$

If we had not noticed that Ae^{7x} is a homogeneous solution and had tried to use it for $y_p(x)$, then we would have gotten

$$y_p'(x) = 7Ae^{7x} \rightarrow y_p''(x) = 49Ae^{7x},$$

$$y_p'' - 5y_p' - 14y_p = 49Ae^{7x} - 35Ae^{7x} - 14Ae^{7x} = 0.$$

Thus trying to equate $(D^2 - 5D - 14)y_p(x)$ to $s(x) = 5e^{7x}$ to find A would have produced the absurdity $0 = 5e^{7x}$, instead of an equation we could solve to find A. To avoid such problems, **always check whether the $y_p(x)$ in the table is a homogeneous solution. If it is, then modify it by a factor of x, or if necessary x^2, before trying to determine its coefficients.**

The next example applies the method of undetermined coefficients to the equation of Example 1.

Example 3

Use the method of undetermined coefficients to solve

$$3\frac{d^2y}{dx^2} - 15\frac{dy}{dx} - 42y = 3e^{7x}\sin x.$$

Solution. First we divide through by 3. Then we have $s(x) = e^{7x}\sin x$. This is of the form $e^{kx}(a\cos cx + b\sin cx)$, where $k = 7$, $a = 0$, $b = 1$, and $c = 1$. So the $y_p(x)$ suggested by Table 1 has the form

$$y_p(x) = e^{7x}(A\cos x + B\sin x).$$

Since the general solution of the associated homogeneous problem is $y_0 = c_1e^{7x} + c_2e^{-2x}$, the suggested $y_p(x)$ can be used *without* modification. (**Note:** *Both* the sine and cosine terms *must* appear in $y_p(x)$ even if $s(x)$ involves only the sine or only the cosine.) The coefficients A and B can be determined from the fact that $y_p(x)$ is a solution to the given equation. We have

$$\begin{aligned} y_p'(x) &= 7e^{7x}(A\cos x + B\sin x) + e^{7x}(-A\sin x + B\cos x)\\ &= e^{7x}(7A + B)\cos x + e^{7x}(-A + 7B)\sin x \end{aligned}$$

and

$$\begin{aligned} y_p''(x) &= 7e^{7x}(7A + B)\cos x - e^{7x}(7A + B)\sin x\\ &\quad + 7e^{7x}(-A + 7B)\sin x + e^{7x}(-A + 7B)\cos x\\ &= e^{7x}(48A + 14B)\cos x + e^{7x}(-14A + 48B)\sin x. \end{aligned}$$

Thus

$$\begin{aligned} y'' - 5y' - 14y &= e^{7x}(48A + 14B)\cos x + e^{7x}(-14A + 48B)\sin x\\ &\quad + e^{7x}(-35A - 5B)\cos x + e^{7x}(5A - 35B)\sin x\\ &\quad + e^{7x}(-14A)\cos x + e^{7x}(-14B)\sin x\\ &= e^{7x}(-A + 9B)\cos x + e^{7x}(-9A - B)\sin x\\ &= e^{7x}\sin x. \end{aligned}$$

Equating coefficients of $e^{7x}\sin x$ and $e^{7x}\cos x$ in this equation, we get

$$\begin{cases} -A + 9B = 0, & (6)\\ -9A - B = 1. & (7) \end{cases}$$

Multiplying (6) by -9 and adding to (7), we find

$$-82B = 1 \rightarrow B = -\frac{1}{82}.$$

From (6) then, $A = 9B = -9/82$. Hence

$$y_p(x) = e^{7x}\left(-\frac{9}{82}\cos x - \frac{1}{82}\sin x\right) = -\frac{e^{7x}}{82}(9\cos x + \sin x).$$

So, as before, we find that the general solution is

$$y = -\frac{e^{7x}}{82}(9\cos x + \sin x) + c_1e^{7x} + c_2e^{-2x}. \blacksquare$$

Besides being conceptually simpler (no substitutions, integrating factors, or integration are required), the method of undetermined coefficients, when it applies, is usually quicker than the reduction-of-order approach. It also has the esthetic appeal of reducing the solution of a *linear* differential equation to the solution of a *system of linear equations* like (6) and (7).

The following simple result makes the method of undetermined coefficients of wider applicability than might be initially supposed.

5.1 THEOREM

The second-order linear differential equation

$$\frac{d^2y}{dx^2} + p\frac{dy}{dx} + qy = s(x) + t(x) \tag{8}$$

has a particular solution $y_1(x) + y_2(x)$, where $y_1(x)$ is a particular solution to $(D^2 + pD + q)y(x) = s(x)$ and $y_2(x)$ is a particular solution to $(D^2 + pD + q)y(x) = t(x)$.

Proof. We have

$$\begin{aligned}
&\frac{d^2(y_1 + y_2)}{dx^2} + p\frac{d(y_1 + y_2)}{dx} + q(y_1 + y_2) \\
&\quad = \left(\frac{d^2y_1}{dx^2} + \frac{d^2y_2}{dx^2}\right) + p\left(\frac{dy_1}{dx} + \frac{dy_2}{dx}\right) + (qy_1 + qy_2) \\
&\quad = \left(\frac{d^2y_1}{dx^2} + p\frac{dy_1}{dx} + qy_1\right) + \left(\frac{d^2y_2}{dx^2} + p\frac{dy_2}{dx} + qy_2\right) \\
&\quad = s(x) + t(x). \quad \boxed{\text{QED}}
\end{aligned}$$

Theorem 5.1 can be a helpful tool in reducing complexity. To find a particular solution of a given linear equation (8), we first find separate particular solutions of the equations

$$(D^2 + pD + q)(y) = s(x) \qquad \text{and} \qquad (D^2 + pD + q)(y) = t(x)$$

and then add them together. (This scheme is called the *method of superposition* in the literature of differential equations.)

Example 4

Find the general solution of

$$\frac{d^2y}{dx^2} - 5\frac{dy}{dx} - 14y = e^{7x}\sin x + 14x^2 - 2.$$

Solution. We have the general solution of the associated homogeneous problem and a particular solution to

$$\frac{d^2y}{dx^2} - 5\frac{dy}{dx} - 14y = e^{7x}\sin x$$

from Examples 1 and 3. So all we need is a particular solution of

$$\frac{d^2y}{dx^2} - 5\frac{dy}{dx} - 14y = 14x^2 - 2. \tag{9}$$

According to Table 1, since $14x^2 - 2$ is a polynomial of degree 2, there is a particular solution to (9) of the form

$$y_p(x) = A_0 + A_1x + A_2x^2,$$

because there is no homogeneous solution of this form. Then $y_p'(x) = A_1 + 2A_2x \rightarrow y_p''(x) = 2A_2$, and so, substituting in the given equation, we have

$$\begin{aligned}\frac{d^2y}{dx^2} - 5\frac{dy}{dx} - 14y &= 2A_2 - 5A_1 - 10A_2x - 14A_0 - 14A_1x - 14A_2x^2 \\ &= (-14A_0 - 5A_1 + 2A_2) - (10A_2 + 14A_1)x - 14A_2x^2 \\ &= 14x^2 - 2.\end{aligned}$$

Equating coefficients of x, x^2, and constants, we get

$$-14A_2 = 14 \rightarrow A_2 = -1;$$

$$-10A_2 - 14A_1 = 0 \rightarrow A_1 = -\frac{10}{14}A_2 = \frac{5}{7},$$

and

$$-14A_0 - 5A_1 + 2A_2 = -2 \rightarrow -14A_0 = -2 + \frac{25}{7} + 2 = \frac{25}{7} \rightarrow A_0 = -\frac{25}{98}.$$

Hence a particular solution to (9) is

$$y_p(x) = -\frac{25}{98} + \frac{5}{7}x - x^2.$$

The general solution of the given problem is then

$$y = -\frac{25}{98} + \frac{5}{7}x - x^2 - \frac{e^{7x}}{82}(9\cos x + \sin x) + c_1e^{7x} + c_2e^{-2x}. \blacksquare$$

While the method of undetermined coefficients is of fairly wide applicability, *it is not universally applicable.* If $s(x)$ does not have one of the forms in Table 1 and is not a sum of such forms, you can resort to the technique of Example 1.

Example 5

Find the general solution of

$$\frac{d^2y}{dx^2} - 8\frac{dy}{dx} + 16y = xe^{-x^2+4x}.$$

Solution. Table 1 has no suggested $y_p(x)$ when $s(x)$ involves e to a power of x other than the first power. So we write

$$[(D-4)\circ(D-4)](y) = xe^{-x^2+4x}$$

and let $z = (D-4)(y)$. We then have

$$(D-4)(z) = xe^{-x^2+4x} \rightarrow \frac{dz}{dx} - 4z = xe^{-x^2+4x}.$$

The first-order linear equation has the integrating factor e^{-4x}. Multiplying through by that, we get

$$\frac{d}{dx}(ze^{-4x}) = xe^{-x^2},$$

so

$$ze^{-4x} = -\frac{1}{2}e^{-x^2} \rightarrow z = -\frac{1}{2}e^{-x^2+4x}.$$

Since $z = (D-4)(y)$, we have

$$\frac{dy}{dx} - 4y = -\frac{1}{2}e^{-x^2+4x}.$$

Again we use the integrating factor e^{-4x}, getting

$$\frac{d}{dx}(ye^{-4x}) = -\frac{1}{2}e^{-x^2} \rightarrow ye^{-4x} = -\frac{1}{2}\int e^{-x^2}\,dx. \tag{10}$$

Even though e^{-x^2} does not have an antiderivative expressible in terms of other elementary functions, the ***error function*** $\operatorname{erf}(x)$ is defined by

$$\operatorname{erf}(x) = \frac{2}{\sqrt{\pi}}\int_0^x e^{-t^2}\,dt$$

and tables of this function exist just as for $\exp(x)$, $\ln x$, and others. So, choosing the right constant of integration enables us to write (10) as

$$ye^{-4x} = -\frac{1}{4}\sqrt{\pi}\operatorname{erf}(x).$$

Hence

$$y_p(x) = -\frac{1}{4}\sqrt{\pi}e^{4x}\operatorname{erf}(x).$$

Thus the given differential equation has general solution

$$y = -\frac{1}{4}\sqrt{\pi}e^{4x}\operatorname{erf}(x) + c_1e^{4x} + c_2xe^{4x},$$

where we have used the result of Example 5 of the last section. ■

Exercises 15.5

In Exercises 1–10, find a particular solution of the given differential equation.

1. $\dfrac{d^2y}{dx^2} + 3\dfrac{dy}{dx} - y = x^2 - 1$

2. $\dfrac{d^2y}{dx^2} + 2\dfrac{dy}{dx} - 3y = 2x^2 - x + 1$

3. $\dfrac{d^2y}{dx^2} - 3\dfrac{dy}{dx} - 4y = 2\sin x$

4. $\dfrac{d^2y}{dx^2} + 2\dfrac{dy}{dx} - 5y = \cos 2x$

5. $\dfrac{d^2y}{dx^2} + \dfrac{dy}{dx} - 6y = 2e^{-x}$

6. $\dfrac{d^2y}{dx^2} + 3\dfrac{dy}{dx} - 5y = 3e^{2x}$

7. $\dfrac{d^2y}{dx^2} + \dfrac{dy}{dx} - 2y = xe^x$

8. $\dfrac{d^2y}{dx^2} + 2\dfrac{dy}{dx} - 4y = -xe^{2x}$

9. $\dfrac{d^2y}{dx^2} - 4y = 3e^{2x}$

10. $\dfrac{d^2y}{dx^2} - 9y = -e^{3x}$

In Exercises 11–26, find the general solution of the given differential equation.

11. $\dfrac{d^2y}{dx^2} - 5\dfrac{dy}{dx} + 6y = xe^x$

12. $\dfrac{d^2y}{dx^2} + \dfrac{dy}{dx} - 6y = xe^{-x}$

13. $\dfrac{d^2y}{dx^2} + 3\dfrac{dy}{dx} - y = \cos x$

14. $\dfrac{d^2y}{dx^2} + 3\dfrac{dy}{dx} - y = \sin x$

15. $\dfrac{d^2y}{dx^2} + 2\dfrac{dy}{dx} + y = 3e^{-x}$

16. $\dfrac{d^2y}{dx^2} + 6\dfrac{dy}{dx} + 9y = 2e^{-3x}$

17. $\dfrac{d^2y}{dx^2} - 2\dfrac{dy}{dx} - 3y = 2e^x - 10\sin x$

18. $\dfrac{d^2y}{dx^2} - 5\dfrac{dy}{dx} - 6y = 2e^x + \cos x$

19. $\dfrac{d^2y}{dx^2} - 2\dfrac{dy}{dx} - 3y = 2e^x - 10\sin x$, where $y(0) = 2$, $y'(0) = 4$

20. $\dfrac{d^2y}{dx^2} - 5\dfrac{dy}{dx} - 6y = 2e^x\cos x$, where $y(0) = 0$, $y'(0) = 1$

21. $y'' - 2y' + y = e^x$, where $y(0) = 1$, $y'(0) = 1$

22. $y'' - 6y' + 9y = e^{3x}$, where $y(0) = 1$, $y'(0) = 1$

23. $y'' - 3y' + 2y = 2x^2 + e^x + 2xe^x + 4e^{3x}$

24. $y'' + 2y' + 2y = x\cos 2x + \sin 2x$

25. $\dfrac{d^2y}{dx^2} - 2\dfrac{dy}{dx} + y = 2xe^{-x^2+x}$

26. $\dfrac{d^2y}{dx^2} - 6\dfrac{dy}{dx} + 9y = xe^{-x^2+3x}$

27. Find the general solution to $d^2y/dx^2 + 9y = \cos 2x$ by finding a complex solution to $y'' + 9y = e^{2ix}$. Use the method of undetermined (complex) coefficients and then take its real part.

28. Use the approach of Exercise 27 to find the general solution of $y'' + y' - 2y = 4\sin 2x$.

29. Use the approach of Exercise 27 to find the general solution of $y'' + y' - 6y = 3\sin x$.

30. Use the approach of Exercise 27 to find the general solution of $y'' + y' - 6y = 3\sin 2x$.

31. By letting $x = e^t$, reduce

$$x^2\frac{d^2y}{dx^2} - 2x\frac{dy}{dx} + 2y = 6\ln x$$

to a second-order linear equation with constant coefficients. Find the general solution on $(0, +\infty)$.

32. Use the method of Exercise 31 to solve

$$x^2\frac{d^2y}{dx^2} + 2x\frac{dy}{dx} - 2y = 6x$$

on $(0, +\infty)$. (The equations in Exercises 31 and 32 are called ***Cauchy–Euler equations.***)

15.6 Applications of Second-Order Linear Differential Equations

Some important applications of second-order linear differential equations to physics and engineering grow out of the study of oscillatory motion and electrical circuits. We will discuss the first of these in some detail and then see that the models for both phenomena are so similar that a great deal of information about the second can be obtained with very little additional work.

We start with the simplest kind of vibrating system, the one pictured in **Figure 6.1.** Here a mass m is attached to a fixed spring. If the mass is moved from its initial position and then released, it oscillates about its equilibrium position. To analyze the motion we need only Newton's law $\mathbf{F} = m\mathbf{a}$ and *Hooke's law*, which was formulated by the English physicist Robert Hooke (1635–1703) in 1676. That law says that the spring exerts a force on the mass proportional to the displacement $\mathbf{x}$ from the equilibrium position 0 in Figure 6.1 and directed oppositely to the motion. That is,

$$\mathbf{F} = -h\mathbf{x}$$

where $\mathbf{x}$ is the directed distance from the equilibrium position to the mass at time t, and h is a positive constant called the *spring constant*. Since the motion of the mass and the force exerted by the spring are both directed along the same line, $\mathbf{F}$ and $\mathbf{x}$ are *one-dimensional* vectors, whose coordinates are then real numbers

FIGURE 6.1

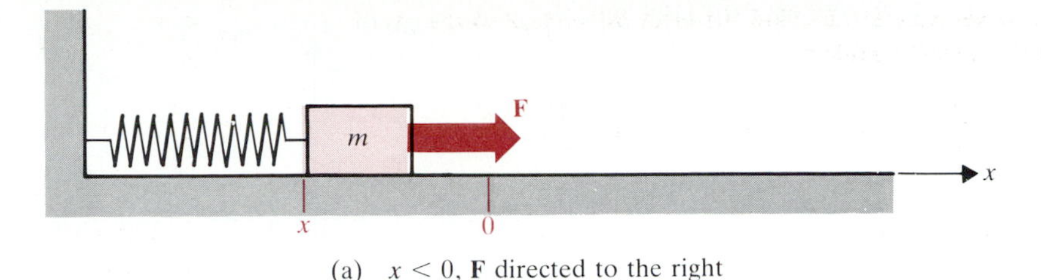

(a) $x < 0$, **F** directed to the right

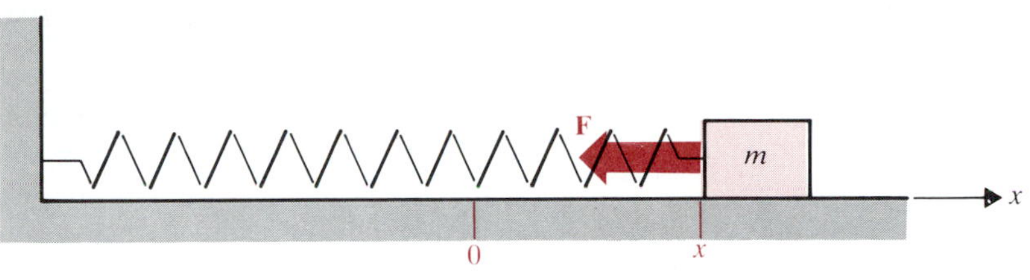

(b) $x > 0$, **F** directed to the left

(see the first part of Section 10.1). Recall that positive real numbers correspond to rightward-pointing vectors, and negative numbers correspond to leftward-directed vectors. Considering F and x as real variables, and using the fact that $F = ma = mx''(t)$, we can write the differential equation describing the vibrating system as

$$m\frac{d^2x}{dt^2} = -hx. \tag{1}$$

According to the following theorem, the motion described by (1) is quite simple.

6.1 THEOREM Under the model (1), the motion of the mass in Figure 6.1 is oscillatory about the equilibrium position: $x(t) = a\cos(\omega t - \phi_0)$, where a, ω, and ϕ_0 are constants.

Proof. To solve (1) for $x = x(t)$, we first put it in standard form:

$$\frac{d^2x}{dt^2} + \frac{h}{m}x = 0 \to \left(D^2 + \frac{h}{m}\right)x = 0.$$

This second-order homogeneous linear equation has an auxiliary equation whose roots are $\pm\sqrt{h/m}i$. Thus the general solution to (1) is

$$x(t) = c_1 \cos\sqrt{h/m}t + c_2 \sin\sqrt{h/m}t. \tag{2}$$

FIGURE 6.2

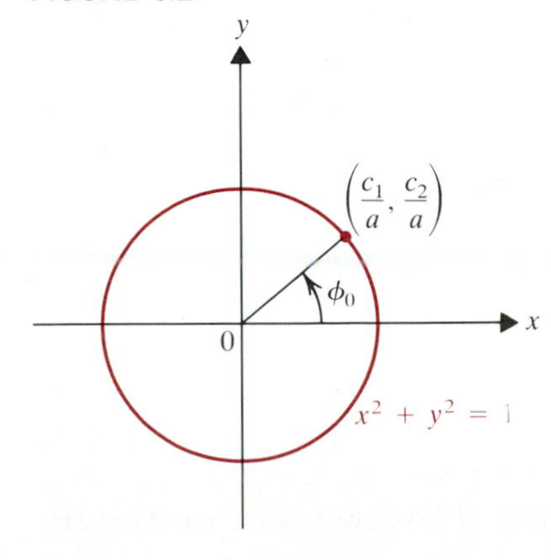

In this form we can't visualize the motion very well, but (2) can be transformed to a formula that is easy to interpret. Let $a = \sqrt{c_1{}^2 + c_2{}^2}$. Then observe that $(c_1/a, c_2/a)$ is on the unit circle $x^2 + y^2 = 1$. Hence by basic trigonometry there is an angle ϕ_0 such that

$$\cos\phi_0 = \frac{c_1}{a} \quad \text{and} \quad \sin\phi_0 = \frac{c_2}{a}.$$

(See **Figure 6.2.**) We can thus rewrite (2) as

$$x(t) = a\left(\frac{c_1}{a}\cos\sqrt{h/m}t + \frac{c_2}{a}\sin\sqrt{h/m}t\right)$$

$$= a(\cos\phi_0\cos\sqrt{h/m}t + \sin\phi_0\sin\sqrt{h/m}t),$$

$$x(t) = a\cos(\sqrt{h/m}t - \phi_0) = a\cos(\omega t - \phi_0) \tag{3}$$

FIGURE 6.3

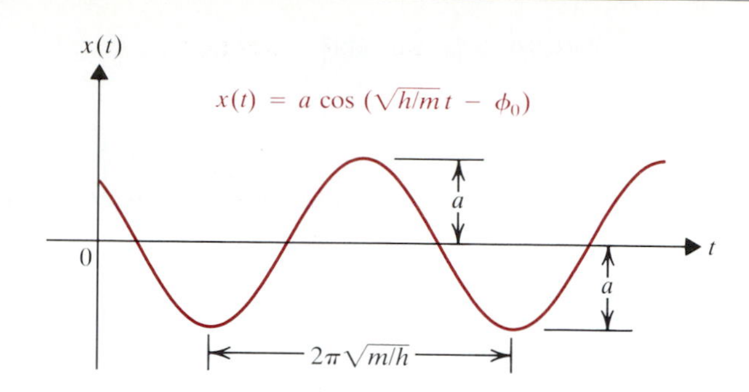

where $\omega = \sqrt{h/m}$. The motion is therefore *periodic*. We graph $x(t)$ versus t in **Figure 6.3.** The mass clearly oscillates between two extremes a units to the left and right of the equilibrium position. QED

6.2 DEFINITION The motion described by (3) is called ***simple harmonic motion.*** The ***amplitude*** is a, the maximum displacement from the equilibrium position. The ***frequency*** is $\omega = \sqrt{h/m}$, measured in radians per second. The ***period*** is $2\pi/\omega$, the time required for one complete cycle. The ***phase angle*** is ϕ_0.

We could express the displacement in terms of the sine function by simply using a different phase angle $\phi_1 = \phi_0 - \pi/2$. Then

$$x(t) = a\cos(\omega t - \phi_0) = a\cos(\phi_0 - \omega t), \qquad \textit{since } \cos\theta = \cos(-\theta)$$

so

$$x(t) = a\sin\left(\frac{1}{2}\pi - (\phi_0 - \omega t)\right) = a\sin(\omega t - \phi_1). \tag{4}$$

For consistency we will use (3) throughout rather than the equivalent formula (4).

The model (1) usually fails to give an accurate representation of oscillatory motion encountered in real-world situations. We have neglected air resistance, friction, and any other kind of damping force that might exist, for example. We thus are led to consider the slightly more sophisticated vibrating system pictured in **Figure 6.4.** We have added a "dashpot" D, which damps the motion and can be thought of as representing air or fluid resistance to the motion of m. Consider, for example, the vertical motion of the front of an automobile after it strikes a bump in the road. The oscillatory motion is damped by the front shock absorbers, and the motion dies down rather than continuing indefinitely as in (4).

FIGURE 6.4

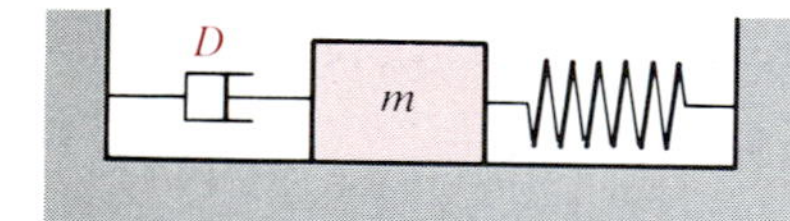

The resisting force to the rightward motion of the mass is often observed to be proportional to the velocity dx/dt of the mass. We thus model this system by the equation

$$m\frac{d^2x}{dt^2} = -hx - k\frac{dx}{dt},$$

where k is a positive constant. In standard form this second-order linear homogeneous equation is

$$\frac{d^2x}{dt^2} + \frac{k}{m}\frac{dx}{dt} + \frac{h}{m}x = 0 \tag{5}$$

As we saw in Theorems 4.7 and 4.8, the solution of (5) depends on whether the auxiliary equation

$$r^2 + \frac{k}{m} r + \frac{h}{m} = 0$$

has real or complex roots. By the quadratic formula, this depends on the sign of

$$b^2 - 4ac = \left(\frac{k}{m}\right)^2 - 4\,\frac{h}{m} = \frac{k^2 - 4mh}{m^2},$$

and hence on the sign of $k^2 - 4mh$. As we might expect, then, the relative sizes of the spring constant h, the damping constant k, and the mass m determine the nature of the motion.

6.3 THEOREM Under the model (5), the system in Figure 6.4 tends to return to its equilibrium position.

Rather than proving this directly, we will give a discussion of the motion that will tell us considerably more than the theorem asserts about the motion of the system in each of the cases $k^2 - 4mh > 0$, $k^2 - 4mh < 0$, and $k^2 - 4mh = 0$. In operator form (5) is

$$\left(D^2 + \frac{k}{m} D + \frac{h}{m}\right)(x) = 0$$

The auxiliary polynomial has roots

$$r_1 = \frac{1}{2}\left(-\frac{k}{m} + \sqrt{\frac{k^2 - 4mh}{m^2}}\right) = \frac{1}{2m}(-k + \sqrt{k^2 - 4mh}),$$

and

$$r_2 = \frac{1}{2m}(-k - \sqrt{k^2 - 4mh}).$$

(a) If $k^2 - 4mh > 0$, then $k > 2\sqrt{mh}$, so the damping constant is more than double the geometric mean of m and h. In this case one says that the motion is *overdamped*. Both roots are real, and $r_2 < r_1$. Since $k > \sqrt{k^2 - 4mh}$, we have

$$-k + \sqrt{k^2 - 4mh} < 0,$$

and thus both roots are negative. The general solution to (5) is

$$x(t) = c_1 e^{r_1 t} + c_2 e^{r_2 t} \tag{6}$$

Hence

$$\lim_{t\to\infty} x(t) = c_1 \lim_{t\to\infty} e^{r_1 t} + c_2 \lim_{t\to\infty} e^{r_2 t} = 0 + 0 = 0,$$

since r_1 and r_2 are negative. So Theorem 6.3 is true. The motion is *not* oscillatory, as can be seen from (6), and dies out rather rapidly if the r_i are not extremely small. Depending on c_1 and c_2, k, m, and h, the mass may pass its equilibrium position once. Two typical graphs of (6) are shown in **Figure 6.5.**

(b) If $k^2 - 4mh < 0$, then $k < 2\sqrt{mh}$. In this case the motion is said to be *underdamped*. The roots r_1 and r_2 are complex,

$$r_1 = \frac{-k}{2m} + i\,\frac{\sqrt{4mh - k^2}}{2m} \quad \text{and} \quad r_2 = \frac{-k}{2m} - i\,\frac{\sqrt{4mh - k^2}}{2m}.$$

FIGURE 6.5

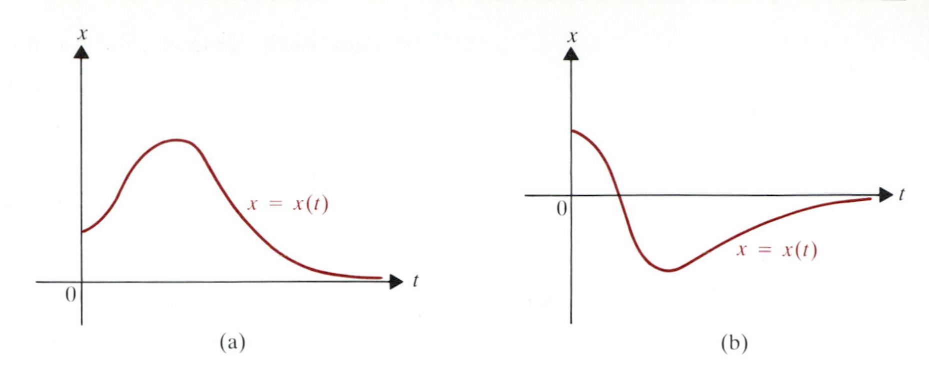

Hence the general solution of (5) is

$$x(t) = e^{-kt/2m}\left(c_1 \cos \frac{\sqrt{4mh - k^2}}{2m} t + c_2 \sin \frac{\sqrt{4mh - k^2}}{2m} t\right).$$

If, as in Theorem 6.1, we let $a = \sqrt{c_1{}^2 + c_2{}^2}$ and choose ϕ_0 so that $\cos \phi_0 = c_1/a$ and $\sin \phi_0 = c_2/a$, then

$$x(t) = ae^{-kt/2m} \cos\left(\frac{\sqrt{4mh - k^2}}{2m} t - \phi_0\right) \tag{7}$$

In this case then we get *damped vibration:* The graph of $x(t)$ resembles a cosine curve whose amplitude gradually decreases toward 0, because

$$\begin{aligned} \lim_{t \to \infty} |x(t)| &= |a| \lim_{t \to \infty} \left(e^{-kt/2m}\left|\cos\left(\frac{\sqrt{4mh - k^2}}{2m} t - \phi_0\right)\right|\right) \\ &\leq |a| \lim_{t \to \infty} e^{-kt/2m} = 0. \end{aligned}$$

So Theorem 6.3 holds here. This time, the mass does oscillate about its equilibrium position, as shown in **Figure 6.6,** which represents a typical graph of (7). We put

$$\omega_1 = \frac{\sqrt{4mh - k^2}}{2m},$$

so $x(t) = ae^{-kt/2m} \cos(\omega_1 t - \phi_0)$.

(c) If $k^2 - 4mh = 0$, then $k = 2\sqrt{mh}$. In this case the motion is said to be *critically damped.* This is an unstable situation in which a small change in either k, m, or h will put us into case (a) or case (b). We have $r_1 = r_2 =$

FIGURE 6.6

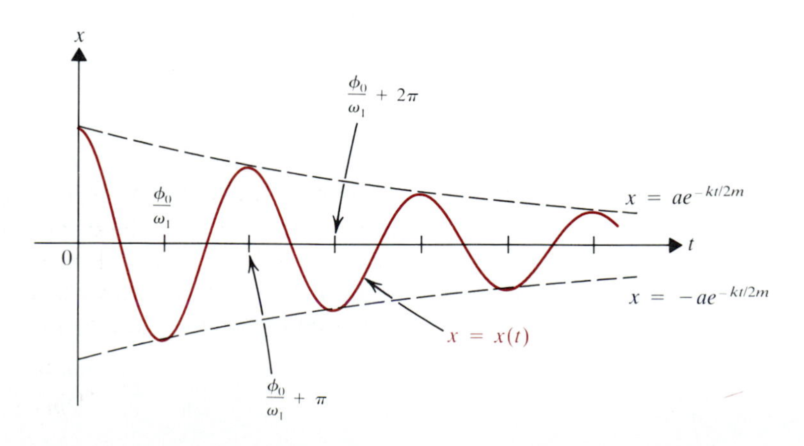

FIGURE 6.7

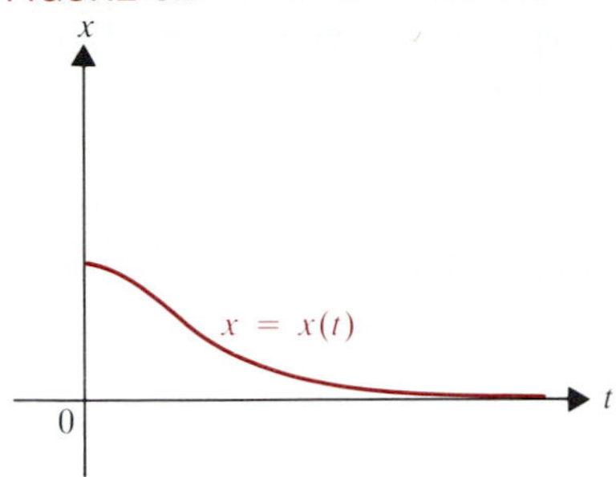

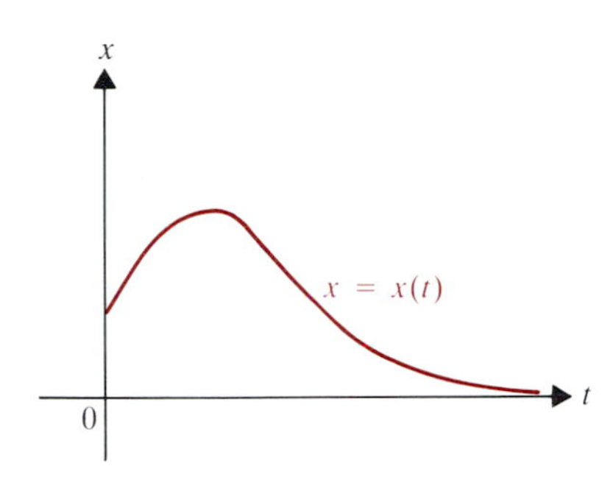

$-k/2m$, so the general solution of (5) is

$$x(t) = c_1 e^{-kt/2m} + c_2 t e^{-kt/2m}. \tag{8}$$

There is no oscillation, and as $t \to \infty$, $x(t)$ again approaches 0. (Use l'Hôpital's rule to see that $c_2 t e^{-kt/2m} = c_2(t/e^{kt/2m})$ goes to 0 as $t \to \infty$.) Two typical graphs of (8) are shown in **Figure 6.7.**

Example 1

In the system in Figure 6.4, suppose that a force $F = 13$ newtons is needed to move the 1-kg mass 1 meter to the right and that the damping constant is 4 kg/s. If the spring is compressed $\frac{1}{4}$ meter to the left and released with zero initial velocity, then find a formula for the displacement $x(t)$ of the mass from its equilibrium position at time t. What kind of motion does the mass exhibit?

Solution. We are given that $k = 4$ and $m = 1$ and that a force of 13 newtons exerted on the spring moves the mass 1 meter to the right. Then Hooke's law gives $13 = h \cdot 1$, so $h = 13$. Then (5) becomes

$$\frac{d^2x}{dt^2} + 4\frac{dx}{dt} + 13x = 0$$

The auxiliary equation is $r^2 + 4r + 13 = 0$, which has roots

$$r = \frac{-4 \pm \sqrt{16 - 52}}{2} = -2 \pm 3i.$$

The motion is thus *underdamped*. We are told that the initial displacement $x(0)$ is $-1/4$ and the initial velocity is

$$\frac{dx}{dt}(0) = 0.$$

Thus $c_1 = -1/4$. Differentiating the expression for $x(t)$, we obtain

$$\frac{dx}{dt} = -2e^{-2t}(c_1 \cos 3t + c_2 \sin 3t) + e^{-2t}(-3c_1 \sin 3t + 3c_2 \cos 3t).$$

Thus when $t = 0$, we have

$$0 = -2c_1 + 3c_2 = \frac{1}{2} + 3c_2 \to c_2 = -\frac{1}{6}.$$

So the solution to the given initial-value problem is

$$x(t) = -\frac{1}{2}e^{-2t}\left(\frac{1}{2}\cos 3t + \frac{1}{3}\sin 3t\right). \quad \blacksquare$$

The motion discussed thus far is called *free motion*, because of the absence of any external force on the system. We now suppose that the mass in Figure 6.4 is acted on by an external force $F(t)$. For instance, a variable-speed fan might blow on the mass to impart a time-dependent force. In place of the homogeneous equation (5), we have the nonhomogeneous equation

$$m\frac{d^2x}{dt^2} + k\frac{dx}{dt} + hx = F(t) \tag{9}$$

as the differential equation of the system. In this case the motion is called ***forced vibration*** or ***oscillation.*** A common type of external force $F(t)$ is $F_0 \cos \omega_0 t$, where ω_0 is different from the system's "natural" frequency $\omega = \sqrt{h/m}$. The following result describes the motion of an undamped system ($k = 0$) modeled by (9).

6.4 THEOREM Under the model (9) with $k = 0$ and $F(t) = F_0 \cos \omega_0 t$, where $\omega_0 \neq \omega$, the position of the mass at time t is

$$x(t) = \frac{2F_0}{m(\omega^2 - {\omega_0}^2)} \sin \frac{1}{2}(\omega - \omega_0)t \sin \frac{1}{2}(\omega + \omega_0)t$$

if the system starts from rest.

Proof. In this case, (9) reduces to

$$m\frac{d^2x}{dt^2} + hx = F_0 \cos \omega_0 t \tag{10}$$

The homogeneous part of the solution was found in Theorem 6.1 to be given by

$$x_0(t) = c_1 \cos \omega t + c_2 \sin \omega t. \tag{2}$$

So all we need is a particular solution $x_p(t)$ of (10). From Table 5.1 there is a particular solution of the form

$$x_p(t) = A \cos \omega_0 t + B \sin \omega_0 t$$

since no such function is a homogeneous solution. Differentiating, we get

$$x_p'(t) = -A\omega_0 \sin \omega_0 t + B\omega_0 \cos \omega_0 t, \tag{11}$$

$$x_p''(t) = -A{\omega_0}^2 \cos \omega_0 t - B{\omega_0}^2 \sin \omega_0 t = -{\omega_0}^2 x_p(t). \tag{12}$$

Thus, substituting into (10), we get

$$-m{\omega_0}^2 x(t) + hx(t) = F_0 \cos \omega_0 t \rightarrow (h - m{\omega_0}^2)(A \cos \omega_0 t + B \sin \omega_0 t) = F_0 \cos \omega_0 t.$$

Hence, since $\omega = \sqrt{h/m}$, we have

$$A = \frac{F_0}{h - m{\omega_0}^2} = \frac{F_0}{m\omega^2 - m{\omega_0}^2} = \frac{F_0}{m(\omega^2 - {\omega_0}^2)} \quad \text{and} \quad B = 0.$$

So the general solution of (10) is

$$x(t) = c_1 \cos \omega t + c_2 \sin \omega t + \frac{F_0}{m(\omega^2 - {\omega_0}^2)} \cos \omega_0 t \tag{13}$$

If the system starts from rest at the equilibrium position, then $x(0) = x'(0)$, so

$$x(0) = 0 = c_1 + 0 + \frac{F_0}{m(\omega^2 - {\omega_0}^2)} \rightarrow c_1 = -\frac{F_0}{m(\omega^2 - {\omega_0}^2)};$$

$$x'(0) = 0 = 0 + \omega c_2 - 0 \rightarrow c_2 = 0.$$

Thus (13) becomes

$$x(t) = \frac{F_0}{m(\omega^2 - {\omega_0}^2)}(\cos \omega_0 t - \cos \omega t)$$

$$= \frac{2F_0}{m(\omega^2 - {\omega_0}^2)} \sin \frac{1}{2}(\omega - \omega_0)t \sin \frac{1}{2}(\omega + \omega_0)t, \tag{14}$$

FIGURE 6.8

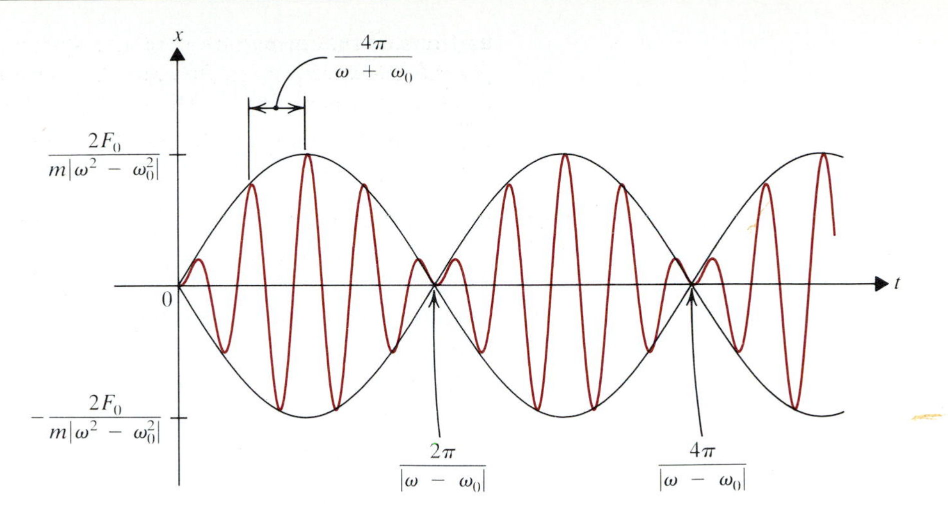

where we used the trigonometric identities

(15) $$\cos(\theta_1 + \theta_2) = \cos\theta_1 \cos\theta_2 - \sin\theta_1 \sin\theta_2$$

(16) $$\cos(\theta_1 - \theta_2) = \cos\theta_1 \cos\theta_2 + \sin\theta_1 \sin\theta_2,$$

where $\theta_1 = \frac{1}{2}(\omega + \omega_0)t$ and $\theta_2 = \frac{1}{2}(\omega - \omega_0)t$. [We subtracted (15) from (16), as you should verify.] QED

Note that if ω_0 is close to ω, then $\omega + \omega_0$ is much larger than $\omega - \omega_0$, so $\sin\frac{1}{2}(\omega + \omega_0)t$ oscillates with small period $4\pi/(\omega + \omega_0)$. The resulting motion is rapidly oscillating with frequency $\frac{1}{2}(\omega + \omega_0)$ and with a slowly changing amplitude influenced by the factor $\sin\frac{1}{2}(\omega - \omega_0)t$. In **Figure 6.8** we show a graph of the motion described in (14), which is said to exhibit *beats*. This phenomenon is used by musicians to tune stringed instruments; the beats are clearly audible when two strings differ in frequency by a few cycles per second.

In case $\omega = \omega_0$ the analysis of Theorem 6.4 changes radically. Since $A\cos\omega_0 t + B\sin\omega_0 t$ is then a *homogeneous* solution, the particular solution $x_p(t)$ must have the form $tA\cos\omega_0 t + tB\sin\omega_0 t$. Thus the motion oscillates *without bound* as $t \to \infty$, a phenomenon known as ***resonance.*** In a physical system this means that the driving force $F(t)$ is supplying energy to the system continuously, causing the amplitude to increase (like a child who swings higher and higher on a playground swing). In practical terms the spring in Figure 6.4 will be stretched to the point of breaking or losing its elasticity, or the chamber walls at one end or the other will be broken down. You can consider this case in more detail in Exercise 3.

We return now to the general (damped) case in (9), with external force $F(t) = F_0 \cos\omega_0 t$, where we no longer restrict ω_0 to be different from $\omega = \sqrt{h/m}$.

6.5 THEOREM

Under the model

(9) $$m\frac{d^2x}{dt^2} + k\frac{dx}{dt} + hx = F_0 \cos\omega_0 t$$

the motion of the system approaches a steady oscillation about the equilibrium point.

Proof. In proving Theorem 6.3 we found that the solution to the associated homogeneous problem

(5) $$\frac{d^2x}{dt^2} + \frac{k}{m}\frac{dx}{dt} + \frac{h}{m}x = 0$$

had one of the forms

(6) $$x_0(t) = c_1e^{r_1t} + c_2e^{r_2t},$$

(7) $$x_0(t) = ae^{-kt/2m}\cos\left(\frac{\sqrt{4mh-k^2}}{2m}t - \phi_0\right),$$

or

(8) $$x_0(t) = c_1e^{-kt/2m} + c_2e^{-kt/2m}.$$

In all three cases we saw that $\lim_{t\to\infty} x_0(t) = 0$, so the motion is ultimately determined by the nature of a particular solution $x_p(t)$ of (9). For this reason, $x_0(t)$ is called a ***transient solution*** and $x_p(t)$ a ***steady state solution.*** The particular solution to (9) has the form

(17) $$x_p(t) = A\cos\omega_0 t + B\sin\omega_0 t,$$

so the motion does approach a steady oscillation, since we can reduce (17) to the form $x_p(t) = a\cos(\omega_0 t - \phi_0)$ by introducing an appropriate phase angle ϕ_0 as in Theorem 6.1. QED

We next take a closer look at $x_p(t)$. The formulas (11) and (12) give $x_p'(t)$ and $x_p''(t)$. Substitution of these into (9) gives

(18) $$(h - m\omega_0^2)(A\cos\omega_0 t + B\sin\omega_0 t) + k(-A\omega_0\sin\omega_0 t + B\omega_0\cos\omega_0 t) = F_0\cos\omega_0 t.$$

Recalling that $\omega = \sqrt{h/m}$, we can write $h - m\omega_0^2 = m(\omega^2 - \omega_0^2)$. Equating coefficients of $\cos\omega_0 t$ and $\sin\omega_0 t$ in (18),

(19) $$\begin{cases} m(\omega^2 - \omega_0^2)A + k\omega_0 B = F_0, \\ -k\omega_0 A + m(\omega^2 - \omega_0^2)B = 0. \end{cases}$$

As you are asked to show in Exercise 2, this system has the solution

(20) $$A = \frac{m(\omega^2 - \omega_0^2)F_0}{m^2(\omega^2 - \omega_0^2)^2 + k^2\omega_0^2}, \qquad B = \frac{k\omega_0 F_0}{m^2(\omega^2 - \omega_0)^2 + k^2\omega_0^2}.$$

So (17) becomes

(17′) $$x_p(t) = \frac{F_0}{m^2(\omega^2 - \omega_0^2)^2 + k^2\omega_0^2}[m(\omega^2 - \omega_0^2)\cos\omega_0 t + k\omega_0\sin\omega_0 t].$$

Now if we let

$$c = \sqrt{m^2(\omega^2 - \omega_0^2)^2 + k^2\omega_0^2},$$

then (cf. Theorem 6.1) we can choose ϕ_0 so that

$$\cos\phi_0 = \frac{m(\omega^2 - \omega_0^2)}{c} \quad \text{and} \quad \sin\phi_0 = \frac{k\omega_0}{c}.$$

We thus have

(21) $$x_p(t) = \frac{F_0}{c^2}(\cos\phi_0\cos\omega_0 t + \sin\phi_0\sin\omega_0 t) = \frac{F_0}{c^2}\cos(\omega_0 t - \phi_0).$$

Here c is never 0, even if $\omega_0 = \omega$. Hence, unlike the undamped case, the motion here is always bounded. If $k < \sqrt{2h}$, then (Exercise 4) c takes on its minimum when $\omega_0^2 = \omega^2 - k^2/(2m^2)$. In this case the amplitude of $x_p(t)$ is maximized.

Seismographs are constructed as vibrating systems of frequency ω such that $\sqrt{\omega^2 - k^2/(2m^2)}$ is located in the middle of the range of frequencies associated with earthquakes. They thus can detect even slight earthquakes by responding strongly to frequencies near ω_0.

Electrical Circuits

FIGURE 6.9

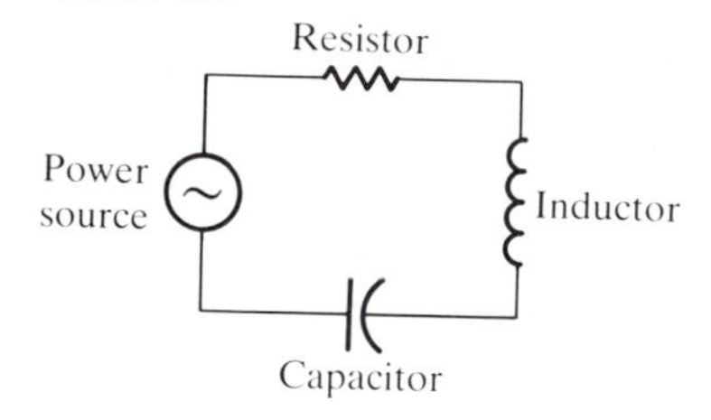

We turn now to a different phenomenon, whose mathematical models are strikingly similar to those of the vibrating systems just considered. The type of circuit we consider is shown in **Figure 6.9.** It has an external power source (EMF), a resistor, an inductor, and a capacitor (condenser). The power source has a time-varying voltage $E(t)$, the resistor a constant resistance R measured in ohms, the inductor a fixed inductance L measured in henrys, and the capacitor a fixed capacitance C measured in farads. (The quantity $1/C$ is called the *elastance.*) Let $q(t)$ be the charge in coulombs on the capacitor at time t, and let $i(t) = dq/dt$ be the current flowing in the circuit at time t, measured in amperes.

We use *Kirchhoff's voltage law*, which says that the sum of the voltage drops around the circuit at any instant is equal to the impressed voltage $E(t)$. From laboratory measurements it is known that the voltage drops across the resistor, inductor, and capacitor are given respectively by $Ri(t)$, $L(di/dt)$, and $(1/C)q(t)$. Thus Kirchhoff's law gives the differential equation

$$L\frac{di}{dt} + Ri(t) + \frac{1}{C}q(t) = E(t). \tag{22}$$

Since $i = dq/dt$, this can be written

$$L\frac{d^2q}{dt^2} + R\frac{dq}{dt} + \frac{1}{C}q(t) = E(t). \tag{23}$$

From a mathematical standpoint, (23) is *indistinguishable* from the equation (9) of a damped vibrating system. We have, in fact, the following table of correspondences. (The frequency ω is measured in hertz (cycles per second).)

If we wish to consider just current and avoid working directly with $q(t)$, then we can do so by differentiating (22). We get the second-order linear equation

$$L\frac{d^2i}{dt^2} + R\frac{di}{dt} + \frac{1}{C}i = E'(t), \tag{24}$$

which again has the same basic form as (9).

Since we have already analyzed (9) thoroughly, we have in effect a complete solution of (23) or (24). All we need to do is translate our earlier results by means of Table 1. Suppose, for example, that $E(t) = (1/\omega_0)E_0 \sin \omega_0 t$. Then $E'(t) =$

TABLE 1

Vibrating System $m\frac{d^2x}{dt^2} + k\frac{dx}{dt} + hx = F(t)$	*Electrical Circuit* $L\frac{d^2q}{dt^2} + R\frac{dq}{dt} + \frac{1}{C}q(t) = E(t)$
Displacement $x(t)$	Charge $q(t)$
Velocity dx/dt	Current $dq/dt = i(t)$
Mass m	Inductance L
Damping constant k	Resistance R
Spring constant h	Elastance $1/C$
External force $F(t)$	Electromotive force $E(t)$
Frequency $\omega = \sqrt{h/m}$	Frequency $\omega = \sqrt{1/CL}$

$E_0 \cos \omega_0 t$ and Theorem 6.5 applies. There is then a *transient current* of the form (6), (7), or (8) and a *steady state current* $i_p(t)$ in the circuit given by the analogue of (17′) or (21). Thus

$$(25) \quad i_p(t) = \frac{E_0}{I^2}\left[L(\omega^2 - \omega_0{}^2)\cos \omega_0 t + R\omega_0 \sin \omega_0 t\right] = \frac{E_0}{I^2}\cos(\omega_0 t - \phi_0),$$

where $I = \sqrt{L^2(\omega^2 - \omega_0{}^2)^2 + R^2\omega_0{}^2}$ is called the *impedance* of the circuit.

Just as mechanical systems have maximal steady-state response (amplitude) when $\omega_0{}^2 = \omega^2 - k^2/2m^2$, here I has a minimum, and the amplitude of $i_p(t)$ is maximized when $\omega_0{}^2 = \omega^2 - R^2/2L^2$. Thus a circuit such as a radio receiver should be designed so that its frequency ω is in the middle of the band of frequencies ω_0 to be received.

Example 2

In the circuit of Figure 6.9, suppose that $E(t) = 400 \sin 200t$, $R = 40$ ohms, $L = 0.25$ henrys, and $C = 4.0 \times 10^{-4}$ farads. Determine the frequency ω, phase angle ϕ_0, impedance, and steady-state current.

Solution. Differentiating the given $E(t)$, we get $E'(t) = 8.0 \times 10^4 \cos 200t$. Comparing this with $E'(t) = E_0 \cos \omega_0 t$, we see that $E_0 = 8.0 \times 10^4$ and $\omega_0 = 200$. Equation (24) thus is

$$0.25\frac{d^2 i}{dt^2} + 40\frac{di}{dt} + (0.25 \times 10^4)i = 8.0 \times 10^4 \cos 200t.$$

In standard form,

$$\frac{d^2 i}{dt^2} + 160\frac{di}{dt} + 10^4 i = 32.0 \times 10^4 \cos 200t.$$

The auxiliary equation is $r^2 + 160r + 10^4 = 0$, whose roots are

$$r = \frac{1}{2}(-160 \pm \sqrt{25600 - 40000}) = -80 \pm \frac{1}{2}\sqrt{-14400} = -80 \pm 60i.$$

Thus the transient current has the form

$$i(t) = e^{-80t}(c_1 \cos 60t + c_2 \sin 60t).$$

From the preceding discussion, the steady-state current is given by (25). We now collect the quantities needed to compute the impedance I: $E_0 = 8.0 \times 10^4$ volts, $L^2 = 6.25 \times 10^{-2}$, $R^2 = 1.6 \times 10^3$, and $\omega_0 = 200$ hertz. From Table 6.1, $\omega = \sqrt{1/CL}$, so $\omega^2 = 1/CL = 1.0 \times 10^4$; thus $\omega = 1.0 \times 10^2$ hertz. Now we can find the impedance

$$\begin{aligned} I &= \sqrt{6.25 \times 10^{-2}(10^4 - 4 \times 10^4)^2 + 1.6 \times 10^3(4 \times 10^4)} \\ &= \sqrt{120.25 \times 10^6} \approx 10.966 \times 10^3. \end{aligned}$$

So the impedance is approximately 1.1×10^4. The phase angle ϕ_0 is given by

$$\cos \phi_0 = \frac{L(\omega^2 - \omega_0{}^2)}{I} = \frac{-7500}{\sqrt{120.25 \times 10^6}} \approx -0.68394,$$

$$\sin \phi_0 = \frac{R\omega_0}{I} = \frac{8000}{\sqrt{120.25 \times 10^6}} \approx 0.72954.$$

So $\phi_0 \approx 2.324$ radians (about 133°). We can conveniently give the steady-state current in the form of the second expression in (25):

$$i_p(t) = \frac{E_0}{I}\cos(\omega_0 t - \phi_0) \approx 7.3\cos(200t - 2.3),$$

where we round off to the two significant figures of accuracy of the measurements of R, L, and C. ■

If we wanted an explicit formula for $i(t)$ at any instant, we would usually find it more convenient to work with the form (23) than with (24).

Example 3

Suppose in Example 2 that $E(t) = 100\sin 200t$, $q(0) = 0$, and $i(0) = 0$. Find a formula for $i(t)$ at any time t.

Solution. The equation in form (23) is

$$0.25\frac{d^2q}{dt^2} + 40\frac{dq}{dt} + (0.25 \times 10^4)q = 100\sin 200t.$$

In standard form we have

$$\frac{d^2q}{dt^2} + 1.6 \times 10^2\frac{dq}{dt} + 1.0 \times 10^4 q = 400\sin 200t. \tag{26}$$

The homogeneous part of the solution is

$$q(t) = e^{-80t}(c_1\cos 60t + c_2\sin 60t).$$

To find a particular solution, let $q_1(t) = A\cos 200t + B\sin 200t$. Then

$$q_1'(t) = -200A\sin 200t + 200B\cos 200t,$$

$$q''(t) = -40{,}000A\cos 200t - 40{,}000B\sin 200t = -40{,}000q_1(t).$$

Substituting into (26) we have

$$-40{,}000(A\cos 200t + B\sin 200t) + 32{,}000(-A\sin 200t + B\cos 200t)$$
$$+ 10{,}000(A\cos 200t + B\sin 200t) = 400\sin 200t.$$

Equating coefficients, we get the system of equations

$$\begin{cases} (-40{,}000 + 10{,}000)A + 32{,}000B = 0, \\ -32{,}000A + (-40{,}000 + 10{,}000)B = 400. \end{cases}$$

Dividing by 2000, we find

$$\begin{cases} -15A + 16B = 0 \\ -16A - 15B = \frac{1}{5}. \end{cases}$$

This system has solution $A = -16/2405 \approx -0.006653$ and $B = -3/481 \approx -0.006237$. Thus the general solution to (27) is

$$q(t) = e^{-80t}(c_1\cos 60t + c_2\sin 60t) - \frac{16}{2405}\cos 200t - \frac{3}{481}\sin 200t.$$

When $t = 0$, $q = 0$. Hence

$$0 = c_1 - \frac{16}{2405}, \text{ so } c_1 = \frac{16}{2405}.$$

Next,

$$i(t) = \frac{dq}{dt} = -80e^{-80t}(c_1 \cos 60t + c_2 \sin 60t)$$
$$+ 60e^{-80t}(-c_1 \sin 60t + c_2 \cos 60t)$$
$$+ \frac{3200}{2405} \sin 200t - \frac{600}{481} \cos 200t.$$
$$= e^{-80t}[(-80c_1 + 60c_2) \cos 60t + (-60c_1 - 80c_2) \sin 60t]$$
$$+ \frac{3200}{2405} \sin 200t - \frac{600}{481} \cos 200t.$$

Since $i(0) = 0$, we have

$$0 = -80c_1 + 60c_2 - \frac{600}{481},$$

$$c_2 = \frac{1}{60}\left(\frac{600}{481} + 80c_1\right) = \frac{10}{481} + \frac{4}{3}c_1 = \frac{10}{481} + \frac{64}{7215} = \frac{214}{7215}.$$

Hence the desired formula for $i(t)$ is

$$i(t) = e^{-80t}\left(\frac{600}{481} \cos 60t - \frac{1112}{481} \sin 60t\right) + \frac{3200}{2405} \sin 200t - \frac{600}{481} \cos 200t$$
$$\approx e^{-80t}(1.2 \cos 60t - 2.3 \sin 60t) + 1.3 \sin 200t - 1.2 \cos 200t. \quad ■$$

Exercises 15.6

1. Show that equation (3) also models the motion of a point revolving with constant angular velocity ω counterclockwise about the circle $x^2 + y^2 = a^2$, which starts at an initial position (x_0, y_0) such that $x_0/a = \cos \phi_0$.
2. Show that the system (19) has solution (20).
3. **(a)** Suppose in Theorem 6.4 that $\omega_0 = \omega$. Find a particular solution to (10).
   (b) Show that if $\omega_0 = \omega$, then the general solution $x(t)$ of (10) tends to infinity as $t \to \infty$, thus establishing the assertion made just before Theorem 6.5.
4. Suppose that $k < \sqrt{2}h$. Then show that $c = \sqrt{m^2(\omega^2 - \omega_0{}^2)^2 + k^2\omega_0{}^2}$ has its minimum, and so $x_p(t)$ in (17) has its maximum amplitude, when
   $$\omega_0{}^2 = \omega^2 - k^2/2m^2.$$
5. A mass of weight 16 pounds, when placed on a 2-foot-long spring, stretches it down to a length of 10 feet. The mass is pulled down 6 inches and then released (with no initial velocity). Find a formula for $x(t)$, the number of feet from the ceiling to the mass, at time t. (Use $g = 32$ ft/s^2.)
6. **(a)** A mass with weight 8 pounds is placed on the lower end of a spring suspended from the ceiling. The weight pulls the spring 6 inches down and comes to rest in its equilibrium position. The weight is then pulled down 3 inches farther and set in motion with initial velocity $v_0 = 1$ ft/s downward. Find a formula in form (2) for the displacement $x(t)$ of the mass from its equilibrium position at time t. (Use $g = 32$ ft/s^2.)
   (b) Express $x(t)$ in the form (3) and determine the amplitude, period, and frequency of the motion.
7. In the system of Figure 6.4, a force of 32 newtons is required to move the 1-kg mass 2 meters to the right. The mass is moved $\frac{1}{2}$ meter to the right and released. Assume that the damping constant is 8 kg/s. Find $x(t)$ at time t. What type of motion does the system exhibit?
8. Repeat Exercise 7 if the damping constant is 10 kg/s.
9. Consider the undamped system in Figure 6.4, in which $m = 1$ kg and a force of 4 newtons moves the mass 1 meter to the right. Suppose that a force $F = 5 \cos t$ is applied to the system. Find a formula for $x(t)$, assuming that the system starts from rest in its equilibrium position.
10. In the system of Figure 6.4, suppose that $m = \frac{1}{2}$ kg. A force of 20 newtons is required to move the mass 2 meters to the right. If the damping constant is 2, and the system begins from rest in its equilibrium position, then find a formula for the displacement $x(t)$ at time t if an external force $F = 5 \cos 2t$ is applied to the system. What is the resonant frequency?
11. A circuit consists of a capacitor and an inductor. Show that the charge $q(t)$ on the capacitor is periodic. What is the frequency? What is the nature of the current?

12. If an electromotive force $E = E_0 \cos \omega t$ (where $\omega = 1/\sqrt{CL}$) is added to the circuit in Exercise 11, then what happens to the charge as $t \to \infty$? What about the current?

13. If a circuit consists of a resistor, capacitor, and inductor (but no impressed electromotive force), then show that as $t \to \infty$, the charge decreases to 0.

14. Discuss the long-term behavior of the charge and current in Exercise 13 if $E(t) = E_0 \cos \omega_0 t$, where $\omega_0 \neq \omega$.

15. A circuit consists of a resistor, inductor, and capacitor with $R = 6.0 \times 10^2$ ohms, $C = 0.5 \times 10^{-5}$ farads, and $L = 4.0 \times 10^{-1}$ henrys. If $q(0) = 1.0 \times 10^{-6}$ coulombs and $i(0) = 0$, then find $q(t)$ and $i(t)$.

16. Repeat Exercise 15 if $R = 2 \times 10^2$ ohms, $C = 5.0 \times 10^{-6}$ farads, and $L = 1.0$ henry.

17. In the circuit of Figure 6.9, suppose that $R = 16$ ohms, $L = 2.0 \times 10^{-2}$ henrys, $C = 2.0 \times 10^{-4}$ farads, and $E = 12$ volts. If $q(0) = 4.8 \times 10^{-3}$ coulombs and $i(0) = 0$, then find formulas for $q(t)$ and $i(t)$.

18. Suppose in the circuit of Figure 6.9 that $E(t) = 110 \sin 120\pi t$, $R = 5.0 \times 10^2$ ohms, $C = 5.0 \times 10^{-6}$ farads, and $L = 3.0$ henrys. Find the steady-state current.

19. Suppose in the circuit of Figure 6.9 that $E(t) = 100 \sin 60t$, $R = 2$ ohms, $L = 0.1$ henrys, and $C = 1/260$ farad. If $q(0) = i(0) = 0$, then find $q(t)$ and $i(t)$. What is the steady-state current?

20. Show that if x is given by (6), then x assumes its maximum value exactly once.

15.7 Power Series and Numerical Solutions*

At this point the only second-order linear differential equations

$$y'' + p(x)y' + q(x)y = s(x) \tag{1}$$

we can solve are those with constant coefficients $p(x) = p$ and $q(x) = q$. The preceding section illustrates that such equations have important applications, but it is still natural to try to solve the more general equation (1), which involves variable coefficients. This section presents two methods for doing so.

The first method applies to equations (1) for which $p(x)$ and $q(x)$ are ***analytic functions*** at a real number $x = a$. By this we mean that the Taylor series expansions of p and q converge to those functions on some interval I about the point a. (Before proceeding, you may want to review Section 9.8, which discussed power series representations of functions.) Our approach will be to find a power series expansion for the solution $y = y(x)$ of (1) on the interval I. That is, we assume that the general solution of (1) is an analytic function

$$y(x) = \sum_{n=0}^{\infty} a_n x^n \tag{2}$$

on I, where the coefficients a_n are real numbers to be determined. The process of determining a_n is very much like the method of undetermined coefficients of Section 5. The theoretical justification of the technique rests on the following result of the German mathematician I. Lazarus Fuchs (1833–1902). The proof is left for more advanced texts.

7.1 THEOREM Suppose that $p(x)$, $q(x)$, and $s(x)$ are analytic functions at $x = a$. Then the general solution of (1) is an analytic function $y = y(x)$ whose Taylor series converges to $y(x)$ on an interval $(a - r, a + r)$. This interval is at least as large as the common interval I on which the Taylor series for $p(x)$, $q(x)$, and $s(x)$ converge to those functions.

To keep matters as simple as possible, we assume that $p(x)$ and $q(x)$ are polynomials and that $s(x)$ is either a polynomial or a function such as e^x, $\sin x$, $\cos x$, or $\ln(1 + x)$ with a well-known or easily computed Taylor series. To solve (1) we first write a power series (2) for the general solution and then determine the coefficients a_n of the series as follows. By Theorem 8.5 of Chapter

* Sections 9.6, 9.7, and 9.8 are prerequisite to the part of this section up to p. 1052.

9, inside the interval I of convergence of (2) we have

$$y'(x) = \sum_{n=1}^{\infty} na_n x^{n-1} \tag{3}$$

Similarly, inside I we also have

$$y''(x) = \sum_{n=2}^{\infty} n(n-1)a_n x^{n-2} \tag{4}$$

If we substitute (2), (3), and (4) into (1), then we get

$$\sum_{n=2}^{\infty} n(n-1)a_n x^{n-2} + \sum_{n=1}^{\infty} na_n p(x)x^{n-1} + \sum_{n=0}^{\infty} a_n q(x)x^n = s(x). \tag{5}$$

Equating the coefficients of x^n on each side of (5) gives a sequence of equations on the coefficients $a_0, a_1, a_2, a_3, \ldots$, which can be solved for the later coefficients in terms of the earlier ones, usually a_0 and a_1.

Some rewriting of summation indices is usually needed to carry out this procedure. The following four examples illustrate the details. In the first two we choose simple linear equations with constant coefficients, which we can easily solve by the methods of Section 4. That enables us to check the results of the power series method against known formulas for $y = y(x)$ and also sheds some light on the significance of the early constants a_0 and a_1.

Example 1

Use the power series scheme to solve $y'' + y = 0$.

Solution. We assume that the solution $y = y(x)$ is given by (2). Then substituting (2) and (4) into the given equation, we obtain

$$\sum_{n=2}^{\infty} n(n-1)a_n x^{n-2} + \sum_{n=0}^{\infty} a_n x^n = 0.$$

The natural impulse is to add these series by adding corresponding coefficients of the powers of x. However, the powers of x are indexed differently in the two series: $x = x^1$ occurs for $n = 3$ in the first but for $n = 1$ in the second; x^2 occurs for $n = 4$ in the first but for $n = 2$ in the second, and so on. To get more readily combinable series, we change the index of summation in the first one to m. That gives

$$\sum_{m=2}^{\infty} m(m-1)a_m x^{m-2} + \sum_{n=0}^{\infty} a_n x^n = 0.$$

To index the powers of x in the same way in these series, we once more change the index of summation in the first series, this time by letting $m - 2 = n$. Then $m - 1 = n + 1$, and $m = n + 2$. Note that when $m = 2$, we have $n = 0$. We thus get

$$\sum_{n=0}^{6} (n+2)(n+1)a_{n+2}x^n + \sum_{n=0}^{\infty} a_n x^n = 0,$$

$$\sum_{n=0}^{\infty} [(n+2)(n+1)a_{n+2} + a_n]x^n = 0. \tag{6}$$

Theorem 8.6 of Chapter 9 says that an analytic function has a *unique* power series. So the only power series for the identically zero function is

$$\sum_{n=0}^{\infty} 0x^n = 0 + 0x + 0x^2 + 0x^3 + \ldots + 0x^n + \ldots.$$

Therefore every coefficient in (6) has to be 0. Hence

$$a_{n+2} = -\frac{1}{(n+2)(n+1)}\, a_n. \tag{7}$$

- When $n = 0$, this says that $a_2 = -\frac{1}{2 \cdot 1}\, a_0$. When $n = 1$, we have $a_3 = -\frac{1}{3 \cdot 2}\, a_1$.
- When $n = 2$, we get $a_4 = -\frac{1}{4 \cdot 3}\, a_2 = \frac{1}{4!}\, a_0$. For $n = 3$, we find $a_5 = -\frac{1}{5 \cdot 4}\, a_3 = \frac{1}{5!}\, a_1$.
- When $n = 4$, we get $a_6 = -\frac{1}{6 \cdot 5}\, a_4 = -\frac{1}{6!}\, a_0$. For $n = 5$, we have $a_7 = -\frac{1}{7 \cdot 6}\, a_5 = -\frac{1}{7!}\, a_1$, and so on.

From (2) then, the general solution takes the form

$$\begin{aligned}
y(x) &= a_0 + a_1 x - \frac{1}{2!}\, a_0 x^2 - \frac{1}{3!}\, a_1 x^3 + \frac{1}{4!}\, a_0 x^4 + \frac{1}{5!}\, a_1 x^5 \\
&\quad - \frac{1}{6!}\, a_0 x^6 - \frac{1}{7!}\, a_1 x^7 + \ldots \\
&= a_0\left(1 - \frac{1}{2!}\, x^2 + \frac{1}{4!}\, x^4 - \frac{1}{6!}\, x^6 + - \ldots\right) \\
&\quad + a_1\left(x - \frac{1}{3!}\, x^3 + \frac{1}{5!}\, x^5 - \frac{1}{7!}\, x^7 + - \ldots\right) \\
&= a_0 \cos x + a_1 \sin x,
\end{aligned}$$

because we recognize the series in parentheses as the Taylor series for $\sin x$ and $\cos x$, which converge for all x. ■

As a simple check, we can put the given constant coefficient equation in the standard form $(D^2 + 1)y = 0$. The auxiliary equation has roots $\pm i$. So the general solution is $y(x) = c_1 \cos x + c_2 \sin x$, in agreement with the result calculated by the power series method. The early coefficients a_0 and a_1, which cannot be found from (7), are actually *arbitrary:* They are the constants c_1 and c_2 that occur in the general solution of second-order homogeneous linear differential equations in Theorem 4.8.

An equation, like (7), which determines later coefficients in terms of earlier ones, is called a *recurrence relation.* Had the equation in Example 1 been a nonhomogeneous one, we still could have proceeded just as we did in the solution.

Example 2

Use the power series scheme to solve $y'' + y = x$.

Solution. Assume as in Example 1 that the solution is given by

$$y(x) = \sum_{n=0}^{\infty} a_n x^n. \tag{2}$$

The reasoning used in Example 1 applies here, but instead of (6) we arrive at

$$\sum_{n=0}^{\infty} [(n+2)(n+1)a_{n+2} + a_n]x^n = x. \tag{6'}$$

Since the only power series for the function $f(x) = x$ is

$$0 + 1x + 0x^2 + 0x^3 + \ldots + 0x^n + \ldots,$$

we have

$$a_{n+2} = -\frac{1}{(n+2)(n+1)} a_n, \qquad \text{for } n \neq 1, \tag{7'}$$

and when $n = 1$,

$$[3 \cdot 2a_3 + a_1] = 1 \rightarrow a_3 = -\frac{1}{3!}(a_1 - 1). \tag{8}$$

When n is even, we get from (7′) the same a_{n+2} as in Example 1:

$$n = 0 \rightarrow a_2 = -\frac{1}{2 \cdot 1} a_0$$

$$n = 2 \rightarrow a_4 = -\frac{1}{4 \cdot 3} a_2 = \frac{1}{4!} a_0$$

$$n = 4 \rightarrow a_6 = -\frac{1}{6 \cdot 5} a_4 = -\frac{1}{6!} a_0, \qquad \text{etc.}$$

However, the *odd* coefficients $a_5, a_7, a_9, \ldots$ differ from the ones in Example 1, because (8) affects them. We find

$$n = 3 \rightarrow a_5 = -\frac{1}{5 \cdot 4} a_3 = \frac{1}{5!}(a_1 - 1),$$

$$n = 5 \rightarrow a_5 = -\frac{1}{7 \cdot 6} a_5 = -\frac{1}{7!}(a_1 - 1), \qquad \text{etc.}$$

So the general solution to this nonhomogeneous problem is

$$\begin{aligned}
y(x) &= a_0 + a_1 x - \frac{1}{2!} a_0 x^2 - \frac{1}{3!}(a_1 - 1)x^3 + \frac{1}{4!} a_0 x^4 \\
&\quad + \frac{1}{5!}(a_1 - 1)x^5 - \frac{1}{6!} a_0 x^6 - \frac{1}{7!}(a_1 - 1)x^7 + \ldots \\
&= a_0\left(1 - \frac{1}{2!}x^2 + \frac{1}{4!}x^4 - \frac{1}{6!}x^6 + - \ldots\right) + x + (a_1 - 1)x \\
&\quad - \frac{1}{3!}(a_1 - 1)x^3 + \frac{1}{5!}(a_1 - 1)x^5 - \frac{1}{7!}(a_1 - 1)x^7 + \ldots \\
&= a_0 \sin x + x + (a_1 - 1)\cos x \\
&= a_0 \sin x + b_0 \cos x + x,
\end{aligned}$$

where $b_0 = a_1 - 1$. This general solution once again is the same one we would get using the approach of Section 5, as you can verify. ■

Naturally, the usefulness of the power series method lies not in affording a more involved way of solving constant coefficient equations, but rather in providing a method to attack *variable* coefficient equations. The next example thus is more illustrative of the practical application of the method.

Example 3

Find a power series representation for the solution of

$$\frac{d^2y}{dx^2} + x\frac{dy}{dx} + (2x^2 + 1)y = 0.$$

Solution. Here $p(x) = x$ and $q(x) = 2x^2 + 1$. Thus both $p(x)$ and $q(x)$ are analytic at 0. As before, we let $y(x) = \Sigma_{n=0}^{\infty} a_n x^n$, so $y'(x) = \Sigma_{n=1}^{\infty} na_n x^{n-1}$ and $y''(x) = \Sigma_{n=2}^{\infty} n(n-1)a_n x^{n-2}$. Then the second term of the given equation is

$$x\frac{dy}{dx} = \sum_{n=1}^{\infty} na_n x^n,$$

and the third term is

$$(2x^2 + 1)y = \sum_{n=0}^{\infty} 2a_n x^{n+2} + \sum_{n=0}^{\infty} a_n x^n.$$

Substituting these and the series for y'' into the given equation, we get

$$\sum_{n=2}^{\infty} n(n-1)a_n x^{n-2} + \sum_{n=1}^{\infty} na_n x^n + \sum_{n=0}^{\infty} 2a_n x^{n+2} + \sum_{n=0}^{\infty} a_n x^n = 0,$$

$$(9) \qquad \sum_{n=2}^{\infty} n(n-1)a_n x^{n-2} + a_0 + \sum_{n=1}^{\infty} (n+1)a_n x^n + \sum_{n=0}^{\infty} 2a_n x^{n+2} = 0.$$

We again need to combine these summations on differently indexed powers of x. In the first sum we change to index of summation m and let $m - 2 = n$. In the third sum we change to index of summation k and let $k + 2 = n$. Then (9) becomes

$$\sum_{m=2}^{\infty} m(m-1)a_m x^{m-2} + a_0 + \sum_{n=1}^{\infty} (n+1)a_n x^n + \sum_{k=0}^{\infty} 2a_k x^{k+2} = 0.$$

Now we use the relations $m = n + 2$ and $k = n - 2$ to convert all these to sums indexed by n:

$$\sum_{n=0}^{\infty} (n+2)(n+1)a_{n+2} x^n + a_0 + \sum_{n=1}^{\infty} (n+1)a_n x^n + \sum_{n=2}^{\infty} 2a_{n-2} x^n = 0.$$

We can combine terms for $n \geq 2$, because the last sum begins only at $n = 2$. Collecting the earlier terms, we have

$$(2) \quad (1)a_2 + (3)(2)a_3 x + a_0 + 2a_1 x + \sum_{n=2}^{\infty} [(n+2)(n+1)a_{n+2} + (n+1)a_n + 2a_{n-2}]x^n = 0,$$

$$(a_0 + 2a_2) + 2(a_1 + 3a_3)x + \sum_{n=2}^{\infty} [(n+2)(n+1)a_{n+2} + (n+1)a_n + 2a_{n-2}]x^n = 0.$$

Thus

$$a_0 + 2a_2 = 0 \rightarrow a_2 = -\frac{1}{2}a_0;$$

$$a_1 + 3a_3 = 0 \rightarrow a_3 = -\frac{1}{3}a_1;$$

and for $n \geq 2$ we have the recurrence relation

$$a_{n+2} = -\frac{1}{(n+2)(n+1)}(2a_{n-2} + (n+1)a_n)$$

- For $n = 2$ this gives $a_4 = -\frac{1}{4 \cdot 3}\left(2a_0 - \frac{3}{2}a_0\right) = -\frac{1}{24}a_0$.
- For $n = 3$ we get $a_5 = -\frac{1}{5 \cdot 4}\left[2a_1 - 4\left(\frac{1}{3}\right)a_1\right] = -\frac{1}{20}\left[\frac{2}{3}a_1\right] = -\frac{1}{30}a_1$.
- For $n = 4$ we find $a_6 = -\frac{1}{6 \cdot 5}\left[2\left(-\frac{1}{2}a_0\right) + 5\left(-\frac{1}{24}\right)a_0\right] = \frac{29}{720}a_0$.
- For $n = 5$ we have $a_7 = -\frac{1}{7 \cdot 6}\left[2\left(-\frac{1}{3}a_1\right) + 6\left(-\frac{1}{30}a_1\right)\right] = \frac{13}{630}a_1$.

Thus the solution is

$$\begin{aligned} y(x) &= a_0 + a_1 x - \frac{1}{2}a_0 x^2 - \frac{1}{3}a_1 x^3 - \frac{1}{24}a_0 x^4 - \frac{1}{30}a_1 x^5 \\ &\quad + \frac{29}{720}a_0 x^6 + \frac{13}{630}a_1 x^7 + \ldots \\ &= a_0\left(1 - \frac{1}{2}x^2 - \frac{1}{24}x^4 + \frac{29}{720}x^6 + \ldots\right) \\ &\quad + a_1\left(x - \frac{1}{3}x^3 - \frac{1}{30}x^5 + \frac{13}{630}x^7 + \ldots\right) \\ &= a_0 y_1(x) + a_1 y_2(x). \end{aligned}$$

Here the even and odd coefficients are too complicated to give a general formula for them. ■

In this example, $y_1(x)$ and $y_2(x)$ are not recognizable as familiar functions. The reason for this is that they are *not* elementary functions! For small values of x, however, we can compute them to any desired accuracy by taking enough terms of the series.

If we were interested in values of $y(x)$ near $x_0 \neq 0$, then instead of (2) we would use

$$y(x) = \sum_{n=0}^{\infty} a_n(x - x_0), \tag{10}$$

the Taylor series expansion near x_0. The next example illustrates that if we change variables by letting $t = x - x_0$, then the calculations are just as in the preceding examples.

Example 4

Find a power series representation for the solution of

$$\frac{d^2y}{dx^2} + (x - 1)\frac{dy}{dx} + y = 0$$

in powers of $x - 1$.

Solution. Let $t = x - 1$. Then

$$\frac{dy}{dx} = \frac{dy}{dt}\frac{dt}{dx} = \frac{dy}{dt} \quad \text{and} \quad \frac{d^2y}{dx^2} = \frac{d^2y}{dt^2}\frac{dt}{dx} = \frac{d^2y}{dt^2}.$$

Thus the original equation is equivalent to

$$\frac{d^2y}{dt^2} + t\frac{dy}{dt} + y = 0.$$

We want a power series solution in powers of t now. So suppose that $y = \Sigma_{n=0}^{\infty} a_n t^n$. Then

$$t\frac{dy}{dt} = \sum_{n=1}^{\infty} n a_n t^n \quad \text{and} \quad \frac{d^2y}{dt^2} = \sum_{n=2}^{\infty} n(n-1)a_n t^{n-2}.$$

Substituting into the given equation, we obtain

$$\sum_{n=2}^{\infty} n(n-1)a_n t^{n-2} + \sum_{n=1}^{\infty} n a_n t^n + \sum_{n=0}^{\infty} a_n t^n = 0,$$

$$\sum_{n=2}^{\infty} n(n-1)a_n t^{n-2} + \sum_{n=1}^{\infty} (n+1)a_n t^n + a_0 = 0.$$

In the first summation we switch to summation index m and then let $m - 2 = n$; so $m = n + 2$ and $m - 1 = n + 1$. We then get

$$\sum_{n=0}^{\infty} (n+1)(n+2)a_{n+2}t^n + \sum_{n=1}^{\infty} (n+1)a_n t^n + a_0 = 0,$$

$$\sum_{n=1}^{\infty} [(n+1)(n+2)a_{n+2} + (n+1)a_n]t^n + a_0 + (1 \cdot 2)a_2 = 0$$

We then have $a_2 = -a_0/2$, and for $n \geq 1$, we find $a_{n+2} = -a_n/(n+2)$. Thus

$$a_3 = -\frac{1}{3}a_1, \qquad a_4 = -\frac{1}{4}a_2 = \frac{1}{8}a_0, \qquad a_5 = -\frac{1}{5}a_3 = \frac{1}{3 \cdot 5}a_1,$$

and

$$a_6 = -\frac{1}{6}a_4 = -\frac{1}{8 \cdot 6}a_0.$$

We therefore have

$$\begin{aligned} y(t) &= a_0 + a_1 t - \frac{1}{2}a_0 t^2 - \frac{1}{3}a_1 t^3 + \frac{1}{8}a_0 t^4 + \frac{1}{3 \cdot 5}a_1 t^5 - \frac{1}{8 \cdot 6}a_0 t^6 + \dots \\ &= a_0\left(1 - \frac{1}{2}t^2 + \frac{1}{2^2 \cdot 2!}t^4 - \frac{1}{2^3 \cdot 3!}t^6 + - \dots\right) \\ &\quad + a_1\left(t - \frac{1}{3}t^3 + \frac{1}{3 \cdot 5}t^5 + - \dots\right) \\ &= a_0 \sum_{i=0}^{\infty} \frac{(-1)^i t^{2i}}{2^i i!} + a_1 \sum_{j=0}^{\infty} \frac{(-1)^j t^{2j+1}}{1 \cdot 3 \cdot 5 \dots (2j+1)}. \end{aligned}$$

Hence

$$\begin{aligned} y(x) &= a_0 \sum_{i=0}^{\infty} \frac{(-1)^i (x-1)^{2i}}{2^i i!} + a_1 \sum_{j=0}^{\infty} \frac{(-1)^j (x-1)^{2j+1}}{1 \cdot 3 \cdot 5 \dots (2j+1)} \\ &= a_0 y_1(x) + a_1 y_2(x). \quad ■ \end{aligned}$$

Again, $y_1(x)$ and $y_2(x)$ are not recognizable as elementary functions (but see Exercise 45). Since they are defined by convergent alternating series, though, they can be approximated to any desired accuracy by Theorem 4.4 of Chapter 9.

Euler's Method and a Refinement

The foregoing may give the impression that the power series method is universally applicable. But we have used it only for equations

(1) $$y'' + p(x)y' + q(x)y = s(x)$$

with polynomial coefficients $p(x)$ and $q(x)$. If $p(x)$ or $q(x)$ is a more complicated function, then finding a power series expansion for y may be prohibitively complicated. Worse still, even if we can find that series, calculating the value of y for some values of x from the power series expansion $y = \Sigma_{n=0}^{\infty} a_n(x - x_0)^n$ would require us to find the sums of several infinite series of constants. Unless these happen to be geometric series, that is a *formidable* task for which the methods of Chapter 9 are of little help. The best way to proceed is to *numerically approximate* $y(x_1)$ for values x_1 of interest by taking a finite partial sum $\Sigma_{n=0}^{k} a_n(x_1 - x_0)^n$.

An often shorter approach is to use a numerical solution method for the given differential equation in the first place. Such a method generates a list of approximate values of the solution $y = y(x)$ at a number of points $x_1, x_2, \ldots, x_n$.

The development of electronic calculators and small computers has enormously increased the practical importance of numerical solution methods, some of which have been known theoretically (but were difficult to apply in concrete examples) for years. You will learn more of these methods if you take further courses in differential equations and numerical analysis. In the rest of this section we will give just an introduction to the idea of numerical procedures for solving differential equations, by considering two of the most elementary of the many numerical solution schemes.

Euler developed a linear approximation technique for the unique solution $y = y(x)$ to the first-order initial-value problem

(11) $$\frac{dy}{dx} = f(x, y), \qquad y(x_0) = y_0.$$

We have already used it in Section 2.8 to approximate y near $x = x_0$ (see Examples 4 and 5 of that section). The method is called ***Euler's method*** or the ***tangent-line method.*** We are given x_0 and y_0, so we know the slope $y'(x_0) =$

FIGURE 7.1

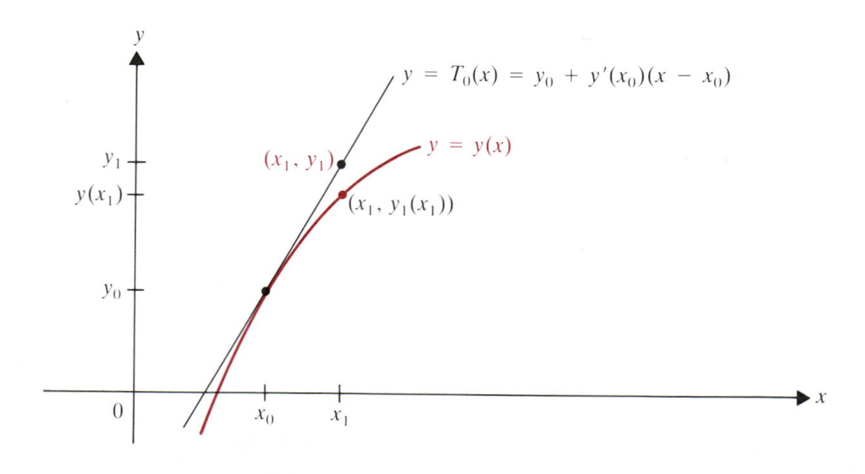

FIGURE 7.2

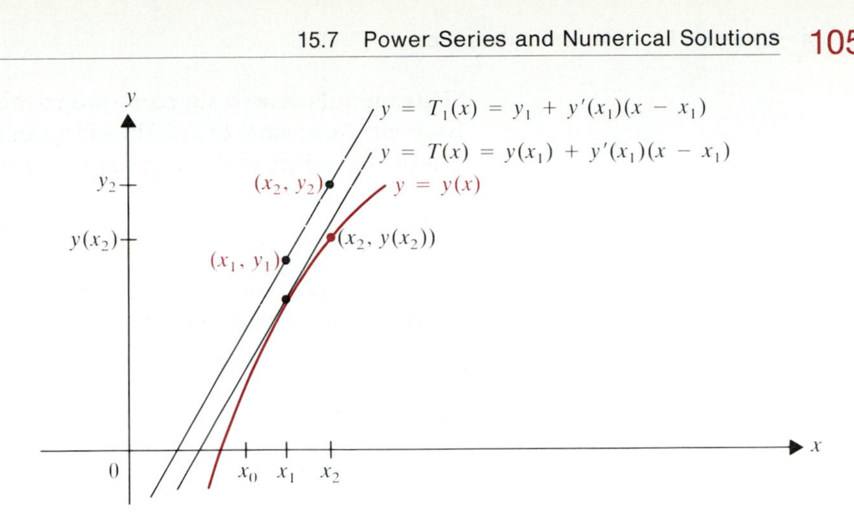

$f(x_0, y_0)$ of the tangent line to the solution $y = y(x)$ at (x_0, y_0). If x_1 is close to x_0, then we can use the tangent approximation $T_0(x) = f(x_0) + f'(x_0)(x - x_0)$ to approximate y at x_1. As **Figure 7.1** illustrates, we get

(12) $$y(x_1) \approx T_0(x_1) = y_0 + y'(x_0)(x_1 - x_0) = y_0 + f(x_0, y_0)(x_1 - x_0) = y_1.$$

Once we have the approximation y_1 for $f(x_1)$, we can approximate $y'(x_1)$ by

$$y'(x_1) \approx f(x_1, y_1).$$

This approximates the slope of the tangent line to $y = y(x)$ at $(x_1, y(x_1))$. Then if x_2 is near x_1, we use the tangent approximation again (see **Figure 7.2**):

$$y(x_2) \approx T_1(x_2) = y_1 + y'(x_1)(x_2 - x_1) \approx y_1 + f(x_1, y_1)(x_2 - x_1) = y_2.$$

We can now continue this scheme, generating in general the approximation $y_{n+1} \approx T_n(x_{n+1})$ given by

(13) $$y(x_{n+1}) \approx y_n + f(x_n, y_n)(x_{n+1} - x_n) = y_{n+1},$$

where y_n is the approximation for $y(x_n)$ obtained at the previous step. (See **Figure 7.3.**) If in (13) we take steps of uniform length $h = x_{i+1} - x_i$, then we have

(14) $$y(x_{n+1}) \approx y_n + hf(x_n, y_n) = y_{n+1}.$$

FIGURE 7.3

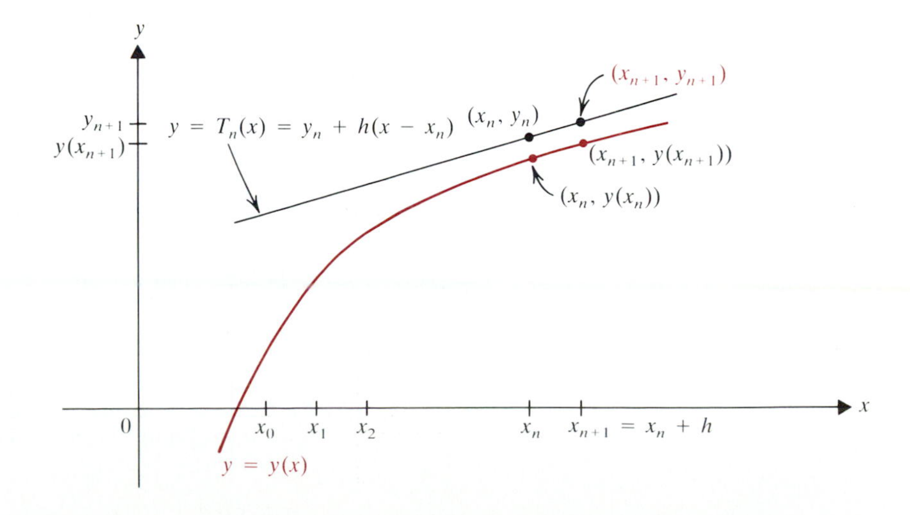

This formula is well suited to programming: It just says that at each iterative step, the new value of y is the old value of y (the one obtained at the last step) plus the product of the step size h and the value of the given function $f(x, y)$ at the last point (x_n, y_n).

To illustrate the method, we consider a simple first-order linear differential equation that we can readily solve for an explicit function $y = y(x)$. (The exact formula for $y(x)$ will enable us to calculate the error in the Euler approximations without recourse to estimation.)

Example 5

Find approximate values of y at x_0, x_1, x_2, x_3, x_4, and x_5 if

$$\frac{dy}{dx} = x - 2y, \qquad x_0 = 0, \qquad y(x_0) = 1,$$

and a uniform step size $h = 0.1$ is used. Compare with the exact solution.

Solution. Here $y(x_0) = y_0 = 1$. From (12) we have

$$y_1 = y_0 + (x_0 - 2y_0)(0.1) = 1 + (-2)(0.1) = 0.8000.$$

From (13) or (14) with $n = 1, 2, 3$, and 4 we find, to four decimal places,

$$y_2 = y_1 + (x_1 - 2y_1)(0.1) = 0.8000 + (0.1000 - 1.6000)(0.1) = 0.6500,$$

$$y_3 = y_2 + (x_2 - 2y_2)(0.1) = 0.6500 + (0.2000 - 1.300)(0.1) = 0.5400,$$

$$y_4 = y_3 + (x_3 - 2y_3)(0.1) = 0.5400 + (0.3000 - 1.0800)(0.1) = 0.4620,$$

and

$$y_5 = y_4 + (x_4 - 2y_4)(0.1) = 0.4620 + (0.4000 - 0.9240)(0.1) = 0.4096.$$

FIGURE 7.4

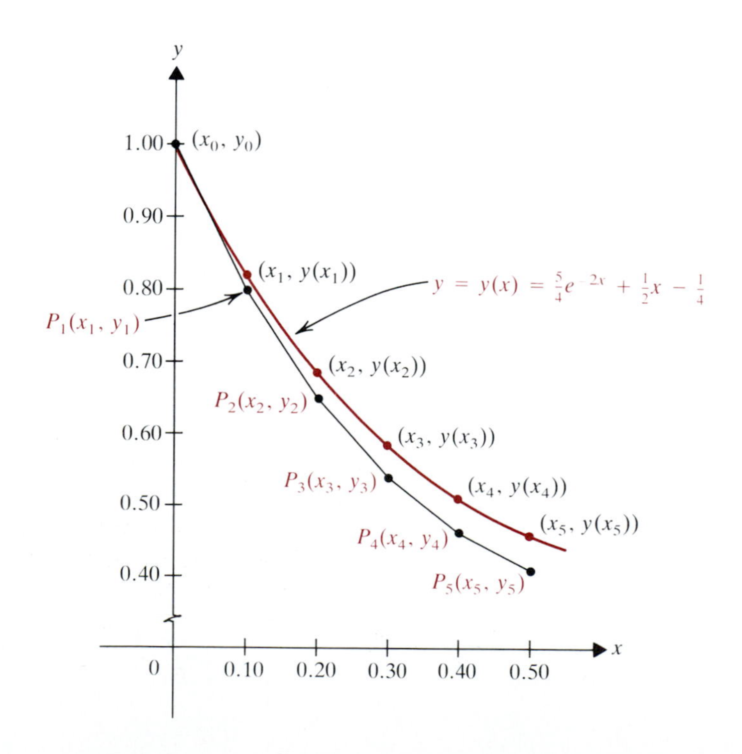

We can easily solve this first-order linear equation by the method of Section 1. Since it has integrating factor e^{2x}, we get

$$\frac{d}{dx}(e^{2x}y) = xe^{2x} \rightarrow y = \frac{1}{2}x - \frac{1}{4} + ce^{-2x}.$$

When $x = x_0 = 0$, $y = 1$, so $c = \frac{5}{4}$. Thus the exact solution is $y = \frac{5}{4}e^{-2x} + \frac{1}{2}x - \frac{1}{4}$. In **Figure 7.4** we illustrate graphically how the points (x_i, y_i) compare with the corresponding points $(x_i, y(x_i))$ on the graph of $y = y(x)$. Connecting those points with the line segments shown produces an approximate graph of $y = y(x)$. In Table 1 we show the approximate values of y_i in comparison with the values of $y(x_i)$ to four decimal places. We show also the percent error in the approximation of $y(x_i)$ by y_i. ■

TABLE 1

x_n	$y(x_n) = \frac{5}{4}e^{-2x_n} + \frac{1}{2}x_n - \frac{1}{4}$	y_n	$\lvert y_n - y(x_n)\rvert$	% *error*
0	1.0000	1.0000	0	0.00
0.1	0.8234	0.8000	0.0234	2.84
0.2	0.6879	0.6500	0.0379	5.51
0.3	0.5860	0.5400	0.0460	7.85
0.4	0.5117	0.4620	0.0497	9.71
0.5	0.4598	0.4096	0.0502	10.92

As the table shows, the error is sizeable. It can be reduced somewhat if a smaller step size is used (see Exercise 30), but the Euler method is not intrinsically highly accurate. We can get a rough idea of why, as well as why the error grows in Example 5, as follows. (More precise discussions are given in numerical analysis or differential equations texts.) Suppose that y'' is continuous and does not change sign on an interval I containing x_0 and all the points $x_1 = x_0 + h$, $x_2 = x_0 + 2h, \ldots, x_n = x_0 + nh$, where we use the Euler approximation formula

$$y_{n+1} = y_n + hf(x_n, y_n) \tag{14}$$

(The function y thus has constant concavity on I.) Then the Lagrange remainder formula (Theorem 6.7 of Chapter 9) says that

$$y(x_{n+1}) = y(x_n) + y'(x_n)h + \frac{1}{2}y''(c_n)h^2 = y(x_n) + f(x_n, y_n)h + \frac{1}{2}y''(c_n)h^2, \tag{15}$$

where c_n lies in the interval (x_n, x_{n+1}). If we assume that $f(x_n, y_n)$ is so close to $y'(x_n) = f(x_n, y(x_n))$ that we can ignore the error, then subtraction of (14) from (15) gives

$$y(x_{n+1}) - y_{n+1} = [y(x_n) - y_n] + \frac{1}{2}y''(c_n)h^2.$$

This equation says that the error in y_{n+1} is the sum of the additional factor $\frac{1}{2}y''(c_n)h^2$ and the error present in y_n. This means that with each iteration, we *add* to the previous error by an amount proportional to h^2, the square of the step size. If M is an upper bound of $y''(x)$ over the interval I, then in using Euler's method n times from $x = x_0$ to $x = x_{n+1}$, we accumulate an error that is roughly proportional to

$$n \cdot \frac{1}{2}Mh^2.$$

Since

$$x_{n+1} = x_0 + nh \rightarrow h = \frac{x_{n+1} - x_0}{n},$$

that approximate error can be expressed as

$$\not{n} \cdot \frac{1}{2} M \cdot \frac{x_{n+1} - x_0}{\not{n}} \cdot h = \frac{1}{2}(x_{n+1} - x_0)h.$$

The accumulated error in x_{n+1} is therefore *roughly* proportional to the step size h.

Because Euler's method is not highly accurate, refinements of it are employed in practice. A commonly used family of refinements comprise the ***Runge–Kutta method.*** We discuss only the *second-order* Runge–Kutta method, which is also known as the ***improved Euler method.*** The idea is to replace $f(x_n, y_n)$ in (14) by the *average* of $f(x_n, y_n)$ and $f(x_{n+1}, y_{n+1})$. Here

$$y_{n+1} = \bar{y}_n + hf(x_n, \bar{y}_n) \tag{16}$$

is computed from the tangent approximation applied at $(x_n, \bar{y}_n)$, where $\bar{y}_n$ is the improved Euler approximation to $y(x_n)$. The formula for the improved Euler approximation $\bar{y}_{n+1}$ is then

$$\bar{y}_{n+1} = \bar{y}_n + \frac{1}{2} h[f(x_n, \bar{y}_n) + f(x_{n+1}, y_{n+1})],$$

where y_{n+1} is computed from (16). Thus

$$\bar{y}_{n+1} = \bar{y}_n + \frac{1}{2} h[f(x_n, \bar{y}_n) + f(x_{n+1}, \bar{y}_n + hf(x_n, \bar{y}_n))], \tag{17}$$

where $h = x_{n+1} - x_n$ is assumed to be constant.

Use of (17) involves more calculation than is needed for Euler's method: Two evaluations of f have to be made at each step. The next example illustrates the fact that while (17) can be used on a calculator with a memory, the amount of calculating it involves makes it better suited to a computer or programmable calculator.

Example 6

Repeat Example 5 using the improved Euler method (17).

Solution. We still have $y(x_0) = \bar{y}_0 = 1$. Continuing from there, we first find y_{n+1} at each stage, and then we use (17) to find $\bar{y}_{n+1}$. As in Example 5, $y_1 = 0.8000$. Then, to four decimal places,

$$\begin{aligned}
\bar{y}_1 &= \bar{y}_0 + \frac{1}{2}(0.1)(x_0 - 2\bar{y}_0 + x_1 - 2y_1) \\
&= 1.000 + (0.05)(-2.0000 - 1.5000) = 0.8250, \\
y_2 &= \bar{y}_1 + 0.1(x_1 - 2\bar{y}_1) = 0.8250 + 0.1(0.1000 - 1.6500) = 0.6700, \\
\bar{y}_2 &= \bar{y}_1 + \frac{1}{2}(0.1)(x_1 - 2\bar{y}_1 + x_2 - 2y_2) \\
&= 0.8250 + 0.05(-1.5500 - 1.1400) = 0.6905, \\
y_3 &= \bar{y}_2 + (0.1)(x_2 - 2\bar{y}_2) = 0.6905 + 0.1(0.2000 - 1.3810) = 0.5724,
\end{aligned}$$

$$\bar{y}_3 = \bar{y}_2 + \frac{1}{2}(0.1)(x_2 - 2\bar{y}_2 + x_3 - 2y_3)$$
$$= 0.6905 + (0.05)(-1.1810 - 0.8448) = 0.5892,$$
$$y_4 = \bar{y}_3 + (0.1)(x_3 - 2\bar{y}_3) = 0.5892 + 0.1(-0.8784) = 0.5014,$$
$$\bar{y}_4 = \bar{y}_3 + \frac{1}{2}(0.1)(x_3 - 2\bar{y}_3 + x_4 - 2y_4)$$
$$= 0.5892 + (0.05)(-0.8784 - 0.6028) = 0.5151,$$
$$y_5 = \bar{y}_4 + (0.1)(x_4 - 2\bar{y}_4) = 0.5151 + 0.1(-0.6302) = 0.4521,$$
$$\bar{y}_5 = \bar{y}_4 + \frac{1}{2}(0.1)(x_4 - 2\bar{y}_4 + x_5 - 2y_5)$$
$$= 0.5151 + (0.05)(-0.6302 - 0.4042) = 0.4634. \blacksquare$$

TABLE 2

x_n	$y(x_n)$	$\bar{y}_n$	$\lvert\bar{y}_n - y(x_n)\rvert$	% error
0	1.0000	1.0000	0	0.00
0.1	0.8234	0.8250	0.0016	0.19
0.2	0.6879	0.6905	0.0026	0.38
0.3	0.5860	0.5892	0.0032	0.55
0.4	0.5117	0.5151	0.0034	0.66
0.5	0.4598	0.4634	0.0036	0.78

Table 2 shows the marked improvement in accuracy that the improved Euler method yields in this example and is suggestive of the gain in accuracy usually afforded by this method. It can be shown that the accumulated error is roughly proportional to h^2 (see Exercises 46–48).

While the error continues to increase as we get farther from x_0 in Example 6, the percent error remains less than 1% throughout. For many computational purposes that is adequate accuracy.

HISTORICAL NOTE

The methods of this section were known, and used by hand, long before the advent of modern computing machines made them convenient. Euler's method dates from the mid-18th century, and the Runge–Kutta method was developed at the end of the last century by the German mathematicians Carl Runge (1856–1927) and M. Wilhelm Kutta (1867–1944).

Exercises 15.7

In Exercises 1–18, find the general solution of the given differential equation in power series form. In Exercises 1–4, compare your solution with the explicit solution obtained using the methods of the preceding sections.

1. $\dfrac{d^2y}{dx^2} - 4y = 0$, about $x = 0$
2. $\dfrac{d^2y}{dx^2} + 4\dfrac{dy}{dx} + 4y = 0$, about $x = 0$
3. $\dfrac{d^2y}{dx^2} + 4y = 0$, about $x = 0$
4. $\dfrac{d^2y}{dx^2} + 9y = 0$, about $x = 0$
5. $\dfrac{d^2y}{dx^2} + x\dfrac{dy}{dx} + 2y = 0$, about $x = 0$
6. $\dfrac{d^2y}{dx^2} - x\dfrac{dy}{dx} - y = 0$, about $x = 0$

7. $y'' + xy' + 2y = x$ about $x = 0$
8. $y'' - xy' - y = x$ about $x = 0$
9. $y'' + xy' + (x^2 + 2)y = 0$, about $x = 0$
10. $y'' + xy' + (3x + 2)y = 0$, about $x = 0$
11. $(2x + 1)y'' + y' + 2y = 0$, near $x = 0$
12. $(1 + x)y'' - y = 0$, near $x = 0$
13. $(x^2 + 1)y'' + x\dfrac{dy}{dx} + xy = 0$, about $x = 0$
14. $y'' - 2(x - 1)y' - y = 0$, about $x = 1$
15. $y'' - xy = 0$, about $x = 1$. [This equation is known as the *Airy equation*, after the English mathematician and astronomer George Airy (1801–1892).]
16. $y'' - 2xy' + \lambda y = 0$, about $x = 0$. [This equation is known as the *Hermite equation*, after the French mathematician Charles Hermite (1822–1901). If λ is an even integer, then one of the basic solutions is a polynomial that gives rise to the *Hermite polynomial*.]
17. $(1 - x^2)y'' - 2xy' + \lambda(\lambda + 1)y = 0$ about $x = 0$. [This is called the *Legendre equation*, after the French mathematician Adrien Legendre (1752–1833). If λ is a positive integer, then one of the basic solutions is a polynomial that gives rise to the *Legendre polynomial*.]
18. $(1 - x^2)y'' - xy' + \lambda y = 0$ about $x = 0$. [This is called the *Tchebysheff equation*, after the Russian mathematician Pafnuti L. Tchebysheff (1821–1894). If λ is the square of a positive integer, then one of the basic solutions is a polynomial that gives rise to the *Tchebysheff polynomial*.]

In Exercises 19–22, find the power series solution about $x_0 = 0$ for the given initial-value problem.

19. $y'' + xy' + 2y = 0$, $y(0) = 0$, $y'(0) = 1$
20. $y'' - xy' - y = 0$, $y(0) = 0$, $y'(0) = 1$
21. $(x^2 + 1)y'' + xy' + xy = 0$, $y(0) = 2$, $y'(0) = 3$
22. $(x^2 - 1)y'' + 3xy' + xy = 0$, $y(0) = 4$, $y'(0) = 6$

In Exercises 23–26, use Euler's method to find approximate values for the solution $y(x)$ to the given initial-value problem at the points $x = 0, 0.1, 0.2, 0.3, 0.4$, and 0.5. In Exercises 23 and 24, find explicit formulas for the solution and compare with your Euler approximations. Carry your calculations to six decimal places.

23. $\dfrac{dy}{dx} - 4y = 1 - x$, $y(0) = 1$
24. $\dfrac{dy}{dx} = 2x + y$, $y(0) = 1$
25. $y\dfrac{dy}{dx} = e^{-x^2}$, $y(0) = 1$
26. $\dfrac{dy}{dx} = \dfrac{1}{3}x^2y^2 + 1$, $y(0) = 1$

In Exercises 27 and 28, (a) use the improved Euler method on the indicated exercise and (b) compare with the exact solution values.

27. Exercise 23
28. Exercise 24

In Exercises 29 and 30, use the improved Euler method on the indicated exercise.

29. Exercise 25
30. Exercise 26

31. Work Exercise 23 using the Euler method and $h = 0.05$. Compare with the results of Exercise 27.
32. Work Example 5 using the Euler method and $h = 0.05$. Compare with the results of Example 6.
33. Euler's method extends to second-order initial-value problems

$$y'' = f(x, y, y'), \qquad y(x_0) = y_0, \qquad y'(x_0) = v_0,$$

as follows. Let $v = y'$. Then the given equation is equivalent to the system of two first-order equations

$$\begin{cases} v = y' \\ v' = f(x, y, v) \end{cases}$$

with initial values $y(x_0) = y_0$ and $v(x_0) = v_0$. Show that Euler's approach leads to

$$\begin{cases} y_1 = y_0 + (x_1 - x_0)v_0 \\ v_1 = v_0 + (x_1 - x_0)f(x_0, y_0, v_0) \end{cases}$$

as second-stage approximations.

34. In Exercise 33 show that the general formula is

$$\begin{cases} y_{n+1} = y_n + hv_n \\ v_{n+1} = v_n + hf(x_n, y_n, v_n), \end{cases}$$

where $h = x_{n+1} - x_n$.

35. Apply the technique of Exercises 33 and 34 to approximate the solution $y = y(x)$ to

$$\frac{d^2y}{dx^2} + \frac{2x}{x^2 - 1}\frac{dy}{dx} - \frac{6}{x^2 - 1}y = 0,$$

where $y(0) = 1$ and $y'(0) = 0$, at $x = 0, 0.1, 0.2, 0.3, 0.4$, and 0.5. Carry calculations to six decimal places.

36. Repeat Exercise 35 for

$$\frac{d^2y}{dx^2} + \frac{1}{x}\frac{dy}{dx} - x^2 - 2y = 0, \qquad y(1) = -2, \qquad y'(1) = 3$$

at $x = 1, 1.1, 1.2, 1.3, 1.4$, and 1.5.

37. In Exercise 11 use Theorem 7.1 to show that the power series solution converges at least on the interval $(-1/2, 1/2)$.
38. In Exercise 12 use Theorem 7.1 to show that the power series solution converges at least on the interval $(-1, 1)$.

PC 39. Rework Exercise 27 on the interval $[0, 1]$ with $h = 0.05$.

PC 40. Rework Exercise 28 on the interval $[0, 1]$ with $h = 0.05$.

PC 41. Rework Exercise 29 on the interval $[0, 1]$ with $h = 0.05$.

PC 42. Rework Exercise 30 on the interval $[1, 2]$ with $h = 0.05$.

PC **43.** Rework Exercise 35 on the interval $[0, 1]$ with $h = 0.05$.

PC **44.** Rework Exercise 36 on the interval $[1, 2]$ with $h = 0.05$.

45. In Example 4 show that $y_1(x) = e^{-(x-1)^2/2}$.

46. Suppose that y'' is continuous and does not change sign on an interval I containing the points $x_0, x_1 = x_0 + h, \ldots, x_{n+1} = x_0 + nh$.

(a) Write the third-order version of (15), the second-degree Taylor polynomial approximation for $y(x_{n+1})$ with Lagrange remainder.

(b) Regard $y'(x_n)$ as so close to $f(x_n, \bar{y}_n)$ that their difference is negligible. Subtract (17) from your result in (a) to show that the error in $\bar{y}_{n+1}$ is the error in $\bar{y}_n$ +

$$\frac{1}{2}[y''(x_n)h^2 - f(x_{n+1}, \bar{y}_n + hf(x_n, \bar{y}_n))h + f(x_n, \bar{y}_n)h] + \frac{y'''(c_n)h^3}{3!}, \tag{18}$$

where c_n is between x_n and x_{n+1}.

47. Use the two-variable Taylor polynomial approximation

$$\begin{aligned} f(x_{n+1}, y) \approx f(x_n, \bar{y}_n) &+ f_x(x_n, \bar{y}_n)h + f_y(x_n, \bar{y}_n)(y - \bar{y}_n) \\ &+ \frac{1}{2}[f_{xx}(x_n, \bar{y}_n)h^2 + 2f_{xy}(x_n, \bar{y}_n)h(y - \bar{y}_n) \\ &+ f_{yy}(x_n, \bar{y}_n)(y - \bar{y}_n)^2] \end{aligned}$$

on the term in color in (18) to show that the bracketed term in (18) is approximately proportional to h^3. [*Hint:* Also use the chain rule on $y'(x) = f(x, y(x))$ to show that $y''(x) = f_x(x, y(x)) + f_y(x, y(x))f(x, y)$.]

48. Conclude from Exercise 47 that the accumulated error after n iterations of the improved Euler method is roughly proportional to h^2.

15.8 Looking Back

This chapter augments the discussion of differential equations found in Sections 3.10, 6.3, and elsewhere. You have learned by now how to solve several of the most important first- and second-order differential equations. In the course of the chapter, several mathematical models involving differential equations were also discussed.

We began in Section 1 with first-order linear differential equations (Definition 1.1). Equation (9) on p. 992 gives the solution of such an equation, although the introduction of an integrating factor makes it unnecessary to deal with that cumbersome general formula for $y(x)$. Section 2 treated a special type of first-order linear differential equation. Exact equations are the differential-equation version of path-independent line integrals and are easily solved once they are recognized as being exact (Theorem 2.2).

Sections 4 and 5 developed the theory needed to solve constant-coefficient second-order linear differential equations. As in the case of first-order linear equations (Theorem 1.2), we found in Theorem 4.2 that the general solution of such an equation is made up of all functions of the form $y_0(x) + y_p(x)$, where $y_0(x)$ is the general solution of the associated homogeneous problem, and $y_p(x)$ is one particular solution. Using the auxiliary equation (Definition 4.5), we derived Theorems 4.7 and 4.8. These show how to find the homogeneous solutions of any second-order linear differential equation with constant coefficients. That may involve the complex exponential function, which was introduced in Definition 3.7, following a review of complex numbers in Section 3.

Section 5 developed methods for finding a particular solution $y_p(x)$ to a second-order linear equation with constant coefficients. One method reduced the problem to solving a first-order linear equation (Examples 1 and 5). The other method was the method of undetermined coefficients, which provides the general form of a particular solution based on the nature of the function $s(x)$ in the equation $y_p'' + py_p' + qy_p = s(x)$ (Examples 2, 3, and 4).

Section 6 was devoted to applications of second-order linear differential equations with constant coefficients to vibrating systems and electrical circuits. We saw that those two apparently quite different phenomena have *identical* mathematical models, and so abstractly they are the same! The final section presented two techniques that can be used when more general second-order differential equations are encountered. The first technique involves finding a power series expansion for the solution $y = y(x)$. The second uses the Euler method, or its improved version, to approximate the solution function at several points $x_1, x_2, \ldots, x_n$.

CHAPTER CHECKLIST

Section 1: first-order linear differential equation; associated homogeneous equation; solution set; integrating factor; unique solution of initial-value problem; falling body models, learning model of Estes.

Section 2: exact first-order differential equations, test for exactness (Theorem 2.2); integrating factors for nonexact equations.

Section 3: complex numbers, real and imaginary parts; complex conjugate, "fundamental theorem of algebra," absolute value (modulus), polar form, De Moivre's theorem, primitive roots of 1; complex exponential function, differentiation and integration of the complex exponential function.

Section 4: second-order linear differential equation, solution set, associated homogeneous equation; linear differential operators D and D^2, polynomial differential operators; auxiliary polynomial, auxiliary equation; general solution of homogeneous equation (Theorems 4.7 and 4.8); initial conditions; boundary conditions.

Section 5: reduction of second-order nonhomogeneous linear equation to first-order linear problem; method of undetermined coefficients; modification of table suggestion when it is a homogeneous solution; method of superposition, error function.

Section 6: Hooke's law, simple harmonic motion; amplitude, frequency, period, phase angles; resistance to motion (damping force), equilibrium, underdamped, overdamped, and critically damped systems; forced vibration, resonance, transient and steady-state solutions; electrical circuits, impedance.

Section 7: analytic functions, power series expansion of solution of a differential equation; recurrence relations on coefficients; Euler's method; improved Euler's method (second-order Runge-Kutta method).

REVIEW EXERCISES 15.8

In Exercises 1–13, solve the given differential equation.

1. $\sin x \dfrac{dy}{dx} + y \cos x = 0$

2. $(x^2 + 1)\dfrac{dy}{dx} - 4xy = x,\ y(0) = 1$

3. $x\dfrac{dy}{dx} - y = x^4y^3,\ y(1) = 2$

4. $(2x^2 - 2y^2)\,dx + (1 - 4xy)\,dy = 0$

5. $(x^2 - 3y^2)\,dx + 2xy\,dy = 0$

6. $\dfrac{d^2y}{dx^2} - 3\dfrac{dy}{dx} + 2y = 0$

7. $\dfrac{d^2y}{dx^2} - 6\dfrac{dy}{dx} + 9y = 0$

8. $\dfrac{d^2y}{dx^2} + \dfrac{dy}{dx} + 2y = 0$

9. $\dfrac{d^2y}{dx^2} - 3\dfrac{dy}{dx} + 2y = 0,\ y(0) = 1,\ y'(0) = 3$. (See Exercise 6.)

10. $\dfrac{d^2y}{dx^2} + \dfrac{dy}{dx} - 6y = xe^{-x}$

11. $\dfrac{d^2y}{dx^2} + 3\dfrac{dy}{dx} - y = \sin 2x$

12. $\dfrac{d^2y}{dx^2} + 5\dfrac{dy}{dx} - 6y = 2e^{3x}$

13. $\dfrac{d^2y}{dx^2} + 5\dfrac{dy}{dx} - 6y = 3e^{2x}$

14. Find a power series about $x_0 = 0$ for $y = y(x)$ if $d^2y/dx^2 - 9y = 0$. Check by solving the equation for y.

15. Find a power series about $x_0 = 0$ for $y = y(x)$ if

$$(x^2 + 2)\frac{d^2y}{dx^2} + 5x\frac{dy}{dx} + 4y = 0.$$

16. Find a power series about $x_0 = 1$ for $y = y(x)$ if $x^2y'' + xy' + y = 0$.

17. Use Euler's method to find approximate values of $y = y(x)$ at $x = 0, 0.1, 0.2, 0.3, 0.4$, and 0.5 if

$$\frac{dy}{dx} = xy^2, \qquad y(0) = 1.$$

18. Redo Exercise 17 using the improved Euler method.

19. An industrial plant starts dumping a pollutant into an initially unpolluted lake at the rate of 10 gallons per day. Each day, the lake loses 1000 gallons of its contents of 10 million gallons by evaporation and receives 990 gallons of pure water from a small stream. Find a formula for the number of gallons of pollutant in the lake after t days. If a concentration of 10 parts pollutant per million parts of fluid is enough to kill fish, then how long will it take for fish in the lake to begin dying?

20. In the undamped system of Figure 6.4, suppose that $m = 1$ kg and a force of 32 newtons is required to move the mass 2 meters to the right. Suppose that a force $F = 5\cos t$ is applied to the system. Find a formula for $x(t)$, assuming that the system starts from its equilibrium position at rest.

21. In the system of Figure 6.4, suppose that the damping constant is 2 and $m = 1/2$ kg. A force of 13 newtons moves the mass 2 meters to the right. If the system begins from rest in its equilibrium position, find a formula for $x(t)$ at time t if an external force $F = 5\cos 2t$ is applied. Find the resonant frequency.

22. A circuit consists of a resistor, inductor, and capacitor with $R = 1.0 \times 10^2$ ohms, $C = 4.0 \times 10^{-6}$ farads, and $L = 1.0 \times 10^{-1}$ henrys, respectively. If $q(0) = 1.0 \times 10^{-6}$ coulombs and $i(0) = 0$, then find $q(t)$ and $i(t)$.

23. Suppose that in the circuit of Figure 6.9, $E(t) = 110\sin 120\pi t$, $R = 3.0 \times 10^3$ ohms, $C = 2.5 \times 10^{-6}$ farads, and $L = 10$ henrys. Find the steady-state current.

24. **(a)** If $z_1 = 2 + 3i$ and $z_2 = -1 + 2i$, then compute $z_1 + z_2$, $z_1 z_2$, and $z_1 \div z_2$.
(b) Describe the geometric effect of multiplying $-3 + 2i$ by $\sqrt{3} - i$.
(c) Compute $(-1 + \sqrt{3}i)^4$.
(d) Find the six complex sixth roots of 1.
(e) Find the sixth roots of -64.

25. **(a)** Compute $e^{\pi i/6}$ and $e^{\ln 3 + \pi i/2}$.
(b) Find the sixth derivative of $e^{x/4}(\cos \frac{1}{4}x + i\sin \frac{1}{4}x)$.

APPENDIX I A Short Table of Integrals

1. $\int u^r \, du = \frac{u^{r+1}}{r+1} + C \qquad \text{if } r \neq -1$

2. $\int \frac{1}{u} \, du = \ln |u| + C$

3. $\int e^u \, du = e^u + C$

4. $\int e^{ku} \, du = \frac{1}{k} e^{ku} + C$

5. $\int b^u \, du = \frac{b^u}{\ln b} + C$

6. $\int \sin u \, du = -\cos u + C$

7. $\int \sin ku \, du = -\frac{1}{k} \cos ku + C$

8. $\int \cos u \, du = \sin u + C$

9. $\int \cos ku \, du = \frac{1}{k} \sin ku + C$

10. $\int \sec^2 u \, du = \tan u + C$

11. $\int \csc^2 u \, du = -\cot u + C$

12. $\int \sec u \tan u \, du = \sec u + C$

13. $\int \csc u \cot u \, du = -\csc u + C$

14. $\int \tan u \, du = -\ln |\cos u| + C = \ln |\sec u| + C$

15. $\int \cot u \, du = \ln |\sin u| + C$

16. $\int \sec u \, du = \ln |\sec u + \tan u| + C$

17. $\int \csc u\,du = \ln|\csc u - \cot u| + C$

18. $\int \frac{du}{\sqrt{1-u^2}} = \sin^{-1} u + C = -\cos^{-1} u + C$

19. $\int \frac{du}{\sqrt{a^2-u^2}} = \sin^{-1} \frac{u}{a} + C$, where $a > 0$

20. $\int \frac{du}{1+u^2} = \tan^{-1} u + C$

21. $\int \frac{du}{a^2+u^2} = \frac{1}{a}\tan^{-1} \frac{u}{a} + C$

22. $\int \frac{du}{u\sqrt{u^2-1}} = \sec^{-1} u + C$

23. $\int \frac{du}{u\sqrt{u^2-a^2}} = \frac{1}{a}\sec^{-1} \frac{u}{a} + C$, where $a > 0$

24. $\int \frac{du}{\sqrt{u^2+1}} = \sinh^{-1} u + C = \ln(u + \sqrt{u^2+1}) + C$

25. $\int \frac{du}{\sqrt{u^2+a^2}} = \sinh^{-1} \frac{u}{a} + C = \ln(u + \sqrt{u^2+a^2}) + C$

26. $\int \frac{du}{\sqrt{u^2-1}} = \cosh^{-1} u + C = \ln|u + \sqrt{u^2-1}| \qquad \text{if } |u| \geq 1$

27. $\int \frac{du}{\sqrt{u^2-a^2}} = \cosh^{-1} \frac{u}{a} + C = \ln|u + \sqrt{u^2-a^2}| \qquad \text{if } |u| \geq a$

28. $\int \frac{du}{1-u^2} = \begin{Bmatrix} \tanh^{-1} u + C & \text{if } |u| < 1 \\ \coth^{-1} u + C & \text{if } |u| > 1 \end{Bmatrix} = \frac{1}{2}\ln\left|\frac{1+u}{1-u}\right| + C$

29. $\int \frac{du}{a^2-u^2} = \begin{Bmatrix} \frac{1}{a}\tanh^{-1} \frac{u}{a} + C & \text{if } |u| < a \\ \frac{1}{a}\coth^{-1} \frac{u}{a} + C & \text{if } |u| > a \end{Bmatrix} = \frac{1}{2a}\ln\left|\frac{a+u}{a-u}\right| + C$

30. $\int \frac{du}{u\sqrt{1-u^2}} = -\operatorname{sech}^{-1} |u| + C = \ln \frac{1-\sqrt{1-u^2}}{|u|} + C \qquad \text{if } |u| < 1$

31. $\int \frac{du}{u\sqrt{a^2-u^2}} = -\frac{1}{a}\operatorname{sech}^{-1} \left|\frac{u}{a}\right| + C = -\frac{1}{a}\ln\left|\frac{a+\sqrt{a^2-u^2}}{u}\right| + C$

32. $\int \frac{du}{u\sqrt{u^2+1}} = -\operatorname{csch}^{-1} |u| + C = \ln\left|\frac{1-\sqrt{1+u^2}}{u}\right| + C$

33. $\int \frac{du}{u\sqrt{a^2+u^2}} = -\frac{1}{a}\operatorname{csch}^{-1} \left|\frac{u}{a}\right| + C = -\frac{1}{a}\ln\left|\frac{a+\sqrt{a^2+u^2}}{u}\right| + C$

34. $\int u^n \cos u\,du = u^n \sin u - n\int u^{n-1} \sin u\,du$

35. $\int u^n \sin u\,du = -u^n \cos u + n\int u^{n-1} \cos u\,du$

36. $\int \sin^m u \cos^n u\,du$

$$= \frac{\sin^{m+1} u \cos^{n-1} u}{m+n} + \frac{n-1}{m+n}\int \sin^m u \cos^{n-2} u\,du \qquad \text{if } m \neq -n$$

37. $\int \sin au \cos bu\, du = -\frac{\cos(a-b)u}{2(a-b)} - \frac{\cos(a+b)u}{2(a+b)} + C$

38. $\int \sin au \sin bu\, du = \frac{\sin(a-b)u}{2(a-b)} - \frac{\sin(a+b)u}{2(a+b)} + C$

39. $\int \cos au \cos bu\, du = \frac{\sin(a-b)u}{2(a-b)} + \frac{\sin(a+b)u}{2(a+b)} + C$

40. $\int \sin^2 u\, du = \frac{1}{2}u - \frac{1}{4}\sin 2u + C$

41. $\int \cos^2 u\, du = \frac{1}{2}u + \frac{1}{4}\sin 2u + C$

42. $\int \sin^m u\, du = -\frac{1}{m}\sin^{m-1} u \cos u + \frac{m-1}{m}\int \sin^{m-2} u\, du$

43. $\int \cos^n u\, du = \frac{1}{n}\cos^{n-1} u \sin u + \frac{n-1}{n}\int \cos^{n-2} u\, du$

44. $\int \tan^n u\, du = \frac{1}{n-1}\tan^{n-1} u - \int \tan^{n-2} u\, du$

45. $\int \sec^m u\, du = \frac{1}{m-1}\tan u \sec^{m-2} u + \frac{m-2}{m-1}\int \sec^{m-2} u\, du$

46. $\int \cot^n u\, du = \frac{-1}{n-1}\cot^{n-1} u - \int \cot^{n-2} u\, du$

47. $\int \csc^n u\, du = \frac{-1}{n-1}\cot u \csc^{n-2} u + \frac{n-2}{n-1}\int \csc^{n-2} u\, du$

48. $\int \ln ax\, dx = x \ln ax - x + C$

49. $\int ue^{au}\, du = \frac{1}{a^2}e^{au}(au-1) + C$

50. $\int u^n e^{au}\, du = \frac{1}{a}u^n e^{au} - \frac{n}{a}\int u^{n-1}e^{au}\, du$

51. $\int e^{au} \sin bu\, du = \frac{e^{au}}{a^2+b^2}[a \sin bu - b \cos bu] + C$

52. $\int e^{au} \cos bu\, du = \frac{e^{au}}{a^2+b^2}[a \cos bu + b \sin bu] + C$

53. $\int \sqrt{u^2+a^2}\, du = \frac{u}{2}\sqrt{u^2+a^2} + \frac{a^2}{2}\ln(u + \sqrt{u^2+a^2}) + C$

54. $\int \sqrt{u^2-a^2}\, du = \frac{u}{2}\sqrt{u^2-a^2} - \frac{a^2}{2}\ln|u + \sqrt{u^2-a^2}| + C$

55. $\int \sqrt{a^2-u^2}\, du = \frac{u}{2}\sqrt{a^2-u^2} + \frac{a^2}{2}\sin^{-1}\frac{u}{a} + C$

56. $\int (a^2-u^2)^{3/2}\, du = \frac{1}{4}u(a^2-u^2)^{3/2} + \frac{3}{8}a^2u\sqrt{a^2-u^2} + \frac{3}{8}a^4 \sin^{-1}\frac{u}{a} + C$

APPENDIX II Proofs of Some Limit Theorems

In this section, we give proofs of some of the results in Section 5 of Chapter 1. They are presented for completeness, to be read (and, it is hoped, understood), but they need not be thoroughly mastered. Skill in constructing proofs such as these is *not* required to become proficient at calculus.

5.2 THEOREM

(a) $\lim_{x \to c} x^2 = c^2$

(b) $\lim_{x \to c} \dfrac{1}{x} = \dfrac{1}{c}$ if $c \neq 0$

Proof. (a) The case $c = 0$ is Exercise 29 of Section 1.4. So we may suppose $c \neq 0$. We give the proof for $c > 0$, leaving the argument for $c < 0$ to Exercise 1. Since $\lim_{x \to c} x^2$ pertains to the behavior of $f(x) = x^2$ near $x = c$, we begin by restricting attention to a fairly small interval around c. Namely, we impose the condition that x be within $c/2$ of the limit point c:

$$|x - c| < \frac{1}{2}c. \tag{1}$$

By Theorem 1.2(c) of Chapter 1, this is equivalent to

$$-\frac{1}{2}c + c < x < c + \frac{1}{2}c \rightarrow \frac{1}{2}c < x < \frac{3}{2}c,$$

that is, to

$$\frac{3}{2}c < x + c < \frac{5}{2}c. \tag{2}$$

Now let $\varepsilon > 0$ be given. We want to make

$$|x^2 - c^2| = |(x + c)(x - c)| = |x + c|\,|x - c| \tag{3}$$

less than ε by taking $|x - c|$ sufficiently small—smaller than some positive number δ. Under restriction (1), we have from (2)

$$|x + c| < \frac{5}{2}c.$$

Hence in the presence of (1), Equation (3) gives

$$|x^2 - c^2| = |x + c|\,|x - c| < \frac{5}{2}c|x - c|.$$

Thus, we will have

$$|x^2 - c^2| < \varepsilon \tag{4}$$

if $|x - c| < 2\varepsilon/5c$. We therefore take δ to be the *smaller* of $c/2$ and $2\varepsilon/5c$. Then if $0 < |x - c| < \delta$, we conclude from (4) that $|x^2 - c^2| < \varepsilon$. This means that $\lim_{x \to c} x^2 = c^2$, according to Definition 4.1 of Chapter 1. QED

(b) We use an approach like the one in part (a). Namely, we first make the restriction that

$$|x - c| < \frac{1}{2}|c|, \tag{5}$$

and use this to replace

$$\left|\frac{1}{x} - \frac{1}{c}\right| = \frac{|c - x|}{|cx|} = \frac{|x - c|}{|c|} \cdot \frac{1}{|x|}. \tag{6}$$

by a *larger* expression of the form $K|x - a|$ for some constant K, like $5c/2$ in part (a). It is easy to make $K|x - a|$ less than any prescribed tolerance $\varepsilon > 0$ by requiring

$$|x - a| < \varepsilon/K.$$

That will then make the smaller expression (6) less than ε also. To obtain K, we use the triangle inequality [Theorem 1.2(d) of Chapter 1] on

$$c = x + (c - x)$$

to get

$$|c| \leq |x| + |c - x| = |x| + |x - c|,$$

which can be rewritten as

$$|c| - |x - c| \leq |x|. \tag{7}$$

Now restriction (5) is equivalent to

$$-\frac{1}{2}|c| < -|x - c|.$$

If we add $|c|$ to both sides of this inequality and use (7), then we obtain

$$|c| - \frac{1}{2}|c| < |c| - |x - c| \leq |x| \rightarrow \frac{1}{2}|c| \leq |x|.$$

Therefore,

$$\frac{1}{|x|} \leq \frac{1}{\frac{1}{2}|c|} = \frac{2}{|c|}.$$

Putting this into (6), we therefore have under restriction (5)

$$\left|\frac{1}{x} - \frac{1}{c}\right| = \frac{|x - c|}{|c|} \cdot \frac{1}{|x|} \leq \frac{|x - c|}{|c|}\frac{2}{|c|} = \frac{2}{c^2}|x - c|.$$

The expression on the right is easy to make less than ε:

$$\frac{2}{c^2}|x - c| < \varepsilon \rightarrow |x - c| < \frac{c^2}{2}\varepsilon.$$

We therefore take δ to be the smaller of $\frac{1}{2}|c|$ and $\frac{1}{2}c^2\varepsilon$. Then if $0 < |x - c| < \delta$, we do have

$$\left|\frac{1}{x} - \frac{1}{c}\right| < \varepsilon.$$

Thus, $\lim_{x \to c} 1/x = 1/c$. QED

5.3 THEOREM Suppose that

(a) $\lim_{x \to c} g(x) = M$ and

(b) $\lim_{t \to M} h(t) = h(M)$.

Then

$$\lim_{t \to c} h(g(x)) = h(M) = h\left(\lim_{x \to c} g(x)\right).$$

Proof. Let $\varepsilon > 0$ be given. From hypothesis (b) there is a $\delta' > 0$ such that

(8) $\qquad |h(t) - h(M)| < \varepsilon \qquad$ for all t satisfying $|t - M| < \delta'$.

(Here we needn't require t to also satisfy $0 < |t - M|$ because h is defined at M and $h(M) = \lim_{t \to M} h(t)$.) Since δ' is a positive number, hypothesis (a) says that there is a $\delta > 0$ such that

(9) $\qquad |g(x) - M| < \delta' \qquad$ whenever $0 < |x - c| < \delta$.

Now restrict attention to t in the range of g, that is, to those t such that $t = g(x)$ for some x. Then (8) and (9) together say that

$$|h(g(x)) - h(M)| < \varepsilon \qquad \text{whenever } 0 < |x - c| < \delta.$$

That is, $\lim_{x \to c} h(g(x)) = h(M)$. QED

5.5 THEOREM Let c be a real number. If n is even, suppose in addition that $c > 0$. Then

$$\lim_{x \to c} \sqrt[n]{x} = \sqrt[n]{c}.$$

Proof. We write $x^{1/n}$ in place of $\sqrt[n]{x}$ for notational simplicity. The case $c = 0$ is Exercise 2, and the case $c < 0$ is Exercise 3. Suppose then that $c > 0$, and $\varepsilon > 0$ is given. We consider $\varepsilon' < c^{1/n}$ in case $\varepsilon \geq c^{1/n}$, and otherwise let $\varepsilon' = \varepsilon$. Our aim is to make

(10) $\qquad |x^{1/n} - c^{1/n}| < \varepsilon' \leq \varepsilon,$

which we do by considering the string of equivalent inequalities

$$-\varepsilon' < x^{1/n} - c^{1/n} < \varepsilon' \leftrightarrow c^{1/n} - \varepsilon' < x^{1/n} < c^{1/n} + \varepsilon',$$

$$(c^{1/n} - \varepsilon')^n < x < (c^{1/n} + \varepsilon')^n,$$

$$-c + (c^{1/n} - \varepsilon')^n < x - c < (c^{1/n} + \varepsilon')^n - c,$$

$$-[c - (c^{1/n} - \varepsilon')^n] < x - c < (c^{1/n} + \varepsilon')^n - c.$$

Let δ be the minimum of the two positive numbers $c - (c^{1/n} - \varepsilon')^n$ and $(c^{1/n} + \varepsilon')^n - c$. Then for $0 < |x - c| < \delta$, we have

$$|x^{1/n} - c^{1/n}| \leq \varepsilon' \leq \varepsilon.$$

Thus, $\lim_{x \to c} \sqrt[n]{x} = \sqrt[n]{c}$. QED

5.7 THEOREM

Sandwich Theorem. Suppose that for all $x \neq c$ in an open interval I containing c, we have

$$g(x) \leq f(x) \leq h(x).$$

If $\lim_{x \to c} g(x) = \lim_{x \to c} h(x) = L$, then $\lim_{x \to c} f(x) = L$ also.

Proof. Let $\varepsilon > 0$ be given. Then there are positive numbers δ_1 and δ_2 small enough that the intervals $(c - \delta_1, c + \delta_1)$ and $(c - \delta_2, c + \delta_2)$ are in I, and furthermore

$$|g(x) - L| < \varepsilon \qquad \text{if } 0 < |x - c| < \delta_1,$$

and

$$|h(x) - L| < \varepsilon \qquad \text{if } 0 < |x - c| < \delta_2.$$

This means that if we let δ be the smaller of δ_1 and δ_2, then

$$-\varepsilon < g(x) - L < \varepsilon \qquad \text{and} \qquad -\varepsilon < h(x) - L < \varepsilon,$$

so

$$L - \varepsilon < g(x) < L + \varepsilon \qquad \text{and} \qquad L - \varepsilon < h(x) < L + \varepsilon. \tag{11}$$

Since $f(x) \leq h(x)$, the second inequality in (11) says that

$$f(x) < L + \varepsilon \qquad \text{for } 0 < |x - c| < \delta. \tag{12}$$

Similarly since $g(x) \leq f(x)$, we have from the first inequality in (11) that,

$$L - \varepsilon < f(x) \qquad \text{for } 0 < |x - c| < \delta. \tag{13}$$

Putting (12) and (13) together, we have

$$L - \varepsilon < f(x) < L + \varepsilon \qquad \text{for } 0 < |x - c| < \delta,$$

that is,

$$-\varepsilon < f(x) - L < \varepsilon \rightarrow |f(x) - L| < \varepsilon$$

for such x. Hence $\lim_{x \to c} f(x) = L$. QED

5.8 THEOREM

Let I be an open interval containing c. Suppose that $\lim_{x \to c} f(x) = L$.

(a) If $f(x) \geq 0$ for all $x \neq c$ in I, then $L \geq 0$.

(b) If $f(x) \leq 0$ for all $x \neq c$ in I, then $L \leq 0$.

Proof. We prove (a) and leave (b) as Exercise 7. To prove that $L \geq 0$ we will show that $L < 0$ is impossible. For suppose that $L < 0$. Then, for $\varepsilon = -\frac{1}{2}L > 0$, there would be some $\delta > 0$ such that

$$|f(x) - L| < -\frac{1}{2}L \qquad \text{for } 0 < |x - c| < \delta.$$

For x in I lying within distance δ of c, we would therefore have

$$-\left(-\frac{1}{2}L\right) < f(x) - L < -\frac{1}{2}L.$$

Hence,

$$f(x) < L - \frac{1}{2}L = \frac{1}{2}L < 0.$$

But by hypothesis $f(x)$ is *never* negative on I. Thus $L < 0$ can't hold. Therefore, $L \geq 0$. QED

Exercises A.II

1. Prove Theorem 5.2(a) for the case $c < 0$.

2. Prove $\lim_{x \to 0} x^{1/n} = 0$. (This is the case $a = 0$ in Theorem 5.5.)

3. If $c < 0$ and n is odd, then
 (a) use the text's proof of Theorem 5.5 to show that $\lim_{-x \to -c} \sqrt[n]{-x} = \sqrt[n]{-c}$.
 (b) use (a) to show that $\lim_{x \to c} x^{1/n} = c^{1/n}$.

4. If $\lim_{x \to c} f(x) = L > 0$, then show there is some $\delta > 0$ such that

$$f(x) > 0 \qquad \text{whenever } 0 < |x - c| < \delta.$$

(*Hint:* Consider $\varepsilon = \frac{1}{2}L$.)

5. Show that $\lim_{x \to c} x^n = c^n$ for any positive integer n.

6. Repeat Exercise 5 for n a negative integer.

7. Prove Theorem 5.8(b).

In Exercises 8–16, show directly from Definition 4.1 of Chapter 1 that the limit has the value asserted.

8. $\lim_{x \to c} \sqrt{x} = \sqrt{c}$. $\left(\textit{Hint:} \text{ Use } \sqrt{x} - \sqrt{c} = \dfrac{x - c}{\sqrt{x} + \sqrt{c}}.\right)$

9. $\lim_{x \to c} x^n = c^n$ for n a positive integer. (*Hint:* Use $x^n - c^n = (x - c)(x^{n-1} + x^{n-2}c + \ldots + xc^{n-2} + c^{n-1})$.)

10. $\lim_{x \to 1} x^2 = 1$. [*Hint:* Use the approach of Theorem 5.2(a).]

11. $\lim_{x \to 1} \dfrac{1}{x} = 1$

12. $\lim_{x \to 2} \dfrac{1}{x + 1} = \dfrac{1}{3}$

13. $\lim_{x \to 2} x^3 = 8$

14. $\lim_{x \to 4} x^2 = 16$

15. $\lim_{x \to 2} (x^2 - x + 1) = 3$

16. $\lim_{x \to -1} (x^2 + 2x - 1) = 2$

APPENDIX III Mathematical Induction

Mathematical induction is a method of proving statements about positive integers n. It is the mathematical version of the domino theory, which you may have encountered in current events or history courses. That theory justifies opposition to an expansionist power by attempting to show that if one country in a certain region falls under the influence of that nation, then inevitably the next neighboring country will also. Once this is accepted, it means that if the expansionist power gains control over even one country in the region, then it will ultimately take control of the entire region. The domino theory takes its name from the simple fact that to knock over an entire set of dominoes standing on edge, it is enough to first check that they are aligned so that whenever any one of them falls it will knock over the next one, and then to knock over the first domino.

For a statement $P(n)$ about positive integers, think of the set $\mathbf{N}$ of positive integers as an infinite set of dominoes. To prove $P(n)$, we show that the set T of all integers n for which $P(n)$ is true is the entire set $\mathbf{N}$. This is done in the following two stages:

(1) Show that $1 \in T$, that is, that $P(n)$ is true for $n = 1$.

(2) Show that whenever any positive integer $k \in T$, then the next positive integer $k + 1 \in T$ also. That is, whenever $P(n)$ is true for the integer $n = k$, then it is also true for the next integer $n = k + 1$.

Step (1) corresponds to knocking over the first domino (the first positive integer). Step (2) corresponds to checking that whenever any domino falls, the next one also falls. The idea is that once we know that the statement P is true for $n = 1$ and that its truth for any integer $n = k$ entails its truth for the next integer $n = k + 1$, then we can be certain that it is true for every positive integer n. Since it holds for $n = 1$, it must also hold for $n = 1 + 1 = 2$; since it holds for $n = 2$, it must also hold for $n = 2 + 1 = 3$; etc.

As a first example of the use of the principle of mathematical induction, we use it to prove Theorem 1.6(a) of Chapter 4, which was proved in another way on p. 234.

Example 1

Use mathematical induction to show that

$$1 + 2 + 3 + \ldots + n = \frac{n(n+1)}{2}. \tag{3}$$

Solution. Following Steps (1) and (2), we first check that (3) holds for $n = 1$. In that case, it asserts that

$$1 = \frac{1 \cdot 2}{2}$$

which is certainly true. Next, we check (2). To do so, we must show that whenever (3) is true for $n = k$, then it is also true for $n = k + 1$. Suppose then that (3) is true for $n = k$:

$$1 + 2 + 3 + \ldots + k = \frac{k(k+1)}{2}. \tag{4}$$

Then we have to show that (3) is also true for the next positive integer $n = k + 1$. For that, we use (4), which is called the ***induction hypothesis.*** We have

$$\begin{aligned} 1 + 2 + \ldots + k + (k+1) &= \frac{k(k+1)}{2} + (k+1) && \textit{by (4)} \\ &= \frac{k^2 + k}{2} + \frac{2k+2}{2} \\ &= \frac{k^2 + 3k + 2}{2} = \frac{(k+2)(k+1)}{2}. \end{aligned}$$

The last expression is $n(n+1)/2$ for $n = k + 1$. Thus, (3) is true for $n = k + 1$ whenever it is true for $n = k$. By the principle of mathematical induction, then, (3) is true for every positive integer n. ■

The next example provides a proof of Equation (1) on p. 38 by mathematical induction.

Example 2

Suppose that $\lim_{x \to c} f_i(x) = L_i$ for $i = 1, 2, \ldots, n$. Then use mathematical induction to show that

$$\lim_{x \to c} [f_1(x) + f_2(x) + \ldots + f_n(x)] = L_1 + L_2 + \ldots + L_n. \tag{5}$$

Solution. By hypothesis, (5) holds for $n = 1$. So we just have to show that it holds for $n = k + 1$ whenever it is true for $n = k$. To do that, suppose that (5) holds for some $n = k$:

$$\lim_{x \to c} [f_1(x) + f_2(x) + \ldots + f_k(x)] = L_1 + L_2 + \ldots + L_k. \tag{6}$$

Then we can use Theorem 5.1(d) of Chapter 1 to show that (5) also holds for $n = k + 1$. We have

$$\begin{aligned} &\lim_{x \to c} [f_1(x) + \ldots + f_k(x) + f_{k+1}(x)] \\ &\quad = \lim_{x \to c} [f_1(x) + \ldots + f_k(x)] + \lim_{x \to c} f_{k+1}(x) \\ &\quad = L_1 + L_2 + \ldots + L_k + L_{k+1}. && \textit{by the induction hypothesis (6) and the hypothesis that } \lim_{x \to c} f_{k+1}(x) = L_{k+1} \end{aligned}$$

Thus (5) does hold for $n = k + 1$ whenever it holds for $n = k$. Therefore, (5) is true for every positive integer n. ■

The idea behind mathematical induction can also be used to show that a given statement $P(n)$ about positive integers holds for all positive integers n beyond some starting point. The next example illustrates how that is done.

Example 3

Show that for every positive integer $n \geq 2$,

(7) $$n^2 > n + 1.$$

Solution. This is not true for $n = 1$, since $1^2 = 1$ is not greater than $1 + 1 = 2$. We therefore apply the principle of mathematical induction starting from $n = 2$, in keeping with the hypothesis of this problem. For $n = 2$, inequality (7) *is* true:

$$2^2 = 4 > 2 + 1 = 3.$$

We next show that (7) is true for $n = k + 1$ whenever it is true for $n = k$. To do so, we first suppose that (7) is true for $n = k$; that is, our induction hypothesis is

(8) $$k^2 > k + 1.$$

Then we have to show that (7) holds for $n = k + 1$. Using (8), we get

$$(k + 1)^2 = k^2 + 2k + 1 > k + 1 + 2k + 1 = k + 2 + 2k > k + 2.$$

Thus, (7) does hold for $n = k + 1$ whenever it holds for $n = k$. Since we have shown that it holds for $n = 2$, we can conclude that it is true that for every positive integer $n \geq 2$. ■

Exercises A.III

In Exercises 1–26, use mathematical induction to prove the given statement $P(n)$ about positive integers.

1. $1 + 3 + 5 + \ldots + (2n - 1) = n^2$.
2. $1 + 4 + 7 + \ldots + (3n - 2) = (3n^2 - n)/2$.
3. $1^3 + 2^3 + 3^3 + \ldots + n^3 = n^2(n + 1)^2/4$ [Theorem 1.6(c) of Chapter 4].
4. $1^3 + 3^3 + 5^3 + \ldots + (2n - 1)^3 = n^2(2n^2 - 1)$.
5. $1 + x + x^2 + \ldots + x^{n-1} = (1 - x^n)/(1 - x)$ for any real number $x \neq 1$.
6. $a + ar + ar^2 + \ldots + ar^{n-1} = a(1 - r^n)/(1 - r)$ if $r \neq 1$ [Equation (1), p. 560].
7. $1 \cdot 2 + 2 \cdot 3 + 3 \cdot 4 + \ldots + n(n + 1) = n(n + 1)(n + 2)/3$.
8. $1 \cdot 2^0 + 2 \cdot 2^1 + 3 \cdot 2^2 + \ldots + n \cdot 2^{n-1} = 1 + (n - 1) \cdot 2^n$.
9. $\dfrac{1}{1 \cdot 2} + \dfrac{1}{2 \cdot 3} + \dfrac{1}{3 \cdot 4} + \ldots + \dfrac{1}{n(n + 1)} = \dfrac{n}{n + 1}$.
10. $\dfrac{1}{2} + \dfrac{2}{2^2} + \dfrac{3}{2^3} + \ldots + \dfrac{n}{2^n} = 2 - \dfrac{n + 2}{2^n}$.
11. 2 is a factor of $n^2 + n$.
12. 3 is a factor of $n^3 + 2n$.
13. 4 is a factor of $5^n - 1$.
14. 5 is a factor of $3^n + 2 \cdot 3^{n-1} - 3 \cdot 2^n + 2^{n-1}$.
15. $3n^2 \geq 2n + 1$.
16. $4^n \geq n^2$.
17. $n! > n^2$ if $n \geq 4$.
18. $n! > n^3$ if $n \geq 6$.
19. $(1 + a)^n > 1 + na$ for $n \geq 2$ and a (a real number) > -1.
20. $\sum_{i=1}^{n} [f(i + 1) - f(i)] = f(n + 1) - f(1)$ [Theorem 1.8(d) of Chapter 4.]
21. If $\lim_{x \to c} f_i(x) = L_i$, for $i = 1, 2, \ldots, n$, then $\lim_{x \to c} [f_1(x) \cdot f_2(x) \cdot \ldots \cdot f_n(x)] = L_1 \cdot L_2 \cdot \ldots \cdot L_n$. Equation (5), p. 40]
22. If $f_i(x)$ is differentiable at $x = c$ for $i = 1, 2, \ldots, n$, then
$$\frac{d}{dx}[f_1(x) + f_2(x) + \ldots + f_n(x)] = f'_1(x) + \ldots + f'_n(x).$$
[Equation (9), p. 73]

23. $\dfrac{d}{dx}[x^n] = nx^{n-1}$.
(Theorem 2.1 of Chapter 2)

24. $\dfrac{d^n}{dx^n}[x^n] = n!$.
(Exercise 44, Section 2.9)

25. If $f(x)$ is differentiable on an interval $I \supseteq [a, b]$ and if $f(x) = 0$ for $n \geq 2$ points $x_1, x_2, \ldots, x_n$ in $[a, b]$, then $f'(x) = 0$ for at least $n - 1$ points in (a, b). [*Hint:* Use Rolle's theorem (Theorem 10.1 of Chapter 2).]

26. If $f^{(n)}(x)$ exists on an interval $I \supseteq [a, b]$ and if $f(x) = 0$ for at least $n + 1$ distinct points in $[a, b]$, then $f^{(n)}(c) = 0$ for at least one point in $[a, b]$.

27. Let $P(n)$ be the statement that

$$1 + 2 + \ldots + n = \frac{1}{8}(2n + 1)^2.$$

(a) Show that $P(n)$ is true for $n = k + 1$ whenever it is true for $n = k$.
(b) Is $P(n)$ true for every positive integer n?

28. Can you find the fallacy in the following "application" of mathematical induction to prove that every blonde girl has blue eyes? Let $P(n)$ be the statement that in any set of n blonde girls, if at least one girl has blue eyes, then every girl in the set has blue eyes. Clearly, $P(n)$ is true for $n = 1$. Suppose that $P(n)$ is true for $n = k$.

(a) To show that $P(n)$ is true for $n = k + 1$, let S be a set of $k + 1$ blonde girls, at least one of whom, say G_1, has blue eyes. Ask one of the girls *other* than G_1 to step aside for a moment. That leaves a set C of k blonde girls, one of whom (G_1) has blue eyes. Then, by the induction hypothesis, all k of the girls in C have blue eyes. Now invite the remaining girl back, and send any member of C out. There are now k blonde girls present, $k - 1$ of whom have blue eyes. So the last (kth) girl must also have blue eyes, by the reasoning just given. Thus, $P(n)$ is true for $n = k + 1$ whenever it is true for $n = k$.

(b) Now let S be the set of all blonde girls. There certainly is at least one blonde girl in S. Then apply $P(n)$ just proved in (a) to conclude that every member of S has blue eyes. That is, every blonde girl must have blue eyes!

APPENDIX IV

Answers to Odd-Numbered Exercises

CHAPTER 1

EXERCISES 1.1, p. 10

1. $(1, +\infty)$ **3.** $[-3/2, +\infty)$ **5.** $(-\infty, 5]$ **7.** $(-\infty, 2)$
9. $(-4/3, 2)$ **11.** $(-\infty, -10] \cup [-1, +\infty)$
13. $(-\infty, -2) \cup (1, +\infty)$ **15.** $[-2, -1] \cup [2, +\infty)$
17. $(-2, -\sqrt{2}) \cup (0, \sqrt{2})$ **19.** $(-\infty, -3) \cup (-1, +\infty)$
21. $(-3, -\sqrt{5}) \cup (\sqrt{5}, 3)$ **23.** $(-\infty, -1] \cup [2/3, +\infty)$
31. *Hint:* Use a formula from algebra. **35.** $H' > 0$, $S' < 0$
37. $T > H'/S'$

EXERCISES 1.2, p. 19

1a. -1 **1b.** $4/3$ **3a.** $x + y = 1$ **3b.** $-4x + 3y = 13$
5a. $y = 6x + 3$ **5b.** $7x + 2y = 10$ **7a.** $2x - 3y = 7$
7b. $y = 5x + 8$ **9.** $F = (9/5)C + 32$ **11a.** 5, (3, 5/2)
11b. 13, (1/2, 3) **13.** $y = 0$ **15a.** $(x + 1)^2 + (y - 3)^2 = 4$
15b. $(x - 2)^2 + (y + 1)^2 = 9$ **17a.** $(x - 5)^2 + (y + 2)^2 = 4$
17b. $(x + 3)^2 + (y - 2)^2 = 9$ **19.** Circle $C(4, -1)$; $r = 5$
21. Circle $C(1, 3)$; $r = 1$ **23.** Point $P(-3, 4)$ **25.** Empty set
27. $v = v_0 + 9.8t$ **29a.** $T = 3x + 1000$ **29b.** \$4000
29c. Slope of graph is 3.
31a. $D = 100{,}000(1 - p)$, $S = 400{,}000(p - 1/2)$ **31b.** $p = \$0.60$
31c. $D = S = 40{,}000$ **33a.**

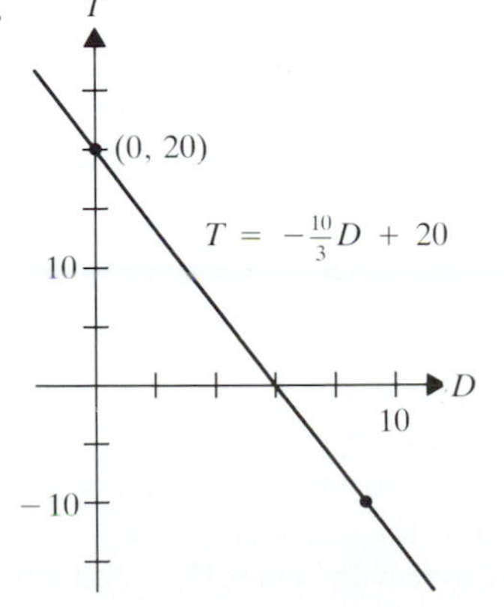

33b. $T = -10x/3 + 20$, where x is distance from inside edge of the batt in cm. and T is temperature in °C **33c.** 10°C
35. $\Delta G = \Delta H - T\Delta S$ **37.** $(1, -2)$ **41.** $(15/2, 9/2)$
45. $x/-1 + y/2 = 1$ **49.** $x + y = 2$ **51a.** -6 **51b.** $2/3$
53. $P(4, 9)$ **55a.** $(a/2, 0)$, $(0, b/2)$, $(a, b/2)$, $(a/2, b)$
57a. $y = y_0 + b(x - x_0)/a$
57b. $P_1\left(\dfrac{b^2x_0 + ac - aby_0}{a^2 + b^2}, \dfrac{-abx_0 + a^2y_0 + bc}{a^2 + b^2}\right)$ **59.** 1

EXERCISES 1.3, p. 27

1. $(-\infty, -2) \cup (-2, +\infty)$ **3.** $(-\infty, 2)$ **5.** $[1, +\infty)$
7. $[3, 4) \cup (4, 6) \cup (6, +\infty)$ **9.** $[-1, 1]$

11.

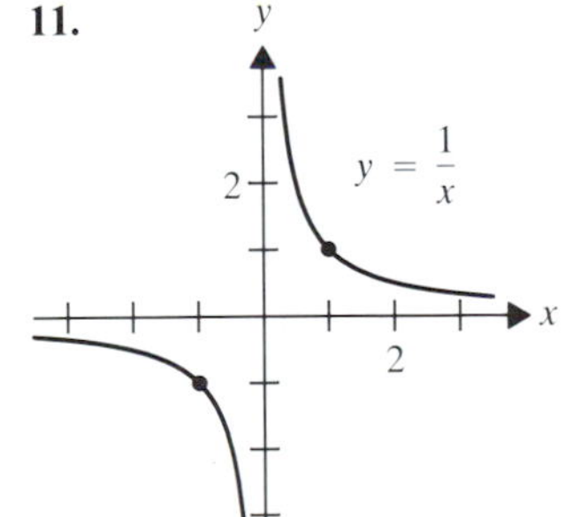

13.

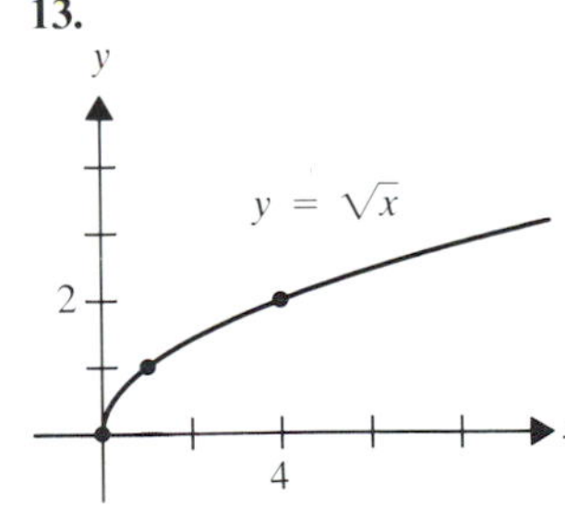

15.

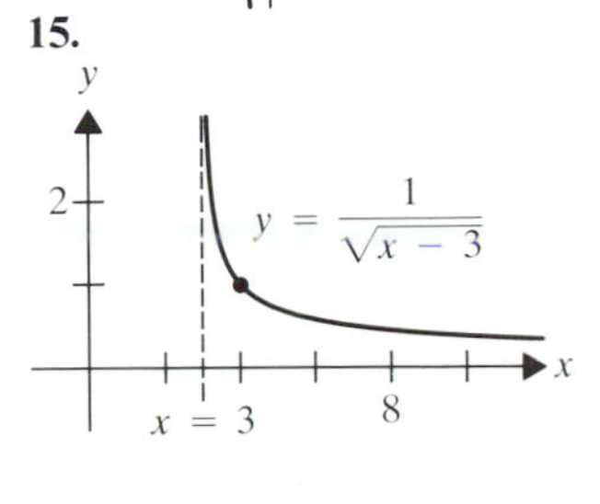

17.

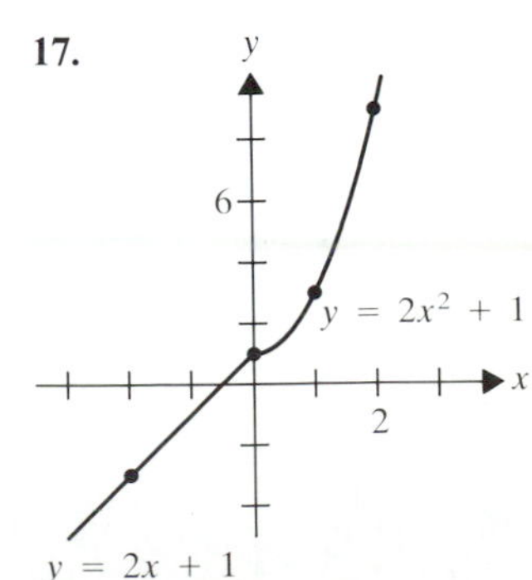

19.

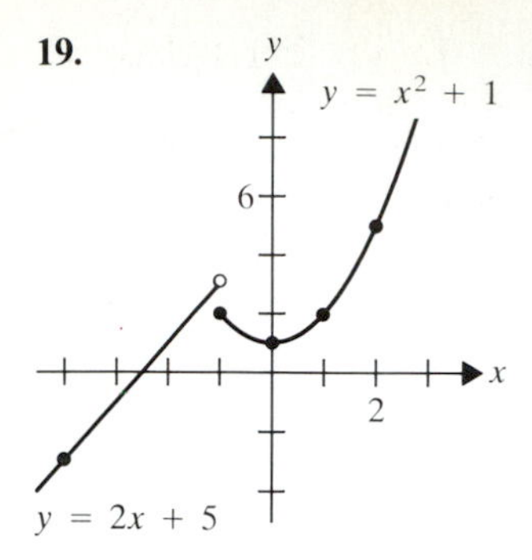

21.

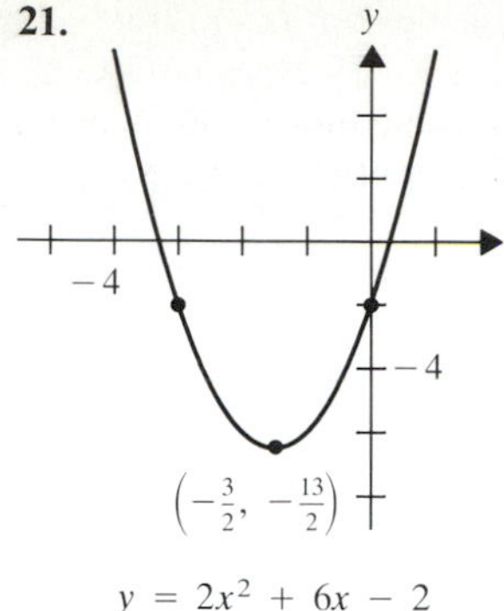

23.

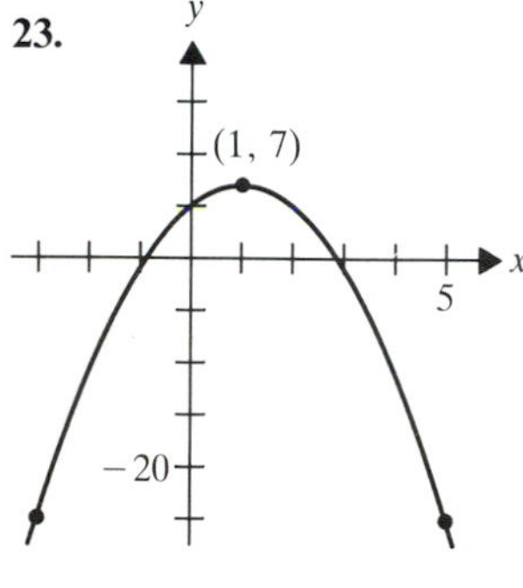

25a. $3/5$ **25b.** -5 **25c.** $\dfrac{4a-1}{4a^2+2a-1}$ **25d.** $\dfrac{2a-a^2}{1+a-a^2}$

25e. $\dfrac{2a+2h-1}{a^2+h^2+2ah+a+h-1}$ **25f.** $\dfrac{2\sqrt{a}-1}{a+\sqrt{a}-1}$

25g. $\dfrac{2a^2-1}{a^4+a^2-1}$ **27a.** a^2-3a+2

27b. $a^2+2ah+h^2-3a-3h+2$ **27c.** $2a+h-3$

29a. x^2+2x-2; $D_{f+g}=D_{f\cdot g}=\mathbf{R}$ **29b.** $2x^3-x^2-2x+1$

29c. $\dfrac{x^2-1}{2x-1}$; $D_{f\div g}=(-\infty, 1/2)\cup(1/2, +\infty)$

29d. $\dfrac{2x-1}{x^2-1}$; $D_{g\div f}=(-\infty, -1)\cup(-1, 1)\cup(1, +\infty)$

29e. $4x^2-4x$; $D_{f\circ g}=\mathbf{R}$ **29f.** $2x^2-3$; $D_{g\circ f}=\mathbf{R}$

31a. $\sqrt{x^2-4}+\dfrac{1}{x+1}$

31b. $\dfrac{\sqrt{x^2-4}}{x+1}$; $D_{f+g}=D_{f\cdot g}=(-\infty, -2]\cup[2, +\infty)$

31c. $(x+1)\sqrt{x^2-4}$; $D_{f\div g}=(-\infty, -2]\cup[2, +\infty)$

31d. $\dfrac{1}{(x+1)\sqrt{x^2-4}}$; $D_{g\div f}=(-\infty, -2)\cup(2, +\infty)$

31e. $\dfrac{\sqrt{-3-4x^2-8x}}{|x+1|}$; $D_{f\circ g}=[-3/2, -1)\cup(-1, -1/2]$

31f. $\dfrac{1}{\sqrt{x^2+4}+1}$; $D_{g\circ f}=(-\infty, -2)\cup(2, +\infty)$

33a. $\dfrac{1}{x-2}+\dfrac{x+1}{x+3}$; $D_{f+g}=(-\infty, -3)\cup(-3, 2)\cup(2, +\infty)$

33b. $\dfrac{x+1}{(x-2)(x+3)}$; $D_{f\cdot g}=(-\infty, -3)\cup(-3, 2)\cup(2, +\infty)$

33c. $\dfrac{x+3}{(x-2)(x+1)}$; $D_{f\div g}=(-\infty, -1)\cup(-1, 2)\cup(2, +\infty)$

33d. $\dfrac{(x+1)(x-2)}{(x+3)}$; $D_{g\div f}=(-\infty, -3)\cup(-3, +\infty)$

33e. $-\dfrac{x+3}{x+5}$; $D_{f\circ g}=(-\infty, -5)\cup(-5, -3)\cup(-3, +\infty)$

33f. $\dfrac{x-1}{3x-5}$; $D_{g\circ f}=\left(-\infty, \dfrac{5}{3}\right)\cup\left(\dfrac{5}{3}, 2\right)\cup(2, +\infty)$

35. $(-\infty, -2)\cup[3, +\infty)$ **37.** $(-\infty, 0)\cup(0, 2)\cup(2, +\infty)$
39. $f(x)=(g\circ h)(x)$ for $g(x)=\sqrt{x}$, and $h(x)=x^2-5x+7$
41. $f(x)=(g\circ h)(x)$ for $g(x)=3$, and $h(x)=1/(2x-3)$
43. 22.5 **47a.** $S=36\pi(25-h)$ **47b.** $h=25-t/4$
47c. $S=36\pi\cdot t/4=9\pi t$ **47d.** $t=100$ **49a.** Even **49b.** Even
49c. Odd **49d.** Even **49e.** Even **49f.** Neither

EXERCISES 1.4, p. 36

1. -14 **3.** -1 **5.** 2 **7.** 1/13 **9.** $-1/9$ **11.** 6 **13.** 1/8
15. -7 **17.** 3 **19.** 1/300 **21.** 0.00049 **23.** 0.00399 **37.** No

EXERCISES 1.5, p. 44

1. 5 **3.** -1 **5.** 1/13 **7.** 8/3 **9.** 1 **11.** $-1/4$ **13.** 2 **15.** 8
17. $-1/4$ **19.** -3 **21.** 0 **23.** 0 **37.** 0
41. No. Consider $f(x)=-x$, $c=1$, $L=1$.

EXERCISES 1.6, p. 50

1. 2 **3.** 0 **5.** Does not exist **7.** Does not exist **9a.** 1
9b. -1 **9c.** Does not exist **11a.** 1 **11b.** 1 **11c.** 1
13a. $-1/10$ **13b.** $-7/18$ **13c.** Does not exist
15a. Does not exist **15b.** 5 **15c.** Does not exist **17a.** 7
17b. 2/5 **17c.** Does not exist **19a.** 3 **19b.** 2
19c. Does not exist **21a.** 1 **21b.** 0 **23a.** 5 **23b.** 27
25a. No **25b.** No

EXERCISES 1.7, p. 58

1. Yes **3.** No **5.** No **7.** No **9.** No **11.** Continuity
13. Continuity **15.** Essential discontinuity **17.** Continuity
19. Removable discontinuity; 1/4 **21.** $[1, 2)\cup(2, +\infty)$
23. $(-\infty, -1)\cup(-1, 1)\cup(1, +\infty)$
25. $(-\infty, -2)\cup(2, +\infty)$
27. $(-\infty, -2)\cup(-2, 1)\cup(1, 2)\cup(2, +\infty)$
29. $(-\infty, -3)\cup(3, +\infty)$ **37.** $r\approx 1.32$; $f(r)\approx -0.0020$
39. $r\approx 0.70$; $f(r)\approx -0.0048$ **43.** None
45. $f+g$ is discontinuous at c.
47. Consider $f(x)=0$, $g(x)=x$ for $x<0$, $g(x)=1$ for $x\geq 0$, and $f(x)=x$, $g(x)=1/x^2$ at $x=0$.

REVIEW EXERCISES 1.8, p. 60

1. $(-1, 4)$, **3.** $(-1, 0)$ **5.** $y=x/3+13/3$ **7.** $y=x/3+7/3$
9a. $\text{TC}=50{,}000+5x$ **9b.** $p=85/7\approx \$12.14$
11. $(x-1)^2+(y+3)^2=34$
13. Circle with center $(2, -3/2)$ and radius $\sqrt{5}/2$

15. $[0, 9) \cup (9, +\infty)$ **17.** $(-\infty, -6] \cup [-2, +\infty)$

19.

x	0	-2	1	2	-1
y	1	1	-2	-7	2

;

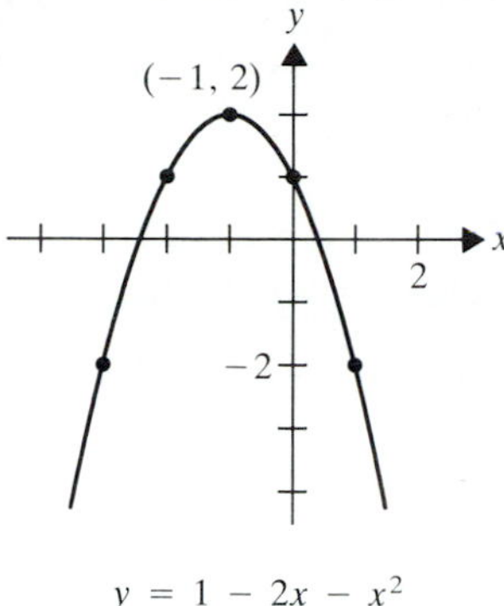

21a. Neither: $f(-1/2) = -1$, but $f(1/2) = 0$ **21b.** Neither
23. 6/7 **25.** Does not exist **27.** $\sqrt{3}$ **29.** 0 **31.** 0
33. Continuous **35.** Removable discontinuity
37. Essential discontinuity **39.** Essential discontinuity
41. $(-\infty, 3) \cup (3, +\infty)$ **43.** $(1, 2) \cup (2, +\infty)$ **51.** No
53. Essential discontinuity at each whole number greater than 0

CHAPTER 2

EXERCISES 2.1, p. 69

1a. $y = 2x$ **1b.** $y = -4x - 3$

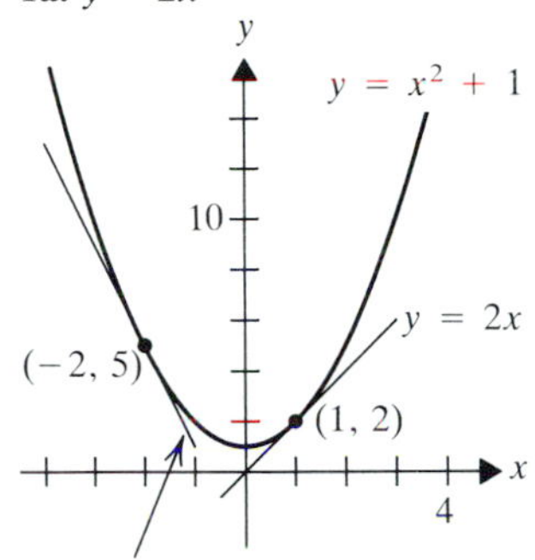

3a. $y = 2 - x$ **3b.** $y = -x/4 - 1$

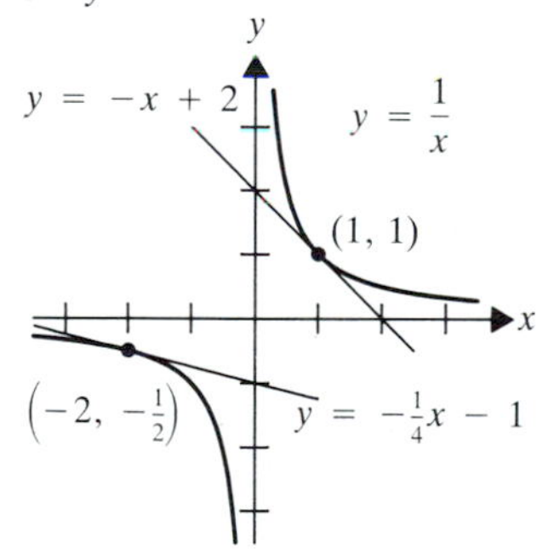

5a. $y = -2x + 5$ **5b.** $y = 4x + 8$

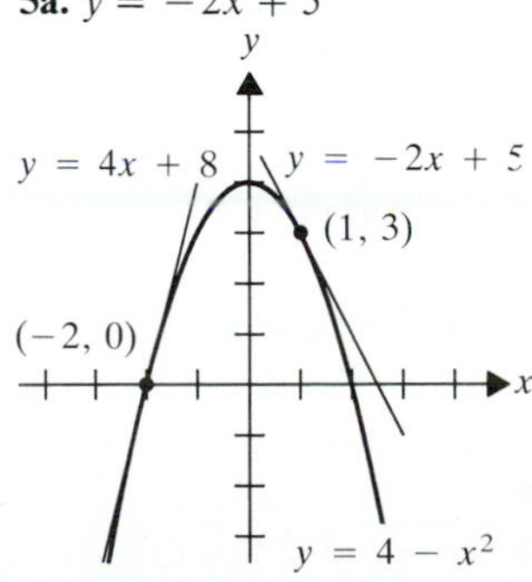

7. $x = -2$

9. 0; $D_{f'} = \mathbf{R}$ **11.** $3x^2$; $D_{f'} = \mathbf{R}$
13. $-2x^{-3}$; $D_{f'} = (-\infty, 0) \cup (0, +\infty)$ **15a.** 1960 **15b.** 3920
17a. -2 **17b.** 0 **17c.** 2
17d. Upward on $(2, +\infty)$; Downward on $[0, 2)$ **19a.** 2
19b. 0 **19c.** -2
19d. Upward on $[0, 2)$; Downward on $(2, +\infty)$
21a. $x + 2y = 5$ **21b.** $x - 4y + 22 = 0$ **23a.** $y = x$
23b. $y = 4x + 15/2$ **25.** They intersect at $(0, -4)$.
37. $D_{f'} = (-\infty, 3) \cup (3, +\infty)$
41a. 0.9983342, 0.8396036, 0.670603, 0.4559019, -0.0499583
41b. 0.99998333, 0.863511, 0.7035595, 0.4956616, -0.005
41c. 0.9999998, 0.8657753, 0.7067532, 0.4995668, -0.000499
41d. 1, 0.8660, 0.7070757, 0.49999565, -0.00004

EXERCISES 2.2, p. 79

1. 2 **3.** $10x - 8$ **5.** $35x^4 - 12x^3 + 14x$

7. $8x - 3 - x^{-2} + 6x^{-3}$ **9.** $\dfrac{8 - 10x}{(5x^2 - 8x + 11)^2}$

11. $(27x^8 - 105x^6 + 22x - 5)(7x^3 - 3x^2 + x + 1) + (21x^2 - 6x + 1)(3x^9 - 15x^7 + 11x^2 - 5x - 2)$
13. $16(3x - 1)(3x^2 - 2x + 1)^7$ **15.** $-24x(2x^2 - 8)^{-7}$

17. $8x(x^2 + 5)^{-2}$ **19.** $\dfrac{12}{(3 - 2x)^2}$ **21.** $10(2x - 3)^4 + \dfrac{70x}{(5x^2 + 1)^8}$

23. $\dfrac{-8x(x^2 + 1)}{(x^2 - 1)^3}$ **25.** $\dfrac{-acx^2 - 2bcx + ae - bd}{(cx^2 + dx + e)^2}$

27. $8x(x^2 + 1)^3 + 40x^3 + 40x$ **33.** $\dfrac{dH}{dt} = \dfrac{dE}{dt} + p\dfrac{dV}{dt}$

35. $\dfrac{1}{75R} \approx 0.162$ **37.** 2.782×10^2 N/s **39a.** $\dfrac{dv}{dS} = \dfrac{K_m V_{\max}}{(K_m + S)^2}$

39b. $\dfrac{dv}{dt} = \dfrac{2K_m V_{\max}}{(K_m + S)^2}$ **41.** $\left[\dfrac{p(x)}{x}\right]' = \dfrac{xp'(x) - p(x)}{x^2}$

EXERCISES 2.3, p. 90

1a. $3\pi/4$ **1b.** $\pi/4$ **1c.** $5\pi/4$ **1d.** $5\pi/3$ **1e.** $7\pi/4$

3.

Quadrant	I	II	III	IV
Sine	+	+	−	−
Cosine	+	−	−	+

5a. $-1/2$ **5b.** $-\sqrt{2}/2$

5c. -1 **5d.** $-1/2$ **5e.** $-1/2$ **5f.** $-1/2$ **5g.** $\sqrt{2}/2$ **7a.** $\sqrt{3}$

7b. -2 **7c.** -1 **7d.** 2 **7e.** $-\sqrt{3}/3$ **7f.** $\sqrt{3}/3$ **7g.** 2

9a. $\sqrt{\dfrac{2-\sqrt{3}}{4}} \approx 0.258819$ **9b.** $\sqrt{\dfrac{2-\sqrt{3}}{4}} \approx 0.258819$

9c. $\sqrt{\dfrac{2+\sqrt{3}}{4}} \approx 0.9659258$ **9d.** $\sqrt{\dfrac{2+\sqrt{3}}{4}} \approx 0.9659258$

13a. $t = 0, \pi/4, 3\pi/4, 5\pi/4, 7\pi/4, \pi, 2\pi$

13b. $t = \pi/6, \pi/2, 5\pi/6, 3\pi/2$ **13c.** $t = 0, \pi/6, 5\pi/6, 2\pi$

17. No, removable discontinuity **19.** 2 **21.** 4/3 **23.** 2/3

25. 0 **27.** 1 **29.** 0 **31.** No limit **33.** 0 **35.** 16/3

37. $\pi/180 \approx 0.0174533$ **39a.** a/b **39b.** 1 **41.** 0

43. Period of $\sin 2x$ is π. Period of $\sin kx$ is $2\pi/k$.

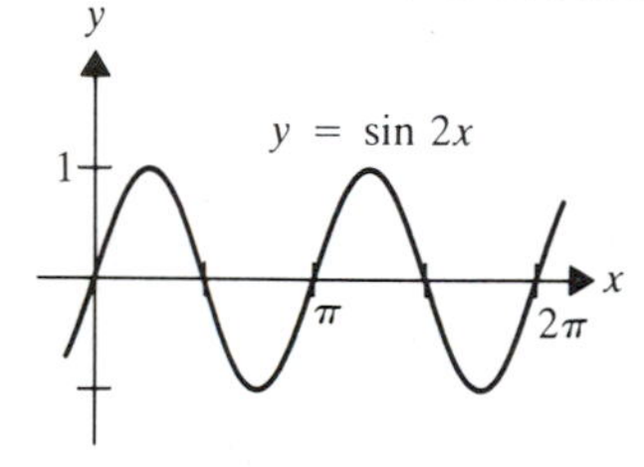

45a. $y = \sin \pi t/6$ **45b.** $y = \cos 2\pi t/p$

47a. $P(0) = 140$, $P(6) = 60$ **47b.** $P(3) = 100 = P(9)$

47c. Yes; the price is cyclical with period 12 months, as is the function.

49. $R(t) = 0.35 + 0.25\cos(\pi t/12)$; t = no. of hours since 6 AM

51. $300 \sin \pi/3 = 150\sqrt{3} \approx 259.8$ ft

EXERCISES 2.4, p. 97

1. $5\cos x + 2\sin x$

3. $x \sin x + 3x^2 \cos x + 5\cos x - \cos^2 x + \sin^2 x$

5. $3\sec^2 x \tan x(\tan x + \sec x)$

7. $(1 + \cos^2 x)(1 + \sin^2 x)^2(\sin 2x)(1 - 2\sin^2 x + 3\cos^2 x)$

9. $(1 + \cos^2 x)(3\sin^2 x \cos x - 4\sin^4 x \cos x + 3\sin^2 x \cos^3 x) - 2\cos x \sin x(\sin^3 x + \sin^3 x \cos^2 x)$

11. $-\dfrac{\cos^3 x + 2\sin^2 x \cos x + 2\sin x \cos x}{(1 + \sin x)^2}$ **13.** $\dfrac{-2\cos x}{(\sin x - 1)^2}$

15. $\dfrac{x\cos x - \sin x}{x^2}$

17. $\dfrac{\sin x \cos x(2\sin^2 x + 3\sin x \cos x - 4\cos^2 x)}{(\sin x + 2\cos x)^2}$

19. $2\cos 2x$ **21.** $3\cos 3x$ **23a.** $y = 1/2 - \sqrt{3}(x - \pi/3)/2$

23b. $y + \sqrt{3}/3 = 4(x - 5\pi/6)/3$

25. $dx/dt = v_0 \cos\alpha$, $dy/dt = -gt + v_0 \sin\alpha$, $dy/dt > 0$ for $t < v_0(\sin\alpha)/g$, $dy/dt < 0$ for $t > v_0(\sin\alpha)/g$

27a. $\dfrac{dF_t}{d\theta} = mg\cos\theta = F_n$, $\dfrac{dF_n}{d\theta} = -mg\sin\theta = F_t$

27b. F_t decreases as θ increases, and F_n decreases as θ increases.

29. $\dfrac{-mg\,\mu(\mu\cos\theta - \sin\theta)}{(\mu\sin\theta + \cos\theta)^2}$; positive when $\theta > \tan^{-1}\mu$; negative when $\theta < \tan^{-1}\mu$

EXERCISES 2.5, p. 104

1a. $y = 1/2 - (x - 1)/2$ **1b.** $y \approx 0.499$ **1c.** 0.499001

3a. $y = -x + \pi/2$ **3b.** $y \approx -0.0174533$ **3c.** $y = -0.0174524$

5. $\Delta y = 0.51$; $dy = 0.50$ **7.** $\Delta y = -0.039601$; $dy = -0.04$

9. $\Delta y = -0.0024814$, $dy = -0.0025$

11. $\Delta y = -0.000202$; $dy = 0$ **13.** 120π **15.** $1/3\ \text{ft}^3$

17. $\pi\sqrt{3}/360 + 1/2 \approx 0.515115$ **19.** $40\pi\ \text{cm}^3$; 3%

21. 60 in.^{2}; 0.41667% **23.** 2; 2% **25.** $\dfrac{0.01}{2\pi\varepsilon_0 r^3}$; $2/r$%

27. $f'(c) = 2$ **29.** $f'(c) = 2$ **31.** $f'(c) = 1/4$ **33.** $f'(c) = -1/4$

35. $f'(c) = 3$ **41.** $C_p = \dfrac{dE}{dt} + p\dfrac{dV}{dt}$

EXERCISES 2.6, p. 111

1. $6(x - 2)(x^2 - 4x + 7)^2$ **3.** $\dfrac{-30x^2 + 8x - 6}{(5x^3 - 2x^2 + 3x - 2)^3}$

5. $-2(x^2 - 5x + 3)^{-3}(2x - 5)(x^3 + x + 1)^3 + 3(x^2 - 5x + 3)^{-2}(x^3 + x + 1)^2(3x^2 + 1)$

7. $\dfrac{2(x^2 + 1)^2(x^3 + 2x^2 - 5x - 1)}{(x^2 + x - 1)^3}$ **9.** $\dfrac{-x^3 + \frac{3}{2}x^2 - x - \frac{3}{2}}{(x^2 + 3)^2\sqrt{x^2 - x + 2}}$

11. $15\cos 3x \cos 2x - 10\sin 3x \sin 2x$ **13.** $6x\sec^2(3x^2 + 1)$

15. $\dfrac{2}{(\cos 2t - \sin 2t)^2}$ **17.** $-2\cos(\cos 2u)\sin 2u$

19. $\dfrac{x\cos\sqrt{1 + x^2}}{\sqrt{1 + x^2}}$ **21.** $\dfrac{3}{2}\sqrt{\sec(3x - 2)}\tan(3x - 2)$

23. $20x\sin 5x^2 \cos 5x^2$ **25.** $\dfrac{x\sqrt{1 + 2x^2} + x}{\sqrt{x^2(1 + 2x^2) + (1 + 2x^2)^{3/2}}}$

27. $\dfrac{4}{x(2x^4 + 5)}$ **29.** $f'(c) = -32$ **31.** 3.025; 3.0250455

33. 0.3328889; 0.3328894 **35.** 0.8346095; 0.83187663

37. $S = 24.4375$; $N = 249.2979$; $\dfrac{dN}{dt} = 4.5820$

39. $S = 25$; $N = 256.25$; $\dfrac{dN}{dt} = 0$ **41.** $A = 10$; yes

43. Falls for $t < 6$, rises for $t > 6$ **45a.** Decreasing

45b. Neither **45c.** Increasing

47a. $W_C = 1136.85$; $T_C \approx -16.25$°C

47b. $T_C = 33 - \dfrac{(10\sqrt{s} + 10.45 - s)(33 - T)}{23.082397}$

47c. $\dfrac{dT_C}{ds} = \dfrac{T-33}{23.082397}\left[\dfrac{5}{\sqrt{s}} - 1\right], \left.\dfrac{dT_C}{ds}\right|_{\substack{T=0\\S=16}}$

$= \dfrac{-33(0.25)}{23.082397} \approx -0.357415$

51a. $\Delta u \neq 0$

EXERCISES 2.7, p. 120

1. Yes; $f^{-1}(x) = (x+1)^{1/3}$; $(f^{-1})'(x) = (1/3)(x+1)^{-2/3}$ $D_{f^{-1}} = \mathbf{R}$

3. Yes; $f^{-1}(x) = \dfrac{1}{x}$; $(f^{-1})'(x) = \dfrac{-1}{x^2}$; $D_{f^{-1}} = \mathbf{R} - \{0\}$

5. No. $f(-1) = f(-2) = 6$; therefore f is not one-to-one.

7. Yes; $f^{-1}(x) = \dfrac{x+1}{1-x}$; $(f^{-1})'(x) = \dfrac{2}{(1-x)^2}$; $D_{f^{-1}} = \mathbf{R} - \{1\}$

9. No. $f(-1) = f(1) = 0$; therefore f is not one-to-one.

11. $D_f = [0, +\infty)$; $f^{-1}(x) = \sqrt{x-1}$ for $x \geq 1$, $(f^{-1})'(x) = \dfrac{1}{2\sqrt{x-1}}$

13. $D_f = [0, 4]$; $f^{-1}(x) = \sqrt{16 - x^2}$; $(f^{-1})'(x) = \dfrac{-x}{\sqrt{16-x^2}}$; $D_{f^{-1}} = [-4, 4]$

15. $\dfrac{3}{2}\sqrt{x^2+3x-1}(2x+3)$ **17.** $\dfrac{10x^2-2}{\sqrt[3]{5x^3-3x+7}}$

19. $(2/3)(x-1)^{-2/3}(x+1)^{-4/3}$

21. $-2(2\cos^2 2x \sin 2x + x^2)(\cos^3 2x - x^3)^{-1/3}$

23. $3x[5x^2 + \sqrt[3]{x^2+1}]^{1/2} \cdot \left[5 + \dfrac{1}{3}(x^2+1)^{-2/3}\right]$ **25.** -1

27a. $\sqrt{2}$ **27b.** sine is 1:1 on $[0, \pi/2]$ **29.** 1/12 **31.** 1 **33.** 1/10

EXERCISES 2.8, p. 124

1. $\dfrac{x}{y}$ **3.** $-\dfrac{y^2+2xy}{x^2+2xy}$ **5.** $-\dfrac{3x^2+4xy+3y^2}{2x^2+6xy-3y^2}$ **7.** $-\sqrt{\dfrac{y}{x}}$

9. $\dfrac{2xy^2+x^2}{-2x^2y+y^2}$ **11.** $-\dfrac{2y^2+x^2y^3}{x^3y^2-3x^2}$

13. $\dfrac{3x^2 - 2(xy^2-x^2y)(y^2-2xy)}{2(xy^2-x^2y)(2xy-x^2)+3y^2}$ **15.** $-\dfrac{y}{x}$

17. $-\dfrac{3y\sin^2 x\cos x + \cos^2 y}{\sin^3 x - 2x\cos y \sin y}$

19. $\dfrac{6x^2y\sin x^3y\cos x^3y + 3y^2\sin^2 xy^2\cos xy^2 + 2xy^2}{-2x^3\cos x^3y\sin x^3y - 6xy\sin^2 xy^2\cos xy^2 - 2x^2y}$

21. 2.014 **23.** 0.993 **25.** 3.0018 **27.** 0.98 **29.** -1.983

31. 2.028, 2.043, 2.058, 2.073

33. 3.003, 3.006, 3.009, 3.012, 3.015 **37.** $x_0x/a^2 + y_0y/b^2 = 1$

EXERCISES 2.9, p. 129

1. $f''(x) = 6x^2 - 9x + 10$ **3.** $f''(x) = 2\sec^2 x\tan x$

5. $f''(x) = 6(x^2-5x+1)(5x^2-25x+26)$

7. $h''(x) = (3/4)x^{-1/2} - (3/2)x^{-5/2}$

9. $m''(x) = -(x^2-x+2)^{-3/2}(2x-1)^2/4 + (x^2-x+2)^{-1/2}$

11. $2\sin 3x + 12x\cos 3x - 9x^2\sin 3x$

13. $6\sin^2(x^2+1)\cos(x^2+1) + 24x^2\sin(x^2+1)\cos^2(x^2+1) - 12x^2\sin^3(x^2+1)$

15. $-4y^{-3}$ **17.** $-16xy^{-5}$ **19.** $\dfrac{-6}{(x+2y)^3}$ **21.** $\dfrac{1}{3x^{4/3}y^{1/3}}$

23. $a = 2$; moves toward the right when $t > 4$, toward the left when $t < 4$, stops at $t = 4$

25. $v < 0$ when $t < 2$; $v > 0$ when $t > 2$; $a > 0$ when $t > 1$; $a < 0$ if $t < 1$; $v = 0$ at $t = 2$; $a = 0$ at $t = 1$

27. $a > 0$ when $t > 7/4$; $v > 0$ when $t < 1/2$ or $t > 3$; $v < 0$ when $1/2 < t < 3$; $v = 0$ at $t = 1/2, 3$; $a = 0$ at $t = 7/4$; $a < 0$ when $t < 7/4$

29. $y' > 0$ for $-2 < t < 0$ or $t > 2$; $y' < 0$ for $t < -2$ or $0 < t < 2$; $y' = 0$ at $t = -2, 0, 2$; $y'' > 0$ on $(-\infty, -2/\sqrt{3}) \cup (2/\sqrt{3}, +\infty)$; $y'' < 0$ for $t \in (-2/\sqrt{3}, 2/\sqrt{3})$; $y'' = 0$ at $t = \pm 2/\sqrt{3}$

31. Oscillates between $x = -1$ and $x = 1$; moves to right on $(0, \pi/2) \cup (3\pi/2, 2\pi) \cup \ldots$; stops at $\pi/2, 3\pi/2, 5\pi/2, \ldots$; moves to left on $(\pi/2, 3\pi/2) \cup (5\pi/2, 7\pi/2) \cup \ldots$; acceleration positive on $(\pi, 2\pi) \cup (3\pi, 4\pi) \cup \ldots$; acceleration negative on $(0, \pi) \cup (2\pi, 3\pi) \cup \ldots$.

33. $f'(x) = 15x^2 - 16x + 1$; $f''(x) = 30x - 16$; $f'''(x) = 30$; $f^{(n)}(x) = 0$ for $n \geq 4$

35. $f'(x) = \cos x$; $f''(x) = -\sin x$; $f'''(x) = -\cos x$; $f^{(4)}(x) = \sin x$, $f^{(5)}(x) = f^{(9)}(x) = \cos x$, etc.

37. $f'(x) = -x^{-2}$, $f''(x) = 2x^{-3}$, $f'''(x) = -6x^{-4}$, $f^{(n)}(x) = (-1)^n n!(x^{-n-1})$, where $n! = n(n-1)\ldots 3\cdot 2\cdot 1$

39. $g''(x)h(x) + 2g'(x)h'(x) + g(x)h''(x)$

41. $g^{(4)}(x)h(x) + 4g'''(x)h'(x) + 6g''(x)h''(x) + 4g'(x)h'''(x) + h^{(4)}(x)g(x)$ **43.** $g''(h(x))[h'(x)]^2 + g'(h(x))h''(x)$

EXERCISES 2.10, p. 135

1. \$25: 2 h 17 min to 2 h 29 min
\$50: 2 h 7 min to 2 h 16 min
\$100: under 2 h 7 min

3. f is differentiable on $(1, 2)$; continuous on $[1, 2]$; $f'(x) = 0$ at $x = 3/2$

5. m is differentiable on $(-2, 4)$; continuous on $[-2, 4]$; $m'(x) = 0$ at $x = 2$

7. f is differentiable on $(-2, 3)$; continuous on $[-2, 3]$; $c = 1/2$

9. m is differentiable on $(1, 4)$; continuous on $[1, 4]$; $c = 2$

11. n is differentiable on $(4, 9)$; continuous on $[4, 9]$; $c = 25/4$

13. f is differentiable on $(-1, 1)$; continuous on $[-1, 1]$; $c = 1 - 2/(3\sqrt{3}) \approx 0.6151$

15. f is neither continuous nor differentiable at $x = 0$

17. f is not differentiable at $x = 3$

19. f is not differentiable at $x = 0$

21. Yes, $c = 1$ works; no, since it isn't differentiable at $x = 0$. The conditions are therefore not necessary.

REVIEW EXERCISES 2.11, p. 137

1. $9x^2 - 8x + 5$

3. $(20x^3 - 6x + 1)(3x^5 - 4x^3 + 2x^2 + 6) + (5x^4 - 3x^2 + x - 1)(15x^4 - 12x^2 + 4x)$

5. $2\sin x\cos^4 x - 3\cos^2 x\sin^3 x - 2x\cos^2 x + 2x^2\cos x\sin x$

7. $\frac{3}{2}[x^2 + 5x - 3]^{1/2}(2x + 5)$

9. $(1/3)(x^3 - 3x^2 + x + 1)^{-2/3}(3x^2 - 6x + 1)$
11. $-(c + 1)^{-2}$ **15.** $y + 1 = -11(x - 1); -1.11$
17. $y - 2 = 5(x - 2)/12; 1.9875$
21. Period is π.

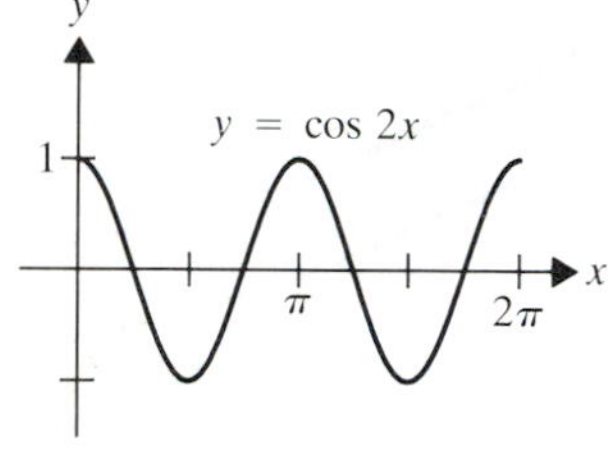

25. $p = 3 + \cos \pi t/2$. Price after four years would be about \$4.00 per pound.
27. 3/4 **29.** 0 **31.** $D_{f'} = (-\infty, 0) \cup (0, +\infty)$ **33.** $6\pi r^2$

35. 1 ft³ **37.** $dh = \dfrac{dV - 2\pi rh\,dr}{\pi r^2}$

39. $1 - 1$; $f^{-1}(x) = \sqrt[3]{x - 1}$; $(f^{-1})'(x) = \dfrac{1}{3(x - 1)^{2/3}}$ for $x \neq 1$

41. Not $1 - 1$: $f(-1) = f(1)$
43. $D_f = [0, +\infty)$, $f^{-1}(x) = (1 - x)^{1/2}$;
$(f^{-1})'(x) = -(1/2)(1 - x)^{-1/2}$ for $x \neq 1$

45. $\dfrac{dy}{dx} = \dfrac{2x - 9x^2y^2}{6x^3y + 2y}$ **47.** $\dfrac{dy}{dx} = \dfrac{\sin y^2 + 2xy \sin x^2}{\cos x^2 - 2xy \cos y^2}$

49. 0.9625
51. $v > 0$ for $t > 5$ or $t < -1$; $a > 0$ for $t > 2$; $a < 0$ for $t < 2$
53. $p'(x) = 24x^3 - 6x + 1$; $p''(x) = 72x^2 - 6$; $p'''(x) = 144x$
$p^{(4)}(x) = 144$, $p^{(n)}(x) = 0$ for $x \geq 5$

55. $\dfrac{4}{3x^{4/3}y^{1/3}}$ **59.** Yes **63b.** $v = \dfrac{V_{max}S_T}{K_m + S_T + E_T}$

CHAPTER 3

EXERCISES 3.1, p. 145

1. 4/3 ft/s **3.** 2000π ft²/min **5.** $\sqrt{3}/30$ ft/min

7. 1.8×10^{-5} cm/min² **9.** 72 in³/s **11.** $\dfrac{dP}{dt} = -0.01 \dfrac{\text{atm}}{\text{min}}$ **13.** $1.5 \dfrac{\text{ft}}{\text{s}}$; $4.5 \dfrac{\text{ft}}{\text{s}}$

15. $225\pi/256$ in.³/s **17.** $340/\sqrt{13} \approx 94.3$ km/h
19. $2400/\sqrt{41} \approx 374.8$ ft/s **21.** 2.4 m/s **23.** $5/4 = 1.25$ ft/s
25. $30\sqrt{10} \approx 94.87$ ft/s **27.** $3/28 \approx 0.1071429$ l/min
29. $-9/4$ units/s **31.** 50 lumens/m² · min. **33.** 12π ft/min

37. $r' = 25\sqrt{30}\, t^{-3/4}/\sqrt{\pi}$

EXERCISES 3.2, p. 154

1a. $R = 50{,}000 + 1500x - 50x^2$ **1b.** $[0, 50]$
1c. $x = 15$, i.e., \$350/month
1d. More, but there still are 75 vacant units.
3. Max. 1 at $x = 0$; min. -3 at $x = 2$
5. Max. 5 at $x = 1$ or 4; min. 1 at $x = 0$ or 3
7. Max. 2 at $x = 8$; min. -2 at $x = -8$
9. Max. 3 at $x = 0$; min. 2 at $x = 1$ or $x = -1$
11. Max. 5 at $x = -1$; min. 1 at $x = 1$
13. Max. 3/4 at $x = 3$; min. 1/2 at $x = 1$
15. Max. 1 at $x = 0$; min. 0 at $x = \pm 1$
17. Max. $3\sqrt{3}/4$ at $x = 1/2$; min. 0 at $x = \pm 1$
19. Max. also occurs at $5\pi/6$, min. also occurs at π.
21. Max. 2 at $x = \pm\pi/3$; min. 1 at $x = 0$
23. Max. $\sqrt{2} - 1$ at $x = \pi/4$; min. $\sqrt{3} - 2$ at $x = -\pi/6$
25. Max. 0 at $x = 0$ **27.** Extreme values are $f(c)$ and $f(d)$.

EXERCISES 3.3, p. 160

1. Increasing on $(3/2, +\infty)$; decreasing on $(-\infty, 3/2)$
3. Increasing on $(-\infty, +\infty)$
5. Increasing on $(1, +\infty)$; decreasing on $(-\infty, 1)$
7. Increasing on the whole real line
9. Increasing on $\ldots \cup (-\pi, -\pi/2) \cup (0, \pi/2) \cup (\pi, 3\pi/2) \cup \ldots$
11. Local min. at $x = 5/2$

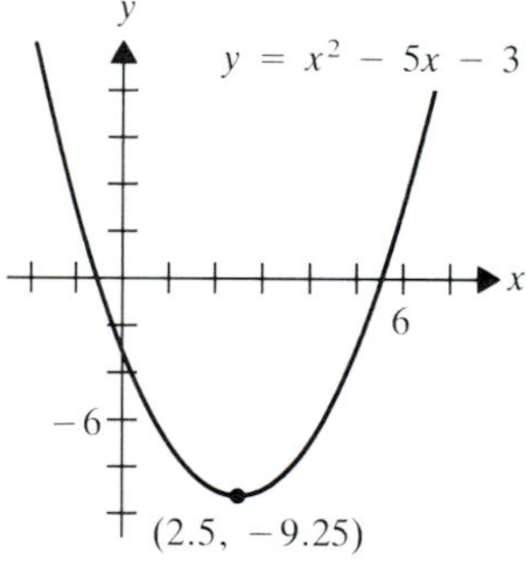

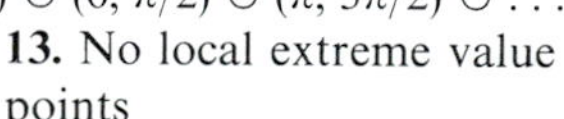
13. No local extreme value points

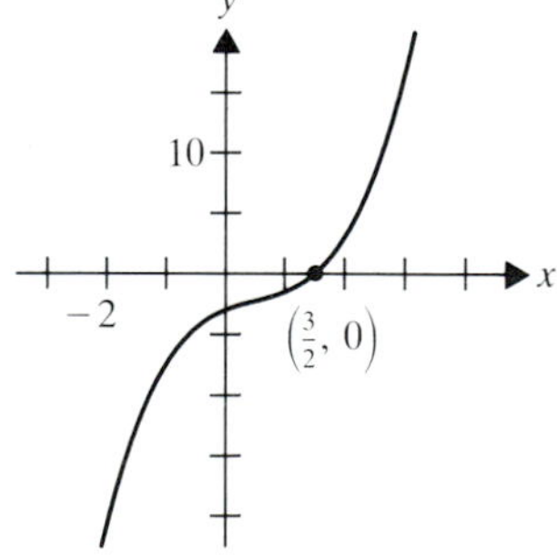

$y = x^3 - \frac{3}{2}x^2 + 2x - 3$

15. Local min. at $x = 1$

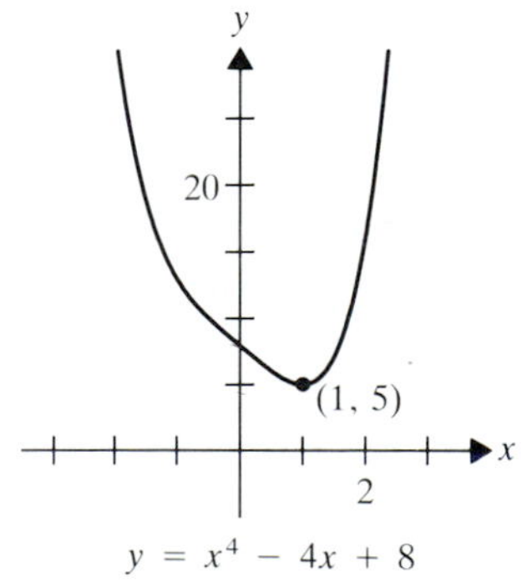

$y = x^4 - 4x + 8$

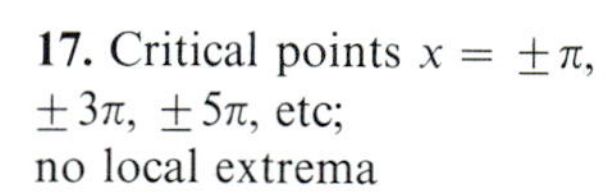
17. Critical points $x = \pm\pi$, $\pm 3\pi$, $\pm 5\pi$, etc; no local extrema

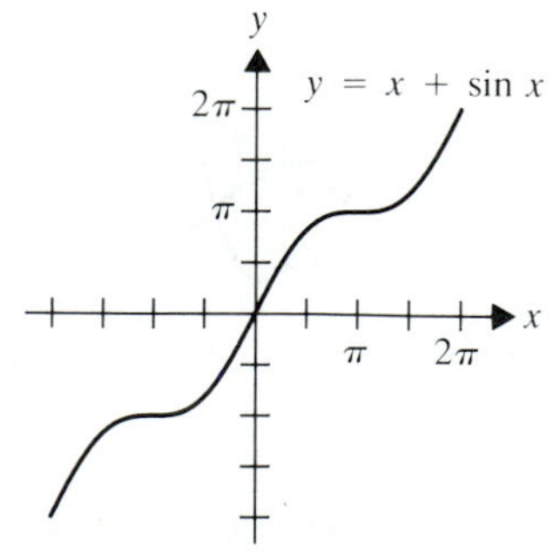

19. Local max. at $x = \pm\pi/2, \pm 3\pi/2, \pm 5\pi/2, \ldots$
Local min. at $x = 0, \pm\pi, \pm 2\pi, \pm 3\pi, \ldots$

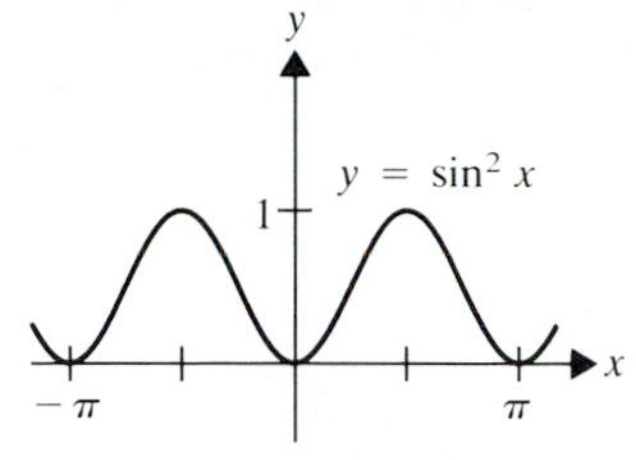

21. Local min. at $x = 4$; local max. at $x = 4/3$; critical point at $x = 0$

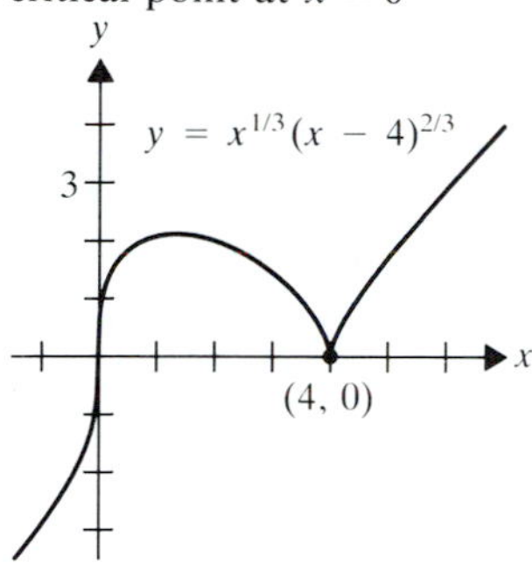

23. Local min. at $x = -2$; critical point at $x = 0$

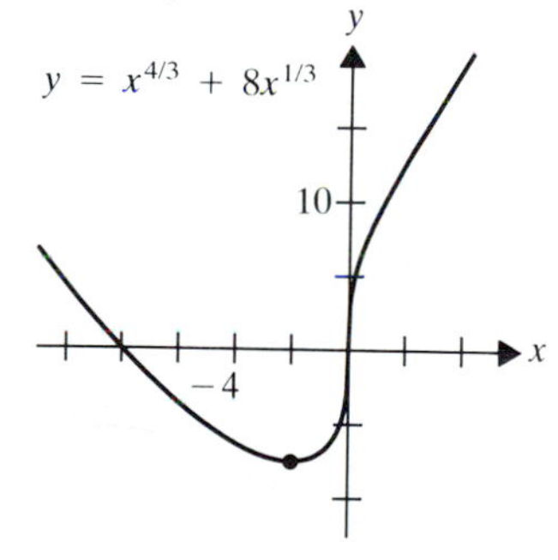

25. Local min. at $x = 2$; local max. at $x = -2$

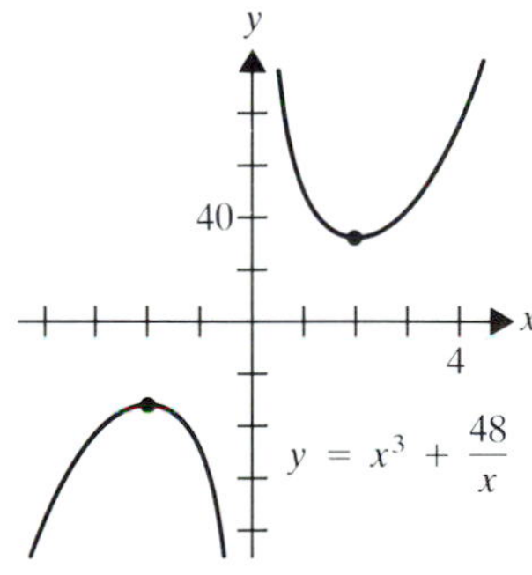

27. Local min. at $x = 1/3$; local max. at $x = -1$

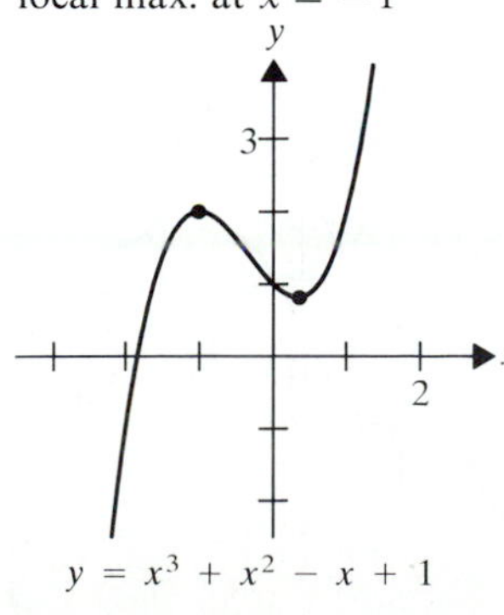

29. Local max. at $x = 0$

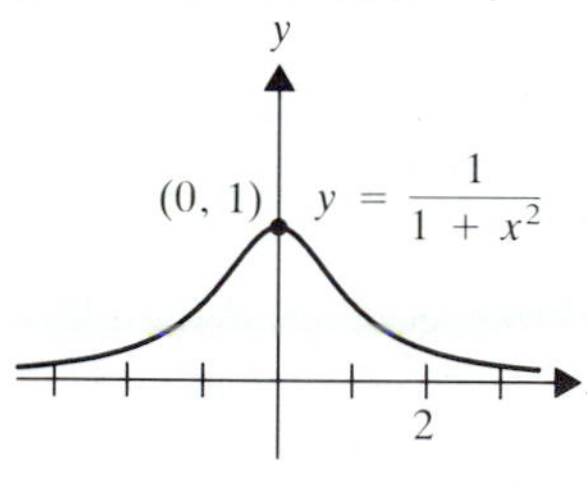

31. Local min. at $x = \pm 1$; local max. at $x = 0$

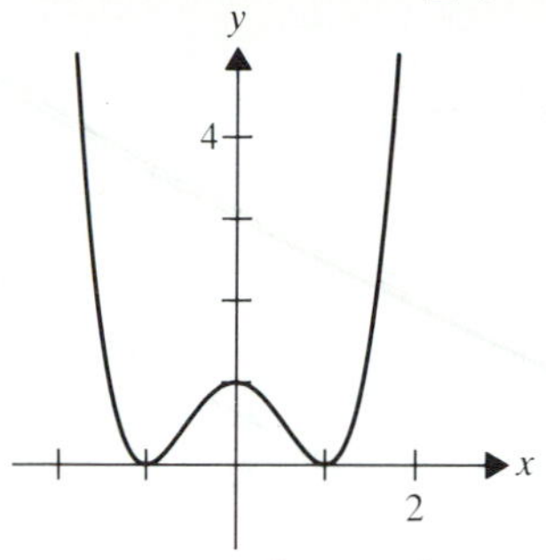

EXERCISES 3.4, p. 168

1. Concave up on $(-\infty, +\infty)$
3. Concave down on $(-\infty, 1/2)$; concave up on $(1/2, +\infty)$; inflection point at $x = 1/2$
5. Concave up on $(-\infty, +\infty)$
7. Concave up on $(-\pi, 0) \cup (\pi, 2\pi) \cup (3\pi, 4\pi) \cup \ldots$; concave down on $(0, \pi) \cup (2\pi, 3\pi) \cup \ldots$; inflection points at all multiples of π
9. Concave up on $\ldots(-5\pi/4, -3\pi/4) \cup (-\pi/4, \pi/4) \cup (\pi/4, 3\pi/4)\ldots$; concave down on $\ldots \cup (-3\pi/4, -\pi/4) \cup (\pi/4, 3\pi/4) \cup \ldots$; inflection points at all odd multiples of $\pi/4$
11. Local and absolute min. at $x = 3/10$
13. Local min. at $x = 6$; local max. at $x = 2$
15. Local min. at $x = -1$ and 2; local max. at $x = 1$
17. Local min. at $x = 0$; local max. at $x = \pm 1$
19. Local min. at $x = 0$ and 2; local max. at $x = 1$
21. Local max. at $x = -1$; local min. at $x = 1$
23. Local min. at $x = 4/5$; local max. at $x = 0$
25. No local extreme value points
27. Local max. at $x = \pm\pi/2, \pm 3\pi/2, \pm 5\pi/2, \ldots$
Local min. at $x = 0, \pm\pi, \pm 2\pi, \pm 3\pi, \ldots$
29. Local min. at $x = \ldots, -7\pi/2, -3\pi/2, \pi/2, 5\pi/2, \ldots$
Local max. at $x = \ldots, -5\pi/2, -\pi/2, 3\pi/2, 7\pi/2, \ldots$
31. Local max. at $x = \ldots, -5\pi/3, -2\pi/3, \pi/3, 4\pi/3, 7\pi/3, \ldots$
Local min. at $x = \ldots, -7\pi/3, -4\pi/3, -\pi/3, 2\pi/3, 5\pi/3, \ldots$

EXERCISES 3.5, p. 178

1. \$26.00 per barrel
3. 65 trees should be planted per acre to maximize yield.
5. $10/3 \times 14/3 \times 70/3$ **7.** 2 ft × 2 ft × 1 ft
9. $r = (158/\pi)^{1/3} \approx 3.69$ cm; $h \approx 7.38$ cm
11. Proceed 250 ft downshore east from A, then lay the rest underwater to B. **13.** Come ashore 4/3 miles downstream.
15. 200 ft^2 **19.** 6.25 in. **21a.** $v = kx(a - x)$ **21b.** $x = a/2$
23a. $t = \dfrac{\sqrt{x^2 + y_1^{\,2}}}{c_1} + \dfrac{\sqrt{(a - x)^2 + y_1^{\,2}}}{c_2}$ **27.** $x = \dfrac{2}{\sqrt{3}}$; $y = 2\sqrt{\dfrac{2}{3}}$

29. $x = y = \dfrac{10}{\pi + 4}$

31. $x = \dfrac{ab}{1 + b}$ where $b = \sqrt[3]{\dfrac{y_1}{y_2}} + x =$ distance from O_1

33. $\dfrac{4}{9}\pi a^3\sqrt{3}$ **35.** $t = \dfrac{6}{305}$ h ≈ 1.1 min, $d = \sqrt{1 - 60t + 1525t^2}$

37. (5/2, −1/2) **39.** 12 in. × 6 in. **41.** $4(4^{2/3} - 1)^{3/2}$
43. $x = M/2$

EXERCISES 3.6, p. 189

1. 1/2 **3.** −5/2 **5.** 0 **7.** $+\infty$ **9.** $-\infty$ **11.** 3/5 **13.** 2/5
15. 1 **17.** 1 **19.** 0 **21.** $-\infty$ **23.** $+\infty$ **25.** $+\infty$ **27.** $+\infty$
29. 0 **31.** 0

33. $y = 0, x = 0$

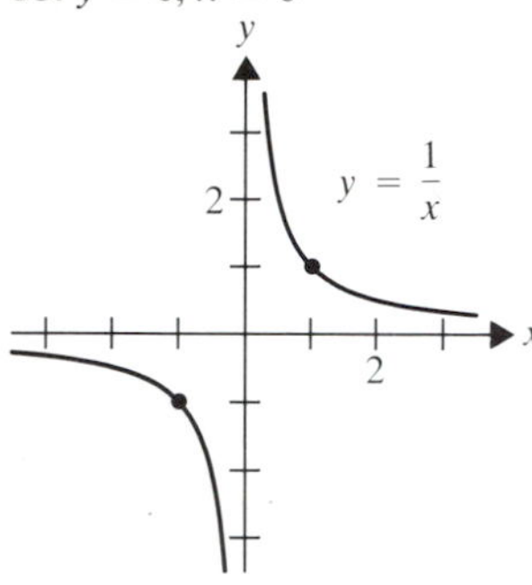

35. $y = 1, x = -1$

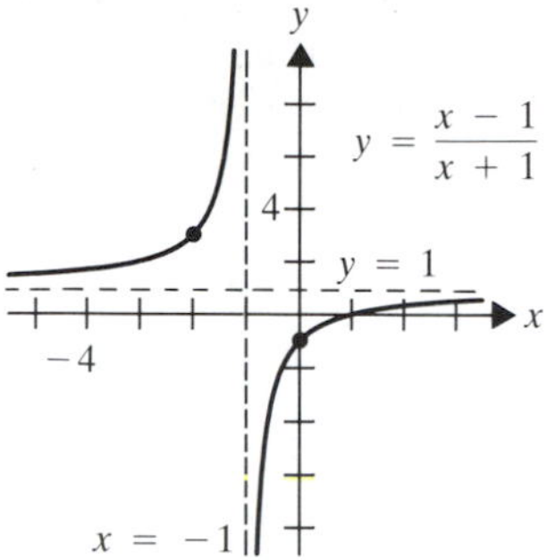

37. $y = 0$

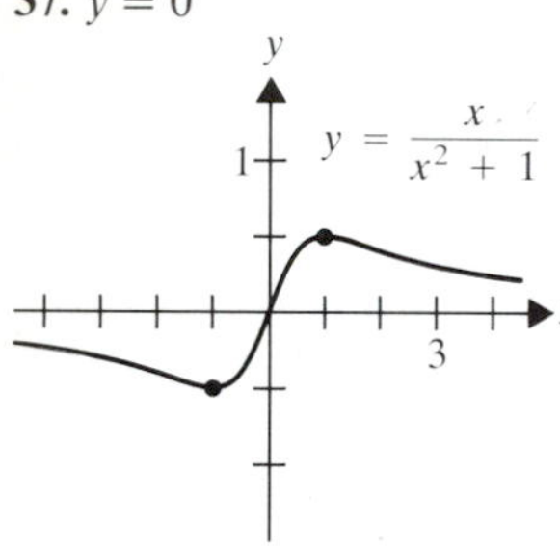

39. $y = 1$

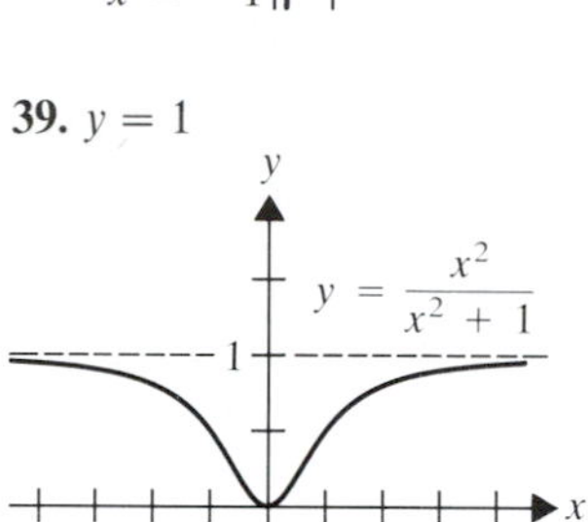

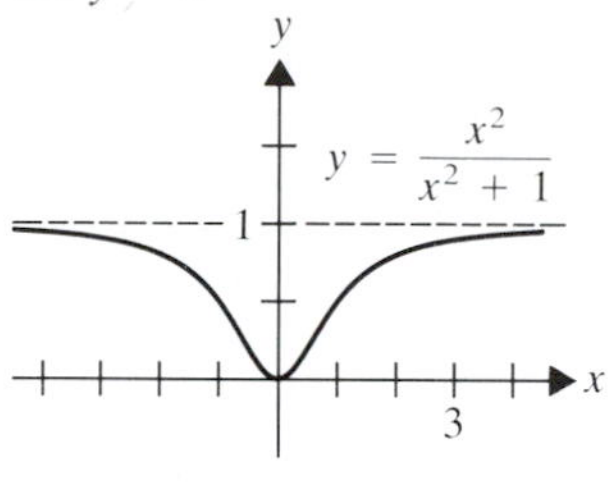

41. $y = 1, x = -3$

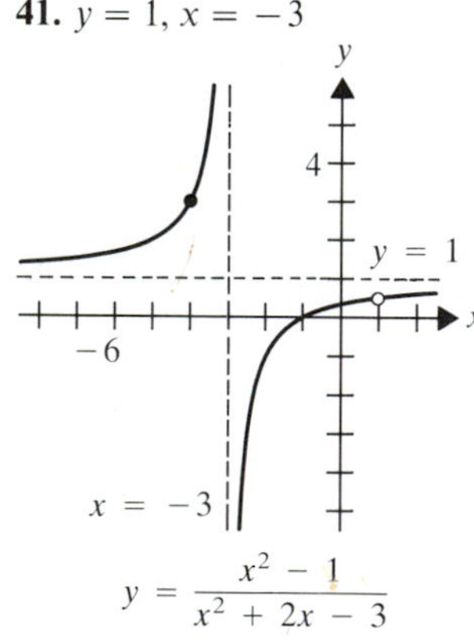

43. $V_{\max}$ **45.** $v \to +\infty$

EXERCISES 3.7, p. 200

1.

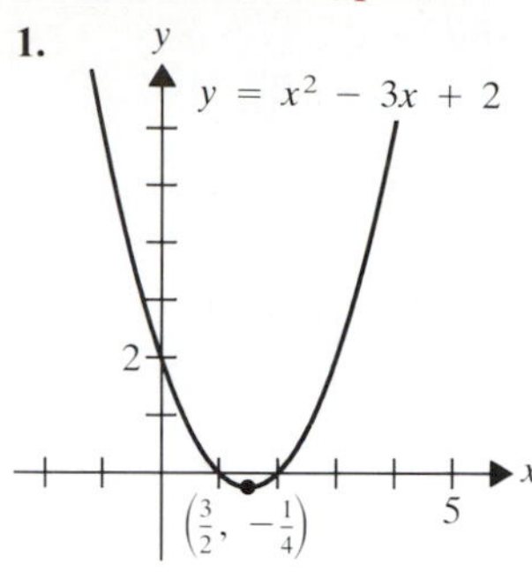

3. See figure for Exercise 13, Section 3.3.

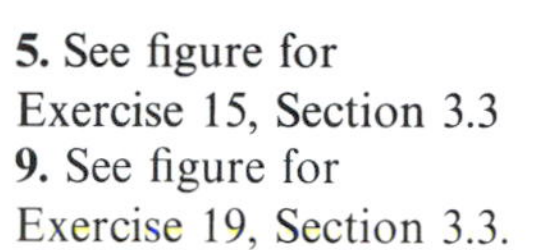

5. See figure for Exercise 15, Section 3.3
9. See figure for Exercise 19, Section 3.3.

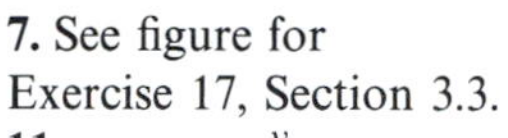

7. See figure for Exercise 17, Section 3.3.

11.

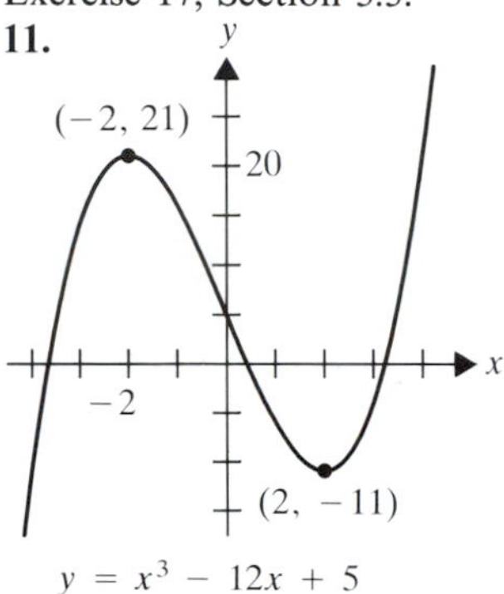

13. See figure for Exercise 43, Section 2.3.
17. See figure for Exercise 21, Section 3.3.

21.

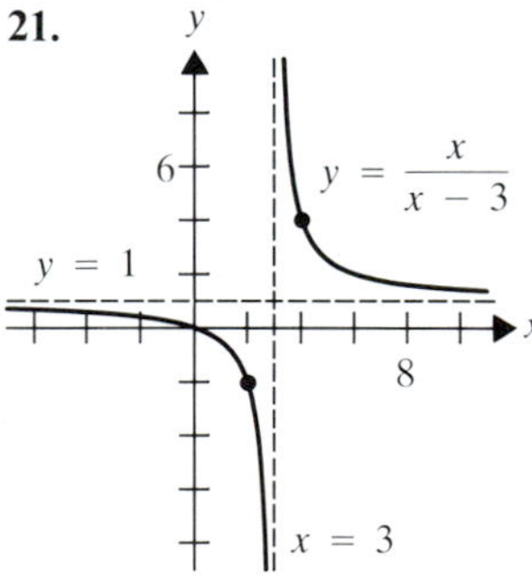

15. See figure for Exercise 19, Section 3.3.
19. See figure for Exercise 23, Section 3.3.
23. See figure for Exercise 29, Section 3.3.

25.

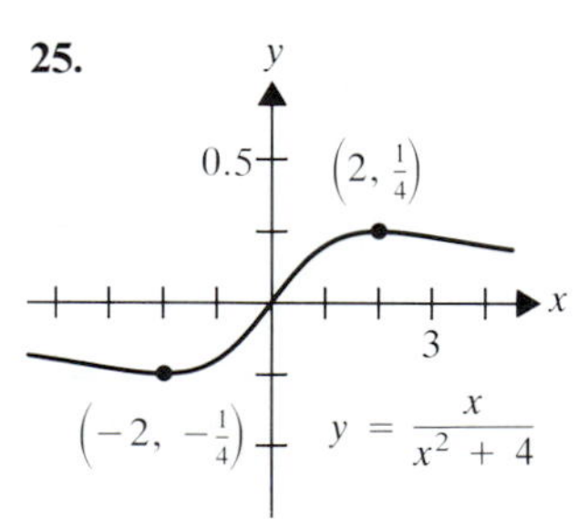

27.

x = −2 y 2 4 x x = 2

$y = \dfrac{x}{x^2 - 4}$

29.

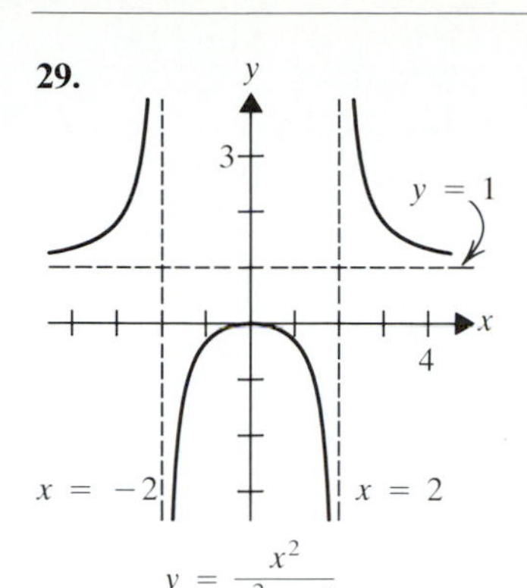

31.

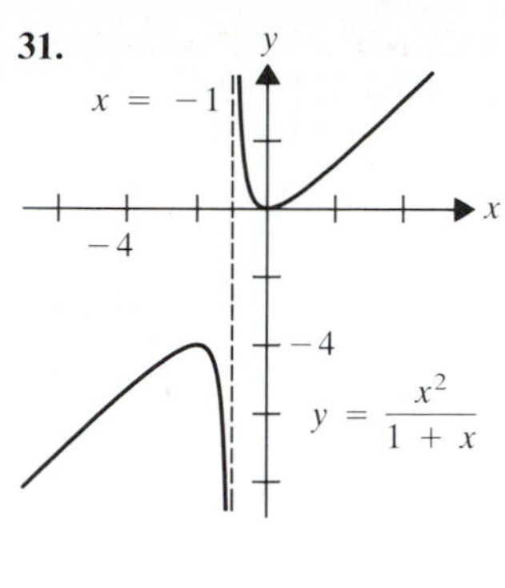

33.

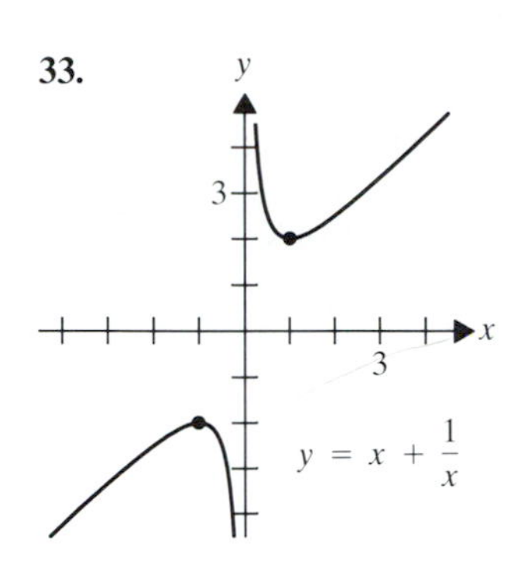

35.

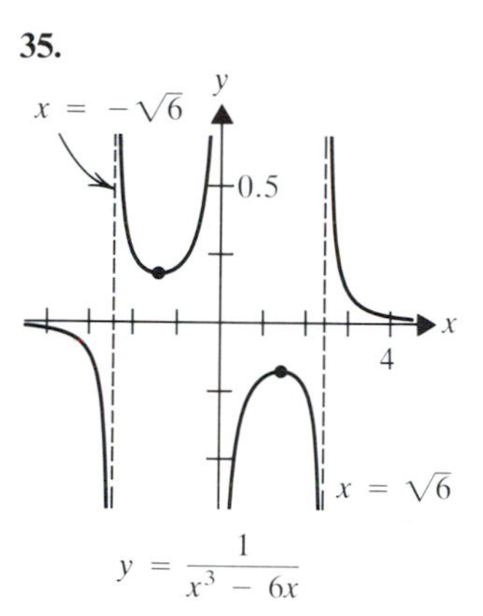

37.

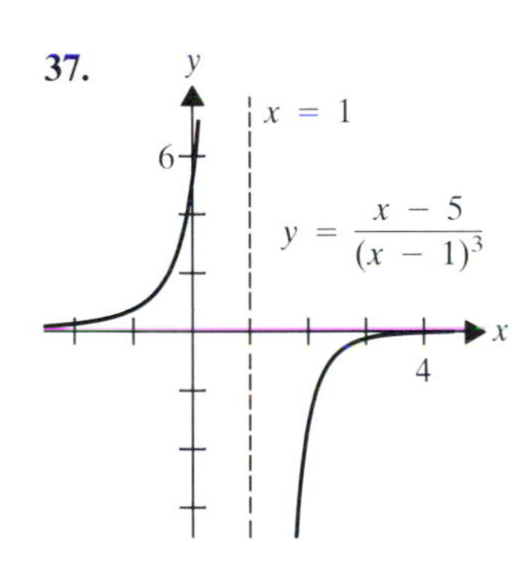

39.

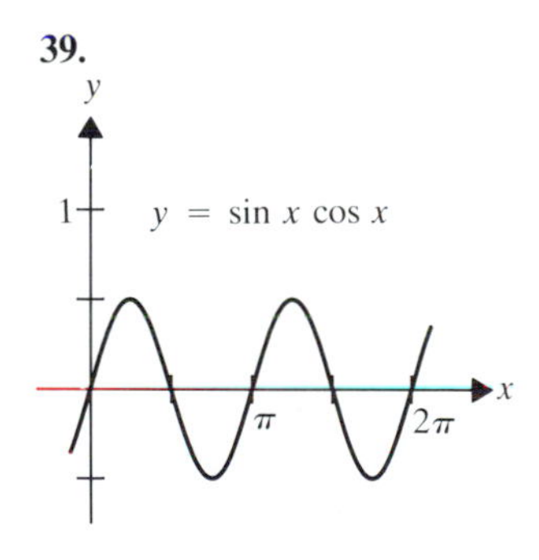

41.

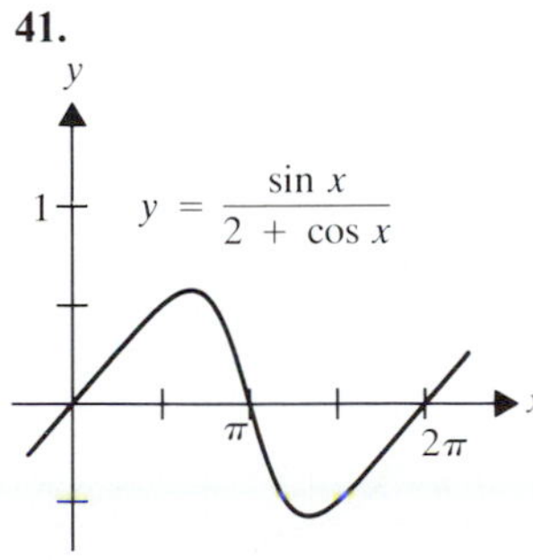

43. See Figure 3.7.10 (p. 198).

EXERCISES 3.8, p. 207

1. 1.7320508 **3.** 1.2599211 **5.** 1.316074 **7.** 1.5848932
9. 2.1544347 **11.** 1.2906488 **13.** 1.517003 **15.** 0.6938225
17. 0.7853982 **19.** 1.8954942 **21.** $x_{n+1} = x_n(2 - xx_n)$

23. $x_{n+1} = \frac{1}{k}\left[(k-1)x_n + \frac{x_1}{x_n^{k-1}}\right]$
25. See answer to Exercise 1. **27.** See answer to Exercise 5.
29. See answer to Exercise 9. **31.** See answer to Exercise 13.
33. See answer to Exercise 17. **35.** -1, -0.51879, 1.29065
37. 1.51700 **39.** -2.061499, -0.396339, 0.693822, 1.764015
41. -18.064158, -14.922565, -11.780973, -8.639380, -5.497787, -2.356195, 0.785398, 3.926991, 7.068583, 10.210176, 13.351769, 16.493361, 19.634954
43. ± 1.89549427, 0 **45.** You get -1, -2, -4, 8, etc.
47. Cycling between 0.5443 and -0.5443

EXERCISES 3.9, p. 216

1. $2\sqrt{x} + C$ **3.** $(3/5)x^5 - (5/2)x^2 + C$
5. $x^3/3 - 3x^2/2 + x + C$ **7.** $-x^{-2}/2 + x^{-1} + 3x + C$
9. $2x^{7/2}/7 + 2x^{-1/2} - 2x^{1/2} + C$
11. $2x^{3/2}/3 - 6x^{1/2} - 2x^{-1/2} + C$ **13.** $\sin^2 2x)/4 + C$
15. $(x^2 - 3)^6/2 + C$ **17.** $-3(x^2 - x + 1)^{-2}/2 + C$
19. $(x^4 - 3x^2 + 5)^{3/2}/3 + C$ **21.** $(1 + 2t^2)^{1/2}/2 + C$
23. $(2x - 1)^{3/2}/3 + C$ **25.** $(\sin^3 x)/3 + C$
27. $-3(\cos 2x)/2 + 2(\cos 3x)/3 + C$ **29.** $(6/5)(\sqrt{x} + 1)^{5/3} + C$
31. $(-1/12)(\cos^6 2x) + C$ **33.** $(\tan 5x)/5 + C$
35. $(-\cot^3 3x)/9 + C$ **37.** $-2(\cot 3x)^{3/2}/9 + C$
39. $(\sec x^3)/3 + C$ **41a.** $(-\cos 2x)/2 + C$ **41b.** $\sin^2 x + K$
41c. $\cos 2x = \cos^2 x - \sin^2 x = 1 - 2\sin^2 x$ **49b.** No
49c. x is not confined to a single interval but rather $x \in (-\infty, 0) \cup (0, +\infty)$, a union of two intervals.
51. $x^2/2 + C$ for $x \geq 0$; $-x^2/2 + C$ for $x < 0$

EXERCISES 3.10, p. 222

1. $v = 21.36$ miles/h; $t = 2.1$ s **3.** $a = 11.73$ ft/s^2; $t = 15$ s
5. $s = -16t^2 + 64t$; $t = 2$; 64 ft; $t = 4$, $v_f = -64$ ft/s
7. $t = 5\sqrt{5}/2 \approx 5.59$ s;
$v_f = -80\sqrt{5} \approx -178.89$ ft/s ≈ 121.97 miles/h
9. $t = 25$; 1/40 mg (20% of original dose) **11.** 1.8 mg
13. $a = g - kv^2/m$
15. $T = -10 + 2[1 - \cos(\pi t/100)]/\pi$; $p = 200$ years
17. 4 joules(N-m) **19.** $y^2 = x^2 + C$
21. $y^2 + 4y = \sin 2x + C$
23. $y^2 - 2y = x^3 + 17x^2 + 2x + C$
25. $y^3 + 3y^{4/3}/4 = \tan x + C$ **27.** $y = -x$
29. $y = -2 + \sqrt{9 + \sin 2x}$ **31.** $y = 1 - \sqrt{x^3 + 2x^2 + 2x + 4}$
33. $y^3 + 3y^{4/3}/4 = \tan x - 1$

REVIEW EXERCISES 3.11, p. 224

1. Max. 8/3 at $x = 2$ or 5; min. -4 at $x = 0$; local min. at $x = 4$
3. Max 268.739 at $x = 0.6$; min. 0 at $x = \pm 3$
5. Max. $\pi + 1$ at $x = \pi$; min. -1 at $x = 0$
7. Decreases on $(-\infty, -1) \cup (0, 1)$; local min. at $x = -1, 1$. Increases on $(-1, 0) \cup (1, +\infty)$; local max. at $x = 0$.
9. Increases on $(-\infty, 4) \cup (6, +\infty)$; decreases on $(4, 6)$. Local max. at $x = 4$; local min. at $x = 6$.
11. Concave down on $(-\infty, 3)$; concave up on $(3, +\infty)$; inflection point at $x = 3$ **13.** \$260; 130 seats
15. $80\sqrt{2}$; $10\sqrt{2} \times 20\sqrt{2}$ **17.** 5/2 **19.** 0 **21.** $+\infty$ **23.** $+\infty$
25. $-\infty$

27. Max. at $x = 2$; min. at $x = 4$; inflection point at $x = 3$

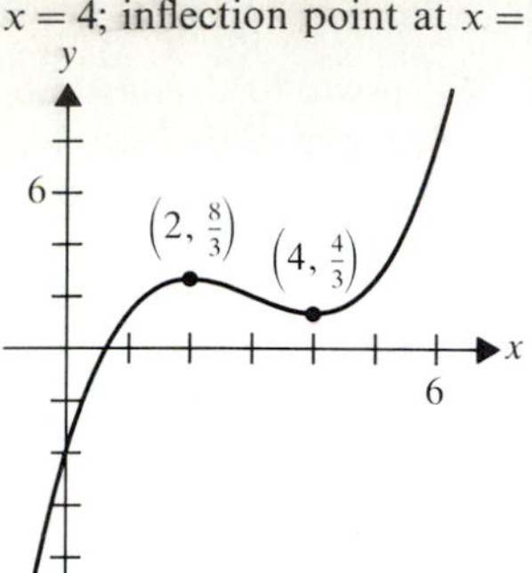

29. Increasing on $(-\infty, +\infty)$; inflection point at $x = 0$

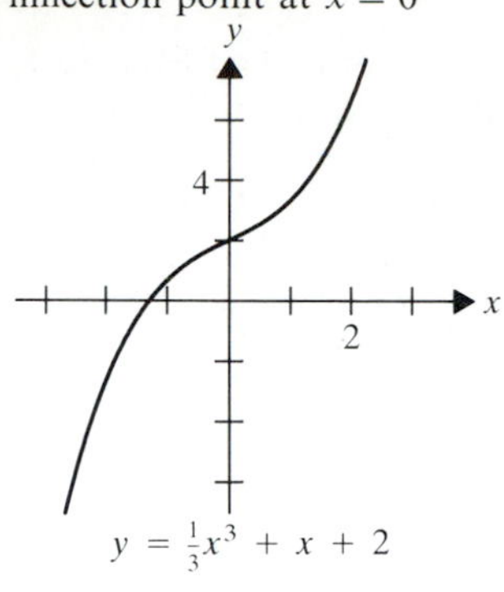

31.

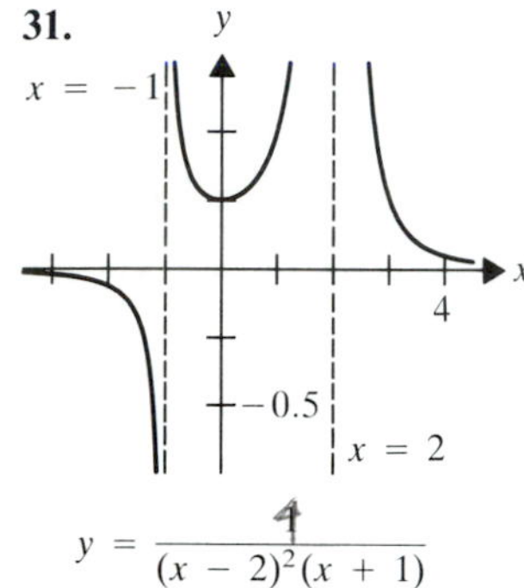

33.

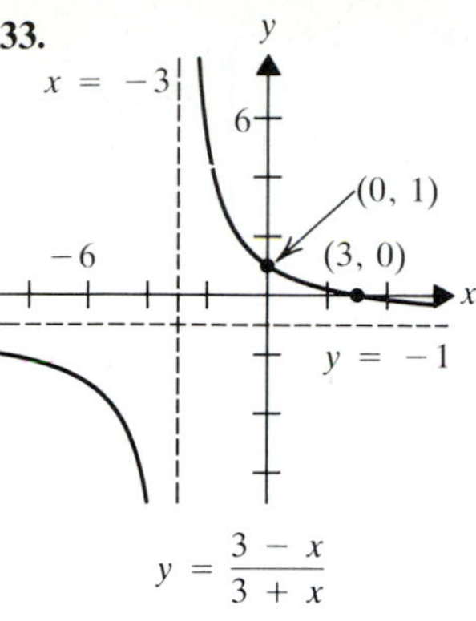

35. 6 ft/s; 10 ft/s **37.** $-2/25\pi$ **39.** -0.543689

41. $6\sqrt{x} - \dfrac{2x^{3/2}}{9} + \dfrac{2(x+1)^{3/2}}{3} + C$ **43.** $(x^2 - 1)^{3/2}/3 + C$

45. $(-\cos x^2)/2 - (\sin^3 x)/3 + C$ **47.** 3 s; 40 ft/s

49. $y^2/2 + 2y = x^2/2 - 2x + C$

51a. $U_A = 12 + (1 + 4r)x - (r + 1)x^2$; maximized when $x = \dfrac{1 + 4r}{2(1 + r)}$

51b. $U_B = 10 + (3 - 2r)x - (1 + r)x^2$; $x = \dfrac{3 - 2r}{2(1 + r)}$

51c. $r = 33\frac{1}{2}\%$ **53d.** 7/8 bar; 7/8 bar

CHAPTER 4

EXERCISES 4.1, p. 238

1. 3.875 **3.** $\pi\sqrt{2}/4 \approx 1.11072$ **5.** 12 **7a.** 2.64 **7b.** 2.66 **9a.** 1.92 **9b.** 2.28 **11a.** 3.52 **11b.** 3.08

13. 12

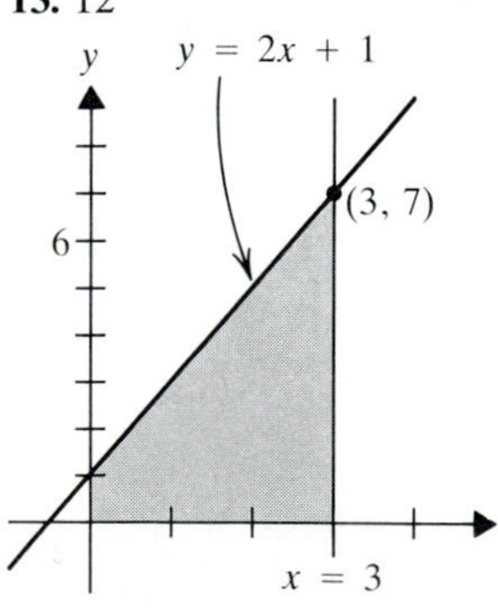

15. 8/3

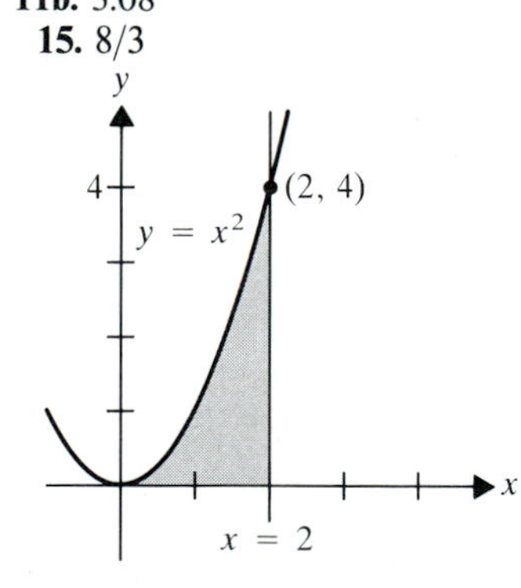

17. 4

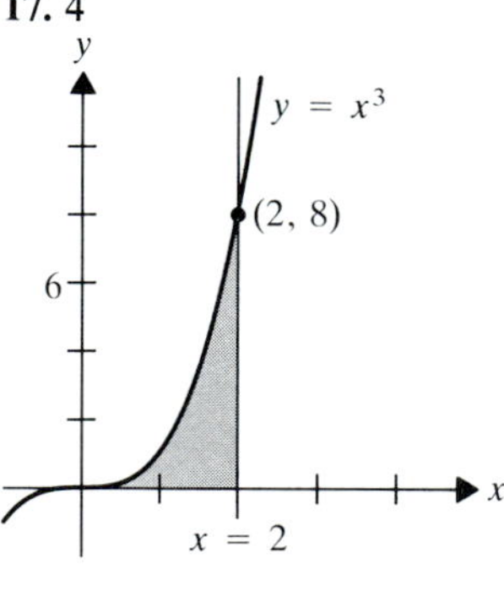

19. 14/3

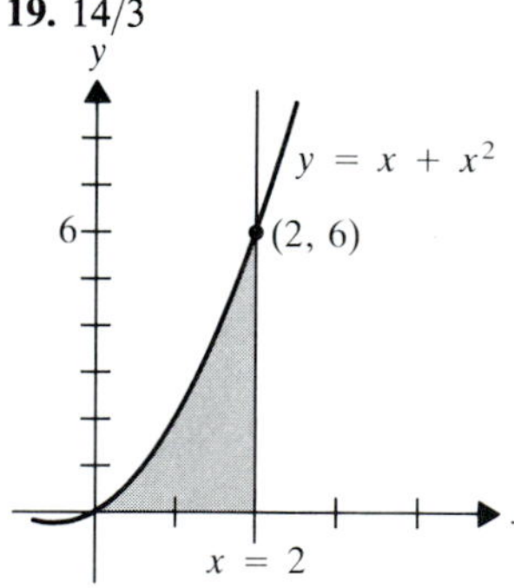

21. 18

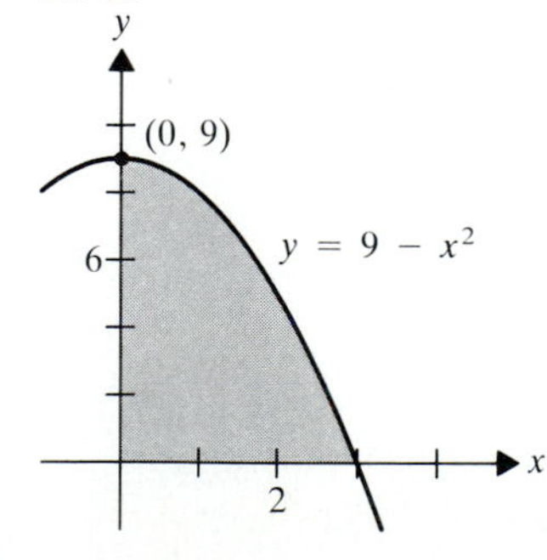

23a. 46 **23b.** 2 **23c.** 3 **23d.** $-43/30$ **25a.** 50/51 **25b.** $2^{101}-2$

37.

n	10	20	40	80	160	320
R_p	8.977500	8.994375	8.998594	8.999648	8.999912	8.999978

39.

n	10	20	40	80	160	320
R_p	2.008248	2.002058	2.000514	2.000129	2.000032	2.000008

41.

n	10	20	40	80	160	320
R_p	0.198336	0.199584	0.199896	0.199974	0.199993	0.199998

43.

n	10	20	40	80	160	320
R_p	2.660000	2.665000	2.666250	2.666563	2.666641	2.666660

45.

n	10	20	40	80	160	320
R_p	0.692850	0.693069	0.693128	0.693142	0.693146	0.693147

47.

n	10	20	40	80	160	320
R_p	1.147499	1.147720	1.147776	1.147789	1.147792	1.147793

21. $F'(x)=\sqrt{x^2-x+1}$ **23.** $\dfrac{1}{\sin^2 x + x^{3/2}}$

EXERCISES 4.2, p. 245

1. $\Delta x_1 = 0.26$; $\Delta x_2 = 0.24$; $\Delta x_3 = 0.25$; $\Delta x_4 = 0.25$; $\Delta x_5 = 0.25$; $|P| = 0.26$
3. $\Delta x_1 = 1/6$; $\Delta x_2 = 1/30$; $\Delta x_3 = 1/20$; $\Delta x_4 = 1/12$; $\Delta x_5 = 1/6$; $\Delta x_6 = 1/2$; $|P| = 1/2$
5. 1/8 **7a.** 0.08 **7b.** -0.28 **9a.** -1.52 **9b.** -1.08
11a. -0.64 **11b.** -0.66 **13.** $\int_0^1 \sqrt{1-x^2}\,dx$
15. $\int_0^2 (x^2+\sin x\cos x)\,dx$ **17.** 9 **19.** $9\pi/4$ **21.** 10/3
23. 28/3 **25.** 68/3 **27.** $\int_{-2}^3 f(x)\,dx$ **29.** $\int_1^5 f(x)\,dx$
31. $\int_{x_0}^{x_0+h} f(x)\,dx$ **33.** 1 **35.** 5/2

EXERCISES 4.3, p. 252

1a. 4 **1b.** $c = 3/2$ **3a.** 5/3 **3b.** $c = \pm 1/\sqrt{3}$
5a. 2 **5b.** $c = \sqrt[3]{2}$ **7a.** 6 **7b.** $c = \sqrt{3}$
9a. 0 **9b.** $c = 0, 1, -1$
11a. $\pi/2$ **11b.** $c = \sqrt{4 - \dfrac{\pi^2}{4}} \approx 1.238$
13. $0 \le \int_0^{\pi/4} \sin x\,dx \le \dfrac{\pi\sqrt{2}}{8}$
15. $\dfrac{\pi}{2} \le \int_0^{\pi/2} (\sin x + \cos x)\,dx \le \dfrac{\pi\sqrt{2}}{2}$
17. $4 \le \int_0^2 (x^3 - 3x + 4)\,dx \le 12$
19. $-506 \le \int \left(x^4 - \dfrac{16x^3}{4} + 1\right)dx \le 44$ **21.** $-g$

EXERCISES 4.4, p. 260

1. 11/6 **3.** 16/3 **5.** 28/3 **7.** 2 **9.** -18 **11.** 0 **13.** 8/3
15. $-1/2$ **17.** $\sqrt{2}-1$ **19.** $-1/\sqrt{3}+\sqrt{3}-\pi/6$
25. $\sqrt{1+(x^2+x)^3}(2x+1)$ **27.** $-(\sin^3 x + x^2)$
29. $-2x\sqrt{1+x^4}+\sqrt{1+(1+x)^2}$
31. $2\sec 2x\tan 2x(\sec^2 2x - 3) - \dfrac{x(x^2+2)}{\sqrt{1-x^2}}$
33. 20/3 **35.** 1

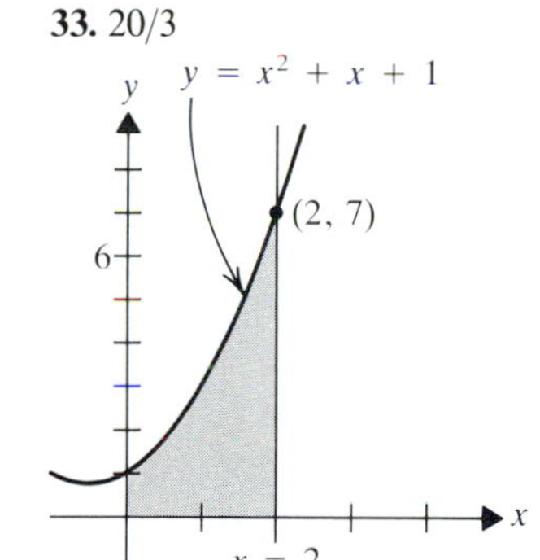

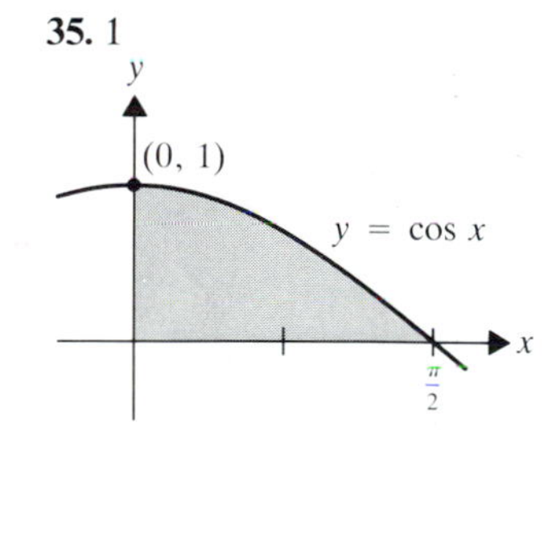

37. $-11/3$ **39.** $\pi/4 + 2/\pi$ **39.** $\pi/4 + 2/\pi$ **43.** $2g$
47. $f(x) = \cos x$; $a = \pi/6$

EXERCISES 4.5, p. 265

1. 14/9 **3.** 78 **5.** $-\sqrt{2}/2$ **7.** 1/8 **9.** 0 **11.** 1/3
13. 8/225 **15.** 3/8 **17.** $(1/3)[\sec(\pi^3/64) - 1]$ **19.** $2-\sqrt{2}$
21. 1 **23.** 2/3

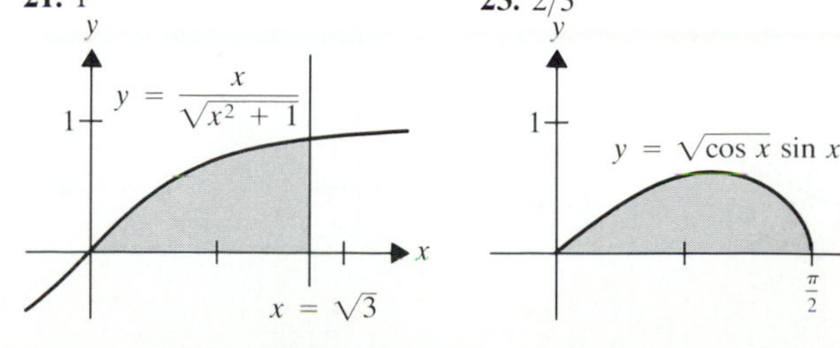

25. 1/6 **27.** $3/4\pi$ **29.** $1500 - 12/\pi$ **31.** $6 - (5^{2/3})/3 \approx 2.273$
33a. 14/5 **33b.** 0.02 L/min **35.** $(8 - 2\sqrt{3})/13$

EXERCISES 4.6, p. 274

1a. 1.0736 **1b.** 1.1564 **1c.** 1.1097 **3a.** 0.9506 **3b.** 1.3433
3c. 1.2131 **5a.** 1.1150 **5b.** 1.1128 **7a.** 0.6956 **7b.** 0.6938
9a. 1.0941 **9b.** 1.0906 **11a.** 1.11136 **11b.** 1.11144
13a. 0.69325 **13b.** 0.69315 **15a.** 1.08941 **15b.** 1.08943
17. 0.54931 **19.** 1.828 **21.** 3.3783 **23.** 3.739 miles
25a. ≤ 0.0125 **25b.** ≤ 0.0489 **27a.** ≤ 0.0034 **27b.** ≤ 0.00084
29a. ≤ 0.00014 **29b.** ≤ 0.00009 **33.** 6 **35.** 19

37.

n	10	20	40	80	160	320
$T_n(f)$	1.185197	1.193564	1.196522	1.197568	1.197938	1.198069
$S_n(f)$	1.193089	1.196354	1.197508	1.197917	1.198061	1.198112

39.

n	10	20	40	80	160	320
$T_n(f)$	1.002052	1.000514	1.000128	1.000032	1.000008	1.000002
$S_n(f)$	1.000016	1.000001	1.000000	1.000000	1.000000	1.000000

REVIEW EXERCISES 4.7, p. 278

1. $-23/8$ **3.** 5.92 **5.** 4/3 **7.** 1/4 **9a.** 165 **9b.** $2n$

9c. $\sum_{i=1}^{16} (2i - 1)$ **11.** $\int_a^b \frac{1 - x^2}{1 + x^2}\,dx$

13a. $\int_{-3}^{5} f(x)\,dx$ **13b.** $\int_{-2}^{1} f(x)\,dx$ **15.** $-\pi^2, \pi^2$ **17.** $9/4\pi$

21. 106/15 **23.** 0 **25.** 1/2 **27.** $\sqrt{97} - 4$ **29.** $\sqrt{\sin^2 x^2 + 2x^{3/2}}$

31. $[(1 - x^2)^{3/2} - 2\sqrt{1 - x^2} + 1] \cdot \frac{-x}{\sqrt{1 - x^2}}$

33. \$333,933.33 **35a.** 1.4683 **35b.** 1.5319 **35c.** 1.4957
37. 1.5001 **39.** 1.4979 **41.** 1.4971 **43.** 1.4972 **45.** 19

CHAPTER 5

EXERCISES 5.1, p. 286

1. 9/2

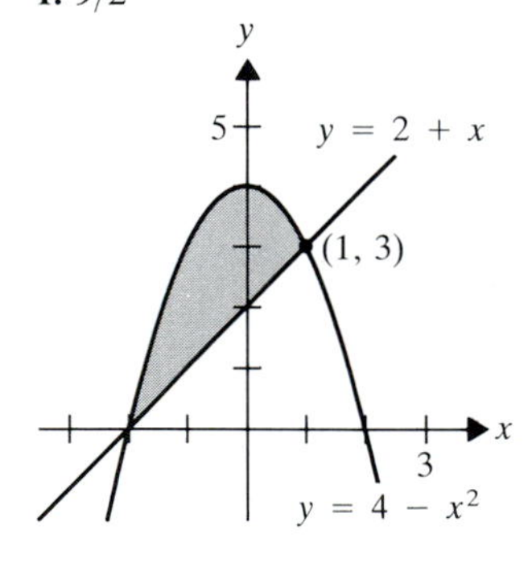

3. 32/3

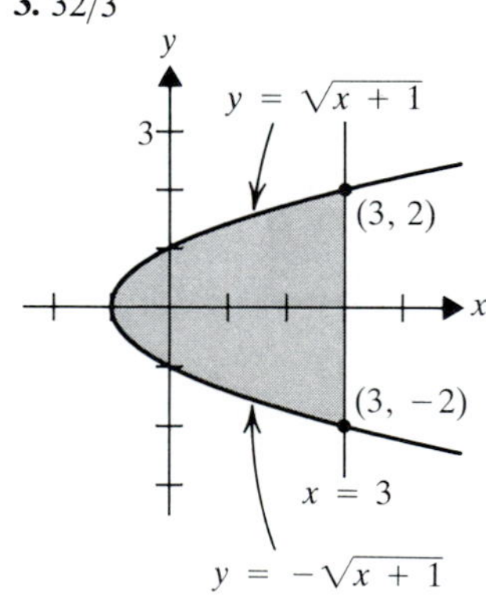

9. 1

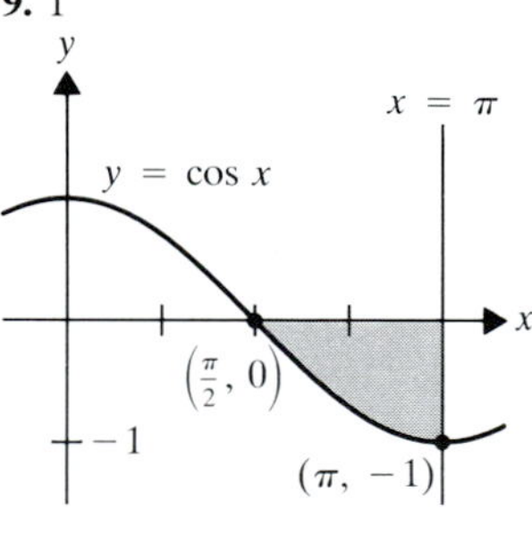

11. $\sqrt{3}$

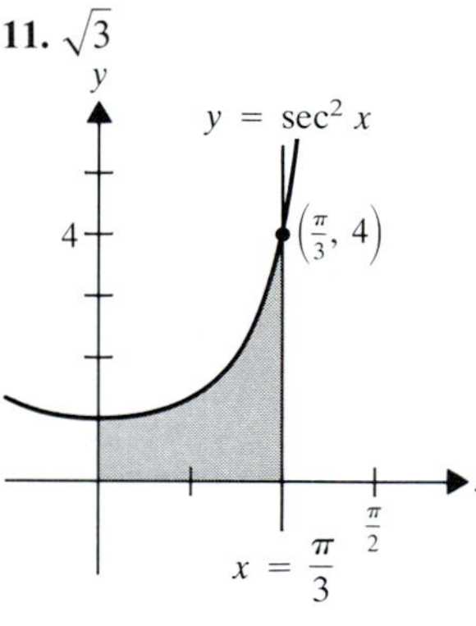

5. 1/3

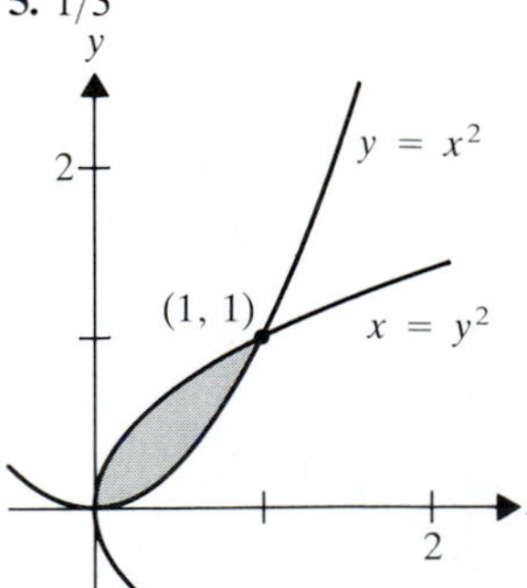

7. 1/12

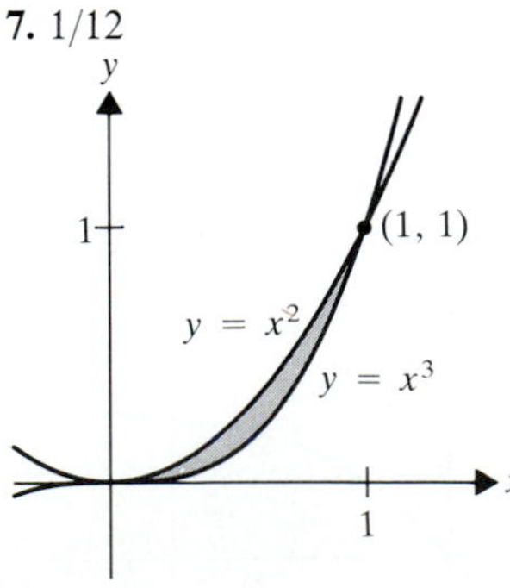

13. $2\sqrt{2}$

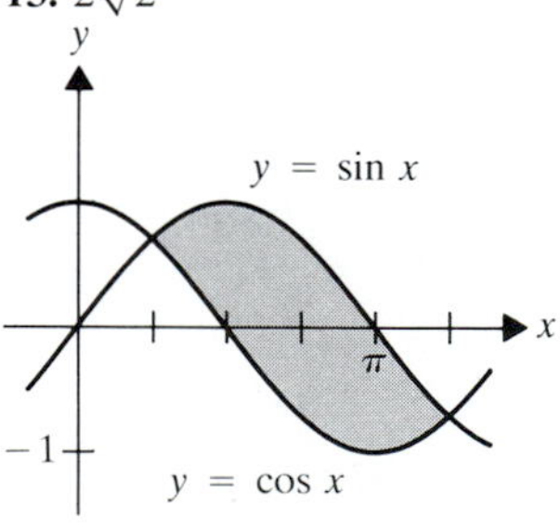

15. 18

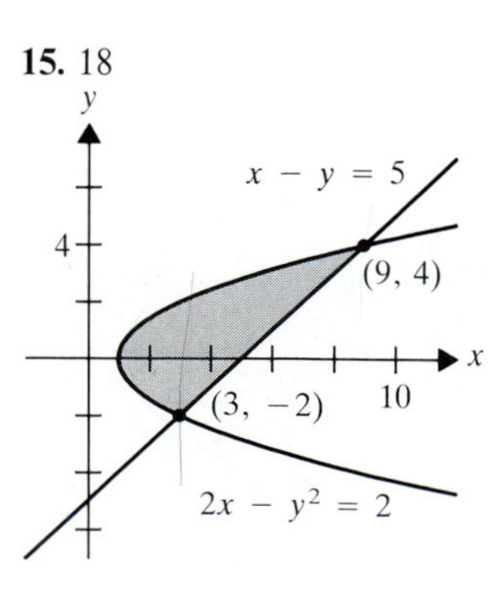

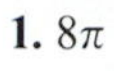

17. 13/6

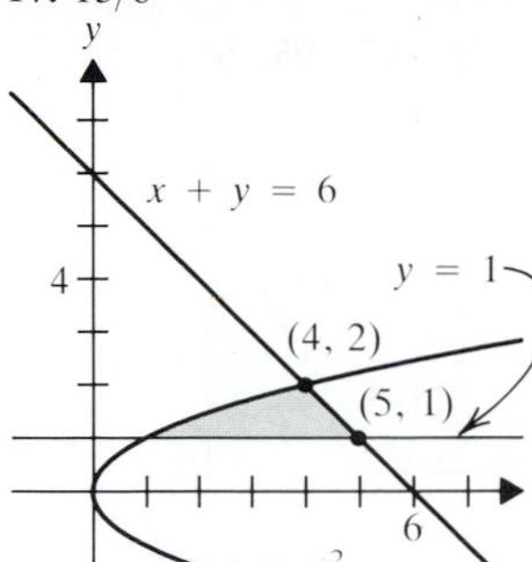

19. 9

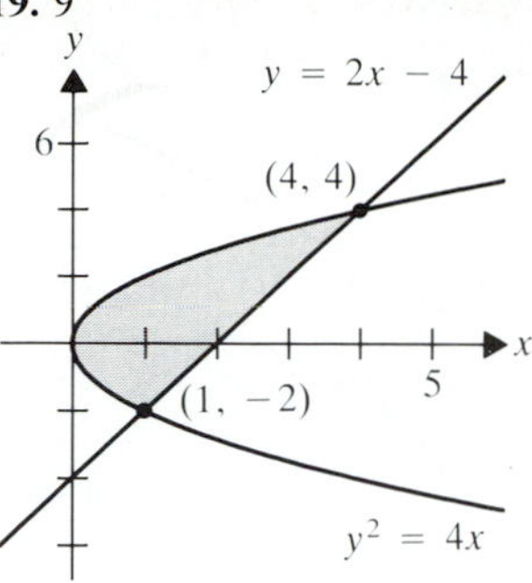

21. 32/3

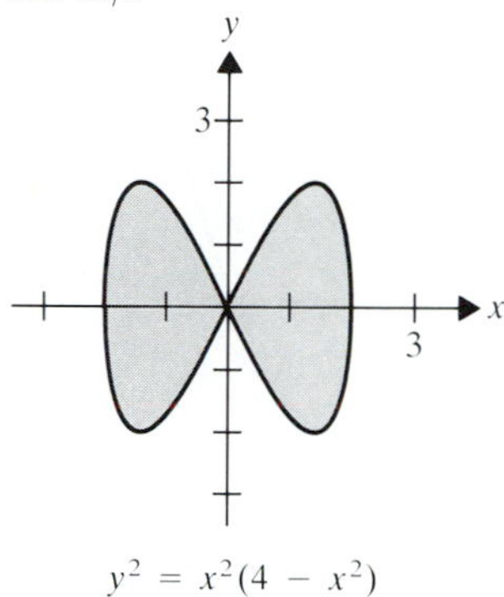

23. 8/3

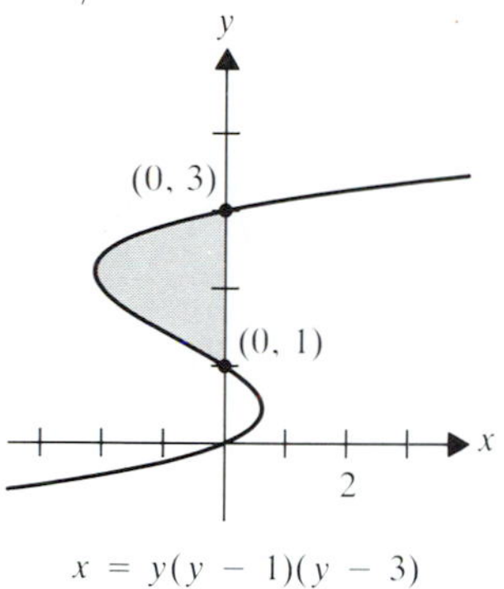

25. 9

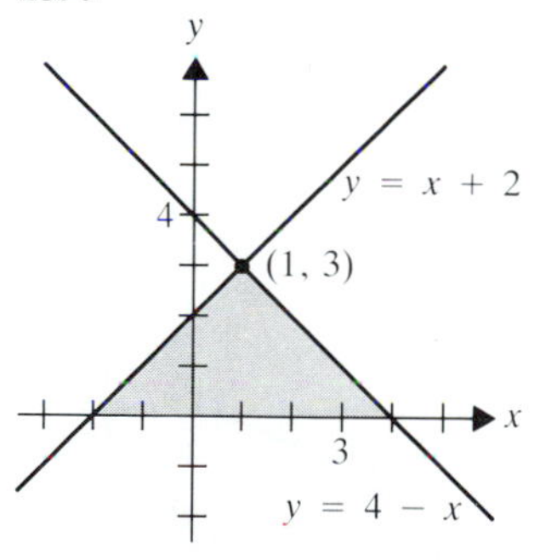

27. 18

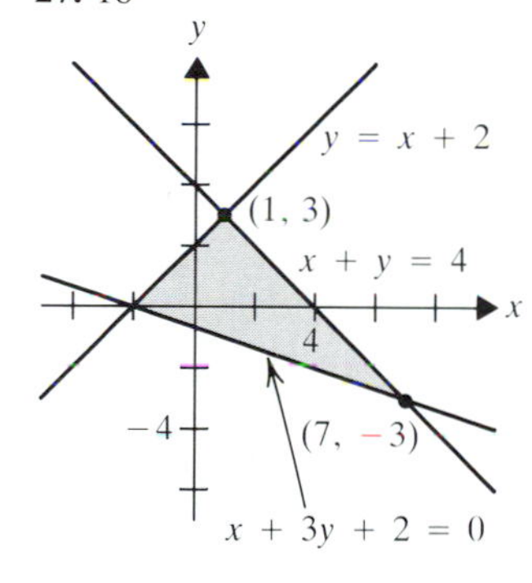

29. 12

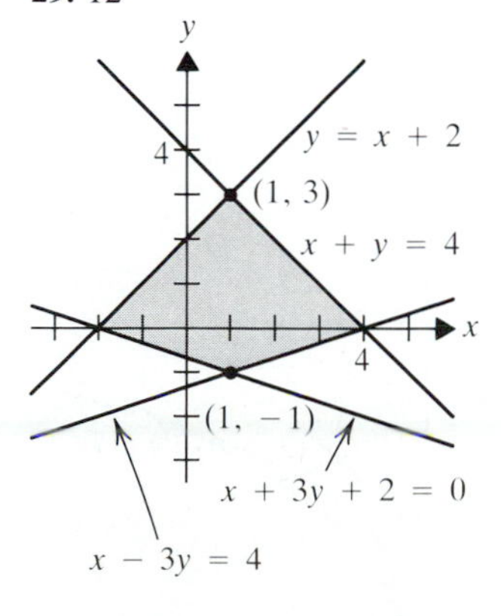

31. 13/6

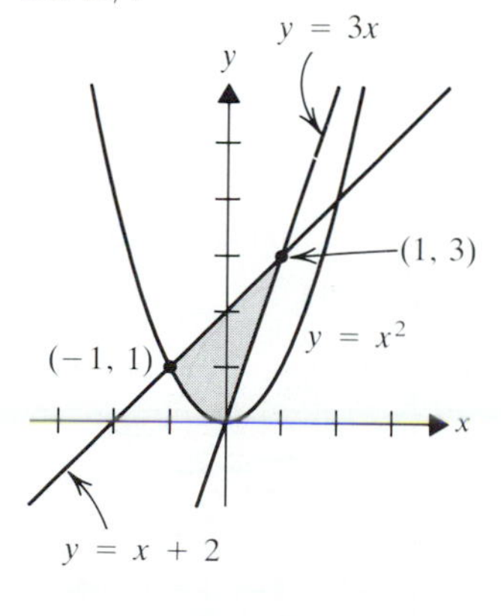

33. \$33,333.33 **35.** \$6666.67 **37.** No

EXERCISES 5.2, p. 295

1. 8π

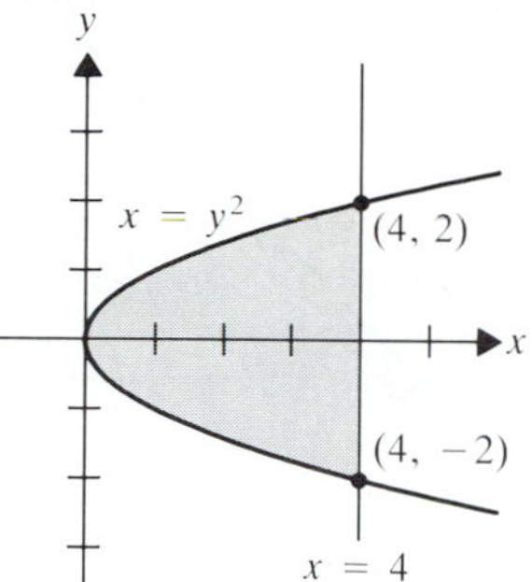

3. π

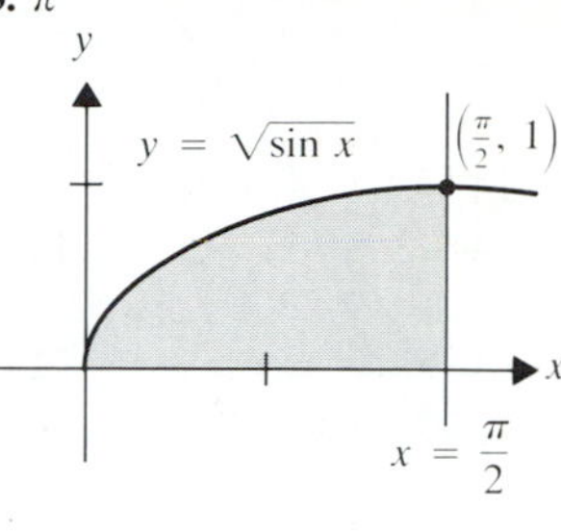

5. $256\pi/5$

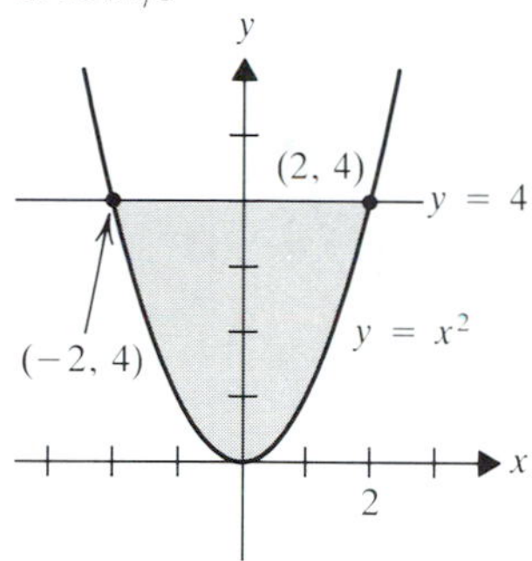

7. $206\pi/15$

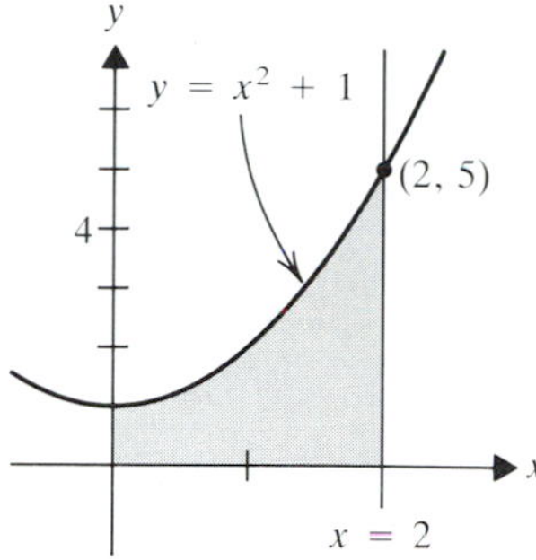

9. $7\pi/3$

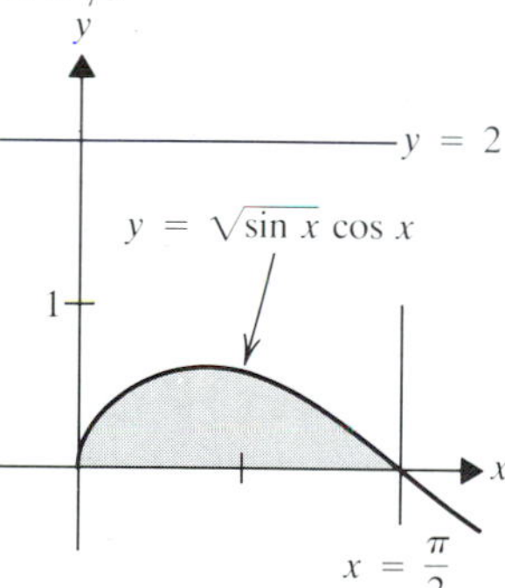

11. π

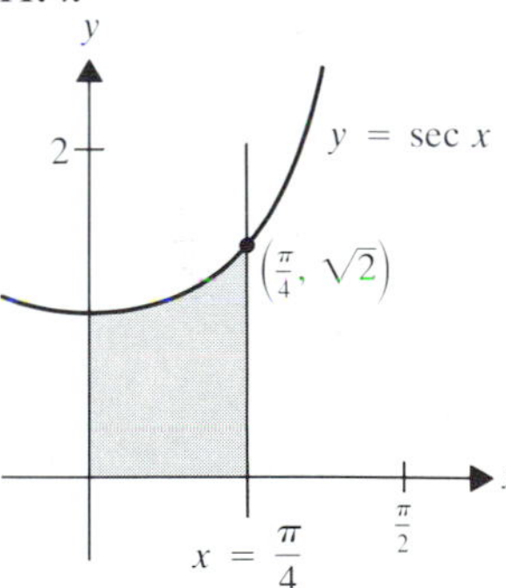

13. $\pi/6$

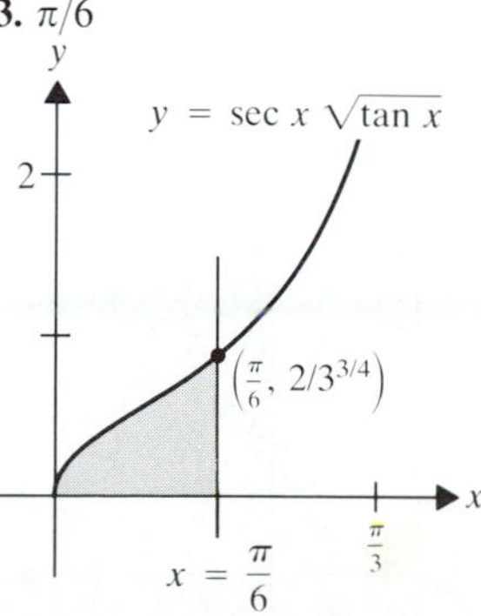

15. $\pi/4$

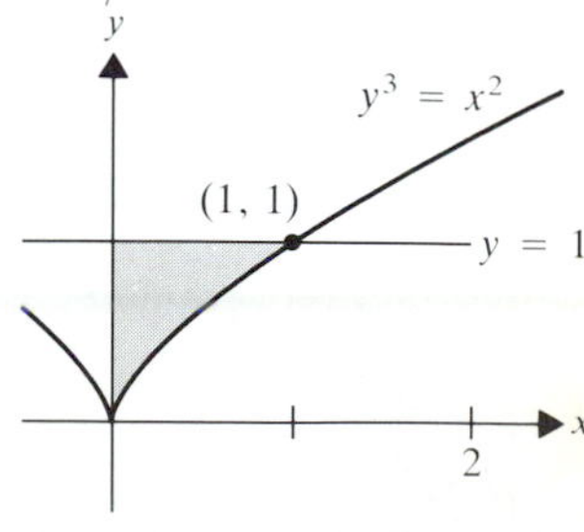

17. $64\pi/15$

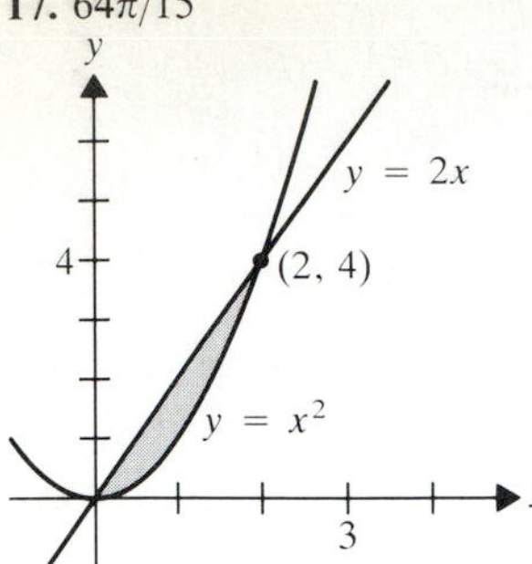

19. $4\pi/15$

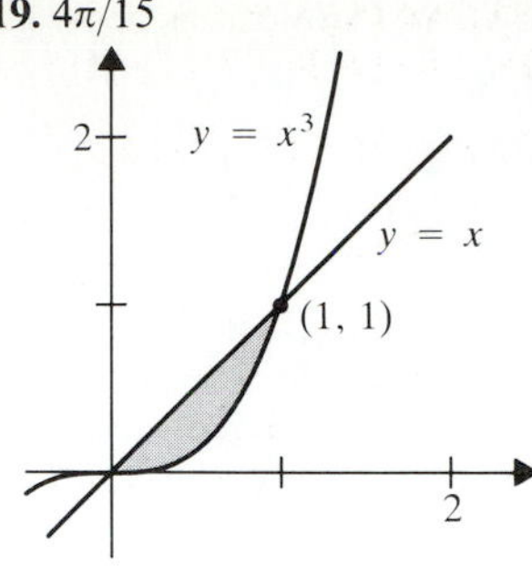

9. $124\pi/5$

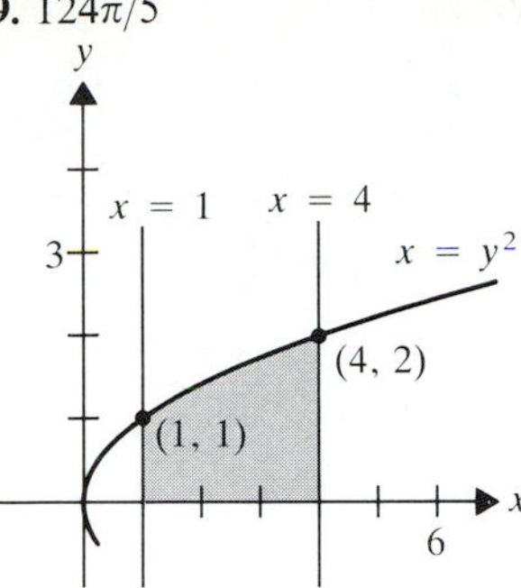

11. $232\pi/5$

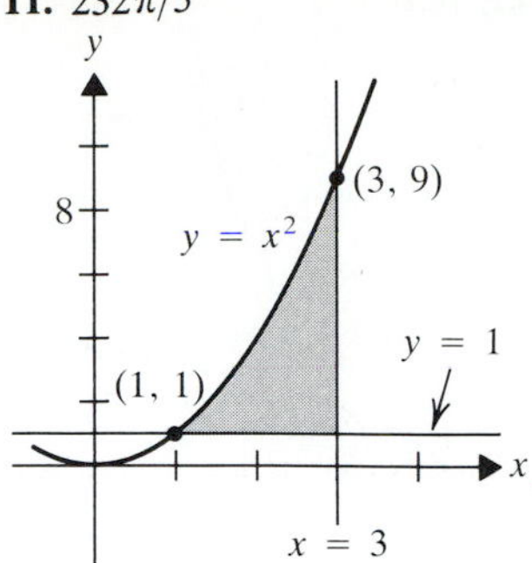

21. $256\pi/5$
(Figure same as for Exercise 1.)

23. 24π

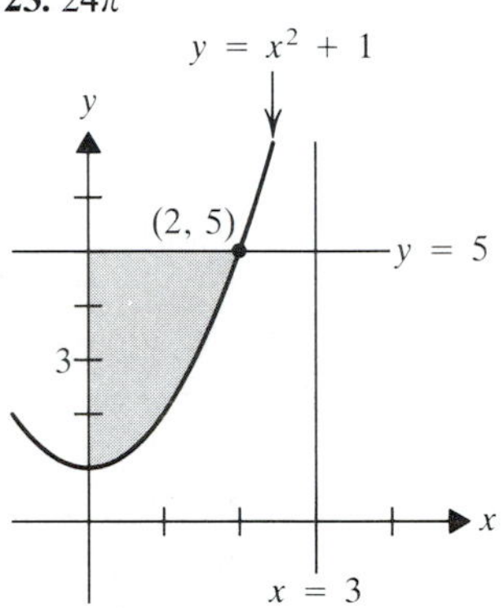

13. 24π

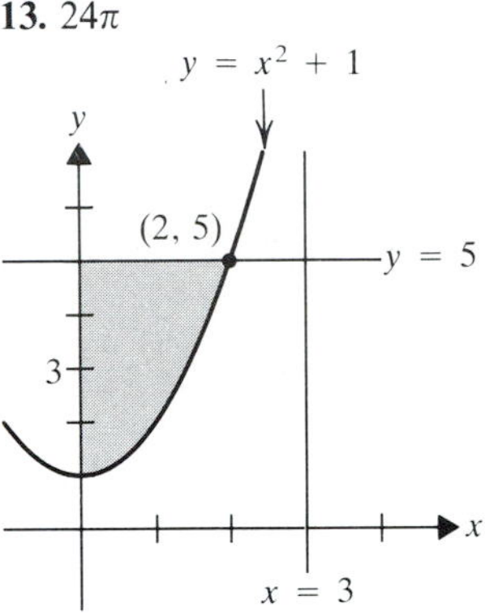

15. $\pi/6$

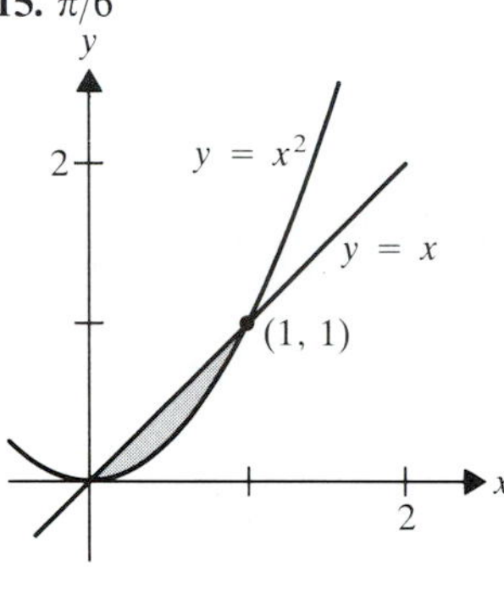

25. $(\pi a^2/h)/3$ **27.** $(4\pi a^3)/3$ **29.** $4(\pi a^2 b)/3$ **31.** $4\pi a^3\sqrt{3}$
33. $a^2h/3$ **35.** $128/3$ **37.** $9\sqrt{3}/2$ **39.** $16\sqrt{3}/9$
41. $\pi h(a^2 + ab + b^2)/3$

EXERCISES 5.3, p. 303

1. 8π

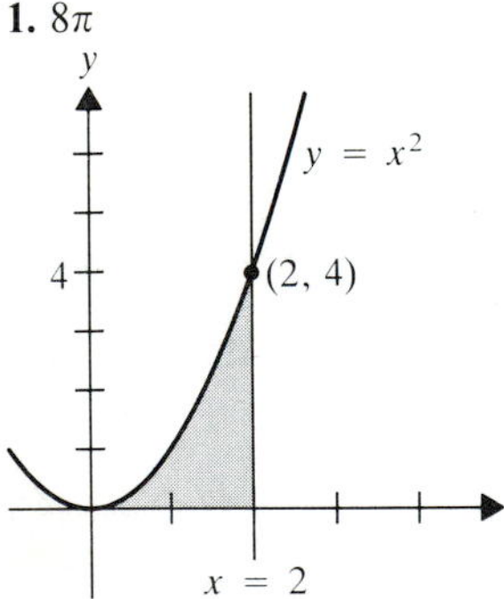

3. 2π

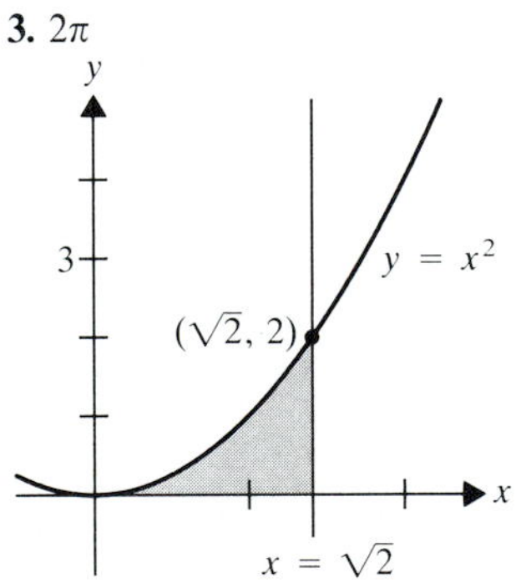

17. $3\pi/5$

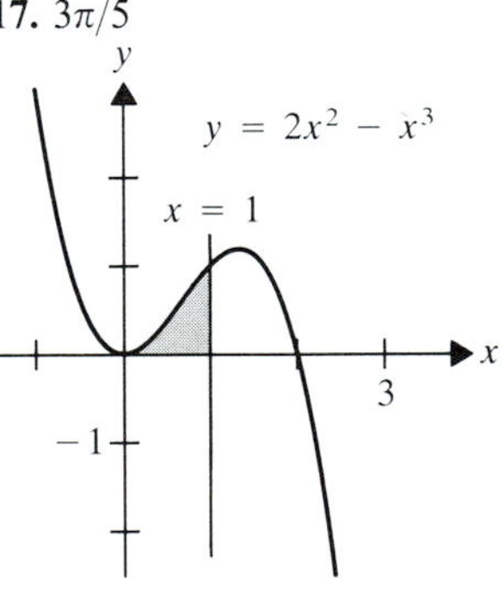

19. $232\pi/5$

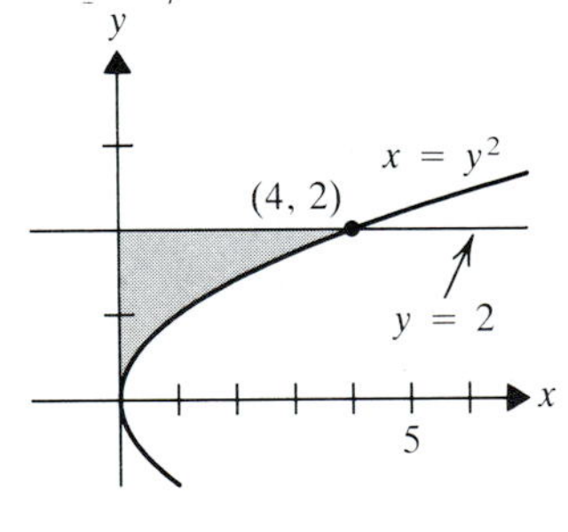

5. π

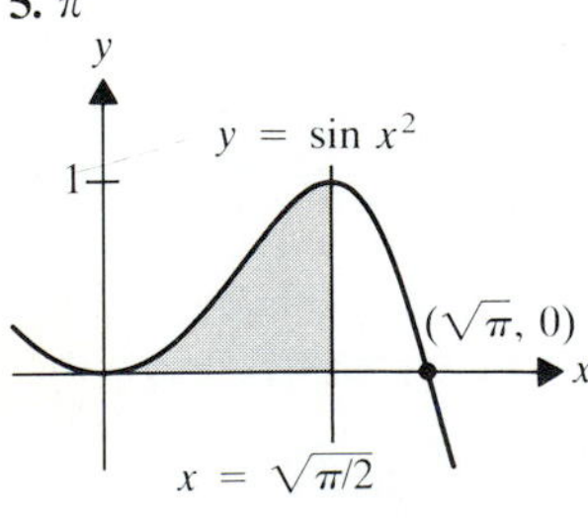

7. $2\pi/3$

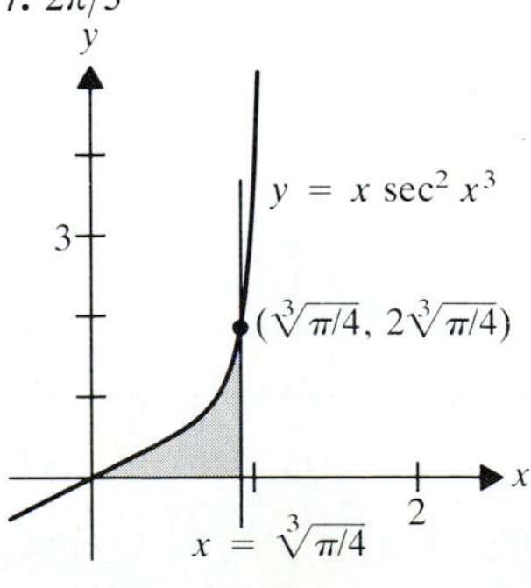

21. $\pi(\sqrt{3} - \pi/3) \approx 2.1515$

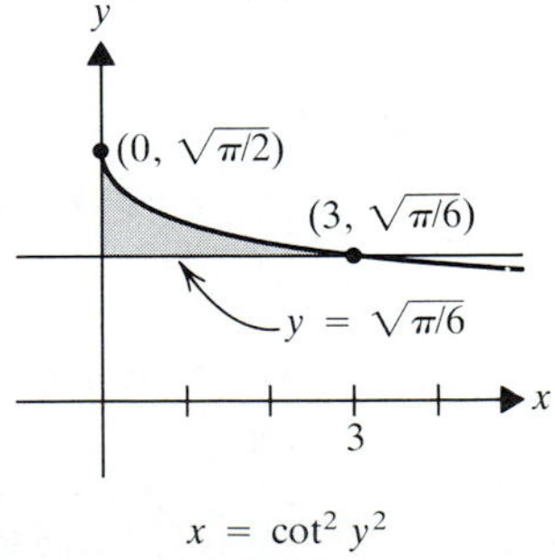

23. $\pi/10$

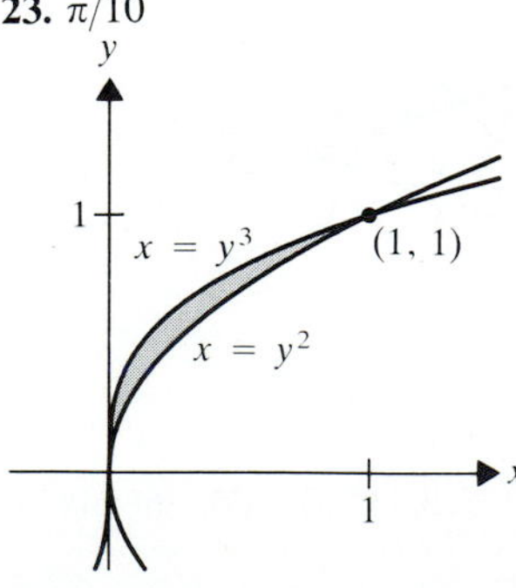

25. $32\pi/3$

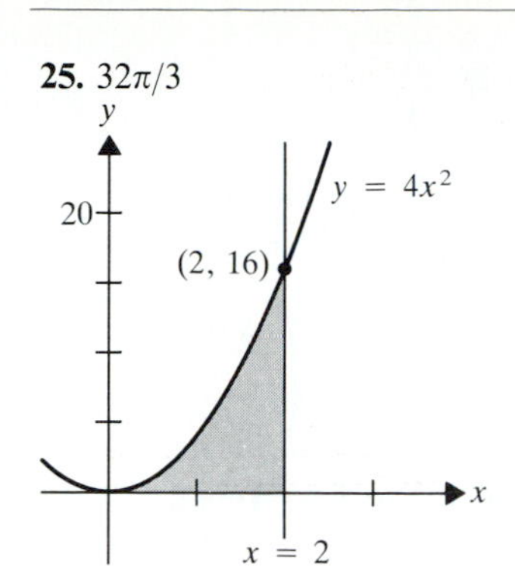

27. 64π

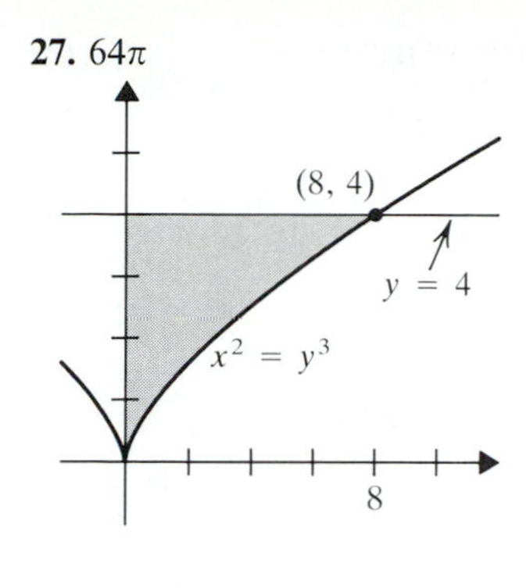

29. $20\pi\sqrt{15}$ **31.** $\frac{4\pi}{3}(a^2 - b^2)^{3/2}$ **33.** $\frac{\pi a^2 h}{3}$ **35.** $4\pi a^3/3$

37. $4\pi a^3\sqrt{3}$

EXERCISES 5.4, p. 309

1. 56/27 **3.** 4/3 **5.** 169/24 **7.** 123/32 **9.** 31/6
11. $10\sqrt{10} - 2\sqrt{2}$ **13.** 80/3 **15.** 22/3 **17.** 3/2 **19.** 2
21. 4/3
23a.

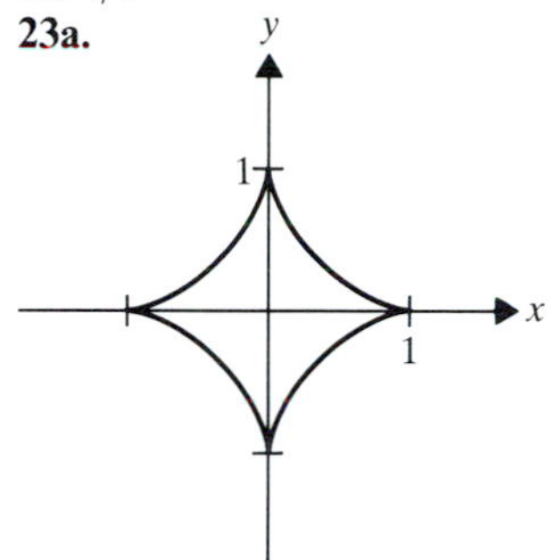

$x^{2/3} + y^{2/3} = a^{2/3}$

23b. 9/8

25. No, because dy/dx fails to exist at $x = 0$.
29. 1.4789424 **31.** 1.9100989 **33.** 1.277978 **37.** No

EXERCISES 5.5, p. 320

1. $\bar{x} = 4$ **3.** (4, 3/5) **5.** (−2, 1) **7.** $\bar{x} = 3$ **9.** $m = 1/3$; $\bar{x} = 3/4$
11. $m = 18$; $\bar{x} = 27/5$ **13.** $m = 3/2$; $\bar{x} = 4/9$ **15.** $m = \pi/2$; $\bar{x} = 0$
17. $m = \ln 3$; $\bar{x} = 2/\ln 3$ **19.** $(\bar{x}, \bar{y}) = (1/3, 1/3)$
21. $m = 2\pi$; $(\bar{x}, \bar{y}) = (0, 8/3\pi)$ **23.** $m = 1/3$; $(\bar{x}, \bar{y}) = (3/4, 3/10)$
25. $m = 16/3$; $(\bar{x}, \bar{y}) = (3/4, 8/5)$ **27.** $m = 9/2$; $(\bar{x}, \bar{y}) = (3/2, 12/5)$
29. $m = 5/12$; $(\bar{x}, \bar{y}) = (12/25, 3/7)$ **31.** $(\bar{x}, \bar{y}) = (1/2, 1/2)$
33. $\bar{x} = \bar{y} = 28/9\pi$
43. 40/7 miles from the center of the smaller; 30/7 miles from the center of the larger.
45. $\bar{x} = 0$; $\bar{y} = 6/\pi$ **47.** $\delta(x) = x^2$ will do.

EXERCISES 5.6, p. 330

1. 70 ft-lbs. **3.** 8.1 ergs **5.** −105/4 ergs **7.** 10 joules
9. −5/2 joules **11.** 1.207×10^9 J **13.** 5.151×10^8 ft-lbs.
15. 6.667×10^{-9} ergs **17.** 300.95 ergs **19.** $28{,}800\pi w$
21. $50{,}400\pi$ J **23.** $\pi a^4 w/4$ **25.** $2000\pi w$
27. 6.24×10^7 lbs. **29.** $4w/3$ N **31.** $125\pi w$ **33.** 1920 lbs.
35. 18,720 lbs. **39a.** $F_b = w(D + h)A$ **39b.** $F_t = wDA$
39c. whA **43.** $a = 6$

REVIEW EXERCISES 5.7, p. 332

1. $16\sqrt{3}/9$ **3.** $4a^3/3$ **5.** 17/6 **7.** $4\pi/15$ **9.** $8\pi/3$ **11.** $32\pi/3$
13. $9\pi/2$ **15.** 8π **17.** $2\pi(1 - \sqrt{3}/2)$ **19.** $24\pi^2$ **21.** 52/3
23. $10\sqrt{10} - 2\sqrt{2}$ **25.** 4 **27.** $m = 36$, $\bar{x} = 63/16$
29. $m = 2\pi$, $\bar{x} = 0$ **31.** $m = 32\delta/3$; $\bar{x} = 0$; $\bar{y} = 8/5$
33. $m = \delta/12$; $\bar{x} = 3/5$; $\bar{y} = 12/35$
35. $m = 58/12$; $\bar{x} = 12/25$; $\bar{y} = 3/7$ **37.** −1/120 joules
39. 444,444 mile-lbs. **41.** $625\pi w \approx 122$ ft/lbs.
43. $608\pi w$ joules **45.** $1.125w \times 10^6 w$
47. $F = 18w \approx 1123$ lbs.

CHAPTER 6

EXERCISES 6.1, p. 345

1. $\frac{2}{2x + 5}$ **3.** $\frac{1}{2x + 5}$ **5.** $\frac{3x^2 - 8x}{3(x^3 - 4x^2 + 6)}$ **7.** $2(\cot x - \tan x)$

9. $-\frac{\sin(\ln(x + 3))}{x + 3}$ **11.** $\frac{1}{x \ln x}$ **13.** $\frac{1 - 4\ln x}{x^5}$ **15.** $\frac{2}{x} + \frac{2}{x}\ln x$

17. $12x\left(\frac{1}{2x^2 + 1} + \frac{x}{x^3 + 1}\right)$ **19.** $\frac{2}{3(x + 1)(x - 1)}$

21. $\frac{2x}{3x^2 - 15} - \frac{16x(3x + 1)}{2x^3 + x^2 + 1}$

23. $xg(x)\left(\frac{1}{x^2 - 1} + \frac{1}{x^2 + 2} + \frac{1}{x^2 + 3}\right)$

25. $g(x)[\cot x - \tan x]$ **27.** $(1/2)\ln|2x + 1| + C$
29. $(1/3)\ln|x^3 + 3x - 2| + C$ **31.** $(1/2)\ln(1 + \sin^2 x) + C$
33. $-\cot(\ln x) - \ln x + C$ **35.** $(\ln x)^4/4 + C$
37. $-1/2(\ln|x|)^2 + C$ **39.** $(1/2)\ln 6$ **41.** 1 **43.** $4\ln 2$
45. $2\pi\ln 2$ **47.** $(3\ln 2)/\pi$
49. $1/x$; $-1/x^2$; $2/x^3$; $(-1)^{n-1}(n - 1)!/x^n$ **51a.** 1.098661
51b. $E_S \le 0.00043$

51c.

n	20	40	80	160	320
S_n	1.0986155	1.0986125	1.0986123	1.0986123	1.0986123

51d. $E_S \le 0.0000001$ **53a.** 1.9485192 **53b.** $E_S \le 0.10368$

53c.

n	20	40	80	160	320
S_n	1.9461369	1.9459262	1.9459112	1.9459102	1.9459102

53d. $E_S \le 0.0000253$ **55a.** -0.69315 **55b.** 1.38629 **55c.** 1.79176 **55d.** 2.30259 **55e.** 2.07944 **57a.** 24 **57b.** 72

61. $\dfrac{C}{V_1 - V_0} \ln \dfrac{V_1}{V_0}$

EXERCISES 6.2, p. 352

1. $(\ln 5)/2$ **3.** $4, -1$ **5.** $2/\ln 5$ **7.** $4, -1$ **9.** $3e^2$ **11.** $x_0 e^{kt}$

13. $-2e^{-2x+1}$ **15.** $2xe^{x^2+1}$ **17.** $(\cos x)e^{\sin x}$ **19.** $\dfrac{e^{\sqrt{x+2}}}{2\sqrt{x+2}}$

21. $e^x \sin 2x + 2e^x \cos 2x$ **23.** $2xe^{x^2} \sec e^{x^2} \tan e^{x^2}$

25. $2xe^{x^2+1}(x^2+1)$ **27.** $e^{-x^2}[1 - 2x - 2x^2]$ **29.** $\dfrac{4}{[e^x + e^{-x}]^2}$

31. $\dfrac{e^x - e^{-x}}{e^x + e^{-x}}$ **33.** $2x$ **35.** $\dfrac{e^{2x-1}}{2} + C$ **37.** $\dfrac{e^{x^2+5}}{2} + C$

39. $2e^{\sqrt{x+1}} + C$ **41.** $e^x - e^{-x} + C$ **43.** $\ln(e^x + e^{-x}) + C$

45. $\dfrac{e^{\tan 2x}}{2} + C$ **47.** $\frac{2}{3}(e - e^{1/2})$ **49.** $\ln[e^2 - e^{-2}] - \ln[e - e^{-1}] = \ln[e + e^{-1}]$

51. $1 - e^{-1}$ **53.** $\pi(e-1)$ **55.** $(e - e^{-1})/2$

59. $250(1 - e^{-18})/9 \approx 250/9$

EXERCISES 6.3, p. 360

1. 12.4244% **3a.** 5.38986% **3b.** 5.39026% **5a.** 17.935% **5b.** 17.939% **5c.** \$1510.59 **5d.** \$1510.11 **7a.** 6.93 years **7b.** 4.62 years **9.** 2.9×10^{-2} gm; 0.43 hours **11.** 9954 years ago **13.** 29% **15.** 9:42 PM **17.** 7.4 min **19.** 1,118,000 **21.** Every four hours **23.** 6.4 billion; 16.4 billion; in the year 2073

25. $t_{1/2} = \dfrac{3}{2k(x_0)^2}$ **27.** After $9\frac{1}{2}$ weeks **31a.** $\Delta V = A\,\Delta x$

31b. $\Delta F = P\,\Delta v/\Delta x \approx P\,dv/dx$, where P is the pressure at some point during compression **31c.** $nRT\ln(V_2/V_1)$ **31d.** negative **35.** 12.607% **37a.** 5.46673% **37b.** 5.46713%

EXERCISES 6.4, p. 367

1. $(3\ln 2)2^{3x}$ **3.** $[(1-2x)\ln 10]\,10^{-x^2+x-1}$

5, 7. $\dfrac{(2x-3)}{(\ln 10)(x^2 - 3x + 1)}$

9. $-2[x^{-3}10^{2/(x-1)} + (x^{-2} - 1)(x-1)^{-2}10^{2/(x-1)}\ln 10]$

11. $(-3\ln 10)(\sin 3x)(10^{\cos 3x})$ **13.** $(6\ln 2)(\tan 3x \sec^2 3x)2^{\tan^2 3x}$

15. $5(10^x + 2^{-x})^4(10^x \ln 10 - 2^{-x}\ln 2)$ **17.** $\dfrac{1}{x \ln x \ln 10}$

19. $(x+1)^{3x}\left(\dfrac{3x}{x+1} + 3\ln(x+1)\right)$ **21.** $x^{\sqrt{x}-1/2}(1 + \ln\sqrt{x})$

23. $\dfrac{21x}{(x^2+1)(x^2+8)\ln 10}$ **25.** $x^{\ln x}\left(\dfrac{2}{x}\ln x\right)$

27. $x^{\tan x}\left((\sec^2 x)(\ln x) + \dfrac{\tan x}{x}\right)$ **29.** $\dfrac{10^{2x-1}}{2\ln 10} + C$

31. $-\dfrac{10^{-3x^2+5}}{6\ln 10} + C$ **33.** $\dfrac{\ln(10^x + 1)}{\ln 10} + C$

35. $\ln|\log_2 x| \cdot \ln 2 + C$ **37.** $\dfrac{2^{\tan^2 x}}{2\ln 2} + C$ **39.** $\dfrac{\ln(5/4)}{\ln 2}$

41. $\dfrac{127}{384\ln 2}$ **47.** $10\log 2 \approx 3.01$ **49.** $10^{1.9} \approx 79.5$ times

EXERCISES 6.5, p. 377

1a. $\pi/4$ **1b.** $-\pi/6$ **1c.** $-\pi/2$ **1d.** 0 **1e.** $\pi/4$ **1f.** $\pi/6$ **3a.** $-\pi/3$ **3b.** $-\pi/4$ **3c.** $\pi/6$ **3d.** 0 **3e.** $-\pi/4$ **3f.** $-\pi/6$ **5a.** $1/(1-x^2)^{1/2}$ **5b.** $(1-x^2)^{1/2}/x$ **5c.** $(1-x^2)^{1/2}/x$ **5d.** $(1+x^2)^{-1/2}$ **7.** $1/2(-x^2-x)^{1/2}$ **9.** $-2x^{-1}(x^4-1)^{-1/2}$ **11.** $(x + x(\ln x)^2)^{-1}$ **13.** $(\cot x)/(\sin^2 x - 1)^{1/2}$

15. $(x\sqrt{x} + \sqrt{x})^{-1}$ **17.** $\dfrac{x\cos^{-1}x - \sqrt{1-x^2}}{(1-x^2)^{3/2}}$

19. $\dfrac{\exp\sqrt{\sin^{-1}x}}{2\sqrt{\sin^{-1}x}\sqrt{1-x^2}}$

21. $[\tan^{-1}x]^{\tan x}\left[\sec^2 x \ln(\tan^{-1}x) + \dfrac{\tan x}{(1+x^2)\tan^{-1}x}\right]$

23. $\pi/18$ **25.** $\sin^{-1}e^x + C$ **27.** $\arctan e^x + C$ **29.** $\pi/8$ **31.** $\sec^{-1}2x + C$ **33.** $(\sec^{-1}x^2)/2 + C$ **35.** $\sin^{-1}(x/a) + C$ **37.** $\pi/8$ **39.** $(1/3)\tan^{-1}[(x-2)/3] + C$

41. $(1/2)\ln(x^2 - 4x + 13) + (2/3)\tan^{-1}[(x-2)/3] + C$ **43.** $\pi/18$ **45.** $\sec^{-1}(x+1) + C$ **47.** $-1/(1+x^2) = D_x \cot^{-1}x$ **53.** $\sqrt{10}$ feet **55.** $x = 3/2$

EXERCISES 6.6, p. 389

1. $2\cosh(2x+5)$ **3.** $\dfrac{x}{\sqrt{x^2+4}}\operatorname{sech}^2\sqrt{x^2+4}$

5. $-2xe^{x^2}\operatorname{sech} e^{x^2}\tanh e^{x^2}$ **7.** $(1/x)\cosh(\ln x)$ **9.** $6x\cosh^2(x^2-2)\sinh(x^2-2)$ **11.** $2x\operatorname{sech} x$ **13.** 0 **15.** $(\sinh x^2)/2 + C$ **17.** $\ln|1 + \sinh x| + C$ **19.** $(2/3)\sinh^{3/2}x + C$ **21.** $(\sinh^5 x)/5 + (\sinh^3 x)/3 + C$ **23.** $-2\operatorname{csch}\sqrt{x+1} + C$ **25.** $(1/3)\sinh^{-1}(3x/5) + C$

27. $\dfrac{1}{4}\ln\left|\dfrac{x+2}{2-x}\right| + C$ **29.** $(1/8)\ln 7$ **31.** $\ln(1/2 + \sqrt{5}/2)$

33. $\ln\dfrac{9}{4+\sqrt{7}}$ **35.** See Figure 6.6, p. 383

43. $\sinh 1 = (e^2 - 1)/2e$ **45.** $(2 + \sinh 2)\pi/4$

47a. $(\cosh t \sinh t)/2 - \displaystyle\int_1^{\cosh t}\sqrt{x^2-1}\,dx$

EXERCISES 6.7, p. 397

3b. ωx **5a.** 1/3 **5b.** 0 **5c.** 8 rad/min **5d.** $\pi/24$ min **7.** $\cos 2t$ **9.** $(-\sqrt{2}/2)\cos(2t - \pi/4)$ **11.** $\omega = 100.0$, $\phi_0 = \tan^{-1}(1/600) \approx 0.001667$, $a = 12.00$ **13.** $\omega = 25.00$, $\phi_0 = \tan^{-1}(1/75) \approx 0.1333$, $a = 6.001$ **23.** $a\sinh t/a$

EXERCISES 6.8, p. 408

1. 1/2 **3.** 1/2 **5.** 1/2 **7.** $+\infty$ **9.** $+\infty$ **11.** 1 **13.** $+\infty$ **15.** 0 **17.** -2 **19.** 0 **21.** 1 **23.** 1 **25.** 0 **27.** 0 **29.** $-\infty$ **31.** 1 **33.** $1/e$ **35.** 1 **37.** 0 **39.** $+\infty$ **41.** 0 **43.** $(x-2)^2$ **45.** $(x-\pi/4)^2/2$ **47.** $e/2$ **49.** 1/2

EXERCISES 6.9, p. 416

3.

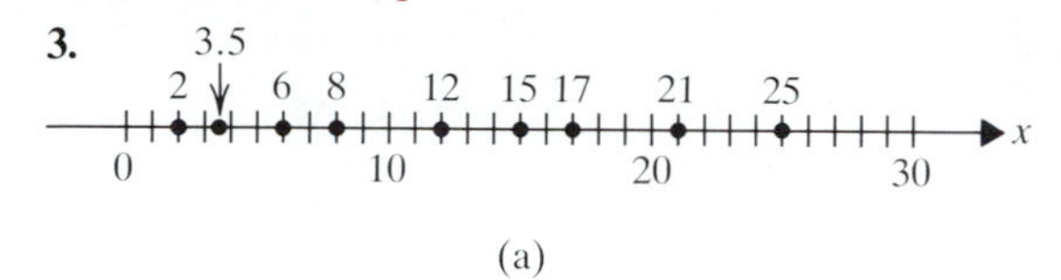

(a)

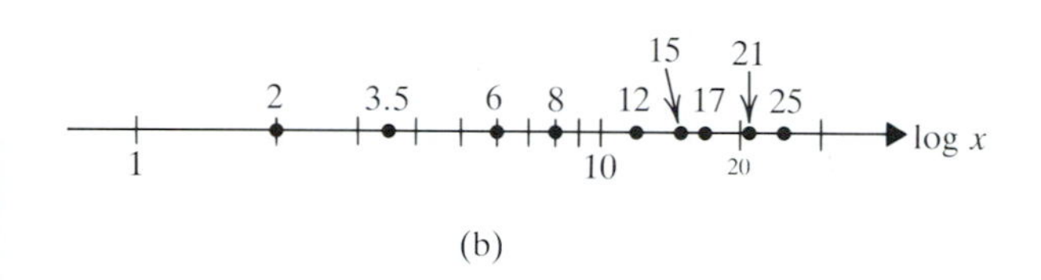

(b)

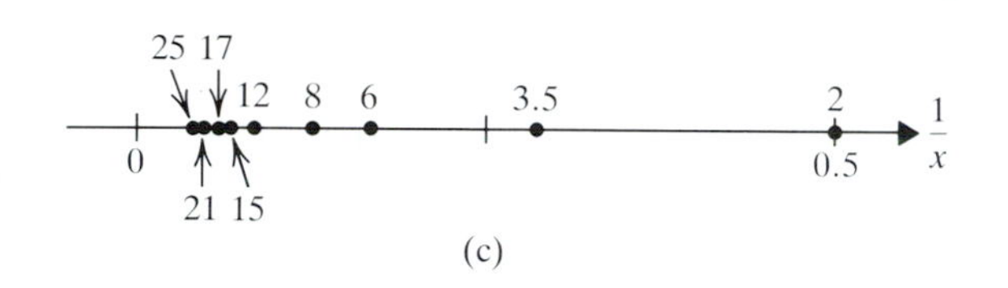

(c)

5,7.

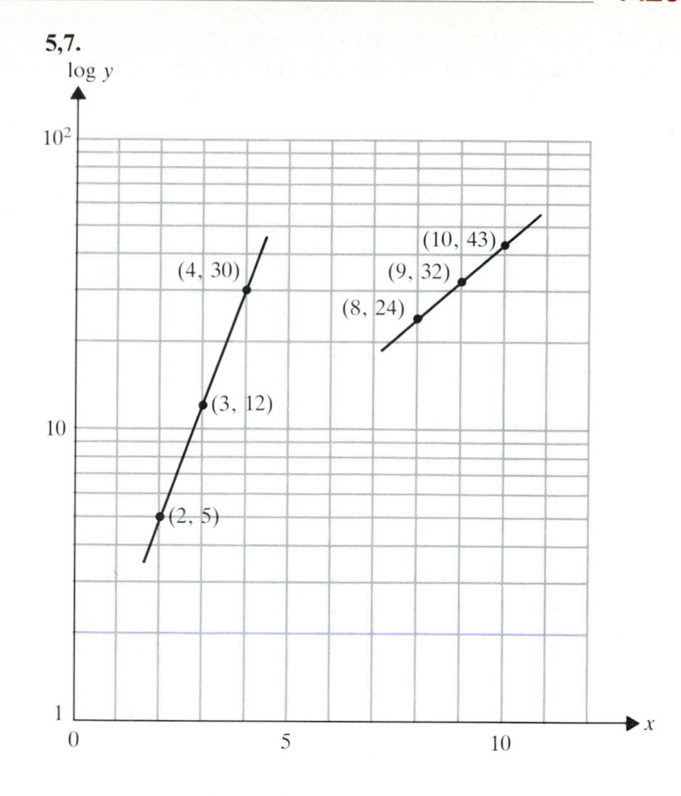

9.

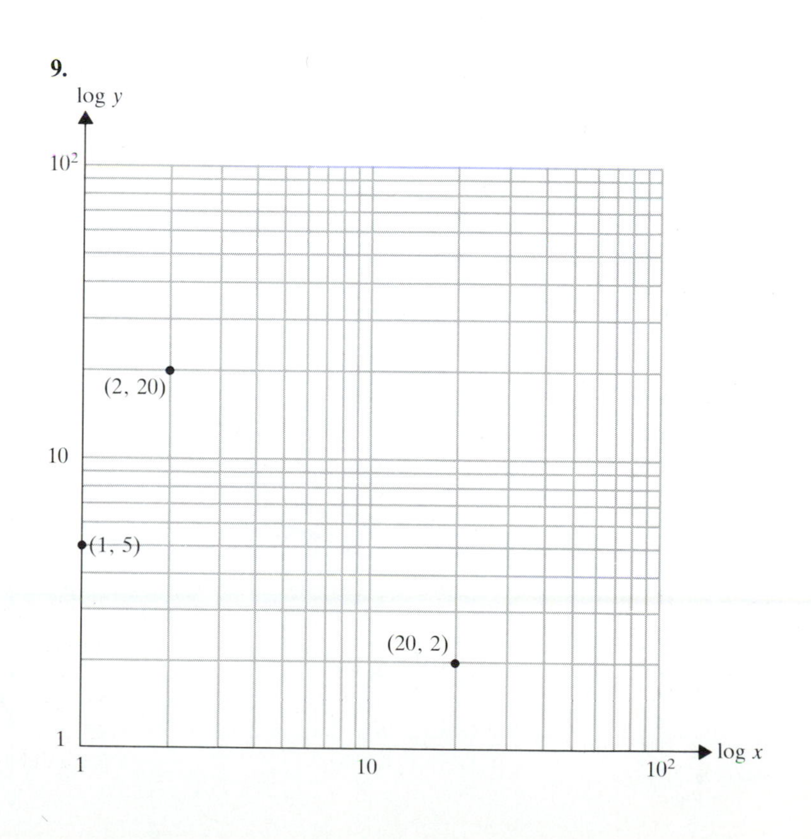

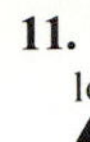

11.

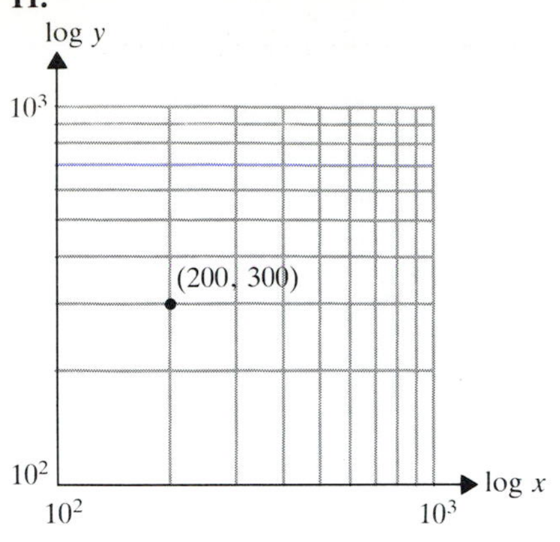

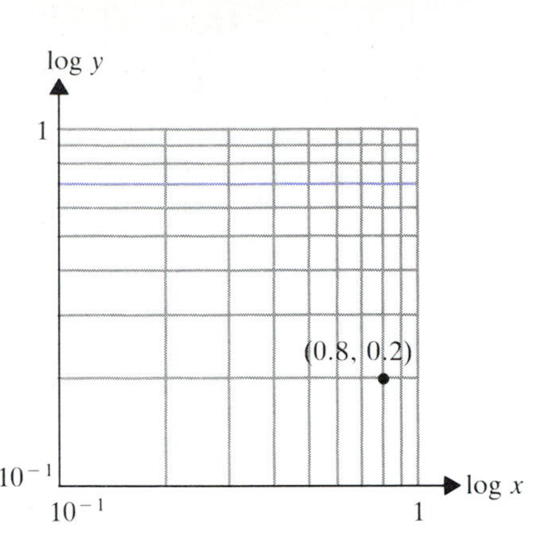

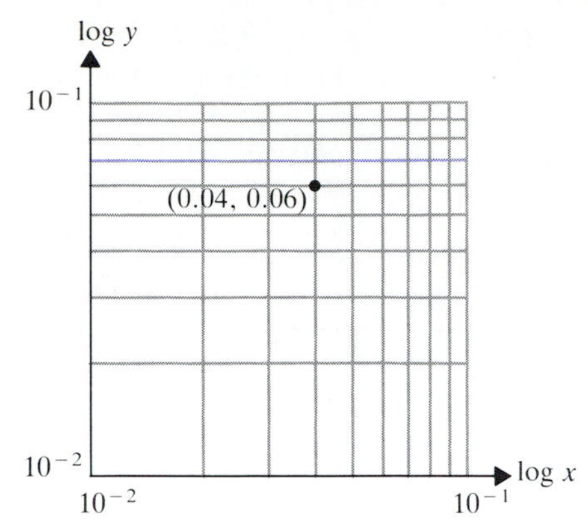

13. $y = 5e^{0.2x}$

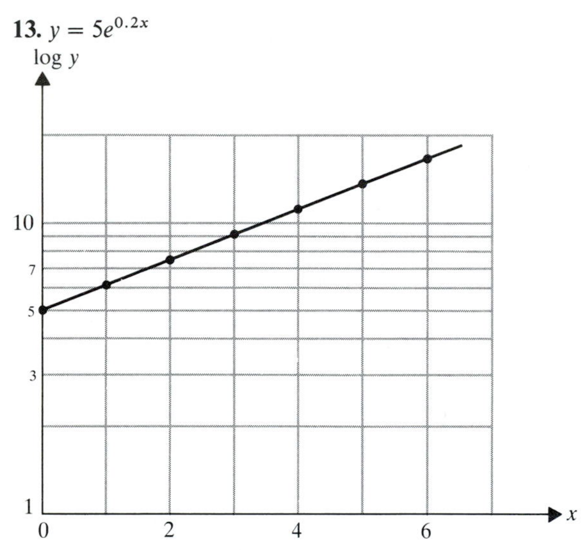

15. $y = -2e^{-0.2x}$

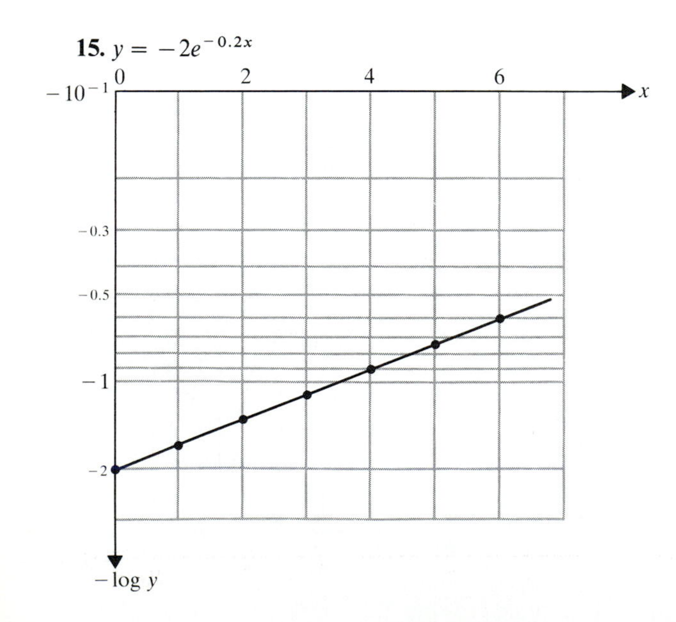

17. $y = 1.6e^{1.2x}$

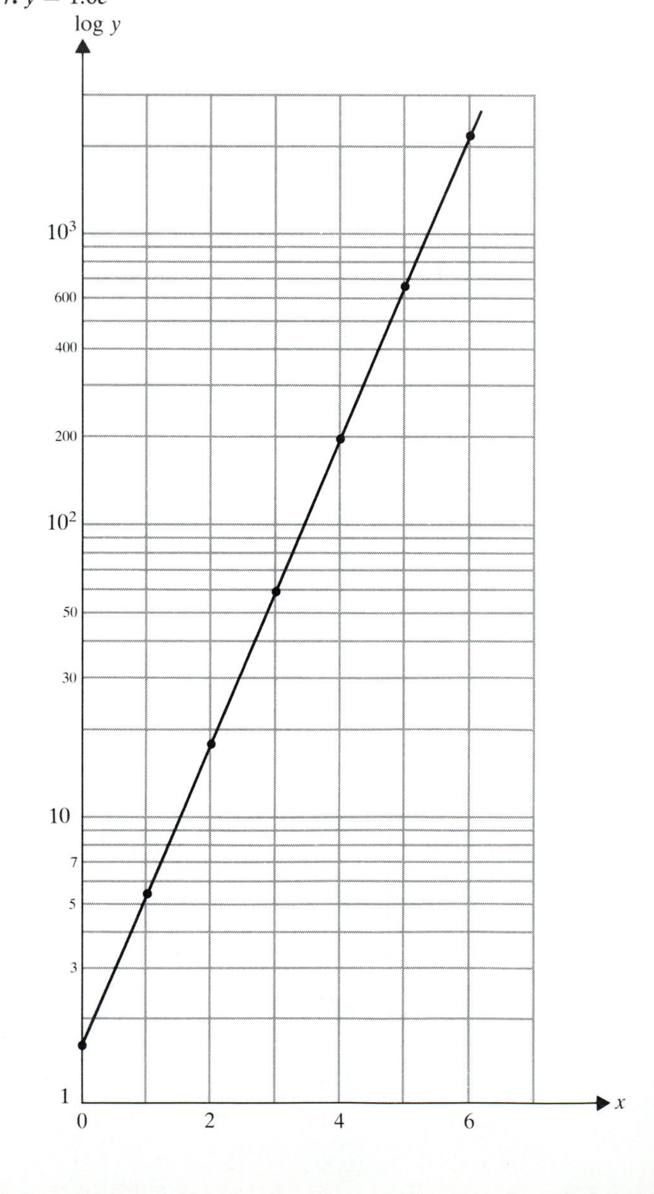

19. $y = 2x^{1/2}$

21. $y = 3x^{-1/2}$

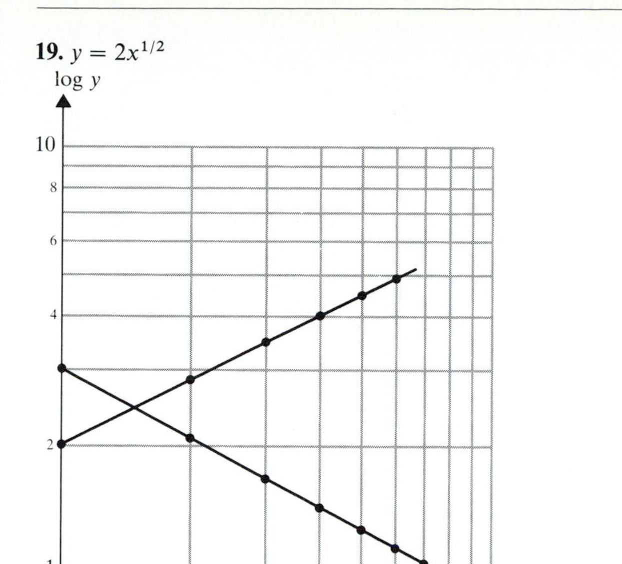

23. $y = -1.5x^{2/5}$

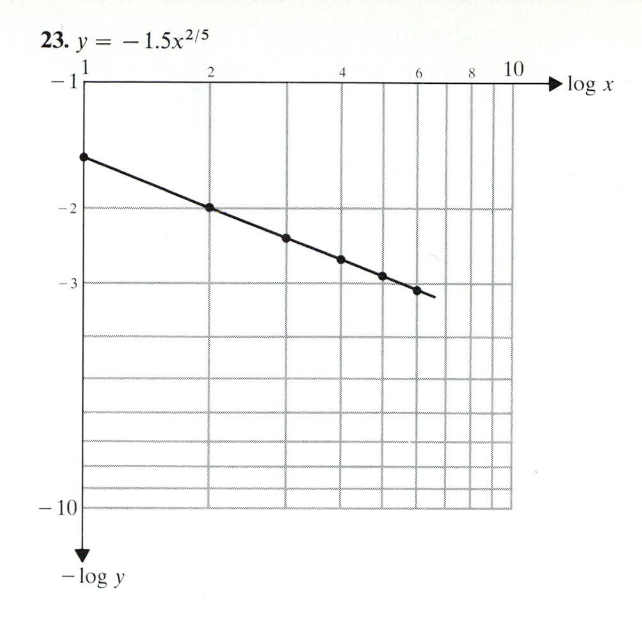

25. $P \approx 9.3V^{-1.4}$

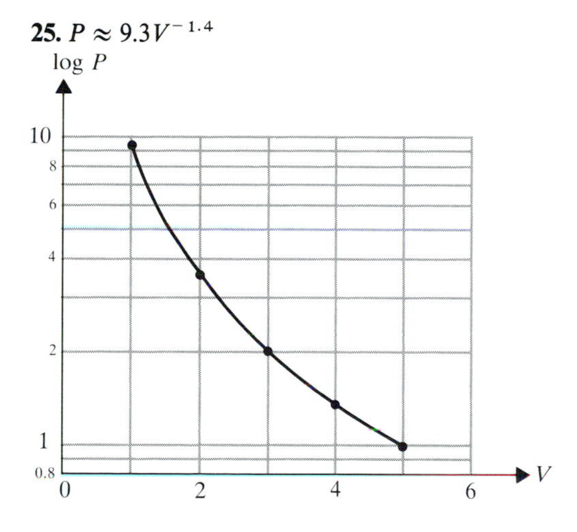

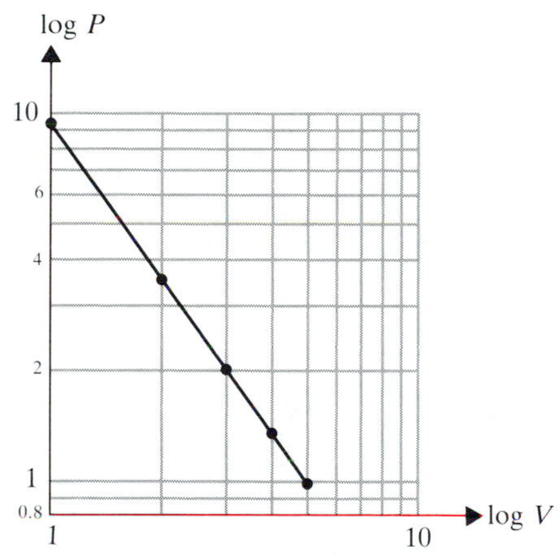

27. $S = 2R^{1/3}$

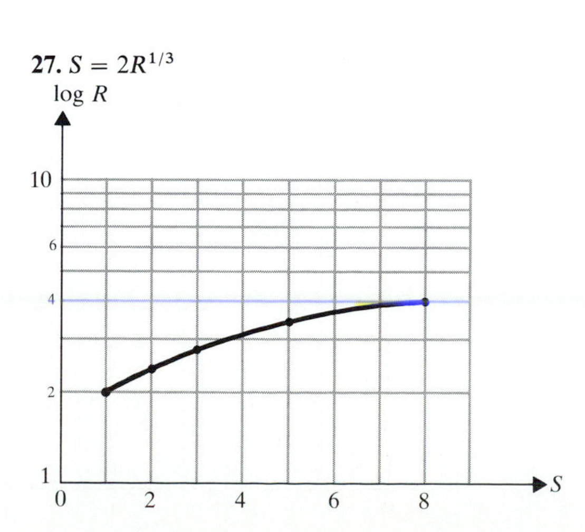

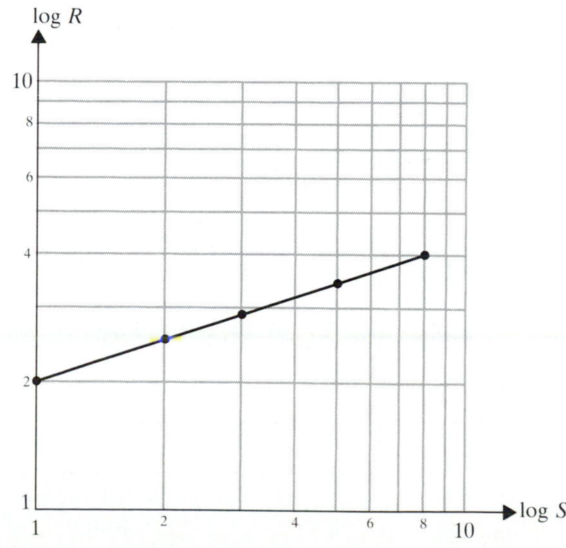

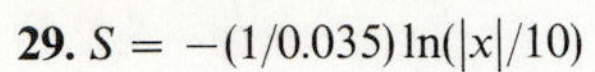

29. $S = -(1/0.035)\ln(|x|/10)$

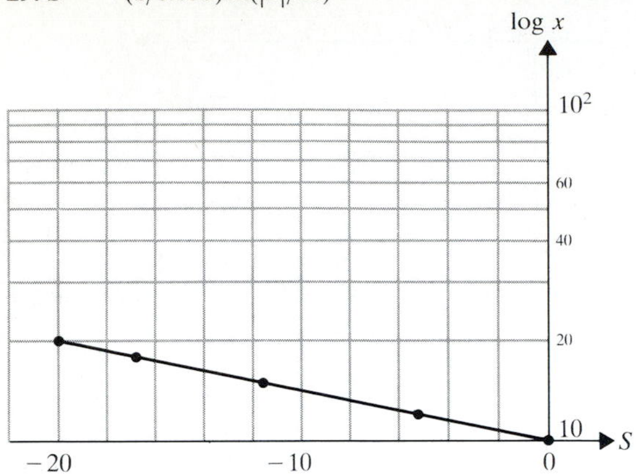

31. $y \approx 0.407D^{3/2}$

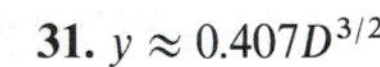

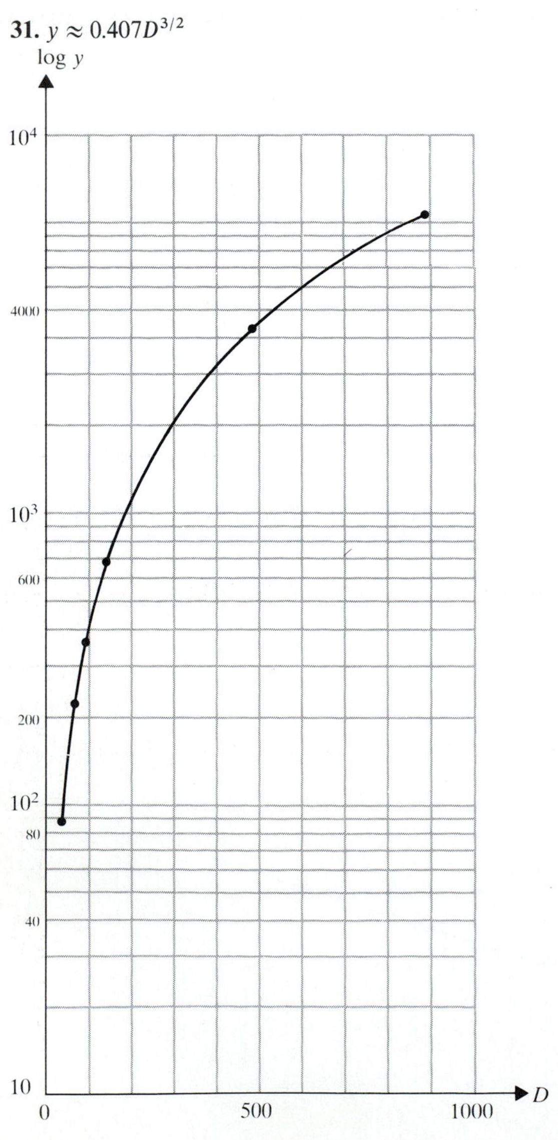

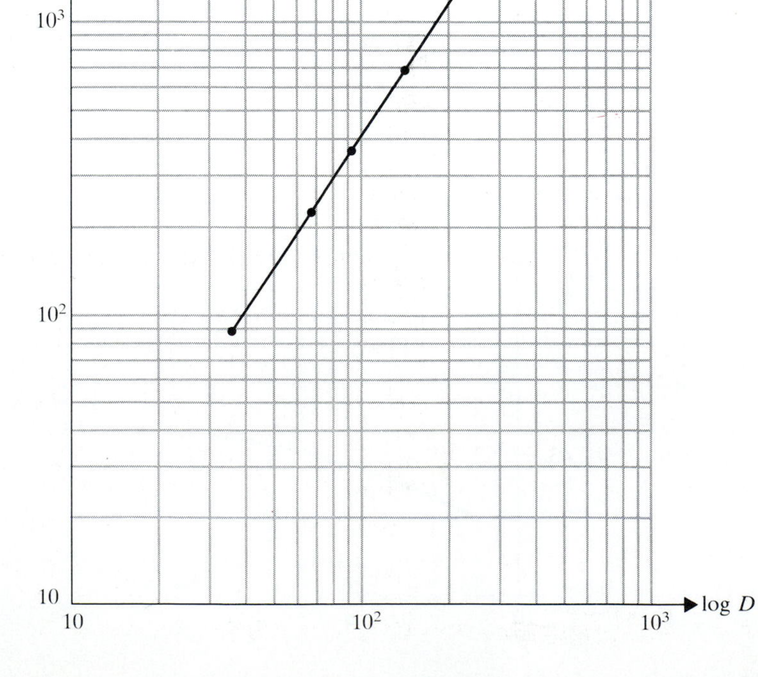

33a. $y = ax/(b + x)$ **33b.** $1/y = (b/a)(1/x) + 1/a$
33c. Straight line **35.** $1/V = (1/v)$ intercept; $-K/v =$ slope; $K = -\text{slope} \div (1/V)$
37. $E_a \approx 4.9 \times 10^4$ joules

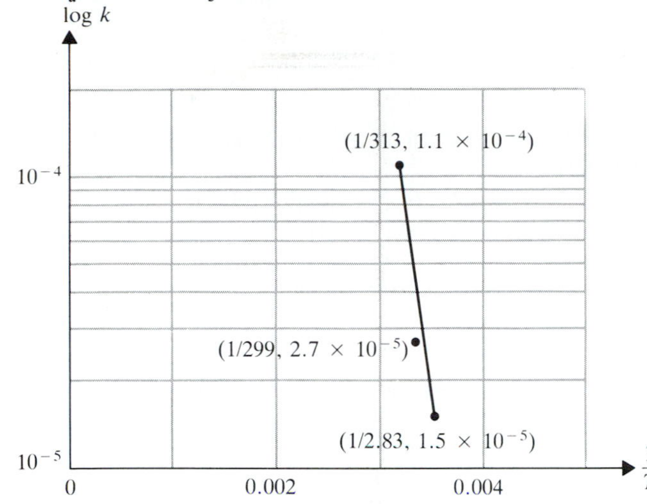

REVIEW EXERCISES 6.10, p. 420

1. $\dfrac{\sin 2x + 3x^2}{2x^2 + 2\sin^2 x}$ **3.** $\dfrac{-(3\sec 3x \tan 3x + 2x)}{\sec 3x + x^2}$

5. $\left(\dfrac{9x^2}{2x^3 - 4} + \dfrac{5x}{x^2 + 5}\right)\sqrt{(x^3 - 2)^3(x^2 + 2)^5}$

7. $\dfrac{2x[e^{x^2} - e^{-x^2}]}{e^{x^2}}$ **9.** $x^{x^2 + \ln x}[x^2 + 2(x^2 + 1)\ln x]$

11. $-2^{\cos x^2}(2x \sin x^2)\ln 2$ **13.** $\dfrac{32x}{(3x^4 - 192)\ln 2}$ **15.** $\dfrac{2xe^{x^2}}{\sqrt{1 - e^{2x^2}}}$

17. $\dfrac{\cosh(\sin^{-1} x)}{\sqrt{1 - x^2}}$ **19.** $\dfrac{-x}{\sqrt{x^2 + 4}} \operatorname{sech}\sqrt{x^2 + 4} \tanh\sqrt{x^2 + 4}$

21. $(1/4)\ln(10x^2 - 4x + 8) + C$ **23.** $(\ln x)^3/3 + C$

25. $t + \ln|t| + C$ **27.** $\ln\left[\dfrac{1}{1 + e^{-x}}\right] + C$ **29.** $\ln|e^x - 1| + C$

31. $\dfrac{-10^{-x^2 + 3}}{2\ln 10} + C$ **33.** $-\sin^{-1}(e^{-x}) + C$ **35.** $\sin^{-1}(x/2) + C$

37. $-\pi/8$ **39.** $(1/2)\tan^{-1}[(x - 3)/2] + C$
41. $\sinh^{-1}(x/3) + C$ **43.** $\cosh^{-1}(x/2) + C$ **45.** -1 **47.** $+\infty$
49. 0 **51.** $-1/2$ **53.** e^2 **55.** 0 **57.** -1 **59.** $e(e - 1)/2$

61. $\dfrac{C}{P_2 - P_1}\ln\dfrac{P_2}{P_1}$ **63.** 12.750% **65.** 12,516.75 years

67. 7.6 min **69a.** 2 **69b.** 0 **69c.** 3 rad/min **69d.** $\pi/9$ min
71. $\omega = 100$, $\phi_0 = \tan^{-1}(1/400) \approx 0.0025$, $a \approx 8.000$
73. $P \approx 10V^{-1.4}$

CHAPTER 7

EXERCISES 7.1, p. 433

1. $3\ln|\sec x^{1/3}| + C$ **3.** $\dfrac{1}{2}\ln|\sec e^{x^2} + \tan e^{x^2}| + C$

5. $\sin x - (\sin^3 x)/3 + C$
7. $\sin x - \sin^3 x + (3/5)\sin^5 x - (\sin^7 x)/7 + C$
9. $-(\cos^3 x)/3 + (\cos^5 x)/5 + C$
11. $-(\cos^5 x)/5 + (\cos^7 x)/7 + C$
13. $2\sin^{1/2} x - (2/5)\sin^{5/2} x + C$
15. $3x/8 - (\sin 2x)/4 + (\sin 4x)/32 + C$
17. $x/16 + (\sin^3 2x)/48 - (\sin 4x)/64 + C$
19. $3\pi/128$ **21.** $(\tan^3 x)/3 + C$
23. $\tan x + (2/3)\tan^3 x + (\tan^5 x)/5 + C$
25. $(\tan^2 x)/2 + (\tan^4 x)/4 + C$ **27.** $-\cot x - (\cot^3 x)/3 + C$
29. $(\tan^4 x)/4 - (\tan^2 x)/2 + \ln|\sec x| + C$
31. $(\tan^4 x)/4 + C$ **33.** $(6\sqrt{3} - 8)/45\sqrt{3}$
35. $(2/5)\sin^{5/2} x - 2\sec^{1/2} x + C$
37. $-(\cot^2 x)/2 + C = -(\csc^2 x)/2 + K$ **39.** $(\sin^2 x)/2 + C$
41. $13/15 - \pi/4$ **43.** $5\pi^2/32$ **47.** $(\cos x)/2 - (\cos 3x)/6 + C$
49. $(\sin 2x)/4 - (\sin 4x)/8 + C$ **53.** 0 if $m \neq \pm n$, π otherwise
57. Use $\tan^n x = (\sec^2 x - 1)^{n/2}$

EXERCISES 7.2, p. 441

1. $xe^x - e^x + C$ **3.** $x\sin x + \cos x + C$
5. $-x^2\cos x + 2x\sin x + 2\cos x + C$ **7.** $(e^2 - 1)/4$

9. $x\tan x - \ln|\sec x| + C$ **11.** $x\ln|x| - x + C$
13. $x\sin^{-1} x + (1 - x^2)^{1/2} + C$
15. $(2/3)x^2(1 + x)^{3/2} - (8x/15)(1 + x)^{5/2} + (16/105)(1 + x)^{7/2} + C$
17. $(1/2)(e^x \sin x - e^x \cos x) + C$
19. $(-1/2)\csc\theta\cot\theta + (1/2)\ln|\csc\theta - \cot\theta| + C$
21. $\pi/4 - 1/2$ **23.** $x^2\cosh x - 2x\sinh x + 2\cosh x + C$
25. $(e/2)(\sin 1 - \cos 1) + 1/2$
27. $[e^{ax}/(a^2 + b^2)][a\cos bx + b\sin bx] + C$
29. $x^3(\tan^{-1} x)/3 - x^2/6 + (1/6)\ln(x^2 + 1) + C$

31. $\dfrac{1}{n - 1}\sec^{n-2}x\tan x + \dfrac{n - 2}{n - 1}\displaystyle\int \sec^{n-2} x\,dx$

33. $x^n e^x - n\displaystyle\int x^{n-1}e^x\,dx$ **35.** $x(\ln x)^n - n\displaystyle\int (\ln x)^{n-1}\,dx$

37. $x^n\cosh x - n\displaystyle\int x^{n-1}\cosh x\,dx$

39. $\dfrac{-e^x}{(n - 1)x^{n-1}} + \dfrac{1}{n - 1}\displaystyle\int e^x x^{-n+1}\,dx$ **41.** 1

43. $(18 + 10(\ln 10)^2 - 20\ln 10)$ **45.** 2π

EXERCISES 7.3, p. 449

1. $\sin^{-1} x + C$ **3.** $x/(1 - x^2)^{1/2} + C$
5. $(1/5)(1 - x^2)^{5/2} - (1/3)(1 - x^2)^{3/2} + C$

7. $11\sqrt{3}/4 + (3/8)\ln|2+\sqrt{3}|$ **9.** $\frac{1}{2}\ln\left|\frac{\sqrt{x^2+4}-2}{x}\right| + C$

11. $-\frac{\sqrt{x^2+1}}{x} + C$ **13.** $\sqrt{5}/2 - 2\ln(\sqrt{5}+1) + 2\ln 2$

15. $\sqrt{x^2-9} - 3\sec^{-1}(x/3) + C$ **17.** $\sec^{-1}x + C$

19. $\ln[(3+\sqrt{5})/4]$ **21.** $\frac{1}{2}\sin^{-1}[(2x+4)/5] + C$

23. $\ln[(3\sqrt{2}+3)/(\sqrt{13}+2)]$

25. $\ln|x-3+(x^2-6x-7)^{1/2}| + C$ **29.** $-\operatorname{sech}^{-1}x + C$

31. $\sqrt{5}/2 + (1/4)\ln(2+\sqrt{5}) \approx 1.479$ vs. 1.4789429

33. $\sqrt{5} - \sqrt{2} + \ln[(\sqrt{5}-1)/(2\sqrt{2}-2)] \approx 1.2220162$

EXERCISES 7.4, p. 458

1. $\frac{1}{2}\ln\left|\frac{x-1}{x+1}\right| + C$ **3.** $\frac{2}{3}\ln|(x-2)^2(x+1)| + C$

5. $(\ln 2)/6 + 7(\ln 4)/6 - (\ln 3)/2 - 2(\ln 5)/3$

7. $2\ln|x| - \ln|x-1| + 3\ln|x+2| + C$ **9.** $\ln 2 + 1/2$

11. $\frac{1}{16}\ln\left|\frac{x-1}{x+3}\right| - \frac{1}{8x-8} - \frac{5}{8x+24} + C$

13. $5\ln\left|\frac{x}{x-1}\right| - \frac{6}{x-1} + C$ **15.** $2\ln(4/3) - 1/6$

17. $x + \ln\left|\frac{x-1}{x+1}\right| + C$

19. $\frac{x^2}{2} + \frac{4\ln|x+2|}{3} + \frac{2\ln|x-1|}{3} + C$

21. $\pi/8 - (5\ln 2)/4$ **23.** $\frac{1}{8}\ln\left|\frac{x-2}{x+2}\right| + \frac{1}{4}\tan^{-1}\frac{x}{2} + C$

25. $\frac{3\ln|2x+1|}{10} + \frac{\ln(x^2+1)}{10} + \frac{2\tan^{-1}x}{5} + C$

27. $\ln|x^3+x| + C$ **29.** $\ln|x^3+1| + K$

31. $\frac{3\ln(x^2+x+1)}{2} - \ln x^2 - \frac{5}{\sqrt{3}}\tan^{-1}\frac{2x+1}{\sqrt{3}} + K$

33. $(1/2)\ln 2 - 1/4$ **35.** $\frac{15}{16}\tan^{-1}\frac{x}{2} - \frac{x+4}{8(x^2+4)} + C$

37. $\frac{\ln(x^2+1)}{2} - 3\tan^{-1}x - \frac{1}{2(x^2+1)} + C$

39. $\frac{\ln(x^2+2x+2)}{2} - \tan^{-1}(x+1) + \frac{2x+2}{x^2+2x+2} + C$

41. $[\ln|(\sin\theta-3)/(\sin\theta+2)|]/5 + C$

43. $x - \ln(1+e^x) + C$ **47.** $u = \frac{at}{t+10} \to a$ as $t \to +\infty$

49. $y = \frac{ab[e^{k(a-b)t}-1]}{ae^{k(a-b)t}-b}$

EXERCISES 7.5, p. 465

1. $2\sqrt{x} - 2\ln(1+\sqrt{x}) + C$ **3.** $3\pi/2 - 152/35$

5. $-2\sqrt{x} + \frac{4}{\sqrt{3}}\tan^{-1}\frac{2x^{1/4}-1}{\sqrt{3}} + 2\ln|x - \sqrt[4]{x} + 1| + C$

7. $\frac{1}{2}\ln\left|\frac{\sqrt{x}-1}{\sqrt{x}+1}\right| + \tan^{-1}\sqrt{x} + C$

9. $(2/7)(x+1)^{7/2} - (4/5)(x+1)^{5/2} + (2/3)(x+1)^{3/2} + C$

11. $(1/6)(1+2x)^{3/2} - (1/2)(1+2x)^{1/2} + C$

13. $(112 - 64\sqrt{2})/15$ **15.** $3(x^2+4)^{5/3}/10 - 3(x^2+4)^{2/3} + C$

17. $(1/5)(x^2-1)^{5/2} + (1/3)(x^2-1)^{3/2} + C$ **19.** $1412/5$

21. $\frac{-2}{1+\tan\frac{x}{2}} + C$ **23.** $-\ln|1-\tan(x/2)| + C$

25. $\pi/(3\sqrt{3})$ **27.** $(1/4)[\tan^2(x/2) + (1/2)\ln|\tan(x/2)|] + C$

29. $\ln\left|x + \frac{3}{2} + \sqrt{x^2+3x}\right| + C$

31. $\frac{1}{\sqrt{2}}\ln|x\sqrt{2} + \sqrt{2x^2-5}| + C$ **33.** $\frac{-19e^{-2}+3}{8}$

35. $(x^3\sin 3x)/3 + (x^2\cos 3x)/3 - (2x\sin 3x)/9 - (2\cos 3x)/27 + C$

37. $35\sqrt{5}/4 + 243(\sin^{-1}\sqrt{5}/3)/8$

43. $2\sqrt{1+e^x} + \ln\left|\frac{\sqrt{1+e^x}-1}{\sqrt{1+e^x}+1}\right| + C$

EXERCISES 7.6, p. 475

1. $\pi/4$ **3.** 0 **5.** Diverges to $+\infty$ **7.** π **9.** Diverges to $+\infty$
11. Diverges to $+\infty$ **13.** 1/4 **15.** Diverges (oscillates)
17. 1/2 **19.** 3/2 **21.** Diverges to $+\infty$ **23.** Diverges to $-\infty$
25. $3[1+2^{1/3}]$ **27.** Does not exist **29.** Diverges to $+\infty$

31. $-1/4$ **33.** $\pi/3$ **35.** 0 **37.** π **39.** $\pi/2$ **41.** $v_1 - \sqrt{\frac{2GM}{R_0}}$

43a. \$50,000 **43b.** \$83,333.33 **45.** \$15.00 per share
51. Converges **53.** Diverges **55.** Diverges

REVIEW EXERCISES 7.7, p. 477

1. $-(\cos^3 x)/3 + 2(\cos^5 x)/5 - (\cos^7 x)/7 + C$ **3.** $\pi/16$

5. $(\tan^5 x)/5 + (\tan^3 x)/3 + C$ **7.** $(\sec^5 x)/5 - (\sec^3 x)/3 + C$

9. $9\sqrt{3}/5 - \pi/3$ **11.** $x^2\sin x + 2x\cos x - 2\sin x + C$

13. $x^2\sqrt{1+x^2} - \frac{2}{3}[1+x^2]^{3/2} + C$

15. $(1/26)[e^{\pi}(3\sqrt{3}+2) + 4]$

17. $x\sec^{-1}x - \ln|x + (x^2-1)^{1/2}| + C$

19. $(x^2\sin^{-1}x)/2 - (\sin^{-1}x)/4 + x(1-x^2)^{1/2}/4 + C$

21. $e^{2x}(2x^2-2x+1)/4 + C$

23. $(x^3\sin^{-1}x)/3 - x^2(1-x^2)^{1/2}/3 + 2(1-x^2)^{3/2}/9 + C$

25. $2\sin^{-1/2}x - x(4-x^2)^{1/2}/2 + C$ **27.** $3\pi/16$

29. $x(a^2+x^2)^{1/2}/2 + (a^2/2)\ln[x + (x^2+a^2)^{1/2}] + C$

31. $\sqrt{2}/4 + \ln[(2\sqrt{2} + 3)/(\sqrt{2} + 1)]$
33. $x(x^2 - a^2)^{1/2}/2 - (a^2/2)\ln|x + (x^2 - a^2)^{1/2}| + C$
35. $-\sqrt{5} - 2\pi + 2\sec^{-1}(-3/2)$
37. $\ln|x - 2 + (x^2 - 4x + 5)^{1/2}| + C$
39. $2\sin^{-1}[(x - 2)/3] - (5 + 4x - x^2)^{1/2} + C$
41. $2\ln|x - 1| - \ln|x| + C$ **43.** $\ln 2 + 1/2$
45. $\ln|x| + \ln[(x - 1)^2(x + 1)^{-2}] + C$
47. $(-1/2)\ln|x| + (5/6)\ln|x + 2| + (2/3)\ln|x - 1| + C$
49. $(1/10)(3\ln 3 + \ln 2 + \pi)$
51. $2\ln|x - 2| - 8/(x - 2) - 9/2(x - 2)^2 + C$
53. $x^2/2 + 2x + (1/4)(27\ln|x - 3| - \ln|x + 1|) + C$
55. $\ln x - (1/2)\ln(x^2 + 1) + 1/(2x^2 + 2) + C$
57. $2x^{1/2} - 3x^{1/3} + 6x^{1/6} - 6\ln(1 + x^{1/6}) + C$
59. $10/3 - 4\ln 2$ **61.** $(2/5)(x - 1)^{5/2} + (2/3)(x - 1)^{3/2} + C$
63. $2e^{\sqrt{2-x}}[(2 - x)^{3/2} + 4\sqrt{2 - x} + 3x - 10] + C$
65. $\dfrac{2}{1 - \tan(\pi/12)} - 2$ **67.** $\dfrac{1}{5}\ln\left|\dfrac{3\tan(x/2) - 1}{\tan(x/2) + 3}\right| + C$ **69.** 2
71. $\pi/8$ **73.** 1 **75.** $\pi/2$ **77.** 3 **79.** $3(2)^{1/3}$ **81.** Diverges
83. π **85.** $\dfrac{x}{(2n - 1)[1 + x^2]^{n-1}} + \dfrac{2n - 3}{2n - 2}\displaystyle\int \frac{dx}{[1 + x^2]^{n-1}}$
87. $\dfrac{2}{b(2n + 3)}\left[x^n(a + bx)^{3/2} - an\displaystyle\int x^{n-1}\sqrt{a + bx}\,dx\right]$
91. $L = \int_0^2 \sqrt{1 + 4x^2}\,dx$ **93.** $\dfrac{abe^{kt(a-b)} - 1}{ae^{kt(a-b)} - b}$ **97.** \$50

CHAPTER 8

EXERCISES 8.1, p. 488

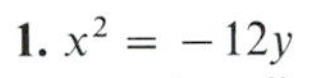

1. $x^2 = -12y$

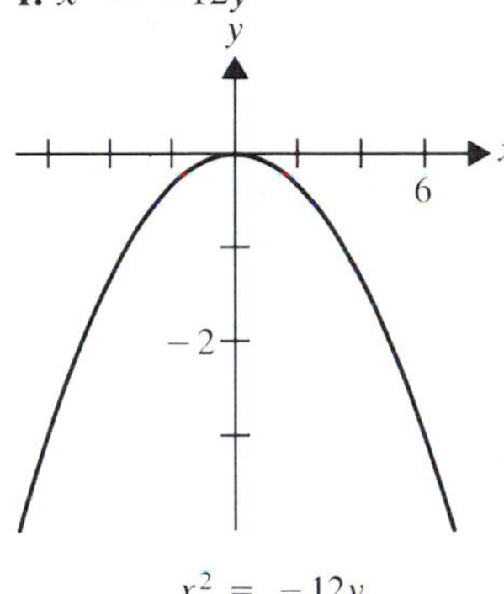

$x^2 = -12y$

3. $y^2 = 8x$

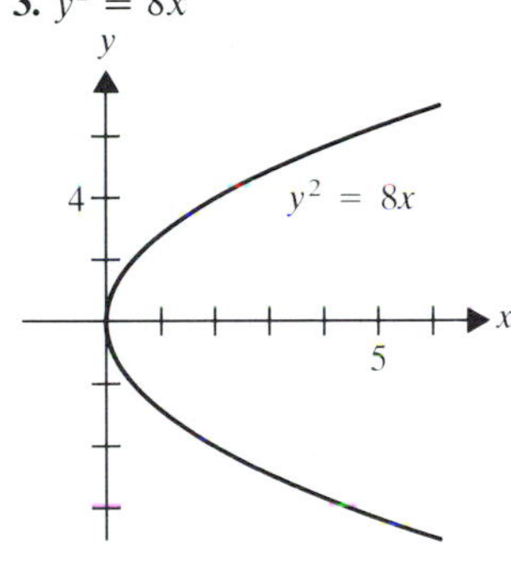

9. $(y - 2)^2 = 12(x - 1)$

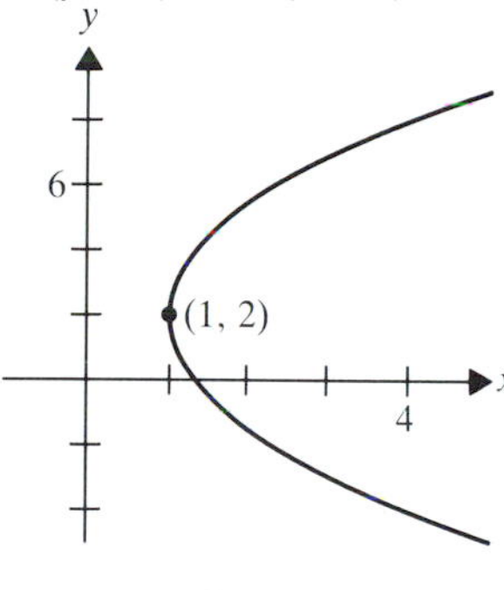

$(y - 2)^2 = 12(x - 1)$

11. $(x - 3)^2 = 12(y - 1)$

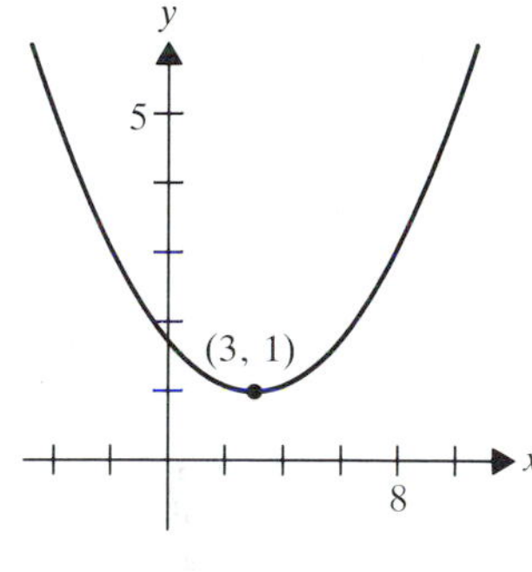

$(x - 3)^2 = 12(y - 1)$

5. $x^2 = -8(y + 1)$

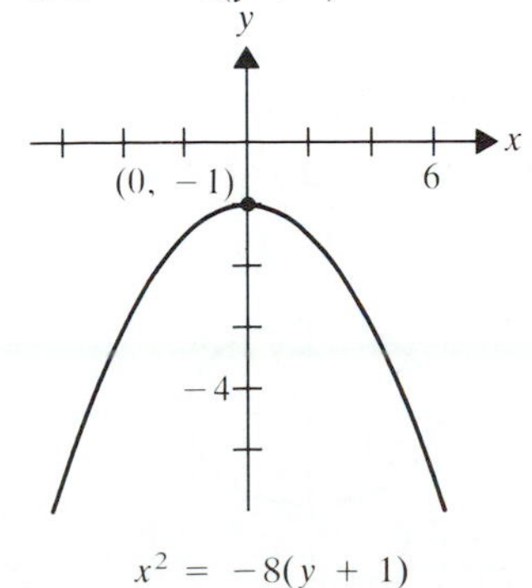

$x^2 = -8(y + 1)$

7. $y^2 = 6(x - 1/2)$

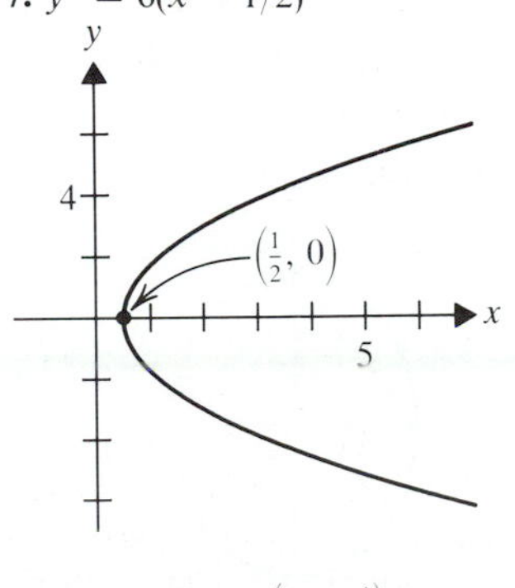

$y^2 = 6\left(x - \frac{1}{2}\right)$

13. $(x + 3)^2 = 12(y - 4)$

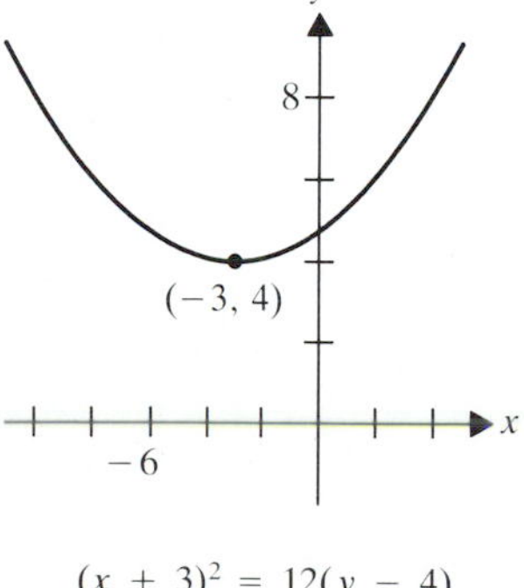

$(x + 3)^2 = 12(y - 4)$

15. $(y - 7)^2 = -8(x + 1)$

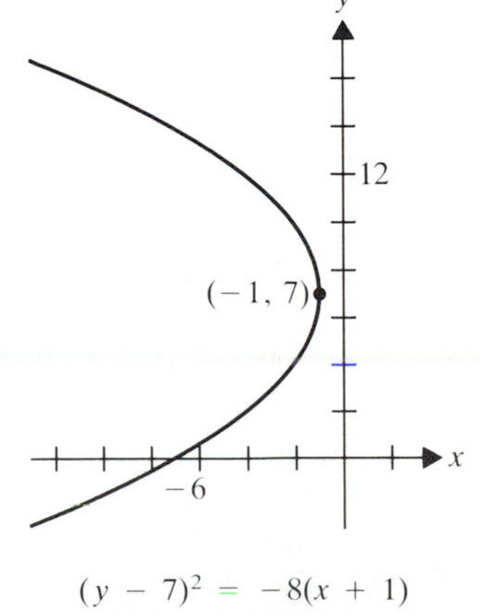

$(y - 7)^2 = -8(x + 1)$

17. $(x-1)^2 = 25(y-1)/2$

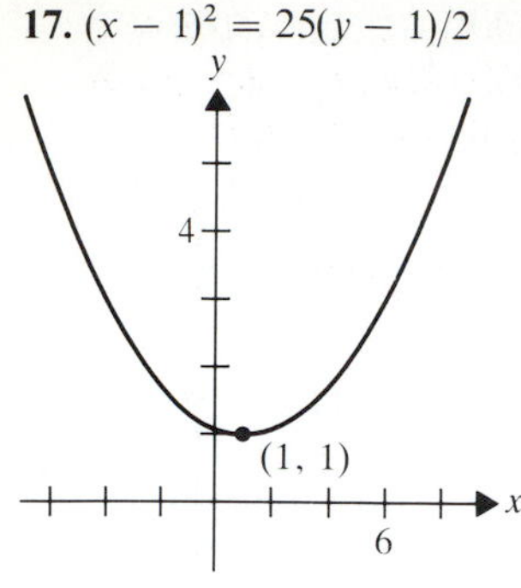

$(x-1)^2 = \dfrac{25(y-1)}{2}$

19. $(y-5)^2 = 2(x+3)$

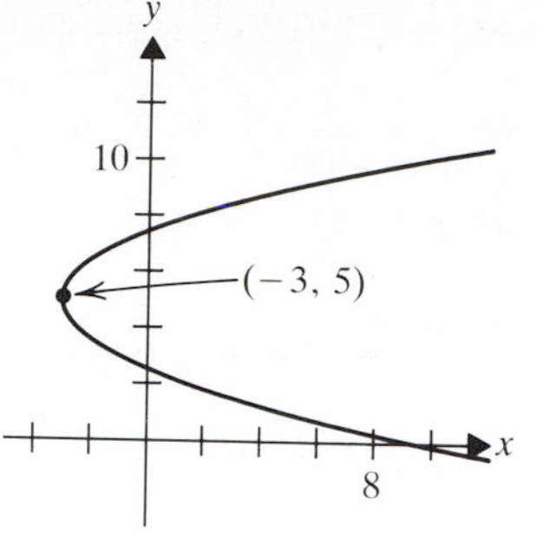

$(y-5)^2 = 2(x+3)$

21. $(x-4)^2 = 2(y-9/2)$

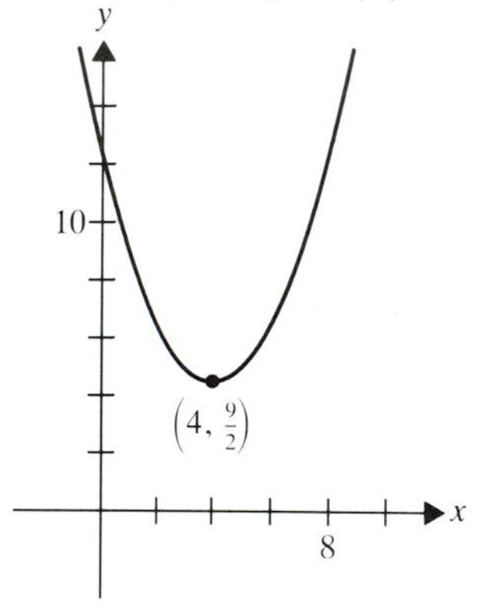

$(x-4)^2 = 2y - 9$

23. $V(0, 0)$; $F(7/4, 0)$; directrix $x = -7/4$

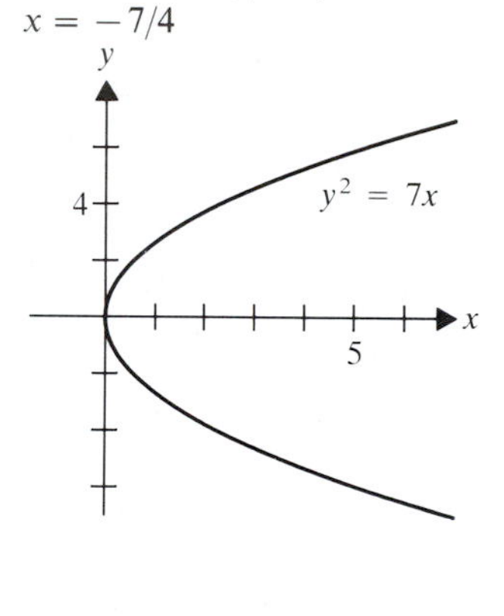

25. $V(0, 0)$; $F(0, -3/2)$; directrix $y = 3/2$

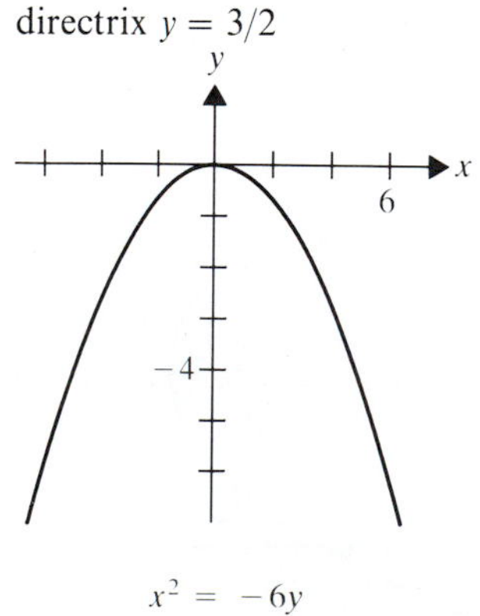

$x^2 = -6y$

27. $V(-1, 3)$; $F(-2, 3)$; directrix $x = 0$

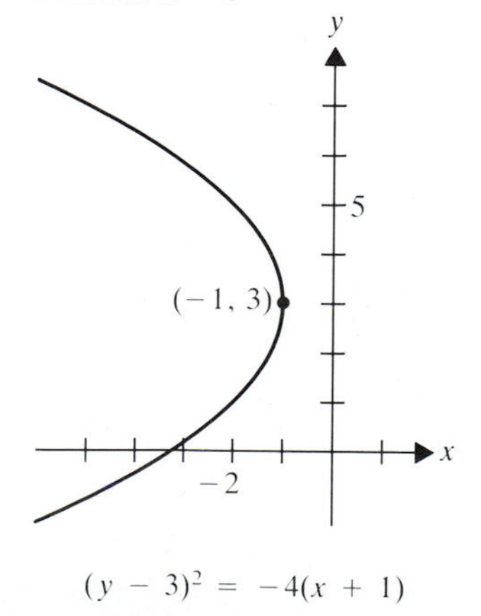

$(y-3)^2 = -4(x+1)$

29. $V(-1, -2)$; $F(-1, -3/2)$; directrix $y = -5/2$

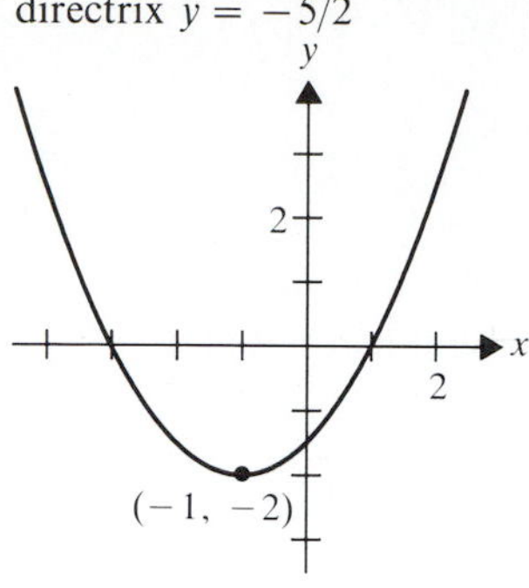

$(x+1)^2 = 2(y+2)$

31. $V(-1, 3/2)$; $F(-1/2, 3/2)$; directrix $x = -3/2$

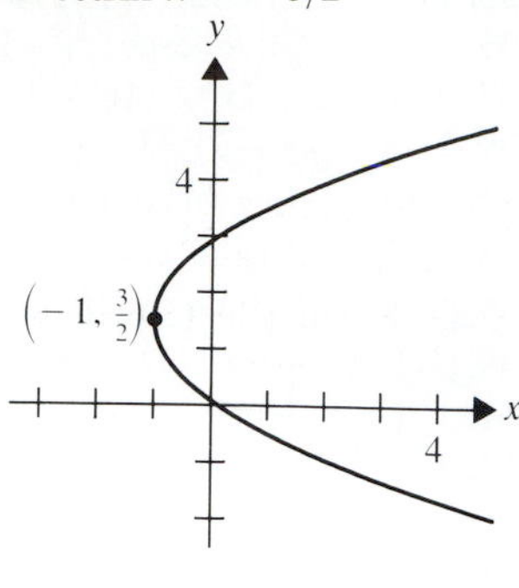

$8x = 4y^2 - 12y + 1$

33. $V(1, 1)$; $F(1, 3/4)$; directrix $y = 5/4$

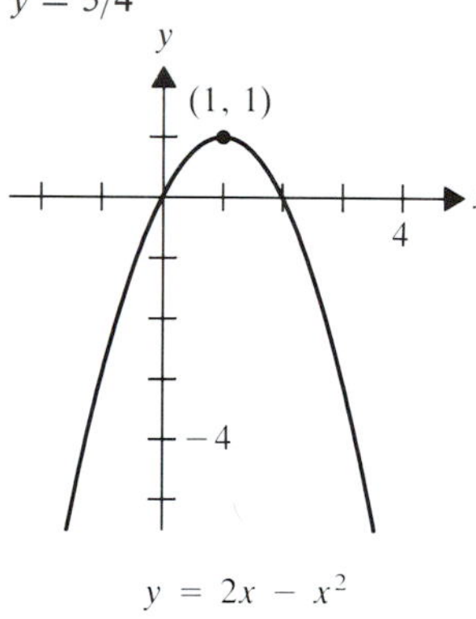

$y = 2x - x^2$

35. $V(-1, -3)$; $F(-1, -3/2)$; directrix $y = -9/2$

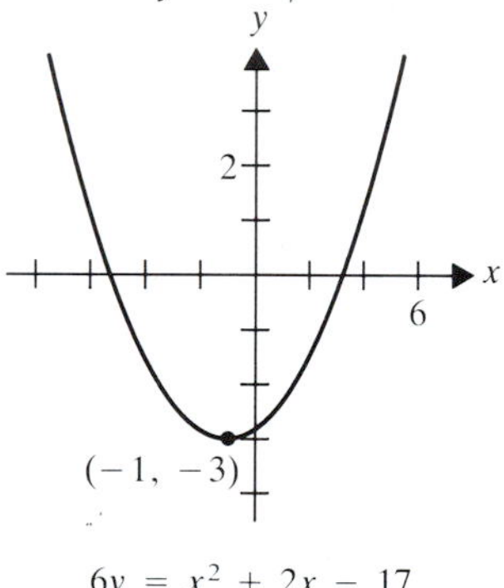

$6y = x^2 + 2x - 17$

41a. $t = (v_0 \sin \alpha)/g$ **41b.** $Y_{max} = (v_0{}^2 \sin^2 \alpha)/2g$

43. 140 m/s **45.** 5/8 in. from the vertex

EXERCISES 8.2, p. 498

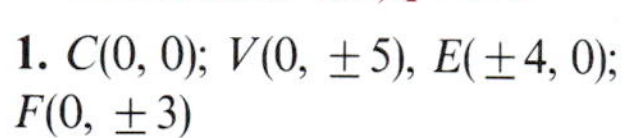

1. $C(0, 0)$; $V(0, \pm 5)$, $E(\pm 4, 0)$; $F(0, \pm 3)$

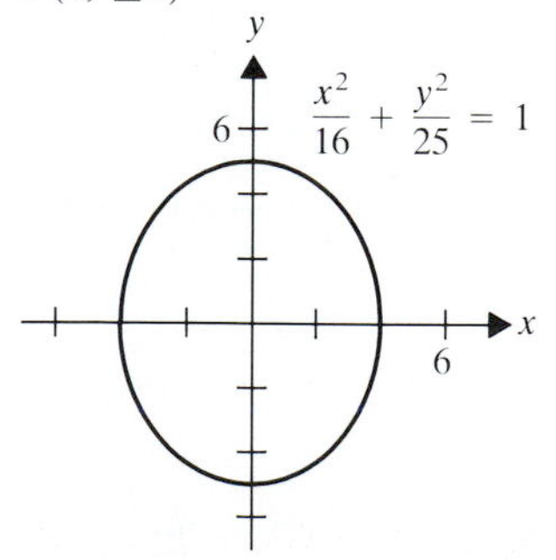

3. $C(0, 0)$; $V(\pm 4, 0)$, $E(0, \pm 3)$; $F(\pm\sqrt{7}, 0)$

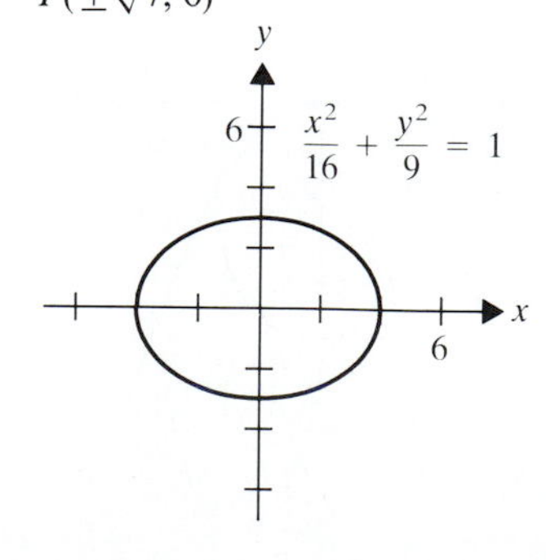

5. $C(0, 0)$; $V(\pm 3, 0)$, $E(0, \pm 2)$; $F(\pm\sqrt{5}, 0)$

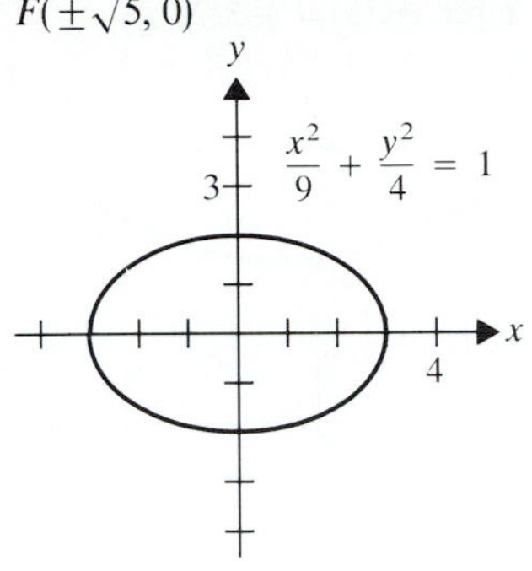

7. $C(0, 0)$; $V(0, \pm 4)$, $E(\pm 2, 0)$; $F(0, \pm 2\sqrt{3})$

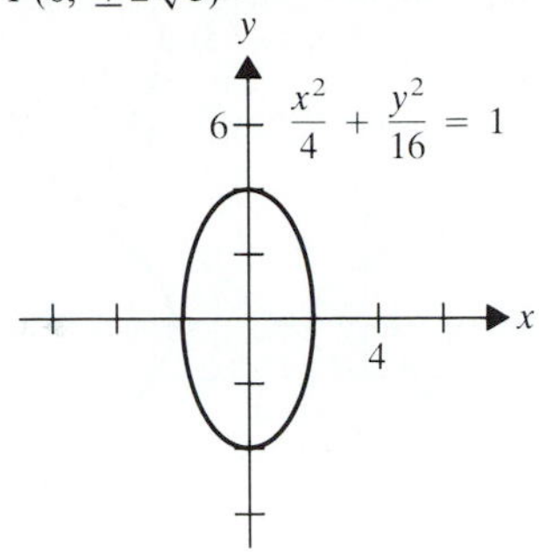

17. $C(1, 0)$; $V_1(-2, 0)$, $V_2(4, 0)$; $F(1 \pm \sqrt{5}, 0)$, $E(1, \pm 2)$

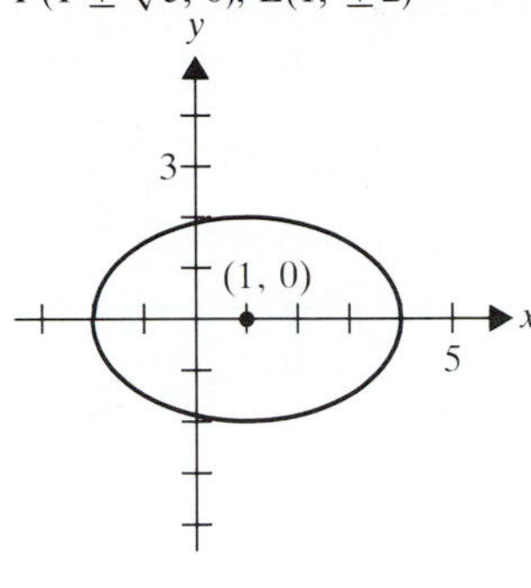

$4x^2 + 9y^2 - 8x = 32$

19. $C(-2, 1)$; $V_1(-2, 4)$, $V_2(-2, 2)$; $F(-2, 1 \pm \sqrt{5})$, $E_1(0, 1)$, $E_2(-4, 1)$

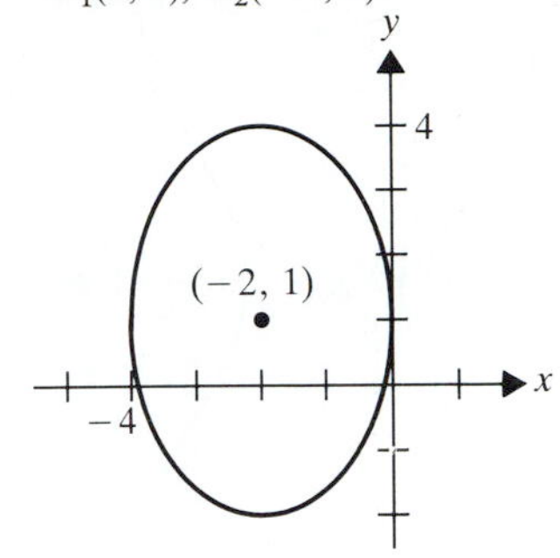

$9x^2 + 36x + 4y^2 - 8y + 4 = 0$

9. $C(0, 3)$; $V(\pm 3, 3)$; $E_1(0, 5)$; $E_2(0, 1)$; $F(\pm\sqrt{5}, 3)$

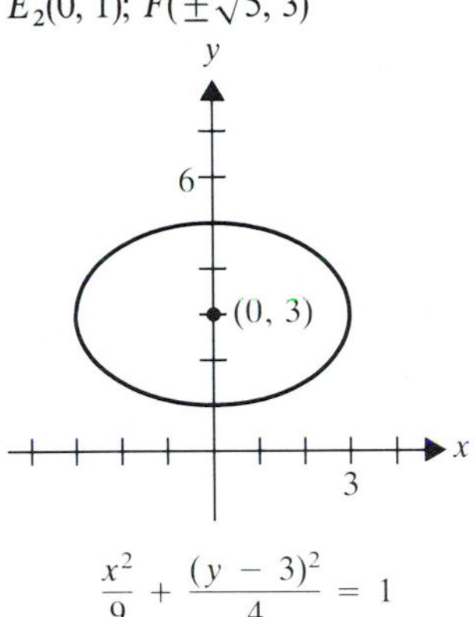

$\frac{x^2}{9} + \frac{(y-3)^2}{4} = 1$

11. $C(-3, 1)$; $V_1(-6, 1)$, $V_2(0, 1)$; $F(-3 \pm \sqrt{5}, 1)$, $E_1(-3, 3)$, $E_2(-3, -1)$

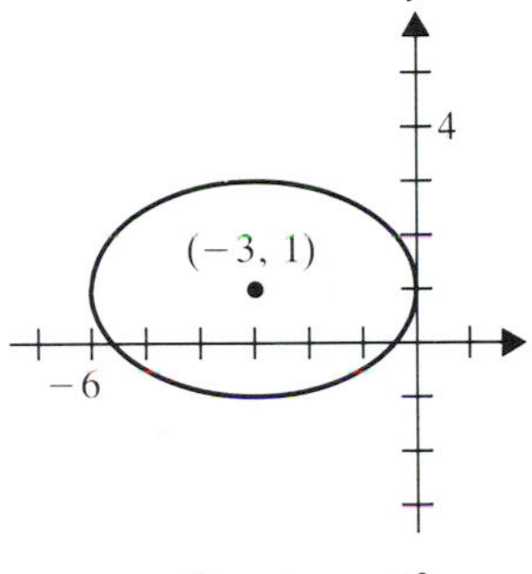

$\frac{(x+3)^2}{9} + \frac{(y-1)^2}{4} = 1$

21. Point $P(-2, 2)$

23. No graph

25. $\frac{x^2}{25} + \frac{y^2}{16} = 1$

27. $\frac{x^2}{36} + \frac{y^2}{100} = 1$

29. $\frac{x^2}{25} + \frac{y^2}{9} = 1$

31. $\frac{x^2}{16} + \frac{y^2}{12} = 1$

33. $\frac{x^2}{16} + \frac{y^2}{25} = 1$

35. $\frac{x^2}{34} + \frac{y^2}{9} = 1$

37. $\frac{(x-2)^2}{20} + \frac{(y-3)^2}{36} = 1$

39. $\frac{(x-2)^2}{9} + \frac{(y-1)^2}{5} = 1$

41. $\frac{(x-4)^2}{25} + \frac{(y+2)^2}{16} = 1$

43. $(x+1)^2 + \frac{(y-2)^2}{4} = 1$

51a. 252,000 miles **51b.** 225,400 miles **51c.** 238,700 miles

53b. $\frac{x^2}{(6800)^2} + \frac{y^2}{(6797)^2} = 1$; $E \approx 0.0294$ **55.** 1.211056

13. $C(3, -2)$; $V_1(3, -5)$, $V_2(3, 1)$; $F(3, -2 \pm \sqrt{5})$, $E_1(5, -2)$, $E_2(1, -2)$

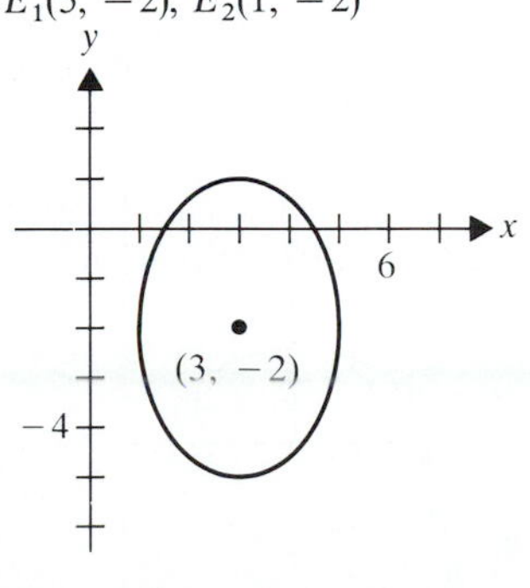

$9(x-3)^2 + 4(y+2)^2 = 36$

15. $C(-1, -2)$; $V_1(-1, -7)$, $V_2(-1, -3)$; $F_1(-1, -6)$, $F_2(-1, 2)$, $E_1(-4, -2)$, $E_2(2, -2)$

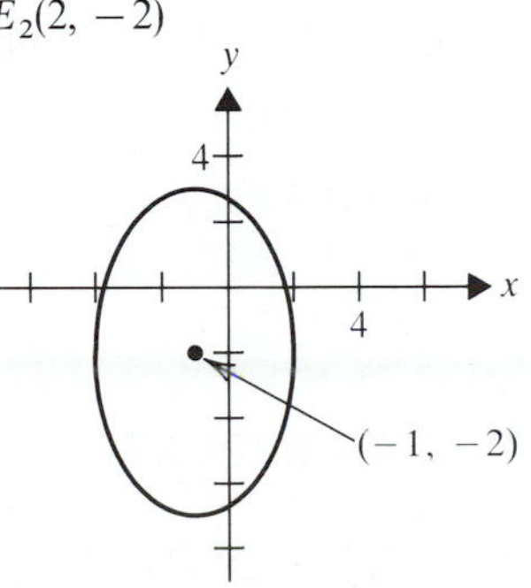

$25(x+1)^2 + 9(y+2)^2 = 225$

EXERCISES 8.3, p. 508

1. $V(\pm 4, 0)$; $F(\pm 5, 0)$; $y = \pm 3x/4$

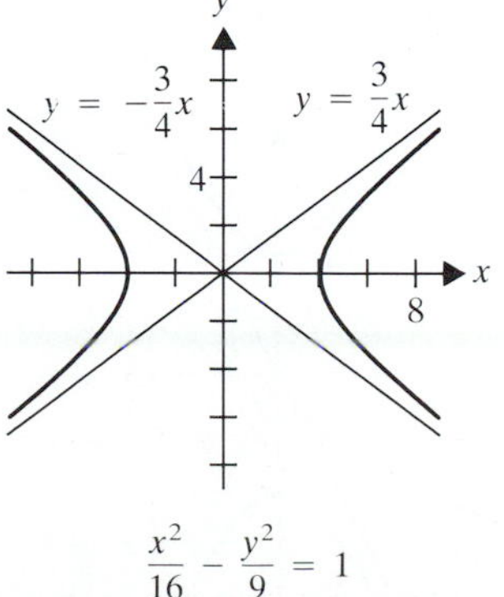

$\frac{x^2}{16} - \frac{y^2}{9} = 1$

3. $V(0, \pm 5)$; $F(0, \pm\sqrt{34})$; $y = \pm 5x/3$

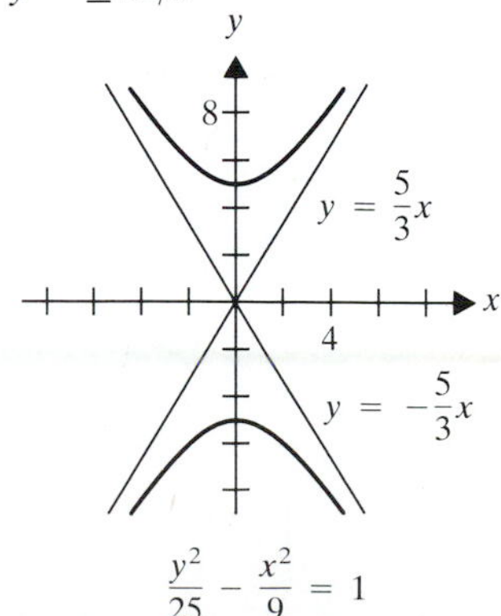

$\frac{y^2}{25} - \frac{x^2}{9} = 1$

5. $V(\pm 3, 0)$; $F(\pm 5, 0)$; $y = \pm 3x/4$

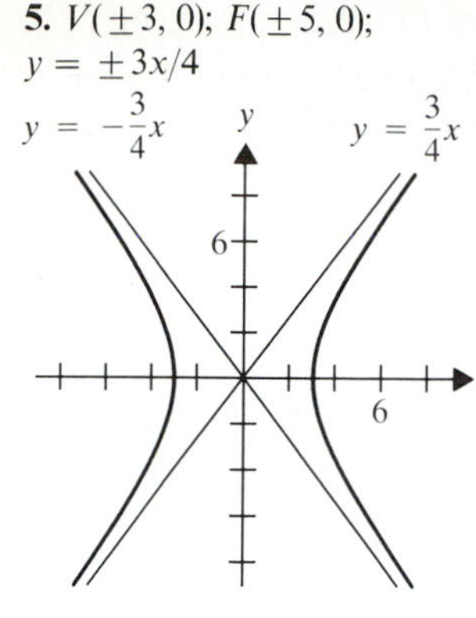

$16x^2 - 9y^2 = 144$

7. $V(0, \pm 5)$; $F(0, \pm 13)$; $y = \pm 5x/12$

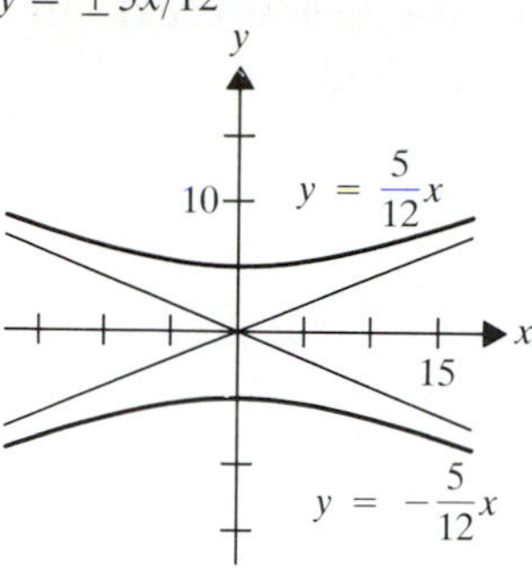

$144y^2 - 25x^2 = 3600$

9. $V(\pm 1, 0)$; $F(\pm\sqrt{2}, 0)$; $y = \pm x$

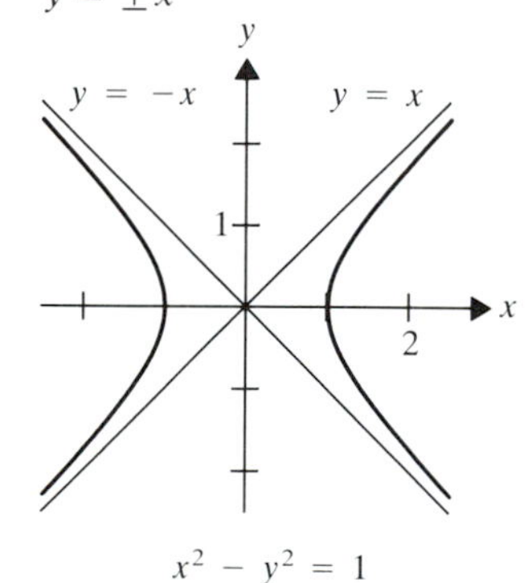

$x^2 - y^2 = 1$

11. $V(\pm 1, 0)$; $F(\pm\sqrt{5}, 0)$; $y = \pm 2x$

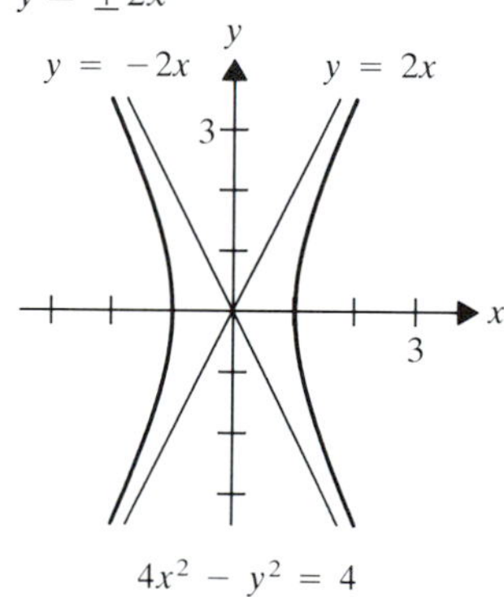

$4x^2 - y^2 = 4$

13. $V(\pm 3, 0)$; $F(\pm 3\sqrt{5}/2, 0)$; $y = \pm x/2$

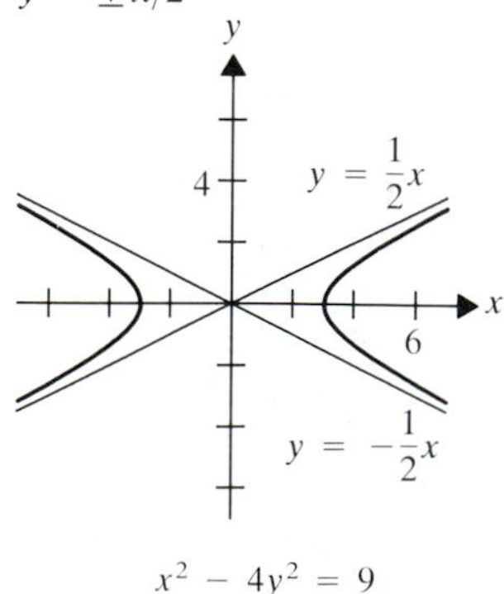

$x^2 - 4y^2 = 9$

15. $V(0, \pm 5)$; $F(0, \pm 5\sqrt{17}/4)$; $y = \pm 4x$

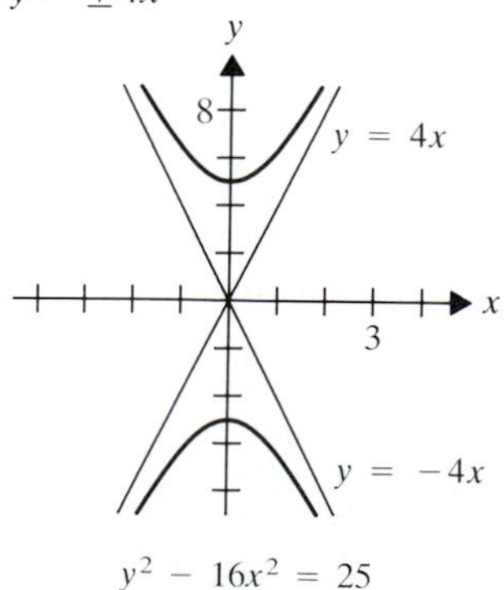

$y^2 - 16x^2 = 25$

17. $V_1(-1, 2)$, $V_2(7, 2)$; $F_1(-2, 2)$, $F_2(8, 2)$; $y - 2 = \pm 3(x - 3)/4$

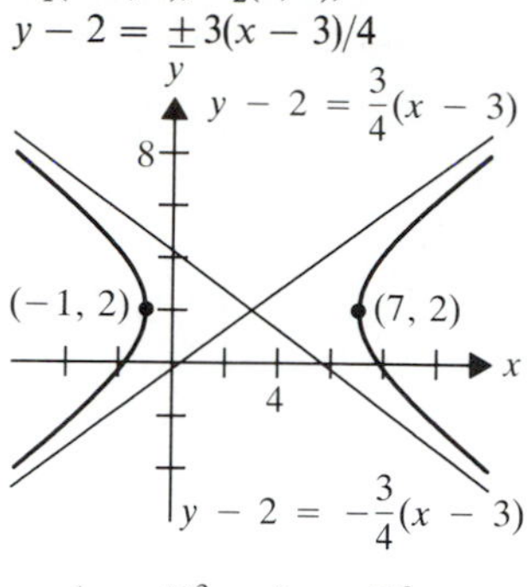

$$\frac{(x-3)^2}{16} - \frac{(y-2)^2}{9} = 1$$

19. $V_1(2, -9)$, $V_2(2, 3)$; $F_1(2, -13)$, $F_2(2, 7)$; $y + 3 = \pm 3(x - 2)/4$

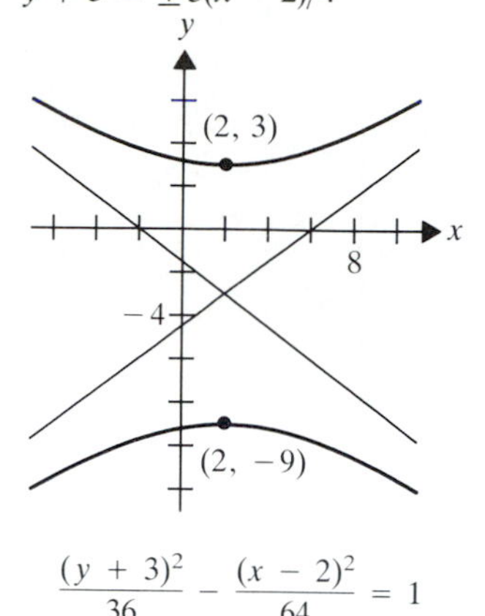

$$\frac{(y+3)^2}{36} - \frac{(x-2)^2}{64} = 1$$

21. $V(-1, \pm 2)$; $F(-1, \pm\sqrt{7})$; $y = \pm 2(x + 1)/\sqrt{3}$

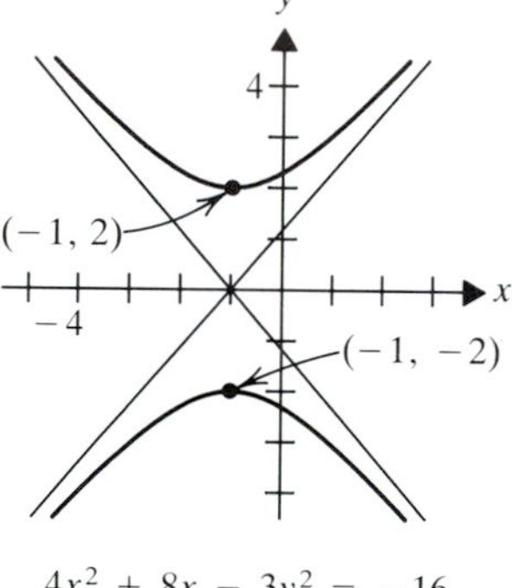

$4x^2 + 8x - 3y^2 = -16$

23. $V_1(-1, -1)$, $V_2(5, -1)$; $F(2 \pm 3\sqrt{5}/2, -1)$; $y + 1 = \pm(x - 2)/2$

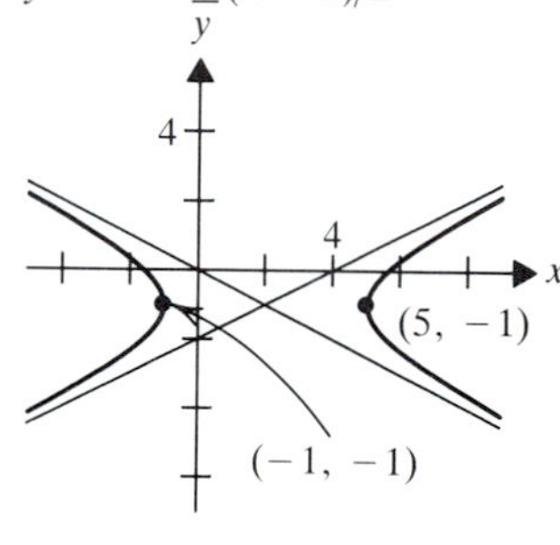

$$\frac{(x-2)^2}{9} - \frac{4(y+1)^2}{9} = 1$$

25. $\dfrac{x^2}{9} - \dfrac{y^2}{16} = 1$ **27.** $\dfrac{y^2}{16} - \dfrac{x^2}{20} = 1$ **29.** $\dfrac{x^2}{36} - \dfrac{y^2}{64} = 1$

31. $\dfrac{y^2}{64} - \dfrac{x^2}{36} = 1$ **33.** $\dfrac{x^2}{16} - \dfrac{y^2}{9} = 1$ **35.** $\dfrac{y^2}{4} - \dfrac{x^2}{4} = 1$

37. $\dfrac{x^2}{9} - \dfrac{4y^2}{9} = 1$ **39.** $\dfrac{(x-2)^2}{4} - \dfrac{(y-2)^2}{5} = 1$

41. $\dfrac{(y-4)^2}{9} - \dfrac{(x+2)^2}{16} = 1$ **43.** $\dfrac{(y+2)^2}{9} - \dfrac{(x-3)^2}{9/4} = 1$

55. 90 km off the coast

EXERCISES 8.4, p. 516

1. $V_1(2, -2)$, $V_2(-2, 2)$; $F_1(-2\sqrt{2}, 2\sqrt{2})$, $F_2(2\sqrt{2}, -2\sqrt{2})$

3. $V_1(1, 1)$, $V_2(-1, -1)$; $F_1(-\sqrt{2}, -\sqrt{2})$, $F_2(\sqrt{2}, \sqrt{2})$

5. $V_1(\sqrt{2}, \sqrt{2})$, $V_2(-\sqrt{2}, -\sqrt{2})$; $F_1(2, 2)$, $F_2(-2, -2)$

7. $(1 + \sqrt{3})X + (1 - \sqrt{3})Y = 4$ **9.** $\dfrac{X^2}{4} + \dfrac{Y^2}{9} = 1$

11. $\dfrac{Y^2}{25} - \dfrac{X^2}{4} = 1$ **13.** $\dfrac{Y^2}{16} - \dfrac{X^2}{4} = 1$ **15.** $Y = 4X^2$

17. $3X^2 + Y^2 = 6$; $\alpha = \pi/4$; ellipse

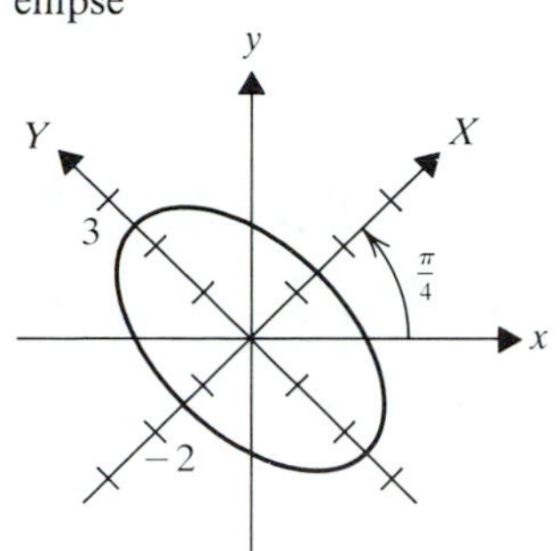

$3X^2 + Y^2 = 6$

19. $X^2 + \dfrac{(Y-1)^2}{4} = 1$; $\alpha \approx 0.644$ rad $\approx 36.9°$; (ellipse)

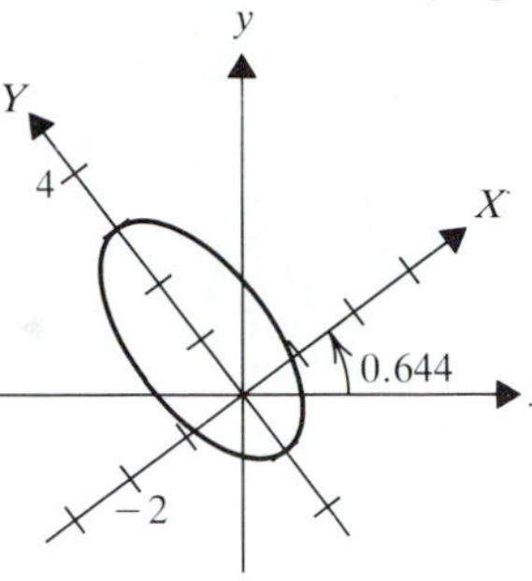
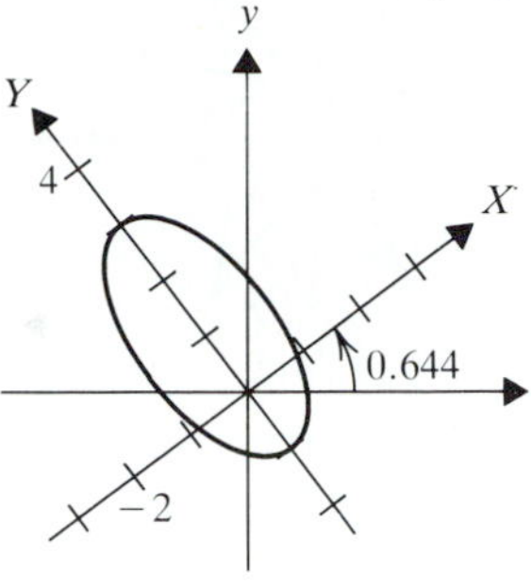

$X^2 + \dfrac{(Y-1)^2}{4} = 2$

21. $Y^2/2 - X^2/10 = 1$; $\alpha = \pi/4$; hyperbola

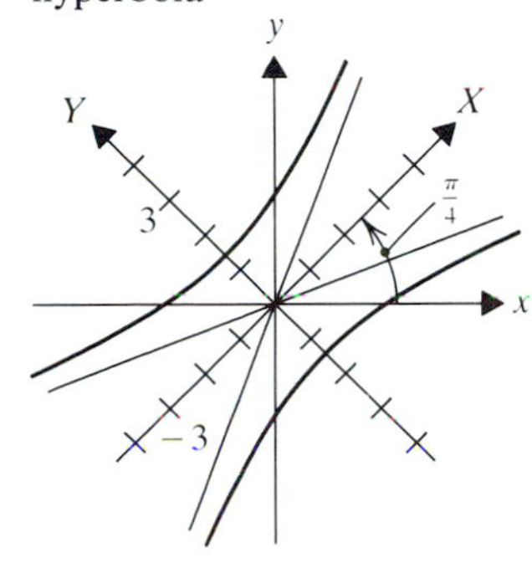

$\dfrac{Y^2}{2} - \dfrac{X^2}{10} = 1$

23. $Y^2/4 - (X+2)^2 = 1$; $\alpha = -\pi/6$; hyperbola

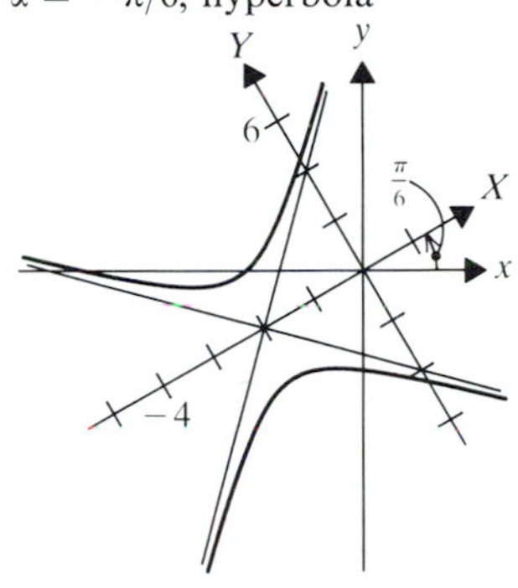

$\dfrac{Y^2}{4} - (X+2)^2 = 1$

25. $\alpha = \pi/4$; $Y - 2 = (X-2)^2$; parabola

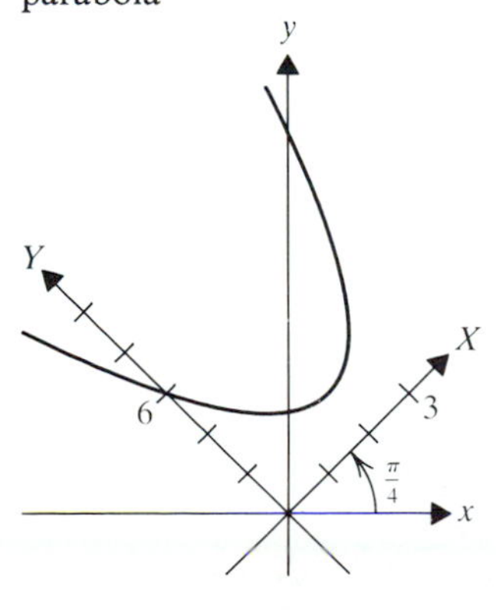

$Y - 2 = (X-2)^2$

27. $\alpha = \pi/4$; $X^2 = 4$; two parallel lines

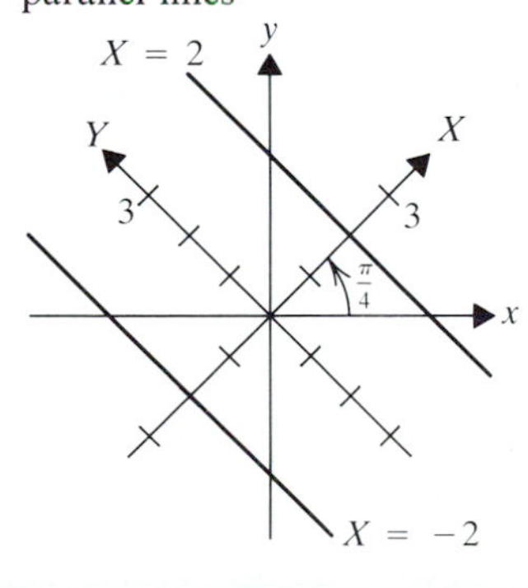

29. Parabola **31.** Hyperbola **33.** Ellipse

EXERCISES 8.5, p. 525

1. Line segment $y = 2x + 3$; $x \in [-1, 3]$

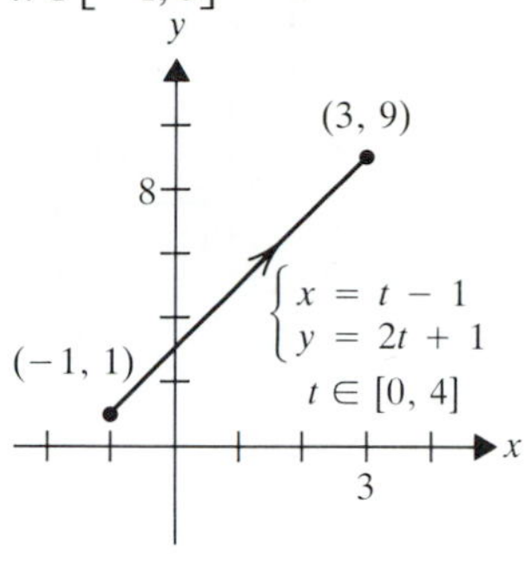

3. Line segment $5x + 6y = 17$; $x \in [-8, 10]$

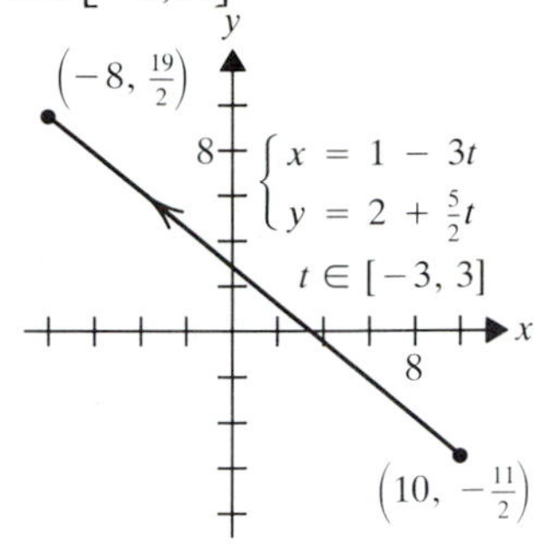

5. Line segment $y = x + 5$; $x \in [-3, -2)$

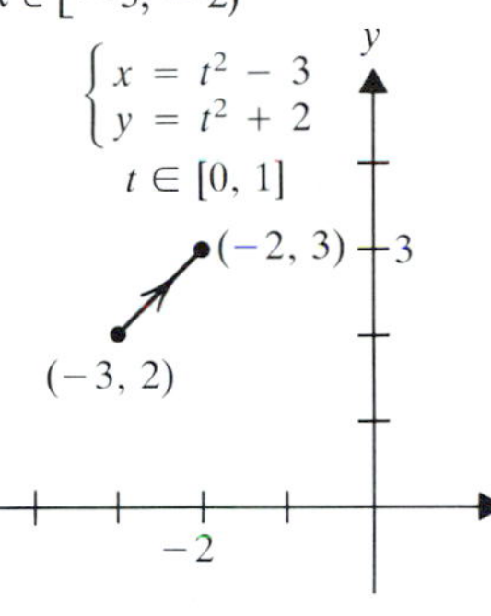

7. $y^2 = 4x/3$; $x \in [0, 12]$

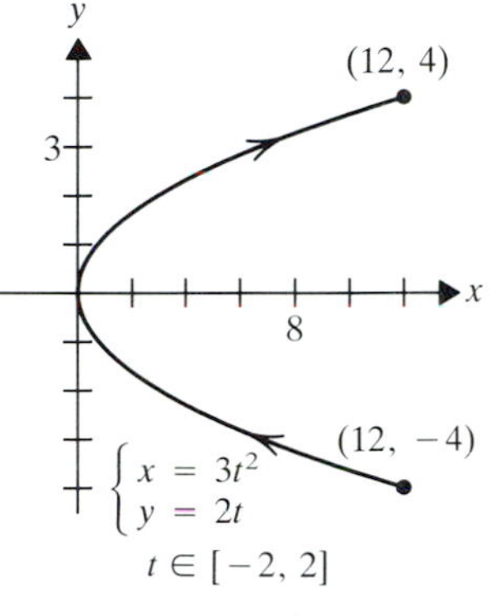

9. $y = (8 - x/2)^{1/2}$; $x \in (-\infty, 8]$

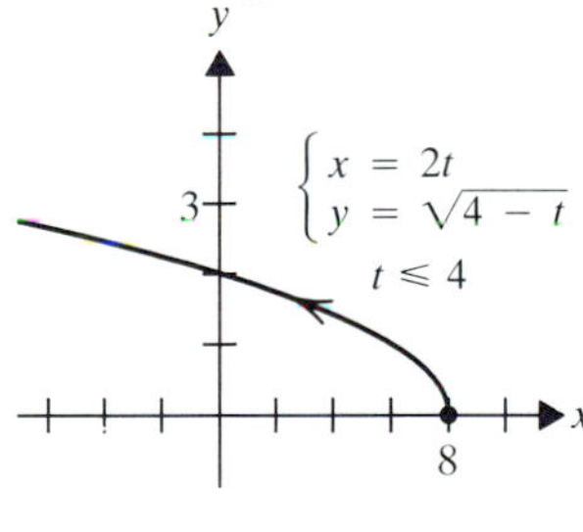

11. $\dfrac{x^2}{9} + \dfrac{y^2}{4} = 1$

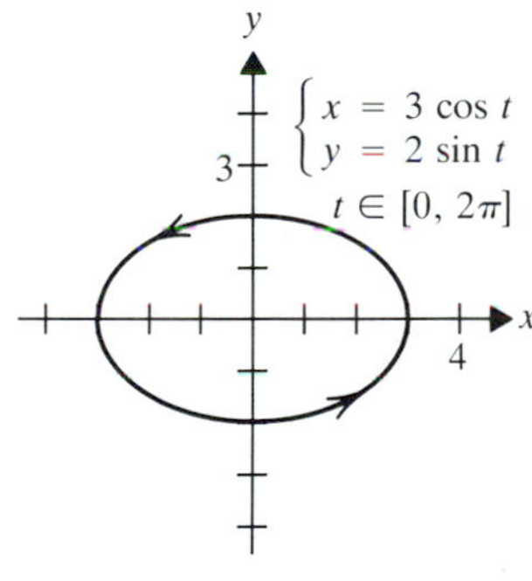

13. $(x+2)^2 + (y-4)^2 = 1$

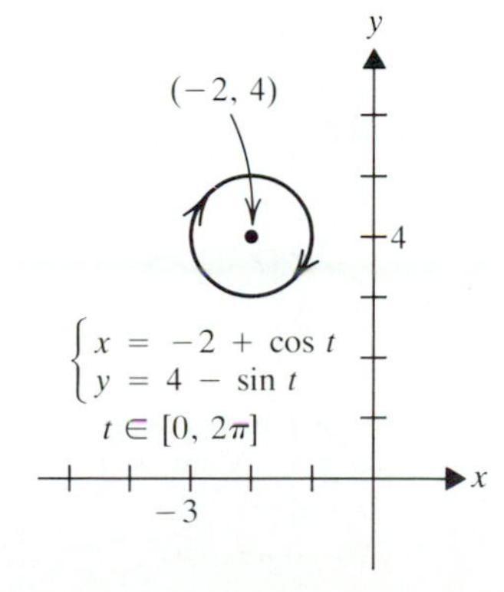

15. $\dfrac{(x-3)^2}{4} + \dfrac{(y-5)^2}{25} = 1$

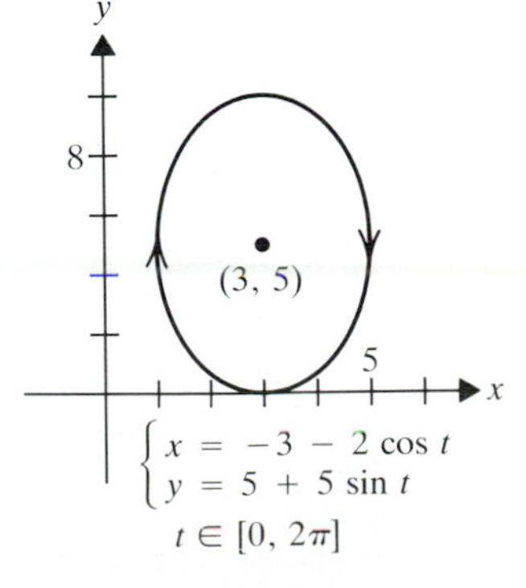

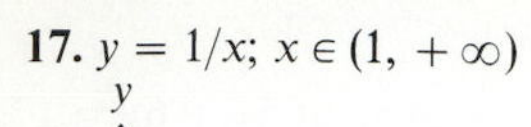
17. $y = 1/x;\ x \in (1, +\infty)$

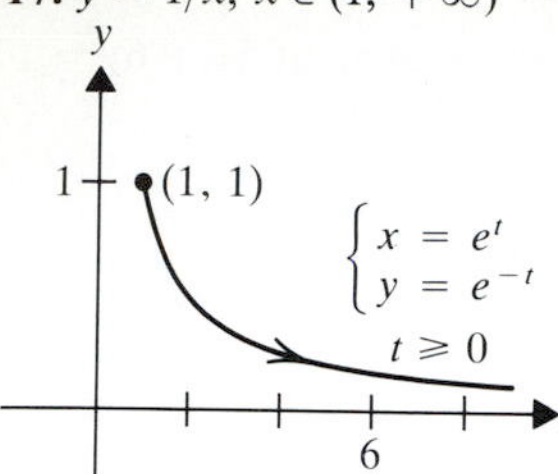

$\begin{cases} x = e^t \\ y = e^{-t} \end{cases}$ $t \geqslant 0$

19. $x^2/4 - y^2/9 = 1$

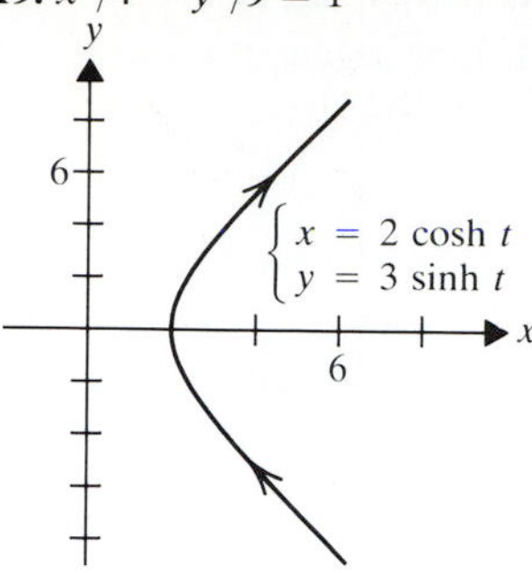

$\begin{cases} x = 2 \cosh t \\ y = 3 \sinh t \end{cases}$

e.

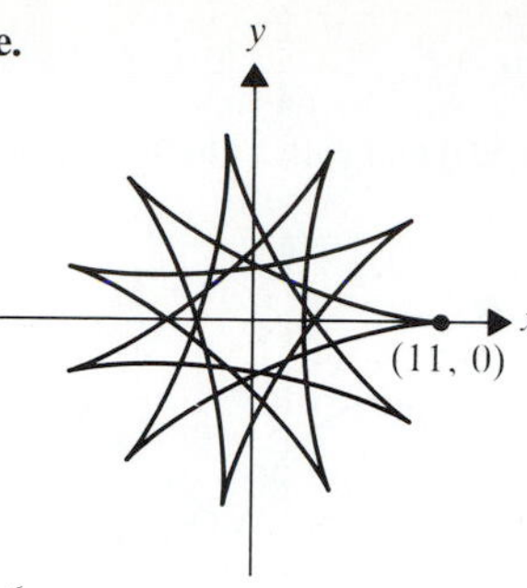

$\begin{cases} x = 7 \cos t + 4 \cos (7t/4) \\ y = 7 \sin t - 4 \sin (7t/4) \end{cases}$
$t \in [0, 8\pi]$

f.

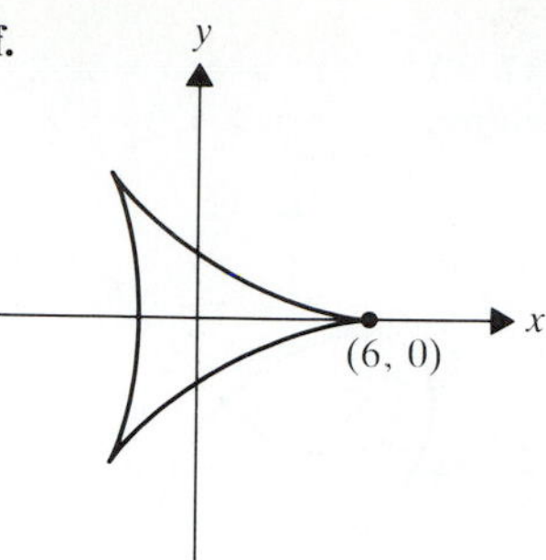

$\begin{cases} x = 4 \cos t + 2 \cos 2t \\ y = 4 \sin t - 2 \sin 2t \end{cases}$
$t \in [0, 4\pi]$

21. 1/3
23. Does not exist (vertical tangent) **25.** $-2/3$
27. $-e^{-4}$ **29.** $4\sqrt{5}$ **31.** 12 **33.** $\sqrt{2}\pi/2$
35. $4(10^{3/2} - 1)/27$ **37.** 6
39. $x = at - b\sin t;\ y = a - b\cos t$ **41c.** No

41d. $6a$ **49.** $\dfrac{d^2y}{dx^2} = \dfrac{\dfrac{dx}{dt}\dfrac{d^2y}{dt^2} - \dfrac{d^2x}{dt^2}\dfrac{dy}{dt}}{\left[\dfrac{dx}{dt}\right]^2}$

g.

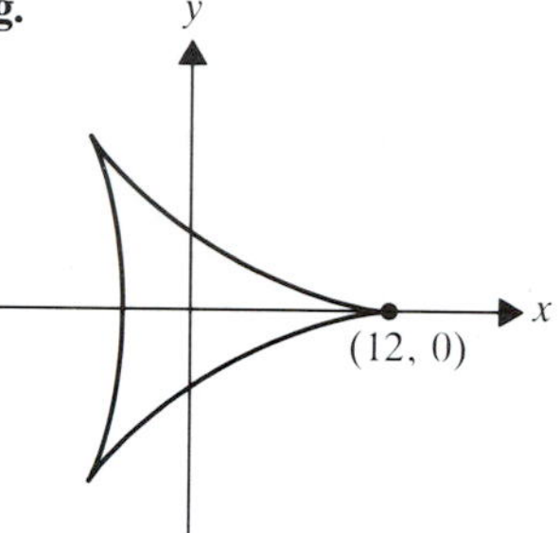

$\begin{cases} x = 8 \cos t + 4 \cos 2t \\ y = 8 \sin t - 4 \sin 2t \end{cases}$
$t \in [0, 8\pi]$

53. a.

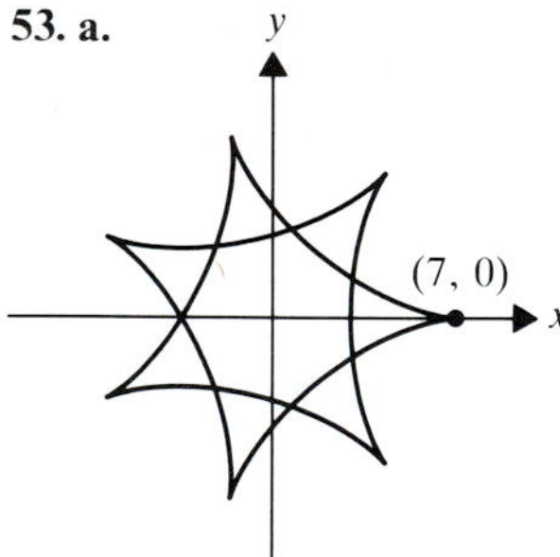

$\begin{cases} x = 5 \cos t + 2 \cos (5t/2) \\ y = 5 \sin t - 2 \sin (5t/2) \end{cases}$
$t \in [0, 4\pi]$

c.

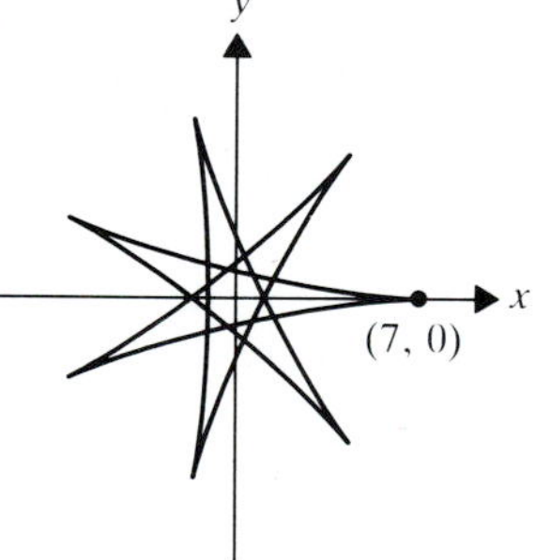

$\begin{cases} x = 4 \cos t + 3 \cos (4t/3) \\ y = 4 \sin t - 3 \sin (4t/3) \end{cases}$
$t \in [0, 6\pi]$

b.

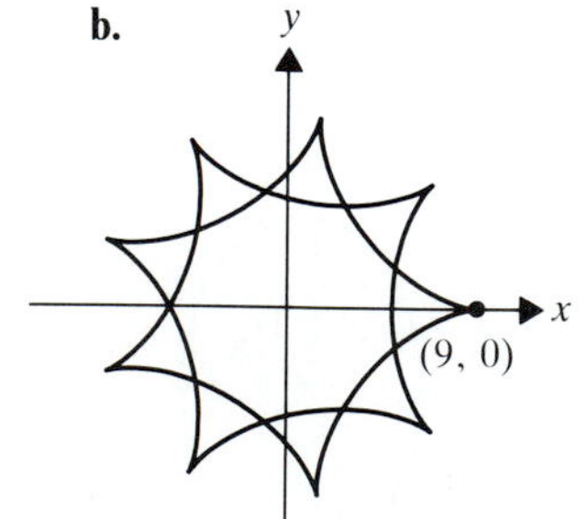

$\begin{cases} x = 7 \cos t + 2 \cos (7t/2) \\ y = 7 \sin t - 2 \sin (7t/2) \end{cases}$
$t \in [0, 4\pi]$

d.

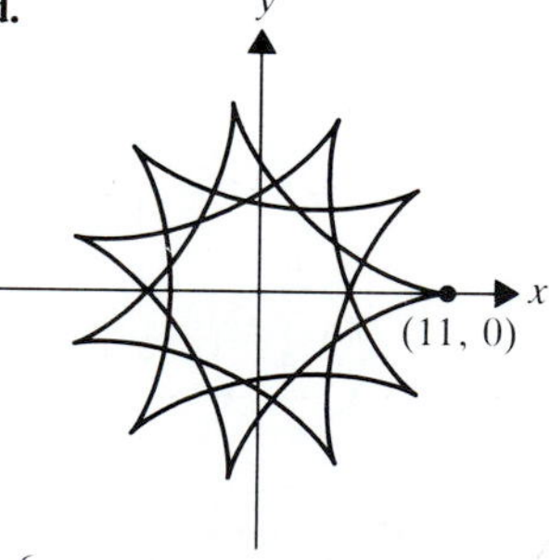

$\begin{cases} x = 8 \cos t + 3 \cos (8t/3) \\ y = 8 \sin t - 3 \sin (8t/3) \end{cases}$
$t \in [0, 6\pi]$

EXERCISES 8.6, p. 535

1.

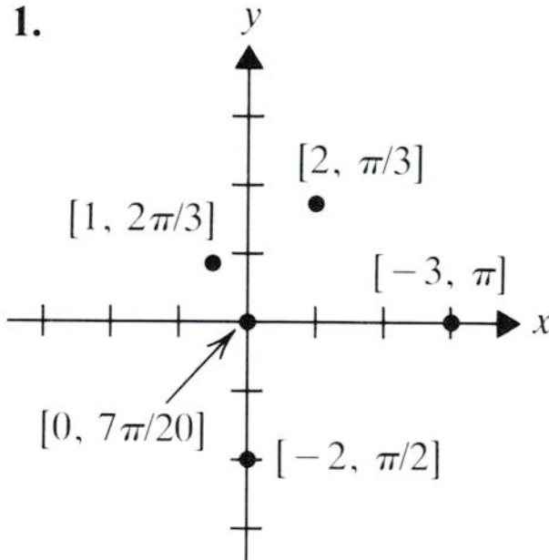

3a. $(1, \sqrt{3})$ **3b.** $(0, -2)$ **3c.** $(3, 0)$ **3d.** $(-1/2, \sqrt{3}/2)$ **3e.** (0
5a. $[\sqrt{5}, \tan^{-1} 2]$ **5b.** $[2, \pi/2]$ **5c.** $[-\sqrt{13}, \tan^{-1} 3/2]$
5d. $[\sqrt{13}, \tan^{-1} 3/2]$ **5e.** $[\sqrt{13}, \tan^{-1}(-3/2)]$ **7.** $r = 3$
9. $r^2 \cos 2\theta = 4$ **11.** $r = 4\cos\theta$ **13.** $r = -8\sin\theta$
15. $r^2 = 16(4\cos^2\theta + \sin^2\theta)^{-1}$ **17.** $y = x/\sqrt{3}$ **19.** $x = 5$
21. $x^2 + y^2 = 36$ **23.** $x^2 + (y - 1)^2 = 1$
25. $x^4 + y^4 + 2x^2y^2 - y^2(1 + 2x) - 2x^3 = 0$ **27.** $x^2 - y^2$ =
29. $2x + y = 1$ **31.** $x^2 = -2y + 1$

33. $r = 1 + 2\cos\theta$

35. $r = 3\cos 2\theta$

37. $r = 3\sin 5\theta$

39. $r = \theta$

41. Cardioid

$r = 1 - \sin\theta$

43. $r = \dfrac{1}{\theta}$

45. Lemniscate

$r^2 = 4\sin 2\theta$

47. Parabola: $x^2/4 = y + 1$

$r = \dfrac{2}{1 - \sin\theta}$

49. Hyperbola: $225(x - 8/15)^2 - 15y^2 = 4$

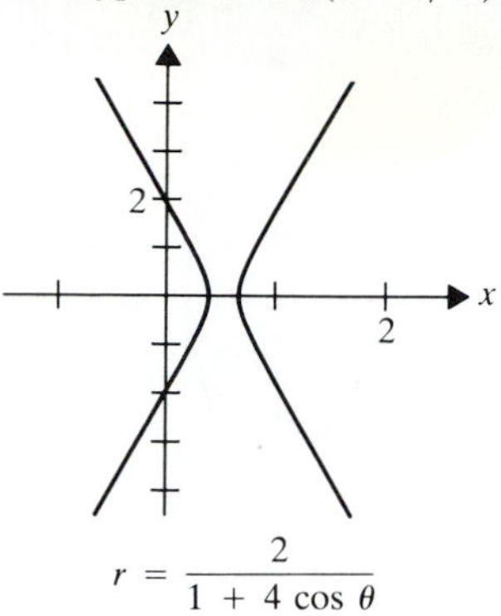

$r = \dfrac{2}{1 + 4\cos\theta}$

51. $r = \dfrac{4}{2 - \cos\theta}$ **53.** $r = \dfrac{12}{1 - 3\cos\theta}$

55. $r = \dfrac{3}{1 - \cos\theta}$ **57.** $r = \dfrac{6}{1 + 2\sin\theta}$

63. Rose with n leaves if n is odd. $2n$ leaves if n is even. As n increases, it becomes star-shaped.

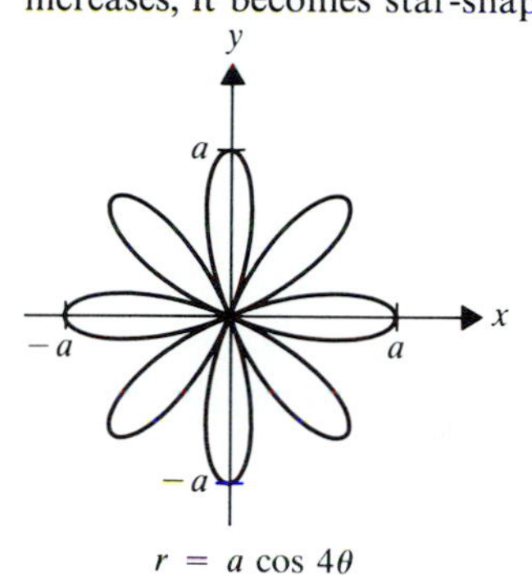

$r = a\cos 4\theta$

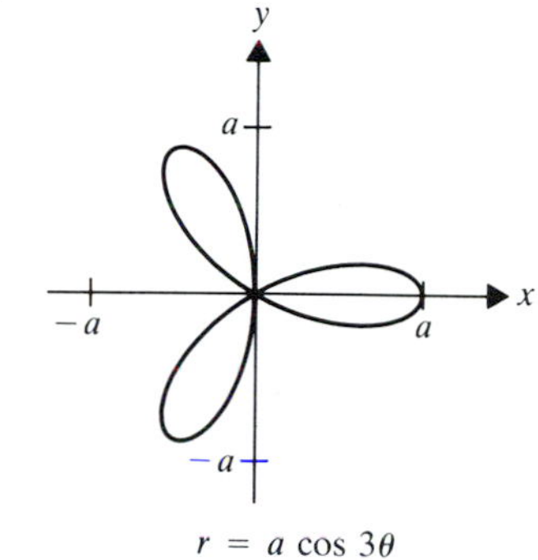

$r = a\cos 3\theta$

EXERCISES 8.7, p. 542

1. 4π **3.** $3\pi/2$ **5.** $9\pi/2$ **7.** 4 **9.** $9\pi/4$ **11.** 2π **13.** π **15.** $\pi/16$ **17.** $9/2$ **19.** $4 + \pi$ **21.** $2 - \pi/4$ **23.** $3\sqrt{3}$ **25.** $8\pi/3 + \sqrt{3}$ **27.** $5\sqrt{3}/2 - \pi/3$ **29.** $\pi - 3\sqrt{3}/2$ **31.** $\pi^3/24$ **33.** $(5\pi - 8)/4$ **35.** $2\pi - 4$ **37.** $\pi/2 - 3\sqrt{3}/4$ **39.** 2 **41.** $8[(\pi^2 + 1)^{3/2} - 1]/3$ **43.** $3\pi/2$ **45.** $\sqrt{5}(e^{\pi} - 1)$ **47.** 4 **49.** 3.141592654 **51.** 7.826465012

REVIEW EXERCISES 8.8, p. 544

1. $y^2 = 12x$

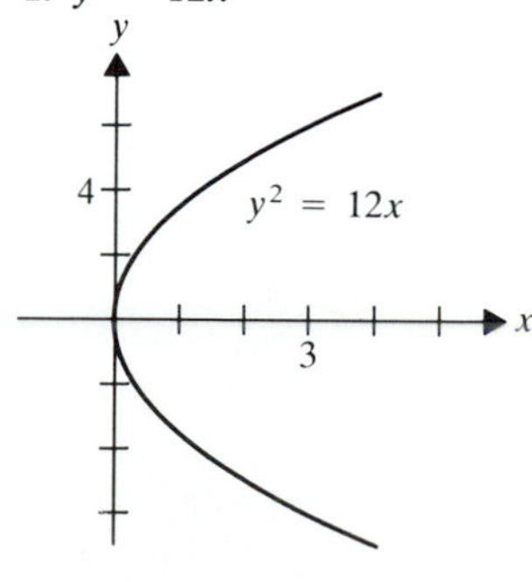

3. $(x - 2)^2 = -4(y - 1)$

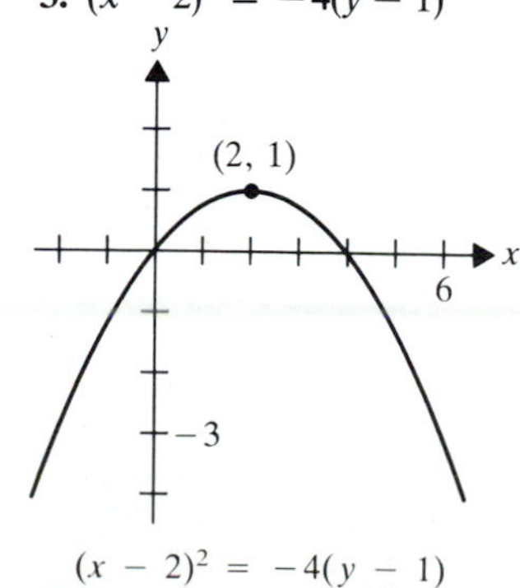

$(x - 2)^2 = -4(y - 1)$

5. $(y-2)^2 = -8(x-2)$

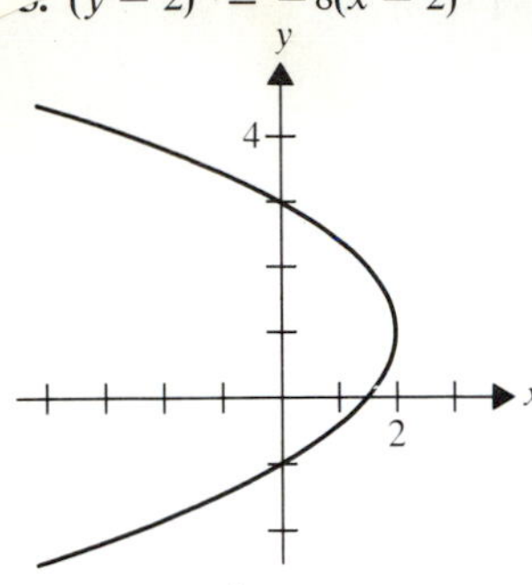

$(y-2)^2 = -8(x-2)$

7. $x^2/9 + y^2/25 = 1$

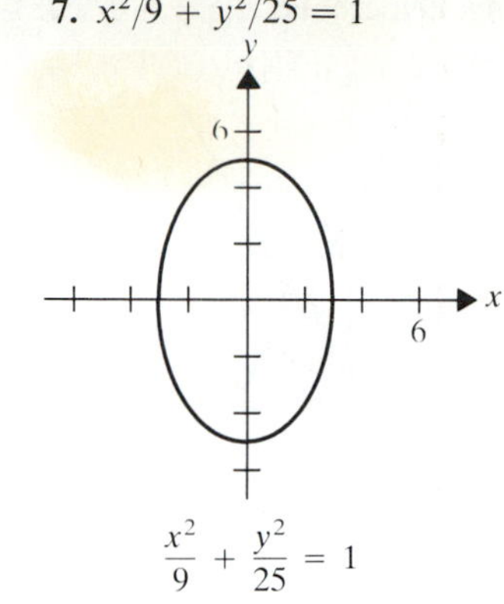

$$\frac{x^2}{9} + \frac{y^2}{25} = 1$$

9. $(x-2)^2/9 + (y+1)^2/25 = 1$

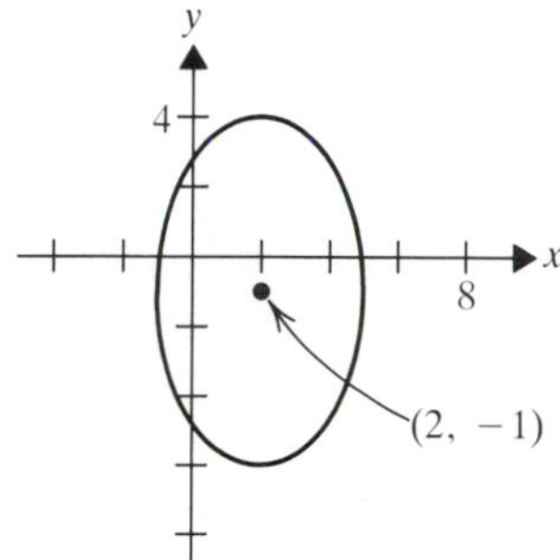

$$\frac{(x-2)^2}{9} + \frac{(y+1)^2}{25} = 1$$

11. $x^2/169 + y^2/25 = 1$

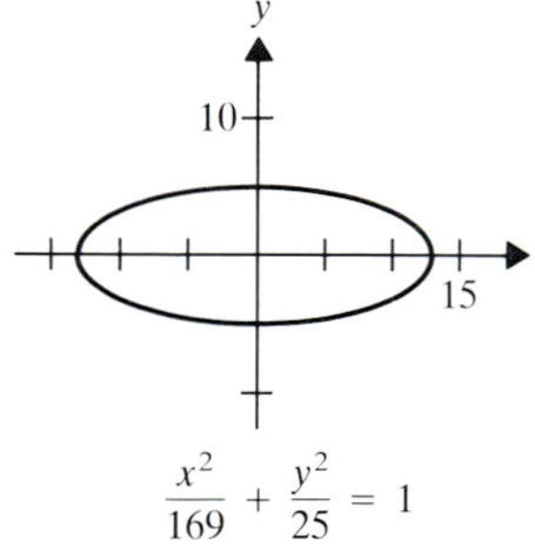

$$\frac{x^2}{169} + \frac{y^2}{25} = 1$$

13. $y^2/9 - x^2/16 = 1$

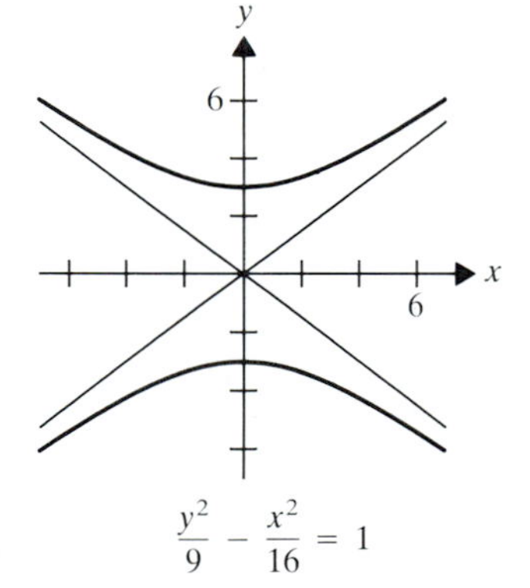

$$\frac{y^2}{9} - \frac{x^2}{16} = 1$$

15. $y^2/9 - 4x^2/9 = 1$

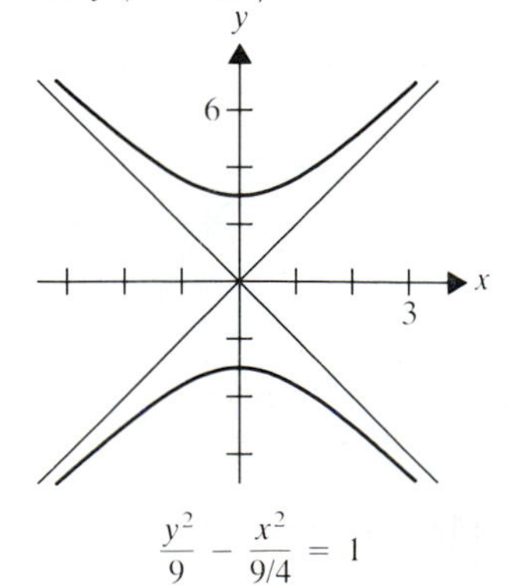

$$\frac{y^2}{9} - \frac{x^2}{9/4} = 1$$

17. Parabola V: $(-2, -1)$; F: $(-29/16, -1)$

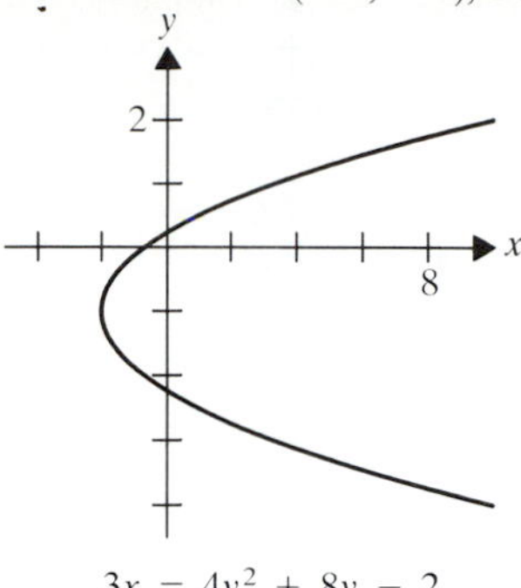

$3x = 4y^2 + 8y - 2$

19. Ellipse V: $(18/5, 0)$, $(2/5, 0)$, $(2, 8/3)$, $(2, -8/3)$;
F: $(2, 32/15)$, $(2, -32/15)$

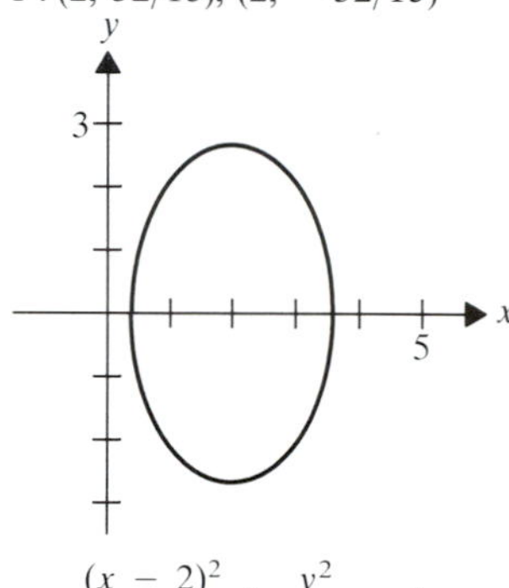

$$\frac{(x-2)^2}{64/25} + \frac{y^2}{64/9} = 1$$

21. Hyperbola V: $(2, 0)$, $(-2, 0)$; F: $(2+\sqrt{7}, 0)$, $(-2-\sqrt{7}, 0)$

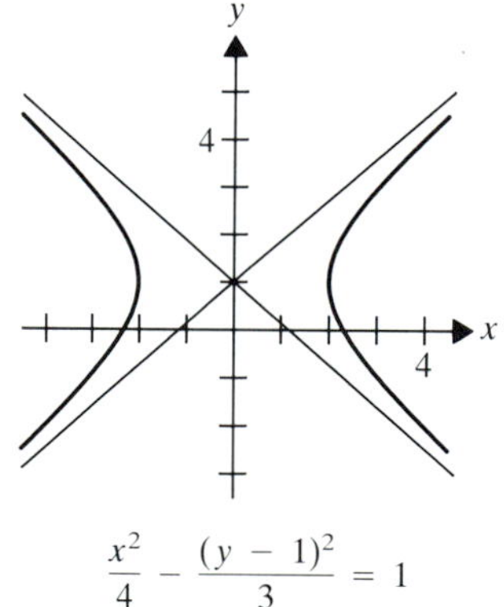

$$\frac{x^2}{4} - \frac{(y-1)^2}{3} = 1$$

23. Hyperbola V: $(3\sqrt{2}, 0)$, $(-3\sqrt{2}, 0)$; F: $(6, 0)$, $(-6, 0)$

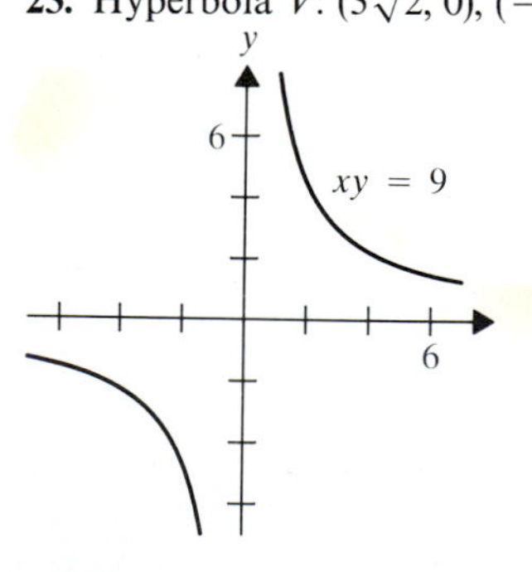

25. $\alpha = \pi/6$; ellipse $X^2/4 + Y^2 = 1$

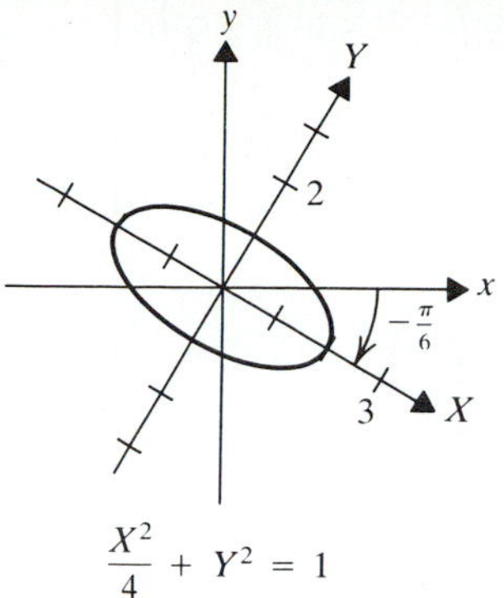

27. $\alpha = \pi/6$; hyperbola $X^2/2 - Y^2/2 = 1$

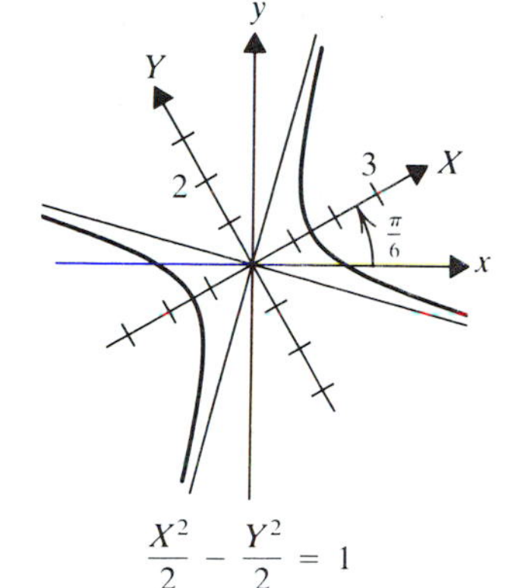

29. Hyperbola **31.** Ellipse **33.** Hyperbola

35. $y = \ln x$; $dy/dx = 1$

$\begin{cases} x = t^2 \\ y = 2 \ln t \end{cases}$ $t > 0$

37. $y = x^2 + 1$; parabola; $dy/dx = 4$ at $(2, 5)$

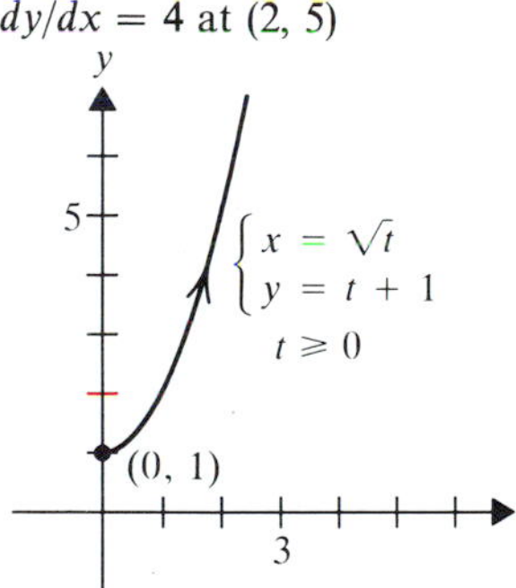

39. $4(x - 2)^2 + 4(y - 3)^2/9 = 1$; ellipse; $dy/dx = -\sqrt{3}$ at $(9/4, 3 + 3\sqrt{3}/4)$

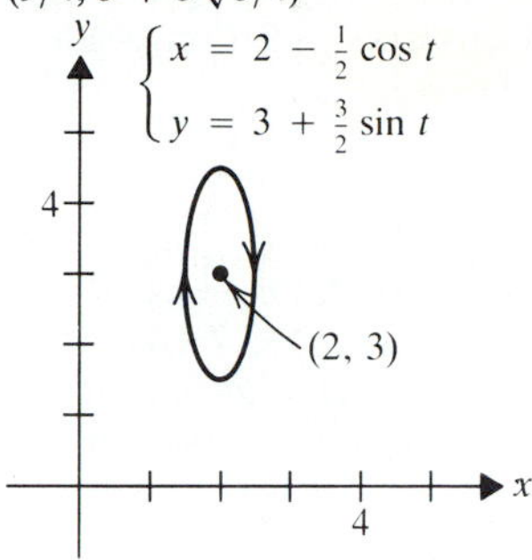

41. $a^2/2 + a$ **43.** $\pi(1 + \pi^2)^{1/2}/2 + (1/2)\ln|\pi + (1 + \pi^2)^{1/2}|$

45. 2 **47.** $r^2 \cos 2\theta + 4 = 0$ **49.** $(x - 1)^2 + (y - 1)^2 = 2$

51.

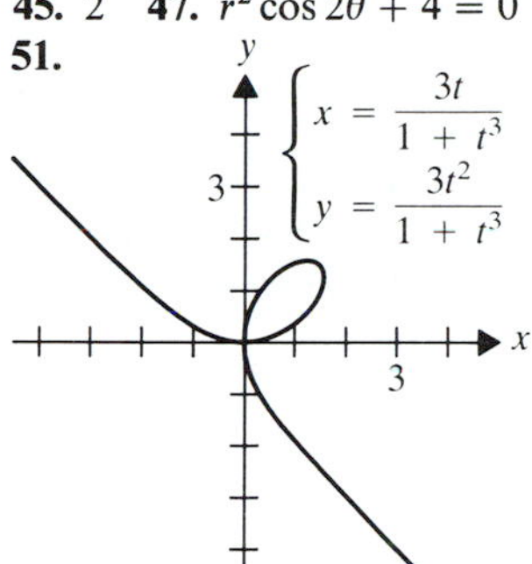

53.

$r = 1 + 2 \sin \theta$

55.

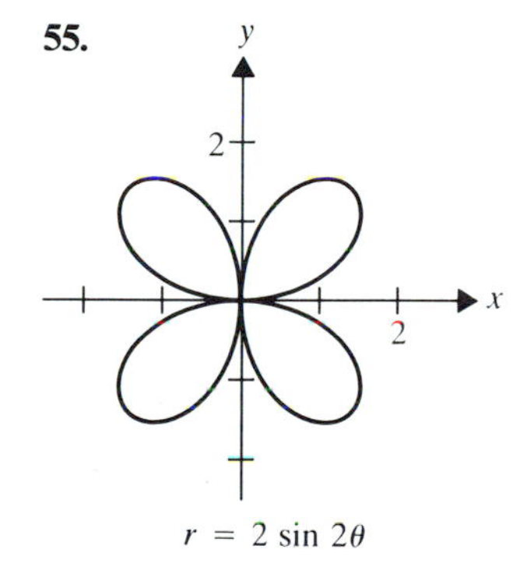

57. 2π **59.** π

61. $\sqrt{3}/4 - \pi/6$ **63a.** Rotated ellipse

63b. $(\sqrt{3}, 2)$, $(-\sqrt{3}, -2)$ at $t = \pi/3, 4\pi/3$

CHAPTER 9

EXERCISES 9.1, p. 557

1. $2^n/n!$ **3.** $(-1)^n$ **5.** $1/n^2$ **7.** Converges to 2/3
9. Diverges to $+\infty$ **11.** Converges to 0 **13.** Converges to 0
15. Diverges **17.** Converges to 1 **19.** Diverges to $+\infty$
21. Converges to $\pi/2$ **23.** Converges to e^a
25. Converges to 1 **27.** Diverges **29.** Converges to 0
47. $(-1)^n/n$
49. 4, 4, 8, 12, 20, 32, 52, 84, 136, 220, 356, 576, 932
55. $f(x) = \sin \pi x$

EXERCISES 9.2, p. 566

1. Divergent **3.** Divergent **5.** Convergent to 20
7. Convergent to 2/3 **9.** Divergent
11. Convergent to $(\sin \theta)/(2 + \sin \theta)$ **13.** Convergent to 8/5
15. Convergent to 1/3 **17.** Divergent **19.** Divergent
21. Divergent **23.** Convergent to 1 **25.** Divergent
31a. 1 **31b.** 1/3
35. $PV = [600e^{-0.08/12}(1 - e^{-0.8})]/[1 - e^{-0.08/12}] \approx$ \$49,395.38

EXERCISES 9.3, p. 575

1. Converges **3.** Converges **5.** Diverges **7.** Diverges **9.** Converges **11.** Diverges **13.** Converges **15.** Converges **17.** Diverges **19.** Converges **21.** Diverges **25.** Yes, it does. **27.** Yes, it does.

EXERCISES 9.4, p. 580

1. Diverges **3.** Converges **5.** Converges **7.** Diverges **9.** Converges **11.** Diverges **13.** Converges **15.** -0.632 **17.** -0.901 **19.** 0.333 **21.** -0.405 **23.** 0.393 **25.** $n = 10$ **27.** $n = 19{,}999$ **31.** $1/21^3 \approx 0.000108$ **33.** 0.631 **35.** 0.69

EXERCISES 9.5, p. 588

1. Absolutely convergent **3.** Absolutely convergent **5.** Absolutely convergent **7.** Absolutely convergent **9.** Conditionally convergent **11.** Divergent **13.** Absolutely convergent **15.** Absolutely convergent **17.** Divergent **19.** Absolutely convergent **21.** Absolutely convergent **23.** Divergent **25.** Conditionally convergent **27.** Absolutely convergent **29.** Absolutely convergent **31.** Absolutely convergent **33.** Divergent **35.** Divergent **37.** Convergent **43.** Divergent

EXERCISES 9.6, p. 600

1. $x - x^3/3! + x^5/5! - x^7/7! + x^9/9!$; $r_9(x) = -(\sin c)x^{10}/10!$, c between 0 and x.
3. $x - x^2/2 + x^3/3 - x^4/4 + x^5/5 - x^6/6$; $r_6(x) = x^7/7(1 + c)^7$, c between 0 and x.
5. $1 - x/2 + x^2/8 - x^3/48 + x^4/(2^4 4!) - x^5/(2^5 5!) + x^6/2^6 6!$; $r_6(x) = -e^{-c/2}x^6/128$, c between 0 and x.
7. $x + x^3/3$; $r_4(x) = [8\sec^6 c\tan^2 c + 36\tan^4 c\sec^3 c - 176\sec^5 c\tan^2 c - 8\sec^7 c]\cdot(x^5/5!)$, c between 0 and x
9. $\sin\pi/8 + (\cos\pi/8)(x - \pi/4)/2 - (\sin\pi/8)(x - \pi/4)^2/8 - (\cos\pi/8)(x - \pi/4)^3/48 + (\sin\pi/8)(x - \pi/4)^4/384 + (\cos\pi/8)(x - \pi/4)^5/3840$; $r_5(x) = (\sin c)(x - \pi/4)^6/46080$, c between x and $\pi/8$
11. $\sqrt{2} + 2\sqrt{2}\left(x - \frac{\pi}{4}\right) - \frac{\sqrt{2}}{2}\left(x - \frac{\pi}{4}\right)^2 - \frac{\sqrt{2}}{3}\left(x - \frac{\pi}{4}\right)^3 + \frac{\sqrt{2}}{24}\left(x - \frac{\pi}{4}\right)^4$; $r_5(x) = \frac{1}{5!}[3\cos c + \sin c]\left[x - \frac{\pi}{4}\right]^5$, c between x and $\frac{\pi}{4}$
13. $\frac{1}{2} - \frac{\sqrt{3}}{2}\left(x - \frac{\pi}{3}\right) - \frac{1}{4}\left(x - \frac{\pi}{3}\right)^2 + \frac{\sqrt{3}}{12}\left(x - \frac{\pi}{3}\right)^3 + \frac{1}{48}\left(x - \frac{\pi}{3}\right)^4 - \frac{\sqrt{3}}{240}\left(x - \frac{\pi}{3}\right)^5 - \frac{1}{1440}\left(x - \frac{\pi}{3}\right)^6$; $r_6(x) = \frac{1}{5040}\left(x - \frac{\pi}{3}\right)^7$, c between x and $\pi/4$
15. $2 + (x - 8)/12 - (x - 8)^2/288 + 5(x - 8)^3/20736 - 5(x - 8)^4/248832$; $r_4(x) = (880c^{-14/3}/243)(x - 8)^5$, c between x and 8
17. $f(x) - P_3(x) = 1 + 2(x - 1) + (x - 1)^2 + (x - 1)^3$; $r_3(x) = 0$
19. $\ln\frac{1}{2} - \sqrt{3}\left(x - \frac{\pi}{3}\right) - 2\left(x - \frac{\pi}{3}\right)^2 - \frac{4\sqrt{3}}{3}\left(x - \frac{\pi}{3}\right)^3$; $r_3(x) = -k(x - \pi/3)^4/4!$ where $k = 4\sec^2 c\tan^2 c + 2\sec^4 c$, c between x and $\pi/3$
21. 0.2079 **23.** 0.0198 **25.** 1.6487 **27.** 0.8829 **29.** 2.04939

EXERCISES 9.7, p. 608

1. Absolute convergence on $(-\infty, +\infty)$
3. Absolute convergence only at $x = 0$
5. Absolute convergence on $(-3/2, 3/2)$; conditional convergence at $x = 3/2$
7. Absolute convergence on $(0, 2)$
9. Absolute convergence on $[-1, 1]$
11. Absolute convergence on $(-2, 0)$; conditional convergence at 0
13. Absolute convergence on $(1, 3)$
15. Absolute convergence on $(1/4, 3/4)$; conditional convergence at $1/4$
17. Absolute convergence on $(1, 3)$; conditional convergence at 1
19. Absolute convergence on $[2, 4]$
21. Absolute convergence on $(-1, 1)$; conditional convergence at -1
23. Absolute convergence on $(-1, 1)$
25. Absolute convergence on $(-8, 8)$
27. 1 **29.** 1

EXERCISES 9.8, p. 617

13. 0.4613 **15.** 0.3103 **17.** 0.1000 **19.** 0.0048
21b. $x - x^3/3 + x^5/5 + x^7/7 + \ldots + (-1)^n x^{2n+1}/(2n + 1) + \ldots$
21c. $f^{(6)}(0) = 0$; $f^{(5)}(0) = 24$
23b. $1 + 2x + 3x^2 + \ldots + nx^{n-1} + \ldots$; $x \in (-1, 1)$
25a. $(-1, 1]$ **25b.** $(-1, 1)$
27. $1 + x/2 + x^2/3! + x^3/4! + \ldots + x^{n-1}/n!$
29. $x + x^2 + x^3/2! + x^4/3! + \ldots + x^{n+1}/n!$
31. $\frac{2}{\sqrt{\pi}}\sum_{n=0}^{+\infty}\frac{(-1)^k x^{2k+1}}{(2k + 1)k!}$ **37a.** At least 10,000

EXERCISES 9.9, p. 625

1. $1 + \sum_{n=0}^{\infty}(-1)^n\frac{(-1)\cdot 1\cdot 3\cdot\ldots\cdot(2n - 3)}{2^n n!}x^n$; $x \in (-1, 1)$
3. $1 + \sum_{n=1}^{\infty}(-1)^n\frac{1\cdot 4\cdot\ldots\cdot(3n - 2)}{3^n n!}x^n$; $x \in (-1, 1)$
5. $1 - \frac{x}{3} + \sum_{n=2}^{\infty}(-1)^{2n-1}\frac{2\cdot 5\cdot\ldots\cdot(3n - 4)}{3^n n!}x^n$; $x \in (-1, 1)$
7. $\sqrt{2}\left[1 + \sum_{n=1}^{\infty}(-1/2)^n\frac{(-1)\cdot 3\cdot\ldots\cdot(2n - 3)}{2^n n}x^n\right]$; $x \in (-1, 1)$
9. $1 + \frac{2x^2}{3} + \sum_{n=2}^{\infty}(-1)^{2n-1}\frac{2\cdot 5\cdot\ldots\cdot(3n - 4)}{3^n n!}[-2x^2]^n$; $x \in (-1, 1)$
11. $1 + x/2 + 3x^2/8 + x^3/16$
13. $1 - x^2k^2/2 + x^4(k^2/6 - k^2/8) + x^6(k^4/12 + k^2/36 - k^6/16)$

15. 5.0990 **17.** 1.8171 **19.** 5.0100 **21.** 10.0033 **23.** 0.3294 **25.** 0.5082 **27.** 0.2505 **29.** 0.485 **31.** $0.4740\sqrt{5\pi} \approx 3.3297$

REVIEW EXERCISES 9.10, p. 626

1a. $a_n = (-1)^n \dfrac{n+2}{3^n}$ **1b.** $a_n = \dfrac{3 \cdot 2^{n-1}}{5^{n-1}}$ **3a.** Increasing, divergent **3b.** Decreasing, convergent **7.** Divergent **9.** Absolutely convergent **11.** Conditionally convergent **13.** Absolutely convergent **15.** Absolutely convergent **17.** Divergent **19.** Conditionally convergent **21a.** 1.6134 **21b.** 1.3956 **29.** $1 - x^2/2 - x^3 - 9x^4/8 - \ldots$

CHAPTER 10

EXERCISES 10.1, p. 636

1a. (5, 7) **1b.** (−1, 3) **1c.** (5, −4) **1d.** (0, 11) **1e.** $\sqrt{29}$; $\sqrt{13}$ **1f.** $\dfrac{1}{\sqrt{29}}(2, 5)$; $\dfrac{1}{\sqrt{13}}(3, 2)$ **3a.** (−1, 1) **3b.** (3, −5) **3c.** (−8, 13) **3d.** (7, −12) **3e.** $|\mathbf{v}| = \sqrt{5}$; $|\mathbf{w}| = \sqrt{13}$ **3f.** $\mathbf{u}_v = (1/\sqrt{5}, -2/\sqrt{5})$; $\mathbf{u}_w = (-2/\sqrt{13}, 3/\sqrt{13})$ **5a.** $4\mathbf{i} + 5\mathbf{j}$ **5b.** $2\mathbf{i} - 3\mathbf{j}$ **5c.** $-3\mathbf{i} + 10\mathbf{j}$ **5d.** $7\mathbf{i} - 5\mathbf{j}$ **5e.** $|\mathbf{v}| = \sqrt{10}$; $|\mathbf{w}| = \sqrt{17}$ **5f.** $\mathbf{u}_v = \dfrac{3}{\sqrt{10}}\mathbf{i} + \dfrac{1}{\sqrt{10}}\mathbf{j}$; $\mathbf{u}_w = \dfrac{1}{\sqrt{17}}\mathbf{i} + \dfrac{4}{\sqrt{17}}\mathbf{j}$ **7a.** $-\mathbf{i} + 3\mathbf{j}$ **7b.** $5\mathbf{i} - 5\mathbf{j}$ **7c.** $-13\mathbf{i} + 14\mathbf{j}$ **7d.** $12\mathbf{i} - 11\mathbf{j}$ **7e.** $|\mathbf{v}| = \sqrt{5}$; $|\mathbf{w}| = 5$ **7f.** $\mathbf{u}_v = \dfrac{2}{\sqrt{5}}\mathbf{i} - \dfrac{1}{\sqrt{5}}\mathbf{j}$; $\mathbf{u}_w = -\dfrac{3}{5}\mathbf{i} + \dfrac{4}{5}\mathbf{j}$ **9a.** $-3\mathbf{i} + 2\mathbf{j}$ **9b.** $-3\mathbf{i} - 2\mathbf{j}$ **9c.** $6\mathbf{i} + 6\mathbf{j}$ **9d.** $-9\mathbf{i} - 4\mathbf{j}$ **9e.** $|\mathbf{v}| = 3$; $|\mathbf{w}| = 2$ **9f.** $\mathbf{u}_v = -\mathbf{i}$; $\mathbf{u}_w = \mathbf{j}$

11. $\mathbf{AB} = (3, 2)$; $|\mathbf{AB}| = \sqrt{13}$

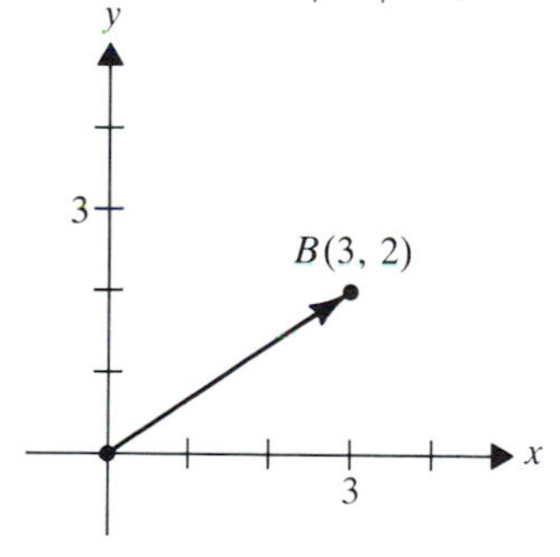

13. $\mathbf{AB} = (3, 4)$; $|\mathbf{AB}| = 5$

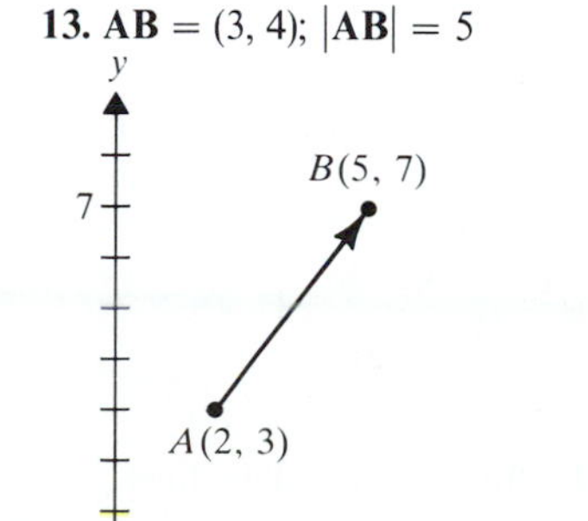

15. $\mathbf{AB} = (-3, -4)$; $|\mathbf{AB}| = 5$

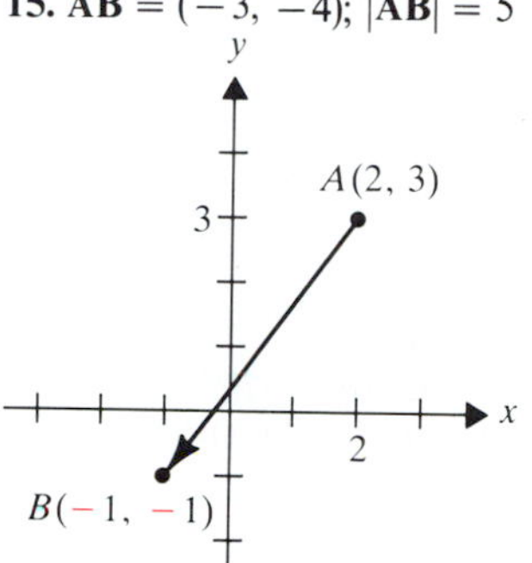

17. $\mathbf{AB} = (5, 12)$; $|\mathbf{AB}| = 13$

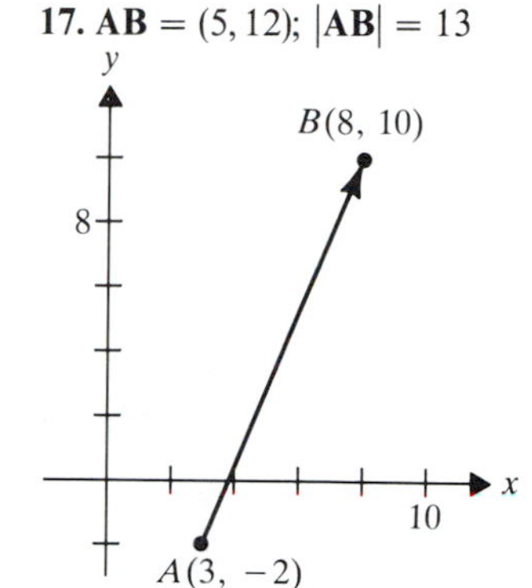

19. $\mathbf{AB} = (1, -1)$; $|\mathbf{AB}| = \sqrt{2}$

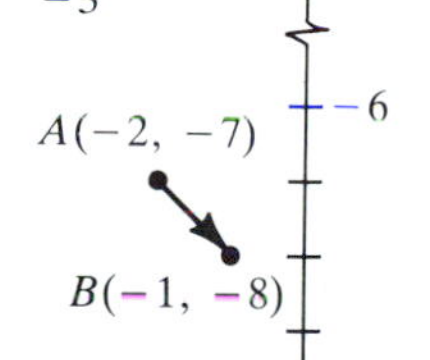

21. Parallel **23.** Nonparallel **25.** Nonparallel **27.** Nonparallel **29.** Nonparallel **31.** Parallel **33.** $-\mathbf{i} - \mathbf{j}$ **35.** $m = 2/3$

EXERCISES 10.2, p. 643

1.

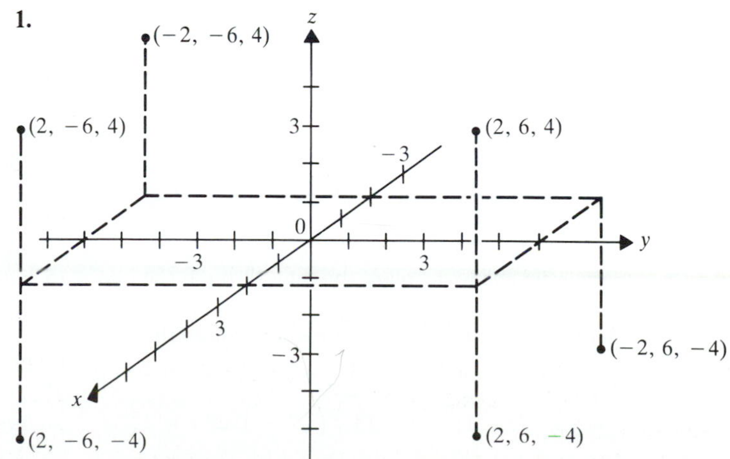

3.

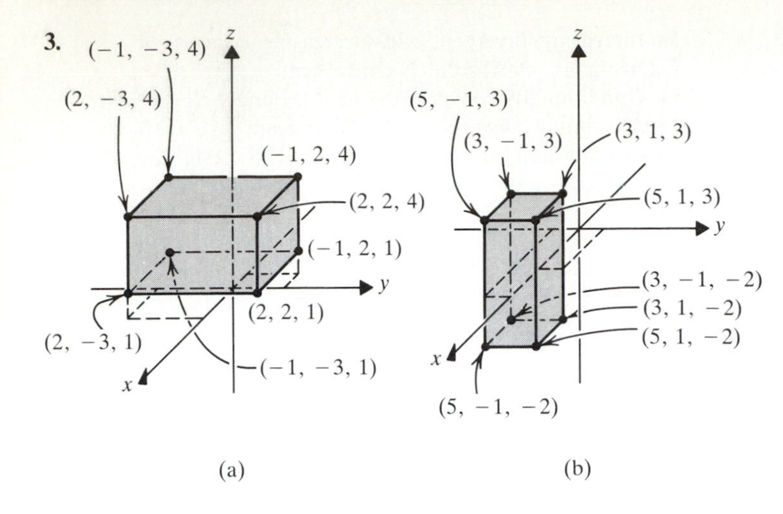

5a. 7; $(-1/2, 1, 4)$ **5b.** 4; $(1, 5/2, 0)$ **7a.** $z = 0$ **7b.** $y = -2$
9a. $y = 0$ **9b.** $x = 3$ **11a.** $(1, 2, 3)$ **11b.** $(-3, 4, 1)$
11c. $(3, -4, -1)$ **11d.** $(-8, 14, 6)$ **11e.** $(10, -10, 0)$
11f. $\sqrt{14}$ **11g.** $\sqrt{6}$ **11h.** $(1/\sqrt{14})(-1, 3, 2)$; $(1/\sqrt{6})(2, -1, 1)$
13a. $(0, -1, 2, 4)$ **13b.** $(2, -3, -2, 2)$ **13c.** $(-2, 3, 2, -2)$
13d. $(6, -10, -4, 10)$ **13e.** $(-6, 8, 8, -2)$ **13f.** $\sqrt{14}$
13g. $\sqrt{7}$ **13h.** $(1/\sqrt{14})(1, -2, 0, 3)$; $1/\sqrt{7}(-1, 1, 2, 1)$
15a. $(-3, 0, -2)$; $\sqrt{13}$
15b. $(1, -3, 2)$; $\sqrt{14}$

17a. $(-2, 0, 4, 2)$; $2\sqrt{6}$ **17b.** $(3, -4, 3, 2)$; $\sqrt{38}$
19a. Plane perpendicular to xy-plane, passing through the line $y = 3x/2 - 3$ in the xy-plane.
19b. Plane perpendicular to yz-plane, passing through the line $y = 3z/2$ in the yz-plane.
21a. $(x - 1)^2 + (y + 2)^2 + (z + 2)^2 = 4$
21b. $(x - 2)^2 + y^2 + (z - 1)^2 = 9$ **23.** $C(3/2, -1, 2)$; $r = 3$

25a. No **25b.** Yes
35a. Orders for East, North, South, and West, respectively
35b. $1.02\mathbf{x}$, $0.978\mathbf{y}$, $1.011\mathbf{z}$, $0.998\mathbf{w}$ **39.** 1.9106 rad $\approx 109.47°$

EXERCISES 10.3, p. 654

1a. $\pi/4$ **1b.** 0.6435 rad **3a.** $5\pi/6$ **3b.** $\pi/3$ **5a.** $\pi/3$
5b. 1.11 rad **7a.** $(1/\sqrt{2})(1, -1, 0)$; $\alpha = \pi/4$, $\beta = 3\pi/4$, $\gamma = \pi/2$
7b. $(1/2)(\sqrt{3}, 0, 1)$; $\alpha = \pi/6$, $\beta = \pi/2$, $\gamma = \pi/3$
9a. $(1/2)(\sqrt{2}, 1, 1)$; $\alpha = \pi/4$, $\beta = \pi/3$, $\gamma = \pi/3$
9b. $(1/5)(4, 2, -\sqrt{5})$, $\alpha = 0.644$, $\beta = 1.159$, $\gamma = 2.034$
11. $(1/\sqrt{13})(3\mathbf{i} + 2\mathbf{j})$; two **13.** $(1/\sqrt{27})(5, 1, 1)$; infinitely many
15. $(1/\sqrt{8})(2, 0, -2, 0)$; infinitely many **17.** $x = -4/3$; no
19. $x = 3$; yes, $x = -5/3$ **21.** $\sqrt{5}$; $(1, 2)$
23. $7/\sqrt{6}$; $(7/6)(\mathbf{i} + \mathbf{j} + 2\mathbf{k})$
25. $(6 + \sqrt{2})/4$; $((6 + \sqrt{2})/16)(-1, 2, 3, \sqrt{2})$ **27.** 15

EXERCISES 10.4, p. 662

1a. $\mathbf{x} = (2, 1) + t(3, -4)$ **1b.** $\mathbf{x} = t(0, 1)$
1c. $\mathbf{x} = (-7, -2) + t(11, 6)$ **3a.** $\mathbf{x} = (1, 2, 3) + t(2, 4, 6)$
3b. $\mathbf{x} = (0, 1, -2) + t(3, 0, 0)$ **3c.** $\mathbf{x} = t(0, 1, 0)$
5a. $\mathbf{x} = t(1, 2, 0)$ **5b.** $\mathbf{x} = (1, 2, -1) + t(1, 2, \sqrt{2})$
7a. $\mathbf{x} = (2, -1, 3) + t(-3, 2, -1)$
7b. $\mathbf{x} = (2, -1, 3) + t(1, -1, 1)$ **7c.** $\mathbf{x} = (1, 0, 0) + t(1, 0, -1)$
9a. $\mathbf{x} = (1, 2, 1, -1) + t(2, -1, 0, 1)$
9b. $\mathbf{x} = (-2, -1, 0, -1) + t(1, -1, -1, 1)$
11. $\mathbf{x} = (1, 2, 3) + t(2, 5, -2)$; $(x - 1)/2 = (y - 2)/5 = (z - 3)/-2$; $x = 1 + 2t$; $y = 2 + 5t$; $z = 3 - 2t$
13. $\mathbf{x} = (-1, 5, 4) + t(1, -1, -2)$; $(x + 1)/1 = (y - 5)/-1 = (z - 4)/-2$; $x = -1 + t$, $y = 5 - t$, $z = 4 - 2t$
15. $\mathbf{x} = (-1, 3) + t(3, 5)$; $(x + 1)/3 = (y - 3)/5$; $x = -1 + 3t$; $y = 3 + 5t$
17. $x = 1 + 3t$; $y = 1 - t$; $z = 2 + 2t$
19. $\mathbf{x} = (-1, 3, 2) + t(1, 0, 1)$ **21.** No, they are skew lines.
23. Yes **25.** No **29.** No
31. Yes, you should slouch down about 0.2 ft.

EXERCISES 10.5, p. 669

1. $3x - 4y + z = 16$

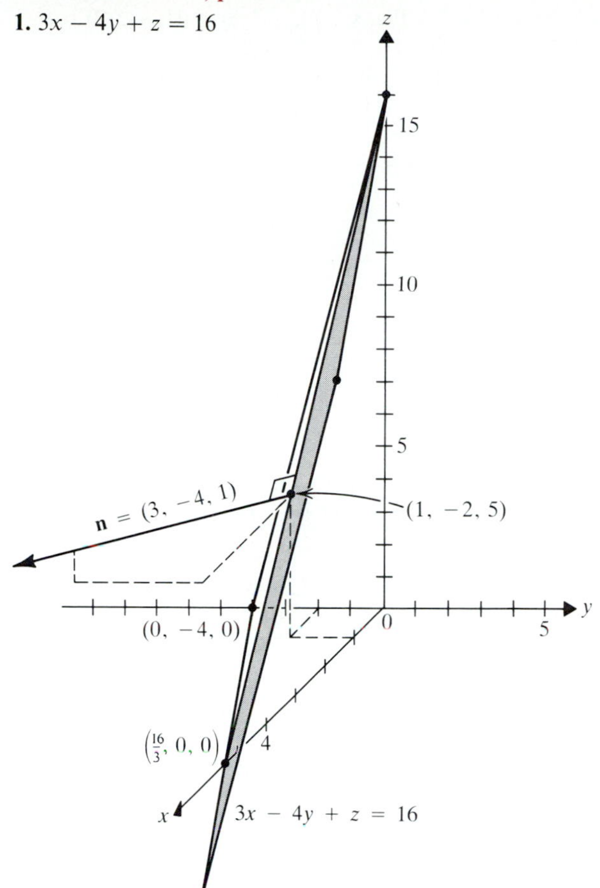

3. $z = 0$, $x = 0$, $y = 0$
5. $3x - 5z + 34 = 0$

7. $3x - 2y = 10$
9. $x + z = 3$ **11a.** $x = 1$ **11b.** $y = -2$
13. $3x - 2y - z = -4$ **15.** $y = -2$
17. (a) and (b) coincide, are parallel to (d), and are perpendicular to (c).
19. $\mathbf{x} = (-2, 0, 3) + t(-4, 1, 3)$; $x = -2 - 4t, y = t, z = 3 + 3t$; $(x + 2)/-4 = y/1 = (z - 3)/3$
21. xy plane: $x = 5 + 5t,\ y = 3t,\ z = 0$
yz plane: $x = 0,\ y = -3 + 4t,\ z = 10t$
xz plane: $x = 5 + 2t,\ y = 0,\ z = -3t$

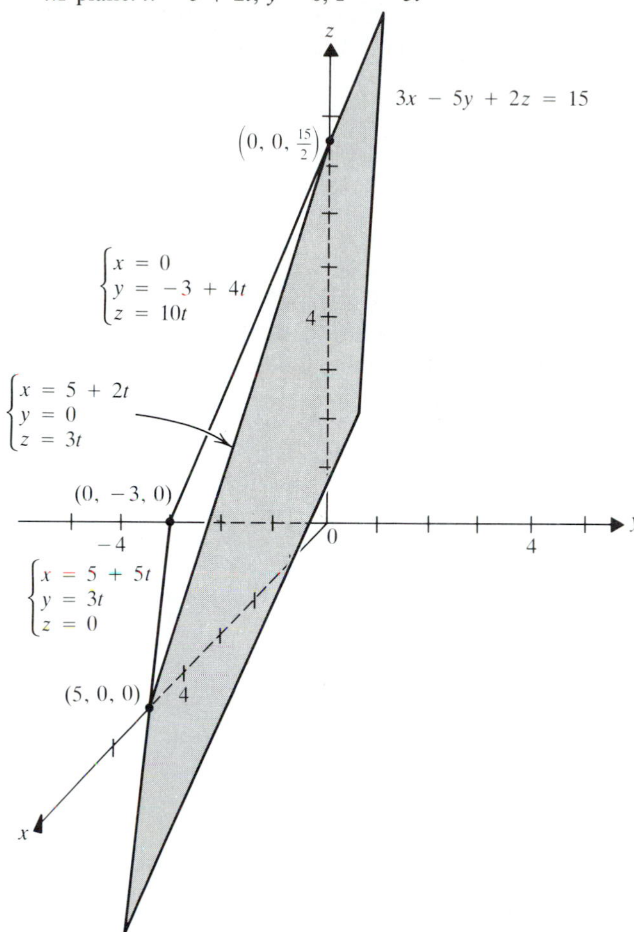

23. (8/5, 1/5, 17/5) **25.** Yes; (3, −2, −1) **31.** $4/\sqrt{33}$
33. $E_4/E_3 = 0.49$ **35.** $5\pi/6$ **37.** $\pi/3$ **39a.** $x + \sqrt{3}y = 2$

EXERCISES 10.6, p. 679

1. $2x + y - z = -2$ **3.** $x + y + z = 4$ **5.** $x + y = 2$
7. $3x - 5y + z = 0$ **9.** $10x - 3y + 5z = 0$ **11.** $2x + y = 1$
13. $3\sqrt{38}$ **15.** 7 **19.** 9/2 **21.** $\sqrt{257}/2$ **23.** Yes, 9 **25.** Yes; 10
27. No **29.** Yes; 3 **31a.** $3\mathbf{i} - \mathbf{j} - 5\mathbf{k}$ **31b.** $\mathbf{i} - 3\mathbf{j} - 3\mathbf{k}$
35. $\pm(1/\sqrt{17})(2, -3, 2)$ **37.** 4/5 **43.** $1/\sqrt{2}$

REVIEW EXERCISES 10.7, p. 681

1a. $4\mathbf{i} + 3\mathbf{j}$ **1b.** $-9\mathbf{i} + 7\mathbf{j}$ **1c.** $\sqrt{10}$ **1d.** $(-1/\sqrt{10})(\mathbf{i} - 3\mathbf{j})$
3a. Neither **3b.** Parallel **3c.** Perpendicular
5a. $(-13, -1, -1)$ **5b.** $(1/15)(46, 7, -2)$ **5c.** 3
5d. $(2/3, -1/3, 2/3)$; $(1/\sqrt{11})(-3, -1, 1)$ **7a.** $\pi/2$ **7b.** $\pi/2$
9a. $(1/\sqrt{13})(2\mathbf{i} - 3\mathbf{j})$; infinitely many
9b. Two vectors: $\pm(1/\sqrt{34})(5\mathbf{i} - 3\mathbf{j})$
11a. $\mathbf{x} = (-1, 1, 3) + t(1, -1, -3)$; $x = -1 + t$; $y = 1 - t$; $z = 3 - 3t$; $(x + 1)/1 = (y - 1)/-1 = (z - 3)/-3$
11b. $\mathbf{x} = (2, -1, 3) + t(1, 2, 2)$; $x = 2 + t$, $y = -1 + 2t$; $z = 3 + 2t$; $(x - 2)/1 = (y + 1)/2 = (z - 3)/2$
13. Yes, at $(-1, 5, 2)$ **15.** No, they are skew.
17. $3y - 4z + 17 = 0$ **19a.** $\mathbf{x} = (1, -2, 3) + t(1, -2, 2)$
19b. $x - 2y + 2z = 20$; $x - 2y + 2z = 2$
19c. $\mathbf{x} = (11, 0, 0) + t(2, 1, 0)$ **21.** $5x - 2y + z + 12 = 0$
23. 0 **25.** $4x + 3y + 3z = 11$
27. $3x - 6y - 7z + 11 = 0$
29. 11 **31.** 14 **33.** $\sqrt{3}s/4$

CHAPTER 11

EXERCISES 11.1, p. 692

1b. The unit circle $x^2 + y^2 = 1$ **1c.** $\mathbf{x}'(0) = (0, 1)$
3a. $\mathbf{x} = (1, 5, 7) + t(1, 4, 12)$ **3b.** $\sqrt{161}$
5a. $\mathbf{x} = (2, 1, 1) + t(4, 1, 3)$ **5b.** $\sqrt{26}$
7a. $\mathbf{x} = (\pi/8, 1, 0) + t(1, 0, -4)$ **7b.** $\sqrt{17}$
9a. $\mathbf{x} = (e, 0, 1) + s(2e, 2, -1)$ **9b.** $(5 + 4e^2)^{1/2}$
11. $\mathbf{i}$; not continuous
13a. $(-4\sin t - 4t\cos t - 3t^2)\mathbf{i} + (2 + 4\cos t - 4t\sin t)\mathbf{j} + (2t\cos t - t^2\sin t - 2\cos t)\mathbf{k}$
13b. $-2\sin t + 2t\sin t + t^2\cos t - 8t$
15a. $[(3t^2 - t^4)\cos t - 5t^3\sin t]\mathbf{i} + [16t^3 - 2t\cos t + t^2\sin t]\mathbf{j} + [2t\sin t + t^2\cos t - 12t^2]\mathbf{k}$
15b. $8t + (3t^2 - t^4)\sin t + 5t^3\cos t$ **17.** $(14/3)\mathbf{i} - (2/3)\mathbf{j}$
19. $-\mathbf{i} + \mathbf{j} + (3\pi^2/8)\mathbf{k}$ **21.** $-2\mathbf{i} + \pi\mathbf{j} + (\pi - \pi^2/2)\mathbf{k}$
23. $\mathbf{x}(t) = (t - 1, t^2/2 + 2, t^3/3 + 2)$

EXERCISES 11.2, p. 700

1. $\mathbf{v}(t) = ([-\sin t/2]/2, [\cos t/2]/2, 1)$, $\mathbf{a}(t) = ([-\cos t/2]/4, [-\sin t/2]/4, 0)$; speed is $\sqrt{5}/2$

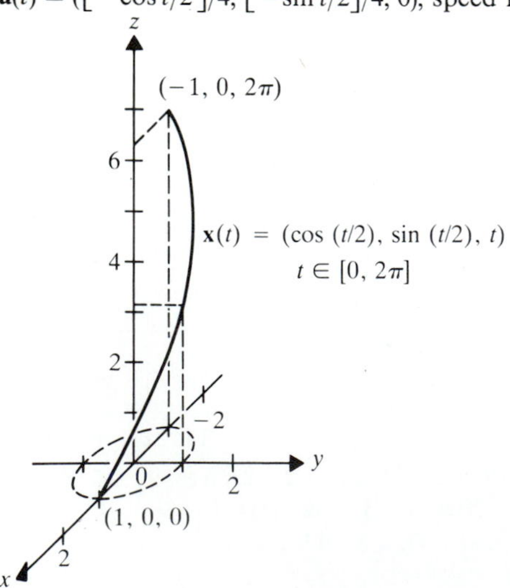

3. $\mathbf{v}(t) = (-t\sin t + \cos t)\mathbf{i} + (t\cos t + \sin t)\mathbf{j} + \mathbf{k}$, $\mathbf{a}(t) = (-t\cos t - 2\sin t)\mathbf{i} - (t\sin t - 2\cos t)\mathbf{j}$; speed is $(t^2 + 2)^{1/2}$

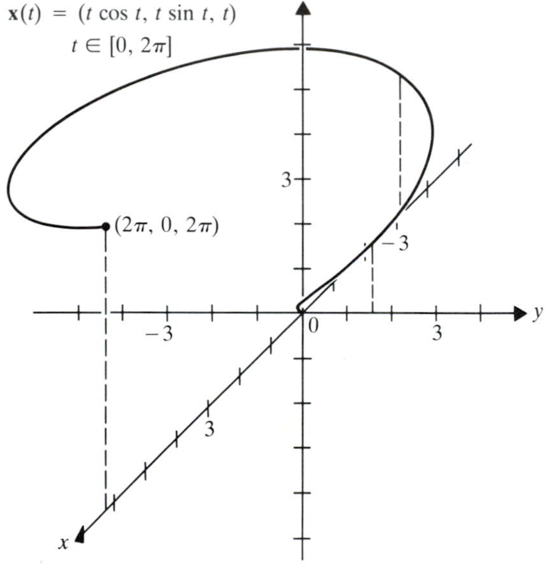

5. $\mathbf{v}(t) = e^t(\mathbf{i} + 2\mathbf{j} + 2e^t\mathbf{k})$; $\mathbf{a}(t) = e^t(\mathbf{i} + 2\mathbf{j} + 4e^t\mathbf{k})$; speed is $e^t(5 + 4e^{2t})^{1/2}$

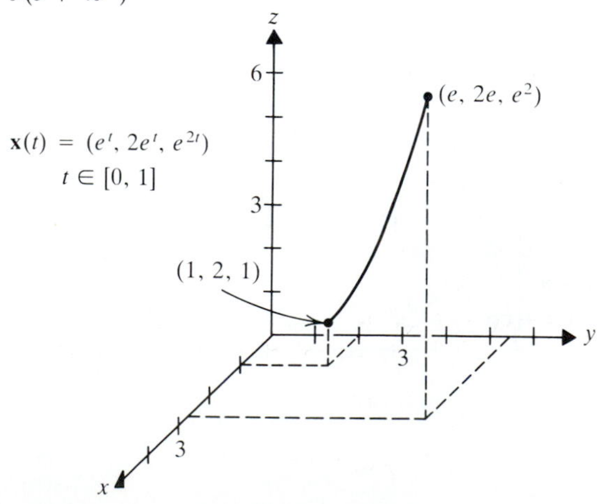

7. $\mathbf{v}(t) = (1, 2\pi\cos 2\pi t, -2\pi\sin 2\pi t)$; $\mathbf{a}(t) = (0, -4\pi^2\sin 2\pi t, -4\pi^2\cos 2\pi t)$; speed $= (1 + 4\pi^2)^{1/2}$
9. $\mathbf{v}(t) = (t\sin t)\mathbf{i} + (t\cos t)\mathbf{j} + 2t\mathbf{k}$; $\mathbf{a}(t) = (\sin t + t\cos t)\mathbf{i} + (\cos t - t\sin t)\mathbf{j} + 2\mathbf{k}$; speed $= \sqrt{5}|t|$
11. $\mathbf{v}(t) = (\sin t + t\cos t)\mathbf{i} + (\cos t - t\sin t)\mathbf{j} + 2t\mathbf{k}$, $\mathbf{a}(t) = (2\cos t - t\sin t)\mathbf{i} - (2\sin t + t\cos t)\mathbf{j} + 2\mathbf{k}$; speed $= \sqrt{1 + 5t^2}$

13. $\frac{\sqrt{5}}{2}(b - a)$ **15.** $2(1 + 4\pi^2)^{1/2}$

17. $e\sqrt{e^2 + 5/4} + \frac{5}{4}\ln\frac{2e + \sqrt{4e^2 + 5}}{5} - \frac{3}{2}$

19. $\frac{\sqrt{3}}{2} + \ln\left[\frac{1 + \sqrt{3}}{\sqrt{2}}\right]$

21. $\mathbf{x}(t(u)) = \frac{u}{\sqrt{17}}\mathbf{i} + \sin\frac{4u}{\sqrt{17}}\mathbf{j} + \cos\frac{4u}{\sqrt{17}}\mathbf{k}$

23. $\mathbf{x}(s) = (2s/\sqrt{6})\mathbf{i} + (s/\sqrt{6})\mathbf{j} + (1 - s/\sqrt{6})\mathbf{k}$

27a. $\mathbf{v}(\theta) = \left[\frac{df}{d\theta}\cos\theta - f(\theta)\sin\theta\right]\mathbf{i} + \left[\frac{df}{d\theta}\sin\theta + f(\theta)\cos\theta\right]\mathbf{j}$

27b. $|\mathbf{v}(\theta)| = \sqrt{\left[\frac{dr}{d\theta}\right]^2 + r^2}$

29a. $\mathbf{x}(t) = (v_0 t\cos\alpha)\mathbf{i} - (gt^2/2 - v_0 t\sin\alpha)\mathbf{j}$; the parabola $y = -gx^2/(2v_0{}^2\cos^2\alpha) + (\tan\alpha)x$
29b. $v_0{}^2\sin 2\alpha/g$ **31.** $\mathbf{x}(t) = t^2\mathbf{c} + t\mathbf{d} + \mathbf{e}$ for $\mathbf{c}, \mathbf{d}, \mathbf{e} \in \mathbf{R}^3$
33. $8\sqrt{2}\pi^3/3$

EXERCISES 11.3, p. 709

1. $K = 1/2$; $\mathbf{T} = (1/\sqrt{2})(-\sin t\mathbf{i} + \cos t\mathbf{j} + \mathbf{k})$; $\mathbf{N} = -\cos t\mathbf{i} - \sin t\mathbf{j}$
3. $K = 3/25$; $\mathbf{T} = (1/5)(3\cos t\mathbf{i} - 3\sin t\mathbf{j} + 4\mathbf{k})$; $\mathbf{N} = -\sin t\mathbf{i} - \cos t\mathbf{j}$
5. $K = 1/5t$; $\mathbf{T} = (1/\sqrt{5})(\sin t, \cos t, 2)$; $\mathbf{N} = (\cos t, -\sin t, 0)$
7. $K = 1/2$; $\mathbf{T} = (-\sqrt{2}/2)\sin 3t\mathbf{i} + (-\sqrt{2}/2)\sin 3t\mathbf{j} + \cos 3t\mathbf{k}$; $\mathbf{N} = (-\sqrt{2}/2)\cos 3t\mathbf{i} + (-\sqrt{2}/2)\cos 3t\mathbf{j} - \sin 3t\mathbf{k}$

9. $K = \frac{\sqrt{19}}{7\sqrt{14}}$; $\mathbf{T} = \frac{1}{\sqrt{14}}(\mathbf{i} + 2\mathbf{j} + 3\mathbf{k})$;
$\mathbf{N} = \frac{1}{\sqrt{266}}(-11\mathbf{i} - 8\mathbf{j} + 9\mathbf{k})$

11. $K = \frac{[5\pi^4 + 16\pi^2 + 8]^{1/2}}{[1 + 5\pi^2]^{3/2}}$; $\mathbf{T} = \frac{1}{[1 + 5\pi^2]^{1/2}}(-\pi\mathbf{i} - \mathbf{j} + 2\pi\mathbf{k})$
$\mathbf{N} = (25\pi^6 + 85\pi^4 + 56\pi^2 + 8)^{-1/2}[-(2 + 5\pi^2)\mathbf{i} + (5\pi^3 + 6\pi)\mathbf{j} + 2\mathbf{k}]$
13. $\mathbf{a_T} = \mathbf{0}$; $\mathbf{a_N} = \mathbf{N} = -\cos t\mathbf{i} - \sin t\mathbf{j} = \mathbf{a}$
15. $\mathbf{T} = (1/5)(3\cos t\mathbf{i} - 3\sin t\mathbf{j} + 4\mathbf{k})$; $\mathbf{a} = -3\sin t\mathbf{i} + 3\cos t\mathbf{j}$; $\mathbf{a_T} = \mathbf{0}$; $\mathbf{a_N} = \mathbf{a}$
17. $\mathbf{a_T} = \sin t\mathbf{i} + \cos t\mathbf{j} + 2\mathbf{k}$; $\mathbf{a_N} = t\cos t\mathbf{i} - t\sin t\mathbf{j}$
19. $\mathbf{a_T} = (11/7)(\mathbf{i} + 2\mathbf{j} + 3\mathbf{k})$; $\mathbf{a_N} = (1/7)(-4\mathbf{i} + 8\mathbf{j} - 9\mathbf{k})$
21. $\mathbf{a_T} = \frac{5\pi}{1 + 5\pi^2}(-\pi, -1, 2\pi)$;
$\mathbf{a_N} = \frac{1}{1 + 5\pi^2}(-2 - 5\pi^2, 6\pi + 5\pi^3, 2)$

23. $K = \frac{\sqrt{5}}{27}$ **29.** 1 **31.** $K = \frac{2}{5\sqrt{5}}$ **33.** 0

EXERCISES 11.4, p. 717

1. ln 2 **3.** No limit **5.** No limit **7.** 0 **9.** No limit
11. Continuous on $\mathbf{R}^2$ except at (0, 0)
13. Continuous on $\mathbf{R}^2$
15. Continuous on $\mathbf{R}^2$ except at (0, 0)
17. Continuous on $\mathbf{R}^2$ except at (0, 0)
19. Continuous except along the x-axis **27.** Yes
31. Yes, limit is -1. **33.** No limit

EXERCISES 11.5, p. 727

1. Parabolic cylinder

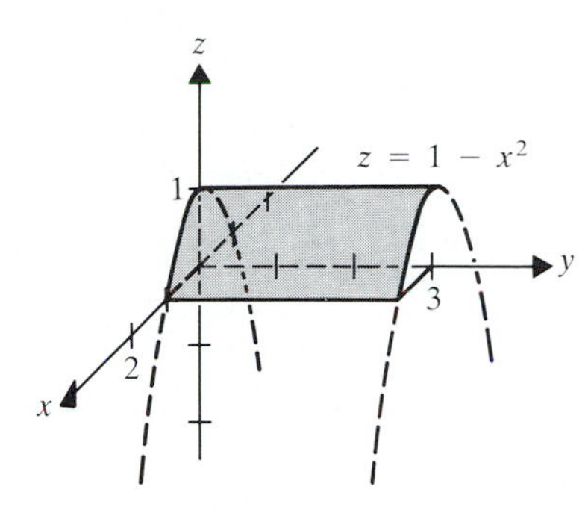

3. Circular cylinder

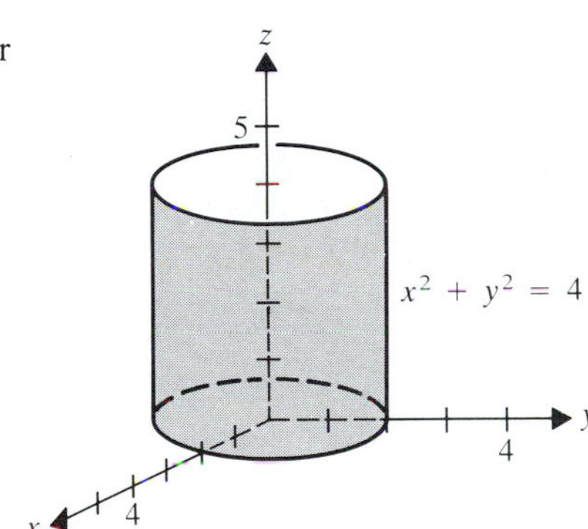

5. Hyperbolic cylinder

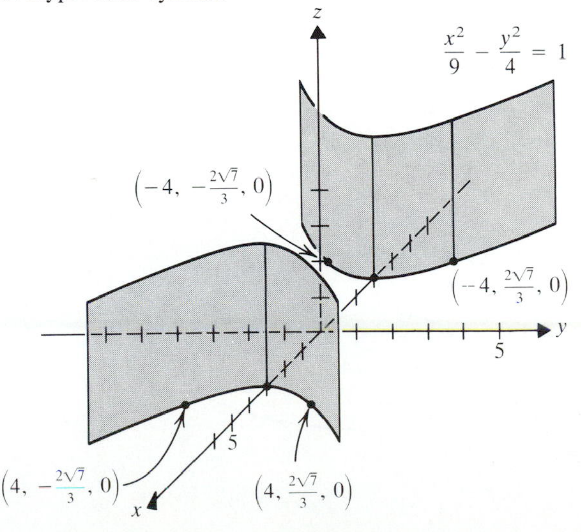

7. Ellipsoid of revolution

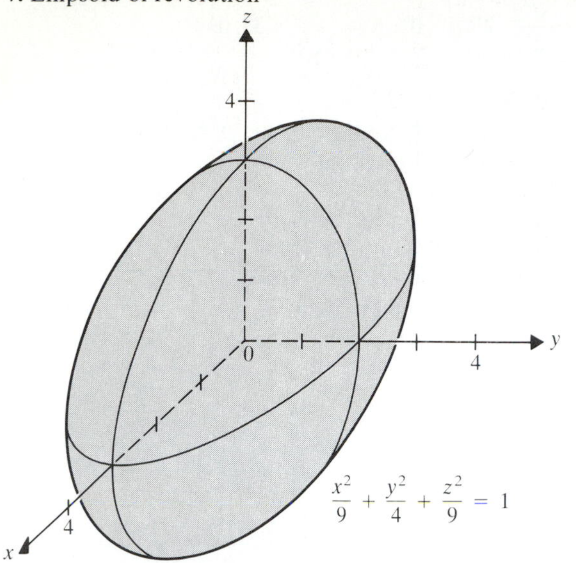

9. $y^2 + z^2 = x^4$ **11.** $x^2 + z^2 = y^4/16$
13. $x^2/4 - y^2/9 - z^2/9 = 1$ **15.** $-x^2/9 + y^2/16 - z^2/9 = 1$
17. $x^2 + y^2 + z^2 = 4$ **19.** $y^2 + z^2 = 9x^2$ **21.** No
23. Yes, of the hyperbola $4x^2 - 3y^2 = 9$ in the xy-plane about the y-axis

25.

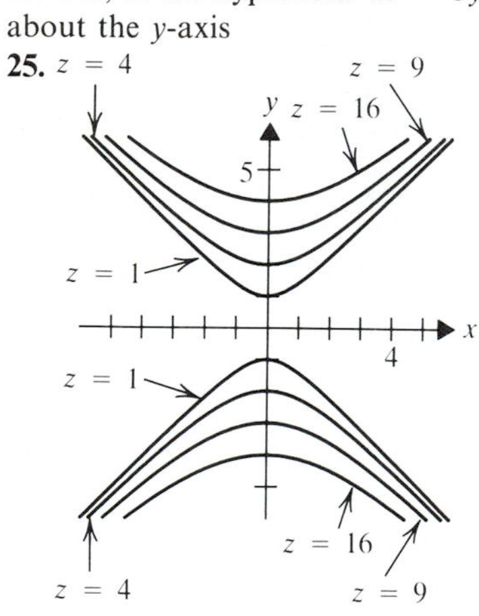

27.

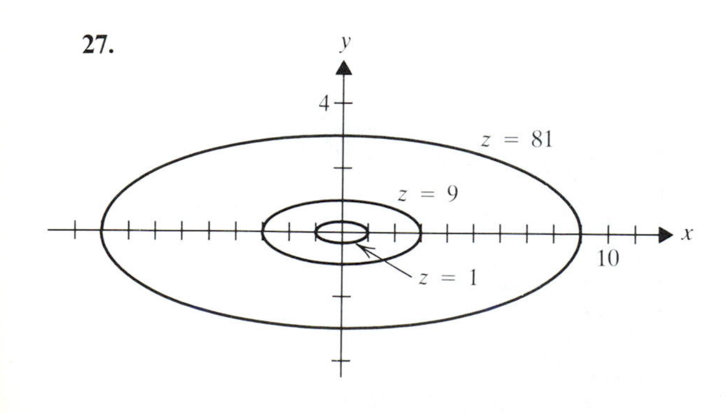

29.

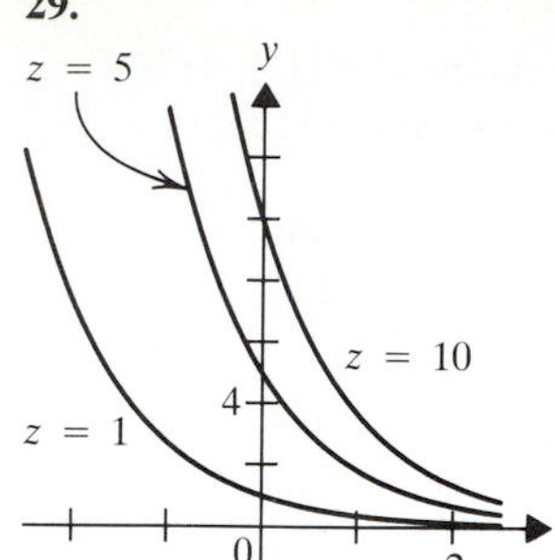

31.

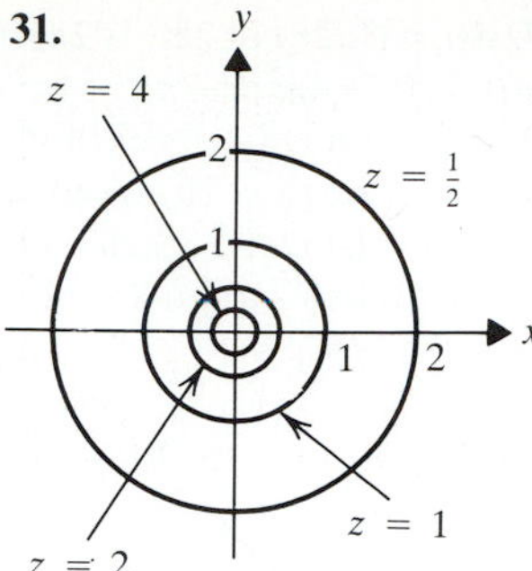

33. Spheres

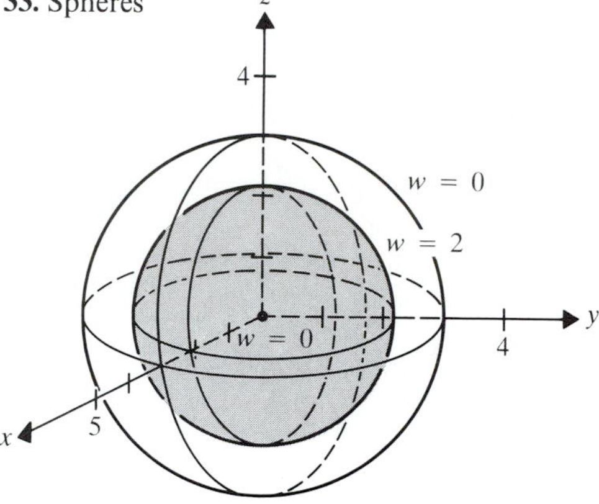

35. Ellipsoids of revolution

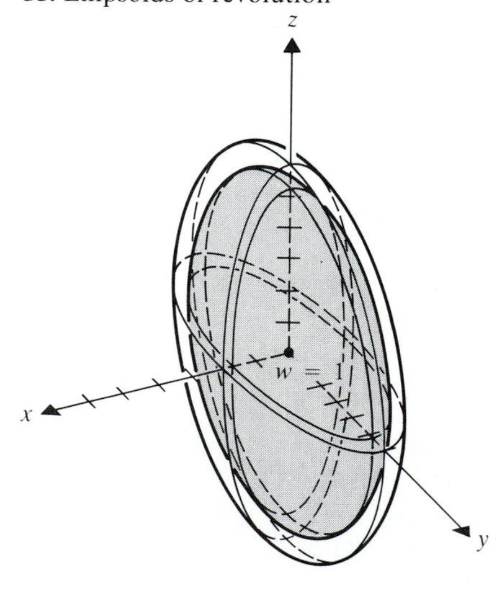

37. Union of planes $y = \pm x$
39. The plane $y = -z$ **41.** The empty set
43. The line $\mathbf{x} = t\mathbf{i} + 2\mathbf{j}$

EXERCISES 11.6, p. 742

1. Elliptical cylinder

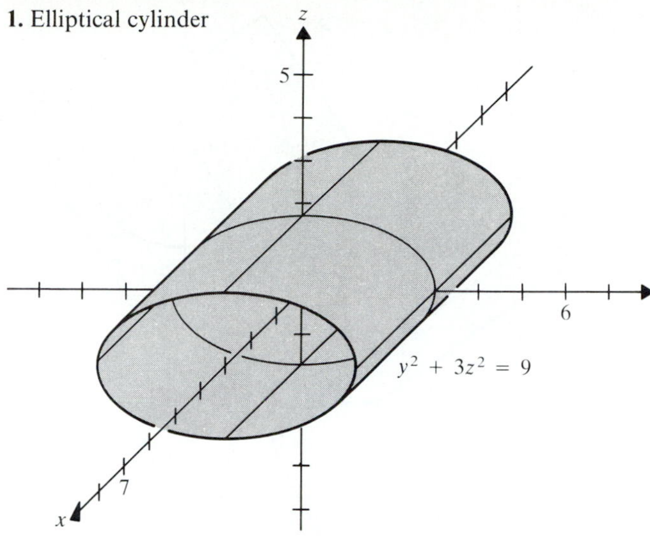

3. Parabolic cylinder

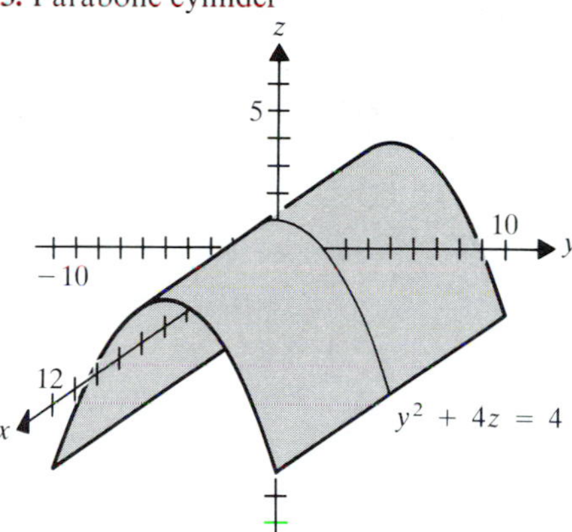

5. Circular cone

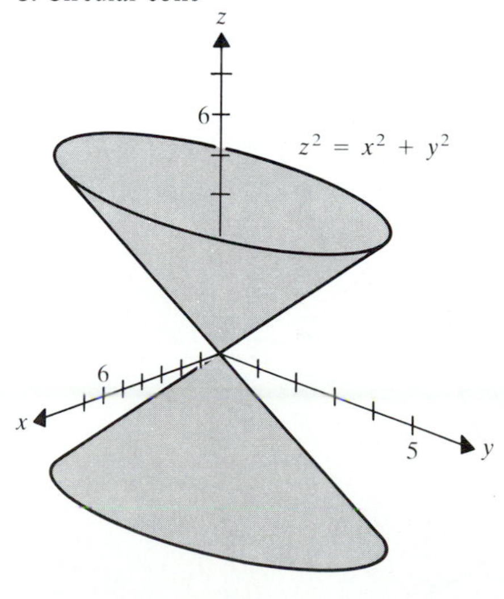

7. Elliptical paraboloid

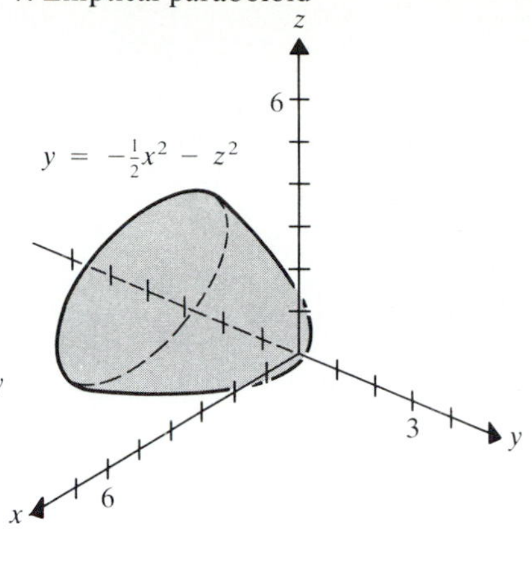

9. Hyperbolic paraboloid

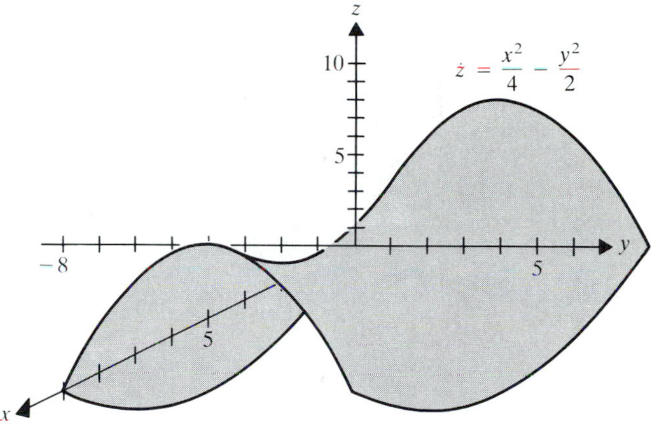

11. Hyperbolic paraboloid

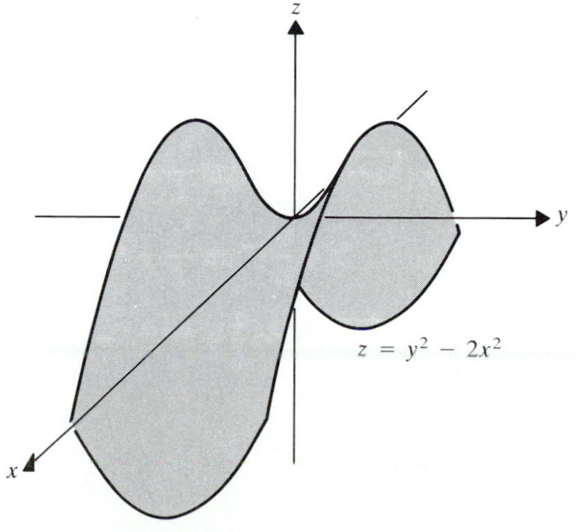

13. Ellipsoid of revolution

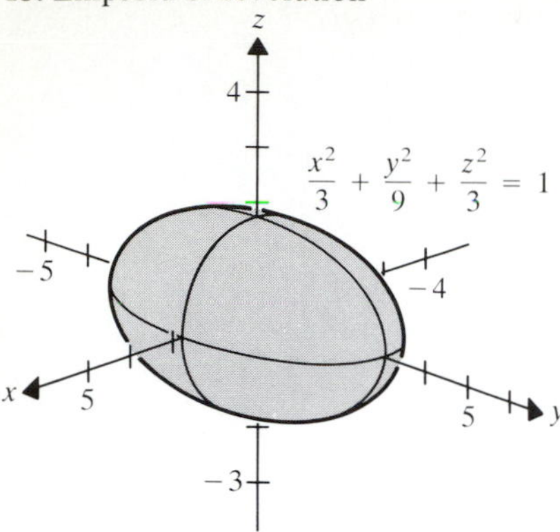

15. Hyperboloid of one sheet

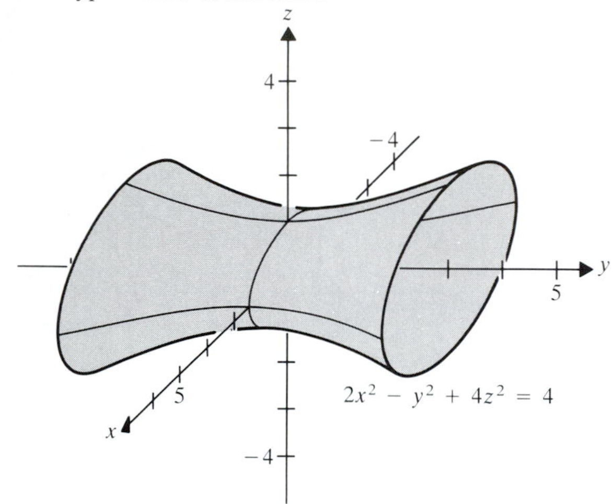

17. Hyperboloid of two sheets

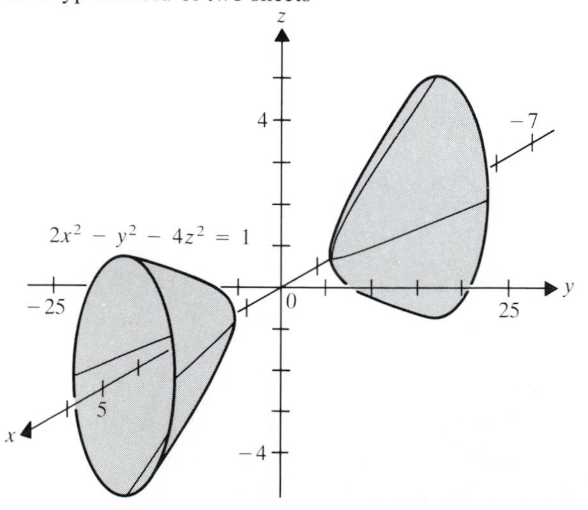

19. Hyperbolic cylinder

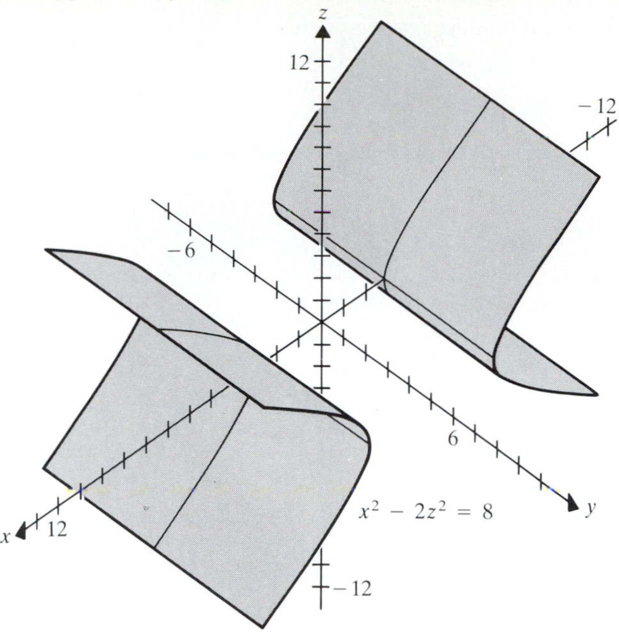

21. Circular cone

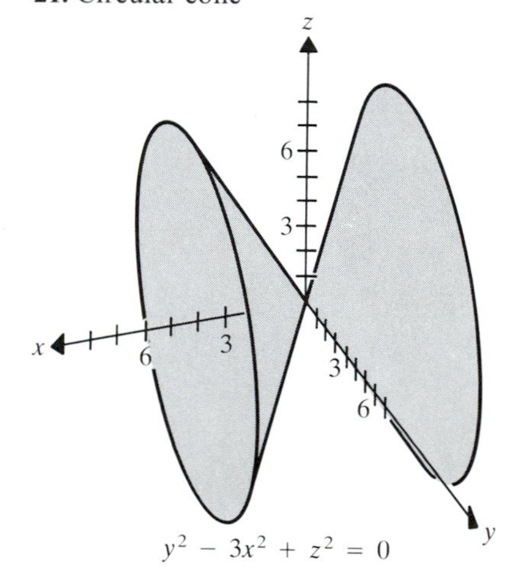

23. Elliptical paraboloid

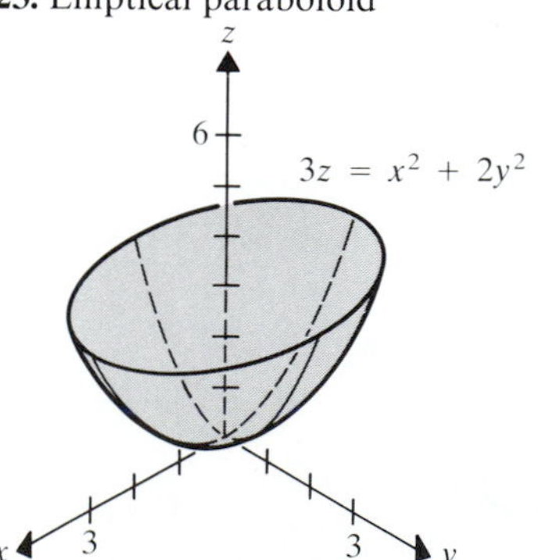

25. Hyperbolic paraboloid

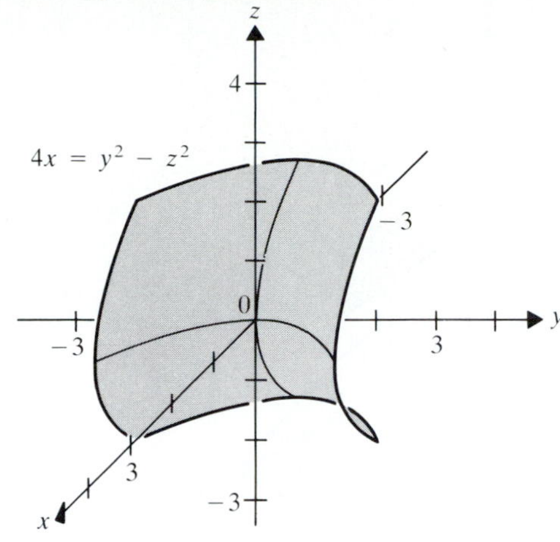

27. The origin

29. Hyperboloid of revolution of one sheet

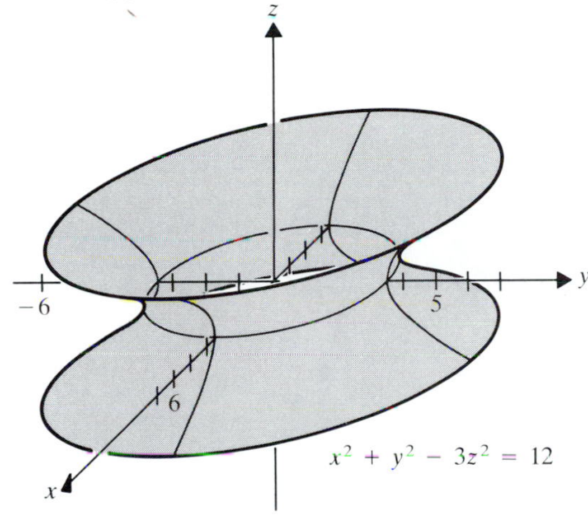

31. Hyperboloid of revolution of one sheet

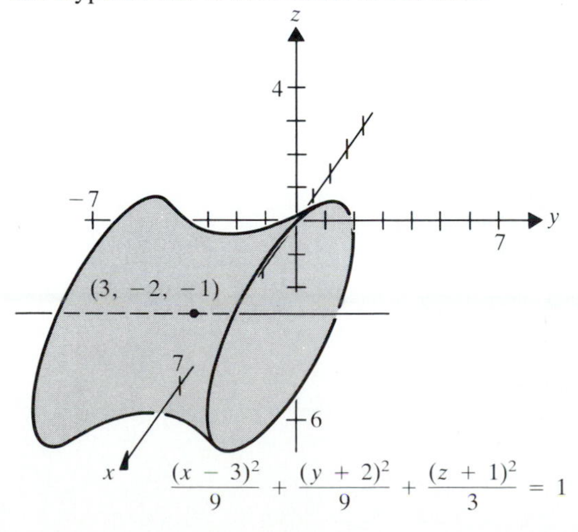

$$\frac{(x-3)^2}{9}+\frac{(y+2)^2}{9}+\frac{(z+1)^2}{3}=1$$

33. Ellipsoid

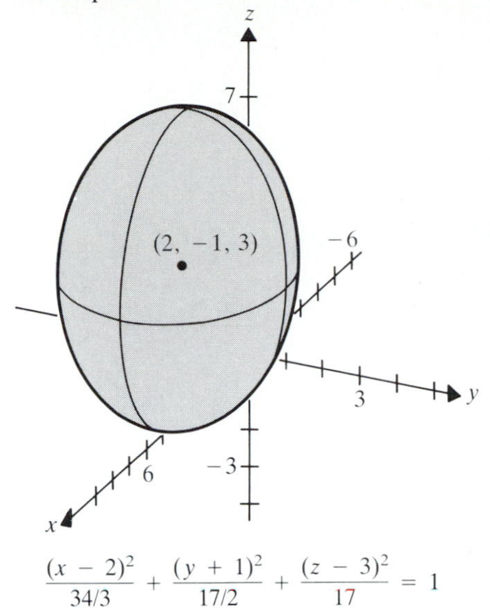

$$\frac{(x-2)^2}{34/3}+\frac{(y+1)^2}{17/2}+\frac{(z-3)^2}{17}=1$$

35. Elliptical paraboloid

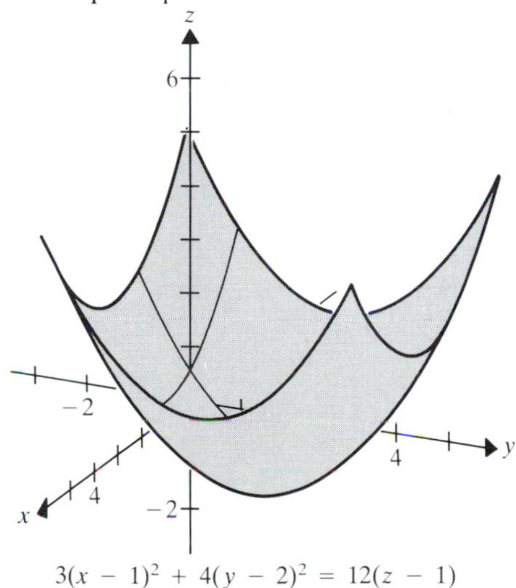

$$3(x-1)^2+4(y-2)^2=12(z-1)$$

REVIEW EXERCISES 11.7, p. 743

1. $\mathbf{x}=\mathbf{j}+(\pi/4)\mathbf{k}+t(-2\mathbf{i}+\mathbf{k})$; speed is $\sqrt{5}$
3. $(1/2)\mathbf{i}+(1/2)\mathbf{j}+(\pi^2/32)\mathbf{k}$
5. $\mathbf{x}(t)=(1/m)(\cos t\mathbf{i}+\sin t\mathbf{j})+(t/m)\mathbf{c}_1+(1/m)\mathbf{c}_2$
9. $K=4/5$; $\mathbf{T}=(1/\sqrt{5})(2\cos 2t,\ 1,\ -2\sin 2t)$;
$\mathbf{N}=(-\sin 2t,\ 0,\ -\cos 2t)$
11a. $\mathbf{v}(t)=(t\cos t+\sin t,\ -t\sin t+\cos t,\ 2t)$;
$\mathbf{v}(\pi)=(-\pi,\ -1,\ 2\pi)$; $ds/dt=(1+5t^2)^{1/2}=(1+5\pi^2)^{1/2}$ $\mathbf{a}(t)=$
$(-t\sin t+2\cos t,\ -t\cos t-2\sin t,\ 2)=(-2,\pi,2)$ at $t=\pi$
11b. $\pi(1+20\pi^2)^{1/2}+(1/2\sqrt{5})\ln[2\pi\sqrt{5}+(1+20\pi^2)^{1/2}]$
11c. $\mathbf{a_T}=\dfrac{(-5\pi^2,\ -5\pi,\ 10\pi^2)}{1+5\pi^2}$; $\mathbf{a_N}=\dfrac{(-2-5\pi^2,\ 6\pi+5\pi^3,\ 2)}{1+5\pi^2}$
13. $\mathbf{R}^2$ except for the line $y=-x$ **15.** $\mathbf{R}^2-\{(0,b)|b\neq 0\}$

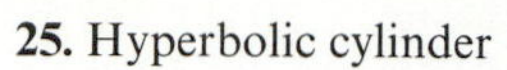

17.

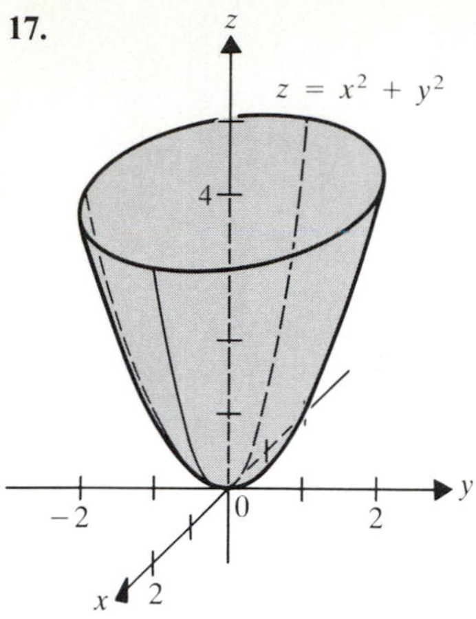

25. Hyperbolic cylinder

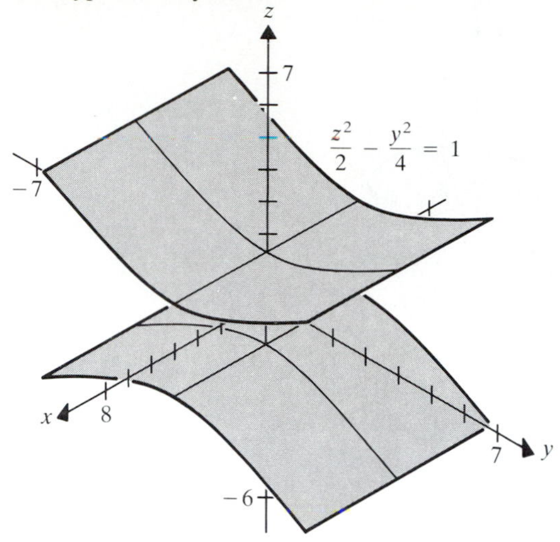

19.

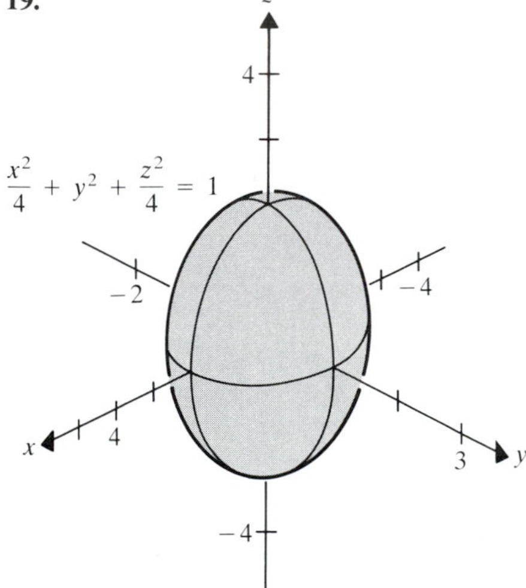

27. Elliptical cone

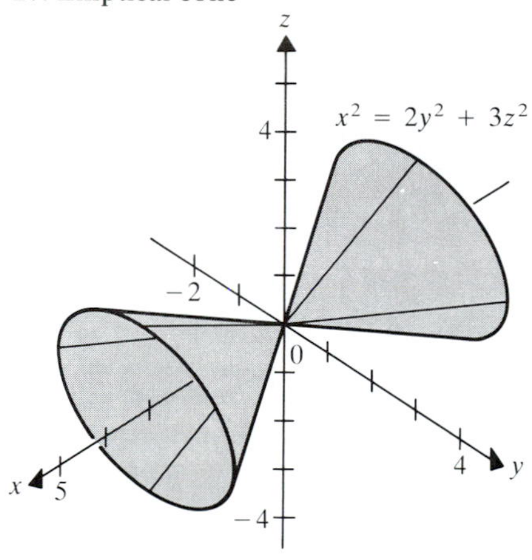

29. Elliptical paraboloid

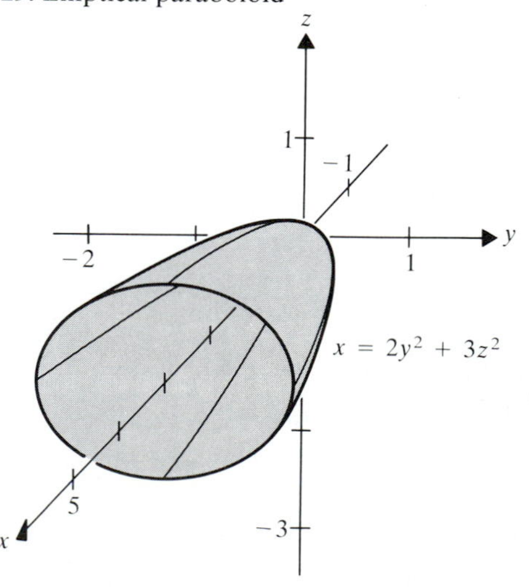

21a. $y^2 + z^2 = (x^2 + 4)^2$ **21b.** $x^2 + z^2 = y - 4$

23.

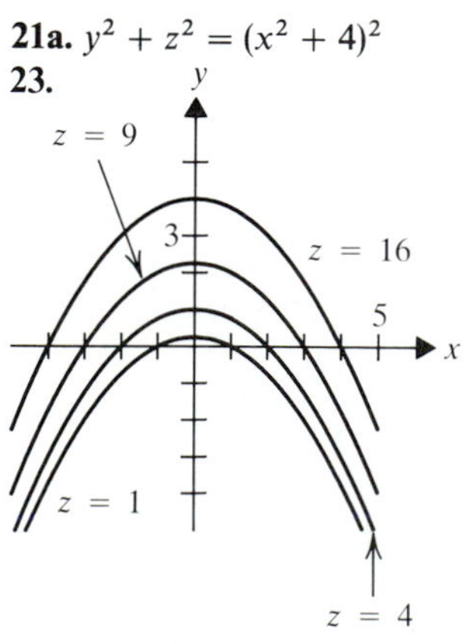

31. Ellipsoid of revolution

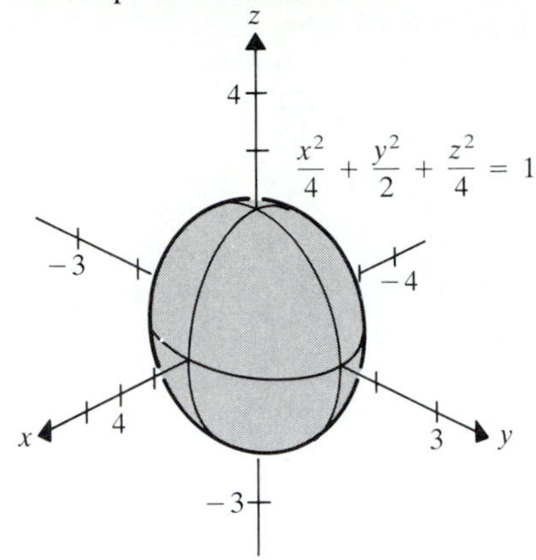

33. Hyperboloid of two sheets

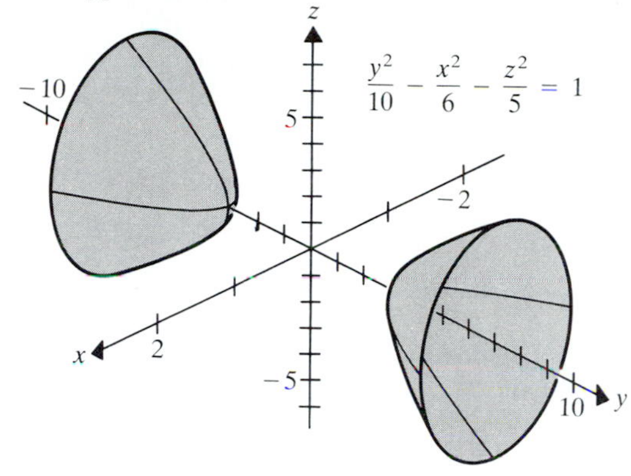

35. Hyperboloid of one sheet

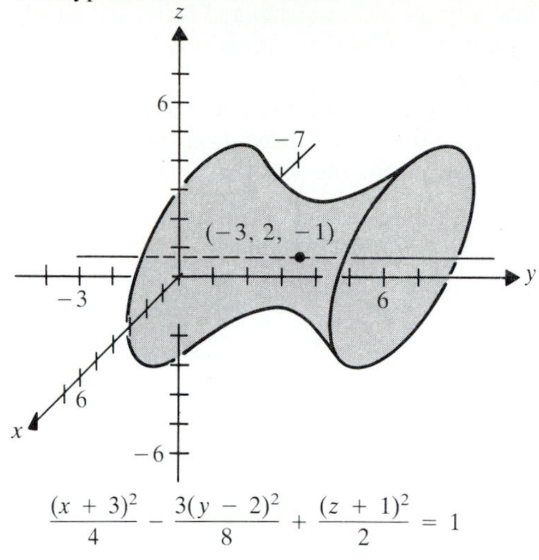

$$\frac{(x+3)^2}{4} - \frac{3(y-2)^2}{8} + \frac{(z+1)^2}{2} = 1$$

CHAPTER 12

EXERCISES 12.1, p. 751

1. $\frac{\partial f}{\partial x} = \frac{x}{\sqrt{x^2+y^2}} = \frac{1}{\sqrt{5}}$ at (1, 2); $\frac{\partial f}{\partial y} = \frac{y}{\sqrt{x^2+y^2}} = \frac{2}{\sqrt{5}}$ at (1, 2)

3. $f_x = -6e^y \sin x - 5ye^{xy}\ln x^2y^2 - 10e^{xy}/x =$ $-6e - 5e^{\pi/2}\ln(\pi^2/4) - (20/\pi)e^{\pi/2}$ at $(\pi/2, 1)$; $f_y = 6e^y\cos x - 5xe^{xy}\ln x^2y^2 - 10e^{xy}/y =$ $-5\pi/2e^{\pi/2}\ln(\pi^2/4) - 10e^{\pi/2}$ at $(\pi/2, 1)$

5. $\frac{\partial f}{\partial x} = \frac{x}{\sqrt{x^2+y^2+z^2}} = \frac{1}{\sqrt{3}}$ at (1, 1, 1);

$\frac{\partial f}{\partial y} = \frac{y}{\sqrt{x^2+y^2+z^2}} = \frac{1}{\sqrt{3}}$ at (1, 1, 1);

$\frac{\partial f}{\partial z} = \frac{z}{\sqrt{x^2+y^2+z^2}} = \frac{1}{\sqrt{3}}$ at (1, 1, 1);

7. $\frac{\partial f}{\partial x} = x^2y^2z\cos xyz + 2xy\sin xyz + yze^{z^2y}\sin xyz = -\pi$ at $(1, 1, \pi)$; $\frac{\partial f}{\partial y} = x^3yz\cos xyz + x^2\sin xyz + xze^{z^2y}\sin xyz -$ $z^2e^{z^2y}\cos xyz = -\pi + \pi^2e^{\pi^2}$ at $(1, 1, \pi)$; $\frac{\partial f}{\partial z} = x^3y^2\cos xyz -$ $2ze^{z^2y}\cos xyz + xye^{z^2y}\sin xyz = -1 + 2\pi e^{\pi^2}$ at $(1, 1, \pi)$

9. $g_x = 2(\tan xy^2z)(\sec^2 xy^2z)(y^2z) = 4$ at $(\pi/4, 1, 1)$ $g_y = 2(\tan xy^2z)(\sec^2 xy^2z)(2xyz) = 2\pi$ at $(\pi/4, 1, 1)$ $g_z = 2(\tan xy^2z)(\sec^2 xy^2z)(xy^2) = \pi$ at $(\pi/4, 1, 1)$

11. $6x + 10y - z = 8$ **13.** $3x - 2y + 3z = 0$

15. $z = (1+e)x + (1+e)y - (e+2)$

17. $(1/\sqrt{137})(6\mathbf{i} + 10\mathbf{j} - \mathbf{k})$ **19.** $(1/\sqrt{22})(-3\mathbf{i} + 2\mathbf{j} - 3\mathbf{k})$

21. $1/\sqrt{6}$

23. $\partial P/\partial T = P/T$; $\partial P/\partial V = -P/V$; P increases by about P/T for each unit increase in T, and decreases by about P/V for each unit increase in V.

25. $\partial d/\partial p = -10$; $\partial d/\partial x = 5$; $\partial d/\partial y = 1$; demand decreases sharply in response to price increase; demand increases sharply in response to price increase by the competition, less in response to costlier aging.
27. $\partial d_1/\partial y$ and $\partial d_2/\partial x$ should be positive. As the price of competition rises, demand for the competition should drop, which should raise demand for the product whose price doesn't rise.
35. (a) 3323 (b) 0.1000, 2.400×10^4

EXERCISES 12.2, p. 763

1. $4\mathbf{i} + 3\mathbf{j}$ **3.** $7\mathbf{i} + 4\mathbf{j}$ **5.** $2e\mathbf{j}$ **7.** $\frac{1}{\sqrt{2}}\mathbf{i} + \frac{1}{\sqrt{2}}\mathbf{j}$

9. $-\frac{4}{25}\mathbf{i} + \frac{3}{25}\mathbf{j}$ **11.** $e^2\left(\sqrt{3}\mathbf{i} + \sqrt{3}\mathbf{j} + \frac{1}{2}\mathbf{k}\right)$ **13.** $-2\sqrt{\pi}\mathbf{k}$

15. $\frac{x}{\sqrt{x^2+y^2}}\mathbf{i} + \frac{y}{\sqrt{x^2+y^2}}\mathbf{j}$; any $(x, y) \neq (0, 0)$

17. $\frac{2x}{1 + x^2 - y}\mathbf{i} - \frac{1}{1 + x^2 - y}\mathbf{j}$; any (x, y) not on the parabola $y = x^2 + 1$

19. $\frac{1}{\sqrt{16 - x^2 - 4y^2 - 4z^2}}(-x\mathbf{i} - 4y\mathbf{j} - 4z\mathbf{k})$; any (x, y, z) inside the ellipsoid $x^2 + 4y^2 + 4z^2 = 16$ **23.** 4.99
25. 32.00 **27.** 2.96 **29.** 17 ft; 0.03 ft **31.** $0.64\pi \approx 2.01$ in.^{3}

EXERCISES 12.3, p. 773

1. $3\sqrt{2}$ **3.** $-3/5$ **5.** $2/3$ **7.** 0 **9.** -4
11. Max. rate of increase: $2\sqrt{5}$ in direction $(1/\sqrt{5})(\mathbf{i} + 2\mathbf{j})$. Max. rate of decrease: $-2\sqrt{5}$ in direction $(-1/\sqrt{5})(\mathbf{i} + 2\mathbf{j})$.
13. Max. rate of increase: 17 in direction $(1/17)(8\mathbf{i} - 15\mathbf{j})$. Max. rate of decrease: -17 in direction $(-1/17)(8\mathbf{i} - 15\mathbf{j})$.
15. Max. rate of increase: $2\sqrt{6}$ in direction $(1/\sqrt{6})(\mathbf{i} + 2\mathbf{j} + \mathbf{k})$. Max. rate of decrease: $-2\sqrt{6}$ in direction $(-1/\sqrt{6})(\mathbf{i} + 2\mathbf{j} + \mathbf{k})$.
17. Max. rate of increase: $2e^{-6}\sqrt{6}$ in direction $(-1/\sqrt{6})(\mathbf{i} + 2\mathbf{j} - \mathbf{k})$. Max. rate of decrease: $-2e^{-6}\sqrt{6}$ in direction $(1/\sqrt{6})(\mathbf{i} + 2\mathbf{j} - \mathbf{k})$.
19. $(-1/3)(\mathbf{i} + 2\mathbf{j} + 2\mathbf{k})$
21. Heat flows outward across the surface and at right angles to it. **23.** $(1/\sqrt{2})(\mathbf{i} + \mathbf{j})$
25. $(-1/z)(x\mathbf{i} + y\mathbf{j})$, when $z \neq 0$.

27a. $\frac{-x}{\sqrt{x^2 + 4y^2}}\mathbf{i} + \frac{2y}{\sqrt{x^2 + 4y^2}}\mathbf{j}$

27b. $$\frac{\frac{dx}{dt}}{\sqrt{\left(\frac{dx}{dt}\right)^2 + \left(\frac{dy}{dt}\right)^2}} = \frac{-x}{\sqrt{x^2 + 4y^2}};$$
$$\frac{\frac{dy}{dt}}{\sqrt{\left(\frac{dx}{dt}\right)^2 + \left(\frac{dy}{dt}\right)^2}} = \frac{2y}{\sqrt{x^2 + 4y^2}}$$

27c. $dy/dx = -2y/x$ **27d.** $y = c/x^2$ **27e.** $y = 1/x^2$
29. $(x_2, y_2) = (0.8, 0.8)$; $(x_3, y_3) = (0.64, 0.64)$; $(x_4, y_4) = (0.512, 0.512)$; $(x_5, y_5) = (0.4096, 0.4096)$
31. $(x_2, y_2) = (1, -0.2)$; $(x_3, y_3) = (1, -0.92)$; $(x_4, y_4) = (1, -1.352)$; $(x_5, y_5) = (1, -1.6112)$
33. $x_2 = y_2 = 0.9729$; $x_3 = y_3 = 0.9436$; $x_4 = y_4 = 0.9118$; $x_5 = y_5 = 0.8773$; $x_6 = y_6 = 0.8396$; $x_7 = y_7 = 0.7986$; $x_8 = y_8 = 0.7540$; $x_9 = y_9 = 0.7056$; $x_{10} = y_{10} = 0.6535$
35. The convergence rate seems to become slower as t decreases.

EXERCISES 12.4, p. 784

1. -1 **3.** $6/5$ **5.** $2\pi/(4 + \pi^2)$ **7.** $(1, 1/2)$
9. $\partial w/\partial s = 2s$; $\partial w/\partial t = 0$
11. $\partial w/\partial s = 2xyz^2t - tx^2z^2/s^2 + 2x^2yz/t$; $\partial w/\partial t = 2xyz^2s + x^2z^2/s - 2x^2yzs/t^2$
13. $\partial w/\partial s = st^2/(1 + s^2t^2)$; $\partial w/\partial t = s^2t/(1 + s^2t^2)$
15. $(4xst - 4syu, 2xs^2 - 2y, 2x - 2ys^2)$; $\mathbf{0}$
17. $(y\cos\theta + x\sin\theta, -y^2 + x^2)$; $(0, -1)$
19. $(tyze^5 + xze^t + txy, yze^s + sxze^t + sxy)$; $e^2(3, 3)$
21. $[1/(x^2 + y^2 + z^2)](x + ytu + ze^{tu}, x + ysu + zsue^{tu}, x + yst + zste^{tu})$; $(2/5, 3/10, 1/2)$ **23.** $4x + y + 2z = 0$
25. $2x - 3y - 2z = 5$ **27.** $3x - 2y + 2z = 1$ **31.** $(1, 0, 0)$

EXERCISES 12.5, p. 793

1. $\frac{dy}{dx} = -\frac{2xy^2 - 3}{2x^2y - 3y^2}$; valid where denominator is nonzero: everywhere but the x-axis and the parabola $y = 2x^2/3$; $y(2.984) \approx 2.014$

3. $\frac{dy}{dx} = -\frac{y + 3x^2}{x + 3y^2}$; valid except on the parabola $x = -3y^2$; $y(-2.005) \approx 3.003$

5. $\frac{dy}{dx} = -\frac{14xy^6 - 5y^5}{42x^2y^5 - 25xy^4 + 9y^2}$; valid when denominator is nonzero; $y(1.005) \approx 0.98$

7. $\frac{dy}{dx} = \frac{x\cos y - 2x^2e^{xy} - x^3e^{xy} + y^2 - 15x^5}{x^2\sin y + x^4e^{xy} - 2xy\ln x}$; valid where denominator exists and is nonzero; $y(0.95) \approx 0.80$
9. 2.028; 2.043, 2.058, 2.073

11. $\frac{\partial z}{\partial x} = \frac{2x\cos xyz - x^2yz\sin xyz + y^3z\cos xyz}{x^3y\sin xyz - xy^3\cos xyz + 2z}$;
$\frac{\partial z}{\partial y} = \frac{-x^3z\sin xyz + 2y\sin xyz + xy^2z\cos xyz}{x^3y\sin xyz - xy^3\cos xyz + 2z}$; valid where denominator is nonzero

13. $\frac{\partial z}{\partial x} = \frac{2x - z}{ye^z + x}$, $\frac{\partial z}{\partial y} = \frac{2y - e^z}{ye^z + x}$; valid where denominator is nonzero

15. $\frac{\partial z}{\partial x} = \frac{-3x^2e^{y+z} + y\cos(x - z)}{x^3e^{y+z} + y\cos(x - z)}$;
$\frac{\partial z}{\partial y} = \frac{-x^3e^{y+z} + \sin(x - z)}{x^3e^{y+z} + y\cos(x - z)}$; valid where denominator is nonzero **17.** 2.00 **19.** 3.12

21. $\frac{\partial x}{\partial y} = \frac{4yz^2 + x^3w^4}{2x - 3x^2yw^4}$; $\frac{\partial z}{\partial w} = \frac{4x^3yw^3 - 3z^3w^2}{-4y^2z + 3z^2w^3}$

23. $\dfrac{\partial y}{\partial w} = \dfrac{x^3y^3z\sin wz}{2xy^3\cos z + 3x^2y^3\cos wz}$

$\dfrac{\partial x}{\partial w} = \dfrac{x^3y^3z\sin wz}{3x^2y^2\cos z + 3x^3y^2\cos wz}$

25. $3x + y + 3z = 1$

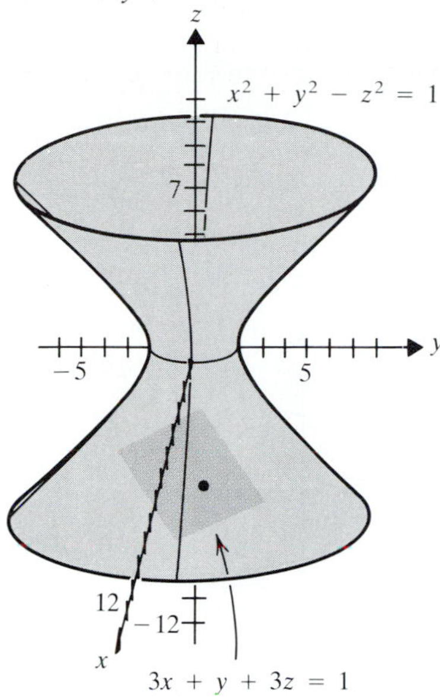

27. $4x + 9y + 2\sqrt{23}z = 36$

$4x + 9y + 2\sqrt{23}z = 36$

$\left(1, 1, \frac{\sqrt{23}}{2}\right)$

$4x^2 + 9y^2 + 4z^2 = 36$

33. $(10, 14, -3)$

EXERCISES 12.6, p. 803

1. $f_x = 6x - 5y$; $f_y = -4y - 5x$; $f_{xx} = 6$; $f_{yy} = -4$; $f_{xy} = f_{yx} = -5$
3. $f_x = 2x\sin xy + x^2y\cos xy$; $f_y = x^3\cos xy$; $f_{xx} = 2\sin xy + 4xy\cos xy - x^2y^2\sin xy$; $f_{yy} = -x^4\sin xy$; $f_{xy} = f_{yx} = 3x^2\cos xy - x^3y\sin xy$

5. $\dfrac{\partial f}{\partial x} = \dfrac{2x}{x^2+y^2}$; $\dfrac{\partial f}{\partial y} = \dfrac{2y}{x^2+y^2}$; $\dfrac{\partial^2 f}{\partial x^2} = \dfrac{2y^2-2x^2}{(x^2+y^2)^2}$; $\dfrac{\partial^2 f}{\partial y^2} = \dfrac{2x^2-2y^2}{(x^2+y^2)^2}$; $\dfrac{\partial^2 f}{\partial y\,\partial x} = \dfrac{\partial^2 f}{\partial x\,\partial y} = \dfrac{-4xy}{(x^2+y^2)^2}$;
7. $f_x = 2x$; $f_y = -2y$; $f_z = 6z$; $f_{xx} = 2$; $f_{yy} = -2$; $f_{zz} = 6$; $f_{xy} = f_{yx} = f_{xz} = f_{zx} = f_{yz} = f_{zy} = 0$
9a. $2x^2/z^3$ **9b.** $2y^2/z^3$

11a. $\dfrac{-72z^2 - 24yz - 26y^2 + 48xy - 24x^2}{(6z+y)^3}$

11b. $\dfrac{384 + 50xy}{(6z+y)^3}$

21. xy **23.** xy **25.** No 2nd-degree Taylor polynomial
27. $2 + (x-1) + 3(y-1) + (x-1)(y-1) + (y-1)^2$
29. $1 + (x^2+y^2)/2$ **31.** y^2
33. $1 - (x-\pi/6)^2/2 - (y-\pi/6)^2/2 - (z-\pi/6)^2/2 - (x-\pi/6)(y-\pi/6) - (x-\pi/6)(z-\pi/6) - (y-\pi/6)(z-\pi/6)$
35. No 2nd-degree Taylor polynomial **37.** 4.99004
39. 17.0312 **41.** $|e(x,y)| \le 0.02$

EXERCISES 12.7, p. 819

1. Absolute min. at $(0, 0)$ **3.** Saddle point at $(1, 2)$
5. Saddle point at $(0, 3)$; local minima at $(2, 3)$ and $(-2, 3)$
7. Local max. at $(1, 2)$; saddle point at $(1, -2)$
9. Saddle points at $(3, 0)$, $(-2, 2)$; local min. at $(3, 2)$; local max. at $(-2, 0)$ **11.** Absolute max. at $(0, 0)$
13. Saddle points at $(\pi/2 \pm n\pi, 0)$ **15.** Local min. at $(0, 0, 0)$
17. Local min. at $(-1/2, -1, 3/2)$ **19.** 2 by 2 by 1; volume 4
21. 2 by 2 by 1
25. $x = y \approx 2.1544$, $z = 43.089$. No; it would be an unattractive four stories.
27a. $P = 2p(7 - p + q) + q(16 + 2p - 6q) - 6(7 - p + q) - 4(8 + p - 3q)$
27b. $p = \$8.75$, $q = \$4.75$; 6 units of A, 5 units of B
29. $y = 62x/35 + 18/35$ **31.** $y = 1.036364x - 0.363636$
33. $(0, 0)$ is a stable equilibrium point.
37. If $D = 0$ and $f_{yy}(a) > 0$ then $f(a)$ is a local min. of $z = f(x_0, y)$.

EXERCISES 12.8, p. 830

1. Max. $\sqrt{5}$, min. $-\sqrt{5}$ **3.** Max. $\sqrt{13}$, min. $-\sqrt{13}$
5. Max. $1/2$, min. $-1/2$ **7.** Max. $\sqrt{3}$, min. $-\sqrt{3}$

9. Max. 3; min. -3 **11.** 2 by 2 by 1, volume 4

13. 2 by 2 by 1 **15.** Max. 6, min. 2 **17.** $1/\sqrt{2}$ at $(1, 1/2, 3/2)$
19. $(-2/\sqrt{3}, 2/\sqrt{3}, -2/\sqrt{3})$ **21.** $16/\sqrt{3}$ **23.** Max. 3, min. 1
25. A square of side 1 **27.** Absolute min. at $(0, 0)$; max. at $(\pm 3, 0)$ **29.** Max. at $(0, \pm 1)$ **31.** Max. at $(0, 0, \pm 2)$
33. $x = 225$; $y = 37.5$

REVIEW EXERCISES 12.9, p. 831

1. $(1/5)(2\mathbf{i} + \sqrt{5}\mathbf{j} + 4\mathbf{k})$, $2x + \sqrt{5}y + 4z = 25$
3. $(9\pi/2)\mathbf{i} + 6\mathbf{j} + (\pi/2)\mathbf{k}$; all entries of ∇f are continuous at $(1, \pi/2, 3)$. **5.** $T = 6.094\,\mathrm{K}$; $E = 0.01767$
7. $160/3$; $(1/\sqrt{61})(6\mathbf{i} + 3\mathbf{j} + 4\mathbf{k})$; $8\sqrt{61}$

9. $(1/3)(5 - 2\sqrt{7}, 1 + 2\sqrt{7}, 2 + \sqrt{7})$ **11.** $-8x + y + 20z = 16$

13. $\dfrac{dy}{dx} = \dfrac{12x^3 - 2xy^3}{3x^2y^2 + 6y^2}$ if $y \neq 0$; $y(0.98) \approx 0.98$

15. $f_{xx} = 2y^3 + 8y^2 - 2$; $f_{xy} = f_{yx} = 6xy^2 + 16xy + 1$; $f_{yy} = 6x^2y + 8x^2$ **17.** 0.066

19. Saddle points at (1, 1) and (1, −1); local max. at (0, 0); Local min. at (2, 0)
21. Saddlepoint at (1, 1) **23.** Max. $2\sqrt{37}$; min. $-2\sqrt{37}$
25. Max. at (±1, 0), min. at (0, 0)

CHAPTER 13

EXERCISES 13.1, p. 845

1. 0.65219907 **3.** 0.75927609 **5.** 0.56174930 **7.** $-27/2$
9. $\ln(4/3)$ **11.** 154 **13.** 2 **15.** 2 **17.** $3 \ln 3 - 4 \ln 2$
19. $e^2 - 2e + 1$ **21.** 0 **23.** 0 **25.** 44/3 **27.** 20/3
29. $e^2/2 - e + 1/2$ **31.** 12 **45.** 2.2081 **49.** 7/6

EXERCISES 13.2, p. 855

1. 8

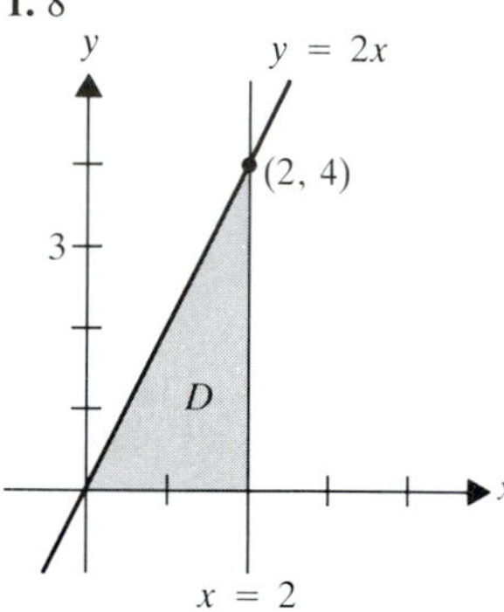

3. 128/15

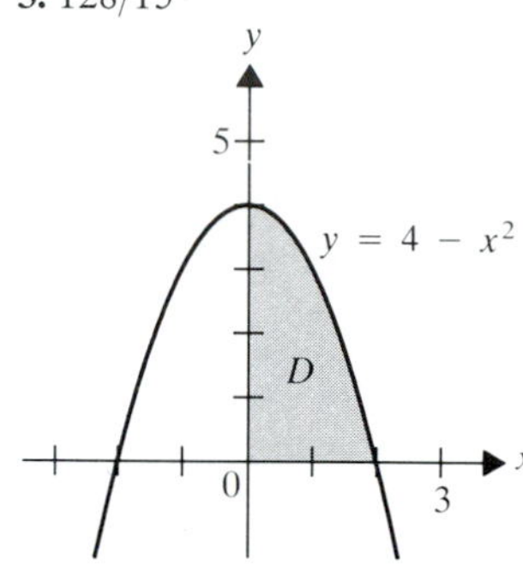

5. 4

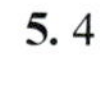

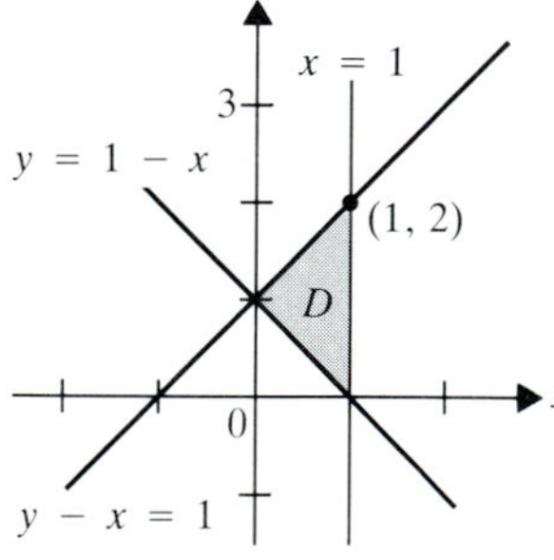

7. $e^{-2}/2 - (3/2)e^{-1} + e^{-1/2}$

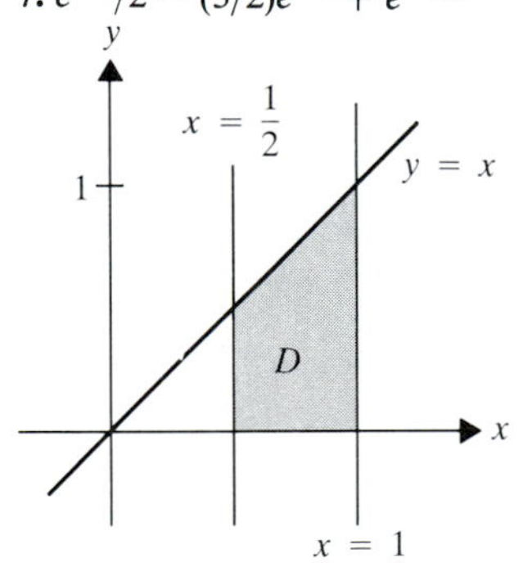

9. 4/3

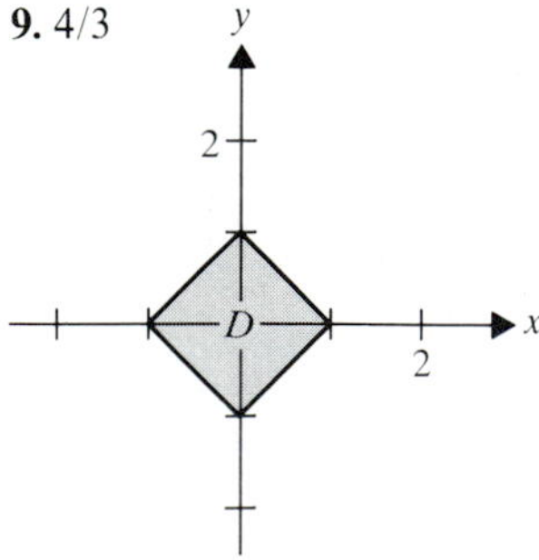

11. $2(27 - 5\sqrt{5})/3$

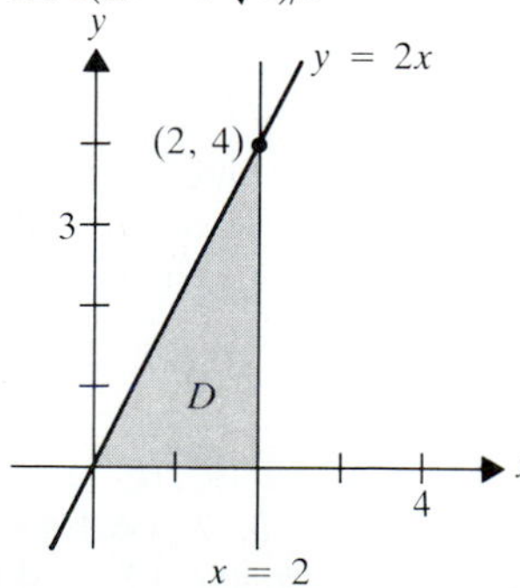

13. 6 **15.** 2/15 **17.** $e - 1$ **19.** $(5/6)\sqrt{5}$ **21.** 5 **23.** 7/6
25. 8/3 **27.** 18

29. 9

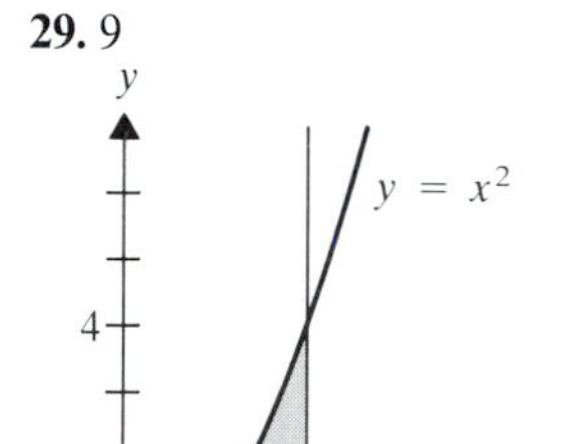

31. $\int_0^4 \int_{\sqrt{y}}^2 f(x, y)\,dx\,dy$

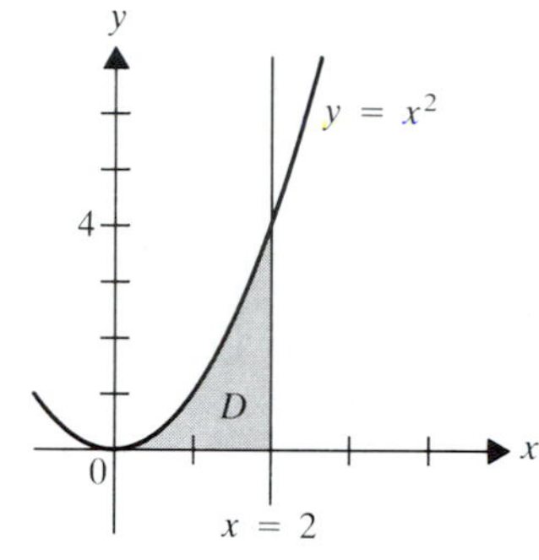

33. $\int_1^e \int_0^{\ln x} f(x, y)\,dy\,dx$

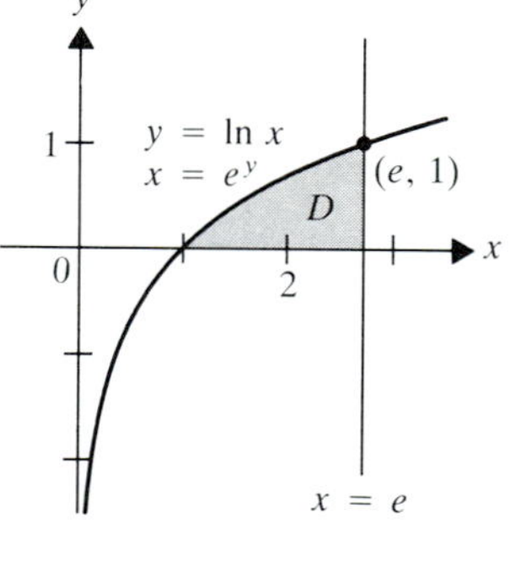

35. $\int_0^1 \int_{-\sqrt{y}}^{\sqrt{y}} f(x, y)\,dx\,dy + \int_1^4 \int_{y-2}^{\sqrt{y}} f(x, y)\,dx\,dy$

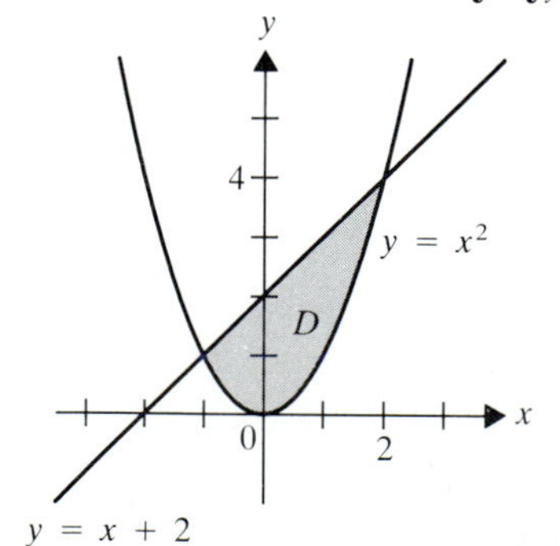

37. $n = 10$: 5.03941082 $n = 80$: 5.03961269
$n = 20$: 5.03959529 $n = 160$: 5.03961283
$n = 40$: 5.03961131 $n = 320$: 5.03961281

39. $n = 10$: 0.180699669 $n = 80$: 0.180672032
$n = 20$: 0.18067376 $n = 160$: 0.180672025
$n = 40$: 0.180672135 $n = 320$: 0.180672029

EXERCISES 13.3, p. 862

1. 5π **3.** $\pi/3$ **5.** $3\pi/2$ **7.** $8\pi/3 - 2\sqrt{3}$ **9.** 8 **11.** $3\sqrt{3}$
13. $2\pi/3$ **15.** $15\pi/2$ **17.** 2π **19.** $2\pi a^3/3$
21. $[4\pi a^3 - 4\pi(a^2 - b^2)^{3/2}]/3$ **23.** $21n(\sqrt{2}+1) - \pi/4$
25. 16/9 **27.** 0 **29.** -40π

EXERCISES 13.4, p. 873

1. 0;

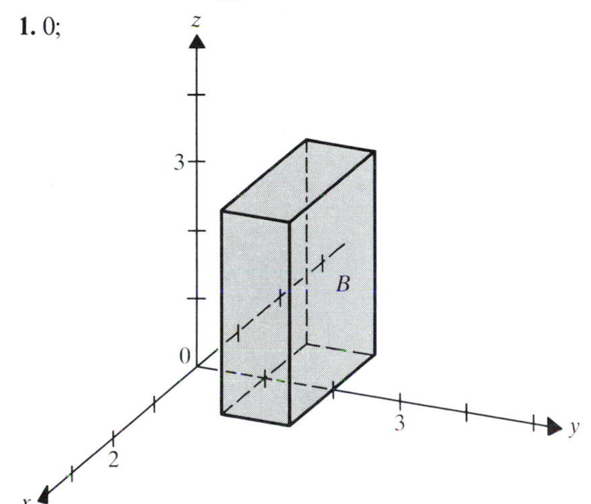

3. 7/30

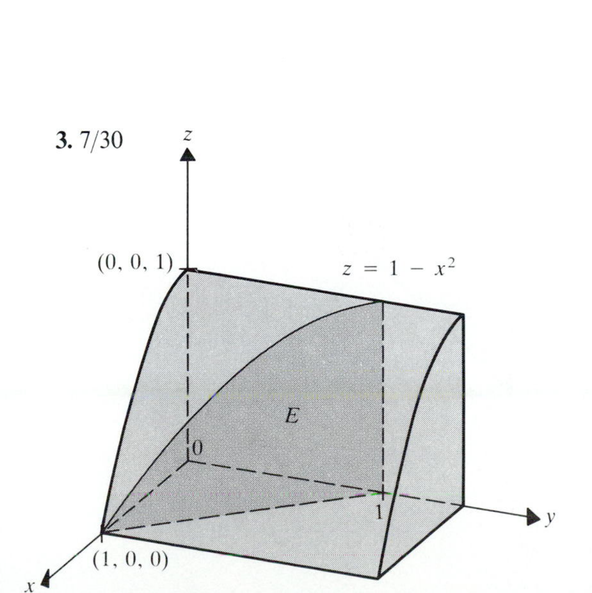

5. 256/195

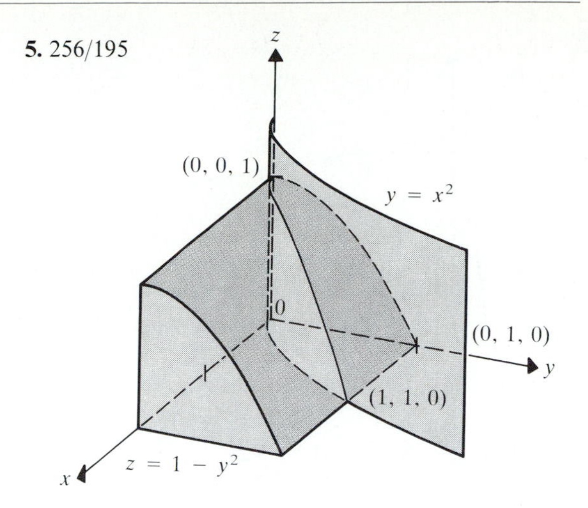

7. $-1/30$

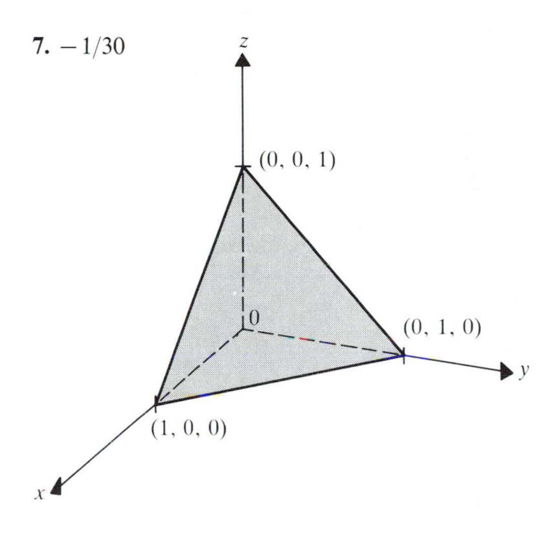

9. 11/120

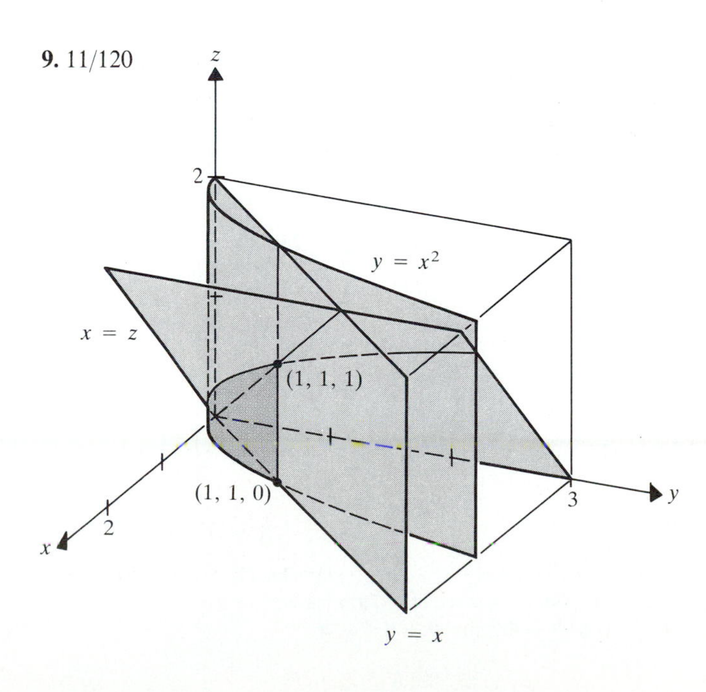

11. 31/20

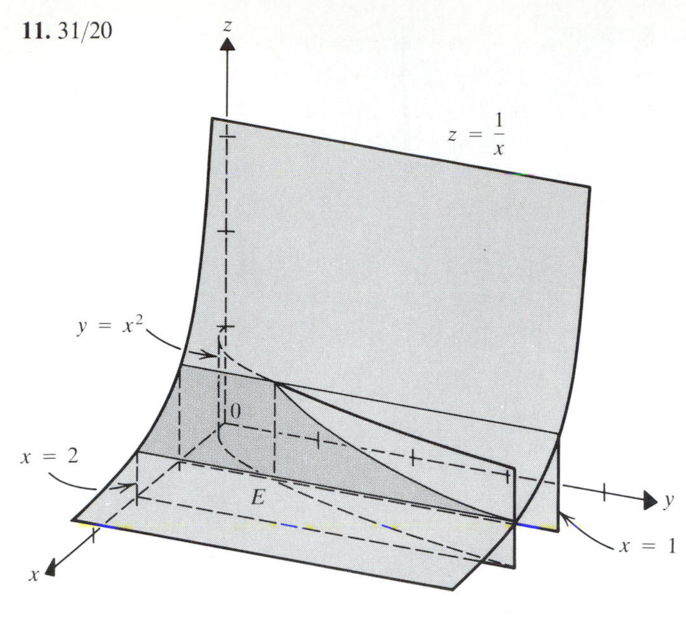

13. 1/3

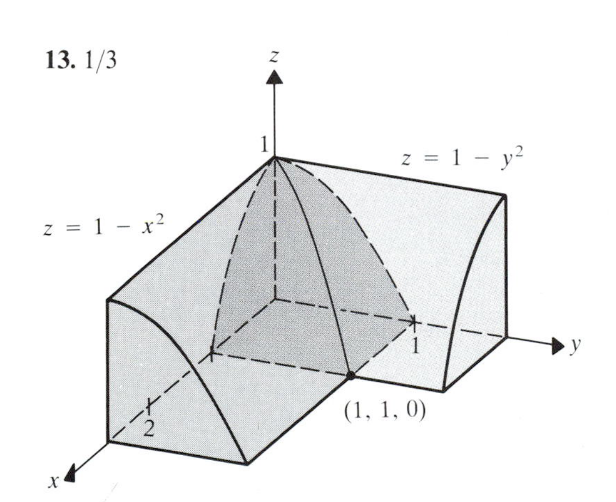

15. $abc/6$
17. 4 **19.** 8π **21.** 8/15 **23.** $8\pi\sqrt{2}$ **25a.** 24
25b. 24

EXERCISES 13.5, p. 880

1a. $[\sqrt{5}, \tan^{-1} 2, 3]$ **1b.** $\left[2, \frac{\pi}{2}, -2\right]$

1c. $\left[\sqrt{13}, \tan^{-1}\left[\frac{3}{2}\right], -1\right]$ **1d.** $\left[-\sqrt{13}, \tan^{-1}\left[\frac{3}{2}\right], 1\right]$

1e. $\left[\sqrt{13}, \tan^{-1}\left[-\frac{3}{2}\right], -1\right]$ **3a.** $(-1, \sqrt{3}, 1)$ **3b.** $(\sqrt{3}, 1, 2)$
3c. $(-1, 0, 2)$ **3d.** $(-\sqrt{3}, -1, -1)$ **3e.** $(1, -\sqrt{3}, -1)$
5a. $r = 4$ **5b.** $r^2 = 9 \sec 2\theta$ **7a.** $r^2 + 9z^2 = 9$
7b. $2r^2 \cos^2\theta - 4r^2 \sin^2\theta + 4z^2 = 4$

9a. $x^2 + y^2 = 9$

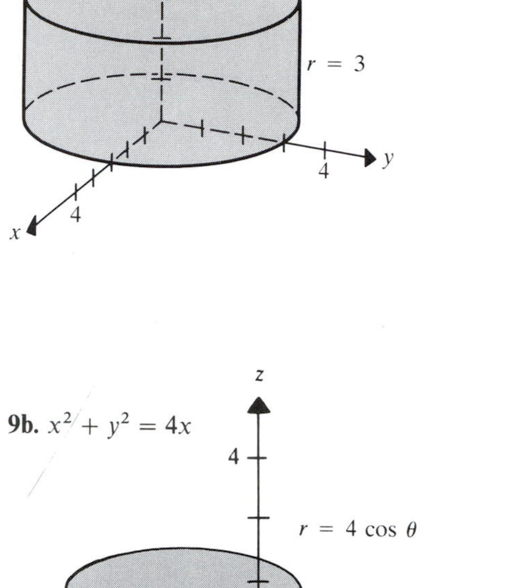

9b. $x^2 + y^2 = 4x$

9c. $y = 2$

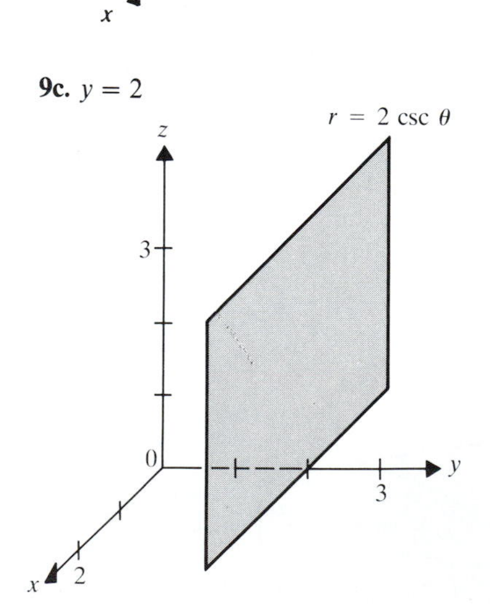

11a. $2x^2 + 2y^2 = (x + \sqrt{6})\sqrt{x^2 + y^2}$

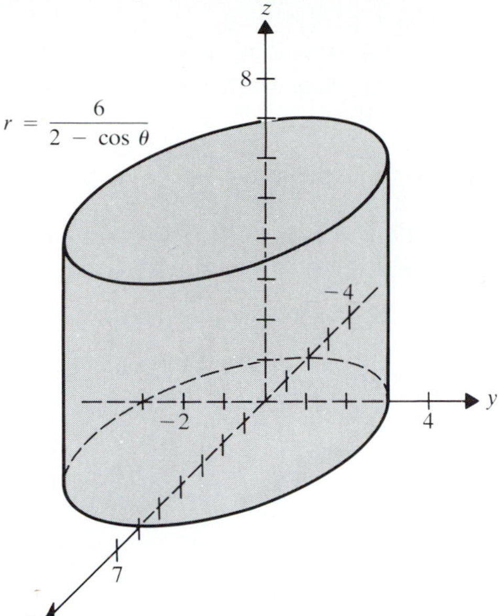

11b. $4x^2 + 4y^2 = z^2$

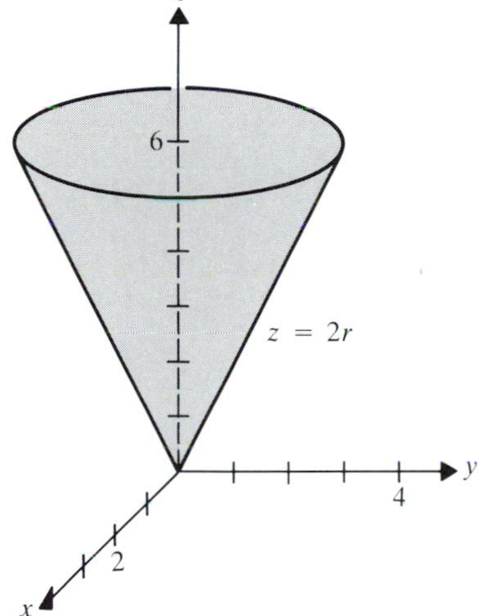

11c. $x^2 + y^2 + z^2 = 9$ **13.** $\pi a^2 h/3$

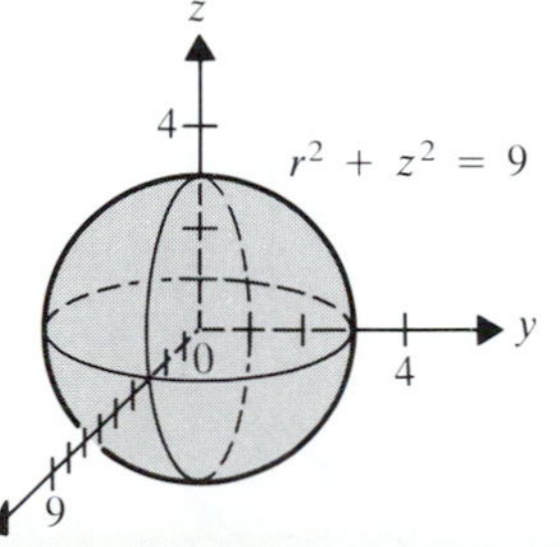

15. 8π **17.** 9π **19.** π **21.** 4π **23.** $2\pi/15$ **25.** a^4 **27.** $\pi/35$
29. $128\pi/15$

EXERCISES 13.6, p. 889

1a. $\{\sqrt{14}, \cos^{-1}(3/\sqrt{14}), \tan^{-1} 2\}$
1b. $\{\sqrt{3}, \cos^{-1}(1/\sqrt{3}), -3\pi/4\}$ **1c.** $\{2\sqrt{2}, 3\pi/4, \pi/2\}$
1d. $\{\sqrt{13}, \cos^{-1}(-3/\sqrt{13}), 0\}$
1e. $\{\sqrt{14}, \cos^{-1}(-1/\sqrt{14}), \pi + \tan^{-1}(-3/2)\}$
3a. $(3\sqrt{2}/4, 3\sqrt{2}/4, 3\sqrt{3}/2)$ **3b.** $(1, \sqrt{3}, 0)$
3c. $(\sqrt{3}/\sqrt{2}, 1/\sqrt{2}, -\sqrt{2})$ **3d.** $(-\sqrt{3} \; -1/\sqrt{2}, 1/\sqrt{2}, -\sqrt{2})$
3e. $(0, 0, -2)$ **5a.** $\rho = 3$ **5b.** $\rho = 6\cos\phi$
5c. $\rho = 4\csc\phi$
7a. $x^2 + y^2 + z^2 = 25$

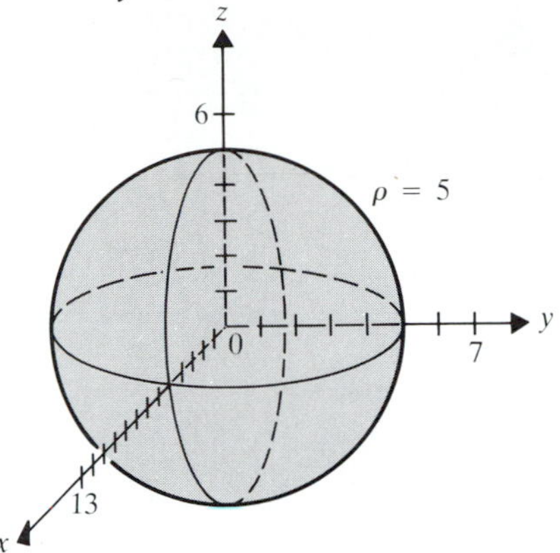

7b. $y = x/\sqrt{3}$

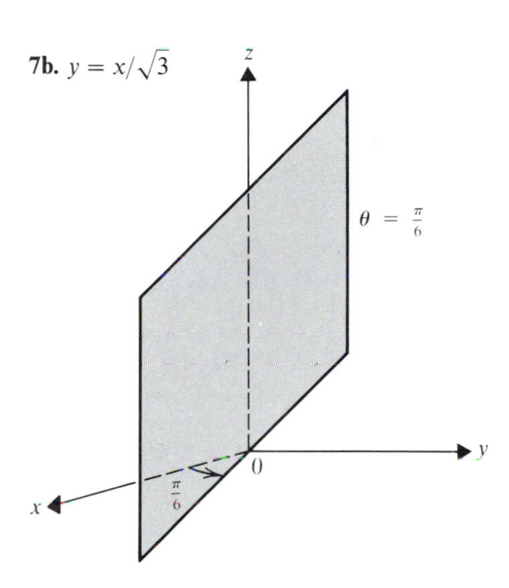

7c. $x^2 + y^2 + (z - 4)^2 = 16$

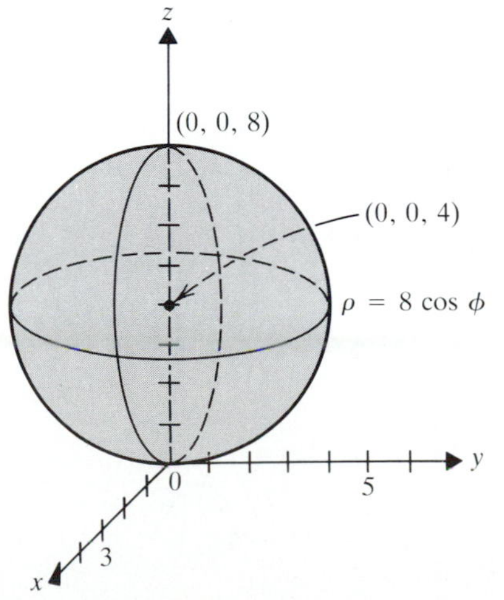

9. $\pi a^2 h$ **11.** $(2\pi/3)(1-\sqrt{3}/2)$ **13.** $28\pi/3$
15. $8\pi/3$

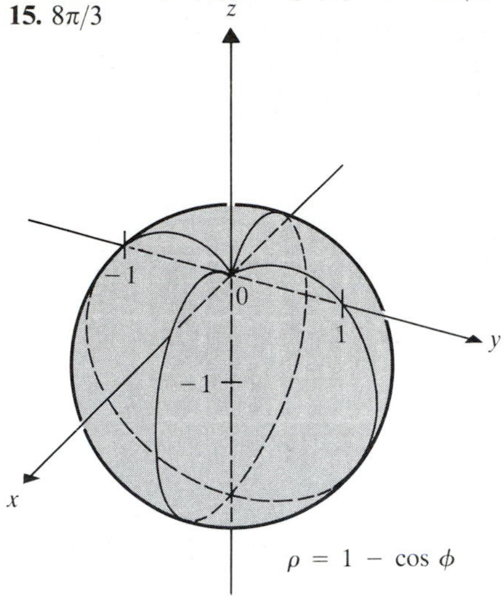

17. π **19.** 4π
21. $\pi/15$ **23.** $(43/3-8\sqrt{3})\pi$
27. $(\pi/2)\rho_0$ where ρ_0 is the radius of the earth
29. $(1/6)(17\sqrt{17}-64)$ **31.** 0

EXERCISES 13.7, p. 897

1. $m=acd/2$; $\bar{x}=(a+b)/3$; $\bar{y}=c/3$
3. $m=6$; $\bar{x}=14/9$; $\bar{y}=13/6$ **5.** $m=3k/20$; $\bar{x}=5/9$; $\bar{y}=4/7$
7. $m=3\pi d/4$; $\bar{x}=\bar{y}=28/9\pi$ **9.** $m=1/2$; $\bar{\mathbf{x}}=(1/2, 1/2, 2/3)$
11. $m=16\pi/3$; $\bar{\mathbf{x}}=(0, 0, 3/4)$

13. $m=4\pi d$; $\bar{x}=\bar{y}=0$; $\bar{z}=9/4$ **15.** $R=\sqrt{\frac{2}{3}}$; $I_x=2\delta$

17. $I=324\pi k$; $R=2$ **19.** $K=648\pi k$ **21.** $K=324\pi\omega^2 dh$, where $\omega=33\frac{1}{3}$ r.p.m. and $h=$ thickness **23.** $(2a/\pi, 2a/\pi)$

25a. $4\pi a^2$ **25b.** $4\pi^2 ab$ **29.** $m=\delta\int_a^b [g(x)-f(x)]\,dx$;

$$M_x=\frac{1}{2}\delta\int_a^b ([g(x)]^2-[f(x)]^2)\,dx;$$

$$M_y=\delta\int_a^b x[g(x)-f(x)]\,dx$$

REVIEW EXERCISES 13.8, p. 899

1. $(e^2-1)/2$ **3.** $4/7$ **5.** $32\pi/15$ **7.** $1/15$ **9.** $9a^4/16$ **11.** 16π
13. $64/3$ **15.** 2π **17.** $18\pi-10\sqrt{5}\pi/3$ **19.** $\pi/2$

21. $\int_1^e \int_0^{\ln x} f(x, y)\,dy\,dx$ **23.** $1-1/e$

25. $m=6$; $\bar{x}=6/5$; $\bar{y}=8/5$ **27.** $I_z=2^{13}\pi/9$; $R=8/3$

CHAPTER 14

EXERCISES 14.1, p. 913

1. Limit is (0, 0, 3); **F** is not continuous at $\mathbf{x}=\mathbf{0}$ because it is not defined there.
3. There is no limit, because **F** is not defined on a deleted open ball around $\mathbf{a}=(0, 0, 2)$: it is undefined on the whole plane $x+y=0$.

5. $J_{\mathbf{F}}(1, \pi, -1)=\begin{bmatrix} 2 & e^{\pi} & 3 \\ -\pi-\pi e & -1-e & \pi e \end{bmatrix}$

7. $J_{\mathbf{F}}(1, 1, 0)=\begin{bmatrix} 1 & 0 & 0 \\ \frac{1}{\sqrt{5}} & \frac{4}{\sqrt{5}} & 0 \\ \pi & \pi & 0 \end{bmatrix}$ **9.** $1+y$; -1

11. $y^2z^2+z^2\cos y$; $4\pi^2-4$ **13.** $y\mathbf{k}$; $4\mathbf{k}$ **15.** **0**; **0**

39. $f_{xx}+f_{yy}=Tr\begin{bmatrix} f_{xx} & f_{yx} \\ f_{xy} & f_{yy} \end{bmatrix}$

41.

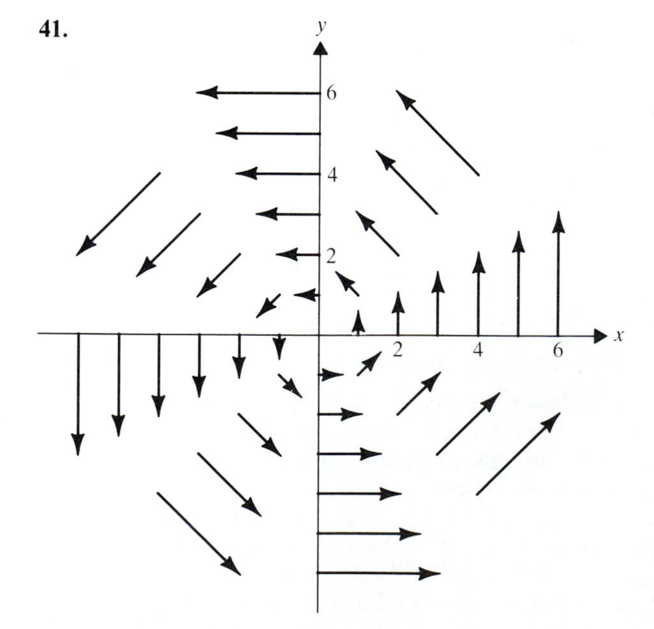

43.

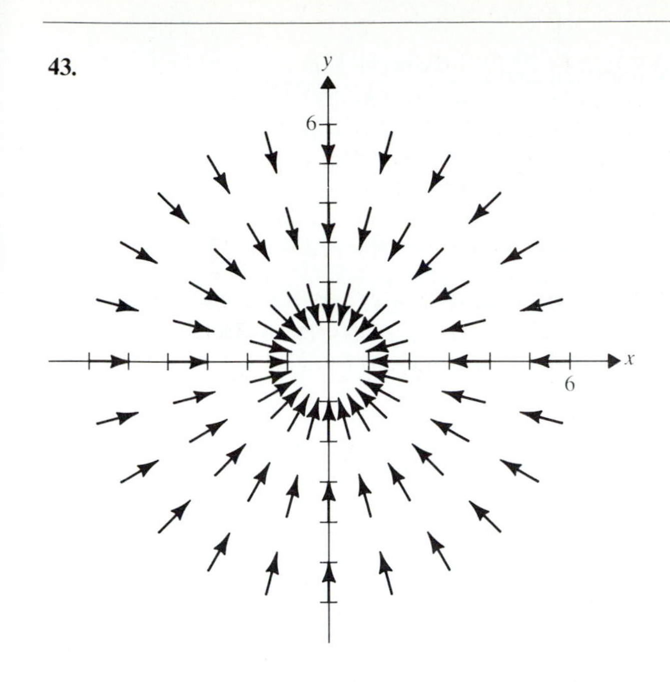

EXERCISES 14.2, p. 923

1. $1/4$ **3.** -2 **5.** $26/15$ **7.** 2π **9.** 1 **11.** 2 **13.** $31/2$
21. -2π **23.** $2\sqrt{2}\pi$ **25.** $(\bar{x}, \bar{y}, \bar{z})$ where $\bar{x}m = \int_\gamma xf(x)\,ds$, $\bar{y}m = \int_\gamma yf(x)\,ds$, $\bar{z}m = \int_\gamma zf(x)\,ds$ **27.** $2\sqrt{2}\pi(1 + 4\pi^2/3)$
29. M **31a.** 3 **31b.** -1

EXERCISES 14.3, p. 933

1. 12 **3.** $-243\pi/2$ **5.** 0 **7.** $12 - \pi$ **9.** $\pi/2$ **11.** 2π **13.** 0

15. 8π **17.** 24π **19.** 7π **23.** $\bar{x} = \dfrac{1}{4A(D)} \oint_{\partial D} x^2\,dy - 2xy\,dx$

35. $2\mathbf{k}$

EXERCISES 14.4, p. 943

1. $5/2$ **3.** $\pi/4$ **5.** 0 **7.** -8 **9.** $-e^{-\pi/2} + e^{9\pi/2}$
11. Yes; $p(x, y) = x^3 + xy + e^y$ **13.** Yes; $p(x, y) = (x^2 + y^2)^{1/2}$
15. Yes; $p(x, y, z) = x^2yz + xz^2 - 2xy^2 + x - 2z$
17. Yes; $p(x, y, z) = x^3 + y^3 - 3xyz$ **23.** $-mgz$
25. $T = m|\mathbf{v}|^2/2$

EXERCISES 14.5, p. 952

1. $\mathbf{X}(u, v) = (u, v, (u^2 + v^2)/2)$, $(u, v) \in [1, 2] \times [2, 5]$
3. $\mathbf{X}(r, \theta) = (r\cos\theta, r\sin\theta, 4 - r)$, $(r, \theta) \in [0, 4] \times [0, 2\pi]$
5. $\mathbf{X}(r, \theta) = (r\cos\theta, r\sin\theta, 3r\cos\theta - 2r\sin\theta)$, $(r, \theta) \in [0, 2] \times [0, 2\pi]$
7. $\mathbf{X}(\phi, \theta) = (\sin\phi\cos\theta, \sin\phi\sin\theta, \cos\phi)$, $(\phi, \theta) \in [0, \pi/4] \times [0, 2\pi]$
9. $\mathbf{X}(u, v) = ((b + a\cos v)\cos u, (b + a\cos v)\sin u, a\sin v)$, $(u, v) \in [0, 2\pi] \times [0, 2\pi]$

11. $\mathbf{X}(u, v) = \left(u, v, \dfrac{1}{c}[d - au - bv]\right)$, $u, v \in \mathbf{R}$.

13. $(\pi/6)(5\sqrt{5} - 1)$ **15.** $(\pi/6)(17\sqrt{17} - 5\sqrt{5})$

17. $(\pi/2)\sqrt{2} + (\pi/2)\ln(1 + \sqrt{2})$ **19.** $\sqrt{14}\int_0^{2\pi}\int_0^2 r\,dr\,d\theta = 4\sqrt{14}\pi$

21. $4\pi^2ab$ **23.** $2\pi ah$ **25.** $(1/2)(a^2b^2 + b^2c^2 + a^2c^2)^{1/2}$
27. $\mathbf{X}(u, v) = (u\cos v, u\sin v, f(u))$, $(u, v) \in [a, b] \times [0, 2\pi]$

EXERCISES 14.6, p. 966

1. $\pi/2$ **3.** 12π **5.** $2\pi[2/3 - 5/6\sqrt{2}]$ **7.** $\pi a^3h + \pi a^4/2$ **9.** 4π
11. $3e - 21$ **13.** $81\pi/2$ **15.** 2π **17.** 3π **19.** $4\pi\delta a$ **21.** 4π
23. -4π

EXERCISES 14.7, p. 976

1. 0 **3.** 8π **5.** 0 **7.** $-\pi$ **9.** $-\pi$ **11.** $-\pi$ **13.** 0 **15.** -4π
19. $(a + b + c)\pi r^2$

EXERCISES 14.8, p. 985

1. $32\pi/3$ **3.** 0 **5.** $11/24$ **7.** $7\pi/4$ **9.** 216 **11.** $7\pi/4$

13. $2\pi = 2\int_0^{\pi}\int_0^{2\cos\theta}\int_{r^2}^{2\cos\theta} r(r\cos\theta + 1)\,dz\,dr\,d\theta$

REVIEW EXERCISES 14.9, p. 988

3. Yes; $\mathbf{F}(x, y) = \nabla((x^2 + y^2)^{-1/2})$ **5.** No **7.** $-3/4$ **9.** -24π
11. 0 **13.** 0 **15.** -4
17a. $\mathbf{X}(r, \theta) = (r\cos\theta, r\sin\theta, r\cos\theta + r\sin\theta)$,

$(r, \theta) \in [0, 1] \times [0, 2\pi]$ **17b.** $\sqrt{3}\pi = \int_0^1\int_0^{2\pi}\sqrt{3}r\,dr\,d\theta$ **19.** 0

21. $\dfrac{1}{3} = \int_0^1\int_0^{1-y}(x, -2y, 1)\cdot(1, 1, 1)\,dx\,dy$

23. $\dfrac{3\pi}{2} = \int_0^1\int_0^1\int_0^{2\pi} r^3\left[\dfrac{1}{2} + \cos^2\theta\right]d\theta\,dr\,dz$ **25.** 0

CHAPTER 15

EXERCISES 15.1, p. 998

1. $y = ce^{\cos x}$ **3.** $y = c\cos x$ **5.** $y = e^{3x}/2 + ce^x$
7. $y = -x^2 - 1/3 + cx^3$ **9.** $y = (-\cos 3x)/3x^2 + cx^{-2}$
11. $y = 1 - x^2$ **13.** $y = -19x^2/3 + x^5/3$
15. $y = 1 + (x + 1)\ln|x + 1|$ **17.** $y = 1/(cx^3 - x^3\ln|x|)$
19. $y^{-2} = x + 1/2 - e^{2x}/4$ **21a.** $L(di/dt) + Ri = E(t)$
21b. $i(t) = E/R + ce^{-Rt/L}$ **23.** $ds/dt = 75 - 3s(t)/10$; $s(t) = 750e^{-0.3t} + 250 \to 250$ as $t \to \infty$
25. $y(t) = 20(1 - e^{-t/5000})$; $2\frac{1}{2}$ min

27. $v = \sqrt{mg/k}\,[(Ce^{Lt} - 1)/(Ce^{Lt} + 1)]$, where $L = 2\sqrt{gk/m}$;

$y = (m/k)\ln\cosh(Lt/2)$; $\lim_{t\to\infty} v(t) = \sqrt{mg/k}$

29a. $k = -gR^2/M$ **29b.** $v_2 = v_0{}^2 - 2gR + 2gR^2/x$
29c. $v_0 > (2gR)^{1/2}$

33. $p(t) = \left[p(0) - \frac{b-c}{l} - \frac{k^2lc}{k^2l^2 + a^2}\right]e^{-klt} + \frac{b-c}{l} + \frac{kc}{k^2l^2 + a^2}(kl\cos at + a\sin at)$

Let $\theta_0 = \tan^{-1}(kl/a)$; then $p(t) \approx \frac{b-c}{l} + \frac{kc\sin(at + \theta_0)}{[k^2l^2 + a^2]^{1/2}}$ for large t; $p(t)$ oscillates about $(b-c)/l$: minimum values occur at $t = [2n\pi - (\pi/2 + \theta_0)]/a$; maximum values occur at $t = [2n\pi + (\pi/2 - \theta_0)]/a$

EXERCISES 15.2, p. 1005

1. $xy + \cos y = \pi - 1$ **3.** $x^3 + 2x^2y + y^2 = c$
5. $x^3/3 + xy^2 = 0$ **7.** $xy + y\cos y - \sin y = c$
9. $x^3y + x^2y^2/2 = c$ **11.** $xy^3 + 2x^2y^2 = c$
17. $x^2 + 4xy + 4y^2 + 6x - 2y = c$
19. The single solution is tangent at each point to some member of the one-parameter family of solutions: $y = cx + c^2$; $y = -x^2/4$

EXERCISES 15.3, p. 1014

3. $z_1 + z_2 = 4 + i$; $z_1z_2 = 5 - i$; $z_1 \div z_2 = 1/13 - 5i/13$
5. $\pm i$, $\pm i\sqrt{2}$
9. The vector $2 - 3i$ is multiplied by $\sqrt{2}$ and rotated 45° counterclockwise.
11. $-8 - 8i\sqrt{3}$
13. $3(\cos\pi/9 + i\sin\pi/9)$, $3(\cos 7\pi/9 + i\sin 7\pi/9)$, $3(\cos 13\pi/9 + i\sin 13\pi/9)$
15. $\sqrt{3}/2 + i/2$, i, $\sqrt{3}/2 + i/2$, $-\sqrt{3}/2 - i/2$, $-i$, $\sqrt{3}/2 - i/2$
19. $e^i = \cos 1 + i\sin 1$; $e^{2\pi i} = 1$
23. $(1 + 2i)e^{x(1+2i)}$, $(-3 + 4i)e^{x(1+2i)}$, $(-11i - 2i)e^{x(1+2i)}$
25. $e^{x/\sqrt{2}}\left[\cos\left(\frac{x}{\sqrt{2}}\right) + i\sin\left(\frac{x}{\sqrt{2}}\right)\right]$
27. $\int e^{ax}\cos bx\,dx = \frac{e^{ax}}{a^2 + b^2}(a\cos bx + b\sin bx) + C$;
$\int e^{ax}\sin bx\,dx = \frac{e^{ax}}{a^2 + b^2}(a\sin bx - b\cos bx) + C$
29. $m = n$ gives π

EXERCISES 15.4, p. 1023

1. $y = c_1e^x + c_2e^{2x}$ **3.** $y = c_1e^{-2x} + c_2e^{-3x}$
5. $y = c_1e^{-3x} + c_2xe^{-3x}$ **7.** $x = e^{-t}(c_1\cos 2t + c_2\sin 2t)$
9. $y = c_1\cos 3x + c_2\sin 3x$ **11.** $y = 2e^{2x} - e^x$
13. $y = \cos 3x - \sin 3x$ **15.** $y = -2(\cos 3x)/3 + \sin 3x$
17. $y = e^{-3x} + 4xe^{-3x}$ **19.** No solution
21. $y = \cos 3x - \sin 3x$ **29.** $y = c$, $x = y - c\ln|y + c| - k$
31. $y = c$, $y = -2x + c$, $y = \frac{1}{c_1}[c_2e^{c_1x+2}]$ **33.** $y = x^2/2 + 3/2$

EXERCISES 15.5, p. 1031

1. $y_p(x) = -x^2 - 6x - 19$ **3.** $(3\cos x - 5\sin x)/17$
5. $-e^{-x}/3$ **7.** $e^x(x^2/6 - x/9)$ **9.** $3xe^{2x}/4$
11. $c_1e^{2x} + c_2e^{3x} + 3e^x/4 + xe^x/2$
13. $c_1e^{r_1x} + c_2e^{r_2x} + \frac{-2\cos x + 3\sin x}{13}$, where $r_1 = (-3 + \sqrt{13})/2$ and $r_2 = (-3 - \sqrt{13})/2$
15. $y = c_1e^{-x} + c_2xe^{-x} + 3x^2e^{-x}/2$
17. $c_1e^{3x} + c_2e^{-x} - e^x/2 + 2\sin x - \cos x$
19. $3e^{3x}/2 + 2e^{-x} - e^x/2 - \cos x + 2\sin x$ **21.** $e^x + x^2e^x/2$
23. $c_1e^x + c_2e^{2x} + x^2 + 3x + 7/2 + 2e^{3x} - x^2e^x - 3xe^x$
25. $-\sqrt{\pi}e^x\,\mathrm{erf}(x)/2 + c_1e^x + c_2xe^x$
27. $c_1\cos 3x + c_2\sin 3x + (\cos 2x)/5$
29. $c_1e^{-3x} + c_2e^{2x} - (21\sin x)/50 - (3\cos x)/50$
31. $y = c_1x + c_2x^2 + 3\ln x + 9/2$

EXERCISES 15.6, p. 1044

3a. $(F_0t\sin\omega_0t)/2m\omega_0$ **5.** $x = 8 + (\cos 2t)/2$
7. $x = (1/2 + 2t)e^{-4t}$; critically damped
9. $x(t) = (10/3)(\sin t/2)(\sin 3t/2) = (5/3)(\cos t - \cos 2t)$
11. $\omega = 1/\sqrt{CL}$; current is also periodic
15. $q = 10^{-6}(2e^{-500t} - e^{-1000t})$ and $i = dq/dt$
17. $i = dq/dt$, where $q = e^{-400t}(2.4 \times 10^{-3}\cos 300t + 3.2 \times 10^{-3}\sin 300t) + 2.4 \times 10^{-3}$
19. $q = (6e^{-10t}/61)(6\sin 50t + 5\cos 50t) - 5(5\sin 60t + 6\cos 60t)/61$;
$i(t) = dq/dt$; $i_p(t) = (-1500\cos 60t)/61 + (1800\sin 60t)/61$

EXERCISES 15.7, p. 1057

1. $y = a_0\left[1 + \frac{(2x)^2}{2!} + \dots + \frac{(2x)^{2n}}{(2n)!} + \dots\right] + \frac{a_1}{2}\left[2x + \frac{(2x)^3}{3!} + \dots + \frac{(2x)^{2n+1}}{(2n+1)!} + \dots\right]$
$= a_0\cosh 2x + a_1\sinh 2x$
3. $y = a_0[1 - (2x)^2/2! + (2x)^4/4! - (2x)^6/6! + - \dots] + (a_1/2)[2x - (2x)^3/3! + (2x)^5/5! - + \dots]$
5. $y = a_0[1 - x^2 + x^4/3 - x^6/(3\cdot 5) + - \dots] + a_1[x - x^3/2 + x^5/(2\cdot 4) - x^7/(2\cdot 4\cdot 6) + - \dots]$
7. $y = a_0\left[1 - x^2 + \frac{1}{3}x^4 - \frac{1}{3\cdot 5}x^6 + \frac{1}{3\cdot 5\cdot 7}x^8 - + \dots\right] + a_1\left[x - \frac{1}{2}x^3 + \frac{1}{2\cdot 4}x^5 - \frac{1}{2\cdot 4\cdot 6}x^7 + - \dots\right]$
9. $y = a_0[1 - x^2 + x^4/4 - + \dots] + a_1[x - x^3/2 + 3x^5/40 - + \dots]$
11. $y = a_0[1 - x^2 + x^3 - 13x^4/12 + - \dots] + a_1[x - x^2/2 + x^3/6 - x^4/8 + - \dots]$
13. $y = a_0[1 - x^3/6 + 3x^5/40 - + \dots] + a_1[x - x^3/6 - x^4/12 + 3x^5/40 + \dots]$
15. $y = a_0[1 + (x-1)^2/2 + (x-1)^3/6 + (x-1)^4/24 + (x-1)^5/30 + \dots] + a_1(x - 1 + (x-1)^3/6 + (x-1)^4/12 + (x-1)^5/120 + \dots]$
17. $y = a_0\left[1 + \sum_{m=1}^{\infty}(-1)^m\frac{\lambda(\lambda-2)(\lambda-49)\cdots(\lambda-4)\times(\lambda+1)\cdots(\lambda+2m-1)}{(2m)!}x^{2m}\right] + a_1\left[x + \sum_{n=1}^{\infty}(-1)^m\{[(\lambda-1)(\lambda-3)\cdots(\lambda-2m+1)(\lambda+2)(\lambda+4)\cdots(\lambda+2m)]/(2m+1)!\}x^{2m+1}\right]$
19. $y = x - x^3/2 + x^5/(2\cdot 4) - x^7/(2\cdot 4\cdot 6) + \dots$

21. $y = 2(1 - x^3/6 + 3x^5/40 + \dots) + 3(x - x^3/6 - x^4/12 + \dots)$
23. 1.000000, 1.500000, 2.190000, 3.146000, 4.474400, 6.324160; *exact solution* $y = x/4 - 3/16 + 19e^{4x}$ gives 1.000000, 1.6090042, 2.505330, 3.830139, 5.7942226

25. 1.000000, 1.100000, 1.190005, 1.270743, 1.342664, 1.406131
27a. 1.000000, 1.595000, 2.463600, 3.737128, 5.609949, 8.369725
27b. See Exercise 23 for exact solution values.
29. 1.000000, 1.095002, 1.180735, 1.257628, 1.325992, 1.386133

31.

x_i	0	0.05	0.10	0.15	0.20	0.25
y_i	1.000000	1.250000	1.547500	1.902000	2.324900	2.829880
$y(x_i)$	1.000000	1.275416	1.609042	2.013766	2.505330	3.102960

x_i	0.30	0.35	0.40	0.45	0.50
y_i	3.433356	4.155027	4.018533	5.018533	6.052239
$y(x_i)$	3.830139	4.715550	5.794226	7.108956	8.712004

35. 1.000000, 1.000000, 0.940000, 0.818182, 0.632538, 0.380708

39.

i	x_i	z_i	$y(x_i)$	% Error
0	0	1	1	0
1	.05	1.27375	1.27541578	.130606445
2	.1	1.604975	1.60904183	.252748472
3	.15	2.0063195	2.01376608	.369783515
4	.2	2.49320979	2.50532985	.483771128
5	.25	3.08446594	3.10295967	.59600281
6	.3	3.80304845	3.83013885	.707295349
7	.35	4.67696911	4.71554996	.818162291
8	.4	5.74040231	5.79422601	.928919456
9	.45	7.03504082	7.10895637	1.03975243
10	.5	8.61174981	8.71200412	1.15076061
11	.55	10.5325848	10.6672035	1.26198752
12	.6	12.8732534	13.052522	1.37344
13	.65	15.7261192	15.9631889	1.48510285
14	.7	19.2038654	19.5155181	1.59694811
15	.75	23.4439658	23.8515751	1.70894102
16	.80	28.6141382	29.1448797	1.82104525
17	.85	34.9189987	35.6073689	1.93322412
18	.90	42.6081783	43.4979035	2.04544378
19	.95	51.9862276	53.1326567	2.15767328
20	1	63.4246977	64.8978032	2.26988505

41.

i	x_i	z_i
0	0	1
1	.05	1.04875007
2	.1	1.09510539
3	.15	1.1391432
4	.2	1.18092148
5	.25	1.22048634
6	.3	1.25787731
7	.35	1.29313121
8	.4	1.32628501
9	.45	1.35737781
10	.5	1.38645227
11	.55	1.41355556
12	.6	1.43873989
13	.65	1.46206274
14	.7	1.48358683
15	.75	1.50337988
16	.80	1.52151421
17	.85	1.53806618
18	.90	1.55311553
19	.95	1.5667447
20	1	1.57903802

43.

i	x_i	y_{Euler}
0	0	1
1	.05	1
2	.1	.985
3	.15	.954887218
4	.2	.909546024
5	.25	.848856057
6	.3	.772690058
7	.35	.68091127
8	.4	.573370157
9	.45	.449900147
10	.5	.31031186
11	.55	.154385046
12	.6	−.0181431268
13	.65	−.20759578
14	.7	−.414384389
15	.75	−.639055809
16	.8	−.882376706
17	.85	−1.14549927
18	.9	−1.43032782
19	.95	−1.74048228
20	1	−2.08463138

REVIEW EXERCISES 15.8, p. 1060

1. $y = k \csc x$ **3.** $y^{-2} = -x^4/3 + 7/12x^2$ **5.** $y^2 = x^2 + cx^3$
7. $y = c_1e^{3x} + c_2xe^{3x}$ **9.** $y = -e^x + 2e^{2x}$
11. $y = c_1e^{(-3+\sqrt{13}x)/2} + c_2e^{(-3-\sqrt{13}x)/2} - 6(\cos 2x)/61 - 5(\sin 2x)/61$ **13.** $y = c_1e^{-6x} + c_2e^x + 3e^{2x}/8$

15. $y = c_0\left[1 + \sum_{m=1}^{\infty} \frac{(-1)^m 2^{m!} x^{2m}}{1 \cdot 3 \cdot 5 \cdot \dots (2m-1)}\right] + c_1\left[\sum_{m=1}^{\infty} (-1)^{m+1} \frac{1 \cdot 3 \cdot 5 \cdot \dots (2m-1)}{4^{m-1}(m-1)!} x^{2m-1}\right]$

17. 1.0000, 1.0000, 1.0100, 1.0304, 1.0623, 1.1074
19. $P = 100{,}000(1 - e^{t/10{,}000})$; after 10 days **21.** $x(t) = (2/87)[3(9\cos 2t + 8\sin 2t) - e^{-2t}(27\cos 3t + 34\sin 3t)$; $\sqrt{13}$
23. $i_p(t) = (13200\pi/I^2)[100(40000 - \omega_0^2)^2 \cos 120\pi t + 3000\omega_0 \sin 120\pi t]$ where $I^2 = 100(40000 - \omega_0^2)^2 + 9 \times 10^6\omega_0^2$ and $\omega_0 = 120\pi$ **25a.** $\sqrt{3}/2 + i/2$; $3i$
25b. $(e^{x/4}/512)[\sin(x/4) - i\cos(x/4)]$

INDEX OF TERMS

Please send me *CALCULUS (Kemeny/Hurley)* formatted for the

☐ IBM-PC ☐ Macintosh ☐ Amiga

at $24.95 per copy. Total: $ ______________.

☐ Check enclosed. Please charge to

☐ VISA ☐ MasterCard ☐ Amex

Card #: ______________________ Expiration Date: ______________

Signature: ______________________________

Please print:

Name: ______________________________

School ______________________________

Street Address ______________________________

City ______________ State ________ Zip ______________

(Or phone 1-800-TR-BASIC to order)

FOLD HERE

CUT PAGE OUT

FOLD HERE

BUSINESS REPLY MAIL

FIRST CLASS MAIL PERMIT NO. 19 HANOVER, NH

POSTAGE WILL BE PAID BY ADDRESSEE

True BASIC Inc.
39 South Main Street
Hanover, NH 03755

See calculus more clearly through the power of your microcomputer

If you have access to a microcomputer, you can take advantage of *CALCULUS (Kemeny/Hurley),* a flexible and easy-to-use software program that is especially designed to supplement this text. It features on-screen generation of examples from the text, plus routines you can use to solve many of the text's computer-oriented problems—even if you have no knowledge of programming!

This software is easy to learn and use, making it ideal for self-study. The program is menu-driven—there's no need to learn "computerese" to make the program run.

With *CALCULUS (Kemeny/Hurley)* you can *see* the behavior of functions graphically, using the program to:

- Choose the topics you want to explore
- Select the examples you want to study
- Input functions of your choice
- Check your homework
- Review for examinations

CALCULUS (Kemeny/Hurley) provides a disk and a handy, 20-page User's Guide. The disk includes:

- **A General–Purpose Routine** that accepts the function you define, takes multiple derivatives, plots the function and its derivatives, finds zeros, and displays tables of values.
- **8 Special–Purpose Routines** for individual topics, including limits, tangents to a curve, maxima and minima, area under a curve, L'Hôpital's rule, parametric equations, Taylor series, and differential equations.
- **Source Code** that can be used to solve many of the text's optional computer problems and serves to introduce you to the powerful True BASIC language.

Hardware Requirements: IBM-PC (or compatible) with graphics adaptor and 192K memory, or the Apple Macintosh, Commodore Amiga, or Atari ST. (Execution of source code requires True BASIC language system, but the menu-driven routines do not.)

John G. Kemeny is professor of Mathematics and Computer Science (and President Emeritus) at Dartmouth College. Noted as an innovator in the use of computers in mathematics education, Dr. Kemeny has defined the undergraduate Finite Mathematics course as it exists today and is the co-inventor of the BASIC computer language.

James F. Hurley is Professor of Mathematics at the University of Connecticut, Storrs. He is the author of three calculus texts, including *CALCULUS* (Wadsworth, 1987).

Your chance to rate

Calculus by James F. Hurley

In order to make this text even more responsive to your needs, it would help us to know what you, the student, thought of *Calculus*. We would appreciate it if you would answer the following questions. Then, cut out the page, fold, seal, and mail it; no postage is required. Thank you for your help.

What Chapters in the text did you skip, or *not* cover in class? (circle)

1 2 3 4 5 6 7 8 9 10 11 12

13 14 15

Prior to taking Calculus, which courses in mathematics had you previously taken?

College Algebra _______ Analytic Geometry _______

Trigonometry _______ Algebra and Trigonometry _______

Precalculus _______ High School Algebra _______ (1 or 2 terms? _______)

Prior to taking Calculus, how long ago did you take your last Algebra course?

Within last 2 years _______ 3–5 years ago _______ Over 5 years ago _______

What is your major course of study?

Engineering _______ Chemistry _______ Physics _______

Biology _______ Mathematics _______ Computer Science _______

Business _______ Social Science _______ Other _______

Would you rather see more applications to your major in college, or more detailed examples and answers?

Applications _______ More detail _______

Did the answers have any typos or misprints? If so, where?

FOLD HERE

CUT PAGE OUT

FOLD HERE

NO POSTAGE NECESSARY IF MAILED IN THE UNITED STATES

BUSINESS REPLY MAIL

First Class Permit No. 34 Belmont, CA

Postage will be paid by
WADSWORTH PUBLISHING COMPANY, INC.
10 Davis Drive
Belmont, California 94002
ATTN: Mathematics Editor

Index of Symbols

$<, \leq$	less than, less than or equal
$>, \geq$	greater than, greater than or equal
$\in$	belongs to
$\{x \mid a \leq x \leq b\}$	set of all x such that $a \leq x \leq b$
$[a, b]$	closed interval
(a, b)	open interval
$(a, b], [a, b)$	half-open intervals
$(-\infty, b), (-\infty, b]$, $(a, +\infty), [a, +\infty)$	half lines
$\mathbf{N}$	set of natural numbers $\{1, 2, 3, \ldots\}$
$\mathbf{Z}$	set of integers $\{0, \pm 1, \pm 2, \ldots\}$
$\mathbf{Q}$	set of rational numbers m/n, $m, n \in Z$, $n \neq 0$
$\mathbf{R}$	set of real numbers
$(-\infty, +\infty)$	real line
$\|x\|$	absolute value of x
$\sqrt{a}$	radical a
$\cup$	set union
$\cap$	set intersection
$\varnothing$	empty set
Δx	increment(al change in) x
$\perp$	perpendicular
$d(P, Q)$	distance between P and Q
$f \circ g$	f composed with g
$\dfrac{\Delta y}{\Delta x}$	difference quotient
$\lim_{x \to c} f(x)$	limit of f as x approaches c
$\lim_{x \to b^-} f(x), \lim_{x \to a^+} f(x)$	one-sided limits of f
$[x]$	greatest integer $\leq x$
$\approx$	approximately equals

Index of Symbols

$f'(c), \left.\frac{dy}{dx}\right\rvert_{x=c}$	derivative of f (or y) at c
$f'(x), D_x f$	derivative function
$\frac{d}{dx}$	differentiation operator
$T(x)$	tangent approximation to $f(x)$
dx, dy	differentials
Å	angstrom
f^{-1}	inverse of the function f
$f''(c), \left.\frac{d^2y}{dx^2}\right\rvert_{x=c}$	second derivative of f (or y) at c
$f^{(n)}(c), \left.\frac{d^ny}{dx^n}\right\rvert_{x=c}$	nth derivative of f (or y) at c
$\lim_{x\to\pm\infty} f(x)$	limit of f as $\lvert x\rvert$ increases without bound
$D_x^{-1}(f(x)), D^{-1}(f)$	antiderivative of f
$\int f(x)\,dx$	indefinite integral of f
ΔA_i	area of approximating rectangle
R_P	rectangular approximation, Riemann sum
$\lvert P\rvert$	norm of a partition
$\sum_{i=k}^{m} f(i)$	sum of $f(i)$ from k to m
$\int_a^b f(x)\,dx$	definite integral of f over $[a, b]$
$\bar{f}$	average value of f
$F(x)\Big\rvert_a^b, F(x)\Big]_a^b$	$F(b) - F(a)$
$I_x(f)$	$\int_a^x f(t)\,dt$
$L_n(f), R_n(f)$	left (right) endpoint approximation of $\int_a^b f(x)\,dx$
$M_n(f)$	midpoint approximation of $\int_a^b f(x)\,dx$
$T_n(f)$	trapezoidal approximation $\int_a^b f(x)\,dx$
$S_n(f)$	Simpson approximation of $\int_a^b f(x)\,dx$
$A(R)$	area of region R
$V(S)$	volume of solid S
$(\bar{x}, \bar{y})$	center of mass of a lamina
M_0	moment with respect to origin
M_x, M_y	moments with respect to x- and y-axes
e	base of natural logarithms
r_{eff}	effective annual interest rate
(X)	concentration of a reactant X
pH	potential of hydrogen
$\int_a^{+\infty} f(x)\,dx, \int_{-\infty}^b f(x)\,dx, \int_{-\infty}^{+\infty} f(x)\,dx$	improper integrals
$[r, \theta]$	polar coordinates
$\lim_{n\to\infty} a_n$	limit of a sequence
$\sum_{n=1}^{\infty} a_n$	infinite series
s_n	nth partial sum of a series
e_n	error in s_n
$p_n(x)$	nth-degree Taylor polynomial
$r_n(x)$	remainder: $f(x) - p_n(x)$
$\sum_{n=0}^{\infty} b_n(x-a)^n$	power series in $x - a$
$\binom{p}{n}$	binomial coefficient
$\mathbf{R}^n$	n-dimensional Euclidean space
$\lvert\mathbf{a}\rvert$	length of a vector $\mathbf{a}$
$\mathbf{i}, \mathbf{j}, \mathbf{k}$	standard basis vectors for $\mathbf{R}^3$